Hexagon Series on Human and Environmental Security and Peace

Volume 10

Series editor

Hans Günter Brauch, Mosbach, Germany

Hans Günter Brauch · Úrsula Oswald Spring
John Grin · Jürgen Scheffran
Editors

Handbook on Sustainability Transition and Sustainable Peace

With Forewords by Heide Hackmann, Executive Director of ICSU;
Hoesung Lee, Chairman, IPCC; and Paul Raskins, President, Tellus
Institute

Editors

Hans Günter Brauch
Department of Political and Social Sciences
Free University of Berlin (ret.)
  AFES-PRESS Chairman
Mosbach
Germany

Úrsula Oswald Spring
Centro Regional de Investigaciones
  Multidiscipinarias (CRIM)
National Autonomous University of Mexico
  (UNAM)
Cuernavaca
Mexico

John Grin
Department of Political Science
University of Amsterdam
Amsterdam
The Netherlands

Jürgen Scheffran
Research Group Climate Change
  and Security (CLISEC) Institute
  of Geography
University of Hamburg
Hamburg
Germany

ISSN 1865-5793   ISSN 1865-5807 (electronic)
Hexagon Series on Human and Environmental Security and Peace
ISBN 978-3-319-43882-5   ISBN 978-3-319-43884-9 (eBook)
DOI 10.1007/978-3-319-43884-9

Library of Congress Control Number: 2016947505

Cover design: deblik, Berlin

Cover illustrations: The top photograph, illustrating Sustainable Peace, was taken from "Annie's New Letters (& notes)" by Anne Selden Annab, a Palestinian-American poet (at: http://anniesnewletters.blogspot.de/2012/09/international-day-of-peace-sustainable.html, posted on the International Day of Peace on 21 September 2012 to show the way to "Sustainable Peace for a Sustainable Future"). The photograph on the left was taken in Bangkok by Hans Günter Brauch in January 2014, the one in the centre by Úrsula Oswald Spring in Yautepec, Morelos, Mexico, and the one on the right showing schoolchildren in Matabeleland in Zimbabwe by Hans Günter Brauch in September 2015. All photographers granted permission for use.

Copyediting: PD Dr. Hans Günter Brauch, AFES-PRESS e.V., Mosbach, Germany

Language editing: Michael Headon, Colwyn Bay, Wales, UK

Typesetting and layout: Thomas Bast, AFES-PRESS e.V., Mosbach, Germany

Printed on acid-free paper

This Springer imprint is published by Springer Nature
The registered company is Springer International Publishing AG Switzerland

Global Environmental Change and Sustainability Transition

Over the past three to four decades, global change research communities across the world have been analysing the dramatic global environmental changes related to increasing fossil fuel consumption, accelerated resource use and a growing world population. The international global environmental change research programmes initiated during this time by the *International Council for Science* (ICSU), often in cooperation with the *International Social Science Council* (ISSC) and relevant UN partners, have made major contributions to our understanding of these processes, and to the complex interlinkages between natural and human systems underlying them.

Yet despite the wealth of scientific knowledge produced, and notwithstanding national and global policy efforts, including UN conferences and international agreements, we have only taken small steps—at best—towards the kind of sustainable development called for back in 1986 in the Brundtland Report *Our Common Future* and in *Agenda 21*, adopted at the first summit in Rio de Janeiro in 1992. Our failure to act and the growing exposure to threats this has generated now mean we face the urgent and immensely challenging task of accelerating transformations to a sustainable, peaceful and just world.

In 2015, the world community agreed on an ambitious set of new commitments to end and, indeed, transform, unsustainable business-as-usual policies and practices towards global socio-economic development. At three high-level meetings in the context of the United Nations, governments took remarkable steps forward in the quest for effective solutions to an integrated global agenda of ending poverty, ensuring social and economic development, and protecting our planet. Three new international agreements were adopted: the 2030 *Agenda for Sustainable Development*, including a set of seventeen *Sustainable Development Goals* (SDGs); the *Paris Climate Agreement*; and earlier in the year the *Sendai Framework for Disaster Risk Reduction*.

The hard work of implementing the goals of these agreements—particularly by national policymakers and practitioners—cuts across the substantive issues addressed and is widely accepted as demanding an integrated, multisector, multi-stakeholder approach within which science has a central role to play. This places a significant new demand on the scientific community: to be more immediately impactful, more effective in making a difference to real-world problems by contributing transformative solutions to a series of interconnected challenges, from poverty eradication to food production for a growing world population, from addressing global financial crises and growing levels of social discontent to managing the looming global water crisis and providing sustainable energy for all. And with this demand comes the challenge to change: to work in ways that most scientists have not been trained to work in, and more often than not, are not supported or rewarded for working in.

Taking science into the so-called 'solutions space' calls not only for a reinvigorated interdisciplinarity, one that takes collaboration across the natural, social, human, medical and engineering sciences to a deeper, more meaningful level in which the collective strengths of disciplines are brought to bear on the framing of problems as much as they are on the production of new research insights. It also calls for the kind of boundary-spanning work (and skills) that enables scientists to engage with other knowledge partners—decision-makers, practitioners, business leaders and citizens—in the co-design and co-production of knowledge, policy and practice.

Solutions-oriented science that is global, inter- and transdisciplinary in nature is increasingly supported at the international level, most notably perhaps by initiatives such as *Future Earth*, a global platform for sustainability research established and now governed by an alliance of international organizations including ICSU and the ISSC (see <http://www.futureearth.org>).

This volume on *Sustainability Transition and Sustainable Peace* shares the Future Earth vision. It makes a timely and important contribution to our understanding of the leading role that science can—and must—play in shaping humanity's future on planet Earth.

Paris, France   23 April 2016

Heide Hackmann
Executive Director
International Council for Science (ICSU)

Climate Change and Sustainable Development Goals

This *Handbook of Sustainability Transition and Sustainable Peace* appears at a particularly timely juncture. In December 2015 the global community came together to reach the Paris Agreement on climate change, more far-reaching than many had dared hope. Three months earlier, world leaders adopted the 2030 Agenda for Sustainable Development, comprising seventeen Sustainable Development Goals to end poverty, protect the planet and ensure prosperity for all. Climate action is one of the seventeen, but in fact all of them are affected one way or another by climate change.

2016 therefore is the year that the world begins a process of implementation of these vital agreements. Never before have so many countries sought to undertake such ambitious goals in the economic and development spheres. Decision-making in a changing world benefits from iterative risk management; the experience of implementing these agreements will be a rich field for researchers in a range of disciplines in the coming years, and policymakers will benefit from their findings. The IPCC too can also contribute, in its coming reports, by assessing the research of the academic community addressed by this Handbook.

The tasks faced by the world and analysed by the Handbook are not to be underestimated: decarbonization, dematerialization, changes in consumption patterns and waste, the restoration of environmental and ecosystem services, the recovery of soils and integrated water management. These are the challenges of the twenty-first century.

One of the key difficulties we face is the multidisciplinary nature of the tasks we have set ourselves. At its simplest this comes down to the apparent conflict between protecting the environment and tackling climate change while ensuring that development continues to lift people out of poverty. I say apparent, because the IPCC has shown that climate (and hence environmental) policy can be profitably mainstreamed into countries' development agendas. Indeed, the mutually reinforcing rather than contradictory nature of development and environmental policy is the essence of what we mean by sustainability.

The potential tension does not lie only between economy and ecology. The IPCC has shown how climate-related hazards exacerbate other stressors, often with negative outcomes for livelihoods, especially for people living in poverty. Climate change impacts can exacerbate poverty in rich and poor countries, and indirectly increase risks of violent conflicts in the form of civil war and inter-group violence by amplifying well-documented drivers of these conflicts such as poverty and economic shocks. By worsening droughts and water shortages, climate change can add to pressures encouraging people to migrate. In other words, climate change and the other sustainable development goals in general have a powerful bearing on peace and human security.

By linking the four social science research areas of peace studies, security studies, development studies and environmental studies, the Handbook makes a welcome contribution to an integrated assessment of these issues. As the IPCC focuses increasingly on solutions in future assessments we too will look increasingly at these areas. In its assessments of the policy response to climate change as well as the causes and impacts, the IPCC aims to be policy-relevant but never policy-prescriptive. But even in a policy-neutral framework, there is a need for a comprehensive, integrated and urgent response. This Handbook provides rich source of insights and information to students and policymakers as they confront these challenges.

Seoul/Geneva   27 April 2016

Hoesung Lee
Chairman
Intergovernmental Panel on Climate Change (IPCC)

The Larger Challenge

Sustainability and security, the foci of this collection, are critical themes of a larger drama now unfolding on the global stage. With threads of connectivity binding people and planet into a single community of fate, a profound historical transition is under way. We have entered the *Planetary Phase of Civilization*,[1] with the outcome indeterminate and contested. Our trajectory in the twenty-first century will depend to a significant degree on the choices we make, or fail to make, now and in the pivotal years ahead. The overarching challenge is to shape a flourishing planetary civilization.

The global future is inherently uncertain, but this much seems clear: the current course will take us helter-skelter into a time of troubles, even calamity. Anachronistic twentieth-century mindsets and institutions—the state-centric political order, growth-driven economies, environmental insensitivity, and consumer culture—are ill-suited to twenty-first-century realities. The gap between the 'is' of the world today and the 'ought' of a civilized tomorrow is a breeding ground for myriad perils best understood as various expressions of a common phenomenon: the disruptive coalescence of one, planet-encompassing social–ecological system.

Emerging conditions in the Planetary Phase demand a new development paradigm, a *Great Transition* centred on the well-being of people and nature in our interdependent, fragile world. With the stakes high and political leadership feeble, a zeitgeist of apprehension is on the rise, bringing fear for the future, despair, and even fatalism. At the same time, antidotes to self-fulfilling pessimism and passivity—social vision and collective action—are gaining traction. Twin motives—necessity and desire—propel the movement for a *Great Transition*: the need to mute looming dangers and the hope for a civilized future. At a fundamental level, the moral imperative to pass on a liveable world resonates with the instincts for survival and empathy that lie deep in the human psyche.

Needless to say, these spurs for deep change confront powerful counterforces, namely, vested interests, institutional inertia, and cultural tenacity. And less noble human impulses—greed, dominance, complacency—too often best the better angels of our nature. Still, despite such resistance and distraction, change is afoot, even in the mainstream where policy reforms are actively pursued, and where the 2015 UN Sustainable Development Goals provide an aspirational framework for guiding further effort. But incrementalism will not suffice: scattershot tweaks to conventional development leave intact the deep drivers of systemic stress.

---

1 The ideas of the *Planetary Phase* and *Great Transition* were introduced in Paul Raskin, Tariq Banuri, Gilberto Gallopín, Pablo Gutman, Allen Hammond, Robert Kates, and Rob Swart: *Great Transition: The Promise and the Lure of the Times Ahead* (Boston: Tellus Institute, 2002), at: <http://www.tellus.org/tellus/publication/great-transition-the-promise-and-lure-of-the-time-ahead>. The *Great Transition Initiative*, an international network devoted to elaborating the framework, publishes scholarship on the theory and practice of global transformation (see: <www.greattransition.org>).

Transformative change entails cultivation of values and institutions consonant with the exigencies and potentialities of the *Planetary Phase*. The expansion of the human project encourages a corresponding expansion of human consciousness to embrace humanity-as-a-whole and its place in the ecosphere, as well as a revised understanding of what constitutes a good life and a good society. Ultimately, the hope for a *Great Transition* correlates with the waxing of a new suite of values—human solidarity, quality of life, and an ecological sensibility—and the waning of the conventional triad of individualism, materialism, and domination of nature.

Each value demarcates a vital domain of strategic action. The concern for human solidarity supports efforts to establish an egalitarian social contract, eradicate poverty, strengthen popular democracy, and cultivate the identity and practice of global citizenship. The emphasis on quality of life inspires initiatives to limit workweeks and foster lifestyles rich in creativity, recreation, relationships, community, spirituality, and other dimensions of a well-lived life. An ecological sensibility motivates endeavours to protect and heal the earth, including advancing the sustainable development agenda, which calls for moving beyond unfettered capitalist growth toward regenerative economies that maintain human impacts within the environmental boundaries of a finite planet.

These values and strategies are emerging as a potent cultural and political influence. The critical uncertainty is whether the forces for a *Great Transition* will mobilize with sufficient speed, scale, and coherence. This question of social agency—Who will change the world?—needs a plausible answer. The myopic concerns of government and business render them poor candidates to lead the way. Civil society holds more potential. Indeed, in recent decades, countless global and local movements and organizations have weighed in on a full spectrum of issues, an eruption of citizen energy that has spread popular awareness, conducted important research, and mounted action campaigns. Much has been achieved, but piecemeal victories have not scaled up to a systemic change of direction. Organizational and conceptual fragmentation has stifled civil society's potential, slicing the integral challenge into a thousand arenas and turfs with no unifying conceptual framework or social vision.

Our pressing task now is to weave disparate efforts into an integrated planetary praxis that can attract and engage widening circles of concerned citizens. Such a 'global citizens movement' is the critical actor for a *Great Transition*, but remains missing from the global stage. Nevertheless, history shows that popular movements arising from widespread grievance and offering a path to rectification can, after a period of slow development, reach a critical mass, then quickly coalesce. Thus, advancing an inclusive and plural global movement towards its tipping point has become a frontline project for shaping a civilization worthy of the name.

In times of social stability, visionaries of a better world can be dismissed, with some justification, as quixotic dreamers. In transformative moments, business-as-usual thinkers are the utopians, and the prophets of 'another world' the pragmatists. If today's generations prove too hidebound to accept the necessity of deep change or too cynical to think such change possible, the legacy could well be a century of decline. Hence, it is urgent but not too late to join together in a new planetary praxis for meeting the larger challenge: building the community of Earth.

Boston, Massachusetts, USA   20 March 2016

Paul Raskin, President
Tellus Institute

Contents

Contents

Acknowledgments

This volume is based on selected papers that were presented at the *First Sustainability Transition and Sustainable Peace Workshop* of *Centre for Regional Multidisciplinary Research* at the *National Autonomous University of Mexico* (CRIM-UNAM) and AFES-PRESS in Yautepec, Morelos, Mexico, on 10-13 September 2012 focusing on "Towards a Fourth Sustainability Revolution and Sustainable Peace: Visions and Strategies for Long Term Transformative Change to Sustainable Development in the 21st Century", funded by the German Foundation for Peace Research (DSF) with logistical support by UNAM/CRIM (Mexico) and AFES-PRESS (Germany).[1]

A second catalytic workshop took place with the financial support of the *International Studies Association* (ISA) on 2 April 2013 in San Francisco. Its theme was "Sustainability Transition and Sustainable Peace: Policy Initiatives of Governments and International Organizations".[2] It continued on 4 April 2013 during the same ISA Convention as a regular panel on the theme of "Sustainability Transition: Theories, Approaches and Perspectives from Europe and Latin America".[3]

Key ideas were also presented by the first two co-editors during their guest professorships at the Political Science Department and the *Social Research Institute* (CUSRI) of Chulalongkorn University in Bangkok, Thailand. This took place on 9-13 December 2013 in the first Winter School during the public protests against the then Thai government.

Furthermore, the first two editors, as co-chairs of the *Ecology and Peace Commission* (EPC) of the *International Peace Research Association* (IPRA), presented selected ideas in November 2012 during the 24th biennial IPRA conference at Mie University in Japan and during the 25th biennial IPRA conference in Istanbul (Turkey). The 25th conference was organized by Sarkarya University in August 2014 and resulted in three peer-reviewed publications.[4]

After intense discussion among the four co-editors, most authors were invited to submit between November 2013 and April 2014. The anonymous double-blind peer review process then started and lasted until February 2016. All editorial decisions were made by the editors in their personal capacity only. The funders had no influence on the themes of this book, on the invitation to the authors, or on the views expressed by its authors. None of the authors and editors receive any personal remuneration from this collaborative research

---

1   This workshop is documented with podcasts at: <http://www.afes-press-books.de/html/sustainability_ workshop_overview.htm>.

2   This workshop is documented with selected powerpoints and papers at: <http://www.afes-press-books.de/ html/workshop_SanFrancisco.htm>.

3   This workshop is documented with abstracts, powerpoints and papers at: <http://www.afes-press-books.de/ html/workshop_SanFrancisco.htm>.

4   Úrsula Oswald Spring; Hans Günter Brauch; Keith G. Tidball (Eds.): *Expanding Peace Ecology: Security, Sustainability, Equity and Peace: Perspectives of IPRA's Ecology and Peace Commission* 1 (Cham–Heidelberg–New York–Dordrecht–London: Springer-Verlag, 2014); Hans Günter Brauch, Úrsula Oswald Spring, Juliet Bennett, Serena Eréndira Serrano Oswald (Eds.): *Addressing Global Environmental Challenges from a Peace Ecology Perspective* (Cham–Heidelberg–New York–Dordrecht–London: Springer-Verlag, 2016); Úrsula Oswald Spring, Hans Günter Brauch, Serena Eréndira Serrano Oswald, Juliet Bennett (Eds.): *Regional Ecological Challenges for Peace in Africa, the Middle East, Latin America and Asia Pacific*. SpringerBriefs in Environment, Security, Development and Peace, vol. 28 (Cham–New York–Heidelberg–Dordrecht–London: Springer International Publishing, 2016).

project. The preparation of this manuscript, its copy- and language-editing and the layout was made possible by a second publication grant by the *German Foundation for Peace Research* (DSF) and by financial support from Springer-Verlag in Heidelberg for the preparation of the camera-ready manuscript and from AFES-PRESS for the preparation of the index.

The editors would like to thank Dr. Christian Witschel, Editorial Director, Earth Sciences, Geosciences Editorial, Springer-Verlag and Dr. Johanna Schwarz, Senior Editor, for their support for this project, and Ms Almas Schimmel, the producer of this book within Springer-Verlag in Heidelberg, for her efficient coordination and implementation of this project, as well as the many other unnamed persons within Springer-Verlag for the care they have devoted to this publication.

All editors are grateful to Mike Headon (Wales/UK) for his extremely careful and perceptive language editing and to Thomas Bast, who in his quiet and efficient way produced several versions of the proofs and the whole layout as well as the index and acted as the webmaster on this book website. The editors are particularly grateful to the sixty authors, who come from eighteen countries on all five continents and represent many scientific disciplines, for their contributions to this book and for their readiness to reflect the critiques and suggestions of the anonymous reviewers who have tried to look beyond the boundaries of their respective disciplines and expertise.

For the forewords, the editors are particularly grateful to Dr. Heide Hackmann, Executive Director, International Council for Science (ICSU), Paris; Prof. Dr. Hoesung Lee, Chairman of the Intergovernmental Panel on Climate Change (IPCC), Seoul/Geneva, and Dr. Paul Raskin, President, Tellus Institute, Boston.

All co-editors are grateful to the anonymous reviewers, from many regions of the world and from many disciplines, for their diligence and their recommendations for improving the quality of the texts in this volume. We thank all book authors who contributed to the review process, as well as the following external reviewers: Alves de Sousa, Célio Alberto (Portugal); Amster, Randall (USA); Arizpe, Lourdes (Mexico); Baker, Lucy (UK); Balaban, Osman (Turkey); Balbo, Andrea (Spain); Basurto Guillermo, Xavier (Mexico/USA); Bharat, Dahiya (Thailand); Bogardi, Janos (Germany); Bosold, David (Germany); Boulding, Russell (USA); Brauch, Hans Guenter [40 formal reviews]; Brizga, Janis (Latvia); Brock, Lothar (Germany); Brzoska, Michael (Germany); Buettner, Thomas (Germany); Cardoso-Castro, Pedro-Pablo (UK/Brazil) [2 reviews]; Castan Broto, Vanesa (UK); Cincotta, Richard (USA); Coenen, Lars (Sweden); Cohen, Maurie (Canada); Conca, Ken (USA); Cornell, Sarah (Sweden); Dalby, Simon (Canada) [8 reviews]; Danielzyk, Rainer (Germany); De Stefano, Lucia (Spain); Deconinck, Heleen (The Netherlands); d'Estree, Tamra (USA); Ehlers, Eckart (Germany); Ehrenberg, John (USA); Espinosa, Angela (UK); Fischer-Kowalski, Marina (Austria); Galtung, Johan (Norway); Garrido, Alberto (Spain); Geels, Frank (The Netherlands/UK); Genus, Audley T. (UK); Gleditsch, Nils-Petter (Norway); Goede Curacao, Miguel (The Netherlands); Greacen, Chris (UK); Grin, John (The Netherlands) [3 reviews]; Guillen-Hanson, Georgina (Germany); Happaerts, Sander (Belgium); Hargroves, Charlie (Australia); Hayden, Anders (Canada); Holsti, Kalevi (Canada); Hope, Ronald (Canada); Jaeger, Jill (Austria) [2 reviews]; Jochem, Eberhard (Germany); Johnson, Twig (USA) [2 reviews]; Karakosta, Chara (Greece); Karvonen, Andy (UK); Kern, Florian (Germany/UK); Kosinski, Leszek (Poland) [2 reviews]; Kumar, Dinesh M. (India); Linden, Bas van der (The Netherlands); Loorbach, Derk (The Netherlands); Lopez Vázquez, Esperanza (Mexico); Lorek, Sylvia (Germany); Lucatello, Simone (Mexico); Lutz, Wolfgang (Austria); Mallett, Alexandra (Canada); Martinez Austria, Polioptro Fortunato (Mexico); Maxton, Graeme

(UK); Meadowcroft, James (Canada) [2 reviews]; Meierding, Emily (Switzerland); Mesjasz, Czeslaw (Poland) [2 reviews]; Naidu, Sirisha (USA); Napoli, Christopher R. (USA); Narkeviciute, Rasa (Denmark); Newman, Peter (Australia); Ometto, Jean (Brazil); Oswald Spring, Úrsula (Mexico) [4 reviews]; Ott, Hermann (Germany); Ozasa, Takao (Japan); Paredis, Erik (Belgium); Penttinen, Elina (Finland); Perez-Floriano, Lorena (Mexico); Perrot, Radhika (South Africa); Peter, Camaren (South Africa) [2 reviews]; Price, Martin (UK); Puppim de Oliveira, Jose (Brazil/Malaysia); Quintana Solórzano, Fausto (Mexico); Reardon, Betty (USA); Rechkemmer, Andreas (Germany/USA); Rohracher, Harald (Sweden); Rosenberg, Mark (Canada); Sahakian, Marlyne (Switzerland) [2 reviews]; Sawdon, John (UK); Schauer, Thomas (Germany); Scheffran, Jürgen (Germany) [7 reviews]; Schroeder, Patrick (Germany); Schulz, Christian (Luxembourg); Simon, Dagmar (Germany); Snihur, Yuliya (France); Swilling, Mark (South Africa) Teitelbaum, Michael (USA); Tongsopit, Jiab (Thailand); Truffer, Bernhard (Switzerland); Uexküll, Nina von (Sweden); Verbong, Geert (The Netherlands); Viola, Eduardo (Brazil); Wieczorek, Anna (The Netherlands); Wungaeo, Surichai (Thailand); Young, Oran (USA) [2 reviews].

Finally, the editors would like to thank the photographers for allowing the use of the cover images to illustrate the topics of this volume. The top photograph, illustrating Sustainable Peace, was taken from "Annie's New Letters (& notes)" by Anne Selden Annab, a Palestinian-American poet (at: <http://anniesnewletters.blogspot.de/2012/09/international-day-of-peace-sustainable.html>, posted on the International Day of Peace on 21 September 2012 to show the way to "Sustainable Peace for a Sustainable Future"). Of the three smaller photographs below, the one on the left was taken in Bangkok by Hans Günter Brauch in January 2014, the one in the centre by Úrsula Oswald Spring in Yautepec, Morelós, Mexico, and the one on the right showing schoolchildren in Matabeleland in Zimbabwe by Hans Günter Brauch in September 2015. All photographers granted permission for use.

Mosbach, Germany – Cuernavaca, Mexico,
Amsterdam, The Netherlands – Hamburg, Germany
in February 2016

Hans Günter Brauch
Úrsula Oswald Spring
John Grin
Jürgen Scheffran

Permissions and Credits

The editors are grateful to the following copyright holders, publishers, authors, and photographers who have granted their permission for the use of copyright material.

The top photograph, illustrating Sustainable Peace, was taken from "Annie's New Letters (& notes)" by Anne Selden Annab, a Palestinian-American poet (at: <http://anniesnewletters.blogspot.de/2012/09/international-day-of-peace-sustainable.html>, posted on the International Day of Peace on 21 September 2012 to show the way to "Sustainable Peace for a Sustainable Future"). The photograph on the left was taken in Bangkok by Hans Günter Brauch in January 2014, the one in the centre by Úrsula Oswald Spring in Yautepec, Morelós, Mexico, and the one on the right showing schoolchildren in Matabeleland in Zimbabwe by Hans Günter Brauch in September 2015. All photographers granted the permission to use these photos for the book cover.

In chapter 1 *Hans Günter Brauch* and *Úrsula Oswald Spring* used the following figures, most of which are in the public domain; for the remaining figures permission was obtained.
- **Figure 1.1:** Atmospheric CO_2 at Mauna Loa Observatory. **Source:** National Oceanic and Atmospheric Administration (NOAA)—Monthly Data for the Atmospheric CO_2 from 1958 to December 2015; at: <http://www.esrl.noaa.gov/gmd/ccgg/trends/full.html> (30 January 2016).
- **Figure 1.2:** The PEISOR Model. **Source:** Brauch 2005: 16; 2007: 28; 2009: 76; Brauch/Oswald Spring 2009: 9.
- **Figure 1.3:** Environmental Quartet: DLDD, Climate Change, Water Degradation and Biodiversity Loss. **Source:** The figure was developed by Úrsula Oswald Spring inspired by MA (2005: 17) and designed by Guillermo A. Peimbert, UNAM/CRIM, Mexico.
- **Figure 1.4:** Reported natural disasters by continent (1950-2015). **Source:** D. Guha-Sapir, R. Below, Ph. Hoyois—EM-DAT: The CRED/OFDA International Disaster Database, <www.emdat.be>, Université Catholique de Louvain, Brussels Belgium. The figures were published in: CRED Crunch, No. 30, January 2013: "Natural disasters in Asia" (Brussels: CRED: 1). Permission was granted on 14 March 2015. The authors are grateful to Régina Below for updating the figure from 1950-2010 to 1950-2015.
- **Figure 1.5:** Interlinkages between, land climate and biodiversity. **Source:** UNCCD (2015: 7).
- **Figure 1.6:** Total renewable water resource per cubic meter and capita (2013). **Source:** UN Water (2015:12).
- **Figure 1.7:** Cascading effects of climate change on food security and nutrition. **Source:** FAO (2016a: 4).
- **Figure 1.8:** Share of global wealth of the top 1 per cent and the bottom 99 per cent respectively. **Source:** Oxfam 2016: 2, based on Credit Suisse data available (2000-2014).
- **Figure 1.9:** The Sustainable Development Goals adopted by the UN in 21015: **Source:** UN; at: <http://www.un.org/sustainabledevelopment/sustainable-development-goals/> (23 February 2016).
- **Figure 1.10:** The three major scenarios and six alternative pathways of a sustainability transition. **Source:** Tellus' Gobal Transition Initiative; at: <http://www.greattransition.org/explore/scenarios> (11 February 2015). Permission was granted by Paul Raskin, President of the Tellus Institute.
- **Figure 1.11:** Multilevel perspective on transitions. **Source:** Geels and Schot (2010: 25) adapted from Geels (2002: 1263) and used with the permission of the author.

- **Figure 1.12:** EU GHG emissions towards an 80 per cent domestic reduction (100 per cent =1990). **Source:** EU (2011b: 5).
- **Figure 1.13:** Total annual anthropogenic GHG emissions by groups of gases 1970–2010. **Source:** IPCC (2014a: 9).
- **Table 1.1:** Sectoral reductions. Source: EU (2011b: 6).

In chapter 2 *Simon Dalby* used a table by the Stockholm Resilience Centre that is in the public domain.
- **Figure 2.1:** Planetary Boundaries for a Safe Operating Space for Humanity. Source: Stockholm Resilience Centre: "Azote Images"; at: <http://www.stockholmresilience.org/21/research/research-programmes/planetary-boundaries.html>.

In chapter 4 *John Grin* compiled two boxes for which no permissions were needed.
- **Box 4.1:** Possible transition pathways according to Geels and Schot (2007: 54–76).
- **Box 4.2:** Different transition patters. **Sources:** Rotmans and Loorbach (2010: 137–139) and De Haan and Rotmans (2011).

In chapter 5 *Uwe Schneidewind, Mandy Singer-Brodowski* and *Karoline Augenstein* used one figure from a report of the *German Advisory Council on Global Change* (WBGU).
- **Box 5.1:** Four Transformative Pillars of the Knowledge Society. **Source:** German Advisory Council on Global Change (WBGU),2011: *World in Transition: A Social Contract for Sustainability, Summary for Policy-Makers* (Berlin: WBGU). This figure is licensed under Creative Commons BY-NC-ND; at: <http://www.wbgu.de/fileadmin/ templates/ dateien/veroeffentlichungen/hauptgutachten/jg2011/wbgu_jg2011_kurz_en.pdf>.

In chapter 8 *Úrsula Oswald Spring* developed a figure and table based on her own research.
- **Figure 8.1:** Sustainable-engendered peace with cultural diversity. **Source:** The author.
- **Table 8.2:** Rise and adaptation of patriarchy in different socio-economic contexts. **Source:** The author.

In chapter 9 *Hans Günter Brauch* used one box (9.1) from the IPCC Report that may be used for scientific purposes without permission, for figure 9.1 he obtained the written permission of SIPRI, figure 9.2 is based on a previous publication co-authored by him and published by Springer, figure 9.3 is licensed under creative commons, and figures 9.4–9.6 are based on a report by IRENA that is in the public domain.
- **Box 9.1:** Excerpts from chapter 12 of "Climate change and human security" by the IPCC, WG II. **Source:** IPCC (2014).
- **Figure 9.1:** World military expenditure (1988–2104). **Source:** © SIPRI (2015) used with the permission of SIPRI, 17 November 2015; at: <http://www.sipri.org/research/armaments/milex/milex-graphs-for-data-launch-2015/World%20military%20expenditure%201988-2014.png>.
- **Figure 9.2:** Five Pillars of Peace Ecology and their four linkage concepts of negative, positive, cultural and engendered peace. **Source:** Oswald Spring, Brauch and Tidball (2014: 19).
- **Figure 9.3:** Battle death rate in state-based conflicts by type (1946–2013) by Max Roser. Annual worldwide battle deaths per 100,000 people. **Source:** PRIO battle deaths dataset (1946–2007) and data provided by Steven Pinker for 2009 and later (based on UCDP and PRIO). Licensed under CC.BYSA by Max Roser; at: <http://ourworldindata.org/ data/war-peace/war-and-peace-after-1945/>.
- **Figure 9.4:** Increase in global wind power capacity (2004–2015). **Source:** REN 21 (2015); at: <http://www.ren21.net/status-of-renewables/global-status-report/>. Permission was granted by the REN 21 Secretariat in Paris on 1 February 2016.

- **Figure 9.5:** Increase in global solar PV capacity (2004-2015). **Source:** REN 21 (2015); at: <http://www.ren21.net/ status-of-renewables/global-status-report/>. The permission was granted by the REN 21 Secretariat in Paris on 1 February 2016.
- **Figure 9.6:** Global new investment in renewable power and fuels, developed and developing countries (2004-2014). **Source:** REN 21 (2015); at: <http://www.ren21.net/status-of-renewables/global-status-report/>. The permission was granted by the REN 21 Secretariat in Paris on 1 February 2016.

In chapter 10 *Hania Zlotnik* has derived the following figures based on the online database released by the Populations Division of the United Nations that is in the public domain.

- **Figure 10.1:** Map displaying the high-fertility, intermediate-fertility and low-fertility countries.
- **Figure 10.2:** Trends in total fertility for the three groups of countries considered, 1950-2010. **Source:** Derived from United Nations (2013a).
- **Figure 10.3:** Trends in under-five mortality for the three groups of countries considered, 1950-2010. **Source:** Derived from United Nations (2013a).
- **Figure 10.4:** Components of future population change in high-fertility countries, medium variant, 2010-2100. **Source:** Derived from United Nations (2013b).
- **Figure 10.5:** Components of future population change in intermediate-fertility countries, medium variant, 2010-2100. **Source:** Derived from United Nations (2013b).
- **Figure 10.6:** Components of future population change in low-fertility countries, medium variant, 2010-2100. **Source:** Derived from United Nations (2013b).

The following tables were derived from publications released by the Populations Division of the United Nations, which are in the public domain.

- **Table 10.1:** Distribution of countries and population among the high-fertility, intermediate-fertility and low-fertility groups. **Source:** Derived from United Nations (2003a).
- **Table 10.2:** Current and projected total fertility for different projection variants, 2005-2010, 2045-2050 and 2095-2100 (children per woman). **Source:** Derived from United Nations (2013a).
- **Table 10.3:** Estimated population in 2010 and projected population in 2050 and 2100 according to different projection variants for the three groups of countries considered. **Source:** Derived from United Nations (2013a).
- **Table 10.4:** Population in 2010 and ratio of the projected population in 2050 and 2100 to the population in 2010 for different projection variants and for the three groups of countries considered. **Source:** Derived from table 10.3.
- **Table 10.5:** Median age in 2010 and according to different projection variants in 2050 and 2100 for the three groups of countries considered. **Source:** Derived from United Nations (2013a).
- **Table 10.6:** Support ratio (population aged 15-64 over population under age 15 plus the population aged 65 or over) in 2010 and according to different projection variants in 2050 and 2100. **Source:** Derived from United Nations (2013a).
- **Table 10.7:** Estimates and projections of contraceptive prevalence and unmet need for contraception for the three groups of countries considered, 1970-2030. **Source:** Derived from United Nations (2013c, 2013d).

In chapter 11 *H. J. Schellnhuber, O. M. Serdeczny, S. Adams, C. Köhler, I. M. Otto, C. F. Schleussner* relied on the following figure by the IPCC which is in the public domain for scientific purposes.

- **Figure 11.1:** The IPCC's Working Group II Assessment: *A global perspective on climate-related risks.* **Source:** IPCC Summary for Policy Makers, Working Group II (2014: 13).

- **Figure 11.2:** Multisectoral hot spots of impacts for two (orange) and three (red) sectors, where at least fifty per cent of climate and impact model combinations agree on the threshold crossing in each sector for a global mean temperature change of up to 4.5°C above 1980–2010. Regions in light grey are regions where no multisectoral overlap is possible at all because of data and assessment restrictions. Dark grey shading shows areas where at least ten per cent of climate and impact model combinations agree on the given threshold crossing. **Source:** Piontek/Müller/Pugh et al. (2013).
- **Figure 11.3:** Schematic mapping of tipping elements of the Earth system and populations density. **Source:** Schellnhuber, pers. comm., after Lenton, Held, Kriegler et al. (2008).

In chapter 12 *Tobias Ide, P. Michael Link, Jürgen Scheffran, Janpeter Schilling* developed their own previously published table further and designed a new map.
- **Figure 12.1:** Overview of the case study areas. **Source:** The authors.
- **Table 12.1:** Central findings of large-N studies on possible pathways connecting climate change to violent conflict. **Source:** Expanded from Ide/Scheffran 2014).

In chapter 13 *Jürgen Scheffran* used 17 figures he had previously developed, partly with co-authors, and adapted them for this chapter.
- **Figure 13.1:** Impact chains and feedbacks in climate-society interaction. **Source:** Adapted and modified from Scheffran, Brzoska, Kominek et al. (2012a: 870).
- **Figure 13.2:** Hot spot regions where climate change impacts are critical for human security and social stability in different regions of the world. **Source:** Scheffran and Battaglini (2011): 33.
- **Figure 13.3:** Micro–macro interaction between local and global levels. **Source:** Scheffran (2015a).
- **Figure 13.4:** The VIABLE adaptive model framework. **Source:** Scheffran, Link and Schilling (2012).
- **Figure 13.5:** Competing transition plans to end nuclear deterrence in the 1980s: cost projections for Reagan's SDI plan (upper part) vs. Gorbachev's plan for elimination of nuclear weapons. **Source:** Adapted from: (top) Pike (1985: 11); (bottom) Spiegel (1986: 98).
- **Figure 13.6:** Simulation of the transition from negative (worst case $w = -1$) to positive perceptions (best case $w = +1$) of the SCX arms race model, with a chaos-like oscillation around $w = 0$. Shown are the security function (upper part) and cost functions for the fluctuating scenario (total and variable cost, lower part). **Source:** Scheffran (1989:245).
- **Figure 13.7:** Tipping point in global nuclear arsenals. **Source:** Scheffran (2015b) based on data from Bulletin of the Atomic Scientists.
- **Figure 13.8:** Security bifurcation diagram of the SCX model for the anti-symmetric case of two agents (here agent 1 is shown). **Source:** Jathe and Scheffran (1992).
- **Figure 13.9:** Dichotomy between winners and losers for a large number of agents. **Source:** Scheffran (2003).
- **Figure 13.10:** Framework for transitions between cycle of violence and cycle of cooperation between two agents. **Source:** Scheffran, Ide and Schilling (2014).
- **Figure 13.11:** Temperature change for parameter variations of: (a) unit reduction cost; (b) max reduction cost; (c) climate sensitivity; (d) initial emission trend. **Source:** Scheffran and Hannon (2007).
- **Figure 13.12:** Simulation of simplified SCX-model dynamics for population, energy consumption, atmospheric CO_2 concentration, gross national product (GNP), risk/GNP and GNP/capita for two actors over fifty years (North: solid line, South: dotted line) *Upper simulation*: Catastrophe scenario if the 1990 situation continues, for high prosperity goals; lower *simulation*: Stabilization scenario for smaller population growth, lower wealth goals, lower emissions, higher efficiency and higher costs per energy unit. **Source:** Scheffran (1999).

- **Figure 13.13:** Range of variables (CO_2 emissions, per capita production, atmospheric concentration, temperature change) over 100 years for low and high emission energy paths. **Source:** Modified from Scheffran (2008b).
- **Figure 13.14:** Value difference between high and low emission energy technology with variation of damage factor (left) and carbon price (right). **Source:** Modified from Scheffran (2008b).
- **Figure 13.15:** Interplay between energy providers and customers for different energy paths (marked here by *A* and *B*). **Source:** Adapted and modified from Scheffran (2006).
- **Figure 13.16:** Simulation of adaptive coalition formation with two energy paths for two energy providers (coalitions), represented by C1 and C2 and six customer A1 to A6 using energy to produce value **Source:** Scheffran (2006).
- **Figure 13.17:** Future climate pathways, tipping points and cascades between conflict and cooperation. **Source:** Modified from Scheffran (2015a).

In chapter 15 *Eckart Ehlers* used material from the IHDP and by Bohle published by the IHDP that is in the public domain in:
- **Figure 15.1:** The IHDP programme: organizational structure, projects and partner institutions. **Source:** IHDP Update 1/1997.
- **Figure 15.2:** Human security as a desirable goal for individuals and societies. **Source:** Bohle (2002).
- **Figure 15.3:** Institutions and their effects on global environmental change/The primary focus of the IDGEC project. **Source:** IDGEC Science Plan.
- **Figure 15.4:** Industrial Transformation: Interactions within the human–environment system with respect to the global carbon cycle. **Source:** IHDP Update 3/2000.
- **Figure 15.5:** Bohle's Conceptual Framework for Vulnerability Analysis. **Source:** Bohle (2001).
- **Figure 15.6:** IHDP's new governance structure. **Source:** IHDP's Strategic Plan 2007-2015, p. 30.

In chapter 16 *Marit Sjovaag* used material from the publications on *The Limits of Growth* with the permission of Jørgen Randers, who was a co-author of all studies.
- **Figure 16.1:** Causal diagram of the World3 model. **Source:** Jørgen Randers presentation material; he also granted permission to use this figure.
- **Figure 16.2:** Causal diagram for the 2052 study—changes from the World Model marked in red. **Source:** Jørgen Randers presentation material; he also granted permission to use this figure.
- **Figure 16.3:** World State of Affairs, 1970-2050. **Source:** Randers (2012: 232). Permission to use this figure was granted by the author.

In chapter 17 *Mark Swilling* used previously published figures and a table of which he is a co-author; material by UNEP's International Resource Panel of which he has been a member and which are in the public domain; and material by close colleagues who granted their permission.
- **Figure 17.1:** Global metabolic rates and incomes (1900-2005), and income. **Source:** Fischer-Kowalski and Swilling (2011: 12).
- **Figure 17.2:** Composite resource price index (at constant prices, 1900-2000). **Source:** United States Geological Survey data cited in: Fischer-Kowalski and Swilling (2011: 13).
- **Figure 17.3:** Commodity price indices. **Source:** Fischer-Kowalski and Swilling (2011: 13).
- **Figure 17.4:** Major contributors to global GHG emissions, including land use and land cover change (measures in CO_2 eqivalents using a 100-year global warming potential). **Source:** UNEP (2010).

- **Figure 17.5:** Distribution of energy use across consumption categories, as identified in different studies, and total energy use measures in kW per capita. **Source:** UNEP (2010).
- **Figure 17.6:** Carbon footprint of different consumption categories (tonnes of CO_2 equivalents per capita in 2001) in 87 countries/regions as a function of expenditure ($ per capita). **Source:** UNEP (2010).
- **Figure 17.7:** Normalized global warming potential of material flows and *Environmentally Weighted Material Consumption* (EMC) for EU-27+1 region. **Source:** UNEP (2010).
- **Figure 17.8:** World Population and Urban Growth Trends (0–2010). **Source:** Angel (2012).
- **Figure 17.9:** Conceptual Framework: Food Systems and Natural Resources. **Source:** Hajer, Westhoek, Ozay et al. (2015): 12.
- **Figure 17.10:** Physical trade according to material composition (1980–2010). **Source:** Fischer-Kowalski, Dittrich, Eisenmenger et al. (2015).
- **Figure 17.11:** Largest net exporters and importers by material composition of net trade in 2010. **Source:** Fischer-Kowalski, Dittrich, Eisenmenger et al. (2015).
- **Figure 17.12:** Growth in production of minerals, 1845–2010. **Source:** Mudd (2009), cited in International Resource Panel Working Group on Global Metal Flows (2013).
- **Figure 17.13:** Computer Chip Elemental Contents. **Source:** Quoted in a Powerpoint presentation by Tom Graedel, November 2013, Stellenbosch University.
- **Figure 17.14:** Agricultural inputs relative to crop yields (1961–2009). **Source:** FAO data compiled by Schutz, cited in Bringezu, Schutz, Pengue et al. (2014).
- **Figure 17.15:** Cereal yields by selected world regions (1961–2011). **Source:** FAO data cited in Bringezu, Schutz, Pengue et al. (2014).
- **Figure 17.16:** High and low estimates of net and gross expansion of agricultural land, 2005–2050. **Source:** Bringezu, Schutz, Pengue et al. 2014—note: for detailed references to the sources of data cited here, see original diagram in this report.
- **Figure 17.17:** Number of people living in water-stressed areas in 2030, by country type. **Source:** UNEP (2011), cited in Urama (2015).
- **Table 17.1:** Increases in Annual Water Demand, 2005– 2030. **Source:** McKinsey (2009), cited in Urama (2015).

In chapter 18 *Czeslaw Mesjasz* developed a figure and several tables based on his own research.
- **Figure 18.1:** Areas and elements of sustainability description and analysis (three/four pillars). **Source:** Own research based upon Adams (2006) and Goodland.
- **Table 18.1:** General definitions of 'sustainability'. **Source:** The author's own research.
- **Table 18.2:** General definitions of 'sustainable development'. **Source:** The author's own research.
- **Table 18.3:** Levels of analysis of sustainability. **Source:** Riley (1992).
- **Table 18.4:** Definitions of sustainability in specific socio-economic domains. **Source:** The author's own research.

In chapter 20 *Mike Hodson, Simon Marvin and Philipp Späth* designed a figure based on their own research:
- **Figure 20.1:** Scalar interrelationships in transitions activities and chains of intermediary spaces. **Source:** The authors.

In chapter 23 *Juan Antonio Le Clercq* developed the following figures and tables based on Mexican sources that are in the public domain:

- **Figure 23.1:** Legislative process concerning climate change in Mexico (1992-2012). **Source:** Information from the Chamber of Deputies, the Senate and the *Legislative Information System* (SIL); at: <diputados.gob.mx/>; <senado.gob.mx/>; and <sil.gobernacion.gob.mx/>.
- **Figure 23.2:** Climate change bills by topic (2006-2012). **Source:** Information from the Chamber of Deputies, the Senate and the *Legislative Information System* (SIL); at: <diputados.gob.mx/>; <senado.gob.mx/>; and <sil. gobernacion.gob.mx/>.
- **Figure 23.3:** Climate change legislative policy calls by topic (1992-2012). **Source:** Based on information from the Chamber of Deputies, the Senate and the *Legislative Information System* (SIL); at: <diputados.gob.mx/>; <senado.gob.mx/>; and <sil.gobernacion.gob.mx/>.
- **Figure 23.4:** Mexican climate policy (2006-2012). **Source:** Information presented in the 4th and 5th National Communications to the UNFCCC (INE: National Institute of Ecology; CICC: Interministerial Commission on Climate Change).
- **Figure 23.5:** Proposed reform for climate regulation in environmental law. **Source:** Information from the legislative project to reform the environmental law (LEEGEPA) approved by Congress (25 April 2012); CICC: Interministerial Commission on Climate Change).
- **Figure 23.6:** Climate change law structure. **Source:** Ley General de Cambio Climático (NSCC: National Strategy on Climate Change; INECC: National Ecology and Climate Change Institute; CICC. Interministerial Commission on Climate Change; C3: Advisory Council on Climate Change).
- **Table 23.1:** Climate policy instruments, objectives and goals (2006-2012). **Source:** Information from PND (2007-2012), ENACC (2007) and PECC (2009).
- **Table 23.2:** Alternative models in the legislative discussion on climate change 2010-2012: **Source:** Information from the National Communications and Legislative Information System (SIL); at: <sil.gobernacion.gob.mx/>.

In chapter 24 *Audley Genus* added a figure and table he developed himself.
- **Figure 24.1:** A discourse-institutional analytical framework. **Source:** The author.
- **Table 24.1:** Discursive legitimation strategies underpinning institutional pillars (rules). **Source:** The author.

In chapter 25 *Jan Jonker and Linda O'Riardon* designed three figures based on their own sources or inspired by selected cited texts by colleagues:
- **Figure 25.1:** The value creation logic behind conventional business models. **Source:** The authors.
- **Figure 25.2:** Process NBMs. **Source:** The authors, based on Shafer, Smith and Linder (2005) and Simanis and Hart (2009).
- **Figure 25.3:** Value creation process NBMs. **Source:** The authors, based on Simanis and Hart (2009).

In chapter 26 *Sylvia Lorek* added a box based on her own research. For the three figures and two tables she obtained the permission of the copyright holders.
- **Box 26.1:** Green Growth and De-growth. **Source:** The author.
- **Figure 26.1:** The Environmental Space Concept. **Source:** Spangenberg (1995).
- **Figure 26.2:** Possible pathways to sustainable consumption from overconsumption. **Source:** Brunner and Urenje (2012).
- **Figure 26.3:** Possible pathways to sustainable consumption from under-consumption. **Source:** Brunner and Urenje (2012).
- **Table 26.1:** Sustainable World Population at Different Consumption Levels. **Source:** Assadourian (2010).

- **Table 26.2:** Examples of enabling mechanisms for sustainable consumption. **Source:** Lebel/Lorek 2010.

In chapter 27 *Hongmin Chen* summarized her research in three boxes, developed three figures and added two tables based on Chinese official and scientific sources that are partly in the public domain.

- **Box 27.1:** Top 1000 Enterprises Energy Conservation Programme. **Source:** NDRC (2006a, 2011).
- **Box 27.2:** China's Carbon Emissions Trading Pilots. **Source:** Zheng (2014); Wilkening, Kachi (2014); Wu, Shi, Xia et al. (2014).
- **Box 27.3:** Project "Energy-saving Products Benefiting the Public". **Source:** MFPRC (2013).
- **Figure 27.1:** MSW treatment by type in China (2004-2010). **Source:** NBS (2014).
- **Figure 27.2:** Household Waste Sorting at Different Stages in Shanghai. **Source:** Huang (2014).
- **Figure 27.3:** Regional Distribution of Per Capita GDP in mainland China in 2013 (104 US dollars). **Source:** NBS (2014).
- **Table 27.1:** Development of Seven Pilot Carbon Trading Schemes in China. **Source:** Zheng Shuang (2014); International Carbon Action Partnership Region; at: <https://icapcarbonaction.com>.
- **Table 27.2:** Development of three types of NGOs in China. **Source:** Website of the Ministry of Civil Affairs, <www. mca.gov.cn>.

In chapter 28 *Philipp Schepelmann, René Kemp and Uwe Schneidewind* developed one figure (28.8) based on their own research and added two maps (figure 28.1) and six photos (figures 28.2-28.7) that are licensed under Wikimedia Commons and based on public sources.

- **Figure 28.1:** Germany, the State of North Rhine-Westphalia and the Ruhr Area. **Sources:** Locator map of the Ruhr District, Germany (TUBS Own work). Licensed under Creative Commons Attribution-ShareAlike 3.0 via Wikimedia Commons; Ruhr District (Ruhr) with all cities with more than 50,000 inhabitants (Threedots (Daniel Ullrich))—Own work. Licensed Regionalverband under Creative Commons Attribution-ShareAlike 3.0 via Wikimedia Commons.
- **Figure 28.2:** View of the city of Gelsenkirchen, to the right the former coal mine, today centre of the Nordstern Parks—view from a former slag heap in the city of Essen. **Source:** Hans-Jürgen Wiese—Own work. Licensed under Creative Commons Attribution-ShareAlike 3.0 via Wikimedia Commons.
- **Figure 28.3:** River Emscher in the Ruhr district near Herne and Herten. **Source:** Arnoldius—Own work. Licensed under Creative Commons Attribution-ShareAlike 2.5 via Wikimedia Commons.
- **Figure 28.4:** Wastewater treatment plant Deusen at the river Emscher. **Source:** Tbachner—Own work. Licensed under Creative Commons Attribution-ShareAlike 3.0 via Wikimedia Commons.
- **Figure 28.5:** Heritage of the IBA Emscher Park at night: Furnace No. 5 at the Landscape Park Duisburg North. **Source:** Tuxyso—own work. Licensed under Creative Commons Attribution-ShareAlike 2.5 via Wikimedia Common.
- **Figure 28.6:** The river Emscher beside the Landscape Park Duisburg. **Source:** DerHexer—own work. Licensed under Creative Commons Attribution-ShareAlike 3.0 via Wikimedia Commons.
- **Figure 28.7:** UNESCO World Heritage Site Zeche Zollverein in Essen. **Source:** Thomas Wolf—own work. Licensed under Creative Commons Attribution-ShareAlike 3.0 via Wikimedia Commons.

- **Figure 28.8:** Main phases and elements of NRW sustainability transition. **Source:** The authors.

In chapter 29 *Belinda Yuen* and *Asfaw Kumssa* summarized their own research in three boxes (29.1-29.3), added six of their own photos (figure 29.1-29.6) for which they own the copyright, and in table 29.1 used data from the United Nations Population Division that are in the public domain.
- **Box 29.1:** Cairo: Africa's Megacity. **Source:** The author.
- **Box 29.2:** Africa's Path to Green Growth. **Source:** The author.
- **Box 29.3:** National Strategy for Green Growth, South Korea. **Source:** The author.
- **Figure 29.1:** A hotel in the slums of Nairobi, Kenya. **Source:** Authors.
- **Figure 29.2:** Rising traffic congestion in Nairobi city, Kenya. **Source:** Authors.
- **Figure 29.3:** Squatter settlement in Metro Manila, the Philippines. **Source:** Authors.
- **Figure 29.4:** Lack of proper solid waste disposal creates an environmental problem in Jakarta, Indonesia. **Source:** Authors.
- **Figure 29.5:** Salt pan barren land being transformed into the Sino-Singapore Tianjin eco-city. **Source:** Authors.
- **Figure 29.6:** Turning waste to recreation land at Semakau, Singapore. **Source:** Authors.
- **Table 29.1:** Urbanization Levels, 1950-2050. **Source:** United Nations, 2012: *World Urbanization Prospects: The 2011 Revision* (New York: United Nations Population Division, Department of Economic and Social Affairs).

In chapter 30 *Gregory Trencher* and *Xuemei Bai* summarized their own research in one box (30.1) and in three tables (tables 30.1-30.3) and designed seven figures (figures 30.1-30.7) based on their own findings.
- **Box 30.1:** Method. **Source:** The authors.
- **Figure 30.1:** Urban systems targeted (n = 15 cases). **Source:** The authors.
- **Figure 30.2:** Types of internal actors (n = 15 cases). **Source:** The authors.
- **Figure 30.3:** Types of external actors (n = 15 cases). **Source:** The authors.
- **Figure 30.4:** Number of external actor types involved (n = 15 cases). **Source:** The authors.
- **Figure 30.5:** Factors motivating partnership formation (n = 15 cases). **Source:** The authors.
- **Figure 30.6:** Mechanisms (n = 15 cases). **Source:** The authors.
- **Figure 30.7:** Barriers and negative factors (n = 13 cases). **Source:** The authors.
- **Table 30.1:** List of partnerships (n = 15). **Source:** The authors.
- **Table 30.2:** Summary of key roles, motivations and potential barriers. **Source:** The authors.
- **Table 30.3:** Overview of Urban Reformation Programme for the Realization of a Bright Low-Carbon Society. **Source:** The authors.

In chapter 31 *Cecilia Tortajada* and *Martin Keulertz* used one figure based on sources from NASA, and in two tables summarized data provided by international organizations that are in the public domain or may be used for scientific purposes without permission.
- **Figure 31.1:** Satellite-based estimates of groundwater depletion in India. **Source:** I. Velicogna, University of California, Irvine; at: <http://www.nasa.gov/topics/earth/features/india_water.html>.
- **Table 31.1:** Regional production and consumption of biofuels, 2011. **Source:** Adapted from HLPE (2013). This information is originally from the US Energy Information Administration/International Energy Statistics, available at: <http://www.eia.gov/>.
- **Table 31.2:** Global water use for energy production in the New Policies Scenario by region (bcm). **Source:** OECD/IEA (2012).

In chapter 32 *Úrsula Oswald Spring* summarized in two tables (32.1-32.2) a survey that was conducted by CRIM/UNAM under her guidance; developed twelve figures based on her own research and on official Mexican sources (figures 32.1-32-3, 32.6-32.13); adapted a figure by Bohle (figure 32.4) and used a photograph taken by Civil Protection, Yautepec that is in the public domain.

- **Figure 32.1:** Location of the study area in the central part of Mexico. **Source:** The author.
- **Figure 32.2:** Location of the study area with municipalities. **Source:** The author.
- **Figure 32.3:** Complexity of the river basin. **Source:** The author.
- **Figure 32.4:** Dual vulnerability. **Source:** Adapted by the author from Bohle (2002).
- **Figure 32.5:** Flood over bridge in the centre of Yautepec. **Source:** Photograph taken by Civil Protection, Yautepec, during the 2010 flood.
- **Figure 32.6:** Part of the Didactic Map of the Yautepec River Basin. **Source:** The author.
- **Figure 32.7:** Multiple risks in the YRB. **Source:** Survey (CRIM/UNAM 2011-2013).
- **Figure 32.8:** Main disasters occurring in the YRB. **Source:** Survey (CRIM/UNAM 2011-2013).
- **Figure 32.9:** Environmental destruction in the YRB. **Source:** Survey (CRIM/UNAM 2011-2013).
- **Figure 32.10:** Dealing with increasing risks in the YRB. **Source:** Survey (CRIM/UNAM 2011-2013).
- **Figure 32.11:** Identity in the YRB (two possible answers). **Source:** Survey (CRIM/UNAM 2011-2013).
- **Figure 32.12:** Income distribution of a peasant household. **Source:** Oswald Spring/Serrano Oswald/Flores Palacio et al. (2014: 343), based on data from SIAP [Statistics from the Ministry of Agriculture].
- **Figure 32.13:** Arenas, agendas, actors and activities of vulnerability and livelihood. **Source:** The author.
- **Table 32.1:** Representative survey in the Yautepec River Basin (YRB). **Source:** Field research (CRIM/UNAM 2010-2013).
- **Table 32.2:** Land use changes in the YRB. **Source:** Field research (CRIM/UNAM 2010-2013).

In chapter 33 *Monica de Andrade* summarized her own research in three tables (33.1-33.3).

- **Table 33.1:** Characterization of sustainability transitions in Brazil: structural changes at local level (landscape level). HD = Hydroelectricity. **Source:** The author.
- **Table 33.2:** Timeline of main policy interventions in Brazil and their impacts on health and the ecosystem, together with their social and economic aspects. **Source:** The author.
- **Table 33.3:** Transition of heath indicators in Brazil. **Source:** The author.

In chapter 34 *Jürgen Scheffran* and *Rebecca Froese* used the following boxes, figures and chapters based on the sources provided below for which no permission was needed. For Figure 34.1 and Table 34.2 permission was granted by Elsevier.

- **Box 34.1:** Aspects of technology transfer and learning in low-carbon development. **Source:** Adapted and condensed from: Lema, Iizuka and Walz (2015).
- **Box 34.2:** Public-Private Partnership for Electricity from Peanut Shells in Senegal. **Source:** Assembled by the authors.
- **Box 34.3:** China-Zambia South-South Cooperation on Renewable Energy Technology Transfer. **Source:** Assembled by the authors.
- **Box 34.4:** Mechanisms for Technology Transfer. **Source:** Assembled from <http://unfccc.int/ttclear/templates/render_cms_page?TTF_home>.
- **Box 34.5:** Solar Home Systems for Bangladesh. **Source:** Assembled by the authors.

- **Box 34.6:** Barriers to technology transfer and investment. **Source:** Based on Santarius/Scheffran/Tricarico 2012.
- **Box 34.7:** Clean Power from the Desert—Concentrated Solar Power (CSP). **Source:** Assembled by the authors.
- **Figure 34.1:** The process of technology transfer. **Source:** Adapted from Karakosta, Doukas, Psarras (2010), with the permission of Elsevier.
- **Figure 34.2:** Innovation Chain. **Source:** Adapted and modified from Grubb (2004) and Tomlinson, Zorlu, Langsley (2008).
- **Figure 34.3:** Global New Investment in Renewable Power and Fuels, Developed and Developing Countries, 2004-2013. **Source:** Adapted from REN 21 (2014).
- **Figure 34.4:** Global New Investment in Renewable Energy by Technology, Developed and Developing Countries, 2013. **Source:** Adapted from REN 21 (2014).
- **Figure 34.5:** FDI inflows, global and by group of economies, 1995-2013 and projections, 2014-2016 in billions of dollars. **Source:** Adapted from UNCTAD FDI-TNC-GVC Information System, FDI/TNC database; at: <www.unctad.org/fdistatistics> (WIR 2014).
- **Figure 34.6:** Estimates of financing for mitigation technologies. **Source:** Adapted from UNFCCC (2009).
- **Table 34.1:** Gaps between actual and required climate finance. **Source:** Authors' representation, data from EC (2015).
- **Table 34.2:** Relative importance of particular types of financial flows to technology transfer pathways for different actors. **Source:** Adapted from Karakosta, Doukas, Psarras (2010), with the permission of Elsevier.
- **Table 34.3:** Current investments, investment needs and gaps, and private sector participation in key SDG sectors in developing countries. **Source:** Adapted from WIR 2014; at: <www.unctad.org/fdistatistics>.
- **Table 34.4:** Summary of actions and incentives for the public and private sectors in the developed and developing countries. **Source:** Adapted from Santarius, Scheffran, Tricarico (2012: 50).

In chapter 35 *Karlson 'Charlie' Hargroves* summarized his joint research with colleagues in table 35.1, designed five figures (35.1-35.5) based on his own research (35.3) and joint research with colleagues (35.1-35.2, 35.4), and adapted a methodology used by colleagues (35.5) for the purpose of his own research.

- **Figure 35.1:** Waves of Innovation. **Source:** Hargroves/Smith (2005).
- **Figure 35.2:** Gross Domestic Product vs. Estimated Environmental Costs (billions) for the United States of America from 1950 to 2004. **Sources:** Smith, Hargroves and Desha (2010), with data reinterpreted by K. Hargroves from Talberth, Cobb and Slattery (2006).
- **Figure 35.3:** Schematic of Key Elements of Carbon Structural Adjustment. **Source:** Hargroves (2014).
- **Figure 35.4:** Whole of Society Approach to Sustainable Development. **Source:** Hargroves and Smith (2005).
- **Figure 35.5:** A method for the prioritization of efforts based on the likely impact on greenhouse gas emissions and the likely willingness to adjust in the area of focus. **Source:** Adapted by K. Hargroves from an adaptation by K. Hargroves and C. Desha of the CBSM Methodology (McKenzie-Mohr/Smith 1999).
- **Table 35.1:** Examples of cost-effective reductions in carbon intensity in various sectors that deliver strong economic multipliers. **Source:** von Weizsäcker, Hargroves, Smith, Desha and Stasinopoulos (2009).

In chapter 36 *Britta Rennkamp* and *Radhika Perrot* summarized their research in two tables (36.1-36.2).

- **Table 36.1:** Overview of the framework for analysis of Wind Energy Technology Drivers and Barriers. **Source:** The authors based on Bergek, Jacobsson, Carlsson et al. (2008).
- **Table 36.2:** Synthesis of the results. **Source:** The authors.

In chapter 37 *Lucy Baker* used a table (37.1) for which written permission was granted by the copyright holders; for the figure (37.1) she used data which are in the public domain.
- **Figure 37.1:** Total envisaged capacity 2030 under policy-adjusted IRP 2010. **Source:** Author's own elaboration using data available from DoE (2011a:17).
- **Table 37.1:** The DoE technical task team. **Source:** McDaid, Austin and Bragg (2010: 5). Permission was granted.

In chapter 38 *Eduardo Viola* and *Larissa Basso* summarized official data that are in the public domain in two tables (38.1-38.2).
- **Table 38.1:** Domestic energy production and domestic energy supply by source 2003-2012 (103 tep (toe)). **Source:** Author's elaboration based on: *Balanço Energético Nacional–séries completas* 1970-2012; at: <https:// ben.epe.gov.br/BENSeriesCompletas. aspx> (12 January 2014).
- **Table 38.2:** UHEs in the Amazon region (Amazon basin and Tocantins-Araguaia basin). **Source:** Author's elaboration, based on data from *Agência Nacional de Energia Elétrica–Banco de Informações sobre Geração*; at: <http://www.aneel.gov.br/aplicacoes/capacidadebrasil/GeracaoTipoFase.asp?tipo=1&fase=3>, and *Programa de Aceleração do Crescimento–PAC*; at: <http://www.pac.gov.br/energia/geracao-de-energia-eletrica/br/30> (26 July 2014). PCHs and UHEs under study or not found on any of the government websites above are excluded from the table.

In chapter 39 *Carl Middleton* compiled two tables adapted from a publication by a British colleague with his permission (39.1-39.2).
- **Table 39.1:** Civil society activity in relation to sustainable electricity transitions. **Source:** The author, adapted from Smith (2012: 189).
- **Table 39.2:** Evolution of Thailand's electricity regime. **Source:** Table structure adapted from Smith (2012: 193-195).

In chapter 40 *Roeland J. in 't Veld* developed two figures to illustrate his own research.
- **Figure 40.1:** The emergence of the knowledge democracy concept. **Source:** The author.
- **Figure 40.2:** Old and new forms coexist and influence each other. **Source:** The author.

In chapter 41 *Sander Happaerts* summarized his own research findings in two boxes (41.1-41.2).
- **Box 41.1:** Transitions in the European Commission's Roadmaps. **Source:** The author.
- **Box 41.2:** Resource efficiency and Europe 2020: the back story. **Source:** The author.

In chapter 42 *Úrsula Oswald Spring, Hans Günter Brauch,* and *Jürgen Scheffran* included the following figures:
- **Figure 42.1:** System approach to natural and anthropogenic factors of global environmental change. **Source:** Oswald Spring (2016), based on data from Rockström, Steffen, Noone et al. (2009).
- **Figure 42.2:** Transition to sustainability. **Source:** Oswald Spring (2016).
- **Figure 42.3:** From diagnosis to remedy: Sustainability Transition and Sustainable Peace: **Source:** Hans Günter Brauch (2016), with many suggestions from Úrsula Oswald Spring.

- **Figure 42.4:** Renewable Energy and Jobs. Renewable Energy Employment by Technology. **Source:** IRENA: *Renewable Energy and Jobs-Annual Review* 2016; at: <http://www.irena.org/DocumentDownloads/Publications/IRENA_RE_Jobs_Annual_Review_2016.pdf.>.
- **Figure 42.5:** Renewable electricity generation in the US by fuel type in the reference case, 2000-2040 (billion kilowatt hours). **Source:** EIA (2015: 41).
- **Figure 42.6:** Matrix of lifestyle differences in the four scenarios: **Source:** SPREAD [Leppänen, Juha; Neuvonen, Aleksi; Ritola, Maria; Ahola, Inka; Hirvonen, Sini; Hyötyläinen, Mika; Kaskinen; Tuuli; Kauppinen, Tommi; Kuittinen, Outi; Kärki, Kaisa; Lettenmeier, Michael; Mokka, Roope, all from Demos Helsinki] (2013): 56-57.
- **Figure 42.7:** How to spread sustainable lifestyles. **Source:** SPREAD: *Scenarios for Sustainable Lifestyles-From Global Champions to Local Loops* (Wuppertal: 2013), based on Geels (2002).

Part I   Moving towards Sustainability Transition

1 Sustainability Transition and Sustainable Peace: Scientific and Policy Context, Scientific Concepts and Dimensions

Hans Günter Brauch[1] and Úrsula Oswald Spring[2]

Abstract

This Handbook links together four social science research programmes—peace studies, security studies, development studies and environmental studies—which have had only limited exchanges on sustainable development, human security and sustainable peace. The Handbook connects these three concepts within the *research paradigm* of 'sustainability transition'. This research paradigm focuses on a large-scale and long-term transformative change of the dominant carbon-intensive development path by addressing the causes of global environmental and climate change. There has been an exponential increase in GHG emissions since the 1950s and a rapid destruction of biodiversity and ecosystem services. These texts can be used in graduate seminars in different scientific disciplines and research programmes and in new transdisciplinary degree programmes. The texts foster longer term proactive strategies and policies and specific measures to realize two policy goals, 'sustainable development' alongside and contributing to a 'sustainable peace', as the possible result of a large-scale transition of the systems of production, consumption, and governance.

Among the key questions in this Handbook are a) whether *business-as-usual* policies and the growing number of climate-induced natural hazards that threaten the survival of millions of people pose threats to international peace and security; b) whether anticipative learning and a forward-looking discourse on long-term transformative changes may contribute to sustainable development and address new dangers to international peace and security in a preventive manner; and c) what lessons may be drawn from the violent consequences of the industrial revolution and used to promote a long-term transformative change towards sustainable development with sustainable peace.

This chapter consists of eight parts. After a brief sketch of opposing scientific and political visions (1.1.), the purpose and objectives of the Handbook are highlighted (1.2.) and a survey (1.3) reviews the challenges posed by global environmental change: population growth, the impacts of climate change, loss of biodiversity, soil erosion and desertification, water scarcity and stress, food scarcity and hunger, and gender implications. It addresses the impacts of different economic development paths (1.4), through integrating the results of global research programmes, of their linkages and their assessment by the IPCC, and through the nexus debates between the fields of water, food and energy security (1.5). The three key concepts of sustainable development, sustainability transition and sustainable peace are introduced (1.6), the evolution of different approaches to sustainability transition is reviewed, the debates on ecosystem restoration, green growth and decarbonization are noted, and six dimensions of the research on 'sustainability transition' are outlined (1.7). The last section introduces the ten parts of this Handbook and offers an overview of its 40 peer reviewed contributions (1.8).

Keywords: Anthropocene, biodiversity, business-as-usual, climate change, cornucopian perspectives, decarbonization, development paths, food, gender, global environmental change, human security, hunger, soil, sustainable development, sustainable peace, sustainability transition, water.

1.1 Introduction and Overview of the Chapter

This *Handbook on Sustainability Transition and Sustainable Peace* follows the *Global Environmental and Human Security Handbook for the Anthropocene*, which focused on the reconceptualization of security (Brauch/Oswald Spring/Grin et al. 2009; Brauch/Oswald Spring/Mesjasz et al. 2008, 2011). In the latter's concluding chapter Oswald Spring and Brauch

© Springer International Publishing Switzerland 2016
H.G. Brauch et al. (eds.), *Handbook on Sustainability Transition and Sustainable Peace*,
Hexagon Series on Human and Environmental Security and Peace 10, DOI 10.1007/978-3-319-43884-9_1

(2011) argued that in the Anthropocene humankind faces two alternative visions and policy strategies:

- *Business-as-usual* (BAU) in a Hobbesian world. Here economic and strategic interests and actions dominate and may lead to a major crisis for humankind, inter-state relations and nature.
- The need for a *transformation* in cultural, environmental, economic and political relations.

Scheffran, Brzoska, Brauch et al. (2012) examined the possible consequences of the *first* alternative and showed, by addressing climate change as a 'threat multiplier', that in the case of no action it might lead to "dangerous climate change" (UNFCCC 1992; Schellnhuber/Cramer/Nakicenovic et al. 2006). This volume deals with the second alternative: 'sustainability transition' that may serve as a sustainable alternative. It would avoid the negative consequences of climate change for human, national and international security (Brauch 2014). Both visions address different coping strategies for this century for *global environmental change* (GEC) and climate change:

- In the first vision, *cornucopian perspectives* or *business-as-usual* suggest technical fixes and the defence of economic, strategic and national interests, with the adaptation and mitigation strategies that are affordable for industrialized countries.
- In the alternative vision of a comprehensive transformation of the global economy, *Politik* (as politics, policy and polity), economics, society and culture, a *sustainable perspective* requires effective new strategies and policies. Their goal should be decarbonization, dematerialization, reduction of the water and environmental footprint, and global cooperation and solidarity. These would contrib-

ute to a sustainable peace with more global equity and social justice.

The consequences of both scientific visions and policy perspectives are:

- The first vision—with minimal reactive adaptation and mitigation strategies—would increase the probability of dangerous global changes in the environment, water, food and climate (IPCC 2013/2014), and there would be linear and chaotic changes in the earth system.
- The sustainability perspective requires a change in *culture* (thinking on the human–nature interface), *world views* (thinking on systems of rule, e.g. democracy vs autocracy, on domestic priorities and policies, and on inter-state relations in the world), *mindsets* (the strategic perspectives of policymakers), and new forms of national and global sustainable *governance* (Oswald Spring/Brauch 2011).

This alternative vision addresses the need for a "transition to [a] much more sustainable global society" (Raskin/Banuri/Gallopin et al. 2002), aimed at peace, freedom, cooperation, care, material well-being and environmental health. Changes in technology, financing and management systems alone will not be sufficient, but "significant changes in governance, institutions and value systems" are needed, resulting in a new major transformation after "the stone age, early civilization and the modern era". These alternative strategies should be "more integrated, more long-term in outlook, more attuned to the natural dynamics of the Earth System and more visionary" (Steffen/Sanderson/Tyson et al. 2004: 291–293).

Taking up this second vision and coping strategy, this Handbook tries to bridge different research programmes in the social sciences, especially environmental studies and ecology, but also sociotechnological perspectives (Grin/Rotmans/Schot 2010), peace and gender studies. It addresses the following questions:

- What are the possible conceptual relationships between a manifold *process* of 'sustainability transition' and the *normative goal* of a 'sustainable peace'?
- Will this long-term transformative process towards sustainability contribute to a more peaceful world or to new forms of violence and war?
- How could global participative governance and solidarity avoid critical tipping points in the Earth and human systems?

---

1   PD Dr. Hans Günter Brauch is chairman of *Peace Research and European Security Studies* (AFES-PRESS), an international scientific NGO (Germany) and editor of five English language book series published by Springer: the Hexagon series <http://www.afes-press-books.de/html/hexagon.htm>, of Springer Briefs series on ESDP <http://www.afes-press-books.de/html/SpringerBriefs_ESDP.htm>, and PSP <http://www.afes-press-books.de/html/SpringerBriefs_PSP.htm>; APESS <http://www.afes-press-books.de/html/APESS.htm> and PAHSEP <http://www.afes-press-books.de/html/PAHSEP.htm>, email: <brauch@afes-press.de>.

2   Prof. Dr. Úrsula Oswald Spring, full-time Professor/Researcher at the *National Autonomous University of Mexico* (UNAM) in the *Regional Multidisciplinary Research Center* (CRIM); email: <uoswald@gmail.com>.

There have been two previous technical revolutions: the *Neolithic and agricultural revolution* (10,000 BP to 6,000 BP) resulted in the formation of states, urbanization, warfare, and religious-ideological control; the *industrial revolution* (since 1750 AD) resulted in a continuing process of scientific and technical innovation that led to the industrialization of warfare (with millions of casualties both among soldiers and the civilian population), billions of poor and hungry people, gender discrimination, and an extreme concentration of wealth in very few hands.

It is at present impossible to answer the question of whether this long-term transformative process aiming at the realization of the *sustainable development goals* (SDG) will contribute to a more peaceful world, a world where peace goes beyond domestic or internal peace and becomes peace among ethnic groups, cultures, gender and religions and nations: peace with nature, with equity and with equality in the Anthropocene.

However, by addressing this linkage and the possible consequences of the process of 'sustainability transition', an endurable peace within society, states and the nations of the globe can be put on the scientific and policy agenda.

The next section highlights the purpose and objectives of this Handbook (1.2). This is followed by a survey (1.3) of the challenges posed by global environmental change. These necessitate a long-term process of sustainability transition in this century because of the silent transition in earth history from the Holocene to the Anthropocene (1.3.1). This has created many challenges caused by severe physical and societal impacts in the twentieth and twenty-first centuries brought about by multiple interactions between the human and earth systems (1.3.2). The section on GEC and its physical and societal impacts (1.3.3) offers a diagnosis—based on the most recent reports by international organizations and regimes—on *population change*, as a major driver for the rising demand for environmental services, food and manufactured goods (1.3.4), on *climate change* impacts and policy responses (1.3.5), on *loss of biodiversity* (1.3.6), on *soil degradation, erosion and desertification* (1.3.7), on *water degradation, scarcity, and stress* (1.3.8), and their joint impacts on *food scarcity and hunger* (1.3.9), as well as on the *gender impacts of global environmental and climate change* (1.3.10).

The next two sections address the impacts of different economic development paths (1.4). It integrates the results of four global research programmes (WCRP, IGBP, DIVERSITAS, IHDP) and their link-

ages. These have been addressed by the ESSP and Future Earth (1.5), and their peer-reviewed research results have been assessed by the *Intergovernmental Panel on Climate Change* (IPCC), with a special focus on human security (1.5.1), and on nexus debates between water, food and energy security (1.5.2).

The sixth section briefly introduces three key concepts of this Handbook: sustainable development, sustainability transition and sustainable peace (1.6), with a special focus on the etymology of the terms and concepts 'sustainable' and 'sustainability' (1.6.1), on transition and transformation (1.6.2), and on different concepts of peace, such as ahimsa, peace with nature, and sustainable peace (1.6.3).

In section seven, the alternative approach to sustainability transition will be reviewed (1.7). The section examines its origins and conceptual evolution in North America and Europe (1.7.1), especially the early US debate (1970s onwards) on sustainability transition (1.7.2), and looks at a Dutch scientific research initiative dating from 2005 (1.7.3), and a German scientific advisory report on a 'Social Contract for Sustainability'(1.7.4). This advisory report put this approach on the policy agenda, thus enriching other policy debates on green growth and decarbonization that had been launched by international organizations and governments (1.7.5). All these debates produced feedback for an emerging new research paradigm in the social sciences (1.7.6), and this has been addressed in at least six dimensions of the research on 'sustainability transition' (1.7.7): the spatial (1.7.7.1), scientific (1.7.7.2), societal (1.7.7.3), economic (1.7.7.4), political (1.7.7.5), and cultural (1.7.7.6) dimensions of sustainability transition.

The last section introduces the ten parts of this Handbook and offers an overview of its 40 peer-reviewed contributions (1.8), examining the move towards sustainability transition (1.8.1), the target of sustainable peace (1.8.2), and meeting the challenges of the twenty-first century, e.g. demographic imbalances, rise in temperature, and the climate–conflict nexus (1.8.3). This is followed by a review of different efforts to initiate research into global environmental change, the limits to growth, and the decoupling of growth and resource needs (1.8.4), as well as through developing theoretical approaches on sustainability and transitions (1.8.5), and analysing national debates on sustainability in North America (1.8.6). In the areas of policy development and implementation, the next sections address preparing transitions towards a sustainable economy and society, and production and consumption and urbanization (1.8.7), examining sustainability transitions in the water, food and health sectors from Latin American and European perspectives (1.8.8), and preparing sustainability transitions in the energy sector (1.8.9). The last part discusses issues of international, regional and national governance in achieving sustainability transition (1.8.10).

1.2   Purpose and Objectives of this Book

Scientifically this Handbook links the four distinct research programmes of peace, security, development and environmental studies, each with their specific goals. Only limited bilateral scientific exchanges have occurred in the past, e.g. between peace and security, and between development and environment, with the aim of promoting sustainable development (WCED 1987), human security (UNDP 1994; CHS 2003) and sustainable peace (see Brauch, chapter 9 in this volume).

The Handbook connects these three scientific concepts and policy goals with the *research paradigm* of 'sustainability transition'. This focuses on socio-technological, socio-economic and societal as well as political processes of a large-scale and long-term transformative change of the dominant carbon-intensive development path. It does this by addressing the specific causes of global environmental and climate change, and the exponential increase in *greenhouse gas* (GHG) emissions and their concentration in the atmosphere since the end of World War II.

Given the enormity of the challenges facing humankind, and the powerful opposition of the industrial and business sectors affected, as well as the deeply anchored consumer society, Clark, Crutzen and Schellnhuber (2004) have called for a *new scientific revolution* (Kuhn 1962), with a change in world view, similar to the 'Copernican Revolution' (Copernicus, Kepler, Galilei, Giordano Bruno et al.) of the fifteenth century that faced violent opposition from the Catholic Church and its inquisition process. Clark, Crutzen and Schellnhuber (2004) have suggested a new world view of 'sustainability' and proposed a "New Social Contract for Sustainability" (WBGU 2011; Pagliani/Bonini, 2011; Mancebo/Sachs 2015).[3]

---

3   UNITAR, Ethos Institute: "Leading International Thinkers Call for New Social Contract", High-level expert panel in Rio de Janeiro, 23 June 2012; at: <https://www.unitar.org/leading-international-thinkers-call-new-social-contract> (23 February 2016).

From the perspective of the *social and political sciences*, multiple theoretical and empirical approaches are presented in this Handbook:

- different notions of *time* in geology, chemistry (Crutzen) and history (Braudel, Kosellek), and of the different duration of processes of transformation, transition and change triggered by 'structure-creating events' (Brauch 2016, chapter 9 in this volume);
- *conceptual theory* by reviewing the evolution (conceptual history) of the changing meanings of the four key political concepts of peace, security, development and environment, and of the three linkage concepts of sustainable development, human security and sustainable peace (see chapter 9 by Brauch);
- from *human geography* (human dimension of global environmental change, see chapters 2 by Dalby and 15 by Ehlers) and *physical geography* or 'geoecology' (Huggett 1995) and anthropology (chapter 14 by Arizpe, Price and Worcester);
- from *scientific breakthroughs* and *sociotechnological* approaches, histories and processes of technological innovations (see chapter 35 by Hargroves);
- from *systems* and *complexity theory* (chapters 13 by Scheffran and 18 by Mesjasz), by addressing the linkages between the earth system and human systems in an era of direct human intervention into nature;
- a.) from theories of *governance* (chapter 40 In 't Veld, and 41 by Happaerts), from local to state and international levels, addressing multiple actors and the three different notions of *Politik* or *politique* as *politics* (process), *policy* (field) and *polity* (norms, legal and regulatory framework).

These new research paradigms and programmes often need to overcome narrow disciplinary approaches and to move to transdisciplinary research designs, both between the different disciplines in the natural sciences, as with *Earth System Science* (ESS) and *Earth Systems Analysis* (ESA), and within the social sciences, as with *sustainability science* (Kates, Clark et al.), but they also need to link the natural and social sciences, e.g. as a *political geoecology* (Brauch/Dalby/Oswald Spring 2011). Understanding and researching individual cases, as well as strategies, policies and specific measures for sustainability transition, requires such a multidisciplinary or 'consilience' approach (Wilson 1998).

The Handbook provides texts that may be used in graduate seminars in different scientific disciplines (anthropology, economics, engineering, geography, history, law, political science, sociology, etc.), in research programmes for development studies, environmental (ecology, sustainability) studies, and peace and security studies, but also in new inter- and transdisciplinary degree programmes, e.g. those in ESS, ESA, geoecology, and transition studies. The Handbook challenges the tendency towards overspecialization of research and knowledge and encourages more holistic thinking in order to facilitate communication between scholars and between scientists and policymakers.

Politically the texts in this Handbook promote longer-term proactive strategies and policies but also shorter-term and specific measures that aim to realize the two complementary policy goals of 'sustainable development' that contributes to 'sustainable peace', as a possible result of a large-scale transition of the systems of production, consumption and governance, both of supply and demand. Two chapters discuss how a process of sustainability transition may foster or endanger the goal of sustainable peace (chapters 9 by Brauch and 13 by Scheffran).

Which peace dividends might a sustainability transition generate, e.g. by countering the prospects of resource conflicts over oil, gas and coal through improvements in energy efficiency and through switching from fossil to renewable energy sources? Which dangers to peace, security and development might be generated by opposition from those industries that are negatively affected by such a transition? These processes focus primarily on unilateral local and national action to reduce GHG emissions, irrespective of global norms and obligations, and on realizing the innovations of these systems, even if the multilateral environment and climate diplomacy remains as paralysed as it has been since 2009 up to the adoption of the Paris Agreement in December 2015.

These changes in the international order differ from the sociotechnical perspective on "the dynamics of transitions", for which Geels and Schot (2010: 11–12) identified five characteristics:

- Transitions are co-evolution processes that require multiple changes in sociotechnical systems and configurations. ...
- Transitions are multi-actor processes, which entail interactions between social groups such as businesses, firms, different types of user groups, scientific communities, policymakers, social movements and special interest groups.

- Transitions are radical shifts from one system or configuration to another. The term 'radical' refers to the scope of change, not to its speed....
- Transitions are long-term processes (40-50 years); while breakthroughs may be relatively fast (e.g. 10 years), the preceding innovation journeys through which new sociotechnical systems gradually emerge usually take much longer (20-30 years).
- Transformations are macroscopic. The level of analysis is that of the 'organizational fields'.

The *Handbook on Sustainability Transition and Sustainable Peace* offers a collection of original peer-reviewed texts by leading scholars from many disciplines that introduce both themes and address their linkages. Both authors and audience are international and the aim is the generation of 'transformative knowledge', with the goal of changing policies towards achieving the dual goal of sustainable development with sustainable peace.

The Handbook examines the impacts of possible 'threat minimizers' through sustainable development policies for international peace and human security (UN 2009a). It emerged from two workshops in Yautepec, Mexico in September 2012 and in San Francisco in April 2013[4] and from a winter school at Chulalongkorn University in Bangkok in December 2013. These addressed possible linkages between scientific discourses and policy debates on global environmental and climate change, on the international, national and human security impacts of climate change issues, on sustainable peace, and on long-term transformative change towards sustainable development. Among the key questions that were addressed in these meetings and which are discussed in the Handbook are:

- Might *business-as-usual* policies that fail to address the challenges posed by global environmental change and the growing intensity of climate-induced natural hazards and disasters threaten the survival of millions of people during this century and pose serious threats to international peace and security?
- Might anticipative learning and a forward-looking public and global discourse on the necessary long-term transformative changes contribute to sustain-

---

4 These workshops are documented with abstracts, Powerpoint presentations and podcasts at: the Yautepec workshop: <http://afes-press-books.de/html/sustainability_ workshop_programme.htm>, and San Francisco Workshop: <http://afes-press-books.de/html/workshop_ SanFrancisco.htm>.

able development and thus address the new dangers to international peace and security in a preventive manner?

- What general policy lessons can be drawn from the violent consequences of the industrial revolution (during the nineteenth century) and the third technical revolution (during the twentieth century) for a long-term transformative change towards sustainable development?

The *Handbook on Sustainability Transition and Sustainable Peace in the Anthropocene* presents a collection of innovative texts written by invited distinguished authors. The Handbook addresses scholars, policymakers and students and may be used across the world for university seminars in environmental studies (sustainability and sustainability transition), development studies, peace and security studies, and international studies. John Grin (see chapter 4 in this volume; Grin/Rotmanns/Schot 2010, 2011), one of the initiators of the sustainability transitions studies discourse, has outlined sustainability transition theory and offered an overview and assessed its gradual evolution:

> The transition perspective on promoting sustainable development recognizes a need for not merely new societal practices, but changes in the structures in which these practices are embedded, and which have co-evolved with earlier practices ('the regime'). This contribution presents insights on the dynamics of sustainable transitions, developed within the context of the Dutch KSI programme (2005–2010), and identifies issues for further research. Three perspectives have been most central to the programme: (i) the sociotechnical approach, primarily rooted in STS [sociotechnological studies] and empirically grounded in historical case studies; (ii) the complex adaptive systems approach, grounded, through action research, in on-going transitions, and (iii) the governance approach, rooted in political science and sociology. I will discuss these approaches as well as two other ones, and indicate some of the main research findings form the KSI programme. For each approach, I will also present the associate governance concepts that translate the theoretical understanding of transitions into prescriptions for shaping them.[5]

Sustainable Peace is a normative goal that aims at 'peace with nature', based on a sustainable development process that has been used in the context of development and peace, and to a lesser extent in environmental studies (Coleman/Deutsch 2012; Oswald Spring/Brauch/Tidball 2014). Alongside negative, positive, cultural and engendered peace, 'sustainable peace' has been recently introduced as one of five con-

ceptual pillars of an emerging 'peace ecology' (Oswald Spring/Brauch/Tidball 2014):

> *Sustainable peace* refers to the manifold links among peace, security and the environment, where humankind and the environment as the two key parts of global Earth face the consequences of destruction, extraction and pollution. ... The sustainable peace concept includes also the processes of recovering from environmental destruction, reducing the human footprint in nature through a less carbon-intensive—and in the long term possibly carbon-free and increasingly dematerialized production processes that future generations may still be able to decide on their own resources and development strategies.

This Handbook suggests linking the goal of achieving sustainable development through sustainable production and consumption with international cooperative multilateral political strategies for achieving international peace and security. It is argued that the possible peace dividends of *sustainability transition* (ST) strategies may reduce future energy resource conflicts (over scarce oil) and the negative consequences of a BAU approach for global environmental and climate change (Oswald Spring 2008, 2009; Oswald Spring/ Brauch 2011).

A key research question of this book is to what extent political strategies aimed at achieving the goal of sustainable development through a gradual process of sustainability transition may reduce the probability of resource conflicts and the security consequences of climate change if policies of *business-as-usual* and a Hobbesian mindset prevail.

To be able to understand the global environmental changes humankind is expected to face during this century, and to be able to create a broad public societal and political awareness in order to create a consensus and political support for proactive coping strategies, a shift from disciplinary to inter- and transdisciplinary as well as to transformative approaches is needed in both the natural and social sciences. This may require a new epistemic community aiming at 'consil-

---

5   John Grin: "Abstract" of: "Theories of Transitions to Sustainable Development: Approach of the Dutch Knowledge Network of Systems Innovation"; see also the podcasts of his presentations to the First Sustainability Transition and Sustainable Peace Workshop, UNAM/CRIM and AFES-PRESS: *Towards a Fourth Sustainability Revolution and Sustainable Peace: Visions and Strategies for Long Term Transformative Change to Sustainable Development in the 21st Century*, 10–13 September 2012, in Morelos, Mexico; at: < http://afes-press-books.de/html/sustainability_workshop _programme.htm>.

ience' (Wilson 1988) in science and closer horizontal cooperation in policy-making.

Proactive coping strategies have been difficult and suboptimal in the past, but they have become essential for understanding and coping with these challenges of developing knowledge about and policies on these complex linkages. Approaches aiming at transformative social science and at developing new knowledge about sustainability transition and its impact on sustainable peace are important tasks for the knowledge sector in universities and research institutes as well as for education and the media.

1.3 The Global Environmental Context

The world faces economic crises, population growth, climate change, water scarcity and pollution, soil depletion, erosion and desertification, food crises, urbanization with slum development, rural–urban and international migration, physical and structural violence, gender, race and ethnic discrimination, youth unemployment, a growing loss of ecosystem services, and an increasing number of resource conflicts. The interaction of these multiple crises may result in extreme outcomes, especially for vulnerable people living in risky places, and may reduce human, gender and environmental security and peace (HUGE; see chapter 8 by Oswald Spring in this volume).

This global environmental context has gradually changed since the agricultural and industrial revolution, but over the last fifty years the survival of Earth and humankind has been more intensively threatened. Scientists only began to be aware of this global environmental change in the 1970s, which is when the scientific discourse began, initially in the natural sciences and from the 1990s also in the social sciences. This process has been referred to as *scientization*. In the late 1980s, the political debate and international negotiations (*politicization*) started, with Gro Harlem Brundtland's concept of sustainability (WCED 1987) and at the *United Nations Conference on the Environment and Development* (UNCED) in 1992 in Rio de Janeiro. The global outcome was the signing of the *United Nations Framework Convention on Climate Change* (UNFCCC) and the *Convention on Biological Diversity* (CBD); adoption of a mandate for the *United Nations Convention to Combat Desertification* (UNCCD) that was signed only in 1994.

Global environmental change is wider than climate change. The term refers to the transformation produced by human beings in the ecosphere, which affects the hydrosphere (the combined mass of water found above, on and under the surface of the planet), the atmosphere (the layer of gases surrounding the surface), the biosphere (the global ecological system where all living beings exist), the lithosphere (the outer layer of the Earth) and the pedosphere (the soil) (see chapter 2 by Dalby; NRC 1999; Oswald Spring 2016).

Changes in the natural system are the result of modifications in agricultural production, of rapid urbanization processes, and of population growth (Oswald 2011). Furthermore, unsustainable productive processes are polluting natural resources and creating health threats for human beings, as well as endangering ecosystems (Elliott 2011). The energy, transportation and production sectors have polluted heavily due to their use of fossil fuels (IEA 2014). In addition, changes in land use and deforestation are reducing the capture of *carbon dioxide* (CO_2) (IPCC 2014a; 2014b). For these reasons, the emissions from *greenhouse gases* (GHG) have increased exponentially (IPCC 2013). In addition, a globalized financial system, unequal credit access, current patterns of consumption and production, and an uneven level of access to resources are also contributing to environmental change. Irrational behaviour has also produced poverty, hunger, and inequality among regions and social groups (Wilkinson/Pickett 2009).

In Earth and human history, drastic changes have gradually occurred since the industrial revolution (1780–1870). Such changes have taken place especially over the last five decades, because of the intensive use of fossil energy; the rapid increase in GHG emissions into the atmosphere; pollution, warming, and acidification of the seas; massive changes in land use, population growth, and an accelerated process of urbanization. The Anthropocene represents a new geological epoch that is changing the history of Earth. Bond, Showers, Cheseby et al. (1997) define it as "the most recent manifestation of a pervasive millennial-scale climate cycle operating independently of the glacial-interglacial climate state". This concept is useful for understanding the transformative negative effects of human activity on the global planet, on its ecosystem services, and on humankind itself. Nevertheless, human agency also has the potential for positive change.

Nobel laureate Paul J. Crutzen claimed in February 2000, during an IGBP meeting in Cuernavaca, that "we are in the Anthropocene" (Moore III 2000; Crutzen/Stoermer 2000). Crutzen (2002) links these changes with the environmental changes induced and produced predominantly by human intervention. The

'Anthropocene' is a term widely used since to denote the present time interval, in which many geologically significant conditions and processes are profoundly altered by human activities. These include changes in erosion and sediment transport associated with a variety of anthropogenic processes, including colonization, agriculture, urbanization, population growth, poverty, inequality (Oxfam 2016), and global warming.

The chemical composition of the atmosphere, oceans and soils has been altered, with significant anthropogenic perturbations of the cycles of elements such as carbon, nitrogen, phosphorus, sulphur, various metals, and emergent pollutants in water and soil (Cortés/Calderón 2011). Environmental and human conditions generated by these perturbations include global warming, ocean acidification, the spreading of oceanic 'dead zones', and biosphere destruction on both land and sea. This has resulted in habitat loss, predation, species invasions, and physical and chemical changes, and has also led to socio-economic changes, with greater inequality, poverty and misery among socially vulnerable groups, together with gender discrimination and violence.

1.3.1 "We are in the Anthropocene"

In the natural and social sciences six historical times (Brauch 2016) may be distinguished. These have triggered multiple contextual changes of different duration: a) *cosmic time* in modern physics and astronomy, that is beyond any human influence; b) *geological time* in contemporary earth sciences; c) the *technical time* of the Neolithic and industrial revolutions that have triggered the change in geological time with the 'silent transition' from the Holocene to the Anthropocene. The French historian Braudel (1969, 1972) distinguished between *three times* in human history: d) *structural time* (e.g. of international order); e) *cyclical time* (e.g. of the government of presidents and prime ministers); and f) the short *time of historical events*. Such historical events (e.g. revolutions, coups, elections, disasters) may trigger changes in governments (cyclical time) or even of international order (structural time, e.g. the Russian revolution of 1917 or the fall of the Berlin Wall in 1989).

Since the industrial revolution a long series of technical inventions, e.g. by Watt, Edison, Curie, Bell, Daimler, Benz, Ford, Röntgen, Zuse, Lederberg, Müller, with indirect intervention into the earth system by humans, has resulted in fundamental changes: the accumulation of CO_2 and the destruction of the ozone layer, biodiversity, soil and water. During the

twentieth century several major changes in international order occurred. Two resulted from global (1919, 1945) and regional wars (Korea, Indochina, Algeria, Angola, Uganda, Liberia, Congo, Iraq, Afghanistan, Yugoslavia, etc.), and military *coups d'état* (Chile, Argentina, Bangladesh, Myanmar, Ghana, Chad, Fiji, etc.). The others were the outcome of liberation struggles (independence of India, Pakistan, Vietnam, African nations) or were peaceful changes (the disintegration of the Soviet Union, which was not a victory for the West (Schweitzer 1994) but the result of a 'learning' (Grunberg/Risse-Kappen 1992) by the new Soviet leadership). Others were related to interventions by the UN Security Council or regional peace bodies (the African Union), some of them with unintended outcomes.

Geologists and Earth scientists use a "*geological time scale*" (GTS) approved by the "International Commission on Stratigraphy" (Cohen/Finney/Gibbard 2012; Gradstein/Ogg/Schmitz 2012). In 2008 the international Commission on Stratigraphy set up a study group "to make the Anthropocene a formal unit of geological epoch divisions", with the goal of reaching a decision by 2016 (Waters/Zalasiewicz/Williams et al. 2014).[6]

With the end of several wars of independence, e.g. the Vietnam War (1975), and of the cold war (1989), a peaceful regional and global transition of the structure, strategies and international politics in the Westphalian sovereignty-based system of nation states resulted. The 'silent' transition from the Holocene[7] to the 'Anthropocene' is more profound, however. The fundamental change from the '*Holocene*' to the '*Anthropocene*' is caused by increasing human interventions in the Earth processes, but there is no agree-

---

6 See the Working Group on The 'Anthropocene', *International Commission on Stratigraphy* (ICS); at: <http://quaternary.stratigraphy.org/workinggroups/anthropocene/>; see also the most recent "Newsletter of the Anthropocene Working Group", September 2014; at: <http://quaternary.stratigraphy.org/workinggroups/anthropo/anthropoceneworkinggroupnewslettervol5.pdf>. The convenor of this working group is Dr. Jan Zalasiewicz (University of Leicester, UK).

7 The Holocene started with the end of the glacial period about 12,000 years ago. This marked the beginning of great civilizations in the Mediterranean, China, India, Egypt, Mesopotamia and Meso-America. The 'Holocene' is a period of geological transition, when a dramatic environmental change took place, with a major rise in sea level as a result of the melting of the huge ice sheets that covered large areas in the northern hemisphere.

Figure 1.1:   Atmospheric CO2 at Mauna Loa Observatory.**Source**: National Oceanic and Atmospheric Administration (NOAA)–Monthly Data for Atmospheric CO_2 from 1958 until December 2015; at: <http://www.esrl.noaa.gov/gmd/ccgg/trends/full.html> (30 January 2016).

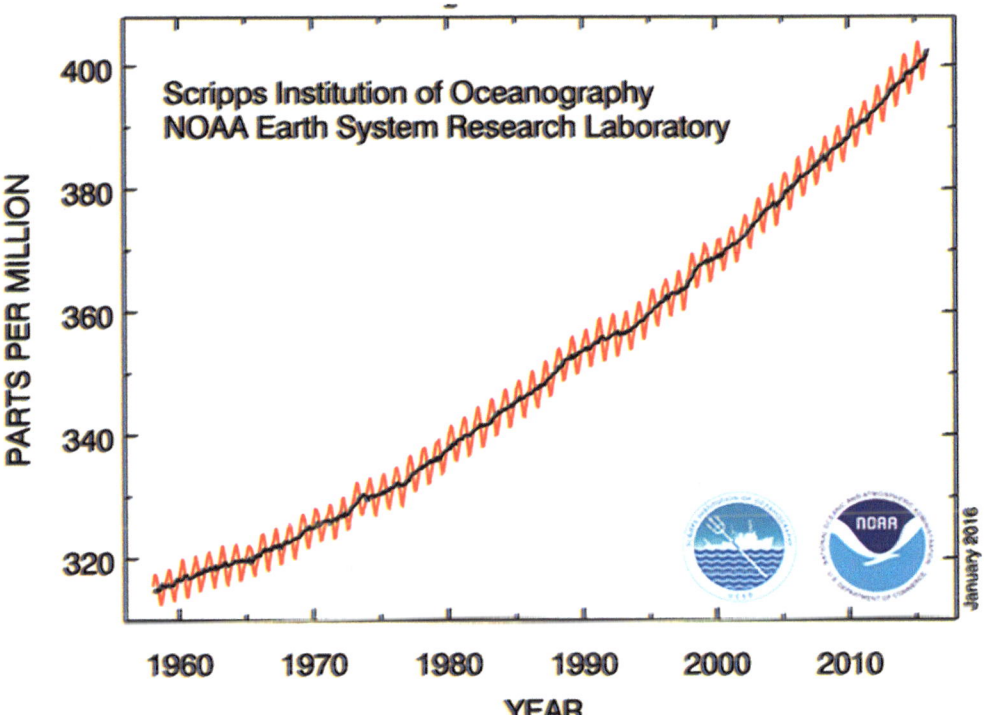

ment on when the 'Anthropocene' actually began. Crutzen initially named the Industrial Revolution as the starting point, while others pointed to the end of World War II, after which greenhouse gas concentrations increased exponentially due to the rapid increase in the burning of oil and natural gas (figure 1.1). But Crutzen (2006: 17) argued that since humankind "will remain a major geological force for many millennia", it is necessary "to develop a world-wide accepted strategy leading to sustainability of ecosystems against human induced stresses", and that this will be "one of the greatest tasks of [hu]mankind, requiring intensive research efforts and wise application of the knowledge".

After the Industrial Revolution the CO2 concentration in the atmosphere rose from about 279 ppm in 1750 to 315 ppm in 1958. But from 1958 to November 2015, CO2 concentration increased to 400.38 ppm.[8] While from 1750 to 1958 CO2 concentration rose by 36 ppm, from 1958 to 2015 it increased by 85 ppm.

8   See "Weekly Data-Atmospheric CO2"; at: <http://co2 now.org/Current-CO2/CO2-Now/weekly-data-atmospheric- co2.html> (2 December 2015), and at: <http://www.esrl.noaa.gov/gmd/ccgg/trends/global.html# global> (30 January 2016).

NOAA noted that "for the past ten years (2005-2014), the average annual rate of increase is 2.11 parts per million (ppm). This rate of increase is more than double the increase in the 1960s".

1.3.2   Changing Interactions between the Human and Earth System in the Anthropocene

The 'silent transition' in earth history from the Holocene to the Anthropocene is the result of multiple interactions between the human and earth systems. Humankind has for the first time directly interfered in nature and has triggered physical effects in the global climate system. This has resulted in small-onset a) temperature rise, b) precipitation change, c) extreme weather events, and d) slow-onset sea-level rise. To analyse these interactions, the PEISOR model (figure 1.2) was developed by Brauch (2005: 16; 2007: 28; 2009: 76). It was inspired by several similar pressure-response models produced by international organizations (OECD 2001; EEA, UNCSD) to securitize GEC issues from an environmental security perspective. This model combines five stages:

Figure 1.2: The PEISOR model. **Source:** Brauch (2009: 76); Brauch and Oswald Spring (2009: 9).

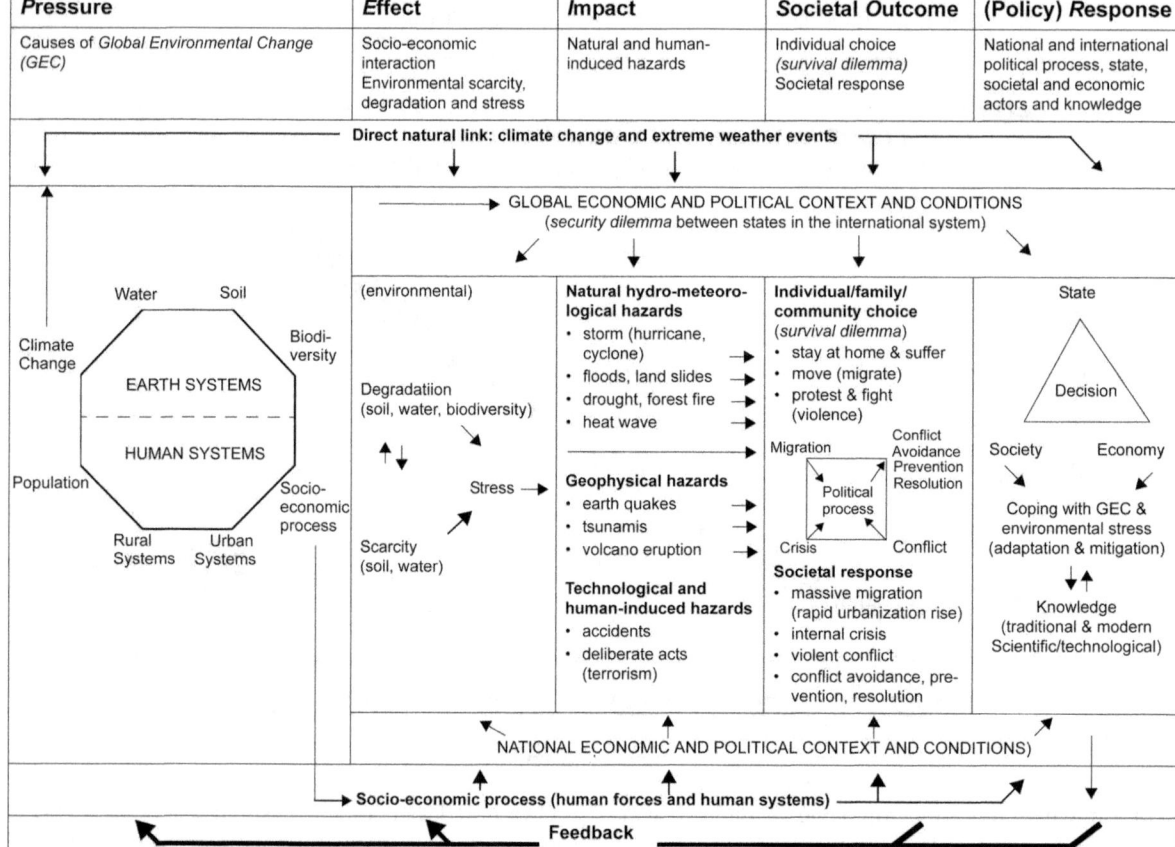

- *P (pressure)* refers to the eight drivers of global environmental change;
- *E* to the *effects* of the interactions on environmental scarcity, degradation, and stress;
- *I* to the extreme *impacts* of human-induced and climate-related natural hazards;
- *SO* to *societal outcomes*: forced migration, slums, crises, conflicts or state failure;
- *R* to the *response* of all stakeholders.

In the PEISOR model, 'Pressure' refers to the eight factors contributing to GEC. They often interact in a non-linear or chaotic way and impact on earth and human systems. The political and societal contexts may affect the socio-economic processes that contribute to anthropogenic environmental degradation or to resource scarcity that may result in environmental stress. Four factors are key drivers of the human system (population, rural systems and agriculture, urbanization, industry and services, and the socio-economic development process), and these affect the 'environmental quartet' of the earth system: climate change (air) (1.3.5), biodiversity (1.3.6), water (1.3.7) and soil (1.3.8).

Figure 1.2 illustrates GEC and the complex interactions and feedbacks between the earth and human systems needed to understand the impact of climate change, water stress and biodiversity loss on *desertification, land degradation and drought* (DLDD). In figure 1.3 the three small cycles with the factors determining climate change, water stress and biodiversity loss each have different effects on DLDD. The wider cycle representing GEC describes the often chaotic interrelations between the earth and human systems and their possible societal outcomes.

1.3.3 Global Environmental Change and its Physical and Societal Impacts in the Twentieth and Twenty-First Centuries

During the twenty-first century, the relationship between the causes and severe societal outcomes of GEC and climate change may result in environmentally-induced massive and forced movements of people, hunger- and famine-induced protests, and small-scale societal violence. It may also possibly result in violent conflicts within and between countries. These

Figure 1.3: Environmental Quartet: DLDD, Climate Change, Water Degradation and Biodiversity Loss. **Source**: The figure was developed by Úrsula Oswald Spring based on MA (2005: 17) and designed by Guillermo A. Peimbert, UNAM/CRIM, Mexico.

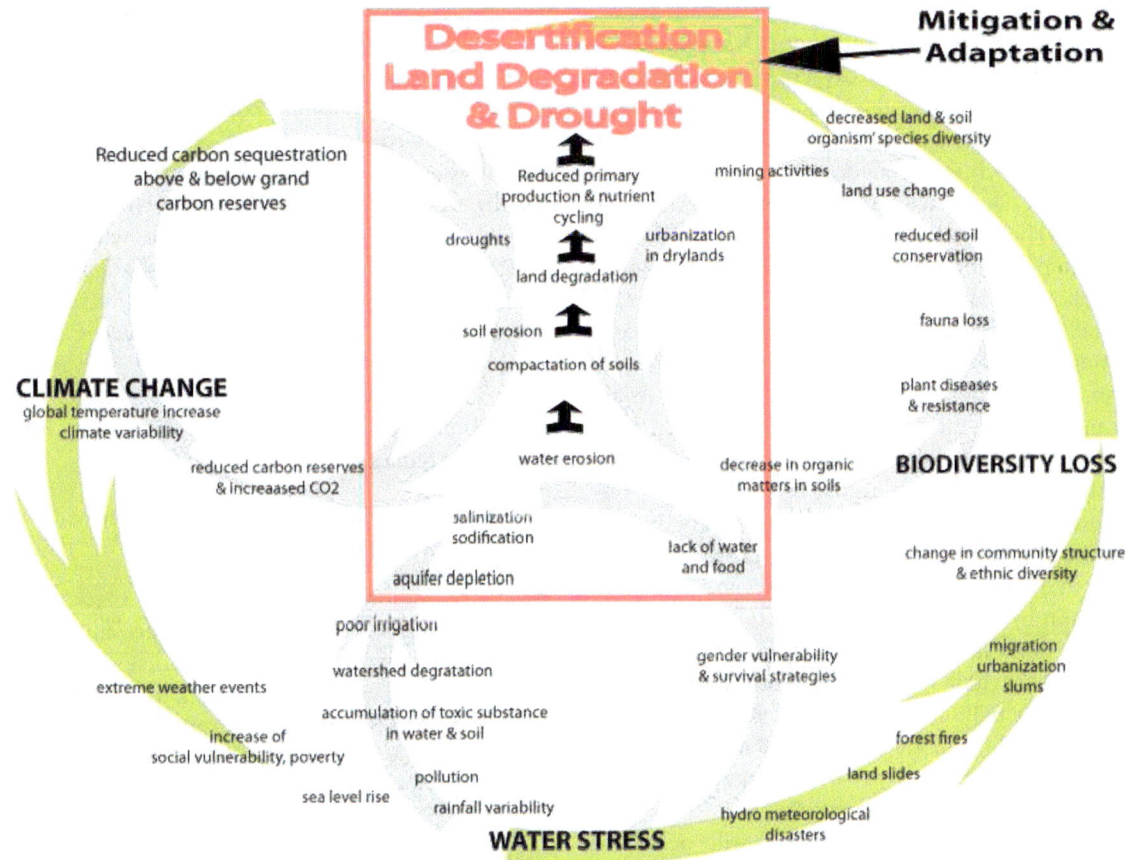

may pose numerous security dangers, and these have increasingly been addressed by governments and international organizations (Brauch 2002, 2009; WBGU 2008; Gleditsch 2012; Scheffran/Brzoska/ Brauch et al. 2012; IPCC 2014a; see chapter 12 by Ide, Link, Scheffran and Schilling; chapter 13 by Scheffran).

While a number of natural scientists had hypothesized on the linkages between the burning of hydrocarbons and an increase of the greenhouse gases in the atmosphere from the nineteenth century onwards (e.g. Tindall, Arrhennius in 1896 et al.), the concept of *global environmental change* and associated research in the natural sciences has gradually emerged from the 1970s onwards (IGBP, WCRP, DIVERSITAS) and in the social sciences from the 1990s (IHDP, chapter 14 by Arizpe/Price/Worcester and chapter 15 by Ehlers). In 1988 the *World Meteorological Organization* (WMO) and the *United Nations Environment Program* (UNEP) established jointly the IPCC and this was endorsed by the *United Nations General Assembly* (UNGA). The role of the IPCC is to assess

on a comprehensive, objective, open and transparent basis the scientific, technical and socio-economic information relevant to understanding the scientific basis of risk of human-induced climate change, its potential impacts and options for adaptation and mitigation.

1.3.4   Demographic Change and Urbanization in the Twenty-First Century

During the twentieth century world population increased threefold, while water use rose sixfold (Steffen/McNeill/Crutzen 2007: 617; Steffen/Grinevald/ Crutzen et al. 2011: 843). The interdependence of technological change, agricultural change (the green revolution with improved seeds and today transgenic seed, irrigation, chemical fertilizers, pesticides, and heavy agricultural machinery), cheap energy, and medical improvement (antibiotics) has resulted in a rapid growth in population worldwide (see chapter 10 by Zlotnik), while birth control, reproductive health for women, and contraception have been limited due to cultural (patriarchy), religious (Christianity, Islam, Buddhism) and social obstacles (poverty, illiteracy, lack of hygienic conditions, public health or medical attention).

The medium variant of the *United Nationa Population Division*'s (UNPD) 2015 Population Revision (2015: 7) projection indicated that a global population of 7.349 billion in 2015 would increase to 9.725 billion in 2050 and reach 11.213 billion by 2100. Most of this growth will take place in the poorest developing countries. The highest increase is forecast for Africa, where the population rose from 133 million in 1950 to 1.186 billion in 2015 and is projected to rise further to 2.478 billion by 2050 and to 4.387 billion by 2100, representing 25 per cent of the global population in 2050 and 39 per cent by 2100 (UNPD 2015: 9). The Asian continent was the most populated region and in 2015 held 4.93 billion of a global total of 7.349 billion people; this figure is projected to rise in 2050 to 5.267 billion and to fall to 4,889 billion by 2100. For North America, population is projected to grow from 358 million in 2015 to 433 million by 2050 and to about 500 million by 2100, while the projected increase for Latin America will be more modest, from 634 million in 2015 to 784 million in 2050, after which it will decline to 721 million by 2100. Only for Europe is a continuous decline projected: from 738 million in 2015 to 734 million in 2030, to 707 million in 2050, and to 646 million by 2100 (UNDP 2015: 7).

Most of this population growth will occur in urban areas. While in 1950 only 30 per cent of the world's population was urban, by 2014 about 54 per cent were residing in an urban context and the projections indicate 66 per cent by 2050. The turning point from rural to urban population was in 2007.

From an occidental view, Eric Hobsbawn (1962: chapter 11) insisted that "urban development in our period [1789–1848] was a gigantic process of class segregation, which pushed the new labouring poor into great morasses of misery outside the centres of government and business and the newly specialized residential areas of the bourgeoisie. The almost universal European division into a 'good' west end and a 'poor' east end of large cities developed in this period." The developing countries could not avoid this segregation process and today the *United Nations Population Fund* (UNFPA) (2012) has stated that informal slum settlements suffer "disproportionately from disease, injury, premature death, and the combination of ill-health and poverty entrenches disadvantage over time" (UNFPA 2012).

In the cities in the developed world, urbanization follows a general pattern of immigration into downtown areas. Most of the impoverished city centres have also exhibited a concentration of human activities and settlements around the so-called 'peripheralization of the core'. Henderson (2002) argued that urbanization in developing countries resulted in very high levels of concentration of people in megacities. Since 1995 an enormous change in cities has occurred globally and today almost half of the world's population lives in big cities, which is also where future growth will mainly occur (Cohen 2006). These megacities create enormous socio-economic, environmental and cultural challenges. This excessive concentration of people has created traffic congestion and air, water and soil pollution. Lack of city planning and management is often combined with corruption, increasing health threats and reducing the quality of life.

1.3.4.1 Impacts of Population Growth and Urbanization

The results are urban heat islands and a lack of safe food, drinking water, drainage systems and sewage treatment plants. Gasoline with lead and a lack of parks and green areas are other additional threats to health and well-being. The eutrophication of water and rain leaches all type of pollutants to aquifers, rivers and seas and has damaged large areas of agricultural land and marine ecosystems (Cortes/Calderón 2011). Nevertheless, planned urban development may also create new environmental opportunities. Land use is concentrated, provision of services is cheaper, food is widely available and increasingly produced even on roof and vertical gardens. Public transport systems often replace private cars, reducing air pollution. Furthermore, parks and green cities are promoting new models of urban settlement and job opportunities. But in poor megacities the absence of job opportunities, food scarcity and escalating poverty may also become a 'threat multiplier'. When good governance is absent and corruption and liberalizing reforms limit the state resources, then public insecurity, organized crime and insurgency could destroy existing development strategies and prevent people's survival.

1.3.4.2 Anthropogenic Global Environmental Change

Projected population growth, the change in hygienic conditions, developments in urbanization, and the increase in or lack of income means that from now until 2100 there will be an increase in the demand for

Figure 1.4: Reported natural disasters by continent (1950–2015). **Source**: D. Guha-Sapir, R. Below, Ph. Hoyois - EM-DAT: The CRED/OFDA International Disaster Database – <www.emdat.be> – Université Catholique de Louvain - Brussels - Belgium, The figures were published in: CRED Crunch, No. 30, January 2013:"Natural disasters in Asia" (Brussels: CRED: 1). Permission of the authors was granted on 14 March 2015. The authors are grateful to Régina Below for updating the figure from 1950-2010 to 1950-2015.

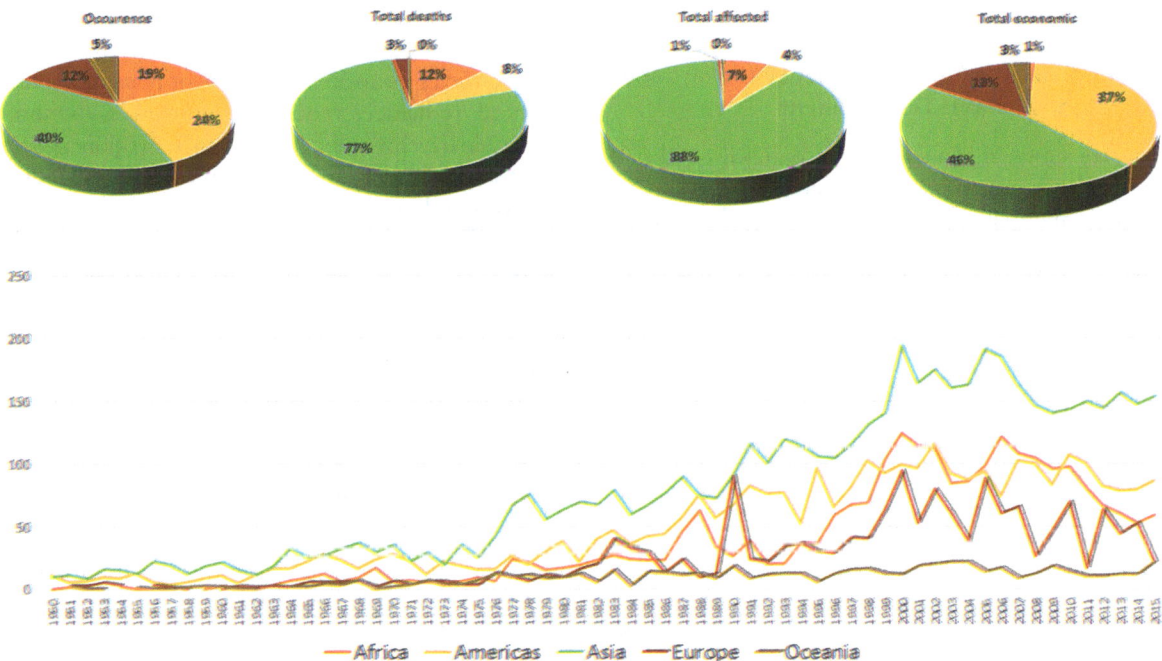

water, food, energy, housing, industrial goods, and public services. Global environmental change has directly affected many natural processes, such as a) climate change—with a rise in temperature, changes in precipitation, extreme weather events (droughts, storms, floods, forest fires, heatwaves) and a rise in sea level; b) loss of biodiversity due to deforestation, changes in land use, land grabbing and industrial and urban development; c) soil degradation, related to the overuse of agrochemicals, inadequate irrigation systems, erosion and desertification; and d) water scarcity, degradation and pollution due to the interaction of population growth, industrial, urban and agricultural pollution, and poor soil and ecosystem management. Confronted with these multiple environmental stressors, humankind can not only destroy but also has the capacity to restore and recover the natural resources and ecosystem services affected.

1.3.5 Climate Change Impacts and Policy Responses

Since the first Report of the IPCC in 1990, scientists have warned that climate change is a 'threat multiplier', which exacerbates other threats to social and natural systems. The existing dual vulnerabilities (Bohle 2001, 2002; Oswald Spring 2013) caused by environmental challenges (posed by GEC and climate change) and social threats (globalization, poverty, violence) have increased the risks that extreme events (figure 1.4) are transformed into disasters.

1.3.5.1 Physical and Societal Impacts of Anthropogenic Climate Change

The physical impacts of anthropogenic climate change include storms, floods, droughts, landslides, forest fires, sea-level rise, changes in precipitation, and extreme temperatures. It places additional burdens on the poorest countries because of their lack of human and financial resources, thus further constraining development paths. In response to these environmental challenges, the IPCC (2014a) has proposed positive interaction between adaptation, mitigation and sustainable development, both within and across regions, social actors, and governmental scales. Mitigation efforts could cause risks to environmental resources and drain financial resources from development priorities, including the alleviation of poverty.

But there are also numerous co-benefits, such as improved air quality and reduced energy and water consumption in urban areas through a greening of cities and from a recycling of water, by shifting to sustainable agriculture and forestry. All these initiatives could contribute to the recovery of ecosystems for carbon storage and other ecosystem services, and also to the enhancement of energy security.

1.3.5.2   Towards Climate-Resilient Pathways and Sustainable Development

Strategies aimed at climate-resilient pathways and sustainable development may contribute to mitigation by reducing GHG emissions and increasing the adaptation and resilience of the people affected. They may simultaneously improve livelihoods and socio-economic well-being, and enhance effective environmental management. Successful implementation depends on policies and cooperation at all levels, where all stakeholders are constructively involved. Usually engineering and technological options are the most common responses; sometimes they are tied together with disaster risk management and water management.[9]

Most assessments of adaptation activities are limited to an analysis of impacts, vulnerabilities, and adaptation planning, primarily from a top-down perspective, where the processes and multiple effects of actions are rarely assessed. Most adaptation policies recognize the value of social, institutional, and ecosystem-based measures and of common enabling factors. The IPCC (2014a) emphasized that these factors should include effective institutions and governance, innovation and investments into environmentally sound technologies and infrastructure, sustainable livelihoods and behavioural and lifestyle changes.

Other global risks of climate change are concentrated in urban areas, due to heat stress, extreme precipitation, inland and coastal flooding, landslides, air pollution, drought, and water scarcity affecting people, assets, economies, and ecosystems. In developing countries the lack of essential infrastructure and services as well as living in flood-prone and highly exposed regions and in poor quality housing increases the vulnerability and exposure of the poor. Inundations due to rapid-onset tropical storms and floods as well as sea-level rise pose risks to the territorial integrity of small island states and to countries with extensive coastlines. Furthermore, changes in sea ice, in shared water resources, and pelagic fish[10] stocks may

increase conflicts between states. Robust national and intergovernmental institutions can enhance cooperation and manage many of these rivalries peacefully.

Integrated policy responses are especially relevant for a) energy planning and implementation; b) interactions between water, food, energy and biological carbon sequestration; and c) urban planning. These integrated policies provide substantial opportunities for enhanced resilience, reduced emissions, and more sustainable development. These policies require financial and economic support from industrialized states for poor developing countries and especially for low-income groups and vulnerable communities. According to the IPCC (2014a), the effectiveness of adaptation may be further enhanced by transparent governance structures, adequate institutional and human capacities, and partnerships with local governments and organized stakeholders or directly exposed social groups.

1.3.5.3   Towards the Paris Agreement

In this complex situation and given the increasing impacts and threats posed by the physical and societal impacts of climate change, the twenty-first *Conference of the Parties* (COP 21) of the *United Nations Framework Convention on Climate Change* (UNFCCC, 1992) took place in Paris in November and December 2015. Its aim was to negotiate and to adopt a new legally binding agreement that would supersede the Kyoto Protocol (1997) to the UNFCCC, which expired in 2012.

After the failure of COP 15 in Copenhagen (2009), the Paris conference was to agree among the member countries of the UNFCCC on a drastic reduction of GHG emissions that would stabilize the "increase in the global average temperature to below 2°C above pre-industrial levels by reducing emissions to 40 gigatonnes or to 1.5°C above pre-industrial levels by reducing to a level to be identified in the special report". The specific measures were to be identified in an IPCC Report in 2018 "on the impacts of global warming of 1.5°C above pre-industrial levels and related global greenhouse gas emission pathways" (Decisions adopted at COP 21).[11]

The resulting "Paris Agreement"[12] was prepared and approved by 196 participating ministers from state parties, who had submitted their *Intended Nationally*

---

9   The role of science and engineering and the technological dimension will be discussed in more detail in the concluding chapter by Oswald Spring, Brauch and Scheffran.

10   See at: "What are pelagic fish?"; at: <http://oceanservice.noaa.gov/facts/pelagic.html> (2 March 2016).

11   See at: <http://unfccc.int/resource/docs/2015/cop21/eng/10a01.pdf> (9 March 2106).

Determined Contributions (INDC) prior to COP 21. It was agreed that this legally binding agreement would enter into force by 2020.

But these proposed INDCs would stabilize the global average temperature at much higher than 3°C above pre-industrial levels by 2100 (FCCC/CP/2015/7). They are insufficient to reduce GHG emissions and certainly would not realize the general target adopted by the same representatives of the state parties.

However, after procedural matters related to the adoption (1-11), the "II. Intended Nationally Determined Contributions" (12-21) are welcome, the "III. Decisions to Give Effect to the Agreement" listed under 'mitigation' (22-41), 'adaptation' (42-47), 'loss and damages' (48-52), 'finances' (53-65), 'technology development and transfer' (66-71), 'capacity building' (72-84), 'transparency of action and support' (85-99), 'global stocktake' (100-102), 'facilitating implementation and compliance' (103-104) and 'final clause' (105) rely exclusively on voluntary actions. The same applies to the "IV. Enhanced Action Prior to 2020" (106-133). The role of "V. Non-Party Stakeholders", especially of "civil society, the private sector, financial institutions, cities and other subnational authorities" is welcomed.

Although the Paris Agreement will be a legally binding international treaty—once it has been ratified by the required number of states—that will succeed the Kyoto Protocol (1997), it avoids any legally binding obligations in quantitative reduction for the state parties and sanctions in the case of lack of implementation. It is a collection of good intentions that "in enhancing the implementation of the Convention, including its objective, aims to strengthen the global response to the threat of climate change, in the context of sustainable development and efforts to eradicate poverty", with the goal of "holding the increase in the global average temperature to well below 2°C above pre-industrial levels and to pursue efforts to limit the temperature increase to 1.5°C above pre-industrial levels" (Article 2). It referred to what state parties 'shall' and 'should' do but not what they 'have to' or 'must' do to achieve these goals.

Article 6 recognized "that some Parties choose to pursue voluntary cooperation in the implementation of their nationally determined contributions". Although the states established in Article 7 "the global goal on adaptation of enhancing adaptive capacity, strengthening resilience and reducing vulner-

ability to climate change, with a view to contributing to sustainable development and ensuring an adequate adaptation response in the context of the temperature goal referred to in Article 2", it was only stated that "each Party shall, as appropriate, engage in adaptation planning processes and the implementation of actions".

Also on "loss and damages" (Art. 8) and on financial support of "developed country parties ... to assist developing country Parties with respect to both mitigation and adaptation in continuation of their existing obligations under the Convention" (Art. 9) no specific legally binding commitments for state parties were adopted. In Art. 10 the "Parties share a long-term vision on the importance of fully realizing technology development and transfer in order to improve resilience to climate change and to reduce greenhouse gas emissions", while in Art. 11 "Capacity-building ... should enhance the capacity and ability of developing country Parties ... [to take] effective climate change action". However, according to Art. 27 "no reservations may be made to this Agreement". Developed countries are committed to provide financial and technological support for developing countries and others are invited to help on a voluntary basis. They "should continue to take the lead in mobilizing climate finance from a wide variety of sources, instruments and channel" (Art. 9) and they are voluntarily committed to provide 100 billion dollars annually from 2020 onwards as a floor. This *burden sharing* is complemented with a *review mechanism* each five years from 2023 on, where each country is "updating and enhancing" (Art. 14) their pledges. The third mechanism is a *recognition of the needs* of vulnerable countries for adverting, minimizing and addressing losses due to climate change. A financial compensation for "loss and damage" (Art. 8) brought up by affected countries was not accepted by industrialized countries. Developed country parties should continue to take the lead by undertaking economy-wide absolute emission reduction targets. Developing country parties should continue enhancing their mitigation efforts, and are encouraged to move over time towards economy-wide emission reduction or limitation targets in the light of different national circumstances.

This achievement was the result of more than twenty-three years of international collective action in the framework of the UN. 'Informal informals' held by bodies such as the G7, G20, OECD, OPEC, and small groups of delegates tried to remove the brackets which denoted disagreements on the text. When no progress was achieved, the conference chair, French

---

12 See at: <http://unfccc.int/files/meetings/paris_nov_2015/application/pdf/paris_agreement_english.pdf> (9 March 2016).

foreign minister Laurent Fabius turned to 'indaba' (Barry 2012). This is a Zulu tradition for discussing and negotiating disputes between groups of elders, which was also used in Durban in 2011 to achieve the outcomes of COP 17. In France this procedure consisted of groups of up to eighty delegates whose task was to overcome the disagreements; the final agreement was then accepted by consensus. Paul Bodnar, the US National Security Council's director of energy and climate change observed: "Very rarely do you get to a moment like that and feel completely that what the world had been looking for was achieved in this particular meeting,"[13] Outside the hall fifty thousand people were waiting. After sleepless nights during the negotiation marathon, when they heard about the success they produced a 'Mexican wave'.[14]

There was much criticism from scientists and social movements. Kumi Naidoo from Greenpeace insisted: "The Paris agreement is only one step on a long road and there are parts of it that frustrate, that disappoint me, but it is progress. The deal alone won't dig us out of the hole that we're in, but it makes the sides less steep."[15] The differences between the developed countries who were doing enough to control GHG emissions and the threshold countries, such as China and India, who should do more, were not resolved. There is an agreement that industrialized nations are responsible for the cumulative emission of GHG since the start of the industrial revolution and that China is now the largest emitter, followed by the US, the twenty-eight countries of the European Union, India, Russia, Indonesia, Brazil, Japan, Canada and Mexico. These ten emitters accounted for 68 per cent of all GHG, have 60 per cent of the population in 2011 and produce 74 per cent of *gross domestic product* (GDP) worldwide (World Resource Insti-

tute).[16] There is also agreement that these GHGs—mainly *carbon dioxide* (CO_2)—are warming up the atmosphere and that global GHG emissions should peak as soon as possible, and fossil fuels, including those used for electricity generation, should be eliminated by 2050, thus realizing a decarbonization.

During the Paris conference new alliances emerged, such as the "Carbon Pricing Leadership Coalition"[17] with seventy-four countries, twenty-three subnational governments and more than a thousand companies committed to putting a price on carbon. President Obama launched "Mission Innovation", by which twenty countries agreed to double research and development spending on clean energy technologies over the next five years.[18] Bill Gates promoted the "Breakthrough Energy Coalition", with twenty-eight of the world's wealthiest people providing funding for clean energy technologies.[19]

Finally, the signatories of the Paris Agreement agreed that climate change is closely intertwined with several *Sustainable Development Goals* (SDG),[20] including gender, water, food and health security, where gender equity (tool 7), the empowerment of the vulnerable, climate-intelligent agriculture and nutrition (Oswald Spring 2016) may increase crop yields and improve the income of small-scale farmers, especially in rain-fed and mountain agriculture.

1.3.6   Biodiversity Loss

Gibbs, Ruesch, Foley et al. (2010) indicate that in the tropics between 1980 and 2000 about fifty-five per cent of pristine forest and twenty-eight per cent of disturbed[21] forest were cleared for new agricultural land. This deforestation represents a 'criminal biodiversity loss' affecting crucial ecosystem services, which are

---

13 See Wilson Center, Brazil Institute, 16 December 2015: "Beyond the Paris Climate Talks: What Was Achieved and What Remains To Be Done"; at: <https://www.wilsoncenter.org/event/beyond-the-paris-climate-talks-what-was-achieved-and-what-remains-to-be-done> (20 February 2016).

14 The Mexican wave was launched during the soccer cup in Mexico in 1973 when 120,000 people in the Stadium Azteca raised their hands and moved them jointly in order to achieve a global wave though a big crowd representing a sea-wave.

15 As cited by Fiona Harvey: "Paris climate change agreement: the world's greatest diplomatic success", in: The Guardian, 14 December 2015; at: <http://www.theguardian.com/environment/2015/dec/13/paris-climate-deal-cop-diplomacy-developing-united-nations> (20 February 2016).

16 See: Mengpin Ge, Johannes Friedrich and Thomas Damassa: "6 Graphs Explain the World's Top 10 Emitters"; at: <http://www.wri.org/blog/2014/11/6-graphs-explain-world%E2%80%99s-top-10-emitters> (20 February 2016). These graphs include the cumulative CO_2 emissions from 1850 to 2011 and from 1990 to 2011.

17 See at: <http://www.carbonpricingleadership.org/> (20 February 2016).

18 See at: <https://www.whitehouse.gov/blog/2015/11/29/announcing-mission-innovation> (20 February 2016); see also at: <http://www.mission-innovation.net/> (20 February 2016).

19 See at: <http://www.breakthroughenergycoalition.com/en/index.html> (20 February 2016).

20 See at: <http://www.un.org/sustainabledevelopment/sustainable-development-goals/> (20 February 2016).

providing food, water, energy, wood, biomass, oxygen and fibre; *supporting* the nutrient cycle, the carbon, nitrogen, sulphur cycle, acting as a biodigestor of waste, and removing naturally different toxins; *regulating* climate, water, waves, and controlling floods, winds and storms; and *offering cultural services* of material and immaterial goods and the patrimony of humanity (MA 2005, 2005a, 2005b).

Human activities, especially in the Anthropocene, have seriously affected Earth's biodiversity. Current extinction rates are a thousand times greater than fossil extinction rates were, and are projected to increase in future to ten times more than the current rate, that is, ten thousand times greater than fossil extinction rates (MA 2005, 2005a, 2005b). This ecological footprint is seriously affecting biocapacity and ecosystem services. During the last four decades many countries, including some of the most biodiverse states (India, China, Mexico, US, Indonesia, Cambodia, Vietnam), have destroyed their natural capital and have created today a bio-debt, which is unable to restore naturally their biodiversity and ecosystem services. The WWF notes that "10,000 representative populations of mammals, birds, reptiles, amphibians and fish, have declined by 52 per cent since 1970" (WWF 2014).[22]

Biodiversity is a cornerstone of developed and developing economies. The *Convention on Biological Diversity* (CBD),[23] as a global agreement, addresses all aspects of biological diversity: genetic resources, species, and ecosystems with their services. It was signed in 1992 in Río and has nearly universal membership, with 196 parties.[24] Two decades later at Rio+20, the 'Strategic Plan' proposed an integrated methodology, where biodiversity was incorporated in economic and social policies. The 'Strategic Plan' (known also as 'Aichi Biodiversity Targets') adopted a long-term vision with five goals, twenty targets, and different tools and mechanisms, able to implement, monitor, review and evaluate achievements.

The five goals address the mainstreaming of biodiversity into other sectors, reducing the pressures on biodiversity, improving the status of biodiversity, enhancing the benefits to all from biodiversity and ecosystem services, and providing for a participatory process of implementation.[25]

This transdisciplinary approach integrates the social, economic and environmental aspects of the SDG, represents a global commitment by all parties and is a flexible national framework for policies that provide mechanisms for monitoring, review and evaluation.

Without healthy conservation of biodiversity, livelihoods, ecosystem services, and natural habitats, food security can be severely compromised. The *Global Biological Outlook* 4 (GBO-4)[26] "finds that reducing deforestation rates have been estimated to result in an annual benefit of US$183 billion in the form of ecosystem services. In addition, many households in developing countries, especially in Asia, derive as much as 50-80 per cent of annual household income from non-timber forest products" (GBO-4 2014: 7). In the UN General Assembly, those countries who were convinced by the economic, environmental and cultural importance of biodiversity declared the years 2011-2020 as the UN Decade on Biodiversity. The *Strategic Plan for Biodiversity* 2011-2020 was adopted in 2010 at the tenth meeting of the *Conference of the Parties to the Convention on Biological Diversity* (CBD).[27]

The 'Nagoya Protocol on Access to Genetic Resources and the Fair and Equitable Sharing of Benefits Arising from their Utilization to the Convention on Biological Diversity'[28] is the second agreement after Cartagena's 'Protocol on Biosafety'.[29] The Nagoya Protocol is a supplementary agreement that was adopted on 29 October 2010 in Nagoya, and entered into force on 12 October 2014, by when seventy-one parties had ratified it (CBD 2011a).

---

21 A disturbed forest or a secondary forest is a "forest that has been logged and has recovered naturally or artificially. Not all secondary forests provide the same value to sustaining biological diversity, or goods and services, as did primary forest in the same location. In Europe, secondary forest is forest land where there has been a period of complete clearance by humans with or without a period of conversion to another land use. Forest cover has regenerated naturally or artificially through planting"; at: <https://www.cbd.int/forest/definitions.shtml> (26 February 2016).

22 See: "Living Planet Report 2014–species and spaces, people and places"; at: <http://wwf.panda.org/about_our_earth/all_publications/living_planet_report/> (20 February 2016).

23 See at: <https://www.cbd.int/> (20 February 2016).

24 The US is not party to the CBD; see at: <https://www.cbd.int/countries/default.shtml?country=us> (20 February 2016).

25 See at: <https://www.cbd.int/sp/targets/> (20 February 2016).

26 See at: <https://www.cbd.int/gbo4//> (20 February 2016).

27 See at: <https://www.cbd.int/sp/> (20 February 2016).

28 See at: <https://www.cbd.int/abs/> (20 February 2016).

29 See at: <https://bch.cbd.int/protocol> (20 February 2016).

The *Global Environment Facility* (GEF) is the international funding mechanism which supports parties in implementing the Convention and the Strategic Plan. The parties have identified and used different additional innovative domestic and international financial resources, and member states have emphasized national priorities and addressed the way they may contribute to the achievement of the SDGs (CBD 2015). National SDG agendas were promoted among the parties to improve their CBD goals.

Five goals were agreed with the aim of achieving the proposed conserved, restored, maintained and evaluated biodiversity with its ecosystem services by 2050 (CBD 2011c):

> Goal A: Address the underlying causes of biodiversity loss by mainstreaming biodiversity across government and society; Goal B: Reduce the direct pressures on biodiversity and promote sustainable use; Goal C: Improve the status of biodiversity by safeguarding ecosystems, species and genetic diversity; Goal D: Enhance the benefits to all from biodiversity and ecosystem services; Goal E: Enhance implementation through participatory planning, knowledge management and capacity building (CBD 2011b: 1).

Both at the local and national levels, legal terms were included into the national SDG plans. Scientifically there are still discussions regarding the monitoring process and national goals that are sometimes in conflict with global ones, such as extractivism[30] from mines and oil in natural protected areas in Latin America, or land grabbing in Africa.

The *National Biodiversity Strategies and Action Plans* (NBSAP) are trying to translate the global goals into national priorities, which must represent the interests of all ministries.[31] Some countries have been able to create an inter-ministerial or inter-agency committee to improve mainstreaming and integration (CBD 2015). Nevertheless, high profits from mines and oil made by private investors are causing them to pressure governments not to adhere to the CBD convention and its additional protocols. But there are also scientists who are concerned about the protocol. They experience many obstacles in their work for the prevention of diseases and in their conservation activities, including the imprisonment of researchers. The national history museums are afraid of limits to maintaining their biological collections and to the interchange among institutions, due to the greater legal certainty and transparency for providers and users. Finally, the Protocol must address the illegal trafficking of wild and especially threatened plants and animals, which has become a profitable business controlled by organized crime.

During studies for the preparation of the sixth report of the *Global Environmental Outlook* (GEO-6),[32] experts found that *Latin American and Caribbean* (LAC) countries contain an important proportion of Earth's natural wealth. The absence of environmental planning, the extraction and transformation of raw materials, population growth together with unsustainable urbanization and consumption patterns have degraded biodiversity, soils, water and air and affected ecosystem services, while pollution and toxic emissions have worsened human health and caused conflicts over land and water use.

1.3.7   Soil Degradation, Erosion and Desertification

From the land that is free of ice only 46.5 per cent is in a natural state. Of the remainder, 38.6 per cent is used for agricultural activities and 14.9 per cent has been eroded and occupied by urban, rural and industrial settlements, mines, dams, railways and roads, or is used for plantations (Leb Hooke 2015). Reducing the human footprint on the environment requires the sustainable use of soil, water and air and the encouragement of biodiversity (MA 2005, 2005a, 2005b). *Integrated water resources management* (IWRM) includes the conservation of forests, soils and riverbeds, the restriction of wind and water erosion, strengthening of the slopes and the reforesting of cleared areas. This lets the rain infiltrate into aquifers and the topsoil remains in the mountain areas.

Soil organic matter (SOM) is crucial for carbon sequestration. SOM is a biogeochemical mixture of organic matter that is in the process of decomposition. Due to microbial degradation and mineralization to CO_2 (and methane (CH_4) in anaerobic environments), the majority of plant litter and compounds added to soil remain for a relatively short time (from a few days to a few years). However, in high perennial tropical forests SOM could stay for millennia and

---

30 See Alberto Acosta: "Extractivism and neoextractivism: two sides of the same curse"; at: <https://www.tni.org/files/download/beyonddevelopment_extractivism.pdf>; and at: <http://wiki.elearning.uni-bielefeld.de/ wiki-farm/fields/ges_cias/field.php/Main/ Unterkapitel53> (20 February 2016).

31 See at: <https://www.cbd.int/nbsap/introduction.shtml> (20 February 2016).

32 See at: <http://www.unep.org/geo/> (20 February 2016).

deforestation could produce an abrupt process in the biogeochemical composition of the tropical soil, leaking the stored carbon to the atmosphere. Water is also essential for adequate microbial activity in soils, as in drier soils the rates of decomposition decrease. But flooded soils have lower amounts of organic matter. A high level of precipitation or intensive wind may produce wind or water erosion, with intensive loss of the natural fertility of soils. Healthy soils support plant growth and provide nitrogen, phosphorous, potassium, calcium, magnesium, sulphur and different trace elements (FAO 2016b).

The causes of land degradation—and in its extreme form, desertification—are massive deforestation, overgrazing of fragile savanna soils, overuse of agrochemicals, use of heavy machinery, bad irrigation practices, lack of soil protection, monocultures, and wind or water erosion. One of the most important effects of soil deterioration is desertification, defined by WWF (2015)[33] as being "characterized by the droughts and arid conditions the landscape endures as a result of human exploitation of fragile ecosystems. Effects include land degradation, soil erosion and sterility, and a loss of biodiversity, with huge economic costs for nations where deserts are growing".

Nkonya, Gerber, Baumgartner et al. (2016) estimated that the overall cost of land degradation per year amounts to US$300 billion. Soil degradation affects 3.2 billion people and thirty per cent of the surface of Earth. Of this, forty-six per cent is related to land use changes at a cost of US$231 billion—equivalent to 0.41 per cent of global GDP—and fifty-four per cent is related to the loss of ecosystem services. Deterioration of the natural soil fertility alone accounts for an annual loss of US$15 billion or 0.07 per cent of global GDP.

The most severely affected region is sub-Saharan Africa, followed by Latin America and the Caribbean. In Latin America, Mexico and Argentina suffer worst, due to severe soil deterioration. Land use has undergone significant change during the last century. Between 1900 and 2000, natural forests and mountain areas decreased from 70.1 per cent to 46.5 per cent; crop land increased from 27.2 per cent to 46.5 per cent; and urban areas from 2.7 to 6.9 per cent (UNCCD 2015: 6). In relation to GEC and climate change, there is an emission gap which requires urgent action to achieve sustainable land management, ecosystem and wetland restoration, climate-smart agriculture, agroforestry, and land use and urban planning to reduce not only environmental costs, but also the threat of disasters. UNCCD (2015) suggested restoring twelve million hectares of degraded land every year to help close the remaining emission gap. This was discussed in Paris and may amount to 25 per cent by 2030.

Soil erosion basically affects the upper layer of fertile soil. Half of the topsoil on Earth has been lost during the last 150 years. Soil quality is further affected by compaction, loss of soil structure, nutrient degradation, and soil salinity, which in extreme cases could result in processes of sodification.[34] From the 'green revolution' of 1950 onwards, agrochemicals have been massively used, increasing atmospheric pollution and GHG (CO_2, N_2O), as well as water and ocean eutrophication, and causing serious risks to human health. The *Food and Agriculture Organization of the United Nations* (FAO) (2016b) has examined global trends in soil management and their effects on ecosystem services as far as supply is concerned, where land use change, decreased water availability and biodiversity, decreased soil saturation and increased biomass have had the most important impacts. As far as regulation of supply is concerned, water storage capacity, decreased filtration capacity and carbon sequestration have suffered losses, while from the cultural aspect, there have been impacts on landscape, with decreased natural and recreational potential.

Figure 1.5 (UNCCD 2015) shows the cycle by which deteriorating soils produce a loss in ecosystem services and decreased mitigation and adaptive capacities, which are further worsened by the poor management of land resources. Further effects are related to *climate change* (CC) and biodiversity losses, where invasive species, diseases, loss of habitat and changes in species abundance are increasing existing land degradation and therefore affecting different ecosystem services. Figure 1.5 also points to linkages among the three Rio agreements of 1992 on climate change (UNFCCC), biodiversity (CBD) and land degradation (UNCCD). It is these linkages that are aggravating GEC and CC, where deterioration in soils increases the impacts of disasters, especially during extreme events, such as drought, floods, landslides, and loss of resource management.

---

33 See WWF (2015) at: <http://www.worldwildlife.org/threats/soil-erosion-and-degradation> (18 February 2016).

34 "Soil salination and sodification is one of the soil degradation processes"; see at: <http://enfo.agt.bme. hu/drupal/en/elearning/11444>; see also: Levy , Fine, Goldstein, Azenkot, Zilberman, Chazan Grinhut (2014).

Figure 1.5: Interlinkages between land, climate and biodiversity. **Source**: UNCCD (2015: 7).

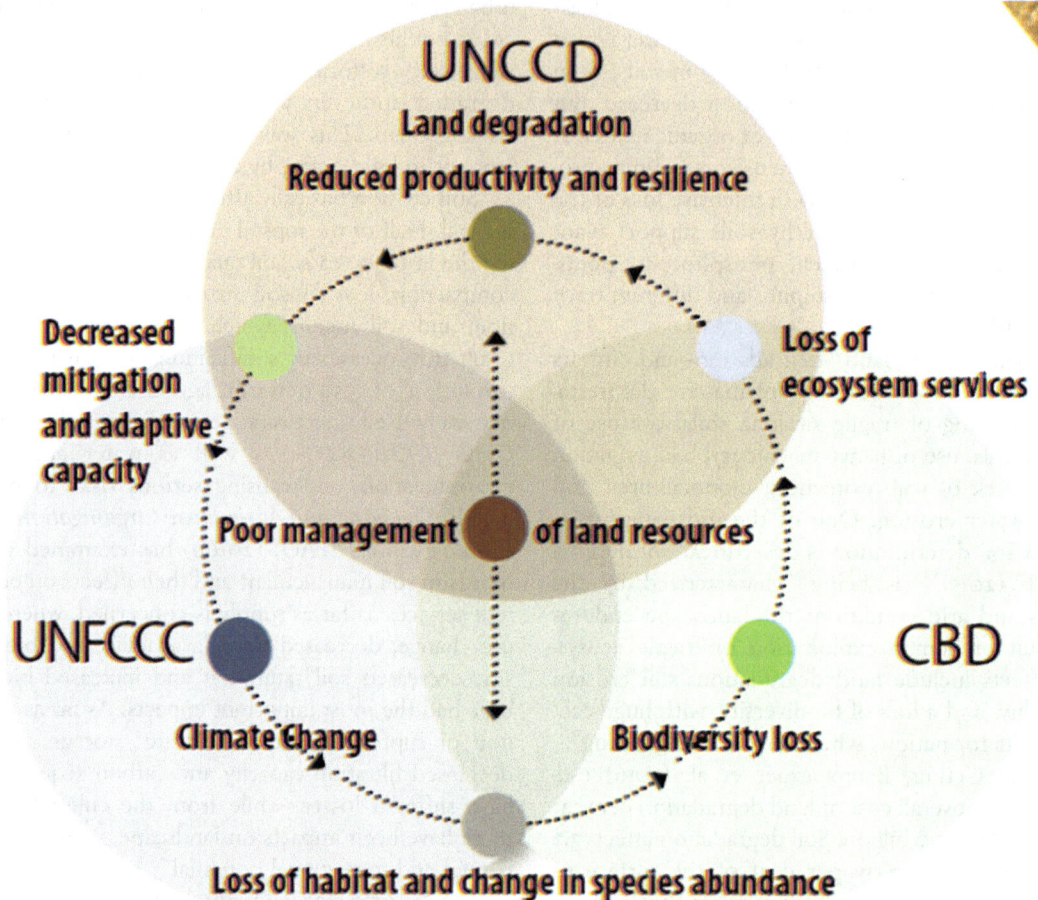

1.3.8   Water Degradation, Scarcity and Stress

Water accessibility, use and management is crucial for everybody in daily life, as well as for food and industrial production. Only recently there have been concerns about the crucial role of ecosystem services and the loss and pollution of water. Clean water is vital for human health, yet still one out of six people does not have access to safe water and 2.5 billion people lack basic sanitation facilities, while large-scale corporate agribusiness consumes seventy per cent of the fresh water on Earth. Ban Ki-moon argued in the introduction to UN Water (2015):

> Water flows through the three pillars of sustainable development–economic, social and environmental. Water resources, and the essential services they provide, are among the keys to achieving poverty reduction, inclusive growth, public health, food security, lives of dignity for all and long-lasting harmony with Earth's essential ecosystems.

In 2014 UN Water recommended five policies: (i) WASH [access to water, sanitation and hygiene]; (ii) water resources; (iii) water governance; (iv) water quality and wastewater management; and (v) water-related disasters.

Water is analysed generally for its environmental, social, and economic impacts. The link between water and ecosystem services, especially wetlands and mangroves, is still under-recognized. Water is also crucial for safe food production, healthy forests, sustainable agriculture, and urban planning, and reduces human and material losses during any disaster. An integrated focus on ecosystems for development and water protection and conservation should be relevant in decision-making and environmental planning. Evaluation of ecosystem services assesses the trade-offs between water conservation, adaptation to more extreme climate events, and safe drinking water.

The social aspect of water is related to poverty and marginalization with a cross-cutting gender under-

Figure 1.6: Total renewable water resources, cubic metres per capita (2013). **Source:** UN Water (2015: 12).

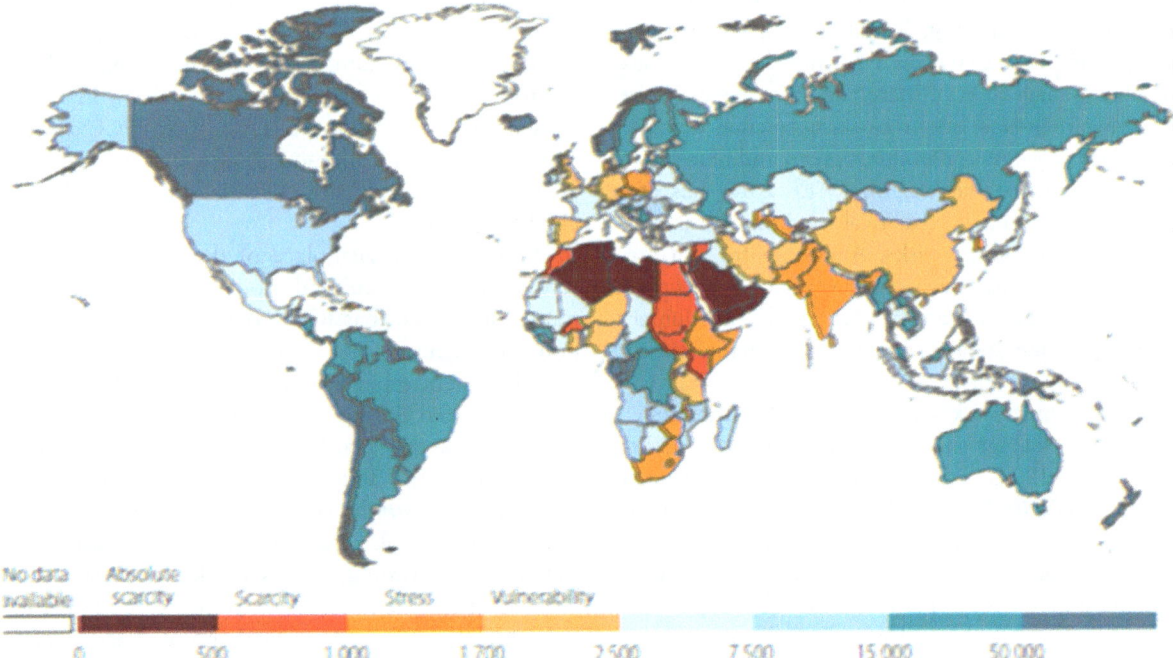

standing. Women are generally responsible for water supply, which is critical for the well-being and health of the family. Clean water reduces water-related illnesses and better hygienic conditions improve well-being. At the household level subsistence food production is crucial to overcoming chronic undernourishment, especially for children under five years. Inequality within and between countries is a key factor in underdevelopment and lack of the basic services of safe water, improved sanitation and healthy food supply. Water is ubiquitous in any productive process and clean water reduces costs in health and loss of working hours. Investments in water facilities increase opportunities for businesses and enable self-sustained small enterprises. Clean water and sewage treatment plants depend on the available technology and trained people to operate these facilities. There exists a wide range of technologies, and depending on the number of people to be served and the financial possibilities they range from the household to big cities. Dispersed settlements may increase the costs of providing basic water and sewage facilities. Nevertheless the impact on social and environmental assets compensates for the financial investment, but also represents a basic human right.

The UN Water Report (UN 2015: 4) estimates that "by 2050, agriculture will need to produce 60 per cent more food globally, and 100 per cent more in developing countries". The water footprint is high in agri-culture (Hoekstra 2011) and requires 1,021 litres of water for every litre of milk; 17,196 litres for 1 kg of chocolate and 15,415 litres for 1 kg of beef. Energy production is also water-intensive and additional energy demand will increase the stress on freshwater resources and cause conflicts with existing water users. Increasing solar, geothermal, and wind energy reduces the demand from the cooling systems of thermal power plants. Manufacturing industry is estimated to increase by 400 per cent between 2000 and 2050 but water and energy-efficient productive processes will save fresh water. Some multinational enterprises are already moving their production chain to water-rich countries.

The water demand is expected to increase in all sectors and climate change will further threaten the water supply. The *UN Water Report* (UN 2015: 11) estimates that with a BAU scenario there will be a global water deficit of forty per cent. Figure 1.6 shows renewable water resources per person in 2013, with North Africa and the Arabian Peninsula threatened by absolute water scarcity; Sudan, Kenya and Ghana by water scarcity; and India, Pakistan, Zimbabwe, Myanmar, Tajikistan, Poland, the Czech Republic, the Horn of Africa, and South Africa by water stress. Several of these water-stressed countries are currently involved in armed conflicts and water scarcity is increasing the risks of political instability.

To reduce water conflicts, several transboundary water basins are globally managed through collaboration (the Nile, Mekong, Danube, etc.), but no legal agreements exist that would provide long-term sustainable management of shared aquifers. Big cities are depleting their aquifers in an unsustainable way and often lack of sanitation pollutes green and blue water. The *World Economic Forum* (WEF 2016) has analysed the economic, geopolitical, environmental, societal and technological risks. At the centre is a profound water crisis, which could produce food crises, profound social instability, large-scale involuntary migration, state collapse and massive unemployment and underemployment.

This complex risk scenario is further threatened by climate change impacts and extreme weather events. The IPCC (2014a) assessed that the *Palmer Drought Severity Index* (PDSI) for 1900 to 2002 has drastically changed, and that the cumulative deficit due to precipitation is affecting surface land moisture. Global estimates suggest a probability of 29.5 per cent of longer dry periods, 42.4 per cent of irregular rainfall; of 15.03 per cent in insufficient rain for crops; and of 13.06 per cent in extreme rainfall. This global prediction especially affects rain-fed agriculture, which still provides most food worldwide.

1.3.9 Food Scarcity and Hunger

Food scarcity is of global importance. In 2015, about 850 million people went hungry. By 2050 the world must feed another two billion inhabitants, especially in the poorest countries. In addition, greater prosperity in China, India and other developing countries has increased the demand for meat, milk and eggs, and animals are fed mainly on cereals, in competition with a vegetarian diet[35] and inflating the price of grains (GHI 2014, 2015). Since 1960, the green revolution has meant that modern agriculture employs mechanization, irrigation and agrochemicals. This has increased productivity at the expense of depletion of the aquifers and soils. Genetically modified seeds now promise pest control, but have affected human health and have not increased productivity (Velázquez 2015). Modern agriculture emits more greenhouse gases (GHG) than all means of transport taken together, especially methane from livestock and rice production, nitrous oxide (N_2O) emitted from fertilized fields, and CO_2 as a result of deforestation for agriculture and ranching (IPCC 2013).

Although 53.5 per cent of primal vegetation has been modified by human activities, more than half of the food produced is still grown organically in home gardens, often by women (FAO 2015; IPCC 2014). Most of the grain from modern agriculture is used for animal feed (FAO 2013b) and transformed into biofuels (Ren21 2014).

Mexico is the birthplace of the so called 'green revolution', an agricultural model using improved or transgenic seeds, massive agrochemicals, mechanization and irrigation. This causes groundwater pollution and abatement especially in the drylands all over the world (Oswald Spring 2011). This model is globally conceived as the 'productive model' of agriculture (Land/Heaseman 2004). The *United Nations Environment Programme*'s (UNEP) GEO-5 report, Gibbs, Ruesch, Foley et al. (2010) and the WWF (2014) have stated that according to this model, and especially because of change in land use from pristine forest to agriculture, intensive agrochemical production modes, and genetically modified organisms, a total of fifteen out of twenty-four ecosystem services have deteriorated (MA 2005). The effects on ecosystems and their services with this 'resource intensive model', aggravated by extensive food subsidies, were further accelerated by the impacts of climate change (IPCC 2013). Particularly in the most affected regions in the South, international organizations and governments have begun to promote adaptation, changes of crop and different seeds to increase the level of resilience in agricultural production.

Because of the increasing changes in food intake, as well as obesity, researchers have included in their models the process of nutrition, access to safe, nutritional and culturally acceptable food for a healthy life, and the permanent local availability of food at reasonable prices. A key issue in 2008 was the increase in the price of basic food. This caused the number of hungry people on the planet to increase to more than one billion. The situation was partially aggravated by drought, by speculation on basic grains in the stock markets, and by the massive use of these grains for biofuel. By 2015, the FAO (2016a) estimated that 800 million people were still chronically undernourished, and 161 million children under five years of age were stunted in their growth;[36] this may result in brain damage and thus affect their future lives (Álvarez/ Oswald Spring 1985). Two billion people lack the essential micronutrients for a healthy life, and more than 500 million are obese.

---

35 A calorie of beef requires ten calories of grain; and a vegetarian diet could feed more people.

Increases in population, income and urbanization, and free trade agreements, have made the situation more complex. Together with increases in the price of food, they have driven changes in the demand for food and the demand for feed. In addition, in industrialized countries subsidies for biofuel and livestock have distorted international grain prices, and financial capital has been used to speculate on basic food staples (De Schutter 2014). Both biofuels and ranching have exacerbated the competition for land, water, and food investment. Since 2011, half of the corn produced has been used for ethanol (biofuel) in the US (REN21 2015: 44). These structural phenomena in food production have been further challenged by the impacts of climate change. Transdisciplinary researchers have linked the food system with global environmental change, gender studies and health impacts. The bridges between the social and natural sciences have played a crucial role. Eriksen (2010) synthesized the empirical evidence and showed that the trade-offs of an integrated food system are often context-specific and differ at different scales, and that climate conditions and changes also play a crucial role.

The IPCC (2014a) identified the following as key factors in the decline in food security: the loss of rural livelihood and income; the deterioration in terrestrial and inland water ecosystems; the deterioration in marine and coastal ecosystems; and food insecurity and the breakdown of regional food systems. Prospective models indicate current and future impacts on water availability and supply, food insecurity, and a deterioration in agricultural income. Drylands especially may suffer from shifts in the areas where food and non-food crops are produced. The FAO (2016a: 4) has developed a scheme of cascading effects of climate change on food and nutrition that includes physical, biological, biophysical, socio-economic, livelihood and health factors. Interactions between natural and social factors, changes in food culture, and economic deprivation are further aggravated by climate change impacts, disasters and land degradation. These affect above all the most vulnerable of the poorest countries in Africa, Asia and Latin America. The fifth assessment report of the IPCC (2013, 2014) mentions co-benefits from mitigation, adaptation and preventive behaviour at the national and local levels.

Such behaviour would guarantee food security and healthy nutrition for the most vulnerable people, mostly women and girls.

1.3.10　Gender Impacts of Global Environmental and Climate Change

GEC and CC are not gender-neutral. Socially determined discrimination and conditions of exclusion have limited access to strategic goods, land, education, and health care (Ímaz/Blazquez/Chao et al. 2014). For millennia, the elimination of women from decision-making processes has made them invisible and prevented their equal participation in productive processes. These structural factors, deeply embedded in the patriarchal system and characterized by authority, exclusion, discrimination, exploitation and violence, have also affected justice and human rights. These historical and socially constructed processes also affect values of solidarity and cultural identity (Serrano Oswald 2014). The sources of threats towards women have developed over thousands of years, influenced by patriarchal institutions (Folbre 2006), religious controls (Jaspers 2013), and the totalitarian exercise of power. MacKinnon (1987) claimed that the distinction between women and men is a not just a difference of sexes, but of powerlessness and power. Nevertheless, Judith Butler (1990) criticized the sex/gender distinction and showed that the socially constructed order upholds current power relationships.

This long-term discrimination has been exercised through violence, and this has affected more vulnerable women and girls and has often threatened their survival in disasters (Fordham 1999; Oswald Spring 2008) and wars. Women's bodies were often made into a battlefield (Hynes 2004). During the past millennia the evolution and consolidation of patriarchy has changed and become regionally differentiated. Nevertheless, there are some common dominant factors: ideological control mechanisms around the world have both sociobiological and social constructivist explanations. Ecofeminists (D'Eaubonne 1974; Mies 1986; Warren 1997) believe that similar power relationships between dominance, greed, exploitation, and violence have been the cause of GEC, CC and the destruction of biodiversity. In this context, the international feminist agenda has emphasized the need for a transverse gender perspective for strategies and actions of mitigation, adaptation and disaster risk reduction, where all types of discrimination against women are eliminated. UNEP recognizes that GEC

36　UNICEF defines stunted growth as "moderate and severe—below minus two standard deviations from median height for age of reference population", at: <http://www.unicef.org/infobycountry/stats_popup2.html> (26 February 2016).

Figure 1.7: Cascading effects of climate change on food security and nutrition. **Source**: FAO (2016a: 4).

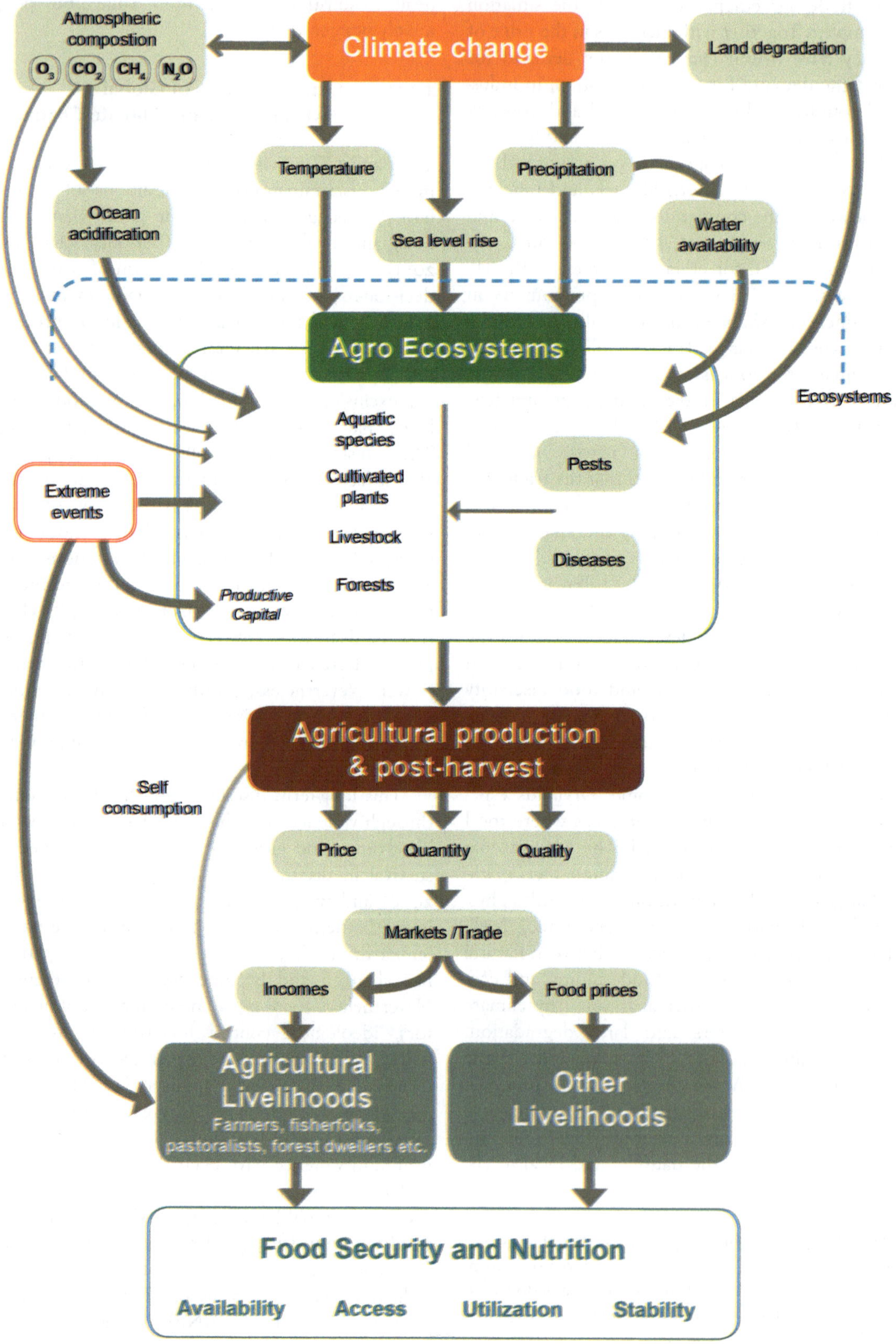

and CC are environmental, technological, social and economic issues, but that they are primarily political and development concerns where gender equality is crucial. The identification of different gender processes in water, food, livelihood, technology and disasters also opens up challenges for collective processes that may create a different qualitative growth, with equality, equity, sustainability and safety for all, men and women.

Traditionally women were always involved in food production, as paid agricultural workers, in the transformation and cooking of food, and in local markets for agricultural and food products. Nevertheless, the modern productive system has made the work of women invisible. Furthermore, modern agriculture is only evaluated in terms of the use of heavy machinery, agrochemicals and large land resources. The IPCC (2014) stated that, especially in developing countries, most of the food is directly produced by women, mostly using positive carbon sequestration processes, reuse and recycling of waste, and the protection of soil, health, and biodiversity. Forests, oceans and soils are crucial for climate mitigation and water capture. People's different capabilities depend on gender, ethnic, age, and social class. This means that governments, international institutions, and social movements need to revise their approaches to vulnerability, risk, prevention, adaptation and resilience.

1.4 Impacts of Different Economic Development Paths

These changes in the global environment have been caused by anthropogenic climate change, loss of biodiversity, processes of soil erosion, degradation and desertification, and processes of water degradation, scarcity and stress. The changes have been facilitated by an increase in the consumption of cheap fossil energy sources (coal, oil and gas). This has emerged slowly since the industrial revolution and has rapidly increased since the end of World War II (1945), and has further intensified since the end of the cold war (1990).

The dominant Western economic development path of *business-as-usual* (BAU) relies on high economic growth rates and uncontrolled financial transaction and accumulation processes. It became globalized with the adoption of state capitalism in China (inspired by Deng Xiaoping's reforms in the 1980s) and the collapse of state communism in the early 1990s. The industrial revolution first emerged in England against the background of the British capitalist system. In the early twentieth century, and especially after 1919, the US replaced Great Britain as the leading world economy. After World War II, the US opted for an active role in economic, political and military world leadership. It was instrumental in setting up and initially controlling the Bretton Woods institutions (World Bank, IMF), the United Nations and regional organizations (OAS), the *World Trade Organization* (WTO), and a system of military alliances (e.g. NATO).

Particularly since the end of the cold war, economic neo-liberalism—influenced by the 'Consensus of Washington'—became the dominant global economic doctrine driving globalized financial processes and markets (Notz 2015). Neo-liberalism had been promoted since the early 1970s and 1980s by neo-liberal economic theories and economists (e.g. the Chicago school of Milton Friedman) and implemented by the economic policies of Margaret Thatcher in the United Kingdom and Ronald Reagan in the United States.

Neo-liberalism means the privatization of public services, state-owned property and banking systems, deregulation, a reduction in governmental spending and public subsidies, economic liberalization, and wage containment (Strahm/Oswald Spring 1990; Stiglitz 2010; Kotz 2015). The World Bank and IMF reinforced this policy with a structural adjustment policy applied to highly indebted countries.

In the dominant development path, nature became an object of exploitation. However, from the 1960s environmental movements and from the 1970s national (ministries, agencies, parties, *non-govern mental organizations* (NGOs)) and international environmental institutions (UNEP, environmental regimes) addressed the negative consequences that were resulting in the anthropogenic processes of GEC, CC, desertification and water scarcity, degradation and stress.

The materialization of ecosystem services and natural resources has also existed in the state socialist system of the Soviet Union (1917–1991), in the People's Republic of China (since 1949), in Eastern Europe (1945–1989), and in selected communist developing countries (Vietnam, North Korea, Cuba, et al.). The environmental impact of socialist development strategies was largely ignored. With the disintegration of the Soviet Union, the transition of former socialist and communist countries in Eastern Europe to a market economy, economic growth, production and international trade were at the centre of new development strategies. In 2007, China became the largest emitter

Figure 1.8:   Share of global wealth of the top 1 per cent and the bottom 99 per cent respectively. **Source**: Oxfam 2016: 2, based on available Credit Suisse data (2000–2014).[a)]

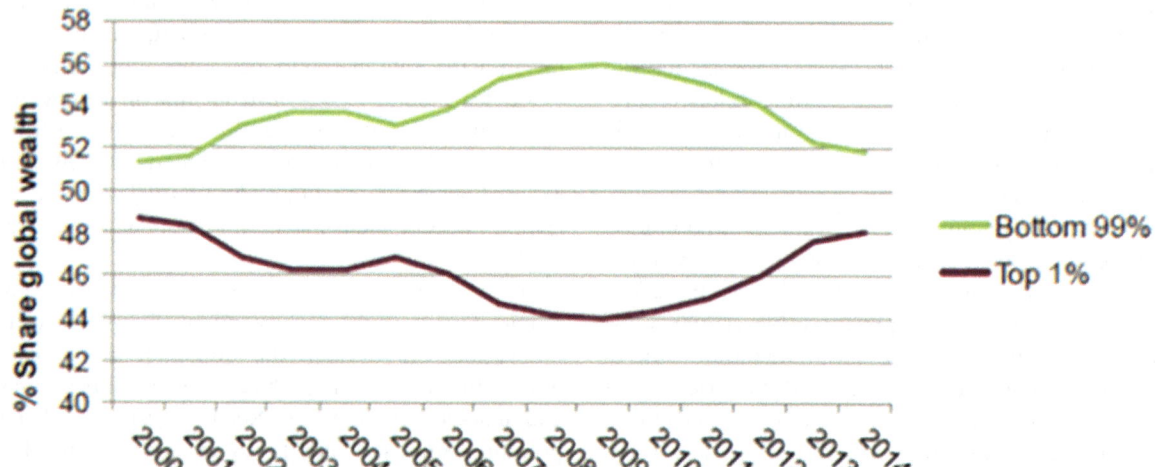

a) OXFAM: An Economy for the 1%. How privilege and power in the economy drive extreme inequality and how this can be stopped (Oxford: January 2016); at: <https://www.oxfam.org/sites/www.oxfam.org/files/file_attachments/ bp210-economy-one-percent-tax-havens- 180116-en_0.pdf> (22 February 2016); for a critique of these data: Chris Giles, "Three reasons to question Oxfam's inequality figures", in: *Financial Times*, 18 January 2016; at: <http:// blogs.ft.com/ftdata/2016/01/18/three-reasons-to-question-oxfams-inequality-figures/> (22 February 2016); Credit Suisse: "Global Wealth in 2015: Underlying Trends Remain Positive" (13 October 2015); at: <https://www.credit- suisse.com/uk/en/about- us/research/research-institute/news-and-videos/articles/ news-and-expertise/2015/10/en/ global-wealth-in-2015- underlying-trends-remain-positive.html> (22 February 2016).

of GHG, producing 15 per cent of global cumulative GHG emissions (US: 16 per cent; EU: 12 per cent; Russia: 6 per cent).[37] In 2014, China emitted 7.4 tonnes of CO_2 per capita, overtaking the average of the twenty-eight states of the EU (6.69 tonnes of CO_2 per capita).[38]

With the neo-liberal 'Washington Consensus',[39] global inequality within and inequity among countries has increased from the early 1990s onwards. Stiglitz (2010) has analysed the impacts of faith in the free market and globalization and has shown that the 'great recession' started in 2008 and produced massive unemployment, poverty and a global crisis in so-

called 'American Capitalism'. The richest country in the world had lived beyond its capacity and low interest rates, due to lax regulations in their own country, caused the mortgage bubble. Stiglitz (2010: 21) argued that the crisis became global because "nearly a quarter of US mortgages had gone abroad" and European financial institutions suffered from these toxic mortgages. Banks, car producers and numerous companies

37 See: Mengpin Ge, Johannes Friedrich and Thomas Damassa: "6 Graphs Explain the World's Top 10 Emitters", World Resources Institute (25 November 21014); at: <http://www.wri.org/blog/2014/11/6-graphs-explain-world%E2%80%99s-top-10-emitters> (2 March 2016).

38   See: EU, Joint Research Centre: "CO_2 time series 1990-2014 per capita for world countries"; at: <http:// edgar.jrc.ec.europa.eu/overview.php?v=CO2ts_pc1990-2014>; *Trends in Global CO2 Emissions 2015 Report* (The Hague: PBL Netherlands Environmental Assessment Agency, 2015); at: <http://edgar.jrc.ec. europa. eu/news_docs   jrc-2015-trends-in-global-co2-emissions-2015-report-98184.pd> (22 February 2016).

39 The International Monetary Fund (IMF), the World Bank and the US Treasury Department imposed ten economic policy prescriptions to stabilize macro-economically the countries affected by economic crises. John Williamson coined the term 'Washington Consensus' in 1989. This forced developing countries in particular to achieve: a disciplined fiscal policy that would avoid a large fiscal deficit in relation to their GDP; reduction of subsidies and reorientation of public spending towards pro-growth and pro-service investments (private education, health care, infrastructure); a broadening of the tax base through tax reforms; market-determined interest rates; competitive exchange rates; liberalization of imports and elimination of trade protections through trade liberalization; openness to foreign investments and free capital flows; privatization of state enterprises; deregulation of norms that restrict international competition, especially in financial institutions; and legal security for private property rights.

collapsed when suppliers went bankrupt. The American government bailout supported businesses generously, while "low-income workers who had to work hard all their life and had done nothing wrong had to take a wage cut, but not the million-dollar-plus financiers who had brought the world to the brink of financial ruin" (Stiglitz 2010: 43).

The outcome is shown in figure 1.8. The richest 1 per cent of people owned 48 per cent of global wealth in 2014.[40] Of the remaining 52 per cent, the richest 20 per cent leave only 5.5 per cent of global wealth for the rest—80 per cent—of the world's population. Thus the 'Washington Consensus' permitted the richest eighty people in the world to accumulate wealth at a rate even faster than in the past, while the wealth of poor people has decreased. Today these eighty people own more than the whole of the bottom fifty per cent of the world's population. Their income has increased rapidly over the last year and grew especially fast after the crisis of 2008, while poor people lost purchasing power.

This inequality is socially constructed and could potentially be overcome with specific policies. Oxfam (2016) proposed that governments should work for their citizens and tackle extreme inequality; promote women's equality and women's rights; adopt a higher and progressive tax burden on financial capital and income; implement free public services and universal social protection by 2020; reorient R&D towards public health services; invest in development finance to promote the integration of poor people; and close international financial and tax loopholes (tax havens).

The abolition of the ten mandates of the Washington Consensus could be added, as well as a reduction in differences of terms of trade between raw materials and manufactured products; the elimination of agricultural subsidies in industrial countries, which are destroying the food supply in poor countries because of an imposed policy of comparative advantage; land-grabbing and extractive industries; corruption and lobbying of politicians; patents and TRIPs (*trade-related aspects of intellectual property rights*); reinvestment of military and arms spending in development projects and equity; reduction of debts for poor countries, and so on.

1.5   Integrating the Global Trends

In the scientific realm, different international efforts have addressed the complex linkages between the four global research programmes (WCRP, IGBP, DIVERSITAS, IHDP). In 2002, the *Earth Systems Science Partnership* (ESSP) emerged from a conference in Amsterdam (2001) and a decade later the *Future Earth Initiative*[41] of the *International Council of Science* (ICSU) succeeded the ESSP and its four research programmes; and the *Global Research Forum on Sustainable Production and Consumption* (GRF-SPaC)[42] emerged from meetings in Rio de Janeiro in June 2012.[43]

1.5.1   The IPCC: Climate Change Impacts for Human Security

So far the direct effect of the four key physical effects of climate change (temperature increase, precipitation change, sea-level rise and increase in extreme events) on security issues (migration, conflicts and wars) is contested (IPCC 2014a, chapter 12; Scheffran/ Brzoska/Brauch et al. 2012; Gleditsch 2012). In its assessment of the peer-reviewed social science literature on climate change and human security, the IPCC (2014a) concluded (see box 9.1 in chapter 9 by Brauch):

- *Climate change will have significant impacts on forms of migration that compromise human security (high agreement, medium evidence). ...*
- *Mobility is a widely used strategy to maintain livelihoods in response to social and environmental changes (high agreement, medium evidence). ...*
- *Some of the factors that increase the risk of violent conflict within states are sensitive to climate change (medium agreement, medium evidence). ...*

---

40 In 2014 1,645 people were listed by Forbes as billionaires. About 30 per cent are US citizens, 85 per cent are aged over fifty and 90 per cent are male (Oxfam 2016: 5).

41 See: <http://www.icsu.org/future-earth/> and: <http://www.icsu.org/future-earth/who>. Future Earth builds on the existing global environmental change programmes (Diversitas, IGBP, IHDP, WCRP and ESSP).

42 See: <http://grf-spc.weebly.com/> and its research agenda: <http://grf-spc.weebly.com/research-agenda.html>.

43 This problem will be discussed in more detail in the concluding chapter by Oswald Spring, Brauch and Scheffran.

- *People living in places affected by violent conflict are particularly vulnerable to climate change (high agreement, medium evidence). ...*
- *Climate change will lead to new challenges to states and will increasingly shape both conditions of security and national security policies (medium agreement, medium evidence)....*

The future impact of climate change on human, national and international security will depend on the success of international diplomatic political efforts to freeze and reduce greenhouse gas emissions, dematerialize the extraction of raw materials, recover ecosystem services, forests, wetlands, and mangroves, protect fresh water, and recover desertified soils.

At the national level, initiatives towards a rigorous application of *Nationally Appropriate Mitigation Actions* (NAMAs) and *Intended Nationally Determined Contributions* (INDCs), and for the protection and restoration of natural areas, climate-smart agriculture and a sustainable economy may counter GEC and CC and conserve the planet for present and future generations.

1.5.2 The Scientific and Political Nexus Debates

With its Risk Report, the *World Economic Forum* (WEF 2011) started a discussion on the security nexus between *water, energy, and food* (WEF), but without including biodiversity. This Forum is an important meeting place for global policy debates between political leaders and the global business elite, who represent multinational corporations and major financial, productive, commercial, and entertainment institutions.

In an early nexus paper the WEF (2011) had adopted a traditional national security approach (Oswald Spring 2016) and ignored the widening, deepening and sectorialization of security (Buzan/Wæver/de Wilde 1998; Brauch/Oswald Spring/Grin 2009). Their nexus approach between *water, energy and food* security reflected the traditional assumptions of a *business-as-usual* strategy.

It ignored a sustainable nexus between *water security* (WS), *energy security* (ENS), *food security* (FS) and *biodiversity security* (BS) from an *environmental* (ES) and *human security* (HS) perspective. Such a HS approach is based on the four pillars of *freedom from fear* and *freedom from want* (UNDP 1994, CHS 2003), *freedom to live in dignity* (Annan 2005) and *freedom from hazard impacts* (Bogardi/Brauch 2005; Brauch/Scheffran 2012).

1.6 Three Key Concepts: Sustainable Development, Sustainability Transition and Sustainable Peace

The key concepts for this book of 'sustainable' and 'sustainability' (1.6.1) and of 'transition' and 'transformation' (1.6.2), the research programme on 'sustainability transition' (1.6.3), and the meanings and uses of 'sustainable peace' (1.6.3) in the humanitarian and scientific community will be introduced here.

1.6.1 Term and Concept: 'Sustainable' and 'Sustainability'

The adjective 'sustainable', 'sustentable', 'nachhaltig' and the noun 'sustainability' have been widely used globally since 1987 in the ecological debate about a preferred development path. The word 'sustainability' is derived from the Latin *sustinere* (*tenere*, to hold; *sub*, up). According to the *Shorter Oxford English Dictionary* (Oxford 2002: 3129), the verb 'sustain' means "1. support the efforts, conduct or cause of (a person); support (a cause or course of action)...; 2. keep (a person, the mind, spirit, etc.) from failing or giving way; 3. cause to continue in a certain state; maintain at the proper level or standard; [and] 4. maintain or keep going continuously (an action or process)..." The adjective 'sustainable' has these meanings: "1. Supportable, bearable; 2. able to be upheld or defended; 3. able to be maintained at a certain rate or level (of economic activity....)." The *Compact Oxford Dictionary of Current English* (Oxford 2002a: 1160) applies the term 'sustainable' to "industry, development or agriculture, avoiding depletion of natural resources". These dictionaries do not include or define separately 'sustainability'.

The German adjective '*nachhaltig*' means "long-enduring, strong" and emerged in the eighteenth century from the noun '*Nachhalt*', indicating "something that is retained for times of distress" (Der Große Duden, volume 7, Etymologie, 1963: 460). The development of the political term 'sustainable development' is reflected in the twenty-first and latest edition of the *Brockhaus Enzyklopädie* (2006, volume 19: 233-237), where the historical roots of the concept are traced to Hans Carl von Carlowitz (1665-1714), who used it in forestry. During the nineteenth century '*Nachhaltigkeit*' or 'sustainability' was used as a principle in forestry, of countering the overuse of wood and guaranteeing a continual supply.

The Club of Rome in its Report *The Limits to Growth* (Meadows/Meadows/Randers et al. 1972;

see chapter 16 by Marino below) used the word 'sustainable' to describe the desirable 'state of global equilibrium'. A report by the *International Union for the Conservation of Nature* (IUCN 1980) included a reference to sustainable development as a global priority. The 'Brundtland Report' (1987) offered the first widely recognized definition of sustainable development as "development that meets the needs of the present without compromising the ability of future generations to meet their own needs". This report contextualized sustainable development within the concept of the 'needs' of the world's poor and "the idea of limitations imposed by the state of technology and social organization on the environment's ability to meet present and future needs" (WCED 1987).

The goal of 'sustainable development' was operationalized in the context of justice, of globalization, of anthropocentric assumptions whose aim was to contribute to securing human existence, and of societal productiveness, while maintaining the possibility of development and action. The key term pointed to *areas of action* (energy, mobility/transportation, construction/living, agriculture/forestry, leisure/tourism, technological development, gender, financing of social systems, ageing population, education/knowledge, development policy), *conflict potentials* (cause of new conflicts, strong vs weak sustainability), and *strategies for implementation* (resource productivity, ecological structural change, resource management, sustainable consumption patterns). Since the first *UN Conference on Environment and Development* (UNCED) in 1972 in Stockholm, the concept of sustainability has been integrated as a joint ecological and development idea, and it was introduced as a political guideline in the Brundtland Commission Report (WCED 1987).

Since the 1980s sustainability has been used in the context of "human sustainability on planet Earth ... as a part of the concept sustainable development..." by the Brundtland Commission (WCED 1987). This Report further stated that "the concept of sustainable development provides a framework for the integration of environment policies and development strategies—the term 'development' being used here in its broadest sense. The word is often taken to refer to the processes of economic and social change in the Third World. But the integration of environment and development is required in all countries, rich and poor. The pursuit of sustainable development requires changes in the domestic and international policies of every nation" (WCED 1987: 3).

Several pillars of sustainability are usually distinguished, the three classic ones being economy, society and environment (UN Millennium Declaration 2000). To these some experts have added 'the next generation' (Brundtland Commission 1987; WCED 1987) as a fourth pillar, and 'politics' (Earth Charter 2000), or 'culture' (Plumwood 2002; Head, Trigger, Mulcock 2005) as a fifth, while others have embedded the economy and society within the environment. The Brundtland concept was later criticized as being arbitrary, an ideological delusion, a utopian hope and a mere illusion.

Agenda 21 was the practical approach to sustainable development. It referred to "four domains of economic, ecological, political and cultural sustainability". Sustainability improves "the quality of human life while living within the carrying capacity of supporting eco-systems". Furthermore, "sustainability implies responsible and proactive decision-making and innovation that minimizes negative impact and maintains balance between ecological resilience, economic prosperity, political justice and cultural vibrancy to ensure a desirable planet for all species now and in the future".

In the social sciences, numerous definitions, research paradigms and programmes have been developed since 1987, when sustainable development was used by policymakers both as a road map and an action plan for achieving sustainability. Sustainability envisions a desirable future state for human societies where living conditions and resource use continue to meet human needs without undermining the 'integrity, stability and beauty' of natural biotic systems. The concept addresses the "concern for the carrying capacity of natural systems with the social, political, and economic challenges faced by humanity".

The *Organization of American States* (OAS), UNEP and the Government of Peru (2001) identified potential conflicts between the concepts of development and sustainability. After the Rio Summit in 1992, the three Rio Conventions on climate change, biodiversity and desertification were not universally implemented, as a few large industrialized countries were unable to ratify them (US) or later withdrew (Canada). Two decades later, at Rio+20 (in June 2012), the governments adopted a more modest and legally nonbinding document, "The future we want", which was a step towards the *Sustainable Development Goals* (SDGs).

The *Earth Charter* (1992) called for the "building of a just, sustainable, and peaceful global society in the 21st century". *Agenda 21*, adopted at UNCED in Rio de Janeiro in June 1992, identified information,

Figure 1.9: The Sustainable Development Goals adopted by the UN in 2015: **Source**: UN; at: <http://www.un.org/sustainabledevelopment/sustainable-development-goals/> (23 February 2016).

integration, and participation as three key goals of sustainable development and as the key task of the *UN Commission on Sustainable Development* (UNCSD). In June 2012, the Rio+20 Conference launched a process to develop *Sustainable Development Goals* (SDGs, see figure 1.9) based on the *Millennium Development Goals*. In the Rio+20 outcome document (2012), member states agreed on ten requirements that these SDGs should comply with.[44] On 25–27 September 2015, the SDGs were adopted by the UN Summit on the post-2015 development agenda.[45]

Sustainable Development Goal 16 "is dedicated to the promotion of peaceful and inclusive societies for sustainable development, the provision of access to justice for all, and building effective, accountable institutions at all levels".[46] Among its twelve key targets are:

- Significantly reduce all forms of violence and related death rates everywhere
- End abuse, exploitation, trafficking and all forms of violence against and torture of children

- Promote the rule of law at the national and international levels and ensure equal access to justice for all
- By 2030, significantly reduce illicit financial and arms flows, strengthen the recovery and return of stolen assets and combat all forms of organized crime
- Substantially reduce corruption and bribery in all their forms
- Develop effective, accountable and transparent institutions at all levels
- Ensure responsive, inclusive, participatory and representative decision-making at all levels
- Broaden and strengthen the participation of developing countries in the institutions of global governance
- By 2030, provide legal identity for all, including birth registration
- Ensure public access to information and protect fundamental freedoms, in accordance with national legislation and international agreements
- Strengthen relevant national institutions, including through international cooperation, for building capacity at all levels, in particular in developing countries, to prevent violence and combat terrorism and crime
- Promote and enforce non-discriminatory laws and policies for sustainable development.

However, in the short-term targets there is no reference to 'sustainability transition' as a process to

---

44 See: <https://sustainabledevelopment.un.org/topics/sustainabledevelopmentgoals> (10 February 2015).

45 See: "UN Sustainable Development Platform"; at: <https://sustainabledevelopment.un.org/> (23 February 2016).

46 See: <http://www.un.org/sustainabledevelopment/peace-justice/> (23 February 2016).

achieve a 'sustainable peace', nor is this term even mentioned. Thus the concept lacks an action component to promote sustainable peace among nations, regions, and people.

In their theoretical introduction to "transitions to sustainable development", Grin, Rotmans and Schot (2010: 2) used a definition by Meadowcroft (2000: 73), where sustainable development aims at "promoting human well-being, meeting the basic needs of the poor and protecting the welfare of future generations (intra- and intergenerational justice), preserving environmental sources and global life-support systems (respecting limits, integrating economics and environment in decision-making, and encouraging popular participation in development processes".

1.6.2   Terms and Concepts: Transition and Transformation

The word 'transition' derives from the two Latin words 'trans' (beyond) and 'ire' (go) and has been used in many disciplines in the natural sciences (geology, genetics, medicine), social sciences (political science, psychology, sociology, anthropology, economics) and humanities (music, painting, sculpture). According to the *Shorter Oxford English Dictionary* (Oxford 2002: 3129) 'transition' means "1. action or process of passing or passage from one condition, action, or place, to another; change"; 2. passage in thought, speech, or writing from one subject to another...; 4. passage from an earlier to a later stage of development or formation ...". The *Longman Dictionary of Contemporary English* (1995: 1537) defines 'transition' as "the act or process of changing from one form or state to another".

The term 'transformation' originates from the Latin verb 'transformare' or 'to transform' and it is used in mathematics, physics, chemistry, hydrodynamics, medicine, genetics, soil science, linguistics, political science, law, military affairs and management. According to the *Shorter Oxford English Dictionary* (Oxford 2002: 3327), 'transformation' means "1. action of changing in form, shape, or appearance; metamorphosis; 2. a complete change in character, nature etc.; ... 6. natural change of form in a living organism; metamorphosis ...".

In the twenty-first (2006) and latest edition of the *Brockhaus Enzyklopädie* 'transition' is mentioned as a concept in molecular genetics as a form of point mutation. The nineteenth edition (1993) of the *Brockhaus Enzyklopädie* (1993, volume 22: 311–312) recognized 'transformation' as a concept in mathematics,

medicine, molecular biology, linguistics, and in economics and politics for the fundamental change of the economic, political and societal system or a state. After the collapse of the communist states of Eastern Europe, the concept was used for the transition from a planned to a market economy.

The twenty-first (2006, volume 27: 660–665) edition of *Brockhaus Enzyklopädie* distinguished 'transformation', 'transformation function', 'transformation society' and 'transformation curve'. Accordingly the transformation concept was used in mathematics, physics, medicine, molecular biology, linguistics, and for political and economic issues as in "system transformation and transformation societies". Here 'transformation society' refers to a concept in political and social science and economics relating to "processes and steering problems of societal change in the context of ... interaction of fundamental political, economic, social and cultural changes". These transformation processes focus on political institutions, political culture, value systems, changes in the economic order, and fundamental changes in societal living together and social structure, as well as changes in ways of life, behavioural patterns and value preferences. This concept assumes that such a transformation may be planned and steered.

Karl Polanyi's book *The Great Transformation* (1944) examined the great social and political upheavals England experienced during the industrial revolution with the rise of the market economy and the modern nation state "as the single human invention he calls the 'market society'". The debate on 'transformation societies' was inspired by the theories of Durkheim and Max Weber on 'social change' and it was rekindled after 1990 by the transformation processes taking place in post-communist countries. Since the mid-1990s the prevailing 'transition' or 'transformation research' schemes in political and social science have analysed past changes induced by industrialization (W. F. Ogburn; K. Mannheim), modernization (T. Parsons; W. W. Rostow), and the collapse of the Soviet Union, focusing on the transition of state-socialist political, economic and societal systems towards Western 'neo-liberal' market economies (v. Beyme/Offe 1995; Merkel 1996).

1.6.3   Ahimsa, Peace with Nature and Sustainable Peace

The English word 'peace' comes from the Latin 'pax', while the Greek 'eirene', the Hebrew 'shalom', and the Arabic 'salaam' mean 'peace with justice', while the Hindi *'ahimsa'* adds the ecological dimension. 'Ahimsa' means 'not to injure', cause no injury, do no harm, and also involves the concept of nonviolence, which according to Jainism, Hinduism, and Buddhism is to be applied to all living beings—including all animals. "The concept reached an extraordinary status in the ethical philosophy of Jainism. Mahatma Gandhi popularized the term, and he strongly believed that the principle of *ahimsa*, of 'cause no injury', includes one's deeds, words, and thoughts" (Oswald Spring 2008).

In *Wege zum Frieden mit der Natur* ('Ways to peace with nature') Klaus Michael Meyer-Abich (1984) suggested an applied philosophy of nature for environmental policy. For Meyer-Abich (1979, 2003), "peace with nature" is both a domestic and an international task, where human behaviour has to be brought back in line with the wholeness of nature, where increasing hazards and disasters are an expression of the disharmony and lack of peace of humankind with nature. 'Peace with nature' has been called a goal of many initiatives.[47]

'Sustainable peace' is a value-oriented concept that has been used by development NGOs and in the social sciences in development and peace studies, including peace psychologists (chapter 6 by Deutsch/ Coleman; chapter 7 by Coleman). Connie Peck (1998: 22-23) introduced the concept into the post-cold war debate on preventive diplomacy and on conflict prevention both as a 'vision' and as a policy programme for conflict prevention. In the conflict prevention and transformation literature, 'sustainable' was primarily used as a synonym for a 'lasting' or 'enduring' peace, without addressing 'peace with nature' (see chapter 9 by Brauch in this volume).

1.7 Alternative Approach to Sustainability Transition

1.7.1 Origins and Conceptual Evolution since the 1970s

The debate on 'sustainability transition' emerged first in the US in the 1970s and was taken up in a report by the US Academy of Science (NRC 1999; see chapter 21 by Twig Johnston in this volume; Raskin/ Banuri/Gallopin et al. 2002). It has looked forward to the processes of a long-term system transformation necessary to contain and reduce the effects of the dominant *business-as-usual* paradigm and to reduce GHG emissions through both *multilateral* quantitative emission reduction obligations and *unilateral* transformations.

From 2005, a specific 'sustainability transition' research paradigm emerged from the *Dutch Knowledge Network on Systems Innovation and Transition* (KSI)[48] and from the Amsterdam Conference in 2009 where the *Sustainability Transition Research Network* (STRN)[49] was founded. This particular approach will be presented below, together with other considerations focusing on long-term transformations in culture, values, behaviour and lifestyle.

The UNFCCC and its Kyoto Protocol did not achieve their goals because of the lack of political will to implement them (Brauch 2012), and the "Paris Agreement" (2015) is based exclusively on voluntary commitments (INDC). For these reasons, unilateral bottom-up strategies to reduce both GHG emissions and energy efficiency will be an important coping strategy for facing the impacts of global environmental and climate change.

---

47 "Make peace with nature" is also the name of a TV show by David F. Surber on WKRC-TV, Channel 12, Cincinnati, Ohio, US; at: <http://www.makepeacewithnature.org/>; see also the "Peace With Nature Initiative" by Costa Rica's former president, Oscar Arias Sanchez; at: <http://www.wegefoundation.com/peacewith-nature-initiative/>; "Pax Natura—Peace with Nature"; at: <http://www.paxnatura.org/> (23 February 2016).

---

48 For KSI see for details: <http://www.ksinetwork.nl/ home>; on its participants: <http://www.ksinetwork. nl/what-is-ksi/participants> and publications: <http:// www.ksinetwork.nl/output/publications>; see also Routledge Studies in Sustainability Transition: <http:// www.routledge.com/books/series/RSST/>

49 For STRN, see: <http://www.transitionsnetwork.org/>; reference list of ST literature: <http://www.transitions-network.org/files/Reference%20list%20to%20transition %20publications.pdf>. STRN has held conferences in Amsterdam (2009), Lund (2011), Copenhagen (2012) Zürich (2013), Utrecht (2014), Brighton (2015), and Wuppertal (2016); selected results are published in: *Journal on Environmental Innovation and Sustainability Transition*; at: <http://www.journals.elsevier.com/ environmental-innovation-and-societal-transitions/>.

1.7.2   Evolution of the US Debate on Sustainability Transition

From the mid-1970s and the late 1990s there have been several initiatives aiming at 'sustainability transition'. The *Tellus Institute*, which was founded in 1976, was an early pioneer of sustainability transition research. Kates (2001: 15325) noted "an effort to re-engage the scientific community around the requirements for a sustainability transition" in the US during the 1990s. This sought "a transition towards a state of sustainable development", which "was studied as a series of interlinked transitions, as a process of adaptive management and social learning, and as a set of indicators and future scenarios". This effort was influenced by "the widespread perception that the continuation of current trends in both environment and development would not provide for the desired state of sustainable development. To the extent that transitions represent breaks in such trends, there was growing interest in identifying the needed or desired transitions in the relationship between society and the environment."

In a first major conceptualization of 'sustainability transition', Speth (1992) specified five "inter-linked transitions (demographic, technological, economic, social and institutional) as collective requirements for a sustainability transition". He proposed "a transition towards technologies that were environmentally benign and reduced sharply the consumption of natural resources, and the generation of waste and pollutants. For the economy, the desired transition was towards one in which prices reflected their full environmental costs. A social transition would move towards a fairer sharing of economic and environmental benefits both within and between countries." As a fifth process, he proposed "a transition in the institutional arrangements between governments, businesses and people that would be less regulation-driven and more incentive-led, less confrontational and more collaborative."

In addition to these approaches, US Nobel Laureate Gell-Mann (1994) "suggested an ideological transition towards a sense of solidarity that can encompass both the whole of humanity and the other organisms of the biosphere and an informational transition that integrates disciplinary knowledge and disseminates it across the society." In order to study these necessary transitions they jointly organized "a major, albeit uncompleted, research effort known as the '2050' project (World Resources Institute, Brookings Institutions, Santa Fe Institute) that sought to track these transitions and understand their interactions".

Kates (2001) noted two different approaches by the US National Academy of Sciences in its report *Our Common Journey: A Transition toward Sustainability* (NRC 1999), and a process of social learning (Social Learning Group 2001). The NRC's research agenda was "based on both current scientific understanding and the requirements of an emerging sustainability science". The report "identified concrete goals and appropriate next steps to accelerate major transitions underway or needed in each sector" and tried "to integrate global and local perspectives to shape a 'place-based' understanding of the interactions between environment and society; and initiate focused research programs on a small set of understudied questions that are central to sustainability transition".

A fourth group focused on long-term scenarios of alternative pathways to transition "about how the global system might unfold". These scenarios reflected "current trends and reform proposals, various forms of social and environmental breakdown, and more fundamental transitions or transformations (Raspin/Gallopin/Gutman et al. 1998, Hammond 1998)". The NRC (1999) report concluded that

> although the future is unknowable a successful transition toward sustainability is possible over the next two generations. This transition could be achieved without miraculous technologies or drastic transformations of human societies. What will be required, however, are significant advances in basic knowledge, in the social capacity and technological capabilities to utilize it, and in the political will to turn this knowledge and know-how into action (Kates 2001: 15328).

The Tellus Institute focuses on "global scenarios, planetary phase of civilization, sustainable development, climate change, eco-efficiency, globalization, information technology", and since 2005 it has addressed

> a *Great Transition* to a sustainable, just, and liveable global civilization. To attain this vision, the world must navigate toward ways of producing, consuming, and living that balance the rights of people today, future generations, and the wider community of life. The prospects for such a transition rest with the ascendance of new values, a planetary consciousness, and a sense of global citizenship. These aims will lie at the heart of the Institute's program of research, education, and network-building in the coming years.[50]

---

50 For Tellus, see at: < http://www.tellus.org/about/> (11 February 2015).

Figure 1.10: The three major scenarios and six alternative pathways of a global transition. **Source**: Tellus' Gobal Transition Initiative; at: <http://www.greattransition.org/explore/scenarios> (11 February 2015).

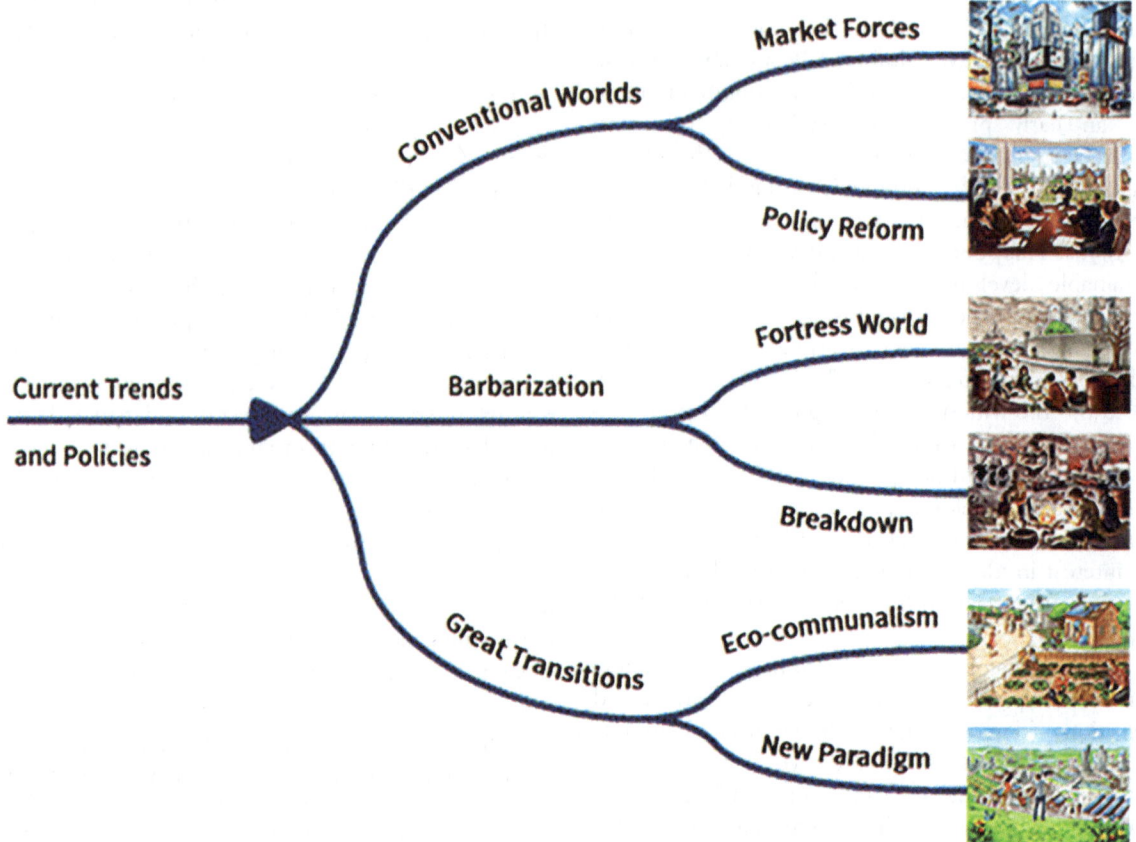

This *Great Transition Initiative* (GTI) coordinates a global network ... [and] spreads the message that a future of enriched lives, global solidarity, and a healthy planet is possible if the citizens of the world join in a vast cultural and political mobilization for change. ... It builds on the ground-breaking work of the international *Global Scenario Group*. With ... over 400 participants from over 40 countries, GTI ... reach[es] out to those seeking greater understanding of global challenges and a positive framework for addressing them.[51]

In 2014, Paul Raskin, the Tellus President, distinguished three global models (figure 1.10): *Conventional Worlds* (business-as-usual), a model which assumes structural continuity of present trends and actors, *Barbarization* (worst case), which assumes "a deluge of instability swamps society's adaptive capacity, leading to a general global crisis and the erosion of civilized norms", and *Great Transitions*, that imagines "how the imperatives and opportunities of the

Planetary Phase might advance more enlightened aspirations", envisioning the new values of "human solidarity, quality-of-life, and an ecological sensibility" instead of "individualism, consumerism, and domination of nature", and aiming for "institutions that support democratic global governance, well-being for all, and environmental sustainability".

Raskin argued (see also his foreword to this book) that at present: "Great Transition precursors announce themselves ... in a rising cosmopolitan consciousness, civil society campaigns, and expanding subcultures seeking more responsible and fulfilling lifestyles." But while the technological feasibility may be easier, changing the cultural and political assumptions is more difficult. He claimed that "the Planetary Phase, by unravelling old patterns and mindsets and urging new ones, opens opportunities for creative social transformation" by fostering "the idea of global citizenship", which "carries both psychological and juridical meanings". But he cautioned that "intergovernmental institutions, transnational corporations, and big civil society organizations are unlikely candidates

---

51 For Tellus' GTI, see at: <http://www.greattransition.org/> (30 January 2016).

Figure 1.11: Multilevel perspective on transitions. **Source**: Geels and Schot (2010: 25), adapted from Geels (2002: 1263) and used with the permission of the author.

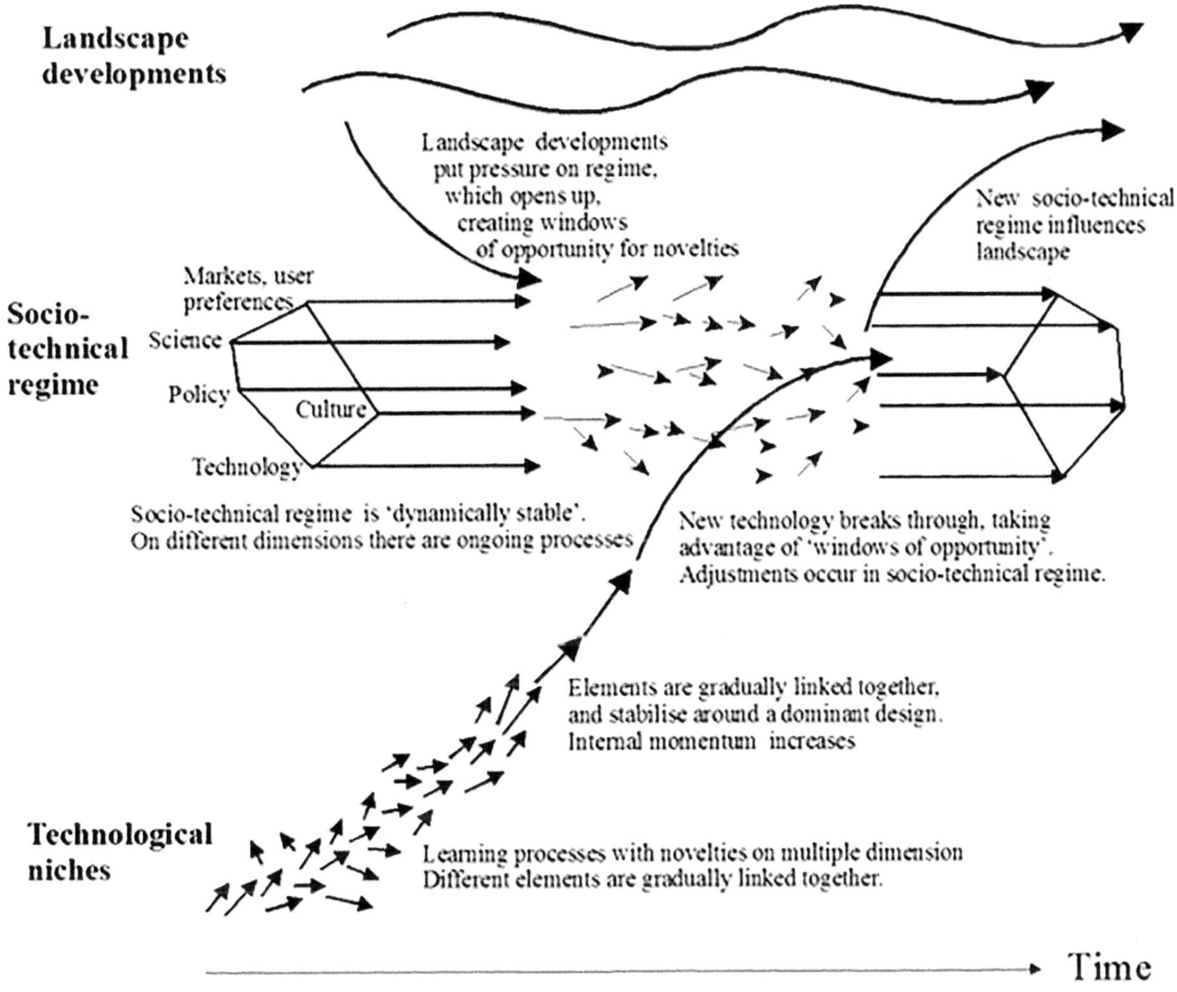

Landscape developments

Landscape developments put pressure on regime, which opens up, creating windows of opportunity for novelties

New socio-technical regime influences landscape

Socio-technical regime

Markets, user preferences

Science

Policy

Culture

Technology

Socio-technical regime is 'dynamically stable'. On different dimensions there are ongoing processes

New technology breaks through, taking advantage of 'windows of opportunity'. Adjustments occur in socio-technical regime.

Elements are gradually linked together, and stabilise around a dominant design. Internal momentum increases

Technological niches

Learning processes with novelties on multiple dimension Different elements are gradually linked together.

Time

for the role of change agent", and hoped that "the natural change agent for a Great Transition would be a vast and inclusive movement of global citizens".[52]

The Great Transition Scenario distinguishes two pathways: "Ecocommunalism" and "New Paradigm". While the first incorporates "the green vision of bioregionalism, localism, face-to-face democracy, small technology, and economic autarky ... [with the] emergence of a patchwork of self-sustaining communities from our increasingly interdependent world seems implausible", the GTI embraces the "New Sustainability Paradigm", which "sees globalization not only as a threat but also an opportunity to construct a planetary civilization rather than rely on the incremental forms of Conventional Worlds or retreat into local-

ism. It envisions the ascendance of new categories of consciousness–global citizenship, humanity-as-whole, the wider web of life, and the well-being of future generations–alongside democratic institutions of global governance".[53]

1.7.3   The Dutch Scientific Research Initiative

A totally different approach to sustainability transition emerged from a large research project by the *Dutch Knowledge Network on Systems Innovation and Transition* (KSI) in the Netherlands, in which eighty-five researchers participated (2005–2010). Grin, Rotmans and Schot (2010) combined "three perspectives on transitions to a sustainable society: complexity the-

---

52 See at: <http://www.greattransition.org/publication/a-great-transition-where-we-stand> (11 February 2015).

53 See at: <http://www.greattransition.org/explore/scenarios> (11 February 2015).

ory, innovation theory, and governance theory". The authors

> seek to understand transitions dynamics, and how and to what extent they may be influenced. ... They do so from the conviction that only through drastic system innovations and transitions it becomes possible to bring about a turn to a sustainable society to satisfy their own needs, as inevitable for solving a number of structural problems on our planet, such as the environment, the climate, the food supply, and the social and economic crisis. Among other things this implies that our world has to overcome the undesirable side effects of the ongoing 'modernization transition', which began around 1750. However, the transition to sustainability has to compete with other developments, and it is uncertain which development will gain the upper hand. In *Transitions to Sustainable Development* the authors ... closely address the need for transitions, as well as their dynamics and design (Grin/Rotmans/Schot 2010: xvii–xix).

A new discourse on 'sustainability transition' has evolved since 2009 within the *Sustainability Transition Research Network* (STRN).[54] It focuses on sustainability problems in the energy, transport, water and food sectors from different scientific perspectives on the ways

> in which society could combine economic and social development with the reduction of its pressure on the environment. A shared idea among these scholars is that due to the specific characteristics of the sustainability problems (ambiguous, complex) incremental change in prevailing systems will not suffice. There is a need for transformative change at the systems level, including major changes in production, consumption that were conceptualized as 'sustainability transitions'.[55]

The STRN defined transitions research as a "new approach to sustainable development" that

> is also developing its own core set of questions and theories. Major research efforts ... have advanced knowledge of transitions to sustainability, particularly in the field of a broad understanding of how major, radical transformations unfold and what drives them. ... Technical changes need to be seen in their institutional and social context, generating the notion of 'socio-technical (s-t) systems', which are often stable and path dependent, and therefore difficult to change. Under certain conditions and over time, the relationships within s-t systems can become reconfigured and replaced in a process that may be called a system innovation or a transition.

The STRN argued that

> transitions to sustainability may ... be strongly context specific. ... It is ... of great interest to explore the varied governance challenges that transitions to sustainability imply in different contexts ... from ... industrial transformation, innovation and socio-technical transitions; integrated assessment; sustainability assessment; governance of SD (political science); policy appraisal community; researchers working on reflexive governance; the resilience community; the ecological economics community; groups of energy-, environment- and sustainability-modelers; and a core sustainability transitions community. ... Research is organized around seven themes: (a) synthesizing perspectives and approaches to transitions; (b) governance, power and politics; (c) implementation strategies; (d) civil society, culture and social movements in transitions; (d) firms and industry; (e) geography of transitions; (e) modelling of transitions.

The STRN defined its mission as coordinating its scientific capacity "towards the production of foresight reports on strategic sustainability policy questions, ... to support the development of a sustainability transitions research community internationally, and provide an independent, authoritative and credible source of analysis and insight into the dynamics and governance of sustainability transitions".[56]

Their multilevel perspective on transitions was influenced by Geels's (2002: 1263) model that distinguished between technological niches, the sociotechnological regime and landscape developments that start with ... interrelated and mutually reinforcing technological innovations in 'niches'. Such multiple events initially face opposition in the respective sociotechnical regime, and once they are overcome they often result in a structural change by exploiting windows of opportunity, and lead to changes in the landscape (figure 1.11).[57]

---

54 See at: <http://www.transitionsnetwork.org/> (22 January 2015).

55 See at: <http://www.transitionsnetwork.org/files/STRN_research_agenda_20_August_2010%282%29.pdf>.

56 See the opening page of the STRN, at: <http://www.transitionsnetwork.org/> (22 January 2015).

57 The *Environmental Innovation and Sustainability Transitions* (EIST) Journal "offers a platform for reporting studies of innovations and socio-economic transitions to enhance an environmentally sustainable economy and thus solve structural resource scarcity and environmental problems, notably related to fossil energy use and climate change." See the 'Environmental Innovation and Sustainability Transitions' (EIST) Journal at: <http://www.journals. elsevier.com/environmental-innovation-and-societal-transitions/>.

1.7.4   A German Scientific Advisory Report on a 'Social Contract for Sustainability'

The research focus of KSI and of the STRN influenced a policy report by the *German Advisory Council on Global Change* (WBGU 2011) on a 'Social Contract for Sustainability' (2011). This report argued that the transformation to a low-carbon society requires that we

> not just accelerate the pace of innovation; we must also cease to obstruct it. ... We must also take into account the external costs of high-carbon (fossil energy-based) economic growth to set price signals, and thereby to provide incentives for low-carbon enterprises. Climate protection is ... a vital fundamental condition for sustainable development on a global level.

The WBGU (2011: 93) report adapted Geels's model and added several megatrends in both the earth system (climate, biodiversity, land degradation, water, raw materials) and the human system (development, democratization, energy, urbanization, food), where innovative changes in the regime may directly affect the megatrends. The report argued that the realization of a low-carbon economy and society is overcoming the many barriers and exploiting the favourable factors (WBGU 2011: 6). The WBGU report stated that "... a low-carbon transformation can only be successful if it is a common goal, pursued simultaneously in many of the world's regions" (WBGU 2011). It discussed (WBGU 2011: 5) the global "remodelling of economy and society towards sustainability as a 'Great Transformation'. Production, consumption patterns and lifestyles in all of the three key transformation fields must be changed in such a way that global greenhouse gas emissions are reduced to an absolute minimum over the coming decades, and low carbon societies can develop."

The transformation towards a climate-friendly society requires that many existing change agents lead to a "concurrence of multiple change" (Osterhammel 2009), which "can trigger historic waves and comprehensive transformations". This transformation aims at a low-carbon society that starts with the decarbonization of energy systems, where "the proactive state and the change agents are the key players". Within the next decade, they should jointly initiate a transformation towards a sustainable energy path. This transformation should permit a major decarbonization of the economy by 2050, avoiding both a rebound effect and a climate crisis.

To achieve a major transformation towards a low-carbon economy and society, the WBGU proposed specific measures for the energy sector, land-use changes and global urbanization that could accelerate and extend the transition to sustainability. The WBGU Report suggested that "research and education are tasked with developing sustainable visions, in co-operation with policy-makers and citizens; identifying suitable development pathways, and realizing low-carbon and sustainable innovations". The WBGU argued "that transformation costs can be lowered significantly if joint decarbonization strategies are implemented in Europe. The transformation also represents a great chance for Europe to make innovation-driven contributions to a globalization process that has a viable future" (WBGU 2011).

A report by the *International Social Science Council* (ISSC 2012: 16) on *Transformative Cornerstones of Social Science Research for Global Change* defined transformation as "a process of altering the fundamental attributes of a system, including in this case structures and institutions, infrastructures, regulatory systems, financial regimes, as well as attitudes and practices, lifestyles, policies and power relations". It argued that the necessary additional social science research will contribute to producing change by calling "for an additional response to global change and to climate change; additional to and building on the enduring focus in this field on adaptation and mitigation; ... a critical questioning of the systems and paradigms that have created climate change and on which climate change rests".

So far all the studies on sustainability transition in North America—by the Tellus Institute, by Speth, Kates and the NRC—and more recently in Europe—by the KSI, STRN and WBGU—and those by the ISSC have not addressed the possible linkage of the suggested transitions—including those of the past—to questions of international and national peace and security. This leads to the second key goal of this Handbook: 'peace with nature' or 'sustainable peace'.

1.7.5   Policy Debates on Green Growth and Decarbonization

UNEP (2011, 2014), OECD,[58] the EU and several governments have promoted an alternative vision and outlined alternative policies for a global green deal, green growth, and a decoupling of economic growth from energy consumption. These policy proposals were partly taken up by the European Commission

---

58 See publications on: "Green growth and sustainable development", at: <http://www.oecd.org/greengrowth>.

Figure 1.12: EU GHG emissions towards an 80 per cent domestic reduction (1990 = 100 per cent). **Source**: EU (2011b: 5).

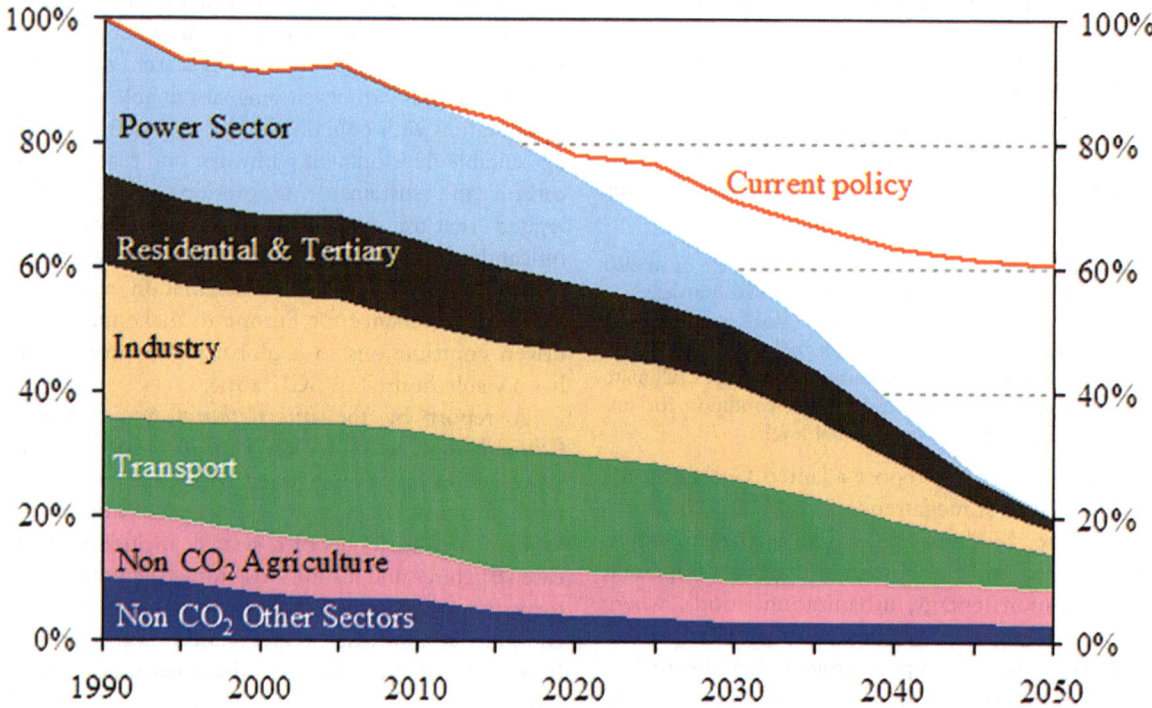

and the European Council in its longer-term goals and policy papers on climate change, its energy (EU 2011), resource (see chapter 41 by Happaerts below, EU 2011c, 2011d) and transport policies[59] (EU 2011a), and in its "Roadmap for moving to a competitive low carbon economy in 2050" (EU 2011b).[60] In this Roadmap the European Commission addressed the goal "of reducing greenhouse gas emissions by 80-95 per cent by 2050 compared to 1990". Based on these goals, the EU's Roadmap outlined milestones with "policy challenges, investment needs and opportunities in different sectors".

The Commission projected the necessary GHG reductions for key sectors (table 1.1). The modelling analysis assumed that "the switch to domestically produced low carbon energy sources will reduce the EU's average fuel costs by between €175 billion and €320 billion per year" (EU 2011: 11-12). The study further assumed that "in 2050, the EU's total primary energy consumption could be about 30 per cent below 2005

levels" and it argued that "without action the oil and gas import bill could instead double compared to today, a difference of €400 billion or more per annum by 2050, the equivalent of 3 per cent of today's GDP" (EU 2011: 12).

Table 1.1: Sectoral reductions. **Source**: EU (2011b: 6).

GHG reductions compared to 1990	2005	2030	2050
Total	-7%	-40 to -44%	-79 to -82%
Sectors			
Power (CO$_2$)	-7%	-54 to -68%	-93 to -99%
Industry (CO$_2$)	-20%	-34 to -40%	-83 to -87%
Transport (incl. CO$_2$ aviation, excl. maritime)	+30%	+20 to -9%	-54 to -67%
Residential and services (CO2)	-12%	-37 to -53%	-88 to-91%
Agriculture (non-CO$_2$)	-20%	-36 to -37%	-42 to -49%
Other non-CO$_2$ emissions	-30%	-72 to -73%	-70 to -78%

The Commission's Roadmap study for moving towards a low-carbon economy indicated

---

59 See "White paper 2011: Roadmap to a Single European Transport Area–Towards a competitive and resource efficient transport system", at: <http://ec.europa.eu/transport/themes/strategies/2011_white_paper_en.htm>.

60 European Commission, Climate Action, "Roadmap for moving to a low-carbon economy in 2050"; at: <http://ec.europa.eu/clima/policies/roadmap/index_en.htm>.

that a cost effective and gradual transition would require a 40% domestic reduction of greenhouse gas emissions compared to 1990 as a milestone for 2030, and 80 per cent for 2050. ... With existing policies, the EU will achieve the goal of a 20 per cent GHG reduction domestically by 2020. ... Deep reductions in the EU's emissions have the potential to deliver benefits in the form of savings on fossil fuel imports and improvements in air quality and public health. ... The Roadmap gives ranges for emissions reductions for 2030 and 2050 for key sectors (EU 2011b: 14).

The Commission planned to use this Roadmap for sector-specific policy initiatives and longer-term funding considerations on how EU funding could support necessary instruments and investments for the transition to a low-carbon economy. Three years later, on 23 October 2014, the European Council agreed that

- the domestic 2030 greenhouse gas reduction target [should be] at least 40 per cent compared to 1990 together with the other main building blocks of the 2030 policy framework for climate and energy. ... This 2030 policy framework aims to make the European Union's economy and energy system more competitive, secure and sustainable and also sets a target of at least 27 per cent for renewable energy and energy savings by 2030.
- The framework presented will drive continued progress towards a low-carbon economy. It aims to build a competitive and secure energy system that ensures affordable energy for all consumers, increases the security of the EU's energy supplies, reduces our dependence on energy imports and creates new opportunities for growth and jobs.
- The 2030 framework ... also takes into account the longer-term perspective set out by the Commission in 2011 in the Roadmap for moving to a competitive low-carbon economy in 2050, the Energy Roadmap 2050 (EU 2011) and the Transport White Paper (EU 2011a). These documents reflect the EU's goal of reducing greenhouse gas emissions by 80–95 per cent below 1990 levels by 2050 as part of the effort needed from developed countries as a group.[61]

By the end of 2015, at COP 21 in Paris, a majority of state parties agreed on legally binding voluntary commitments as defined by the member countries themselves. They would reduce their GHG emissions in the decades to come. Moreover, they would fully

implement these commitments by moving gradually towards a low-carbon economy and by realizing a sustainability transition in the energy and production sectors and adopting national policies that supported sustainable consumption by their citizens. However, no global targets and no sanctions were approved.

While the EU, UN, UNEP and OCED have suggested a transition towards a green economy, so far only very few countries have initiated detailed policy programmes aiming at a sustainable energy transition (e.g. Germany,[62] the UK,[63] France,[64] Austria, Denmark and Japan).[65] As major producers of solar panels and of wind power, China,[66] India,[67] Mexico (see chapter 23 by Le Clercq)[68] and many other OECD and threshold countries have also adopted major energy transition laws and policies.

The *International Renewable Energy Agency* (IRENA) argued in its Roadmap for a Renewable Energy Future (IRENA 2016: 23) that to move towards

---

61 See: "2030 framework for climate and energy policies", at: <http://ec.europa.eu/clima/policies/2030/index_en.htm> (22 January 2015).

62 Federal Ministry for the Environment (29 March 2012). *Langfristszenarien und Strategien für den Ausbau der erneuerbaren Energien in Deutschland bei Berücksichtigung der Entwicklung in Europa und global [Long-term Scenarios and Strategies for the Development of Renewable Energy in Germany Considering Development in Europe and Globally]* (PDF) (Berlin: Federal Ministry for the Environment, BMU).

63 HM Government: *The UK Low Carbon Transition Plan–National strategy for climate and energy* (London: HM Government, 15 July 2009).

64 See: "France launches energy transition"; at: <http://www.developpement-durable.gouv.fr/-France-launches-its-energy-> (24 February 2016).

65 See: "The Energy Transition in Europe: initial lessons from Germany, the UK and France", at: <http://www.cerre.eu/events/energy-transition-europe-initial-lessons-germany-uk-and-france> (24 February 2016); see also the "Berlin Energy Transition Dialogue"; at: <http://www.energiewende2015.com/> and <http://www.dena.de/veranstaltungen/berlin-energy-transition-dialogue-2016.html> (24 February 2016).

66 See: Wang, Tao; Watson, Jim, n.d: *China's Energy Transition Pathways for Low Carbon Development* (Falmer: SPRU).

67 Reddy, B. Sudhakara, 2014: *India's Energy Transition–Pathways for Low-Carbon Economy* (New Delhi: Indira Gandhi Institute of Development Research, Mumbai, July); at: <http://www.igidr.ac.in/pdf/publication/WP-2014-025.pdf> (24 February 2016).

68 See: "Mexico Energy Forum"; at: <http://www.renewableenergymexico.com/>; Sonal Patel: "Nieto: Mexico's Energy Transition Will Persevere despite Dismal Oil Prices", 22 February 2016; at: <http://www.powermag.com/nieto-mexicos-energy-transition-will-persevere-despite-dismal-oil-prices/> (24 February 2016).

a 2°C goal the global share of renewables would have to double by 2030 what would require "a six-fold increase".

Both the Roadmaps of the EU for moving to a low-carbon economy by 2050 and of IRENA for realizing a *Renewable Energy Future* by 2030 are model projections that do not imply any binding legal obligations. Rather, policy decisions and their subsequent full implementation into national policies are needed.

The fossil and nuclear energy industries and the car and highway lobbies have attacked the IPCC as a key messenger by supporting the campaign of climate critics. In the US Congress, they have succeeded in blocking all major climate bills proposed by the Obama Administration.[69] The tar sand lobby in Canada (Dalby 2016) and the fracking industry in the US have invested heavily in new fossil technologies and supported policies to prevent legal obligations by not ratifying the Kyoto Protocol in US or by withdrawing from it in Canada (in December 2011).

The weak performance and implementation of quantitative GHG emissions reductions will most severely affect the highly socially and environmentally vulnerable developing and least developed countries. These have already seen the highest number of deaths and people affected. The economic losses from climate-induced hazards and health impacts were highest in developed countries, because of insurance. It is projected that many countries with a continuing high level of population growth and a high level of people below the poverty line will also have a low level of resilience and limited capabilities for adaptation and mitigation during the twenty-first century.

1.7.6 Towards a New Research Paradigm in the Social Sciences

Both the scientific discourse and the policy debates have closely interacted. While the policy debate since the publication of the Brundtland Report (1987) has partly triggered funding for new scientific institutions and research projects, the scientific debate has since moved much further from developing an approach to zero growth, to a reduction of the overuse of nature and the recuperation of ecosystem services that are

essential for humans and nature. While natural scientists (Clark/Crutzen/Schellnhuber 2004) have called for a 'second Copernican revolution in science' and the development of a new scientific world view and a new sustainability paradigm, in the social sciences several approaches to 'sustainability research' exist:

- the '*sociotechnical approach*' of sociologists and historians who examine technical innovations (inventions, breakthroughs and setbacks) in their specific national political, economic and societal contexts with the aim of drawing generalizable *lessons from past long-term transformative innovation processes* for the necessary transition to sustainable development;
- the '*empirical approach*' of *policy analysis* that observes and assesses ongoing processes of sustainability transition, i.e. of discussion, planning, steering and implementation of processes of energy transition or change ("*Energiewende*");
- '*discourse analysis*', that reviews and interprets scientific discourses, and the societal and political debates of multiple actors;
- '*constraint analysis*', that analyses systemic (mindset), technical (laws of physics, status of innovation), ideological (e.g. cornucopian, Hobbesian), and interest-driven (lobbies of affected industries and trade unions) obstacles to strategies and policies aiming at sustainability transition.

Besides these numerous scientific approaches to analysing strategies, policies and measures that aim at processes of 'sustainability transition' with the goal of realizing a sustainable development path, the debate is taking place in many scientific disciplines and thus several dimensions may be distinguished, among them, besides the historical or *temporal* dimension, *spatial, scientific, societal, economic, political* and *cultural* dimensions, which will not be dealt with in detail in this multidisciplinary Handbook.

1.7.7 Dimensions of Research into 'Sustainability Transition'

The emerging research and publications on 'sustainability transition' have focused on spatial (1.7.5.1), scientific (1.7.5.2), societal (1.7.5.3), economic (1.7.5.4), political (1.7.5.5), and cultural (1.7.5.6) dimensions.

1.7.7.1 The Spatial Dimension of Sustainability Transition

Within the evolving discourse on sustainability transition, the proposal by Coenen, Benneworth and

---

69 Naomi Klein: "Capitalism vs. the Climate—Denialists are dead wrong about the science. But they understand something the left still doesn't get about the revolutionary meaning of climate change", in: *The Nation*, 28 November 2011; at: <http://www.thenation.com/article/164497/capitalism-vs-climate> (25 January 2014).

Truffer (2010, 2012) of a spatial dimension[70] was more limited. They argued that "an explicit analysis of the geography of transitions contributes to the extant transitions literature in a variety of ways. Firstly it provides a contextualization and reflection on the limited territorial sensitivity of existing transitions analysis". Coenen and Truffer (2012: 1) claimed that "environmental innovations and sustainability related initiatives have received increasing attention in the recent economic geography and regional studies literature". They suggested future research should combine both traditions in sustainability-related research in regional studies. They distinguished between two trends in sustainability transition studies: a focus on a) the *technological innovation systems* (TIS) *approach* and b) a focus on the *multilevel perspective* (MLP); both rely on innovation and technology studies (Coenen/Diaz Lopez, 2010).

Coenen and Truffer (2012: 8) noted that the emerging sustainability transitions research is still "nascent and partially immature" and lacks a specific geographic or spatial dimension. They reviewed forty-seven papers that had appeared in the *Journal of Regional Studies* since 1993 and which referred explicitly to sustainability concerns. Some focused on "assessing and quantitatively measuring the state of sustainable development in regions", while others "highlighted the interdependence of sustainable regional development with natural resources and ecosystems" or submitted "conceptually and theoretically programmatic contributions" that focused on "a) ecological modernization and regulationist approaches, b) industrial ecosystems and, c) what can be considered as the most prominent thread in regional studies, the development of a framework to analyse policy processes that help shape 'sustainable regions'", including regional aspects of sustainable production and consumption. Coenen and Truffer (2012: 8) concluded that

> the major weaknesses being that technologies and sectoral (trans-)formation processes rarely receive very explicit consideration. Either there is a strong focus on institutional change at the expense of technological

change, regional production structures at the expense of consumer and citizen related processes or alternatively a strong but singular focus on (experimental) policies for regional sustainability. The complementarities between regional studies and transitions studies therefore warrant some further scrutiny.

However, regional studies usually only look at the lower level of the geographical scale, while international relations address the more abstract level of the relations among states, societies and economies, thus linking international with transnational relations, including negotiations towards achieving policy declarations on decarbonization of the economy and a shift towards green growth. The discussion on 'sustainability transition' has so far primarily focused on the micro-level of socio-economic and societal and technological innovations, and has not addressed the impacts of strategies and policies within a business-as-usual world view or mindset and an alternative sustainability perspective on international peace and security (see chapter 2 by Dalby and chapter 20 by Hodson/Marvin/Späth in this volume).

To simplify, a continuation of the present trend in the consumption of fossil fuels (coal, oil, gas) will not only raise the level of GHG emissions but also increase the demand for these non-renewable energy sources and increase their price, and may result in military conflicts over access to and control of hydrocarbons, especially in the Middle East. What policy scenarios and strategies are foreseeable whose strategies, policies and measures are oriented towards long-term transformative change towards sustainable development, and may enhance the prospects for international cooperation, peace with security and a long-term vision of a positive or sustainable peace?

1.7.7.2   Scientific Dimension of Sustainability Transition

The development of new scientific and technological knowledge is crucial for initiating processes that call for multiple transitions towards sustainability. In "Science for Global Sustainability: Toward a New Paradigm", Clark, Crutzen and Schellnhuber (2004: 3) provided the conceptual context for the Dahlem Workshop on "Earth Systems Science and Sustainability" (2003). They pointed to "the need for harnessing science and technology in support of efforts to achieve the goal of environmentally sustainable human development in the Anthropocene". They noted the great transformation during the twentieth century that had resulted in an increase of cropland by a factor of two, of world population by a factor of

---

70 See Teis Hansen and Lars Coenen, 2013: *The Geography of Sustainability Transitions: A Literature Review* (Lund: Circle); at: <http://wp.circle.lu.se/upload/CIRCLE/workingpapers/201339_Hansen_Coenen.pdf>; Teis Hansen and Lars Coenen, 2014: "The Geography of Sustainability Transitions: Review, Synthesis and Reflections on an Emergent Research Field"; at: <http://www.regionalstudies.org/uploads/fortaleza.pdf> (24 February 2016).

four, water use by a factor of eight, energy use by a factor of 16 and industrial output by a factor of 40 (based on McNeill 2000, 2009). They discussed both the opportunities and challenges in facing and coping with the impacts of GEC and CC. These were addressed in the Amsterdam Declaration (2001) that called for the *Earth System Science Partnership* (ESSP) that has evolved during the past decade (Leemans/Rice et al. 2011), until it was replaced by *Future Earth* in 2012.

They considered earth systems science as a key promoter of such a transition, which requires a change in the scientific world view and orientation that recognizes that sustainable development is a knowledge-intensive activity. They pointed to a growing consensus "that management systems for a sustainability transition need to be systems for adaptive management and social learning". In conclusion, they suggested "A New Contract for Planetary Stewardship" linking science and society. This resulted in 2011 in the WBGU Report "A Social Contract for Sustainability".

From the perspective of international relations, ecology and geography this may require a readiness from the natural sciences to bring international and peace and security considerations into their analysis, as well as a readiness from the social sciences to consider the results of ESS and ESA in their own analyses. This has stimulated Brauch, Oswald Spring and Dalby (2011) to call for a new "Political Geoecology for the Anthropocene". The scientific discourse on 'sustainability transition' must be broadened from its narrow initial focus as it has evolved since the Amsterdam conference in 2009 towards a wider scope that comprises all the dimensions of 'sustainability transition'.

1.7.7.3   Societal Dimension of Sustainability Transition

Political, economic and societal strategies for 'sustainability transition' cannot be implemented against the wishes, values and preferences of the people concerned. Such a long-term and global transformative change requires not only 'hard' changes in the production, energy and transportation systems and in human settlements and habitats, but also many 'soft' changes in human values, belief systems, world views and mindsets. The societal dimension of the discourse on sustainability transition has so far focused on the changes needed in human values, perception and behaviour that would result in new lifestyles, ways of life and patterns of consumption.

Ingelhart's (1977, 1998) work on value change addressed the emergence of post-materialist values since 1945. These have resulted in an "increasing power of new social movements ... as the expression of a wider cultural value change (Inglehart 2008)" (cited in WBGU 2011: 69). However, this value change and the global contextual change since 1989 has not affected the prevailing world view in US society and the mindset of its policymakers in the US Congress. While during the fifth wave of the *World Value Survey* (WVS 2010), close to eighty per cent of the US population saw global warming, or the greenhouse effect, as serious or very serious, President Obama failed to get any climate change legislations adopted in either house of the US Congress or with the Supreme Court (2016) (Klein 2011). This is a clear indication of the high volatility of these WVS and demonstrates that it hardly mattered politically, given the strong economic and ideological interests of climate change opponents and sceptics.

For a behavioural change towards sustainability transition a temporary change in public preferences and attitudes is not sufficient. Rather, fundamental changes in human behaviour are needed, with major changes in lifestyles and consumptive preferences and patterns. This would result in a lower ecological footprint and in a reduction of individual carbon emissions. However, this cannot be achieved by changes only on the demand side. It also requires major changes on the supply side with regard to green and renewable energy systems, public and low-carbon transport systems and products with a much lower carbon footprint.

New social movements and political parties may contribute to creating both awareness of and positive political frameworks for a change in the lifestyles and the preferred way of life of a majority of the people. Thus, changing the 'soft' human and societal side of 'sustainability transition' may be as difficult if not even more difficult than changing the sociotechnological framework on which most research is so far focused. The WBGU (2011: 78-79) argued that

> for the transformation of economy and society towards sustainability, the political, economic and technological path dependency is also a significant barrier (Liebowitz/ Marjolis 1995; Pierson 2004). An existing system of institutions (norms, contracts negotiating and decision-making modi, etc.), but also of technologies and infrastructures, can hinder far-reaching social changes. ... In politics and the economy, path dependent processes and developments frequently result in mistakes becoming the established norm, and the continued absence of learning effects. Individual and political decision-mak-

ers, as well as the public, are led by crisis management routines that were developed for past problem cases. Clinging to past thought and action patterns can lead to an 'objective' pressure to change, which merely results in modification, rather than transformation of the status quo, delaying the replacement of the fossil-nuclear energy system with a sustainable energy system even further.

While new scientific results and new publicly shared knowledge does not change values, attitudes, preferences or behaviour, the enduring change of these soft factors requires simultaneous changes in the hard factors of the economic system and production and consumption processes, as well as in the policy process.

1.7.7.4 Economic Dimension of Sustainability Transition

According to the IPCC's (2011) *Special Report on Renewable Energy Sources and Climate Change Mitigation* and the WBGU's (2011: 119) assessment, "the sustainable potential of renewable energies is fundamentally sufficient to provide the world with energy". The IPCC (2011: 15) stated that: "There are multiple pathways for increasing the shares of RE [*renewable energy*] across all end-use sectors." This applies to the transportation, building and agricultural sectors and requires long-term integration efforts including "investment in enabling infrastructure; modification of institutional and governance frameworks; attention to social aspects, markets and planning; and capacity building in anticipation of RE growth".

The IPCC (2011) argued that "historically, economic development has been strongly correlated with increasing energy use and growth of GHG emissions, and RE can help decouple that correlation, contributing to sustainable development (SD)." Renewables can also make a significant contribution to global mitigation efforts, given that "a significant increase in the deployment of RE by 2030, 2050 and beyond is indicated in the majority of the 164 scenarios reviewed" in the IPCC's *Special Report on Renewable Energy Sources and Climate Change Mitigation* (SRREN). The IPCC argued that "individual studies indicate that if RE deployment is limited, mitigation costs increase and low GHG concentration stabilizations may not be achieved". Also "a transition to a low-GHG economy with higher shares of RE would imply increasing investments in technologies and infrastructure".

As fossil fuels are the major cause for the increase of GHG emissions in the atmosphere, an increase of RE is a necessary but not sufficient policy for coping with GHG emissions. A gradual and comprehensive decarbonization of the economy requires major improvements in energy efficiency and a great transformation in many economic sectors that may be opposed by economic financial and labour interests representing the old sectors.

The WBGU Report (2011) also proposed an intensification of policies of sustainable production and consumption and major initiatives in buildings, living and land-use planning, in mobility and communication and in food; this would require both climate-compatible agricultural management and a change in dietary habits.

Initiating sustainable transformation in cities with the highest energy growth potential can become a major force for innovation and investment in new infrastructure. This requires new governance actors (Corfee-Morlot/Kamal-Chaoui/Donovan et al. 2009) who can reduce traffic by a "spatial integration of urban functions", thus "achieving a high quality of life for inhabitants". Further, "energy infrastructure integration (CHP [*combined heat and power*] technology, heating and cooling systems, smart grids, electromobility, etc.) can benefit considerably from the spatial density" (WBGU 2011: 173). While "land-use systems cannot become completely emissions-free", "a significant contribution from land use" is needed, including "stopping deforestation and switching to sustainable forest management, as well as the promotion of climate-friendly agriculture and dietary habits" (WBGU 2011: 173).

FAO (2013a) has proposed a climate-smart agriculture, fishery and forestry, which includes the recovery of ecosystems, ecosystem services, organic agriculture, protection of seed diversity, local and regional marketing, and sustainable food patterns. FAO (2015) stated that, worldwide, half of the food is lost as waste in the whole cycle of production, transformation, marketing, and consumption. "Without accounting for GHG emissions from land use change, the carbon footprint of food produced and not eaten is estimated to 3.3 Gtonnes of CO_2 equivalent: as such, food wastage ranks as the third top emitter after US and China. Globally, the blue water footprint (i.e. the consumption of surface and groundwater resources) of food wastage is about 250 km^3, which is equivalent to the annual water discharge of the Volga river, or three times the volume of lake Geneva. Finally, produced but uneaten food vainly occupies almost 1.4 billion hectares of land; this represents close to 30 percent of the world's agricultural land area" (FAO 2013b: 6)

Figure 1.13: Total annual anthropogenic GHG emissions by groups of gases 1970–2010. **Source**: IPCC (2014d: 9).

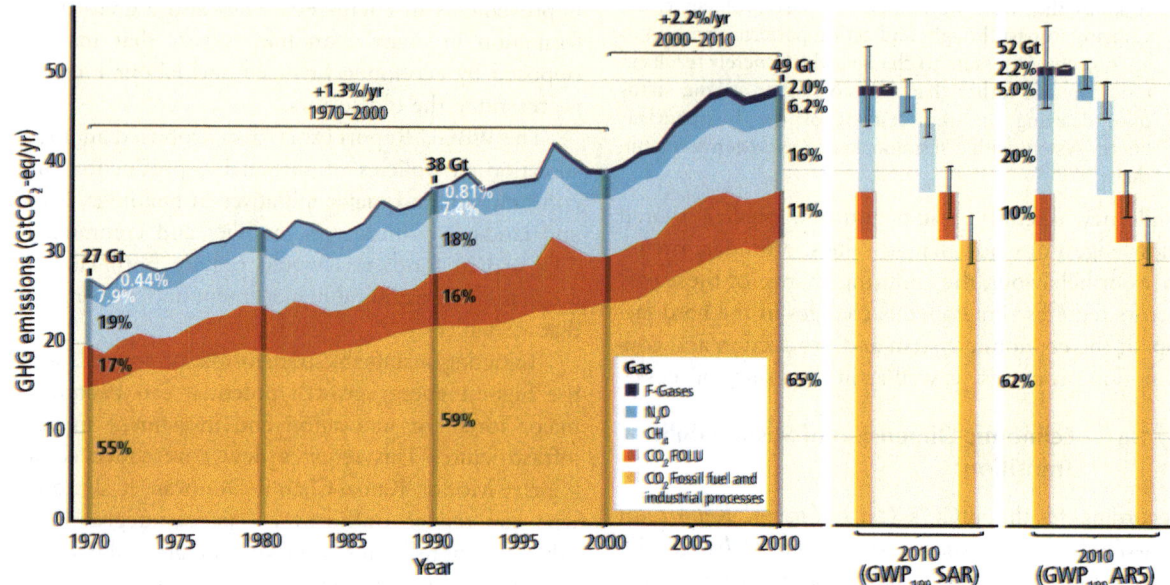

Working Group 3 of the Fifth Assessment Report of the IPCC[71] concluded in its "Summary for Policymakers" (2014: 8–9) that: "CO2 emissions from fossil fuel combustion and industrial processes contributed about 78% of the total GHG emission increase from 1970 to 2010, with a similar percentage contribution for the period 2000–2010" (figure 1.13) and "about half of cumulative anthropogenic CO2 emissions between 1750 and 2010 have occurred in the last 40 years". Furthermore, "annual anthropogenic GHG emissions have increased by 10 GtCO2eq between 2000 and 2010, with this increase directly coming from energy supply (47%), industry (30%), transport (11%) and buildings (3%) sectors".

To counter these trends, the IPCC's WG 3 assessed several "mitigation pathways and measures in the context of sustainable development". They argued that "mitigation scenarios in which it is likely that the temperature change caused by anthropogenic GHG emissions can be kept to less than 2°C relative to pre-industrial levels are characterized by atmospheric concentrations in 2100 of about 450ppm CO2eq". However, to reach this target would include "substantial cuts in anthropogenic GHG emissions by mid-century through large-scale changes in energy systems and potentially land use" (IPCC 2014d: 12). But "delaying mitigation efforts beyond those in place today through 2030 is estimated to substantially increase the

difficulty of the transition to low longer-term emissions levels and narrow the range of options consistent with maintaining temperature change below 2°C relative to pre-industrial levels" (IPCC 2014d: 14). The IPCC agreed that "mitigation scenarios reaching about 450 to about 500 ppm CO2eq by 2100 show reduced costs for achieving air quality and energy security objectives, with significant co-benefits for human health, ecosystem impacts, and sufficiency of resources and resilience of the energy system" (IPCC 2014d: 16) and that "mitigation scenarios reaching around 450 ppm CO2eq concentrations by 2100 show large-scale global changes in the energy supply sector" (IPCC 2014d: 18).

Working Group WG 3 of IPCC AR5 also noted the high relevance of soft factors, for example: "[e]fficiency enhancements and behavioural changes, in order to reduce energy demand compared to baseline scenarios without compromising development, are a key mitigation strategy in scenarios reaching atmospheric CO2eq concentrations of about 450 to about 500 ppm by 2100" and "behaviour, lifestyle and culture have a considerable influence on energy use and associated emissions, with high mitigation potential in some sectors, in particular when complementing technological and structural change" (IPCC 2014d: 22).

For the IPCC, "decarbonizing (i.e. reducing the carbon intensity of) electricity generation is a key component of cost-effective mitigation strategies in achieving low-stabilization levels (430–530ppm CO2eq); in most integrated modelling scenarios, decarbonization hap-

---

71 See at: <http://www.ipcc.ch/pdf/assessment-report/ar5/wg3/ipcc_wg3_ar5_summary-for-policymakers.pdf>.

pens more rapidly in electricity generation than in the industry, building, and transport sectors". The IPCC suggested replacing "coal-fired power plants with modern, highly efficient natural gas combined-cycle power plants or combined heat and power plants", while increasing reliance on bioenergy may imply "challenges and risks" (IPCC 2014d: 22).

In the transport sector, "technical and behavioural mitigation measures for all transport modes, plus new infrastructure and urban redevelopment investments, could reduce final energy demand in 2050 by around 40% below the baseline" (IPCC 2014d: 23) and in the building sector: "lifestyle, culture and behaviour significantly influence energy consumption in buildings", and "the energy intensity of the industry sector could be directly reduced by about 25% compared to the current level through the wide-scale upgrading, replacement and deployment of best available technologies, particularly in countries where these are not in use and in non-energy intensive industries" (IPCC 2014d: 25).

In *Agriculture, Forestry and Other Land Use* (AFOLU) the IPCC saw "the most cost-effective mitigation options in forestry are afforestation, sustainable forest management and reducing deforestation, with large differences in their relative importance across regions. In agriculture, the most cost-effective mitigation options are cropland management, grazing land management, and restoration of organic soils" (IPCC 2014d: 26). With regard to human settlements and urbanization, "the largest mitigation opportunities ... are in rapidly urbanizing areas where urban form and infrastructure are not locked in, but where there are often limited governance, technical, financial, and institutional capacities" (IPCC 2014d: 28).

In discussing the financial aspects, the IPCC argued that "in many countries, the private sector plays central roles in the processes that lead to emissions as well as to mitigation. Within appropriate enabling environments, the private sector, along with the public sector, can play an important role in financing mitigation" (IPCC 2014d: 31). *Working Group* (WG) 3 of the IPCC concluded that "policy linkages among regional, national, and sub-national climate policies offer potential climate change mitigation and adaptation benefits" that can "be established between national policies, various instruments, and through regional cooperation" (IPCC 2014d: 32).

1.7.7.5 Political Dimension of Sustainability Transition

The political dimension of 'sustainability transition' has been extensively discussed and many approaches,

analysis and proposals have been submitted so far. Grin (2010: 223) suggested that the transition to sustainable development can no longer rely on centralized government institutions of political administrative steering, given the "more prominent role of the interactions between the state, market, and society". He argued that a governance perspective "allows us to consider transition management, strategic niche management and interrelated processes in the real world".

Grin (2010: 237) reviewed the contemporary processes of institutional change in modern societies with regard to a) structural changes affecting the polity and dealing with institutional transformations between the four key actors: state, market, society and science; b) structural changes in innovative systems relating to the development and use of new technologies, and c) the emergence of new, often transnational, arrangements for corporate governance.

This perspective lacks an analysis of the transformation of international politics of peace and security that compares the international, regional and global impacts of a continuation of *business-as-usual* policies and of the sustainability paradigm. Based on the first three levels, Grin (2010: 247) argued that "at the regime level, major processes of transformations go on in the institutions of state, market, civil society and knowledge, and their mutual alignment."

Grin interpreted these changes as the result of two processes resulting from: a) "influences on the regime from landscape-level trends, such as globalization, individualization, Europeanization and the politicization of side effects, as well as derived trends such as privatization and liberalization;" and b) the responses to "the challenges these practices have come to face during the past two decades as a consequence of feedback processes."

A different approach was taken by Roeland J. in 't Veld (2011: xv; see chapter 40 in this volume). He suggested decision-makers in politics, business, science, civil society and the media should create governance arrangements beyond traditional borders; "sustainability requires transgovernance", where action is based on thinking: a) beyond classical governance style and towards a culturally sensitive meta-governance for sustainable development; b) beyond disciplinary scientific research, towards more transdisciplinarity.

These studies by Grin (2010) and in 't Veld (2011) link the intensive scientific debate on global environmental and climate governance to the process of sustainability transition. Other studies have addressed governance aspects and perspectives of sustainability transition (Loorbach 2007). Governance aspects were also discussed prior to the Rio+20 summit but no proposals for international governance for sustainability transition, e.g. the upgrading of UNEP from a programme to a specialized agency, were adopted in its outcome document.

1.7.7.6 The Cultural Dimension of Sustainability Transition

While many studies of sustainability transition have focused on technological innovation in the relevant industrial sectors, especially energy, and on governance aspects, the societal and cultural dimension has been less prominent. Myles (2012) found a deep cultural transition was needed in the British army and Morrow, Rothwell, Burford et al. (2013) inquired into the cultural dimension of transition for overseas medical staff in the UK.

In the social sciences there has been an intensive debate on postmodern values and value changes (Inglehart 1977) and on changes of attitudes and preferences towards sustainability (Raskin/Banuri/Gallopin et al. 2002; Leiserowitz/Kates/Parris et al. 2006). The WBGU used values as "a shared perception of something worth striving for", where cultural values refer "to something that has evolved socioculturally, something that exists independent of individuals". It stated that "attitudes relate to certain objects, people (groups), ideas, and ideologies, or specific situations" (Häcker/Stapf 1994). In contrast to short-term intentions and long-term value systems, attitudes "represent evaluations and action tendencies with regard to attitude objects, and are usually stable over the medium term", while opinions are understood as "verbalizations of attitudes and values".

The WBGU (2011: 77) argued that various barriers prevent "value systems from impacting on behaviour, at both individual and social or structural level", and that a change in behaviour requires "a material and cognitive basis". A transition towards sustainability is structurally constrained by the prevailing path dependence, as well as the high level of carbon infrastructure and its political and electoral influence on decision-makers in parliaments and in the executive sector.

Analysis of the so-called soft aspects of sustainability transition, e.g. of the constraints, obstacles and barriers to a change in opinion, attitudes, value systems and behaviour, requires the expertise of not only sociologists, social psychologists and anthropologists, but also of political scientists. It should include an analysis of cognitive perceptual and evaluative barriers created by the established traditional world views of scientists and the mindsets of policymakers (Oswald Spring/Brauch 2011).

1.8 Structure of this Handbook and its Contributions

The sixty authors of this Handbook are citizens of or work in eighteen different countries from all five continents. They come from *or* work in Africa (3, all from South Africa); the Americas (16): Brazil (3), Canada (1), Mexico (5), and the US (7); Asia (5): China (2), Singapore (2) and Thailand (1); Australia (3); and Europe (33 *or* 43): particularly Belgium (1), France (1), Germany (21), Ireland (2), Norway (1), Poland (1), Sweden (1), the Netherlands (7), and the United Kingdom (8). Three-fifths of them are male and two-fifths female.

The authors were trained or are working in the following scientific disciplines:

- *Anthropology* (3): Lourdes Arizpe, Twig Johnson and Úrsula Oswald Spring;
- *Climate Studies* (6): Sophie Adams, Rebecca Froese Hans Joachim Schellnhuber, Carl-Friedrich Schleussner, Olivia Maria Serdeczny, and Simon Dalby;
- *Demography* (2): Hongmin Chen and Hania Zlotnik;
- *Development Studies* (4): Sophie Adams, Lucy Baker, Asfaw Kumssa and Carl Middleton;
- *Ecology and Environmental Studies* (9): Úrsula Oswald Spring, Philipp Schepelmann, Monica de Andrade, Philipp Späth, Bonno Pel, Belinda Yuen, Sander Happaerts, Cecilia Tortajada and Gregory Trencher;
- *Economics* (11): Karoline Augenstein, Audley Genus, Jan Jonker, Flor Avelino, Sylvia Lorek, Czeslaw Mesjasz, Linda O'Riordan, Ilona Magdalena Otto, Uwe Schneidewind, Mike Hodson and Hongmin Chen;
- *Engineering* (3): Karlson 'Charlie' Hargroves, Carl Middleton and Martin Keulertz;
- *Geography* (8): Simon Dalby, Eckart Ehlers, Peter Michael Link, Simon Marvin, Jürgen Scheffran, Janpeter Schilling, Rebecca Froese and Martin F. Price;
- *Philosophy* (3): Claudia Köhler, Bonno Pel, Olivia Maria Serdeczny and Úrsula Oswald Spring;
- *Physics* (4): John Grin, Czeslaw Mesjasz, Jürgen Scheffran and Hans Joachim Schellnhuber;
- *Political Science* (13): Karoline Augenstein, Hans Günter Brauch, John Grin, Juan Antonio Le

Clercq, Radhika Perrot, Britta Rennkamp, Marit Sjøvaag, Carolyn M. Stephenson, Eduardo Viola, Mike Hodson, Larissa Basso, Tobias Ide and Lucy Baker;

- *Psychology* (2): Coleman, Morton Deutsch, and Úrsula Oswald Spring;
- *Sociology* (3): Flor Avelino, Claudia Köhler and Sir Robert Worcester;
- *Sustainable Development* (5): Sander Happaerts, Roeland J. in 't Veld, Mark Swilling, Mike Hodson and René Kemp; and
- *Urban Studies* (3): Xuemei Bai, Belinda Yuen and Shivant Jhagroe.

The book reflects a methodological pluralism and several chapters aim for a transdisciplinary and transformative approach. All chapters in this Handbook have been peer-reviewed, and the book is organized into ten parts.[72]

1.8.1   Moving towards Sustainability Transition

Simon Dalby (CIGI Chair in the Political Economy of Climate Change at the Balsillie School of International Affairs and Professor of Geography and Environmental Studies at Wilfrid Laurier University, Waterloo, Canada) argues in chapter 2, "Contextual Changes in Earth History", that human activities have changed the parameters of the Holocene so much that we now live in the Anthropocene. Studies in earth system science suggest that sustainable development and plans for transitions to a sustainable peace must consider the possibility of rapid phase shifts in the biosphere. Constraining human activities to ecological boundaries in the earth system is key to sustainability, but rapid shifts may be coming. Sustainability planning must think beyond national security and recognize that human actions are shaping the future of the planet and are changing the geopolitical context. Adopting the perspective of a geopolitical ecology focusing on global economic production rather than only on environmental protection is key to the future if planetary stewardship of the Anthropocene is to be successful.

In chapter 3, "Paradigm and Praxis Shifts: Transitions to Sustainable Environmental and Sustainable Peace Praxis", *Carolyn Stephenson* (Associate Professor of Political Science at the University of Hawaii at

Manoa and affiliate faculty member in its Matsunaga Institute for Peace, US) claims that the paradigms of both sustainable environmental practice and sustainable peace practice have rapidly diffused in past decades, but unevenly across time and space, across regions, and within countries. Until recently, researchers in environmental studies and peace studies have often ignored each other. Cooperation between them is a significant change. This chapter reviews the paradigms of sustainable environment and sustainable peace as an incomplete engagement with each other.

John Grin (Professor at the Department of Political Science of the University of Amsterdam, The Netherlands) introduces in chapter 4 on "Transition Studies: Basic Ideas and Analytical Approaches" different approaches to (i) understanding and (ii) shaping transition dynamics. He distinguishes among three key approaches: 1) A *socio-technical approach*, with the multi-level perspective as its main concept, and strategic niche management as its governance concept; 2) A *complex (adaptive) system based approach*, using the concept of transition patterns, with transition management as its governance concept; 3) A *governance approach*, which has elaborated the politics of transition dynamics, with reflexive design as the core of its approach to shaping transition. Further, work based on the social theory of practice have helped to bring in consumption practices into transition theory, and an approach focusing on innovation systems has helped to understand how different arrangements may become interconnected so as to produce a transition.

In chapter 5, "Transformative Science for Sustainability Transitions", *Uwe Schneidewind* (President of the Wuppertal Institute for Climate, Environment, Energy and Professor for Sustainable Transition Management at the University of Wuppertal), *Mandy Singer-Brodowski* (scientific coordinator of the Center for Transformation Research and Sustainability at the University of Wuppertal) and *Karoline Augenstein* (research fellow at the Wuppertal Institute for Climate, Environment, Energy) (all from Germany) argue that sustainability transitions require a knowledge production that contributes actively to the Grand Challenges of twenty-first-century societies. In their view, scientific institutions play a key role in the transformation towards sustainability and peace. A 'transformative science' is needed that not only analyses processes of transformation, but also actively supports and accelerates them. Drawing from the German science system, they introduce the concept of 'transformative science' and its implications for (1) the methodologies of transdisciplinary and transformative

---

72 The summaries of the chapters are based on condensed abstracts written by the authors.

research, (2) institutional capacity-building for facilitating such a research approach, and (3) the national science systems and national science policies that will enable this new mode of knowledge production.

1.8.2   Aiming for Sustainable Peace

The second part of the book addresses sustainable peace and includes two chapters on the use of this concept in peace psychology, from a gender perspective and in its environmental dimension as peace with nature. In chapter 6, *Morton Deutsch* (E. L. Thorndike Professor and director emeritus of the Morton Deutsch International Center for Cooperation and Conflict Resolution (MD-ICCCR) at Teachers College, Columbia University, US) and *Peter T. Coleman* (Professor of Psychology and Education at Columbia University at Teachers College and The Earth Institute and who teaches courses in Conflict Resolution, Social Psychology, and Social Science Research; Director of the Morton Deutsch International Center for Cooperation and Conflict Resolution (MD-ICCCR) at Teachers College, Columbia University, US) introduce "The Psychological Components of a Sustainable Peace", based on the introductory chapter of their book (Coleman/Deutsch 2012), They enhance the understanding of sustainable peace by supplementing the standard approach of studying the prevention of destructive conflict, violence, war and injustice with the equally important investigation of the promotion of the basic conditions and processes conducive to lasting peace. In addition to addressing the pervasive realities of oppression, violence and war, peace requires us to understand and envision the alternatives we wish to construct.

In chapter 7, "The Essence of Peace? Toward a Comprehensive and Parsimonious Model of Sustainable Peace", *Peter T. Coleman* (Columbia University, US) highlights the basic commonalities of a "harmonious sustainable peace", summarizes the main components of sustainable peace, presents a sketch of a more parsimonious model of sustainable peace informed by dynamic systems theory and dynamic minimalism, and outlines an agenda for future study and education in this area

In chapter 8, "Development with Sustainable-Engendered Peace: A Challenge during the Anthropocene", *Úrsula Oswald Spring* (Research Professor at the *National Autonomous University of Mexico* (UNAM) in the *Regional Multidisciplinary Research Center* (CRIM) in Cuernavaca, Mexico) examines the evolution of the concept of peace from a negative towards a positive, sustainable, and engendered concept, with reference to long-term violence, embedded in the patriarchal system and characterized by authoritarianism, exclusion, discrimination, exploitation, and violence. These threats are consolidated by patriarchal institutions, religious controls, self-identified beliefs and social representations, and totalitarian exercise of power affecting natural resources and human development. The chapter explores the concept of a sustainable-engendered peace, attempts to understand the links to patriarchy and its war system, and explores the potential for a holistic and cosmopolitan peace that may challenge the root causes of violence and destruction.

In chapter 9, "Sustainable Peace in the Anthropocene: Towards Political Geoecology and Peace Ecology", *Hans Günter Brauch* (Privatdozent (Adj. Prof.) at the Faculty of Political Science and Social Sciences, Free University of Berlin (ret.) and chairman of *Peace Research and European Security Studies* (AFES-PRESS, Germany) conceptualizes possible linkages between the 'sustainability transition' research paradigm and the rethinking of peace, security, development and the environment or ecology since 1990. With the transition from the Holocene to the Anthropocene the threat to the survival of humankind has changed. We are now the threat due to the increase in our burning of hydrocarbons and the accumulation of greenhouse gases in the atmosphere. This new threat cannot be countered with traditional military strategies any longer but requires a long-term transformative change of our economy with its production and consumption and of the energy, transportation, agricultural and housing sectors towards a low-carbon economy that would be the result of a process of a transition to sustainability. Such a long-term transformative change can avoid climate-induced and resource-scarcity-driven violent conflicts. This chapter recommends the development of possible linkages between sustainable development, human security and sustainable peace in the context of a political geoecology and a peace ecology. There is a need for more conceptual, theoretical and empirical research on the possible linkages between peace studies and ecology, that takes into account the changed human and environmental conditions in the Anthropocene. The added value is to sensitize the research on 'sustainability transition' to reflect on the impact of its realization on sustainable peace and human security.

1.8.3 **Meeting the Challenges of the Twenty-First Century: Demographic Imbalances, Temperature Rise and the Climate–Conflict Nexus**

In chapter 10, "Population imbalances: Will they continue over the rest of the century?", *Hania Zlotnik* (consultant and former director of the United Nations Populations Division, US/Mexico) states that current population dynamics vary markedly between countries. In 2010, forty-six per cent of the world's population lived in countries with below-replacement fertility, while nineteen per cent lived in countries with such high fertility that their combined populations are likely to double by 2050. The remaining thirty-five per cent lived in countries where fertility levels had declined considerably but where the potential for continued population growth remains high. Both population momentum and long-standing differences in population dynamics between these groups of countries are expected to continue in the medium term. Reducing population imbalances by the end of the century requires that the high-fertility countries accelerate the reduction of fertility to reach below-replacement levels before 2100.

In chapter 11, "The Challenge of a 4°C World by 2100", *Hans Joachim Schellnhuber* (Director of the *Potsdam Institute for Climate Impact Research* (PIK), Germany), *Olivia Maria Serdeczny* (research analyst at *Climate Analytics*, Germany), *Sophie Adams* (research analyst at *Climate Analytics*, from Australia), *Claudia Köhler* (Personal Executive Advisor to the Director, PIK, Germany), *Ilona Magdalena Otto* (researcher at PIK, Germany), and *Carl Friedrich Schleussner* (climate scientist with *Climate Analytics*, Berlin and PIK, Germany) argue that our understanding of the earth system and the climate change impacts expected in the coming decades is developing at a rapid pace. They argue that we need to take into account the different degrees of vulnerability not only across but also within nation states. The ramifications of non-linear impacts and their uneven distribution are likely to be deleterious to the stability and well-being of our societies. If we wish to understand the challenges associated with a 4°C world, such a world needs to be imagined.

In chapter 12, "The climate-conflict nexus: Pathways, regional links and case studies", *Tobias Ide* (head of the research field Peace and Conflict at the *Georg Eckert Institute* in Braunschweig, Germany), *P. Michael Link* (postdoctoral scientist at the *Research Group Climate Change and Security* (CLISEC), Insti-tute of Geography, University of Hamburg, Germany), *Jürgen Scheffran* (professor at the Institute of Geography and the *Center for Earth System Research and Sustainability* (CEN), Germany, head of CLISEC in the CliSAP Cluster of Excellence, Hamburg University, Germany), *Janpeter Schilling* (Associate at International Alert in London, postdoctoral researcher with CLISEC at Hamburg University, Germany) claim that the role of climate change as a potential cause of violent conflict has been debated in the scholarly and policy communities for years. They review the most recent quantitative and qualitative literature and find that research on the issue has produced little consensual findings so far. Further, they discuss major theoretical, conceptual and empirical issues and describe possible pathways linking climate change to violent conflict. They analyse the climate-conflict nexus in different world regions and present three qualitative case studies: in north-western Kenya, the Nile Basin, and Israel/Palestine. They see possible reasons for the lack of scientific consensus in the difficulties existing approaches have in adequately capturing the complex links between climate change, vulnerability, and violent conflict.

In chapter 13, "From a Climate of Complexity to Sustainable Peace: Viability Transformations in the Anthropocene", Jürgen Scheffran (see above, Germany) argues that in an increasingly interconnected climate of complexity, the stabilization of human-environment interactions is a major challenge in international relations that demands the integration of complexity science with global governance. This chapter highlights several cases of complex crises where cascading events have affected international stability. Climate change is considered as a risk multiplier which disturbs the balance between natural and social systems and amplifies the consequences through complex impact chains that affect the functioning of critical infrastructures and supply networks; intensify the nexus of water, energy and food; lead to production losses, price increases and financial crises in other regions through global markets; undermine human security, social living conditions and political stability; and trigger or aggravate migration movements and conflict situations. An integrative framework of human-environment interaction is used to analyse destabilizing developments, tipping elements and cascading risks, as well as concepts of resilience, viability and sustainable peace.

1.8.4 Initiating Research on Global Environmental Change, Limits to Growth, Decoupling of Growth and Resource Needs

The next four chapters deal with international efforts to initiate research on global environmental change, on the 'Limits to Growth', and on a decoupling of economic growth from resource needs. In chapter 14, *Lourdes Arizpe* (Professor, Regional Center for Multi-disciplinary Research, National Autonomous University of Mexico), *Martin F. Price* (Professor, Director of the *Centre for Mountain Studies* (CMS), Perth College, *University of the Highlands and Islands* (UHI), Scotland, UK), and *Sir Robert Worcester* (Emeritus Chancellor of the University of Kent, Visiting Professor and Honorary Fellow at the London School of Economics, UK) document "The first decade of initiatives for research on the Human Dimensions of Global (Environmental) Change (1986–1995)". They present and discuss the emergence of initiatives for research on the human dimensions of global change up to the 1996 launch of the *International Human Dimensions Programme on Global Environmental Change* (IHDP). In 1990, *International Social Science Council* (ISSC) launched the *Human Dimensions of Global Environmental Change Programme* (HDP), identifying seven broad research topics. Simultaneously various related national and regional research initiatives emerged.

In chapter 15, "From HDP to IHDP: Evolution of the International Human Dimensions of Global Environmental Change Programme (1996–2014)", *Eckart Ehlers* (Professor emeritus, Bonn University, Germany) reviews as a process of maturation of social science research the challenges of climate and global environmental change that became part of the *Earth System Science Partnership* (ESSP) and in 2015 were fully integrated into the Future Earth initiative.

In chapter 16, "From the Limits to Growth to 2052", *Marit Sjovaag* (political scientist working on climate policy, Norway) summarizes the content of and the debate around the 'Limits to Growth' and its updates, from 1972 until 2012. The chapter presents the background and political context to the original 1972 study, and follows the debate on the four books. The methodology of systems analysis is briefly introduced and the analyses are presented, as are the reactions from academia and political and economic interests. The chapter addresses the research questions of why the debate around the 'Limits to Growth' became so polarized, and how the original scenarios have been updated over the last four decades. The chapter concludes with Randers' forecast (2012) for the next forty years.

Finally, in chapter 17, "Preparing for Global Transition: Implications of the Work of the International Resource Panel", *Mark Swilling* (Professor of Sustainable Development, School of Public Leadership, University of Stellenbosch, South Africa) briefly describes the establishment of the *International Resource Panel* (IRP) by UNEP in 2007 as a panel of scientific experts. In his review of the IRP's activity he uses a material flow analysis perspective, as the primary focus of the IRP is on global resource use and potential alternatives. The notion of a 'third great transformation' is deployed to suggest that the work of the IRP is documenting the endgame of the industrial socio-metabolic regime. Three clusters of reports are addressed: (a) global resource perspectives, with special reference to decoupling rates of economic growth from rates of resource use by focusing on the importance of resource productivity; (b) nexus themes, including cities, food, trade, and GHG mitigation technologies; and (c) specific resource challenges with respect to two clusters of issues, metals and ecosystem services. He concludes that a resource-use perspective adds to an understanding of the unsustainability of the current global system, complementing the outputs of climate science on the effects of anthropogenic carbon emissions and ecosystem science on the implications of biodiversity degradation. The IRP anticipates the possibility of a more resource efficient socio-metabolic order but it has not yet addressed the dynamics and modalities of transition from an institutional and macro-economic perspective.

1.8.5 Developing Theoretical Approaches to Sustainability and Transitions

In the next three chapters, an economist from Poland (Czeslaw Mesjasz), three political scientists from the Netherlands (B. Pel, Flor Avelino and S. S. Jhagroe), and three geographers from the UK and Germany (Mike Hodson, Simon Marvin and Philipp Späth) present different theoretical approaches to sustainability and transitions.

In chapter 18, "Sustainability and Complexity: A Few Lessons from Modern Systems Thinking", *Czeslaw Mesjasz* (Associate Professor, Faculty of Management, Cracow University of Economics, Poland) argues that sustainable development, sustainability and sustainability transition are associated with the growing complexity of sociopolitical and ecological systems and their interactions. The chapter identifies and assesses the applications of the concepts deriving from complexity studies to the discourse on sustainability and sustainable development and on related terms. Its main hypothesis is that a sophisticated language dealing with complexity and applied in the narratives about sustainability and related ideas requires profounder clarification so as to provide new insights concerning the description, explanation of causal links, prediction, normative approach and influence on societal phenomena.

In chapter 19, "Critical Approaches to Transitions Theory", *Bonno Pel* (postdoctoral researcher, *Centre d'Études du Développement Durable, Université Libre de Bruxelles*, Belgium), *Flor Avelino* (researcher and lecturer, DRIFT, Erasmus University Rotterdam, the Netherlands), and *Shivant Jhagroe* (PhD researcher, Eindhoven University of Technology, The Netherlands) argue that since its emergence as a theory of sustainability transformation, transitions theory has attracted both policymakers and researchers. As transitions approaches become established in research and policy, a process of institutionalization can be witnessed. Yet notwithstanding this mainstreaming, transitions theory continues to be controversial. Questions have been raised about its theorization of agency and transformation dynamics, and especially about the normative assumptions underlying its intervention strategies. There are repeated calls for 'critical approaches' to transitions theory. This contribution explores these, guided by a constructive attitude. The argument starts from the consideration that transitions theory harbours distinctly 'critical' elements: What are the critical contents of transitions theory? How can the critical contents of transitions theory be retained and developed further? These questions are answered through a historical comparison with the critical-theoretical project as initiated by Marx, Horkheimer and Adorno, amongst others. As with transitions studies, this project was meant to diagnose the social problems of its time, and to articulate corresponding remedial strategies. It ran into various internal contradictions, however, and these provide useful insights for the further development of critical transitions. The main conclusion is that transitions theory is well equipped to deal with these critical-theoretical paradoxes, but also displays tendencies towards relapsing into the pitfalls.

In chapter 20, *Mike Hodson* (Research Fellow, University of Manchester, *Sustainable Consumption Institute* (SCI) and *Manchester Institute of Innovation Research* (MIoIR), UK), *Simon Marvin* (Direc-

tor, Urban Institute and Professor, Department of Geography, Sheffield University, UK) and *Philipp Späth* (who heads a research project on Smart EcoCities, Institute of Environmental Social Sciences and Geography, Albert-Ludwigs-University Freiburg, Germany) address "Sub-national, Inter-scalar Dynamics: The Differentiated Geographies of Governing Low-Carbon Transitions—With Examples from the UK" with the aim of improving the analytical understanding of low-carbon transitions at and in between multiple geographical scales, particularly 'below' the national level. Taking the *Multilevel Perspective* (MLP) as their starting point, they show that it offers tools for thinking through the institutional and technological conditions and rules through which regimes reproduce or change over time. But it is not well equipped to facilitate the study of transitional dynamics as they unfold in space. The chapter sets out several levels on a spatial scale and activities that are relevant to transition activity but with which the MLP so far engages only partially. The chapter explicitly identifies very different, yet coexisting, scales of transition activity that aim at low-carbon transitions in the UK. It demonstrates the possibilities of attributing different spaces and levels to particular transition activities on a geographical scale and highlights the dynamics between these scaled activities. It initiates debate on the 'appropriate' mechanisms for dialogue 'between' governance levels and differently scaled transition activities.

1.8.6   Analysing National Debates on Sustainability in North America

The next three chapters by Twig Johnson (US), Simon Dalby (Canada) and Juan Antonio Le Clercq (Mexico) offer three case studies on aspects of national debates and policies in the three *North American Free Trade Agreement* (NAFTA) countries.

In chapter 21, "Policy, Politics and the Impact of Transition Studies", *Twig Johnson* (Writer/Consultant, Kingston, Massachusetts, US) argues that transition studies are undertaken in a variety of contexts and at various levels of society. Researchers hope and in some cases expect that their studies will be used to address challenges to sustainability transition. The relationship between research and effective action poses a challenge that transition studies must address. As part of an effort to study the different contexts in which sustainability transition studies have developed, Johnson discusses a 1999 report by the US National Research Council (NRC), *Our Common Journey: A*

Transition toward Sustainability (NRC 1999). The study was begun in 1996 during the Clinton and Gore administration. Sustainable development and environment were high on their list of priorities. The time seemed ripe for looking at ways in which science could better support US policy efforts to transition to sustainability. It was the best of times. Shortly after publication (December 1999) the political climate changed radically, especially after the 9/11 terrorist attacks. Johnson reviews this study and related efforts to address the sustainability transition challenge fifteen years since its publication.

In chapter 22, "Geopolitics, Ecology and Stephen Harper's Reinvention of Canada", *Simon Dalby* (see above, Canada) analyses the impact of the election to power of the Conservative Party in Canada in 2006 with a vision of the world that was much more competitive than previous Liberal or earlier conservative visions. Dalby interprets this as an attempt to reinvent Canada as a player in a world of competitive geopolitics rather than as a good citizen in a shared biosphere. Foreign and domestic policy have been shaped by this new view, and this has led to the abrogation of the Kyoto protocol. Since Canada is an energy superpower and oil exporter, it has also led to substantial attacks by the government on environmental science and regulatory processes, as these might obstruct resource company projects. What is being sustained in this process is a vision of Canada antithetical to what in most parts of the world would be considered sustainable. The lessons to be learnt for sustainable transitions are many, most notably the importance of thinking carefully about conventional politics and the dangers of narrowly-cast nationalist and populist attacks on environmental policies and sustainability initiatives.

In chapter 23, "Regime Change, Transition to Sustainability and Climate Change Law in México", *Juan Antonio Le Clercq* (professor, International and Political Sciences Department, *Universidad de las Américas, Puebla* (UDLAP), Mexico) examines the creation of the Climate Change Law in Mexico as a governance regime, where the legislative deliberations oriented towards creating the Climate Change Law between 2005 and 2012 are understood as a pathway to a sustainable transition. The chapter discusses the principal elements of Mexican climate policy and identifies three competing alternatives for climate legislation in 2012. The chapter argues that the energy sector limits the possibilities for a sustainable transition in Mexico.

1.8.7 Preparing Transitions towards a Sustainable Economy and Society, Production and Consumption and Urbanization

In part seven, chapters 24 to 30 deal with "Preparing Transitions towards a Sustainable Economy and Society, Production and Consumption and Urbanization". In chapter 24, "Sustainability Transitions: A Discourse-Institutional Perspective", *Audley Genus* (Professor, Small Business Research Centre, Kingston University, Kingston upon Thames, UK) addresses the complex web of activities and actors necessary for the transition to sustainability. He reviews various contributions, with a concern for the role that language and institutional arrangements play in related developments, in order to improve the understanding of the problems involved, the issues at stake, and the implications for policy and practice. He presents a framework for a discussion of sustainability transitions, neo-institutional theory, and critical discourse analysis by drawing on neo-institutional theory and critical discourse analysis on different factors and actors implicated in (un)sustainable patterns of production and consumption and the (in)effective governance of environmental sustainability-related science, technology and other phenomena. He considers potential insights from a discourse-institutional perspective and the progress that has been made using this approach.

In chapter 25, "New Business Models: Examining the Role of Principles Relating to Transactions and Interactions", *Jan Jonker* (Professor of Sustainable Entrepreneurship, Nijmegen School of Management, Radboud University Nijmegen and Chaire d'Excellence Pierre de Fermat, Toulouse Business School in Toulouse (France), The Netherlands) and *Linda O'Riardon* (Professor of Business Studies and International Management, Director of the *KompetenzCentrum for Corporate Social Responsibility* (KCC), FOM University of Applied Sciences in Essen, Germany) claim that our current economic ideas no longer function and that sustainability concerns are often central. They present the results of exploratory research initiated by Radboud University Nijmegen into *new business models* (NBMs). Their study focuses on business models that create so-called 'multiple value(s)', a way of organizing that focuses not just on organization between organizing entities. This approach generates social and ecological as well as economic value. For this research, a series of interviews was conducted in order to gain insight into the phenomenon of NBMs and to combine this fresh empirical evidence with theoretical underpinnings from previous scholarship. This data was then used to explore the field and discover the nature of NBMs, their features, and how they function in (micro-)practice.

In chapter 26, "Sustainable Consumption", *Sylvia Lorek* (head, *Sustainable Europe Research Institute* (SERI) Germany e.V.) offers key arguments from the sustainable consumption literature. The chapter introduces 'environmental space' as an early concept which embedded sustainable consumption within natural and social boundaries. It explains why a floor as well as a ceiling for the environmental space has to be considered and reflects on the space itself, its size and how to share it. Various possible paths of transition to reach the environmental space from a position of over- and under-consumption are linked to various schools in sustainability research. Lorek emphasizes the need for a detailed analysis of the concepts of 'green growth' and 'de-growth'. Relating these concepts to sustainable consumption research and politics, she distinguishes between strong and weak sustainable consumption, and outlines enabling mechanisms for sustainable consumption.

In chapter 27, "Sustainable Consumption and Production in China", *Hongmin Chen* (Associate Professor, Department of Environmental Science and Engineering and researcher, Fudan Urban Environmental Management Research Center and Fudan Tyndall Center, all at Fudan University, Shanghai, China) states that the increasing problem of severe resource scarcity, environmental pollution and ecological degradation has become a great constraint on economic development in China. *Sustainable consumption and production* (SCP), as a primary way to decouple economic development from environmental degradation, has been undertaken in China with energy-saving and environmental protection policies. She discusses the main practices of SCP, including the adjustment of the industrial structure, the promotion of a circular economy and clean production, the Energy Conservation and Emission Reduction Project, green procurement, and progress in waste management. Though China has made some achievements in SCP, it still faces serious challenges.

In chapter 28, "The Eco-restructuring of the Ruhr Area as an Example of a Managed Transition", *Philipp Schepelmann* (acting director, Research Group for Material Flows and Resource Management, Wuppertal Institute for Climate, Environment and Energy, Germany), *René Kemp* (Professor of innovation and sustainable development, *International Centre for Integrated Assessment and Sustainable Development* (ICIS), Maastricht University and professorial fellow, UNU–MERIT, The Netherlands), and *Uwe Schneidewind* (see above, Germany) present the eco-restructuring of the Ruhr area as a remarkable case of a managed transition, bringing out the importance of regional actors and factors alongside external stimuli. The evidence in support of the authors' hypothesis of a managed transition in the Ruhr district is: (i) the vision of blue skies above the Ruhr, (ii) the reconstruction of the Emscher river system, and (iii) the emerging energy transition. The ongoing transition of the Ruhr district requires further analysis which the authors plan to undertake. The chapter provides a first attempt at offering an evidence-based narrative of a region as a real-world laboratory for ecological modernization.

In chapter 29, "Transition towards Sustainable Urbanization in Asia and Africa", *Belinda Yuen* (Lee Kuan Yew Centre for Innovative Cities, Singapore University of Technology, Singapore) and *Asfaw Kumssa* (Chief Technical Advisor, Office of the United Nations Resident Coordinator's Office, Nairobi, Kenya) show that both continents are the world's least urbanized regions. But they are fast urbanizing; eighty per cent of the world's projected urban growth to 2050 is expected to take place in Africa and Asia. Even though cities in Africa and Asia are increasingly recognized as engines of growth, they are sites of extremely high population density, congestion, informal housing and the concomitant expansion of slums and squatter settlements, infrastructure shortages and environmental degradation. As rapid urbanization continues, how Africa and Asia manage the urban transition will not just affect urban economic efficiency but will also define their greenhouse gas footprint. They compare the urbanization of both regions and discuss the challenges and opportunities for transition towards sustainable urbanization.

In chapter 30, "The Role of University Partnerships in Urban Sustainability Experiments: Evidence from Asia", *Gregory Trencher* (Assistant Professor in Environmental Science and Policy, Clark University, US) and *Xuemei Bai* (Professor of Urban Environment and Human Ecology, Fenner School of Environment and Society, Australian National University, Australia) argue that university-driven partnerships and experiments for advancing urban sustainability are flourishing across the globe. Responding to drivers such as calls for stakeholder engagement in research, tangible social and economic contributions, and government funding incentives, Asian research universities are forming cross-sector partnerships and implementing various sociotechnical experiments. The authors examine the role of university partnerships in knowledge co-production and the implementation of urban sustainability experiments in industrialized Asian nations. By examining fifteen cases from Singapore, Japan, Hong Kong and Korea, they highlight common attributes and investigate the functions, motivations, barriers and significance of the roles assumed by different societal sectors. A case study from the University of Tokyo illustrates these attributes in context.

1.8.8 Examining Sustainability Transitions in the Water, Food and Health Sectors from Latin American and European Perspectives

The next three chapters, in part eight, examine sustainability transitions in the water, food and health sectors from Latin American and European perspectives. In chapter 31, "Future global water, food and energy needs", *Cecilia Tortajada* (Senior Research Fellow, Institute of Water Policy, Lee Kuan Yew School of Public Policy, Singapore) and *Martin Keulertz* (postdoctoral research associate, Department for Agricultural and Biological Engineering, Purdue University in Indiana, US) argue that if present trends continue, it is unlikely that increasingly overused ecosystems will meet global water, food and energy needs. Scarcity, pollution, mismanagement and misallocation of natural resources will impact every sector on which humankind depends for survival. In a globalized economy with increasingly free movement of commodities and financial and human capital, a poor understanding of the most pressing issues and their interconnectedness and interdependences will cause irreparable damage to Earth and its people. This is a challenging context for global development, and to understand and manage the interdependencies between the various sectors and their global impact will require comprehensive planning and policy implementation, institutional resilience, partnerships across economic sectors, and innovation in development in order to sustainably manage resources in the wake of population growth and climate change.

In chapter 32, "Sustainability Transition in a Vulnerable River Basin in Mexico", *Úrsula Oswald Spring* (see above, Mexico) examines a transition process that is greatly affected by climate change, social deterioration and the drugs war. The Yautepec River Basin in Central Mexico is particularly prone to climate impacts because of its abrupt slopes, numerous affluents, and a high population density in its floodplain, which is frequently exposed to the impacts of hurricanes. Taking laissez-faire policies, illegality, government corruption, and public insecurity into account in a region with high levels of dual vulnerability (environmental and social), and which is subject to extreme events, the chapter reviews the constraints on the process of transition towards sustainability, especially when the goals of the people and government are not the same. People organize themselves to increase public security and to reduce the impacts of extreme events which directly affect their lives and livelihoods, while the government promotes a wide infrastructure. The study examines the risks perceived by the people and the ways in which they reduce these risks of extreme events. The transition process requires an integration of numerous social, economic, political, environmental, judicial, cultural and mental factors. These will help to develop processes and resilience from the bottom up to deal with this complex emergency, as well as to force the different levels of government to enhance the human security of the people.

In chapter 33, "Sustainability Transition in the Health Sector in Brazil", *Monica de Andrade* (Health Promotion Graduate Program, University of Franca, Brazil) analyses the dynamic and the forces influencing political, sociotechnical and cultural features. The aim of her study is to identify transition in health in a Brazil facing globalization, demographic, migration, environmental and climate change effects and inequities. Brazil has improved its capacity to formulate, implement, monitor and evaluate multisectoral and universal public policies in the last twenty years, with the implementation of universal health care and conditional cash transfers. These policies have resulted in the improvement of the health indicators of extreme poverty and hunger, under-five mortality, maternal health, infectious diseases, and primary education coverage. Since the frequency of extreme weather events has been increasing globally and in Brazil cardiovascular diseases are a leading cause of death, the consequences for health systems must be considered. Mortality due to traffic accidents has also increased, despite policy efforts and restrictive laws, overwhelming health services and public financial resources.

1.8.9 Preparing Sustainability Transitions in the Energy Sector

The six chapters in part nine examine "Preparing Sustainability Transitions in the Energy Sector". *Jürgen Scheffran* (see above, Germany) and *Rebecca Froese*

(Research Assistant, CLISEC, Hamburg University, Germany) argue in chapter 34, "Enabling Environments for a Sustainable Energy Transformation: The Diffusion of Know-How, Technologies and Innovations in Low-Carbon Societies", that a sustainability transition requires transformation processes and innovations in different fields, including new technologies, products and infrastructures as well as new social rules and norms. Greening the economy rests on the rapid and effective dissemination of climate-friendly technologies, in particular of renewable and efficient energy systems. A high level of finance and smart governance are required in this process in order to involve the economic and technological capacities of all countries. Within this progression, the international diffusion of know-how, investments and technologies needs to be strengthened in order to build up production capacities and the demand for low-carbon goods. Furthermore, business and government actors have to establish international cooperation so that they can adapt technologies to the local context and support the transformation towards a low-carbon society.

In chapter 35, "Considering a Structural Adjustment Approach to the Low-Carbon Transition", *Karlson Hargroves* (Postdoctoral Fellow, Curtin University, Perth, Australia) raises a range of questions about how to effectively transition economies to low-carbon operation over the coming decades as the world comes to grips with the need to significantly reduce greenhouse gas emissions. Increasing pressure is now being felt across a range of sectors to reduce emissions, in particular carbon-related fuel consumption, and this is leading to autonomous emissions reduction efforts—typically ad hoc and business-led. However, in order to meet the ambitious targets for the reduction of greenhouse gas emissions now set by the world's largest economies, a structural adjustment approach may be needed, and this needs to be effectively underpinned and appropriately expedited at an economy-wide level. Hargroves introduces key lessons from structural adjustment programmes for the low carbon transition and how the willingness to adjust structures of the economy to deliver low-carbon outcomes can be increased.

Britta Rennkamp (researcher, Energy Research Centre and fellow, African Climate and Development Initiative, University of Cape Town, South Africa) and *Radhika Perrot* (senior researcher, *Mapungubwe Institute for Strategic Reflection* (MISTRA), Johannesburg, South Africa) examine, in chapter 36, "Drivers and Barriers to Wind Energy Technology Transitions

in India, Brazil and South Africa", by analysing actors and institutions that have played a role in the development and diffusion of wind energy technologies, from a *technological innovation systems* (TIS) perspective. They introduce innovation capabilities, the enabling environment and policy-independent strategies as the main drivers for the development of TIS. Their aim is to contribute to improving the understanding of drivers of and barriers to innovation relevant to successful transitions towards cleaner technologies and less carbon-intensive economies.

In chapter 37, "Sustainability Transitions and the Politics of Electricity Planning in South Africa", *Lucy Baker* (research fellow, Science Policy Research Unit, University of Sussex and visiting fellow, Energy Research Centre, University of Cape Town; UK) shows that after decades of cheap, abundant coal-fired electricity, from which large international mining and energy conglomerates and wealthy households have benefited disproportionately, South Africa is experiencing a supply-side crisis. In 2011, the country's first *integrated resource plan for electricity* (IRP) was promulgated after a prolonged consultation process throughout 2010. This plan anticipates that renewable energy will constitute twenty per cent of installed generation capacity by 2030, and that this will deliver approximately nine per cent of supply. Coal will retain the greatest share alongside a potential yet currently uncertain nuclear sector. She examines electricity governance in South Africa and the highly politicized policy-making process for IRP in which vested interests have played a major role; and considers the extent to which IRP has facilitated a low-carbon transition.

In chapter 38, "Low Carbon Green Economy: Brazilian Policies and Politics of Energy, 2003–2014", *Eduardo Viola* (Full Professor, International Relations at the University of Brasília and senior researcher, *Brazilian Council for Scientific Research* (CNPQ), Brazil) and *Larissa Basso* (PhD candidate, University of Brasília and researcher, *Brazilian Federal Agency for Support and Evaluation of Graduate Education* (CAPES), Brazil) argue that achieving sustainability is complex. A low-carbon green economy is introduced as a good paradigm for guiding policy-making towards respecting these boundaries; it includes the objective of the decarbonization of energy systems in order to mitigate climate change. The authors show how Brazilian energy policies and politics evolved from 2003 to 2014, and they identify an upward and downward trend in the production of energy from low-carbon sources and a stable trend in energy efficiency. They conclude that Brazil has not yet embraced the low-carbon green economy even though significant forces are promoting it, while powerful political forces are opposed.

In chapter 39, "Sustainable Electricity Transition in Thailand and the Role of Civil Society", *Carl Middleton* (lecturer, *MA in International Development Studies* (MAIDS) Program and Deputy Director for International Research, *Center for Social Development Studies* (CSDS), Faculty of Political Science, Chulalongkorn University, Bangkok, Thailand) explains the creation and resistance to change of Thailand's centralized and fossil-fuel-intensive electricity regime through a Sustainability Transition and Multilevel Perspective lens, with an emphasis on the sector's political economy (actors, interest, power). The incumbent electricity industry has evolved from a state-owned monopoly to a partially-privatized industry dominated by the state utility and several large independent power producers, and is increasingly shaped by the sector's financialization. Middleton's analysis demonstrates how important global landscape shifts articulate with the sector's domestic political economy, including a shifting global development paradigm from a developmentalist state to liberal market principles, as well as the impact of waves of global economic crises.

1.8.10 International, Regional and National Governance towards Sustainability Transition

In chapter 40, "Governance of Sustainable Development in Knowledge Democracies—Its Consequences for Science", *Roeland J. in 't Veld* (UNESCO Professor of Governance and Sustainability at Tilburg University, The Netherlands) claims that governance is the way in which a society organizes decision-making. Advanced societies turn into knowledge democracies where the relationships between politics, media and science intensify and change continuously. He argues that the quest for sustainable development takes place within knowledge democracies, where sustainability covers economic, social and ecological issues, whose governance is complex. The future direction and content of sustainable development depend on uncertain future determinants such as technological innovation and the evolution of social values. The multidimensional nature of sustainability requires integration and a recognition of a multilevel, multiscale, multidisciplinary character. Development is about change, transitions and transformations. Governance of sustainable development has to cope with complex dynamics. Governance that furthers transitions focuses on the interaction between representative and participatory democracy, and the optimization of the contribution of science. He addresses the specific consequences for science relating both to disciplinary research and transdisciplinarity.

Sander Happaerts (European Commission, Brussels and part-time assistant professor, Leuven International and European Studies, University of Leuven, Belgium) argues in chapter 41, "Discourse and Practice of Transitions in International Policy-Making on Resource Efficiency in the EU", that for several years, 'transitions' has been the new buzzword in international policy-making. The emergence of an international transitions discourse is linked to debates on the green economy and a low-carbon society, and is manifested in the UN system and in EU discussions. Happaerts analyses international policy initiatives from the perspective of sustainability transitions, with the aim of moving towards a better understanding of the potential role of international policy-making in current and future transitions. A case study of the European Commission's policy initiatives on resource efficiency demonstrates a strong awareness of transitions thinking, consistently integrated into the discourse with the intention of creating a sense of urgency and convincing other actors of the need for fundamental change. Furthermore, transitions thinking has served as the inspiration for several principles, goals and instruments, such as the European Resource Efficiency Platform.

In the concluding chapter, Úrsula Oswald Spring (Mexico), Hans Günter Brauch (Germany), and Jürgen Scheffran (Germany) summarize the key messages of this book and offer suggestions for future research on sustainability transition and sustainable peace.

References

Álvarez, Enrique; Oswald Spring, Úrsula, 1993: *Desnutrición Crónica o Aguda Materno Infantil y Retardos en el Desarrollo*, Aporte de Investigación No. 59 (Cuernavaca: CRIM-UNAM).

Annan, Kofi A., 2005: *In Larger Freedom: Towards Security, Development and Human Rights for All. Report of the Secretary General for Decision by Heads of State and Government in September 2005*. A/59/2005 (New York: United Nations, Department of Public Information, 21 March).

Barry, Azamiou, 2012: *El Impacto de los Conflictos Regionales en el Proceso de Integración en África* (PhD thesis, UNAM, Faculty of Political and Social Sciences, Mexico, D.F.).

Beyme, Klaus von; Offe, Claus (Eds.), 1995: *Politische Theorien in der Ära der Transformation*, as: *PVS Sonderheft*, 26 (Wiesbaden: Westdeutscher Verlag)

Bogardi, Janos; Brauch, Hans Günter, 2005: "Global Environmental Change: A Challenge for Human Security—Defining and conceptualising the environmental dimension of human security", in: Rechkemmer, Andreas (Ed.): *UNEO—Towards an International Environment Organization—Approaches to a sustainable reform of global environmental governance* (Baden-Baden: Nomos): 85-109.

Bohle, Hans-Georg, 2001: "Vulnerability and Criticality: Perspectives from Social Geography", in: *IHDP Update*, 2/01: 3-5.

Bohle, Hans-Georg, 2002: "Land degradation and Human Security", in: Plate, Erich (Ed.): *Environment and Human Security, Contributions to a workshop in Bonn* (Bonn).

Bond, Gerard; Showers, William; Cheseby, Maziet; Lotti, Rusty; Almasi, Peter; de Menocal, Peter; Priore, Paul; Cullen, Heidi; Hajdas, Irka; Bonani, Georges, 1997. "A Pervasive Millennial-Scale Cycle in North Atlantic Holocene and Glacial Climates", in: *Science* 278, 14 (November): 1257-1266.

Brauch, Hans Günter, 2002: "Climate Change, Environmental Stress and Conflict—AFES-PRESS Report for the Federal Ministry for the Environment, Nature Conservation and Nuclear Safety", in: Federal Ministry for the Environment, Nature Conservation and Nuclear Safety (Ed.): *Climate Change and Conflict. Can climate change impacts increase conflict potentials? What is the relevance of this issue for the international process on climate change?* (Berlin: Federal Ministry for the Environment, Nature Conservation and Nuclear Safety): 9-112; at: <http://www.afes-press.de/pdf/Brauch_ClimateChange_BMU.pdf>.

Brauch, Hans Günter, 2005: *Environment and Human Security. Freedom from Hazard Impact*, InterSecTions, 2/2005 (Bonn: UNU-EHS); at: <http://www.ehs.unu.edu/file.php?id=64>.

Brauch, Hans Günter, 2007c: "The Model: Global Environmental Change, Political Process and Extreme Outcomes", in: Oswald Spring, Úrsula (Ed.): *Encyclopedia of Life Support System* (Oxford: Oxford-EOLSS Publisher), vol. 39; at: <http://www.eolss.net/EI-39B-toc.aspx>.

Brauch, Hans Günter, 2009: "Securitizing Global Environmental Change", in: Brauch, Hans Günter; Oswald Spring, Úrsula; Grin, John; Mesjasz, Czeslaw; Kameri-Mbote, Patricia; Behera, Navnita Chadha; Chourou, Béchir; Krummenacher, Heinz (Eds.), 2009: *Facing Global Environmental Change: Environmental, Human, Energy, Food, Health and Water Security Concepts* (Berlin—Heidelberg—New York: Springer-Verlag): 65-102.

Brauch, Hans Günter, 2012: "Climate Paradox of the G8: legal obligations, policy declarations and implementation gap", in: *Revista Brasileira de Política Internacional*. Special issue edited by Eduardo Viola & Antonio Carlos Lessa on: *Global Climate Governance and Transition to a Low Carbon Economy* (Brasilia: Instituto Brasileira de Realacoes Internacionais): 30-52.

Brauch, Hans Günter, 2014: "From Climate Change and Security Impacts to Sustainability Transition: Two Policy Debates and Scientific Discourses", in: Oswald Spring, Úrsula; Brauch, Hans Günter; Tidball, Keith G. (Eds.): *Expanding Peace Ecology: Peace, Security, Sustainability, Equity and Gender: Perspectives of IPRA's Ecology and Peace Commission*. (Heidelberg—Dordrecht—London—New York: Springer-Verlag): 33-61.

Brauch, Hans Günter, 2016: "Historical Times and Turning Points in a Turbulent Century: 1914, 1945, 1989 and 2014?", in: Brauch, Hans Günter; Oswald Spring; Ursula; Bennett, Juliet; Serrano Oswald, Serena Eréndira (Eds.): *Addressing Global Environmental Challenges from a Peace Ecology Perspective* (Cham—Heidelberg—Dordrecht—London—New York: Springer-Verlag, forthcoming).

Brauch, Hans Günter; Dalby, Simon; Oswald Spring, Úrsula, 2011: "Political Geoecology for the Anthropocene", in: Brauch, Hans Günter; Oswald Spring, Úrsula; Mesjasz, Czeslaw; Grin, John; Kameri-Mbote, Patricia; Chourou, Béchir; Dunay, Pal; Birkmann, Jörn (Eds.), 2011: *Coping with Global Environmental Change, Disasters and Security—Threats, Challenges, Vulnerabilities and Risks* (Berlin—Heidelberg—New York: Springer-Verlag): 1453-1486.

Brauch, Hans Günter; Oswald Spring, Úrsula, 2009: *Securitizing the Ground—Grounding Security* (Bonn: UNCCD, 2009); at: <http://www.unccd.int/knowledge/docs/dldd_eng.pdf>.

Brauch, Hans Günter; Oswald Spring, Úrsula; Grin, John; Mesjasz, Czeslaw; Kameri-Mbote, Patricia; Behera, Navnita Chadha; Chourou, Béchir; Krummenacher, Heinz (Eds.), 2009: *Facing Global Environmental*

Change: Environmental, Human, Energy, Food, Health and Water Security Concepts (Berlin–Heidelberg–New York: Springer-Verlag).

Brauch, Hans Günter; Oswald Spring, Úrsula; Mesjasz, Czeslaw; Grin, John; Dunay, Pal; Behera, Navnita Chadha; Chourou, Béchir; Kameri-Mbote, Patricia; Liotta, P. H. (Eds.), 2008: *Globalization and Environmental Challenges: Reconceptualizing Security in the 21ˢᵗ Century* (Berlin–Heidelberg–New York: Springer-Verlag).

Brauch, Hans Günter; Oswald Spring, Úrsula; Mesjasz, Czeslaw; Grin, John; Kameri-Mbote, Patricia; Chourou, Béchir; Dunay, Pal; Birkmann, Jörn (Eds.), 2011: *Coping with Global Environmental Change, Disasters and Security–Threats, Challenges, Vulnerabilities and Risks* (Berlin–Heidelberg–New York: Springer-Verlag)

Brauch, Hans Günter; Scheffran, Jürgen, 2012: "Introduction: Climate Change, Human Security, and Violent Conflict in the Anthropocene", in: Scheffran, Jürgen; Brzoska, Michael; Brauch, Hans Günter: Link, Michael; Schilling, Janpeter (Eds.), 2012: *Climate Change, Human Security and Violent Conflict: Challenges for Societal Stability* (Berlin–Heidelberg: Springer-Verlag): 3-40.

Braudel, Fernand, 1969: "Histoire et science sociales. La longue durée" ; in: *Écrits Sur l'Histoire* (Paris: Flammarion): 41-84.

Braudel, Fernand, 1972: *The Mediterranean and the Mediterranean World in the Age of Philip II,* 2 volumes (New York: Harper & Row).

Brockhaus Enzyklopädie, 19ᵗʰ ed., 1991, vol. 15 (Wiesbaden–Mannheim: Brockhaus).

Brockhaus Enzyklopädie, 21ˢᵗ ed., 2006, vol. 27 (Wiesbaden–Mannheim: Brockhaus).

Brundtland Commission (World Commission on Environment and Development), 1987: *Our Common Future. The World Commission on Environment and Development* (Oxford–New York: Oxford University Press).

Butler, Judith, 1990: *Gender Trouble: Feminism and the Subversion of Identity* (London: Routledge).

Buzan, Barry; Wæver, Ole; Wilde, Jaap de, 1998: *Security. A New Framework for Analysis* (Boulder–London: Lynne Rienner).

CBD (Secretariat of the Convention on Biological Diversity) 2009: *Connecting Biodiversity and Climate Change Mitigation and Adaptation: Report of the Second Ad Hoc Technical Expert Group on Biodiversity and Climate Change* (Montreal: Technical Series No. 41).

CBD, 2011a: "Nagoya Protocol on Access to Genetic Resources and the Fair and Equitable Sharing of Benefits Arising from their Utilization to the Convention on Biological Diversity", at: <https://www.cbd.int/abs/doc/protocol/nagoya-protocol-en.pdf>.

CBD, 2011b: "Aichi Biodiversity Targets", at: <https://www.cbd.int/sp/targets/>.

CBD, 2011c: *Strategic Plan for Biodiversity 2011-2020 and the Aichi Targets "Living in Harmony with Nature"*, at:

<https://www.cbd.int/doc/strategic-plan/2011-2020/Aichi-Targets-EN.pdf>.

CBD, 2015: *National Biodiversity Strategies and Action Plans (NBSAPs)*, at <https://www.cbd.int/nbsap/>.

CHS (Commission on Human Security), 2003, 2005: *Human Security Now, Protecting and empowering people* (New York: Commission on Human Security); at: <http://www.humansecurity-chs.org/finalreport/>.

Clark, William C.; Crutzen, Paul J.; Schellnhuber, Hans Joachim, 2004: "Science and Global Sustainability: Toward a New Paradigm", in: Schellnhuber, Hans Joachim; Crutzen, Paul J.; Clark, William C.; Claussen, Martin; Held, Hermann (Eds.): *Earth System Analysis for Sustainability* (Cambridge, MA; London: MIT Press): 1-28.

Coenen, Lars; Benneworth, Paul; Truffer, Bernhard, 2010: *Towards a spatial perspective on sustainability transitions.* CIRCLE paper no. 2010/08 (Lund: Lund University, Center for Innovation, Research and Competence in the Learning Economy [CIRCLE]).

Coenen, Lars; Benneworth, Paul; Truffer, Bernhard, 2012: "Towards a spatial perspective on sustainability transitions", in: *Research Policy,* 41,6: 968-979.

Coenen, Lars; Lopez, Diaz, 2010: "Comparing system approaches to innovation and technological change for sustainable and competitive economies: An explorative study into conceptual commonalities, differences and complementarities", in: *Journal of Cleaner Production,* 18: 1149-1160.

Coenen, Lars; Truffer, Bernhard, 2012: "Environmental Innovation and Sustainability in Regional Studies", in: *Regional Studies,* 46,1: 1-21.

Cohen, Barney, 2006: "Urbanization in developing countries: Current trends, future projections, and key challenges for sustainability", *Technology in Society* 28: 63-80.

Cohen, K. M.; Finney, S.; Gibbard, P. L., 2012: "International Chronostratigraphic Chart: International Commission on Stratigraphy", at: <www.stratigraphy.org> (Chart reproduced for the 34th International Geological Congress, Brisbane, Australia, 5-10 August 2012.)

Coleman, Peter T.; Deutsch, Morton (Eds.), 2012: *Psychological Components of Sustainable Peace* (New York–Heidelberg–Dordrecht–London: Springer, 2012).

Compact Oxford Dictionary of Current English (2002).

Corfee-Morlot, J.; Kamal-Chaoui, L.; Donovan, M. G.; Cochran, I.; Robert, A.; Teasdale, P.-J., 2009: *Cities, Climate Change and Multi-Level Governance.* OECD Environmental Working Paper (Paris: OECD).

Cortes, Juana E.; Calderón, César G., 2011: "Potable Water Use From Aquifers Connected to Irrigation of Residual Water", in: Oswald Spring, Úrsula (Ed.): *Water Resources in Mexico* (Heidelberg – Berlin: Springer): 189-200.

Crutzen, Paul J., 2002: "Geology of Mankind", in: *Nature,* 415,3 (January): 23.

Crutzen, Paul J., 2006: "The Anthropocene", in: Ehlers, Eckart; Krafft, Thomas (Eds.): *Earth System Science in the Anthropocene* (Berlin–Heidelberg–New York: Springer): 13-18.

Crutzen, Paul J.; Stoermer, Eugene F., 2000: "The 'Anthropocene'", in: *IGBP Newsletter 41–The Global Change Newsletter*, 17-18.

D'Eaubonne, Françoise 1974: *Le Féminisme ou la Mort* (Paris: Pierre Horay).

De Schutter, Olivier, 2014: *Report of the Special Rapporteur on the right to food*, A/HRC/25/57 (New York: UNGA).

Der Große Duden, 1963: *Etymologie*, vol. 7 (Mannheim: Bibliographisches Institut).

Earth Charter, 2000: "The Earth Charter", at: <http://earthcharter.org/discover/the-earth-charter>.

Elliott, John E.; Bishop, Christine A.; Morrissey, Christy A., 2011: *Wildlife Ecotoxicology: Forensic Approaches* (New York: Springer).

EM-DAT (Guha, Sapir D.; Hoyois, Ph.; Below, R.), 2014: *Annual Disaster Statistical Review 2013. The numbers and trends* (Brussels: CRED); at: <http://cred.be/sites/default/files/ADSR_2013.pdf> (20 February 2016).

Ericksen, Polly J., 2007: "Conceptualizing food systems for global environmental change research", in: *Global Environmental Change*, at: <http://greenconsensus.com/education/food/materials/03_due_september26/02_eriksen07-foodsystems.pdf>.

Eriksen, Thomas H., 2010: *Small Places, Larges Issues An Introduction to Social and Cultural Anthropology* (New York: Palgrave Macmillan).

EU (European Commission), 2011: *Energy Roadmap 2050* (Brussels: European Commission).

EU (European Commission), 2011a: *White Paper. Roadmap to a Single European Transport Area–Towards a competitive and resource efficient transport system* (Brussels: European Commission).

EU (European Commission), 2011b: *A Roadmap for moving to a competitive low carbon economy in 2050* (Brussels: European Commission).

EU (European Commission), 2011c: *A resource-efficient Europe–Flagship initiative under the Europe 2020 Strategy* (Brussels: European Commission).

EU (European Commission), 2011d: *Analysis associated with the Roadmap to a Resource Efficient Europe. Part II* (Brussels: European Commission).

European Commission, Climate Action, "Roadmap for moving to a low-carbon economy in 2050"; at: <http://ec.europa.eu/clima/policies/roadmap/index_en.htm>.

FAO, 2013a: *Climate-smart agriculture sourcebook* (Rome: FAO).

FAO, 2013b: *Food wasting footprint. Impacts on natural resources* (Rome: FAO).

FAO, 2015: *Global guidelines for the restoration of degraded forests and landscapes in drylands: building resilience and benefiting livelihoods*, Forestry Paper No. 175 (Rome: FAO).

FAO, 2016a: *Climate change and food security: risks and responses* (Rome: FAO).

FAO, 2016b: *Status of the World's Soil Resources* (Rome: FAO).

Folbre, Nancy (2006), "Rethinking the Child Care Sector", in: *Journal of the Community Development Society*, 37, 2 (Summer): 38-52.

Fordham, Maureen, 1999: "The Intersection of Gender and Social Class in Disaster: Balancing Resilience and Vulnerability", in: *Intern. J. of Mass Emergency and Disasters* 17,1 (March): 15-36.

GBO-4, 2014: *Global Biodiversity Outlook*, at: <https://www.cbd.int/gbo/gbo4/publication/gbo4-en-hr.pdf>.

Geels, Frank W., 2002: "Technological Transitions as evolutionary reconfiguration processes: A multi-level perspective and a case study", in: *Research Policy*, 31,8-9: 1257-1274.

Geels, Frank W.; Schot, Johan, 2010: "The Dynamics of Transitions. A Socio-Technical Perspective", in: Grin, John; Rotmans, Jan; Schot, Johan (Eds.): *Transitions to Sustainable Development. New Directions in the Study of Long Term Transformative Change* (New York: Routledge): 11-104.

Gell-Mann, Murray, 1994: *The Quark and the Jaguar: Adventures in the Simple and in the Complex* (New York: Freeman).

German Brockhaus Encyclopedia (16th Ed.), 1955, vol. 8: 243.

GHI [Global Hunger Index], 2014: *2014 Global Hunger Index: The Challenge of Hidden Hunger* (Bonn: Welthungerhilfe; Washington DC: International Food Policy Research Institute).

GHI [Global Hunger Index], 2015: *2015 Global Hunger Index: Armed Conflict and the Challenge of Hunger* (Bonn: Welthungerhilfe; Washington DC: International Food Policy Research Institute).

Gibbs, H.K.; Ruesch, A.S.; Foley, J.A., 2010: "Tropical Forests Were the Primary Sources of New Agricultural Land in the 1980s and 1990s", in: *Proc. Natl. Acad. Sci*, 107, 38: 16732-7.

Gibbs, H.K.; Salmon, J.M. "Mapping the world's degraded lands", in: *Appl. Geogr*, 2015, 57, 12-21.

Gleditsch, Nils-Petter, 2012: "Whither the weather? Climate change and conflict", in: *Journal of Peace Research, special issue: Climate change and conflict*, 49,1 (January-February): 9-18.

Gradstein, Felix M.; Ogg, James G.; Schmitz, Mark D.; Ogg, Gabi, 2012: *The Geologic Time Scale 2012* (Boston: Elsevier).

Grin, John, 2010: "Understanding Transitions from a Governance Perspective", in: Grin, John; Rotmanns, Jan; Schot, Johan, 2010: *Transitions to Sustainable Development. New Directions in the Study of Long Term Transformative Change* (New York, NY–London: Routledge): 223-319.

Grin, John; Rotmanns, Jan; Schot, Johan, 2010: *Transitions to Sustainable Development. New Directions in the Study of Long Term Transformative Change* (New York, NY–London: Routledge);

Grin, John; Rotmanns, Jan; Schot, Johan, 2011: "On patterns and agency in transition dynamics: some key insights from the KSE programme, in: *Environmental Innovation and Societal Transitions*, 1 (2011): 76-81.

Grunberg, Isabelle; Risse-Kappen, Thomas, 1992: "A Time for Reckoning? Theories of International Relations and the End of the Cold War", in: Allan, Pierre; Goldmann, Kjell (Eds.): *The End of the Cold War* (Dordrecht: Martinus Nijhoff Publishers): 104-46.

Häcker, H.; Stapf, K. H., 1994: *Dorsch Psychologisches Wörterbuch* (Bern–Göttingen–Toronto–Seattle: Hans Huber).

Hammond, A., 1998: *Which World? Scenarios for the 21st Century* (Washington, DC: Island Press).

Head, Lesley; Trigger, David; Mulcock, Jane, 2005: "Culture as Concept and Influence in Environmental Research and Management", in: *Conservation and Society*: 3,2 (December): 251-264.

Henderson, Vernon, 2002: "Urbanization in Developing Countries", in: *World Bank Research Observer*, 17,1, March: 89-112.

HM Government: *The UK Low Carbon Transition Plan–National strategy for climate and energy* (London: HM Government, 15 July 2009).

Hobsbawm, Eric, 1962: *The age of revolution: 1789-1848* (London: Hachette): chapter 11.

Hoekstra, Arjen, 2011: *The water footprint assessment manual: Setting the global standard* (London: Earthscan).

Huggett, Richard John, 1995: *Geoecology. An Evolutionary Approach* (London–New York: Routledge).

Hynes, H. Patricia, 2004: "On the Battlefield of Women's Bodies: An Overview of the Harm of War to Women", in: *Women's Studies International Forum*, 27: 431-445

IEA, 2014: *World Energy Outlook 2014* (Paris: OECD/IEA).

Ímaz, Mireya; Blazquez, Norma; Chao, Verania; Castañeda, Itzá; Beristain, Ana (Eds.), 2014: *Cambio climático. Miradas de género* (Mexico: PUMA, CEIICH, CRIM, FC, FCPS, FM, FP, PINCC-UNAM/UNDP).

In 't Veld, Roeland J., 2011: *TRANSGOVANCE. The Quest for Governance of Sustainable Development* (Potsdam: Institute for Advanced Sustainability Studies [IASS]).

Inglehart, R., 1977: *The Silent Revolution. Changing Values and Political Styles among Western Publics* (Princeton, N.J.: Princeton University Press).

Inglehart, R., 1998: *Modernisierung und Postmodernisierung. Kultureller, wirtschaftlicher und politischer Wandel in 43 Gesellschaften* (Frankfurt/M.–New York: Campus).

Inglehart, R., 2008: "Changing values among western publics", in: *West European Politics*, 31,1-2: 130-146.

IPCC, 2007: *Climate Change 2007. The Physical Science Basis*. Working Group I Contribution to the Fourth Assessment Report of the IPCC (Cambridge–New York: Cambridge University Press); also at: <http://www.ipcc-wg1.org/>.

IPCC, 2007a: *Climate Change 2007. Mitigation and Climate Change*. Working Group III Contribution to the Fourth Assessment Report of the IPCC (Cambridge–New York: Cambridge University Press); at: <http://www.ipcc.ch/publications_and_data/ar4/wg3/en/figure-spm-2.html> (9 February 2015).

IPCC, 2011: *Renewable Energy Sources and Climate Change Mitigation* [SRREN] (Geneva: IPCC).

IPCC, 2013: *Climate Change 2013–The Physical Science Basis. Working Group I Contribution to the Fifth Assessment Report of the Intergovernmental Panel on Climate Change* (Cambridge–New York: Cambridge University Press); at: <http://www.ipcc.ch/report/ ar5/wg1/>.

IPCC, 2014: *Climate Change 2014–Synthesis Report, Summary for Policymakers* (Geneva: IPCC); at: <http://www.ipcc.ch/pdf/assessment-report/ar5/syr/SYR_AR5_SPMcorr2.pdf>.

IPCC, 2014a: "Human Security", in: *Climate Change 2014–Impacts, Adaptation, and Vulnerability–Part A: Global and Sectoral Aspects. Working Group II Contribution to the Fifth Assessment Report of the Intergovernmental Panel on Climate Change* (Cambridge–New York: Cambridge University Press): 755-701; at: <https://www.ipcc.ch/pdf/assessment-report/ar5/wg2/WGIIAR5-Chap12_FINAL.pdf>.

IPCC, 2014b: *Climate Change 2014–Impacts, Adaptation, and Vulnerability–Part A: Global and Sectoral Aspects. Working Group II Contribution to the Fifth Assessment Report of the Intergovernmental Panel on Climate Change* (Cambridge–New York: Cambridge University Press).

IPCC, 2014c: *Climate Change 2014–Impacts, Adaptation, and Vulnerability–Part B .... Working Group II Contribution to the Fifth Assessment Report of the Intergovernmental Panel on Climate Change* (Cambridge–New York: Cambridge University Press).

IPCC, 2014d: *Climate Change 2014–Mitigation of Climate Change: Working Group III Contribution to the Fifth Assessment Report of the Intergovernmental Panel on Climate Change* (Cambridge–New York: Cambridge University Press).

IRENA (International Renewable Energy Agency), 2016: *REmap: Roadmap for a Renewable Energy Future. 2016 Edition* (Abu Dhabi: IRENA).

ISSC (International Social Science Council), 2012: *Transformative Cornerstones of Social Science Research for Global Change* (Paris: UNECO, International Social Science Council).

IUCN [International Union for Conservation of Nature and Natural Resources], 1980: *World Conservation Strategy* (Gland: IUCN).

Jaspers, Karl, 2013: *One Century of Karl Jaspers' General Psychopathology* [edited by Giovanni Stanghellini and Thomas Fuchs] (Oxford: Oxford Medicine on Line).

Kates, Robert W., 2001: "Sustainability Transition: Human-Environment Relationship", in: Smelser, N. J.; Baltes, Paul B. (Eds.): *International Encyclopedia of the Social and Behavioral Sciences (New York:* Pergamon): 15325-15329.

Kates, Robert W.; Clark, William C.; Corell, Robert; Hall, J. Michael; Jaeger, Carlo C.; Lowe, Ian; McCarthy, James J.; Schellnhuber, Hans Joachim; Bolin, Bert; Dickson, Nancy M.; Faucheux, Sylvie; Gallopin, Gilberto C.; Grubler, Arnulf; Huntley, Brian; Jager, Jill; Jodha, Narpat S.; Kasperson, Roger E.; Mabogunje, Akin; Matson, Pamela Harold Mooney, Berrien Moore III, Timothy O'Riordan, Uno Svedin, 2001: "Sustainability science", in: *Science,* 292: 641-642.

Klein, Naomi: "Capitalism vs. the Climate—Denialists are dead wrong about the science. But they understand something the left still doesn't get about the revolutionary meaning of climate change", in: *The Nation,* 28 November 2011; at: <http://www.thenation.com/article/164497/capitalism-vs-climate> (25 January 2014).

Kotz, David M., 2015: *The Rise and Fall of Neoliberal Capitalism* (Cambridge: Harvard University Press).

Kuhn, Thomas, 1962: *The Structure of Scientific Revolutions* (Chicago: University of Chicago Press).

Land, Tim, Michael Heaseman, 2004: *Food Wars. The Global Battle for Mouths, Minds and Markets* (London: Earthscan).

Leb Hooke, Roger, 2015: University of Maine, cited in: Bourne, Joel K.: "The End of Plenty. Special Report the Global Food Crisis", in: *National Geographic* 215,6 (June): 26-59.

Leemans, Rik; Rice, Martin; Henderson-Sellers, Ann; Noone, Kevin, 2011: "Research Agenda and the Policy Input of the Earth Systems Science Partnership for Coping with Global Environmental Change", in: Brauch, Hans Günter et al. (Eds.): *Coping with Global Environmental Change, Disasters and Security—Threats, Challenges, Vulnerabilities and Risks* (Berlin–Heidelberg–New York: Springer-Verlag): 1205-1220.

Leiserowitz, A. A.; Kates, R. W.; Parris, T. M., 2006: "Sustainability values, attitudes, and behaviors: a review of multinational and global trends", in: *Annual review of environment and resources,* 31: 413-444.

Lenton, Timothy; Held, Hermann; Kriegler, Elmar; Hall, Jim W.; Lucht, Wolfgang; Ramstorf, Stefan; Schellnhuber, Hans Joachim, 2008: "Tipping elements in the Earth's climate system", in: *Proceedings of the National Academy of Science* (PNAS), 105,6 (12 February): 1786-1793.

Levy , Guy J.; Fine, Pinchas; Goldstein, Dina; Azenkot, Asher; Zilberman, Avraham; Chazan, Amram; Grinhut, Tzfrir, 2014: "Long term irrigation with treated wastewater (TWW) and soil sodification", in: *Biosystems Engineering,* 128 (December): 4-1.

Liebowitz, S.J.; Marjolis, S.E., 1995: "Path dependence, lock-in and history", in: *Journal of Law, Economics and Organization,* 11: 205-226.

Longman's Dictionary of Contemporary English (1995).

Loorbach, Derk Albert, 2007: *Transition Management* (PhD thesis, Rotterdam: Erasmus University Rotterdam).

MA, 2005: *Ecosystems and Human Well-Being: Synthesis* (Washington DC: Island Press); at: <http://www.millenniumassessment.org/documents/document.300.aspx.pdf>.

MA, 2005a: *Ecosystems and Human Well-Being: Desertification Synthesis* (Washington DC: World Resources Institute).

MA, 2005b: "Dryland Systems", in: Millennium Ecosystem Assessment (Ed.): *Ecosystems and Human Well-Being: Current State and Trends. Findings of the Condition and Trend Working Group* (Washington DC: Island Press): 623-662.

MacKinnon, Catherine, 1987: *Feminism Unmodified: Discourses on Life and Law* (Boston: Harvard University Press).

Mancebo, François; Sachs, Ignacy (Eds.), 2015: *Transitions to Sustainability* (Dordrecht: Springer);

Marsh, George P., 1864 [1965]: *The Earth as Modified by Human Action* (Cambridge, MA: Belknap–Harvard University Press).

McCright, A. M.; Dunlap R. E., 2000: "Challenging global warming as a social problem: An analysis of the conservative movement's counter-claims", in: *Social problems,* 47,4: 499-522.

McKechnie, Jean L. (Ed.), 1983: *Webster's New Universal Unabridged Dictionary* (New York: Dorset & Baber).

McNeill, John R., 2009: "The International System, Great Powers, and Environmental Change since 1900", in: Brauch, Hans Günter; Oswald Spring, Úrsula; Grin, John; Mesjasz, Czeslaw; Kameri-Mbote, Patricia; Behera, Navnita Chadha; Chourou, Béchir; Krummenacher, Heinz (Eds.), 2009: *Facing Global Environmental Change: Environmental, Human, Energy, Food, Health and Water Security Concepts* (Berlin–Heidelberg–New York: Springer-Verlag): 43-52.

McNeill, John R., 2000: *Something New Under the Sun: An Environmental History of the 20th-Century World* (New York: Norton).

Meadowcroft, James, 2000: "Sustainable development: A new(ish) idea for a new century?", in: *Political Studies,* 48: 270-387.

Meadows, Donella H.; Meadows, Dennis; Randers, Jørgen; Behrens III, William W., 1972: *The Limits to Growth: A Report for the Club of Rome's Project on the Predicament of Mankind* (New York: Universe).

Merkel, Wolfgang (Ed.), 1996: *Systemwechsel. Theorien, Ansätze und Konzeptionen* (Opladen: Leske + Budrich).

Meyer-Abich, Klaus M. (Ed.), 1979: *Frieden mit der Natur* (Freiburg–Basel–Wien: Herder).

Meyer-Abich, Klaus M. [KMA], 2003: "Frieden mit der Natur", in: Simonis, Udo E. (Ed.): *Ökolexikon* (München: Beck): 84.

Meyer-Abich, Klaus Michael, 1984: *Wege zum Frieden mit der Natur: Praktische Naturphilosophie für die Umweltpolitik* (München: Carl Hanser).

Mies, Maria, 1986: *Patriarchy and Accumulation on a World Scale* (Melborne: Zed Book).

Moore III, Berrien, 2000: "Sustaining Earth's life support systems—the challenge for the next decade and beyond", in: *IGBP Newsletter 41—The Global Change Newsletter*: 1-3.

Morrow, G.; Rothwell, C..; Burford, B.; Illing, J., 2013: "Cultural dimensions in the transition of overseas medical graduates to the UK workplace", in: *Med Teach*, 35,10 (October): 1537-45.

Myles, Arthur Tilney Angus, 2012: *The Cultural Dimension of Army Transition* (Fort Leavenworth: School of Advanced Military Studies)

Nkonya, Ephraim; Gerber, Nicolas; Baumgartner, Philip.; Braun, Jochim von; de Pinto, Alex; Graw, Valerie; Kato, Edward; Kloos, Julia; Walter, Teresa, 2011: *The Economics of Land Degradation—Towards an Integrated Global Assessment* (Frankfurt am Main—Berlin—Bern—Bruxelles—New York—Oxford—Wien: Peter Lang).

NRC (National Research Council), 1999: *Global Environmental Change: Research Pathways for the Next Decade* (Washington DC: National Academy Press).

OAS/UNEP/Government of Peru, 2001: *Case Study of Environmental Management: Integrated Development of An Area in the Humid Tropics—The Selva Central of Peru* Washington: OAS).

OECD, 2001: *Environmental Indicators—Towards sustainable development* (Paris: OECD).

Osterhammel, Jürgen, 2009: *Die Verwandlung der Welt. Eine Geschichte des 19. Jahrhunderts* (München: C. H. Beck).

Oswald Spring, Úrsula, 2008: "Peace and Environment: Towards a Sustainable Peace as Seen from the South", in: Brauch, H. G. et al. (Eds.): *Globalization and Environmental Challenges: Reconceptualizing Security in the 21st Century* (Heidelberg—New York: Springer): 113-126;

Oswald Spring, Úrsula, 2009: "Sustainable Development", in: De Rivera, Joe (Ed.): *Handbook on Building Cultures of Peace* (New York: Springer): 211-227.

Oswald Spring, Úrsula (Ed.), 2011: *Water Resources in Mexico* (Berlin—Heidelberg: Springer).

Oswald Spring, Úrsula, 2013: "Dual vulnerability among female household heads", in: *Acta Colombiana de Psicología*, 16, 2: 19-30.

Oswald Spring, Úrsula, 2016: "Global Environmental Change", in: Sosa-Nunez, Gustavo; Atkins, Ed (Eds.): *Environment, Climate Change and International Relations*, at: <http://www.e-ir.info/publications/>.

Oswald Spring, Úrsula; Brauch, Hans Günter, 2011: "Coping with Global Environmental Change—Sustainability Revolution and Sustainable Peace", in: Brauch, Hans Günter; Oswald Spring, Úrsula; Mesjasz, Czeslaw; Grin, John; Kameri-Mbote, Patricia; Chourou, Béchir; Dunay,

Pal; Birkmann, Jörn, 2010: *Coping with Global Environmental Change, Disasters and Security—Threats, Challenges, Vulnerabilities and Risks* (Berlin—Heidelberg—New York: Springer-Verlag): 1487-1504.

Oswald Spring, Úrsula; Brauch, Hans Günter; Tidball, Keith G. (Eds.): *Expanding Peace Ecology: Security, Sustainability, Equity and Peace: Perspectives of IPRA's Ecology and Peace Commission* 1 (Cham—Heidelberg—New York: Springer, 2014).

Oxfam, 2016: *An Economy for the 1%. How privilege and power in the economy drive extreme inequality and how this can be stopped* (Oxford: January 2016).

Oxford, 2002a: *Compact Oxford English Dictionary* (Oxford: University Press); Online version at: <www.askoxford.com>.

Oxford, 52002: *Shorter Oxford English Dictionary* (Oxford—New York: Oxford University Press).

Pagliani, Paola; Bonini, Astra, 2011: *A New Social Contract For Sustainable Human Development* (New York: UNDP).

Peck, Connie, 1998: *Sustainable Peace: The Role of the UN and Regional Organizations in Preventing Conflicts* (Lanham—Boulder—New York: Carnegie Commission on Preventing Deadly Conflict).

Pierson, P., 2004: *Politics in Time—History, Institutions and Social Analysis* (Princeton—New York: Princeton University Press).

Plumwood, Val (Ed.), 2002: *Environmental Culture: The Ecological Crisis as Reason* (New York—London: Routledge)

Polanyi, Karl, 1944: *The Great Transformation: The Political and Economic Origins of Our Time* (Boston: Beacon Press).

Raskin, P.; Banuri, T.; Gallopin, G.; Gutman, P.; Hammond, A.; Kates, A.; Swart, R., 2002: *Great transition. The promise and the lure of the times ahead. Report of the Global Scenario Group*. Pole Start Series Report 10 (Stockholm: Stockholm Environment Institute).

Raskin, P.; Gallopin, G.; Gutman, P.; Hammond, A.; Swart, R., 1998: *Bending the Curve: Toward Global Sustainability. A Report of the Global Scenario Group* (Boston: Stockholm Environment Institute).

Reddy, B. Sudhakara, 2014: *India's Energy Transition—Pathways for Low-Carbon Economy* (New Delhi: Indira Gandhi Institute of Development Research, Mumbai, July); at: <http://www.igidr.ac.in/pdf/publication/WP-2014-025.pdf> (24 February 2016).

Scheffran, Jürgen; Brzoska, Michael; Brauch, Hans Günter: Link, Michael; Schilling, Janpeter (Eds.), 2012: *Climate Change, Human Security and Violent Conflict: Challenges for Societal Stability* (Berlin—Heidelberg: Springer-Verlag).

Schellnhuber, Hans Joachim; Cramer, Wolfgang; Nakicenovic, Nebojsa; Wigley, Tom; Yohe, Gary (Eds.), 2006: *Avoiding Dangerous Climate Change* (Cambridge: Cambridge University Press).

Schweitzer, Peter, 1994: *Victory. The Reagan Administration's Secret Strategy That Hastened the Collapse of the Soviet Union* (New York: Atlantic Monthly Press).

Serrano Oswald, Eréndira Serena, 2009: "The Impossibility of Securitizing Gender vis a vis Engendering Security", in: Brauch, Hans Günter; Oswald Spring, Úrsula; Grin, John; Mesjasz, Czeslaw; Kameri-Mbote, Patricia; Behera, Navnita Chadha; Chourou, Béchir; Krummenacher, Heinz (Eds.): *Facing Global Environmental Change: Environmental, Human, Energy, Food, Health and Water Security Concepts* (Berlin–Heidelberg: Springer): 1151-1164.

Serrano Oswald, Eréndira Serena, 2010: *La Construcción Social y Cultural de la Maternidad en San Martín Tilcajete, Oaxaca* (PhD Thesis, UNAM: Instituto de Antropología, Mexico. D.F.).

Serrano Oswald, Eréndira Serena, 2014: "Migration, woodcarving and engendered identities in San Martín Tilcajete, Oaxaca", in: Truong, Thanh-Dam; Gasper, Des; Handmaker, Jeff, Bergh, Sylvia (Eds.): *Migration, Gender and Social Justice. Perspectives on Human Insecurity* (Heidelberg: Springer): 173-192.

Speth, J., 1992: "The transition to a sustainable society", in: *Proceedings of the National Academy of Sciences of the United States of America*, 89: 870-872.

Steffen, Will; Sanderson, Angelina; Tyson, Peter D.; Jäger, Jill; Matson, Pamela A.; Moore III, Berrien; Oldfield, Frank; Richardson, Katherine; Schellnhuber, Hans Joachim; Turner II, B. L.; Wasson, Robert J., 2004: *Global Change and the Earth System. A Planet under Pressure. The IGBP Series* (Berlin–Heidelberg–New York: Springer-Verlag).

Steffen, Will; Grinevald, Jacques; Crutzen, Paul J.; McNeill, John, 2011: "The Anthropocene: conceptual and historical perspectives", in: *Phil. Trans. R. Soc. A* 369: 843.

Steffen, Will; McNeill, John; Crutzen, Paul J., 2007: "The Anthropocene: Are Humans Now Overwhelming the Great Forces of Nature?", in: *Ambio*, 36: 614-621.

Stiglitz, Joseph E., 2010: *Freefall. America, Free Markets, and the Sinking of the World Economy* (New York: Norton).

Strahm, Rudolf; Oswald Spring, Úrsula, 1990: *Por esto somos tan pobres* (Cuernavaca: CRIM-UNAM).

The Social Learning Group, 2001: *Learning to Manage Global Environmental Risks, vol. 1: A Comparative History of Social Responses to Climate Change, Ozone Depletion, and Acid Rain. The Social Learning Group* (Cambridge, MA: The MIT Press).

The Social Learning Group, 2001a: *Learning to Manage Global Environmental Risks, vol. 2: A Functional Analysis of Social Responses to Climate Change, Ozone Depletion, and Acid Rain. The Social Learning Group* (Cambridge, MA: The MIT Press).

UN (United Nations), Department of Economic and Social Affairs, Population Division, 2015: *World Population Prospects: The 2015 Revision, Key Findings and Advance Tables*. Working Paper No. ESA/P/WP.241 (New York: United Nations).

United Nations, 2000: "Millennium Declaration"; at: <http://www.un.org/en/development/devagenda/millennium.shtml >

UN, 2009a: *Climate change and its possible security implications. Report of the Secretary-General.* A/64/350 of 11 September 2009 (New York: United Nations).

UN, 2015: *UN Water Report. Water for a Sustainable World*, at: <http://unesdoc.unesco.org/images/0023/002318/231823E.pdf>.

UNCCD, 2015: *Land Matters for Climate. Reducing the Gap and Approaching the Target* (Bonn: UNCCD).

UNDP, 1994: *Human Development Report 1994. New Dimensions of Human Security* (New York–Oxford–New Delhi: Oxford University Press); at: <http://hdr.undp.org/reports/global/1994/en/pdf/hdr_1994_ch2.pdf>.

UNEP, 2011: *Green Economy Report: Towards a Green Economy: Pathways to Sustainable Development and Poverty Eradication* (Nairobi: UNEP); at: <http://www.unep.org/greeneconomy/GreenEconomyReport/tabid/29846>.

UNEP, 2012: *Global Environmental Outlook; Environment for the Future We Want, GEO-5* (Valletta: UNEP).

UNEP, International Resource Panel, 2011: *Decoupling Natural Resource Use and Environmental Impacts from Economic Growth* (Paris: UNEP).

UNEP, International Resource Panel, 2014: *Decoupling 2: Technologies, Opportunities and Policy Options* (Paris: UNEP).

UNFCCC, 1992: "United Nations Framework Convention on Climate Change"; at: <http://unfccc.int/resource/docs/convkp/conveng.pdf>.

UNFPA, 2012: *Urbanization, gender and urban poverty: Paid work and unpaid carework in the city* (New York: UNFPA).

UNPD (UN Populations Division), 2015: *World Population Prospects. 2015 Revision: Key findings & advance tables* (New York: UN) at: <http://esa.un.org/unpd/wpp/publications/files/key_findings_wpp_2015.pdf>.

Velázquez, Juana María, 2014: *La emergencia de los nuevos actores de las relaciones internacionales. Estrategias de defensa para los recursos genéticos del maíz frente a los organismos genéticamente modificados* (MSc Thesis, UNAM, Faculty of Political and Social Sciences [FCPS UNAM], Mexico, D.F.).

Wang, Tao; Watson, Jim, n.d.: *China's Energy Transition Pathways for Low Carbon Development* (Falmer: SPRU).

Waters, C. N.; Zalasiewicz, J. A.; Williams, M.; Ellis, M. A.; Snelling, A. M. (Eds.): *Stratigraphical Basis for the Anthropocene* (London: Geological Society of London, 2014)

WBGU, 2008: *World in Transition–Climate Change as a Security Risk* (London: Earthscan); at: <http://www.wbgu.de/wbgu_jg2007_engl.html >.

WBGU, 2011: *World in Transition—A Social Contract for Sustainability* (Berlin: German Advisory Council on Global Change, July 2011).

WCED (World Commission on Environment and Development), 1987: *Our Common Future. The World Commission on Environment and Development* (Oxford–New York: Oxford University Press).

WEF [World Economic Forum], 2011: *Water Security. The Water-Food-Energy-Climate Nexus* (Davos: WEF).

WEF [World Economic Forum], 2016: *The Global Risk Report* 2016 (Davos: WEF).

Wilkinson, Richard; Pickett, Kate, 2010: *The Spirit Level: Why Equality is Better for Everyone* (London: Penguin).

Wilson, Edward O., 1998: *Consilience* (New York: Knopf).

WVS [World Value Survey], 2010: *Values Change the World. World Values Survey* (Stockholm: World Values Survey Association).

WWF 2014: *Living Planet Report* 2014, at: <wwf.or.jp/activities/lib/lpr/WWF_LPR_2014.pdf>.

WWF (Hoegh-Guldberg, O. et al.), 2015: *Reviving the Ocean Economy. The Case For Action—2015* (Gland: WWF-International).

WWF, 2015: "Threats: soil erosion and desertification", at: <http://www.worldwildlife.org/threats/soil-erosion-and-degradation>.

2   Contextual Changes in Earth History: From the Holocene to the Anthropocene — Implications for Sustainable Development and for Strategies of Sustainable Transition

Simon Dalby[1]

Abstract

Human activities have changed many of the key parameters of the Holocene geological epoch of the recent past so much that we now live in the Anthropocene. New perspectives in earth system science suggest that sustainable development and plans for transitions to a sustainable peace now have to consider the possibilities of rapid phase shifts in the biosphere. Constraining human activities to within a safe operating space defined by key ecological boundaries in the earth system is key to sustainability but planning has to recognize that rapid shifts may be coming. The implications of this suggest that sustainability planning has to think beyond notions of national security and recognize that human actions are shaping the future configuration of the planet and hence changing the geopolitical context. Adopting a perspective of geopolitical ecology with a focus on global economic production rather than only on traditional ideas of environmental protection is key to the future if planetary stewardship of the Anthropocene is to be successful.

Keywords: Anthropocene, Earth System Science, ecological phase shift, Holocene, sustainable development, Planetary Boundaries, Planetary Stewardship, political geoecology, safe operating space.

2.1   Earth History and Sustainable Development

Discussions about peaceful transitions to a sustainable society are driven by an often-implicit understanding that humanity ought to live in a planetary system that is at least broadly similar to the geological circumstances of the last ten thousand years.[2] Geologists and Earth system scientists usually call this period the Holocene. It provided the ecological conditions that facilitated the emergence of human civilization. Now recent research into the earth system, and a growing recognition of the sheer scale of human transformation of many environments, suggests that the assumption of a relatively stable geological context for humanity is at best misleading, and at worst a dangerous failure to think carefully about the new context that humanity is creating for itself in the new epoch driven by human actions, this new epoch of the Anthropocene.

If rapid ecological change accelerates in the next few decades, as all indications are that it will, then rapid adaptations to new circumstances have to be part of the planning for transitions to more sustainable modes of life. Sustainable peace strategies must consider the possibility of rapid and unexpected ecological phase shifts. If peace is to prevail these will have to be lived through without major powers resorting to military action in attempts to deal with at least some of the consequences of environmental and social disruptions. This is the key implication that arises from juxtaposing discussions of peace, transitions and sustainable development with the new insights into geological and ecological sciences.

---

1   Simon Dalby is CIGI Chair in the Political Economy of Climate Change at the Balsillie School of International Affairs, and Professor of Geography and Environmental Studies at Wilfrid Laurier University, Waterloo, Canada; at: <sdalby@gmail.com>.
2   My thanks to Aleksandra Szaflarska for most helpful research assistance and editing and to the Social Sciences and Humanities Research Council for support through a 'Borders in Globalization' partnership grant.

© Springer International Publishing Switzerland 2016
H.G. Brauch et al. (eds.), *Handbook on Sustainability Transition and Sustainable Peace*,
Hexagon Series on Human and Environmental Security and Peace 10, DOI 10.1007/978-3-319-43884-9_2

2.1.1 Formulating Sustainable Development

The *World Commission on Environment and Development* (WCED), chaired by Gro Harlem Brundtland, popularized the phrase 'sustainable development' in its widely cited report *Our Common Future*. The famous definition at the beginning of the second chapter of the report reads: "Sustainable development is development that meets the needs of the present without compromising the ability of future generations to meet their own needs" (WCED 1987: 43). This has become the standard definition of one of the most widely used terms in contemporary international politics.

The authors of *Our Common Future* went on to elaborate on the definition, stating that it involved two key concepts. The first is "the concept of 'needs', in particular the essential needs of the world's poor, to which overriding priority should be given" (WCED 1987: 43). The second concept is "the idea of limitations imposed by the state of technology and social organization on the environment's ability to meet present and future needs" (WCED 1987: 43). But the authors went on to emphasize that the social and the physical are inextricably interconnected. "Even the narrow notion of physical sustainability implies a concern for social equity between generations, a concern that must logically be extended to equity within each generation" (WCED 1987: 43). How to do this is not exactly easy; claims to inter-generational and intra-generational equity persist in discussions of sustainable development, but the dramatic trajectory of economic change since the Brundtland Report was published a generation ago has apparently not operated on the report's principles despite the repeated invocation of the term 'sustainable development'.

Bluntly put, the term was at best a compromise. It was an attempt to incorporate Northern concerns with environment with Southern concerns about development. Fifteen years after Indira Gandhi called poverty the worst kind of pollution at the United Nations Conference on the Human Environment in Stockholm, the necessity of dealing with rapid environmental change and with impoverishment in many parts of the world required that some compromise between development and environmental protection be articulated in international forums. At the time, alarm about the depletion of stratospheric ozone and concerns about industrial accidents—with Chernobyl and Bhopal very much on people's minds—was coupled with worries about deforestation and the limited possibilities of expanding agricultural production.

Ozone depletion in particular made it clear that some environmental vulnerabilities were in fact widely shared and some sense of global cooperation was necessary to deal with these matters.

Southern leaders were adamant that Northern environmental issues should not be used as a method for constraining what they saw as essential Southern economic growth (Kjellen 2008). Given that most of the big environmental problems of the time were caused by Northern activities, simple matters of justice required that those who had caused the problems be the ones to pay for the solution. Where ozone depleting substances were a problem, Southern leaders insisted that Northern economies help provide technological alternatives to compensate for what they portrayed as foregone development opportunities. Such principles linking environment to development have subsequently been key to much of the diplomatic discussions about aid and development. More recently, these themes have been key to international discussions of climate change where technology transfer and development aid are part of the negotiations under the rubric of common but differentiated responsibilities (Brunee/Streck 2013). This terminology has become the taken-for-granted language for discussing many international political matters, not just obviously and immediately 'environmental' matters.

This has more recently been complemented by attempts to reconfigure economies and societies in ways that are more obviously 'sustainable' in that they reduce energy and resource use and hence put less pressure on ecological systems. They remain key themes in the more recent scholarly discussions of sustainable transitions (Grin/Rotmans/Schot 2010) and policy-oriented documents focusing on innovations necessary for global governance (WGBU 2011). The growing recognition in at least some states, and European ones in particular, that de-carbonizing energy supplies is key to dealing with climate change is linked to a recognition that development cannot be equated with economic growth as traditionally understood. Ever larger appropriation of resources to feed increased material production is anathema to serious attempts to think about sustainability and further disrupts the livelihoods of many of earth's poorer peoples (Nixon 2011). Coupled with new measurement metrics, only most obviously the notion of an ecological footprint that calculates the amount of land surface needed for each economic or social entity to be at least carbon-neutral, such strategies attempt to dramatically increase efficiencies and recycle materials. They also attempt to limit the extraction of new resources from fields, forests and mines while simultaneously trying to deal with questions of global justice (Sachs/Santarius 2007).

2.1.2   Sustainability in the Anthropocene

Given the scale of the changes already caused by humanity, earth system scientists are now suggesting that we live in a new geological epoch, one commonly called the Anthropocene following Paul Crutzen's (2002) popularization of the term. There are complicated technical discussions about when the Anthropocene might have started, related to which ecological functions of the system are defined as the most important (Ruddiman 2005), and scepticism on the part of at least some geologists as to whether this is a useful formulation for solving scientific questions in stratigraphy (Autin/ Holbrook 2012). The most commonly accepted view is that the Anthropocene began with the industrial revolution and the growth of the use of fossil fuels, first coal and subsequently petroleum and natural gas (Steffen/Crutzen/McNeill 2007). Steam power was key to the industrial revolution period, both as a source of industrial power and as a mode of locomotion that rapidly connected parts of the global economy and greatly facilitated the extraction of resources and the spread of commercial agriculture.

While this changed many aspects of the global system, starting in the aftermath of the Second World War the global economy began what is now called a period of "great acceleration" (Steffen/Grinewald/Crutzen/McNeill 2011). Powered by an increasing use of petroleum in addition to the coal use, mass consumption economies grew rapidly, first in the United States, Europe and Japan. Subsequently, the rise of Asian economies–in Korea, Singapore and elsewhere–extended this pattern, with China in particular adopting a capitalist-driven consumption model of 'development' in the 1980s. India too has joined the race to consume. Carbon dioxide in the atmosphere has risen rapidly so that it is now present in quantities unknown in the geological history not only of the Holocene, since the last ice age, but back through previous ice ages and interglacial periods stretching back at least 800,000 years.

In earth system science terms, what comes next is a matter for humanity to decide. A 'planetary stewardship' would seem to be the desirable next phase of the Anthropocene. But if the earth system is to be sustained in something roughly approximating Holocene conditions, many things will have to change, not least

the understandings at beginning of environment' and humanity's place 'in it'. In the terms of international relations, what should be secured to facilitate a sustainable earth is rather different from what has been seen as essential until very recently; geopolitics can no longer operate on the assumption that the 'playing field' of international politics is a given (Hommel/Murphy 2013). The key point about the Anthropocene perspective is that climate change and other ecological changes are remaking the context of international politics (Dalby 2014).

2.1.3   Recontextualizing Peace and Sustainability

To fill in some of the details for recontextualizing geopolitics in these terms, this chapter first looks to the discussions of earth system science and how the current geological situation is understood. Considerable caution is needed in invoking this particular scientific view of present circumstances; science is not unrelated to attempts to govern human affairs, and the political implications of attempting to see the earth as a whole are not trivial (Lovbrand/Stripple/Wiman 2009). Nonetheless, insofar as environmental contexts are part of the larger considerations of peace and sustainability in coming decades, the earth system science perspective provides a contextualization that distances analysis from an undue focus on states and demands an engagement with the specific material contexts of vulnerability in an innovative way that makes it difficult to avoid the key issues of the politics of security (Dalby 2009). The larger engagement between humanities, social and natural sciences that has often been bypassed by disciplinary foci on one or the other is in urgent need of engagement, and earth sciences provide an especially productive way to link environmental change to history (Hornborg/McNeill/Martinez-Alier 2007) as well as to a wider range of humanities scholarship (Palsson/Szerszynski/Sörlin et al. 2012). Insofar as peace is to be linked to sustainability, such intellectual conversations are simply essential; these frameworks are increasingly being used to discuss innovative development policies as well as climate adaptations (Pisano/Berger 2013; Raworth 2012).

While earth system science cannot provide a blueprint for a sustainable future, it has developed a loose framework for what is called a 'safe operating space' for humanity in light of key ecological functions of the biosphere. The chapter reviews these prior to returning to the questions of what is needed in terms of transitional strategies and how international security needs to be rethought if a sustainable earth is to be produced in coming generations. In the words of the unofficial report to the United Nations Conference on the Human Environment in Stockholm in 1972, there is "Only One Earth" (Ward/Dubos 1972). How we think about it is now rather different from that early environmental view of what needs to be done, not least because we have recently come to understand humanity as a geological-scale actor in the earth system. While earth system science does not provide answers to the key political questions facing humanity, it does provide a framing of the options that is increasingly influential (Dalby 2013a).

The rest of this chapter argues that, whatever the finer points of transition and peace strategies engaging sustainability, they all now have to be considered in light of these new insights into the new geological circumstances that humanity is creating for future generations. The chapter first turns to earth system science and the discussion of phase shifts, tipping points and the key question of the boundaries of a safe operating space for humanity in the earth system. These boundaries involve more than climate change that gets most contemporary attention; it is important to consider other ecological changes that humanity is making if the context for sustainability is to be adequately formulated. Later sections of the chapter emphasize that notions of stewardship and transitions have to be understood in light of these new global ecological understandings. The final section suggests that any consideration of peaceful transitions or sustainability now has to include both a recognition that any proposed transition involves decisions about what kind of ecology its strategies imply and, crucially, that rapid ecological change may be the context in which the transitions happen. Any plans for a transition to a new less rapacious mode of economy will also have to include thinking about how to peacefully cope with rapid and sometimes unanticipated ecological change. Earth system science has profound implications for how social sciences now understand their task (Schellnhuber/Crutzen/Clark et al. 2004); taking these seriously is essential for all strategies for economic sustainability.

2.2   Earth System Science

Human actions are often viewed as external drivers of ecosystem dynamics; examples include fishing, water extracting, and polluting. Through such a lens the manager is an external intervener in ecosystem resilience. However, many of the serious, recurring problems in natural resource use and environmental management stem precisely from the lack of recognition that ecosystems and the social systems that use and depend on them are inextricably linked. It is the feedback loops among them, as interdependent social–ecological systems, that determine their overall dynamics and sustainability (Folke/Jansson/Rockström et al. 2011: 722).

With much more attention now placed on the question of sustainability, and given increasing concerns regarding a wide range of large-scale ecological changes and climate change in particular—well beyond the concerns in *Our Common Future*—the possibilities of transitioning to a sustainable mode of economic life on the part of developed economies extend the conceptual framework of sustainable development further. Earth system sciences have emphasized how difficult it is to clearly define the parameters of 'physical sustainability' while also confirming the necessity of understanding social considerations as an essential part of the biosphere. The planet is 'under pressure' from widespread human activities (Steffen/Sanderson/Tyson et al. 2004). While in the 1970s environmentalists had often looked to the discussion of "the limits to growth" in terms of pollution and resource availability   (Meadows/Meadows/Randers/Behrens 1972, 1974), now the earth system science literature nuances these matters by looking to a more wide-ranging series of boundaries to what has been called the "safe operating space" for humanity (Rockström/Steffen/Noone et al. 2009a, 2009b). Climate change in particular has raised questions about how we might now understand 'physical sustainability' given that human actions are already changing some of the key parameters of the biosphere.

This is a profound shift in understanding of humanity's place in the larger cosmological ordering of things. Just as astronomy's proofs that the earth orbited the sun rather than the other way round shook human conceptions profoundly as modern science began its investigations, now earth system sciences are making clear that the planet is not a given stable context into which humanity was recently added, but rather a dynamic system that humanity is now profoundly and rapidly changing. 'Physical sustainability' is not a stable given context for humanity; human systems are actively shaping the future geology of the planet, directly altering terrestrial ecosystems and indirectly changing many other aspects of the biosphere, and need to be contextualized that way in any serious thinking about how to address the needs of future generations (Ellis 2011). We are in this new epoch of the Anthropocene, one where human actions are leaving traces in the sedimentary record in many remote places, a geomorphological footprint as it were of the age of humanity (Brown/Tooth/Chiverrell et al. 2013). These actions may yet leave a distinctive geological footprint on the history of the planet (Clark 2012). Even if the geological legacy we leave may not be this epochal when viewed from millions of years in the future, the rapidly changing context is more than enough to raise profound questions for societal stability and with it human security in coming decades (Scheffran/Brzoska/Brauch et al. 2012).

2.2.1 Ecological Phase Shifts

As a result of the enormous complexity of the system as a whole, it is not possible to predict the outcomes of rapidly increasing human pressures on the Earth System, but it is clear that thresholds have been or are being reached, beyond which abrupt and irreversible changes occur. These changes will affect the basic life-support functions of the planet (UNEP 2012: 210).

In earth system science terms, the current transformations that humanity has set in motion amount in some accounts to an approaching phase shift in how the biosphere functions (Barnosky/Hadly/Bascompte et al. 2012). Ecological thresholds have either already been crossed or are in danger of being crossed with the consequence that ecosystems will likely operate in new and potentially unpredictable ways (Huggett 2005).

The shift from one state to another can be caused by either a 'threshold' or 'sledgehammer' effect. State shifts resulting from threshold effects can be difficult to anticipate, because the critical threshold is reached as incremental changes accumulate and the threshold value generally is not known in advance. By contrast, a state shift caused by a sledgehammer effect—for example the clearing of a forest using a bulldozer—comes as no surprise. In both cases, the state shift is relatively abrupt and leads to new mean conditions outside the range of fluctuation evident in the previous state (Barnosky/Hadly/ Bascompte et al. 2012: 52).

These shifts can occur at various scales, and while the overall effect may be global, it is important to emphasize that the cumulative effects of many small changes may cross thresholds at larger scales.

In the context of forecasting biological change, the realization that critical transitions and state shifts can occur on the global scale, as well as on smaller scales, is of great importance. One key question is how to recognize a global-scale state shift. Another is whether global-scale state shifts are the cumulative result of many smaller-scale events that originate in local systems or instead require global-level forcings that emerge on the planetary scale and then percolate downwards to cause changes in local systems. Examining past global-scale

state shifts provides useful insights into both of these issues (Barnosky/Hadly/Bascompte et al. 2012: 53).

Those past events suggest that the current transition is more rapid than previous dramatic changes in the earth system, the most recent of which was the transition from the last ice age.

While transitions happen very quickly relative to the fairly stable states that precede them, the pace of human adaptation of numerous aspects of the biosphere may be unprecedented. "Global-scale forcing mechanisms today are human population growth with attendant resource consumption, habitat transformation and fragmentation, energy production and consumption, and climate change. All of these far exceed, in both rate and magnitude, the forcings evident at the most recent global-scale state shift, the last glacial-interglacial transition" (Barnosky/Hadly/Bascompte et al. 2012: 53) which gave rise to the geological period of the Holocene. This is one of the worrisome factors in our present circumstances: there are few clear geological analogies to draw upon to anticipate how the earth system will respond to the new forcing mechanisms humanity has created.

However, the question of whether there are global tipping points that will mean the earth system in total will rapidly tip into some new format is disputed, and much remains to be studied on potential linkages between different drivers of system change (Hughes/ Carpenter/Rockström et al. 2013). Brook, Ellis, Perring et al. (2013) suggest that for at least four of the main drivers a phase shift in the immediate future is unlikely at least for terrestrial ecosystems.

Our examination of the evidence suggests that four principal drivers of terrestrial ecosystem change—climate change, land use change, habitat fragmentation, and biodiversity loss—are unlikely to induce planetary-scale biospheric tipping points in the terrestrial realm. Criteria that would increase the likelihood of such a global-scale tipping point—homogeneity of response over space at a short timescale, interconnectivity, and homogeneity of a causative agent across space—are not met for any of these drivers. Instead, terrestrial ecosystems are likely to respond heterogeneously to these variable forcings and, with a few exceptions, show limited interconnectivity" (Brook, Ellis, Perring, et al. 2013: 399-400).

In part this is because humanity has already transformed so much of the terrestrial ecosystem that there is not a natural state that might tip in terms of land use. All of which makes the case for great caution in predicting global ecological consequences of further changes. It also emphasizes the key point that ecological change is highly geographically variable in the earth system and human vulnerabilities are dependent on this and the increasingly artificial circumstances in which most of us live (Dalby 2009).

The most worrisome dimension to all this is not that the world will gradually change as a result of human activities, but that the earth system will be rapidly changed in ways that are not conducive to human flourishing. This may happen if the whole earth system enters a rapid phase of non-linear change that results in a new relatively stable configuration, but one very different from that so far familiar to human societies. While this may be unlikely in the immediate future, and some of the earlier concerns about rapid change may have been exaggerated (Committee on Understanding and Monitoring Abrupt Climate Change and its Impacts 2013), exactly how many ecosystems will respond to accelerating climate change is a crucial unknown (Hughes/Carpenter/Rockström et al. 2013). The Anthropocene presents us with a new set of questions concerning the cumulative consequences of human actions, and possible interactions among them, that are key to any discussion of what kind of future we may be creating. The question of the Anthropocene as posed by Rockström, Steffen, Noone et al. (2009a) is nothing less than "What are the non-negotiable planetary preconditions that humanity needs to respect in order to avoid the risk of deleterious or even catastrophic environmental change at continental to global scales?"

2.2.2   Boundaries, Thresholds and Tipping Points

Such 'planetary preconditions' are not easy to establish, not least because ecological matters rarely work in simple linear processes. They often have very considerable abilities to function while key drivers of important facets vary considerably. Many are resilient too, being able to bounce back after serious disruptions. Not all ecosystems function in patterns that are immediately obvious and they are sometimes interconnected over distances in ways that are hard to clearly analyse. Many have threshold values that, once surpassed, lead to systems changing dramatically as they cross a "tipping point" (Lenton/Held/Kriegler et al. 2008). In the case of the global ecosystem, human and ecological systems are so interconnected and enmeshed that they now have to be considered together if any discussion of sustainability is to make sense. "Because ecosystems are variable, one must focus on the risk, not the certainty, of exceeding an objectively defined target or threshold" (Bennett/Car-

penter/Cardille 2008: 132). Calculating such things is rarely easy, but clearly many earth scientists are convinced that some boundaries have already been crossed, and others may well be in the next few decades, with consequences that are potentially disastrous to the contemporary modes of human existence.

More specifically, ecological change has to be understood in terms of potential non-linearities, thresholds, and tipping point responses to stresses that drive systems in particular ways.

> Ecosystem attributes such as species abundance or biological carbon sequestration can respond in three (stylized) ways to biotic and abiotic drivers. The first type of response is characterized by being consistently proportional to the magnitude of the driver, thus exhibiting a 'smooth response' pattern, where no single critical point can be determined. In the second class of ecosystem change, the response, at some critical level of forcing, is amplified by internal synergistic feedbacks and thus becomes nonlinear in relation to the driver, changing the slope of the response curve. The third class similarly involves nonlinearity, but exhibits hysteresis, in which at least two stable states exist, implying limited reversibility. The term tipping point applies to the second and third class of ecosystem change and refers specifically to the inflection point or threshold at which the ecosystem response becomes nonlinear or the rate of change alters steeply (Brook/Ellis/Perring et al. 2013: 397).

Rapid and unpredictable change is what worries most political decision-makers: the potential for drastic disruptions to increasingly artificial social-ecological systems is what has stimulated seemingly endless invocations of environmental security since the Brundtland Commission explicitly raised concerns that environmental disruptions could potentially cause military conflict (Floyd/Matthews 2013). Such considerations have become all the more urgent because scientific evaluations of current transformations are identifying thresholds in many systems.

> Theoretical or empirical evidence of tipping points, manifesting on decadal to centennial time scales, exists at local and regional scales for many subsystems of the Earth system, including the cryosphere, ocean thermohaline circulation, atmospheric circulation, and marine ecosystems. In the terrestrial biosphere, tipping points involve ecosystem attributes such as species abundance or carbon sequestration responding nonlinearly and potentially irreversibly to proximate drivers like habitat loss or climate change (Brook/Ellis/Perring et al. 2013: 396).

The interconnected nature of these suggests very clearly that any attempt to think carefully about sustainability, and strategies for transitions towards more sustainable human systems, will face fraught interpretive tasks in terms of the science. These will be just as fraught as potential governance arrangements if in the next few decades humanity seriously tries to shape a functional biosphere for future generations.

2.2.3   Multiple System Stressors

All of this is made more complicated by the simple but unavoidable point that multiple stressors are working simultaneously on most systems.

> In a related area of concern, we are struggling conceptually with how to propose robust boundaries for issues that are spatially distributed heterogeneously around the world. Part of the answer relates to the potential geographic specificity of process and function—the primary concern is not the physical intervention in the structure itself. Thus for instance, deforesting the equatorial/tropical Amazon basin really might be more of a planetary cause for concern than land use change over an equivalent area elsewhere, not because of what it materially consists of nor the area involved, but because of the interplay of that particular patch of vegetation with the processes influencing global water and energy balance (Cornell 2012: 2).

Thus, while global boundaries are suggestive, specific ecosystems in particular places matter and trying to ascertain which of these are most important in terms of ecological function is not easy, although it is essential for earth system governance (Bierman 2012).

To even think in such terms requires a conceptual shift from modern notions of a nature external to humanity that provides the environmental context for humanity, to formulations that understand at least the affluent fossil fuel powered part of humanity as a key ecological actor in what effectively are geological processes. Such considerations suggest that humanity itself be understood in geological terms given the scale of its actions, a discussion that has given rise to various prior formulations of present times in geological terms before science settled on the informal use of 'the Anthropocene' in the last decade (Davis 2011). What is also clear is the relative novelty of major human interventions in the biosphere, although on closer examination the question of when humanity started to have a noticeable impact on the biosphere and which impact is most important turns out to be very complicated (Ruddiman 2005). Nonetheless, what is clear is that human actions are now transforming the biosphere and the concern is that unless great care is taken not to cross crucial thresholds, we may change it in ways that threaten human civilization profoundly. Not crossing key thresholds in the biosphere is a key part of any strategy that aims to transition from current economic practices to ones that remain safely within the planetary operating space (Rockström/Steffen/Noone et al. 2009a).

2.3   Planetary Boundaries

> Industrial human systems, in just two centuries, have already introduced at least three clearly novel biospheric processes: the use of fossil energy to replace biomass fuel and human and animal labour, revolutionizing human capacity for ecosystem engineering, transport and other activities; the industrial synthesis of reactive nitrogen to boost agro-ecosystem productivity; and, most recently, genetic engineering across species (Ellis 2011: 1013).

In attempting to provide at least a preliminary answer to questions about how far such transformations can be taken while keeping essential biospheric processes working in more or less the conditions humanity is familiar with, Rockström, Steffen, Noone et al.'s (2009a) formulations of a safe operating space for humanity suggest that nine planetary boundaries need to be especially carefully monitored (see figure 3.1). While these are obviously not precisely definable technical measures, they are postulated as conditions short of those that might plausibly be thresholds that, if crossed, might shift ecological conditions from the present desirable state into one much less desirable from the human point of view. Thresholds are defined in terms of coupled human natural systems and nonlinear transitions, and the example given is the recent unanticipated retreat of Arctic ice caused by anthropogenic global warming (2007 was a year of especially dramatic reduction in the Arctic Ocean ice cover).

Given the complexity of earth system processes, such simple definitions are very difficult to operationalize in terms of practical metrics. "Some Earth System processes, such as land use change, are not associated with known thresholds at the continental to global scale, but may, through continuous decline of key ecological functions (such as carbon sequestration), cause functional collapses, generating feedbacks that trigger or increase the likelihood of a global threshold in other processes (such as climate change)" (Rockström/Steffen/Noone et al. 2009b). While these may occur at smaller scales (in particular biomes), they may become a matter of global concern when aggregated if their occurrence is widespread. Determining the boundaries prior to such thresholds is not easy and depends on judgments in the face of numerous uncertainties.

Figure 2.1: Planetary Boundaries for a Safe Operating Space for Humanity. **Source:** Stockholm Resilience Centre: "Azote Images"; at: <http://www.stockholmresilience.org/21/research/research-programmes/planetary-boundaries.html>.

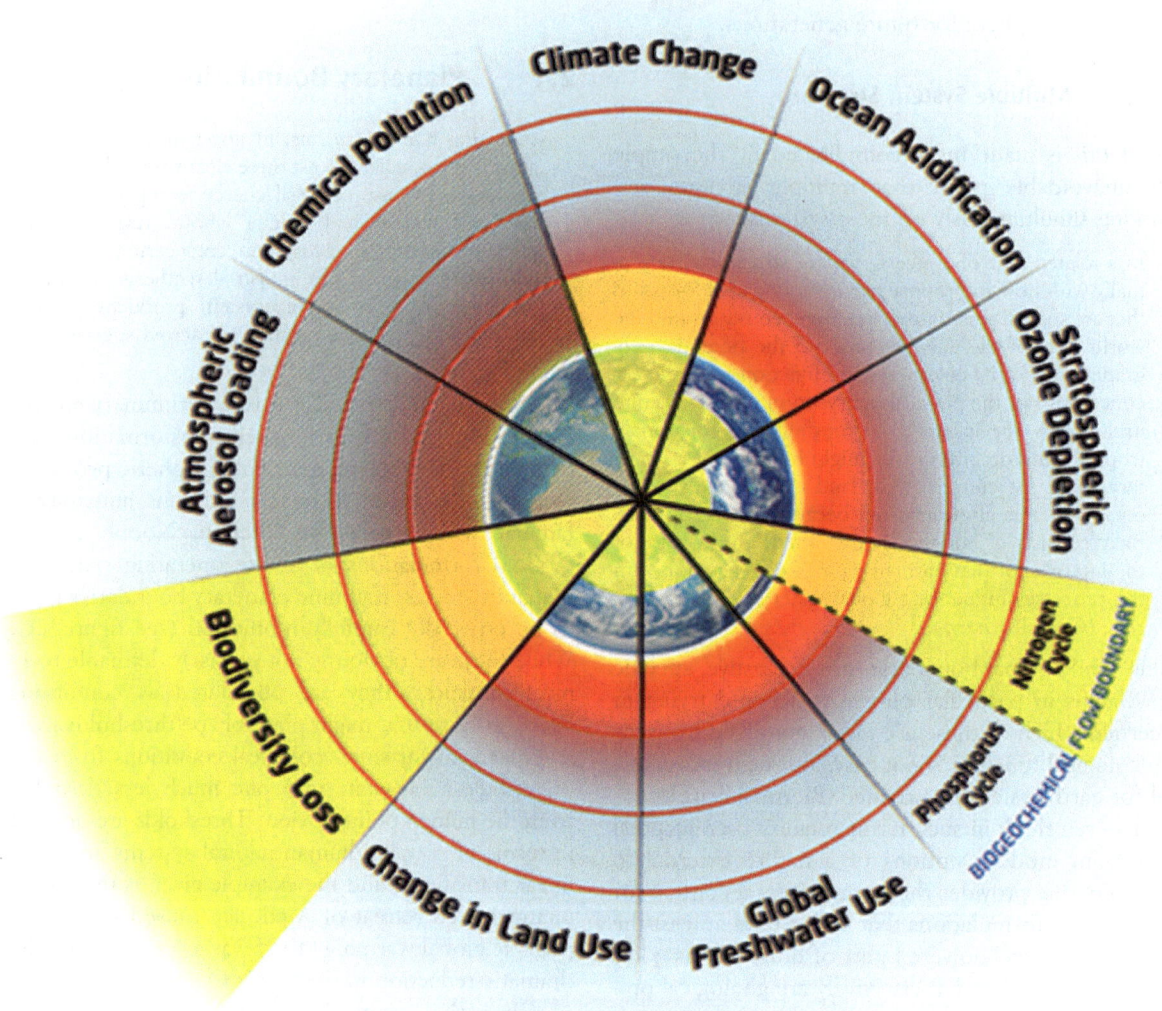

With such caveats carefully noted, the Rockström team identified nine boundaries to a safe operating space for humanity about which they are especially concerned. Three of these are systemic processes at the planetary scale, namely, climate change, ocean acidification, and stratospheric ozone depletion. While the first two are processes with global-scale thresholds, in the case of stratospheric ozone this is less clear. The other boundaries deal with aggregated processes from local and regional changes. The associated thresholds here are less clear in the case of global phosphorous and nitrogen cycles, atmospheric aerosol loading, freshwater use, and land use change. Biodiversity loss and chemical pollution are also listed: they are clearly slower processes than the others and lack obvious global-scale thresholds. From such categorizations, climate change and ocean acidification, both of which are predominately caused by the accumulation of carbon dioxide due to human use of fossil fuels, as well as deforestation, are the immediate cause for concern, being both planetary processes and ones that are happening quickly. All the others listed matter as well, as they are important parts of the life support systems for humanity even if there is no way to establish precise thresholds yet. Even more complicated is that these processes interplay and connect in numerous ways; changes in one may cause other processes to cross boundaries.

The initial planetary boundaries framework was updated and extended in 2015 (Steffen/ Richardson/ Rockström et.al 2015). The new version added updates to the initial scientific estimates, and added an additional discussion of what they termed "novel entities" a category that encompasses the new products of industrial civilization which has added numerous new things to the biosphere, the consequences of which are at least so far less than clear. The revised formulation also emphasized the geographical diversity of boundaries, noting that some boundaries are transcended in some regions, but not elsewhere, a matter that makes the framework more precise, but adds difficulty to the task of aggregating the local boundaries into global calculations. Johan Rockström (Rockström and Klum 2015) at least argues that it is still possible to have a civilization based on abundance within these revised planetary boundaries; but critics are doubtful whether there is the necessary clarity concerning what needs to be done to stay within the climate change boundary in particular (Anderson 2015).

2.3.1 Climate Change

The most high-profile theme in the discussion of the Anthropocene is climate change: the body of science related to this topic is now huge. The consensus, widely adopted at the Copenhagen climate negotiations in 2009, is that any warming above 2 degrees Celsius is to be avoided as it will be dangerously disruptive. What is less clear is the long-term level of carbon dioxide in the atmosphere that will keep the climate below this threshold. Concentrations reached 400ppm briefly in 2013, and while environmental activists suggest that perhaps 350ppm is the maximum level that should be maintained in the long run—approximately the level when *Our Common Future* was written—there are as yet few serious suggestions as to how the atmosphere can be brought back to such a level, one already reached after the first few decades of the great acceleration. With the Arctic Ocean ice cap already melting and warming the northern hemisphere as a result of increased albedo, one of the positive feedbacks that concerns climate science is clearly already in operation. Enhanced methane emissions from melting northern permafrost also suggest accelerated warming in this region. The trend to ever-greater emissions of carbon dioxide from combustion and further destruction of forests suggests rapid climate change. Considerations of rapid climate change are now part of the scenario planning exercises for the future (Anderson/Bows 2011) and increasingly a matter of concern to financial planners who have belatedly begun to consider the consequences of climate system disruptions (Potsdam Institute 2012; see the chapter by Schellnhuber et al. in this volume).

In terms of the boundary debate, climate is one of the key potential drivers that might cross significant tipping points this century with potentially serious disruptions to human systems. Lenton, Held, Kriegler et al.'s (2008) summary suggested that Arctic Sea Ice, the Greenland Ice Sheet, the Atlantic Thermohaline Circulation, the West Antarctic Ice Sheet, the El Nino-Southern Oscillation, Indian Summer Monsoon, Sahara/Sahel and West African Monsoon, Amazon forest dieback, and possible Boreal forest dieback were all contenders for major changes as a consequence of accelerating climate change. While this is far from encompassing the whole earth system, these components are substantial parts of it and they raise the alarm about potential climate security risks very clearly, especially as the monsoons are key to feeding much of Asia and Africa. But as Lenton's (2013) subsequent investigation of environmental shocks makes clear, all this change also matters in human terms: as a matter of how vulnerable people are in particular places and, related to that, a matter of institutional preparedness—or the lack thereof—in particular societies. All this is especially important because, despite repeated warnings that this boundary is one that humanity is on track to transcend, some key scientists are warning that scientific projections of what is needed to prevent global heating, are still not being taken anything like seriously enough by politicians (Anderson 2015).

2.3.2 Ocean Acidification

One of the so-called 'carbon sinks' that removes carbon dioxide from the atmosphere is the ocean where the gas is absorbed in surface waters. But this process is itself a matter with profound ecological consequences.

> Nearly one third of the carbon dioxide released by anthropogenic activity is absorbed by the oceans. But for this fact, current atmospheric CO_2 concentrations would be higher than they already are. However, CO uptake lowers the pH and alters the chemical balance of the oceans, in particular the solubility of calcium salts. This phenomenon is called ocean acidification, and is occurring at a rate faster than at any time in the last 300 million years (Gillings/Hagan-Lawson 2014: 3).

The solubility of calcium salts is a key factor in the success of coral reefs and other marine creatures dependent on shells; if the water is too acidic, coral skeletons or shells may dissolve and cause reefs to stop functioning and shellfish to die. The ecosystem consequences of this phenomenon and other disruptions of marine life may be crucial to the future of the biosphere:

> In the final analysis, protection of the ocean may be more important than protection of atmosphere or land because it stores more carbon, mediates climate variability and provides essential ecosystem services (Gillings/Hagan-Lawson 2014: 4).

This point is especially important given the obviously global dimensions to the oceans. Despite the large-scale implications of shifts in oceanic functions, they frequently remain a low priority for environmentalists whose focus is on terrestrial systems, which are perceived as more immediately a matter of human experience. Indeed, the focus on 'greening' things in environmental politics and the formulation of 'green' policies suggests a focus on chlorophyll that is key to growing plants. But the earth system perspective and a focus on Anthropocene life suggests that, given the importance of the atmosphere, oceans, and ozone layer, a focus on blue formulations, following from oxygen in ozone and in water in particular, might be more appropriate. Correcting this inherent 'terrestrial bias' is one of the key implications of earth system thinking.

2.3.3   Stratospheric Ozone Depletion

If there is a success story in global environmental management and international cooperation, it is clearly the Montreal Protocol and its subsequent additional amendments and extensions (Benedick 1991). These mandated the end to the production of *chlorofluorocarbons* (CFCs) and the gradual reduction of the use of other halocarbons. While they remain in the atmosphere and will for decades more as they gradually decay, the annual ozone holes over the poles are not increasing, although in 2014 detection of new chemicals in the atmosphere in small quantities raised concern about the efficacy of this regime (Laube/Newland/Hogan et al. 2014). Coincidentally, the reduction in CFCs in the atmosphere has been a useful climate change measure given their potency as greenhouse gases. While other chemicals such as nitrous oxides will still have some detrimental effects on ozone levels in the stratosphere, the immediate danger of removing the essential ultraviolet filter from the upper atmosphere, something essential to terrestrial life, has been removed.

However, while this success is worth emphasizing, so too is the point that the combination of easily identifiable dangers, the availability of technical replacements for the outlawed gases, and the relative simplicity of the technical issues allowed for a relatively rapid evolution of policy. Likewise, financial compensation to Southern states for difficulties they might have encountered in making the transition was forthcoming and was not a prohibitive cost to signatories to the agreements. But this regime, frequently invoked as a model for dealing with other 'global' environmental matters, and a prominent part of the discussions about climate change in particular, may not be a good fit for more complicated issues where numerous technologies operate and where substitutions are much more difficult to identify and implement (Hoffman 2005).

2.3.4   Phosphorous and Nitrogen Cycles

One of the keys to the rapid transformation of rural landscapes has been the availability of artificial fertilizers that have, when coupled with tractors and other farm machinery, facilitated industrial-scale monoculture farming. While the fossil fuel 'subsidy' to natural systems has boosted productivity dramatically in the so-called 'green revolution', it has done so by disrupting rural social systems and ecologies and by dramatically increasing the circulation of nitrogen and phosphorous through the biosphere. "Nitrogen flux through the biosphere is primarily a biological process, while phosphorus availability arises slowly through geological weathering. Humans sidestep the phosphorus bottleneck by mining and distribution of fertilizer onto agricultural lands, thus inadvertently increasing the flow of phosphorus into the oceans" (Gillings/Hagan-Lawson 2014: 5). Eutrophication and other ecological disruptions result from the addition of artificial fertilizers into aqueous environments; ocean anoxic events that have caused large-scale die-offs may have been caused by phosphorous being washed into the ocean.

However, the boundary on this ecological change is distant when considered at a global scale despite the fact that phosphorous run-off rates may be significant for some coastal waters. An important consideration with managing phosphorous is that it is geographically heterogeneous: while the application of fertilizers in some places may have eutrophication effects on terrestrial waterways, phosphorous deficiencies elsewhere limit ecosystem productivity. This means that a global boundary for phosphorous is very difficult to calculate even if specific ecosystems have transcended boundary conditions (Carpenter/Bennett 2011). Nitrogen pollution also has regional effects but as yet does not seem to be close to any global threshold (Vries/Kros/Kroeze/Seitzinger 2013). The nitrogen cycle is more complicated to assess as, in the form of nitrous oxide, it is an important greenhouse gas and hence also counts as part of the climate change calculations.

2.3.5 Atmospheric Aerosols

Combustion of fossil fuels, and in particular their inefficient combustion, leads to particulate matter and various chemicals in the atmosphere that have numerous effects (Tsigaridis/Krol/Dentener et al. 2006). These have been of particular concern recently in terms of their immediate pollution effects in Asia and the possible effects global warming may have on disrupting the Asian monsoon, the key rainfall pattern that supplies agricultural systems in the region with water, thus helping feed a large portion of the human population (Kitoh/Endo/Kumar et al. 2013). Ironically, aerosols also act as cooling agents in the atmosphere, shading the ground from sunlight. This effect is clearly visible in studies of the consequences of volcanic eruptions and has become one of the proposed ideas for geoengineering to artificially cool the planet in coming decades if climate change becomes an immediately hazardous phenomenon. As such, aerosols are both pollution and a health hazard but also potentially an artificial sunshade, if practical attempts to cool the planet are undertaken. Hence the great difficulty in assessing the total impact of aerosols on global heating as well as other related human effects.

2.3.6 Freshwater Use

Human activities divert large quantities of surface and ground water for farming, industrial use, as well as for basic human needs such as drinking, bathing, cooking, and household use, making discussions of water security for humanity very complicated indeed (Grey/Garrick/Blackmore et al. 2013). Potable water is key to basic hygiene and disease prevention and, as such, a key dimension to human functioning in many increasingly artificial ecosystems. Water supplies are both a matter of 'natural' supply (as in rain and snow), but also, given the extensive plumbing systems that now supply cities in particular, very much a matter of artificial hydrology. Many rivers no longer flow all the way to their estuaries due to the volume of water diverted en route. Groundwater aquifers are also being pumped dry in many places. While this may not have many direct ecological effects beyond the locations where aquifers feed water springs and hence provide water for ecosystems and human use, the effects matter once the water is depleted and the unsustainable activities dependent on that water source have to be discontinued.

Climate change may alter rain and snow patterns, causing droughts and forcing ecosystems and farming arrangements into new configurations. The California drought emergency of 2014 suggests difficult political choices concerning the allocation of remaining water supplies. Such decisions regarding prioritization have practical, social and philosophical implications for the communities and actors affected. Water is an essential part of human politics, contested and used in numerous ways that defy easy categorization, but an unavoidable necessity in all human activities no matter how humans try to govern its use (Linton 2010). It is clear that human use of fresh water is rapidly increasing due to food cultivation requirements in particular. As such, governance issues will be an important part of sustainability transitions. "This indicates that the remaining safe operating space for water may be largely committed already to cover necessary human water demands in the future" (Rockström/Steffen/Noone et al. 2009b: 16).

2.3.7 Land Use Change

Humanity cleared land for agricultural purposes throughout much of the Holocene period. Indeed, part of the argument that the Anthropocene started long before the industrial revolution is related to the release of methane from agricultural activities (Ruddiman/ Vavrus/Kutzbach/He 2014). The scale of the transformations that have already taken place are such that ecologists have suggested that traditional classifications of the world's large geographical designations of natural areas in terms of biomes now needs to be updated with the addition of various "anthromes" (Ellis/Goldewijk/Siebert et al. 2010). Deforestation reduces carbon sink capabilities at least in forests where trees do not decay quickly and return their carbon to the atmosphere. The albedo of bare land is very different from that of a tree canopy. Forests' water retention functions also affect other hydrological functions. Clearly terrestrial land cover is key to many ecological matters and it is very hard to aggregate these into any one meaningful global threshold.

That said, humanity is already using much of the most fertile parts of the planet's land surface and we will need to implement many changes in how land is used if the ecological footprint of agriculture is to be reduced while its efficiency is simultaneously increased to feed a still-growing population (Foley/Ramankutty/Brauman et al. 2011). Part of the concern about land use change is the question, "Can a threshold for habitat clearance effects on biodiversity be defined on a global scale?" (Brook/Ellis/Perring et al. 2013: 399). The answer would seem to be negative not least because "thresholds are deeply context dependent" and "tipping points might differ between scales" (Brook/Ellis/Perring et al. 2013: 399). Beyond that, it is also worth emphasizing again that, given the diversity of terrestrial land cover, it is unlikely that terrestrial ecosystems will universally respond to a global tipping point crossing some other boundary; they are more likely to react heterogeneously in the event of major disruptions.

2.3.8   Biodiversity Loss

The huge changes to landscapes, including deliberate land clearing as well as inadvertent habitat disruption, in combination with hunting and fishing, has already led to the extinction of many species. The rates of extinction are much greater than the normal background rates of species disappearance in the geological record, suggesting that we are living through the sixth global extinction event in the planet's history (Kolbert 2014). While many species have disappeared, other artificial species, such as farm animals, have expanded greatly. These, however, are dependent on human systems and not a replacement for the diverse species that make up 'natural' ecosystems. The rate of extinction is the key consideration: the alarming pace of extirpation has been driven by most of the other ecological processes in addition to direct human predation on particular species. Given the diversity of species in tropical rainforests, many of them with very limited geographical ranges, forest clearing is an especially damaging human activity in terms of reducing species diversity. Further complicating efforts to stem biodiversity loss, conservationists warn that assuming that these ecosystems have already been radically disrupted may undercut important conservation efforts that can still protect many species, especially in tropical areas less immediately susceptible to climate change playing out more intensely in polar regions (Caro/Darwin/Forrester et al. 2012).

While possible food and pharmaceutical derivatives may be being destroyed, the large concern is that unknown future possibilities for life are being pre-

cluded. In the very long run, the planet will no doubt replenish life forms, but humanity faces an impoverished range of life forms in the centuries ahead. It is important to note that not all species are equally important in ecosystem function: removing "top predators and structurally important species such as corals and kelp, results in disproportionately large impacts on ecosystem dynamics" (Rockström/Steffen/Noone et al. 2009b: 15). As such, while an overall reduction in the rate of extinction by several orders of magnitude is needed to push biodiversity loss to a level within the safe operating space, specific species may have a disproportionate effect on particular ecosystems and their functioning. It consequently becomes clear that managing the global scale must be balanced with micro-scale interventions.

The overall pattern of biodiversity loss in comparison to previous mass extinction events is not clear (Condamine/Rolland/Morlon 2013), even if the trend in particular places due to 'sledgehammer' clearing effects frequently is observable and better understood. This remains the case in the updated version of the planetary boundaries framework where matters of biodiversity loss are nuanced by dealing with them in terms of biosphere integrity, and focusing on functional and genetic diversity in specific biomes (Steffen/Richardson/Rockström et.al 2015). The lack of clarity about baselines in terms of species numbers and extinction rates remains a measurement problem in terms of the precise location of the boundary even if the trajectory of rapid extinction is clear.

2.3.9   Chemical Pollution

While pesticides were a central driver in the rise of environmentalism in the United States in particular, inspired by Rachel Carson's book *Silent Spring*, other forms of industrial pollution have long been a problem both for ecosystems and as a direct cause of human health issues. Recent smog events in China are reminiscent of the situation in London sixty years earlier when the death of thousands as a result of smog caused by coal fires finally led to comprehensive efforts to reduce smoke. Many of the environmental campaigns of the 1960s and 1970s in the developed world led to technological innovations that removed pollutants from smokestacks and effluent pipes. Ecological modernization provided numerous technical fixes to pollution problems but only rarely led to more fundamental social change. With the rise of globalization, industrial production frequently moves to states with less rigorous regulations, effectively outsourcing pollution rather than permanently and directly addressing the problem. Of great concern are the very low-level toxic substances that may have effects on particular species and thus alter whole ecosystems indirectly. These include such substances as endocrine-disrupting chemicals. Where the thresholds on such activities might be is not yet clearly known.

2.3.10 Boundary Priorities

The most immediate concerns are with atmospheric greenhouse gases. Nonetheless, as Rockström and colleagues emphasize, the nine factors identified as safe operating space boundaries interact and interconnect in complicated ways. It is worth remembering that chlorofluorocarbons, the most obvious cause of stratospheric ozone depletion, are also powerful greenhouse gases. The regime to curtail their production is effectively also an agreement to deal with climate change even if it is not designed explicitly to do this! Crossing one boundary may have many serious and unpredictable consequences for others: this simple but difficult point is key to any serious consideration of how to facilitate transitions to sustainability. Sustainability usually implies a fairly stable context for humanity but, as the earth system science analyses briefly summarized here suggest, in the present context sustainability of societies has to be considered in terms of the rapidly changing context for humanity and the simple fact that societies have been changing rapidly as rural transformations and urbanization interact.

Viewed as a totality, the planetary boundaries perspective suggests that humanity has effectively taken its own fate into its hands in terms of the future configuration of the planetary system. However, while we have clearly crossed the boundaries in terms of biodiversity loss, the artificial production of nitrogen in the atmosphere, and climate change, it appears that human action has at least halted the dangerous trend of ozone depletion and limited it to the areas within the high-altitude polar vortex wind systems. Although depletion will remain a problem until at least the middle of the current century, given the existing inventory of ozone depleting substances already released, it is important to remember that the boundary has not been crossed nor does it seem likely that this will happen given the widespread agreement that chlorofluorocarbons and related chemicals are too dangerous for more than very limited use. Dealing with this boundary has been relatively easy given that the components necessary for a solution were practical and political opposition to agreements was relatively weak. Other boundaries are much more difficult to deal with, even if it is clear where they might be and how close we are coming to some of the thresholds.

2.4 Planetary Stewardship for a Sustainable Earth?

Implied but rarely spelled out in the discussion of sustainable development is the assumption that the planetary conditions inherited from the Holocene are essential for future generations to meet their needs. The discussion presupposes that the baseline condition of the planet is one given by the Holocene parameters that facilitated the emergence of human civilization and that the planetary boundaries are effectively guard-rails beyond which humanity should not venture. Sustainable development is about economic change while effectively maintaining Holocene conditions, ones that are presumably the optimal state for humanity. The extraordinary growth in human numbers and associated increase in economic activity over the last half century of the great acceleration have been powered by the extraction and combustion of fossil fuels. This process has reversed the long-term ecological pattern of life sequestering carbon from the atmosphere, and effectively made industrial humanity a new geological force in the planetary system. Constraining the further expansion of these activities and planning to reduce the carbon dioxide levels in the atmosphere to something approaching pre-industrial levels is key to most policies of transition to a sustainable future. However, the very awkward assumption about this formulation is that it presupposes precisely what current processes of unsustainable development have started to fundamentally change:

> Environmentalist traditions have long called for a halt to human interference in ecology and the Earth system. In the Anthropocene, the anthropogenic biosphere is permanent, the legacy of our ancestors, and our actions as human systems a force of nature, making the call to avoid human interference with the biosphere irrelevant. The implication is clear; the current and future state of the terrestrial biosphere is up to us, and will be determined by human systems of one form or another, whether it is the momentum of our past or new pathways we are able to achieve in the future (Ellis 2011: 1027).

It is abundantly clear that decisions about human economic activities are now central to constructing the future of the planet and instrumental in determining whether key boundaries will be crossed. Preventing these transgressions while simultaneously working back to levels within the boundaries in terms of nitrogen and climate change, as well as drastically reducing the rate of loss of biodiversity, is key to any transition to a human condition that lives within the safe operating space. The alternative requires a transition to a very different configuration of the biosphere, one impossible to predict precisely but, given the phase shifts already looming, one very different from the conditions that have given rise to human civilization. The implicit assumption in juxtaposing transitions and a peace that can be sustained is that ecological changes will not be so drastic or so quick that major powers resort to military force in attempts to control the human consequences of the disruptions.

In Steffen, Persson, Deutsch et al.'s (2011) terms, the next phase of the Anthropocene requires planetary stewardship, a complex matter of global governance that will require numerous social innovations if the earth is to be kept within a safe operating space loosely analogous to Holocene conditions. While production of chlorofluorocarbons and other ozone-depleting substances has largely ceased due to international cooperation, this has occurred despite a remarkable amount of foot-dragging on the part of specific sectors of some economies, notably strawberry producers in California, reluctant to give up their particular mode of soil sterilization (Gareau 2013). The larger lesson of this case is the necessity of thinking in terms of how humanity produces things, and how these processes are governed. Phasing out the use of ozone-depleting substances emphasizes that merely regulating the use of substances is not enough for at least some of the planetary boundaries; prohibition of certain activities may be required, and that, in turn, is a matter of global governance where industrial corporations as well as state governments must be involved in determining production priorities.

Looking further ahead, it is clear that decisions about such things as the continued production of coal-powered electricity generation stations are matters of industrial policy that have global consequences. If climate boundaries are not to be yet further transcended, then thinking in terms of the political economy of energy systems is essential to future planetary governance. But this is more than a matter of governments regulating some detrimental environmental consequences of economic activity; it is about production decisions and, quite literally, *who* decides *what* gets made. While there is a wide diversity of economic decision-making authorities currently operating, the neo-liberal modes of letting markets make such decisions seem unlikely to constrain the use of fossil fuels

quickly enough to begin reducing carbon dioxide emissions any time soon, even if they do focus attention on how markets might work as governance mechanisms (Newell/ Paterson 2010). The climate boundary has been breached and the larger patterns of the great acceleration suggest that this trajectory is not likely to change quickly despite repeated, although as yet limited, efforts to restrict emissions under the *United Nations Framework Convention on Climate Change* (UNFCCC).

This has in turn raised the question of geoengineering, attempting to artificially manipulate the global climate to at least temporarily mask the effects of elevated levels of greenhouse gases (Hamilton 2013). However, as the discussion of the other planetary boundaries emphasizes, dealing with just one facet of boundary crossing, such as the failure to cope with rapidly rising greenhouse gases, needs to take into consideration other facets of the earth system. As one of the leading proponents of experimenting with climate engineering technologies is keen to emphasize, while artificial sulphate aerosol use may be feasible as a method of "solar radiation management", it will not address the major issue of ocean acidification—hence, reducing emissions of carbon dioxide must remain the priority for global environmental politics (Keith 2013).

The implications of such considerations are profound. While arguments for a transition to a sustainable future suggest that rapidly reducing the disruptions to the natural arrangements that humanity has known since the end of the last ice age approximately ten millennia ago are essential, it is no longer the case that this future will occur in the given circumstances implied by the invocation of physical sustainability. The assumption built into environmental concerns through the discussion of the limits to growth in the 1970s, the 1980s discussions of sustainable development, and subsequently through the initial formulations of the UNFCCC in the 1990s is that the planet ought to be kept in more or less the configuration that has so far nurtured civilization. This is implied in the formulation of 'physical sustainability' in *Our Common Future*, and spelled out clearly in the formulation of avoiding dangerous human interference with the climate system that is the key operant phrase in the UNFCCC.

But all this is very new in human affairs. "At the first major global environmental governance conference—the 1972 Stockholm Conference on the Human Environment—none of the major earth system challenges that we discuss today was on the agenda. And this was merely forty years ago. Hardly anybody talked then about ozone depletion, climate change, desertification, or the mass extinction of species" (Biermann 2012: 9). Given these novel circumstances, it is hardly surprising that humanity does not have institutions, much less governance structures, to effectively deal with such issues. However, as other chapters in this book suggest, such structures are now urgently necessary, and need to be formulated so that warfare is precluded as an adaptive mechanism, both because it will make other adaptations more difficult and, in the event of major weapon use, add yet further unpredictable perturbations to the earth system. In the words of the German Advisory Committee on Climate Change, a United Nations 2.0 is needed, a new structure whose purpose "… would be to take the planetary guard rails into account as a guiding principle that governs all UN actions, the pursuit of which would guarantee the protection of climate and environment in order to stabilize the Earth system as much as peace, security and development" (WGBU 2011: 316). While this remains an aspiration, the speed of current ecological changes make thinking about new habits of cooperation to deal with coming transformations a crucial part of any transition strategy.

2.5   Sustainable Transitions in a Rapidly Changing Future

Most of the ideas about mitigating climate change, the predominant emphasis in climate discussions until relatively recently, epitomized by the UNFCCC, have either implicitly or explicitly assumed that the Holocene condition is the optimal biospheric arrangement for humanity. Given the patterns of agricultural activity at the moment, taking advantage of most of the temperate humid biomes to grow crops using petroleum-powered industrial production systems is not surprising, but it is important to note that the assumption that this is how the social organization of food production ought to be, or that extensive monocultures of grain are the only way to feed humanity in the long run, are simply that: assumptions. If these systems are not optimal, then perhaps other ecological possibilities open up and new ways of feeding, fuelling, and housing humanity may be possible. Thinking in these terms makes it clear that sustainable development strategies need to operate in ways that do not foreclose such possibilities for future generations; a flexible interpretation of future needs is a crucial component of such policies.

Many of the ecological ideas that structure thinking about environmental strategies also presume some form of stability, or at least a notion of homoeostasis as the desirable state toward which policy should direct human activity. The last generation of ecological management thinking has drawn heavily on notions of resilience, as well as the frameworks of panarchy and non-linear changes (Gunderson/ Holling 2001). Earlier assumptions of stable ecosystems and ecological transitions back to a climax condition following disruptions have not been entirely abandoned, but ecological thinking is now clearly much more complicated than visions built upon simplistic assumptions of a stable nature disrupted by human economic activity. Even in arguments about resilience, however, it is clear that the assumptions of stability are integral to subsequent 'bounce back' strategies after a major disruption. Policies using such thinking usually postulate a given relatively stable situation which, after facing disruption, can return to more or less the situation prior to the disruption.

However, as the growing awareness of the sheer scale of the human transformation of the biosphere becomes clear and the failure of humanity to curb the use of fossil fuels in particular ensures at least some climate change is inevitable, policy is frequently turning to questions of how to adapt to these new circumstances. In Bangladesh, where poor people are especially vulnerable to flooding and storms moving up from the Bay of Bengal, adaptation is the order of the day. Flood shelters and rebuilding coastal mangrove forests are necessary tools for dealing with rising sea levels and inundations. In such states there is limited choice in terms of policy; they have done little to cause climate change and can do little to change the global energy mix that is accelerating the process. What sustainability and the transition to it might mean in such changing circumstances suggests that the most important aspect of sustainability is the flexibility to adapt to new circumstances as they arise. How to make social systems that can change quickly without disruptions, social breakdown, or the use of organized coercion is not easy but it is key to any serious attempt to link peace with sustainability.

The clear implication to be drawn from earth system science is that such questions are in need of immediate scholarly attention. "These are admittedly huge tasks, but are vital if the goal of science and society is to steer the biosphere towards conditions we desire, rather than those that are thrust upon us unwittingly" (Barnosky/Hadly/Bascompte et al. 2012: 57). Sustainable development implies that economic transitions to modes of industrial activity that do not exceed the parameters of the conditions inherited from the Holocene are key to maintaining a relatively stable biosphere, the *sine qua non* for future generations being able to supply their own needs.

But if some of those key parameters have already been exceeded, as the discussion of planetary boundaries suggests, then what kind of transition is needed to peacefully move to what kind of future must be much more carefully deliberated. Nature can no longer be taken for granted as some sort of given context for humanity; the Anthropocene discussion makes it clear that the future of humanity and whether that future is peaceful or not depends on much more than traditional discussions of the causes of war. Now humanity it shaping its context in novel ways, and that too has to be a key part of any discussion of transitions to new modes of economy and life.

2.6   Implications for Contextualizing Sustainable Peace

Focusing on the insights of earth system science and the clear understanding that humanity is shaping the future in ways that are much more profound than has been understood until very recently requires that social scientists and policymakers interested in thinking through both strategies of sustainable development and peaceful modes of transition to more ecologically benign economic modes of human life now have to incorporate at least four key themes in their work. These are: a notion of security very different from cold war versions; a recognition that geopolitical contexts are changing; a perspective on political geoecology; and, crucially, a focus on economic production rather than just on environmental protection as key to the next phase of the Anthropocene.

2.6.1   Beyond National Security

We are currently on track for a much warmer world, one where ecological disruptions are inevitable (Anderson/Bows 2011). The potential for human conflict caused by these disruptions to undo the economic progress and improvements in human welfare made in recent generations is a key part of the logic of sustainable development spelled out in *Our Common Future*. Indeed, this is the logic that underpins the whole discussion of environmental security even if the empirical evidence on the small scale repeatedly suggests that environmental change does not directly

cause violence (Theisen/Holtermann/Buhaug 2011; O'Loughlin/Witmer/Linke et al. 2012). But larger-scale ecological phase shifts might present the world system with much greater disruptions than small-scale rural displacements and agricultural failure in specific regions. Transitions to sustainability that focus on the complexity of social change in particular parts of the world are important, but how particular places play a part in shaping the overall configuration of the future matters greatly (Grin/Rotmans/Schot 2010). Planning for these in advance rather than relying on force to try to deal with some of the disruptions is now key; it requires a move towards policies for global human security.

Nonetheless, much of the recent literature that links climate change to security persists with traditional American formulations wherein instabilities in distant parts of the world are understood as potential threats, whether through terrorism or violence spilling over from conflict zones. More specifically, climate change has been formulated as a threat enhancer or 'multiplier' implying that related policies are essential in preventing future conflicts (CNA 2007; Campbell/Gulledge/McNeill et al. 2007; UN 2009). The necessity of dealing with climate change is specified as key in these formulations but what is threatening in terms of security is defined in terms of political instabilities in peripheral areas, not the disruptions caused by the global economy. Recent attempts to think through the practicalities of environmental insecurity by focusing on water and climate change suggest that structural inequalities in the global economy and political discrimination are more important immediate drivers of conflict if not interstate war (Zografos/Goulden/Kallis 2014). This key finding links back to the initial WCED (1987) formulations of sustainable development that insist that matters of equity are key to sustainable development that is peaceful.

More recently, as numerous institutions and governments start to come to terms with the fact that some climate change is inevitable, policy questions about how to adapt to changing circumstances are coming to the fore (Dalby 2013b). But now adaptations themselves are causing further environmental transformations. States are buying or leasing land in other states to ensure supplies of food in future, leading to further displacements of peoples to make way for plantations and commercial farming arrangements. These trends add to land use transformations and, in some cases, chemical pollution that put stress on existing ecosystems. Extending the modern agricultural development model that has already caused dramatic disruptions to many ecosystems may just accelerate environmental change; these systems need innovations as part of a larger effort to rethink global governance in terms of planetary boundaries (Galaz/ Biermann/Crona et al. 2011). Such policies have 'backdraught' effects that need to be considered in terms of unanticipated consequences of trying to deal with climate change while ignoring other aspects of environmental change (Dabelko/Herzer/Null et al. 2013). Attempts to adapt to climate change have consequences in terms of how landscapes are remade: such changes need to be thought of in those terms if earth system analysis is to be worked into policy considerations.

2.6.2 Changing Geopolitical Contexts

This has international repercussions in terms of trade and the international flows of investment that shape how one state impacts another's ecology. All of this emphasizes decisions about what is being made, and how, rather than discussions of protecting an environment that has already been transformed many times over: this is the context for twenty-first century geopolitics. A further extension of this point is that the geopolitical rivalries of the present are frequently being played out in the arcane details of trade negotiations (Wang/Gu/Li 2012). Who will write the rules on technological standards related to new generations of energy technologies now matters, and here the mostly neglected dimension of geopolitics in discussions of sustainable transitions needs to be engaged directly (Markand/Raven/Truffer 2012). How what gets produced in future will also quite literally shape the planetary ecology of the future.

Understanding that geopolitics is no longer a simple matter of military rivalry but a matter of shaping the technological future to the short-term advantage of the rule-writers—with long term implications for how the biosphere is shaped—is crucial to linking earth system thinking to matters of geopolitics. This becomes even more important where discussions of possible geoengineering experiments enter the discussion (Galaz 2012; Luke 2010; Humphreys 2011) as a contribution to attempts to keep the planet's climate within a safe operating space. The related point is that environmental changes have already transformed the terrain of great power rivalry; nineteenth-century assumptions about the given context of politics are no longer tenable premises for serious political discussion (Hommel/Murphy 2013).

2.6.3 The Perspective of Political Geoecology

Thinking about the point that nature is no longer the given context for humanity in terms of international peace and security requires a very different approach from that of traditional geopolitics, with its attendant focus on territorial states and their rivalries as key to war and peace. Thinking in terms of a 'political geoecology' of human actions at the global scale shaping the future global context, rather than taking it for granted, is now key to any notion of a sustainable peace if serious consideration is to be given to ensuring that climate and other ecological disruptions do not trigger violent policy responses (Brauch/Dalby/ Oswald Spring 2011). A geopolitical imagination of competing territorial states as what is most important in human affairs is simply out of date as the premise for either policy prescription or academic analysis. While states may still write many of the rules for international trade, and as such are still a key institution in shaping the future, understanding geopolitics in these terms is now much more important than traditional discussions of elite military rivalries and struggles to dominate the planet.

While technical details of trade arrangements may be important in the shaping of the future, so too are the cities and towns in which the majority of us now live. The Anthropocene has also been about the urbanization of humanity and the construction of increasingly large artificial environments of concrete and asphalt. The commodities and resources that are extracted often at great distance in rural hinterlands supply these consumption spaces, spaces that also function as centres of political innovation in many ways (Magnusson 2011). How cities are rebuilt and governed to simultaneously reduce their carbon footprint and make them less vulnerable to extreme events will matter greatly in coming decades. But as Naomi Klein (2014) has ruefully noted, thinking about practical ways to green cities and building sensible public transit systems, communal green spaces, and local food systems has not been a priority among many political movements. If members of such movements and communities were thinking seriously about the artificial landscapes of the present, these considerations should certainly be prioritized.

Even more important is to think about how suburban sprawl, which is so dependent on automobiles and uses large amounts of energy very inefficiently, can be reworked with ecological principles in mind. Likewise, informal solutions to the huge challenges of urbanization in the global South that take ecological issues seriously also present possibilities for adaptation that improve the lot of residents of the new cities there (Robertson 2012). All of this urbanization is dependent on the integration of the global economy which is in turn shaping the life chances of people in the new cities of our time, and hence producing new geopolitical circumstances that make old imperial struggles to gain colonies or zero sum games to directly control agricultural territories increasingly anachronistic. Struggles over economic activity are now shaping national policies as well as city strategies; the task for a political geoecology is think through these interconnections as they shape the increasingly artificial habitats of the future.

2.6.4 Producing the Anthropocene

Thinking in these terms makes it clear that production—quite literally *what* humanity is making—is a key consideration in understanding and constructing our future; getting that clearly in focus is integral to a sustainable transition (Harris 2012). The Anthropocene will be shaped by decisions taken both by community planners and executives in boardrooms who decide which commodities will be made and how they will be produced. It makes a big difference if new electrical systems are powered by solar panels or coal-burning power stations. If automobile manufacturers stop making gasoline-powered cars in favour of other propulsion systems, this too will have all sorts of ramifications for climate and other ecological changes. Global cooperation on such matters is a key component for transitions to sustainability.

In part, this point also relates to where they are produced. Territorial strategies to 'green' some societies by outsourcing the production of energy- and pollution-intensive industries to less regulated societies, an accusation frequently levelled at European states,

do not solve problems when viewed in earth system terms. Authority over ecological processes necessitates deeper action than state territorial strategies of rule, especially when territorial arrangements for such policies as emissions trading quotas and ecological offsets are involved. Outsourcing pollution may satisfy some limited state-based counting methods, but it is not an ecologically sensible strategy if more than short term geographically-specific spaces are considered. This new geography of connection in the increasingly globalized economy, and the possibilities of commodity chain governance also suggest that the traditional assumptions of state-based national security are insufficient frameworks for governance in the next phase of the Anthropocene.

All of which becomes ever more pressingly urgent if the earth system discussion is engaged precisely because assumptions of a stable ecological context for humanity can no longer be taken for granted. Peace is a condition to be struggled for but the struggle will be carried out in future in ecological and social conditions that are rapidly changing. Any strategy of transition from present consumption-based extractive modes of economy to ones that can be sustained in the long run have to recognize that this transition will happen against the backdrop of dramatic ecological change. Strategies that link development and sustainability now have to factor in the possibilities of ecological phase shifts if the earth system boundaries are further transgressed in the next few decades. This makes planning more difficult, but also emphasizes the fact that humanity is making the future of the planetary system as well as its own economic and social future. They are but two sides of the same coin, a matter requiring a transition to new ways of thinking about economics and politics if peaceful human societies are to be created as the next phase of the Anthropocene.

References

Anderson, Kevin; Bows, Alice, 2011: "Beyond Dangerous Climate Change: Emission Scenarios for a New World", in: *Philosophical Transactions of the Royal Society A*, 369: 20–44.

Anderson, Kevin 2015: "Duality in Climate Science" *Nature Geoscience* Online 12 October.

Autin, Whitney J.; Holbrook, John M., 2012: "Is the Anthropocene an issue of stratigraphy or pop culture?", in: *GSA Today*, 22,7: 60–61.

Barnosky, Anthony, D.; Hadly, Elizabeth A.; Bascompte, Jordi; Berlow, Eric L.; Brown, James H.; Fortelius, Mikael; Getz, Wayne M.; Harte, John; Hastings, Alan;

Marquet, Pablo A.; Martinez, Neo D; Mooers, Arne; Roopnarine, Peter; Vermeij, Geerat; Williams, John W.; Gillespie, Rosemary; Kitzes, Justin; Marshall, Charles; Matzke, Nicholas; Mindell, David P.; Revilla, Eloy: Smith, Adam B., 2012: "Approaching a state shift in Earth's biosphere", in: *Nature*, 486: 52–58.

Benedick, Richard E., 1991: *Ozone Diplomacy* (Cambridge, MA: Harvard University Press).

Bennett, Elena M.; Carpenter, Stephen R.; Cardille, Jeffrey A., 2008: "Estimating the Risk of Exceeding Thresholds in Environmental Systems", in: *Water, Air, and Soil Pollution*, 191: 131–138.

Biermann, Frank, 2012: "Planetary boundaries and earth system governance: Exploring the links", in: *Ecological Economics*, 81: 4-9.

Brauch, Hans Günter; Dalby, Simon; Oswald Spring, Úrsula, 2011: "Political Geoecology for the Anthropocene", in: Brauch, Hans Günter; Oswald Spring, Úrsula; Mesjasz, Czeslaw; Grin, John; Kameri-Mbote, Patricia; Chourou, Béchir; Dunay, Pal; Birkmann, Jörn (Eds.), 2011: *Coping with Global Environmental Change, Disasters and Security–Threats, Challenges, Vulnerabilities and Risks*. Hexagon Series on Human and Environmental Security and Peace, vol. 5 (Berlin Heidelberg–New York: Springer-Verlag): 1453-1486.

Brook, Barry W.; Ellis, Erle C.; Perring, Michael P.; Mackay, Anson W.; Blomqvist, Linus, 2013: "Does the terrestrial biosphere have planetary tipping points?", in: *Trends in Ecology & Evolution*, 28: 396-401.

Brown, Anthony G.; Tooth, Steven ; Chiverrell, Richard C.; Rose, James; Thomas, David S.G.; Wainwright, John; Bullard, Jonna E.; Thorndycraft, Varyl R.; Aalto, Rolf; Downs, Peter, 2013: "The Anthropocene: is there a geomorphological case?", in: *Earth Surface Processes and Landforms*, 38: 431-434.

Brunnée, Jutta; Streck, Charlotte, 2013: "The UNFCCC as a negotiation forum: towards common but more differentiated responsibilities", in: *Climate Policy*, 13,5: 589-607.

Campbell, Kurt M.; Gulledge, Jay; McNeill, J.R.; Podesta, John; Ogden, Peter; Fuerth, Leon; Woolsey, R. James; Lennon, Alexander T.J.; Smith, Julianne; Weitz, Richard; Mix, Derek, 2007: *The Age of Consequences: The Foreign Policy and National Security Implications of Global Climate Change* (Washington, DC: Center for Strategic and International Studies).

Caro, Tim; Darwin, Jack; Forrester, Tavis; Ledoux-Bloom, Cynthia; Wells, Caitlin, 2012: "Conservation in the Anthropocene", in: *Conservation Biology*, 26: 185-188.

Carpenter, Stephen R.; Bennett, Elena M., 2011: "Reconsideration of the planetary boundary for phosphorus", in: *Environmental Research Letters*, 6: 014009.

Carson, Rachel, 1962: *Silent Spring* (Boston: Houghton Mifflin Harcourt).

Clark, Nigel, 2012: "Rock, Life, Fire: Speculative Geophysics and the Anthropocene", in: *Oxford Literary Review*, 34: 259-276.

CNA [CNA Corporation], 2007: *National Security and the Threat of Climate Change* (Alexandria: VA: Center for Naval Analysis [CNA]).

Committee on Understanding and Monitoring Abrupt Climate Change and its Impacts, 2013: *Abrupt Impacts of Climate Change: Anticipating Surprises* (Washington: National Academies Press).

Condamine, Fabien L.; Rolland, Jonathan; Morlon, Helene, 2013: "Macroevolutionary perspectives to environmental change", in: *Ecology Letters*, 16: 72-85.

Cornell, Sarah, 2012: "On the System Properties of the Planetary Boundaries", in: *Ecology and Society*, 17.

Crutzen, Paul J., 2002: "Geology of Mankind–The Anthropocene", in: *Nature*, 415: 23.

Dabelko, Geoff; Herzer, Lauren; Null, Schuyler; Parker, Meaghan; Stiklor, Russell (Eds.), 2013: "Backdraft: The Conflict Potential of Climate Change Adaptation and Mitigation", in: Woodrow Wilson Center Environmental Change and Security Program, *Report*, 14,2: 1-60.

Dalby, Simon, 2009: *Security and Environmental Change* (Cambridge: Polity).

Dalby, Simon, 2013a: "Human security in the Anthropocene: the implications of earth system analysis", in: Sygna, Linda; O'Brien, Karen; Wolf, Johanna (Eds.): *A Changing Environment for Human Security. Transformative approaches to research, policy and action* (London–New York: Routledge–Earthscan): 27-33.

Dalby, Simon, 2013b: "Climate Change: New Dimensions of Environmental Security", in: *RUSI Journal*, 158,3 (June/July): 34-43.

Dalby, Simon, 2014: "Rethinking Geopolitics: Climate Security in the Anthropocene", in: *Global Policy*, 5,1: 1-9.

Davis, Robert, 2011: "Inventing the Present: Historical roots of the Anthropocene", in: *Earth Sciences History*, 30,1: 63-84.

Ellis, Erle C.; Goldewijk, Kees K.; Siebert, Stefan; Lightman, Deborah, Ramankutty, Navin; 2010: "Anthropogenic transformation of the biomes, 1700 to 2000", in: *Global Ecology and Biogeography*, 19: 589-606.

Ellis, Erle C., 2011: "Anthropogenic transformation of the terrestrial biosphere", in: *Philosophical Transactions of the Royal Society A*: 369, 1010-1035.

Floyd, Rita; Matthew, Richard (Eds.), 2013: *Environmental Security: Approaches and Issues* (London: Routledge).

Foley, Jonathan A.; Ramankutty, Navin; Brauman, Kate A.; Cassidy, Emily S.; Gerber, James S.; Johnston, Matt; Mueller, Nathaniel D.; O'Connell, Christine; Ray, Deepak K.; West, Paul C.; Balzer, Christian; Bennett, Elena M.; Carpenter, Stephen R.; Hill, Jason; Monfreda, Chad; Polasky, Stephen; Rockström, Johan; Sheehan, John; Siebert, Stefan; Tilman, David; Zaks, David P.M., 2011: "Solutions for a Cultivated Planet", in: *Nature*, 478: 337-342.

Folke, Carl; Jansson, Asa; Rockström, Johan; Olsson, Per; Carpenter, Stephen R.; Chapin III, F. Stuart; Crépin, Anne-Sophie; Daily, Gretchen; Danell, Kjell; Ebbesson, Jonas; Elmqvist, Thomas; Galaz, Victor; Moberg, Fredrik; Nilsson, Mans; Osterblom, Henrik; Ostrom, Elinor; Persson, Asa; Peterson, Garry; Polasky, Stephen; Steffen, Will; Walker, Brian; Westley Frances; 2011: "Reconnecting to the Biosphere", in: *Ambio*, 40: 719-738.

Galaz, Victor, 2012: "Geo-engineering, Governance, and Social-Ecological Systems: Critical Issues and Joint Research Needs", in: *Ecology and Society*, 17;1; at: <http://www.ecologyandsociety.org/vol17/iss1/>.

Galaz, Victor; Biermann, Frank; Crona, Beatrice; Loorbach, Derk; Folke, Carl; Olsson, Per; Nilsson, Mans; Allouche, Jeremy; Persson, Åsa; Reischl, Gunilla, 2012: "Planetary boundaries–exploring the challenges for global environmental governance", in: *Current Opinion in Environmental Sustainability*, 4: 80-87.

Gareau, Brian J., 2013: *From Precaution to Profit: Contemporary Challenges to Environmental Protection in the Montreal Protocol* (New Haven: Yale University Press).

Gillings, Michael R.: Hagan-Lawson, Elizabeth L., 2014: "The cost of living in the Anthropocene", in: *Earth Perspectives*, 1,2.

Grey D.; Garrick D.; Blackmore, D.; Kelman, J.; Muller, M.; Sadoff, C., 2013: "Water security in one blue planet: twenty-first century policy challenges for science", in: *Philosophical Transactions of the Royal Society* A 371: 20120406; at: <http://dx.doi.org/10.1098/rsta.2012.0406>.

Grin, John; Rotmans Jan, Schot, Johan, 2010: *Transitions to Sustainable Development. New Directions in the Study of Long term transformative change* (New York–London: Routledge).

Gunderson, Lance; Holling, C.S., 2001: *Panarchy: Understanding Transformations in Systems of Humans and Nature* (New York: Island Press).

Hamilton, Clive, 2013: *Earthmasters: The Dawn of the Age of Climate Engineering* (New Haven: Yale University Press).

Harris, Steven R., 2012: *Pushing the Boundaries: The Earth System in the Anthropocene.* (Bristol: Schumacher Institute for Sustainable Systems Report).

Hoffmann, Matthew J., 2005: *Ozone Depletion and Climate Change: Constructing a Global Response* (Albany: State University of New York Press).

Hommel, Demian; Murphy, Alec B., 2013: "Rethinking geopolitics in an era of climate change", in: *GeoJournal*, 78: 507-524.

Hornborg, Alf; McNeill, John R.; Martinez-Alier, Joan (Eds.), 2007: *Rethinking Environmental History: World System History and Global Environmental Change* (Lanham: Altamira).

Huggett, Andrew J., 2005: "The concept and utility of 'ecological thresholds' in biodiversity conservation", in: *Biological Conservation*, 124: 301-310.

Hughes, Terry P.; Carpenter, Stephen; Rockström, Johan; Scheffer, Marten; Walker, Brian; 2013: "Multiscale Regime Shifts and Planetary Boundaries" in: *Trends in Ecology and Evolution* 28,7: 389-395.

Humphreys, David, 2011: "Smoke and Mirrors: Some Reflections on the Science and Politics of Geoengineering", in: *Journal of Environment and Development*, 20,2: 99-120.

Keith, David, 2013: *A Case for Climate Engineering* (Cambridge, MA: MIT Press).

Kitoh, A.; Endo, H.; Kumar, K.K.; Cavalcanti, I.F.A.; Goswami, P.; Zhou, T., 2013: "Monsoons in a changing world: A regional perspective in a global context", in: *Journal of Geophysical Research: Atmospheres*, 118: 3053-3065; at: <doi:10.1002/jgrd.50258>.

Kjellen, Björn, 2008: *A New Diplomacy for Sustainable Development: The Challenge of Global Change* (New York: Routledge).

Klein, Naomi, 2014: *This Changes Everything: Capitalism vs. The Climate* (Toronto: Knopf).

Kolbert, Elizabeth, 2014: *The Sixth Extinction: An Unnatural History* (New York: Henry Holt).

Laube, Johannes C.; Newland, Mike J.; Hogan, Christopher; Brenninkmeijer Carl A.; Fraser, Paul J.: Martinerie, Patricia.; Oram, David E.; Reeves, Claire E.; Rockmann, Thomas; Schwander, Jakob; Witrant, Emmanuel; Sturges, William T., 2014: "Newly detected ozone-depleting substances in the atmosphere", in: *Nature Geoscience*, 7: 266-269; at: <doi:10.1038/ngeo2109>.

Lenton, Timothy M., 2013: "What early warning systems are there for environmental shocks?", in: *Environmental Science & Policy*, 27: S60-S75.

Lenton, Timothy; Held, Hermann; Kriegler, Elmar; Hall, Jim W.; Lucht, Wolfgang; Ramstorf, Stefan; Schellnhuber, Hans Joachim, 2008: "Tipping elements in the Earth's climate system", in: *Proceedings of the National Academy of Science* (PNAS), 105,6 (12 February): 1786-1793.

Linton, Jamie, 2010: *What is Water?: The History of a Modern Abstraction* (Vancouver: University of British Columbia Press).

Lövbrand, Eva; Stripple, Johannes; Wiman, Bo, 2009: "Earth System governmentality", in: *Global Environmental Change* 19: 7-13.

Luke, Timothy, 2010: "Geoengineering as global climate change policy", in: *Critical Policy Studies*, 4,2; 111-126.

Magnusson, W., 2011: *Politics of Urbanism: Seeing like a City* (London: Routledge).

Markard, Jochen; Raven, Rob; Truffer, Bernhard, 2012: "Sustainability transitions: An emerging field of research and its prospects", in: *Research Policy*, 41: 955-967.

Meadows, Donella H.; Meadows, Dennis L.; Randers, Jørgen; Behrens III, William W., 1972: *The Limits to Growth: A Report for the Club of Rome's Project on the Predicament of Mankind* (New York: Universe–Earth Island).

Meadows, Donella H.; Meadows, Dennis L.; Randers, Jørgen; Behrens III, William W., 1974: *The Limits to Growth* (London: Pan).

Newell, Peter; Paterson, Matthew, 2010: *Climate Capitalism* (Cambridge: Cambridge University Press).

Nixon, Rob, 2011: *Slow Violence and the Environmentalism of the Poor* (Cambridge, MA: Harvard University Press).

O'Loughlin, J.; Witmer, Frank D.W.; Linke, Andrew M.; Laing, Arlene; Gettleman, Andrew; Dudhia, Jimy, 2012: "Climate variability and conflict risk in East Africa 1990-2009", in: *Proceedings of the National Academy of Sciences*, 109,45. 18344-18349.

Palsson, Gisli; Szerszynski, Bronislaw; Sörlin, Sverker; Marks, John; Avril, Bernard; Crumley, Carole; Hackmann, Heide; Holm, Poul; Ingram, John; Kirman, Alan;

Buendía, Mercedes P.; Weehuizen, Rifka, 2013: "Reconceptualizing the 'Anthropos' in the Anthropocene: Integrating the social sciences and humanities in global environmental change research", in: *Environmental Science & Policy*, 28: 3-13.

Pisano, Umberto; Berger, Gerald, 2013: *Planetary Boundaries for Sustainable Development* (Vienna: European Sustainable Development Network Quarterly Report).

Potsdam Institute for Climate Impact Research and Climate Analytics, 2012: *Turn Down the Heat: Why a 4°C Warmer World Must be Avoided* (Washington, DC: The World Bank).

Raworth, Kate, 2012: *A safe and just space for humanity: can we live within the doughnut?* (Oxford: Oxfam).

Robertson, Melanie, 2012: *Sustainable Cities: Local Solutions in the Global South* (Ottawa: International Development Research Centre).

Rockström, Johan; Steffen, Will; Noone, Kevin; Persson, Åsa; Chapin, III, F. Stuart; Lambin, Eric; Lenton, Timothy M.; Scheffer, Marten; Folke, Carl; Schellnhuber, , Hans Joachim; Nykvist, Björn; De Wit, Cynthia A.; Hughes, Terry; Leeuw, Sander van der; Rodhe, Henning; Sörlin, Sverker; Snyder, Peter K.; Costanza, Robert; Svedin, Uno; Falkenmark, Malin; Karlberg, L.; Corell, Robert W.; Fabry, Victoria J.; Hansen, James; Walker, Brian; Liverman, Diana M.; Richardson, Katherine; Crutzen, Paul; Foley, Jonathan A., 2009a: "Planetary boundaries: exploring the safe operating space for humanity", in: *Ecology and Society*, 14,2,: 32; at: <http://www.ecologyandsociety.org/vol14/iss2/art32/>.

Rockström, Johan; Steffen, Will; Noone, Kevin; Persson, Åsa; Chapin, F. Stuart; Lambin, Eric F.; Lenton, Timothy M.; Scheffer, Marten; Folke, Carl; Schellnhuber, Hans Joachim; Nykvist, Björn; de Wit, Cynthia A.; Hughes, Terry; van der Leeuw, Sander; Rodhe, Henning; Sörlin, Sverker; Snyder, Peter K.; Costanza, Robert; Svedin, Uno; Falkenmark, Malin; Karlberg, Louise; Corell, Robert W.; Fabry, Victoria J.; Hansen, James; Walker, Brian; Liverman, Diana; Richardson, Katherine; Crutzen, Paul; Foley, Jonathan A., 2009b: "A safe operating space for humanity", in: *Nature*, 461,24 (September): 472-475.

Rockström, Johan; Kulm, Mattias 2015; *Big World Small Planet: Abundance within Planetary Boundaries* (Stockholm: Max Strom).

Ruddiman, William; Vavrus, Steve; Kutzbach John; He, Feng, 2014: "Does pre-industrial warming double the anthropogenic total?", in: *The Anthropocene Review*; at: <DOI: 10.1177/2053019614529263>.

Ruddiman, William F., 2005: *Plows, Plagues, and Petroleum: How Humans Took Control of Climate* (Princeton. NJ: Princeton University Press).

Sachs, Wolfgang; Santarius, Tilman, 2007: *Fair Future: Resource Conflicts, Security and Global Justice* (London: Zed).

Scheffran, Jürgen; Brzoska, Michael; Brauch, Hans Günter; Link, Peter Michael; Schilling, Janpeter (Eds.), 2012: Climate Change, Human Security and Violent Conflict: Challenges for Societal Stability (Heidelberg–Dordrecht–London–New York: Springer).

Schellnhuber, Hans Joachim; Crutzen, Paul J.; Clark, William C.; Claussen, Martin; Held, Hermann (Eds.), 2004: *Earth System Analysis for Sustainability* (Cambridge, MA; London: MIT Press).

Steffen, Will; Sanderson, Angelina; Tyson, Peter D.; Jäger, Jill; Matson, Pamela A.; Moore III, Berrien; Oldfield, Frank; Richardson, Katherine; Schellnhuber, Hans Joachim; Turner II, B.L.; Wasson, Robert J., 2004: *Global Change and the Earth System. A Planet under Pressure. The IGBP Series* (Berlin–Heidelberg–New York: Springer-Verlag).

Steffen, Will; Crutzen, Paul; McNeill, John R., 2007: "The Anthropocene: are humans now overwhelming the great forces of nature?", *Ambio*, 36,8: 614-21.

Steffen, Will; Grinevald, Jacques; Crutzen, Paul; McNeill, John, 2011: "The Anthropocene: conceptual and historical perspectives", in: *Philosophical Transactions of the Royal Society A*, 369: 842-867.

Steffen, Will; Persson, Åsa; Deutsch, Lisa; Zalasiewicz, Jan; Williams, Mark; Richardson, Katherine; Crumley, Carole; Crutzen, Paul; Folke, Carl; Gordon, L.; Molina, Mario; Ramanathan, Veerabhadran; Rockström, Johan; Scheffer, Marten; Schellnhuber, Hans Joachim; Svedin, Uno, 2011: "The Anthropocene: From global change to planetary stewardship", in: *Ambio*, 40: 739-761.

Steffen, Will; Richardson, Katherine; Rockström, Johan; Cornell, Sarah E.; Fetzer, Ingo; Bennett, Elena M.; Biggs, Reinette; Carpenter, Stephen R.; de Vries, Wim; de Wit, Cynthia A.; Folke, Carl; Gerten, Dieter; Heinke, Jens; Mace, Georgina M.; Persson, Linn M.; Ramanathan, Veerabhadran; Reyers, Belinda; Sörlin, Sverker; 2015: Planetary boundaries: Guiding human development on a changing planet *Science* 347(6223).

Theisen, Ole M.; Holtermann, Helge; Buhaug, Halvard, 2011: "Climate Wars?: Assessing the Claim that Drought Breeds Conflict", in: *International Security*, 36,3: 79-106.

Tsigaridis, K.; Krol, M.; Dentener, F. J.; Balkanski, Y.; Lathière, J; Metzger, S.; Hauglustaine, D. A.; Kanakidou, M., 2006: "Change in global aerosol composition since preindustrial times", in: *Atmospheric Chemistry and Physics*, 6: 5143-5162.

UNEP [United Nations Environment Programme], 2012: *Global Environmental Outlook 5: Environment for the future we want* (Nairobi: UNEP).

United Nations Secretary General, 2009: *Climate Change and its Possible Security Implications* (New York: United Nations, 11 September, A/64/350).

Vries, Wim de; Kros, Johannes; Kroeze, Carolien; Seitzinger, Sybil P., 2013: "Assessing planetary and regional nitrogen boundaries related to food security and adverse environmental impacts", in: *Current Opinion in Environmental Sustainability*, 5,3-4: 392-402.

Wang, Limao; Gu, Mengchen; Li, Hongqiang, 2012: "Influence path and effect of climate change on geopolitical pattern" *Journal of Geographical Sciences*, 22,6: 1117-1130.

Ward, Barbara; Dubos, Rene, 1972: *Only One Earth* (Harmondsworth: Penguin).

WCED [World Commission on Environment and Development], 1987: *Our Common Future* (Oxford: Oxford University Press).

WGBU [German Advisory Council on Global Change], 2011: *A World in Transition: A Social Contract for Sustainability* (Berlin: WGBU).

Zografos, Christos; Goulden, Marisa C.; Kallis, Giorgios, 2014: "Sources of human insecurity in the face of hydro-climatic change", in: *Global Environmental Change* 29: 327-336.

3   Paradigm and Praxis Shifts: Transitions to Sustainable Environmental and Sustainable Peace Praxis

Carolyn M. Stephenson[1]

Abstract

The diffusion of both paradigms of sustainable environmental practice and sustainable peace practice has quickened in the last thirty to forty years, but has occurred unevenly across time and space, across regions, and even within individual countries and subregions of countries. Some disciplines have been more hospitable to one or the other. In large part, until recently, environmental studies has not found peace issues relevant, nor peace studies environmental issues. The beginning of the coming together of these paradigms and their practice is a significant change. This chapter examines the evolution of the separate paradigms of sustainable environment and sustainable peace, and their gradual but as yet incomplete engagement with each other. It also examines texts at the level of global governance, particularly at the United Nations, with respect to the same issues, asking how and why UN and other documents and conceptualizations in the 1970s have increasingly begun to reflect the linkages between these issues.

Keywords: Environmental security, human security, international environmental studies, peace studies, spaceship earth, stable peace, sustainable development, sustainable peace.

3.1   Transitions in Paradigms and Praxis

While there has developed a somewhat technical literature in transitions, especially sustainability transitions, this paper relies on the ordinary everyday meaning of the term, in the belief that it may be useful to speak political and social science in a way that can be comprehended by ordinary people, especially if one is trying to change paradigms and praxis of the general population. Since much of the work on sustainability transitions has focused especially on environmental sustainability, this work will place more emphasis on the development of work on sustainable peace, and on the development of peace studies as a field, as well as on the interrelationships between peace studies and environmental studies.

In 1949 Aldo Leopold published, posthumously, *A Sand County Almanac* and, within it, "The Land Ethic". He argued that ethics has extended, in "a process of ecological evolution", from relations between individuals to the relation between the individual and society, and that this needs to expand to a societal ethic on "man's relation to land" (202-3). "All ethics so far evolved rest upon a single premise: that the individual is a member of a community of interdependent parts"(203). "The land ethic simply enlarges the boundaries of the community to include soils, waters, plants, and animals, or collectively: the land" (204). Leopold was both a game manager and an advisor on conservation to the United Nations. This extension of ethics can be observed in the fields of both peace studies and environmental studies, but it is not yet complete in either. The linkage of paradigm to praxis was evident in Leopold's life and work and there are attempts, in both peace and environmental studies, to make such linkages and to enlarge the size of the community. Yet there is no real epistemic community that includes scholars in both peace studies and environmental studies, even within the social sciences. There are primarily different journals, different organizations or sections of organizations, different schools. Linkage of concerns over international

---

1   Prof. Dr. Carolyn M. Stephenson, Associate Professor of Political Science at the University of Hawaii at Manoa and affiliate faculty member in its Matsunaga Institute for Peace; Email: <cstephen@hawaii.edu>.

© Springer International Publishing Switzerland 2016
H.G. Brauch et al. (eds.), *Handbook on Sustainability Transition and Sustainable Peace*,
Hexagon Series on Human and Environmental Security and Peace 10, DOI 10.1007/978-3-319-43884-9_3

peace and security with concerns over sustainable environment and development is an important step in linkages between paradigms and praxis in both these fields. The development of a paradigm inclusive of the concerns of both fields could well improve praxis as well.

This chapter examines the early historical development of the fields of international environmental studies and peace studies, including both conceptual and institutional changes.[2] It looks first at the academic fields of international environmental studies and the various conceptual waves of peace studies, and then the beginnings of convergence of the fields, especially as exemplified by Kenneth Boulding. It then looks at global governance contributions to sustainable environment and sustainable peace, incorporating in particular work on the Culture of Peace, the Nobel Peace Prize, the evolution of concepts of security, and the Millennium and Sustainable Development Goals, concluding that the convergence has progressed but remains incomplete.

3.2   Sustainable Environment and Environmental Studies

Environmental studies as a field of social science research really had its beginnings in the 1960s, and began to emerge in earnest in the 1970s. Early emphases included simply recognition of the physical and

---

2   *Editor's comment*: There is a rich literature that covers the more recent scientific discussion on environmental studies and peace research and their emerging linkages, including environmental peacemaking (Conca/Dabelko 2002) and peace ecology (Amster 2014; Oswald Spring/Brauch/Tidball 2014; Brauch/Oswald Spring/Bennett et al. 2016); see e.g. the previous reviews of the state of the scientific debate in previous volumes of this Hexagon Book Series on *Human and Environmental Security and Peace* (HESP) that may be easily accessed at: <http://www.afes-press-books.de/html/hexagon.htm>; see also the three volumes of the *Ecology and Peace Commission* (EPC) of IPRA in the *SpringerBriefs on Environment, Security, Development and Peace* (ESDP), at: <http://www.afes-press-books.de/html/SpringerBriefs_ESDP.htm>. Older texts by pioneers of peace and ecology studies are also included in the *SpringerBriefs on Pioneers in Science and Practice* (PSP); at: <http://www.afes-press-books.de/html/SpringerBriefs_PSP.htm>. The websites on the books in these three book series offer references and links to the state of the global scientific discussion on these linkages. See also many recent references in chapters 1-2, 6-9 and in the concluding chapter of this book.

biological environment as part of the environment of international politics, as well as concerns over the harm being done by humans to the environment, the dangers of increases in world population, notions of resource scarcity and the limits to growth, and the idea of the environment as a commons. Harold and Margaret Sprout brought considerations of the environment into the study of international relations in their 1962 textbook *Foundations of International Politics*. That their work was regarded as pivotal by the international studies community was confirmed by the naming of the International Studies Association's Environmental Studies Section annual book award in their name. The year 1962 also saw the publication of Rachel Carson's *Silent Spring*, which warned of the dangers of pesticide pollution, and awoke an environmental movement against DDT.

Garrett Hardin's essay on the "Tragedy of the Commons", in the December 1968 issue of *Science*, was a part both of the emerging literature on the dangers of world population growth to the environment and the emerging approach of looking at the world as a commons. Paul and Anne Ehrlich's works on population stressed and popularized the dangers of population growth, characterizing it as *The Population Bomb* (Ehrlich 1968; Ehrlich/Ehrlich 1970). While Ehrlich's use of the word "bomb" was clearly related to the dangers of nuclear proliferation that were beginning to be recognized in that period leading up to the development of the 1970 Nuclear Non-Proliferation Treaty, his analogy to the enormity of the nuclear bomb did not carry through to an analysis of the linkages between population and war. Hardin's approach to the "tragedy of the commons" was controversial, as he advocated "mutual coercion mutually agreed upon" to curb population growth. But his reconceptualization of the commons as an area of joint decision-making as related to population and the environment led to a burgeoning literature on common property resource regimes, much of it in disagreement with his original thesis. Elinor Ostrom's 1990 work *Governing the Commons* argued against Hardin's logic of the necessity of coercion and for the possibility of cooperative action in governing fisheries and other common property resource regimes. Her work was awarded the Nobel Prize in Economic Science in 2009.

The series of books that began with the Club of Rome reports on the limits to growth also included population as a limiting factor (see chapter by Slovaag Marino in this volume), but added into their MIT-based computer modelling the factors of resources, food output per capita, industrial output per capita, and pollution, stressing the interlinkages between these factors and subcategories of each of the factors.

Donella and Dennis Meadows, Jorgen Randers and William Behrens's 1972 *The Limits to Growth* models stressed the dangers that would be likely to result from the interactions of these factors. War, however, was not one of the factors considered. While there was praise for the systematic global modelling in the book, there was scepticism about the conclusions. The book, however, did inform the analysis at the UN's 1972 Stockholm Conference on the Human Environment, which helped to put the issue of environment on the global agenda. Mesarovic and Pestel's 1974 *Mankind at the Turning Point,* the second report to the Club of Rome, carried further the analysis on limits to growth by disaggregating the world into ten regions with very different streams of development. When Meadows and Randers in 1992, at the time of the Rio "Earth Summit", the UN Conference on Environment and Development, updated their findings in the new book, *Beyond the Limits,* they were able to answer their critics with a resounding clarity that in fact, things were in fact worse than they had originally stated (see chapter 16 by Marino in this vol.).

Kenneth Boulding, economist and peace researcher, was also an early pioneer in thinking about global environmental limits. In his 1966 article "The Economics of the Coming Spaceship Earth", he argued that we needed to proceed from the paradigm of unlimited resources in a "cowboy economy" to a paradigm recognizing that the earth has more the limits of a spaceship. In his 1968 book *Ecodynamics,* he advocated a systems approach to the environment. Among the limits Boulding saw, however, were not only population but also rising inequality.

Several other books were significant in advancing the focus on resource scarcity, William Ophuls, in his 1977 *Ecology and the Politics of Scarcity: Prologue to a Theory of the Steady-State,* argued that "man and nature" were "on a collision course", and that industrial man was rapidly exceeding the earth's carrying capacity. He argued that our thinking needed to be changed toward a theory of the steady-state economy either by teaching or by coercion, i.e. the development of law. Ophuls also argued for the necessity of a new paradigm. This emphasis on changing the patterns of consumption was later followed up by the work of Princen (2005).

Another influential work was E. F. Schumacher's *Small is Beautiful: Economics as if People Mattered,* first published at the time of the oil/energy crisis in 1973. Schumacher argued that the dehumanization of the economic system, as it grew larger and larger, would harm the provision of human needs, and that the furtherance of economic growth was not the be-all and end-all of politics. In the face of the expansion of what would become the *European Union* (EU), he argued that local production might put economic control back into the hands of people. He argued the importance, not so much of the tenet "small is beautiful", but of the need for things to be done at the level appropriate to that activity, some small, some medium, some large. His emphasis on subsidiarity, that things should be done at the lowest level at which they could, ultimately became a principle that was adopted—albeit not quite in the same sense as Schumacher conceptualized it—by the EU, where it is currently interpreted in different ways by different countries, with the UK holding that this means that the EU itself should not take on what states can do, and Germany going further, holding that this principle goes even further down, and that sometimes this means that sub-units of states should hold primary responsibility.

If one looks at the development of education in environmental studies, one can go back to the development of what were originally schools of forestry. The New York State College of Forestry was founded at Cornell in 1898, the Yale Forest School in 1900, and the SUNY College of Environmental Science and Forestry at Syracuse in 1911. But the original emphasis did not fit so much with Leopold's land ethic; this was forestry for human consumption, not forestry for the sake of forests. It was only with the rise of environmental concerns in the early 1970s that educational institutions focusing on the broader issues of environment were formed. Among the most significant were the School of Public and International Affairs at Indiana, under the influence of Lynton Caldwell, and the renamed Yale School of Forestry and Environmental Studies, both created in 1972.

Also very early on, many of the international organizations which remain significant in environmental issues were formed, before many of the peace organizations we will examine below. The *International Council of Scientific Unions* (ICSU), now the International Council for Science, was created in 1931. The *International Union for the Conservation of Nature* (IUCN) was formed in 1948. The Environmental Studies Group was formed within the International Studies Association (ISA) in 1974, but there was little overlap between either the initial or the present membership of that and the Peace Studies Section of ISA.

Intellectually, one could go back even further to the work of Thomas Malthus, whose 1798 *Essay on the Principle of Population* began the emphasis on the problem of population growth that was later continued by Hardin, Ehrlich and others mentioned above. Malthus argued that there are different ratios for the increase of population and food, with population growing in a geometric ratio and food an arithmetic ratio, that thus population will exceed the food supply, also that the checks on human population, primarily the food supply and famine, epidemic disease, and war, do not have much effect on the size of populations. Interestingly enough, he does raise some of the questions of inequality, including the inequality of women and men, that later came to concern peace studies and to be significant in understanding the relationships between gender, population, development and peace. Again, however, Malthus was concerned with the social system, not with the environment for environment's sake.

3.3 Sustainable Peace and Peace Studies

The origins of peace research could be judged to be as far back as Thomas Hobbes's *Leviathan* (1651) or even Plato or Thucydides. Hobbes's characterization of war as originating from the combination of the greed and equality of human beings, with the solution being a social contract dependent on the creation of authoritative institutions, has been a thread in both realist and idealist strands of peace research ever since. The Lockean claim of the right of revolution against the tyranny of authoritarian institutions has been a second important strand. Yet the origins of peace research as a separate field of inquiry are probably better traced only within the twentieth century. Elsewhere this author (Stephenson 2008) has argued that there have been three primary waves of peace studies worldwide since its beginnings between the world wars, a) a first wave consisting of research on the causes of war and conditions of peace; b) a second wave on peace education and critical peace research; and c) a third wave dealing with the institutionalization of peace research and education, with those waves intermingling with each other and with waves from other related fields in the period after the end of the cold war and 9/11.[3]

In what might be called the first wave of peace studies, which ran roughly 1945 to the late 1960s, the quantitative study of war was the primary focus. The work of US political scientist Quincy Wright and UK (Quaker) physicist Lewis Richardson looked at the causes of war over the long term, with Richardson in particular finding that homogeneity among states did not preclude arms races and war. Together with the socio-cultural theories of Pitrim Sorokin, these works revolutionized the emerging field of international relations. Peace research institutes were first established in this period, among them the *Institut Français de Polémologie* (1945), the Center for Conflict Research and David Singer's *Correlates of War Project* at the University of Michigan (1957), the *Peace Research Institute Oslo* (PRIO, originally 1957, independent in 1966), and the *Stockholm International Peace Research Institute* (SIPRI, 1966). Significant journals included the *Journal of Conflict Resolution* (1957) and the *Journal of Peace Research* (1964). Organizations included the 1959 *Lancaster Peace Research Centre/ Richardson Institute*, the 1963 *Conference* (now *Council*) on *Peace Research in History*, and the 1964 *International Peace Research Association* (IPRA).

The second wave of peace studies, roughly 1970 to the late 1980s, focused primarily on the development of peace education in the US and on critical peace research in Europe. The first undergraduate peace studies programme in the US, at Manchester College, established in 1948, was joined by many others, including the first chair in peace studies at Colgate in 1971, along with the first chair in the UK at Bradford in 1973. Uppsala (Sweden) began the *Department of Peace and Conflict Research* in 1971. Other research institutes were founded: the *Tampere Peace Research Institute* (TAPRI, in Finland) in 1970, and the Peace Research Institute Frankfort, also 1970, as the first of the German peace research institutes. In 1970 IPRA added its *Peace Education Commission* (PEC). Regional organizations were formed: the *Canadian Peace Research and Education Association* (CPREA) in 1966, the *Consortium on Peace Research, Education and Development* (COPRED) in the US in 1970 (which became the Peace and Justice Studies Association in 2001), and the *Asian Peace Research Association* (later the *Asia-Pacific Peace Research Association*, or APPRA) in 1974; the First *Latin American Conference on International Relations and Peace*

---

3 A fuller version of this argument can be found in Stephenson (2008: 1534–1548). This chapter is available for download at: <http://www.sciencedirect.com/science/article/pii/B9780123739858001306>. For a conceptual overview of the evolution of peace studies, see also Stephenson (2010: 5579–5603).

Research was held in 1994 (CLAIP). Surprisingly, Europe, although a center of peace research, did not form a regional organization until the late 1980s. The *European Peace Research Association* (EUPRA) began in 1988, and held its first conference in 1991. The *African Peace Research and Education Association* began in 2000. Critical peace research began especially at the Peace Research Institute Oslo, where Johan Galtung was then located, and where he put forth the ideas that peace was more than the absence of war and direct violence, and included the presence of social justice, or the absence of structural violence (1969, 1971). He termed these two kinds of peace 'negative' and 'positive peace', terms that had been used earlier by Martin Luther King in much the same way. Galtung's conceptualization of peace revolutionalized the field. It was criticized especially by Kenneth Boulding (1977), who argued that while justice was essential for peace, conceptual clarity required that one not be defined in terms of the other.

The third wave of peace studies, roughly during the 1980s, saw considerable institutionalization of peace studies at the national and international levels, especially in government funded institutes. These included: the *University for Peace* (UPEACE, whose establishment was approved by UN General Assembly Resolution 35/55 of 5 December 1980) in Costa Rica, the *Canadian Institute for Peace and International Security* (1984-1992), the *Peace Research Centre at Australian National University* (1984-1994), the *US Institute of Peace* (1984), the *Copenhagen Peace Research Institute* (1985-2005), and the *Scientific Research Council on Peace and Development* in the USSR (1979). New journals included *Negotiation Journal* (1985), *Peace Review* (1989), and *Security Dialogue* (1992, renamed from the earlier *Bulletin of Peace Proposals*). While the instigation for much of this was expanded peace activism, especially the anti-nuclear movements that developed in the early 1980s, some of the field narrowed a bit to focus on nuclear war and international security.

At the same time, peace studies also began to focus not only on the causes of war and conditions of peace, as well as on structural violence, but also on peaceful processes that might serve as alternatives to war. Two important fields that could be seen either as subfields of peace studies or as intersecting with it were conflict resolution and nonviolent action. Conflict resolution practice emerged primarily in the domestic and local arena, but was later reinvigorated by the work on peacemaking, peacekeeping and peacebuilding that came with the UNs *Agenda for Peace* in 1992. The Hewlett Foundation in the US was instrumental in advancing research on both the theory and practice of conflict resolution, first funding the *Program on Negotiation* at Harvard (1983) and then many other universities from 1984 to 2004. Fisher and Ury's book *Getting to Yes* and the many sequels established principled negotiation as a key approach to conflict resolution. The Institute for *Conflict Analysis and Resolution* at George Mason University established the first M.S. in conflict resolution in 1982, later expanding to a Ph.D. as well, while the *Kroc Institute* at Notre Dame, begun in 1986, focused its M.A. in peace studies more broadly. The study of nonviolent action, which had received its initial energy from Gene Sharp's landmark three-volume *Politics of Nonviolent Action* (1973) and his *Program on Nonviolent Sanctions in Conflict and Defense* at Harvard (to 2005), received new energy from the creation of the *Einstein Institution* in 1983. The field of nonviolence studies continued to expand with the PBS TV series and book by Ackerman and DuVall on *A Force More Powerful* (2001) and their foundation of the *International Center on Nonviolent Conflict* (ICNC) in 2002. Most recently, the award-winning *Why Civil Resistance Works* (2011) by Chenoweth and Stephan has further deepened research in this area, as has the Roberts and Ash book *Civil Resistance and Power Politics* (2011), the result of an important conference at Oxford that linked theory and case studies in many regions of the world. However, although there were social movements using nonviolent action to deal with environmental conflicts, such as the Chipko movement in India, the anti-dam movement, and the rubber tappers in Brazil, the use of nonviolent action for environmental goals was not highlighted by most researchers focusing on nonviolent action. Other peace researchers focused on integrating these various sets of methods in the hope of leading toward a system that would provide alternatives to war. One of the first conferences attempting to integrate alternative methods for international security was COPRED's 1979 annual meeting, which resulted in an edited book (Stephenson, 1982).

What might either be characterized as a fourth wave of peace studies, or as a spreading out and commingling of earlier waves, occurred in the aftermath of the bombings of the World Trade Center in New York City and the Pentagon in Washington DC on 11 September 2001. The consolidation of US power that had occurred especially in the decade after the end of the cold war had challenged all of the fields concerned with the international system.

While many who study or comment on international relations argue that the world changed after both the end of the Cold War and 9/11, most of those in the field of peace studies see the same fundamental issues and structures continuing to challenge us. Ethical issues around peace, power, violence and justice came to the forefront of the field again. The democratic peace hypothesis that began with Kant's work received new attention. International law and organization were studied both as conservative structures which promote and maintain the interests of the powerful, and as liberating structures which promote human rights, justice, peace and a clean environment. The role of sanctions and of other various forms of intervention in the prevention of genocide, war crimes and crimes against humanity and the promotion of human rights were debated and researched. Additional attention went to the role of smart, focused sanctions, and to the role of post-conflict transitional justice mechanisms, with debates over the relative effectiveness of punishment mechanisms like the International Criminal Court versus reconciliation mechanisms like South Africa's Truth and Reconciliation Commission. The reconceptualization of security begun by feminists and alternative security researchers in the 1970s, and taken up by more conventional political elites during the 1980s, expanded to bring the concept of human security into centre focus following UNDP's *Human Development Report* (HDR 1994), the *Report of the Commission on Human Security* (CHS 2003) and the *Human Security Report* (2005).

A development strain has always linked peace studies to international relations. With the development of peacebuilding as a major subfield in conflict resolution, examination of the linkages between war and violent conflict, development, justice issues, and various forms of intervention has become a central focus of study linking all three fields, as well as linking them to security studies. Two new peacebuilding journals, the *Journal of Peacebuilding and Development* in 2002 and *Peacebuilding* in 2013, emerged. This has long been a focus at Bradford, perhaps the premier peace studies programme. In contrast, however, the linkages between environment, development, and conflict, while studied intensively in certain parts of the field of peace studies, for example at PRIO, have not been central to the field.

3.4 Beginnings of Convergence in Sustainable Environment and Sustainable Peace

Choucri and North (1972) argued that the emphasis on the study of crises in international relations needed to be supplemented with the study of the dynamics of competition and conflict in the longer run. This included the study of the impact of population growth, technology and resources. They argued that it was important for a policymaker to know whether population growth or growth in technology was more likely to influence conflict outcomes, but that since such variables are not very much manipulable in the short term, policymakers tend to ignore them. Their basic proposition was that "differential rates of population growth in combination with differential rates of technological growth contribute to international competition and sometimes to conflict…" (Choucri/North 1972: 83) They argued that:

> the demands and specialized capabilities of a society combine multiplicatively to produce what might be called lateral pressure. This amounts to a tendency to undertake activities farther and farther from the original boundaries of the society, to acquire some degree of influence or control over a wider extent of space or among a larger number of people (Choucri/North 1972: 90).

Their research focused especially on the major powers during the period 1870-1914. While this research did not deal with the issue of environment as we conceive of it today, the concern with population growth, harkening back to Malthus's hypothesis, was a predecessor to later work on environment, conflict and security. While Choucri and North's work was seen as significant in international studies, it was never really picked up in peace studies.

Attention began to be given to environmental issues within the framework of peace research and peace education in the late 1980s and early 1990s. The first issue of the Five College Program on Peace and World Security's *PAWSS Perspectives on Teaching Peace and World Security Studies* focused on "Peace and the Environment: Making the Connections in the Classroom" (1990). Drawing on the expertise of those in environmental studies, the issue argued for integrating environmental concerns into peace and international relations courses. Ake Bjerstedt of the University of Malmö, Sweden, School of Education, who served as Executive Secretary of the *International Peace Research Association's Peace Education Commission* (IPRA/PEC), edited "Peace, Environment and Education", which served as the newsletter for PEC during the period beginning in 1990. While environment was a concern of some, it was not an overriding concern of the field. It was not integrated widely into peace studies courses at that time, and a quick perusal of the issues of *Peace, Environment and Education* shows only a few issues with a focus on environment.

The argument was put forward in the 1980s that it was important to integrate environmental concerns into the concept of security. With the publication of the *World Commission on Environment and Development*'s (WCED 1987) *Our Common Future*, the concepts of both sustainable development and environmental security came into currency in both policy and academic circles. Many had by then called for the redefinition of security to include environmental security. Among the earliest was Lester Brown (1977). Others included Richard Ullman (1983) and Arthur Westing (1986). Jessica Matthews (1989) also argued that security needed to be redefined to include environmental security. Sverre Lodgaard (1990) stated that the concept of environmental security "conveys a message that environmental problems have a legitimate claim for status as military problems have".

Critical environmental studies scholars found problems with the concept of environmental security, arguing that a focus on security marginalizes concerns over environmental justice. Simply adding in environmental problems as a security threat is incompatible with the basic requirements of solving environmental problems. Some scholars, such as Barry Buzan, Ole Wæver, and Jaap de Wilde (1998) of the Copenhagen School, have argued that "securitization" of the environment would lead to inappropriate policy choices. Dan Deudney (1990) also critiqued the concept of environmental security, arguing that the security discourse is a discourse of militarism incompatible with an environmental ethic. The structures developed for national security rely on violence, which is of little use in dealing with environmental problems. Dalby has provided a comprehensive overview of the problems of environmental security conceptualization and policy (2002). Later work began to focus not only on the potential for conflict, but also on that of cooperative responses to scarcity (Ronnefeldt 1997).

Thomas Homer-Dixon (1994), then Chair of the Peace and Conflict Studies Program at the University of Toronto, and researchers in the Project on Environmental Change and Acute Conflict began systematic case study research on the relationships between environmental scarcity, population growth, depletion and

degradation of resources, unequal access to resources and violent conflict. Finding relationships in the developing world, and hypothesizing that scarcity would become more acute, he predicted that states' inability to deal with environmental problems would lead them to fragment or become more authoritarian. Robert Kaplan (1994) in *The Atlantic Monthly*, drawing on Homer-Dixon's research and arguing against Huntington's "clash of civilizations" thesis, cited evidence that scarcity, crime, overpopulation, tribalism, and disease were destroying our social fabric and bringing about anarchy, especially in West Africa, but in other regions as well.

There continues to be a lack of agreement on the conceptualization of environmental security. Most treat environmental security as meaning security for humans, or states, from environmental problems; only a few base their conceptualization on security for the environment. This has been especially true as environmental security has been incorporated into national defence thinking; for example, as the US Department of Defense established the Deputy Under Secretary of Defense for Environmental Security in 1993. If one returns to Aldo Leopold's argument in "The Land Ethic" about the expansion of the community as a tenet of ecological evolution, perhaps a solution to the debate over the utility of the concept of environmental security might be its expansion to include security both for the environment itself and for all people in the international system, thus overcoming the narrowness of the traditional concept of national security.

The *Peace Research Institute Oslo* (PRIO) was one of the first to embark on research into the possible relationships between environment, especially resource scarcity and degradation, and conflict. Early research concluded that there was not good evidence of any past relationship between water and armed conflict. However, Peter Gleick of the Pacific Institute made important arguments on the relationships between fresh water and conflict (1993). He also provided a comprehensive look at freshwater resources and a database on water and conflict (2009). Nils Petter Gleditsch (1998), in a critique of the literature on armed conflict and the environment, citing scarce resources as a source of armed struggle, and evaluating the argument that environmental degradation would increase scarcity and therefore contribute to an increase in armed struggle, concluded that there had been little systematic study on these issues. He identified a number of problems in the literature, among them lack of conceptual clarity, neglect of significant variables, and questions about the direction of causality.

PRIO has continued its work on the examination of the relationships between environment and conflict. Gleditsch introduced a special issue of the *Journal of Peace Research* (2012) on climate change and conflict. Among the works included were articles by Bergholt and Lujala on climate-related natural disasters, economic growth, and armed civil conflict, Bernauer and Siegfried on climate change and international water conflict, Slettebak on climate-related natural disasters and civil conflict, and a number of articles on specific conflicts in various regions. To its credit, the issue included conflicting viewpoints, and also focused on the role of institutions in mitigating or resolving such conflicts. Others have also taken up the issues of the relationships between climate change, water and conflict, including Scheffran, Brzoska, Kominek and Link (2012) and Carter and Veale (2015).

Another significant area of convergence was the development of environmental peacemaking and environmental peacebuilding. Significant academic work began in the 1980s, with Larry Susskind's 1980 *Breaking the Impasse* focusing on the use of dispute resolution in public disputes. Bacon and Wheeler published their *Environmental Dispute Resolution* text in 1984. Gail Bingham reviewed a decade of experience with environmental dispute resolution in 1986. Drawing on the beginnings of this work, and on Conca and Dabelko's (2002) environmental peacemaking work, which made the case for environmental peacemaking, including case studies in different world regions, both theory and practice further developed after the turn of the century. Both UNEP and UNDP developed programmes focusing on environmental peacebuilding in war-torn or conflict regions. Carius published Environmental Peacebuilding in 2006 and Conca and Wallace an article on UNEP's work in 2009.[4]

3.5   Kenneth Boulding—The Exception

One of the very few exceptions to the lack of attention given to the environment in peace studies and to peace and security in environmental studies was Kenneth Boulding, an economist who is generally regarded as one of the founders of the field of peace

studies. Boulding (1910-1993) was both economist and ecologist.

Among the books for which Kenneth Boulding was most widely recognized were his books *The Image* (1956) and the *Conflict and Defense* (1962). *The Image* represented Boulding's attempt to show a single thread running through the diverse academic disciplines "to produce something like a general theory" (Boulding 1956: 139). In *Conflict and Defense* (1962), rooted in game theory, Boulding puts forward a general theory of conflict ranging from the individual, through economic and industrial-organizational, to the international level, focusing especially on the dangers of international conflict, and concluding with approaches to conflict resolution and control.

It has been argued that Boulding actually laid the foundations of ecological economics in his book *A Reconstruction of Economics* (1950). Robert Scott (2015) argues in his new biography of Boulding that "Many people (including me) think *Reconstruction* was Boulding's most underappreciated work" (Scott 2015: 66). Boulding (1966) was far ahead of either economics or peace studies when in 1965 he first spoke of earth as a spaceship, characterizing in his classic article "The economics of the coming Spaceship Earth" the "cowboy economy" as a more open system and the "spaceman economy" as the more closed system which would need to be the economy of the future. In *Ecodynamics*, employing an evolutionary perspective, and seeing the world and universe as a "total system of interacting parts" (Boulding 1978: 31), he argued for a "more accurate image of the future" (Boulding 1978: 373). Among the limits to growth he saw the rise of world inequality and the necessity for population control. Several chapters also foreshadow his threefold conceptualization of power.

In his *Stable Peace* (1978a), Boulding spoke of the continuum of war and peace, rather than seeing them as diametrically opposed. He argued that we can move back and forth between stable war, unstable war, unstable peace, and stable peace, with the goal being to move to more stable peace. But Boulding also elaborated on a more complex conceptualization of power itself in his *Three Faces of Power* (1989). He argued not only for the carrot-and-stick dichotomy, but also argued that what he termed social power was an equally prevalent mode of power in human life, including in international relations. Boulding argued that political-military power was primarily destructive power, based on threat, while economic power was primarily productive, based on exchange, and social power was primarily integrative, based on the power

---

4   The website <www.environmentalpeacebuilding.org> provides other useful links to much of this work.

of love. He argued both that society, including international society, is primarily reliant on social or integrative power, and that we would all be better off if we continued to move society further in that direction by reliance on less coercive forms of power.

Boulding's work on the development of general systems theory and his later work on 'human betterment' framed an understanding of both the dynamics of power and peace and of ecodynamics. Yet his work on peace was read primarily by peace studies scholars, and his work on environment by environmental studies scholars. Very few made the linkages he made. The work of this book begins to follow that lineage.

3.6   Global Governance, Sustainable Environment and Sustainable Peace

In some ways, it can be argued that the institutions of global governance have done more to incorporate the areas of sustainable environment and sustainable peace than have the academic fields of environmental studies and peace studies. In a number of multilateral and other forums, conceptualization of peace and security, for whatever reasons, began to link environment and development issues to traditional peace issues. Among these were the Culture of Peace programme in UNESCO, the Nobel Peace Prize, the series of commissions on changing concepts of security, and the Millennium Development Goals, which by 2015 morphed into the Sustainable Development Goals. The final documents from the series of UN conferences that began with the 1972 UN Conference on the Human Environment in Stockholm, the food and population conferences in Rome and Bucharest in 1974, the International Women's Year Conference in Mexico City in 1975, and carrying through to their follow-on conferences in 1992, 1994, and 1995, also began to reflect the linkages between war, peace, food, population, gender, and the environment.

3.6.1   Conceptualizing Peace: the Culture of Peace

Peace studies began to examine and incorporate the concept of a "culture of peace", although this was done primarily within UNESCO and the United Nations rather than in academia. The Seville Statement on Violence, drafted by natural and social scientists during the UN International Year for Peace in 1986, foreshadowed the work that was begun at UNE-

SCO in the early 1990s. UN work on the culture of peace was furthered by UN General Assembly resolution 52/15 of 20 November 1997, by which it proclaimed the year 2000 as the "International Year for the Culture of Peace", and its resolution 53/25 of 10 November 1998, by which it proclaimed the period 2001–2010 as the "International Decade for a Culture of Peace and Non-Violence for the Children of the World". The General Assembly adopted the Declaration on a Culture of Peace on 13 September 1999, and the Programme of Action on a Culture of Peace in General Assembly resolution 53/243 of 6 October 1999. The Programme defined eight domains of action:

- a culture of peace through education
- sustainable economic and social development
- respect for all human rights
- equality between women and men
- democratic participation
- understanding, tolerance and solidarity
- participatory communication and the free flow of knowledge
- international peace and security.

All of these areas of work have been incorporated into peace studies, and many into more conventional work in international relations as well. What is significant about the Culture of Peace framework, however, is that it highlights the interrelationships between each of the domains of a culture of peace. One of the links between paradigms of sustainable development and sustainable peace is found in the section on Sustainable Economic and Social Development, which links a culture of peace to both environmental sustainability and gender equality.

There appears to be a reciprocal relationship between sustainable economic and social development and gender equality. This domain of the Culture of Peace combines the importance of traditional economic development, as well as social or human development, and concern with the protection of the environment. The environment vs. development debate that came about with the first UN Conference on the Human Environment in Stockholm in 1972, over whether the South had the right to destroy their environment in the course of development, as the North had done, resulted in the Brundtland Commission's 1987 popularization of the term "sustainable development". Sustainable development, defined by the Commission as "development that meets the needs of the present while not compromising the rights of future generations", incorporated the view that, if development did not pay attention to environmental protection, it would not be sustainable (WCED 1987). We will examine both the development side and the environment side of this conceptualization.

Many of the measures of economic as well as social and human development are highly correlated with gender equality. Women's basic literacy improves the chances of child immunization and reduces under-five child mortality. Two measures of gender empowerment, female education, and lifetime exposure to employment, appear to increase the probability of contraceptive use, reducing family size, which in turn seems to improve child survival rates, especially for girl children. While the relationships are complex, improvements in gender equality appear to lead to improvements in health and well-being both at the level of the family and that of the society (Kabeer 1999). Research also suggests that in societies where more than thirty per cent of the legislators are female, there is a greater emphasis on social and human development concerns such as health and education. Gender equality also appears to be positively related to many of the Millennium Development Goals adopted by the UN in 2000, including especially the reduction of infant and maternal mortality.

Second, there are also linkages between women and the environment. Women, in many societies, are the drawers of water and the hewers of wood; their approaches to consumption are key in determining environmental sustainability. There are clear links between gender and land distribution, biodiversity, and water management (UNEP 2004). The Women's Environment and Development Organization, today one of the primary women's NGOs, was founded in 1991 to link and to show the linkages between gender and environmental concerns. The Green Belt Movement was founded in 1977 by Kenyan Wangari Maathai, who received the 2004 Nobel Peace Prize for her work on the intersection of the issues of democracy, environmental protection and restoration, human rights, and especially women's rights.

3.6.2 Conceptualizing Peace: The Nobel Peace Prize

While the Nobel Peace Prize is not strictly peace research, the diversity of its awards can also be seen as a fertile source of theorizing on the relationships of the conditions of peace. One of the ways in which the general public becomes aware of the conceptualization of peace is the annual awarding of the Prize. Within peace research, there has been a small strain of solid historical research on the Prize. Alfred Nobel, a

Swede and the inventor of dynamite, created the Nobel Prizes to be awarded "to those who, during the preceding year, shall have conferred the greatest benefit on mankind", with the award divided into five equal parts, including one part to the person who shall have done: "the most or the best work for fraternity between the nations, for the abolition or reduction of standing armies and for the holding and promotion of peace congresses".[5]

In 1895, shortly before his death, while Norway and Sweden were under common governance (until 1905), and Norwegian liberals were working for self-governance, he decreed that the prize for champions of peace was to be awarded by a committee of five persons to be elected by the Norwegian Storting. Since the first Nobel Peace Prize was awarded in 1901, divided between Jean Henri Dunant, founder of the International Committee of the Red Cross, and Frederic Passy, founder of the first French peace society, Nobel Peace Prizes have been awarded to persons and institutions who represent a wide variety of conceptualizations of peace.

Over the course of the 1997-2007 decade the Nobel Peace Prize broadened further, being awarded to both governmental and non-governmental individuals and institutions; and in 2004 and 2007 environmental work was added, when the Nobel Peace Prize was awarded to Wangari Maathai (2004) and to the *Intergovernmental Panel on Climate Change* (IPCC) and Albert Arnold (Al) Gore, Jr. (2007).[6]

The Prize's conception of peace has always been broad. Within that decade alone, it included awards for efforts to ban weapons, find a peaceful solution to the Northern Ireland conflict, provide humanitarian assistance, promote democracy and human rights, promote reconciliation in divided societies such as South and North Korea, work for a more organized and peaceful world, find peaceful solutions to conflicts, work for democracy and human rights (especially women's rights), contribute to sustainable development, democracy and peace, prevent nuclear energy from being used for military purposes, create economic and social development through microcredit, and build up and disseminate knowledge about man-made climate change. The awarding of the Peace Prize to Wangari Maathai for her development of the

---

5   See: "Alfred Nobel's Will"; at: <nobelpeaceprize.org>.
6   See for an overview of all Nobel Peace Prizes at: <http://www.nobelprize.org/nobel_prizes/peace/>. A list of all Nobel Peace Prize recipients is at: <http://www.nobel-prize.org/nobel_prizes/peace/laureates/index.html>

Green Belt Movement in Kenya and her advancement of environment, women, and democracy was a significant broadening of the notion of peace.

The October 2007 announcement of the award of the Prize to the United Nations' *Intergovernmental Panel on Climate Change* (IPCC) for its work on the science of global warming and to Al Gore for his publicizing of the issue, particularly through the documentary "An Inconvenient Truth", was contested immediately from various political positions. The most common criticism of the 2007 Nobel award appears to have been that environmental issues do not fall within the province of an award for peace. Certainly environmental issues were not specified in Nobel's will, but neither were they recognized as important on their own at the turn of the century. Today it is clear that there are many important linkages between environmental problems and peace, as there are between peace and justice. The Nobel citation said that the IPCC and Gore had emphasized "the processes and decisions that appear to be necessary to protect the world's future climate, and thereby reduce the future threat to the security of mankind".

In the peace research community, another criticism was that the Prize has often tended to reward government-related rather than grassroots work. This presages the question of peace research's independence from or integration with government, and whether it should serve the needs of government or civil society and, as a derivative argument, whether it should or should not accept support from government. This relates as well to the arguments in international relations as to who are to be seen as the actors in international relations, only states, as realists argue, or also civil society actors including non-governmental organizations, as liberal institutionalists have argued in putting forward the concepts of transnational relations and complex interdependence (Keohane/Nye 2001). These debates, in addition to the positive peace/negative peace debates, have been central in framing the development of peace research and peace studies. Similar debates have arisen over the concept of environmental security, a part of the evolving debate both in peace studies and even more so in environmental studies.

3.6.3   Evolving Concepts of Security

The traditional concept of security developed at the time of the peace of Westphalia in 1648, with the idea of nation-state sovereignty, backed up by the balance of power. This can be contrasted with the development of

concepts of international security, which came along with the development of multilateral security institutions (Brauch/Oswald Spring/Grin et al. 2009; Brauch/Oswald Spring/Mesjasz et al. 2008, 2011).

First, the concept of 'collective security' came about with the League of Nations in 1919, failed when the League of Nations failed in stopping aggression, but came back stronger under the United Nations in 1945 when the UN Charter in articles 2.3 and 2.4 said that "all nations shall settle their international disputes by peaceful means" and "refrain in their international relations from the threat or use of force against the territorial integrity or political independence of any state". Collective security is backed up by enforcement. If any nation violates that ban against the use of force, the UN under Chapter VII can enforce action upon it by either non-military (article 41) or military (article 42) sanctions.

A step backward toward national security was 'collective defence' under the alliances of the NATO Pact in 1949 and the Warsaw Pact in 1955. Collective security was based on the idea that an attack on one was an attack on all but this, again, was still based on the use of force.

A somewhat more interesting concept is that of 'common security'. Common security came about with the Brandt Commission in 1980, which focused on North-South issues and economic security and presaged human security. After that the *Independent Commission on Disarmament and Security Issues* (Palme Commission 1982) brought about the concept of 'common security', in the context of nuclear war. The Palme Commission defined common security as the notion that you couldn't have a victory in nuclear war and therefore security had to be mutual, and that the implications of that are that you have to reduce military budgets.

Of all of this series of evolving concepts of security, only the Brundtland Commission's work addressed environmental issues. The World Commission on Environment and Development in 1987 talked about the concept of 'environmental security' and their report *Our Common Future* publicized the notion of sustainable development, the notion that you couldn't have development unless you took account of the environment.

Japan's 1983–84 Defence White Paper introduced the concept of *comprehensive security*, including traditional national self-defence, but also introduced access to food and access to energy as security issues and talked about the use of development aid as a security mechanism as well. While there was no mention of environmental security, the issues of access to food and energy intersected with the scarcity issues encountered in environmental studies.

3.6.4   Human Security

The concept of human security was introduced in the *Human Development Report* (HDR) in 1994 and was then taken up by the Commission on Human Security that ran from 2001 to 2003 and issued its report, *Human Security Now*, in 2003, with Sadako Ogata and Amartya Sen as primary authors. It was funded by the government of Japan and the Rockefeller Foundation and a number of other groups. It followed on from the concept introduced in the *Human Development Report* 1994. The *Commission on Human Security* (CHS) defined human security in the following way: "to protect the vital core of all human lives in ways that enhance human freedoms and human fulfilment".

To the Commission, human security meant protecting fundamental freedoms, freedoms that are the essence of life. It meant protecting people from critical, severe and pervasive threats and situations. It meant using processes that build on people's strength and aspirations. It meant creating political, social, environmental, economic, military and cultural systems that together give people the building blocks of survival.

Human security in its broader sense embraces far more than the absence of violent conflict, building on the *Human Development Report's* conceptualization. It is security based on people, individual security rather than state security, though both of these argue very much that these are complementary and not different. It talks about protecting people in cases of violent conflict but also about empowering people, and so it refers to a number of different aspects: protecting people in violent conflict, protecting people from the proliferation of arms, protecting the security of "people on the move", working to provide minimum living standards everywhere, prioritizing universal access to basic health care, empowering all people with universal basic education.

This particular report focused on the broad concept of human security. There is a big debate in the field as to whether the concept should be broad and essentially include development and human rights, or whether it should still be focused narrowly and involve protection from direct threats of violence.

On the other hand, the *Human Security Report* (2005) that came out of the Liu Institute in Vancouver, Canada, funded by the government of Canada and others, adopted the narrow concept of human security. They decided to focus on threats based on violent conflict, and to talk also about some of the indirect effects of those, like refugees and disease, but to focus largely on the threat of war and other factors that involve human violence. The *Human Security Report* focuses on wars: both inter-state wars, which have reduced considerably, and intra-state wars, which have actually been increasing, as well as on genocide, on the violence that comes with crime, on political repression, on terrorism, on human rights abuse, particularly the kind that kills people, and on the displacement that comes with wars and also the disease that often comes with wars.

As with the concept of environmental security, there has been much criticism of human security. One of the first and most cited critics is Roland Paris (2001). There continues to be disagreement on what human security is, with 21 comments on the subject published in a special issue of *Security Dialogue* in 2004. While the concept of human security certainly broadened the concept of security itself, it did little to incorporate environmental security or environmental issues.

3.6.5   From the Millennium Development to the Sustainable Development Goals

The *Millennium Development Goals* (MDGs), adopted by world leaders at the UN's Millennium Summit in September 2000 and reaffirmed at the World Summits in September 2005 and 2010, followed on from the conceptualization of human security, although they did not make the explicit linkages between development and peace and security, nor even include peace. The MDGs, however, did explicitly include environmental issues, although only in Goal 7. Intended to be achieved by 2015, the MDGs stressed the interconnectedness of all goals.

Following on from the many UN Decades of Development, beginning in 1961, and the conceptualization of human development in 1970 and human security in 1974 by the UN Development Programme, as well as from the series of UN conferences on environment, food, population, and women beginning in the 1970s, the MDGs explicitly included a focus on ensuring environmental sustainability in Goal 7: "Ensure environmental sustainability." By this time the environment had made it onto the world agenda, although the linkages between war, peace, gender, population, environment and development were still not always explicitly addressed. As these goals morphed into the seventeen *UN Sustainable Development Goals (SDGs)*, adopted in 2015 to be accomplished by 2030, the concept of sustainable development publicized by the Brundtland Commission in 1987 became an even stronger part of our way of thinking.[7]

The SDGs continued many of the MDGs, but this time eleven goals explicitly addressed issues of sustainability (Goals 2, 6 (which was a target in the MDGs), 7, 8, 9, 11, 12, 13, 14, 15, and 17). A notable integrative innovation was goal 16, which explicitly linked peace and the environment, reading: "Promote peaceful and inclusive societies for sustainable development, provide access to justice for all and build effective, accountable and inclusive institutions at all levels." But sustainable development may well be an oxymoron; development, over the long term, may not be sustainable, particularly if it includes the connotation of growth, as in Goal 8. Attention to the values of environment may mean not developing the fossil fuels now accessible rather than exploiting them.

3.7   Conclusions

While the conceptualization of 'sustainability transitions' (Brauch, chapter 1 in this volume) may be an important contribution both to the fields of peace studies and environmental studies, it may also be important to consider this in the light of systems thinking. Boulding's work on the development of systems theory and that of the *Limits to Growth* project continue to be important here. Systems evolve; though they may exist in equilibrium at any one point in time, the equilibrium is likely to evolve also with the introduction of step-level functions that change any of the basic conditions. Stability is unlikely over time. Sustainability suggests stability. The environment has been changing over millennia, and will continue to change. The question may not so much require transitions to environmental sustainability as simply to acknowledge the importance of the environment, both to human society and on its own. This implies acknowledgment of the detrimental role human beings have played with respect to the health of the environment, and a commitment to a transition to protecting the environment.

Whether the concept 'sustainability transitions' best captures this may be debated. Among the key maxims toward environmental protection is the precautionary principle put forward by the UN in the 1992 Rio Declaration, and earlier in climate change and other negotiations, that: "Where there are threats of serious or irreversible damage, lack of full scientific certainty shall not be used as a reason for postponing cost-effective measures to prevent environmental degradation." The adoption of a precautionary principle with respect to peace and human rights could also be important. A transition to an expanded concept of our human community would improve our work both on peace and the environment. We can come back to Aldo Leopold's dictum: the evolution of human ethics has involved the expansion of our understanding of community, and the adoption of a "land ethic" would be an important transition in our ethical development.

---

7   On the *Millennium Development Goals* (MDGs) of 2000 see: <http://www.un.org/millenniumgoals/>; and on the *UN Sustainable Development Goals* (SDGs) of 2015 see: <www.un.org/sustainabledevelopment/sustainable-development-goals>.

References

Ackerman, Peter; DuVall, Jack, 2000: *A Force More Powerful: A Century of Nonviolent Struggle* (London - New York: Palgrave/St. Martin's).

Amster, Randall, 2015: *Peace Ecology* (Boulder, CO: Paradigm).

Bacon, L.S.; Wheeler, M. 1984: *Environmental Dispute Resolution* (New York: Plenum Press).

Bergholt, Drago; Lujala, Paivi, 2012: "Climate-related natural disasters, economic growth, and armed civil conflict", in: *Journal of Peace Research*, 49,1 (January): 147-162.

Bernauer, Thomas; Siegfried, Tobia, 2012: "Climate change and international water conflict in Central Asia", in: *Journal of Peace Research*, 49,1 (January): 227-239.

Bingham, Gail, 1986: *Resolving Environmental Disputes: A Decade of Experience* (Washington, DC: Conservation Foundation).

Boulding, Kenneth E., 1956: *The Image: Knowledge in Life and Society* (Ann Arbor: University of Michigan Press).

Boulding, Kenneth E., 1962: *Conflict and Defense: A General Theory* (New York: Harper & Brothers).

Boulding, Kenneth E. 1966: "The Economics of the Coming Spaceship Earth", in: Jarrett, H. (Ed.): *Environmental Quality in a Growing Economy* (Baltimore: Johns Hopkins Press).

Boulding, Kenneth E., 1977: "Twelve Friendly Quarrels with Johan Galtung", in: *Journal of Peace Research*, 14: 75-86.

Boulding, Kenneth E., 1978: *Ecodynamics* (Beverly Hills, CA: Sage).

Boulding, Kenneth E., 1978a: *Stable Peace* (Austin: University of Texas Press).

Boulding, Kenneth E., 1989: *Three Faces of Power* (Newbury Park: Sage).

Brauch, Hans Günter; Oswald Spring, Úrsula; Bennett, Juliett; Oswald Serrano, Serrena Erendira (Eds.), 2016: *Addressing Global Environmental Challenges from a Peace Ecology Perspective* (Cham–Heidelberg–New York–Dordrecht–London: Springer).

Brauch, Hans Günter; Oswald Spring, Úrsula; Grin, John; Mesjasz, Czeslaw; Kameri-Mbote, Patricia; Behera, Navnita Chadha; Chourou, Béchir; Krummenacher, Heinz (Eds.), 2009: *Facing Global Environmental Change: Environmental, Human, Energy, Food, Health and Water Security Concepts* (Berlin - Heidelberg - New York: Springer-Verlag).

Brauch, Hans Günter; Oswald Spring, Úrsula; Mesjasz, Czeslaw; Grin, John; Dunay, Pal; Behera, Navnita Chadha; Chourou, Béchir; Kameri-Mbote, Patricia; Liotta, P.H. (Eds.), 2008: *Globalization and Environmental Challenges: Reconceptualizing Security in the 21st Century* (Berlin - Heidelberg - New York: Springer-Verlag).

Brauch, Hans Günter; Oswald Spring, Úrsula; Mesjasz, Czeslaw; Grin, John; Kameri-Mbote, Patricia; Chourou, Béchir; Dunay, Pal; Birkmann, Jörn (Eds.), 2011: *Coping with Global Environmental Change, Disasters and Security - Threats, Challenges, Vulnerabilities and Risks* (Berlin - Heidelberg - New York: Springer-Verlag).

Brown, Lester, 1977: "Redefining Security", Worldwatch Paper 14 (Washington DC: Worldwatch Institute).

Buzan, Barry; Wæver, Ole; de Wilde, Jaap, 1998: *Security: A New Framework for Analysis* (Boulder: Lynne Rienner Publishers).

Carson, Rachel, 1962: *Silent Spring* (Greenwich, Conn.: Fawcett Publications).

Caldwell, Lynton, 1984: *International Environmental Policy* (Durham, N.C.: Duke University Press).

Carter, Timothy Allen; Veale, Daniel Jay, 2015: "The timing of conflict violence: Hydraulic behavior in the Ugandan civil war", in: *Conflict Management and Peace Science*, 32 (September): 370-394.

Carius, Alexander, 2006: "Environmental Cooperation as an Instrument of Crisis Prevention and Peacebuilding: Conditions for Success and Constraints". Paper submitted to the 2006 Berlin Conference on the Human Dimensions of Global Environmental Change "Resource Policies: Effectiveness, Efficiency, and Equity", Berlin, 17-18 November 2006.

Chenoweth, Erica; Stephan, Maria J., 2011: *Why Civil Resistance Works* (New York, NY: Columbia University Press).

Choucri, Nazli; North, Robert C., 1972: "Dynamics of International Conflict: Some Policy Implications of Population, Resources, and Technology", in: *World Politics*, 24 (Spring): 80-122.

Commission on Human Security, 2003: *Human Security Now* (New York: Commission on Human Security); at: <http://www.un.org/humansecurity/sites/www.un.org.humansecurity/files/chs_final_report_-_english.pdf>.

Conca, Ken; Dabelko, Geoffrey D. (Eds.), 2002: *Environmental Peacemaking* (Washington, DC: Woodrow Wilson Center Press—Baltimore: Johns Hopkins University Press).

Conca, Ken; Wallace, Jennifer, 2009: "Environment and Peacebuilding in War-torn Societies: Lessons from the UN Environment Programme's Experience with Postconflict Assessment", in: *Global Governance*, 15,4, 485-504.

Dalby, Simon, 2002: *Environmental Security* (Minneapolis: University of Minnesota Press).

Deudney, Daniel H., 1990: "The Case Against Linking Environmental Degradation and National Security", in: *Millennium*, 19: 461-476.

Ehrlich, Paul, 1968: *The Population Bomb* (New York: Ballentine).

Ehrlich, Paul R.; Ehrlich, Anne, 1970: *Population, Resources, Environment* (San Francisco: W. H. Freeman & Co.).

Fisher, Roger; Ury, William, 1981: *Getting to yes: Negotiating agreement without giving in* (New York: Penguin).

Galtung, Johan, 1969: "Violence, Peace, and Peace Research", in: *Journal of Peace Research*, 6,3: 167-191.

Galtung, Johan, 1971: "A Structural Theory of Imperialism", in: *Journal of Peace Research*, 8,2: 81-117.

Galtung, Johan, 1975: *Essays in peace research*, Vol. I. (Copenhagen: Christian Ejlers).

Gartzke, Erik, 2012: "Could climate change precipitate peace?", in: *Journal of Peace Research*, 49,1 (January): 177-192.

Gleditsch, Nils Petter, 1998: "Armed Conflict and the Environment: A Critique of the Literature", in: *Journal of Peace Research*, 35:3 (May): 381-400.

Gleditsch, Nils Petter, 2012: "Whither the weather? Climate change and conflict", in: *Journal of Peace Research*, 49,1 (January): 3-9.

Gleick, Peter, 1993: "Water and Conflict", in: *International Security*, 18,1: 79-112.

Gleick, Peter H., 1998: *The World's Water 1998-1999*. (Washington DC: Island Press).

Gleick, Peter H. et al., 2013: *The World's Water, Volume 8. The Biennial Report on Freshwater Resources* (Washington, DC: Island Press).

Hardin, Garrett, 1968: "Tragedy of the Commons", in: *Science*, 162,13 (December): 1243-1248.

Homer-Dixon, Thomas F., 1994: "Environmental Scarcities and Violent Conflict", in: *International Security*, 19,1 (Summer): 5-40.

Human Security Centre, University of British Columbia, 2005: *Human Security Report 2005: War and Peace in the 21st Century* (New York: Oxford University Press); at: <http://www.humansecurityreport.info>.

Independent Commission on Disarmament and Security Issues, 1982: *Common Security* (New York: Simon and Schuster).

Independent Commission on International Development Issues, 1980: *North-South: A Programme for Survival* (Cambridge, MA: M.I.T. Press).

Kaplan, Robert D., 1994: "The Coming Anarchy", in: *The Atlantic Monthly*, 273,2 (February).

Lodgaard, Sverre, 1990: "Environmental Conflict Resolution", paper presented at UNEP meeting on Environmental Conflict Resolution (March): 18.

Mack, Andrew, 1985: *Peace research in the 1980s* (Canberra: Australian National University).

Malthus, Thomas Robert, 1798 [1993]: *An Essay on the Principle of Population as It Effects the Future Improvement of Society* (Oxford: Oxford University Press).

Matthews, Jessica, 1989: "Redefining Security", in: *Foreign Affairs*, 68,2 (Spring): 162-177.

Meadows, Donella H.; Meadows, Dennis L.; Randers, Jorgen; Behrens, William W. III, 1972: *The Limits to Growth* (New York: Universe Books).

Meadows, Donella H.; Meadows, Dennis L.; Randers, 1992: *Beyond the Limits* (Post Mills, VT: Chelsea Greens Publishing).

Mesarovic, Mihaljo; Pestel, Eduard, 1974: *Mankind at the Turning Point* (New York: E. P. Dutton)

Ophuls, William, 1977: *Ecology and the Politics of Scarcity: Prologue to a Political Theory of the Steady State* (San Francisco: W. H. Freeman).

Ostrom, Elinor, 1990: *Governing the Commons* (Cambridge: Cambridge University Press).

"Peace and the Environment: Making the Connections in the Classroom", in: *PAWSS Perspectives on Teaching Peace and World Security Studies*, 1,1 (April 1990).

Oswald Spring, Úrsula; Brauch, Hans Günter; Tidball, Keith G. (Eds.), 2014: *Expanding Peace Ecology: Peace, Security, Sustainability, Equity and Gender–Perspectives of IPRA's Ecology and Peace Commission* (Cham–Heidelberg–New York–Dordrecht–London: Springer).

Paris, Roland, 2001: "Human Security–Paradigm Shift or Hot Air?", in: *International Security*, 26,2: 87-102.

Princen, Thomas, 2005: *The Logic of Sufficiency* (Cambridge, MA: MIT Press).

Princen, Thomas, 2010: *Treading Softly: Paths to Ecological Order* (Cambridge, MA: MIT Press).

Princen, Thomas; Maniates, M.; Conca, Ken (Eds.), 2002: *Confronting Consumption* (Cambridge, MA: MIT Press).

Richardson, Lewis F., 1960: *Arms and Insecurity* (Pittsburgh, PA: Boxwood Press).

Richardson, Lewis F., 1960: *Statistics of deadly quarrels* (Pittsburgh, PA: Boxwood Press).

Roberts, Adam and Timothy Garton Ash (eds.), 2011: *Civil Resistance and Power Politics: The Experience of Nonviolent Action from Gandhi to the Present*. Oxford: Oxford University Press.

Ronnefeldt, Carsten F., 1997: "Three Generations of Environmental Security Research", in: *Journal of Peace Research*, 34,4 (November): 473-482.

Scheffran, Jurgen; Brzoska, Michael; Kominek, Jasmin; Link, P. Michael, 2012: "Climate Change and Violent Conflict", in: *Science*, 336,6083 (18 May): 869-871.

Schumacher, E. F., 1976: *Small is Beautiful: Economics as if People Mattered* (New York: Harper and Row–London: Blond and Briggs, 1973).

Scott, Robert, 2015: *Kenneth Boulding: A Voice Crying in the Wilderness* (Basingstoke–New York: Palgrave Macmillan).

Sharp, Gene, 1973: *The Politics of Nonviolent Action* (Boston: Porter Sargent Publishers).

Slettebak, Rune T., 2012: "Don't blame the weather! Climate-related natural disasters and civil conflict", in: *Journal of Peace Research*, 49,1 (January): 163-176.

Sorokin, Pitrim, 1937: *Social and Cultural Dynamics*, vol. 3. (New York: American Book Co.)

Sprout, Harold; Sprout, Margaret, 1962: *Foundations of International Politics* (Princeton: Princeton University Press).

Stephenson, Carolyn M. (Ed.), 1982: *Alternative methods for international security* (Washington DC: University Press of America).

Stephenson, Carolyn M., ²2008: "Peace Studies, Overview", in: Kurtz, Lester (Ed.): *Encyclopedia of Violence, Peace, & Conflict* (Oxford: Academic Press), 1534-1548.

Stephenson, Carolyn M., 2010: "Peace Research/Peace Studies: a Twentieth Century Intellectual History", in: Denmark; Robert A. (Ed.): *The International Studies Encyclopedia*. Vol. IX (Chichester, UK: Wiley-Blackwell): 5579-5603.

Stevis, Dimitris, 2010. "International Relations and the Study of Global Environmental Politics: Past and Present", in: *The International Studies Encyclopedia*, vol. VII. (Oxford: Wiley-Blackwell): 4476-4507.

Susskind, Lawrence: Cruickshank, Jeffrey, 1987: *Breaking the Impasse: Consensual Approaches to Resolving Public Disputes* (New York: Basic Books).

Tir, Jaroslav; Stinnett, Douglas M., 2012: "Weathering Climate Change: Can institutions mitigate international water conflict?", in: *Journal of Peace Research*, 49,1 (January): 211-225.

Ullman, Richard H., 1983: "Redefining Security", in: *International Security*, 8,1 (Summer): 129-153.

United Nations Development Programme, 1994: *Human Development Report 1994: New Dimensions of Human Security* (New York: UNDP).

United Nations, 1999: Programme of Action on a Culture of Peace. General Assembly Resolution 53/243 of 6 October 1999.

UNESCO, World Commission on the Ethics of Scientific Knowledge and Technology. *The Precautionary Principle* (Paris: UNESCO); at: <unesdoc.unesco.org/images/0013/001395/139578e.pdf>.

UN Secretary-General, 1992: *An Agenda for Peace*. A/47/277- S/24111 (New York: United Nations); at: <http://www.un-documents.net/a47-277.htm>.

Westing, Arthur H., 1986: "An Expanded Concept of International Security", in: Westing, Arthur H. (Ed.): *Global Resources and International Conflict* (Oxford: Oxford University Press).

WCED (World Commission on Environment and Development), 1987: *Our Common Future*. (Oxford: Oxford University Press).

"What is 'Human Security'? Comments by 21 Authors", (2004): in *Security Dialogue*, 35,3 (September): 347-87.

Wright, Quincy, 1942, 1965. *A study of war* (Chicago: University of Chicago Press).

4 Transition Studies: Basic Ideas and Analytical Approaches

John Grin[1]

Abstract

As a background to later contributions, this chapter provides a concise introduction to different approaches to (i) understanding and (ii) shaping transition dynamics:
1. A *sociotechnical approach*, with the multilevel perspective as its main concept, and strategic niche management as its governance concept;
2. A *complex (adaptive) system-based approach*, using the concept of transition patterns, with transition management as its governance concept;
3. A *governance approach*, where the politics of transition dynamics are examined, with reflexive design as the core of its approach to shaping transition.

In addition, work based on the social theory of practice has helped to bring consumption practices into transition theory, and an approach focusing on innovation system has helped to explain how different arrangements may become interconnected so as to produce a transition.

Keywords: Sociotechnical transitions, multilevel perspective, strategic niche management, complex adaptive systems, transition management, reflexive governance, reflexive design, power, legitimacy.

4.1 Introduction: Rationale and Nature of Transitions and the Scope of Transition Studies

Transitions may be *defined* as profound societal transformations in that they involve changes in both multiple, interacting societal practices and the institutional, structural and discursive structures in which these are embedded. The *rationale* for transition studies lies in particular diagnosis of persistent societal problems—problems that appear to resist potential solutions, such as the depletion of specific resources, continued greenhouse gas emissions, and flooding in river flow areas. Central to that diagnosis is the presumption, well-rooted in evolutionary economics and sociology, *science and technology studies* (STS) and social theory, that the definitions of societal problems, the

practices that tend to be used to deal with them, and societal structures themselves co-evolve with each other, meaning that the evolution of each over time co-shapes the evolution of the other two. If problem definitions that are fundamentally novel become dominant, a call for fundamentally different practices may result. These are often not well served by incumbent structures. This may generate inertia, and lead to a more fertile ground for resistance to transformative practices than acceptance of them. As a result, such innovative practices may be difficult to establish, may easily fall back into normality, or may not survive in the long run.

This in a nutshell is why such 'novel type' problems may be so persistent, and why in transition studies persistent problems are seen as providing a rationale for a transition (Grin/Rotmans/Schot 2010: 1–5). From that key point of departure the two main research questions follow: how may transition dynamics be understood, and how may they be shaped so as to deal with persistent problems?

Transition studies has its intellectual roots in work by many scholars from a variety of fields, most notably STS (*science and technology studies*) and innovation studies, and the field of complex systems studies

1 John Grin (The Netherlands) is a full Professor in Policy Science, in particular of System Innovation, in the Department of Political Science, University of Amsterdam, and Co-Director of the Programme Group Transnational Configurations, Conflicts and Governance, Amsterdam Institute for Social Science Research (AISSR); email: <j.grin@uva.nl>.

and socio-ecological system studies (Smith/Voß/Grin 2010; Grin 2008). These strands of work were stimulated by a global sense of urgency concerning climate change and other sustainability problems, and by programmes like the *International Human Dimensions Programme on Global Environmental Change* (IHDP). Transition studies as a field received a major boost when Dutch scholars were fortunate enough to get a major research programme funded, the *Dutch Knowledge Network on System Innovations and Transitions* (KSI, 2005-2010). This five-year programme (Rotmans/Schot/Grin 2004) comprised eighty-five researchers, including thirty-eight PhD students, at eight Dutch universities and at TNO, a Dutch *Great Technological Institute* (GTI); they collaborated with some 750 transition practitioners in several dozen transition projects, some of which had been established by KSI.[2]

Normal, healthy academic routines of critical scrutiny were applied to novel strands of research, and KSI spent ten per cent of its 22 million budget on internationalization. As a result, important exchanges took place with scholars from a host of other countries who had been working, in their own ways, on transitions or related themes; these scholars began to critically scrutinize the Dutch approaches. A series of international workshops, together with sessions at major conferences, special issues of scholarly journals and so on helped to improve and further develop original theorizing, and added important new themes and modes of theorizing. This also helped to overcome the field's original Eurocentric focus (e.g. Berkhout/Angel/Wieczorek 2009; Berkhout/Verbong/Wieczorek 2010). KSI also initiated a transition journal (*Environmental Innovation and Sustainability Transitions, EIST*)[3] as well as a book series with Routledge.[4] At present, the field counts more than thousand scholars from all continents, assembled in the *Sustainability Transitions Research Network* (STRN),[5] founded at the first international transitions conference in Amsterdam in 2009. They meet at the annual conferences organized by STRN as well as during sessions at other academic conferences, and frequently undertake internationally collaborative pro-

jects. By now a vast, global literature has evolved (Markard/Raven/Truffer 2012; Chappin/Ligtvoet 2014). Lessons from and for practitioners are being exchanged.[6]

This chapter seeks to introduce the approaches included in the original KSI programme. While these are primarily those based on the sociotechnical (4.2), complex system (4.3) and governance theory (4.4) approaches, in section 4.5 we will briefly discuss two other approaches, focusing on social practices and technical innovation systems respectively.

The objective of this chapter is to provide a background to the following chapters by discussing some of the main concepts and insights that were developed at the cradle of the field. For each of the approaches the chapter will describe (i) how it conceives of transition dynamics and on the basis of what theoretical ideas, and (ii) which governance concept it proposes. While some recent research which further develops these original concepts will be discussed in the next sections, many other recent lines of research will not be discussed. The most important ones will be briefly indicated in the final section, however.

4.2   The Sociotechnical Approach and Strategic Niche Management

The first approach is the *sociotechnical approach*. It focuses on the interactions of the practices and objects of social, economic and technological innovation with each other and with incumbent sociotechnical systems. The central concept it proposes for mapping and understanding transition dynamics is the *multilevel perspective* (MLP), which has become a core notion in transition studies more generally.

4.2.1   Understanding Transition Dynamics

This approach is rooted in STS, evolutionary theory and the 'contextual history' of technology. One seminal source of the MLP is a major study by Rip and Kemp (1998), which seeks to contribute to a better understanding of how, and under what conditions, deliberate technical change can contribute to dealing

---

2   See for details at: <www.ksinetwork.nl>.

3   See for details at: <http://www.journals.elsevier.com/environmental-innovation-and-societal-transitions/>.

4   See for details on the Routledge Studies on Sustainability Transitions; at: <https://www.routledge.com/series/RSST>; and at: <http://www.sustainabilitytransitions.com/en/background>.

5   See for details at: <www.strn.org>.

6   See at: <http://www.transitiepraktijk.nl/en/>; see videos on: "Making Transitions Happen Platform"; at: <https://www.youtube.com/watch?v=GEzXMNc7yGk&index=4&list=PLu8ruQWIRJIGPwxo5i_aOj_A7QPrK-MusP> and on "The journey of sustainable business"; at: <https://www.youtube.com/watch?v=RSDnAVkdaAM&feature=related> (31 March 2016).

with climate change. It begins by reviewing the bodies of literature on large technical systems, innovation system approaches, evolutionary theories, actor network theory, and the tradition of the social construction of technology. The authors then organize their findings into a two-dimensional scheme (Rip/Kemp 1998: 339). Along one axis, 'configurations that work' gradually merge into structuring entities and, eventually, 'seamless webs' that provide even stronger structuration. The other axis indicates scope. At the meso-level, there are regimes defined as "the rule-set or grammar embedded in a complex of engineering practices, production process technologies and product characteristics, skills and procedures, ways of handling relevant artefacts and persons, ways of defining problems—all of them embedded in institutions and infrastructures" (Rip/Kemp 1998: 338).

Usually, innovations will be shaped by the incumbent regime. However, the regime is not static, but may change, partly under the influence of its linkages to innovations at the micro-level and (changes at) the landscape level. On the basis of these findings, the authors claim that "to understand and explain technical change, a combination of economic and sociological theory is necessary" (Rip/Kemp 1998: 356). While no single explanation may be sensibly given, it is crucial to note path dependencies and irreversibilities as major factors in technical change.

A second seminal source of the MLP sheds more specific light on the potential of change beyond the incumbent regime. Drawing on a vast empirical basis of studies of the development of technology in the Netherlands in the nineteenth century (for an English summary volume of the Dutch book series, see Schot/Rip/Lintsen 2010), Johan Schot (1998) observes that in many cases, the development of technology appears to follow established patterns. Theoretically guided by a range of evolutionary theories, he explains this by pointing out that the development of technology is structured by a *technological regime,* which combines "the rules embedded in engineering practices (cognitive structuring) with the rules following from the embodiment of a given set of technologies in specific production and consumption patterns" (Schot 1998: 190). In some cases, however, regime shifts appeared to take place, especially when the efforts of interested parties coincided with favourable exogenous ('landscape') developments. Under such conditions, a niche may be formed: "a limited and local alliance (network) between the party that produces the new technology and the party that uses it (the sponsor)" (Schot 1998: 193), which protects

technologies under development from the incumbent regime. These niches create links between variation and selection, contributing to *quasi-evolution.*

Crucial in this connection is the finding that actors' expectations are not only a source of new technical options (variation), but also a "guiding force in the process of technological development" and "the development and articulation of expectations gradually improve the definition of the technology and the environment in which it is supposed to function" (Schot 1998: 196). Such dynamics of expectations and articulation in niches, formed by change agents under favourable exogenous conditions, may contribute to regime change, especially when constructive interference may be brought about between niche innovations, changes or instabilities in the regime, and landscape developments. This point has been further developed by Frank Geels (2005). Building upon earlier work as well as on a wide variety of literatures, concerning for instance the role of users and the role of increasing returns in innovation processes, he develops and employs a more detailed version of the MLP. The three levels are considered as a 'nested hierarchy', with an increasing degree of structuration from the niche level to the landscape level (Geels 2005: chapter 3). Drawing upon a set of case studies, he then (Geels 2005: 246-279) reconfirms and adds analytical accuracy to the claims of the MLP that transitions occur through interlinkage and interaction between developments at the three levels, and that they often involve multiple technologies (cf. figure 1.11, in this volume). Van Driel and Schot (2005) have interpreted this as interferences between changes occurring at different scales of time (Braudel 1980).

At that stage, the MLP had become a relatively well-developed heuristic framework, yielding the following insights on transition dynamics. First, its most central claim is that regime shifts occur through interlinkage and interaction between multiple developments at the three levels. Second, and more precisely, in this connection, other crucial insights are that this requires

- the *creation of niches*, locations that are actively protected against part of the influence of the existing regime; and
- *strategic action* in the sense of creating linkages to overcome inertia through connecting dynamics at all three levels.

Subsequent work by Schot, Geels and collaborators has produced several dozen case studies. These have yielded detailed accounts of the mechanisms through

Box 4.1: Possible transition pathways according to Geels and Schot (2007: 54–76).

- *Reproduction pathway*: in the absence of landscape pressure, the regime is likely to remain dynamically stable.
- *Transformation pathway*: if there is moderate landscape pressure at a time when niche innovations have not yet been sufficiently developed, then regime actors will respond by modifying the direction of the development path and innovation activities.
- *De-/re-alignment pathway*: if landscape change is divergent, large and sudden, then increasing regime problems may lead to erosion. This then creates space

for the emergence of multiple niche innovations that co-exist and compete for attention and resources. Eventually a new dominant core for realignment of a new regime emerges.
- *Substitution pathway*: if there is substantial landscape pressure at a point when niche innovations have developed sufficiently, then these will break through and replace the existing regime.
- *Reconfiguration pathway*: niche innovations are initially adopted in the regime to solve local regime.

which transitions and system innovations unfold—including the role of outsiders, the importance of policy support, the role of the market mechanism, the role of wide cultural visions and promises, and the importance of strategic games and competition throughout. Also, based on this body of case studies, a typology has been developed for transition pathways through which interlinkage and interaction may occur (Geels/Schot 2007; Schot/Geels 2010). These pathways differ from each other in two dimensions: the nature of landscape influences (continuing pressure on incumbent regime; a disruptive effect, hyper-turbulence, an avalanche effect, or a specific shock effect) and the degree of development of niche innovations. Five proposed pathways are summarized in box 4.1.

4.2.2   Shaping Transition Dynamics

Following their understanding of transition dynamics, Rip and Kemp state two crucial questions about how those dynamics are shaped are: "how to increase the chances for better path dependencies (…[and]) how to identify and realize transition paths from the present situation to a more desirable one" (Rip/Kemp 1998: 371). The authors emphasize the need to use the model as a heuristic device, which can guide, but not replace, contingent analysis. At the core of their ideas is their proposal that governance should be seen as a matter of the 'modulation' of ongoing dynamics within and between the three levels. They draw attention to the need to develop the perspective so as to take into account normative goal-setting, power asymmetries and strategic action, as well as to diagnose the course contemporary industrial societies are taking.

The sociotechnical approach has drawn on earlier work on strategic niche management (Schot/Hoogma/Elzen 1994; Kemp/Schot/Hoogma 1998; Weber/Hoogma/Lane et al. 1999; Hoogma/Kemp/Schot et al. 2002; Truffer/Metzner/Hoogma 2004) as

a basis for shaping transition dynamics in a way that draws strongly on experiments. In *strategic niche management* (SNM), novelties are able to develop within a niche that protects them from the adverse influences of the regime and provides helpful structural conditions not prevalent in the regime. Importantly, a niche must be understood as an analytical construct: a segment of the market with relatively interested consumers and helpful supply chains may be supported by a tax exemption or by government-funded research.

In much early work on SNM, a niche was understood more or less as an incubator, nurturing a premature novelty until it was strong enough to enter the normal world, through processes of learning, network formation and the shaping of expectations, and taking place within and between niche experiments. Later, Smith and Raven (2012) proposed and elaborated three functions: shielding, nurturing and empowerment.

Another limitation of early work was that much less attention was paid to the question of how the novelty could eventually help to fundamentally change the world. That problem has been addressed by later research. Elzen, Geels, Hofman et al. (2004) have proposed the *Sociotechnical Scenario* approach. Scenarios may provide guidance for the development of niche experiments and their connection to changes at the regime and landscape levels. They should be designed such they are both sustainable for a specific set of actors and attainable in the sense that their realization may build on ongoing changes at the regime and landscape levels. This kind of analysis has been used in different fields, including car mobility (Marletto 2014) and energy (Foxon/Hammond/Pearson 2010; Verbong/Geels 2012). Along similar lines, Raven (2005, 2006) has explicitly situated SNM in ongoing multilevel dynamics. Thus the capacity of niche experiments to contribute to regime change is understood as co-determined by the ways in which they interact with ongoing changes at the regime and landscape levels. Smith (2007) has proposed the concept of translation (from actor network theory) to theorize the agency involved in bringing about such connections. In later work, Raven and others (Raven 2007; Raven/Verbong 2007, 2009; Raven/Verbong/ Schilpzand et al. 2011) have elaborated this understanding so as to comprise the dynamic interaction between niche experiments and multiple regimes, a key issue as many transitions involve the reordering of 'systems' (e.g. biofuel implies the emergence of a system reconnecting elements from the energy and agriculture systems). Amongst many other findings, such work reveals on the one hand additional difficulties in reconciling experiments with two different regimes, and on the other, more opportunities for strategically operating agents to alternately draw on elements from different regimes. Schot and Geels (2008) have synthesized this and other work, seeing SNM as a form of reflexive governance (Grin 2006), and proposing a new version of SNM, where niches facilitate sustainable innovation journeys in which the internal and external dynamics of niches are deliberately connected. Seyfang and Haxeltine (2012) have added other drivers of dynamics, most notably group (identity) formation.

Other important work has been done, for example, focusing on better understanding the internal dynamics of niches in bringing about protection (Ulmanen/Verbong/Raven 2009; Ulmanen 2013; Boon/Moors/Meijer 2014), and on the specific nature of 'grassroots' niches, rooted in civil society (Seyfang/Hielscher/Hargreaves et al. 2014).

4.3 The Complex Systems Approach and Transition Management

4.3.1 Understanding Transition Dynamics

The second approach to explaining sustainability transitions (Rotmans/Loorbach 2010) has been inspired by two strands of systems theory in particular. In the first, *integrated assessment*, systems are depicted as comprising mutually interacting sociocultural, economic and physical stocks and flows (Rothman/Robinson 1997; cf. Pahl-Wostl 2002; Weaver/Jansen/van Grootveld et al. 2008). At any specific moment in time, the system may be characterized by its stocks. These stocks change relatively slowly compared to their own volume. Relatively rapid changes may occur

through flows, associated with ongoing socio-economic and natural processes. This depiction involving stock flow and agents is of obvious use in conceptualizing the basic notion of sustainable development as involving the closing of substance cycles.

The second body of theory involves *complex adaptive systems* (CAS) theory (Gunderson/Holling 2002; Holland 1996). Compared with more traditional systems theory, it adds the possibility of adaptation through (complex, non-linear) responses to external (e.g. 'landscape', in MLP terms) perturbations. Under normal conditions, when changes in the environment are moderate at most, the system maintains a particular dynamic equilibrium of stocks, flows and agents (an 'attractor'). This is due to an internal structure which emerges over time from processes at a lower or higher level of aggregation. Attractor and associate structure are the CAS depiction of the regime. In the case of stronger external perturbations, agents will induce non-standard responses, through a constant stream of self-reinforcing changes in flows and agent behaviour—this is how a niche may be represented in CAS terms). In specific circumstances this may bring the system into a different state of dynamic equilibrium, centring its dynamics on a different attractor. This transformative change from one dynamic system equilibrium to another, involving a change in the deep structure of the system, constitutes a transition. Systemic adaptability and resilience may also be seen as intrinsic features of sustainable development.

Following this understanding of niche and regime, De Haan and Rotmans (2011; cf. Haxeltine/Whitmarsh/Bergman et al. 2008) have developed a typology of 'transition patterns'—basic modes of transition dynamics, understood as interactions between the niche and regime levels and involving an intermediate level: the niche-regime. The basic patterns are listed in box 4.2. Compared with the transition pathways that have resulted from the sociotechnical approach, the 'niche to niche-regime' and 'niche-regime to regime' patterns form two steps in a process in which regime changes gradually result from changes at the niche level, with the niche-regime as a key element. The main difference is its analytical nature (Grin/Marijnen 2011). Sociotechnical pathways, derived from historical analysis, follow in more or less phenomenological terms the process of (re-, de-)alignment. They thus relate relatively clearly to everyday terms. CAS patterns take a more abstract perspective, providing insight into the underlying systemic causation in terms of stocks, flows and agents. Patterns may be

> **Box 4.2: D**ifferent transition patters. **Sources**: Rotmans and Loorbach (2010: 137–139), and De Haan and Rotmans (2011).
>
> • A 'niche to niche-regime' pattern, where niches emerge, cluster and form a niche-regime which starts to undermine the incumbent regime.
> • A 'niche-regime to regime' pattern, in which the niche is absorbed or combined with the incumbent regime,
>
> which thus evolves into a new regime.
> • A 'regime to niche-regime' pattern, in which a massive change at the landscape level induces regime change through significant landscape changes, or competition from niches and niche-regimes.

analytically related to core notions of sustainable development.

4.3.2   Shaping Transition Dynamics

Associated with and rooted in this (complex) system understanding of transition dynamics, *Transition Management* (TM) has been proposed as a governance concept for transitions (Rotmans/Kemp/van Asselt 2001; Loorbach 2007; Kemp/Loorbach/Rotmans et al. 2009; Rotmans/Loorbach 2010). A key point of departure is the idea that the actor (alliance) that seeks to deliberately influence a system towards a desirable transition is bound, in and through doing so, to become part of the system. Even if we initially conceive of such action as an external intervention, as such it will provoke a (complex) response from the system, partly unforeseen. The steering agent, who has already become part of the web of causation, can do little better than make her next move informed by that response, and the awareness that that move will draw her even further into the system (Van der Brugge/Van Raak 2007). This, in essence, is the notion of self-governance: the governing system essentially is part of the system to be governed (Loorbach 2007: 175). As Rotmans and Loorbach (2010) point out, the methodological implication is that action research may play a key role in both developing the theoretical understanding of transition dynamics and in shaping and promoting transitions in actual contexts. This has led researchers in this tradition to develop ways of analysing transition dynamics, in order to be able to monitor their progress as a basis for learning (Van der Brugge 2009; Taanman 2014).

Transition Monitoring is one of the four key activities, mutually related in a dynamic, iterative mutual way, that co-constitute TM. It gathers insights into the dynamics and (desired and undesired) progress of a second activity—performing transitions experiments—as inputs into the other two activities: the development of visions, and the structuring of strategic problems, the development of a transition agenda and the derivation of associated transition paths. These take

place in a so-called transition arena, composed on the basis of the idea of selective participation: a critical mass of 'front runners', visionary thinkers and entrepreneurial actors complement regime actors with a certain openness to change.

There is no fixed order for undertaking these activities. The major Dutch governmental Energy Transition programme started with problem structuring. It then involved visioning, agenda-building and the defining of tracks. For each track a set of experiments was then defined (Rotmans/Loorbach 2010: 180-199). But the Transition Programme on Long-Term Care, for instance, started with existing, relatively radical experiments and helped them to 'transitionize' (Van den Bosch/Neuteboom forthcoming). In both cases, subsequent programme evolution comprised a wide set of different relations between the four core activities, with demands for accountability and funding, monitoring activities, and an emphasis on experimentation and external feedbacks as the main drivers.

When explained in detail (Loorbach 2007, 2010; Rotmans/Loorbach 2010: 140-159), TM has been underpinned by principles derived from complex adaptive system theory. For instance, the idea of emergence translated into creation niches as a structural context that may nurture novel practices, around which further structures may develop. The principle of focusing on front runners is based on the idea that they, by connecting experiments to exogenous changes, may promote what is called in CAS dissipative structures, i.e. (regime) structures that lose energy to the environment. Simultaneously, they may develop novel visions that provide guidance to distributed practices, yielding a novel attractor in the system: the principle of promoting radical change through small, interconnected steps.

Through such correspondences between CAS and TM, transition management may, generically as well as in specific cases, be developed further on the basis of modelling (Haxeltine/Whitmarsh/Bergman et al. 2008) in ways that are still being further developed (De Haan/Ferguson/Adamowicz et al. 2014; Holtz/Alkemade/de Haan et al. 2015). Other recent work on transition management focuses on more detailed understanding of specific aspects, such as the role of agency (Brown/Farrelly/Loorbach 2013; Rauschmayer/Bauler/Schäpke 2015), urban transition labs (Nevens/Frantzeskaki/Gorissen et al. 2013) and partnerships in TM (Frantzeskaki/Wittmayer/Loorbach et al. 2014). There is also work applying it to novel domains, such as the financial crisis (Loorbach/Lijnis-Hüffenreuter 2013) and Agenda 21 (Wittmayer/van Steenbergen/Rok et al. 2015), as well as work which focuses more on particular places, like cities, than on a domain (Wittmayer/van Steenbergen/Rok et al. forthcoming).

4.4   Reflexive Governance and Reflexive Design

4.4.1   Understanding Transition Dynamics

The final approach to transition studies is the governance approach (Grin/van de Graaf/Vergragt 2003; Grin 2010, see also Voß/Bauknecht/Kemp 2006; Avelino 2009, 2011). This designation may be somewhat confusing, given that the two approaches described in the preceding two sections also discuss the governance of transitions (the second central question in this chapter). This third approach is nevertheless so called as it has its home base in governance studies. It differs from both the sociotechnical and the complex systems approaches in how the answers to the questions that guide this chapter (how to understand and how to shape transitions) are related. In the sociotechnical approach, the second question was answered by discussing strategic niche management, a governance concept derived from MLP-based theory on transition dynamics. In the second approach, the answers to both questions are derived from complex systems theory, while methodologically, the further development of either informs the other.

The governance approach draws substantially on results of studies of transition dynamics from both other approaches (especially MLP insights into transition dynamics, and the notion of adaptation in CAS), but opens them up from the governance question. Basically, this requires a dual vision of the dynamics of transition: iterating between a helicopter view and the perspective of a performing, situated actor. This is because the essence of governance is in the interaction between context and agency; the art of governance is thus in skilfully employing dual vision, like a person wearing Google glasses. Moreover, as we will

see, it is an insight into this relationship which enables us to understand, unlike with both other approaches, the politics evoked by transitions as part and parcel of transition dynamics, thus opening up new ways of dealing with such politics.

To develop such a dual vision, several points of departure are needed. First, structuration theory is used to understand how the agency involved in transitions depends on the structural context (Grin/van de Graaf/Vergragt 2003; cf. Grin 2006). While niche and regime are both sets of practices and associated structures, it is stressed that the regime may also been seen as structuring niche practices. Regular practices are, by definition, privileged by the incumbent regime, while niche practices may face inertia as the regime tends to 'draw' them back into regular practices, and does not offer proper rules and resources. This means that there is little support for their development, while actors resisting those niche practices may be better served by the regime (Roep/van der Ploeg/Wiskerke 2003; Grin/van de Graaf/Vergragt 2003; Grin/Felix/Bos et al. 2004; Bos/Grin 2008; Geels 2014). While TM acknowledges the inertia and resistance it may encounter, the governance approach offers opportunities for a more fine-grained, anticipatory analysis.

Second, transition dynamics is explicitly situated in the real world, i.e. in the context of ongoing long-term ('landscape') trends such as Europeanization, individualization and the politicization of side effects, and their interaction with the practices and structures of various societal domains. Regarding structure, attention is drawn to the nature of the four key institutions of modern society (market, state, civil society and knowledge) and their mutual alignments in various arrangements: the market system, the governance system and the innovation system. These institutions and their mutual alignment (together designated the 'institutional rectangle') have co-evolved with dominant practices and therefore differ between domains. Yet it is possible to discern some more or less generic trends (Grin 2010: 237-248), such as the changing/increasing role of civil society in the innovation system (the societization of science and technology since the 1970s), the governance system (in the wake of the emergence of the new social movements) and the market system (e.g. partnerships between NGOs and firms). This has led to less rigid, pre-given functional differentiation, making practices more contingent upon contextual agency. A second generic trend is re-spatialization of these systems due to globalization and Europeanization: the locus of activities has become not just the context, but also the outcome of agency, which may more or less strategically choose between a variety of venues (see also Späth/Rohracher 2010, 2012, 2014; Hodson/Marvin 2010).

Within the governance approach, such developments are an important part of the helicopter view implied in the dual vision. As such, they acquire additional meaning when understood against the backdrop of modernization theory, especially the versions espoused by Beck (1997) and Giddens (1991). From this perspective, transitions essentially become a matter of (1) redirecting the co-evolution of structure and agency towards (2) an orientation which goes beyond the control-mode orientation characterizing 'first' or 'simple' modernity (Beck 1997), as well as beyond the categorizations and identities of the 'emancipatory politics' of that era (Giddens 1991), and takes sustainable development as a normative orientation. This has to be achieved (3) amidst the turbulence of a variety of exogenous trends in the real world, as discussed above.

Crucial in the process of re-orientation is reflexivity, understood as what Voß and Kemp (2006) have called 'second-order reflexivity.' While 'first-order reflexivity' captures the unconscious and unintended, 'reflex-like', consequences (side effects and risks) of early modernization processes, second-order reflexivity is about self-critical and self-conscious reflection on the processes of modernity. It evokes a sense of agency, intention and change. Here actors reflect on and confront not only the self-induced problems of modernity, but also the approaches, structures and systems that reproduce them (Grin/Felix/Bos et al. 2004; Stirling 2006). It is this emphasis on reflexivity that turns this approach into a form of reflexive governance (Voß/Bauknecht/Kemp 2006).

4.4.2   Shaping Transition Dynamics

Crucial to shaping transitions in this approach is that changes at the level of practice and the regime inform each other. Niche practices central to SNM need regime changes in order to develop and spread. Conversely, regime changes undertaken to promote novel practices need to be accompanied by experiments at the practice level to determine whether they work out well or need to be adapted: a central element of TM. While direct interaction between agents working to achieve change at the two levels is important and elaborated in this 'dual-track governance', intermediaries may also play a pivotal role (Grin 2006, 2010). Research into care farming (the therapeutic use of farming practices) has shown that the embedding of

these intermediaries into one or more regimes is a crucial influence on their effectiveness (Hassink/ Hulsink/Grin 2014; Grin 2014). These insights into three different kinds of 'governance work', all in their own way requiring a dual vision, have been deepened on the basis of three different strands of planning theory, focusing on the design of local practices, the shaping of such practices through the adaptation of their structural embedding, and the interconnecting of 'local' and 'global' changes (Grin 2010).

A key aspect of the governance of transitions are its politics, including *powering*. The lack of an adequate conception of power in early versions of the sociotechnical and complex system approaches has been rightly criticized by a variety of authors, who have pointed out that there are a priori grounds for expecting (i) resistance from incumbent actors (Meadowcroft 2007: 308); (ii) inertia from the incumbent regime (Meadowcroft 2005: 488-489); and (iii) competing processes of change (associated with e.g. liberalization or Europeanization), which may or may not help to promote a transition towards a particular, desired direction (Shove/Walker 2007: 767; Meadowcroft 2005: 490). Others have raised the question of how transitions may be influenced in a situation of distributed power (Meadowcroft 2007: 308-312).

Fundamental to addressing this shortcoming is the realization (Grin 2010) that power is not only a quality of actors and their interaction per se. Power is also embedded in structure, something that Arts and Van Tatenhove (2004) have called 'dispositional power.' Transitions by their nature involve structural change, due both to changes in practice and to *longue durée* trends at the MLP's 'landscape' level (which thus also embodies power—structuring power, as Arts and Van Tatenhove call it). Therefore, transition dynamics may as much shape power dynamics as the other way around—in fact, they may be seen as implied in each other. As a corollary, the more fruitful question about power in transition becomes: how may power dynamics and transition dynamics reinforce each other over time?

This opens up novel perspectives on 'powering' in transitions. For example, *longue durée* trends do not determine transition dynamics, but may be key to it since agents may strategically draw upon them in order to interfere in that interaction. Thus, while it has often been claimed that globalization and liberalization by their very nature work against sustainable development, the achievement of the MacSharry reforms in the EU's agricultural policy (Grin/Marijnen 2011) has been analysed as a counter-example, in which EU commissioner MacSharry strategically drew upon these trends to push for a structural change in the common agricultural policy. Similarly, then minister Mansholt drew on post-war inclinations towards 'Americanization' to promote the modernization of Dutch agriculture, 1945-1970 (Grin 2012). Stirling (2014) has recently argued that contemporarily there seems to be a window of opportunity for a transition to a non-fossil energy regime.

Hoffman has further developed this understanding of powering as part and parcel of transition dynamics, and significantly advanced the point theoretically, framing transitions as intertwined processes of creativity and powering. Inertia and resistance comprise powering and provoke creativity, which in its turn may help open up novel modes of powering. Creativity and powering are understood in relational terms, i.e. the work of creativity and powering is conceived as relating innovative practices to structural elements in relevant fields, while simultaneously adapting these practices to these fields (Hoffman 2013; Hoffman/Loeber forthcoming).

We may conceive of legitimization in a similar way. As Hendriks and Grin (2007) have shown, legitimization often arises in and through the contestation occurring at the interfaces between transition practices and the spheres in which they are embedded. And legitimization too may be understood as relating a novel practice to resources embedded at the regime level (Hassink/Hulsink/Grin 2014).

Seeing and exploiting opportunities for reconnecting practices, regimes and landscape trends involves second-order reflexivity. As we have seen, this may arise from the confrontation between transition practices and a nearby regime. Yet proposals have also been developed within the governance approach for more deliberate instances of the designing of both visions and practices. This has been called (Grin/Felix/Bos et al. 2004) reflexive design or reflexive interactive design (Bos/Groot Koerkamp 2009). It has been epistemologically grounded (Bos 2008) and further developed through practical experience (with actual societal impact) in the area of livestock systems (for an overview of this evolution, see Bos/Grin 2012). Its micro-dynamics have been analysed in terms of how to shift from a perspective rooted in past experience towards one putting a 'different future' centre stage (Lissandrello/Grin 2011). Schuitmaker (2012) has developed a method for identifying which structural regime elements need to change by analysing barriers in transition practices from the perspective of a historical-critical analysis of the regime (Schuit-

maker 2012). Inghelbrecht (2016) has explored how the results of reflexive design exercises may be written into material designs.

The governance approach has thus strongly drawn on the two other approaches' theorizing of the dynamics of transitions, but at the same time added conceptual starting points for understanding the politics of such dynamics, and translated this into proven methods for shaping transitions. By now, this more accurate political understanding is being developed both for the MLP-based sociotechnical approach (e.g. Geels 2014; Raven/Kern/Verhees et al. forthcoming) and for transition management, where the early work by Avelino (2009, 2011) and Kern (Kern/Howlett 2009; Smith/Kern 2009) has now been joined by, for example, Brown/Farrelly/Loorbach et al. (2013), Frantzeskaki/Wittmayer/Loorbach et al. (2014), Voß (2014) and Wittmayer/van Steenbergen/Rok et al. (forthcoming). Apart from its place in these three original approaches, the politics of transitions has become an object of study in and of itself (e.g. Seyfang/Haxeltine 2012; Pel/Teisman/Boons et al. 2012; Jørgensen 2012; Goulden/Bedwell/Rennick-Egglestone et al. 2014; McGuirk/Bulkeley/Dowling 2014; Pesch 2015; Avelino/Grin/Jhagroe/Pel forthcoming).

4.5   The Social Practice Theory and Technical Innovation System Approaches

While, as has just been implied, a range of new approaches has emerged by now, two other approaches emerged simultaneously with the three just discussed. The first is *social practice theory* (Shove/Pantzar 2005; Spaargaren/Martens/Beckers 2006). It seeks to fill two important gaps (Shove/Walker 2007; Spaargaren/Oosterveer/Loeber 2012) in MLP theorizing: the lack of properly accounting for agency, and the reduction of consumption to one dimension of the regime of a system defined by its supply side, thus analytically excluding the possibility of change (co-)caused by changes in consumption. Recognizing that agency should be understood not individually, as rational choice theory would, but as social and embedded, the proponents of this approach listed above have sought to address both shortcomings through bringing into transition theory the concept of social practices, i.e. everyday practices (such as cooking, laundering, getting the children to school), as they are typically and habitually performed in society. In a seminal article, Reckwitz (2002) has pointed out that such practices bring together, in a particular way, heterogeneous elements such as bodily and mental activities, knowledge and skills, financial resources, material artefacts, and so on. Showering, for instances, brings together things as different as particular standards of cleanliness, a particular bathroom design, a water supply infrastructure, and particular soaps (Shove 2004).

Spaargaren, Oosterveer and Loeber (2012) have connected to Grin's (2010) understanding of the MLP as depicting different levels of structuration of practices and integrated the notion of embedded social practices into the MLP. This introduces as part of the regime objects like refrigerators, intermediary infrastructures like (globalizing) retail firms, and (induced by individualization and cultural globalization) a diversity of lifestyles (Spaargaren/Oosterveer 2010). This approach was used as an analytical framework in an edited volume on transitions in the area of food in OECD countries, and led to a variety of theoretical contributions (Spaargaren/Loeber/Oosterveer 2012). In addition to a wider understanding of the regime, as just exemplified, it also demonstrates different methods of interaction between consumer niche practices and the regime, as well as interactions between the practices of consumption, retailing and production. Such insights may well be developed into an extension of the typology of transition pathways discussed in section 4.2. While the volume's analyses make clear that there is a gross power differential between consumers and retailers, there are also interesting examples of how critical consumers may, eventually, make a difference.

Shove and Walker (2010) provide a thoughtful analysis of how the London Traffic Congestion Zone Scheme has affected citizens' practices of going to work, taking children to school, and so on. In complex ways, the agency of consumers and the complex interference between different practices may yield (partly unintended) patterns, in addition to attempted interventions. In this finding resonate notions like self-organization and emergence, from the complex systems approach to transitions (4.3), which thus may be expanded to better account for the everyday practices of consumers or citizens.

The second approach is the *technical innovation system* approach. A central idea in this approach is that technical innovation requires an adequate innovation system to support the functions needed to nurture a sociotechnical innovation. These functions, derived from Jacobsson and Johnson (2000) and further elaborated by Hekkert, Suurs, Negro et al. (2007:

426-427), are: entrepreneurial activities, knowledge development, knowledge diffusion, guidance of the search, market formation, resource mobilization, and creation of legitimacy. It should be understood that 'innovation system' in this literature focuses on a specific technology, while what was called 'innovation system' in the previous section fulfils similar functions, but on the level of a particular domain, where it is part of the regime.

Rooted, like the sociotechnical and complex system approaches, in evolutionary theory, a basic assumption is that more fundamental innovations, like those involved in a transition, may often not be embedded in an appropriate innovation system. In such cases, so-called 'systemic instruments' (Smits/ Kuhlman 2004) may play important roles: platforms situated in between the institutions of state, market and society, and that bring about the structural connections between the elements needed to support innovation functions. Wieczorek (2014) has contributed to the further development of the notion of a system instrument by elaborating how structural system deficits and systemic problems can be related, so as to inform the design of system instruments.

Hekkert, Suurs, Negro et al. (2007) stress that innovations entail many interactions between these functions, and they mention two 'virtuous cycles' of functions that often occur. One motor driving such virtuous cycles is entrepreneurial action for better structural conditions. Another motor is driven by the 'guidance of the search' function. These are the engines that may drive the emergence and transformation of innovation systems. Meanwhile, empirical studies (e.g. Suurs/Hekkert 2009; Negro/Alkemade/ Hekkert 2012; Gabaldon-Estevan/Hekkert 2013) have yielded significant insight into the driving mechanisms.

From the approaches to transition studies discussed above, the sociotechnical approach is obviously closest to the technical innovation system approach, as both have a sociotechnical focus. Reviews of the relationship between these two conceptualizations (Bergek/Hekkert/Jacobsson 2008; Bergek/Hekkert/ Jacobsson et al. 2015; Markard/Truffer 2008) agree that a *key difference* between the two is that the technical innovation system approach has an individual technology as a focus, and sees the system as comprised of those elements needed for that specific technical innovation, while the sociotechnical approaches focus on the transformation of a system that will fulfil a particular societal need (like food or energy) or, in later work, serve a particular place (like a region or city). A key point that hitherto has remained unclear is the nature and dynamics relationship between innovation system and regime. While in particular more recent work on technical innovation systems (e.g. Bergek/Hekkert/Jacobsson et al. 2015; Markard/Wirth/Truffer 2016) pays significant attention to their interaction with 'context', the nature of that context remains under-theorized, comprising both 'regimes' (explicitly understood as in MLP literature), and other elements (like legal systems) that in MLP theory would be included under the 'regime'. Also, the dynamics of that context is much less theorized than in MLP-based literature. There seems little interest in integrating insights on, for example, transition pathways, or in an understanding of the differences in temporal scales involved in (slow) changes in the incumbent structural context, the latter's day-to-day impact on innovative practices, and the way in which very slow changes in the wider context ('landscape' in MLP terms) may affect direct structural context.

Yet this literature furnishes an important understanding of the micro-dynamics of building a proper innovation system around a technical innovation, that is, into how a niche-regime may emerge and/or niche experiments may induce regime change. In addition, these insights may be of wider significance: it seems that it is possible to understand regime-building in similar terms—in a wider sense around innovations of a primarily non-technical nature (Hassink/Hulsink/Grin 2016). In order to fully benefit from this contribution to understanding the dynamics and shaping of transitions, such insights should firstly be seen as relevant to the work involved in creating or connecting to new structural elements while drawing on existing ones, and secondly be integrated with MLP theory, including its understanding of pathways.

4.6 Concluding Reflections

In the preceding sections, the three main approaches to transition studies were reviewed. In terms of the first guiding question, they share a portrayal of transitions as interactions between experiments, regime changes and exogenous trends. The sociotechnical approach, empirically rooted in historical case studies, has yielded transition pathways that enable discussion of possible patterns of interaction between these levels in more or less everyday terms. The complex systems approach has generated a more analytical insight into the underlying mechanisms, summarized in various types of transition patterns that underlie the soci-

otechnical pathways. Further conceptual integration would probably promote further development of either. The third, governance theory based, approach draws on these insights while adding an understanding of how the 'politics' (powering, legitimization) of transitions are part and parcel of transition dynamics—thus yielding new insight into both transition dynamics and its politics.

Social practice theory has contributed insight into the agency in the embedded practices which shape transition dynamics, including the previously under-emphasized agency of users and consumers. It can, and should, be better integrated into the MLP. Technical innovation systems theory has a lot to offer in terms of insight into the drivers of transition patterns; this too should be further explored in future research.

Further development of these various approaches and their relations may be both used in and informed by emerging research into novel objects (the financial crisis, the circular economy, the transformation of the welfare state, etc.) as well as novel units of analysis—the unit of analysis of transition research (Grin 2008) has expanded from consisting of mainly societal domains, such as energy, agrifood, mobility, water management and health care, to also including, for example, cities or regions. These relatively recent strands of research may offer novel insights into these original approaches.

Another more recent development in transition research is an increasing awareness of the role of place and space in all five approaches discussed above (Smith/Voß/Grin 2010; Coenen/Raven/Verbong 2010; Truffer/Coenen 2012; Raven/Schot/Berkhout 2012; Wieczorek/Raven/Berkhout 2015; Hansen/Coenen 2015). Places and connections shape regime-niche dynamics for all units of analysis; the governance and social practices approaches have highlighted the importance of proper analysis of the interaction between local, embedded, practices and globalization; and place and space are obviously particularly important when studying regions or cities.

The critical debate about the approaches discussed above may increase over the next few years. Work from about 2008 onwards owes a lot to such criticism, and the debate is still proving useful for further theoretical development (Avelino/Grin/Jhagroe et al. forthcoming). A final ongoing trend concerns the increasing amount of attention being paid to non-developed countries (e.g. Patankar/Patwardhan/Verbong 2010; Vreugdenhil/Taljaard/Slinger 2012; Fatimah/Raven/Arora 2015; Jolly/Raven 2015; Pant/Adhikari/Bhattarai 2015—and many chapters in this volume). Such studies may help to further develop the various theoretical approaches, not only because novel contexts mean a more varied empirical basis, but also because comparison of European with non-European cases may shed new light on the former—they may, for instance, reveal how much regime

dynamics is intertwined with the development of the capitalist welfare state; or how (reflexive) modernization takes different forms in different places. All in all, transition studies is here to stay for a while, and for good reasons.

References

Arts, Bas; van Tatenhove, Jan, 2005: "Policy and power: A conceptual framework between the 'old' and 'new' policy idioms", in: *Policy Sciences*, 37,3-4: 339-356.

Avelino, Flor, 2009: "Empowerment and the challenge of applying transition management to ongoing projects", in: *Policy Sciences*, 42,4: 369-390.

Avelino, Flor, 2011: *Power in Transition: Empowering Discourses on Sustainability Transitions* (PhD dissertation, Erasmus University Rotterdam).

Avelino, Flor; Grin, John; Jhagroe, Shivant; Pel, Bonno, forthcoming 2016: Special issue of *Journal of Environmental Policy and Planning* on 'The Politics of Sustainability Transitions: connecting theories and dispersed struggles'.

Beck, Ulrich, 1997: *The re-invention of politics. Rethinking Modernity in the Global Social Order* (Cambridge: Polity Press).

Bergek, Anna; Hekkert, Marko; Jacobsson, Staffan, 2008: "Functions in innovation systems: a framework for analysing energy system dynamics and identifying goals for system-building activities by entrepreneurs and policy makers", in: Foxon, Timothy; Köhler, Jonathan; Oughton, Christine (Eds.): *Innovation for a Low Carbon Economy: Economic, Institutional and Management Approaches* (Cheltenham: Edward Elgar).

Bergek, Anna; Hekkert, Marko; Jacobsson, Staffan; Markard, Jochen; Sandén, Björn; Truffer, Bernhard, 2015: "Technological innovation systems in contexts: Conceptualizing contextual structures and interaction dynamics", in: *Environmental Innovation and Societal Transitions*, 16,3: 51-64.

Berkhout, Frans; Angel, David, Wieczorek Anna J. (Eds.), 2009. "Sustainability transitions in developing Asia: Are alternative development pathways likely?", Special issue of *Technological Forecasting and Social Change*, 76,2: 215-290.

Berkhout, Frans; Verbong, Geert, Wieczorek, Anna J. (Eds.), 2010: "Socio-technical experiments in Asia—a driver for sustainability transitions?", Special issue of *Environmental Science and Policy*, 13,4: 261: 338.

Boon, Wouter P. C.; Moors, Ellen H. M.; Meijer, Albert J., 2014: "Exploring dynamics and strategies of niche protection", in: *Research Policy*, 43: 792-803.

Bos, Bram, 2008: "Instrumentalization Theory and Reflexive Design in Animal Husbandry", in: *Social Epistemology*, 22,1: 29-50.

Bos, Bram; Grin, John, 2008: "'Doing' Reflexive Modernization in Pig Husbandry: The Hard Work of Changing the Course of a River", in: *Science, Technology & Human Values*, 33,4: 480-507.

Bos, Bram; Grin, John, 2012: "Reflexive interactive design as an instrument for dual track governance", chapter 7 in: Barbier M.; Elzen B. (Eds.) *System Innovations, Knowledge Regimes, and Design Practices towards Transitions for Sustainable Agriculture* (Inra [online], posted 20 November 2012).

Bos, Bram; Groot Koerkamp, Peter W. G., 2009: "Synthesizing needs in system innovation through methodical design. A methodical outline on the role of needs in Reflexive Interactive Design (RIO)", in: Poppe, Krijn J.; Termeer, Kathrien J. A. M.; Slingerland, Maja (Eds.) *Transitions towards sustainable agriculture and food chains in peri-urban areas* (Wageningen: Wageningen Academic Publishers): 219-238.

Braudel, Fernand, 1980: *On History* (Chicago: University of Chicago Press).

Brown, Rebecah R.; Farrelly, Megan; Loorbach, Derk A., 2013: "Actors working the institutions in sustainability transitions: The case of Melbourne's stormwater management", in: *Global Environmental Change*, 23,4: 701-718.

Chappin, Emile J. L.; Ligtvoet, Andreas, 2014: "Transition and transformation: A bibliometric analysis of two scientific networks researching socio-technical change", in: *Renewable and Sustainable Energy Reviews*, 30: 715-723.

Coenen, Lars; Raven, Rob P. J. M.; Verbong, Geert P. J., 2010: "Local niche experimentation in the energy transitions: a theoretical and empirical exploration of proximity and disadvantages", in *Technology in Society*, 32,4: 295-302.

de Haan, Fjalar J.; Ferguson, Briony C.; Adamowicz, Rachelle C.; Johnstone, Phillip; Brown, Rebekah R.; Wong, Tony H. F., 2014: "The needs of society: A new understanding of transitions, sustainability and liveability", in: *Technological Forecasting and Social Change*, 85: 121-132.

de Haan, J.; Rotmans, Jan, 2011: "Patterns in transitions: Understanding complex chains of change", in: *Technological Forecasting and Social Change*, 78,1: 90-102.

Elzen, Boelie; Geels, Frank W.; Hofman, Peter S.; Green, Ken, 2004: "Socio-technical scenarios as a tool for transition policy: an example from the traffic and transport domain'", in: Elzen, B.; Geels, Frank W.; Green, K. (Eds.): *System Innovation and the Transition to Sustainability. Theory, Evidence and Policy* (Cheltenham: Edward Elgar): 251-281.

Fatimah, Yuti Ariani; Raven, Rob P. J. M.; Arora, Saurabh, 2015: "Scripts in transition: Protective spaces of Indonesian biofuel villages", in: *Technological Forecasting and Social Change*, 99: 1-13.

Foxon, Timothy J.; Hammond, Geoffrey P.; Pearson, Peter J. G., 2010: "Developing transition pathways for a low carbon electricity system in the UK", in: *Technological Forecasting and Social Change*, 77,8: 1203-1213.

Frantzeskaki, Niki; Wittmayer, Julia; Loorbach, Derk, 2014: "The role of partnerships in 'realizing' urban sustainability in Rotterdam's City Ports Area, the Netherlands", in: *Journal of Cleaner Production*, 65: 406-417.

Gabaldon-Estevan, Daniel; Hekkert, Marko P., 2013: "How Does the Innovation System in the Spanish Ceramic Tile Sector Function?", in: *Boletin de la Sociedad Espanola de Ceramica y Vidrio*, 52,3: 151-158.

Geels, Frank W., 2005: *Technological Transitions and System Innovations: A Co-Evolutionary and Socio-Technical Analysis* (Cheltenham: Edward Elgar).

Geels, Frank W., 2014: "Regime Resistance against Low-Carbon Transitions: Introducing Politics and Power into the Multi-Level Perspective", in: *Theory Culture & Society*, 31,5: 21-40.

Geels, Frank W.; Schot, Johan W., 2007: "Typology of socio-technical transition pathways", in: *Research Policy*, 36,3: 399-417.

Giddens, Anthony, 1991: *Modernity and self-identity: self and society in the late modern age* (Cambridge: Polity Press).

Goulden, Murray; Bedwell, Ben; Rennick-Egglestone, Stefan; Rodden, Tom; Spence, Alexa, 2014: "Smart grids, smart users? The role of the user in demand side management", in: *Energy Research & Social Science*, 2: 21-2.

Grin, John, 2006: "Reflexive modernization as a governance issue—or: designing and shaping *Re*-structuration", in: Voß, Jan-Peter; Bauknecht, Dierk; Kemp, René (Eds.): *Reflexive Governance for Sustainable Development* (Cheltenham: Edward Elgar): 54-81.

Grin, John, 2008: "The Multi-Level Perspective and the design of system innovations", chapter 3, in: van den Bergh, Jeroen C. J. M.; Bruinsma, Frank R. (Eds.): *Managing the transition to renewable energy: theory and macro-regional practice* (Cheltenham: Edward Elgar): 47-80.

Grin, John, 2010: "Understanding Transitions from a Governance Perspective", Part III,) in: Grin, John; Rotmans, Jan; Schot, Johan, 2010: *Transitions to Sustainable Development. New Directions in the Study of Long Term Structural Change* (New York: Routledge): 223-314.

Grin, John, 2012: "The politics of transition governance in Dutch agriculture. Conceptual understanding and impli-

cations for transition management", in: *Int. J. Sustainable Development*, 15,2: 72-89.

Grin, John, 2014: "Verandering bewerken in een veranderende context: lessen uijt de transitie van de Nederlandse landbouw", in: *Res Publica*, 56,4: 455-480.

Grin, John; Felix, Francisca; Bos, Bram; Spoelstra, Sierk, 2004: "Practices for reflexive design: lessons from a Dutch programme on sustainable agriculture", in: *Int. J. Foresight and Innovation Policy*, 1,1-2: 126-149.

Grin, John; Marijnen, Esther, 2011: "Global Threats, Global Changes and Connected Communities in the Agrofood system", in: Brauch, Hans Günter; Oswald Spring, Úrsula; Mesjasz, Czeslaw; Grin, John; Kameri-Mbote, Patricia; Chourou, Béchir; Dunay, Pal; Birkmann, Jörn (Eds.), 2011: *Coping with Global Environmental Change, Disasters and Security–Threats, Challenges, Vulnerabilities and Risks* (Berlin–Heidelberg–New York: Springer-Verlag): 1005-1018.

Grin, John; Rotmans, Jan; Schot, Johan (with contributions by Geels, Frank; Loorbach, Derk), 2010: *Transitions to Sustainable Development. New Directions in the Study of Long Term Structural Change* (New York: Routledge).

Grin, John; van de Graaf, Henk; Vergragt, Philip, 2003: "Een derde generatie milieubeleid: Een sociologisch perspectief en een beleidswetenschappelijk programma", in: *Beleidswetenschap*, 17,1: 51-72.

Gunderson, Lance H.; Holling, C. S. (Eds.), 2002: *Panarchy: understanding transformations in human and natural systems* (Washington DC: Island Press).

Hansen, Teis; Coenen, Lars, 2015: "The geography of sustainability transitions: Review, synthesis and reflections on an emergent research field", in: *Environmental Innovation and Societal Transitions*, 17: 92-109.

Hassink, Jan; Hulsink, Willem; Grin, John, 2016: "Entrepreneurship in agriculture and healthcare: Different entry strategies of care farmers", in: *Journal of Rural Studies*, 43,2: 27-39.

Hassink, Jan; Hulsink, Wim; Grin, John, 2014: "Farming with care: the evolution of care farming in the Netherlands", in: *NJAS–Wageningen Journal of Life Sciences*, 68,7: 1-11.

Haxeltine, Axel; Whitmarsh, Lorraine; Bergman, Noam; Rotmans, Jan; Schilperoord, Michel; Köhler, Jonathan, 2008: "A Conceptual Framework for transition modelling", in: *International Innovation and Sustainable Development*, 3,1-2: 93-114.

Hekkert, Marko; Suurs, Roald; Negro, Simona; Smits, Ruud; Kuhlman, Stefan, 2007: "Functions of innovation systems: a new approach for analysing technological change", in: *Technological Forecasting and Social Change*, 74,4: 413-432.

Hendriks, Carolyn M.; Grin, John, 2007: "Contextualising Reflexive Governance: The politics of Dutch transitions to sustainability", in: *Journal of Environmental Policy & Planning*, 9,3-4: 1-17.

Hodson, Mike; Marvin, Simon, 2010: "Can cities shape socio-technical transitions and how would we know if they were?", in: *Research Policy*, 39: 477-485.

Hoffman, Jesse G., 2013: "Theorizing power in transition studies: the role of creativity in novel practices in structural change", in: *Policy Sciences*, 46,3: 257-275.

Hoffman, Jesse; Loeber, Anne, forthcoming: "Exploring the Micro-politics in Transitions from a Practice Perspective: The Case of Greenhouse Innovation in the Netherlands", in: *Journal of Environmental Policy and Planning*.

Holland, John H., 1996: *Hidden Order: How Adaptation Builds Complexity* (New York: Basic Books).

Holtz, Geor; Alkemade, Floortje; de Haan, Fjalar; Köhler, Jonathan; Trutnevyte, Evelina; Luthe, Tbias; Halbe, Johannes; Papachristos, George; Chappin, Emile; Kwakkel, Jan; Ruutu, Smpsa, 2015: "Prospects of modelling societal transitions: Position paper of an emerging community", in: *Environmental Innovation and Societal Transitions*, 17: 41-58.

Hoogma, Remco; Kemp, René; Schot, Johan; Truffer, Bernhard, 2002: *Experimenting for sustainable transport: the approach of strategic niche management* (London: Spon Press).

Inghelbrecht, Linde, 2016: *'GM Crops in the EU' as a Wicked Problem. On Technology, Morality and a Polarised Debate* (PhD dissertation, University of Ghent).

Jacobsson, Staffan; Johnson, Anna, 2000: "The diffusion of renewable energy: an analytical framework and key issues for research", in: *Energy Policy*, 28,9: 625-640.

Jolly, S.; Raven, R. P. J. M., 2015: "Collective institutional entrepreneurship and contestations in wind energy in India", in: *Renewable and Sustainable Energy Reviews*, 42: 999-1011.

Jørgensen, Ulrik, 2012: "Mapping and Navigating Transitions—The Multi-level Perspective compared with Arenas of Development", in: *Research Policy*, 41: 996-1010.

Kemp, René; Loorbach, Derk; Rotmans, Jan, 2007: "Transition management as a model for managing processes of co-evolution towards sustainable development", in: *International Journal of Sustainable Development & World Ecology*, 14,10: 78-91.

Kemp, René; Schot, Johan; Hoogma, Remco, 1998: "Regime Shifts to Sustainability through Processes of Niche Formation. The Approach of Strategic Niche Management", in: *Technology Analysis and Strategic Management*, 10,2: 175-195.

Kern, Florian; Howlett, Michael, 2009: "Implementing transition management as policy reforms: a case study of the Dutch energy sector", in: *Policy Sciences*, 42,4: 391-408.

Lissandrello, Enza; Grin, John, 2011: "Reflexive planning as design and work: lessons from the Port of Amsterdam", in: *Planning Theory and Practice*, 12,2: 223-248.

Loorbach, Derk A.; Huffenreuter, R. Lijnis, 2013: "Exploring the economic crisis from a transition management perspective", in: *Environmental Innovation and Societal Transitions*, 6: 35-46.

Loorbach, Derk, 2007: *Transition Management: New Mode of Governance for Sustainable Development* (Utrecht: International Books).

Loorbach, Derk, 2010: "Transition Management for Sustainable Development: A Prescriptive, Complexity-Based Governance Framework", in: *Governance: An International Journal of Policy, Administration, and Institutions*, 23,1: 161-183.g

Markard, Jochen; Raven, Rob; Truffer, Bernhard, 2012: "Sustainability transitions: An emerging field of research and its prospects", in: *Research Policy*, 41: 955-96.

Markard, Jochen; Truffer, Bernhard, 2008: "Technological innovation systems and the multi-level perspective: Towards an integrated framework", in: *Research Policy* 37,4: 596-615.

Markard, Jochen; Wirth, Steffen; Truffer, Bernhard, 2016: Institutional dynamics and technology legitimacy—A framework and a case study on biogas technology, in: *Research Policy* 45,1: 330-344.

Marletto, Gerardo, 2014: "Car and the city: Socio-technical transition pathways to 2030", in: *Technological Forecasting and Social Change*, 87 (September): 164-178.

McGuirk, Pauline; Bulkeley, Harriet; Dowling, Robyn, 2014: "Practices, programs and projects of urban carbon governance: perspectives from the Australian city", in: *Geoforum*, 52: 137-147.

Meadowcroft, James, 2005: "Environmental political economy, technological transitions and the state", in: *New Political Economy*, 10,4: 479-498.

Meadowcroft, James, 2007: "Who is in charge here? Governance for Sustainable Development in a Complex World", in: *Journal of Environmental Policy and Planning*, 9,3: 299-314.

Negro, Simona O.; Alkemade, Floortje; Hekkert, Marko P., 2012: "Why does renewable energy diffuse so slowly? A review of innovation system problems", in: *Renewable and Sustainable Energy Reviews*, 16,6: 3836-3846.

Nevens, Frank; Frantzeskaki, Niki; Gorissen, Leen; Loorbach, Derk, 2013: "Urban Transition Labs: co-creating transformative action for sustainable cities", in: *Journal of Cleaner Production*, 50,1: 111-112.

Pahl-Wostl, Claudia, 2002: "Participative and stakeholder-based policy design, evaluation and modeling processes", in: *Integrated Assessment*, 3,1: 3-14.

Pant, Laxmi P.; Adhikari, Bhim; Bhattarai, Kiran Kumari, 2015: "Adaptive Transition for Transformations to Sustainability in Developing Countries", in: *Current Opinion in Environmental Sustainability*, 14: 206-212.

Patankar, Mahesh; Patwardhan, Anand; Verbong, Geert, 2010: "A promising niche: waste to energy project in the Indian dairy sector", in: *Environmental Science & Policy*, 13,4: 282-290.

Pel, Bonno; Teisman, Geert; Boons, Frank, 2012: "Transition by translation: The Dutch travel information innovation cascade", in: Geels, Frank; Kemp, René; Dudley, Geoff; Lyons, Glenn (Eds.), 2012: *Automobility in transition?* (London: Routledge).

Pesch, Udo, 2015: "Tracing discursive space: Agency and change in sustainability transitions", in: *Technological Forecasting & Social Change*, 90: 379–388.

Rauschmayer, Felix; Bauler, Tom; Schäpke, Niko, 2015: "Towards a thick understanding of sustainability transitions–Linking transition management, capabilities and social practices", in: *Ecological Economics*, 109: 211-221.

Raven, Rob P. J. M., 2005: *Strategic niche management for biomass* (PhD thesis, Eindhoven University of Technology, the Netherlands).

Raven, Rob P. J. M., 2006: "Towards alternative trajectories? Reconfigurations in the Dutch electricity regime", in: *Research Policy*, 35,4: 581-595.

Raven, Rob P. J. M., 2007: "Co-evolution of waste and electricity regimes: multi-regime dynamics in the Netherlands (1969-2003)", in: *Energy Policy*, 35,4: 2197-2208.

Raven, Rob P. J. M.; Verbong, Geert P. J., 2007: "Multi-regime interactions in the Dutch energy sector. The case of combined heat and power in the Netherlands 1970-2000", in: *Technology Analysis and Strategic Management*, 19,4: 491-507.

Raven, Rob P. J. M.; Verbong, Geert P. J., 2009: "Boundary crossing innovations: case studies from the energy domain", in: *Technology in Society*, 31,1: 85-93.

Raven, Rob P. J. M.; Verbong, Geert P. J.; Schilpzand, Wouter F.; Witkamp, Marten J., 2011: "Translation mechanisms in socio-technical niches: a case study of Dutch river management", in: *Technology Analysis and Strategic Management*, 23,10: 1063-1078.

Raven, Rob; Kern, Florian; Verhees, Bram; Smith, Adrian (forthcoming): "Niche construction and empowerment through socio-political work. A meta-analysis of six low-carbon technology cases", in: *Environmental Innovation and Societal Transitions*.

Raven, Rob; Schot, Johan; Berkhout, Frans, 2012: "Space and scale in socio-technical transitions", in: *Environmental Innovation and Societal Transitions*, 4: 63-78.

Reckwitz, Andreas, 2002: "Toward a Theory of Social Practices", in: *European Journal of Sociology*, 5,2: 243-263.

Rip, Arie; Kemp, René, 1998: "Technological change", in: Rayner, Steve; Malone, Elizabeth L. (Eds.), *Human choice and climate change* (Columbus, Ohio: Battelle Press: 327-399.

Roep, Dirk; van der Ploeg, Jan Douwe; Wiskerke, Han S. C., 2003: "Managing technical-institutional design processes: some strategic lessons from environmental co-operatives in the Netherlands", in: *Netherlands Journal of Agrarian Studies*, 51,1-2: 195-217.

Rothman, Dale; Robinson, John, 1997: "Growing pains: a conceptual framework for considering integrated assessments", in: *Environmental Monitoring and Assessment*, 46,1-2: 23-43.

Rotmans, Jan, Schot, Johan; Grin, John (in collaboration with Derk Loorbach and Ruud Smits), 2004: *The KSI Knowledge Project. Knowledge Network for System Innovation: Transitions to a Sustainable Society*. Pro-posal for the Bsik programme of the Dutch Government.

Rotmans, Jan; Kemp, René, van Asselt, Marjolein; 2001: "More evolution than revolution: transition management in public policy", in: *Foresight*, 3,1: 15-31.

Rotmans, Jan; Loorbach, Derk, 2010: "Towards a Better Understanding of Transitions and Their Governance: A Systemic and Reflexive Approach", in: Grin, John; Rotmans, Jan; Schot, Johan (with contributions by Geels, Frank; Loorbach, Derk), Part II; *Transitions to Sustainable Development. New Directions in the Study of Long Term Structural Change* (New York: Routledge): 105-221.

Schot, Johan W.; Geels, Frank, W., 2008: "Strategic niche management and sustainable innovation journeys: theory, findings, research agenda, and policy", in: *Technology Analysis & Strategic Management*, 20,5: 537-554.

Schot, Johan W.; Geels, Frank W., 2010: "The dynamics of transitions: a sociohisotrical perspective", in: Grin, John; Rotmans, Jan; Schot, Johan (with contributions by Geels, Frank; Loorbach, Derk), Part I; *Transitions to Sustainable Development. New Directions in the Study of Long Term Structural Change* (New York: Routledge): 11-101.

Schot, Johan W.; Rip, Arie; Lintsen, Harry (Eds.), 2010: *Technology and the making of the Netherlands: The age of contested modernization, 1890-1970.* (Cambridge, Mass.: MIT Press–Zutphen: Walburg Pers).

Schot, Johan, 1998: "The usefulness of evolutionary models for explaining innovation. The case of the Netherlands in the nineteenth century", in: *History and Technology*, 14: 173-200.

Schot, Johan; Hoogma, Remco; Elzen, Boelie, 1994: "Strategies for shifting technological systems. The case of the automobile system", in: *Futures*, 26,10: 1060-1076.

Schuitmaker, Tjerk Jan, 2012: "Identifying and unravelling persistent problems", in: *Technological Forecasting & Social Change*, 79: 1021-1031.

Seyfang, Gill Y.; Haxeltine, Alex, 2012: "Growing Grassroots Innovations: Exploring the Role of Community-Based Initiatives in Governing Sustainable Energy Transitions", in: *Environment and Planning C Government & Policy*, 30,3: 381-400.

Seyfang, Gill; Hielscher, Sabine; Hargreaves, Tom; Martiskainen, Mari; Smith, Adrian, 2014: "A grassroots sustainable energy niche? Reflections on community energy in the UK", in: *Environmental Innovation and Societal Transitions*, 13: 21-44.

Shove, Elisabeth, 2004: "Sustainability, system innovation and the laundry", in: Elzen, Boelie; Geels, Frank; Green, Ken (Eds.): *System Innovation and the Transition to Sustainability: Theory, Evidence and Policy* (Cheltenham: Edward Elgar): 76-94.

Shove, Elisabeth, 2010: "Beyond the ABC: climate change policy and theories of social change", in: *Environment and Planning A*, 42,6: 1273-1285.

Shove, Elisabeth; Pantzar, Mika, 2005: "Consumers, Producers and Practices", in: *Journal of Consumer Culture*, 5,1: 43-64.

Shove, Elisabeth; Walker, Gordon, 2007: "CAUTION! Transitions ahead: politics, practice, and sustainable transition management", in: *Environment and Planning A*, 39,4: 763-770.

Shove, Elisabeth; Walker, Gordon, 2008: "Transition Management™ and the politics of shape shifting", in: *Environment and Planning A*, 40: 1012-1014.

Shove, Elisabeth; Walker, Gordon, 2010: "Governing transitions in the sustainability of everyday life", in: *Research Policy*, 39,4: 471-476.

Smith, Adrian, 2007: "Translating sustainabilities between green niches and sociotechnical regimes", in: *Technology Analysis & Strategic Management*, 19,4: 427-450.

Smith, Adrian; Kern, Florian, 2009: "The transitions storyline in Dutch environmental policy", in: *Environmental Politics*, 18,1: 78-98.

Smith, Adrian; Raven, Rob, 2012: "What is protective space? Reconsidering niches in transitions to sustainability", in: *Research Policy*, 41,6: 1025-1036

Smith, Adrian; Voß, Jan-Peter; Grin, John, 2010: "Innovation studies and sustainability transitions: the allure of adopting a broad perspective, and its challenges", in: *Research Policy*, 39: 435-448.

Smits, Ruud; Kuhlman, Stefan, 2004: "The rise of systemic instruments in innovation policy", in: *Int. J. Foresight and Innovation Policy*, 1,1-2: 4-32.

Spaargaren, Gert, 2003: "Sustainable Consumption: A Theoretical and Environmental Policy Perspective", in: *Society & Natural Resources*, 16,8: 687-701.

Spaargaren, Gert; Loeber, Anne M. C.; Oosterveer, Peter J. M., 2012: "Food futures in the making", in: Spaargaren, Gert; Oosterveer, Peter J. M.; Loeber, Anne M. C. (Eds.), 2012: *Food Practices in Transition—Changing Food Consumption, Retail and Production in the Age of Reflexive Modernity* (New York—London: Routledge).

Spaargaren, Gert; Martens, Susan; Beckers, Theo, 2006: "Sustainable technologies and everyday life", in: Verbeek, Peter-Paul; Slob, Adriaan (Eds.): *User Behaviour and Technology Development: Shaping Sustainable Relations between Consumers and Technologies* (Dordrecht: Springer): 107-118.

Spaargaren, Gert; Oosterveer, Peter J. M., 2010: "Citizen-Consumers as Agents of Change in Globalizing Modernity: The Case of Sustainable Consumption", in: *Sustainability*, 2,7: 1887-1908.

Spaargaren, Gert; Oosterveer, Peter J. M.; Loeber, Anne M. C., 2012: "Sustainable Transitions in Food Consumption, Retail and Production", in: Spaargaren, Gert; Oosterveer, Peter J. M.; Loeber, Anne M. C. (Eds.), 2012: *Food Practices in Transition—Changing Food Consumption, Retail and Production in the Age of Reflexive Modernity* (New York—London: Routledge).

Späth, Philipp; Rohracher, Harald, 2010: "'Energy regions': The transformative power of regional discourses on socio-technical futures, in: *Research Policy*, 39,4: 449-458.

Späth, Philipp; Rohracher, Harald, 2012: "'Energy regions': The transformative power of regional discourses on socio-technical futures", in: *Research Policy*, 39: 449-458.

Späth, Philipp; Rohracher, Harald, 2014: "Beyond Localism: The Spatial Scale and Scaling in Energy Transitions". in: Padt, Frans J. G.; Opdam, Paul F. M.; Polman, Nico B. P.; Termeer, Kathrien J. A. M. (Eds.): *Scale-sensitive Governance of the Environment* (Oxford: John Wiley): 106-121.

Stirling, Andrew, 2006: "Precaution, foresight and sustainability", in: Voß, Jan-Peter; Bauknecht, Dierk; Kemp, René (Eds.): *Reflexive governance for sustainable development* (Cheltenham: Edward Elgar): 225-272.

Stirling, Andrew, 2014: "Transforming power: Social science and the politics of energy choices", in: *Energy Research & Social Science*, 1: 83-95.

Suurs, Ruud A. A., Hekkert, Marko P., 2009: "Cumulative causation in the formation of a technological innovation system: the case of biofuels in the Netherlands", in: *Technol. Forecasting Social Change*, 76: 1003-1020.

Switzer, Andrew; Bertolini, Luca; Grin, John, 2013: "Transitions of Mobility Systems in Urban Regions: A Heuristic Framework", in: *J. Environmental Policy and Planning*, 15,2: 141-160.

Taanman, Mattijs, 2014: *Looking for Transitions: a monitoring approach for sustainable transition programmes* (PhD dissertation, Erasmus University Rotterdam).

Taanman, Matthijs; Wittmayer, Julia; Diepenmaat, Henk, 2012: "Monitoring on-going vision development in system change programmes", in: *Journal on Chain and Network Science*, 12,2: 125-136.

Truffer, Bernhard, Coenen, Lars, 2012: "Environmental Innovation and Sustainability Transitions in Regional Studies", in: *Regional Studies*, 46,1: 1-21.

Truffer, Bernhard; Metzner, André; Hoogma, Remco, 2004: "The coupling of viewing and doing: strategic niche management and the electrification of individual transport", in: *Greener Management International*, 37: 111-124.

Ulmanen, Johanna Heleena, 2013: *Exploring policy protection in biofuel niche development: a policy and strategic niche management analysis of Dutch and Swedish biofuel development, 1970-2010* (Eindhoven: Boxpress); at: <http://alexandria.tue.nl/extra2/762655.pdf>.

Ulmanen, Johanna Heleena; Verbong, Geert P. J.; Raven, Rob P. J. M., 2009: "Biofuel developments in Sweden and the Netherlands. Protection and socio-technical change in a long-term perspective", in: *Renewable and Sustainable Energy Reviews*, 13: 1406-1417.

Van den Bosch, Suzanne; Neuteboom, Jord (forthcoming): "The Making of a Transition Program in the Dutch Care Sector", in: Broerse, Jacqueline E. W.; Grin, John

(Eds.): *Towards system innovations in health systems. Understanding historical evolution, innovative practices and opportunities for a transition in healthcare* (New York: Routledge).

Van der Brugge, Rutger, 2009: *Transition Dynamics in Social-Ecological Systems. The Case of Dutch Water Management* (PhD dissertation, Erasmus University, Rotterdam).

Van der Brugge, Rutger; van Raak, Roel, 2007: "Facing the adaptive management challenge: insights from transition management", in: *Ecology and Society*, 12,2: 33 [online].

Van Driel, Hugo; Schot, Johan W., 2005: "Radical innovation as a multilevel process: introducing floating grain elevators in the Port of Rotterdam", in: *Technology and Culture*, 46,1: 51-76.

Verbong, Geert P. J.; Geels, Frank W., 2012: "Future electricity systems: visions, scenarios and transition pathways", in: Loorbach, Derk; Verbong, Geert P. J. (Eds.): *Governing the energy transition: reality, illusion or necessity?* (New York–Oxford: Routledge): 203-219.

Voß, Jan-Peter, 2014: "Performative policy studies: realizing 'transition management'", in: *Innovation: The European Journal of Social Science Research*, 27,4: 317-343.

Voß, Jan-Peter; Bauknecht, Dierk; Kemp, René (Eds.), 2006: *Reflexive governance for sustainable development* (Cheltenham: Edward Elgar).

Voß, Jan-Peter; Kemp, René, 2006: "Sustainability and reflexive governance: Introduction", in: Voß, Jan-Peter;

Bauknecht, Dierk; Kemp, René (Eds.), 2006: *Reflexive governance for sustainable development* (Cheltenham: Edward Elgar): 3-28.

Vreugdenhil, Heleen; Taljaard, Susan; Slinger, Jill H., 2012: "Pilot projects and their diffusion: A case study of integrated coastal management in South Africa", in: *International Journal of Sustainable Development*, 15,1-2: 148-172.

Weaver, Paul; Jansen, Leo; van Grootveld, Geert; van Spiegel, Egbert; Vergragt, Philip, 2000: *Sustainable technology development* (Sheffield: Greenleaf Publishing).

Weber, Matthias; Hoogma, Remco; Lane, Ben; Schot, Johan, 1999: "Experimenting with Sustainable Transport Innovations: A Workbook for Strategic Niche Management" (Wierden: Promotioneel Drukwerk Service).

Wieczorek, Anna J., 2014: *Towards sustainable innovation. Analysing and dealing with systemic problems in innovation systems* (PhD dissertation, Utrecht University).

Wieczorek, Anna J.; Raven, Rob; Berkhout, Frans, 2015: "Transnational linkages in sustainability experiments", in: *Environmental Innovation and Societal Transitions*, 17: 149-165.

Wittmayer, Julia; van Steenbergen, Frank van; Rok, Ania; Roorda, Chris (forthcoming): "Governing sustainability: a dialogue between Local Agenda 21 and transition management. Local Environment", in: *The International Journal of Justice and Sustainability*.

5 Transformative Science for Sustainability Transitions

Uwe Schneidewind[1], Mandy Singer-Brodowski[2]and Karoline Augenstein[3]

Abstract[4]

Sustainability Transitions require a knowledge production that contributes actively to the Grand Challenges of twenty-first-century societies. Scientific institutions play a key role in this domain in the transformation towards sustainability and peace. Against this background a Transformative Science is needed: a mode of science that not only analyses processes of transformation, but also actively supports and accelerates them. This chapter will introduce the concept of Transformative Science and its implications for (1) the methodologies of transdisciplinary and transformative research, (2) institutional capacity-building for facilitating such a research approach, and (3) the national science systems and national science policies that enable this new mode of knowledge production. The case of the German science system is introduced to describe an ongoing science system transition with special regard to the role of civil society organizations.

Keywords: Transformative science, science system transition, transformative research, transdisciplinarity, governance of science, civil society participation, Wuppertal Institute[5].

5.1 Need for and Definition of 'Transformative Science'

The concept of transformative science is closely connected to international debates about the 'Great Transformation' (WBGU 2011) or the 'grand challenges' announced by the EU (Reid/Chen/Goldfarb et al. 2010). The meaning of this concept and the new role it assigns to science and academia can only be fully comprehended against the background of these debates. Humanity in the twenty-first century is faced with massive upheaval and the challenge of guaranteeing a good life for nine billion people. Planetary boundaries (Rockström/Steffen/Noone et al. 2009) have been discovered that set clear limits for resource-intensive economies, for political systems that are not adequately oriented *towards* social welfare for all, and for the carefree continuation of today's lifestyles and

---

1 Prof. Dr. Uwe Schneidewind is president of the Wuppertal Institute for Climate, Environment and Energy and Professor for Sustainable Transition Management at the University of Wuppertal; Email: <uwe.schneidewind@wupperinst.org>.

2 Mandy Singer-Brodowski is a research fellow at the Wuppertal Institute for Climate, Environment and Energy and a founding member of the German network for student initiatives in sustainability; Email: <mandy.singer-brodowski@wupperinst.org>.

3 Karoline Augenstein is a research fellow at the Wuppertal Institute for Climate, Environment and Energy; Email: <karoline.augenstein@wupperinst.org>.

4 We would like to thank the reviewers for helpful comments and suggestions that significantly improved this chapter.

5 The Wuppertal Institute for Climate, Environment and Energy was founded in 1991, when decision-makers around the world became aware of global climate change caused by humankind as a new global challenge. For more than twenty years now, the Wuppertal Institute has undertaken research and developed models, strategies and instruments for transitions to a sustainable development at local, national and international level. Sustainability research at the Wuppertal Institute focuses on resource-, climate- and energy-related challenges and their relation to economy and society. Special emphasis is put on analysing and stimulating innovations that decouple economic growth and wealth from natural resource use. The overall research focus is on transition processes towards sustainable development. Scientific research towards this end combines its approaches to generate practical and actor-oriented solutions. An overview of current projects of the Wuppertal Institute can be found at: <http://wupperinst.org/en/projects/>.

© Springer International Publishing Switzerland 2016
H.G. Brauch et al. (eds.), *Handbook on Sustainability Transition and Sustainable Peace*,
Hexagon Series on Human and Environmental Security and Peace 10, DOI 10.1007/978-3-319-43884-9_5

patterns of consumption. The *German Advisory Council on Global Change* (WBGU) therefore claims in its flagship report *World in Transition* that a "great transformation towards a decarbonized society" is needed:

> Adding together all of these challenges involved in the transformation to come, it becomes clear that the upcoming changes go far beyond technological and technocratic reforms: the business of society must be founded on a new 'business basis'. *This is, in fact, all about a new global social contract for a low-carbon and sustainable global economic system* (WBGU 2011: 1).

At the international level, momentum is currently created by the UN's *Sustainable Development Goals* (SDGs). On the one hand, these address the slow progress towards more sustainable development (e.g. in the context of international climate negotiations about a binding agreement) and identify key fields of action. On the other hand, the SDGs serve as a new narrative that describes the necessary change processes; they offer a comprehensive framework for coordinating sustainable development efforts and strategies at regional, national and international levels, and for integrating them in a shared vision of a globally just and much less resource-intensive world society.

The emergence and steady growth of new fields of research, such as transition studies and related approaches focusing on a better understanding of complex system innovations, can also be explained as a phenomenon in this broader context. Current transformation challenges require radical change in infrastructures, institutions and lifestyles, and these have to be dealt with by different sectors and societal subsystems at different but interrelated levels.

In this transformation challenge, science and science policy plays a central—while often underestimated—role. Over the past years, science policy has come to be equated with innovation policy (Martin 2012) and this has contributed to the generation of unrestrained economic growth and the development and diffusion of technologies that have often caused severe and harmful side effects for society and the environment (Beck 1992). Therefore, achieving sustainable development crucially depends on the kind of knowledge that is produced in modern societies. Meeting the key societal challenges of today demands socially robust knowledge that can be applied under diverse, uncertain and unforeseeable conditions. The term *mode-2 science* (Nowotny/Scott/Gibbons 2001) emerged at the turn of the millennium and captures a new mode of scientific knowledge production, which facilitates and argues for a pluralization of the places where relevant knowledge is produced and of actors involved in the production of knowledge. A central claim in this debate is that a 're-contextualization' (Rip 2011: 5) of science in society is needed in order to accommodate the increasing demand for participation by non-scientific stakeholders and the growing complexity of knowledge production in the age of reflexive modernity.

In this context, transformative science is understood as a mode of science that not only analyses processes of transformation but also actively contributes to them: by developing new methodological approaches, and by explicitly focusing on the institutions shaping scientific knowledge production and science policy at the level of (national) science systems. The concept builds on successful experiences in the field of sustainability science, which has been established in the academic sphere from the early 2000s onwards (e.g. Clark/Dickson 2003; Kates/Clark/Corell et al. 2001). The WBGU has introduced the term 'transformative research' in its flagship report and defines it as research that analyses "transformation processes with regard to their causes, conditions and development" (WBGU 2011: 373) and actively contributes to "transformation processes through specific innovations in the relevant sectors" (WBGU 2011: 373).

Transformative research aims not only at delivering analyses of complex and sustainability-related systems but also at supporting transformational changes in sustainability transitions (Wiek/Ness/Schweizer-Ries et al. 2012). This transformational agenda goes back to the beginnings of the debate about sustainability science (Clark/Dickson 2003: 8059), but in the practice of carrying out sustainability research it is confronted with many challenges (Wiek/Ness/Schweizer-Ries et al. 2012). For instance, engaging an extended peer community for research processes in post-normal science, as suggested by Funtowicz and Ravetz (1993: 752ff.), or concrete user engagement in sustainability research (Talwar/Wiek/Robinson 2011) require a completely new understanding of the role of researchers in the process of knowledge production (Wittmayer/Schaepke 2014). Researchers aiming at participatory knowledge production in complex und uncertain systems and transition processes (Loorbach/Frantzeskaki/Thissen 2011) will need a whole range of new competencies.

Building on strategies of 'transformative research' and introducing the broader concept of 'transformative science' highlights the crucial fact that knowledge production in a transformative mode is always embedded in the institutional context of the established sci-

> **Box 5.1:** Four Transformative Pillars of the Knowledge Society. **Source**: German Advisory Council on Global Change (WBGU),2011: World in Transition: A Social Contract for Sustainability, Summary for Policy-Makers (Berlin: WBGU). This figure is licensed under Creative Commons BY-NC-ND; at: <http://www.wbgu.de/fileadmin/templates/dateien/veroeffentlichungen/hauptgutachten/jg2011/wbgu_jg2011_kurz_en.pdf>.
>
> The German Advisory Council on Global Change (WBGU 2011) distinguishes four transformative pillars of the knowledge society:
> - *Transformation research* focuses on "the basic principles, conditions and progression of transformation processes" (WBGU 2011: 351).
> - *Transformative research* "supports transformation processes in practical terms", e.g. by developing relevant technological and social innovations, applying methods for facilitating inter- and transdisciplinary research processes, and by actively including non-academic stakeholders (WBGU 2011: 351f.).
> - *Transformation education* "makes the findings of transformation research available to society" (WBGU 2011: 352).
> - *Transformative education* encourages changes in actual social practices and focuses on creating awareness of concrete options for action and solution approaches (WBGU 2011: 352).
>
>
> Typification of transformation research and education.

ence system, and this influences and bounds this specific quest for new forms of knowledge production.

Transformative science is not limited to research and education that focuses on analysing sustainability challenges and relevant systems; it goes beyond systems analysis and aims at catalysing and supporting transformation processes towards sustainable development through suitable forms of knowledge production and transfer. Consequently, the concept of transformative science has massive implications for (1) methodologies of transdisciplinary and transformative research, (2) institutional capacity-building for facilitating such a research approach in the field of sustainability science, and (3) national science systems, where this type of institutional change is the subject of controversial debate.

With regard to the methodological challenges discussed in the first part of this chapter, transformative research refers to the discourse on transdisciplinarity, which focuses on the development of research designs suitable for addressing sustainability-oriented research questions. A key aspect is the methodologically robust integration of the different forms of knowledge of different actors and stakeholders. Basic principles of transdisciplinary research have been adopted by global science programmes, such as Future Earth, by introducing concepts of 'Co-Design' and 'Co-Production' of knowledge, which explicitly provide a new role for non-scientific actors.

In addition, transformative science focuses on the institutional dimension regarding the places where knowledge is traditionally produced and reflects on the impact institutional structures have on the science–society interface. An institutional perspective is crucial because inter- and transdisciplinary research approaches have been faced with substantial barriers over the past years. Establishing such approaches requires institutional capacity-building that goes beyond project-based funding (Lyall/Fletcher 2013). In contrast to the methodological issues, the institutional dimension of the change envisaged in the science system remains understudied so far and will be discussed in the second part of the chapter.

Furthermore, the concept of transformative science is at the core of the political controversies that emerge when science actively assumes responsibility for societal developments. A number of controversial issues appear in this debate, for instance, the role of academic freedom versus the societal responsibility of academia, university autonomy versus university management, academic excellence versus transdisciplinarity, fostering innovation and ethics. Tensions in these fields are increased by the demands of external actors (e.g. civil society organizations) for a more open science system. By including a perspective on these issues, transformative science is positioned in the context of new theories of the governance of science, and the focus is explicitly on these negotiation and interaction processes. In the third part of this chapter, a discussion of these issues will be presented for the case of the German science system.

5.2   Methodological Challenges of a Transformative Research

A more detailed elaboration of the concept of 'transformative science' has to start by outlining transformative research, which is deeply intertwined with the discourse of transdisciplinarity and the basis of 'transformative science'.

Transformative research takes as its starting point various societal transformation challenges in the context of urban transitions, transitions to sustainable energy and transport systems, dealing with resource scarcity and the pressures caused by unlimited economic growth. The complexity of such transition processes can be illustrated by the example of the German energy transition ('Energiewende'), i.e. the nuclear phase-out and restructuring process of the entire German energy system, oriented towards increased energy efficiency and the development of renewable energies: This transformation not only includes the substitution of traditional technology but also supports social innovations between the affected stakeholders (companies, decision-makers, users etc.).

The energy transition is a typical example of a complex problem, in the sense that the object of study can hardly be separated from its context (Scholz/Lang/Wiek et al. 2006: 228). Such problems are deeply embedded in a complex system (Scholz/Tietje 2002). Most sustainability challenges can be described as 'wicked problems' which do not fit the disciplinary logic of academic problem definitions. Against the background of global unsustainable developments such as climate change, the term 'super-wicked problems' has been recently introduced. 'Super-wicked problems' are characterized by four key features: "time is running out; those who cause the problem also seek to provide a solution; the central authority needed to address them is weak or non-existent; and irrational discounting occurs that pushes responses into the future" (Levin/Cashore/Bernstein et al. 2012: 124). The urgent need to address these super-wicked problems using scientific approaches underlines once again the need for a transformational research agenda.

Nevertheless, transformative research basically draws on the methods developed in the field of transdisciplinary sustainability science, which emerged as a distinct field of research around the turn of the millennium (Clark 2007).

Transdisciplinarity can be defined as "a reflexive, integrative, method-driven scientific principle aiming at the solution or transition of societal problems and concurrently of related scientific problems by differentiating and integrating knowledge from various scientific and societal bodies of knowledge. This definition highlights the need for transdisciplinary research to comply with the following requirements: (a) focusing on societally relevant problems; (b) enabling mutual learning processes among researchers from different disciplines (from within academia and from other research institutions), as well as actors from outside academia; and (c) aiming at creating knowledge that is solution-oriented, socially robust (cf., for example, Gibbons 1999), and transferable to both scientific and societal practice." (Lang et al. 2012: 26f.).

This quotation summarizes the state of the art of the debate about transdisciplinarity and it shows that disciplinary approaches of defining relevant aspects of a broader research question and suitable methods for addressing these remain the core of good scientific practice. However, it is imperative that different disciplinary perspectives are connected and related to each other early in the research process, in order to be able to gain a comprehensive understanding of complex problems—and to cultivate a much deeper appreciation for the approaches and methods of other disciplines. In the case of transdisciplinary research processes, non-academic stakeholders are integrated as well, in order to provide relevant practical and transformation knowledge. Roland Scholz coined the simple yet ambitious phrase "disciplined interdisciplinarity in transdisciplinary processes" (Scholz 2011: XVII), which captures the central idea of the whole endeavour.

Thus, a disciplinary approach remains the essential starting point for transdisciplinary and transformative research. However, since sustainability science has to cope with complex systems and transitions processes, characterized by huge uncertainties and ambiguity (Kates/Clark/Corell et al. 2011, 641), there is an additional need for (1) cooperation in interdisciplinary teams and (2) participatory involvement of affected stakeholders (Loorbach/Frantzeskaki/Thissen 2011: 80f.). The basic idea is that such an approach leads to the production of socially robust knowledge.

5.2.1 Status Quo of Transdisciplinary Science?

An important milestone for transdisciplinary research was the conference "Transdisciplinarity: Joint Problem Solving among Science, Technology and Society" held in Zürich in 2000 (Klein/Grossenbacher-Mansuy/Häberli et al. 2001), which stimulated an intensive debate on the value, goals and processes of transdisciplinary research.

In transdisciplinary research processes, non-scientific actors are ideally integrated into all stages of a research process, i.e. from the formulation of the research question and the selection of methods to the discussion of findings and results. The basic aim is to arrive at a shared identification and systematization of the problem that is to be studied, and to ensure a continuous feedback process between researchers and non-scientific actors throughout the research process and the subsequent up-scaling of results (Hirsch Hadorn/Biber-Klemm/Grossenbacher-Mansuy et al. 2008). Following Jahn, Bergmann and Keil (2012: 7f.), this requires an integration of the perspectives of both actor groups at three levels:

1. at an epistemic level, the different kinds of knowledge (scientific knowledge and practical knowledge) of the involved actors need to be integrated;
2. at a social-organizational level, the varying interests and activities of involved actors need to be integrated;
3. at a communicative level, the different (professional) languages and forms of expression of the involved actors need to be integrated, in order to arrive at a shared understanding of the problem, the research process and the results; this also needs to be expressed in some form of language conversion.

The three levels of integration show that transdisciplinary knowledge production requires organizational framework conditions that facilitate cooperation between researchers and non-scientific actors at eye level. At the same time, creating acceptance and the conditions for epistemological pluralism and reflexivity is a central aim of institutional change for sustainability-oriented knowledge production (Miller/Munoz-Erickson/Redman 2011: 188f.). Researchers who have been successful in current science systems for a considerable amount of time have to develop this kind of reflexivity, because otherwise they tend to rely on their academic routines and action strategies (e.g. using highly specialized professional languages), making communication with non-scientific stakeholders extremely difficult.

Discourses on transdisciplinarity have gained substantial momentum over the past ten to fifteen years. Important impulses have been created by the discussion of suitable quality criteria for this new mode of science. Quality criteria have been defined for the different phases of a transdisciplinary research process: "problem identification and systematization, participative generation of solution-oriented and compatible knowledge, re-integration and application of the generated knowledge" (Vilsmaier/Lang 2014: 101). Most experts in the field of transdisciplinary research (Jahn 2008; Lang/Wiek/Bermann et al. 2012) have defined similar quality criteria for the different phases of the research process.

Most recently, the idea of transdisciplinarity has gained a prominent position in the redefinition of 'Global Change Research'. In the newly constituted programme of Future Earth Science, the concepts of Co-Design of research questions and processes and Co-Production of knowledge play a key role. The discourse on transdisciplinarity has successfully entered the field of science and science policy.

However, an overview of transdisciplinary research projects shows that there is still a large gap between the aspirations of an ideal-typical transdisciplinary sustainability science and the reality of actual research projects: "there is a gap between 'best practice', transdisciplinary research as advocated, and transdisciplinary research as published in scientific journals" (Brandt/Ernst/Gralla et al. 2013: 5). Furthermore, transdisciplinary research is currently still a niche discipline in the science system as a whole. The academic mainstream is firmly based on traditional disciplinary quality criteria. Central questions thus remain: what institutional framework conditions are needed to facilitate transdisciplinary research, and what are the differentiation criteria delineating the boundaries between transdisciplinary research and transformative science?

5.2.2 From Transdisciplinary Research to Transformative Research

The outline of the basic principles of transdisciplinary research presented above (cf. also Jahn/Bergmann/Keil 2012; Lang/Wiek/Bermann et al. 2012) shows that a new relationship between researchers and non-scientific actors is central. This new relationship is built on the acceptance of the different epistemological backgrounds of the involved scientific and non-scientific research partners and it aims at the generation of knowledge that is socially robust and solution-oriented. The concept of transformative research further stresses this aspiration: it contributes to the generation of different forms of knowledge and actively catalyses concrete transformation processes.

In addition to the principles of knowledge integration developed in the field of transdisciplinary research, a central element of transformative research is active intervention in a specific field of research. Such an interventionist character is described by the WBGU in its conceptualization of transformative research:

> It supports transformation processes in practical terms through the development of solutions and technical and social innovations, including diffusion processes in economy and society, and opportunities for their acceleration, and demands, at least in part, systemic perspectives and inter- as well as transdisciplinary procedure methods, including stakeholder participation (WBGU 2011: 351f.).

The aim of transformative research is to actively generate impulses for change in society and to intervene in concrete transformation processes by scientific means and in the course of a research process. It contributes to an Experimental Turn in the social sciences, which has been observed in political science as a "significant change in perspective" (Morton/Williams 2010: 3) towards a focus on experiments as a suitable method for studying causality in complex real-life settings (Greenberg/Shroder 2004). This experimental turn includes a move away from abstract modelling approaches that are independent of specific contexts and introduces a new focus on analysing systems through 'real-world experiments' (Groß/Hoffmann-Riem/Krohn 2005). The aim of real-world experiments is to identify characteristic patterns in transformation processes and to generate 'pattern languages', which provide orientation to concrete actors involved in transformation processes (Schneidewind/Singer-Brodowski 2014: 73).

The interventionist character of transformative research also requires a more explicit focus on (sectoral) societal subsystems and on catalysing, accompanying, analysing and reflecting on complex system innovations in these systems. In this way, an active contribution to transformation processes is generated. Knowledge generated in these types of research processes is socially and scientifically robust. From a methodological point of view, this research mode is similar to the field of 'action research'. Together with the focus on societal systems, this makes research in real-world laboratories a suitable strategy (cf. Schneidewind/Scheck 2013).

However, the interventionist character of transformative research is a controversial issue in the field of transdisciplinary research, because it implies a redefinition of the role of science in society, where science is no longer just a provider of objective and neutral knowledge to politics, economy and society. Mittelstraß stresses that transdisciplinary research is in fact a new 'scientific and research principle', but that it does not have a direct impact on scientific standards of rationality (2003: 22). Jahn, Bergmann and Keil (2012: 2) also stress that transdisciplinarity is a "research approach, not a theory, methodology or institution". Nonetheless, the institutional consequences of transdisciplinarity 'between mainstreaming and marginalization' are spelled out clearly: "the true challenges of transdisciplinary collaboration are underestimated and [that] those who take them seriously become marginalized" (Jahn/Bergmann/Keil 2012: 1). Yet still, the necessity of associated forms of institutional change is only hinted at. A discussion of the institutional consequences of a new mode of sustainability science expands the focus of transformative research and calls for developing a broader notion of transformative science.

5.3   Institutional Challenges of a Transformative Science

It has been shown that, in principle, transdisciplinary research is possible and it has developed fruitfully over the past years as regards methods and fields of application. However, why has its diffusion in actual research practice and across concrete research projects been relatively slow? It is argued here that this is mainly due to the lack of a focus on an institutional perspective. In order to facilitate inter- and transdisciplinary as well as transformative research, new institutional framework conditions in universities and other scientific organizations are needed. A number of arguments are relevant in this context:

1. Transdisciplinary and transformative research is in conflict with the established disciplinary organization of science. A brief look at the field of science studies shows that the science system follows a distinct logic of functioning which is detrimental to transdisciplinary and transformative aspirations: scientific communities can be described in terms of epistemic communities that tend to develop within disciplinary boundaries and focus on specialization within their specific fields of research. They withdraw from non-scientific perspectives, from the

political regulation of science and thus also from societal expectations (cf. e.g. Gläser/Lange 2007: 441). "The autopoietic nature of disciplines assures their perennial and, in principle, their unlimited regeneration, growth and differentiation" (Weingart 2014: 163). Efforts towards the establishment of transdisciplinary and transformative research thus remain an uphill battle, if they do not explicitly include an institutional perspective, because they are in conflict with the deep structural principles of scientific communities. The disciplinary organization of science (especially within universities) limits the possibilities for designing research projects that are motivated by concrete societal challenges from the outset.

2. A closely connected argument is related to established reputation-building and qualification mechanisms in the science system, which is firmly placed within the disciplinary logic of academia. Over the past decades, scientific impact has emerged as the key criterion for assessing the quality of research as well as of individual researchers. Science policy instruments are also geared towards increasing scientific impact—together these factors prove to be a disincentive for engaging in transdisciplinary research. Scientists focusing on transdisciplinary research early in their career face major barriers in their further academic career. Therefore, identifying measurable indicators of societal impact (e.g. Bornmann 2013) are decisive for the further development of sustainability science. Additionally, institutional framework conditions need to be established within universities (specific research groups, institutes, faculties) that offer a protected space for discussing societal quality criteria for transdisciplinary research and developing career pathways in accordance with such criteria.

3. Adopting a perspective on real-world problems that guides the research process and organizing research based on transdisciplinary and transformative principles questions the self-conception of science and individual researchers. Tensions emerge in the related debates about autonomy versus freedom of research and lead to massive controversies in the realm of science policy. It can safely be assumed that scientific self-conceptions represent the most prominent barriers to institutional change because they touch upon the issue of freedom of research, which is argued to be at risk where science focuses more explicitly on societal challenges. It is feared that science as a whole will be increasingly politicized (Shinn 2002), if

external expectations are entering the safe haven of academia. A common accusation thus centres on the 'normativity' of transdisciplinary and transformative research. However, this criticism in fact applies to any field of research that avoids making inherent norms and values transparent and hides behind seemingly 'objective' data. Again, the field of science studies provides an important contribution to this debate, showing that social negotiation processes within science have always shaped the generation of new (and only allegedly objective) knowledge (Felt/Nowotny/Taschwer 1999: 136ff.). A crucial example in this context is the current state of economic research, which has focused on objective calculations and thereby failed to predict dramatic developments during the financial crisis, and up until now has ignored substantial foundations of economic activity, such as natural capital, which are not included in common economic models and calculation methods. In contrast, transdisciplinary and transformative research make different forms of knowledge (e.g. the target knowledge of societal stakeholders regarding desirable futures) explicit and integrate it into a cooperative research process—in this way being fully transparent and thus complying with the most fundamental requirement of good scientific practice (Schneidewind/Singer-Brodowski 2014: 380). While debates about normativity and academic freedom are often avoided in the sustainability science community, they are a central element in the concept of transformative science.

A final reason for the slow development of transdisciplinary research over the past years has to do with the lack of 'transformative' infrastructures, i.e. the arenas in which sustainability scientists can discuss such fundamental questions and at the same time have the opportunity to develop new methods for and approaches to a more interventionist research process. The decisive impact of favourable structural conditions can be shown for the case of the Leuphana University of Lüneburg, where the foundation of the Faculty of Sustainability has allowed for the integration of approaches from the natural and social sciences. As a result, the Faculty of Sustainability has not only emerged as an important think tank in the German field of sustainability science, it also plays an important role in the education of a new generation of young scientists thinking and working in a transdisciplinary fashion.

Transformative science calls for an institutional revolution if it is to contribute to the mainstreaming of sustainability science. This distinguishes transformative science from other concepts (post-normal science, action research, intervention research, transdisciplinary research), because they call for an alternative form of doing research (for sustainability) and the changing roles and competencies of researchers in the concrete research processes. However, they lack a perspective on the institutional framing, which is needed to provide additional space and resources for transformative research, as well as suitable incentives and structures, which reduce the risk for sustainability researchers of being permanently overburdened.

The necessary institutional change depends not only on the various academic institutions themselves (universities, non-university research institutes, departmental research in government ministries, that should enable research processes across single disciplines and in cooperation with non-academic stakeholders), but also on those agencies shaping the framework conditions for academic knowledge production, such as administrative agencies, scientific policy advisors and other relevant political authorities (that are called to facilitate these non-conventional research approaches with a focus on a societal impact). Finally, change is needed at the deeper level of implicit routines, self-conceptions and paradigms in the science system, which has been shaped by the institutional framework conditions.

Transformative science actively contributes to societal transformations towards sustainable development and is itself subject to a continuous transformation process, where new interrelations between specific national or regional science systems and other societal subsystems are forged. This affects also the level of the curriculum and higher education, which should be oriented more towards societal challenges and foster education for sustainable development.

The interplay of change towards transformative science approaches and the related institutional change has been defined in terms of a mode-3 science (Schneidewind/Singer-Brodowski 2014: 103ff.). It builds on the concept of mode-2 science (Nowotny/Scott/Gibbons 2001) and further develops it as a concept of continuous self-reflexion of science and an opening-up of the science system with a view to societal transformation challenges. Mode-3 science is further based on the concept of third-order change or higher-order learning as developed in system theory (Sterling 2003: 127 ff.), which goes beyond the reflection of implicit routines during learning processes, and conceptualizes epistemic change, "a corrective change in the system of alternatives from which choice is made" (Bateson 1972: 293). This type of change can be fostered by actively including civil society actors, because civil society organizations can function as an external corrective with a critical perspective on the 'blind spots' in the science system. Defining transformative science in terms of mode-3 science includes basic principles such as "plurality, heterodox thinking and inclusion of civil society. At the same time the normal-science-mode is not negated but transcended" (Schneidewind/Singer-Brodowski 2014: 123).

Sophisticated approaches to delineating a mode-3 knowledge production have described it in terms of an advancement of the triple helix (i.e. the interplay of science, politics and industry) to a quadruple helix that also includes civil society (Carayannis/Campbell 2012). "The Mode 3 Knowledge Production System architecture focuses on and leverages higher order learning processes and dynamics that allow for both top-down government, university and industry policies and practices and bottom-up civil society and grassroots movements initiatives and priorities to interact and engage with each other towards a more intelligent, effective, and efficient synthesis" (Carayannis/Campbell 2012: 3).

Mode-3 science places academic knowledge production in a context of societal challenges and broadens the set of relevant scientific actors to also include civil society, and it discusses the related institutional implications with regard to relevant forms of knowledge, scientific organizations and quality criteria for research.

5.4   Achieving Institutional Self-transformation: Towards a 'New Governance of Science'

5.4.1   From Science Policy to Governance of Science

A focus on institutional mechanisms is instructive with regard to the science system as a whole, and especially in the context of its political framework conditions. The way in which political framework conditions for a specific field of research have developed is regularly discussed, particularly by researchers as the concerned parties who have to react to new emphases in research funding programmes. Apart from this, these issues have been analysed in the field of 'science policy research', where the political steering processes of the science system are studied. Relevant questions in this field deal with issues of the effectiveness of science policy, science policy as innovation policy (Martin 2012), and the changing role of steering committees. New science policy instruments have internationally contributed to a strengthening of decision-making power by university boards and thus to greater flexibility and responsiveness towards external demands (Jansen 2010: 47). Science policy research also studies the changing interplay between industry, politics and science, i.e. the triple-helix model (Etzkowitz/Leydesdorff 2000).

Based on the current state of research in the field of science studies, it can be shown that the science system is confronted with dynamic change processes and shifting boundaries, where new demands are voiced by industry or civil society actors and media penetration is increasing. These changing conditions also have a massive impact on science policy itself, as well as on research carried out in this field. "These transformations require … a fundamental change in perspective in science policy: from traditional 'science-policy making' to the 'governance of science' which is currently taking shape" (Grande/Jansen/Jarren et al. 2013: 19, translation by authors).

In general, a governance approach provides a perspective on negotiation processes and the interdependencies of actors in a specific field. Governance theories adopt a social science perspective and analyse patterns of handling interdependencies between actors (Schimank 2007: 29). The term governance depicts the interplay of "all co-existing forms of collectively regulating societal issues: from the institutionalized self-regulation of civil society, to the different kinds of interaction between public and private actors, and the sovereign actions of state actors" (Mayntz 2003: 72, translation by the authors). The following characteristics of governance approaches are identified by Grande, Jansen, Jarren et al. (2013: 20, translation by authors):

1. 'Emphasis on non-hierarchical forms of providing public goods,
2. a critical perspective on the nation state as the exclusive provider of public goods,
3. a non-hierarchical integration of private actors in the provision of public goods,
4. the complexity of political action in a world of blurred boundaries and, following from this,
5. the 'necessity of coordination (of action) and, beyond that, of cooperation (between different actors)'.

The governance debate has over the past years spread out into an increasingly differentiated field. A prominent strand of research has developed around the concept of reflexive governance, which describes governance processes and policy analyses as "interlinked with and open to feedback from broader social, technological and ecological changes, both in terms of innovative action and structural change" (Voß/Smith/Grin 2009: 280). The field of transition studies has also built on the concept of governance (Chappin/Ligtvoigt 2014: 717) and explicitly emphasizes the importance of reflexive approaches.

The theoretical positioning of transformative science in a context of 'governance of science' emphasizes a move away from top-down steering processes; it takes up the impulses from science politicians and statistically analyses their effects. The theoretical approach of science governance includes a more diverse set of actors and external demands (e.g. by civil society) in its analyses, since they have begun to influence academic knowledge production (Jansen 2007).

The role of civil society as an increasingly active agent in science policy shows the value of such a broadened perspective. In current discussions of science policy and in the field of science studies, the analysis of negotiation processes and the handling of interdependencies between researchers and civil society actors play only a minor role. In the following section, the case of Germany shows the increase in dynamics caused by the more active involvement of civil society organizations in science policy.

5.4.2 The Role of Civil Society in the Governance of Science

Organized civil society has not entered the field of science policy as an independent stakeholder so far. Apart from a few exceptions, theoretical analyses of civil society as a stakeholder in the field of science policy are hardly to be found (Wehling/Viewhöver 2013). At the same time, civil society organizations do play a key role in fostering societal transformation processes. However, they can only fulfil this role adequately if they can rely on scientific expertise in sustainability-related challenges.

Transfer of knowledge and expertise has traditionally worked well in specific cases in Germany, where civil society organizations such as the large environmental associations were well connected with academic partners in specific types of universities. "Together with the growing environmental movement, a large number of reform universities were founded in the 1970s. In these universities, controversially discussed issues such as environmental protection played an important role and many professors were appointed who focused on research questions close to the concerns of environmental associations" (Schneidewind/Singer-Brodowski 2014: 309). Such a pattern of transferring knowledge between civil society and science has not been broadly diffused beyond the issue of environmental protection throughout the science system as a whole, but this is nonetheless a valuable example of successful cooperation.

This type of close cooperation deteriorated with the changes in the science systems sketched above. "The generation of professors appointed in the 1970s is resigning. Many of their professorships are not reappointed or with severely changed denominations. Environmental associations suddenly experience the ... results of a self-reducing science system that solely focuses on inner-academic disciplinary expertise" (Schneidewind/Singer-Brodowski 2014: 309). Pressing questions voiced by civil society are not matched with sufficient amounts of research funding. Therefore, the large environmental associations, such as BUND,[6] have published their own science-political position papers over the past few years (e.g. BUND 2012). These position papers highlight the discrepancies between the funding priorities of the Federal Ministry of Education and Research and the research questions that, from the perspective of civil society, are decisive for achieving sustainable development: for instance, instead of spending billions of euros on technological innovation and the development of e-mobility, research projects should rather focus on the development of new concepts of mobility.

Another large environmental association in Germany, NABU,[7] has intensively dealt with the science-political strategy of the German government in the field of 'bio-economy' and has, in 2011, published a study on the specific focus of the research programme on 'bio-economy', worth two billion euros (NABU 2011). The study showed that the adopted research strategy focuses almost exclusively on technological solutions and basically ignores the position of civil society. NABU is also the first environmental

association that has established among its staff the official position of an advisor for research and science policy, whose task it is to integrate the different policy fields within NABU with the field of science policy.

The various science-political activities of the German environmental associations have increasingly emerged as an important catalyst for a more sustainability-oriented science policy (similar initiatives at an international level are presented in box 5.2). In this way, they support the role of other large stakeholders, such as the independent sustainability research institutes and students, who are actively involved in the transformation of the science system and of universities. In 2012, concerted activities resulted in the foundation of a common platform, 'Forschungswende' ('Research Transition'), which independently voices their demands, such as the integration of civil society issues in the conceptualization and calls for new research programmes. These demands were taken up in the coalition agreement of the new German Government (2013 – 2017). Organized civil society has thus become established as an important new player in the field of science policy and its participation demands have contributed to an opening-up of science policy towards the great societal challenges. Such initiatives are not limited to environmental lobby groups but also include stakeholders in other fields directly and personally affected by (a lack of) research, such as patient associations and consumer organizations (cf. Wehling 2012). Ober (2014) argues that involving civil society actors can contribute to a democratization of science policy, because the increasing tendency to make science-political decisions in expert commissions that are not democratically elected has led to an exclusion of citizens in this policy field.

Including civil society in science policy can develop an additional potential. Wehling and Viehöver (2013) describe a double participation of civil society in science. On the one hand, it can offer welcome contributions in agenda-setting processes, which often play a role in politically sensitive fields (e.g. nuclear technology, genetic engineering) and where civil society involvement can contribute to greater societal acceptance of new technologies. On the other, it can also offer "unwelcome civil society participation", for instance, in cases where patient organizations successfully demand research on rare diseases and through continuous lobbying achieve the establishment of new research programmes. It is argued that "civil society organizations can play an important role for the governance of science, especially when they proactively and in self-organized ways contribute to the

---

6 Friends of the Earth Germany (Bund für Umwelt und Naturschutz Deutschland, BUND) is an environmental NGO with currently more than 480,000 members and supporters in Germany. More information can be found at: <http://www.bund.net/ueber_uns/bund_in_english/>.

7 The *Nature and Biodiversity Conservation Union* (Naturschutzbund Deutschland, NABU) is one of the oldest environmental associations in Germany with currently more than half a million members and supporters (see at: <http://www.nabu.de/en/nabu/>).

Box 5.2: Global Efforts towards Transformative Research and Civil Society Participation. **Source:** The authors.

'Future Earth' is a global research programme and coordinating platform for inter- and transdisciplinary research on transformations towards sustainable development. It was launched by the UN (including UNESCO, UNEP, and UNU), the *International Council for Science* (ICSU), the *International Social Science Council* (ISSC), and the Belmont forum of funding agencies. It is not only a platform for connecting scientists; its explicit aspiration is to generate knowledge together with societal partners in co-design and co-production processes. For more information see: <http://www.futureearth. org>.

The ISSC has developed a global social science research agenda on global environmental change: *The Transformative Cornerstones of Social Science Research for Global Environmental Change* (see chapter 14 by Arizpe/Price/Worcester in this volume). This agenda originates from the idea that co-designing research processes and co-producing knowledge together with civil society is imperative for addressing sustainability challenges in a solution-oriented way and for achieving actual societal impact. It also calls for institutional change in the global science system, since there are no adequate funding structures for this type of research (see at: <https://igfagcr.org/sites/default/files/news/issc_transformative_cornerstones_report.pdf>).

development and design of research and technology" (Wehling/Viehöver 2013: 213, translation by authors).

Thus, external demands by civil society can be included in theoretical analyses in the field of the governance of science—in contrast to research approaches with a more traditional focus on science policy—and the concrete impact of civil society on the science system as a whole can be studied in theoretically differentiated ways. Furthermore, a governance approach is suitable for capturing all the relevant negotiation and interaction processes that have an impact on the science system. In this context, the demands for participation and the necessary processes play an important role. A perspective on the science system and science policy against the background of the century's major challenge of achieving sustainable development can be captured in terms of 'governance of science' as a comprehensive concept that is close to reality and that facilitates the analysis of the complex interdependencies in the science system. Transformative science is positioned in this exact context.

5.5 Conclusion

Humanity is facing massive challenges. The important role of science in contributing to sustainability transitions is so far only partially being recognized. This chapter has introduced the concept of transformative science, which aims at catalysing the necessary processes through suitable forms of knowledge production. The concept of transformative science emerges from three specific strands of thought:

1. Transformative science is based on debates about transdisciplinary/transformative research and emphasizes the aspirations of scientists to inter-

vene in complex systems and adopt a new mode of research carried out in real-world laboratories.

2. Transformative science not only focuses on the problem dimensions of sustainability science, but also adopts a perspective on the necessity for institutional change, in order to build the framework conditions for better sustainability science.

3. Transformative science also focuses on the science system as a whole, which is itself facing massive transformations. Building on theoretical approaches of the 'governance of science', it argues for non-hierarchical forms of organization in science and the acceptance of external actors (such as organized civil society), which are playing an increasingly important role in national science policy.

Change processes that have contributed to an opening-up of the science system have briefly been sketched for the case of the German science system (Schneidewind/Augenstein 2012). Science system transformations, in the larger context of transitions to sustainability, require a process of reflection on the institutional conditions for a broadening and a quality enhancement of sustainability sciences as a whole. A science system transition presents a complex challenge, but it is not a lost cause. The German example shows how reform processes extending over twenty years have prepared the ground. Even though the German case cannot be directly compared with other countries, it can be assumed that structural similarities can be observed in other cases. This has been illustrated by examples of global initiatives such as the Future Earth programme and the global change research agenda of the International Social Science Council.

References

Avelino, Flor, 2009: "Empowerment and the challenge of applying transition management to ongoing projects", in: *Policy Science*, 42,4: 369-390.

Bateson, Gregory, 1972: *Steps to an Ecology of Mind* (San Francisco: Chandler).

Beck, Ulrich, 1992: *Risk Society: Towards a New Modernity* (London: Sage).

Bornmann, Lutz, 2013: 'What Is Societal Impact of Research and How Can It Be Assessed? A Literature Survey", in: *Journal of the American Society for Information Science and Technology*, 64,2: 217-233.

Brandt, Patric; Ernst, Anna; Gralla, Fabienne; Luederitz, Christopher; Lang, Daniel J.; Newig, Jens; Reinert, Florian; Abson, David J.; von Wehrden, Henrik, 2013: "A review of transdisciplinary research in sustainability science", in: *Ecological Economics*, 92: 1-15.

BUND (Bund für Umwelt- und Naturschutz Deutschland), 2012: *Nachhaltige Wissenschaft. Plädoyer für eine Wissenschaft für und mit der Gesellschaft. Discussion Paper* (Berlin: BUND).

Carayannis, Elias G.; Campbell, David F. J., 2012: *Mode 3 Knowledge Production in Quadruple Helix Innovation Systems. 21st Century Democracy, Innovation and Entrepreneurship for Development* (New York: Springer).

Chappin, Emile J. L.; Ligtvoet, Andreas, 2014: "Transition and transformation: A bibliometric analysis of two scientific networks researching socio-technical change", in: *Renewable and Sustainable Energy Reviews*, 30: 715-723.

Clark, William C., 2007: "Sustainability Science: A room of its own", in: *Proceedings of the National Academy of Science of the United States of America*, 104,6: 1737-1738.

Clark, William C.; Dickson, Nancy M., 2003: "Sustainability science: The emerging research program", in: *Proceedings of the National Academy of Sciences of the United States of America*, 100,14: 8059-8061.

Etzkowitz, Henry; Leydesdorff, Loet, 2000: "The dynamics of innovation: From national systems and 'Mode 2' to a Triple Helix of university-industry-government relations", in: *Research Policy*, 29: 109-123.

Felt, Ulrike; Nowotny, Helga; Taschwer, Klaus, 1999: *Wissenschaftsforschung—Eine Einführung* (Frankfurt a.M.—New York: Campus).

Funtowicz, Silvio O.; Ravetz, Jerome R., 1993: "Science for the Post-Normal Age", in: *Futures*, 25,7 (September): 739-755.

Gibbons, Michael, 1999: "Science's new social contract with society", in: *Nature*, 402 (6761 Suppl): C81-C84.

Gläser, Jochen; Lange, Stefan, 2007: "Wissenschaft", in: Benz, Arthur; Lütz, Susanne; Schimank, Uwe; Simonis, Georg (Eds.): *Handbuch Governance: Theoretische Grundlagen und empirische Anwendungsfelder* (Wiesbaden: VS): 237-251.

Grande, Edgar; Jansen, Dorothea; Jarren Ottfried; Schimank, Uwe; Weingart, Peter, 2013: "Die neue Governance der Wissenschaft. Zur Einleitung", in: Grande, Edgar; Jansen, Dorothea; Jarren Ottfried; Rip, Arie; Schimank, Uwe; Weingart, Peter (Eds.): *Neue Governance der Wissenschaft. Reorganisation—Externe Anforderungen—Medialisierung* (Bielefeld: Transcript): 15-48.

Greenberg/Shroder 2004

Groß, Matthias; Hoffmann-Riem, Holger; Krohn, Wolfgang, 2005: *Realexperimente. Ökologische Gestaltungsprozesse in der Wissensgesellschaft* (Bielefeld: Transcript).

Hirsch Hadorn, Gertrude; Biber-Klemm, Susette; Grossenbacher-Mansuy, Walter; Hoffmann-Riem, Holger; Joye, Dominique; Pohl, Christian; Wiesmann, Urs; Zemp, Elisabeth, 2008: "The emergence of transdisciplinarity as a form of research", in: Hirsch Hadorn, Gertrude; Hoffmann-Riem, Holger; Biber-Klemm, Susette; Grossenbacher-Mansuy, Walter; Joye, Dominique; Pohl Christian; Wiesmann, Urs; Zemp, Elisabeth (Eds.): *Handbook of Transdisciplinary Research* (Berlin: Springer): 19-42.

Jahn, Thomas, 2008: "Transdisziplinarität in der Forschungspraxis", in: Bergmann, Matthias; Schramm, Engelbert (Eds.), *Transdisziplinäre Forschung. Integrative Forschungsprozesse verstehen und bewerten* (Frankfurt a.M.—New York: Campus): 21-37.

Jahn, Thomas; Bergmann, Matthias; Keil, Florian, 2012: "Transdisciplinarity—between Mainstreaming and marginalisation", in: *Ecological Economics*, 79: 1-10.

Jansen, Dorothea (Ed.), 2007: *New Forms of Governance in Research Organizations. From Disciplinary Theories towards Interfaces and Integration* (Dordrecht: Springer).

Jansen, Dorothea, 2010: "Von der Steuerung zur Governance: Wandel der Staatlichkeit?", in: Simon, Dagmar; Knie, Andreas; Hornbostel, Stefan (Eds.): *Handbuch Wissenschaftspolitik* (Wiesbaden: VS): 39-50.

Kates, Robert W.; Clark, William C.; Corell, Robert; Hall, Michael J.; Jaeger, Carlo C.; Lowe, Ian; McCarthy, James J.; Schellnhuber, Hans Joachim; Bolin, Bert; Dickson, Nancy M.; Faucheux, Sylvie; Gallopin, Gilberto C.; Gruebler, Arnulf; Huntley, Brian; Jäger, Jill; Jodha, Narpat S.; Kasperson, Roger E.; Mabogunje, Akin; Matson, Pamela; Mooney, Harold; Moore III, Berrien; O'Riordan, Timothy; Svedin, Uno, 2001: "Sustainability Science", in: *Science*, 292,5517: 641-642.

Klein, Julie; Grossenbacher-Mansuy, Walter; Häberli, Rudolf; Bill, Alain; Scholz, Roland W.; Welti, Myrtha (Eds.), 2001: *Transdisciplinarity: Joint problem solving among science, technology and society. An effective way for managing complexity* (Basel—Boston—Berlin: Birkhäuser).

Knie, Andreas; Simon, Dagmar, 2010: "Stabilität und Wandel des deutschen Wissenschaftssystems", in: Simon,

Dagmar; Knie, Andreas; Hornbostel, Stefan (Eds.): *Handbuch Wissenschaftspolitik* (Wiesbaden: VS): 26-38.

Lang, Daniel J.; Wiek, Arnim; Bermann, Matthias; Stauffacher, Michael; Martens, Pim; Moll, Peter; Swilling, Mark; Thomas, Christopher J., 2012: "Transdisciplinary research in sustainability science: practice, principles, and challenges", in: *Sustainability Science*, 7,1: 25-43.

Levin, Kelly; Cashore, Benjamin; Bernstein, Steven; Auld, Graeme, 2012: "Overcoming the tragedy of super wicked problems: constraining our future selves to ameliorate global climate change", in: *Policy Sciences*, 45,2: 123-152.

Loorbach, Derk; Frantzeskaki, Niki; Thissen, Will, 2011: "A Transition Research Perspective on Governance for Sustainability", in: Jaeger, C.C. et al. (Ed.): *European Research on Sustainable Development* (Berlin, Heidelberg: Springer): 73-89.

Lyall, Catherine; Fletcher, Isabel, 2013: "Experiments in interdisciplinary capacity-building: The successes and challenges of large-scale interdisciplinary investments", in: *Science and Public Policy*, 40,1: 1-7.

Martin, Ben R., 2012: "The evolution of science policy and innovation studies", in: *Research Policy*, 41,7: 1219-1239.

Mayntz, Renate, 2009: *Über Governance: Institutionen und Prozesse politischer Regelung* (Frankfurt a.M.: Campus).

Miller, Thaddeus R.; Muñoz Erickson, Tischa; Redman, Charles L., 2011: "Transforming knowledge for sustainability: towards adaptive academic institutions", in: *International Journal of Sustainability in Higher Education*, 12,2: 177-192.

Mittelstraß, Jürgen, 2003: *Transdisziplinarität—wissenschaftliche Zukunft und institutionelle Wirklichkeit. Konstanzer Universitätsreden* (Konstanz: Universitätsverlag).

NABU, 2011: "Bioökonomie. Können neue Technologien die Energieversorgung und die Welternährung sichern?"; at: <http://www.nabu.de/imperia/md/content/nabude/gentechnik/nabu-bio konomie.pdf> (31 May 2014)

Nowotny, Helga; Scott, Peter; Gibbons, Michael, 2001: *Re-Thinking Science. Knowledge in the Public in an Age of Uncertainty* (Cambridge: Polity Press).

Ober, Steffi, 2014: "Wissenschaftspolitik nachhaltiger gestalten", in: *GAIA*, 23,1: 11-13.

Reid, Walter V.; Chen, Daici; Goldfarb, Leah; Hackmann, Heide; Lee, Yuan Tshe; Mokhele, Khotso; Ostrom, Elinor; Raivio, Kari; Rockström, Johan; Schellnhuber, Hans Joachim; Whyte, Anne, 2010: "Earth System Science for Global Sustainability: Grand Challenges", in: *Science*, 330,6006: 916-917.

Rip, Arie, 2011: "Science Institutions and Grand Challenges of Society: A Scenario", in: *Asian Research Policy*, 2,1: 1-9.

Rockström, Johan; Steffen, Will; Noone, Kevin; Persson, Åsa; Chapin, F. Stuart; Lambin, Eric F.; Lenton, Timothy M.; Scheffer, Marten; Folke, Carl; Schellnhuber, Hans Joachim; Nykvist, Björn; de Wit, Cynthia A.; Hughes, Terry; van der Leeuw, Sander; Rodhe, Henning; Sörlin, Sverker; Snyder, Peter K.; Costanza, Robert; Svedin, Uno; Falkenmark, Malin; Karlberg, Louise; Corell,

Robert W.; Fabry, Victoria J.; Hansen, James; Walker, Brian; Liverman, Diana; Richardson, Katherine; Crutzen, Paul; Foley, Jonathan A., 2009b: "A safe operating space for humanity", in: *Nature*, 461,24 (September): 472-475.

Schimank, Uwe, 2007: "Elementare Mechanismen", in: Benz, Arthur; Lütz, Susanne; Schimank, Uwe; Simonis, Georg (Eds.): *Handbuch Governance: Theoretische Grundlagen und empirische Anwendungsfelder* (Wiesbaden: VS): 29-55.

Schneidewind, Uwe; Augenstein, Karoline, 2012: "Analyzing a transition to a sustainability-oriented science system in Germany", in: *Environmental Innovation and Societal Transitions*, 3: 16-28.

Schneidewind, Uwe; Scheck, Hanna, 2013: "Die Stadt als Reallabor für Systeminnovationen", in: Rückert-John, Jana (Ed.): *Soziale Innovation und Nachhaltigkeit* (Wiesbaden: Springer): 229-248.

Schneidewind, Uwe; Singer-Brodowski, Mandy, 22014 (updated edition): *Transformative Wissenschaft. Klimawandel im deutschen Wissenschafts- und Hochschulsystem* (Marburg: Metropolis).

Scholz, Roland W.; Daniel J. Lang; Arnim Wiek; Walter, Alexander I.; Stauffacher, Michael, 2006: "Transdisciplinary case studies as a means of sustainability learning: Historical framework and theory", in: *International Journal of Sustainability in Higher Education*, 7,3: 226-251.

Scholz, Roland; Tietje, Olaf, 2002: *Embedded Case Study Methods* (Thousand Oaks: Sage Publications).

Scholz, Roland, 2011: *Environmental Literacy in Science and Society. From Knowledge to Decisions* (Cambridge: Cambridge University Press).

Shinn, Terry, 2002: "The triple helix and new production of science. Prepackaged thinking on science and technology", in: *Social Studies of Science*, 32,4: 599-614.

Sterling, Stephen, 2003: "Whole systems thinking as a basis for paradigm change in education: explorations in the context of sustainability"; at: <http://www.bath.ac.uk/cree/sterling/index.htm> (31 May 2014).

Talwar, Sonia; Wiek, Arnim; Robinson, John, 2012: "User engagement in sustainability research", in: *Science and Public Policy*, 38,5: 379-390.

Vilsmaier, Ulli; Lang, Daniel, 2014: "Transdisziplinäre Forschung", in: Heinrichs, Harald; Michelsen, Gerd (Eds.): *Nachhaltigkeitswissenschaften* (Berlin—Heidelberg: Springer): 87-114.

Voß, Jan-Peter; Smith, Adrian; Grin, John, 2009: "Designing long-term policy: rethinking transition management", in: *Policy Science*, 42,4: 275-302.

WBGU (German Advisory Council on Global Change), 2011: *World in Transition—A Social Contract for Sustainability* (Berlin: WBGU); at: <http://www.wbgu.de/en/flagship-reports/fr-2011-a-social-contract/>.

Wehling, Peter, 2012: "From invited to uninvited participation (and back?): rethinking civil society engagement in technology assessment and development", in: *Poiesis Prax*, 9: 43-60.

Wehling, Peter; Viehöver, Willy, 2013: "'Uneingeladene' Partizipation der Zivilgesellschaft. Ein kreatives Element der Governance von Wissenschaft", in: Grande, Edgar; Jansen, Dorothea; Jarren, Ottfried; Rip, Arie; Schimank, Uwe; Weingart, Peter (Eds.): *Neue Governance der Wissenschaft. Reorganisation–externe Anforderungen–Medialisierung* (Bielefeld: Transcript): 213-234.

Weingart, Peter, 2014: "Interdisciplinarity and the New Governance of Universities", in: Weingart, Peter; Padberg, Britta (Eds.): *University Experiments in Interdisciplinar-ity. Obstacles and Opportunities* (Bielefeld: Transcript): 151-174.

Wiek, Arnim; Ness, Barry; Schweizer-Ries, Petra; Brand, Fridolin S.; Farioli, Francesca, 2012: "From complex systems analysis to transformational change: a comparative appraisal of sustainability science projects", in: *Sustainability Science*, 7,2: 5-24.

Wittmayer, Julia M.; Schäpke, Niko, 2014: "Action, research and participation: roles of researchers in sustainability transitions", in: *Sustainability Science* (21 August): 1-14.

Part II   Aiming at Sustainable Peace

6 The Psychological Components of a Sustainable Peace: An Introduction

Morton Deutsch and Peter T. Coleman

Abstract

The purpose of The Psychological Components of a Sustainable Peace, a book edited by Peter Coleman and Morton Deutsch (Coleman & Deutsch, 2012), is to enhance understanding of sustainable peace by supplementing the standard approach of studying the prevention of destructive conflict, violence, war and injustice with the equally important investigation of the promotion of the basic conditions and processes conducive to lasting peace. For in addition to addressing the pervasive realities of oppression, violence and war, peace requires us to understand and envision what alternatives we wish to construct. Recognizing the ultimate need for multidisciplinary frameworks to best comprehend and foster sustainable peace, we hoped to elicit what contemporary psychology might have to contribute to such a framework. This chapter provides a brief historical and conceptual context for the many fine scholarly chapters that follow in The Psychological Components of a Sustainable Peace (Coleman/Deutsch 2012).[1]

Keywords: Sustainable peace, war prevention, cooperation, conflict resolution, social justice, power, needs and emotions, psychodynamics, creative thinking, reconciliation

6.1 Introduction

William James, the first peace psychologist, was a most distinguished scholar and also an insistent public voice on issues of war and peace. He was deeply opposed to imperialism and the war fever with which it was associated. He was at one time the vice president of the Anti-Imperialist League, and he published articles and letters in newspapers as well as making many speeches against the Monroe Doctrine, the Spanish-American War, the colonization of the Philippines and Cuba, and so forth (Perry 1948).

James was opposed to war but he admired the heroic and courageous actions associated with the military. For James, the appeal of war and the military

did not come primarily from people's negative predispositions, but from their desire to face challenge and adversity and, in so doing, to realize their potentials in such virtues as fidelity, cohesiveness, tenacity, and heroism. In his famous paper, *The Moral Equivalent of War* (James 1917), he sought to articulate how the manly virtues associated with the military and war could find expression in the midst of a pacific civilization and thus be a moral substitute for war. This chapter takes a different orientation than that of James and much of psychological writings related to issues of war and peace. Their focus has mainly been on what psychological theory and research can contribute to the very important concern, *the prevention of war*. *The Psychological Components of a Sustainable Peace* (Coleman/Deutsch 2012) is concerned with what psychological theory and research can contribute to the promotion of a *harmonious, sustainable peace*.

Underlying this orientation is our belief that promoting the ideas and actions which can lead to a sustainable, harmonious peace can not only contribute to the prevention of war, but will also lead to more positive, constructive relations among people and nations and to a more sustainable planet. This chapter has

---

1 This chapter was first published as the introductory chapter by the authors, in: Coleman, Peter T.; Deutsch, Morton (Eds.): *Psychological Components of* Sustainable Peace. *Series: Peace Psychology Book Series* (New York: Springer 2012). The permission to include this text was granted by Springer in Heidelberg in 2013. The text has been updated and the most recent scientific literature has been added by the authors.

three brief sections: (1) Psychological contributions to the prevention of war and violent, destructive conflicts; (2) The nature of a sustainable, harmonious peace; and (3) The psychological components of a sustainable, harmonious peace.

6.2 Psychological Contributions to the Prevention of War and Violent, Destructive Conflicts

6.2.1 Debunking the Inevitability of War

One of the earliest and most important contributions of psychologists and other social scientists was to debunk the myth that war was inevitable because of mankind's innate aggressiveness. As early as 1945, the Society for the Psychological Study of Social Issues published a book, *Human Nature and Enduring Peace* (Murphy 1945), which included a statement endorsed by the leading psychologists of that time, "If man can live in a society which does not block and thwart him, he does not tend to be aggressive; and if a society of men can live in a world order in which the members of the society are not blocked or thwarted by the world arrangements as a whole, they have no intrinsic tendency to be aggressive" (Murphy 1945: 20).

On 16 May 1986 a multinational and multidisciplinary group of scientists, organized by David Adams (a psychologist), issued the *Seville Statement on Violence,* which was subsequently adopted by UNESCO on 16 November 1989. The statement was designed to refute "the notion that organized human violence is biologically determined". The statement contains five core ideas. These ideas are:

1. It is scientifically incorrect to say that we have inherited a tendency to make war from our animal ancestors.
2. It is scientifically incorrect to say that war or any other violent behaviour is genetically programmed into our human nature.
3. It is scientifically incorrect to say that in the course of human evolution there has been a selection for aggressive behaviour more than for other kinds of behaviour.
4. It is scientifically incorrect to say that humans have a 'violent brain'.
5. It is scientifically incorrect to say that war is caused by 'instinct' or anysingle motivation.

The statement concludes: "Just as 'wars begin in the minds of men', peace also begins in our minds. The same species who invented war is capable of inventing peace. The responsibility lies with each of us" (Adams/Barnett/Bechtereva et al. 1990).

Another myth that has been debunked is that there are no peaceful societies. Much work by anthropologists has demonstrated the existence of many peaceful societies, large as well as small. Some excellent books about peaceful societies are: Fry (2006), *The Human Potential for Peace: An Anthropological Challenge to Assumptions about War and Peace,* Howell and Willis (1989) *Societies at Peace: Anthropological Perspectives,* and Kemp and Fry (2004) *Keeping the Peace: Conflict Resolution and Peaceful Societies around the World.*

6.2.2 Psychology and the Prevention of War

After the end of World War II, stimulated by the development of nuclear weapons, the emergence of the United Nations, and the development of the Cold War between the Soviet Union and the United States, a significant number of psychologists began to become active in applying psychology to the prevention of war. Such psychologists as Ed Cairns, Leila Dane, Joseph de Rivera, Morton Deutsch, Daniel Druckman, Ronald Fisher, Susan Fiske, Jerome Frank, Irving Janus, Herbert Kelman, Paul Kimmel, Evelin Lindner, Susan McKay, Susan Opotow, Charles Osgood Dean Pruitt, Ann Sandon, Milton Schwebel, Ervin Staub, Richard Wagner, Michael Wessels, Ralph White, and many others were very active in writing papers, giving talks, and participating in conferences with citizen groups as well as with officials from the US State and Defense Departments. They wrote about: motivations and misperceptions which led to war; such processes as 'autistic hostility', 'self-fulfilling prophecies', and 'unwitting commitments' that perpetuate destructive conflicts; they analysed and criticized the psychological assumptions involved in 'nuclear deterrence'; they considered processes for reducing tension and hostility such as mediation and GRIT (the *graduated reduction in tension*); they identified 'groupthink' which, in tense situations, limits the alternatives of interpretation and action available to the group; they identified the conditions which give rise to destructive rather than constructive resolution of conflict; they analysed current international hostilities such as the Cuban Missile Crisis and the Vietnam War in terms of how psychological factors affected their development and course. Scholars from other disciplines (political science, economics, sociology, law, etc.) often participated with psychologists in multidisciplinary books and conferences; most notably Andrea Bartoli, Jacob Bercovitch, Kenneth and Elise Boulding, Roger Fisher, Mary Parker Follett, Johan Galtung, Ted Gurr, Robert Jervis, Debra Kolb, Victor Kremenyuk, Louis Kriesberg, Jean Paul Lederach, Chris Mitchell, Robert Mnookin, Linda Putnam, Anatol Rapaport, David Riesman, Harold Saunders, Thomas Schelling, Gene Sharp, Larry Suskind, William Ury, and William Zartman.

They wrote about such topics as: arms control and disarmament; non-physical methods of disarmament; economic steps toward peace; East and West; military defence; reducing international tensions; building a world society; international cooperation and the rule of law, ethnic conflicts, and negotiation and mediation.

6.2.3 Modern Peace Psychology

With the end of the Cold War, the break-up of the Soviet Union, and the dissolution of the pro-Soviet Eastern Bloc during the 1980s, the attention of Western peace psychology became less focused on preventing war between the United States and the Soviet Union. As Christie, Tint, Wagner et al. (2008: 542) point out:

> The focal concerns of post-Cold War peace psychology have become more diverse, global, and shaped by local geohistorical contexts in part because security concerns are no longer organized around the U.S.-Soviet relationship. For example, countries aligned with the Global South and developing parts of the world tend to associate peacebuilding efforts with social justice, in part because political oppression and the unequal distribution of scarce resources diminish human well-being and threaten survival. In geohistorical contexts marked by deeply divisive intractable conflicts and oppositional social identities, such as the conflicts in Northern Ireland, the Middle East and parts of Africa, research and practice often focus on the prevention of violent episodes through the promotion of positive intergroup relations. In the West, the research agenda is dominated by efforts to more deeply understand and prevent terrorism.

During the Cold War, but especially afterwards, not only were there many psychological articles and workshops aimed at psychological intervention into specific violent conflicts, whether at the international, intergroup, or interpersonal levels; there was also much psychological work to develop theory that might improve psychologically based interventions. Galtung's (1969) important distinctions between direct and structural violence provides useful distinctions between much of the early and more recent work of psychologists concerned with issues of peace, conflict, and violence. Structural violence is embedded in the values, social norms, laws, social structures, and procedures within a society or community which systematically disadvantage certain individuals and groups so that they are poorer, sicker, less educated, and more harmed than those who are not disadvantaged. Much of the early work was focused on direct violence; on the causes and conditions which give rise to aggression and physical violence. More recent work has often been concerned with the bidirectional relationship between conflict and social injustice (structural violence).

The literature and contributions to the modern fields of peace psychology and conflict resolution have grown so large that no summary will be presented here. However, in a number of recent books there are excellent presentations and summaries of this work. They include: Christie, Tint and Wagner et al. et al. (2001); Blumberg, Hare and Costin (2007); Deutsch, Coleman and Marcus (2006); Fisher (1990, 1997); Kriesberg (2006); Lederach (1994, 1997); Pruitt and Kim (2004).

In a book of essays on preventing World War III (Wright/Evan/Deutsch 1962), Quincy Wright, a distinguished historian, wrote:

> A world society capable of settling international disputes and preventing war is possible, and without such a society the maintenance of peace in the shrinking world will be increasingly difficult. The basic problem in preventing World War III is, therefore, the building of such a society. Observation of the history of groups merging into supersocieties indicated that such a development normally proceeds through four stages which may considerably overlay. They are (1) the establishment of *communication* and trade among independent groups; (2) the process of *acculturation* through mutual borrowing of technologies and syntheses of values; (3) the emergence of common cultural standards and techniques, inducing *cooperation* to maintain norms, achieve goals, and promote common interests in the developing culture; and (4) the increase of the efficiency of such cooperation by the establishment of a central *organization* with authority to recommend, guide, or even compel appropriate action, at first by the component groups and eventually by individuals.

Similarly, in the foreword to the important book, *Building Peace: Sustainable Reconciliation in Divided Societies* (Lederach 1997), Richard Solomon, President of the United States Institute of Peace, offered this image:

> Sustainable peace requires that long-time antagonists not merely lay down their arms but that they achieve profound reconciliation that will endure because it is sustained by a society-wide network of relationships and

mechanisms that promote justice and address the root causes of enmity before they can regenerate destabilizing tensions (Solomon 1997: ix).

We agree with Wright and Solomon that a sustainable world peace will require the building of such a society imbued with such mechanisms and relationships. Below, we stress what we consider to be the psychological requirements of such a society.

1. A strong sense of positive interdependence among the units composing the greater society. They should feel as well as believe that the units are so linked that they "sink or swim together". Such common bonds are most prevalent in societies organized around cross-cutting structures, where members of different ethnic groups play, work, and socialize together (LeVine/Campbell 1972; Varshney 2002).

2. A strong sense of global, as well as local, patriotism and loyalty. Their sense of identity is strongly linked to the global as well as their local community. Such phrases as 'Irish American', 'Jewish American', and 'Italian American' indicate the possibility of such dual or multiple identities.

3. The sharing of such basic common values as recognition that all human beings despite differences or disagreements have the right to be treated with respect, dignity, and justice as well as to have their basic needs fulfilled. The United Nations Universal Declaration of Human Rights, adopted by the United Nations General Assembly on 10 December 1948, is a much fuller statement of these basic values.

4. Mutual understanding, which is fostered by the freedom to be informed as well as the freedom to communicate and by the ability to have the message being communicated expressed or translated so that it is mutually understood by the sender and receiver of the messages. Quick, accurate computer translation of different languages may become a substitute for a common, universal language.

5. A sense of fair recourse. Inevitably, conflicts between people and between groups will occur and experiences of injustices and even oppression will arise. When such problems develop, the presence of fair and efficient means of recourse goes a long way in decreasing the probability that they will culminate in either criminal or political forms of violence (Gurr 2000). Of course, history is filled with instances of the opposite, where unmet needs combined with a limited sense of recourse

have resulted in extraordinary episodes of violence, revolution and human suffering.

6. Social taboos against the use of violence to solve problems. The biggest single predictor of spikes in violence in Western society is the presence of international wars (Gurr). There are similar correlations to be found between incidents of local ethno-political violence and the normalization of violence as a legitimate method of communal problem-solving, as well as between experiences of domestic abuse as a child and the perpetration of domestic abuse as an adult. In contrast, anthropological research has documented the central importance of social taboos against violence for fostering more internally and externally peaceful societies (Fry 2006).

These six psychological requirements constitute a set of basic building-blocks for fostering a harmonious, sustainable peace. No one aspect would be sufficient, nor would the presence of all six necessarily be adequate. However, the more that a society invests in each of these components, the more they will decrease the prevalence of destructive conflict and the more they will increase the probability that peaceful relations will be sustained.

6.3 The Psychological Components of a Sustainable Peace

Below, we characterize briefly what we consider to be key psychological components; these were shown to the contributors as we invited their contributions. Individual chapters address these components as the distinguished contributors see fit. The chapters do not exhaust the potential contributions of psychological theory and research to the development of a sustainable peace, nor do they cover what other disciplines (e.g. economics, political science, sociology, international relations, history, the physical and biological sciences) can contribute to the development of a sustainable peace. Their aim is to stimulate other psychologists to make further contributions and to inform educated citizens and public officials as well as other social scientists of existing and potential psychological contributions to this area of knowledge.

The key psychological components discussed in *The Psychological Components of a Sustainable Peace* (Coleman/Deutsch 2012) are:

1. *Effective Cooperation* At the international level, the developmental of harmonious peaceful rela-

tions among nations will require effective cooperation in dealing with such issues as climate change, proliferation of weapons of mass destruction, pandemics of contagious diseases, global economic development, failed states, and so on. Similarly, in interpersonal relations such as marriages, if a couple is unable to cooperate effectively on matters that are central to their identities whether these be religious concerns, sexual relations, political views, economic relations, lifestyles, child-raising or in-laws, it will be difficult for them to have a harmonious, peaceful marriage. Much research has been done on the conditions which give rise to successful cooperation and to its effects (see Johnson/Johnson 2005; Deutsch 2006, 2011).

2. *Constructive Conflict Resolution*

Among extended relations of all sorts—whether at the interpersonal, intergroup, or international levels—it is inevitable that conflict will arise. Some of the conflicts are not central to the relationship and may persist and be mainly ignored without harming the relationships. Other conflicts which threaten the well-being or identity of one or more of the participants in the relationship cannot be suppressed or ignored without harming the involved parties and their relationship. How such conflicts are resolved—constructively or destructively—is critical in determining whether harmonious, cooperative relationships will persist and be strengthened or will deteriorate into bitter, hostile relations.

During the past several decades, there has been extensive theoretical and research investigation of the effects of constructive and destructive processes of conflict resolution as well as of the conditions which give rise to each process (for summaries, see for instance Deutsch/Coleman/Marcus 2006; Bercovitch/Kremenyuk/Zartman 2009). There is also a growing literature of useful, practical, advice in how to manage conflict (see e.g. Moore 1996; Gottman/Silver 1999; Schneider/Honeyman 2006; Thompson 2008).

3. *Social Justice*

Relationships that are just foster effective cooperation and constructive conflict resolution. Injustice and oppression, on the other hand, foster and are fostered by destructive conflict. Similarly, effective cooperation is inhibited or destroyed by injustice and oppression.

It is useful to make a distinction between *injustice* and *oppression*. Oppression is the experience of repeated, widespread, systemic injustice. It need not be extreme and involve the legal system

(as in slavery, apartheid, or the lack of the right to vote) nor violent (as in tyrannical societies). Harvey (1999) has used the term 'civilized oppression' and Sue, Capodilupo, Torino et al. (2007) the term 'micro aggression' to characterize the everyday processes of oppression in normal life. Civilized oppression "is embedded in unquestioned norms, habits, and symbols, in the assumptions underlying institutions and rules, and the collective consequences of following those rules. It refers to the vast and deep injustices some groups suffer as a consequence of often unconscious assumptions and reactions of well-meaning people in ordinary interactions which are supported by the media and cultural stereotypes as well as by the structural features of bureaucratic hierarchies and market mechanisms" (Young 1990: 41).

There is an extensive literature dealing with overcoming injustice and oppression which is too extensive to present here. The main themes are: *Awakening the Sense of Injustice; Persuasion Strategies for Changing Oppression; Relationships and Power Strategies for Change* (see Deutsch 2006, for more elaboration).

4. *Power and Equality*

The distribution of power, the equality or inequality of the parties involved in any relationship, plays a critically important role in determining the characteristics of the relationship. For instance, Adam Curle (1971), a mediator working with ethnic conflicts in Africa in the 1960s and 1970s, observed that as conflicts moved from unpeaceful to peaceful relationships, their course could be charted from one of relative inequality between the groups to relative equality. He described this progression toward peace as involving four stages. In the first stage, conflict was 'hidden' to the lower-power parties because they remained unaware of the injustices that affected their lives. Here, any activities or events resulting in conscientization (erasing ignorance and raising awareness of inequalities and inequities) moved the conflict forward. An increase in awareness of injustice led to the second stage, confrontation, when demands for change from the weaker party brought the conflict to the surface. Under some conditions, these confrontations resulted in the stage of negotiations, which were aimed at achieving a rebalancing of power in the relationship in order for those in low power to increase their capacities to address their basic needs. Successful negotiations moved the conflicts to the final stage of sustainable peace, but only if they led to a restructuring of the relationship that addressed effectively the substantive and procedural concerns of those involved.

5. *Human Needs and Emotions*
Neither effective cooperation, constructive conflict resolution, nor social justice is likely when basic human needs are unsatisfied. Maslow (1954) has identified the basic human needs as: physiological, safety, belongingness and love, esteem, and self-actualization. Frustration of these needs lead to diverse emotional consequences such as apathy, fear, depression, humiliation, rage, and anger. These emotions are not conducive to effective cooperation, constructive conflict resolution, or any other psychological component of a harmonious, sustainable peace. The view that the frustration of one's needs is purposeful and unjust gives rise to intense feelings of humiliation which Lindner (2006) has described as the 'nuclear bomb of emotions'.

6. *The Psychodynamics of Peace*
From Freud on psychodynamic theorists have been interested in how individual and group psychodynamics have contributed to constructive, peaceful, or destructive, violent relationships at the international as well as interpersonal levels. The psychodynamic approach emphasizes the interdependence between internal conflicts and external conflicts. Thus, internal conflict between a socially prohibited desire (e.g. desire for homosexual contact) and guilt feelings may lead to anxiety and such defence mechanisms against anxiety as projection where the struggle in yourself is denied and is projected on to or attributed to another.

External conflict can also give rise to internal conflict. Psychodynamic approaches also emphasize the importance of understanding how the past and development of an individual, group, or society play a critical role in forming self-identity, as well as the values, symbolic meanings, attitudes, and predispositions to behaviour.

7. *Creative Problem-Solving*
Betty Reardon, a noted peace educator, once said, "The failure to achieve peace is in essence a failure of imagination" (pers. comm.). The freedom and ability to imagine new possibilities as well as the capacity to select judiciously from these possibilities what is novel, interesting, and valuable (Simon 2001) are central to creative problem-solving. The conditions which foster the freedom and ability to create novel and valuable solutions not only are conditions in the problem-solver (individual or

group), but also are in conditions in the social context which affects the problem-solver. Creative problem-solving is necessary to overcome the obstacles which block effective cooperation and the impasses which hinder constructive conflict resolution.

8. *Complex Thinking*
Simple thinking is directed at the here-and-now and, often, has an 'either/or' quality. It does not take into account the future or past or what is occurring in different locales and remote places, nor that solutions to problems often involve the integration of apparently opposed alternatives and the creation of new alternatives. At the international level such problems as climate change, depletion of basic resources, worldwide economic recession, terrorism, and weapons of mass destruction require the ability to think of the future as well as of the past, to think globally as well as locally. Similarly, in married couples such issues as college tuition for one's children, retirement income, care for elderly parents, and maintaining the positive in marital relations require complex thinking.

9. *Persuasion and Dialogue*
As Ledgerwood, Chaiken and Gruenfeld (2006) have pointed out: "Persuasion is distinct from coercion in that persuasion is influence designed to change people's minds, whereas coercion involves influence designed to change people's behaviours (with little regard for whether they have actually changed their minds)." Lasting change is more likely to result from persuasion than coercion.

Persuasion involves communication by a *source* of a *message,* through a *medium,* designed to *reach* and influence a *recipient.* Whether the recipient will be persuaded by the message is a function of the characteristics of each of the foregoing elements as well as the characteristics of the relationship between the source and the recipient. Sustainable, harmonious peaceful relations require the mutual ability to persuade one another. Without this ability, a convergence of values, information, and actions as well as mutual satisfaction of needs is not likely to occur.

Dialogue, unlike persuasion, is not unilateral. It is a mutual process in which the interaction parties openly communicate and actively listen to one another with mutual respect and a feeling of mutual equality. Each communicates what is important and true for her without derogating what is true and important for others. They seek to learn together and to find common meaning by exploring the assumptions underlying their individual and collective beliefs. Dialogue is a collaborative and creative process in which the participants are open to change as they seek common ground and mutual understanding.

10. *Reconciliation*

After destructive conflicts in which the conflicting parties have inflicted grievous harm (humiliation, destruction of property, torture, assault, rape, murder) on one another, the conflicting parties may still have to live and work together in the same communities. This is often the case in civil wars, ethnic and religious conflicts, gang wars and even family disputes that have taken a destructive course. Consider the slaughter that has taken place between Hutus and Tutsis in Rwanda and Burundi (Staub 2012); between blacks and whites in South Africa; between the 'Bloods' and 'Crips' of Los Angeles; the Protestants and Catholics in Northern Ireland; and among Serbs, Croats and Muslims in Bosnia. Is it possible for forgiveness and reconciliation to occur under such conditions? If so, what fosters these processes? Recently, a considerable psychological literature has emerged in response to this question (see Lederach 1997, 2002, 2003, 2005).

After bitter destructive conflict, it can be expected that reconciliation will be achieved, if at all, after a slow process with many setbacks as well as advances. The continuous and persistent help and encouragement of powerful and respected third parties is often necessary to keep the reconciliation process moving forward and to prevent its derailment by extremists, misunderstandings or harmful actions by either of the conflicting parties. The help and encouragement must be multifaceted. It must deal not only with the social-psychological issues addressed so well in this volume, but also, justly, with such institutions as the economic, political, legal, educational, health care and security, whose effective functioning is necessary for a sustained reconciliation.

11. *Education*

One of the most important things that educators can do to foster each of the psychological components discussed above is to exemplify these components in their own behaviour in and out of the classroom and also in the pedagogy, curricula, and organizational functioning of the school. To achieve these objectives will require changes in the education and training of school personnel, particularly teachers and administrators, as well as new requirements in the hiring of school personnel.

In recent years, it has been increasingly recognized that schools have to change in basic ways if we are to educate children so that they are for rather than against one another, so that they develop the ability to resolve their conflicts constructively rather than destructively and are prepared to live in a peaceful world. This recognition has been expressed in a number of interrelated movements: cooperative learning, conflict resolution, and education for peace. In our view, there are several key components in these overlapping movements: cooperative learning; conflict resolution training; the constructive use of controversy in teacher subject matters; the creation of dispute resolution centres in the schools; and development of knowledge of and a commitment to human rights and social justice. Students should also acquire, at the appropriate age level, substantive knowledge in such fields as political science, international relations, arms control and disarmament, economic development, the global environment, and world trade, fields which are also important to world peace, together with other substantive knowledge and skills necessary to function as responsible adults. They should also become informed and sensitized to the many injustices that exist globally as well as locally so that they can be intelligently active in bringing about social change.

12. *Norms for Policy*

Psychological principles play a central role in the development of policies and norms that support sustainable peace, where peace is defined comprehensively to include the prevention and mitigation of episodes both of direct violence and structural violence. Sustainable peace requires changes at the level of norms and policies and psychologically-informed principles and activism have played a role in changing policies and/or norms. Some potential examples can be found in research and activism/practice that created: (a) a climate that made the Oslo Accords possible; (b) a movement that led to the removal of secrecy clauses from the Truth and Reconciliation Act, thereby making some of the testimony public; (c) serial dramas that have been used to change norms in regard to intergroup relations; and (d) emancipatory agendas that have increased voice and representation among the oppressed throughout Latin America.

13. *The Practice of Sustainable Peace*

Peace is never achieved, but rather is a process that is fostered by a variety of cognitive, affective, behavioural, structural, institutional, spiritual, and cultural components. Accordingly, there are wide arrays of ideas and methods that can be learned, practised and mastered to help bolster and sustain peace. This chapter will detail some of these practices.

The preceding discussion of psychological components of a sustainable, harmonious peace is meant to be an introduction, not a substitute for the excellent chapters which follow. It represents our preliminary thinking which gave rise to *The Psychological Components of a Sustainable Peace* (Coleman/Deutsch 2012) and stimulated our desire to have an expert in each area write each of the various chapters. We have asked the authors of the chapters to describe where possible:

1. The nature of the psychological component which is the focus of the chapter, originally appearing in *The Psychological Components of a Sustainable Peace* (Coleman/Deutsch 2012).
2. The conditions which give rise to it (Provide research evidence as well as theory).
3. Its effects, positive and negative (Provide research evidence as well as theory).
4. Generalize the implications of the preceding for the development of a harmonious, sustainable peace at the interpersonal, intergroup, and international levels.
5. Indicate what further development of theory and research is needed.

We have encouraged the authors to discuss the psychological components that are the focus of their chapters in *The Psychological Components of a Sustainable Peace* (Coleman/Deutsch 2012) in the interaction of different types of social actors: the interpersonal, intergroup, and international. We believe it is fruitful to take a social-psychological approach to all types of social interaction. Several key notions in a social-psychological approach are:

1. Each participant in a social interaction responds to the other in terms of his/her perceptions and cognitions of the other; these may or may not correspond to the other's actualities.
2. Each participant in a social interaction, being cognizant of the other's capacity for awareness, is influenced by his/her own expectations concerning the other's actions as well as by his/her perceptions of the other's conduct. These expectations may or may not be accurate; the ability to take the role of the other and to predict the other's behav-

iour is not notable in either interpersonal or international crises.
3. Social interaction is not only initiated by motives by also generates new motives and alters old ones. It is not only determined but also determining. In the process of rationalizing and justifying actions that have been taken and effects that have been produced, new values and motives emerge. Moreover, social interaction exposes one to models and exemplars which may be identified with and imitated. Thus, a child's personality is shaped largely by the interactions he/she has with his parents and peers and by the people with whom he/she identifies. Similarly, a nation's institutions may be considerably influenced by its interrelations with other nations and by the existing models of functioning that other nations provide.
4. Social interaction takes place in a social environment—in a family, a group, a community, a nation, a civilization—that has developed techniques, symbols, categories, rules, and values that are relevant to human interactions. Hence, to understand the events that occur in social interactions one must comprehend the interplay of these events with the broader social context in which they occur.
5. Even though each participant in a social interaction, whether an individual or a group, is a complex unit composed of many interacting subsystems, the individual or group can act in a unified way toward some aspect of their environment. Decision-making within the individual as within the nation can entail a struggle among different interests and values for control over action. Internal structure and internal process, while less observable in individuals than in groups, are characteristic of all social units.

References

Adams, David, Barnett, Scott A.; Bechtereva, Nataliya P.; Carter, Bonnie F.; Rodriquez Delgado, Jose M.; Diaz, Jose Luis; Eliasz, Andrzej, 1990: "The Seville statement on violence", in: *American Psychologist,* 45,10: 1167-1168.

Bercovitch, Jacob; Kremenyuk, Victor; Zartman, William, 2009: "Introduction: The nature of conflict and conflict resolution", in: Bercovitch, Jacob; Kremenyuk, Victor; Zartman, William (Eds.): *The Sage handbook of conflict resolution* (Thousand Oaks: Sage Publishers).

Blumberg, Herbert H.; Hare, Paul A.; Costin, Anna (Eds.), 2007: *Peace psychology: A comprehensive introduction* (Cambridge: Cambridge University Press).

Christie, Daniel J.; Tint, Barbara S.; Wagner, Richard V.; Winter, Deborah D., 2001: *Peace, conflict and violence: Peace psychology for the 21st century* (Upper Saddle River: Prentice Hall).

Christie, Daniel J.; Tint, Barbara S.; Wagner, Richard V.; Winter, Deborah D., 2008: "Peace psychology for a peaceful world", in: *American Psychologist,* 63: 540-552.

Coleman, Peter T.; Deutsch, Morton (Eds.), 2012: *Psychological Components of Sustainable Peace.* (New York: Springer).

Curle, Adam, 1971: *Making peace* (London: Tavistock).

Deutsch, Morton, 22006: "Cooperation and competition", in: Deutsch, Morton; Coleman, Peter T.; Marcus, Eric C. (Eds.): *The handbook of conflict resolution: Theory and practice* (San Francisco: Jossey-Bass): 23-42.

Deutsch, Morton, 2011: "Cooperation and competition", in: Coleman, Peter T. (Ed.): *Conflict, interdependence, and justice: The intellectual legacy of Morton Deutsch* (New York: Springer).

Deutsch, Morton; Coleman, Peter T.; Marcus, Eric C. (Eds.), 22006: *The handbook of conflict resolution: Theory and practice* (San Francisco: Jossey-Bass).

Fisher, Ron J., 1990: *The social psychology of intergroup and international conflict resolution* (New York: Springer).

Fisher, Ron J., 1997: *Interactive conflict resolution* (Syracuse: Syracuse University Press).

Fry, Douglas P., 2006: The human potential for peace: An anthropological challenge to assumptions about war and violence (New York: Oxford University Press).

Galtung, Johan, 1969: "Violence, peace and peace research", in: *Journal of Peace Research,* 6,3: 167-191.

Gottman, John M.; Silver, Nan, 1999: *The seven principles for making marriage work* (New York: Three Rivers Press).

Gurr, Ted R., 2000: *Peoples versus states* (Washington, DC: United States Institute of Peace Press).

Harvey, Jean, 1999: *Civilized oppression* (Lanham: Rowman & Littlefield).

Howell, Signe; Willis, Roy (Eds.), 1989: *Societies at peace: Anthropological perspectives.* (London–New York: Routledge).

James, William, 1917: "The moral equivalent of war", in: *Memoirs and studies* (London: Longmans).

Johnson, David W.; Johnson, Roger T., 2005: "New developments in social interdependence theory", in: *Psychology Monographs,* 131,4: 285-358.

Kemp, Graham; Fry, Douglas P. (Eds.), 2004: Keeping the peace: Conflict resolution and peaceful societies around the world (New York: Routledge).

Kriesberg, Louis, 32006: Constructive conflicts: From escalation to resolution (Lanham: Rowman & Littlefield).

Lederach, John Paul, 1994: Beyond prescription: Perspectives on conflict, culture and training (Syracuse: Syracuse University Press).

Lederach, John Paul, 1997: Building peace: Sustainable reconciliation in divided societies (Washington, DC: United States Institute of Peace Press).

Lederach, John Paul, 1999: *The journey toward reconciliation* (Scottsdale: Herald Press).

Lederach, John Paul, 2002: A handbook of international peacebuilding: Into the eye of the storm (San Francisco: Jossey-Bass).

Lederach, John Paul, 2003: "Cultivating peace: A practitioner's view of deadly conflict and negotiation", in: Darby, John; MacGinty, Roger (Eds.): *Contemporary peacemaking: Conflict, violence, and peace processes* (New York: Palgrave Macmillan).

Lederach, John Paul, 2005: The moral imagination: The art and soul of building peace (New York: Oxford University Press).

Ledgerwood, Alison; Chaiken, Shelly; Gruenfeld, Deborah H.; Judd, Charles M., 2006: "Changing minds: Persuasion in negotiation and conflict resolution", in: Deutsch, Morton; Coleman, Peter T.; Marcus, Eric C. (Eds.): *The handbook of conflict resolution: Theory and practice* (San Francisco: Jossey-Bass Publishers).

LeVine, Robert A.; Donald T. Campbell, 1972: Ethnocentrism: Theories of conflict, ethnic attitudes, and group behavior (New York: Wiley).

Lindner, Eveline, 2006: Making enemies: Humiliation and international conflict (London: Praeger Security International).

Maslow, Abraham Harold, 1954: *Motivation and personality* (New York: Harper).

Moore, Christopher W., 1996: *The mediation process* (San Francisco: Jossey-Bass).

Murphy, Gardner (Ed.), 1945: *Human nature and enduring peace* (Oxford: Houghton Mifflin).

Perry, Ralph Barton, 1948: *The thought and character of William James* (Cambridge: Harvard University Press).

Pruitt, Dean G.; Kim, Sung Hee, 32004: Social conflict: Escalation, stalemate, and settlement (New York: McGraw-Hill).

Schneider, Andrea Kupfer; Honeyman, Christopher (Eds.), 2006: *The negotiator's fieldbook* (Chicago: American Bar Association).

Simon, Herbert A., 2001: "Creativity in the arts and the sciences", in: *The Canyon Review and Stand,* 23: 203-220.

Staub, Ervin, 2012: "Reconciliation Between Groups, the Prevention of Violence, and Lasting Peace", in: Coleman, Peter T.; Deutsch, Morton (Eds.): *Psychological Components of Sustainable Peace* (New York: Springer): 245-264.

Sue, Derald Wing; Capodilupo, Christina M.; Torino, Gina C.; Bucceri, Jennifer M.; Holder, Aisha; Nadal, Kevin L.; Esquilin, Marta, 2007: "Racial microaggressions in everyday life: Implications for clinical practice", in: *American Psychologist,* 62,4: 271-286.

Thompson, Michael, 2008: "New game teaches peaceful conflict resolution";

Varshney, Ashutosh, 2002: *Ethnic conflict and civic life: Hindus and Muslims in India* (New Haven: Yale University Press).

Wright, Quincy; Evans, William M.; Deutsch, Morton (Eds.), 1962: *Preventing World War III: Some proposals* (New York: Simon & Schuster).

Young, Marion Iris, 1990: *Justice and the politics of difference* (Princeton: Princeton University Press).

7   The Essence of Peace? Toward a Comprehensive and Parsimonious Model of Sustainable Peace

Peter T. Coleman

Abstract

This concluding chapter has four sections. First, it highlights the basic commonalities in the discussion of the construct of a "harmonious sustainable peace". Second, it offers a summary of the main components of sustainable peace. Third, it presents a sketch of a more parsimonious model of sustainable peace informed by dynamical systems theory and dynamic minimalism. Finally, it outlines an agenda for future study and education in this area

Keywords: Sustainable peace, commonalities, a parsimonious model, dynamic systems theory, research education.

> *For every thousand pages published on the causes of war, there is less than one page directly on the causes of peace.*
>
> Historian Geoffrey Blainey (1988)

The purpose of developing the book *The Psychological Components of a Sustainable Peace* (Coleman/ Deutsch 2012) was to achieve three main objectives. First, to enhance our understanding of sustainable peace by supplementing the standard approach of studying the *prevention* of destructive conflict, violence, war and injustice with the equally important investigation of the *promotion* of the basic conditions and processes conducive to lasting peace. For in addition to addressing the pervasive realities of oppression, violence and war, peace requires us to understand and envision what alternatives we wish to construct. Second, in the context of this new inquiry, we hoped to help clarify and better specify the meaning of *sustainable peace*. Third, with respect to the ultimate need for multidisciplinary frameworks to best comprehend and foster sustainable peace, we hoped to elicit what contemporary psychology might have to contribute to such a framework.

The good news is that through their many excellent chapters, the contributors to *The Psychological Components of a Sustainable Peace,* edited by Coleman and Deutsch (2012) have helped make great progress toward meeting the three objectives; identifying

a wide variety of factors at different levels of analysis associated with the promotion of peace. However we are now left with an embarrassment of riches. The seventeen chapters which constitute *The Psychological Components of a Sustainable Peace* (Coleman/ Deutsch 2012) offer a vast array of psychosocial conditions and processes which have been linked to sustainable peace. While a critical first step, this bounty of information leaves us with a rather cluttered state of understanding. Therefore, this concluding chapter from *The Psychological Components of a Sustainable Peace* (Coleman/Deutsch 2012) will attempt to offer a synthesis of the research and ideas presented in the book. It has four sections. First, it returns to the discussion of the meaning of a harmonious, sustainable peace broached in the introduction, and highlights the basic commonalities of the construct that underlie the many aspects of peace described in the chapters of the book. Second, it offers a brief summary of the main components of sustainable peace presented in the book, organized within a nested, multi-level framework. Third, it offers a sketch of a more parsimonious model of sustainable peace, informed by dynamical systems theory (Nowak/Vallacher 1998) and dynami-

© Springer International Publishing Switzerland 2016
H.G. Brauch et al. (eds.), *Handbook on Sustainability Transition and Sustainable Peace,*
Hexagon Series on Human and Environmental Security and Peace 10, DOI 10.1007/978-3-319-43884-9_7

cal minimalism (Nowak 2004), which conceptualizes the effects of the many component parts on the probabilities of stable dynamics of destructive conflict and peace. And finally, it outlines an agenda for future study and education in this area.

7.1 The Meaning of Sustainable Peace Revisited

Peace is both complicated and simple. For example, a search of the *Thomson Reuters Web of Knowledge* database on articles published in English since 2000 with 'peace' in the their title reveals over forty terms distinguishing different types or aspects of peace (see table 7.1). This is more than a matter of semantics. Peace can differ in a variety of ways, including by level (interpersonal to international to global peace), direction (internal and external peace), durability (from fragile to enduring peace), source or conditions (peace through coercion, democratic participation, economic incentive, etc.), type (negative, positive and promotive peace) and scope (local to global peace).

Table 7.1: Types and components of peace. **Source:** Thomson Reuters web of knowledge database 2000–2011.

Agonistic peace	Movable peace
Armed peace	Negative peace
Capitalist peace	Nuclear peace
Cold-warm peace	Overt peace
Commercial institutional peace	Partial peace
Democratic peace	Peace-building
Durable peace	People's-civil peace
Enduring peace	Perpetual peace
Feminist peace	Positive peace
Fragile peace	Post-liberal peace
Global-world peace	Precarious peace
Hegemonic peace	Realistic peace
Holistic Gaia peace	Relative peace
Holistic Inner Peace	Republication peace
Holistic intercultural peace	Sustainable peace
Hybrid peace	Technological peace
Imperfect peace	Tyrannical peace
Kantian peace	Uneasy peace
Lasting peace	Unqualified peace
Liberal peace	Virtual peace
Monadic peace	Dynamic-static peace

This vision of sustainable peace narrows the focus of our discussion somewhat, although even this type of peace can still differ by level, direction, source, type and scope, and these differences affect the nature of the facilitating and inhibiting conditions associated with them. For example, the psychological conditions conducive to holistic inner peace are likely to differ dramatically from the conditions which foster sustained international peace.

Nevertheless, all forms of sustainable peace share some basic underlying qualities, reflecting a relative absence of destructive conflict, tension and violence and a presence of constructive conflict, harmony and well-being. Therefore, building on Boulding (1978), we define *sustainable peace* as existing in a state where the probability of using destructive conflict and violence to solve problems is so low that it does not enter into any party's strategy, while the probability of using cooperation and dialogue to promote social justice and well-being is so high that it governs social organization and life.

Thus, the many factors, conditions and processes related to peace presented in the chapters of *The Psychological Components of a Sustainable Peace* by Coleman and Deutsch (2012) can be understood in the context of their relative effects on (1) decreasing probabilities for destructive conflict, violence and injustice and (2) increasing probabilities for promotive peace. They are summarized below.

7.2 A Nested Model of the Psychosocial Components of Sustainable Peace

The psychosocial components from chapters 1 to 17 in the book *The Psychological Components of a Sustainable Peace* (Coleman/Deutsch 2012), where this text was first published, are summarized briefly and organized below by level (micro, meso and macro) and by *orientation* (prevention of destructive conflict or promotion of sustainable peace; see table 3.1 in Coleman 2012a). Micro-level components include those involving individuals; meso-level are those within families, schools, organizations and communities; and macro-level those involving policies and institutions of societies, states and the international community. Many of the factors associated with preventing destructive conflict may also be necessary for promoting positive relations, and vice versa. However they are each categorized here as oriented toward where they are most commonly employed.

Each of the factors presented can operate in isolation, but typically is nested within a communal system of interlacing forces which affect the relative probabilities of destructive conflict and peace. Whether components operate at higher macro-levels or lower micro-levels will affect the rate and scope of their impact (Klein/Kozlowski 2000). Of course, the probabilities of maintaining a culture of constructive conflict and peace increase considerably when multiple factors are operating and aligned across levels. Exactly how best to operationalize this, however, is highly dependent on the particulars of the local situation.

7.3   Micro-level Factors (Prevention of Destructive Conflict)

Individuals, particularly when acting in concert, are key agents in processes of transforming and preventing conflict. Of course, exceptional leaders such as Mahatma Gandhi, Martin Luther King, Nelson Mandela, and the recent Liberian Nobel Laureate Leymah Gbowee can have a disproportionate impact on conflict resolution and prevention (Disney/Gbowee 2012). However, as many of the chapters suggest, destructive conflict is most likely to be mitigated when all individuals in a society adopt and internalize the following components:

- Awareness of the causes, consequences and escalatory tendencies of destructive conflict and violence.
- Moderately high levels of self-monitoring, restraint and regulation of internal impulses for destructive or violent acts.
- Satisfaction of basic human needs including physiological needs, safety and dignity.
- Values, attitudes, skills and behaviours supporting non-violence.
- Moderate levels of tolerance for uncertainty.
- High levels of tolerance for and openness to difference.
- capacity for forgiveness.

7.4   Micro-level Factors: Promotion of Sustainable Peace

Individuals also play a foundational role in increasing the probabilities for sustainable peace. They increase when individuals in a society adopt and internalize the following:

- Recognition of the interdependence of all people, similar and different, local and global.
- A strong self-transcendent value orientation committed to the welfare of others and society, with a sufficient self-enhancement orientation to mitigate individual resentment.
- A healthy balance of openness to change *and* conservatism, responsive to changing times and circumstances.
- Values, attitudes, skills and behaviours promoting cooperation and trust.
- Knowledge, attitudes and skills for constructive conflict resolution.
- Higher levels of integrative, emotional, behavioural and social identity complexity.
- Capacities for tolerance, humanization, realistic empathy (understanding how a situation looks to someone else) and compassion for members of one's in-groups and out-groups.
- An appreciation of environmental stewardship and equitable sharing of the earth's resources among its members and with all human beings.
- Language for peace: a large lexicon for all aspects of cooperative and peaceful relations and sufficient use of such terms to foster automaticity.
- A strong sense of global identity with a concrete understanding of the steps that need to be taken locally to act as a global citizen.

7.5   Meso-level Factors: Prevention of Destructive Conflict

Some scholar-practitioners suggest that mid- or meso-level factors play a disproportionate role in determining community dynamics related to conflict, as these forces are situated between both micro and macro forces and thus have an important mediating effect between, for instance, governance and policy and individual beliefs, values and behaviour (Dugan 1996; Lederach 1997; Kriesberg 1999). Some of the meso-level factors relevant to conflict prevention and mitigation identified by our contributors include:

Social taboos against corporal punishment and other forms of violence in the home, schools, workplace and public spaces.

Norms of gender equity and equality in the home, schools and the workplace. Early access to peace education and multicultural tolerance programs in pre-school, elementary and middle school.

Opportunities for peaceful sublimation of aggression through competitive or extreme sports, occupations, creative arts, etc.

Functional and accessible venues for constructive, non-violent action to seek recourse and address perceived injustices and other harms.

Strong norms for procedural and distributive justice in schools, workplaces, marketplaces, and elsewhere in the community.

7.6 Meso-level Factors: Promotion of Sustainable Peace

With regard to increasing the probabilities for fostering and sustainingpeace,the following mid-level components were identified as critical:

- Strong norms valuing and nurturing children.
- Early socialization of children oriented toward mutual care and nurturance.
- Cross-cutting structures fostering common interests, activities and bonds across different ethnic and religious groups.
- Structures of cooperative task, goal and reward interdependence in schools, workplaces, and politics.
- Programs and workshops in constructive conflict resolution and creative problem-solving for children, adults, parents and leaders of schools, businesses and politics.
- Shared, accurate and transparent collective memories of past events, conflicts and relationships between groups.
- Common use of peaceful language in popular media and normal daily discourse.
- Strong emphasis on both local and superordinate identities at the ethnic, communal, national and global levels.

7.7 Macro-level Factors: Prevention of Destructive Conflict

Conditions, mandates, regulations and processes operating at the macro-level have the paradoxical effect of being the most distant from individual-level behaviours and yet influencing these behaviours most rapidly once implemented (Klein/Kozlowski 2000). And although what our leaders do and say at this level may often seem to not have a major impact on our day-to-day lives, they can have a substantial symbolic effect on us as well as significantly alter the social conditions in which our conflicts take place. Thus, our contributors rightly identified the following macro-level factors as important to the prevention and mitigation of destructive conflict.

- Recognition and understanding of the inordinately strong salience of threat and tendencies towards inequality and competition in many societies across the globe.
- Established national political and social institutions that ensure the implementation and follow-through of negotiated settlements.
- Well-coordinated early warning systems operating through local governments and NGOs networked locally, regionally and globally for efficient communication.
- Use of crisis-mapping: an open-source platform for collecting and plotting local cell phone accounts of the commission of violent atrocities to inform the international community of emerging crises in a timely manner.
- Use of the internet and other social technologies to mobilize broad non-violent movements for social justice and corporate responsibility.
- Coordination between local governments, civil society and international organizations to prevent violent conflict.
- Well-functioning global organizations and institutions such as the United Nations, the International Criminal Courts, Interpol, and the Universal Declaration of Human Rights.
- Developing awareness and knowledge of the Dilemma of the Commons and how to overcome it.

7.8 Macro-level Factors: Promotion of Sustainable Peace

Finally, chapters of the book *The Psychological Components of a Sustainable Peace* by Coleman and Deutsch (2012) also identified these important influences for promoting and maintaining a state of sustainable peace:

- A societal idea of peace that includes an ethic of interethnic unity, care and nurturance of others, which is as strong (or stronger) as the view of peace as something that need be secured and defended.
- Societies that define themselves as internally and externally peaceful.

- A transcultural elite with shared norms of tolerance, cooperation, and creative problem-solving, who model for all the efficacy and value of constructive, non-violent action.
- National governance structures tending towards egalitarianism and democracy.
- A strong community of global citizens engaged locally in initiatives fostering global citizenry and addressing shared global concerns (climate change, poverty, etc.).
- Political and business ethics that are in harmony with nature and environmental stewardship.
- Institutions which reflect and uphold self-transcendent values.
- Gender parity with a proportional number of women in the highest positions of leadership in business, politics and the military.
- Use of the internet and other social technologies to mobilize broad social movements for humanitarian works and global peace.
- Strong communications, trade, and cultural and civilian exchanges between nations
- Peace-mapping: an open-source platform for collecting and plotting local cell phone accounts of nations.the commission of peaceful and humanitarian acts to inform and inspire other potential 'third-siders'.
- The establishment of peace parks: natural parks located at the borders between disputing nations where development and use of the parks are offered as superordinate goals.

These multiple factors operating across three levels together constitute a system of sustainable peace, distinguishing such communities and societies from those locked in systems of dominance and destructive conflict. The question then becomes, can we conceptualize how these systems develop, stabilize and change, and how groups and communities can move from one to the other?

7.9   A Dynamical-Minimal Model of Sustainable Peace

In Cultures of Peace: The Hidden Side of History, Elise Boulding (2000) points out that our warlike culture is accompanied by a concurrent culture of peace. This is the view we have taken in this book: that all communities and societies inflicted by destructive conflict and war have a latent potential for peace, and that societies at peace often harbour a latent potential for hostilities. For instance, in Islands of Agreement:

Managing Enduring Armed Rivalries (2007<XREF>), Gabriella Blum describes the many examples of cooperation and exchange operating in the context of long-enduring armed rivalries such as between India and Pakistan, Greece and Turkey, and Israel and Lebanon. These havens of cooperation in the context of enmity effectively reduce suffering and loss and allow mutually beneficial exchanges to take place, and are evidence of the latent potential for peace inherent in all societies, even those currently engulfed by war. On the other hand, if we examine the current state of Northern Ireland, we see a somewhat fragile state of peace and often hear the rumblings of what could once again become a dynamic of violence. Both potentials exist.

The trick then is not to simply be able to move from one state (war) to the other (peace) constructively. In fact the international community has got quite good at this, seeing a dramatic increase in the number of wars ending through negotiation rather than through unilateral military victory. In fact, these numbers have flipped since the end of the Cold War, with today twice as many wars ending through negotiation as through military victory (Mason/Crenshaw/McClintock et al. 2007). Incredibly, from 1988 to 2003, more wars ended through negotiation than had in the previous two centuries (United Nations 2004). After peaking in 1991, the number of civil wars had dropped by roughly forty per cent by 2003 (United Nations 2004). This indicates that local, regional and international peacemakers have an increasing positive impact in peace mediation and transitions to systems of peace.

However, over twenty-five per cent of the wars ended through negotiations relapse into violence within 5 years (Suhrke/Samset 2007). In some cases, such as in Rwanda and Angola, more people were harmed and died after peace agreements were ratified by the parties and then failed (Stedman et al. 2002). And these failed-peace states seem to begin a new downward spiral. States with civil wars in their history are far more likely to experience renewed violence (Mason/Crenshaw/McClintock et al. 2007). And the longer such conflicts last, the greater the chances of recurrence of war (Collier 2000). Thus, the priority focus today should fall on sustainability—how to increase the probabilities that once societies transition to peace, they will be able to remain there and navigate the many challenges to peace that can accompany its implementation and maintenance.

A new theoretical approach to understanding and promoting sustainable peace is informed by the efforts of a multidisciplinary research team working to apply insights and methods from complexity science to understanding peace (see Nowak/Bui-Wrzosinska/Vallacher/Coleman 2012 and Coleman/Bui-Wrzosinska/Vallacher et al. 2006; Coleman/Vallacher/Nowak et al. 2007; Coleman/Hacking/Stover et al. 2008; Coleman/Vallacher/Nowak et al. 2011; Nowak/Vallacher/Bui-Wrzosinska et al. 2006; Vallacher/Coleman/Nowak et al. 2010). They suggest that qualitative differences in the dominant patterns of social behaviour (such as those found in peaceful societies versus hostile or warring societies) can be accounted for by a few basic factors. Accordingly, their research attempts to identify, from scholarship and practice, the fundamental factors that promote sustained peaceful dynamics in communities or, put another way, that make societies immune to prolonged destructive or violent conflict.

The basic model centres on the concept of *attractor*, a concept from applied mathematics. In a dynamical system composed of many parts or 'elements', an attractor is a relatively stable state or pattern of behaviour that coordinates or integrates the elements (see Nowak and Vallacher 1998). In a mental system, an attitude or a belief functions as an attractor if it integrates and provides common meaning for different events, memories, and pieces of information, even if these mental 'elements' by themselves might be interpreted in very different ways. In a social system (e.g. a group or society), an ideology functions as an attractor if it provides a shared reality and frame of reference for collective action, even if the members of the group or society each have divergent needs and interests. Metaphorically, an attractor 'attracts' the system's elements to a common state or pattern, providing coherence and stability in the face of new and confusing experiences (e.g. ambiguous information, unexpected events). Once a system is governed by an attractor, it actively resists threats that would change the way the elements (e.g. thoughts, individuals) are organized. From a dynamical perspective, then, attempts to challenge a person's firmly held attitude or a group's ideology are likely to backfire, strengthening rather than weakening the attractor, and thus may intensify rather than reduce antagonism and violence in a situation characterized by conflict (see Staub 2012 for a more detailed discussion of attractor properties and dynamics).

Research on attractors has found that groups (e.g. communities, gangs, societies) typically have more than one attractor governing the way they think about and behave toward other groups. This means that hostile and destructive interaction patterns between groups may coexist with the potential for peaceful interactions between such groups. At any one time, however, only one attractor (e.g. negative) is likely to manifest, with the other attractors (e.g. positive) virtually invisible to observers, or even to the participants themselves.

The existence of *latent attractors* suggests that under the right conditions, the groups may demonstrate a sudden and dramatic change in their thoughts, feelings, and actions vis-à-vis one another. Thus, the interactions within a community can move from one manifest attractor (such as peace) to another previously latent attractor (such as war), sometimes even in response to a rather minor incident that triggers the latent pattern of thought, feeling, and action. This scenario of *nonlinear change* is evident both in sudden outbreaks of group violence in situations of relative peace (such as has occurred in Northern Ireland) and in sudden outbreaks of peace in situations of protracted conflict (such as occurred in the 1990s in Mozambique after sixteen years of civil war). Recognition that the current state of communal life can coexist with other potential but latent patterns of interaction (each with differing degrees of 'attracting' power) underlies an ambitious research agenda and provides the foundation for the following set of recommendations for promoting sustainable peace.

7.9.1   Be Aware That War and Peace Potentials Can Coexist

As the attractor landscape in figure 7.1 indicates, groups and communities typically hold the potential for dramatically different types of interaction patterns simultaneously. One attractor may capture the state of the system for extended periods of time (as is seen during protracted periods of conflict). However, this does not mean that peace-building initiatives (peace education, dialogue groups, intergroup cultural exchanges, common community projects, etc.) during this period are for naught. Here, the idea of *latent attractors* provides an important new perspective for understanding peace. In this view, the malignant thoughts, feelings, and actions characterizing a group's dynamics during conflict represent only the most salient and visible attractor for the group.

Particularly if there is a long history of interaction with the out-group, there may be other potential patterns of mental, affective, and behavioural engagement vis-à-vis members of the out-group, including those for positive relations. With this in mind, identi-

Figure 7.1: A dynamical system with two attractors corresponding to constructive relations (A) and destructive relations (B).

fying and reinforcing latent (positive) attractors, not simply disassembling the manifest (negative) attractors, should be the aim of conflict prevention in service of sustainable peace. In other words, in addition to attempts at achieving *negative peace* (an end to destructive conflict and violence), and the goal of *positive peace* (establishing fair systems of opportunity and justice) we must also strive to enhance *promotive peace*—the establishment of strong attractors for positive, constructive social relations. These objectives can be accomplished by implementing many of the initiatives summarized in the above multi-level framework.

7.9.2 'Reverse Engineer' Negative, Destructive Attractors

When conflicts do arise, the most obvious need is to quell any violence and contain actively destructive processes. This is often done by introducing police support, peacekeeping troops, or other forms of regional or international military interventions. However, even when systems de-escalate and appear to return to a state of peace, the potential for destructive interactions (destructive attractors) still exists. It is important, then, that we work actively to deconstruct and dismantle the negative attractors.

In generic terms, the deconstruction of an attractor entails focusing on the elements comprising the pattern of behaviour rather than focusing on the pattern itself. In the context of conflict, this means calling attention to specific actions, events, and pieces of information without noting their connection to the pattern in which they are embedded. When decoupled in this fashion, the lower-level elements may become reconfigured into an entirely different pattern (e.g. a positive view of the out-group and a benign or peaceful interaction pattern). The important point is this: attacking the pattern itself is likely to intensify rather than weaken the pattern because of the tendency for attractors to resist change, so one should focus instead on isolating elements and thereby weakening or eliminating the reinforcing feedback loops among

them. The chapters in the book *The Psychological Components of a Sustainable Peace* edited by Coleman and Deutsch (2012) present a variety of ways in which this can be accomplished in real-world settings, including: *introducing negative feedback loops (early-warning systems, cross-cutting structures, international monitoring, etc.); institutionalizing more nuanced, alternative conflict narratives (through media, textbooks, official accounts, etc.); and limiting the pervasive spread of conflict by allowing movement of the parties.*

7.9.3 Increase Complexity for Peace

Research has also shown that constructive social relations are characterized by relatively high levels of cognitive, emotional, behavioural, and structural complexity (see Nowak/Bui-Wrzosinska/Vallacher/Coleman 2012). Such complexity is advantageous when groups face problems or conflicts with other groups. As conflicts intensify, there is a strong tendency for the parties' thoughts, feelings, and behaviours to become more simple and black and white (which is evidence of strong attractor dynamics). If the conflict spreads to the community level and persists, then we see the same polarization occur in social networks, groups, and institutions. However, communities and groups who maintain more complex cross-cutting (intergroup) structures and social networks, who hold more complex (multiple group) social identities, and who display more complex cognitive, emotional, and behavioural (adaptive) patterns, have been found to be more tolerant of out-groups, display less violence when conflicts spark, and engage in a more constructive manner when conflicts become difficult. Thus, *sustainable peace requires structures and processes that foster increased contact and complexity.*

7.9.4 Increase Movement for Peace

The findings from research support the basic idea that peace is associated with movement (Bartoli et al. 2010). When people and groups get trapped in narrow attractors for social relations, whether in patterns of destructive conflict, oppressor-oppressed dynamics, or even in patterns of isolation and disengagement from others, their well-being tends to deteriorate and their level of resentment tends to build. These traps may be constituted by physical structures such as segregated spaces, or by social-psychological constraints such as norms, attitudes and ideologies. When trapped in such a well, people can be creative at becoming ever more destructive, oppressed, independent, etc., which acts to deepen the attractor and makes it less likely they will be able to escape its pull. Of course, any pattern of behaviour may be functional in certain situations; a destructive orientation fits very well in times of armed conflict. But these patterns can become dominant and pervasive, so that when the current situation changes, or when people move to different situations, it is critical for people to *adapt*—to take up different patterns of behaviours that are appropriate to the varied situations they face. From this perspective, *sustainable peace requires the establishment of conditions that allow for movement and adaptation.* At times, even 'jiggling' the system—almost random movement—can break patterns and restore flexibility.

7.9.5 Peace is Associated with a Sufficient yet Tolerable Rate of Movement toward Justice

Decades of research on the psychology of justice has found that movement is also central to justice and peace. First, a sense of relative deprivation has been found to be a fundamental source of ethno-political conflict and instability in otherwise peaceful communities. This is the injustice felt when people experience a gap in what they feel their group deserves and what it can achieve, in comparison to similar groups. This experience is typically triggered by change—shifts in the status quo that affect what groups expect, what they can get, and with whom they compare themselves. However, it is the need for procedural justice, or the sense that there exist fair processes for the allocation of goods and for recourse against grievances, that has been shown to be critical to addressing injustice, even more so than actually receiving fair outcomes. Furthermore, the rate at which justice is achieved is also critical. Peace scholars have found that minority groups who feel that the channels for fair recourse are blocked or too unresponsive are more likely to revolt (Gurr 2000). However, they have also found that when particular minority groups ascend to justice and equal treatment very rapidly, this can raise the aspirations, envy, and resistance of other groups (including those in power), and thus destabilize communities (Gurr 2000; Lederach 1997). Thus, procedures of justice that provide a sufficiently steady response to the grievances of all stakeholders are a necessary condition for sustainable peace.

7.9.6 Foster Repellors for Violence

Anthropological research summarized by Fry and Miklikowska (2012) has shown that communal taboos against violence have existed for the bulk of human history, and were a central component of our ancestors, the prehistoric nomadic hunter-gatherer bands. Indeed, a key characteristic of peaceful societies, both historically and in the contemporary world, is the presence of non-violent values, norms, ideologies, and practices. Although non-violent norms are practised in many communities around the globe, they are often overwhelmed by more violent ideologies, messages, and social modelling. There are a wide-variety of parenting and educational methods for fostering more non-violent, pro-social attitudes and skills in children, such as violence-prevention, tolerance, cooperative learning, conflict resolution, and peace education curriculum, to name just a few. However, sustainable peace will require a much more concerted effort to teach non-violent values, norms, and practices to young people and to better limit exposure of youth to gratuitous forms of violence and to destructive social modelling by adults and public leaders.

7.9.7 Realize that Peace is never Achieved

Peace is a dynamic process, not an outcome. It requires a set of fair processes and procedures that allow all stakeholders to negotiate for their needs and rights, in order to create unity out of diversity. Indeed, peace initiatives uninformed by an ongoing process of reading feedback are destined to do more harm than good. Research has found that the most effective decision-makers are those who are able to continually adapt; by remaining open to feedback, they can reconsider their decisions and alter their course if necessary (Dorner 1996). These leaders make more, not fewer, decisions as their plans unfold, and ultimately are able to enhance the well-being of the communities with which they work. Thus, effectiveness comes from flexibility not rigidity. In this way, we can work to increase the probabilities that peace will emerge and be sustained.

7.10 Conclusion: An Agenda for Sustainable Peace

Today, very few scholars study peace. However, the few that do tell us that today there are approximately eighty societies worldwide who could be categorized as having low levels of internal aggression, and seventy societies who are peaceful in their relations with other groups and communities in their regions (Fry 2006). Unfortunately, our understanding of such groups, and of the conditions that foster pro-social relations, is extremely limited. There is often an unarticulated assumption in research on war and conflict that a thorough understanding of the problem of destructive conflict will provide insight into conditions and processes which foster and sustain peace. This assumption has been found by researchers to be unfounded and incorrect (see Gottman/Murray/Swanson et al. 2002; Losada 1999; Losada/Heaphy 2004). Destructive conflict and peace are not end points of a single dimension but rather often coexist as separate dimensions. It is clearly time to champion the study of peace in its own right. In dynamical-systems terms, the coexistence of malignant and (potential) peaceful possible relations is tantamount to the coexistence of two attractors constraining the dynamics of the parties to a conflict. Although effort should be devoted to the deconstruction of the negative attractor, attention should also be devoted to strengthening the positive attractor for intergroup relations. There may be little immediate effect of fostering opportunities for positive relations between the groups, but such efforts plant the seed for a possible transformation should conditions change in a way that destabilizes existing mental, affective, and behavioural patterns.

If such a seed is not planted, it cannot take root even if the negative attractor is somehow discredited or otherwise destabilized. A dynamical system does not change unless it has a new space to occupy. A latent attractor essentially represents a new space for intergroup relations.

Therefore, the approach to the study of sustainable peace we advocate includes the following components:

- Movement beyond the primary focus on destructive conflict, violence and war (problems) to the equally important study of sustainable peace *(solutions).*
- Movement away from simple, linear models of cause-and-effect toward more complex, *holistic* models of sustainable peace situated within constellations of ecological, biological, psychological, social, economic, and other structural forces.
- An enhanced capacity to work collaboratively across a variety of disciplines to better understand and foster sustainable peace through *multiple perspectives* and complementary approaches.

- A shift in emphasis away from achieving particular short-term outcomes (peace treaties, agreements, etc.) toward establishing and maintaining the conditions for *sustainable* peace processes in communities over time.
- An enhanced capacity to communicate and build *partnerships* from science to policy/practice and from policy/practice to science. The establishment of local, regional, and global *networks* of support and information on best practices for increasing the probabilities of sustainable peace.

What is required at this stage is an investment in a concerted effort to bring together scholars, practitioners, and policymakers from a variety of disciplines to work to understand sustainable peace beyond the level of case-based descriptions, to get at the essence of their underlying dynamics. This could include initiatives such as:

- *Support for the development of basic theory and research on sustainable peace.* There are few scholars conducting basic research on the fundamental (necessary and sufficient) conditions and processes for sustainable peace (Doug Fry and Marta Miklikowska are exceptions). However, it is critical that the applied frameworks which inform practice be informed by basic, sound, empirically-tested theoretical models, in order to foster peace most effectively.
- *Graduate-level, multi-disciplinary, theory-practice courses on sustainable peace.* Courses which involve a core group of faculty from different disciplines that are committed to working together to weave and develop the ideas and practices of sustainable peace. These courses could move from basic theory through applied models to strategies and tactics for intervention, and it is vital that we involve academics and UN/NGO practitioners and policymakers as guest lecturers on the course.
- *The development of a data-based index for annual reporting on state and regional levels of sustainable peace.* This project could build of off the Global Peace Index and other such resources (FAST International, International Crisis Group, Human Security Report) and could involve the business and academic communities in developing a comprehensive methodology (beyond early warning and violence prevention) for measuring and reporting on sustainable peace worldwide. This initiative could be informed by such initiatives as the Gross National Happiness Index (Yones 2006), the eight bases of a Culture of Peace (UN Resolu-

Figure 7.2: A nested model of the psychological components of sustainable peace.

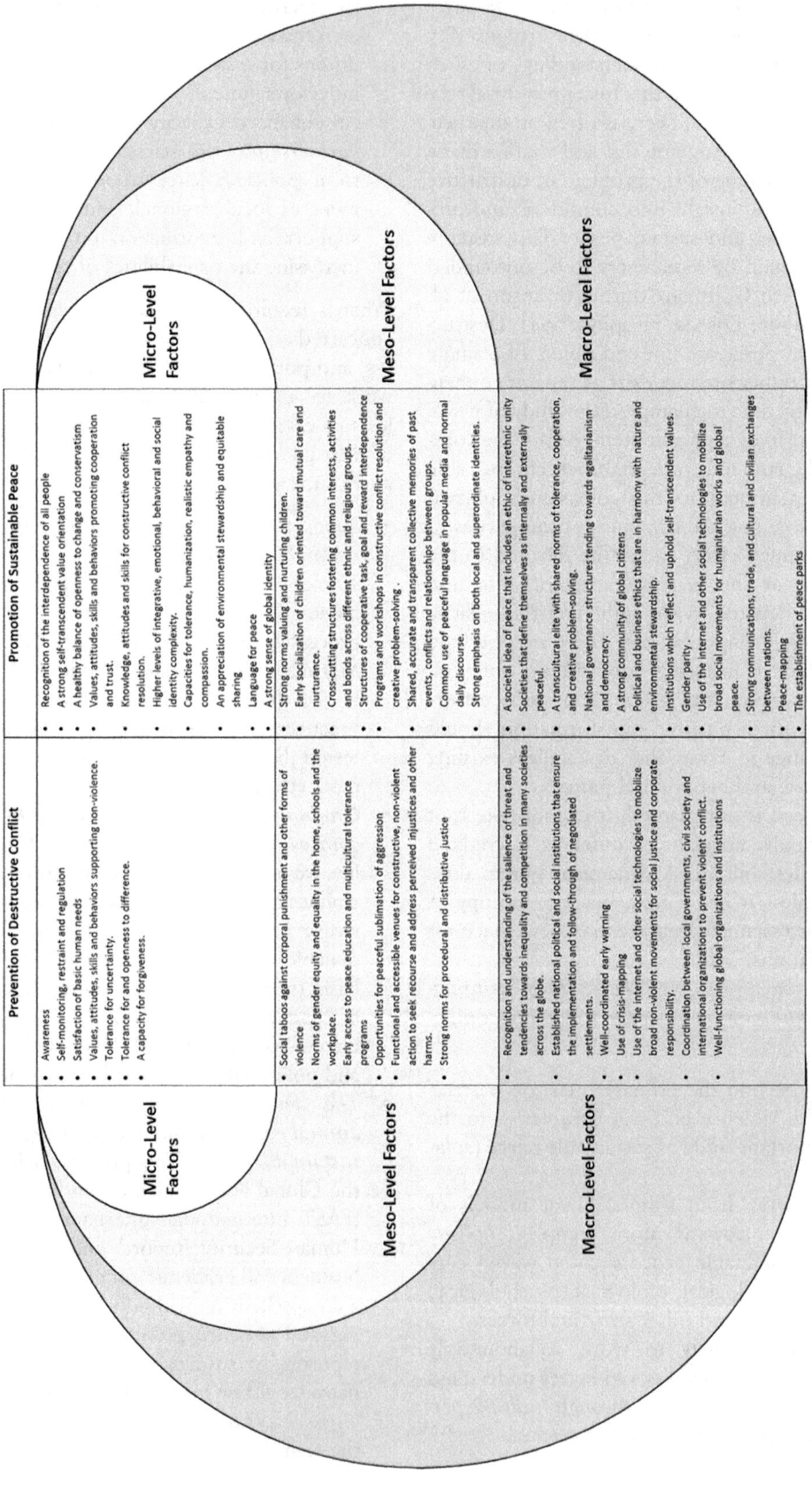

	Prevention of Destructive Conflict	Promotion of Sustainable Peace
Micro-Level Factors	• Awareness • Self-monitoring, restraint and regulation • Satisfaction of basic human needs • Values, attitudes, skills and behaviors supporting non-violence. • Tolerance for uncertainty. • Tolerance for and openness to difference. • A capacity for forgiveness.	• Recognition of the interdependence of all people • A strong self-transcendent value orientation • A healthy balance of openness to change and conservatism • Values, attitudes, skills and behaviors promoting cooperation and trust. • Knowledge, attitudes and skills for constructive conflict resolution. • Higher levels of integrative, emotional, behavioral and social identity complexity. • Capacities for tolerance, humanization, realistic empathy and compassion. • An appreciation of environmental stewardship and equitable sharing • Language for peace • A strong sense of global identity
Meso-Level Factors	• Social taboos against corporal punishment and other forms of violence • Norms of gender equity and equality in the home, schools and the workplace. • Early access to peace education and multicultural tolerance programs • Opportunities for peaceful sublimation of aggression • Functional and accessible venues for constructive, non-violent action to seek recourse and address perceived injustices and other harms. • Strong norms for procedural and distributive justice	• Strong norms valuing and nurturing children. • Early socialization of children oriented toward mutual care and nurturance. • Cross-cutting structures fostering common interests, activities and bonds across different ethnic and religious groups. • Structures of cooperative task, goal and reward interdependence • Programs and workshops in constructive conflict resolution and creative problem-solving. • Shared, accurate and transparent collective memories of past events, conflicts and relationships between groups. • Common use of peaceful language in popular media and normal daily discourse. • Strong emphasis on both local and superordinate identities.
Macro-Level Factors	• Recognition and understanding of the salience of threat and tendencies towards inequality and competition in many societies across the globe. • Established national political and social institutions that ensure the implementation and follow-through of negotiated settlements. • Well-coordinated early warning • Use of crisis-mapping • Use of the internet and other social technologies to mobilize broad non-violent movements for social justice and corporate responsibility. • Coordination between local governments, civil society and international organizations to prevent violent conflict. • Well-functioning global organizations and institutions	• A societal idea of peace that includes an ethic of interethnic unity • Societies that define themselves as internally and externally peaceful. • A transcultural elite with shared norms of tolerance, cooperation, and creative problem-solving. • National governance structures tending towards egalitarianism and democracy. • A strong community of global citizens • Political and business ethics that are in harmony with nature and environmental stewardship. • Institutions which reflect and uphold self-transcendent values. • Gender parity. • Use of the internet and other social technologies to mobilize broad social movements for humanitarian works and global peace. • Strong communications, trade, and cultural and civilian exchanges between nations. • Peace-mapping • The establishment of peace parks

tion A/RES/52/13), the Peace Scale (Klein/Goertz/ Diehl et al. 2008), and the Universal Declaration of Human Rights. Another possible step would be to develop a dynamical computational model with variables from multiple disciplines that have been shown to predict violence and peace and then try to keep a 'Violence Watch' as well as a 'Peace Watch' on countries by plugging data in to see if we can identify nations susceptible to outbreaks of violence and outbreaks of peace (latent attractors).

An annual theory-practice-policy forum on sustainable peace. There is currently a need for an annual gathering of policymakers, peace practitioners and scholars, where leading-edge research on sustainable peace could be translated and provided to policymakers. Collaborative, multidisciplinary work of this nature requires a common language based on an integrative platform to facilitate communication and coordination across the legendary disciplinary and theory-practice divides. Theapproach of dynamical systems, a scientific paradigm widely employed across scientific disciplines, provides such a platform.

References

Bartoli, Andrea; Bui-Wrzosinska, Lan; Nowak, Andrzej, 2010: "Peace is in movement: A dynamical systems perspective on the emergence of peace in Mozambique", in: *Peace and Conflict: Journal of Peace Psychology,* 16,2: 211-230.

Blum, Gabriella, 2007: Islands of agreement: Managing enduring armed rivalries (Cambridge, MA: University Press)

Boulding, Kenneth, 1978: *Stable peace* (Austin: University of Texas Press).

Boulding, Elise, 2000: *Cultures of peace: The hidden side of history* (Syracuse: Syracuse University Press).

Coleman, P. T., 2006: "Conflict, complexity, and change: A meta-framework for addressing protracted, intractable conflicts–III", in: *Peace and Conflict: Journal of Peace Psychology,* 12,4: 325-348.

Coleman, Peter T.; Bui-Wrzosinska, Lan; Vallacher, Robin R.; Nowak, Andrzej, 2006: "Protracted conflicts as dynamical systems", in: Schneider, A. K.; Honeyman, C. (Eds.): *The negotiator's field book: The desk reference for the experienced negotiator* (Chicago: American Bar Association Books): 61-74.

Coleman, Peter T.; Vallacher, Robin R.; Nowak, Andrzej; Bui-Wrzosinska, Lan, 2007: "Intractable conflict as an attractor: Presenting a model of conflict, escalation, and intractability", in: *American Behavioral Scientist,* 50: 1454-1475.

Coleman, Peter T.; Hacking, Antony; Stover, Mark; Fisher-Yoshida, Beth; Nowak, Andrzej, 2008: "Reconstructing ripeness I: A study of constructive engagement in protracted social conflicts", in: *Conflict Resolution Quarterly,* 26,1: 3-42.

Coleman, Peter T.; Vallacher, Robin R.; Nowak, Andrzej; Bui-Wrzosinska, Lan; Bartoli, Andrea, 2011: "Navigating the landscape of conflict: Applications of dynamical systems theory to protracted social conflict", in: Ropers, Norbert (Ed.): *Systemic thinking and conflict transformation* (Berlin: Berghof Foundation for Peace Support).

Coleman, Peter T.; Deutsch, Morton (Eds.), 2012: *Psychological Components of Sustainable Peace* (New York: Springer).

Collier, Paul, 2000: Economic causes of civil conflict and their implications for policy (Washington, DC: World Bank).

Disney, Abigail; Gbowee, Leymah, 2012: "Gender and Sustainable Peace", in: Coleman, Peter T.; Deutsch, Morton (Eds.): Psychological Components of Sustainable Peace (New York: Springer): 197-204.

Dorner, Dietrich, 1996: The Logic of Failure: Recognizing and avoiding error in complex situations (New York: Metropolitan Books).

Dugan, 1996: "A Nested theory of conflict", in: Women in Leadership, 1: 9-20.

Fry, Douglas P., 2006: The human potential for peace: An anthropological challenge to assumptions about war and violence (New York: Oxford University Press).

Fry, Douglas P.; Miklikowska, Marta, 2012: "Culture of Peace", in: Coleman, Peter T.; Deutsch, Morton (Eds.): Psychological Components of Sustainable Peace (New York: Springer): 227-244.

Gottman, John M.; Murray, James; Swanson, Catherine C.; Tyson, Rebecca; Swanson, Kristin R., 2002: The mathematics of marriage (Cambridge, MA: The MIT Press).

Gurr, Ted R., 2000: *Peoples versus states* (Washington, DC: United States Institute of Peace Press).

Klein, James; Kozlowski, Steve W. J. (Eds.), 2000: *Multilevel theory, research, and methods in organizations: Foundations, extensions, and new directions.* Society for industrial and organizational psychology frontiers series (San Francisco: Jossey-Bass).

Klein, James; Goertz, Gary; Diehl, Paul, 2008: "The peace scale: Conceptualizing and operationalizing nonrivalry and peace", in: *Conflict Management and Peace Science,* 25: 67-80.

Harrison, Selig S.; Kreisberg, Paul H.; Kux, Dennis, 1999: *India and Pakistan: The first fifty years* (Cambridge-New York: Woodrow Wilson Center Press–Cambridge University Press).

Lederach, John Paul, 1997: Building peace: Sustainable reconciliation in divided societies (Washington, D.C.: United States Institute of Peace Press).

Losada, Marcial, 1999: "The complex dynamics of high performance teams", in: *Mathematical and Computer Modelling*, 30: 179-192.

Losada, Marcial; Heaphy, Emily, 2004: "The role of positivity and connectivity in the performance of business teams: A nonlinear dynamics model", in: *American Behavioral Scientist*, 47: 740-765.

Mason, T. David.; Crenshaw, Martha; McClintock, Cynthia; Walter, Barbara, 2007: *How political violence ends: Paths to conflict de-escalation and termination*. APSA task force on political violence and terrorism, Group 3. Presented at the annual meeting of the American Political Science Association, Chicago.

Nowak, Andrzej, 2004: "Dynamical minimalism", in: *Personality and Social Psychology Review*, 8: 138-145.

Nowak, Andrzej; Vallacher, Robin, 1998: *Dynamical social psychology* (New York: Guilford Press).

Nowak, Andrzej; Vallacher, R.; Bui-Wrzosinska, Lan; Coleman, Peter T., 2006: "Attracted to conflict: A dynamical perspective on malignant social relations", in: Golec, Agnieszka; Skarzynska, Krystyna (Eds.): *Understanding social change: Political psychology in Poland* (Haauppague: Nova Science Publishers Ltd.).

Nowak, Andrzej; Bui-Wrzosinska, Lan; Vallacher, Robin: Coleman, Peter T., 2012: "Sustainable Peace: A Dynamical Systems Perspective", in: Coleman, Peter T.; Deutsch. Morton (Eds.): Psychological Components of Sustainable Peace (New York: Springer): 265-282.

Seager, Thomas, 2008: "The sustainability spectrum and the sciences of sustainability", in: Business Strategy and the Environment, 17: 444-453.

Staub, Ervin, 2012: "Reconciliation Between Groups, the Prevention of Violence, and Lasting Peace", in: Coleman, Peter T.; Deutsch, Morton (Eds.): Psychological Components of Sustainable Peace (New York: Springer): 245-264.

Stedman, Stephen. J.; Rothchild, Donald; Cousens, Elizabeth M., 2002: *Ending civil wars: The implementation of peace agreements* (Boulder: Lynne Rienner).

Suhrke, Astri; Samset, Ingrid, 2007: "What's in a figure? Estimating recurrence of civil war", in: *International Peacekeeping*, 14,2: 195-203.

United Nations, 2004: A more secure world: Our shared responsibility. Report of the United Nations High-Level Panel on threats, challenges and change (New York: United Nations).

Vallacher, Robin; Coleman, Peter T.; Nowak, Andrzej; Bui-Wrzosinska, Lan, 2010: "Rethinking intractable conflict: The perspective of dynamical systems", in: American Psychologist, 65,4: 262-278.

Yones, Med, 2006: *The American pursuit of unhappiness: Gross National Happiness (GNH), a new socioeconomic policy. Executive white paper* (Las Vegas: International Institute of Management).

8   Development with Sustainable-Engendered Peace: A Challenge during the Anthropocene

Úrsula Oswald Spring[1]

Abstract

This chapter examines the evolution of the peace concept from its understanding as a negative concept towards a positive, structural, sustainable, and engendered peace. The concept of a 'sustainable-engendered peace' refers to the structural factors related to long-term violence, deeply embedded in the patriarchal system and characterized by authoritarianism, exclusion, discrimination, exploitation and violence. This dominant social structure affects values such as equity, equality and justice, and often even threatens the survival of individuals and social groups. Further, this dominant system has also concentrated the wealth of earth within a small group of oligarchs who manage multinational enterprises. The sources of threats have been consolidated over thousands of years by patriarchal institutions, religious controls, self-identified beliefs and social representations, and totalitarian exercise of power, and have also affected natural resources. Faced with these global threats, the chapter explores the potential of the concept of a sustainable-engendered peace, and attempts to reach an understanding of the deeply anchored links to patriarchy and its war system that are related to the physical, social and cultural threats of the dominant values and behaviour in the Anthropocene. The text also explores the potential for a concept of holistic and cosmopolitan peace that can challenge the root causes of violence and destruction, and it discusses the goal of just and equal power structures for human beings and nature.

Keywords: Sustainable development, sustainable-engendered peace, Anthropocene, peace concept, patriarchy, civilization, empowerment, equity, values, cultural identity.

8.1   Introduction

This chapter examines[2] the evolution of the peace concept from negative (UN Charter 1947) towards a positive peace (Galtung 1967) where structural (Galtung 1968), environmental (Conca 1994; Oswald Spring 2008), and engendered peace elements (Oswald Spring/Brauch/Tidball 2014; Oswald Spring 2004) are integrated in a diversified culture of peace (Boulding 2000). A 'sustainable-engendered peace'

explores the structural factors related to long-term violence, which are deeply embedded in the patriarchal system (Mies 1968; Reardon/Snauwaert 2015a, 2015b).

Over thousands of years, patriarchy has developed a complex system of power, exploitation and control, where economy, politics, wars, culture, religious beliefs, identity, and psychosocial roots have been adapted to historical and regional differences. The dominant factors of this system are authoritarianism, control, violence, exclusion, discrimination, exploitation, and concentration of wealth. The reference object refers to all people with no power or limited access to power at the global and local level, such as women, the indigenous, youth, the elderly, the unemployed, the poor, etc. (Calva 2012a, 2012b, 2012c). The values at risk are established gender relations (Lagarde 1990; Lamas 1996), deeply rooted in the dominant social representations of gender (Jodelet 1991; Serrano Oswald 2010), which are manipulated by an oligarchy (Stiglitz

1   Prof. Dr. Úrsula Oswald Spring, full-time Professor/Researcher at the *National University of Mexico* (UNAM) in the *Regional Multidisciplinary Research Center* (CRIM); email: <uoswald@gmail.com>.

2   I want to express my sincere thanks for the creative comments I received from Betty Reardon, Hans Günter Brauch, and from three anonymous peer reviewers. They helped to strengthen the arguments; however, any errors are my own responsibility. I am also grateful to Mike Headon for his careful English revision.

2010; Yiamouyiannis 2013) and reinforced by religious fundamentalism, hierarchical churches (Gutiérrez 2013), and schools (Gramsci 1975, 1998). This social structure affects values such as equity, equality, solidarity, justice (Truong/Gasper/Handmaker et al. 2014), cultural identity (Arizpe 2015, Serrano Oswald 2014) and often even the survival of individuals and social groups (Oswald Spring 1994). The sources of threats have been consolidated over thousands of years by patriarchal institutions (Folbre 2006), religious controls (Jasper 2013), and the totalitarian exercise of power (Held 2004).

Nevertheless, today not only humanity as a whole, but even the survival of the planet itself are threatened by an aggressive exploitation of natural resources, including mining, oil extraction, and the exploitation of timber (UNEP 2012; Crutzen 2011). Further, a small oligarchic global elite has imposed patterns of production that deplete and pollute natural resources (Westing 2013) and create a wasteful and unequal consumerism (MA 2005), precarious livelihoods, and corporate production patterns (Ellis 2003). Monopolized finances, unequal terms of trade, a consumerist propaganda worldwide and an unequal control of resources (UN/DESA 2015, Calva 2012a, 2012b) dominate these global environmental changes (Brauch/Oswald Spring/Mesjasz et al. 2008, 2011; Brauch/Oswald Spring/Grin 2009).

Faced with these deeply-rooted global threats, this chapter addresses the following research questions: how may the widened concept of a 'sustainable-engendered peace' provide a deeper understanding and new conceptual tools so that alternatives may be proposed for a transition to sustainable development? How may a sustainable-engendered peace cope with and overcome the present economic structure and power structure and the global threats to the survival to humankind and planet Earth in the era of the Anthropocene (ISSC 2013)?

The chapter starts with definitions of the 'Anthropocene' (Crutzen 2002; 2011), of 'structural peace' (Webel/Galtung 2007), of a 'culture of peace' (Boulding 2000, 2016; UNESCO 2000), and of 'sustainable development' (Brundtland 1987). Then it explores the emergence, history and adaptation of patriarchy from its origins until today. It shows how this socio-economic, military and political system of organization could develop from a decentralized village structure to monarchies, slavery, colonialism, feudalism, capitalism, and neo-liberalism, and produce an economic oligarchy based on a symbolic relation with militarism and the war system (Reardon 1996; Reardon/Snau-

waert 2015a, 2015b). It analyses the structural elements that limit the sustainable development of humankind in the present unequal and unsustainable globalized world and explains the threats emanating from the production and consumption process for both humankind and planet Earth. Section 8.4 explores the potential of the concept of 'sustainable-engendered peace', a concept that will help in the analysis of patriarchy and its war system, which are related to the physical, socio-economic and cultural threats of dominant values and behaviour in the Anthropocene. In the final section, the chapter discusses critically the research question and explores the potential for a sustainable-engendered peace. The conclusion proposes an integrated negative, positive, structural, cultural, engendered, and sustainable peace. This holistic and cosmopolitan shift requires a change to the root causes of economic monopoly, violence and destruction in order to move towards just and equal economic and power structures that can nurture both human beings and nature.

8.2   Conceptual Definitions: Global Environmental Change, Anthropocene, Structural Peace, Culture of Peace, Sustainable-Engendered Peace and Development

8.2.1   Global Environmental Change

Global environmental change (GEC) is wider than *climate change* (CC). The term refers to the transformation produced by human beings in the ecosphere and that affects the hydrosphere, the atmosphere, the biosphere, the lithosphere and the pedosphere (Brauch/Oswald Spring/Mesjasz et al. 2008, 2011; Brauch/Oswald Spring/Grin 2009). Changes in the natural system are the result of numerous changes in systems of agricultural production, of a rapid urbanization process, and of population growth—the human population tripled during the last century, but water consumption increased sixfold (Oswald 2011). Further, unsustainable productive processes are polluting natural resources and creating health threats for human beings, as well as endangering ecosystems (Elliott 2011). The energy, transportation and production sectors have consumed increasing amounts of fossil fuel (IEA 2014), and in addition deforestation is reducing the capture of CO_2 (IPCC 2013, 2014a,

2014b, 2014c, 2014d). The emissions from greenhouse gases (IPCC 2013) have increased exponentially. Together with changes in land use, the intensive use of chemical fertilizers and pesticides in agriculture (FAOSTAT 2013), water pollution, deforestation and unsustainable waste production and management, both natural and human systems have seriously deteriorated. Finally, a globalized financial system, unequal credit, patterns of consumption and production, and an uneven level of access to material and immaterial resources are leading contributors to GEC.

8.2.2 Concept of the Anthropocene

Paul Crutzen (2002) proposed the idea of the Anthropocene. This concept relates to the environmental changes produced, predominantly by human intervention and corporate enterprises, in the earth system since the industrial revolution, but especially during the last five decades, because of the intensive use of fossil energy, the rapid increase in greenhouse gas emissions into the atmosphere, the pollution and warming of the seas, and an accelerated process of urbanization. The Anthropocene is a new geological epoch that is changing the earth history of the Holocene, it is defined as "the most recent manifestation of a pervasive millennial-scale climate cycle operating independently of the glacial-interglacial climate state" (Bond/Willima/Maziet et al. 1997). This concept is useful for understanding the transformative negative effects of human activity on the planetary ecology, but this chapter supports the argument that human agency also has a positive potential for change. It is precisely the dominant patriarchal global order that is limiting human choice and agency. By choosing peace actions that lead towards an envisioned political, social, economic and cultural transformation, a sustainable-engendered peace paradigm may open ways for new socio-economic orders, which will transform the present exploitation of humanity and earth into equal, peaceful and sustainable behaviours.

8.2.3 Structural Peace and Globalization

Quincy Wright ([1942] 1965z: 1090) had already distinguished between 'negative peace', defined as the absence of war, and a dynamic understanding of peace. This involves political, social, cultural and technological aspects and "the conception of positive peace ... [that] encroaches upon many established conceptions and interests". Wright (1965: 1093) argued that if it "is to be realized, peace must be accepted not merely in symbols and myths but also in personalities and cultures".

Galtung (1967, 1968) distinguished between 'negative peace' (absence of physical or personal violence, reduction of arms or a state of non-war) and 'positive peace' (absence of structural violence, repression, discrimination, class conflicts and injustice). The lack of a 'structural peace' limits the full potential of human beings or of nations and is often related to institutional racism and sexism, where women and girls especially are discriminated (Anttila-Hughes/Hsiang/Hsiang 2013) and underpaid (Truong/Gasper/Handmaker et al. 2014).

A sustained peacebuilding process needs to address these root causes of social and physical conflict and discrimination. The present system of unequal access to natural and social resources has aggravated the structural imbalance (Stiglitz 2007). The World Bank (2014) has indicated that the very richest became wealthier during and after the financial crises of 2008, while among the majority of the world population many lost their well-being, their jobs and their income. 'Structural peace' means equity in economic terms, well-being in social interactions, and equality and lack of discrimination at the personal, family, social, regional and international levels. Senghaas (1997) proposed five conditions among nations to achieve a lasting peace: positive interdependence, symmetry of interdependence, homology, and entropy, together with common softly-regulating institutions that are able to promote a 'civilisatory hexagon'. Peacekeeping and its supervision can be considered a civilising project. His civilisatory hexagon consists of six elements: the disarmament of citizens, the rule of law, democratic participation, social justice, a culture of constructive conflict, and interdependencies and control of emotions (Senghaas 1997; 2013).

On the other hand, a 'destructive globalization' (Woodward 2000) is increasing the differences between poor and rich in North and South (Howell 2007). Globalization means the international integration of raw materials, products, services, finances, and knowledge through a free-market system (Calva 2012a, 2012b). In theoretical terms, there should be no limits to this trade in merchandise, people and ideas. The South has opened its borders and its trade; nevertheless, the industrialized countries have imposed increasing controls, subsidies and limits on their economic and human interchanges.

Militarized borders (Schomerus/de Vries 2014), the brain drain (Tigau 2012), *trade-related aspects of intellectual property rights* (TRIPs), protection by patents, and commercial or non-commercial barriers bring disadvantages for the poorer countries. This *business-as-usual* or neo-liberal system has widened the gap not only between industrialized and developing countries, but also between social classes inside the industrialized countries. Ideologically, it has strengthened a global world view of consumerism, fashion, trademarks, violence and fake modernity, where cultural diversity is subsumed into the monopolized market system with a single goal: to create a unique system of dominance or a hegemony controlled by transnational enterprises (Calva 2012a, 2012b, 2012c) and supported by the remaining superpower.

Nevertheless, structural peace has not focused sufficiently on the essential 'gender apartheid' of the global corporate economy. This system includes various forms of violence against women that have become corporately organized. Frequently human trafficking, sweatshops and the arms trade are business of the same corporation. However, all this money obtained illegally from organized crime, tax evasion, unequal terms of trade, and so on is laundered in the same corporate global financial system.

8.2.4   A Culture of Peace and International Institutions

The concept of a 'culture of peace' includes a diverse set of values, traditions, behaviours, attitudes and ways of life with respect for life on earth and for human beings (UNESCO 2002). Conflicts between human beings, communities and states should be managed through conflict resolution, where discriminative social representations and violence are ended through negotiation that can lead to win-win conditions for all those involved. In this positive sense, a 'culture of peace' should promote peacebuilding education (Reardon/Snauwaert 2015a, 2015b) in order to deepen the dialogue and increase cooperation between races, genders and ages. The existence of two concepts, peace and culture, has also created tensions. Groff and Smoker (1995) insisted that the wider understanding of culture must include symbols, rituals, heroes and values, but the crucial issue of an integrated culture of peace is related to shared values, including relationships with others, with nature, and with God. A culture of peace has a visible and a hidden dimension of culture, both of which influence the peacebuilding process.

Declaration 53/243A on a 'Culture of Peace' was accepted by the UN General Assembly in September 1999. It is based on the United Nations Charter, the constitution of UNESCO, and the Universal Declaration of Human Rights. This Declaration was launched just when the cold war had ended, but many violent conflicts were still going on and an intensive trade in weapons still taking place. The declaration is based on nine principles:

1. Respect for life, the ending of violence and promotion and the practice of non-violence through education, dialogue and cooperation;
2. Full respect for the principles of sovereignty, territorial integrity and political independence of states and non-intervention into matters which are essentially within the domestic jurisdiction of any state, in accordance with the Charter of the United Nations and international law;
3. Full respect for and promotion of all human rights and fundamental freedoms;
4. Commitment to peaceful settlement of conflicts;
5. Efforts to meet the developmental and environmental needs of present and future generations;
6. Respect for and promotion of the right to development;
7. Respect for and promotion of equal rights and opportunities for women and men;
8. Respect for and promotion of the right of everyone to freedom of expression, opinion and information;
9. Adherence to the principles of freedom, justice, democracy, tolerance, solidarity, cooperation, pluralism, cultural diversity, dialogue and understanding at all levels of society and among nations; fostered by an enabling national and international environment that itself fosters peace.

In the year 2000 the United Nations (UN Resolution A/RES/52/15) launched an "International Decade for a Culture of Peace and Non-violence for the Children of the World (2001–2010)". The goal is to "create a 'grand alliance' of existing movements that unites all those already working for a culture of peace in its eight domains of action", of which UNESCO became the "Focal Point for the International Year for the Culture of Peace (UN Resolution E/1997/47) and lead agency for the Decade (UN Resolution A/55/47)".[3]

Despite all these efforts, no peace dividend was realized at the end of the cold war. Instead, exclusive globalization, greed for natural resources, exploitation of the labour force (ILO 2014), violence (Mies 1986) and injustice (Truong/Gasper/Handmaker et al. 2014; Fraser 1994) have prevented such a 'culture of peace'.

While a culture of peace has been embraced by some civil society organizations, it is poorly understood in terms of acknowledging the deep psychosocial roots in patriarchy of the culture of violence and the war system. The UN Charter lays claim to a commitment to equality between men and women, but the basic document on the *Conference of Parties* (COP) Declaration lacks gender awareness. It is precisely the term "respect for" instead of "acting towards" that permits continued resistance to specific policies for achieving that equality on the basis of culture.[4]

Rather, a 'culture of war' has prevailed, a culture that was intensified during 2014 with the centenary of the start of World War I, the seventy-fifth anniversary of the German attack on Poland that resulted in World War II, and the twenty-fifth anniversary of the fall of the Berlin Wall. These events triggered some peaceful changes: the reunification of Europe and other parts of the world, such as South East Asia, and a peaceful expansion of both the *European Union* (EU) and of the *Association of Southeast Asian Nations* (ASEAN), involving former opponents in a bipolar world as new partners.

8.2.5 Sustainable Development in the Era of the Anthropocene

There are many definitions of 'sustainable development', with many of them based on the Brundtland Report (1987): "Development that meets the needs of the present without compromising the ability of future generations to meet their own needs". It includes well-being and quality of life in the present, but with an intergenerational approach that looks to future generations.

This definition includes several conflicts: clean air for people or polluted air so that businesses can make high profits. We also need to define which level of needs we are talking about —basic or luxurious. We must also take the long term into account. Meeting the needs of the future we want (UN 2012) depends

on how well we are able to balance social, economic, environmental and cultural goods in the present. For example, are we able to tackle the threats of GEC and CC today by reducing the emissions of greenhouse gases, by reorganizing the processes of production, extraction and urbanization, and by conserving forests and jungles (ISSC 2013)? How can people balance household needs with industrial growth and the productive use of labour? How can humankind improve equity through participative involvement, empowerment, social mobilization and cultural preservation without creating failed states and an anoxic environment? How can humanity allow ecosystem services to function and restore biodiversity and clean air, soil and water, when natural resources are being extracted for industrial activities (Willaarts/Garrido/Llamas 2014)?

Finally, there is also the problem of the territorial ownership of resources (often located in the South), and the necessity of negotiating prices that are just and not only promote the conservation of existing resources but also lead to integrated development for everyone. As the deterioration in natural resources has shown, the paradigm of sustainable development is a benchmark but not a reality in the present system of production (Lelé 1991; The Great Debate Schools Programme 2011). When the ongoing processes of development between 1980 and 2005 are compared, there is no sustainability: resource intensity was reduced by twenty-five per cent, population increased by forty per cent, extraction of raw material increased by fifty per cent, and global *gross domestic product* (GDP) augmented by as much as 125 per cent (SERI 2010).

Our ecological footprint in 2014 was equivalent to the depletion of 1.5 planets (WWF 2014), even though we have only one that has favourable conditions for human life. In the distant past, fossil records show that for every thousand mammals, only one became extinct. The current extinction rate is a thousand times greater than the fossil record, and the future is expected to be ten times more destructive still than the present (MA 2005). During the last six decades of the Anthropocene in hand of multinational enterprises, nine million km^2 of soil (an area roughly the size of China) were moderately degraded and three million km^2 were severely degraded, losing their original biological function. We know that healthy soil is a basic precondition for living organisms (FAO 2012). All these data provoke the following questions: what are the obstacles to a sustainable and peaceful future for humankind and for nature? What violent system

---

3 See: UNESCO; at: <http://www.unesco.org/new/en/brasilia/about-this-office/prizes-and-celebrations/inter-national-decade-for-a-culture-of-peace-and-non-violence-for-the-children-of-the-world/>; for UNESCO's official website see at: <http://www3.unesco.org/iycp/uk/uk_sum_decade.htm> (15 September 2014).

4 This international 'soft' language was probably necessary to make it possible for the document to achieve a successful passage through the General Assembly.

lies behind the concentration of natural and economic goods that is capable of destroying both humankind and nature?

8.3 History and Adaptation of Patriarchy: A Structural Impediment to Equity

In the European tradition, the term and concept of 'patriarchy' emerged from Latin (*patriarchia*) and Greek (πατριάρχης) roots and is widely used in the contemporary social sciences. Etymologically, the *Shorter Oxford English Dictionary* (52002, vol. 2: 2122) defines 'patri-' as a combinatory form that refers in anthropology and sociology to a "social organization defined by male dominance or relationship through the male line". It defines 'patriarch' in contemporary British English as: "1 The male head or ruler of a family or tribe...; 2 (Ecclesiastical) An honorific or official title given to a bishop of high rank ... 3 (A person regarded as) the father or founder of an institution, tradition, science, school of thought etc.; 4 A man or thing which is the oldest or most venerable of a group; esp[ecially] the oldest man in a village or neighbourhood; a veteran, a grand old man". 'Patriarchy' is defined in the same dictionary as a "patriarchal system or society or government; rule by the eldest males of a family; a family, tribe, or community so organized", and 'patriarchalism' as a "patriarchal rule or government", while 'patriachate' is "the position or office of patriarch; the jurisdiction of a patriarch; the residence of a patriarch". The German term 'Patriarchalismus' refers in economic sociology to the unlimited dominance of men in the family, tribe or society.[5]

From an American perspective, for *Webster's Third New International Dictionary* (2002: 1656) 'patriarchy' refers to a "social organization marked by the supremacy of the father in the clan or family in both domestic and religious functions, the legal dependence of wife or wives and children, and the reckoning of descent and an inheritance in the male line; a society so organized".

Originally, the term 'patriarchy' was understood as a system of society or government, where the father or the oldest male is the head of the family and descent and heritage is traced through the male line (Silverblatt 1987; Federici 1999; Meillassoux 1981). This means that the social structure of living is patrilocal and inheritance is patrilineal. These definitions describe a situation, but lack any analysis, which explains when the change from cooperation to exploitative patriarchy occurred among humans and what the structural elements and social relations were that favoured it. The following structural, functional, and diachronic analysis may help to deconstruct this long-term pattern of violence and exclusion.

Feminists (Reardon/Snauwaert 2015a, 2015b; Hartsock 1983, 1996; Young 1992) have analysed patriarchal structures and found that the social structure is male-dominated; for example, the most powerful roles in all sectors of society are held by men and the least valued by women (Lagarde 1990). A second hypothesis is that powerful men control the social structure because they consider that they are the only ones able to exert functions of control, and therefore women need male control, supervision, and protection. The control functions are often reinforced by violence, discrimination and exclusion. This division of labour and power has produced social values and male and female roles and social norms. Males generally work outside, are providers, and perform economic and political activities: they are the *homo sapiens*, while the work of women is devalued and 'made invisible'. Women are traditionally confined to the interior of their houses in order to care for the family, and they take on the role of being-for-others as *homo domesticus*.

Benett (2006: 54) argued that "Patriarchy might be everywhere, but it is not everywhere the same, and therefore patriarchy, in all its immense variety, is something we need to understand, analyse, and explain". It is precisely this variety and diversity which allows the present androgenic world to be overcome in different cultural contexts where power is exercised through control, oppression and the exploitation of nature and human beings. In order to deepen the understanding of the concept of an 'engendered peace', a paradigm without violence, control and destruction, it is crucial to understand the origin and process of consolidation of patriarchy, as well as the inherent power structure within which patriarchy developed and was consolidated.

---

5   "Unbeschränkte Herrschaft des Mannes in der Familie, der Verwandtschaftsgruppe, der Gesellschaft, sei es durch geltende Abstammungs- und Nachfolgeregeln, durch anerkannte Herrschaftsbeziehungen oder durch organisierte Unterdrückung der Frauen in einer Schicht oder einer Gesellschaft." See at: <http://www.wirtschaftslexikon.co/d/patriarchalismus/patriarchalismus.htm>.

8.3.1 Origin and Emergence of Patriarchy

During the first phase of the Neolithic and the initial agricultural revolution, social organization changed drastically, due to a fundamental change in the method of food supply. In place of hunting and gathering, humans began to domesticate some of the basic food items, such as millet in Africa and China some 10,000 years BP[6] (Lu/Zhang/Liu et al. 2009; Colledge 2007). Genetic evidence from rice shows that rice originates from a single domestication in the Pearl River valley in China between 13,500 and 8,200 BP (Huang/Kurata/Wei et al. 2012). Wheat originated from the wild *emmer* and was first cultivated in the Levant as early as 11,600 BP. Heun, Schäfer-Pregl, Klawan et al. (1997) used DNA to identify the earliest wheat where *einkorn* was found on a site near the Karacadag Mountain Range, in modern Turkey. Maize was found to have grown around 8,700 BP in the Balsas river valley of Mexico (Ranere/Piperno/Holst et al. 2009) and was transformed from *teocinte* to the present cob of corn.

The long period of experimentation, including failures and adaptation, with these basic food crops required knowledge and cooperation among family members, regardless of their sex. Archaeologists and anthropologists agree that in this early phase of the agricultural revolution social groups had a better chance of survival when they cooperated. Goddesses and gods protected humans from natural catastrophes and there was no formal system of family relationship between men and women. Smaller or larger human groups lived together, and fertility and reproduction were at the centre of interest, since they provided a sufficient labour force for agricultural activities. The people transferred their agricultural knowledge to the next generation, which developed it further. The preservation of food was crucial, especially in extreme climate conditions. In cold climates, meat and fish were frozen on the ice, and in tropical climates food was dried in the sun. Later the food staple was improved and was fermented, pickled, cured, canned, and transformed into jam and jelly (Zerzan 2010).

The earliest human lifestyle (hunting and gathering) and later independent rural economic units in agricultural and pastoral settlements (8,000 BP) required a high level of collaboration between men and women (Federici 2004). Female fertility was crucial for the survival of these agricultural societies, and sexual promiscuity and group marriage were common (Morgan 1881). Goddesses symbolized the ruthless fertility of nature, and sexual regulations were progressively introduced to avoid incest (Malinowsky 1929). Human populations increased as each of these processes was able to improve conditions of food and life, and technologies progressed slowly. The leap forward occurred with the invention of plough, irrigation and flood control, so that in warmer climates several harvests could be produced each year.

With the domestication of food crops and livestock, technological improvements, better methods of conservation and especially with irrigation, a food surplus allowed food to be accumulated within productive units. Scientists agree that this important innovation began in Mesopotamia (Adams 1965). People in the hills were threatened by food scarcity and hunger and migrated from the mountains to the floodplain. However, these valleys between the Euphrates and the Tigris were often flooded and then later dried out. To overcome these arid climate conditions, the people built and maintained irrigation systems for villages. Around 3,500 BP villages grew into larger communities (Neugebauer 1993). Neighbouring food-scarce communities often attacked these cities. The situation of rural villages also changed drastically, with violent invasions by male hordes from the eastern steppes (del Valle 1993), and later different tribes occupied particular territories, although they were often driven out by poor climatic conditions.[7]

To find protection, Sumeria people organized themselves into larger cities, which represented small independent countries surrounded by agricultural areas. Each city had its own ruler and the irrigated farmland provided people with different sorts of food. Society became socially differentiated and people specialized in specific activities. Disputes over water became so intense that they led to armed conflicts, the military was created, and arms were produced to deal with internal conflicts and to protect the city against foreign invaders. The citizens built walls and dug moats outside the city to prevent ene-

---

6 BP = "Before Present" (used in archaeological dating for prehistory; "Present" is taken as 1950 CE). For historical dates, and to avoid a Eurocentric cultural approach, this chapter uses BCE and CE rather than BC and AD.

---

7 The wild Germanic tribes from Central and Northern Europe (Vandals, Goths, Langobards and others) brought the Roman Empire down; later the nomadic pastoralist Huns invaded southeastern Europe; there is archaeological evidence of Olmec expansion as far as Guatemala, and in China stronger dynasties conquered weaker ones.

mies from entering and to protect their people within the walls (Arnold 2004). Groups of men were trained for both attack and defence, and a social class of soldiers emerged in order to defend wealth in urban settlements. These armies were not just defensive but also invaded neighbouring cities and other kingdoms and towns, and stole goods, women and technological knowledge from them. Successive conquests of smaller regional cultures strengthened the regional dominance of an ethnic group. Slowly, greater kingdoms and empires were established which were better equipped militarily, and the division of labour was consolidated. Power was concentrated in the hands of a powerful king who ruled over his subjects and slaves. Women were successfully eliminated from decision-making and the exercise of power, and disappeared inside their houses. Violent conquest was a crucial ingredient of this patriarchal expansion, led by men and controlled by male armed forces.

Ideologically, fertility goddesses were replaced by male gods (the Sumerian Ishkurthe, who held supreme power, or Enki, the god of fresh water and male fertility, the Greek Zeus, the Meso-American Quetzalcoatl, etc.) and later by the religious control of one male father god (in the Jewish, Christian and Muslim religions). This ideological shift further diminished the role of women. They were confined to the house, their fertility controlled by a husband, and their labour was considered unproductive. This 'gender blindness' took away the highly visible efforts of women, because they did not fit the preferred patriarchal paradigm. Fathers gained authority over women, children and property, based on institutions of male rules and privileges that meant female subordination and expropriation. Patriarchy is thus closely related to gender roles, which are socially constructed and which generate norms through habits, costumes, violence and repression. With the historical evolution of societies, the division of labour became more diversified (Reilly 2010). Goods, including women, were traded first in local and regional markets, and later long-distance market structures were established during the great empires of Egypt, Meso-America, Inca, China and India (Meillasoux 1981; Hoodfar 1997; Silverblatt 1987).

Patriarchy includes therefore polity, religious belief, political economy, social discrimination, sexual control, and women's loss of power in decision-making processes. The origins of the political dimension of patriarchy, which has existed through all stages of the consolidation of political structures, included a war system to reinforce and consolidate the political system (see table 8.1).

In the beginning of patriarchy, the king further reinforced his legitimacy and took on the attributes of a divine supreme being. This king as God at the summit of the civil and religious hierarchy reigned over the city-state. The belief that patriarchy is part of the 'natural order' or 'ordained' by God has been invoked by fundamentalists from the majority of religious traditions and by the economic elite. Human coexistence was regulated by social rules that clearly discriminated against women and girls. The origins of patriarchy are also linked with a process of primitive accumulation, with social stratification, and with a hierarchical system of rule reinforced by religious fundamentalist beliefs. The first document carved in stone was the Hammurabi code of 1754 BP. This code contains 282 laws, which organized all aspects of life in the city-kingdom. The code distinguished between free men and slaves and is based on the punishment philosophy of 'an eye for an eye, a tooth for a tooth'. One-third of the code addressed issues of family relationships, household, inheritance, divorce, paternity and sexual behaviour. Women were clearly discriminated against and oppressed by this code. After marriage, a women's sexuality became the property of her husband, and adultery by women was punished by the death penalty. Female slaves belonged to their master. Girls from higher classes had to remain virgins until their marriage, and inheritance was patrilineal. This code is the first written evidence of patriarchal relationships in an agricultural society (table 8.1).

Similar processes occurred in Greece, Egypt, China, Meso-America and the Roman Empire, and widened the scope of patriarchy. Goddesses were successively replaced by male gods and later often by one dominant and vindictive male god. During the past 4,000 to 5,000 years these initial patriarchal relationships were transformed and have shown a tremendous capacity to adapt to different systems of rule (e.g. king as God, city-kingdoms, monarchies, slavery colonialism, republics, autocracies, communism, liberal democracy), economic systems (slavery, feudalism, mercantilism, capitalism, socialism, neo-liberalism), and processes (barter, conquest, piracy, interchange, market). Nevertheless, during these thousands of years of evolution, patriarchy also adapted to regional and social differences, even though its basic roots continue to be violence, discrimination, domination, exploitation, and oppression in household and society, in short, a system of waging war in order to maintain the dominant power relationships. Its vertical structure integrated cultural and ritual elements into its system of rule in Mesopotamia, China, Meso-America, the Roman Empire, and later in the European conquest of Asia, America, and Africa. It replaced the traditional legal forms of social coexistence and imposed worldwide a single system of rule controlled by Bretton Wood: the International Monetary Fund, the World Bank and the World Trade Organization. Within these patriarchal social representations and exercises of power, empires, feudalism, conquest, monarchies, mercantilism, capitalism, and socialism emerged and laid the basis for the current neo-liberal globalization at the hands of corporate enterprises, international legal support (Serrano 2013), and military control by the US.

In the mid-twentieth century a postmodern globalization emerged that is largely influenced and controlled by neo-liberal multinational oligarchies, authoritarian governments and hierarchical structured and exclusive religions. The state of law has developed an independent system of rule divided between the executive, the legislative and the judiciary branches of government at the national level. Nonetheless, international treaties and normative systems of operation have been inefficient when leading states and domestic political forces or interests have opposed them and tried to block or constrain such changes in the systems of operation (Diehl/Ku/Zamora 2003). All legal international frameworks have largely reflected occidental interests and have replaced the original tribal or indigenous community norms and rules in the southern hemisphere (Bartra 2012). The global interests of a transnational oligarchy (Lander 2011), supported by hegemonic governments, represent the present patriarchal system, where warfare, the corporate economy and political power determine the present neo-liberal process of globalization.

With this historical construction of gender differentiation of male-ness and women-ness (Mies 1986), together with the glass ceiling for women, patriarchy spread from within the family structure up to the level of the kingdom, e.g. in the Greek democracy, the Roman Empire, in feudalism, mercantilism and capitalism (Reardon/Snauwaert 2015a, 2015b). In England in the seventeenth century, Robert Filmer developed the political theory of "Patriarcha" for the kingdom and compared the *paterfamilias* with the king. The origin of patriarchy preceded the institution of private property (Engels 1940 [1884]), and goes back to the

earliest establishment of cities in the river valleys of the Jordan, the Euphrates and Tigris, the Nile, and the Indus (more than 3,000 years BCE). Similar conditions were found a thousand years later in the cities of Meso-America and China. These urban processes were consolidated in a similar way among regions, people, and social developments. Vertical relationships of rule, with labour specialization, class distinction, social hierarchies, kingship, priests, writers, artisans and the military, thus a clear division of labour, characterized the goal of an unequal appropriation of the surplus. Without any doubt, the productive work of women was essential to maintaining society, but women's power structures and processes of decision-making were systematically devaluated and obscured.

Undoubtedly, during the last millennia the evolution and consolidation of patriarchy has changed and become differentiated regionally. Nevertheless, there are some common dominant factors: the ideological control mechanisms around the world have both socio-biological and social constructivist explanations. While the first uses genetics and biology to explain male control, the second focuses on socially constructed gender roles, which have changed over time, could change further, and must be abolished in order to allow the creation of an engendered security and peace. Anthropologists have suggested that most hunter-gatherer and early agricultural societies were egalitarian and cooperative between genders (Hughes/ Hughes 2001). Following social and technological innovations (irrigation, cities, technology), social stratification, fatherhood, and sexual control over women and girls became dominant (Kraemer 1991).

In economic terms, Hartmann (1976) locates the interrelationship between patriarchy and female subordination at the material level, where the control of women's labour, in both the private and public spheres, is in the hands of men. Today the reduction in the purchasing power of wages obliges an increasing number of families to have two incomes. Usually, job segregation by gender means that "economic class differences and patriarchal social control are maintained. Job segregation and the wage differential tend to keep women connected to and partially dependent on men, so that even though we have a high divorce rate, the practice of marriage is perpetuated" (Hartmann 1976: 147) and therefore so is the concept of reproduction.

Table 8.1: Rise and adaptation of patriarchy in different socio-economic contexts. **Source:** The author.

Type of society	Era	Activities	System of Rules	Goals	Ideology, faith	Adaptation to risks	Gender relations
Neolithic and agricultural revolution	8000–3000 BCE	hunting, gathering, pastoralism, subsistence agriculture, millet in China (5000 BCE)	emerging rules against incest	subsistence, survival, cooperation among humans	fertility cult, goddesses, Mother Earth as archetype, protecting gods, stone carving, painting of caves	migration due to climate variability, lack of resilience, survival strategies,	equality, equity, cooperation, participation in labour and local rule, control of incest
Centralized god-state (Babylonia) China: Liangzhu dynasty (3400–2250 BP)	3000–50 BCE	crop development, livestock, agriculture with irrigation & aquaculture, handicraft, division of labour, military, writing on stone	paterfamilias, monarchy with god-kings, legal written rules of social behaviour, slaves	subsistence, territorial expansion, kingdoms, elite, social stratification, slaves, improvement of agricultural technologies (plough, irrigation, flood defences)	king as god, absolutism, Buddha, evaluation of cultural expressions, one male god (Judaism)	gods, goddesses, migration, invasion, army, militarized defences, protected settlements, women are turned invisible	women submissive to husbands, central male power, control of virginity of girls and sexuality of women, complex societal hierarchies, castes, slaves, discrimination, sanctions against women
Slavery, Xia dynasty (1600–1050 BCE); Shang dynasty (1700–1046 BCE), Zhou dynasty (1046–256 BCE), Mayan kingdoms	3000 BCE - 1772 CE	household productive activities (mines, plantations), rural villages and emergence of towns	monarchy, state with senate (Greece), empires in China and Meso-America, political entities and mandate of heaven; written laws and clear social rules	labour force, free service, exploitation of slaves and women, territorial expansion	master, dominance, private, public and state property	buy freedom (bail-out) with money or service, escape, struggle	patriarchal violence exclusion, control, rape, discrimination
Roman Empire	27 BCE–476 CE	market institutions, global commerce, conquest	empire with senate, protection of citizens within the *limes*, laws and social rules	territorial expansion, conquest, control, *pax romana*	male gods, one father god, expansion of Christianity and Buddhism	war outside *limes*, emergence of democratic institutions and division of power	patriarchal patrilineal patrilocal discrimination
Feudalism, Inca, Aztec, Chinese empires, European monarchies	9th–15th century CE	lords, vassal fiefs, agriculture, wars, recognition and protection by monarch	religious and civil authorities, kingdoms, dynasties, cities, lords, tributes, written law and personal agreements	reciprocal legal and military obligation, individual states, monarchies, original accumulation	one male father god, Christianity, Occidentalism, Confucianism in China, Buddhism in S, SE and NE Asia, Hinduism in India and Nepal	ceremonies agreements protection, alliance, migration, wars, military, exploitation, renting land	patriarchal patrilineal patrilocal discrimination against women, burning of witches
Conquest	15th–19th century CE	control of market in natural resources and products, exploitation of mines, food products, and natural resources	empires, kingdoms, lords, absolutism, colonization, occidental laws and customs imposed in colonies	colonies, gold, territorial expansion, raw material, land, new food products, medicinal herbs	dominant Catholic church, one male father god destruction of traditional beliefs and cultures	war, military, destruction of autochthonous cultures, ideological and religious control, colonial control	patriarchal patrilineal patrilocal, genocide, rape, discrimination, violence

Type of society	Era	Activities	System of Rules	Goals	Ideology, faith	Adaptation to risks	Gender relations
Mercantilist capitalism	17th–18th century CE	market partially coordinated, exploitation, trade	emergence of nation states, trade rules and consolidation of trade routes, Westphalian peace agreement (1648)	government monopoly with private investments	one male father god, goods	production and distribution toward market forces, violence, respect for internal differences of states	patriarchal patrilineal patrilocal, extended families, exploitation, discrimination, violence, exclusion, women and child labour
From commercial to labour capitalism	1750–1930 CE	industrial revolution, proletariat, labour unions	division of rules (executive, legislative, judicial), trade unions	paid labour force, proletariat, oil, GHG, resource scarcity, repression of any opposition	one male father god, money, class struggle	exploitation urbanization, laws agreements, poverty, migration, hunger, Red Cross	patriarchal patrilineal small families, female and child labour, exploitation, discrimination, individualism
Socialism–Marxism	1848 (1917) CE – present	state-organized economy, social welfare, education, militarization, Eastern bloc, bureaucracy, a one party state	USSR and Chinese Communist Party, economy planned by the state, Supreme Soviet, central committee, democratic centralism, Warsaw Pact, massive famine, repression	bureaucratic centralized economy, scarcity of basic goods, high level of military expenditure, technological improvement	social equality, proletariat, party elite, socialism, atheism, cold war, representative wars in the South and East	education, social welfare, military warfare, Warsaw Pact equality of labour (men–women), women's emancipation in labour system	patriarchal small family under state control, female political participation, gender discrimination, glass ceiling for women, violence, party hierarchy
From industrial to state capitalism and social capitalism	1930–1980 CE	national sovereignty, control of capital by the state, taxes, redistribution of some income, economic blocs	welfare state, division of rules (executive, legislative, judicial), UN Charter, human rights, NATO, US dominance, Bretton Wood organizations	protection of labour force, trade, and finances; equality, resource abuse, pollution, scarcity, science and technology	one male father god, capitalism, welfare state, cold war, independence movements in India, Africa and Asia	welfare state, pensions, social security, militarization, military warfare, feminist groups, anti-war movements	patriarchal patrilineal small family, female suffrage, some redistribution of wealth, support of female labour force, housewifization, subordination, inequality
Neo-liberalism, internationalization of finance, globalization of markets, fashion, consumerist culture, superpower military control, business-as-usual	1980 CE –present	financial and market control, privatization, lack of social protection, globalization, liberalization, deregulation, FTA, financial crises, youth unemployment, military and police control, illegal activities (drugs, arms, human trafficking), money laundering, emerging countries, BRICS	Hegemonic world order, military interventions, division of power (executive, legislative, judicial) systems, small government with large security sector, global oligarchy with multinational enterprises, binding agreements, Washington Consensus (IMF, World Bank, WTO), NATO, structural adjustment policy	economic crises, laissez-faire, free market, globalization and deregulation of financial markets, unemployment, increase of poverty and hunger, development of productive forces, R&D, innovation, resource exploitation, GEC, climate change, ideological domination, massive global consumerism	one male father god, money and economic totalitarianism, elite, poverty, oligarchy with governmental legitimization, social consciousness, risk perception, horoscope	informal labour force, migration, poverty, resource wars: oil, mines, water, social conflicts, warfare and economic organizations (e.g. OECD, EU, ASEAN, NAFTA, BRICS and other economic associations for cooperation and disaster risk reduction, urbanization, horoscope, massive tourism, human rights groups, feminists, alternative media	patriarchal, patrilineal small family, individual violence, incipient female political participation, discrimination, glass ceiling, but also women's and civil rights movements in the US, empowerment, resistance, power-from-within, self-subjugation, contraception, solidarity, peace building, ecofeminism

Edholm, Harris and Young (1972) distinguish three forms of the concept of reproduction: 1. social reproduction that includes the reproduction of the total conditions of production; 2. the reproduction of the labour force, and 3. biological reproduction. This distinction helps to combat the Marxist emphasis on the importance of the economy for reproduction, and to understand that patriarchy goes far beyond the rise of capitalism. Various feminists have defined the evolution of this dominant social system, which can be found globally in different temporal and spatial frameworks, as patriarchy. It is "the power of the fathers: a familial-social, ideological, political system in which men—by force, direct pressure, or through ritual, tradition, law and language, customs, etiquette, education, and the division of labour, determine what part women shall or shall not play, and in which the female is subsumed under the male   " (Benett 2006: 55).

Maria Mies (1986) asks why the exploitation and oppression of women are not only accidental phenomena but an intrinsic part of the systems that have controlled women throughout thousands of years of humanity and its evolution, and why such a brutal system could not be overthrown with modernity. She insists that capitalism cannot function without patriarchy and that the "struggle against all capitalist–patriarchal relation ... [means that we] have to extend our analysis to the system of accumulation on a world scale, the world market or the international division of labour" (Mies 1986: 39). How did this happen?

These brief summaries are an attempt to explain in general terms how patriarchy emerged and became a global archetype of male and female roles and behaviours, which are culturally embedded in order to support an unequal, exploitative, violent and destructive system of rules. A crucial requirement related to the consolidation of patriarchy worldwide is to understand the manifest and hidden power relations, which enabled minority social groups to accumulate power and wealth at the cost of the majority. This basic system of patriarchy evolved and created the potential for further development and greater consolidation. Different social systems emerged, but all of them reinforced a system where power and wealth were concentrated in a minority of male leaders at the cost of the majority. The present oligarchic system of neo-liberal control, where a small number of corporate enterprises dominate the world economy, is the historical result of patriarchy consolidated through different economic and political processes such as slavery, feudalism, colonialism, and capitalism.[8]

While economic factors are the basis of any social system and they are crucial to understanding the consolidation of patriarchy worldwide in the past and the present, it is existing power combined with violence, war, conquest, ideological superiority and religious belief that has permitted this sophisticated dominant system of rule. The global arena is today one of neoliberal globalization, where Margaret Thatcher could insist that 'TINA': "there is no alternative". The next chapter reviews these power relations from a feminist point of view.

8.3.2   Feminist Analysis of Power Relations

Most feminist theories try to explain women's subordination by analysing the intersections between sexism, racism, heterosexism, and ethnic and class oppression, while Marxist feminists include economic domination. Ecofeminists believe that similar power relationships between dominance, greed, exploitation, and violence (D'Eaubonne 1974; Mies 1986; Warren 1997) cause environmental destruction. Other approaches envision the possibility of overcoming such subordination through both individual and collective resistance, and through empowerment. Allen (1998, 1999, 2008, 2011), Hartsock (1983, 1996), and Young (1992) have analysed the concept of power as a central element in understanding this subordination and in discovering the limits to and potential for a sustainable-engendered peace and worldview.

Adrienne Rich (1976) gazes into the fissure that separates modern women's link to maternity and the "theories, ideals, archetypes, descriptions" of the patriarchal culture that is replacing this relationship. "All human beings are born of women therefore all have a connection to motherhood, more fundamental than tribalism or nationalism ... [But patriarchal culture] has created images of the archetypal Mother, which reinforce the conservatism of motherhood and convert it to an energy for the renewal of male power". Rich insists that art should evoke in its cultural scope "rape and its aftermath ... marriage as economic dependence ... the theft of childbirth from women ... laws regulating contraception and abortion

---

8   The development of capitalism in Britain destroyed most domestic production and forced women, men and children into the factory system. Similar processes occurred in Latin America, Asia and Africa, where rapid urbanization was a result of rural-urban migration in order to improve income and to obtain better services in the cities, although the outcome was shantytowns.

... the absence of social benefits for mothers". This archetype is charged with patriarchal elements, such as rape, violence, and submission. The active role of women is visible in neither society nor peacebuilding.

There are four different approaches to analysing power relations. The first relates to the exercise of 'power-to' in the traditional way of Thomas Hobbes, who understood the concept as power-to "obtain some future apparent good" (Hobbes 1985 [1641]: 150). Max Weber added 'power-over': "the probability that one actor within a social relationship will be in a position to carry out his own will despite resistance..." (Weber 1978: 53). Michel Foucault developed this approach in his analysis of institutions of control. He affirmed that "if we speak of the structures or the mechanisms of power, it is only insofar as we suppose that certain persons exercise power over others" (Foucault 1983: 217). It was Hannah Arendt who understood power as "the human ability not just to act but to act in concert" (Arendt 1970: 44). In her sense, power is the capacity to act, although she states that "power springs up between men when they act together and vanishes the moment they disperse" (Arendt 1958: 200). Lukes also defined power also as a possibility when he affirms that power "is a potentiality, not an actuality—indeed a potentiality that may never be actualized" (Lukes 2005: 69).

The fourth way of understanding power is as a systemic conception where power is seen as "the ways in which given social systems confer differentials of dispositional power on agents, thus structuring their possibilities for action" (Haugaard 2010: 425), and precisely this type of power is able to promote an engendered sustainable peace. This systemic conception highlights the ways in which "historical, political, economic, cultural, and social forces enable some individuals to exercise power over others, or inculcate certain abilities and dispositions in some actors but not in others" (Allen 2001: 3). The systemic approach goes back to Weber's understanding as 'power-over'. Feminists are interested in understanding the gender-based relationships of domination and subordination, and how women could empower, and then how such relations could be generalized in the whole of society. In a liberal feminist approach, power is understood as a resource oriented towards a positive social good ('power-to'). Okin found that in the distribution "between husbands and wives of such critical social goods as work (paid and unpaid), power, prestige, self-esteem, opportunities for self-development, and both physical and economic security, we find socially

constructed inequalities between them, right down the list" (Okin 1989: 136).

This approach is underlain by a broader social, cultural, institutional and structural context in which power relations are organized. Young (1990) focused on the relationship of domination, understood as oppression, patriarchy and subjection. She considered power to be a special 'power-over' relation, which is unjust and illegitimate. She was inspired by Simone de Beauvoir (1974), who understood power as domination or oppression. Beauvoir recognized that women are partly responsible for this power oppression by submitting to the status of the 'Other' in order to avoid the anguish of authentic existence. From a radical point of view, MacKinnon (1987: 123) insisted that "women/men is a distinction not just of difference, but of power and powerlessness. ... Power/powerlessness is the sex difference".

Judith Butler (1990) and other critics of the sex/gender distinction think that sex differences or gender differences are socially constructed in order to maintain the existing relationships of power. Economic and political power within the family is thus concentrated in the hands of husbands or dominant male figures and supports the system of private property and patrilineal heritage. Frye (1983) compares this male domination to a master/slave relationship, and she defines oppression as "a system of interrelated barriers and forces which reduce, immobilize, and mould people who belong to a certain group, and effect their subordination to another group (individually to individuals of the other group, and as a group, to that group)" (Frye 1983: 33).

In the case of historical materialism, feminists have criticized Marxism because of its lack of understanding of gender. Young (1990: 21) went behind the dual systems theory: women's oppression arises from two distinct and relatively autonomous systems. "The system of male domination, most often called 'patriarchy', produces the specific gender oppression of women; the system of the mode of production and class relations produces the class oppression and work alienation of most women". She proposed an authentic feminist historical materialism in which she identified five types of oppression: economic exploitation, socio-economic marginalization, lack of power or autonomy over one's work, cultural imperialism, and systematic violence (Young 1992: 183-193).

Hartsock understood that the dominant group, such as men and the ruling class or oligarchy, is the one carrying out the analysis of power. She is concerned with "(1) how relations of domination along lines of gender are constructed and maintained and (2) whether social understandings of domination itself have been distorted by men's domination of women" (Hartsock 1983: 1). Patricia Hill Collins (2002) proposed the concept of "interlocking systems of oppression" where racial, gender, and class-based subordination are intertwined. Foucault (1977) influenced radical feminists, although Bartky (1990: 65) criticizes him because he did not understand that disciplinary practices are gendered due to the fact that women's bodies are rendered more docile than the bodies of men. "[I]t is women themselves who practise this discipline on and against their own bodies ... The woman who checks her make-up half a dozen times a day to see if her foundation has caked or her mascara run, who worries that the wind or rain may spoil her hairdo, who looks frequently to see if her stockings have bagged at the ankle, or who, feeling fat, monitors everything she eats, has become, just as surely as the inmate in the Panopticon, a self-policing subject, a self committed to relentless self-surveillance. This self-surveillance is a form of obedience to patriarchy" (Bartky 1990: 80). Butler notes that subjection is a paradoxical form of power. It has an element of domination and subordination. She writes: "if, following Foucault, we understand power as forming the subject as well as providing the very condition of its existence and the trajectory of its desire, then power is not simply what we oppose but also, in a strong sense, what we depend on for our existence and what we harbour and preserve in the beings that we are" (Butler 1997: 2). Fraser (1989) insisted that the normative notions (autonomy, legitimacy, sovereignty) are themselves effects of modern power, and Hartsock (1990) claimed that the deeper analysis of power can be performed only from the standpoint of the dominated, exploited and subordinated.

From an analytical feminist viewpoint, Cudd (2006: 21-23) understood oppression as "an objective social phenomenon". She mentioned four conditions: 1) the group condition, which states that individuals are subjected to unjust treatment because of their (ascribed) membership; 2) the harm condition, where individuals are systematically and unfairly harmed due to such membership; 3) the unjustified coercion condition of the harms; and 4) the privilege condition, which states that such coercive, group-based harms count as oppression only when there exist other social groups who derive a reciprocal privilege or benefit from that unjust harm.

Confronted with these different approaches, feminists from various theoretical backgrounds have reconceptualized power as the capacity to empower and therefore to transform oneself, others and social conditions. These feminists understand power not as 'power-over' but as 'power-to'. On the other hand, Wartenberg (1990) argues that this transformative power is a type of 'power-over', but one distinct from domination because it permits the empowerment of those over whom power is exercised. Miller insisted that women's perspective on power takes a different approach: "there is enormous validity in women's not wanting to use power as it is presently conceived and used. Rather, women may want to be powerful in ways that simultaneously enhance, rather than diminish, the power of others" (Miller 1992: 247-248). Therefore, for Hoagland, "power-from-within" is "the power of ability, of choice and engagement. It is creative; and hence it is an affecting and transforming power but not a controlling power" (Hoagland 1988: 118).

Hartsock (1983: 226) took the view that male power relations established over thousands of years "should allow us to understand why the masculine community constructed ... power, as domination, repression, and death, and why women's accounts of power differ in specific and systematic ways from those put forward by men ... such a standpoint might allow us to put forward an understanding of power that points in more liberatory directions". Years before, Hannah Arendt had rejected the command–obedience model of power and had insisted on "the human ability not just to act but to act in concert" (Arendt 1970: 44). Her understanding of power brings together individual empowerment with a focus on community or collective empowerment, which is also the key to the understanding of resistance by all oppressed groups, including indigenous peoples, racial groups, and minorities. The question that remains for the creation of a sustainable-engendered peace is then, how can this transforming power be used to overcome patriarchy, the war system, and human and environmental destruction?

While the hopes of women and other vulnerable groups suffered a backlash from the dominant structures, after World War II independence movements in Asia and Africa challenged imposed colonial structures. In the US, women's and civil rights movements changed the most discriminative behaviours and produced a social consciousness of social and political processes. In Japan and the USSR women left their homes to join what the government called 'the war effort'. In the global South,[9] the lives of women remained mostly unchanged, although the dictatorships in Latin America presented women with the possibility of fighting against military repression and for their disappeared children and grandchildren.

The 'pill' and other modes of contraception gave women—and not only in the North—more control over their fertility and a sense of autonomy that fed into the women's movement. Issues of health and social costs were transferred to the South, where women were used as human guinea pigs—as an uneducated market, they became victims of the promotion of formula baby milk, while the rest of the world returned to breastfeeding for the sake of their children's health. Immigrant women from the South were also increasingly used as cheap household labour with little or no rights. While Betty Friedan's (1963) rallying cry against the 'feminine mystique' allowed women to put limits on 'housewifization' and 'domesticity', the differences worldwide are enormous.

All these analytical and practical reflections have challenged the constructed self-discipline and internalization of gender roles in most societies, although the enemy —corporate power based on patriarchy— is immense. None of the advances made in women's empowerment could halt the progress of the neo-liberal model, where patriarchy has developed an almost perfect model of exclusion, concentration of wealth, exploitation, discrimination and violence. A common struggle involving both men and women to defeat this corporate power requires global empowerment and a social pact without patriarchal relations. This change will simultaneously produce an economic, environmental, cultural and ideological restoration of dignity and respect for humans and nature. Only the transformation of these deeply embedded patriarchal structures will be able to overcome thousands of years of consolidation and perfection of patriarchy and the present-day inhuman corporations. This change to the roots of the system will also open the way to a sustainable-engendered peace for South and North, women and men, the elderly and children.

---

9   There is no doubt that the economic and social conditions of women worldwide are very different and the different phases of the evolution of patriarchy and later capitalism have left a large population of men and women, the elderly and children bereft of the basic requirements of life. The present chapter tries to overcome the occidental approach that sees the Western rise of capitalism as the key reason for the exploitation of human beings and nature. The text focuses on the root causes of violence, which began thousands of years ago and through the consolidation of neo-liberalism culminated in a system of exclusion.

8.4   Engendering Sustainable Peace and Security by Transforming Patriarchy and its War System

Through this conceptual review, patriarchy can be seen to be understood historically as being imposed on women through violence (via conquest, spoliation, rape and feminicide), discrimination (by laws and rules), subordination (through economic and sexual control), hierarchy (by the notion of paterfamilias, today represented by the global oligarchy), inequality (in education, income, leisure and political power), through exclusion (patrilineal and patrilocal inheritance and exclusive globalization), and through social representations (constructed self-discipline and internalization of gender roles).

As Burton (2013) maintained, during the agricultural revolution and the consolidation of god-cities, motherhood was bereft, fertility goddesses abandoned, and reproduction shut away into the private sphere. This spoliation of motherhood has given rise to gender inequalities not only among women, but also among men with less power (the indigenous, the unemployed, youth and the elderly). Together with discrimination, it has limited opportunities for employment, education and art, and undermined solidarity because of the dominant individualistic values that have been imposed by the Western value system. The same mechanisms that were used for the oppression of women were employed in the destruction of Mother Earth. In the past, humanity depended directly on ecosystem services for its survival, although there were fewer humans on earth. The pressure on soil, water, air and biodiversity was less. Through the processes of industrialization, the massive use of fossil energy, the exponential growth of population, and the wasteful consumerist society, during the last six decades earth history has moved from the Holocene to an era called by Crutzen (2002) the Anthropocene. Patriarchy has destroyed nature through the same mechanisms as it has done with women, changing the face of the earth by producing a global environmental change.

On the social side, the *information technology* (IT) revolution after World War II gave women a new hope that they might empower and overcome the patriarchal structures. Many women and also men thought that the integration of the three revolutions (agricultural, industrial and IT) would allow women and socially vulnerable people to get rid of backward, feudal and patriarchal relationships. However, what actually happened was that the structures of dominance were consolidated as the neo-liberal patriarchy transformed women into housewives. These new social arrangements also increased the establishment of nuclear/small families, where children and especially the elderly lack space because of small living-spaces. This *housewifization* (Mies 1986) affects not only women and family relationships, but also male labour and especially youth.[10] Since 2008, most countries have lost incomes and jobs during the latest financial crisis; poverty has grown, and the anorexic state, because of earlier massive processes of privatization, is unable to balance monopolized capital, the needs of the people and the needs of jobless workers.

There is no doubt of the consolidation of the present neo-liberal system by a market economy, globalization, liberalization, deregulation, privatization, and free trade agreements. Selective World Bank investments, the structural adjustment policy of the International Monetary Fund, and the regional and global treaties of free trade agreements under the jurisdiction of the World Trade Organization with TRIPs have created a billon hungry people (mostly children), three billion people without safe water and improved sanitation, and fifty-eight corporate multinational oligarchs which own half the wealth on earth (UNDP 2014: 21).

Therefore a key question is how the embedded system of patriarchy can be transformed. Betty Reardon (1985) looked into the deep fear-based psychological dynamics that underlie the symbiotic relationship between patriarchy and the war system: "I continue to insist [that] … the oppression of women and the legitimation of coercive force which perpetuates war, the two major pillars of patriarchy, are mutually dependent conditions…" [Reardon 1985:98]. In the same book she insists that "… militarism and militarization [are] the bastions and bulwark of patriarchy" (cited in Reardon 2015a: 74). She actively worked for peace education and linked it with peacekeeping, peacemaking and peacebuilding, and with Agenda 21, thus promoting an agenda of conflict reduction and resolution and an agenda for development and for restoring the environment.

To achieve a sustainable peace she listed four basic transformations needed to overcome patriarchy, militarism and injustice:

1. The general adoption of a feminist, holistic, gender-equal perspective.

2. A fundamental change in worldview, which includes the widespread inclusion of feminist values in all levels of society, including the public domain and government.

3. Shifting the concept of security from national security to human security, and a cosmopolitan ethic.

4. A widespread increase in self-awareness among the population (Snauwaert 2015: XVI).

Combining this with Young's proposal we may include in the concept of a sustainable-engendered peace the reduction of economic exploitation, global policies to overcome socio-economic marginalization, training for increased power or autonomy over one's own work, the eradication of cultural imperialism, and the eradication of systematic violence. Ecofeminists also draw attention to the relationship between patriarchy and the plundering of natural resources: "there are important connections between how one treats women, people of colour, and the underclass on one hand and how one treats the nonhuman natural environment on the other" (Warren 1997: XI). An engendered peace requires the healing of the environment through a massive reduction in fossil energy and greenhouse gas emissions, a full recycling of all materials, and a systematic restoration of all ecosystems and renewable natural resources.

8.4.1 Transition to a Sustainable-Engendered Peace

The analysis presented above has described the development of at least five thousand years of patriarchy and the complexity of embedded power relations from the family to the global level. In addition, awareness is growing that conflicts and post-conflict situations have different implications for women and men. There have been few efforts at international and government levels to propose gender-sensitive policies. This comes partly from a lack of gender knowledge, but also demonstrates the continuing resistance to losing small benefits from the patriarchal system. A third problem is the lack of money available for gender-sensitive peace efforts, as most of the money is used internationally for UN peacekeeping forces, who have limited opportunities to promote a stable peace as their task is oriented towards the maintenance of a negative peace. Their modes of action are militarized and the UN operates under the dominant patriarchal state paradigm. Their troops have been among the perpetrators of sexual violence against women and have often failed to protect the vulnerable (Ruanda).

---

10 Unemployment rates among youth are extremely high, especially in South Africa (53.6%), Libya (50.8%), Spain (57.3%), Greece 58.4%, and Croatia (51.5%), to mention only those countries with the highest unemployment rates on earth (ILO 2014).

It is difficult to promote an engendered peace when in the recent history of any conflict—but also during so-called times of peace—women and girls have been the victims of sexual violence and torture carried out to express domination along gender lines, as well as to humiliate the male enemy. Arendt (1979) distinguishes between the space of appearance, as a space of political freedom and equality, constructed by citizens through the medium of speech and persuasion, and the space of the common world. The second is a shared and public human world with institutions, agendas, and actors, which provides a relatively stable environment for activities. Both spaces are essential for the consolidation of citizenship, and hence essential for a change in the dominant power relationships. Arendt insists that the recovery of a common and shared world will reactivate a mode of citizenship where individuals and groups can establish relationships of support, reciprocity and solidarity. It is precisely in the common space where collective ideas can emerge and group preferences are strengthened. It is in this space where gender-sensitive practices and peacebuilding can be developed.

There have been many movements worldwide that have achieved major political goals without arms and through non-violent activities. There was Gandhi's non-violent movement for India's independence; the Civil Rights movement in the US against racism; the ending of apartheid in South Africa and the construction of a rainbow democracy; the Velvet Revolution in Czechoslovakia that overcame authoritarianism; and the fall of the Berlin wall and the reunification of both Germanies. There is also the interesting case of Costa Rica, a country that abolished its army in 1948 and invested the savings in education and public health. Today it is the country with the highest Human Development Index and the lowest crime rate in Latin America (UNDP 2014).

Worldwide there is a growing concern about environmental and social deterioration, as a result of the current process of neo-liberal globalization. Most of these efforts occur in isolation, sometimes through leadership struggles. A systemic process is still lacking where first a collective thinking emerges, then a readiness to act, and later a process of judgement, which is necessary, as Hannah Arendt argued, for efficiency in the common space and for achieving goals through peaceful means. This means that first of all it is necessary to design an agenda for combating patriarchy from a clearly gender-equal perspective. This means equal salaries for equal work and making visible unpaid household work and caring activities, which

are mostly carried out by women. The revalorization of household and caring activities is needed to produce a more just distribution of the unpaid workload at the household level (Parvin/Bélanger 1996). At the local level, all types of discrimination and oppression in jobs, public spaces, and the media must be made known and prosecuted at law. At the regional, national and international levels, there is the need for a rigorous assessment together with an adjustment of all legal frameworks towards gender equality, with clear guidelines for a policy of equity and equality. This includes the abolition of the military and of warfare.

In the second phase, young people and the alternative media can promote a worldview where feminist values are anchored at all levels of society, including in the public domain and in government. In the third phase, there should be a shift at national and international levels from the narrow militaristic view of security to human security (UNDP 1994) and a "widespread increase in self-awareness among the population" (Snauwaert 2015: XVI), where the environment, sustainability, and gender security (Oswald 2013) are key values.

8.4.2 Engendering a Sustainable Peace: Imagining the Possible

Thousands of years of written and institutional woman-hating activities have limited the understanding of how harmonious relations between nature, humankind, and gender might be achieved. Violence, authoritarianism, discrimination, and exclusion have not only affected the environment, but also limited democracy and gender equity, and increased risks for both men and women (Beck 2009, 2011). "The patriarchal culture of control and domination is the root of all social and ecological violence. It corrupted the original unity of man and woman and is now corrupting the unity between humanity and the human habitat" (Burton 2013). Peace and security were conceived within a male-dominated system and conceptual evolution is imbued with these patriarchal biases. For this reason, Serrano Oswald (2009) noted that security and peace must be engendered in conceptual terms to overcome the underlying prejudices.

A transition to a sustainable-engendered peace requires an ethical framework and a clear strategy for change (Grin/Rotman/Schot 2010). The current method of reaching a holistic peace is seen as a step-by-step procedure through international organizations, governments, social movements, and *non-governmental organizations* (NGOs). This may be too slow to achieve the goal, as environmental and social conditions are deteriorating so fast that the traditional diplomatic way may be too late to transform thousands of years of oppression and violence into a sustainable-engendered peace. Disasters tripled between 2000 and 2009 compared with their number between 1980 and 1989. About eighty per cent of this increase is climate-related, and greenhouse gases have risen above 400 ppm. Since 1990, extreme events that resulted in disasters have affected about 217 million people yearly. Time and the speed of change are becoming crucial variables in changing the paradigm of patriarchy to a sustainable-engendered peace.

Facing this reality, most feminists propose a more radical way, where a cultural change in worldview and mindset is necessary in order to overcome the present system of patriarchal oppression (Oswald/Brauch 2011). This means a change in the globally organized way of life and changes in values, norms, beliefs, institutions and productive processes, and also requires the human-oriented development of science and technology. It means a collective learning process, one which is not based on the so-called 'natural laws' of patriarchy where women are invisible, but one which understands that discrimination, violence, and exploitation are socially constructed and follow hegemonic power interests. To defeat patriarchy, deeply anchored structures and mechanisms of control must be demolished. As Simone de Beauvoir argued and Frye (1983) emphasized, there are not just external factors of control, but also internal ones, which have created a system of articulated barriers and forces that are able to immobilize people. As Hannah Arendt stressed, there are also social cognition processes, which are able to overcome the deep structures of violence and empower groups of people to promote holistic and complex sustainable change.

Finally, the difference in the public response to the fourth and fifth *Intergovernmental Panel on Climate Change* (IPCC) assessment reports (IPCC 2007, 2007a, 2007b, 2013, 2014a, 2014b, 2014c, 2014d), greater awareness, and protest marches (for example, in New York in September 2014) are social indicators for change. More and very severe extreme events (IPCC 2012), increasing financial costs, and a loss of life and livelihood have persuaded people to get rid of the present system of exploitation and destruction of the environment and human beings. A radical change towards sustainability and equity in the management of earth is required (Conca 1994). It is exactly greater knowledge about power structures, better communication about world disasters, and global solidarity

that are opening new ways to move away from Margaret Thatcher's TINA (there is no alternative) to the TAMA (there are many alternatives) of ecofeminists.

8.5   Some Initial Conclusions

Returning to the research question, a widened concept of sustainable-engendered peace may offer a deeper understanding of the historical development of patriarchy, the destruction of the environment, and the power relations and economic interests intrinsic to both processes. Women have powerful ways of enhancing rather than diminishing the power of others. This means that a sustainable-engendered peace is a 'power-from-within', which leads first to a preference for change and later to a commitment to action. A sustainable-engendered peace starts from a positive understanding of power, which will lead to new actions to overcome all types of oppression, including self-subjugation. The concept of a sustainable-engendered peace focuses on a process of empowerment, primarily to overcome systemic violence, economic oppression, socio-economic marginalization, a lack of autonomy in decision-making, cultural imperialism and environmental destruction. Figure 8.1 shows how the different elements of peace, such as negative, positive, sustainable, cultural, structural and engendered peace may produce feedback to and solid bases for a process of empowerment.

The first step is to overcome the narrow concept of military security and promote policies aimed at human security, defined as freedom from fear, freedom from want, freedom from hazard impacts, and freedom to live in dignity within a state of law (Brauch/Oswald Spring/Mesjasz et al. 2008, 2011; Brauch/Oswald Spring/Grin et al. 2009). It also begins to make visible the existing harmony among gender relations, generations, races and classes, as well as the solidarity and the gift economy that exist between all human beings (Vaughan 2007). The gift from nature through ecosystem services is another starting point for understanding that sustainability is vital to living well for both present and future generations. Present and future generations need a healthy environment with plenty of resources for a dignified livelihood; especially taking into account that the world population is still growing and that by 2050 we may reach ten billion.

The concept of diversity is crucial in a world with numerous ethnic, religious, cultural, and social groups. The integration of traditional and modern knowledge

Figure 8.1:   Sustainable-engendered peace with cultural diversity. **Source**: The author.

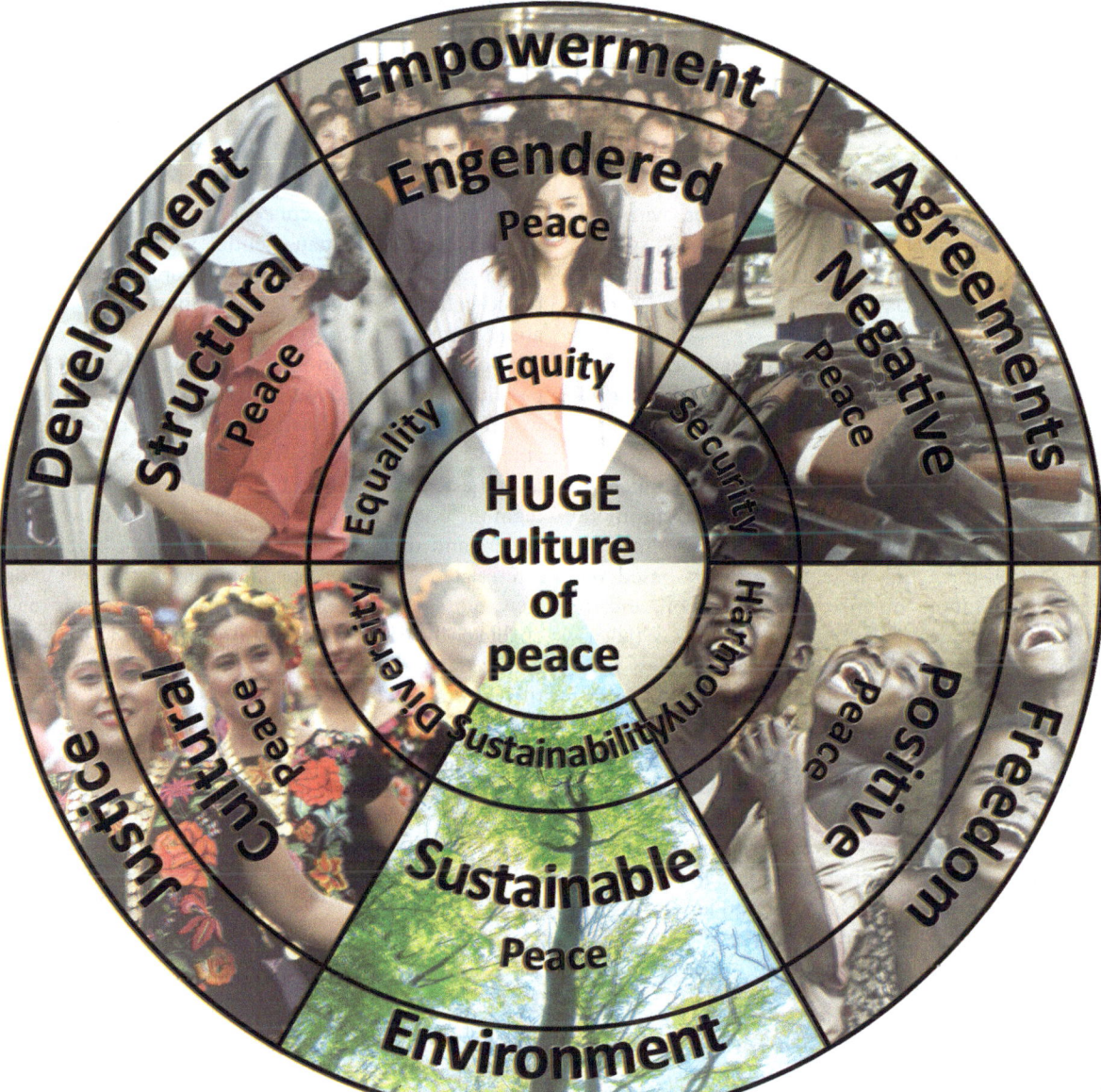

from all over the world may offer new ways of dealing differently with approaching conflicts (Oswald Spring 2004), but will especially offer new ways of promoting equality.

Wilkinson and Pickett (2010) have shown that countries with fewer internal levels of difference have the longest life expectancy and the lowest rates of violence. This means that the redistribution of wealth will also offer rich people better conditions of life than the present concentration of wealth with widespread violence. There are enough resources on earth to grant each world citizen an adequate quality of life

and the ability to live with dignity. Finally, numerous oppressive mechanisms at global level are limiting the development of the largest part of world society. These subordinated groups could achieve greater equity and empowerment among women, races, classes, ethnic groups and religions. Link(age)s, feedbacks, and reinforcements among all these factors may promote greater personal and collective self-awareness, which is crucial to understanding and taking action towards a peaceful and sustainable transition.

The *United Nations Environment Programme* (UNEP 2011) has promoted a new sustainable policy called the "green economy", designed to overcome local resource problems. It integrates holistic approaches from different theoretical positions such as ecofeminism, postmodernism, ecology, peace movements, green politics and parties, anti-globalization movements, and anarchists. The proposed "green economy" encourages an economic future with renewable energy, green buildings, sustainable transport systems, water, waste and land management, with soil restoration and food sovereignty that includes subsistence agriculture mostly in the hands of women (Bennholdt-Thomsen/Faraclas/Werlhof 2001; Oswald Spring 2009).

From a feminist, holistic, and gender-equal perspective, feminist values are deeply embedded at all levels of society, and this could be realized by governments and the business community. A paradigm shift towards a sustainable-engendered peace would include pathways to negotiate agreements and to promote actions for change. All these efforts together may be able to transform the present worldview based on greed, exploitation, violence, and oppression, and to promote a fundamentally different worldview that is inclusive, just, participative, and equal. This new feminist paradigm, as suggested in this chapter, focuses on a globally holistic, cosmopolitan, ethically oriented, sustainable-engendered peace with clean natural resources and safe food (Bennholdt-Thomsen/Mies 1999). This shift may help overcome the patriarchal system and achieve an ethnically-oriented, gender-sensitive, religiously diverse and just society. If the root causes of violence and destruction in world society are transformed, the environment is provided with an opportunity to recover its crucial natural processes and to continue to offer clean resources such as water, air, soil, and biodiversity. A different way of managing the earth will enable the environment to deliver the crucial ecosystem services that provide, regulate, support, and create cultural benefits for humans and nature.

A radical understanding of justice will allow us to move from the narrow conception of military security to an integrated human, gender and environmental security, a HUGE security (Oswald Spring 2009) and peace—a holistic peace—, where opportunities to care about nature and other human beings are created. Actions based on bottom-up initiatives supported by policies of equity and mechanisms of redistribution of wealth from the top down may offer the most vulnerable people alternatives for empowerment and self-awareness. This sustainable-engendered peace paradigm includes a holistic approach that will allow us to understand that life on earth and among humans is interrelated and interdependent. From this point of view, more just and more equal power structures may enable civil society to restructure global and local societal arenas and to educate and train actors to promote the societal values of

> love, genuine caring for others, fairly sharing all that is available to the group; and empowerment, helping group members to achieve fulfilment, cooperation and maturity—making together for mutual fulfilment. Feminism is profoundly transformational, for it calls for fundamental changes in personal values and human relationships as well as in structures and systems (Reardon 1980: 14).

References

Adams, Richard M., 1965: *Heartland of Cities, Surveys of Ancient Settlement and Land Use on the Central Floodplain of the Euphrates* (Chicago: University of Chicago Press).

Allen, Amy, 1998: "Rethinking Power," in: *Hypatia*, 13: 21–40.

Allen, Amy, 1999: *The Power of Feminist Theory: Domination, Resistance, Solidarity* (Boulder: Westview Press).

Allen, Amy, 2008: *The Politics of Our Selves: Power, Autonomy, and Gender in Contemporary Critical Theory* (New York: Columbia University Press).

Allen, Amy, 2011: "Feminist Perspectives on Power", in: *Stanford Encyclopedia of Philosophy*; at: <http://plato.stanford.edu/entries/feminist-power/>.

Anttila-Hughes, Jesse; Hsiang, Keith; Hsiang, Solomon M., 2013: "Destruction, Disinvestment, and Death: Economic and Human Losses Following Environmental Disaster"; at: <http://papers.ssrn.com/sol3/papers.cfm?abstract_id=2220501>.

Arendt, Hannah, 1979: *The Recovery of the Public World*, ed. Hill, Melvyn A. (New York: St. Martin's Press).

Arendt, Hannah, 1970: *On Violence* (New York: Harcourt Brace & Co.).

Arendt, Hannah, 1958: *The Human Condition* (Chicago: University of Chicago Press).

Arizpe, Lourdes, 2015: *Vivir para crear Historia. Antología de Estudios sobre Desarrollo, Migración, Género e Indígenas* (Cuernavaca: CRIM-UNAM; M.A. Porrúa).

Arnold, Bill T., 2004: *Who were the Babylonians?* (Atlanta: Society of Biblical Literature).

Bartky, Sandra, 1990: *Femininity and Domination: Studies in the Phenomenology of Oppression* (New York: Routledge).

Bartra, Armando, 2012: *Los nuevos herederos de Zapata. Campesinos en movimiento 1920–2012* (Mexico: CNPA, Circo Maya, Cámara de Diputados).

Beauvoir, Simone de, 1974: *Gender Trouble: Feminism and the Subversion of Identity* (New York: Routledge).

Beck, Ulrich, 2011: "Living in and Coping with World Risk Society", in: Brauch, Hans Günter; Oswald Spring, Úrsula; Mesjasz, Czeslaw; Grin, John; Kameri-Mbote, Patricia, Chourou, Béchir; Dunay, Pal; Birkmann, Jörn (Eds.): *Coping with Global Environmental Change, Disasters and Security–Threats, Challenges, Vulnerabilities and Risks* (Berlin–Heidelberg–New York: Springer): 11-16.

Beck, Ulrich, 2009: *World at Risk* (Cambridge: Polity Press).

Benett, Judith, 2006: *History Matters. Patriarchy and the Challenge of Feminism,* (Philadelphia. PA: Pennsylvania University Press).

Bennholdt-Thomsen, Veronika; Faraclas, Nicholas; von Werlhof, Claudia (Eds.), 2001: *There is an Alternative: Subsistence and Worldwide Resistance to Corporate Globalization* (London: Zed).

Bennholdt-Thomsen, Veronika; Mies, Maria, 1999: *The Subsistence Perspective: Beyond the Globalised Economy* (London: Zed).

Bond, Gerard; Showers, William; Cheseby, Maziet; Lotti, Rusty; Almasi, Peter; de Menocal, Peter; Priore, Paul; Cullen, Heidi; Hajdas, Irka; Bonani, Georges, 1997: "A Pervasive Millennial-Scale Cycle in North Atlantic Holocene and Glacial Climates", in: *Science*, 278 (14 November): 1257-1266.

Borras, Saturnino M.; Franco, Jennifer C.; Kay, Cristobal; Spoor, Max, 2011: *Land grabbing in Latin America and the Caribbean in broader international perspectives. A paper prepared for and presented at the Latin America and Caribbean seminar: 'Dinámicas en el mercado de la tierra en América Latina y el Caribe',* 14-15 November (Santiago, Chile: FAO Regional Office): 1-54.

Boulding, Elise, 2000: *Cultures of Peace. The Hidden Side of History* (Syracuse: Syracuse University Press).

Boulding, J. Russell, 2016: *Elise Boulding: A Pioneer in Peace Research, Peace-making, Feminism and the Family: From a Quaker Perspective* (Cham: Springer).

Brauch, Hans Günter; Oswald Spring, Úrsula; Mesjasz, Czeslaw; Grin, John; Kameri-Mbote, Patricia, Chourou, Béchir; Dunay, Pal; Birkmann, Jörn (Eds.), 2011: *Coping with Global Environmental Change, Disasters and Security–Threats, Challenges, Vulnerabilities and Risks* (Berlin–Heidelberg–New York: Springer).

Brauch, Hans Günter; Oswald Spring, Úrsula; Grin, John; Mesjasz, Czeslaw; Kameri-Mbote, Patricia; Behera, Navnita Chadha; Chourou, Béchir; Krummenacher, Heinz (Eds.), 2009: *Facing Global Environmental Change: Environmental, Human, Energy, Food, Health and Water Security Concepts* (Berlin–Heidelberg–New York: Springer).

Brauch, Hans Günter; Oswald Spring, Úrsula; Mesjasz, Czeslaw; Grin, John; Dunay, Pal; Behera, Navnita Chadha; Chourou, Béchir; Kameri-Mbote, Patricia; Liotta, P. H. (Eds.), 2008: *Globalization and Environmental Challenges: Reconceptualizing Security in the 21st Century* (Berlin–Heidelberg–New York: Springer).

Brundtland Commission, 1987: *Our Common Future* (Oxford: Oxford University Press).

Burton, Bruce A., 2013: *The Three D's: Democracy, Divinity and Drama: An Essay on Gender and Destiny* (Columbia TN: SynergEbooks).

Butler, Judith, 1990: *Gender Trouble: Feminism and the Subversion of Identity* (New York: Routledge).

Butler, Judith, 1997: *Excitable Speech: Toward a Politics of the Performative* (New York: Routledge).

Calva, José Luis (Ed.), 2012a: *Análisis estratégico para el Desarrollo: Estrategias económicas exitosas en Asia y en América Latina,* vol. 2 (México: Juan Pablos).

Calva, José Luis (Ed.), 2012b: *Análisis estratégico para el Desarrollo: Políticas macroeconómicas para el desarrollo sostenido,* vol. 2 (México: Juan Pablos).

Calva, José Luis (Ed.), 2012c: *Análisis estratégico para el Desarrollo: Derechos sociales y desarrollo incluyente,* vol. 12 (México: Juan Pablos).

Colledge, Sue, 2007: *The origins and spread of domestic plants in southwest Asia and Europe* (London: University College, Institute of Archaeology).

Collins, Patricia Hill, Maldonado, Lionel A.; Maldonado, Dana Y.; Barrie Thorne; Weber, Lynn, 2002: "Symposium on West and Fenstermaker's 'Doing Difference'", in: Fenstermaker, Sarah; West, Candace (Eds.): *Doing Gender, Doing Difference* (New York: Routledge).

Conca, Ken, 1994: "In the name of sustainability: Peace studies and environmental discourse", in: *Peace and Change,* 19: 91-113.

Crutzen, Paul J., 2011: "The Anthropocene: Geology of Mankind", in: Brauch, Hans Günter; Oswald Spring, Úrsula; Mesjasz, Czeslaw; Grin, John; Kameri-Mbote, Patricia, Chourou, Béchir; Dunay, Pal; Birkmann, Jörn (Eds.): *Coping with Global Environmental Change, Disasters and Security–Threats, Challenges, Vulnerabilities and Risks* (Berlin–Heidelberg–New York: Springer): 3-4.

Crutzen, Paul J., 2002: "Geology of Mankind", in: *Nature,* 415,3 (January): 23.

Cudd, Ann, 2006: *Analysing Oppression* (Oxford: Oxford University Press).

d'Eaubonne, Françoise, 1974: *Le Féminisme ou la Mort* (Paris: Pierre Horay).

del Valle, Teresa (Ed.), 1993: *Gendered Anthropology* (London: Routledge).

Diehl, Paul F.; Ku, Charlotte; Zamora, Daniel, 2003: "The Dynamics of International Law: The Interaction of Normative and Operating Systems", in: *International Organization,* 57,1 (Winter): 43-75.

Edhom, Felicity, Harris, Olivia; Young, Kate, 1977: "Conceptualizing Women", in: *Critique of Anthropology,* 3: 9-10.

Elliott, M. Lynn, 2011: "First report of Fusarium wilt caused by Fusarium oxysporum f. sp. palmarum on Canary Island date palm in Florida", in: *Plant Disease,* 95: 356.

Ellis, F.; Kutengule, M.; Nyasulu, A., 2003: "Livelihoods and rural poverty reduction in Malawi", in: *World Development*, 31,9 (September): 1495-1510.

Engels, Friedrich, 2004 [1884]: *The Origin of the Family, Private Property and the State*, (Chippendale: Resistance Books).

FAOSTAT, 2013: "Pesticides", at: <http://faostat.fao.org/site/424/default.aspx#ancor>.

Federici, Silvia, 2007: "Women, Land-Struggles and The Valorization of Labor"; at: <http://www.commoner.org.uk/10federici.pdf>.

Federici, Silvia, 2004: *Caliban and the Witch: Women, the Body, and Primitive Accumulation* (Brooklyn: Autonomedia).

Federici, Silvia, 1999: "Reproduction and Feminist Struggle in the New International Division of Labor", in: Dalla Costa, Mariarosa; Dalla Costa, Giovanna F. (Eds.): *Women, Development and Labor of Reproduction. Struggles and Movements* (Trenton, NJ: Africa World Press).

Folbre, Nancy, 2006: "Rethinking the Child Care Sector", in: *Journal of the Community Development Society*, 37,2 (Summer): 38-52.

Foucault Michel, ²1983: "Afterword: The Subject and Power", in: Dreyfus, Hubert; Rabinow, Paul (Eds.) *Michel Foucault: Beyond Structuralism and Hermeneutics* (Chicago: University of Chicago Press).

Foucault, Michel, 1977: *Discipline and Punish: The Birth of the Prison* (New York: Vintage).

Fraser, Nancy, 1994: "After the Family Wage: Gender Equity and the Welfare State", in: *Political Theory*, 22,4: 591-618.

Fraser, Nancy, 1989: *Unruly Practices: Power, Discourse and Gender in Contemporary Social Theory* (Minneapolis: University of Minnesota Press).

Frye, Marilyn, 1983: *The Politics of Reality: Essays in Feminist Theory* (Freedom: The Crossing Press).

Friedan, Betty, 1963: *The Feminine Mystique* (New York: W. W. Norton and Co.).

Galtung, Johan, 1968: "Peace", in: *International Encyclopedia of the Social Sciences* (London–New York; Macmillan): 487-496.

Galtung, Johan, 1967: "Peace Research: science, or politics in disguise?" in: *International spectator*, 21,19: 1573-1603.

Gramsci, Antonio, 1998: *Escritos Políticos* (Mexico: Siglo XXI Eds.).

Gramsci, Antonio, 1975: *Notas sobre Maquiavelo, Política y el Estado moderno* (México: Juan Pablos).

Grin, John; Rotmanns, Jan; Schot, Johan, 2010: *Transitions to Sustainable Development. New Directions in the Study of Long Term Transformative Change* (New York–London: Routledge).

Groff, Linda; Smoker, Paul, 1995: "Creating Global-Local Cultures of Peace", at: <http://www.gmu.edu/programs/icar/pcs/smoker.htm>.

Gutiérrez, Luis, 2013: *Mother Pelican. A Journal of Solidarity and Sustainability*; at: <www.pelicanweb.org>.

Hartman, Heidi, 1976: "Women and the Workplace: The Implications of Occupational Segregation", in: University of Chicago (Ed.): *Capitalism, Patriarchy, and Job Segregation by Sex*, 1,3 (Spring) (Chicago: University of Chicago Press): 137-169.

Hartsock, Nancy, 1996: "Community/Sexuality/Gender: Rethinking Power," in: Hirschmann, Nancy J.; Di Stefano, Christine (Eds.): *Revisioning the Political: Feminist Reconstructions of Traditional Concepts in Western Political Theory* (Boulder: Westview Press).

Hartsock, Nancy, 1990: "Foucault on Power: A Theory for Women?", in: Nicholson, Linda (Ed.): *Feminism/Postmodernism* (New York: Routledge).

Hartsock, Nancy, 1983: *Money, Sex, and Power: Toward a Feminist Historical Materialism* (Boston: Northeastern University Press).

Haugaard, Mark, 2010: "Power: A 'Family Resemblance' Concept", in: *European Journal of Cultural Studies*, 13,4: 419-438.

Held, David, 2004: *Global Covenant: The Social Democratic Alternative to the Washington Consensus* (Cambridge: Polity Press).

Heun, Manfred; Schäfer-Pregl, Ralf; Klawan, Dieter; Castagna, Renato; Accerbi, Monica; Borghi, Basilio; Salamini, Francesco, 1997: "Site of Einkorn Wheat Domestication Identified by DNA Fingerprinting", in: *Science*, 278: 1312-4; at: <doi:10.1126/science.278.5341.1312>.

Hoagland, Sarah Lucia, 1988: *Lesbian Ethics: Toward a New Value* (Palo Alto: Institute of Lesbian Studies).

Hobbes, Thomas, 1985 [1641]: *Leviathan* (New York: Penguin Books).

Hoodfar, Homa, 1997: *Between Marriage and the Market. Intimate Politics and Survival in Cairo* (Berkeley: University of California Press).

Howell, George, 2007: "The North-South Environmental Crisis: An Unequal Ecological Exchange Analysis", in: *New School Economic Review*, 21: 77-99.

Huang, Xuehui; Kurata, Nori; Wei, Xinghua; Wang, Zi-Xuan; Wang, Ahong; Zhao, Qiang; Zhao, Yan; Liu, Kunyan et al. 2012: "A map of rice genome variation reveals the origin of cultivated rice", in: *Nature*, 490,7421: 497-501.

Hughes, Sarah Shaver; Hughes, Brady, 2001: "Women in Ancient Civilizations", in: Adas, Michael (Ed.): *Agricultural and pastoral societies in ancient and classical history* (Philadelphia: Temple University Press): 118-119.

IEA, 2014: *World Energy Investment Outlook 2014. Special Report*, at: <http://www.iea.org/publications/freepublications/publication/weio2014.pdf>.

ILO, 2014: "Trends Econometric Models, April 2014", at: <http://www.ilo.org/global/about-the-ilo/multimedia/maps-and-charts/WCMS_244259/lang-en/index.htm>.

IPCC, 2014a: *Climate Change 2014: Impacts, Adaptation and Vulnerability. Part A: Global and Sectoral Aspects. Contribution of Working Group II to the Fifth Assessment Report of the Intergovernmental Panel on Cli-*

mate Change (Cambridge–New York: Cambridge University Press).

IPCC, 2014b: *Climate Change 2014: Impacts, Adaptation and Vulnerability. Part B: Regional Aspects. Global and Sectoral Aspects. Contribution of Working Group II to the Fifth Assessment Report of the Intergovernmental Panel on Climate Change* (Cambridge–New York: Cambridge University Press).

IPCC, 2014c: *Climate Change 2014: Mitigation of Climate Change. Working Group III Contribution to the Fifth Assessment Report of the Intergovernmental Panel on Climate Change* (Cambridge–New York: Cambridge University Press).

IPCC, 2014d: *Climate Change 2014. Synthesis Report. Contribution to the Fifth Assessment Report of the Intergovernmental Panel on Climate Change* (Cambridge–New York: Cambridge University Press).

IPCC, 2013: *Climate Change 2013: The Physical Science Basis. Contribution of Working Group I to the Fifth Assessment Report of the Intergovernmental Panel on Climate Change* (Cambridge–New York: Cambridge University Press).

IPCC, 2012: *Managing the Risks of Extreme Events and Disasters to Advance Climate Change Adaptation. A Special Report of Working Groups I and II of the Intergovernmental Panel on Climate Change* (Cambridge–New York: Cambridge University Press).

IPCC, 2007: *Climate Change 2007. The Physical Science Basis,* Working Group I Contribution to the Fourth Assessment Report of the IPCC (Cambridge–New York: Cambridge UP); at: <http://www.ipcc-wg1.org/>.

IPCC, 2007a: *Climate Change 2007. Impacts, Adaptation and Vulnerability.* Working Group II Contribution to the Fourth Assessment Report of the IPCC (Cambridge–New York: Cambridge University Press); at: <http://www.ipcc-wg2.org/>.

IPCC, 2007b: *Climate Change 2007. Mitigation and Climate Change.* Working Group III Contribution to the Fourth Assessment Report of the IPCC (Cambridge–New York: Cambridge University Press); at: <http://www.ipcc-wg3.org/>.

ISSC, 2013: *World Social Science Report 2013: Changing Global Environments* (Paris: ISSC).

Jasper, Allison, 2013: "Religion, Feminism, and Gender-making Theory", in: *Mother Pelican,* 9,1: 6-8.

Jodelet, Denise, 1991: *Madness and Social Representation* (London: Harvester/Wheatsheaf).

Kraemer, Sebastian, 1991: "The Origins of Fatherhood: An Ancient Family Process", in: *Family Process,* 30,4: 377-392.

Lagarde y de los Ríos, Marcela, 1990: *Los cautiverios de las mujeres. madresposas, monjas, putas, presas y locas* (México: PUEG/UNAM).

Lamas, Marta (Ed.), 1996: *El género. La construcción cultural de la diferencia sexual* (Mexico: PUEG-Porrúa).

Lander, Edgard, 2011: *The Green Economy: The Wolf in Sheep's Clothing* (Amsterdam: Transnational Institute,

November); at: <http://www.tni.org/sites/www.tni.org/files/download/green-economy.pdf>.

Lelé, Sharachchandra M., 1991: "Sustainable Development. A Critical Review", in: *World Development,* 19,6: 607-621.

Lu, Houyuan; Zhang, Jianping; Liu, Kam-biu; Wu, Naiqin; Li, Yumei; Zhou, Kunshu; Ye, Maolin; Zhang, Tianyu; Zhang, Haijiang; Yang, Xiaoyan; Shen, Licheng: Xu, Deke; Li, Quan, 2009: "Earliest domestication of common millet (*Panicum miliaceum*) in East Asia extended to 10,000 years ago", in: *Proceedings of the National Academy of Sciences of the United States of America,* 106,18: 7367-7372.

Lukes, Steven, ²2005: *Power: A Radical View* (London: Macmillan).

MA, 2005: *Ecosystem and Human Wellbeing. Millennium Ecosystem Assessment* (Washington, DC: Island Press).

MacKinnon, Catharine, 1987: *Feminism Unmodified: Discourses on Life and Law* (Cambridge: Harvard University Press).

Malinowsky, Bronislaw, 1929: *The Sexual Life of Savages in North-West Melanesia: An Ethnographic Account of Courtship, Marriage and Family Life among the Natives of the Trobriand Highlands, British New Guinea* (Boston: Beacon Press).

Meillassoux, Claude, 1981: *Maidens, Meal and Money: Capitalism and the Domestic Community* (Cambridge: Cambridge University Press).

Mies, Maria, 1986: *Patriarchy and Accumulation on a World Scale* (Melbourne: Zed Books).

Miller, Jean Baker, 1992: "Women and Power", in: Wartenberg, Thomas (Ed.), *Rethinking Power* (Albany: SUNY Press).

Morgan, Lewis H., 2003 [1881]: *Houses & House–Life of the American Aborigines* (Salt Lak City: University of Utah Press).

Neugebauer, O., ²1993: *The Exact Sciences in Antiquity* (New York: Barnes and Noble).

Okin, Susan Moller, 1989: *Justice, Gender and the Family* (New York: Basic Books).

Oswald Spring, Úrsula, 2013: "Seguridad de género", in: Flores, Fátima (Ed.): *Representaciones Sociales y contextos de investigación con perspectiva de género* (Cuernavaca: CRIM-UNAM): 225-256.

Oswald Spring, Úrsula (Ed.), 2011: *Water Resources in Mexico. Scarcity, Degradation, Stress, Conflicts, Management, and Policy* (Berlin - Heidelberg - New York: Springer).

Oswald Spring, Úrsula, 2009: "A HUGE Gender Security Approach: Towards Human, Gender, and Environmental Security", in: Brauch, Hans Günter; Oswald Spring, Úrsula; Grin, John; Mesjasz, Czeslaw; Kameri-Mbote, Patricia; Behera, Navnita Chadha; Chourou, Béchir; Krummenacher, Heinz (Eds.): *Facing Global Environmental Change: Environmental, Human, Energy, Food, Health and Water Security Concepts* (Berlin - Heidelberg - New York: Springer): 1165-1190.

Oswald Spring, Úrsula, 2008: *Gender and Disasters: Human, Gender and Environmental Security: A HUGE Challenge*. SOURCE ('Studies of the University: Research, Counsel, Education'), Publication Series of UNU-EHS, No. 8/2008 (Bonn: United Nations University, Institute for Environment and Human Security).

Oswald Spring, Úrsula, 2004: *Resolución noviolenta de conflictos en sociedades indígenas y minorías* (México: Coltlax, Böll Foundation).

Oswald Spring, Úrsula 1994: *Estrategias de supervivencia en la Ciudad de México* (Cuernavaca: CRIM-UNAM).

Oswald Spring, Úrsula; Brauch, Hans Günter; Tidball, Keith G. (Eds.), 2013: *Expanding Peace Ecology. Peace, Security, Sustainability, Equity and Gender. Perspectives of IPRA's Ecology and Peace Commission* (Cham–Heidelberg–New York–Dordrecht–London: Springer).

Oswald Spring, Úrsula; Brauch, Hans Günter, 2011: "Coping with Global Environmental Change–Sustainability Revolution and Sustainable Peace", Brauch, Hans Günter; Oswald Spring, Úrsula; Mesjasz, Czeslaw; Grin, John; Kameri-Mbote, Patricia, Chourou, Béchir; Dunay, Pal; Birkmann, Jörn (Eds.): *Coping with Global Environmental Change, Disasters and Security–Threats, Challenges, Vulnerabilities and Risks* (Berlin–Heidelberg–New York: Springer): 1487-1504.

Parvin, Ghorayshi; Bélanger, Claire (Eds.), 1996: *Women, Work, and Gender Relations in Developing Countries: A Global Perspective* (Westport: Greenwood Press).

Ranere, Anthony J.; Piperno, Dolores R.; Holst, Irene; Dickau, Ruth; Iriarte, José, 2009: "The cultural and chronological context of early Holocene maize and squash domestication in the Central Balsas River Valley, Mexico", in: *Proceedings of the National Academy of Sciences*, 106,13: 5014-5018.

Reardon, Betty A., 1996: *Sexism and War System* (Syracuse: Syracuse University Press).

Reardon, Betty A., 1985: "Civic Responsibility for a World Community", in: Conrad, D.; Thomas, T. M. (Eds.), *Images of an Emerging World: From a War System to Peace System* (Katyam, India: Prakasam Publications).

Reardon, Betty A., 1980: "Moving to the Future", in: *Network* 8,1.

Reardon, Betty A.; Snauwaert, Dale, 2015a: *Betty A. Reardon: A Pioneer in Education for Peace and Human Rights*. Springer Briefs on Pioneers in Science and Practice No. 26–presented by Dale Snauwaert (Cham–Heidelberg–New York–Dordrecht–London: Springer).

Reardon, Betty A.; Snauwaert, Dale, 2015b: *Betty A. Reardon: Key Texts in Gender and Peace*. Springer Briefs on Pioneers in Science and Practice No. 27–Texts and Protocols No. 13 (Cham–Heidelberg–New York–Dordrecht–London: Springer).

Reilly, Kevin, [4]2010: Worlds of History, Vol. I: To 1550: A Comparative Reader (Bedford: Bedford Books).

Rich, Adrienne, 1976: *Of Women Born. Motherhood as Experience and Institution* (New York: W.W. Norton).

Schomerus, Mareike; de Vries, Lotje, 2014: "Improvising border security: 'A situation of security pluralism' along South Sudan's borders with the Democratic Republic of the Congo", in: *Security Dialogue*, 45: 279-294.

Senghaas, Dieter, 2013: *Dieter Senghaas: Pioneer of Peace and Development Research*. Springer Briefs on Pioneers in Science and Practice No. 6 (Heidelberg: Springer).

Senghaas, Dieter, 1997: "Frieden–ein mehrfaches Komplexprogramm", in: Senghaas, Dieter (Ed.): *Den Frieden machen* (Frankfurt am Main: Suhrkamp): 560-575.

SERI, 2010: "SERI Global Material Flow Database"; at: <http://www.materialflows.net/fileadmin/docs/materialflows.net/MFA_technical_report_May_2013.pdf>.

Serrano, Omar, 2013: *The Domestic Sources of European Foreign Policy: Defence and Enlargement* (Amsterdam: Amsterdam University Press).

Serrano Oswald, Serena Eréndira, 2009: "The Impossibility of Securitizing Gender vis a vis Engendering Security", in: Brauch, Hans Günter; Oswald Spring, Úrsula; Grin, John; Mesjasz, Czeslaw; Kameri-Mbote, Patricia; Behera, Navnita Chadha; Chourou, Béchir; Krummenacher, Heinz (Eds.): *Facing Global Environmental Change: Environmental, Human, Energy, Food, Health and Water Security Concepts* (Berlin–Heidelberg–New York: Springer): 1151-1164.

Serrano Oswald, Serena Eréndira, 2010: *La Construcción Social y Cultural de la Maternidad en San Martín Tilcajete, Oaxaca* (PhD Thesis, UNAM: Instituto de Antropología, Mexico. D.F.).

Serrano Oswald, Serena Eréndira, 2014: "Migration, woodcarving and *engendered identities* in San Martín Tilcajete, Oaxaca", in: Truong, Thanh-Dam; Gasper, Des; Handmaker, Jeff, Bergh, Sylvia (Eds.), *Migration, Gender and Social Justice. Perspectives on Human Insecurity* (Heidelberg: Springer): 173-192.

Silverblatt, Irene, 1987: *Moon, Sun, and Witches: Gender Ideologies and Class in Inca and Colonial Peru* (Princeton: Princeton University Press).

Snauwaert, Dale, 2015: "Preface", in: Reardon, Betty A.; Snauwaert, Dale, *Betty A. Reardon: Key Texts in Gender and Peace*. Springer Briefs on Pioneers in Science and Practice No. 27–Texts and Protocols No. 13 (Cham–Heidelberg–New York–Dordrecht–London: Springer): ix-xx.

Stiglitz, Joseph, 2010: *Freefall: America, Free Markets, and the Sinking of the World Economy* (New York: W.W. Norton).

Stiglitz, Joseph, 2007: *Globalisation and its Discontent* (New York: W.W. Norton).

The Great Debate Schools Programme, 2011: Newcastle University; at: <http://www.thegreatdebate.org.uk/>.

Tigau, Camelia 2012: *¿Fuga de Cerebros o nomadismo Académcio?* (Mexico, D.F.: UNAM, Centro de Investigaciones sobre América del Norte).

Truong, Thanh-Dam; Gasper, Des; Handmaker, Jeff; Bergh, Sylvia (Eds.), 2014: *Migration, Gender and Social Jus-*

tice. Perspectives on Human Insecurity (Heidelberg–New York–Dordrecht–London: Springer).

UN, 2012: *The Future We Want*, G.A. Res. 66/288, 84, U.N. Doc. A/RES/66/288 (11 October 2012); at: <http://sustainabledevelopment.un.org/futurewewant.html>.

UN, 2000: UN Resolution A /RES/52/15) International Decade for a Culture of Peace and Non-violence for the Children of the World (2001-2010).

UN, 1999: *The Declaration 53/243 A on a Culture of Peace was accepted in September* 1999 (New York: UN).

UN Charter, 1947: *United Nations Charter* (New York: United Nations Press).

UN/DESA, 2015: *World Economic Situation and Prospects 2015* (New York: United Nations).

UNDP, 2014: *Human Development Report. Sustaining Human Progress: Reducing Vulnerabilities and Building Resilience* (New York: UNDP).

UNEP, 2011: *Towards a Green Economy. Pathways to Sustainable Development and Poverty Eradication* (Nairobi: UNEP); at: <http://www.unep.org/greeneconomy/ Portals/88/documents/ger/ger_final_dec_2011/ Green%20EconomyReport_Final_Dec2011.pdf>.

UNEP, 2012: *Global Environmental Outlook* 5 (Nairobi: UNEP).

UNESCO, 2002: *Universal Declaration on Cultural Diversity and Culture of Peace* (Paris: UNESCO).

UNESCO, 2000: *World Education Report* (Paris: UNESCO).

Vaughan, Genevieve, 2007: *Women and the Gift Economy: A Radically Different Worldview Is Possible* (Toronto: Inanna Publications and Education).

Vickers, Katherine, 2011: *Ghanaian Women. Creating Economic Security: An Analysis of Gender, Development, and Power in the Volta Region of Ghana, West Africa.* MA thesis in Applied Anthropology (Corvallis: Oregon State University).

Warren, Karen J., 1997: "Introduction", in: Warren, Karen (Ed.): *Ecofeminism, Women, Culture, Nature* (Bloomington: Indiana University Press): XI-XVI.

Wartenberg, Thomas, 1990: *The Forms of Power: From Domination to Transformation* (Philadelphia: Temple University Press).

Webel, Charles; Galtung, Johan, 2007: *Handbook of Peace and Conflict Studies* (London: Routledge).

Weber, Max, 1978: *Economy and Society: An Outline of Interpretive Sociology* (Berkeley: University of California Press).

Westing, Arthur H., 2013: *Pioneer on the Environmental Impact of War* (Heidelberg–New York–Dordrecht–London: Springer).

Wilkinson Richard; Pickett, Kate, 2010: *The Spirit Level: Why Equality is better for everyone* (Dexter: Bloomsbury Press).

Willaarts, Bárbara A.; Garrido, Alberto; Llamas, M. Ramón, 2014: *Water and Food Security in Latin America and the Caribbean* (London: Routledge).

Woodward, Susan, 2000: *The Postmodern State and the World Order* (London: Demos).

World Bank, 2014: *Risk and Opportunity, World Development Report* (Washington, DC: World Bank).

World Wildlife Fund, 2014: *Living Planet Report. Species and Spaces, People and Places* (Gland: WWF).

Wright, Quincy, 1965 [1942]: *A Study of War* (Chicago: University of Chicago Press).

Yiamouyiannis, Zeus, 2013: *Transforming Economy: From Corrupted Capitalism to Connected Communities* (Kindle-Amazon).

Young, Iris Marion, 1990: *Throwing like a Girl and Other Essays in Feminist Philosophy and Social Theory* (Bloomington: Indiana University Press).

Young, Iris Marion, 1992: "Five Faces of Oppression" in: Wartenberg, Thomas (Ed.): *Rethinking Power* (Albany: SUNY Press).

Zerzan, John, 2010: *Patriarchy, Civilization, and the Origins of Gender,* at: <http://theanarchistlibrary.org/ library/John_Zerzan__Patriarchy__Civilization__And_ The_Origins_Of_Gender.pdf>.

9 Sustainable Peace in the Anthropocene: Towards Political Geoecology and Peace Ecology

Hans Günter Brauch[1]

Abstract

This chapter attempts to conceptualize possible and plausible linkages between the emerging 'sustainability transition' research paradigm and the conceptual debate on a rethinking of peace, security, development and the environment or ecology, within the context of four research programmes carried out since the end of the Cold War. Within the framework of a shift in earth history from the Holocene to the Anthropocene during the past sixty years, the threat to the survival of humankind has fundamentally changed. No longer are 'others' the threat, but 'we' are, due to the exponential increase in the burning of hydrocarbons and the resulting accumulation of greenhouse gases in the atmosphere. This new anthropogenic threat can no longer be countered with traditional military strategies and means. In the twenty-first century, there needs to be a long-term transformative change towards a low-carbon economy, in production and consumption, and in the energy, transportation, agricultural and housing sectors. Only thus can dangerous climate change and chaotic tipping points in the climate system be avoided. Such a low-carbon economy should be the result of a transition to sustainability, necessitating not just sociotechnical changes but changes in perception, values, behaviour and lifestyles. Such a long-term transformative change to sustainability may possibly prevent two types of conflicts: climate-induced violent conflicts, and those driven by resource scarcity. On the conceptual level, this chapter suggests possible linkages that may be developed in the Anthropocene between sustainable development, human security and sustainable peace in the context of both a *political geoecology*—between the natural and social sciences—and a *peace ecology*—between peace, security, development and environmental studies. Its key message is the need for more conceptual, theoretical and empirical research into possible linkages between peace studies and ecology that takes into account the changed human and environmental conditions in the framework of the Anthropocene. The added value is to sensitize research on 'sustainability transition' so that it reflects on the impact of its realization on sustainable peace and human security.

Keywords: Anthropocene, sustainability transition, peace, security, development, ecology, environment, sustainable development, human security, sustainable peace, political geoecology, peace ecology, avoidance of resource and climate conflicts.

9.1   Introduction[2]

Since the mid-twentieth century, a fundamental change in earth history has occurred, brought about by humankind's intervention into nature and the Earth System through the massive consumption and burning of fossil energy sources. Now, "we are in the Anthropocene", as Paul J. Crutzen stated at a conference in Cuernavaca in late February 2000.[3] At about the same time Hans Joachim Schellnhuber (1999) called for a 'second Copernican revolution' and William C. Clark contributed to the NRC Study (1999) *Our Common Journey. A Transition towards Sustainability* (Johnson 2016, in this volume).

In May 2003, during a Dahlem workshop on "Earth Systems Analysis for Sustainability" William C.

1   PD Dr. Hans Günter Brauch is chairman of *Peace Research and European Security Studies* (AFES-PRESS), an international scientific NGO (Germany) and editor of five English language book series published by Springer: the *Hexagon series* <http://www.afes-press-books.de/html/hexagon.htm> and two Springer Briefs series on *ESDP* <http://www.afes-press-books.de/html/SpringerBriefs_ESDP.htm> and *PSP* <http://www.afes-press-books.de/html/SpringerBriefs_PSP.htm>; and the two new series: The Anthropocene (APESS) and Pioneers in Arts, Humanities,Science,Engnieering, Practice (PAHSEP); email: <brauch@afes-press.de>.

Clark, Paul J. Crutzen and Hans Joachim Schellnhu- ber (2004: 1) spoke of "the extraordinary revolution in our understanding of the Earth System that is now underway". They called for a "new paradigm aiming at sustainability" and a "new contract between science and society", calls prompted by "the urgent need to harness science and other forms of knowledge in pro- moting a worldwide sustainability transition that enhances human prosperity while protecting the Earth's life-support systems and reducing hunger and poverty". Clark, Crutzen and Schellnhuber (2004: 24) summarized their observations in this statement:

> We are currently witnessing the emergence of a new sci- entific paradigm that is driven by unprecedented plane- tary-scale challenges, operationalized by transdiscipli- nary centennium-scale agendas, and delivered by multi- scale co-production based on a new contract between science and society.

These insights have emerged gradually since 1896, when the Swedish physicist Svante Arrhenius hypoth- esized a linkage between the burning of hydrocarbon energy sources (coal, oil, gas) and the accumulation of greenhouse gases in the atmosphere. It was seventy- five years until this hypothesis was taken up by natural scientists in the early 1970s; climate change then became a recognized scientific issue, and formed a major research objective for four international scien- tific programmes (*scientization*).[4]

In the late 1980s, issues of global environmental change became political problems. For example, in December 1988 the IPCC was established and a man- date for the negotiation of the Rio conventions (1992) was approved by the UN General Assembly (*politici- zation*). Since 2000, the possible security impacts of global environmental change and climate change have been addressed by policymakers and scientists, that is, they have been *securitized* (Brauch 2002, 2009; Gled- itsch 2012; Scheffran/Brzozka/Brauch et al. 2012; Scheffran/Brzoska/Kominek et al. 2012; Salehyan 2014; chapter by Ide/Link/Scheffran et al. in this volume).

Since the end of World War II, in the policy-ori- ented social sciences, four research fields have evolved: a) *security studies*, b) *peace studies* or *peace research*, c) *development studies* and d) *environmen- tal studies* or *ecology*. There has been only limited sci- entific exchange between them, especially between peace studies and environmental studies (chapter 3 by Stephenson in this volume).[5] The linguistic and con- ceptual evolution of these four key guiding concepts has been discussed previously in the framework of a 'conceptual quartet' (Brauch 2009).

The study of issues raised by global (environmen- tal and climate) change has gradually emerged since the mid-1980s, through four global research pro- grammes. Only one, IHDP, which focused on the

2   This author is grateful for many critical and constructive comments and valuable suggestions to Mr Russell Boul- ding, Bloomington, Indiana, USA; Prof. Dr. Lothar Brock, J.W. Goethe University, Frankfurt, Germany; Prof. Dr. Michael Brzoska, director of the ISFH, Ham- burg University, Germany; Prof. Dr. Ken Conca, Ameri- can University, Washington DC, USA; Prof. Dr. Simon Dalby, Wilfrid Laurier University, Waterloo, Canada; Prof. Dr. Johan Galtung, Norway; Prof. Dr. Nils-Petter Gleditsch, PRIO, Oslo and University of Trondheim, Norway; Prof. Dr. Kalevi Holsti, University of British Columbia, Vancouver, Canada; Prof. Dr. Czeslaw Mes- jasz, Economic University of Cracow, Cracow, Poland; Prof. Dr. Ursula Oswald Spring, UNAM, CRIM, Cuer- navaca, Mexico; Prof. Dr. Jürgen Scheffran, Hamburg University, Germany. I am also grateful to Nobel laure- ate Prof. Dr. Paul J. Crutzen (2016), who taught me more about the human predicament with his work on the nuclear winter, the Anthropocene and the need for a 'sustainability revolution' than my own discipline. All shortcomings are solely those of the author.

3   On the history of the 'Anthropocene' concept, see Stef- fen, Grinevald, Crutzen and McNeill (2011); for an over- view of the public debate, see: Ellis, Erle, 2013: "Anthropocene", in: *Encyclopedia of Earth*; at: <http:// www.eoearth.org/view/article/150125>. The term was first used by Crutzen and Stoermer (2000); Paul J. Crutzen (2002, 2012, 2016) independently reinvented and popularized it. Crutzen has explained, "I was at a conference where someone said something about the Holocene. I suddenly thought this was wrong. The world has changed too much. So I said: 'No, we are in the Anthropocene.' I just made up the word on the spur of the moment. Everyone was shocked. But it seems to have stuck." (Stoermer as cited at: <http://en.wikipe- dia.org/wiki/Anthropocene> (12 Nov. 2015).

4   See the *World Climate Research Programme* (WCRP), which was set up in 1980 (Church, Asrar, Busalacchi et al. 2011); the *International Geosphere-Biosphere Pro- gramme* (IGBP) in 1986 (Noone, Nobre, Seitzinger 2011); DIVERSITAS followed in 1991 and modified in 2002 (Walther, Larigauderie, Loreau 2011); and IHDP in 1996 (*von* Falkenhayn, Rechkemmer, Young 2011; see chapter 14 by Arizpe, Price, Worcester and chapter 15 by Ehlers in this volume).

5   This list of area-specific research programmes is not com- prehensive and many issue-specific research paradigms and programmes have also emerged, such as globaliza- tion and migration research, technology assessment and sociotechnical innovations; and many theory-specific approaches, such as systems and complexity theory.

"international human dimensions" of global change, had a social science focus, and only one of its many projects on *Global Environmental Change and Human Security* (GECHS) addressed 'human' security issues (Barnett/Matthew/O'Brien 2008; Sygna/O'Brien/Wolf 2013).[6] These discussions were evaluated for the first time by the IPCC (2014, II, chapter 12) in its fifth assessment report (for a critique, see Gleditsch/Nordas 2014). As systematic research on GEC and peace issues has hardly emerged so far,[7] one goal of this chapter is to inspire more innovative research of a bridge-building nature in the years to come.

Since the nuclear age began, with the development and deployment of nuclear and hydrogen weapons, massive use of such weapons would have been unable to act as a defence and could only have destroyed the areas to be defended (von Weizsäcker 1972). Crutzen and Birks (1982) argued that a big nuclear war could cause a 'nuclear winter'; this challenged the views of some US civilian nuclear strategists, who argued that wars using nuclear weapons could be fought and won (Gray/Payne 1980) in Central Europe. While during the Cold War, when a nuclear doomsday was being discussed, the perceived enemy was always the 'other' superpower (e.g. a war between East and West) or nuclear nation (Pakistan vs. India), in the Anthropocene the threat to human survival has fundamentally changed from the 'other' to 'us'.

'We' have become the major threat to our own survival. This threat is no longer posed by the 'other' through 'weapons of mass destruction' but by our consumptive behaviour and our use of 'fossil energy sources' that has resulted in an exponential increase in greenhouse gases (Brauch 2016, 2016a). The impact of the 'we' can be measured by the exponential increase in greenhouse gas concentration. This rose from between 260 and 279 ppm during the Holocene (the last 12,000 years) to 405 ppm in early 2015, with an increase of 90 ppm since 1958 (315 ppm) alone. This threat is not posed by weapons but by our unsustainable production and consumption patterns. It has spread rapidly from the industrialized to the developing threshold countries, who between 1990 and 2015

have increased their CO_2 emissions by 50 to 250 per cent (Brauch 2016b).

Between 2005 and 2010, a new research paradigm, 'sustainability transition', was developed by the *Dutch Knowledge Network on Systems Innovations and Transitions* (KSI), using "complex systems analysis, a socio-technical ... and a governance perspective" (Grin/Rotmans/Schot 2011: xvii). This Handbook aims to examine possible conceptual linkages between the proposed 'sustainability transition' paradigm, which is an *analytical concept*, and the primarily *normative debate* on 'sustainable peace'. So far, the conceptual linkages between a *process* of 'sustainability transition' and the *goal* of a 'sustainable peace' have not been systematically addressed.

Global research into 'sustainability transition' has been discussed at the annual conferences of the *Sustainability Transition Research Network* (STRN) since 2009 in Amsterdam,[8] while as a normative concept 'sustainable peace' is being used and addressed in UN circles (UN Women;[9] UNU [Keating/Knight 2004]) and by NGOs that focus on development (Busumtwi-Sam 2002), peacebuilding (Sending 2010) and conflict resolution (Peck 1998) and post-conflict situations (Roeder/Rothchild 2005). However, it has hardly been addressed in the framework of ecological, environmental and global change problems (Oswald Spring/Brauch/Tidball 2014).

This chapter attempts to contribute to the debate on conceptual bridge-building between 'peace research' and 'environmental studies' in general, and on the narrower assumed linkages between the impacts of GEC and GCC for humankind and its security and the process of 'facing' and 'coping' with those impacts through a process of 'sustainability transition' with the goal of avoiding conflict (Galtung's 'negative peace') and of achieving a 'sustainable peace' (Galtung's positive peace with justice). One long-term policy goal of such a 'sustainability transition' is a "low-carbon global economy", and this was endorsed in June 2015 by the heads of states and governments of the G7 in Elmau (Germany):

---

6 O'Brien, Sygna, Wolf (2013); O'Brien, St. Clair, Kristoffersen (2010); Matthews, Barnett, McDonald et al. (2010).

7 Only a few chapters in the so far three books of IPRA's *Ecology and Peace Commission* (EPC) have discussed these specific linkages: Oswald Spring, Brauch, Tidball (2014); Brauch, Oswald Spring, Bennett et al. (2016); Oswald Spring, Brauch, Serrano et al. (2016).

8 For the STRN, see at: <http://www.transitionsnetwork.org/>; and for the annual conferences at: <http://www.transitionsnetwork.org/events/conferences>; the *STRN-Newsletter* is available at: <http://www.transitionsnetwork.org/newsletter> and the STRN manifest is at: <http://www.transitionsnetwork.org/manifest>.

9 See UN Women at: <http://www.unwomen.org/en/what-we-do/peace-and-security> (1 December 2015).

We commit to doing our part to achieve a low-carbon global economy in the long-term including developing and deploying innovative technologies striving for a transformation of the energy sectors by 2050 ... To this end we also commit to develop long term national low-carbon strategies.[10]

This 'long-term' policy goal has been analysed in many publications by the OECD[11] and the UNEP.[12] The UNEP's International Resource Panel[13] has called for a 'decoupling' of sustainable economic growth from an increase in fossil energy consumption and in GHG emissions. This requires a better understanding of the concepts that influence scientists and policymakers who address problems of 'sustainable development' (WCED 1987), as well as a better understanding of 'green growth' and of a transition towards a 'low-carbon' economy and society during the twenty-first century.

According to the Summary for Policymakers of the Synthesis Report of the Fifth Assessment Report (AR5) of the *International Panel on Climate Change* (IPCC 2014: 1,4,6–7), there is general agreement among scientists and policymakers:

> Human influence on the climate system is clear, and recent anthropogenic emissions of greenhouse gases are the highest in history. Recent climate changes have had widespread impacts on human and natural systems. ... Warming of the climate system is unequivocal, and since the 1950s, many of the observed changes are unprecedented over decades to millennia. The atmosphere and ocean have warmed, the amounts of snow and ice have diminished, and sea level has risen. ... Anthropogenic greenhouse gas emissions have increased since the pre-industrial era, driven largely by economic and population growth, and are now higher than ever. This has led to atmospheric concentrations of carbon dioxide, methane and nitrous oxide that are unprecedented in at least the last 800,000 years.

This assessment supports Crutzen's claim that "we are now in the Anthropocene". The effects of this change in earth history caused by human interventions into the earth system have been observed, and they are gradually being understood by scientists and policymakers and by the population at large:

> Their effects, together with those of other anthropogenic drivers, have been detected throughout the climate system and are *extremely likely* to have been the dominant cause of the observed warming since the mid-20th century. ... In recent decades, changes in climate have caused impacts on natural and human systems on all continents and across the oceans. Impacts are due to observed climate change, irrespective of its cause, indicating the sensitivity of natural and human systems to changing climate. ... Changes in many extreme weather and climate events have been observed since about 1950. Some of these changes have been linked to human influences, including a decrease in cold temperature extremes, an increase in warm temperature extremes, an increase in extreme high sea levels and an increase in the number of heavy precipitation events in a number of regions.

Empirical evidence from the observed security impacts of climate change is still limited, and the linkage is contested among social scientists (Burke/Miguel/Satyanath et al. 2009; Hsiang/Burke/Miguel 2013; Buhaug 2010, 2014; Bernauer/Böhmelt/Koubi 2013; Theisen/ Gleditsch/Buhaug 2013); however, the strategic, military[14] and intelligence community (NIC 2008, 2012, 2012a) has taken into account considerations of climate in planning their medium- and long-term postures, so that they will be able to protect their infrastructure and operate their forces under these changed conditions.

Policymakers and science advisers in the UK have argued that, as a consequence of these assumed and projected impacts on security, humankind may face two new and closely related security threats during this century—from the impact of climate change and from declining resources. British Defence Secretary John Reid and Sir David King, a former science adviser to Prime Minister Tony Blair, mentioned the dual challenge of climate-induced and scarcity-driven violent resource conflicts. On 28 February 2006, John Reid warned, in a speech at Chatham House:

> that global climate change and dwindling natural resources are combining to increase the likelihood of violent conflict over land, water and energy. Climate

---

10 See G7 Leaders' Declaration, Schloss Elmau, Germany, 8 June 2015; at: <https://www.whitehouse.gov/the-press-office/2015/06/08/g-7-leaders-declaration> (14 August 2015).

11 See OECD on "green growth", at: <http://www.oecd.org/greengrowth/> (13 Nov. 2015); for publications on green growth, see at: <http://www.oecd. org/greengrowth/green-growth-key-documents.htm> (13 Nov. 2015).

12 See all recent UNEP publications, at: <http://www.unep.org/publications/> (13 Nov. 2015); see on the green economy at: <http://www.unep.org/greeneconomy/> (13 Nov. 2015).

13 See on this panel and its publications, at: <http://www.unep.org/resourcepanel/> (13 Nov. 2015).

14 Brzoska (2015); at: <http://www.emeraldinsight.com/doi/abs/10.1108/IJCCSM-10-2013-0114?journalCode=ijccsm&>.

change, he indicated, 'will make scarce resources, clean water, viable agricultural land even scarcer'–and this will 'make the emergence of violent conflict more rather than less likely'.[15]

Sir David King claimed in 2004 that during the twenty-first century climate-induced conflicts will be a bigger threat than terrorism;[16] in 2009 he argued that the US intervention in Iraq in 2003 was primarily driven by energy security concerns, given the reliance of the US "on foreign oil from unstable states".

> 'Unless we get to grips with this problem globally, we potentially are going to lead ourselves into a situation where large, powerful nations will secure resources for their own people at the expense of others.' ... 'I went into the White House in 2001 to persuade them that de-carbonising their economy was the way forward. I didn't get much shrift at that time. What I can tell you is that, if I had managed to persuade the government of America that investing (instead of going into Iraq) in de-carbonising their economy with roughly a tenth of [the estimated \$3 trillion the US spent on the war], they would have managed it.' ... King summed up by saying that with growing population and dwindling resources, fundamental changes to the global economy and society were necessary.[17]

In the aftermath of the dual natural geophysical hazard (an earthquake triggering a tsunami) that hit Fukushima on 11 March 2011 and caused the meltdown of several nuclear reactors, the 'cheap' answer to anthropogenic climate change–by moving to nuclear energy–was challenged. This had a negative effect on German politics, and a few months later the German government decided to move out of nuclear energy. Meanwhile Japan–after a temporary shutdown of most of its reactors–only three years later, during the Abe administration, resumed the use of its nuclear reactors for electricity production.

The reaction of society and of the political world to the scientific assessments of the IPCC and the

hydro-meteorological hazards triggered by both climate change and by geophysical natural disasters has been significantly different in the major democracies. Australia, New Zealand, Spain, Canada, Portugal, Japan and the US all missed their national reduction targets under the UNFCCC's Kyoto Protocol. On the other hand, Russia, all the former Socialist countries, the UK, Germany, the EU, France and Italy stuck to their commitments and even over-performed, for different reasons.[18] So far, only the twenty-eight EU countries have agreed on a joint strategy aiming at a low-carbon economy and society by 2050 (EU 2011; Brauch 2016: see chapter 42 by Happaerts in this volume); but at least the goal was endorsed by the US President and by the prime ministers of Japan and Canada during the G8 conferences between 2007 and 2011.

Is there a linkage between the two perceived new security threats and the fundamental transformation of the economy and society suggested by the proponents of the 'sustainability transition' research paradigm, with the normative goal of a sustainable peace? The observed anthropogenic changes in the earth's climate system have triggered physical impacts (temperature increase, sea-level rise, precipitation changes and increases in extreme weather events) and societal outcomes (migration, crises and conflicts), and these have increasingly posed environmental challenges that may become 'human' and also 'national' and 'international security problems and concerns' (Brauch 2002, 2009; WBGU 2008; UNSG 2009; Scheffran/Brzoska/Brauch et al. 2012; Gleditsch 2012, IPCC 2014). This anthropogenic change in earth history (Steffen/Grinevald/Crutzen et al. 2011) requires a reconceptualization of the four key concepts of peace, security, development and environment, as well as of the linkages between 'sustainable development' (WCED 1987), 'human security' (UNDP 1994; CHS 2003) and 'sustainable peace' (Peck 1998), in the framework of the Anthropocene.

A heuristic research question is whether and to what extent a transition towards sustainability may–in some future scenarios–help prevent the two related climate-induced and resource scarcity-driven violent

---

15 Michael T. Klare: "The Coming Resource Wars–America's closest ally has announced that climate change has ushered in an era of violent conflict over energy, water and arable land", in: *Alternet* (9 March 2006); at: <http://www.alternet.org/story/33243/the_coming_resource_wars>.

16 Steve Connor: "US Climate Policy Bigger Threat to World than Terrorism", in: *Global Policy Forum* (9 January 2004); at: <https://www.globalpolicy.org/global-taxes/45219.html>.

17 David King, 2009: "UK's ex-science chief predicts century of 'resource' wars", in: *The Guardian*, 133 February; at: <http://www.theguardian.com/environment/2009/feb/13/resource-wars-david-king>.

18 See: "Changes in greenhouse gas emissions excluding Land Use, Land Use Change and Forestry (LULUCF) from 1990 to 2012 in percent". Source: UNFCCC (2014), at: <http://unfccc.int/ghg_data/ghg_data_unfccc/items/4146.php>; with data without LULUCF <http://unfccc.int/files/inc/graphics/image/jpeg/total_excl_2014.jpg> and with LULUCF <http://unfccc.int/files/inc/graphics/image/jpeg/total_incl_2014.jpg> (13 November 2015).

conflicts (as suggested by Sir David King) and whether such a process may thus be beneficial for achieving the goal of sustainable peace in the Anthropocene. Since our experience is that the two previous agricultural and industrial revolutions have resulted in a revolution in warfare and more deadly conflicts (Brauch 2016), such a thought experiment is necessary so that the context of the research on sustainability transitions can be broadened in order to take 'peace and security' considerations in the Anthropocene into account.[19]

This chapter is structured in ten parts. After this introduction, the Anthropocene is framed in the context of historic times (9.2), and the dual thesis is developed that we are both the 'threat' but also the 'solution', if we can aim for policy constellations that prevent new environmental and climate conflicts from occurring (9.3). After a brief summary of the conceptual quartet (9.4), the linkages between these four key concepts are discussed as 'sustainable development' (9.5), 'human security' (9.6), and as 'sustainable peace' in the Anthropocene (9.7). These conceptual considerations are further explored in the framework of two new proposed scientific approaches: a wider 'political geoecology' (9.8) and a 'peace ecology' (9.9). The chapter finally offers hypothetical considerations on linkages between 'sustainability transition' and 'sustainable peace' in the Anthropocene (9.10).

---

19 In a comment to an earlier draft of this chapter Kalevi Holsti argued that the author wants "to marry peace research with environmentalist thought in order to provide a conceptual framework for developing systematic thought about 'sustainability transition'. ... The road to sustainability transition is blocked by traditional Hobbesian security thinking, by continued blockage of meaningful GHG-emission reduction policies, by subsidies to oil and other carbon industries, by lack of political will, by industry lobbies, and the like." Holsti is "somewhat more pessimistic ... on future possibilities" noting that most of the literature ignores "the demand side of the equation". He claims that our epoch is dominated by the philosophy of possessive individualism", where "we focus on the consequences of this philosophy rather than its causes". And he continues "the optimists argue that technological advances can resolve the incompatibility between our prevailing definitions of the 'good life' and its consequences on the biosphere. ... The other sources of international conflict seem to be much stronger."

9.2   Historical Times: We are Living in the Anthropocene

In the natural and social sciences six historical times are distinguished. These have triggered numerous contextual changes of different duration. Elsewhere, the author has examined past long-term transformations, turning points and transition processes, focusing specifically on the Anthropocene (Brauch 2016). In earth and human history, five historical times may be distinguished. This argument is inspired by Braudel's (1969, 1972) three historical times: the history of structures, of repetitive cycles and of events, as well as by Crutzen's reference to the Anthropocene.

As has been argued elsewhere, the peaceful global change of 1989 was not a victory for the military superiority of the West (Schweitzer 1994), but rather the result of a 'learning' (Grunberg/Risse-Kappen 1992) by the new Soviet leadership that triggered unintended results (Brauch 2016). Since the end of the Cold War, 'transition studies' has emerged in political science, focusing on the transition of state socialist political, economic and societal systems towards Western 'neoliberal' market economies (Wolfgang Merkel 1996).

However, the discourse on 'sustainability transition' fundamentally differs from that on 'transition studies'. It addresses transformations in scientific, societal, economic, and political systems, together with a radical cultural transformation that can prevent catastrophic changes in climate, soil and water and major losses in biodiversity. The 'industrial revolution' and the many technological innovations led to a major human intervention into the earth system with an exponential increase in *greenhouse gases* (GHG) in the atmosphere, something that had not occurred in millions of years of climate variability (IPCC 1990, 1996, 2001, 2007, 2013).

The emerging scientific debate on 'sustainability transition'[20] addresses the many needs to reduce GHG emissions. There are both *multilateral* obligations to reduce emissions quantitatively, and a need for *unilateral* transformations. But since the UNFCCC and its Kyoto Protocol did not achieve their goals, due to a lack of political will to implement these obligations (Brauch 2012), unilateral strategies to reduce both GHG emissions and energy intensity[21] have become more important as a coping strategy for facing the impacts of global environmental and climate change. (see chapter 1 on the outcome of COP 21 in Paris).

Taking a sociotechnical perspective on "the dynamics of transitions", Geels and Schot (2010: 11-12) described five characteristics of transitions:

- Transitions are co-evolution processes that require multiple changes in sociotechnical systems and configurations. ...
- Transitions are multi-actor processes, which entail interactions between social groups such as businesses, firms, different types of user groups, scientific communities, policymakers, social movements and special interest groups.
- Transitions are radical shifts from one system or configuration to another. The term 'radical' refers to the scope of change, not to its speed....
- Transitions are long-term processes (40–50 years); while breakthroughs may be relatively fast (e.g. 10 years), the preceding innovation journeys through which new sociotechnical systems gradually emerge usually take much longer (20–30 years).
- Transformations are macroscopic. The level of analysis is that of the 'organizational fields'.

In Geels' and Schot's view, Braudel's historical times offer a general heuristic "for studying long-term processes, multi-causality, co-evolution, lateral thinking, anti-reductionism, patterns, context and the use of different time scales" (Geels/Schot 2010: 15). In their dynamic model they distinguish between *niches*, where the radical innovations occur; the sociotechnical *landscape* that offers the exogenous context; and the middle level of the sociotechnical *regime*, where the innovations have to succeed and overcome multiple obstacles.[22]

This Dutch research project influenced a policy report from the *German Advisory Council on Global Change* (WBGU 2011) on a 'Social Contract for Sustainability' (2011). This argued that the transformation

to a low-carbon society requires an 'accelerated pace of innovations' but there is also a need to overcome the hurdles, including cutting subsidies for fossil energy sources and making use of favourable trends in society. "Sustainable development means more than climate protection [and] include[s] many other natural resources, such as fertile soil and biological diversity". The WBGU (2001:93) adapted Geels's model and added several megatrends in both the Earth System (climate, biodiversity, land degradation, water, raw materials) and the Human System (development, democratization, energy, urbanization, food), where innovative changes in the regime may directly affect the megatrends. The WBGU report stated that

> [t]he transformation into a sustainable society requires a modern framework to allow ... almost nine billion people to lead 'the good life', both in terms of living with each other, and living with nature: a new *Contrat Social* ... [that] represents a special agreement between science and society. ... It is also about a new culture of democratic participation. ... A low-carbon transformation can only be successful if it is a common goal, pursued simultaneously in many of the world's regions. Therefore, the social contract also encompasses new ways of shaping global political decision-making and cooperation beyond the nation state.[23]

The WBGU (2011: 5) discussed the global "remodelling of economy and society towards sustainability" with the goal of changing "production, consumption patterns and lifestyles in all three key transformation fields ... in such a way that global greenhouse gas emissions are reduced to an absolute minimum over the coming decades, and low carbon societies can develop." The social dynamics for a change in the direction of climate protection requires *a knowledge-based strategy*, a *proactive state* to establish a relevant framework aiming for structural change by guaranteeing the implementation of climate-friendly innovations, and it also counts on the *cooperation of the international community* and on global governance

---

20 See the Sustainability Transition Research Network (STRN); at: <http://www.transitionsnetwork.org/> (22 January 2015), which focuses on sustainability problems in the energy, transport, water and food sectors from different scientific perspectives on the ways: "in which society could combine economic and social development with the reduction of its pressure on the environment. A shared idea among these scholars is that due to the specific characteristics of the sustainability problems (ambiguous, complex) incremental change in prevailing systems will not suffice. There is a need for transformative change at the systems level, including major changes in production, consumption that were conceptualized as 'sustainability transitions'."

21 According to a Special Report by *The Economist* on Energy and Technology (17 January 2015), China had the highest increase in the consumption of fossil energy but also the highest decline in energy intensity between 1990 and 2012 and spent over $56 billion on renewables in 2013, more than all of Europe.

22 The figure is available in Geels and Schot (2007), at: <http://www.sciencedirect.com/science/article/pii/S0048733307000248> and at: <https://www.google.com/search?q=Geels%2C+Schot+2007&ie=utf-8&oe=utf-8>; and Geels (2012); at: <http://community.eldis.org/.5ad501d7/Geels%202012%20corrected%20proofs.pdf>.

23 See WBGU, Press Release on: "Low-Carbon Economy and Sustainable Development: A Social Contract for Sustainability"; at: <http://www.wbgu.de/fileadmin/templates/dateien/presse/presseerklaerungen/downloads/wbgu_presse_20110407_engl.pdf>.

structures as the indispensable driving force for the transformation (WBGU 2011: 5-6).

Both the Dutch project and the WBGU Report (2011) avoided any contextualization of their theoretical approach and their policy proposal in the international foreign and security context and there was no discussion of the two related concepts of 'human security' and 'sustainable peace' in the Anthropocene era of earth history.

9.3   Not 'They' but 'We' are the Threat and Solution

In strategic studies, the 'other' country, with its military forces, economic potential and ideological attractiveness, was perceived as 'the threat'. With globalization and the emergence of 'new wars' (Kaldor 1999, 2002, 2012, 2013) with an 'asymmetric' and often 'invisible' opponent in the form of terrorists, drug cartels, or organized crime, this logic has not changed during the US-led 'war on terror'. But with the emergence of the global debate on the security implications of global environmental and climate change, a totally different type of threat has been identified, where human beings as producers and consumers pose a threat that cannot be defeated by military means. Georg Boomgarden (2007), a state secretary of the German foreign ministry, argued:

> 'If we ask ourselves who the enemy is in climate change, using the concepts of classic security policy, we must conclude that we are turning nature itself into an enemy'. ... 'And with this enemy, neither deception nor deterrence is going to be of any use. The later we adapt, the greater the cost will be'. ... Avoiding security-relevant cataclysms of global extent required the course to be set today. The time window for possibly irreversible processes to occur as a result of global temperatures rising by more than two degrees compared to pre-industrial days was about to close.

The *German Advisory Council on Global Change* (WBGU 2008) argued in a previous report on *Climate Change as a Security Risk* that "without resolute counteraction, climate change will overstretch many societies' adaptive capacities within the coming decades. This could result in destabilization and violence, jeopardizing national and international security to a new degree". But a positive outcome may also be possible if the international community "recognizes climate change as a threat to humankind and soon sets the course for the avoidance of dangerous anthropogenic climate change by adopting a dynamic and globally coordinated climate policy".

The report described probable new conflict constellations "as typical causal linkages at the interface of environment and society, whose dynamic can lead to social destabilization and, in the end, to violence": a) climate-induced degradation of freshwater resources; b) climate-induced decline in food production; c) climate-induced increase in storm and flood disasters; and d) environmentally-induced migration. The WBGU identified "six key threats to international security and stability which will arise if climate change mitigation fails": 1) a possible increase in the number of weak and fragile states as a result of climate change; 2) risks to global economic development; 3) risks of growing international distributional conflicts between the main drivers of climate change and those most affected; 4) the risk to human rights and the industrialized countries' legitimacy as global governance actors; 5) the triggering and intensification of migration; and 6) the overstretching of classic security policy. In the WBGU's view, "climate policy ... becomes preventive security policy, for if climate policy is successful in limiting the rise in globally averaged surface temperatures to no more than 2°C relative to the pre-industrial value, the climate-induced threat to international security would likely be averted".

When the Norwegian Nobel Committee awarded the Nobel Peace Prize of 2007 to both the IPCC and to Al Gore it justified its choice by stating:

> Extensive climate changes may alter and threaten the living conditions of much of mankind. They may induce large-scale migration and lead to greater competition for the earth's resources. ... There may be increased danger of violent conflicts and wars, within and between states. ... The Norwegian Nobel Committee is seeking to contribute to a sharper focus on the processes and decisions that appear to be necessary to protect the world's future climate, and thereby to reduce the *threat to the security of mankind*.

In its Fifth Assessment Report (AR5), the IPCC (2014) assessed the literature on climate change and human security (Gleditsch 2015; Brauch 2016).

In the Anthropocene, with the intervention of humankind into the earth system and into nature, 'we are the threat' that is posed by 'our' economic behaviour and by the prevailing production, transportation and consumption processes that rely heavily on the burning of hydrocarbons for the production of our food, goods and services and for our movement.

If 'we are the threat', then a sustainable peace policy in the Anthropocene has to address these multiple causes that are unrelated to classic security considerations. Such a 'sustainable peace' policy must deal with the obstacles to policies that aim for a sustainability transition or a great transformation of major economic sectors; but it must also address our values, perceptions, world views, and mindsets, all of which influence our economic, societal, political and consumptive behaviour. It also requires major changes in our agricultural, economic, housing, transportation, and environmental policies, with the aim of a gradual dematerialization and decarbonization of the energy and other key economic sectors that have been major producers of greenhouse gases. This challenges the Hobbesian thinking that relies on military solutions whose aim is the control of those regions and countries that have the largest reserves of fossil energy sources.

9.4 'Conceptual Quartet' in the Anthropocene: Peace—Security—Development—and the Environment

Of the four concepts of a 'conceptual quartet' (Brauch 2008), peace and security are key goals of the UN Charter. 'Development' and 'environment' were added to the political agenda in the 1950s and 1970s. While these concepts have been widely used in the social sciences, systematic conceptual analyses of these concepts and their linkages are rare.

In the Anthropocene there is a need to reconceptualize these four key concepts and the related linkage concepts of 'sustainable development' (1987), 'human security' (1994) and 'sustainable peace' (1998). These three linkage concepts are considered as conceptual pillars of two proposed research programmes of a "political geoecology" and of a "peace ecology" under the specific conditions of the Anthropocene, the new era of earth history we have silently entered.

9.4.1 The Peace Concept

The English word 'peace' originates from the Latin 'pax'. The German word 'Frieden' indicates a 'condition of quietness, harmony, resolution of warlike conflicts', and in Russian 'mir' means both 'peace' and 'the world'. While the Greek *eirene*, the Hebrew *shalom*, and the Arab *salām* all imply 'peace with justice', the Hindi *ahimsa* adds the ecological dimension.[24]

Plato argued that war and conflicts should be avoided within the *polis*. Aristotle stressed that all political goals can only be realized under conditions of peace, and war is only acceptable as a means for the defence of the polis. During the Roman period, 'pax' was closely tied to law and contracts, where the

pax Romana relied on subjugation under the emperor in exchange for protection against external intruders. Besides 'peace within the state' that was achieved through its monopoly of the means of force, 'peace between and among states' has been a major concern of international law since the sixteenth century. Authors have still considered war a legitimate means for the realization of interests between states (*ius ad bellum*), while calling for constraints during war (*ius in bello*). In his treatise for an *eternal peace* (1795), Kant proposed a ban on war itself and developed a legal framework for a permanent peace based on six preliminary and three definite articles that called for a democratic system of rule, a league of nations, and respect for human rights.

After World War I, the Kantian tradition influenced the creation of the League of Nations, while after World War II, the United Nations gained teeth with the Security Council. However, during the Cold War a bipolar power system dominated, with nations relying on military alliances instead of on the 'collective security' of the UN Charter. With the end of the Cold War, 'new wars' emerged as resource, ethnic, and religious conflicts, primarily within states but also as pre-emptive wars. During the 1990s proposals for a new international order were gradually replaced by power-driven concepts of preventive wars and the 'war on terror'.

Detached from these political contexts, peace has been defined as a 'basic value' and as a 'goal of political action', as a situation of non-war, or as a utopia of a just and sustainable world. Galtung (1967, 1968, 1969; Galtung/Fischer 2013) distinguished between 'negative peace' (absence of physical or personal violence) and 'positive peace' (absence of structural violence, repression, and injustice), taking into account the methods of "economic exploitation and/or political repression in intra-country and inter-country class relations". He distinguished negative, indifferent and positive relations that often result in *negative peace* (absence of violence, cease-fire, indifferent relations) or *positive peace* (harmony).[25]

In the UN Charter of 1945, the 'concept of peace' is noted as its key mission in Art. 1,1: "to maintain

---

24 For a discussion of the "meanings of peace" that includes the philosophical debate in China (including Lao Tzu, Confucius, Mo Tzu), and in Buddhist, Hindu, and Judaeo-Christian thinking with the goal of "achieving positive peace", see Webel, Barash ([2]2008), part IV on "Building Positive Peace", in a chapter on "Ecological Well-Being" (397–418); see also Oswald Spring (2008a).

international peace and security", and "to take effective collective measures for the prevention and the removal of the threats to the peace, and for the suppression of acts of aggression or other breaches of the peace", as well as peaceful conflict settlements. Wolfrum (1994: 50) pointed to both narrow and wide interpretations of peace in the UN Charter. In Art. 1(2) and 1(3) the Charter uses a wider and more positive peace concept when it calls for the development of "friendly relations among nations" and for "international cooperation in solving international problems of an economic, social, cultural, or humanitarian character." In chapter VI on the Pacific Settlement of Disputes, Art. 33 uses a 'negative' concept of peace that is "ensured through prohibitions of intervention and the use of force" (Tomuschat 1994: 508). In Chapter VII of the UN Charter dealing with "Action with Respect to Threats to the Peace, Breaches of the Peace, and Acts of Aggression", in Art. 39, a 'negative' concept of peace dominates, with a reference to "the absence of the organized use of force between states". In the framework of Chapter IX on "International Economic and Social Cooperation", Art. 55 (3) refers to "universal respect for, and observance of, human rights and fundamental freedoms." It has been suggested that "the right of self-determination, to peace, development, and to a sound environment" (Partsch 1994: 779) should be incorporated as "human rights of the third generation" (Vasak 1984: 837).

Thus, in the UN Charter of June 1945, a narrow or 'negative' concept of peace has been at the centre with a few direct references to 'positive' aspects to be achieved by 'friendly relations among nations', and by 'international cooperation'. The 'positive peace' concept indicates peaceful social and cultural beliefs and norms, the presence of economic, social and political justice and a democratic use of power including nonviolent mechanisms of conflict resolution. 'Sustainable peace' or 'peace with nature' was added later to the debate in the UN.

Inspired by the 'ahimsa' concept, Mahatma Gandhi's thinking had a significant impact on Arne Næss's

(1973, 1986, 1989) environmental philosophy and his 'deep ecology', and on Schumacher's (1999) 'small is beautiful' philosophy. Within peace research, only a few scholars addressed environmental challenges, among them Kenneth and Elise Boulding.[26] In *The Economics of the Coming Spaceship Earth* (1966, 1970),[27] Kenneth Boulding used the symbol of cowboys arguing that the world of unlimited resources is coming to an end, and in *Ecodynamics* (1978)[28] he combined evolutionary biology, ecology, peace research and Keynesian, socio- and environmental economics (Khalil 1996). Later, Boulding (1983) developed his concept of the 'empty niche' in biology, societal evolution and artefacts, biological catastrophes, ecological and human interactions, social ecosystems and finally, evolutionary economics. Elise Boulding co-edited a book on climate change (Chen/Boulding/ Schneider 1983) and linked peace to ecology, insisting that there is no true peace without ecological links, respect for nature and human ecology (Morrison 2005). The concepts of 'environmental peacebuilding' (Conca 2002) and 'peace ecology' (Kyrou 2007; Oswald Spring/Brauch/Tidball 2014; Amster 2014) later contributed to this bridge-building.

In the US peace research and international relations community Patrick M. Regan (2014) and Paul F. Diehl (2016), in their presidential addresses to the Peace Studies Society (2013) and the International Studies Association (2015), called for "Bringing peace

---

25 Johan Galtung commented on an earlier draft: "To me both sustainable peace and environment are guided by two basic deep structures: diversity and symbiosis; hence very dynamic. To impose one culture, Western, one structure, capitalism will lead to the collapse of both; cultures-structures in partnership and species, abiota and biota in symbiosis will lead to higher complexity, evolution–that then has to be watched but is promising. What I read seems to me very compatible with these simple propositions."

---

26 Russell Boulding, the oldest son of Elise and Kenneth Boulding, wrote to this author on 10 February 2015 that the oldest comment of his father on environmental issues "was his poem *A Conservationist's Lament; The Technologist's Reply* which he wrote during the ... 1955 symposium on *Man's Role in Changing the Face of the Earth.* [His] 1966 paper *Economics of the Coming Spaceship Earth* is probably the most often cited paper ..., but he laid some theoretical foundations for an approach to sustainable economics in *A Reconstruction of Economics* (Boulding 1950) and even earlier critiques of mainstream economic theory. ... Robert Scott's [2015] biography ... discuss[es] the 1950 book and also the 1966 paper and its significance/impact on the development of ecological economics. ... Scott notes that [his] first actual reference to Spaceship Earth was in a speech given in May 1965 titled *Earth as a Space Ship* presented to the Committee on Space Sciences and Washington State University. [In] April 1965 he presented a paper *Economics and Ecology* at a Conservation Foundation Conference on "Future Environments in North America" (1966).

27 See: <http://www.jayhanson.us/page160.htm>.

28 For a summary by Tanya Glaser, see at: <http://www.colorado.edu/conflict/peace/example/boul7525.htm>.

back in" and for "Looking beyond war and negative peace", thus returning to suggestions put forward by Quincy Wright (1942, 1954) and Johan Galtung (1964, 1967, 1985, 2012) decades earlier. Neither Regan nor Diehl addressed environmental challenges, nor did they try to recontextualize peace in the framework of the Anthropocene.

Environmental issues were not discussed in Peter Wallensteen's (2015) book on *quality peace*, but he mentioned climate change twice, writing about the present post-Cold War period as "a time that soon will be described as an imprudent spending spree before the Great Global Climate Crisis, similarly to how the 1920s is now seen as the irresponsible period before the Great Depression" (p. 61) and as a source of new cooperation over dwindling resources, arguing that scarcity may give rise not only to conflict but also to cooperation (p. 193).[29]

In the framework of the 'silent transition' in earth history from the Holocene to the Anthropocene, the peace concept must be reconceptualized, and the goals of peace studies and peace research courses and degrees must be developed further so that they can reflect on the impacts of our own productive and consumptive behaviour as a cause of global environmental change and thus also of its potential violent consequences. The normative goal of 'peace with nature' in a process of transition towards more sustainable development strategies of decarbonization and dematerialization must be addressed. The first step is one of agenda-setting.

9.4.2 The Security Concept

The reconceptualization of security in the twenty-first century[30] has gradually evolved since the end of the East-West conflict (1989-1991) and has been influenced by the end of the Cold War, processes of globalization, and the emerging impacts of global environmental change. Since the late 1970s, an expanded concept of security has been discussed in academia (Krell 1981; Buzan 1983). In the policy debate, the 'security concept' has gradually widened since the late

1980s. Ullman (1983), Mathews (1989) and Myers (1989, 1993) put environmental concerns on the US national security agenda. Since the 1990s, many European governments have adopted an extended concept of security.

Buzan, Wæver and de Wilde (1998) distinguished between wideners and traditionalists, focusing on the primacy of a narrow military concept of security (Walt 1991; Chipman 1992). The Copenhagen School (Wæver 1995, 1997; Buzan/Wæver/de Wilde 1998) distinguished five dimensions (military, political, economic, societal and environmental) and five referent objects or levels of analysis (international, regional, national, domestic groups, individual). But they did not review the sectorialization of security (Brauch 2009a) from the perspective of national, international and human security (Brauch 2009). Others referred to five vertical levels of security analysis: global or planetary (Steinbruner 2000), regional (Buzan/Wæver 2003), national (Tickner 1995), societal (Møller 2003; Wæver 2008) and human (UNDP 1994; CHS 2003) security.

In the post-Cold War era, within the UN, NATO and the EU, different security concepts have coexisted: a state-centred political and military concept, and an extended security concept with economic, societal, and environmental dimensions, while some countries have adhered to a narrow concept of national security that emphasizes the military dimension.

Although since the Westphalian order (1648)–and especially since the nineteenth century–the key security 'actor' has been the 'state' (Holsti 1991, 2016), it has not necessarily been the major 'referent object', which is often referred to as 'the people', whose survival is at stake. A major debate has evolved since the late 1980s over whether the state as the key referent object ('national security') should be extended to include the people (individuals, humankind) as 'human security'. Buzan (1991) distinguished between the international, state and individual levels of analysis and emphasized the inherent tension between the latter two, but he remained critical of the human security approach (Buzan 2004a).

While the classical means and instruments of a narrow security policy have remained the prerogative of the military and diplomacy, in the EU this classical *domaine réservé* of the nation state is changing with common policies and strategies, and increased common voting in international institutions (UN, OSCE). In many international regimes (food, climate, desertification), the EU is a full member alongside its twenty-eight member states. Its common *European Foreign*

---

29 The author is grateful to Peter Wallensteen for this reference (by email on 18 November 2015).

30 This section relies on Brauch, Oswald Spring, Grin et al. (2009) and on Brauch, Oswald Spring, Mesjasz et al. (2009, 2011). For an overview of the debate on the reconceptualizing of security, see Brauch (2005, 2008, 2009a, 2011, 2012), and from a peace research perspective, see Albrecht and Brauch (2008).

and Security Policy (CFSP) and *Security and Defence Policy* (ESDP) have affected the traditional national military and diplomatic leverage.

Within international organizations sector-specific security concepts are widely used, such as 'environmental security', 'food security', 'global health security', 'energy security', and 'livelihood security'. The drivers of the theoretical security discourse and the centres of conceptual innovation have shifted away from the United States. During the 1980s, conceptual thinking on 'alternative security' in Europe searched for alternatives to mainstream deterrence doctrines and nuclear policies (Weizsäcker 1972; Brauch/Kennedy 1990, 1992, 1993; Møller 1991). By 2008, discourses on security were no longer a primarily American social science (Crawford/Jarvis 2001; Hoffmann 2001). The American national security concept was challenged by alternative security experts in Europe from the 1970s onwards and also by new national perspectives during the 1990s, when geopolitics re-emerged in southern Europe (France, Italy, Spain) and elsewhere. Since 1990, new centres of conceptual innovation have emerged and new journals on security problems have been launched.

- In Europe, *Aberystwyth*, *Paris*, and *Copenhagen* are associated with critical 'schools' of security theory (Wæver 2004).
- The *human security concept* was promoted by Mahub ul Haq (UNDP 1994), and developed further by the *Human Security Commission* (CHS 2003) co-chaired by Sadako Ogata and Amartya Sen.
- Civil society organizations in South Asia developed the concept of *livelihood security*.
- International organizations introduced the sectoral concepts of *energy, food, water, health,* and *soil security*.
- In the US, Canada and Switzerland the *environmental security* concept and research into it developed during the 1990s.
- The *Earth System Science Partnership* (ESSP) and its four programmes: IHDP (*International Human Dimensions Programme*), IGBP (*International Geosphere-Biosphere Programme*), WCRP (*World Climate Research Programme*) and *Diversitas* resulted in global scientific networks that are addressing new security problems.

These new centres of conceptual innovation challenged the narrow state-focused security concept of realists during the Cold War. But the world view of experts working on war, security and strategic studies still determines the political mindsets of decision-makers in the post-Cold War era (Albrecht/Brauch 2008). From 1989, in environmental security research, three phases have been distinguished (Dalby/Brauch/Oswald Spring 2009) and a fourth phase has been suggested (Oswald Spring/Brauch/Dalby 2009).

In the Anthropocene, the new era of earth history, the goals of the security concept must again be reconceptualized and the tasks of security policy must be redefined. If the 'other' nation, its weapons and ideologies are not the 'sole' or 'major' threat to the survival of humankind any longer and 'we' have become a new non-military threat to our own survival, security must increasingly reflect on these new challenges besides the traditional threats and those associated with the 'new wars' and 'terrorism' of the post-Cold War era. Simon Dalby[31] has pioneered the debate on a reconceptualization of security in the Anthropocene, which implies for him that:

> We have to shift our understanding of the human place in the cosmos to one of us as an active agent in remaking the biosphere, not a passive spectator in environmental matters. Threats to modernity are internal, generated by fossil fueled modes of economy, not by distant peripheral 'others' with evil intentions. Anthropocene security has to start from this reversal of traditional geopolitical premises and think how urban civilization can be quickly rebuilt so we can all live well without burning things to do so, and before we disrupt the climate system to such an extent that civilization can't cope.[32]

Addressing, understanding, facing and coping with the new anthropogenic causes of insecurity requires major scientific and political efforts. However, national and international politics are increasingly influenced by a revival of nationalist ideologies and radical fundamentalist religious belief systems and ideologies that are being exploited by terrorist groups who want to re-establish an Islamic State in a new caliphate in Syria and Iraq, and in parts of Africa. Hence security in the Anthropocene must analyse traditional threats, those posed by the 'new wars', and also long-term human-induced environmental challenges triggered, caused or influenced by human interventions into the Earth system during the Anthropocene.

---

31 Dalby (2007, 2009, 2013, 2013a, 2014, 2015, 2015a, 2016); Brauch/Dalby/Oswald Spring (2011).

32 This brief summary was provided to this author by Simon Dalby (in an email on 14 November 2015) and is cited here with his permission.

9.4.3 The Development Concept

The concept of 'development' is defined as "1. the act or process of growing or developing; 2. the product of developing; 3. a fact or event, especially one that changes a situation; 4. an area of land that has been developed" (McLeod 1985: 305). The *Shorter Oxford English Dictionary* (⁵2002: 662) adds: "Economic advancement or industrialization. In economics, development is defined as a synonym for economic growth. The term is also used for the improvement of the living conditions that includes besides the standard of living also social indicators, and aspects of distribution (income, public goods, and infrastructure)".

Ake (1993: 239–243) stated that after World War II, development theory emerged during decolonization as a variant of modernization theory. However, these theories "were at best heuristic devices" that were "too general and too vague to be taken seriously as scientific theories and paradigms" because "their major concepts could not be operationalized and their empirical referents were unclear". Toye (1996: 212–215) observed that by 1965 "prolonged and steady increase of national income" was identified as an indicator of economic development. It is accompanied by rapid population growth due to declining mortality, longer life expectancy, rapid urbanization, and improved standards of literacy and education. The identification of these indicators has been criticized if the distribution of income remains unequal and if the population majority remains impoverished. Some have claimed that "indicators of economic growth and structural change must be complemented by indicators of improvement in the quality of everyday life for most people". Sen (1981, 1984, 1994, 1999, 2000, 2000a) argued that distribution of income should be complemented by a fair distribution of entitlements to food, shelter, clean water, clothing and household utensils. These definitions excluded environmental factors contributing to and constraining economic development, especially natural hazards and disasters.

The policy goals of development differed between the industrial (OECD, G7, G8) and developing countries (G77 and China). During the Cold War, in a bipolar world, these goals were closely associated with economic systems. The goals differed on import-substitution or export-led industrialization, and capital- or labour-intensive strategies. Stallings (1995) used this concept primarily for economic development, i.e. for growth and equity of distribution. He pointed to five new elements in the new international context for development since 1990: "the end of the Cold War, new relations among advanced capitalist powers, increased globalization of trade and production, shifting patterns of international finance, and new ideological currents" (Stallings 1995: 2).

Decolonization and global competition between rival systems and modes of production predominated during the Cold War where development aid was an instrument of global strategic policy–the geo-strategic and geo-economic importance of developing countries was rewarded with economic and military assistance. Development assistance was supplied by governments, the EU, international organizations, financial institutions (e.g. the World Bank Group, the EIB) and development banks (Asian, African, Latin American and Islamic development banks), and by non-governmental economic, societal, and humanitarian organizations. Since 1990, development assistance from OECD countries as a percentage of their GDP has dropped. There was no peace dividend after the end of the Cold War and the geo-strategic importance of several developing countries declined, as did the security-motivated economic aid, and this contributed to weak, failing or failed states.

Development research emerged after World War II as an objective of social and political science that focused on the preconditions for and features of development processes, especially the economic, social, political and cultural factors that enhance or restrain development. Later the goals of development and the causes of underdevelopment were added. Two main theories of modernization emerged, to be used by scientists in OECD countries, and critical approaches were influenced by theories of imperialism, *dependencia*, self-reliance, and autocentric development.

With the end of the Cold War a crisis in development theories was noted (Boeck 1995: 69–80). Development theories and strategies for poverty eradication, and social and sustainable development are linked to the state, market, community, and civil society (Kothari/Minougue 2002: 1–15).

During the 1950s and 1960s most development experts stressed 'economics first' through investment-driven economic development strategies with a focus on industrialization. Since 1980, the focus has shifted to poverty and development and a basic human needs approach (Boserup 1970; Sen 1981; W. Arthur Lewis and Richard Jolly;[33] Burton 1998[34]). This was reflected in an upgrading of poverty eradication programmes, but until 1985 there was no emphasis on issues of governance, development of social capital, or institution-building and capacity-building for self-reliance. During the 1990s there was a gradual shift to agriculture, gender issues, and participatory community development so that people were put first, as reflected in the Human Development Reports, which introduced 'human security' (UNDP 1994) as a complement to 'human development'.

With the concept of 'sustainable development', the Brundtland Commission (WCED 1987) has initiated a global scientific and political debate that has reflected on the impacts of economic development on the environment (see section 9.5 below; chapters 1 and 18 by Mesjasz). With the emergence of global environmental and climate change as major human challenges from 1988 and on through the 1990s and since 2000, in the Anthropocene era of earth history, the scope of the debate on sustainable development has narrowed to green growth, green economy, dematerialization, and decarbonization aiming at a low-carbon economy resulting from a process of sustainability transition.

9.4.4   The Environment and Ecology Concept

In English dictionaries the terms 'environment' and 'ecology' have been given many different meanings. The definition of 'ecology' in the *Shorter Oxford English Dictionary* (⁵2002: 789) is: "1. The branch of biology that deals with organisms' relations to one another and to the physical environment in which they live; (the study of) such relations as they pertain to a particular habitat or a particular species; also human ecology; 2. The political movement that seeks to protect the environment, esp[ecially] from pollution." According to *Webster's Third New International Dictionary* (2002: 720) 'ecology' is: "1. a branch of science concerned with the interrelationship of organisms and their environments especially as manifested by natural cycles and rhythms, community development and structure, interaction between different kinds of organisms, geographic distributions, and population alterations; 2. the totality or pattern of relations between organisms and their environment; 3. human ecology."

While the term 'environment' has many meanings, the scientific concept has been more specific. The *Encyclopaedia Britannica* (1998, IV: 512) has defined

'environment' as: "the complex of physical, chemical, and biotic factors that act upon an organism or an ecological community and ultimately determine its form and survival". Aspects of the natural environment of human beings are covered in that work under *atmosphere, hydrosphere, biosphere,* and *geosphere.*

The concept of ecology has been developed by many scientists from different disciplines and world regions, in part using observations of indigenous cultures in the Americas, China, India and the Middle East, where knowledge of the uses and dangers of plants and animals is crucial for human survival and cultural development. According to Ellen (1996: 207), the concept of ecology "has been centrally concerned with the concept of adaptation and with all properties having a direct and measurable effect on demography, development, behaviour and [the] spatio-temporal position of an organism." *Human ecology* is used in human geography, urban sociology and anthropology. Ellen stated that "the other major impact of ecological concepts in the social sciences has been in the relation of political environmentalism, and to environment and development. ... Increasing attention is also being paid to the cultural construction of nature, indigenous technological knowledge, the management of collectively owned resources, and environment history" (Ellen 1996: 208).

Vladimir Vernadsky (1926) defined the biogeochemical cycles as the sum of all ecosystems. Arthur Tansley (1935) described an interactive system between living things and their environment and ecology, and from this he created a science of ecosystems that was crucial for the development of ecology as a modern systems science. Odum (1953, 1975, 1977, 1998) defined ecology as the study of the linkages of organisms and of groups of organisms with their environment and of their structure and functions (Nentwich et al. 2004: 1). The concept of ecology was used in biology chiefly in terms of 'autoecology' and 'synecology', as 'population ecology', as 'community ecology' and as 'systems ecology', and in physical geography as 'landscape ecology' (Troll 1968) or 'geoecology' (Huggett 1995, 2000). The ecology concept was initially related to the biophysical sciences and only after World War II was the concept used in the social sciences and the humanities. In the *Encyclopedia of Global Environmental Change*, Munn (2002a: xi, xiv) wrote:

In the 1960's, the scientific community began to use the word *environment* in this new non-specialist sense. ... In the ensuing decades, the world community has come to see the 'environment' in many different ways, as a

---

33 See: Polanyi Levitt (2008); at: <http://unchronicle.un.org/article/w-arthur-lewis-pioneer-development-economics/>; Emmerij/Jolly/Weiss, Thomas (2005); Emmerij (2005); at: <http://www.rrojasdatabank.info/widerconf/Emmerij.pdf>.

34 Burton (1998); at: <http://www.gmu.edu/academic/ijps/vol3-1/burton.htm>.

life-support system, as a fragile sphere hanging in space, as a problem, a threat and a home. ... In the 1970's and 1980's; ... *global environmental change* acquired a popular currency. ... Another vital insight began to emerge about 1980: the inescapably interlinked nature of these many environmental changes. ... Thus, the term *global environmental change* has come to encompass a full range of globally significant issues relating to both nature and human-induced changes in the Earth's environment, as well as their socio-economic drivers.

Fleming (2002, II: 290), writing on the environment, distinguished between *abiotic* (climate, minerals, soil, sunlight, water) and *biotic* (organisms) factors that are linked by "the flow of energy and the cycling of nutrients". Lovelock (1975, 1979, 1986, 1988, 1992), in cooperation with Margulis (1974, 1974a), expressed the complicated physical, chemical, and biological processes that maintain life on earth in the *Gaia hypothesis*, which claims "that the entire range of living matter on Earth defines the material conditions needed for its survival, functioning as a vast organism ... capable of modifying the biosphere, atmosphere, oceans, and soil to produce the physical and chemical environment that suits its needs" (Oxford 1998).

From an international relations perspective, Ronald Mitchell (2002: 500-516) focused on a) agenda-setting, b) policy formulation, and c) policy implementation and effectiveness and policy evolution and social learning. For the analysis of national and international environmental governance and regime formation, all three stages of the policy process are relevant. On environmental issues, especially on population growth and resource constraints, two opposite traditions have evolved (Kennedy 1992; Gleditsch 2003):

- a *pessimist* or *Neo-Malthusian view*, stimulated by Malthus's *Essay on Population* (1798) that stresses the limited carrying capacity of the Earth to feed the growing population;
- an *optimist* or *Cornucopian view* that believes an increase in knowledge, human progress, and breakthroughs in science and technology can cope with these challenges.

These two *ideal type* positions have dominated the environmental debate since the Club of Rome's *Limits of Growth* (Meadows/Meadows/Randers et al. 1972; Meadows/Meadows/ Randers 1982, 2004; chapter by Marino in this volume) and Lomborg's (2001) *Skeptical Environmentalist* (Gleditsch 2003). Rayner and Malone (2002, V: 109-123) pointed to a descriptive vs. an interpretative tradition in social science analyses of global environmental change. While the *descriptive tradition* relies on quantitative methods

"of tracing stocks and flows of social data through time and space", using methods and models from the natural sciences, the interpretive tradition tries "to understand motivations, ideas, and values". But both are essential for research, e.g. descriptive approaches "have revealed much what would happen under various scenarios of climate change", while "interpretive approaches can provide value-oriented parameters as a basis for choosing among candidate policies".

In the social sciences, the analysis of issues of global environmental changes and of human–nature relationships (Glaeser 1995, 2002: 11-24) are polarized between epistemological idealism and realism, or between *social constructivism* and *neo-realism*. The *neo-idealist orientation* has highlighted two aspects: a) the uncertainty of scientific knowledge and claims; and b) the attempt to explain the scientific and public recognition of environmental change as influenced by political and historical forces (Rosa/Dietz 1998). Glaeser calls for a combination of the strongholds of both positions, i.e. for a critical analysis of the assumptions and models of the natural scientists and of their inherent interests.

Within the scientific discipline of international relations, the analysis of problems of global environmental change has been pursued from different theoretical or practical orientations. Paterson (2000: 5) distinguished between six basic positions: a) *liberal institutionalism*, b) *realism*, c) *eco-authoritarianism*, d) *eco-socialism* (Pepper 2002: 224-225), e) *social ecology* (Pepper 2002a: 484) and f) *deep ecology* (Pepper 2002b: 211), to which one may add g) *ecofeminism* (Warren 2002: 218-224). These all differ on perceived causes and appropriate responses.

Several ideologies have been distinguished: *ecocentrism* "that centres on and prioritizes the whole planetary ecosystem", reflecting a *biocentric* focus on the biosphere, and *Gaiacentric* that focuses on the Earth as one living system. Homer-Dixon (1999: 28-46) distinguished between *neo-Malthusians* (biologists, ecologists); *economic optimists* (economic historians, neoclassical economists, agricultural economists); and *distributionists* (poverty, inequality, misdistribution of resources).

Paelke (2002: 49-61) distinguished two waves in the establishment and institutionalization of environmental politics: a) the early environmental movement with an often apocalyptic and apolitical dimension that focused on pollution and global sustainability concerns, and b) a second wave with "the re-emergence of conservationist and biodiversity concerns". Mostafa Tolba (2002: 1-13) noted eight trends in the national and international responses of industrialized countries to environmental problems: a) the inclusion of environmental impacts into sectoral policies; b) an increase in cross-sectoral policies; c) the replacement of a reactive approach to pollution control by a preventive one; d) a growing interest in economic instruments as incentives to energy and pollution control; e) the promotion of energy efficiency, energy conservation, and environmentally sound processes in industry, transport and domestic environments; f) the recognition of the international, and often regional, nature of many environmental problems; g) increased public information and participation; and h) better environmental science and monitoring (Tolba 2002: 5).

Since the 1960s and 1970s many new governmental and non-governmental institutions have been set up, and an increasing number of environmental laws and regulations have been adopted in OECD countries. The developing countries have followed this pattern "but with a different range of concerns and on a different time-scale" (Tolba 2002: 8), with a primary focus on land and freshwater management and food production. While for them development is crucial "to improve the quality of life, eliminate poverty and support the infrastructure needed in order to deliver the health care, education and other institutions essential to the national future", many countries have prepared national conservation strategies.

International environmental regimes and institutions have gradually evolved since the end of World War II within the framework of the UN. In 1948 the IUCN (World Conservation Union) was founded by state and non-governmental members to protect natural areas and species. The decision at the Stockholm Conference (1972) to set up the *UN Environment Programme* (UNEP) in Nairobi and the adoption of the Agenda 21 and of several environmental regimes at the *United Nations Conference on the Environment and Development* (UNCED) in Rio in 1992 were major steps towards international responses.

It has been argued above that in the Anthropocene a reconceptualization of these four basic concepts of the 'conceptual quartet' is needed. As well as this, a reframing of the research programmes on the environment and ecology and a reformulation of environmental policies may be necessary. This would proactively address the causes of global environmental change and its projected consequences. This would then make it less likely that these new environmental challenges will present security threats in the decades ahead.

This will be discussed below in the context of the three established concepts, scientific goals and policies of 'sustainable development' (9.5), 'human security' (9.6) and 'sustainable peace' (9.7).

9.5 Moving From Sustainable Development to a Long-Term Transformation to Sustainability Transition

Since the UN Conference on the Human Environment in Stockholm (1972), many representatives of developing countries have called for 'additional' funding by the North to deal with the global environmental issues caused by the industrialized nations. While the debate between modernization and critical theories of development has excluded the environment, since the late 1980s controversies between proponents of sustainable development and neoclassical modernization and critical theories of development have increased.

The concept of 'sustainable development' was politically introduced—fifteen years later—in the Brundtland Report (1987: 8) that defined it as meaning "to ensure that it meets the needs of the present without compromising the ability of future generations to meet their own needs". Sustainable development was understood as "a process of change in which the exploitation of resources, the direction of investments, the orientation of technological development, and institutional change are made consistent with future as well as present needs" (Brundtland 1987: 9). 'Sustainable development' combines two key concepts: of needs, in particular the essential needs of the world's poor, to which overriding priority should be given; and the idea of limitation imposed by the state of technology and social organization on the environment's ability to meet present and future needs (Brundtland Report 1987: 43).

For the Brundtland Commission, "sustainable development is a process of change in which the exploitation of resources, the direction of investments, the orientation of technological development; and institutional change are all in harmony and enhance both current and future potential to meet human needs and aspirations". Their report calls for a 'sustainable development' path, necessitating "a concern for social equity between generations, a concern that must logically be extended to equity within each generation". Since the publication of the Brundtland Report in 1987, sustainable development has become the key concept guiding both policy and scientific debates over the past nearly three decades.

The UNCED conference in June 1992 in Rio de Janeiro resulted in the signing of the *UN Framework Convention on Climate Change* (UNFCCC) and of the *UN Convention on Biodiversity* (CBD) and the adoption of Agenda 21 and of a mandate for negotiating a *UN Convention to Combat Desertification* (UNCCD). The *United Nations Commission on Sustainable Development* (CSD) was to ensure effective follow-up of UNCED by reviewing progress in the implementation of Agenda 21 and of the Rio Declaration on Environment and Development and by providing policy guidance for the *Johannesburg Plan of Implementation* (JPOI) at the local, national, regional and international levels.

In 2000 a summit meeting of the *United Nations General Assembly* (UNGA) in New York adopted the Millennium Declaration, with eight *Millennium Development Goals* (MDG) to be achieved by 2015. Goal 7 focused on ensuring 'environmental sustainability' (UN 2011: 50). The UN 2005 World Summit Outcome Document refers to the "interdependent and mutually reinforcing pillars" of sustainable development as economic development, social development, and environmental protection.[35]

The *UN Conference on Sustainable Development* (UNCSD or Rio+20) in June 2012 adopted a 'legally non-binding outcome document', *The Future We Want*, calling for a green economy in the context of *sustainable development* (SD) and poverty reduction, an institutional arrangement for SD, and a framework for action and follow-up (UNGA 2012). It was proposed to develop a set of *Sustainable Development Goals* (SDGs), guidelines on green economy policies, a ten-year framework on sustainable consumption and production[36] which would include guidelines on *green economic policies*, and a ten-year framework for sustainable consumption and production.

The SDGs defined in a report by an intergovernmental committee of experts on sustainable development financing, together with other documents, were adopted at the UN Summit on 26 September 2015 by heads of state and government and high representatives,[37] in a document entitled *Transforming our world: The 2030 Agenda for Sustainable Develop-*

ment. In the policy debate at the UN General Assembly during the *UN Summit on Sustainable Development* 2015 (25-27 September 2015):

> Climate change was repeatedly stressed as an existential threat that jeopardizes hard-won development gains and undermines efforts for future sustainable development. Without addressing climate change, achieving the 2030 Agenda and the other SDGs will not be possible. Its negative impacts are putting at risk the potential for progress in the economic, social and environmental areas. ... The transition to renewable energy sources was underscored as particularly important, while improving the efficiency of energy and natural resources.[38]

Among the interactive dialogues, one dealt with "Fostering sustainable economic growth, transformation and promoting sustainable consumption and production" and stressed the need for the "economies ... to undertake structural and technological transformation in order to attain higher technological intensity, greater value added and productivity and full integration in international trade". It was further argued that: "economic growth must be decoupled from environmental degradation and combat climate change to protect future generations. Accelerating a real shift towards more sustainable lifestyles will require changing how we consume and use valuable resources".[39]

The seventeen SDGs "with 169 associated targets" include a "transformational vision" of "a world in which every country enjoys ... sustainable economic growth and decent work for all", where "consumption and production patterns and use of all natural resources ... are sustainable". The SDGs are general policy guidelines for a process of sustainability transition and call for "access to affordable, reliable, sustainable and modern energy for all" (goal 7); "sustain-

35 United Nations, 2005: "2005 World Summit Outcome Document"; at: <http://www.who.int/hiv/ universalaccess2010/worldsummit.pdf>.

36 See at: <https://sustainabledevelopment.un.org/rio20> (22 January 2015).

37 See at: <https://sustainabledevelopment.un.org/rio20> (22 January 2015); see: "Post-2015 Development Agenda" with access to all adopted documents at: <https://sustainabledevelopment.un.org/post2015> (20 August 2015), including: "Transforming our World: the 2030 Agenda for Sustainable Development".

38 See at: <https://sustainabledevelopment.un.org/content/documents/8521Informal%20Summary%20-%20UN%20Summit%20on%20Sustainable%20Development%202015.pdf>; <https://sustainabledevelopment.un.org/post 2015/summit>; <http://www.un.org/en/ga/info/meetings/69schedule.shtml> (1 December 2015).

39 UNGA: "70th Session of the General Assembly United Nations Summit on Sustainable Development 2015. Informal summary"; at: <https://sustainabledevelopment.un.org/content/documents/8521Informal%20Summary%20-%20UN%20Summit%20on%20Sustainable%20Development%202015.pdf> (14 November 2015).

able economic growth" (goal 8); "inclusive and sustainable industrialization and ... innovation" (goal 9); "sustainable consumption and production patterns" (goal 12); and a "global partnership for sustainable development" (goal 17).

Goal 12, "sustainable consumption and production patterns", includes "implement the 10-year framework of programmes on sustainable consumption and production" (12.1), so that it will be possible by 2030 to achieve "the sustainable management and efficient use of natural resources" (12.2), "encourage companies ... to adopt sustainable practices and to integrate sustainability information into their reporting cycle" (12.6), and "support developing countries to strengthen their scientific and technological capacity to move towards more sustainable patterns of consumption and production" (12a), and to "rationalize inefficient fossil-fuel subsidies that encourage wasteful consumption by removing market distortions" (12c).

Global climate governance has been paralysed due to the failure of the *Conference of Parties* since Copenhagen (COP 15, 2009) to adopt a legally binding post-Kyoto climate regime. At the political level several OECD states have failed to implement their legal obligations under the UNFCCC and the Kyoto Protocol (1990–2012) and to adopt a post-Kyoto regime (Brauch 2012). The Durban outcome "included a decision by Parties to adopt a universal legal agreement on climate change as soon as possible, and no later than 2015".

Democratic governance is not relevant for distinguishing the different climate performance of the G7/G8 and G20 countries. An implementation gap exists among democracies between EU countries (leaders) and large OECD countries in North America and in the Asia–Pacific region (laggards) with legally binding reduction obligations.

Among the G20 countries different strategies towards a long-term transformative change to sustainability can also be observed. During the G20 summit in Antalya on 15–16 November 2015 their leaders were more constrained than the G7 had been in Elmau (May 2015) recognizing "that actions on energy, including improving energy efficiency, increasing investments in clean energy technologies and supporting related research and development activities will be important in tackling climate change and its effects" without calling for decarbonization. They supported a move "to adopt a protocol, another legal instrument or an agreed outcome with legal force under the UNFCCC that is applicable to all Parties".[40]

During the Anthropocene it is necessary to move beyond declaratory national policy statements and international agreements with limited implementation and to adopt binding national policies with economic incentives, with the aim of moving towards a gradual sustainable transformation of the national and global economy towards a low-carbon economy, especially of an energy sector that relies heavily on fossil fuels.

In the Anthropocene, 'sustainability transition' is both a scientific concept and a research programme and paradigm. It also refers to a policy process that aims for a long-term transformative change of policies, the economy, society and science. As Clark, Crutzen and Schellnhuber (2004: 2–3) argued, 'sustainability' is "a normative concept regarding not merely what *is*, but also what *ought to be* the human use of the Earth (Kates 2001)". They noted that "while the goal of environmentally sustainable development in the Anthropocene was generally recognized", its implementation has largely failed since the Johannesburg summit (2002) and at Rio II (2012), where no legally binding commitments were adopted.

The US National Academy of Science claimed in 1999 that "a successful transition toward sustainability is possible over the next two generations", with "significant advantage in knowledge, in the social capacity and the technical capabilities to utilize it, and in the political will to turn this knowledge into action" (NRC 1999: 160; Johnson 2016).

While our scientific knowledge and technical capabilities have significantly progressed since 1999, in the context of the Bush Administration's 'war on terror' from 2001 onwards there was a return to a military agenda driven by Hobbesian power rationales, with new wars in Afghanistan (1991), Iraq (2003), Libya (2011) and Syria (2013). At the same time the environmental and global change agenda was partly disputed, ignored or downgraded. Not surprisingly, billions of US dollars and euros have been spent since the mid-1990s and especially since 9/11 on these wars and their humanitarian consequences. Global military expenditure in constant US dollars has increased again since 1997, and since 2008 it has reached its highest level since the end of the Cold War in 1989 (figure 9.1).

Three of the G7 countries failed to achieve their goals under the *Kyoto Protocol* (KP) and many devel-

---

40 See the "G20 Leaders' Communiqué Antalya Summit, 15–16 November 2015"; at: <http://www.consilium. europa.eu/press-releases-pdf/2015/11/40802205150_en_63583293780000000.pdf>.

Figure 9.1: World military expenditure (1988–2104). **Source:** © SIPRI (2015) with the permission of SIPRI. 17 November 2015; at: <http://www.sipri.org/research/armaments/ milex/milex-graphs-for-data-launch-2015/ World%20military%20expenditure%201988-2014.png>.

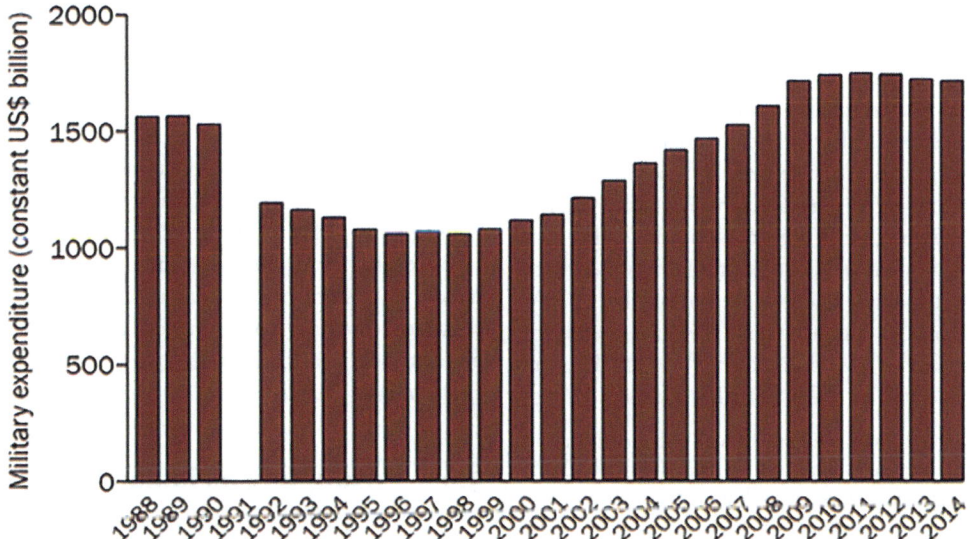

oping and threshold countries (China, India) totally ignored the goals of the UNFCCC that they had endorsed, and increased their GHG and emissions by up to 250 per cent between 1990 and 2014 (Brauch 2016a). At COP 15 of the UNFCCC in Copenhagen (1999) and at the Rio II Summit (2012), there was a lack of political will to act decisively to meet the recognized, declared and endorsed challenges that face humankind in the Anthropocene, among both major industrialized countries and rapidly developing threshold countries. Most industrialized countries have also failed to meet their declared commitments to spend 0.7 per cent of their GNP on development aid. The humanitarian crisis—with millions of war refugees and distressed migrants—facing Europe in autumn 2015 was partly driven or influenced by Western 'wars' in the Balkans, in Afghanistan and in the MENA region.

While the 'political will' has often been lacking since the start of the twenty-first century, scientific research, knowledge and discourse on the problems of global environmental and climate change have progressed significantly, as Clark, Crutzen and Schellnhuber (2004:15) have pointed out: "the Anthropocene crisis forces humanity to manage consciously a transition toward sustainable use of the Earth". However, the hopes they put in the proposal of the NRC (1999) that a sustainable development could be achieved over two generations have not yet been realized. They called for a new Copernican revolution, a new paradigm for sustainability and a new 'social contract'

between science and society for planetary stewardship (Clark/Crutzen/Schellnhuber 2004: 23-24).

The WBGU (2011: 1-2) took up the proposal for *A Social Contract for Sustainability* aiming at a "low-carbon and sustainable global economic system" and it stated that

> One key element of such a social contract is the 'proactive state', a state that actively sets priorities for the transformation, at the same time increasing the number of ways in which its citizens can participate, and offering the economy choices when it comes to acting with sustainability in mind. The social contract also encompasses new forms of global political will formation and cooperation.

The WBGU used this concept for its call "for the transformation towards sustainability...as an analogy to the emergence of the industrialised societies during the course of the 19th century", following Karl Polanyi's (1944) analysis in his *Great Transformation* that described the embedding of "the uncontrolled market dynamics and innovation processes into a constitutional state, democracy and the creation of the welfare state". The WBGU stated (2011: 8) that this idea:

> refers to the necessity of humankind taking collective responsibility for the avoidance of dangerous climate change and other dangers to the planet. ... This needs a voluntary capping of the usual options for economic growth in favour of giving the people in those parts of the world already suffering the consequences of our irresponsible behaviour, and particularly future generations, room to manoeuvre. For another, the transforma-

tion needs a powerful state, counterbalanced by extended participation on the part of its citizens.

The WBGU report was discussed during an international symposium in Berlin in May 2012 where Chancellor Angela Merkel argued that "the two crucial elements ... must ... be ... switching to renewables, and dealing more efficiently with energy and the resources we have." She stressed that her government "decided to raise the proportion of renewables in our overall energy consumption to sixty per cent by 2050. For electricity consumption, that figure is to be eighty per cent." Merkel concluded that "sustainability needs to become a central tenet in every area of our lives."[41] At the G7 meeting in Elmau (Germany), the heads of major industrialized countries called for a legal instrument to be adopted at COP 21 in Paris, which

> should enable all countries to follow a low-carbon and resilient development pathway in line with the global goal to hold the increase in global average temperature below 2°C. ... In order to incentivize investments towards low-carbon growth opportunities we commit to the long-term objective of applying effective policies and actions throughout the global economy.[42]

The key challenge in the years ahead will be whether the agreements adopted at COP 21 will be implemented by all major contributors to the rapidly rising GHG and CO2 concentrations in the atmosphere. The 'rationale' of the research on 'sustainability transition' is that these changes should occur nationally and regionally, e.g. within the EU, irrespective of international commitments, and that a creative scientific and investment climate must be created such that sustainable niches and changes in the regime and the landscape will occur globally and become as competitive as wind power and increasingly photovoltaics have done since 1990.

The gradual shift from the normative goal of *sustainable development* (SD) towards an analytic concept and policy process of *sustainability transition* (ST) requires a reframing of the social science agenda, including the four key programmes of security, peace, development and environmental studies, as well as of the courses offered and existing degree programmes

(see below, section 9.8). In the context of the Anthropocene, the many related scientific and political concepts of green growth, green economy, low-carbon economy, dematerialization, decarbonization and decoupling must be discussed and developed further.

Prior to this recent debate on 'sustainability transition' and peace, Scheffran (1997, 1998, 1999, 2011) analysed multiple linkages between "peace and sustainable development" where he discussed the linkages between economic growth and violent conflicts, the consequences of resource scarcity, and between sustainable development and the limits of satisfying human needs. As a step towards sustainable peace, Scheffran suggested moving from a negative towards a positive linkage (see also chapter 34 by Scheffran/Froese and chapter 13 by Scheffran in this volume).

9.6 Adopting a Human Security Approach

After the end of the Cold War 'human security' emerged as a new security concept that shifted the focus from the 'state-centred' perspectives of national, international and global security to a 'people-centred' or 'human-centred' security.[43] Before 1989, a few scholars had already called for a human security approach (Westing 1989, 1989a, 2013). After UNDP (1994) had launched the concept of human security, the traditional narrow understanding of military and political security was widened to include economic, societal and ecological security dimensions, as well as deepened to include human and gender security. As a result, its reference objects changed from sovereignty and territory to human beings and humankind. The values at risk shifted to identity, sustainability and survival, while the sources of threats changed from other states to global environmental change, globalization and financial crises. Some expanded the human security discourse to the environmental dimension and to interactions between the individual and humankind as a cause and victim of global environmental change (Brauch 2005, 2008a, 2009).

Since the late 1990s two different concepts have existed in the political realm. While Canada, Norway and the European members of the *Human Security Network* (HSN) preferred a narrow concept that

---

41 See the text documentation from the symposium of 9 May 2012 in Berlin; at: <http://www.wbgu.de/fileadmin/templates/dateien/symposium2012/Documentation_Symposium.pdf> (23 August 2015).

42 See "G7 Leaders' Declaration, Schloss Elmau, Germany, June 8, 2015"; at: <https://www.whitehouse.gov/the-press-office/2015/06/08/g-7-leaders-declaration> (14 August 2015).

43 This section relies on previous texts by this author (Brauch 2005, 2005a, 2009a) and a co-authored chapter (Brauch/Scheffran 2012), with the permission of the co-author.

focused on 'freedom from fear' and dealt with humanitarian initiatives and human rights, Japan opted for a wider concept that included the human development agenda, with 'freedom from want'. In the social sciences, several proponents of a narrower concept of human security (Mack 2004, 2004a, 2005; Krause 2004, 2004a, 2008, 2008a; MacFarlane/Khong 2006) were opposed to including development and environment policy.

In the social sciences, the concept has been widely discussed in political science, international relations, and especially in *development studies* (Picciotto/Olonisakin/Clarke 2007; Ulbert/Werthes 2008) and *peace studies* (Tadjbakhsh/ Chenoy 2006; Thakur 2006/2007) and to a lesser extent in *environmental studies* (Page/Redclift 2002; Dodds/Pippard 2005) and *security studies* (Dannreuther 2007; Booth 2005, 2007), where many realists ignored this discourse (Kolodziej 2005). It has also been discussed in geography (Lonergan 1999, Bohle/O'Brien 2007), in international law (von Tigerstrom 2007), in education (Nelles 2003), in philosophy (Fabra Mata 2007), in theology (Eisen 2008), and in the health sciences (Leaning 2009). Human security has been discussed in relation to *national* (urlevic-Lukic 2004), *international* (Dannreuther 2007) and *global security* (Stoett 1999), as well as in the field of sectoral security concepts, including water, food, health security and livelihood security (Bohle 2009).

Human security requires a fundamental shift in thinking on security. It addresses different policy requirements and needs, with horizontally integrated political coping strategies where the role of development and environment policies is vital, and the mission of the military changes from fighting wars to *protecting* the people against genocide and natural hazards; society and social movements should *empower* people to build resilience and to enhance their coping capacities. Human security concepts and agendas enhance societal groups and knowledge-based scientific epistemic communities as new *securitizing actors* that can put global environmental and climate change on national and international policy agendas.

Within international relations, the concept of human security has remained controversial. While many realists and the strategic studies community, as well as 'state-centred' peace researchers, have rejected the human security concept, authors with liberal and constructivist perspectives and from peace research have rallied behind it. Uvin (2004, 2008) used the concept as a "conceptual bridge between the ... fields of humanitarian relief, development assistance, human

rights advocacy, and conflict resolution" (Owen 2004). For Hampson (2004), human security gives voice to the politically marginalized, while Acharya (2004) interpreted it as a response to the globalizing of international policy, and for others, human security is a response to genocide and the limits of sovereignty that can justify humanitarian interventions.

Thomas and Tow (2002, 2002a) distinguished general human security 'threats' such as hunger and disease, and specific ones, such as "single actions that have an immediate effect on the safety or welfare of victims and demand immediate remedy", to which 'peacekeeping' emerges as a major response along with peace enforcement measures. They argued that human security could be considered "a valid paradigm for identifying, prioritizing and resolving emerging transnational security problems", and that it offers ways to respond to these challenges by "safeguarding and improving the quality of life" for individuals and groups.

Owen (2004) suggested combining UNDP's (1994) wide definition with a threshold-based approach. He argued that "human security is the protection of the vital core of all human lives from critical and pervasive environmental, economic, food, health, personal and political threats", regardless of whether people are affected by floods, communicable disease, or war. Rather, all those threats would be included "that surpass a threshold of severity [and] would be labelled threats to human security" (2004).

Khagram, Clark and Raad (2003: 107-135) discussed both environmental threats (and their impacts on human survival, well-being and productivity) and environmental opportunities for human security. Environmental change can have direct and immediate effects on well-being and livelihoods, it can also impact on health, economic productivity and political instability. Environmental threats can affect "individuals, families, communities, social organizations, identity groups (women, children), diasporas, governments and biological species". Also, a single environmental threat "can have potentially adverse effects at multiple scales from the household to the planetary". On the other hand, environmental protection, cooperation and peacemaking can improve human security.

Mary Kaldor (2007: 182-187) who headed the *Study Group on Europe's Security Capabilities* (Kaldor 2004; Glasius/Kaldor 2005) argued that "it includes both civilian and military elements; it offers a way to act, a set of principles for crisis management." In order to put the concept into practice she suggested five principles (as both ends and means) for the implementation of human security "in a continuum of varying degrees of violence that always involve[d] elements of both protection and reconstruction:" 1) the primacy of human rights; 2) legitimate political authority; 3) multilateralism; 4) the bottom-up approach; and 5) a regional focus. Kaldor (2007: 190-191) claimed that even after new wars have been stabilized, "individuals still experience high levels of physical insecurity as a consequence of high crime rates, high human rights violations and high levels of violence against women." She suggested that a human security approach "would aim both to stabilize conflicts and address the sources of insecurity." This requires public security based on the rule of law and effective law enforcement. The implications of this for international organizations are an expanded international presence, new human security forces, and a legal framework.

The economic and social priorities for conflict prevention include: a) combining humanitarian and development assistance; b) creation of legitimate employment and self-sustaining livelihoods; c) institution-building, including the rule of law; d) attention to the importance of infrastructure and public works; e) education and social services; and f) generating tax revenues. She concluded that a human security approach could benefit development by: a) "providing the conditions (physical safety, rule of law and sustainable institutions) that are integral to development; b) focusing on human development and strengthening weak institutions; and c) stressing the needs of individuals and communities" (Kaldor 2007: 193-195).

There has been an intensive scientific discourse on human security concepts and issues among and with scholars from developing countries, in Latin America (Neff 2003; Palma 2003; Kornblith 2003; Bonilla 2003; Lopez 2003; De Lombaerde/Norton 2009; Rojas Aravena 2009; Singh 2009), in South Asia (Ariyabandu/Fonseka 2009; Chari/Gupta 2003, 2003a; Abdus Sabur 2003, 2009; Bajpai 2000, 2003, 2004, 2005; Najam 2003, 2003a), in South East Asia (Othman 2009; Wun'Gaeo 2009), in Japan (Shinoda 2009), in Africa (Hendricks 2008; Goucha/Cilliers 2001; Naidoo 2001; Hussein/Gnisci/Wanjiru 2004; Mulongo/Kibasomba/Njeri Kariri 2005; Mutesa/Nchito 2005 Mpagala/Lwehabura 2005; Mataure 2005; Ngoy Kaungulu 2005; Correira de Barros 2005; Mlambo 2005; Dzimba/Matooane 2005; Poku/Sandkjaer 2009), and in the Arab world (Chourou 2005, 2009).

The human security concept has also been widely debated in Africa. On the political level the debate on human security was triggered by a *conference on security, stability, development and cooperation* (CSS-DCA) in Kampala in 1991 where five hundred African politicians and leaders adopted the Kampala Document that formed the basis for the comprehensive security agenda of the African Union. The human security concept influenced the *New Partnership for Africa's Development* (NEPAD 2001) and the adoption of a *Common African Defence and Security Policy* (2004). However, there was disagreement on the policy priorities for achieving its aims (Hendricks 2008: 137-142).

In June 2001, the *Institute for Security Studies* (ISS) in Pretoria hosted UNESCO's regional conference on *Peace, Human Security and Conflict Prevention in Africa* (Goucha/Cilliers 2001: vi-vii). This addressed: a) promoting human capacity-building in the member states of the region; b) helping African countries to establish a strategy for the prevention of HIV/AIDS and other contagious diseases; c) mobilizing and acting as a catalyst for international cooperation in support of initiatives by African Member States; and d) promoting the active participation of communities and representatives of civil society in the planning and implementation of development programmes.

This selective review of definitions of human security and of threats for human security, as well as of areas of relevance for human security issues, documents both the diversity and the lack of a clear consensus. This last has hindered communication with policymakers and their efforts to move from declaratory statements to concrete policy initiatives and actions. And so after two decades the conceptual discourse remains inconclusive and the definition of human security being used depends on the approach and preferences of the respective author. In the policy and scientific debate on human security, environmental security problems have been mostly ignored.

Kofi Annan argued (2005) that human security should be based on the three pillars of 'freedom from want' (human and sustainable development agenda), 'freedom from fear' (the traditional peace and security agenda) and 'freedom to live in dignity' (democracy, rule of law and human rights agenda). Bogardi and Brauch (2005; Brauch 2005, 2005a, 2009) suggested a fourth conceptual pillar, 'freedom from hazard impact', which would be brought about by reducing the vulnerability and enhancing the coping capabilities of societies confronted with natural and human-induced hazards, through empowering people and enhancing their resilience.

Box 9.1: Excerpts of chapter 12 on "Climate change and human security" by the IPCC, WG II. **Source**: IPCC (2014).

- *Climate change will have significant impacts on forms of migration that compromise human security (high agreement, medium evidence)*. Some migration flows are sensitive to changes in resource availability and ecosystem services. Major extreme weather events have in the past led to significant population displacement, and changes in the incidence of extreme events will amplify the challenges and risks of such displacement. Many vulnerable groups do not have the resources to be able to migrate to avoid the impacts of floods, storms and droughts. ... Migrants themselves may be vulnerable to climate change impacts in destination areas, particularly in urban centres in developing countries ...

- *Mobility is a widely used strategy to maintain livelihoods in response to social and environmental changes (high agreement, medium evidence)*. ... Expanding opportunities for mobility can reduce vulnerability to climate change and enhance human security ...

- There is insufficient evidence to judge the effectiveness of resettlement as an adaptation to climate change. *Some of the factors that increase the risk of violent conflict within states are sensitive to climate change (medium agreement, medium evidence)*. The evidence on the effect of climate change and variability

on violence is contested ... Although there is little agreement about direct causality, low *per capita* incomes, economic contraction, and inconsistent state institutions are associated with the incidence of violence ... These factors can be sensitive to climate change and variability. Poorly designed adaptation and mitigation strategies can increase the risk of violent conflict ...

- *People living in places affected by violent conflict are particularly vulnerable to climate change (high agreement, medium evidence)*. Evidence shows that large-scale violent conflict harms infrastructure, institutions, natural capital, social capital and livelihood opportunities. Since these assets facilitate adaptation to climate change, there are strong grounds to infer that conflict strongly influences vulnerability to climate change impacts ...

- *Climate change will lead to new challenges to states and will increasingly shape both conditions of security and national security policies (medium agreement, medium evidence)*. Physical aspects of climate change, such as sea level rise, extreme events and hydrologic disruptions, pose major challenges to vital transport, water, and energy infrastructure ...

These four pillars of a wide concept of human security match four different human security agendas that were addressed by the UN member states during the two thematic discussions on human security on 22 May 2008 and on 14 April 2011.[44] UN Secretary-General Ban Ki-Moon (UNSG 2009) proposed in his report on *Climate change and its possible security implications* "several 'threat minimizers', ... [to] lower the risk of climate-related insecurity. ... Accelerated action at all levels is needed to bolster these threat minimizers". The daily survival problems of a few billion people, their social vulnerability, and their physical exposure to climate change are creating additional dangers for human security.

In his Human Security report of 2010, the Secretary-General applied the human security concept to "Climate change and the increase in the frequency and intensity of climate-related hazard events". A *policy-focused human security approach* to climate change prioritizes the climate-induced security threats

that humankind will *face* during the twenty-first century. Its task is to develop policies for coping better with the human security impacts of climate change by measures of mitigation, adaptation and resilience-building to *protect* and to *empower* the people affected. 'Freedom from Hazard Impact' requires a proactive environmental strategy for implementing the Rio Conventions. It requires a close cooperation between those agencies working on the global environmental change and the hazard agenda (early warning, disaster response, disaster preparedness, resilience building and reduction of social vulnerability).

In its fifth Assessment Report (AR5), the IPCC (2014) for the first time assessed social science literature on the human security impacts of climate change (box 9.1). Gleditsch and Nordas (2014) commented critically on the report, pointing to the lack of a consensus in the scientific community on the linkage between climate change and war (Nordas/Gleditsch 2013).

This IPCC assessment of the social science literature requires that the environmental and the climate change dimension of human security and its fourth conceptual pillar of 'Freedom from Hazard Impact' should be taken up in the debates in the UN General

---

44 The following section is based on the author's presentation at this meeting and on his background paper, accessible at: <http://www.afes-press-books.de/html/hexagon_05_PressConf_Presentations.htm#NY2>.

Assembly and in the discussions of the UN Security Council as well as in future reports from the UN Secretary-General (2009, 2010, 2012) on both human security and climate change. So far, these have treated both as two separate policy issues. In the social sciences and especially in peace, security, development and environmental studies, these observed linkages need more emphasis in conceptual, theoretical, and empirical research.

Simon Dalby is one of the very few scientists who work on security (see chap. 2 in this volume) and human security in the Anthropocene. "How we respond to these changes will be at the heart of what security means in the coming decades; how we think about linking ecological change and what we feel is important enough to need securing will soon shape future human prospects profoundly" (Dalby 2013a: 28). He argues that "security thinking needs to be substantially updated to be relevant to the new era of the Anthropocene" (Dalby 2013a: 28), in terms of what future is to be secured and which challenges must be contained, including the "increasing severity of hazards" (Dalby 2013a: 30) and the highly vulnerable urban infrastructure. Human security in the Anthropocene is, for Dalby, "a matter of what we choose to make, build and use. It is about investment decisions, only most obviously in terms of energy choices and transportation modes" (Dalby 2013a: 32). And he indicates that "The shift from a perspective that takes rival territorial states as the key to global order to one that looks to low carbon modes of economy providing basic needs for all the world's people is dramatic" (Dalby 2013a: 32).

9.7 Developing Sustainable Peace Further

'Sustainable peace' is a value-oriented and idealist concept that has been used by development NGOS and IGOs and in the social sciences in development and peace studies, especially by peace psychologists (chapter 6 by Deutsch/Coleman; chapter 7 by Coleman). In her book on *Sustainable Peace*, Connie Peck (1998: 22-23) introduced this concept into the post-Cold War debate on preventive diplomacy and on conflict prevention both as a 'vision' and as a 'policy programme' for conflict prevention, She argued that

> the twin concepts of sustainable development and sustainable peace could provide a full, and more focused and acceptable agenda for conflict prevention. ... The search for sustainable peace will therefore need to be

based on the establishment of the rule of law (a rights-based approach) and the institutionalization of problem solving (an interest-based approach) to replace violent conflict (a power-based approach).

Peck linked 'sustainable peace' and 'sustainable development' to 'human security', stating that

> sustainable peace is dependent upon addressing human security needs through the development of a fair process that can foster and maintain that security. Assistance in the creation of sustainable peace must therefore be based first on a thorough understanding of human security needs at the local level, and second on knowledge about how these might be best addressed through appropriate institutional and structural mechanisms (Peck 1998: 225).

Deutsch and Coleman have discussed above (chapter 6) "what psychological theory and research can contribute to the promotion of a *harmonious, sustainable peace*". The purpose of their book (Coleman/Deutsch 2012) was to enhance the "understanding of sustainable peace by supplementing the standard approach of studying the *prevention* of destructive conflict, violence, war and injustice with the equally important investigation of the *promotion* of the basic conditions and processes conducive to lasting peace."

Following Solomon (1997: ix), Deutsch and Coleman stated that: "Sustainable peace requires that long-time antagonists not merely lay down their arms but that they achieve profound reconciliation that will endure because it is sustained by a society-wide network of relationships and mechanisms that promote justice and address the root causes of enmity before they can regenerate destabilizing tensions". They argued that "a sustainable world peace will require the building of such a society" and they outline ... the "psychological requirements of such a society".

Coleman defined "sustainable peace as existing in a state where the probability of using destructive conflict, oppression and violence to solve problems is so low that it does not enter into any party's strategy, while the probability of using cooperation, dialogue and collaborative problem-solving to promote social justice and well-being is so high that it governs social organization and life", a definition which he elaborated for peace education.[45]

In a World Bank study, Lie, Malmin, and Gates (2007) discussed 'sustainable' or 'durable' peace in the context of 'post-conflict justice', while Lederach (1998)

---

45 Peter Coleman, "The Missing Piece in Sustainable Peace"; at: <http://blogs.ei.columbia.edu/2012/11/06/the-missing-piece-in-sustainable-peace/> (15 Nov. 2015).

proposed building sustainable peace "beyond violence". The UN Secretary-General in his 2001 report on conflict prevention argued that "effective conflict prevention is a prerequisite for achieving and maintaining sustainable peace, which in turn is a prerequisite for sustainable development".[46] In a similar direction the activities of the Department for Peacekeeping Operations (DPKO) "are designed to support peaceful transitions towards the building of sustainable peace in post-conflict environments" (Pampell/Sen 2005).

The Advanced Consortium on Cooperation, Conflict and Complexity (2015)[47] at Columbia University produced an expert survey report on *The Sustainable Peace Mapping Initiative: What is Sustainable Peace?* which concluded that among the seventy-four scholars there was "no clear consensus… on the definition of key elements associated with sustainable peace", but some points of convergence:

> All … respondents either defined peace as the presence of processes and conditions for preventing *negative, destructive* dynamics and outcomes (violence, war, injustice, exclusion etc.), as the presence of processes and conditions promoting more *positive, constructive* dynamics and outcomes (mutual respect, cooperation, justice, harmony, environmental sustainability etc.), or as a combination of both (AC4 2015: 9).

The authors of the expert survey took four components into account in their working definition of sustainable peace: "*dynamic processes, positivity* and *negativity, systemic context* and *sustainability*", and proposed a

> working definition of sustainable peacefulness as *a set of dynamics that result in a high probability of robust patterns of constructive interactions between stakeholders and communities and a low probability of destructive interactions. Such dynamics establish and are established by a robust, enabling, and self-perpetuating context of peacefulness* (AC4 2015: 10–11).

The authors distinguished six thematic categories: 1) well-being, 2) quality of relations, cooperation and interdependence, 3) conflict management and resolution, 4) access to resources, equality, and human security, 5) institutional capacity and governance, and 6)

violence, non-violence and security. For the purpose of our own analysis the fourth is most relevant; the participants associated it with "freedom of speech, access to information, economic opportunities, access to healthcare, and education, social protection, arts, freedom of religion, social and income inequalities, environmental protection, [and] public goods provision", which corresponds with the definition of human security as freedom from fear and want.

In a proposal for a "systems-based applications of sustainable peace", Coleman, Burns, Fisher-Yoshida and Fisher (2015) noted that for many policymakers sustainable peace is elusive; for some it means

> a stable, long-lived peace that is associated with the preservation of the status quo through adequate security and protection from outside influence. For others, sustainability in the context of peace is associated with adaptation and renewal—a creative-adaptive peace—which recognizes that all social systems are in flux and progress through multiple states or stages over time (AC4 2015: 1).

Coleman, Burns, Fisher-Yoshida et al. (AC4 2015a: 1) argue that for decades peace research "has primarily studied the pathologies of war, violence, aggression and conflict" and only few researchers have studied peace "as a positive state". They deplore the fact that, so far, "interdisciplinary approaches are limited, and [a] comprehensive review of the scientific contributions across disciplines is yet to be undertaken. As a result, the complexity, multidisciplinarity, dynamism, and sustainability of peace are not well understood."

However, Peck, Deutsch and Coleman and most texts on sustainable peace have often ignored environmental challenges and their possible consequences for new types of conflict during the Anthropocene. This theme has therefore been underexplored, and conceptual, theoretical and empirical research is needed to develop the concept of a sustainable peace in the Anthropocene.

The respondents to the expert survey did not contextualize 'peace' in the context of 'peace with the environment' and the 'silent transition' to the Anthropocene that has not yet been socially constructed by most political scientists, international relations experts and peace researchers. Moving towards 'sustainability transition' may be conceived as the positive, creative-adaptive cooperative peace initiative that helps to contain or even prevent new causes of distress migration, conflict and, in the worst and most unlikely case, of wars to emerge. It requires shifting resources from the 'war system' to sustainable development, more equality, economic opportunities, and justice (Oswald 2008).

---

46 See: *Prevention of Armed Conflict—Report of the Secretary-General* (A/55/985-S/2001/574). Para.10.

47 Advanced Consortium on Cooperation, Conflict and Complexity, 2015: *The Sustainable Peace Mapping Initiative: What is Sustainable Peace? Expert Survey Report* (New York: The Earth Institute, Columbia University); see also at: <http://ac4.ei.columbia.edu/ac4-supported-initiatives/sustainable-human-develop-ment/sustainable-peace-systems-mapping-initiative/> (18 November 2015).

9.8   Conceptualizing an Emerging Political Geoecology

In the Anthropocene a rethinking of the relationship between humankind and nature, including the political realm and the social sciences, is needed, and it will make geopolitical approaches in the Hobbesian tradition obsolete.[48] In the Anthropocene, the old spatializations of international political, strategic and economic relations as 'geopolitics', 'geostrategy' and 'geoeconomy' have been primarily power- and interest-based and were partially ideologically legitimated. The ideas of many of these spatializations can be traced to the writings of Thomas Hobbes and to the pioneers of geopolitics and geostrategy: Kjellén (1917/1924), Mackinder (1904, 1905, 1918), Mahan (1890, 1897), Haushofer (1932). But these scientific and political approaches are unable to cope with the new environmental security problems posed by issues of global environmental change during the Anthropocene.

Instead of these Hobbesian spatializations of international political, strategic and economic relations, that totally ignore the environmental dimension and issues of global environmental change, a new concept of a *'political geoecology'* was developed. Geopolitics, geostrategy and geoeconomics have understood space as 'territory', as a component of national sovereignty, and have ignored the environment. But several approaches from the natural sciences, such as Lovelock's 'Gaia hypothesis', Edward O. Wilson's (1998, 1998a) 'consilience', Huggett's (1995) 'geoecology', and the 'earth systems analysis' approach (Schellnhuber/Wenzel 1998; Steffen/Sanderson/Tyson et al. 2004) have ignored politics.

The specific goal of a 'political geoecology' is to bring in politics, and especially peacebuilding and widened security issues (Dalby 2009, 2009b), and to outline a policy vision for the twenty-first century that aims for a sustainable peace (Brauch/Oswald Spring 2009a; Oswald 2009). Political geoecology was introduced as a new concept for research and teaching that links the approaches in physical and human geography with those of other social sciences. This requires a political dimension that moves from 'anticipatory knowledge' to 'proactive action' inspired by a political geoecology, with guidelines for a cooperative environ-

mental policy in the Anthropocene. The environment is not external to the human enterprise, because we are part of the earth.

With their proposed 'political geoecology for the Anthropocene', the authors want to launch a debate among scholars in the natural and social sciences. The old geopolitical and geostrategic research and policy debates were unaware of environmental issues, while the new research programmes in physical geography on 'geoecology' ignored 'politics', that is, the fields, processes, institutions, and legal frameworks that are needed for the implementation of sustainable development. The proposed perspective of a 'political geoecology for the Anthropocene' aims for a long-term global cooperative political strategy in a multilateral context. This requires a rethinking of one of the key goals of the UN Charter, "to maintain international peace and security", to include the new challenges humankind faces in the twenty-first century, and to develop a new strategy for a sustainable peace based on a gradual decarbonization of the global economy that can cope with the impacts of global environmental change.

The proposed new approach of a political geoecology combines the necessary *political orientation* with a *spatial focus* that combines different scales by implementing the many visions, perspectives and programmes for a sustainable development, so that both the linear projections of 'dangerous climate change' and the possible chaotic tipping points in the earth system are prevented from becoming reality. These new security dangers do not differentiate between countries, and they cannot be countered by military superiority.

Security policy in the Anthropocene requires primarily non-military instruments, and it also needs a gradual reallocation of scarce resources to be used for strategies of adaptation and mitigation that address the impacts of GEC and global and regional climate change. Purely technical solutions are bound to fail if there is no comprehensive political strategy for a progressive global green economy and no change in human lifestyles and social responsibility.

The cooperation between natural scientists working on geoecology and those working on programmes on *earth systems science* (ESS) addressing issues of global environmental change and climate change has not been easy. By adding a 'political dimension' to spatially sensitive 'geoecology' Brauch, Dalby and Oswald Spring (2011) call on the natural sciences to bring in perspectives from the policy sciences, and call

---

48 This chapter relies on previous work by the author (Brauch 2003) and on a co-authored chapter (Brauch, Dalby, Oswald Spring 2011), where this concept was developed further and discussed in detail. Some text is used here with the permission of the co-authors.

on the social sciences to reflect the new knowledge produced in the natural sciences.

As a result of linking 'ecology' as a scientific concept with the normative ethical, political and scientific goal of 'peace', scientific analysis must be broadened, as must be action-oriented political thinking and associated strategies, policies and measures to achieve peace, with its different features of a 'negative', 'positive', 'cultural', an 'engendered' and a 'sustainable peace'. Peace ecology should be conceptualized in relation to its five key pillars, embracing peace, security, equity, gender and sustainability.

9.9   Conceptualizing an Emerging Peace Ecology

Since the 1990s, environmental security research has shifted from environmental scarcity, degradation and conflict to the dangers posed by global environmental and climate change (Sygna/O'Brien/Wolf 2013). With the direct impacts of humans upon ecosystems in the Anthropocene (Crutzen 2002, 2011) and with the progressive securitization of issues of GEC during the past decade, these anthropogenic changes are increasingly threatening human lives and livelihoods. Worldwide, the destruction of key ecosystem services, the pollution of air, water and soil, land use change and extreme events are creating new 'anthropogenic challenges' for humankind, although these do not pose the threat of violent conflict and war.

Conca (1994: 20) suggested an "environmental agenda for peace studies" and a discussion on whether "ecologically desirable futures include concerns for peace and justice". He argued that it is not enough "to place 'sustainable development' and 'ecological security' alongside peace or social justice as 'world-order values'" but that scholars must ask whether "not only their formal definitions, but also their metaphorical and institutional associations, further the purposes of peace, justice, and community". Conca (2002: 9) fundamentally challenged a core premise of the debate on environmental (in)security and conflict by asking "whether environmental cooperation can trigger broader forms of peace defined as a continuum ranging from the absence of violent conflict to the inconceivability of violent conflict" by also addressing "problems of structural violence and social inequality" and by "building an imagined security community" based on peaceful conflict resolution.

Christos N. Kyrou (2007) introduced the concept of 'peace ecology'[49] as an "integrative, multi-contextual,

and case sensitive approach in identifying resources for conflict and violence transformation" with as its goal "to include issues of conflict analysis and peacebuilding" in environmental studies. Peace ecology calls philosophically for "peace with nature" (Meyer-Abich 1979, 2003), both a domestic and an international task where human behaviour has to be brought back into line with the wholeness of nature, where increasing hazards and disasters are an expression of the disharmony and absence of humankind's peace with nature.

How humans respond to these new threats to their survival depends not only on the world view of the scientists but also on the mindset of the elites. Business-as-usual prevails when the political, economic and military elites are unwilling or unable to act to address the root causes of GEC and climate change.

Peace ecology in the Anthropocene was conceptualized by Oswald Spring, Brauch and Tidball (2014: 18-19) within the framework of six conceptual pillars: peace, security, equity, sustainability, culture and gender, where *negative peace* (non-war) is defined by the linkages between peace and security, while for the relationship between peace and equity the concept of *positive peace* is defined by peace with social justice and global equity; for interactions between peace, gender and environment the concept of *cultural peace* was proposed, and for the relations between peace, equity and gender the concept of an *engendered peace* was suggested.

These five pillars of peace ecology point to different conceptual features of peace. The classic relationship between 'international peace and security' in the UN Charter refers only to a narrow *negative peace* without war and violent conflict. It aims for the prevention, containment and resolution of conflicts and violence and the absence of 'direct violence' in wars and repression. To achieve peace with equity or *positive peace* requires the absence of 'structural violence'. This is achieved by overcoming social inequality, discrimination, marginalization and poverty where there is no access to adequate food, water, health or educational opportunities.

Sustainable peace involves the manifold links between peace, security and the environment, where humankind and the environment as the two interdependent parts of global Earth face the consequences of destruction, extraction and pollution (Oswald Spring 2008). The concept of sustainable peace also includes

---

49 This section partly relies on a previous co-authored chapter (Oswald Spring, Brauch, Tidball 2014), with the permission of the two co-authors.

Figure 9.2: Five Pillars of Peace Ecology and their four linkage concepts of negative, positive, cultural and engendered peace. **Source:** Oswald Spring, Brauch and Tidball (2014: 19).

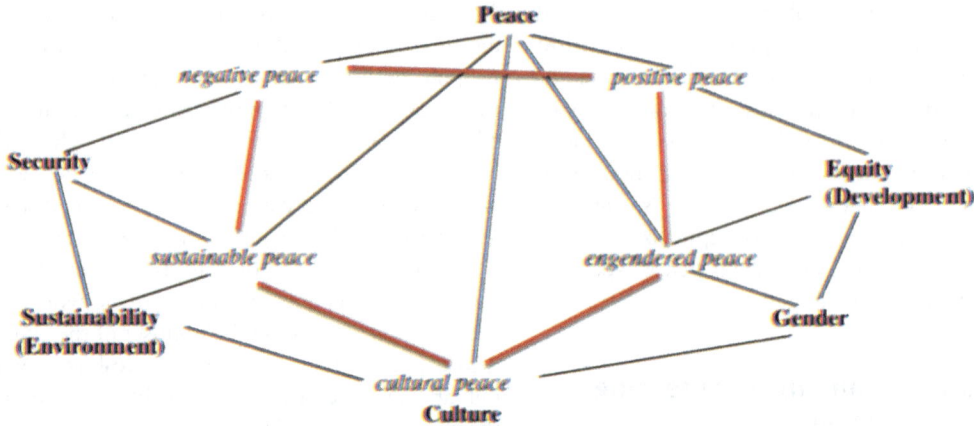

the processes of recovering from environmental destruction, reducing the human footprint in ecosystems through less carbon-intensive processes, and in the long term possibly carbon-free and dematerialized production, so that future generations may still decide on their own resource and development strategies.

Policies aiming for 'sustainability transition' are thus part of a positive strategy that addresses possible new causes of instability, crises, conflicts and in the worst case even war. These causes may be either the scarcity of fossil energy sources or the possible security consequences of anthropogenic global environmental and climate change, either of which may be triggered by linear trends as well as by chaotic tipping points (Lenton/Held/Kriegler et al. 2006).

The relation between peace, the environment and gender may result in *cultural peace*. *Cultural peace* facilitates the creation of peace in the minds and actions of humankind. It socializes people so that religious and social discrimination can be overcome, by establishing human rights granted equally to all people. This will enable them to develop the ability to negotiate solutions to present and future conflicts peacefully and to share political, economic, social and cultural powers. The rights also respect different ecosystems, by taking into account their vulnerability to human actions.

Finally the links between peace, equity and gender introduce an *engendered peace*, creating relations between men and women, children, young people and elderly people based on equity. This will build up the capacity for negotiation, exchange, sharing, bargaining and developing tools, where the most vulnerable are empowered through development processes, non-violent conflict management, and disaster risk reduction (see chapter 8 by Oswald Spring in this volume). For Amster (2015)

peace ecology is more than merely a conceptual synthesis of peace and ecology. In essence, it contemplates the ways in which the same environmental processes that often drive conflict—e.g. resource depletion, anthropogenic climate change, food and water shortages—can also become opportunities for peaceful engagement. ... When confronted with a crisis ... recent research [argues] that people are actually more likely to cooperate than they are to compete, and that instances of violence in such situations are the exception rather than the norm (e.g. Solnit 2009). ... People around the world who strive to manage scarce essential resources (as with sensitive watersheds) often find that their mutual reliance on the resource transcends even profound cultural and political differences—and in some cases even warring parties have found ways to work together positively on such issues. Informed by schools of thought including *social ecology* ... and *deep ecology*, peace ecology is concerned equally with the human–human and human-environmental interfaces as they impact the search for peace at all levels (Amster 2015: 8-9).

Amster (2015: 143ff.) reviewed different concepts of and approaches to environmental cooperation, the concepts of *environmental dispute resolution* (Caplan 2010), *environmental peacemaking* (Conca/Dabelko 2002; Ali 2007, 2009) and *environmental peacebuilding* (Ecopeace 2016). However, he did not frame his argument in the context of the Anthropocene, nor did he discuss issues of sustainability transition or of the long-term global transition and transformation of the national and global economies that needs to be achieved through strategies aiming at a gradual decarbonization and dematerialization of the economy.

9.10   Sustainability Transition with Sustainable Peace

This concluding section focuses on possible interrelations between the research paradigm of a 'sustainability transition' and the goal of 'sustainable peace'. According to Geel's model, sociotechnological innovations emerge from small niches, and they often face major opposition within society (sociotechnical regime) from coalitions of special interest groups, business representatives and policymakers, who would lose from these innovations. This process will only succeed if new coalitions of societal, economic and political actors can successfully overcome opposing interests before these innovations can succeed globally (landscape). Globally successful innovations must take external factors—both earth systems and global economic and social megatrends—into account. However, both the DKI (Grin/Rotmans/Schot 2010) and the WBGU (2011) have not yet reflected on the implications for peace and security.

The two major technical revolutions in earth history have both resulted in technical and strategic revolutions and in more violent warfare. During the two world wars the total mobilization and militarization of technological innovations resulted in between 7,750,919 and 15 million deaths in World War I[50] and between 50 and 85 million deaths in World War II—if those who died from war-related famine and diseases are included[51]—and in more than 50 million deaths since 1945.[52]

The initial question was whether a long term transformative change towards sustainability —as in the aftermath of the Neolithic and the industrial revolutions—might result in a 'higher form of killing' or whether this transformation would foster more cooperative forms of multilateral cooperation and conflict resolution and peacebuilding. No definite answer is possible because future political constellations, decisions and events cannot be predicted.

However, different scenario outcomes may be assumed:

- A pessimist or *worst-case scenario* that assumes that—given the present global and national economic, political and societal power structures—such a major change in human and state behaviour is unforeseeable and that new technological innovations will be absorbed by the military, resulting in new revolutions in warfare.

- An optimist or *best-case scenario* assumes that a 'sustainability transition' to a low-carbon economy may foster the realization of a utopian vision of a global peaceful order, where war and patriarchy (Oswald Spring, chapter 8) must be overcome before a lasting, eternal or 'sustainable peace' in the Kantian tradition can prevail.

- A *pragmatic scenario* combines features of both the pessimist and the optimist scenarios and assumes that the behaviour of human beings and their interest in wealth, power and domination are unlikely to change. But sociotechnical innovations and a new scientific revolution towards sustainability (Clark/Crutzen/Schellnhuber 2004) will fundamentally change the world view of scientists and partly change the mindset of policymakers. Whether such a 'sustainability transition' might lead to a more peaceful and cooperative international order is an open question that cannot be scientifically answered.

Only this third *pragmatic* scenario will be discussed, from a pragmatic 'Grotian perspective' (Bull 1977; Bull/Kingsbury/Roberts 1992; Dean 2008, 2014) on

---

50 The figures for WWI differ: The American *Public Broadcasting System* (PBS) estimated a total number of deaths for all conflict parties as 8,528,831, of wounded as 21,189,154, and of prisoners and missing as 7,750,919, amounting to a total number of casualties of 37,466,904; at: <https://www.pbs.org/greatwar/resources/casdeath_pop.html>; a similar figure was offered by: <http://www.historylearningsite.co.uk/world-war-one/world-war-one-and-casualties/first-world-war-casualties>; according to Wikipedia for World War I, estimates of total deaths range from 9 million to over 15 million; at: <https://en.wikipedia.org/wiki/World_War_I_casualties>, based on Matthew White, Source List and Detailed Death Tolls for the Primary Megadeaths of the Twentieth Century: <http://necrometrics.com/20c5m.htm>.

51 The figures for WW II differ even more: Wikipedia offers a total for military deaths of between 21 and 25,5 million, of civilian deaths between 29 and 30,5 million, and of civilian deaths due to war-related famine and diseases of between 19 and 28 million, amounting to a total of 70 to 85 million; at: <https://en.wikipedia.org/wiki/World_War_II_casualties>; another source estimates 20,858,800 military deaths, 20,858,800 civilian deaths and 48,231,700 total deaths; at: <http://warchronicle.com/numbers/WWII/deaths.htm>.

52 The Worldmapper claimed that 51 million people were killed in wars between 1945 and 2000; at: <http://www.worldmapper.org/display.php?selected=287> (19 November 2015); see Max Roser, 2015: "War and Peace after 1945"; in: *OurWorldInData.org;* at: <http://ourworldindata.org/data/war-peace/war-and-peace-after-1945/> (19 November 2015).

Figure 9.3: Battle death rate in state-based conflicts by type (1946–2013) by Max Roser. Annual worldwide battle deaths per 100,000 people. **Source**: PRIO battle deaths dataset (1946–2007) and data provided by Steven Pinker for 2009 and later (based on UCDP and PRIO). Licensed under CC.BYSA by Max Roser; at: <http://ourworldindata.org/data/war-peace/war-and-peace-after-1945/>.

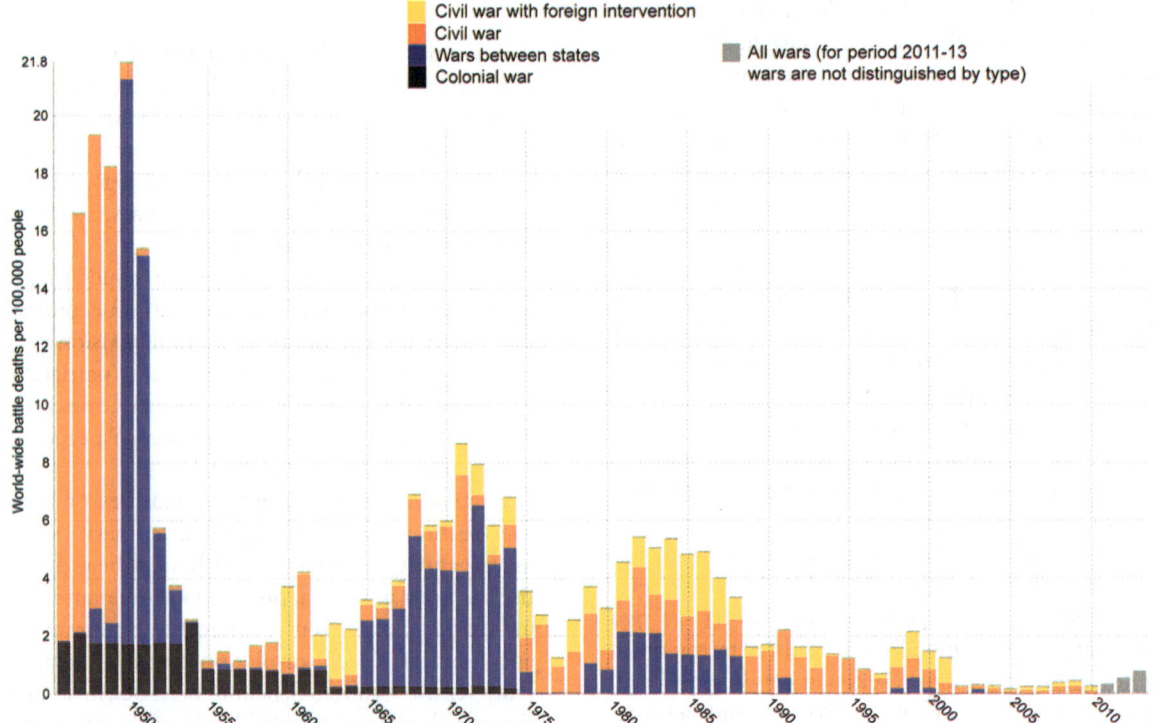

peace and security issues. A 'sustainability transition' towards a low-carbon industry challenges the old carbon-based industry, energy utilities, and the highway and car transportation lobbies whose economic interests would be negatively affected. Since 2007 these industries have been major sponsors of climate change critics (Klein 2011) and of their aggressive campaign against the IPCC. They have significantly changed public opinion about climate change issues in the US, Canada and Australia, where politicians who supported their economic interests have been successful.

Powerful interest groups in the primary exporting countries of coal, oil and natural gas have countered the debates on global climate change and the arguments for a decarbonization of the economy, while countries that import from fossil energy sources (e.g. the EU, but also China and India) have increasingly relied on renewable energy sources (wind, solar) for electricity.[53]

The global capacity of wind power has risen from 48 GW (2004) to 370 GW (2014) and of solar PV from 3.7 GW (2004) to 177 GW (2015), and global new investment in renewable power and fuels in both developed and developing countries increased from

US$45 billion to US$270 billion between 2004 and 2014 (figure 9.4, 9.5, 9.6). While both new fossil energy sources and renewables may counter resource conflicts, only the latter may avoid the negative security consequences of the physical effects of anthropogenic climate change.

As a 'decoupling of growth from energy consumption' (UNEP 2014; von Weizsäcker 2014) is possible if there is an energy efficiency improvement of a factor 4, 5 or 10 (von Weizsäcker/Lovins/Lovins 1997; von Weizsäcker/Hargroves/Smith et al. 2009; von Weizsäcker 2014) and through a replacement of fossil fuels with renewable energy sources, dependence on energy imports will gradually decline and resource wars may

---

53 See: REN 21 [Renewable Energy Network for the 21st Century], 2015: *Renewables 2015, Global Status Report* (Paris: REN21 c/o UNEP); REN 21, at: <http://www.ren21.net/wp-content/uploads/2015/07/REN12-GSR2015_Onlinebook_low1.pdf>; IRENA: *Renewable Energy Capacity Statistics* (Masdar City, Abu Dhabi: IRENA); at: <http://www.irena.org/DocumentDownloads/Publications/IRENA_RE_Capacity_Statistics_2015.pdf> (19 November 2015).

Figure 9.4: Increase in global wind power capacity (2004–2015). **Source:** REN 21 (2015); at: <http://www.ren21.net/status-of-renewables/global-status-report/>.

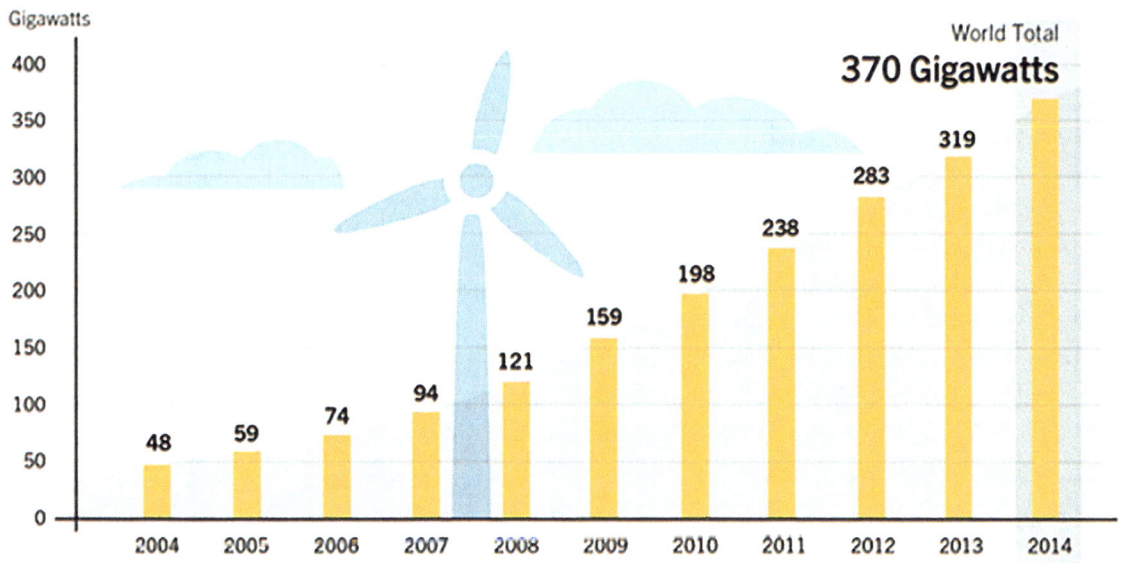

Figure 9.5: Increase in global solar PV capacity (2004–2015). **Source:** REN 21 (2015); at: <http://www.ren21.net/status-of-renewables/global-status-report/>.

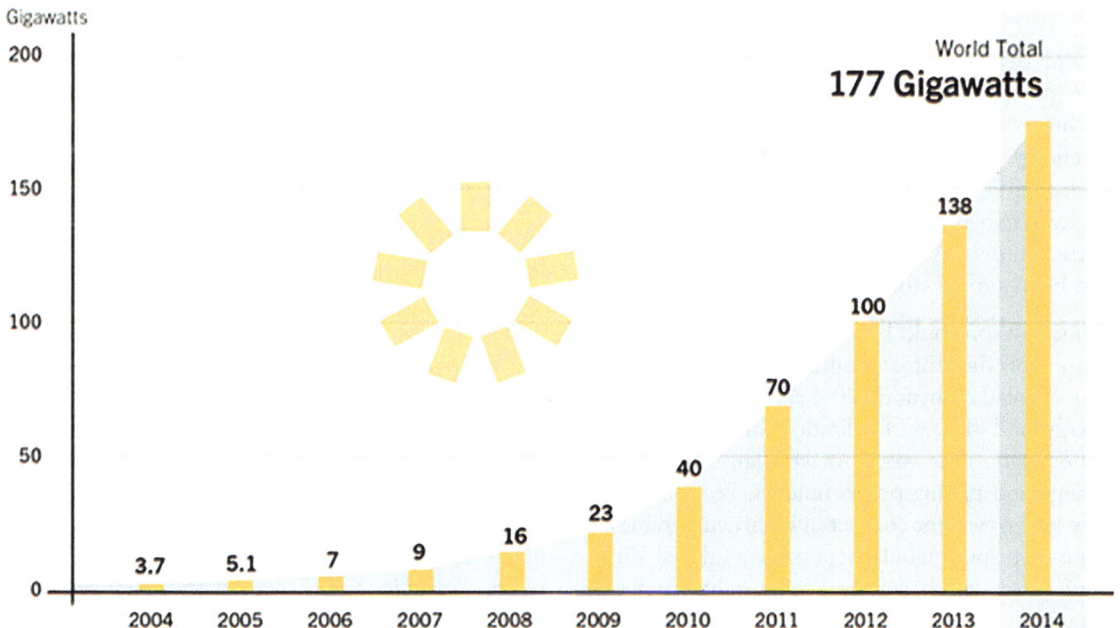

be less likely. This is a declared goal of the EU's roadmap for a low-carbon economy by 2050 (EU 2011b; Brauch 2016a).

Policymakers who favour a transformation of the economy towards a low-carbon society will face major and well-funded opposition. So far only the European Union (EU 2011, 2011a, 2011b, 2011c, 2011d; Hap-

paerts in this volume; Brauch 2016) and a majority of its twenty-eight countries have favoured such a strategy. Both the European Commission and the Council have argued that a sustainability transition in the energy and transport sector as the result of an efficiency revolution and a gradual replacement of fossil with renewable energy sources would significantly reduce both energy

Figure 9.6: Global new investment in renewable power and fuels, developed and developing countries (2004–2014). **Source:** REN 21 (2015); at: <http://www.ren21.net/status-of-renewables/global-status-report/>.

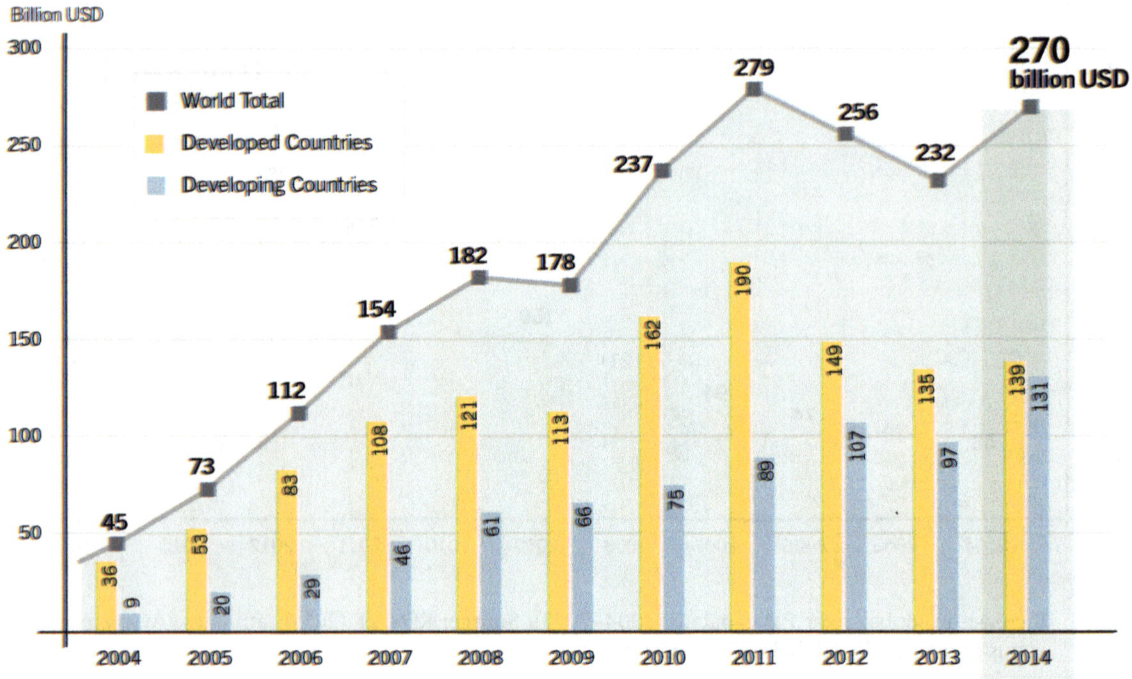

costs and dependence on imports of fossil energy sources from unstable regions such as the Middle East. This would not only decouple economic growth from energy consumption but possibly also from dependence on contested regions.

At least two peaceful outcomes of such a transition process towards a low-carbon economy and society can be assumed with some plausibility:

1. Climate wars would become highly unlikely if the causes of this threat could be countered by a major global reduction in the consumption of fossil energy and thus a stabilization of GHGs in the atmosphere. This would make dangerous climate change and tipping points unlikely. Preventing this new type of violent conflict in highly vulnerable climate hot spots would represent a global diplomatic strategy that fosters a sustainable peace (Brauch 2002; WBGU 2008; Scheffran/Brzoska/ Brauch et al. 2012; IPCC 2014).

2. For those countries that adopt and implement such a long-term transformative change towards sustainability in the energy, production, transportation, housing and agricultural sectors, their dependence on a scarce supply and rising energy prices from contested regions would decline. This would reduce the pressure to intervene militarily in order to guarantee access to and transportation of these

energy sources and to prevent their economy from collapsing. A second major beneficial outcome would be that classical resource conflicts, especially over oil and gas and other scarce strategic minerals, might be reduced (Klare 2001, 2012).[54]

Combining both challenges, Michael T. Klare (2013) argued: "Two nightmare scenarios—a global scarcity of vital resources and the onset of extreme climate change—are already beginning to converge and in the coming decades are likely to produce a tidal wave of unrest, rebellion, competition and conflict."[55]

The probable reduction of the likelihood of old 'resource conflicts' and the avoidance of new 'climate conflicts' through economic transformation towards sustainability would by itself not change the international order. It could, however, possibly foster new political coalitions of those countries who could invest in concrete projects that might result in such a dual

---

54 Stephen Kinzer: "America's next security threat: resource wars—This century's clashes will be conflicts over access to food, water, and energy", in: *Boston Globe*, 31 August 2014.

55 See Klare (2013); on this dual linkage, see also Leggett (2001, 2005, 2005a), Homer-Dixon (2006); and Lovins, Datta, Bustness et al. (2005).

decoupling of many highly vulnerable countries situated in climate change hot spots.

Thus, a sustainability transition towards a low-carbon economy by reducing the probability of two types of conflicts might foster policy strategies that aim for a more sustainable peace with more human security, where freedom from the impacts of climate-induced hazards would also be more likely. This linkage between the three new policy goals of a 'sustainable development' with 'human security' and 'sustainable peace' would draw basic lessons from the security implications of the Anthropocene, where 'we' have become the threat to the survival of humankind and only 'we' as part of humankind itself can offer a solution to the consequences of human intervention into the earth system.

These conceptual considerations have suggested the contextualization of the 'sustainability transition paradigm' and policy strategies for a 'low-carbon economy' as a sustainable peace goal within a human security strategy, as Sir David King (UK) has suggested. As only 'we' can offer the solutions to the impact of human interventions into the earth system it is crucial to bring sustainable peace considerations into the mainstream of conceptual thinking and policy action that aims for a long-term transformative change towards a low-carbon economy with high energy efficiency during this century.

The ten key messages of this chapter[56] are:

1. During the twentieth century humankind has for the first time significantly interfered in nature and the earth system and has silently moved as a result of multiple anthropogenic interventions—especially during the past sixty years—from the Holocene to the Anthropocene (Crutzen 2002, 2011; Crutzen/Brauch 2016), a new era of earth history.

2. This silent transition requires a "Second Copernican Revolution" (Schellnhuber 1999) with the emergence of a new 'scientific paradigm' aiming at sustainability and a new 'social contract for sustainable development' (Clark/Crutzen/Schellnhuber 2004; WBGU 2011) that can be advanced through a process of 'sustainability transition'.

3. This means that during the twenty-first century, the policy-focused social sciences must rethink and reconceptualize the key concepts of, and the design of research programmes on, peace, secu-

rity, development and the environment, including their cultural contexts.

4. Irrespective of the outcome of multilateral global climate diplomacy and of the adoption of climate treaties and their implementation, unilateral national and regional strategies aiming at a 'sustainability transition' in key economic sectors (energy, transportation, production and consumption, urbanization and housing, agriculture and forestry with water and soil management) are needed; they must be based on economic incentives and rationales.

5. A strategy that aims for a 'sustainability transition' towards a low-carbon economy may contribute to the realization of the triple goal of 'sustainable development' with 'human security' and 'sustainable peace' by reducing the probability of conflicts over energy, other non-renewable resources (fossil fuels, minerals) and renewable resources (including water and arable soil), and caused by the physical and societal impacts of climate change.

6. The two suggested multidisciplinary scientific research programmes of a 'political geoecology' and of a 'peace ecology' in the Anthropocene may contribute to the development of a new sustainability revolution (Oswald Spring/Brauch 2011) with a new science paradigm aiming for sustainability.

7. The goal of the proposed 'political geoecology' is to add to the 'geoecology' paradigm used in physical geography the 'political' dimension, including issues of peace and security, while sensitizing policy-oriented social scientists for new knowledge generated by scholars in earth system sciences and geoecology.

8. The proposed 'peace ecology' paradigm will link environmental and peace studies by addressing both the negative consequences of pollution and global environmental change and also the opportunities created by 'environmental peacemaking' and 'sustainability transition', within the framework of the *Sustainable Development Goals* (SDGs) in the Anthropocene.

9. Both research paradigms and approaches, when they are developed and discussed, must take the specific framework of the 'Anthropocene' fully into account. 'Sustainable peace in the Anthropocene' requires a rethinking of this 'normative concept' that is widely used for conflict prevention and post-conflict peacebuilding, focusing also on 'peace with nature'.

10. The proposed 'sustainability revolution' requires a *transformation* of global cultural, environmental,

---

56 This summary has benefitted from critical comments by Ursula Oswald Spring, based on a joint publication.

economic and political relations (Oswald Spring/ Brauch 2011: 1487ff.). This must replace the dominant vision and strategies of *business-as-usual* in a Hobbesian world where economic and strategic interests and behaviour predominate. Such Hobbesian interests and behaviour may possibly lead to a major crisis for humankind in inter-state relations, where the survival of the vulnerable is put at risk. If this happens, Earth as a habitat for humans and ecosystems will be destroyed.

In the dominant vision of *business-as-usual* a *cornucopian perspective* predominates that suggests primarily technical fixes (geoengineering) and the defence of economic, strategic and national interests in a way that only OECD countries can afford. The alternative vision requires a *sustainable perspective* based on global equity and social justice. The first vision with its minimal reactive strategies will most likely increase the probability of 'dangerous climate change' (Schellnhuber/Cramer/Nakicenovic et al. 2006) with both linear and chaotic changes in the climate system and their sociopolitical consequences. For Oswald Spring and Brauch (2011), the sustainability perspective

requires a change in *culture* (thinking on the human-nature interface), *world views* (thinking on systems of rule, e.g. democracy vs. autocracy, on domestic policies and inter-state relations), *mindsets* (the strategic perspectives of policymakers) and new forms of national and global *governance*.

This alternative vision identifies the need for a "new paradigm for global sustainability" (Clark/Crutzen/Schellnhuber 2004), for a "transition to [a] much more sustainable global society" (Raskin/Banuri/Gallopin et al. 2002), aimed at peace, freedom, material well-being and environmental health. Changes in technology and management systems alone will not be sufficient, but "significant changes in governance, institutions and value systems" are needed. These alternative strategies should be "more integrated, more long-term in outlook, more attuned to the natural dynamics of the Earth System and more visionary" (Steffen/Sanderson/Tyson et al. 2004: 291-293) and should take planetary boundaries into account (Rockström/Kulm 2015; see chapter 4 by Dalby).

References

Abdus Sabur, A.K.M., 2003: "Evolving a Theoretical Perspective of Human Security: the South Asian Context", in: Chari, P.R.; Gupta, Sonika (Eds.): *Human Security in South Asia* (New Delhi: Social Science Press): 35-51.

Abdus Sabur, A. K. M., 2009: "Theoretical Perspective on Human Security: A South Asian View", in: Brauch, Hans Günter; Oswald Spring, Úrsula; Grin, John; Mesjasz, Czeslaw; Kameri-Mbote, Patricia; Behera, Navnita Chadha; Chourou, Béchir; Krummenacher, Heinz (Eds.), 2009: *Facing Global Environmental Change: Environmental, Human, Energy, Food, Health and Water Security Concepts* (Berlin–Heidelberg–New York: Springer-Verlag): 1003-1011.

AC4 (Advanced Consortium on Cooperation, Conflict and Complexity), 2015: *The Sustainable Peace Mapping Initiative: What is Sustainable Peace? Expert Survey Report* (New York: Columbia University, The Earth Institute, AC4).

AC4 (Advanced Consortium on Cooperation, Conflict and Complexity by Coleman, Peter; Burns, Danny; Fisher-Yoshida, Beth; Fisher, Joshua D.), 2015a: "Systems-based Applications of Sustainable Peace: A stakeholder-informed approach to designing interactive policy support tools for policy making (New York: Columbia University, The Earth Institute, AC4).

Acharya, Amitav, 2004: "A Holistic Paradigm", in: *Security Dialogue*, 35,3 (September): 355-356.

Ake, Claude, 1993: "Development and Underdevelopment", in: Krieger, Joel (Ed.): *The Oxford Companion to Politics of the World* (New York–Oxford: Oxford University Press): 239-243.

Albrecht, Ulrich; Brauch, Hans Günter, 2008: "Security in Peace Research and Security Studies", in: Brauch, Hans Günter; Oswald Spring, Úrsula; Mesjasz, Czeslaw; Grin, John; Dunay, Pal; Behera, Navnita Chadha; Chourou, Béchir; Kameri-Mbote, Patricia; Liotta, P.H. (Eds.): *Globalization and Environmental Challenges: Reconceptualizing Security in the 21st Century* (Berlin–Heidelberg–New York: Springer-Verlag): 501-523.

Ali, Saleem H. (Ed.), 2007: *Peace Parks: Conservation and Conflict Resolution* (Cambridge, MA: MIT Press).

Ali, Saleem H., 2009: "Extractive Industries and the Environmental Aspects of International Security", in: Brauch, Hans Günter; Oswald Spring, Úrsula; Grin, John; Mesjasz, Czeslaw; Kameri-Mbote, Patricia; Behera, Navnita Chadha; Chourou, Béchir; Krummenacher, Heinz (Eds.), 2009: *Facing Global Environmental Change: Environmental, Human, Energy, Food, Health and Water Security Concepts* (Berlin–Heidelberg–New York: Springer-Verlag): 335-342.

Amster, Randall, 2014, 2015: *Peace Ecology* (Boulder, CO: Paradigm).

Annan, Kofi A., 2005: *In Larger Freedom: Towards Security, Development and Human Rights for All. Report of the Secretary General for Decision by Heads of State and Government in September* 2005. A/59/2005 (New York: United Nations, Department of Public Information, 21 March).

Ariyabandu, Madhavi Malalgoda; Fonseka, Dilrukshi, 2009: "Do Disasters Discriminate? A Human Security Analysis of the impact of the Tsunami in India, Sri Lanka and of the Kashmir Earthquake in Pakistan", in: Brauch, Hans Günter; Oswald Spring, Úrsula; Grin, John; Mesjasz, Czeslaw; Kameri-Mbote, Patricia; Behera, Navnita Chadha; Chourou, Béchir; Krummenacher, Heinz (Eds.), 2009: *Facing Global Environmental Change: Environmental, Human, Energy, Food, Health and Water Security Concepts* (Berlin–Heidelberg–New York: Springer-Verlag): 1215-1226.

Arizpe, Lourdes; Price, Martin F.; Worcester, Robert, 2016: "The First Decade of Initiatives for Research on the Human Dimensions of Global (Environmental) Change (1986-1995)", in: Brauch, Hans Günter; Oswald Spring, Úrsula; Grin, John; Scheffran; Jürgen (Eds.), 2016: *Handbook on Sustainability Transition and Sustainable Peace* (Cham–Heidelberg–New York–Dordrecht–London: Springer).

Babcock Gove, Philip, 2002: *Webster's Third New International Dictionary* (Springfield, MA: Merriam Webster).

Bajpai, Kanti, 2000: "Human Security: Concept and Measurement", Occasional Paper 19 (Notre Dame, IN: Kroc Institute for International Peace Studies, University of Notre Dame); at: <http://kroc.nd.edu/ocpapers/op_19_1.PDF>.

Bajpai, Kanti, 2003: "The Idea of Human Security", in: *International Studies*, 40,3 (July-September): 195-228.

Bajpai, Kanti, 2004: "An Expression of Threats Versus Capabilities Across Time and Space", in: *Security Dialogue*, 35,3 (September): 360-361.

Bajpai, Kanti, 2005: "Human Security: Concept and Measurement", in Bajpai, Kanti; Mallavarapu Siddharth (Eds.): *International Relations in India: Bringing Theory Back Home* (New Delhi: Orient Longman): 275-332.

Barnett: Jon; Matthew, Richard A.; O'Brien; Karen, 2008: "Global Environmental Change and Human Security", in: Brauch, Hans Günter; Oswald Spring, Úrsula; Mesjasz, Czeslaw; Grin, John; Dunay, Pal; Behera, Navnita Chadha; Chourou, Béchir; Kameri-Mbote, Patricia; Liotta, P.H. (Eds.): *Globalization and Environmental Challenges: Reconceptualizing Security in the 21st Century* (Berlin–Heidelberg–New York: Springer-Verlag): 355-362.

Bellamy, Alex J.; McDonald, Matt, 2002: "'The Utility of Human Security': Which Humans? What Security? A Reply to Thomas & Tow", in: *Security Dialogue*, 33,3 (September): 373-377.

Bernauer, Thomas; Böhmelt, T.; Koubi, Valery, 2012: "Environmental changes and violent conflict", in: *Environmental Research Letters*, 7,1: 1-8.

Boeck, Andreas, 1995: "Entwicklungstheorien", in: Nohlen, Dieter; Schultze, Rainer-Olaf (Eds.): *Lexikon der Politik, vol. 1: Politische Theorien* (München: C.H. Beck): 69-80.

Bogardi, Janos; Brauch, Hans Günter, 2005: "Global Environmental Change: A Challenge for Human Security–Defining and conceptualising the environmental dimension of human security", in: Rechkemmer, Andreas (Ed.): *UNEO–Towards an International Environment Organization–Approaches to a sustainable reform of global environmental governance* (Baden-Baden: Nomos): 85-109.

Bohle, Hans-Georg, 2009: "Sustainable Livelihood Security. Evolution and Application", in: Brauch, Hans Günter; Oswald Spring, Úrsula; Grin, John; Mesjasz, Czeslaw; Kameri-Mbote, Patricia; Behera, Navnita Chadha; Chourou, Béchir; Krummenacher, Heinz (Eds.), 2009: *Facing Global Environmental Change: Environmental, Human, Energy, Food, Health and Water Security Concepts* (Berlin–Heidelberg–New York: Springer-Verlag): 521-528.

Bohle, Hans-Georg; O'Brien, Karen, 2007: "The Discourse on Human Security: Implications and Relevance for Climate Change Research. A Review Article", in: *Climate Change and Human Security*. Special issue of *Die Erde*, 137,3: 155-163.

Bonilla, Adrian, 2003: "Human Security in the Andean Region", in: Goucha, Moufida; Rojas Aravenna, Francisco (Eds.), 2003: *Human Security, Conflict Prevention and Peace* (Paris: UNESCO): 337-351.

Boomgarden, George, 2007: "Opening speech", in: GTZ (Ed.): *Desertification: a security threat? Analysis of risks and challenges. A conference on the occasion of the World Day to Combat Desertification* 2007 Federal Foreign Office, Berlin, 26 June 2007 (Berlin: GTZ, August 2008): at: <http://www.auswaertiges-amt.de/diplo/de/Aussenpolitik/InternatOrgane/VereinteNationen/VN-Engagements/DesertificationAndSecurity-Download.pdf>: 4-5.

Booth, Ken (Ed.), 2005: *Critical Security Studies and World Politics* (Boulder, CO–London: Rienner).

Booth, Ken, 2007: *Theory of World Security* (Cambridge–New York: Cambridge University Press).

Boserup, Ester, 1970: *Women's Role in Economic Development* (New York: St. Martin's Press).

Boulding, Kenneth E., 1950: *Reconstruction of Economics* (New York: John Wiley & Sons).

Boulding, Kenneth E., 1966: "The Economics of the Coming Spaceship Earth", in: Jarrett, Henry (Ed.): *Environmental Quality in a Growing Economy, Essays from the Sixth RFF Forum on Environmental Quality held in Washington, March 8 and 9, 1966* (Baltimore, Johns Hopkins Press): 3-14.

Boulding, Kenneth E., 1970: "The Economics of the Coming Spaceship Earth", in: Boulding, Kenneth E., (Ed.): *Beyond Economics: Essays on Society, Religion, and Ethics* (Ann Arbor: University of Michigan Press): 275-287.

Boulding, Kenneth E., 1978: *Ecodynamics: A new theory of societal evolution* (Beverly Hills: Sage).

Boulding, Kenneth E., 1983: "Ecodynamcis", in: *Interdisciplinary Science Reviews*, 8,2 (June): 108–113.

Brauch, Hans Günter, 2002: "Climate Change, Environmental Stress and Conflict–AFES-PRESS Report for the Federal Ministry for the Environment, Nature Conservation and Nuclear Safety", in: Federal Ministry for the Environment, Nature Conservation and Nuclear Safety (Ed.): *Climate Change and Conflict. Can climate change impacts increase conflict potentials? What is the relevance of this issue for the international process on climate change?* (Berlin: Federal Ministry for the Environment, Nature Conservation and Nuclear Safety, 2002): 9–112; at: <http://www.afes-press.de/pdf/ Brauch_ClimateChange_BMU. pdf>.

Brauch, Hans Günter, 2003: "Towards a Fourth Phase of Research on Human and Environmental Security and Peace: Conceptual Conclusions", in: Brauch, Hans Günter; Liotta, P.H; Marquina, Antonio; Rogers, Paul; Selim, Mohammed El-Sayed (Eds.): *Security and Environment in the Mediterranean. Conceptualising Security and Environmental Conflicts* (Berlin–Heidelberg: Springer 2003): 919–954.

Brauch, Hans Günter, 2005: *Environment and Human Security. Freedom from Hazard Impact*, InterSecTions, 2/2005 (Bonn: UNU-EHS); at: <http://www.ehs.unu. edu/file.php?id=64>.

Brauch, Hans Günter, 2005a: *Threats, Challenges, Vulnerabilities and Risks in Environmental Human Security*. Source, 1/2005 (Bonn: UNU-EHS); at: <http://www. ehs.unu.edu/index.php?module=overview&cat=17&menu =36>.

Brauch, Hans Günter, 2008: "Introduction: Globalization and Environmental Challenges: Reconceptualizing Security in the 21st Century", in: Brauch, Hans Günter; Oswald Spring, Úrsula; Mesjasz, Czeslaw; Grin, John; Dunay, Pal; Behera, Navnita Chadha; Chourou, Béchir; Kameri-Mbote, Patricia; Liotta, P.H. (Eds.): *Globalization and Environmental Challenges: Reconceptualizing Security in the 21st Century* (Berlin–Heidelberg–New York: Springer-Verlag): 27–43.

Brauch, Hans Günter, 2008a: "Conceptual Quartet: Security and its Linkages with Peace, Development and Environment", in: Brauch, Hans Günter; Oswald Spring, Úrsula; Mesjasz, Czeslaw; Grin, John; Dunay, Pal; Behera, Navnita Chadha; Chourou, Béchir; Kameri-Mbote, Patricia; Liotta, P.H. (Eds.): *Globalization and Environmental Challenges: Reconceptualizing Security in the 21st Century* (Berlin–Heidelberg–New York: Springer-Verlag): 65–98.

Brauch, Hans Günter, 2009: "Securitizing Global Environmental Change", in: Brauch, Hans Günter; Oswald Spring, Úrsula; Grin, John; Mesjasz, Czeslaw; Kameri-Mbote, Patricia; Behera, Navnita Chadha; Chourou, Béchir; Krummenacher, Heinz (Eds.), 2009: *Facing Global Environmental Change: Environmental,* *Human, Energy, Food, Health and Water Security Concepts* (Berlin–Heidelberg–New York: Springer-Verlag): 65–102.

Brauch, Hans Günter, 2009a: "Introduction: Facing Global Environmental Change and Sectorialization of Security", in: Brauch, Hans Günter; Oswald Spring, Úrsula; Grin, John; Mesjasz, Czeslaw; Kameri-Mbote, Patricia; Behera, Navnita Chadha; Chourou, Béchir; Krummenacher, Heinz (Eds.), 2009: *Facing Global Environmental Change: Environmental, Human, Energy, Food, Health and Water Security Concepts* (Berlin–Heidelberg–New York: Springer-Verlag): 27–44.

Brauch, Hans Günter, 2009b: "Human Security Concepts in Policy and Science", in: Brauch, Hans Günter; Oswald Spring, Úrsula; Grin, John; Mesjasz, Czeslaw; Kameri-Mbote, Patricia; Behera, Navnita Chadha; Chourou, Béchir; Krummenacher, Heinz (Eds.): *Facing Global Environmental Change: Environmental, Human, Energy, Food, Health and Water Security Concepts*. (Berlin–Heidelberg–New York: Springer-Verlag): 965–990.

Brauch, Hans Günter, 2012: "Climate Paradox of the G-8: legal obligations, policy declarations and implementation gap", in: *Revista Brasileira de Politica Internacional*. Special issue edited by Eduardo Viola & Antonio Carlos Lessa on *Global Climate Governance and Transition to a Low Carbon Economy* (Brasilia: Instituto Brasileira de Realacoes Internacionais): 30–52.

Brauch, Hans Günter, 2014: "From Climate Change and Security Impacts to Sustainability Transition: Two Policy Debates and Scientific Discourses", in: Oswald Spring, Úrsula; Brauch, Hans Günter; Tidball, Keith G. (Eds.): *Expanding Peace Ecology: Peace, Security, Sustainability, Equity and Gender: Perspectives of IPRA's Ecology and Peace Commission* (Heidelberg–Dordrecht–London–New York: Springer-Verlag): 33–61.

Brauch, Hans Günter, 2016: "Historical Times and Turning Points in a Turbulent Century: 1914, 1945, 1989 and 2014?", in: Brauch, Hans Günter; Oswald Spring; Úrsula; Bennett, Juliet; Serrano Oswald, Serena Eréndira (Eds.): *Addressing Global Environmental Challenges from a Peace Ecology Perspective* (Cham–Heidelberg–Dordrecht–London–New York: Springer International Publishing).

Brauch, Hans Günter, 2016a: "Building Sustainable Peace by Moving towards Sustainability Transition", in: Brauch, Hans Günter; Oswald Spring, Úrsula; Bennett, Juliet (Eds.): *Addressing Global Environmental Challenges from a Peace Ecology Perspective* (Cham–Heidelberg–Dordrecht–London–New York: Springer International Publishing).

Brauch, Hans Günter; Dalby, Simon; Oswald Spring, Úrsula, 2011: "Political Geoecology for the Anthropocene", in: Brauch, Hans Günter; Oswald Spring, Úrsula; Mesjasz, Czeslaw; Grin, John; Kameri-Mbote, Patricia; Chourou, Béchir; Dunay, Pal; Birkmann, Jörn (Eds.), 2011: *Coping with Global Environmental Change, Disasters and Secu-*

rity–Threats, Challenges, Vulnerabilities and Risks (Berlin–Heidelberg–New York: Springer International Publishing): 1453-1486.

Brauch, Hans Günter; Kennedy, Robert (Eds.), 1990: *Alternative Conventional Defense Postures for the European Theater. Vol. 1: The Military Balance and Domestic Constraints* (New York–Bristol–Washington–London: Crane Russak: Taylor & Francis Group).

Brauch, Hans Günter; Kennedy, Robert (Eds.), 1992: *Alternative Conventional Defense Postures for the European Theater. Vol. 2: The Impact of Political Change on Strategy, Technology and Arms Control* (New York–Philadelphia–Washington–London: Crane Russak: Taylor & Francis Group).

Brauch, Hans Günter; Kennedy, Robert (Eds.), 1993: *Alternative Conventional Defense Postures for the European Theater. Vol. 3: Force Posture Alternatives for Europe after the Cold War* (Washington–Philadelphia–London: Crane Russak: Taylor & Francis Group).

Brauch, Hans Günter; Oswald Spring, Úrsula, 2009: *Securitizing the Ground. Grounding Security* (Bonn: UNCCD).

Brauch, Hans Günter; Oswald Spring, Úrsula, 2009a: "Towards Sustainable Peace for the 21st Century", in: Brauch, Hans Günter; Oswald Spring, Úrsula; Grin, John; Mesjasz, Czeslaw; Kameri-Mbote, Patricia; Behera, Navnita Chadha; Chourou, Béchir; Krummenacher, Heinz (Eds.): *Facing Global Environmental Change: Environmental, Human, Energy, Food, Health and Water Security Concepts* (Berlin–Heidelberg–New York: Springer-Verlag): 1295-1310.

Brauch, Hans Günter; Oswald Spring, Úrsula; Grin, John; Mesjasz, Czeslaw; Kameri-Mbote, Patricia; Behera, Navnita Chadha; Chourou, Béchir; Krummenacher, Heinz (Eds.), 2009: *Facing Global Environmental Change: Environmental, Human, Energy, Food, Health and Water Security Concepts* (Berlin–Heidelberg–New York: Springer-Verlag).

Brauch, Hans Günter; Oswald Spring, Úrsula; Mesjasz, Czeslaw; Grin, John; Dunay, Pal; Behera, Navnita Chadha; Chourou, Béchir; Kameri-Mbote, Patricia; Liotta, P.H. (Eds.), 2008: *Globalization and Environmental Challenges: Reconceptualizing Security in the 21st Century* (Berlin–Heidelberg–New York: Springer-Verlag).

Brauch, Hans Günter; Oswald Spring, Úrsula; Mesjasz, Czeslaw; Grin, John; Kameri-Mbote, Patricia; Chourou, Béchir; Dunay, Pal; Birkmann, Jörn (Eds.), 2011: *Coping with Global Environmental Change, Disasters and Security–Threats, Challenges, Vulnerabilities and Risks* (Berlin–Heidelberg–New York: Springer-Verlag).

Brauch, Hans Günter; Oswald Spring; Úrsula; Bennett, Juliet (Eds.), 2016: *Addressing Global Environmental Challenges from a Peace Ecology Perspective* (Cham–Heidelberg–Dordrecht–London–New York: Springer International Publishing).

Brauch, Hans Günter; Scheffran, Jürgen, 2012: "Introduction", in: Scheffran, Jürgen; Brzoska, Michael; Brauch, Hans Günter; Link, Peter Michael; Schilling, Janpeter (Eds.): *Climate Change, Human Security and Violent Conflict: Challenges for Societal Stability* (Berlin–Heidelberg–New York: Springer-Verlag, 2012): 3-40.

Braudel, Fernand, 1969: "Histoire et science sociales. La longue durée", in: *Écrits Sur l'Histoire* (Paris: Flammarion): 41-84.

Braudel, Fernand, 1972: *The Mediterranean and the Mediterranean World in the Age of Philip II,* 2 volumes (New York: Harper & Row).

Brockhaus, 21 2006: *Brockhaus Enzyklopädie,* 30 vols (Leipzig–Mannheim: FA Brockhaus).

Brundtland Commission (World Commission on Environment and Development), 1987: *Our Common Future. The World Commission on Environment and Development* (Oxford–New York: Oxford University Press).

Brzoska, Michael (2015): "Climate Change and Military Planning", in: *International Journal of Climate Change Strategies and Management,* 7,2, <doi:10.1108/IJCCSM-10-2013-0114>.

Buhaug, Halvard, 2010: "Climate not to blame for African civil wars", in: *PNAS,* 107: 16477-16482.

Buhaug, Halvard, 2014: "Concealing agreements over climate-conflict results", in: *PNAS,* 111,6: E636.

Bull, Hedley, 1977: *The Anarchical Society. A Study of Order in World Politics* (New York: Columbia University Press–London: Macmillan).

Bull, Hedley; Kingsbury, Benedict; Roberts, Adam (Eds.), 1992: *Hugo Grotius and International Relations* (Oxford: Clarendon Press).

Burke, M.B.; Miguel, E.; Satyanath, S.; Dykema, J.A.; Lobell, D.B., 2009: "Warming increases the risk of civil war in Africa", in: *PNAS,* 106,49: 20670-20674.

Burton, John W., 1998: "Conflict Resolution: The Human Dimension", in: *International Journal of Peace Studies,* 3,1; at: <http://www.gmu.edu/academic/ijps/vol3-1/burton.htm>.

Busumtwi-Sam, James, 2002: "Development and Human Security: Whose Security and from What?", in: *International Journal,* 57,2: 253-272.

Buzan, Barry, 1983: *People, States & Fear. The National Security Problem in International Relations* (Brighton: Harvester Books - Chapel Hill: University of North Carolina Press; 2nd ed. 1991; reprint with new preface 2007).

Buzan, Barry, 2004: "A Reductionist, Idealistic Notion that Adds Little Analytical Value", in: *Security Dialogue,* 35,3 (September): 369-370.

Buzan, Barry, 2 1991: *People, States and Fear. An Agenda for International Security Studies in the Post-Cold War Era* (London: Harvester Wheatsheaf–Boulder, Co.: Lynne Rienner).

Buzan, Barry; Wæver, Ole, 2003: *Regions and Powers: The Structure of International Security* (Cambridge: Cambridge University Press).

Buzan, Barry; Wæver, Ole; de Wilde, Jaap, 1998, 2 2004: *Security. A New Framework for Analysis* (Boulder–London: Lynne Rienner).

Caplan, James A., 2010: *The Theory and Principles of Environmental Dispute Resolution* (San Bernardino, CA: edrusa.com).

Chari, P. R.; Gupta, Sonika (Eds.), 2003a: *Human Security in South Asia: Gender, Energy, Migration and Globalisation* (New Delhi: Social Science Press).

Chari, P.R.; Gupta, Sonika, 2003: "Introduction", in: Chari, P.R.; Gupta, Sonika (Eds.). *Human Security in South Asia* (New Delhi, Social Science Press): 1-21.

Chen, Richard S.; Boulding, Elise; Schneider, Stephen H. (Eds.), 1983: *Social science and climate change* (Dordrecht: Reidel).

Chipman, John, 1992: "The Future of Strategic Studies: Beyond Grand Strategy", in: *Survival*, 34,1: 109-131.

Chourou, Bechir, 2005: *Promoting Human Security: Ethical, Normative and Educational Frameworks in the Arab States* (Paris: UNESCO).

Chourou, Béchir, 2009: "Human Security in the Arab World: A Perspective from the Maghreb", in: Brauch, Hans Günter; Oswald Spring, Úrsula; Grin, John; Mesjasz, Czeslaw; Kameri-Mbote, Patricia; Behera, Navnita Chadha; Chourou, Béchir; Krummenacher, Heinz (Eds.), 2009: *Facing Global Environmental Change: Environmental, Human, Energy, Food, Health and Water Security Concepts* (Berlin–Heidelberg: Springer-Verlag): 1021-1035.

CHS [Commission on Human Security], 2003: *Human Security Now* (New York: Commission on Human Security); at: <http://www.humansecurity-chs.org/finalreport/>.

Church, John A.; Asrar, Ghassem R.; Busalacchi, Antonio J.; Arndt, Carolin E., 2011: "Climate Information for Coping with Environmental Change: Contributions of the World Climate Research Programme", in: Brauch, Hans Günter; Oswald Spring, Úrsula; Mesjasz, Czeslaw; Grin, John; Kameri-Mbote, Patricia; Chourou, Béchir; Dunay, Pal; Birkmann, Jörn, 2010: *Coping with Global Environmental Change, Disasters and Security–Threats, Challenges, Vulnerabilities and Risks* (Berlin–Heidelberg–New York: Springer-Verlag): 1257-1270.

Clark, William C.; Crutzen, Paul J.; Schellnhuber, Hans Joachim, 2004: "Science and Global Sustainability: Toward a New Paradigm", in: Schellnhuber, Hans Joachim; Crutzen, Paul J.; Clark, William C.; Claussen, Martin; Held, Hermann (Eds.): *Earth System Analysis for Sustainability* (Cambridge, MA; London: MIT Press): 1-28.

Coleman, Peter T., "The Essence of Peace? Toward a Comprehensive *and* Parsimonious Model of Sustainable Peace", in: Brauch, Hans Günter; Oswald Spring, Úrsula; Grin, John; Scheffran; Jürgen (Eds.), 2016: *Handbook on Sustainability Transition and Sustainable Peace* (Cham–Heidelberg–New York–Dordrecht–London: Springer).

Coleman, Peter T.; Deutsch, Morton (Eds.), 2012: *Psychological Components of Sustainable Peace* (New York–Heidelberg–Dordrecht–London: Springer, 2012).

Conca, Ken, 1994: "In the Name of Sustainability: Peace Studies and Environmental Discourse", in: Kakonen, Jyrki (Ed.): *Green Security or Militarized Environment* (Dartmouth: Aldershot): 7-24.

Conca, Ken, 2002: "The Case for Environmental Peacemaking", in: Conca, Ken; Dabelko, Geoffrey (Eds.): *Environmental Peacemaking* (Baltimore: Johns Hopkins University Press): 1-22.

Conca, Ken; Dabelko, Geoffrey (Eds.), 2002: *Environmental Peacemaking* (Baltimore: Johns Hopkins University Press).

Correira de Barros, Manuel, 2005: "Profiling youth involved in the informal markets of Luanda", in: Mulongo, Keith; Kibasomba, Roger; Kariri, Jemina Njeri (Eds.): *The Many Faces of Human Security. Case Studies of Seven Countries in Southern Africa* (Pretoria: Institute for Security Studies): 201-224; at: <http://www.reliefweb.int/rw/RWFiles2005.nsf/FilesByRWDocUNIDFileName/KKEE-6HZQQQ-iss-saf-10nov.pdf/$File/iss-saf-10nov.pdf>.

Crawford, Robert M.A.; Jarvis, Darryl S. L. (Eds.), 2001: *International Relations–Still an American Social Science? Toward Diversity in International Thought* (Albany: State University of New York Press).

Crutzen, Paul J., 2002: "Geology of Mankind", in: *Nature*, 415,3 (January): 23.

Crutzen, Paul J., 2011: "The Anthropocene: Geology by Mankind", in: Brauch, Hans Günter; Oswald Spring, Úrsula; Mesjasz, Czeslaw; Grin, John; Kameri-Mbote, Patricia; Chourou, Béchir; Dunay, Pal; Birkmann, Jörn (Eds.), 2011: *Coping with Global Environmental Change, Disasters and Security–Threats, Challenges, Vulnerabilities and Risks* (Berlin–Heidelberg–New York: Springer-Verlag): 3-4.

Crutzen, Paul J.; Birks, John W., 1982: "The atmosphere after a nuclear war: Twilight at noon", in: *Ambio*, 2/3: 114-125.

Crutzen, Paul J.; Brauch, Hans Günter (Eds.), 2016: *Paul J. Crutzen: A Pioneer on Atmospheric Chemistry, Biosphere, and Climate in the Anthropocene* (Cham–Heidelberg–New York–Dordrecht–London: Springer).

Crutzen, Paul J.; Stoermer, Eugene F., 2000: "The Anthropocene", in: *IGBP Newsletter*, 41: 17-18.

Dalby, Simon, 2015a: "Anthropocene Formations: Environmental Security, Geopolitics and Disaster", in: *Theory, Culture and Society*. Special issue on *Geosocial Formations and the Anthropocene*: in: OnlineFirst (August); at: <http://tcs.sagepub.com/content/early/recent>.

Dalby, Simon, 2007: "Ecology, Security, and Change in the Anthropocene", in: *Brown Journal of World Affairs*, 13,2: 155-164.

Dalby, Simon, 2009: *Security and Environmental Change* (Cambridge: Polity).

Dalby, Simon, 2009a: "Peacebuilding and Environmental Security in the Anthropocene", in: Péclard, Didier (Ed.): *Environmental Peacebuilding: Managing Natural*

Resource Conflicts in a Changing World. Conference Paper 2009-1 (Bern: Swisspeace): 8-21.

Dalby, Simon, 2013: "Biopolitics and climate security in the Anthropocene", in: *Geoforum* (October): 184-192.

Dalby, Simon, 2013a: "Human Security in the Anthropocene: The Implications of Earth Systems Analysis", in: O'Brien, Karen; Wolf, Johanna; Sygna, Linda (Eds.): *The Changing Environment for Human Security: Transformative Approaches to Research, Policy, and Action* (London: Earthscan): 27-33.

Dalby, Simon, 2014: "Rethinking Geopolitics: Climate Security in the Anthropocene", in: *Global Policy*, 5,1 (February): 1-9.

Dalby, Simon, 2015: "International Security in the Anthropocene", in: *E_International Relations* (23 February); at: <http://www.e-ir.info/2015/02/23/international-security-in-the-anthropoce-ne/>.

Dalby, Simon, 2016: "Climate Security in the Anthropocene: 'Scaling up' the Human Niche", in: Wapner, Paul; Elver, Hilal (Eds.): *Reimagining Climate Change* (New York: Routledge): 29-48.

Dalby, Simon; Brauch, Hans Günter; Oswald Spring, Úrsula, 2009: "Environmental Security Concepts Revisited During the First Three Phases (1983-2006)", in: Brauch, Hans Günter; Oswald Spring, Úrsula, Grin, John; Mesjasz, Czeslaw; Kameri-Mbote, Patricia; Behera, Navnita Chadha; Chourou, Béchir; Krummenacher, Heinz (Eds.), 2009: *Facing Global Environmental Change: Environmental, Human, Energy, Food, Health and Water Security Concepts* (Berlin–Heidelberg–New York: Springer-Verlag): 781-790.

Dannreuther, Roland, 2007: *International Security. The Contemporary Agenda* (Cambridge: Polity).

De Lombaerde, Philippe; Norton, Matthew, 2009: "Human Security in Central America", in: Brauch, Hans Günter; Oswald Spring, Úrsula; Grin, John; Mesjasz, Czeslaw; Kameri-Mbote, Patricia; Behera, Navnita Chadha; Chourou, Béchir; Krummenacher, Heinz (Eds.), 2009: *Facing Global Environmental Change: Environmental, Human, Energy, Food, Health and Water Security Concepts* (Berlin–Heidelberg–New York: Springer-Verlag): 1063-1076.

Dean, Jonathan, 2008: "Rethinking Security in the New Century—Return to the Grotian Pattern", in: Brauch, Hans Günter; Oswald Spring, Úrsula; Mesjasz, Czeslaw; Grin, John; Dunay, Pal; Behera, Navnita Chadha; Chourou, Béchir; Kameri-Mbote, Patricia; Liotta, P.H. (Eds.), 2008: *Globalization and Environmental Challenges: Reconceptualizing Security in the 21st Century* (Berlin–Heidelberg–New York: Springer-Verlag): 3-6.

Dean, Jonathan, 2014: "Rethinking Security in the New Century—Return to the Grotian Pattern", in: Brauch, Hans Günter; Grimwood, Teri (Eds.): *Jonathan Dean: Pioneer in Détente in Europe, Global Cooperative Security Arms Control and Disarmament* (Cham–Heidelberg–New York–Dordrecht–London: Springer-Verlag, 2015): 207-212.

Deutsch, Morton; Coleman, Peter T., 2016: "The Psychological Components of a Sustainable Peace: An Introduction", in: Brauch, Hans Günter; Oswald Spring, Úrsula; Grin, John; Scheffran; Jürgen (Eds.): *Handbook on Sustainability Transition and Sustainable Peace* (Cham–Heidelberg–New York–Dordrecht–London: Springer).

Diehl, Paul F., 2016: "Exploring Peace. Looking Beyond War and Negative Peace", in: *International Studies Quarterly* (forthcoming).

Dodds, Felix; Pippard, Tim, (Eds.) 2005: *Human and Environmental Security: An Agenda for Change* (London–Sterling, VA: Earthscan).

urdevic-Lukic, Svetlana, 2004: "Broadening Security Concept–From 'National' to 'Human Security'", in: *Izvorni Naucni rad*, 56, 4 (December): 397-408.

Dzimba, J.; Matooane, Matsolo, 2005: "The impact of stock theft on human security: Strategies for combating stock theft in Lesotho", in: Mulongo, Keith; Kibasomba, Roger; Kariri, Jemina Njeri (Eds.): *The Many Faces of Human Security. Case Studies of Seven Countries in Southern Africa* (Pretoria: Institute for Security Studies). 265-300; at: <http://www.reliefweb.int/rw/RWFiles 2005.nsf/FilesByRWDocUNIDFileName/KKEE-6HZQQQ-iss-saf-10nov.pdf/$File/iss-saf-10nov.pdf>.

Ecopeace [Kool, Jeroen; Bamya, Saeb; Talozi, Samer] 2016: *Sustainable Development in the Jordan Valley–Final Report of the Regional NGO Master Plan* (Cham–Heidelberg–New York–Dordrecht–London: Springer, forthcoming).

Ehlers, Eckhart, 2016: "From HDP to IHDP: Evolution of the International Human Dimensions of Global Environmental Change Programme", in: Brauch, Hans Günter; Oswald Spring, Úrsula; Grin, John; Scheffran; Jürgen (Eds.), 2016: *Handbook on Sustainability Transition and Sustainable Peace* (Cham–Heidelberg–New York–Dordrecht–London: Springer).

Eisen, Robert, 2008: "Human Security in Jewish Philosophy and Ethics", in: Brauch, Hans Günter; Oswald Spring, Úrsula; Mesjasz, Czeslaw; Grin, John; Dunay, Pal; Behera, Navnita Chadha; Chourou, Béchir; Kameri-Mbote, Patricia; Liotta, P.H. (Eds.): *Globalization and Environmental Challenges: Reconceptualizing Security in the 21st Century* (Berlin–Heidelberg–New York: Springer-Verlag): 253-261.

Ellen, R.F., 1996: "Ecology": in: Kuper, Adam; Kuper; Jessica (Eds.): *The Social Science Encyclopedia* (London–New York: Routledge): 207-209.

Ellis, Erle, 2013: "Anthropocene", in: *Encyclopedia of Earth*; at: <http://www.eoearth.org/view/article/150125>.

Emmerij, Louis, 2005: "Turning points in development thinking and practice", WIDER Jubilee Conference on WIDER Thinking Ahead: The Future of Development Economics, Helsinki, 17-18 June 2005: at: <http://www.rrojasdatabank.info/widerconf/Emmerij.pdf>.

Emmerij, Louis; Jolly, Richard; Weiss, Thomas G., 2005: "Economic and Social Thinking at the UN in Historical Perspective", in: *Development and Change*, 36,2: 211-235.

Encyclopaedia Britannica, [1974] 1991, 1998: *The New Encyclopaedia Britannica* (Chicago: Encyclopaedia Britannica).

EU (European Commission), 2011: *Energy Roadmap 2050. Communication from the Commission to the European Parliament, the Council, the European Economic and Social Committee and the Committee of the Regions*, COM(2011) 885/2 (Brussels: European Commission).

EU (European Commission), 2011a: *White Paper: Roadmap to a Single European Transport Area—Towards a competitive and resource efficient transport system.* COM(2011) 144 final (Brussels: European Commission, 28.3.2011).

EU (European Commission), 2011b: Communication from the Commission to The European Parliament, The Council, The European Economic and Social Committee and The Committee of the Regions: *A Roadmap for moving to a competitive low carbon economy in 2050.* COM(2011) 112 final (Brussels: European Commission, 8.3.2011).

EU (European Commission), 2011c: *Roadmap to a Resource Efficient Europe* (Brussels: European Commission).

EU (European Commission), 2011d: *White Paper. Roadmap to a Single European Transport Area—Towards a competitive and resource efficient transport system* (Brussels: European Commission).

Fabra Mata, Javier, 2007: "Playing the Right Game? Human Security Research and its Implications in Policymaking". Paper presented at the annual meeting of the International Studies Association, 48th Annual Convention, Chicago, 28 February; at: <http://www.allacademic.com/me-ta/p180091_index.html>.

Fleming, Richard A., 2002: "Environment", in: Mooney, Harold A.; Canadell, Joseph G. (Eds.): *Encyclopedia of Global Environmental Change*, vol. 2: *Biological and Ecological Dimensions of Global Environmental Change* (Chichester: John Wiley): 290.

Galtung, Johan, 1964: "An editorial", in: *Journal of Peace Research*, 1,1.

Galtung, Johan, 1967: "Peace Research: science, or politics in disguise?", in: *International Spectator,* 21,19: 1573-1603.

Galtung, Johan, 1968: "Peace", in: *International Encyclopedia of the Social Sciences* (London—New York; Macmillan): 487-496.

Galtung, Johan, 1969: "Violence, Peace and Peace Research", in: *Journal of Peace Research*, 3: 167-191.

Galtung, Johan, 1985: "Twenty-Five Years of Peace Research: Ten Challenges and Some Responses", in: *Journal of Peace Research*, 22,2: 141-158.

Galtung, Johan, 2012: "Positive and Negative Peace", in: Webel, Charles P.; Johansen, Jørgen (Eds.): *Peace and Conflict Studies: A Reader* (New York: Routledge): 75-78.

Galtung, Johan; Fischer, Dietrich: *Pioneer of Peace Research* (Heidelberg—New York—Dordrecht—London: Springer, 2013).

Geels, Frank W., 2002: "Technological transitions as evolutionary reconfiguration processes: A multi-level perspective and a case study", in: *Research Policy*, 31,8-9: 1257-1274.

Geels, Frank W., 2012: "A socio-technical analysis of low-carbon transitions: introducing the multi-level perspective into transport studies", in: *Journal of Transport Geography*, 24 (September): 471-482; at: <http://community.eldis.org/.5ad501d7/Geels%202012%20corrected%20proofs.pdf>.

Geels Frank W.; Schot, Johan, 2007: "Typology of sociotechnical transitions pathways", in: *Research Policy*, 36: 399-417; at: <http://www.sciencedirect.com/science/article/pii/S0048733307000248> and at: <https://www.google.com/search?q=Geels%2C+Schot+2007&ie=utf-8&oe=utf-8>.

Geels, Frank W.; Schot, Johan, 2010: "The Dynamics of Transition: A Socio-Technical Perspective", in: Grin, John; Rotmans, Jan; Schot, Johan, 2010: *Transitions to Sustainable Development. New Directions in the Study of Long Term Transformative Change* (New York, NY—London: Routledge): 11-101.

Glaeser, Bernhard, 1995: *Environment, Development, Agriculture, Integrated Policy through Human Ecology* (London: UCL).

Glaeser, Bernhard, 2002: "The Changing Human-Nature Relationships (HNR) in the Context of Global Environmental Change", in: Timmerman, Peter (Ed.): *Encyclopedia of Global Environmental Change*, vol. 5: *Social and Economic Dimensions of Global Environmental Change* (Chichester: John Wiley): 11-24.

Glasius, Marlies; Kaldor, Mary, 2005: "Individuals First: A Human Security Strategy for the European Union", in: *Internationale Politik und Gesellschaft—International Politics and Society*, 8,1 (Spring): 62-83.

Gleditsch, Nils Petter, 2003: "Environmental Conflict: Neomalthusians vs. Cornucopians", in: Brauch, Hans Günter; Liotta, P.H; Marquina, Antonio; Rogers, Paul; Selim, Mohammed El-Sayed (Eds.): *Security and Environment in the Mediterranean. Conceptualising Security and Environmental Conflicts* (Berlin—Heidelberg: Springer): 477-486.

Gleditsch, Nils Petter, 2015: *Nils Petter Gleditsch: Pioneer in the Analysis of War and Peace* (Cham - Heidelberg—New York—Dordrecht—London: Springer).

Gleditsch, Nils-Petter, 2012: "Whither the weather? Climate change and conflict", in: *Journal of Peace Research, special issue: Climate change and conflict,* 49,1 (January-February): 9-18.

Gruenewald, David A, 2003: "The best of both worlds: a critical pedagogy of place," in: *Educational Researcher*, 32, 4: 3-12.

Gleditsch, Nils-Petter; Nordas, Ragnhild, 2014: "Conflicting Messages? The IPCC on conflict and human security", in: *Political Geography*, 43: 82-90.

Goucha, Moufida; Cilliers, Jakkie (Eds.), 2001: *Peace, Human Security and Conflict Prevention in Africa* (Paris: UNESCO).

Gove, Philip Babcock, 2002: *Webster's Third New International Dictionary of the English Language. Unabridged* (Springfield. MA: Merriam-Webster Inc.).

Gray, Colin S.: Payne, Keith B., 1980: "Victory is Possible", in: *Foreign Policy*, 39: 14-28.

Grin, John; Rotmans, Jan; Schot, Johan, 2010: *Transitions to Sustainable Development. New Directions in the Study of Long Term Transformative Change* (New York, NY–London: Routledge).

Grunberg, Isabelle; Risse-Kappen, Thomas, 1992: "A Time for Reckoning? Theories of International Relations and the End of the Cold War", in: Allan, Pierre; Goldmann, Kjell (Eds.): *The End of the Cold War* (Dordrecht: Martinus Nijhoff Publishers): 104-146.

Hampson, Fen Osler, 2004: "A Concept in Need of a Global Policy Response", in: *Security Dialogue*, 35,3 (September): 349-350.

Happaerts, Sander, 2016: "Discourse and Practice of Transitions in International Policy-making on Resource Efficiency in the EU", in: Brauch, Hans Günter; Oswald Spring, Úrsula; Grin, John; Scheffran, Jürgen (Eds.): *Handbook on Sustainability Transition and Sustainable Peace* (Cham–Heidelberg–New York–Dordrecht–London: Springer).

Hargroves, Karlson 'Charlie', 2016: "Considering a Structural Adjustment Approach to the Low Carbon Transition", in: Brauch, Hans Günter; Oswald Spring, Úrsula; Grin, John; Scheffran; Jürgen (Eds.), 2016: *Handbook on Sustainability Transition and Sustainable Peace* (Cham–Heidelberg–New York–Dordrecht–London: Springer).

Hargroves, Karlson J.; Smith, Michael H. (Eds.), 2005: *The Natural Advantage of Nations*, The Natural Edge Project (London: Earthscan).

Haushofer, Karl, 1932: *Jenseits der Großmächte* (Leipzig-Berlin: B.G. Teubner).

Hendricks, Cheryl, 2008: "Die Weiterentwicklung der Agenda der menschlichen Sicherheit. Afrikanische Perspektiven", in: Ulbert, Cornelia; Werthes, Sascha (Eds.): *Menschliche Sicherheit. Globale Herausforderungen und regionale Perspektiven* (Baden-Baden: Nomos): 137-148.

Hoffmann, Stanley, 2001: "An American Social Science. International Relations", in: Crawford, Robert M.A.; Jarvis, Darryl S. L. (Eds.), 2001: *International Relations–Still an American Social Science? Toward Diversity in International Thought* (Albany: State University of New York Press): 27-51.

Holsti, Kalevi J., 1991: *Peace and War: Armed Conflicts and International Order 1648-1989* (Cambridge: Cambridge University Press).

Holsti, Kalevi J., 1996: *The state, war, and the state of war* (Cambridge: Cambridge University Press).

Holsti, Kalevi J., 2016: *Kalevi Holsti: A Pioneer in International Relations Theory, Foreign Policy Analysis, History of International Order, and Security Studies* (Cham–Heidelberg–New York–Dordrecht–London: Springer International Publishing).

Homer-Dixon, Thomas F., 1999: *Environment, Scarcity, and Violence* (Princeton, NJ: Princeton University Press).

Homer-Dixon, Thomas F., 2006: *The Upside of Down: Catastrophe, Creativity, and the Renewal of Civilization* (Knopf, Island Press).

Homer-Dixon, Thomas F.; et al., 2010: *How Peak Oil and the Climate Crisis Will Change Canada (and Our Lives)* (Toronto: Vintage, Random House).

Hsiang, S.M.; Burke, M.; Miguel, E., 2013: "Quantifying the influence of climate on human conflict", in: *Science*, 341,6151: 1-14.

Huggett, Richard John, 1995: *Geoecology. An Evolutionary Approach* (London–New York: Routledge).

Huggett, Richard John, 2000: "Geoecology", in: Hancock, Paul; Skinner, Brian J. (Eds.): *The Oxford Companion to the Earth* (Oxford: Oxford University Press); at: <http://www.encyclopedia.com/doc/1O112-geoecology.html>.

Hussein, Karim; Gnisci, Donata; Wanjiru, Julia, 2004: *Security and Human Security: an Overview of Concepts and Initiatives. What Implications for West Africa?* SAH/D(2004)547 (Paris: Sahel and West Africa Club, December).

Ide, Tobias; Link, P. Michael; Scheffran, Jürgen; Schilling, Janpeter, 2016: "The Climate-Conflict Nexus: Pathways, Regional Links, and Case Studies", in: Brauch, Hans Günter; Oswald Spring, Úrsula; Grin, John; Scheffran; Jürgen (Eds.): *Handbook on Sustainability Transition and Sustainable Peace* (Cham–Heidelberg–New York–Dordrecht–London: Springer).

IPCC, 1990: *Climate Change. The IPCC Impacts Assessment* (Geneva: WMO, UNEP; IPCC).

IPCC, 1996: *Climate Change 1995. The Science of Climate Change. Contributions of Working Group I to the Second Assessment Report of the Intergovernmental Panel on Climate Change* (Cambridge: Cambridge University Press).

IPCC, 2001: *Climate Change 2001. The Scientific Basis* (Cambridge–New York: Cambridge University Press).

IPCC, 2007: *Climate Change 2007. The Physical Science Basis*, Working Group I Contribution to the Fourth Assessment Report of the IPCC (Cambridge–New York: Cambridge University Press); also at: <http://www.ipcc-wg1.org/>.

IPCC, 2013: *Climate Change 2013 - The Physical Science Basis. Working Group I Contribution to the Fifth Assessment Report of the Intergovernmental Panel on Climate Change* (Cambridge–New York: Cambridge University Press); at: <http://www.ipcc.ch/report/ar5/wg1/>.

IPCC, 2014: *Climate Change 2014 - Synthesis Report, Summary for Policymakers* (Geneva: IPCC); at: <http://

www.ipcc.ch/pdf/assessment-report/ar5/syr/SYR_AR5_SPMcorr2.pdf>.

IPCC, 2014a: "Human Security", in: *Climate Change 2014–Impacts, Adaptation, and Vulnerability - Part A: Global and Sectoral Aspects. Working Group II Contribution to the Fifth Assessment Report of the Intergovernmental Panel on Climate Change* (Cambridge–New York: Cambridge University Press): 755-701; at: <http://www.ipcc.ch/pdf/assessment-report/ar5/wg2/WGIIAR5-Chap12_FINAL.pdf>.

IRENA: *Renewable Energy Capacity Statistics* (Masdar City, Abu Dhabi: IRENA); at: <http://www.irena.org/DocumentDownloads/Publications/IRENA_RE_Capacity_Statistics2015.pdf> (19 November 2015).

Johnson, Twig, 2016: "Policy, Politics and the Impact of Transition Studies", in: Brauch, Hans Günter; Oswald Spring, Úrsula; Grin, John; Scheffran; Jürgen (Eds.): *Handbook on Sustainability Transition and Sustainable Peace* (Cham–Heidelberg–New York–Dordrecht–London: Springer).

Kaldor, Mary, 1999: *New and Old Wars: Organized Violence in a Global Era* (Cambridge: Polity–Stanford: Stanford University Press).

Kaldor, Mary, 2002. *New and Old Wars: Organized Violence in a Global Era* (Cambridge: Polity).

Kaldor, Mary, 2007: *Human Security. Reflections on Globalization and Intervention* (Cambridge: Polity).

Kaldor, Mary, 2012: *New and Old Wars: Organised Violence in a Global Era* (Cambridge: Polity).

Kaldor, Mary, 2013: "In Defence of New Wars", in: *Stability*, 2,1 (4):1-16.

Kaldor, Mary (convenor) et al., 2004: *A Human Security Doctrine for Europe. The Barcelona Report of the Study Group on Europe's Security Capabilities.* Presented to EU High Representative for Common Foreign and Security Policy Javier Solana (London: LSE, 15 September): at: <http://www.lse.ac.uk/Depts/global/Publications/HumanSecurityDoctrine. pdf>.

Kant, Immanuel [1795], 1968: "Zum ewigen Frieden. Ein Philosophischer Entwurf", in: Kant, *Werke in 10 Bänden*, Wilhelm Weischedel (Ed.) (Darmstadt: Wissenschaftliche Buchgesellschaft): 195-251.

Kant, Immanuel, [1795], 1992: *Perpetual Peace and Other Essays* [transl. by Ted Humphrey] (Indianapolis–Cambridge: Hackett Publishing Co.).

Kates, R.W., 2001: "Queries on the human use of the Earth", in: *Annual Review of Energy and the Environment*, 26: 1-26.

Keating, Tom; Knight, W. Andy (Eds.), 2004: *Building Sustainable Peace* (Tokyo, UNU–Edmonton: The University of Alberta Press).

Kennedy, Paul, 1992: *Preparing for the Twenty-First Century* (New York: Random House).

Khagram, Sanjeev; Clark, William C., Firas Raad, Dana, 2003: "From the Environment and Human Security to Sustainable Security and Development", in: Chen, Lincoln; Fukuda-Parr, Sakiko; Seidensticker, Ellen (Eds),

2003: *Human Insecurity in a Global World* (Cambridge: Harvard University Press): 107-136.

Khalil, Elias L., 1996: "Kenneth Boulding: Ecodynamicist or Evolutionary Economist?", in: *Journal of Post Keynesian Economics*, 19,1 (Autumn): 83-100.

Kjellén, Rudolf, 1917, 1924: *Der Staat als Lebensform* (Leipzig: S. Hirzel).

Klare, Michael T., 2001: *Resource Wars: The New Landscape of Global Conflict* (New York: Henry Holt–Metropolitan Books).

Klare, Michael T., 2012: *The Race for What's Left: The Global Scramble for the World's Last Resources* (London: Picador).

Klare, Michael T., 2013: "How Resource Scarcity and Climate Change Could Produce a Global Explosion", in: *The Nation*, 22 April); at: <http://www.thenation.com; article; 173967; how-resource-scarcity-and-climate-change-could-produce-global-explosion>.

Klein, Naomi, 2011: "Capitalism vs. the Climate–Denialists are dead wrong about the science. But they understand something the left still doesn't get about the revolutionary meaning of climate change", in: *The Nation* (28 November); at: <http://www.thenation.com/article/164497/capitalism-vs-climate>.

Kolodziej, Edward A., 2005: *Security and International Relations* (Cambridge: Cambridge University Press).

Kornblith, Miriam, 2003: "Human Security: Definition and Challenges for Latin America and the Caribbean", in: Goucha, Moufida; Rojas Aravenna, Francisco (Eds.), 2003: *Human Security, Conflict Prevention and Peace* (Paris: UNESCO): 317-336.

Kothari, Uma; Minougue, Martin, 2002: "Critical Perspectives on development: an Introduction", in: Kothari, Uma; Minougue, Martin (Eds.): *Development Theory and Practice* (Basingstoke–New York: Palgrave): 1-15.

Krause, Keith, 2004: "Is Human Security 'More than Just a Good Idea'?", in: BICC (Ed.): Brief 30: *Promoting Security: But How and for Whom?* (Bonn: BICC): 43-46.

Krause, Keith, 2004a: "The Key to a Powerful Agenda, if Properly Delimited", in: *Security Dialogue*, 35,3 (September): 367-368.

Krause, Keith, 2008: "Kritische Überlegungen zum Konzept der menschlichen Sicherheit", in: Ulbert, Cornelia; Werthes, Sascha (Eds.): *Menschliche Sicherheit. Globale Herausforderungen und regionale Perspektiven* (Baden-Baden: Nomos): 31-50.

Krause, Keith, 2008a: "Building the agenda of human security: policy and practice within the Human Security Network", in: *Rethinking Human Security. International Social Science Journal Supplement*, 57 (Paris: UNESCO): 65-80.

Krell, Gert, 1981: "The Development of the Concept of Security", in: Jahn, Egbert; Sakamoto, Yoshikazu (Eds.): *Elements of World Instability: Armaments, Communication, Food, International Division of Labour* (Frankfurt: Campus): 238-254.

Kyrou, Christos N., 2007: "Peace Ecology: An Emerging Paradigm in Peace Studies", in: *The International Journal of Peace Studies*, 12,2 (Spring/Summer): 73-92.

Leaning, Jennifer, 2009: "Health and Human Security in the 21st Century", in: Brauch, Hans Günter; Oswald Spring, Úrsula; Grin, John; Mesjasz, Czeslaw; Kameri-Mbote, Patricia; Behera, Navnita Chadha; Chourou, Béchir; Krummenacher, Heinz (Eds.): *Facing Global Environmental Change: Environmental, Human, Energy, Food, Health and Water Security Concepts* (Berlin–Heidelberg–New York: Springer-Verlag): 541-552.

Lederach, John Paul, 1998: "Beyond Violence: Building Sustainable Peace", in: Weiner, Eugene (Ed.): *The Handbook of Interethnic Coexistence* (New York: Continuum Publishing): 236-245.

Leggett Jeremy K., 2001: *The Carbon War: Global Warming and the End of the Oil Era* (London: Routledge).

Leggett Jeremy K., 2005: *The Empty Tank: Oil, Gas, Hot Air, and the Coming Financial Catastrophe* (New York: Random House).

Leggett, Jeremy K., 2005a: *Half Gone: Oil, Gas, Hot Air and the Global Energy Crisis* (London: Portobello Books).

Lenton, Timothy; Held, Hermann; Kriegler, Elmar; Hall, Jim W.; Lucht, Wolfgang; Ramstorf, Stefan; Schellnhuber, Hans Joachim, 2008: "Tipping elements in the Earth's climate system", in: *Proceedings of the National Academy of Science* (PNAS), 105,6 (12 February): 1786-1793.

Lie, Tove Grete; Binningsbø, Helga Malmin; Gates, Scott, 2007: "Post-Conflict Justice and Sustainable Peace", in: The World Bank, WP S4191; at: <http://web.worldbank.org/archive/website01241/WEB/IMAGES/WPS4191.PDF>.

Lomborg, Bjørn, 2001: *The Skeptical Environmentalist. Measuring the Real State of the World* (Cambridge–New York: Cambridge University Press).

Lonergan, Steve, 1999: *Global Environmental Change and Human Security Science Plan*. IHDP Report No 11 (Bonn: IHDP).

Lopez, Ernesto, 2003: "Human Security Agenda: The Case of MERCOSUR", in: Goucha, Moufida; Rojas Aravenna, Francisco (Eds.), 2003: *Human Security, Conflict Prevention and Peace* (Paris: UNESCO): 353-363.

Lovelock, James E., 1975: "Thermodynamics and the Recognition of Alien Biosphere", in: *Proceedings of the Royal Society* (London): B, 189: 167-181.

Lovelock, James E., 1979: *Gaia. A New Look at Life on the Earth* (Oxford: Oxford University Press).

Lovelock, James E., 1986: "Geophysiology: A New Look at Earth Science", in: *Bulletin of the American Meteorological Society*, 67: 392-397.

Lovelock, James E., 1988: *The Ages of Gaia: A Biography of our Living Earth* (Oxford: oxford University Press).

Lovelock, James E., 1992: *Gaia: The Practical Science of Planetary Medicine* (Stroud: Gaia Books).

Lovelock, James E.; Margulis, L., 1974: "Atmospheric Homeostasis By and For the Biosphere. The Gaia Hypothesis", in: *Tellus*, 26: 1-10.

Lovelock, James E.; Margulis, L., 1974a: "Homeostasis Tendencies of the Earth's Atmosphere", in: *Origins of Life*, 5: 93-103.

Lovins, Amory B.; Datta, E. Kyle; Bustness, Odd-Even; Koomey, Jonathan G.; Glasgow, Nathan J., 2005: *Winning the Oil Endgame: Innovation for Profit, Jobs and Security* (London: Earthscan).

MacFarlane, S. Neil; Khong, Yuen Foong, 2006: *Human Security and the UN. A Critical History* (Bloomington–Indianapolis: Indiana University Press).

Mack, Andrew, 2004: "The Concept of Human Security", in: BICC (Ed.): *Promoting Security: But How and for Whom?* Brief 30 (Bonn: BICC): 47-50.

Mack, Andrew, 2004a: "A Signifier of Shared Values", in: *Security Dialogue*, 35,3 (September): 366-367.

Mack, Andrew, 2005: "The Concept of Human Security", in: Helsinki Process Secretariat, (Ed.), 2005b: *Helsinki Process Papers on Human Security* (Helsinki. Foreign Ministry Publications): 9-16.

Mackinder, Halford J., 1904: "The Geographical Pivot of History", in: *Geographical Journal*, 23: 421-444.

Mackinder, Halford J., 1905: "Man-Power as a Measure of National and Imperial Strength", in: *National and English Review*, 45: 143.

Mackinder, Halford J., 1918: *The Teaching of Geography and History: A Study of Method* (London: George Philip).

Mahan, Alfred, 1890: *The Influence of Sea Power upon History, 1660–1783* (Boston: Little Brown)

Mahan, Alfred, 1897: *The Interest of America in Seapower* (London: Sampson Law).

Malthus, Thomas Robert, 1798 [1993]: *An Essay on the Principle of Population as It Affects the Future Improvement of Society* (Oxford: Oxford University Press).

Marino, Marit Sjøvaag, 2016: "From *The Limits to Growth* to 2052", in: Brauch, Hans Günter; Oswald Spring, Úrsula; Grin, John; Scheffran; Jürgen (Eds.): *Handbook on Sustainability Transition and Sustainable Peace* (Cham–Heidelberg–New York–Dordrecht–London: Springer).

Mataure, Michael M., 2005: "Individual confidence and personal security in the 2005 Zimbabwean elections", in: Mulongo, Keith; Kibasomba, Roger; Kariri, Jemina Njeri (Eds.): *The Many Faces of Human Security. Case Studies of Seven Countries in Southern Africa* (Pretoria: Institute for Security Studies): 97-158; at: <http://www.reliefweb.int/rw/RWFiles2005.nsf/FilesByRWDocUNIDFileName/KKEE-6HZQQQ-iss-saf-10nov.pdf/$File/iss-saf-10nov.pdf>.

Mathews, Jessica Tuchman, 1989: "Redefining Security", in: *Foreign Affairs*, 68,2 (Spring): 162-177.

Matthews, Richard A.; Barnett, Jon; McDonald, B.; O'Brien, Karen L. (Eds.), 2010: *Global Environmental*

Change and Human Security (Cambridge, MA: MIT Press).

McLeod, William T., 1985, 1986: *The New Collins Concise English Dictionary* (London: Guild Publishing).

Meadows, Donella H.; Meadows, Dennis L.; Randers, Jorgen, 1982: *Beyond the Limits* (White River Jct., Vt: Chelsea Green Publishing).

Meadows, Donella H.; Meadows, Dennis L.; Randers, Jørgen, 1992: *Beyond the Limits* (Post Mills, Vt.: Chelsea Green Publishing).

Meadows, Donella H.; Meadows, Dennis L.; Randers, Jorgen, 2004: *The Limits to Growth–the 30-Year Update* (White River Jct., Vt: Chelsea Green Publishing).

Meadows, Donella H.; Meadows, Dennis L.; Randers, Jorgen; Behrens III, William W., 1972: *The Limits to Growth* (New York: Universe Books).

Merkel, Wolfgang (Ed.), 1996: *Systemwechsel. Theorien, Ansätze und Konzeptionen* (Opladen: Leske + Budrich).

Meyer-Abich, Klaus M. (Ed.), 1979: *Frieden mit der Natur* (Freiburg–Basel–Wien: Herder).

Meyer-Abich, Klaus M. [KMA], 2003: "Frieden mit der Natur", in: Simonis, Udo E. (Ed.): *Ökolexikon* (München: Beck): 84.

Mitchell, Ronald, 2002: "International Environment", in: Carlsnaes, Walter; Risse, Thomas; Simmons, Beth A. (Eds.): *Handbook of International Relations* (London–Thousand Oaks–New Delhi: Sage): 500-516.

Mlambo, Norman, 2005: "Perceptions of human security in democratic South Africa: Opinions of students from tertiary institutions", in: Mulongo, Keith; Kibasomba, Roger; Kariri, Jemina Njeri (Eds.): *The Many Faces of Human Security. Case Studies of Seven Countries in Southern Africa* (Pretoria: Institute for Security Studies): 225-262; at: <http://www.reliefweb.int/rw/RWFiles 2005.nsf/FilesByRWDocUNIDFileName/KKEE-6HZQQQ-iss-saf-10nov.pdf/$File/iss-saf-10nov.pdf>.

Møller, Bjørn, 1991: *Resolving the Security Dilemma in Europe. The German Debate on Non-Offensive Defence* (London: Brassey's).

Møller, Bjørn, 2003: "National, Societal and Human Security: Discussion–A Case Study of the Israeli-Palestine Conflict", in: Brauch, Hans Günter; Liotta, P.H; Marquina, Antonio; Rogers, Paul; Selim, Mohammed El-Sayed (Eds.): *Security and Environment in the Mediterranean. Conceptualising Security and Environmental Conflicts* (Berlin-Heidelberg: Springer 2003): 277-288.

Morrison, Mary Lee, 2005: *Elise Boulding: A Life in the Cause of Peace* (Jefferson, NC: McFarland).

Mpagala, Gaudens P.; Lwehabura, Jonathan M.K., 2005: "Zanzibar: conflict resolution and human security in the 2005 elections", in: Mulongo, Keith; Kibasomba, Roger; Kariri, Jemina Njeri (Eds.): *The Many Faces of Human Security. Case Studies of Seven Countries in Southern Africa* (Pretoria: Institute for Security Studies): 39-96; at: <http://www.reliefweb.int/rw/RWFiles2005.nsf/FilesByRWDocUNIDFileName/KKEE-6HZQQQ-iss-saf-10nov.pdf/$File/iss-saf-10nov.pdf>.

Mulongo, Keith; Kibasomba, Roger; Njeri Kariri, Jemina (Eds.), 2005: *The Many Faces of Human Security. Case Studies of Seven Countries in Southern Africa* (Pretoria: Institute for Security Studies).

Munn, Ted (Ed.), 2002: *Encyclopedia of Global Environmental Change (Egec)*, 5 vols (Chichester, UK: John Wiley).

Mutesa, Fredrick; Nchito, Wilma, 2005: "Human security, popular participation and poverty reduction in Zambia", in: Mulongo, Keith; Kibasomba, Roger; Kariri, Jemina Njeri (Eds.): *The Many Faces of Human Security. Case Studies of Seven Countries in Southern Africa* (Pretoria: Institute for Security Studies): 7-38; at: <http://www.reliefweb.int/rw/RWFiles2005.nsf/FilesByRWDocUNIDFileName/KKEE-6HZQQQ-iss-saf-10nov.pdf/$File/iss-saf-10nov.pdf>.

Myers, Norman, 1989: "Environment and Security", in: *Foreign Policy*, 74 (Spring): 23-41.

Myers, Norman, 1993: *Ultimate Security: The Environmental Basis of Political Stability* (New York–London: Norton).

Næss, Arne (translated by D. Rothenberg), 1989: *Ecology, Community and Lifestyle: Outline of an Ecosophy* (Cambridge: Cambridge University Press).

Næss, Arne, 1973: "The Shallow and the Deep, Long-Range Ecology Movement. A Summary", in: *Inquiry*, 16: 95-100.

Næss, Arne, 1986: "The Deep Ecological Movement Some Philosophical Aspects", in: *Philosophical Inquiry*, 8: 1-2.

Naidoo, Sagaren, 2001: "A Theoretical Conceptualization of Human Security", in: Goucha, Moufida; Cilliers, Jakkie (Eds.): *Peace, Human Security and Conflict Prevention in Africa* (Paris: UNESCO): 1-9.

Najam, Adil (Ed.), 2003: *Environment, Development and Human Security: Perspectives from South Asia* (Lanham, Md: University Press of America).

Najam, Adil, 2003a: "The Human Dimensions of Environmental Insecurity: Some Insights from South Asia", in: Woodrow Wilson International Center for Scholars (Ed.): *Environmental Change and Security Project Report Issue No. 9* (Washington DC: Wilson Center, ECSP): 59-73; at: <http://wwics. si.edu/topics/pubs/najam. pdf>.

Neff, Jorge, 2003: "Human Security and Mutual Vulnerability", in: Goucha, Moufida; Rojas Aravenna, Francisco (Eds.), 2003: *Human Security, Conflict Prevention and Peace* (Paris: UNESCO): 29-60.

Nelles, Wayne (Ed.), 2003: *Comparative Education, Terrorism and Human Security. From Critical Pedagogy to Peacebuilding* (New York–Houndmills: Palgrave Macmillan).

Nentwich, Wolfgang; Bacher, Sven; Beierkuhnlein, Carl; Brandl, Roland; Grabherr, Georg, 2004: *Ökologie* (Heidelberg: Spektrum der Wissenschaft).

Ngoy Kaungulu, Huert, 2005: "Urban security in Kinshasa: A socio-demographic profile", in: Mulongo, Keith; Kibasomba, Roger; Kariri, Jemina Njeri (Eds.): *The Many Faces of Human Security. Case Studies of Seven Countries in Southern Africa* (Pretoria: Institute for Security

Studies): 97-158; at: <http://www.reliefweb.int/rw/RWFiles2005.nsf/FilesByRWDocUNIDFileName/KKEE-6HZ QQQ-iss-saf-10nov.pdf/$File/iss-saf-10nov.pdf>.

NIC [National Intelligence Council], 2008: *Global Trends 2025: A Transformed World*. NIC 2008-003 (Washington DC: US Government Printing Office, November); at: <www.dni.gov.nic/NIC_2025_project.html>.

NIC, 2012: *Global Trends 2030: Alternative Worlds* (Washington DC: US Government Printing Office); at: <http://www.dni.gov/files/documents/GlobalTrends_2030.pdf>.

NIC, 2012a: "Global Water Security"; at: <http://www.dni.gov/files/documents/Special%20Report_ICA%20Global%20Water%20Security.pdf>.

Noone, Kevin J.; Nobre, Carlos; Seitzinger, Sybil, 2011: "The International Geosphere-Biosphere Programme's (IGBP) Scientific Research Agenda for Coping with Global Environmental Change", in: Brauch, Hans Günter; Oswald Spring, Úrsula; Mesjasz, Czeslaw; Grin, John; Kameri-Mbote, Patricia; Chourou, Béchir; Dunay, Pal; Birkmann, Jörn, 2010: *Coping with Global Environmental Change, Disasters and Security–Threats, Challenges, Vulnerabilities and Risks* (Berlin–Heidelberg–New York: Springer-Verlag): 1249-1256.

Nordäs, Ragnhild; Gleditsch, Nils-Petter, 2013: "The IPCC, human security, and the climate conflict nexus", in: Redclift, M.; Grasso, M. (Eds.): *Climate change and human security* (London: Elgar): 67-88.

NRC (US National Research Council, Policy Division, Board on Sustainable Development), 1999: *Our Common Journey: A Transition toward Sustainability* (Washington DC: National Academy Press); at: <http://www.nap.edu/openbook.php?record_id=9690&page=1>.

O' Brien, Karen L.; St. Clair, A.; Kristoffersen, B. (Eds.), 2010: *Climate Change, Ethics, and Human Security* (Cambridge: Cambridge University Press).

O'Brien, Karen L.; Sygna, L.; Wolf, J. (Eds.), 2013: *A Changing Environment for Human Security: Transformative Approaches to Research, Policy and Action* (London: Earthscan).

Odum, Eugene P., 1953: *Fundamentals of Ecology* (Philadelphia: W.B. Saunders).

Odum, Eugene P., 1975: *Ecology, the link between the natural and the social sciences* (New York: Holt, Rinehart and Winston).

Odum, Eugene P., 1977: "The emergence of ecology as a new integrative discipline", in: *Science*, 195: 1289-1293.

Odum, Eugene P., 1998: *Ecological Vignettes: Ecological Approaches to Dealing with Human Predicaments* (Amsterdam: Harwood).

Oswald Spring, Ùrsula, 2008: "Peace and Environment: Towards a Sustainable Peace as Seen from the South", in: Brauch, Hans Günter; Oswald Spring, Úrsula; Mesjasz, Czeslaw; Grin, John; Dunay, Pal; Behera, Navnita Chadha; Chourou, Béchir; Kameri-Mbote, Patricia; Liotta, P.H. (Eds.): *Globalization and Environmental Challenges: Reconceptualizing Security in the 21ˢᵗ Cen-*

tury (Berlin–Heidelberg–New York: Springer-Verlag): 113-126.

Oswald Spring, Ùrsula, 2008a: "Oriental, European and Indigenous Thinking on Peace in Latin America", in: Brauch, Hans Günter; Oswald Spring, Úrsula; Mesjasz, Czeslaw; Grin, John; Dunay, Pal; Behera, Navnita Chadha; Chourou, Béchir; Kameri-Mbote, Patricia; Liotta, P.H. (Eds.): *Globalization and Environmental Challenges: Reconceptualizing Security in the 21ˢᵗ Century* (Berlin–Heidelberg–New York: Springer-Verlag): 175-194.

Oswald Spring, Úrsula, 2009: "A HUGE Gender Security Approach: Towards Human, Gender and Environmental Security", in: Brauch, Hans Günter; Oswald Spring, Úrsula; Grin, John; Mesjasz, Czeslaw; Kameri-Mbote, Patricia; Behera, Navnita Chadha; Chourou, Béchir; Krummenacher, Heinz (Eds.), 2009: *Facing Global Environmental Change: Environmental, Human, Energy, Food, Health and Water Security Concepts* (Berlin–Heidelberg: Springer): 1165-1190.

Oswald Spring, Úrsula, 2016: "Development with Sustainable-Engendered Peace: A Challenge during the Anthropocene", in: Brauch, Hans Günter; Oswald Spring, Úrsula; Grin, John; Scheffran; Jürgen (Eds.), 2016: *Handbook on Sustainability Transition and Sustainable Peace* (Cham–Heidelberg–New York–Dordrecht–London: Springer).

Oswald Spring, Úrsula; Brauch, Hans Günter, 2011: "Coping with Global Environmental Change–Sustainability Revolution and Sustainable Peace", in: Brauch, Hans Günter; Oswald Spring, Úrsula; Mesjasz, Czeslaw; Grin, John; Kameri-Mbote, Patricia; Chourou, Béchir; Dunay, Pal; Birkmann, Jörn, 2010: *Coping with Global Environmental Change, Disasters and Security–Threats, Challenges, Vulnerabilities and Risks* (Berlin–Heidelberg–New York: Springer-Verlag): 1487-1504.

Oswald Spring, Úrsula; Brauch, Hans Günter; Tidball, Keith G. (Eds.), 2014: *Expanding Peace Ecology: Peace, Security, Sustainability, Equity and Gender–Perspectives of IPRA's Ecology and Peace Commission* (Cham–Heidelberg–New York–Dordrecht–London: Springer).

Oswald Spring, Úrsula; Brauch, Hans Günter; Tidball, Keith G., 2014: "Expanding Peace Ecology: Peace, Security, Sustainability, Equity and Gender", in: Oswald Spring, Úrsula; Brauch, Hans Günter; Tidball, Keith G. (Eds.), 2014: *Expanding Peace Ecology: Peace, Security, Sustainability, Equity and Gender - Perspectives of IPRA's Ecology and Peace Commission* (Cham–Heidelberg–New York–Dordrecht–London: Springer): 1-32.

Othman, Zarina, 2009: "Human Security Concepts, Approaches and Debates in Southeast Asia", in: Brauch, Hans Günter; Oswald Spring, Úrsula; Grin, John; Mesjasz, Czeslaw; Kameri-Mbote, Patricia; Behera, Navnita Chadha; Chourou, Béchir; Krummenacher, Heinz (Eds.), 2009: *Facing Global Environmental Change: Environmental, Human, Energy, Food, Health and*

Water Security Concepts (Berlin–Heidelberg–New York: Springer-Verlag): 1037-1048.

Owen, Taylor, 2004: "Human Security: Conflict, Critique and Consensus: Colloquium Remarks and a Proposal for a Threshold-Based Definition, Special Section: What is Human Security?" in: *Security Dialogue, September* 35,3 (Autumn): 345-387.

Oxford University Press, ⁵2002: *Shorter Oxford English Dictionary on Historical Principles* (Oxford–New York: Oxford University Press).

Oxford, 1998: *Reference Encyclopedia* (Oxford–New York: Oxford University Press).

Paelke, Robert, 2002: "Environmental Politics", in: Timmerman, Peter (Ed.): *Encyclopedia of Global Environmental Change*, vol. 5: *Social and Economic Dimensions of Global Environmental Change* (Chichester: John Wiley): 49-61.

Page, Edward A.; Redclift, Michael (Eds.), 2002: *Human Security and the Environment: International Comparisons* (Cheltenham: Edward Elgar).

Palma, Hugo, 2003: "Peace, Human Security and Conflict Prevention in Latin America and the Caribbean", in: Goucha, Moufida; Rojas Aravenna, Francisco (Eds.), 2003: *Human Security, Conflict Prevention and Peace* (Paris: UNESCO): 103-114.

Pampell, Camille; Sen, Anjalina, 2005: *Beyond Conflict Prevention: How Women Prevent Violence and Build Sustainable Peace* (New York: Global Action to Prevent War–Women's International League for Peace and Freedom, October).

Partsch, Karl-Josef, 1994: "Art. 55 (c)", in: Simma, Bruno (Ed.): *The Charter of the United Nations. A Commentary* (Oxford: Oxford University Press): 776-793.

Paterson, Matthew, 2000: *Understanding Global Environmental Politics. Domination, Accumulation, Resistance* (Basingstoke–New York: Palgrave).

Peck, Connie, 1998: *Sustainable Peace: The Role of the UN and Regional Organizations in Preventing Conflicts* (Lanham–Boulder–New York: Carnegie Commission on Preventing Deadly Conflict).

Pepper, David, 2002: "Ecosocialism", in: Timmerman, Peter (Ed.): *Encyclopedia of Global Environmental Change*, vol. 5: *Social and Economic Dimensions of Global Environmental Change* (Chichester: John Wiley): 224-225.

Pepper, David, 2002a: "Social Ecology", in: Timmerman, Peter (Ed.): *Encyclopedia of Global Environmental Change*, vol. 5: *Social and Economic Dimensions of Global Environmental Change* (Chichester: John Wiley): 484.

Pepper, David, 2002b, "Deep Ecology", in: Timmerman, Peter (Ed.): *Encyclopedia of Global Environmental Change*, vol. 5: *Social and Economic Dimensions of Global Environmental Change* (Chichester: John Wiley): 211.

Picciotto, Robert; Olonisakin, Funmi; Clarke, Michael, 2007: *Global Development and Human Security* (New Brunswick – London: Transaction Publishers).

Poku, Nana; Sandkjaer, Bjorg, 2009: "Human Security in Sub-Saharan Africa", in: Brauch, Hans Günter; Oswald Spring, Úrsula; Grin, John; Mesjasz, Czeslaw; Kameri-Mbote, Patricia; Behera, Navnita Chadha; Chourou, Béchir; Krummenacher, Heinz (Eds.), 2009: *Facing Global Environmental Change: Environmental, Human, Energy, Food, Health and Water Security Concepts* (Berlin–Heidelberg–New York: Springer-Verlag): 1049-1062.

Polanyi Levitt, Kari, 2008: "W. Arthur Lewis: Pioneer of Development Economics", in: *UN Chronicle*, 45,1 (March), at: <http://unchronicle.un.org/article/w-arthur-lewis-pioneer-development-economics/>.

Polanyi, Karl, 1944: *The Great Transformation: The Political and Economic Origins of Our Time* (Boston: Beacon Press).

Raskin, P.; Banuri, T.; Gallopin, G.; Gutman, P.; Hammond, A.; Kates, A.; Swart, R., 2002: *Great transition. The promise and the lure of the times ahead.* Report of the Global Scenario Group. Pole Start Series Report 10 (Stockholm: Stockholm Environment Institute).

Rayner, Steve; Malone, Elizabeth L., 2002: "Social Science and Global Environmental Change", in: Munn, Ted. (Ed.): *Encyclopedia of Global Environmental Change*, vol. 5: Timmerman, Peter (Ed.): *Social and Economic Dimensions of Global Environmental Change* (Chichester: John Wiley): 109-123.

Regan, Patrick, 2014: "Bringing Peace Back in: Presidential Address to the Peace Science Society, 2013", in: *Conflict Management and Peace Science*, 31,4: 345-356.

REN 21 [Renewable Energy Netzwork fort he 21st Century], 2015: *Renewables 2015, Global Status Report* (Paris: REN21, c/o UNEP).

REN 21, 2015a: *Renewables 2015, Global Status Report* (Paris: REN21, c/o UNEP), at: <http://www.ren21.net/wp-content/uploads/2015/07/REN12-GSR2015_Onlinebook_low1.pdf>.

Rockström, Johan; Kulm, Mattias, 2015; *Big World Small Planet: Abundance within Planetary Boundaries* (Stockholm: Max Strom).

Roeder, Philip G.; Rothchild, Donald S. (Eds.), 2005: *Sustainable peace. Power and democracy after civil wars* (Ithaca, NY: Cornell University Press).

Rojas Aravenna, Francisco, 2009: "Human Security: a South American Perspective", in: Brauch, Hans Günter; Oswald Spring, Úrsula; Grin, John; Mesjasz, Czeslaw; Kameri-Mbote, Patricia; Behera, Navnita Chadha; Chourou, Béchir; Krummenacher, Heinz (Eds.): *Facing Global Environmental Change: Environmental, Human, Energy, Food, Health and Water Security Concepts* (Berlin–Heidelberg–New York: Springer-Verlag): 1085-1094

Rosa, E.A.; Dietz, T., 1998: "Climate Change and Society: Speculation, Construction and Scientific Investigations", in: *International Sociology*, 13: 421-455.

Salehyan, Idean (Ed.), 2014: *Political Geography–Special Issue: Climate Change and Conflict*, 43 (November): 1-90.

Scheffran, Jürgen, 1997: "Frieden und nachhaltige Entwicklung. Interdisziplinäre Betrachtungen zum Verhältnis von Umwelt, Entwickung und Frieden", in: Vogt, Wolfgang; Jung, Eckhard (Eds.): *Kultur des Friedens. Wege zu einer Welt ohne Krieg* (Darmstadt: Wissenschaftliche Buchgesellschaft): 43-49.

Scheffran, Jürgen, 1998: "Wege zu einer nachhaltigen Entwicklung des Friedens", in: Scheffran, Jürgen; Vogt, Wolfgang R. (Eds.): *Kampf um die Natur* (Darmstadt: Wissenschaftliche Buchgesellschaft): 291-301.

Scheffran, Jürgen, 1999: "Environmental Conflicts and Sustainable Development: A Conflict Model and its Application in Climate and Energy Policy", in: Carius, Alexander; Lietzmann, Kurt M. (Eds.): *Environmental Change and Security: A European perspective* (Berlin– Heidelberg: Springer): 95-218.

Scheffran, Jürgen, 2011: "Frieden und nachhaltige Entwicklung", in: Giessmann, H.J.; Rinke, B. (Eds.): *Handbuch Frieden* (Wiesbaden: VS Verlag): 310-323.

Scheffran, Jürgen, 2016: "From a Climate of Complexity to Sustainable Peace: Viability Transformations in the Anthropocene", in: Brauch, Hans Günter; Oswald Spring, Úrsula; Grin, John; Scheffran; Jürgen (Eds.): *Handbook on Sustainability Transition and Sustainable Peace* (Cham–Heidelberg–New York–Dordrecht–London: Springer).

Scheffran, Jürgen; Brzoska, Michael; Brauch, Hans Günter; Link, Peter Michael; Schilling, Janpeter (Eds.): *Climate Change,Human Security and Violent Conflict: Challenges for Societal Stability* (Berlin–Heidelberg–New York: Springer-Verlag, 2012)

Scheffran, Jürgen; Froese, Rebecca, 2016: "Enabling Environments for a Sustainable Energy Transformation: The Diffusion of Know-how, Technologies and Innovations in Low-Carbon Societies", in: Brauch, Hans Günter; Oswald Spring, Úrsula; Grin, John; Scheffran; Jürgen (Eds.): *Handbook on Sustainability Transition and Sustainable Peace* (Cham–Heidelberg–New York–Dordrecht–London: Springer).

Schellnhuber, Hans Joachim, 1999: "'Earth System' analysis and the Second Copernican Revolution", in: *Nature*, 402,2: C19-C23.

Schellnhuber, Hans Joachim; Cramer, Wolfgang; Nakicenovic, Nebojsa; Wigley, Tom; Yohe, Gary (Eds.), 2006: *Avoiding Dangerous Climate Change* (Cambridge: Cambridge University Press).

Schellnhuber, Hans Joachim; Serdeczny, Olivia Maria; Adams, Sophie; Köhler, Claudia; Otto, Ilona Magdalena; Schleussner, Carl-Friedrich, 2016: "The Challenge of a 4°C World by 2100", in: Brauch, Hans Günter; Oswald Spring, Úrsula; Grin, John; Scheffran; Jürgen

(Eds.): *Handbook on Sustainability Transition and Sustainable Peace* (Cham–Heidelberg–New York–Dordrecht–London: Springer).

Schellnhuber, Hans-Joachim; Wenzel, Volker (Eds.), 1998: *Earth System Analysis. Integrating Science for Sustainability* (Berlin–Heidelberg: Springer).

Schott, Max, 2009: "Human Security: International Discourses and Local Reality–Case of Mali", in: Brauch, Hans Günter; Oswald Spring, Úrsula; Grin, John; Mesjasz, Czeslaw; Kameri-Mbote, Patricia; Behera, Navnita Chadha; Chourou, Béchir; Krummenacher, Heinz (Eds.): *Facing Global Environmental Change: Environmental, Human, Energy, Food, Health and Water Security Concepts* (Berlin–Heidelberg–New York: Springer-Verlag): 1105-1114.

Schumacher, Ernst Friedrich, 1973, 1989, 1999: *Small Is Beautiful: A study of economics as if people mattered* (London: Blond and Briggs; New York, N.Y.: Harper Perennial; Vancouver, B.C.: Hartley & Marks).

Schweitzer, Peter, 1994: *Victory. The Reagan Administration's Secret Strategy That Hastened the Collapse of the Soviet Union* (New York: Atlantic Monthly Press).

Scott, Robert, 2015: *Kenneth Boulding* (Basingstoke–New York: Macmillan Palgrave).

Sen, Amartya, 1981: *Poverty and Famines: An Essay on Entitlement and Deprivation* (Oxford: Oxford University Press).

Sen, Amartya, 1984: *Resources, Values, and Development* (Cambridge, Mass.: Harvard University Press).

Sen, Amartya, 1994. "Freedom and Needs", in: *New Republic*, 10 and 17 (January): 31-38.

Sen, Amartya, 1999: *Development as Freedom* (New York: Alfred A. Knopf).

Sen, Amartya, 2000: *Why human security*, Paper presented at the International Symposium on Human Security, Tokyo, 28 July 2000; at <http://www.humansecurity-chs.org/activities/outreach/Sen2000.pdf>, 21 March 2005.

Sen, Amartya, 2000a: *Development as Freedom* (New York: Anchor).

Sending, Ole Jacob (Ed.), 2010: *Learning to Build a Sustainable Peace–Ownership and Everyday Peacebuilding* (Oslo: Norwegian Institute of International Affairs–Bergen: Chr. Michelsen Institute).

Shinoda, Hideaki, 2009: "Human Security Initiatives of Japan", in: Brauch, Hans Günter; Oswald Spring, Úrsula; Grin, John; Mesjasz, Czeslaw; Kameri-Mbote, Patricia; Behera, Navnita Chadha; Chourou, Béchir; Krummenacher, Heinz (Eds.): *Facing Global Environmental Change: Environmental, Human, Energy, Food, Health and Water Security Concepts* (Berlin–Heidelberg–New York: Springer-Verlag): 1097-1104.

Singh, Joseph G., 2009: "Relevance of Human and Environmental Security Concepts for the Military Services: A Perspective of a Former Chief of Staff", in: Brauch, Hans Günter; Oswald Spring, Úrsula; Grin, John; Mesjasz, Czeslaw; Kameri-Mbote, Patricia; Behera, Navnita Chadha; Chourou, Béchir; Krummenacher, Heinz

(Eds.): *Facing Global Environmental Change: Environmental, Human, Energy, Food, Health and Water Security Concepts* (Berlin–Heidelberg–New York: Springer-Verlag): 1245-1252.

SIPRI, 2015: *SIPRI Yearbook 2015: Armaments, Disarmament and International Security* (Oxford –New York: Oxford University Press); at: <http://www.sipri.org/researcharmaments/milex/milex-graphs-for-data-launch-2015/World%20military%20expenditure%201988-2014.png>.

Solnit, Rebecca, 2009: *A Paradise Built in Hell–The Extraordinary Communities that Arise in Disaster* (New York: Penguin).

Stallings, Barbara (Ed.), 1995: *Global change, regional response* (Cambridge: Cambridge University Press).

Steffen, Will; Sanderson, Angelina; Tyson, Peter D.; Jäger, Jill; Matson, Pamela A.; Moore III, Berrien; Oldfield, Frank; Richardson, Katherine; Schellnhuber, Hans Joachim; Turner II, B.L.; Wasson, Robert J., 2004: *Global Change and the Earth System. A Planet under Pressure. The IGBP Series* (Berlin–Heidelberg–New York: Springer-Verlag).

Steffen, Will; Grinevald, Jacques; Crutzen, Paul; McNeill, John, 2011: "The Anthropocene: conceptual and historical perspectives", in: *Phil. Trans. R. Soc. A*: 369: 843.

Steinbruner, John D., 2000: *Principles of Global Security* (Washington DC: Brookings).

Stoett, Peter, 1999: *Human and Global Security. An Exploration of Terms* (Toronto–Buffalo–London: University of Toronto Press).

Suess, Eduard, 1875: *Die Entstehung Der Alpen* [*The Origin of the Alps*] (Vienna: W. Braunmüller).

Sygna, Linda; O'Brien, Karen; Wolf, Johanna (Eds.), 2013: *A Changing Environment for Human Security. Transformative Approaches to research, policy and action* (London–New York: Routledge–Earthscan).

Tadjbakhsh, Shahrbanou; Chenoy, Anuradha, 2006: *Human Security: Concepts and Implications* (London: Routledge).

Tansley, Arthur G., 1935: "The use and abuse of vegetational terms and concepts", in: *Ecology*, 16,3: 284-307.

Thakur, Ramesh, 2006, 2007: *The United Nations, Peace and Security* (Cambridge–New York: Cambridge University Press).

Theisen, O.M.; Gleditsch, Nils-Petter; Buhaug, Halvard, 2013: "Is Climate Change a driver of armed conflict?", in: *Climatic change*, 117,3: 971-996.

Thomas, Caroline, 1999: "Introduction", in: Thomas, Caroline; Wilkins, Peter (Eds). *Globalization, Human Security and African Experience* (Boulder, CO: Lynne Rienner): 1-22.

Thomas, Nickolas; Tow, William T., 2002: "The Utility of Human Security: Sovereignty and Humanitarian Intervention", in: *Security Dialogue*, 33,2 (Summer): 177-192.

Thomas, Nickolas; Tow, William T., 2002a: "Gaining Security by Trashing the State? A Reply to Bellamy &

McDonald", in: *Security Dialogue*, 33,3 (Autumn): 379-382.

Tickner, J. Ann, 1995: "Re-visioning security", in: Booth, Ken; Smith, Steve (Eds.): *International relations theory today* (University Park: Pennsylvania State University Press).

Tigerstrom, Barbara von, 2007: *Human Security and International Law. Prospects and Problems*. Studies in International Law 14 (Oxford–Portland, OR: Hart Publishing).

Tolba, Mostafa Kamal, 2002: "Envrionmental Responses: An Overview", in: Munn, Ted: *Encyclopedia of Global Environmental Change*, vol. 4: (Ed.): Tolba, Mostafa K. (Ed.): *Responding to Global Environmental Change* (Chichester, UK: John Wiley): 1-13.

Tomuschat, Christian, 1994: "Chapter VI. Pacific Settlement of Disputes, Art. 33", in: Simma, Bruno (Ed.): *The Charter of the United Nations. A Commentary* (Oxford: Oxford University Press): 505-514.

Toye, John, 1996: "Economic Development", in: Kuper, Adam; Kuper; Jessica (Eds.): *The Social Science Encyclopedia* (London–New York: Routledge): 212-215.

Troll, Carl, 1968: "Landschaftsökologie", in: Tuxen, R. (Ed.): *Pflanzensoziologie und Landschaftsökologie. Berichte das 1963 Internalen Symposium der Internationalen Vereinigung für Vegetationskunde* (Essen: IVF).

Ulbert, Cornelia; Werthes, Sascha, 2008: "Menschliche Sicherheit–Der Stein des Weisen für globale und regionale Verantwortung? Entwicklungslinien und Herausforderungen eines usmtrittenen Konzepts", in: Ulbert, Cornelia; Werthes, Sascha (Eds.): *Menschliche Sicherheit. Globale Herausforderungen und regionale Perspektiven* (Baden-Baden: Nomos): 13-27.

Ullman, Richard, 1983: "Redefining Security", in: *International Security* 8,1 (Summer): 129-153.

UN. 2011: *The Millennium Development Goals Report 2011* (New York: UN).

UNDP, 1994: *Human Development Report 1994. New Dimensions of Human Security* (New York–Oxford–New Delhi: Oxford University Press); at: <http://hdr.undp.org/reports/global/1994/en/pdf/hdr_1994_ch2.pdf>.

UNEP, 2011: *Green Economy Report: Towards a Green Economy: Pathways to Sustainable Development and Poverty Eradication* (Nairobi: UNEP); at: <http://www.unep.org/greeneconomy/GreenEconomyReport/tabid/29846>.

UNEP, 2014: *Decoupling 2: Technologies, Opportunities and Policy Options- Report to UNEP's International Resource Panel* (Nairobi: UNEP).

UNGA, 2012: *Draft resolution: The Future we Want*, A/66/L.56 (New York: UN, 24 July).

UNSG, 2001: *Prevention of Armed Conflict–Report of the Secretary-General. A/55/985-S/2001/574* (New York: United Nations).

UNSG, 2009: *Climate change and its possible security implications. Report of the Secretary-General. A/64/350* of 11 September 2009 (New York: United Nations).

UNSG, 2010: *Human Security–Report of the Secretary-General. A/64/701* of 8 May 2010 (New York: United Nations).

UNSG, 2012: *Follow-up to General Assembly resolution 64/291 on human security. Report of the Secretary-General. A/66/763* of 5 April 2012 (New York: United Nations); at: <http://www.un.org/humansecurity/sites/www.un.org.humansecurity/files/n1228537.pdf>.

Uvin, Peter, 2004: "A Field of Overlaps and Interactions", in: *Security Dialogue*, 35,3 (September): 352.

Uvin, Peter, 2008: "Development and Security: Genealogy and Typology of an Evolving International Policy Area", in: Brauch, Hans Günter; Oswald Spring, Úrsula; Mesjasz, Czeslaw; Grin, John; Dunay, Pal; Behera, Navnita Chadha; Chourou, Béchir; Kameri-Mbote, Patricia, Liotta, P.H.; (Eds.): *Globalization and Environmental Challenges: Reconceptualizing Security in the 21st Century* (Berlin–Heidelberg–New York: Springer-Verlag): 151-164.

Vasak, K., 1984: "Pour une troisième génération des droits de l'homme", in: Swinarski, Christophe (Ed.): *Studies in Honour of Jean Pictet* (The Hague: Martinus Nijhoff Publishers): 837

Vernadsky, V. I., 1998 [1926]: *The Biosphere* (New York: Copernicus, Springer).

von Falkenhayn, Louise; Rechkemmer, Andreas; Young, Oran R., 2011: "The International Human Dimensions Programme on Global Environmental Change–Taking Stock and Moving Forward", in: Brauch, Hans Günter; Oswald Spring, Úrsula; Mesjasz, Czeslaw; Grin, John; Kameri-Mbote, Patricia; Chourou, Béchir; Dunay, Pal; Birkmann, Jörn (Eds.): *Coping with Global Environmental Change, Disasters and Security–Threats, Challenges, Vulnerabilities and Risks* (Berlin–Heidelberg–New York: Springer-Verlag): 1221-124.

von Weizsäcker, Carl Friedrich von (Ed.), 1972: *Kriegsfolgen und Kriegsverhütung* (München, Wien: Hanser, 1972).

Wæver, Ole, 1995: "Securitization and Desecuritization", in: Lipschutz, Ronnie D. (Ed.): *On Security* (New York: Columbia University Press): 46-86.

Wæver, Ole, 1997: *Concepts of Security* (Copenhagen: Department of Political Science).

Wæver, Ole, 2004: "Aberystwyth, Paris, Copenhagen: New 'Schools in Security Theory and their Origins between Core and Periphery". Paper for 45th International Studies Association Convention. Montreal, 17-20 March.

Wæver, Ole, 2008: "Peace and Security: Two Evolving Concepts and their Changing Relationship", in: Brauch, Hans Günter; Oswald Spring, Úrsula; Mesjasz, Czeslaw; Grin, John; Dunay, Pal; Behera, Navnita Chadha; Chourou, Béchir; Kameri-Mbote, Patricia; Liotta, P.H. (Eds.): *Globalization and Environmental Challenges: Reconceptualizing Security in the 21st Century* (Berlin–Heidelberg–New York: Springer-Verlag): 99-111.

Walt, Stephen, 1991: "The Renaissance of Security Studies", in: *International Studies Quarterly*, 35,2 (June): 211-239.

Walther, Bruno A.; Larigauderie, Anne; Loreau, Michel, 2011: "DIVERSITAS: Biodiversity Science Integrating Research and Policy for Human Well-Being", in: Brauch, Hans Günter; Oswald Spring, Úrsula; Mesjasz, Czeslaw; Grin, John; Kameri-Mbote, Patricia; Chourou, Béchir; Dunay, Pal; Birkmann, Jörn (Eds): *Coping with Global Environmental Change, Disasters and Security–Threats, Challenges, Vulnerabilities and Risks* (Berlin–Heidelberg–New York: Springer-Verlag): 1235-1248.

Warren, Karen J., 2002: "Ecofeminism", in: Timmerman, Peter (Ed.): *Encyclopedia of Global Environmental Change*, vol. 5: *Social and Economic Dimensions of Global Environmental Change* (Chichester: John Wiley): 218-224.

WBGU, 2008: *World in Transition–Climate Change as a Security Risk* (London: Earthscan); at: <http://www.wbgu.de/wbgu_jg2007_engl.html>.

WBGU, 2011: *World in Transition–A Social Contract for Sustainability* (Berlin: German Advisory Council on Global Change, July 2011).

WCED (World Commission on Environment and Development), 1987: *Our Common Future. The World Commission on Environment and Development* (Oxford–New York: Oxford University Press).

Webel, Charles P.; Barash, David P., ²2008: *Peace and Conflict Studies* (London: Sage),

Weizsäcker, Carl Friedrich von (Ed.), 1972: *Kriegsfolgen und Kriegsverhütung* (München: Hanser).

Weizsäcker, Ernst Ulrich von (Ed.), 2014: *Ernst Ulrich von Weizsäcker: A Pioneer on Environmental, Climate and Energy Policies* (Cham–Heidelberg–New York–Dordrecht–London: Springer).

Weizsäcker, Ernst Ulrich von; Hargroves, Charlie; Smith, Michael H.; Desha, Cheryl J.K.; Stasinopoulos, Peter, 2009: *Factor Five. Transforming the Global Economy through 80 % Improvements in Resource Productivity* (London: Earthscan).

Weizsäcker, Ernst Ulrich von; Lovins, Amory; Lovins, Hunter, 1997: *Factor Four. Doubling Wealth, Halving Resource Use* (London: Earthscan).

Westing, Arthur H. (Ed.), 1989: *Comprehensive Security for the Baltic: An Environmental Approach*, (London: Sage).

Westing, Arthur H., 1989a: "The Environmental Component of Comprehensive Security", in: *Bulletin of Peace Proposals*, 20,2: 129-134.

Westing, Arthur H., 2014: *Texts on Environmental and Comprehensive Security* (Cham–Heidelberg–New York–Dordrecht–London: Springer-Verlag).

Wilson, Edward O., 1998: "Integrated science and the coming century of the Environment". in: *Science*, 279: 2048-2049.

Wilson, Edward O., 1998a: *Consilience* (New York: Knopf).

Wolfrum, Rüdiger, 1994: "Chapter 1. Purposes and Principles, Art. 1", in: Simma, Bruno (Ed.): *The Charter of the*

United Nations. A Commentary (Oxford: Oxford University Press): 49-56.

Wright, Quincy, 1942, 1965: *A Study of War* (Chicago—London: University of Chicago Press).

Wright, Quincy, 1954: "Criteria for judging the relevance of researches on the problems of peace", in: *Research for Peace* (Oslo: Institute for Social Research): 3-98.

Wun'Gaeo, Surichai, 2009: "Environment as an Element of Human Security in Southeast Asia: Case Study on the Thai Tsunami", in: Brauch, Hans Günter; Oswald Spring, Úrsula; Grin, John; Mesjasz, Czeslaw; Kameri-Mbote, Patricia; Behera, Navnita Chadha; Chourou, Béchir; Krummenacher, Heinz (Eds.): *Facing Global Environmental Change: Environmental, Human, Energy, Food, Health and Water Security Concepts* (Berlin—Heidelberg—New York: Springer-Verlag): 1131-1142.

10   Population Imbalances: Their Implications for Population Growth over the Twenty-first Century

Hania Zlotnik[1]

Abstract

Current population dynamics vary markedly among countries. In 2010, 46 per cent of the world population lived in countries with below-replacement fertility, whose populations are experiencing rapid population ageing and will likely decrease in the future. In sharp contrast, 19 per cent of the world population lived in countries with such high fertility that their combined populations are likely to double by 2050 and to keep growing over the rest of the century. The remaining 35 per cent lived in countries where fertility levels had declined considerably since their peak but where the potential for continued population growth over the medium term remains high. Owing to both population momentum and the long-standing differences in population dynamics between those groups of countries, these diverse population trends are expected to continue over the medium term, thus maintaining today's population imbalances and producing a high potential for marked population growth over the course of this century, especially in some of the world's poorest countries. Reducing those population imbalances by the end of the century and achieving the stabilization of the world's population requires that the high-fertility countries accelerate the reduction of fertility so that it reaches below-replacement level before 2100.

Keywords: Below-replacement fertility, population ageing, population dynamics. population growth, population imbalances, population momentum.

10.1   Introduction

In 2014, as governments debate which goals they are to adopt to spur sustainable development between 2015 and 2030 (United Nations 2014a), the world population continues to grow, having reached 7 billion in 2011 and being poised to reach 8 billion between 8 and 12 years from now (that is, between 2022 and 2026). Crucially, the earlier that 8th billion is reached, the more likely it is that the world population will add several additional billions by the end of the century. Because most of the people who will be alive over the next twelve years have already been born, the exact moment at which the next billion is reached depends mostly on how many children are born over the next few years. That is, fertility trends are the main determinant of population increases and, because

of the multiplicative nature of population growth, small changes in global fertility now and in the immediate future can have a major impact on the long-term potential for further population growth.

Although the rapid population growth that the world has experienced since 1950 has been accompanied by improvements in the standard of living and the quality of life of the majority of the world population, it is becoming increasingly clear that the planet is under stress because of human activity and that ensuring better livelihoods for future generations poses many challenges. Thus, attaining universal food security by mid-century, a basic goal to ensure future well-being, may be feasible but is far from certain. The need to increase the global food supply by at least 70 per cent to be able to feed the nine billion people expected to inhabit the Earth by 2040 requires intensifying the use of existing agricultural land, particularly in the populous countries of Asia, and extending the land under cultivation in many of the world's least developed countries where population growth is expected to be the fastest (Butler and Dixon 2012;

---

1   Dr. Hania Zlotnik is an independent consultant and was Director, Population Division, Department of Economic and Social Affairs, United Nations Secretariat (2005-2012); Email: <hania.zlotnik@hotmail.com>.

© Springer International Publishing Switzerland 2016   239
H.G. Brauch et al. (eds.), *Handbook on Sustainability Transition and Sustainable Peace*,
Hexagon Series on Human and Environmental Security and Peace 10, DOI 10.1007/978-3-319-43884-9_10

West et al, 2014). Achieving those changes while simultaneously reducing the deleterious effects of agriculture on global warming and water quality and reducing water use will demand large additional investments in agriculture and the judicious use of technology. Expanding access to clean energy is another daunting task that has just started and that is still progressing too slowly to meet the needs of the rapidly growing populations of the developing world.

These and other challenges are all scaled up by population numbers and will be more difficult to meet the larger the population becomes. Yet intergovernmental and even expert discussions about how to attain a sustainable future rarely consider future population numbers as a factor subject to modification. Because policymakers usually focus on the short term and changes in population dynamics usually take time to have a significant effect on overall population numbers, the need to modify current population dynamics is often disregarded.

This chapter, by focusing on the long-term implications of changes in population dynamics, shows that, if the world population is to remain below 10 billion by the end of this century, there is an urgent need to accelerate the reduction of fertility in the countries that still have high fertility levels, and it is also necessary that the countries with intermediate levels of fertility continue reducing them. According to the most recent estimates of fertility levels, referring to 2005–2010, 62 countries, accounting for 19 per cent of the world population in 2014, had high fertility; another 64 countries, accounting for 35 per cent of the world population, had intermediate fertility, and a further 75 countries, representing 46 per cent of the world population, had low fertility.

Analysis of past trends shows that the high-fertility countries of today have had, on average, higher fertility levels than the rest since 1950 and, although their levels of mortality in childhood have declined substantially, their fertility has declined very slowly. Therefore, even though they account for just 19 per cent of today's population, their population is increasing rapidly and may grow to represent 43 per cent of the world population in 2100 even if their fertility declines at a moderate pace in future. Because most of the high-fertility countries are low-income or lower-middle-income countries, they are the least well prepared to satisfy the basic needs of rapidly growing populations. In contrast, the most prosperous or rapidly growing economies are those of countries whose fertility has reached low levels and whose overall population is likely to decrease during this century, so that

by 2100 they may account for 28 per cent of the world population, down from 46 per cent today.

Past trends reveal that such 'population imbalances', meaning the marked differences in population dynamics that over the long run change the regional distribution of the world population, have been at play since at least 1950. However, the expectation has been that population trends will converge so that all countries will eventually achieve a state of low fertility and low mortality that, aside from producing a world population that changes very slowly, would also contribute to improve human well-being by maximizing longevity and facilitating higher investments in the human capital of each child. Yet, in view of developments over the past two decades, one may question whether that convergence will actually occur and, if it does, how rapidly it will be achieved. Considering the number and characteristics of the countries that continue having high fertility, there is greater uncertainty today than just two decades ago about whether convergence will happen. By exploring the consequences of a delayed convergence, this chapter provides evidence that should prompt those interested in sustainability to advocate a sustained and well-funded effort to accelerate the reduction of fertility in high-fertility countries through the promotion of a smaller family norm and the uptake of voluntary family planning in countries where still too many children are born and too many die from preventable causes.

10.2   The Demographic Transition as the Rationale for Convergence

By any standards, world population trends changed dramatically during the second half of the twentieth century. The reductions of mortality that had started in the more advanced countries in the nineteenth century expanded rapidly to the developing world particularly after 1950 as better nutrition, improved sanitation and modern medical interventions successfully controlled or treated the most deadly infectious and parasitic diseases that were endemic in those countries. Plummeting infant and child mortality coupled with fertility levels that remained high resulted in accelerated population growth. However, with the expanded adoption of family planning and the use of modern contraceptives, fertility eventually declined also, sometimes very rapidly and often to unprecedentedly low levels. By the end of the twentieth century, most developed countries and an increasing number of developing countries were experiencing levels of fertility so low that, if maintained for a generation or more, they would lead to decreasing populations.

These developments have led demographers to develop a model of demographic change known as 'the demographic transition'. In the 1950s, this model provided a stylized description of the changes in fertility and mortality trends that took place as populations passed from a state where both mortality and fertility were high and population growth was low to one where both mortality and fertility were low and population growth would again be very low. According to the demographic transition, mortality would decline first while fertility remained high until people realized that, because of improved child survival, it was no longer necessary to have so many children in order to achieve the number of adult children they desired. That realization would prompt people to reduce their fertility and, as mortality continued to decline, fertility would drop further until once more fertility levels would counterbalance mortality and the natural increase of the population would reach zero. By the 1990s, noting that many populations were completing the demographic transition not by reaching a balance between fertility and mortality but rather by overshooting the fertility decline and ending in a stage where fertility was too low to counterbalance mortality, demographers modified the original description of the last stage of the transition by recognizing that it could produce not an unchanging population with zero population growth but rather a decreasing population.

By 2014, changes in fertility and mortality trends consistent with the demographic transition had taken place in all countries in the world. In every country, mortality is currently lower today than it was in the 1950s and fertility has also declined with respect to the peak level it attained after 1950. However, the pace of mortality and fertility reductions as well as their timing has varied considerably among countries. Currently, countries that are most advanced in the demographic transition are characterized by having very low fertility and low mortality. Those that are the least advanced still have relatively high fertility levels. The rest have advanced in the path to low fertility but have not yet reached the very low levels that characterize the last stage of the demographic transition. Consequently, those three groups of countries have today different population profiles and different prospects for future population change. Countries with low fertility have rapidly ageing populations and, if their populations are not decreasing already, expect to experi-ence population reductions in the future. Countries with high fertility still have young populations that are growing fast and have a high potential for future population increases. The rest are experiencing moderate rates of population growth and have moderately ageing populations but their future hinges on whether or not they complete the demographic transition by reaching very low fertility levels. These differences in current population dynamics are at the root of the population imbalances that characterize world population today and are likely to remain a feature of the world population over the next four or five decades at least. Furthermore, developments over those decades will determine whether there is any chance for those imbalances to start waning during the second half of this century as the fertility and mortality of all countries converge to low levels.

10.3   Classifying Countries into Three Groups According to Their Net Reproduction Rates

The countries of the world can be classified into three groups according to their fertility levels. The classification used in this chapter is based on the value of the *net reproduction rate* (NRR) for each country. NRR is a measure of the reproductive potential of a population. It takes into account both the overall level of fertility of a country (that is, a country's total fertility[2]) and the level of mortality during the first 25 or 30 years of life. NRR indicates the average number of daughters expected to survive to the mean age at childbearing that a cohort of women would have by exact age 50 were those women to experience from exact age 15 onwards the age-specific fertility rates observed during a given period without being subject to mortality. The *United Nations Population Division* (UNPD) estimates NRR for the 201 countries[3] that had at least 90,000 inhabitants[4] in 2010 (United Nations 2013a). In the majority of those 201 countries,

---

2   Total fertility is a measure of the level of fertility in a population. Total fertility is the average number of children a hypothetical cohort of women would have at the end of their reproductive period if they were subject during their whole lives to the fertility rates of a given period and if they were not subject to mortality. It is expressed as children per woman.

3   The term 'countries' is used here to encompass also areas that are dependent territories but whose data are reported separately from that of the country from which they depend.

Figure 10.1: Map displaying the high-fertility, intermediate-fertility and low-fertility countries.

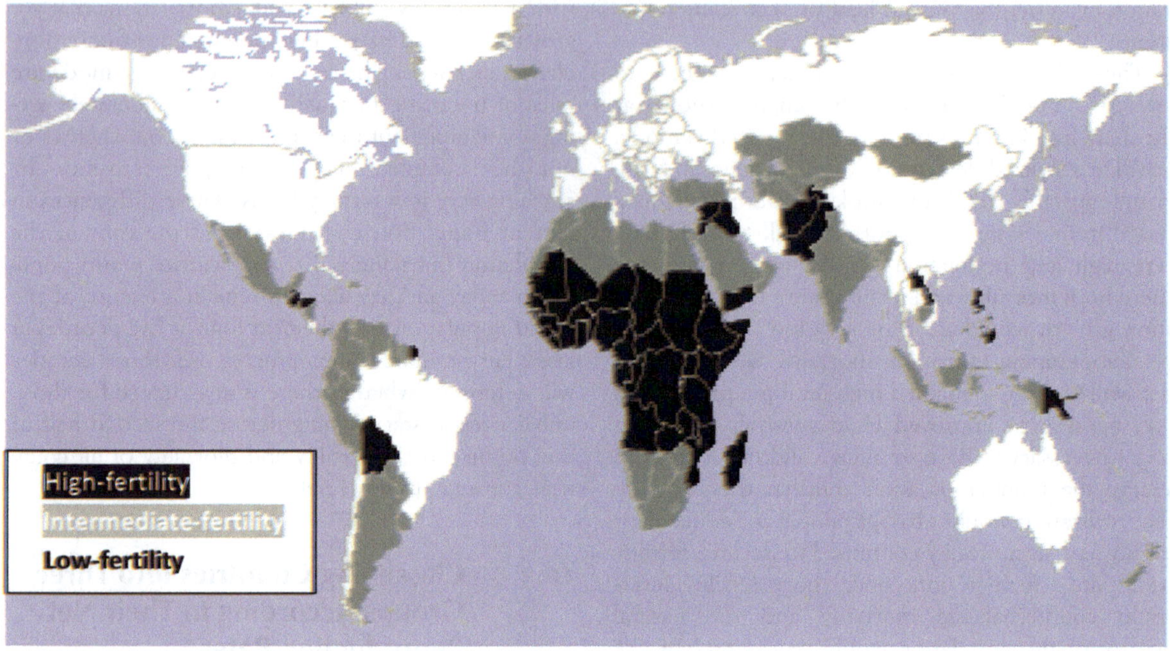

fertility and mortality have been the major components of past population change.

A net reproduction rate of 1.0 daughter per woman indicates that every woman of reproductive age is eventually replaced by a daughter who survives to an age where she herself may have a child. Therefore, when NRR = 1.0, the population is on track to 'replace itself' and fertility is said to be at 'replacement level'.[5] When NRR is higher than 1.0 daughter

per woman, the current level of fertility is higher than it needs to be to counterbalance the current level of mortality and, therefore, the population is poised to increase from generation to generation and fertility is said to be 'above replacement level'. When NRR is lower than 1.0 daughter per woman, the reverse holds, that is, the current level of fertility is not sufficient to counteract the current level of mortality and, therefore, the population is poised to decrease from generation to generation. In this latter case, fertility is said to be 'below replacement level'.

Estimates of NRR referring to the period 2005-2010 have been used to classify the 201 countries with estimates available into three groups. The *high-fertility countries* are those with an NRR equal to or higher than 1.5 daughters per woman, that is, they are countries whose most recently measured fertility and mortality levels, if maintained, would lead to an increase of at least 50 per cent between one generation and the next. The second group of countries, denominated *intermediate-fertility countries*, are those whose NRR in 2005-2010 varies from 1.0 daughter per woman to just under 1.5 daughters per woman. Lastly, the *low-fertility countries* are those whose NRR is below 1.0 daughter per woman, that is, their fertility is below replacement level. If an NRR below 1.0 daughter per woman is sustained over a period of a generation or more, the population will decrease.

According to the above classification, in 2005-2010, 62 countries[6] had high fertility levels, 64 had

---

4   The United Nations Population Division does not produce estimates of the components of population change for countries with fewer than 90,000 inhabitants, and that is why they cannot be included in the analysis presented here. In 2010, 32 countries or areas had fewer than 90,000 inhabitants each and their joint population amounted to approximately one million. Because of their small populations, in many of those countries migration is a very important component of population change.

5   Since NRR depends on both the level of fertility and that of mortality, 'replacement level' fertility varies according to the level of mortality: the higher the probability of dying before age 30, the higher total fertility needs to be to assure that a daughter survives to the age of reproduction. When mortality is low, a total fertility of around 2.10 children per woman is usually sufficient to ensure replacement and that number is often described as 'replacement-level fertility'. When mortality is high, however, replacement-level fertility can be much higher, sometimes amounting to values well above 3 children per woman.

Table 10.1: Distribution of countries and population among the high-fertility, intermediate-fertility and low-fertility groups. **Source**: Derived from United Nations (203a).

		Number of countries			Percentage		
Region	Total	High-fertility countries	Intermediate-fertility countries	Low-fertility countries	High-fertility countries	Intermediate-fertility countries	Low-fertility countries
Total	201	62	64	75	31	32	37
Developed countries	44	0	2	43	0	2	98
Developing countries	157	62	62	32	39	39	20
Africa	57	41	14	2	72	25	4
Asia	50	11	21	18	22	42	36
Latin America and the Caribbean	38	4	22	12	11	58	32
Oceania	12	6	5	0	50	50	0

		Population in 2014 (billions)			Percentage of population		
Total	7,24	1,36	2,52	3,37	19	35	46
Developed countries	1,26	0,00	0,00	1,25	0	0	100
Developing countries	5,99	1,36	2,52	2,11	23	42	35
Africa	1,14	0,89	0,24	0,01	78	21	1
Asia	4,22	0,43	1,93	1,86	10	46	44
Latin America and the Caribbean	0,62	0,04	0,35	0,24	6	55	39
Oceania	0,01	0,01	0,00	0,00	84	16	0

intermediate fertility levels[7] and 75 had low fertility levels[8] (see table 10.1). Most of the developed countries[9] (98 per cent) have below-replacement fertility levels and are classified as low-fertility countries. Only two developed countries, Iceland and New Zealand, had intermediate-fertility levels in 2005–2010 and both had fertility close to replacement level. Among developing countries, 32 (20 per cent) had below-replacement fertility, including two of the most populous countries: Brazil and China. Ninety per cent of the low-fertility countries are high-income or upper middle-income countries according to the World Bank's classification.

The remaining 144 developing countries are equally divided among the high-fertility and the intermediate-fertility countries. In Africa, the majority of countries had high fertility levels in 2005–2010 (72 per cent). In developing Oceania, half of the countries belonged to the high-fertility group. High-fertility countries are less common in Asia, where they account for 22 per cent of the total, and in Latin America and the Caribbean, where they account for just 11 per cent. Nevertheless, all developing regions still have significant numbers of countries maintaining high fertility. Ninety per cent of the high-fertility coun-

6   *High-fertility countries:* Afghanistan, Angola, Benin, Bolivia, Burkina Faso, Burundi, Cameroon, Central African Republic, Chad, Comoros, Congo, Côte d'Ivoire, Democratic Republic of the Congo, Djibouti, Equatorial Guinea, Eritrea, Ethiopia, Federated States of Micronesia, French Guiana, Gabon, Gambia, Ghana, Guatemala, Guinea, Guinea-Bissau, Honduras, Iraq, Jordan, Kenya, Lao People's Democratic Republic, Liberia, Madagascar, Malawi, Mali, Mauritania, Mayotte, Mozambique, Niger, Nigeria, Pakistan, Papua New Guinea, Philippines, Rwanda, Samoa, Sao Tome and Principe, Senegal, Sierra Leone, Solomon Islands, Somalia, South Sudan, State of Palestine, Sudan, Syrian Arab Republic, Tajikistan, Timor-Leste, Togo, Tonga, Uganda, United Republic of Tanzania, Vanuatu, Yemen and Zambia.

7   *Intermediate-fertility countries:* Algeria, Antigua and Barbuda, Argentina, Bahrain, Bangladesh, Belize, Bhutan, Botswana, Brunei Darussalam, Cambodia, Cape Verde, Colombia, Dominican Republic, Ecuador, Egypt, El Salvador, Fiji, French Polynesia, Grenada, Guadeloupe, Guam, Guyana, Haiti, Iceland, India, Indonesia, Israel, Jamaica, Kazakhstan, Kiribati, Kuwait, Kyrgyzstan, Lesotho, Libya, Maldives, Mexico, Mongolia, Morocco, Namibia, Nepal, New Caledonia, New Zealand, Nicaragua, Oman, Panama, Paraguay, Peru, Qatar, Réunion, Saint Vincent and the Grenadines, Saudi Arabia, Seychelles, South Africa, Sri Lanka, Suriname, Swaziland, Turkey, Turkmenistan, United States Virgin Islands, Uruguay, Uzbekistan, Venezuela, Western Sahara and Zimbabwe.

tries are low-income or lower middle-income countries according to the World Bank's classification.

Intermediate-fertility countries are mostly concentrated in Asia and in Latin America and the Caribbean where they account for 42 per cent and 58 per cent of the countries respectively. Among countries in developing Oceania, half had intermediate fertility in 2005–2010 and so did a quarter of those in Africa. India, the most populous country in the world, is among the intermediate-fertility countries, as are Indonesia, Bangladesh, Mexico and Egypt (in order of population size). Those five countries account for 75 per cent of the population of all intermediate-fertility countries. Sixty-three per cent of intermediate-fertility countries are high-income or upper-middle-income countries, according to the World Bank's classification, and 9 per cent are low-income countries.

In 2014, 46 per cent of the world population lived in low-fertility countries, 35 per cent in intermediate-fertility countries and 19 per cent in high-fertility countries (table 10.1). The populations of low-fertility and intermediate-fertility countries are highly concentrated in just a few countries. As just noted, five countries account for 75 per cent of the 2.5 billion people living in the intermediate-fertility countries. Among the 3.4 billion people living in the low-fertility countries, 75 per cent live in just nine countries of which the most populous are, in order of population size, China, the United States, Brazil, the Russian Federation and Japan. In comparison, the 1.4 billion people living in high-fertility countries are less concentrated in a few populous countries given that 17 countries are needed to account for 75 per cent of their overall

population. The most populous among them are: Pakistan, Nigeria, the Philippines and Ethiopia, in order of population size.

10.4 Past Trends in Total Fertility and Under-five Mortality[10]

Although the three groups of countries defined above were determined on the basis of a synthetic measure of reproductive potential (the net reproduction rate) referring to 2005–2010, it turns out that the three groups have had quite different trends in total fertility and under-five mortality[11] since at least 1950. That is, their current differences are not only the result of diverging trends since 1950 but reflect also long-standing differences that have been more persistent than expected. In particular, the high-fertility countries have, as a group, consistently had higher fertility and higher mortality in childhood than the other two groups. In 2005–2010, one in every ten children born in the high-fertility countries would die before age five. Nevertheless, that probability of dying was two-thirds lower than it had been in the early 1950s, a reduction that in many other countries has been sufficient to trigger a major reduction of fertility. Therefore, the slow fertility decline observed among the high-fertility countries is unlikely to be mainly a response to the persistence of high childhood mortality.

Another development of note relates to the continued decline of total fertility among the low-fertility countries, a group that reached replacement-level fertility in the early 1990s but continued to see its fertility drop, so that it has been below-replacement level since at least 1995. Such a sustained fertility decline has meant that, rapid as the reduction of fertility has been in the intermediate-fertility countries, their fertility has still not 'caught up' with that of the low-fertility countries. Furthermore, the convergence of the fertility of the two groups, if it happens at all, may still take several decades, a period over which the population

---

8 *Low-fertility countries*: Albania, Armenia, Aruba, Australia, Austria, Azerbaijan, Bahamas, Barbados, Belarus, Belgium, Bosnia and Herzegovina, Brazil, Bulgaria, Canada, Channel Islands, Chile, China, Costa Rica, Croatia, Cuba, Curaçao, Cyprus, Czech Republic, Democratic People's Republic of Korea, Denmark, Estonia, Finland, France, Georgia, Germany, Greece, Hong Kong, Hungary, Iran, Ireland, Italy, Japan, Latvia, Lebanon, Lithuania, Luxembourg, Macao, Malaysia, Malta, Martinique, Mauritius, Montenegro, Myanmar, Netherlands, Norway, Poland, Portugal, Puerto Rico, Republic of Korea, Republic of Moldova, Romania, Russian Federation, Saint Lucia, Serbia, Singapore, Slovakia, Slovenia, Spain, Sweden, Switzerland, Taiwan, Thailand, The Former Yugoslav Republic of Macedonia, Trinidad and Tobago, Tunisia, Ukraine, United Arab Emirates, United Kingdom, United States and Viet Nam.

9 Developed countries include all the countries in Europe plus Australia, Canada, Japan, New Zealand and the United States.

---

10 All population estimates and projections cited in this chapter were derived from the 2012 *Revision* of *World Population Prospects* prepared by the Population Division of the Department of Economic and Social Affairs of the United Nations Secretariat in New York (United Nations 2013a). Calculations of the relevant indicators for the groups of countries used in this chapter were made by the author.

11 Under-five mortality is the probability of dying between birth and exact age 5. It is expressed as number of deaths per 1,000 births.

Figure 10.2: Trends in total fertility for the three groups of countries considered, 1950–2010. **Source:** Derived from United Nations (2013a).

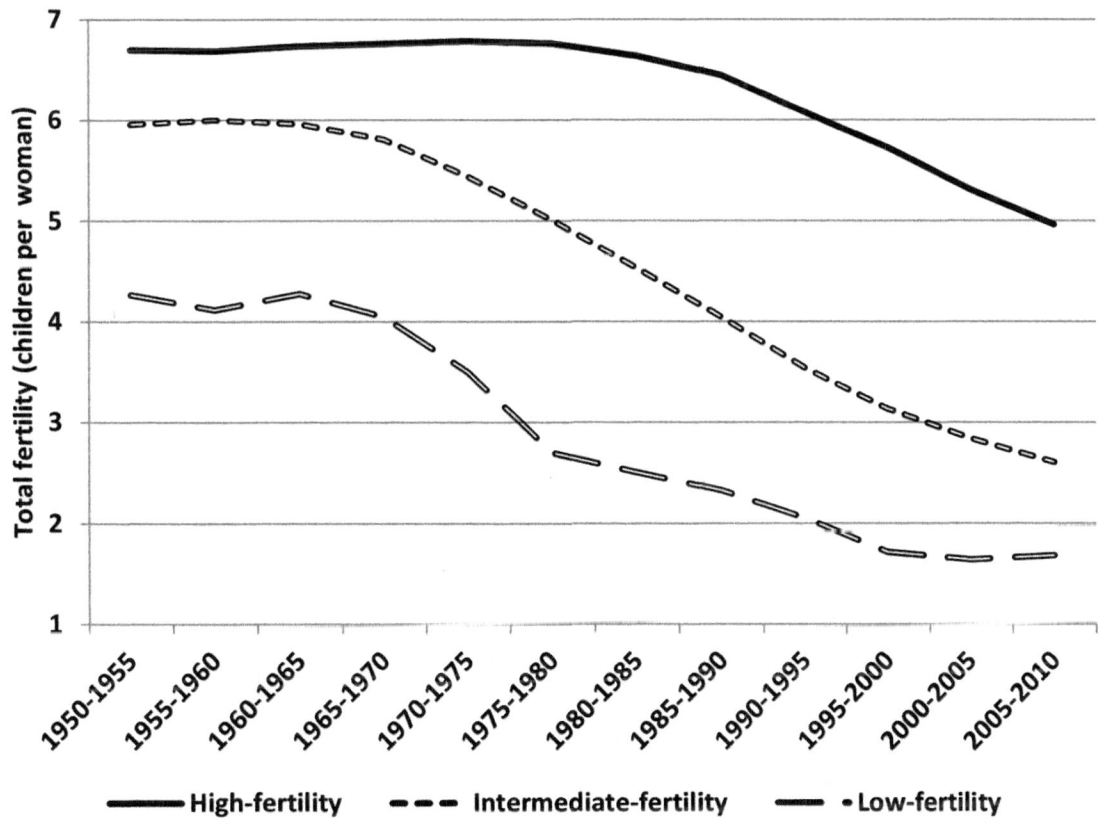

━━━ High-fertility   ━ ━ ━ Intermediate-fertility   ━━━ • Low-fertility

of the intermediate-fertility countries will continue to increase.

10.4.1   Trends in Total Fertility

In 1950–1955, total fertility among the high-fertility countries averaged 6.70 children per woman, 0.76 children higher than the 5.96 children per woman estimated for the intermediate-fertility countries, which, in turn, surpassed the total fertility of the low-fertility countries (4.26 children per woman) by 1.70 children per woman. The large difference between the total fertility of the intermediate-fertility countries and that of the low-fertility countries was to be expected because the latter group includes nearly all developed countries, the vast majority of which had experienced major reductions of fertility since the late nineteenth century. Thus, although many of them had experienced a 'baby boom' during the 1950s and early 1960s, their average total fertility was a low 2.83 children per woman in 1950–1955.

More remarkable is the difference between the total fertility of the high-fertility and the intermediate-

fertility countries in the early 1950s, since there was no reason to expect that their past experience would have differed so markedly at the early stages of the demographic transition. Also remarkable is the difference in the timing of the fertility decline in the high-fertility and the intermediate-fertility countries. Thus, whereas in the group of intermediate-fertility countries average total fertility began to decrease between 1960–1965 and 1965–1970, that of the high-fertility countries started to decline ten years later, between 1970–1975 and 1975–1980. Furthermore, during the 35 years following peak total fertility, the high-fertility countries saw their total fertility decline, on average, by 0.5 children per decade, whereas in the intermediate-fertility countries the decline averaged 0.8 children per decade. That is, the fertility decline in the high-fertility countries has been significantly slower than in the intermediate-fertility countries (figure 10.2). Consequently, by 2005–2010, total fertility in the intermediate-fertility countries was 56 per cent lower than at its peak, averaging 2.6 children per woman in 2005–2010, whereas that of the high-fertility countries averaged 4.96 children per woman in 2005–2010 and had declined by just 26 per cent from its peak value of 6.79 children per woman in 1970-1975.

The sharp decline in fertility achieved by the intermediate-fertility countries since 1965 brought them closer to the low-fertility countries by 2005-2010 but, remarkably, their total fertility of 2.6 children per woman still surpassed by 0.9 children per woman that of the low-fertility countries, which averaged 1.7 children per woman in 2005-2010. Indeed, the low-fertility countries had recorded an almost equally rapid reduction of fertility as the intermediate-fertility countries, averaging a reduction of 0.7 children per decade during the 35 years following their peak fertility level. In addition, the total fertility of the low-fertility countries had kept on falling even after reaching replacement level with the result that they had maintained, as a group, below-replacement fertility since at least 1995.

In sum, as of 2005-2010, the differences in fertility levels between the three groups of countries considered were remarkable, with the high-fertility countries having, on average, a total fertility of nearly 5 children per woman as a result of consistently higher fertility since 1950, a late onset of the fertility decline and a slow pace of fertility reductions. The intermediate-fertility countries, with an average total fertility in 2005-2010 of 2.6 children per woman, had, as a group, recorded major and rapid reductions of fertility but were, nevertheless, still far from achieving the low levels characterizing the low-fertility countries, whose total fertility averaged a low 1.7 children per woman. As this chapter will show, future prospects for stopping population growth hinge on the eventual convergence of all countries to low fertility levels similar to those already attained by the group of low-fertility countries.

10.4.2 Trends in Under-Five Mortality

Trends in childhood mortality have also been distinct for the three groups of countries considered. The high-fertility countries have consistently had the highest mortality under age five since 1950 and although it has dropped by nearly two-thirds, their under-five mortality was still a high 109 deaths under age 5 per 1,000 births in 2005-2010. Under-five mortality was high for the intermediate-fertility countries in the 1950s, but lower than in the high-fertility countries (272 vs 307 deaths under age 5 per 1,000 births in 1950-1955). Since then, the under-five mortality of the intermediate-fertility countries has dropped by more

than 80 per cent, reaching 51 deaths per 1,000 births in 2005-2010.

Under-five mortality was markedly lower for the low-fertility countries in the 1950s being, at 156 deaths under age 5 per 1,000 births, about half what it was in the high-fertility countries at that time. Yet its decline has been even steeper than that achieved by the intermediate-fertility countries, so that its level in 2005-2010, at 18 deaths per 1,000 births, is lower by 88 per cent than what it was in 1950-1955.

Interestingly, the decline in fertility in the intermediate-fertility countries started when their average under-five mortality was higher than it was in the high-fertility countries when their own fertility started to decline and, in both cases, under-five mortality was high when fertility began to decrease (at 187 and 177 deaths under age 5 per 1,000 births, respectively). In 1985-1990, when the under-five mortality of the intermediate-fertility countries stood at 105 deaths per 1,000 births, their total fertility was 4.04 children per woman, 0.9 children below the total fertility in the high-fertility countries (4.96 children per woman) in 2005-2010, when the under-five mortality of the latter group was 109 deaths under age 5 per 1,000 births. That is, similar levels of under-five mortality have been associated with higher fertility in the high-fertility countries than in the intermediate-fertility countries indicating that there is not an automatic and tight relationship between declining mortality in childhood and declining fertility and that factors other than the reduction of mortality are hindering rapid reductions in fertility in the high-fertility countries.

Because the reduction of under-five mortality has been faster in the intermediate-fertility countries than in the high-fertility countries, by 2005-2010 the under-five mortality of the former was closer to that of the low-fertility countries, although it had been closer to that of the high-fertility countries in the 1950s (figure 10.3). In the high-fertility countries, the slow decline in under-five mortality between 1980 and 2000 owes much to the impact of the HIV/AIDS epidemic on the highly affected countries that are members of that group. Therefore, as in the case of fertility, convergence in the levels of under-five mortality among the three groups considered is still in the future and, especially in the case of the high-fertility countries, the gap between their under-five mortality and that of the other groups will likely take decades to close.

Figure 10.3: Trends in under-five mortality for the three groups of countries considered, 1950–2010. Source: Derived from United Nations (2013a).

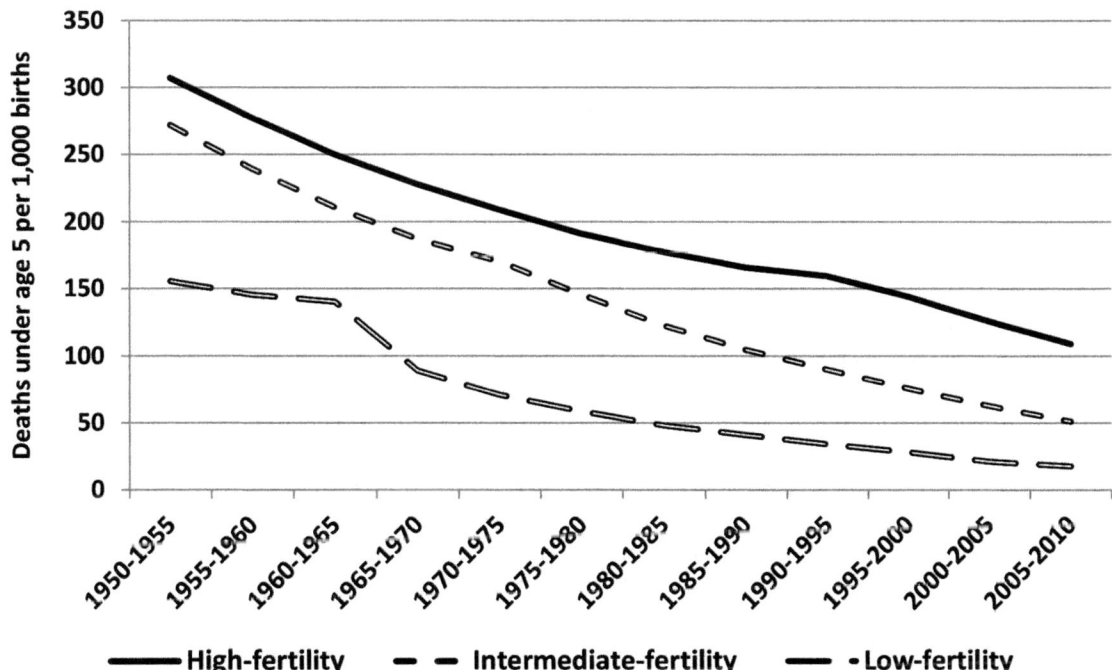

10.4.3   Consequences of Past Fertility and Mortality Trends

Differences in past fertility and mortality trends among the three groups of countries considered have produced major differences in population growth rates that, in turn, have given rise to the population imbalances described earlier in the chapter. Despite their higher mortality, the persistence of high fertility in the high-fertility countries has produced a very high rate of annual population growth, which peaked at 2.84 per cent in 1977 and is still a high 2.42 per cent in 2014. The intermediate-fertility countries also reached a peak annual growth rate of 2.36 per cent in 1977 but it has since declined to an estimated 1.22 per cent in 2014, still far from zero but declining at a good pace. The most remarkable change has occurred among the low-fertility countries, whose annual growth rate peaked at 1.83 per cent in 1967 and is now a low 0.5 per cent per year and declining fast.

Another source of population imbalances are differences in the extent of population ageing experienced by the different groups of countries considered. The speed of the fertility decline and its starting point determine the speed of population ageing, that is, of the increasing concentration of the population at adult ages at first and then at older ages. The best

overall indicator of population ageing is the median age of the population, that is, the age that divides the population into two equal parts, with half the population being younger that the median age and half being at the median age or older.

At the start of the demographic transition, decreases in mortality and the increases in fertility that sometimes accompany them produce increasing numbers of surviving children and, therefore, lead to a rejuvenation of the population as the proportion of people in the younger ages increases. Such rejuvenation produces a decreasing median age. However, once fertility starts to decline, its reduction eventually leads to decreasing numbers of children being born and thus triggers population ageing, reflected in an increasing median age.

As expected, between 1950 and 2010, the median ages of the three groups of countries considered declined at first and eventually started to increase. In the low-fertility countries, the median age declined from 25.7 years in 1950 to 23.9 in 1970 and in the intermediate-fertility countries, it dropped from 20.9 to 18.9 years over the same period before it began to increase. In contrast, in the high-fertility countries the median age declined for 40 years, from 19.1 years in 1950 to 17.4 years in 1990. By 2010, the median age for the low-fertility countries had risen to 35.5 years, that of the intermediate-fertility countries to 25.7 years and the one for the high-fertility countries to 18.8 years.

Population imbalances relative to the population age structure were already evident in the 1950s, when the low-fertility countries had an older population than the intermediate-fertility countries and the latter an older one than the high-fertility countries. Furthermore, those differences have become starker over time, with the median age of the low-fertility countries increasing faster than that of the intermediate-fertility countries (by 48 per cent from 1970 to 2010, as compared to 36 per cent over the same period in the intermediate-fertility countries). Population ageing has been slowest in the high-fertility countries, whose median age rose by a mere 8 per cent since 1990. Furthermore, the median age of the high-fertility countries was lower in 2010 than it had been in 1950.

In view of these trends, although the populations of all three groups of countries are ageing, a convergence of the shape of their age distributions is not expected to occur soon. The population of the low-fertility countries, which is already far advanced in the ageing process, will continue to have a rapidly growing older population. That of the intermediate-fertility countries is poised to enter a phase of rapid ageing, caused by their rapid fertility decline, whereas the population of the high-fertility countries will remain relatively young. Furthermore, provided their fertility decline continues or accelerates, the high-fertility countries will soon enter a phase where their rising proportions of persons of working age have the potential to boost economic growth provided they can be productively employed (Bloom/Canning/Sevilla 2003).

One way of gauging the potential economic impact of the age distribution of a population is to consider the ratio of the number of persons of working age (usually considered to be those aged 15 to 64) to the number of children (persons under age 15) and the number of elderly persons (those aged 65 years or over). That indicator is known as the 'support ratio' and, although it does not reflect the actual number of producers or workers in a population relative to the dependants in the population, it is a useful indicator of the relative size of the number of potential workers to the number of potential dependants. For the three groups of countries considered, the support ratio has always been highest for the low-fertility countries as a group, having reached a minimum of 1.44 potential workers per dependant in 1965 and increased to 2.39 potential workers per dependant in 2010. The support ratio for the intermediate-fertility countries also

reached a minimum in 1965, at 1.18 potential workers per dependant, and then rose to 1.83 in 2010. The lowest support ratio corresponds to the high-fertility countries where a minimum of 1.07 potential workers per dependent was reached in 1990, a value that has since increased to 1.21 in 2010.

These trends in the support ratio confirm that, owing to their early and sustained reduction of fertility, the low-fertility countries have had a more favourable age distribution since 1950 than the other two groups of countries. In the intermediate-fertility countries, the age distribution has become more favourable to economic growth since 1965 but it has yet to reach a support ratio above 2, let alone the 2.39 that the low-fertility countries had in 2010. In the high-fertility countries, the persistently young age distribution has been a drag rather than a boost for the economy because of the high child dependency burden it implies. Hence, although their support ratio has started to increase, unless their fertility decline accelerates, the high-fertility countries may never reach a support ratio of 2 potential workers per dependant.

10.5   The Future of Fertility and its Implication for Future Population Trends

Differences in the past fertility trends experienced by the three groups of countries considered have major implications for future population trends. The high-fertility countries are expected to continue having higher fertility and higher mortality than the rest and to continue experiencing rapid population growth for some time. The key questions regarding these countries are whether and when their fertility will drop to replacement level or below. The sooner their fertility reaches replacement level, the sooner their fast population growth will be curbed and the smaller their peak population will be. Alternatively, the slower their fertility decline is in future, the larger their population will become. As this section will show, a variety of population projections indicate that the high-fertility countries of today are very likely to be the major contributors to future population growth at the global level.

For the intermediate-fertility countries, the key questions are whether and when their fertility will drop below replacement level. If it does so within the next few decades, their contribution to overall population growth over the rest of the century will be moderate. The slower this group is in reaching below-replacement fertility, the more it will contribute to future population growth and the higher their peak population will be.

Lastly, the low-fertility countries, whose average fertility has been below replacement level since 1995, are expected to continue having below-replacement fertility until 2100. The key questions in this case are whether the overall fertility of this group will increase over the century and whether it will remain below replacement level until 2100. If the total fertility of the low-fertility countries remains at or close to the low level it has already reached, their overall population will decrease markedly by 2100. A slowly increasing fertility that remains below replacement level will slow down the contraction of their population but, unless their fertility increases above replacement level, their population will decrease, therefore contributing to moderate global population growth.

The different projection variants and scenarios produced by the United Nations Population Division as part of the most recent set of official United Nations population estimates and projections (United Nations 2013a) will be used to explore the implications of future fertility trends for the three groups of countries considered.

10.5.1 The United Nations Projection Variants and Scenarios

In the official United Nations population projections, the 'medium-fertility variant' is the reference for the preparation for other projection sets. For a given country, its future fertility path is derived using stochastic simulations of future fertility paths that take account of past fertility trends in that country and of the pattern and variability of fertility trends in all other countries since 1950. The medium variant therefore reflects a central path informed by all the data available on fertility change since 1950.

The medium variant is the basis for deriving two other variants, namely, the 'high-fertility' and 'low-fertility' variants, that allow an assessment of the sensitivity of population projections to small changes in projected fertility. The high-fertility variant (also called 'high variant', for short) projects, for each country, a total fertility that remains half a child above the total fertility projected by the medium variant during the projection period (2010 to 2100). The low-fertility variant (also called 'low variant', for short) projects, for each country, a total fertility that remains half a child below that projected by the medium variant over the projection period.

The United Nations Population Division also prepares a number of 'scenarios', that is, projections aimed to answer 'what-if' questions. Two of those scenarios are considered here. The first, called the 'no-change scenario', illustrates what would happen if fertility and mortality were to remain constant in each country at the level they had in 2005-2010. This scenario provides a useful basis on which to judge whether the most recently estimated fertility and mortality levels, if maintained indefinitely, could lead to future population numbers that may be 'sustainable'.

The second scenario of interest is the 'instant-replacement scenario', which shows what would happen if each country were to attain and maintain replacement-level fertility immediately after 2010, when the projections start. Therefore, this scenario projects the total fertility[12] of each country in such a way that it produces an NRR that remains constant at 1.0 daughter per woman during the projection period (that is, from 2010 to 2100). Demographic theory predicts that a population that maintains replacement-level fertility over lengthy periods and whose mortality stops changing reaches eventually a stationary state where both its size and age distribution remain constant. In the instant-replacement scenario, no country reaches the stationary state by 2100, partly because mortality is projected to keep on decreasing over the rest of the century. Nevertheless, for countries with fertility above replacement level, the instant-replacement scenario provides a measure of the minimum population increase that one would expect if current and future generations of reproductive age had only the number of children necessary to replace themselves.

The rest of this section will focus on fertility trends and their implications for future population growth based on the results of the different projection variants and scenarios just described. Table 10.2 displays the total fertility levels estimated for 2005-2010 and those projected for 2045-2050 and 2095-2100 by the different projection variants and scenarios for the three groups of countries considered. Table 10.3 shows the population size of those three groups of countries as of 2010 and as projected to 2050 and 2100 according to the same projections.

---

12 Under low mortality conditions, a total fertility close to 2.10 children per woman usually implies an NRR of 1.0 daughter per woman. Therefore, that level of total fertility is often considered to indicate that fertility is at replacement level. However, when mortality is high or moderate, replacement level will be reached at higher levels of total fertility, as the instant-replacement scenario shows.

Table 10.2: Current and projected total fertility for different projection variants, 2005–2010, 2045–2050 and 2095–2100 (children per woman). **Source:** Derived from United Nations (2013a).

Group	2005-2010	No-change scenario	High variant	Medium variant	Instant-replacement scenario	Low variant
				2045-2050		
High-fertility countries	4,96	5,20	3,47	2,98	2,18	2,50
Intermediate-fertility countries	2,60	2,62	2,39	1,89	2,12	1,39
Low-fertility countries	1,68	1,71	2,29	1,80	2,08	1,30
				2095-2100		
High-fertility countries	4,96	5,50	2,59	2,10	2,10	1,62
Intermediate-fertility countries	2,60	2,64	2,34	1,84	2,09	1,34
Low-fertility countries	1,68	1,77	2,39	1,89	2,07	1,39

Table 10.3: Estimated population in 2010 and projected population in 2050 and 2100 according to different projection variants for the three groups of countries considered. **Source:** Derived from United Nations (2013a).

Group	2010 Population (billions)	Projected population (billions)				
		No-change scenario	High variant	Medium variant	Instant-replacement scenario	Low variant
				2050		
High-fertility countries	1,2	3,5	3,2	2,8	1,9	2,5
Intermediate-fertility countries	2,4	3,6	3,7	3,3	3,3	2,8
Low-fertility countries	3,3	3,1	4,0	3,5	3,9	3,1
Total	**6,9**	**10,2**	**10,9**	**9,5**	**9,0**	**8,3**
				2100		
High-fertility countries	1,2	13,0	6,8	4,7	2,1	3,1
Intermediate-fertility countries	2,4	4,9	5,0	3,1	3,6	1,8
Low-fertility countries	3,3	2,1	4,8	3,0	4,2	1,8
Total	**6,9**	**19,9**	**16,6**	**10,9**	**9,9**	**6,7**

10.5.2 Future Fertility and its Implications for Population Size

10.5.2.1 The High-Fertility Countries

According to all the projection variants and scenarios considered, the high-fertility countries are expected to maintain the highest fertility among the three groups of countries considered over the whole projection period. Of particular interest are the results of the no-change scenario for the high-fertility countries. In that scenario future fertility remains constant for each country at the level it had in 2005–2010, but for groups of countries total fertility under the no-change scenario rises because the countries with higher fertil-

ity gain weight within each group. Thus, for the high-fertility countries, total fertility in the no-change scenario increases by slightly more than half a child between 2005–2010 and 2095–2100, passing from 4.96 children per woman to 5.50 children per woman by century's end. This increasing fertility results in accelerated population growth, which yields a population of 13 billion for the high-fertility countries alone in 2100. Such a huge number of people, were it to materialize, would undoubtedly prevent the attainment of a sustainable future in the high-fertility countries of today and would very likely impose heavy constraints on the sustainability of the world population as a whole. These results indicate that the current lev-

Table 10.4: Population in 2010 and ratio of the projected population in 2050 and 2100 to the population in 2010 for different projection variants and for the three groups of countries considered. **Source**: Derived from table 10.3.

Group	2010 Population (billions)	Ratio of projected population to 2010 population				
		No-change scenario	High variant	Medium variant	Instant-replacement scenario	Low variant
				2050		
High-fertility countries	1,2	2,8	2,6	2,3	1,5	2,0
Intermediate-fertility countries	2,4	1,5	1,6	1,4	1,4	1,2
Low-fertility countries	3,3	0,9	1,2	1,1	1,2	0,9
Total	**6,9**	**1,5**	**1,6**	**1,4**	**1,3**	**1,2**
				2100		
High-fertility countries	1,2	10,6	5,5	3,8	1,7	2,6
Intermediate-fertility countries	2,4	2,0	2,1	1,3	1,5	0,8
Low-fertility countries	3,3	0,6	1,5	0,9	1,3	0,5
Total	**6,9**	**2,9**	**2,4**	**1,6**	**1,4**	**1,0**

els of fertility and mortality in the high fertility countries are not sustainable.

The high variant produces the next largest population for the high-fertility countries in 2100. Under that variant, the average total fertility of the group declines slowly, passing from 4.96 children per woman in 2005-2010 to 3.47 children per woman in 2045-2050 and to 2.59 children per woman in 2095-2100. Hence, according to the high variant, the total fertility of the high-fertility countries remains well above replacement level over the projection period and leads therefore to robust population growth. Thus, their overall population increases from 1.2 billion in 2010 to 6.8 billion in 2100, implying that by century's end the high-fertility countries would have a population nearly as large as that of the whole world in 2010. That is, if the total fertility of the high-fertility countries declines in future as slowly as projected in the high variant, they alone could add more than 5 billion inhabitants to the Earth.

According to the medium variant, whose assumptions about future fertility and mortality trends are derived from all the information available on past trends of fertility and mortality change, the total fertility of the high-fertility countries declines more rapidly than in the high variant, but still not as fast as the medium variant projections prepared in the late 1990s had forecast (United Nations 2000). Thus, their fertility drops from 4.96 children per woman in 2005-2010 to 2.98 children per woman in 2045-2050 and to 2.10

children per woman in 2095-2100, meaning that it reaches replacement level only towards the end of the century. Such a prolonged decline to replacement level leads to a significant increase in the population of the high-fertility countries, which reaches 4.7 billion in 2100. Yet, this total is 2.1 billion lower than that projected under the high variant and although it represents almost a quadrupling of the population of the high-fertility countries between 2010 and 2100 (table 10.4), it implies that their population growth rate would be well on the way of reaching zero in the early decades of the twenty-second century if they were to maintain low fertility from 2100 onward.

If the high-fertility countries managed to reduce their fertility faster than in the medium variant, they might approach the path set by the low variant, according to which their fertility declines from 4.96 children per woman in 2005-2010 to 2.50 children per woman in 2045-2050 and reaches a low 1.62 children per woman in 2100. Thus, under the low variant, the fertility of the high-fertility countries drops below replacement level during the second half of this century and reaches a very low level towards its end. Such a path of fertility decline results in a more moderate population increase compared to that in the medium variant, with the population of the high-fertility countries rising from 1.2 billion in 2010 to a peak of 3.2 billion in 2093 and then declining slowly to 3.1 billion in 2100. Hence, under the low variant, the overall population of the high-fertility countries would increase just about two and a half times between 2010 and 2100 instead of nearly quadrupling as in the medium variant (table 10.4).

As already noted, the high-fertility countries still have a young population, therefore their potential for future population growth is high because the current and future generations of potential parents are large. The instant-replacement scenario provides an indication of the minimum increase that might be expected in the high-fertility countries because of both their youthful age distribution and the projected decline of mortality. According to the instant replacement scenario, the total fertility of the high-fertility countries would need to drop from 4.96 children per woman in 2005-2010 to 2.42 children per woman[13] in 2010-2015 in order to reach replacement level immediately and it would then decline in response to declining mortality until it reached 2.10 children per woman in 2095-2100. Under those conditions, the population of the high-fertility countries would increase by just 0.9 billion between 2010 and 2100, to reach 2.1 billion at century's end. This increase can be interpreted as the minimum likely to occur in the high-fertility countries provided their mortality continues to decline.

In sum, depending on the speed of future fertility declines in the high-fertility countries, their population may vary over a relatively wide range in 2100, from a high of 6.8 billion as projected in the high variant to a low of 2.6 billion as projected in the low variant. The results of the no-change and the instant-replacement scenarios are not considered realistic. Nevertheless, the no-change scenario shows that the continuation of current levels of fertility and mortality in the high-fertility countries would lead to a very large population by century's end and suggest that, were high population growth to continue in those countries, it would very likely prevent the attainment of sustainability. The instant-replacement scenario indicates that, at a minimum, the high-fertility countries will add nearly another billion people to the world population by 2100.

Among the three groups of countries considered and in all the projections discussed, except for those associated with the instant-replacement scenario, the high-fertility countries make the largest contribution to world population growth during 2010-2100.

Hence, the future increase of the world population depends highly on what happens in the high-fertility countries and, in turn, their future population growth hinges on the path they follow to achieve low fertility. It is important, therefore, to support policies and programmes that contribute to accelerate the reduction of fertility in those countries.

Given the significant and rapid fertility decline that has already occurred in the rest of the world (that is, among the 81 per cent of the population living in the low-fertility and the intermediate-fertility countries), the sense of urgency that prevailed in the 1960s and 1970s about the need to curb population growth has disappeared. For at least the past twenty years, the expectation has been that the countries lagging behind in reducing fertility would follow the path of their predecessors almost automatically. Yet the expected rapid declines in fertility have failed to materialize and recent evidence suggests that, instead of declining, total fertility levels in some high-fertility countries may be stalling if not increasing. Furthermore, the case for reducing fertility to moderate population growth and increasing the chances of achieving sustainability is complicated by the fact that, with their youthful populations, the high-fertility countries are very likely to experience rapid population growth over the next few decades even if their fertility declines precipitously. Thus, the projections to 2050 produce a relatively narrow range of variation for their overall population, ranging from 2.5 billion in the low variant to 3.2 billion in the high, with the medium variant producing a population of 2.8 billion. Only by extending the projection horizon can one grasp the big difference that accelerating or delaying the reduction of fertility can make over the long run. As the projections shows, the higher the population size the high-fertility countries have by mid-century, the more likely they will be to add several billion people to the world population by 2100.

10.5.2.2   The Intermediate-Fertility Countries

The intermediate-fertility countries, as a group, have achieved success in reducing fertility. However, with an average total fertility of 2.60 children per woman in 2005-2010, they still have some way to go to reach low fertility. Furthermore, encompassing 35 per cent of the world's population, they still have considerable potential to add a few billions to the world population over this century. Thus, if their fertility were to remain constant at the level it had in 2005-2010 and their mortality remained unchanged, the no-change scenario shows that their population could double by

---

13 Their current replacement level is higher than 2.10 children per woman because the high-fertility countries still have relatively high mortality, especially in childhood. As their mortality declines, so will their replacement-level fertility.

2100, rising from 2.4 billion in 2010 to 4.9 billion by century's end (table 10.3). These results suggest that the continued reduction of fertility in the intermediate-fertility countries is still necessary to avoid the increasing demands that a doubling of their population would place on the environment.

The high variant illustrates what would happen if the fertility of the intermediate-fertility countries failed to decline to below-replacement level and remained, instead, slightly above replacement level until 2100.[14] Even relatively small positive deviations from replacement level accompanied by declining mortality result in significant population growth over the long run, so that the population of the intermediate-fertility countries reaches 5 billion in 2100 under the high variant, a number higher than that projected by the no-change scenario because the latter does not incorporate a reduction of mortality.

In the medium variant, the total fertility of the intermediate-fertility countries is projected to fall below replacement level in 2030–2035 and to remain at below-replacement level thereafter, so that by 2045–2050 it would be at 1.89 children per woman and it would further decline to 1.84 children per woman by 2095–2100. This trend in fertility reduces their potential for continued population growth, leading to an overall population increase of just 0.9 billion between 2010 and 2065 (from 2.4 billion in 2010 to 3.3 billion in 2065) when their population starts decreasing to reach 3.1 billion in 2100. That is, between 2010 and 2100, the medium variant projects that the population of the intermediate-fertility countries would have increased by just 0.7 billion people.

That small increase compares favourably with the increase projected under the instant-replacement scenario where the total fertility of the intermediate-fertility countries remains at replacement level instead of dropping below that level as it does in the medium variant. That trend in fertility coupled with declining mortality produces an increasing population for the intermediate-fertility countries, which reaches 3.6 billion in 2100, a figure half a billion higher than that projected under the medium variant.

In contrast, the low variant, by projecting very low fertility levels for the intermediate-fertility countries (below 1.50 children per woman after 2035), reduces their overall population by a fourth: from 2.4 billion in 2010 to just 1.8 billion in 2100.

In sum, although the intermediate-fertility countries still have the potential of experiencing sustained population growth if their total fertility does not fall below replacement level over the next few decades, their already moderate fertility levels ensure that such potential is also moderate. In comparison with the high-fertility countries whose population is very likely to nearly quadruple or more, that of the intermediate-fertility countries might at most double by the end of the century. Nevertheless, because of their large current population, the intermediate-fertility countries are expected to make, as a group, the second largest contribution to world population growth among the three groups of countries considered. Only if their fertility were to fall below 1.5 children per woman and remain at such low levels during most of the century would they contribute to a net reduction of the world population by 2100.

Although it is generally expected that the total fertility in the intermediate-fertility countries will continue to fall, it is by no means certain that it will or that all countries in that group will reach below-replacement levels soon or at all. In India, for instance, a significant proportion of the population continues to have high fertility and, if such high levels are maintained, they may slow down the national fertility decline or even reverse it. In other countries, the successful programmes that accelerated the reduction of fertility have been dismantled or are no longer getting the financial and public support that was such an important factor in ensuring their effectiveness. These developments may slow down or even prevent further fertility reductions and make it more likely that the intermediate-fertility countries would continue to add to the world population long after the end of the century.

10.5.2.3   The Low-Fertility Countries

In contrast to the other two groups, the group of low-fertility countries is projected to have an increasing total fertility in almost every scenario and projection variant, excepting only the low variant. For this group of countries the importance of achieving a sustained, even if small, increase in total fertility over the long run is underscored by the results of the no-change scenario which, by maintaining fertility and mortality constant for each country at the low levels reached in 2005–2010, produces a 2100 population of just 2.1 billion, nearly one-third lower than the population that the low-fertility countries had in 2010 (3.3 billion).

---

14 It remains about half a child above replacement level until 2030 and about a quarter of a child above replacement level after 2045, varying between 2.72 to 2.32 children per woman over the projection period.

In contrast with the sharp population reduction produced by the maintenance of recent levels of low fertility coupled with unchanging mortality, the medium variant projects a small decrease in the population of the low-fertility countries by assuming that their total fertility increases from 1.68 children per woman in 2005-2010 to 1.80 in 2045-2050 and then to 1.89 children per woman in 2095-2100. Such a modest increase in fertility, coupled with declining mortality, is sufficient to prevent a rapid reduction of the population, producing an overall population of 3.0 billion in 2100, down from 3.3 billion in 2010.

Because the low-fertility countries already have below-replacement fertility, they constitute the only group for which the instant-replacement scenario produces an increasing population. Thus, were their total fertility to increase immediately to replacement level, rising to 2.18 children per woman in 2010-2015 and then declining to 2.07 children per woman toward the end of the century as mortality decreases, their population would increase from 3.3 billion in 2010 to 4.2 billion in 2100.

An even larger population increase would occur if the fertility of the low-fertility countries were to follow the path projected by the high variant whose total fertility increases from 2.11 children per woman in 2010-2015 to 2.39 in 2095-2100. That is, under the high variant the fertility of the low-fertility countries would not only remain well above replacement level but would also increase over time. That trend produces a population that rises steadily from 3.3 billion in 2010 to 4.8 billion in 2100.

In contrast with the other projections for the low-fertility countries, the low variant projects a declining total fertility until 2050 and a very slight increase thereafter. The result is a total fertility that remains well below 1.5 children per women until 2100 and leads to a sharp reduction of the population: from 3.3 billion in 2010 to 1.8 billion in 2100.

Because the low-fertility countries, as a group, accounted for 46 per cent of the world population in 2010, their potential for future population growth or decline is of considerable significance for world population trends. Given that below-replacement fertility has been a feature of virtually all the low-fertility countries for at least two decades and the measures to increase fertility adopted by countries that have been concerned about the persistence of low fertility have generally not resulted in large increases, it is expected that the low-fertility countries as a group will maintain below-replacement fertility over the foreseeable future and, consequently, that their overall population will

decrease eventually, thus contributing to counterbalance the population growth expected to occur in other countries. Nevertheless, as the results of the different projections described above indicate, the increase in total fertility needed to prevent the population of the low-fertility countries from decreasing is small by historical standards and may not be beyond the realm of possibility especially if the total fertility of China, the world's most populous country, were to increase in future.

10.5.2.4   Implications of the Different Projections for World Population Size

According to the medium variant, which is the one most commonly used as reference, the world population would increase from 7.2 billion in 2014 to 9.5 billion in 2050 and further to 10.9 billion in 2100. By 2100, according to this variant, the population of both the low-fertility and the intermediate-fertility countries would be decreasing slowly whereas that of the high-fertility countries would still be growing. Furthermore, because of the rapid growth of their population over the rest of the century, the high-fertility countries would account in 2100 for 43 per cent of the world population, up from the 19 per cent in 2014. Partly because of this expansion, the projected population growth in the high-fertility countries at the end of the century would more than counterbalance the decline in the other two groups of countries so that the world population would still have the potential of continuing to increase after 2100. Thus, according to the medium variant, the relatively slow reduction of fertility projected for the high-fertility countries coupled with their declining mortality would lead to population increases not only sufficient to counterbalance the population reductions projected in the low-fertility countries, which account for 46 per cent of the world population today, but also large enough to dominate future population growth at the world level.

Further evidence that rapid population growth in a small proportion of the world population can not only counterbalance population decreases in a major part of the population but also surpass the growth in the intermediate-fertility countries, which constitute 35 per cent of today's world population, is provided by the no-change scenario. According to that scenario, the population of the low-fertility countries decreases by 0.2 billion between 2010 and 2050, while that of the intermediate-fertility countries rises by 1.2 billion and that of the high-fertility countries increases by 2.3 billion. Clearly, the potential decreases in the

46 per cent of the world population that lived in the low-fertility countries in 2014 is no match for the huge potential increases in the rest of the population and, especially, for those projected for the 19 per cent of the population that still has high fertility.

As already noted, in order to moderate the expected population growth in the high-fertility countries, their fertility would have to follow a path closer to that projected under the low variant, which projects their total fertility falling below replacement level by 2065–2070. If the high-fertility countries managed to follow the low variant while the other two groups followed the medium variant, the world population would increase from 7.2 billion in 2014 to 9.2 billion in 2050, reach a peak of 9.6 billion around 2075 and decline thereafter to 9.3 billion in 2100. Indeed, the world population might reach a peak over this century if it were possible to achieve an accelerated reduction of fertility in the high-fertility countries of today and if the fertility of the intermediate-fertility countries would drop below replacement level over the next few decades.

In comparison with the results of the medium variant, those produced by the high variant are sobering. If fertility were to remain just half a child above that projected by the medium variant for each country, the world population in 2100 would reach 16.6 billion, more than doubling today's population. According to the high variant, the population of the high-fertility countries alone would reach 6.8 billion by 2100, equivalent to the size of the world population in 2010 and representing more than a five-fold increase with respect to the population that the high-fertility countries had in 2010.

In sum, if the world population is to remain within manageable limits, it is imperative for the high-fertility countries to reduce their fertility at least as fast as and preferably faster than projected in the medium variant. It is also necessary for the intermediate-fertility countries to reach below-replacement fertility as soon as possible and for all countries to maintain below-replacement fertility for a long time after they reach it in order to counterbalance the high potential for continued growth that large populations with favourable age distributions have.

How likely are these changes to be realized? In the past, actions taken by governments, non-governmental organizations and the international community have facilitated and contributed to spur and accelerate the social and behavioural changes leading to lower fertility. However, over the past two decades, the attention and funding being invested in providing peo-ple with the means and the social support that they require in order to have only the children they desire and no more have not kept up with the growth of population and the increasing demand for family planning services. Therefore, many countries, especially those having low average incomes and suffering from poor governance, are unlikely to experience soon the changes necessary to reduce fertility. Even in countries with better governance where fertility levels remain high, the focus of key actors on other pressing issues has meant that the challenge of reducing population growth through the achievement of lower fertility ranks low on the governments' development agenda. Hence, the chances of achieving rapid reductions of fertility in those countries also seem low. These conclusions, however, are based on purely qualitative assessments. Can quantitative methods be brought to bear on the issue?

Recently, the United Nations Population Division has issued the second version of a set of probabilistic projections that include, for the first time, probabilistic projections for groups of countries (United Nations 2014b). The groups used by the United Nations do not include the three groups considered in this chapter. Nevertheless, the probabilistic projections for some world regions provide useful insights into the uncertainty surrounding future outcomes. The United Nations probabilistic projections have the advantage of being data-driven, that is, their future variability is modelled using estimates of variability derived from data on the trends experienced since 1950 and therefore allow a purely quantitative assessment of the probability of projected trends.

Future developments in sub-Saharan Africa are considered here as a good representation of what may happen in the high-fertility countries as a whole. In 2014, 93 per cent of the population of sub-Saharan Africa lived in high-fertility countries and they constituted 65 per cent of the population in all high-fertility countries. The probabilistic projections for sub-Saharan Africa show that the 80 per cent confidence interval for its population in 2100 ranges between 3.13 billion and 4.7 billion. Given that the population of sub-Saharan Africa was 0.83 billion in 2010, those results imply that there is a 90 per cent probability that the population of that region will nearly quadruple by the end of the century (be multiplied by a factor of 3.7, to be exact). Since the 95 per cent confidence interval for the population of sub-Saharan Africa in 2100 ranges from 2.82 billion to 5.30 billion, there is almost a certainty that its population will at least triple by 2100. These results suggest that, in view of past trends, the probability that the high-fertility countries may follow a low-fertility path such as that embodied by the low variant is virtually nil. This conclusion is consistent with the qualitative assessment made above on the basis of the low priority that reducing fertility has today on the development agenda.

There is no ideal region to represent the intermediate-fertility countries, since they are distributed among all developing regions. However, Southern Asia, which includes India, the most populous country with an intermediate fertility level, is a region where 88 per cent of the population lives in countries with intermediate fertility and where that population accounts for 59 per cent of the overall population of the intermediate-fertility countries. According to the probabilistic projections, the population of Southern Asia, which rose to 1.7 billion in 2010, has an 80 per cent probability of being between 2.10 billion and 2.55 billion in 2050, indicating that the region still has a significant potential for continued population growth over the medium term. Yet depending on the path the population takes over the next few decades, that potential may decrease during the second half of this century, as indicated by the fact that the 80 per cent confidence interval for the 2100 population of Southern Asia ranges from 1.6 billion to 2.9 billion in 2100, with a median value of 2.20 billion. That is, there is more than a 50 per cent probability that the population of Southern Asia will be higher in 2100 than it was in 2010, with a 50 per cent chance of surpassing 2.20 billion by century's end. These findings are congruent with the analysis presented earlier about the prospects for future population growth in the groups of intermediate-fertility countries. That group may also today almost be certain of seeing their population increase by 2050 and have a greater than even chance of having a larger population in 2100 than they had in 2010. Yet, depending on the path the intermediate-fertility countries follow over the next few decades, they also have a substantial chance of experiencing a population reduction before the end of the century.

Two regions will be taken as representative of the low-fertility countries: Eastern Asia and Europe, in both of which virtually all the population lived in low-fertility countries in 2014. Eastern Asia, which includes China, accounted for 48 per cent of the population of the low-fertility countries in 2014 and Europe accounted for a further 22 per cent. The probabilistic projections for both regions confirm that it is very unlikely that their populations will be larger in 2100 than they were in 2010. For Eastern Asia, whose population in 2010

was 1.6 billion, the 95 per cent confidence interval for the population projected to 2100 ranges from 0.8 billion to 1.6 billion, implying that the probability of having a population higher than 1.6 billion in 2100 is a low 2.5 per cent. For Europe, whose population in 2010 amounted to 0.74 billion, the 95 per cent confidence interval for 2100 ranges from 0.55 billion to 0.74 billion, implying again that the chances of experiencing a population increase over the long run are slim. These findings confirm the general expectations that experts have about the persistence of low fertility in the countries that have attained it and reflect the trends observed since 1950. Yet a word of caution may be in order: those trends do not reflect properly the moderate increases in fertility experienced by developed countries during the 'baby boom' because, by starting in 1950, they exclude data on the low fertility levels that developed countries had reached in the 1930s and 1940s. At the time it occurred, the 'baby boom' was unexpected and caught experts by surprise. Hence, especially in the case of the low-fertility countries, one must bear in mind that unexpected events may occur even if their probability of occurrence is low. Furthermore, it bears underscoring that the low-fertility countries would benefit from small increases in fertility that would prevent a too precipitous decrease in population size and moderate somewhat population ageing.

For the world as a whole, the probabilistic projections produce a 95 per cent confidence interval for the 2050 population ranging from 9.0 billion to 10.1 billion and for the 2100 population from 9.0 billion to 13.3 billion. These ranges of variation are significantly narrower than those suggested by the low and high variants, which are: 8.3 billion to 10.9 billion in 2050 and 6.7 billion to 16.6 billion in 2100. The narrow intervals produced by the probabilistic projections result from the nature of the random processes that mix countries following different projection paths, whereas the low and high variants impose similar paths, if not to all countries, at least to those in the three groups of countries distinguished in this chapter. The probabilistic projections indicate that it is very unlikely that the world population will fall below 9 billion over the course of this century but it is also unlikely that it will rise above 13.3 billion. However, for the world population not to rise above this level, it is necessary for all regions to follow fertility paths that are well below that of the high variant.

Interpreting the results of the probabilistic projections is difficult for at least two reasons. First, any single future path has theoretically a zero probability of

Figure 10.4: Components of future population change in high-fertility countries, medium variant, 2010–2100. **Source:** Derived from United Nations (2013b).

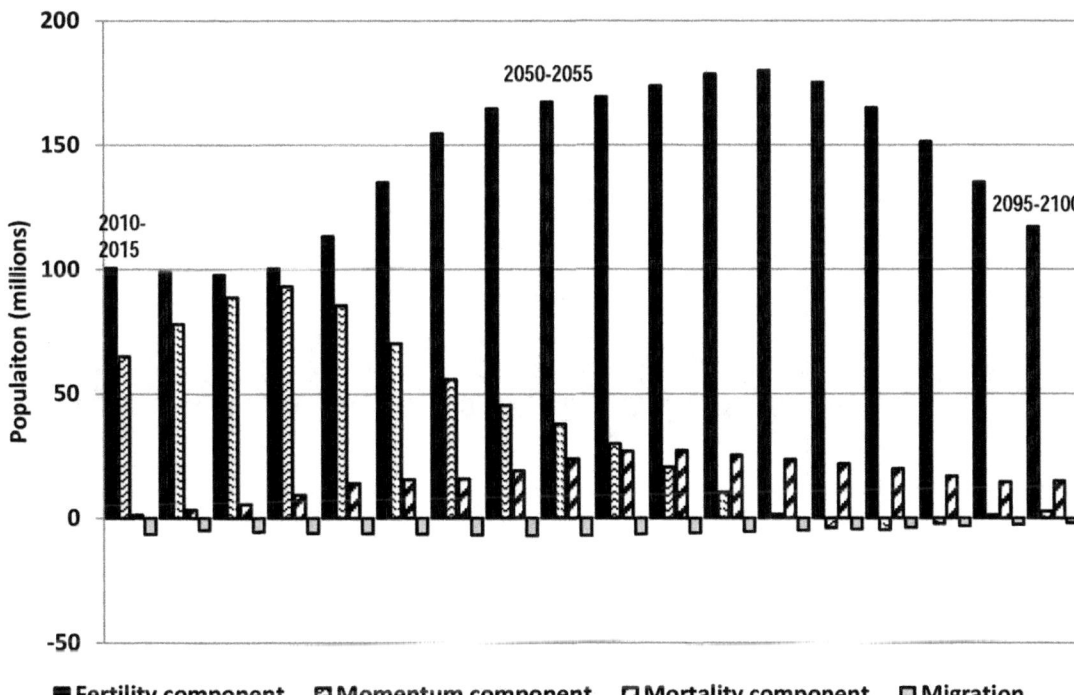

■ Fertility component   ▨ Momentum component   ▫ Mortality component   ▫ Migration

being realized. Second, except for the median path, which corresponds to the medium variant, there is no way of relating the results of a probabilistic path to the fertility and mortality levels that generated it. For instance, the combination of fertility and mortality trends that would produce the population path generated by the upper limit of the 95 per cent confidence intervals for future population size are not known. Therefore, it is useful to refer to the results of the deterministic projections discussed earlier to assess the types of changes in fertility that would produce outcomes within the confidence limits yielded by the probabilistic approach. In the case of the probabilistic projections for the world as a whole, in order for the population to remain below 13.3 billion as projected, the future fertility of most countries and especially of the high-fertility countries would have to remain well below that projected in the high variant. That is, the results of the probabilistic projections are not 'predictions': they will not be realized automatically. They indicate that, if past experience could be replicated, outcomes within the confidence interval range could be achieved. It is important to stress, therefore, that past experience includes political and financial commitments that were successful, in many instances, in generating the voluntary behavioural changes necessary to reduce fertility.

10.6   The Components of Future Population Change in the Medium Variant

As fertility declines, there is at first a greater concentration of the population in adult ages and, among women in particular, in the reproductive ages. Such a distribution is favourable to the maintenance of population growth because, even if women reduce on average the number the children they bear, the increasing numbers of women bearing children ensure that the number of children born remains high and, consequently, the population keeps on growing. The effect of the age distribution in maintaining or accelerating population growth is known as 'population momentum'. The United Nations Population Division has recently produced a report providing estimates of the future contribution of population momentum to the projected population trends under the medium variant (United Nations 2013b). The report distinguishes the contribution of four distinct components of population change: fertility, mortality, net migration and population momentum. Calculation of these components for the three groups of countries considered shows remarkable differences between them (figures 10.4, 10.5 and 10.6).

Figure 10.5: Components of future population change in intermediate-fertility countries, medium variant, 2010–2100. **Source**: Derived from United Nations (2013b).

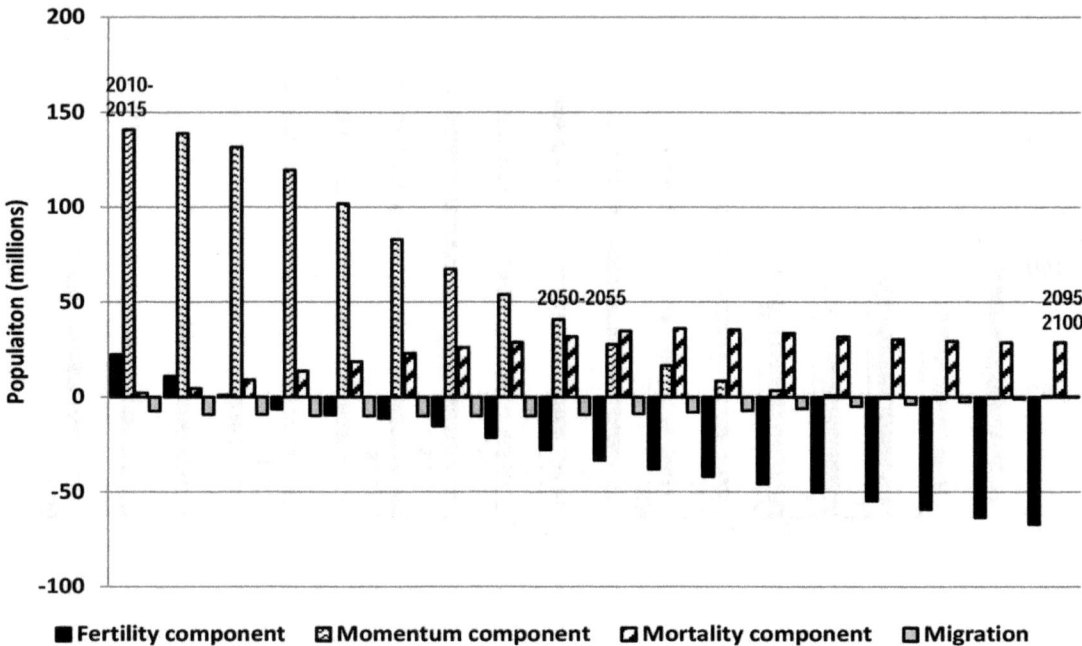

Figure 10.6: Components of future population change in low-fertility countries, medium variant, 2010–2100. **Source**: Derived from United Nations (2013b).

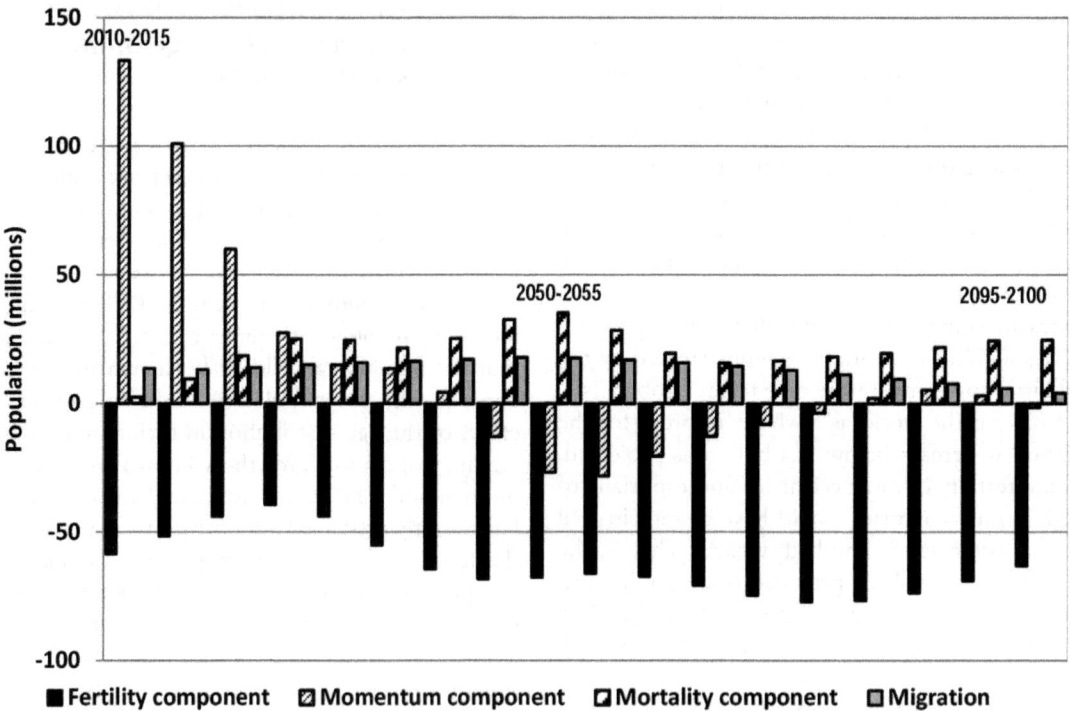

In the high-fertility countries, fertility is by far the major component of population change but population momentum makes an important contribution to population growth, especially before 2050. After 2050, the fertility contribution is dominant.

In the intermediate-fertility countries, population momentum is the major component of population change until about 2050 and fertility is the major component thereafter, but fertility contributes then to population reductions rather than to population growth because, after 2050, the total fertility of the intermediate-fertility countries remains below replacement level in the medium variant.

In the low-fertility countries, population momentum is still making a major contribution to population growth but will not do so for long. After 2025, the contribution of population momentum is small and, even as soon as 2020, population momentum is not sufficient to counteract completely the negative contribution of below-replacement fertility. From 2030 onward, that negative contribution dominates the components of population change and is reinforced between 2045 and 2080 by a smaller but significant negative contribution of population momentum. It is noteworthy that the group of low-fertility countries is the only one for which the contribution of net migration remains positive during the whole projection period. However, it is small and insufficient to counteract the negative contribution of low fertility.

These results reinforce the conclusions derived from considering the different projection variants and scenarios: the high-fertility countries have the highest potential for continued growth driven mainly by the continuation of relatively high, even if declining, fertility. The population of the intermediate-fertility countries also has considerable potential for future growth but, because that growth comes mostly from population momentum, that potential drops markedly by mid-century. Hence, provided their fertility falls below replacement level by 2035–2040, as projected in the medium variant, the contribution of fertility to population change in the intermediate-fertility countries will be largely in the negative direction after 2050, thus counteracting any tendency to population growth.

Lastly, the low-fertility countries can still expect to experience population growth driven by population momentum over a decade or so. Beyond 2025, however, population momentum will not be sufficient to counteract the negative contribution of below-replacement fertility to population change in the low-fertility countries. Therefore, unless their total fertility increases to replacement level or above, their overall population will decrease.

10.7   Future Population Ageing

Population ageing results primarily from declining fertility, although decreasing mortality at adult and older ages also contributes to it. Because every country in the world has already experienced some reduction of fertility, all populations today are ageing and, as discussed in section 10.4.3 above, population ageing is most advanced in the low-fertility countries, less so in the intermediate-fertility countries and has barely started in the high-fertility countries. Since all the projection variants and scenarios considered in this chapter, with the sole exception of the no-change scenario, project further fertility reductions for the high-fertility and the intermediate-fertility countries, their projected populations under those variants are sure to continue ageing. For the low-fertility countries, several variants project fertility increases which will slow down population ageing.

Table 10.5 shows the values of the median age estimated for 2010 and projected according to different projection variants and scenarios to 2050 and 2100. In all projection variants and scenarios the future population of the high-fertility countries remains younger than that of the intermediate-fertility countries and this one, in turn, remains younger than the population of the low-fertility countries. The no-change scenario produces the lowest median age for the high-fertility and intermediate-fertility countries in both 2050 and 2100, but at the cost of major increases in population size, especially in the case of the high-fertility countries whose population by 2100 would increase more than tenfold with respect to 2010 (table 10.3). The high variant produces the next lowest set of median ages, which do not surpass age 40 even for the low-fertility countries but at the cost of marked increases in population size. The low-fertility countries, for instance, would have to increase their population by about 50 per cent by 2100 to prevent their median age from rising above 40 years, and for the high-fertility countries, the population would have to increase more than fivefold to prevent the median age from rising above 31 years.

The instant-replacement scenario provides a useful standard against which to judge the performance of other variants because, by maintaining fertility at replacement level for nearly a century, by 2100 it approaches the age distribution of a population that is close to experiencing zero population growth. That is, the results of the instant-replacement scenario may be thought as indicative of the minimum degree of ageing that the population would experience in order

Table 10.5: Median age in 2010 and according to different projection variants in 2050 and 2100 for the three groups of countries considered. Source: Derived from United Nations (2013a).

Group	2010 Median age (years)	No-change scenario	High variant	Medium variant	Instant-replacement scenario	Low variant
				2050		
High-fertility countries	18,8	19,6	23,4	25,5	33,8	28,2
Intermediate-fertility countries	25,7	31,8	33,1	37,6	36,9	42,5
Low-fertility countries	35,5	44,8	39,7	45,4	40,2	50,8
				2100		
High-fertility countries	18,8	18,8	30,9	35,6	38,9	41,4
Intermediate-fertility countries	25,7	31,8	38,0	45,4	41,3	54,7
Low-fertility countries	35,5	44,5	39,5	47,1	43,6	56,6

Table 10.6: Support ratio (population aged 15–64 over population under age 15 plus the population aged 65 or over) in 2010 and according to different projection variants in 2050 and 2100. **Source**: Derived from United Nations (2013a).

Group	Support ratio in 2010	No-change scenario	High variant	Medium variant	Instant-replacement scenario	Low variant
				2050		
High-fertility countries	1,21	1,23	1,52	1,70	2,22	1,92
Intermediate-fertility countries	1,83	1,79	1,77	1,98	1,88	2,21
Low-fertility countries	2,39	1,64	1,44	1,51	1,48	1,56
				2100		
High-fertility countries	1,21	1,17	1,69	1,80	1,60	1,85
Intermediate-fertility countries	1,83	1,72	1,49	1,41	1,46	1,19
Low-fertility countries	2,39	1,58	1,36	1,26	1,35	1,05

to get to a stationary state. For the intermediate-fertility countries, the instant-replacement scenario produces a median age that is about 4 years lower than that produced by the medium variant (41.3 years vs 45.4 years) and that difference is similar for the low-fertility countries (43.6 years vs 47.1 years). Yet, to attain those lower median ages, the joint populations of the intermediate-fertility and the low-fertility countries would have to increase by an additional 1.7 billion by 2100.

For the high-fertility countries, the medium variant produces a lower median age than the instant-replacement scenario because the former maintains a higher fertility for the high-fertility countries during most of the projection period. Consequently, reducing the median age of the high-fertility countries by 3.4 years with respect to that projected in the instant-replacement scenario requires an increase of their 2100 pop-

ulation by 2.6 billion (the difference between the population in 2100 projected by the medium variant and that projected by the instant-replacement scenario).

The many examples cited above show that slowing population ageing can be achieved only by accommodating the higher population growth resulting from higher fertility. Such an unavoidable trade-off makes it imperative that societies adapt to population ageing since it is unlikely that ever-growing populations will prove to be sustainable. Furthermore, the initial stages of population ageing have beneficial aspects. The support ratio, for instance, increases when the proportion of the population in the working ages rises. As table 10.6 shows, the support ratio for the high-fertility countries is projected to increase according to every variant that projects declining fertility and it increases more rapidly the faster and sharper the decline in fertility. From that perspective, following the fertility path associated with the low variant would bring about the most beneficial changes to the high-fertility countries and would have the advantage of limiting their population growth more than in the medium variant.

The support ratio for the intermediate-fertility countries is already high and is projected to increase at least until 2050 in the medium variant, the low variant and in the instant-replacement scenario, but will then decline considerably. That is, after 2050, the beneficial changes in the age distribution associated with declining fertility would have run their course in the intermediate-fertility countries and their population would enter the final stage of population ageing. Yet, if the intermediate-fertility countries follow the medium variant, their support ratio by 2100 will not be less than it had been before their fertility started declining and it will be qualitatively different from what it was in the 1960s because a higher proportion of the 'dependent' population will consist of older people than it did in the 1960s. In comparison with children, who are real 'dependants' because they require care and services and are generally not allowed to engage in productive activities in modern societies, people aged 65 or over may remain productive, continue to engage in the labour force and be likely to have accumulated savings that will prevent them from being fully dependent on transfers. Appropriate policies and institutions will be necessary to promote the accumulation of savings for use after retirement but there is increasing evidence that older people do not fit the traditional 'dependent' mould.

As for the low-fertility countries, because they are already in the last stage of population ageing and the process is accelerating, they are projected to see a sharp decline in their support ratio in all projection variants and scenarios. They all indicate that the most rapid decline of the support ratio in low-fertility countries will likely occur over the next four decades. Beyond 2050, their support ratio would continue to decline but more slowly. By 2100, the low-fertility countries would have the lowest support ratio among the three groups considered. In the medium variant, the support ratio for low-fertility countries is projected to be 1.26 potential workers per dependant in 2100, low by historical standards but still higher than 1.0. The no-change scenario, which projects lower fertility for the low-fertility countries than the medium variant, produces a higher support ratio (1.58 potential workers per dependant) because it maintains mortality constant at the level it had in 2005–2010 and therefore fewer people survive to advanced ages by 2100 than according to the medium variant. If the mortality reductions projected by the medium variant are real-

ized, it is to be hoped that such rising longevity may be accompanied by healthier and more productive lives beyond age 65.

10.8   The Persistence of Population Imbalances

The projection results presented thus far provide ample reasons to conclude that current population imbalances will continue to shape the population dynamics of the world over the next four or five decades. The population of the high-fertility countries will continue to grow and is likely to double by mid-century. Whether it keeps on increasing rapidly thereafter depends in great measure on how fast their fertility levels decline from now to 2050. The persistence of high fertility in the past has meant that the population of the high-fertility countries still has a young age distribution with high numbers of children under age 15. Therefore, their support ratio is low but is starting to increase. The more rapid the reduction of fertility that the high-fertility countries achieve in future, the faster their support ratio will grow, increasing its potential to boost their economic growth.

At the other extreme, the low-fertility countries are already advanced in the process of population ageing and their low fertility, which remains below replacement level, will likely produce a decreasing population at some point during the coming two or three decades. The period of most rapid population ageing will also occur during the coming four decades. By 2050, the support ratio of the low-fertility countries as a group is likely to have fallen by about a third from its 2010 level but changes thereafter will depend on how much mortality declines at older ages and whether fertility continues to rise while remaining below replacement level. The median age of the low-fertility countries, which was 35.5 years in 2010, will probably rise to 45 years or higher by mid-century. By adapting to a decreasing and ageing population, the countries in this group will make a major contribution to the eventual stabilization of world population and set the example for countries that are still far from reaching below-replacement fertility.

As for the intermediate-fertility countries, they have, as a group, an average total fertility sufficiently close to replacement level that if, as projected in the medium variant, it attains replacement level and goes below it over the next three or four decades, their population will grow mainly because of population momentum until 2050. Consequently, the intermediate-fertility countries appear likely, as a group, to add about a billion to the world population by 2050 unless their fertility drops below replacement level much earlier. Although the intermediate-fertility countries still have a relatively young population, with a median age of 25.7 years, it is ageing more rapidly than that of low-fertility countries did and will likely have a median age well above 40 years by 2050. Their support ratio, which stood at 1.83 potential workers per dependant in 2010, seems likely to change little by 2050, perhaps increasing slightly if their fertility continues to fall, but it is projected to decline significantly thereafter as population ageing continues and the period with below-replacement fertility lengthens. Provided the total fertility of intermediate-fertility countries does not stall at levels above replacement, by the end of the century their population would share similar traits with that of the low-fertility countries and the major imbalances remaining then would be those between the high-fertility countries of today and the rest.

One scenario that would reduce markedly the population imbalances prevalent today among the three groups of countries considered would be that combining the medium variant for the low-fertility and the intermediate-fertility countries with the low variant for the high-fertility countries. Under that combination, the three groups of countries would have below-replacement fertility by the end of the century and their populations would be declining. Each group of countries would account for about a third of the world population, which would total 9.3 billion. All three groups of countries would have median ages above 41 years, though the high-fertility countries would still have a younger population than the intermediate-fertility countries and these, in turn, would have a lower median age than the low-fertility countries. The high-fertility countries would have a more advantageous support ratio than the rest, with 1.85 potential workers per dependant and it would be decreasing slowly. Realizing this type of convergence over the course of this century is feasible even if it may seem unlikely given the trends followed by the high-fertility countries in the recent past. The more reason, therefore, to concentrate effort and support in promoting the reduction of fertility in the high-fertility countries of today, especially in the ways discussed below.

10.9   Concluding Thoughts: What Will it Take for the Changes Projected to Happen?

Population projections are not predictions and though they provide elements to judge how much current population dynamics need to change in order to realize particular outcomes, their realization is not automatic. This chapter has stressed that future population growth and prospects depend crucially on the path that future total fertility takes, particularly in today's high-fertility countries. Among the proximate determinants of fertility change, the level of contraceptive use among women subject to the risk of childbearing is a key one. Demographers consider that women aged 15 to 49 who are married or in a stable union represent the main group of women at risk of childbearing. The percentage of those women using contraception at a particular point in time is a measure of contraceptive prevalence. It is possible to distinguish between women who use the most effective contraceptive methods (usually denominated 'modern methods') and those who use traditional methods whose effectiveness in preventing pregnancy is lower. Therefore, contraceptive prevalence can be measured with respect to all methods or just taking into account the women using modern methods.

A measure indicative of how well the need for contraception is being satisfied is the 'unmet need for contraception'. This measure is calculated as the percentage of women aged 15 to 49, who are married or in union, who are not pregnant or sterile and who do not wish to become pregnant soon or do not wish to have another child ever but who are not using any method of contraception. The unmet need for modern contraception includes women who are using a traditional method of contraception among those having an unmet need for contraception.

Table 10.7 shows the estimated and projected levels of contraceptive prevalence and unmet need among women in the three groups of countries considered. These estimates were derived from the set of country estimates produced by the United Nations Population Division (United Nations 2013c, 2013d; Alkema/Kantorova/Menozzi/Biddlecom 2013). The latter are based on a set of observations that are incomplete (not all countries have the required data at every point in time) and subject to a variety of errors. Therefore, the country estimates and those for the groups considered here should be taken as indicative of trends and differences but should not be interpreted too strictly. Nevertheless, the estimates pre-

Table 10.7: Estimates and projections of contraceptive prevalence and unmet need for contraception for the three groups of countries considered, 1970–2030. **Source:** Derived from United Nations (2013c, 2013d).

Group	Percentage of women aged 15-49 who are married or in union					
	1970	**1980**	**1990**	**2000**	**2010**	**2030**
	Using a method of contraception (any method)					
High-fertility countries	6	9	14	22	29	47
Intermediate-fertility countries	15	32	45	54	60	66
Low-fertility countries	53	64	72	78	78	76
	Using a modern method of contraception					
High-fertility countries	2	4	8	15	21	38
Intermediate-fertility countries	10	25	38	47	53	59
Low-fertility countries	36	52	64	72	73	72
	With an unmet need for contraception					
High-fertility countries	28	28	28	27	25	20
Intermediate-fertility countries	27	25	20	16	14	11
Low-fertility countries	17	12	9	7	6	7
	With an unmet need for modern contraception					
High-fertility countries	31	33	34	35	33	28
Intermediate-fertility countries	32	32	27	23	21	18
Low-fertility countries	34	24	17	13	11	11

sented in table 10.7 show that clear differences exist between the high-fertility, the intermediate-fertility and the low-fertility countries in terms of contraceptive prevalence and unmet need for contraception. For all time points, contraceptive prevalence is lowest in the high-fertility countries and highest in the low-fertility countries. Conversely, unmet need for contraception is highest in the high-fertility countries and lowest in the low-fertility countries. Although such differences were to be expected, it is important to stress that the estimates of contraceptive prevalence and unmet need for contraception presented here were derived independently from the estimates of total fertility discussed earlier in this chapter.

According to table 10.7, just 22 per cent of women aged 15-49 who were married or in union were using some method of contraception in 2010 in high-fertility countries and the percentage using a modern method was a low 15 per cent. Although both these levels were higher than they had been in 1970, contraceptive prevalence in high-fertility countries had increased very slowly between 1970 and 1990 and, although it increased more rapidly after 1990, its speed of change between 1990 and 2010 was moderate at best. This experience contrasts with that of the intermediate-fertility countries whose level of contraceptive prevalence in 2010 had reached 60 per cent, having

increased very fast between 1970 and 1990, when it attained 45 per cent, and increasing at a more moderate pace between 1990 and 2010, although still faster than in the high-fertility countries after 1990.

In low-fertility countries, contraceptive prevalence was already high in 1970 (at 53 per cent) although there was room to increase the use of modern contraceptive methods since only 36 per cent of women of reproductive age who were married or in union were using them. Over the next 20 years, the use of modern methods in low-fertility countries nearly doubled and overall contraceptive prevalence increased to 72 per cent. That is, by 1990 nearly 3 out of every 4 women of reproductive age who were married or in union in low-fertility countries were using some method of contraception and by 2010 that same proportion was using a modern method of contraception.

These past trends in contraceptive use are congruent with the changes in total fertility experienced by each group of countries. Because projected levels of future contraceptive prevalence are available, one may consider whether they can guide action to achieve the fertility reductions embodied by the population projections considered earlier in this chapter. The projected levels of contraceptive prevalence to 2030 are derived by taking account exclusively of past trends in contraceptive prevalence in each country, trends in countries in the same region and those in all countries. They are obtained, therefore, independently of projections of future fertility. Consequently, it is not certain that the projected levels of contraceptive prevalence will necessarily lead to the total fertility levels projected under, for instance, the medium variant. However, a comparison of the level of fertility experienced by the intermediate-fertility countries in 1990 at the time when their contraceptive prevalence was 45 per cent suggests that the projected level of contraceptive prevalence of 47 per cent in 2030 for high-fertility countries may be sufficient to reduce fertility as projected by the medium variant. Assuming that to be the case, it will be necessary for high-fertility countries to accelerate the uptake of contraceptive use and, in particular, to expand the use of modern contraceptives during 2010–2030 if the projected contraceptive prevalence is to be attained by 2030. Moreover, the projected increase in contraceptive prevalence needs to occur even as the number of women of reproductive age in high-fertility countries increases rapidly as a result of the high fertility levels of the past.

Although achieving a contraceptive prevalence of 47 per cent and a prevalence of modern contraceptive use of 38 per cent over the next 15 years in high-fertility countries should be feasible based on the experience of intermediate-fertility countries, those levels are unlikely to be achieved without serious commitment from governments, non-governmental organizations and the international community to support family planning in the high-fertility countries. Experts have noted that since the mid-1990s, that support has waned (Cleland/Bernstein/Ezeh et al. 2006), but recent efforts by private donors to galvanize support for family planning are expected to make a difference. Bongaarts, Cleland, Townsend et al. (2012) provide an overview of changing perspectives about the effectiveness of family planning programmes and describe strategies to reduce barriers to the use of contraception in low-income countries, such as those included in the group of high-fertility countries considered here.

In the case of the intermediate-fertility countries, increasing contraceptive prevalence from 60 per cent in 2010 to 66 per cent in 2030 and the use of modern contraception from 53 per cent in 2010 to 59 per cent in 2030 seems feasible in view of past experience. Nevertheless, achieving those goals requires sustained support for family planning, especially considering that the number of women of reproductive age in intermediate-fertility countries will be increasing over the next two decades. It is of concern that several countries in this group, after having successfully

expanded contraceptive use and reduced fertility, have been reducing their support for family planning both in financial terms and in terms of commitment to the provision of services. If those changes make fertility stall at current levels, the potential for sustained population increases in intermediate-fertility countries may become a reality.

In both the high-fertility and the intermediate-fertility countries, unmet need for contraception is projected to decline from 2010 to 2030. In the intermediate-fertility countries, the decline in unmet need is slower than it has been in the past, but for the high-fertility countries it is faster than in the past. Furthermore, the projected reduction of unmet need in both groups of countries still leaves a substantial proportion of women of reproductive age with an unmet need for contraception in 2030. That is to be expected, especially in the high-fertility countries where, as fertility declines and increasing numbers of women wish to reduce their fertility, the demand for contraception will likely grow faster than the capacity to meet it. Furthermore, as the existence of a significant level of unmet need for contraception in the low-fertility countries suggests, even when contraceptive use is high, some unmet need will remain unsatisfied.

To sum up, current population imbalances are to a large extent the result of different trends in the reduction of fertility among countries. The use of contraception, particularly modern contraception, is the major immediate means of reducing fertility. Given the stark differences in levels and trends of fertility that already exist between the high-fertility, the intermediate-fertility and the low-fertility countries, their influence cannot be erased over the medium-term future. However, if those imbalances are to diminish over the long run, it is necessary to accelerate the transition to low fertility in the high-fertility countries and ensure that the intermediate-fertility countries reach below-replacement levels of fertility over the coming two or three decades. To do so, a key strategy is to expand contraceptive use, especially in the high-fertility countries where the unmet need for contraception remains especially high. Continuing to support and facilitate access to modern contraception, especially among low-income groups, is also necessary in the intermediate-fertility countries. If the high-fertility countries could achieve the reductions of fertility projected by the low variant while all other countries followed fertility paths close to those projected by the medium variant, the world might witness the beginning of population stabilization by the end of this century.

References

Alkema, Leontine; Kantorova, Vladimira; Menozzi, Clare; Biddlecom, Ann, 2013: "National, Regional, and Global Rates and Trends in Contraceptive Prevalence and Unmet Need for Family Planning between 1990 and 2015: A Systematic and Comprehensive Analysis", in: *The Lancet* (online: 13 March; at: <http://dx.doi.org/10.1016/S0140-6736(12)62204-1>).

Bloom, David E.; Canning, David; Sevilla, Jaypee, 2003: *The Demographic Dividend: A New Perspective on the Economic Consequences of Population Change* (Santa Monica, CA: Rand Corporation).

Bongaarts, John; Cleland, John; Townsend, John W.; Bertrand, Jane T.; Das Gupta, Monica, 2012: *Family Planning Programs for the 21st Century: Rationale and Design* (New York: The Population Council).

Butler, Colin D.; Dixon, Jane, 2012: "Plentiful Food? Nutritious Food?", in: Rosin, Christopher; Stock, Paul; Campbell, Hugh (Eds.): *Food Systems Failure: The Global Food Crisis and the Future of Agriculture* (New York–London: Routledge–Earthscan): 98–113.

Cleland, John; Bernstein, Stan; Ezeh, Alex; Faundes, Anibal; Glasier, Anna; Innis, Jolene, 2006: "Family Planning: The Unfinished Agenda", *The Lancet*, 368, 9549 (18 November): 1810–27.

West, Paul C.; Gerber, James S.; Engstrom, Peder M.; Mueller, Nathaniel D.; Brauman, Kate A.; Carlson, Kimberley M.; Cassidy, Emily S.; Johnston, Matt; MacDonald, Graham K.; Ray, Deepak K.; Siebert, Stefan, 2014: "Leverage points for improving food security and the environment", in: *Science*, 345, 6194 (18 July): 325–328.

United Nations, 2000: *World Population Prospects: The 1998 Revision*. Volume 3: Analytical Report (New York: United Nations).

United Nations, 2013a: *World Population Prospects: The 2012 Revision*. Comprehensive Data Set (Excel Format) (New York: United Nations); at: <http://esa.un.org/unpd/wpp/index.htm>.

United Nations, 2013b: *Demographic Components of Population Growth*, Technical Paper No. 2013/3 (New York: United Nations); at: <http://www.un.org/en/development/desa/population/publications/technical/2013-3.shtml>.

United Nations, 2013c: *Model-based Estimates and Projections of Family Planning Indicators: 2013 Revision* (New York: United Nations), at: <http://www.un.org/en/development/desa/population/theme/family-planning/cp_model.shtml>.

United Nations, 2013d: *Estimates and Projections of the Number of Women Aged 15–49 Who Are Married or in a Union: 2013 Revision* (New York: United Nations); at: <http://www.un.org/en/development/desa/population/theme/marriage-unions/marriage_estimates.shtml>.

United Nations, 2014a: *Outcome Document. Open Working Group on Sustainable Development Goals*; at: <http://sustainabledevelopment.un.org/focussdgs.html>.

United Nations, 2014b: *Probabilistic Population Projections based on World Population Prospects: The 2012 Revision*; at: <http://esa.un.org/unpd/ppp/>.

11   The Challenge of a 4°C World by 2100

Hans Joachim Schellnhuber, Olivia Maria Serdeczny, Sophie Adams,
Claudia Köhler, Ilona Magdalena Otto and Carl-Friedrich Schleussner[1]

Abstract

As evidenced by the Fifth Assessment Report of the IPCC, our understanding of the Earth System and the climate change impacts expected in the coming decades is developing at a rapid pace. Contributing to this progress, the first ever Inter-Sectoral Impact Model Intercomparison *(ISI-MIP)* has helped to paint a clearer picture of potential impacts at different levels of global mean warming. However, along with such advances the limitations of our understanding become more apparent. A number of processes are scarcely or not at all reflected in current assessments of the risks associated with significant levels of warming. These include critical thresholds in the Earth system which, once breached, can give rise to non-linear impacts. Recent insights from West Antarctica indicate that we have already 'tipped' several large glacier systems there, suggesting that the risk of crossing such thresholds might be much greater than previously thought. Also excluded from a sectoral perspective are the intricate interdependencies between systems and the potential for an initial impact to cascade into a chain of impacts, or for impacts to occur simultaneously and interact in complex ways. Finally, we need to take into account the different degrees of vulnerability not only across but also within nation states. The ramifications of non-linear impacts and their uneven distribution are likely to be deleterious to the stability and well-being of our societies and will, we hope, never be realized. However, if we wish to understand the challenges associated with a 4°C world[2], such a world needs to be imagined.

Keywords: 4°C World, non-linearity, social vulnerability, cascading impacts, intersectoral hotspots, thresholds, tipping elements, uncertainty, complex systems.

11.1   Introduction

A scenario in which global mean temperature reaches 4°C above pre-industrial levels by the end of the century constitutes a real risk. Current policies would lead us to a warming of 3.7°C above pre-industrial levels by 2100, with about a 40% chance of exceeding 4°C by that time (climateactiontracker.org). What this means for the livelihoods, welfare and security of future generations remains highly uncertain. However, research on climate change impacts is progressing rapidly and our understanding of the implications of high emission scenarios continues to deepen.

Although we are not able to project future impacts with the precision that policymakers might hope for, we now know enough to build a picture of a 4°C world and some of the risks that it will entail. For example, as analysed in Schellnhuber/Hare/Serdeczny et al. (2012, 2013), projections from five bias-corrected general circulation models (Hempel/Frieler/Warszawski et al. 2013) indicate a severe increase in the frequency and intensity of extreme heat in a 4°C world. Heatwaves of unprecedented magnitude—five standard deviations outside the historical norm—are

---

1   Prof. Dr. Hans Joachim Schellnhuber is director of the Potsdam Institute for Climate Impact Research (PIK); Olivia Maria Serdeczny is research analyst at Climate Analytics; Sophie Adams is a doctoral candidate at the University of New South Wales, Australia; Claudia Köhler is executive advisor to the director of the Potsdam Institute for Climate Impact Research; Dr. Ilona Magdalena Otto is a postdoctoral researcher at the Potsdam Institute for Climate Impact Research and a postdoctoral researcher at Zhejiang University, Department of Land Management, Hangzhou, China; Dr. Carl-Friedrich Schleussner is a research analyst at Climate Analytics.
2   The term '4°C world' is used as shorthand for an increase of 4°C in global mean temperature above pre-industrial levels by the end of the century.

Figure 11.1: The IPCC's Working Group II Assessment: *A global perspective on climate-related risks.* The colour shading indicates the additional risk due to climate change when a temperature level is reached and then sustained or exceeded. Undetectable risk (white) indicates no associated impacts are detectable and attributable to climate change. Moderate risk (yellow) indicates that associated impacts are both detectable and attributable to climate change with at least *medium confidence*, also accounting for the other specific criteria for key risks. High risk (red) indicates severe and widespread impacts, also accounting for the other specific criteria for key risks. Purple shows that very high risk is indicated by all specific criteria for key risks. Temperatures on the left are given for the baseline period 1986–2005 and above pre-industrial levels on the right. **Source:** IPCC Summary for Policy Makers, Working Group II (2014: 13).

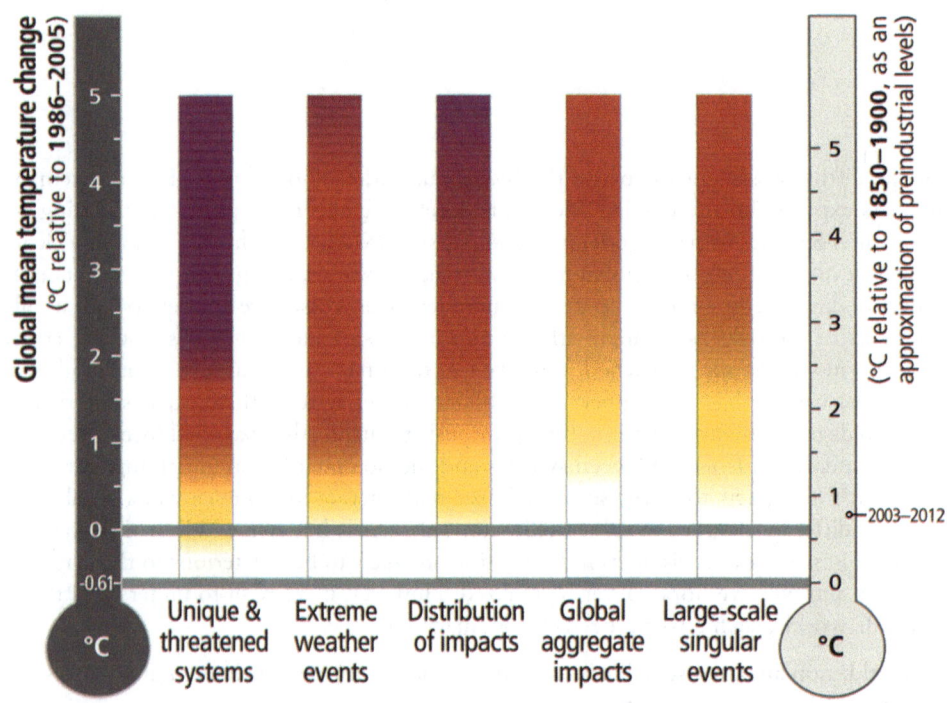

expected to affect 60 per cent of the global land area by the end of the century. Today's extremes, like the Central European heatwave of 2010, would be the new normal, affecting 80 per cent of the surface. Rainfall patterns would be amplified with decreases of up to 30 per cent during the dry season and increases of up to 30 per cent in the wet season projected for South Asia (Schellnhuber/Hare/Serdeczny et al. 2012, 2013). Sea-level rise poses another severe risk, particularly to coastal populations at low latitudes. Semi-empirical projections of regional sea-level rise indicate that sea-level rise at tropical and subtropical locations will be up to 20 per cent higher than the global mean, reaching 1.2 m in the Western Pacific

(Schellnhuber/Hare/Serdeczny et al. 2012, 2013). It is clear that these risks pose enormous obstacles to human development and security (German Advisory Council on Climate Change 2007). That climate change will hit the poorest hardest is now widely accepted. But climate change impacts are not confined to the developing world, and indeed climate change is expected to also threaten the more comfortable ways of life of the industrialized regions of the world. Based on a common understanding of shared responsibility for present and future generations, the challenge of climate change requires no less than a global transformation towards sustainability (German Advisory Council on Climate Change 2011).

Box 11.1: The Inter-Sectoral Impact Model Intercomparison Project (ISI-MIP)

The *ISI-MIP*-Project is a community-driven modelling effort and the first intercomparison of climate impacts to bring together several sectors expected to be affected by climate change: agriculture, water, ecosystems, human health (malaria) and coastal infrastructure. The project aims to quantitatively assess global climate impacts at different stages of global warming using a consistent setting. It is based on earlier impact model comparisons and helps to identify research gaps as well as to provide robust results on societally relevant climate impacts. ISI-MIP's core element is a public data archive, which is driven by the motivation to create transparency and to lead to model improvement. Based on the quantitative analysis of intermodel variations, the project also aims to estimate uncertainties for *General Circulation Models* (GCMs) as well as for *Global Impact Models* (GIMs). ISI-MIP aims to initiate a continuous process of coordinated improvement of impact modelling and assessment that involves the entire scientific community. In this sense, ISI-MIP pursues a dual purpose: that of facilitating the process of understanding and of model development within the research community.

In the following we provide an outline of the challenges humanity could be faced with in a 4°C world. We begin with a brief overview of projections of climate change impacts in a number of sectors of particular importance to society. The results are derived from scientific publications based on the first Inter-Sectoral Impact Model Intercomparison Project (ISI-MIP; see box 11.1; Warszawski/Frieler/Huber et al. 2013). In light of the uncertainties associated with climate change projections, we then review the literature to discuss some of the 'known unknowns' that have emerged in recent research—those risks that at present remain inadequately captured in modelling exercises and that require further investigation. These include the potential for 'tipping points' in the Earth system and human sectors, as well as for complex interactions between impacts, all of which could have substantial adverse consequences for the human populations inhabiting a 4°C world.

In this discussion we investigate some of the key risks as identified by the Second Working Group of the Intergovernmental Panel on Climate Change (see figure 11.1). Based on expert judgment, five areas of emergent risks of climate change and its societal impacts indicate how risks mount with increasing global mean temperature. These areas include the distributional aspect of impacts as well as risks related to large-scale singular events. While our chapter cannot provide a detailed understanding of those key risks that societies will be faced with in a 4°C world, it does illustrate the scope of the challenge both for scientists to advance our understanding and for policymakers to act even in conditions of uncertainty.

11.2 Sectoral Impact Estimates Derived from ISI-MIP

In order to study the differences between warming scenarios of a 1, 2, 3 and 4 °C temperature rise above pre-industrial levels, the *Inter-Sectoral Impact Model Intercomparison Project* (ISI-MIP) brought together thirty research teams from twelve countries involved in the assessment of climate impacts in multiple sectors (Warszawski/Friend/Ostberg et al. 2013). In the first project phase, the researchers compared 28 impact models across five different sectors within the consistent setting provided by the ISI-MIP modelling protocol (see box 11.1). The sets of climate data used were derived from five climate simulations, each simulating four *Representative Concentration Pathways* (RCPs) (Hempel/Frieler/Warszawski et al. 2013). The results of this interdisciplinary exercise allow for a better understanding both of the differences between low and high warming scenarios and of the remaining research gaps and the main sources of the uncertainty that hamper climate impact projections. Seemingly paradoxically, a model comparison such as this increases the range of results, and the results therefore appear less sharp than those of a single simulation. At the same time, however, the results become more reliable compared to the results of just one simulation. Some of the key insights of the first fast-tracked phase, including for the water and agricultural sectors, are presented below.

11.2.1 Sectoral Results

Changes in precipitation patterns due to climate change pose threats to human populations through both increased risk of direct water scarcity and associated agricultural impacts. Drought conditions and extreme precipitation and flood events are also hazardous to human lives and livelihoods. The ISI-MIP results for the water sector published in Schewe/ Heinke/Gerten et al. (2013) focus on the impacts of climate change on water resources and availability. Although population change accounts for the larger part of the overall change in water scarcity, climate change significantly increases the number of people at risk: a warming of approximately 2.7°C above pre-industrial levels is projected to expose an additional 15 per cent of the global population to a severe decrease in water resources. The number of people living with absolute water scarcity (<500 m3 per capita per year) is expected to grow by another 40 per cent compared with a scenario without climate change. The study shows that the greatest increase in water scarcity could occur at a warming of 2–3°C above pre-industrial levels. At 3°C warming, 10 out of 100 people would be affected by absolute water scarcity—compared to one to two people per 100 people today. Significant regional differences appear: while water availability in the Mediterranean region, the Middle East, the southern USA and southern China decreases, it is projected to increase in southern India, western China and parts of East Africa. Increasing drought conditions are likely to place affected populations under immense pressure. According to Prudhomme/Giuntoli/Robinson et al. (2013), as warming approaches 4°C drought conditions are expected to deteriorate in the Americas, large parts of tropical and southern Africa, the Mediterranean region, south-east China and Australia. Overall, an increase in drought conditions is projected for 40 per cent of all land area analysed in nearly half of all simulations of a 4°C world used in the study. Minor changes or decreased drought conditions are found only in northern Canada, north-east Russia, the Horn of Africa and parts of Indonesia.

With its high exposure and strong dependence on rainfall, agricultural productivity is particularly vulnerable to climate change. Rosenzweig/Elliott/Ruane et al. (2013) bring together and compare the results of seven crop models within the frameworks of the Agricultural Model Intercomparison and Improvement Project, and ISI-MIP. Despite strong variation across models, overall results show a robust negative trend of 50 per cent decreases in yield for important crops such as wheat and maize in the tropics in a scenario roughly corresponding to a 4°C world. The response is stronger if the uncertain CO_2 fertilization effect is not taken into account. The authors stress that while higher latitudes are expected to see increases in yields, important soil properties in these regions are understudied and add another factor of uncertainty to projections and another risk to global food production.

A risk analysis of ecosystem shifts under climate change by Warszawski/Friend/Ostberg et al. (2013) aggregates changes in the biochemical ecosystem state as a proxy for the risk of ecosystem changes at different levels of global warming. The main result shows a risk for 5–19 per cent of the naturally vegetated land surface at 2°C of global warming above 1980–2010 levels. The area at risk of severe changes is projected to grow as global mean temperature rises, approximately doubling for a warming of between 2°C and 3°C and reaching a median value of 35 per cent of all naturally vegetated land surface at approximately 4°C warming. Some regions characterized by exceptional biodiversity coincide with those found to be at risk of severe ecosystem changes, such as the montane grass and shrublands of the Tibetan Plateau Steppe, and several moist forest regions, such as the Guayanan Highland moist forests of South America. The authors identify strong divergence in results across models, with discrepancies in the globally aggregated level of risk of ecosystem change and its geographical distributions resulting from the diversity of modelling approaches and their respective sensitivities to climate change and increasing CO_2 concentrations. Overall, however, there is agreement within the model ensemble regarding the increasing risk of severe change on a global scale under all warming scenarios considered.

11.2.2   Climate Impact Hotspots

According to the analysis provided by Piontek/ Müller/Pugh et al. (2013), every tenth person lives in a place on Earth that can become a hotspot of unrestrained climate change. With increasing global mean temperature, the risk of overlapping impacts grows and leads to higher adaptation pressure. Within the ISI-MIP framework, Piontek/Müller/Pugh et al. (2013) define hotspots as regions of multisectoral exposure, where metrics in two or more sectors indicate severe change from average conditions. In order to identify such hotspots, the authors analyse data across a range of impact areas including agriculture, malaria, water resources and ecosystems. They find that following the strict assessment criteria (climate and impact model agreement of at least 50 per cent), overlapping of multisectoral impacts starts at 3°C warming above the mean temperature of the 1980–2010 period, with at least two of the four impact sectors overlapping at 4°C. The most prominent hotspot is the southern Amazonian Basin, which is affected by impacts in three areas: agricultural yields, ecosystems,

Figure 11.2: Multisectoral hotspots of impacts for two (orange) and three (red) sectors, where at least fifty per cent of climate and impact model combinations agree on the threshold crossing in each sector for a global mean temperature change of up to 4.5°C above 1980–2010. Regions in light grey are regions where no multisectoral overlap is possible at all because of data and assessment restrictions. Dark grey shading shows areas where at least ten per cent of climate and impact model combinations agree on the given threshold crossing. **Source:** Piontek/Müller/Pugh et al. (2013).

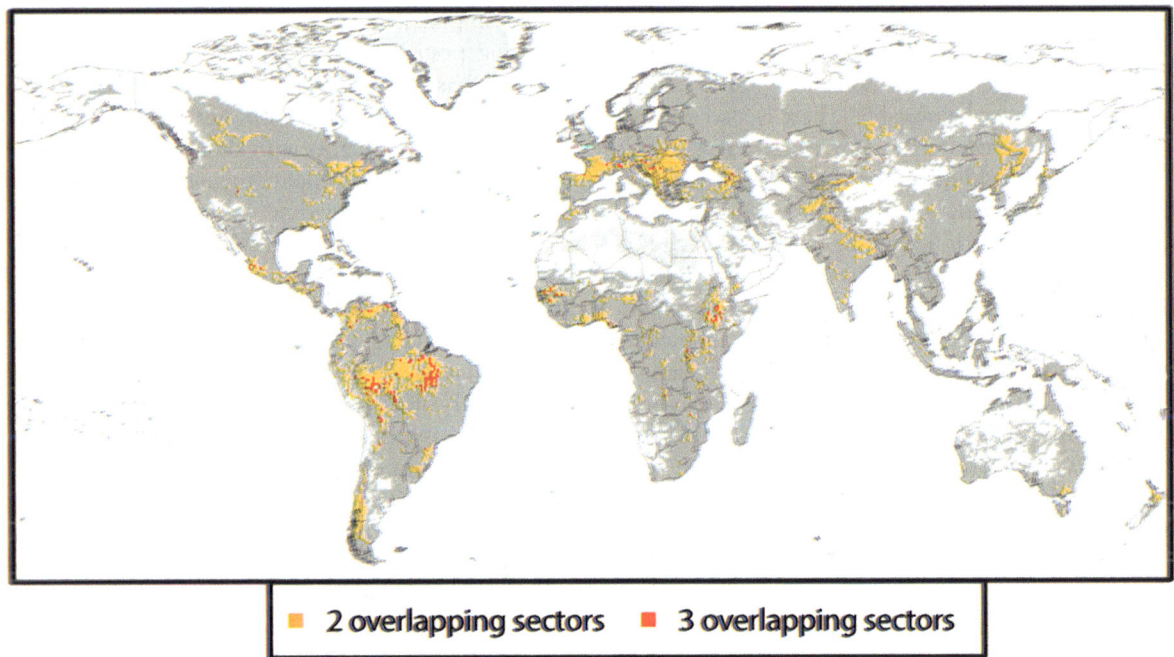

■ **2 overlapping sectors**   ■ **3 overlapping sectors**

and river discharge (see figure 11.2). The second largest hotspot identified is southern Europe, as this region is expected to experience pressures in water resources and ecosystems. A further region at high risk of multiple impacts is the Ethiopian Highlands, where malaria extension, crop yield reduction and ecosystem change are projected to coincide. Similarly, northern regions of Africa are expected to be affected by either reductions in discharge and crop yields or by crop yield reduction and ecosystem change. Where impacts in different sectors overlap they increase exposure, will likely lead to interaction between impacts, and most certainly will compound adaptation pressures.

Critically, loosening the strict criteria to 10 per cent agreement between climate and impact models in order to include a worst-case scenario substantially increases the area at risk of overlap between multisectoral climate impacts (dark grey areas in figure 11.2). Areas at risk cover most of the global land area, with the exception of the high northern latitudes, the Sahara, and individual areas in Australia and China. In particular the exception of the Arctic from a worst-case assessment indicates that this assessment is limited to the sectors in question and does not capture

impacts such as cultural loss and health impacts other than malaria, which have both already been observed in high-latitude northern regions (e.g. Crate 2008). Integrating further sectors into this analysis would thus likely lead to a different, albeit not more optimistic picture of climate impacts.

11.3 Critical Thresholds, Cascading Impacts and Sectoral Interactions within Human Systems

The analysis outlined above provides an important indication of the risks arising from multiple impacts occurring in one location. Particularly in cases in which such impacts are experienced simultaneously, they would place enormous pressure on affected populations. Over the last decade or so the scientific community has been developing conceptual approaches and models that reflect the interactions and positive feedbacks that can occur between natural and societal systems (e.g. Brauch 2007). Agent-based models also allow for a better understanding of societal responses to changes in climatic conditions (e.g. Smith 2014).

However, climate impact modelling exercises for the most part do not capture potential interactions between impacts, as well as the ways in which climate change impacts can influence and be influenced by non-climatic factors and processes such as urbanization and environmental degradation. Current assessments also tend to not be able to capture the significance of critical thresholds within human sectors, beyond which the scale of impacts grows non-linearly and adaptation may no longer be possible. There is also a possibility that an initial impact can trigger a series of impacts across multiple sectors or different parts of the globe—particularly given the interconnectedness of today's world. Such dynamics, which could lead to abrupt, non-linear or potentially irreversible changes within sectors, with significant consequences for human populations, are illustrated with examples in this section. While the overall damage of climate impacts will be strongly shaped by societal developments and the state of readiness in the face of large-scale environmental change, a comprehensive risk assessment needs to incorporate the possible ways in which climate impacts may unfold to affect societies. Some of these are illustrated in the following examples.

11.3.1 Non-linearities Associated with Critical Thresholds within Sectors

Responses to rising temperatures can be abrupt if critical system thresholds are crossed. Such thresholds are likely to exist not only in components of the Earth system (discussed below in section 11.5), but also in human sectors such as agriculture and health care. While modelling development in individual impact sectors is currently under way (e.g. Rosenzweig/Jones/Hatfield et al. 2013), the processes underlying non-linear responses within individual sectors are typically not yet represented in climate impact projections.

One such potential non-linear response has been found in agriculture. Observations point to threshold behaviour in the growth of important crops including maize, wheat, soya, and cassava. When exposed to extreme temperatures or drought conditions during critical growth periods, plant development can be severely inhibited, leading to substantive yield losses. Non-linear temperature effects on these crops have been identified in several regions of the world, including the United States, Africa, India and Europe. For example, in the USA, significant non-linear effects have been observed when local temperature rises above 29 °C for corn, 30 °C for soybeans, and 32 °C for cotton (Schlenker/Roberts 2009). In South Asia, the upper temperature sensitivity threshold for current cultivars of rice is 35-38 °C and of wheat is 30-35 °C (Wassmann/Jagadish/Sumfleth et al. 2009). Further factors exacerbating the risk of non-linear effects include inhibited plant growth due to increasing salinity in deltaic regions, the spread of pests and diseases, declines in pollinators associated with habitat loss, and increasing pesticide use (Potts/Biesmeijer/Kremen et al. 2010; Wassmann/Jagadish/Sumfleth et al. 2009). Most current crop models do not account for the interaction of such factors with climatic changes (Warren 2011) and their net impact remains a considerable uncertainty associated with a warming of 4 °C.

Critical thresholds may also exist in social systems, beyond which climatic pressures produce rapid, unforeseen and unprecedented effects. However, they are not well understood. Archaeological studies indicate that natural disasters can cause significant social and cultural shifts—through mass migration and economic and political upheaval—in instances in which the effects of the disaster make existing ways of life untenable (Riede 2013). It is plausible that present institutions could similarly be pushed beyond their capacity to maintain normal operation when faced with multiple unprecedented pressures from climate change impacts. For example, various stresses on human health, such as heatwaves, malnutrition, reduced quality of drinking water due to salt water intrusion and the spread of vector-borne diseases, could overburden health care systems to the point that coping with given stresses is no longer possible. The long-term repercussions remain unclear. Alternative means of coping are likely to be developed by those affected, including migration to more secure locations; however, the human and cultural costs of such changes are likely to be high.

11.3.2 Non-linearities Due to the Interaction of Impacts

Most model-based climate impact studies present results for individual natural or human systems. However, as global mean temperature rise approaches 4 °C, it can be expected that impacts will more frequently coincide with one another and interact. Such interactions could produce a combined effect on human populations that is greater than the sum of the first-order impacts.

Interactions could occur between multiple impacts within a single sector, or between impacts across different sectors. For example, productivity in the agricultural sector could be severely undermined by a combination of direct impacts associated with changing temperature and precipitation patterns and indirect climate change impacts resulting from the expansion of weeds and invasive species (Potts/Biesmeijer/Kremen et al. 2010). An example of how impacts within the agricultural sector might interact with impacts in other sectors is that of the repercussions of a large shock to agricultural production following an extreme weather event such as extreme temperatures, drought or flooding. The resulting risk of malnutrition could significantly compound the health impacts associated with a possible concurrent outbreak of transmittable disease or heat stress (WHO 2009). In such a scenario the human costs could be considerable and very likely greater than in a scenario of agricultural impacts only.

Importantly, interactions between climate change impacts would occur against a backdrop of non-climatic pressures on natural and human systems. In effect, climatic factors interact with non-climatic pressures. The latter, which include population growth and environmental degradation, will influence the nature and scale of the ultimate repercussions of climate change for populations. Declining agricultural productivity along with a simultaneous increase in demand may, for example, lead to an expansion of agricultural area at the cost of natural ecosystems and the ecosystem services on which some people depend. Further, adaptation measures in one sector may exacerbate impacts in another—for example, if dam construction impedes the resilience of natural ecosystems (Warren 2011). However, the outcome of these interactions is highly uncertain, given the contingency and complexity of factors involved.

11.3.3   Non-linearities Due to Cascading Impacts

It is becoming increasingly apparent as research progresses that climate change impacts can produce many indirect effects. In other words, an initial impact can set off a chain of further impacts, which tend to not be taken into account in climate impact assessments. Such cascading impacts present another layer of risk to be integrated into a comprehensive understanding of the challenges of 4°C of warming by 2100.

For example, the impacts of climatic changes and extreme events on critical physical infrastructure or goods traded in international markets can ripple through the economy, leaving their mark well beyond the initial point of impact. This could occur as a result of damage to seaports from impacts including sea-level rise, more intense tropical cyclone activity, higher

storm surges, severer river flooding, and altered wave regimes, with consequent effects on the supply chain (Becker/Acciaro/Asariotis et al. 2013). Another example is that of the long-term economic and welfare effects of damage to transport infrastructure due to a severe flooding event in Mozambique in 2000. Road connections between the capital Maputo and the rest of the country, along with the railway line to Zimbabwe, were destroyed for nearly a year. According to Chinowsky/Arndt (2012), this caused per capita economic growth to decline to the slowest growth rate in two decades. Economic growth rebounded when the primary north-south road link was repaired in 2001. Further, linkages between infrastructure systems can cause impacts in one to echo throughout others. For example, disruptions to wastewater treatment could increase flood losses due to contaminated floodwaters and decrease availability of cooling water for power generation (Kirshen/Ruth/Anderson 2007).

Similarly, impacts on production sites and distribution networks of widely traded goods can reverberate across the globe. Based on data collected by Lenzen/Kanemoto/Moran/Geschke (2012), Levermann/Clark/Marzeion et al. (2014) estimate that for receiving countries the knock-on effects of the cessation of exports by the country of origin may well exceed the direct economic damages. For example, six per cent of the US economy relies on imports from the Philippines. It would thus be directly affected if all Philippine exports ceased in the wake of a catastrophic weather event such as a tropical cyclone. However, indirect effects—mainly through the channels of retail trade—could resonate throughout the wider US economy, of which 21 per cent is calculated to be affected. Similarly, extreme events expose the market's susceptibility to weather-related production shocks. When coal exploration was put on hold in Queensland, Australia, following the combined effects of heavy rainfall and Cyclone Yasi in 2010-11, coking coal prices rose by 25 per cent the following year (Levermann 2014).

11.4   The Differential Social Impacts of Climate Change

The overall effect of climate impacts will be determined in large part by the patterns of capital and commodity flows and the degree of complexity in social networks. Another decisive factor driving the nature and scale of impacts and impact cascades and human vulnerability to them is future socio-economic development. While its trajectory remains uncertain, it is becoming increasingly clear that socio-economic development itself will be negatively affected by the impacts of climate change.

The ways in which climate change will affect human development and efforts to alleviate poverty are increasingly receiving research attention. Although there are as yet few studies which fully capture and quantify the socio-economic repercussions of a global mean warming of 4°C, it is apparent that climate change impacts have the potential to impede practically all aspects of human life and threaten long-term development goals. Climate change acts as a multiplier on other environmental and social stressors, further increasing the vulnerability of the poor and those already disadvantaged (see chapter by Ide/Link/Scheffran/Schilling in this volume). It is therefore considered likely that climate change will widen the divide between rich and poor, or between those populations that are generally more resilient and those that are less so.

11.4.1   Differential Social Vulnerability

Social vulnerability has been defined by Füssel (2012), following Blaikie, Cannon, Davis et al. (1994) and Adger and Kelly (1999), as the inadequate capacity of individuals, groups or communities to cope with and adapt to external stress placed on their livelihoods and well-being. This is determined by the availability of resources and by the capacity of individuals and groups to call on these resources. A complex set of characteristics influence a person's vulnerability, including initial well-being and health; livelihood circumstances and qualifications; capacity and willingness of individuals to minimize risks to themselves and their property; the forms of hazard preparedness provided by society and the availability of public services (e.g. disaster relief, shelters); and social and political networks and institutions (e.g. people's right and capacity to express their needs; see also Brauch 2011; Cannon/Twigg/Rowell 2003).

Marginalized and disadvantaged social groups and individuals are therefore often more vulnerable to the impacts of climate change than others (Warner/Erhart 2009). Accordingly, a wide range of literature reports that those people that are disadvantaged in society due to their age, gender, or ethnicity are among those most vulnerable to negative climate change impacts (e.g. Blaikie/Cannon/Davis et al. 1994; WHO 2009; Mearns/Norton 2010; FAO 2013; Wang/Otto/Lu 2013). This includes marginalized groups and individuals in middle- and high-income countries (e.g. Sti-

vers 2007; Napier/Mandisodza/Andersen et al. 2006; Vandentorren/Bretin/Zeghnoun et al. 2006).

In this context we use the term 'differential social vulnerability' to refer to the varying degrees of adverse effect that different individuals and social groups in one location may suffer from the climate stressors they are exposed to. Critically, patterns of differential vulnerability to climate change impacts have the potential to shape future patterns of inequality across the globe and within regions. For example, children, women and the elderly are generally expected to be more affected by negative health impacts related to the transmission of diseases and pathogens, increase of diarrhoeal and respiratory illnesses, under- and malnutrition, and heat stress (WHO 2009, 2012; EACC Synthesis World Bank Group 2010; Reyburn/Kim/Emch et al. 2011; UN Habitat 2011). The inhabitants of Small Island Developing States and low-lying areas are particularly vulnerable to the salinization of fresh water and arable land as well as to the impacts of storm surges (WHO 2009). Indigenous groups in the Arctic region may suffer a decrease in food sources and severe cultural impacts as reduced sea ice causes the animal populations on which they depend to decline, affecting their hunting and food-sharing cultures (Bronen 2008; ACIA 2004; Crowley 2010). Ethnic minorities often work outdoors in sectors such as construction and agriculture and tend to have few income alternatives. As they receive inadequate protection by labour law and trade unions, they have been identified as being particularly prone to dangerous working conditions, including exposure to heat stress (Vásquez-Léon 2009; Hanna/Kjellstrom/Bennett et al. 2011; Aw 2010). Urban populations are more vulnerable to increased temperatures due to a combination of higher inner-city temperatures, high population density, and inadequate sanitation and freshwater services (WHO 2009).

Although vulnerability and poverty are not synonymous, they tend to coincide. Indeed, it is clear that the poorest people in any given community are often disproportionately affected by environmental stressors such as disasters, particularly over the long term (Shepherd/Mitchell/Lewis et al. 2013). Recent trend analysis shows that poverty will be increasingly concentrated in particular areas in the future (Shepherd/Mitchell/Lewis et al. 2013). Such locations tend to be those in which people are most vulnerable to climate-related stressors as economic and political forces confine the poor to living in high-risk landscapes (Warner/Erhart 2009). For example, Amarasinghe/Samad/Anputhas (2005) show that in Sri Lanka the poorest households are located in dry areas where small-sized agricultural holdings depend on rain-fed production and where income diversification opportunities are scarce due to the large distance from roads and cities. Similarly, the urban poor are expected to be among the most vulnerable groups to the impacts of climate change, due to the location of many cities in coastal areas as well as an increasing trend towards urbanization. Urbanization often drives urban migrants into informal settlements, which tend to be located in areas most prone to extreme weather, such as steep slopes or alluvial plains. They also lack basic services and infrastructure, such as waste collection and sanitation, and this can increase the prevalence of transmittable disease. In addition, as the urban poor are the net buyers of food, they are particularly vulnerable to increases in food prices following climate-induced production shocks and declines (Hein/Metzger/Leemanns 2009; Kumssa/Jones 2010; Smit/Parnell 2012; McMichael/Montgomery/Costello 2012).

11.4.2   The Potential of Climate Change to Exacerbate Vulnerability

Climate change impacts can affect people in multiple ways and with long-lasting effects. The channels through which extreme weather events, for example, can impact upon populations and economies include death and disability, sudden loss of income, depletion of assets, loss of public infrastructure, and macroeconomic shocks (UNDP 2007; Shepherd/Mitchell/Lewis et al. 2013). As for the multifarious changes to their lives and lived surroundings, many of the critical impacts on the poor cannot be measured in economic terms. These include degradation or loss of ecosystem services and social networks, which in some cases, particularly in developing countries, may in fact be of greater consequence than economic losses (UNFCCC 2013). Where there is inadequate capacity to recover, these impacts can place people in an ever more vulnerable position in the face of future impacts. This can lead to a cycle of losses and maladaptive strategies including divestment of productive assets such as livestock, exchange of land for food, and the sale of firewood, leading to increased deforestation (UNDP 2007). Particularly detrimental to human resilience in the face of climate change are consecutive disasters that do not allow enough time for societies and households to recover.

Climate change impacts such as droughts and extreme storms carry the potential to push vulnerable households beyond an observed threshold at which they fall into a poverty trap (Carter/Little/Mogues/Negatu 2007). This makes them acutely vulnerable to further impacts, effectively preventing recovery and impeding economic growth and human development in the longer term. Climate-related impacts on food availability, for example, may lead to long-term setbacks in areas such as health, education and future employment opportunities for children in those households that are unable to avoid the detrimental effects of malnourishment, stunting, and missed schooling (Shepherd/Mitchell/Lewis et al. 2013).

Social and economic asset endowments largely determine the capacity of a household or individual to withstand shocks, with the likelihood of impoverishment decreasing as household prosperity rises (Little/Stone/Mogues et al. 2006). There appears to be no clear threshold, but at an income of more than US$4 per day, the risk of falling into poverty has been shown to be greatly reduced (Shepherd/Mitchell/Lewis et al. 2013). Insofar as climate change threatens to cause people to fall into poverty, it can be seen as a threat to human development and the advances in poverty alleviation made in recent years.

Climate change also has the potential to influence broad social processes and trends that are driven by non-climatic factors. For example, there are strong indications that climate extremes as well as ongoing negative climate change impacts on water availability and food production are contributing to population movements in sub-Saharan Africa (Chaves/Koenraadt 2010) and South East Asia (Marks 2011; Warner 2010; World Bank 2010). Recent studies based on household surveys show that climate events may account for approximately 6–12 per cent of current levels of migration in North Africa and the Middle East (Grant/Burger/Wodon 2014). Indeed, the role that weather patterns play in such dynamics could increase in the future as climate conditions continue to deteriorate. The poorest migrants—those lacking the social networks and resources to go through formal migration procedures—may choose life-threatening escape channels. These include so-called 'boat people', who are at risk of trafficking, crime, and death (Murray/Williamson 2011; Düvell 2012). As a consequence of population movements, migrants can be exposed to the risks associated with informal settlements, as outlined above. In their new locations they may also face tensions across ethnic identities and political and legal restrictions as well as competition and limitations to access to land (Tacoli 2009).

To sum up, those people who are initially the most vulnerable to climate change are made all the more vulnerable by climate change. This could cause resilience to be progressively degraded. The relatively less vulnerable parts of the world, on the other hand, would be less affected by this feedback loop. In this way, climate change has the potential to drastically deepen inequality across the world.

11.5 (In)Stability of Earth System Components

Climate change may interact with several large-scale components of the Earth system in a non-linear way. Due to the complex nature of these systems their response to increasing global mean temperature may be highly amplified by internal dynamics and in some cases even lead to regime shifts. The human consequences of such major perturbations are still little understood. This means that their integration into a comprehensive climate risk assessment is still for the most part lacking.

11.5.1   The El Niño Southern Oscillation

One large-scale component of the Earth system of high societal relevance is the *El Niño Southern Oscillation* (ENSO), which is expected to be affected by and interact with anthropogenic climate change. The ENSO is the dominant mode of natural variability of the Earth climate system and consists of irregular warm (El Niño) and cold (La Niña) sea-surface patterns in the central Pacific Ocean. Anomalous, El Niño-type conditions are related to disastrous flooding events in Latin America on the one hand and droughts in Australia and large areas of South East Asia on the other, and can have far-reaching effects on Atlantic hurricane activity and the global monsoon system (see e.g. Donnelly/Woodruff 2007; Kumar/Rajagopalan/Hoerling et al. 2006).

Additionally, the ENSO affects the global heat budget as it strongly alters ocean heat uptake in the Pacific. In fact, the recent 'hiatus' decade of slowed global warming has been related to the absence of strong El Niño events and a corresponding increase in ocean heat uptake (Balmaseda/Trenberth/Källén 2013; Meehl/Arblaster/ Fasullo et al. 2011; Rahmstorf/Foster/Cazenave 2012).

Given its socio-economic importance and possible growing future impacts, short-term prediction of ENSO evolution is a crucial, albeit challenging, scientific task. Based on an analysis of the network of teleconnections of the El Niño basin with the surrounding Pacific Ocean, Ludescher/Gozolchiani/Bogachev et al. (2013) have developed a new forecasting scheme for the occurrence of El Niño events. While predictions have to date been limited to approximately six months, the method introduced by Ludescher/Gozolchiani/Bogachev et al. (2013) allows a robust twelve-month forecast and thus enables adaptation in, for example, the agricultural sector in advance of potential ENSO related climatological extremes.

As natural ENSO variability both temporally and spatially is very large, the last IPCC report remained inconclusive about future ENSO evolution (Collins et al. 2013). Recent model intercomparison studies, however, reveal a trend towards more extreme El Niño events over the twenty-first century (Cai/Borlace/Lengaigne et al. 2014; Power/Delage/Chung et al. 2013).

11.5.2   The Indian Monsoon

Palaeoclimatological records provide evidence that the Indian monsoon system has undergone abrupt transitions in the past that might have been the result of a regime shift in the underlying moisture-advection feedback (Levermann/Schewe/Petoukhov/Held 2009; Schewe/Levermann 2012). While such dramatic shifts are not projected for the twenty-first century, increases in absolute rainfall and in the number of monsoon failure events are projected. An increase in the interannual variability of the Indian monsoon may further pose a risk to agricultural productivity patterns which are highly adapted to the timing of monsoon rainfall (Menon/Levermann/Schewe et al. 2013). Thus, the monsoon circulation may become less reliable and potentially more extreme, which represents a serious risk for agriculture and the hydrological cycle in South East Asia.

11.5.3   The Amplification of Planetary Waves

An increase in the frequency of climatological extremes, both warm and cold, has been observed in the Northern Hemisphere. This might be related to a weakening of the zonal mean jet stream as well as quasi-resonant amplification of planetary waves in the Northern Hemisphere jet stream (Petoukhov/Rahmstorf/Petri/Schellnhuber 2013). Under such quasi-resonant conditions, the amplitude of these planetary waves—their longitudinal extension—is enhanced, which leads to uncommon northward (southward) penetration of warm (cold) air masses. At the same time, the planetary waves that normally meander around the

Figure 11.3: Schematic mapping of tipping elements of the Earth system and populations density. **Source:** Schellnhuber, pers. comm., after Lenton, Held, Kriegler et al. (2008).

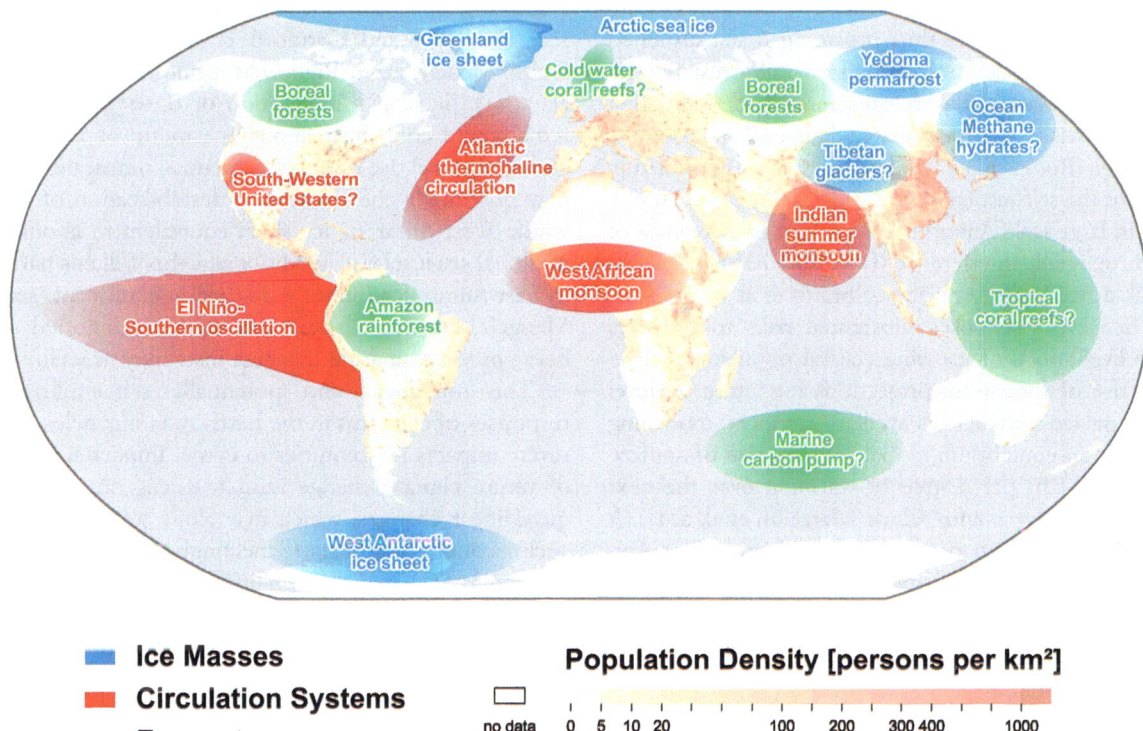

globe become quasi-stationary, which can lead to persistent extreme weather conditions over several weeks.

Stationary wave patterns have been observed in connection to severe flooding events in central Europe, heatwaves in Russia and the US, as well as cold winters in central Europe and the US over the last decade. Additionally, quasi-resonant amplification could translate into extremes occurring simultaneously over the Northern Hemisphere.

The role of climate change in the recent increase in occurrence of such events has not been conclusively established. However, a decreased temperature gradient between the subtropics and the polar regions is favourable to the establishment of quasi-resonant conditions. During the last decade, we have seen a dramatic loss in Artic sea-ice accompanied by a strong warming signal in the Arctic region (Cowtan/Way 2013), and it might be that this loss is partially responsible for the increase in mid-latitude extreme events over the same period (Coumou/Petoukhov/Rahmstorf et al. 2014; Petoukhov/Semenov 2010; Tang/Zhang/Francis 2013). While evidence is growing that this phenomenon might be responsible for a variety of recent weather extremes, the resolution of state-of-the-art General Circulation Models is too low to cor-

rectly resolve it, which means that future Northern Hemisphere mid-latitude climate extremes may potentially be underestimated.

11.5.4 Tipping Elements

The three examples outlined above illustrate the importance of considering the non-linear responses of large-scale components of the climate system to anthropogenic interference. In addition, a variety of these components may show self-amplifying dynamics that can lead to system-immanent regime shifts. These are the so-called "tipping elements" of the climate system (Lenton/Held/Kriegler et al. 2008, depicted in figure 11.3), which include ecosystems such as the Amazon rainforest and parts of the cryosphere such as the Greenland and West Antarctic ice sheets.

Moisture recycling, that is, the re-evaporation of precipitated water, over the Amazon rainforest is a crucial part of the climate of South America, where in some regions more than 70 per cent of the moisture that precipitates is of continental origin (van der Ent/Savenije/Schaefli et al. 2010). While a critical threshold of annual precipitation exists below which tropical rainforests are unlikely to persist (Hirota/ Holmgren/

Nes et al. 2011; Malhi/Aragão/Galbraith et al. 2009), the extent of evaporation depends on the vegetation type and is considerably higher for rainforest than grassland or pasture, thus providing a self-sustained positive feedback. As the Amazon rainforest is critically endangered by deforestation and increasing drought risk due to climate change (Prudhomme/Giuntoli/Robinson et al. 2013), there is a risk of disruption to the self-sustaining hydrological processes, resulting in large-scale Amazon dieback as a consequence of anthropogenic interference (Davidson/Araújo/Artaxo et al. 2012; Malhi/Aragão/Galbraith et al. 2009).

Sea-level rise poses substantial risks to the lives and livelihoods of growing coastal populations. Irrespective of short-term projections for future sea-level rise, palaeo-evidence as well as ice-sheet modelling suggest an equilibrium global sea-level rise of approximately 2.3 m per degree of warming over the next millennia (Levermann/ Clark/Marzeion et al. 2013). A non-linear reaction of the Greenland and West Antarctic ice sheets to warming ocean and atmospheric temperatures may significantly contribute to this rise. The total ice volume of the Greenland ice sheet corresponds to 7 m of sea-level rise. The ice sheet rests nearly fully on bedrock above sea level, and its elevation, as well as the albedo feedback, will positively amplify melting induced by climate change . As the ice sheet starts to melt, elevation decreases and thus the temperature at the ice sheet's surface rises. At the same time, the melting ice sheet covers less and less land area, which changes the local surface albedo and thus changes absorbed radiation. Given these two dominant feedbacks, the critical threshold for the Greenland ice sheet is estimated at between 0.8 °C and 3.2 °C global mean temperature rise above the pre-industrial temperature, with a best estimate of 1.6 °C (Robinson/Calov/Ganopolski 2012). Recent insights about the bed topography of the ice sheet reveal that about 88 per cent of total ice discharge from Greenland into the ocean is channelled through deeply incised submarine glacial valleys (Morlighem/Rignot/Mouginot et al. 2014). This indicates that the Greenland ice sheet is more sensitive to oceanic melt than previously thought, suggesting a higher degree of sensitivity in the tipping dynamics of the ice sheet.

An even more dramatic picture emerges for the mostly marine West Antarctic ice sheet, which rests on bedrock below sea level. Increased sub-ice-shelf oceanic melting underneath the floating ice shelves fringing the Antarctic coast can potentially destabilize the ice sheet and lead to rapid disintegration. Recent evidence from observations and numerical model simulations indicate that such a destabilization is already under way for several large glacier systems in West Antarctica (including the Pine Island and Thwaites glaciers, see: Favier/Durand/Cornford et al. 2014; Joughin/Smith/Medley, 2014; Rignot/Mouginot/Morlighem et al. 2014). Such a disintegration of these glacier systems would contribute about one metre or more to global sea-level rise on multi-centennial timescales and may potentially contribute to a destabilization of the whole West Antarctic ice sheet equivalent to about 4 m global sea-level rise. Additionally, the Wilkens basin in East Antarctica (about 3-4 sea-level equivalent, see: Mengel/Levermann 2014) was recently identified as being prone to marine ice-sheet instability as well.

The non-linear and potentially self-amplifying responses of elements in the Earth system may lead to severe impacts for centuries to come. Impact analyses of future climate change tend to focus on gradual, quasi-linear changes, which may alone lead to severe socio-economic impacts (Schellnhuber/Hare/Serdeczny et al. 2012), and the non-linear responses of large-scale climate components to anthropogenic perturbations are rarely taken into account (Lenton/Ciscar 2012). However, the self-amplifying nature of the responses of many of these systems may represent an at least equally significant threat to humanity, as they can neither be controlled nor easily reversed by future generations. While the assessment of such critical tipping points is not a trivial task, the recent alarming results from West Antarctica illustrate the risks posed by interference with these systems at levels of warming already experienced. In the light of these post-IPCC deadline insights, the IPCC assessment of the climate risk posed by "large-scale singular events" (see figure 11.1, high risk assessed above 3 °C warming) seems to be far too optimistic.

11.5.5   Concluding Remarks

Climate change impacts do not occur in a vacuum. The negative impacts of increasing temperatures and changing rainfall patterns on human sectors can be expected to interact with each other or cascade into potentially far-reaching chains of impacts. These cumulative effects will be mediated and shaped by societal developments including population and economic growth, which in turn affect and are affected by climatic changes. Impacts projected for ecosystems, agriculture, and water supply in the twenty-first century could, for example, lead to large-scale displacement of populations, with manifold consequences for human security, health and economic and trade systems. While each of these sectors has been the subject of research, researchers do not fully understand the wide-ranging and concomitant consequences for society which appear to be increasingly likely with higher levels of warming. Given the intricate complexity of the Earth system, impacts may also occur in a non-linear manner, significantly exacerbating the challenge for precautionary risk reduction and other adaptation measures. As the world warms towards 4°C above pre-industrial levels, the risk of complex and abrupt impacts increases to combine with the direct effects of unprecedented heat extremes and asymmetric sea-level rise.

Important sectoral model development is under way and the research community continues to fill in the picture of what a 4°C world would look like. While this will with time help to ameliorate our lack of knowledge, there will never be full certainty. In light of this, adaptation policy-making and planning needs to find ways to minimize the risks associated with climate change impacts both expected and unexpected. The safest way forward for humanity, however, is to simply avoid rather than to attempt to be prepared for the unexpected. Mitigating climate change remains our safest bet for the lives and well-being of present and future generations.

References

Adger, W. Neil; Kelly, P. Mick, 1999: "Social Vulnerability to Climate Change and the Architecture of Entitlements", in: *Mitigation and Adaptation Strategies for Global Change*, 4, 3-4 (September): 253-266.

Amarasinghe, Upali A.; Samad, Madar; Anputhas, Markandu, 2005: *Locating the poor: Spatially disaggregated Poverty Maps for Sri Lanka* (Colombo: International Water Management Institute, Research Report 96).

Aw, Tar-Ching, 2010: "Global Public Health and the United Arab Emirates", in: *Asia Pacific Journal of Public Health*, 22.3: 19S-24S.

Balmaseda, Magdalena A.; Trenberth, Kevin E.; Källén, Erland, 2013: "Distinctive Climate Signals in Reanalysis of Global Ocean Heat Content", in: *Geophysical Research Letters*, 40 (February): 1754-59.

Becker, Austin H.; Acciaro, Michele; Asariotis, Regina; Cabrera, Edgard; Cretegny, Laurent; Crist, Philippe; Esteban, Miguel; Mather, Andrew; Messner, Steve; Naruse, Susumu; Ng, Adolf K. Y.; Rahmstorf, Stefan; Savonis, Michael; Song, Dong-Wook; Stenek, Vladimir; Velegrakis, Adonis, F., . 2013: "A Note on Climate Change Adaptation for Seaports: A Challenge for Global Ports, a Challenge for Global Society", in: *Climatic Change*, 120, 4: 683-95.

Blaikie, Piers; Cannon, Terry; Davis, Ian; Wisner, Ben, 1994: *At Risk, Natural Hazards, Peoples' Vulnerability and Disasters* (Abingdon: Routledge).

Brauch, Hans Günter, 2007: "Landscape Ecology and Environmental Security: Basic concepts and regional applications for the Mediterranean in the 21st Century", in: Petrosillo, Irene; Müller, Felix; Jones, Bruce K.; Zurlini, Giovanni; Krauze, Kinga; Victorov, Sergey; Li, Bai-Lian; Kepner, William G. (Eds.): *Use of Landscape Sciences for the Assessment of Environmental Security* (Dordrecht: Springer): 21-42.

Brauch, Hans Günter, 2011: "Concepts of Security Threats, Challenges, Vulnerabilities and Risks", in: Brauch, Hans Günter; Oswald Spring, Úrsula; Mesjasz, Czeslaw; Grin, John; Kameri-Mbote, Patricia; Chourou, Béchir; Dunay, Pal; Birkmann, Jörn (Eds.), 2011: *Coping with Global Environmental Change, Disasters and Security—Threats, Challenges, Vulnerabilities and Risks* (Berlin–Heidelberg–New York: Springer-Verlag): 61-106.

Cai, Wenju; Borlace, Simon; Lengaigne, Matthieu, van Rensch, Peter ; Collins, Mat; Vecchi, Gabriel; Timmermann, Axel; Santoso, Agus; McPhaden, Michael, J.; Wu, Lixin; England, Matthew H.; Wang, Guojian; Guilyardi, Eric; Jin, Fei-Fei, 2014: "Increasing Frequency of Extreme El Niño Events due to Greenhouse Warming", in *Nature Climate Change*, 5,1: 1-6.

Cannon, Terry; Twigg, John; Rowell, Jennifer, 2003: *Social Vulnerability, Sustainable Livelihoods and Disasters*. Report to DFID (London: DFID, Conflict and Humanitarian Assistance Department and Sustainable Livelihoods Support Office): 1-63.

Carter, Michael R.; Little, Peter D.; Mogues, Tewodaj; Negatu, Workneh, 2007: "Poverty traps and natural disasters in Ethiopia and Honduras", in: *World Development*, 35, 5: 835-856.

Chaves, Luis Fernando; Koenraadt, Constantianus J.M., 2010: "Climate change and highland malaria: fresh air for a hot debate", in: *The Quarterly Review of Biology*, 85, 1: 27-55.

Chinowsky, Paul; Arndt, Channing, 2012: "Climate Change and Roads: A Dynamic Stressor-Response Model", in: *Review of Development Economics*, 16, 3: 448-62.

Collins, Matthew; Knutti, Reto; Arblaster, Julie; Dufresne, Jean, Louis; Fichefet, Thierry; Friedlingstein, Pierre; Gao, Xuejie; Gutowski, William J.; Johns, Tim; Krinner, Gerhard; Shongwe, Mxolisi; Tebaldi, Claudia; Weaver, Andrew J.; Wehner, Michael, "Chapter 12 : Long-Term Climate Change : Projections, Commitments and Irreversibility", in: Stocker, Thomas, F.; Qin, Dahe; Plattner, Gian-Kasper; Tignor, Melinda M. B.; Allen, Simon, K.; Boschung, Judith; Nauels, Alexander; Xia, Yu; Bex, Vincent; Midgley, Pauline M. (Eds.): *Climate Change 2013: The Physical Science Basis. Contribution of Working Group I to the Fifth Assessment Report of the Intergovernmental Panel on Climate Change* (Cambridge; New York: Cambridge University Press): 1029-1136.

Coumou, Dim; Petoukhov, Vladimir; Rahmstorf, Stefan; Petri, Stefan; Schellnhuber, Hans Joachim; 2014: "Quasi-Resonant Circulation Regimes and Hemispheric Synchronization of Extreme Weather in Boreal Summer" in: *Proceedings of the National Academy of Sciences Early Edit*, 111, 34: 12331-12336.

Crate, Susan A., 2008: "Gone the Bull of Winter? Grappling with the Cultural Implications of and Anthropology's Role(s) in Global Climate Change", in: *Current Anthropology*, 49, 4: 569-95.

Davidson, Eric A.; de Araújo, Alessandro C.; Artaxo, Paulo; Balch, Jennifer K.; Brown, I. Foster; Bustamante, Mercedes M. C.; Coe, Michael T.; DeFries, Ruth S.; Keller, Michael; Longo, Marcos; Munger, J. William; Schroeder, Wilfrid; Soares-Filho, Britaldo S.; Souza, Carlos M.; Wofsky, Steven C.; 2012: "The Amazon Basin in Transition", in: *Nature*, 481, 7381: 321-28.

Donnelly, Jeffrey P; Woodruff, Jonathan D, 2007: "Intense Hurricane Activity over the Past 5,000 Years Controlled by El Nino and the West African Monsoon", in: *Nature*, 447, 7143: 465-68.

Dutton, Andrea; Lambeck, Kurt, 2012: "Ice volume and sea level during the last interglacial", in: *Science*, 337, 6091: 216-219.

Düvell, Franck, 2012: "Transit Migration: A Blurred and Politicised Concept", in: *Population, Space and Place*, 18, 4: 415-427

FAO (Food and Agriculture Organization of the United Nations), 2013: *The state of food insecurity in the World. The multiple dimensions of food security* (Rome: Food and Agriculture Organization of the United Nations).

Favier, Lionel; Durand, Gael; Cornford, Stephen L.; Gudmundsson, G. Hilmar; Gagliardini, Olivier; Gillet-Chaulet, Fabien; Zwinger, Thomas; Payne, Anthony; Le Brocq, Anne M., 2014: "Retreat of Pine Island Glacier controlled by marine ice-sheet instability", in: *Nature Climate Change*, 4, 2: 117-121.

Friend, Andrew D.; Lucht, Wolfgang; Rademachera, Tim T.; Keribina, Rozenn; Betts, Richard; Cadulee, Patricia; Ciais, Philippe; Clark, Douglas B.; Dankers, Rutger; Falloon, Pete D.; Itoh, Akihiko; Kahana, Ron; Kleidon, Axel; Lomasj, Mark R.; Nishinah, Kazuya; Osterberg, Sebastian; Pavlick, Ryan; Peylin, Philippe; Schaphoff, Sibyll; Vuichard, Nicolas; Warszawski, Lila; Wiltshired, Andy; Woodwardj, F. Ian, 2013: "Carbon residence time dominates uncertainty in terrestrial vegetation responses tofuture climate and atmospheric CO2", in: Proceedings of the National Academy of Sciences, 111, 9: 3280-3285.

Füssel, Hans-Martin, 2012: "Vulnerability to Climate Change and Poverty", in: Edenhofer, Ottmar; Wallacher, Johannes; Lotze-Campen, Hermann; Reder, Michael; Knopf, Brigitte; Müller, Johannes (Eds.) *Climate Change, Justice and Sustainability* (Dordrecht: Springer): 9-17.

German Advisory Council on Climate Change, 2007: *World in transition: climate change as a security risk* (London: Earthscan).

German Advisory Council on Climate Change, 2011: *World in transition—a social contract for sustainability* (Berlin: WBGU).

Grant, Audra; Burger, Nicholas; Wodon, Quentin, 2014: "Climate-Induced Migration in the MENA Region: Results from the Qualitative Fieldwork", in: Wodon, Quentin; Liverani, Andrea; Joseph, George; Bougnoux, Nathalie (Eds.) *Climate Change and Migration: Evidence from the Middle East and North Africa* (Washington DC: The World Bank): 163-190.

Hanna, Elisabeth G.; Kjellstrom, Tord; Bennett, Charmian; Dear, Keith, 2011: "Climate Change and Rising Heat: Population Health Implications for Working People in Australia", in: *Asia Pacific Journal of Public Health*, 23,2: 14S-26S.

Hein, Lars; Metzger, Marc, J.; Leemans, Rik, 2009: "The Local Impacts of Climate Change in the Ferlo, Western Sahel", in: *Climatic Change*, 93: 465-483.

Hempel, Sabrina; Frieler, Katja; Warszawski, Lila; Schewe, Jacob; Piontek, Franziska, 2013: "A Trend-Preserving Bias Correction—the ISI-MIP Approach", in *Earth System Dynamics*, 4: 219-36.

Hirota, Marina; Holmgren, Milena; Van Nes, Egbert H.; Scheffer, Martin, 2011: "Global Resilience of Tropical Forest and Savanna to Critical Transitions", in: *Science*, 232: 232-235.

IPCC, 2014: "Summary for Policymakers", in: Field, Christopher B; Barros, Vicente R.; Dokken, David J.; Mach, Katharine J.; Mastrandrea, Michael D.; Bilir, Eren; Chatterjee, Monalisa; Ebi, Kristie L.; Estrada, Yuka; Genova, Robert C.; Girma, B.; Kissel, Eric S.; Levy A.N., MacCracken, Sandy; Mastrandrea, Patricia R.; White, Leslie L. (Eds.) *Climate Change 2014: Impacts, Adaptation and Vulnerability. Part A: Global and Sectoral Aspects. Contribution of Working Group II to the Fifth Assessment Report of the Intergovernmental Panel on Climate Change* (Cambridge; New York: Cambridge University Press): 1-32.

Joughin, Ian; Smith, Benjamin E.; Medley, Brooke, 2014: "Marine Ice Sheet Collapse Potentially Underway for the Thwaites Glacier Basin, West Antarctica", in: *Science*, 344: 735-38.

Kirshen, Paul; Ruth, Matthias; Anderson, William, 2007: "Interdependencies of Urban Climate Change Impacts and Adaptation Strategies: A Case Study of Metropolitan Boston USA", in: *Climatic Change*, 86, 1-2: 105-22.

Kumar, K Krishna; Rajagopalan, Balaji; Hoerling, Martin; Bates, Gary; Cane, Mark, 2006: "Unraveling the Mystery of Indian Monsoon Failure during El Niño", in: *Science*, 314, 5796: 115-19.

Kumssa, Asfaw; Jones, John F., 2010: "Climate Change and Human Security in Africa", in: *International Journal of Sustainable Development and World Ecology*, 17,6: 453-461.

Lenton, Timothy M.; Ciscar, Juan-Carlos, 2012: "Integrating tipping points into climate impact assessments", in: *Climatic Change*, 117: 585-597.

Lenton, Timothy; Held, Hermann; Kriegler, Elmar; Hall, Jim W.; Lucht, Wolfgang; Ramstorf, Stefan; Schellnhuber, Hans Joachim, 2008: "Tipping elements in the Earth's climate system", in: *Proceedings of the National Academy of Sciences of the United States of America*, 105, 6: 1786-1793.

Lenzen, Manfred; Kanemoto, Keiichiro; Moran, Daniel; Geschke, Arne, 2012: "Mapping the Structure of the World Economy", in: *Environmental Science & Technology*, 46, 15: 8374-81.

Levermann, Anders. 2014: "Make Supply Chains Climate-Smart", in: *Nature, 506*: 27-29.

Levermann, Anders: Clark, Peter U.; Marzeion, Ben; Milne, Glenn A.; Pollard, David; Radic, Valentina; Robinson, Alexander, 2013: "The Multimillennial Sea-Level Commitment of Global Warming", in: *Proceedings of the National Academy of Sciences of the United States of America*, 110, 34: 13745-50.

Levermann, Anders; Schewe, Jacob; Petoukhov, Vladimir; Held, Hermann, 2009: "Basic Mechanism for Abrupt Monsoon Transitions", in: *Proceedings of the National Academy of Sciences of the United States of America*, 106, 49: 20572-77.

Little, Peter D.; Stone, M. Priscilla; Mogues, Tewodaj; Castro, A. Peter; Negatu, Workneh, 2006: "'Moving in place': Drought and poverty. Dynamics in South Wollo, Ethiopia", in: *Journal of Development Studies*, 42, 2: 200-225.

Ludescher, Josef; Gozolchiani, Avi; Bogachev, Mikhail I.; Bunde, Armin; Havlin, Shlomo; Schellnhuber, Hans Joachim, 2013: "Correction for Ludescher et Al., Improved El Nino Forecasting by Cooperativity Detection", in: *Proceedings of the National Academy of Sciences of the United States of America, 110*, 47: 19172-73.

Malhi, Yadvinder; Aragão, Luiz E. O. C.; Galbraith, David; Huntingford, Chris; Fisher, Rosie; Zelazowski, Przemyslaw; Sitch, Stephen; McSweeney, Carol; Meir, Patrick, 2009: "Exploring the Likelihood and Mechanism of a Climate-Change-Induced Dieback of the Amazon Rainforest", in: *Proceedings of the National Academy of Sciences of the United States of America, 106*, 49: 20610-15.

Marks, Danny, 2011: "Climate change in Thailand: Impact and Response", in: *Contemporary Southeast Asia: A Journal of International and Strategic Affaires*, 33, 2: 229-258.

McMichael, Tony; Montgomery, Hugh; Costello, Anthony, 2012: "Health Risks, Present and Future, from Global Climate Change", in: *BMJ*, 344: 1359.

Mearns, Robin; Norton, Andrew (Eds.), 2010: *Social Dimensions of Climate Change. Equity and Vulnerability in a Warming World* (Washington, DC: The World Bank).

Meehl, Gerald A.; Arblaster, Julie M.; Fasullo, John T.; Hu, Aixue; Trenberth. Kevin E., 2011: "Model-Based Evidence of Deep-Ocean Heat Uptake during Surface-Temperature Hiatus Periods", in: *Nature Climate Change, 1*, 7: 360-64.

Mengel, Matthias; Levermann, Anders, 2014: "Ice plug prevents irreversible discharge from East Antarctica", in: *Nature Climate Change, 4*: 451-455.

Menon, Arathy; Levermann, Anders; Schewe, Jacob; Lehmann, Jascha; Frieler, Katja, 2013: "Consistent Increase in Indian Monsoon Rainfall and Its Variability across CMIP-5 Models", in: *Earth System Dynamics Discussions, 4*, 1: 1-24.

Morlighem, Mathieu; Rignot, Eric; Mouginot, Jeremie; Seroussi, Helene; Larour, Eric: "Deeply incised submarine glacial valleys beneath the Greenland ice sheet"; in: *Nature Geoscience, 7*: 18-22.

Murray, Royce Bernstein; Williamson, Sarah Petrin, 2011: *Migration as a Tool for Disaster Recovery: A Case Study on U.S. Policy Options for Post-Earthquake Haiti*, Working Paper 255 (Washington, DC: Center for Global Development).

Napier, Jaime L.; Mandisodza, Anesu N.; Andersen, Susan M.; Jost, John T., 2006: "System Justification in Responding to the Poor and Displaced in the Aftermath of Hurricane Katrina", in: *Analyses of Social Issues and Public Policy*, 6, 1: 57-73

Müller, Christoph.; Robertson, Richard D., 2013: "Projecting Future Crop Productivity for Global Economic Modeling", in: Agricultural Economics, 45, 1: 37-50.

Petoukhov, Vladimir; Rahmstorf, Stefan; Petri, Stefan; Schellnhuber, Hans Joachim, 2013: "Quasiresonant Amplification of Planetary Waves and Recent Northern Hemisphere Weather Extremes", in: *Proceedings of the National Academy of Sciences of the United States of America, 110*, 14: 5336-41.

Petoukhov, Vladimir; Semenov, Vladimir A., 2010: "A Link between Reduced Barents-Kara Sea Ice and Cold Winter Extremes over Northern Continents", in: *Journal of Geophysical Research*, 115: D21111.

Piontek, Franziska; Müller, Christoph; Pugh, Thomas A. M.; Clark, Douglas B.; Deryng, Delphine; Elliott, Joshua; Colón González, Felipe de Jesus; Florke, Martina; Folberth, Christian; Franssen, Wietse; Frieler, Katja; Friend, Andrew D.; Gosling, Simon N.; Hemming, D.; Khabarov, N.; Kim, Hyungjun; Lomas, Mark R.; Masaki, Yoshimitsu; Mengel, Matthias; Morse, Andrew; Neumann, Kathleen; Nishina, Kazuya; Ostberg, Sebastian; Pavlick, Ryan; Ruane, Alex C.; Schewe, Jacob; Schmid, Erwin; Stacke, Tobias; Tang, Qiuhong; Tessler, Zachary D.; Tompkins, Adrian M.; Warszawski, Lila; Wisser, Dominik; Schellnhuber, Hans Joachim, 2013: "Multisectoral Climate Impact Hotspots in a Warming World", in: *Proceedings of the National Academy of Sciences of the United States of America*, 111, 9: 3233-3238.

Potts, Simon G; Biesmeijer, Jacobus C.; Kremen, Claire; Neumann, Peter; Schweiger, Oliver; Kunin, William E,. 2010: "Global Pollinator Declines: Trends, Impacts and Drivers", in: *Trends in Ecology & Evolution*, 25, 6: 345-53.

Power, Scott; Delage, François; Chung, Christine; Kociuba, Greg; Keay, Kevin, 2013: "Robust Twenty-First-Century Projections of El Nino and Related Precipitation Variability", in: *Nature*, 502: 541-545.

Prudhomme, Christel; Giuntoli, Ignazio; Robinson, Emma L.; Clark, Douglas B.; Arnell, Nigel W.; Dankers, Rutger; Fekete, Balázs M.; Franssen, Wietse; Gerten, Dieter; Gosling, Simon N.; Hagemann, Stefan; Hannah, David, M.; Kim, Hyungjun; Masaki, Yoshimitsu; Satoh, Yusuke; Stacke, Tobias; Wada, Yoshihide; Wisser, Dominik, 2013: "Hydrological Droughts in the 21st Century, Hotspots and Uncertainties from a Global Multimodel Ensemble Experiment", in: *Proceedings of the National Academy of Sciences of the United States of America*, 111, 9: 3262-3267. 1222473110-.

Rahmstorf, Stefan; Foster, Grant; Cazenave, Anny, 2012: "Comparing Climate Projections to Observations up to 2011", in: *Environmental Research Letters*, 7, 4: 044035.

Reyburn, Rita; Kim, Deok Ryun; Emch, Michael; Khatib, Ahmed; von Seidlein, Lorenz; Ali, Mohammad, 2011: "Climate variability and the outbreaks of cholera in Zanzibar, East Africa: a time series analysis", in: *The American Journal of tropical medicine and hygiene*, 84, 6: 862-869.

Riede, Felix, 2013: "Towards a Science of Past Disasters", in: *Natural Hazards*, 71: 335-362.

Rignot, Eric; Mouginot, Jeremy; Morlighem, Mathieu; Seroussi, Helene; Scheuchl, Bernd, 2014: "Widespread, Rapid Grounding Line Retreat of Pine Island, Thwaites, Smith and Kohler Glaciers, West Antarctica from 1992 to 2011", in: *Geophysical Research Letters*, 41, 10: 3502-3509.

Robinson, Alexander; Calov, Reinhard; Ganopolski, Andrey, 2012: "Multistability and Critical Thresholds of the Greenland Ice Sheet", in: *Nature Climate Change* 2, 6: 429-32.

Rosenzweig, Cynthia; Elliott, Jushua; Deryng, Delphine; Ruane, Alex C.; Müller, Christoph; Arneth, Almut; Boote, Kenneth J., Folberth; Christian; Glotter, Michael; Khabarov, Nikolay; Neumann, Kathleen; Piontek, Franziska; Pugh, Thomas, A. M.; Schmid, Erwin; Stehfest, Elke; Yang, Hong; Jones, James W., 2013; "Assessing Agricultural Risks of Climate Change in the 21st Century in a Global Gridded Crop Model Intercomparison", in: *Proceedings of the National Academy of Sciences of the United States of America*, 111, 9: 3268-3273.

Rosenzweig, Cynthia; Jones, James W.; Hatfield, Jerry L.; Ruane, Alexander C.; Boote, Kenneth J.; Thorburn, Peter; Antle, John M., Nelson, Gerald C.; Porter, Cheryl; Janssen, S.; Asseng, Senthold; Basso, Bruno; Ewert, Frank; Wallach, Daniel; Baigorria, Guillermo; Winter, Jonathan M., 2013: "The Agricultural Model Intercomparison and Improvement Project (AgMIP): Protocols and Pilot Studies", in: *Agricultural and Forest Meteorology*, 170: 166-82.

Schaeffer, Michiel; Gohar, Laila; Kriegler, Elmar; Lowe, Jason; Riahi, Keywan; van Vuuren, Detlef, 2013: "Mid- and Long-Term Climate Projections for Fragmented and Delayed-Action Scenarios", in: Technological Forecasting & Social Change, in press.

Schellnhuber, Hans Joachim; Hare, William L.; Serdeczny, Olivia; Adams, Sophie; Coumou, Dim; Frieler, Katja; Marin, Maria; Otto, Ilona M.; Perrette, Mahé; Robinson, Alexander; Rocha, Marcia; Schaeffer, Michiel; Schewe, Jacob; Wang, Xiaoxi; Warszawski, Lila, 2012: *Turn Down the Heat: Why a 4°C Warmer World Must Be Avoided* (Washington DC: The World Bank).

Schellnhuber, Hans Joachim; Hare, William L.; Serdeczny, Olivia; Adams, Sophie; Baarsch, Florent; Schwan, Sunsanne; Coumou, Dim; Robinson, Alexander; Vieweg, Marion; Piontek, Franziska; Donner, Reik; Runge, Jakob; Rehfeld, Kira; Rogelj, Joeri; Perette, Mahé; Menon, Arathy; Schleussner, Carl-Friedrich; Bondeau, Alberte; Svirejeva-Hopkins, Anastasia; Schewe, Jacob; Frieler, Katja; Warszawski, Lila; Rocha, Marcia, 2013: *Turn Down the Heat: Heat: Climate Extremes, Regional Impacts, and the Case for Resilience* (Washington DC: The World Bank).

Schewe, Jacob; Heinke, Jens; Gerten, Dieter; Haddeland, Ingjerd; Arnell, Nigel W.; Clark, Douglas B.; Dankers, Rutger; Eisner, Stephanie; Fekete, Balazs M.; Colon-Gonzalez, Felipe J.; Gosling, Simon N.; Kim, Hyungjun; Liu, Xingcai; Masaki, Yoshimitsu; Portmann, Felix T.; Satoh, Yusuke; Stacke, Tobias; Tang, Qiuhong; Wada, Yoshihide; Wisser, Dominik; Albrecht; Frieler, Katja; Piontekm Franziska; Warszawski, Lila; Kabat, Pavel, 2013: "Multimodel Assessment of Water Scarcity under Climate Change", in: *Proceedings of the National Academy of Sciences of the United States of America*, 111, 9: 3245-3250.

Schewe, Jacob; Levermann, Anders, 2012: "A Statistically Predictive Model for Future Monsoon Failure in India", in: *Environmental Research Letters*, 7, 4: 044023.

Schlenker, Wolfram; Roberts, Michael J., 2009: "Nonlinear Temperature Effects Indicate Severe Damages to U.S. Crop Yields under Climate Change", in: *Proceedings of the National Academy of Sciences of the United States of America*, 106, 37: 15594-98.

Shepherd, Andrew; Mitchell, Tom; Lewis, Kirsty; Lenhardt, Aamanda; Jones, Lindsye; Scott, Lucy; Muir-Wood, Robert, 2013: *The geography of poverty, disasters and climate extremes in 2030* (London: Overseas Development Institute).

Smit, Warren; Parnell, Susan, 2012: "Urban sustainability and human health: an African perspective", in: *Current Opinions on Environmental Sustainability*, 4: 443-450.

Smith, Christopher S., 2014: "Modelling Migration Futures: Development and Testing of the Rainfalls Agent-Based Migration Model—Tanzania", in: *Climate and Development*, 6, 1: 77-91.

Stivers, Camilla, 2007: "So poor and so black: Hurricane Katrina, public administration, and the issue of race", in: *Public Administration Review*, 67, 1: 48-56.

Tacoli, Cecilia, 2009: "Crisis or Adaptation? Migration and Climate Change in a Context of High Mobility", in: *Environment and Urbanization*, 21,2: 513-525.

Tang, Qiuhong; Zhang, Xuejun; Francis, Jennifer A., 2013: "Extreme Summer Weather in Northern Mid-Latitudes Linked to a Vanishing Cryosphere", in: *Nature Climate Change*, 4, 1: 45-50.

UNDP (United Nations Development Programme), 2007: *Fighting climate change: Human solidarity in a divided world*, United Nations Human Development Report 2007/2008 (New York: UNDP).

UNFCCC (United Nations Framework Convention on Climate Change), 2013: *Technical Paper: Non-economic losses in the context of the work programme on loss and damage* (Bonn: UNFCCC).

UN Habitat (United Nations Human Settlements Programme), 2011: *Cities and Climate Change: Global Report on Human Settlements* (New York: UN Habitat).

Van der Ent, Rudi J.; Savenije, Hubert H. G.; Schaefli, Bettina; Steele-Dunne, Susan C., 2010: "Origin and Fate of Atmospheric Moisture over Continents", in: *Water Resources Research*, 46, 9: W09525

Vandentorren, Stéphanie; Bretin, Philippe; Zeghnoun, Abdelkrim; Mandereau-Bruno, Laurence; Croisier, Alice; Cochet, Christian; Ribéron, Jacques; Siberan, Isabelle; Declercq, Beatrice; Ledrans, Martine, 2006: "August 2003 heat wave in France: risk factors for death of elderly people living at home", in: *European Journal of Public Health*, 16, 6: 583-591.

Vásquez-Léon, Marcela, 2009: "Hispanic farmers and farmworkers: Social networks, institutional exclusion and climate vulnerability in Southeastern Arizona", in: *American Anthropologist*, 111, 3: 289-301.

Wang, Xiaoxi; Otto, Ilona M.; Lu, Yu, 2013: "How physical and social factors affect village-level irrigation: An institutional analysis of water governance in northern China", in: *Agricultural Water Management* 119: 10-18.

Warner, Koko, 2010: "Global environmental change and migration: Governance challenges", in: *Global Environmental Change*, 20,3: 402-413.

Warner, Koko; Erhart, Charles, 2009: *In Search of Shelter: Mapping the Effects of Climate Change on Human Migration and Displacement* (Bonn: UNU-EHS).

Warren, Rachel, 2011: "The role of interactions in a world implementing adaptation and mitigation solutions to climate change", in: *Philosophical Transactions. Series A, Mathematical, Physical, and Engineering Sciences*, 369, 1934: 217-41.

Warszawski, Lila; Frieler, Katja; Huber, Veronika; Piontek, Franziska; Serdeczny, Olivia; Schewe, Jacob, 2013: "The Inter-Sectoral Impact Model Intercomparison Project (ISI-MIP): Project Framework", in: *Proceedings of the National Academy of Sciences of the United States of America*, 111, 9: 3228-3232.

Warszawski, Lila; Friend, Andrew; Ostberg, Sebastian; Frieler, Katja; Lucht, Wolfgang; Schaphoff, Sibyll; Beerling, David; Cadule, Patricia; Ciais, Philippe; Clark, Douglas B.; Kahana, Ron; Ito, Akihiko; Keribin, Rozenn; Kleidon, Axel; Lomas, Mark; Nishina, Kazuya; Pavlick, Ryan; Rademacher, Tim Tito; Buechner, Matthias; Piontek, Franziska; Schewe, Jacob; Serdeczny, Olivia; Schellnhuber, Hans Joachim, 2013: "A Multi-Model Analysis of Risk of Ecosystem Shifts under Climate Change", in: *Environmental Research Letters* 8, 4: 044018

Wassmann, Reiner; Jagadish, S. V. Krishna; Heuer, Sigrid; Ismail, Abdel; Redona, Edilberto; Serraj, Rachid; Singh, Rakesh Kumar; Howell, Gregory; Pathak, Himanshu; Sumfleth, Kay, 2009: "Climate Change Affecting Rice Production: The Physiological and Agronomic Basis for Possible Adaptation Strategies", in: *Advances in Agronomy*, 101: 59-122.

WHO [World Health Organization], 2009: *Protecting health from climate change: connecting science, policy and people* (Geneva: WHO).

WHO [World Health Organization], 2012: *Malaria Fact Sheet* (Geneva: WHO).

World Bank, 2010: *The Social Dimension of Adaptation to Climate Change in Bangladesh* (Washington, DC: The World Bank).

World Bank, 2010: *Economics of Adaptation to Climate Change: Social Syntheses Report* (Washington, DC: The World Bank).

12 The Climate-Conflict Nexus: Pathways, Regional Links, and Case Studies

Tobias Ide, P. Michael Link, Jürgen Scheffran and Janpeter Schilling[1]

Abstract

The role of climate change as a potential cause of violent conflict has been debated in the scholarly and policy communities for several years. We review the most recent quantitative and qualitative literature and find that research on the issue has produced little consensual findings so far. Further, we discuss major theoretical, conceptual and empirical issues and describe possible pathways linking climate change to violent conflict. To illustrate these issues, we analyse the climate-conflict nexus in different world regions and present three qualitative case studies in north-western Kenya, the Nile Basin, and Israel/Palestine. We find that possible reasons for the lack of scientific consensus may be the difficulties of existing approaches to adequately capture the complex links between climate change, vulnerability, and violent conflict.

Keywords: Climate change, environment, violence, conflict, pathways, Kenya, Nile Basin, Israel, Palestine.

12.1 Introduction

Climate change caused by human activities such as the emission of greenhouse gases is likely to cause an overall worldwide increase in the number and/or intensity of drought conditions, heatwaves, floods, and other natural disasters,[2] with important regional variations (IPCC 2013; see chapter 11 in this volume by Schellnhuber/Serdeczny/Adams et al.). Failure to apply far-reaching mitigation and adaptation measures is likely to cause reduced freshwater availability, increased food insecurity, biodiversity loss, infrastructure destruction, a reduction of economic growth, poverty, non-voluntary migration, and increased stress on health systems (IPCC 2014). Those already poor and marginalized will be affected most severely by climate change and its impacts on biophysical and social systems (see chapter 11 in this volume by Schellnhuber/Serdeczny/Adams et al.). The impacts of current and future climate change are aggravated by rapid population growth (see chapter 10 in this volume by Zlotnik), poverty, large-scale land acquisitions by foreign investors, and similar processes of resource appropriation by the rich and powerful (Cotula 2012; Houdret 2012).

Climate change is increasingly seen as a threat to several dimensions of security, including human security (Brauch/Scheffran 2012). Since at least 2007, the possible implications of climate change and its consequences for peace, conflict, and security have received the attention of researchers, think tanks, and policymakers alike. In the same year, the IPCC and Al Gore received the Nobel Peace Prize for their efforts to mitigate climate change. The Nobel Prize Committee (2007) explained its decision as follows: "Extensive climate changes may alter and threaten the living conditions of much of mankind. They may induce large-scale migration and lead to greater competition for the earth's resources [...]. There may be increased danger of violent conflicts and wars, within and between states." In the same year and again in 2011, the UN Security Council discussed the issue of climate

---

1 Tobias Ide is head of the research field Peace and Conflict at the Georg Eckert Institute and associated member of CLISEC. P. Michael Link is a postdoctoral scientist at the Research Group Climate Change and Security (CLISEC), Institute of Geography, University of Hamburg. Jürgen Scheffran is the head of CLISEC and professor at the Institute of Geography, University of Hamburg. Janpeter Schilling is a postdoctoral scientist at CLISEC and an Associate at International Alert (London).

2 The term natural disasters in this chapter is used to refer to natural hazards (potential geo- or biophysical threats) which become active and negatively influence a humanitarian crisis.

change. And very recently, the CNA Military Advisory Board (2014: 2)[3] warned that "in many areas, the projected impacts of climate change [...] will serve as catalysts for instability and conflict". Are such claims justified by the scientific literature? Recently, a meta-study by Hsiang et al. (2013: 1) found "strong causal evidence linking climatic events to human conflict" (see also Hsiang/Burke 2014). This conclusion was heavily criticized (Buhaug 2014; Buhaug/Nordkvelle/ Bernauer et al. 2014). Other reviews of the literature could not detect any scientific consensus on the relationship between climate change and violent conflict (Ide/Scheffran 2014; Scheffran/Brzoska/Kominek et al. 2012a; Theisen/Gleditsch/Buhaug 2013). Mixed results were also presented in the fifth assessment report of the IPCC (2014), particularly in chapters 12 and 19 (Gleditsch/Nordås 2014).

The debate on climate change and (violent) conflict is highly relevant for the debate on sustainability transitions and sustainable peace. In order to realize a sustainable transition towards a world characterized by low CO2 emissions, a small ecological footprint and a strong role for democracy, stable, reliable and accountable institutions are necessary. The possible adverse impacts of climate change on societal stability could erode the development of such institutions, and this might trigger a downward spiral of violent conflict and environmental degradation (UNEP 2009). Similarly, violent conflict, especially within states, is known to have net negative effects on the environment (Reuveny/Mihalache-O'Keef/Li 2010). On the other hand, recent studies suggest that if the impacts of adverse environmental changes on conflict are adequately handled, they can facilitate cooperation and reconciliation between rival groups and contribute to sustainable peace (Ide/Scheffran 2014; Scheffran 2011; Streich/Mislan 2014; see chapter 13 in this volume by Scheffran).

The aim of this chapter is to give the reader an overview of the current academic debate on the possible link between climate change and violent conflict and to highlight the complexity of the issue by addressing conceptual and empirical perspectives. In the next section, we discuss major theoretical and conceptual issues in the research on the nexus between climate change and violent conflict (12.2). Afterwards, the results of qualitative and quantitative empirical

research on possible climate-conflict links are evaluated for possible pathways and world regions (12.3). We then illustrate the theoretical issues and causal pathways discussed by presenting the findings of three qualitative case studies (12.4). Each of these studies was conducted in a region that is widely perceived to be a hot spot of climate change and conflict, namely north-western Kenya (12.4.1), the Nile River Basin (12.4.2), and Israel/Palestine (section 12.4.3). Finally, a conclusion is drawn (12.5).

12.2   Conceptual Considerations on the Nexus of Climate Change and Violent Conflict

As in many other social science research fields, the literature on climate change and (violent) conflict does not use its core concepts and variables in a standardized manner (Salehyan 2014). Climate change is commonly defined as "a statistically significant variation in either the mean state of the climate or in its variability, persisting for an extended period (typically decades or longer)" (IPCC 2001), while climate itself refers to the average patterns of meteorological variables such as temperature and precipitation over a long period of time. But empirical studies operationalize climate change in many different ways, ranging from monthly and annual changes (often compared with long-term averages or trends) to climate processes that take place over decades and even centuries (Buhaug/Nordkvelle/Bernauer et al. 2014). In terms of their dependent variable, most studies on climate change and conflict focus on the reciprocal use of direct, physical violence against humans, typically resulting in at least one fatality and including the state as a party (Scheffran/Ide/Schilling 2014). Some scholars use wider definitions of violence as dependent variables, ranging from small-scale, non-lethal communal violence (and sometimes even baseball disputes) to full-blown civil and inter-state wars (Hsiang/Burke/ Miguel 2013). Forms of structural violence as well as genocide have so far been largely absent in the research on climate change and conflict (Zimmerer 2014).

In this chapter, we define conflict broadly as a process in which at least two social groups perceive their respective interests as contradictory and undertake actions in order to enforce or articulate these interests (Baron 1990; Dietz/Engels 2014). Conflicts can catalyse positive societal change. However, if differences in interests cannot be overcome or at least mediated, tensions increase, pushing parties in conflict

---

3   The Center for Naval Analyses at the Institute for Public Research is a nonprofit research and analysis organization located in Arlington, VA; see at: <http://www.cna.org/centers/ipr#sthash.73vdGBvB.dpuf>.

towards extreme actions such as the use of violence. When at least one of these social groups uses direct, physical violence in order to articulate or enforce its interests, the conflict is considered to be violent (Bonacker/Imbusch 2010: 83-87).

Climate change is often conceived as a "threat multiplier" (e.g. Buhaug/Theisen 2012: 51; Johnstone/ Mazo 2011: 11; Njiru 2012: 520). This metaphor is used to make the point that if there are no pre-existing conditions favourable to violent conflict onset (i.e. factors that climate change can multiply), then the implications of climate change for societal stability and violent conflict will not be significant. Such factors include, inter alia, population growth, livelihood insecurity, increased demand and unequal distribution of resources, pre-existing social, ethnic, or religious tensions, and the lack of political legitimacy of governments (Carius/ Tänzler/Winterstein 2006; Dixon 2009).

From an actor-centred approach, two main causal mechanisms are relevant for the potential links between climate change and violent conflict. In order to start, perpetuate, or join a violent conflict, individuals and social groups must have both the motivation and the capability or opportunity to do so (Scheffran/ Ide/Schilling 2014; Scheffran/Link/Schilling 2012). Motivations may be relevant if climatic changes (higher temperatures, prolonged droughts, or more intense storm events) negatively affect the availability of key resources such as food, land, water, or income. The resulting competition for these resources can take place along pre-existing cleavages and escalate into violence (Homer-Dixon/Blitt 1998). If certain social groups are more strongly affected by natural disasters than others or receive less post-disaster support, relative deprivation can increase grievances and facilitate their violent escalation (Berrebi/Ostwald 2011; Nel/Righarts 2008). Similarly, resource scarcity can lead people to migrate, causing tensions over scarce resources or identity conflicts in the receiving area (Reuveny 2007). Regarding capability and opportunity, climate change may reduce tax revenues due to its negative impact on economic growth and instead increase the amount of money that a state has to spend on infrastructure and disaster relief because of extended damage caused by disasters. Hence, climate change may weaken already fragile states, making them an easier target for rebel groups and reducing their capacity to prevent communal violence (Kahl 2006). Furthermore, climate change potentially reduces recruitment costs for fighters. If, for instance, resource scarcity and natural disasters lead to lower yields and lower incomes, the opportunity costs for people to join a conflict group

decline (Barnett/Adger 2007). Climate change may also diminish the capacity to fight, e.g. by limiting resource abundance, destroying military infrastructure or restricting the mobility of rebel forces (Adano/ Dietz/Witsenburg et al. 2012; Salehyan/Hendrix 2014).

One of the main deficits in the literature on the effects of climate change and violent conflict is the absence of a theoretical basis that would take into consideration the complexities of these issues. One promising approach is to represent the relationship between climate change and violent conflict within a comprehensive framework that allows the integration of the environmental and societal conditions of violent conflict, as well as quantitative and qualitative factors (for a comprehensive analysis see Scheffran/Battaglini 2011; Scheffran/Link/Schilling 2012). In particular, the degree of vulnerability is important in determining the conflict potential of climate change (Ide/Schilling/Link et al. 2014). Vulnerability is defined here as the "predisposition to be adversely affected" (IPCC 2012: 32) by climate change and can be broken down into three factors: i) exposure to climate change, ii) sensitivity to climate change, and iii) adaptive capacity (Adger 2006; IPCC 2007a). Exposure to climate change concerns the frequency and intensity of extreme weather events, the variability of weather parameters such as temperature and precipitation, cloud and wind patterns, and long-term changes in such parameters. These changes act as stressors on natural systems and human societies, which may have direct or indirect impacts on natural resources and related infrastructure (e.g. soil and water, ecosystems, agriculture and land use, forests and biodiversity, energy and economic systems, networks, and infrastructures) that are important for human livelihood, well-being, and survival (IPCC 2014).

Sensitivity is "the degree to which a system is modified or affected by perturbations" (Adger 2006: 270). For instance, a region in which a large percentage of the population depends on rain-fed agriculture is more sensitive to more intense/frequent droughts (as a consequence of climate change) than a region characterized by a strong tertiary sector. Furthermore, vulnerability critically depends on adaptive capacity and how people actually respond to climate stress. Some responses can help adapt and minimize the risks, others may cause new problems, e.g. through mitigation risks, maladaptation, forced migration, or climate engineering (Maas/Scheffran 2012; Scheffran/Cannaday 2013; Tänzler/Maas/Carius 2010). For existential threats the spectrum of responses may be restrained, making non-legal and violent acts more likely, but these could also force people to work together to improve their chances for survival, e.g. through risk reduction, strengthening resilience, and improving sustainability (Feil/Klein/Westerkamp 2009).

Vulnerability and conflict affect each other, implying that the climate-conflict link is not a one-way street. Considering the regional exposure to climate change and violent conflict, there are countries that are sensitive to both violent conflict and future climate change. A comparison of the number of deaths from natural disasters and battle-related deaths in the past shows that both are highest in countries with a low human development index (Scheffran/Brzoska/Kominek et al. 2012b). Inhabitants of these countries already experience increased threats to their lives and health that undermine human development. If climate change adds to these risks and vulnerabilities, it can augment humanitarian crises and aggravate existing conflicts without directly causing them. Thus, it can be argued that violent conflict has a devastating effect that makes societies also more vulnerable to climate change. In turn, climate change can undermine social stability, which increases vulnerability to violent conflict (Scheffran/Ide/Schilling 2014).

While theoretical considerations can help to structure and assess the complex climate-conflict nexus, it is essential to examine in more detail the potential pathways between climate change and violent conflict as well as the regionally specific conditions. To distinguish facts from fiction it is essential to analyse the empirical state of the art and the research deficits, and to make an in-depth assessment of pathways and regional conditions for selected case studies.

12.3   State of the Art

12.3.1   Pathways and Quantitative Evidence

In the research on climate change and conflict, qualitative single-case and quantitative large-N studies are the dominant methods. A few studies have made efforts to explore the middle ground between them and concluded that environmental stress is related to violent conflict, but only under very specific conditions (Bretthauer 2014; Ide 2015). Quantitative large-N studies are currently the most widely accepted methodological approach in the research on climate change and violent conflict, although they face severe problems regarding the quality of their data sets and their ability to capture complex human-nature interac-

tions (Bernauer/Böhmelt/Koubi 2012; Detges 2014; Ide/Scheffran 2014; Scheffran/Brzoska/Kominek et al. 2012a; Vivekananda/Schilling/Smith 2014). While some recent studies suggest a link between water scarcity and military inter-state disputes (Devlin/Hendrix 2014; Hensel/Mitchell/Sowers 2006; Tir/Stinnett 2012), climate change is unlikely to have an impact on inter-state wars in the short to middle term (Gleditsch/Furlong/Hegre et al. 2006; Wolf/Yoffe/Giordano 2003). Regarding intra-state violent conflict, more studies are available but there is less consensus. Particular attention has been paid to the degradation of freshwater resources, the decline in food production, increasing storm and flood disasters, and environmentally-induced migration (Scheffran/Battaglini 2011; WBGU 2008). In the following, we discuss the degree of consensus among quantitative studies on the six pathways connecting climate change and violent conflict. The pathways are structured along the physical impacts of climate change and include higher temperatures, reduced precipitation, more precipitation extremes, lower freshwater availability, land degradation, and climate-related natural disasters (see table 12.1).

12.3.1.1   Temperature Changes

Higher temperatures are suspected of increasing the risk of violent conflict onset because they aggravate water scarcity (through evaporation and the melting of glaciers) or crop failure (through heat and water stress). Several studies in social psychology indicate "that in many settings hot temperatures cause increases in aggression" (Anderson 2001: 37), which may escalate into conflict via spirals of violence or an increased preparedness to fight (Anderson/DeLisi 2011). However, the empirical link between higher temperatures and violent conflict remains disputed, as can be observed in table 12.1. Burke, Miguel, Satyanath et al. (2009) claim a strong link between higher temperatures and more violent conflict but the robustness of their findings has been subject to controversy (Buhaug 2010; Hsiang/Meng 2014). Existing empirical evidence does not allow us to draw clear or universal conclusions on the relationship between higher temperature and violent conflict. Such a link would be quite indirect and is therefore hard to support by qualitative evidence. We are not aware of any concrete empirical cases that justify a significant temperature-violence link. There are some studies that claim a strong and direct link between colder temperatures and war for long historical periods (Tol/Wagner 2010; Zhang/Brecke/Lee et al. 2007) but these studies

Table 12.1: Central findings of large-N studies on possible pathways connecting climate change to violent conflict. **Source:** Expanded from Ide/Scheffran 2014).

Pathway	Increases the risk of violent conflict significantly	Does not significantly increase the risk of violent conflict
higher temperatures	(Burke/Miguel/Satyanath et al. 2009) (Hsiang/Meng/Cane 2011) (Landis 2014) (Maystadt/Ecker 2014) (O'Loughlin/Witmer/Linke et al. 2012) (O'Loughlin/Linke/Witmer 2014) (Yeeles 2015)	(Buhaug 2010) (Couttenier/Soubeyran 2013) (Koubi/Bernauer/Kalbhenn et al. 2012) (Salehyan/Hendrix 2014) (Theisen 2012) (Wischnath/Buhaug 2014)
reduced precipitation	(Bohlken/Sergenti 2010) (Couttenier/Soubeyran 2013) (Ember/Adem/Skoggard et al. 2012) (Fjelde/von Uexkull 2012) (Hendrix/Glaser 2007) (Hendrix/Salehyan 2012) (Hsiang/Meng/Cane 2011) (Maystadt/Ecker 2014) (Miguel/Satyanath/Sergenti 2004) (Raleigh/Choi/Kniveton 2015) (Raleigh/Kniveton 2012) (von Uexkull 2014)	(Buhaug 2010) (Buhaug/Theisen 2012) (Brückner/Ciccone 2010) (Miguel, Edward; Satyanath et al. 2009) (Koubi/ Bernauer/Kalbhenn et al. 2012) (Meier/Bond/Bond 2007) (Nel/Righarts 2008) (O'Loughlin/Witmer/Linke et al. 2012) (O'Loughlin/Linke/Witmer 2014) (Salehyan/Hendrix 2014) (Theisen 2012) (Theisen/Holtermann/Buhaug 2012) (Wischnath/Buhaug 2014) (Witsenburg/Adano 2009) (Yeeles 2015)
more precipitation extremes	(Hendrix/Salehyan 2012) (Raleigh/Kniveton 2012)	(Koubi/Bernauer/Kalbhenn et al. 2012) (O'Loughlin/Linke/Witmer 2014) (Wischnath/Buhaug 2014)
lower freshwater availability	(Gizelis/Wooden 2010) (Hauge/Ellingsen 1998) (Raleigh/Urdal 2007)	(Couttenier/Soubeyran 2013) (Hendrix/Glaser 2007) (Salehyan/Hendrix 2014) (Theisen 2008)
land degradation	(Biermann/Petschel-Held/Rohloff 1998) (Esty/Goldstone/Gurr et al. 1999) (Hauge/Ellingsen 1998) (Raleigh/Urdal 2007) (Theisen 2008)	(Hendrix/Glaser 2007) (Rowhani/Degomme/Guha-Sapir et al. 2011)
climate-related natural disasters	(Berrebi/Ostwald 2011) (Besley/Persson 2011) (Drury/Olson 1998) (Hsiang/Meng/Cane 2011) (Nardulli/Peyton/Bajjalieh 2015) (Nel/Righarts 2008)	(Bergholt/Lujala 2012) (Ghimire/Ferreira/Dorfman 2015) (Omelicheva 2011) (Slettebak 2012)

focus on the pre-industrial and pre-globalization era and are therefore of limited relevance for violent conflicts in the twentieth and twenty-first centuries.

12.3.1.2 Precipitation Reductions

Lower precipitation is associated with a lower availability of water and a lower availability of food due to crop failure. Lack of these resources can stimulate grievances and provide opportunities for violent behaviour as discussed in section 12.2. Table 12.1 indicates that the majority of large-N studies find no link between precipitation reduction and violent conflict onset. However, five recent and methodologically advanced studies do detect such a link (Ember/Adem/Skoggard et al. 2012; Fjelde/von Uexkull 2012; Hendrix/Salehyan 2012; Raleigh/Kniveton 2012; von Uexkull 2014). The debate continues. This is especially the case since many of the studies of precipitation and violence have problems interpreting their results in a meaningful way. To give an example: for Kenya, both Theisen (2012) and Witsenburg and Adano (2009) find that violent conflicts occur primarily in regions and time periods with above-average rainfall. But this does not necessarily support their conclusion that low precipitation and the associated scarcity of pasture and water are unrelated to violent conflict. From the perspective of spatial theory, one would expect people who are negatively affected by reductions in precipitation to migrate into wetter areas and to compete for the remaining resources there (Detges 2014; Rutten/Moses 2014). Similarly, during wetter periods opportunity structures are more favourable for violent cattle raids: cattle are strong, vegetation for hiding and labour surplus are available, rain washes away tracks etc. But that does not exclude the possibility that these raids become necessary in the first place due to the need to restock herds after a severe drought (Schilling/Opiyo/Scheffran et al. 2014).

12.3.1.3 More Precipitation Extremes and Increased Precipitation Variability

For many regions (e.g. northern and eastern Africa) climate models do not necessarily predict an absolute decrease in precipitation, but rather an increase in precipitation variability. This means that instances of heavy rainfall, which may cause floods, landslides, or crop failure, alternate with longer periods of very little rainfall. The potential consequences of such alternating climate patterns include a reduced ability for anticipatory planning (e.g. of stockpiling, sowing, or infrastructure work), an increase in food, water, economic, and cultural insecurity, increased grievances as well as a weakening of the state, and a reduction of opportunity costs for violence. Social scientists have reacted to this discussion and investigated whether more rainfall extremes increase the risk of violent conflict. However, research on this issue is very recent and has so far not produced unambiguous results, as can be deduced from table 12.1.

12.3.1.4 Freshwater Availability

Earlier studies on environmental change, resource scarcity, and violent conflict also focus on freshwater scarcity, which can be caused by climate change as well. However, the degree of scientific consensus on the link between water scarcity and violent conflict is limited, as table 12.1 indicates. Most data on water scarcity are also not exogenous to socio-economic variables such as level of development, resource management practices, or population size, and are usually only available at a national level. In contrast, many recent studies use subnational administrative areas or grid cells as the unit of analysis. This makes it even harder to draw causal inferences from studies on freshwater scarcity and violent conflict (Koubi/Bernauer/Kalbhenn et al. 2012; Meierding 2013).

12.3.1.5 Land Degradation

With the conflict relevance of land degradation, there seems to be the greatest degree of scientific consensus. Five out of seven statistical studies find a link between land degradation and violent conflict onset. The existence of such a link is also claimed by various qualitative studies, which emphasize a link between land/soil degradation, land scarcity, the political instrumentalization of the related grievances, and violent conflict (e.g. Homer-Dixon/Blitt 1998; Kahl 2006). However, four out of the five studies detecting a correlation between soil degradation and violent conflict use the criticized GLASOD data set (Benjaminsen 2008), while the most recent study using a more comprehensive land degradation indicator is unable to confirm the existence of such a link (Rowhani/Degomme/Guha-Sapir et al. 2011). Furthermore, Fearon (2010) and Urdal (2005) find no correlation between per capita cropland scarcity and violent conflict. Per capita cropland scarcity may be a particularly appropriate operationalization since resource scarcity is the supposed mediating mechanism between land degradation and violent conflict, while even a high level of land degradation may be unproblematic if the affected area is sparsely populated.

12.3.1.6 Climate-related Natural Disasters

Natural disasters can destabilize societies with weak economies, mixed political regimes, and pre-existing conflicts. The hypothesized links between natural disasters and climate change are well summarized by Bhavnani (2006: 38):

> Natural disasters in general contribute to conflict because they create competition for scarce resources, exacerbate inequality with the unequal distribution of aid, change power relationships between individuals, groups, and the organizations that serve them, and can create power vacuums and opportunities for warlords to usurp power.

However, table 12.1 indicates a lack of consensus on the link between climate-induced natural disasters and violent conflict. Nearly all of these studies rely on the EM-DAT database (CRED 2014). To be recorded by

EM-DAT, a natural disaster must either cause at least ten fatalities, affect at least a hundred people, cause the declaration of a state of emergency, or cause a call for international assistance. This again raises concerns with regard to endogeneity. Strong states with a wealthy population are less likely to experience violent conflict (Dixon 2009) and they can often also cope better with natural disasters, thus reducing the number of events EM-DAT registers for such countries.

12.3.1.7   Other Pathways

There are no quantitative studies on the link between environmental or climate change-induced migration and violent conflict, presumably because of inadequate or incomplete data (Reuveny 2008). Recent studies suggest considering migration as an adaptive response to climate change that includes the development of capabilities of migrant networks (Black et al. 2011; Scheffran/Marmer/Sow 2012). There is wide agreement in the literature that low economic growth, which is predicted to be a consequence of climate change (IPCC 2014), increases the risk for violent conflict onset (Dixon 2009; Hegre/Sambanis 2006). However, explicit quantitative tests have so far only provided weak support for this pathway (Koubi/Bernauer/Kalbhenn et al. 2012). Finally, climate change is also likely to contribute to food insecurity (IPCC 2014). Several recent empirical studies suggest that higher food prices increase the risk of riots and other forms of violent conflict events (e.g. Rowhani/Degomme/Guha-Sapir et al. 2011; Hendrix/Haggard 2015; Raleigh/Choi/Kniveton 2015; Smith 2014; Wischnath/Buhaug 2014b). However, Sneyd/Legwego/Fraser (2013) show that such forms of violence are triggered by a complex set of political and economic factors rather than by higher food prices alone.

12.3.2   The Climate-Conflict Nexus in Different World Regions

Several regions have emerged as focal points of the qualitative debate on climate change and conflict in recent years, each of which has vulnerabilities specific to itself (Scheffran/Battaglini 2011; WBGU 2008):

a.) Middle East: In early 2011, large protests against the regimes in Egypt, Libya, and Tunisia marked the beginning of the 'Arab Spring'. Some analysts consider rising food and especially bread prices as an important cause of these protests, especially in Egypt. Price spikes were caused by, among other factors, extreme droughts in Russia (Egypt's largest wheat supplier) and China, the world's largest wheat consumer and producer (Johnstone/Mazo 2011; Sternberg 2012). Gleick (2014) recently argued that an exceptional drought in important agricultural regions from 2006 to 2010 fuelled grievances and the protests that escalated into the Syrian civil war. Other researchers remain sceptical and point to political mismanagement as the key reason for the humanitarian crisis succeeding the Syrian drought (de Châtel 2014). The impact of climate change on the Nile (Link et al. 2012) and the Israeli–Palestinian conflict (Brown/Crawford 2009; Feitelson/Tamimi/Rosenthal 2012) is also frequently discussed.

b.) East Africa: The role of drought in causing violent conflict between pastoral groups, or between pastoral and sedentary groups, is also intensely debated for Ethiopia, Kenya, Tanzania, and Uganda. While some studies conclude that such a link exists (Ember/Adem/Skoggard et al. 2012; Schilling/Opiyo/Scheffran 2012; Temesgen 2010), others remain skeptical (Adano et al. 2012; Benjaminsen/Maganga/Abdallah 2009; Eaton 2008; see also 12.4.1). The ethno-political violent conflict in western Sudan is sometimes considered to be "the world's first climate war" (Faris 2007). However, most researchers agree that drought-induced livelihood insecurities, grievances, and migration patterns were at best one among several causes of the civil war in Darfur (Scheffran/Ide/Schilling 2014; Selby/Hoffmann 2014). Other assessments even conclude that the years preceding the onset of violent conflict in 2003 were unusually wet (Brown 2010; Kevane/Gray 2008).

c.) West Africa: In the northern regions of Nigeria, more frequent and/or intense droughts have caused a southward migration of pastoralist groups, and this has contributed to livelihood or cultural conflicts with sedentary farmers in the receiving regions (Nyong/Fiki/McLeman 2006; Obioha 2008). Similar conflict dynamics are described for Mali, although their link to climate change remains disputed (Benjaminsen/Ba 2009; Benjaminsen/Alinon/Buhaug et al. 2012). Other studies focus on the conflict relevance of land tenure systems in Cameroon (Gausset 2007) and food price spikes in Burkina Faso (Engels 2014).

d.) Central Asia: Irrigated agriculture is of great importance in many central Asian states. The water for irrigation is mainly extracted from large rivers (e.g. Syr Darya, Amu Darya), whose flow is likely to be reduced in the long run by glacial melting. The resulting water scarcity has the potential to worsen already existing international and inter-communal tensions in this ethnically heterogeneous and in parts politically unstable region. However, the relevance of water for the dynamics of (violent) conflict in central Asia has been challenged (Bernauer/Siegfried 2012; Bichsel 2009; de Martino/Carlsson/Rampolla et al. 2005; Sehring/Giese 2011).

Other potential climate conflict hot spots include North Africa (Schilling/Freier/Hertig et al. 2012), South Asia (Wischnath/Buhaug 2014), Latin America (Stark/Guillén/Brady 2012), and the Pacific (Weir/Virani 2011). However, more in-depth research on these regions is necessary. Following our review of the qualitative and quantitative evidence in this section, we now present in-depth qualitative analyses of three conflict hot spots and assess how the dynamics of (possible) disputes are influenced by climate change.

12.4   Case Studies

In order to illustrate the conceptual and empirical points above and to complement the extensive discussion of the quantitative literature with further qualitative evidence, we now discuss the existence and relevance of a potential climate-conflict nexus for three case studies. In order to maximize the insights gained, we choose three areas that are quite different with regard to their location, the actors involved, the means of conflict, and the level of violence. Turkana (Kenya) is characterized by frequent armed violence between non-state pastoralist groups, while the disputes about water from the Nile are carried out non-violently between the riparian states. The Israeli–Palestinian conflict cannot be clearly characterized as either intra- or inter-state. It has also experienced, at least in recent years, high levels of structural violence and the erratic use of direct, physical violence. All three cases are also frequently discussed as potential climate security hot spots (Brown/Crawford 2009; O'Loughlin/Witmer/Linke et al. 2012; Piontek 2010; WBGU 2008).

12.4.1   Turkana (Kenya)

Turkana, located in north-western Kenya, is among the country's poorest and economically and politically most marginalized administrative Districts (UNDP 2006, 2010). The dominant livelihood is pastoralism, which is well adapted to the semi-arid to arid climate. Most of the rainfall is received during the 'long rains'

between March and May and the 'short rains' between October and December (McSweeney/New/Lizcano 2008). Because of the region's strong dependence on rain-fed water and pasture resources and the parallel occurrence of violent conflict between pastoral groups, the media tend to draw direct links between drought and conflict in Turkana. In response to the drought of 2009, the Guardian (2009) for example announced "the first climate change conflicts". Scientific studies on pastoral conflict in Kenya suggest that matters are more complex while at the same time acknowledging that climate-influenced resource availability plays a role in pastoral conflicts (Campbell/Dalrymple/Rob et al. 2009; Mkutu 2008; Schilling/Opiyo/Scheffran et al. 2014; Temesgen 2010; Theisen 2012; Witsenburg/Adano 2009). However, the importance of environmental changes and particularly of global climate change as a conflict driver varies across studies. Several studies (Campbell/Dalrymple/Rob et al. 2009; Ember/Adem/Skoggard et al. 2012; Eriksen/Lind 2009; Meier/Bond/Bond 2007; Mkutu 2008; Njiru 2012; Omolo 2010; Temesgen 2010) stress that climate-related and climate-unrelated resource scarcity is a conflict driver while others associate a higher level of violence with more rainfall and hence increased resource availability (Adano/Dietz/Witsenburg et al. 2012; Eaton 2008; Theisen 2012; Witsenburg/Adano 2009). In contrast, it is undisputed that the Turkana engage in violent conflict with other groups and particularly with the Pokot of Kenya and Uganda (Mkutu 2006; Schilling/Opiyo/Scheffran 2012). These conflicts mostly articulate themselves through violent livestock thefts called 'raids' and direct attacks on individuals and even villages (Schilling/Opiyo/Scheffran 2012). Between 2006 and 2009 the local non-governmental organization *Turkana Pastoralist Organization* (TUPADO) recorded 592 raid-related deaths in Turkana (Schilling/Opiyo/Scheffran 2012; TUPADO 2011).

Over the past fifty years Kenya has experienced a warming of 1°C, which is well above the global average (IPCC 2007b). Local ground measures from the meteorological station in Lodwar suggest that the temperature increase is even stronger in Turkana (Schilling/Opiyo/Scheffran et al. 2014).

Precipitation data for the past fifty years show no clear trend for Kenya (Eriksen/Lind 2009; Tutiempo 2011). However, the proportion of rain falling in heavy rainfall events has increased (McSweeney/New/Lizcano 2008). Some authors (e.g. Mude/Barrett/Carter et al. 2009) argue that the frequency of droughts has increased in northern Kenya. For Tur-

Figure 12.1: Overview of the case study areas. **Source:** The authors.

Case studies ■ Israel and Palestine   ■ Nile Basin States   ■ Turkana (Kenya)

kana, Opiyo, Wasonga, Nyangito et al. (2014) find that the occurrence of extreme droughts has increased since the 1990s. While climate projections for eastern Africa are less reliable than for other parts of Africa (IPCC 2013; Klehmet 2009), the majority of models suggest a strong increase in temperature and rainfall variability (IPCC 2007b, 2013; Schilling/ Opiyo/Scheffran et al. 2014). Both trends are likely to increase the drought frequency and possibly also the severity of droughts.

While droughts have the potential to significantly decrease herds on occasion, the increased rainfall variability is likely to decrease the predictable availability of water and pasture in general. Increased drought risk and fewer reliable resources are likely to affect the pastoral conflicts over water, pasture, land, and territory. Based on interviews with members of the Turkana and Pokot communities and an analysis of raiding records and climate data, Schilling, Opiyo, Scheffran et al. (2014) have developed the *Resource Abundance and Scarcity Threshold* (RAST) hypothesis, which suggests that droughts increase the pressure to compensate own livestock losses through raiding. But the raiding to restock is mostly conducted in and around the rainy seasons because the raided animals are healthier, they can travel longer distances, and raiders find cover in the higher vegetation. In contrast, raiding during dry seasons is only conducted when a certain threshold of resource scarcity is reached. During this time, raiding serves the purpose of gaining or securing control over scarce pasture and water resources rather than restocking. The RAST hypothesis hence suggests that climatic changes have and are likely to continue to aggravate the violent conflicts between the Turkana and the Pokot.

However, this conclusion does not mean that climate change is an actual cause of conflict. Political and economic marginalization, including the limitation of pastoral movement and the inability of the central government to provide security for the pastoral communities in northern Kenya, are key conflict causes that are then exacerbated by climate-related resource scarcities (Schilling/Opiyo/Scheffran 2012; Schilling/Opiyo/ Scheffran et al. 2014). Recently, major oil reserves have been found in the conflict zone between the Turkana and the Pokot (Tullow 2014). While oil exploitation has aggravated pastoral conflicts in Sudan (Chavunduka/Bromley 2011), it is too early to tell if oil is not only driving global climate change but also local conflicts in Turkana.

12.4.2   Nile River Basin

The Nile River Basin is a region with a cultural history that dates back to the fourth millennium BCE. Since then it has served as a lifeline for the entire region. As there are no other water sources, riparians are often critically dependent on the Nile waters, particularly in Egypt which, together with Sudan, Ethiopia, and Uganda, is the key player in the Nile Basin when it comes to the sharing of the waters of the river. The balance of power that can be observed today is the result of a long set of developments in the past that cemented Egypt's dominance in the area. However, this dominance is challenged by the other riparians of the Nile who are economically developing quickly. In conjunction with a continued strong population increase in the region and progressing climate change (IPCC 2013), there is a distinct possibility that conflicts may erupt over the distribution of the limited Nile waters among the riparians. On the other hand, there are basin-wide efforts to cooperate on water issues, and

these have led to the formation of the *Nile Basin Initiative* (NBI) (Peichert 2003). Here, we want to address the boundary conditions that govern the distribution of this scarce resource among the many riparians.

Current disputes in the Nile Basin about water allocation are based on the historic asymmetry between the downstream country Egypt, which can be considered the hydro-hegemon (Cascão 2009; Zeitoun/Warner 2006) of the basin, on the one side, and the upstream states on the other side. Hydro-hegemony rests on three pillars: the location of the country, its military, economic, ideational, political, and bargaining power, and its exploitation potential. Even though Egypt's location far downstream puts it at a disadvantage, it is the dominant country when it comes to power and exploitation potential. This status of Egypt has grown historically and is based on the extensive external support it has received from Great Britain in colonial times and from other countries in subsequent decades, e.g. for the construction of the Aswan High Dam (Allan 2009; Link/Piontek/Scheffran et al. 2012; Piontek 2010). Egypt is highly dependent on the Nile waters since there are practically no other sources of renewable water in the country (Schilling/Freier/Hertig et al. 2012; Waterbury 1979). Consequently, it has a strong interest in preserving its 'historic' rights to the river's water in treaties.

The two fundamental treaties initially governing the water allocation of the Nile date back to 1929 and 1959. The first was a treaty between the United Kingdom, Egypt, and Sudan, and institutionalized "the Egyptian belief in 'natural and historic rights' to the Nile water" (Cascão 2009: 245). An adjustment was necessary because of the construction of the Aswan High Dam in 1959. The resulting agreement implicitly left no water to the upstream countries of the Nile. Therefore, this treaty manifested a strong division between the upstream and downstream regions of the Nile Basin (Waterbury 1979).

Since the beginning of the twenty-first century, there have been extensive efforts by all riparians to overcome this division and to foster cooperation. The NBI has become an institutional framework for many cooperative projects (Mekonnnen 2010). Furthermore, a Cooperative Framework Agreement was negotiated and signed by the upstream countries Ethiopia, Uganda, Rwanda, and Tanzania, followed by Kenya (Al Masry/Al Youm 2010). With the signature of Burundi in February 2011 as the sixth country, the ratification of the agreement now could proceed despite Egypt's and Sudan's continued opposition (World Politics Review 2011). This development highlights the upstream countries' increasing emancipation and their challenge to Egypt's status as hydro-hegemon. In addition, the Arab Spring that has triggered considerable political change in Egypt since 2011 has created the opportunity to alter the Egyptian approach towards regional integration and cooperation over the Nile waters. Also, the independence of South Sudan in July 2011 has the potential of influencing upstream-downstream relations by possibly leading to new coalitions.

Current struggles for adequate water allocation schemes in the Nile Basin occur in the wake of large challenges regarding demographic and socio-economic development, which steadily increase the pressure on the Nile as a water resource, while there is no basin-wide agreement on Nile water use yet and a situation of low-level conflict remains (Cascão 2009). Under these circumstances, climate change imposes an additional pressure on water resources and augments the effects of other current challenges that could possibly trigger conflict and social destabilization (Goulden/Conway/Persechino 2009; Kim/Kaluarachchi 2009). One important issue resulting from altered temperature and precipitation patterns is the changed agricultural yield (Agrawala/Moehner/El Raey et al. 2004; Müller/Cramer/Hare et al. 2011). Agriculture is likely to become even more dependent on irrigation, which in turn reduces the possibilities of allocating the water to other purposes. Therefore, measures need to be taken to offset this higher vulnerability; one possibility is the sharing of benefits from water use among riparians instead of sharing the water itself (Rahman 2013). However, this appears to be difficult to achieve because of lack of political will to cooperate and differences in socio-economic backgrounds among the riparians.

So even without altered environmental conditions due to climate change, the chances for cooperation among riparians are already hard to assess. The additional influence of climate change on cooperation in the Nile Basin is still uncertain (Kim/Kaluarachchi 2009; Link/Piontek/Scheffran et al. 2012). This has to do with the fact that it is not yet clear how climate change will affect the water availability in the basin in the future. Climate models disagree on whether the overall water availability in the Nile will increase or decrease or whether merely the variability of precipitation will increase. The key area for the water supply of the Nile Basin is the Ethiopian highlands (Kim/Kaluarachchi 2009). This makes it hard to define adequate measures to increase the adaptive capacity of the region. Also, the riparians' actions may be guided by short-term political considerations despite the fact that effective climate change adaptation is likely contingent on strong cooperation. Reduced water availability may increase vulnerability, conflict, and barriers to cooperation but simultaneously presses riparians to cooperate for the most efficient use of the available water (Solomon 2014). And there is hope that in the long run cooperation will prevail, as it is much more likely in transboundary watersheds than conflict (Wolf 1998; Yoffe/Wolf/Giordano 2003). In contrast, the short-term impact of changing environmental conditions on the riparian relations is likely to result in an enhanced securitization of water. This would lead to increased distrust among the riparians (Stetter/Herschinger/Teichler et al. 2011) and is thus a threat to basin-wide cooperation as it emphasizes unilateral action rather than joint efforts (Link/Piontek/Scheffran et al. 2012).

In addition to water allocation issues, Egypt is potentially exposed to and adversely affected by another kind of climate change impact: sea level rise. Some of the coastal zones of Egypt are artificially protected, and large parts of the coast rely on natural protective measures. Nonetheless, some of the low-lying areas are hardly protected and thus highly vulnerable (Frihy/El-Sayed 2013; Link/Kominek/Scheffran 2013). Already, a moderate increase in sea level can cause extensive economic damage (Dasgupta/Laplante/Murray et al. 2011) and force millions of people to adapt or relocate, particularly in metropolitan areas such as Alexandria (El Raey 1997). The current state of Egyptian coastal protection is insufficient to counter the expected future rise in sea level in the eastern Mediterranean (Dasgupta/Laplante/Meisner et al. 2009).

For this reason, coastal protection should become a more important issue on the agenda of the country (Elsharkawy/Rashed/Rached 2009), but the current political instability in Egypt actually causes environmental protection to be often disregarded in comparison with civil rights and economic development (Dasgupta/Laplante/Meisner et al. 2009). Also, coastal protection necessitates cooperation at all administrative levels and an increase in civil awareness (Malm 2013), both of which are aspects that are far from being realized (Link/Kominek/Scheffran 2013). All in all, the multiple facets of climate change will place a significant burden on the adaptive capacities of the riparians of the Nile Basin. It will take a joint cooperative effort to effectively deal with these challenges as failures of states would result in increased basin-wide conflict.

12.4.3   Israel–Palestine

The Middle East in general and Israel/Palestine in particular has often been characterized as a region in which climate change has the potential to worsen existing tensions and to create new ones (e.g. Amery 2002; Brown/Crawford 2009; Scheffran/Battaglini 2011). In general, the dispute between Israel and Palestine is a political conflict (Moore/Guy 2012). However, the conflict has a water dimension as well. Water has been a contested issue in past Israeli-Palestinian negotiations and it is indeed one of the few issues on which no final agreement could be reached during the peace process in the 1990s (Lautze/Reeves/Vega et al. 2005). Water distribution between both sides is also highly unequal and disputed. According to the 1995 Oslo II agreement between the government of Israel and the *Palestinian Liberation Organization* (PLO), Israel is allowed to withdraw 912 million cubic metres of water per year (MCM/y) from the shared mountain and coastal aquifers. Palestinians are only entitled to withdraw 253 MCM/y (Zeitoun 2008: 48, see below for a per capita estimate). Furthermore, Palestinians have no access to the River Jordan at all.

Due to the ongoing Israeli military occupation of the West Bank and the legal framework established by the Oslo II accord, the Israeli administration can block the development of water infrastructure or the increase of well extraction quotas in the West Bank (Selby 2003). It frequently uses this right among other reasons because it is the lower riparian with regard to the mountain aquifer shared between Israel and the West Bank (Selby 2013). Regarding the heavily overexploited coastal aquifer shared between Israel and the Gaza Strip, Israel is the upper riparian. Altogether, fresh water availability per capita is approximately four times higher in Israel than in Palestine and the gap is even larger for agricultural water. Consequently, many Palestinians have to survive with less than the 100 litres per day recommended by the *World Health Organization* (WHO) (Daoudi 2009).

Israel and Palestine experienced unusually little rainfall during the first decade of the twenty-first century when compared with historical records. After two very wet years in 2012 and 2013, a major drought hit the region again in early 2014. Climate models suggest that temperatures in the region will rise; this is one of the factors contributing to the region's vulnerability. Rainfall will become less predictable and more erratic, although it is not clear yet whether the absolute amount of precipitation will increase or decrease. Salt water intrusion into the coastal aquifer is expected due to sea level rise, which will cause a further decrease of the already poor quality of Gaza's only water source. However, most experts agree that climate change is unlikely to cause tremendous changes to the regional water quality or quantity before 2030 (Feitelson/Tamimi/Rosenthal 2012; Messerschmid 2012).

Several analysts predict that a decrease in water availability and quality will worsen existing water-related grievances and tensions and thus contribute to the escalation of the Israeli-Palestinian conflict. Based on the above discussion, we cannot deny this possibility. But it is important to emphasize that the Israeli-Palestinian water conflict is not primarily based on the physical scarcity of water as a resource. The amount of water available in the region has increased considerably in recent years due to ambitious Israeli seawater desalination and wastewater recycling programs (Aviram/Katz/Shmueli 2014; Feitelson/Rosenthal 2012; Fischhendler/Dinar/Katz 2011). Similarly, virtual water can be and is imported in large quantities in the form of food (Shuval 2007). In spite of these developments the water conflict did not de-escalate but rather became fiercer (Selby 2013). This is the case because confrontative discourses and the unequal power structures are the major drivers of the Israeli-Palestinian water conflict.

In the dominant Israeli discourse, the water resources of the region are considered limited although the overall water availability can be increased through technological measures. Water is also perceived as essential because of its importance for agriculture, which is strongly linked to the Zionist ideas of settling the land, making the desert bloom, and forming a secure and self-sufficient Jewish homeland. Furthermore, water is securitized in the sense that giving up control over the region's water resources is unthinkable in the dominant Israeli discourse, mainly because Israel considers itself to be surrounded by enemies (Fröhlich 2012; Lipchin 2007). In the dominant Palestinian discourse, exactly this Israeli control over water resources is constructed as an existential threat. Water is perceived to be intrinsically tied to land and without control over sufficient land/water resources a viable Palestinian state is considered impossible. According to the dominant Palestinian discourse, water scarcity in the region is not natural but politically induced by unfair Israeli policies (Daoudi 2009; Fröhlich 2012). These contradictory and rather confrontative discursive constructions of the same physical water reality are a major driver of the Israeli-Palestinian water conflict. However, this symbolic confrontation does not take place in a vacuum. Rather, Israel is the dominant military and economic power in the region and its depiction of the regional water situation is internationally much more accepted (Zeitoun 2008). On the national as well as on the international level, the costs for the Israeli government of maintaining the highly unequal water distribution and governance are rather low. Israel is likely to benefit from this situation, e.g. through higher water availability and political support by settler groups (Messerschmid 2012).

In summary, water is certainly an important although not the most important factor in the Israeli-Palestinian conflict while climate change is likely to aggravate the regional water situation. The main drivers of the water conflict, however, are not physical resource scarcity and declining water quality but confrontative discourses and asymmetric power and incentive structures. The potential of climate change to act as a threat multiplier in the Israeli–Palestinian conflict is thus rather limited. These findings also highlight the importance of considering factors such as discourses, identities, and power structures as crucial elements of vulnerability vis-à-vis climate change and violent conflict (Adger/Dessai/Goulden et al. 2009; Jabri 1996).

12.5   Conclusion

The links between climate change and violent conflict remain disputed. Depending on the spatial and temporal scales and the data sets used, large-N studies provide different and at times opposing conclusions regarding the possible impacts of higher temperatures, less precipitation, larger precipitation extremes, lower freshwater availability, soil degradation, or natural disasters on violent conflict onset. Qualitative studies also provide no clear picture. This also holds true for the cases presented in this chapter. Climate change in combination with several other political, social, and economic trends seems to influence the intensity of pastoralist violence in north-western Kenya. Such a nexus has not yet manifested in the Nile Basin although it will take coordinated efforts by the major riparian states to adapt to climate change in order to avoid increasing conflict in the future. With regard to water in Israel and Palestine, it can be said that confrontational discourses and asymmetric power structures are much more important for the course of the conflict than the amount of water physically available, which is predicted to be reduced due to climate change.

Possible reasons for this lack of scientific consensus may be the difficulties of existing approaches to adequately capture the complex links between climate

change vulnerability and violent conflict in theory and to operationalize them in empirical research. Particularly the different operationalizations of the dependent variables (onset or incidence of civil war, armed conflict, or social violence) and the independent variables, and the different units of analysis (countries, administrative units, or grid cells spatially, and decades, years, or months temporally) require nuanced and tailor-made theoretical approaches (Landis 2014; Salehyan 2014). The onset of civil war at the country level with an annual temporal resolution, for instance, is very likely to be driven by other factors and dynamics than the persistence of communal violence on the administrative unit level with a monthly temporal resolution. Other important issues include the poor availability and quality of data sets and deficits regarding the development and combination of adequate methodological approaches (Ide/Scheffran 2014; Selby 2014). However, it is also possible that the existence and nature of a climate-conflict nexus may vary within short time periods and geographically at the local level, so that general conclusions on the issue are hardly possible. And finally, as climate change progresses, the relevance of conflict over climate change adaptation and mitigation measures (Dabelko/Herzen/Null et al. 2013) and of geoengineering (Maas/Scheffran 2012) increases, while the consequences of climate change for violent conflict in (peri-)urban areas (Saha 2012) and the impact of high-impact non-linear events may become more likely (see chapter 11 in this volume by Schellnhuber/Serdeczny/Adams et al.). These issues are far from being understood in detail. The possible links and interactions between climate change and violent conflict will continue to be an important issue in scientific as well as in policy debates. In order to allow the readers to better keep track of these debates, box 12.1 provides an overview of the websites we consider most relevant in this research field.

Finally, we think it is important to mention the growing literature on environmental peace (Ide/Scheffran 2014), environmental peacemaking (Conca/Dabelko 2002), disaster diplomacy (Kelman 2012), and natural resource management (Ratner/Meinzen-Dick/May et al. 2013). This research suggests that the predicted impacts of climate change can under certain circumstances provide incentives for social groups to overcome their differences and cooperate in order to cope with and adapt to common threats such as storms, floods, or droughts. Not much is known about this climate-cooperation nexus. However, it provides additional reasons that with progressing climate change we are not necessarily heading towards a

---

Box 12.5: Relevant websites for the discussion about climate change, the environment, conflict and cooperation.

http://www.newsecuritybeat.org/
The blog of the Wilson Center's Environmental Change and Security Program. Probably the most comprehensive website on environmental change, population, conflict, and security.

http://www.ecc-platform.org/
Website of the Environment, Conflict and Cooperation platform.

http://www.climateandsecurity.org
Website of the Center for Climate & Security.

http://www.environmentalpeacebuilding.org/
Website of the Environmental Peacebuilding Initiative.

http://www.clisec-hamburg.de/
Website of the Research Group Climate Change and Security (CLISEC), to which the authors of this chapter belong.

http://www.g-feed.com/p/about-us_22.html
A blog on the interaction between global food, environment and economic dynamics. Two of the leading climate-conflict scholars, Solomon Hsiang and Marshall Burke, regularly write for the website.

http://www.prio.org/Projects/Project/?x=905
Home page of the project Security Implications of Climate Change at the Peace Research Institute Oslo (PRIO).

http://www.wiso.uni-tuebingen.de/faecher/ifp/lehrende/ipol/forschungsprojekte/climasec.html
Website of the ClimaSec Project, which investigates the securitization of climate change.

http://www.clico.org/
Website of the research project Climate Change, Hydro-conflicts and Human Security

http://www1.american.edu/ted/ice/subic.htm
Website of the Inventory of Conflict & Environment, which contains around 200 case studies on environmental conflict from all regions of the world.

http://worldwater.org/water-conflict/
Contains an extensive chronology of water conflicts, generated by Peter Gleick.

http://www.springer.com/series/8090
An overview of the Hexagon Series on Human and Environmental Security and Peace, which contains many publications on environmental/climate change and conflict.

---

more violent world, but that severe environmental challenges can instead have the potential to facilitate transitions towards sustainable livelihoods and sustainable peace.

References:

Adano, Wario; Dietz, Ton; Witsenburg, Karen M.; Zaal, Fred, 2012: "Climate change, violent conflict and local institutions in Kenya's dryland", in: *Journal of Peace Research*, 49,1: 65-80.

Adger, W. Neil, 2006: "Vulnerability", in: *Global Environmental Change*, 16,3: 268-281.

Adger, W. Neil; Dessai, Suraje; Goulden, Marisa; Hulme, Mike; Lorenzoni, Irene; Nelson, Donald R.; Naess, Lars Otto; Wolf, Johanna; Wreford, Anita, 2009: "Are there social limits to adaptation to climate change?", in: *Climatic Change*, 93,3-4: 335-354.

Agrawala, Shardul; Moehner, Annett; El Raey, Mohamed; Conway, Declan; van Aalst, Maarten; Hagenstad, Marca; Smith, Joel, 2004: *Development and climate change in Egypt: focus on coastal resources and the Nile* (Paris: OECD).

Al Masry; Al Youm, 2010: "Kenya signs Nile basin deal, rules out discussion with Mubarak"; at: <http://www.masress.com/en/almasryalyoumen/42678> (29 November 2014).

Amery, Hussein, 2002: "Water wars in the Middle East: a looming threat", in: *The Geographical Journal*, 168,4: 313-323.

Anderson, Craig, 2001: "Heat and violence", in: *Current Directions in Psychological Science*, 10,1: 33-38.

Anderson, Craig; DeLisi, Matt, 2011: "Implications of global climate change for violence developed and developing countries", in: Forgas, Joseph; Kruglanski, Arie; Wil-

liams, Kipling (Eds.): *The Psychology of Social Conflict and Aggression* (New York: Psychology Press): 249-265.

Barnett, Jon; Adger, W. Neil, 2007: "Climate change, human security and violent conflict", in: *Political Geography*, 26,6: 639-655.

Baron, Robert A., 1990: "Conflict in organizations", in: Murphy, Kevin R.; Saal, Frank E. (Eds.): *Psychology in organizations: integrating science and practice* (Hillsdale: Lawrence Erlbaum): 197-216.

Benjaminsen, Tor A., 2008: "Does supply-induced scarcity drive violent conflicts in the African Sahel? The case of the Tuareg rebellion in northern Mali", in: *Journal of Peace Research*, 45,6: 819-836.

Benjaminsen, Tor A.; Ba, Boubacar, 2009: "Farmer-herder conflicts, pastoral marginalisation and corruption: a case study from the inland Niger delta of Mali", in: *Geographical Journal*, 175,1: 71-81.

Benjaminsen, Tor; Alinon, Koffi; Buhaug, Halvard; Buseth, Jill Tove, 2012: "Does climate change drive land-use conflict in the Sahel?", in: *Journal of Peace Research*, 49,1: 97-111.

Bergholt, Drago; Lujala, Paivi, 2012: "Climate-related natural disasters, economic growth, and armed civil conflict", in: *Journal of Peace Research*, 49,1: 147-162.

Berrebi, Claude; Ostwald, Jordan, 2011: "Earthquakes, hurricanes, and terrorism: do natural disasters incite terror?", in: *Public Choice*, 149,3: 383-403.

Besley, Timothy; Persson, Torsten, 2011: "The logic of political violence", in: *The Quarterly Journal of Economics*, 126,3: 1411-1445.

Bhavnani, Rakhi, 2006: "Natural disasters conflicts"; at: <http://rakhibhavnani.ca/ bhavnanisummary.pdf> (30 October 2011).

Biermann, Frank; Petschel-Held, Gerhard; Rohloff, Christian 1998: "Umweltzerstörung als Konfliktursache? Theoretische Konzeptionalisierung und empirische Analyse des Zusammenhangs von 'Umwelt' und 'Sicherheit'", in: *Zeitschrift für Internationale Beziehungen*, 5,2: 273-308.

Black, Richard; Bennett, Stephen; Thomas, Sandy M.; Beddington, John R., 2011: "Migration as adaptation", in: *Nature*, 478,7370: 447-449.

Bohlken, Anjali Thomas; Sergenti, Ernest John, 2010: "Economic growth and ethnic violence: an empirical investigation of Hindu-Muslim riots in India", in: *Journal of Peace Research*, 47,5: 589-600.

Bonacker, Thorsten; Imbusch, Peter, 2010: "Zentrale Begriffe der Friedens- und Konfliktforschung: Konflikt, Gewalt, Krieg, Frieden", in: Imbusch, Peter; Zoll, Ralf (Eds.): *Friedens- und Konfliktforschung: eine Einführung* (Wiesbaden: VS Springer): 67-142.

Brauch, Hans Günter; Scheffran, Jürgen, 2012: "Introduction: climate change, human security, and violent conflict in the Anthropocene", in: Scheffran, Jürgen; Brzoska, Michael; Brauch, Hans Günter; Link, Peter Michael; Schilling, Janpeter (Eds.): *Climate change, human security and violent conflict: challenges for societal stability* (Berlin–Heidelberg: Springer): 3-40.

Bretthauer, Judith M., 2015: "Conditions for peace and conflict: applying a fuzzy-set Qualitative Comparative Analysis to cases of resource scarcity", in: *Journal of Conflict Resolution*, 59,4: 593-616.

Brown, Ian A., 2010: "Assessing eco-scarcity as a cause of the outbreak of conflict in Darfur: a remote sensing approach", in: *International Journal of Remote Sensing*, 31,10: 2513-2520.

Brown, Oli; Crawford, Alec, 2009: *Rising temperatures, rising tensions: climate change and the risk of violent conflict in the Middle East.* (Winnipeg: International Institute for Sustainable Development).

Brückner, Markus; Ciccone, Antonio 2010: "International commodity prices, growth, and the outbreak of civil war in sub-Saharan Africa", in: *The Economic Journal*, 120, 5: 519-534.

Buhaug, Halvard, 2010: "Climate not to blame for African civil wars", in: *PNAS*, 107,38: 16477-16482.

Buhaug, Halvard; Nordkvelle, Jonas; Bernauer, Thomas Bernauer; Böhmelt, Tobias; Brzoska, Michael; Busby, Joshua W.; Ciccone, Antonio; Fjelde, Hanne; Gartzke, Erik; Gleditsch, Nils Petter; Goldstone, Jack A.; Hegre, Håvard; Holtermann, Helge; Koubi, Vally; Link, P. Michael ; Link, Jasmin S. A.; Lujala, Päivi; O'Loughlin, John; Raleigh, Clionadh; Scheffran, Jürgen; Schilling, Janpeter; Smith, Todd G.; Theisen, Ole Magnus; Tol, Richard S. J.; Urdal, Henrik; von Uexkull, Nina, 2014:

"One effect to rule them all? A comment on climate and conflict", in: *Climatic Change*, 127,3-4: 391-397.

Buhaug, Halvard; Theisen, Ole Magnus, 2012: "On environmental change and armed conflict", in: Scheffran, Jürgen; Brzoska, Michael; Brauch, Hans Günter; Link, Peter Michael; Schilling, Janpeter (Eds.): *Climate Change, Human Security and Violent Conflict: Challenges for Societal Stability* (Berlin: Springer): 43-55.

Burke, Marshall B.; Miguel, Edward; Satyanath, Shanker; Dykema, John A.; Lobell, David B., 2009: "Warming increases the risk of civil war in Africa", in: *PNAS*, 106,49: 20670-20674.

Bush, Ray, 2010: "Food riots: poverty, power and protest", in: *Journal of Agrarian Change*, 10,1: 119-129.

Campbell, Ivan; Dalrymple, Sarah; Rob, Craig; Crawford, Alec, 2009: *Climate change and conflict: lessons from community conservancies in northern Kenya* (Winnipeg: IISD).

Carius, Alexander; Tänzler, Dennis; Winterstein, Judith, 2006: *Weltkarte von Umweltkonflikten* (Berlin: Adelphi).

Cascão, Ana Elisa, 2009: "Changing power relations in the Nile river basin: unilateralism vs. cooperation?", in: *Water Alternatives*, 2,2: 245-268.

Chavunduka, Charles; Bromley, Daniel, 2011: "Climate, carbon, civil war and flexible boundaries: Sudan's contested landscape", in: *Land Use Policy*, 28, 4: 907-916.

CNA Military Advisory Board, 2014: *National security and the accelerating risks of climate change* (Alexandria: CNA Corporation).

Conca, Ken; Dabelko, Geoffrey D. (Eds.), 2002: *Environmental peacemaking* (Baltimore: Johns Hopkins University Press).

Cotula, Lorenzo, 2012: "The international political economy of the global land rush: a critical appraisal of trends, scale, geography and drivers", in: *Journal of Peasant Studies*, 39,3-4: 649-680.

Couttenier, Mathieu; Soubeyran, Raphael, 2013: "Drought and civil war in sub-Saharan Africa", in: *The Economic Journal*, 124,1: 201-244.

CRED, 2014: "EM-DAT: the OFDA/CRED international disaster database", <http://www.emdat.be/> (17 July 2014).

Dabelko, Geoffrey D.; Herzen, Lauren; Null, Schuyler; Parker, Meaghan; Sticklor, Russell (Eds.), 2013: *Backdraft: the conflict potential of climate change adaptation and mitigation.* (Washington DC: Woodrow Wilson Center).

Daoudi, Mohammed Dajani, 2009: "Conceptualization and debate on environmental and human security in Palestine", in: Brauch, Hans Günter; Oswald Spring, Úrsula; Grin, John; Mesjasz, Czeslaw; Kameri-Mbote, Patricia; Chadha Behera, Navnita; Courou, Béchir; Krummenacher, Heinz (Eds.): *Facing global environmental change: environment, human, energy, food, health and water security concepts* (Berlin–Heidelberg: Springer): 873-883.

Dasgupta, Susmita ; Laplante, Benoit; Meisner, Craig; Wheeler, David; Yan, Jianping, 2009: "The impact of sea level rise on developing countries: a comparative analysis", in: *Climatic Change*, 93,3-4: 379-388.

Dasgupta, Susmita; Laplante, Benoit; Murray, Siobhan; Wheeler, David, 2011: "Exposure of developing countries to sea-level rise and storm surges", in: *Climatic Change*, 106,4: 567-579.

de Châtel, Francesca, 2014: "The role of drought and climate change in the Syrian uprising: untangling the triggers of the revolution", in: *Middle Eastern Studies*, 50,4: 521-535.

de Martino, Luigi; Carlsson, Annica; Rampolla, Gioanluca; Kadyrzhanova, Inkar; Svedberg, Peter; Denisov, Nickolai; Novikov, Viktor; Rekacewitcz, Philippe; Simonett, Otto; Skalvik, Janet Fernandez; del Pietro, Dominique; Rizollio, Diana; Palosaari, Marika, 2005: *Transforming risks into cooperation: Central Asia, Ferghana/Osh/Khujand area* (Geneva: ENVSEC).

Detges, Adrien, 2014: "Close-up on renewable resources and armed conflict: the spatial logic of pastoralist violence in northern Kenya", in: *Political Geography*, 42,1: 57-65.

Dietz, Kristina; Engels, Bettina, 2014: "Immer (mehr) Ärger wegen der Natur?—Für eine gesellschafts- und konflikttheoretische Analyse von Konflikten um Natur", in: *Österreichische Zeitschrift für Politikwissenschaft*, 43, 1: 73-90.

Dixon, Jeffrey, 2009: "What causes civil war? Integrating quantitative research findings", in: *International Studies Review*, 11,4: 707-735.

Drury, A. Cooper; Olson, Richard Stuart, 1998: "Disasters and political unrest: an empirical investigation", in: *Journal of Contingencies and Crisis Management*, 6,3: 153-161.

Eaton, David, 2008: "The business of peace: raiding and peace work along the Kenya-Uganda border (part I)", in: *African Affairs*, 107,426: 89-110.

El Raey, Mohamed, 1997: "Vulnerability assessment of the coastal zone of the Nile delta of Egypt to the impacts of sea level rise", in: *Ocean & Coastal Management*, 37,1: 29-40.

Elsharkawy, H.; Rashed, H.; Rached, I., 2009: "Climate change: the impacts of sea level rise on Egypt", Paper for the 45th ISOCARP Congress, Porto, 18-22 October 2009.

Ember, Carol C.; Adem, Teferi Abate; Skoggard, Ian; Jones, Eric C., 2012: "Livestock raiding and rainfall variability in northern Kenya", in: *Civil Wars*, 14,2: 159-181.

Engels, Bettina, 2014: "Contentious politics of scale: the global food price crisis and local protest in Burkina Faso", in: *Social Movement Studies*, online first. DOI: 10.1080/14742837.2014.921148.

Eriksen, Siri; Lind, Jeremy, 2009: "Adaptation as a political process: adjusting to drought and conflict in Kenya's drylands", in: *Environmental Management*, 43,5: 817-835.

Esty, Daniel; Goldstone, Jack; Gurr, Ted Robert; Harff, Barbara; Levy, Marc; Dabelko, Geoffrey; Surko, Pamela; Unge, Allan, 1999: "State failure task force report: phase II findings", in: *Environmental Change and Security Project Report*, 5,1: 49-72.

Faris, Stephen, 2007: "The real roots of Darfur", in: *The Atlantic Monthly*, 2007, April: 2.

Fearon, James D., 2010: Governance and civil war onset, WDR 2011 background paper (Washington DC: World Bank).

Feil, Moira; Klein, Diana; Westerkamp, Meike, 2009: *Regional cooperation on environment, economy and natural resource management: how can it contribute to peacebuilding?* (Brussels: Initiative for Peacebuilding).

Feitelson, Eran; Tamimi, Abdelrahman; Rosenthal, Gad, 2012: "Climate change and security in the Israeli-Palestinian context", in: *Journal of Peace Research*, 49,1: 241-257.

Fjelde, Hanne; von Uexkull, Nina, 2012: "Climate triggers: rainfall anomalies, vulnerability and communal conflict in sub-Saharan Africa", in: *Political Geography*, 31,7: 444-453.

Frihy, Omran E.; El-Sayed, Mahmoud Kh., 2013: "Vulnerability risk assessment and adaptation to climate change induced sea level rise along the Mediterranean coast of Egypt", in: *Mitigation and Adaptation Strategies for Global Change*, 18,8 : 1215-1237.

Fröhlich, Christiane, 2012: "Security and discourse: the Israeli-Palestinian water conflict", in: *Conflict, Security & Development*, 12,2: 123-148.

Gausset, Quentin, 2007: "Land tenure and land conflicts among the Adamawa, Northern Cameroon", in: Derman, Bill; Odgaard, Rie; Sjaastad, Espen (Eds.): *Conflicts over land and water in Africa* (Oxford: James Currey).

Ghimire, Ramesh; Ferreira, Susana; Dorfman, Jeffrey H., 2015: "Flood-induced displacement and civil conflict", in: *World Development*, 66,1: 614-628.

Gizelis, Theodora-Ismene; Wooden, Amanda E., 2010: "Water resources, institutions, & intrastate conflict", in: *Political Geography*, 29,8: 444-453.

Gleditsch, Nils Petter; Furlong, Kathryn; Hegre, Håvard; Lacina, Bethany; Owen, Taylor, 2006: "Conflicts over shared rivers: Resource scarcity or fuzzy boundaries?", in: *Political Geography*, 25,4: 361-382.

Gleditsch, Nils Petter; Nordås, Ragnhild, 2014: "Conflicting messages? The IPCC on conflict and human security", in: *Political Geography*, 43,1.

Gleick, Peter, 2014: "Water, drought, climate change, and conflict in Syria", in: *Weather, Climate, and Society*, 6,3: 331-340.

Goulden, Marisa; Conway, Declan; Persechino, Aurelie, 2009: "Adaptation to climate change in international river basins in Africa: a review", in: *Hydrological Sciences*, 54, 5: 805-828.

Guardian, 2009: "The first climate change conflicts", <http://www.guardian.co.uk/journalismcompetition/

professional-climate-change-conflicts/print> (23 June 2011).

Hauge, Wenche; Ellingsen, Tanja, 1998: "Beyond environmental scarcity: causal pathways to conflict", in: *Journal of Peace Research*, 35,3: 299-317.

Hegre, Håvard; Sambanis, Nicholas, 2006: "Sensitivity analysis of empirical results on civil war onset", in: *Journal of Conflict Resolution*, 50,4: 508-535.

Hendrix, Cullen S.; Glaser, Sarah M., 2007: "Trends and triggers: climate, climate change and civil conflict in sub-Saharan Africa", in: *Political Geography*, 26,6: 695-715.

Hendrix, Cullen S.; Salehyan, Idean, 2012: "Climate change, rainfall, and social conflict in Africa", in: *Journal of Peace Research*, 49,1: 35-50.

Hendrix, Cullen; Haggard, Stephan, 2015: "Global food prices, regime type, and urban unrest in the developing world", in: *Journal of Peace Research*, 52,2: 143-157.

Homer-Dixon, Thomas; Blitt, Jessica (Eds.), 1998: *Ecoviolence: links among environment, population, and security* (Lanham: Rowman & Littlefield).

Houdret, Annabelle, 2012: "The water connection: irrigation, water grabbing and politics in southern Morocco", in: *Water Alternatives*, 5,2: 284-303.

Hsiang, Solomon; Burke, Marshall, 2014: "Climate, conflict, and social stability: what does the evidence say?", in: *Climatic Change*, 123,1: 39-55.

Hsiang, Solomon; Burke, Marshall; Miguel, Edward, 2013: "Quantifying the influence of climate on human conflict", in: *Science*, 341, 6151: 1-14.

Hsiang, Solomon M.; Meng, Kyle C., 2014: "Reconciling disagreement over climate–conflict results in Africa", in: *PNAS*, 111, 6: 2100-2103.

Hsiang, Solomon M.; Meng, Kyle C.; Cane, Mark A., 2011: "Civil conflicts are associated with the global climate", in: *Nature*, 476,7361: 438-441.

Ide, Tobias, 2015: "Why do conflicts over scarce renewable resources turn violent? A qualitative comparative analysis", in: *Global Environmental Change*, 33,1: 61-70.

Ide, Tobias; Scheffran, Jürgen, 2014: "On climate, conflict and cumulation: suggestions for integrative cumulation of knowledge in the research on climate change and violent conflict", in: *Global Change, Peace & Security*, 26,3: 263-279.

Ide, Tobias; Schilling, Janpeter; Link, Jasmin S. A.; Scheffran, Jürgen; Ngaruiya, Grace; Weinzierl, Thomas, 2014: "On exposure, vulnerability and violence: spatial distribution of risk factors for climate change and violent conflict across Kenya and Uganda", in: *Political Geography*, 43,1: 68-81.

IPCC, 2001: "TAR, working group I: the scientific basis, appendix I–glossary", <http://www.ipcc.ch/ipccreports/tar/wg1/518.htm> (24 July 2014).

IPCC, 2007a: *Climate change 2007: synthesis report* (Geneva: IPCC).

IPCC, 2007b: *Climate change 2007: The physical science basis* (Geneva: IPCC).

IPCC, 2012: *Managing the risks of extreme events and disasters to advance climate change adaptation* (New York: Cambridge University Press).

IPCC, 2013: *Climate change 2013: the physical science basis* (Cambridge: Cambridge University Press).

IPCC, 2014: *Climate change 2014: impacts, adaptation, and vulnerability* (Geneva: IPCC).

Jabri, Vivienne, 1996: *Discourses on violence: conflict analysis reconsidered* (Manchester: Manchester University Press).

Johnstone, Sarah; Mazo, Jeffrey, 2011: "Global warming and the Arab Spring", in: *Survival: Global Politics and Strategy*, 53,2: 11-17.

Kahl, Colin H., 2006: *States, scarcity, and civil strife in the developing world* (Princeton: Princeton University Press).

Kelman, Ilan, 2012: *Disaster diplomacy: how disasters affect peace and conflict* (London: Routledge).

Kevane, Michael; Gray, Leslie, 2008: "Darfur: rainfall and conflict", in: *Environmental Research Letters*, 3,3: 1-10.

Kim, Ungtae; Kaluarachchi, Jagath J., 2009: "Climate change impacts on water resources in the upper Blue Nile River Basin, Ethiopia", in: *Journal of the American Water Resources Association*, 45,6: 1361-1378.

Klehmet, Katharina, 2009: *Klima in Ostafrika—Modellvalidierung und Untersuchung regionaler Charakteristika* (Bonn: Rheinische Friedrich-Wilhelms-Universität).

Koubi, Vally; Bernauer, Thomas; Kalbhenn, Anna; Spilker, Gabriele, 2012: "Climate variability, economic growth, and civil conflict", in: *Journal of Peace Research*, 49,1: 113-127.

Landis, Steven T., 2014: "Temperature seasonality and violent conflict: the inconsistencies of a warming planet", in: *Journal of Peace Research*, 51,5: 603-618.

Lautze, Jonathan; Reeves, Meredith; Vega, Rosaura; Kirshen, Paul, 2005: "Water allocation, climate change, and sustainable peace: the Israeli proposal", in: *Water International*, 30,2: 197-209.

Link, P. Michael; Piontek, Franziska; Scheffran, Jürgen; Schilling, Janpeter, 2012: "On foes and flows: vulnerabilities, adaptive capacities and transboundary relations in the Nile river basin in times of climate change", in: *L'Europe en formation*, 365, 3: 99-138.

Link, Peter Michael; Kominek, Jasmin; Scheffran, Jürgen, 2013: "Impacts of accelerated sea level rise on the coastal zones of Egypt", in: *Mainzer Geographische Studien*, 55,1: 79-94.

Lipchin, Clive, 2007: "Water, agriculture and Zionism: exploring the interface between policy and ideology", in: Lipchin, Clive; Pallant, Eric; Saranga, Danielle; Amster, Allyson (Eds.): *Integrated water resources management and security in the Middle East* (Dordrecht: Springer): 251-268.

Maas, Achim; Scheffran, Jürgen Scheffran, 2012: "Climate conflicts 2.0? Geoengineering as a challenge for international peace and security", in: *Security & Peace*, 30,4: 193-200.

Malm, Andreas, 2013: "Sea wall politics: uneven and combined protection of the Nile Delta coastline in the face of sea level rise", in: *Critical Sociology*, 39,6: 803.832.

Maystadt, Jean François; Ecker, Olivier, 2014: "Extreme weather and civil war: does drought fuel conflict in Somalia through livestock price shocks?", in: *American Journal of Agricultural Economics*, 96,4: 1157-1182.

McSweeney, C.; New, M.; Lizcano, G., 2008: "UNDP Climate Change Country Profiles–Kenya"; at: <http://country-profiles.geog.ox.ac.uk/> (25 June 2011).

Meier, Patrick; Bond, Doug; Bond, Joe, 2007: "Environmental influences on pastoral conflict in the Horn of Africa", in: *Political Geography*, 26,6: 716-735.

Meierding, Emily, 2013: "Climate change and conflict: avoiding small talk about the weather", in: *International Studies Review*, 15,2: 185-203.

Mekonnnen, Dereje Zeleke, 2010: "The Nile Basin cooperative framework agreement negotiations and the adoption of a 'water security' paradigm: flight into obscurity or a logical cul-de-sac?", in: *The European Journal of International Law*, 21,2: 421-440.

Messerschmid, Clemens, 2012: "Reality and discourses of climate change in the Israel-Palestinian conflict", in: Scheffran, Jürgen; Brzoska, Michael; Brauch, Hans Günter; Link, Peter Michael; Schilling, Janpeter (Eds.): *Climate change, human security and violent conflict: challenges for societal stability* (Berlin–Heidelberg: Springer): 423-459.

Miguel, Edward; Satyanath, Shanker; Sergenti, Ernest, 2004: "Economic shocks and civil conflict: an instrumental variables approach", in: *Journal of Political Economy*, 112,41: 725-753.

Mkutu, Kennedy, 2006: "Small arms and light weapons among pastoral groups in the Kenya-Uganda border area", in: *African Affairs*, 106,422: 47-70.

Mkutu, Kennedy, 2008: *Guns and governance in the Rift Valley: pastoralist conflict and small arms* (Bloomington: Indiana University Press).

Moore, Dahlia; Guy, Anat, 2012: "The Israeli-Palestinian conflict: the sociohistorical context and the identities it creates", in: Landis, Dan; Albert, Rosita (Eds.): *Handbook of ethnic conflict: international perspectives* (New York: Springer): 199-240.

Mude, Andrew; Barrett, Christopher B.; Carter, Michael R.; Chantarat, Sommarat; Ikegami, Munenobu; McPeak, John, 2009: *Index based livestock insurance for northern Kenya's arid and semi-arid lands: the Marsabit pilot* (Nairobi: International Livestock Research Institute).

Müller, Christoph; Cramer, Wolfgang; Hare, William L.; Lotze-Campen, Hermann, 2011: "Climate change risks for African agriculture", in: *PNAS*, 108,11: 4313-4315.

Nardulli, Peter F.; Peyton, Buddy; Bajjalieh, Joseph, 2015: "Climate change and civil unrest: the impact of rapid-onset disasters", in: *Journal of Conflict Resolution*, 59, 2: 310-335.

Nel, Philip; Righarts, Marjolein, 2008: "Natural disasters and the risk of violent civil conflict", in: *International Studies Quarterly*, 52,1: 159-185.

Njiru, Beth Njeri, 2012: "Climate change, resource competition, and conflict amongst pastoral communities in Kenya", in: Scheffran, Jürgen; Brzoska, Michael; Brauch, Hans Günter; Link, Peter Michael; Schilling, Janpeter (Eds.): *Climate change, human security and violent conflict: challenges for societal stability* (Berlin–Heidelberg: Springer): 513-527.

Nobel Peace Prize Committee, 2007: "The Nobel Peace Prize for 2007", <http://www.nobelprize.org/nobel_prizes/peace/laureates/2007/press.html> (19 August 2014).

Nyong, Anthony; Fiki, Charles; McLeman, Robert, 2006: "Drought-related conflicts, management and resolution in the West African Sahel: considerations for climate change research", in: *Die Erde*, 137, 3: 223-248.

O'Loughlin, John; Linke, Andrew M.; Witmer, Frank D. W., 2014: "Effects of temperature and precipitation variability on the risk of violence in sub-Saharan Africa, 1980-2012", in: *PNAS*, 111,47: 16712-16717.

O'Loughlin, John; Witmer, Frank D.W.; Linke, Andrew M.; Laing, Arlene; Gettelman, Andrew; Dudhia, Jimy, 2012: "Climate variability and conflict risk in East Africa, 1990-2009", in: *PNAS*, 109,45: 18344-18349.

Obioha, Emeka E., 2008: "Climate change, population drift and violent conflict over land resources in northeastern Nigeria", in: *Journal of Human Ecology*, 23,4: 311-324.

Omelicheva, Mariya Y., 2011: "Natural disasters: triggers of political instability?", in: *International Interactions*, 37,4: 441-465.

Omolo, Nancy A., 2010: "Gender and climate change-induced conflict in pastoral communities: case study of Turkana in northwestern Kenya", in: *African Journal on Conflict Resolution*, 10,2: 81-102.

Opiyo, Francis; Wasonga, Oliver Vivian; Nyangito, Moses; Schilling, Janpeter, 2014: "Pastoralists' adaptation and coping response to drought stresses in Turkana of northern Kenya", in: *Climate and Development* (forthcoming).

Peichert, Henrike, 2003: "The Nile Basin Initiative: a catalyst for cooperation", in: Brauch, Hans Günter; Liotta, Peter H.; Marquina, Antonio; Rogers, Paul F.; El-Sayed Selim, Mohammed (Eds.): *Security and environment in the Mediterranean: conceptualising security and environmental conflicts* (Berlin–Heidelberg: Springer): 761-774.

Piontek, Franziska, 2010: The impact of climate change on conflict and cooperation in the Nile basin. MA thesis (Hamburg: University of Hamburg).

Rahman, Majeed A., 2013: "Water security: Ethiopia-Egypt transboundary challenges over the Nile River Basin", in: *Journal of Asian and African Studies*, 48,1: 35-46.

Raleigh, Clionadh; Kniveton, Dominic, 2012: "Come rain or shine: An analysis of conflict and climate variability in East Africa", in: *Journal of Peace Research*, 49,1: 51-64.

Raleigh, Clionadh; Urdal, Henrik, 2007: "Climate change, environmental degradation and armed conflict", in: *Political Geography*, 26,6: 674-694.

Raleigh, Clionadh; Choi, Hyun Jin; Kniveton, Dominic, 2015: "The devil is in the details: an investigation of the relationships between conflict, food price and climate across Africa", in: *Global Environmental Change*, 32,1: 187-199.

Ratner, Blake D.; Meinzen-Dick, Ruth; May, Candace; Haglund, Eric, 2013: "Resource conflict, collective action, and resilience: an analytical framework", in: *International Journal of the Commons*, 7,1: 183-208.

Reuveny, Rafael, 2007: "Climate change-induced migration and violent conflict", in: *Political Geography*, 26,6: 656-673.

Reuveny, Rafael, 2008: "Ecomigration and violent conflict: case studies and public policy implications", in: *Human Ecology*, 36,1: 1-13.

Reuveny, Rafael; Mihalache-O'Keef, Andreea S.; Li, Quan, 2010: "The effect of warfare on the environment", in: *Journal of Peace Research*, 47,6: 749-761.

Rowhani, Pedram; Degomme, Olivier; Guha-Sapir, Debarati; Lambin, Eric, 2011: "Malnutrition and conflict in East Africa: the impacts of resource variability on human security", in: *Climatic Change*, 105,1: 207-222.

Rutten, Marcel; Moses, Mwangi, 2014: "How natural is natural? Seeking conceptual clarity over natural resources and conflicts", in: Bavinck, Maarten; Pellegrini, Lorenzo; Mostert, Erik (Eds.): *Conflicts over natural resources in the global south: conceptual approaches* (Boca Raton: CRC Press): 51-70.

Saha, Sujan, 2012: "Security implications of climate refugees in urban slums: a case study from Dhaka, Bangladesh", in: Scheffran, Jürgen; Brzoska, Michael; Brauch, Hans Günter; Link, Peter Michael; Schilling, Janpeter (Eds.): *Climate change, human security and violent conflict: challenges for societal stability* (Berlin–Heidelberg: Springer): 595-611.

Salehyan, Idean, 2014: "Climate change and conflict: making sense of disparate findings", in: *Political Geography*, 43,1: 1-5.

Salehyan, Idean; Hendrix, Cullen, 2014: "Climate shocks and political violence", in: *Global Environmental Change*, 28, 1: 239-250.

Scheffran, Jürgen, 2011: "The security risks of climate change: vulnerabilities, threats, conflicts and strategies", in: Brauch, Hans Günter; Oswald Spring, Úrsula; Mesjasz, Czeslaw; Grin, John; Kameri-Mbote, Patricia; Chourou, Béchir; Dunay, Pal; Birkmann, Jörn (Eds.): *Coping with global environmental change, disasters and security* (Berlin–Heidelberg: Springer): 735-756.

Scheffran, Jürgen; Battaglini, Antonella, 2011: "Climate and conflicts: the security risks of global warming", in: *Regional Environmental Change*, 11,1: S27-S39.

Scheffran, Jürgen; Brzoska, Michael; Kominek, Jasmin; Link, P. Michael; Schilling, Janpeter, 2012a: "Climate change and violent conflict", in: *Science*, 336,6083: 869-871.

Scheffran, Jürgen; Brzoska, Michael; Kominek, Jasmin; Link, P. Michael; Schilling, Janpeter, 2012b: "Disentangling the climate-conflict nexus: empirical and theoretical assessment of vulnerabilities and pathways", in: *Review of European Studies*, 4,5: 1-15.

Scheffran, Jürgen; Cannaday, Thomas, 2013: "Resistance against climate change policies: the conflict potential of non-fossil energy paths and climate engineering", in: Maas, Achim; Bodó, Balázs; Burnley, Clementine; Comardicea, Irina; Roffey, Roger (Eds.): *Global environmental change: new drivers for resistance, crime and terrorism?* (Baden-Baden: Nomos): 261-292.

Scheffran, Jürgen; Ide, Tobias; Schilling, Janpeter, 2014: "Violent climate or climate of violence? Concepts and relations with focus on Kenya and Sudan", in: *International Journal of Human Rights*, 18,3: 366-387.

Scheffran, Jürgen; Link, Peter Michael; Schilling, Janpeter, 2012: "Theories and models of the climate-security interaction: framework and application to a climate hot spot in North Africa ", in: Scheffran, Jürgen; Brzoska, Michael; Brauch, Hans Günter; Link, Peter Michael; Schilling, Janpeter (Eds.): *Climate change, human security and violent conflict: challenges for societal stability* (Berlin–Heidelberg: Springer): 91-131.

Scheffran, Jürgen; Marmer, Elina; Sow, Papa, 2012: "Migration as a contribution to resilience and innovation in climate adaptation: social networks and co-development in Northwest Africa", in: *Applied Geography*, 33,1: 119-127.

Schilling, Janpeter; Freier, Korbinian P.; Hertig, Elke; Scheffran, Jürgen, 2012: "Climate change, vulnerability and adaptation in North Africa with focus on Morocco", in: *Agriculture, Ecosystems & Environment*, 156,1: 12-26.

Schilling, Janpeter; Opiyo, Francis; Scheffran, Jürgen, 2012: "Raiding pastoral livelihoods: motives and effects of violent conflict in north-eastern Kenya", in: *Pastoralism*, 2,25: 1-16.

Schilling, Janpeter; Opiyo, Francis; Scheffran, Jürgen; Weinzierl, Thomas, 2014: "On raids and relations: climate change, pastoral conflict and adaptation in north-western Kenya", in: Bob, Urmilla; Bronkhorst, Salomé (Eds.): *Conflict-sensitive adaptation to climate change in Africa* (Berlin: BWV): 241-267.

Selby, Jan, 2003: "Dressing up domination as 'cooperation': the case of Israeli–Palestinian water relations", in: *Review of International Studies*, 29,1: 121-138.

Selby, Jan, 2013: "Cooperation, domination and colonisation: the Israeli–Palestinian joint water committee", in: *Water Alternatives*, 6,1: 1-24.

Selby, Jan, 2014: "Positivist climate conflict research: a critique", in: *Geopolitics*, 19,4: 829-856.

Selby, Jan; Hoffmann, Clemens, 2014: "Beyond scarcity: rethinking water, climate change and conflict in the Sudans", in: *Global Environmental Change*.

Shuval, Hillel, 2007: "'Virtual water' in the water resource management of the arid Middle East", in: Shuval, Hillel; Dweik, Hassan (Eds.): *Water resources in the Middle East: Israel-Palestinian water issues–from conflict to cooperation* (Berlin–Heidelberg: Springer): 133-139.

Slettebak, Rune T., 2012: "Don't blame the weather! Climate-related natural disasters and civil conflict", in: *Journal of Peace Research*, 49,1: 163-176.

Smith, Todd, 2014: "Feeding unrest: disentangling the causal relationship between food price shocks and sociopolitical conflict in urban Africa", in: *Journal of Peace Research*, 51, 6: 679-695.

Sneyd, Lauren Q.; Legwegoh, Alexander; Fraser, Evan D.G., 2013: "Foot riots: media perspectives on the causes of food protest in Africa", in: *Food Security*, 5,4: 485-497.

Solomon, Hussein, 2014: "Potential for cooperation rather than conflict in the face of water degradation: the cases of the Nile River and Okavango River basins", in: *Journal for Contemporary History*, 39,1: 69-94.

Stark, Jeffrey; Guillén, Sergio; Brady, Cynthia, 2012: *Follow the water: emerging issues of climate change and conflict in Peru* (Washington DC: USAID).

Sternberg, Troy, 2012: "Chinese drought, bread and the Arab Spring", in: *Applied Geography*, 34,1: 519-524.

Stetter, Stephan; Herschinger, Eva; Teichler, Thomas; Albert, Mathias, 2011: "Conflicts about water: securitization in a global context", in: *Conflict and Cooperation*, 46,4: 441-459.

Streich, Philip A.; Mislan, David Bell, 2014: "What follows the storm? Research on the effect of disasters on conflict and cooperation", in: *Global Change, Peace & Security*, 26,1: 55-70.

Tänzler, Dennis; Maas, Achim; Carius, Alexander, 2010: "Climate change adaptation and peace", in: *Wiley Interdisciplinary Reviews: Climate Change*, 1,5: 741-750.

Temesgen, Amsale K., 2010: *Climate change to conflict? Lessons from southern Ethiopia and northern Kenya* (Oslo: Fafo).

Theisen, Ole Magnus, 2008: "Blood and soil? Resource scarcity and internal armed conflict revisited", in: *Journal of Peace Research*, 45,6: 801-818.

Theisen, Ole Magnus, 2012: "Climate clashes? Weather variability, land pressure, and organized violence in Kenya, 1989-2004", in: *Journal of Peace Research*, 49,1: 81-96.

Theisen, Ole Magnus; Gleditsch, Nils Petter; Buhaug, Halvard, 2013: "Is climate change a driver of armed conflict?", in: *Climatic Change*, 117,3: 613-625.

Theisen, Ole Magnus; Holtermann, Helge; Buhaug, Halvard 2012: "Climate wars? Assessing the claim that drought breeds conflict", in: *International Security*, 36,3: 79-106.

Tullow, 2014: "Operational update–Kenya", <http://www.tullowoil.com/index.asp?pageid=137&category=&year=Latest&month=&tags=84&newsid=878> (23 April 2014).

TUPADO, 2011: *Turkana pastoralist organisation incident register 2000-2010* (Lodwar: TUPADO).

Tutiempo, 2011: "Kenya", <http://www.tutiempo.net/en/Weather/Kenya/KE.html> (18 August 2011).

UNDP, United Nations Development Programme, 2006: *Kenya National Human Development Report 2006* (Nairobi: UNDP).

UNDP, United Nations Development Programme, 2010: *Kenya National Human Development Report 2009* (Nairobi: UNDP).

UNEP, United Nations Environment Programme, 2009: *From conflict to peacebuilding: the role of natural resources and the environment* (Nairobi: UNEP).

Urdal, Henrik, 2005: "People vs. Malthus: Population Pressure, Environmental Degradation, and Armed Conflict Revisited", in: *Journal of Peace Research*, 42,4: 417-434.

von Uexkull, Nina, 2014: "Sustained drought, vulnerability and civil conflict in Sub-Saharan Africa", in: *Political Geography*, 43,1: 16-26.

Waterbury, John, 1979: *Hydropolitics of the Nile basin* (Syracuse: Syracuse University Press).

WBGU, 2008: *World in transition–climate change as a security risk* (London: Earthscan).

Weir, Tony; Virani, Zahira, 2011: "Three linked risks for development in the Pacific Islands: climate change, disasters and conflict", in: *Climate and Development*, 3,3: 193-208.

Wischnath, Gerdis; Buhaug, Halvard, 2014: "On climate variability and civil war in Asia", in: *Climatic Change*, 122,4: 709-721.

Witsenburg, Karen M.; Adano, Wario R., 2009: "Of rain and raids: violent livestock raiding in northern Kenya", in: *Civil Wars*, 11, 4: 514-538.

Wolf, Aaron T.; Yoffe, Shira B.; Giordano, Mark, 2003: "International waters: identifying basins at risk", in: *Water Policy*, 5,1: 29-60.

World Politics Review, 2011: "Global insider: Nile basin water rights", <http://www.worldpoliticsreview.com/trend-lines/8520/global-insider-nile-basin-water-rights> (18 August 2014).

Yeeles, Adam, 2015: "Weathering unrest: the ecology of urban social disturbances in Africa and Asia", in: *Journal of Peace Research*, 52,2: 158-170.

Zeitoun, Marc, 2008: *Power and water in the Middle East: the hidden politics of the Palestinian-Israeli water conflict* (London: Tauris).

Zeitoun, Marc; Warner, Jeroen, 2006: "Hydro-hegemony: a framework for analysis of trans-boundary water conflicts", in: *Water Policy*, 8,5: 435-460.

Zhang, David; Brecke, Peter; Lee, Harry; He Yuan-Qing; Zhang, Jane, 2007: "Global climate change, war, and population decline in recent human history", in: *PNAS*, 104,49: 19214-19219.

Zimmerer, Jürgen, 2014: "Climate change, environmental violence and genocide", in: *International Journal of Human Rights*, 18,3: 265-280.

13   From a Climate of Complexity to Sustainable Peace: Viability Transformations and Adaptive Governance in the Anthropocene

Jürgen Scheffran[1]

Abstract

In an increasingly interconnected climate of complexity, the stabilization of human-environment interactions is a major challenge in international relations and demands the integration of complexity science with global governance. This chapter highlights several cases of complex crises where cascading events affect international stability. Climate change is considered as a risk multiplier which disturbs the balance between natural and social systems and amplifies the consequences through complex impact chains that affect the functioning of critical infrastructures and supply networks; intensify the nexus of water, energy and food; lead to production losses, price increases and financial crises in other regions through global markets; undermine human security, social living conditions and political stability; and trigger or aggravate migration movements and conflict situations. An integrative framework of human-environment interaction is used to analyse destabilizing developments, tipping elements and cascading risks, as well as concepts of resilience, viability and sustainable peace. Whether climate stress fuels a cycle of violence or climate policy drives a transition towards a cycle of cooperation and sustainable peace depends on the human and societal responses. Strategies for viability transformations and adaptive governance range from climate mitigation and adaptation and the building of social networks to new capabilities of disaster management, crisis prevention and conflict resolution. Several examples are presented showing how transition and transformation processes can be analysed with an agent-based model framework.

Keywords: Adaptive governance, anthropocene, climate, complexity, conflict, cooperation, crisis, migration, social networks, sustainability transition, sustainable peace, transformation, viability.

13.1   Introduction

Since the report to the Club of Rome on *The Limits to Growth* was published in 1972, human-environment interactions have become ever more complex (see chapter 16 by Sjovaag in this volume). The report played a prominent role in helping us understand the world as a complex dynamic system where natural boundaries are violated by uncontrolled human interventions. It has been one of the drivers in promoting the concept of complexity, which evolved in the 1980s as a new paradigm in science. With the chaotic breakdown of the East-West conflict in 1989, complexity also served as a framework for transitions in society. While established scientific concepts based on linear systems, rational choice and simple two-player games were the dominant paradigms of the cold war, they reached their limits in the new emerging world order characterized by rapid changes and transitions in the international system that were marked by shifting pathways and transformation processes (see chapter 1 by Brauch/Oswald Spring in this volume). Although complexity science (see chapter 18 by Mesjasz in this volume) is challenging the dominant rationalist, realist and reductionist framework, the new understanding still needs to move from the academic realm into the day-to-day world of policy-making (Geyer/Pickering 2011). The world's complexity is apparently growing and there is a need to develop new practical approaches to address this complexity.

The 1989 turning point in world history was followed by a period of disorder and a transformation

---

1   Prof. Dr. Jürgen Scheffran, Institute of Geography, CliSAP Research Group Climate Change and Security, Center for Earth System Research and Sustainability, University of Hamburg; email: <juergen.scheffran@uni-hamburg.de>.

towards a fractal and fragile international landscape that continues to be unstable and full of surprises. Numerous factors and actors are interrelated in complex networks involving national, subnational and transnational actors in complex crises and conflicts. When everything is interconnected, seemingly small changes in one part of the world could have significant impacts in other parts. The tight couplings may lead to cascading crises that spiral out of control (Kominek/Scheffran 2012, Scheffran 2015a), such as the global economic crisis of 2008, the Arab Spring of 2011, the wars in Iraq, Afghanistan, Libya and Syria, the division and civil war in Ukraine in 2014, the Greek debt crisis from 2010 to 2015, the European refugee crisis of 2015, and recent terror attacks in Beirut, Paris, Brussels and elsewhere. These events are interconnected in many ways and through multiple channels that are often invisible (Scheffran 2016b).

In this complex network of crises, environmental change is connected with other problem areas through multiple linkages from local to global levels. Irrespective of when, where and how the 'limits to growth' take effect, the continued expansion of human activities in the anthroposphere has become a driving force that transforms the earth system into a new geological epoch, the 'Anthropocene', with multiple consequences such as climate change, land degradation, resource scarcity and biodiversity loss. In this nexus of overlapping and interconnected problem areas that creates a "climate of complexity" (Rothe 2015), the world may continue on a slippery slope of escalating risk cascades of environmental destruction, poverty and violent conflict, running full speed into natural boundaries and their forces. The challenge of the twenty-first century is whether humankind can anticipate and avoid hazardous pathways by counteracting forces that slow down and change course towards a more sustainable, peaceful and viable world, while at the same time taking into account natural boundaries (Scheffran 2015a). Along these lines, this chapter is based on the following structure:

- Drawing on various research activities by the author over three decades, this chapter will focus on the systemic issues of the current expansive growth-oriented pathways and related tipping points and risk cascades, in the context of complexity science (13.2).
- The next section highlights linkages between climate change as a potential risk multiplier in the world's complex crises, emergencies and disasters through interactions between climate stress, environmental change, human responses and social

conflicts (13.3). Basic human needs and values (such as the availability of water, food, energy, health and wealth) will be used to illustrate the vulnerability to environmental changes, which may lead to social disruption through instability events (such as forced displacement, riots, insurgencies, urban violence and war).

- The author will further discuss the role of social dynamics in transformation (13.4) and conceptual issues relevant to transformation (13.5), including stability, resilience, sustainable development and peace, in order to derive the concept of viability as a framework for the analysis and governance of transition and transformation processes (13.6).
- Finally, the author will introduce the VIABLE model as a framework for the analysis of transformation processes, including examples from the cold war arms race, the complex security landscape after the cold war, climate change and related risks, and coalition formation in energy transformations (13.7).

13.2   Tipping Points and Risk Cascades in Complex Security Landscapes

13.2.1   The Role of Complexity Science

The complexity of a system reflects the difficulty of understanding, describing or representing the system in an appropriate language. This has led to numerous attempts to define complexity (for an early survey, see Scheffran 1983). Key terms in complexity theory are: chaos, bifurcation, attractor, fractal and self-organization (Mesjasz 2010; see his chapter 18 in this volume). Complexity theory is (in)famous for popularizing statements like 'everything is connected to everything' and 'small causes can have big effects'. The interplay between the complexity and stability of dynamic systems has shaped ecosystem research since the 1970s (May 1972; Casti 1979; Scheffran 1983). Evolving complex systems (e.g. biological organisms, ecosystems, social systems) are generally more robust against variations in their environment, while systems that do not use learning skills to adapt their complexity to changing environmental conditions are exposed to the risk of instability (Scheffran 2015a).

In transition processes that change the qualitative structure of systems, the complexity–stability relationship is particularly relevant, and the methods of nonlinear dynamics and self-organization can be used to analyse phase transitions. Key questions are: under which conditions small-scale micro-level events lead to qualitative changes at the macro level, and whether there are thresholds and tipping points beyond which chain reactions and risk cascades are triggered that propagate in space and time and induce a qualitative system change. These include complex social interactions such as stock market crashes, revolutions, mass exoduses and violent conflicts (Kominek/Scheffran 2012). While systems are often robust against disturbances in the core region of stability (Held/Schellnhuber 2004), close to critical thresholds between regions of stability and instability small variations and uncertainties can decide whether systems break down or new ones are formed. This is symbolized by the famous butterfly effect in chaos theory, which may occur when a system is already 'on the edge', driven by other processes.

At tipping points the dynamics of change accelerates and induces a qualitative switch of behaviour which is sensitive to triggering events such as natural disasters, mass migrations and social movements, leading to a chain reaction of cascading sequences, e.g. when actions by one actor provoke more intense actions by other actors that undermine the stability of the whole system (Scheffran 2016b). A crucial phenomenon is path dependency (Kominek/Scheffran 2012), when social actors are locked in certain pathways of action that are self-enforcing and hard to change individually, e.g. when agents follow other agents or organizations, leading to a collective dynamics. A related term is the sensitivity of couplings between variables, which determines how changes spread in the network of interconnections (Scheffran/Link/Schilling 2012). As a result of non-linear effects, an increase in environmental variables such as global temperature and emissions above a threshold may trigger instabilities, tipping points and cascading sequences that could undermine the viability of natural and social systems. A governance challenge is to develop new approaches that stabilize the interaction, e.g. with the help of concepts such as adaptive complexity and stability (Scheffran 2015a).

One particular form of social instability is conflict emerging from incompatible actions, values, behaviours and priorities of agents who undercut each other's values and provoke responses that generate further losses and responses, thus leading to conflict escalation and waste of resources if the conflict is not resolved. The transition from conflict to cooperation requires adaptation towards common positions and mutually beneficial actions in order to stabilize the interaction.

13.2.2 The Complex Security Landscape

Human history has been shaped by complex social transformation processes, such as revolutions and the collapse of civilizations. More recent events include the end of the cold war followed by a new world (dis)order of high complexity and instability and a securitization of environmental, economic, social and human dimensions (Wæver 1997; Buzan/Wæver/de Wilde 1998; Scheffran 2008c). Initially the hostile relationship between the former superpowers was replaced by a more cooperative relationship but the positive consequences of the abandoned East-West conflict were challenged by countering trends. The unipolar dominance of the United States and NATO provoked resistance from Russia, China and other powers; conflicts in the Balkans, Africa, the Middle East and other parts of the world triggered foreign military interventions; the proliferation of nuclear weapons and missiles continued, and new arms races emerged, including that in outer space; new technologies accelerated a 'revolution in military affairs' (Neuneck 2008). 'New wars' (Kaldor 2012) and terrorism demonstrate that individuals and small groups can have a huge impact and contribute to the cycles of hatred and violence.

That small differences can matter was demonstrated by the 2000 presidential election of George W. Bush in the United States, when a few individual votes in the State of Florida changed the course of history. The 9/11 terror attacks by a small terrorist group triggered a chain reaction of violence, from the invasion of Afghanistan and Iraq to the civil war in Syria, which feeds terror attacks and refugee movements affecting the Middle East, Europe and other parts of the world.

13.2.3 The Nexus of Growth, Power and Violence

More 'tipping points' may occur in the future, in particular when the risks of environmental degradation, climate change, poverty, and hunger affect living conditions in many parts of the world; this could result in severe security threats. Especially vulnerable are societies on the edge of instability, such as 'states in transition', where the transition is from authoritarian to democratic regimes, and states that lack legitimacy and cannot protect citizens from harm. Human insecurity and personal instability could trigger social and political instability, and vice versa. The marginal impact of environmental change could undermine our ability to solve problems and further dissolve state structures, possibly leading to their collapse. Particularly critical is the situation in fragile and failing states with social fragmentation, weak governance structures and inadequate management capacities (Milliken/Krause 2003; Rotberg 2003; Starr 2008).

Economic growth, political power differences and violent force may become drivers of crises (Scheffran 1996). While growth is a major feature of life, it can be destructive when out of balance. A population grows exponentially (proportionately to the size of the population) if the birth rate exceeds the death or mortality rate (on demographic factors of population growth, see chapter 10 in this volume by Zlotnik). Balancing birth and death rates is a precondition for preserving life. If this balance is disturbed and is not stable, then births may exceed deaths or vice versa, leading to exponential expansion or decline, with a thin dividing line between both. Growth can continue in a cancer-like chain reaction that will cause pressure, stress and destruction in its environment, until limits are reached that do not permit further growth.

Over history, the human population has grown through increasing birth rates and falling death rates. The growth principle has also applied to the expansion of capital, investments, income, technology, energy and resource flows, and to political power, as well as to violent means of destruction. Malthus's concern was that human exponential growth would result in increasing death rates once the population reached its resource limits. However, humans were able to overcome constraints on resources and to expand into new spaces by using their problem-solving capabilities and developing technical and social innovations that stretched the limits to growth, allowing more wealth to be generated on a shrinking base of natural resources. Continued expansion of the human sphere puts further pressure on natural resources and ecosystems. As we reach planetary boundaries (Rockström/ Steffen/Noone et al. 2009; see chapter 2 in this volume by Dalby) in the Anthropocene, the question arises of whether a balance will be established by an increase in death rates or by the reduction of birth rates. Both represent different pathways: the first means crisis, disaster and death, while the second is associated with a global 'demographic transition', wealth creation and sustainability within natural boundaries. Before the second is discussed, the next section will associate the first pathway with the climate of complexity, focusing on possible linkages between climate change as a risk multiplier and the world's complex crises.

13.3   Climate Change as a 'Risk Multiplier' in Complex Crises

Although much is known about the effects of climate change on the components of the earth system, the interaction between the subsystems is still poorly understood. Since changes in one system can have direct or indirect effects on other systems, local events can propagate through complex causal chains and feedbacks on various spatial and temporal scales. Accordingly, climate change has been called a 'risk amplifier' and a 'threat multiplier' (CNA 2007, EU 2008, WBGU 2008).

Some critical questions are relevant in this context. What happens if climate impacts exceed the adaptive capacity of natural and social systems? Are there ranges of tolerance where systems remain stable, or outside of which turmoil, destabilization and 'tipping points' leading to qualitatively different states of the earth system are likely? Is it possible that climate change triggers regional or global risk cascades? Will critical infrastructures which are essential for the economy and society become dysfunctional? How does the risk-multiplying effect of climate change connect various problem areas? Will the stressor climate change lead to a destabilization of human-environment interaction or set reactions in motion which weaken the effects of the disturbance?

Some aspects are highlighted in the following section, based on general considerations about the complexity and stability of human-environment interaction, in particular the interaction between climate change and society[2]. The complex relationships in highly networked systems are illustrated by various types of risk, including instabilities in the climate system; hot spots of climate change and human insecurity; vulnerable infrastructures and networks; economic and financial crises; social and political instability; environmental migration; and climate change and violent conflict. As became clear during the Arab Spring, such processes have primary and secondary consequences for Europe which are mediated through mechanisms at global, regional and local levels.

---

2   This section was presented in a modified form at the Planetary Security Conference in The Hague, 2–3 November 2015 (Scheffran 2015b).

13.3.1   Instabilities in the Climate System

Weather and climate are considered primary examples of complex systems, and the Lorenz equations (a simplified mathematical model for atmospheric convection dynamics) became one of the roots of chaos theory (Sparrow 1982). Complex dynamic processes are characteristic of the climate system and are difficult to predict from knowledge of individual factors and equations. As is known from climate history, the planet went through different climate extremes, from the ice ages to warm periods, with variations occurring over short periods of time. Compared with previous abrupt climatic changes, climatic conditions in recent periods of human history have been relatively stable. Little is known about how stable the current world climate is, given the massive interference from anthropogenic greenhouse gas emissions. Furthermore, there are significant uncertainties regarding climate sensitivity and the estimation of critical stability limits, where the climate could experience a 'tipping' into unknown regions of the climate system, posing one of the gravest dangers facing mankind.

One focus of climate change research is weather extremes such as hurricanes, droughts, forest fires, floods and heatwaves, which can be described through processes of non-linear dynamics such as phase transitions, critical thresholds, and chaos; these have become paradigms in complex systems theory and in many application areas (Scheffran 1983; Bunde/Kropp/Schellnhuber 2002; Kurths/Maraun/Zhou et al. 2009; Kavalski 2015). In terms of their frequency and intensity, extreme events represent phenomena at the boundaries of the climate system. They are outside the 'normal' range and are associated with extreme circumstances that can burden the functionality and stability of the natural and social systems affected, and overwhelm their resilience and viability. The number and intensity of extreme weather events is likely to increase in the future (IPCC 2012; Rahmstorf/Coumou 2011).

In addition to single local events the climate system itself can become unstable if critical tipping points are reached, for example by exceeding certain thresholds in global mean temperature that trigger amplifying effects (Lenton/Held/Kriegler et al. 2008). These include the weakening of the North Atlantic thermohaline circulation, the rapid melting of ice shelves in Greenland and west Antarctica, the release of greenhouse gases such as methane from frozen soils in Siberia or Canada, and the change in the Asian monsoon. These phenomena and related chains of events can lead to a global and lasting transformation in the earth system. Massive and abrupt climate change could also overwhelm the ability of the strongest states and societies to deal with the problems. Less time-critical, although globally more hazardous in the long term, is the rise in sea level that creates stress on many coastal regions and threatened islands. Given the large number of uncertainties, it is a risky experiment to move into unknown areas of the climate system, where amplification of impacts and tipping elements opens up the possibility of global destabilization.

Figure 13.1: Impact chains and feedbacks in climate–society interaction. **Source**: Adapted and modified from Scheffran, Brzoska, Kominek et al. (2012a: 870).

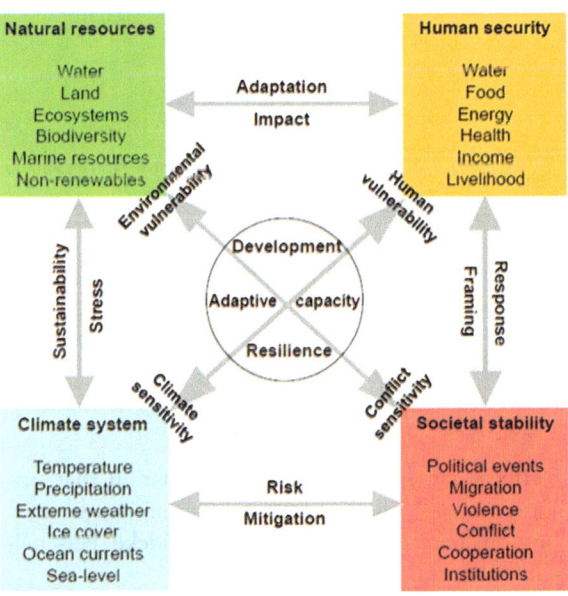

Figure 13.1 shows an integrative framework describing the complex interactions in the earth system (Scheffran/Brzoska/Kominek et al. 2012a, 2012b). The couplings in this network can be characterized by sensitivities which each represent the impact that a change in one variable has on another variable. In the context of the climate, sensitivity is "the degree to which a system is affected, either adversely or beneficially, by climate variability or climate change" (IPCC 2007: 881). Changes in the climate system affect the functioning of ecological systems and natural resources (e.g. soil, forests, biodiversity). Depending on the vulnerability, this can have an impact on human security, e.g. by affecting the supply of water, energy, food or economic goods. Even for uncertain sensitivities the sign

Figure 13.2: Hot spot areas where climate change impacts are critical for human security and social stability in different regions of the world. **Source**: Scheffran and Battaglini (2011): 33.

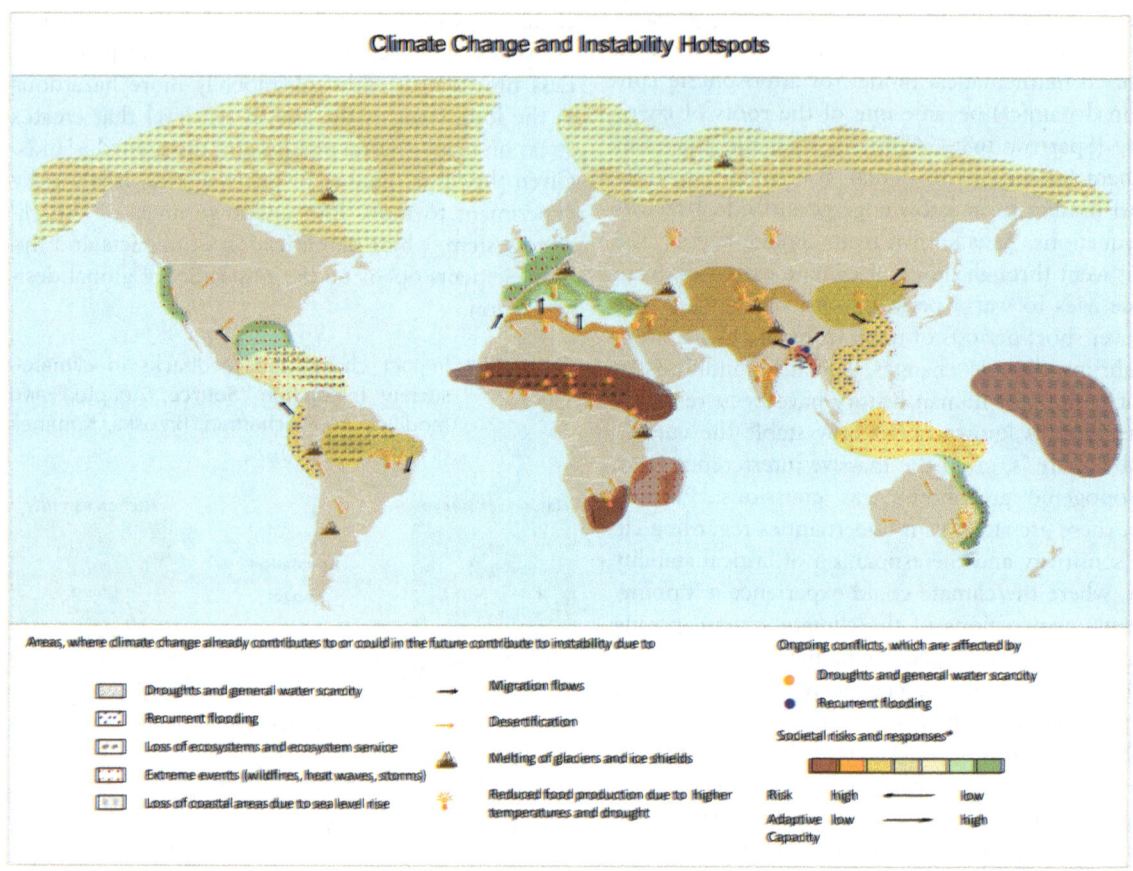

of effects can provide valuable information that can be used to classify relevant patterns or 'syndromes' using qualitative analysis (Eisenack/Lüdeke/Petschel-Held et al. 2007).

Human reactions to environmental change can affect the stability of societal structures. This can cause conflicts and social destabilization in regional climate hot spots, and in a networked world transfer mechanisms can propagate in a domino effect. On the other hand, cooperative and sustainable response strategies for mitigation and adaptation can be initiated to weaken and mitigate the causes and consequences of climate change. The problem is to develop practical strategies that are adequate to the level of complexity and that address the challenges of climate change in order to prevent dangerous instabilities, as well as to facilitate the necessary system changes. If reached, the goals agreed in the *UN Framework Convention on Climate Change* (UNFCCC) will stabilize the climate system at a non-dangerous level. To achieve this, the underlying interactions need to be

better understood, and an anticipatory-adaptive policy framework is needed that avoids or circumvents risky pathways, as well as allowing for a timely and qualitative system transformation that takes the form of a self-organized stabilization. Some types of complex interactions are discussed below, where climate change may act as a 'risk multiplier' and trigger social destabilization in complex crises.

13.3.2 Hot Spots of Climate Change and Human Insecurity

In regions strongly affected by climate change (hot spots), there are multiple stresses on human security (figure 13.2). Hydro-meteorological disasters (storms, floods and droughts) are an immediate danger to the life and health of the most affected people (Germanwatch 2014) in both developing countries (e.g. the Indus flood in Pakistan in 2010, drought in China in 2010–11, typhoon in the Philippines in 2013) and industrialized countries (e.g. the European heatwave of 2003; the Elbe floods in Germany in 2002 and 2013; tropical storms in the US in 2005 and 2012; wildfires in Russia in 2010). The consequences can be so severe that in the regions of concern adequate assistance is not possible and social systems become overloaded. For instance, in 2005 hurricane Katrina caused enormous damage to the south coast of the US and led to more than 1,800 deaths; it displaced hundreds of thousands of people and overwhelmed disaster management. The heatwave of 2003 in Europe left behind tens of thousands of casualties and damage to agriculture worth tens of billions of euros. The Indus flood in 2010, the worst for more than eighty years, flooded a fifth of the land area of Pakistan with consequences for twenty million people; it led to about 2,000 deaths and destroyed 1.7 million homes and a large part of the infrastructure (Gemenne/Brücker/Glasser et al. 2011).

On the other hand, climate change affects the long term availability of natural resources, and this can contribute to shortages and an unbalanced distribution of resources. Examples include the degradation of fresh water, forests and farmlands, shortage of nutrition, the threat to biodiversity, and overfishing. Whether people are able to cope with the consequences and to limit the risks depends on their vulnerability and adaptive capacity. These are influenced by their access to resources, information and technologies, as well as by the stability and effectiveness of institutions (Adger/Lorenzoni/O'Brien 2009).

In the hot spots, climate stress causes great human suffering and significant economic and social losses, and this undermines human security. A large proportion of the risks are not only or primarily related to climate change, but are often also caused by pre-existing local problems in the affected areas. These include the destruction of ecosystems, a high level of poverty, political instability, overuse of land, and the absence of early warning systems and disaster protection. Most vulnerable to climate stress are those regions whose economies are dependent on climate-sensitive resources and where infrastructures are particularly exposed to climate change. These include developing countries with a high level of dependence on agriculture; coastal areas and river basins; and hot and dry regions. In these hot spots human security is at stake, if natural resources which are of fundamental importance for the existence of people and the satisfaction of their needs are depleted or degraded.

In many cases, the primary consequences are confined to the affected areas, but they may be connected to remote regions through humanitarian aid, civil protection or other direct interventions (including military operations). In many cases, however, indirect and secondary effects going beyond the hot spot region occur. Some reactions may further aggravate the situation, e.g. when people in need enforce the overexploitation of resources, move to other risk areas, or use violence against competitors, in order to ensure their own survival. For instance, rising land prices induce a search for cheaper land; this problem is often found in developing countries, where it leads to environmental and supply risks and displaces local users. Another example is the impact of land scarcity on the availability of water and on related crop losses. Where there are droughts in parts of Asia, Africa and America, global food markets can cause a fall in prices elsewhere. In this context, various trade-offs and exchange processes between different resources need to be taken into account. This is expressed in the nexus of water, energy and food (see the following sections).

13.3.3 Vulnerability of Infrastructure, Technical Systems and Supply Networks

Climate change may affect the viability of critical economic and social infrastructures and supply networks. These include systems for the supply of water, food and energy, goods and services, systems for the provision of communication, health, transportation and security, and human settlements and political institutions. It is not only the failure of important subsystems that matters, but also the possibility that disruption of certain system components spreads across couplings and leads to the collapse of the entire system. This is true for developing countries which depend directly on ecosystem services and agriculture and to a different degree for developed countries which rely on highly networked technical systems but have more sophisticated protection and response mechanisms. The stronger the impacts are and the more subsystems are affected, the harder it is for societies to absorb the consequences.

Corresponding relationships are examined in risk research for the failure of complex technical systems in which the combination of different events results in the loss of control mechanisms. This examination started with the debate on various spectacular accidents in high-risk technologies (Bhopal, Challenger, Chernobyl), which demonstrated that large-scale technology (chemical and nuclear engineering, biotechnology and genetic engineering, aerospace and defence technology) is not completely manageable and contains a 'residual risk', comparable to natural disasters. Since not all contingencies are predictable in complex systems, it is often enough that a minor event can initiate a chain of events which appears as a 'normal accident' but leads to catastrophes in tightly coupled human-machine systems (Perrow 1984). In a globalized world these tight couplings occur not only in technical systems but in other fields as well.

A spectacular example of a risk cascade of natural and technological disasters was the earthquake in Japan on 11 March 2011, which triggered a chain of events with global effects. The tsunami flooded parts of the Japanese coast, cost many lives, and triggered the accident at Fukushima that destroyed several nuclear reactors, spreading radioactivity globally through the atmosphere and the ocean. Because of the consequences of this large-scale accident, the Japanese power grid, the nuclear industry, stock markets, oil prices and the global economy were all affected. Automobile manufacturers and electronics companies worldwide cut back production because important components were no longer being delivered from Japan. The shock waves from the nuclear disaster triggered the energy transition in Germany. This disaster impressively demonstrates how a single event can set into motion different chains of processes and can change the political environment (Kominek/Scheffran 2012; Scheffran/Burroughs/Leidreiter et al. 2015; Scheffran 2016b).

In addition to earthquakes or technological accidents, weather extremes can also hit critical nodes in the networks of business, technology and society. For industrialized countries such as Germany, whose economies and societies depend on a functioning infrastructure, the stability of a sustainable supply system in response to extreme weather events is of great importance. Weather extremes such as the 2003 heatwave, the storm surge in 2013 or the Elbe flood in 2013 lead to temporary impairments of transport or energy supply systems. From the perspective of climate research, it is important to assess whether extreme weather events that exceed the limits of adaptation of supply networks can occur over a given period. It is important to identify the critical nodes and links in the global supply network and to understand how local failures of infrastructure components affect the global supply. Network research analyses what happens in the case of the failure of network elements, how shocks propagate into the power grid, and whether chain reactions and cascading effects which can cause a collapse of the network can occur.

As an example we discuss here the failure of power grids. Since virtually all other supply networks depend on a well-functioning power grid, their failure affects all parts of society. Most hydro-meteorological disasters affect parts of the national electricity grid only temporarily, and power is then restored. However, there have also been cases in which the power supply fell victim to a large area blackout caused by minor events. In the biggest blackout in history in July 2012, over 600 million people in northern and eastern India were affected due to an overloading of the power grid. In November 1965, about thirty million people in the north-eastern United States and in many parts of Canada remained without electricity for about six days. In California, there were regular power outages caused by insufficient generation capacity and market manipulation (Brand/Scheffran 2005). During a major power blackout in Europe in November 2006, parts of Germany, France, Belgium, Italy, Austria and Spain were temporarily disconnected from the power supply. While in these cases there were various underlying causes, in other cases weather events were the trigger. In November 2005, after a heavy snowfall in North Rhine-Westphalia and Lower Saxony, one of the largest power outages in German history occurred, and some 250,000 people were without power for several days, resulting in a financial loss of about €100 million (Deutschländer/ Wichura 2005). More recently, the snowstorm in North America at the turn of 2013 and 2014 caused major power cuts for hundreds of thousands of people, leading to partial failure of the communication and transport systems.[3]

If the supply system for a particular resource is hit, this often has effects on other resources, in particular the nexus of water, energy and food (Beisheim 2013; IEA 2012). Energy is needed for irrigation and for the production of food, water, and energy supply, in particular for the extraction, transportation and processing of fossil and nuclear energy. The development of unconventional gas and oil reserves (fracking, oil sands and oil shales) has led to an increasing need for water and land. In addition, renewable energy sources such as hydropower, biofuels and geothermal power need large amounts of water. Regions with low rainfall are dependent on artificial irrigation for the cultivation of plants for food and energy. Even with large solar power plants in desert areas, water supply is a

---

3 "Ice storm blackouts frustrate tens of thousands"; at: <http://www.cbc.ca/news/canada/ice-storm-blackouts-frustrate-tens-of-thousands-1.2476866>.

critical issue. Overall, the water demand for energy generation is estimated to rise twice as much as the demand for energy (Beisheim 2013).

Climate change affects this nexus in many ways and increases the competition between water, energy and food. For instance, nuclear power plants are vulnerable because they are dependent on the flow of cooling water. A warming of the waters or long periods of drought or floods affect power generation in developed countries, leading to critical situations when cooling water is no longer available or water in power plants is below the critical level (Beisheim 2013: 24). Through hurricane Katrina in 2005 in the Gulf of Mexico region, more than a quarter of total offshore oil production, almost a fifth of total natural gas production and almost half of available refining capacity were temporarily disabled, as well as important oil pipelines, thousands of oil rigs, and a large proportion of rail and sea transport (for an overview, see Kumins/Bamberger 2005).

Typhoon Haiyan in the Philippines 2013 destroyed part of the country's supply of renewable energy (Bradsher 2013). Since renewable energy sources (bio, hydro, wind, solar) depend on certain meteorological conditions, they are affected by climate change..

The forced cultivation of bioenergy plants as part of a climate change strategy has intensified a global resource competition as large amounts of water and land are needed that are therefore no longer available for food production (Scheffran 2010). With a shortage of resources due to climate change, market prices increase, making the expansion of agricultural production more attractive. This may lead to a rise in the demand for production factors such as water, energy, pesticides and fertilizers, which in turn are associated with increasing environmental pollution and a growing demand for land (Beisheim 2013). On the other hand, rising food prices have adverse effects on poor populations (see below). To some extent competition can be mitigated by synergistic effects, e.g. when water power plants achieve an optimal trade-off between water and energy use, solar energy is used for water desalination, organic waste is used in food production for energy purposes, and new jobs are created that contribute to development in rural areas.

It is difficult to make supply networks more resilient to climate change impacts if disruptive events occur in rapid succession and with multiple effects which hit a system simultaneously, with short time delays or in a narrow geographical area. With the increasing intensity and frequency of climate-related events, the question arises as to when the limits of capacity and resilience of infrastructure are reached and whether existing safeguards and adaptation measures are sufficient.

13.3.4 Economic and Financial Crises

Assets and economic processes such as global freight, trade and financial markets that regulate the exchange between supply and demand are also subject to climate change. Financial transactions and pricing information represent virtual transfer mechanisms which link the events with each other globally and within a very short period of time. If there is a disturbance in one place as a result of climate-related events, such as production losses, bankruptcies of companies or a fall in the stock market, it is propagated across global networks and markets.

The economic crisis of 2008 demonstrated the instability of the complex interconnected global economy. Powered by the reckless speculation and lending practices of financial institutions and short-sighted human behaviour, local events and individual responses escalated, pushing the global financial system to the brink of collapse. After the critical limit was exceeded, self-reinforcing mechanisms were triggered, leading to losses of hundreds of billions of dollars and euros worldwide. Public investment and regulatory policies were initially unable to compensate for the short-term fluctuations. The interaction between rating agencies and government responses led to an explosive situation. In Europe, the global economic crisis was followed by a crisis in southern Europe, most dramatically in Greece.

Although other factors were at work here, climate and economic crises may interact and lead to a downward spiral. According to the Stern review, abrupt and extensive changes in the climate system could wreak havoc in global trade and financial markets (Stern 2006). Since there are various links between climate change and financial markets, risk cascades can be analysed and classified (Haas 2010; Onischka 2009). In addition to the risk of direct economic damage, global impacts are possible through the loss of production, supply shortages and price increases. Thus, extreme events in one country can induce production losses in another country, and these can spread through global supply chains (Levermann 2014). While the direct damages and costs of weather extremes have been recognized for several years, the indirect economic consequences are still poorly understood. Some issues have been raised in the context of the indirect effects of bioenergy, particularly as in other parts of the world the price and cultivation of cereals has been affected by bioenergy production (Scheffran 2010), e.g. in the so-called 'tortilla crisis' in Mexico in 2007.

In the energy sector various risks (natural disasters, infrastructure problems, strikes, riots, wars, political interventions) may lead to constraints on supply and strong market variations. High oil prices, as in 2008, are a driver for recession, social risks and willingness to accept a high level of environmental risk (Beisheim 2013). Relevant questions can be raised if production losses are observed through extreme events in a country that is a food supplier. These events include the heatwaves and related fires lasting several weeks in Russia in the summer of 2010, which resulted in an export ban on wheat (FAZ 2010). The droughts in the US in 2011 and 2012 and in China in 2010–2011 were associated with price increases in food commodities. For poorer countries the consequences of integration into the globalized economy can be as momentous as the direct effect of domestic local events. Even in developed countries such as Germany, the impacts of extreme events on the consumer are noticeable. Europe is not immune to the consequences if negative developments in the Mediterranean lead to a spiral of escalation. An economically weakened southern Europe is more vulnerable to climate-related risks and would have a lower potential for adaptation. Here problems of water and food supply could hit tourism and agriculture, lead to conflict and migration, affect neighbouring countries, and spread across continents. Some examples are discussed below.

1. *Floods in Australia* 2010–2011: In the wake of tropical cyclone Tasha, Queensland and New South Wales were affected by heavy rainfall in 2010 and 2011. The worst flooding in fifty years inundated an area the size of Germany and France combined; it included seventy cities, thirty-five people lost their lives and 200,000 people were evacuated, including from parts of the metropolitan area of Brisbane. According to media reports the damage was of the order of AUS$1 billion and the loss to GDP stood at AUS$13 billion, which had a significant impact on the economic performance of Australia. Furthermore, due to flooding about forty coal mines were temporarily closed or operated at reduced power, so that the production capacity of the largest coal exporter in the world was severely impaired. Coal mining in Queensland fell by thirty per cent; coal production fell from 471 million tonnes in the previous year to 405 mil-

lion tonnes. At times, the domestic coal industry was losing more than €70 million per day (Oldag/ Walterlin 2011). Since the cost of raw materials is more than eighty per cent of the production cost of steel, this triggered rising prices and supply bottlenecks in the steel industry. The expanding Australian chain reaction was also felt in Germany and had an impact on the automotive industry, on mechanical engineering, and on other industries (Spiegel 2011).

2. *Flood in Thailand* 2011: As a result of a violent and unusually long-lasting monsoon Thailand was hit in October and November 2011 by the worst floods in fifty years, which affected nearly twelve per cent of the country. The consequences were almost 400 deaths, property damage of more than €11 billion, substantial loss of economic growth, temporary drops in tourist numbers, and massive crop losses (a quarter below the previous year). The neighbouring countries of Cambodia and Laos were also affected. In addition to the regional consequences, the disaster had an impact on the world economy. Supply failures in electronic components led to bottlenecks in the international electronics and computer industry and a large increase in hard drive prices in Germany (Feddern/Schnurer 2011). German companies such as Volkswagen had problems with the delivery of important parts. Japan's automobile companies suffered repeated losses in production, in particular shortly after the Fukushima disaster. Although the Thai electrical and electronics industry was put back by the flooding, the industry has recovered faster than expected. With the stronger shift of production to Asia, plagued by earthquakes and floods, the consultant A. T. Kearney noted that eighty per cent of companies were adversely affected (Gärtner 2011).

3. *Drought in China in* 2010 *and* 2011: In November 2010, a once-in-a-century drought in China's eastern wheat belt threatened the winter wheat crop, which accounts for twenty-two per cent of the harvest of the world's largest producer and consumer of wheat. An area of 1.6 million hectares and more than 300 million people were affected. The severe drought hit the domestic and agricultural water supply and led to the closure of parts of the Yangtze River for navigation, as well as to the drying-up of water resources and to reduced hydropower generation. In early 2011, more than 2.2 million people and 2.73 million units of livestock nationwide suffered from a lack of water. Utilizing expe-

rience from past famines (1958-1961), the Chinese government has taken measures to reduce the risk of crop failure. They have invested in new hydro projects and bought wheat on the international market to compensate for the losses from the drought (Sternberg 2013). As a significant proportion (between six and eighteen per cent) of annual global wheat production is traded across borders (Lampietti/Michaels/Magnan et al. 2011), the decline in supply led to an increase in wheat prices and serious economic impacts in the import-dependent countries of North Africa and the Middle East (Sternberg 2013; see the following section).

13.3.5 Social and Political Destabilization

Directly or indirectly through the integration of physical, economic and geopolitical risks in a globally networked world, the impact of climate-related events could also undermine social and political stability in the regions concerned and possibly cause global turmoil. Due to globalization, combined with rapid developments in computer technology and in communication and transportation systems, people are increasingly globally connected and able to respond collectively and rapidly to local changes. Accordingly, social and political changes in one region can have significant impacts in other regions. It is possible that small groups and individuals can set in motion global chains of events that have an influence on international relations. This became clear in a spectacular way with the end of the cold war, the terrorist attacks of 11 September 2001, the Arab Spring in 2011, and the 2015 refugee crisis, where each had a significant impact on Europe. Individuals can also put solution processes in motion, as in the cases of whistle-blowers from Daniel Ellsberg to Edward Snowden, who brought sensitive information to the public in order to draw attention to deficits and problems.

The spirals of hate, terror and violence have their own dynamics across different levels. Environmental destruction, poverty and hunger affect social conditions in many parts of the world. Particularly sensitive are fragile and weak states with social fragmentation and poor governance and management capacity who cannot guarantee the core functions of government, such as law and public policy, the state's monopoly on force, welfare, participation, and basic public services in infrastructure, health and education (WBGU 2008).

Climate change may contribute to destabilization, especially where societies are in transition, for instance from authoritarian to democratic regimes. On the edge of instability, natural disasters can undermine the legitimacy and ability of states to protect their citizens from harm. If the agriculture of a developing country is severely damaged, the livelihood and existence of many people is at stake. The loss of life, income, wealth, jobs, health, family or friends provokes opposition and unrest that threaten the social contract and undermine the political order. Some of these processes act slowly and contribute to the erosion of social and political stability, others run quickly and overwhelm the problem-solving and adaptive capacity of communities. Various destabilizing processes may increase in climate hot spots and spread into neighbouring regions. With the decay of the social and political order, non-state actors (private security companies, terrorist groups, warlords) penetrate into domains opened up by the power vacuum. Countries with a low income and a low level of adaptive capacity are particularly at risk, while richer societies have more potential capacity for adaptation. However, in view of global interdependence, they cannot simply ignore the destabilization in other parts of the world which may affect them as well through complex chains.

Various natural disasters have been associated with a temporary collapse of law and order. Looting and criminal acts have occurred after heavy storms, for instance after hurricane Katrina in the US in 2005 and after the typhoon in the Philippines in 2013. After some storms and floods in southern Asia and Central America, the distribution of aid and relief goods was subject to disputes that were partly conducted violently (WBGU 2008). In addition to storms and floods with usually temporary and local impacts on the food supply, droughts have a direct and lasting impact on global food markets because of their wide spatial and temporal scale. Many people who are highly dependent on agricultural production and the local availability of water resources are affected. In contrast, it is usually the indirect influence of climate-related events and disasters on water, food and population that gives such type of disasters an international dimension.

The most significant consequences include food shortages and a consequent increase in food prices, which undermine living conditions for poor social groups. Throughout history 'bread protests' and food riots have contributed time and again to political and social change, as in the French and Russian Revolutions. This includes recent global supply crises such as those of 2008 and 2011, where food prices quickly multiplied and the number of hungry people increased by 100 million to 1 billion (Beisheim 2013). For instance, in 2008 uprisings related to food crises provoked a change of government in Haiti, while in Cameroon twenty-four people were killed during protests and about 1,500 were arrested (Sternberg 2012).

A spectacular example is provided by the social and political upheavals in the *Middle East and North Africa* (MENA) since 2011. The series of protests and uprisings in the Arab world led to a conflagration that affected the entire region and provoked a change of regime in several countries (Johnstone/Mazo 2011). Starting with the unrest in Tunisia in early 2011 which forced the president to flee, the revolutionary impulse spread to Libya, Egypt, Syria and other MENA countries. It was accelerated and multiplied by electronic media and social networks (Kominek/Scheffran 2012), which enabled the rapid spread of protest and motivated others to join the protest movement. An example is seen in the demonstrations in Tahrir Square in Cairo and in other countries. Although supporters of the regime responded with force against the demonstrators, the self-organized resistance remained largely peaceful, forcing the resignation of President Mubarak. In the following years the situation turned violent in some countries, especially in Libya and Syria.

Which role rising food prices played here and to what proportion climate change and extreme weather events might have affected these processes is still the subject of scholarly debate. At the beginning of the revolt some media reports suggested a link with the sharp rise in food prices at the turn of 2010-2011. A collection of papers published by the Center for American Progress examined the impact of climate change on the upheavals in the Arab world (Werrell/Femia 2013). The argument was not so much that climate change was a primary cause, but rather that in an explosive political crisis the effects of climate change can act as an additional stressor that can exceed a 'tipping point'. One of the factors that occurred before the crisis was the drought in China in 2010 and 2011 (described above), which exerted pressure on the international market price of wheat and influenced the availability of food products. This coincided with other factors that further increased world food prices, including high oil prices, the development of bioenergy, and speculation on the global food markets (Johnson 2011).

The consequences affected much of the MENA region where the world's nine largest importers of food are located, seven of which saw political pro-

tests. Many households spend, on average, more than a third of their income on food (Sternberg 2013), while people in Western countries spend less than ten per cent. The dependence of Arab states on imported food makes them vulnerable to fluctuations in global commodity markets. Low incomes and high levels of resource imports and spending on food taken together affect food security. Reinforced by the sharp rise in bread prices, the existing public discontent with the government was magnified. In Egypt, the largest wheat importer in the world with a rapidly growing population, three per cent of the national income was spent on wheat subsidies (Sternberg 2013). As early as 1977 there was the 'bread intifada' in Egypt in which seventy-seven people died, and in 2008 there were bread riots. However, no protests took place in Israel or the United Arab Emirates, which have a high per capita income, a small food share of income, and better adaptive capacities.

The chain of events before, during and after the Arab Spring illustrates how in the networked world extreme events can affect international relations, mediated through economic, social and political processes. In this complex pattern of overlapping stressors (Werz/Hoffman 2013), climate change was not the main cause, but contributed to triggering the chain. The political upheavals happening today affect the stability of the Mediterranean region and coincide with the economic crisis in southern Europe. For Europe these events are quite significant, because of the civil wars in Libya and Syria and increasing migration from North Africa, the Middle East and sub-Saharan Africa (see the following sections).

13.3.6 Environmental Migration

As early as 1990, the first IPCC Assessment Report warned that changes in temperature and precipitation "could initiate large migrations of people, leading over a number of years to severe disruptions of settlement patterns and social instability in some areas" (IPCC 1990: 55). The IPCC Special Report of 2012 on extreme events and disasters stated that in the future climate extremes would have a larger impact on migration (IPCC 2012). And the most recent (fifth) Assessment Report found evidence for increased mobility or displacement in seventeen cases, while decreased mobility was found in six cases and socially differentiated mobility outcomes in five (IPCC 2014: 769).

According to the European Commission's position paper, Europe must be prepared for substantially increased pressure from migration (EU 2008). In regions of poverty, violence, injustice and social insecurity, climate-induced stress could induce migratory movements. The *International Organization for Migration* (IOM) cites as reasons for environmental migration sudden or progressive changes in the environment that affect life or living conditions so that people are forced or choose to leave their homes temporarily or permanently (IOM 2008). In a narrow sense, Biermann/Boas (2008) define 'climate refugees' as people who leave their habitat as a result of rising sea levels, droughts and water shortages. By contrast, the *United Nations High Commissioner for Refugees* (UNHCR) (2011) argues against the use of the terms 'environmental refugee' and 'climate refugee', as they are imprecise and the term 'refugee' should be limited to political persecution and threats. As long as there are no international agreements, the legal status of climate-induced migration remains unregulated.

According to the *Internal Displacement Monitoring Centre* (IDMC), 32.4 million people have been displaced in 2012 by natural disasters (mostly weather-related), more than twice as many as in 2011. In two flood disasters in 2012 alone, 6.9 million people in Nigeria were displaced and 6.1 million people in India (IDMC 2013). The global numbers declined to 22.3 million in 2013 and to 19.3 million in 2014 (IDMC 2015). In contrast, the number of internally displaced people showed an opposite trend, from 22.8 (15.4) million in 2012, to 33.3 (16.7) million in 2013 and 38 (19.5) million in 2014 (the figures in parentheses indicate refugees). Estimates of future climate migrants in the literature vary widely, from fifty million to a billion people; the figure is disputed (Jakobeit/Methmann 2012; Foresight 2011).

Climate change is one among several possible drivers for human migration. Given a large number of possible reasons for flight and complex relationships, the impact of climate on migration is difficult to determine. Changes in the environment can not only promote but also impede migration by increasing the poverty of the rural population and thereby limiting the opportunities to escape (trapped populations). Environmental impacts and vulnerabilities can increase if people migrate to ecologically fragile and conflict-affected regions, including coastal cities which are affected by storms and sea level rise. In Europe and the United States climate migration is seen as a security problem and conflict factor, possibly leading to ethnic, religious and political tensions between the local population and immigrants. One contributing factor is competition for scarce resources such as arable and pasture land, housing, water, jobs and social services. Media coverage of events such as the drought in Somalia, boat people in the Mediterranean and refugee movements along the Balkan route reinforce threat perceptions in Europe. With the establishment of the *European Agency for the Management of Operational Cooperation at the External Borders* (FRONTEX) the 'defence' of and against refugees—including environmental and climate migrants—is to be expanded. One response to the refugee crisis of 2015 was to increase border controls in southern Europe.

So far climate or environmental factors have not been identified as major contributions to international South–North migration. The majority of people affected by precarious environmental conditions remain in their home region or migrate to nearby urban areas. For weak and marginalized people, it is more difficult to overcome long distances or other barriers (e.g. language and cultural barriers) than it is for privileged classes. How far migration can be proved to trigger political instability and conflict is debated (Barnett/Adger 2007; Reuveny 2007). Response patterns in security policy narrow the scope of action to combatting symptoms, with the risk of a 'chain reaction' between increasing migration pressure and countermeasures (such as enhanced border protection, as observed during the 'refugee crisis' of 2015-16). Adaptation strategies and international cooperation can help overcome risks and even develop migration into an important measure of adaptation to climate change (Foresight 2011), strengthening the resistance and resilience of the affected communities. Migration networks can contribute to resilience and stable structures between source and destination countries, such as the transfer of payments, knowledge and technology (Adger/Kelly/Winkels et al.2002; Scheffran/Marmer/Sow 2012).

Industrialized countries are not spared from environmental migration. The debate became more important when hurricane Katrina in 2005 forced hundreds of thousands of people to flee New Orleans, among them a number of refugees who never returned. Risk zones vulnerable to flooding in coastal or river areas can also become uninhabitable in Europe and lead to migration, even if a larger number of domestic environmental migrants are not expected in the foreseeable future. In contrast, the debate on the immigration of refugees from conflict areas can provoke massive internal social conflicts;

this has become an issue in Europe, particularly in Germany, since the summer of 2015.

A large number of immigrants come from the MENA region (especially from Syria and Iraq), Afghanistan and the Sahel. In addition to the destabilizing remote impacts that have been mentioned in the previous section, these regions are directly affected by climate change, and this potentially increases the pressure for migration (BW 2012). Because of high population growth rates, climate change and resource depletion in large parts of Africa, availability of drinking water and arable and pasture land is expected to decline, affecting hundreds of millions of people (UNPD 2015, IPCC 2007; Schilling/Freier/Hertig et al. 2012, Busby/Raleigh/Salehyan 2013). Environmental variations increase the pressure on agriculture, fisheries and pastoralism (Werz/Hoffman 2013). Water availability in some countries is already below the threshold for water scarcity of 1,000 cubic metres per person per year. In Libya in 2009 this was only about 95.8 cubic metres and in Syria 356 cubic metres per capita, significantly lower than the figures for 2002 and well below the world average (World Bank 2013).

In the years before the rebellion Syria suffered devastating droughts (Kelley/Mohtadi/Cane et al. 2015) that hit the main growing areas of the country hardest, and many people were driven from the countryside to the cities. In 2010 alone, 50,000 Syrian families were affected. In 2002 more than thirty per cent of Syrians worked in agriculture; this had fallen to less than fifteen per cent by 2010 (Werz/Hoffman 2013). Syria has suffered from the presence of more than a million Iraqi refugees who fled after the US invasion of 2003. In Tunisia, due to urbanization, the rural population is declining despite the growth of the total population. In Algeria there have been repeated violent protests in cities, especially by unemployed immigrants from sub-Saharan Africa. Yemen suffers from the struggle over dwindling water resources and illegal water sources. Because of its growing population, the capital Sana'a could use up its groundwater in a few years. Agriculture in Morocco is particularly vulnerable to climate change (Schilling/Freier/Hertig et al. 2012). Water and food supply in Egypt depends heavily on the Nile, which is increasingly being used by upper riparian states (Link/Piontek/Scheffran et al. 2012).

Due to the growth in population and the flow of people into the cities, ethnic, religious and social tensions are increasing. Health and social services are under pressure, infrastructure and utility networks are overloaded. Frustration sparks protests, especially among the young population, and increases their will-

ingness to leave the country to travel to the north. The problems of North Africa are linked in complex ways with those of the Sahel, which is also affected by climate change. Libya has been the destination of many migration routes from the south because of the oil revenues and related jobs. Nearly 700,000 immigrants constitute more than ten per cent of the total population. After the government was overthrown, tensions with immigrants increased, and many of those who were expelled were suspected of collaborating with Mu'ammer al-Gaddafi. Since the border became more permeable, some of the armed mercenaries who went to Mali and other countries of the Sahel have contributed to destabilization there. In sub-Saharan Africa, climate change, desertification and scarcity of resources have become connected with economic and social marginalization, political instability and violent conflict; this has undermined the livelihoods of farmers and herders and increased pressure for migration (Gemenne/Brücker/Ionesco 2013).

13.3.7   Climate Change and Violent Conflict

By altering natural and social livelihoods in many regions of the world, climate change represents a potential driver for conflict and related acts of violence. These include civil wars and military interventions that in turn are associated with various negative consequences such as famine, economic crises, refugees, resource exploitation and environmental degradation (WBGU 2008). The joint vulnerability to environmental change and violent conflict in hot spots may multiply and lead to a spiral of escalation that is difficult to control and can spread to other regions (see figure 13.2).

Different fields of conflict are relevant in the context of climate change (see chapter 12 by Ide/Link/Scheffran et al. in this volume). There is a widely held assumption that the growing consequences of global warming will increase the likelihood of conflicts that are aggravated by the destruction of human livelihoods and resources. In addition, there are potential differences over strategies for avoiding climate change as well as over the financing of these strategies, which can involve damages, costs, and the resistance of other actors. Examples include the controversy over the use of nuclear power as a contribution to CO_2 abatement or the debate about the consequences of bioenergy, which also determine the German discourse (Webersik 2010; Scheffran/Cannaday 2013). The same applies to differences on adaptation to climate change and its security implications, such as alternative farming practices, protection measures such as dam building, and military operations in disaster management. Particularly prone to conflict are technical interventions into the climate system (geoengineering or climate engineering) in order to remove CO_2 from the atmosphere or to influence the earth's radiation budget. Such measures raise critical issues of technical and economic feasibility, risks and responsibilities at global, national and local levels (Brzoska/Link/Maas et al. 2012). In all these consequences and responses there are concerns about justice when it comes to the distribution of the costs, benefits and risks of climate change, and this can complicate cooperative solutions.

The potential contribution of environmental change and resource use to violent conflict has been the subject of scientific controversy for more than two decades. While some studies claim that natural disasters and resource scarcity put social systems under stress, threaten their stability and make violent conflict more likely, others see no clear causality for historical events that is detectable by statistical methods, and emphasize the ability of human societies to deal with resource issues through collaboration and innovation (see reviews in Brauch 2002, 2009). So far, most environmental conflicts have been regional in scope and have presented no threat to international security (Carius/Tänzler/Winterstein 2006). The connections are regionally quite different and depend on the resource type. While for renewable resources (water, food, biodiversity) scarcity is more likely to be a conflict factor, for non-renewable resources (fossil fuels, uranium, diamonds, coltan) abundance is more likely to lead to conflict. In both cases, violent conflicts consume resources, and this can drive or restrain a spiral of violence (Scheffran/Ide/Schilling 2014).

More recent debate has addressed the links between climate change and violent conflict. This issue was raised in the fifth IPCC Assessment Report (IPCC 2014; Gleditsch/Nordas 2014). Some studies looking at long historical periods have found significant correlations between climate variability and violent conflicts, particularly in the Little Ice Age in Europe between the fifteenth and the nineteenth centuries. Research on more recent periods has produced mixed results, which depend in a complex way on the regional context and on the conflict situation (see reviews in Scheffran/Brzoska/Kominek et al. 2012a, 2012b). Studies using data selectively and studies aiming to prove the relationship between climate change and violence over all historical periods, regions of the world, forms of violence and causal mechanisms

(Burke/Miguel/Satyanath et al. 2009; Hsiang/Burke/Miguel et al. 2013) have exacerbated the controversy (Buhaug 2010; Buhaug/Nordkvelle/Bernauer et al. 2014).

Regardless of the interpretation of historical data, the impact of future climate change goes beyond previous experiences, leaving space for scenarios, plausibility considerations and speculations. It is indeed possible that societies have been able to adapt to moderate climate change over historical periods of time, but they may be overwhelmed in the future by rapid and strong climatic change that exceeds their adaptive capacities. There is a wide range of possible conflict constellations (WBGU 2008) associated with the effects of climate change on rainfall and water scarcity, land use and food security, migration and refugee movements, extreme weather events and natural disasters, and vegetation and biodiversity. All these processes can become conflict factors individually or in conjunction (see chapter 12 by Ide/Link/Scheffran et al. in this volume). In addition, the effects of climate change on infrastructure and social destabilization (see section 13.5) may become crucial and trigger societal 'tipping points', leading to social unrest, riots, violence, crime and armed conflict.

Particularly pronounced is the vulnerability of agrarian societies with a high level of population growth and a low level of development, or that are already suffering from violent conflicts (Raleigh/Urdal 2007). Whether climate change acts as a 'threat multiplier' and creates a 'climate of violence' depends largely on how people and societies respond to change, and on whether their adaptive capacities and institutional structures are adequate for maintaining stability and supporting viable solutions. While the rich industrialized countries are not spared by climate change, they may benefit from more advanced economic and institutional conditions for problem-solving and conflict management. Potential issues of conflict in Europe include tensions over territorial claims and natural resources in the Arctic and the Mediterranean (figure 13.2). The melting of the polar ice sheets affects the strategic interests of Europe, Russia and North America. Efforts between Europe, the Middle East and Northern Africa to build a power grid based on renewable energy open up the possibility of converting the Mediterranean from a region dominated by oil interests towards a region of cooperative security (Brauch 2012; Scheffran/Brauch 2014), provided that the utilization of energy is sustainable, and promotes development, peace and justice.

As an example, the importance of climate change for conflict in North and East Africa is discussed below. In the wake of the Arab spring, Syria and Libya experienced bloody unrest that led to a coup in the case of Libya, and in the case of Syria to a bloody civil war that has spread the cycle of violence to other regions. In both countries, climate change plays a dual role. With a rise in temperature and decreased precipitation water supply and agriculture are hit, affecting people's lives. The indirect consequences of conflict are mediated by the political unrest in which international food markets play a role. As in other MENA countries, both countries are facing water problems that compromise the supply of this elementary good (Schilling/Freier/Hertig et al. 2012).

For several decades, Sudan has been characterized by political instability and violent conflicts, reinforced by national power games, regional struggles and global geopolitics. Peripheral regions such as Darfur are characterized by marginalization and exclusion, and this leads to disintegration and secession. The complex nexus of problems includes population pressure, unsustainable exploitation of land and forests, declining agricultural productivity, food insecurity, and the spread of diseases such as malaria. Associated problems are environmental changes and resource degradation, which cause water shortages and the deterioration of pasture land in the northern Sahel following drought and desertification. This exacerbates competition for resources between herders and sedentary farmers. The expansion of mechanized agriculture continues to deprive nomadic peoples of their traditional migration routes, to dispossess peasants, and possibly lead to serious tensions

Attempts by the Sudanese government to set up new administrative structures have weakened traditional mechanisms of conflict resolution and resource allocation between tribal groups. This issue has become intermingled with the conflict between the government and the rebels. The resulting expulsions, killings and destructions, as well as the inflow of weapons, have increased uncertainty and fuelled the spiral of violence in Darfur. In the vicinity of crowded refugee camps further environmental and health problems have been caused by firewood collection and by infectious diseases (Scheffran/Ide/Schilling 2014).

The role of climate change as a 'conflict amplifier' in Darfur is controversial. While some observers classify Darfur as a "tragic example of a social collapse as a result of an ecological collapse" (UNEP 2007: 12–13), others are concerned about the oversimplification of the Darfur conflict (Butler 2007). They criticize the

government of Sudan for exploiting the climate argument to distract from its own responsibility (Verhoeven 2011). Overall, climate change is one of many conflict factors that reinforce each other in a complex way (for other regional case studies, see Scheffran/Brzoska/Brauch et al. 2012c; also chapter 12 by Ide/Link/Scheffran et al. in this volume).

13.4 Social Dynamics in Transformative Interactions

The previous section highlights how climate change is challenging the balance between natural and social systems and interacting with other phenomena in complex crises. It is crucial that we understand how systems respond and interact to ensure their viability and survival against destabilizing consequences and avoid 'tipping points' and 'cascading breakdowns'. Natural and social systems have evolved over long historic times and developed the ability to cope with changing environmental conditions by adapting their pathways to new circumstances. Some of these pathways may aggravate problems (maladaptation), others could help to diminish harm and develop new opportunities. A key question is to what degree of climate change societies can adapt and how effective and creative they are in developing strategies that are appropriate, fast enough and sufficiently complex to cope with the speed and intensity of global change (Scheffran 2015a). Thus, to understand a 'sustainability transition' it is essential to go beyond the analysis of interactions in crisis events and to investigate social dynamics, such as social networks and collective behaviour that would support a transformation. Some initial considerations are presented in the following section, with a few examples to illustrate social transformation, without in-depth analysis.

13.4.1 Social Interactions and Networks

There are numerous agents and networks in sustainability or low-carbon transformations and they include regions, countries, businesses and citizens, each acting across multiple levels according to their interests, capabilities and rules of behaviour. At local levels, citizens and consumers are the key players; at the global level of decision-making, the main actors are the governments of nation states, or groupings among them. Local and global decision-making are connected in both directions via multiple layers of aggregation, with the intermediate national level connecting individual citizens, governments and multinational organizations (figure 13.3). Each layer has its decision procedures for setting targets and implementing actions. To manage social dilemmas and complex crises, new rules, norms, innovations and institutions need to evolve (Ostrom 2000). In a world where national governments are still key actors, the challenge is to balance competing processes across different levels, from citizen empowerment and privatization to the formation of multinational organizations and institutions of global governance.

Figure 13.3: Micro–macro interaction between local and global levels. **Source:** Scheffran (2015a).

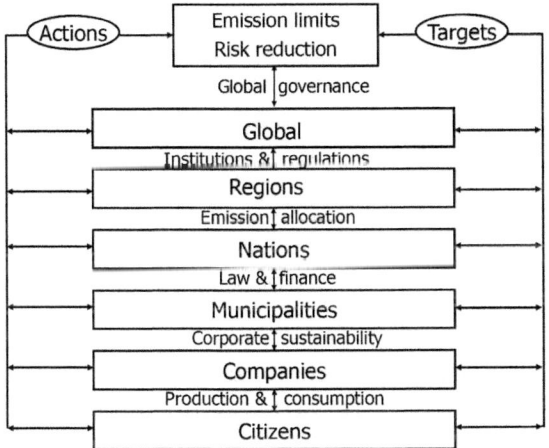

Social networks can promote the development of democratic and participative structures and open up opportunities for cooperation, peacekeeping and conflict resolution at local, regional and global levels. Each individual agent is a node in the network and can use its structures as power amplifiers, for destructive as well constructive purposes, to support or oppose a transformation. Social networks existed long before the Internet, though it has tremendously accelerated the formation of networks. Leaving aside climate change, there is a long history of social movements, from the workers' movement to the student, peace and environmental movements, which have played important roles in social change and political revolutions. An example is the peace movement of the early 1980s, which shaped public debate around the world and contributed to the momentous transformation that saw the end of the cold war.

In social networks, different forms of communication have been developed to deal with risks and conflicts, and these have facilitated discussions, discourses and disputes about the direction collective

action should take. Protest, resistance and whistle-blowing can bring negative developments to public and political attention, help avoid unacceptable consequences and risks, and build win-win solutions. Potential linkages between climate change and the Arab Spring of 2011 have been discussed in section 13.3.5, considering food price and social media to be key factors (Kominek/Scheffran 2012). In some cases events have induced a peaceful transition, in others they have led to civil war. A large amount of research has analysed the conditions for violence and conflict. There is less research on preventing the causes of social instability and violent conflict and building social networks of resource use, conflict resolution and cooperation (Ngaruiya/Scheffran/Yang et al. 2015). In the context of 'sustainability transitions', new research is emerging on natural boundaries (Rockström/Steffen/Noone et al. 2009); forerunners in transitions (Hasselmann/Cremades/Filatova et al 2015); and the limits to the Anthropocene (Lüthje/Schäfer/Scheffran et al. 2011).

13.4.2 Collective Action and Governance

As we face common problems like climate change, without coordination and governance there is the risk that the world's more than seven billion people (2010) will waste resources, act against each other and run into conflicts. Every country, industry and citizen contributes to greenhouse gas emissions, and to driving the world into a climate of complexity. The question is whether and how quickly agents and stakeholders are learning to use their steering wheels in a coordinated manner. This is needed to develop goals and adaptive strategies that can drive the planet through the complex and foggy landscapes of climate change, where information is limited and uncertain, but continuously updated. A lack of agreement on the underlying causes, on the risks to be expected and on the actions required is impeding progress. To agree on a path that keeps the cumulative emissions and temperature changes within tolerable limits so that dangerous climate change is averted is a collective action problem. Part of the solution is to translate goals and implement agreements into rules, structures and institutions that prevent or minimize the causes, consequences and conflicts in the complex crises described above. Any substantial progress has been hampered by the expected costs and risks of transformation, by the constraints placed and the resistance by lobbying and partial interests that undermine the required cooperation.

To overcome the hurdles associated with the 'tragedy of the commons' (Hardin 1968) a multilevel process of negotiation is required (figure 13.3) that involves citizens, firms, institutions and states (Scheffran 2008). In negotiations, agents mutually restrain and adjust their actions in order to achieve joint benefits, to reduce costs, or to diminish risks. This process requires binding and verifiable agreements such as the Paris Agreement of 2015, an important stepping stone that needs to be implemented. Identifying individual emission commitments for each agent and ensuring their compliance remains a challenge (Scheffran 2016a). To address diverging preferences, some degree of cooperation is indispensable. Climate policy could help to unite the international community towards dynamic and globally coordinated actions and to achieve co-benefits with other policy fields (Ipsen/Rösch/Scheffran 2001).

To manage the technical and socio-economic transformation for sustainable development demands collective decision-making, learning and negotiation among multiple stakeholders, so that solutions can be found that bridge the differences, avoid conflicts and establish cooperative structures. Besides climate change, the transitions between conflict and cooperation have been studied extensively for fisheries management. In that domain, collective decision-making, learning and negotiation have been demonstrated to be more beneficial for both natural and social systems than individual actions of rational choice (Eisenack/Kropp/Scheffran 2007b; Bendor/Scheffran 2009). Different concepts can complement each other and enable joint action to reduce the burden on humans and nature and strengthen social skills. Governance mechanisms that strengthen human creativity and citizens' participation in democratically legitimated decisions, balance interests in stakeholder dialogues and mediation. Peaceful coexistence, cooperation and innovation are instrumental for a viable society and important for satisfying human needs and protecting livelihoods. To speed up emission reductions and technological change in the energy, agricultural and other sectors, an unprecedented degree of international action and cooperation is required. This must include the international transfer of investments and technologies to shift the composition and structure of the energy system towards emission reductions (see chapter 34 by Scheffran/Froese in this volume).

A challenge in the transition is to bridge micro-macro linkages in collective action across multiple levels of society through the formation of coalitions that acquire capabilities from individuals to generate joint values (Göbeler/Scheffran 2003; Scheffran 2006). In these coalitions agents can adapt their climate targets or merge their resources and investments to be better off by acting together rather than acting alone (Eisenack/Kropp/Scheffran 2007). Coalitions are reasonable if individual action is insufficient or inefficient, e.g. in acquiring a critical number of votes or a critical mass of investment to realize projects. Coalition formation describes the transition from individual to collective action as a bargaining process where the probability of joining a coalition increases with the values actors expect from it (Göbeler/Scheffran 2003; Scheffran 2006).

1. In resource-based coalitions actors merge their investments to generate joint benefits which are then distributed to the individual actors (Scheffran 2006). With the entry into force of the Kyoto Protocol in 2005, the international community has established a first set of cooperative instruments that address the problem of global warming. Market mechanisms are assumed to provide an efficient and cost-effective allocation between regions and businesses. Emission trading is designed to achieve emission reductions in regions and business sectors where they are least costly. The Paris Agreement of 2015 opens up opportunities for a wide range of strategies and instruments in climate policy where each state commits to individual emission reductions and related investment, so that the joint global temperature constraints can be achieved over the coming decades (see Scheffran 2016a).

2. In value-based coalitions actors seek agreement by adapting their positions and values to each other. Coalition formation is more likely among actors with similar positions. For instance, countries tend to cooperate more closely if they prefer similar levels of carbon concentration or temperature change, as can be seen in the formation of country groups in the *United Nations Framework Convention on Climate Change* (UNFCCC) and the Kyoto process. Moving beyond Kyoto has been a challenge for the policy process that is supposed to implement the longer-term objectives and manage the potentially severe implications in case of failure. While global warming affects the integrity of natural and social systems, with severe risks to their stability and security, UNFCCC Article 2 demands the stabilization of atmospheric greenhouse gas concentrations at levels that "prevent dangerous anthropogenic interference with the climate system". In Copenhagen in 2009 (COP 15)

and Paris 2015 (COP 21), this was translated into a common global goal of a global rise in temperature of less than 2°C. In this context, the Paris summit and the agreement adopted at COP 21 was a major expression and achievement of a joint global climate policy and of the emerging world civil society. More than 30,000 participants acted together, despite conflicting interests and power structures.

These issues pose a challenge for the scientific community, which is increasingly becoming involved in normative issues, value judgments and soft science; these all require innovative integrated approaches at the science-policy interface. Science can also contribute to assessment of the decision-making process by developing analytical concepts and the application of decision methods and tools (Sprinz 2005). In the following section, we discuss alternative approaches and concepts for developing a new framework appropriate for analysing social interaction, collective decision-making and governance across multiple levels.

13.5 Concepts of 'Sustainability Transitions'

Analytical concepts evaluate the evolution of complex dynamic systems according to criteria and indicators to avoid destabilizing consequences, tipping points and cascading breakdowns, and to facilitate a switch to a transformation of human–environment interaction that balances natural and social systems. After a short discussion of the concepts of stability, resilience, sustainability and peace, we focus on the concept of viability as an emerging integrative framework.

13.5.1 Stability and Resilience

The stability concept is a well-established paradigm in system dynamics and has played a significant role in international relations during the cold war (Scheffran 2011, 2015a; Scheffran/Link/Schilling 2012). In general terms stability means the ability to dampen the effect of disturbances in order to keep the essential characteristics of a system state (often an equilibrium) within boundaries. If this fails, there is the possibility of destabilization associated with systemic breakdowns, phase transitions and transformation processes. Examples include the collapse of natural and social systems under existential stress, transitions between war and peace, and the transformation from the exploitation of natural resources to their sustaina-

ble use. In the climate context it is relevant whether anthropogenic global warming leads to a destabilization of human-environment interaction, as described above, or is contained by self-stabilizing mechanisms.

Resilience often refers to the capability of a system to handle and survive external challenges and disturbances such as climate change. A 'resilient' system is able to withstand external shocks and to maintain or restore the conditions of its own existence (Holling 1973). Social resilience implies that a community is able to respond appropriately to disturbing changes (Adger/Kelly/Winkels et al. 2002). In a resilient social environment, actors cope with and withstand the disturbances caused by climate change in a dynamic way that preserves, rebuilds, or transforms their livelihood. Resilience depends on multiple abilities and characteristics, such as connectivity, innovation and creative thinking. Key resilience strategies include the building of networks, the cultivation of diversity, and the maintenance of flexibility (Scheffran 2015a). In the context of crisis management, resilience describes the "capacity over time of a system, organization, community, or individual to create, change, and implement multiple adaptive actions, procedures, and processes in the face of various types of crises" (Penuel/Statler/Hagen 2013: 811). In a changing world, individuals, communities, and organizations that are willing and able to innovate have a greater potential to confront a crisis that challenges the existing structure and to relate to it as an opportunity for further development.

13.5.2 Sustainable Development

Concepts of 'sustainable development' seek to balance economic, social and ecological issues of concern for present and future generations and integrate the human sphere (socio-sphere) into the boundaries of the natural environment (eco-sphere) (WCED 1987; Scheffran/Stoll-Kleemann 2002; see chapters 1 by Brauch/Oswald Spring and 18 by Mesjasz in this volume). The challenge is to make apparently conflicting objectives compatible (Scheffran 1997): 1. *sustain*[4] refers to *preservation* and *upholding* (in German "Erhaltung") of natural resources as the life-enabling base of society and precondition for human existence; 2. *development* means the *unfolding* of opportunities and abilities to improve human well-being and promote societal progress.[5] While the first task is to keep

---

4 The term *sustain* goes back to the Latin *sustinere* (*sub* = up, *tenere* = hold) which translates into *upholding*.

given abilities, development seeks to expand them in a progressive direction.

These goals were already suggested in the definition by the World Commission on the Environment and Development (1987): "sustainable development is development that meets the needs of the present without compromising the ability of future generations to meet their own needs". Thus, the task is to balance the human needs and abilities of current and future generations. While 'needs' are based on human preferences and values, the term 'ability' is related to the opportunities created by natural resources (Scheffran 2011) and the capacity to act, which is a precondition for achieving needs. In a nutshell, sustainability can be defined as the ability to sustain ability. Ability has many facets related to human action, from capacity-building and Amartya Sen's (1986) 'capability approach' to the various dimensions of natural, physical, human and social capital (Scheffran/Remling 2013). While natural and physical capital determine the fundamental role of nature and technology for human capability and productivity, human capital comprises knowledge, skills and capabilities that enable human beings to act in new ways, while social capital refers to the institutions, relationships and norms that shape the quality and quantity of a society's interactions.

Integrated sustainability strategies seek to balance the two ends of the relationship between natural resource input and human value output, taking into consideration several contextual and intermediate factors (Scheffran/Stoll-Kleemann 2002). Sustainability indicators apply to human impacts on natural resources as a source (extraction for human production and consumption) and as a sink (impacts of human activity on nature). The impact of resources on humans is also twofold: resources as a production factor needed for human well-being and as a risk factor for human security. The 'Kaya identity' combines the drivers of human impact on carbon emissions and on environmental change (Kaya/Yokuburi 1997; IPCC 2014). Affecting these drivers represents different sustainability strategies: changing labour productivity, technical innovation, investment in human and social capital, resource intensity, renewable energy and sustainable lifestyles and consumption patterns, environmental protection, risk reduction and conflict management. Instead of destroying boundaries through the growing extraction and destruction of material,

energetic and biological resources, development into a multidimensional space of values would ensure the preservation of natural conditions and boundaries. The implementation of sustainable development has met various obstacles which are in conflict with competing interests and the world's power structures.

13.5.3   Peace and Sustainable Peace

In human history, visions of peace have a long tradition and have served as a model for conceptions of alternative futures of societies. Peace is often mainly associated with its negative definition as the absence of war (Galtung 1967), or in a broader sense the absence of violence (intentional destruction). Peace strategies are based on the principles of non-violence, preventing and excluding the use of violence and diminishing its potential through disarmament, transparency and confidence-building. The absence of immediate physical violence is, however, not necessarily evidence for peace. A situation could hardly be called peaceful if stronger actors were to suppress the weaker ones, who then surrender for fear of repression. This would correspond to what Johan Galtung (1969) has called 'structural violence', as a situation that encompasses severe constraints on possibilities for development. The criticism has been levied that giving violence such a broad meaning could lose focus and see anything that is interpreted as a loss compared to an ideal or desired world as violent (Scheffran/Ide/Schilling 2014). Renouncing violence as an illegitimate means of power is a more viable pathway in the pursuit of peace and sustainable development than 'just wars'. While negative peace is characterized by 'freedom from fear' of violence and hostility, positive peace rests on the freedom to develop opportunities for a satisfying and healthy life. This includes basic human rights and justice, as well as participation and conviviality in social interactions, reconciliation, healthy social relationships, prosperity in matters of social or economic welfare, and a working political order that balances the interests of all.

Like sustainable development, peace rests on two key principles with regard to the existence and development of human rights (Czempiel 1995: 170): 1. preservation and protection of the existence, integrity and identity of each individual by excluding violence; 2. self-fulfilment and unfolding of the individual through equal distribution of development opportunities. Upholding and unfolding are thus essential categories of both sustainable development and peace.

---

5   This dual role of holding and unfolding (*Erhaltung und Entfaltung*) has been recognized in Bender (1991, 1998).

Common features facilitate the joint conception of *sustainable peace* (see chapters 1 by Brauch/ Oswald Spring and 9 by Brauch in this volume). The idea of peace is accepted in political conflicts and international relations, often as a normative concept for the avoidance of violence and war. After the cold war, however, sustainable peace became a framework for global and planetary security, that would protect the earth against environmental threats and harmonize the conflicting goals of environmental and security policy. Sustainable peace addresses multiple linkages in this nexus. On the one hand, it assesses the negative interactions between armed conflict, environmental destruction and low levels of development (Scheffran/Vogt 1998). Both the lack of sustainability and peace are associated with the loss of development opportunities. If the world is more violent and less peaceful, sustainable development may also be threatened, and this could further aggravate conflicts. Degradation of sustainability undermines living conditions. This further undermines the conditions for peace and leads to violence, running into a 'vicious cycle' of a non-peaceful and unsustainable world, where the principles of non-violence, justice and preservation of life are violated, thus distracting attention from fair and equitable sustainable development.

On the other hand, positive linkages can emerge between human development, environmental protection and peace-building, leading to a 'virtuous cycle'. While sustainable development may contribute to peace, the preservation of peace may be an essential precondition for the cooperative implementation of sustainable development. As suggested above, sustainability connects the development opportunities of society to the preservation of the environment, and the development of individuals depends on the preservation of their existence in times of peace. To avoid conflicts and to strengthen the positive links between natural and social capital, it is essential that both the natural and social sciences jointly contribute to integrated sustainability strategies and that they find co-benefits, e.g. through consensus, arrangement and conflict resolution in environmental and security issues. It is important to understand how environmental degradation, low levels of development, and violence mutually influence each other; but also whether sustainable development can be achieved, even under non-peaceful conditions; and whether peace is possible, even under unsustainable conditions.

Whether positive (cooperative) or negative (conflictive) linkages will prevail depends on circumstances and human responses as well as on their speed and adequacy. While both pathways may develop in parallel, it is possible that there is a bifurcation or tipping point beyond which either pathway could prevail, following mutually enforcing mechanisms that trigger cycles of insecurity and violence or cycles of cooperation and peace (see section 13.7 below). For the time being the world may enter an undefined und uncertain path on the brink between conflict and cooperation, where small changes may drive regional development in one or the other direction. We may not know the boundary between stability and instability, how far away we are from it, how fractal its shape is, or how robust against changes it is. For the time being, the world may be moving along the edge close to a tipping point.

If seriously implemented in the mechanisms of local and global governance, the goal of sustainable peace could be an important contribution to a preventive strategy of global risk avoidance and reduction rooted in the satisfaction of basic human needs without destroying the essential conditions of life. People who are satisfied with themselves, their natural environment and their social livelihood are assumed to have less incentive to use violence or violate the social norms essential for their wealth. In this respect, sustainable peace would be an important contribution to the preservation of peace, more effective than attempts to control violence in a non-peaceful world.[6]

In addition to the two aspects of preservation and development (upholding and unfolding) already mentioned, the constructive and transformative aspects of sustainable peace can be added by reference to a third task: shaping and designing a future world. In the German language the word 'Gestaltung' can be represented by the English term 'configuring' to fit something into a proper shape, form or design, thus creating a harmonious relationship between the real and the desired world.[7] To summarize our conceptual excursion, the triangular relationship between sustainability, development and peace can be represented by the three tasks of upholding, unfolding and configuring (in German: *Erhaltung, Entfaltung, Gestaltung*, following a suggestion by Bender 1991, 1998). In short: upholding current abilities serves as a basis for unfolding enabling opportunities and configuring environmental transformation towards new pathways and spaces. The challenge is to find viable pathways that

---

6 In German, the word for 'satisfaction' is 'Zufriedenheit', which includes the word 'Frieden' meaning peace.

7 An alternative term is 'immoulding' which however is obsolete in modern English language.

transform current human–environment interactions into the desired future ones, taking into account the complex non-linear dynamics of the respective systems.

13.6 Viability as an Integrative Framework of Transformation Processes

The conceptual considerations in the previous section have laid the ground for developing the overarching integrative framework of 'viability' as a basis for analysing future transformation processes and pathways. This goes beyond established approaches, such as forward system dynamics simulations and integrated assessment modelling based on rational choice and optimal control of dynamic interactions between natural and socio-economic systems under climate change.

13.6.1 Tolerable Windows and Adaptive Control

Optimal control methods seek to maximize time-discounted utility functions (welfare), which include expected benefits, potential damages and the costs invested. Integrated assessment models use optimal growth of production functions with capital, labour and technology to design climate policies (Nordhaus 1993; Edenhofer/Lessmann/Kemfert et al. 2006), usually assuming complete knowledge utilized to select time-discounted control paths. The merits and limits of optimization-based integrated assessment have been widely recognized (see Füssel 2006). The requirements of rational choice methods such as optimal control and game theory, e.g. perfect foresight and complete information, are hardly fulfilled in decisions taken under conditions of deep uncertainty and complexity. Going beyond optimization over long-term horizons, adaptive decision-making and cooperative management allow actions and targets to be adjusted to the limited knowledge of the climate state and the capabilities of decision-makers, as well as to the complex socio-economic interactions that undermine predictability (Scheffran 2008a).

Forward-looking approaches determine a set of emission trajectories based on computer simulations in order to project future system dynamics from initial conditions by variation of scenario-dependent parameters. An alternative method is to use inverse approaches that calculate admissible 'channels' of emission trajectories compatible with a tolerable or targeted temperature range. An example is the *Tolerable Windows Approach* (TWA) and related concepts (safe landing, guard rails) that restrain and adjust the path of GHG emissions to keep average temperature change within viable bounds of natural and social systems and avoid critical risk levels (Bruckner/Petschel-Held/Toth et al. 1999; Petschel-Held/Schellnhuber/Bruckner et al. 1999). The space to manoeuvre for future climate policy is provided by assessments and judgments of climate change that take into account vulnerabilities and adaptive capacities as well as critical thresholds for phase transitions and extreme events that may cause qualitative system changes. Once the tolerable boundaries are found, control variables and regulation mechanisms for staying within these boundaries are identified.

An extension of optimal control and TWA is adaptive control, where actions are taken according to rules that respond to the actual state of a system in a prescribed direction (Scheffran 2008b). Adaptive control approaches constrain and adjust GHG emission pathways towards a target or towards staying within a viability domain of the climate system (e.g. a certain range of carbon concentration or global-average temperature change). Actors decide and act on the basis of incomplete knowledge, usually limited to a spatial and temporal window of attention. An important rule in climate policy is to stay below critical atmospheric carbon concentration and temperature thresholds. Other boundaries have been identified by Rockström, Steffen, Noone et al. (2009). Once boundaries are defined, the distance from the current position as well as the rate of change constitute drivers that are used to adjust the correcting response in order not to exceed the threshold.

Adaptive rule-based approaches are useful in complex and deeply uncertain situations where actors do not know enough about the future to calculate long-term optimization (Scheffran 2008a, 2008b). Within the range of uncertainty, a forward-looking actor can estimate whether future pathways reach the target or enter 'forbidden' areas, something which requires actions within given constraints. Continuously updated scientific information is essential for estimating whether expected future trends are tolerable within given limits or corrections are needed, just like steering and speed controls in a motor vehicle which are adapted by sight, destination and events. Similar adaptive control strategies are implemented in other contexts via decision rules and responses to a changing environment, taking into account actions taken by other actors.

Timing matters, as rapidly changing environments require faster action than long-term decision-making. For instance, impact-driven responses are too late in climate policy because of the considerable time lag of several decades between emission reductions and the visible impact of climate change (such as sea level change). In addition, the effect of policies will be delayed: for instance, the replacement of infrastructure and technology such as buildings, power stations and transport systems as part of a low-carbon transformation can take several decades. Another issue is time discounting by decision-makers, where the long-term future is given a lower value than the short-term future; this leads to 'wait and see' policies that favour short-term investment priorities. More appropriate are anticipative and adaptive strategies that are precautionary and seek to avoid risky pathways.

13.6.2 The Viability Framework of Anticipative and Adaptive Governance

Adaptive control and guard rails approaches can be embedded in a broader management framework of viability in order to analyse complex human–environment interactions and transformation processes. Since the 1980s 'viability' has emerged as a paradigm in complexity and sustainability science (Aubin 1991; Grossmann/Watt 1992; Bossel 2001; Scheffran 2015a). Originating from the Latin word 'viabilis', which combines the words *via* and *abile*, viability refers to 'able and enabling pathways' and the ability of a system, such as an organism, ecosystem or social system, to live, grow, and develop within the limits of its specific environmental conditions and boundaries. These boundaries may be transformed themselves by expansion or reduction of the available space of possibilities and opportunities.[8] If as a result of changing environmental conditions and human actions tolerable domains of viability are left, key functions of the system can be lost, and this threatens its identity and existence. Catastrophic changes and cascading events may lead to the loss of valuable functions and a breakdown of the system. To maintain viability it is essential for a system to stay within critical tolerance thresholds of its ability to exist and set up control and regulation mechanisms in time to avoid exceeding these limits (Aubin/Saint-Pierre 2007).

Viability theory provides methods and tools for adaptive management of dynamic systems within given boundaries. Mathematical viability theory studies algorithmic methods in complex uncertain systems (e.g. the biosphere, societies, markets, rural and urban areas, etc.) which develop in neither a completely deterministic nor a completely stochastic way, in conjunction with their viability constraints (Aubin/Saint-Pierre 2011).

A crucial but often neglected issue of viability is identifying the 'essential' attributes that characterize a system. While an individual system is represented by a large set of attributes, a class of systems requires only a limited set of attributes. For instance, the term 'human' is characterized by a few attributes that are typical for all humans, while a particular human being can be described in detail by numerous specific characteristics such as size, weight, location, age, name, etc. Interactions in the real world, such as moving through time and space, continuously change attributes that are less essential while existential attributes are preserved (such as the attributes that identify humans in general). Transformations change attributes along certain pathways, some of which may be essential and lead to qualitatively new systems, such as the ones that have created humans in biological evolution or that identify them in social evolution. Conceptions and constructions of identity play an essential role, as these define the key attributes and functions of the system—those that are to be maintained or achieved.

Viability combines quantitative and qualitative indicators that are partly subject to measurable data and scientific criteria (e.g. threshold for breakdown of ecosystems, number of conflicts) and partly subject to value-based judgements for the 'essential' attributes of identity or liveability of socio-ecological units (organisms, humans, societies) which can decide according to their own constructions and follow rules about when and why they change behaviour. Viability translates judgments based on value and capability into system states and trajectories that are to be achieved or avoided.

To preserve the attributes essential for a system's identity, certain capabilities are required to compensate for disturbances. Where the permissible limits are and when the viability of a system is compromised are things that are difficult to determine for complex systems. Exceeding limits means losing system functions and capabilities that are essential for their identity. If identity is based on a large number of key attributes, viability is harder to achieve if any of these attributes

---

8 Wikipedia refers to viability as the ability of a system to maintain itself or recover its potentialities. This resembles elements from the definition of resilience in a wider context.

is threatened. If critical thresholds are passed, the system is more difficult to stabilize. In particular, this is the case once a loss of crucial capabilities is experienced (such as human disabilities in walking or seeing) or pathways are pursued that provoke hostile responses from others that exceed the system's own capacity to deal with. On the other hand, additional capabilities can be gained along these pathways, e.g. by growing income or cooperative connections in social networks, thus creating mutual benefits. To meet the 'precautionary principle', early (self-)regulation mechanisms are required in order to prevent instability and maintain a system without compromising the viability of interconnected systems.

In the environmental field, viability-related approaches have been applied, for example, to the design of tolerable windows for fish harvests (Ben-Dor/Scheffran/Hannon 2009; Eisenack/Scheffran/Kropp 2006) and to guard rails for greenhouse gas concentration (Petschel-Held/Schellnhuber/Bruckner et al. 1999). Like any type of control this approach selects certain future pathways in dynamic systems and excludes others, by following normative concepts. Viability allows other concepts such as stability, resilience, sustainable development and peace to be integrated as specific representations for particular contexts, each having specific mechanisms to tackle climate change (Redman 2014). Since climate change may threaten the viability of systems and push them towards non-viable dynamics, it is important to diminish climate vulnerability and strengthen adaptation, to avoid the creation of a dangerous chain of events before it develops and becomes unstoppable.

The viability concept is closely related to the adaptive management of transformation processes in human–environment interactions (Scheffran/Link/Schilling 2012; Scheffran 2015a). Viable adaptive control is a useful instrument for designing and adjusting the complex interaction between the socio-economic, environmental and political spheres in natural resource management (see, for example, Aubin/Saint-Pierre 2007). Once political requirements or goals are defined, they can be translated into boundary conditions that allow stakeholders to identify the controls necessary to stay within sustainable limits and maintain resilience, e.g. for resource use or carbon emissions, and to avoid non-viable regions of collective human action in rural–urban interactions. Viable control and adaptive management are needed to implement actions based on locally updated information and decision rules for each actor that respond to the changing state of a system (Scheffran 2015a). In this

context a conceptual framework and a meta-model of multi-stakeholder interaction processes will be appropriate to implement the viability concept (see the VIABLE model in the next section).

Viability assessment seeks to identify non-viable areas and critical thresholds where the identity or existence of the system is threatened, for instance critical levels of water scarcity, food insecurity, poverty, mass movements, conflict escalation and ecosystem stability. The next task is to analyse dynamic pathways and consequences in the indicator space, identifying key patterns and trends in the earth system and how they could be affected by climate change projections. Using the classification of typical non-viable patterns and pathways in hot spots, possible destabilizing and conflicting pathways and responses that exceed critical thresholds and lead to non-viability would be explored, and their likely consequences discussed. Assessments can evaluate past trends and projected scenarios regarding the key indicators, including agent-based modelling of potential pathways and response patterns.

Finally, key capabilities, opportunities, rules and institutional mechanisms are identified in order to regulate and control the viability of pathways in the indicator space, avoiding or moving away from non-viable areas or crossing thresholds in either direction. Specifically, this includes variables that affect the pathways towards increasing viability of human–environment interaction; it also includes the discussion of opportunities for viable regulation and control. Examples are water, food and energy infrastructure and efficiency; possible social and technical innovations in resource use; social networks and remittances from migrants; cooperation, collective action, institutions and governance mechanisms; and strategies for disaster management, protection and resilience. The impacts of alterations in crucial couplings can be studied in order to improve viability in resource or migrant networks and possibly allow the resolution of conflicts between agents. Viability theory can be used for analysing and testing boundary conditions for various human–environment interactions.

The viability concept can be expanded to conflict studies in the context of viable adaptive networks (Scheffran/Link/Schilling 2012; Scheffran 2015a). While conflict escalation may violate essential viability conditions (e.g. through the use of violence), institutional mechanisms and regulations can maintain viability through conflict resolution and cooperation. The viability concept can also be used for analysing transformation processes between opposing pathways of sustainability and non-sustainability, conflict and cooperation, or peace and war. In the next section, a framework is described for modelling such transitions.

13.7 The VIABLE Model Framework of Societal Transformation and Adaptive Governance

To study the complexity and stability of interactions between systems and agents, various methods and tools have been developed, including system dynamics, dynamic game theory, agent-based modelling and social network analysis. Particularly relevant for complexity thinking are adaptive networks that switch dynamically between alternative pathways in response to changing internal and external conditions. Understanding the emergence of collective behaviour and the evolution of cooperation is a dynamic field of current interdisciplinary research. Of great interest is the dynamic spread of processes and patterns across social networks, such as the diffusion of social behaviour or technical innovations (e.g. Kempe/Kleinberg/Tardos 2005; Angourakis/Santos/Galán et al. 2014). Social networks evolve through processes of interaction such as conflict, cooperation, competition, and coalition formation among a large number of participants (Flint/Diehl/Scheffran et al. 2009; Maoz 2010).

Figure 13.4: The VIABLE adaptive model framework. **Source:** Scheffran, Link and Schilling (2012).

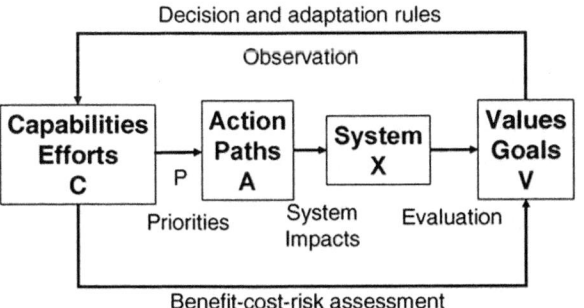

13.7.1 The VIABLE Model of Human Action and Social Interaction

In the remainder of this chapter, example cases of transformation processes are highlighted in the framework of a dynamic agent-based model of social interaction that has been developed by the author. The model framework VIABLE (*V*alues and *I*nvestments in *A*gent-*B*ased interaction and *L*earning for *E*nviron-

mental systems) describes the dynamic behaviour and interaction of agents who apply (invest) proportions of their available capabilities (C) according to priorities (P) to given action pathways (A) that change their environment (X). The observed impacts are evaluated based on the agent's values and goals (V), which are a function of the benefits, costs, and risks of the actions taken. Over time, an agent's actions are adjusted through repeated feedback and learning cycles according to decision and adaptation rules in response to environmental changes, the actions of other agents, and the assessment of expected benefits, costs and risks associated with each action path. These actions are adjusted according to the agent's ability to evaluate the natural and social environment around them and implement their decision rules.[9] Since the choice between different action pathways and the emerging social interaction patterns (such as conflict and cooperation, social networks and coalition formation) are core elements of the VIABLE model, it offers a framework for the modelling of transformation processes. Rather than explaining the structure of the model in detail (which has been done elsewhere: Scheffran 2003; Scheffran/Hannon 2007; Bendor/Scheffran 2016), a few examples and results are given in fields where the model has actually been applied to transformation processes.

13.7.2 Ending Nuclear Deterrence: Missile Defence and Nuclear Disarmament

The VIABLE model had its origins in the analysis of the cold war arms race between the superpowers in the 1980s, starting from the question of how to end nuclear deterrence, and moving from *Mutually Assured Destruction* (MAD) to *Mutually Assured Survival* (MAS). Alternative approaches have been presented: missile defence (the Strategic Defense Initiative SDI launched by US President Ronald Reagan in March 1983) or complete nuclear disarmament (suggested by the Soviet General Secretary Mihkail Gorbachev's plan in January 1986) (see figure 13.5). Concerns about destabilization led to the failure of the Reykjavik summit in October 1986, when both strategies became incompatible.

To study the complex intricacies of these issues, the SCX Model of security, cost and armament dynamics was developed in a PhD thesis by the author between 1986 and 1989, a year of momentous transi-

---

9   Model variants such as SCX, VCX, and VCAPS have been used in different application fields.

Figure 13.5: Competing transition plans for ending nuclear deterrence in the 1980s: cost projections for Reagan's SDI plan (upper part) vs. Gorbachev's plan for the elimination of nuclear weapons. **Source:** Adapted from: (top) Pike (1985: 11); (bottom) Spiegel (1986: 98).

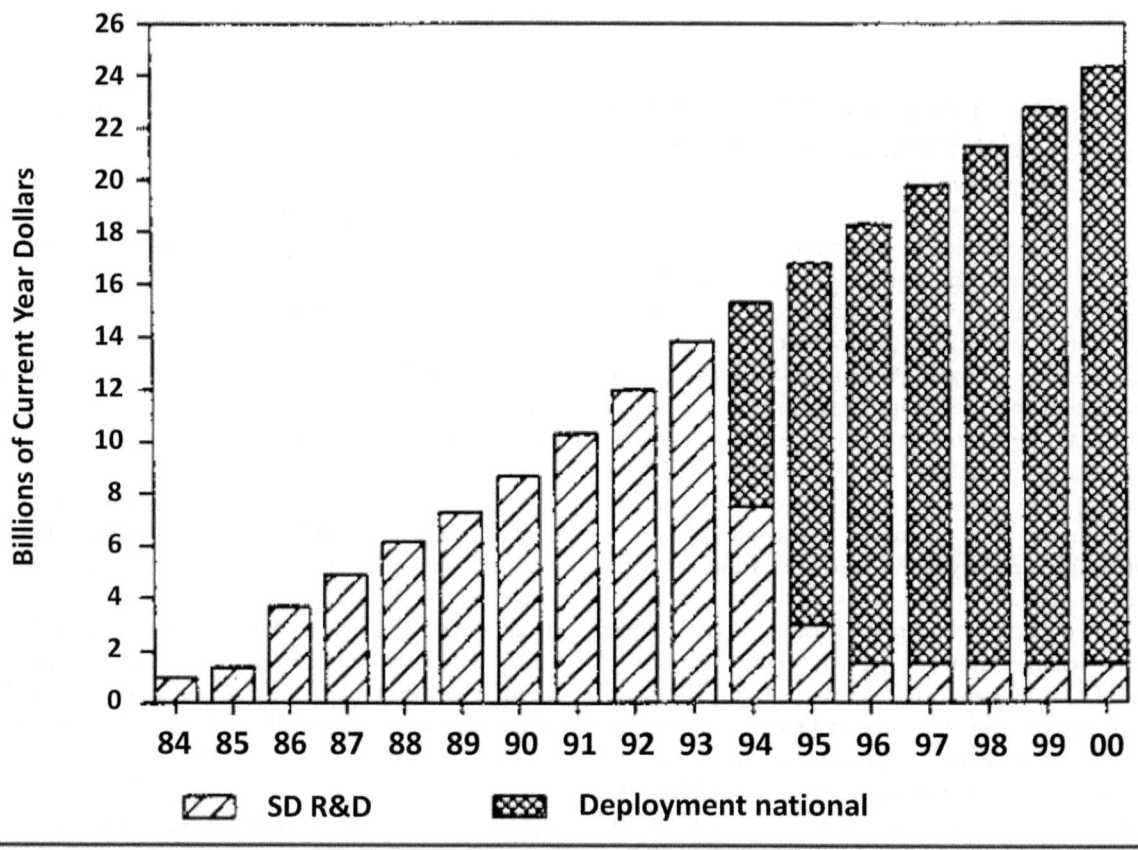

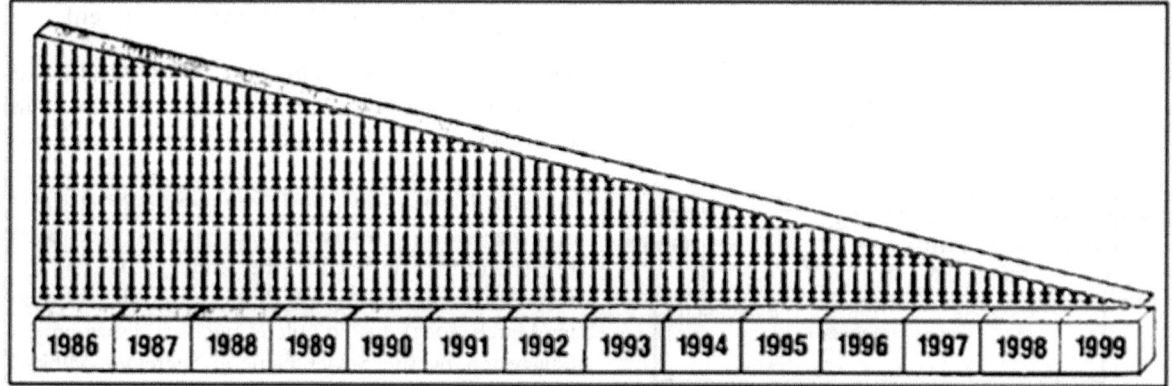

tions that led to the end of the cold war (Scheffran 1989).[10] Figure 13.6 shows fifty-year model runs for two selected variables (Security and Cost functions) for one of the scenarios. The results show variation of

the so-called 'worst-case parameter' between worst-case (w = -1) and best-case (w = +1) assumptions within a range of uncertainty of variables of strategic warfare (e.g. number of nuclear-armed missiles, capabilities of missile defence and space weapons, explosive yields, multiple warheads, missile accuracy, costs, etc.). The middle of the uncertainty range is represented by w = 0. It is striking that in the transition between the extreme cases a qualitative change of cascading events beyond a tipping point occurs. Moving

---

10 The thesis was submitted in early October 1989 and the final defence took place on 21 December 1989. In between these dates, the Berlin Wall fell, and with it the East European political regimes of the Warsaw Pact, except for the Soviet Union, which disappeared in 1991.

Figure 13.6: Simulation of the transition from negative (worst case $w = -1$) to positive (best case $w = +1$) perception of the SCX arms race model, with a chaos-like oscillation around $w = 0$. The figure shows the security function (lower part) and cost functions for the fluctuating scenario (total and variable cost, upper part) $\delta = 0$ implies no time delay in responses. **Source**: Scheffran (1989: 245).

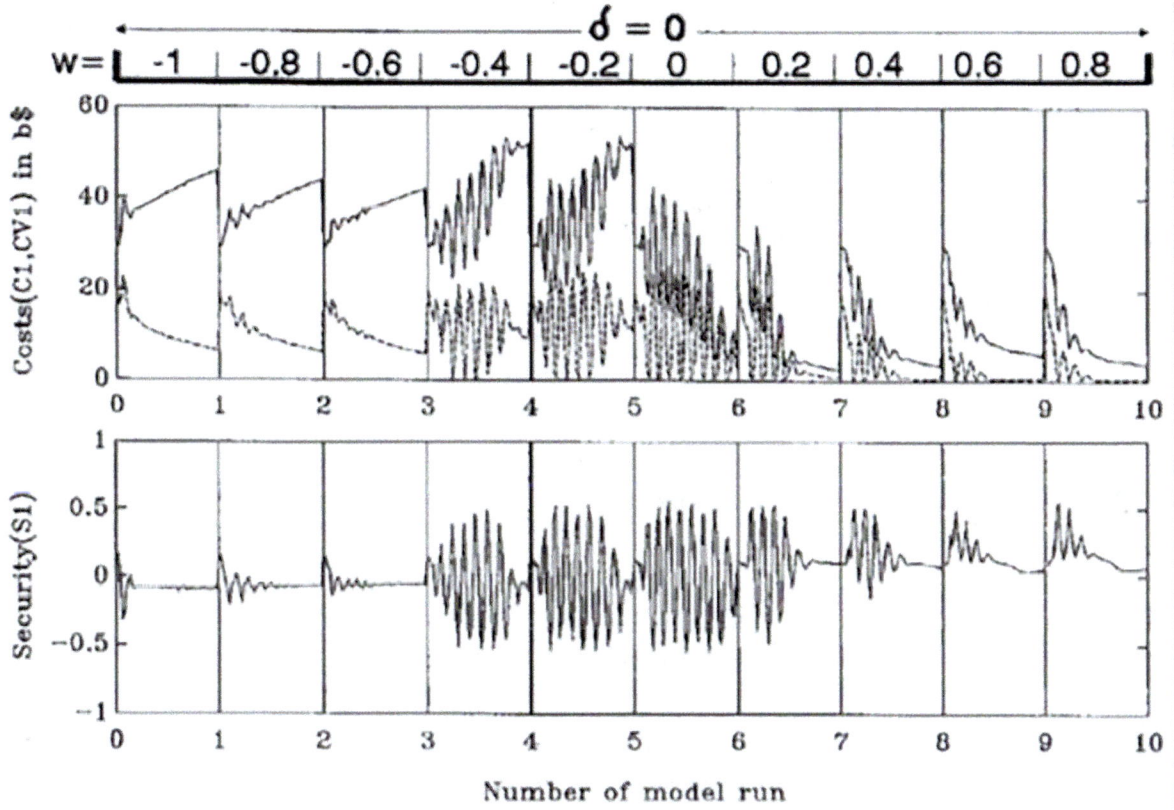

Figure 13.7: Tipping point in global nuclear arsenals. **Source**: Scheffran/Burroughs/Leidreiter et al. 2015, based on data from the Bulletin of the Atomic Scientists.

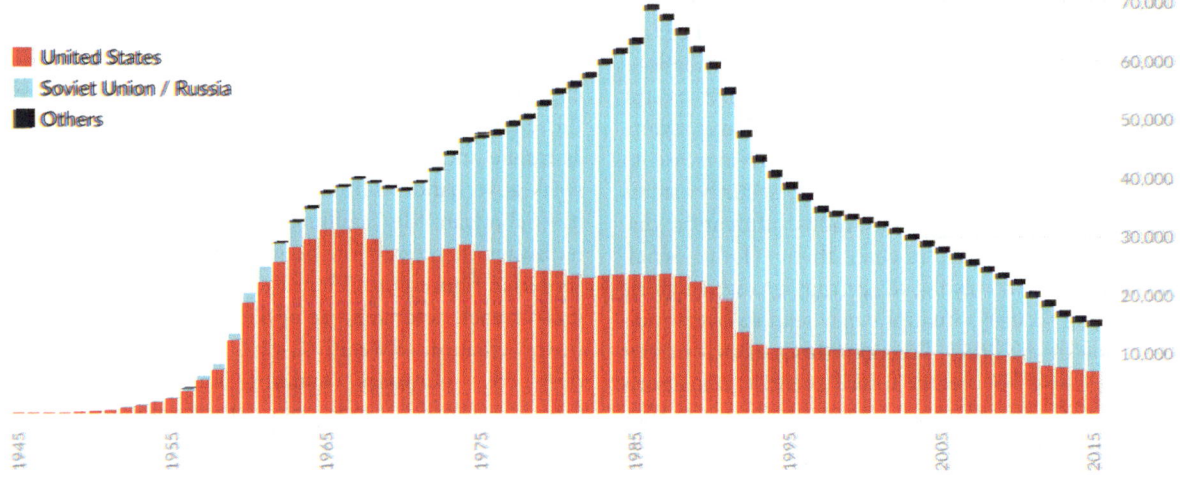

from negative perceptions (negative security and high costs) to positive perceptions (positive security and low costs) there is a transition zone around $w = 0$

with strong chaos-like fluctuations. Thus, for the same initial strategic situation, completely different outcomes result, depending on interpretation from hostile to friendly attitudes on both sides.

Such a shift from hostile to friendly attitudes and perceptions actually occurred between 1986 and 1989 when Gorbachev's policy of glasnost and perestroika opened the way for a new relationship between the United States and Russia. Threat perceptions declined after 1989 when the Soviet Union disappeared in 1991 and both sides significantly reduced their nuclear arsenals (figure 13.7) and even considered the complete abolition of nuclear weapons. At the same time, the United States continued to push for missile defence. This led to a hostile reaction from Russia and is one of the factors in the new climate of hostility that has slowed down progress in nuclear disarmament. This points to the fact that these issues are still relevant today.

13.7.3 Modelling Complex Conflict Landscapes and Cycles of Violence after the Cold War

After the chaotic ending of the cold war, the new conflict environment became increasingly complex, with a growing number and diversity of agents and multiple overlapping security dimensions. Within the SCX model variant of the VIABLE framework, the fractal security landscape can be represented by the bifurcation diagram in figure 13.8, which shows the range of anti-symmetric security dilemma situations for an increasing reaction (given by the so-called memory parameter) which indicates the strength of the agents' responses to an experienced change of security (for instance a strong response to security losses). For low reactivity the system shows periodic oscillations around the satisfying security state (here $S = 0$); for growing reaction the oscillation is damped and asymptotically approaches zero. With a further increase in reactivity a cascading sequence of security states emerges, which matches the famous bifurcation diagram of the logistic map in chaos theory (Jathe/Scheffran 1992). It is interesting to see that strong reactions to changing security situations may lead to overshooting in one or the other direction; this makes it hard if not impossible to predict future security transitions beyond the edge of chaos, since small events may drive the situation in one or the other direction. In a tightly coupled world of complex crises, rapidly changing security situations can emerge (as described in earlier sections of this chapter).

What adds to the complexity is the interaction of multiple agents. Going beyond two-player games, a hypothetical situation of thirty agents each is considered, with asymmetrically distributed initial conditions of security and costs (marked by crosses +) and impacts on each other in their social networks (figure 13.9). Following the dynamics of the SCX model the agents separate into those diminishing security and increasing costs (running into conflictive interactions) and those who benefit from the interaction by improving security at declining costs. Thus, the world falls apart (bifurcates) into 'winners' and 'losers', provided there are no behavioural changes, represented by shifting impact factors (Scheffran 2003). Who belongs to either group largely depends on the initial conditions and the stability of the interaction matrix, measured by its eigenvalues. While most agents who initially are in the negative or positive security domain stay there, several agents cross the divide as a result of the interaction (they either win or lose from the efforts of others).

Many conflicts are related to the 'security dilemma', where threats to the security of one agent provoke reactions that threaten the security of other agents, contributing to the 'cycle of violence' mentioned above (Scheffran/Ide/Schilling 2014). As a prominent example, this inherently unstable interaction led to cascading threats in World War 1 (Flint/Diehl/Scheffran et al. 2009). The dynamics of the conflict is driven by the interaction between motivation (incentives for survival and countering opponents) and capability (increasing losses and needs for violent force). Violence can transform (e.g. from intercommunal violent conflict to insurgencies or interstate wars) and spread to neighbouring states or regions, e.g. through (cross-border) migration, ethnic links, natural resource flows, black markets or arms exports. Societies prone to spirals of violence are those in transition or on the edge of instability, such as fragile and failing states with social fragmentation, weak governance and inadequate management capacity (Scheffran/Ide/Schilling 2015). Key variables and pathways of interaction between two actors are presented in figure 13.10. Here the resources and capabilities of each actor may be applied for productive purposes (creation of wealth assets) in cooperation or for destructive purposes (building and use of violent force) in conflict. Once critical thresholds of insecurity and violence have been passed, a self-enforcing spiral of violence may emerge when violent acts provoke more violent acts; similarly, a self-enforcing cycle of cooperation and peace can be induced (see Scheffran/Ide/Schilling 2014). Agents can choose to switch from productive to destructive purposes (and vice

Figure 13.8: Security bifurcation diagram of the SCX model for the anti-symmetric case of two agents (here agent 1 is shown). **Source**: Jathe and Scheffran (1992).

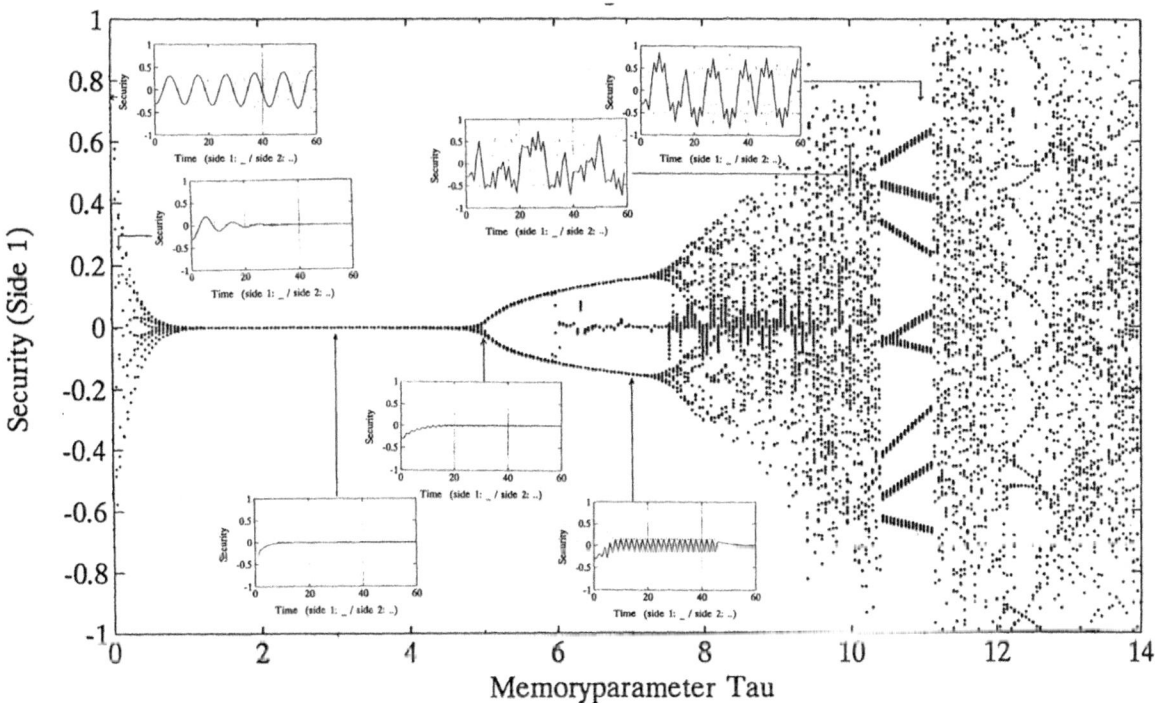

Figure 13.9: Dichotomy between winners and losers for a large number of agents. **Source**: Scheffran (2003).

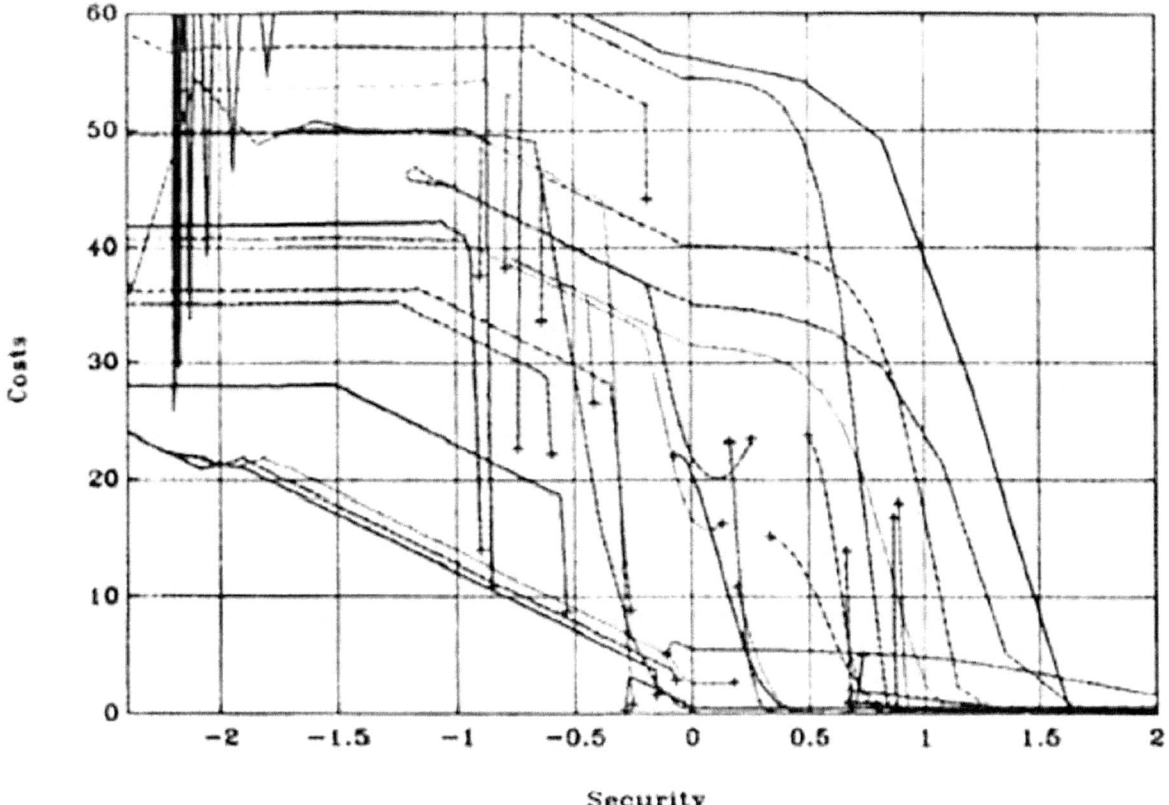

Figure 13.10: Framework for transitions between cycle of violence and cycle of cooperation between two agents. **Source:** Scheffran, Ide and Schilling (2014).

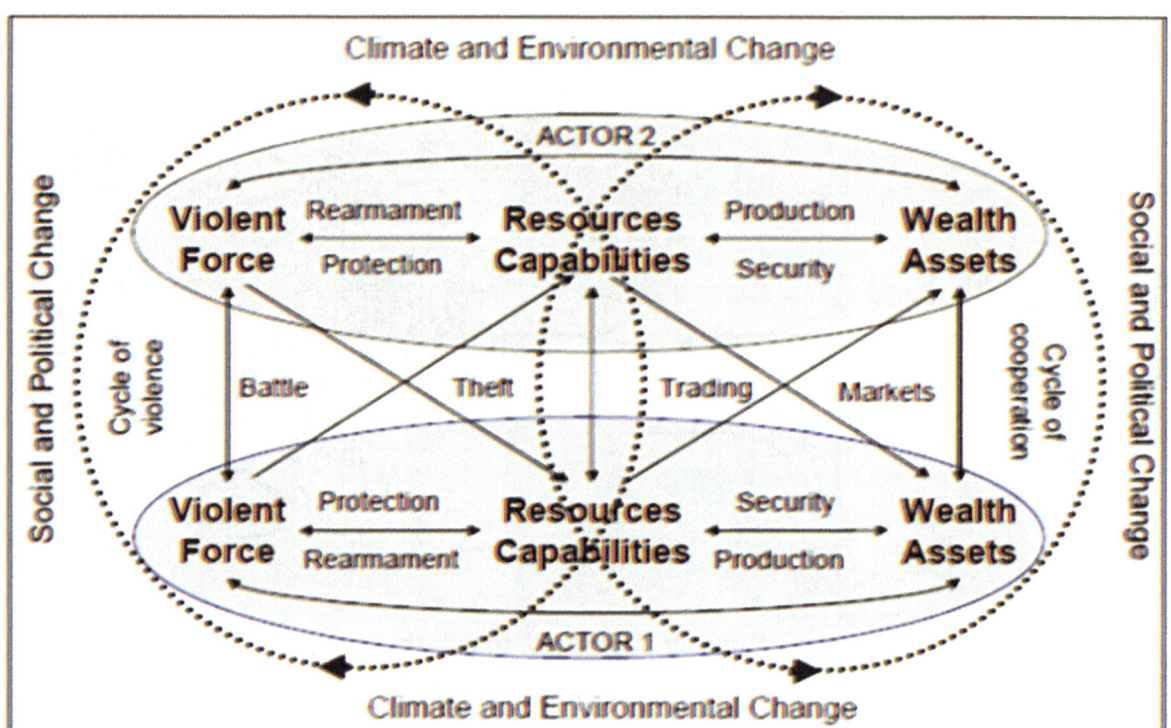

versa) according to their motivations; collective actions could drive agents into self-enforcing cycles of violence or cooperation.

13.7.4   From Climate Risk to Emission Reduction

Human-environment interaction has been extensively simulated within the VIABLE model framework (and its VCX variant for value *V*). Besides other examples, e.g. cooperative and sustainable management of fisheries (Scheffran 2000; Bendor/Scheffran 2009), modelling of energy and climate change has played an important role. In a simplified model version with adaptive control strategies the impact of parameter variations is demonstrated so that the conditions can be visualized that lead to the transition from high-emission to low-emission pathways, starting from a baseline business-as-usual scenario and a 2°C temperature limit (figure 13.11). The model is less a prediction than a demonstration tool that describes the fundamental dynamic relationship between actual and cumulative carbon emissions, atmospheric carbon concentration and global temperature change beyond pre-industrial levels for a particular parameter choice.

Figure 13.11a shows variation of the abatement costs per emission unit, which demonstrates that the target is nearly achievable for a unit cost of $20–50 billion, which will reduce the level of carbon emissions by 1 Gt (with some overshooting). For larger unit costs, temperature exceeds the limit. Variation of the maximum reduction cost shows the relevance of the total investment that the world is willing to spend on emission reductions each year (figure 13.11b). For the baseline case, $100 billion per year may stop temperature growth in about 100 years and reverse it but still misses the target, while $200 billion per year would prevent temperature from exceeding the limit and bring it back below the threshold. For high climate sensitivity (above 3.5°C) the temperature target is missed, while a lower sensitivity of 2.5°C starting from the baseline allows the temperature increase to be reversed below the threshold after temporarily overshooting (figure 13.11c). Finally, the model is sensitive to the initial trend of emission growth as a main driver of global warming (figure 13.11d).

Going beyond the world as a single agent, other simulations over fifty years are presented for two agents, called 'North' (industrialized countries: Actor 1) and 'South' (developing countries: Actor 2), modifying a basic integrated assessment model (Nordhaus

Figure 13.11: Temperature change for parameter variations of: (a) unit reduction cost; (b) maximum reduction cost; (c) climate sensitivity; (d) initial emission trend. **Source:** Scheffran and Hannon (2007).

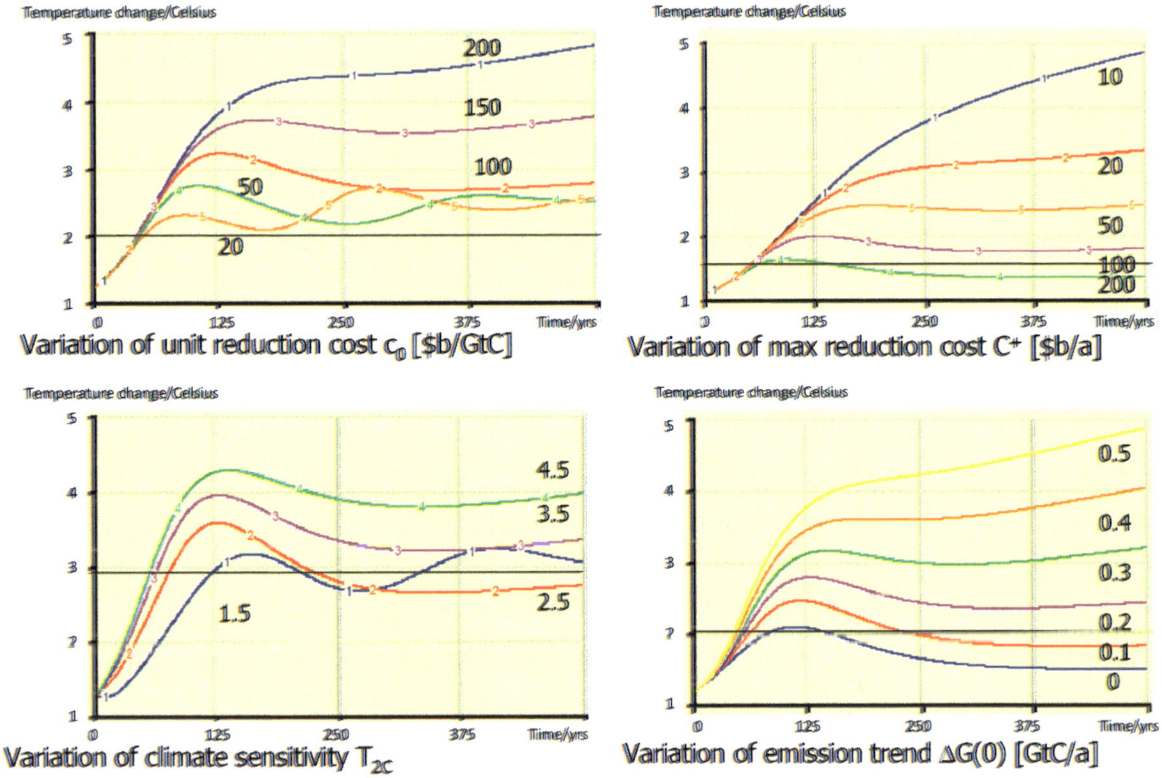

Variation of unit reduction cost c_0 [$b/GtC]

Variation of max reduction cost C^+ [$b/a]

Variation of climate sensitivity T_{2c}

Variation of emission trend $\Delta G(0)$ [GtC/a]

1993) with adaptive elements from the VCX model (Scheffran/Jathe 1995, Scheffran 1999). Both regions are striving for a certain level of wealth, slightly above the present average for the North and much higher than in the South. Both invest in energy production which increases wealth but also increases the potential environmental damage (risk) from climate change, which may affect GNP negatively. The choice of the other model parameters (population growth; emissions, costs, wealth and risk effects per unit of energy; initial conditions for energy consumption, GNP, CO_2 content of the atmosphere) represent a standardized reference case, meant to correspond to the initial year 1990.

As expected, in the numerical simulation the attempt by the developing region to reach the level of wealth per capita of the industrialized region, despite a doubling of the population, leads to a dramatic increase in energy consumption and GNP to more than fifty times the present level. After about ten years, the South outstrips the North's GNP, but the North achieves the goal of GNP per capita sooner. The CO_2 content of the atmosphere increases about

fivefold, which is associated with high risks (a few per cent of GNP).

To avoid this pessimistic catastrophe scenario, a few parameters are altered: a linear decline in the population growth rate, lower and asymmetric wealth targets in North and South, technologies with fifty per cent lower CO_2 emissions, fifty per cent higher energy efficiency, and higher energy costs. Under these conditions, energy consumption in the South only grows about fourfold, while it declines by half in the North. The CO_2 concentration in the atmosphere increases to a slightly higher level than today but much lower than in the first scenario. The lower wealth goals are achieved, despite drastic increases in energy prices, with substantially less environmental damage (below one per cent of GNP).

13.7.5 Technical Innovation and Coalition Formation in Sustainable Energy Transitions

One of the greatest challenges is the transformation of the energy system from fossil and nuclear power to renewable energy sources. The key questions are

Figure 13.12: Simulation of the simplified SCX model dynamics for population, energy consumption, atmospheric CO_2 concentration, gross national product (GNP), risk/GNP and GNP/capita for two actors over fifty years (North: solid line, South: dotted line). *Upper simulation*: Catastrophe scenario if the 1990 situation continues, for high prosperity goals; *lower simulation*: Stabilization scenario for smaller population growth, lower wealth goals, lower emissions, higher efficiency and higher costs per energy unit. **Source**: Scheffran (1999).

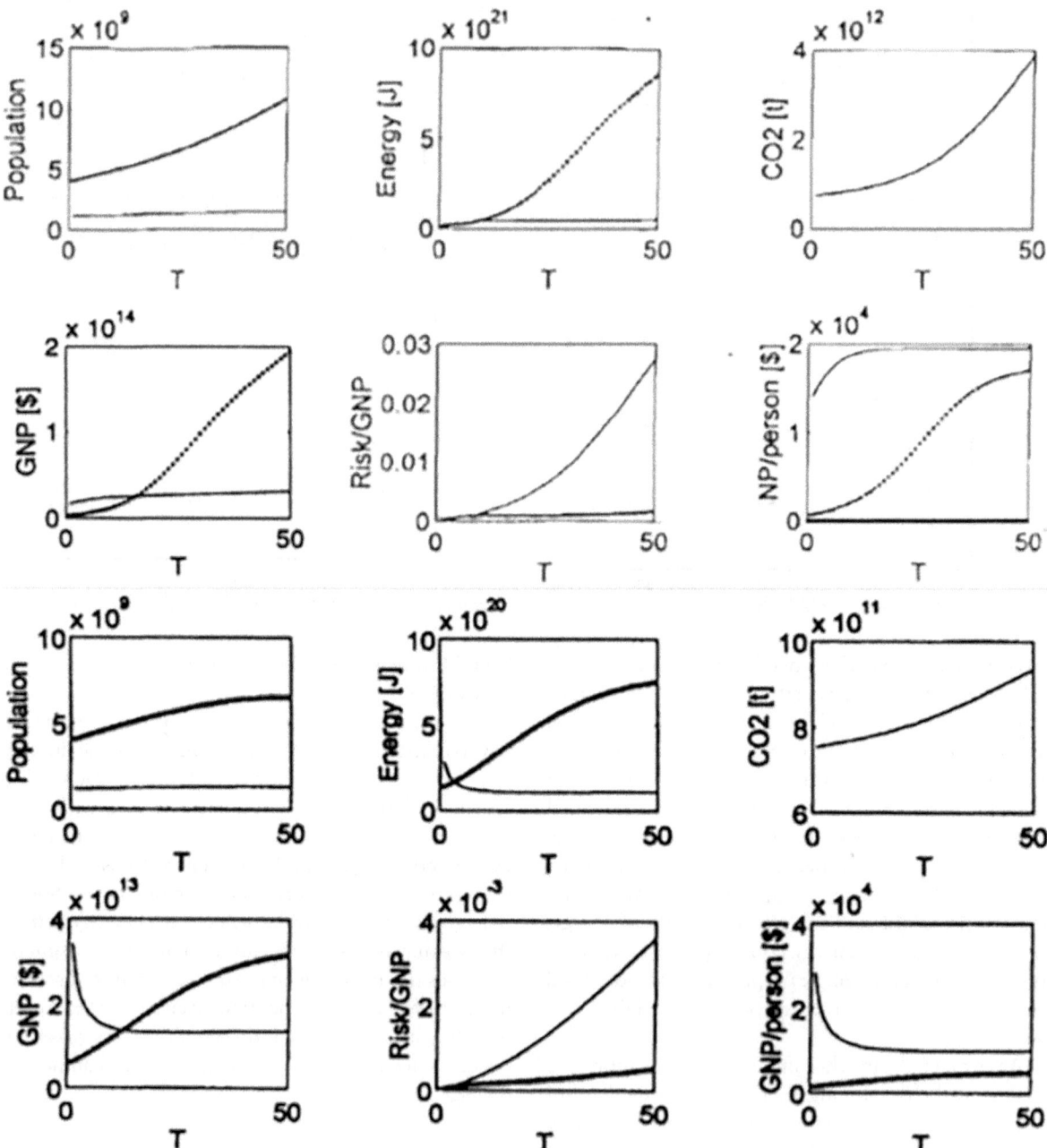

focusing on the conditions for switching to low-carbon energy pathways, and building societal infrastructures and coalitions to support an energy transition. Integrated assessment in the VIABLE model framework couples a basic climate model to economic production with energy as a production factor, which is controlled by the allocation of investments to alternative energy technologies. Investment strategies are shaped by value functions, including utility, costs and climate damages for a given future time horizon, and these are translated into permissible emission limits for keeping atmospheric carbon concentrations and

Figure 13.13: Range of variables (carbon emissions, per capita production, atmospheric CO_2 concentration, temperature change) over 100 years for low- and high-emission energy paths. **Source:** Modified from Scheffran (2008b).

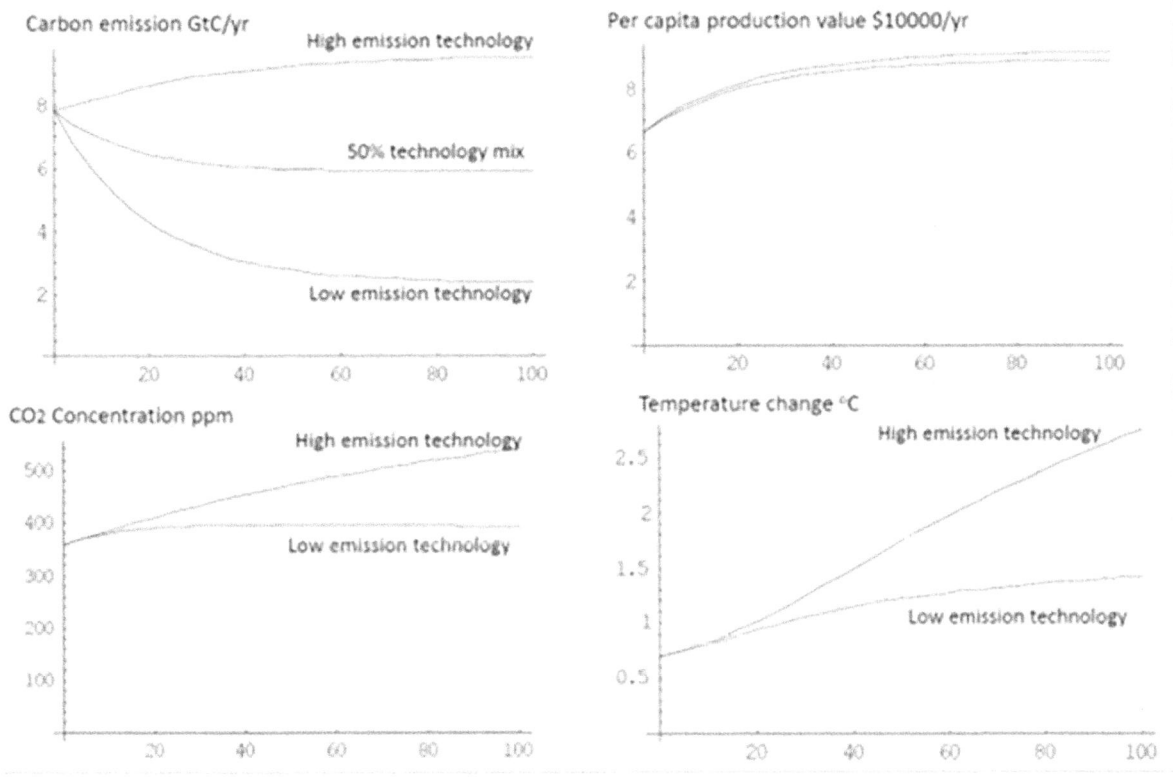

Figure 13.14: Value difference between high and low emission energy technology with variation of damage factor (left) and carbon price (right). **Source:** Modified from Scheffran (2008b).

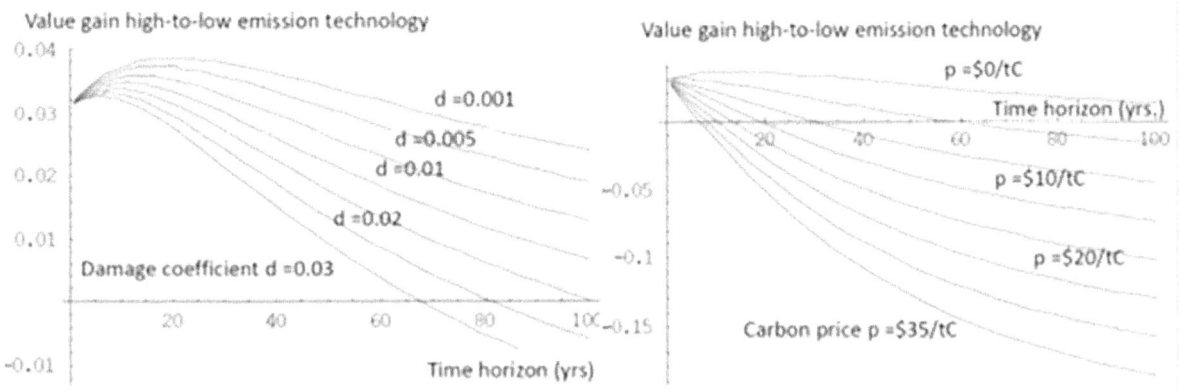

global mean temperature below a given threshold. Conditions for switching between energy paths with different costs and carbon intensities are identified, and the sensitivity of the results to variation in crucial parameters is discussed, in particular time discounting, climate damage, taxes, and time horizons for decision-making (Scheffran 2008b).

Two energy paths are compared: the established path, and a new path with twice that cost and half the emissions per energy unit. The variables in figure 13.13 are carbon emissions, per capita production, atmospheric CO_2 concentration and temperature change over a period of 100 years for high- and low-carbon pathways. While emissions show the largest variation, between 10 GtC/year for high-emission technology and 2 GtC/year for low-emission technology, per capita production varies only by a small margin of 1–2 per

Figure 13.15: Interplay between energy providers and customers for different energy paths (marked here by *A* and *B*). **Source**: Adapted and modified from Scheffran (2006).

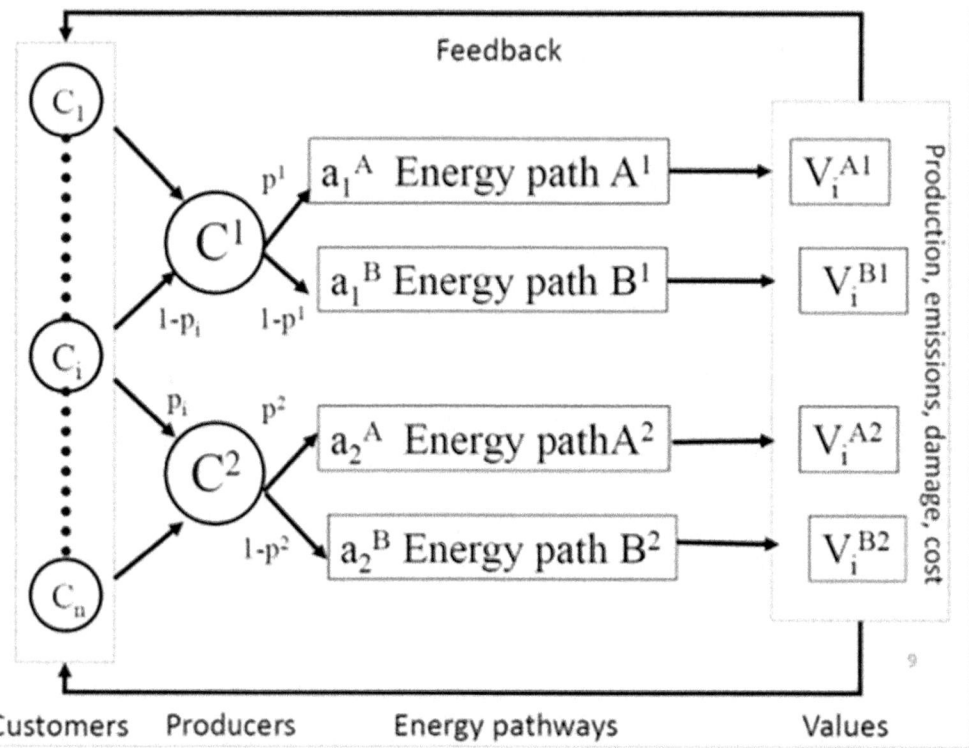

Figure 13.16: Simulation of adaptive coalition formation with two energy paths for two energy providers (coalitions), represented by C1 and C2, and six customers A1 to A6 using energy to produce value. **Source**: Scheffran (2006).

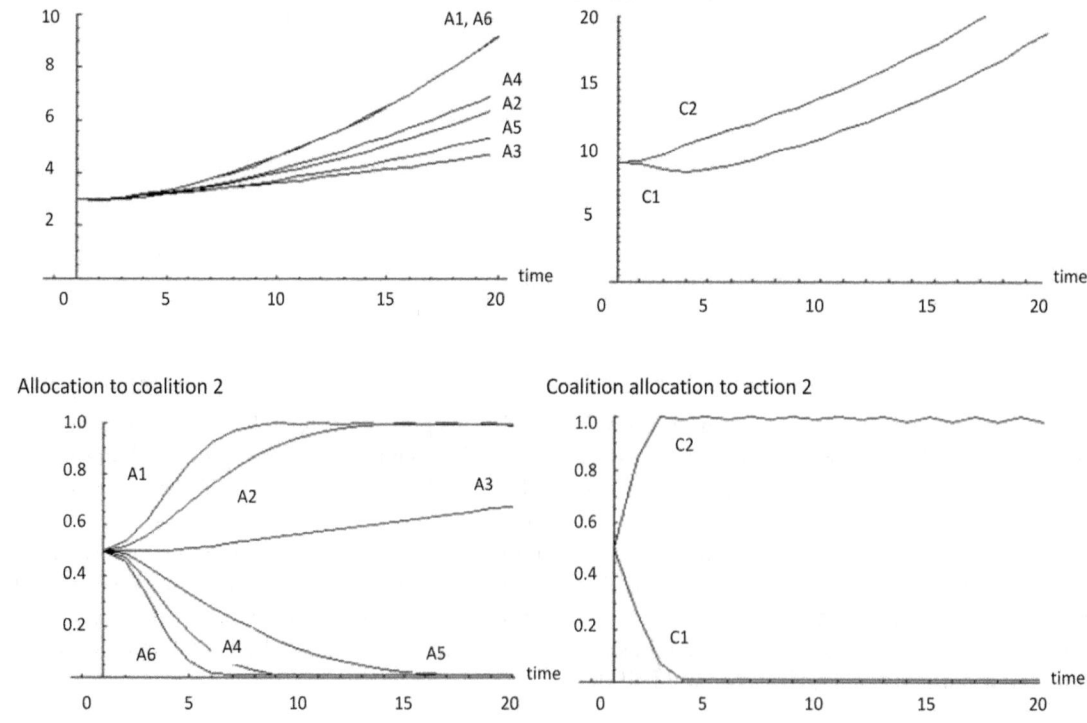

cent. Concentration varies between around 400 and 550 ppm after 100 years, while temperature varies between 1.5°C and more than 2.5°C.

Figure 13.14 plots the annual value differences between high- and low-emission energy technologies for different time horizons of decision-making. It plots value differences against time horizons for varying damage factors (figure 13.14a) and carbon taxes (figure 13.14b). There is an initial value barrier against switching towards the low-carbon energy path, which declines with increasing damage and tax. Increasing climate damage lowers the barrier until it becomes negative. Accordingly, a tax will reduce the barrier, but much faster. A tax creates a relative incentive to switch to new technology in the near term. The example case demonstrates that timing matters. Will emission reductions and investment be delayed to a later stage, running the risk of severe climate change, or will the required action be taken earlier, at the cost of diminished near-term economic benefits? Even though the overall value may be still the same, both paths differ considerably with regard to the distribution of risks and costs in time and space.

As suggested earlier, energy transitions are driven by multi-agent interactions. These have been simulated in the VIABLE framework in order to analyse the formation of adaptive coalitions among energy providers and consumers, where the coalitions have been formed to allocate investment to different energy pathways to generate value (for model structure, see Scheffran 2006 and figure 13.15). Each coalition of energy producers receives payments from individual consumers to supply energy through different energy paths to produce value for the consumers. Here we have a triple decision situation: allocation of investment from consumers to producers and from producers to energy paths, and distribution of value to consumers. Each of the decision variables follows adjustment rules.

Figure 13.16 presents the hypothetical interplay between two energy providers and six customers who have the choice between two energy paths: the old technology path with high emissions and low costs, and the new one that cuts emissions by up to fifty per cent, but where unit costs are up to fifty per cent higher (Scheffran 2006). Customers differ in energy efficiency of production, where the efficiency for energy 1 declines from customer 1 to 6 and the efficiency for energy 2 increases from customer 1 to 6. The range for unit cost, efficiency and carbon emissions is a factor of two, and damages per emission unit gradually increase from the first customer to the

sixth customer. The generated value of the agents can be reinvested. The model runs in figure 13.16 show a growth in the resources of agents and coalitions, where agents A1 and A6 and coalition C2 have the strongest growth rates. Coalition C1 specializes in the old energy path and coalition C2 in the new energy path, where both have cost advantages. Customer A3 is largely indifferent between the coalitions. Customers A1 and A2 emerge as early adopters of the new energy path and prefer coalition C2 while customers A4 to A6 still prefer coalition C1 which represents the established cheaper energy path with higher emissions. In another study, the adaptive model has been applied to emission trading in eleven world regions (Scheffran 2004; Scheffran/Leimbach 2006).

13.8   Conclusions

In an increasingly interconnected world where systems are tightly coupled across different scales, stabilization of human–environment interactions under conditions of climate change is a major challenge in international relations that requires the integration of complexity science with global governance. This chapter has highlighted several cases of complex crises, where interconnected systems and cascading events have magnified individual events to the degree that they affect the stability of large systems. Climate change is considered a 'risk multiplier', which disturbs the balance between natural and social systems and amplifies the consequences through complex impact chains in interconnected systems. Through multiple pathways, climate change can affect the functioning of critical infrastructures and supply networks; intensify the nexus of water, energy and food; lead to production losses, price increases and financial crises in other regions through global markets; undermine human security, social living conditions and political stability; and trigger or aggravate migration movements and conflict situations.

Whether these linkages will materialize ultimately depends on the strengths and sensitivities of each coupling. Some effects could act over long distances, for instance through interventions or humanitarian aid in remote regions affected by violent conflict or large migration movements. Climate change as a threat multiplier could affect different security dimensions and provoke 'security dilemmas' with an increasing level of threat perceptions and armed conflicts. In the most affected states the erosion of social order and state failure could be aggravated, leading to a spiral of

Figure 13.17: Future climate pathways, tipping points and cascades between conflict and cooperation. **Source:** Modified from Scheffran (2015a).

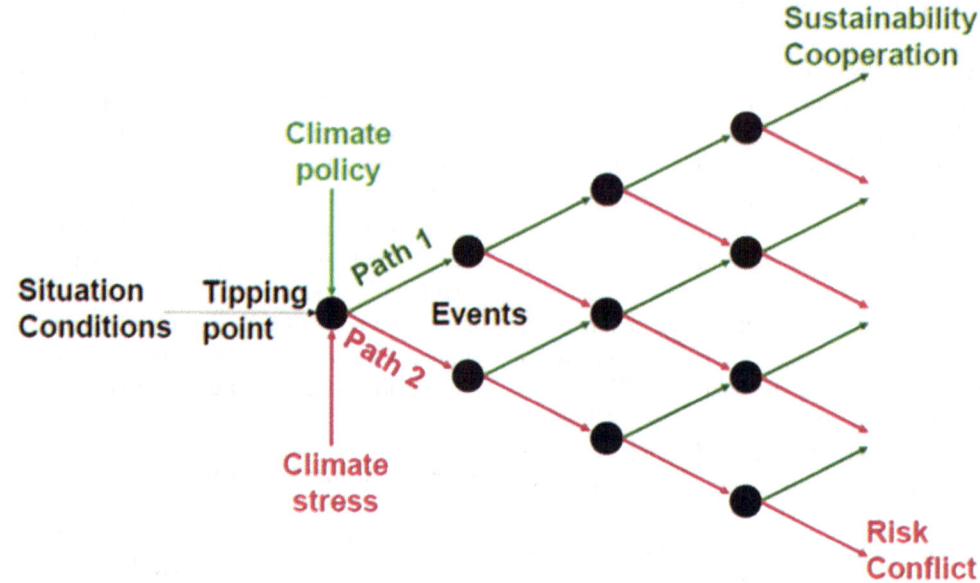

corruption, crime and violence. To stabilize the interaction, actors could move towards mutually beneficial solutions (win–win), e.g. by resource sharing and risk management. Whether climate stress fuels a cycle of violence or climate policy drives a transition towards a cycle of cooperation and sustainable peace depends on the human and societal responses (figure 13.17).

An integrative framework of human-environment interaction can help us to analyse destabilizing developments, tipping elements and cascading risks. To survive destabilizing consequences, affected systems need to adapt to the changing circumstances to ensure their viability. Concepts of anticipative and adaptive governance influence critical decision points and adjust actions along multiple causal chains to protect human security, develop social livelihood and strengthen societal resilience. The goal is to avoid risky pathways and facilitate a qualitative and self-organized system transformation towards sustainability. Nevertheless, given the complexity and interconnectedness of affected systems, significant uncertainties exist regarding consequences and related damages, and this underscores the need for new approaches to decision-making under conditions of uncertainty and novel concepts for addressing the nexus of multiple problem areas that lead to complex crises.

To some degree, social systems have the ability to cope with the magnitude of climate change and develop possible alternative pathways of human responses and actions. To succeed, responses need to be timely and adequate compared to the speed and intensity of climate change. A key question is to what degree of climate change societies can adapt, and how effective and creative they are in developing coping strategies that are complex enough to deal with the challenges. One task is to translate environmental change into new social and political structures and institutions that avoid or minimize social instability. Rules and regulations that guarantee peaceful coexistence are characteristic of a viable society and this in turn is important for satisfying human needs.

Concepts of resilience, viability and sustainable peace can strengthen people's social capability in their effective, creative and collective efforts to handle the problems of climate change. In a resilient social environment, actors are able to cope with and withstand the disturbances caused by climate change in a dynamic way that will enable them to preserve, rebuild, or transform their livelihood. Key viability strategies, supporting a "new climate for peace" (Rüttinger/Stang/Smith et al. 2015), include climate mitigation and adaptation; the building of networks, the cultivation of diversity and the maintenance of flexibility; migrant networks that facilitate the exchange of knowledge, income and other resources; new capabilities to manage disasters; arms control, non-proliferation and disarmament; regional security concepts, crisis prevention, conflict resolution and confidence-building; and innovative institutional frameworks and legal mechanisms. The 2015 Paris Agreement offers a framework of opportunities that need to be realized.

References

Adger, W. Neill; Kelly, P Mick; Winkels, Alexandra; Huy Luong Quang; Locke, Catherine; 2002: "Migration, Remittances, Livelihood Trajectories, and Social Resilience", in: *Journal of the Human Environment*, 31,4: 358-366.

Adger, W. Neill; Lorenzoni, Irene; O'Brien, Karen (Eds.), 2009: *Adapting to Climate Change: Thresholds, Values, Governance* (Cambridge: Cambridge University Press).

Angourakis, Andreas; Santos, José Ignacio; Galán, José Manuel; Balbo, Andrea L., 2014: "Food for all: An agent-based model to explore the emergence and implications of cooperation for food storage", in: *Environmental Archaeology*, 20,4 (November): 349-363.

Aubin, Jean-Pierre, 1991: *Viability Theory* (Boston–Basel–Berlin: Birkhäuser).

Aubin, Jean-Pierre; Saint-Pierre, Patrick, 2007: "An introduction to viability theory and management of renewable resources", in: Kropp, Jürgen; Scheffran, Jürgen (Eds.): *Decision Making and Risk Management in Sustainability Science* (New York: Nova Science): 43-80.

Barnett, Jon; Adger, W. Neill, 2007: "Climate change, human security and violent conflict", in: *Political Geography*, 26: 639-655.

Beisheim, Marianne (Ed.), 2013: *Der "Nexus" Wasser-Energie-Nahrung–Wie mit vernetzten Versorgungsrisiken umgehen?* (Berlin: Stiftung Wissenschaft und Politik, Deutsches Institut für Internationale Politik und Sicherheit).

Bender, Wolfgang, 1991: *Erhaltung und Entfaltung als Kriterien für die Gestaltung von Wissenschaft und Technik.* IANUS Working Paper 9/1991 (Darmstadt: IANUS).

Bender, Wolfgang, 1998: "Erhaltung und Entfaltung–Leitbilder für Frieden und nachhaltige Entwicklung", in: Scheffran, Jürgen; Vogt, Wolfgang R. (Eds.): *Kampf um die Natur* (Darmstadt: Wissenschaftliche Buchgesellschaft): 249-266.

BenDor, Todd; Scheffran, Jürgen; Hannon, Bruce, 2009: "Ecological and economic sustainability in fishery management: A multi-agent model for understanding competition and cooperation", in: *Ecological Economics*, 68,4 (15 February): 1061-1073.

BenDor, Todd; Scheffran, Jürgen, 2016: "Agent-Based Modeling of Environmental Conflict and Cooperation", book manuscript (in preparation for publication).

Biermann, Frank; Boas, Ingrid, 2008; "Protecting Climate Refugees: The Case for a Global Protocol", in: *Environment*, 50,6 (November/December): 8-16.

Bossel, Hartmut 2001: "Assessing viability and sustainability: a systems-based approach for deriving comprehensive indicator sets", in: *Conservation Ecology*, 5,2: 12; at: <http://www.consecol.org/vol5/iss2/art12/>.

Bradsher, Keith, 2013: "Font of Natural Energy in the Philippines, Crippled by Nature", in: *The New York Times*, 21 November.

Brand, Ruth, Scheffran, Jürgen, 2005: *The price to pay for deregulation of electricity supply in California*", in: von Weizsäcker, Ernst Ulrich; Young, Oran R.; Finger, Matthias, Beisheim, Marianne (Eds.): *Limits To Privatization–How to Avoid Too Much of a Good Thing.* Report to the Club of Rome (London: Earthscan): 79-83.

Brauch, Hans Günter, 2002: "Climate Change, Environmental Stress and Conflict - AFES-PRESS Report for the Federal Ministry for the Environment, Nature Conservation and Nuclear Safety", in: Federal Ministry for the Environment, Nature Conservation and Nuclear Safety (Ed.): *Climate Change and Conflict. Can climate change impacts increase conflict potentials? What is the relevance of this issue for the international process on climate change?* (Berlin).

Brauch, Hans Günter, 2009: "Securitizing Global Environmental Change", in: Brauch, Hans Günter; Oswald Spring, Úrsula; Grin, John; Mesjasz, Czeslaw; Kameri-Mbote, Patricia; Behera, Navnita Chadha; Chourou, Béchir; Krummenacher, Heinz (Eds.), 2009: *Facing Global Environmental Change: Environmental, Human, Energy, Food, Health and Water Security Concepts* (Berlin - Heidelberg - New York: Springer-Verlag): 65-102.

Bruckner, Thomas; Petschel-Held, Gerhard; Toth, Ferenc L.C.; Helm, Carsten; Leimbach, Marian; Schellnhuber, Hans Joachim, 1999: "Climate change decision-support and the tolerable windows approach", in: *Environmental Modeling and Assessment*, 4: 217-234.

Brzoska, Michael; Link, Peter M.; Maas, Achim; Scheffran, Jürgen (Eds.), 2012: "Geoengineering: An Issue for Peace and Security Studies?", in: *Sicherheit & Frieden/Security & Peace*, Special Issue, 30,4.

Buhaug, Halvard, 2010: "Climate not to blame for African civil wars", in: *Proceedings of the National Academy of Sciences (PNAS)*, 107: 16477-16482.

Buhaug, Halvard; Nordkvelle, Jonas; Bernauer, Thomas; Böhmelt, Tobias; Brzoska, Michael; Busby, Joshua W.; Ciccone, Antonio; Fjelde, Hanne; Gartzke, Erik; Gleditsch, Nils-Petter; Goldstone, Jack A.; Hegre, Håvard; Holtermann, Helge; Koubi, Valley; Link, Jasmin S. A.; Link, Peter M.; Lujala, Paivi; O'Loughlin, John; Raleigh, Clionadh; Scheffran, Jürgen; Schilling, Janpeter; Smith, Todd G.; Theisen, Ole Magnus; Tol, Richard S.J.; Urdal, Henrik; von Uexkull, Nina, 2014: "One effect to rule them all? A comment on climate and conflict", in: *Climatic Change*, 127,3-4: 391-397.

Bunde, Armin; Kropp, Jürgen, Schellnhuber, Hans Joachim (Eds.), 2002: *The Science of Disasters–Climate Disruptions, Heart Attacks, and Market Crashes* (Berlin-London: Springer).

Bundeswehr, 2012: *Klimafolgen im Kontext–Implikationen für Sicherheit und Stabilität im Nahen Osten und Nordafrika*, Teilstudie 2 (Strausberg: Dezernat Zukunftsanalyse des Planungsamtes der Bundeswehr).

Burke, Marshall B.; Miguel, Edward; Satyanath, Shanker; Dykema, John A.; Lobell, David B.; 2009: "Warming increases the risk of civil war in Africa", in: *Proceedings of the National Academy of Sciences*, 106: 20670-20674.

Busby, Joshua W.; Raleigh, Clionadh; Salehyan, Idean, 2013: "The Political Geography of Climate Vulnerability, Conflict and Aid in Africa", in: Huth, Paul K.; Wilkenfeld, Jonathan; Backer, David A. (Eds.): *Peace and Conflict 2014* (Boulder, Co: Paradigm Publishers).

Butler, Declan, 2007: "Darfur's climate roots challenged", in: *Nature*, 447: 1038.

Buzan, Barry; Wæver, Ole; Wilde, Jaap de, 1998: *Security: A New Framework for Analysis.* (Boulder, Co.: Lynne Rienner).

Carius, Alexander; Tänzler, Dennis; Winterstein, Judith, 2006: "Weltkarte von Umweltkonflikten" (Berlin: Adelphi); at: <http://www.wbgu.de/fileadmin/templates/dateien/ veroeffentlichungen/hauptgutachten/jg2007/wbgu_jg2007_ex02.pdf>.

Casti, John, 1979: *Connectivity, complexity and catastrophe in large-scale systems* (Vienna: IIASA).

CNA Corporation, 2007: *National security and the threat of climate change* (Alexandria, Va.: The CNA Corporation).

Czempiel, Ernst-Otto, 1995: "Der Friede—sein Begriff, seine Strategien", in: Senghaas, Dieter (Ed.): *Den Frieden denken* (Frankfurt am Main: Suhrkamp): 165-176.

Deutschländer, Thomas, Wichura, Bodo, 2005: "Das Münsterländer Schneechaos am 1. Adventswochenende 2005", in: *Klimastatusbericht* 2005: 163-167; at: http://www.dwd.de/DE/leistungen/klimastatusbericht/publikationen/ksb2005_pdf/15_2005.pdf?__blob=publicationFile&v=1.

Edenhofer, Ottmar; Lessmann, Kai; Kemfert, Claudia; Grubb, Michael; Köhler, Jonathan, 2006: "Induced Technological Change: Exploring its Implications for the Economics of Atmospheric Stabilization", in: *Energy Journal*. Special Issue, Endogenous Technological Change and the Economics of Atmospheric Stabilisation (March): 57-108.

Eisenack, Klaus; Lüdeke, Matthias K. B.; Petschel-Held, Gerhard; Scheffran, Jürgen; Kropp, Jürgen P., 2007: "Qualitative Modelling Techniques to Assess Patterns of Global Change", in: Kropp, Jürgen, Scheffran, Jürgen (Eds.): *Advanced Methods for Decision Making and Risk Management in Sustainability Science* (New York: Nova Science): 99-146.

Eisenack, Klaus; Scheffran, Jürgen; Kropp, Jürgen P., 2007: "Viability Analysis of Management Frameworks for Fisheries", in: *Environmental Modeling and Assessment*, 11,1: 69-79.

EU, 2008: Rat der Europäischen Union: *Klimawandel und Internationale Sicherheit* (Brüssel: European Union); at: <http://register.consilium.europa.eu/pdf/de/08/st07/st07249.de08.pdf>.

FAZ, 2010: "Russlands Exportverbot treibt den Weizenpreis", in: *Frankfurter Allgemeine Zeitung*, 6 August.

Feddern, Boi; Schnurer, Georg, 2011: "Teure Terabytes—Überschwemmung in Thailand führt zu Lieferengpässen bei Festplatten", in: *c't magazin*, 24 November; at: <http://www.heise.de/ct/artikel/Teure-Terabytes-1370334.html>.

Flint, Colin; Diehl, Paul; Scheffran, Jürgen; Vasquez, John; Chi, Sang-hyun, 2009: "Conceptualizing Conflict Space: Toward a Geography of Relational Power and Embeddedness in the Analysis of Interstate Conflict", in: *Annals of the Association of American Geographers*; 99(5-5):827-835.

Foresight, 2011: *Migration and Global Environmental Change*. Final Project Report (London: Government Office for Science).

Füssel, Hans-Martin, 2007: "Methodological and empirical flaws in the design and application of simple climate-economy models", in: *Climatic Change*, 81: 161-185.

Galtung, Johan, 1967: *Theories of Peace: A Synthetic Approach to Peace Thinking* (Oslo: International Peace Research Institute, September).

Galtung, Johan, 1969: "Violence, Peace, and Peace Research", in: *Journal of Peace Research*, 6,3: 167-191.

Gärtner, Markus, 2011: "Thailand: Flutdesaster bedroht globale Lieferketten", in: *Manager Magazin*, 14 November 2011; at: <http://www.manager-magazin.de/politik/weltwirtschaft/a-797205.html>.

Gemenne, François; Brücker, Pauline; Glasser, Joshua (Eds.), 2011: *The State of Environmental Migration 2010* (Paris: Institute for Sustainable Development & International Relations, International Organization for Migration).

Gemenne, François; Brücker, Pauline; Ionesco, Dina, 2013: *The State of Environmental Migration 2013* (Paris: Institute for Sustainable Development and International Relations, International Organization for Migration).

Gennies, Sidney, 2013: "Wie Deutschland nach der Flut auf die Beine kommt", in: *Der Tagesspiegel*, 18 July 2013; at: <http://www.tagesspiegel.de/politik/hochwasser-2013-wie-deutschland-nach-der-flut-auf-die-beine-kommt/8510988.html>.

Germanwatch, 2014: *Global Climate Risk Index 2015* (Bonn: Germanwatch); at: <http://germanwatch.org/en/7659>.

Geyer, Robert; Pickering, Steve; 2011: "Applying the tools of complexity to the international realm: from fitness landscapes to complexity cascades", in: *Cambridge Review of International Affairs*, 24,1: 5-26.

Gleditsch, Nils Petter, Nordås, Ragnhild, 2014: "Conflicting messages? The IPCC on conflict and human security", in: *Political Geography*, 43 (November): 82-90.

Göbeler, Frank; Scheffran, Jürgen, 2003: "Extended power values and dynamic coalition formation", in: Petrosjan, Leon; Zenkevich, Nikolay A. (Eds.): *Proceedings of the 10th International Symposium on Dynamic Games and Applications, St. Petersburg, Russia* (St. Petersburg), 11/2002: 338-340.

Grossmann, Wolf-Dieter; Watt, Kenneth E.F.., 1992: "Viability and Sustainability of Civilizations, Corporations, Institutions and Ecological Systems", in: *Systems Research*, 9,1: 1-41.

Haas, Armin, 2010: "Klimawandel und Finanzmärkte", in: *UmweltWirtschaftsForum (uwf)*, 18,1 (March): 3-9.

Hardin, Garrett, 1968: "The Tragedy of the Commons", in: *Science*, 162,3859: 1243-1248.

Hasselmann, Klaus; Cremades, Roger; Filatova, Tatiana; Hewitt, Richard; Jaeger, Carlo; Kovalevsky, Dmitry; Voinov, Alexey; Winder, Nick, 2015: "Free-riders to forerunners. Commentary", in: *Nature Geoscience*, Advance Online Publication, 1-4.

Held, Hermann; Schellnhuber, Hans Joachim, 2004: "Evolution of Perturbations in Complex Systems", in: Steffen, Will; Sanderson, Angelina; Tyson, Peter D.; Jäger, Jill; Matson, Pamela A.; Moore III, Berrien; Oldfield, Frank; Richardson, Katherine; Schellnhuber, Hans Joachim; Turner II, B.L.; Wasson, Robert J., 2004: *Global Change and the Earth System. A Planet under Pressure. The IGBP Series* (Berlin - Heidelberg –New York: Springer-Verlag): 145-147.

Holling, C.S., 1973: "Resilience and stability of ecological systems", in: *Annual Review of Ecology and Systematics*, 4, 1-23.

Hsiang, Solomon M.; Burke, M.; Miguel, Edward, 2013: "Quantifying the influence of climate on human conflict", in: *Science Express* (2 August); doi:10.1126/science.1235367.

IDMC, 2013: "Global Overview 2012" (Geneva: Internal Displacement Monitoring Centre); at: <http://www.internal-displacement.org/publications/global-overview-2012>.

IDMC, 2015: "Global Estimates 2015–People displaced by disasters" (Geneva: Internal Displacement Monitoring Centre); at: <http://www.internal-displacement.org/publications/2015/global-estimates-2015-people-displaced-by-disasters>.

IEA, 2012: *World Energy Outlook 2012* (Paris: OECD/IEA): 501-528.

IOM, 2008: *World Migration Report 2008* (Geneva: International Organization for Migration), at: <http://www.iom.int/jahia/Jahia/cache/offonce/pid/1674?entryId=20275>.

IPCC, 1990: *First Assessment Report, Overview* (Geneva: IPCC); at: <https://www.ipcc.ch/ipccreports/1992 %20IPCC%20Supplement/IPCC_1990_and_1992_Assessments/English/ipcc_90_92_assessments_far_overview.pdf>

IPCC, 2007: *Climate Change 2007. Climate Change Impacts, Adaptation and Vulnerability. Intergovernmental Panel on Climate Change* (Cambridge: Cambridge University Press).

IPCC, 2012: *Managing the Risks of Extreme Events and Disasters to Advance Climate Change Adaptation (SREX.* (Geneva: IPCC).

IPCC, 2014: *Climate Change 2014–Impacts, Adaptation, and Vulnerability–Part A: Global and Sectoral Aspects.*

Working Group II Contribution to the Fifth Assessment Report of the Intergovernmental Panel on Climate Change (Cambridge–New York: Cambridge University Press).

Ipsen, Dirk; Rösch, Roland; Scheffran, Jürgen, 2001: "Cooperation in Global Climate Policy: Potentialities and Limitations", in: *Energy Policy*, 29,4: 315-326.

Jakobeit, Cord; Methmann, Chris, 2012: "Climate Refugees' as Dawning Catastrophe? A Critique of the Dominant Quest for Numbers", in: Scheffran, Jürgen; Brzoska, Michael; Brauch, Hans Günter; Link, Peter Michael; Schilling, Janpeter (Eds.) 2012: *Climate Change,Human Security and Violent Conflict: Challenges for Societal Stability* (Berlin–Heidelberg–New York: Springer-Verlag): 301-314.

Jathe, Markus; Scheffran, Jürgen, 1992: *Security, Stability and Costs in the Armament Dynamics: The SCX Model Framework. Paper for 4th World Peace Science Congress, Rotterdam, May 1992;* "Abstract", in: *Conflict Management and Peace Science,*; 12,2 (05/1992):98-99.

Jathe, Markus, Scheffran, Jürgen, 1995: "Modelling International Security and Stability in a Complex World", in: Bergé, P.; Conte, R; Dubois, M.; van Thran Thanh, J. (Eds.): *Chaos and Complexity* (Editions Frontières).

Johnson, Toni, 2011: "Food Price Volatility and Insecurity." CFR.org. Council on Foreign Relations, 24 June 2011. Web. 1 May 2016.

Johnstone, Sarah; Mazo, Jeffrey, 2011: "Global warming and the Arab spring", in: *Survival*, 53,2 (April): 11-17.

Kaldor, Mary, 2012: *New and Old Wars: Organized Violence in a Global Era* (Cambridge: Polity).

Kavalski, Emilian (Ed.), 2015: *World Politics at the Edge of Chaos* (Albany, NY: State University of New York Press).

Kaya, Yoichi; Yokoburi, Keiichi, 1997: *Environment, energy, and economy: strategies for sustainability* (Tokyo: United Nations University Press).

Kelley, Colin P.; Mohtadi, Shahrzad; Cane, Mark A.; Seager, Richard; Kushnir, Yochanan, 2015: "Climate change in the Fertile Crescent and implications of the recent Syrian drought", in: *PNAS*, 112,11: 3241-3246.

Kempe, David; Kleinberg, Jon; Tardos, Éva, 2005: "Influential Nodes in a Diffusion Model for Social Networks", in: [L. Caires, et al., Eds.] *ICALP 2005, LNCS*, 3580 (Heidelberg: Springer): 1127-1138.

Kominek, Jasmin; Scheffran, Jürgen, 2012: "Cascading Processes and Path Dependency in Social Networks", in: Soeffner, Hans-Georg (Ed.): *Transnationale Vergesellschaftungen*, CD ROM (Wiesbaden: VS Springer).

Kumins, Lawrence; Bamberger, Robert, 2005: *Oil and Gas Disruption from Hurricanes Katrina and Rita*. Congressional Research Service, 21 October 2005; at: <http://fpc.state.gov/documents/organization/55824.pdf>.

Kurths, J.; Maraun, D.; Zhou, C.S.; Zamora-Lopez, G.; Zou, Y., 2009: "Dynamics in complex systems", in: *European Review*, 17,2: 357-370.

Lampietti, Julian A.; Michaels, Sean; Magnan, Nicholas; McCalla, Alex F.; Saade, Maurice; Khouri, Nadim, 2011:

"A strategic framework for improving food security in Arab countries", in: *Food Security*, 3: S7-S22.

Lenton, Timothy; Held, Hermann; Kriegler, Elmar; Hall, Jim W.; Lucht, Wolfgang; Ramstorf, Stefan; Schellnhuber, Hans Joachim, 2008: "Tipping elements in the Earth's climate system", in: *Proceedings of the National Academy of Science* (PNAS), 105,6 (12 February): 1786-1793.

Levermann, Anders, 2014: "Make supply chains climate-smart", in: *Nature* 506: 27-29 (6 February).

Link, P. Michael; Piontek, Franziska; Scheffran, Jürgen; Schilling, Janpeter, 2012: "On Foes and Flows: Vulnerabilities, Adaptive Capacities and Transboundary Relations in the Nile River Basin in Times of Climate Change", in: *L'Europe en Formation*, 365: 99-138.

Lüthje, Corinna; Schäfer, Mike; Scheffran, Jürgen, 2011: *Limits to the Anthropocene. What are the challenges and boundaries of science for the post-normal age?* EGU General Assembly 2011, in: *Geophysical Research Abstracts*, 13: 11795.

Maoz, Zeev, 2010: *Networks of Nations: The Evolution, Structure, and Impact of International Networks, 1816-2001* (New York: Cambridge University Press).

May, Robert M., 1972: "Will a large complex system be stable?", in: *Nature*, 238: 413-414.

Mesjasz, Czeslaw, 2010: "Complexity of Social Systems", in: *Acta Physica Polonica A*, 117,4 (April), 706-715.

Milliken, Jennifer; Krause, Keith, 2003: "State failure, state collapse and state reconstruction: concepts, lessons, and strategies", in: Miliken, Jennifer (Ed.): *State Failure, Collapse and Reconstruction,*. (London: Blackwell): 753-774.

Neuneck, Götz, 2008: "The revolution in military affairs: Its driving forces, elements, and complexity", in: *Complexity*, 14,1: 50-61.

Ngaruiya, Grace W.; Scheffran, Jürgen; Yang, Liang, 2015: "Social Networks in Water Governance and Climate Adaptation in Kenya", in: Leal Filho, Walter; Sümer, Vakur (Eds.): *Sustainable Water Use and Management* (Berlin–Heidelberg: Springer): 151-167.

Nordhaus, William D., 1993: "Rolling the DICE: An optimal transition path for controlling greenhouse gases", in: *Resource and Energy Economics*, 15: 27-50.

Oldag, Andreas; Wälterlin, Urs, 2011: "Kettenreaktion mit Kohle", in: *Süddeutsche Zeitung*. 6 January 2011; at: <http://www.sueddeutsche.de/wirtschaft/ueberschwemmung-in-australien-kettenreaktion-mit-kohle-1.1043437>.

Onischka, Mathias, 2009: *Definition von Klimarisiken und Systematisierung in Risikokaskaden*, Diskussionspaper (Wuppertal: Wuppertal Institut für Klima, Umwelt, Energie, September).

Ostrom, Elinor, 2000: "Collective Action and the Evolution of Social Norms", in: *Economic Perspectives*, 14,3: 137-158.

Penuel, K. Bradley; Statler, Matt; Hagen, Ryan, 2013: *Encyclopedia of Crisis Management*. (London: Sage).

Perrow, Charles, 1984: *Normal Accidents: Living with High-Risk Technologies* (Princeton, N.J.: Princeton University Press).

Peter, Camaren; Swilling, Mark, 2014: "Linking Complexity and Sustainability Theories: Implications for Modeling Sustainability Transitions", in: *Sustainability*, 6, 1594-1622.

Petschel-Held, Gerhard; Schellnhuber, Hans Joachim; Bruckner, Thomas; Toth, Ferenc L.; Hasselmann, Klaus, 1999: "The Tolerable Windows Approach: Theoretical and Methodological Foundations", in: *Climatic Change*, 41,3-4: 303-331.

Pike, John, 1985: "The Star Wars Budget", in: *F.A.S. Public Interest Report*, 38,3 (March): 10-11.

Rahmstorf, Stefan; Coumou, Dim, 2011: "Increase of extreme events in a warming world", in: *PNAS*: 17905-17909.

Raleigh, Clionadh; Urdal, Henrik, 2007: "Climate change, environmental degradation and armed conflict", in: *Political Geography*, 26,6: 674-694.

Redman, Charles, 2014: "Should Sustainability and Resilience Be Combined or Remain Distinct Pursuits?", in: *Ecology and Society*, 19,2 (January).

Reuveny, Rafael, 2007: "Climate change-induced migration and violent conflict", in: *Political Geography*, 26: 656-673.

Rockström, Johan; Steffen, Will; Noone, Kevin; Persson, Åsa; Chapin, III; F. Stuart;Eric F. Lambin, Timothy M. Lenton, Marten Scheffer, Carl Folke, Hans Joachim Schellnhuber, Björn Nykvist, Cynthia A. de Wit, Terry Hughes, Sander van der Leeuw, Henning Rodhe, Sverker Sörlin, Peter K. Snyder, Robert Costanza, UnoSvedin, Malin Falkenmark, Louise Karlberg, Robert W. Corell, Victoria J. Fabry, James Hansen, Brian Walker, Diana Liverman, Katherine Richardson, Paul Crutzen, Jonathan A. Foley, 2009: "A safe operating space for humanity", in: *Nature*, 461, 472-475.

Rotberg, Robert I., 2003: "Failed states, collapsed states, weak states: causes and indicators", in: Rotberg, Robert I. (Ed.): *State Failure and State Weakness in a Time of Terror* (Washington DC: Brookings Institution Press): 1-25.

Rothe, Delf, 2015: *Securitizing Global Warming: A Climate of Complexity* (London: Routledge).

Rüttinger, Lucas; Stang, Gerald; Smith, Dan; Tänzler, Dennis; Vivekananda, Janani; et al., 2015: *A New Climate for Peace—Taking Action on Climate and Fragility Risks* (Berlin: Adelphi—London: International Alert—Washington: Wilson Center—Paris: EUISS).

Scheffran, Jürgen, 1983: *Komplexität und Stabilität von Makrosystemen mit Anwendungen* (Marburg: Universität Marburg, Department of Physics).

Scheffran, Jürgen, 1989: *Strategic Defense, Disarmament, and Stability—Modelling Arms Race Phenomena with Security and Costs under Political and Technical Uncertainties*. PhD thesis (Marburg: IAFA).

Scheffran, Jürgen, 1996: "Leben bewahren gegen Wachstum, Macht, Gewalt—Zur Verknüpfung von Frieden und nachhaltiger Entwicklung", in: *Wissenschaft und Frieden*, 3/1996: 5-9.

Scheffran, Jürgen, 1997: "Frieden und nachhaltige Entwicklung. Interdisziplinäre Betrachtun-gen zum Verhältnis von Umwelt, Entwickung und Frieden", in: Vogt, Wolfgang; Jung, Eckhard (Eds.): *Kultur des Friedens. Wege zu einer Welt ohne Krieg* (Darmstadt: Wissenschaftliche Buchgesellschaft): 43-49

Scheffran, Jürgen, 1998: "Wege zu einer nachhaltigen Entwicklung des Friedens", in: Scheffran, Jürgen; Vogt, Wolfgang R. (Eds.): *Kampf um die Natur* (Darmstadt: Wissenschaftliche Buchgesellschaft): 291-301.

Scheffran, Jürgen, 1999: "Environmental Conflicts and Sustainable Development: A Conflict Model and its Application in Climate and Energy Policy", in: Carius, Alexander; Lietzmann, Kurt M. (Eds.): *Environmental Change and Security: A European perspective* (Berlin- Heidelberg: Springer): 95-218.

Scheffran, Jürgen, 2000: "The dynamic interaction between economy and ecology. Cooperation, Stability and Sustainability for a Dynamic-Game Model of Resource Conflicts", in: *Mathematics and Computers in Simulation*, 53,4: 371-380.

Scheffran, Jürgen, 2003: "Calculated Security? Mathematical Modelling of Conflict and Cooperation", in: Booß-Bavnbek, Bernhelm; Høyrup, Jens (Eds.) Mathematics and War, (Basel: Birkhäuser): 390-412.

Scheffran, Jürgen, 2006: "The Formation of Adaptive Coalitions", in: Haurie, Alain; Muto, Shigeo; Petrosjan, Leon A. (Eds.): *Advances in Dynamic Games*, Volume 8, Annals of the International Society of Dynamic Games (Boston: Birkhäuser): 163-178.

Scheffran, Jürgen, 2008a: "Preventing Dangerous Climate Change", in: Grover, Velma I. (Ed.): *Global Warming and Climate Change* (New York: Science Publishers), vol. 2: 449-482.

Scheffran, Jürgen, 2008b: "Adaptive management of energy transitions in long-term climate change", in: *Computational Management Science*, 5,3: 259-286.

Scheffran, Jürgen, 2008c: "The Complexity of Security", in: *Complexity*, 14,1: 13-21.

Scheffran, Jürgen, 2010: "Criteria for a Sustainable Bioenergy Infrastructure and Lifecycle", in: Mascia, P. N.; Scheffran, Jürgen; Widholm, Jack (Eds.), 2010: *Plant Biotechnology for Sustainable Production of Energy and Co-products* (Berlin: Springer): 409-443.

Scheffran, Jürgen, 2011a: "The Security Risks of Climate Change: Vulnerabilities, Threats, Conflicts and Strategies", in: Brauch, Hans Günter; Oswald Spring, Úrsula; Mesjasz, Czeslaw; Grin, John; Kameri-Mbote, Patricia; Chourou, Béchir; Dunay, Pal; Birkmann, Jörn (Eds.): *Coping with Global Environmental Change, Disasters and Security* (Berlin: Springer): 735-756.

Scheffran, Jürgen, 2011b: "Frieden und nachhaltige Entwicklung", in: Giessmann, Hans Joachim; Rinke, Bernhard (Eds.): *Handbuch Frieden* (Wiesbaden: VS Verlag): 310-323.

Scheffran, Jürgen, 2015a: "Complexity and Stability in Human-Environment Interaction: The Transformation from Climate Risk Cascades to Viable Adaptive Networks", in: Kavalski, Emilian (Ed.): *World Politics at the Edge of Chaos* (Albany, NY: State University of New York Press): 229-252.

Scheffran, Jürgen, 2015b: "Climate Change as a Risk Multiplier in a World of Complex Crises", Planetary Security Conference, The Hague, Netherlands; November 2015.

Scheffran, Jürgen, 2016a: "Der Vertrag von Paris: Klima am Wendepunkt?", in: *WeltTrends—Das außenpolitische Journal*, 112,24 (February): 4-9.

Scheffran, Jürgen, 2016b: "Kettenreaktion außer Kontrolle: Vernetzte Technik und das Klima der Komplexität", in: *Blätter für deutsche und internationale Politik*, 3: 101-110.

Scheffran, Jürgen; Battaglini, Antonella, 2011: "Climate and conflicts—The security risks of global warming", in: *Regional Environmental Change*, 11 (Suppl. 1): 27-39.

Scheffran, Jürgen; Brauch, Hans Günter, 2014: "Conflicts and Security Risks of Climate Change in the Mediterranean Region", in: Goffredo, Stefano; Dubinsky, Zvy (Eds.): *The Mediterranean Sea* (Dordrecht: Springer): 625-640.

Scheffran, Jürgen; Brzoska, Michael; Brauch, Hans Günter; Link, Peter Michael; Schilling, Janpeter (Eds.) 2012: *Climate Change, Human Security and Violent Conflict: Challenges for Societal Stability* (Berlin—Heidelberg—New York: Springer-Verlag).

Scheffran, Jürgen; Brzoska, Michael; Kominek, Jasmin; Link, P. Michael; Schilling, Janpeter, 2012a: "Climate change and violent conflict", in: *Science*, 336: 869-871.

Scheffran, Jürgen; Brzoska, Michael; Kominek, Jasmin; Link, P. Michael; Schilling, Janpeter, 2012b: "Disentangling the Climate-conflict Nexus", in: *Review of European Studies*, 4,5: 1-13.

Scheffran, Jürgen; Burroughs, John; Leidreiter, Anna, van Riet, Rob; Ware, Alyn, 2015: *The Climate-Nuclear Nexus: Exploring the linkages between climate change and nuclear threats*. Affiliation (Hamburg: World Future Council).

Scheffran, Jürgen; Cannaday, Thomas, 2013: "Resistance Against Climate Change Policies: The Conflict Potential of Non-fossil Energy Paths and Climate Engineering", in: Maas, Achim; Balazs, Bodó; Burnley, Clementine; Comardicea, Irina; Roffey, Roger (Eds): *Global Environmental Change: New Drivers for Resistance, Crime and Terrorism?* (Baden-Baden: Nomos).

Scheffran, Jürgen; Hannon, Bruce, 2007: "From complex conflicts to stable cooperation: Cases in environment and security", in: *Complexity*, 13,2: 78-91.

Scheffran, Jürgen; Ide, Tobias; Schilling, Janpeter, 2014: "Violent Climate or Climate of Violence? Concepts and Relations with Focus on Kenya and Sudan", in: *International Journal of Human Rights*, 18,3: 369-390.

Scheffran, Jürgen; Jathe, Markus, 1995: "Modelling the Impact of the Greenhouse Effect on International Stability", in: Kopacek, Peter (Ed.): *Supplementary Ways for Improving International Stability, Proceedings from the IFAC conference, Vienna, Austria,* 29 September–1 October 1995 (Oxford: Pergamon): 31-38.

Scheffran, Jürgen; Leimbach, Marian, 2006: "Policy-Business Interaction in Emission Trading between Multiple Regions", in: Antes, Ralf; Hansjürgens, Bernd; Letmathe, Peter (Eds.): *Emissions Trading and Business* (Heidelberg: Physica-Verlag): 353-367.

Scheffran, Jürgen; Link, Peter Michael; Schilling, Janpeter, 2012: "Theories and Models of Climate-Security Interaction: Framework and Application to a Climate Hot Spot in North Africa", in: Scheffran, Jürgen; Brzoska, Michael; Brauch, Hans Günter; Link, Peter Michael; Schilling, Janpeter (Eds.) 2012: *Climate Change, Human Security and Violent Conflict: Challenges for Societal Stability* (Berlin–Heidelberg–New York: Springer-Verlag): 91-132.

Scheffran, Jürgen; Marmer, Elina; Sow, Papa, 2012: "Migration as a contribution to resilience and innovation in climate adaptation", in: *Applied Geography,* 33: 119-127.

Scheffran, Jürgen; Remling, Elise, 2013: "The social dimensions of human security in climate change", in: Redclift, Michael R.; Grasso, Marco (Eds.): *Handbook on Climate Change and Human Security* (Chichester: Edward Elgar): 137-163.

Scheffran, Jürgen; Stoll-Kleemann, Susanne, 2002: "Implementation of Integrated Sustainability Strategies in Europe–Multi-level Participation and Conflict Management in Climate and Biodiversity Regimes", in: Biermann, Frank; Brohm, Rainer; Dingwerth, Klaus (Eds.): *Global Environmental Change and the Nation State. Proceedings of* 2001 *Berlin Conference on the Human Dimensions of Global Environmental Change* (Potsdam: PIK): 329-339.

Schilling, Janpeter; Freier, Korbinian P.; Hertig, Elke; Scheffran, Jürgen, 2012: "Climate change, vulnerability and adaptation in North Africa with focus on Morocco", in: *Agriculture, Ecosystems and Environment,* 156: 12-26.

Sen, Amartya, 1985: *Commodities and Capabilities* (Amsterdam: North-Holland).

Sparrow, Colin, 1982: *The Lorenz Equations: Bifurcations, Chaos, and Strange Attractors.* Springer.

Spiegel, 1986: Abrüstung: "Die Sache bringt Bewegung", in: *Der Spiegel,* 4/86: 97-98.

Spiegel, 2011: "Flut in Australien: Wichtige Kohleminen noch wochenlang unbrauchbar", in: *Spiegel online* 6 January 2011; at: <http://www.spiegel.de/wirtschaft/flut-in-australien-wichtige-kohleminen-noch-wochenlang-unbrauchbar-a-738054.html>.

Starr, Harvey (Ed.), 2008: "Failed States", Special Issue. *Conflict Management and Peace Science,* 25,4.

Stern, Nicholas, 2006: *The Economics of Climate Change–The Stern Review* (Cambridge: Cambridge University Press).

Sternberg, Troy, 2012: "Chinese drought, bread and the Arab Spring", in: *Applied Geography,* 34: 519-524.

Sternberg, Troy, 2013: "Chinese Drought, Wheat, and the Egyptian Uprising: How a Localized Hazard Became Globalized", in: Werrell, Caitlin E.; Femia, Francesco (Eds.): *The Arab Spring and Climate Change.* A Climate and Security Correlations Series (Washington, D.C.: Center for American Progress, Stimson Center, February): 7-14.

UNEP, 2007: *Sudan: post-conflict environmental assessment* (Nairobi: United Nations Environment Programme).

UNHCR, 2011: "Expert Meeting on Climate Change and Displacement", at: <http://www. unhcr.org/cgi-bin/texis/vtx/search%5C?page=&comid=4e01e63f2&keywords=Bellagio-meeting>.

Verhoeven, Harry, 2011: "Climate Change, Conflict and Development in Sudan: Global Neo-Malthusian Narratives and Local Power Struggles", in: *Development and Change,* 42: 679-707.

Wæver, Ole, 1997: *Concepts of Security* (Copenhagen: Department of Political Science, University of Copenhagen).

Walker, Brian; Holling, C. S.; Carpenter, Stephen R.; Kinzig, Ann, 2004: "Resilience, adaptability and transformability in social-ecological systems", in: *Ecology and Society,* 9,2:5; at: < http://www.ecologyandsociety.org/vol9/iss2/art5/>.

WBGU, 2008: *Climate Change as a Security Risk* (Berlin: Wissenschaftlicher Beirat der Bundesregierung Globale Umweltveränderungen/German Advisory Council on Global Change).

Webersik, Christian, 2010: *Climate Change and Security* (New York: Prager).

Werrell, Caitlin E.; Femia, Francesco (Eds.), 2013: *The Arab Spring and Climate Change.* A Climate and Security Correlations Series (Washington, D.C.: Center for American Progress, Stimson Center, February).

Werz, Michael; Hoffman, Max, 2013: "Climate Change, Migration, and Conflict", in: Werrell, Caitlin E.; Femia, Francesco (Eds.): *The Arab Spring and Climate Change.* A Climate and Security Correlations Series (Washington, D.C.: Center for American Progress, Stimson Center, February): 33-40.

World Bank, 2013: "Renewable internal freshwater resources per capita (cubic meters)"; at: <http://data.worldbank.org/indicator/er.h2o.intr.pc/countries/tnLY-sY-1A-1W?display= graph>.

World Commission on Environment and Development, 1987: *Our Common Future* (Oxford: Oxford University Press).

14   The First Decade of Initiatives for Research on the Human Dimensions of Global (Environmental) Change (1986–1995)

Lourdes Arizpe[1], Martin F. Price[2] and Robert Worcester[3]

Abstract

By the end of the 1980s, very different meanings of 'global change' existed, promoted by different constituencies in the social and natural sciences (Price 1989). One could be described as anthropocentric, emphasizing the interactions between people and their institutions, primarily at scales extending to decades. This chapter presents and discusses the emergence of initiatives for research on the human dimensions of global change until the 1996 launch of the International Human Dimensions Programme on Global Environmental Change (IHDP). In 1987, the International Social Science Council (ISSC) joined with IFIAS and UNU to develop a Human Response to Global Change Programme. In 1990, ISSC launched the Human Dimensions of Global Environmental Change Programme (HDP), based on the "Framework for Research on HDGEC" (Jacobson/Price 1990), identifying seven broad areas in which research should be done. The first half of the 1990s were also characterized by the emergence and development of various national and regional (e.g., European) initiatives for research on HDGEC. In the subsequent two decades, as described elsewhere in this book, substantial advances have been made; many of them emerged from the initiatives described in this chapter.

Keywords: Social science, global environmental change, International Social Science Council, human dimensions of environmental change, land use, deforestation, industrial metabolism, social sustainability, culture and sustainability, perceptions and opinions on environmental change.

14.1   Introduction[4]

As several chapters in this volume clearly show, our planet is experiencing changes at the global scale.

The volume's title refers to 'global environmental change', but the biophysical environment is only part of the complex world in which we live, and many other parts of this are also experiencing changes at the global scale. These include social, economic and political systems and, until the early 1980s, the predominant use of the term 'global change' was by social scientists, to refer to changes in these systems. For example, the philosopher Robert Heilbroner stated that "Everybody senses that our age is one of profound turmoil, a time of deep change" (Heilbroner/Campbell 1975). He identified the proliferation of nuclear weapons, population growth and industrial growth as fundamental problems, linked to significant concerns about pol-

---

1   Lourdes Arizpe, is former President of the International Social Science Council, held the position of Assistant Director General for Culture at Unesco and is Professor at the National University of Mexico (la2012@correo.crim.unam.mx).

2   Martin F. Price, Professor, University of the Highlands and Islands, UK, UNESCO Chair in Sustainable Mountain Development, Perth College, Perth, Scotland, UK; Adjunct Professor, University of Bergen, Norway; Email: <martin.price@perth.uhi.ac.uk>.

3   Sir Robert Worcester KBE DL is the Founder of MORI (Market & Opinion Research International), London, former President of the World Association for Public Opinion Research (WAPOR) and former Senior Vice President of the International Social Science Council. Sir Robert is Emeritus Chancellor of the University of Kent and Visiting Professor of Government at LSE, as well as Deputy Chairman of the Magna Carta Trust.

---

4   Of this chapter, section 14.1 and 14.2 rely heavily on Price, M.F. 1989: "Global change: defining the ill-defined", in: *Environment*, 31 (8), 18-20, 42-44, and Price, M.F. 1990: "Humankind in the biosphere: the evolution of interdisciplinary research", *Global Environmental Change; Human and Policy Dimensions*, 1, 3-13.

lution, inflation, energy and national security, and the possible loss of affluent lifestyles. Through the late 1970s and well into the 1980s, 'global change' continued to signify social and political change associated with similar challenges, such as international insecurity, unequal development, and decreases in the quality of life (Väyrynen 1979; Gazzo 1979; Prebisch 1980; Gati 1983; Price 1989).

In the 1980s, the concept of 'global change', now more commonly referred to as 'global environmental change', emerged, first in a study sponsored by the National Aeronautics and Space Administration (NASA), which effectively identified that we are now in the Anthropocene, referring to "changes that may affect the habitability of the earth" (Goody 1982: 3). In the following years, publications resulting from a study concluded by the US National Research Council (NRC, 1983) and a symposium sponsored by the International Council of Scientific Unions (ICSU) in 1984 (Malone/Roederer 1985) specifically referred to global change and led to the establishment of the International Geosphere-Biosphere Programme (IGBP), launched in 1986 as "A study of global change". Over the next two years, a plan for action was developed, outlining a research programme with four main components: terrestrial biosphere-atmosphere chemistry interactions; marine biosphere-atmosphere interactions; biospheric aspects of the hydrological cycle; effects of climate change on terrestrial ecosystems (IGBP 1988).[5]

Thus, by the end of the 1980s, very different meanings of 'global change' existed, promoted by different constituencies in the social and natural sciences (Price 1989). One could be described as anthropocentric, emphasizing the interactions between people and their institutions, primarily at scales extending to decades. The other could be described as geocentric, emphasizing the interacting processes of the Earth's atmosphere, biosphere, geosphere and hydrosphere, typically over far longer timescales. While people were recognised as driving forces, and also as affected by climate change and the productivity of ecosystems, they were not at all a focus, as encapsulated in a 'conceptual model of global system' published by NASA's Earth System Science Committee (ESSC 1988) in which 'human activities' were restricted to three 'bubbles' around the edge of the diagram. In this context, the goal of this chapter is to present and discuss the

emergence of initiatives for research on the human dimensions of global change—incorporating elements of both meanings—until the 1996 launch of the International Human Dimensions Programme on Global Environmental Change (IHDP).[6]

14.2   The Institutional Framework

As described below, the International Social Science Council (ISSC) launched the Human Dimensions of Global Environmental Change Programme (HDP) in 1990. This was the outcome of a long process involving a considerable number of institutions, introduced briefly in this section.

The ISSC [7] was created in 1953 as UNESCO set out to establish international non-governmental organizations (NGOs), including scientific organizations. In the aftermath of World War II, scientists were to develop their arts and sciences to reconstruct new and peaceful countries (Platt 2002: 12). Claude Lévi-Strauss became the first Secretary General of the ISSC. In accordance with its first two aims, "advance the quality, novelty and utility of the social sciences worldwide [and] advance social science research across national and regional boundaries",[8] the establishment of disciplinary organizations and national social science associations was encouraged. By the end of the 1980s, the ISSC had more than sixty member organizations and member associations, and took the initiative to create several important research programmes.

In 1969, ICSU established the Scientific Committee on Problems of the Environment (SCOPE); its objectives included "to advance knowledge of the influence of humans on their environment, as well as the effects of these environmental changes upon people, their health and welfare".[9] While ICSU's member unions are from the natural sciences, during the 1980s, two, from the disciplines of geography and psychology, were also members of the ISSC.[10] The establishment of SCOPE derived particularly from increas-

---

5   The IGBP closed at the end of 2015. Many of its activities are now part of Future Earth: see <http://www.futureearth.org/>.

6   For a previous assessment see: Louise von Falkenhayn, Andreas Rechkemmer and Oran R. Young and at: <http://www.ihdp.unu.edu/>.

7   At: <http://www.worldsocialscience.org/> (12 January 2015), and at: <http://ngo-db.unesco.org/r/or/en/1100053199> (12 January 2015).

8   At: <http://www.worldsocialscience.org/about/history-mission/>, accessed 12 January 2015.

9   At: <http://www.scopenvironment.org/downloadpubs/scope5/foreword.html> (21 December 2014).

ing awareness of the interactions of people and the biosphere, as manifested particularly by the 1972 UN Stockholm Conference on the Global Environment, for which SCOPE prepared the first report (SCOPE 1971). While many projects until the end of the 1980s focused almost exclusively on natural sciences, others involved significant numbers of both natural and social scientists (Price 1990).

A further four relevant global institutions emerged in the early 1970s, showing the increasing recognition of the need for international and interdisciplinary collaboration with regard to the challenges of global change. UNESCO's Man and the Biosphere (MAB) programme began in 1971. Its general objective was similar to those of SCOPE, but it explicitly aimed to "develop the basis within the natural and social sciences for the rational use and conservation of the resources of the biosphere and for the improvement of the global relationship between man and the environment; to predict the consequences of today's actions on tomorrow's world".[11] By 1984, MAB had some success in integrating natural and social scientists—and sometimes local decision-makers and people—in policy-relevant research (di Castri 1985). Nevertheless, the involvement of social scientists was limited (Spooner 1984).

In 1972, representatives of the Soviet Union, the US, and ten other countries from the Eastern and Western blocs established the International Institute for Applied Systems Analysis (IIASA). It created international interdisciplinary teams that used advanced systems analysis to study global challenges. In the context of this chapter, a key programme was on sustainable development of the biosphere (Clark/Munn 1986). Also in 1972, the International Federation of Institutes for Advanced Study (IFIAS) was founded; among its original objectives were "to promote and carry out joint transdisciplinary and transnational research on world problems with a special emphasis on social, ethical and humanistic aspects".[12] In the same year, the UN General Assembly adopted the decision to establish the United Nations University (UNU). Academic activities began in 1975; the first

priority programme areas were world hunger, natural resources, and human and social development.[13]

14.3 Emerging Issues in Social Science Research on Global Change

Until the 1980s, relatively few social scientists, with the exception of some anthropologists and geographers, had considered the biophysical environment as a critical constraint on the survival of human societies (Dunlap 1980). However, as the decade progressed, and building on the interdisciplinary initiatives such as those outlined above, social scientists began taking up questions of global change, and the related, but also contested, concept of sustainable development,[14] according to different disciplinary perspectives. In addition to the IIASA programme mentioned above, which included perspectives from economics, energy, geography, history, law, and management and policy studies, a few key examples are mentioned below.

In geography, researchers looked at human displacements in territorial or ecological systems, applying concepts such as risk and vulnerability (Kates et al. 1985). What kinds of micro-social models could be built in the framework of global change? Robert Kates had already developed an approach that defined decisions of social movements which decisively alter the course of a process as 'events'. 'Initial conditions' are those existing immediately prior to a major event that caused a sufficiently significant alternation of these conditions that their rippling effects can still be discerned in the present condition.

One early issue was how to define the processes under study in a global context. García (1986) proposed a three-tiered structure linking micro- to macro-processes. To analyse the dynamics of such processes, he stressed that the property of the system is not found in its components but in the relationships between them. Decision-making then becomes important to understand, for example, the emergence of famines. Researchers led by Gilberto Gallopin at the Bariloche Foundation gave a different perspective on global reality, proposing the concept of 'global impoverishment' to encompass both ecological and economic impoverishment as the central process of global change (Gallopin/Guttman/Maletta 1989; Leff, 1986). A counterpoint from more industrialized coun-

---

10 Others have subsequently joined, from anthropology/ethnography, and sociology.

11 <http://www.unesco.org/new/en/natural-sciences/environment/ecological-sciences/man-and-biosphere-programme/mab40/press/history/> (10 January 2015).

12 <http://www.ifias.ca/IFIASinfo/IFIASinfohist.html> (12 January 2015).

---

13 <http://unu.edu/about/unu/history#overview>, (21 December 2014).

14 As early as 1989, 190 definitions existed (Pezzey 1989).

tries was provided by Udo Simonis (Simonis 1989), describing the relationships between economic structure and environmental impacts, and identifying both deficiencies of environmental policy and needs to integrating ecological dimensions into economic policy.

Sociologists and anthropologists, many of them members of research committees of the International Sociological Association and the International Union of Anthropological Sciences (both belonging to ISSC) focused on 'actions that generate social structures' through the choices made by individuals or groups. Choices are indeed made by individuals, but all such choices are embedded in exchanges. Thus, in terms of the relationship of human populations to the environment, choice is not constrained by an individual's psychological motivation but by the given range of options which that individual or group has in a given economic and social structure. Additionally, it was argued that "...human beings also have the capacity to adapt to rapidly changing environments, and so room must be left for new knowledge. Yet this gift of adaptation is based on the one single ability that people have....and that is the ability to learn from experience" (Arizpe 1989/2013: 39).

Psychologists also worked on various aspects of global change and human action. Lennart Sjöberg (Sjöberg 1989) suggested some key points with regard to: the use of cognitive and attitudinal approaches for behaviour modification; cognitive biases which tend to cause overconfidence and exaggerated belief in environmental stability; the difficulty of giving priority to the collective good and the preservation of the environment; environmental attitudes, and the difficulty of using mass media to change them; and that, while the individual perspective is important for a significant share of pollution and resource consumption, civic behaviour and organizational psychology also have potential. 'Cultural sustainability' was also taken up as perceptions of resource depletion and environmental change—not as psychological states but as expressions of cultural values in a context of accelerated social change—emerged as a major research challenge. This fundamental stratum of cultural values is the basis of assessments of options and choices made (Clark 1988). Nevertheless, the heuristic boundaries of units under analysis mean "deciding whether one culture should be considered a single unit in spite of internal diversity or whether its subcultures should be considered discrete entities" (Arizpe 1991:49). The interpretive nature of such methodological decisions in the heuristics of boundaries creates theoretical

problems and methodological issues of statistical measurement.

The examples above are all drawn from individual disciplines, but the 1980s were also a decade in which a particular aspect of global change—climate change—became an increasing focus of research attention from various social science disciplines, increasingly working together to address this complex phenomenon (e.g., Kellogg/Schware 1981; Chen et al. 1983; NRC 1983; Kellogg 1987; Glantz 1988; Parry et al. 1988). In an interdisciplinary effort, Lourdes Arizpe, Wolfgang Lutz and Robert Constanza argued that "To achieve a sustainable pattern of resource use and population we must understand and control the interactions of population and per capita resources as mediated by technology, culture and values"(Arizpe/Lutz/Constanza 1992: 61). Finally, this was a period in which boundaries between disciplines became blurred, and new hybrids emerged. To some extent, this resulted from interdisciplinary programmes and initiatives such as those mentioned above. Mattei Dogan and Robert Pahre (Dogan and Pahre 1989) identified various examples of hybridization, noting that sociology "is perhaps the most open of all social sciences... [and] interacts with all the other social sciences" (op cit. p 459), and also describing hybridization in history, political science, economics, and geography. They concluded that hybridization, through borrowing concepts, theories, and methods as well as the exchange of findings, would continue to stimulate innovate throughout the social sciences—and that this requires courage.

14.4   Towards a Research Programme on the Human Dimensions of Global (Environmental) Change

In 1986, the Sixteenth General Assembly of the ISSC adopted a resolution creating an ad hoc committee to explore the possibility of developing an international social science programme to parallel and complement the emerging IGBP, recognizing that human activities had become a significant force in global change. This decision was based in two imperatives: first, increasing recognition among social scientists that they had to advance in developing new theoretical perspectives and models to encompass all dimensions of global change; and, second, the fact that natural scientists had already begun developing the IGBP.

Even in the early discussions in the Committee, 'global change' was considered too broad a term given the many ongoing global processes, often considered as global change in the anthropogenic sense, including various processes sometimes described as 'globalization', especially in economics, political science and sociology. However, the committee also recognised the emergence of the geocentric perspective espoused by the IGBP, and the emphasis for the research programme in the social sciences increasingly focused on this perspective.

In 1987, ISSC joined with IFIAS and UNU to develop a Human Response to Global Change Programme, which soon became the Human Dimensions of Global Change Programme (HDGCP). This renaming resulted from a second conceptual discussion regarding the 'social' in the formulation of research boundaries. The members of the ad hoc committee were unwilling to emphasize disciplinary boundaries at a time when both the social sciences were insisting on revising disciplinary boundaries and the research field under discussion itself required interdisciplinary collaboration among scientists. Consequently, the terms traditionally used did not seem to be appropriate. The 'social aspects' of global environmental change would need to include in its formulation, at the very least, the social, economic, political, geographical, and cultural aspects of global change: hence, the decision to adopt 'human dimensions'. This seemed to capture the specificity of the topics that the social sciences would take up in relation to the environment, yet at the same time implied an openness to interdisciplinary work that not only left out none of the social sciences but could also include research by natural sciences which considered anthropogenic actions.

Over a one-year period, eleven meetings were held around the world in order to define the scope of the programme. These were followed by a symposium in Tokyo in September 1988 to define the programme's objectives and research initiatives (IFIAS/ISSC/UNU 1989). This initiative continued, with UNESCO joining the steering committee of the HDGCP in 1989, and a number of workshops took place under the auspices of the four (and other) organizations. Some developed themes for the proposed ISSC programme (see below), while the majority developed the research initiative derived from the Tokyo symposium, most with a policy emphasis (Price 1990).

In December 1988, the ISSC established a Standing Committee on the Human Dimensions of Global Change. The initial fourteen members of the committee included Elza Berquo (Brazil), Rene Passet (France), Ademola Salau (Nigeria), Krishnamurthy Srinivasan (India), Kerry Turner (UK), Björn Wittrock

(Sweden), and Zhang Pei Yang (China). However, these were not able to continue, and the eventual committee consisted of nine individuals, each from a specific discipline and country: Harold Jacobson (political science US, chair); Lourdes Arizpe (anthropology, Mexico); Daniel Bertaux (sociology, France); Ashish Bose (demography, India); Takashi Fuji (economics, Japan); Leszek Kosinski (geography, Canada); Kurt Pawlik (psychology, Germany); Renat Perelet (systems analysis, Russia); Robert Worcester (public opinion, UK). They met three times in 1989 and 1990, with the main task of drafting an action plan for research on what they now referred to as the human dimensions of global environmental change (HDGEC).

These ongoing processes were also influenced by the publication of 'Our Common Future' by the World Commission on Environment and Development (WCED 1987). In 1989, the NRC held a forum on 'Global change and our common future' at which the chair of the WCED, Gro Harlem Brundtland, clearly stated the convergence of the two types of global change, explicitly linking what was now becoming more widely referred to as 'global environmental change' (and particularly climate change) with sustainable development (Brundtland 1989). At the same meeting, Roberta Miller of the US National Science Foundation noted the need for social science research programmes that should not only "feed into the physical and natural science activities" (i.e, the emerging IGBP), but also "be concerned with those elements of the global change research agenda that are purely social and economic but that ultimately are as powerful determinants of environmental change as ongoing physical and biological processes" (Miller 1989: 86 87).

In November 1990, after two years of debates on research carried out in social science and in consultation with the member associations and organizations of the ISSC as well as those of ICSU, and in collaboration with UNESCO's Social Science Sector, the ISSC Standing Committee presented their output. Particularly because of the very diverse perspectives represented on the Committee, this was not an action plan, but a framework for research on HDGEC (Jacobson/Price 1990). It identified seven broad areas in which research should be carried out:

- social dimensions of resource use;
- perception and assessment of global environmental conditions and change;
- impacts of local, national, and international social, economic and political structures and institutions;
- land use;

- energy conversion and consumption;
- industrial growth; and
- environmental security and sustainable development.

The first three topics encompass the central issues involved in understanding HDGEC (Jacobson 1992). The dynamic interactions among populations, resources and technology set fundamental parameters on the anthropogenic contribution to global change. Two areas were especially signalled out for research: economics and demographic change. At that time, concerns over population growth and the need to bring about a demographic transition were priorities on the agenda of the social sciences. Importantly, in seeking a social science perspective on global environmental change, the Standing Committee emphasized that the ways in which individuals and government officials, industry managers, and the general public perceive and assess environmental changes will affect how they behave toward sources of global change and will determine steps they might take to adapt or mitigate the effects of such changes. In a forward-looking strategy, the Standing Committee was keen to draw attention to the advances in cognitive sciences that detailed the structure and function of information processing, on the one hand; and on the other hand, the understanding of the relationship between the mass media and public perception that could provide a strong base for this research.

Other members of the Standing Committee, particularly those living in what were then called 'Third World' countries, were more interested in discussing political institutions and economic policies that had direct and indirect, as well as intended and unintended, effects on global environmental change. It must be remembered that, at the time, dictatorships and *coups d'état* in Latin America and Africa and the effects of rapid modernization were the focus of attention of social scientists in these regions. Thus, they began to perceive environmental questions as the secondary objective after the development concerns of democratization and poverty alleviation, and became deeply involved in these debates. The Standing Committee, however, was able to build a broad framework that encompassed both environmental depletion as well as development needs.

Each of the first three topics demanded integrated analyses from a range of disciplines, including anthropology, economics, political science and sociology. Much discussion was devoted to debating the institutional and programmatic form that such interdisciplinary work would take. A key concern was the realiza-

tion that social science studies conducted within the borders of nation-states were not producing the data needed to understand the social aspects of global processes. As an example, a specific study group on demography was created; and this showed how national population statistics were very difficult to use in identifying global trends. Although some studies based on population statistics could show broad international trends, such as world population growth processes, they could rarely be used for comparative studies between given countries, since each established different definitions and categories of basic census data. Understanding global processes required the standardization and formalization of statistical data across countries. This became the central objective of the Human Development Project of the United Nations Development Programme which began to consolidate during the same period. Lourdes Arizpe participated in both projects, and ensured that the building of concepts and models, and the same major concerns were discussed and exchanged.

The next three themes of the framework for research on HDGEC were recognised as major proximate causes of global change—land use, energy conversion and consumption, and industrial growth—jointly responsible for the majority of greenhouse gas emissions. Understanding the dynamics of these activities was considered crucial to fully understanding the dynamics of global change. The framework explicitly linked the last topic, environmental security and sustainable development, to the report of the WCED; this was seen as embodying the normative tradition in the social science, namely, exploring the values involved in human actions. These investigations should involve description, analysis and prescription and would be an essential component of social science research on global environmental change. The costs and benefits of various strategies to adapt to global change or to mitigate its effects should be explored, and norms for dealing with issues of international and intergenerational equity should also be taken into account.

14.5 Initiatives on the Human Dimensions of Global Environmental Change, 1990–1995

The framework for research on HDGEC (Jacobson/Price 1990) was published in November 1990. A draft work programme for the implementation of the HDGEC programme (HDP) was reviewed at a scientific symposium on HDGEC held in Palma de Mallorca in the same month, in conjunction with the Eighteenth General Assembly of the ISSC, and then adopted by the Standing Committee. From this point onwards, its aim was to stimulate and encourage research among the members of the scientific organizations. Several unions already had research committees to study the processes that were changing the environment in geographical, economic and social terms. Many had already highly developed research activities on the interaction of humans and the natural environment, for example, the International Union of Anthropological and Ethnological Sciences which had a long tradition of studies of 'Man's Habitat'. The establishment of the HDP led many more member associations to create committees, study groups and research networks to study global environmental change. The then President of ISSC, Luis Ramallo, proposed that the Secretariat of the HDP be established in Barcelona, with funding from Spanish sources. It subsequently moved to Geneva, with Swiss government support, with an office and a scientific directorate which continued until August 1996 (Platt 2002: 34).

Following the lead of the IGBP and the World Climate Research Programme (WCRP), the HDP promoted interdisciplinary and international research organized on the basis of 'work programmes' to cover the two-year periods between ISSC General Assemblies (Jacobson, 1992). Given that the availability of relevant data was highlighted as crucial for research in this field, the 1991-1992 work programme (ISSC 1990) concentrated heavily on assessing available data and establishing data requirements to generate new databases in this field. Six working groups on data were therefore established. The themes, chairs, and outputs of those that published reports are listed below:

- demographic data (John Clarke, professor of geography, Durham University, UK: Clarke/Rhind 1992);
- economic data (Gary Yohe, professor of economics, Wesleyan University, US: Yohe/Segerson 1992);
- social science survey data (Robert Worcester, Senior Vice President of ISSC and former President of the World Association for Public Opinion Research (WAPOR): Worcester/Barnes 1991).

The work programme also stated that social science research on HDGEC required conceptual and methodological innovations. Seven working groups on the design and development of research, in the seven broad areas identified in the framework for research,

were therefore established, of which two published reports:

- land-use/land-cover change (jointly with the IGBP) (Billie Lee Turner, professor of geography, Clark University, US: Turner/Moss/Skole 1993);
- perceptions and assessments of global environmental conditions and change (Kurt Pawlik, professor of psychology, Hamburg University, Germany: Pawlik 1991).

In addition, most of the other working groups organized workshops. The outcomes of much of this work were reported at the second scientific symposium of the HDP, on 'Creating the Database' in 1992 and, subsequent to this, a Steering Committee, including an Executive Committee, was established. A third scientific symposium took place in 1995. By this time, four research programmes had emerged within the HDP, on attitudes and perceptions; demographic and social dimensions of resource use, industrial transformation and energy use, and land-use and land-cover change (jointly with the IGBP). The HDP had also identified two more fields for further study: institutions and environmental security; and three other areas on which further exploration was needed: human health and global environmental change, trade and environment, and the vulnerability of human populations. The papers and posters presented at the symposium addressed these very diverse themes (HDP 1996).

The late 1980s and first half of the 1990s were also characterized by the emergence and development of various national and regional (e.g. European) initiatives for research on HDGEC. In reality, the majority of funding for this research was and continues to be provided at the national level (and also, the European level). In Canada, the national global change programme included social scientists from its inception in 1985 (Braybrooke/Paquet 1987). In the US, NRC's Committee on Global Change proposed research into six areas of human dimensions in 1988 (Clark 1988), and the National Academy of Sciences and the NRC established a Committee on the Human Dimensions of Global Change in 1989. This published a substantial report in 1992 which presented the state of knowledge and proposed a national research programme on the human dimensions of global change (Stern/Young/Druckman 1992). In the same year, the Social Science Research Council appointed a Committee for Research on Global Environmental Change to foster interdisciplinary research, and the Centre for International Earth Science Information Network (CIESIN) was established as an independent NGO to provide information that would help scientists, decision-makers, and the public better understand the changing relationship between human beings and the environment. In 1991, it organized a workshop which produced a 'social process diagram' showing the interactions between different elements of social systems, as a tool for designing research agendas concerning the causes and effects of, and responses to, different types of global change (Kuhn/Luterbacher/Wiegandt 1992). In Europe, scientific interest emerged early in Sweden, with a conference in 1988 (Svedin/Heurling 1988) and a national committee on human dimensions of global change in 1991. By 1994, various initiatives had also developed in Belgium, France, Germany, Italy, the Netherlands, Norway, Switzerland and the United Kingdom. These initiatives were summarized by Price (Price 1994) in a report designed to assist the European Commission in defining linkages between existing initiatives and its own activities in research and development programmes. The Academy of the Social Sciences in Australia began to address such topics with a conference in 1990 (Brookfield/Doube 1990).

Many of these activities were linked, to one extent or another, not only to the HDP, led by the ISSC, but to the activities of other international organizations (Price 1994). These included the European Science Foundation's project on 'Environment, Science and Society', IIASA projects on global environmental change, global economic and technological transitions, and systems methods for the analysis of global issues; various projects of the UNU and the Organization for Economic Cooperation and Development (OECD); and working groups II and III of the Intergovernmental Panel on Climate Change (IPCC).

14.6   Conclusions

The decade from 1986 to 1995 was one in which scientists from both natural and social science disciplines, as well as an increasing number of politicians and members of the public around the world, recognised the increasingly rapid rates of change in not only the biophysical, but also the human, systems of our planet. Two themes, climate change and the imperative of sustainable development, were particularly important drivers. From the scientific perspective, this implied a critical need for new ways of thinking, new types of data, and new approaches to research, often requiring collaboration across disciplinary boundaries. As Miller (Miller 1994) noted, this requires new conceptualization of research problems, agreements on measurement, and time to learn to work together. These challenges certainly characterized the development of strategies for research on the HDGEC, both in the development of the ISSC's framework for research and in its subsequent implementation. However, this was only one of many international initiatives for research on HDGEC and, equally, much of the research that needs to be undertaken in relation to HDGEC has been, and will continue to be, done by individuals and groups within and across disciplines, typically at the local, subnational, or national level. In the subsequent two decades, as described elsewhere in this book, substantial advances have been made; many of them emerged from the initiatives described in this chapter.

References:

Arizpe, Lourdes, 1989: "On the Cultural and Social Sustainability of World Development", in Emmerj, Louis, (Ed.): *One World or Several?* (Paris: OECD Development Centre): 45-61. Republished in *Development*, 1997:43-57. Republished in: Arizpe, Lourdes, 2014: "On the Cultural and Social Sustainability of World Development", in *Lourdes Arizpe: A Mexican Pioneer in Anthropology* (Heidelberg: Springer-Verlag), 2014: 31-42.

Arizpe, Lourdes, 1991: "The Global Cube" in: Global Environmental Change. International Social Science Journal, XLIII,4 (Blackwell Publishers, Unesco, 1991), 599-608. Republished in *Lourdes Arizpe: A Mexican Pioneer in Anthropology* (Heidelberg: Springer-Verlag) 2014: 43-55.

Arizpe, Lourdes; Lutz, Wolfgang; Constanza, Robert, 1992: "Population and Natural Resource Use", in: Dooge, J.C.I.; Goodmanm GT; la Riviere, J.W.M. (Eds.): *Agenda for Science on Environment and Development*

into the 21st Century. (Cambridge: University of Cambridge): 61-68.

Baybrooke, D.; Paquet, G., 1987: "Human dimensions of global change: the challenge to the humanities and social sciences", in: *Transactions of the Royal Society of Canada*, 5: 271-291.

Brookfield, Harold; Doube, Loene (Eds.), 1990: *Global change: The Human Dimensions*. (Canberra: Academy of the Social Sciences in Australia).

Brundtland, Gro Harlem, 1989: "Global change and our common future: The Benjamin Franklin lecture", in: Ruth S. DeFries and Thomas F. Malone (Eds.), *Global Change and our Common Future*. Washington: National Academy Press: 10-18.

Chen, R.S.; Boulding, E.; Schneider, S.H., 1983: *Social science and climate change*. (Dordrecht: Reidel).

Clark, William C., 1988: "The Human Dimensions of Global Environmental Change", in: Committee on Global Change, *Toward an Understanding of Global Change* (Washington DC: National Academy Press): 134-200.

Clark, W.C.; Munn, R.E. (Eds.), 1986: *Sustainable Development of the Biosphere.* (Cambridge: Cambridge University Press).

Clarke, John I.; Rhind, David, D.W., 1992: *Population data and global environmental change.* (Paris: International Social Science Council/UNESCO).

Di Castri, Francisco, 1985: "Twenty years of international programmes on ecosystems and the biosphere", in: Malone, Thomas F.; Roederer, J.G. (Eds.), 1985: *Global Change.* Cambridge: Cambridge University Press: 314-331.

Dogan, Mattei; Pahre, Robert, 1989: "Hybrid fields in the social sciences"; in: *International Social Science Journal*, 121: 457-470.

Dunlap, R.E., 1980: "Paradigmatic change in social science"; in: *American Behavioral Scientist* 24(1): 5-14.

Earth System Science Committee (ESSC), 1988: *Earth System Science: A Closer View.* (Washington DC: National Aeronautics and Space Administration).

Gallopin, Gilberto C.; Gutman, Pablo; Maletta, Héctor, 1989: "Global impoverishment, sustainable development and the environment: a conceptual approach", in: *International Social Science Journal*, 121: 375-397.

García, Rolando, 1986 : "Conceptos básicos para el estudio de sistemas complejos", in: Leff, Enrique (Ed.), *Los Problemas del Conocimiento y la Perspectiva Ambiental del Desarrollo.* (Mexico: Siglo XXI).

Gati, T.T. (Ed.), 1983: *The U.S., the U.N., and the Management of Global Change.* (New York: New York University Press).

Gazzo, L., 1979: "Ten years' respite: The Club of Rome's Campaign for Global Change", in: *Atlas World Press Review*, January 1979: 44.

Glantz, Michael H. (Ed.), 1988: *Societal Responses to Regional Climatic Change: Forecasting by Analogy.* (Boulder: Westview).

Goody, J.M., 1982: *Global change: Impacts on habitability.* Report JPL D-95. Pasadena, California: Jet Propulsion Laboratory, California Institute of Technology.

Heilbroner, R.; Campbell, C., 1975: "Global Change: From Psychology Today, February '75", in: *Current*, March 1975: 24-34.

HDP, 1996: *Global Change, Local Challenge* (2 vols.). Report No. 8. (Geneva: HDP).

IFIAS/ISSC/UNU, 1989: *Report of the Tokyo International Symposium on the Human Response to Global Change* (Toronto: IFIAS).

IGBP, 1988: *The International Geosphere-Biosphere Programme: A Study of Global Change (IGBP): A Study of Global Change. A Plan for Action.* (Stockholm: IGBP Secretariat).

International Social Science Council, 1990: *Human Dimensions of Global Change Work Program, 1991-1992* (Paris: UNESCO/ISSC).

Jacobson, Harold K.; Price, Martin F., 1990: *A Framework for Research on the Human Dimensions of Global Environmental Change* (Paris: ISSC/UNESCO).

Jacobson, Harold, 1992: "Human Dimensions of Global Environmental Change Programme", in: *Environment*: 34, 5: 44.

Kates, Robert W; Ausubel, Jesse H.; Berberian, Mimi (Eds.), 1985: *Climate Impact Assessment: Studies of the Interaction of Climate and Society.* (New York: John Wiley and Sons).

Kellogg, William W., 1987: "Human impact on climate: the evolution of an awareness", in: *Climatic Change* 10: 113-136.

Kellogg, William W.; Schware, Robert, 1981: *Climate Change and Society.* (Boulder: Westview).

Kuhn, William R.; Luterbacher, Urs.; Wiegandt, Ellen, 1992: *Pathways of Understanding: The Interactions of Humanity and Global Environmental Change.* (Saginaw: CIESIN).

Leff, Enrique (ed.), 1986: *Los Problemas del Conocimiento y la Perspectiva Ambiental del Desarrollo.* (Mexico: Siglo XXI, eds).

Malone, Thomas F.; Roederer, J.G. (Eds.), 1985: *Global Change.* Cambridge: Cambridge University Press.

Miller, Roberta B., 1989: 'Human Dimensions of Global Environmental Change', in: Ruth S. DeFries and Thomas F. Malone (Eds.), *Global Change and our Common Future.* (Washington: National Academy Press: 84-89).

Miller, Roberta B., 1994: "Interactions and collaborations in global change across the social and natural sciences", in: *Ambio*, 23(1): 19-24.

National Research Council, 1983: *Toward an International Geosphere-Biosphere Program: A Study of Global Change.* (Washington DC: Commission on Behavioral and Social Sciences and Education, NAS/NRC).

Parry, M.L.; Carter, T.R.; Konijn, N.T. (Eds.), 1988: *The impact of climatic variations on agriculture.* (2 vols.) (Dordrecht: Kluwer).

Pawlik, Kurt, 1991: *Perception and assessment of global environmental conditions and change (Pagec): Report 1.* (Barcelona: HDP Secretariat).

Pezzey, J., 1989: *Economic Analysis of Sustainable Growth and Sustainable Development.* Environmental Working Paper, No.15. (Washington, DC: World Bank).

Platt, Jennifer, 2002: *Fifty Years of the International Social Science Council* (Paris: International Social Science Council).

Prebisch, R., 1980: Towards a Theory of Global Change. *UNCLA Review.*

Price, Martin F., 1989: 'Global Change: Defining the Ill-defined', in: *Environment*, 31 (8): 18-20, 42-44.

Price, Martin F., 1990: "Humankind in the biosphere: the evolution of international interdisciplinary research", in:

Global Environmental Change: Human and Policy Dimensions 1: 3-13.

Price, Martin F., 1994: *Options for EC-level Activities on the Human Dimensions of Global Change.* (Brussels: European Commission Directorate-General for Science, Research and Development).

SCOPE, 1971: *Global Environmental Monitoring.* (Stockholm: SCOPE).

Simonis, Udo E., 1989: "Ecological modernization of industrial society: three strategic elements", in: *International Social Science Journal*, 121: 347-361.

Sjöberg, Lennart, 1989: "Global change and human action: psychological perspectives", in: *International Social Science Journal*, 121: 413-432.

Spooner, B., 1984: "The MAB approach: problems, clarifications and a proposal", in: Di Castri, F.; Baker, F.W.G.; Hadley, M. (Eds.): *Ecology in Practice, Part II: The Social Response.* (Dublin: Tycooly: 324-339).

Stern. Paul C.; Young, Oran R.; Druckman, D. (Eds.), 1992: *Global Environmental Change: Understanding the Human Dimensions.* (Washington DC: National Academy Press).

Svedin, Uno; Heurling, Bo (Eds.), 1988: *Swedish Perspectives on Human Response to Global Change.* (Stockholm: Swedish Council for Planning and Coordination of Research).

Turner, B.L.; Moss, R.H.; Skole, D.L. (Eds.), 1993: *Relating land use and global land-cover change: A proposal for an IGBP-IHDP core project.* (Stockholm: IGBP and HDP).

Väyrynen, Raimo, 1979: "East-west relations and global change: The foreign policy ideology of Zbigniew Bzrezinski", in: *Current Research on Peace and Violence,* 2 (1), 20-37.

Worcester, Robert M.; Barnes, Samuel H., 1991: *Dynamics of Societal Learning about Global Environmental Change* (Paris: ISSC/UNESCO).

World Commission on Environment and Development, 1987: *Our Common Future.* (Oxford: Oxford University Press).

Yohe, G.W; Segerson, K., 1992: *Economic Data and the Human Dimensions of Global Environmental Change: Creating a Data Support Process for an Evolving Long Term Research Program.* (Barcelona: HDP).

15 From HDP to IHDP: Evolution of the International Human Dimensions of Global Environmental Change Programme (1996–2014)

Eckart Ehlers[1]

Abstract

IHDP's performance between 1996 and 2014 can be described as a process of maturation of social science research vis-à-vis the challenges of climate and global environmental change. It is characterized by a number of independent research projects and an increasing embeddedness into joint programme developments with the natural sciences. Incorporation of IHDP into the *Earth System Science Partnership* (ESSP) and its full integration into the Future Earth initiative bear witness to these facts. The emergence of the concept of the Anthropocene, the discussions around a 'geology of mankind', and the increasingly undisputed conclusions of IPCC reports about the role of human footprints in shaping our planet by stimulating global warming and exerting deep impacts on the atmosphere, geosphere, hydrosphere and biosphere—all three observations, assumptions and related theories have greatly influenced the steadily increasing importance of the social sciences in research into global change. In 2014, IHDP joined IGBP and DIVERSITAS in merging under the umbrella of Future Earth.

Keywords: Anthropocene, Future Earth, Earth System Science Partnership, integration, Institutional Dimensions of Global Environmental Change, Industrial Transformation, human security.

15.1 Introduction

This chapter, a direct continuation of the previous one (Arizpe/Price/Worcester 2016), is a reflection of IHDP's development since its transfer from its Geneva (Switzerland) headquarters to Bonn (Germany). This transfer and reorganization falls into a period of worldwide changes, politically and ecologically. In 1989 the Berlin Wall dividing West and East Germany came down. The collapse of the former Soviet Union seemed to open a new era of cooperation between West and East and to replace the political confrontations that had dominated international politics since the end of World War II. The year 1992 saw the 'Earth Summit' in Rio de Janeiro (Brazil), a turning point for global change research. The Rio conference not only opened up insights into the global dimension of ecological and environmental problems, but also demonstrated the necessity of increased international cooperation in order to tackle them. The 1992 Earth Summit was preceded by an increasing 'scientization' (Brauch 2009) of specific fields of climate research and its impacts on the geosphere and the biosphere. The creation of the *World Climate Research Programme* (WCRP) in 1980 (Church/Asrar/Busalacchi et al. 2011) and the establishment of the *International Geosphere-Biosphere Programme* (IGBP) in 1986 (Noone/Nobre/Seitzinger 2011) bear witness to these developments. Thus Rio 1992 and its follow-ups (Rio+10 in Johannesburg (South Africa) in 2002 and Rio+20 in 2012, again in Rio de Janeiro) became important cornerstones (Hackmann/St. Clair 2012) for the development of not only interdisciplinary research endeavours but also of international research cooperation in the fields of research into global change.

Social science research, although organizing itself from 1986 onwards, remained in a somewhat reserved and ambiguous position (see Arizpe/Price/Worcester 2016 [in this volume]). HDP/IHDP's early years are

---

1 Prof. Dr. Eckart Ehlers, University of Bonn, Department of Geography, Prof. Emeritus; email: <ehlers@giub.uni-bonn.de>.

© Springer International Publishing Switzerland 2016
H.G. Brauch et al. (eds.), *Handbook on Sustainability Transition and Sustainable Peace*,
Hexagon Series on Human and Environmental Security and Peace 10, DOI 10.1007/978-3-319-43884-9_15

characterized by an intensive search for a solid theo-retical and methodological foundation for its specific role in research into global environmental change and for the identification of relevant themes and topics. It should, however, also be noted that the acceptance of the social science approaches by the representatives of the natural science disciplines was not always encourag-ing. The great challenge was to find common grounds, that is common language, common themes and topics and common projects—a tedious yet in the end suc-cessful approach. In 2014 IHDP (von Falkenhayn/ Rechkemmer/Young 2011), together with IGBP (Noone/Nobre/Seitzinger 2011) and DIVERSITAS (Walther/Larigauderie/Loreau 2011), became one of the founding programmes of *Future Earth*,[2] the new programme of integrative sustainability science.

15.2   Moving from HDP to IHDP (1995–1996)

The Third HDP Scientific Symposium on "Global Change, Local Challenge", held in Geneva on 20-22 September 1995 was surely the climax of the 'old' *Human Dimensions Programme* (HDP). It was at the same time, however, the starting point of its continu-ation as the *International Human Dimensions of Global Environmental Change Programme* (IHDP). Led by Harold K. Jacobson (1986-1994, including its preparatory stage) and by Martin Parry (1995-96), HDP operated from its headquarters in Geneva under Ellen Wiegandt as its scientific director. Prior to HDP's 'grand finale' in September 1995, eight reports and a number of occasional papers had identified a number of social science deliverables and set the arena for IHDP's future challenges.

The Geneva symposium 1995 and its participants reads like a 'Who's Who?' of social science research on global environmental change, reflecting the broad range of themes and topics of that meeting. Plenary sessions focused on

- Risk, Uncertainties and Human Dimensions of *Global Environmental Change* (GEC);
- Population, Resources and GEC;
- Political Institutions and Global Change;
- National Programmes and International Global Change Research;
- GEC Research and Implications for Policy;
- Data Systems Demonstrations.

Plenaries were complemented by a number of focused discussion groups outlining lacunae and future needs of social science research vis-à-vis the challenges of cli-mate and environmental changes and as a collabora-tor in the *World Climate Research Programme* (WCRP), the *International Geosphere-Biosphere Pro-gramme* (IGBP), and/or DIVERSITAS, the pro-gramme established to address the complex scientific questions posed by the loss in biodiversity and ecosys-tem services and to offer science-based solutions. The issues at stake were (among others):

- Data for Global Change;
- Developing Regionally Sensitive Energy Models;
- Environmental Security and Local Sustainability;
- Global Change and Vulnerable Regions: Coastal Zones and Small Islands;
- Integrating Science for Global Change: Research Collaboration between Natural and Social Sciences (HDP, WCRP, IGBP);
- Local Land Use Patterns and their Global Impacts;
- Individual Perceptions and National and Interna-tional Environmental Policy;
- Industrial Transformation in Developed, Transi-tional and Developing Countries: Scenarios for Global Change;
- Regional Freshwater Resources and Global Change;
- Human Health and GEC.

As a result of this admittedly ambitious 'tour de force', the participants agreed upon a kind of manifesto in which the vision and mission of a new HDP became visible. In view of its importance as a milestone docu-ment of the human dimensions programme and as a link between the early stages of HDP and the new IHDP (from 1996 onwards), it may be appropriate to present it in full (box 15.1).

It would be going beyond the limits of this histor-ical retrospective to dive into the details of the some-what engaged and sometimes controversial discus-sions of this seminal meeting. It is, however, worth mentioning that until 1995 the situation of HDP was far from ideal in a number of ways.

15.3   IHDP 1996–2014: In Search of Stability and Sustainability

The year 1996 finally brought about substantial changes in the programme, based on two decisive decisions concerning the overall organizational struc-ture of the programme.

---

2   See "Future Earth", at: <http://www.futureearth.org/>.

The Third Scientific Symposium of the *Human Dimensions of Global Change Programme* (HDP) of the *International Social Science Council* (ISSC) was held in Geneva from 20 to 22 September 1995. Close to 200 people from nearly 40 countries from all major regions of the world attended the meeting. Representatives of national research programmes and regional networks, universities and research institutes, government funding agencies, and multinational organizations (including UNEP, WHO and EU) participated in the meeting. The purpose of the symposium was to provide an international and interdisciplinary forum for discussion of the crucial scientific challenges related to the human dimensions of global environmental change. Plenary speakers addressed a wide range of issues, including archaeological evidence of human responses to climate change in earlier periods, the relationship between population and environmental change, human health effects of global change, international environmental accords, and regional differences in vulnerability. Open discussion groups covered topics such as energy use, environmental security, data needs and policies, and the integration of social and natural science research, and developed a number of specific suggestions for future IHDP activities based on these discussions.

The symposium demonstrated a widespread and growing interest in research questions related to human dimensions of global environmental change. An increasing number of national HDP committees are being established to coordinate research at the national level. The meeting agreed that this development should be encouraged, and urged that communication be strengthened among national committees and between these committees and the HDP. Participants strongly affirmed the need for the HDP programme to promote the development of an integrated framework to facilitate inter- and transdisciplinary research, to help coordinate efforts undertaken through national, regional and disciplinary programmes, and to focus research on important gaps or particularly promising fields of study. There was a consensus that the agenda of the HDP programme covers some of the research questions considered to be most important by scientists working in this area, and that the HDP should do its best to respond also to the concerns of governmental as well as non-governmental entities trying to cope with the very complex political, economic, and social problems of global environmental change.

Geneva, 22 September 1995

First: HDP, so far under the umbrella of the *International Social Science Council* (ISSC), became co-sponsored by the *International Council of Scientific Unions* (ICSU), as ICSU was called at that time. HDP was renamed IHDP.

It would be short-sighted not to acknowledge that the inclusion of the human dimension in both ISSC and ICSU had an intellectual 'prehistory'. As early as 1991, the *International Social Science Journal* published a special issue on "Global Environmental Change. Concepts, data, methods, modelling, cooperation with natural sciences". It contains a number of very basic and fundamental articles, emphasizing the overdue cooperation between natural and social sciences. Imperatives and scientific necessities are clearly identified. Lourdes Arizpe argues in favour of a "joint agenda with the natural sciences" (Arizpe 1991). Even more explicit is J. W .M. La Rivière (1991), who stresses the need for a more or less indispensable cooperation between natural and social scientists. J. B. Robinson (1991) even proposes very specific approaches to model natural and human system interactions by combining physical flows and actor systems views. Worth mentioning is G. C. Gallopin's (1991) contribution, in which the problem of scale is outlined: how can global and local processes be combined? As a matter of fact, scaling issues were to

become one of the major methodological and practical challenges in combining natural and social science research—a problem taken up in the early years of the new IHDP (see section 15.3.2.1).

It is beyond the scope of this review to reflect the sometimes highly controversial discussions about the role of the social sciences in future global change research. The fact is that from the early 1990s onwards these discussions gained momentum and became a decisive challenge for the newly-established IHDP. The demand for much stronger participation by the social sciences in many more aspects of global change issues and the deliverance of reliable and robust data to be included in global change modelling became almost mandatory (Kates/Chen 1993).

Second: IHDP and its headquarters were transferred from Switzerland to Bonn in Germany. Thanks to generous support from German government sources as well as from the University of Bonn, IHDP moved into a spacious office right in the centre of the university's natural science departments and could be staffed in such a way that restructuring and reorganization became possible under adequate framework conditions. The author of this review was appointed jointly by the councils of ISSC and ICSU as the new chair of IHDP; Larry Kohler, a political scientist from the US who was on leave from the *International*

Labour Office (ILO) in Geneva, was appointed as IHDP's new executive director.

The official opening of the new IHDP Secretariat took place on 31 January 1997. The opening ceremony and the related expectations are described in the *ISSC Newsletter 76* as follows: "About thirty people from the 'IHDP family' attended, including the Secretary-General of the ISSC, Professor Leszek A. Kosinski, and the executive director of the ICSU, Ms Julia Merton-Lefèvre". The report concludes modestly but also with determination "… the new IHDP, still in its 'adolescent' phase, is eager to develop its potential as the social science equivalent to the other major international programmes dealing with global environmental change, the *International Geosphere-Biosphere Programme* (IGBP) and the *World Climate Research Programme* (WCRP), and thus soon [to] become an 'adult' member of the global change research family".

15.3.1   IHDP's New Mission and Organizational Restructuring

IHDP's mission and vision as well as its objectives were defined as follows:

> IHDP is an international, interdisciplinary, social science programme to promote and co-ordinate research aimed at describing, analysing and understanding the human dimensions of global environmental change, i.e. (a) the way people and societies contribute to global environmental change; (b) the way global environmental change affects people and societies; and (c) ways and means for people and societies to mitigate and adapt to global environmental change.

Thus, the change from HDP to IHDP meant both a continuation and a new beginning. IHDP could build on the foundations laid by its HDP predecessors (Arizpe/Price/Worcester 2016 [in this volume]). On the other hand, IHDP had not only to establish entirely new organizational structures, but also to create a scientific profile of its own and develop joint ventures with existing ICSU-sponsored activities. The new IHDP *Scientific Committee* (SC) was formed, covering a broad range of social sciences and ensuring global internationality. Its members were:

- Prof. Eckart Ehlers, chair, geography, Germany;
- Prof. Arild Underdal, vice-chair, political scientist, Norway;
- Dr. Peter de Janosi, Treasurer, economics (IIASA), USA;
- Prof. Hans B. Opschoor, economics, The Netherlands;

- Prof. Mauricio T. Tolmasquim, economics, Brazil;
- Dr. Anne V. Whyte, geography, Canada.

Ex-officio members were:

- Dr. David Skole, ecology, *Land Use and Land Cover Change* (LUCC);
- Prof. Kurt Pawlik, psychology (ISSC);
- Prof. J.W. Maurits la Rivière, microbiology (ICSU);
- Prof. Peter S. Liss, environmental chemistry (IGBP);
- Prof. W.L. Gates, meteorology, climate modelling (WCRP);
- Prof. Joao M. Morais, IGBP Secretariat, Stockholm, social science liaison officer.

The overall structure of the new IHDP is summarized in figure 15.1.

The following years saw regular changes within IHDP's Scientific Committee as well as the statutory changes in the chairholders and executive directors of the programme:

Chairs:

- Eckart Ehlers, Germany (1996-1999)
- Arild Underdal, Norway (1999-2002)
- Coleen Vogel, South Africa (2002-2006)
- Oran R. Young, USA (2006-2011)
- Sir Partha Dasgupta, India (2012-2014)

Executive Directors:

- Larry Kohler, USA (1998-1999)
- Jill Jaeger (1999-2002)
- Barbara Goebel (2002-2005)
- Andreas Rechkemmer, Germany (2005-2009)
- Anantha Kumar Duraiappah, India (2010-2014)

Until the very end of IHDP in 2014 the chairs and executive directors were always accompanied by outstanding scientists and academic advisors as members of IHDP's SC, among them (to mention just a few) William C. Clark, Geoffrey Dabelko, Carl Folke, Gilberto C. Gallopin, Heinz Gutscher, Carlo Jaeger, Ruth Khasaya Oniango, Tatyana Kluvankova-Oravska, Elena Nikitina, Elinor Ostrom, Xizhe Pen, P. S. Ramakrishnan, M. A. Mohamed Salih, Roberto A. Sanchez-Rodriguez, Akilagpa Sawyer, and Paul L.G. Vlek.

While IHDP's Scientific Committee, its appointed and ex-officio members, and the representatives of IHDP's core and joint projects ensured the constant rejuvenation of the programme and its research strategies, regular 'Open Meetings' of the IHDP community took place in different parts of the world. While the early conventions attracted between 250 and 400 participants, Bonn in 2005 had more than a thousand registered social scientists from all parts of the world

Figure 15.1: The IHDP programme: organizational structure, projects and partner institutions. **Source**: IHDP Update 1/ 1997.

GECHS: Global Environmental Change & Human Security
ICSU: International Council of Scientific Unions
IDGEC: Institutional Dimensions of Global Environmental Change
IT: Industrial Transformation
IGBP: International Geosphere-Biosphere Programme
ISSC: International Social Science Council
LUCC: Land Use and Land Cover Change
START: Global Change System for Analysis, Research and Training
WCRP: World Climate Research Programme

Source: IHDP Update 1/1997

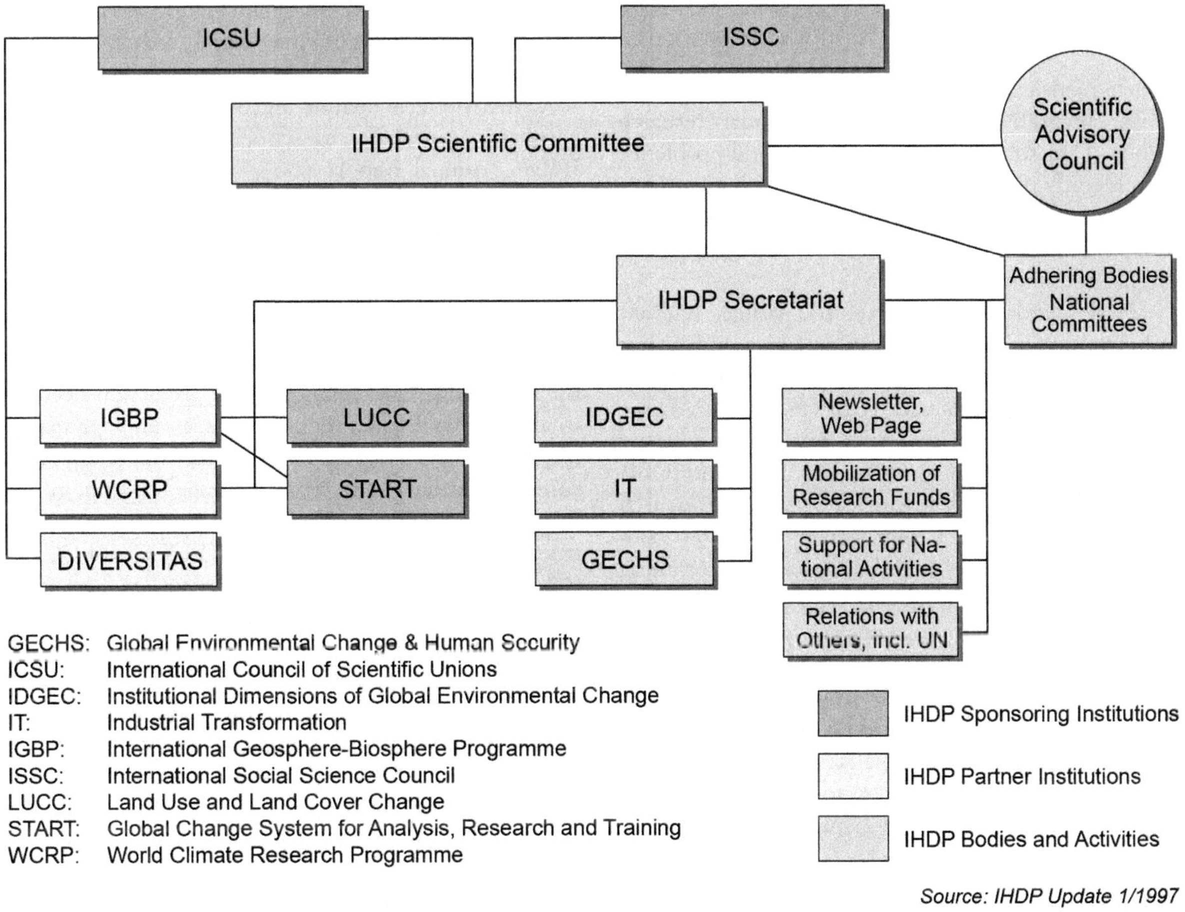

(IHDP-Update 01/2006). These Open Science Meetings proved themselves valuable and almost indispensable inputs into IHDP's research portfolio and were also innovative for cooperation beyond disciplinary and cultural boundaries.

Altogether, one is probably justified in characterizing the transfer from Geneva to Bonn as a kind of rejuvenation of social science research within the GEC community. The installation of a viable administration and its equipment with highly committed staff resulted in the indicated organizational restructuring and the forceful development of research projects that soon became visible flagships of IHDP.

15.3.2 IHDP Becomes a Fully-Fledged Scientific Programme: The Decade 1996–2005/6

15.3.2.1 Project Developments

Against this background, the years from 1996 onwards saw the steady and systematic growth of IHDP and its

development into a fully-fledged programme of its own. Immediately after its establishment in Bonn and as a result of intensive discussions within the newly-established SC, it was agreed that IHDP should select three core projects to be developed as flagship activities. These were:(a) the *Institutional Dimensions of GEC* (IDGEC, led by Oran Young and Arild Underdal), (b) *Industrial Transformation* and GEC (IT, led by Pier Vellinga), and (c) *Global Environmental Change and Human Security* (GECHS, led by Steve Lonergan). These were placed on a 'priority fast track', with a view to building upon previous work in these areas by the former HDP. Based on the outcomes of preceding workshops, internal scoping reports, draft science plans and their external reviews, in 1999 IHDP was able to publish three Science Plans:

- *Science Plan: Institutional Dimensions of Global Environmental Change (IDGEC)*. IHDP Report 9, 1999.

- *Science Plan: Global Environmental Change and Human Security (GECHS)*. IHDP Report 11, 1999.
- *Science Plan: Industrial Transformation (IT)*. IHDP Report 12, 1999.

Science plans are the first step in developing an innovative, integrative and also international research agenda. They formulate research needs and problems. They identify research foci and related activities. They also discuss models and methods to be applied and consider a priori programmatic links to other projects and programmes.

IHDP's endeavour to develop a visible, relevant and cross-cutting scientific profile of its own was not restricted to these three core projects alone. It also inherited joint projects with IGBP and WCRP. These included the *Land Use and Land Cover Change (LUCC)* project jointly sponsored by IHDP and IGBP. The START project [*System for Analysis, Research and Training*], another joint activity (WCRP/IGBP/IHDP), continued to pursue its goal of promoting regional research, training and capacity-building support for GEC scientists, and it is the only joint project that has maintained its independence until today.

Almost simultaneously with the three science plans of IDGEC, GECHS and IT, the LUCC project published its Implementation Strategy, specifying in detail those activities that were to fulfil the mandate outlined in its earlier science plan:

- *Implementation Strategy: Land-Use and Land-Cover Change (LUCC)*. IGBP Report 48/IHDP Report 10, 1999 (see also Lambin/Geist 2006).

However, the development of science plans and implementation strategies was only part of IHDP's new agenda. In 1998, IHDP saw the publication of two working papers, dealing with the intricate problems of "Scaling issues in the social sciences", co-authored by the later Nobel Laureate Elinor Ostrom (Gibson/Ostrom/Ahn 1998), and with "The problem of fit between ecosystems and institutions" (Pritchard/Colding/Berkes et al. 1998). Both papers were to become basic contributions not only to the newly developed core projects, but also to the general discussion on coordination between natural and social science paradigms.

This survey and reconstruction of project developments would be incomplete if one did not include those projects that developed in the later years of IHDP's first decade. They include additional core projects of IHDP such as *Urbanization and Global Environmental Change* (UGEC), focusing on the impacts of rapid global urbanization processes and

their environmental and socio-economic causes and consequences. It is worth mentioning that as a by-product of this programme, nine editions of a journal-like publication "UGEC Viewpoints" (1/2008–9/2013) were produced.

Other projects include a growing number of joint activities, especially in cooperation with IGBP. IHDP's Annual Report 2006/07 also mentions the LOICZ (*Land–Ocean Interactions in the Coastal Zone*) project and the *Global Land Project* (GLP) as core projects. Both, however, were joint activities, in operation since 1993 under the auspices of IGBP. They became a joint IGBP/IHDP initiative in 2004 and their cross-cutting aim and goal was "to provide the knowledge, understanding and prediction needed to allow coastal communities to assess, anticipate and respond to the interaction of global and local pressures which determine coastal change"–a truly multidisciplinary challenge. The same, by the way, holds true for the GLP, successor to the LUCC and GCTE (*Global Change and Terrestrial Ecosystems*) project, inaugurated as a joint IGBP/IHDP project in 2006. Other joint activities included the continuation of the *Global Carbon Project* (GCP), the *Global Environmental Change and Food Systems* (GECAFS) project launched in 2001, and regional studies such as MAIRS (*Monsoon Asia Integrated Regional Study*) and support for capacity-building. Special mention should be made of the IHOPE (*Integrated History of People on Earth*) project, which got under way in 2005. The aim of this project, again co-sponsored by IGBP and IHDP, is to analyse co-evolution of ancient societies with their environments on different time scales (millennial, centennial, decadal)–a truly cross-cutting approach with a strong historical aspect.

15.3.2.2 From Science Plans to Syntheses

While it is impossible to go into details of IHDP's core project activities, an attempt may be made to point to the central themes, aims and goals of the three projects mentioned above and their foci within the sustainability discourse.

GECHS (Global Environmental Change and Human Security), dating back in its preliminary phase to HDP times, formulated the first draft of its Science Plan in 1997, finalizing it in 1999. It ended, after ten years of intensive research, in June 2010 under the leadership of Karen O'Brien, then GECHS-SC. She summarized GECHS's performance as follows: "GECHS research has conceptualized and characterized vulnerability, analysing what it means for the future of marginal populations and living conditions.

GECHS research has also explored the physical dimensions of challenges to human security, for example, in the context of water scarcity and food. Underlying all these challenges are the individual and social mechanisms that affect how a lack of security actually takes hold and manifests" (IHDP Update 2/2009, p. 4). Themes and topics indicate foci of GECHS research: "Sustainability in water and sanitation", "Resilience as agency", "Ethics of global environmental change", "Managing extremes" and "Sustainable adaptation: emphasizing local and global equity and environmental integrity".

Figure 15.2: Human security as a desirable goal for individuals and societies. **Source**: Bohle (2002).

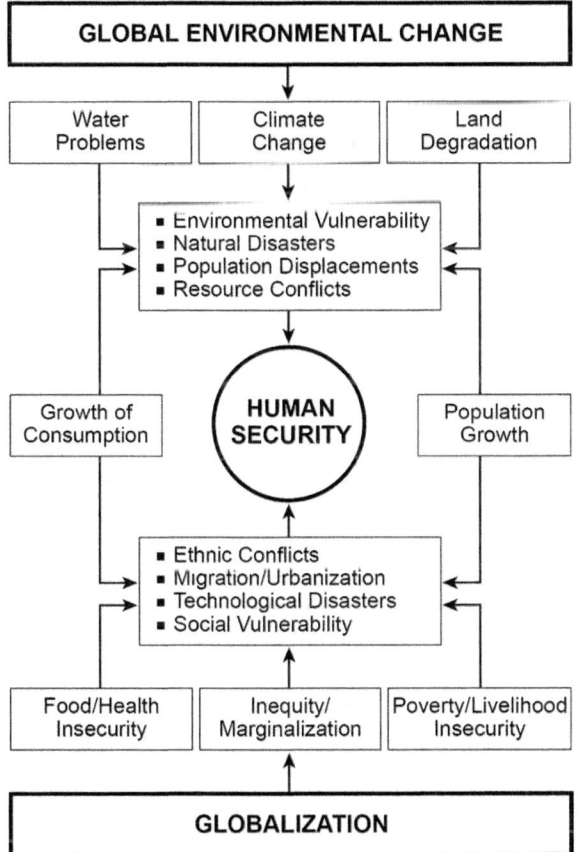

In fact, human security in the broadest sense of the word and as an ultimate goal of social and environmental stability became GECHS's main research interest. The complexity of this object in view becomes apparent in figure 15.2. Human security, embedded in and endangered by the consequences of globalization and the deterioration of our natural environment, is recognized as a central and crucial aspect of humanity

on its pathway towards sustainability. The summarizing presentation in figure 15.2 shows the intricate interactions of natural and societal impacts on human (in-)security. At the same time it indicates the core problems of vulnerabilities that will be dealt with in more detail in section 15.3.2.3.

IDGEC (Institutional Dimensions of Global Environmental Change): this project is an extremely important part of IHDP's scientific activities. It is by necessity the most cross-cutting topic, since institutions are indispensable instruments for the regulation and management of disputes; and disputes are manifold, not only over environmental and resource regimes but also over conflicting interests in social and political issues, and of course over juxtaposing views from natural and social perspectives. The IDGEC Science Plan (1999) reads like an introduction to institutional effects on GEC and the foci of IDGEC's projects (figure 15.3)

As mentioned earlier, IDGEC activities were continuously accompanied by reflections on the problems of fit, interplay and scale—problems of importance also to other IHDP projects. In view of the crucial role that institutions play in GEC as determinants of human-environment interactions, it is not surprising that several issues of IHDP Update were specifically devoted to aspects of governance and institutions: IHDP Update 3-4/2006; IHDP Update 1/2007 (with emphasis on the IDGEC Synthesis Conference), and IHDP Update 3/2009 as a special issue on Governance as a cross-cutting theme in human dimensions science.

The IDGEC Synthesis Conference in Bali (Indonesia) at the end of 2006 marked the end of a highly successful project initiative. Oran Young, long-time chair of IDGEC and for many years chair of the IHDP-SC, could sum up by saying that the project "has produced results that add to our understanding of the roles that institutions play wherever they occur ..." (IHDP Update 1/2007, p. 2), and the project achieved new insights into the roles that institutions play in steering human-environment interactions.

In view of the outstanding relevance of institutions to almost all aspects of research, IDGEC's continuation beyond its termination was not surprising. The publication of the IDGEC synthesis volume in 2008 (Young/King/Schroeder 2008) marks the almost logical outset to promoting *"Earth System Governance"* (ESG). In October 2008 IHDP formally approved the "Earth System Governance Project" (ESGP) as a new core project. IHDP Update 3/2009 is a comprehensive tribute to this new activity. It contains a number

Figure 15.3: Institutions and their effects on global environmental change/The primary focus of the IDGEC project. **Source**: IDGEC Science Plan.

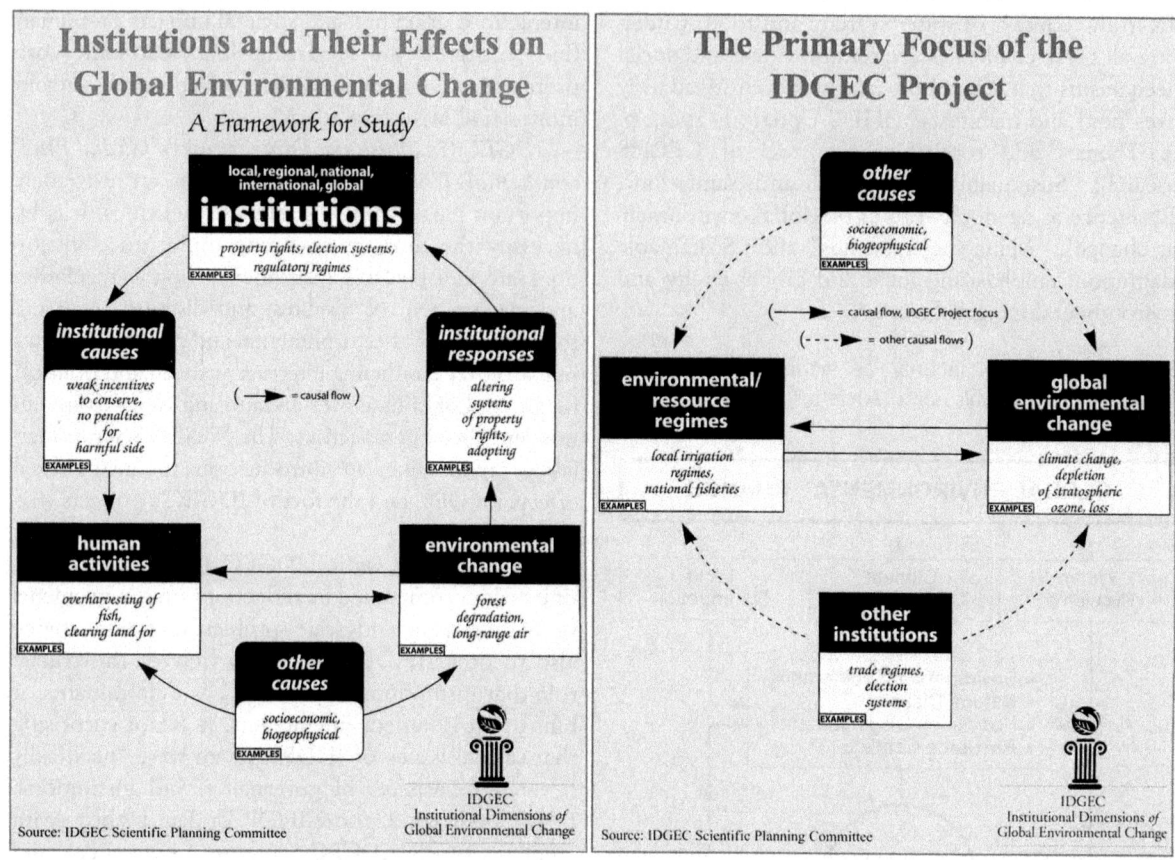

of basic articles in fields such as urban areas, health, food and their governance structures.

IT (Industrial Transformation), the third core project established in 1999, started off with three objectives: to understand complex society-environment interactions; to identify driving forces for change; and to explore development trajectories that have a significantly smaller burden on the environment (Vellinga/ Herb 1999).

Its outset was a focus on industrial production and industrial outputs that, in the past, had been completely unsustainable. The researchers' analysis, initially focused on work in industrialized countries, made it clear that technological innovations are scarcely sufficient for creating less resource-intensive and less polluting systems. Institutional and social interventions are also necessary to achieve what has been called a "social-technical system". The transfer and application of the test phase to the rapidly industrializing countries of Asia discovered the importance of "sustainability experiments", which were obvious important preconditions for more sustainable and long-term pathways.

Like GECHS and IDGEC, IT cannot be viewed independently of other environmental and social factors. Urbanization, resource extraction, energy consumption and mobility: these and other phenomena are closely connected with industrial transformation. Cross-cutting activities and analyses alongside other IHDP projects and joint activities with IGBP and WCRP were practically mandatory. From the very beginning, IT has been closely connected with the Carbon Joint Project, renamed as the Global Carbon Project (GCP), of IGBP/IHDP/WCRP. In addition, IHDP's own Global Carbon Cycle Research initiative served as a basis for an international Carbon Research Framework (Gupta/Lebel/Vellinga 2001). The interactions of the IHDP projects in pursuit of this initiative are presented in figure 15.4 (see also IHDP Update 3/2000 for a special edition on the carbon cycle and its relevance for society—a collection of six articles/statements).

Figure 15.4: Industrial Transformation: Interactions within the human–environment system with respect to the global carbon cycle. **Source**: IHDP Update 3/2000.

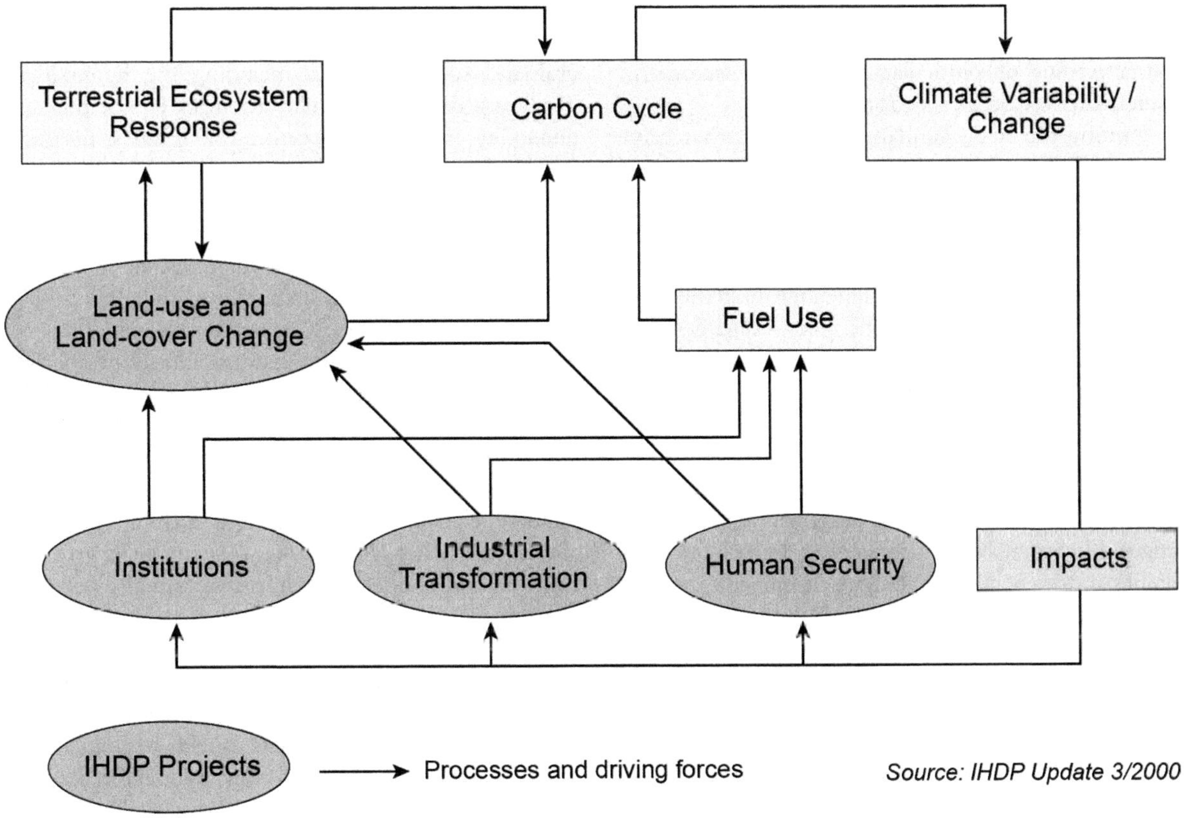

15.3.2.3 IHDP and the Sustainability Discourse

The development of all GEC research programmes under the auspices of ICSU and ISSC are characterized by a steadily growing tendency to proceed from observation and data collection via modelling to scenario developments. The early reports of the IPCC (*Intergovernmental Panel on Climate Change*) are good examples of this kind of research and its conclusions in terms of predictions.

The turn of the century/millennium may also be called a turning point for GEC research. These years saw not only a number of seminal publications (Kates/Clark/Corell et al. 2001; Kasperson/Kasperson 2001; Lubchenko 1998; NRC 1999—to name just a few), but also the propagation of the Anthropocene (Crutzen/Stoermer 2000) and of a geology of mankind (Crutzen 2002). All this had a deep impact on IHDP's sustainability discussions and its discourses on how individuals and societies cope or are forced to cope with climate change, risks and disasters. Human security proved to be a crucial issue not only for IHDP's core projects, but also for the programmes' methodological and theoretical discussions in general. It is impressive to see just how intensive these discussions were and to what extent they also began to influence joint programmes, the activities of the ESSP, endorsed networks and capacity development initiatives. To a great extent derived from IHDP's core projects, the sustainability discourse and all its facets nevertheless developed their own agenda and produced remarkable insights in fulfilment of IHDP's original vision and mission.

It would be going far beyond the aims and scope of this report to delve into the details of these analyses. The fact is that the sustainability discourses of IHDP are closely connected with the concepts of vulnerability, risks and hazards, resilience, adaptation, mitigation, susceptibility, and coping capacity. And these are the foci of continuous academic and application-oriented discussions and further strategic developments up to today. They all deal more or less explicitly with the attempt to overcome the deep gulfs in methods, theories and practice between the natural and social sciences and to bridge the fundamental differences between them. Sustainability science as perceived by an international group of experts led them to argue: "A new field of sustainability science is emerging that seeks to understand the fundamental character of interactions between nature and society. Such an understanding must encompass the interaction of global processes with the ecological and social characteristics of particular places and sectors ..." (Kates/Clark/Corell et al. 2001; 641).

Among the many fundamental papers that came out of IHDP's core projects, a rather short but essential and complex paper on vulnerability is Hans-Georg Bohle's "Conceptual Model for Vulnerability". This combines different social science approaches and theories as explanatory causes of vulnerability together with coping strategies to overcome the challenges and threats caused by natural or social hazards and risks. It is to a certain extent a very comprehensive approach. Bohle's external side of vulnerability covers both natural and social risks and hazards as part of human exposures. Coping strategies (e.g. adaptation or mitigation) are part of the social response to such challenges. However, Bohle's conceptual framework goes far beyond mere factual analysis. It embeds the specific cases/objects of research into a set of methodological and theoretical considerations, thus transferring the special case to a general problem. It is probably fair to say that these theoretical as well as conceptual considerations mark the increasing reliance of social science research on global environmental change.

Figure 15.5: Bohle's Conceptual Framework for Vulnerability Analysis. **Source:** Bohle (2001).

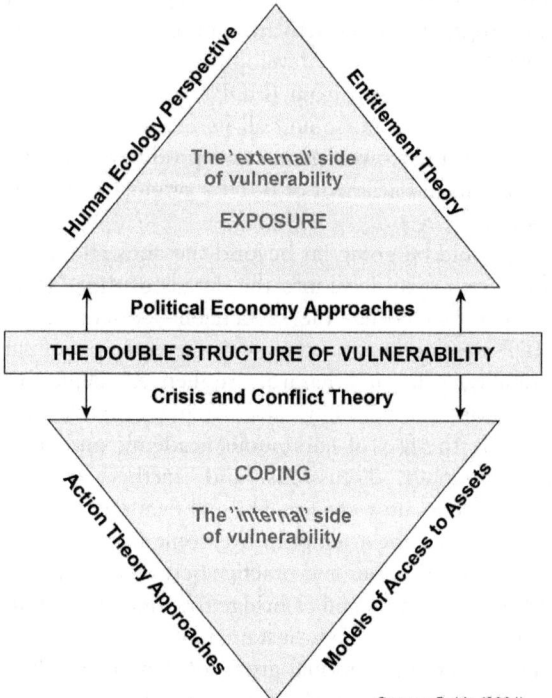

Source: Bohle (2001)

Bohle's comprehensive approach, published in IHDP Update's special issue on vulnerability (IHDP Update 2/2001), is just one example of IHDP's attempts to serve as a bridge-builder between the natural and social sciences, including the humanities. There are other and similar attempts to put the sustainability issue at the centre of IHDP's research agenda (see, for example, figure 15.2). Besides these developments, mentioned in the review of IHDP's core projects, there has been an intensive and inspiring conceptual discussion on many facets connected with the discourse on sustainability. While it is probably fair to say that IDGEC ensured, independently of its own research foci, the institutional frame for the discourse on sustainability in general, it was GECHS in particular that inspired the discussion on sustainability with a wide range of methodological and theoretical inputs. Key themes and topics were, for instance, "climate change vulnerability and adaptation, water governance issues, rural livelihoods and food security, conflicts and cooperations in trans-boundary resource management, and interactions between global environmental change and globalization" (IHDP Annual Report 2008, p. 24). GECHS's Synthesis Conference on "Human Security in an Era of Global Change" in 2009 also put the multifaceted aspects of human (in-)security including natural as well as social risks and hazards at the centre of their final discussions.

It would be going far beyond the scope of this article to try to document the scientific outcomes in detail. It must therefore suffice to point to IHDP's Updates and Annual Reports. At least until 2007, the projects' progress reports are (partly) equipped with lists of selected publications and are suggested as further reading.

15.3.3 IHDP's Second Decade (2006/7–2014): New Foci, New Directions and New Pathways towards "Future Earth"

In 2007 IHDP published a new 'Strategic Plan 2007-2015' in which, besides continuing cutting-edge research, special emphasis was given to capacity development, community building, and outreach, as well as placing a stronger focus on science–policy interactions. Figure 15.6 reflects this new strategy and the new organizational structure.

Both the programme structure of the new IHDP in 2007 and the composition of IHDP's Scientific Committee reflect the new approach. While the core of the SC covers exclusively social science disciplines and economics, its ex-officio members are mediators

Figure 15.6: IHDP's new governance structure. **Source:** IHDP's Strategic Plan 2007–2015, p. 30.

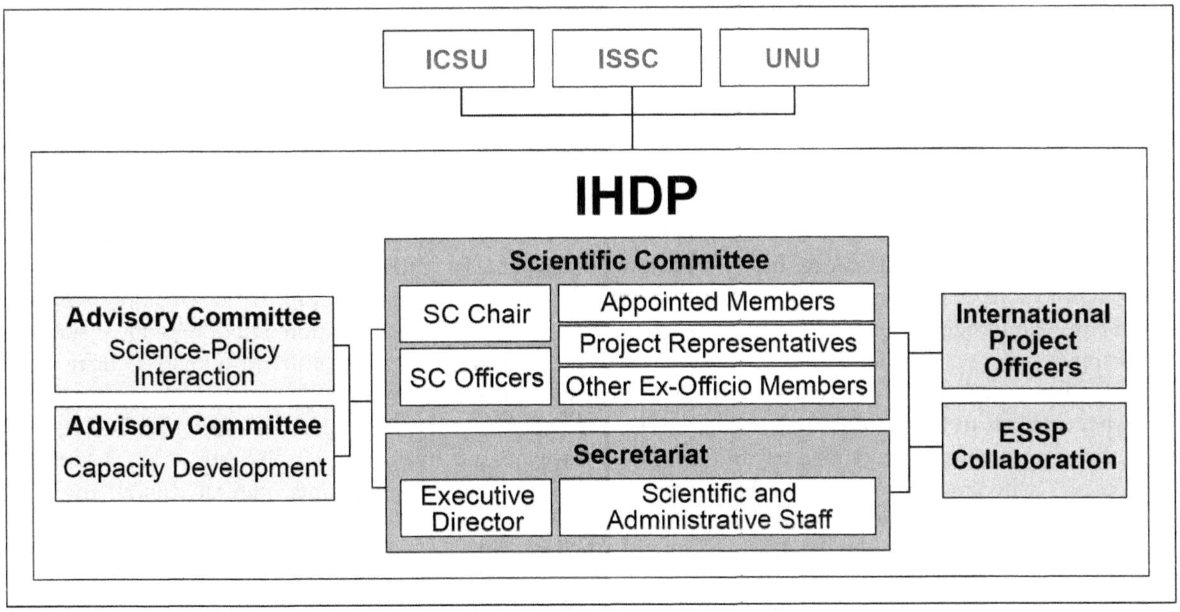

Source: IHDP's Strategic Plan 2007-2015

Box 15.6: IHDP Project Activities in 2008. **Source:** IHDP Annual Report 2008.

Core projects: IHDP currently has six core projects. Core projects are conceptualized and supported by IHDP, sometimes in partnership with various other programmes to identify and generate new, cutting-edge research activities, promote international cooperation, and build linkages between policy-makers and researchers.

- Earth System Governance (ESG)
- Global Environmental Change and Human Security (GECHS)
- Global Land Project (GLP)
- Industrial Transformation (IT)
- Land-Ocean Interactions in Coastal Zones (LOICZ)
- Urbanization and Global Environmental Change (UGEC).

ESSP Joint Projects: IHDP is currently conducting four projects in conjunction with the Earth System Science Partnership (ESSP). These projects are formed and run in a similar way to the core projects, but with support from all partners of ESSP. In addition, IHDP cooperates with ESSP to endorse other networks.

- Global Environment Change and Food Systems (GECAFS)
- Global Carbon Project (GCP)
- Global Water Systems Project (GWSP)
- Global Environmental Change and Human Health (GECHH)
- Monsoon Asia Integrated Regional Study (MAIRS)
- System for Analysis, Research and Training (START).

Initiatives: Initiatives are potential IHDP projects. These projects are in the process of preparing and reviewing a science plan to present to the IHDP Scientific Committee.

- Integrated Risk Governance (IRG)
- Integrated History of People on Earth (IHOPE)
- Knowledge, Learning and Societal Change (KLSC).

Endorsed Networks:

- Mountain Research Initiative (MRI)
- Population-Environment Research Network (PERN)
- Young Human Dimensions Researchers (YHDR).

---

of IHDP's obligations towards the natural sciences. The same holds true for IHDP's programme structure, its sponsoring institutions and its partnership with existing IGBP and WCRP activities.

The Strategic Plan 2007-2015 provides valuable background information. While some of IHDP's core projects had already terminated by 2007, others were in the final stages of the preparation of synthesis reports. Almost at the same time, in October 2008, a broadly-

based international research initiative, the *Earth System Governance Project* (ESGP), got under way and became a new core project within IHDP. ESSP also implemented four global-scale joint projects on water, food, carbon and human health, all with the strong involvement of IHDP. IHDP's extensive portfolio in 2008 is summarized in box 15.2.

All this went hand in hand with new directions in regard to IHDP's new strategies. Besides its continued involvement in research, the Strategic Plan 2007-2015 set distinctly new standards and declared "science-policy interaction a top priority for IHDP in its second decade". Science-policy interaction meant, above all, capacity development, community building and outreach. Sustainable capacity development designed as a close interaction between eminent scientists, policymakers and practitioners, and other stakeholders and beneficiaries, was to be accompanied not only by conferences and publications, but also by new major activities: specific workshops, training seminars, mentorship programmes for emerging scientists, scholarships, etc.

As indicated earlier, the last years of IHDP are somewhat difficult to reconstruct and to judge. The fact is that the programme expanded beyond its traditional cores. It became a kind of cohesive centre of the new and integrative GEC projects. It was especially under the far-sighted IHDP chairmanship of Oran Young, one of the founding fathers of the programme, that IHDP and its IDGEC programme ensured an urgently needed cohesion among IHDP's core projects and its joint activities (Young/King/Schroeder 2008). Its continuation under the umbrella of ESG is therefore a logical consequence (see: From IDGEC to ESG—Passing the Torch. 1st Synthesis book of the former IDGEC project, Delhi 2008; see also Biermann 2007). It is obvious that ESG, formally inaugurated as a new core project within IHDP in 2008, had to fulfil crucial functions and tasks for the overall GEC community, as IDGEC was now in the past. ESG is challenged by "the development of theories to understand Earth System governance, as well as strategies to advance it" (IHDP Annual Report 2008, 18). In order to achieve its interdisciplinary, international and multi-scale aims and goals, ESG's comprehensive Science and Implementation Plan was published in 2009 (Biermann/Betsill/Gupta et al. 2009).

ESSP, first discussed in Amsterdam 2001 and officially launched as an outcome of the ESSP Open Science Conference "Global Environmental Change: Regional Challenges" in Beijing in November 2006, must be seen as an important step in further fostering cooperation between WCRP, IGBP, IHDP and DIVERSITAS. This partnership is based on the conviction that the complexity and dynamics of highly sensitive socio-ecological systems needs cooperation between different programmes of GEC research and, if possible, the bottom-up development of joint projects such as GCP, GECAFS, GWSP and GECHH in order to tackle the causes and consequences of air and water pollution, water scarcity, food problems, risks and

hazards to human health—and all this under aspects such as vulnerability, resilience, and adaptation and coping strategies. Thus, and finally, DIVERSITAS, IGBP and IHDP became united under one umbrella organization, ending their experiences as ESSP and merging into the new programme called "Future Earth". Designed as a ten-year international research programme, Future Earth was launched in June 2012 at the UN Conference on Sustainable Development (Rio+20). Its initial design report (2013) includes the following statement: "Future Earth will address issues critical to poverty alleviation and development such as food, water, energy, health and human security, and the nexus between these areas and the over-arching imperative of achieving global sustainability. It will provide and integrate new insights in areas such as governance, tipping points, natural capital, the sustainable use and conservation of biodiversity, lifestyles, ethics and values. It will explore the economic implications and social transformations towards a low-carbon future. Future Earth will explore new research frontiers and establish new ways to produce research in a more integrated and solutions-oriented way."

Is it too critical or even arrogant to note that language, vision and mission as well as programme details sound rather familiar? This "familiarity" becomes even more striking if one looks at Future Earth's conceptual framework, which stresses "that humanity is an integral part of the dynamics and interactions of the Earth System and that this has important implications for global sustainability …". It is probably not an exaggeration to state that IHDP's handwriting has had a major impact on the initial design of this executive summary.

Its ultimate aims and goals also sound somewhat familiar. Future Earth will be—in its own words—a global platform to deliver:

- *Solution-oriented* research for sustainability, linking environmental change and development challenges to satisfy human needs for food, water, energy, health;
- *Effective interdisciplinary collaboration* across natural and social sciences, humanities, economics, and technology development to find the best scientific solutions to multifaceted problems;
- *Timely information for policymakers* in generating the knowledge that will support existing and new global and regional integrated assessments;
- *Participation* of policymakers, funders, academics, and other sectors of civil society in co-designing and co-producing research agendas and knowledge; and

- *Increased capacity-building* in science, technology, and innovation, especially in developing countries, and engagement of a new generation of scientists.

It looks as if IHDP in its almost twenty years of existence must be seen as a pacemaker in bridging the gaps between the natural and social sciences and the humanities. It has done so through the cohesive power of its core and joint projects and especially through its intensive contributions in the fields of institutional and governance structures. ESSP projects and their successful implementation have been major steps towards the pursuit of serious large-scale integrative and cross-disciplinary research initiatives. The conscious inclusion of START (*Global Change System for Analysis, Research and Training*) supports IHDP's advertised interest in science-policy interactions, outreach and knowledge dissemination (IHDP Strategic Plan 2007-2015. 2007; IHDP Annual Reports 2006-2007, 2008). Thus, and for the time being, Future Earth is an important additional step forward in global (environmental) change research or, as Future Earth calls it, "a step-change in Earth system research". Its main message is that IGBP, DIVERSITAS, and IHDP have ceased to exist. Together they are now "Future Earth".

In June 2014, the IHDP headquarters and offices closed. In doing so IHDP as representative of the social sciences and the humanities within climate and global change research finished its specific mission. The legacy of IHDP will live on in Future Earth.

15.3.4   IHDP 1996–2014: A Retrospective Summary

Looking back at IHDP's institutional and organizational reforms between the late 1990s and its amalgamation with IGBP and DIVERSITAS into the new Future Earth programme in 2014, it may be appropriate to put special emphasis on the formal concomitants of IHDP's development between 1996 and 2014. Science and research would have been impossible without framework conditions, by which is meant functioning and effective communication and information structures. Equally important and basically dependent on those structures are the scientific networks beyond programme and disciplinary boundaries that emerge over time.

Central to the coherence of the international human dimensions community and its connections with its partner organizations WCRP, IGBP and DIVERSITAS was the IHDP newsletter *Update*. Over almost twenty years, *Update* developed from a meagre eight-pager to an impressive journal-like publica-

tion, consequently changing its newsletter status into that of a *Magazine of the International Human Dimensions Programme on Global Environmental Change*. The newsletter version of the *IHDP Update* in particular contains a wealth of organizational details of the programme's various activities and details of the development of the core projects and of meetings and cooperation—and, of course, policy statements and scientific articles. A reconstruction of IHDP's history from 1996/7 until early 2014 is impossible without the *IHDP Update*. The same holds true for IHDP's Annual Reports, the first of which was also published in 1997 and whose last edition was published in 2013.

However, IHDP's indispensable communication aids also underwent important organizational and structural changes. In 2007 IHDP came under the umbrella of the *United Nations University* (UNU) as an additional sponsor alongside ISSC and ICSU. The UN University and its Bonn branch has a special focus on *Environment and Human Security* (UNU-EHS),[3] and the themes and topics of UNU-EHS proved to be so close to those of IHDP that cross-cutting cooperation with IHDP was just a matter of time. In addition, UNU-EHS also agreed to host the IHDP Secretariat at its headquarters at the UN Campus in Bonn. This new affiliation with an internationally visible partner and its focus on the increasingly close interaction between nature and society granted IHDP access to many other UN agencies and secretariats located in Bonn, such as the *United Nations Framework Convention on Climate Change* (UNFCCC),[4] the *United Nations Convention to Combat Desertification* (UNCCD),[5] and others.

Formal changes in the second decade of its existence included a new format for the IHDP Update. From October 2007 onwards, this functioned as a "Magazine". As a consequence, it focused on specific themes or projects; at the same time, however, it lost its well-established functions as a newsletter in the proper sense of the word. In 2012, these functions were taken over by a short-lived publication called "The Networker". In 2012 "Update" was entirely replaced by a new journal-like publication called "Dimensions".

Taking into account the science and research in its core programmes, its growing embeddedness in joint activities, and its involvement in multidisciplinary networks, one can distinguish three distinct phases in IHDP's twenty years of existence, as follows.

---

3   See at: <http://www.ehs.unu.edu/>.
4   See at: <http://unfccc.int/2860.php>.
5   See at: <http://www.unccd.int/en/Pages/default.aspx>.

15.3.4.1 The Period 1996–2001

This initial period was characterized not only by the endeavour to establish the programme logistically, but also by the development of central projects and their science plans (see 15.3.2) and by (as one of the reviewers has called it) its "success ... as a real-science driven enterprise". Both executive directors, Larry Kohler and Jill Jaeger, played crucial roles in these activities. As already indicated, and as will be addressed in further detail in chapter 15.4, these years saw the establishment of IT, GECHS and IDGEC as core projects and the development of LUCC as a joint project in cooperation with IGBP, as well as increased involvement in START activities. The Global Change Open Science Conference in Amsterdam in July 2001 may probably be considered as a turning point, since it marks the first really integrated Global Environmental Change Conference (Steffen/Jaeger/Bradshaw 2002). The "Amsterdam Declaration on Global Change" pays tribute to this integrative approach and focuses beyond doubt on the critical role of humans. The joint statement by WCRP, IGBP, IHDP, and DIVERSITAS notes, for example (Steffen/Jaeger/Bradshaw 2002: 206-7):

- The Earth System behaves as a single, self-regulating system made up of physical, chemical, biological and human components.
- Human activities significantly influence Earth's environment in many ways in addition to greenhouse gas emissions and climate change.
- Earth System Dynamics are characterized by critical thresholds and abrupt changes. Human activities could inadvertently trigger such changes with severe consequences for Earth's environment and inhabitants.

On the basis of these (and other) findings, the Declaration concludes:

- An ethical framework for global stewardship and strategies for Earth System management are urgently needed.
- A new system of global environmental science is required.

Amsterdam 2001 was thus a decisive event for IHDP. The role and importance of human interferences in and impacts on nature and on the environments of societies was broadly acknowledged. IHDP's science plans and implementation strategies and the rapid progress of its projects were a timely answer to these challenges.

15.3.4.2 The Period 2001–2009

This second period may—again in line with a reviewer's suggestion—be called the "real heyday of IHDP" or its "golden era". It is marked by IHDP's increasing endeavours to pursue research in line with a highly integrative and social-science-based "sustainability science" (Kates/Clark/Corell et al. 2001). Almost parallel to this short but seminal article, a number of similarly influential publications opened the pathways for the increased involvement of social science in research into global environmental change. Following a comprehensive stocktaking by the US National Research Council on "Human Dimensions of Global Environmental Change" (NRC 1999), the Stockholm Environment Institute organized an "International Workshop on Vulnerability and Global Environmental Change" and published its results (SEI 2001). Almost simultaneously, IHDP published a special issue on vulnerability (IHDP Update 2/2001).

The period from 2001 onwards is characterized by the successful progress of the IHDP research initiatives mentioned above (IDGEC, IT and GECHS), as well as by IHDP's joint projects with IGBP. In addition, new IHDP projects as well as joint projects were begun. In 2001, GECAFS (Global Environmental Change and Food Systems) was launched as a project focusing on the linkages between food security, food systems and global change; a joint project with IGBP and WCRP. As another example, in 2006 the project on Urbanization and Global Environmental Change (UGEC) got under way. LUCC merged into a new project, the Global Land Project (GLP), co-sponsored by IGBP, and the second phase of the Land-Ocean Interactions in the Coastal Zone project (LOICZ 2) became a joint effort by IHDP and IGBP.

Other outstanding achievements of this period are the publication of IHDP's Strategic Plan 2007-2015 under the chairmanship of Oran B. Young (IHDP 2007), the continued engagement in connection with the development of an Earth System Science Partnership (ESSP) inaugurated in an Open Science Conference in Beijing in November 2006; the achievements of this period culminated in the IHDP Open Meeting in Bonn 2009, with more than a thousand participants.

15.3.4.3 The Final Period 2009–2014

The third period is less easy to characterize. 'Restructuring' may be an appropriate term to describe this rather incoherent phase, but restructuring towards what? As one of the critical reviewers of this paper noted, there were obvious "efforts to build IHDP Bonn as a research unit independent of the core projects", but there were also tendencies by the secretariat to work under the auspices of the ICSU in pursuit of the *Future Earth* (FE) programme. IHDP's decision to relocate its headquarters under the umbrella of the UNU (United Nations University) and its Bonn branch on *Environment and Human Security* (EHS) also falls into this period. Finally, in early 2012 IHDP started a new magazine *Dimensions*, which (in the words of IHDP's Executive Director Anantha Duraiappah) "will be IHDP's premier dissemination outlet for key findings from its portfolio of projects" (Dimensions 2012, p. 3)—a clear difference from IHDP's earlier and well-established information policies, and with only four issues, very short-lived. It was complemented by an equally short-lived "Quarterly Newsletter" between April 2011 and May 2014, replacing the former IHDP Update.

15.4 An Outlook: GEC Research and "Future Earth"

It is more than a mere coincidence that IHDP in its almost twenty years of existence really succeeded in bringing humankind, as one of the main triggers of both climate change and global environmental change, into the foreground of global change research. In doing so, IHDP not only fulfilled its own mission and vision (see above), but it also helped to pave the way for a basically new understanding of the intricate and dense interactions between nature and societies. The increasingly alarming results from IPCC, the postulate of a new geological era, the Anthropocene (Crutzen/ Stoermer 2000; Ehlers 2008), and the propagation of a "geology of mankind" (Crutzen 2002) are milestones of a new interpretation of society—nature relationships and interactions. IHDP can take pride in the fact that it has participated in the development of these new paradigms.

Both IPCC and its warnings about the climate change impacts by human activities, and the general discussion about the 'Anthropocene', its characteristics and consequences, are durable indicators of the increasing role of the human dimension in shaping global environmental changes. The questions as to whether this is really "a new epoch of geological time" (Zalasiewicz/Williams/Haywood et al. 2012; see also Steffen/Grinewald/Crutzen et al. 2011) and whether it is "an epoch of our making" (Syvitski 2012) have greatly enhanced and fostered reflection on the importance of human interference into the nature of Planet Earth. It is especially Syvitski's ultimate statement

> Interestingly, while portions of society still refuse to acknowledge the role of humans in affecting global climate, they appear more willing to accept that the modern world is anything but pristine and strongly under the influence, if not control, of society...Humans have changed the Earth in a number of fundamental ways, many of which are far less known than global warming (Syvitski 2012: 12-13)

that underlines even stronger commitments from the social science community to cope with the increasing number of threats to our fragile Planet Earth. IHDP 1996-2014 and its projects (IDGEC, IT, GECHS, UGEC and many others) can take pride in the fact that it has been an essential promoter of these insights. IHDP was successful in raising awareness and expanding our consciousness of the causes and consequences of climate and environmental change. And, above all, it has taken up the serious and timely demands for the pursuit of a new "sustainability science" (Kates/Clark/ Corell et al. 2001) and for science's responsibility for society and its well-being (Lubchenco 1998; Gibbons 1989; Lövbrand/Stripple/Widman 2009).

For millennia, humanity has lived under the threat of nature. Today, nature exists under the threat of humankind. It may therefore be appropriate to finish this review of IHDP 1996-2014 by quoting a few conclusions from a Nobel Prize Laureate symposium on Global Sustainability held in Stockholm in May 2011. Some of its key messages stress the need for new ways of thinking and include an urgent call for a new alliance between nature and society. The Nobel Prize Laureates' advice for new forms of independent social-ecological systems of global change research is as follows:

> It is time for a new social contract for global sustainability rooted in a shift of perception—from people and nature seen as separate parts to interdependent social-ecological systems. ... Most current economic and technological solutions are ecologically illiterate and too linear and single problem-orientated. ... We need a new type of 'social-ecological' innovation and technologies that work more directly for social justice, poverty alleviation, environmental sustainability and democracy ... (Third Nobel Laureate Symposium on Global Sustainability 2011).

IHDP has surely contributed to meeting these challenges. It remains to be seen how Future Earth will proceed in its efforts to achieve transformations towards sustainability.

References

Arizpe, Lourdes, 1991: "The global cube", in: *International Social Science Journal, 130: Global Environmental Change*: 599-608.

Arizpe, Lourdes; Price, Martin F.; Worcester, Robert, 2016: "The First Decade of Initiatives for Research on the Human Dimensions of Global (Environmental) Change (1986-1995)", in: Brauch, Hans Günter; Oswald Spring, Úrsula; Grin, John; Scheffran, Jürgen (Eds.): *Sustainability Transition and Sustainable Peace Handbook*. Hexagon Series on Human and Environmental Security and Peace 10 (Heidelberg–New York–Dordrecht–London: Springer-Verlag, 2016): pages [in this volume].

AVISO (Information Bulletin on Global Environmental Change and Human Security), 1999- (Victoria, B.C.: GECHS).

Balstad Miller, R., 1991: "Social science and the challenge of global environmental change", in: *International Social Science Journal 130: Global Environmental Change*: 609-627.

Biermann, Frank, 2007: "'Earth System Governance' as a Cross-cutting Theme of Global Change Research." Global Environmental Change. Human and Policy Dimensions 17, 3-4: 326-337.

Biermann, Frank; Betsill, Michelle M.; Gupta, Joyeeta; Kanie, Norichika; Lebel, Louis; Livermann, Diane; Schröder, Heike; Siebenhüner, Bernd, 2009: "Earth System Governance: People, Places and the Planet", Science and Implementation Plan of the Earth System Governance Project. ???

Bohle, Hans-Georg, 2001: "Vulnerability and Criticality. Perspectives from Social Geography", in: *Update IHDP*, 2/01: 1-5.

Bohle, Hans-Georg, 2002: "Land Degradation and Human Security", Paper presented to the UNU/RTC Workshop on 'Environment and Human Security', Bonn, 23-25 October 2002.

Brauch, Hans Günter, 2009: "Securitizing Global Environmental Change", in: Brauch, Hans Günter; Oswald Spring, Úrsula; Grin, John; Mesjasz, Czeslaw; Kameri-Mbote, Patricia; Behera, Navnita Chadha; Chourou, Béchir; Krummenacher, Heinz (Eds.), 2009: *Facing Global Environmental Change: Environmental, Human, Energy, Food, Health and Water Security Concepts* (Berlin–Heidelberg–New York: Springer-Verlag): 65-102.

Church, John A.; Asrar, Ghassem R.; Busalacchi, Antonio J.; Arndt, Carolin E., 2011: "Climate Information for Coping with Environmental Change: Contributions of the World Climate Research Programme", in: Brauch, Hans Günter; Oswald Spring, Úrsula; Mesjasz, Czeslaw; Grin, John; Kameri-Mbote, Patricia; Chourou, Béchir; Dunay, Pal; Birkmann, Jörn, 2010: *Coping with Global Environmental Change, Disasters and Security–Threats, Challenges, Vulnerabilities and Risks* (Berlin–Heidelberg–New York: Springer-Verlag): 1257-1270.

Clark, William C.; Dickson, Nancy; Jäger, Jill; Eijndhoven, Joree van (Eds.), 2001: "Learning to Manage Global Environmental Risks. The Social Learning Group".
Volume 1: A Comparative History of Social Responses to Climate Change, Ozone Depletion, and Acid Rain (361 p.)
Volume 2: A Functional Analysis of Social Responses to Climate Change, Ozone Depletion, and Acid Rain (222 p.)

Crutzen, Paul J., 2002: "Geology of Mankind", in: *Nature*, 415,3 (January): 23.

Crutzen, Paul J., 2006: "The Anthropocene", in: Ehlers, Eckart; Krafft, Thomas (Eds.): *Earth System Science in the Anthropocene* (Berlin–Heidelberg–New York: Springer): 13-18.

Crutzen, Paul J.; Stoermer, Eugene F., 2000: "The Anthropocene", in: *IGBP Newsletter*, 41: 17-18.

Dimensions, 2012: "The human factor in the global environmental debate", Magazine of the International Human Dimensions Programme on Global Environmental Change, Issue 1 Bonn.

Ehlers, Eckart, 1994: "ICSU and the Social Sciences", in: *Science International*, 56: 17-20.

Ehlers, Eckart, 2008: *Das Anthropozän. Die Erde im Zeitalter des Menschen* (Darmstadt: Wissenschaftliche Buchgesellschaft).

Ehlers, Eckart; Kosinski, Leszek A., 1998): "From HDP to IHDP", in: *Global Environmental Research*, 1,1-2: 95-96.

Future Earth, 2013: *Future Earth Initial Design Report of the Transition Team. Paris: International Council for Science* (Paris: ICSU).

Gallopin, Gilberto C., 1991: "Human dimensions of global change: linking the global and the local processes", in: *International Social Science Journal 130: Global Environmental Change*: 707-718.

Geist, H.J.; Lambin, E.F., 2001: *What Drives Tropical Deforestation? A meta-analysis of proximate and underlying causes of deforestation based on subnational case study evidence*. LUCC Report Series No. 4 (Louvain-la Neuve: publisher).

Gibbons, M., 1999: "Science's new social contract with society", in: *Nature*, 402, Suppl. 2: C81-C84.

Gibson, C.; Ostrom, Elionore; Ahn, T.K., 1998: *Scaling Issues in the Social Sciences*. IHDP Working Paper No. 1 (Bonn: IHDP, May).

GLP, 2005: *Science Plan and Implementation Strategy*. IGBP Report No. 53 / IHDP Report No. 19 (Stockholm: IGBP Secretariat).

Gupta, Joyeeta; Lebel, Louis; Vellinga, Pier; Young, Oran (IHDP Secretariat), 2001: *IHDP Global Carbon Cycle Research. International Carbon Research Framework* (Bonn: IHDP).

Hackmann, Heide; St. Clair, Asunción Lera, 2012: *Transformative Cornerstones of Social Science Research for*

Global Change (Paris: International Social Science Council, May).

IHDP Update 2/2001: *Special Issue on Vulnerability Research* (7 short articles on specific vulnerability aspects).

IHDP (International Human Dimensions Programme on Global Environmental Change) (2007): *Strategic Plan 2007-2015. Framing Worldwide Research on Human Dimensions of Global Environmental Change* (Bonn: IHDP).

IHDP Update 2009: *GECHS Synthesis. Human Security in an Era of Global Change* (Special Update Issue).

IPCC, 2001: *Climate Change 2001. The Scientific Basis of Working Group I to the Third Assessment Report of the Intergovernmental Panel on Climate Change* (Cambridge, UK–New York: Cambridge University Press).

IPCC, 2007: *Climate Change 2007. The Physical Science Base. Contribution of Working Group I to the Fourth Assessment Report of the Intergovernmental Panel on Climate Change* (Cambridge, UK–New York: Cambridge University Press).

ISSC (International Social Science Council), 1997: "IHDP Office Inaugurated", in: *Newsletter*, 67: 1-2.

Kasperson, Jeanne X.; Kasperson, Roger E., 2001: *International Workshop on Vulnerability and Global Environmental Change. SEI Risk and Vulnerability Programme Report 2001-01* (Stockholm: Stockholm Environmental Institute).

Kates, Robert W.; Clark, William C.; Corell, Robert; Hall, J. Michael; Jaeger, Carlo C.; Lowe, Ian; McCarthy, James J.; Schellnhuber, Hans Joachim; Bolin, Bert; Nancy M. Dickson; Sylvie Faucheux; Gilberto C. Gallopin; Arnulf Grübler; Brian Huntley; Jäger, Jill; Jodha, Narpat S.; Kasperson, Roger E.; Mabogunje, Akin; Matson, Pamela; Mooney, Harold; Moore III, Berrien; O'Riordan, Timothy; Svedin, Uno, 2001: "Sustainability Science", in: *Science*, 292,5517 (27 April): 641-642

Kates, Robert W.; Chen, R.S., 1993: "Poverty and global environmental change", in: *IGU Bulletin*, 43: 5-14.

Lambin, Eric F.; Geist, Helmut (Eds.), 2006: *Land-Use and Land-Cover Change. Local Processes and Global Impacts.* IGBP Series (Berlin–Heidelberg: Springer).

Leemans, Rik; Rice, Martin; Henderson-Sellers, Ann; Noone, Kevin, 2011: "Research Agenda and Policy Input of the Earth System Science Partnership for Coping with Global Environmental Change", in: Brauch, Hans Günter; Oswald Spring, Úrsula; Mesjasz, Czeslaw; Grin, John; Kameri-Mbote, Patricia; Chourou, Béchir; Dunay, Pal; Birkmann, Jörn, 2010: *Coping with Global Environmental Change, Disasters and Security–Threats, Challenges, Vulnerabilities and Risks* (Berlin–Heidelberg–New York: Springer-Verlag): 1205-1220.

Lövbrand, Eva; Stripple, Johannes; Widman, Bo, 2009: "Earth System Governmentality. Reflections on Science in the Anthropocene", in: *Global Environmental Change*, 19: 7-13.

Lonergan, Stephen, (no year/1994?): *The Role of Environmental Degradation in Population Displacement. Global Environmental Change and Human Security Project.* International Human Dimensions Programme on Global Environmental Change. Research Report 1. (Victoria, B.C.: GECHS).

Lonergan, Stephen, 1999: *Global Environmental Change and Human Security (GECHS). Science Plan.* IHDP Report No. 11 (Bonn: IHDP).

Lubchenko, J., 1998: "Entering the Century of the Environment: A New Social Contract for Science", in: *Science*, 279: 491-497.

National Research Council et al., eds., 1999: *Human Dimensions of Global Environmental Change. Research Pathways for the Next Decade.* Committee on the Human Dimensions of Global Change, Commission on Behavioral and Social Sciences and Education; Committee on Global Change Research, Board of Sustainable Development. Policy Division, National Research Council. Washington DC: National Academy Press.

Noone, Kevin J.; Nobre, Carlos; Seitzinger, Sybil, 2011: "The International Geosphere-Biosphere Programme's (IGBP) Scientific Research Agenda for Coping with Global Environmental Change", in: Brauch, Hans Günter; Oswald Spring, Úrsula; Mesjasz, Czeslaw; Grin, John; Kameri-Mbote, Patricia; Chourou, Béchir; Dunay, Pal; Birkmann, Jörn, 2010: *Coping with Global Environmental Change, Disasters and Security–Threats, Challenges, Vulnerabilities and Risks* (Berlin–Heidelberg–New York: Springer-Verlag): 1249-1256.

Nunes, C.; Augé, J.I., (Eds.), 1999: *Land-Use and Land-Cover Change. Implementation Strategy.* IGBP Report 48/IHDP report 10 (Stockholm: IGBP–Bonn: IHDP).

Pritchard Jr., L.; Colding, J.; Berkes, F.; Svedin, U.; Folke, C., 1998: *The Problem of Fit between Ecosystems and Institutions.* IHDP Working Paper No. 2 (Bonn: IHDP, May).

Rivière, J.W.M. La, 1991: "Co-operation between natural and social scientists in global change research: imperatives, realities, opportunities", in: *International Social Science Journal 130: Global Environmental Change*: 619-627.

Robinson, J.B., 1991: "Modelling the interactions between human and natural systems", in: *International Social Science Journal 130: Global Environmental Change*: 629-646.

Steffen, Will; Jaeger, Jill; Carson, David J.; Bradshaw, Clare (Eds.), 2002: *Challenges of a Changing Earth.* Proceedings of the Global Change Open Science Conference Amsterdam/NL, 10-13 July 2001 (Berlin–Heidelberg–New York: Springer-Verlag)

Steffen, Will; Grinevald, J.; Crutzen, Paul; McNeill, John, 2011: "The Anthropocene: conceptual and historical perspectives", in: Special issue of: *Philosophical Transactions of the Royal Society A*, 2011: 842-867.

Syvitski, J., 2012: "Anthropocene: An Epoch of our Making", in: *Global Change. International Geosphere-Biosphere Programme*, Issue 78: 12-15.

Third Nobel Laureate Symposium on Global Sustainability, 2011: *Transforming the World in an Era of Global Change. Stockholm*. Executive Summary of Scientific Background -Reports (Stockholm: Stockholm Resilience Centre, Stockholm University).

Turner II, Billie Lee; Meyer, N.B., 1991: "Land use and land cover in global environmental change: considerations for study", in: International Social Science Journal 130: Global Environmental Change: 669-679.

Turner II, Billie Lee; Skole, D.; Sanderson, St.; Fischer, G.; Fresco, L.; Leemans, Rik, 1995: *Land-Use and Land-Cover Change. Science/Research Plan*. IGBP Report No. 35 – HDP Report No. 7 (Stockholm: IGBP–Geneva: HDP).

Vellinga, Pier; Herb, Nadia (Eds.), 1999: "Industrial Transformation Science Plan, International Human Dimensions Programme", in: *IHDP Report*, No.12; at: <http://www.uni-bonn.de/ihdp/ITSciencePlan/>.

von Falkenhayn, Louise; Rechkemmer, Andreas; Young, Oran R., 2011: "The International Human Dimensions Programme on Global Environmental Change–Taking Stock and Moving Forward", in: Brauch, Hans Günter; Oswald Spring, Úrsula; Mesjasz, Czeslaw; Grin, John; Kameri-Mbote, Patricia; Chourou, Béchir; Dunay, Pal; Birkmann, Jörn, 2010: *Coping with Global Environmental Change, Disasters and Security–Threats, Challenges, Vulnerabilities and Risks* (Berlin–Heidelberg–New York: Springer-Verlag): 1221-1234.

Walther, Bruno A.; Larigauderie, Anne; Loreau, Michel, 2011: "DIVERSITAS: Biodiversity Science Integrating Research and Policy for Human Well-Being", in: Brauch, Hans Günter; Oswald Spring, Úrsula; Mesjasz, Czeslaw; Grin, John; Kameri-Mbote, Patricia; Chourou, Béchir; Dunay, Pal; Birkmann, Jörn, 2010: *Coping with Global Environmental Change, Disasters and Security– Threats, Challenges, Vulnerabilities and Risks* (Berlin– Heidelberg–New York: Springer-Verlag): 1235-1248.

Young, Oran R., 1999: *Institutional Dimensions of Global Environmental Change (IDGEC). Science Plan*. IHDP Report No. 9 (Bonn: IHDP).

Young, Oran R., 2009: "Institutions, Governance, and the Evolution of the IHDP", in: *IHDP Update*, Issue 3 (November): 7-8.

Young, Oran R.; King, L.A.; Schroeder, H. (Eds.), 2008: *Institutions and Environmental Change. Principal Findings, Applications and Research Frontiers* (Cambridge: MIT Press).

Zalasiewicz, Jan; Waters, Colin N.; Williams, Mark; Barnosky, Anthony D.; Cearreta, Alejandro; Crutzen, Paul J.; Ellis, Erle; Fairchild, Jan J.; Grinevald, Jacques; Haff, Peter K.; Hajdas, Irka; Leinfelder, Reinhold; McNeill, John; Odada, Eric O.; Poirier, Clement; Richter, Daniel; Steffen, Will; Summerhayes, Colin; Syvitski, James P.M.; Vidas, Davor; Wagreich, Michael; Wing, Scott L.; Wolfe, Alexander P.; Zhisheng, An; Oreskes, Naomi; 2014: "When did the Anthropocene begin? A mid-twentieth century boundary level is strategically optimal", in: *Quaternary International*, details at: <http://dx.doi.org/10.1016/quaint2014.11.045>.

Zalasiewicz, Jan; Williams, Mark; Haywood, Alan; Ellis, Michael, 2012: "The Anthropocene: a new epoch of geological time". Special issue of *Philosophical Transactions of the Royal Society* A: 369.

16   From *The Limits to Growth* to 2052

Marit Sjøvaag

Abstract

This chapter summarizes the content of and the debate around *The Limits to Growth* and its updates, from 1972 until 2012. It presents the background and political context to the original 1972 study, and follows the debate in and around the four books on whether a 'fair and free market' can provide sufficient benefits for us all. The methodology of systems dynamics is briefly introduced. The analyses are presented in some detail, as are the reactions from both academia and political and economic interests following the books' publication. The two main research questions are why the debate around *Limits to Growth* became so polarized, and what we have learned about the original scenarios over the last four decades. The chapter therefore includes a synthesis of research on how the original World3 Standard run has compared to subsequent reality. Randers' forecast for the next forty years constitutes the fitting end point for this analysis.

Keywords: Limits to Growth, economic growth, polarization, system dynamics, modelling.

16.1   Introduction[1]

In 1972 a small book appeared that was to have an impact far above what its modest size indicated. It presented a set of possible scenarios for the future. *The Limits to Growth* (hereafter: LTG) (Meadows/ Meadows/Randers et al., 1972) kick-started a debate about the pros and cons of economic growth, and gave inspiration to the growing environmental movement in the industrialized world.

The hard and bitter debate that ensued is not yet concluded, but it has over the years changed in volume, intensity and character. If it is at all possible to identify one fundamental issue in a debate that has been going on for over four decades, it is first and foremost a question of whether our decision-making (mainly economic) models are capable of integrating sufficient information to ensure that the effect of human behaviour remains within the planet's carrying capacity—in other words, can we trust a free market to produce ultimate benefits for all?

Limits to Growth did not appear from nowhere. It followed an intellectual tradition that started at least

a decade previously with Rachel Carson's (year) *Silent Spring*, which highlighted the dangers to our environment of using new technologies (in her case pesticides, in particular DDT) whose consequences are unknown. But it was not only the intellectual tradition of questioning human activity's impact on the environment that was followed from *Silent Spring*. As early as the 1960s powerful industrial interests had spent large sums of money on discrediting the kind of research in *Silent Spring*, and subsequently in LTG. The battlefronts of the conflict were therefore already set and hard before LTG was published, at least in the US, where the debate was the most polarized. Once the message had been sent that the authors were 'doomsayers', 'neo-Malthusians' and 'anti-growth', it proved almost impossible to alter the impression in the public mind.

Two questions are of particular interest in this context. Firstly, why did the debate raised by the *Limits to Growth* become so polarized? And secondly, what have we learned over the last four decades about the essential questions posed by the LTG team?

In order to answer these questions this chapter will synthesize the main message from the Limits to Growth study (hereafter the LTG) as well as the two updates: *Beyond the Limits* (BLT), from 1992, and *Limits to Growth—the 30-Year Update* from 2004.

---

1   I wish to thank Prof. Jørgen Randers for helpful information and moral support throughout the writing process. Any errors are mine alone.

© Springer International Publishing Switzerland 2016
H.G. Brauch et al. (eds.), *Handbook on Sustainability Transition and Sustainable Peace*,
Hexagon Series on Human and Environmental Security and Peace 10, DOI 10.1007/978-3-319-43884-9_16

Finally the main findings from 2052—*A Global Forecast for the Next 40 Years* will be presented. This book can in some ways be seen as a sequel to the others, but unlike the LTG series, '2052' is presented as a forecast, not as a collection of possible futures. It will therefore be possible to empirically validate (or not) its accuracy over the next forty years.

The chapter is organized in four main parts. Part one presents the Limits to Growth from 1972 (16.2), its methodological foundations, its context, and the immediate reactions. Part two reviews the two update studies from 1992 and 2004 (16.3). Part three looks at the message and conclusions from the book 2052: *the next 40 years* (16.4), while part four offers conclusions (16.5).

16.2   The Limits to Growth—A System Dynamic Analysis of Planet Earth (1972)

The LTG was written as a 'report for the Club of Rome's project on the predicament of mankind'. The Club of Rome[2] traces its origins back to a meeting in 1968 between thirty individuals who believed that 'the major problems facing mankind are of such complexity and are so interrelated that traditional institutions and policies are no longer able to cope with them, nor even to come to grips with their full content' (Meadows/Meadows/Randers et al. 1972: 9-10).

The Club of Rome commissioned a group of young researchers at MIT under the direction of Professor Dennis Meadows to carry out 'Phase One of the Project on the Predicament of Mankind'. The intellectual inspiration for their work came from Professor Jay Forrester, an engineer who had turned his attention to management systems and to using computers in the simulation of social systems (Forrester 1989). He is the founder of system dynamics.

16.2.1   Key Concepts and Methodology

Before summarizing the findings of the LTG, it is useful to give a brief presentation of the methodological foundations upon which the study was conducted. So much has been said and opined about the LTG results that it is worth clarifying the methodological ambitions from the outset.[3]

Models[4] are 'a formalization of our assumptions about a system' (Hall/Day 1977, cited in Bardi 2011: 16). We often refer to individuals having 'mental models'. Such mental models are a representation of the surrounding world, and are crucial to our ability to make sense of our observations. Scientific models, however, should start from clear, explicit assumptions, so that others can validate their appropriateness and usefulness in different contexts. Scientific models allow us to make explicit what we assume about a system, and when the models are quantitative (i.e. mathematically formulated expressions define how various system parameters are related), they allow us to calculate how the parameters—and hence the system—will vary with time.

The assumptions may be more or less right or wrong. Large discrepancies between predictions and observations can lead us to conclude that the model is wrong, but the opposite is not true. A model is not necessarily right even if it predicts what we go on to observe. However, the more empirical data that agree with simulated results, the more we tend to 'trust' the model—we have a higher degree of confidence that it will be useful for predicting the future behaviour of the system.

System dynamics[5] emerged as a discipline in the 1950s. Initially it was used for industrial purposes, but from the 1960s it was used to model social and eco-

---

2   The Club of Rome is a "global network of independent and renowned thinkers [...that...] analyses today's challenges facing the world, their root causes and the possible futures, in a systematic and holistic manner. The Club of Rome encourages global debate in order to set in motion actions that by the middle of the century will ensure a more secure, equitable, and prosperous world." See at: <http://www.clubofrome.org/> and <http://www.clubofrome.org/?p=324>

3   This is not meant as an exhaustive introduction to system dynamics. For those interested, see for example Ford 2009, <http://www.public.asu.edu/~kirkwood/sysdyn/SDIntro/SDIntro.htm> or <http://www.systemdynamics.org/what-is-s/>.

4   For an excellent overview of the debate surrounding *Limits to Growth* see Ugo Bardi (2011), *Limits to Growth Revisited*. Ugo Bardi's blog can be found at: <www.cassandralegacy.blogspot.com>.

5   One typically distinguishes between linear and non-linear systems. In linear systems there are well-defined and well-known causal relationships, whereas non-linear systems involve parameters that interact, and they are often very difficult to predict over longer time spans. Cause and effect are often not well understood in non-linear systems. They are central to the group of systems called 'complex', i.e. systems that involve feedback relations.

nomic systems. System dynamics looks at several fundamental elements in a system: stocks, flows and feedback.[6] These elements vary over time and are interdependent. The stock is something's actual amount (e.g. the number of individuals in a population), whereas the flow is the variation over time of the stock (e.g. births and deaths). Feedback occurs when the intensity of a flow into a stock depends on the size of the stock. The feedback can be positive (for example, a larger population increases the number of births, if the birth rate is constant) or negative (a larger population increases the number of deaths, if the death rate is constant). However, "the fact that a system is dominated by feedback effects (...) does not imply that it is completely random. On the contrary, these systems often follow well-defined patterns" (Bardi 2011: 8), and this allows us to build models that help us analyse possible trajectories of the system's behaviour.

16.2.2 Key Scientific Approach and Messages of *The Limits to Growth* Study

LTG was a study of the Earth system using system dynamics. The research group identified what they saw as the five most central system parameters, and built a quantitative model expressing the relationship between these parameters (and some others) using mathematical expressions.

The ambition of the LTG research group was to study the interrelatedness of five basic factors that determine growth in population and ecological footprint on Earth. The question they asked was simple: how do changes in one of these factors impact on the other factors?

The five factors were:

- population
- agricultural production
- natural resources
- industrial production
- pollution.

Defining and explaining the nature of the growth of the system was a central part of the LTG study. The book goes to some length to explain the characteristics of 'exponential growth', and to emphasize how exponential growth differs from linear growth. It explains the concepts of 'doubling time' and 'positive

feedback loops'. It is hard to overestimate how difficult the human mind finds it to think about exponential growth, particularly in relation to processes that take a long time. The classic illustration used in the book is the old French riddle that visualizes the suddenness with which the 'end' is approaching:

> Suppose you own a pond on which a water lily is growing. The lily plant doubles in size each day. If the lily were allowed to grow unchecked, it would completely cover the pond in 30 days, choking off the other forms of life in the water. For a long time the lily plant seems small, and so you decide not to worry about cutting it back until it covers half the pond. On what day will that be? On the twenty-ninth day, of course. You have one day to save your pond (Meadows/Meadows/ Randers et al. 1972: page).

The LTG research group built a system dynamic model of the world, and called it World3. Like all models, it was a simplified representation of the *real* world system this is one of the purposes of modelling—and in this case many constraints were obvious: there was insufficient knowledge about the physical processes specifying the relationships between the parameters, and it was (and is) difficult to quantify the effects of socio-psychological factors (for example, of the population's potential value change on the consumption of both industrial goods and services). The authors grouped these sources of uncertainties into three categories:

1. the relationship between economic variables that are relatively well understood,
2. the effect of socio-psychological factors that are difficult to quantify, and
3. the biological variables which are largely unknown (Meadows/Meadows/Randers et al. 1972: 103-104).

The authors were therefore aware of the shortcomings in knowledge about the system they wanted to simulate. However, rather than waiting for more research about the exact effects of socio-psychological factors or biological variables, which would take a long time in coming, they proceeded to develop the World Model, representing the world system to the best of their ability. A visual presentation of the causal links in the World Model is given in figure 16.1.

In 1972, the recent development of sufficiently sophisticated computer technology had profoundly changed our ability to handle large amounts of data simultaneously. We have become nonchalant when it comes to computing capacity, but in 1972 this was still in its infancy, and while the World Model used in the

---

6   For a highly readable introduction to the basics of system dynamics as used in the LTG study, see Bardi (2011), chapters 3 and 4.

Figure 16.1: Causal diagram of the World3 model. **Source**: Jørgen Randers presentation material; he also granted permission to use this figure.

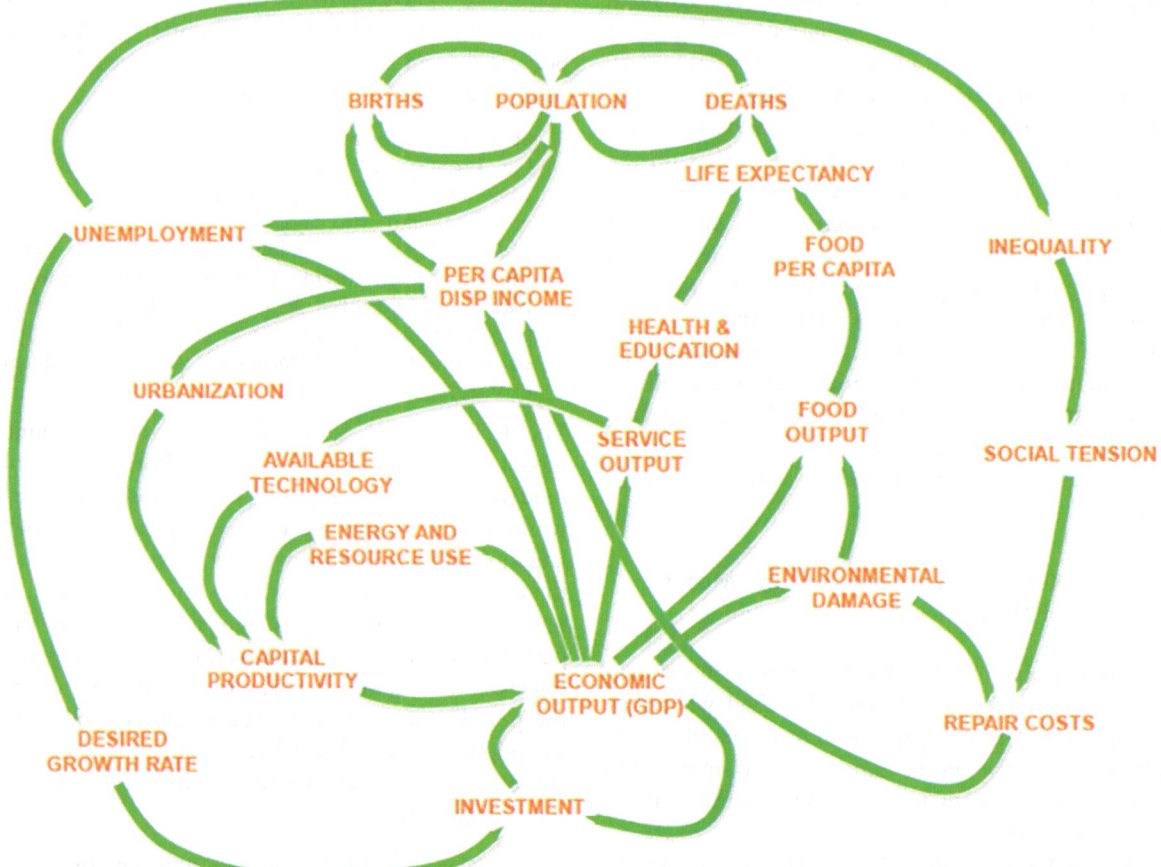

Limits to Growth is simple from our current perspective, it was the most sophisticated in its time.

The discussion in the LTG study started from a 'standard run' of World3, where one assumed 'no significant change'[7] in the physical, economic or social relationships that were seen to govern the world system. The standard run was concurrent with historical data for the period 1900-1970. This is not surprising; the parameters of the model were chosen to ensure such a match, so the match is in itself not a sign of the correctness of the model. The model was built empirically, in some cases using historical relationships between the five basic factors to specify the equations used in the model. The standard run showed that if

continued unchanged, the exponential growth in population, food and industrial output would result in a rapidly diminishing resource base that again would lead to a slowdown in industrial growth, and subsequently (because of delays in the system) to a rapid decline in food production and population.

To investigate the effect of various policies in World3, the researchers used the 'standard run' as a baseline and simulated policy implementation by changing one or more of the input parameters. 'Policies' should here be understood as behavioural changes at the system level, rather than as a political process or pronounced ambitions (for example about recycling rates or reduction of ozone-depleting gases). The policy changes included unlimited access to nuclear energy (enabling extensive recycling and resource substitution), and controls on population and pollution. The overwhelming majority of scenarios produced futures with overshoot and collapse.

Having observed collapse, i.e. a sudden and dramatic decline in population or production, the

---

7   This important fact is often overlooked. The model was built so that it concurred with observed data in the period 1900-1970, and the standard run therefore assumed that the interdependent processes simply would continue as they had done for the previous seven decades.

researchers set themselves the task of identifying policies that would stabilize the system so that uncontrollable collapses would be avoided, and that at the same time would be capable of satisfying the basic material requirements of all people.

The input parameters were thus changed with the explicit goal of identifying a stable solution for world population with an adequate food supply and industrial output. Of course, several input combinations would be able to produce a stable system, and one set was presented in the LTG study. It consisted of:

- Stabilizing the population at 1975 levels (by setting the birth rate equal to the death rate)
- Stabilizing industrial capital at 1990 levels (by setting the investment rate equal to the depreciation rate)
- Setting resource consumption per unit of industrial output to one-fourth of its 1970 value (introduced in 1975)
- Shifting society's consumption preferences more towards services and less towards manufactured goods (starting in 1975)
- Setting pollution generation per unit of industrial and agricultural output to one-fourth of its 1970s value (introduced in 1975)
- Diverting some capital from industrial to agricultural production even if it was 'uneconomic' (to ensure proper nourishment for all)
- Maintaining soil quality through altered use of agricultural capital
- Increasing average lifetime for industrial capital.

These policies were, according to LTG, one (of several) possible sets that it was physically possible to maintain on Earth. Resource depletion would still occur, but at a rate sufficiently slow to allow for technology to adjust to changes in resource availability.

To reach this sustainable model, however, one would have to implement the above assumptions, which were highly unlikely. Simulations with less unrealistic assumptions[8] avoided collapse in the period up to 2100, whereas a simulation where policy implementation was delayed until 2000 showed collapse before 2100.

Thus, of the twelve model runs that were presented in the book, only two represented a system

that would not overshoot and collapse before 2100. Two conclusions stand out. Firstly, a free-market-oriented, liberal, individualistic world will not deliver the set of measures needed for the system to stabilize. Secondly, the later growth-reducing policies are implemented, the smaller the chances of avoiding overshoot and collapse.

The basic message from LTG can be summarized thus:

a.) The planet we live on is finite, and as the population and its ecological footprint increase in size, the relative size of the Earth system on which we depend becomes smaller.
b.) If developments in population, economic activity and use of nature's resources continue unchanged, overshoot of global carrying capacity is likely within the twenty-first century.
c.) Once in overshoot, physical contraction is inevitable. This may happen as a sudden contraction and collapse, or as a 'managed decline'.

16.2.3 Immediate Reactions—Academic and Political Debate During the 1970s

When looking at the debate that followed the publication of LTG, one should be excused for believing that the book had prophesied apocalypse and the end of humankind as we know it. However, as is often the case with highly polarized debates, the basic text seems to have been little read and even less understood. Far from being apocalyptic in its approach, the LTG study was presented as a warning signal and an invitation to think about the possible futures if the growth model we live by are to continue unchanged.

The title of the book, *Limits to Growth*, was provocative in an age where economic growth was hailed as the panacea for the world's ills. Economists, therefore, would be expected to take issue with the LTG conclusions, and they did. This academic debate was, at least initially, relatively civil in form. The LTG conclusions were out of sync with the consensual wisdom among mainstream economists and public policymakers in the 1970s, which led to questioning both of model and methodology.

One of many criticisms of the LTG came from a research team in Sussex University's *Science Policy Research Unit* (SPRU), who in 1973 published their own run of World3. Their results differed significantly from those of the LTG team. The Sussex team abolished the basic assumption of absolute resource limits, and replaced them with 'ongoing exponential increases in available resources' (Ekins 1993: 270). Thus, they

---

8 The less unrealistic assumptions were 'perfect birth control', meaning that only wanted children would be born; an average family size of two children; and industrial output stabilized at the level of average per capital output in 1975.

effectively embraced the idea that human ingenuity would develop good substitutes for non-renewable resources, and that we would learn to use existing resources much more effectively. "To postpone collapse indefinitely these rates of improvement must obviously be competitive with growth rates of population and consumption so that even if the overall growth is rapid, it is also balanced" (Cole/Freeman/Jahoda et al. 1973).

How could different scientists reach such widely different conclusions when they used the same model? Lecomber identified three points of contention that could explain the modellers' differences: their optimism regarding the rate of technical progress, how they saw future changes in the composition of output, and the possibilities of substitution. He took care, however, to emphasize that "this establishes the *logical* conceivability, not the certainty, probability or even the possibility in practice, of growth continuing indefinitely" (Lecomber, cited in Cole 1999: 89).

The economist William Nordhaus has been a vociferous participant in the debate about limits to growth since the early 1970s. In a 1973 article he criticized the World3 model, mainly on the grounds that it in his view was built on (mostly) plausible hypotheses about relationships between parameters, but that none of these relationships were empirically tested. His criticism was strongly worded:

> the dynamic theory put forward in the work represents no advance over earlier work. [...T]he economic theory put forth in *World Dynamics* is a major retrogression from current research in economic growth theory. [...W]ithout the scantest reference to economic theory or empirical data, Forrester predicts that the world's material standard of living will peak in 1990 and then decline. *Sic transit gloria* (Nordhaus 1973: 1182-1183).

Nordhaus gave voice to a widely held view among economists, namely that the researchers working with the World3 model had neglected economic research, and that they did so at their own peril. Apart from the factual error that the World3 did not *predict* the future (it merely simulated and produced scenarios based on a simplified model of the world system),[9] Nordhaus can be excused for wanting to defend economists' intellectual territory. With hindsight we may regret this lack of curiosity and mutual understanding between system dynamics and conventional economics—and conventional economists.

If reactions from the academic community were strong, they remained civil compared to some of the reactions in the public political debate. The LTG's

conclusions were in many ways stark and depressing, despite the non-alarmist language used in the book. However, it is reported that none of the researchers involved in the study envisaged the uproar their publication would cause. The book's American publishers, Potomac Associates, "were arranging to present a copy of the book to every senator, representative, governor and UN ambassador" (AtKisson 2011: 10), in addition to organizing an event at the Smithsonian institute. The publication thereby received high-level political attention, reported in the American press with headlines such as "Mankind warned to curb growth or face catastrophe" (Chicago Sun Times), "Scientists warn of global catastrophe", and "Panel on growth strives to stave off world ruin" (AtKisson 2011). The subsequent debate was often hostile, polarized, and increasingly dominated by 'public legends', perceived as facts, but without root in the LTG study.

The most important of these legends is that the LTG study predicted a crisis. Far from predicting anything at all, the LTG presented twelve different scenarios based on the World3 model. The fact that an overwhelming majority of the scenarios they produced led to overshoot and collapse some time during the twenty-first century was clearly threatening to all those with a stake in the existing system. Moreover, the 'Cassandra effect', that we are reluctant to listen to bringers of bad news, may explain why many dug in to their trenches rather than participating in constructive debate about how to preserve the benefits of the existing model while attempting to change it to avoid its negative effects.

The authors of LTG were unprepared for the massive reaction the publication would create, and for the polarization that followed. If they had known the stir they would cause, maybe they would have taken greater care to emphasize that their scenarios were not predictions, but an invitation to start a debate about

---

9   The difference between a prediction and a scenario is not always obvious, and the two terms are often used interchangeably in daily speech. However, they differ both in intent and in level of precision. Whereas a 'prediction' is a statement about something that will happen in the future, and is therefore empirically testable, a scenario is one (of several possible) internally consistent stories about the future. Therefore, whilst a prediction can be said to be right or wrong once the future that was predicted has arrived, a scenario can be neither right nor wrong. Rather, the scenario is produced to inform and encourage thinking and debate about the choices we face. See also <http://www.ipcc-data.org/guidelines/pages/definitions.html>.

the paradigm within which we live our lives. This, at least, seems to be one lesson they have learned in the updated versions of the World3 simulations, which appeared in 1992 and 2004, and to which we now turn our attention.

16.2.4   Legacy: Long-Term Impact of *The Limits to Growth Study*

The LTG study was hotly debated in the 1970s, but attention receded throughout the 1980s. The general mood lifted after a decade of economic difficulties (remember the world oil crisis in 1973), and '[i]f the 1973 oil crisis had looked to many as the crisis predicted by the LTG, the end of the crisis, in the mid 1980s, seemed to be the refutation of the same predictions' (Bardi 2011: 87).

16.3   The Updates

Despite being in the business of mapping possible futures, none of the authors of the LTG had anticipated the ferocity and acerbity of the debate that followed its publication. They were scientists, not politicians or demagogues, and had not prepared for a public exchange of words on the scale of what actually happened. Two updates were published twenty and thirty years after the LTG study respectively. This part of the chapter will take a closer look at the results from these updates, and inscribe them in the political context they appeared in.

16.3.1   *Beyond the Limits*

Three of the authors of the LTG decided to publish an updated version of the World3 simulations with new numbers twenty years after the original publication. They called the new book *Beyond the Limits* (hereafter: BTL) (Meadows/Meadows/Randers 1992). In the preface they wrote:

> Much has happened in twenty years to bring about technologies, concepts, and institutions that can create a sustainable future. (...) When we began working on the present book, we simply intended to document those countervailing trends in order to update The Limits to Growth for its reissue on its twentieth anniversary. We soon discovered that we had to do more than that. As we compiled the numbers, reran the computer model, and reflected on what we had learned over two decades, we realized that the passage of time and the continuation of many growth trends had brought the human

society to a new position relative to its limits (Meadows/Meadows/Randers 1992: page).

The rerun of the World3 model with updated numbers led the authors to conclude that we had already passed the point of sustainable use of many resources and generation of many pollutants, that is, we were in 'overshoot'. Having learned from the debate following the LTG publication, they took great care to spell out that theirs was a message of hope and opportunity to do something that would improve the possibilities for choice in the future, although the latter was relatively rapidly vanishing.

The structure of BTL is very similar to the structure of the LTG. It starts with an explanation of the phenomenon of exponential growth, followed by a presentation of the limits to growth in the form of sources (renewable and non-renewable) and sinks (nature's capacity to absorb the wastes of human activity, in particular the atmosphere's reaction to greenhouse gases and some other air pollutants). The book subsequently explains the drivers of growth in the world system, including feedback loops, and finally presents a discussion of technological possibilities for solving the problems identified.

This logic underlines one important fact about their basic message, namely that the economy is a subsystem of the physical world, and therefore constrained by it. This might seem uncontroversial today, but it still proved a hard sell to many economists. The idea has been further developed by many within the field of ecological economics.[10]

Beyond the Limits was published in the year of the United Nations Conference on Environment and Development, better known as the Rio Earth Summit. This conference resulted in Agenda 21, which brought the 'sustainable development' concept on to the global political agenda.[11] Many environmentalists at the time were optimistic that world society finally would rise to the task and bring about sustainable living. But at the same time the awareness grew that the tools we had for managing these challenges were inad-

---

10 See Constanza and Daly (1987). For further information about ecological economics, see the website of the International Society of Ecological Economics: <http://www.isecoeco.org/>.

11 The concept's most common definition, 'sustainable development is development that meets the needs of the present without compromising the ability of future generations to meet their own needs', first appeared in the report of the World Commission on Environment and Development, also known as the Brundtland Commission.

equate. To the extent that agreement was achieved on UN documents and policies, these remained vague and non-binding on the member states. This reflected the underlying old conflict about which role the state should have in the economy: on the one hand, states who were positive to global, binding regulations, and on the other, states who advocated as little state intervention as possible, in the interest of individual freedom of choice and free, unregulated markets.

BTL, like its predecessor, presented scenarios resulting from the World3 model. The 1991 run was for technical reasons carried out in a different computer language from the 1972 run, but the model itself remained unchanged. Some of the parameters were slightly altered to take into account historical data from 1972 to 1990. For example, World3 had underestimated both the (negative) impact of erosion on arable land and the (positive) effect of technological development in agriculture, and World3/91 was changed to make a better fit with observed data (Meadows/Meadows/Randers 1992: 245).[12]

However, *Beyond the Limits* is not only a product of a new run of an existing system dynamic model with adjusted numbers. It is also in many ways a communications exercise, inasmuch as it is an attempt to explain more clearly what had been misunderstood in the debate after the *Limits to Growth*. The authors emphasize that they are aware of the shortcomings of their model, and that they do not want to make predictions, but to present a set of possible futures. In the spirit of providing not only hope, but also guidance on preferred action (to answer the question 'but what can we *do*?'), they provide an 'action plan' for transition to a sustainable system. This plan consists of six elements that will improve our chances of building a sustainable society: improving the signals; speeding up response times; minimizing use of non-renewable resources; preventing erosion of renewable resources; using all resources with maximum efficiency; and slowing and eventually stopping exponential growth in population and physical capital.

William Nordhaus, one of the loudest academic critics of the 1972 book, was not convinced of the model's appropriateness despite the result from the new data. Rather, in his 1992 paper "Lethal Model 2: The Limits to Growth Revisited", he clearly ridicules what he terms the 'anti-growth movement'[13] on the grounds that the 'predicted' resource shortages never

materialized (despite this being clearly a wrong interpretation of the 1972 study). He is, however, open to the possibility that there might be physical limits to human activity. Having presented his own model for studying the future of economic growth, he concedes that "our ignorance is vast [...but...] I will hazard the guess that resource constraints are likely to be a small but noticeable impediment to economic growth over the next few decades in advanced industrial countries—although an obstacle that will continue to be surmounted by technological advance" (Nordhaus 1992: 39).

During the 1990s, not long after the publication of BTL, Mathias Wackernagel and William Rees developed a metric for measuring the ecological impact of human consumption, the now well-known 'ecological footprint'.[14] The concept of ecological footprint was clearly useful in the debate about limits to growth, as it provided a clear link between the physical environment on our planet and human consumption. It has since become a household staple in almost all debates about environmental policy in its broadest interpretation, and is incorporated in subsequent LTG studies.

16.3.2   The 30-Year Update

Limits to Growth: The 30-Year Update (Meadows/Meadows/Randers 2004) was 'the third edition in a series'. The World3 model was updated slightly. The most important changes were (1) that the cost of technology was estimated to decline more rapidly than in previous analyses, and (2) that the desired family size was set to respond more strongly to growth in industrial output, i.e. that as material welfare grew, women would choose to have fewer children, and this process was also assumed to be speedier than in previous analyses.

The preceding thirty years had seen many important developments in the public debate about sustainability and the difficulties with unlimited growth. The Brundtland Commission had brought the concept of sustainable development to the fore, and inspired much academic research around questions of nature's

---

12 Other similar changes were made. For a complete overview, see BTL appendix: Research and Teaching with World3.

13 The authors of LTG, as ecological economists, had never purported to be 'anti-growth'. They did, however, try to distinguish between material throughput and societal welfare.

14 The ecological footprint is a measure of how much land and water area is required for human activity. It is estimated that we currently (2015) use 1.5 Earths. For more information and updates, see at: <http://www.footprintnetwork.org/en/index.php/GFN/>.

constraints on human activity. This was obviously important in and of itself, but in our context it provided the authors with new concepts in their communication. In particular, the 'ecological footprint', both concept and measurements, was incorporated in their text. It also proved an invaluable tool in visualizing the meaning of 'overshoot', so central to their findings. Their new Human Welfare Index proved an effective tool for visualizing whether the future would be better or worse for those living there.

The 30-year update therefore included the variables 'human welfare' and 'ecological footprint'. The Human Welfare Index in World3 was inspired by the UN Human Development Index, and was calculated using *gross domestic product* (GDP), life expectancy and education. The Ecological Footprint Indicator in World3 was inspired by Wackernagel's 'ecological footprint' and was built using arable land used for crop production in agriculture, urban land used for urban-industrial-transportation infrastructure, and the amount of absorption land required to neutralize the emission of pollutants (Meadows/Meadows/Randers 2004: appendix 2).

The message from the third book in the series does not make cheerful reading. The language is still both clear and accessible, but the message is stark. The authors also concede that this third book is not so much about making new contributions to the debate as about providing the old debate with new numbers. Their review of the initial World3 scenarios shows that the actual development of the central parameters follow closely the standard run in World3 from 1972. The previous decade had shown that the lofty ambitions of the Earth Summit in 1992 were not being implemented.[15] Therefore, the 'action plan' from BTL was substituted by a list of 'the tools we don't yet know how to use'. These tools were: visioning; networking; truth-telling; learning; and loving.

16.3.3   Reactions to the Updates

Debate about the human condition within the physical boundaries of planet Earth had also taken on new forms since the 1970s. The strands had multiplied, and it is almost impossible to map them all. But two issues stand out: the question of limits to economic growth, and the debate about climate change.

The old question of whether there are limits to economic growth, so much ridiculed by both academics and politicians after the publication of LTG, saw a sea-change with the publication of an article by economist Matthew R. Simmons in 2000 entitled "Revisiting the Limits to Growth—could the Club of Rome have been right, after all?"[16] He compared some of the LTG scenarios within the energy sector with what had happened in the succeeding quarter of a century, and concluded that a shortage of non-renewable energy sources was a real worry. He then says: "Is there time to begin the thoughtful work which the Club of Rome hoped would take place post 1972? I would hope so. But, another 10 years of neglect to these profound issues will probably leave any satisfying solutions too late to make a difference. In hindsight, The Club of Rome turned out to be right. We simply wasted 30 important years by ignoring this work."

Graham Turner compared the last thirty years of historical developments with the LTG scenarios in 2008 (Turner 2008), and found that the LTG scenario with the closest match to historical data was the standard run of World3. His paper was another example of a re-appreciation of the LTG study in the twenty-first century, also exemplified by Hall and Day (2009) who note that "[i]f we are to resolve these issues, including the important one of climate change, in any meaningful way, we need to make them again central to education at all levels of our universities, and to debate and even stand up to those who negate their importance" (Hall/Day 2009: 237). The lower intensity of the debate in the 1980s and 1990s had allowed the hard fronts to soften, and a new generation of researchers appeared, with little or no stake in the debate from the 1970s.

This softening of the fronts in the conflict seem also to have spilt over on the mainstream macroeconomists, who are increasingly participating in the debate about environmental policy and the place of economic policy instruments in it. Moreover, social scientists of different hues have recently turned their attention to environmental problems, in particular climate change. The new voices in the debate contribute to broadening the discussions wider than simply for or against economic growth.

Since the turn of the century, and particularly after 2007, the climate policy debate has become more institutionalized and better anchored in the political

---

15 This is of course in itself a complex story, and it can be argued that some important developments took place in the global arena in the 1990s, particularly with the signing of the Kyoto Protocol in 1997.

16 See at: <http://www.resilience.org/stories/2000-09-30/ revisiting-limits-growth-could-club-rome-have-been-correct-after-all-part-one>.

process both nationally and globally, and scientific knowledge about nature's processes is significantly greater than in 1972. Extreme weather events are increasingly seen as evidence of a changing climate, and the loss of biodiversity, pollution problems, and rising energy prices have all contributed to a general acceptance (among many if not all mainstream economists) that there are physical limits to what nature can sustain. This might seem rather a small achievement, but the hope is that the seed contained in this fundamental truth can flourish into a real change in how we use economic modelling, and subsequently in how modern societies arrive at decisions for collective action.

It may, of course, be a case of too little too late. As was argued both in BTL and in the 30-year update, time is running out. Humankind is already in overshoot and unless large structural changes take place, collapse seems the likely scenario. This was the background against which the book 2052—A Global Forecast for the Next Forty Years appeared.

16.4   2052—A Global Forecast for the Next Forty Years

There are two important differences between 2052—A Global Forecast for the Next Forty Years (Randers 2012) and the three other books that were discussed above. Firstly, it has only one author. Secondly, '2052' is not a scenario study, but a *forecast* of what Randers believes will actually happen by 2052. This means that the author himself has identified one scenario among several possible ones that fulfil three criteria:

- the described future must be internally consistent, i.e. the development in different variables must not cause them to contradict one another;
- the trend lines are developed based on existing data, which means that the basic assumption is *no major change* in any of the large global trends;
- the described future is what the author finds the most likely when the two first conditions are met.

The central global trends that form the basis of the forecast are GDP (based on population growth and productivity), investment, consumption, energy use, climate impacts, food production, and land use. The story is based on the same 'fairly simple' set of cause-and-effect relationships as in the World3 models, with the following important amendments:

- Urbanization leads to smaller families. This will not be because of lack of food or ill health, but

because more people will wish to have fewer children, and can better control the number of offspring because of access to education, health services and contraception.
- Labour productivity will continue to grow, but ever more slowly, because of resource depletion, pollution, climate change, and rising inequity.
- CO2 emissions will at first increase with energy use (although at a slower rate than in many other forecasts, because of increased energy efficiency), but then gradually disconnect as the share of renewable energy accelerates. The emissions will lead to higher concentrations of greenhouse gases in the atmosphere, higher temperatures, and more climate damage to planet Earth (Randers 2012: 55ff).

A visual representation of the causal mechanisms are given in figure 16.2, where the changes from the World model are marked in red.

After defining the fundamental causal relationships, detailed forecasts are presented about both physical and non-physical variables. The first of the physical variables is population growth, which, according to Randers' forecast, will peak in 2040, at just above eight billion. This is lower than most official UN estimates,[17] and the most important driver in Randers' forecast is that women will choose to have fewer children as urbanization increases. GDP will continue to grow, but not as rapidly as hitherto, and with a significantly different distribution from what we have seen to date. The big, emerging economies will continue to grow much faster (particularly China), and the US and Europe will remain relatively stable at around 2010 levels. The rest of the world "will stay unpleasantly near their current GDP per person" (Randers 2012: 77).

When it comes to energy use, Randers forecasts that use of fossil energy sources will peak around 2030, and that renewable energy use will increase throughout the period. This will lead among other things to lower climate intensity, a process that will increase in pace after 2030 because of the higher visibility of climate change impacts on the economy. Because of the peak in fossil use, emissions from energy will also peak around 2030. This, however, is not enough to meet the two-degree target. According

---

17 But there are other institutions that forecast along the same lines, see e.g. <http://www.businessinsider.com/deutsche-population-will-peak-in-2055-2013-9>.

Figure 16.2: Causal diagram for the 2052 study-changes from the World Model marked in red. **Source**: Jørgen Randers presentation material; he also granted permission to use this figure.

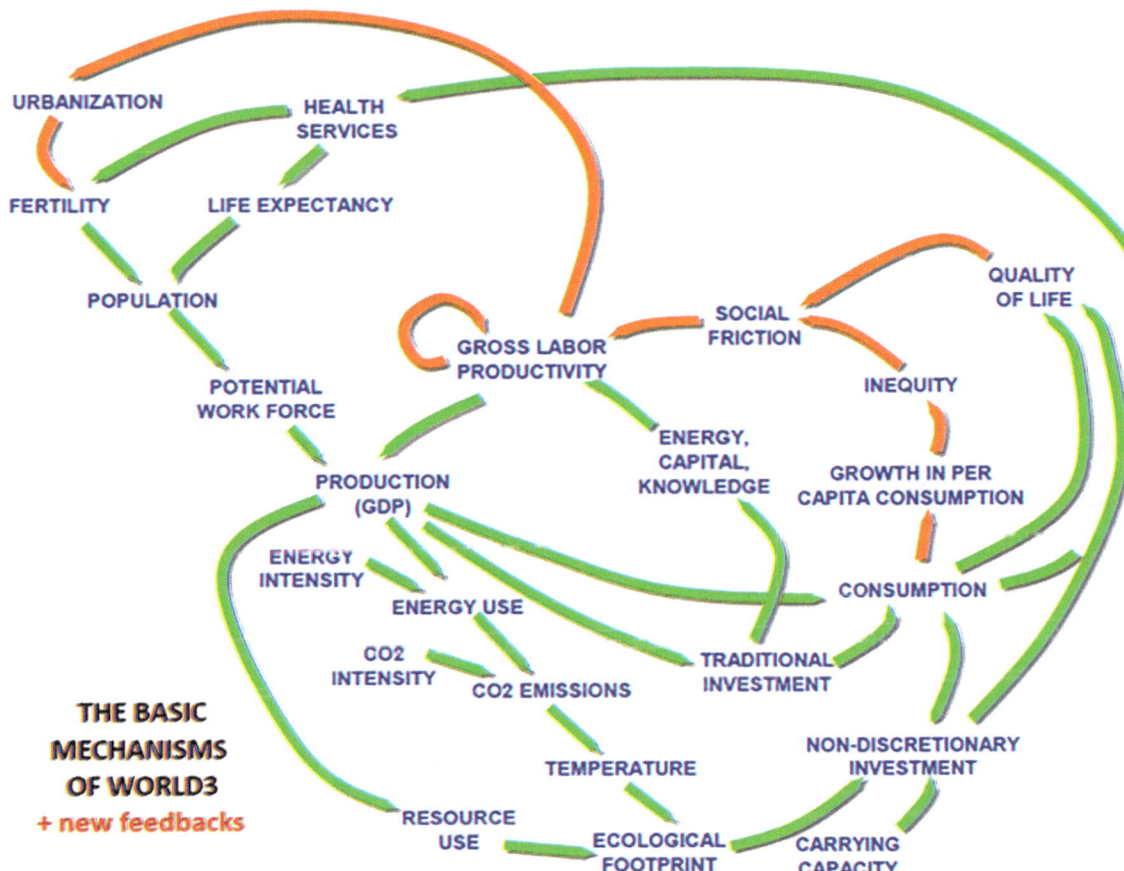

to this forecast, global warming will exceed two degrees compared to pre-industrial times around the middle of the twenty-first century.

Food production will increase and peak around 2040. Food per person will follow a similar curve, but the decline in population will mean that the 'food per person' curve will decrease slower than the 'food production' curve. "Grossly simplified, my forecast to 2052 says that there will be enough energy, grain, and chicken, plus some fish–with some exception for the poor" (Randers 2012: 143).

Important elements in the 'non-material future' include stronger government, and urbanization that leads to megacities and the omnipresence of the Internet. The 'zeitgeist' will, according to Randers, focus on fragmentation, local solutions, slum urbanization, and the Internet providing fertile ground for human creativity.

The general forecast is followed by forecasts for five regions (US, OECD minus US, China, BRISE[18] (the fourteen largest 'emerging' economies), and

ROW (the *rest of the world*). The maybe most striking feature of these regional forecasts is their diversity. This underlines the large differences that will be experienced in the different regions of the world–not only because 30-cm sea level rise will have very different impacts on society depending on whether one lives on a low Pacific island or on the top of a rugged cliff, but also because the forecast describes some states that will become stronger (but have to invest more to cope with the effects of climate change), and other states that will remain unstable, unable to provide poor inhabitants with basic security and resources.

The language and style in the 2052 book is even more personal and direct than in the LTG series. The big picture depicted in 2052 is a world that is less bad than the author had feared before starting the forecasting work. It is far from his 'ideal' world, and it is a world where people are poorer than he (together

18 Please explain BRISE here.

Figure 16.3: World State of Affairs, 1970–2050. **Source:** Randers (2012: 232). Permission to use this figure was granted by the author.

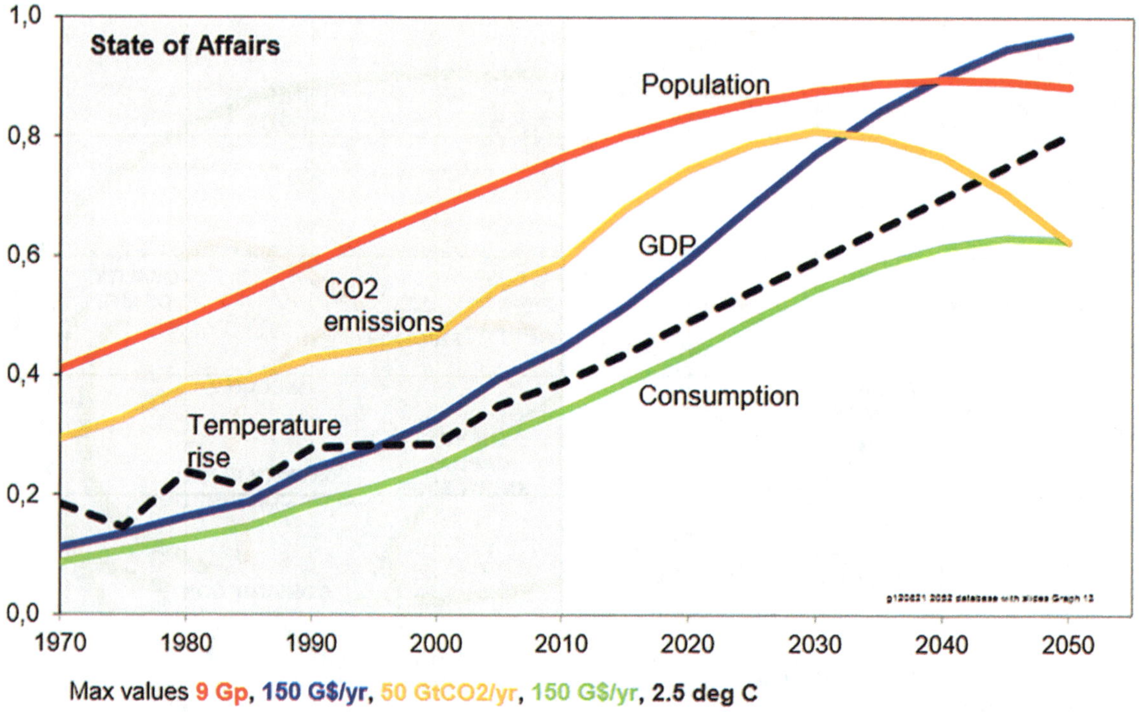

Max values **9 Gp**, **150 G$/yr**, **50 GtCO2/yr**, **150 G$/yr**, **2.5 deg C**

with many economists) originally expected, but it is also a world where no major collapse will occur before the end-date of the forecast.

Randers closes his book with a list of twenty 'pieces of personal advice'. It is impossible to quote them all here, so here is a selection:

- Focus on satisfaction rather than income
- Invest in great electronic entertainment and learn to prefer it
- Live in a place that is not overly exposed to climate change
- Stop believing that all growth is good
- Remember that your fossil-based assets—suddenly one day—will lose their value
- In politics, remember that the future will be dominated by physical limits.

'2052' is the work of someone who is clearly disappointed with the world and the lost opportunities for change over the last four decades. The real disappointment is that it has proved impossible to beat short-termism, both in economics and in politics. Solving the climate challenge will mean choosing the slightly more expensive solution in the short term, and making (and keeping) electoral promises about something further than one electoral cycle away does not ensure re-election.

16.4.1   Further Resources

In spring 2014 the book was translated into seven languages. The material and analysis is rendered accessible through a website dedicated to the project (<www.2052.info>), where readers can find some of the experts' glimpses into the future and make their own forecasts based on the model.

'2052' has formed the basis for a long series of lectures and presentations,[19] and has contributed to debate in an impressive number of countries around the world. Among the many themes covered in the debates are population growth and birth control policies, environmental degradation, the future of the energy sector, and short-termism and challenges for democracy.[20]

16.5   Concluding Remarks

In the introduction two questions were raised: firstly, why did the debate about the *Limits to Growth* become so polarized, and secondly what have we

19   See <http://www.2052.info/presentations/>.
20   See <http://www.2052.info/articles/>.

learned about the fundamental questions posed by the LTG team?

The answer to the first question can be found in the interests that felt threatened by the publication. Large industrial interests (in particular producers of fossil energy and other scarce resources) would clearly have a hard time believing that the fundamental credo and the resources from which they had created their wealth and large fortunes were to run out any time soon. Economists and politicians, who believed economic growth to be the panacea for increasing human welfare, found their world view questioned. And the population at large generally wanted and needed work and an income, and saw economic growth as the safest option for meeting these needs.

Criticism is often met with hostility. What now seems a relatively uncontroversial message about planetary boundaries was radical when it first appeared. The authors of the LTG were not prepared for the animosity their message would entail, and despite their attempt at framing their message as one of 'hope' because they presented 'possible futures', the necessary changes to bring the world to a sustainable development path were threatening for many of the established elites, economic, political and industrial.

We should also remember that the LTG was not the only environmental message that was met with hostility. The 1960s and the 1970s were decades of controversy and large societal conflict in many parts of the world, but particularly in the US.

Subsequent updates of the LTG study were also clearly presented in different lights, with more emphasis on 'hope' and 'love' (in 1992), and 'doom and gloom' (in 2004). It seems reasonable to assume that the authors chose different strategies because of their previous experiences, and because they fundamentally wanted to engage the world in a debate about where we are heading.

What has humankind learned over the last four decades?

Graham Turner has shown that the standard run of World3 seemed to be a rather good description of reality as observed thirty years after the scenario's publication.[21] Growth has continued, both in global population, production, and ecological footprint. Humankind is in a situation of overshoot with a current consumption of approximately 1.5 Earths. Every year twice as much CO2 is emitted as is being absorbed in oceans and forests, which will lead to temperature rise. Biodiversity loss and ocean acidification is happening.

The global governance system under the UN has not been able to produce a globally binding *greenhouse gas* (GHG) emission reduction treaty during the past forty years. Whether a comprehensive global agreement will occur in 2015 is still an open question, but it is highly unlikely that it will be sufficient for the world to meet its 2 °C target.

This review of the four books has shown that they have all concentrated on the development of the relatively uncontroversial 'physical' variables of human development and its effect on the world around us. Discussions about wider security and social development issues have been largely non-existent. This must be characterized as one of the major shortcomings of these books. The analysis and discussion about the effects of physical changes (lower food production, climate change, biodiversity loss), i.e. how they will be felt by humans and how they will impact on social structures, stability and security, are pivotal to our possibility of grasping the real challenges. In these books it is often assumed that resource shortages may lead to social unrest, but no closer analysis or causal mechanisms are discussed.

The 2052 book only partially puts right this issue. Discussions are included about the importance of focusing on quality of life ('happiness') rather than economic growth, but the analysis is rather superficial. This is also the case for political processes, which, to the extent that they are mentioned, are criticized for not delivering policies that bring about sustainable living. Democracy is particularly vulnerable, whereas the Chinese political system gets a bit more credit for delivering on the necessities of preventing the worst causes of pollution. There is no further discussion about whether or how the political process, democratic or otherwise, can contribute to solving the *social* problems resulting from climate change.

The story of *Limits to Growth* and its aftermath is at least two different stories. One is about how global governance has been unable to take sufficient action to prevent many of the challenges that were highlighted even as early as the early 1970s. This is a story that easily leaves us feeling helpless in face of humanity's destiny. But there is another story that is sometimes forgotten. That is the story about how one slim volume managed to incite debate, raise deep ethical questions, spur a search for deeper knowledge about the world in which we live, and remain relevant for several generations.

---

21 The real test on the LTG scenarios will come over the next couple of decades, when the scenarios start to deviate.

References

AtKisson, Alan, 22011: *Believing Cassandra. How to be an optimist in a pessimist's world. EarthScan* 2010 (London–Washington DC: Earthscan); at: <www.earthscan.co.uk/atkisson>.

Cole, Henry S.D.; Freeman, Christopher; Jahoda, Marie; Pavitt, Keith E.R. (Eds.), 1973: *Thinking about the Future: a Critique of the Limits to Growth* (London: Chatto and Windus for Sussex University Press).

Cole, Matthew A., 1999: "Limits to Growth, Sustainable Development and Environmental Kuznets Curves: An Examination of the Environmental Impact of Economic Development", in: *Sustainable Development*, 7: 87-97.

Constanza, Robert; Daly, Herman E., 1987: "Toward and Ecological Economics", in: *Ecological Modelling*, 38: 1-7.

Daly, Herman E., 1996: *Beyond Growth* (Boston: Beacon Press).

Ekins, Paul, 1993: "'Limits to Growth' and 'Sustainable Development': grappling with ecological realities", in: *Ecological Economics*, 8: 269-288.

Ford, Andrew, 22009: *Modelling the Environment* (Washington: Island Press).

Forrester, Jay, 1989: "The Beginning of System Dynamics". Banquet talk at the international meeting of the System Dynamics Society, Stuttgart, Germany, 13 July; at: <web.mit.edu/sysdyn/sd-intro/D-4165-1.pdf>.

Hall, C.A.S.; Day, J.W., 2009: "Revisiting the Limits to Growth after Peak Oil", in: *American Scientist*, 97: 230-237.

Meadows, Donella H.; Meadows, Dennis L.; Randers, Jorgen; Behrens III, William W., 1972: *The Limits to Growth* (New York: Universe Books).

Meadows, Donella H.; Meadows, Dennis L.; Randers, Jorgen, 1982: *Beyond the Limits* (White River Jct., Vt: Chelsea Green Publishing).

Meadows, Donella H.; Meadows, Dennis L.; Randers, Jorgen, 2004: *The Limits to Growth–the 30-Year Update* (White River Jct., Vt: Chelsea Green Publishing).

Nordhaus, William D., 1973: "World Dynamics: Measurement Without Data", in: *The Economic Journal*, 83,332: 1156-1183.

Nordhaus, William D., 1992: *Lethal Model 2: The Limits to Growth Revisited*, Brookings Papers on Economic Activity, 2 (Washington DC: Brookings Institution).

Randers, Jorgen, 2012: *2052–A Global Forecast for the Next 40 Years* (White River Jct., Vt: Chelsea Green Publishing).

Turner, Graham, 2008: "A comparison of *The Limits to Growth* with 30 years of reality", in: *Global Environmental Change*, 18,3: 397-411.

17 Preparing for Global Transition: Implications of the Work of the International Resource Panel

Mark Swilling

Abstract

The International Resource Panel (IRP) was established as an expert scientific panel by UNEP in 2007. By using a material flow analysis perspective, the primary focus of the IRP is on global resource use and potential alternatives. The notion of a 'third great transformation' was deployed to suggest that the work of the IRP is documenting the endgame of the industrial socio-metabolic regime. Three clusters of reports were addressed: (a) global resource perspectives, with special reference to decoupling rates of economic growth from rates of resource use by focusing on the importance of resource productivity; (b) nexus themes, including cities, food, trade, and GHG mitigation technologies; and (c) specific resource challenges with respect to two clusters of issues, namely metals and ecosystem services .The conclusion reached is that a resource-use perspective adds to our understanding of the unsustainability of the current global system, complementing the outputs of climate science on the effects of anthropogenic carbon emissions and ecosystem science on the implications of biodiversity degradation. To this extent, the work of the IRP anticipates the possibility of a more resource efficient socio-metabolic order. However, the IRP has to date not addressed specifically the dynamics and modalities of transition from an institutional and macro-economic perspective.

Keywords: Decoupling, transformation, sustainable development, sustainability science, material flow analysis, long-wave theory.

17.1 Introduction

This chapter will analytically review the work of the *International Resource Panel* (IRP) from the perspective of global transition theory.[1] It will be argued that the IRP can be understood as a collaborative effort by a diverse group of researchers to document the metabolic case for why the industrial epoch has effectively reached the end of its 250-year historical cycle. Although this documentary evidence suggests that the necessary conditions are in place for a transition to a more sustainable long-term development cycle, this by no means implies that the IRP has developed a view on whether sufficient conditions exist for such a transition to happen. Now that the Sustainable Development Goals have been approved, this may provide the context for such a task. The IRP has yet to pay attention to the key factors that will determine the nature of such a transition, namely social actors, their networks and the highly complex dynamics of the institutions that embody the intentions of organized historic and current socio-economic interests.

The *International Resource Panel* (IRP) was established by the *United Nations Environment Programme* (UNEP) in 2007 (see <http://www.unep.org/resourcepanel/>). It is currently co-chaired by Dr. Ernst Ulrich von Weiszäcker and Dr. Ashok Khosla and has twenty-four members from twenty-six countries. It is not constituted like the *Intergovernmental Panel on Climate Change* (IPCC) as an inter-*governmental* expert panel. Instead, it is a panel of experts funded by governments and UNEP that also has a Steering Committee made up of government representatives who consider the scientific reports of the Panel members, but without the requirement that reports must first be approved by the Steering Committee before they are published. The Steering Com-

---

1 The arguments presented in this chapter have been developed exclusively by the author and do not in any way reflect the views of the IRP, UNEP or individual members of the IRP.

© Springer International Publishing Switzerland 2016
H.G. Brauch et al. (eds.), *Handbook on Sustainability Transition and Sustainable Peace*,
Hexagon Series on Human and Environmental Security and Peace 10, DOI 10.1007/978-3-319-43884-9_17

mittee, however, does have the power to approve the initiation of reports. The Panel members come from a wide range of scientific disciplines and intellectual traditions, with some closely allied to their respective governments while others are thoroughly independent and even oppositional within their domestic policy environments.

The objectives of the International Resource Panel are to:

- provide independent, coherent and authoritative scientific assessments of policy relevance on the sustainable use of natural resources and their environmental impacts over the full life cycle;
- contribute to a better understanding of how to decouple economic growth rates from the rate of resource use and environmental degradation.

17.2   Contextualizing the Work of IRP

There is growing acceptance across a wide range of audiences that 'modern society' is currently facing historically unprecedented challenges. The advent of the 'Anthropocene' comes with an all-pervasive sense that landscape pressures like climate change, resource depletion and ecosystem breakdown threaten the conditions of existence of human life as we know it (Crutzen 2002). The onset of the global economic crisis in 2007/8 has resulted in a realization that we may have come to the end of the post-World War II long-term development cycle (Gore 2010; Swilling 2013b), and there is little understanding of what will come next. Simultaneously, there are those who argue that we may have reached a metabolic turning point that marks the endgame of the industrial era (Fischer-Kowalski 2011; German Advisory Council on Global Change 2011: 81; Haberl/Fisher-Kowalski/Krausmann et al. 2011). The result of these converging industrial and metabolic crises is an interregnum Edgar Morin has usefully called a 'polycrisis' (Morin 1999: 73).

Reflecting the thought patterns and influence of Schumpeterian long-wave theory (Foxon 2011; Freeman & Louca 2001; Köhler 2012; Perez 2002; Swilling 2013b), the *German Advisory Council on Global Change* (GACGC or WBGU) has argued that we should anticipate the third 'great transformation' comparable in its historical significance to the first two 'great transformations': the Neolithic revolution some 13,000 years ago and the industrial revolution some 250 years ago (German Advisory Council on Global Change 2011). Both can be defined as great transformations because they both resulted in funda-

mental shifts in the metabolic foundations of society: for the Neolithic transformation this entailed a shift to permanently occupied land, cultivated soils, harvested biomass, animal power, clay, rocks and the basic implements of pre-industrial agriculture; and then 250 years ago a shift to fossil fuels, metals, construction minerals and massive increases in biomass use and water use with the onset of the industrial revolution (Fischer-Kowalski/ Haberl 2007). For the GACGC, the third great transformation must be about radical decarbonization and resource efficiency to "provide wealth, stability and democracy within the planetary boundaries" (German Advisory Council on Global Change 2011: 81). However, all those who use long-wave theory recognize that these transitions are by no means linear and therefore cannot be easily predicted: they are highly complex processes that manifest differently across geographical scales and historical time. Key events can coalesce unexpectedly with accumulated macro-level structural shifts and the dynamics of conjunctural realignments to open up hitherto unlikely future trajectories.

It has been argued elsewhere that the year 2009 might be such a tipping point (Swilling 2013b): the collapse of Lehman Brothers at the end of 2008 was the key event, the conjunctural realignment was the ending of the post-World War II long-term development cycle (represented by the fact that 2009 was the first year since 1945 that the global economy shrank) and the structural shifts were reflected in the gradual realization that we could be breaching planetary boundaries in dangerous ways for human survival (marked by events such as the G20 adoption of the 'green economy' concept in 2009, following on from the awarding of the Nobel Prize to the IPCC, shock events like hurricane Katrina, heatwaves and economic shifts such as the rapid acceleration of investments in renewable energy technologies that were no longer confined to innovation niches).

Following Geels (2002), it is preferable to use the notion of 'transition' to describe these major metabolic shifts rather than the notion of 'transformation', on the understanding that it is specific transformations along the way that drive an overall transition. We can thus refer to a sustainability-oriented just transition as an alternative to the notion of a 'third great transformation' (Swilling/Annecke 2012), although these notions will be used interchangeably in this chapter.

Three conditions make this transition unique, only one of which is given sufficient emphasis by the GACGC Report. The first, recognized by the GACGC, is the fact that it is probably going to depend on the collective intent of specific constellations of actors who will need to collaborate at global, national and local levels. It is for this reason that the GACGC Report argues as follows:

"The imminent transition must gain momentum on the basis of the scientific findings and knowledge regarding the risks of continuation along the resource intensive development path based on fossil fuels, and shaped by policy-making to avoid the historical norm of a change in direction in response to crises and shocks" (German Advisory Council on Global Change 2011: 84).

This statement clearly defines the historic role of anticipatory science as key driver of the next great transformation (Poli 2014). This is why the work of the IRP, the IPCC, Future Earth and many other global scientific initiatives is significant. If they can contribute to the translation of anticipatory science into an anticipatory culture, then accumulated evidence about the risks we face and the potentials that can be exploited might just tilt the balance in favour of human survival (Poli 2014).

However, what needs greater emphasis is the implications of the *information and communication technology* (ICT) revolution that has transformed our organizational capabilities for learning and knowing—what Perez calls the fifth developmental surge that has taken place since the start of the industrial era (Perez 2009). Following Castells, just as the technologies of the combustion engine, electricity, telephony and mass production made possible the industrial revolution and its associated organizational arrangements (nation state, joint stock company, etc), so has the ICT revolution resulted in a new mode of organization, namely the network (Castells 2009). This new package of technologies and organizational modes has resulted in 'self-managed mass communications' with major implications for knowledge dissemination, innovation and collective action. For Brynjolfsson and McAffee, this provides the basis for the 'second machine age'—a new era of highly networked mutually interdependent activities that will unleash extraordinary creativity and productive potential (Brynjolfsson/McAffee 2014).

The internal informational recomposition of the industrial age that has taken place since the early 1970s has major implications for the anticipated sustainability-oriented transition. Following earlier work to address this question (Swilling 2013b), there is sufficient evidence that innovations made possible through network organization powered by ICTs will

be stimulated by the financial returns that can be made by repairing the future of the planet (see discussion of Decoupling 2 below). If the 'second machine age' makes possible low carbon and resource-intensive modes of consumption and production, then a key driver of the third great transformation begins to come into view.

The third condition that makes this transition unique is the fact that we are living in an *urbanized* Anthropocene. The majority of the world's population now lives in cities and the next 2-2.5 billion people who are expected to be living on the planet by 2050 when the population is expected to hit the 9-9.5 billion mark will land up in African and Asian cities. Demographic projections suggest that we will be constructing in cities over the four decades to 2050 more material infrastructures than what we have built over the past 400 years (Angel 2012). This harsh spatial reality has major implications for how networks coalesce to rapidly translate anticipatory science into fundamental changes in the geographies of everyday consuming and producing. What remains almost certain is that these networks will not only emerge from the networks of innovation-oriented cities that are flowering around the world, they will also be using the challenges faced by cities as the laboratories for testing the technologies that will be deployed during the next great transformation.

The co-evolutionary dynamics of anticipatory science, the network mode of organization and learning made possible by the ICT revolution, and the reconfiguration of spaces of agglomeration caused by accelerated urbanization create conditions that make it possible to consider the end of the industrial epoch and how the long-term sustainability-oriented transition—or third great transformation—could emerge (Swilling/Annecke 2012). This provides the context for understanding the enormous significance of the rapidly expanding body of work that has been generated by the IRP since 2007. At the simplest level, the IRP is providing documented evidence across a range of fields that it is no longer possible to conceive of a future for modern society that rests on the assumption that there are unlimited resources available for ensuring the well-being of over nine billion people on a finite planet by 2050. In other words, the IRP is documenting the end of the industrial socio-metabolic era and by implication anticipates a more sustainable era in the future. However, the IRP has also put in place within the global policy community a way of thinking that is different to the two other mainstream bodies of sustainability science, namely climate science and ecosystem science. By thinking of sociotechnical and socio-economic systems as socio-metabolic systems that consume, transform and dispose of resources extracted from natural systems, the IRP has put in place a key conceptual framework for imagining the dimensions and modalities of a more sustainable epoch. The notion that we need to decouple economic development and well-being from the rising rate of resource consumption is potentially a very radical idea, especially if this implies massive reductions in resource use per capita for people living in rich countries and a redefinition of development for those policy-makers in poorer countries committed to poverty eradication. How we build a more equitable world of over nine billion people by 2050 without destroying the planet will not only depend on the mainstreaming of an appropriate political economy to replace neo-liberalism (Picketty 2014); it will also mean that a political economy that is appropriate for imagining the next great transformation will have to recognize that it will be necessary to fundamentally reconfigure the flow of non-renewable and renewable resources through our sociotechnical and socio-economic systems. When read together from the transition perspective outlined here, the research assessments generated by the IRP since 2007 provide a significant starting point and partial foundation for achieving such a synthesis.

17.3   Overview of the Work of the IRP

Unlike the IPCC, the IRP does not produce an integrated report at specific points in time. Instead, the IRP publishes reports as and when they have been produced by one or more members of the IRP and their respective research teams. This means there is no integrated synthesis of the IRP's body of knowledge. For the purposes of this chapter, the work of the panel has been divided into the following categories:[2]

- *global resource perspectives*, with special reference to decoupling rates of economic growth from rates of resource use by focusing on the importance of resource productivity (Decoupling 1 and Decoupling 2), the environmental impacts of products and materials, and the beginnings of scenario thinking;

- *nexus themes*, including cities, food, trade, and *greenhouse gas* (GHG) mitigation technologies;
- *specific resource challenges* with respect to two clusters of issues, namely metals (both stocks in use and recycling) and ecosystem services (including water and land-use/soils).[3]

The global resource perspectives define the IRP's commitment to focus on the resource inputs into the global economy, and therefore how future economic trajectories (whether growth-oriented or not) can be decoupled from the prevailing rising level of resource use over time. Without this kind of decoupling, a transition will be unlikely. Nexus themes are about specific spheres of action constituted by highly complex sociotechnical systems where the potential for decoupling exists. Specific resource challenges are about resource regimes that are both under threat from, for example, rising demand and prices, and can also be potential threats to larger systems that are dependent on them.

17.4   Global Resource Perspectives

The environmental science of pollution, climate science and ecosystem science have traditionally been the three underlying bodies of science that have supported the claims of the environmental movement. In recent years material flow analysis has emerged as the fourth body of science, with roots in industrial ecology, resource economics and political economy (Fischer-Kowalski 1998, 1999). Major historical reinterpretations of agricultural and industrial economic transitions have now been written that are clearly extremely useful for anticipating the dynamics of future transitions (Fischer-Kowalski/Haberl 2007; Giampietro/Mayumi/Sorman 2012; Smil 2014). The focus has shifted from the negative environmental impacts of the outputs of industrial processes to the material inputs into a global economy that depends on a finite set of material resources. This is the discursive framework within which the work of the IRP should be located.

One of the first reports produced by the IRP (generally referred to as 'Decoupling 1') entitled *Decoupling Natural Resource Use and Environmental Impacts from Economic Growth* presented evidence on the use of four categories of resources: biomass (everything from agricultural products, to clothing

---

2   All completed reports referred to below are available on the IRP website at: <http://www.unep.org/ resource-panel/>.

3   The IRP has also produced reports on biofuels and forests, but these have not been included in this analysis.

Figure 17.1: Global metabolic rates and incomes (1900–2005), and income. **Source:** Fischer-Kowalski and Swilling (2011: 12).

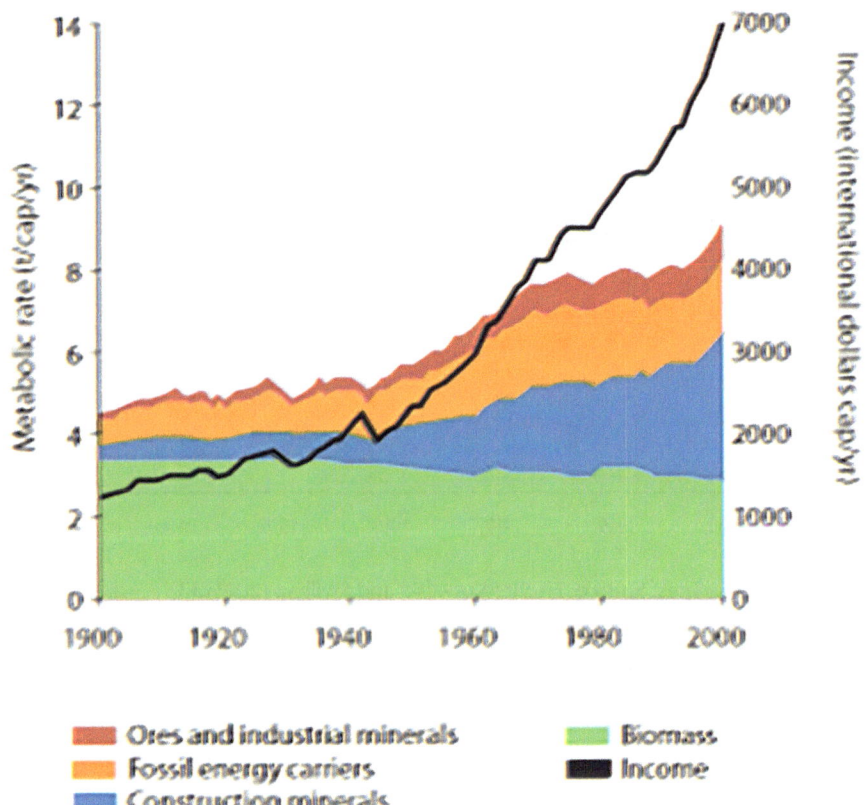

material like cotton, to forest products), fossil fuels (oil, coal and gas), construction minerals (cement, building sand, etc.) and ores and industrial minerals (Fischer-Kowalski/Swilling 2011). The Decoupling 1 Report shows that by the start of the twenty-first century the global economy consumed between 47 and 59 billion metric tons of resources (which is equal to half what is physically extracted from the crust of the earth). Between 1900 and 2005 total material extraction increased over this period by a factor of 8, while GDP increased by a factor of 23 for the same period. As reflected in figure 17.1, the result is relative decoupling between rates of resource use and global growth rates.

As the Decoupling 1 Report shows, rising global resource use during the course of the twentieth century (including the metabolic shift that took place from mid-century onwards as non-renewables grew and dependence on renewable biomass declined in relative terms) corresponded with declining real resource prices—a trend that came to an end in 2000–2002. Since 2000–2002, the macro-trend in real

resource prices has been upwards (notwithstanding dips along the way).

The McKinsey Global Institute report (which was published after the IRP report) generally confirms the trends identified by the Decoupling 1 Report, demonstrating that resource prices have increased by 147 per cent in the decade since 2000. As a result investments in resource productivity over the long-term can generate returns of 10 per cent, more if the US$1.1 trillion "resource subsidies" are removed (McKinsey Global Institute 2011).

A key conclusion of the Decoupling 1 Report is that a transition to a more sustainable global economy will depend on absolute resource reduction in the developed world, and relative decoupling of economic growth rates from rates of resource use in the developing world. If this is not achieved, the Report argues, the result may well be an increase in total resource use from 60 *billion tonnes* (Bt) in 2005 to 140 Bt by 2050 if all nine billion people living on the planet by then consume the equivalent of the average European (i.e. 16 *tonnes per annum per capita* (t/cap), which is half what the average American con-

Figure 17.2: Composite resource price index (at constant prices, 1900–2000). **Source**: United States Geological Survey data cited in Fischer-Kowalski and Swilling (2011: 13).

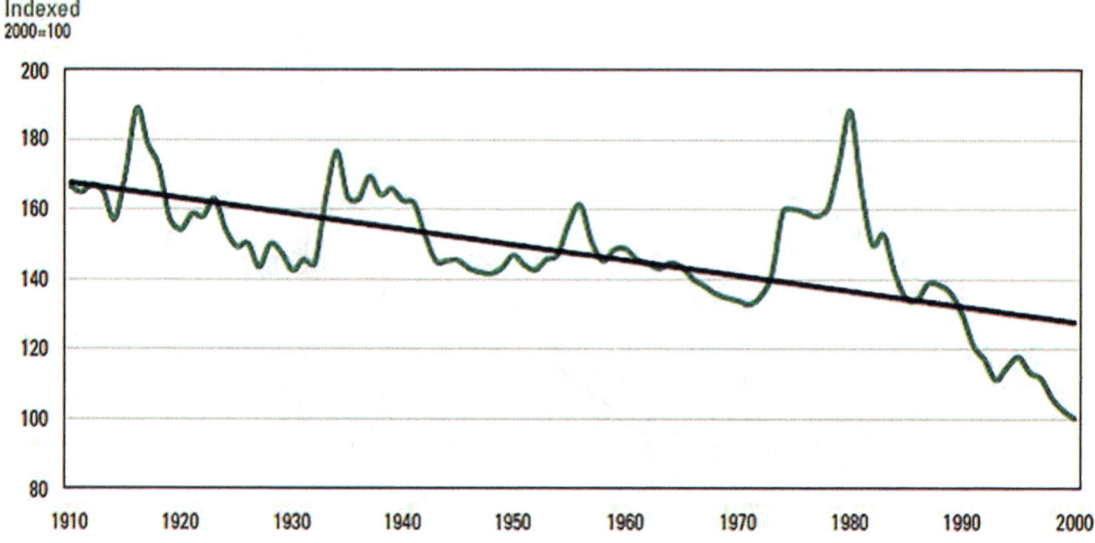

Figure 17.3: Commodity price indices. **Source**: Fischer-Kowalski and Swilling (2011: 13).

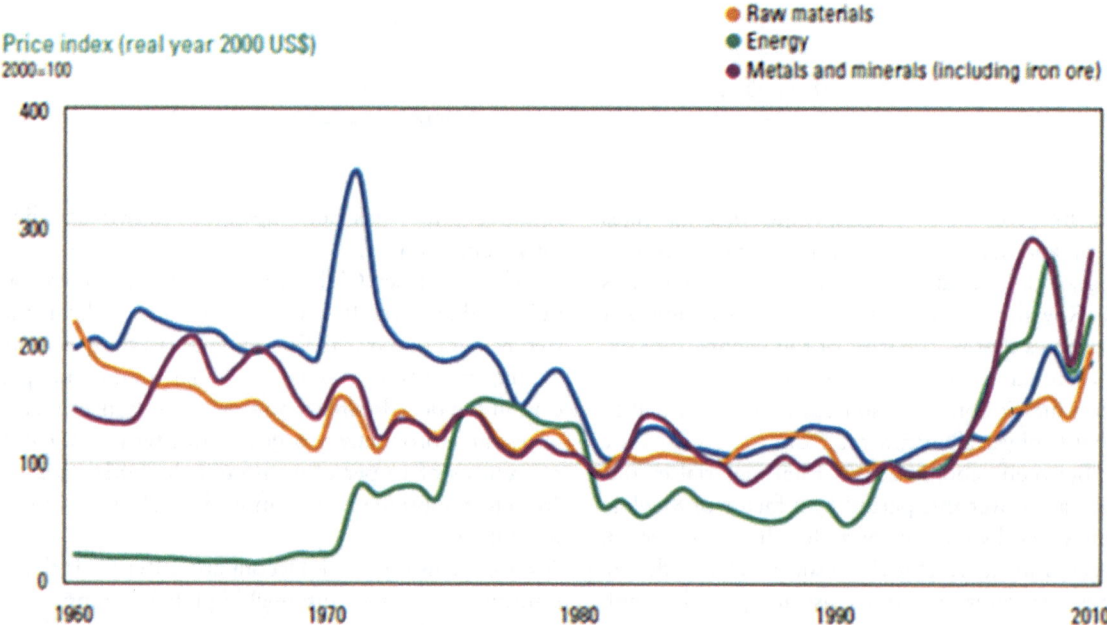

sumes). However, if the convergence point is 8 t/cap, the total material requirement would be 70 Bt by 2050 on a planet of nine billion people. The Decoupling 1 Report suggests that the material equivalent of living in ways that will result in the emission of 2 tonnes of CO_2 per annum per capita by 2050 on a planet of nine billion people (as recommended by the IPCC) may well be 60 Bt or 6 t/cap for everyone. Although the latter is the logical consequence of the science of

the IPCC that all countries approved, it implies a 'great transformation' equal in significance to the metabolic transformations that resulted in the Neolithic and Industrial Revolutions.

Reinforcing the argument of the Decoupling 1 Report, another early IRP report entitled *Assessing the Environmental Impacts of Consumption and Production: Priority Products and Materials* (referred to as the Priority Materials Report) addressed key ques-

Figure 17.4: Major contributors to global GHG emissions, including land use and land cover change (measures in CO_2 eqivalents using a 100 year global warming potential).**Source**: UNEP (2010).

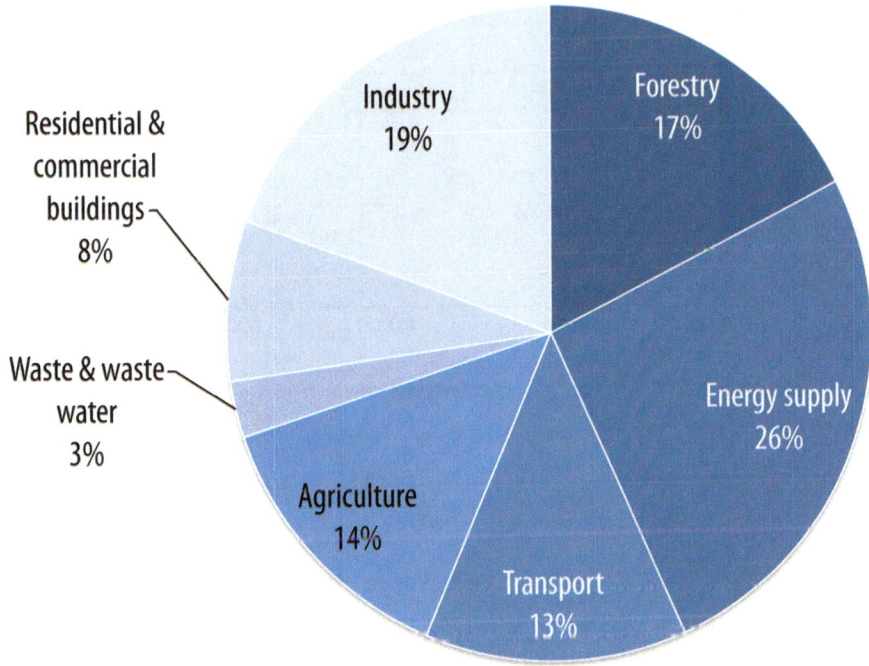

Figure 17.5: Distribution of energy use across consumption categories, as identified in different studies, and total energy use measures in kW per capita. **Source**: UNEP (2010).

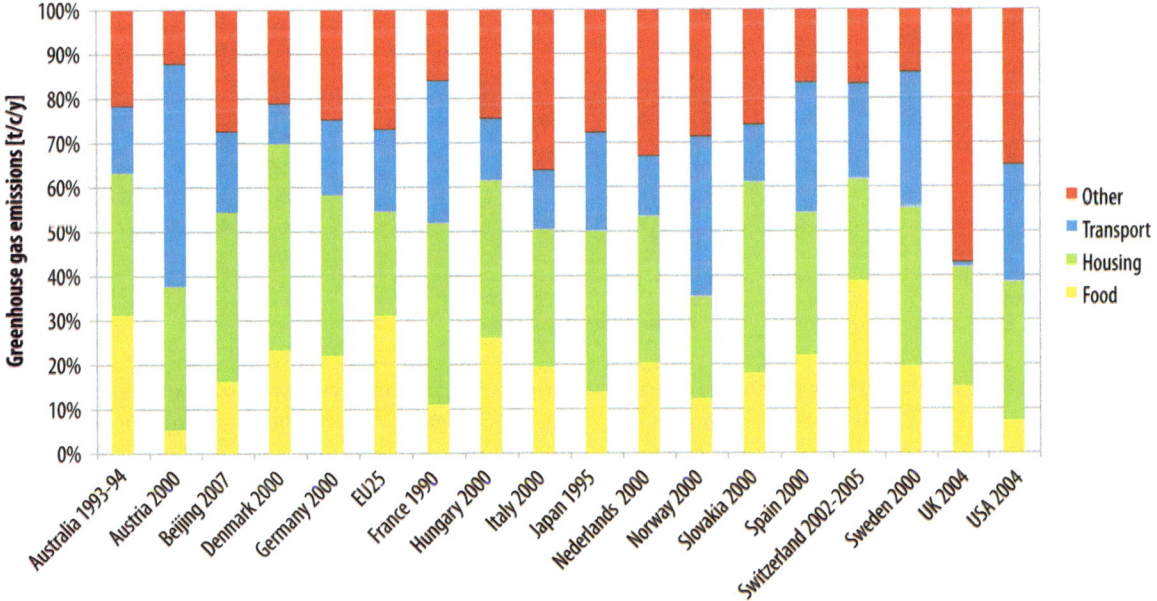

tions of relevance to this review, only three of which are addressed here: Which industries are the most responsible for contributing to environmental and resource pressures? What products and services have the greatest environmental impacts? Which materials

have the greatest environmental impacts across their respective life cycles? (UNEP 2010)

As indicated in figure 17.4 and related information in the Priority Materials Report, the energy industry, followed by industry and forestry (through deforesta-

Figure 17.6: Carbon Footprint of different consumption categories (tonnes of CO₂ Equivalents per capita in 2001) in 87 countries/regions as a function of expenditure ($ per capita). **Source:** UNEP (2010).

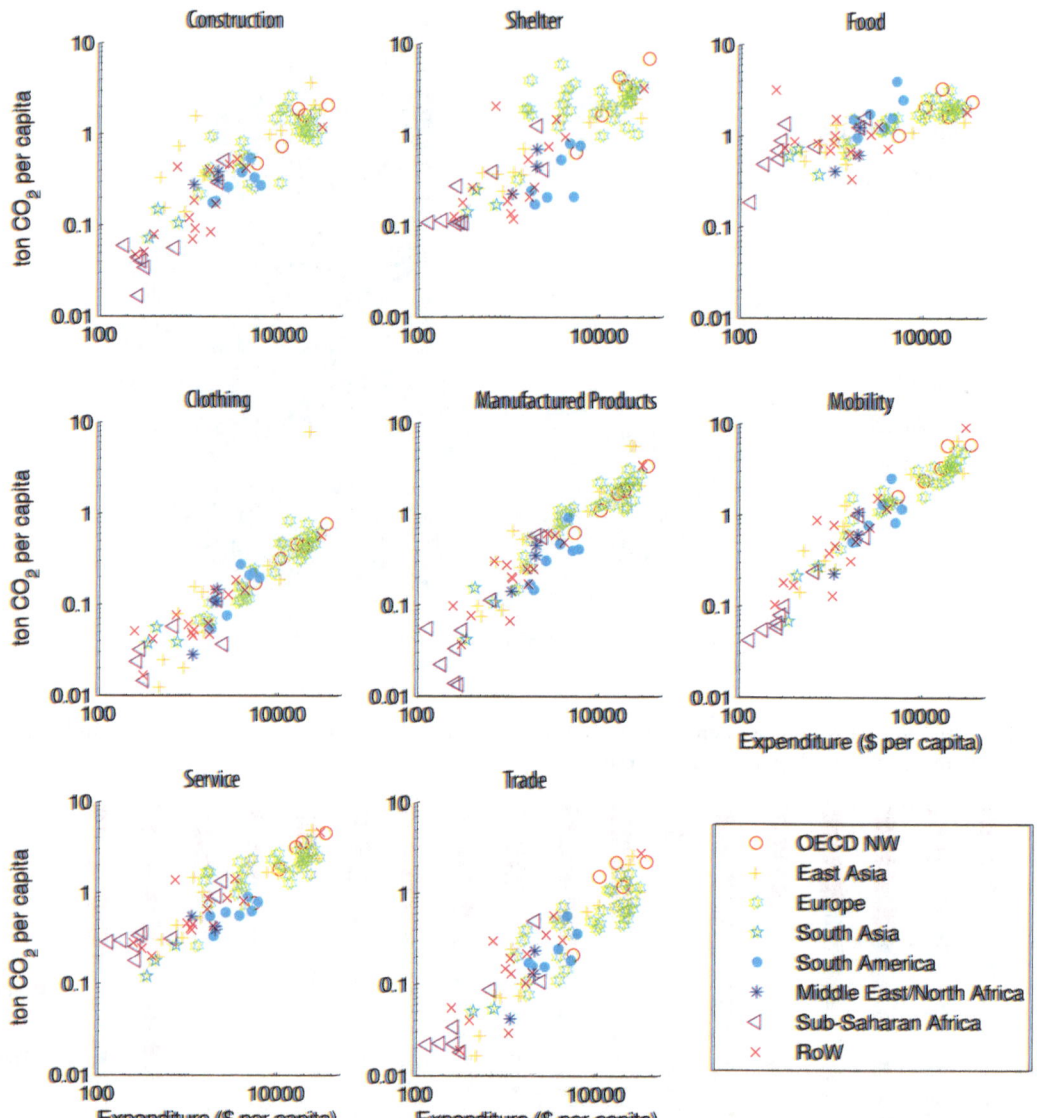

tion), are the greatest contributors to climate change, abiotic resource depletion, and sometimes eutrophication, acidification and toxicity.

As far as consumption is concerned, the Report shows that transport, housing and food are responsible for 60 per cent of all impacts (see figure 17.5). Given that these are overwhelmingly configured by urban systems, this prioritization reinforces the argument of the City-Level Decoupling Report (discussed later) that interventions that address these priorities will have to take into account their spatial contextuality.

However, even more important is the unsurprising fact that as incomes go up so do the enviromental

impacts. Figure 17.6 clearly shows that there is no decoupling when it comes to rising incomes and related environmental impacts.

As far as the environmental impacts of materials are concerned, the Priority Materials Report shows that animal products, fossil fuels and key metals (iron, steel and aluminium) have the greatest impacts. However, only integrated data for Europe exists.

The first bar in figure 17.7 shows resource use in kilograms per capita, while the second and third show the environmental impacts weighted according to their environmental impacts over the life cycle (the so-called *environmentally-weighted material consumption*, EMC). The second gives the EMC for the *global*

Figure 17.7: Normalized global warming potential of material flows and *Environmentally weighted Material Consumption* (EMC) for EU-27+1 region. **Source**: UNEP (2010).

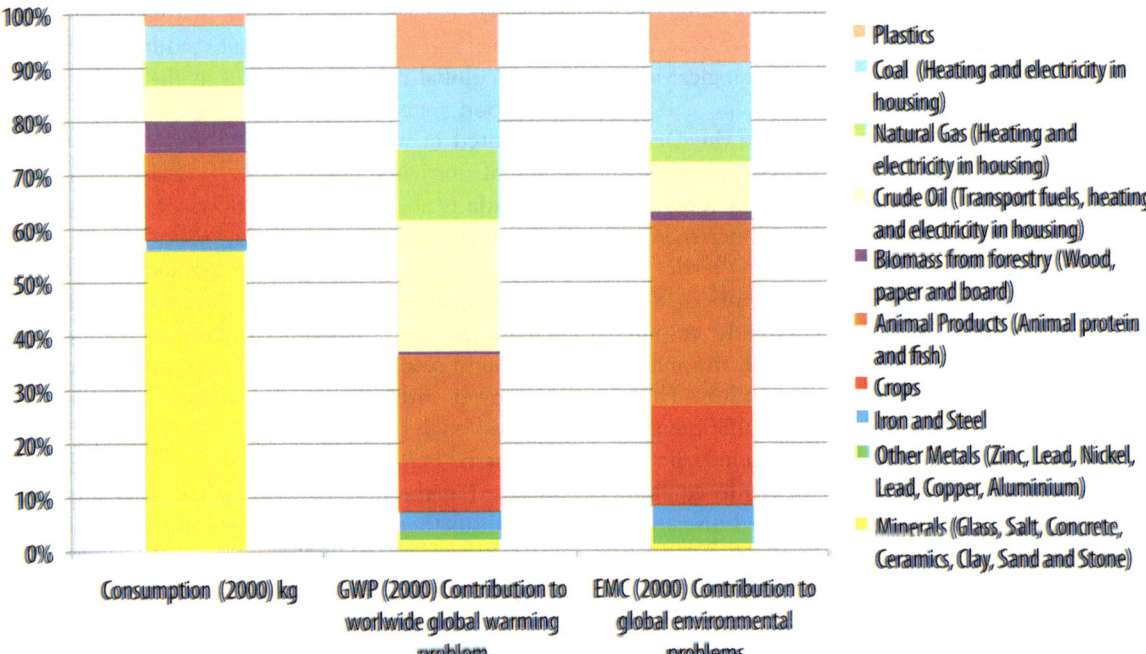

warming potential (GWP), while the third adds up a large number of environmental impact categories such as global warming, acidification, land use competition plus others with equal weightings.

The *Priority Materials Report* concludes that future economic growth and development on a business-as-usual basis will exacerbate these trends. The impact of fossil fuels and agricultural activities are seen as the top two priorities that must be addressed if a transition to a more sustainable order is to be achieved.

In a follow-up to the Decoupling 1 Report, the IRP report entitled *Decoupling 2: Technologies Opportunities and Policy Options* (launched at Green Week, Brussels, in June 2014) argued that there are three types of decoupling (UNEP 2014):

- decoupling through maturation: found mainly in developed countries, this is a natural process caused by saturated demand, levelling off or even decline of populations, minimal new construction and a shift towards services;
- decoupling through burden shifting to other countries: by off-shoring the resource extraction and related impacts to other countries and then excluding this reality from material use calculations, it is possible for many countries to create the appearance of decoupling—in reality, as recent research has shown, if the ecological rucksacks are

attributed to the consumer and not producer, this apparent decoupling disappears (Wiedman/ Schandl/Lenzen et al. 2013);

- decoupling by intentionally improving resource productivity: as a "paradigm shift", this type of decoupling "requires technological and institutional innovations, resource-efficient infrastructure, low-material-intensity manufacturing, public awareness and appropriate attitudes and behaviours" (UNEP 2014:5).

The Decoupling 2 Report demonstrated that since 2000 metal prices have risen by 176%, rubber by 350%, energy by 260% and food by 22.4% (with some projecting an increase for food of 120–180% by 2030). Demand for water by 2030 is expected to have risen by 40%, exceeding existing capacity by 60%. Possibly even more important than price increases is price volatility and related supply shocks (UNEP 2014). The Decoupling 2 Report documents a wide range of emerging alternatives that are made possible by these price increases and argues the case for replicating radical resource productivity improvement on a global scale. Many examples are provided, including the potential to reduce energy and water demand in developed economies by 50–80% using existing energy and water efficiency technologies; how developing countries investing in new energy infrastructure could reduce energy demand by half over the next twelve years if energy efficiency and renewable energy technologies were adopted now rather than later; and that decoupling technologies could result in resource savings equal to US$2.9 to 3.7 trillion each year until 2030 if the policy, regulatory and technological innovations are put in place (UNEP 2014).

The most significant contribution of the Decoupling 2 Report is the suggestion that radical resource productivity can be achieved by introducing a resource tax system that is used to gradually and incrementally increase real resource prices over the long term. This tax could be used to ameliorate rising prices when these occur, and to counteract declining resource prices when these occur, thus providing the market with a level of certainty over the long term. This is crucial for counteracting what is inevitably going to happen if nothing of this kind is done, namely increasing price volatility that will tend to reinforce short-term investment perspectives with limited investment in innovation. Long-term innovation-driven investments will not thrive if prices remain volatile.

Informed by the thinking reflected in the above Reports and what has emerged in the nexus and specific resource challenge reports, during the course of 2013 and 2014 IRP member Professor Tom Graedel led a group that has started to consider the formulation of future scenarios. The group has decided to adopt the GEO 4 scenarios that will be used to frame future resource use storylines. These scenarios are as follows:

- markets first—a "business-as-usual" scenario;
- policy first—a "make the world greener" scenario;
- security first—an isolationist scenario of rising social, environmental, and economic tension;
- sustainability first—a new environment and development paradigm with more equitable outcomes and institutions that is, in effect, consistent with GACGC's 'third great transformation'.

If the IRP succeeds in integrating the various strands of its work into some credible scenarios using the GEO4 (or similarly well-accepted) storylines, this will contribute a new perspective to the already cluttered scenario-building scene. More importantly, it will help to bring the IRP's research firmly into the realm of the emerging science of anticipation (Poli 2014).

17.5   Nexus Themes

Each nexus theme can be defined as a complex of interrelated resource use and environmental impact issues that can be analysed by reference to a particular cross-cutting process. Cities concentrate in space in particular context-specific ways all the resource use and environmental impact issues addressed by the general global reports discussed in the previous section. Food systems are globally, regionally and locally constituted in ways that connect incredibly complex flows of nutrients, energy, water, wastes and materials. Trade is about the global flows of resources and their associated ecological rucksacks that can be attributed to the producing or consuming countries with drastically differing results. And GHG mitigation technologies are massive composites that require energy and resource inputs that are intended to produce lower carbon and more resource-efficient outputs. Although the IRP's work on these nexus themes is discussed below, only the report on cities had been published at the time of writing. This may mean the other nexus themes are dealt with more superficially.

17.5.1   City-Level Decoupling

The main aim of the Cities Working Group of the IRP is to apply the insights generated by the new literature on urban metabolism (Barles 2009; Barles 2010; Costa/Marchettini/Facchini, 2004; Farrao/Fernandez 2013; Heynen/Kaika/Swyngedouw 2006; Kennedy/Pincetl/ Bunje 2011; Ramaswami/Weible/Main et al. 2012; for the most comprehensive overview of the literature see Robinson/Musango/Swilling et al. 2013; Swilling/Robinson/ Marvin et al. 2013; Swyngedouw 2006; Weisz/Steinberger 2010) to the challenge of designing, building and operating more sustainable urban infrastructures.

The first urbanization wave took place between 1750 and 1950, and resulted in the urbanization of about 400 million people in what is now the developed world. The second urbanization wave between 1950 and 2030 is expected to result in the urbanization of close to four billion people in the developing world in less than a century. By 2007 just over fifty per cent of the global population lived in cities. Hence we should be talking about the 'urban Anthropocene'.

Based on UN population data for 1950-2050 (Department of Economic and Social Affairs, United Nations 2012), the total global urban population is expected to increase from 3.5 billion in 2010 (of which 73 per cent were in cities in developing countries) to 7.3 billion in 2050 (by which time 83 per cent will be living in cities in developing countries). This means that by 2010 the global process of urbanization that began in earnest in 1800 (see figure 17.8) had only

Figure 17.8: World Population and Urban Growth Trends (0-2010). **Source:** Angel (2012).

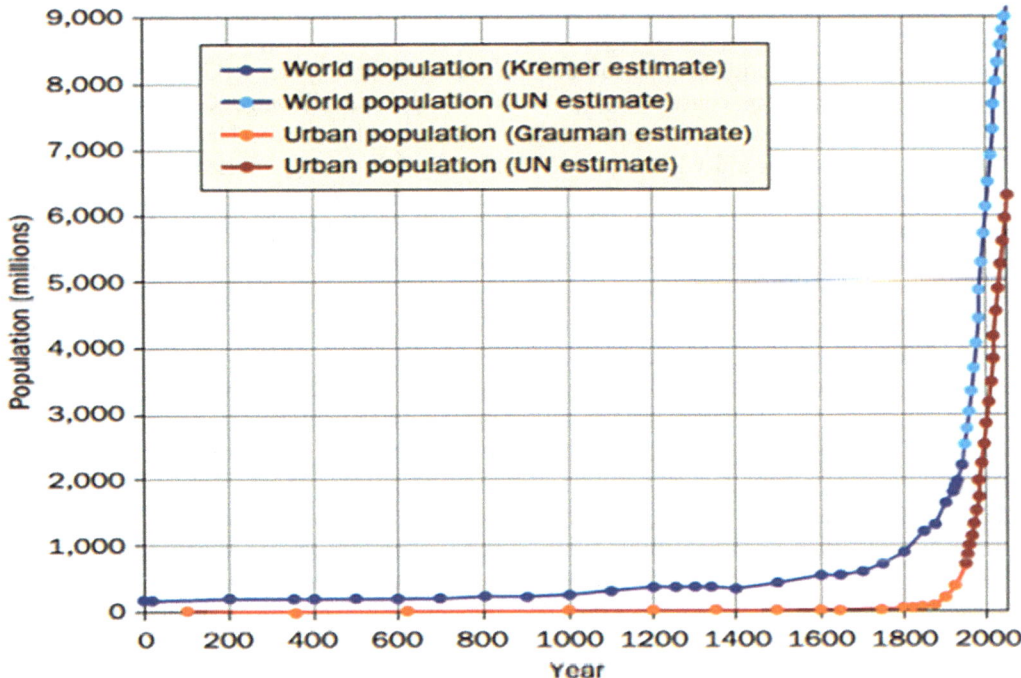

resulted in the urbanization of 48 per cent of households that are expected to live in cities by 2050.

Furthermore, according to the groundbreaking UN Habitat report *Challenge of Slums* (United Nations Centre for Human Settlements 2003), of the 3.5 billion who were living in cities by 2010, one billion lived in slums. In other words, 210 years of urbanization had created a decent quality of life for only two-thirds of all urban dwellers. Resolving this problem must, therefore, be seen as integral to a just urban transition by 2050.

It follows, therefore, that 52 per cent of the urban fabric that is expected to exist by 2050 must still happen over the four decades to 2050. The significant proportion of the additional urban population of nearly four billion people will end up in developing country cities, in particular Asian and African cities. If we include the one billion people who live in slums, then it follows that material infrastructures of one kind or another will need to be assembled for an additional five billion new urban dwellers by 2050.

This raises an obvious and vitally important question: what will the resource requirements of future urbanization be if business-as-usual sociotechnical systems are deployed to assemble built environments? What are the resource implications of more sustainable sociotechnical systems? Unfortunately, no one has attempted to answer these questions yet, which is why the IRP decided at its meeting Santiago in May 2014

to approve a study by its Cities Working Group that aims to address these questions.

This new study will build on the first IRP Report on cities entitled *City-Level Decoupling: Urban Resource Flows and the Governance of Infrastructure Transitions* (referred to as the City Decoupling Report) (Swilling/Robinson/Marvin et al. 2013). Noting the proliferation of reports on sustainability in cities, the core argument of this Report was that insufficient attention was paid in these reports and in the academic literature to the strategic importance of urban infrastructures. It was noted that urban infrastructures conduct resource flows (e.g. energy, waste, water and sanitation) through urban systems. It follows, therefore, that in order to transition from linear unsustainable urban metabolisms to more circular sustainable urban metabolisms it will be necessary to reconfigure urban infrastructures.

Thirty case studies of urban infrastructure transitions across all world regions were documented in order to demonstrate that there is plenty of evidence that various initiatives are under way. Furthermore, it was noted that intermediaries play a crucial role as facilitators of change. *Global technology companies* (GTCs) have emerged as new major players within the urban policy space. Companies like IBM, Siemens, Cisco, Altech and others have all mounted 'smart city' programmes, with Songdo city in South Korea the poster child for what this means in practice

(Kuecker 2013). The last time public discourse was inundated with visions of grand city-wide transformations was in the late 1800s with respect to sanitation infrastructure and in the post-World War I period during the lead-up to the highway construction programme that transformed cities around the world (Hajer/Dassen 2014). The result today is the promotion of massive global escalations in investments in urban infrastructures in cities in both developed and developing countries, sometimes but not always within a 'smart city' framework (Airoldi/Biscarini/Saracina 2010; Doshi/Schulam/Gabaldon 2007; Siemens/ PWC/Berwin Leighton Paisne 2014; World Business Council for Sustainable Development 2014). The algorithmic urbanism promoted by the GTCs holds many dangers, including the consolidation of new wealth-based digital divides and greenwashing (Hajer 2014; Luque/ Marvin/ McFarlane 2013).

Unfortunately, very little data were available about the metabolic flows before and after a given intervention. The City Decoupling Report therefore concluded that in order to assess whether urban infrastructure innovations do, indeed, result in more sustainable outcomes it would be necessary in future to use the tools of urban metabolic flow analysis in combination with systems dynamics modelling. These tools will make it possible to evaluate and, therefore, model the effect of a given set of sociotechnical innovations at two levels: at the level of design with respect to a given sectorial intervention (e.g. a public transit system or sewage treatment plant) and at the level of city-wide planning with respect to the overall resource productivity of the entire urban system.

Given that over fifty per cent of the global population now lives in cities and given the extent of future urban growth, retrofitting and building new urban infrastructures that reproduce sustainable urban metabolisms may well be the single most important driver of the third great transformation.

17.5.2   GHG Mitigation Technologies

Given that the energy transition is going to be the most important driver of the third great transformation, it follows that more needs to be known about the environmental implications of the suite of renewable energy technologies that are regarded as the cornerstone of this transition. In a draft highly-detailed 500-page report entitled *Green Energy Choices: The Benefits, Risks and Trade-offs of Low-Carbon Technologies for Electricity Production* (referred to as the Green Energy Report) the following technologies

were assessed using Life Cycle Assessment: wind power, hydropower, *photovoltaics* (PV), *concentrated solar power* (CSP), geothermal power, *natural gas combined cycle power* (GCCP) with and without CO_2 *capture and storage* (CCS), and coal-fired power with and without CCS (Hertwich/Gibon/Arveson et al. 2015).[4] Bioenergy, nuclear energy, oil-fired power plants and ocean energy were not assessed.

The Green Energy Report found that wind, PV, CSP, hydro and geothermal power generate GHG emissions over the life cycle of less than 50gCO2e/ kWh. This compares favourably to coal-fired power plants that generate 800-1,000gCO2e/kWh over the life cycle and GCCP (without CCS) that generates 500-600gCO2e/kWh over the life cycle. CCS can reduce emissions of fossil power plants by only 200-300gCO2e/kWh. As far as pollution and related health impacts are concerned, renewables reduce impacts by seventy to ninety per cent. Similarly, impacts of renewables on ecosystems are a factor of three to ten lower than fossil power plants (Hertwich/Gibon/Arveson et al. 2015).

By contrast, the Report shows, a global transition to renewables (with some GCCP for peak loading and some coal power plants) would require an increased use of steel, cement and copper in comparison to the continuation of the business-as-usual fossil-fuel-based system. Furthermore, renewables depend on various rare earth metals like indium and tellurium, as well as silver (Hertwich/Gibon/Arveson et al. 2015). There is no consensus in the literature on the security of supply of these materials. However, their concentration in China is well known.

In short, from a purely technical perspective (which of course ignores institutional change, financing and learning) the environmental impacts of renewables are substantially reduced compared to fossil-fuel-based energy supply. However, the resource inputs with respect to steel, cement and copper may be greater if alternative technologies for these aspects of the clean energy infrastructure are not found. Increased requirements of bulk materials such as steel, cement and copper can easily be met with current production rates.

17.5.3   Food Systems

As argued by the *Priority Materials* Report, the food system is a major user of resources and a major con-

---

4   Note that at the time of writing this report had not gone through the UNEP peer-review process.

Figure 17.9: Conceptual Framework: Food Systems and Natural Resources. **Source**: Hajer, Westhoek, Ozay et al. (2015: 12.

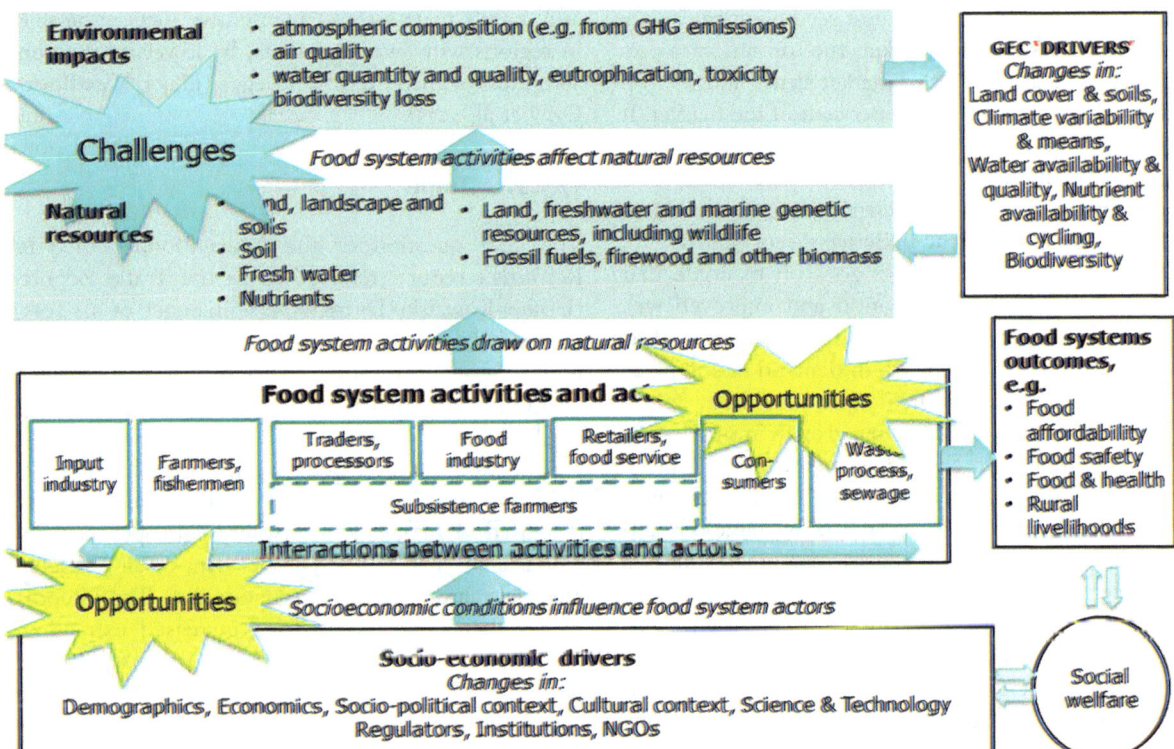

tributor to negative environmental impacts. Food systems are highly complex global-local systems that are currently in deep crisis as several long-term megatrends accumulate into a perfect storm. Breaking from the dominant tendency to see food insecurity as a problem of production, the Food Working Group of the IRP adopts a food system perspective that, in turn, makes it possible to see food insecurity as a direct and persistent symptom of a flawed global food system (Hajer/Westhoek/Ozay et al. 2015). Food security is defined as a situation where all people, at all times, have physical, social and economic access to sufficient, safe and nutritious food which meets their dietary needs and food preferences for an active and healthy life (Food and Agriculture Organization 1996). Considered in terms of the distribution of dietary energy supply, 868 million people around the world were considered chronically undernourished in 2013 (FAO 2013: ix). In addition, a further two billion people experienced the negative health consequences of micronutrient deficiencies (FAO et al. 2012: 4). About 850 million of the people estimated to be undernourished live in developing countries (FAO A/B et al. 2012: 8). Food insecurity is one of the key indicators of a system incapable of responding to the pressures

that it faces. The capacity of the food system to ensure food availability and thus food security is shaped by a wide variety of factors, but the increase in population, urbanization and improved welfare are the most important (Food and Agriculture Organization 2013: ix) drivers of food system change. The conceptual framework captured in figure 17.9 represents this complex set of actors and networks.

The IRP's report on the food system essentially mounts the following argument (Hajer/Westhoek/Ozay et al. 2015). The global food system is now dominated by large-scale modern systems that have replaced localized family-farm-based food economies with large-scale globalized processing and retail activities, long value chains, regulatory standards and transnational companies. One result of neo-liberalization since the 1980s has been the shift in food governance from largely localized upstream governance systems to the big global downstream players, in particular the food processors and retailers. The result is that the food system is now primarily configured for short-term profit rather than the long-term continuity of farming systems and the ecosystems they used to depend on. Global and national governance systems tend to reinforce this orientation because it is perceived to be more 'efficient'. As a result concentration in the off-farm sectors of the food value chain are high and rising: the three largest seed companies control 50 per cent of the market; the top ten agro-processors have 28 per cent of market share; and the top ten food retailers control 10 per cent of the market. It is this shift in power that is contested by the agro-ecological movements who want a return to local food bio-economies where sufficiency ensures the long-term sustainability of the underlying ecosystems.

The Food System Report goes on to argue that population growth, urbanization and improved welfare imply a ten per cent increase in food demand by 2025, with the fastest growth in demand taking place where logistical infrastructures are weakest. Given that urbanization and economic growth in developing countries implies an expanding middle class, a nutrition transition is under way from calorie-rich diets (cereals) to energy-rich diets (meat, vegetable oils and sugars). Energy-rich food requires far greater natural resource inputs, including the fact that instead of being consumed directly by humans grains are used as inputs for livestock production. This, in turn, increases the demand for land for cereal production and grazing. Furthermore, now that supermarkets have become the dominant food delivery systems in all regions where middle-class consumers are significant, energy-rich food is transported over longer distances, requires more packaging and depends on vast globally structured networks of interconnected specialized companies and value chains. The combined impact of these processes includes soil degradation as land is overexploited, depletion of acquifers and fish stocks, eutrophication due to nutrient losses (rising by twenty per cent over the next forty years) and diminished biodiversity. Climate change is expected to reduce crop production in key regions of the world (Hajer/Westhoek/Ozay et al. 2015).

The Food System Report concludes that there are significant opportunities to increase resource use efficiency in the food system, while simultaneously reducing environmental impacts. On the supply side, important options include increasing yields in certain low-yield regions with higher potential using a more balanced mixed of natural resources (including agro-ecological systems and higher input of minerals), leading to an increase in the output per unit of land, water and human labour; increased nutrient use efficiency in the food chain, and consequent reduction of nutrient losses to the environment; development of resource efficient aquaculture systems; and sustainable land and water management using agro-ecological approaches. On the demand side, the two key strategies would be reduction of food losses and wastes, and a shift to less resource-intensive diets, especially in regions with 'western' diets, by lowering the consumption of meat, dairy and eggs (Hajer/Westhoek/Ozay et al. 2015).

17.5.4   Trade

The core question of the *International Trade in Resources* report (referred to as the Trade Report) (Fischer-Kowalski/Dittrich/ Eisenmenger et al. 2015) is whether or not the global trading system contributes to greater resource efficiency and diminished environmental impacts.

The Trade Report clearly shows that although trade in volume increased by a factor of 2.5 between 1980 and 2010, trade measured in monetary terms increased dramatically to twenty-eight per cent of global GDP in 2010. Fifteen per cent of all extracted materials are traded internationally. The Report argues that while incentivizing increased extraction, one key result was increased financial revenues for poor resource-rich countries that rapidly became major exporters. In theory this should have positive developmental consequences that would need to be weighed against the environmental costs; however, as Collier has argued, in reality the more dependent on resource rents economies become the less likely they are to have the governance mechanisms to translate resource rents into developmental benefits—a dynamic known as the resource curse (Collier 2010).

The Trade Report goes on to argue that a closer look at physical trade (see figure 17.10) reveals that the total volume of materials traded between 1980 and 2010 more than doubled, with fossil fuels making up around fifty per cent of the total volume traded. However, reflecting the levelling-off of oil production and trade globally since 2005 (Murray/King 2012), growth rates in trade in oil have declined since 2005.

Figure 17.11, sourced from the Trade Report, reflects who the largest exporters and importers were in 2010. Representing a shift from twentieth-century trends, by 2010 only thirty per cent of all countries were net material suppliers while seventy per cent of all countries had become net importers—and now this was not neatly split along North-South lines. South American countries, Scandinavia, and west and central Asian countries, together with Canada, Australia and the south-east Asian islands, have become the largest suppliers of materials. The largest importers were the US, Japan and Western Europe.

Figure 17.10: Physical trade according to material composition (1980-2010). **Source:** Fischer-Kowalski, Dittrich, Eisenmenger et al. (2015).

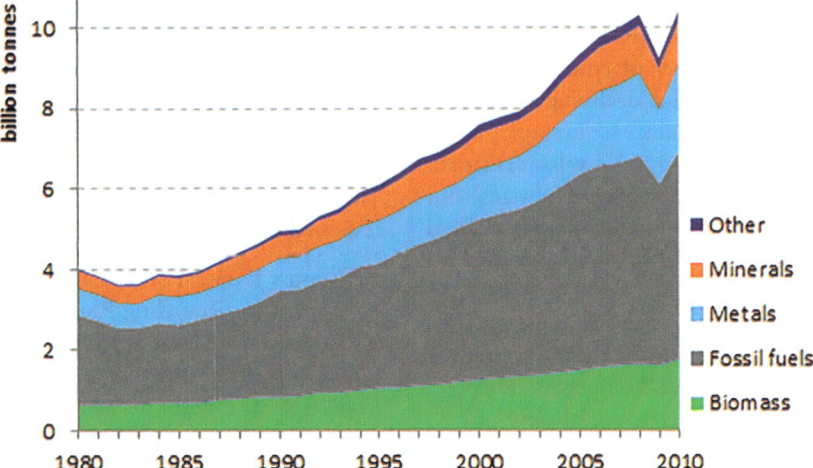

Figure 17.11: Largest net exporters and importers by material composition of net trade in 2010. **Source:** Fischer-Kowalski, Dittrich, Eisenmenger et al. (2015).

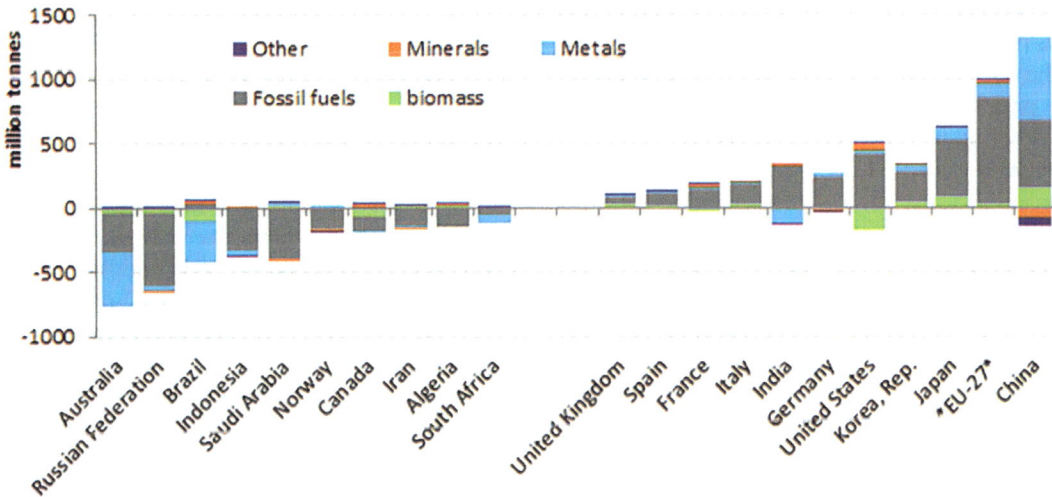

The strategic significance of the mode of analysis in the Trade Report is that it adds another perspective to the assumptions made in the Decoupling 1 Report. Using calculations of 'domestic material extraction' (domestic extraction minus exports plus imports), Decoupling 1 effectively employed a producer perspective that allocated the ecological rucksack (i.e. materials used to produce exports) of imported goods to the exporting country. If, however, the ecological rucksack is attributed to the importing country, apparent decoupling by burden shifting is no longer possible (Wiedman/Schandl/Lenzen et al. 2013). Indeed, Wiedman, Schandl, Lenzen et al. (2013) calculated that forty per cent of the resources extracted were used to enable the exports of goods and services to other countries (Wiedman/Schandl/

Lenzen et al. 2013). The map below reflects the material footprint of nations (in t/cap) where ecological rucksacks are attributed to the consumer and not the producer. This new perspective has become the basis for the *International Trade in Resources* report.

The Trade Report argues that the twentieth-century international division of labour was characterized by declining resource prices that in general made it possible for Northern industrialized countries to act as importers of primary resources and exporters of manufactured goods, with Southern countries as the exporters of primary resources and importers of manufactured goods. The Trade Report confirms that this picture is rapidly changing in the twenty-first century. In a context of rising resource prices, some fast industrializers in the global South have become both importers of primary resources and exporters of manufactured goods, and some industrialized countries like Canada and Australia have become increasingly important exporters of primary resources. In general, there are an increasing number of countries dependent on resource imports and a declining number of countries that are providing an ever-greater proportion of resource exports (Fischer-Kowalski/Dittrich/ Eisenmenger et al. 2015). Trade makes physical burden-shifting possible, but this is unmasked if ecological rucksacks are attributed to the consumer and not the producer.

The Trade Report ends by saying that a conclusive answer to the core question about the role of trade in resource use and environmental impacts is not possible at this stage, especially if a balanced view of environmental and developmental factors is taken into account. This, however, is not the question that guides the primary concerns of this chapter—this chapter is interested in the dynamics of transformation. From this perspective, the declining number of countries providing more and more primary resources within the context of a long-term super-cycle of rising resource prices is clearly the most important limiting factor. The rise of 'resource nationalism' in Africa (together with rising labour costs in China which makes manufacturing through beneficiation increasingly viable in Africa (Swilling 2013a)) suggests that the rising state of resource prices is unlikely to be reversed in the near future, and the adoption by the EU of a Resource Efficiency strategy suggests that rich resource-importing countries will start to find ways of reducing their dependence on resource imports (European Commission 2011). Both signal new directions of change with potentially transformative implications.

17.6   Specific Resource Challenges

The series of IRP reports that deal with specific resource challenges have addressed metals (four reports), water (two reports), land-use and soils, and forests.[5] They all recognize that these resources will in one way or another be required by society irrespective of whether there is a sustainability-oriented transition or not. It therefore follows that it is necessary to understand the complex dynamics that will shape the availability of these resources over time and what actions will be required to ensure that these resources are managed and used in more sustainable ways as part of a wider 'great transformation' process.

17.6.1   Metals

The Metals Working Group has published four peer-reviewed reports and one working paper:[6]

- Report 1: Metal Stocks in Society (2010);
- Report 2a: Recycling Rates of Metals (2011);
- Report 2b: Metal Recycling: Opportunities, Limits, Infrastructure;
- Report 3: Environmental Risks and Challenges of Anthropogenic Metals Flows and Cycles (2013);
- Working Paper: Estimating Long-Term Geological Stocks of Metals.
- Future reports include:
- Report 4: Future Demand Scenarios for Metals;
- Report 5: Critical Metals and Policy Options.

All economies, no matter their level of development, depend on metals of various kinds. The rise of the information age and related increased demand for hi-tech electronic goods has resulted in rapid increases in demand for speciality (or rare earth) metals like lithium and indium. Simultaneously, the accelerated growth and rapid urbanization in the BRICS (Brazil, Russia, India, China, South Africa) -plus countries (e.g. Turkey, Mexico, Nigeria and Indonesia)—especially China—has resulted in massive increases in demand for base metals. Figure 17.12 depicts the growth in demand in recent decades. In combination, these two driving forces of demand have deepened the criticality of a wide range of metals. For example, as the Metals e-Book makes clear, the future demand for zinc, copper, nickel and aluminium just for the expansion of the global energy system are in each case several magnitudes greater than current demand (e.g. demand for aluminium is expected to grow from 500 *gigagrams per year* (Gg/y) to over 5,500 Gg/y by 2050 just for non-fossil fuel infrastructure).

The increasing complexity of electronic goods is a major driver of demand for a wide range of metals to

---

5   The Reports on biofuels and forests have not been included, partly because the implications of biofuels is incorporated into the land and soil group, and the forests report was compiled in a way that does not quite fit into the overall orientation of the IRP.

6   A summary is contained in an e-book available on the IRP website at <http://www.unep.org/resourcepanel>. Unless alternative sources are specified, the data referred to in this section is taken from this e-book (International Resource Panel Working Group on Global Metal Flows 2013).

Figure 17.12: Growth in production of minerals, 1845–2010. **Source:** Mudd (2009), cited in International Resource Panel Working Group on Global Metal Flows (2013).

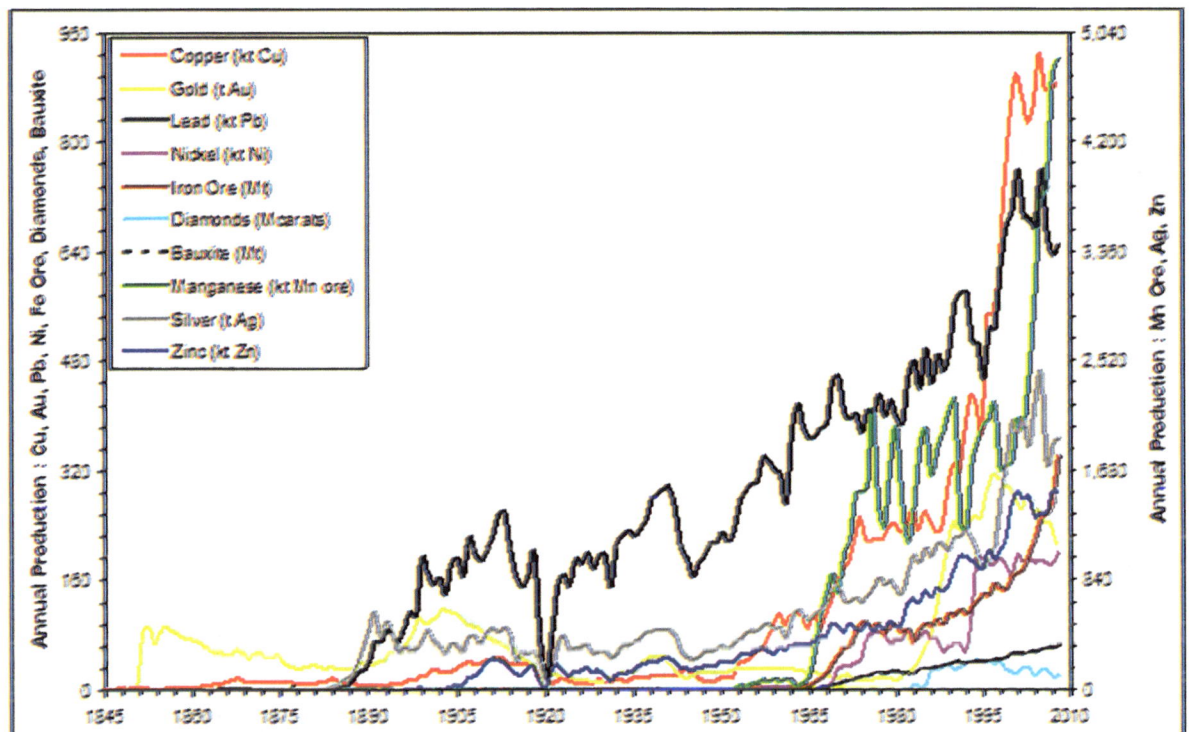

produce the compounds required by these goods. Figure 17.13 reveals the growth in the number of elements needed to make a microchip.

Although a lack of information prevents high-confidence estimations about resource depletion (Smil 2014), what is clear from the work of the Metals Working Group is that there are also other factors that increase supply risks. These include, according to reports of this Working Group, challenging technological conditions (depth, composition of ore as ore grades decline), economic variables (adequacy of infrastructure, size of deposit), environmental constrains (natural habitats, ecosystem services), and geopolitical dynamics (trade barriers, political instability, weak states) (International Resource Panel Working Group on Global Metal Flows 2013).

Global metals production is a major contributor to environmental pollution and energy demand. The Working Group's reports shows that no less than seven to eight per cent of global energy use and therefore energy-related GHG emissions can be attributed to metals production. Whereas 20 MJ of energy is needed to make a kilogram of steel, 200,000 MJ is needed to make a kilogram of platinum (International Resource Panel Working Group on Global Metal Flows 2013). A major driver of increased future energy

demand of metals production is the decline in ore quality—three times more material must be moved today to extract a kilogram of ore compared to a century ago.

Report 1 estimated the quantity of metals being used by society for the period 2000–2006. The average for aluminium was 80 kg per capita, with a range of 350–500 kg/cap for developed economies and 35 kg/cap for the least developed economies. Similarly for copper: 35-55 kg/cap is the global average, ranging from 140–300 kg/cap to 30–40kg/cap; and 2200kg/cap for iron, ranging from 7,000–14,000kg/cap to 2,000kg/cap. Obviously, the same pattern replicates itself for each metal (International Resource Panel Working Group on Global Metal Flows 2013), the implication being that global development targets aimed at eradicating poverty and achieving greater equity will entail significant increases in metals consumption in developing countries.

To diminish the environmental impact and energy requirements of metal production it will be necessary to increase the recycling rates of metals. Report 2a demonstrated that the *end of life-recycling rates* (EoL-RR) for metals are very low: EoLRR of above fifty per cent can be found for only eighteen metals.

Figure 17.13: Computer Chip Elemental Contents. **Source**: Quoted in a Powerpoint presentation by Tom Graedel, November 2013, Stellenbosch University.

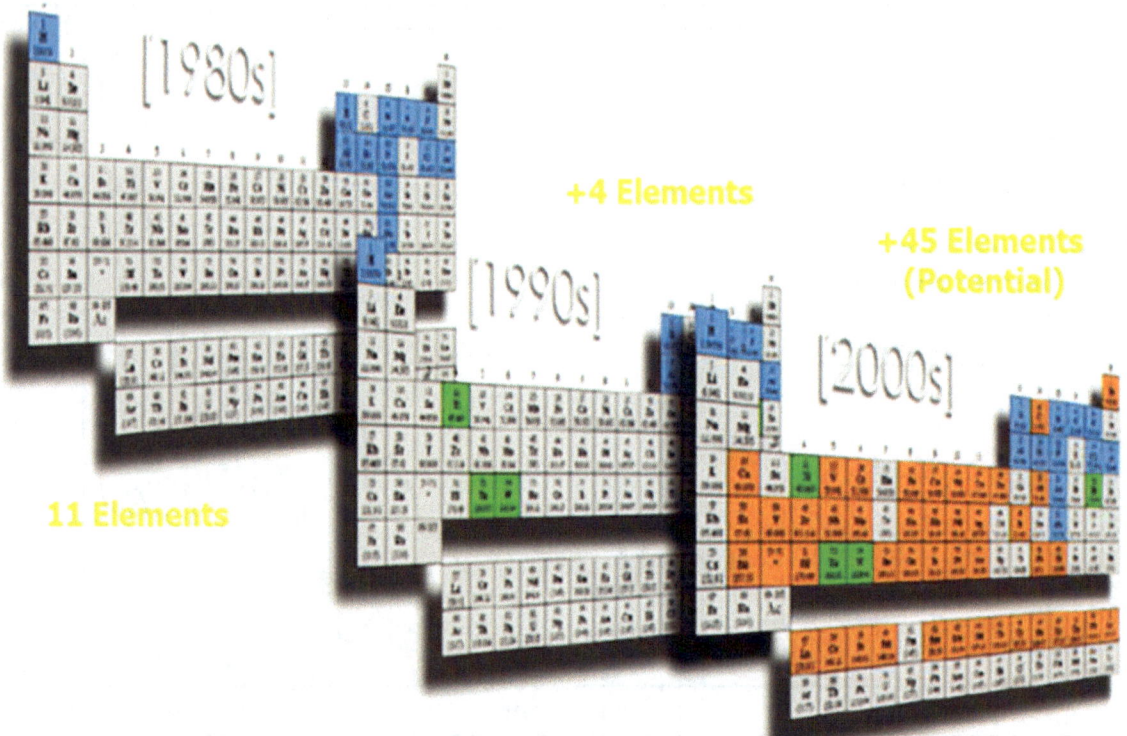

The Report shows that part of the explanation for low EoLRRs is rising demand and the long in-use life of metals. However, another more important explanation is that the design of products has not hitherto taken into account the need for end-of-life recovery and reuse. Disassembly and metals recovery is not what designers have been required to do. To increase EoLRRs to fifty per cent or more for all metals as part of a wider sustainability-oriented transition, it will be necessary, the Report argues, to radically change the way products are designed (i.e. design for disassembly) and substantial investments in new collection infrastructures will be necessary (International Resource Panel Working Group on Global Metal Flows 2013). As resource prices continue to rise as demand continues to grow, driven mainly by the requirements of the information age and urbanization, the financial viability of design for disassembly will more than likely improve. This will be a crucial driver of a more fundamental transformation.

17.6.2   Land-Use and Soils

A century of steadily declining food prices came to an end at the turn of the millennium. Since then food prices have been rising steadily and so has the number of large-scale land transactions (including so-called 'land grabbing'). Except for the first decade of the twenty-first century, there has been no decade since 1900 where there is evidence of steadily rising food prices. This pattern is expected to continue with major implications for land use and food security (Swilling/Annecke 2012: Chapter 6).

Following Scherr, the total ice-free land surface of the Earth is 13 *billion hectares* (Bha) of which 1.5 Bha is unused 'wasteland' and an additional 2.8 Bha is unused and inaccessible. This leaves 8.7 Bha which humans in the Anthropocene can choose to 'use' for a wide variety purposes, including pasture, forests and cropland. Of this, 3.2 Bha are potentially arable, the rest being marginal land from a cultivation perspective and covered by forest, grassland and permanent vegetation.[7] Of the potentially arable land, only 1.3 Bha is deemed to be moderate to highly productive. Just under half of the 3.2 Bha of potentially arable land (1.47 Bha) is cultivated as cropland. This means that just over ten per cent of the ice-free land surface of the Earth is the resource on which humans depend

---

7   Lambin and Meyfroidt estimate that there are approximately 4 Bha available for 'rain-fed agriculture' (Lambin/Meyfroidt 2011: 3466).

Figure 17.14: Agricultural inputs relative to crop yields (1961–2009). **Source:** FAO data compiled by Schutz, cited in Bringezu, Schutz, Pengue et al. (2014).

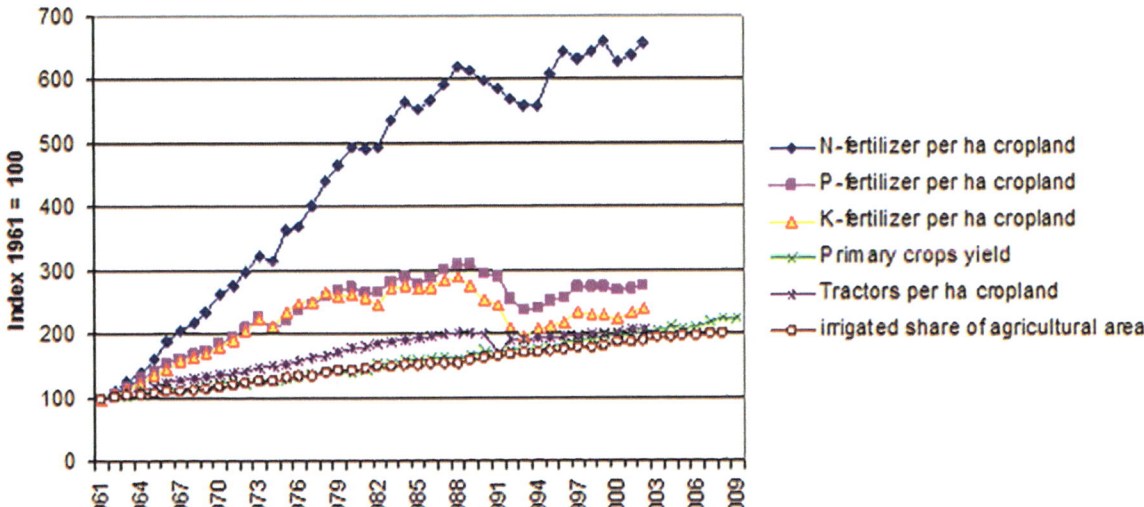

for the bulk of their food. This 1.47 Bha of cropland, plus approximately 3.2 Bha of permanent pasture and 4 Bha of permanent forest and woodland, is what makes up the 8.7 Bha of 'usable' land (Scherr 1999). Half of the developing world's arable and perennial cropland is in just five countries—Brazil, China, India (with twenty-two per cent), Indonesia and Nigeria. It is noticeable that the only African countries with very extensive or moderately extensive arable land resources are Nigeria, Ethiopia, South Africa and Sudan. The majority of African countries have limited arable land resources with high population pressures and it is estimated that sixty-five per cent of Africa's agricultural land is degraded (Scherr 1999). Yet African countries are earmarked by all the models of the future for substantial yield increases—it is also where most of the land grabs are taking place (Cotula/Vermeulen/Leonard et al. 2009).

Global land use, rising food prices, soil degradation and accelerated land transactions (as countries scramble to secure food supplies) provides the context for the IRP Report entitled *Assessing Global Land Use: Balancing Consumption with Sustainable Supply* (generally referred to as the *Land and Soils Report*) (Bringezu/Schutz/Pengue et al. 2014). Launched at the World Economic Forum in January 2014, the report raises very serious questions about the sustainability of expanding agricultural production in a world dominated by a resource inefficient food system that does not cater for the needs of the nearly billion or so people who are undernourished.

Figure 17.14 reveals that since 1961, inputs (nitrogen, phosphorus, potassium, tractors) have tended to rise at a faster rate than crop yields, and this within the context of a doubling of irrigated land area.

At the same time, soil degradation continues, with twenty-three per cent of soils degraded by 1990. Two to five million hectares are degraded per annum.

The international division of agricultural labour is clearly reflected in figure 17.15, which reveals the gap between yields in Europe and North America, where high external input intensive industrial farming is prevalent, and the yields in developing countries where soil degradation levels are high, infrastructures are poor and farming is still dominated by 400 million small farmers (only forty per cent of whom use chemical inputs).

Expanding agricultural land use is driven in part by rising demand for food and non-food biomass which cannot be compensated by higher yields. This net expansion occurs at the expense of grasslands, savannahs and forests. However, expansion is also driven by the need to compensate for expanding cities and soil degradation. This plus net expansion results in gross expansion. Based on an assessment of a wide range of studies (for sources of data cited see original diagram in *Land and Soils Report*), the dimensions of the net and gross expansion of agricultural land are represented in figure 17.16.

The Land and Soils Report shows the future land requirements to meet food supply (after exhausting yield growth potential) are estimated to be between 71 *million hectares* (Mha) and 300 Mha. The rapidly

Figure 17.15: Cereal yields by selected world regions (1961-2011). **Source**: FAO data cited in Bringezu, Schutz, Pengue et al. (2014).

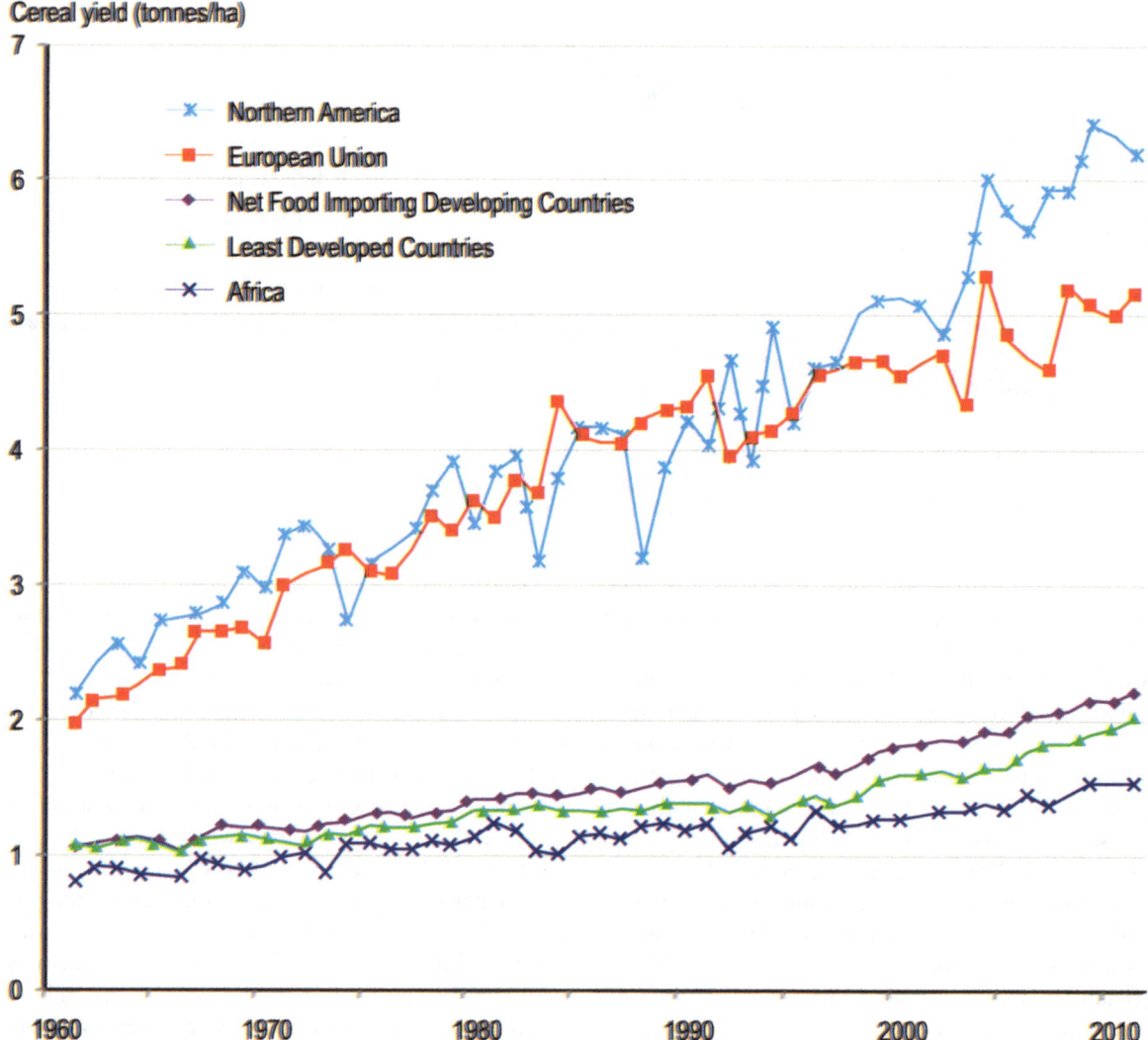

Cereal yield (tonnes/ha)

- Northern America
- European Union
- Net Food Importing Developing Countries
- Least Developed Countries
- Africa

expanding demand for land to grow biofuel crops is estimated to be 48 MHa to 80 Mha, and the requirements for additional biomaterials (wood, textile crops, etc) is estimated to be be 4 MHa to 115 Mha. To compensate for the expanding built environment (that tends to destroy the most valuable agricultural land), between 107 Mha and 129 Mha may be needed. Assuming that a significant proportion of degraded soils cannot be restored,[8] estimates of the requirements to compensate for degradation range from 90 Mha to 225 Mha. This means that the estimates for

gross additional agricultural land requirements to meet growing needs range between 320 Mha and 849 Mha. This suggests that the needs are much greater than FAO's estimate of 120 Mha (Bringezu/Schutz/Pengue et al. 2014).

It needs to be recognized that land-use change in favour of agriculture is one of the primary drivers of rising CO_2 emissions and biodiversity loss. We need, therefore, to accept that there are absolute limits to the quantity of global land that can be used for agriculture. Taking into account various factors, the *Land and Soils Report* proposes that the expansion of global cropland should be halted by 2020, at which point it is estimated global cropland will have expanded from about 1.5 Bha to 1.64 Bha. In other words, although an additional 140 Mha of cropland is

_____

8 Although seriously degraded soils are difficult and costly to restore, there are still about 300 Mha of lightly degraded soils that can be restored mainly by changing management practices.

Figure 17.16: High and low estimates of net and gross expansion of agricultural land, 2005-2050. **Source**: Bringezu, Schutz, Pengue et al. 2014—note: for detailed references to the sources of data cited here, see original diagram in this report.

Business-as-usual expansion	Low estimate (Mha)	High estimate (Mha)
Food supply	71	300
Biofuel supply	48	80
Biomaterial supply	4	115
Net expansion	**123**	**495**
Compensation for built environment	107	129
Compensation for soil degradation	90	225
Gross expansion	**320**	**849**

bound to have very negative environmental effects, the *Land and Soils Report* nevertheless estimates that it may be possible to remain within this 'safe operating space' and thus avoid the far more negative consequences of an expansion in the range 320 Mha to 849 Mha, as suggested by the sum of existing research. To achieve this reduction in future requirements, the *Land and Soils Report* recommends the following (Bringezu/Schutz/Pengue et al. 2014):

- massively increase the existing land potential by restoring degraded soils and using existing soils optimally—how to do this is the focus of the next report on land and soils that is currently under way;
- ensure that national governments have the capacity to control expansion of agricultural land in order to avoid uncontrolled destruction of biodiversity, forests and pastures;
- limit meat/dairy consumption and change the way the food system works—again, as mentioned earlier in this chapter, this is the focus of the forthcoming IRP report.

The Report recommends the following specific sets of interventions (Bringezu/Schutz/Pengue et al. 2014):

- reducing the demand for meat/dairy products and reducing the levels of food waste could save between 96 and 135 Mha;
- halving the global biofuel targets could save between 24 and 40 Mha;
- controlling the demand for biomaterials could save up to 57 Mha;

- limiting the expansion of cities into productive agricultural by just ten per cent of the expected impact could save between 11 and 13 Mha;
- restoring a third of degraded soils could save between 30 and 74 Mha.

In short, a mix of strategies and measures to reduce overconsumption of certain foods, reduce food waste and limit the consumption of non-food biomass products while at the same time improving land management could save between 160 and 320 Mha by 2050. Cropland area would still expand to meet, in particular, the demand for increased food production to meet the needs of those who have enough, but not as much.

17.6.3   Water

According to the Water Decoupling Report (Urama 2015), integrated water resources management faces two closely interlinked obstacles—one, on the supply side, of unpolluted freshwater resources for a growing world population, and the other, on the demand side, of water for increased agricultural output, water-intensive industries and domestic use. The problems associated with the supply and demand of water, such as significant increases in water pollution and freshwater withdrawals, are driven by population increase, urbanization, rising living standards, unsustainable water governance (which includes inefficient supply and demand management), agricultural land uses (specifically irrigation), industrial production, ecosystem degradation and climate change.

The as yet unpublished but completed *Water Decoupling Report* addresses the challenge of water availability and use in light of mounting global challenges to security of supply (Urama 2015). Water withdrawals on a global scale have increased at a rate almost double the human population growth rate, from 600 billion cubic meters in 1,900 to 4,500 billion cubic meters in 2010. This could grow to between 6,350 and 6,900 billion cubic meters by 2030 if an average economic growth scenario and efficiency gains are assumed. This represents a forty per cent demand gap above currently accessible water resources, including return flows.

Table 17.1 illustrates the expected increases in water withdrawal demand for human activities by 2030. The highest incremental demand is expected to occur in sub-Saharan Africa at 283 per cent, while the lowest is expected in North America at 43 per cent.

In terms of global freshwater use to support human activities, currently 70% is used in agriculture

Table 17.1: Increases in Annual Water Demand, 2005–2030. **Source:** McKinsey (2009), cited in Urama (2015).

Region	Projected Change from 2005
China	61%
India	58%
Rest of Asia	54%
Sub-Saharan Africa	283%
North America	43%
Europe	50%
South America	95%
Oceania	109%

(output of which is estimated to increase by another 65% by 2030) of which 15–35% is considered unsustainable, especially in cases where groundwater is extracted faster than it can be recharged. An additional 22% of fresh water is used in industries (estimated to grow by an additional 25% by 2030), but this can range from as high as 60% in industrialized countries to as low as 10% in some developing countries. Lastly, 8–11% is for domestic use (estimated to grow by another 10% by 2030), at an average of about 50 litres per person per day, although also with great variability (International Water Management Institute 2007 and Gleick 2010, both cited in Urama 2015).

On the supply side, it is estimated that over the next twenty years, water supply would need to be 140% higher than in the past twenty years to meet the increasing demand and ensure accessible, reliable and sustainable provision of existing water supplies (Urama 2015). The obstacles are as follows. Readily available sources of fresh water are under significant stress already, with the shrinking of many freshwater lakes, the drying up of rivers that subsequently never reach the ocean, and the overuse of groundwater resources, something that is already occurring in many regions. Further limiting these water resources, the Water Report argues, are increasing rates of pollution with over 405 dead zones currently on record globally in coastal waters. Lastly, further water is lost due to inefficiencies in technologies, the most pertinent example being the loss of drinking water from municipal distribution systems before it even reaches the consumer, where on average thirty per cent (and in extreme cases up to eighty per cent) of water is lost. This is equivalent to over US$18 billion worth of water per year that does not generate revenue, indicat-

ing the need for efficiency and productivity gains (Urama 2015).

Figure 17.17 shows the number of people living in water-stressed areas between 2005 and 2030, as calculated by UNEP (UNEP 2011, cited in Urama 2015). The OECD estimates that nearly 3.9 billion people will experience severe water stress by 2030.

One of the greatest issues relevant to the supply of fresh water is the level of pollution through human activities. The most relevant sources include mining activities, agriculture, landfill, and industrial and urban wastewater effluents. The main pollutants from agriculture, for example, include pesticides, organic compounds and nutrients from fertilizer that end up in water bodies, causing eutrophication and ultimately leading to "dead zones" (Urama 2015). Furthermore, pollution from industrial activities is in the form of seventy per cent of untreated industrial wastes being dumped into waters (UN-Water 2009 cited in Urama 2015).

In many developing countries sanitation and wastewater treatment causes major water pollution, and scenarios have been found where as much as eighty-five to ninety-five per cent of sewage is discharged directly into rivers, lakes and coastal areas, causing large amounts of revenue to be spent dealing with waterborne diseases instead of generating new wealth (Tropp 2010 cited in Urama 2015). .

Lastly, the number of people vulnerable to water-related disasters, particularly flooding as a result of climate change, deforestation, population growth, rising sea levels, and human settlement in flood-prone lands, may reach two billion by 2050 (Urama 2015).

All these obstacles make a strong case for water decoupling, that is, using fewer units of water resources per unit of economic output, while also reducing other adverse socio-economic and environmental trade-offs downstream, such as rates of water pollution, known as impact decoupling.

Relative decoupling is shown to be beneficial to human well-being, environmental flows in river basins, and to economic growth (Urama 2015). Achieving sustainable decoupling in the water sector will require innovative structural transformations in economic pathways; integrated water management policy and practices at local, national, river-basin, and global scales; and substantive investments in improved technologies and innovations for improving water efficiency and productivity at the appropriate temporal and spatial scales. Improving technical and allocative efficiency and resource productivity in the key water use sectors could offset up to sixty per cent of the

Figure 17.17: Number of people living in water-stressed areas in 2030, by country type. **Source**: UNEP (2011), cited in Urama (2015).

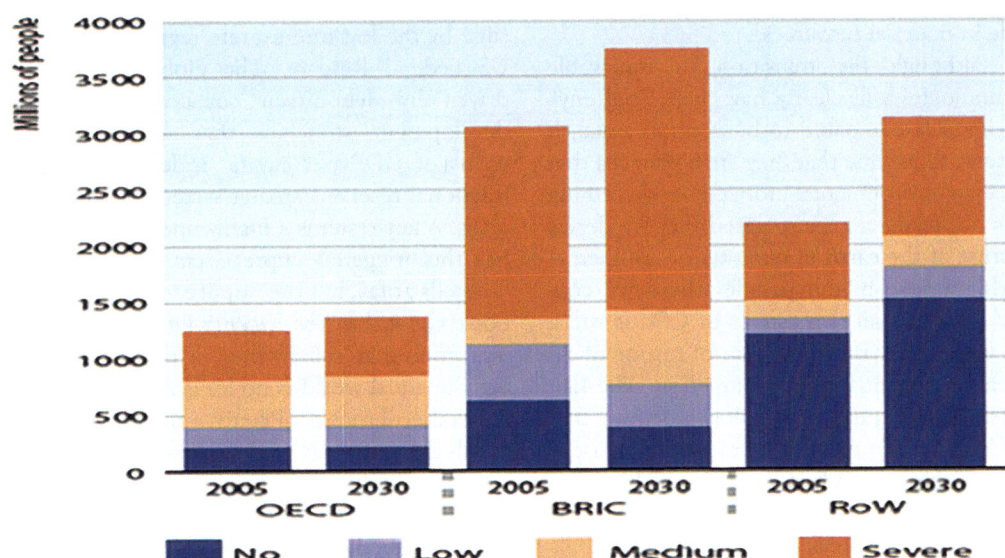

anticipated growth in demand for water by 2030 (Urama 2015).

17.7   Discussion: Implications of the Work of the IRP for Global Transition Thinking

It was argued at the outset that the work of the IRP can be understood within the wider context of the so-called 'third great transformation' or more precisely a 'sustainability-oriented transition'. Following the GACGC, it was argued that a structural metabolic shift would be the distinctive feature of this transition. It should be clear from the discussion thus far that although the IRP does not address the question of transition directly, when the completed and ongoing work is read together it does provide an extraordinary body of rich empirical evidence that confirms the notion that we are experiencing the endgame of the industrial socio-ecological regime.

Although the notion of decoupling is contested on the grounds that it implies that fundamental structural change can be avoided while greening consumption (Jackson 2009), the global resource perspectives provided in the Decoupling 1, Decoupling 2 and Environmental Impacts reports all confirm that unless the global systems of production and consumption are in fact radically transformed it will not be possible to build a world without poverty where average consumption (of around six t/cap) is consistent with

what available planetary systems can provide on a sustainable basis. This message goes way beyond the carbon-centred argument of the IPCC that has succeeded in establishing the notion of a low carbon transition within the global policy community. From the perspective of the IRP, this will not be sufficient. A low carbon destruction of planetary resources is not an appropriate future trajectory.

The conclusions of the work on the various nexus themes confirm that superficial modifications to the sociotechnical systems that we depend on will not suffice. To double the extent of the world's urban settlements, it will be necessary to fundamentally change the way we design, build and operate cities. Many aspects of urban living that are taken for granted will clearly have to be replaced with information-rich alternatives that embed cities in sustainable technological and ecological cycles.

The crisis of the current highly complex tightly-coupled global food system poses a major risk to the survival of the global population, in particular the poor. Although this might be the most difficult socio-technical system to change, fundamental changes to this system might well be driven by the social and health consequences of deepening food insecurity.

The global trade in material resources is already changing rapidly as fewer and fewer countries become increasingly important exporters of primary resources to an increasing number of industrializing and industrialized resource-importing countries. The rise of resource nationalism in many resource-rich developing countries and the resource efficiency movement in many resource-importing developed countries suggests future trajectories that will have major implications for global trade in material resources.

Finally, although the transition to renewable energy technologies will clearly have beneficial environmental impacts compared to business as usual, it would be naïve to assume that they are a panacea that will produce an environmental utopia. Like everything we humans do, resources are required that we derive from the crust of the earth in one way or another. A future world of nine billion people where we consume six t/cap and emit two tonnes of CO_2 is still a world that will require the extraction of resources on a scale equal to current levels of extraction. The IRP work on renewable energy technologies shows that the demand for certain materials may well increase if we implement the low carbon transition. This is an early warning that the sustainability-oriented transition is not a simplistic break that miraculously heals the planet.

Finally, the IRP work on the specific resource challenges in the metals and ecosystems sectors clearly shows that a sustainability-oriented transition will depend on extraordinary efforts to change the way we use the three most basic ingredients of contemporary modern living: metals (in particular for the global electronic infrastructure), water and land. However, no matter what we do, there is no way we can do without these three key resources. Indeed, it is clear from the evidence that we will need more of them all and that this will have to be done in a way that ensures that those who currently live in poverty gain greater access to these resources. The challenge, however, is to trigger new consumption and production systems that create new economic opportunities out of the need, for example, to 'design for disassembly' when it comes to metals, or to design and build decentralized urban water and sanitation systems that can use water more efficiently and recycle all waste water, or replicate on a massive scale the agro-ecological farming methods that have proved to be able to increase yields by restoring the soils and ecosystems.

This discussion raises the question about whether the IRP should go beyond its current mandate and become the global body that addresses in an integrated way the challenge of making the sustainability-oriented transition happen. For this to be possible, however, it will be necessary to embed the research done to date on resource flows within an analysis of the dynamics of the global economy. During the quarter-century leading up to the great contraction of 2009, the neo-liberalization of the global economy resulted in the rise to dominance of finance capital on the back of a wave of financial innovations made possible by the low-interest-rate regime advocated by the US Federal Reserve. The global economy is now driven by debt-driven consumption and national development strategies that measure progress in terms of GDP per capita. Indeed, without growth, fractional reserve banking systems will face a serious crisis. A key result is a highly unequal world. Not only has this triggered unprecedented social movements (Castells 2012), but the popularity of Thomas Piketty's book *Capital in the Twenty-First Century* is a clear indication that 'rule by the one per cent' of an increasingly unequal world is no longer a legitimate mode of societal governance (Piketty 2014). It is arguable that this highly complex and increasingly unstable global economic architecture is incompatible with the metabolic dynamics and requirements of the third great transformation. If this argument is correct, it will be necessary to merge the body of work thus far developed by the IRP with the emerging body of work by progressive ecological and institutional economists who have begun to consider the details of alternative future economic arrangements and trajectories that will be fundamentally different to what mainstream macro-economic theory assumes is the natural order of things.

17.8   Conclusion

In this first academic review of the IRP's body of work, it has been argued that it may be appropriate to use long-wave theories of transition to understand the contribution of the IRP to the wider field of anticipatory science. What is anticipated by those who use this perspective is that in some way the metabolic, sociotechnical and institutional regimes of the industrial epoch will be replaced over time by an alternative more sustainable epoch characterized by more sustainable metabolic flows made possible by reconfigured and transformed sociotechnical and institutional regimes. Although the IRP's work does not directly address transition per se, when read together the various strands of thought and evidence in the completed and current work do suggest that it is highly unlikely that the industrial epoch can continue into the medium- and long-term future if it depends on the continuous increase in consumption of natural resources. There are elements, of course, across the reports that could be woven into a more robust and systematic conception of transition: the types of decoupling envisaged in the Decoupling 2 Report, the recommended dietary and land-use changes in the Land and Soils Report, the key role of cities in the City-Decoupling Report, and the unintended consequences of a transition to clean energy, to cite only a few examples.

Three broad conclusions flow from this analysis.

Firstly, when collected together, the totality of evidence mobilized by the IRP supports the notion that future well-being and development (whether growth-based or not) will have to be decoupled from rising rates of resource use. Relative decoupling is not sufficient. Absolute reductions in resource use will be necessary. To implement this idea, however, a fundamental restructuring of prevailing modes of production and consumption will be necessary. Decoupling is not simply sophisticated greenwash. It will mean significant changes for consumers in developed economies, and in developing countries committed to poverty eradication it will be necessary to replace resource-intensive development pathways with resource-efficient development pathways that end up delivering to more people a fairer deal resulting in less inequality and therefore more long-term democratic stability.

Secondly, the IRP's work reveals the futility of naïve assumptions about what will be attainable in a sustainable world populated by over nine billion people, most of whom will be living in cities. All past human activity has depended on the exploitation of natural resources in one way or another. Humans currently have technical and institutional capabilities to exploit these resources on an unprecedented scale. Anticipatory science is needed to show that this cannot continue. However, if the results from IRP research are anything to go by, massive reductions in resource use are possible, but they cannot be eliminated or reduced to insignificant levels. A world of

over nine billion people without poverty may well need what was extracted from the earth in 2000. The finding made in the Green Energy Choices Report that more of certain materials might be needed is highly significant. The only question is how this will be done. Will these sociotechnical processes become part of closed-loop techno-industrial and ecological cycles or not? That will become the key longer-term question, not simply a zero-sum calculation based on how much less can be consumed. This, in turn, might make it possible to make a shift from only focusing on 'resource limits' to focusing more on 'resource potential'. This shift is already under way in a number of reports.

Thirdly, the IRP work on resource limits and potentials needs to be integrated into a wider holistic theory of economic development that is not GDP-centred. The gradual dismantling of neo-liberalism is already under way as intolerance of poverty and inequality reaches new heights in the wake of the global economic crisis and states in the developing world disassociate themselves from the hegemonies of Western thinking. This is clearly a positive movement. However, if an alternative is constructed that anticipates the great transformation that once again assumes that there is an unlimited supply of natural resources, then a major opportunity will have been missed. The economic theories informing the developmental states run the risk of making this mistake. However, this is unlikely to become a mainstream habit because the century-long decline in resource prices ended in 2002. If those who have predicted a long-term supercycle of forty to sixty years of rising resource prices prove to be correct, then we can safely anticipate that the economic theory that replaces the reductionist simplicities of neo-liberalism will indeed need to come to terms with the expanding body of work produced by the IRP. This will surely justify the efforts by those who have made this work possible.

References

Airoldi, Marco; Biscarini, Lamberto; Saracina, Vito, 2010: *The Global Infrastructure Challenge: Top Priorities for the Public and Private Sectors* (Milan: Boston Consulting Group).

Angel, Shlomo, 2012: *Planet of Cities*. (Hollis, NH: Puritan Press).

Barles, Sabine, 2009: "Urban Metabolism of Paris and its Region". *Journal of Industrial Ecology*, 13,6: 898-913.

Barles, Sabine, 2010: "Society, Energy and Materials: The Contribution of Urban Metabolism Studies to Sustainable Urban Development Issues", in: *Journal of Environmental Planning and Management*, 53,4: 439-455.

Bringezu, Stefan; Schutz, Helmut; Pengue, Walter; O'Brien, Maghan; Garcia, Fernando; Sims, Ralph; Howarth, Robert; Kauppi, Lea; Swilling, Mark; Herrick, Jeffrey, 2014: *Assessing Global Land use: Balancing Consumption with Sustainable Supply* (Nairobi: United Nations Environment Programme).

Brynjolfsson, Erik; McAffee, Andrew, 2014: *The Second Machine Age: Work, Progress, and Prosperity in a Time of Brilliant Technologies* (New York; London: W. W. Norton Co.).

Castells, Manuel, 2009: *Communication Power* (Oxford: Oxford University Press).

Castells, Manuel, 2012: *Networks of Outrage and Hope* (Cambridge, UK: Polity).

Collier, Paul, 2010: "The Political Economy of Natural Resources", in: *Social Research*, 77,4: 1105-1132.

Costa, Alessandro; Marchettini, Nadia; Facchini, Angelo, 2004: "Developing the Urban Metabolism Approach into a New Urban Metabolic Model", in: Marchettini, Nadia; Brebbia Carlos; Tiezzi, Enzo; Wadhwa, Lal C. (Eds.): *The Sustainable City III* (London: WIT Press).

Cotula, Lorenzo; Vermeulen, Sonja; Leonard, Rebeca; Keeley, James, 2009: *Land Grab or Development Opportunity? Agricultural Investment and International Land Deals in Africa* (London: International Instituted for Environment and Development; FAO; International Fund for Agricultural Development).

Crutzen, Paul J., 2002: "The Anthropocene: Geology and Mankind", in: *Nature*, 415: 23.

Department of Economic and Social Affairs, United Nations, 2012: *World Urbanization Prospects: The 2011 Revision*. New York: United Nations.

Doshi, Viren, Schulam, Gary; Gabaldon, Daniel, 2007: "Lights! Water! Motion!", in: *Strategy and Business*, 47: 39-53.

European Commission, 2011: *Life and Resource Efficiency: Decoupling Growth from Resource Use* (Luxembourg: European Commission).

Ferrao, Paulo; Fernandez, John. E., 2013: *Sustainable Urban Metabolism* (Cambridge, MA: MIT Press).

Fischer-Kowalski, Marina, 1998: "Society's Metabolism: The Intellectual History of Materials Flow Analysis, Part I, 1860-1970". *Journal of Industrial Ecology*, 2,1: 61-78.

Fischer-Kowalski, Marina, 1999: "Society's Metabolism: The Intellectual History of Materials Flow Analysis, Part II, 1970-1998". *Journal of Industrial Ecology*, 2,4: 107-136.

Fischer-Kowalski, Marina, 2011: "Analysing Sustainability Transitions as a Shift between Socio-Metabolic Regimes". *Environmental Transition and Societal Transitions*, 1: 152-159.

Fischer-Kowalski, Marina; Dittrich, Monika; Eisenmenger, Nina; Ekins, Paul; Fulton, Julian; Kastner, Thomas; Schandl, Heinz; West, Jim; Wiedman, Thomas, 2015: *International Trade in Resources: A Biophysical Assessment*. Report for the International Resource Panel (Nairobi: UNEP).

Fischer-Kowalski, Marina; Haberl, Helmut, 2007: *Socioecological Transitions and Global Change: Trajectories of Social Metabolism and Land use* (Cheltenham, UK: Edward Elgar).

Fischer-Kowalski, Marina; Swilling, Mark, 2011: *Decoupling Natural Resource use and Environmental Impacts from Economic Growth*. Report for the International Resource Panel. Paris: United Nations Environment Programme.

Food and Agriculture Organization, 1996: *World Food Summit: Rome Declaration on World Food Security* (Rome: Food and Agriculture Organization); at: <http://www.fao.org/WFS> (19 January 2014).

Food and Agriculture Organization, 2013: *The State of Food and Agriculture 2013: Better Systems for Better Nutrition* (Rome: Food and Agriculture Organization).

Foxon, Timothy J., 2011: "A Coevolutionary Framework for Analysing a Transition to a Sustainable Low Carbon Economy", in: *Ecological Economics*, 70: 2258-2267.

Freeman, Chris; Louca, Francisco, 2001: *As Time Goes by: From Industrial Revolutions to the Information Revolution* (Oxford: Oxford University Press).

Geels, Frank W., 2002: "Technological Transitions as Evolutionary Reconfiguration Processes: A Multi-Level Perspective and a Case-Study", in: *Research Policy*, 31,8-9: 1257-1274.

German Advisory Council on Global Change, 2011: *World in Transition: A Social Contract for Sustainability* (Berlin: German Advisory Council on Global Change).

Giampietro, Mario; Mayumi, Kozo; Sorman, Alvegul. H., 2012: *Metabolic Pattern of Societies: Where Economists Fall Short*. Abingdon; New York: Routledge.

Gore, Charles, 2010: "Global Recession of 2009 in a Long-Term Development Perspective", in: *Journal of International Development*, 22: 714-738.

Haberl, Helmut; Fischer-Kowalski, Marina; Krausmann, Fridolin; Martinez-Alier, Joan; Winiwarter, Verena, 2011: "A Socio-Metabolic Transition Towards Sustainability? Challenges for another Great Transformation", in: *Sustainable Development*, 19: 1-14.

Hajer, Maarten, 2014: "On being Smart about Cities: Seven Considerations for a New Urban Planning and Design", in: Maarten, Hajer (Ed.): *Smart About Cities: Visualizing the Challenge for 21st Century Urbanism* (Amsterdam: Forthcoming).

Hajer, Maarten; Dassen, Ton, 2014: *SMART about Cities: Visualising the Challenge for 21st Century Urbanism* (Rotterdam: NAI and PBL).

Hajer, Maarten; Westhoek, Henk; Ozay, Leyla; Ingram, James; van Berkum, Siemen, 2015: *Food Systems and Natural Resources*. Report for the International Resource Panel (Paris: UNEP).

Hertwich, Edgar G; Gibon, Thomas; Arveson, Anders; Bayer, Peter; Bouman, Evert; Bergersen, Joe; Heath, Garvin; Aloisi de Larderel, Jacqueline S.; Ramirez, Andrea; Suh, Sangwon, 2015: *Green Energy Choices: The Benefits, Risks, and Trade-Offs of Low Carbon Technologies for Electricity Production—Technical Summary*. Report for the Intertional Resource Panel (Paris: UNEP).

Heynen, Nikolas; Kaika, Maria; Swyngedouw, Erik, 2006: *In the Nature of Cities: Urban Political Ecology and the Politics of Urban Metabolism* (London - New York: Routledge).

International Resource Panel Working Group on Global Metal Flows, 2013: *E-Book: International Resource Panel Work on Global Metal Flows*. Report for the International Resource Panel (Nairobi: UNEP).

Jackson, Tim, 2009: *Prosperity without Growth? The Transition to a Sustainable Economy.* (United Kingdom: Sustainable Development Commission).

Kennedy, Christopher; Pincetl, Stephanie; Bunje, Paul, 2011: "The Study of Urban Metabolism and its Applications to Urban Planning and Design", in: *Environmental Pollution,* 159: 1965-1973.

Köhler, Jonathan, 2012: "A Comparison of the Neo-Schumpeterian Theory of Kondratiev Waves and the Multi-Level Perspective on Transitions", in: *Environmental Innovation and Societal Transitions,* 3: 1-15.

Kuecker, Glen David, 2013: *Building the Bridge to the Future: New Songdo City from a Critical Urbanism Perspective.* SOAS University of London Centre of Korean Studies Workshop on New Songdo City and South Korea's Green Economy: An Uncertain Future: London, 5 June 2013.

Lambin, Eric F.; Meyfroidt, Patrick, 2011: "Global Land use Change, Economic Globalization, and the Looming Land Scarcity", in: *Proceedings of the National Academy of Science,* 108,9: 3465-3472.

Luque, Andres; Marvin, Simon; McFarlane, Colin, 2013: Smart Urbanism: Cities, Grids and Alternatives?, in: Hodson, Mike; Marvin, Simon (Eds.): *After Sustainable Cities?* (London - New York: Routledge).

McKinsey Global Institute, 2011: *Resource Revolution: Meeting the World's Energy, Materials, Food, and Water Needs* (McKinsey Global Institute); at: <http://www.mckinsey.com/client_service/sustainability.aspx> (6 December 2011).

Morin, Edgar, 1999: *Homeland Earth* (Cresskill, NJ: Hampton Press).

Murray, James; King, David, 2012: "Oil's Tipping Point has Passed", in: *Nature,* 481: 433-435.

Perez, Carlota, 2009: "The Double Bubble at the Turn of the Century: Technological Roots and Structural Implications", in:. *Cambridge Journal of Economics,* 33: 779-805.

Perez, Carlota, 2002: *Technological Revolutions and Financial Capital: The Dynamics of Bubbles and Golden Ages* (Cheltenham, UK: Elgar).

Piketty, Thomas, 2014: *Capital in the Twenty-First Century* (Cambridge, Mass.: Belknap Press).

Poli, Roberto, 2014: "Anticipation: A New Thread for the Human and Social Sciences?", in: *CADMUS,* 2,3: 23-36.

Ramaswami, Anu; Weible, Christopher; Main, Deborah; Heikkila, Tanya; Siddiki, Saba; Duvall, Andrew; Pattison, Andrew; Bernard, Meghan, 2012: "A Social-Ecological-Infrastructural Systems Framework for Interdisciplinary Study of Sustainable City Systems", in: *Journal of Industrial Ecology,* 16,6: 801-813.

Robinson, Blake; Musango, Josephine; Swilling, Mark; Joss, Simon; Mentz-Lagrange, Sasha, 2013: *Urban Metabolism Assessment Tools for Resource Efficient Cities.* Report commissioned by Global Initiative for Resource Efficient Cities hosted by UNEP. Stellenbosch: Sustainability Institute; at: <http://www.sustainabilityinstitute.net/newsdocs/document-downloads/cat_view/23-research-project-outputs> (1 July 2014).

Scherr, Sara, 1999: *Soil Degradation: A Threat to Developing-Country Food Security by 2020?.* Food, Agriculture and the Environment Discussion Paper 63 (Washington, DC: International Food Policy Research Institute); at: <http://www.env-edu.gr/Documents/Soil %20Degradation%20-%20A%20Threat%20to%20Developing-Country%20-%20Food%20Security%20by%202020.pdf> (12 April 2008).

Siemens, PWC; Berwin Leighton Paisner, 2014: *Investor Ready Cities: How Cities can Create and Deliver Infrastructure Value* (city: Siemens - PWC - Berwin Leighton Paisner); at: <http://www.pwc.com/en_GX/gx/psrc/publications/assets/pwc-investor-ready -cities- vi.pdf > (6 April 2015).

Smil, Vaclav, 2014: *Making the Modern World: Materials; Dematerialization* (Chichester, UK: Wiley).

Swilling, Mark; Robinson, Blake; Marvin, Simon; Hodgson, Mike, 2013: *City-Level Decoupling: Urban Resource Flows and the Governance of Infrastructure Transitions.* Paris: United Nations Environment Programme; at: <http://www.unep.org/resourcepanel/ Publications/ AreasofAssessment/Cities/tabid/106447/Default.aspx>.

Swilling, Mark, 2013a: "Beyond the Resource Curse: From Resource Wars to Sustainable Resource Management in Africa", in: Minderman, G.; Raman, V.; Cloete, F.; Woods, G. (Eds.): *Good, Bad and Next in Public Governance* (The Hague: Elevent International Publishing).

Swilling, Mark, 2013b: "Economic Crisis, Long Waves and the Sustainability Transition: An African Perspective", in: *Environmental Innovations and Societal Transitions,* 6: 95-115.

Swilling, Mark; Annecke, Eve, 2012: *Just Transitions: Explorations of Sustainability in an Unfair World* (Cape Town - Tokyo: UCT Press; United Nations University Press).

Swyngedouw, Erik, 2006: "Metabolic Urbanization: The Making of Cyborg Cities". in: Heynen, Nikolas; Kaika, Maria; Swyngedouw, Erik (Eds.): *In the Nature of Cities* (London: Routledge).

UNEP, 2010: *Assessing the Environmental Impacts of Consumption and Production: Priority Products and Materials.* A Report of the Working Group on the Environmental Impacts of Products and Materials to the International Panel for Sustainable Resource Management (Paris: UNEP).

UNEP, 2014: *Decoupling 2: Technologies Opportunities and Policy Options* (Paris: UNEP).

United Nations Centre for Human Settlements, 2003: *The Challenge of Slums: Global Report on Human Settlements* (London: Earthscan).

Urama, Kevin, 2015: *Water Decoupling Report.* Unpublished report for the International Resource Panel (Paris: UNEP).

Weisz, Helga; Steinberger, Julia. K., 2010: "Reducing Energy and Materials Flows in Cities", in: *Current Opinion in Environmental Sustainability*, 2: 185-192.

Wiedman, Thomas O.; Schandl, Heinz; Lenzen, Manfred; Moran, Daniel; Suh, Sangwon; West, James; Kanemoto, Keiichiro, 2013: "The Material Footprint of Nations", in: *Proceedings of the National Academy of Science*, vol.,issue (date): pages; at: <www.pnas.org/cgi/doi/10.1073/pnas.1220362110>.

World Business Council for Sustainable Development, 2014: *The Urban Infrastructure Initiative* (Geneva: World Business Council for Sustainable Development).

Part V Developing Theoretical Approaches on Sustainability and Transitions

18   Sustainability and Complexity: A Few Lessons from Modern Systems Thinking

Czesław Mesjasz[1]

Abstract

Sustainable development, sustainability and sustainability transition are associated with the growing complexity of sociopolitical systems and ecological systems, and of their interactions. These concepts are becoming even more intricate since in academic considerations and in policy-making the concept of complexity is not precisely defined. The term complexity usually relates to a special class of non-linear mathematical models which are relevant to phenomena described with characteristics measurable in the ratio scale which occur in nature and in social systems. It is also used as analogy and metaphor. The aim of this chapter is to identify and to assess the applications of the concepts deriving from complexity studies in the discourse on sustainability and sustainable development and on related terms—transition to sustainability and management of transition to sustainability. The main hypothesis of the chapter is that a sophisticated language dealing with complexity and applied in the narratives on sustainability and related ideas requires a profounder clarification so as to provide new insights concerning description, explanation of causal links, prediction, normative approach and influence upon societal phenomena.

Keywords: Complexity, sustainable development, sustainability transition, transition management.

18.1   Introduction

The idea of sustainable development, viewed as an action plan for achieving sustainability of the world society and of its components, has undoubtedly become one of most important social, economic and political ideas at the turn of the twenty-first century. Treated as a response to the environmental threats caused by economic development and increasing population, sustainability is viewed as a kind of ultimate objective. However, a closer look at the discourse on sustainability leads to the conclusion that very often it is defined in too general terms. In consequence of its broad meaning, 'sustainability', 'sustainable development' and associated terms can be perceived as fuzzily defined metaphors, or in other words, as a kind of linguistic variables (Zadeh 1975).

Sustainable development, sustainability and sustainability transition are associated with the growing complexity of sociopolitical systems, ecological systems, and their interactions. The concepts are becoming even more intricate since in academic considerations and in policy-making the concept of complexity is not precisely defined. The term 'complexity' is usually related to a special class of non-linear mathematical models which are relevant to phenomena described with characteristics measurable on the ratio scale and which occur in nature and in social systems. Such models, in turn, are depicted by well-known terms such as bifurcation, chaos, emergent properties, 'butterfly effect', and so on. Complexity can be also viewed as a qualitative feature of systems. In the latter case, the meaning of 'complexity' is intersubjective and is expressed with analogies and metaphors which are not always useful in scientific discourse (see chapter 13 by Scheffran).[2]

---

1   Associate Professor dr habil., Cracow University of Economics, Cracow, Poland; Email: <mesjaszc@uek.krakow.pl>.

---

2   Although metaphor is a type of analogy, in this chapter metaphor and analogy are treated separately since analogy can be of a more universal character, for example a simulated mathematical model of a social system, while metaphor is a purely linguistic vehicle.

© Springer International Publishing Switzerland 2016
H.G. Brauch et al. (eds.), *Handbook on Sustainability Transition and Sustainable Peace*,
Hexagon Series on Human and Environmental Security and Peace 10, DOI 10.1007/978-3-319-43884-9_18

This means that in studying the links between the complexity of socio-economic systems and of ecological systems, the Wittgensteinian language games of several concepts have to be considered (Wittgenstein 2002). Without delving into such games, it is necessary to strive for more precise definitions of such terms as 'complexity', 'sustainable development', 'sustainability', 'transition to sustainability' and 'management of transition to sustainability'. Numerous attempts have been already made to define the concepts examined in this chapter. In only a few of these attempts, however, have the links between the deeper meaning of sustainability and its associated terms, and the ideas taken from complexity studies/theory, been investigated, e.g. Jenks/Smith 2006; Loorbach 2007; Cilliers 2008; Picard 2010; Espinosa/Walker 2011. The concepts of sustainability transition and transition management are built upon various attributes of complexity, e.g. Rotmans/Kemp/van Asselt et al. 2001; Rotmans 2005; Loorbach 2007, 2008; Camaren/Swilling 2014.

The aim of the chapter is to identify and to assess the applications of the concepts deriving from complexity studies in the discourse on sustainability, sustainable development and on terms related to them—transition to sustainability and management of transition to sustainability.[3] The main hypothesis of the chapter is that despite a sophisticated language related to complexity and applied in the narratives on sustainability and related ideas, there is a need to maintain the basic classical interpretations of the characteristics of complexity as far as description, explanation of causal links, prediction, the normative approach and influence upon societal phenomena are concerned. This is of particular importance since in numerous works on the complexity of social systems and of sustainability, the narratives are based upon carelessly used analogies and metaphors, the main

reason being that the subjective (intersubjective) character of social systems is not sufficiently recognized.

The analysis presented in this chapter is focused upon conceptual and linguistic considerations. It is of special importance for three reasons. First, a closer look at the concepts and definitions used in the social sciences allows for a profounder understanding and more precise operationalization. Second, the rhetoric of sustainability-related scientific discourse is often applied in political considerations on development and growth. This means that the weaknesses and imprecision of the ideas could be transferred into policy practice where they can become perceived as void 'pseudoscientific' slogans. The third element of the proposed approach is methodological. The chapter is treated as an introduction to a more rigorous quantitative research in which 'text mining' and similar methods could be used to produce a kind of 'inventory' of the applications of complexity characteristics to narratives on sustainability, which, in turn, would allow the building of typologies, the deepening of interpretations, the creation of a ground for the phenomenology of the complexity of sustainability, and subsequently, the enhancement of the effectiveness of applications in practice.

In the first part of the chapter, interpretations of sustainability and sustainable development are presented. The second part includes a survey of interpretations of the ideas of sustainability transition and transition management which have been developed in the early twenty-first century. In the third part, barriers to defining complexity and the complexity of social systems are analysed. The reasons for and examples of abuses and misuses are described. It is stressed that social systems are of subjective (intersubjective) character. Since prediction is the key element of the discourse on sustainable development, the barriers of prediction stemming from the complexity of social systems are described. In the fourth part, the links between ideas taken from complexity studies and sustainability, sustainable development, sustainability transition and transition management are surveyed and a preliminary assessment is made. The final part contains conclusions and recommendations for future research.

---

3   Because of the vagueness of the idea of complexity, in my earlier works I have refrained from using the terms 'complexity theory' and 'complexity science', although the ideas of 'emerging sciences or science of complexity' had already been proposed by Waldrop (1992) and Richardson and Cilliers (2001), and several authors use the term 'complexity theory'. Taking into account the numerous examples of the application of the terms 'complexity science' and 'complexity theory', the latter term is used in this chapter, sometimes interchangeably with 'complexity studies'. However, it would be going too far to claim that complexity theory is a developed scientific theory. The same applies to complexity science.

18.2   Systemic Characteristics of Sustainability

18.2.1   Definitions of Sustainable Development and Sustainability

18.2.1.1   Sustainable Development and Sustainability: Main Questions

The origins of the concept of sustainable development and sustainability at all levels and types of human systems, starting with local and ending with global systems, can be traced in the economic thought of the eighteenth century. In 1798, Malthus worried about how Britain's apparently inexorable rise in population could be sustained from a finite amount of land. In 1865, Jevons wondered how Britain's ever-increasing consumption of energy could be sustained from finite supplies of coal (Pezzey/Toman 2002: iv). At that time such considerations were rather a part of general reflections upon the human condition than any part of policy-oriented discussion.

In the modern era, the initial ideas associated with sustainable development emerged in the 1960s (IISD 2009). The first significant intellectual impulse for considering the sustainability of world society was delivered by the authors of the first Report to the Club of Rome in 1972 (Meadows/Meadows/Randers et al. 1972). In the Report, a relative simple computer simulation model was used. Five variables were examined in the original model: world population, industrialization, pollution, food production, and resource depletion. The authors explored the possibility of a sustainable feedback pattern for the world that would be achieved by altering growth trends among the five variables under simulated scenarios. In most of the discussions after the Report's publication its simplicity and the pessimistic conclusions partly stemming from it were often noted.

However, the meaning of both terms proved to be remarkably difficult to define, and it was similarly difficult to use the terms more precisely. The problem lies in their systemic/structural properties. This means that, when discussing sustainable development and sustainability, three questions have to be answered:

1. How can sustainability and sustainable development be defined?
2. What are the objects of reference of sustainable development and sustainability?
3. How can sustainability and sustainable development be operationalized and measured?

18.2.1.2   Survey of Definitions of Sustainable Development and Sustainability

Discussions among scholars and policymakers on sustainable development and sustainability were initiated after the publication of the report *Our Common Future* prepared under the supervision of Gro Harlem Brundtland. In the report, sustainable development was defined as development that meets the needs of the current generation, without compromising the needs of future generations (WCED 1987: 37).

Since then, an increasing wave of interpretations and applications of the concepts relating to sustainable development and sustainability has emerged. Although sometimes redundant, multiple ways of defining sustainability are useful for different situations and different purposes.

Taking into account the number and sometimes the generality and superficiality of the definitions of the two terms, it may be concluded that in some instances they have become 'buzzwords' of contemporary sociopolitical rhetoric, very often abused and misused in science and in policy-making.

There have been numerous attempts to define both terms and to provide a critical analysis of their interpretations. Obviously it would be a simplification to focus solely on that kind of popular use of the concepts of sustainability and sustainable development. Therefore, only a brief survey of the main ideas along with synthetic definitions of both terms are presented below.

18.2.1.3   Domains, Levels and Issues of Sustainability

There are several typologies of areas of social life where sustainability is taken into account—see tables 18.3 and 18.4. The most popular one embodies three 'pillars': social, environmental and economic (Adams 2006). There are two typologies which include various conceptions of 'four pillars'. The first specifies human, social, economic and environmental pillars, the second environmental responsibility, social equity, economic health, and cultural vitality. The following areas of sustainability and sustainable development

Table 18.1: General definitions of 'sustainability'. **Source:** The author's own research.

Definition
Sustainable: able to be used without being completely used up or destroyed; involving methods that do not completely use up or destroy natural resources; able to last or continue for a long time.   Full definition of sustainability: Capable of being sustained: a: of, relating to, or being a method of harvesting or using a resource so that the resource is not depleted or permanently damaged,   b: of or relating to a lifestyle involving the use of sustainable methods.[a]
A 'sustainable society' lives by the nine principles...   1. Respect and care for the community of life.   2. Improve the quality of human life.   3. Conserve the earth's vitality and diversity.   4. Minimize the depletion of non-renewable resources.   5. Keep within the earth's carrying capacity.   6. Change personal attitudes and practices.   7. Enable communities to care for their own environments.   8. Provide a national framework for integrating development and conservation.   9. Create a global alliance (IUCN/UNEP/WWF 1991: 10).
The duty imposed by sustainability is to bequeath to posterity not any particular thing with rare exceptions such as Yosemite, for example, but rather to endow them with whatever it takes to achieve a standard of living at least as good as our own and to look after their next generation similarly. We are not to consume humanity's capital, in the broadest sense (Robert Solow, quoted in Anand/Sen 2000: 2033).
Sustainability is an economic state where the demands placed upon the environment by people and commerce can be met without reducing the capacity of the environment to provide for future generations. It can also be expressed in the simple terms of an economic golden rule for the restorative economy: Leave the world better than you found it, take no more than you need, try not to harm life or the environment, make amends if you do (Hawken 1993: 139).
Sustainability is about stabilizing the currently disruptive relationship between earth's two most complex systems—human culture and the living world (Hawken 2007: 172).
A development is said to be weakly sustainable if the development is non-diminishing from generation to generation. This is by now the dominant interpretation of sustainability.   "The second interpretation, known as 'strong sustainability', sees sustainability as non-diminishing life opportunities. This should be achieved by conserving the stock of human capital, technological capability, natural resources and environmental quality" (Brekke 1997: 91).
The core of mainstream sustainability thinking has become the idea of three dimensions: environmental, social and economic sustainability. These have been drawn in a variety of ways, as 'pillars', as concentric circles, or as interlocking circles. ... The IUCN Programme 2005–8, adopted in 2005, used the interlocking circles model to demonstrate that the three objectives need to be better integrated, with action to redress the balance between dimensions of sustainability (Adams 2006: 2).
Four pillars of sustainability: Human, Social, Economic and Environmental: *Human sustainability* means maintaining human capital. Human capital is a private good of individuals, rather than between individuals or societies. The health, education, skills, knowledge, leadership and access to services constitute human capital. Investments in education, health, and nutrition of individuals have become accepted as part of economic development.   *Social sustainability* means maintaining social capital. Social capital is investments and services that create the basic framework for society. It lowers the cost of working together and facilitates cooperation: trust lowers transaction costs. Commonly shared rules, laws, and information promote social sustainability.   The widely accepted definition of *economic sustainability* is maintenance of capital, or keeping capital intact. Thus Hicks's definition of income—the amount one can consume during a period and still be as well off at the end of the period—can define economic sustainability, as it devolves on consuming value-added (interest), rather than capital.   *Environmental Sustainability* itself seeks to improve human welfare by protecting Natural Capital. As contrasted with economic capital, Natural Capital consists of water, land, air, minerals and ecosystem services, hence much is converted to manufactured or economic capital.[b]
The four pillars include Environmental Responsibility, Social Equity, Economic Health, and Cultural Vitality. While it is useful to organize sustainability in terms of these four pillars, it is the integration between them that will drive sustainability, highlight opportunities for innovation and reduce duplication of efforts.[c]

a)Merriam-Webster Dictionary (no year); at <http://www.merriam-webster.com/dictionary/sustainability> (6 March 2013).
b) Goodland, Robert (no year): Sustainability: Human, Social, Economic and Environmental (Washington DC: World Bank); at: <http://www.earthfutureaustralia.org/-4-pillars-of-sustainability.html> (18 September 2013).
c) Sustainable Kingston (no year): "Four Pillars of Sustainability", Sustainable Kingston Corporation; at: <http://sustainablekingston.ca/community-plan/four-pillars-of-sustainability> (11 February 2014).

Table 18.2: General definitions of 'sustainable development'. **Source:** The author's own research.

Definition
Sustainable development is development that meets the needs of the present without compromising the ability of future generations to meet their own needs. It contains within it two key concepts: - the concept of needs, in particular the essential needs of the world's poor, to which overriding priority should be given; and   - the idea of limitations imposed by the state of technology and social organization on the environment's ability to meet present and future needs (WCED 1987: 37).
Principles for Sustainable Development: "Business as usual is no longer an option—for government, private sector or individual citizens. Our soils, waters, forests and minerals are not inexhaustible. Farms, industries, homes and lifestyles must become more sustainable, in every community on our planet. To be sustainable, development must improve economic efficiency, protect and restore ecological systems, and enhance the well-being of all peoples".[a]

a) IISD (International Institute for Sustainable Development), 2009: "The Sustainable Development Timeline"; at: <http://www.iisd.org/ pdf/2009/sd_timeline_2009.pdf> (13 February 2014).

Table 18.3: Levels of analysis of sustainability. **Source:** Riley (1992).

Analysis of sustainability		
Level of analysis	**Typical characteristics of sustainability (cumulative)**	**Typical determinants of sustainability**
Field/production unit	Productive crops & animals; conservation of soil & water; low levels of crop pests & animal diseases	Soil & water management; Biological control of pests; use of organic manure; fertilizers; crop varieties & animal breeds
Farm	Awareness by farmers; economic & social needs satisfied; viable production systems	Access to knowledge, external inputs and markets
Country	Public awareness; sound development of agro-ecological potential; conservation of resources	Policies for agricultural development; population pressure; agricultural education, research & extension
Region/continent/world	Quality of the natural environment; human welfare & equity mechanisms; international agricultural research & development	Control of pollution; terms of trade; distribution

relating to various sociopolitical and socio-ecological systems (areas of reference) can be distinguished:

- sociopolitical sustainability,
- sustainable use of resources (environmental sustainability),
- sustainable agriculture,
- economic sustainability,
- regional sustainability,
- sustainable business and production.

The hierarchy of sustainability areas embodies the following levels: global, macro-regional, transnational, micro-regional, market-oriented corporations, agricultural units and non-market organizations. It might be worth considering the proposal to add individual persons as the fundamental (nuclear) units of sustainability-oriented studies. Definitions of sustainability for every area and every level have been proposed in the literature. As an example, the levels of analysis of sustainability proposed by Riley (1992) are shown in table 18.3.

The selection of examples presented in tables 18.1, 18.2, 18.3 and 18.4 embodies the main interpretations which can be helpful in delving into the complexity of the idea of sustainability and the multitude of its extensions and derivative concepts.

Sustainability can be described and analysed within the framework of the four pillars. All pillars are equivalent but undoubtedly it is human economic activity which determines the present and the future of the fundamental conditions of human living that are treated as desired and sustainable. Viewed from

Table 18.4: Definitions of sustainability in specific socio-economic domains. **Source:** The author's own research.

Definition
Triple bottom line of sustainability—social, environmental (ecological), economic (financial) (three pillars, TBL, 3BL, three P's—People, Planet, Profit) (Elkington 1994).[a]
It must be emphasized that the actor of sustainability is not an individual human being, but rather a generation of human beings making decisions. The generation metaphorically is made to function as a single being attempting to sustain a flow of imputed benefits analogous to but more inclusive than the income of the individual human being. It is this generation's ethical responsibility to 'maintain' a 'broadly defined capital stock' to sustain a 'broadly defined income' for the benefit of future generations (Brätland 2006: 10).
Corporate Sustainability is a business approach that creates long-term shareholder value by embracing opportunities and managing risks deriving from economic, environmental and social developments. Two guiding principles of corporate sustainability: – sustainable business practices are critical to the creation of long-term shareholder value in an increasingly resource-constrained world, – sustainability factors represent opportunities and risks that competitive companies must address.[b]
Sustainable businesses are defined as follows: 1. Replace nationally and internationally produced items with products created locally and regionally. 2. Take responsibility for the effects they have on the natural world. 3. Do not require exotic sources of capital in order to develop and grow. 4. Engage in production processes that are human, worthy, dignified, and intrinsically satisfying. 5. Create objects of durability and long-term utility whose ultimate use or disposition will not be harmful to future generations. 6. Change consumers to customers through education (Hawken 1993: 144).
Sustainable Production is the creation of goods and services using processes and systems that are: – non-polluting, – conserving of energy and natural resources, – economically viable, – safe and healthful for workers, communities, and consumers, – socially and creatively rewarding for all working people. If production is sustainable, then the environment, employees, communities, and organizations—all benefit. These conditions can lead, always in the long term, and often in the short term, to more economically viable and productive enterprises.[c]

a) Triple Pundit (no year); at: <http://www.triplepundit.com/> (24 July 2014).
b) Dow Jones Sustainability Index; at: <http://www.sustainability-indices.com/sustainability-assessment/corporate-sustainability.jsp> (18 February 2014).
c) Lowell Center for Sustainable Production (no year); at: <http://www.sustainableproduction.org/about.what.php> (18 September 2013).

the economic perspective, the main challenge in thinking about sustainability and sustainable development is to balance the present and future needs of humanity. In economic terms this means that the following issues have to be dealt with.

The first is to define the future desired process of development, which can be viewed as self-sustainable. The process should be described with a set of parameters (variables—continuous and punctuated (discrete)). Then the interactions between subsystems must be depicted. An important element in understanding sustainable development is to capture the mechanisms of intertemporal transfer of resources, which can be seen as an element of the integration of human progress with environmental conservation.

As with other areas of sustainability theory and policy, the ideas taken from complexity studies are to a large extent a conceptual foundation of the economic facets of sustainability. They especially concern prediction and mechanisms for the transfer of resources.

The time horizon of the transfer covers one or two generations (twenty-five to fifty years) since a longer perspective in economic (and political) planning is not reasonable. A very long-term perspective is sometimes needed in ecological analyses, although a hundred-year forecast often seems very inappropriate.

As an example of analyses of inter-temporal, sustainability-oriented economic mechanisms, the ideas developed by Anand and Sen (2000) can be used as a source of inspiration. The most important issues of

Figure 18.1: Areas and elements of sustainability description and analysis (three/four pillars). **Source**: Own research based upon Adams (2006) and Goodland.

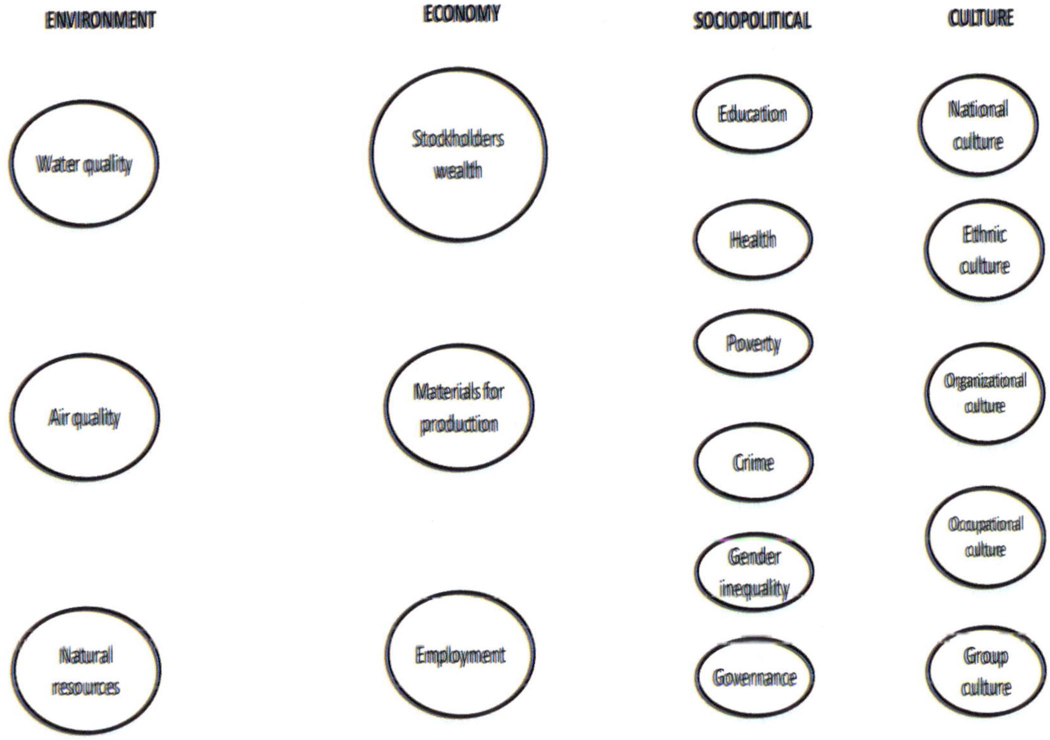

economic aspects of sustainability and sustainable development are the following:

1. The needs of future generations.
2. The volume and kind of capital which has to be left to the following generations.
3. Adequate models of economic growth.
4. Preservation of capital inherited from past generations.
5. Definition of human capital and of its growth.
6. Dilemma—to preserve or to enhance productive capabilities (broad stock capital) by the present generation with a perspective looking towards future generations.
7. Is the continuous increase in standards of living for all people achievable and justifiable?
8. Dilemma—support for the poor today vs. capital left for the future.

18.2.2   Operationalization and Measures of Sustainability and Sustainable Development

Studies of the links between the complexity of socioeconomic systems and sustainability also need to explain the methods of creating the qualitative and quantitative characteristics of sustainability. The number and variety of the definitions shown above is an obstacle to the operationalization and development of the quantitative characteristics (indicators) of sustainability and sustainable development measurable on two scales, ratio and interval. Indicators of sustainability must capture the interdependence between three, or four, 'pillars'. In addition, there is a dilemma concerning the number and quality of indicators. A large number of indicators usually leads to low precision, overlaps and correlations of indicators. Too few indicators do not allow a collection to be prepared which could be representative for description and analysis. As with other methods of assessing social systems, this dilemma remains unsolvable, and specific collections of indicators depend upon the decisions of the authors of the methods. Traditional indicators measure changes in one part of a social entity as if they were entirely independent of the other parts. Sustainability indicators reflect the fact that the elements of three (or four) different pillars are very tightly interconnected directly and indirectly. In figure 18.1, four areas and their elements are presented, and in order to make the scheme clear, no relations are modelled using lines.

There may be different methods of preparing indicators of sustainability. One of them has been proposed by the Sustainable Measures Company. Such a methodology usually derives from assumptions and questions. These are the questions proposed by this company:[4]

1. What is an indicator of sustainability?
2. What are the characteristics of effective sustainability indicators?
3. Are there any checklists that can be used to evaluate sustainability indicators?
4. How can the indicators be organized?
5. How many and what kind are needed?
6. What data sources are available for indicators?
7. Can we find examples/benchmarks of good sustainability indicators?
8. Are there any education/training materials that explain indicators and sustainability?

A survey and assessment of sets of sustainability partial indicators usually labelled as "indices" (macro-indicators) remains beyond the scope of this chapter. In order to show how different sets of indicators help to deal with the complexity of sociopolitical/socio-economic and socio-ecological systems, only the most commonly used indicators are listed. Gender-related indices are also included in this collection since a proper gender policy helps to maintain sustainability at all levels of society.

1. *American Human Development Report* (AHDR),
2. *Ecological Footprint* (EF).
3. *Gender-related Development Index* (GDI).
4. *Genuine Progress Indicator* (GPI).
5. *Gross National Happiness* (GNH).
6. *Human Development Index* (HDI).
7. *National Human Development Report* (NHDR).
8. *The Happy Planet Index* (HPI).
9. *The Index of Sustainable Economic Welfare* (ISEW).

The indicators from the above proposals have been widely discussed and assessed both positively and negatively (McGillivray 1991; Klasen 2006). From a methodological point of view they share several common features.

First, they go beyond traditional economic indicators by adding other characteristics—social, environmental, human, and political.

Second, more attention is paid to non-economic indicators, and in some cases (Ecological Footprint and gender-oriented indicators) other characteristics are considered as crucial for sociopolitical and economic development.

Third, their authors attempt to make them 'objective' by applying indicators measurable with the ratio or interval scales. Wherever possible, the basic idea behind all these approaches is to cast doubt on wealth development at any price, with the inclusion of non-market commodities, positive and negative, to yield an aggregated macro-indicator in monetary terms. This leads to a situation in which a critical approach to results stemming from the intersubjective creation of those indicators is neglected. Although their authors do not declare it openly, they are aiming at making those partial indicators measurable, not correlated and not redundant with the other indicators within the same set (index, macro-indicator).

Fourth, the development of numerous indices may contribute to confusion in policy-oriented discourse because of their lack of precision, their too high level of detail, and incommensurability of partial indicators within the indices and between the aggregate indices.

Fifth, the extension of factors and subsequently the number of indicators of sustainability and sustainable development may lead to stretching their meaning beyond the possibility of theoretical usefulness and practical applicability. It may create a situation like that of the deepening and broadening of security, which may sometimes lead to its being seen as becoming too broad and too universal and subsequently losing its theoretical and practical usefulness.

Sixth, there may also be problems resulting not only from overlaps of indicators and indices of different concepts of sustainability and sustainable development but also from overlaps with indicators applied in defining other characteristics such as resilience and robustness.

18.3 Sustainability Transition as a Theoretical and Policy-Making Concept

18.3.1 Sustainability Transition

Changing society at all levels, beginning with small groups and ending on a global scale, has always been

---

4 The questions are a modified version taken from Sustainable Measures Company (no year); at: <http://www.sustainablemeasures.com/company> (24 March 2014). The website of the Sustainable Measures Company states that it ceased its activities in 2014.

an important issue for theorists. It is not just a matter of description and understanding; there is also the challenge of how to influence the change. This leads to discussions about goals, norms and methods of change. At the micro-level, i.e. companies and non-market institutions, influencing change is mirrored in change management or project management. Theory and practice show that influencing change at that level is achievable although the expected results are never guaranteed (Kotter 1995).

There are more significant challenges in attempting to influence or even control change at higher levels of the social hierarchy—industry, a domain of social activity such as a health service or innovation, a region in a country, a country, a multi-country region, and global. In such cases the systems and the changes become more complex and policy-making requires the coordination of goals and of the activities of numerous actors. Problems at the macro-scale include security, macro-economic policy, social policy, international finance, trade, and environment.

The term transition, deriving its metaphorical sense from the phase transition in physics and chemistry, has been applied to changes at societal levels above the micro-scale, e.g. a company, and may concern such areas as energy, agriculture or regions. Usually it is equivalent to regime transformation, technological revolutions, technological transitions, system innovation, and transition management (Geels/Schot 2007).

Societal transitions are defined as processes of change that structurally alter the culture, structure and practices of a societal system. As with the case of sustainable development, these processes take a long time (one to two generations) at the level of societal systems, although partial processes (for example, fundamental changes in thinking or radical innovation) can happen almost overnight. A societal transition results from interacting changes in all societal domains (e.g. economy, ecology, institutions, technology and welfare) (Loorbach 2007: 17).

Changes in all these areas are especially significant when sustainability and sustainable development are taken into account. There are plenty of ways of assessing the characteristics of socio-economic and environmental systems, but the recommended policies, processes and methods of achieving or approaching the desired states are usually defined at a very high level of generality. One of the ways to specify more precisely not only what sustainability and sustainable development are, but how to achieve them, is the idea of sustainability transition (transition to sustainabil-

ity), or even transition to sustainable development (Grin/Rotmans/Schot 2010) and transition management.

Conceptual and policy problems resulting from difficulties in defining sustainability and sustainable development also come to the fore when studying the meaning of sustainability transition and transition management. These concepts were first put forward at the beginning of the twenty-first century by theoreticians and policymakers in the Netherlands (Rotmans/Kemp/van Asselt et al. 2001; Rotmans/Kemp/van Asselt 2001). The authors of the ideas treat them as new interdisciplinary, multidisciplinary and even transdisciplinary "transition studies", begetting transition theory and even a new paradigm. These lead to new types of research: sustainability science, integrated assessment, post-normal science, and action research (Loorbach 2007: 33-34).

These have emerged over the past few years as a new approach to dealing with complex societal problems and governance in the context of such problems. In the Netherlands, the UK and Belgium, efforts have been made to develop transition policies in areas such as energy, building, health care, mobility, and water management (Loorbach/Rotmans 2010).

Transition is a process of structural societal change from one relatively stable system state to another via a co-evolution of markets, networks, institutions, technologies, policies, individual behaviour and autonomous trends. The complexity of a transition means that it has a multitude of driving factors and impacts. A transition can be accelerated by one-time events, such as big accidents or crises (such as the oil crisis). Slow changes in the external environment determine the undercurrent for a fundamental change; superimposed on this undercurrent are events such as disasters, which may accelerate the transformation process. Transitions are thus multi-causal, multi-level, multi-domain, multi-actor and multi-phase processes. In this new research field, transition processes are studied from a variety of system perspectives: sociotechnical systems, innovation systems and complex adaptive systems.

In the most general sense, the transition can be treated as a continuous process of societal change, whereby the structure of society (or a subsystem of society) changes fundamentally. This societal transformation process has the following characteristics (Loorbach 2007: 18):

- it concerns large-scale technological, economical, ecological, sociocultural and institutional developments that influence and reinforce each other,

- it is a long-term process that covers at least one generation (twenty-five years), and
- there are interactions between different scale levels (niche, regime, landscape).

The process-oriented character of the transitions approach can be depicted with the following propositions (Loorbach, 2007; Loorbach/Frantzeskaki/Thissen 2011):

Proposition 1: Sustainability transitions are long-term processes of fundamental societal change that incorporate processes of societal, ecological, economic, cultural and technological evolution.

Proposition 2: Enabling societal processes of change (transitions) requires an integrated understanding of the dynamics of change and deliberate and reflexive strategies so as to allow for self-orientation of society towards a sustainable development pathway.

Proposition 3: Innovation and sustainable development are interlinked: to develop sustainably means to continuously innovate and redefine existing culture, structures and practices in an evolutionary manner. More specifically, a focus on sustainability could trigger innovations that comply with sustainability values as well as such innovations becoming the stimuli for the initiation of multi-domain processes for societal transitions to sustainability.

Proposition 4: Sustainability transitions are continuous open-ended processes of societal innovation. Governance for sustainability transitions has thus to secure sustainability values such as long-term orientation and intergenerational justice.

Summing up the above, it may be stated that sustainability transition constitutes a specific case of transition which results from narrowing the meaning of transition by setting a point of departure—an unsustainable state (static), or perhaps, more frequently, a more or less stable process. The purpose of the transition is to achieve a desired/expected but difficult to define state(?), unstable equilibrium, stable equilibrium(?), or process(?).

18.3.2   Transition Management

The concept of transition sustainability provides a theoretical background for the set of theoretical concepts and activities described as transition management. The goal of transition management is to enable, facilitate and guide transitions to sustainability. It tries to do so through structuring processes of governance based on the principles of complexity, transition and sustainability.[5]

Definitions of transition management have evolved since 2001 (Rotmans/Kemp/van Asselt et al. 2001; Rotmans/Kemp/van Asselt 2001; Loorbach 2007, 2008). The concept of transition management is a broad one, meaning that its definitions have many dimensions. According to Lorbach (2008: 12):

> ...transition management is a deliberative process to influence governance activities in such a way that they lead to accelerated change directed as sustainability ambitions. Transition management is thus defined as meta-governance: how do we influence, coordinate and bring together actors and their activities in such a way that they reinforce each other to such an extent that they can compete with dominant actors and practices? Transition management is thus about creating space (in a sense: governance niches) for innovative governance at all levels, as a strategy to develop alternatives to the regime. Transition management anticipates increasing pressures on the regime level (e.g. predevelopment phase) or tries to provide a more fundamental reflection and long term orientation while the process of change is underway (e.g. acceleration phase).

Transition management can be depicted with several sets of characteristics relating to its framework, governance—in Kooiman's sense (1993), levels and cycle. Loorbach (2008: 13-14) proposes the following tenets of complexity governance to be treated as a part of transition management:

1. Multi-domain, multi-actor and multi-level systems thinking is necessary.
2. Long-term thinking (at least twenty-five years) as a framework for shaping short-term policy in the context of persistent societal problems.
3. Objectives should be flexible and adjustable at the system level. The complexity of the system is at odds with the formulation of specific objectives.
4. Creating space for niches in transition arenas and transition experiments. A niche is a new structure, a small core of agents, that emerges within the system and that aligns itself with a new configuration. The new alignment is often the emergent property of the system. An emergent structure is formed around niches to stimulate the further development of these niches and the emergence of niche regimes.
5. A focus on front-runners. In this context, by front-runners we mean agents with particular competen-

---

5   The complexity of broadly defined sociopolitical governance (Jessop 2004) and of corporate governance (Goergen/Malline/Mitleton-Kelly et al. 2010) has already been studied, albeit with a predominantly qualitative approach.

cies and qualities: creative minds, strategists and visionaries.

6. Guided variation and selection.

7. Radical change in incremental steps. Radical, structural change is needed to erode the existing deep structure (incumbent regime) of a system and ultimately dismantle it. Immediate radical change, however, would lead to maximal resistance from the deep structure, which cannot adjust to radical change that happens too fast.

8. Learning-by-doing and doing-by-learning. Social learning is a pivotal aspect of societal transition processes, aimed at 'reframing', changing the perspective of the actors involved.

9. Anticipation and adaptation. Anticipating future trends and developments, taking account of weak signals and seeds of change acting as the harbingers of the future, is a key element of a proactive, long-term strategy for transition management, which has emerged over the past few years as a new approach to dealing with complex societal problems and governance in the context of these problems.

To complete this partial characterization of transition management, it should be added that four different types of governance activities can be distinguished when observing the behaviour of actors in the context of societal transitions: strategic, tactical, operational and reflexive. It is also important from the point of view of applicability that transition management can be implemented through a specific methodology (cycle) that includes the following stages (Loorbach 2008: 15):

- problem structuring, establishment of the transition arena and envisioning;
- developing a transition agenda, a vision of sustainability, and developing and deriving the necessary transition paths;
- establishing and carrying out transition experiments and mobilizing the resulting transition networks;
- monitoring, evaluating and learning.

This necessarily simplified description of transition management provides the necessary background for discussion of the links between this concept and complexity theory. More details can be found in the growing body of literature on this and associated topics; see, for example, Rotmans/Kemp/van Asselt et al. 2001; Rotmans/Kemp/van Asselt 2001; Loorbach 2007, 2008; Grin/Rotmans/Schot 2010; and Loorbach/Frantzeskaki/Thissen 2011.

18.4 The Meaning of Complexity

18.4.1 Definitions of Complexity: Models, Analogies and Metaphors

Complexity is undoubtedly one of most popular concepts used in contemporary science and policy-making. Studies of complexity are rooted in cybernetics and systems thinking.[6] The first attempts to define and study complex entities go back to the works of Weaver (1948) ("disorganized complexity" and "organized complexity"), Simon (1962)–the Architecture of Complexity, and Ashby (1963)–the Law of Requisite Variety. In his search for explaining the meaning of complexity, Lloyd (2001) identified forty-five different ways of describing complexity. A very convincing picture of the intricacy of the field of complexity science can be found in the scheme proposed by Castelani (2014). In other writings, numerous definitions of complexity have been formulated and scrutinized– Prigogine and Stengers (1984), Waldrop (1992), Gell-Mann (1995), Kauffman (1993, 1995), Holland (1995), Bak (1996), Bar-Yam (1997), Haken (1997), Prigogine (1997, 2003) and Biggiero (2001).

Unequivocal distinction of complex systems from "classical" systems is not possible. In the works by Wiener (1948/1961), Ashby (1963, defining "first-order cybernetics" and 'hard' systems thinking), and Bertalanffy (1968), complexity was treated as one of the important features of systems without considering the role of the observer. Their works were the first to list the systemic/cybernetic characteristics of systems: system, element, relation, subsystem, environment, input, output, feedback, black box, equilibrium, stability, synergy, turbulence.

Taking a preliminary approach, the complexity of systems derives from the number of elements and of their interactions. It can be also characterized by a multitude of such traits as adaptability, adaptation, attractor, autopoiesis, chaos, bifurcations, butterfly effect, closed system, co-evolution, complex adaptive systems, dynamic systems, edge of chaos, emergent properties, far-from-equilibrium states, fitness landscape, fractals, non-linearity, open system, path

---

6 There are various interpretations of the relationship between cybernetics and systems thinking but, following Bertalanffy (1968), it can be agreed that the former can be regarded as a part of the latter. To avoid unnecessary typological considerations, it is also assumed that studies of complex systems are treated as a part of systems thinking (Mesjasz 1988; Midgley 2003).

dependence, power law, reflexivity, scale-free networks, self-organization, self-organized criticality, self-reflexivity, synergy, synergetics, and turbulence. These ideas are extensively depicted in a large number of writings of which only a small fraction are quoted in this chapter.[7]

Impossibility of decomposition and incomprehensibility are also treated as important facets of complexity. Gell-Mann (1995) shows that complexity can be treated as a function of the number of interactions between the elements in a system. Nicolis and Prigogine (1989) prefer measures of complexity based on system 'behaviour' rather than on any description of system interactions. Similarly, behaviour is also fundamental to the analysis and description of *Complex Adaptive Systems* (CAS) by Holland (1995).

Ideas originating in systems thinking and complexity studies are used in the social sciences as models, analogies and metaphors. According to this distinction, the term 'model' is used only for mathematical structures. Mathematical models in complexity studies can be applied in three areas: computer-based experimental mathematics, high-precision measurements made across various disciplines and confirming the 'universality' of complexity properties, and rigorous mathematical studies embodying new analytical models, theorems and results.

Models, analogies and metaphors are instruments of theory in the social sciences and are used for description, the explanation of causal relations, prediction, anticipation, the normative approach, prescription, retrospection, retrodiction, control and regulation, or, in a modern approach, influence upon the system. It is also worthwhile adding that models, analogies and metaphors deriving from systems thinking/complexity studies are gaining a special significance in the social sciences. They are treated as 'scientific' and hence obtain supplementary political influence resulting from 'sound' normative (more specifically, prescriptive) legitimacy in any debate on security theory and policy[8].

It must be mentioned that unlike physics, chemistry and biology, where only mathematical models are used in prediction, in the social sciences it is also qualitative considerations that are used in prediction. Therefore, the role of analogies and metaphors taken from complexity studies must be taken into account with sufficient care.

One of most influential ideas of complex research are the scale-free networks developed by Barabási and Albert (1999; Barabási 2003). After finding that various networks, including some social and biological networks, had heavy-tailed degree distributions, Barabási and collaborators coined the term 'scale-free network' to describe the class of networks that exhibit a power-law degree distribution, which they presumed could describe all real-world networks of interest.

The above ideas can be called 'hard' complexity research by analogy with 'hard' systems thinking, and to some extent, with 'first-order cybernetics'. It includes the mathematical modelling of systems with well-defined and measurable characteristics in physics, chemistry, and the natural sciences, as well as in society.[9]

'Soft' complexity research, a term coined by analogy with 'soft' systems thinking and 'second-order cybernetics', includes the ideas of complexity developed in other areas—in cybernetics and systems thinking, in the social sciences, and in psychology. Its ideas can be divided into two groups. The first includes those based upon analogies and metaphors drawn from 'hard' complexity studies; these dominate in social sciences theory and practice, and are very often abused and misused (Gleick 1987; Castelani 2014). The second group includes the indigenous qualitative concepts of complexity such as those developed by Luhmann (1995).

Subjectivity is the first aspect of complexity in the 'soft' approach. Following this line of reasoning, from the point of view of second-order cybernetics, or in a broader approach, constructivism (Glasersfeld 1995; Biggiero 2001), complexity is not an intrinsic property of an object but depends on the observer. Usually it is said that "complexity, like beauty, is in the eye of the beholder".

---

7   It may be observed that the titles of the writings that include concepts such as chaos, edge of chaos, complexity, turbulence, etc., draw the additional attention of non-specialists in the field, as well as social scientists who do not have a sufficient background in the mathematical aspects of these terms, and last but not least, the general public. This extra 'appeal' of works with titles and narratives embodying these utterances is likely to be one of the reasons for the simplifications, misuses and abuses.

8   The differences between prescriptive approach based upon "common sense", and qualitative reasoning, and normative approach based upon ideal, mathematical models of rationality in decision making has been described by (Bell/Raiffa/Tversky 1988).

9   The distinction between 'soft' and 'hard' complexity was also introduced by Richardson and Cilliers (2001), though with a slightly different meaning.

As to identifying a genuine epistemological meaning of complexity, Biggiero (2001: 3), basing his approach on some properties of the relationships between observers (human or cognitive systems) and observed systems (all kind of systems), treats predictability of the behaviour of an entity as the fundamental criterion for distinguishing various kinds of complexity. He proposes three classes of complexity: (a) objects not deterministically or stochastically predictable at all; (b) objects predictable only with infinite computational capacity; (c) objects predictable only with a transcomputational capacity.

Coming from this typology, he defines '*observed irreducible complexity* (OIC)' as those states of unpredictability which allow an object to be classified in one of the three classes. This definition allows complexity to be distinguished semantically in the new sense.

The typologies presented by Biggiero lead to two conclusions that are important in studying social systems. Firstly, the first class is characterized by self-reference, relating to the many forms of undecidability and interactions between observing systems (Foerster 1982). This property, being a foundation of 'second-order cybernetics', in some sense favours the subjective interpretations of complexity. Secondly, human systems are characterized by the presence of all sources and types of complexity (Biggiero 2001: 4-6). One can summarize this by stating that human systems are the 'complexities of complexities'.

In the social sciences, and particularly in sociology, attention is given to the concepts of complexity of systems proposed by Luhmann. This is the main idea of 'soft' complexity, akin to 'second-order cybernetics'. Luhmann is one of only a few authors to have made an attempt to provide a comprehensive definition of social systems based solely on communication and on the concept of the autopoiesis (self-creation) of biological systems. According to Luhmann, a complex system is one in which there are more possibilities than can be actualized. Complexity of operations means that the number of possible relations becomes too large with respect to the capacity of elements to establish relations. It means that complexity enforces selection. The other concept of complexity is defined as a problem of observation. Now, if a system has to select its relations itself, it is difficult to foresee what relations it will select, for even if a particular selection is known, it is not possible to deduce which selections would be made (Luhmann 1990: 81).

Complexity of social systems as developed by Luhmann is strongly linked to self-reference, since reduction of complexity is also a property of the system's own self-observation, although no system can possess total self-insight. This phenomenon is representative for the epistemology of modern social sciences, where observation and self-observation, reflexivity and self-reflexivity, self-reference and subsequently intersubjectivity play an important role. According to this interpretation, social systems become self-observing, self-reflexive entities trying to solve any problems that arise through the process of adaptation (learning).

Taking an epistemological stance which might be called moderate constructivism, it should be emphasized that definitions of all categories do not have any 'objective' character, independent from the observer. This is a basic epistemological assumption in modern social sciences. Therefore, in systems thinking dealing with both 'hard' and 'soft' complexity, intersubjective interpretations of concepts are the point of departure for investigations.

The links between intersubjectivity and complexity of all kinds of systems need to be further investigated. The challenges stemming from intersubjectivity are of special importance in the discourse on 'soft' complexity, where reflexivity, self-reflexivity and self-reference are taken into consideration. This issue is thus crucial for social systems, in which the systemic properties are always the constructs of participants/observers. Paradoxically, with a few exceptions, these links have not yet been comprehensively analysed.

Intersubjectivity is one of the key issues of modern psychology and sociology, although there is no commonly accepted definition of the concept (Gillespie/ Cornish 2009). Bednarz (1984) made an initial attempt to investigate how intersubjectivity is connected with Luhmann's concept of the social system. In Bednarz's approach, Husserl's (1970) idea of transcendental intersubjectivity is treated as the source of all objectivity and meaning—including that of the world itself. Intersubjectivity should be always be borne in mind when defining and studying social systems.

18.4.2 Non-linearity, Intersubjectivity, Prediction and Complexity

There are three important aspects of complexity which have to be explained in a more detailed way in the discourse on sustainability and sustainable development: non-linearity and the limits of prediction resulting from the complexity of social systems. The main reason for choosing these characteristics is that they can be treated as the fundamental facets of the complexity of socio-economic and ecological systems which are applied in the discourse on sustainability.

Complex systems exhibit non-linear behaviour that is called positive feedback where internal or external changes to a system produce amplifying effects. Non-linear systems can generate a specific temporal behaviour which is called chaos. Chaotic behaviour can be observed in time series as data points that appear random and devoid of any pattern but which demonstrate a deeper, underlying effect. During unstable periods such as chaos, non-linear systems are susceptible to shocks (sometimes very small). This phenomenon, called 'sensitivity to initial conditions' and popularized as Lorenz's 'butterfly effect', exemplifies the cases where a small change may generate a disproportionate change (Gleick 1997). The major lesson of non-linear dynamics is that a dynamic system does not have to be 'complex' or be described by a large set of equations in order for the system to exhibit chaos. All that is needed is three or more variables and some embedded non-linearity (Ilachinski 1996: 57).

When considering non-linearity and its consequences in complex systems two phenomena should be borne in mind. Firstly, as Stanislaw Ulam once remarked, discerning non-linear phenomena and their mathematical models was "like defining the bulk of zoology by calling it the study of 'non-elephant animals'". His point, clearly, was that the vast majority of mathematical equations and natural phenomena are non-linear, with linearity being the exceptional, but important, case (Campbell 1997: 218). Secondly, it should be also mentioned that to distinguish between the statements that linear is predictable and non-linear is not predictable is a simplification. For instance, Newton's equations for the two-body Kepler problem (the Sun and one planet) are non-linear and yet explicitly solvable. It means that non-linearity does not always lead to chaos. At the same time the fundamental equation of quantum mechanics, Schrödinger's equation, is absolutely linear (Sokal/Bricmont 1998: 144-145).

Prediction is the key factor in discussions about sustainability, sustainable development and sustainability transition. It has been widely discussed in numerous works ranging from philosophy to mathematical modelling.[10] Here only a few barriers to prediction will be discussed. Leaving aside the analysis of the sense of causality and explanation, it may be stated that prediction is the key element of scientific reasoning (Popper 1974). Prediction is associated with such terms as forecasting, anticipation, predictability and prophecy. The basic meaning of prediction is a statement or claim

that a particular event will occur in the future. Narrowing the sense, it may be added that the place and time of the event are known as well.

Prediction can be interpreted as the selection of one world from an infinite number of 'potential worlds' which can emerge from any given set of circumstances. This interpretation brings in the question of causality. First of all it must be stressed that causality may be interpreted as objective if proved with logical or empirical evidence. At the same time causality has an intersubjective character. Observers (participants) agree that a specific course of events has led to a specific outcome although it is not certain whether both cause and effect are unique and cannot be replaced. Prediction is also associated with anticipation, i.e. expectation or making decisions based upon the predicted states of the future, beliefs, etc.

Since social systems are 'complexities of complexities', understanding the barriers to the prediction of the behaviour of social systems requires a broader survey, including obviously the barriers of mathematical modelling.

In the most fundamental sense, in which the impact of postmodernism/post-structuralism is not incidental, social phenomena (social systems) are a mixture of a tangible component (physical and biological processes) with an intangible intersubjective component which emerges in the discourse. In a linguistically-oriented approach, the discourse on social systems in economics, management and finance includes metaphors, analogies and mathematical models. In such an approach, the barriers of prediction begin not only at the epistemological level (the future is unknown) but also at the ontological level (what is the object of which the future is to be predicted and what is the initial state of that object?). There are multiple possible futures created in the social discourse and they are shaped by various subjective factors determined by independent, 'natural' processes as well as by the actions and utterances of conscious individuals (participants/observers).

Reference to social learning brings in the main source of the complexity of social systems i.e. reflexivity and self-reflexivity. Regularities generated by social systems can be brought about by homeostatic causal loops, self-regulation through feedback and reflexive self-regulation.

The above interpretation of the prediction of the behaviour of social systems is very rudimentary but should be sufficient for further discussion on sustainability and on related ideas.

---

10 This discussion of prediction draws partly on Mesjasz (2008a).

18.5 Determinants and Main Characteristics of Complexity of Sustainability

18.5.1 Assumptions of the Survey

A closer look at the conceptual barriers to defining vague notions of sustainability, sustainable development, sustainability transition, transition management and, last but not least, complexity, allows us to conclude that paradoxically, they may be perceived as representative of both postmodernist and constructivist trends, as represented, for example, by Baudrillard (1994), Cilliers (1998), Foucault (2007), and Glasersfeld (1995), as well as for "reflexive modernity" (Beck/ Giddens/Lash 1994) and "liquid modernity" (Baumann 2000). Therefore it is natural that they require a thorough, multidimensional analysis. In the theoretical assessment it is necessary to take into consideration the value of the concepts from the point of view of the already-mentioned functions of analogies, metaphors and mathematical models, i.e. description, explanation of causal links, prediction, normative approach, and the possibility of influencing societal phenomena (regulatory approach).

The above general framework of the description and assessment of the application of complexity ideas in research into and the practice of sustainability-related concepts is perfectly reflected in a self-assessment by the authors of the idea of sustainability transition (Lorbach 2007: 23):

> Sustainable development becomes rather complex when one tries to operationalize it in terms of governance strategies. After the initial optimism during the 1990s about win–win opportunities, it is increasingly understood that there are trade-offs between different values and interests in any type of development (at least in the short term) and that each development raises new problems for society. Sustainable development should therefore be considered as a continuous process in which these values and interests are discussed, negotiated and balanced. This means that sustainability in itself can never be defined objectively beforehand, but that process-conditions and contextual factors can be formulated which ensure an equal representation, pluriform debate and informed discussion. Approaching sustainable development as a continuous process of change means that it cannot be translated into a blueprint or a defined end state from which criteria could be derived and unambiguous decisions are taken to get there. Rather, it is a multi-dimensional, dynamic and plural concept that can neither be translated into the narrow terms of static optimization nor be conducive to strate-

gies based on direct control, fixed goals and predictability.

This opinion, according to which there is almost nothing constant in theoretical ideas and praxis associated with sustainability, would require a separate discussion. Looked at from the point of view of the identification and assessment of the links between complexity studies and sustainability-related theory and applications, such an approach presents a specific logical challenge. As was shown in earlier parts of the chapter, complexity cannot be defined unequivocally. This is quite common with a multitude of social ideas. However, complexity's meaning also mirrors the vagueness and impossibility of identifying and defining other concepts. There are some exceptions—more precise definitions of 'hard' complexity, according to which complex systems are simply mathematically well-defined systems with 'emergent properties', depicted as *Complex Adaptive Systems* (CAS) (Gell-Mann 1994; Holland 1995). While according to these and other authors who deal with 'hard' complexity, Complex Adaptive Systems can be precisely defined using mathematical models, other authors use the concept as a metaphor which is closer to 'soft' complexity, e.g.: "Most fundamentally, ecological and socio-economic systems are complex, adaptive systems, integrating phenomena across multiple scales of space, time and organizational complexity" (Levin 2006: 328). A similar approach is fundamental to the theory of sustainability transition, e.g. (Rotmans 2005: 32: Loorbach 2007: 55).

Remaining in the area of 'soft' complexity, it must be emphasized that it is necessary to maintain a certain 'core' set of concepts that allow for the description of complexity in a classical way, i.e. with 'classical' definitions and not just with the arbitrary use of metaphors. Otherwise, the discourse beginning "complexity is something difficult to define" is tautological and non-productive in theory and practice.

Therefore, the following point of departure for further consideration is proposed. Despite the vagueness of the ideas, methods and interpretations presented, especially in the works on sustainability transition and transition management, the traditional questions and determinants of complexity relating to description, explanation of causal links, prediction, normative approach and influence upon societal phenomena should be applied in description and assessment of the links between complexity theory and sustainability, and its related terms. It would allow for further less precise discourse and the main reason for this is simply effectiveness of communication, which in turn increases the probability of effective action.

Referring to the above survey of interpretations of complexity, it should be stressed that with the exception of precisely-defined models of social phenomena described with measurable parameters (ratio scale, and perhaps, interval scale), all other applications are based upon analogies and metaphors, or in other words, linguistic variables. It is thus necessary to consider the following factors that determine the discourse on complexity and sustainability and on related ideas:

- consequences of the intersubjective character of the systems under consideration,
- intersubjective character of complexity, especially in the 'soft' interpretation,
- limited applicability of analogies and metaphors taken from complexity theory,
- epistemological and methodological impact of contemporary trends in the social sciences (sustainability science, integrated assessment, post-normal science and action research),
- excessive relativization of ideas, objects of study, and methods,
- social/political biases of the stakeholders, especially the leaders and scientific advisors,
- cognitive/cultural biases of the stakeholders, especially the leaders and scientific advisors,
- biases of stakeholders and scientific advisors leading to tendencies of domination and politicization of the discourse,
- the necessity to prove, and at the same time the limitations to proving, what the theoretical and empirical advantages and disadvantages of narratives and operationalization dealing with complexity studies are, compared with other approaches.

The interpretations of sustainability and complexity and of all the above concepts will be scrutinized. Their relevance will then be analysed in relation to their components, operationalization and applications. First and foremost, any a priori definitions and interpretations will be assessed from the point of view of their relevance and validity. Obviously it is not possible to study all aspects of the meaning of these terms, and so a selection of characteristics and of criteria must be made.

There are significant epistemological limitations to conducting such an assessment. First and foremost, as is often stressed in transition theory and in sustainability science, which are seen as sciences without uncertainty, everything can be treated as a matter of discourse, subjectivity and intersubjectivity. Using more sophisticated terminology, the main epistemological message of all such considerations relating to integrated assessment, post-normal science and action research can be and should be reduced to a few rigorously defined concepts (Loorbach 2007: 33-35).

Another epistemological limitation of the discourse on sustainability and related ideas is that it is not possible to prove empirically whether, for example the implementation of the concept of transition by the application of the narratives and methods of complexity analogies, metaphors and mathematical models is more effective than the application of narratives and methods which do not utilize such tools. Obviously, in the light of the theories proposed, it does not seem possible or necessary to conduct that kind of research, but at the same time the results of the applications *eo ipso* confirm the validity of such an approach (Lorbach/Rotmans 2010).

In consequence, the empirical scope of the description and assessment conducted in this chapter is limited and is predominantly based upon selected theoretical publications and descriptions of case studies concerning sustainability and related concepts that are available in the literature. It should be treated solely as an introduction to further, more systematic studies.

18.5.2 Complexity and Sustainability of Modern Society

Since the onset of the debate about sustainable development, complexity, more or less precisely defined, has been an indispensable element of the debate. The complexity of socio-economic and ecological systems had already been spelled out in *The Limits to Growth* in 1972 (Meadows/Meadows/Randers et al. 1972: 9, 190). In the following years, complexity became one of most important issues in sustainability discourse (Capra 1982). It was viewed as the fundamental attribute of relations between society, technology and the environment, though it was not defined precisely but rather used as a kind intuitive term referring to the number of elements, the number of their relationships, non-linearity and emergent properties. Complexity is considered at every level of the societal hierarchy and in every domain of sustainability and sustainable development.

1. Challenges of sustainability and complexity concerning the development of global systems, regions and states (Diamond 1997, 2005; Wallerstein 2000).
2. Indicators of sustainability (Meadows 1998).
3. Increasing and uncontrollable complexity of socio-economic systems (civilizations) as the reason for their collapse (Tainter 1988; 2000).

4. Holistic qualitative approach to sustainability with the use of ideas drawn from complexity studies (Capra 1982; Urry 2005; Jenks/Smith 2006; Levin 2006; Homer-Dickson 2002; Espinosa/Harnden/Walker 2008; Espinosa/Walker 2011).

5. Sustainability as a hierarchic recursive system based upon the 'Viable System Model' (Beer 1979; Schwaninger 2006; Espinosa/Harnden/Walker 2008; Espinosa/ Walker 2011).

6. Sustainability, complexity and learning systems—the emergence of a "global and sustainable learning society as an effect of coevolution of society and its environment" (Laszlo 2003: page).

7. Sustainability interpreted using a postmodernist approach based on Foucault's radical relational agency and complexity. Non-linearity as one of the main tenets of complexity in ecological considerations (Picard 2010).

In this survey of the applications of complexity-related analogies, metaphors and mathematical models in theory and policy-making concerning sustainability and sustainable development, the following functions of theory can be used as the background to a point of departure: description, identification of causal links, prediction, the normative approach, and influence upon societal phenomena (change management).

Defining in detail all the characteristics of complexity presented in this section and in the subsequent sections would require extensive description and explanation. The characteristics can be treated solely as a preliminary set of guidelines which can help to focus the discussion. There are many other questions which can be added here and in the sections concerning sustainability transition and transition management.

It must be borne in mind that research on sustainability, sustainability transition and transition management are a part of sustainability science (Kates/Clark/Corell et al. 2001), a field closely related to *Integrated Assessment* (IA) by Rotmans (2005). 'Sustainability science' is a more or less general term for a development in science as a whole towards more multi- and interdisciplinary research into complex societal issues. Sustainability science mainly deals with the field of global environmental and sustainability research, and also emphasizes the importance of the involvement of stakeholders in the knowledge development process. While IA offers concrete tools and methods for complexity and sustainability research, sustainability science redefines the role of research and researchers at an abstract level. For transition research this is relevant because the ambitions behind transition research are similar to those behind sustainability science: scientific and societal impact based on an active and participatory role by researchers (Loorbach 2007: 34-35).

18.5.2.1 Description

The ideas of sustainability and sustainable development are to a large extent built upon the language borrowed from both 'hard' and 'soft' complexity discourse. Analogies and metaphors dominate at higher levels of generalization. In the operationalization of more precisely defined ideas, mathematical models are applied. In the complexity-based narratives and models used to describe sustainability and sustainable development, the following issues need to be considered:

1. Identification of the systems and processes which have to be sustainable.
2. Identification of the processes leading to sustainability (methods of identification—qualitative, quantitative).
3. General definitions of sustainability.
4. Holistic approach to sustainability.
5. Relations between socio-economic systems and their environment.
6. Domains of sustainability.
7. Sustainable systems and states in the domains.
8. Methods of identification of the goals of sustainability.
9. Character of sustainability—a final stable state or just the perpetually changing state of the world (punctuated equilibria)?
10. Sustainable development as a perpetual learning process.
11. Essential variables of sustainability and methods for their definition and measurement.
12. Types of relations between sustainability of elements (subsystems) and sustainability of the systems/processes defined on any scale—economic unit (enterprise, other organizations), regional/local, national, international, regional/international, global.

18.5.2.2 Causal Relations

Causal relations within the systems constitute a foundation for studying and influencing the dynamics of sustainable development. Their identification constitutes at the same time a part of description. Identification of all monocausal and multicausal relations in a complex system is impossible. They constitute the very definition of complexity, including emergent properties. For this reason, only some example relationships can be pointed out.

1. Relations in the processes leading to sustainability (constituting sustainable development).
2. Types of causality.
3. The causal links between domains of sustainability.
4. Hierarchical and heterarchical relationships within sustainable systems.
5. Structure-agency relations in sustainable systems.
6. Self-organization as the process leading to and maintaining sustainability.
7. Loops of reflexivity in sustainability systems and processes.
8. Causal relations between sustainability of elements (subsystems) and sustainability of the systems/processes defined at any scale–economic unit (enterprise, other organizations), regional/local, national, international, regional/international, global.
9. Characteristics of the learning process–learning loops, norms of learning.

18.5.2.3 Prediction

This is the most important and at the same time the most controversial facet of the application of complexity ideas in sustainability discourse since it determines the sense of the discourse. The definition of the future sustainable state(s), processes or series of punctuated equilibria which can be representative of sustainable development is undoubtedly the key issue in the whole of sustainability discussion. Examples of determinants of the value of ideas and methods drawn from complexity studies are:

1. Expected (desired) characteristics of future sustainable states/processes.
2. Time horizon of predictions.
3. Methods and validity of predictions of long-time horizons necessary in multigenerational sustainability processes.
4. Types of causality and prediction.
5. Methods of prediction and validity of methods of prediction.
6. Quantitative and qualitative methods of prediction.
7. Prediction, learning and reflexivity.

18.5.2.4 Normative Approach

As in the case of prediction, the norms and goals of sustainability are very difficult to establish. Even if they are not defined as a stable state, the goals of sustainable development need to be determined. Usually both norms and goals are depicted as multigenerational (one to two generations) phenomena. This case echoes a classical challenge in such normative social sciences as economics, finance, management and security studies. Theoreticians and practitioners are aware that their predictive capabilities are limited, but at the same time they have to determine the goals and norms. While in classical management at the micro-level, goals resulting from the interplay between market and stakeholders are relatively easier to develop, at the higher levels of the social hierarchy the phenomena are much more difficult to grasp because of a multiplicity of elements, factors and interactions. Some examples of determinants of complexity-related normative aspects of sustainability and sustainable development are:

1. Contemporary qualitative and quantitative norms of sustainability.
2. Future norms of sustainability.
3. Patterns of selection/emergence of the norms of sustainability.
4. Norms as the result of learning process.

18.5.2.5 Influence upon Societal Phenomena

This function of theory, as well as of analogies, metaphors and mathematical models, can be understood in a twofold way. First, when traditional 'hard' systems thinking is applied, it can be reduced to simple regulation and control rooted in 'first-order cybernetics', where the observer is outside the system. Second, when 'soft' systems thinking or 'second-order cybernetics' is taken into account, the role of participant/observer and the interpretative/constructivist aspects of theory and practice must be borne in mind. However, in both approaches mathematical models drawn from 'hard' complexity as well as analogies and metaphors dealing with 'hard' and 'soft' complexity discourse should be taken into account.

1. Control parameters of processes leading to sustainability.
2. Methods of identification of control parameters.
3. Methods of direct and indirect influence.
4. Learning processes and their characteristics.
5. Influence upon learning and self-organization.
6. Balance between spontaneity and self-organization, and external influences.
7. Links between the goals and the processes of change.

18.5.3 Complexity Studies, Transition Theory and Transition Management

Concepts associated with complexity, or complexity theory or science, are the key foundation of sustainability transition theory. The narratives of sustainability transition theory and transition theory management are completely permeated with the terminology of complex systems research. Complexity studies are viewed as a 'unifying principle' (Camaren/Swilling 2014).

The following passage perfectly mirrors the complexity-related assumptions of sustainability transition and transition management.

> The basic mechanisms that underlie change in complex adaptive systems are coevolution, emergence and self-organization (Holland 1995). Societal systems can be considered as complex adaptive systems. Societal sectors consist of numerous interlinked elements (e.g. actors and institutions), there is a high degree of uncertainty about their interactions and feedback and they have an open and nested character in terms of different levels of organization. From this perspective, typical complex system behaviour can be recognized, as for example emerging structures, co-evolving (policy) domains and self-organising processes can be observed. One of the possible patterns distinguished is that of transition: a system in a relatively stable equilibrium is (suddenly) going into a phase of rapid change through a process in which self-organization and coevolution play an important role before a new equilibrium is found (Loorbach/Frantzeskaki/Thissen 2011: 77).

It must be stressed that the authors of the concepts of sustainability transition and transition management are aware of subjectivity and of the arbitrary character of their ideas. "A societal system" does not exist in reality, nor does a "regime". They clearly state that any analysis of a system is arbitrary and only valid as long as it is supported or recognized by actors operating within it. In other words, a system definition is a product of social construction and any model for analysis should support this process.... The concepts of transition theory should therefore not be regarded as goals, but rather as a means for analysis. Possibly, there are many alternative system, scale or phase models that could be of as much use. Acknowledging the limitations stemming from subjectivity and intersubjectivity, they claim that

> ...in its short existence, transition theory has proved to be of value for integrative, long term analysis of complex societal processes, and has presented a promising starting point for redefining governance in the context of these transitions.... The objective of transition research therefore is not to achieve objective analysis,

but to develop coherent, integrative and long-term analytical tools that provide a basis for societal debate, policy and reflection on future development. Transition analysis should stimulate and support the necessary problem structuring processes, reflexive capacity and social learning that create the conditions for change to occur (Loorbach 2007: 22).

This reflection is essential in the analysis of the links between complexity studies and sustainability transition and transition management. It shows that applications of complexity ideas are conceptually well-grounded and the assessment will not be reduced to a simple criticism of simplifications and even ignorance, something which is, alas, too frequent in applications of complexity ideas in the social sciences. However, it also may provide grounds for justification of considerations that are too broad and superficial, especially when the statement of possible alternative systems is taken into account.

In the case of the assessment of sustainability transition, evaluation deals partly with an already-made assessment of the links between sustainability and sustainable development, and complexity. Taking into account the similar assumptions concerning analogies, metaphors, mathematical models, intersubjectivity, and so forth, a ranking of the determinants of the complexity of the systems undergoing transition through transition management can be presented.

18.5.3.1 Description

1. Identification of the systems which are the subject of transition.
2. Methods of identification of transition undergoing transitions.
3. Properties of the unsustainable initial state (stage of the process).
4. Properties of the sustainability transition which derive from the complexity of the initial state.
5. Properties of the final state (process) of the sustainability transition which derive from its complexity.
6. Characteristics of the transition process relating to its complexity.

18.5.3.2 Causal Relations

Causal relations in sustainability transition both reflect interactions within the system and emerge between various activities in the process of transition. As in in the case of sustainability, identification of all causal and multicausal relations in a complex system is impossible; only some examples of determinants of causal relationships can be pointed out.

1. Interactions in the systems in the transition process.
2. Causal links between subprocesses of transition.
3. Causal links between domains of sustainability.
4. Causal links in hierarchies and heterarchies.
5. Feedback loops.

18.5.3.3   Prediction

In predicting the results of sustainability transition it is the future state/process which has to be at least partly determined. In the case of a specific area where the sustainability transition is designed, the following determinants that concern prediction should be taken into account:

1. Characteristics of unsustainable states/processes.
2. Expected states of parameters describing the future sustainable states/processes.
3. Time horizon of predictions.
4. Methods of prediction and validity of methods of prediction.
5. Objective limitations of prediction.
6. Subjective limitations of prediction—individual specific cognitive, and collective - social, political and cultural.
7. Characteristics of the learning process—basic learning mechanisms, assumptions, norms?

18.5.3.4   Normative Approach

As explained in the earlier discussion, at the beginning of the sustainability transition process the norms and goals are unclear. Even if they are not defined as a stable state it is necessary to determine what they could be. Owing to the variety of transition projects, their time horizon may be more difficult to determine. The basic determinants of complexity are as follows.

1. Goals and norms of the transition.
2. Methods of development of the goals or methods of influence of emergence of the goals.
3. Spontaneity vs. control over development/emergence of goals and norms.
4. Characteristics of the learning process leading to emergence of goals and norms.

18.5.3.5   Influence upon societal phenomena

Determinants of complexity of description, identification of causal links, prediction and the normative approach have been described in relation to sustainability transition. Theoretical considerations and discussion concerning applications of transition management focus upon the influence of societal phenomena

(events, processes, spontaneous/non-spontaneous). The dominant characteristics of complexity of the context and of the goals of sustainability transition management have been described earlier; for sustainability transition management, only one function of analogies, metaphors and mathematical models remains: influence upon societal phenomena. Bearing in mind 'hard' and 'soft' complexity, the following questions can be asked:

1. Qualitative characteristics and quantitative control parameters of processes/phenomena   leading to sustainability.
2. Methods of direct/indirect influence upon characteristics and parameters.
3. Links between the goals and the processes of change.
4. Degrees of spontaneity and external influence upon the subprocesses of transition.

18.5.4   Understanding the Complexity of Sustainability, Sustainable Development, Sustainability Transition and Transition Management

18.5.4.1   Assumptions and Criteria of Assessment

The above considerations may constitute a point of departure for preliminary identification and assessment of the relevance of applications of complexity-related concepts to the ideas of sustainability, sustainability development, and sustainability transition and management. The assessment is by definition of an introductory character and further systematic studies are needed. It is even possible that because of the scope and intricacy of the research area, such a comprehensive evaluation is not achievable. In the assessment the following criteria should be taken into account:

1. Awareness of the subjective (intersubjective) and reflexive character of social systems and of definitions of their complexity.
2. Comprehensive understanding of the ideas of complexity ideas—'hard' and 'soft' complexity.
3. Comprehensive understanding of the meaning of analogies and metaphors drawn from complexity studies which are applied in the social sciences, including sustainability theory and policy, sustainability transition theory and sustainability management.
4. Relevance of applications of metaphors, analogies and mathematical models relating to the concepts drawn from complexity theory.

5. Comparison of advantages and disadvantages of complexity-based theory and applications of sustainability transition and transition management.
6. Misuses and abuses of complexity ideas.

When studying and assessing the applications of the utterance "complexity" in the social sciences, it is first essential to identify which characteristics of complexity are being applied and how they are used—as analogies, metaphors or mathematical models. In the majority of applications, metaphors and analogies can be found.

A deeper analysis of the uses of metaphors in theory and in practice has been conducted in several areas of the social sciences (Lakoff/Johnson 1980, 1989; Dobuzinskis 1992; Ortony 1993; Mirowski 1989, 1994; Morgan 1996). When studying the uses of metaphors, the first step is to distinguish their types and their specificity. Then it is necessary to be aware of the negative consequences of the reification of metaphors (Vandenberghe 2014). Fulfilling these two demands makes it possible to evaluate the effectiveness of the application of metaphors and analogies. Evaluation can be performed on the basis of their validity in the five functions of theory described above.

Operationalization of the characteristics of complexity is the key component of the 'hard' approach to complexity. It is usually performed in two stages. In the first stage, the verbal categories are defined in such a way as to make possible their quantification, that is, they are described using ratio or interval scales, and then either statistical processing is performed or other types of mathematical models are used. In the second stage, the model, be it statistical or of another type, is applied, and then the relevance of the model to reality is assessed.

18.5.4.2 Complexity of Sustainability and of Sustainable Development

In the considerations linking sustainability and sustainable with various characteristics of complexity all types of applications can be found: analogies, metaphors and mathematical models. The following general regularities can be observed.

Complexity is treated as an indispensable part of modern socio-economic systems and ecological systems. In a majority of cases, complexity is introduced without any deeper reflection concerning the difference between 'hard' and 'soft' complexity and its metaphorical character. In particular, no reflection on the subjectivity of the meaning is provided. The examples below are obviously not to be treated as absolutely representative but, as mentioned earlier, can be viewed as examples that can be used to shape future, more systematic studies on the occurrence of ideas taken from complexity theory/complexity studies in the discourse on sustainability and sustainable development.

The degree of reflection on the meaning of complexity in sustainability discourse varies. Some works include a deeper reflection on complexity; examples are Espinoza/Porter (2011) and Espinosa/Walker (2011). Elsewhere, a more arbitrary approach is applied. As already mentioned, social and ecological systems and their interactions (co-evolution) are depicted as complex adaptive systems, e.g. Levin (2006). Another method of application, in which a postmodernist interpretation of ecological problems is analysed with the use of Foucault's idea of radical relational agency and complexity, employs a profounder qualitative understanding of complexity as well as an interpretation of complexity theory:

> Complex systems, however, "are not 'things' in the noun like sense but processes; nor are they 'things' in the categorical sense because nothing underwrites the linguistic academic habit of collecting them together" (Jenks/Smith 2005: 23). Complexity theory does not take complex systems to be an object of study. Rather, complexity theory describes the *processes* that occur within complexly arranged systems. That complex systems are defined by processes is consistent with a radical relational agency because it underscores the *interactions* that occur within these systems. A complex system is not defined by any innate characteristics, but by the processes enacted by the social and ecological participants that make it up. In other words, complex systems cannot be 'things' because they are *by definition* the set of complex interactions that take place between humans, nonhumans and environments (Picard 2010: 8).

The metaphor of the 'edge of chaos' seems particularly attractive, not only in security-oriented discourse and management but also in the ecological consideration. Although attempts at operationalization are made, e.g. in Carroll and Burton (2000), in most cases this phrase is used as a metaphor. A rigorous model of the 'edge of chaos' state must include a system defined with a set of measurable variables where changes to some of them represent the dynamics of the system. Otherwise, it is just a metaphor and users should be aware of its limitations.

> Ecological systems are on the edge of chaos without a 'natural' tendency towards equilibrium [...]. Indeed, many ecological systems themselves depend not upon stable relationships but upon massive intrusions, of extraordinary flows of species from other parts of the globe and of fire, lightning, hurricanes, high winds, ice storms, flash floods, frosts, earthquakes and so on. The

'normal' state of nature is not one of balance and repose; the normal state of nature is to be recovering from the latest disaster (Urry 2005: 6).

In the study of broadly defined sustainability on the global scale and on the scale of nations, the concept of complexity is applied in its basic form, even without stressing the role of emergent properties:

> Complexity [of society] is generally understood to refer to such things as the size of the society, the number and distinctiveness of its parts, the variety of specialized social roles that it incorporates, the number of distinct social personalities, and the variety of mechanisms for organizing these into a coherent, functioning whole. Augmenting any of these dimensions increases the complexity of the society (Tainter 1988: 23).

Although such an approach is a simplification which reduces complexity solely to "objective" categories in the case of systems which can be described using a mechanistic analogy, in the case of global systems it is highly valid. The system of states or other entities (regional institutions, global organizations) can and should be perceived in such a way. It both allows the application of mathematical models and provides a basis for further analyses, which may include deeper considerations on subjectivity, intersubjectivity, and the constructivist character of such collectivities.

Looking at the above examples, which obviously may not be fully representative, and confronting them with the demands of theory and policy-making, as well as with the above criteria, the following remarks can be made. First and foremost, the application of the ideas of complexity theory requires a deep and thorough investigation if it is to achieve results that will allow a comprehensive assessment. A preliminary evaluation leads to the conclusion that none of the criteria are fully applied or fulfilled in work on the complexity of sustainability and sustainable development. It is obviously impossible to achieve this in every work on sustainability, but at least some reflections need to be included. As an example of an exception that meets that requirement, the works quoted above may be mentioned (Espinosa/Harnden/Walker 2008; Espinosa/Walker 2011). Such a situation may have a number of negative consequences of a theoretical, practical and political nature. As far as theoretical considerations are concerned, it means that the validity of the discussion on sustainability can be easily weakened. Additionally, if an extended argument about sustainable development is presented without explaining what is to be sustainable and how the desired, possibly temporary, state (temporary equilib-

rium) is to be defined, then the argument is easy to bring into question.

Summarizing the above remarks and sketching a framework for future studies, it can be concluded that in almost every application of ideas relating to complexity in the discourse on sustainability, a more or less concealed mechanistic analogy drawn from physics, mechanics or biology is being applied. Such an approach is the only one possible, even when modern critical, constructivist postmodernist paradigms are taken into account. What is always obviously different is the degree to which authors are aware of the limitations of their approach, deriving as it does from the intersubjective character of all the categories applied in the discourse. Even in the case of mathematical models derived from operationalization, the choice of the model built upon the central metaphor (Lakoff 1993) is never objective, as is often claimed by the models' creators.

18.5.4.3 Complexity of Sustainability Transition and Transition Management

While discussion about the links between complexity and sustainability and sustainable development is at a very early stage and by necessity built upon only a selection of very different examples, consideration of the impact of complexity theory on sustainability transition and transition management involves two coherent theoretical and interrelated concepts.

As was shown earlier in this chapter, both the above ideas are a kind of synthesis of modern social thought on constructivism, reflexivity, holistic thinking, self-organization, governance, and a very broadly defined complexity of social and ecological systems. Undoubtedly these two concepts reflect the limits of relevant modern social thinking. Their relationship with complexity depends on the criteria formulated in the assumptions of this section.

First, as reflected in work quoted above, the reflection of the subjectivity of ideas of complexity is embedded in the theory and application of transition theory and transition management. It concerns the understanding of the systems and, to a lesser extent, of the ideas deriving from complexity studies. Second, and this can be viewed as a weakness of both ideas, is the lack of profounder interpretations of social systems and their characteristics. Although it must be emphasized that systems and their characteristics are perpetually constructed by actors during the transition process through reflection, no sufficiently precise details of such characteristics are provided. This is partly compensated for in the examples of applications where the learning processes and the emergence of systems are demonstrated (Loorbach/Rotmans 2010). It would be helpful if the norms and loops of learning were depicted in a more detailed manner. These are simply general remarks concerning ideas about systems, and they indicate further directions for reflecting upon transition to sustainability and sustainability transition.

The main focus of attention is the assessment of the application of complexity ideas in the transition to sustainability and sustainability management. In this case it must be admitted that despite the declarations of awareness of subjectivity, the ideas of complexity are sometimes applied without sufficiently profound reflection. Several examples are quoted below.

To start with, we can consider complex adaptive systems, one of the core ideas in all sustainability transition discourses. A rigorous and epistemologically valuable reference to complex adaptive systems should be supported with a detailed mapping of the ideas of operationalization, as described by Holland (2006) and others. Otherwise, the applications are simply metaphors and would require a deeper reflection on their cognitive validity. Obviously, the operationalization of some characteristics is always acceptable, but then an open question remains as to what extent such operationalization allows a full-fledged systems approach to be maintained and a system to be properly identified.

Similar reflections concern another complexity-related idea—self-organization. It is one of the main features of the complexity of social systems. It is regarded as a common trait of complex systems in both 'hard' and 'soft' approaches, and numerous examples of its use can be found in the works quoted above. However, as in all other cases, a proper identification of self-organization requires more precision in defining the system, the relationships and the determinants and patterns of *eigendynamik*, including attractors, emergence, and so forth. Taking into account the differences between understanding complexity in 'hard' interpretations and in the interpretation proposed by Luhmann, five ways of bringing self-organization into social theory are proposed. First, operationalization, i.e. identifying the 'hard' part of the system. This is measurable either on a ratio or an interval scale. Mathematical models exhibiting emergence can then be prepared. Examples of such an approach when dealing with sustainability transition have been already developed (Timmermans 2008; Timmermans/de Haan/Squazzioni 2008; Faber/Alkemade 2011). The second method, proposed by Luh-

mann (1990) and developed by Leydesdorff (2001) among others, was built upon the concept of an autopoietic social system of meaningful communication. Having defined the social system in a relatively rigorous way, it is possible to identify the processes of self-organization in a qualitative way, optionally supported by operationalization. The fourth method is to use analogies and metaphors from the 'hard' complexity approach, without any reference to Luhmann's social autopoiesis. The fifth method, which could be seen as the most advanced approach, differs from the others by parallel applications of ideas from 'hard' and 'soft' complexity science in order to identify and analyse self-organization.

The last example reminds us that there may be references to non-linearity in the discourse on sustainability transition. As was stated above, non-linearity is a typical feature of the behaviour of different systems, and linearity is only the special case (Campbell 1997: 218). Moreover, non-linearity is not always a source of computational barriers. In order to identify a non-linear disordered behaviour, a set of well-defined parameters must be applied and relevant mathematical models developed. In other cases, as is often the case in the discourse on sustainability transition and transition management, non-linearity is used as metaphor, and so a detailed analysis of its meaning should be performed before this term is employed in theory or practice. Especially intriguing is the question of how non-linearity can be captured in a qualitative manner by the participant-observers. Obviously, positive feedback loops can be identified without mathematical formalism, but appropriate operationalization would be difficult, if not impossible, to accomplish.

Applications of other ideas drawn from complexity theory in the transition to sustainability and transition management, such as adaptation, attractor, chaos, edge of chaos, learning, and punctuated equilibria, could be scrutinized in the same manner. The methodology of assessment could be the same as in the above cases. A thorough description and evaluation of the links between complexity theory and the discourse on sustainability transition and transition management is beyond the scope of this study. However, even an evaluation made at a certain level of generality can lead to a better understanding of the complexity of sustainability transition and transition management. It is also important that the proposed level of generality enables the researcher to avoid delving into the sophisticated language of modern sociology which is used in the narratives concerning these two concepts.

The main conclusions stemming from a preliminary assessment of the applications of complexity ideas to transition to sustainability and transition management are as follows. Transition to sustainability and transition management undoubtedly constitute the most comprehensive concepts of large-scale societal change, developed on the basis of state-of-the-art studies (or theory) of differently defined complexity. As is not the case with a multitude of other applications of the various characteristics of complex systems in the social sciences, the authors of both concepts are to a large extent aware of the limitations of these applications that stem from intersubjectivity and from the constructivist character of social systems. However, while such limitations are taken into account in reflecting upon the systems, their properties and their functioning, the idea of complexity and its attributes are used with insufficiently deep reflection. In several works quoted above in this chapter, complexity and its characteristics are treated as almost a priori categories. Insufficient attempts are made to explain what the system is, how it is constructed or how it is emergent. A similar set of questions can be applied to the patterns of self-organization. By the same token, a deeper explanation is required for another feature of complex systems—the edge of chaos.

Application of the ideas taken from complexity theory in sustainability transition and transition management is affected by another weakness resulting from the need for knowledge of the genesis of the ideas connected with complexity. Without a fuller explanation, it is not possible to apply Luhmann's concept of the complexity of social systems using other ideas of complexity taken from 'hard' complexity—thermodynamics, (e.g. Prigogine/Stengers 1984; Nicolis/Prigogine 1989) or mathematical modelling of systems complexity (e.g. Waldrop 1992; Holland 1995; Bak 1996). Any attempts at operationalization or joint qualitative interpretation require a very thorough and deep understanding of both domains of the discourse on complexity.

The role of metaphors and analogies is another issue in the assessment of the links between complexity theory and sustainability transition and transition management. Although applications of metaphors are included in the ontological and epistemological foundations of sustainability transition and transition management, e.g. Loorbach (2007: 86–87) and the references to Stacey (1996) in Loorbach's work, they do not seem sufficient. Any action-oriented theories dealing with so many ideas from complexity studies should include much more profound analysis of the terms applied as analogies and metaphors. This analysis seems to be insufficient in the entire body of knowledge on sustainability transition and transition management.

There is another advantage of the use of analogies and metaphors, independent of analysis of their cognitive validity. Metaphors and analogies relating to complexity may be viewed as heuristic instruments that enhance creativity in description, explanation, prediction, setting norms and influencing social systems in the theory and practice of sustainability transition and transition management. This role of complexity theory could be utilized in prediction. Emergent phenomena in modelling systems with Complex Adaptive Systems can only serve as predictions to some extent, since they are not recurrent. However, models of CAS can deliver patterns which were 'unimaginable', and in this way they could be applied in prediction.

When choosing theoretical ideas, be they paradigm, theory or model for application in practice, it is always necessary to demonstrate their advantages over other ideas. There may be different norms and criteria for proving the superiority in an objective manner, or at least for persuading others about their advantages. In the theory of sustainability transition and transition management, the 'added value' of the complexity-based approach is insufficiently advertised. It would be worthwhile to learn what additional practical results have been and can be achieved thanks to the proposed concepts in comparison with other approaches, say, for example, systems analysis and design or traditional methodologies drawn from large-scale project management. Otherwise the proponents of new ideas are under the threat of Occam's razor.

A final remark concerns the overall character of both concepts discussed. They are built upon a reinforced ontological and epistemological rule that says—nothing is constant. Everything can be accommodated, adapted, changed after the reflections of the participant-observers. One may even pose the question: are the reflection patterns and/or loops constant or variable? Obviously, this stands in agreement with the conditions of our modern fast-changing society in flux. A question thus arises. Does changeability of the theory and practice of sustainability transition and transition management result from the fundamental ontological and epistemological limits of the objects (systems) under study, or perhaps, is it the result of the weaknesses of a theory which is not able to answer questions concerning applicability? Of course, we know that *panta rei* and we have to go 'from being to becoming', but is it possible to capture that reality in such a comprehensive way as it is presented in the theory of transition to sustainability theory and transition management, assuming that nothing can be stable and there are no patterns which could at least make classical measurement possible? This problem was discussed by Anand and Sen (2000: 2037), who used the polarity overspecification vs. underspecification of sustainable development. The ideas of sustainability transition and transition management are somewhat unspecified, partly because of the insufficiently analysed applications of ideas drawn from complexity theory. This may lead to such high levels of relativization of ideas that it even undermines the processes of communication treated as the transfer of meaning. It may also provide arguments for the criticism that ideas, objects and methods are fuzzy to such an extent that narratives based upon broad linguistic variables, instead of discovering the very sense of the problem, may lead to its misreading and even concealment.

18.6   Conclusions

The conclusions and recommendations stemming from the above considerations can be divided into two groups. The first relates to sustainability and sustainable development and the second concerns transition and transition management.

The complexity of sustainability and sustainable development was somehow discovered at a very early stage of the development of these ideas. Initially it took the form of fuzzily described labels, but soon, together with the development of complexity science, it became one of the core issues in the discourse on sustainability. As with other vague ideas concerning social life, the terms sustainability and sustainable development cannot be defined in a precise way, and thus their meaning emerges in intersubjective exchange. The problems in defining sustainability become even more profound when ideas associated with complexity are included in research. In the works linking complexity and sustainability, applications may vary, beginning with loosely defined metaphors and analogies, and ending with various attempts at operationalization. The main advantage of the application of complexity-related concepts in the theory and practice of sustainability is the enhancement of possibilities for description, explanation of causal links, prediction and influence on social phenomena. It can be concluded that at present the theory and policy of sustainable development is to a large extent dependent

upon ideas drawn from complexity theory, broadly defined.

The preliminary research on the application of complexity-related ideas in sustainability research and practice shows that, with the exception of the operationalization of some ideas taken from complexity theory, analogies and metaphors borrowed from complex systems dominate the discourse. In the majority of cases they are used without a deeper reflection on the intersubjective character of social systems and of their complexity. Social systems are the product of intersubjective discourse and the same characteristic of the discourse on complexity makes studies and actions very intricate. In many cases, the terms associated with complexity theory are misused and abused in discussions about sustainability, and the vague meaning of that and of utterances associated with it are applied as a kind of decoration, adding a 'scientific value' to the discourse.

It is obviously only a margin of applications of complexity concepts since even a purely but well-thought use of those concepts would contribute to a better understanding of the fast-changing modern world, and in consequence, should lead to the development of more effective policy measures.

While the complexity of societal and environmental phenomena has been included in the theory and policy of striving for sustainable development, ideas deriving from complexity studies have constituted the backbone of the ideas of sustainability transition and transition management. The assessment of the application of complexity-related ideas to the theoretical deliberations on those two concepts and their application leads to the following conclusions.

First, sustainability transition and transition management are to a large extent based upon ideas drawn from a scientific trend which in a more or less justifiable way is called complexity theory or complexity science. This scientific trend is based upon a set of phenomena, processes, names and mathematical models whose proponents aim to study the complexity of natural and social phenomena with a set of simple descriptions, explanations and predictions. Since social systems are an effect of tangible and intangible components synthesized into a common meaning in intersubjective discourse, the authors of the ideas of sustainability transition and transition management are well aware of the limitations stemming from the very character of these systems. However, the ideas of complexity used in the rhetoric dealing with these two ideas is not always sufficient. Although a broad spectrum of problems are properly described and explained using such ideas as adaptation, edge of chaos, learning, self-organization, and so forth, the main conclusion is that the sense of ideas drawn from complexity studies has not been sufficiently reflected on and scrutinized in the context of the applications. This particularly concerns the multitude of interpretations of the characteristics of complexity and the intersubjective process leading to the creation or emergence of the meaning of the term.

Second, such an approach leads to an impression of conceptual redundancy over practical needs, and also, over the sense of a theory which may not be too supportive in the development of new ideas and new methods. For example, how can one identify and then stimulate the process of self-organization among the reflexive and self-reflexive members of an autopoietic social system which is described metaphorically(?) as being at the edge of chaos? It looks as if such a rhetoric may not be too productive.

The main directions for further research are common to applications of complexity-related ideas in sustainability and sustainable development, as well as to sustainability transition and transition management.

The most important direction of research is to clarify the 'language of complexity' used in both areas. Obviously, it is neither necessary nor possible to always provide the deeper analyses of the meaning of the terms used in discussions about sustainability-related problems, but at the same time it is necessary to maintain a deeper analysis of the terms connected with complexity in the theories which are built on the concept of complexity. It is especially important when the idea has some fifty or more interpretations.

The second direction of study is to develop conceptual and methodological frameworks, skeletons of ideas and methods drawn from complexity studies, which should then be applied in the theory and practice of sustainability and sustainability development, and in sustainability transition and transition management. This set should embody advanced conceptual proposals that have been thoroughly analysed from the point of view of their relevance to complexity theory and to social reality. Both qualitative methodologies and operationalization should be included. Perhaps as a source of inspiration though not as a benchmark, the ideas of agile project management could be used, although these are relevant rather at the micro-scale. Such a methodological toolbox would not be closed, that is, it would not include only a limited number of precisely defined ideas. It should be open, but at the same time, any ideas and methods, before inclusion, should be tested for their conceptual and cognitive validity as well as for their applicability. In this way, even broad metaphors could be included, on the condition that those who would spell them out interpret them properly. The development of sets of indicators for sustainable development is the first step in this direction. In the case of transition management, where broad methodologies are already applied, it would only be necessary to raise the demands for refining the definitions.

References

Adams, William M., 2006: "The Future of Sustainability: Rethinking Environment and Development in the Twenty-first Century", *Report of the IUCN Renowned Thinkers Meeting*, 29-31 January; at: <http://cmsdata.iucn.org/downloads/iucn_future_of_sustanability.pdf> (27 August 2014).

Anand, Sudhir; Sen Amartya, 2000: "Human Development and Economic Sustainability", in: *World Development*, 28,12: 2029-2049.

Ashby, W. Ross, 1963: *An Introduction to Cybernetics* (New York: Wiley).

Bak, Per, 1996: *How Nature Works: The Science of Self-Organized Criticality* (New York: Springer).

Barabási, Albert-László, Albert Réka, 1999: "Emergence of Scaling in Random Networks", in: *Science*, 286,5439 (October): 509-512.

Barabási, Albert-László, 2003: *Linked. How Everything is Connected to Everything Else and What It Means for Business, Science, and Everyday Life* (New York: Penguin).

Bar-Yam, Yaneer, 1997: *Dynamics of Complex Systems* (Reading, MA: Addison-Wesley).

Baudrillard, Jean, 1994: *Simulacra and Simulation* (Ann Arbor, MI: University of Michigan Press).

Bauman, Zygmunt, 2000: *Liquid Modernity* (Cambridge: Polity Press).

Beck, Ulrich; Giddens, Anthony; Lash, Scott, 1994: *Reflexive Modernization. Politics, Tradition and Aesthetics in the Modern Social Order* (Stanford: Stanford University Press).

Bednarz, John, jr., 1984: "Complexity and Intersubjectivity: Towards the Theory of Niklas Luhmann", in: *Human Studies*, 7,1/4: 55-69.

Bell, David E.; Raiffa, Howard; Tversky, Amos (Eds.), 1988: *Decision Making: Descriptive, Normative, and Prescriptive Interactions* (Cambridge-New York: Cambridge University Press).

Bertalanffy, Ludvig von, 1968: *General Systems Theory* (New York: Braziller).

Beer, Stafford, 1979: *The Heart of Enterprise* (London–New York: John Wiley).

Biggiero, Lucio, 2001: "Sources of Complexity", in: *Human Systems, Nonlinear Dynamics, Psychology and Life Sciences*, 5,1 (January): 3-19.

Brekke, Kjell Arne, 1997: *Economic Growth and the Environment: On the Measurement of Income and Welfare* (Cheltenham: Edward Elgar).

Brätland, John, 2006: "Toward a Calculational Theory and Policy of Intergenerational Sustainability", in: *The Quarterly Journal of Austrian Economics*, 9,2 (Summer): 13-45.

Camaren, Peter; Swilling, Mark, 2014: "Linking Complexity and Sustainability Theories: Implications for Modeling Sustainability Transitions", in: *Sustainability*, 6,3 (March), 1594-1622.

Campbell, David K., 1987: "Nonlinear Science. From Paradigms to Practicalities", in: *Los Alamos Science*, 15, Special Issue, Stanislaw Ulam: 218-262; at: <http://library.lanl.gov/cgi-bin/getfile?00285753.pdf > (17 June 2007).

Capra, Fritjof, 1982: *The Turning Point. Science, Society, and the Rising Culture* (New York: Bantam Books).

Carroll, Tim; Burton, Richard M., 2000: "Organizations and Complexity: Searching for the Edge of Chaos", in: *Computational & Mathematical Organization Theory*, 6,4 December): 319-337.

Castellani, Brian, 2014: "Brian Castellani on the Complexity Sciences", *Theory, Culture & Society*, blog, 9 October; at: <http://theoryculturesociety.org/brian-castellani-on-the-complexity-sciences/> (20 November 2014).

Chaitin Gregory J., 2001: *Exploring Randomness* (London: Springer).

Chichilnisky, Graciela, 1997: "What is sustainable development?", in: *Land Economics* 73,4: 467-491.

Cilliers, Paul, 2008: "Complexity Theory as a General Framework for Sustainability Science", in: Burns, Michael; Weaver, Alex (Eds.): *Exploring Sustainability Science: A Southern African Perspective* (Stellenbosch: Sun Press): 39-57.

Cowan, George A.; Pines, David; Meltzer, David (Eds.), 1994: *Complexity, Metaphors, Models, and Reality*. Santa Fe Institute Studies in the Sciences of Complexity Proceedings, vol. 19 (Reading, Mass.: Addison-Wesley).

Diamond, Jared, 1997: *Guns, Germs and Steel: The Fates of Human Societies* (New York: W. W. Norton).

Diamond, Jared, 2005: *Collapse: How Societies Choose to Fail or Succeed* (New York: Viking Press).

Dobuzinskis, Laurent, 1992: "Modernist and Postmodernist Metaphors of the Policy Process: Control and Stability vs. Chaos and Reflexive Understanding", in: *Policy Sciences*, 25,4 (November): 355-380.

Dow Jones Sustainability Index; at: <http://www.sustainability-indices.com/sustainability-assessment/corporate-sustainability.jsp> (18 February 2014).

Elkington, John, 1997: *Cannibals with Forks: The Triple Bottom Line of Twenty-First Century Business* (Oxford: Capstone).

Espinosa, Angela; Harnden, Roger; Walker Jon, 2008: "A Complexity Approach to Sustainability–Stafford Beer Revisited", in: *European Journal of Operational Research*, 187,2 (June): 636-651.

Espinosa, Angela; Porter, Terry, 2011: "Sustainability, Complexity and Learning: Insights from Complex Systems Approaches", in: *The Learning Organization*, 18,1: 54-72.

Espinosa, Angela; Walker, Jon, 2011: *A Complexity Approach to Sustainability: Theory and Application*. Series on Complexity Science (London: Imperial College Press).

Faber, Albert; Alkemade, Floortje, 2011: "Success or Failure of Sustainability Transition Policies. A Framework for the Evaluation and Assessment of Policies in Complex Systems", Paper for the DIME (Dynamics of Institutions & Markets) Final Conference, Maastricht, 6-8 April.

Figuières, Charles; Guyomard, Hervé; Rotillon, Gilles, 2010: "Sustainable Development: Between Moral Injunctions and Natural Constraints", in: *Sustainability*, 2,11: 3608-3622.

Foucault, Michel, 2007: *Security, Territory, Population. Lectures at the Collège de France* 1977-1978 (New York: Picador/Palgrave Macmillan).

Foerster, von Heinz, 1982: *Observing Systems* (Seaside, CA: Intersystems Publications).

Geels, Frank W., 2010: "Ontologies, Socio-Technical Transitions (to Sustainability), and the Multi-Level Perspective", in: *Research Policy*, 39,4, 495-510.

Geels, Frank W., Schot Johan, 2007: "Typology of Sociotechnical Transition Pathways", in: *Research Policy*, 36,3: 399-417.

Gell-Mann, Murray, 1994: "Complex Adaptive Systems", in: Cowan George A.; Pines, David; Meltzer, David (Eds.): *Complexity, Metaphors, Models, and Reality, Santa Fe Institute Studies in the Sciences of Complexity Proceedings*, vol. 19 (Reading, Mass.: Addison-Wesley): 17-45.

Gell-Mann, Murray, 1995: "What is Complexity?", in: *Complexity*, 1,1: 16-19.

Gillespie, Alex; Cornish, Flora, 2009: "Intersubjectivity: Towards a Dialogical Analysis", in: *Journal for the Theory of Social Behaviour*, 40,1: 19-46.

Glazersfeld, Ernst, von, 1995: *Radical Constructivism: A New Way of Knowing and Learning*, (London: The Farmer Press).

Gleick, James, 1987: *Chaos: The Making of a New Science* (New York: Viking Press).

Goergen, Marc; Malline, Christine; Mitleton-Kelly, Eve; Al-Hawamdeh, Ahmed; Hse-Yu Chiu, Iris, 2010: *Corporate Governance and Complexity Theory* (Cheltenham: Edward Elgar).

Gold, Mary V., 2007; "Alternative Farming Systems Information Center, National Agricultural Library", in: US Department of Agriculture; at: <http://afsic.nal.usda.gov/sustainable-agriculture-definitions-and-terms>.

Grin, John; Rotmans, Jan; Schot Johan, in collaboration with Geels, Frank; Loorbach, Derk, 2010: *Transitions to*

Sustainable Development. New Directions in the Study of Long Term Transformative Change (New York: Routledge).

Haken, Hermann, 2004: *Synergetics. Introduction and Advanced Topics* (Berlin: Springer).

Hawken, Paul, 1993: *The Ecology of Commerce. A Declaration of Sustainability* (New York: Harper Business).

Hawken, Paul 2007: *Blessed Unrest: How the Largest Movement in the World Came Into Being and Why No One Saw It Coming* (New York: Penguin).

Heal, Geoffrey, 1998: *Valuing the Future: Economic Theory and Sustainability* (New York: Columbia University Press).

Holland, John D., 1995: *Hidden Order. How Adaptation Builds Complexity* (New York: Basic Books).

Holland, John D., 2006: "Studying Complex Adaptive Systems", in: *Journal of Systems Science and Complexity,* 19,1: 1-8

Homer-Dixon, Thomas, 2002: *The Ingenuity Gap. Facing the Economic, Environmental, and Other Challenges of an Increasingly Complex and Unpredictable World* (New York: Vintage Books).

Husserl, Edmund, 1970: *The Crisis of European Sciences and Transcendental Phenomenology* (Evanston, IL: Northwestern University Press).

Ilachinski, Andrew, 1996: *Land Warfare and Complexity, Part I: Mathematical Background and Technical Sourcebook* (Alexandria, VA: Center for Naval Analyses); at: <http://www.cna.org/isaac/lw1.pdf> (4 September 2006).

IISD (International Institute for Sustainable Development), 2009: "The Sustainable Development Timeline"; at: <http://www.iisd.org/pdf/2009/sd_timeline_2009.pdf> (13 February 2014).

IUCN/UNEP/WWF, 1991, *Caring for the Earth: A Strategy for Sustainable Living* (Gland, Switzerland: IUCN: The World Conservation Union–UNEP: United Nations Environment Programme–WWF: Worldwide Fund for Nature); at: <https://portals.iucn.org/library/efiles/documents/CFE-003.pdf> (16 April 2012).

James, Paul, 2015: *Urban Sustainability in Theory and Practice. Circles of Sustainability* (London: Routledge).

Jenks, Chris; Smith, John, 2006: *Qualitative Complexity: Ecology, Cognitive Processes and the Re-Emergence of Structure in Post-Humanist Social Theory* (London–New York: Routledge).

Jessop, Bob, 2004: "Multi-level Governance and Multi-level Meta-governance", in: Bache Ian; Flinders Matthew (Eds.): *Multi-Level Governance* (Oxford: Oxford University Press): 49-74.

Kahneman, Daniel, Tversky, Amos, 1979: "Prospect Theory: An Analysis of Decision under Risk", in: *Econometrica,* 47,2: 263-292.

Kates, Robert W.; Clark, William C.; Corell, Robert; Hall, J. Michael; Jaeger, Carlo C.; Lowe, Ian; McCarthy, James J.; Schellnhuber, Hans Joachim; Bolin, Bert; Dickson, Nancy M.; Faucheux, Sylvie; Gallopin, Gilberto C.; Grü-

bler, Arnulf; Huntley, Brian; Jäger, Jill; Jodha, Narpat S.; Kasperson, Roger E.; Mabogunje, Akin; Matson, Pamela; Mooney, Harold; Berrien, Moore III; O'Riordan, Timothy; Svedin, Uno, 2001: "Environment and Development–Sustainability Science", in: *Science* 292,5517 (April): 641-642.

Kauffman, Stuart A., 1993: *The Origins of Order: Self-Organization and Selection in Evolution* (Oxford: Oxford University Press).

Kauffman, Stuart A., 1995: *At Home in the Universe. The Search for Laws of Self-Organization and Complexity* (Oxford: Oxford University Press).

Klasen, Stephan, 2006: "UNDP's Gender-Related Measures: Some Conceptual Problems and Possible Solutions", in: *Journal of Human Development,* 7,2 (July): 243-274.

Kooiman, Jan, (2003): *Governing as Governance* (Newbury Park, CA: Sage).

Kotter, John P., 1995: "Leading Change: Why Transformation Efforts Fail", in: *Harvard Business Review,* 73,2 (March-April): 59-67.

Lakoff, George; Johnson, Mark, 1980, 1999: *Metaphors We Live By* (Chicago: University of Chicago Press).

Lakoff, George, 1993: "The Contemporary Theory of Metaphor", in: Ortony, Andrew (Ed.): *Metaphor and Thought* (Cambridge: Cambridge University Press): 202-251.

Laszlo, Aleksander, 2003: "Evolutionary Systems Design: A Praxis for Sustainable Development", in: *Organisational Transformation and Social Change,* 1,1 (March): 29-46.

Levin, Simon A., 2006. "Learning to Live in a Global Commons: Socioeconomic Challenges for a Sustainable Environment", in: *Ecological Research,* Special Feature, 21,3: 328-333.

Leydesdorff, Loet, 2001: *A Sociological Theory of Communication: The Self-Organization of the Knowledge-Based Society* (Universal Publishers/uPUBLISHCOM, USA).

Lloyd, Seth, 2001: "Measures of Complexity: A Nonexhaustive List", in: *IEEE Control Systems Magazine,* 21,4: 7-8.

Loorbach, Derk A., 2007: "Transition Management. New Mode of Governance for Sustainable Development" (PhD thesis, Erasmus Universiteit, Rotterdam); at: <repub. eur.nl/pub/10200/proefschrift.pdf> (14 June 2013).

Loorbach, Derk, 2008: "Why and How Transition Management Emerges", Paper for the Berlin Conference on the Human Dimensions of Global Environmental Change, Berlin 22-23 February; at: <http://userpage.fu-berlin.de/ffu/akumwelt/bc2008/papers/bc2008_393_Loorbach.pdf> (24 March 2013).

Loorbach, Derk; Rotmans, Jan, 2010: "The Practice of Transition Management: Examples and Lessons from Four Distinct Cases", in: *Futures,* 42: 237-246.

Loorbach, Derk; Frantzeskaki, Niki; Thissen, Wil, 2011: "A Transition Research Perspective on Governance for Sustainability", in: Jaeger, Carlo C.; Tàbara, J. David; Jaeger, Julia (Eds.): *European Research on Sustainable Development, vol. 1: Transformative Science Approaches for Sustainability* (Berlin–Heidelberg: Springer): 73-89.

Lowell Center for Sustainable Production (no year); at: <http://www.sustainableproduction.org/abou.what.php> (18 September 2013).

Luhmann, Niklas [Bednarz, Jr., John; Baecker, Dirk (translators)], 1995: *Social systems* (Palo Alto: Stanford University Press); originally published in German in 1984.

Luhmann, Niklas, 1990: *Essays on Self-Reference* (New York: Columbia University Press).

Map of Complexity Science (no year); at: <http://www.art-sciencefactory.com/complexity-map_feb09.html> (18 January 2014).

McGillivray, Mark, 1991: "The Human Development Index: Yet Another Redundant Composite Development Indicator?", in: *World Development*, 19,10: 1461-1468.

Meadows, Donella H.; Meadows, Dennis L.; Randers, Jorgen; Behrens, William W. III, 1972: *The Limits to Growth* (New York: New American Library).

Meadows, Donella H., 1998: *Indicators and Information Systems for Sustainable Development* (Hartland Four Corners, VT: Sustainability Institute).

Meadows, Donella H.; Randers, Jorgen; Meadows, Dennis L., 2004: *The Limits to Growth: The 30 Year Update* (London: Chelsea Green Publishing).

Mesjasz, Czesław, 1988: "Applications of Systems Modelling in Peace Research", in: *Journal of Peace Research*, 25,3: 291-334.

Mesjasz, Czesław, 2008: "Security as Attributes of Social Systems", in: Brauch, Hans Günter; Oswald Spring, Úrsula; Mesjasz, Czeslaw; Grin, John; Dunay, Pal; Behera, Navnita Chadha; Chourou, Béchir; Kameri-Mbote, Patricia; Liotta, P.H. (Eds.): *Globalization and Environmental Challenges: Reconceptualizing Security in the 21st Century* (Berlin–Heidelberg–New York: Springer): 45-62.

Mesjasz, Czesław, 2008a: "Prediction in Security Theory and Policy, in: Brauch, Hans Günter; Oswald Spring, Úrsula; Mesjasz, Czeslaw; Grin, John; Dunay, Pal; Behera, Navnita Chadha; Chourou, Béchir; Kameri-Mbote, Patricia; Liotta, P.H. (Eds.): *Globalization and Environmental Challenges: Reconceptualizing Security in the 21st Century* (Berlin–Heidelberg–New York: Springer): 889-900.

Mesjasz, Czesław, 2010: "Complexity of Social Systems", in: *Acta Physica Polonica A*, 117,4 (April), 706-715; at: <http://przyrbwn.icm.edu.pl/APP/PDF/117/a117z468.pdf> (29 October 2011).

Midgley, Gerard (Ed.), 2003: *Systems Thinking*, vols I-IV (London: Sage).

Mirowski, Philip, 1989: *More Heat than Light: Economics as Social Physics, Physics as Nature's Economics* (Cambridge: Cambridge University Press).

Mirowski, Philip (Ed.), 1994: *Natural Images in Economic Thought: 'Markets Read in Tooth and Claw'* (Cambridge: Cambridge University Press).

Morgan, Gareth, 1996: *Images of Organization* (London: Sage).

Nicolis, Gregoire; Prigogine, Ilya, 1989: *Exploring Complexity: An Introduction* (New York: W. H. Freeman).

Ortony, Andrew (Ed.), 1993: *Metaphor and Thought* (Cambridge: Cambridge University Press).

Pezzey John C. V., Toman Michael A., 2002: "The Economics of Sustainability: A Review of Journal Articles", in: *Discussion Paper* 02-03 (Washington DC: Resources for the Future).

Picard, Kezia E., 2010: "A Radical Relational Agency: Foucault, Complexity Theory and Environmental Resistances" (PhD thesis, University of Nottingham); at: <http://eprints.nottingham.ac.uk/11450/1/Thesis.pdf> (7 May 2014).

Popper, Karl R., 1974: *The Poverty of Historicism* (London: Routledge & Kegan Paul).

Prigogine, Ilya; Stengers Isabelle, 1984: *Order Out of Chaos* (New York: Bantam).

Prigogine, Ilya, 1997: *End of Certainty* (New York: The Free Press).

Prigogine, Ilya, 2003: *Is Future Given?* (Singapore: World Scientific Publishers).

Richardson, Kurt; Cilliers, Paul, 2001: "Special Editors' Introduction. What Is Complexity Science? A View from Different Directions", in: *Emergence*, 3,1: 5-23.

Riley, Ralph, 1992: "The Challenge to Science in Food and Agriculture", in: *Towards Sustainable Crop Production Systems*, Monograph Series, 11 (Stoneleigh: Royal Agricultural Society of England): 13-15.

Rotmans, Jan, 2005: *Societal Innovation: Between Dream and Reality Lies Complexity* (Rotterdam: Erasmus Research Institute of Management); at: <http://repub.eur.nl/pub/7293/EIA-2005-026-ORG%20905892 1050%20ROTMANS.pdf> (22 July 2014).

Rotmans, Jan; Kemp, René; van Asselt, Marjolein, Geels, Frank; Verbong, Geert; Molendijk, Kirsten, 2001): *Transitions & Transition Management: The Case For a Low Emission Energy Supply* (Maastricht: International Centre for Integrated Assessment and Sustainable Development (ICIS)).

Rotmans, Jan; Kemp, René; van Asselt, Marjolein, 2001: "More Evolution Than Revolution: Transition Management in Public Policy", in: *Foresight*, 3,1: 15-31.

Rotmans, Jan; Loorbach, Derk, 2009: "Complexity and Transition Management", in: *Journal of Industrial Ecology*, 13,2 (April): 184-196.

Ruckelshaus, William D., 1989: "Toward a Sustainable World", in: *Scientific American*, 261,3 (September): 166-174.

Schwaninger, Markus, 2006: "The Quest for Ecological Sustainability: A Multi-Level Issue", in: Trappl, Robert (Ed.): *Cybernetics and Systems*, vol. 1, Proceedings of the Eighteenth European Meeting on Cybernetics and Systems Research (Vienna: University of Vienna and Austrian Society for Cybernetic Studies): 149-154.

Sen, Amartya, 2013: "The Ends and Means of Sustainability", in: *Journal of Human Development and Capabili-*

ties: A Multi-Disciplinary Journal for People-Centered Development, 14,1: 6-20.

Senge, Peter M., 1990: *The Fifth Discipline. The Art and Practice of the Learning Organization* (New York: Doubleday).

Simon, Herbert A., 1962: "The Architecture of Complexity", in: *Proceedings of the American Philosophical Society*, 106,6 (December): 467-482.

Simon, Herbert, 1997: *Models of Bounded Rationality: Empirically Grounded Reason* (vol. 3) (Cambridge, MA: MIT Press).

Sokal, Alan; Bricmont, Jean, 1998: *Fashionable Nonsense. Postmodern Intellectuals' Abuse of Science* (New York: Picador).

Stacey, Ralph D., 1996: *Complexity and Creativity in Organizations* (San Francisco: Berrett-Koehler).

Stacey, Ralph D.; Griffin, Douglas; Shaw, Patricia, 2000: *Complexity and Management. Fad or Radical Challenge to Systems Thinking?* (New York: Routledge).

Tainter, Joseph, 1988: *The Collapse of Complex Societies* (Cambridge: Cambridge University Press).

Tainter, Joseph, 2000: "Problem Solving: Complexity, History, Sustainability, Population and Environment", in: *A Journal of Interdisciplinary Studies*, 22,1 (September): 3-41.

Timmermans, Jos, 2008: "Punctuated Equilibrium in a Non-Linear System of Action", in: *Computational and Mathematical Organization Theory*, 14: 350-375; at: <http://link.springer.com/article/10.1007%2Fs10588-008-9031-5> (10 June 2014).

Timmermans, Jos; de Haan, Hans; Squazzoni, Flaminio, 2008: "Computational and Mathematical Approaches to Societal Transitions", in: *Computational and Mathe-matical Organization Theory*, 14: 391-414; at: <http://link.springer.com/article/10.1007%2Fs10588-008-9035-1> (10 June 2014).

Triple Pundit (no year); at: <http://www.triplepundit.com/> (24 July 2014).

Urry, John, 2005: "The Complexity Turn", in: *Theory, Culture & Society*, 22,5: 1-14.

Vandenberghe, Frédéric, 2014: *What's Critical about Critical Realism?: Essays in Reconstructive Social Theory* (New York: Routledge).

Waldrop, M. Mitchell, 1992: *Complexity: The Emerging Science at the Edge of Order and Chaos* (New York: Simon & Schuster).

Wallerstein, Immanuel, 2000: *The Essential Wallerstein* (New York: The New Press).

Weaver, Warren, 1948: "Science and Complexity", in: *American Scientist*, 36,4: 536-544.

Wiener, Norbert, 1948/1961: *Cybernetics: Or Control and Communication in the Animal and the Machine* (Paris: Hermann & Cie—Cambridge, MA: MIT Press).

Wittgenstein, Ludwig, 2002: *Philosophical Investigations* [German text with a revised English translation] (Oxford: Blackwell).

WCED (World Commission on Environment and Development), 1987: *Our Common Future*, Report of the World Commission on Environment and Development (United Nations); at: <http://www.un-documents.net/our-common-future.pdf> (16 April 2007).

Zadeh, Lotfi, 1975: "The Concept of a Linguistic Variable and its Application to Approximate Reasoning", in: *Information Sciences,* part I,8: 199-249; Part II,8: 301-357; Part III,9: 43-80.

Other Literature

A Portal on Sustainability (no year): "Definitions of Sustainability"; at: <http://www.cap-lmu.de/fgz/portals/sustainability/definitions.php> (27 April 2013).

Goodland, Robert, (no year): Sustainability: Human, Social, Economic and Environmental (Washington, DC: World Bank); at: <http://www.earthfutureaustralia.org/-4-pillars-of-sustainability.html> (18 September 2013).

Merriam-Webster Dictionary (no year); at: <http://www.merriam-webster.com/dictionary/sustainability> (6 March 2013).

Sustainable Kingston (no year), *Four Pillars of Sustainability*, Sustainable Kingston Corporation; at: <http://sustainablekingston.ca/community-plan/four-pillars-of-sustainability> (11 February 2014).

Sustainable Measures Company (no year); at: <http://www.sustainablemeasures.com/company> (24 March 2014).

19   Critical Approaches to Transitions Theory

Bonno Pel[1], Flor R. Avelino[2] and Shivant S. Jhagroe[3]

Abstract

Since its emergence as a theory of sustainability transformation, transitions theory has started to gain currency with both policymakers and researchers. As transitions approaches become established in research and policy, a process of institutionalization can be witnessed. Yet notwithstanding this mainstreaming, transitions theory continues to be controversial. Questions have been raised about its theorization of agency and transformation dynamics, and especially about the normative assumptions underlying its intervention strategies.

Arguably, these recurring questions call for 'critical approaches' to transitions theory. This contribution explores these, guided by a constructive attitude. The argument starts from the consideration that transitions theory harbours distinctly 'critical' elements, and that polemical juxtapositions between critical and uncritical transitions approaches are unnecessary: *What are the critical contents of transitions theory? How can the critical contents of transitions theory be retained and developed further?* These questions are answered through a historical comparison with the critical-theoretical project as initiated by Marx, Horkheimer and Adorno, amongst others. As with transitions studies, this project was meant to diagnose the social problems of its time, and to articulate corresponding remedial strategies. It ran into various internal contradictions, however, and these provide useful insights for the further development of critical transitions. The main conclusion is that transitions theory is well equipped to deal with these critical-theoretical paradoxes, but also displays tendencies towards relapsing into the pitfalls.

Keywords: Transitions, critical theory, politics, transformation paradoxes, normative foundations.

19.1   Introduction: Transitions Theory as a Critical Theory

As can be seen throughout this volume, it is probably fair to say that transitions theory is coming of age. It has started to gain currency amongst both researchers and policymakers as an integrative framework with which to grasp and manage processes of structural societal transformation towards sustainability. Recently Markard, Raven and Truffer (2012) provided an overview of transition studies in which they distinguish four main perspectives in the field: 1) the *Multilevel Perspective* (MLP; Geels 2005a, 2010); 2) *Strategic Niche Management* (SNM) (Kemp/Schot/Hoogma 1998; Schot/Geels 2008; Smith/Raven 2012); 3) *Technological Innovation Systems* (TIS; Hekkert/Suurs/Negro et al. 2007; Wieczorek/Hekkert 2012); and 4) *Transition Management* (TM; Rotmans 2006; Loorbach 2007, 2010). Alongside and in conjunction with these main strands, various other themes and approaches are developed in direct relation to socio-technical transition studies. This is further evidence that the field is becoming more mainstream (see also chapter 4 by John Grin in this volume).

Notwithstanding this institutionalization of transitions theory, however, its maturity as a theory of soci-

---

1   Bonno Pel, PhD, Institut de Gestion de l'Environnement et d'Aménagement du Territoire (IGEAT), Centre d'Etudes du Développement Durable, room DB.6.250, Université Libre de Bruxelles, 50 avenue FD Roosevelt, 1050 Bruxelles, Belgium; Email: <Bonno.Pel@ulb.ac.be>.

2   Flor Avelino, PhD, Dutch Research Institute For Transitions (DRIFT), Faculty of Social Sciences, Erasmus University Rotterdam, Burgemeester Oudlaan 50, T-building, Room 36, Rotterdam, The Netherlands; Email: <avelino@drift.eur.nl>.

3   Shivant Jhagroe (MSc., MA), Faculty of Social Sciences, Erasmus University Rotterdam, Burgemeester Oudlaan 50, T-building, room T17-45, Rotterdam, The Netherlands, jhagroe@fsw.eur.nl.

otechnical systems transformation remains subject to debate. Not surprisingly, the different transition-theoretical strands and origins make for continuous debate on the dynamics and management of sustainability transitions. Yet apart from this theoretical diversity, typical of multidisciplinary fields of study more generally, there are also more specific concerns. Ever since its inception as a framework for transformations in sociotechnical systems, transitions theory has met with doubts as to whether it can live up to its particularly high ambitions. Various questions have been raised regarding its theorization of agency and transformation dynamics, for example (Berkhout/Smith/Stirling 2004; Genus/Coles 2008; Smith/Voß/Grin 2010; Jørgensen 2012): Is the niche-regime-landscape heuristic really adequate vis-à-vis the complex reality that it is supposed to describe? Moreover, the transition-theoretical repertoire of intervention and management has met with sometimes harsh reactions, especially regarding the normative assumptions underlying these—intentionally profound—interventions in society (Berkhout 2006; Smith/Stirling 2008; Hendriks 2009; Stirling 2011; Røpke 2013): What counts as sustainable development? What and whose 'systems' are in need of transformation, and in what direction? Who decides, and through which procedures and on which grounds?

Especially these interrogations of its underlying normative assumptions indicate a need for 'critical approaches' to transitions theory (and its associated practices). The above questions provide a strong reminder that the implicit choices in transitions diagnoses and remedial strategies need to be accounted for. However, this is not to side with, or argue against, earlier allusions and allegations that transitions theory amounted to a novel version of (typically uncritical) social engineering (as expressed in e.g. Shove/Walker 2007 2008; van Duineveld/Beunen/van Ark et al. 2008). This contribution seeks to avoid adversarial debates on the critical contents of transitions theory, and the attendant false dichotomies that prevent these contents from being developed further. The key consideration from which to proceed further is that transitions theory harbours distinctly 'critical' elements. In some transition-theoretical accounts these may only operate in the distant background, while being more explicit in others. Still, notwithstanding the differences in approaches that exist within this field of research, transitions theory as a whole can be considered as a 'critical' theory from the outset. Crucially, it aims to interrogate the dominant social structures of its time, a critical attitude that is captured in the cen-

tral notion of sociotechnical 'regimes'. Likewise, the different strands share a dedication to the development of alternative development paths and practical solutions, i.e. to critical remedies, where these dominant structures are held to be no longer sustainable (Geels 2005a; Grin/Rotmans/Schot 2010; Zijlstra/Avelino 2012).

To consider *transitions theory* (TT) as a *critical theory* (CT) is only one possible understanding of it, of course. As other chapters in this volume bring out, transitions theory is also in many ways a positive, solution-oriented theory, and not in the first place a line of questioning and reflection. Still, considering that the 'regime' is a key analytical concept through which current practices are relentlessly questioned for their sustainability, the proposed understanding of transitions theory as critical theory seems plausible. A consequence of this particular understanding is that this chapter tends to focus on the *multilevel perspective* (MLP) and *transition management* (TM), and pays less attention to other strands such as technological *innovation systems* (TIS) and *strategic niche management* (SNM) that are less explicitly critical in their approaches. First and foremost, this chapter is intended to provide an overview of recent developments towards more emphatically critical transitions approaches, as they are developed in different transition-theoretical strands. Closely related to this overview is the task of systematizing these somewhat dispersed advances. *What are the critical contents of transitions theory? And considering the suggestions that these have somehow become lost, how can they be retained and developed further?*

This chapter will proceed as follows. First, we explain in what sense transitions theory is considered a 'critical' theory. Its unmistakeable similarities with the critical-theoretical project, as initiated by Marx, Horkheimer, Adorno and Habermas, amongst others, allow us to answer research questions through historical comparison. As with transitions studies, this critical project was meant to diagnose the social problems of its time, and to articulate corresponding remedial strategies. Crucially, it ran into various internal contradictions, however. We shall show how these contradictions resurface in contemporary debate on transitions theory (19.2). Next, three of these paradoxes are discussed, in order to structure the debate on critical transitions theory: after discussions on normative foundations (19.3), theorization of social evolution (19.4), and agency/structure (19.5), conclusions are drawn about the critical content of transitions theory and its further development (19.6).

19.2 Critical Content of Transitions Theory: Revisiting Critical Theory and its Paradoxes

As introduced, transitions theory can be considered to harbour distinctly 'critical' elements. It should be borne in mind that the latter term is used in a specific sense. It does not imply a negative or adversarial attitude (as suggested by colloquial understandings of 'criticism' or 'being critical of'), nor should it be confused with being important or crucial (as in 'critical infrastructures' or 'critical paths'). Instead, transitions theory is considered critical in the sense that it aims to interrogate the dominant social structures of its time, and to envisage alternative development paths in cases where these dominant structures are held to be undesirable, or un*sustainable*.

This specific philosophical usage of the term originally stems from Kant, who famously argued that knowledge is necessarily mediated by subjective categories. Following this line of reasoning, Hegel and Marx argued that knowledge must therefore be inherently historical, i.e. shaped by the prevailing social conditions under which human subjects live and act. Kant's critical self-understanding allowed for all sorts of philosophical elaborations, and it was Marx who has been most influential in developing it into a basis for transformative social theory. As he witnessed how the industrial revolution created profound social changes and serious social cleavages, he considered it imperative to move beyond detached philosophizing. Instead, the critical and modern self-understanding should become relevant mainly by articulating how people's thinking and living are determined by social structures, and how these typically operate 'behind their backs'. To Marx, the key structure in need of such critical unmasking was the system of capital accumulation, with its inherent tendencies towards exploitation and alienation.

Later, Horkheimer and Adorno (1947) dissociated themselves somewhat from Marx's materialistic focus on industrial production and labour. They did seek to maintain and develop his spirit and mission, however: their Frankfurt School promoted a social theory that was likewise set to articulate the social conditions of its time, so as to unveil and counter the dominant structures impeding true self-determination. In the Germany of the 1930s, they considered, this Enlightenment promise of self-determination had become a misleading caricature of itself. Their theoretical conclusion was that the rise of Nazism signalled new structures of domination to have emerged, next to

and in conjunction with those of capitalism. In its later post-World War II stages, the critical-theoretical project continued through such debates on society's crucial structures of domination. Marxist suspicion was mobilized against 'consumerism', against the homogenizing tendencies of 'mass society' and against the 'culture industry'. The various critical accounts of society led to particularly vehement debate, often revolving around alternative interpretations of Kant, Hegel, Marx, and Horkheimer and Adorno (Honneth 1991). The most influential theoreticians in this respect have been Habermas and Foucault. Whereas the first sought to establish a normative grounding to diagnoses of domination and system pathologies (Habermas 1998), namely his ideal speech construct, the latter rather showed how any such fundament would be prone to becoming a source of domination itself (Foucault 1998). To Foucault, Kant's critical spirit could be best applied through constant questioning of thoughts and practices—and this understanding of critique emphatically included self-questioning (Kelly 1998). On the other hand, whilst agreeing with most of this critical understanding, Habermas sought to safeguard the possibility of rational critique and social diagnosis. In the process of and alongside these philosophical debates, the critical project also inspired activists, artists and political parties. And as can be witnessed from research strands such as critical sociology, critical geography, gender studies, and political economy, critical theory also had a considerable following in the social sciences (Rasmussen 1996).

As indicated earlier, transitions theory bears several fairly obvious similarities with this critical tradition, and these similarities account for its critical content. First of all, transitions theory is similar in spirit, as it is equally motivated by a sense of urgency about the social problems of its time. Even when this sense of urgency or engagement tends to be somewhat less pronounced in current transition-theoretical work, early contributions such as Elzen, Geels and Green (2004), Rotmans (2006) and Loorbach (2007) were very explicit about the authors' sustainability concerns. Second, transitions theory is premised on the assumption that these problems have structural roots: current sustainability challenges are considered to be so particularly persistent due to the societal structures that operate 'behind people's backs'. In a similar way to Marx's work on the laws of capital, transitions theory conceives of sociotechnical 'regimes' as seldom questioned rules or grammar that shape thinking, acting and producing (Kemp/Schot/Hoogma 1998; Geels 2005a). In strategic niche management, for example (Smith/Raven 2012), the viability of an innovation is not considered to be a fixed and objective given. What is 'realistic' precisely depends on regime structures that are man-made, and—at least in principle—open to change. Third, transitions theory shares the critical-theoretical commitment to emancipation. Transitions theory refuses to limit itself to the diagnostics of system lock-in, but involves the task of devising corresponding remedies ('system lock-out', as expressed in Geels 2005a) as well. This practical orientation not only speaks from the various accounts of transition management, strategic niche management and technological innovation systems, all providing tools to empower sustainability actions, but the practical orientation also speaks from the commitments to trans-disciplinary, inclusive and socially engaged science (e.g. Grin/Rotmans/Schot 2010) and the inclusion of 'lay' expertise in experimenting and visioning processes (Loorbach 2007). Fourth and finally, both transitions theory and critical theory are historical, evolutionary theories: they co-evolve with the societal structures that they are intended to diagnose. Whilst critical theory grounded its theorization of social evolution mainly in Hegelian-Marxist dialectics, transitions theory is underpinned by a broad synthesis of social-scientific insights.

In summary, transitions theory displays strong similarities with its critical-theoretical predecessors—even when it is tapping quite different sources of knowledge. The latter fact reminds us that there are also significant differences to consider, and even suggests that transitions theory may have learned from its critical-theoretical predecessors. Indeed, detailed accounts of such lesson-drawing have been put forward, regarding both diagnostics (Geels 2005a; Geels/Schot 2007) and remedial strategies (Grin 2010; Loorbach 2007). Still, these accounts of theoretical origins mainly reveal that transitions theory is hardly directly informed by critical theory. The apparent similarities seem to stem mostly from the various 'critical' branches of social science that have developed through, and alongside, the sociological-philosophical 'critical project': one may think of science and technology studies (STS), evolutionary economics and neo-institutional theory as typically heterodox approaches that constitute essential critical ingredients of transitions theory.

So even when transitions theory can be appreciated as a contemporary form of critical theory, it cannot be considered an actual successor. The similarities between transitions theory and critical theory are quite clear, but there is only distant kinship. It is there-fore all the more interesting to see how transitions theory and critical theory seem to coincide in the counterarguments and objections they have received. For the development of critical approaches to transitions theory, these similarities in particular merit consideration. The theoretical impasses of critical theory can be considered well rehearsed, and as such they provide ready available opportunities for learning and reflection.

Throughout its development, critical theory has met with strong criticisms, ranging from principled objections and attempts at refutation to rather adversarial and crudely formulated allegations. All in all, the critical project has lost much of its earlier appeal: to begin with, Marxism has come under a cloud not only due to the experiences with communism, but also due to a certain theoretical inertia. Class-based struggle is only one dimension of contemporary political cleavages, for example, and 'capitalism' does not capture the variety of possible political economies under that heading. These Marxist categories have become rather blunt instruments of critique. In addition, the analyses of the Frankfurt School, whose diagnoses inspired many to criticize or to fight 'the System', now appear as rather outdated and one-sided. As their cultural-critical notions of 'mass society' have been surpassed by social diversification, the associated diagnoses of systemic problems and identification of culprits have become likewise problematic. Now these assertions of outdated critique remind us that it is easy to ridicule earlier critical theory with the benefit of hindsight. Its less refined and exaggerating manifestations making for excellent 'straw man' targets, critical theory is vulnerable to easy criticisms and caricature. It is therefore important to focus on the more precise formulations amongst the many objections that have been raised. Apart from incidental misconceptions and empirical inaccuracies, internal contradictions have been identified as flaws that seem to be inherent in critical theory. These paradoxes are most relevant to this volume, as they resurface in recent transitions theory debate. In the next sections, the following paradoxes will be dealt with:

1. Normative foundations (when norms are admitted to be socially constructed).
2. Keeping up with evolution (in a theory with universal aspirations).
3. Balancing lock-in and lock-out (when sustainability challenges are considered to be both persistent and soluble).

Confronting transitions theory with these paradoxes, its critical contents can be specified, as can how these can be retained and developed further.

19.3   Critical Theoretical Paradoxes (I): Normative Foundations

The central mission of the critical-theoretical project was to interrogate and lay bare the social conditions under which the Enlightenment promise of self-determination was to be realized (McCarthy 1998). The seminal 'Dialectic of Enlightenment' by Horkheimer and Adorno (1947) was to show how this ideal was far from realized—instead, it had turned into an ideological smokescreen, serving the powers that be. Raising people's awareness about their silent subjugation was to be the critical-theoretical remedy towards achieving *true* self-determination. This line of reasoning has been most influential, and not only in the academic sense: the neo-Marxist cultural critique inspired counter-cultural movements around the world, and continues to inform new social movements. Moreover, the classical Marxist appeals to class-based struggle have even manifested themselves in many revolutions and armed struggles. Unsurprisingly, the critical-theoretical line of thinking was scrutinized ever more fiercely, the more influential it became within and outside academia. These objections often concerned the adequacy of the system diagnoses, but were particularly frequently raised against the overly confident ivory tower perspective presupposed by the objections themselves: How could the critical theorists see what the common man couldn't? How could they themselves free their diagnoses from the dominant discourses of their time? And how could the diagnoses of alienated, enslaved people hold, when the alleged patients themselves claimed to feel fine?

Through such questions, directly targeting the ways in which the critical diagnoses served as legitimizations for revolutionary action or armed resistance, the critical-theoretical project was questioned for its own implicit assumptions. Popper challenged the critical theorists to scientifically account for their historicist prophecies, for example. Other critics rather went along with the critical line of reasoning to some extent, only to turn it against the ivory tower aspirations and undemocratic tendencies that often accompanied it: especially Foucault (1998) and Luhmann (1990 2002) have relentlessly confronted the critical theorists with their lack of *self*-critique (Kelly 1998). Critically interrogating the critical accounts of system pathologies and false beliefs themselves, the normative assumptions underlying these diagnoses often proved difficult to account for (Honneth 1996). Crucially, these investigations of normative yardsticks showed the 'ivory tower' lines of critical theory to be rather uncritical forms of critique.

Similar critiques-of-the-critical have been raised against transitions theory: especially Duineveld, Beunen and van Ark et al. (2008) and Shove and Walker (2007, 2008) echoed the Foucauldian inquiries after the distanced, 'voycuristic' system diagnoses through the MLP, and their deployments as apparently self-evident bases for action. In particular they took issue with the notion of transition management and its suggestion of control and steering. Moreover, as transition advocates were seen to remain relatively silent on the politics and power play involved, the question was suspiciously posed as to whether there were dirty hands to hide. The issue of normative assumptions also evoked cogent inquiries by Smith and Stirling (2008, 2010) and Berkhout (2006). Following up on these critiques, Pel and Boons (2010) and Pel (2012b) addressed the practical implications of system demarcations and 'sustainability' understandings in transitions theory (see also Ulrich 2003 on this critical systems thinking). Furthermore, various authors have stressed the directionality of transitions (Stirling 2009, 2011; Cohen 2010; Røpke 2012). The key message is then that once transitions theory starts to obscure the diversity of possible transition pathways and the attendant political choices, it will lose its critical contents. These contributions usefully specify earlier inquiries after the choices and demarcations implied with transitions theory and management. The emphasis on the directionality of transition processes helps substantiate how the normative yardstick of sustainability is less than clear-cut, yet is often mistaken as 'post-political' (Kenis/Mathijs 2014). Finally, there have been critiques that focus on the politics and choices involved with transitions-based remedial strategies, i.e. with transitions practice. In a special issue on transitions-in-the-making, Voß, Grin and Smith (2009) observed the tendency of transitions approaches to be captured and co-opted by regime structures. This tragic fate of critical remedies, their proneness to reproducing the very dominant structures they were set to transform, is a classical critical-theoretical theme.

The above inquiries after its normative foundations have prompted various attempts to safeguard and bolster the critical contents of transitions theory. First of all, allegations of technocratic 'management' from the ivory tower have led to an altogether greater awareness that the term transition *management* bears misleading connotations. This has led in turn to a stronger focus on critical contents such as negotiated visions, emphatically inclusive governance processes, and participatory future scanning. These advances help avoid critical-theoretical one-sidedness, and the associated 'totalitarian temptation' of one-sided action. Second, and related to the first issue, there are several contributions that seek to develop transitions theory more explicitly in the context of reflexive modernization (Beck/Bonns/Lau 2003; Voß/Bauknecht/Kemp 2006; Grin 2010). Avelino and Grin (2014) argue that transitions theory should embrace the tradition of critical deconstruction, while at the same time reaching for reconstruction. This work on reflexive modernization strongly reminds us of Habermas' attempts to safeguard a self-aware but constructive critical theory. Third, the attention given to the practices and politics of transitions-in-the-making has become an integral part (or branch) of transitions theory, as can be witnessed from Paredis 2008; Avelino 2009, 2011; Kern/Howlett 2009; Meadowcroft 2009; Hendriks 2009; Kemp/Rotmans 2009; Loorbach/Rotmans 2010; Smith/Voß/Grin 2010; Farla/Markard/Raven/Coenen 2012; Geels 2014; and Jhagroe/Loorbach 2014. This reflexive attention to the diversity of interests and perspectives involved, the choices made and the uncertainties and ambiguities encountered, can be appreciated as a definitive step away from the overly self-assured, elitist critical theory of earlier days.

Finally, there is the question of whether sustainability can be maintained as a firm, foundational and unproblematic normative yardstick. Considering the pluralistic, diversity-oriented turn that can be witnessed, transitions theory is approaching the Foucauldian point at which 'sustainability' is also acknowledged to be a potentially oppressive regime-structure—an unquestioned societal 'grammar' with considerable material implications. In a way, the reification of social problems is counteracted by the analytical attention to shifting and multifaceted problems (see Penna/Geels 2012, for example). Moreover, some of the ANT-inspired work in transitions research seems to part with the 'persistent sustainability problems' (Schuitmaker 2012) as objective grounds for action. On that account, the struggle between different sustainability understandings is precisely what is at issue in transition processes. Still, such radical constructivism may be a reflexive bridge too far for most transitions theorists. The reality of a class of persistent systemic problems can be considered constitutive for transitions theory as such (Rotmans 2006), in the first place.

Moreover, one could also consider how the critical-theoretical toolbox could be perverted into interest-driven climate change scepticism (Latour 2004), i.e. into a theoretical smokescreen that undermines transitions theory. So there may be only a fine line between transitions theory and reflexive governance (Rip 2006), but it is an important line. Transitions theory is in many ways reflexive, but it is also guided by strong engagement and the desire to steer clear of relativism. Transitions theory is quite Habermasian in its normative orientation. However reflexive, it is inclined to consider relativist accounts of critique as distracting and 'quietist' (Habermas 1998; Pel 2014), i.e. as no longer critical for the normative silence on the unsustainable conditions of their time.

In other words, there are clear developments towards a transitions theory that is more critical about its normative assumptions. Still, it tends to remain somewhat less critical about the reality status of the 'persistent problems'. These operate as normative anchors for the proposed strategies for 'regime' change.

19.4 Critical Theoretical Paradoxes (II): Keeping up with Evolution

Critical theory revolves around the understanding that knowledge is shaped by social conditions, and therefore, by their evolution. Still, a recurring internal contradiction has been the very tendency to seek to escape from this predicament, and build a universally valid critical theory that somehow transcends the context in which it is formulated. One of the most well-known examples of this is Marx's universalist understanding of class struggle. In the advanced Western societies, however, the juxtaposition between the labour class and the capitalist class has to a significant extent been surpassed by the emergence of a large middle class and general increases in wealth. Moreover, the class struggle frame has not only become less adequate to the diagnosis of contemporary political economy, it can also be criticized for drawing attention away from other, newly emerging societal issues (Honneth 1991; Rosa 2012). Like 'class struggle', earlier diagnoses of 'mass society' and 'consumer society' have arguably failed to (entirely) keep up with societal evolution. The first has largely been surpassed by societal differentiation, for example, and the second can be considered negligent of global inequality: in contemporary North-South dialogues, it is increasingly acknowledged that consumer society may be something to critique and remedy in some contexts, whilst remaining a legitimate societal aspiration in others. Moreover, earlier critical theorists have been reproached for staying silent about, and therefore for sustaining, gender and ethnic inequalities (Young 1990). And finally, it is particularly relevant for transitions theory how the critical attention given to societal power relations has often overshadowed the emergence of sustainability challenges. The critical-theoretical paradox is then that the aspirations towards universality did not fortify the critique, but often even diminished the accuracy of the diagnoses. As Foucault liked to demonstrate, the foundational and presumed trans-historical critiques often turned into less than critical mental straightjackets—with oppressing tendencies themselves.

In the face of this struggle with temporality and evolution, transitions theory seems much better equipped than its predecessors (if only through the benefit of hindsight). An often-cited and telling example is the account of evolving mobility problems: the current automobility system, with its greenhouse gas emissions and other system pathologies, may be considered an unsustainable regime in need of transformation. Still, there is the transition-theoretical awareness that the era of the horse-drawn carriage wasn't ideal either. The associated pollution from horse excrement and the traffic dangers were no less persistent problems (Geels 2005b). Thanks to these technological-historical accounts, transitions theory is equipped with a strong awareness that systems, their pathologies and also the underlying ways of diagnosis evolve. Its critical contents also speak from the nuanced phrasing of regimes being *no longer* sustainable (Rotmans 2006). This avoids the pitfall of hypostasis, i.e. of maintaining critiques of systems that have transformed in the meantime. Next to the history of technology, transitions theory owes its relatively advanced understanding of system dynamics to evolutionary economics, *science and technology studies* (STS), neo-institutional theory and insights on *complex adaptive systems* (CAS). The very notion of a 'regime' implies a dynamic, evolutionary understanding of dominant systems, which crucially bypasses any reification of 'the system'. Similarly, CAS theory helps articulate how systems can be solid *and* adaptive, or 'dynamically stable' (Geels/Schot 2007; Grin/Rotmans/Schot 2010). Moreover, it offers a world view of nested and co-evolving systems, which allows for accordingly layered, differentiated and therefore *nuanced* diagnoses and remedies.

Notwithstanding these impressive theoretical resources, there is still intense debate about the transition-theoretical capacity to keep up with evolution. The debate on the different types of transition pathways (Geels/Schot 2007) is ongoing, there have been moves towards the analysis of multiple niches and regimes (Schot/Geels 2008; Raven/Verbong 2007, 2009), and there have been systematic comparisons and syntheses between the frameworks for sociotechnical and socio-ecological transitions and technological innovation systems (van der Brugge/van Raak 2007; Smith/Stirling 2010; Westley/Olssen/Folke et al. 2012). Alongside this, there are the advances in the modelling of system dynamics (e.g. Haxeltine et al. 2008; De Haan 2010). The sheer volume of work performed in that area brings out how transitions theory strongly taps into formal methods and systems science. More generally, one can consider how the combined work on the *multilevel perspective* (MLP), *strategic niche management* (SNM), *technological innovation systems* (TIS), *transition management* (TM), and *socioecological systems theory* (SES) is developing towards a transitions science, and towards a management of transitions that is thoroughly underpinned by scientific insight. Like Marx's attempts to underpin his critical theory with science-based 'laws of capital', these predominantly functionalistic approaches indicate how transitions theory is generally intended as a scientific approach to the diagnosis of systemic problems.

Arguably, the developments towards a transitions science described above do significant work in helping transitions theory to keep up with evolution. In this respect transitions theory diverges somewhat from the humanities orientation of much earlier critical theory. What is more, the very tendencies towards transitions science go against the frequent critical-theoretical charges against positivist 'scientism', which was mistrusted for its altogether system-reproducing tendencies. In that respect, it should be mentioned that there are also transition-theoretical advances that do not so much seek to solidify the accounts of system evolution, but rather take direct issue with transition-theoretical aspirations towards universality. ANT-inspired scholars continue to challenge the tendency to reify systems and the associated diagnoses of pathologies (Maassen 2012; Jørgensen 2012; Hodson/Marvin 2012; Chilvers/Longhurst 2013), stressing the dynamic and fluid character of systemic structures. Especially relevant in this regard is the trend that has recently emerged towards geographical concreteness and differentiation (Coenen/Benneworth/Truffer

2012; Raven/Schot/Berkhout 2012; Späth/Rohracher 2012; Eshuis et al. 2012). This rapidly expanding field of research takes into account that transitions theory has been very much shaped by the (North-)Western European context in which it was initially developed, and deliberately takes it to other places and spaces. This particular transition-theoretical development can be appreciated as a direct response to the above paradox: transitions theory is acknowledged to be a situated type of knowledge.

19.5   Critical Theoretical Paradoxes (III): Balancing Lock-in and Lock-out

As has become clear already, critical theory has faced many profound counter-arguments that have seriously compromised this project of research-driven emancipation. Another often-heard objection is simply that it is so depressing. The eloquent and compelling accounts of systematic oppression, inert systems and counter-productive resistance tend to dash all hopes of remedies—an issue taken up in Ernst Bloch's 'Principle of Hope', and more recently in Unger (1987). The latter especially has focused on a paradox that critical theorists have painstakingly sought to deal with: if social structures are indeed decisive for the formation of thought and action, there can be no independent position from which to articulate a critique that challenges them. Conversely, all too bright perspectives on transformation cast doubt on the underlying societal diagnoses, which apparently have underestimated conditions of system lock-in. This transformation paradox has been a recurring theme in the Habermas versus Foucault debate (Kelly 1998; Honneth 1991, 1996). More generally, the difficulty of developing a balanced perspective on both system lock-in and system lock-out continues to challenge the various practical attempts at transformation: should transformation come from a 'march through the institutions', or through revolutionary action from outsiders? Is transformation possible in the first place? Or does a proper analysis of the state of affairs rather warrant resignation, ironical distance or cautiously reflexive action?

This third paradox is of course very familiar to transitions theory. It indicates the notorious structure-agency debate, for which transitions theory has developed and assembled its own tools. Regarding the paradox, Geels (2005a) has formulated quite clearly the idea that transitions theory is set to articulate both

system lock-in and its inverse of lock-out. He acknowledges this to be a particularly challenging task, as the two tend to be split over different disciplines. Hence the elaborate theoretical construct of the Multilevel Perspective on transitions, which draws on several bodies of knowledge to achieve the desired balance. Compared with earlier critical theory, this synthesis has several features that leave transitions theory better prepared to deal with the above-stated transformation paradox: regimes are dynamically stable but not deterministic, niches are not complete outsiders in opposition to systemic structures but also co-constitutive of sociotechnical systems, and the addition of the landscape level constitutes a changing background rather than a deterministic force. Crucially, landscape factors are considered to be developments that intermittently open and close windows of opportunity. The latter alone is a significant reframing of the traditional dichotomous framing of power relations, in which revolutionaries are pitted against the 'system'. As discussed in the previous section, this dynamic, CAS-based representation of regimes beats earlier critical-theoretical accounts of monolithic and inert systems.

Not surprisingly however, the MLP synthesis has been and continues to be subject to discussion. As Geels (2010) indicated, this is a matter of maintaining the critical contents of transitions theory. The MLP cannot be a 'truth machine', he explains (see also Genus/Coles 2008; Smith/Voß/Grin 2010), as it constitutes a synthesis of different perspectives on transitions that depart from diverging, and possibly even incommensurable, paradigms. This emphatically critical contribution to transitions studies, raising attention to the diversity of its underpinning theories, clarifies much of the continuous debate on the balance between lock-in and lock-out.

First, it draws attention to the striking fact that transitions theory is a form of critical theory that goes without, and even inverts, the pessimistic, dystopian undertone that is traditional in the genre. In fact, there have been critiques of the great confidence in emerging niches as sources of transformation from the very beginning (Berkhout/Smith/Stirling 2004; Schot/Geels 2008). Second, this draws attention to the divide between 'global' and 'local' models of transitions (Geels 2010). Even when these types of transitions theory share the dynamic understanding of structure-agency relations, the first typically emphasizes structure and the second agency. This has spawned various attempts at theoretical integration and cross-fertilization, such as practice theory (Seyfang/Haxeltine 2012; Hargreaves Haxeltine/Longhurst 2011; see earlier Shove (2004) and ANT (Maassen 2012; Pel 2012a; Jørgensen 2012). Third, there have been various developments towards more dynamic representations of the levels distinguished: dynamic niche-regime interactions and translation processes between the two (Smith 2007; Raven/Verbong/Schilpzand et al. 2011), alternative perspectives on the exogenous status of the societal 'landscape' (Avelino 2011; Riddell/Westley 2013), the phenomenon of active regime resistance (Geels 2014; Hess 2014); and the equally easily forgotten transformation potentials of endogenous regime renewal (Wells/Nieuwenhuis 2012).

Finally, the structure-agency debate within transitions studies has received a new impulse that relates directly to its often implicit inclination towards innovation. Indeed, this orientation may silently introduce an optimistic view on the scope for agency and change, even when the inertias of regime structures are acknowledged. Shove (2012) therefore argues that the 'shadowy side' of innovation, i.e. its relapse into earlier system states, merits more analytical attention. Similarly, Turnheim and Geels (2013) have turned transition-theoretical attention towards processes of discontinuation—the darker side of creative destruction that is easily forgotten in transitions idealism. These attempts in particular to balance innovation and decline can be appreciated as impulses towards a more emphatically critical transitions theory. After all, they question the unwarranted embraces of innovation which transitions theory may share with the societal conditions that it is set to diagnose.

19.6 Conclusion: Critical Approaches to Transitions Theory

Through these three confrontations with critical-theoretical paradoxes, the notion of 'critical approaches' to transitions theory has been clarified. In this conclusion, the two research questions can be answered: *What are the critical contents of transitions theory? And considering the suggestion that these have somehow become lost, how can they be retained and developed further?*

As regards the first question, we can conclude that transitions theory is particularly well-equipped to deal with the critical-theoretical paradoxes of 'keeping up with evolution' (19.4) and 'balancing lock-in and lock-out' (19.5). This can largely be attributed to the various dynamic, evolutionary perspectives and associated bodies of knowledge that have been assembled in the

different transition-theoretical strands. Many of these resources were not available to the earlier critical theorists, who also tended to uphold rather crude understandings of systems compared to the transition-theoretical differentiation between sociotechnical subsystems. On the other hand, transitions theory is less convincingly critical regarding the 'normative foundations' paradox (19.3). Definite advances in this respect have been made already, such as a greater attention to the politics of transitions and the methodological shift towards transitions-in-the-making. Still, much of the critical questioning remains hidden under the presumed self-evidence of persistent sustainability challenges, and the objectivity of the 'regime' structures to which these challenges are traced. This apparent relapse into objectivism and ivory tower critiques, and the more general negligence of the normative assumptions underlying transitions diagnoses and remedies, seems to be the flip side of the aspiration towards transitions science. In this process of fortification and scientization, transitions theory is somewhat uprooted from its origins in the humanities.

Regarding the second question, there are nevertheless considerable developments towards a more emphatically critical transitions theory. A great many recent advances can be witnessed that take direct issue with the paradoxes discussed: in order to keep up with evolution, not only new systemic problems but also new diagnostics are being explored. Moreover, the transition-theoretical critique is increasingly acknowledged to be situated, as the thrusts towards geographical differentiation nicely illustrate. Furthermore, the transformation paradoxes around lock-in seem to be more and more acknowledged as inevitable, and as permanent challenges to transitions theory. In this regard it is telling how the transitions research community manages to reconcile numerous disciplinary backgrounds, and how there is a general awareness of the different perspectives and theoretical assumptions at play. Finally, even when transitions theory is mainly critical in terms of 'regime' analysis, there is the self-examination that helps avoid overconfident and one-sided critique. Reflecting on ontological assumptions, and also the recent confrontations of its 'innovation-bias', can be appreciated as thrusts towards self-critical transitions theory. Interestingly, many cases of the refinements discussed can even be traced to earlier critical-theoretical discussions. We could consider, for example, how ANT scholars bring in distinctly postmodernist-Foucauldian perspectives to counter a 'transitions science' that is too confident about its systems diagnoses. Likewise, the many efforts to provide transitions theory with a stronger scientific underpinning reflect how it is a distinctly modernist kind of critical theory–more in the spirit of Marx and Habermas than of Adorno and Foucault. Meanwhile, there is the work that develops transitions theory as a kind of reflexive modernization, which could be considered a third, intermediate position. All in all abundant advances are being made towards retaining the critical contents of transitions theory, and these can be seen to constitute different modes of critique.

Finally, we can consider whether transitions theory will develop in greater or lesser continuity with its critical predecessors. As discussed, there are clear developments towards an emphatically critical transitions theory. On the other hand, there is also the trend towards transitions science that rather drives transitions theory away from the generally more reflective, humanities-driven critical theory. In their turn, these opposite trends remind us of other differences between transitions theory and critical theory. They remind us that transitions theory is also more practical in its outlook, more future-oriented and more professionalized than most of its critical predecessors.

In these respects it goes beyond the largely deconstructive diagnostics (Avelino 2011) of its predecessors, or is at least very different from such awareness-raising activity. The existence of repertoires such as transition management, strategic niche management and the technological innovation systems framework shows how transitions theory is strongly dedicated to the development of remedies (Pel 2014). Interestingly, this very instrumental quality seems to account for at least part of its current institutionalization. Typically, such mainstreaming was treated with suspicion by most critical theorists, who had a sixth sense for the ensuing 'capture' (Voß/Smith/Grin 2009) and neutralization of critique. On the other hand, the instrumental attitude of transitions theory could also be considered very much in line with the critical theoretical project, in as far as it continues to aim for engaged research. In other words, transitions theory is likely to continue the critical-theoretical project, but only in some of its aspects. Yet however it may develop, the critical-theoretical legacy could come in useful: it reminds us that transitions theory too is shaped by the prevailing social conditions of its time.

References

Avelino, Flor, 2009: "Empowerment and the challenge of applying transition management to ongoing projects", in: *Policy Science*, 42,4: 369-390.

Avelino, Flor, 2011: "Power in Transition: Empowering Discourses on Sustainability Transitions" (PhD dissertation, Erasmus University Rotterdam, Dutch Research Institute for Transitions).

Avelino, Flor; Grin, John, 2014: "Beyond Deconstruction. A Reconstructivist Perspective on Sustainability Transition Governance", paper for the Fifth International Conference on Sustainability Transitions, 27-29 August 2014, Utrecht.

Beck, Ulrich; Bonns, Wolfgang; Lau, Christoph, 2003: "The theory of reflexive modernization: problematic, hypotheses and research programme", in: *Theory, culture & society*, 20, 2: 1-33.

Berkhout, Frans; Smith, Adrian; Stirling, Andy, 2004: "Socio-technological regimes and transition contexts", in: Elzen, Boelie; Geels, Frank; Green, Ken (Eds.), 2004: *System Innovation and the Transition to Sustainability: Theory, Evidence, and Policy* (Cheltenham: Edward Elgar): 48-75.

Berkhout, Frans, 2006: "Normative Expectations in Systems Innovation", in: *Technology Analysis & Strategic Management*, 18,3-4: 299-311.

Chilvers, Jason; Longhurst, Noel, 2013: *Participation in Transition(s): Emergent engagement, politics and actor dynamics in low carbon energy transitions* (University

of East Anglia, Norwich: Science, Society and Sustainability Research Group).

Cohen, Maurie, 2010: "Destination Unknown: Pursuing Sustainable Mobility in the Face of Rival Societal Aspirations", in: *Research Policy*, 39,4: 459-470.

Coenen, Lars; Benneworth, Paul; Truffer, Bernhard, 2012: "Toward a spatial perspective on sustainability transitions", in: *Research Policy*, 41,6: 968-979.

De Haan, Hans, 2010: "Towards Transition Theory" (PhD dissertation, Erasmus University Rotterdam, Dutch Research Institute For Transitions).

Duineveld, Martijn; Beunen, Raoul; van Ark, R.; van Assche, Kris; During, Roel, 2008: *The difference between knowing the path and walking the path: een essay over het terugkerend maakbaarheidsdenken in beleidsonderzoek* (Wageningen: Leerstoelgroep Sociaal-ruimtelijke Analyse Wageningen Universiteit).

Elzen, Boelie; Geels, Frank; Green, Ken (Eds.), 2004: *System Innovation and the Transition to Sustainability: Theory, Evidence, and Policy* (Cheltenham: Edward Elgar).

Eshuis, Jasper; Spekkink, Wouter; Loorbach, Derk; Roorda, Chris; Stuiver, Marian; van Steenbergen, Frank, 2012: "Challenges and Tensions in Area Based Transitions", Paper for the Seventh International Conference in Interpretive Policy Analysis, Tilburg, NED, 5-7 July.

Farla, Jacco; Markard, Jochen; Raven, Rob; Coenen, Lars, 2012: "Sustainability transitions in the making: A closer look at actors, strategies and resources; in: *Technological Forecasting & Social Change*, 79,6: 991-998.

Foucault, Michel, 1998: "Two Lectures", in Kelly, Michael (Ed.): *Critique and Power: recasting the Foucault/Habermas debate* (Cambridge, Mass.: MIT Press): 17-46.

Geels, Frank, 2005a: *Technological Transitions and System Innovations: A Co-evolutionary and Socio-Technical Analysis* (Cheltenham: Edward Elgar).

Geels, Frank, 2005b: "The dynamics of transitions in socio-technical systems: A multi-level analysis of the transition pathway from horse-drawn carriages to automobiles (1860-1930)", in: *Technology Analysis & Strategic Management*, 17,4: 445-476.

Geels, Frank; Kemp, Rene; Dudley, Geoff; Lyons, Glenn (Eds.), 2012: *Automobility in transition? A sociotechnical analysis of sustainable transport* (Routledge: London).

Geels. Frank; Schot, Johan, 2007: "Typology of sociotechnical transition pathways", in: *Research Policy*, 36,3: 399-417.

Geels, Frank, 2010: "Ontologies, socio-technical transitions (to sustainability), and the multi-level perspective", in: *Research Policy*, 39,4: 495-510.

Geels, Frank; Penna, Caetano 2014: "Societal problems and industry reorientation: Elaborating the Dialectic Issue LifeCycle (DILC) model and a case study of car safety in the USA (1900-1995)", in: *Research Policy*, 2014.

Geels, Frank, 2014: "Regime resistance against low-carbon transitions: Introducing politics and power into the multi-level perspective", in: *Theory, Culture & Society*, 31,5: 21-40

Genus, Audley; Coles, Anne-Marie 2008: "Rethinking the multi-level perspective of technological transitions", in: *Research Policy* 37, 9: 1436-1445.

Grin, John, 2010: "Understanding Transitions from a Governance Perspective" in: Grin, John; Rotmans Jan; Schot, Johan (Eds.), *Transitions to Sustainable Development: New Directions in the Study of Long Term Transformative Change* (New York: Routledge): 221-319.

Grin, John; Rotmans, Jan; Schot, John (Eds.), 2010: *Transitions to Sustainable Development: New Directions in the Study of Long Term Transformative Change* (New York: Routledge).

Habermas, Jürgen, 1998: "Some Questions Concerning the Theory of Power: Foucault again", in: Kelly, Michael (Ed.): *Critique and Power: recasting the Foucault/Habermas debate* (Cambridge, Mass.: MIT Press): 79-107.

Hargreaves, Tom; Haxeltine, Alex; Longhurst, Noel; Seyfang, Gill., 2011: *Sustainability transitions from the bottom-up: Civil society, the multi-level perspective and practice theory*, Norwich: CSERGE Working Paper 2011-0.1.

Haxeltine, Alex; Whitmarsh, Lorraine; Bergman, Noam; Rotmans, Jan; Schilperoord, Michel; Kohler, Jonathan, 2008: "A Conceptual Framework for transition modelling", in: *International Journal of Innovation and Sustainable Development*, 3,1: 93-114.

Hess, David J., 2014: "Sustainability transitions: A political coalition perspective", in: *Research Policy*, 43,2: 278-283.

Hekkert, Marko; Suurs, Roald; Negro, Simona; Kuhlmann, Stefan; Smits, Ruud, 2007: "Functions of innovation systems: A new approach for analysing technological change", in: *Technological Forecasting and Social Change*, 74,4: 413-432.

Hendriks, Carolyn, 2009: "Policy design without democracy? Making democratic sense of transition management", in: *Policy Science*, 42,4: 341-368.

Hodson, Mike; Marvin, Simon, 2010: "Can cities shape socio-technical transitions and how would we know if they were?", in: *Research Policy*, 39,4: 77-485.

Honneth, Axel, 1991: *The critique of power: Reflective Stages in a Critical Social Theory* (Cambridge, Mass.: MIT Press).

Honneth, Axel, 1996: "Pathologies of the Social: The Past and Present of Social Philosophy", in: Rasmussen, David (Ed.): *The Handbook of Critical Theory* (Cambridge, Mass.: Blackwell): 369-396.

Horkheimer, Max; Adorno, Theodor, 1947, 2002: *Dialectic of Enlightenment: Philosophical Fragments* (Stanford: University Press).

Jhagroe, Shivant; Loorbach, Derk, 2014: "See no evil, hear no evil: The democratic potential of transition management", in: *Environmental Innovation and Societal Transitions*; at: <doi:10.1016/j.eist.2014.07.001>.

Jørgensen, Ulrik, 2012: "Mapping and navigating transitions—The multi-level perspective compared with arenas of development", in: *Research Policy*, 41: 996-1010.

Kelly, Michael. (Ed.), 1998: *Critique and Power: recasting the Foucault/Habermas debate* (Cambridge, Mass.: MIT Press).

Kemp, Rene; Schot, Johan; Hoogma, Remco, 1998: "Regime shifts to sustainability through processes of niche formation: The approach of strategic niche management", in: *Technology Analysis & Strategic Management*, 10,2: 175-198.

Kemp, Rene; Rotmans, Jan, 2009: "Transitioning policy: co-production of a new strategic framework for energy innovation policy in the Netherlands", in: *Policy Sciences*, 42,4: 303-322.

Kenis, Anneleen; Mathijs, Erik, 2014: "(De)politicising the local: The case of the Transition Towns movement in Flanders (Belgium)", in: *Journal of Rural Studies*, 34: 172-183.

Kern, Florian; Howlett, Michael, 2009: "Implementing transition management as policy reforms: a case study of the Dutch energy sector", in: *Policy Sciences*, 42,4: 391-408.

Latour, Bruno, 2004: "Why Has Critique Run out of Steam? From Matters of Fact to Matters of Concern", in: *Critical Inquiry*, 30,2: 225-248.

Loorbach, Derk, 2007: *Transition Management: New Mode of Governance for Sustainable Development* (Utrecht: International Books).

Loorbach, Derk; Rotmans, Jan, 2010: "The practice of transition management: Examples and lessons from four distinct cases", in: *Futures*, 42: 237-246.

Loorbach, Derk, 2010: "Transition Management for Sustainable Development: A Prescriptive, Complexity-Based Governance Framework", in: *Governance*, 23,1: 161-183.

Luhmann, Niklas, 1990, "Ich sehe was, was Du nicht siehst", in: *Soziologische Aufklärung*, 1990,5: 228-234.

Luhmann, Niklas, 2002: *Theories of distinction: redescribing the descriptions of modernity* (Stanford, CA: Stanford University Press).

Maassen, Anne, 2012: "Heterogeneity of Lock-In and the Role of Strategic Technological Interventions in Urban Infrastructural Transformations", in: *European Planning Studies*, 20,3: 441-460.

Markard, Jochen; Raven, Rob; Truffer, Bernhard, 2012: "Sustainability transitions: an emerging field of research and its prospects", in: *Research Policy*, 41,6: 955-967.

McCarthy, Thomas, 1998: "The Critique of Impure Reason: Foucault and the Frankfurt School", in: Kelly, Michael (Ed.): *Critique and Power: recasting the Foucault/Habermas debate* (Cambridge, Mass.: MIT Press): 243-282.

Meadowcroft, John, 2009: "What about the politics? Sustainable development, transition management, and long term energy transitions", in: *Policy Science* 42,4: 323-340.

Paredis, Erik, 2008: *Transition management in Flanders. Policy context, first results and surfacing tensions* (Ghent: Ghent University, Centre for Sustainable Development and Flemish Policy Research Centre on Sustainable Development).

Pel, Bonno; Boons, Frank, 2010: "Transition through subsystem innovation? The case of traffic management", in: *Technological Forecasting & Social Change*, 77,8: 1249-1259.

Pel, Bonno, 2012a: "System innovation as Synchronization: innovation attempts in the Dutch traffic management field" (PhD dissertation, Erasmus University Rotterdam, Department of Public Administration).

Pel, Bonno, 2012b: "Reflection on transition management: mobility policy between integration and differentiation", in: Shiftan, Yoram; Stead, Dominic; Geerlings, Harry (Eds.), *Transition towards Sustainable Mobility: the Role of Instruments, Individuals and Institutions* (Farnham: Ashgate): 53-69.

Pel, Bonno, 2014: "Interactive Metal Fatigue: a conceptual contribution to Social Critique in Mobilities Research", in: *Mobilities*, No.,Issue: 1-19.

Rasmussen, David (Ed.), 1996: *The Handbook of Critical Theory* (Cambridge, Mass.: Blackwell).

Raven, Rob; Verbong, Geert, 2007: "Multi-Regime Interactions in the Dutch Energy Sector: The Case of Combined Heat and Power in the Netherlands 1970-2000", in: *Technology Analysis & Strategic Management*, 19,4: 491-507.

Raven, Rob; Verbong, Geert, 2009: "Boundary Crossing Innovations: Case studies from the energy domain", in: *Technology in Society*, 31: 85-93.

Raven, Rob; Verbong, Geert; Schilpzand, Wouter; Witkamp, Marten, 2011: "Translation mechanisms in socio-technical niches: a case study of Dutch river management", in: *Technology Analysis & Strategic Management*, 23,10: 1063-1078.

Raven, Rob; Schot, Johan; Berkhout, Frans, 2012: "Space and scale in socio-technical transitions", in: *Environmental Innovation and Societal Transitions*, 2012,4: 63-78.

Riddell, Darcy; Westley, Frances, 2013: "Mutual Reinforcement Dynamics and Sustainability Transitions: Civil Society's Role in Influencing Canadian Forest Sector Transition", Paper presented at the 4th International Conference on Sustainability Transitions, Zurich, SUI, 19-21 June 2013.

Rip, Arie, 2006, "A co-evolutionary approach to reflexive governance—and its ironies", in: Voß, Jan-Peter; Bauknecht, Dierk; Kemp, Rene (Eds.): *Reflexive Governance for Sustainable Development* (Cheltenham: Edward Elgar): 82-100.

Røpke, Inge, 2012: "The unsustainable directionality of innovation—The example of the broadband transition", in: *Research Policy*, 41,9: 1631-1642.

Rosa, Hartmut, 2012: *Weltbeziehungen im Zeitalter der Beschleunigung—Umrisse einer neuen Gesellschaftskritik* (Berlin: Suhrkamp).

Rotmans, Jan, 2006: *Societal Innovation: Between dream and reality lies complexity* (Rotterdam: Erasmus University).

Schot, Johan; Geels, Frank, 2008: "Strategic niche management and sustainable innovation journeys: theory, findings, research agenda, and policy", in: *Technology Analysis and Strategic Management*, 20,5: 537-554.

Schuitmaker, Tjerk Jan, 2012: "Identifying and unravelling persistent problems", in: *Technological Forecasting & Social Change*, 79,6: 1021-1031.

Seyfang, Gill; Haxeltine, Alex, 2012: "Growing Grassroots Innovations: Exploring the role of community-based social movements in sustainable energy transitions", in: *Environment and Planning C*, 30,3: 381-40

Shove, Elisabeth, 2004: "Sustainability, system innovation and the laundry", in: Elzen, Boelie; Geels, Frank; Green, Ken (Eds.), 2004: *System Innovation and the Transition to Sustainability: Theory, Evidence, and Policy* (Cheltenham: Edward Elgar): 76-94.

Shove, Elisabeth; Walker, Gordon, 2007: "CAUTION! Transitions ahead: politics, practice, and sustainable transition management", in: *Environment and Planning A*, 39,4: 763-770.

Shove, Elisabeth; Walker, Gordon, 2008: "Transition Management ™ and the politics of shape shifting" in: *Environment and Planning A*, 40, 4: 1012-1014.

Shove, Elisabeth, 2012: "The Shadowy Side of Innovation: Unmaking and Sustainability", in: *Technology Analysis & Strategic Management*, 24,4: 363-375.

Smith, Adrian; Voß, Jan-Peter; Grin, John, 2010: "Innovation studies and sustainability transitions: The allure of the multi-level perspective and its challenges", in: *Research Policy* 39, 4: 435-448.

Smith, Adrian; Stirling, Andy, 2008, *Social-ecological resilience and socio-technical transitions: critical issues for sustainability governance* (Brighton: STEPS Centre).

Smith, Adrian; Stirling, Andy, 2010: "The politics of social-ecological resilience and sustainable socio-technical transitions", in: *Ecology and Society*, 15,1, article 11. [we need the pages]

Smith, Adrian; Raven, Rob, 2012: "What is protective space? Reconsidering niches in transitions to sustainability", in: *Research Policy*, 41: 1025-1036.

Späth, Philipp; Rohracher, Harald, 2012: "'Energy regions': The transformative power of regional discourses on socio-technical futures", in: *Research Policy*, 39,4: 49-458.

Stirling, Andy, 2009: *Direction, Distribution and Diversity! Pluralising Progress in Innovation, Sustainability and Development* (Brighton: STEPS Centre).

Stirling, Andy, 2011: "Pluralising progress: From integrative transitions to transformative diversity", in: *Environmental Innovation and Societal Transitions*, 1,1: 82-88.

Turnheim, Bruno; Geels, Frank, 2013: "The destabilisation of existing regimes: Confronting a multi-dimensional framework with a case study of the British coal industry (1913-1967)", in: *Research Policy*, 42,10: 1749-1767.

Ulrich, Werner, 2003: "Beyond methodology choice: critical systems thinking as critically systemic discourse", in: *Journal of the Operational Research Society*, 54,3: 325-342.

Unger, Roberto M., 1987: *False Necessity: Anti-necessitarian Social Theory in the Service of Radical Democracy* (Cambridge: University Press).

Van der Brugge, Rutger; van Raak, Roel, 2007: "Facing the adaptive management challenge: insights from transition management", in: *Ecology & Society*, 12,2: 33.

Voß, Jan-Peter; Bauknecht, Dierk; Kemp, Rene (Eds.), 2006: *Reflexive Governance for Sustainable Development* (Cheltenham: Edward Elgar).

Voß, Jan-Peter; Smith, Adrian; Grin, John, 2009: "Designing long-term policy: Rethinking transition management", in: *Policy Sciences*, 42,4: 275-302.

Wells, Peter; Nieuwenhuis, Paul, 2012: "Transition failure: Understanding continuity in the automotive industry", in: *Technological Forecasting & Social Change*, 79,9: 1681-1692.

Westley, Frances; Olssen, Per; Folke, Carl; Homer-Dixon, Thomas; Vredenburg, Harrie; Loorbach, Derk; Thompson, John; Nilsson, Måns; Lambin, Eric; Sendzimir, Jan; Banerjee, Banny; Galaz, Victor; Van der Leeuw, Sander, 2011, "Tipping towards sustainability: Emerging pathways of transformation", in: *AMBIO, A journal of the human environment*, 40,7: 762-780.

Wieczorek, Anna; Hekkert, Marko, 2012: "Systemic instruments for systemic innovation problems: A framework for policy makers and innovation scholars", in: *Science and Public Policy*, 39: 74-87.

Young, Iris, 1990, *Justice and the Politics of Difference* (Princeton: University Press).

Zijlstra, Toon; Avelino, Flor, 2012: "A Socio-Spatial Perspective on the Car Regime", in: Geels, Frank; Kemp, Rene; Dudley, Geoff; Lyons, Glenn (Eds.): *Automobility in transition? A sociotechnical analysis of sustainable transport* (Routledge: London): 153-170.

20 Subnational, Inter-scalar Dynamics: The Differentiated Geographies of Governing Low Carbon Transitions— With Examples from the UK

Mike Hodson[1], Simon Marvin[2] and Philipp Späth[3]

Abstract

This chapter aims to improve our analytical understanding of low carbon transitions at and in between multiple geographical scales, particularly 'below' a national level. Taking the Multilevel Perspective (MLP) as our starting point we show that it offers tools for thinking through the institutional and technological conditions and rules through which regimes reproduce or change over time. But it is not very well equipped to study transitional dynamics as they unfold in space. This chapter sets out a range of levels on a spatial scale and activities that are relevant to transition activity but with which the MLP so far engages only partially. The chapter explicitly identifies very different, yet coexisting, scales of transition activity which aim at low carbon transitions in the UK. It demonstrates the possibilities of attributing particular transition activities with different spaces and levels on a geographical scale and highlights the dynamics between these scaled activities. It does this to open up debate about the 'appropriate' mechanisms for dialogue 'between' governance levels and differently scaled transition activities.

Keywords: Geography of transitions; low carbon transitions; sustainability transitions; multilevel perspective (MLP); scaling; multilevel governance; subnational climate policy; United Kingdom; urban governance.

20.1 Introduction

This chapter aims to improve our analytical understanding of low carbon transitions at and in between multiple geographical scales, particularly 'below' a national level. Activities that contribute to a transition, whether that is in electricity systems, heat and buildings, mobility systems, food systems or whatever have effects of different spatial reach, and are connected with a wide range of socially constructed spaces. From our geographical perspective, it is important to acknowledge that both spaces (e.g. cit-

ies, regions, neighbourhoods) and scales (the city scale, the regional scale) are continually being made, negotiated and remade.

One very prominent approach to understanding sustainability or low carbon transitions, and more generally to explain stability and change in sociotechnical systems like energy, mobility and so on is the so-called *multilevel perspective* (MLP). This approach provides an important heuristic by distinguishing three levels of structuration (niche, regime, landscape) and paying attention to ways in which dynamics at these distinct levels interconnect (Geels 2002; Geels/Schot 2007; Smith/Stirling/Berkhout 2005; Verbong/Geels 2010), Through the interrelated concepts of niche, regime and landscape the MLP has worked to bring together the ways in which sociotechnical systems are organized and relatively stabilized but also change gradually and more radically. Potential for agency and change is located mostly in relatively fluid 'sociotechnical niches', while the particularly long-lasting and more structural elements are attributed to the 'landscape level'. The regimes are located at an intermediary level

1 Dr. Mike Hodson, Research Fellow, Sustainable Consumption Institute and Manchester Institute of Innovation Research, University of Manchester; email: <michael.hodson@mbs.ac.uk>.

2 Dr. Simon Marvin, Carillion Chair of Low Carbon Cities and Communities, Department of Geography, University of Durham; email: <simon.marvin@durham. ac.uk>.

3 Dr. Philipp Späth, Assistant Professor at Freiburg University; email: <spaeth@envgov.uni-freiburg.de>.

© Springer International Publishing Switzerland 2016
H.G. Brauch et al. (eds.), *Handbook on Sustainability Transition and Sustainable Peace*,
Hexagon Series on Human and Environmental Security and Peace 10, DOI 10.1007/978-3-319-43884-9_20

of 'structuration'. The value of the MLP is that it offers tools for thinking through the institutional and technological conditions and rules through which regimes reproduce or change over time. It also offers insights into how regimes are reconfigured a) through internal pressures; b) as a response to broader exogenous landscape pressures; and c) through the development of sociotechnical niches of protected spaces. In doing this it allows us to assess transition dynamics and activities which aim to bring about radical or incremental systemic change. While focusing on levels of structuration and developments over time, the MLP so far has not been very well equipped to study transitional dynamics as they unfold in space, i.e. in particular places and at different levels of spatial reach (Raven/Schot/Berkhout 2012).

This chapter sets out a range of levels on a spatial scale and activities that are relevant to transition activity but with which the MLP so far engages only partially. The chapter explicitly identifies very different, yet coexisting, scales of transition activity which aim at low carbon transitions in the UK. There are two contributions in particular that the chapter makes: (1) it highlights the possibilities of attributing particular transition activities with different spaces and levels on a geographical scale, and considers the links between transition activities and spatial attributes not to be eternally fixed properties but to be constructed and contestable; and (2) it highlights the dynamics between these scaled activities and sets out the need to undertake further work to refine understanding of these dynamics, as well as the intermediary activities that operate in particular between governance levels. It does this to open up debate about the 'appropriate' mechanisms for dialogue 'between' governance levels and differently scaled transition activities.

Taking the multilevel perspective as our departure point, we engage with transition activities of different 'scaling'. We illustrate this through sections on: national political-economic configurations; urban transition; 'hyper-local' transition activity; and—as a cross-cutting issue—the role of intermediary spaces in coordinating transition activity across different scales. The chapter sets out a need for mechanisms and relationships for negotiating and coordinating activities across transition scales and governance levels. It is not intended to be totalizing in its approach but to address key issues in relation to scale. We do this because many transitions are possible. There are many ways of organizing transition activity and many scales of action—hence there is a need to understand them, their organiza-

tion, their implications and their possible points of interconnection.

The chapter is organized in six sections. In section 20.2 we review some key concepts of the MLP and the role of geography in it and how a renewed round of globalization in the last four decades has resulted in a more differentiated multilevel governance. The next three sections (20.3-20.5) identify the role of the 'national', the 'urban' and then the 'hyper-local' transition activities. Section 20.6 concludes with a synthesis of the argument and discusses the implications for future research.

20.2   The Role of Scaling in Transition Activities: Complementing the MLP

Globally the issues of climate change, resource constraints and economic crises have raised critical questions about how to respond to ecological and resource constraints. It is clear that responding to these issues requires a significant and 'systemic' transformation in energy, mobility, water, waste and other systems.

However, when it comes to specifying the extent to which systems have to be transformed, the direction of such transformations and how a transition could or should unfold, people tend to have different views. The nature of responses that are to be mobilized and in particular the scale at which responses are constructed naturally reflect different views of the nature and scaling of the problems to be addressed. There is a significant issue, therefore, that needs addressing: in system transition, how do practitioners and we as analysts understand the interplay of transition activities of different scale and in different spaces of activity? We need to ask: to what extent are the objectives, motivations and priorities of differently scaled transition activities congruent, and to what extent do they contradict each other? Are (for example) national priorities around carbon emissions reduction and the promotion of green growth compatible with local priorities of more local democracy or more localized and decentralized energy systems, and does this match well with wider public and householder views of comfort, convenience and reducing bills?

The MLP has developed three key concepts: landscape, regime and niche. These have been discussed and described in detail elsewhere (Geels 2002; Geels/Schot 2007). Here, we note that understanding sociotechnical transitions means recognizing their non-linearity, the constitution of regimes, which are comprised of heterogeneous elements (contexts of rules, established practices and stability) just as niches are (sites of radical innovation), and exogenous landscape pressures and the interplay of these different levels.

Analysis of regimes requires understanding their configurations of technology, culture, science and political and institutional interests. Understanding configurations of regimes provides insight into the reproduction and possible frictions or pressures and hence points of regime transformation. The concept of landscape allows us to link exogenous political, ideological, economic and ecological pressures to regime stabilization, reconfiguration or substitutability.

Pressures can also emerge from niches that are intentionally created as 'protected spaces' in order to experiment with and grow alternative sociotechnical configurations. Such niche activities can be but not necessarily are responses to landscape pressures and may aim to put pressure on different regimes in more or less strategic ways. Processes of experimental configuration, strategy and learning are important to niche construction as is the extent to which niches generate momentum and, from this, the relationship of niches to regimes.

Most empirical analyses of transition processes, however, define the system under transformation—hence their entity of study—with reference to national boundaries (e.g. the Dutch or Danish electricity systems; see Markard/Raven/Truffer 2012). This widespread methodological nationalism has deprived Transition Studies of some of its conceptual richness.

But recently, there has been a developing body of work which addresses, for example, the urban and regional geographies of transition and questions of scale and space explicitly (Hodson/Marvin 2010; Bulkeley/Castan-Broto/Hodson/Marvin 2011; Coenen/Truffer 2012; Geel, 2010; Rohracher/Spaeth 2014, Spaeth/Rohracher 2010, 2014; Raven/Schot/Berkhout 2012).

Since around 2009, many geographers, urbanists and some scholars of Transitions and Science and Technology Studies (STS) have constructively engaged with the MLP to add some spatial sensitivity. In their work, they shed light on many scales and spaces of transitions activity that potentially contribute to processes of transition. We are convinced that the MLP provides very important inroads to understanding the stability and dynamics of sociotechnical systems and that it can be improved to also reflect the spatiality of transitions. We therefore engage with its key concepts and try to develop an understanding of how scales

interrelate and are constructed. The MLP, we conclude, would benefit particularly from a geographical and programmatic view on scale, governance and organization. In the following sections we set out different national, urban and hyper-local scales of transitions activities. We do this in terms of a selective set of issues and to highlight the critical point of this chapter which is the need to build effective mechanisms for dialogue and exchange between scales in transition activities.

20.3 The Scaling of Transition Activities in Systems of Multilevel Governance

Concurrent transition activities differ in their spatial reach and relate to very different governance levels. Despite the great importance that national territories and related legislative frameworks still have today, it is not satisfactory to focus our analysis just on nationally defined activities. There are particular reasons to relocate the 'national' in transition activity since the development of a new phase of globalization since the 1970s, when the role of the national state as a Keynesian, welfarist guarantor of mixed social democratic economies and as being necessary for redistributing and ameliorating capitalism's more inequitable impacts was challenged. The state now was seen rather as impeding the effective operation of markets.

Related to this shift, various new modes of governance emerged as national states, to varying degrees, were themselves meant to promote the benefits of privatization, liberalization, competition and markets; responsibilities were relocated 'upwards' to global and supranational bodies moved 'horizontally' to non-state organizations e.g. by contracting out, and 'downwards' through devolutions and/or decentralizations. This in fact often strengthened the authoritarian functions of the national state (Gamble, 1994). However, it also opened up the field of governing to a much wider array of actors and scales. Cultures of governing became organizationally fragmented within and across multiple scales. This also enabled the definition of an array of 'low carbon' scales and spaces and initiated a struggle over their cognitive, regulative and normative formation.

Particularly when considering the significant challenges posed by climate change, resource constraints and economic crises, many people wish there were the capacity and capability to develop effective action on transitions and in that context to coordinate action across governance levels (Hodson/Marvin 2013). Low carbon transition will probably unfold, if at all, through mutually supportive activities at different scales and in many spatial configurations. Important scales and arenas for negotiations about decarbonization activities are not pre-given but are, as we will show, the result of very dynamic processes of scaling and contestation. There is also contestability and negotiation between differently scaled actors in the construction of low carbon spaces. For the MLP to really help us as a heuristic in understanding transitions and how they unfold, we need to think through different scales and how they relate to the basic categories of the MLP. In the following, we characterize four scalar responses in relation to low carbon transitions activity in the UK, beginning with national low carbon strategies.

20.3.1 National Low Carbon Strategy and Institutional Configurations

As briefly discussed above, the national scale has been very important in the delineation of regime contexts of most transition studies undertaken in the past. In addition, the MLP is often—but not always—applied to national systems and regimes and focuses on the nationwide creation of conditions for niches. What in fact is important here are the ways in which national institutions, particularly in the economic realm, set the parameters and priorities of what low carbon transition will entail. Institutions at a national level transform wider-landscape economic, ecological and political pressures into policies and, through their historically generated priorities, set conditions that enable, favour or disenable particular forms of low carbon activity at other levels, mostly through economic priorities, standards and regulation. This focus on economic priorities and institutions at a national level allows us to link the concept of landscape from the MLP with low carbon transition activities in the sense that it contextualizes landscape pressures in relation to national low carbon political-economic institutional frameworks. Institutional configurations, in this way, are both producers and responders at the national level. They are the historically specific institutional conditions which shape the parameters of systems' adapting.

A key issue is how national institutions condition regime and niche activities. They have significantly modified the way in which, for example, offshore wind has developed in the UK or likewise why battery vehicles, congestion charging and road pricing, domestic

retrofit, and new grid technologies are developing in the ways that they are by promoting and mobilizing particular policies, skills, finance, research, etc.

We hence need to study influential governance mechanisms with regard to (a) the coordination of national institutions with each other, (b) direct ways in which national institutions influence the reshaping of particular regimes and niches, and (c) importantly, how other scalar activities are shaped by national institutions and vice versa. This is important because it tells us something about the ways in which particular institutional configurations (coordinated in, for example, interventionist or market-oriented ways) and histories condition particular regimes and niches but also other scalar activities and the sorts of strategies that are generated through these configurations.

This poses two challenges: in order to understand institutional interrelationships and the ways in which they condition particular approaches to transition, we need to analyse their different cognitive, regulative and normative dimensions (Scott, 2008); longitudinal approaches can address the dynamics and relationships of these configurations. Institutions are often seen as being about reproduction and stability, but it is both this and their transformative potential that we are interested in for transitions and the types of institutional innovations that may be required and that actors may be experimenting with.

Such a conceptualization of institutions allows us to analyse the role of particular institutions in transitions but also the sorts of changes in institutional configurations that are necessary for effective transition activity. This means it is important to understand the mechanisms, resources and instruments through which institutional configurations are 'internally' lubricated. In terms of coordination, what is important is what type of state approach there is to low carbon (here taking the example of the UK and low carbon): a state-coordinated approach or one that promotes more market and less state

Therefore it is key to ask: how does 'the state' at the national level predominantly develop and operationalize low carbon policy/priorities? Which (legacies of) rules are involved in this and how? To what extent is this (for example) an economic agenda? What forms of organization are important in this? In terms of an analytical framework there needs to be a bringing together of three things: elements, modes and orders of governing. Loosely following Kooiman (2003), we ask:

1. *How is national low carbon governance problematized, responded to and acted upon?*—that is, the

framing of problems, strategies, visions, instruments and actions.

2. *What is the national institutional context of low carbon activities?*—that is, the governance of interactions between different national institutions of relevance to low carbon actions.

3. *What norms and principles shape ways of governing themselves?*—that is, the predominant ways of achieving political objectives, through, for example, markets and hierarchies.

20.3.2   UK Low Carbon Problem/Response

The significance of the 2008 UK Climate Change Act was in its positioning of the UK as the first country in the world to have a legally binding framework for cutting carbon emissions. The UK state is playing a central role in setting out the boundaries that will shape the parameters of low carbon futures. Key to this is the development of new forms of carbon accounting and their translation into sectoral and territorially based carbon reduction targets.

The dominant approach to implementation in many low carbon plans in the UK has been to prioritize low carbon technology-based responses (see Committee on Climate Change 2010; Skea/Ekins/Winskel 2011). This has produced a range of strategies that set out the technological options and possibilities—in the form of pathways or scenarios—to a specified time in the future, whether that is for example 2025 or 2050. In the case of national reports (Committee on Climate Change 2010), the issue that is addressed through technology pathways is how UK emissions reduction targets can be reduced and the economic costs and opportunities of different options. Often this means that technology-based responses are better understood as techno-economic responses.

These responses have produced a UK agenda which promotes a range of market-based technology-led responses including: the marketized construction of new offshore wind production systems; the promotion of low carbon vehicles and associated infrastructures; a market-based mechanism, the Green Deal, for retrofitting the UK's housing stock; and the reconfiguration of the electricity grid to facilitate and be compliant with these and other new forms of electricity generation and consumption (Hodson/Marvin 2013).

20.3.3   UK Low Carbon Institutional Configurations

Technology-focused approaches to low carbon futures have much to say about the technological possibilities and economic costs of different low carbon pathways and scenarios. This then leads to certain sectors and specialisms being promoted, for example the development of UK offshore wind capacity and capability. The UK is favourably positioned to develop offshore wind capacity and capability given the shallow water depth and consistent high wind speeds around a significant area of its coast. The Crown Estate has, in recent years, begun to shift from a relatively benign role in managing the marine estate. In doing this it has moved to playing a more active role which marries the technological possibilities afforded by the UK's position with a shifting view of its role as a sustainable manager of the marine estate that considers economic possibilities as well as environmental and social considerations. Its development planning is explicitly orientated around meeting UK targets.

In doing this the Crown Estate, working closely with UK government departments, has designated leases for offshore wind farms in three rounds from 2000. A third round of development (announced in December 2007) is estimated at between three and five times the scale of the first two rounds combined and involves the development of nine zones off the coast of the UK. This is part of a national government attempt to position the UK as a 'leader' in this field and to potentially create the world's largest offshore wind *market*. The zones are widely drawn, where the developer, chosen through competition, can develop in the parts of the zone they consider to be most fruitful. The process of development is long-term (up to four years) and requires addressing and developing an effective response to a wide range of issues. These include constructing supply chains in relation to skills, infrastructure installation, manufacturing and component supply, and so on. This has particular geographies to it related to those areas of coastline encountering beneficial wind speed and water depth, but also to the adjacent, land-based existing R&D capacity and capability.

The story here is one where there is a primacy of market-making with a secondary role for industrial interventionism. The making of offshore wind markets is based on a developing configuration of government targets and zoning, a shifting role for the Crown Estate from benign manager to active market-maker, encouraging investors and developers but on the basis of a pre-existing culture of light touch regulation propounded through pre-existing regulatory structures. Industrial interventionism is thus largely seen as a means of creating the supply chains required in supporting the making of markets (Hodson/Marvin 2013).

The financing of low carbon futures is being constructed through experimentation that is designed to create a stable 'regime' of low carbon financing. These experiments involve attempts to develop new policy instruments, forms of financial products and innovation and new institutions, and they often and primarily involve policy and private sector interests although they do extend beyond this. The levels of investment are huge, with around £200 billion investment in the energy system alone by 2020 estimated to be required. This of course ignores investment required post-2020, investment needed in other utility networks such as water, waste and transport, and also investment in adapting to climate change through the building of defences and relocation (Vivid Economics, 2011).

20.3.4 Norms and Principles of Governing

Within this dominant market-making response the focus of discussion and action in political and policy terms concerns low carbon economies and technologies to the notable neglect of low carbon societies. The assumed techno-economic benefits of low carbon economies are laid out, celebrated and presented as the dominant 'solutions'. These solutions claim to offer not only 'new', accelerated forms of economic growth but also secure 'clean' resource flows necessary literally to power growth and—so the claim goes—to reduce human impact on global ecological change. In short, techno-economic 'low carbon market solutions' underpin and facilitate claims of widespread economic and ecological transformation to society.

This appears to add up to efforts to construct low carbon Britain as a market opportunity, where the role of the state is to govern the imposition of a low carbon transition through the preparation of subnational territories for low carbon accumulation, based on a notion of transition largely constituted by political, regulatory and business elites (Hodson and Marvin, 2013). The emerging dominant transition pathway is as an economic strategy designed to provide a context for another wave of growth-orientated development. This is a profoundly conservative and limited approach dressed within a transformational rhetoric of challenging targets, new possibilities and economic opportunity. Exclusive sets of techno-economic inter-

ests are preparing Britain for a narrowly conceived low carbon economic transition.

20.4   Urban Transition

A key issue is the extent to which these nationally defined problems, responses and institutional configurations also shape subnational transition activities. Here we examine the question: what is the role of the urban in transitions? In recent years there has been an increasing engagement with this question from the perspective of urban studies (Hodson/Marvin 2010; Bulkeley/Castan-Broto/Hodson/Marvin 2011), social studies of technology (Rohracher/Spaeth 2014), and those central to the development of the MLP (Geels 2010).

There are many ways in which transitional activities framed in national orders can interact with those framed in urban orders of governance (Späth/Rohracher 2012). Here, we are particularly interested in two typical ways in which institutional configurations at a national scale inform urban framings of problems, responses and actions. Key to understanding the role of the urban in transition is whether the urban is a site for the cascading down of national transition activity—through niche activities or efforts to reconfigure regimes—or whether urban orders of governing are mobilized to develop distinctly urban responses. In order to illustrate such dynamics between national and urban orders of governance, we now take one metropolitan area—Greater Manchester— into focus and one critical sector of transition: the energy efficiency of the built environment.

20.4.1   Retrofit as Urban Economic Competition

The dominant representation of retrofit in Greater Manchester positions the city-region externally as a low carbon first mover to attract inward investment. Underpinning this is a logic of low carbon 'entrepreneurialism' and a focus on the economic opportunities and benefits this affords. In one estimate, effectively addressing climate change in the city-region over the five-year period of Greater Manchester being a Low Carbon Economic Area could contribute to saving six million tonnes of CO_2, create or support 34,800 jobs, and be a demonstrable exemplar for the wider region and for the UK. This broad view is promoted by a number of plans and strategy documents from the Mini-Stern to the Sustainable Energy Action

Plan and the *Low Carbon Economic Area* (LCEA) for the Built Environment.

Greater Manchester plans for a retrofitting agenda are set out in the draft *Greater Manchester Low Carbon Housing Retrofit Strategy* (LCHRS) published in 2011. Retrofitting is viewed as an emissions reduction strategy in relation to Greater Manchester's carbon reduction emissions targets in a broader national context. It is also seen as a way of achieving 'first mover' economic status and positioning Greater Manchester as leader in an emerging UK retrofit market. In doing so the development of a retrofit agenda is seen as a way to attract private investment to the city-region. Retrofit is therefore being positioned as being about green growth, about job creation, skills development and product innovation. In following this narrative, Greater Manchester is positioned as a "national test-bed"—a means of aligning with national retrofit programmes such as the Green Deal, implementing national programmes, and doing so to achieve national standards such as Energy Performance Certificate (EPC) ratings and to access associated national resources and subsidies such as through the *Feed-in Tariff* (FiT), *Renewable Heat Incentive* (RHI) and the *Energy Company Obligation* (ECO). With this in mind, retrofit is seen as requiring the demonstration of tangible products such as large-scale technologies, the reconfiguration of show houses and exemplar properties.

In addition to these dominant narratives of retrofit there is also a view that retrofit can address fuel poverty and improve well-being through improving existing homes, streets and neighbourhoods and the effectiveness and efficiency of the existing building stock; and that it also requires forms of behavioural change. The dominant message of retrofit in Greater Manchester is that it is about the making of new markets, that is it is about governing Greater Manchester to make it amenable to the market opportunities afforded by retrofit.

20.4.2 Urban Institutions and Governance in Making Urban Retrofit Markets

Greater Manchester's formal low carbon response has been configured through institutions that have developed a view of Greater Manchester over the last two decades as an 'entrepreneurial' metropolitan area. This seeks to position the city-region in ways that are favourable to testing the feasibility of national priorities and to attracting inward investment.

The dominant response in Greater Manchester has involved rebundling the types and sites of retrofit buildings through a threefold hierarchy of responses that involves prioritizing: (1) improving the built fabric of homes through measures such as insulation, glazing and airtightness and also of improving appliances and fittings within buildings such as lighting; (2) more 'carbon literate' home owners, involving the monitoring of energy in homes and the installation of switches; and then (3) filling any energy gap with investment in new forms of renewable energy generation. This hierarchy has been developed as a framework for addressing the application of Basic, Intermediate and Whole House bundles of retrofit measures to different 'types' of owner-occupied and privately rented households.

This packaging of retrofit measures raises the issue of where the finance for such measures will come from. The Strategy estimates that domestic retrofit in Greater Manchester over a ten-year period could require up to £27 billion of investment. Yet the broader context is one where a long period of austerity governance is likely to predominate in the UK and where public finance to underpin such developments is limited. In this context there are numerous attempts to build relationships between Greater Manchester organizations and providers of finance and investment. The Low Carbon Housing Retrofit Strategy outlines potential sources of raising finance that includes bond issue, institutional finance, government finance, bank debt, social financing and private equity. But given the limited discretion of city-regional decision-makers in framing standards and obtaining finance there are pressures to incorporate or align with the priorities of the Green Deal and other national programmes such as the *Technology Strategy Board's* (TSB) 'Retrofit for the Future' programme.

20.4.3 Shared and Implicit Understanding of the Urban Governance of Making Markets

Though there are ongoing relationships between national government departments and institutional interests in Greater Manchester it is the shared and often implicit understanding at the level of meta-governance that is important, especially in terms of the primacy of markets, the necessity for market policy instruments, and the need for economic positioning of cities in a competitive race.

This positions the city-region as a receiver of the priorities of the state in a hierarchical relationship that seeks to promote the benefits of markets for achieving buildings retrofit. This is a way of viewing the relationship between the framing of national low carbon problems, instruments and actions as one where the transition activity is *on* the city-region. Why retrofit is an issue that has been prioritized in Greater Manchester requires an understanding of the changing governance context of Greater Manchester and its dynamics with national government, particularly since the 1970s, and the ways in which this has contributed to shaping contemporary action on retrofit.

Greater Manchester was established in 1974 and operated on the basis of two-tier governing arrangements where the strategic-level Greater Manchester County Council shared power with the ten metropolitan boroughs that constituted it. The Greater Manchester County Council was abolished in 1986 by the Thatcher government and subsequently many powers were devolved back to the ten boroughs while other powers, including transport and emergency services, operated at the metropolitan level. There was also the emergence in the post-1986 landscape of an urban growth coalition within Greater Manchester, which particularly promoted the urban core of the metropolitan area and a series of infrastructure projects and mega-events. Within the metropolitan area there has been a concentration of political and governing power in the hands of agencies and coalitions of political elites and business. This has meant a strategic tier in Greater Manchester that is opaque to the Greater Manchester public, where significant focus is on the urban core. There has been a further step-change in the second half of the 2000s with Greater Manchester being designated a Statutory City-Regional Pilot by the UK government in 2009, and subsequently with the designation and establishment of the Greater Manchester Combined Authority in 2011 constituted by ten (indirectly) elected members of the ten local authorities.

The result of this has been the emergence of a new metropolitan governance at a metropolitan scale but one where the embedded capacity to act is limited, where national priorities remain an important shaper of metropolitan priorities, and where the financial crisis post-2008 has created the conditions for an era of austerity within which efforts to constitute the capacity to shape retrofitting strategies needs to be understood.

20.5   Hyper-Local Transition Activity: Letting a Thousand Flowers Bloom?

In contrast to the dominant market making perspective there are also very many transition activities within urban areas that frame low carbon problems, instruments and actions in very different ways. These frequently mobilize framings of problems that may be about carbon emissions but also may be about a wider set of issues including local democracy, self-sufficiency, empowerment, building local economies and so on. We can characterize these as transitions in rather than on urban areas.

For example, there are many projects and schemes that emanate from neighbourhoods, community groups and settings (Hodson 2014; see also Burrai 2014; Barlow 2014). These may be community energy schemes (hydro, solar, energy efficiency), community food growing schemes, local eco-house exemplars and so on. That is not to say that there is no formal policy involvement—as there may be through, for example, national or city-regional funding schemes—but it is to say that these initiatives are largely developed by neighbourhood or place-based groups, organizations, businesses and collections of individuals to address the motivations of groups of local interests and people embedded in local contexts.

Given this, the motivations for involvement in hyper-local retrofit activities in Greater Manchester include those that seek to: promote economic development through carbon reduction; reduce the carbon footprint of a town, and promote 'sustainability'; and use the retrofit agenda as vehicle for education, outreach and building refurbishment, as part of wider processes of building community engagement.

20.5.1   Building Transitions Activity IN Places Rather Than ON Them

Motivations for local, bottom-up retrofit activity in Greater Manchester are clearly diverse and manifold. What is clear is that these cannot be understood in terms of a singularity. Local retrofit responses take place in a diversity of contexts. The different level of capacities that they can draw upon can be very wide-ranging. This means that some groups are starting out from scratch, some just have a concept, and other groups may have many years of experience. They may also have financial, planning and other forms of expertise that newcomers may not have. Some are small-scale micro-energy generation activities, other relate to behavioural change initiatives, whilst others still work across energy, food and community resilience at a local scale.

Despite the differences across a range of projects there is often a common view in hyper-local retrofit initiatives about making communities relevant again— this is a view of trying to recover something that has been felt to be lost. In that respect this can often be seen as an antidote to the prevalence of top-down initiatives. Hyper-local retrofit has in many ways been about giving voice to community and empowering them through the development of, for example, renewable energy as a real opportunity for members to shape their own future, not only in a low carbon sense but in an economic sense as well. This means that local retrofit is often about both energy generation and conservation, and the generation of income through micro-generation projects in village halls and community buildings and income generation through larger wind or hydro turbines.

But there are frequently significant development costs with these initiatives. Constraints of capacity can mean that a challenge for local retrofit is tightly limited in resources, and this requires finding new ways of working. This highlights something of the limits to community responses with communities often actively having to cultivate networks of financial and other forms of expertise and knowledge, including national and EU funding and other forms of support including emotional support.

20.5.2 Alternatives to Top-Down Market-Making Governance: Remaking the City in Practice

The meta-governance of these hyper-local activities can be understood as an alternative to dominant, top-down approaches to the governance of urban transition activities. Specifically, drawing on thirty hyper-local retrofit projects in Greater Manchester (Hodson 2014; Hodson/Marvin/Bulkeley 2013; Burrai 2014; Barlow 2014), the following issues are important aspects in understanding the constitution of an alternative approach to governing transition activity in the city.

Rather than a singularity of vision, there are very many attempts to remake the city in practice across a range of diverse sites. To take a few examples: this has meant the transformation of a museum roof into a roof garden, an experiment in lived sustainability in a suburban house, installing a ground-source heat pump in a cathedral, and revitalizing old industrial infrastructures such as canals, reservoirs and railway lines.

This involves changing the use and value of sites from below. These initiatives aim at many different ends, whether generating local energy, growing local food, pooling collective buying power, or building local decision-making structures. These all focus on promotion of forms of localism and people taking ownership of their own energy, food and building services and infrastructures, often in ways that are not driven primarily by a profit imperative. Bottom-up and collective forms of organization are often mobilized. These are often organized in ways that bring together community energy/food generation and community ownership through forms of cooperative organization, Industrial Provident Societies, Community Interest Companies and so on.

Yet many of these organizations also connect with the city and the national levels. In many initiatives national, EU and metropolitan scales remain important. One can see this through the labyrinthine funding relationships that projects have to build. In terms of community energy in Greater Manchester this has often involved UK Department of Energy and Climate Change funding. Community renewables projects have also factored in payments from national government for generating surplus energy (through the Feed-in Tariff) in their business plans. This creates a dynamic between the priorities of community projects and the priorities of national government. There are dangers at either extreme: on the one hand community projects require funding, but on the other hand they also have their own priorities which often differ from national priorities. At one extreme this may result in an inability of community groups to act on their aims and plans while on the other they may become delivery agents for national government. This is not a fixed situation and requires those who lead initiatives to actively work to hold together and negotiate these priorities. This ongoing search for funding creates pressure but it has also resulted in new collaborations and social innovation. This raises issues about the boundedness (Vegragt/Brown 2008) or the degree of protection (Raven/ Smith 2012) of these 'niche' spaces.

20.6 Conclusion

This chapter has set out ways of thinking about different scales in relation to transition and the MLP. In doing this it also demonstrates that processes of transition are multi-scalar, with activities simultaneously occurring at multiple scales. It has done so to illustrate the ways in which scalar spatial configurations, and the priorities they mobilize, make contributions to processes of transition. We have shown how transition activities are defined on national, urban, hyper-local and even smaller scales and where coalitions of interests emerge. Coalitions define spaces in particular ways. A coalition, for example, which is tightly linked to the space of the national state is promoting a view of the primacy of markets in low carbon transitions and the need for effective political coordination of institutions to shape low carbon transitions; And urban contexts are viewed and portrayed by some in a way that turns them into recipients of national transition initiatives, while others emphasize hyper-local experiments in urban contexts that contribute to a very different view of urban transitions.

With this context in mind, the diagram below shows the balance between top-down and bottom-up ways of conceiving transition activity and the degree to which transitions seek to reconfigure both systems and spaces.

There are two different axes to the diagram. The vertical axis is concerned with two conceptions of transition. Transitions 'on' conceives of the transition through more 'top-down' or 'outside-in' processes of, for example, market segmentation based on the public's propensity to engage in different forms of environmentally beneficial action. In this case activity is targeted at activating such behaviours through the transition process. Transition 'in' instead conceives of action as being embedded within particular routines, practices and cultural contexts, resulting more in a bottom-up or inside-out movement.

The horizontal axis is concerned with the role of publics in different forms of change. Spatial configuration is primarily concerned with constructing spatially or contextually embedded priorities for change. These configurations can develop contingent priorities but the responses may be consistent with the low carbon transition, or they can potentially develop competing imaginaries of transition. System configuration is primarily concerned with the purposive vision of low carbon transitions and with ensuring that the public complies in playing their role as a delivery mechanism by adopting the new roles assigned to users.

The loci that we highlight in this two-dimensional space may be seen to represent 'ideal types' of the scalar issues discussed above, but most importantly they show us that there are many different elements and aims of scalar transition activity. There are also ongoing dynamics and negotiations between different scalar interests and issues in the 'intermediary' space

between scales. This results in chains of intermediary spaces that facilitate the dynamics between scales and between different interests and priorities.

What emerges from this landscape of scalar transition activity are critical gaps between scales that require processes of intermediation between the differing conceptions of the role of publics in transition activities. We can use our understanding of the specificity of the UK transition landscape to begin to speculate about which existing instruments, activities and organizations may currently form the 'chains' linking these different conceptions of transition across scales.

Looking at these gaps—labelled 1 to 5—we can start to see how they are currently populated. In 1 the national government environment department (DEFRA) has taken a lead in working with consultancies, market research agencies and research institutions in constructing profiles of environmental behaviour through segmented demographics (DEFRA 2008). In 2 there are a series of specific instruments such as the Green Deal and Feed-in Tariffs through which the market segmentation approach to publics is passed down to urban authorities. In 3 we can see how the national energy and climate change department (DECC) has played a critical role in seedcorn funding and supporting diverse hyper-local experiment and activities that are based on particular embedded local priorities rather than demographics. Similarly urban authorities in 4 have funded community groups to work on activities in contexts when the demographic segmentation approach has failed to stimulate market activity in retrofit (Hodson/Marvin 2011). We can also see in 5 how groups developing local initiatives have attempted to upscale and aggregate activity through roadshows, working across neighbourhoods, and other attempts to embed activity in particular spatial contexts.

Our overview here shows that the chains of intermediaries are haphazard, not always well stabilized, often not interactive, and poorly coordinated. It is not clear how they work on each other or whether they can effectively produce some form of spatial reconfiguration or system change. These various scalar relationships to transition and the dynamics and intermediary spaces between them pose a question which is fundamental to transitions activities and interrelationships of scales: to what extent could they or should they be coordinated, and if so, how? This is an issue of institutional innovation and new forms of governance. The reorientation of the state and the horizontal and vertical movement of many of its functions mean that increasingly over recent decades new actors

Figure 20.1: Scalar interrelationships in transitions activities and chains of intermediary spaces. **Source:** The authors.

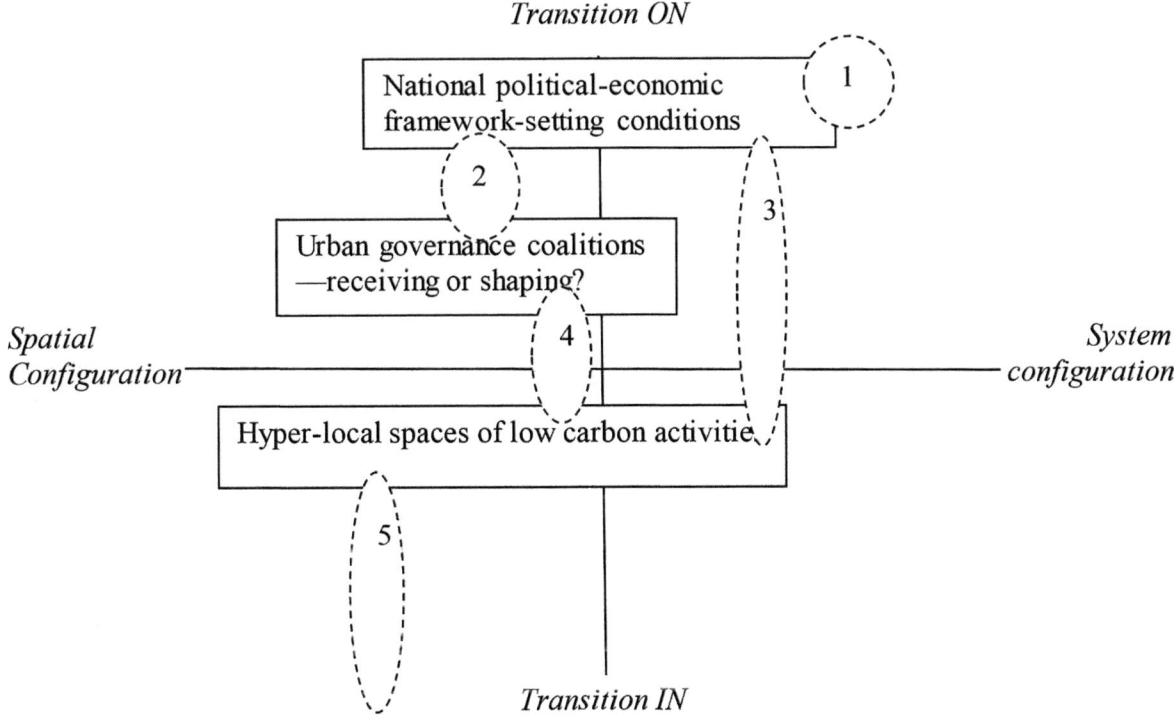

and institutions have become key to effectively addressing societal problems. Yet they are coordinated in various ways—in spaces of intermediation—but not in a way where this is joined up in a systemic approach.

These scalar interactions with existing systems and transition activity therefore pose significant challenges and indicate the need for further debate with the MLP. These engagements are likely to be many as issues related to scale become more commonplace in debates around transition and the MLP. There are, however, three issues that we have identified in this paper:

1. We have illustrated the ways in which political-economic institutional configurations have shaped a particular low carbon response at UK national level. It is important to comparatively understand the ways in which different national political economies mediate landscape pressures and set conditions for regime reproduction/reconfiguration and niche activities. What follows from this is the need to understand the relationship between national and urban scales. Are landscape pressures mediated through national frameworks on to urban centres? Or are landscape pressures interpreted differently through political-economic institutions

at urban scales? Following from this in terms of system transition analysis is the issue of whether system priorities can be scaled down to the urban scale from the national or (and) whether distinct systems can be constituted at an urban scale.

2. The literature on sociotechnical niches can tell us much about how, why and the extent to which hyper-local spaces are 'protected'. Hyper-local spaces can also talk back to the literature by showing the limits of geographical protection where national and local priorities may differ. Thus seeking to 'protect' a local transition space, where national and local priorities differ, may result in the local space missing out on national funding and other resources or alternatively may result in these national priorities overriding local priorities and fundamentally changing the priorities of these spaces.

3. Furthermore, there is the issue of how scalar transition activity works on one dimension to promote top-down low carbon activity or alternatively bottom-up activity; and on another dimension the ways in which scalar activity works on system (re-)configuration on the one hand or the (re-)configuration of space on the other. Of course, these are extremes and there will be scalar activity that seeks

to reconfigure both space and system although the balance may be in favour of one or the other.

Future research would benefit from addressing three issues in particular. First, further development of theoretical and conceptual contributions on scale, space and their interrelationships in processes of transition; second, work around scales of transition activity and the ways in which they can and should be organized; third, longitudinal case studies of the role of the global, international, national, urban, hyper-local and publics in transition processes. Furthermore, the analytical categories developed here may inspire and support comparative analyses of how scaled transition activities are interplaying differently in diverse polities of multilevel governance. In particular, our understanding of the important governance mechanisms that we find *in between* these scales may be improved on the basis of such comparative research.

References

Accenture, 2011: *Carbon Capital: Financing the Low Carbon Economy* (London: Barclays).

Barlow, Catherine, 2014: *Retrofit Alternatives in five UK Cities*, Working Paper EPSRC Retrofit (Cardiff: Cardiff University).

BIS, 2011: *Update on the design of the Green Investment Bank* (London: HM Government).

Brown, Halina; Vergragt, Philip, 2008: "Bounded Socio-technical experiments as agents of systemic change: the case of zero-energy residential building", in: *Technological Forecasting and Social Change*, 75: 107-130.

Bulkeley, Harriet; Castan-Broto, Vanessa; Hodson, Mike; Marvin, Simon (Eds), 2011: *Cities and Low Carbon Transitions* (London: Routledge).

Burrai, Elisa, 2014: *Retrofit Alternatives in Greater Manchester*, Working Paper EPSRC Retrofit 2050 project (Cardiff: Cardiff University).

Coenen, Lars; Truffer, Bernhard, 2012: "Places and Spaces of Sustainability Transitions: Geographical Contributions to an Emerging Research and Policy Field", in: *European Planning Studies*, 20,3: 367-374.

Committee on Climate Change, 2010: *Building a low-carbon economy—the UK's innovation challenge* (London: HM Government).

Gamble, Andrew, 1994: *The Free Economy and the Strong State: The Politics of Thatcherism*, (Basingstoke: Macmillan).

Geels Frank and Schot Johan, 2007: Typology of sociotechnical transition pathways, *Research Policy*, 36, 399-417.

Geels, Frank, 2011: The role of cities in technological transition: analytical clarifications and historical examples, in: Bulkeley, Harriet; Castan-Broto, Vanesa; Hodson, Mike; Marvin, Simon (Eds.), 2011: *Cities and Low Carbon Transitions* (London: Routledge).

Geels, Frank, 2002: "Technological transitions as evolutionary reconfiguration processes: a multi-level perspective and a case study", in: *Research Policy*, 31: 1257-1274.

GMLCHRS, 2011: *Greater Manchester Low Carbon Housing Retrofit Strategy: Discussion draft* (Manchester: GMLCHRS, 30 September.); at: <http://www.envirolink.co.uk/wp-content/uploads/2011/10/GM-Low-Carbon-Housing-Retrofit-Strategy-discussion-draft.pdf>.

Hodson, Mike; Burrai, Elisa; Barlow, Catherine, 2013: "Reshaping the Material Fabric of the City: Low Carbon Spaces of Transformation or Continuity?" Paper for international workshop on 'Constructing and contesting spaces for low-carbon energy innovation', 26-28 November 2013, Eindhoven, Eindhoven University of Technology, School of Innovation Sciences, The Netherlands.

Hodson, Mike; Marvin, Simon, 2013: *Low Carbon Nation?* (London: Earthscan).

Hodson, Mike; Marvin, Simon, 2011: "Governing the Reconfiguration of Energy in Greater London: Practical Public Engagement as 'Delivery'", in: Devine-Wright, Patrick (Ed.), *Renewable Energy and the Public* (London: Earthscan).

Hodson, Mike; Marvin, Simon, 2010: "Can cities shape socio-technical transitions and how would we know if they were?", in: *Research Policy*, 39: 477-485.

Hodson, Mike; Marvin, Simon; Bulkeley, Harriet, 2013: "The Intermediary Organisation of Low Carbon Cities: A Comparative Analysis of Transitions in Greater London and Greater Manchester", in: *Urban Studies*, 50,7 (May): 1403-1422.

Hodson, Mike, 2014: *Remaking the Material Fabric of the City? Why it matters, how it is being done, and what this tells us*. Report for the Greater Manchester Local Interaction Platform of Mistra Urban Futures (Salford: SURF).

Hollingsworth, J. Rogers; Boyer, Robert, 1997: "Coordination of economic actors and social systems of production", in: Hollingsworth, J. Rogers; Boyer, Robert (Eds.): *Contemporary capitalism: the embeddedness of institutions* (Cambridge: Cambridge University Press).

Jessop, Bob, 2002: *The Future of the Capitalist State* (Cambridge: Polity Press).

Kooiman, Jan, 2003: *Governing as Governance* (Sage: London).

Markard, Jochen; Raven, Rob; Truffer, Bernhard, 2012: "Sustainability transitions: An emerging field of research and its prospects", in: *Research Policy*, 41,6: 955-967.

McMeekin, Andy; Southerton, Dale, 2012: "Sustainability transitions and final consumption: practices and socio-technical systems", in: *Technology Analysis and Strategic Management*, 24,4: 345-361.

Raven, Rob; Schot, Johan; Berkhout, Frans, 2012: "Space and scale in socio-technical transitions", in: *Environmental Innovation and Societal Transitions*, 4,0: 63-78.

Rohracher, Harald; Späth, Philipp, 2014: "The Interplay of Urban Energy Policy and Socio-technical Transitions: The Eco-cities of Graz and Freiburg in Retrospect", in: *Urban Studies* 51,7: 1415-1431.

Scott, William Richard, 2008: *Institutions and Organizations: Ideas and Interests, 3rd Edition* (London: Sage).

Shove, Elisabeth; Pantzar, Mika; Watson, Matthew, 2012: *The Dynamics of Social Practice: Everyday life and how it changes* (Los Angeles: Sage).

Skea, Jim; Ekins, Paul; Winksel, Mark, 2011: *Energy 2050: Making the transition to a secure low carbon energy system* (London: Earthscan).

Smith, Adrian; Raven Rob, 2012: "What is protective space? Reconsidering niches in transitions to sustainability", in: *Research Policy*, 41: 1025-1036.

Smith, Adrian; Stirling, Andy; Berkhout, Frans, 2005: "The governance of sustainable socio-technical transitions", in: *Research Policy* 34,10: 1491-1510.

Späth, Philipp; Rohracher, Harald, 2010: "'Energy Regions': The transformative power of regional discourses on socio-technical futures", in: *Research Policy*, 39,4: 449-458.

Späth, Philipp; Rohracher, Harald, 2012: "Local Demonstrations for Global Transitions—Dynamics across Governance Levels Fostering Socio-Technical Regime Change Towards Sustainability", in: *European Planning Studies*, 20,3: 461-479.

Späth, Philipp; Rohracher, Harald, 2014: "Beyond Localism: The Spatial Scale and Scaling in Energy Transitions", in: Padt, Frans J. G.; Opdam, Paul F. M.; Polman, Nico B. P.; Termeer, Catrien J. A. M. (Eds.): *Scale-sensitive Governance of the Environment* (Oxford: John Wiley): 106-121.

Verbong, Geert; Geels, Frank, 2010: "Exploring sustainability transitions in the electricity sector with socio technical pathways", in: *Technological Forecasting and Social Change*, 77,8: 1214-1221.

Vivid Economics, 2011: *The economics of the Green Investment Bank: costs and benefits, rationale and value for money: report prepared for The Department for Business, Innovation & Skills* (London: Vivid Economics).

Part VI   Analysing National Debates on Sustainability in North America

Policy, Politics and the Impact of Transition Studies

Twig Johnson[1]

Abstract

Transition studies are undertaken in a variety of contexts and at various levels of society. Virtually all of them entertain the hope and in some cases the expectation of being used to address challenges to the sustainability transition. The relationship between research and effective action poses a challenge that transition studies must address. As part of an effort to study the different contexts in which sustainability transition studies have developed, the author will discuss a 1999 report by the US National Research Council (NRC) entitled *Our Common Journey: A Transition Toward Sustainability* (NRC 1999). The study began in 1996 during the Clinton/Gore administration. Sustainable development and environment were high on the list of priorities again. The time seemed ripe for looking at ways in which science could better support US policy efforts to transition to sustainability. It was the best of times. Shortly after publication (December 1999) a contested election was held and the political climate changed radically. Then there were the 9/11 terrorist attacks. It was the worst of times. The author will look at this study and related efforts to address the sustainability transition challenge in the fifteen years since publication.

Keywords: Sustainability science, social science, politics (US), sustainable development, Earth Summits, Bush, Gore, Obama, sustainability transition, political gridlock and policy impact of transition studies, global climate change.

21.1 Introduction

As part of an effort to study the different contexts in which sustainability transition studies have developed, this chapter discusses a December 1999 report by the US National Research Council (NRC) entitled *Our Common Journey: A Transition Toward Sustainability* (NRC 1999). Why was such a study undertaken during the Clinton administration? What did it seek to do? What impact did it have on scientific and policy debates? Why did this discussion have so little impact during the Bush administration? Given the polarized, gridlocked and much reported paralysis in Washington, what are the prospects for the US regaining some of its lost leadership in addressing, for example, global climate change?

Addressing these questions will include observations and understandings based on decades of experience working at the interface between research and action in many of the institutions and on many of the issues central to US and global efforts to transition to equitable and sustainable development.[2] My training and fieldwork as a cultural anthropologist, while including tools from many disciplines, emphasized participant observation above all. Margaret Mead, one of my professors, in answering a question on a

---

1 Twig Johnson, PhD, is a writer and consultant formerly serving as a senior policymaker and administrator for the US Government (USAID, Peace Corps), the United Nations (UNICEF) and NGOs (WWF, NRC). Email: <twigjohnson@gmail.com>.

2 Including my efforts to develop the Science and Technology for Sustainability programme in the Policy and Global Affairs Division of the US *National Research Council* (NRC). Though trained as a cultural anthropologist specializing in ecological and applied anthropology, most of my career and participant observations have been spent in government (Peace Corps, *US Agency for International Development* (USAID), private non-profit organizations (*World Wildlife Fund* (WWF), *International Institute for Environment and Development* (IISD)) and international organizations (United Nations Children's Fund (UNICEF), and the donor's *Forest Advisory Group* (FAG).

© Springer International Publishing Switzerland 2016
H.G. Brauch et al. (eds.), *Handbook on Sustainability Transition and Sustainable Peace*,
Hexagon Series on Human and Environmental Security and Peace 10, DOI 10.1007/978-3-319-43884-9_21

particular methodology, told me to "never forget that you are your most important methodology". I hope in this essay to be a useful "key informant".

21.2   Background of the Study

Well before the Clinton administration, interest in the US and global scientific community in supporting environmentally sustainable development efforts had grown. The 1980s had seen a steady growth in the number, strength, technical capacity and political sophistication of *non-governmental organizations* (NGOs) and civil society in general. Many countries had created or strengthened Ministries of the Environment in response to growing problems of transboundary air and water pollution, acid rain, drought and massive fires in remaining tropical forests and increases in cancer-causing UV radiation due to the weakening of the atmospheric ozone layer.

The 1990s in Washington DC continued to be an exciting time for those working on global issues of sustainable development.[3] Preparing for, attending and following up on the 'Earth Summit' [the 1992 United Nations Conference on Environment and Development (UNCED)] held in Rio had generated additional momentum, thanks, I observed, in large part, to non-governmental organizations, many of which had been preparing for a decade. Senator Al Gore was also on the very large US Delegation. His book *Earth in the Balance* (Gore 1992) had recently been published and he was a hero to many, especially to members of the large number of NGOs in attendance, some serving on governmental delegations, others participating in the Global NGO Forum.

Gore had long been intellectually engaged in the science and technology dimensions of global climate change and had been, along with Senators Tim Wirth of Colorado (Democrat) and John Chafee of Rhode Island (Republican), a knowledgeable and enthusiastic advocate on environmental issues. In January of 1993 he was sworn in as Vice President and wrote an introduction to the 1994 edition of Rachel Carson's *Silent Spring* (Carson 1994). There was reason to be optimistic that sustainable development would be high on the government's list of priorities. Early on, the State Department was reorganized with the crea-

tion of a Global Bureau and Tim Wirth was appointed as Under Secretary of State for Global Affairs.

In addition to the improved political climate, an activist civil society and increases in the scientific understanding of international environmental threats and opportunities, there was the upcoming end of the century, for which institutions were preparing. The *National Research Council* (NRC), the operating arm of the *National Academy of Sciences* (NAS),[4] had created a *Board on Sustainable Development* (BSD) to consider how science and engineering could best support efforts to achieve sustainable development. A consensus had formed among the leadership of the Academies that the transition to sustainability would be the defining challenge of the upcoming twenty-first century.

21.3   The Study

In this context, the Board on Sustainable Development began a study[5] aimed at strengthening and energizing the connections between science and efforts to achieve the goals of sustainable development. Unlike most NRC studies, this one was not in response to a request by government, but rather a product of the interest and support of the philanthropist George Mitchel and the NRC leadership.

The study *Our Common Journey: Toward a Transition to Sustainability* set out to "reinvigorate the essential strategic connections between scientific research, technological development, and societies' efforts to achieve environmentally sustainable improvements in human well-being" (NRC 1999: 2). Throughout the preparations for the 1992 Earth Summit, support from science had been included as important, but was treated as one of many major groups, including women, children and youth, indigenous people, non-governmental organizations, local authorities, workers and trade unions, business and industry, and

---

3   During this period I served as Director of the Office of Forestry, Environment and Natural Resources, in the Science and Technology Bureau, at *the US Agency for International Development* (USAID).

4   The National Academy of Sciences is a "private, non-profit, self-perpetuating society of distinguished scholars engaged in scientific and engineering research, dedicated to the furtherance of science and technology and to their use for the general welfare. Upon the authority of the charter granted to it by the Congress in 1863, the Academy has a mandate that requires it to advise the federal government on scientific and technical matters" (NRC 1999: iii).

5   NRC reports are produced by the unpaid efforts of scientists who volunteer significant amounts of their time over several years to participate in a study group or task force. A small staff supports their work.

farmers needing support.[6] In the enthusiastic spread of the sustainable development concept in the aftermath of the Earth Summit and many subsequent international conferences, 'science', along with other 'major groups', was marginalized. What, then, the Board asked, would a serious effort by science and technology to achieve the goals of sustainable development look like?

The study began by reviewing the internationally agreed goals, objectives and indicators for sustainable development. "To meet the needs of a much larger but stabilizing human population, to sustain the life support systems of the planet, and to substantially reduce hunger and poverty." While there had developed "a broad consensus about minimal goals and targets, ... there is seldom analysis of these goals' implications, their potential interactions with one another, or their competing claims on scarce resources" (NRC 1999: 14).

Progress in achieving results had been both weak and uneven, as had many of the efforts to effectively link science and decision-making. Robust support from science would need to deal with the inevitable tensions that emerge "between broadly based and highly focused research strategies; between integrative, problem-driven research and research firmly grounded in particular disciplines; and between the quest for generalizable scientific understanding of sustainability issues and the localized knowledge of environment-society interactions that give rise to those issues and generate the options for dealing with them" (NRC 1999: 10).

Three priority actions were proposed to advance the research agenda of "what might be called sustainability science": an integrated place-based framework to understand society–environment interactions; focused research on a small set of questions needed to better understand these interactions; and better utilization of existing approaches to link knowledge and action. In addition, new ways of learning from large-scale, long-term efforts to meet major challenges were required (NRC 1999: 10-14, 279-288).

Our Common Journey was uncommon, not only in terms of its non-governmental origin, but also in the depth and breadth of its exploration of what a science in support of the sustainability transition would require. To those interested in the transition to sustainability, sustainable development,[7] and the quality

of sound, scientific information to help guide advocacy and action, the release of this study was a welcome event. Here was a serious effort to link separate academic disciplines in order to solve urgent global problems. In December of 1999 *Our Common Journey* was published.

21.4   What Was the Impact on Scientific Debates?

As *Our Common Journey* was not a typical NRC demand-driven study requested by government there was little precedent for following up. Also, as expected, not everyone agreed that a new independent "sustainability science" was necessary. Following its 1999 release, a series of seminal articles were published over the next few years.[8] These were central to the development of sustainability science and its gradual acceptance in the US and wider scientific community. In addition, the co-chairs of the Study Group, Robert Kates and William C. Clark, and many of the group members participated in the various international scientific conferences linked to preparations for the 2002 *World Conference for Sustainable Development* (WCSD).[9] The *International Conference of Scientific Unions* (ICSU), the *Third World Academy of Science* (TWAS), and the *World Federation of Engineering Organizations* played key roles in providing science and technology input into WCSD preparations leading to a parallel Side Event.

The National Research Council followed up creating a Science and Technology for Sustainability Program and using its convening power to establish a Roundtable on Science and Sustainability in order to engage leaders of government, science, business and civil society on priority topics.[10]

The most important impact of this report and the journal articles published in the next few years has

---

6   See Section III on Major Groups, chapter 31 Scientific and technological community. 31.1-31.12 of Agenda 21 at: <http://sustainabledevelopment.un.org/content/documents/Agenda21.pdf>.

7   "... to ensure that [sustainable development] meets the needs of the present without compromising the ability of future generations to meet their own needs" (WCED 1987); at: <http://upload.wikimedia.org/wikisource/en/d/d7/Our-common-future.pdf> and <http://www.un-documents.net/wced-ocf.htm>.

8   See especially: Kates, Clark, Corell et al. (2001); Clark and Dickson (2003); Clark (2007).

9   See for example the Freiburg Workshop report; at: <http://sustainabilityscience.org/content;html?contentid=774>

10   See at: <http://sites.nationalacademies.org/PGA/sustainability/PGA_048724>.

been the further development and rapid spread of a new sustainability science. Over the last fifteen years it has moved from being an interdisciplinary scientific approach to become a recognized scientific programme institutionalized in academic journals, undergraduate and graduate courses, advanced degree programmes and research centres at major universities.[11] Robert Kates (2010, 2011) has produced a freely available, carefully edited online source[12] that provides an excellent introduction to further development of sustainability science in the context of sustainable development. A recent textbook, *Sustainability Science* (Bert de Vries, 2013) builds a framework for his students at Utrecht University to use to both understand and act in relation to (un)sustainable development.[13]

21.5   Impact on Policy Debates: The Case of Climate Change

Looking at US Climate change policy is instructive. Of the sustainability challenges highlighted in *Our Common Journey*, global climate change stands out as one of the most serious for a number of reasons. It affects multiple sectors of society to move ahead toward the transition and it has cumulative or delayed consequences that are felt over a long time. In addition, its impacts are irreversible or difficult to change, and they interact with each other to damage earth's life support systems. It also provides an opportunity to learn from a uniquely large-scale, long-term effort to solve a global problem that requires new forms of linkage between the natural and social sciences. New approaches to providing policy-relevant research are

necessary, as are the means of communicating them effectively. There are also lessons to be learned from various policy and programme choices made around the world.

Some background on this issue in the United States is necessary. The year 1988 was when many woke up to the fact that the climate was changing because of our actions. Forty-five per cent of the US experienced the worst drought since the 'dust bowl years' of the 1930s. Media coverage of the 'ozone hole' caused by our use of chlorofluorocarbons (CFCs) was widespread and was now linked scientifically to cancer. Televised satellite photographs of the massive burning of tropical forests in the Amazon and South East Asia fed concerns that we were damaging the life-support systems of 'spaceship Earth'. Tropical forest burning in Central America darkened the skies of Texas if the wind was right. Dust storms following massive deforestation in China were making it impossible for Los Angeles to meet its mandated air quality targets. In Florida, insects blown in from Africa were discovered, and they had survived the trip. *Time* Magazine designated "Endangered Earth" as "Planet of the Year" (Dressler/Parson 2011: 22-30).

Global concern about what was happening to the climate led to the creation in that year of the *Intergovernmental Panel on Climate Change* (IPCC), established by the United Nations Environment Programme (UNEP) and the World Meteorological Organization (WMO) and subsequently endorsed by the UN General Assembly. It regularly produces consensus reports on the status of scientific understanding of the causes, consequences, and possible responses to global climate change caused by increasing concentrations of greenhouse gases in the atmosphere. Based on the work of thousands of volunteer scientists from around the world, it continues to be the most creditable source of information for policy-makers and policy-shapers on the current state of scientific knowledge on global climate change and its potential environmental and socio-economic impacts.[14]

National and international organizations increased their capacity to deal with environmental challenges, though often starting from a low base. Non-governmental environmental organizations sprang up alongside proliferating Ministries of the Environment (and environmental units in most other government ministries). NGOs had been negotiating among themselves and by the time diplomatic negotiations about a framework convention began they were in place and prepared.

11 For a general overview see "Sustainability science" (31 December 2013), in: *Wikipedia, The Free Encyclopedia*; at: <http://en.wikipedia.org/w/index.php?title=Sustainability_science&oldid=588503013>.   See also <https://en.wikipedia.org/w/index.php?title=Sustainability_science&oldid=588503013> for an overview of various approaches to sustainability science (5 February 2014).

12 Available at: <http://www.hks.harvard.edu/centers/cid/publications/faculty-working-papers/cid-working- paperno.-213>.

13   Bert J. M. de Vries has taught a course on sustainability science at Utrecht University for many years, in connection with his research at the *Netherlands Environmental Assessment Agency* (PBL). This textbook is based on that course. The book provides a historical introduction into patterns of past (un-)sustainable development and into the emergence of the notion of sustainable development (de Vries 2012: 11-30).

The involvement of civil society and especially the part played by NGOs was historic. In her award-winning *Foreign Affairs* article "Power Shift", Jessica Tuchman Mathews put it this way:

> Whether from developing or developed countries, NGOs were tightly organized in a global and half a dozen regional Climate Action Networks, which were able to bridge North-South differences among governments that many had expected would prevent an agreement. United in their passionate pursuit of a treaty, NGOs would fight out contentious issues among themselves, and then take an agreed position to their respective delegations. When they could not agree, NGOs served as invaluable back channels, letting both sides know where the other's problems lay or where a compromise might be found.

> As a result, delegates completed the framework of a global climate accord in the blink of a diplomat's eye— 16 months—over the opposition of the three energy superpowers, the United States, Russia, and Saudi Arabia. The treaty entered into force in record time just two years later. Although only a framework accord whose binding requirements are still to be negotiated, the treaty could force sweeping changes in energy use, with potentially enormous implications for every economy (Mathews 1997).

The result was the first international agreement on climate change, *the UN Framework Convention on Climate Change* (UNFCCC). Under the Framework Convention, countries agreed on a goal of stabilizing greenhouse gas concentrations in the atmosphere at a level that would prevent dangerous anthropogenic interference with the climate system. The Framework Convention was signed by US President George H. W. Bush and ratified unanimously by the US Senate. In the two decades since the signing of the Frame-

work Convention, earth scientists have confirmed the basic climate science through both modelling and observations. Warming and other changes to the earth system are under way, and the dangers look even graver than they did two decades ago.

The UNFCCC was, however, only a framework for further negotiations. By 1995, countries realized that the provisions for emission reductions in the Convention were inadequate. They launched negotiations to strengthen the global response to climate change, and, two years later, adopted the Kyoto Protocol that legally commits developed countries to emission reduction targets.

The Clinton administration had signed the treaty over the written objections of the Senate, but had not proceeded to seek the advice and consent of the Senate necessary to ratify the treaty, knowing it would be defeated. In the foreign policy arena, the Senate has the authority of "advice and consent" regarding treaty-making. Given the difficulty of winning the two-thirds vote necessary for treaty ratification in the US, the Clinton administration, as others have done in the past, was considering ways of achieving some of the goals of the treaty without formal ratification.

Yet preliminary progress had been made on this major threat to planetary sustainability and through the IPCC, scientists from around the world were playing a critically important role. *Our Common Journey* was published in October of 1999. The time seemed ripe for an important contribution from sustainability science to US policy. *Then the political climate changed radically.*

21.6 Bush: A Radical Change in Context

In 2000, shortly after publication of *Our Common Journey*, there was the disputed election of George Bush, ultimately decided by a conservative US Supreme Court. Two months after taking office in January 2001, the Bush administration announced that the US would not ratify the Kyoto Protocol for reducing greenhouse gas emissions, claiming scientific uncertainty in addition to negative effects on the US economy as two of the reasons. The "lack of scientific knowledge" argument was retracted later in the face of several NRC reports and criticisms from the larger scientific community, including many scientists working in the federal government.

Then came the September 11, 2001 attacks on the World Trade Center, the Pentagon, and the attempted

---

14 The Fifth Assessment Report (AR5) of the IPCC released in 2013 and 2014 provides a clear view of the current state of scientific knowledge relevant to climate change; see <http://www.ipcc.ch/The Fifth Assessment Report>. It comprises three *Working Group* (WG) reports and a *Synthesis Report* (SYR). The IPCC *Working Group I* (WG I) assesses the physical scientific aspects of the climate system and climate change. *Working Group II* (WG II) assesses the vulnerability of socioeconomic and natural systems to climate change, negative and positive consequences of climate change, and options for adapting to it. It also takes into consideration the interrelationship between vulnerability, adaptation and sustainable development. *Working Group III* (WG III) assesses options for mitigating climate change through limiting or preventing greenhouse gas emissions and enhancing activities that remove them from the atmosphere.

attack on the White House [foiled by the heroic passengers crashing their own plane]. All was televised, recorded in heartbreaking detail, and replayed over and over. Shock, anger, and fear rocketed through modern communications media and triggered an angry patriotism not seen since the attack on Pearl Harbor that launched a previously reluctant America into World War II. When an aide whispered into the ear of President George Bush that the United States was under attack, he sat frozen in a tiny chair in a classroom where he had been reading a story to the children. He looked stunned, 'gobsmacked', momentarily paralyzed, as were millions of others around the globe, by this unexplained, horrendous event. There followed a great deal of understandable confusion. After Pearl Harbor it had been clear what America needed to do. President Roosevelt had seen the necessity for some time. In addition to his thoughts on the matter, he had a large number of ideas brilliantly and frequently articulated by the British prime minister Winston Churchill. President Bush did not know what to do. It was not clear what was happening and why. The who-what-when-where-why-and-how questions had no answers. And Bush was no Roosevelt nor was he a Churchill.

A large new bureaucracy, Homeland Security, was created and an ill-defined 'War on Terror' was launched. As the administration prepared to invade Iraq (never linked to the 9/11 attacks), the style, tone and rhetoric of US foreign policy seemed to come out of a sort of Clint Eastwood 'spaghetti western'. In the process, the G. W. Bush administration not only failed in its ill-conceived objectives, it also undermined US credibility, squandering the immense global sympathy and goodwill that followed the 9/11 attacks. The follies of the occupation of Iraq and 'nation-building' in Afghanistan followed, with an agenda dominated by claims of keeping America safe from real or suspected enemies, foreign and domestic. Having inherited a rare balanced budget from the Clinton administration, failing to pay for the immense costs of his policies G. W. Bush plunged the country into debt.

In her 10 October 2007 Congressional Testimony before the House National Security and Foreign Affairs Subcommittee, Jessica Tuchman Mathews stated that:

> The events of 9/11 have had far less of an effect on the real world than that day had on the American psyche. Iraq is a very different matter. The war's monopoly on our political energy has now stretched to 5 years—an eon in a time of fast-moving global change. One of its greatest—as yet uncounted—costs is the degree to which it has sucked the oxygen from almost every other issue. A dra-

matically changing global climate might as well not be happening. The reappearance of huge federal budget deficits is hardly noticed. The need for change in an unsustainable energy policy has barely surfaced. And, in these five years a number of international security problems have grown, from neglect, into full-blown crises (Mathews 2007: 2).

After the mess made during the first term by going to war and not paying for it, and after the failure of promised results was revealed to be based on (wilful?) ignorance, avoiding available information by knowledgeable people, and after his actions were executed with all the bravado, arrogance and sensitivity of a Hollywood gunslinger, George W. Bush was re-elected.

The electorate does not pay attention to many details. Many in the Washington foreign affairs community working on a wide range of global issues, including some colleagues working on security and terrorism issues, were shocked. Friends in the foreign press and diplomatic services expressed incredulity. I recall a social scientist talking about an analysis of voting patterns reporting that the pattern in this election was largely unchanged, with the exception of young mothers in the South, who had changed their former vote to support Bush this time. The reason most given for the switch was to "avoid changing Commanders-in-Chief in time of war". So it went.

The negative results of the Bush administration environment policies have been well documented. For example, on 16 January 2009, four days before he left office the White House released a bulletin describing Bush as a careful steward of the environment. The same day the *Natural Resources Defense Council* (NRDC), the Sierra Club and many others took vigorous exception. Suzanne Goldenberg's article "The worst of times: Bush's environmental legacy examined" in *The Guardian* of 16 January 2009 details their near universal outrage. They saw the Bush years as a concerted assault on environmental progress, from the administration's undermining of the science of climate change to its dismantling of environmental safeguards and its uncritical support for mining and oil interests. Frequently mentioned was not only his failure to act on global climate change but his administration's covert attempt to silence the science alerting us to the urgency of the problem.[15]

In terms of the very real but less immediately salient threat of global climate change and the more gen-

---

15 See at: <theguardian.com>, Friday 16 January 2009, 10.45 EST.

eral challenge of the transition to sustainability, the administration seemed happy to abdicate both domestic and global environmental leadership, minimize the problem, and roll back much hard-won progress (see for example Kennedy 2005).

Following the Kyoto announcement, the group responsible for coordinating and integrating US government research on climate, the *Global Change Research Program* (GCRP) was renamed with the weaker title of the *Climate Change Science Program* (CCSP). The Bush administration was to be regularly charged with suppressing scientific information and its implications for society.

21.7   Challenges of the UNFCCC Framework

Independently of the actions of the Bush administration, criticism of the UNFCCC process had been growing. In the winter of 2000, international talks on the implementation of planned emissions standards again faltered, a resolution again postponed (Victor 2011, 2011a). Outside the environmental community, and even within it, support for Kyoto weakened. Indeed Earth Summit agreements on climate, biodiversity and desertification had not produced any concrete results on climate, biodiversity or desertification.

Many reasons for the repeated failure of these attempts to forge global environmental agreements were offered (Victor 2011; Bell 2011). The most persuasive related to the architecture of negotiations under UN auspices—an inclusive process, involving 194 countries—polluters, victims and everyone in between—all of whom officially have equal weight in the proceedings and any single nation might block the will of all the others. Goals and objectives produced this way will be too broad and vague because governments in that setting are unwilling to do anything more concrete. Negotiations now include a bigger basket of issues than most multinational talks—including weaning whole economies off coal and oil, protecting disappearing forests, saving the small island states that are likely to be underwater within a few decades, and managing all the related costs.

Suggestions for improvements have included encouraging bottom-up initiatives at national, regional, and global levels and leveraging national self-interest rather than wishful thinking (Bell 2011).

> Moving to a variety of nimbler negotiating vehicles supplementing the UNFCCC is worth a try, as is disentangling goals for emissions reductions from debates about

legal structures and venues. The large docket of issues could be broken up, narrowing specific negotiations by issue or region, by greenhouse gas emissions contribution, or by tools and methods to achieve greenhouse gas reductions. Finally, climate change talks, which involve a problem with the potential to disrupt the life of every human on earth, must graduate into the elite arena. They should be directly guided by powerful ministries, which can make commitments and sell them at home in ways most environmental authorities cannot.

> Controlling greenhouse gas emissions is a messy real-world challenge that will most likely require an … evolving progression of messy real-world solutions. But if the climate establishment can acknowledge more variation of approaches and entry points, it increases its chances of realizing its most important goals, even if it does not tie a nice, neat bow on them (Bell/Blechman/Zigler 2011: 4).

However, giving talks top-level attention, segmenting issues, and diversifying negotiating arenas all seem promising. It is precisely because climate disaster is impending that the world should get creative about addressing the problem. Within the UN negotiating framework, however, it will be hard to get them to consider approaches that might challenge the rule of consensus, which many equate with equity (Bell/Blechman/Zigler 2011: 3).

In fact, a number of countries, including the US, have been diversifying their approaches for some time. And there are some bright spots. In the absence of federal leadership, and in some cases because of it, much progressive action was led by states and cities. California, the sixteenth largest emitter of carbon dioxide worldwide, for example, is among several US states that have entered into partnerships or passed laws for controlling greenhouse gases ahead of, even in spite of, the federal government.

In August 2006, not long after the publication of the 2006 Stern report on global warming that he had requested, Prime Minister Tony Blair made history by travelling to and signing, with Republican Governor Arnold Schwarzenegger, an agreement with the state of California to cooperate on a range of global environmental issues. "California will not wait for our federal government to take strong action on global warming", said Governor Schwarzenegger, a Republican, who would also send a wake-up call to US automakers by creating emissions standards for automobiles.[16] The Bush administration attempted to halt this action, claiming federal responsibility. They were overruled by the Supreme Court on the grounds of States' Rights.

---

16 See at: <http://gov.ca.gov/news.php?id=2770>.

In August 2008, Massachusetts moved forward with a comprehensive regulatory program to address climate change. The *Global Warming Solutions Act* (GWSA) requires Massachusetts to set economy-wide *greenhouse gas* (GHG) emission reduction goals to achieve reductions of as much as twenty-five per cent below 1990 state-wide levels by 2020, and eighty percent below those levels by 2050.[17]

A month before the *19th Conference of Parties* (COP19) in November 2013 in Warsaw, California launched a regional pact to harmonize and intensify the efforts of California, Oregon, Washington, and British Columbia to reduce greenhouse gases. "California isn't waiting for the rest of the world before it takes action on climate change", said Governor Brown. "Today, California, Oregon, Washington and British Columbia are all joining together to reduce greenhouse gases". Recently the California Air Resources Board announced an agreement with Quebec outlining steps and procedures to fully harmonize and integrate their cap and trade programmes. In September 2013, Governor Brown joined China's top climate official, National Development and Reform Commission Vice Chairman Xie Zhenhua, to sign a first-of-its-kind agreement on climate change between the commission and a subnational entity. This followed landmark partnerships established earlier this year on the governor's Trade and Investment Mission to China, including agreements signed with China's Minister of Environmental Protection Zhou Shengxian to improve air quality and with Jiangsu province to promote renewable energy.[18]

These actions by California and New England are meant as examples of the large number of actions taken by states, municipalities, businesses, voluntary organizations, schools and churches, homeowners, and consumers of automobiles and appliances.

21.8 Obama—Beginning the Turnaround?

On 20 January 2009, Barack Hussein Obama became the forty-fourth president of the United States. Within the week he issued two presidential memoranda on energy. One directed the Department of Transportation to increase fuel efficiency standards to 35 miles to the gallon (15 km/l) and the other directed the *Environmental Protection Agency* (EPA) to allow individual states to set stricter tailpipe emissions regulations than the federal standard. In February he provided $54 billion in funds to encourage domestic renewable energy production, make federal buildings more energy-efficient, improve the electricity grid, repair public housing, and weatherize modest-income homes.

Presidential memoranda and proclamations do not require Congressional approval. This is critically important. The US government was designed, quite deliberately, not to work very well. The design works, perhaps too well. Separations of powers, checks and balances, make it very difficult to do much in many areas. It is extremely easy to halt actions. There is a long list of actions requiring Senate confirmation, but the presidential ability to negotiate an international agreement stands out. Any agreement must receive the advice and consent of the Senate by a two-thirds majority. This is why the Clinton/Gore administration never sent the Kyoto Protocol they had signed forward to the Senate. It had no chance of passage.

In 2010 the EPA issued rules restricting greenhouse gas emissions from cars and trucks by 2012. This was based on the Agency's finding that greenhouse gasses harm human health. As in the case with CFCs, the scientific link to a threat to health created motivation and powerful justifications for action.

Obama re-engaged positively in international climate negotiations, attending the December 2009 Copenhagen COP, which left many with very reduced expectations of what was possible through the UNFCCC process.

In April of that year he launched the Major Economies Forum on Energy and Climate, to facilitate a candid dialogue among major developed and developing economies on climate change and the challenge of clean energy. The seventeen member economies are: Australia, Brazil, Canada, China, the European Union, France, Germany, India, Indonesia, Italy, Japan, Korea, Mexico, Russia, South Africa, the United Kingdom, and the United States. Their Forum on Energy and Climate works to drive transformational low-carbon technologies in the energy sector. The Climate and Clean Air Coalition to Reduce Short-Lived Climate Pollution focused on such short-lived pollutants as methane, black carbon, and *hydrofluorocarbons* (HFCs). Together these account for one-third of current global warming, and addressing them can prevent more than two million premature deaths a year, and avoid the annual loss of over thirty million tons of

---

17 See at: <http://www.mass.gov/eea/pr-pre-p2/pr-2010/ press-release-re-clean-energy-and-climate-plan.html> and <http://www.mass.gov/eea/agencies/massdep/air/climate>.

18 See at: <http://gov.ca.gov/news.php?id=18205>.

crops. Partnership has expanded beyond the founding partners (Bangladesh, Canada, Ghana, Mexico, Sweden, and the UN Environment Programme) to include over thirty countries and the European Commission.

At the 2011 *Asia-Pacific Economic Cooperation* (APEC) Summit, leaders agreed to eliminate non-tariff barriers to environmental goods and services, thus lowering costs and increasing the dissemination of clean technologies. Leaders further committed to phasing out inefficient fossil fuel subsidies and aimed to reduce the energy intensity of APEC economies by forty-five per cent by 2035. In June 2013, President Obama and Chinese President Xi agreed to work together and with other countries to use the Montreal Protocol to phase down HFCs, a critical step forward towards a global agreement. In June 2013 President Obama launched a comprehensive Climate Action Plan.[19]

His Climate Action Plan, his 25 June 2013 speech on climate change[20] and his 2014 State of the Union Speech to Congress[21] leave no doubt as to the seriousness of his intentions and the effort he has made to do what he can using the powers of the Presidency. However, a polarized, gridlocked Congress dominated by right-wing extremists hampers him.

While Obama certainly represents a turnaround in the Presidency, he is in many areas limited to actions that do not require Congressional action. He must deal with a Congress that demonstrates little capacity or interest in solving problems and has dedicated much of its time to defeating anything he proposes.

21.9   Polarization, Gridlock and a Trickle of Contrarians

The evidence from hundreds of scientific and economic studies finds serious to grave impacts on agriculture, coastlines and associated settlements, and ecosystems, as well as increasing acidification of the oceans and threats to many species around the world. Science progresses, the earth warms, glaciers melt, the oceans become increasingly acidic, but the climate contrarians change not (Nordhaus 2012: 85).

Contrarians are becoming fewer in number, less and less credible, and not nearly as well funded as in the past. People look around at storm damage, floods, and droughts and know the climate is changing. The process had been similar in the case of regulating tobacco and controlling chlorofluorocarbons.

The insurance industries in the US as in Europe were early believers in the problem of climate change as they saw the effects of the increases in the frequency and severity of storms consistently predicted by the scientists. This cost them money so they paid attention. The support funds from industries threatened by efforts to reduce greenhouse gas emissions are much reduced. Even the US Chamber of Commerce, with much of industry, has now accepted the reality of climate change caused by greenhouse gas emissions. The press has belatedly become more responsible in looking at the credibility of its sources.

The political gridlock in Washington these days is, however, different. Two of Washington's most respected and even-handed analysts[22] wrote *It's Even Worse than It Looks: How the American Constitutional System Collided with the New Politics of Extremism* which became a *New York Times* bestseller. The Republican Party today, they report, has little in common even with Ronald Reagan's GOP, or with earlier versions that believed in government. It has become "an insurgent outlier—ideologically extreme; contemptuous of the inherited social and economic policy regime; scornful of compromise; unpersuaded by conventional understanding of facts, evidence and science; and dismissive of the legitimacy of its political opposition" (Mann/Ornstein 2013: [Kindle Edition] 178).

They also charge the press with using false equivalence to explain outcomes, when Republican obstructionism and Republican rejection of science and basic facts have no Democratic equivalents. It's much easier to write stories that convey a false impression that the two sides are equally implicated.

Obama's re-election in 2012 was a setback for Republicans and a number of Republicans interested in solving problems are beginning to speak out. But too many remain worried about having to face well-funded right-wing challengers. Recent polls show a declining popularity of the Republicans, seen as too

---

19 See at: <http://www.whitehouse.gov/share/climate-action-plan>.

20 See at: <http://www.whitehouse.gov/the-press-office/2013/06/25/remarks-president-climate-change>.

21 See at: <http://www.whitehouse.gov/the-press-office/remarks-president-state-union-address>.

22 Norman Ornstein of the American Enterprise Institute and Thomas Mann of the Brookings Institution are political scientists with decades of providing even-handed analysis of Washington politics, and especially of Congress. Their 2006 volume *The Broken Branch* focused on the institutional shortcomings of Congress while this volume examines the broader current political dysfunction.

extreme and out of touch. The polls also revealed new lows for the federal government but record highs for State and municipal government.[23]

21.10   Conclusions

The NRC Study *Our Common Journey: Toward The Transition to Sustainability* contributed to the process of strengthening and energizing the connections between science, technology and engineering and their efforts to support the transition to sustainability. It did this through the normal scientific channels of publishing to a largely scientific audience, and in addition through the vigorous and determined international outreach efforts of co-chairs Kates and Clark and the extraordinary scientists in the Study Group as well as Sherburne Abbott, the lead NRC staff person responsible for organizing, editing and frequently in some cases writing. They are themselves movers and shakers in influential institutions and one could track the marks they have left in intellectual contributions, institutional capacities, and undergraduate and graduate programmes. They participated in many of the preparations for the 1992 *World Conference for Sustainable Development* (UNCED) in Johannesburg, South Africa, which provided rare opportunities for strengthening international scientific cooperation in developing sustainability science. The World Conference failed to meet expectations.

Efforts aimed at global and national policy-making mostly failed not for want of trying but because of the collapse of global environmental summitry and the Bush administration's disinterest in, ignorance of, and animosity toward what many regard as the greatest strategic threat to our security we have yet faced. The high level governmental demand for support and cooperation from scientists working on sustainability issues that one could expect from the Clinton/Gore administration had been replaced with outright hostility and suppression of research results.

Fortunately, failure is mostly confined to the federal government in Washington. The front line of problem-solving has shifted to states and municipalities, to educational institutions and research centres, to the many partnerships between NGOs, scientists, engi-

neers and corporations who see a decarbonizing future as an opportunity.

The feedbacks keep on coming. Storms, wind, drought, and crazy weather are, as predicted, becoming stronger and more frequent. Dramatic and sometimes disastrous events can influence policymakers by demanding their attention. The global science assessment of the IPCC becomes ever more solid and sensitive and can now give better information on actions that work.

In an interview on 26 January 2013 during the World Economic Forum, Lord Nicholas Stern reflecting on his 2006 Stern Report said he actually underestimated the risks posed by global climate change. At the same meeting *International Monetary Fund* (IMF) Director Christine Lagarde, former Conservative Finance Minister of France, noted the "Increasing vulnerability from resource scarcity and climate change, with the potential for major social and economic disruption". She called climate change "the greatest economic challenge of the twenty-first century".[24]

Global climate change is just one of the many threats to a sustainability transition along with extreme poverty, the growing gap between rich and poor within as well as between countries, terrorism, crime, pollution, disease, weapons of mass destruction, environmental protection, non-tariff trade barriers, intellectual property, and the challenge of cooperative governance where states are challenged by an ever growing number of non-state single-interest actors. Overall there is a rapidly expanding poorly-regulated market economy that continues on a collision course with our planet. These problems are all global. They must be solved and managed globally. This will require institutional capacities and systems of governance that are currently lacking. Sustainability transition studies have begun to address this need.

---

23 See Pew Research Center; at: <http://www.people-press.org/2013/02/26/gop-seen-as-principled-but-out-of-touch-and-too-extreme/2/> and at: <http://www.people-press.org/2013/04/15/state-govermnents-viewed-favorably-as-federal-rating-hits-new-low/2/>.

---

24 Nicholas Stern: "I got it wrong on climate change—it's far, far worse"; at: <http://www.theguardian.com/environment/2013/jan/27/Nicholas-stern-climate-change-davos/print>.

References

Bell, Ruth Greenspan; Blechman, Barry; Zigler, Micah, 2011: "Beyond the Durban Climate Talks", in: *Foreign Affairs* (30 October 2011); at: <http://www.foreignaffairs.com/articles/136627/ruth-greenspan-bell-barry-blechman-and-micah-ziegler/beyond-the-durban-climate-talks> (5 May 2014).

Carson, Rachel, 1962: *The Silent Spring* (Boston: Houghton Mifflin Harcourt Publishing Company).

Carson, Rachel, 1994: *Silent Spring* (Boston: Houghton Mifflin Company).

Clark, William C., 2007: "Sustainability Science: A Room of its Own"; in: *Proceedings of the National Academy of Science,* 104,6: 1737-1738.

Clark, William C.; Dickson, Nancy, M., 2003: "Sustainability Science: The emerging research program"; in: *Proceedings of the National Academy of Science,* 100,14: 8059-8061.

De Vries, Bert J. M., 2012: *Sustainability Science* (Cambridge: Cambridge University Press [Kindle Edition]).

Dressler, Andrew; Parson, Edward A., 22011: *The Science and Politics of Global Climate Change: A Guide to the Debate* (Cambridge: Cambridge University Press).

Gore, Al, 1992: *Earth in the Balance: Ecology and the Human Spirit* (New York: Plume).

Gore, Al, 1994: "Introduction", in: Carson, Rachel: *Silent Spring* (Boston: Houghton Mifflin Company).

IPCC (Intergovernmental Panel on Climate Change), 2013, 2014: *The Fifth Assessment Report* (Geneva: IPCC).

Kates, Robert W., (Ed.), 2010: *Readings in Sustainability Science and Technology.* CID Working Paper No. 213. Center for International Development, Harvard University. Cambridge, MA: Harvard University, December 2010; at: <http://www.hks.harvard.edu/centers/cid/publications/faculty-working-papers/cid-working-paperno.-213>.

Kates, Robert W., 2011: *Readings in Sustainability Science and Technology* (Cambridge, Mass.: Harvard University, Centre for International Development).

Kates, Robert W.; Clark, William C.; Corell, Robert; Hall, J. Michael; Jaeger, Carlo C.; Lowe, Ian; McCarthy, James J.; Schellnhuber, Hans Joachim; Bolin, Bert; Dickson, Nancy M.; Faucheux, Sylvie; Gallopin, Gilberto C.; Gruebler, Arnulf; Huntley, Brian; Jäger, Jill; Jodha, Narpat S.; Kasperson, Roger E.; Mabogunje, Akin; Matson, Pamela; Mooney, Harold; Moore III, Berrien;

O'Riordan, Timothy; Svedin, Uno, 2001: "Sustainability science", in: *Science,* 292,5517 (27 April): 641-642.

Kennedy, Robert F. Jr., 2005: *Crimes Against Nature How George W. Bush and His Corporate Pals Are Plundering the Country and Hijacking Our Democracy* (New York: HarperCollins [Kindle Edition]).

Mann, Thomas; Ornstein, Norman, 2013: *It's Even Worse than It Looks: How the American Constitutional System Collided with the New Politics of Extremism* (New York, NY: Basic Books, Paperback Edition [Kindle Edition]).

Mathews, Jessica Tuchman, 2007: "Six Years Later: Assessing Long-Term Threats, Risks and the U.S. Strategy for Security in a Post-9/11 World", Testimony (Washington DC: US Congress, House Committee on Oversight and Government Reform, National Security and Foreign Affairs Subcommittee, 10 October).

Mathews, Jessica Tuchman, 1997: "Power Shift", in: *Foreign Affairs,* 76,1 (January/February): 50-66.

Nordhaus, William D., 2012: "Why the Global Warming Skeptics Are Wrong", in: *The New York Review of Books* (22 March); at: <http://www.nybooks.com/articles/archives/2012/mar/22/why-global-warming-skeptics-are-wrong/>.

NRC (US National Research Council, Policy Division, Board on Sustainable Development), 1999: *Our Common Journey: A Transition Toward Sustainability* (Washington DC: National Academy Press); at: <http://www.nap.edu/openbook.php?record_id=9690&page=1>.

Ornstein, Norman; Mann, Thomas, 2006: *The Broken Branch: How Congress Is Failing America and How to Get It Back on Track* (Oxford–New York: Oxford University Press).

Victor, David G., 2011: *The Collapse of the Kyoto Protocol and the Struggle to Slow Global Warming* (Princeton, N.J.: Princeton University Press).

Victor, David, 2011a: *Global Warming Gridlock: Creating More Effective Strategies for Protecting the Planet* (New York: Cambridge University Press [Kindle Edition]).

WCED (United Nations World Commission on Environment and Development), 1987: *Our Common Future* (Oxford: Oxford University Press).

22   Geopolitics, Ecology and Stephen Harper's Reinvention of Canada

Simon Dalby[1]

Abstract

The election of the Conservative Party to power in Canada in 2006 brought with it a vision of the world that was much more competitive than previous Liberal or much earlier conservative visions. Key to all this, and the focus of this chapter, is an attempt to reinvent Canada as a player in a world of competitive geopolitics rather than as a good citizen in a shared biosphere. Foreign and domestic policy have been shaped by this new view, leading to the abrogation of the Kyoto protocol and, given the identification of Canada as an energy super-power and oil exporter, substantial attacks by the government on environmental science and regulatory processes, apparently because these might obstruct resource company projects. What is being sustained in this process is a vision of Canada antithetical to what in most parts of the world would be considered sustainable. The lessons to be learnt for sustainable transitions are many, most notably the importance of thinking carefully about conventional politics and the dangers of narrowly-cast nationalist and populist attacks on environmental policies and sustainability initiatives.

Keywords: Canada, geopolitics, Harper Government, neoconservative, political strategy.

22.1   Whatever Happened to Canada?

In the 1980s Canada had a reputation as a leader in developing environmental policy. The establishment of the *International Development Research Centre* (IDRC) with its international network of activists and academics made development a key theme in policy discussions. The *Canadian International Development Authority* (CIDA) delivered aid, at least some of which was environmentally sensitive. Much of the background work on the *World Development Commission on Environment and Development* (WCED) in the 1980s was done in Ottawa. The Montreal Protocol, a key part of the international framework for controlling and subsequently stopping the production of ozone-depleting chemicals, was named after a Canadian city where some of the key negotiations were conducted. The Global Atmosphere Conference of 1988 was held in Toronto, and issued statements that explicitly linked climate change to global security in terms that are now familiar a generation later.

Focusing more explicitly on the domestic scene, environmental regulations were a fairly widespread matter for the Federal Government in the 1970s, although the division of powers in the confederation does give the ten Provinces responsibility for overseeing the exploitation of natural resources. Nonetheless given Federal responsibilities of fisheries, oceans and water issues, and the system of national parks, there was an active Federal presence on environmental matters. Major research efforts to develop plans for a "Conserver Society" were undertaken in the 1970s (Solomon 1978). By the early 1990s, coincident with the Earth Summit in Rio, 'Green Plans' were in place to deal with many matters of sustainability. Connecting the domestic and the international spheres, Canadian businessman Maurice Strong oversaw both the 1972 Conference on the Human Environment in Stockholm and the United Nations Conference on Environment and Development twenty years later in Rio de Janeiro.

Another twenty years later the Canadian situation is very different. While both the 1980s government and the Harper government in power from 2006 include the term 'conservative' in their name, they

---

1   Simon Dalby PhD is CIGI Chair in the Political Economy of Climate Change, Balsillie School of International Affairs, and Professor of Geography and Environmental Studies, Wilfrid Laurier University, Waterloo, Ontario, Canada; Email: <sdalby@gmail.com>.

H.G. Brauch et al. (eds.), *Handbook on Sustainability Transition and Sustainable Peace,*
Hexagon Series on Human and Environmental Security and Peace 10, DOI 10.1007/978-3-319-43884-9_22

were very different political entities. Canada abrogated its Kyoto protocol commitments, walking away from the agreement rather than dealing with the consequences of its failure to abide by its obligations. By 2014 CIDA effectively no longer existed. The Tar Sands in Alberta were a cause célèbre among environmentalists looking for a symbol of everything that is wrong with contemporary economic trajectories. The huge tailing ponds and massive infrastructure tearing up the boreal forest while using natural gas to power the transformation of bitumen into petroleum is the classic example of carboniferous capitalism run amok. Prime Minister Stephen Harper proudly boasted of Canada as an energy superpower while defunding environmental research and silencing scientists. Some of his ministers hinted darkly at sinister foreign interests behind environmental objections to the extraction of Canadian resources and the operation of Canada-based mining companies in many other parts of the world. Canada's environmental performance was rated with Kazakhstan and Saudi Arabia in the international environmental hall of shame! This was a very different Canada (Kaplan 2011) and one with a foreign policy that was narrowly focused and little interested in multilateral arrangements (McLeod Group 2012).

Given the ideologically driven nature of the Harper government's silencing of dissent and abandonment of much environmental science, activists and scientists ruefully joked that Canada has moved from "reality-based decision-making" to "decision-based reality-making". Watching the Federal political scene in recent years in Canada there is much truth to this assertion; the agenda for Canada's Conservative government (2006-2015) was driven by the resource sector and short-term extractive industry profitability. The mining boom stretched round the world; it wasn't just a domestic matter (Gordon 2010). The electoral support the Conservative Party drew from rural and resource extractive industries was coupled to an ideological argument that this is the route to material wealth, and hence environmentalists who would regulate the industry and examine its project proposals too closely in regulatory hearings were a danger to the prosperity and suburban lifestyle that is portrayed as the envy of much of the world. After all, immigrants supposedly come to Canada precisely to get the benefits of this mode of consumption, based on Canada's traditional economic priorities of providing resources to the North American industrial system, if not to the global economy.

While the environmentally-friendly image of Canada was never the whole story, and there are histories of industrial pollution and resource exploitation across the country that are by no stretch of the imagination 'sustainable' (Baldwin/Dalby 2010), nonetheless the abandonment of anything more than the pretence to be good environmental stewards on the part of the current Federal government is noteworthy. The rhetoric of environmental protection persists, when necessary, as in making the case that Tar Sands petroleum is extracted in a manner that is supposedly environmentally benign, but the reduction of monitoring and research and the effective muzzling of federal scientists makes it clear that this is only public relations.

There are some important lessons to be learned from the sad example of Canadian federal politics that need to be remembered in thinking about strategies for sustainable transitions in other places. The first lesson is that environmentalists forget about the finer points of politics at their and the planet's peril. As Australians with the Howard and more recently Abbott governments have also learned, the departure from power of parties sympathetic to at least some environmental causes can set sustainability back dramatically. It makes getting sustainability back on track all the more difficult by entrenching the power of certain sectors of the economy and scrapping research and monitoring that allow environmental matters to be assessed and debated in a reasonable manner. This is precisely what the Federal government in Canada set about doing in 2006 with very worrisome consequences for both Canada and the planet. All this matters because a politics very different from the policies of the Conservative Party in Canada are needed if sustainability is to be taken seriously.

22.2   Canada's Conservative Government

After a couple of minority parliaments from 2006 to early 2011, and some political manoeuvrings that raised the ire of many Canadians, the Conservative Party formed a parliamentary majority government in May 2011. Given the peculiarities of the "first past the post" Westminster-style electoral system, this was achieved with the support of only about forty per cent of those who actually voted. While the name Conservative is still there, this was a corporate agenda led government with a populist right-wing rhetoric (drawn from its 1990s predecessor "The Reform Party") determined to remake the country, and one apparently more obsessed with partisan motivation than has usually been the case in Canadian federal politics (Nadeau 2010). The result was an ideological agenda supporting the oil industry in Alberta and Saskatchewan mainly, and defining energy security in terms of fossil fuel production. Canada is supposedly an energy superpower, and the supplier to the North American market, and possibly in the future elsewhere further afield if the Keystone pipeline to Texas or the Northern Gateway one across the north of British Columbia get built in coming years.

Stephane Dion, the former Liberal environment minister who had been a key part of the Kyoto process, was ridiculed when, as subsequently leader of the Liberal Party he introduced a set of policy suggestions for a green shift in Federal taxation to tax carbon and compensate by reducing income taxes. Populist rhetoric against taxes was used to ridicule the proposed policy and ensure that it never got reasonably discussed. Dion lost the 2008 election when he ran on a "green shift" electoral platform. This has made carbon taxes more difficult to implement in Canada, although the province of British Columbia has one that has been used to fund popular public services which might have potential for wider application should political entrepreneurs emerge willing to promote such a system elsewhere (Elgie/Mackay 2013). Michael Ignatieff who mostly ignored climate change in his subsequent unsuccessful attempt to remove Stephen Harper in the 2011 election subsequently replaced Dion as leader.

The Conservatives long made the argument that they will follow American policies when it comes to climate change so as not to impinge on Canadian business, but in the process abandoned any attempt to take the initiative and develop a comprehensive 'made in Canada' policy. Despite useful reductions in coal-burning power station emissions, mostly as a result of Provincial initiatives in Ontario in particular, it is clear that even the modest Federal promises for reductions in fossil fuel emissions are unlikely to be met. No Federal strategies on renewable energy development emerged; windmills and solar panels were left to the provinces concerned with their own energy grids and sustainability issues. While this makes sense in that these are mostly provincial responsibilities in Canada, an energy superpower means in this case one that supplies natural gas and petroleum, not one that thinks seriously about the long term and about how sustainable futures are to be powered. It was a policy that effectively locks Canadian policy into a carbon trap

dependent on markets for high-priced marginal petroleum supplies long into the future (Haley 2011).

Abandoning much of the earlier legislation and regulations that required environmental reviews of developments, the omnibus budget bill of April 2012 eviscerated Federal environmental regulations and cut government science capabilities, while censoring scientists' public statements and refusing to allow them to do media interviews without prior clearance from government political overseers (Turner 2013). This led to unprecedented protests by scientists in the summer of 2012 focused on the theme of the "death of evidence". This theme also captures other government changes such as the evisceration of census information gathering and reducing the functions of Statistics Canada as part of the overall reduction in public knowledge about Canada. Further cuts to Federal oversight of lakes and waterways followed in 2013 and the trend to abandon environmental protection continued; the mantra of business self-regulation was related to the abandonment of data collection and monitoring across the country.

Kyoto protocol abrogation fitted into this agenda too. While it was clear that Canada was never going to make cuts in energy use that would get overall emissions down to the 1990-based targets, "Canada's New Government", as the Conservative Party wished to be known in its early years in power, abandoned even the pretence that Canada would try. There was certainly no intention of spending money on offsets to pay for non-compliance. Arguing that remaining in Kyoto would unreasonably hamper Canadian oil companies and that it was obviously an ineffective agreement; the decision to abandon it was entirely in keeping with a government focused only on resource development. The lack of concern with the precedent, or the abrogation of any attempt at political leadership on a matter of global importance, was in keeping with the Conservative ideological agenda that showed scant concern for the wider world (Bosold/Hynek 2010). Unless, of course, there were domestic constituencies whose votes might count in the electoral calculus that kept the government in power by a narrow margin; foreign policy was understood here as important in so far as it generates electoral benefits.

In turn this was related to what at least initially was seen as a much more militaristic foreign policy by Harper, who was happy to lend Canadian forces to imperial actions in Afghanistan and subsequently Libya, although the total amount of resources channelled to the military was not in keeping with a major shift to a more muscular foreign policy (Lang 2012). A persistent series of failures in the procurement process for the F-35 fighter jet programme undermined the credibility of at least some of this focus on the military (Nossal 2012/13); subsequent budget constraints in 2014 severely limited military purchases and caused outrage among veterans' groups where health benefits are limited while money is spent on commemorating past military actions. But the tone of foreign policy is clearly much more militarist than Canadian policy in previous governments; peacekeeping is a thing long abandoned by Canadian governments despite its earlier useful contributions to international affairs. NATO actions in support of American initiatives have been undertaken, including in Afghanistan and Libya; UN peacekeeping missions mostly have not (Charbonneau/Cox 2010).

The ideological orientation of the government was one where the world is understood in competitive terms, one where international action is a matter of self-interested action, a matter of "enlightened sovereignty" in foreign minister John Baird's (2011) terms where temporary coalitions of interest matter, and international institutions don't. Unless, that is, Israel was concerned, where by specifying it as a democracy surrounded by non-democracies, unabashed support for whatever it does was apparently the obviously morally correct foreign policy (Martin 2012). The abrogation of a nuanced foreign policy in favour of moralistic slogans and the periodic denunciation of Iran was quite clear. Some of this was plainly driven by domestic electoral considerations appealing to immigrant groups, but a confrontational geopolitics underlay this explicitly.

Domestically the removal of many environmental regulations and the effective carte blanche given to resource companies violated the provisions of many of the treaties signed with the indigenous groups who were conquered and dispossessed in the process of colonization by European settlers (see Grant 2014). This in turn caused protest movements and continued political opposition to the Conservative government, who disregarded their obligations to meaningfully consult with many native peoples about environmental changes and developments that take place on their territories. This may turn out to be a major source of conflict should the decision to build the Northern Gateway pipeline from Alberta to the coast of British Columbia near Kitimat be taken despite the vehement opposition of many native groups whose land would be crossed by the pipeline.

In so far as lip-service to environmental matters was served, it was entirely in environmental moderni-

zation mode; technological innovations ensure "safe" production. In the case of the Tar Sands, industry groups repeatedly invoked improved techniques in television advertising campaigns in particular, supposedly making products both more secure and environmentally benign. The argument that this is ethical oil, in contrast to petroleum imported from conflict-ridden areas or non-democracies, was simply added on to this formulation (Levant 2010). Nowhere was it countenanced that leaving the bitumen in the ground might be sound strategy and that renewable energy development would be a better priority. This was a policy devoted to resource extraction rather than building an industrial base in Canada; rent from resources rather than innovation in production reprised much of the history of Canadian development but is also a policy that, not least by inflating the Canadian dollar, undercut industrial exports too (Stanford 2008). In the Arctic where the consequences of climate change are most obvious the Harper Government's northern strategy has been to enhance resource extraction and extend property rights of corporations to facilitate their access even if this comes at the cost of indigenous inhabitants' environments and food supplies (Medalye/Foster 2012). Whatever this may be it wasn't Canadian leadership on anything that matters in terms of sustainability or responsible environmental stewardship. When the price of oil collapsed in mid 2014 the folly of an economic policy so dependent on one sector of the economy became abundantly clear.

22.3   Political Ontologies

Running through all this was a simple will to power, an agenda that saw disagreement as opposition and a view of politics that was both narrowly focused and concerned with winning rather than with the consequences that are not immediately reflected in the proverbial bottom line. The bottom lines that mattered were both the corporate balance sheets and the electoral calculus that ensured the persistence of Conservative rule. This was a view of the world that assumed competition as the given context for human activity, and one that was unconcerned about matters of environment. Indeed environmentalists were the enemy, given that they apparently obstructed corporate plans for resource extraction and hence they had to be silenced or denigrated. Therefore their expressions of alarm about climate change were to be either dismissed, or finessed with arguments that suggested economic activity is a far more important priority in a harsh competitive world.

Such formulations relied on a worldview that discounted the consequences of Canadian actions while simultaneously boasting of being a superpower. They operated as though Canada, being a sovereign country, and an enlightened one apparently, was in some key senses simply separate from the rest of the world—foreign policy action was frequently about domestic electoral advantage, not it seemed about shaping the international order beyond pressing Canadian business advantages abroad and issuing moral condemnations of regimes outside the Western world. In such an ontological universe winning and ruling is what matters; having a sustainable biosphere isn't important given that as winners, and rulers, if environmental disruptions happen 'we' will apparently be either wealthy of powerful enough to evade the consequences. Anyway Canada is only a small player on the world scene, with half a percentage of the world's population and only a few per cent of the world's oil production so what we might do on the matter of environment is unimportant, so why pay attention if it will deprive our corporations of profits in the immediate future?

This contradiction-ridden thinking runs directly in opposition to an ecological worldview of any sort where the interconnections between things and the common placement of all beings in one biosphere is the starting point for discussions of sustainability. It assumes a given context that simply provides the opportunity to make money, and failure to take these opportunities suggests moral and political inadequacies that should be swept aside by those who obviously know better. Ecological science simply gets in the way of what is important and so should be silenced, dismissed or ignored. This was politics in the raw, a matter of power first and foremost, as Steven Harper's political biographers make very clear (Martin 2010; Wells 2013). A strategy for a sustainable future simply has to deal with these political challenges from a corporate sector armed with both a neo-liberal ideology and a populist rhetoric if it is to be effective. This is no small challenge especially when, as is the case with the Canadian Conservative Party, it explicitly tried to use political divisions and wedge issues as a mode of political rule.

The other important point in all this is that environmental gains are not necessarily permanent. As the American environmental movement learned to its cost in the 1980s when the Reagan administration set about dismantling at least some of the legislative and administrative systems set in place in the previous dec-

ades, getting laws made is only the initial step in a sustainability agenda. Keeping the political pressure on to ensure that they are enforced and that the long-term benefits are forthcoming requires a permanent political and cultural effort. The danger of formulating matters in terms of a sustainability transition is that there is an assumption of a stable end point. Another lesson from the Canadian case in the last decade is that environmental progress can be overturned if corporate agendas driven by resource extraction priorities in particular regain the political initiative. Politics is an ongoing process, not an end point, and discussing sustainable transitions cannot afford to operate on the assumption that there is a stable end point where sustainability is ensured in perpetuity.

22.4   Neo-Conservative Rhetoric

Looking to the Canadian example it seems that at least six key sources of political thinking fed into the ideological mélange that supported the right wing coalition that constituted the Conservative party. Given the distance that they have travelled ideologically from the earlier Progressive Conservatives who were in power in the 1980s when the Montreal Protocol was negotiated, the Global Atmosphere conference was convened, and national Green plans formulated, it's perhaps appropriate to call this neo-conservatism, and not only because of its militarist inclinations. Backed by media outlets that were sympathetic to both business interests as well as populist stories of moral clarity and victimhood in the face of supposedly big government, think tanks including the Fraser Institute have shaped the predominately corporate policy agenda. The lessons of the last decade in Canada suggest clearly that policies for sustainable futures need to directly tackle all these ideological components, albeit in different combinations in particular places around the world.

The first political configuration is the larger neo-liberal logic that prioritizes the market as that which is the primary object of wealth creation, the focus for government promotional efforts and the supposed provider of welfare for the population. This notion of prosperity is the key to the good life and that agenda was key to the re-articulation of Canadian identity in terms of winners on the global stage in comparison to other states, and the provider of a commodity and real estate vision of the good life at home. Most obviously this translates into aspirations for a suburban lifestyle, one that developers are happy to provide in the sprawling automobile suburbs that now house so many Canadians (Blais 2010). This has apparently entrenched the power of both petroleum companies and property developers in shaping the political economy of Canada. The shift of power has been both to the West, where most of the oil industry is based, with its headquarters in Calgary, and in electoral terms into the suburbs of most Canadian cities.

Second is the ambiguous impact of religion on public policy. While many of the Conservative party are what are called 'social conservatives' there are clearly networks of religious institutions especially in the Western provinces that draw on 'fundamentalist' doctrines, not least those that don't take things like climate change or evolution seriously. While it isn't clear that Prime Minister Harper's religious proclivities actually shaped his policy preferences, it is certainly a matter of discussion among those who have traced the rise of new religious organizations and their connection to the Conservative Party (Macdonald 2010). What is more important than the specifics of religious dogma is that these religious themes support the larger cultural politics of moral rectitude and the focus on personal salvation and economic success that underlies the assertion of conservative values. The alternative much 'greener' engagements between religion and environmental advocacy in North America (Wilkinson 2012), with their notions of environmental stewardship and responsibility for caretaking the earth were noticeably absent from the social conservative discourse.

Third, the rhetoric of the Conservative Party frequently used the figure of the persecuted outsider in appealing to numerous perceived grievances among Canadian voters. The regional dimensions of this were clear in the slogan used repeatedly by the Western-based Reform Party as in 'the West wants in'. Supposedly excluded from the halls of power in Ottawa, the Federal capital, and in Toronto where the financial heartland of Canada resides, the absence of parliamentary representation was a longstanding source of complaint. The social aspirations of Westerners and the much older social hierarchies in Central Canada supplied a very simple if misleading rhetoric of them and us that populist politicians used to good effect in election campaigns.

This social aspiration and sense of entitlement denied fed, fourth, into simplistic geopolitical formulations and imperial throwbacks in the formulation of Canadian foreign policy. Naïve assumptions of moral superiority were used to justify involvement in the bombing of Libya as well as the presence of Canadian troops in Afghanistan. Language that contrasted democracies with dictators and terrorists painted the world as a dangerous place requiring moral certainties and the necessity to use force to ensure that Canada and the West is triumphant in a series of violent confrontations with evil and dangerous Others. The necessity of using virtuous violence is a given once the geopolitical formulations of distant danger and imminent threat dominate the script. It all fit with the Prime Minister Harper's proclivities for controlling things too.

Such Manichean formulations with their differential attribution of virtue and threat also spilled over into the highly charged partisan attacks on domestic political opponents. This fifth factor, the ad hominem politics, focused on the personalities and their supposed failings much more than it did on the policy prescriptions that those politicians or activists might be espousing; messengers are more important than the content of messages. While this was a widespread tactic, not least in the vitriolic campaigns against climate change activists and scientists which spilled over into the Canadian political scene, its prevalence as a way of silencing serious political discussion was especially dangerous when linked to 'gotcha' journalism and the personality politics in the talk show format of political commentary. Winning the argument was what apparently mattered. Reasoned political discussion and simple matters of factual accuracy are difficult to use effectively in such venues, a matter of considerable difficulty for sustainability advocates who unwisely assume a common acceptance of environmental values and reasoned debate as the basis for a democratic politics.

Sixth is the link between prosperity, threat and security. A key part of the Harper government's political rhetoric was the protection of Canadians from threats to their prosperity, and environmentalists who might object to pipelines and tar sands exploitation were of course linked to dangerous foreign funding arrangements. Where external threats are formulated to suburban prosperity the task for sustainable transition advocates becomes especially difficult. The insidious cultural messages celebrate conspicuous consumption as praiseworthy rather than excessive, irresponsible and eminently taxable! The inverted quarantine of suburban living where private provision of everything from security systems to bottled water is premised on a spatial division of safe internal domestic spaces from external threats, and the related cultural politics of family, masculinist safety provision and the necessity to rely on force in a dangerous world, presents a geopolitics of division and violence that is antithetical to an ecologically sane mode of life for most of humanity (Szasz 2007).

These assumptions, powerfully reprised in conservative thinking of the Canadian variety, are now precisely the problems that need to be confronted. The claims to moral superiority, and the implicit assumptions of competition as the ontological given for humanity, likewise need once again to be confronted by a more complex geopolitical ontology that neither accepts the territorial assumptions of modern states as the final word on human organization, nor the ever-larger consumption of materials and energy as the aspirational motivation for politics. In short, the growth dynamics of capitalism and the dominance claims of extractive resource sector corporations and their political allies has to be confronted by both an alternative set of ontological premises and a political strategy that offers plausible alternative modes of governance.

22.5   Anthropocene Geopolitics?

Calls for a sustainable transition suggest the necessity of limiting consumption precisely because we live in an interconnected biosphere in which we are linked fairly directly to one another. The conservative political rhetoric and the ontological assumptions of autonomy and competition are antithetical to all this. Reinforcing assumptions of sovereignty and rivalry, of a world divided into competing states, with morality related to our success, even if that is part of a zero-sum game, the conservative invocation of the necessity of virtuous violence to ensure dominance of the current suburban mode of consumption is precisely what needs to be transcended. But that is not enough intellectually or politically now in the epoch of the Anthropocene (Brauch/Dalby/Oswald Spring 2011). Spelling out some of the implications of the Anthropocene, as the next few paragraphs do here, emphasizes how radically different a sustainable politics has to be from the kind of thinking epitomized by the Canadian Conservative Party.

The ecological premises for sustainability frequently assume a stable situation; life lived in ways that doesn't compromise the possibilities of future generations, to finesse the classic definition of sustainable development in *Our Common Future* (World Commission 1987). What the earth system analyses are making increasingly clear is that the geopolitical premises of modernity, of competing territorial states and expanding industrial power, don't work as the tools for thinking sustainably about the future. Neither should state boundaries be simply assumed (Fall 2010), although they are much more stable since the United Nations system has become the established institutional framework for international politics.

Ecological geopolitics has yet to come to terms with these conundrums; much hard thinking needs to be done on these matters. The irony that governance and hence political rivalry is understood through a series of categories specified in territorial terms precisely when what matters apparently crosses those boundaries is key to understanding the current dilemmas and to suggesting modes of analysis that can more effectively grapple with what is coming (Bulkeley/ Andonova/Betsill et al. 2014). Shifting from a physics model of power, one of competing autonomous entities, of surveillance, territorial demarcation and military enforcement to an ecological sensibility that recognizes interconnection and change rather than permanence and fixity as key to flourishing life, is a fundamental ontological challenge to conventional understandings of politics and society. On the largest scale, that of the biosphere as a whole, this is exactly what now has to be brought into political discussion.

Many of the more thoughtful analyses of climate change have tried to tackle the matter beyond the conventional formulations that apply resource management or state administrative apparatus to the problem. The multifaceted nature of climate change, coupled with the urgency of addressing it, makes this a 'super-wicked' problem that defies a simple solution (Levin/Cashore/Bernstein et al. 2012). There are many technological innovations that are less carbon-intensive, but no technical fix that can resolve the issue in engineering terms. Climate change is part of the larger transformation of the global biosphere and as such touches on the most basic conditions of human existence. How it is tackled, or not, goes to the heart of politics, and to the big questions of world order that are the key matters of geopolitics.

Put most simply, humanity has changed the composition of the planet's atmosphere and raised the level of carbon dioxide close to 400 parts per million, well above levels that the planet has known in the last few hundred thousand years. This will inevitably set off disruptions to how weather systems function and do so in unpredictable ways. Unpredictable precisely because there is no analogous state in the recent history of the planet to which we can refer for indications as to how things might play out. Humans are part of this picture and will either suffer the consequences of disruptions or reap some benefits from new opportunities dependent on the political and economic circumstances they find themselves in.

In so far as people avoid the worse consequences of climate disruptions, it will be because governments make reasonable preparations, but at least in terms of formal state structures so far many don't seem to be adapting quickly to the new realities of climate change. The Canadian Conservative government simply ignored climate realities in its rush to promote petroleum production and eviscerate the scientific knowledge systems that monitor changes and investigate ecological responses. Making these less susceptible to the vagaries of partisan politics is clearly something that needs further attention from all those concerned about sustainability. European states are further ahead but their record is patchy at best. Attempts to negotiating binding arrangements under the umbrella agreement of the widely adopted *UN Framework Convention on Climate Change* (UNFCC) have generated huge conferences and numerous other meetings but the carbon dioxide levels in the atmosphere continue to rise at an accelerating rate.

It is important not to focus on climate alone. The global economy has been constructed by large-scale environmental change in terms of the extension of agriculture into most of the world's ecosystems that can support crops (Ellis/Goldewijk/Siebert 2010). The loss of habitat for other species has led to the extinction of a sizeable percentage of the life forms that the planet supported until recently. Fishing combined with pollution and now increased acidification of the oceans due to rising carbon levels has transformed aquatic life systems too, with untold future consequences for many species apart from humanity. Human industrial systems 'fix' more nitrogen for fertilizers than natural processes do. Phosphorous likewise is increasingly an artificial cycle. Most large rivers have been dammed, diverted and modified, some to the extent that the waters in them don't reach their estuaries. Artificial urban habitats have been built that transform regional ecosystems as they provide the basic necessities of life for the burgeoning global population.

This is the new context for the human drama, one of our own making (UNEP 2012). It is within this new context of an increasingly artificial world that human vulnerabilities, and all sorts of insecurities, play out now. Given these new circumstances it seems fitting to many earth system scientists to designate the current period in terms of the Anthropocene, literally the age of humanity. In the Anthropocene, artificial circumstances define our existence and infrastructure,

markets and politics matter much more in terms of who lives and who dies than the immediate consequences of weather events, however severe or dramatic (see chapter 3 by Dalby in this volume). This is a global urban system that stretches beyond the actual boundaries of individual cities enmeshing people everywhere in the economic and ecological linkages that are globalization. This is a view of the world nearly entirely at odds with that of the Canadian Conservative Party.

22.6   Rethinking Politics

While the ontological premises for thinking intelligently about sustainability are easy enough to outline there remain major difficulties in challenging the dominant understandings of 'environment' that are no longer very helpful in thinking about the future. Most obviously what climate change and the discussion of the Anthropocene make clear is that what needs to be focused on much more clearly than in the past is matters of what is now being made. The point of the Anthropocene is just that; we are making the future (Brauch/Dalby/Oswald Spring 2011). The commodities, houses, roads and energy systems that power our constructions are the new ecological context, not an addition to an environment, parts of which have to be protected in the sense of preserved. Obviously crucial ecosystems need to be protected from some aspects of development, but now they need to serve also as migratory pathways for species set in motion by climate change. Stability is over and ecological responses are happening simultaneously with human changes to environments. Thinking in terms of ecological sustainability now has to work in terms of mobile ecologies.

This is antithetical to much of the territorial thinking that traditionally structured parks and preservation. Ecological planners understand this well; the questions are whether the territorial administrative tools we use to control the mobility of those considered undesirable, both human and other species, can now be adapted appropriately to facilitate migrations. A static cartographic imagination of virtuous locals facing threatening foreigners, the nationalistic impulse all too readily mobilized in the face of changes that are rendered dangerous, is precisely the wrong geographical framework for dealing with what is coming. But it is implicit in many of the formulations that contemporary 'conservative' thinking uses. If the projections of dangerous climate change that are increasingly appearing in both climate science and now in the international investment analyses of future risks (PIK 2012), come to pass, rapid adjustments to changing circumstances will be needed, not militaristic attempts to prevent change by force.

In the Canadian case there has been considerable attention paid to the rapidly changing configuration of the Arctic as the ice retreats and the prospects of both new trade routes as well as oil and gas production in the newly accessible waters loom. While alarmist stories of imminent conflict there, and arguments about whose jurisdiction applies where, make good headlines, they are not what matters most (Kraska 2011). Arguing about who owns which island or what agency gets jurisdiction over sea-lanes isn't grappling with the bigger story. These are however familiar tropes that allow politicians photo opportunities and nationalist sound bites and journalists easy story lines that don't require much thinking about the causes of these changes. But this is dealing with symptoms of climate change, not dealing with the causes. It's adaptation after a fashion rather than mitigation, and while that is necessary now it's a focus that once again displaces attention from the more important issues of how to decarbonize capitalism quickly to slow the changes to many ecosystems.

Focusing only on the resource extraction sector deals with part of the problem of an unsustainable political economy, rather than the whole Canadian story. The consumption landscapes being built around the major urban centres without effective public transport, uneconomical because of the low density, and requiring extensive infrastructure construction as well as on-going energy consumption because of automobile use and the large houses that are key to the whole lifestyle, are the other side of the unsustainable ecology of this exemplary North American profligacy. The 'lock-in' of energy consumption for coming generations is part of the problem but the larger cultural politics of privatized consumption makes a politics of solidarity more difficult. Mobilizing around keeping taxes low, given the large expenditures involved in suburban houses, plays into the rhetoric of small government, not into innovative governance arrangements and the need to construct ecologically sustainable infrastructure with an eye to the long term in a changing climate.

22.7 Conclusions

At least six not particularly novel conclusions can be drawn from the Canadian federal political story of the last decade for those who wish to think seriously about sustainabilities and transitions to a post-carboniferous economy and society.

Most obviously is the simple lesson that activists concerned to make a more sustainable world ignore politics at their peril. Focusing only on narrow technical matters of environment, and assuming that legislation to protect various things will remain intact in the long run, are two dangerous tendencies that may appeal to academics in particular. The dynamics of multiple parties in the Canadian federal parliament are another factor that is less pronounced in other democratic states that have more equitable seat allocation systems than is the case in the Canadian 'first past the post' electoral arrangement. Nonetheless the rise of reactionary parties in times of crisis is a danger that sustainability campaigners need to anticipate. We do after all live in a political world, not a reasonable one. Thinking carefully how to make sustainability initiatives survive political storms is as important as building infrastructure that can survive extreme weather events.

The second lesson is that political economy matters! Confusing the rhetorics that legitimize states as the provider of services to all citizens with how power actually operates is a political trap. The rise of neo-liberal ideologies involves both numerous modes of rule that involve markets and management techniques that reassert the power of states, and does so in ways that underscore their primary function in facilitating capital accumulation. Rising inequality has been the result in most states where the business ideologies and managerial arrangements have been reasserted in the name of efficiency and prosperity. Power has also slipped away from states as they become enmeshed in complicated international trade agreements where foreign companies have the rights to sue governments trying to initiate social and environmental regulations deemed a challenge to corporate profits. In these situations states become dependent on external economic logics rather than domestic political priorities. Power doesn't reside simply with states; inter-governmental arrangements increasingly matter.

Which suggests, third, assuming that policy-making is actually coherent within states is also sometimes a mistake. Not only do different parts of state apparatuses have different priorities, forces outside the normal assumptions that modern states are coherent unitary entities often drive them. This is frequently exacerbated by nationalist rhetoric and the assertion of sovereignty in perverse ways that obscures the international economic situation. In the Canadian case provincial powers are often in conflict with federal priorities and the solutions are complex political compromises, or at least they were until the Stephen Harper conservatives became the majority government and set about limiting cooperation and consultation and imposing solutions where they could.

Fourth, is a reminder that governance by market is frequently a chimera. Markets can't decide many things, not least the most important things like what the future of the biosphere should be! They may be good at some distributional issues, but the profit motive and the failure of markets to signal long-term environmental dangers have brought us to the climate crisis. In Nicholas Stern's terms (2006), in his crucial economic analysis of the issue, climate change is the biggest market failure the world has ever seen. The early 2013 crisis in European carbon markets only emphasizes the point that market mechanisms are only as good as the rules that run them and these are unavoidably political issues (Paterson 2012). Focusing on production decisions and national policies about what gets produced rather than assuming that markets will make the things we need without being constructed to do so is important. Feed-in tariffs and other financial measures have been key to getting renewable energy industries started in many places; there is a politics to this that forces innovation. But such innovations are apparently easier in states without fossil fuel producers who understand such innovations as the competition.

That said, fifth is the simple but important point that in the globalized world economy of the present authority as well as climate governance initiatives (Bulkeley/Andonova/Betsill et al. 2014) is diffusing to new sites—provinces, cities, corporations and international arrangements. Sustainability strategies need to bear in mind that cooperation in all sorts of venues is going to be needed, and in the aftermath of the global financial crisis of the last few years and the exposure of the flawed assumption that financial systems work reliably or that banks are best left to their own devices, new opportunities to use the tools that they have developed for other ends may arise. Leo Panitch and Sam Gindin (2012) close their recent overview of the rise of global capitalism by musing on the political conditions that might yet turn international financial institutions into public utilities devoted to very different priorities from profit-making through

ever more abstract derivative trading schemes. This point also needs to be remembered in coming discussions of sustainable institutions.

Finally, to return to the ontological premise point made earlier in this chapter. The assumptions that the world is an intrinsically competitive place and the only option is to play to win have to be confronted directly in any discussion of sustainable futures. The task ahead for scholars concerned about sustainable transitions is to change the assumption that politics is necessarily only about dominating a divided world. Instead we need to focus on matters of how to share a crowded one (Dalby 2014). This is going to require a much larger discussion of inter-generational and intra-generational equity and a focus on making useful things that do not foreclose future ecological flexibilities. It is going to require thinking hard about how to adapt to changing circumstances in cooperative ways, rather than trying to resist change by using old-fashioned borders and the threats of force. Such geopolitical premises have no place in a sustainable future once it is realized that we are all part of a single biosphere that we are collectively remaking at something close to breakneck geological speed. Above all it is going to require much more careful thinking about political strategy where, based on short-term economic and parochial nationalist premises, there is intense opposition to sustainability initiatives.

While the Canadian Conservative Party was defeated in a Federal Election in October 2015, their actions over a period of nine years to dramatically reduce the capacity of the Canadian state to both monitor environmental change, and facilitate policies of transition to sustainability, is a warning to transition advocates that pre-empting such political programmes has to be part of any strategy that looks to the long-term future. The Liberal Government sworn in on November 4th 2015 with Justin Trudeau as Prime Minister has much work to do to repair the damage to Canada and its international reputation. It will probably help that Stephane Dion will serve as foreign affairs minister in that new government; his record on climate will probably ensure that at least Canada will no longer be an obstacle in climate negotiations. But a possible transition to a more sustainable Canadian future has been tragically delayed by at least a decade.

References

Baird, John, 2011: Address to the United Nations, New York, September 2011; at: <http://www.international.gc.ca/media/aff/speeches-discours/2011/2011-030.aspx-lang=eng&view=d>.

Baldwin, Andrew; Dalby, Simon, 2010: "Canadian Middle Power Identity, Environmental Biopolitics and Human Insecurity", in: Bosold, David; Hynek, Nikola (Eds.): *Canada's New Foreign and Security Policy Strategies* (Oxford: Oxford University Press): 121-137.

Blais, Pamela, 2010: *Perverse Cities: Hidden Subsidies, Wonky Policy and Urban Sprawl* (Vancouver: University of British Columbia Press).

Bosold, David; Hynek, Nikola (Eds.), 2010: *Canada's New Foreign and Security Policy Strategies* (Oxford: Oxford University Press).

Brauch, Hans Günter; Dalby, Simon; Oswald Spring, Úrsula, 2011: "Political Geoecology for the Anthropocene", in: Brauch, Hans Günter; Oswald Spring, Úrsula; Mesjasz, Czeslaw; Grin, John; Kameri-Mbote, Patricia; Chourou, Béchir; Dunay, Pal; Birkmann, Jorn (Eds.), 2011: *Coping with Global Environmental Change, Disasters and Security—Threats, Challenges, Vulnerabilities and Risks* (Berlin Heidelberg—New York: Springer-Verlag), 1453-1486.

Bulkeley, Harriet; Andonova, Lilliana B.; Betsill, Michele M.; Compagnon, Daniel; Hale, Thomas; Hoffmann, Matthew J.; Newell, Peter; Paterson, Matthew; Roger, Charles; Vandeveer, Stacy D., 2014: *Transnational Climate Change Governance* (Cambridge: Cambridge University Press).

Charbonneau, Bruno; Cox, Wayne S. (Eds.), 2010: *Locating global order: American power and Canadian security after 9/11* (Vancouver: University of British Columbia Press).

Dalby, Simon, 2014: "Environmental Geopolitics in the Twenty First Century", in: *Alternatives: Local, Global, Political* 39: 3-16.

Elgie, Stuart; McClay, Jessica, 2013: "BC's Carbon Tax Shift Is Working Well after Four Years (Attention Ottawa)", in: *Canadian Public Policy*, 39: S1-S10.

Ellis, Erle C.; Goldewijk, Kees K.; Siebert, Stefan; Lightman, Deborah; Ramankutty, Navin; 2010: "Anthropogenic transformation of the biomes, 1700 to 2000", in: *Global Ecology and Biogeography,* 19: 589-606.

Fall, Juliet, 2010: "Artificial states? On the enduring geographical myth of natural borders", in: *Political Geography,* 29: 140-147.

Gordon, Todd, 2010: *Imperialist Canada* (Winnipeg: Arbeiter Ring).

Grant, Elizabeth, 2014: "What's in a Name: A lot if you're talking Aboriginal Title", in: *Open Democracy* (2 July); at: <https://www.opendemocracy.net/5050/elizabeth-grant/whats-in-name-lot-if-you%E2%80%99re-talking-aboriginal-title>.

Haley, Brendan, 2011: "From Staples Trap to Carbon Trap: Canada's Peculiar form of Carbon Lock In", in: *Studies in Political Economy*, 88: 97-132.

Kaplan, Gerard, 2011: "Be Very Afraid: Harper is Inventing a new Canada", in: *Globe and Mail* (16 December).

Kraska, James (Ed.), 2011: *Arctic Security in and Age of Climate Change* (Cambridge: Cambridge University Press).

Lang, Eugene, 2012: "The Harper Doctrine", in: *National Post* (16 April); at: <http://fullcomment.nationalpost.com/2012/04/16/eugene-lang-the-harper-doctrine/>.

Levant, Ezra, 2010: *Ethical Oil: The Case for Canada's Oil Sands* (Toronto: McLelland and Stewart).

Levin, Kelly; Cashore, Benjamin; Bernstein, Steven; Auld, Graeme, 2012: "Overcoming the tragedy of super wicked problems: constraining our future selves to ameliorate global climate change", in: *Policy Sciences, 45,*2: 123-152.

Martin, Lawrence, 2010: *Harperland: The Politics of Control* (Toronto: Viking).

Martin, Patrick, 2012: "Baird sticks to party line—Israel's Likud party", in: *Globe and Mail* (3 February).

McDonald, Marci, 2010: *The Armageddon Factor: The Rise of Christian Nationalism in Canada* (Toronto: Random House).

McLeod Group, 2012: *The Cowboy Way: Or Canadian Foreign Policy under a Majority Conservative Government* (Ottawa: McLeod Group); at: <http://www.mcleodgroup.ca/>.

Medalye, Jacqueline; Foster, Ryan, 2012: "Climate Change and the Capitalist State in the Canadian Arctic: Interrogating Canada's Northern Strategy", in: *Studies in Political Economy*, 90: 87-114.

Nadeau, Christian, 2010: *Rogue in Power: Why Stephen Harper is Remaking Canada by Stealth* (Toronto: James Lorimer).

Nossal, Kim R., 2012/13: "Late learners: Canada, the F-35, and lessons from the New Fighter Aircraft program", in: *International Journal*, 68,1: 167-184.

Panitch, Leo; Gindin, Sam, 2012: *The Making of Global Capitalism: The Political Economy of American Empire* (London: Verso).

Paterson, Matthew, 2012: "Who and what are carbon markets for? Politics and the development of climate policy", in: *Climate Policy*, 12: 82-97.

PIK [Potsdam Institute for Climate Impact Research and Climate Analytics], 2012: *Turn Down the Heat: Why a 4°C Warmer World Must be Avoided* (Washington, DC: The World Bank).

Solomon, Lawrence, 1978: *The Conserver Solution* (Toronto: Doubleday).

Stanford, James, 2008: "Staples, Deindustrialisation and Foreign Investment: Canada's Economic Journey Back to the Future", in: *Studies in Political Economy*, 82: 7-34.

Stern, Nicholas, 2006: *The Economics of Climate Change* (Cambridge: Cambridge University Press).

Szasz, Andrew, 2007: *Shopping our way to Safety* (Minneapolis: University of Minnesota Press).

Turner, Chris, 2013: *The War on Science* (Vancouver: Greystone Books).

UNEP [United Nations Environment Programme], 2012: *Global Environmental Outlook GEO5: Environment for the Future we Want* (Nairobi: United Nations Environment Programme).

WCED [World Commission on Environment and Development], 1987: *Our Common Future* (Oxford: Oxford University Press).

Wells, Paul, 2013: *The Longer I am Prime Minister: Stephen Harper and Canada, 2006–* (Toronto: Random House).

Wilkinson, Katherine K., 2012: *Between God and Green: How Evangicals are Cultivating a Middle Ground on Climate Change* (Oxford: Oxford University Press).

23   Regime Change, Transition to Sustainability and Climate Change Law in México

Juan Antonio Le Clercq[1]

Abstract

This chapter analyses the creation of the Climate Change Law in Mexico as a governance regime where horizontal and vertical interplay became key elements. The legislative deliberations oriented towards creating the Climate Change Law, mainly between 2005 and 2012, are understood as a preliminary and as take-off phases of a pathway to a sustainable transition. The principal elements of Mexican climate policy are discussed. Three competing alternatives for climate legislation in 2012 are identified, as are the reasons for opting for a specific climate regime. The chapter argues that even though the Mexican climate regime represents a very ambitious model in its design, and although a preliminary and take-off phase can be identified, problems of fit with other policy objectives, especially the features of the energy sector, limit the possibilities for a sustainable transition in Mexico.

Keywords: Regime change, sustainable transitions, transitions phases, climate change law, institutional change, Mexico.

23.1   Introduction: Regime Change as a Pathway to Sustainable Transition

Climate change is a global collective action problem that no country can solve by itself. Its complex characteristics and differentiated impacts demand international cooperation for the distribution of rights, duties and the sharing of financial burdens (Barret 2007; Gardiner 2011). Without a more efficient and fairer global political architecture it will be extremely difficult, in fact almost impossible, for humankind to start stabilizing CO_2 emissions in the next decade in order to contain the impact of global warming around the globe (Helm/Hepburn 2009; Biermann/Pattberg 2012; Dryzek/Norgaard 2013).[2]

But it is mistaken to conceive of climate change as an exclusively international issue. Negotiations for a new global climate deal will make a difference if the deal defines more effective criteria for the parties to design their national policies from a polycentric or multilevel perspective and to develop human and organizational capabilities according to their particular needs (Paavola 2008; Ostrom 2009).

Institutional change is a prerequisite for the integration of the climate agenda into the policy framework of developing countries. It is also a priority task for the improvement of more efficient multilevel coordination mechanisms. Institutional weaknesses lead to reduced capabilities to fulfil goals and responsibilities as defined in the *United Nations Framework Convention on Climate Change* (UNFCCC), such as emissions inventories, cost-benefit analysis of available options for emissions reduction, systematic research, vulnerability assessment, and the design of adaptation strategies.

The environmental governance perspective points to a relationship between the quality of institutional design, regime interplay and effectiveness in dealing

1   Juan Antonio Le Clercq is a full-time Professor of International Relations and Political Science at the Universidad de las Américas Puebla (UDLAP); Sta. Catarina Mártir. Cholula, Puebla. C.P. 72810. México; Email: <juan.leclercq@udlap.mx>.

2   The author is grateful to Úrsula Oswald Spring, Celeste Cedillo and Ronald Guy Emerson for their comments and criticisms on different versions of this document and to the anonymous reviewers of a previous draft.

with global challenges such as biodiversity loss, the degradation of ecosystem services or the impacts of climate change (Young 2002; Young/King/Schroeder 2008; Paavola 2007; Biermann/Betsill/Gupta et al. 2010; Bechtel/Scheve 2013; Van de Graaf/De Ville 2013; Young 2013). Reports from agencies such as the *Organisation for Economic Co-operation and Development* (OECD), the World Bank and the *United Nations Environment Programme* (UNEP) also tend to focus on the importance of strategies for overcoming institutional inertia and maximizing the benefits of institutional change in developing countries (WB 2010; OECD 2010).

Institutions matter in dealing with climate change. However, even when developing countries recognize the importance of institutional change, the development of new capabilities and the creation of specific regulatory frameworks for sustainable development, these are especially complex problems. Designing reportable mitigation and adaptation policies and simultaneously fostering institutional change is a difficult combination when economic growth is a top priority, there is a shortage of financial resources, and socioeconomic inequalities are consequently widened.

The discussion about the importance of institutional change and its different paths in the face of global climate change is especially relevant for Mexico after its adoption of a Climate Change Law in 2012. This law, which has been internationally recognized as ambitious flagship legislation, formalizes a national climate strategy and a programme developed during the first decade of this century and establishes them in a long-term framework for coping with climate change.[3]

Between 2010 and 2012, during the legislative process that resulted in the *Climate Change Law* (CCL), it became evident that Mexico required new and different institutional mechanisms for three main reasons: 1) even as a Non-Annexe country without miti-

gation obligations according to the Kyoto Protocol, its energy matrix depends ninety per cent on fossil fuels and, as a by-product, is the major contributor to *greenhouse gas* (GHG) emissions from energy generation and consumption in Latin America;[4] 2) its territory is highly vulnerable to droughts, floods, ecosystem loss, heatwaves, changes in precipitation patterns, infrastructure destruction, damage caused by natural disasters and decline in agricultural production, with additional effects for the section of the population living in poverty;[5] 3) since the Earth summit of 1992, Mexico has played an important role in setting up the global climate architecture and has supported emission reduction goals for emerging economies.

This chapter tries to respond to three related questions: Why did legislators in Mexico opt for a climate regulation model that resembles a governance regime? Does the climate regime define a clear path to a sustainable transition? To what extent will the specific features of the *Climate Change Law* (CCL) have an effective impact on Mexico's climate policy?

The theoretical approach to answering these questions uses an environmental governance perspective that integrates the conceptual elements developed in the institutional dynamics literature (particularly the contributions of Oran R. Young, Jouni Paavola and Frank Biermann), and the theory of transition to sustainability (the works of John Grin, Frank Geels, René Kemp, Jan Rotmans and Johan Schot). In discussing the emergence of the CCL, the analysis focuses par-

---

3 The Climate Change Law (*Ley General de Cambio Climático*) was approved on 19 April 2012 after two years of legislative debate (see at: <http://www.diputados. gob.mx/LeyesBiblio/ref/lgcc/LGCC_orig_06jun12.pdf>). In 2007 the Calderon Administration defined the first National Climate Change Strategy (*Estrategia Nacional de Cambio Climático*), together with a programme with specific mitigation and adaptation goals and objectives (*Programa Especial de Cambio Climático* 2008-2012). In this chapter these documents are referred to as ENACC and PECC, and they can be consulted at: <http://cambioclimatico.inecc.gob.mx/ccygob/ccygob. html> and at: <http://dof.gob.mx/nota_detalle.php? codigo=5107403&fecha= 28/08/2009>.

4 Mexican GHG emissions grew by 33.4 per cent between 1990 (561,035.2 Gg CO2e) and 2010 (8,748,252.2 Gg CO2e) due to an energy balance that depends ninety per cent on fossil fuels (SEMARNAT 2012; Fifth National Communication to the UNFCCC (2012, 195-197); at: <http://www2.inecc.gob.mx/publicaciones/ download/685.pdf>. According to the *International Energy Agency* (IEA) (2014: 48-57), Mexico's emissions for 2012 were 435.79 MtCO2 from fuel combustion only. These figures are far lower than those of other Non-Annexe 1 Parties such as China (8,205.86 MtCO2) or India (1,954.02 MtCO2), but higher than any other Latin American country except Brazil (440.24 MtCO2). The World Resource Institute identifies Mexico as the ninth highest global emitter, with a 1.67% contribution to global CO2 emissions, at the same level as Indonesia, Iran, Canada, South Korea and Australia; see at: <http:// www.climatecentral.org/news/greenhouse-gas-emissions-by-country-19167>.

5 The PECC (2014-2018) includes an updated diagnosis of the risks and vulnerabilities facing Mexico; at: <http:// www.dof.gob.mx/nota_detalle.php?codigo=5342492& fecha=28/04/2014>.

ticularly on how a national climate regime model, a vertical and horizontal interplay design, was imposed over two less ambitious legislative alternatives, and the way this process fosters a sustainable transition to some degree.

The rest of this chapter is organized in six parts. First we discuss the relevance of some concepts central to the environmental governance perspective, such as regime and vertical and horizontal interplay, and then we distinguish the different phases of a sustainable transition. Second, the influence of international-national interplay is highlighted for the case of Mexico. Third, the main features of the Mexican climate policy implemented between 2006 and 2012 are identified. In the following two sections the analysis centres on the Mexican legislative process understood as a preliminary phase and on the approval of the CCL as a take-off moment for a sustainable transition. In the last part we conclude.

23.2   A Preliminary Phase of a Sustainable Transition

Following Paavola (2008), Ostrom (2009) and Young (2013), 'interplay' is defined as an institutional interaction with two dimensions: a) vertical or horizontal institutional levels; or b) between different arenas (environment, energy, urban planning, social or economic development, etc.). Horizontal interplay refers to an institutional or agent interaction within the same policy or organizational level (local, national, regional or international), while vertical interplay refers to multilevel or polycentric interplay. Both dimensions necessarily relate regime-building or institutional change to the problem of fit (Young/King/Schroeder 2008; Folke/Pritchard/Berkes et al. 2007; Paavola/Gouldson/Kluvánkova-Oravská 2009; Biermann/Betsill/Gupta et al. 2010; Keohane/Victor 2010).

Governance regimes are institutional compounds within the same or different levels of organization, with a non-rigid integration that produces interaction, mutual influence and feedback effects (Keohane/Victor 2010; Oberthür/Gehring 2011; Paavola/Gouldson/Kluvánkova-Oravská 2009; Biermann/Betsill/Gupta et al. 2010; Young 2012). Interpreting the case of Mexico from the perspective of governance regimes has the analytical advantage of enabling institutional dynamics to be understood as nested interactions and overlaps instead of reducing it to the work of a single institution or a law.[6]

A first argument is that Mexican legislators involved in the creation of the CCL assumed that a central feature of the law design had necessarily to be a nested interaction or 'transversality'. They understood that dealing with the complexities of climate change required a new framework—a regime, for example, capable of solving the problem of interplay and fit. A second argument is that once they promoted a climate regime rather than alternatives oriented towards reforms of individual laws, indirectly they also linked institutional change with a transition to sustainability.

A 'transition' is a deep transformation that affects the structures, practices and value systems of a society, an open process involving multiple organizational levels and different phases. More than an instrument to operationalize governance models or a change towards sustainable development (Parris/Kates 2003; Kemp/Parto/Saed 2005; Geels 2005; Geels/Schot 2007; Grin/Rotmans/Schot 2011), this concept refers to the actors' capability to lead and determine the transformation processes in a society's sociotechnical subsystems.[7] A transition means a "radical transformation toward a sustainable society as an answer to a number of persistence problems facing modern societies" (Grin/Rotmans/Schot 2011: 11-12).

According to Kemp, Parto and Saed (2005), a transition towards sustainability requires a deliberate attempt to transform the social structure in a series of gradual stages. Geels and Schot (2007) stress the need for multilevel changes in sociotechnical regimes with the goal of establishing a new equilibrium and alignment among their constitutive elements. Grin, Rotmans and Schot (2011) differentiate the change processes involved in transitions towards sustainability using five characteristics: 1) co-evolutionary processes requiring changes or reconfigurations in the sociotechnical systems; 2) changes involving multiple actor

---

6   The idea of regime change and genesis has been central in international theory and global governance literature following the influence of Krasner 1983, Keohane 1982, Young 1986 and Haggard/Simmons 1987. Krasner (1983: 19) defines 'regime' as "implicit or explicit principles, norms, rules and decision-making procedures around which actors' expectations converge in a given area of international relations".

7   The concept 'sociotechnical system' refers to a "level of societal domains or functions such as transport, energy, housing agriculture and food, communication and healthcare". Sustainable transitions are "shifts from one socio-technical system to another" (Grin/Schot/Geels 2011: 6).

interactions; 3) radical transformations of a system or configuration towards a new one; 4) long-term gradual processes; and 5) macroscopic transformations at the organizational level.

Regardless of the approach, a transition towards sustainability involves the multilevel interaction of three core variables (Geels 2005; Geels/Schot 2011; Grin/Rotmans/Schot 2011):

1. The existence of 'niches' for innovation. This implies that a small group of innovative actors are capable of identifying and fostering changes that stem from their expectations and visions regarding a sustainable future, thus starting broader transformations in a society's sociotechnical subsystems;
2. 'Regime level' changes. In this case the impact of niche-level innovations opens windows of opportunity to change the configuration of sociotechnical regimes and guide them towards new balances that are more favourable for the development of new innovations;
3. The 'landscape change' or innovations in the regime that impact the sociotechnical picture as a whole, thus spreading the new kinds of practices and consolidating the process of transition towards sustainability.

While the degree to which political and social actors can design and control this kind of transformation is arguable, much less clear is the intention of the Mexican Congress to activate a transition; if the pre-assumptions of these approaches are correct, then by choosing a climate regime they put on the agenda the implicit need to foster some kind of transition towards sustainability. In the statement of purpose of the main initiative and the preliminary draft of the law, Congress assumed that as a consequence of the CCL a major social transformation guided by sustainability principles would be promoted. In other words, the Mexican case represents an attempt to frame a transition towards sustainability through a planned regime change.

The fact that the Mexican climate regime involves a transition towards sustainability does not necessarily imply that this objective will be achieved in the near future. On the contrary, a transition in Mexico entails conflictive and contradictory coexistence between practices oriented towards sustainable principles and the reproduction of an economic growth model based on the exploitation of hydrocarbons. Nor does the existence of a regime as a formal framework mean that an effective capability will be available to regulate individual practices or conduct. The Mexican case illustrates the problem of fit[8] between different

regime types, economic development and social objectives encompassed within the general objective of transition towards sustainability.[9]

From a multiphase perspective, a transition towards sustainability can be divided into four clearly differentiated phases: 1) a pre-development phase "in which the status quo of the system changes in the background, but these changes are not visible"; 2) a take-off phase in which the "process of structural change picks up momentum"; 3) an acceleration phase when "structural changes become visible"; 4) and a stabilization phase "where a new dynamic state of equilibrium is achieved" (Grin/Rotmans/Schot 2011 2011: 4-5).

The creation of specific law for climate change in Mexico may be interpreted as the definition of a transitional agenda that encompasses two phases inherent in the transition process: first, a pre-development phase in which actors identify the problem and begin to define a public response to it; and second, a take-off phase which actually begins with the approval and publishing of the Law and with it the legislative decision on establishing a climate regime.

When we understand the creation of a climate regime as an arena for transition it is possible to identify a pre-development phase, a moment where changes in the background are still not visible as part of a sustainable transition path in which: 1) political actors and social organizations identify and diagnose the problem and its consequences for Mexico; 2) political actors acknowledge the importance of developing an agenda that involves institutional changes and the formulation of public policies; 3) the government defines national goals and adaptation and mitigation objectives and assumes international commit-

---

8 The problem of fit refers to (in)compatibility, (mis)match or contradictions between a set of social institutions or between the properties of a biophysical system and the features of an institutional framework. See Young/King/Schroeder 2008 and Folke/Pritchard/Berkes et al. 2007.
9 The particularity of climate change in Mexico is discussed in Julia Martínez and Adrián Fernández Breamauntz: *Cambio climático: una visión desde México* (Mexico, D.F.: SEMARNAT-INE, 2004); Simone Lucatello and Daniel Rodríguez Velázquez: *Las dimensiones globales del cambio climático. Un panorama desde México. ¿Cambio social o crisis ambiental?* (Mexico, D.F.: Instituto Mora-UNAM, 2011); and José Luis Calva: *Cambio Climático y políticas de desarrollo sustentable* (Mexico, D.F.: Análisis Estratégico para el Desarrollo vol. 14, Consejo Nacional de Universitarios, 2012).

ments; 4) government authorities establish programmes and specific objectives as pressure increases from *non-governmental organizations* (NGOs) for higher commitments and the private sector lobbies to avoid impacts or costs on their activities; and 5) once the agenda is defined, it is possible to evaluate and verify the level to which the programme's goals have been fulfilled.

23.3   National–International Interplay

The design of the Mexican climate legislation is framed by a context characterized by a high level of participation in international climate change negotiations. Mexico ratified the UNFCCC on 7 May 1993 and the Kyoto Protocol on 24 November 2000; promotes at international commitment level the national goals established in the *Climate Change Special Programme* (PECC) to reduce emissions of *greenhouse gases* (GHG), and became one of the first countries to define voluntary obligations in compliance with the agreements achieved in the Copenhagen Agreement.[10] Additionally, Mexico is the only country that has presented up to five national communications as a Non-Annexe Party of the Kyoto Protocol.[11]

The interplay of national politics with climate change and international negotiations was important for Ernesto Zedillo and Vicente Fox's governments, though it was Felipe Calderón who became the leading promoter of Mexican climate policy between 2006 and 2012. During this administration, international negotiation became an interplay element that influenced the definition of national policies. The

international commitments that were taken on, the activism in international negotiations, and the growing media and political impact of climate summits all played an important role in justifying the need for action in the national arena. In addition, advances in the definition of the *National Climate Change Strategy* (ENACC) and PECC, the development of methodologies to implement plans at state level, and the deliberations to create the CCL were all used at an international level as proof of Mexico's commitment to a global climate architecture.

The organization of the Sixteenth *Conference of the Parties* (COP) in Cancun in 2010, and the incorporation of a climate agenda into G20, signalled the importance of national–international interplay in this period. The organization of the COP, which involved political risks after its previous failure in Copenhagen, was translated into a success for the Calderon administration in many senses: a) it relaunched a stalled international negotiation by opening a parallel way based on a portfolio of projects and mechanisms that did not necessarily depend on a great binding agreement; 2) it integrated the proposal outlined in the PECC to create a Green Fund, oriented towards guaranteeing more resources to finance mitigation and adaptation projects in non-Annexe 1 countries; 3) it consolidated Mexican foreign policy on international negotiations and allowed Felipe Calderon's leadership to be promoted through presenting him as a central figure in global climate policy; and 4) no less importantly, it put the national climate policy, created during this government period, into the national and international spotlight (De Alba 2010; Hedegaard 2010; Le Clercq 2010).

Strengthing international cooperation and Mexico's active participation in climate negotiations were important issues during the legislative deliberations. Nevertheless, the different projects discussed do not incorporate specific mandates related to international-national interplay. Congress acknowledges the importance of defining national commitments through the law and of reinforcing the Mexican position in the negotiations, but a Non-Annexe Party condition was assumed and, therefore, the national climate policy was understood as a strictly voluntary effort.[12]

Alternatively, it would be an error to think that the climate legislation design was an inevitable conse-

---

10 The National Mitigation Actions of Mexico are included in the Copenhagen Agreement Appendix 2: "Mexico adopted its Special Climate Change Programme in 2009, including a set of nationally appropriate mitigation and adaptation actions to be undertaken in all relevant sectors. The full implementation of the Programme will achieve a reduction in total annual emissions with respect to a business-as-usual scenario of 51 million tonnes CO_2e by 2012. Mexico aims at reducing its GHG emissions by up to 30 per cent with respect to the business-as-usual scenario by 2020, provided that adequate financial and technological support from developed countries is provided as part of a global agreement"; see at: <http://unfccc.int/meetings/cop_15/copenhagen_ accord/items/5265.php>.

11 The National Communications were presented in 1997, 2001, 2006, 2009 and 2012. They can be consulted at: <http://unfccc.int/national_reports/non-annex_i_natcom/submitted_natcom/items/653.php>.

12 Mexican authorities have consistently stated in their programmes and official documents that the national climate policy represents a voluntary and not an internationally binding effort.

quence of Mexican activism in international negotiations. Even if Congress identified international benefits for the Mexican position in approving the CCL, the process ran on a different track and was driven by a deliberation about the internal benefits of approving the climate legislation and providing an impulse to some form of transition towards sustainability. In other words, the approval of the CCL must be understood as a legislative outcome that responded to national objectives and as an endogenous equilibrium of the political parties' positions, the federal government's objectives, and the agendas of the social actors involved in the process.

23.4   The Definition of a National Climate Change Policy

Between 2006 and 2012 national climate programmes were designed with measurable and verifiable goals and objectives. This model of national climate policy defines a coordination scheme for the federal government agencies to fulfil mitigation and adaptation goals and objectives, and establishes a methodological definition for elaborating local climate plans that correspond to the national strategy.

The main elements of the Mexican policy (table 23.1) are linked in an extensive planning chain comprising:

1. The existence of an inter-ministerial commission (CICC) as a coordination device for the different agencies of the federal government involved.
2. Objectives defined in the *National Development Programme* (PND 2007-2012). Besides taking a Sustainable Human Development approach to the planning of public policies, this includes two objectives and eight specific strategies for mitigating GHG and the goals for adaptation to face the inevitable effects of climate change.[13]
3. Strategic guidelines defined in 2007 through the ENACC. Even though this is only binding at the federal level, it proposes a diagnosis of the national challenges, specific action lines regarding

energy generation and use, vegetation and land use, and vulnerability and adaptation, as well as Mexico's general position in international negotiations.

4. Implementation of the strategic guidelines and objectives defined in the PND 2007-2012 and the ENACC through the PECC for the period 2008-2012. This consists of 100 mitigation, adaptation and interplay objectives, translated into 294 measurable, verifiable and reportable goals.
5. Methodologies and criteria for producing state climate action plans coordinated by the *National Institute of Ecology* (INE) and the voluntary register of the private sector's emissions through the Mexican Programme on Greenhouse Gases (Programa GEI México).[14]

Regardless of its virtues and limitations, the main problem of the climate policy implemented between 2006 and 2012 was its inability to guarantee its own continuity after a change in administration. While the PECC was designed as a planning instrument that set immediate (51 million tCO_2 by 2012) and long-term (indicative objective to reduce fifty per cent of total emissions by 2050) mitigation goals, and that defined an adaptation trajectory divided into stages of vulnerability assessment and evaluation of economic implications (from 2008 to 2012), capability strengthening (from 2013 to 2030) and consolidation (from 2031 to 2050), it could not legally extend its objectives beyond 2012. Even though the definitions of mitigation goals and obligations were based on a CO_2 emissions trending scenario established on the Base Emissions Line, the PECC define its goals as "an indicative objective or aspirational goal", which, in addition, depended on "the industrialized countries giving the financial and technological support to complement the national efforts".[15]

From a multilevel governance perspective (Paavola 2008; Ostrom 2009; Young 2013), it became evident that this model was not capable of solving vertical national interplay or the coordination problems between government levels, and neither was it capable of establishing a fit between conflicting sectorial objectives. While it sets measurable and verifiable goals and objectives, the problems of correspondence

---

13 The *National Development Programme* (PND) is the general planning document for each government administration. It establishes national objectives, strategies and government priorities for a six-year period. The PND 2007-2012 included for the first time a sustainability axis and climate change goals. See PND (2007: 21-37, 231-265); at: <http://dof.gob.mx/nota_detalle.php?codigo=4989401&fecha=31/05/2007>.

14 The GEI Programme is a voluntary emissions registration scheme with more than 120 participating companies from different economic sectors. It is coordinated by the private sector and the environment authorities (SEMARNAT); see at: <http://www.geimexico.org/>.

15 See PECC (2009: vii, 14, 20).

Table 23.1: Climate policy instruments, objectives and goals (2006–2012). **Source:** Information from PND (2007–2012), ENACC (2007) and PECC (2009).

Instrument	Energy Mitigation	Land Change Mitigation	Adaptation	Institutional Capacity
PND 2007–2012	Four strategies: efficiency and clean technology for generation; efficient use; international standards for vehicle emissions; emission recuperation from waste	Objectives for biodiversity and forests	Four strategies: integration of adaptation and planning, climate scenarios, sector vulnerability assessment, information diffusion	PECC links its objectives with the PND 2006–2012 thematic axes
ENACC 2007	14 mitigation opportunities equivalent to +/- 106.8 MtCO2e/year 17 Guidelines	Mitigation potential: Conservation: 11,000–21,000 MtCO2 Storage: 18.08–42.16 MtCO2 Substitution 2.5 MTCO2 Mitigation agriculture and livestock: 9.7 MtCO2e 7 Guidelines	Vulnerabilities identification and capability development: risk management, water management, biodiversity and environmental services, agriculture and livestock, coastal areas, human settlements, energy generations and use 13 Guidelines	National definition for international climate negotiations
PECC 2009–2012 100 objectives, 294 goals 2012 Mitigation goal 2012 50.7 MtCO2e	70% of emissions reduction goal *Generation*: 10 objectives, 23 goals for 18.0 MtCO2e *Use*: 17 objectives, 30 goals for 11.9 MtCO2e *Waste*: 3 objectives, 7 goals for 5.5 MtCO2e Identification of opportunities for the private sector but without definition of objectives or goals.	*Agriculture, Forest and other land use*: 8 objectives and 36 goals for a 30% contribution to the emissions reduction goal	Implementation phase of vulnerability assessment and valuation economic impacts. *Risk management*: 5 objectives, 12 goals; *Water resources*: 5 objectives, 30 goals; *Agriculture, Livestock and fishery*: 9 objectives, 30 goals; *Ecosystems*: 8 objectives, 39 goals; *Energy Industry and services*: 4 objectives, 9 goals; *Zoning and urban development*: 1 objective, 6 goals; *Public Health*: 1 objective, 6 goals	Transversal policy guidelines *Foreign Policy; Institutional development*: 4 objectives, 11 goals; *Climate Change economics*: 8 objectives, 22 goals; *Education, Training, Information and Communication*: 7 objectives, 30 goals; *Research and technological development*: 10 objectives, 12 goals
Reported Advance: 80%/ 40.69 MtCO2e	Reported Advance Generation: 89% Use: 49% Waste: 19%	Reported Advance. 115%	Reported Advance. 65%	Reported Advance: 53%

and fit with other policies also made the possibility of fostering climate policy within the broader objective of a sustainable transition difficult. Even when climate policy focused on setting the basis for encouraging the decarbonization of the economy and increasing capabilities for adaptation, the truth is that this aspiration was in conflict with an energy model based on hydrocarbon production, a tax regime with few incentives for generating renewable energy and with a high level of subsidies for fuel and electricity, and the definition

Figure 23.1: Legislative process concerning climate change in Mexico (1992–2012). **Source**: Information from the Chamber of Deputies, the Senate and the *Legislative Information System* (SIL); at: < diputados.gob.mx/>; <senado.gob.mx/>; and <sil.gobernacion.gob.mx/>.

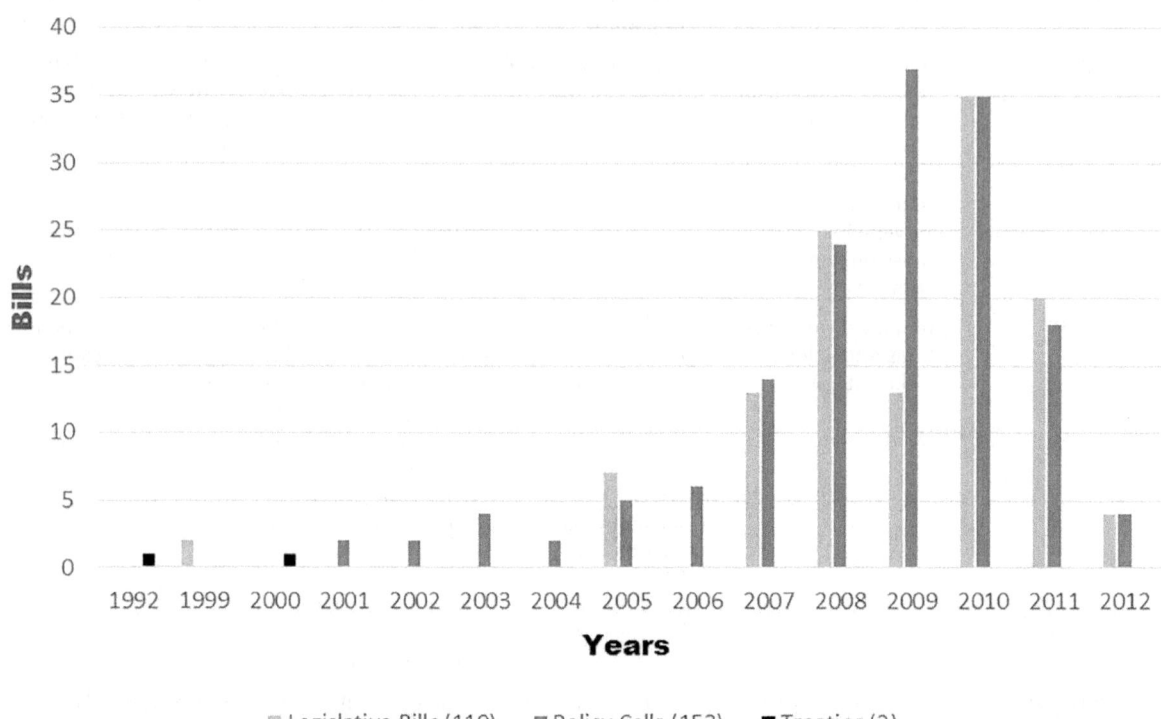

23.5 The Mexican Legislative Process and Climate Change: the Preliminary Phase

While the process of creating the Climate Change Law began in 2010, climate change had been a relevant issue in Congress since 2000, with the inclusion of reform bills and policy calls in legislative debates.[16] Interest in climate change legislation in Mexico began early in the first decade of the twenty-first century, although it was between 2005 and 2010 when the number of amendment projects (initiatives) regarding

climate change increased. At this stage the need for an adequate legal framework was acknowledged and possible alternatives for new regulation modalities, incentives systems and policy instruments were identified. Between 1992 and 2012, 119 legislative bills and 154 policy calls were presented specifically on climate change topics in Congress by the different parties (figure 23.1).[17]

Legislative work concerned with climate change gains relevance in the Mexican Congress up to the LX and LXI legislatures (2006–2012), not to mention the nine initiatives and eighteen policy calls that were presented in the previous years. This period involved two different groups of deputies (elected in 2006 and

---

16 For explanatory purposes, the term 'policy calls' is used to translate the legislative mechanism "Proposición con Punto de Acuerdo", which refers to the capacity of Congress to exhort governmental authorities. Congress can ask the government to do or stop doing certain things in the policy arena but this is a non-binding power. Nevertheless these exhortations are important in deliberations concerning national policy.

---

17 194 bills and 244 policy calls were identified that included the word 'climate change' between 1992 and 2012. In some cases the reference was circumstantial, so in the end 119 bills and 154 policy calls were included in the sample. All the legislative information used in this article was gathered and classified by year, political party, chamber of origin and topic from the available data in *the Chamber of Deputies, the Senate* and the *Legislative Information System* (SIL); at: <diputados. gob.mx/>; <senado.gob.mx/>; and <sil.gobernacion. gob. mx/>.

Figure 23.2: Climate change bills by topic (2006–2012). **Source:** Information from the Chamber of Deputies, the Senate and the *Legislative Information System* (SIL); at: <diputados.gob.mx/>; <senado.gob.mx/>; and <sil. gobernacion.gob.mx/>.

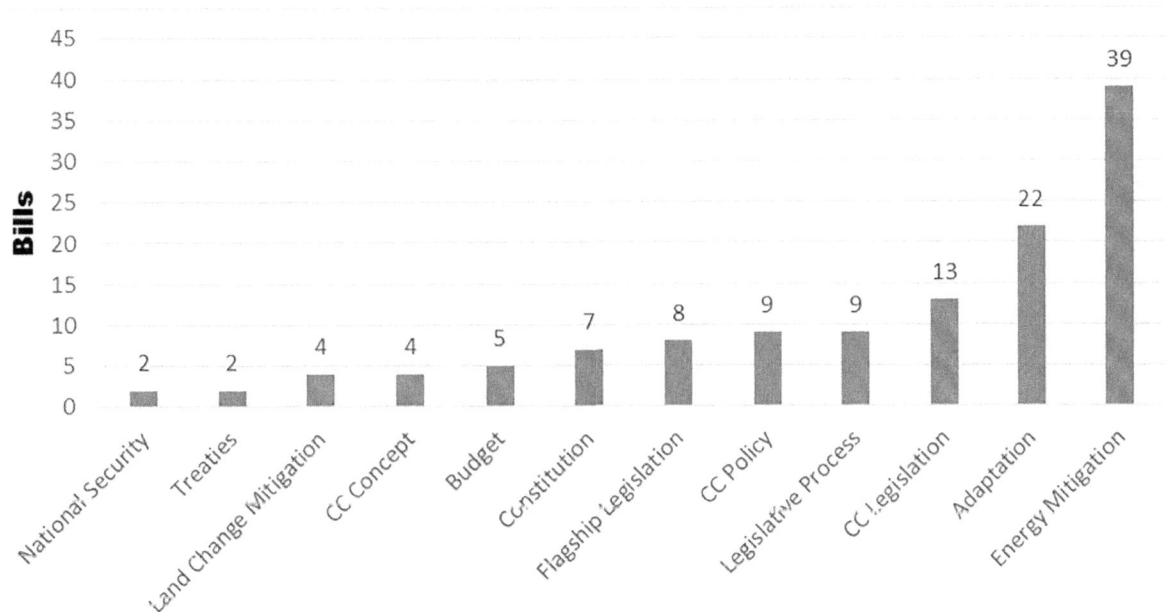

2009, respectively), but the same group of senators (elected in 2006). From the perspective of the climate change debate, these administrations are characterized by encompassing the beginning and failure of international discussion on the need for a period of post-Kyoto obligations, the formulation of a national policy on climate change, and the outbreak of the greatest international economic crisis of the post-war period. This group of legislators also participated in the 2008 energy reform process, approved legislation centred on promoting renewable energies, and presented the main proposals that shaped the CCL in 2012.

The absence of projects from the Executive Power during any given period stands out, but it is especially evident during the LX and LXI legislatures, which concurred temporarily with the issuing of the ENACC, the PECC, the organization of COP 16 in Cancun and the beginning of deliberations regarding the creation of a general law to address the climate change problem. Felipe Calderon's legislative agenda focused on security and economic issues, followed at a great distance by education and energy.

If we break down the relevant law initiatives and policy calls presented by each parliamentary group since 1999 by theme, we can understand the way in which legislators and parties have tried to simplify the

complexity of climate change effects and to define criteria for an institutional response. When we analyse the content of the different legislative bills presented on climate change since 1992, it can be seen that legislators have understood that the definition of the legal basis for the mitigation of GHG derived from the generation and consumption of energy (thirty-one per cent) must be seen as a top priority, despite the fact that Mexico is a Non-Annexe 1 country and therefore not obliged to define punctual goals for emissions reduction. The second important issue is the diversity of proposals centred on problems related to adaptation (eighteen per cent). We have to add eight initiatives of a comprehensive nature that incorporate both mitigation and adaptation objectives and nine that incorporate new faculties for authorities and criteria for national strategies and programmes (figure 23.2).

It never ceases to be amazing that most initiatives centred on adaptation are only half the number of those centred on mitigation, especially if the high vulnerability that characterizes the Mexican territory and the high frequency of natural disasters and periods of drought are taken into account, as well as the fact that the legislative discourse has insisted on setting adaptation as a priority in the definition of policies and resource allocation in the face of climate change. In this case a reactive view of the inevitable conse-

Figure 23.3: Climate change legislative policy calls by topic (1992–2012). **Source:** Based on information from the Chamber of Deputies, the Senate and the *Legislative Information System* (SIL); at: <diputados.gob.mx/>; <senado.gob.mx/>; and <sil.gobernacion.gob.mx/>.

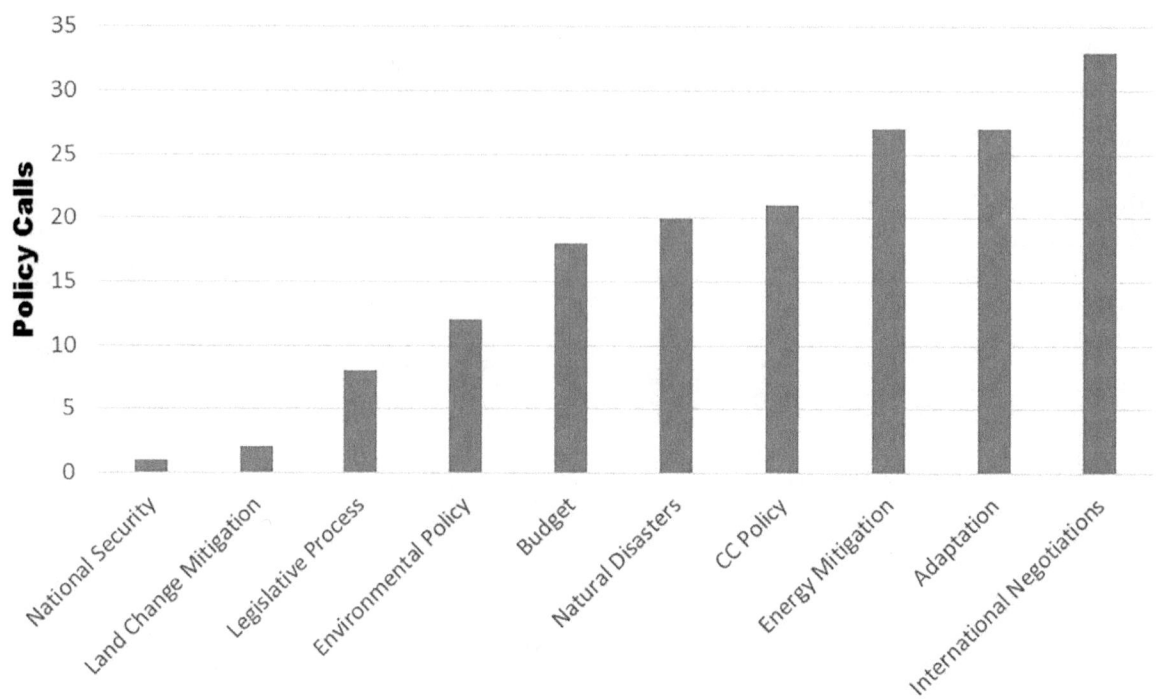

quences of climate change prevails, and this is often translated into pleas for the federal authorities to act once disasters have occurred and to allocate greater financial resources to local authorities.

Regarding the 169 policy calls that have been presented since 2001 on climate change, policies for energy mitigation and adaptation have been given high priority (16 per cent) but have come second as the main concern to the conduct of foreign policy (20 per cent). Other issues that have gained relevance are the definition of the goals and objectives of government policy (13 per cent), political discussion centred in natural disasters (12 per cent), and the demand for budgetary resources to face the effects of the rise in temperature and of natural disasters (11 per cent).

During the last decade there has been growing legislative interest in creating a new institutional framework for dealing with climate change in Mexico. Nonetheless, most bills limit their purpose to including minimal amendments and conceptual changes to current laws. It is from the legislative initiatives presented in 2010, a set of "flagship legislation" proposals or pieces oriented towards creating a new legal framework or, at least, towards modifying substantial layers of pre-existing legislation (Townsend/Fankhauser/

Matthews et al. 2011), that the idea of a climate regime took off.

23.6 The Climate Change Law as a Take-Off Phase

2010 represented a turning point in the legislative deliberation on climate change in Mexico, because a group of legislators presented a set of much more ambitious initiatives, 'flagship legislation' initiatives, that changed the terms of the debate and activated a process of deliberation involving the different parties represented in Congress, NGOs, private sector organizations, specialists and federal government authorities. From the perspective of a transition to sustainability and its division into separate phases (Grin/Rotmans/Schot 2011), the legislative bills and policy presented between 1990 and 2010 can be understood as part of a 'preliminary' phase, but the adoption of the CCL in 2012 and the creation of a climate regime are part of a 'take-off' phase towards some modality of sustainable transition.

While legislators acknowledged the relevance of the ENACC and PECC in their proposals, and almost all the legislative drafts took as a reference the policy instruments developed between 2006 and 2012, it was also evident that: 1) the policy did not have elements to guarantee its continuity in the long term and it was binding only at federal level, and therefore depended on the goodwill and personal commitment of future governments; 2) there were no criteria for coordination with local governments, apart from the methodologies defined by the INE; 3) it was necessary to design new institutions, planning and market instruments, and participation and information devices, to meet the mitigation objectives beyond the established 2012 goal and to advance to a second phase in the adaptation policy;[18] 4) while the PECC clearly defines mitigation goals and fosters capabilities to face adaptation problems, it does not place them in the broader context of a transition to sustainability; 5) even though the PECC established mitigation goals for 2012 and a longer mitigation path was signalled, there were no objectives or strategies to fulfil these in the long term, and it was also difficult to foster the goals without a transition towards a sustainable low carbon economy; 6) finally, the programme consisted of public mitigation efforts with no mandatory commitments on the side of the private sector following the trends indicated by the National Emissions Inventory.

The social organizations and experts involved in the legislative deliberation also questioned the sense and scope of the national policy, something that became increasingly evident in different forums and working groups that were organized between 2010 and 2012 by Congress. For the main environmental organizations, the objectives and goals set for 2012 were insufficient; they were not ambitious enough to tackle Mexican challenges and they were unrelated to a sustainability transition approach.[19]

In the case of the process of creating the Mexican climate change law, the identification of a window of opportunity (the emergence of a novel issue on the national agenda), the acknowledgement of the seriousness of the risks involved (the vulnerability of the nation in the face of the phenomenon's conse-

quences), and the recognition of institutional weakness (the non-existence of a framework legislation), all explain why between 2010 and 2011 a group of legislators presented a set of legislative bills substantively different from all those related to previous projects, and began legislative negotiations oriented towards creating a specific institutional regime for dealing with climate change.[20]

The legislative deliberation on climate change began formally with the bill presented by Senator Cardenas from the *National Action Party* (PAN), and was followed by proposals from other parties, and gradually representatives of the private sector, social organizations, experts and government authorities became involved in the process, influencing the agenda for deliberation and draft documents of the legislative commissions. It should be pointed out that, given the diversity of the actors involved, the creation of a climate regime reflected a kind of bottom-up governance deliberation, where an equilibrium was not automatically imposed.

During the legislative committees' deliberations and in the different public forums organized by Congress, five major problems were identified as key elements of the future law, issues unsolved by the national climate policy: 1) more effective coordination or horizontal interplay between federal agencies in order to fulfil strategic goals and objectives; 2) polycentric verti-

---

18 PECC defines a first phase focused on capacity-building, p 62.

19 PECC advances were evaluated both by the government—see "*Informe de Avances del Programa Especial de Cambio Climático 2009–2012*, at: <http://www.conagua.gob.mx/conagua07/contenido/documentos/pecc12.pdf>; and by social organizations and experts—see "*Evaluación del Programa Especial de Cambio Climático*", at: <http://imco.org.mx/wp-content/uploads/2013/2/studie_2_pecc_web_ok4.pdf>.

20 The most important projects for fostering a climate regime were presented in 2010 and 2011: the proposal to create an "Instituto Nacional de Cambio Climático", from Senator Silvano Aureoles Conejo, PRD, 23 February 2010, see at: <http://www.senado.gob.mx/?ver=sp&mn=2&sm=2&id=24011>; the proposal to create the "Ley General de Cambio Climático", from Senator Alberto Cárdenas Jiménez, PAN, 25 March 2010, see at: <http://www.senado.gob.mx/index.php?ver=sp&mn=2&sm=2&id=24481>; the proposal to create the "Ley General de Adaptación y Mitigación al Cambio Climático", from Deputy Araceli Vázquez Camacho, PRD, 23 November 2010; see at: <http://sil.gobernacion.gob.mx/Busquedas/Basica/ResultadosBusquedaBasica.php?SID=74ee29a3226f14629c8ba8796e90a168&Serial=9f745acc1d1d575f4c238b92c994b3fa&Reg=44&Origen=BB&Paginas=15>; the proposal to create the "Ley General de Sustentabilidad y Cambio Climático", from Senator Ricardo Monreal Ávila of the Labour Party (PT), 8 June 2011; see at: <http://www.senado.gob.mx/index.php?ver=sp&mn=2&sm=2&id=30130>; and the proposal to create the "Ley General de Cambio Climático y Desarrollo Sustentable", from Senator Francisco Labastida, PRI, 11 October 2011; see at: <http://www.senado.gob.mx/index.php?ver=sp&mn=2&sm=2 id=32204>.

Figure 23.4: Mexican climate policy (2006–2012). Source: Information presented in the 4th and 5th National Communications to the UNFCCC (INE: National Institute of Ecology; CICC: Inter-Ministerial Commission on Climate Change).

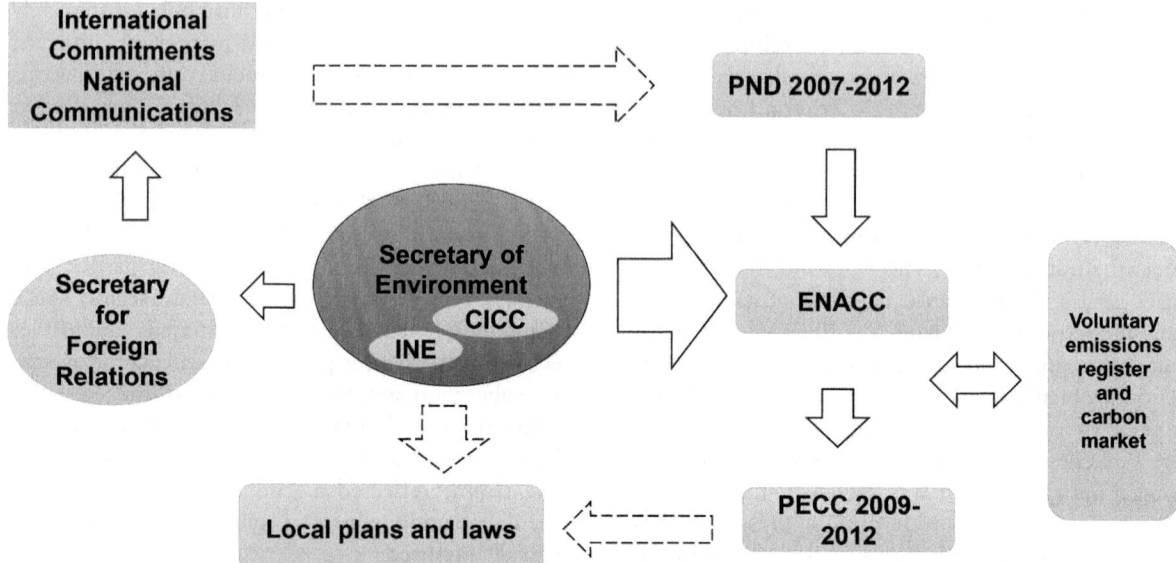

cal interplay between different government levels; 3) the problem of the fit between conflicting sectorial objectives; 4) the continuity and coherence of the climate policy in the long term; and 5) the need to define a sustainable transition agenda and the design of the proper instruments and goals to achieve it.

These deliberations were translated into an argument over three possible alternatives for a new climate-related institutional design to frame national climate policy (Keohane 2014, Aklin/Uperlainen 2013a), two of which necessarily required Congressional approval:

The first option for the Mexican Congress was to support the climate policy in the non-binding terms established between 2006 and 2012 (figure 23.4). The approach was made up of the climate change national policy and its instruments, and the ENACC, PECC and voluntary coordination agreements with local governments, with no changes to the law.[21] The main problem with this option was the impossibility of guaranteeing its continuity beyond Felipe Calderon's administration, because as there was no obligation in law, it depended exclusively on the priorities defined

in the PND by the next government. This perspective represented the preferred alternative for the federal government, including environmental authorities, and for the private sector. In institutional terms it consisted of an intervention in existing public programmes and policies, a "governance framework" approach (Paavola/Gouldson/Kluvánkova-Oravská 2009), or executive action within the existing legal framework, and did not legally require the design of further instruments or mechanisms to foster a transition towards sustainability.

The second alternative was to approve amendments only to the environment law (*Ley General del Equilibrio Ecológico y La Protección al Ambiente*, LEGEEPA) and include specific climate change mandates, and consequently strengthen its capability to regulate mitigation and adaptation programmes. In this case, the model sought to keep climate change definitions as an issue of a fundamentally environmental nature (figure 23.5). This perspective had also been developed in previous law project; in fact, a total of seventeen bills presented prior to 2010 had as their main proposals the modification of this particular law as the criterion for articulating public action to combat climate change. Nevertheless, during the negotiations, this variant was shaped specifically from projects presented by Green Party (*Partido Verde Ecologista de México*, PVEM) legislators and was understood as a more pragmatic approach, mainly because it took into account the resistance that the

---

21 PECC's objective 4.2.2 makes a review of the climate policy mandatory (p. 76), something that was not carried out by the government. Even though Alberto Cardenas was a member of the ruling party, his proposal must be understood as an independent legislative effort.

Figure 23.5: Proposed reform for climate regulation in environmental law. **Source:** Information from the legislative project to reform the environmental law (LEEGEPA) approved by Congress (25 April 2012); CICC: Inter-Ministerial Commission on Climate Change).

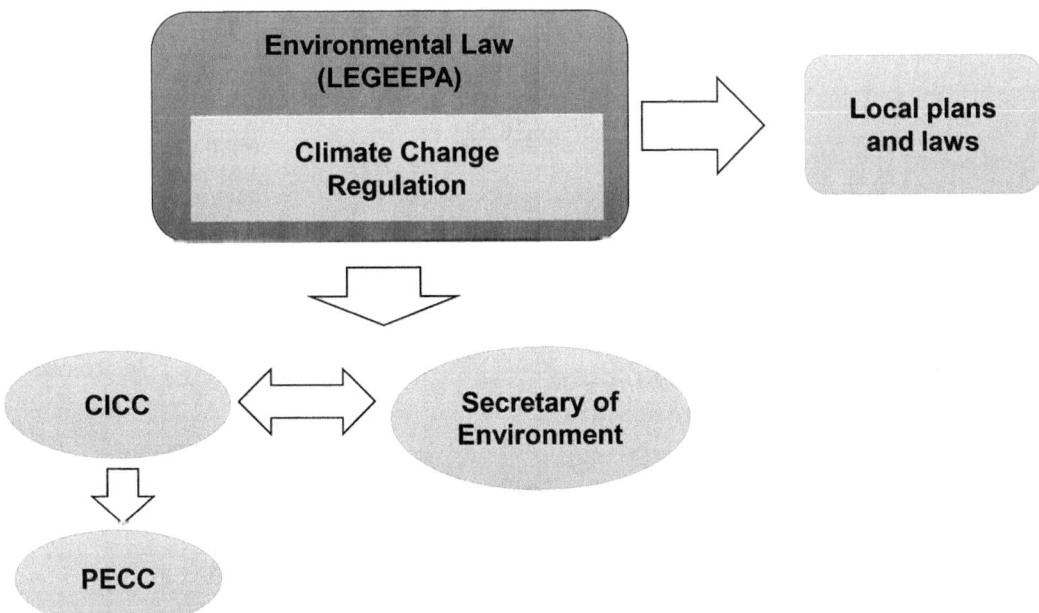

creation of a climate regime generated at different stages of the legislative process (especially from the private sector, government authorities and some legislators).

This option, which was approved in both chambers but was returned modified to Chamber of Deputies by the Senate, was dismissed only once the CCL was approved in April 2012. For some legislators, this reform was capable of balancing the agenda of the public administration, the interests of the private sector, and some of the demands of NGOs. As this reform includes minimal modifications to the environmental law, it represented a legal foundation for the climate policy implemented during the Calderon administration and a guarantee of its continuity.

Finally, Congress discussed the creation of the CCL as a more ambitious approach that included principles, institutions, coordination mechanisms between government levels and specific policy instruments, following the criteria contained in the flagship initiatives (figure 23.6). This reform involved, first, the establishment of a proper climate regime, capable of transversally integrating different realms of the policy arenas and defining criteria for horizontal and vertical interplay between the different government levels; and, second, an understanding of the legislative process as the way to foster some kind of transition towards sustainability through regime change. This model had extensive support from NGOs and experts, but faced

opposition from the private sector and some agencies of the federal government at different stages of the legislative process.

These three models coexisted as possible scenarios during the legislative process between March 2010 and April 2012. None of them imposed itself automatically nor inevitably as the best response for institutional change. The continuation of the same policy was a possibility by default in case no agreement was reached between the political forces represented in Congress, which did not necessarily have priority in legislating on climate change. The reform of the environment law stood as a viable legislative alternative until April 2012, when the climate change law gained support in the Chamber of Deputies, and the framing of the climate policy in the regime model became irreversible.

The approved CCL reflects an equilibrium derived from the negotiations between political, social and economic actors who participated in the legislative process (Greif/Kingston 2001). It did not necessarily represent the most efficient or effective response to the threat of global warming in Mexico, but as a legislative equilibrium it reflects the relative strength of the political actors in using their right of veto (Tsebelis 1992), the response of political and social actors to the unequal distributive effects of institutional change (Knight 1992), and the discourses reflecting different senses of what was understood as an appropriate kind

Figure 23.6: Climate change law structure. **Source:** Ley General de Cambio Climático (NSCC: National Strategy on Climate Change; INECC: National Ecology and Climate Change Institute; CICC. Inter-Ministerial Commission on Climate Change; C3: Advisory Council on Climate Change).

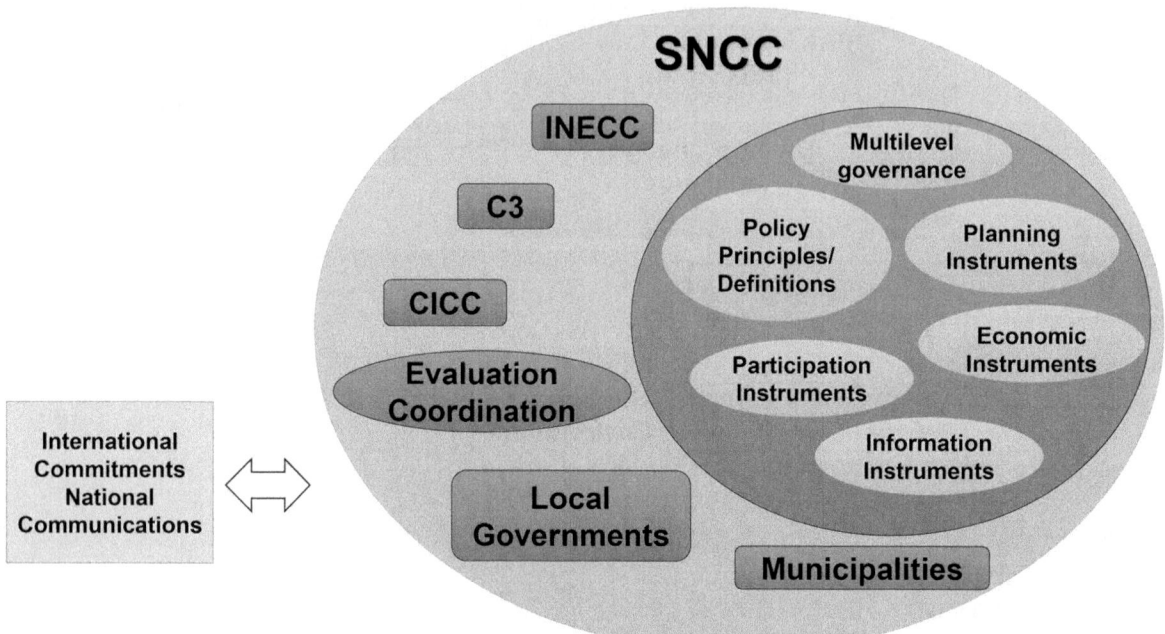

of regulation (March/Olsen 1989). In this sense, while exogenous factors account for the need for institutional change, the framing of a climate regime is explained as a process endogenous to legislative deliberation and its negotiation dynamics.

Why was the resulting equilibrium a climate regime, when the government would have preferred the minimum modifications to legislation and the private sector wanted a voluntary action model and not binding obligations supported by law? First, the core group of legislators involved in the legislative process always kept the regime option as the dominant alternative and gradually imposed this model on other legislators who were superficially involved in the discussions. Second, by presenting their own project, the legislators of the *Partido Revolucionario Institucional* (PRI, Institutional Revolutionary Party), the main veto power player, became actively involved in the process. Third, while some representative elements of the private sector, such as steel and carbon producers and manufacturing industry, tried to stop the passing of the CCL, and government authorities kept on presenting critical observations to the project even a few weeks before the final vote in 2012, in the end they were able to insert changes that contributed towards their own agenda, especially with respect to the definition of market instruments such as green taxes and the creation of a trade and cap system. Finally, the

NGOs, regardless of the particularity of their own agendas, operated as a real counterweight during legislative deliberation, and they contributed to limiting the effects of rejection by other actors. In this sense, the creation of the CCL was truly a bottom-up legislative process. The central elements and differences between the legislative models are summarized in table 23.2.

23.7   Conclusion: Implementation Challenges and the Risk of Failure of the Sustainable Transition

The concepts of climate regime, regime complex, interplay and fit, derived from an environmental governance approach, have the virtue of allowing us to understand the creation of the CCL in Mexico as a different process from the simple approval of an isolated piece of law. Environmental governance concepts are useful for structuring an analysis that integrates the deliberate design of a regime with the conflictive negotiation between actors who promote different institutional framing.

The construction of a climate regime was imposed as an equilibrium between three clearly differentiated models to regulate climate change at a national level: 1) the continuation of goals and objectives defined in

Table 23.2: Alternative models in the legislative discussion on climate change 2010–2012: **Source**: Information from the National Communications and Legislative Information System (SIL); at: <http://sil.gobernacion.gob.mx/portal>.

Elements	Climate Governance Framework based on CC Policy	Changes to the Environmental Law (LEGEEPA)	Climate Change Law as a Climate Regime
Type	Public Policy Programme	Environmental Law amendments	Creation of a specific climate regime
Legal Fundament	Constitution, Planning Law, Federal Administration Law, Environmental Law, Environmental Law regulations	Reforms articles 5, 7, 19, 23 of the Environmental Law and creates a Chapter II Bis: "On Climate Change".	New law with 116 articles It is defined as a guarantee of the constitutional right to a healthy environment
Features	PND, ENACC, PECC, CICC, international activism	Specific mitigation and adaptation mandates in the Environmental Law	SNCC, governing principles, policy definitions, multilevel governance, planning, information, participation and economic instruments Transparency rules and sanctions
Key Feature	Special programme with objectives and goals for 2012	Specific chapter on Climate Change	National System on Climate Change
Institutions	CICC created in 2005	Formalizes CICC in the Environmental Law	SNCC, formalizes CICC, reforms INECC
Policy continuity	Non-binding at the federal level beyond 2012	Mandate to create and update the PECC	Force to create and update ENACC, PECC and local plans
Multilevel governance	CICC formally coordinates the federal government's climate actions Non-mandatory coordination with local governments for the design of local climate plans	CICC established in Environmental Law Defines federal obligations and limits local responsibilities to formulation, implementation and evaluation of local programmes	Distributes competences between the levels of government in a national system (SNCC) CICC becomes the government coordination device inside SNCC
Bottom-up Governance	ENACC and PECC subject to public consultation	None	Particular chapter and social participation and creation of advisory councils with social representation
Planning Instruments	ENACC and PECC binding at federal level Voluntary local plans and emissions registration	Mandate to create and update the PECC	ENACC, PECC, state and municipal plans, mandatory inventories, emissions registration and risk atlas; integrates information system, evaluation and transparency mechanisms and sanctions
Market instruments	PECC budget	Doesn't create new instruments besides those included in the Environmental Law	Specific chapter that includes instruments from the Environmental Law Gives authority to establish financial mechanisms and a carbon market National Climate Fund
Mitigation goals	PECC defines a short-term 50.7 MtCO2e mitigation goal by 2012; a 30% reduction by 2020 and 50% by 2050 National commitment to COP15	Mandate to fulfil PECC goals	National aspirational mitigation goals defined in 2nd temporary article; 35% renewable electricity generation in 2024

the national climate change policy without reforms to the law; 2) minimal modifications to the Environmental Law to include mitigation and adaptation objectives; and 3) the creation of a specific climate regime for solving problems of national horizontal interplay (coordination between federal departments), national vertical interplay (concurrency between government levels), fit between sectorial objectives, continuity of the climate change policy in the long term, and the impetus towards a sustainability transition.

We argued that the policy definitions, Mexico's participation in international negotiations, and the set of initiatives and policy calls presented between 1990 and 2010 can all be identified as part of the 'preliminary' phase of a sustainability transition, while the approval of the CCL in 2012 constitutes the 'take-off' phase.

Unlike the other legislative options, the creation of the CCL represented the implicit intention of promoting some kind of sustainability transition that would then enter properly into a stage characterized by the need to implement its mandates and during which the feasibility of a sustainable transition would be genuinely at stake.

Since its approval in April 2012, the mandates of the Climate Change Law have begun to be implemented by federal and local authorities. In early 2013 the federal government installed the CICC, in accordance with its new legal structure. Later, the ENACC 10-20-40 and the PECC 2014-2018 were published.[22] According to information from the *National Ecology and Climate Change Institute* (INECC), local governments are starting to develop their local climate action plans, but in most cases they are still at a very early stage of development, and overall progress is limited compared with the results reported in 2012.[23]

Even though strategy lines, mitigation and adaptation goals and objectives are defined in these new instruments, thus guaranteeing the continuity of climate change policies, those obligations that have been implemented stray from the idea of a climate regime developed by the legislators. First, the design of the ENACC and the PECC for the period 2013-2018 does not include a detailed or at least a general

evaluation of what had been achieved by 2012. In fact, there is no mention whatsoever of the previous strategy and programme. This represents an important clash with the objectives of the original legislators, who understood the continuity of environment policy as the gradual and incremental development of mitigation and adaptation objectives and goals, not as a simple six-yearly reinvention of the programmes.

Secondly, and related to the previous point, the definition of climate policy instruments in itself does not mean more effective mitigation and adaptation actions. On the contrary, the new programmes show some signs of regression. According to the PECC 2014-2018, and in spite of the mitigation goals set by 2012, "Should this trending scenario continue it is calculated that in 2020 the national GHG emissions would be a billion tons, that is to say twenty-eight per cent more than those in 2010. It is important to point out that the recent energy reform will involve an increase in the hydrocarbon production activities and it is likely that this sector's GHG emissions will increase as well".[24] In other words, the GHG reduction goals are transformed into a twenty-eight per cent increase in emissions as a consequence of the energy reforms fostered by the Peña Nieto administration, which focus on boosting hydrocarbon exploitation, especially deep-water reserves and non-conventional natural gas, as part of the government's definition of 'clean energy'.[25]

The mitigation goal is also conditional on international financing. According to the PECC 2014-2018,

> For the challenge that the mitigation of emissions of compounds and GHG poses, Mexico has made the commitment of reducing by 30 per cent in relation to the 2020 baseline as well as by 50 per cent by 2050, compared to 2000 emissions, through indicative objectives and aspirational goals as contained in the LGCC (Climate Change Law). This is why this Programme, as well as including the action lines related to the Federal Expenditure Budget, includes others, marked with an asterisk (*), that are subject to obtaining national or

---

22 For ENACC 2013, see at: <http://dof.gob.mx/nota_detalle.php?codigo=5342492&fecha=28/04/2014> and for PECC 2014-2018, see at: <http://www4.unfccc.int/submissions/indc/Submission%20Pages/submissions.aspx http://www.dof.gob.mx/nota_detalle.php?codigo=5301093&fecha=03/06/2013>.

23 Advances in local climate action plans can be consulted at: <http://www2.inecc.gob.mx/sistemas/peacc/>.

24 PECC (2014-2018), 2014. This reference can be found in part 1.3, "Las emisiones de compuestos y gases de efecto invernadero en México (Greenhouse Gases and Compounds in Mexico)". The official electronic document lacks page references.

25 The official definition of 'clean energy' goals in the Electric Industry Law, includes renewables sources and those defined by the authorities according to official emission norms and parameters, at: <http://www.cfe.gob.mx/ConoceCFE/1_AcercadeCFE/MarcoLegalyNormativo/Lists/Leyes1/Attachments/26/Leydelaindustriaelectrica11ago.pdf>

international financial and technological support, both private and public.[26]

On 30 March 2105, Mexico submitted to the UNFCCC its Intended National Determined Contribution with the aim of achieving an international legally binding agreement in order to keep the increase in the global average atmospheric temperature below 2°C. Mexico assumes conditional and non-conditional reduction goals related to its emissions baseline:

Unconditional Reduction

Mexico is committed to reduce unconditionally 25 per cent of its greenhouse gases and short lived climate pollutants emissions (below BAU) for the year 2030. This commitment implies a reduction of 22 per cent of GHG and a reduction of 51 per cent of Black Carbon.This commitment implies a net emissions peak starting from 2026, decoupling GHG emissions from economic growth: emissions intensity per unit of GDP will reduce by around 40% from 2013 to 2030.

Conditional Reduction

The 25% reduction commitment expressed above could increase up to a 40% in a conditional manner, subject to a global agreement addressing important topics including international carbon price, carbon border adjustments, technical cooperation, access to low-cost financial resources and technology transfer, all at a scale commensurate to the challenge of global climate change. Within the same conditions, GHG reductions could increase up to 36%, and Black Carbon reductions to 70% in 2030.[27]

Mexico's emissions goals were well received by the international community because it was the first Non-Annexe 1 country to submit its Intended Nationally Determined Contribution. It contains an ambitious GHG emissions target even if it decreases its previous unconditional national commitments.[28] Its main problem remains the contradiction of a very ambitious formal climate agenda and a national energy matrix dependent on fossil fuels and whose emissions are expected to grow by at least twenty-eight per cent due to the energy reform.

This leads us to a central problem in the design of the Mexican climate regime: it does not matter if the modality chosen by the actors was the most complex framing possible to face climate change and foster mitigation goals or a decarbonization of the economy, if the broader objectives of energy production and consumption do not solve the problem of fit with energy policies, urban development plans, conservation objectives and fiscal priorities. The CCL represents a very ambitious design on paper, but it will become irrelevant if it is not able to transform other policy arenas as part of a transition to a sustainability agenda.

The concept of transition towards sustainability has relevance since it allows the sense of the climate regime definition to be framed before the possible alternatives are examined. However, the inconsistency between the objectives of climate policy and those of national economic development warns us that it is not enough to promote a large-scale social transformation through regime change (sustainability-oriented changes), if the contradictory practices of an economic growth model based on hydrocarbon exploitation continue. The implementation phase of a sustainability transition in developing countries means a difficult conflict between different types of regimes, policy arenas and social objectives. Even though a preliminary and a take-off stage take place as part of a transition to sustainability, solving the problem of fit is central to fostering a sustainable model through regime change. This means that sustainable transitions are not only technical problems, but politically contentious issues (Meadowcroft 2009; Akin/Urpelainen 2013b; Keohane 2014; Javeline 2014).

---

26 PECC 2014-2018, 2014: This reference can be found in part 1.3, "Las emisiones de compuestos y gases de efecto invernadero en México (Greenhouse Gases and Compounds in Mexico)". The official electronic document lacks page references.

27 The Mexican Intended Nationally Determined Contribution can be consulted at: <http://www4.unfccc.int/submissions/indc/Submission%20Pages/submissions.aspx>.

28 On December the 10th 2015 the Mexican Congress approved the Energy Transition Law. This law, a complement to a larger energy reform, defines new institutional basis for an energy transition and establishes obligations and goals for emissions mitigation, clean energy generation and energy efficiency. It mandates a participation of clean energy in the electric sector of at least 25% to 2018, 30% to 2021 and 35% to 2024. With its approval, the Congress finally recognizes the importance of fostering clean energy generation as a way to fulfill Intended Nationally Determined Contributions submitted to the UNFCCC and as a path to a low carbon economy. As an instrument for a transition to sustainability, the Energy Transition Law can become a game changer for the energy sector in the next years, but its success largely depends on its implementation process and the willingness of national authorities and the private sector to fulfill its mandates.

References

Adger, Neil; Andrew Jordan, 2009: *Governing Sustainability* (New York: Cambridge University Press).

Aklin, Michael; Urpelainen, Johannes, 2013a: "Debating Clean Energy: Frames, counter frames, and audiences", in: *Global Environmental Change*, 23,5 (October):1225-1232.

Aklin, Michael; Urpelainen, Johannes, 2013b: "Political Competition, Path Dependence, and the Strategy of Sustainable Energy Transitions", in: *American Journal of Political Science*, 57,3: 643-658.

Barret, Scott, 2007: *Why Cooperate? The Incentive to Supply Global Public Goods.* (Oxford: Oxford University Press).

Bechtel, Michael M.; Scheve, Kenneth F., 2013: "Mass support for global climate agreements depends on institutional design", in: *Proceedings of the National Academy of Sciences (PNAS)*, 110,34: 13763-13768.

Biermann, Frank; Betsill, Michel M.; Gupta, Joyeeta; Kanie, Norichika; Lebel, Louis; Liverman, Diana; Schroeder, Heike; Siebenhüner, Bernd; Zondervan, Ruben, 2010: "Earth system governance: a research framework", in: *International Environmental Agreements*, 10: 277-298.

Biermann, Frank; Pattberg, Phillipp (Eds.), 2012: *Global Environmental Governance Reconsidered* (Cambridge, Mass: MIT Press).

Delmas, A. Magali; Young, Oran R. (Eds.), 2009: *Governance for the Environment* (New York: Cambridge University Press).

Dryzek, John S.; Norgaard, Richard B.; Schlosberg, David, 2013: *Climate-Challenged Society* (Oxford: Oxford Univertity Press).

Evans, James P., 2012: *Environmental Governance* (New York: Routledge).

Folke, Carl; Pritchard, Lowell Jr.; Berkes, Fikret; Colding, Johan; Svedin Uno, 2007: "The Problem of Fit Between Ecosystems and Institutions: Ten Years Later", in: *Ecology and Society*, 12,1: 30.

Gardiner, Stephen M., 2011: *A Perfect Moral Storm* (Oxford: Oxford University Press).

Geels, Frank W., 2005: "Processes and patterns in transitions and systems innovations: Refining the co-evolutionary multi-level perspective", in: *Technical Forecasting & Social Change*, 72: 681-696.

Geels, Frank W.; Schot, Johan, 2007: "Typology of sociotechnical transition pathways", in: *Research Policy*, 36: 399-417.

Greif, Avner; Christopher Kingston, 2011: "Institutions: Rules or Equilibria?", in: Schofield, Norman; Christopher Kingston (Eds.) 2011: *Political Economy of Institutions, Democracy and Voting* (New York: Springer): 13-53.

Grin, John; Rotmans, Jan; Schot, Johan, 2011: *Transitions to Sustainable Development: New Directions in the Study of Long Term Transformative Change* (New York: Routledge).

Haggard, Stephan; Simmons, Beth A., 1987: "Theories of international regimes", in: *International Organization* 41,3: 491-517.

Helm, Dieter; Hepburn, Cameron, 2009: *The Economics and politics of Climate Change* (Oxford: Oxford University Press).

International Energy Agency (IEA), 2014: "Key World Energy Statistics", at: <http://www.iea.org/publications/freepublications/publication/KeyWorld2014.pdf>.

Javelin, Debra, 2014: "The Most Important Topic Political Scientists are not Studying: Adaptation to Climate Change", in: *Perspectives on Politics*, 12: 420-434.

Kemp, Rene; Parto, Saed; Gibson, Robert, 2005: "Governance for sustainable development: moving from theory to practice", in: *International Journal of Sustainable Development*, 8,1/2: 13-30

Keohane, Robert O, 1982; "The Demand for International Regimes", in: *International Organizations*, 36,2, on: *International Regimes* (Spring 1982): 325-355.

Keohane, Robert O.; David Victor, G., 2010: "The Regimen Complex for Climate Change", in: *The Harvard Project on International Climate Agreements*, Discussion Paper 10-33.

Keohane, Robert O., 2014: "The Global Politics of Climate Change: Challenge for Political Sciense. The 2014 James Madison Lecture", in: *PS: Political Science & Politics*, 48,1 (January 2015): 19-26.

Kingston, Chistopher; Caballero, Gonzalo, 2009: "Comparing Theories of Institutional Change", in: *Journal of Institutional Economics*, 5,2: 151-180.

Krasner, Stephen 1983: "Structural Causes and Regime Consequences: Regimes as Intervening Variables", in: Krasner, Stephen: *International Regimes* (Cambridge: Cornell University Press): 1-21.

Le Clercq, Juan Antonio, 2011: "Cambio climático: políticas nacionales y bases institu-cionales", in: *Diálogo Político*, 3/2011 (Buenos Aires: Konrad-Adenauer-Stiftung): 97-115.

March, James G.; Olsen, Johan P., 1989: *Rediscovering Institutions. The Organizational Basis of Politics* (New York: The Free Press).

Meadowcroft, James, 2009: "What about the politics? Sustainable Development, transition management, and long term energy transitions", in: *Policy Science*, 42: 323-340.

OECD, 2010: *Cities and Climate Change*: (Paris: OECD).

Ostrom, Elinor, 2009: "A Polycentric Approach for Coping with Climate Change", Policy Research Working Paper 5095.

Paavola, Jouni, 2007: "Institutions and environmental governance: A reconceptualization", in: *Ecological Economics*, 63: 93-103.

Paavola, Jouni, 2008: *Explaining Multi-Level Environmental Governance* (University of Leeds: Sustainability Research Institute).

Paavola, Jouni; Gouldson, Andrew; Kluvánkova-Oravská, Tatiana, 2009: "Interplay of Actors, Scales, Frameworks and Regimes in the Governance of Biodiversity", in: *Environmental policy and Governance*, 19: 148-158.

Parris, Thomas M.: Kates, Robert W., 2003: "Characterizing a sustainability transition: Goals, targets, trends, and driving forces", in: *Proceedings of The National Academy of Sciences* (PNAS), 100,14 (8 July): 8068-8073.

Townshend, Terry; Fankhauser, Sam; Matthews, Adam; Féger, Clement; Liu, Jin; Narciso, Thais, 2011: *The Globe Climate Change Legislation Study* (London: Globe International, Grantham Research Institute on Climate Change and the Environment).

Townshend, Terry; Fankhauser, Sam; Aybar, Rafael; Collins, Murray; Landesman, Tucker; Nachmany, Michal; Pavese, Carolina, 2013: *The Globe Climate Change Legislation Study* (London: London School of Economics, Globe International, Grantham Research Institute on Climate Change and the Environment, World Summit of Legislators).

Tsebelis, George, 1992: *Veto Players* (New York: Russell Sage Foundation).

The World Bank, 2010: *Development and Climate Change. World Development Report* 2010 (Washington D.C.: The World Bank).

Young, Oran R., 1986: "International Regimes: Toward a New Theory of Institutions", in: *World Politics*, 39,1 (October): 104-122.

Young, Oran R. 2002: *The Institutional Dimensions of Environmental Change* (Cambridge, Mass: MIT Press).

Young, Oran R.; King, Leslie A.; Schroeder, Heike 2008: *Institutions and Environmental Change* (Cambridge, Mass: MIT Press).

Young, Oran R., 2011: "Effectiveness of international environmental regimes: Existing knowledge, cutting-edge themes, and research strategies", in: *Proceedings of the National Academy of Sciences (PNAS)*, 108,50: 19853-19860.

Young, Oran R., 2013: *On Environmental Governance* (London: Paradigm Publishers).

Zeijl-Rozema, Annemarie Van; Cövers, Ron; Kemp, René; Martens, Pim, 2008: "Governance for Sustainable Development: A framework", in: *Sustainable Development*, 16: 410-421.

24   Sustainability Transitions: A Discourse-institutional Perspective

Audley Genus[1]

Abstract

This chapter addresses the complex web of activities and actors necessary to achieve the much vaunted yet elusive transition to sustainability. The chapter reviews diverse contributions which have in common a concern for the role that language and institutional arrangements play in related developments. These, it is argued, have a capacity that has not been fully realized to improve our understanding of the problems involved, the issues at stake and implications for policy and practice. The chapter presents a framework which is employed to provide orientation for a discussion of these contributions which may otherwise be considered in an isolated manner or with limitations on cross-disciplinary conversation. At the heart of the discussion are sustainability transitions, neo-institutional theory, and critical discourse analysis. The chapter draws on neo-institutional theory and critical discourse analysis to highlight the role of different types of factors and actors implicated in (un)sustainable patterns of production and consumption and the (in)effective governance of environmental sustainability related science, technology and other phenomena. It considers the potential insights to be gained from application of a discourse-institutional perspective and progress that has been made with the development of such an approach.[2]

Keywords: Neo-institutional theory; sustainable production; sustainable consumption; rules; innovation; critical discourse analysis; public engagement; legitimacy.

24.1   Introduction

There remains a great deal of consternation about the damaging effects of anthropogenic climate change, how such impacts may be mitigated and the policies and strategies that might be implemented to facilitate adaptation thereto. In the UK, to take one example, there appears to be a great deal of anxiety regarding whether and how related commitments are going to be achieved but also a lingering controversy over the 'facts' of climate change and the (over-) ambitious nature of targets set.[3] The framing of the debate tends to revolve around the discourse of competitiveness and the rhetoric of 'keeping the lights on', much to the chagrin of 'deep greens'.

Academic research has investigated some of the fundamental issues in a manner which highlights the potential contribution of a range of disciplines and sub-disciplines, sometimes in combination but often in disciplinary isolation. The chapter seeks to build on these contributions, noting that to gain a richer, more comprehensive insight into the issues at stake and implications for policy and practice it is necessary to recognize and to learn from a wide range of findings. The chapter thus reflects on research on: the processes and mechanisms which embed or reproduce pro-environmental or unsustainable values and actions; the role of language in substantiating the status quo and in 'carrying' institutional change; the role of firms, citizens/consumers and others (e.g. academic researchers) as agents of institutional change or inertia, or as catalytic intermediaries. The chapter is broad in scope, taking in where and how (un)sustainability is talked and written about or symbolized, the

---

1   Prof. Audley Genus, Small Business Research Centre, Kingston University, Kingston Hill, Kingston upon Thames KT2 7LB; Email: <a.genus@kingston.ac.uk>.

2   This chapter is a revised version of a paper published in: *Sustainability*, 6,1 (2014): 283-305.

3   In January 2014 proposals were announced by the European Commission to abandon national targets for renewable energy generation; see at: <http://ec. europa.eu/ energy/doc/2030/com_2014_15_en.pdf> (24 January, 2014): 6.

---

creation, disappearance and re-emergence of institutions, and the role of universities and public engagement in the governance of science, innovation, technology and environmental sustainability.

At the heart of the discussion are neo-institutional theory and critical discourse analysis, focusing on attempts to develop a combined discourse-institutional approach, which might be applied to generate insights into the realization of the transition to sustainability. This is a topic of great contemporary interest, since 'discourse institutionalism' is the 'newest' of the 'new institutionalisms' (Schmidt 2010: 303). The chapter proceeds first by reviewing salient contributions from each of these areas, interweaving what may be gained conceptually from these contributions with a discussion of the difficulties experienced in relation to policy-making, technological innovation and use, and public discourse of environmental sustainability. The discussion is organized into the following sections on: (a) the nature of institutions and the relevance of institutional analysis for sustainability transitions; (b) the nature and relevance of critical discourse analysis to institutional inertia and innovation; and (c) a review of research connected with sustainability transitions which to varying degrees adopt a 'discourse-institutional' approach. Finally, the concluding section reflects on what has been gained from the work reviewed for improving our thinking and practices regarding the facilitation of a more environmentally sustainable society.

24.2   Neo-institutional Theory

This section of the chapter provides an overview of work in the broad topic area of institutional theory which bears on our understanding of the transition to sustainability, particularly that research associated with the 'new institutionalism' or neo-institutional theory. This is concerned with developments in institutional economics, institutional sociology, and institutional theory in political science, which point the way toward adopting a deeper understanding of the role of institutions—and to some extent language—in the transformations necessary to mitigate or adapt to climate change. In economics the 'new' institutionalism refers to contributions from transactions costs theory (Coase 1937; North 1990; Williamson 1985), information economics (Simon 1972), and industrial organization (Katz/Shapiro 1985), for example. An institutional economist (Ostrom/Gardner/Walker 1994), developed an IAD (institutional analysis and develop-

ment) framework, which made it clear that all (institutional) rules—defined as 'shared understandings' regarding what actions are 'desirable, prohibited or permitted'—are formulated in language. It is interesting to note that in political science March and Olsen (1984, 1989) pointed to but did not analyse the importance of language in institutions. A significant contribution has come from Rydin (1999, 2003), whose research on environmental planning has led to an 'institutional discourse' approach which builds on this earlier work. Developments connected with new institutional economics which have informed subsequent work on systems innovation and thence sustainability transition is the work of Nelson and Winter (1982) on organizational routines and technological regimes.

In organizational sociology the institutionalization of neo-institutionalist approaches was assisted by certain landmark publications. These include articles by Meyer and Rowan (1977) and Zucker (1977), the book co-edited by Powell and DiMaggio (1991), and Scott's book *Institutions and Organizations* (Scott 1995, 2008a). These drew on the earlier work of Berger and Luckmann (1967), Hughes (1936), Selznick (1957) and Parsons (1960). The foundational contributions emphasize the properties of institutions, and how institutional environments imposed 'requirements and/or constraints' on rather passive organizations, specifically in relation to the formal structure of individual organizations (Scott 2008b: 429). They observe that institutionalization is partly a matter of instilling value in an organization, which subscribes to beliefs which it considers to be legitimate. Later, researchers focused on institutional change (Dacin/Goodstein/Scott 2002) in fields of interacting organizations, in which there could be competing values, beliefs and regulative and normative rules (Scott 1987). Over time the kinds of organizations and fields studied has widened to encompass business firms in addition to the public sector or professional bodies which were the focus of much of the earlier work (such as art museums and trade unions). Delbridge and Edwards (2007) conclude that early neo-institutionalist research: (a) initially assumed the 'continuity and stability' of social systems and the conformity of organizational field-level practices, then later (b) unduly emphasized processes of institutional change, whilst failing to 'engage meaningfully' with the importance of 'action in maintaining institutional conditions'. Some researchers have sought to avoid these pitfalls by addressing the coexistence of institutionalization processes and processes of de-institutionalization, thus attending both to stability, continuity and sharing of practices within a field and to 'atrophy or change' (Dacin/Dacin 2013; Greenwood/Suddaby/ Hinings 2002; Mohr/White 2008). There is a concern to introduce and extend a discursive approach to institutionalization (Phillips/Malhotra 2008).

A branch of institutional theory involves work on institutional entrepreneurship on the innovation of new rules and practices, and in the past emphasized "the actions of a single or small number of actors" (Lounsbury/Crumley 2007: 993). Over time the problem of realizing successful institutional entrepreneurship has turned its attention to the depth of prevailing institutional logics, the power of existing governance mechanisms to structure the behaviour of incumbents, and the prevalence of structuration processes linked to the interaction of diverse actors, which constrain non-isomorphic change (i.e. that which does not merely copy what other organizations are doing). Of relevance here are the role of transnational organizations in institution-building (e.g. supra-national treaties and legally enforceable commitments on climate change) and the empowering (not merely constraining) aspects of prevailing norms and beliefs (see Garud/ Hardy/Maguire 2007).

Scott (2008a: 48) defines 'institutions' as "comprised of regulative, normative and cultural-cognitive elements that, together with associated activities and resources, provide stability and meaning to social life". He also identifies three types of institutional rules, which guide actions in a *regularized* way within organizational 'fields'. Organizational fields are comprised of the actors which supply and consume a given set of goods and services but also those which exert influence over these, such that 'field' includes users, consumer groups, government bodies and professional associations, and others. 'Field' is thus a broader notion that 'industry' (DiMaggio/Powell 1991) and is a common object of investigation for researchers seeking to understand the nature of institutions and institutionalization.

The three types of institutional rules are briefly described as follows. First, there are *formal/regulative* rules. Typically, manifestations of these rules appear as legislation and regulations, state grants and subsidies, and technical standards. These partly structure innovative and entrepreneurial activities connected with the governance, design, production, and use of renewable energy or other sustainable technologies and the attractiveness and growth of these sectors. Their effectiveness, though, is in part due to the interpretations that actors have of the credibility and legitimacy of incentives and sanctions associated with adherence to the rules in question. Next, *normative*

rules may be identified. These relate to the norms, responsibilities and obligations, which need to be adhered to or performed. They involve norms regarding the proper role and conduct of industry, government, communities, and consumers/users, for example, in relation to prevailing ways of generating or using electricity or heat, for example, and to the promotion or inhibition of novel approaches and practices for doing so. This author argues that this should include the institutionalization of new business models and the de-institutionalization of existing ones (Genus, 2012). Lastly, *cultural-cognitive* rules are implicated in the values, beliefs and guiding assumptions, which are held regarding the effectiveness of policies for promoting sustainable technologies and practices. They also refer to beliefs about the benefits of related technology, such as those which negatively view wind turbines as dangerous to wildlife or ugly, compared with the view which sees them as a vital part of our energy future. The guiding principles may be implicated in the heuristics which routinely are employed to evaluate investments in sustainable technologies in practice.

Institutional rules are transmitted by different types of carriers. Four types of carriers have been identified (Jepperson 1991; Scott 2003, 2008a): (a) symbolic systems including rules, laws, values, expectations, categories, typifications and schema; (b) relational systems including governance systems, power systems, regimes, authority systems, structural isomorphism and identities; (c) routines including protocols, standard operating procedures, jobs, roles, obedience to duty and scripts; and (d) artefacts including objects complying with mandated specifications, objects meeting conventions, standards and objects possessing symbolic value.

One may also identify three processes of institutionalization: habitualization, objectification and sedimentation; and three mechanisms by which institutionalization occurs: coercion, normative commitment and imitation (Scott 2008a). Habitualization refers to the process, pre-institutionalization, by which new ideas and practices assume a taken-for-granted quality in the thoughts and actions of actors in a limited domain and begin to be recognized as such by others outside it. Objectification concerns the process by which the nascent institutional rules are articulated; it is the means by which institutions begin to solidify, a development which is necessary for sedimentation to occur. Sedimentation is the process through which rules seep into society, and is a matter of deep and extensive diffusion. The chapter will in section 24.4 consider the employment of these institutional elements and processes in studies of sustainability transition which are relevant to a discourse-institutional approach. First, however, the following section provides some necessary background on discourse analysis, specifically on approaches to critical discourse analysis, to give orientation to subsequent discussions in the chapter.

24.3   Critical Discourse Analysis

Inspired by the work of Michel Foucault (1972) discourse analysis is an approach—or, rather, a collection of approaches—that emphasizes relations between language and power. *Critical discourse analysis* (CDA) is a branch of discourse analysis which provides methods—some would say a methodology—for conducting social research based on the analysis of language to expose sources of influence, persuasion and dominance in social life. The political programme of CDA brings into focus those individuals, groups or organizations excluded or marginalized from—in this case—the governance and achievement of environmental sustainability, and practices implicated in the making of or adherence to institutional rules. As a leading proponent of CDA states: "it often makes sense to use discourse analysis in conjunction with other forms of analysis, for instance ethnography or forms of institutional analysis" (Fairclough 2003: 2), to address the effects of the social construction of discourse in terms of the generation and maintenance of institutions connected with the governance of environmental sustainability.

Fairclough's version of CDA explicitly aims to expose social inequalities and to account for social change. Such a critical approach to discourse analysis, in addressing what it is about political systems (such as 'new capitalism') that constrains or facilitates our ability to develop our potential as members of society, attends to a social, rather than an individualized notion of speaking and acting. It also emphasizes the need for an honest recognition and reflection on values, including those of researchers. CDA examines connections among three levels of analysis: texts (which may be written, spoken, or symbolic), discursive practices, and social practice, the wider social 'context' that bears upon texts and discursive practices. A critical approach to discourse analysis seeks to link text with underlying power structures in society and the discursive practices upon which the text was drawn. CDA focuses on how social relations, identity, knowledge, and power are constructed through discourse. Social

structures (or 'social practice') are abstract, highly stable entities which define a set of possibilities. Social events occur within these set of possibilities. However, there is no simple causal relation between social events and structures. Their relationship is mediated by a network of discursive practices. Discursive practices are more or less durable forms of social activity, which are articulated together to constitute social fields, institutions and organizations, e.g. the field of technology assessment. Discourse analysts refer to 'fields' as networks of social practice (Chouliaraki/Fairclough 2010; c.f. Bourdieu 1983).

Discourses 'cannot be studied directly'; one needs to scrutinize the "texts that constitute them" (Phillips/Lawrence/Hardy 2004: 636). Such analysis focuses upon the processes of text production, distribution and consumption (Fairclough 2001) and upon bodies of texts, since it is the way in which individual texts draw on other texts and discourses that creates meanings which are significant for social relations and institutionalization (Phillips/Lawrence/Hardy 2004). Further, institutionalization involves more than imitation of actions taken locally. Such actions are shared and given meaning through texts about them which diffuse through an organizational field, drawing the attention of researchers to the search for actions which generate texts and for texts which might affect action and institutionalization (Phillips/Lawrence/Hardy 2004). Meanwhile an 'order of discourse' is a network of discursive practices in its language aspects. The elements of orders of discourse are genres, discourses and styles. A 'genre' is a way of acting in social life, examples including interviews, speeches or meetings, whereas the term 'style' refers to ways of being, which connote the identities of protagonists and those they apply to others. Both of these may be distinguished from 'representation', which refers to different perspectives applied to the comprehension of reality. Discourses do not stand alone but may be interconnected. Hence one speaks of intertextuality or interdiscursivity—which are phenomena through which one text or discourse draws upon, combines or influences others. These elements select certain possibilities within the set of possibilities provided by the language, and exclude others. Fairclough (2001) distinguishes between the discursive and the non-discursive, arguing that while all language is social, not all social activity is linguistic—he cites economic production as an example. This is at odds with more staunchly post-structuralist views of discourse, such as that advanced by Laclau and Mouffe (1985, 2001), which do not make a distinction between discursive

Table 24.1: Discursive legitimation strategies underpinning institutional pillars (rules). **Source:** The author.

Institutional pillar	Corresponding underpinning discursive legitimation strategy	Mechanism by which discursive legitimation occurs
Formal/regulative rules	Authorization	Recognition and acceptance of authority claims (e.g. of person(s), customs or laws)
Normative rules	Moral evaluation	Justification with reference to value systems
Cultural-cognitive rules	Rationalization	Justification based on beliefs about utility of action
(Institutional carrier)	Mythopoesis	Narrative (storytelling)

and non-discursive elements of social life such that social reality is wholly constituted in discourse. For Fairclough, discourse (in the sense of semiotics and language) is an element of the social at all the three levels of social practice, discursive practice and social events. Thus language and semiotics correspond to social practice, the orders of discourse correspond to discursive practice, and texts correspond to social events.

Ideas about legitimation provide a cross-over point for discourse analysis and neo-institutional analysis. Strategies and practices for legitimation call for institutional analysis but to elicit greater insight should be founded on an understanding of the role of discursive strategies of legitimation in giving effect to prevailing and new patterns of thinking and behaviour. At this point it is helpful to review these various strategies of legitimation, indicating their relevance to the chapter. From an institutional perspective legitimation occurs as a result of the operation of a variety of mechanisms and processes (see table 24.1). As mentioned above, the former are coercive, mimetic or normative in character and have the effect of making formal, normative and cultural-cognitive rules appear solid (Scott 2008a). Institutionalizing processes may involve the objectification and thence diffusion of novel practices across organizational fields. Zucker (1987) has identified a number of indicators of institutionalization, which include changes in language use.

Now, with the role of language in institutionalization in mind, consider a discursive standpoint on legitimation, which would involve scrutiny of a body of texts pertaining to the research questions and social context at issue. (Remember that legitimation is only one element of the analysis of text within CDA—there is also 'intertextual' analysis, which is concerned with how one text draws on other texts. Also, textual analysis is but one element of CDA—the others are the analysis of discursive practice and the analysis of broader social practice.) Discursive legitimation strat-

egies could be said broadly to underpin the aforementioned institutional pillars in the following way. Thus rather than regulative rules demanding and enforcing compliance in any straightforward direct manner, their mandating or coercive effect may rely on the discursive legitimation strategy of authorization. Here, legitimation is accorded on the basis of authority and custom, including recognition and acceptance of the authority claims of those who make or enforce laws and regulations so as to constrain the actions of those subjected thereto. Normative institutions may be said to be underpinned discursively through moral evaluation, in which legitimation is granted by reference to value systems. Cultural-cognitive institutional rules may be underpinned by the discursive strategy of rationalization, by which legitimation involves reference to beliefs about the utility of institutionalized action. This applies, for example, to beliefs about the nature and effectiveness of government policies in a particular arena (and indeed of government per se) and to heuristics and tools for analysis (e.g. cost-benefit analysis, or triple bottom-line accounting). In the discursive strategy of mythopoesis legitimation is accomplished through narrative. This storytelling is a mechanism which can carry all the types of institutional pillar referred to above. Discursive-institutional phenomena occur across somewhat conflicting discourses, which are comprised of constituent story sets characterized by certain types of discursive practices ('genres', 'styles', discourse coalitions and members/non-members), in which protagonists identify themselves or others. As mentioned above, discursive practice mediates wider social practice and specific texts in particular social events.

In addition to attending to the legitimation and carriage of existing and new institutional rules, a discourse-institutional approach may address other elements of institutionalization through language. Figure 24.1 recognizes this by suggesting that important

Figure 24.1: A discourse-institutional analytical framework. **Source:** The author.

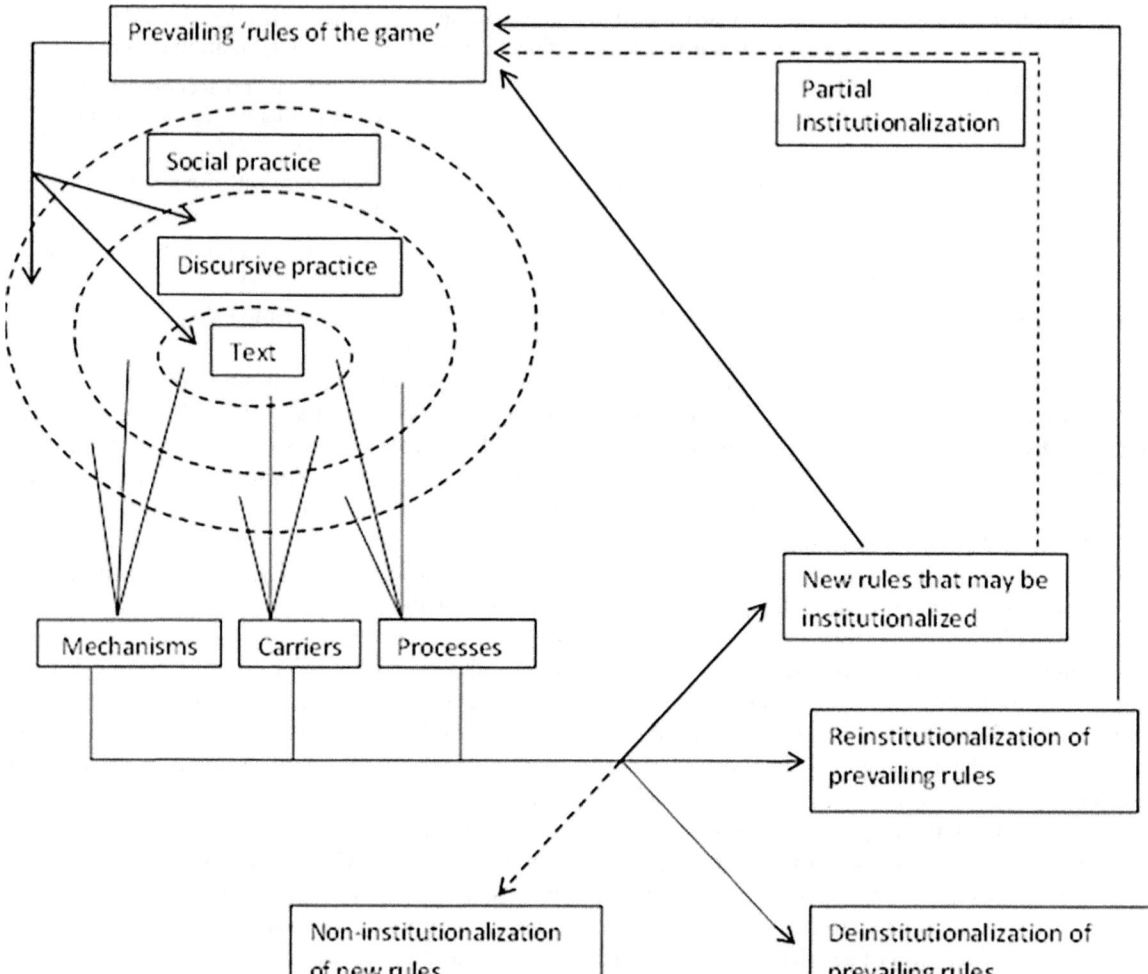

social phenomena, in this case the institutionalization of sustainability, may be depicted as regularities in text and discursive practice; these have a dialectical relationship with social practice, the prevailing historical context and social structures (or overarching rules of the game) which shape and which are partly shaped by those discursive regularities.

New institutional rules are created in speech and writing and visual 'events' (different 'modalities' of text), and in discursive practice, as well as by non-discursive phenomena. Material conditions will be of significance, such as funding and other resource allocations and facilities. Prospective rules may (or may not) be institutionalized depending on the strength and effectiveness of the coercive, obligatory and mimetic mechanisms underpinning the three types of institutional rules in question, and the carriers via which they are transmitted. The extent to which these new

rules are pervasive is a matter of their sedimentation, connected with processes of diffusion and their legitimacy. Such diffusion may be partial and limited to 'habitualization' or 'objectification' at a level below that of the organizational field or wider society. Institutionalization may fail (non-institutionalization) and prevailing rules may be reinforced (re-institutionalization) or undone (de-institutionalization). Thus the section has presented a framework which draws on both neo-institutional and critical discourse analysis, which may be employed to inform an assessment of available discourse-institutionalist contributions to the topic of sustainability transitions. It is to this assessment that the chapter now turns.

24.4 Discourse-institutional Analysis of Sustainability Transition

This section reviews literature on discourse-institutional approaches connected with the transition to sustainability. The approach to reviewing literature taken here is now described. First, during the period from December 2013 to August 2014 the author searched for published research by entering permutations of the key words 'discourse' and 'institution', combined with 'sustainability transition', 'sustainable consumption', and 'sustainable production' into the Web of Knowledge database. Results were selected to be shown for 'all years' and with no limitation on subject area. Searches were thus conducted by topic and title. To give an indication of the extent of scholarly activity in the topic area, it may be noted that seventeen results were obtained from Web of Knowledge when searching by title words for articles on 'discourse, institution', and sustainable transition but 178 were obtained when employing the broader topic search using the key words 'discourse' and 'institution'. To capture papers published in journals not covered by Web of Knowledge and those contributions appearing in monographs or edited works searches in the SCOPUS database and Google Scholar were conducted using similar search terms and subject and date selections as above. Other relevant contributions were suggested by colleagues and contacts working in the field. The core sample of articles reviewed here was achieved after reading the abstracts of papers listed in the search results. Those which were concerned with phenomena of low relevance to the interests of chapter were eliminated from further consideration. The remaining papers were read in full and reviewed in relation to the following themes:

- disciplinary basis
- concept and analysis of institutions
- analysis of text and discursive practice
- contribution to understanding sustainable transitions.

24.4.1 Disciplinarity

There is a variety in the disciplinary basis of contributions, which affects what researchers choose to focus on and in what ways they conduct their work. With regard to institutions it is apparent that a number of projects draw inspiration from and are informed by notions of institution advanced within the field of political science or environmental sociology, as dis-

tinct from treatments of institutions found in sociology or economics. Of the core outputs found in the search conducted in preparing this chapter the majority of contributions cite the work of Hajer or Schmidt, whilst few cite foundational neo-institutionalist sociologists such as Scott or Powell and DiMaggio, or institutional economists such as Ostrom. For instance, Arts and Buizer (2009) conduct a 'discursive-institutionalist' analysis of global forest governance and Palmer (2010) one of UK policy on biofuels for transport, and Den Besten, Arts and Verkoojien (2014) apply a 'discursive-institutionalist' approach to the analysis of the *Reducing Emissions from Deforestation and Forest Degradation* (REDD+) policy, based on developments that took place over the period 2004-2011. In these examples institutional analysis is informed by the work of political scientists (Marsh/Stoker 2002; Schmidt 2008, 2010), whilst ideas regarding the combination of discourse and institutional analysis draw on Hajer (1995; Hajer/Versteeg 2005). Another example is that of Kern, whose work also draws on Hajer and Schmidt to analyse the "discursive politics of governing transitions towards sustainability" and to identify connections among ideas in conflicting discourses, interests and institutional contexts which shape policy (Kern 2011; 2012: 90). Other contributions, such as that by Young (2006), draw on neo-institutionalists such as March and Olsen (1998) and Ostrom (1990) but in general, in terms of references cited or not cited, are less clearly connected with prior work on (critical) discourse analysis. This does not mean that such work fails to provide interesting insights; indeed it sheds light on how discourse coalitions are formed and the implications of these for certain actors being able to impose their views of 'the problem' and what to do about it on others. Young's (2006) work, for example, remarks on the role of 'duelling' discourses in fragmentation of governance arrangements, dominant discourses in systems inertia, and 'blocking' coalitions in creating institutional dysfunction and uncertainty. What are called 'cognitive transitions' refers to the synthesis of previously disintegrated groups of actors, which may culminate in novel arrangements for, in this case, managing marine protection across species. It is more that this kind of research is not fully informed by an approach which systematically analyses text and discursive practices in social relations to effectively analyse particular institutional mechanisms, carriers or processes.

Readers wishing to develop an appreciation of what discursive-institutionalism can contribute to knowledge and practice regarding sustainable transition may find that a number of publications to which they may be directed as result of searching online for relevant literature are only partially or limitedly pursuing such an approach. Either the discursive or institutional aspect may be under-developed, something which a multi- or trans-disciplinary approach could address. For example, Fairclough (2012) suggests that a 'critical discourse analyst' could work with concepts from other research disciplines, and that studies may be conducted by a project team in which the various members have expertise from a range of disciplines. However, caution is warranted. Pursuing an *inter*disciplinary approach, to build a single combined framework, may marginalize some problem definitions and interpretations of research relevance to policy (Shove 2011), or undervalue differences in approach and emphasis (Scott 2008a). Genus (2004) sees potential in *multi*disciplinary approaches which facilitate 'contact points' which enable researchers to debate and learn about diverse perspectives and methods. Such an approach will be ineffective to the extent that some researchers seek to colonize other disciplines or are insufficiently versed in the literature or methods with which they seek to engage.

24.4.2   Analysis of Institutions

In the literature on systems innovation, neo-institutional theory has been invoked by Geels (2004), for example, to analyse the 'deep structure' of sociotechnical systems. Here, there is argued to be a set of regimes which comprise sociotechnical systems; the stability of the regimes and ultimately of the focal sociotechnical system relies in part on the rules which regime members have in common and which connect regimes with each other. More specifically within studies of technological and sustainability-related transitions and transition management, neo-institutional theory has been applied to analyse the diffusion (or rather lack of diffusion) of micro-generation technologies in the UK (Genus 2012), renewable energy (e.g. Reddy/Painuly 2004; Szarka 2006), and bioenergy (e.g. Söderberg 2011) or biofuels (German/Schoneveld 2012). Wolsink (2012) has done so to examine social acceptance of smart electricity grids for distributed electricity generation. There are other streams of work which lie beyond the systems innovation/transitions domain but which share certain concerns of that literature, such as that pertaining to the democratic governance of environmental policy-making and conservation, forestry and biodiversity. Here one also finds appeals to

institutional change and neo-institutional analysis. As mentioned above in section 24.2, Rydin (2003) draws upon the work of March and Olsen (1989) and Ostrom, Gardner and Walker (1994) in attempting to fill the gap in institutional analysis of environmental planning—though not 'transition' as such—with fuller attention to "language use in communicative situations" (Rydin 2003: 49). Den Besten, Arts and Verkooijen (2014) see institutions and discourses as a 'spiral' in which embedded rules, conventions and norms are susceptible to change by ideas which are the carriers of discourses and convey interests. At the same time stabilized institutions may enable or constrain discourses. They argue that the 'spiral' helps to understand stability and change and how discourses open up or close down in relation to temporarily stable institutions.

Some general remarks may be made regarding the analysis of institutions in prior work related to the transition to sustainability. First, Coenen and Dìaz Lòpez (2010) note that the treatment of institutions in sociotechnical systems research is more coherent and consistent than that within technological innovation systems and sectoral systems of innovation branches of innovation studies, and emphasize the role of learning processes in transitions. However, a criticism of this work is that, in general, there has been insufficient or incomplete attention to ways in which these rules are created, persist or are undone. Moreover, there has not been enough explicit attention to 'carriers', mechanisms, and processes associated with institutionalization, or de- or non-institutionalization. In this and in other work, 'institution' has been used merely as a synonym for 'organization', or governmental arrangements (German/Schoneveld 2012; Reddy/Painuly 2004). Much of the research on discourses of environmental sustainability addresses institutions as organizations, particularly focusing on governmental organizations, and what contributors variously refer to as the 'institutional set-up' or 'institutional context'. Fischer (1993), for example, considers how the over-reliance on experts and a technical framing of policy issues contribute to weakening the influence of citizens over environmental politics in the US, a phenomenon which is accomplished by the growth of powerful economic and governmental institutions (which he uses to mean 'organizations'). If not employed to mean a (usually governmental) organization, 'institution' is commonly employed in a general or everyday sense without theoretical underpinning (Söderberg 2011; Szarka 2006).

Other contributions develop a line of analysis which conceives of institutions in ways which discourse analysts would term genres- and styles-related discursive practice, i.e. they draw attention to where and how sustainability is discussed, and by whom and with what effect. For example Hajer states that his focus is on analysis of institutional practices 'within which discourse is produced' and which might afford more democratic debates about environmental issues (Hajer 2003: 103-4). Rydin (1999) considers the potential of 'new' environmental policy-making institutions (or 'genres') such as citizen juries or group evaluation exercises in bringing together 'polyphonic' communities whom formal authorities must learn to trust. Smith and Kern (2009) consider the implications for more radical transition 'storylines' of the capture of environmental discourse by mainstream interests (or 'institutional priorities'). However, in addition to seeing 'institution' as ministerial priorities, Smith and Kern depict as institutional innovations the transition platforms, themes, pathways and experiments created by the Dutch economics ministry as the embodiment in practice of a new energy transition policy. In certain contributions 'discourse' is separated analytically from 'institutions'. In the work of Arts and Buizer (2009) there is on the one hand an analysis of discursive 'frames' of meaning applied by various protagonists regarding forestry biodiversity or economic management, whilst on the other the institutionalization of practices for governance are examined. Rydin considers that previous attempts to devise a synthetic approach to institutional and discourse analysis have insufficiently addressed the role of language, a shortcoming she argues may be remedied by bringing in a discursive element to the analysis. Thus she develops an institutionalist framework in which regularities in language use, actors' discursive strategies, linguistic resources and interactions in communication, analysed using a rhetorical method, take centre stage.

24.4.3   Analysis of Text and Discursive Practice

A number of contributions offer a view of discourse institutions which emphasize policy-making practices and institutional contexts, though not through textual analysis per se. Examples of this pertain to studies with disparate foci, including forestry and biodiversity (Arts/Buizer 2009; Den Besten/Arts/Verkoojien 2014; Li 2007), food (Zwartkruis/Moors/ Faria 2012), water management (Schmidt 2014), sustainable employment (Köves/Király/Pataki 2013), building policy (Melchert 2007), energy (Hisschemoller/Bode/

van de Kerkhof 2006; Jiusto and McCauley 2010; Späth 2012), urban sustainability (Mazza/Rydin 1999; Rydin 2003), national and international environmental policy-making (Young 2006; Den Besten/Arts/Verkoojien 2014; Koning/Mol 2009) and the role of universities in promoting sustainability (Dlouhá/Huisingh/Barton 2013). Where language is central the emphasis of the work reviewed in relation to discourse is not on textual analysis, nor of the interrelation of text, discursive practice and social practice but rather the latter two elements. Hajer, for example, states that he is more concerned with these last two elements than with linguistic analysis (Hajer 2003). Indeed it has been argued that a focus on language (as text) is not as helpful to understanding institutional practices 'which are an expression of discursive power too' (Arts/Buizer 2009: 342).

Connected with the governance of sustainability transitions there have been notable contributions to our understanding of urban sustainability discourses (e.g. Mazza/Rydin 1997) and the politics of environmental discourse (e.g. Hajer 1995). Mazza and Rydin point to three constraints on the integration of environmental sustainability concerns into urban planning policy: (a) the dominant role of economic interests in shaping policy discussions; (b) the degree to which the admission of environmental issues into policy discussions relies upon the involvement of environmental groups and their representatives; and (c) the (un)availability of local tools and resources which limit discussion of sustainability themes. They base their analysis on identification of actors and their roles in urban policy-making in designated research sites, the assessment of local networks of significance to policy-making and a definition of multiple, possibly overlapping policy-making discourses. The latter they refer to as 'images' of urban sustainability. Employing rhetorical analysis, Rydin's (2003) approach to discourse and institutions makes a valuable contribution to understanding the significance of language in social organization. Yet compared with the version of critical discourse analysis discussed in this chapter, it does less to expose connections among texts, their production and consumption, and a critique of society at a general level.

Hajer (1993) combines analysis of discourse through which reality is produced with analysis of social practices from which constructs emerge and in which actors who make statements engage. At the heart of his work are discourse coalitions, which are groups of actors sharing a social construct in relation to environmental debates and policy-making. Hajer marks up the role of storylines in (more or less) synthesizing different discourses into a (somewhat) coherent whole, whilst masking the complexity of discourse. For Hajer, domination and institutionalization of discursive space produce 'solidified' social practices and reasoning. In this way, practice is the medium by which certain actors impose their views of sustainability on others. Hajer (2009) develops this thinking by bringing in a dramaturgical perspective to highlight connections among the legitimacy of governance processes and the symbolism attached to the performance of certain policy-making routines. If the performance (e.g. presentation of the national budget or state-organized public debate) is not as expected or in accordance with 'the rules', the legitimacy accorded to the policy-making process and policymakers may erode. Where there is no consensus on what 'the rules' are for the process and substance of politics, this is described as an 'institutional void', in which politics carries on in multiple spaces and state policy actors need both to negotiate with plural and ad hoc networks (of NGOs, business, citizen movements) *and* conventional governmental systems and practices.

Smith and Kern (2009) analyse storylines in Dutch environmental policy-making, concerned that co-option of environmental storylines into incumbent interests and prevailing discourses impacts negatively on the subsequent institutionalization of ideas and thinking about environmental sustainability. Smith and Kern (2009) give an account of the successful diffusion of the 'flexible storyline' of Dutch environmental transition, which owes much to collaboration among researchers of sustainable transitions and policymakers in the Netherlands. The openness to reinterpretation of this storyline enabled a range of governmental actors to 'buy into' the developing environmental policy. However, this came at a price, since the coalition had to develop an approach which could 'work with the grain' of the commitments of other government ministries (other, that is, than the Ministry of Housing, Spatial Planning and Environment, which was the government department responsible for developing the new policy). Hence the transition approach in the Netherlands, fearful of remaining a 'sideshow', incorporated 'market-based instruments', the new commitment to liberalize the Dutch energy market and a concern to develop the 'knowledge economy'.

24.4.4 Contribution to Understanding Sustainable Transitions

The contributions reviewed here have added a wealth of knowledge to a wide range of sustainability-related phenomena. However, these are more or less distant from the concerns of those who wish to fully understand or intervene to realize sustainability transitions, conceived of as pervasive and environmentally sustainable transformations in the way we live. Recall that such transitions are said to occur over decades, perhaps forty to ninety years in the making, requiring innovation or revival of sustainable practices on the part of heterogeneous actors over wide geographical space and economic sectors. Reflecting on the contributions from the literature, what does one observe? First, 'discourse-institutional' sustainability research is not necessarily about 'transitions' as just outlined. There is considerable variety in the works reviewed here. Thus some albeit very interesting and thought-provoking contributions focus on relatively discrete and short-term case studies, in which the period under scrutiny is less than ten years (Den Besten/Arts/Verkoojien 2014; Genus/Theobald, in press; Palmer 2010; Zwartkruis/Moors/Faria 2012). Others consider changing phenomena focusing on a ten to twenty year period, such as Mol's (2005) study of the promotion of cleaner production in China from the early 1990s to the early 2000s. Shove and Walker (2010) refer to changes that have taken place over 'a generation' with regard to weekly baths and daily showering. Another variant is the future-orientated study, in which the researchers hypothesize (or encourage project participants to create) visions of the future. Here the researchers (and sometimes the participants) engage in 'backcasting' (Köves/Király/Pataki 2013) to understand the nature of local guiding visions and how they may link up with arguments being made more broadly in the society in question (Späth 2012), or consider the potential impact of an imagined future event on sub-sectors of a given society (such as van Koppen/Mol/van Tatenhove 2010) do in relation to the implications of a collapse in the Gulf Stream on urban infrastructure and rural planning in the Netherlands). Temporally, at least, other contributions account for deep changes taking place over thirty years or more, for example with regard to Dutch sustainable building policy (Melchert 2007) and forestry management (Li 2007).

Recent discourse-institutional contributions vary greatly in the extent to which they address core concerns of the founding literature on sustainability transitions, such as the definition and analysis of transition pathways, and systemic interactions between landscape, regimes and niches in major subsystems of society. There are those which continue the pursuit of such lines of inquiry (e.g. Hargreaves/Hielscher/Seyfang et al. 2013; Jiusto/McCauley 2010; Smith/Seyfang 2013). Others, however, fundamentally question foundational thinking about sustainable transition, as represented in the body of knowledge that has emerged over the past twenty years on systems innovation and transition management. Until relatively recently much attention was focused on technologies of energy production. An example of work with a different empirical focus is Shove's work on the conventions inherent in users' everyday unsustainable practices (such as daily resource-intensive showering) and the emergence, disappearance and reappearance of sustainable practices, such as cycling and road congestion charging in the UK (Shove 2012; Shove/Walker 2010). Other contributions focus on demodernization and ecological modernization (Koning/Mol 2009; Melchert 2007); the discursive politics and democracy of transition (Hendriks 2009; Hendriks/Grin 2007; Kern 2011; 2012; Smith/Kern 2009); and governmentality (Hajer/Versteeg 2005; Hajer 2009; Li 2007). Unsurprisingly, authors working from such standpoints do not conceive of their work in the same way as transition researchers inspired by systems innovation, nor do they draw on the same 'discursive repertoire' (Rumpala 2013). Thus, for example in the work of Mol and Spaargaren one sees the language of ecological modernization, which is advanced as a 'transhistorical' concept but not articulated in the terminology of transition, transition pathways or transition management (Mol/Spaargaren, 2000; Spaargaren/Mol, 2008; Li/Bluemling/Mol et al. 2014).

Spatially, and in terms of actor focus, the studies reviewed vary in scope. There is a distinct subtopic of research on grassroots niches and environmental activism and their relation to incumbent regimes and systems innovation (Hargreaves/Hielscher/Seyfang et al. 2013; Smith/Seyfang 2013). Other work focuses on particular regime actors or intermediaries, such as universities (e.g. Dlouhá/Huisingh/Barton 2013), national sectors such as egg production in the Netherlands (Zwartkruis/Moors/Faria 2012), Dutch sustainable building policy (Melchert 2007) and sustainable employment in Hungary (Köves/Király/Pataki 2013), international governance (Koning/Mol 2009; Arts/Buizer 2009; Den Besten/Arts/Verkooijen 2014) and disputes about scale (Lindseth 2006). In terms of methodological approach there is an interesting distinction to make between an understanding of transi-

tion as the outcome of collaborative or antagonist relations among different groups in society and as produced by discourse coalitions, domains or networks constituted by the stories and actions which become institutionalized (c.f. Knox/Savage/Harvey 2006). The former approach is apparent in contributions which emphasize the need for more inclusive engagement and communication among multiple stakeholders (Dlouhá/Huisingh/Barton 2013; van Koppen/Mol/van Tatenhove 2010). It may be characterized as being concerned with network structures, asking the question: who knows (or who talks with) whom? The latter conception draws attention to texts and practices in which certain actors are included and others excluded and which domains and story sets hold sway whether or not they enact sustainability, or particular versions of an environmental agenda. Such an approach prizes and critiques coalition-building (Hajer 1995). The building of coalitions is a potential vehicle for enabling institutional change though one in which the political role of 'boundary-crossing' scientists represents both an opportunity for science and a threat (to the autonomy and objectivity of science and scientists) (Kemp/Rotmans 2009). Such engagement may well involve actors sharing very different interpretations of what is meant by 'transition' and how it should be achieved, as well as different perspectives of what this 'sharing' should entail. The definition and making through institutionalized discursive practices of the policy context and of the boundary separating active protagonists from excluded but likely affected parties is one area that has developed and should continue to develop in improving our understanding of sustainability transitions. Another less developed area concerns explicit analysis of 're-contextualization', an aspect of discourse analysis which addresses the translation or repackaging of ideas and practices from one field or locality to others (Fairclough 2012; van Leeuwen 1993a, 1993b) where they may be legitimated, possibly at different social scales (local, subnational, national, supranational). Fairclough argues that his approach to investigating re-contextualization requires analysis of relations between existing and new discourses, including ways in which primary 'nodal' and secondary discourses are articulated together in the formulation of strategic problems, decision process, decision options and plans for their implementation (Fairclough 2012).

24.5   Conclusion

This chapter has aimed to develop our understanding of research on the transition to sustainability, based on a review of contributions which to varying degrees recognize the complementarity of neo-institutional analysis and (critical) discourse analysis. There is great diversity in this work, invoking discourse and institutions, which addresses directly and indirectly, partially and more fulsomely the core concerns of the wider body of research on sustainable transitions and transitions management. The contributions to this literature are insightful in a number of ways. Fundamentally, contributors such as Schmidt (2008; 2010), Arts and Buizer (2009), Rydin (2003) and Hajer (1995; Hajer and Versteeg 2005) have developed combined discourse-institutional frameworks which draw on both discourse analysis and institutional thinking. For example, Hajer's (2003) approach to discourse institutions has highlighted how the discourse of 'ecology' has provided a language which has 'disciplined' the debate about nature in the Netherlands, leading to new and 'enduring policy practices'. Others draw attention to the persistence of unsustainability practices and the reinstatement of once pervasive sustainable practices (Shove 2012). Further, a number of contributions highlight the potential or marginalization of certain actors, such as NGOs and citizens at large, in grassroots social movements (Hargreaves/Hielscher/ Seyfang et al. 2013; Smith/Kern 2009; Smith/Seyfang 2013). Yet appreciative insights could be obtained regarding a discourse-institutional perspective if the conceptual underpinnings of such an approach were more fully articulated. This applies to the definition

of the types of institutions that may be of significance to governance and innovation linked to the transition to sustainability. It applies also to a more explicit analysis of the carriers, mechanisms and processes by which institutionalization, non-institutionalization or de-institutionalization of arrangements, norms and cultures occur. Furthermore, such neo-institutional analysis could be strengthened by addressing in a more rounded and critical way the role of discourse in substantiating the institutions and institutional processes identified. This would necessitate an approach transcending discourse as merely linguistics or purposive use of language. Previous work has addressed the nature of discourses and discursive strategies implicated in the embedding of certain perspectives relating to sustainability, and the social and political machinations connected with this and the problem of effectively challenging incumbent interests and arguments. From a critically discursive point of view, what earlier contributions have lacked is a comprehensive treatment of specific *texts*, practices and the social conditions governing these, which together (re)produce the institutional reality of governance in relation to the transition to sustainability. To do so whilst attending comprehensively to the issues of institutionalization and de-institutionalization is no small undertaking—more a research programme than a one-off project (cf. Wodak/Meyer 2008). This chapter suggests that to define and implement a research agenda informed by discourse-institutional thinking, but also mindful of the above observations regarding the opportunities and limitations of inter- and multidisciplinary research, could be beneficial to improving understanding of the transition to sustainability.

References

Arts, Bas; Buizer, Marleen, 2009: "Forests, Discourses, Institutions: a Discursive-institutional Analysis of Global Forest Governance", in: *Forest Policy and Economics,* 11,5-6: 340-347.

Berger, Peter L; Luckmann, Thomas, 1967: *The Social Construction of Reality* (New York: Doubleday).

Bourdieu, Pierre, 1983: "The Field of Cultural Production, or the Economic World Reversed", in: *Poetics,* 12: 311-56.

Chouliaraki, Lilie; Fairclough, Norman, 2010: "Critical Discourse Analysis in Organizational Studies: Towards an Integrationist Methodology", in: *Journal of Management Studies,* 47: 1213-1218.

Coase, Richard, 1937: "The Nature of the Firm", in: *Economica,* 4: 386-405.

Coenen, Lars; Dìaz Lòpez, Fernando, 2010: "Comparing Systems Approaches to Innovation and Technological Change for Sustainable and Competitive Economies: an

Explorative Study into Conceptual Commonalities, Differences and Complementarities", in: *Journal of Cleaner Production,* 18,12: 1149-60.

Dacin, M. Tina; Dacin, Peter A., 2013: "Traditions as Institutionalized Practice: Implications for Deinstitutionalization", in: Greenwood, Peter; Oliver, Christine; Sahlin, Kerstin; Suddaby, Roy (Eds.): *The Sage Handbook of Organizational Institutionalism* (London: Sage): 327-351.

Dacin, M. Tina; Goodstein, Jerry; Scott, W. Richard, 2002: "Institutional Theory and Institutional Change: Introduction to the Special Research Forum", in: *Academy of Management Journal,* 45,1: 45-54.

Delbridge, Rick; Edwards, Tim, 2007: "Reflections on Developments in Institutional Theory: Toward a Relational Approach", in: *Scandinavian Journal of Management,* 23,2: 191-205.

Den Besten, JanWillem; Arts, Bas, Verkoojien, Patrick, 2014: "The Evolution of REDD+: an Analysis of Discursive-institutional Dynamics", in: *Environmental Science and Policy*, 35,1: 40-48.

DiMaggio, Paul J; Powell, Walter W, 1991: "The Iron Cage Revisited: Institutional Isomorphism and Collective Rationality in Organizational Fields", in: DiMaggio, Paul J.; Powell, Walter W. (Eds.): *The New Institutionalism in Organizational Analysis* (Chicago, IL: University of Chicago Press): 63-82.

Dlouhá, Jana; Huisingh, Donald; Barton, Andrew, 2013: "Learning networks in higher education: universities in search of making effective regional impacts", in: *Journal of Cleaner Production*, 49,1: 5-10.

Fairclough, Norman, 2001: *Language and Power* (London: Longman).

Fairclough, Norman, 2003: *Analyzing Discourse: Textual Analysis for Social Research* (London: Routledge).

Fairclough, Norman, 2012: "Critical Discourse Analysis", in: *International Advances in Engineering and Technology (IAET)*, 7: 452-82.

Fischer, Frank, 1993: "Policy Discourse and the Politics of Washington Think Tanks", in: Fischer, Frank; Forrester, John, (Eds.): *The Argumentative Turn in Policy Analysis and Planning* (Durham, NC: Duke University Press): 21-42.

Foucault, Michel, 1972: *The Archaeology of Knowledge* (Pantheon: New York).

Garud Raghu: Hardy, Cynthia; Maguire, Steve, 2007: "Institutional Entrepreneurship as Embedded Agency: an Introduction to the Special Issue", in: *Organization Studies*, 28,7: 957-969.

Geels, Frank, 2004: "From Sectoral Systems of Innovation to Socio-Technical Systems: Insights about Dynamics and Change from Sociology and Institutional Theory", in: *Research Policy*, 33: 897-920.

Genus, Audley, 2004: "Understanding Inertia: Developing a Multi-disciplinary Perspective?", in: Ghobadian, Abby; O'Regan, Nicholas; Gallear, Duncan; Viney, Howard (Eds.): *Strategy and Performance: Achieving Competitive Advantage in the Global Market Place Selected Papers from the 22nd Annual British Academy of Management Conference* 2002 (London: Palgrave MacMillan): 203-220.

Genus, Audley, 2012: "Changing the Rules? Institutional Innovation and the Diffusion of Microgeneration", in: *Technology Analysis and Strategic Management*, 24,7: 711-27.

Genus, Audley; Theobald, Kate (in press): "Creating Low Carbon Neighbourhoods: a Critical Discourse Analysis", in: *European Urban and Regional Studies*.

German, Laura; Schoneveld, George, 2012: "Biofuel Investments in sub-Saharan Africa: A Review of the Early Legal and Institutional Framework in Zambia", in: *Review of Policy Research*, 29,4: 467-91.

Greenwood, Royston; Suddaby, Roy; Hinings, Bob, 2002: "Theorizing Change: the Role of Professional Associations in the Transformation of Institutionalized Fields", in: *Academy of Management Journal*, 45,1: 58-80.

Hajer, Maarten, 1993: "Discourse Coalitions and the Institutionalization of Practice: The Case of Acid Rain in Britain", in: Fischer, Frank; Forrester, John (Eds.): *The Argumentative Turn in Policy Analysis and Planning* (Durham, NC: Duke University Press): 43-76.

Hajer, Maarten, 1995: *The Politics of Environmental Discourse: Ecological Modernization and the Policy Process* (Oxford: Oxford University Press).

Hajer, Maarten, 2003: "A Frame in the Fields: Policy-making and the Reinvention of Politics", in: Hajer, Maarten; Wagenaar, Hendrik (Ed.): *Deliberative Policy Analysis: Understanding Governance in the Network Society* (Cambridge: Cambridge University Press) 88-110.

Hajer, Maarten, 2009: *Authoritative Governance: Policy-making in the Age of Mediatization* (Oxford: Oxford University Press).

Hajer, Maarten; Versteeg, Wytske, 2005: "A Decade of Discourse Analysis of Environmental Politics: Achievements, Challenges, Perspectives", in: *Journal of Environmental Policy and Planning*, 7,3: 175-84.

Hargreaves, Tom; Hielscher, Sabine; Seyfang, Gill; Smith, Adrian, 2013: "Grassroots Innovations in Community Energy: the Role of Intermediaries in Niche Development", in: *Global Environmental Change*, 23: 868-880.

Hendriks, Carolyn, 2009: "Policy Design without Democracy? Making Democratic Sense of Transition Management", in: *Policy Sciences*, 42: 341-368.

Hendriks Carolyn; Grin, John, 2007: "Contextualizing reflexive governance: the politics of Dutch energy transitions", in: *Public Administration*, 86,4: 1009-1031.

Hisschemoller Matthijs; Bode, Ries; van de Kerkhof, Marleen, 2006: "What Governs the Transition to a Sustainable Hydrogen Economy? Articulating the Relationship between Technologies and Political Institutions, in: *Energy Policy*, 34,11: 1227-1235.

Hughes, Everett, 1936: "The Ecological Aspects of Institutions", in: *American Sociological Review*, 1,2: 180-189.

Jepperson, Ronald, 1991: "Institutions, Institutional Effects, and Institutionalization", in: DiMaggio, Paul J.; Powell, Walter W. (Eds.): *The New Institutionalism in Organizational Analysis* (Chicago, IL: University of Chicago Press): 143-163.

Jiusto Scott; McCauley Stephen, 2010: "Assessing Sustainability Transition in the US Electrical Power System", in: *Sustainability*, 2,2: 551-575.

Katz, Michael L.; Shapiro, Carl, 1985: "Network Externalities, Competition, and Compatibility", in: *American Economic Review*, 75,3: 424-440.

Kemp, Rene; Rotmans, Jan, 2009: "Transitioning Policy: Co-production of a new Strategic Framework for Energy Innovation Policy in the Netherlands", in: *Policy Sciences*, 42,4: 303-322.

Kern, Florian, 2011: "Ideas, institutions, and interests: explaining policy divergence in fostering 'system innovations' towards sustainability", in: *Environment and Planning* C: 29,6: 1116-1134.

Kern, Florian, 2012: "The discursive politics of governing transitions towards sustainability: the Carbon Trust", in: *International Journal of Sustainable Development*, 15,1-2: 90-106.

Knox, Hannah; Savage, Mike; Harvey, Penny, 2006: "Social networks and the study of relations: networks as method, metaphor and form", in: *Economy and Society*, 35,1: 113-140.

Koning, Niek; Mol, Arthur, 2009: "Wanted: Institutions for Balancing Global Food and Energy Markets", in: *Food Security*, 1,3: 291-303.

Köves, Alexandra; Király, Gabor; Pataki, György; Balázs, Bálint, 2013: "Backcasting for sustainable employment: a Hungarian experience", in: *Sustainability*, 5,7: 2991-3005.

Laclau Ernesto; Mouffe, Chantal, 1985: *Hegemony and Socialist Strategy: Towards a Radical Democratic Politics* (London: Verso).

Laclau Ernesto; Mouffe, Chantal, 2001: *Hegemony and Socialist Strategy, 2nd edition* (London: Verso).

Li, Jia; Bluemling, Bettina; Mol, Arthur P. J.; Herzfeld, Thomas, 2014: "Stagnating Jatropha Biofuel Development in Southwest China: An Institutional Approach", in: *Sustainability*, 6: 3192-3212.

Li, Tania Murray, 2007: "Practices of Assemblage and Community Forest Management", in: *Economy and Society*, 36,2: 263-293.

Lindseth, Gard, 2006: "Scalar Strategies in Climate Change Politics: debating the Environmental Consequences of a Natural Gas Project", in: *Environment and Planning C*, 24: 739-754.

Lounsbury, Michael; Crumley, Ellen, 2007: "New Practice Creation: an Institutional Perspective on Innovation", in: *Organisation Studies*, 28: 993-1012

March, James; Olsen, Johan, 1984: "The New Institutionalism: Organizational Factors in Political Life", in: *American Political Science Review*, 78,3: 734-749.

March, James; Olsen, Johan, 1989: *Rediscovering Institutions: The Organizational Basis of Politics* (New York: The Free Press).

March, James; Olsen, Johan, 1998: "The Institutional Dynamics of International Political Orders", in: *International Organization*, 52: 943-969.

Marsh, David; Stoker, Gerry, 2002: Theories and Methods in Political Science (Basingstoke: Palgrave).

Mazza, Luigi; Rydin, Yvonne, 1997: "Urban Sustainability: Discourses, Networks and Policy Tools", in: *Progress in Planning*, 41: l-74.

Melchert, Luciana, 2007: "The Dutch Sustainable Building Policy", in: *Building and Environment*, 42,2: 893-901.

Meyer, John W.; Rowan, Brian, 1977: "Institutionalized Organizations: Formal Structure as Myth and Ceremony", in: *American Journal of Sociology*, 83: 340-363.

Mohr, John W.; White, Harrison C., 2008: "How to Model an Institution", in: *Theory and Society* 37,5: 485-512.

Mol, Arthur, 2005: "Institutionalising Cleaner Production in China: the Cleaner Production Law", *International Journal of Environmental and Sustainable Development*, 4,3: 227-245.

Mol, Arthur P J; Spaargaren, Gert, 2000: "Ecological Modernisation in Debate: A Review", *Environmental Politics*, 9,1: 17-49.

Nelson, Richard R; Winter, Sidney G., 1982: *An Evolutionary Theory of Economic Change* (Cambridge, MA: Harvard University Press).

North, Douglas C., 1990: *Institutional Change and Economic Performance* (Cambridge: Cambridge University Press).

Ostrom, Elinor, 1990: *Governing the Commons: The Evolution of Institutions for Collective Action* (New York: Cambridge University Press).

Ostrom, Elinor; Gardner, Roy; Walker, James, 1994: *Rules, Games and Common-Pool Resources* (Ann Arbor, MI: University of Michigan Press).

Palmer, James, 2010: "Stopping the unstoppable? A discursive-institutionalist analysis of renewable transport fuel policy", in: *Environment and Planning C*, 28: 992-1010.

Parsons, Talcott, 1960: *Structure and Process in Modern Societies* (Glencoe: Free Press).

Philips, Nelson; Lawrence, Thomas; Hardy, Cynthia, 2004: "Discourses and Institutions", in: *Academy of Management Review*, 29,4: 635-652.

Philips, Nelson; Malhotra, Namrata, 2008: "Taking Social Construction Seriously: extending the Discursive Approach in Institutional Theory", in: Greenwood, Peter; Oliver, Christine; Sahlin, Kerstin; Suddaby, Roy (Eds.): *The Sage Handbook of Organizational Institutionalism* (London: Sage): 702-720.

Powell, Walter W.; DiMaggio, Paul J. (Eds), 1991: *The New Institutionalism in Organizational Analysis* (Chicago, IL: University of Chicago Press).

Reddy, Sudakhar; Painuly, Jyoti Prasad, 2004: "Diffusion of Renewable Energy Technologies—Barriers and Stakeholders' Perspectives", in: *Renewable Energy*, 29: 1431-1447.

Rumpala, Yannick, 2013: "The Search for "Sustainable Development" Pathways as a New Degree of Institutional Reflexivity", in: *Sociological Focus*, 46,4: 314-336.

Rydin, Yvonne, 1999: "Can we talk Ourselves into Sustainability? The role of Discourse in the Environmental Policy Process", in: *Environmental Values*, 8: 467-484.

Rydin, Yvonne, 2003: *Conflict, Consensus and Rationality and Environmental Planning* (Oxford: Oxford University Press).

Schmidt, Jeremy, 2014: "Water Management and the Procedural Turn: Norms and Transitions in Alberta", in: *Water Resources Management*, 28,4: 1127-1141.

Schmidt, Vivien, 2008: "Discursive Institutionalism: the Explanatory Power of Ideas and Discourse", in: *Annual Review of Political Science*, 11: 303-326.

Schmidt, Vivien, 2010: "Taking ideas and discourse seriously: explaining change through discursive institutionalism as the fourth 'new institutionalism'", in: *European Political Science Review*, 2,1: 1-25.

Scott, W. Richard, 1987: "The Adolescence of Institutional Theory", in: *Administrative Science Quarterly*, 32,4: 493-511.

Scott, W. Richard, 1995: *Institutions and Organizations* (Thousand Oaks, CA: Sage).

Scott, W Richard, 2003: "Institutional Carriers: Reviewing Modes of Transporting Ideas over Time and Space and Considering their Consequences", *Industrial and Corporate Change*, 12: 879-894.

Scott, W. Richard, 2008a: *Institutions and Organizations* (Thousand Oaks, CA: Sage).

Scott, W. Richard, 2008b: "Approaching adulthood: the maturing of institutional theory", in: *Theory and Society*, 37: 427-442.

Selznick, Phillip, 1957: *Leadership in Administration: A Sociological Interpretation* (Evanston, Ill: Row, Peterson).

Shove, Elizabeth, 2011: "On the Difference between Chalk and Cheese—a Response to Whitmarsh et al's Comments on 'Beyond the ABC: Climate Change Policy and Theories of Social Change'", in: *Environment and Planning A*, 43: 262-264.

Shove, Elizabeth, 2012: "The Shadowy Side of Innovation: Unmaking and Sustainability", in: *Technology Analysis and Strategic Management*, 24,4: 363-375.

Shove, Elizabeth; Walker, Gordon, 2010: "Governing transitions in the sustainability of everyday life", in: *Research Policy*, 39: 471-476.

Simon, Herbert A., 1972: "Theories of bounded rationality", in: McGuire, Charles Bartlett; Radner, Roy (Eds.): *Decision and Organizations* (London: North Holland): 161-176.

Smith, Adrian; Kern, Florian, 2009: "The Transitions Storyline in Dutch Environmental Policy", in: *Environmental Policy*, 18: 78-98.

Smith, Adrian; Seyfang, Gill, 2013: "Constructing Grassroots Innovations for Sustainability", in: *Global Environmental Change*, 23,5: 827-829.

Söderberg, Charlotta, 2011: "Institutional Conditions for Multi-sector Environmental Policy Integration in Swedish Bioenergy Policy", *Environmental Politics*, 20,4: 528-546.

Spaargaren, Gert; Mol, Arthur P J, 2008: "Greening Global Consumption: Politics and Authority", *Global Environmental Change*, 18: 350-359.

Späth, Philipp, 2012: "Understanding the Social Dynamics of Energy Regions—The Importance of Discourse Analysis", in: *Sustainability*, 4: 1256-1273.

Szarka, Joseph, 2006: "Wind Power, Policy Learning and Paradigm Change", *Energy Policy*, 34: 3041-3048.

van Koppen, Kris; Mol, Arthur; van Tatenhove, Jan P. M., 2010: "Coping with Extreme Climate Events: Institutional Flocking", in: *Futures*, 42,7: 749-758.

van Leeuwen, Theo, 1993a: "Genre and Field in Critical Discourse Analysis: A Synopsis", in: *Discourse and Society*, 4,2: 193-223.

van Leeuwen, Theo, 1993b: "Language and representation : the Recontextualisation of Participants, Activities and Reactions" (PhD dissertation, University of Sydney, Department of Linguistics).

Williamson, Oliver, 1985: *The Economic Institutions of Capitalism* (New York: Macmillan).

Wodak, Ruth; Meyer, Michael, 2008: "Critical Discourse Analysis: History, Agenda, Theory, and Methodology", in: Wodak, Ruth; Meyer, Michael (Eds.): *Methods of Critical Discourse Analysis* (London: Sage): 1-33.

Wolsink, Maarten, 2012: "The Research Agenda on Social Acceptance of Distributed Generation in Smart Grids: Renewable as Common Pool Resources", in: *Renewable and Sustainable Energy Reviews*, 16: 822-835.

Young, Oran, 2006: "Vertical Interplay among Scale-dependent Environmental and Resource Regimes", in: *Ecology and Society*, 11: 1-27.

Zucker, Lynne, 1977: "The Role of Institutionalization in Cultural Persistence", in: *American Sociological Review*, 42: 726-743.

Zucker, Lynne, 1987, "Institutional Theories of Organization", in: *Annual Review of Sociology*, 13: 443-464.

Zwartkruis, Joyce; Moors, Ellen; Faria, Jacco: van Lente, Harro, 2012: "Agri-food in Search of Sustainability: Cognitive Interactional and Material Framing", in: *Journal on Chain and Network Science*, 12,2: 99-110.

25 New Business Models: Examining the Role of Principles Relating to Transactions and Interactions

Jan Jonker and Linda O'Riordan[1]

Abstract

Different sources indicate signals that our current economic ideas no longer function. New ways of organizing are emerging in which sustainability is often central. This chapter presents the results of exploratory research initiated by Radboud University Nijmegen on new business models (NBMs). The research demonstrates that NBMs appear to be 'hot' and 'happening'. But what is a business model and in what sense is it sustainable? This study focuses on business models that create so-called 'multiple value(s)', which refers to a way of organizing that not only focuses on the task of organization itself, but also on organization between organizations—or better: organizing entities. This approach to organizing generates social and ecological value, in addition to economic value. For the purpose of this research, a series of interviews were conducted in order to gain insight into the phenomenon of NBMs. The aim was to combine this fresh empirical evidence with theoretical underpinnings from previous scholarship in order to explore the field, discover the nature of NBMs, their features, and how they function in (micro-)practice. Ultimately, this examination revealed the phenomenon of an altered balance between the simultaneous organization of different values such as nature, care, attention, and money. While many roads lead to interesting discoveries with respect to these aspects, and the research is still at an early stage, the first results from the study indicate some initial clear common denominators emerging from this journey. These preliminary findings suggest that early NBMs can be generally categorized into different streams based on the practice of sharing, trading, and creating. Most significantly, the results indicate that the ability to connect holds increasing social and economic value, and that these connections create all sorts of new consortia and constituent configurations.

Keywords: Business models, multiple value creation, new economy, sustainable development, collaborative advantage.

25.1 Introduction

This chapter presents the results of exploratory qualitative research initiated by the Radboud University Nijmegen on *new business models* (NBMs). The research study was triggered by various developments (e.g. resource depletion, environmental degradation, rise in commodity prices, grass-roots developments) which indicate that the current economic construct is under duress, suggesting that new ways of organizing are essential in which sustainability[2] might play a more central role (Braungart/McDonough 2009). The study is based on the premise that NBMs could offer part of a potential solution to address a number of these issues. However, the extant research in this field

---

1 Prof. Dr. Jan Jonker, Radboud University Nijmegen, The Netherlands; email: <j.jonker@fm.ru.nl> and Prof. Dr. Linda O'Riordan, FOM University of Applied Sciences, Essen, Germany; email: <linda.oriordan@fom. de>.

2 We consider the 'root' definition of sustainability the one provided by the *World Commission on Environment and Development* (WCED), also known as the Brundtland Commission (1987): "a development which meets the needs of current generations without compromising the ability of future generations to meet their own needs". Core to the thinking underpinning this definition is to consider sustainability as a process, as 'work in progress' and not as a definable entity. While it can be argued that this is a rather 'old' definition, it still offers the best possible option for making sense of sustainability given a specific context, issue, and setting.

---

is still in its infancy. As a result, many analytical questions arise with respect to NBMs regarding their conceivable value as a vehicle to enhance sustainable development. Consequently, their role in fostering multiple value(s) creation as a new way of organizing[3] in order to generate social and ecological, as well as mere exclusive, economic value, requires systematic examination. Significantly, the novel approach to organizing proposed in this chapter not only focuses on the task of organization itself, but also on organization *between* organizations—or better: organizing entities.

The purpose of this research is therefore to explore this field, to discover the nature of NBMs, their features, and how they function in (micro-)practice, with the aim of understanding how they can be effectively organized. To answer these questions, the chapter commences with compact desk research to gain insight from existing literature. Following this review of the literature, a first exploratory study was carried out in the Netherlands, which resulted in the development of a research protocol, identifying a collection of thirty cases based on this protocol, and undertaking a total of twelve interviews.

In a second round of fieldwork a total of 274 cases were collected by different teams of researchers from fourteen knowledge institutes.[4] Based on the quality of the entries (assessed on the basis of the collectively-used protocol), each team selected a number of cases for interviews. These interviews were conducted in twelve countries across Europe. To ensure comparable results, the respective research teams employed the same interview protocol that was developed during the first study, which was based on seventeen 'open' questions. Part of this protocol was that each respondent was invited to draw his or her own business model. These two exploratory qualitative studies were undertaken in the spring of 2012 and of 2013. They are the main empirical sources of this chapter.

To present the results of this study, the remainder of this chapter is structured as follows. The next section sets the context in which the requirement for NBMs is established by undertaking a literature review to examine the major overall developments from which NBMs, as a form of alternative organizing construct, emerge. This includes an appraisal of the key relevant terms, their related underpinning theoretical concepts, and definition. The research design which was employed to collect the data presented in this chapter is then explained, and the preliminary fresh empirical findings which were obtained via the in-depth interviews are revealed. The chapter concludes with a summary of the phenomenon of NBMs, and an epilogue which addresses how it is envisaged that NBMs will develop in the near future.[5, 6]

25.2 Why Think Differently about Business Models?

The world as we think we know it no longer exists. Because we don't have a detailed image of the future, we stick to a world-view based on the ideas of the second Industrial Revolution, which led to enormous growth in prosperity in the years following the Second World War. There are several clues to support the prem-

---

3 Here we define the verb 'organizing' as a dynamic system (as opposed to a structural dimension) which reflects the undertakings in and between organizations. In this definition, organizations are groups of people (or systems) organized for a particular purpose (e.g. their business proposition). In this interpretation, the business undertaking is an organization [system] (as opposed to has an organization [structure]) (See Kutschker/Schmid 2008: 1084-1085 for further details).

4 The word 'knowledge institute' is used here since the fourteen participating teams were linked to 'conventional' universities, universities of applied science, and private educational institutes.

5 Conducting research is mainly a 'people' job. Valid, let alone good, research is impossible without the respondents' time, without dialogue, and without a university's entire infrastructure, consisting of libraries, search systems, employees, and students. Accordingly, we owe a great deal of thanks to everyone who has helped, either directly or indirectly, towards realizing this study. Special credit is due in particular to the interviewees. This research would never have seen the light of day without their time and willingness to share open-hearted insights into their organizational practices regarding new business models. We hope we have lived up to all their expectations, even though it is not always possible to recognize each point of view at an individual level.

6 The original working paper for this chapter was written in Dutch in collaboration with Marloes Tap and with the support of Tim van Straaten. We are indebted to them for their help and backing. The research on which the chapter is based was initiated by the *Nijmegen School of Management* (NSM) of Radboud University Nijmegen in the Netherlands in 2012, and subsequently expanded and repeated in Europe with the help of fourteen scientific partners in 2013. This expansion led to rich and valuable sources of information that is only partly incorporated here. Materials from various sources, together with a series of interviews based on a common protocol, are the basis for this chapter. The results summarized here are entirely the authors' responsibility.

ise that we are currently at such a turning point. Everywhere in the world we can see a combination of crises: a financial crisis, an energy crisis, and a climate crisis. These are global problems of size and complexity that we have never experienced before, let alone solved. There are no guidelines to follow. One thing is clear: we cannot keep heading in the direction we have chosen (Gunning 2011).

This quote from Tex Gunning (former CEO of the global paint maker Akzo Nobel) indicates that our world, our society, our way of organizing and of acting within the economy is changing. Ominously, a growing amount of research suggests that a 'linear' economic-organizational approach is no longer sustainable (see e.g. Meadows/Meadows/Randers et al. 1972; then Brundtland 1987; the UN Millennium Assessment 2001, 2005; and the many reports that followed that initiative), which means that we are, in every possible way, looking for new solutions to shape an alternative way of organizing. Ban Ki-moon, the Secretary-General of the United Nations, recently stated that our current model of economic development and growth is 'suicidal'. "Things really need to go differently, we just don't realize yet that we need to radically turn the wheel," said John Elkington (2012), the inventor of the triple bottom line concept,[7] not so long ago. Such a radically different approach to organizing is often referred to as the 'green', 'blue' or 'circular' economy.[8] The exact similarities or differences between these emerging and therefore not always

clearly demarcated concepts are not relevant here. What is however important is that these approaches strive for a different way of organizing based on the ambition of an economy that is designed in a new way. Of further significant relevance is that this different way of thinking might lead to a transition towards a circular economy, which requires a fundamental change in organizing. This development activates the need for innovative mechanisms that are gradually being termed *new business models* (NBMs).

In general, business models are structured concepts of the interlinkage between resources and competencies linked to specific needs, thus providing a logic for value creation in a specific field or sector. More specifically, they have been defined in previous research (and additionally in greater detail below) as models which illustrate both the network of parties and the different capabilities involved in creating, commercializing, and delivering value (e.g. Osterwalder/Pigneur 2010; Jonker 2012: 14). Here, we make a distinction between conventional and new business models. The logic of conventional business models leads to a purely financially-driven cost-benefit analysis. Hence, in conventional business models, only one (economic) value is central. Thus the strategic purpose of the business is ultimately focused on providing benefit based on the (narrow) interests of the organization as *the* priority, and typically, with precedence, its owners. This logic defines the exclusive economic intent, driven by a profit maximization objective, behind the creation of value in conventional business models. An enhancement to this approach is the NBM, the essence of which is to collectively create more than one value by considering a broader range of values in the cost-benefit analysis that is more broadly shared by a group of people. Significantly, the reasoning underlying this augmented approach leads to different so-called multiple value creation logic (Elkington 1997; Jonker 2012; O'Riordan 2010), one that is community-driven instead of organization-centred. Below we will present some examples of organizing multiple value creation in a collective and shared manner. To position the examples, we will examine some overall societal and organizational developments from which NBMs evolve.

Connecting these overall developments leads to a multitude of movements, pilot projects, and transitions which take the form of so-called 'organizing activities' with an eye to multiple value(s) creation based on collective organizing. Examples include initiatives in which waste becomes food; sewage water becomes a new source of income; or elderly care is

---

7   The triple bottom line concept stands for an approach to value creation where companies are supposed to simultaneously create social, ecological and financial value in a balanced way. This concept is globally used and serves for many companies as a starting point to develop a so-called sustainability strategy.

8   Reference is made to various schools of thought here, all concentrating on exploring a different economic 'design'. The 'green' economy is one that results in improved human well-being and social equity, while significantly reducing environmental risks and ecological scarcities (see <www.unep.org/greeneconomy>). The 'blue' economy was a notion launched by Gunter Pauli. Key to this approach is a circular economy based on ideas of biomimicry. This evolves from a core business based on a core competence to a portfolio of businesses that generate multiple benefits for business and society (<www.theblueeconomy.org>). A circular economy is an alternative to a traditional linear economy (make, use, dispose) in which resources are kept in use for as long as possible, the maximum value is extracted from them whilst in use, and then they are recovered and regenerated into new products and materials at the end of each life cycle (<www.wrap.org.uk>).

organized via a street-based co-operative which generates care, guidance, and education. In such co-operatives, the street residents are the shareholders and care coins serve as the currency. Further examples include a feed-in policy which enables city agriculture to flourish and perhaps in the near future, a regional 'Robin Hood' tax will emerge that offers a constant source of finance for innovative projects in which sustainability as a social and ecological task is central. All of these examples comprise illustrations of organizing between organizations and citizens, between communities and citizens, and so on, and all are based on co-creation.

Underpinning the 'fuzzy' range of activities in the examples presented above are two identifiable movements with respect to a development towards a sustainable mindset, and a transition towards organizing not only within but most significantly between organizations. To elaborate, sustainable thinking and acting is gradually beginning to find a position in business. As a result, nowadays the debate about sustainable development and about organizing in a sustainable manner has gained a solid foothold in almost every economic sector. At the same time, the position of sustainable organizing is also shifting. This means that the focus is no longer only on what happens *inside* organizations. Instead, the challenge of organizing in a sustainable manner increasingly lies in the way in which organizations develop common approaches. As a consequence, sustainability is no longer exclusively a challenge for (existing) companies or innovative entrepreneurs (Jonker 2003). Instead, it appears and has become relevant on several 'aggregation levels' in society, not only within organizations, but also within broader social settings where all kind of actors such as *non-governmental organizations* (NGOs), citizens, networks, and co-operations play a role.

This implies that sustainability must be organized not only within (existing) organizations but also and simultaneously *between* organizations, (new) parties and a whole range of stakeholders. This changing organizational approach indicates that sustainability is moving from a position as a side issue to being a central theme—at the heart of the business proposition.

Conventional business models, like all business models, provide a logic for value creation, but are not designed to create multiple value(s). This is primarily due to the fact that the value of sustainability, like so many other societal and/or environmental values, is not expressed in terms of money, and thus is not taken into account in cost-benefit analysis. The challenge of designing and implementing NBMs lies in

looking beyond financial aspects to a focus on making money *in combination* with aims, such as taking care of one another, creating safety, protecting the environment, and generating social capital. Significantly, if money is no longer the all-decisive central element, it becomes necessary to think differently about money, transactions, trade, and in particular about what exactly is of value in these. In such an approach, sustainability becomes inclusively embedded in this new way of emergent thinking. Different parties (e.g. citizens, companies, NGOs and others) in society have intuitively and deliberately 'sensed' this new development and as a result have begun to design and implement NBMs. These NBMs can be described in a variety of ways such as innovative, risky, special, or entrepreneurial, but the fact remains that a 'widening' group of people—not just entrepreneurs—are proposing these models. Sometimes this happens as a deliberate search for new business approaches, and in other cases it occurs accidentally (certainly not less valuably), simply because people wish to make changes in society. The research upon which this paper is based provides a basis for this observation.

It is therefore not surprising that NBMs appear in totally different settings and that different parties work on NBMs, sometimes purposefully and sometimes unintentionally, almost as a 'by-product'. This happens not just in the 'running of business' as a result of a transition or transformation, but also 'on the street' as a consequence of people's doing business with each other by creating co-operations within the context of a certain function (e.g. nutrition, care, or mobility). This is often also an outcome of innovative collaboration between parties, which was not previously self-evident. Gradually, whether driven by design or chance, this form of collaboration has become known as the 'sharing' or 'collaborative' economy. In this sense, NBMs can be described as the micro-translation of a value-creating logic that underpins this line of economic thinking. While multiple value creation (Elkington 1997) has marked the start of this new way of thinking about values, it also appears that the development of different ways of collaborating is crucial within this change. As a consequence, the principles of collaborative and shared value creation come into play.

Characteristic of the process of developing and experimenting with NBMs is the absence of any kind of central plan or control by a national or local government. Nobody is really steering this process, even though a lot of different initiators and 'players' can be distinguished. While it can be said to hold true for conventional business models as well, the distinguishing feature really is the co-collective creation of a business model based on a configuration of people and companies. In general, it appears that transition movements are often decentralized, driven, and shaped by bottom-up initiatives. Altogether, a beautifully muddy yet rich 'field' of co-operative organizing comes to surface, based on the new and often unusual configurations adopted by the different parties involved, leading to multiple, shared and collective value creation. It is to this topic of new ways of organizing to create multiple value(s) that the next section now turns, and from there to the target of defining NBMs more precisely.

25.3 Confusion of Tongues and the Theoretical Essence of NBMs

Organizations were created and developed to help realize different kinds of collective value(s). To fulfil this task, organizing is not a goal in itself, but rather a means of realizing what is of value for, with, and by each other. Keeping this in mind, three common values include social, economic, and ecological components. Within this context, sustainability can be interpreted as a general (overarching) value in which these common values are embedded. Accordingly, sustainability is not in itself something that needs to be organized. Instead, in essence, sustainability is about organizing that which has value from a societal point of view (based on Jonker/Diepstraten/Kieboom 2011; Jonker 2011). In this vision of value creation, sustainability is embedded in a systematic and coherent way of thinking. Based on this new mindset, sustainability is achieved through collaboration by collectively working on that which is of value. Within this approach, multiple value creation then becomes more than a new way of working—it is a new collaborative ability or a fresh strategy, a novel philosophy which takes the form of an innovative approach for realizing value creation. Consequently it profoundly impacts companies' *raison d'être*. In return, it also impacts the meaning of sustainability which transitions from being a responsibility of governments or companies to becoming a shared endeavour. Ultimately, such a transition might even lead to a new approach to developing business models which could become the norm. In such a scenario, this concept of an NBM could comprise a construct that may itself serve to generate a more sustainable organization.

The following views support this claim. "Value co-creation is not efficient based on the traditional approach to the value creation process. Instead, it requires a complete re-consideration of how a company operates and cannot be approached within the context of a traditional value creation system" (Tanev/Knudsen/Gerstlberger 2009). Moreover, both Simanis and Hart (2009) as well as Porter and Kramer (2011) consider the idea of organizing based on the idea of multiple values to comprise a step towards a new, more inclusive form of capitalism. "Economic growth is an overly restricted concept. It's time to exchange the idea of economic welfare for the idea of total welfare, being about physical, intellectual, social, and spiritual value creation" (Gunning 2011).

Having reviewed a selection of the major overall developments from which NBMs, as a form of alternative organizing construct, emerge, and having examined new ways of organizing to create value, we now turn to the task of attempting to define NBMs.

25.3.1 Definitions

Many definitions of business models are in circulation. While this is not the place to sum them all up and make an extensive comparison, many of the definitions share some common features. A simple standard working definition is: a business model describes the organizational logic of the process of value creation and delivery. A second definition of a business model is: the main organizational logic for creating value (Linder/Cantrell 2000). As a result, a business model provides insight into the value that an organization or co-operation can offer to different parties. In doing so, it depicts the different abilities and network partners needed for creating, marketing, and delivering that value (Osterwalder/Pigneur 2010).

What can be derived from the different definitions of business models (Bertens/Statema,2011; Houtgraaf/Bekkers 2010; Osterwalder/Pigneur 2010) is the implicit assumption of organizing *within* an organization. Organizing is something organizations 'do'. This presumes an identifiable organizational entity that intentionally translates the value proposition (the value promise) into the value that is to be delivered. The organization is then the primary subject in organizing that proposition through which stakeholders and customers, in an implicit monetary transactional relationship, express their perceived value for that proposition by buying it. It is clear that organizations are rarely capable of organizing the provision of value propositions on their own; they practically always work within the context of a (value) chain. Many authors agree that the essence of value creation lies in what Osterwalder c.s (2010) termed "the architecture of the organization and the network of partners". Collaborating on value creation is therefore, by definition, a collective task of value chains and/or networks. Yet the consequence of this line of reasoning is that in the course of creating value, customers often do not have a choice in shaping their own logic; they are 'framed' in the value chain as predefined by the organization. This does not however mean they might not have some degree of actionable logic with which to partially create their own value. Still, if that is the case, this individually-driven part of value creation is trapped and limited within the restraints of the overall predefined logic as designed by the organization 'in charge'.

Figure 25.1: The value creation logic behind conventional business models. **Source:** The authors.

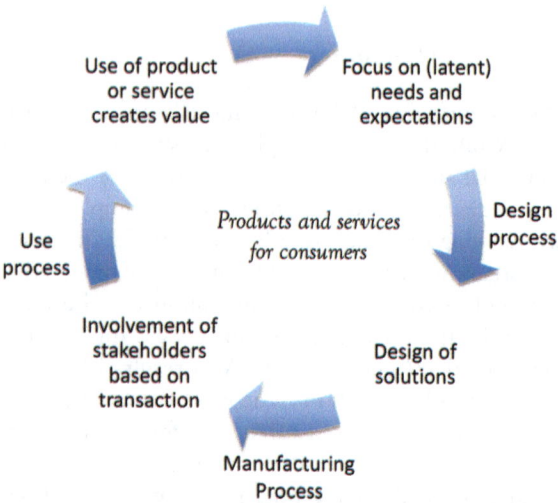

The thinking sustaining current, conventional business models is the idea of delivering products and services that are better, faster, more economical in their use, a bit 'greener', and preferably cheaper compared with their competitors; this has been the leading and often implicit thinking behind creating a 'competitive edge'. Adversely, this is the basis for a continuous improvement mindset which leads to incremental change. In this approach, the key to doing business remains in a conventional transaction model that is based on the value of money. Accordingly, sustainability is threatened within the boundaries of the organization or together with other partners in the value chain. The underpinning profit paradigm is not called into question, leading to a business case that tries to

'greenify' the actual business proposition. This, in turn, leads to strategies that opt for 'less' (less water, less oil, less energy, etc.). In such an approach, it is common to term the resulting type of strategy as eco-efficiency. All in all, it leads to tactics based on first-generation thinking about sustainability, in which the business model—let alone the nature of the transaction or the value creation—is not fundamentally questioned.

Figure 25.2: Process NBMs. **Source:** The authors, based on Shafer, Smith, and Linder (2005) and Simanis and Hart (2009).

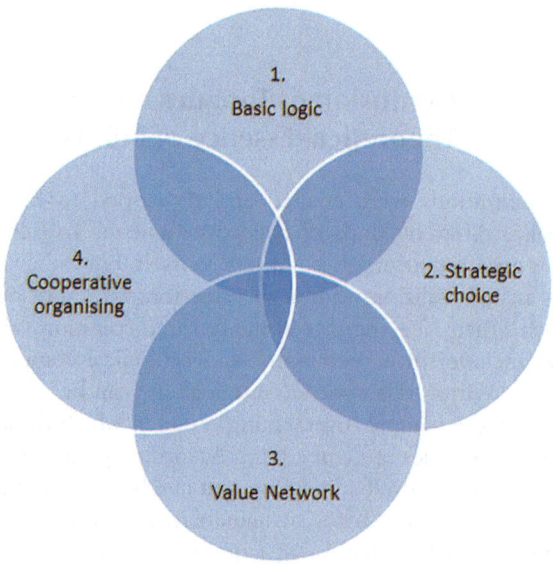

Bertens and Statema (2011) lament the fact that they have not been able to find 'earning' models that were not based on traditional economic thinking in their explorative research on business models. That observation calls into question what needs to be 'earned' within the construct of a business model. One could propose that in the quest to qualify earnings for NBMs, values such as a sense of belonging, attention, security, enjoyment, safety, care, etc. might prove beneficial. Fundamentally, this debate critically questions the nature of the value created in the conventional model.

It is important that NBMs are not only sustainable because of the changing logic of earnings based on principles leading to a different kind of transaction, but that they additionally intrinsically possess an innovative value proposition. A good example of such a model is the Dutch construction company Dijkhuis. This company migrated from cost leader (a traditional strategy in the contractor scene) to product leader.

Sustainability, according to their idea, is about considering buildings from a fundamentally different point of view, one which involves selling a concept rather than just a building. In their approach, the value proposition is not just about the sustainable construction of the building, it is about selling and managing that building's life-cycle—as a company in co-operation with parties in and around that building.

Figure 25.3: Value creation process NBMs. **Source:** The authors, based on Simanis and Hart (2009).

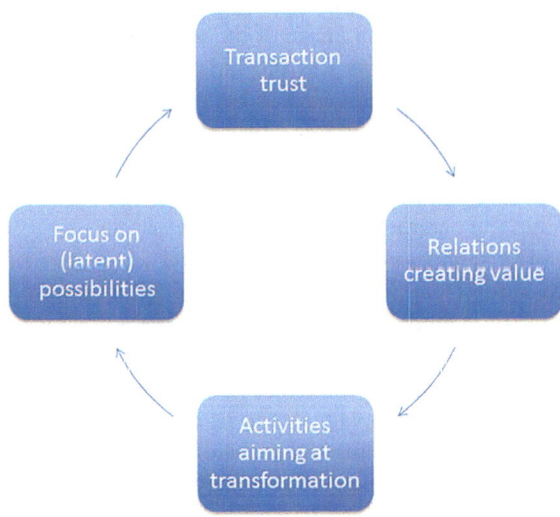

Preliminary research suggests that developing an NBM is about organizing something new not just within an organization, but within the entire value network. Characteristic of a value network, compared with a classical value chain, is the high level of equality among the parties involved. In this approach, the organization(s) and the different stakeholders involved work together on an equal basis. This vision of sustainable business models further provides insight into the way in which the output of some initiatives serves as input for other parties within the value network, enabling abundant amounts of surplus value. Therefore, it is not the thinking of one party on how to organize sustainably that counts, but rather how the entire value network can collectively participate in the process of value creation.

The rationale underpinning NBMs (see figures 25.2 and 25.3) is the fact that companies *and* communities build a collaborative relationship based on recognized and enacted mutual responsibility. The following four building blocks can be distinguished:

1. a basic logic (principles and ideas which the parties share);

2. strategic choices (which the parties make);
3. ways of organizing that lead to co-operation;
4. developing and maintaining a value network.

Together, these building blocks which enable mutual value creation simultaneously serve as an entry barrier to competitors. The embedded earning model leads to collective property. Outcomes do not belong to one organization, party, or individual. Thus, a new form of advantage takes shape based on collaboration instead of competition, leading to what we term 'collaborative advantage'.

Following this review of the literature presenting the key relevant terms and their related underpinning theoretical concepts, the next section explains the research design employed to collect the fresh empirical findings presented in this chapter. The focus of that research design was to answer the research questions that arose from the gaps identified in the above review of past scholarship with respect to the nature of NBMs, their features, and how they are, or can be, organized.

25.4   Research Design: Searching for NBMs

Given the melange of debates on the theme of NBMs and organizing to achieve sustainable development identified from the literature review presented above, empirical investigation was undertaken in order to attempt to shed more light on the subject. In overview, based on the information established via the compact literature study carried out to investigate the nature of business models, two qualitative exploratory studies were conducted in the spring of 2012 and of 2013. In addition, at different stages of the research, a series of interviews were conducted with people working on NBMs. Overall, however, this study does not claim to be complete, since not all of the material collected over the years during which the research was undertaken has been used.

To elaborate, in order to carry out the empirical research to address the issues identified in the previous section, the actual search for NBMs necessitated great entrepreneurial creativity. During the summer of 2010, two databases were initially assembled which were focused on the question of *how people organize sustainability*. The research soon provided 300 Dutch examples and 400 British examples. These databases (one based on Dutch and one based on British examples) were created on the basis of (Internet) desk research guided by the characteristics of NBMs revealed in the initial 2011 study. We hoped to find some good examples of NBMs among these, but their identification was not so obvious. More specifically, many websites provide little or no information on their business model, let alone make it possible to use this public information to make a distinction between transactions, earning money, and (multiple) value creation. Moreover, during the course of this endeavour, we found that when new entrepreneurs were asked about their business models, the answers varied according to a wide range of notions regarding the meaning of the term 'business model' itself. Following several analytical attempts, and due to the unsatisfactory results obtained, this approach was abandoned.

Subsequently, in a new approach, a search was initiated for people who claimed to work with NBMs (see below for further details). This initially led to the identification of various names, which led to the unearthing of new names at our request. Methodologically, this is a fine example of the so-called "snowball method" (e.g. Goodmann 1960; Robson 2002: 265). However, it was not clear in advance whether the potential respondents are actually working on NBMs—and if so, to what degree. As a result, the research aimed to discover what these targets were developing, while the fundamental explorative nature of the research did not prescribe in advance what the result of that research would be. Therefore, this methodology is a fine example of the classical dilemma of Baron Munchausen, who once pulled himself metaphorically out of the swamp. More specifically, resulting from this approach, a list of people who say they work on NBMs was generated in the research design, but since the notion of what an NBM denotes is only 'vaguely' known, it is not possible to identify the result *ex ante*. Consequently, it is only possible to know that we have found an NBM when this subsequently occurs (i.e. *ex post*).

Based on the literature study and several (internal) conversations, an interview protocol was constructed in an attempt to bring structure into the approach for conducting the interviews. Key questions included:

1. Can you tell us something about your business model and why it might be new?
2. What is meant by 'sustainability' within that model?
3. When and why did you start developing an NBM?
4. Which requirements should be met by an NBM which is also focused on sustainability?
5. How would you categorize the business model and what is your particular 'logic' for creating value?
6. What kinds of value does the business model create?
7. Did you start new collaborations or are previous collaborations strengthened for or by the new model? (If so, with which actors, and how?)
8. Did you encounter new products and markets by means of the NBM?
9. What results do you think have been accomplished by the NBM?

These questions were employed in the interviews, during which respondents were also asked to actually sketch a graphic illustration of their model. Although this is a relatively simple request, it is valuable to grasp on paper the internal image of the respondent's 'entrepreneurial activity' via this diagrammatic visualization. While it would have been even better to have additionally asked and identified via a process of reasoning how that 'virtual image' is consistent with the interview data, unfortunately there was not enough time to do so within the parameters of the interview setting.

The target sample included a total of a series of 274 cases from which a selection of 50 interviews[9] were conducted during 2013 by two different teams of researchers from fourteen knowledge institutes[10] spanning twelve countries. Now that the research design has been explained, the next section presents the preliminary findings from the data collected using the interview method outlined above.

---

9 Further details of the raw data are available upon request by email, should they be required.

10 The knowledge institutes involved included: Austria (University of Graz—Institut für Systemwissenschaften, Innovations- & Nachhaltigkeitsforschung), Belgium (Free University Brussels—Solvay Business School), Croatia (University College for Economics, Entrepreneurship and Management "Nikola Subic Zrinski"), France (ESC Toulouse Business School), France (Université Paris-Dauphine & Institut d'Administration des Entreprises de Paris-IAE), Germany (Universität Kassel—Wirtschaftswissenschaften—Nachhaltige Unternehmensführung), Germany (FOM University of Applied Sciences, Hochschule für Ökonomie & Management gemeinnützige GmbH, KompetenzCentrum for Corporate Social Responsibility), Ireland (University of Limerick, Kemmy Business School, Department of Accounting & Finance), Lithuania (Kaunas University of Technology, Faculty of Economics and Management, Department of Business Economics), Netherlands (Radboud University Nijmegen—Nijmegen School of Management), Poland (University of Łodz, Faculty of Management), Portugal (ISMAI—Instituto Superior da Maia), Switzerland (University of Applied Sciences and Arts Northwestern Switzerland FHNW, School of Business), and Turkey (Yaşar University, School of Economics and Administrative Sciences).

25.5   Sorting the Outcomes

During the pre-exploration and construction phase of the two databases for the study, we first attempted to discover several 'streams' or 'categories' of new sustainable business models. The purpose was not so much to categorize the models, but much more to discover where the common denominators lay, as well as to discern the value creation logic underlying NBMs. This step led to the discovery of three value-creating patterns of interaction: *sharing, trading,* and *creating*. During the interviews, most respondents appeared to place their model in one or more of these categories. The respondents also had a couple of interesting 'variations' to these streams such as mesh-working, bargaining, and collaborative use. However, the decision was made to initially focus on the following three categories.

25.5.1   Sharing

The sharing of social capital, time, and ability emerged as a recurring element in many NBMs. Sharing, in this context, means to collectively use specific assets in various degrees. The investigated models were frequently based on a variety of collaborations such as sharing people, ideas, equipment, property, data, and transport. Many of these models resemble co-operations, i.e. a form of organizing that appears to be re-emerging in popularity. This contemporary turn to the concept of 'conventional' co-operations could be interpreted as an attempt to tackle the various malfunctioning issues in the current way of organizing. Key characteristics of this approach include, for example, sharing private property, insuring differently, and collective financing or collective purchase discounts. Sharing knowledge and networks also seems to be an important basis for conducting business in many of the models that were investigated; it leads to a *raison d'être* and growth. However, as Ronald van den Hoff (Seats2Meet) put it: "How do you create a value network? How do you even create a network? It's the challenge a lot of organizations are confronted with. We believe the characteristic of a value network compared to a value chain is the high level of equality." The term 'mutuality', which was mentioned several times during the interviews, is related to this idea. Consequently, this evidence suggests that sharing of tangible and intangible assets between different parties is the essence of such NBMs.

25.5.2   Trading

Trading, meaning transactions in the form of barter based on the transfer of previously-negotiated value, emerged as a second stream of value-creating logic NBMs. Many NBMs entail transactions with alternate payment methods, such as points, credits, advertisements, tweets, time, and savings systems. Deploying these alternative means could be a way of stimulating certain demand and achieve (micro-)behavioural change. The logic of value creation in a business model is based on realizing transactions without any means of payment. The focus then shifts to trading services; for example, 'If you maintain that website for us, we'll take care of your administration' or 'If I can borrow your roof to put solar panels on it, I'll pay your rent'. The models are characterized by basing the transactions not only on money, but additionally on the social value(s) of capital, networks and attention, as well as on the sources of the capital being organized. The underlying logic is that if people deploy these in practice, they can generate value for themselves and their environment other than profits.

25.5.3   Creating

Creating multiple value(s) leading to win-win situations via a set of varying methods that are used simultaneously is a further characteristic of NBMs. As Tom Vroemen (CrowdAboutNow) put it: 'The business model generates the values of knowledge, social value (such as involvement), security, and trust.' For example, a business model can save energy and reduce CO_2 emissions while concurrently creating economic profits. The transaction model is then aligned with the mutual goals of a range of different interests. There are other ways to create a win-win situation, such as allowing the customer to determine the value, relocating funds, or shifting the concept of ownership. With respect to allowing the customer to determine the value, Google's organization of its pay-per-click[11] approach to advertising, where the final charge to the advertiser is partly determined by the client themselves or via a form of 'bidding process', is a case in point (Farris/ Bendle/Pfeifer et al. 2010). Such approaches were aptly described by Ronald van den

---

11 Also labelled cost-per-click, this is an Internet advertising model used to direct traffic to websites, in which advertisers pay the publisher (typically a website owner) when the ad is clicked. It is defined simply as "the amount spent to get an advertisement clicked".

Hoff (Seats2Meet), who stated: "We're going from value chains towards value networks. We facilitate the unexpected relevance of the meeting. Because that's apparently the surplus value we as producers have to offer." These approaches lead to the creation of economic, social, and ecological values which are often overlooked in a traditional business model. Simone Veldema (GeenbizzStartup) put it this way: "What's new about it, is many companies now looking for ways to create value. It's about sharing value. If you map it, you can look at new models more and more. It's what we know as value mapping."

These three streams share the common characteristic that connecting and reciprocity are central to the value network. This suggests that without linkages and collaborations, nothing can be traded, shared, or created. Therefore, the art and ability to connect is interpreted as an important aspect of the new way of doing business. Through those connections, all kinds of new consortia and configurations of parties come to the surface, including collaborations of different and often not-so-obvious partners or 'unlikely' or uncommon alliances. Private individuals, local governments, neighbourhood initiatives, and large commercial companies are thereby becoming connected with each other in newly-created structures.

Several values lie behind this development. The condition for the transformation is that sustainable business models need to realize, facilitate, and maintain these connections; after all, organizing sustainability emerges *between* companies, rather than just *within* them. Interviewees agreed on this point and stressed the fact that social embedding—or anchoring—is an important aspect of NBMs in society. Assuming this condition as an intrinsic premise of the new models for organizing, it then becomes difficult to think in terms of ownership—that is, who possesses the network? Such constructs can only function in terms of *access*. Ultimately, this suggests that significance appears to lie in *making connections* and *creating involvement*.

25.5.4   Other Categorizations

Besides the sharing, trading, and creating streams discussed above, the respondents were additionally asked to be creative and come up with additional categories themselves. Among the many categories which the respondents mentioned appeared so-called 'freemium' models, collective purchase models, matching models, models based on copying nature (biomimicry), circle and life cycle thinking, boosters,

thinking in shifts, and collaborators. In the words of Simone Veldema (GreenBizStartUp): "If you look at the Earth's functioning, it has been working well for so long, it's a self-providing system. Businesses or models should work the same way." A lot of these concepts and underlying ideas scratch the surface of the categorizations we have used, namely sharing, trading, and creating. All of these alternatives further demonstrate that business models generate multiple value(s) in their own way. Taking a more analytical stance, the 'building blocks' that constitute these models appear to be always the same. Apparently, a sustainable business model must meet the following four criteria: (1) sharing knowledge, (2) making connections between various constituents (more than just companies), (3) being aware of the potential for collaborating in networks, and finally (4) constituents must share the idea of multiple value(s) creation. The organizing task is then about the specific configuration of those elements in a certain context, which gives the model its unique character.

25.5.5   Aggregation Levels

Apart from a categorization based on the streams that we have chosen, NBMs can additionally be positioned at different 'aggregation' levels. This means that models can be positioned at a street, neighbourhood, or village level. To elaborate, this could be based on location, property, or function. Examples include a location level on a street (e.g. the location of the energy generation), a building (e.g. building and maintaining a school, including safety), or a function (e.g. a care co-operation). A business model could accordingly have the organizational logic of the value creation and delivery process as its target, or it could be focused on a region or even on a country level.

The research further revealed a number of business models that can be positioned at an over-arching level. These are focused on making the entire chain, market, or sector more sustainable. These models try to achieve local change by altering a sector or value network on a higher level. Lucas Simons of SCOPEinsight aims to accomplish this in the agricultural sector by focusing on the lack of access to financial backing experienced by farmers in East Africa. Simons attempts to solve this issue by increasing the transparency between farmers and the banks by means of a rating system which is designed to map the farmers' credibility. In this way, the model is focused on market transformation. Another example of a business model on this level is the *Equal Opportunity Model*

(EOM) of Diederik van Duijn. The purpose of this model is to redesign the traditional value chain model by eliminating some of its steps. This particular model tries to realize a new way of directly involving the end-users in the production chain. This way, the farmer receives a fair price, the consumer is not overcharged, and sustainable products move into the mainstream.

The various streams and tentative distinctions in the levels of NBMs that were developed in this research and presented in the findings above are anything but complete. It is therefore entirely possible that some factors may have been overlooked. Accordingly, due to the relatively short amount of time within which this research was conducted, and the large variety of NBMs existing in the field, it is likely that more categorizations will be found. In that regard, for clarification, due to the requirement to present concise results, not all of the possible categorizations and variations which emerged in the collected data are elaborated upon, and/or illustrated in detail in this chapter. Nevertheless, supplementary categorization definitely deserves added attention in further research that could well reveal organization mechanisms which do not currently receive much, or indeed any, attention.

25.5.6 Selected Inspiring Examples of NBMs

As noted previously, it is not particularly easy to find NBMs. Even when we do have a 'hit', the nature of the model's conceptualization, development, or practical experience is not immediately apparent. For instance, this may frequently arise due to a lack of background experience; websites may have been built, but may not have been maintained, and as a result detailed information is missing. This can make it difficult to understand the characteristics and mechanisms that lie behind a particular NBM, as well as creating challenges when talking about a business model in progress. Nonetheless, we did manage to find a number of good examples of NBMs, which are briefly addressed in this section.

In overview, it is evident that each model exhibits its own 'way' of sharing, creating, or trading. Many examples of the NBMs investigated in this research can be placed into the *sharing* group, a category to which almost every respondent unfailingly reacted. According to Koen Sieben, the founder of the *Centre for Young Entrepreneurship* (CvJO), sharing is in his organization's 'genes'. The business model is based on the idea of sharing knowledge and networks among affiliated entrepreneurs without presenting a bill.

Sharing knowledge also appears to be important to the consultancy organizations interviewed that were focused on sustainable entrepreneurship. Boukje Vastbinder of Enyini indicates that the innovative aspect of their concept is the focus on 'passing on' knowledge, rather than keeping it hidden internally, such as in the conventional approach adopted, for example, by McKinsey. The 'sharing' category accordingly demonstrates that organizations with sustainable business models definitely do not strive to be protective; instead, as Camille van Gestel of Off Grid Solutions says: "They should have an open attitude". However, knowledge and networks are not the only items which can be shared. An example of collective funding is CrowdAboutNow, which transforms borrowing money into a peer-to-peer process for entrepreneurs by taking the banks out of the equation. In this initiative, entrepreneurs borrow money from a large group of people instead of one party, which spreads the risk.

There are also some interesting examples of *creating* within the stream. Boukje Vastbinder and Esther Blom of Enyini believe that their business model can lead to social and ecological, as well as economic, profits. These profits include values such as involvement, stimulation, and inspiration, but also the spread of risk, improved quality of air, and biodiversity. Nils Roemen of the Waarmakerij has identified that his customers determine the value of a service in retrospect. This is because the system of trading time for money, which stems from the industrial age, no longer applies. Apart from that, business models are based on giving new life to social surplus value, which is over-abundant, and will not be used when it is located in the wrong place. Another phenomenon within the creating stream is the movement of property. The process is less about transactions of property and more about transactions of values, as stated by Douwe Jan Joustra of TurnToo. The process involves using products rather than owning them, and this touches on the concepts denoted in the *trading* stream outlined above. For example, people do not need a lamp when they desire light; they do not want chairs, they want to sit. TurnToo is an initiative which helps to enable this process by mediating between the supplier and end-user. There is also a transition from consumer to producer and a shift of funding. Vincent Mooij has provided a good example of this with his concept of MyEnergy: the essence of the model is that there are no more fixed payments to the energy supplier; instead, you turn this payment into a bank loan and then invest in solar panels so you can generate your own energy.

25.5.7   Financial Sustainability

In addition to the three above-mentioned categories, the interview analysis revealed a further essential aspect. Almost every respondent believed financial sustainability was inseparably linked with a new sustainable business model. This highlights a focus on the requirement for a business model to generate money, and that focus emphasizes that the returns should ultimately outweigh the costs. After all, guaranteeing continuity is another form of being sustainable. This evidence suggests that NBMs do not exist just to 'contribute towards a better world'. Organizations are increasingly beginning to see NBMs as both a prerequisite for and a means of taking a stand in society, possibly because they appreciate that there is no alternative.

25.6   Key Features of NBMs

This provisional list registers seven features of organizing NBMs.

1. NBMs can be described as a form of co-operative collaboration, as a central element in which doing business is the art of the new collaboration, and where connecting increasingly drives social and economic value.

2. Deliberately creating multiple value(s) is a key attribute which aims to achieve a balance between values such as nature, care, attention, and money. Part of this involves cultivating perception or community-building regarding a product and/or service.

3. Money is no longer the only means of trade; time, energy, or care can also be earned, deployed or exchanged. This can be extended to sharing profits with participants.

4. The development of an economy based on needs and uses (now and in the future) which consequently employs credit books for energy, warmth, vegetables, or care, for example.

5. Ownership of property or the means of production is no longer central—access to these resources is perhaps more important. This could mean just paying for use, not for ownership, or 'relocating' the ownership (for example, such that the producer remains the owner).

6. Parties expressing and securing long-term commitment to each other: for example, if I am now earning 'care credits' that I will not need for another ten years, there needs to be a large amount of trust

in the relationship (i.e. governance and confidence) to ensure that these credits will be 'kept safely' and subsequently available when the need arises.

7. The use of alternative 'money' (time, care, points, etc.). Experimenting with complementary local money such as the approach taken by the *local exchange trading system* (LETS) (also known as local employment and trading system or local energy transfer system) in addition to 'care points', time dollars, and others.

Having presented the research findings which detailed the nature and features of NBMs, the next section discusses the future development of NBMS; subsequent sections then summarize and conclude the chapter.

25.7   Future Development of NBMs

How these NBMs will further develop in the near future can be perceived as an exciting journey of discovery in which conventional and new thinking will merge into innovative and not-yet-imagined forms of 'business' based on transactions with regard for a broader range of value(s). Here and there we can even see cautious attempts towards a paradigm shift (and the current system really does need to change), but it is far too early as yet to speak with any degree of confidence about such developments. However, as Tom Vroemen (CrowdAboutNow) put it: "In order to make a true leap forward with sustainable business models, there needs to be a component expressing the intrinsic motivation of the entrepreneur and the others involved to be truly sustainable." In another interview, Matthijs Sienot said: "If I wanted to make a lot of money, I would have chosen a different road. Let me put it differently: if I wanted to make a lot of money *fast*. But we chose to primarily make a difference, make a useful contribution to a transition which we believe has to take place with the talents you've got."

It is important to note that the generation of actual new 'business' models does not use the word 'business' in its classical meaning. Consequently, this new phenomenon can be viewed within the context of transaction models and of trade relations based on alternative forms of value(s). It is clear that there is a movement in society, often bottom-up, in which people try to reinvent transaction models with an altered view of value(s). While this movement is not surprising, given the social and emotional impact of the many institutional failures of the conventional monetary systems which continue to occur without interruption, ultimately it is still early days. Many initiatives are fragile and need time to demonstrate their viability in the longer run. To become a more interesting experiment, NBMs require engagement and governance in the long run. Significantly, however, it is not yet clear how this can be secured, either socially, legally, or institutionally—to name just a few of the system requirements in which it needs to be embedded.

25.7.1   What is Standing in the Way of a Breakthrough for NBMs?

This study has shown that NBMs are emerging. They emerge in different sectors and streams and at different levels of aggregation. While they cannot yet be pronounced fully mainstream, they are becoming increasingly easy to find and to recognize. Nonetheless, there are two factors preventing a major breakthrough by NBMs.

Firstly, many respondents feel that governments have not yet joined the transition towards the so-called green or blue economy. Many financial and legal systems do not work in favour of NBMs, primarily because these systems do not recognize value other than economic profit; they thereby prevent innovation because they are still based on traditional transaction thinking, with money as the central unit of trade. The government could make things much clearer—especially in these times of crisis and uncertainty—by stimulating and facilitating a clear transition towards a green economy in every possible way. Fiscal law could play an important role here. Unblocking these barriers is crucial for the speedy creation of a new level playing field.

Another obstacle preventing the breakthrough of NBMs is that there needs to be a change of paradigm. We need to start thinking differently and to dare to do things differently, but we fail to do this, because we are waiting for the next generation of technology, the right simulating measures, the right moment, or sufficient security. Many of these factors are still missing. Waiting for that 'next-generation technology' is precisely one of the reasons for the lack of such a growth in volume so that prices can quickly decline. Another aspect here is the so-called 'rebound effect': once people adopt measures which lead to more eco-efficient behaviour, they often afterwards undertake behaviour which counteracts this activity (e.g. they tend to leave the lights on longer since it does not cost more). As a result, without the necessary aware-

ness, such behaviour ultimately proves counterproductive to what we were actually trying to achieve in the first place.

A potential limiting factor in the enablement of NBMs is the lack of conventional data, something which might impede the development of thinking and acting based on the circles of trust noted above. The explorative research presented in this chapter does not provide data about the volume of actual returns that this new generation of business model is generating. No insights are given either by a conventional ruler (money) or an alternative ruler (value), let alone any visions into the potential they could generate. In other words: is this work on NBMs a marginal phenomenon (which will always be there), or are there new developments that, despite not yet being part of the discussion, will quickly grow in size and become an important part of the game?

It is tempting to simply give the classic academic answer 'more research is needed'. While this is true, it is worth asking whether the phenomenon of NBMs should continue to develop further. The answer to this question might be akin to when Henry Ford was asked if he had done market research before launching the successful and groundbreaking Model T Ford. His answer was: "No, because people would only have asked for a faster horse." In order to encourage progress in the right direction, it may consequently not be the best idea to 'ask the frogs about whether to drain the pool', as they say in Germany. However, an optimist would at this point note that the examples of NBMs presented in this work indicate that there is at least some light at the end of the tunnel. Or, as they say in Africa, 'not even frowning frogs can stop the cows from drinking from the pond.'

25.8   Summary

To gain greater insight into the phenomenon of NBMs, this research has explored their features, how they function in (micro-)practice, and possibly most crucially, their conceivable value as a vehicle for enhancing sustainable development and entrepreneurship via new ways of organizing in order to generate multiple value(s). Ultimately, this study has examined the phenomenon of an altered balance between the simultaneous organization of different values such as nature, care, attention, and money.

The first results from this investigation suggest that early models can be generally categorized into three different streams, based on the practices of sharing, trading, and creating. Furthermore, the findings indicate that key components of almost all of the models under investigation include activities related to working to create an experience or a community or both that has to do with combining a product or products with a service or services. The actual value proposition is, as a result, and almost by definition, a product-service construct. Within this context, co-operative collaboration emerges as a central element. Organizing then becomes the art and the capability of creating and maintaining these new forms of collaboration. Most significantly, the results suggest that the ability to connect holds increasing social and economic value, and that these connections create all sorts of new consortia and configurations of parties.

25.9   Conclusion

Right at the interface of connecting, community-building, and co-operation is the location where renewal, innovation, and thus NBMs come into existence. An interesting feature of this new scenario is that money no longer predominates as the only means of trade. Economic traffic is based on 'exchanging and satisfying' needs, which among other effects, means that having 'access to' the means of production becomes more important than actually owning them. Ownership and control make way for use and employment.

However, it is only possible to organize such an economy co-operatively based on long-term commitment. This renders securing trust in relations via collaboration a necessary condition. How to actually shape this scenario raises a whole series of new questions.

Consequently, this research merely represents the beginning of a journey of discovery with respect to NBMs in which sustainability is central in several ways. Further exploration will probably reveal more streams, categories, features, and values. Accordingly, this chapter does not pretend to be complete; questions for follow-up research based on these conclusions are already waiting. Nonetheless, this work clearly demonstrates that a transition has been set in motion which is rapidly changing the conventional practice of merely economic business models.

We live in a time of great transition, in which we are searching together for new transaction models as mechanisms for achieving collaborative advantage in a new (sustainable) economy. But because everything we do and do not do is based on a financially-driven transaction model, it is difficult or sometimes even impossible to escape reality in such a way that brings new transaction models to the surface and, by using them, demonstrate their feasibility and worth. Hopefully, the insights provided in this chapter have contributed towards demonstrating the significance, viability, and value of NBMs.

References

Bertens, Coen; Statema, Hidde, 2011: "Business models of eco-innovations: an explorative study into the value network of the business models of eco-innovations and some Dutch case studies"; at: <http://www.rvo.nl/sites/default/files/bijlagen/Business%20models%20of%20eco-innovations%20-%20december%202011%20.pdf> (20 December 2014).

Braungart Michael; McDonough, William, 2009: *Re-Making the Way We Make Things* (London: Vintage Press).

Brundtland, Gro Harlam, 1987: *Our Common Future* (Oxford, UK: Oxford University Press).

Elkington, John, 2012: "Het onmogelijke duurt altijd iets langer", in: *The Optimist*; at: <http://theoptimist.nl/het_onmogelijke_duurt_altijd_iets_langer/> (15 December 2014).

Elkington, John, 1997: *Cannibals with Forks: The Triple Bottom Line of 21st Century Business* (Gabriola Island, CA: New Society Publishers).

Farris, Paul W.; Bendle, Neil T.; Pfeifer, Philip E.; Reibstein, David J., 2010: *Marketing Metrics: The Definitive Guide to Measuring Marketing Performance* (Upper Saddle River, NJ: Pearson Education, Inc.).

Goodman, Lenard A., 1960: *Snowball Sampling: The Annals of Mathematical Statistics* (Chicago: University of Chicago).

Gunning, Tex, 2011: "Weten wat van waarde is 2", in: *The Optimist*; at: <http://theoptimist.nl/weten_wat_van_waardeis/> (15 December 2014).

Hart, Stuart L.; Milstein, Mark B., 2007: *In Search of Sustainable Enterprise*; at: <http://www.stuartlhart.com/sites/stuartlhart.com/files/VALUE_INCLUSIVE_p36-43.qxd_.pdf/> (date).

Houtgraaf, Dirk; Bekkers, Marleen, 2010: *Business modellen: Focus en samenhang en organisaties* (Culemborg, NL: Van Duuren Media).

Jonker, Jan, 2012: *New business models. An exploratory study of changing transactions creating multiple value(s)* (Nijmegen, Netherlands: Radboud University, Nijmegen School of Management).

Jonker, Jan; Diepstraten, Frans.; Kieboom, Jos., 2011: *Inleiding in maatschappelijk verantwoord en duurzaam ondernemen; Chronologisch overzicht en Verklarende woordenlijst* (Deventer, NL: Kluwer).

Jonker, Jan, 2011: *Duurzaam Denken en Doen: Inspiratieboek voor onze gezamenlijke toekomst* (Deventer, NL: Kluwer).

Jonker, Jan, 2003: "In Search of Society: Redefining Corporate Social Responsibility, Organisational Theory and Business Strategies", in: Batten, John A.; Fetherston, Thomas A. (Eds.): "Social Responsibility: Corporate Governance Issues", in: *Research in International Business and Finance*, 17: 423-441.

Kutschker, Michael; Schmid, Stefan, 62008: *Internationales Management* [International management] (Munich: Oldenbourg)

Linder, Jane; Cantrell, Susan, 2000: *Changing Business Models: Surveying the Landscape.* Accenture Working Paper, Institute for Strategic Change; at: <http://course.shufe.edu.cn/jpkc/zhanlue/upfiles/edit/201002/20100224120954.pdf/> (15 December 2014).

Meadows, Donella H.; Meadows, Dennis L.; Randers, Jørgen; Behrens III, William W., 1972: *The Limits to Growth: A Report for the Club of Rome's Project on the Predicament of Mankind* (New York: Universe—Earth Island).

O'Riordan, Linda, 2010: *Perspectives on corporate social responsibility (CSR): Corporate approaches to stakeholder engagement in the pharmaceutical industry in the UK and Germany.* Doctoral thesis, Bradford University School of Management, Bradford, UK; at: <http://ethos.bl.uk/OrderDetails.do?uin=uk.bl.ethos.545644> (15 December 2014)

Osterwalder, Alex; Pigneur, Yves, 2010: *Business Model Generation: A Handbook for Visionaries, Game Changers & Challengers* (New Jersey, NJ: John Wiley & Sons).

Porter, Michael; Kramer, Marc 2011: "Creating Shared Value: How to Reinvent Capitalism and Unleash a Wave of Innovation and Growth", in: *Harvard Business Review*, 89,1-2: 62-77.

Robson, Colin, 22002: *Real World Research* (Oxford, UK: Blackwell Publishing).

Shafer, Scott M.; Smith, Jeff H.; Linder, Jane C., 2005: "The power of business models", in: *Business Horizons*, 48,3. 199-207.

Simanis, Erik; Hart, Stuart, 2009: "Innovation from the inside out", in: *MIT Sloan Management Review* (Cambridge, MA: MIT Press); at: <http://www.stuartlhart.com/sites/stuartlhart.com/files/50414.pdf> (15 December 2014).

Tanev, Stoyan; Knudsen, Mette; Gerstlberger, Wolfgang, 2009: "Value co-creation as part of an integrative vision for innovation management", in: *Technology Innovation Management Review*; at: <http://timreview.ca/article/309> (15 December 2014).

UN Millennium Eco-system Assessment, 2005: *Living Beyond Our Means: Natural Assets and Human Well being* (New York: UN).

UN Millennium Assessment, 2001: "Guide to the Millennium Assessment Reports"; at: <http://www.millenniumassessment.org> (5 January 2015).

26 Sustainable Consumption

Sylvia Lorek[1]

Abstract

This chapter elaborates on sustainable consumption and provides key arguments from the sustainable consumption literature. It introduces 'environmental space' as one of the early concepts which embedded sustainable consumption within natural and social boundaries. It explains why a floor as well as a ceiling for the environmental space has to be considered and reflects on the space itself, its size and how to share it. Various possible paths of transition to reach the environmental space from a position of overconsumption as well as from underconsumption are described and linked to various schools of thought in sustainability research. Specific emphasis is given to a more detailed analysis of the two concepts of 'green growth' and 'de-growth'. Relating these concepts to sustainable consumption research and politics, the chapter distinguishes between strong and weak sustainable consumption and outlines some enabling mechanisms for sustainable consumption.

Keywords: Sustainable consumption, environmental space, green growth, de-growth, enabling mechanisms, strong sustainable consumption.

26.1 Introduction

Consumption is a vital element within the economic system. The economy in turn is a social construction through which society structures the exchange of goods and services. It is embedded in the natural system of planet Earth. As the rich literature on bioeconomics or ecological economics explains (Daly 1996; Martínez-Alier/Healy/Temper et al. 2013), the way the economy is structured has led to fundamental problems because it overstretches the limits of the Earth's carrying capacity, insofar as human activities inevitably alter the face of the system Earth (Steffen/Crutzen/McNeill 2007).

There are valid reasons to make consumption sustainable. The *UN Conference on Environment and Development* (UNCED) in its 'Agenda 21' identified for the first time unsustainable consumption and production patterns, *particularly in industrialized countries,*[2] as the major cause of the continued deterioration of the global environment (United Nations 1992). The resulting call for sustainable consumption and production is based on the ideas that, firstly, plane-

tary boundaries set the limits in the long run (Costanza 1989; Georgescu-Roegen 1966), and, secondly, that societies work better if they are based on democracy and score highly on equality (Pickett/Wilkinson 2009). It can be expected that such sustainable consumption requires more than just changing a few unsustainable habits and products here and there. It requires changes in the economy, the infrastructures serving our daily lives, and the dominant culture, as well as the institutions and power relationships which drive these (Vergragt/Akenji/Dewick 2014).

Such a broad approach was—and still is—not always reflected in the use of the term 'sustainable consumption'—neither in the academic or political literature, nor in classroom discussions on sustainability, nor in concerned statements by environmentally engaged neighbourhoods. For quite some time, mainly in the late 1990s, 'sustainable consumption' was limited to the decision-making of private consumers, following the UNCED idea that consumers are not only

---

1 Dr. Sylvia Lorek, Sustainable Europe Research Institute; Email: <sylvia.lorek@seri.de>.

2 At present, more than half of the 'consumer society' is based in (former) developing countries. Nevertheless, leadership towards more sustainable consumption patterns is still required from developed countries. Hence, examples in this article are taken mainly from a Northern, and especially European, context.

© Springer International Publishing Switzerland 2016
H.G. Brauch et al. (eds.), *Handbook on Sustainability Transition and Sustainable Peace*,
Hexagon Series on Human and Environmental Security and Peace 10, DOI 10.1007/978-3-319-43884-9_26

the victims but also the cause of environmental problems. As a result, sustainable consumption was reflected in the concepts of sustainable household consumption or sustainable consumption behaviour (Georg 1999; Haake/Jolivet 2001; Noorman/Uiterkamp 1998; Zacarias-Farah/Geyer-Allély 2003). At that time, emphasis was placed on case studies and single-product advice about what consumers could do to contribute to sustainability. The picture has developed since then, as least in scientific writing and debate. It is now increasingly recognized that such product- and/or service-oriented approaches might lead to some less unsustainable consumption here and there (Brunner/Urenje 2012), but they are hardly a contribution towards consumption pattern that can be called sustainable.

To distinguish sustainable consumption explicitly from such fragmented considerations, this chapter takes sustainable *resource* consumption as a starting point and takes the complete life cycle of products as well as lifestyles into account. In this context, sustainable consumption stands for

1. limiting the consumption of depletable resources via
    a) more efficient use,
    b) as well as their substitution by renewable resources,
    c) as well as reducing consumption levels in general, and
2. limiting the use of renewable resources to their reproduction rate.

Sustainable (resource) consumption thus involves the consumption patterns of industries, governments, households and individuals (United Nations 1992). In addition to this approach in physical terms, strong sustainable consumption also takes into account the social aspects of, for example, labour rights (ILO 2001), sustainable livelihoods (Lebel/Lorek 2008), and unequal access to and distribution of resources, as reflected in the literature on environmental rights and environmental justice (Martínez-Alier/Healy/Temper et al. 2013; Martínez Alier 1997).

To date, the physical, material flow aspects dominate research on sustainable consumption. In the late 1990s, researchers began to ask which kinds of consumption cause the most environmental impact. They found that the greatest environmental impact per capita is caused by food, housing and mobility (Gatersleben/Steg/Vlek 2002; Lorek/Spangenberg 2001; Noorman/Uiterkamp 1998; Tukker/Jansen 2006). More recent data are provided, for example, by the *European Environmental Agency* (EEA 2013). They point out that only a few product groups contribute significantly (between 30 and 50 per cent) to environmental pressures:

- construction works, that is, buildings and infrastructures;
- food products, beverages and alcohol;
- products of agriculture, forestry and fishing (also food products bought directly from farms rather than via food manufacturers);
- electricity, gas, steam and hot water (the majority of which is contributed by electricity).

These four product groups contribute 42 per cent to greenhouse gas emissions, 52 per cent to acidifying emissions, 37 per cent to ground ozone precursors, and 55 per cent to the *Total Material Requirements* (TMR) embodied in all consumed products in 2005. Interestingly, they only represent 17 per cent of total consumption expenditure (EEA 2013).

So sustainable consumption research broadly agrees on the main sources of the environmental burdens of consumption. Now, key questions to be addressed are, 'how can the impacts be reduced?', 'who should start changing?', 'with which tools?', 'in which time frame? Still, the consideration of social or development aspects is neglected in many cases. This chapter addresses selected lines of argument from the broad literature on sustainable consumption.

Section 26.2 introduces the concept of environmental space. It explains why a floor as well as a ceiling for environmental space has to be considered and reflects on the space itself, its size and how to share it. Section 26.3 outlines some possible means of transition to reach environmental space from a position of overconsumption as well as from one of under-consumption. Section 26.4 then sharpens the focus for a more detailed analysis of two strains of concepts in sustainable consumption research and politics, distinguishing between strong and weak sustainable consumption, and section 26.5 outlines some enabling mechanisms for sustainable consumption. Section 26.6 concludes the arguments.

26.2   Consuming Within Limits

One of the early approaches towards sustainable consumption embedded it in the concept of *environmental space*. This concept was developed in the early 1990s in the Netherlands (Opschoor/Reinders 1991; Weterings/Opschoor 1994), and further elaborated in

Figure 26.1: The Environmental Space Concept. **Source:** Spangenberg (1995).

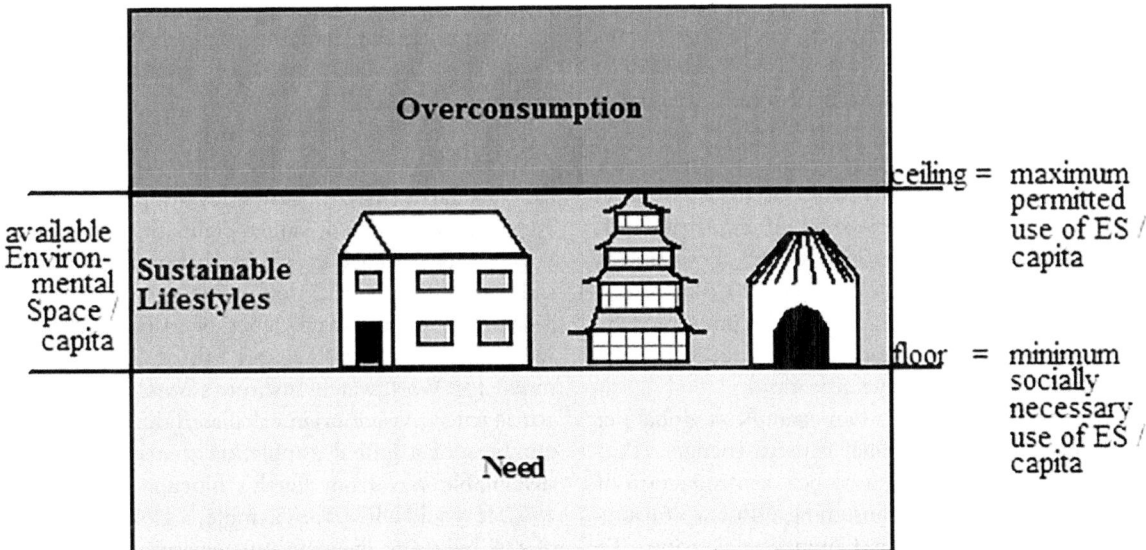

a European context (Spangenberg 1995). It distinguishes a space for free choice of consumption patterns from two zones of unsustainability: the domain of environmentally unsustainable overconsumption and one of socially unsustainable under-consumption. It is based on the insight of a limitation of various resources available for human consumption such as fossil energy, timber, and copper, as well as building and agricultural land.

The two border lines as well as the environmental space itself all deserve a closer look.

26.2.1   The Floor

The lower level, called the 'floor of the environmental space' represents the minimum requirements of material (resource) consumption necessary to live a dignified life. In Latin America this line is known as *linea de dignidad*. It necessarily applies to every citizen because no one should live below it. A dignified life should not only satisfy the physiological needs of nutrition and shelter. The level also includes the availability of the resources needed for actively participating in politics, culture, and the other processes of society (Cruz/Stahel/Max-Neef 2009; Max-Neef/Elizalde/Hopehayn 1989). The contributions of Nussbaum and Sen (1993) on the quality of life, especially their work on the capability approach (Nussbaum 2001; Sen 1999), are influential contributions to this context. Human well-being, living a good life, universal human rights (including social rights), and the extended definition of health by constituting a right

to a minimum income—see the *Preamble to the Constitution of the World Health Organization* (United Nations 1948) are further elements that help to create the idea of a floor with concrete requirements. To operationalize the floor, the concept of a *social protection floor*, developed and promoted by the ILO in collaboration with the WHO, plays a similarly important role (ILO 2011). This suggests measures and institutional reforms to achieve both basic income security and universal access to affordable essential social services. At the UN level, the Human Development Index published annually by the *United Nations Development Programme* (UNDP 2014) since 1990 reflects such considerations.

It is important to note that such a concept is not alien to economic thinking in general. Adam Smith, as early as 1776, stressed the necessity of providing all people with the means of leading 'a life without shame'. However, its shortcoming is that it plays no role in current neoclassical economics (Lorek/Spangenberg 2014).

26.2.2   The Ceiling

The upper limit of the environmental space, called the 'ceiling', is the maximum use of resources as well as the maximal tolerable amount of environmental damage that the Earth can accept without destroying ecological systems in such a way that natural systems—and with them social and economic systems—collapse. Various studies have calculated upper thresholds on a per capita basis, including those for (auto-)mobility, water use and meat consumption, but also, for example, for the use of mineral resources (Buitenkamp/Venner/Wams 1992; Spangenberg 1995). The formal agreement to limit global warming to 2°C can be seen as such a ceiling. In their article in *Nature*, Rockström and twenty-eight colleagues took a broader approach and defined a 'safe operating space for humanity' based on ten criteria (Rockström/Steffen/Noone et al. 2009). This provides an extended empirical basis, emphasizing the key dimensions to care about since a significant (biodiversity, nitrogen cycle) or slight (climate) transgression of the acceptable limits to damage has already occurred, or is soon about to happen (phosphorus cycle, ocean acidification).

Taking CO2 emissions as an example, a global per capita level respecting global climate change reduction targets would call for a 90 per cent reduction of CO2 emissions for overconsuming affluent consumers (Chakravarty/Chikkatur/Coninck et al. 2009). To demonstrate what this reduction target could mean in practice, Swiss researches and practitioners developed the concept of the '2000 Watt Society' (Jochem 2004). According to their concept, 2000 watts per person would be a fair share of energy consumption for the global population. Interestingly, a comparison of energy use and the Human Development Index showed that the current energy supply is sufficient for a high level of human development for everyone if it is fairly shared. The issue is not the resource availability, but its distribution (Steinberger/Roberts 2009).

26.2.3   Size of the Environmental Space

What remains between floor and ceiling are specific consumption corridor(s) (Di Giulio/Fuchs 2014). The crucial question is what size these corridors are. Let us first consider the overall environmental space. At first sight it seems to be quite fixed, as planet Earth has a relatively constant surface. Nevertheless, to some degree it can vary. Research indicates, for example, that the available productive land is shrinking—due to desertification, as a result of global warming and rising sea water levels, or simply through the overuse of renewable resources. As well as this, the amount of non-renewable resources is of course constantly depleting. On the other hand the volume could increase again—for example, if new energy sources were discovered and the energy problem were solved. This would allow land which is currently used for producing energy crops to be reallocated to food production. In addition, increasing effort is being spent to optimize technological solutions to sustain-

ing and maximizing ecosystem services and the biocapacity of the Earth—but these include some debatable attempts such as geoengineering (Anshelm/Hansson 2014) and the development of *genetically modified organisms* (GMOs).

The overall size of the environmental space is of concern because it influences what the concept calls the 'fair Earth share' of resources per person. Accepting the general idea of similar rights for every human, environmental space should be shared as fairly as possible among the global population. So, after resource availability the fair share depends on population size. In his chapter "The Rise and Fall of Consumer Cultures" for Worldwatch Institute's *State of the World 2010* (2010), Assadourian calculated different possible numbers of a global population that could live in a sustainable way from Earth's biocapacity (see table 26.1). It would allow, for example, a global population of 13.8 billion to live within its environmental space if every person lived on a biocapacity of one global hectare. At the other extreme, the calculation demonstrates that our planet only could tolerate 1.4 billion people if everyone consumed in similar way to the average US citizen.

Table 26.1: Sustainable World Population at Different Consumption Levels. **Source**: Assadourian (2010).

Consumption level	Per capita income (2005) (GNI, PPP, 2008 dollars)	Biocapacity/ person (global hectares)	Sustainable population at this level (billion)
Low-income	1,230	1.0	13.6
Middle-income	5,100	2.2	6.2
High-income	34,690	6.4	2.1
United States	45,580	9.4	1.4
Global average	9,460	2.7	5.0

This indicates that both population and affluence—taking technology into consideration—have to be targeted to limit consumption to a size that fits the available biocapacity.

26.2.4   Critical Reflections

The more detailed the calculation of environmental space appears—at a country, city or personal level—the

Figure 26.2: Possible pathways to sustainable consumption from overconsumption. **Source**: Brunner and Urenje (2012).

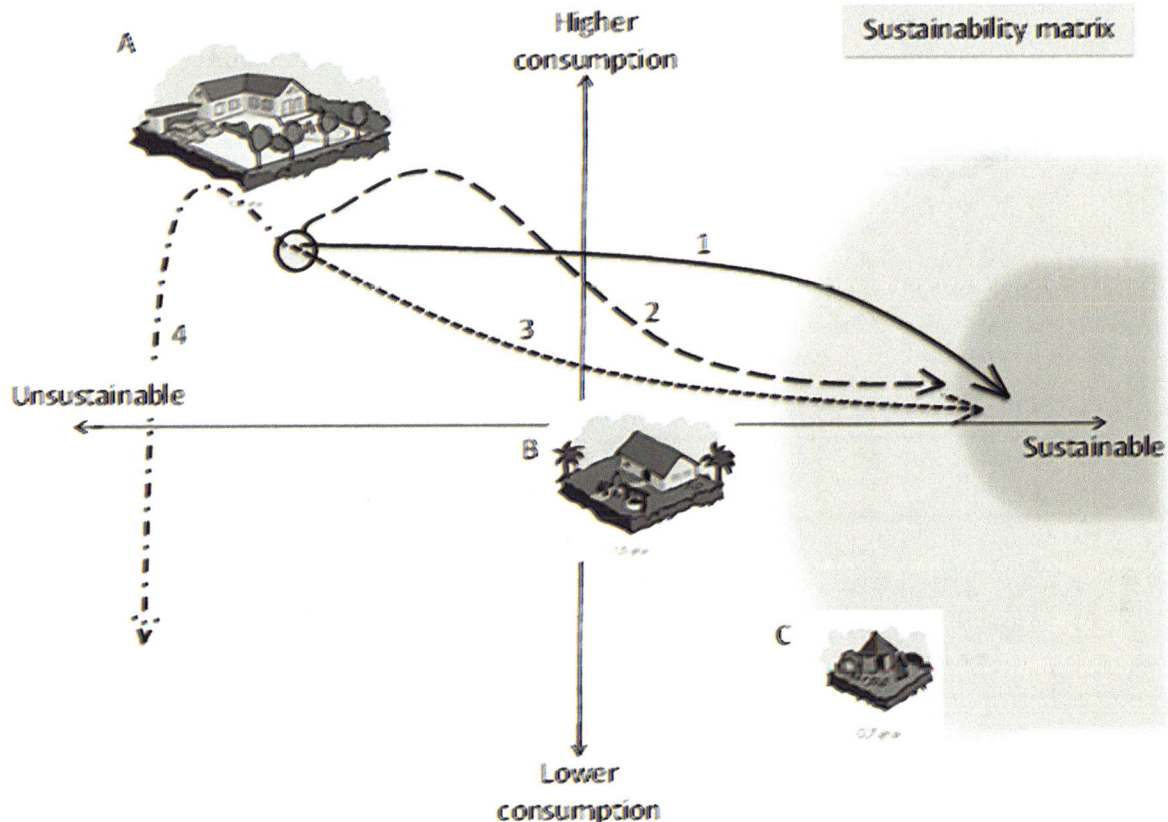

more difficult it is to develop precise recommendations in units of measure of specific resources. Nevertheless, the concept is relatively explicit in pointing towards both situations: where consumption is too high and where it is too low. Its main advantage, therefore, is its inherent call to increase equity of consumption opportunities (UNEP 2001). Accordingly, the concept is often used in the context of environmental justice (McLaren 2003; Rice 2007). The question of how to redistribute resources, however, is generally left open.

26.3 Structuring Transition Paths towards Sustainable Consumption

The crucial question is how to reach the environmental space from a situation of overconsumption as well as from one of under-consumption.

So far neither researchers, nor policymakers, nor members of civil society can agree on what a sustainable future might look like, despite a huge amount of modelling, visioning, and backcasting and forecasting

exercises and scenarios. To illustrate some of the main lines of argument, this section uses a simplified matrix indicating over- and under-consumption on the y axis and sustainable/unsustainable stages on the x axis. As in table 26.1, the environmental space in this illustration is represented through the *ecological footprint*, a resource accounting framework for measuring human demand on the biosphere measured in *global hectares* (gha) (Wackernagel/Rees 1996). The available environmental space for one person is assumed in this matrix to be around one gha on average. This stage of sustainability is illustrated by the dark grey field on the right-hand side of the sustainability axis. For the purpose of better visualization overconsumption is characterized by the big house in the upper left corner using 10 gha, while under-consumption is characterized by the small cottage using 0.7 gha. Figure 26.2 indicates some possible pathways through which overconsumption could shrink to a sustainable stage. Figure 26.3 outlines possible pathways by which under-consumption might be increased to a sustainable level. The arrows reflect possible pathways which can be related to typical strains of scientific and political argument.

Figure 26.3: Possible pathways to sustainable consumption from under-consumption. **Source:** Brunner and Urenje (2012).

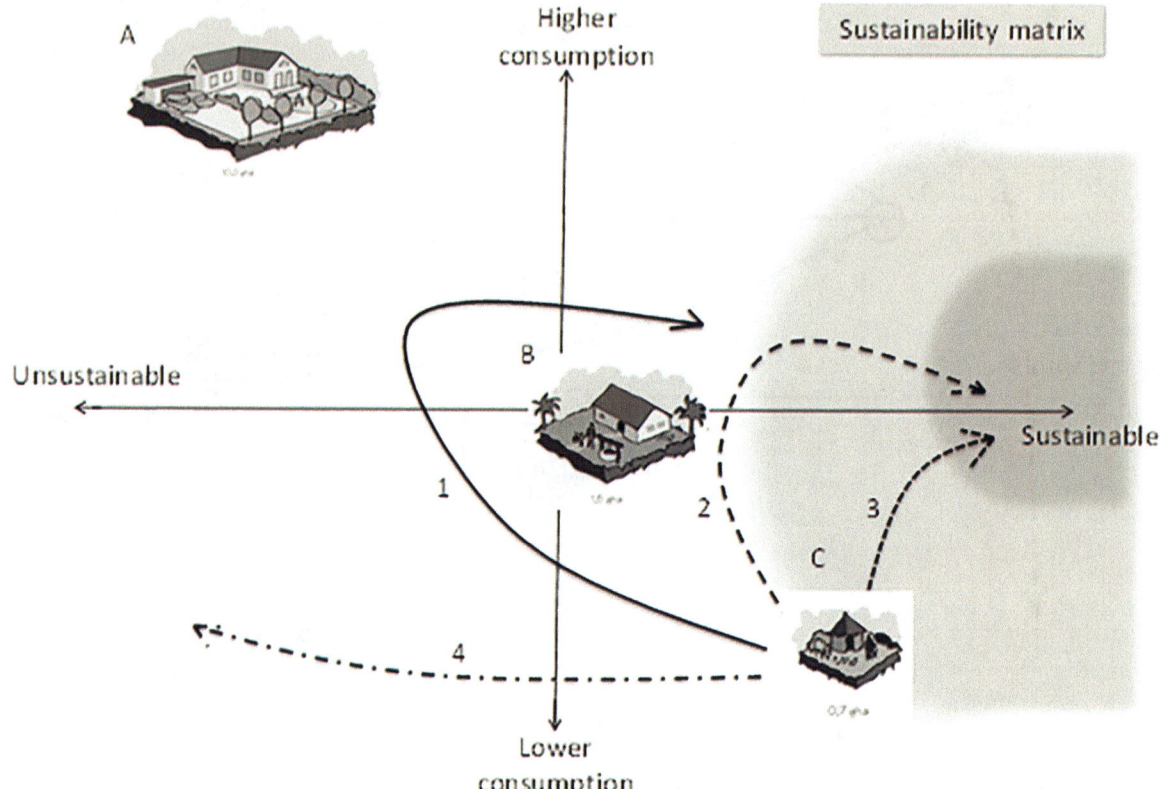

Arrow 1 in figure 26.2 describes how proponents of green growth (OECD 2011) and of the green economy (UNEP 2011) debate the possible transition towards sustainability. They argue that resource consumption can stabilize through improvements in efficiency and investment in green technology and will then slowly reduce. Others, such as the proponents of ecological modernization (Ayres/Simonis 1993; Ayres/Ayres/Frankl 1996), promote the development represented by arrow 2, where some further growth in consumption is still accepted in the expectation that technological innovation and resource productivity will soon reduce environmental burdens quite rapidly. The development of a new source of energy often plays a crucial role in such an argument. The de-growth researchers (Demaria/Schneider/Sekulova 2013; Lorek/Fuchs 2013) argue instead that due to rebound effects economic growth always goes hand in hand with an increase in resource consumption and energy use—which is impossible in a finite world. As it is recognized that economies in developing countries need to grow to achieve equity, material throughput and energy use in developed countries need to de-grow immediately. This argument is illustrated by arrow 3.

Finally, arrow 4 indicates what will probably happen under a business-as-usual scenario: continuous growth will eventually lead to an ecological and economic collapse.

Despite being an important political challenge the question of how to raise under-consumption towards a sustainable level is rarely the focus of sustainable consumption research (Lorek/Barber/Onthank 2013). It may be roughly considered that arrows 2 and 3 in figure 26.3 characterize some kind of leapfrogging. This term describes the immediate uptake of new, less resource-intensive technologies to raise living standards (Tukker 2005). Here, arrow 1 indicates a green growth model. This promotes high growth rates for underdeveloped countries in the expectation that in line with the Environmental Kuznet Curve (see for example Stern/Common/Barbier 1996) the environmental problems related to growth will decline again as soon as a specific level of income is gained and clean technology can be afforded. Then the path turns towards more sustainability. It has to be recognized, however, that it is mainly the arrow 4 path which is promoted by current development models and is the one most likely to be preferred by many

---

Box 26.1: Green Growth and De-growth. **Source:** The author.

Green growth means fostering economic growth and development while ensuring that natural assets continue to provide the resources and environmental services on which our well-being depends. It focuses on the synergies and trade-offs between the environmental and economic aspects of sustainable development, but also considers the social aspect on the basis that without good governance, transparency, and equity, no transformative growth strategy can succeed. One important component in promoting green growth is the provision of the right economic incentives for influencing household decisions. Soft measures, however, such as labelling and public information campaigns also are recognized as having a significant complementary role. Therefore, the desired change in behaviour will require a mixture of these instruments

De-growth means a deliberate downscaling of production and consumption in order to increase human well-being and enhance ecological conditions and equity on the planet. It calls for a future where societies live within their ecological means, with open but mainly localized economies, and where resources are more equally distributed through new forms of democratic institutions. A de-growth understanding of innovation focuses on new social and technical arrangements that enable individuals and societies to live convivially and frugally. In this context, de-growth challenges the centrality of GDP as an overarching policy objective. De-growth explicitly seeks a sustainable shrinking and differentiates itself from the involuntary and painful processes of austerity or recession.

---

people in developing countries. This path therefore represents the business-as-usual scenario.

As indicated already, most academic and political debate about sustainable consumption is split between analysing and developing approaches and instruments for the pathways characterized by arrow 1 and arrow 3 in figure 26.2. The next section will therefore further elaborate on the green growth/de-growth arguments.

26.4 The Search for the Best Way: Curing Consumption Impacts or Changing Consumption Patterns?

The core of the sustainable consumption discourse revolves around the question of whether sustainable consumption needs green growth, hence a *differentiation* of the actual growth model, or a de-growth path to shrink it towards a sustainable consumption level. De-growth proponents argue using the logic of ecological economics and perceive nature (or the environment) as the framing dimension of sustainability. In their argument, human society is perceived as a subsystem of nature and economics as a constructed subsystem of societies (Schneider/Kallis/Martinez-Alier 2010). According to these scholars the ceiling is so overstretched (Rockström/Steffen/Noone et al. 2009) that it needs a substantial reduction in resource consumption first before something like a steady-state economy can be reached (Kerschner 2010; O'Neill 2012). Proponents of green growth logic appear to be more pragmatic. They see economic power as strongest in societies, and think that if it is wisely and strategically channelled it can best cure the impact of unsustainable consumption within the current economic

system (OECD 2014b, 2014a). Proponents of both lines of thought, however, argue that the path they are suggesting will lead to a higher level of well-being for a larger share of the global population (box 26.1).

Translating green growth and de-growth into the terminology of sustainable consumption, the relevant literature distinguishes in this context between strong sustainable consumption closer to the path of de-growth and weak sustainable consumption via a green growth path (Berg 2011; Fuchs/Lorek 2005; Hobson 2013; Lorek/Fuchs 2013).

Weak sustainable consumption proponents more or less take the current consumption levels as given and so encourage every effort to satisfy them with fewer resources (European Commission 2008). They see the prevailing market economy, based on free choice and consumer sovereignty, as a powerful vehicle in this context. Governments, research organizations and civil society organizations who follow this path stress their respect for individual lifestyle choices, and point out that market economy systems (as actually structured) need to constantly increase consumption in order to sustain the economy and especially to sustain full employment. Thus—in the name of sustainable consumption—interventions are carefully calibrated to address environmental problems while not slowing down the economy. They accordingly promote and foster initiatives through which sustainable consumption can be achieved via an increase in the efficiency of products, production processes, services, and the provision of these services.

Others, however, argue that this reduces sustainable consumption to sustainable consumer procurement only (Fedrigo/Hontelez 2010). In fact, in this weak form the concept of sustainable consumption is often used to argue that it is consumers who should shift the market towards sustainability and that interventions should focus on urging consumers to improve the sustainability characteristics of their consumption choices as well as on enabling consumers to do so (Fuchs 2013). Akenji (2014) blames this attitude for making consumers the scapegoats of unsustainability.

Strong sustainable consumption researchers—and thus de-growth advocates—in their turn perceive consumers as locked in to consumption practices through their habits and routines, as well as to the structural constraints resulting from the technological, socio-economic, political and cultural environments they are living in (Cohen/Murphy 2001; Maniates 2001; Røpke 1999; Sanne 2002; Shove 2010). According to their logic, the change of structure has to be the focus. These scholars also support and appreciate the availability of environmentally or socially superior products in markets and the provision of relevant information to consumers as necessary preconditions for change. However, they claim this can only be a starting point because most decisions about sustainable or unsustainable consumption paths take place hidden from the consumers' sphere of influence. Changes in communication technologies, global finance and trade have developed a remarkable influence on the sustainability of consumption long before the consumer ever makes a choice. A similar situation is found with demographic and gender roles which, for example, induce shifts in job situations and time allocation with remarkable consequences for patterns of household consumption (Fuchs/Lorek 2002; Ropke/Godskesen 2007). What seem to be individual decisions or individual lifestyle characteristics are in fact systematic societal shifts which make individual sustainable lifestyles less and less possible.

Although this dichotomy between weak and strong sustainable consumption describes two divergent streams of thought and practice, there are conceptual and practice-based spaces where they intersect (Hobson 2013). As soon as it comes to detailed activities weak (because technology-based and/or efficiency-based) instruments might build a necessary condition into a strong sustainable consumption scenario, at least to win time. The uptake of e-mobility through e-bikes might serve as an example. As a technological innovation they can pave the way for a modal shift in mobility—as long as they indeed replace cars and not ordinary bicycles (Rose 2012). The line between weak and strong sustainable consumption is sharply drawn insofar, however, as the search for transition paths are either restricted to solutions possible within the current system or actively envision demand for and support for a change in the system(s).

A widely neglected aspect in both strains of discussion so far about why consumption habits and structures have developed and seem to be developing further in the direction of unsustainability is the question of power. Power plays a central role in creating structural barriers to sustainable consumption and in delimiting opportunities for intervention. Actors typically mentioned in regard to executing power over consumption structures are administrations from global to local level, business companies and organizations, civil society organizations, media, and last but not least individuals, whether acting as consumers in their daily household context or as engaged citizens. However, the relationship between power and sustainable consumption is only just beginning to be analysed (Fuchs/Di Giulio/Glaab et al. forthcoming). Research so far highlights the close relationship between sustainable consumption and fundamental questions of democracy: how can one assure equality in participation in today's democracies, characterized as they are by large asymmetries in resources and access to institutions and decision-makers? (Fuchs 2013).

26.5   Enabling Mechanisms for Strong (or Weak) Sustainable Consumption?

At the present time, researchers and practitioners are exploring and proposing imperatives and implementation mechanisms for fostering the sustainability of consumption. A literature review has identified eleven main ways in which sustainable production-consumption systems could be made possible (Lebel/Lorek 2010). These are presented in table 26.2, and range from initiatives which emphasize production activities to those which are more consumption-related. At first view, some of the mechanisms appear to follow a weak sustainable consumption approach, in that they try to optimize the system from within, through 'greening the supply chain' or 'certify and label'. Others indicate a rethinking of the system, through 'use less' or 'service rather than sell'. Here we can already see that strong sustainable consumption approaches are not limited to those parts of the product chain which are traditionally linked to consumer activities. In the end, however, to stay within the environmental space all the mechanisms will have to be developed further:

Table 26.2: Examples of enabling mechanisms for sustainable consumption. **Source:** Lebel and Lorek (2010).

Enabling mechanism	Short description	Concerns, constraints, or challenges
Produce with less	Innovations in production process reduce the impact per unit made.	Rebound effects occur through which gains are wiped out by increases in the number of units or how they are used.
Green supply chains	Firms with leverage in a chain impose standards on their suppliers to improve environmental performance.	There may be unfair control of small producers.
Co-design	Consumers are involved in design of products to meet functions with less environmental impact.	Incentives are not adequate to involve consumers.
Produce responsibly	Producers are made responsible for waste from the disposal of products at the end of their life.	Incentives for compliance without regulation may be too low for many types of products.
Service rather than sell	Producers provide service rather than sell products; this reduces the number of products made while still providing to consumers the functions they need.	This is a difficult transition for firms and consumers to make as it requires new behaviours and values.
Certify and label	Consumers buy labelled products. As labels are based on independent certification, producers with good practices increase their market share.	Consumers are easily confused by too much information or by a lack of transparency and credibility in competing schemes.
Trade fairly	Agreements are made with producers that may include minimum price and other investments or benefits. Consumers buy products labelled as or sold through fair trade channels while producers get a better deal.	Mainstream trade still dominates. It is hard to maintain fair trade benefits to producers when a product becomes mainstream.
Market ethically	Reducing unethical practices in marketing and advertising would reduce wasteful and overconsumption practices.	There is reluctance by policymakers to tackle very powerful private sector interests with regulation.
Buy responsibly	Campaigns that educate consumers about impacts of individual products, classes of products and consumption patterns change behaviour overall.	Converting intentions and values into actions in everyday life is often difficult for consumers. Issues of convenience, flexibility, and function still matter a lot.
Use less	Consumption may be reduced for a variety of reasons, for example, as a consequence of working less. There are many potential environmental gains from less overall consumption.	There is a dominant perception that using less means sacrifice. Less income and less consumption may not automatically translate into better consumption impacts.
Increase wisely	Increasing consumption of under-consumers can be done in ways that minimize environmental impacts as economic activity expands.	Wealthy developed countries need incentives and goodwill to assist the poor and those in developing countries, for example, by leaving adequate space and natural resources for them to develop.

26.6   Conclusion

Our consumption is embedded in the boundaries of our natural system but shaped by economic, societal and cultural structures. To direct consumption towards sustainability, the concept of *environmental space* can provide useful orientation. It helps us to visualize how consumption can be unsustainable if it exceeds an upper as well as a lower limit. What exactly these upper and lower limits are, however, has not yet been sufficiently well formulated. Even less well formulated are concrete suggestions about how to reach the safe operating space which is appropriate for the consumption of countries, regions or individuals. What is being debated is what the best pathway is that such a development should take in order to achieve sustainable consumption. This paper has developed arguments for a green growth path that relies on the transformative capacity of the economic system and respects and highlights consumer sovereignty. These conditions are recognized as a necessary but probably insufficient condition for (global) consumption patterns that remain within environmental limits and so characterize a weak sustainable consumption path. The second strain of argument reflected is for a degrowth path which demands a shrinking of economic activities, investing in research and directing policies into a change of systemic structures, including the

framing conditions for the economy. Such a development would require strong sustainable consumption governance. Research into this path still has to convince its opponents that the idea envisioned of a sustainable shrinking does not lead to the collapse of economic and social security systems, for example, and is completely different from austerity or recession situations. Various enabling mechanisms to foster sustainable consumption have been identified. They are ready to be used in any way.

References

Akenji, Lewis, 2014: "Consumer scapegoatism and limits to green consumerism", in *Journal of Cleaner Production,* 63: 16-23.

Anshelm, Jonas; Hansson, Andres, 2014: "Battling Promethean dreams and Trojan horses: Revealing the critical discourses of geoengineering", in: *Energy Research & Social Science,* 2: 135-144.

Assadourian, Erik, 2010: "The Rise and Fall of Consumer Cultures", in: Worldwatch Institute (Eds): *State of the World 2010 - Changing Cultures* (W.W. Norton: New York): 3-20.

Ayres, Robert U.; Simonis, Udo Ernst, 1993: *Industrial metabolism: Restructuring for sustainable development* (Tokyo, New York: UN University Press).

Ayres, Robert U.; Ayres, Leslie W., 1996: *Industrial ecology: towards closing the materials cycle* (Cheltenham: Edward Elgar).

Berg, Annuka, 2011: "Not roadmaps but toolboxes: Analysing pioneering national programmes for sustainable consumption and production", in: *Journal of Consumer Policy,* 34,1: 9-23.

Brunner, Wolfgang; Urenje, Shepherd, 2012: *The sustainability matrix* (Visby Swedish International Centre of Education for Sustainable Development); at: <http://conferences.chalmers.se/index.php/tlhe/nu2012/paper/download/149/43>.

Buitenkamp, Maria; Venner, Henk; Wams, Theo, (Eds.), 1992: *Sustainable Netherlands,* (Amsterdam: Milieu Defense; Friends of the Earth Netherlands).

Chakravarty, Shoibal; Chikkatur, Ananth; de Coninck, Heleen; Pacala, Stephen; Socolow, Robert; Tavoni, Massimo, 2009: "Sharing global CO2 emission reductions among one billion high emitters", in: *Proceedings of the National Academy of Sciences,* 106,29: 11884-11888.

Cohen, Maurie .J.; Murphy, Joseph, 2001: *Exploring Sustainable Consumption: Environmental Policy and the Social Sciences* (Oxford: Elsevier).

Costanza, Robert, 1989: "What is Ecological Economics?", in: *Ecological Economics,* 1,1: 1-7.

Cruz, Ivonne; Stahel, Andri; Max-Neef, Manfred, 2009: "Towards a systemic development approach: Building on the Human-Scale Development paradigm", in: *Ecological economics,* 68,7: 2021-2230.

Daly, Herman E., 1996: *Beyond Growth: The Economics of Sustainable Development* (Boston, MA: Beacon Press).

Demaria, Federico; Schneider, Francois; Sekulova, Filka; Martinez-Alier, Joan, 2013: "What is Degrowth? From an Activist Slogan to a Social Movement", in: *Environmental Values,* 22: 191-215.

Di Giulio, Antonietta; Fuchs, Doris, 2014: "Sustainable Consumption Corridors: Concept, Objections, and Responses", in: *GAIA,* 23,1: 184-192.

EEA, 2013: "Environmental pressures from European consumption and production", in: *EEA Technical report No 2/2013* (Copenhagen: European Environment Agency).

European Commission, 2008: *Sustainable Consumption and Production and Sustainable Industrial Policy Action Plan* (Brussels: European Commission, COM (2008) 397 final).

Fedrigo, Doreen; Hontelez, John, 2010: "SCP: An Agenda Beyond Sustainable Consumer Procurement", in: *Journal of Industrial Ecology,* 14,1: 10-12.

Fuchs, Doris, 2013: "Sustainable consumption", in: *The Handbook of Global Climate and Environment Policy,* 215-230.

Fuchs, Doris; Lorek, Sylvia, 2002: "Sustainable Consumption Governance in a Globalizing World", in: *Global Environmental Politics,* 2,1: 19-45.

Fuchs, Doris, 2005: "Sustainable Consumption Governance—A History of Promises and Failures", in: *Journal of Consumer Policy,* 28: 261-288.

Fuchs, Doris; Di Giulio, Antonietta; Glaab, Katharina; Lorek, Sylvia; Maniates, Michael; Princen, Tom; Ropke, Inge, forthcoming: "Power: what's missing in consumption and absolute reductions research and action", in: *Journal of Cleaner Production.*

Gatersleben, Brigitta; Steg, Linda; Vlek, Charles, 2002: "Measurement and determinants of environmentally significant consumer behavior", in: *Environment and Behavior,* 34,3: 335-362.

Georg, Susse, 1999: "The social shaping of household consumption", in: *Ecological Economics,* 28,3: 455-466.

Georgescu-Roegen, Nicholas, 1966: *Analytical Economics: Issues and Problems* (Cambridge, MA: Harvard University Press).

Haake, Julia; Jolivet, Patrick, 2001: "Some reflections on the link between production and consumption for sustainable development", in: *International Journal of Sustainable Development*, 4,1: 22-32.

Hobson, Kersty, 2013: "'Weak' or 'strong' sustainable consumption? Efficiency, degrowth, and the 10 Year Framework of Programmes", in: *Environment and Planning C: Government and Policy*, 31,6: 1082-1098.

ILO [International Labour Office] (Ed.), 2011: *Social Protection Floor for a fair and inclusive globalization* (Geneva: Report of the Adisory Group).

Jochem, Eberhard, 2004: "R&D and innovation policy—preconditions for making steps towards a 2000 watt/cap society", in: *Energy & Environment*, 15,2: 283-296.

Kerschner, Christian, 2010: "Economic de-growth vs. steady-state economy", in: *Journal of Cleaner Production*, 18,6: 544-551.

Lebel, Louis; Lorek, Sylvia, 2008: "Enabling Sustainable Production-Consumption Systems", in: *Annual Review of Environment and Resources*, 33: 241-275.

Lebel, Louis; Lorek, Sylvia, 2010: "Production-consumption systems and the pursuit of sustainability", in: Lebel, L.; Lorek, S.; Daniel, R. (Eds.): *Sustainable production consumption systems: knowledge, engagement and practice* (Springer: Dordrecht): 1-12.

Lorek, Sylvia; Spangenberg, Joachim H., 2001: "Indicators for environmentally sustainable household consumption", in: *Int. J. Sustainable Development*, 4,1: 101-120.

Lorek, Sylvia; Fuchs, Doris, 2013: "Strong Sustainable Consumption Governance—Precondition For A Degrowth Path?", in: *Journal of Cleaner Production*, 38: 36-43.

Lorek, Sylvia; Spangenberg, Joachim H., 2014: "Sustainable consumption within a sustainable economy—beyond green growth and green economies", in: *Journal of Cleaner Production*, 63: 33-44.

Lorek, Sylvia; Barber, Jeffrey; Onthank, Karen (Eds.), 2013: *Global and Regional Research on Sustainable Consumption and Production Systems: Achievements, Challenges and Dialogues*, (Boston: Workshop Report of the Global Research Forum on Sustainable Production and Consumption, 13-15 June 2012, Rio de Janeiro).

Maniates, Michael, 2001: "Individualization: Plant a Tree, Buy a Bike, Save the World?", in: *Global Environmental Politics*, 1,3: 31-52.

Martínez-Alier, Joan; Healy, Hali; Temper, Leah; Walter, Mariana, 2013: *Ecological economics from the ground up* (London: Routledge).

Martínez Alier, Joan, 1997: "Environmental justice (Local and Global)", in: *Capitalism Nature Socialism*, 8,1: 91-107.

Max-Neef, Manfred; Elizalde, Antonio; Hopehayn, Martin, 1989: "Human Scale Development. An Option for the Future", in: *Development Dialogue*, 1989,1: 7-80.

McLaren, Duncan, 2003: "Environmental space, equity and the ecological debt", in: Agyeman, J.; Bullard, R. D.; Evans, B. (Eds.): *Just sustainabilities: Development in an unequal world* (London: Earthscan): 19-37.

Noorman, Klaas Jan; Uiterkamp, Ton Schoot, 1998: *Green Households?: Domestic Consumers, the Environment and Sustainability* (London: Earthscan).

Nussbaum, Martha, 2001: *Women and human development: The capabilities approach* (Cambridge: Cambridge University Press).

Nussbaum, Martha; Sen, Amartya, 1993: *The quality of life* (Oxford: Oxford University Press).

O'Neill, Daniel W., 2012: "Measuring progress in the degrowth transition to a steady state economy", in: *Ecological Economics*, 84: 221-231.

OECD, 2011: *Towards Green Growth* (Paris: OECD).

OECD, 2014a: *Greening Household Behaviour—Overview from the 2011 Survey—Revised edition* (Paris: OECD).

OECD, 2014b: *Green Growth Indicators 2014* (Paris: OECD Green Growth Studies).

Opschoor, Hans; Reinders, Lucas, 1991: "Towards sustainable development indicators", in: Kuik, O.; Verbruggen, H. (Eds.): *In search of indicators of Sustainable Development* (Dordrecht: Kluwer Academic Publishers): 7-27.

Pickett, Kate; Wilkinson, Richard, 2009: *The Spirit Level: Why more equal societies almost always do better* (London: Allan Lane).

Rice, James, 2007: "Ecological unequal exchange: international trade and uneven utilization of environmental space in the world system", in: *Social Forces*, 85,3: 1369-1392.

Rockström, Johan; Steffen, Will; Noone, Kevin; Persson, Åsa; Chapin, F. Stuart; Lambin, Eric F.; Lenton, Timothy M.; Scheffer, Marten; Folke, Carl; Schellnhuber, Hans Joachim; Nykvist, Björn; de Wit, Cynthia A.; Hughes, Terry; van der Leeuw, Sander; Rodhe, Henning; Sörlin, Sverker; Snyder, Peter K.; Costanza, Robert; Svedin, Uno; Falkenmark, Malin; Karlberg, Louise; Corell, Robert W.; Fabry, Victoria J.; Hansen, James; Walker, Brian; Liverman, Diana; Richardson, Katherine; Crutzen, Paul; Foley, Jonathan A., 2009b: "A safe operating space for humanity", in: *Nature*, 461,24 (September): 472-475.

Ropke, Inge; Godskesen, Mirjam, 2007: "Leisure activities, time and environment", in: *International Journal of Innovation and Sustainable Development*, 2,2: 155-174.

Røpke, Inge, 1999: "The dynamics of willingness to consume", in: *Ecological Economics*, 28,3: 399-420.

Rose, Geoffrey, 2012: "E-Bikes and Urban Transportation: Emerging issues and unresolved questions", in: *Transportation*, 39,1: 81-96.

Sanne, Christer, 2002: "Willing consumers—or locked-in? Policies for a sustainable consumption", in: *Ecological Economics*, 42,1-2: 273-287.

Schneider, Francois; Kallis, Girogos; Martinez-Alier, Joan, 2010: "Crisis or opportunity? Economic degrowth for social equity and ecological sustainability", in: *Journal of Cleaner Production*, 18,6: 511-518.

Sen, Amartya, 1999: *Commodities and capabilities* (Oxford: Oxford University Press).

Shove, Elisabeth, 2010: "Beyond the ABC: climate change policy and theories of social change", in: *Environment and Planning A,* 42,1273-1285.

Spangenberg, Joachim H. (Ed.), 1995: *Towards Sustainable Europe* (Luton: Friends of the Earth Publications).

Steffen, Will; Crutzen, Paul J.; McNeill, John R., 2007: "The Anthropocene: are humans now overwhelming the great forces of nature", in: *Ambio: A Journal of the Human Environment,* 36,8: 614-621.

Steinberger, Julia K.; Roberts, J. Timmons, 2009: *Across a Moving Threshold: energy, carbon and the efficiency of meeting global human development needs* (Klagenfurt: Alpen-Adria Universität).

Stern, David I.; Common, Michael S.; Barbier, Edward B., 1996: "Economic growth and environmental degradation: the environmental Kuznets curve and sustainable development", in: *World Development,* 24,7: 1151-1160.

Tukker, Arnold, 2005: "Leapfrogging into the future: developing for sustainability", in: *Int. J. Innovation and Sustainable Development,* 1,1-2: 65-84.

Tukker, Arnold; Jansen, Bart, 2006: "Environmental impacts of products: a detailed review of studies", in: *Journal of Industrial Ecology,* 10,3: 159-182.

UNDP, 2014: *Human Development Report* [online text]; at: <http://hdr.undp.org/en/global-reports>.

UNEP, 2001: *Consumption opportunities. Strategies for change—A report for decision makers* (Paris: United Nations Environmental Programme).

UNEP, 2011: *Towards a GREEN Economy—Pathways to Sustainable Development and Poverty Eradication. A Synthesis for Policy Makers* (New York: United Nations).

United Nations, 1948: *Preamble to the Constitution of the World Health Organization; entered into force on 7 April 1948* (New York: United Nations).

United Nations 1992: *Agenda 21: Results of the World Conference on Environment and Development* (New York: United Nations).

Vergragt, Philip; Akenji, Lewis; Dewick, Paul, 2014: "Sustainable production, consumption, and livelihoods: global and regional research perspectives", in: *Journal of Cleaner Production,* 63: 1-12.

Wackernagel, Mathis; Rees, William E., 1996: *Our ecological footprint: reducing human impact on the earth* (Gabriola Islands, BC, Canada: New Society Publishers).

Weterings, Rob; Opschoor, Johannes (Hans). B., 1994: *Towards environmental performance indicators based on the notion of environmental space.* Report to the Advisory Council for Research on Nature and Environment (Rijswijk, The Netherlands: Advisory Council for Research on Nature and Environment).

Zacarias-Farah, Adriana; Geyer-Allély, Elaine, 2003: "Household consumption patterns in OECD countries: trends and figures", in: *Journal of Cleaner Production,* 11,8: 819-827.

27 Sustainable Consumption and Production in China

Hongmin Chen[1]

Abstract

The rising problem of severe resource scarcity, environmental pollution and ecological degradation has become a great constraint on economic development in China. *Sustainable consumption and production* (SCP), as a primary way to decouple economic development from environmental degradation, has been undertaken via various energy-saving and environmental protection policies in China. The main practices for SCP in China, including the adjustment of the industry structure, the promotion of a circular economy and cleaner production, the Energy Conservation and Emission Reduction Project, green procurement, and progress in waste management are discussed. Though China has made some achievements in SCP, it is still facing serious challenges. The extensive economic growth model, strengthened by local governments with the incentive of the expectation of promotion and financial reward, aggravates the problems of resources and the environment. The imbalanced development in China, including uneven development among regions, imbalanced expenditure structure between consumption and investment, and the rising inequality in income resulting in imbalanced consumption with insufficient consumption accompanied by overconsumption, is challenging the coordination of development and protection. The weak capacity for scientific and technological innovation and the underdevelopment of non-governmental organizations in China leave the government as the main player in SCP, and this is inefficient for the development of SCP. It is suggested that SCP should be put forward as a development strategy and be conducted more systematically using existing policies in China so as to promote SCP more efficiently. China's further reform is expected to lift the institutional barriers to SCP and promote economic transformation, balanced development and social equity. In addition, the long-term capacity for SCP calls for the improvement of independent technical innovation in enterprises and extensive social participation.

Keywords: Industrial restructuring, cleaner production, circular economy, energy conservation and emission reduction, green procurement, technical innovation, economic transformation.

27.1 Introduction

China has achieved great economic progress in the past three decades. Its gross domestic product (GDP) has continuously increased by an annual average growth rate of 9.83 per cent between 1978 and 2012 (NBS 2014) and it has overtaken Japan as the world's second largest economy. But the rapid economic growth has been accompanied by severe resource depletion, environmental pollution and ecological destruction (Hu/Zhang/Lin et al. 2014; Zhang 2014). The country's energy consumption has risen from 571.44 million *tonnes of coal equivalent* (tce) in 1978 to 3617.32 million tce in 2012, with an average annual growth rate of 5.58 per cent (NBS 2014). China has become the world's largest emitter of CO_2 (IEA 2014). Meanwhile, China's per capita ecological footprint rose to 2.1 global hectares, 2.5 times its ecological carrying capacity in 2008 (WWF–China 2012). Its per capita domestic material consumption as well as its per capita GHG emissions have exceeded the world average (IEA 2014; UNEP 2013). China's accumulated air pollution has led to severe haze covering almost the whole country. It is argued that the rising amount of environmental and health damage, as well as the rising costs of energy supply, waste disposal, tackling pollution and eco-remediation, offset the achievements of development (Wang/Mauzerall

---

1 Dr. Hongmin Chen, Associate Professor, Department of Environmental Science and Engineering, Fudan University, Shanghai, China; Email <swingboat77@gmail.com>.

© Springer International Publishing Switzerland 2016
H.G. Brauch et al. (eds.), *Handbook on Sustainability Transition and Sustainable Peace*,
Hexagon Series on Human and Environmental Security and Peace 10, DOI 10.1007/978-3-319-43884-9_27

2006; World Bank 2007; Wang 2010; Matus/Nam/ Selin et al. 2012). It is widely recognized that the growing conflict between economic growth, energy and environment in China has restrained the country's further development (Hu/Zhang/Lin et al. 2014; Li/Bao/Xiu et al. 2010; Wu/Shi/Xia et al. 2014). Natural resources and ecosystems will no longer support 'business-as-usual' development while China is still in the process of industrialization and urbanization. It is estimated that by 2020, some 850 million people, about 60 per cent of the total population, will be living in China's urban areas, up from about 650 million in 2010 (Atsmon/Magni/Li et al. 2012). Thus, how to decouple economic development from resource depletion and environmental degradation (Wu 2014) and how to build a well-off society with less pressure on resources and the environment has become a crucial challenge for China.

Sustainable development, defined as development that meets the needs of the present without compromising the ability of future generations to meet their own needs (WCED 1987), has already become a global consensus since the 1992 United Nations Conference on Environment and Development. Since then unsustainable consumption and production patterns have been identified as the major cause of global environmental deterioration (United Nations 1992). *Sustainable Consumption and Production* (SCP), which aims to do "more and better with less" by minimizing the use of natural resources, toxic materials and emissions of waste and pollutants over the life cycle while increasing the quality of life for all (Oslo Symposium 1994), has been heralded as a primary way of implementing sustainable development. In order to accelerate the shift to SCP, the 2002 World Summit on Sustainable Development called on governments to promote a 10-*Year Framework of Programmes* (10YFP) in support of regional and national initiatives to boost SCP (UNEP 2011). The multi-stakeholder Marrakech Process, launched in 2003, has supported the implementation of SCP and provides inputs for the development of the 10-Year Framework. At Rio+20, a 10-Year Framework Plan on SCP was decided on, and this is now in the process of being implemented by national governments and various regional groups in cooperation with UNEP.

China was one of the first few countries to propose and implement sustainable development strategies. In March 1994, the Chinese government issued "China's Agenda 21: White Paper on China's Population, Environment and Development in the 21st Century" and then in 1996 incorporated sustainable devel-

opment into national strategies. In 2003, China proposed the scientific outlook on development as the guiding principle which enriched the connotations of sustainable development in China (NDRC 2012). So far SCP, as a concept, has not yet been incorporated into China's national development plans or laws. Various policies, regulations and pilot projects promoting sustainable development have reflected the idea and practices of SCP in China. This chapter introduces and discusses the development of SCP in China. The remainder of this text is structured as follows: section 27.2 describes the main policies and practices for promoting *Sustainable Production* (SP) in China and its existing problems, followed by an introduction to China's *Sustainable Consumption* (SC) development in section 27.3, as SC has been paid less attention to than SP in China. Section 27.4 discusses the key challenges China is still facing in the promotion of SCP. Section 27.5 provides a brief summary of SCP development in China and some suggestions for its further development.

27.2   Sustainable Production in China

Unsustainable production has been recognized as the major cause of various resource and environmental problems in China. Sustainable production, which requires minimizing the impact of production on natural resources and the environment, has not yet been clearly introduced as a principle for industrial development. Many policies and practices which contribute to a shift to sustainable production have been conducted in China.

27.2.1   Adjustment of Industry Structure

China is undergoing an accelerated industrialization and urbanization process. The Chinese economy, with a high proportion of resource-intensive industries as well as severe overcapacity in a number of industrial sectors, places great pressure on resources and environment. Thus industrial restructuring, which aims at shifting from highly polluting heavy industries to a more service-oriented, high-tech economy, has become an important part of sustainable production in China.

China is the world's biggest producer of steel, aluminium and cement, producing about 48.5 per cent of the world's crude steel and 57.5 per cent of the world's cement in 2013.[2] Official statistics indicate that only 72 per cent, 73.7 per cent, 71.9 per cent, 73.1 per cent and 75 per cent of capacity was used in the steel, cement, electrolytic aluminium, glass and shipbuilding sectors, respectively, in 2012, far less than that of other countries. China has issued a large number of policies to tackle excess capacity. The *State Council* (SC) issued a "Guidance on Resolving Serious Excess Capacity" in 2013 which forebade the construction of new projects in industrial sectors with excess capacity and halted new investment in iron and steel, cement and other industries with serious excess capacity (SC 2013). In the meanwhile, the elimination of outdated production capacity has also been widely carried out. Since 2011, the *Ministry of Industry and Information Technology* (MIIT) has mandated eight batches of enterprises, 7,885 enterprises in total, to eliminate outdated production capacity. The number of sectors involved has increased from thirteen in the Eleventh Five-Year Plan (FYP) period (2006–2010) to twenty-one in 2012. In addition, since 2010, the MIIT has publicized its annual progress in eliminating outdated production capacity (CDP 2014).

Besides resolving excess and outdated production capacity, efforts have also been made to promote the development of modern services and strategic emerging industries. Since the beginning of the twenty-first century, a series of policies has been issued to support and regulate the development of the service sector. The share of the service industry in the economy has increased from 34.8 per cent in 2000 to 46.8 per cent in 2013 (NBS 2014). In October 2010, China issued "the SC Decision on Speeding up the Cultivation and Development of Strategic Emerging Industries", in which seven industries such as the renewable energy industry and the energy-saving and environmental industries were identified as strategic emerging sectors (NDRC 2012).

The adjustments to the industry structure aim at broadening employment channels and reducing the pressure of economic development on resources and the environment. But existing regional protectionism in China has slowed the process. Despite the calls by the central government to curb excess capacity, local governments are reluctant to eliminate local laggard and excess capacity for the sake of GDP and employment. The achievements in eliminating laggard capacity have largely been offset by the expansion of new capacity. According to the National Bureau of Statistics, crude steel production capacity was reduced by 76 million tonnes during 2006–2012, while at the

same time the new production capacity totalled 440 million tonnes.[3] On the other hand, competition among local governments for the development of strategic emerging industries has created artificial demand for these industries, leading to another round of excess capacity. The renewable energy industry is most typical in the new round of excess capacity. Excess capacity in both traditional manufacturing and emerging sectors has made industrial restructuring more complicated and difficult to cope with. Prohibiting overlapping investment and the closure of outdated production facilities may only ease the problem, but not cure the disease.

27.2.2 Promotion of a Circular Economy and Cleaner Production

A *circular economy* (CE), proposed by two British environmental economists in 1990 as a closed loop of material flows in the economy (Pearce/Turner 1990), was first introduced into China in the late 1990s (Zhu 1998). Since China's environmental protection strategy embarked on a gradual transition from *end-of-pipe* (EOP) pollution treatment to pollution prevention in the 1990s (Shi/Zhang 2006), CE has aroused great interest and it has been regarded as a new development strategy to help China leapfrog into a more sustainable economic structure rather than being regarded as an incrementally improving environment management policy (Zhu 2008; Geng/Doberstein 2008a). Thus, a lot of policies and regulations were issued to promote a CE since 2000, including the "Circular Economy Promotion Law". It has been determined as a key strategy for China's national development plan since 2005 (Geng/Fu/Sarkis et al. 2012). The main focus of the CE has been expanded from a narrow system of waste recycling to the broad efficiency-oriented control of all stages of production, distribution and consumption (Su/Heshmati/Geng et al. 2013).

Current CE practices in China are carried out at three levels, the enterprise level, the industrial parks level, and the regional level, all of which are the basic manifestations of CE at the micro-, meso-, and macro-levels, respectively (Hong/Li 2013). CE pilot programmes have been launched in twenty-eight provinces (municipalities, autonomous regions), 133 cities (districts and counties), 256 industrial parks and 1,352 enterprises (NDRC 2012). Su, Heshmati, Geng et al.

---

2 See at: <http://www.chinaccm.com/39/20140124/392202_1678542.shtml> and at: <http://minerals.usgs. gov/ minerals/pubs/commodity/cement/mcs-2014-cemen. pdf>.

3 See at: <http://www.chinairn.com/news/20130805/153103234.html>.

(2013) pointed out that practices at micro- and meso-levels are more vibrant than those at the macro-level because the complexity of practices increases when the scale level increases.

A large number of studies have been conducted to evaluate the CE performance at different levels with different evaluation indicator systems (Wang/Huang/Chen 2006; Zhu/Zhu 2007; Yang/Gao/Chen 2011; Geng 2011; Zhao/Christensen/Lu et al. 2011; Li/Su 2012). However, only a few studies exist on cycle flows at the national level. Li, Zhang, and Liang (2013) used the reutilization-extended *economy-wide material flow analysis* (EW-MFA) to investigate comprehensive resource reutilization in China from 2000 to 2010, and they found that about sixty per cent of overall solid waste generation had been reutilized, and more than twenty per cent of the total resource requirement was reutilized resource by the year 2010. Official data indicated that one-fifth to one-third of raw materials for steel, non-ferrous metals and paper pulp came from the recycling of resources; twenty per cent of raw materials for cement originated from solid waste; and the overall utilization rate of industrial solid waste had reached sixty-nine per cent (NDRC 2012); while in Japan, the reutilization rate for iron and steel and aluminium in 2012 was 90.8 per cent and 99.9 per cent, respectively (Japan Business Federation 2014).

Several widely recognized and repeatedly emphasized challenges may slow down the implementation of the CE in China, including weak economic incentives, lack of reliable information, lax enforcement of regulations, shortage of advanced technology, poor leadership and management, lack of stakeholder participation, and lack of a standard system for performance assessment (Peng/Li/Shi 2005; Wen/Yuan 2005; Wang/Chen/Chen et al. 2007; Su/Heshmati/Geng et al. 2013; Liu/Bai 2014). There exists a striking gap between awareness and actual behaviour in developing a circular economy.

Cleaner production, which could be recognized as CE practice at the enterprise level, has also been widely promoted by the Chinese government. CP in China has progressed through the formation stage (1973–1992), the legislative stage (1993–2002), and the institutionalization stage (2003) (Duan/Zhou 2007). The Law of the People's Republic of China on Promotion of Cleaner Production, issued in 2002 and enacted in 2003, established the legal basis for CP in China. A number of governmental documents have been issued to guide and supervise enterprises in implementing CP (Dan/Yu/Yin et al. 2013). The

Cleaner Production Audit (CPA) has become the main measure in promoting CP for enterprises. Compared with foreign companies that conduct CPA voluntarily and actively (Howgrave-Graham/van Berkel 2007; Chavalparit/Ongawandee 2009), the promotion of CP in China has been confronted with many obstacles, such as a lack of economic incentive policies, a lack of corporate responsibility, financial obstacles, and a shortage of technology in companies (Shi/Peng/Liu et al. 2008).

In China, three categories of enterprises are required to conduct compulsory CPA regularly, covering those that exceed the national or local discharging standards or the total amount control targets for pollutants; exceed the energy consumption limits for unit product; and use toxic and harmful materials in production or discharge toxic and harmful substances. Other enterprises are encouraged to take voluntary CPA. Since 2003, many industry sectors, such as metallurgy, chemical, non-ferrous metals, and machinery have established CP centres and more than 760 CPA advisory services have been set up. Capacity-building activities, such as CPA training and relevant education programmes, are being undertaken (Geng/Haight/Zhu 2007). Over 20,000 enterprises carried out CPA, accounting for nine per cent of all industrial enterprises above a designated size in China. As a result, sulphur dioxide, ammoniacal nitrogen, chemical oxygen and energy demand were estimated to have been reduced by 939,000 tonnes, 56,000 tonnes, 2.46 million tonnes and 56.14 million tce respectively since the implementation of CP between 2003 and 2010 (MIIT 2012).

Although some progress has been made (Duan 2001), the universal adoption of CP is far from satisfactory (Mol/Liu 2005). First, there were obvious differences between regions in the promotion of compulsory CPA in key enterprises, since the provincial government plays a leading role in enforcing the related CP regulations, supervising the implementation of CP and providing technical support and capacity-building (Geng/Wang/Zhu et al. 2010). Second, *small and medium-sized enterprises* (SMEs), which account for ninety-nine per cent of all Chinese enterprises, could get little technical and economic support for their CP efforts, since governments pay more attention to big companies and provide more funds to cover their CP expenses (Wang/Shu/Wang 2014; Zhang/Zhang 2008). Thus, the application and effectiveness of CP in these SMEs is doubtful (Geng/Wang/Zhu et al. 2010). Considering the reality that most SMEs are confronted with difficulties such as obsolete equipment and technology, untrained and inexperienced workers, and inadequate financial resources, their production has caused significant negative environmental impacts (Shi/Zhang 2006). Third, most CP in Chinese enterprises aims at savings and efficiency gains through process, technology or management improvements. It was estimated that only 44.6 per cent of the technical renovation suggestions in CPA reports were implemented (MIIT 2012). Other CP practices, such as shifting to eco-design or delivering new products and services with lower environmental impact across their life cycle, which require initiatives from enterprises, were rarely conducted in China. Though according to the "Twelfth FYP on industrial CP development", eco-design of industrial products, improvement of CP technology and substitution of toxic and harmful materials or products are taken as the main tasks of CP in the Twelfth Five-Year period (MIIT 2012), they are scarcely being conducted by enterprises under the current circumstances.

27.2.3 Comprehensively Promoting Energy Conservation and Emission Reduction

Energy Conservation (EC) strategy has always been noted as the key measure for achieving sustainable development in China, and improving energy efficiency has been on the policy agenda since the early 1980s (Hou/Zhang/Tian et al. 2011; Yuan/Kang/Yu et al. 2011). With continuous efforts, China's energy intensity, calculated as the energy consumption per unit of GDP, fell roughly fifty-five per cent between 1990 and 2010 (NBS 2014). China's CO_2 intensity declined by almost 58 per cent compared with the average rate of decrease of 30 per cent for developed countries and the global average of 15 per cent during these two decades (IEA 2014).

China has set an EC target in its FYP for National Economic and Social Development for many years, but it was in the Eleventh FYP (2006–2010) that a binding target, reducing energy intensity by twenty per cent, and the main pollutant of sulphur dioxide(SO_2) and chemical oxygen demand (COD) by ten per cent by 2010, was first set as of 2005 (NPC 2006; Yuan/Kang/Yu et al.2011), and this was referred to as the *Energy Conservation and Emissions Reduction* (ECER) target. In order to ensure the accomplishment of the target, all government divisions at different levels were required to make an effort. The SC approved and distributed a scheme disaggregating the national EC target to each province, and provincial governments disaggregated their targets to cities and

counties (Zhou 2006). The head of the local government was responsible for the accomplishment of the targets in each administrative area. Following this requirement, a hierarchical administration system for EC ranging from the nation to the province and city level was developed during the Eleventh FYP period. Besides this, various EC policies, including legislation, government regulation, technology standards, finance and tax, as well as price policy, were released at the national level, and this laid a sound foundation for the accomplishment of the target (Yuan/Kang/Yu et al. 2011; SC 2007).

With strong determination and vast inputs, China's national energy consumption per unit of GDP was reduced by 19.1 per cent between 2005 and the end of 2010, which was 95.5 per cent of the target. Sulphur dioxide and chemical oxygen demands were reduced by 14.29 per cent and 12.45 per cent respectively, far beyond the pollution reduction targets (SC 2011). It was estimated that the amount of energy saved by the achievement of a 19.1 per cent improvement in energy intensity were about 630 million tce, taking the baseline condition as no improvement in energy intensity from 2005 levels. The carbon dioxide emissions savings were estimated at 1,550 million tonnes (Tsinghua University 2012).

The promotion of ECER during the Eleventh FYP period was conducted in a top-down manner. Command-and-control policies played the most important role. Administrative approaches, such as the Ten Key Energy Conservation Projects, the Top 1000 Enterprises Energy Conservation Programme, and the closure of small plants and outdated production capacity (Price/Levine/Zhou et al. 2011) were implemented as the main measures.

The widely adopted administrative approaches helped ensure governments at all levels implemented the ECER work. But the disadvantages of relying on administrative approaches were also apparent. Firstly, they created a great burden for the government. During the Eleventh FYP period, the government was not only the regulator, but at the same time the provider of finance and technology for EC. Secondly, administrative instruments may be effective when applied to a relatively small number of large enterprises or institutions, but they are ineffective in the case of numerous small-scale industrial plants. The success of the Top 1000 programme and the inefficiency of phasing out inefficient capacity in the steel industry is such a case (Yuan/Kang/Yu et al. 2011). Last but not least, the lack of EC initiatives by energy users made EC unsustainable. The targets were hard to meet, while the

---

Box 27.1: Top 1000 Enterprises Energy Conservation Programme. **Source**: NDRC (2006a, 2011).

China's Top 1000 Enterprises Energy Conservation Programme (Top 1000 Programme), as one of the key energy conservation promotion measures, was launched in April 2006. The programme required the largest 1,000 energy consuming industrial enterprises in nine major energy-consuming industries, each consuming a minimum of 180,000 tce in 2004, to set and implement energy-saving targets.[a]

The 1,008 enterprises on the final list of the Top 1000 Programme consumed 670 million tce in 2004, accounting for one-third of China's total energy consumption. The Top 1000 Programme aimed at reducing energy consumption by 100 million tce via energy efficiency improvement per unit product. Local energy conservation authorities were responsible for tracking, guiding and supervising the work on energy-saving in these enterprises (NDRC 2006a). The National Development and Reform Commission signed target-setting agreements with local governments, and local governments signed agreements with enterprises on their energy conservation targets (NDRC 2011). Each enterprise received an EC target which was defined by energy consumption per unit product. In order to reach

their target, the Top 1000 enterprises are expected to establish an EC working group, formulate an EC plan, establish EC incentive mechanisms, invest in energy efficiency improvement options, conduct energy audits, establish an energy utilization reporting system, enhance EC publicity and training, etc. These enterprises are required to report their quarterly energy consumption by fuel to the NBS (NDRC 2006a). It is worth noting that the Top 1000 Programme requires the integration and implementation of a series of policies and the actual boundary between different kinds of policy is not clear-cut in practice.

According to official data, the comprehensive energy intensity per unit product of aluminium, ethylene and caustic soda decreased more than thirty per cent and the comprehensive energy intensity of oil refining, electrolytic aluminium and cement more than ten per cent, while the energy intensity of coal-fired electricity generation declined by almost ten per cent. Total energy conservation through the Top 1000 Programme reached 150 million tce compared with a frozen 2005 energy intensity baseline (NDRC 2011).

---

a)   See at: <http://iepd.iipnetwork.org/policy/top-1000-energy-cosuming-enterprnises-program>.

---

administrative emphases changed. In fact, energy intensity increased rather than declined during the first half of 2010, since in response to the unexpected global financial crisis in 2008 the government focused on providing vast investments to stimulate the economy. Central government's rush to implement a very strict restructuring policy through strong administrative means did not change the situation (Yuan/Kang/Yu et al. 2011).

The Twelfth FYP for National Economic and Social Development (2011–2015) issued by the Chinese government continued to devote considerable attention to ECER and established new targets for the period 2011 to 2015. It called for energy intensity to decline by a further sixteen per cent and carbon dioxide emissions per unit of GDP to be reduced by seventeen per cent in 2015 compared with 2010. The carbon intensity target was first addressed in the Twelfth FYP in response to the forty to forty-five per cent reduction in carbon intensity from 2005 levels by 2020 which was first announced during the Copenhagen negotiations in 2009.

Using experience and lessons from the Eleventh FYP, the Top 10,000 Enterprises Energy Conservation Programme was created, modelled on the Top 1000 Programme. It covered industrial enterprises that consumed more than 10,000 tce per year and transport

enterprises or public buildings that consumed more than 5,000 tce per year. The Top 10,000 Programme covered sixty per cent of total energy consumption and the total number of enterprises reached about 17,000. According to data released by NDRC, the Top 10,000 Programme had achieved sixty-nine per cent of the EC targets during 2011–2012 (NDRC 2013). But as the number of enterprises grows, so do the challenges of collecting accurate data and enforcing targets. Price, Levine, Zhou et al. (2011) pointed out that both national and local government need to enhance their capacities to implement, monitor and evaluate these programmes.

During the Twelfth FYP period, market-oriented approaches are expected to take a more important role in promoting ECER. The Twelfth FYP established the goal of promoting the development of energy-saving and environmental industries and gradually establishing a carbon trading market. China is currently implementing seven regional pilot *emissions trading systems* (ETSs) which are expected to serve as a testing ground for a national ETS (Carbon Market Watch 2013).

It is recognized that the pilots have made important progress in experimenting with how to use ETS in a Chinese context. Local and national officials are already studying how to potentially go beyond pilot

---

Box 27.2: China's Carbon Emissions Trading Pilots. **Source**: Zheng (2014); Wilkening, Kachi (2014); Wu, Shi, Xia et al. (2014).

In October 2011, the NDRC designated five cities (Beijing, Shanghai, Tianjin, Shenzhen and Chongqing) and two provinces (Guangdong and Hubei) as regions for carbon emissions trading pilot projects. The seven pilot regions account for eighteen per cent of the population and thirty per cent of GDP in China.

From 2013 to 2014, the seven pilot projects have developed carbon emissions trading schemes, set out management and trading rules, and launched the carbon emission trading system. Focusing only on carbon dioxide, the seven pilot projects covered roughly forty to sixty per cent of a city or province's total emissions and the aggregate of all emissions regulated were around 0.8 billion tonnes of CO2, which became the second largest emissions trading market in the world, after the European Union. The total number of enterprises or institutions covered reaches over 2,200, which includes heavy manufacturing sectors such as steel, cement, and petrochemicals, and non-industrial high-emitting sectors such as airports, hotels and financial service providers. Besides direct emissions generated by activities within the area/sector covered, the seven pilot projects also cover indirect emissions linked to electricity or thermal energy used in production, due to the regulated

electricity prices. Most allowances in these pilot projects were allocated freely, though the Guangdong pilot project requires companies to buy a portion of their allowances through an auction (three per cent of the total in 2013 increasing to ten per cent by 2015; Zheng 2014). In July 2014, the Beijing, Shanghai, Tianjin, Guangdong and Shenzhen pilot projects completed their first compliance audit after regulated emitters surrendered their allowances. In these five pilot regions, the average compliance rate was 98.53 per cent. Though it showed a high degree of compliance, trading volumes were generally low and many pilot regions like Beijing, Tianjin and Guangdong postponed their compliance deadline due to the relatively low participation of enterprises in the carbon trading market. A tendency of allocating too many permits may have made emissions trading inefficient. But it should also be noted that the pilot schemes are not a long-term project, but rather a way to experiment with various emissions trading design elements, such as the threshold for inclusion, sector coverage, cap setting, allowance allocation and the *monitoring, reporting and verification* (MRV) system, while helping companies get used to the instrument before a national programme is rolled out (Wilkening/Kachi 2014; Wu/Shi/Xia et al. 2014).

---

a) <http://www.chinalawinsight.com/2014/09/articles/dispute-resolution/chinas-pilot-carbon-markets-at-a-glance/>.

---

programmes (Song/Lei 2014). Although the Chinese government has been ambitious to establish a national carbon market, there are still numerous challenges, including legal and regulatory barriers, lack of a mature market, interactions with existing energy and climate policies, equity concerns relating to the initial allocation where there is an uneven development background, information disclosure and the *monitoring, reporting and verification* (MRV) system, risk of carbon leakage, etc. (Han/Olsson/Hallding et al. 2012; Zhang/Karplus/Cassisa et al. 2014a; Zhang/Springmann/Karplus 2014b; Wu/Shi/Xia et al. 2014). These challenges need to be seriously considered while preparing for a national ETS in China.

27.3 Sustainable Consumption in China

Although China acknowledged the need to establish a sustainable consumption pattern in its Agenda 21 in 1994, the concept of sustainable consumption has not yet been formally defined and sustainable consumption has not received as much attention as sustainable

production. But there are a few policies and practices that reflect the concept of sustainable consumption in China.

27.3.1 Eco-labelling and Green Procurement

Information provision has been regarded as an effective tool for changing consumption patterns towards sustainable consumption through a kind of "voluntary approach" (Russell/Krarup/Clark 2005). Eco-labelling, which provides environmental product information for consumers, is one of the most promising forms of environmental information policy (Loureiro/McCluskey/Mittelhammer 2002). Since the German Blue Angel was first released in 1977, a number of eco-labelling programmes have been developed all around the world (Shen/Saijo 2009).

The Chinese government issued a mandatory and comparative energy information labelling programme known as the China Energy Label in 2005, following the legal provisions of the Energy Conservation Law and supporting regulations (Zhou/Khanna/Fridley et al. 2013). So far, forty-six national energy efficiency

Table 27.1: Development of Seven Pilot Carbon Trading Schemes in China. **Source**: Zheng Shuang (2014); International Carbon Action Partnership Region; at: <https://icapcarbonaction.com>.

	Covered CO_2 emissions Mtonnes）	Share of total emissions	Number of covered entities	Emissions threshold for coverage (t CO_2/year)	Launch date	Deadline for the first compliance audit	Compliance rate
Beijing	70	40%	490	>10,000 for industry and other sectors	28 Nov 2013	27 June 2014	97.10%
Shanghai	150	57%	191	>20,000 for industry; >10,000 for other sectors	26 Nov 2013	30 June 2014	100.00%
Tianjin	150	60%	114	>20,000 for industry and other sectors	26 Dec 2013	25 July 2014	96.50%
Chongqing	100	39.5%	240	>20,000 for industry	19 June 2014	-	-
Shenzhen	30	40%	635	>5,000 for industry and other sectors	18 June 2013	1 July 2014	99.40%
Guangdong	350	58%	202	>20,000 for industry	19 Dec 2013	15 July 2014	98.90%
Hubei	120	33%	153	>60,000 for industry	2 April 2014	-	-

Box 27.3: Project "Energy-saving Products Benefiting the Public". **Source:** MFPRC (2013).

The "Energy-saving Products Benefiting the Public" project provided a promotion list of energy-efficient products which covered three major categories including household appliances, vehicles and industrial products, fifteen varieties and about 100,000 types of energy-efficient products. The central government has allocated more than 40 billion yuan and released more than twenty detailed regulations for the implementation of the project.

It was reported that since the implementation of the project, public awareness of energy-saving products had been greatly improved, and the market share of energy-efficient products rose rapidly, which promoted technological progress and industrial upgrading. Taking the air conditioner as an example, the average price for energy-efficient air conditioners fell, while their market share rose from five per cent before the project to more than seventy per cent. New energy efficiency standards were implemented more smoothly and the production of inadequately energy-efficient air conditioners was halted. The entire energy efficiency of the air conditioner industry was increased by twenty-four per cent. Though the project "Energy-saving Products Benefiting the Public" ended in June 2013, it was concluded that the combination of subsidy policies, mandatory standards for products and the market mechanism was 'a good example of promoting the upgrading of products' (MFPRC 2013). But it should also be noted that subsidy policies may lead to excessive consumption.

---

standards, covering electronic products for home, commercial and industrial use, have been issued. The China Energy Label rates the energy efficiencies of appliances in five classes ranking from 1 to 5, where 1 is the most energy-efficient and 5 the least (Shen/Saijo 2009). It was expected to improve consumers' awareness and their preference for energy-saving products so as to ease the contradiction of boosting both domestic consumption and energy-saving targets in China. Besides providing energy information, China had also been issuing subsidies for energy-efficient lighting products since 2007. In addition, in June 2009, the Chinese government began implementing the project 'Energy-saving Products Benefiting the Public', which aimed at promoting energy-efficient products with energy labels of rank 1 or 2 through financial subsidies.

Similarly, the China environment label, which covers products from automobiles, building materials, textiles, electronics, chemicals and furniture to packaging has formed a complete system with standards, certifications, auditing and quality assurance. More than 30,000 products within 1,500 enterprises have been given the China environmental label. Annual output reaches 100 billion yuan (Lou 2008).

As government itself is a giant consumer, whose procurement accounts for roughly fifteen to twenty per cent of the country's GDP (European Commission 2004), by using its purchasing power to opt for energy-saving or environmentally friendly goods and services it can make a great contribution to SCP (Cao/Yan/Zhou 2009). Thus, *Green Public Procurement* (GPP) has been recognized as an effective way to develop capacity for green supplies and markets (Ho/Dickinson/Chan 2010) by encouraging manufacturers to implement energy saving and environmentally friendly design and production of products (Parikka-Alhola 2008).

Following developed countries' GPP practices, the Chinese government first adopted a policy of preferential purchasing of energy-saving products in 2004. Then in 2007 a programme of compulsory government procurement of energy-saving products began. In the meantime, government procurement began to give priority to products with environmental labels. So far, fifteen issues of government procurement lists with energy-saving products and thirteen issues of lists with environment label products have been released. Both lists are updated each year, and the past two years have seen the rapid expansion of product types and quantities in both lists.

Yet effective GPP promotion is still a major challenge for Chinese governments at different levels. Firstly, limited budgets mean that costs are still key barriers (Zhu/Geng/Sarkis 2013). Although GPP encourages government agents to purchase green products even if the cost is higher than non-green alternatives (Marron 1997), this is not true in many cases where local and government ordinances require choosing the lowest price bid (Zhu/Geng/Sarkis 2013). Secondly, the individual behaviour of procurement officials has been identified as another key factor in implementing GPP practices successfully (Meehan/Bryde 2011). The low environmental awareness of procurement personnel and long-standing relationships between local procurement officials and local suppliers have been observed to hinder the promotion of effective GPP (Geng/Doberstein 2008b). Thirdly, supplier monitoring is another important dimension for GPP practices (Wan/Lu 2009). It was revealed that the Chinese consumers are confronted by an overload of certifications and claims which they could not trust (McNeil/Hathaway 2005). Of great importance is the need to enhance the credibility and independence of China's product certification and

labelling systems. Additionally, the quality and performance standard of environmental products could be improved (Philipps/Espert/Eichhorst 2011). Last but not least, the current procurement laws and regulations do not implement a concept of sustainable procurement and a green supply chain. It is suggested that the best green procurement is no procurement and therefore it is crucially important for the government to carefully review and control government demand to avoid improper demand and procurement (Cao/Yan/Zhou 2009).

27.3.2 The Promotion of Waste Management

With the great improvement in the Chinese people's living conditions, the staggering increase in waste production throughout urban and rural China has become a critical problem. It is estimated that China annually generates 250 million tonnes of *municipal solid waste* (MSW), or a quarter of the world's total waste (Balkan 2012). The daily amount of MSW per capita in urban areas is about 1.0–1.18kg (World Bank 2005; Zhang/Tan/Gersberg 2010; Zheng/Song/Li et al. 2014), and the average annual growth rate for MSW from 1979 to 2009 was approximately seven per cent (Zou 2011). According to World Bank (2012) estimates, China will produce at least 480 million tonnes of MSW by 2030.

The increasing amount of solid waste has already overtaken the treatment capacity such as landfill and incineration, posing a great threat to the health of the people. By the year 2009, only 66.8 per cent of waste was being treated, leaving almost one-third of the total waste either not properly disposed of or untreated (Lybæk/Klemmensen/Li et al. 2010). About seventy-nine per cent of the treated waste goes to landfills. Incineration, which converts MSW to energy, is becoming an attractive option in China (Figure 27.1), while nearly two-thirds of China's cities are afflicted with chronic "waste siege" syndrome (Wang/Pei/Huang et al 2009). But increasing *"Not in My Backyard"* (NIMBY) protests (Wu/Dai 2014) have made the siting and operation of incineration plants a big challenge. The lack of reliable data and transparency on the incineration industry in China leaves a gap of trust between the citizens and the government, and public awareness of the health risks of incineration has also increased (Vanacore 2012). Thus, reducing, reusing and recycling of MSW based on waste classification has become a key solution to alleviating the city's waste disposal dilemma.

Figure 27.1: MSW treatment by type in China (2004–2010). **Source:** NBS (2014).

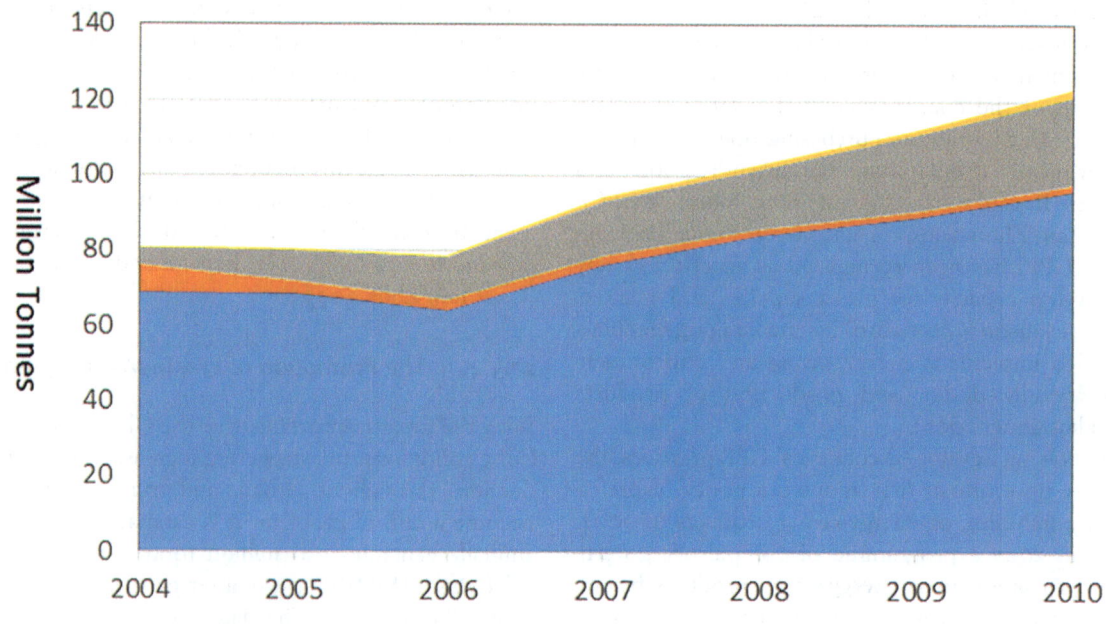

Therefore, in recent years, Beijing, Shanghai, Guangzhou and other big cities in China have carried out large-scale residential garbage sorting programmes. Actually, garbage sorting is not new to China's big cities. Many big cities, such as Beijing, Shanghai, Guangzhou, Shenzhen and Hangzhou, had carried out garbage sorting campaigns as trial projects in selected communities since the late 1990s. However, this has yet to be effective due to the lack of related back-end processing facilities and public participation. In the new round of garbage sorting programmes that started around 2010, these cities are aiming to gradually promote garbage sorting all around the city. More efforts have been put into the establishment of separate collecting, transport, and processing facilities and the implementation is expected to come with strict regulations and penalties. In Guangzhou, where 18,000 tonnes of waste are produced daily, city officials have implemented a fee-for-service programme to encourage residents to separate waste (Wang 2012). In Guangzhou, household garbage is required to be sorted into four categories, namely recyclable garbage, harmful garbage, wet garbage and dry garbage. Other cities have produced similar waste classification requirements. Since many Chinese families have the habit of recycling recyclable wastes such as papers and plastics which will be collected by informal waste collectors, and harmful wastes are relatively small in number, the biggest challenge for residents is separating

wet and dry waste. Wet waste covers kitchen waste and other organic waste, which accounts for between forty and seventy per cent of residential waste in different cities (Zhang/Tan/Gersberg 2010). Dry waste covers all the other waste except recyclable, harmful and wet waste. Volunteers in each community have been working to help guide and encourage the residents' garbage sorting.

Containers of different colours and instructions for different wastes showed that most people did not separate their garbage or failed to separate their garbage correctly, though most residents recognized the need for waste sorting and claimed that they knew how to sort waste. An Internet survey carried out jointly by Minyi China and Sohu.com indicated that only 18.2 per cent of the interviewees confirmed that they carried on separating the wastes as time went on, while more than 90.5 per cent support the promotion of waste separation.[4] A survey of 2,000 respondents carried out by the Shanghai Statistics Bureau revealed that 98.9 per cent of respondents were willing to sort their garbage. However, the actual sorting rate at household level in pilot regions was lower than twenty per cent.[5] Similarly, the separation ratio of MSW in

4   See at: <http://zqb.cyol.com/html/2011-04/19/nw.
     D110000zgqnb_20110419_1-07.htm>.
5   See at: <http://www.chinadaily.com.cn/china/2014-04/
     11/content_17425951.htm>.

Beijing was only approximately fifteen per cent (Xiao/ Bai/Ouyang et al. 2007). Since so many people failed to classify wastes correctly, related regulations and punishment mechanisms[6] have not been implemented to date.

With a low rate of waste sorting at the household level, many community organizations in these big cities began to hire one or more people to do on-site garbage sorting near the litter bins (called "secondary sorting"), so as to meet the waste sorting and reduction requirement. According to a survey of 182 communities in Shanghai (Huang 2014), 74 per cent of the communities relied on secondary sorting, while 23 per cent of the communities carried out source sorting at the household level with secondary sorting as a supplementary activity (figure 27.2). Secondary sorting shows a good performance level of waste classification, but the labour cost is greatly increased and it will also destroy residents' enthusiasm for waste sorting.

Figure 27.2: Household Waste Sorting at Different Stages in Shanghai. **Source**: Huang (2014).

sorted by waste collection and transportation company 3%

sorted at household level 23%

sorted by specific employees at community level 74%

This phenomenon indicates the deficiency in China's government-dominated work with residents. Garbage sorting, as a project promoted by the government, may require a visible and good performance over a short time. On the other hand, garbage sorting, as a daily activity related to residents' behavioural changes, is a long-term process. The demand for the short-term effects of government projects conflicts with the need for long-term behavioural change. Cultivating the

habit of waste sorting at household level has proved to be difficult when implemented as a top-down policy. The government policy may raise people's awareness or increase their willingness to carry out waste sorting, but what matters more is how to keep them constantly practising it. As government policies do not instantly lead to behavioural changes, it calls for the transformation of the governmental function in such activity and the introduction of non-government forces to promote long-term change. Moreover, education and campaigns and clearly clarified responsibilities for different stakeholders with mandatory regulations and punishment mechanisms are needed to achieve these goals.

27.4 Challenges Facing Sustainable Consumption and Production in China

Although the Chinese government has invested great manpower and resources in promoting the transition towards a more resource-conserving and environment-friendly society, China still faces several critical challenges that hinder the transition towards SCP.

27.4.1 The Extensive Growth Model

China's economy has expanded at a staggering pace since the opening-up and the economic reforms. But China's extensive growth model, which relies on high resource inputs and results in broad resource depletion, environmental pollution and ecological degradation, is now widely agreed to be exhausted and unsustainable (Zhang 2012; Dorrucci/Pula/Santabárbara 2013).

China's extensive growth model, which successfully fires the enthusiasm of local governments at different levels to promote economic growth, is enhanced by the GDP-oriented performance evaluation and fiscal system (Oi 1992; Zhou 2004; Wang/ Zhang/Zhang et al. 2007). Although central government has not formally regarded the GDP growth rate as a performance evaluation criterion, it is a prevailing phenomenon that local officials delivering punchy growth rates can get rapid promotion. This has created a political atmosphere favouring high speed (Zhou 2004). On the other hand, the existing fiscal decentralization system forces local governments at each level to gain fiscal revenue via economic growth (Wang/Zhang/Zhang et al. 2007). Although the central government provides fiscal transfers for less devel-

---

6   According to the regulation coming into force in Guangzhou from 1 April 2011, those who are caught failing to put their garbage into the right bin will be fined up to 50 yuan.

oped regions, most of the transfers require matching inputs from local governments. Thus the higher the economic growth rate, the more fiscal revenue local governments can receive, and this also covers the wages of government officials. And with more fiscal revenue, officials can achieve better performance and so find more opportunities for promotion.

Influenced by incentives regarding their political futures and financial rewards, government officials' fever for the growth rate leads to some phenomena that aggravate the problems of resources and the environment. Firstly, local governments can easily rely on investment to maintain a high rate of growth. Local government-dominated investment mania in a wide range of industries forms an investment bubble which boosts the upper reaches of industries that consume high amounts of energy, such as steel and cement. This is a major source of China's sharp increase in energy consumption and environmental pollution. Secondly, local governments compete against each other to gain investments, and this leads to a serious overlapping of investment and industries, as well as to redundant construction projects. This is an important cause of rebounding overcapacity in China. Fierce competition between local governments also leads to China's low labour costs, low land costs, and low environmental costs (Li/Gong 2013). Finally, those areas which are not suitable for industrial development and less competitive in inviting investment, including many towns and villages, also undertake responsibility for promoting GDP growth. This results in the small-scale, low-tech or highly polluting industrial enterprises spreading throughout China, and this hinders the country's economic restructuring and industrial upgrading.

Considering the extensive growth model as unsustainable, the Chinese government has sought to shift to an intensive development model, from investment-led to consumption-driven growth. But little progress has been made until now due to little progress in shifting the incentive structure of performance evaluation and the fiscal system. Central government has stipulated a rather conservative but balanced goal of seven per cent GDP growth for the Twelfth FYP period, while only five provinces (Beijing, Shanghai, Hebei, Zhejiang and Guangdong) have formulated growth goals of less than ten per cent for the year 2011, with all the rest between twelve and thirteen per cent (Yuan/Kang/Yu et al. 2011). If unchecked, this trend would ultimately lead to over-investment and aggravate pressure on resources and the environment.

27.4.2 Imbalanced Development

The pronounced imbalanced development in various fields of China has made the integration of development and protection particularly difficult. These imbalances include pronounced differences between the developed eastern coastal provinces and the less developed central and western provinces (Keidel 2007; Feng/Siu/Guan et al. 2012), the structural imbalance in expenditure and the increasing inequality in income (Dorrucci 2013).

Apart from the eastern coastal areas, most regions of China are still in the middle or even the early stages of industrialization and urbanization (NDRC 2012). In 2013, the developed eastern coastal provinces, with forty-nine per cent of the total population, accounted for sixty-four per cent of China's GDP, and the less developed central and western provinces contributed thirty-six per cent of GDP though they held fifty-one per cent of the population. The per capita GDP in the inland regions is less than half of that in the coastal regions on aggregate (Fan/Kanbur/Zhang 2011) and the differences between individual provinces and municipalities can reach up to a factor of four (Figure 27.3). It is recognized that regional inequality has been rising since the late 1980s (Kanbur/Zhang 2005), with only modest signs of decline (Zhang/Zou 2012). In the meantime, the imbalanced development between urban and rural areas is increased by industrialization and urbanization. There are still over 642 billion people living in the rural areas (NBS 2014), where living conditions and public services have lagged far behind urban levels. This great spatial inequality in education, health care and social security is a huge threat to China's SCP. China still faces a tough task in improving people's livelihood, and so top priority has been given to economic growth in most regions.

In recent years, the eastern developed regions in China, like Beijing and Shanghai, have attempted to say "goodbye" to giving priority to GDP and pursuing a transformation in economic development. Some industries with high levels of energy consumption and pollution have begun to move out of these places. But at the same time, the less developed regions in central and western China have competed to attract these industries. It should be noted that the ecological environment of many central and western regions is more vulnerable than that of the eastern regions. This kind of inter-regional transfer of industry may increase pollution while contributing little to GDP and employment in the whole country.

Figure 27.3: Regional Distribution of Per Capita GDP in mainland China in 2013 (10^4 US dollars). **Source**: NBS (2014).

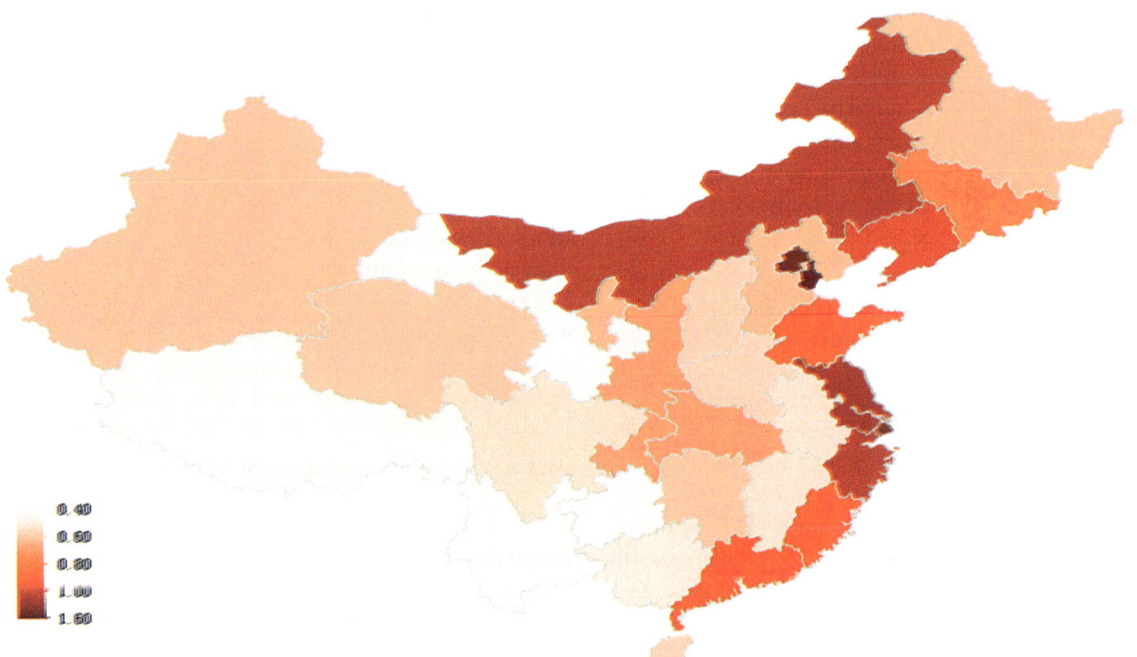

The imbalanced development also reflects on the expenditure structure, which features in high investment and low consumption (Li/Gong 2013). According to China's NBS, final consumption as a share of China's GDP fell from 62.3 per cent to 49.5 per cent between 2000 and 2012, while the investment share of GDP rose from 35.3 per cent to 47.8 per cent, and the private consumption rate declined from 46.4 per cent to 36.0 per cent (NBS 2014). However, China's private consumption has grown at more than 8 per cent per annum (Ma/McCauley/Lam 2013). Thus, the imbalance rather consists in an abnormally high contribution of resource-extensive investment to growth (Dorrucci/Pula/Santabárbara 2013), fuelled by the extensive growth model.

Rising income inequality is another reflection of uneven development. Xie and Zhou (2014) indicated that China's income inequality, measured using the Gini coefficient, has been in the range 0.53–0.55 since 2005, which was very high both from the perspective of China's past and in comparison with other countries at similar stages of economic development. These figures were also higher than has been acknowledged in government statistics, with the Gini coefficient in the range 0.47–0.49 between 2003 and 2012. Income inequality creates the problem of insufficient consumption (Li 2013) combined with overconsumption. At the end of 2012, China still had a population of over 100 million people living in poverty, according

to the new poverty line set in 2011 (rural residents' per capita annual net income being 2,300RMB, about US$350). Most of these people live in regions with harsh natural conditions and poor access to public services, making the task of poverty alleviation extremely difficult (NDRC 2012). Besides this, China's vast rural population and poor households in the urban areas face a liquidity constraint–although they have strong incentives to consume, they are unable to do so. On the other hand, overconsumption and luxury consumption is expanding rapidly. According to the World Luxury Association report, China has become the world's largest consumer of luxury goods (Li/Wang 2012). A growing middle class, affected by the global wave of consumerism, is expanding rapidly. It is projected that by 2030 China will have the largest population of consumers in the world, with an urban middle class of over half a billion people (Wetzel 2012). Consumption has become a symbol of success and social status. Overconsuming is in vogue nowadays. Traditional Chinese frugal consumption has become a byword for smallholder consciousness, belittled by the younger generation. The general public, especially the younger generation, aspire to and pursue western lifestyles, characterized by automobiles, big houses, and gated communities, and which have a deleterious effect on the environment. Therefore, China is on the way to a high-consumption society, although the current overall level of consumption is not high. Given the large population under insufficient consumption and a rising population pursing overconsumption, the conflicts and trade-offs between building a consumption-driven economy and a resource-saving and environmentally friendly society should be explicitly addressed instead of ignored (Dai/Chen 2010). Increasing domestic consumption creates an urgent need to ensure that the emphasis is on sustainable consumption (CCICED 2013).

27.4.3 Weak Scientific and Technological Innovation Capacities

China is at a critical point for its economic transformation, in which scientific and technological innovation plays an important role. China has come a long way in encouraging domestic technological innovation. The past few years have seen a great improvement in China's innovation capacity, but there is still a big gap compared with developed countries. The independent innovation capacity of Chinese enterprises is still too weak. Though according to the "2013 world intellectual property indicator" report released by the *World Intellectual Property Organization* (WIPO), China accounted for the largest number of applications filed throughout the world for the four types of intellectual property (WIPO 2013), China has only a limited number of invention patents and few influential patents. In the 2013 Global Innovation Index list, China ranked at number 35 (Cornell University/INSEAD/WIPO 2013). It is argued that the expansion of innovation activity instead of the growth of innovation efficiency and effectiveness is the main contributor to the development of China's innovation capacity (Mu/Ren/Song et al. 2010).

The Chinese government released the "Outline of Medium and Long-term Plan for National Science and Technology Development (2006–2020)" in 2006 (SC 2006), with an ambitious goal of becoming an innovative country by 2020 (Mu/Ren/Song et al. 2010). The Twelfth FYP for National Economic and Social Development announced an "innovation target" of 3.3 patents per 10,000 people (NPC 2011). But most Chinese enterprises are not fully ready for independent innovation. Most Chinese private industrial firms are small in scale and with relatively low capital intensity and simple technology, and they also have difficulty in getting bank loans. Thus these firms' capacity for technology absorption, adaptation, and creation are often limited. Large and medium-sized *state-owned enterprises* (SOEs), who find it much easier to get bank loans, are the leading performers in

R&D activities, but the innovation efficiency of SOEs is relatively low. Meanwhile, the market ecosystem is not functioning well enough to promote innovation. Underdevelopment of market institutions, such as distortions in pricing, weak enforcement of regulations especially the weak protection of intellectual property rights, and barriers to entry, exit, and fair competition, tends to discourage firms from investing in innovation (Zhang/Zeng/Mako et al. 2009). Weak technological innovation capacities leave most Chinese enterprises in the manufacturing and assembling stage without core technologies, and their competitiveness rests more on price and quantity, thus exacerbating the pressure on resources and the environment.

27.4.4 Underdevelopment of Non-governmental Organizations

Promoting SCP continuously requires a multi-stakeholder approach which involves the government, business and civil society. In China, the government is viewed as the leader in fostering SCP. But relying on the government alone will inevitably lead to high costs and low efficiency, and with the declining public trust in government, such costs will rise rapidly. Meanwhile, an increasing number of environmental pollution events and the rising environmental awareness of residents are continuously calling for more social participation in organized activities related to environmental protection. In this way, *non-governmental organizations* (NGOs) need to and can play a vital role in many fields of SCP, such as engaging civil society, businesses, and the public sector; conducting research to facilitate policy development; building institutional capacity; taking on environmental innovation activities; carrying out environmental education; and helping people live more sustainable lifestyles (Lyuba 2000; Adeel 2003).

There are three types of NGOs operating in China: social organizations, private non-enterprise units, and foundations or branches of international NGOs (Xu/Zhao 2010). Most of the social organizations are top-down NGOs, and are usually product of government reform and in some way affiliated to the government. NGOs of this type are also referred to as government-organized NGOs (GONGOs), and are not considered non-governmental organizations in the real sense of the word (Li 2012). Private non-enterprise units are bottom-up NGOs, and do not have as close a relationship with the government, hence they are normally referred to as grass-roots NGOs (Li 2012). Statistics from the Ministry of Civil Affairs

Table 27.2: Development of three types of NGOs in China. **Source:** Website of the Ministry of Civil Affairs, <www.mca.gov.cn>.

	2006	2007	2008	2009	2010	2011	2012	2013
Social Organizations($\times 10^4$)	19.2	21.2	23	23.9	24.5	25.5	27.1	28.9
Private Non-enterprise Units ($\times 10^4$)	16.1	17.4	18.2	19	19.8	20.4	22.5	25.5
Foundations	1144	1340	1597	1843	2200	2614	3029	3549

show that officially registered NGOs including GON-GOs in China numbered 547,549 (Table 27.2), a relatively small number for China's large population.

These years saw the number of NGOs grow rapidly in China, but several key barriers, widely recognized as hampering the development of NGOs, still exist, including restricted legislative environment, shortage of human resources and funds, lack of public trust and poor management capability (Lu 2007; Ru/Ortolano 2009; Li 2012; Liu 2013; Huang 2014). Registration of NGOs in China is strictly controlled and difficult. Only when an organization finds a related government department as a sponsor and supervisor government body is it eligible to submit an application for registration to the Ministry of Civil Affairs. Many grass-roots NGOs can hardly find a government agency to act as their sponsor/supervisor, as this often means a higher workload and even political trouble for the sponsoring agency (Xu/Zhao 2010). The difficulty of registering NGOs leaves many grass-roots NGOs unregistered, which means they cannot raise funds, and they face potential legal risks. Others register as business entities or use other registration strategies (Lu 2007), which means they cannot enjoy preferential tax policies. Hence the total number of bottom-up or grass-roots NGOs is unknown (Chen 2006). Even though some grass-roots NGOs succeed in registering as social organizations or private non-enterprise units, they still suffer from insufficient and unstable funding problems, and this is a crucial problem for their operation and growth (Li 2012). In China, corporate or personal donations for public welfare must be made through a few qualified NGOs,[7] in order for these donations to receive a pre-tax deduction. This means that most NGOs including grass-roots NGOs as well as some GONGOs find it almost impossible to obtain the qualification and will have difficulty in raising funding. Due to the policy environment and low rates of pay as well as low social status, NGOs in China, especially grass-roots NGOs, lack sufficient human resources. Few highly skilled professionals are willing to work for NGOs on a full-time basis (Li 2012). For this reason, most NGOs lack skilled workers in the legal, tax and accounting fields and in their respective working fields such as health, education, environment, etc. (Liu 2013). Insufficient funds and human resources also lead to poor management capacity in NGOs. With the majority of grass-roots NGOs unrecognized by the government, their internal governance is often problematic, characterized by irregularity and a lack of transparency (Xu/Zhao 2010). They also have difficulty in establishing and organizing medium- to large-size projects as these projects are largely dependent on government contacts and connections (Hasmath/Hsu 2008). In a social environment where government traditionally manages everything (Liu 2013), public understanding and trust in NGOs and their work is still very low. In addition, NGOs without legal recognition find it difficult to earn trust from society (Ru/Ortolano 2009), while the financial misdeeds and scandals that have been disclosed in some NGOs have left a very damaging influence on the reputation of NGOs in China. The low level of public trust in NGOs also explains why they cannot find enough funding from society.

It is quite encouraging that some restrictions on the development of NGOs are beginning to be lifted as reforms take place. Since 2011, four types of groups, namely industrial associations, science and technology organizations, charities, and organizations providing social services, have been able to register directly in a number of provinces where they do not need to find a government office to supervise their operation. The changes are expected to apply nationwide with the amendment in related laws or regulations (The Economist 2014). On the other hand, the Internet is also changing the situation of NGOs. In recent years, with the help of the Internet, Chinese NGOs have begun to earn public understanding and to rally public support for their activities. They have initiated a large number of nationwide environmentally friendly activities, such as the "clean your plate campaign", which aims at reducing food waste. The

---

7   The Ministry of Finance, the Administration of Taxation and the Ministry of Civil Affairs jointly issue the list of qualified non-government organizations (NGOs). In 2012, only 148 qualified NGOs were on the list.

campaign, initiated by NGOs, was first launched on Weibo (a Chinese version of Facebook) and was soon joined by millions of netizens across China. Based on the network, a strong NGO community is expected to play a more important role in promoting SCP in the future.

27.5   Prospect of Sustainable Production and Consumption in China

As a developing country with a large population and inadequate per capita resources, China is confronted with serious problems in dealing with development and protection. SCP, as a primary measure for implementing sustainable development and integrating development and protection, has been carried out in various areas in China. Since SCP has not been clearly put forward as a strategy in China, SCP is not conducted in a systematic pattern, but features in China's energy-saving and environmental protection-related policies and practices, such as the adjustment of industry structure, the promotion of a circular economy and cleaner production, the promotion of the Energy Conservation and Emission Reduction Project, and progress in green procurement and waste management. With great efforts and inputs, China has made some achievements in SP, especially in improvements to production efficiency. But it seems that the pace of improvements in efficiency is inadequate to offset the rapid increases in consumption (CCICED 2013), with a growing middle class affected by the global wave of consumerism. Actually, SC in China has not received as much attention as SP from the government and the public because of the relatively low proportion of consumption in the GDP expenditure structure. However, China's consumption has grown at a rapid pace, and the imbalance in the expenditure structure is rather due to an abnormally high contribution of investment to growth. Considering China's rapidly growing middle class and its shift to a consumption-driven economy, SCP should be clearly identified in a Chinese context and put forward as a national strategy in laws and in China's medium- and long-term plans for economic and social development.

An SCP framework should be developed at the national level to provide instructions for the promotion of SCP. As mentioned above, China has already introduced many SCP-related policies and practices. The key problem lies in their effective implementation. Existing SCP-related concepts, practices and policies should be straightened out and integrated into the SCP framework, so as to reduce duplication of effort and minimize conflicts between policies. In addition, some crucial challenges, including the extensive economic growth model, imbalanced development, weak scientific and technological innovation capacities, and underdevelopment of non-government organizations, are recognized as stemming from institutional problems and hinder the effective implementation of SCP-related policies and practices. Thus, China's further reform is expected to lift institutional barriers, including the optimization of fiscal and tax systems, the adjustment of the performance evaluation system, the transition of government functions and the expansion of market and social forces spaces and the construction of the rule of law, so as to promote economic transformation, balanced development, and social equity, and to guarantee the smooth and efficient implementation of SCP policies. Though government will still play an important role in the promotion of SCP, the long-term capacity for SCP lies mainly in the improvement of independent technical innovation in enterprises and extensive social participation in SCP. As well as this, the institutional arrangements need to give full play to the role of markets and civil society.

References:

Adeel, Zafar (Eds.), 2003: *East Asian Experience in Environmental Governance: Response in a Rapidly Developing Region* (New York: United Nations University Press).

Atsmon, Y.; Magni, M.; Li, L.; Liao, W., 2012: "Meet the 2020 Chinese Consumer", McKinsey Consumer & Shopper Insights, at: <http://www.mckinseychina.com/meet-the-2020-chinese-consumer/>.

Balkan, Elizabeth, 2012: "Waste-to-Energy in China", Presentation for Energizing China's Waste China Environment Forum roundtable, Woodrow Wilson Center, Washington DC, 14 June, at: <http://www.wilsoncenter.org/sites/default/files/Elizabeth%20Balkan%20PowerPoint.pdf>.

Carbon Market Watch, 2013: "China's Pilot Emissions Trading Systems", at: <http://carbonmarketwatch.org/zh-hans/chinas-pilot-emissions-trading-systems/>.

Cao, Fuguo; Yan, Yuying; Zhou, Fen, 2009: "Towards Sustainable Public Procurement in China: Policy and Regulatory Framework, Current Developments and the Case for a Consolidated Green Public Procurement Code", at: <http://www.ippa.org/IPPC4/Proceedings/07 GreenProcurement/Paper7-7.pdf>.

CCICED (China Council for International Cooperation on Environment and Development), 2013: "Sustainable Consumption and Green Development", CCICED 2013 Annual General Meeting, 13-15November, at <http://www.cciced.net/encciced/event/AGM_1/2013agm/speeches2011/ 201311/P020131105353646289008.pdf>.

CDP, 2014: "CDP China 100 Climate Change Report 2014", at: http://www.cdpchina.net/media/cdp/file/CDP%20China%202014%20Re-port_EN_25%20Sept_no%20appendix4.pdf>.

Chavalparit, O.; Ongawandee, M., 2009: "Clean Technology for the Tapioca Starch Industry in Thailand", in: *Journal of Cleaner Production*, 17: 105-110.

Chen, Jie, 2006: "The NGO Community in China: Expanding Linkages With Transnational Civil Society and Their Democratic Implications", in: *Political science*, 29: 40.

Cornell University; INSEAD; WIPO, 2013: "The Global Innovation Index 2013: The Local Dynamics of Innovation", Geneva, Ithaca, at: <http://www.globalinnovationindex.org/content.aspx?page=gii-full-report-2013>.

Dai, Xingyi; Chen, Hongmin, 2010: "Cracking the Conflict between Resource Conservation and Domestic Demand Expansion" (in Chinese), in: *Journal of China University of Geosciences (Social Sciences Edition)*, 10,1: 43-47.

Dan, Zhigang; Yu, Xiuling; Yin, Jie; Bai, Yanying; Song, Danna; Duan, Ning, 2013: "An Analysis of the Original Driving Forces behind the Promotion of Compulsory Cleaner Production Assessment in Key Enterprises of China", in: *Journal of Cleaner Production*, 46: 8-14.

Dorrucci, Ettore; Pula, Gabor; Santabárbara, Daniel, 2013: "China's Economic Growth and Rebalancing", European Central Bank, occasional paper series, No 142, at: <http://www.ecb.europa.eu/pub/pdf/scpops/ecbocp142.pdf>.

Duan, N., 2001: "Cleaner Production, Eco-industry and Circular Economy" (in Chinese), in: *Research of Environmental Sciences*, 14,6: 1-4.

Duan, N.; Zhou, C.B., 2007: "Research and Analysis of the Formation Process of Compulsory Regulation and Policy of Cleaner Production Audit" (in Chinese), in: *China Population Resources and Environment*, 17,4: 107-110.

European Commission, 2004: "A Report on the Functioning of Public Procurement Markets in the EU: Benefits from the Application of EU Directives and Challenges for the Future", at: <http://ec.europa.eu/internal_market/publicprocurement/docs/public-proc-market-final-report_en.pdf>.

Fan, S.; Kanbur, R.; Zhang, X., 2011: "China's Regional Disparities: Experience and Policy", in: *Review of Development Finance*, 1,1: 47-56.

Feng, K.; Siu, Y.; Guan, D.; Hubacek, K., 2012: "Analyzing Drivers of Regional Carbon Dioxide Emissions for China", in: *Journal of Industrial Ecology*, 16:600-611.

Geng, Yong, 2011: "Eco-indicators: Improve China's Sustainability Targets", in: Nature, 477: 162.

Geng, Yong; Doberstein, Brent, 2008a: "Developing the Circular Economy in China: Challenges and Opportunities for Achieving 'Leapfrog Development'", in: *International Journal of Sustainable Development and World Ecology*, 15,3: 231-239.

Geng, Yong; Doberstein, Brent, 2008b: "Greening Government Procurement in Developing Countries: Building Capacity in China", in: *Journal of Environment Management*, 88: 932-938.

Geng, Y.; Haight, M.; Zhu, Q.H., 2007: "Empirical Analysis of Eco-industrial Development in China", in: *Sustainable Development*, 15,2: 121-133.

Geng, Yong; Fu, J.; Sarkis, J.; Xue, B., 2012: "Towards a National Circular Economy Indicator System in China: An Evaluation and Critical Analysis", in: *Journal of Cleaner Production*, 23:216-224.

Geng, Yong; Wang, Xinbei; Zhu, Qinghua; Zhao, Hengxin, 2010: "Regional Initiatives on Promoting Cleaner Production in China: a Case of Liaoning", in: *Journal of Cleaner Production*, 18, 1502-1508.

Han, Guoyi; Olsson, Marie; Hallding, Karl; Lunsford, David, 2012: "China's Carbon Emission Trading: An Overview of Current Development", FORES Study, Stockholm Environment Institute, at: <http://www.sei-international.org/mediamanager/documents/Publications/china-cluster/SEI-FORES-2012-China-Carbon-Emissions.pdf>.

Hasmath, Reza; Hsu, Jennifer, 2008: "NGOs in China: Issues of Good Governance and Aaccountability", in: *Asia Pacific Journal of Public Administration*, 30,1: 1-11.

He, Jiankun; Yu, Zhiwei; Zhang, Da, 2012: "China's Strategy for Energy Development and Climate Change Mitigation", in: *Energy Policy*, 51: 7-13.

Ho, L.W.P.; Dickinson, N.M.; Chan, G.Y.S., 2010: "Green Procurement in the Asian Public Sector and the Hong Kong Private Sector", in: *Natural Resources Forum*, 34: 24-38.

Hong, Jinglan; Li, Xiangzhi, 2013: "Speeding up Cleaner Production in China through the Improvement of Cleaner Production Audit", in: *Journal of Cleaner Production*, 40:129-135.

Hou, Jian; Zhang, Peidong; Tian, Yongsheng; Yuan, Xianzheng; Yang, Yanli, 2011: "Developing Low-carbon Economy: Actions, Challenges and Solutions for Energy Savings in China", in: *Renewable Energy*, 36: 3037-3042.

Howgrave-Graham, A.; van Berkel, R., 2007: "Assessment of Cleaner Production Uptake: Method Development and Trial with Small Businesses in Western Australia", in: *Journal of Cleaner Production*, 15: 787-797.

Hu, He; Zhang, Xiaohong; Lin, Lili, 2014: "The Interactions between China's Economic Growth, Energy Production and Consumption and the Related Air Emissions during 2000-2011", in: *Ecological Indicators*, 46: 38-51.

Huang, Jian, 2014: "Study on the Development of Non-government Organization in China from the Perspective of Democratic Politics" (in Chinese), PhD thesis, The Central School of the Chinese Communist Party.

Huang, Wenfang, 2014: "Study on Municipal Solid Waste Classification and Reduction Model in Shanghai" (in Chinese), Working paper, Department of Environment Science and Engineering, Fudan University.

IEA (International Energy Agency), 2014: "CO2 Emissions from Fuel Combustion Highlights", at: <http://www.iea.org/publications/freepublications/publication/CO2EmissionsFromFuelCombustionHighlights2014.pdf>.

Japan Business Federation, 2014: "Results of Fiscal 2013 Follow-up Survey of Voluntary Action Plan on the Environment: Section on the Establishment of a Sound Material-Cycle Society", at: <https://www.keidanren.or.jp/policy/2014/034_sokatsu.pdf>.

Kanbur, R.; Zhang X., 2005: "Fifty Years of Regional Inequality in China: A Journey through Central Planning, Reform, and Openness", in: *Review of Development Economics*, 9,1: 87-106.

Li, Gan, 2013: "Income Inequality and Consumption in China", at: <http://international.uiowa.edu/files/international.uiowa.edu/files/file_uploads/incomeinequalityinchina.pdf>.

Li, H.Q.; Bao, W.J.; Xiu, C.H.; Zhang, Y.; Xu, H.B., 2010: "Energy Conservation and Circular Economy in China's Process Industries" in: *Energy*, 35: 4273-4281.

Li, J.J.; Wang, W., 2012: "Tax Policy Research on the Construction of a Sustainable Consumption" (in Chinese), in: *Driven Economy*, 330,11: 14-18.

Li, Nan; Zhang, Tianzhu; Liang, Sai, 2013: "Reutilisation-extended Material Flows and Circular Economy in China", in: *Waste Management*, 33: 1552-1560.

Li R.H.; Su C.H., 2012: "Evaluation of the Circular Economy Development Level of Chinese Chemical Enterprises", in: *Procedia Environmental Sciences*, 13: 1595-1601.

Li, Wenpu; Gong, Min, 2013: "China's Growth Model and Structural Imbalance in the Open Economy", in: *Procedia—Social and Behavioral Sciences*, 77: 37-54.

Li, Yuejin, 2012: "The Role and Development of Grassroots NGOs in Eastern China", Master thesis, Norwegian University of Science and Technology, at: <https://www.ntnu.no/documents/10443/21424885/Yuejin+Li.pdf/44323d8d-ae07-4d7d-8add-17e33f4e175a>.

Li, Xia; Peng, Ning; Zhou, Ye, 2014: "International Practice and Policy Implications of Sustainable Consumption" (in Chinese), in: *China Population, Resources and Environment*, 24,5: 46-50.

Liu, Dongwei, 2013: "NGOs Emerge in China, but Face More Challenges", Thomson Reuters Foundation, at: <http://www.trust.org/item/20131007121609-glxo2/>.

Liu, Yong; Bai, Yin, 2014: "An Exploration of Firms' Awareness and Behavior of Developing Circular Economy: An Empirical Research in China", in: *Resources, Conservation and Recycling*, 87: 145-152.

Lou Y.B., 2008: "The Establishment of Sustainable Production and Consumption Patterns", Chinese Environment News, 11 June: 001.

Loureiro, M.L.; McCluskey, J.J.; Mittelhammer, R.C., 2002: "Will Consumers Pay a Premium for Eco-labeled Apples?", in: *Journal of Consumer Affairs*, 36: 203-219.

Lu, Y., 2007: "Environmental Civil Society and Governance in China", in: *International Journal of Environmental Studies*, 64,1: 59-69.

Lybæk, Rikke Bak; Klemmensen, Børge; Li, Ruofei; Liu, Sibei, 2010: "Municipal Solid Waste Management in China", Roskilde University, at: <http://diggy.ruc.dk/bitstream/1800/5513/1/Municipal%20Solid%20Waste%20Management%20in%20China.pdf>.-

Lyuba, Zarsky, 2000: "From Bystanders to Collaborators: New Roles for Civil Society in Urban-industrial Environmental Governance in Asia", in: Angel, David P.; Rock, Michael T. (Eds.): *Asia's Clean Revolution, Industry Growth and the Environment* (Sheffield: Greenleaf Publishing).

Ma, Guonan; McCauley, Robert; Lam, Lillie, 2013: "The Roles of Saving, Investment and the Renminbi in Rebalancing the Chinese Economy", in: *Review of International Economics*, 21,1: 72-84.

Marron, D.B., 1997: "Buying Green: Government Procurement as an Instrument of Environmental Policy", in: *Public Finance Review*, 25: 285-305.

Matus, K.; Nam, K.M.; Selin, N.E., et al., 2012: "Health Damages from Air Pollution in China", in: *Global Environment Change*, 22: 55-66.

McNeil, Gary; Hathaway, David, 2005: "Green Labeling and Energy Efficiency in China", in: *China Environment Series Issues*, 7: 72-73.

Meehan, J.; Bryde, D., 2011: "Sustainable Procurement Practice", in: *Business Strategy Environment*, 20: 94-106.

MFPRC (Ministry of Finance of the People's Republic of China), 2013: "The Project of 'Energy efficient Products Benefiting the Public' Achieve Remarkable Results" (in Chinese), at: <http://jjs.mof.gov.cn/zhengwuxinxi/diaochayanjiu/201307/t20130711_960347.html>.

MIIT, 2012: "12th Five-Year-Plan for Industry Cleaner Production Development" (in Chinese), at: <http://www.miit.gov.cn/n11293472/n11295091/n11299314/14484311.html>.

Mol, A.P.J.; Liu, Ying, 2005: "Institutionalising Cleaner Production in China: the Cleaner Production Promotion Law", in: *International Journal of Environment and Sustainable Development*, 4,3: 227-245.

Mu, Rongping; Ren, Zhongbao; Song Hefa; Chen, Fang, 2010: "Innovative Development and Innovation Capacity-building in China", in: *International Journal of Technology Management*, 51,2-4: 427-452.

NBS (National Bureau of Statistics), 2014: China Statistical Yearbook Data (in Chinese), at: <http://data.stats.gov.cn>.

NDRC (National Development and Reform Commission), 2006a: "Notice on the Implementation Plan for Top-1000 Enterprises Energy Conservation Program" (in Chinese), at: <http://bgt.ndrc.gov.cn/zcfb/200604/t20060414_499304.html>.

NDRC, 2006b: "Implementation Suggestions of Ten Key Energy-Conservation Projects during the Eleventh Five-Year Plan" (in Chinese), at: <http://www.gov.cn/zwgk/2006-08/02/content_352716.htm>.

NDRC, 2011: "Review on the Energy Conservation and Emission Reduction in the 11th FYP" (in Chinese), at: <http://www.gov.cn/gzdt/2011-09/30/content_1960586.htm>.

NDRC, 2012: "The People's Republic of China National Report on Sustainable Development" (in Chinese), at: <http://www.china-un.org/eng/zt/sdreng/P020120608816288649663.pdf>.

NDRC, 2013: "Public Notice of Evaluation Results of Top-10000 Program Energy Saving Targets in 2012" (in Chinese), at: <http://www.sdpc.gov.cn/zcfb/zcfbgg/201401/t20140103_574473.html>.

NPC (The National People's Congress), 2006: "Program Outline of the 11th Five-Year Plan of National Economy and Social Development of the People's Republic of China" (in Chinese), at: <http://www.gov.cn/gongbao/content/2006/content_268766.htm>.

NPC, 2011: "Program Outline of the 11th Five-Year Plan of National Economy and Social Development of the People's Republic of China" (in Chinese), at: <http://www.gov.cn/2011lh/content_1825838.htm>.

Oi, J., 1992: "Fiscal Reform and the Economic Foundations of Local State Corporatism in China", in: *World Politics*, 45: 99-126.

Oslo Symposium, 1994: *Sustainable Consumption*. Oslo, Norway; 19-20 January.

Parikka-Alhola, K., 2008: "Promoting Environmentally Sound Furniture by Green Public Procurement", in: *Ecological Economy*, 68: 472-485.

Pearce, David W.; Turner, R. Kerry, 1990: *Economics of Natural Resources and the Environment* (Baltimore: Johns Hopkins University Press).

Peng, S.Z.; Li, Y.; Shi, H.; Zhong, P., 2005: "Studies on Barriers for Promotion of Clean Technology in SMEs of China" (in Chinese), in: *China Population Resources and Environment*, 3,1: 9-17.

Philipps, Sebastian; Espert, Valentin; Eichhorst, Urda, 2011: "Advancing Sustainable Public Procurement in Urban China", Sustainable Public Procurement in Urban Administrations in China—An action under EuropeAid's SWITCH-Asia Programme, Paper No. 14.

Price, L.; Levine, M.D.; Zhou, N.; Fridley, D.; Aden, N.; Lu, H.; McNeil, M.; Zheng, N.; Qin, Y.; Yowargana, P., 2011: "Assessment of China's Energy-Saving and Emission- Reduction Accomplishments and Opportunities during the 11th Five Year Plan", in: *Energy Policy*, 39: 2165-2178.

Ru, Jiang; Ortolano, Leonard, 2009: "Development of Citizen-Organized Environmental NGOs in China", in: *Voluntas*, 20: 141-168.

Russell, C.S.; Krarup, S.; Clark, C.D., 2005: "Environment, information and consumer behaviour: an introduction", in: Krarup, S.; Russell, C.S. (Eds.): *Environment, Information and Consumer Behaviour* (Cheltenham: Edward Elgar): 1-30.

SC (The State Council), 2007: "Comprehensive Working Plan for Energy Conservation and Emission Reduction" (in Chinese), at: <http://www.gov.cn/xxgk/pub/govpublic/mrlm/200803/t20080328_32749.html>.

SC, 2011: "Comprehensive Working Plan for Energy Conservation and Emission Reduction during the 12th Five-Year Plan Period" (in Chinese), at: <http://www.gov.cn/zwgk/2011-09/07/content_1941731.htm>.

SC, 2013: "Guidance on Resolving the Serious Excess Capacity" (in Chinese), at: <http://www.gov.cn/zwgk/2013-10/15/content_2507143.htm>.

Shen, Junyi; Saijo, Tatsuyoshi, 2009: "Does an Energy Efficiency Label Alter Consumers' Purchasing Decisions? A Latent Class Approach Based on a Stated Choice Experiment in Shanghai", in: *Journal of Environmental Management*, 90: 3561-3573.

Shi, H.; Peng, S.Z.; Liu, Y.; Zhong, P., 2008: "Barriers to the Implementation of Cleaner Production in Chinese SMEs: Government, Industry and Expert Stakeholders' Perspectives", in: *Journal of Cleaner Production*, 16: 842-852.

Shi, Han; Zhang, Lei, 2006: "China's Environmental Governance of Rapid Industrialization", in: *Environmental Politics*, 15,2: 271-292.

Song, Ranping; Lei, Hongpeng, 2014: "Emissions Trading in China: First Reports from the Field", WRI, at: <http://www.wri.org/blog/2014/01/emissions-trading-china-first-reports-field>.

Su, Biwei; Heshmati, Almas; Geng, Yong; Yu, Xiaoman, 2013: "A Review of the Circular Economy in China: Moving from Rhetoric to Implementation", in: *Journal of Cleaner Production*, 42: 215-227.

The Economist, 2014: "Chinese Civil Society: Beneath the Glacier", 12 April, at: <http://www.economist.com/news/china/21600747-spite-political-clampdown-flourishing-civil-society-taking-hold-beneath-glacier>.

Thomson Reuters, 2013: "Top 100 Global Innovators", at: <http://top100innovators.com/>.

Tsinghua University, 2012: "Annual Review of Low-Carbon Development in China (2011- 2012)", at: <http://www.climatepolicyinitiative.org/generic_datas/view/publication/123>.

United Nations, 1992: *Agenda 21: Results of the World Conference on Environment and Development* (New York: United Nations).

UNEP, 2011: "Global Outlook on SCP Policies", at: <http://www.unep.fr/shared/publications/pdf/DTIx1387xPA-GlobalOutlookonSCPPolicies.pdf>.

UNEP, 2013: "Recent Trends in Material Flows and Resource Productivity in Asia and the Pacific", at: <http://www.unep.org/pdf/RecentTrendsAP%28FinalFeb2013%29.pdf>.

Vanacore, Tara Sun, 2012: "Refusing to Waste Away: China's Tale of Trash Cities and the Incinerator Boom", China Environment Forum, at: <http://www.wilsoncenter.

org/sites/default/files/China_Incineration_SunVanacore_Part1_2.pdf>.

Wan, J.K.; Lu, X.F., 2009: *Government Green Procurement and Its Supplier Selection* (in Chinese), (Wuhan: Wuhan University Technology Press).

Wang, Liao; Pei, Tingquan; Huang, Chuan; Yuan, Hui, 2009: "Management of Municipal Solid Waste in the Three Gorges Region", in: *Waste Management*, 29,7: 2203-2208.

Wang D.; Shu X.W.; Wang D., 2014: "Status and Incentives of Cleaner Production in China" in: *Environmental Science and Management*, 39(3): 179-181

Wang, Haotong, 2012: "Guangzhou's Rubbish Charge Struggle", at: <http://www. chinadialogue.net/article/show/single/en/5057-Guangzhou-s-rubbish-charge-struggle>.

Wang, Ji, 1999: "China's National Cleaner Production Strategy", in: *Environment Impact Assessment Review*, 19: 437-456.

Wang, S.; Huang, X.J.; Chen, Y., 2006: "On Evaluating Regional Cycling Economy: A Case Study of Jiangsu Province" (in Chinese), in: *Journal of Jiangxi Agricultural University*, 5,1: 110-113.

Wang, X.; Mauzerall, D. L, 2006: "Evaluating Impacts of Air Pollution in China on Public Health: Implications for Future Air Pollution and Energy Policies", in: *Atmospheric Environment*, 40: 1706-1721.

Wang, Y., 2010: "The Analysis of the Impacts of Energy Consumption on Environment and Public Health in China", in: *Energy*, 35: 4473-4479.

Wang, Y.H; Chen, W.; Chen, J.L.; Duan, X.J., 2007: "Analysis of the Environmental Pressures on Industrial Firms with LISREL Model in the Zone along the Yangtze River of Jiangsu Province" (in Chinese), in: *Geographical Research*, 226,4: 705-711.

Wang, Yongqin; Zhang, Yan; Zhang, Yuan; Chen, Zhao; Lu, Ming, 2007: "On China's Development Model: The Costs and Benefits of China's Decentralization Approach to Transition" (in Chinese), in: *Economic Research Journal*, 1: 4-16.

Wen, Liang; Yuan, Jinyu, 2005: "Research on the Policy and Barrier of SME's Transformation to Circular Economy" (in Chinese), in: *Journal of Changsha University of Science and Technology*, 20,2: 14-16.

Wetzel, Jonathan; Li, Xiujun Lillian; Cheng, William, 2012: "What's Next for China?", McKinsey Insights China, at: <https://www.fccihk.com/files/dpt_image/5_committees/library/General%20Library/Whats-next-for-China-Jan-22-v2.pdf>.

Wilkening, Kristian; Kachi, Aki, 2014: "What next for Chinese carbon trading?", at: <https://www.chinadialogue.net/article/show/single/en/7096-What-next-for-Chinese-carbon-trading->.

WIPO, 2013: "2013 World Intellectual Property Indicators", at: <http://www.wipo.int/export/sites/www/freepublications/en/intproperty/941/wipo_pub_941_2013.pdf>.

World Bank, 2005: "Waste Management in China: Issues and Recommendations", at: <http://siteresources.

worldbank.org/INTEAPREGTOPURBDEV/Resources/China-Waste-Management1.pdf>.

World Bank, 2007: "Cost of Pollution in China: Economic Estimates of Physical Damages", at: <http://documents.worldbank.org/curated/en/2007/02/7503894/cost-pollution-china-economic-estimates-physical-damages>.

World Bank, 2012: "What a Waste: A Global Review of Solid Waste Management", at: <http://siteresources.worldbank.org/INTURBANDEVELOPMENT/Resources/336387133485261076/FrontMatter.pdf>.

WCED (World Commission on Environment and Development), 1987: *Our Common Future*, (Oxford: Oxford University Press).

Wu, Dan, 2014: "Evaluation and Prospect on the Decoupling Trend of Economic Development and Water Resource Utilization in China" (in Chinese), in: *Journal of Natural Resources*, 29,1: 46-54.

Wu, Huaqing; Shi, Yan; Xia, Qiong; Zhu, Weidong, 2014: "Effectiveness of the Policy of Circular Economy in China: A DEA-based Analysis for the Period of 11th Five-year-plan", in: *Resources, Conservation and Recycling*, 83: 163-175.

Wu, Libo; Qian, Haoqi; Li, Jin, 2014: "Advancing the experiment to reality: Perspectives on Shanghai pilot carbon emissions trading scheme", in: *Energy Policy* (online), at: <http://dx.doi.org/10.1016/j.enpol.2014.04.022i>.

Wu, Ying; Dai, Xuezhen, 2014: "China's Not-in-My-Backyard Protest in the Process of Urbanization", in: *Urban China in the new era: market reforms, current state, and the road forward* (Berlin: Springer).

WWF-China, 2012: *China Ecological Footprint 2012 Report* (Beijing: WWF-China).

Xiao, Yi; Bai, Xuemei; Ouyang, Zhiyun; Zheng, Hua; Xing, Fangfang, 2007: "The Composition, Trend and Impact of Urban Solid Waste in Beijing", in: *Environmental Monitoring and Assessment*, 135: 21-30.

Xie, Yu; Zhou, Xiang, 2014: "Income Inequality in Today's China", in: *PNAS*, 111,19: 6928-6933.

Xu Ying; Zhao Liao, 2010: "China's rapidly growing non-governmental organizations", EAI Background Brief No. 514, at: <http://www.eai.nus.edu.sg/BB514.pdf>.

Yang, Q.; Gao, Q.Q.; Chen, M., 2011: "Study and Integrative Evaluation on the Development of Circular Economy of Shanxi Province", in: *Energy Procedia*, 5:1568-1578.

Yuan, Jiahai; Kang, Junjie; Yu, Cong; Hu, Zhaoguang, 2011: "Energy Conservation and Emissions Reduction in China—Progress and Prospective", in: *Renewable and Sustainable Energy Reviews*, 15: 4334-4347.

Zhang C.L.; Zeng D.Z.; Mako W.P.; Seward J., 2009: "Promoting Enterprise-Led Innovation in China" (Washington DC: The International Bank for Reconstruction and Development/The World Bank); at: <http://siteresources.worldbank.org/CHINAEXTN/Resources/318949-1242182077395/peic_full_report.pdf>.

Zhang, Da; Karplus, V.J.; Cassisa, C.; Zhang, Xiliang, 2014a: "Emissions Trading in China: Progress and Prospects", in: *Energy Policy* (online), at: <http://www.sciencedi-rect.com/science/article/pii/S0301421514000275>.

Zhang, Da; Springmann, Marco; Karplus, Valerie, 2014b: "Equity and Emissions Trading in China", The MIT Joint Program on the Science and Policy of Global Change, TSINGHUA–MIT China Energy & Climate Project Report No. 257, at: <http://globalchange.mit.edu/files/document/MITJPSPGC_Rpt257.pdf>.

Zhang, Dongqing; Tan, S.K.; Gersberg, R.M., 2010: "Municipal Solid Waste Management in China: Status, Problems and Challenges", in: *Journal of Environmental Management*, 91:1623-1633.

Zhang, Haiyan; Lahr, Michael L., 2014: "China's Energy Consumption Change from 1987 to 2007: A Multi-regional Structural Decomposition Analysis", in: *Energy Policy*, 67: 682-693.

Zhang, Junjie, 2012: "Delivering Environmentally Sustainable Economic Growth: The Case of China"; at: <http://asiasociety.org/files/pdf/Delivering_Environmentally_Sustainable_Economic_Growth_Case_China.pdf>.

Zhang, L.L.; Zhang, K.L., 2008: "Cleaner Production in China's Small and Medium-sized Enterprises" (in Chinese), in: *Consumer Guide*, 3: 70-71.

Zhao, Y.; Christensen, T.H.; Lu, W.; Wu, H.; Wang, H., 2011: "Environmental Impact Assessment of Solid Waste Management in Beijing City, China", in: *Waste Management*, 31: 793-799.

Zheng, Lijun; Song, Jiancheng; Li, Chuanyang; Gao, Yunguang; Geng, Pulong; Qu, Binni; Lin, Linyan, 2014: "Preferential Policies Promote Municipal Solid Waste (MSW) to Energy in China: Current Status and Prospects", in: *Renewable and Sustainable Energy Reviews*, 36: 135-148.

Zheng, Shuang, 2014: "Investigation Report on the Seven Pilot Carbon Trading Scheme" (in Chinese), in: *China Energy*, 36,2: 23-27.

Zhou, Dadi, 2006: "Allocation of the National Energy Intensity Target to the Provinces and Sectors: Policy Recommendations", Presentation at the Energy Foundation forum on implementing China's 2010 Twenty-Per-Cent Energy Efficiency Target, 9 November.

Zhou, Lian, 2004: "The Incentive and Cooperation of Government Officials in the Political Tournaments: An Interpretation of the Prolonged Local Protectionism and Duplicative Investments in China" (in Chinese), in: *Economic Research Journal*, 6: 33-40.

Zhou, Nan; Khanna, Nina; Fridley, David; Romankiewicz, John, 2013. "Development and Implementation of Energy Efficiency Standards and Labeling Programs in China: Progress and Challenges", at: <http://www.escholarship.org/uc/item/9sd0p3hj>.

Zhu, Dajian, 1998: "Sustainable Development Appeals for Cycle Economy" (in Chinese), in: *Science and Technology Review*, 9: 39-42.

Zhu, Dajian; Qiu, Shoufeng, 2008: "Eco-efficiency Indicators and Their Demonstration as the Circular Economy Measurement in China" (in Chinese), in: *Resources and Environment in the Yangtze Basin*, 17,1, 1-5.

Zhu, Qinghua; Geng, Yong; Sarkis, Joseph, 2013: "Motivating Green Public Procurement in China: An Individual Level Perspective", in: *Journal of Environmental Management*, 126: 85-95.

Zhu, Y.; Zhu, D.J., 2007: "A Revised Circular Economy Model and Its Application Based on Objectivity-process-subjectivity Analysis" (in Chinese), in: *Shanghai Environmental Science*, 26.1: 14-18.

Zou, Xiaolong, 2011: "Municipal Solid Waste Management in China with Focus on Waste", at: <http://r-cube.ritsumei.ac.jp/bitstream/10367/3651/1/51209626.pdf>.

28   The Eco-restructuring of the Ruhr District as an Example of a Managed Transition

Philipp Schepelmann[1], René Kemp[2] and Uwe Schneidewind[3]

Abstract

The eco-restructuring of the Ruhr area presents a remarkable case of a managed transition, bringing out the importance of regional actors and factors, alongside external stimuli. The evidence in support of the authors' hypothesis of a managed transition in the Ruhr district is: (i) the vision of blue skies above the Ruhr, (ii) the reconstruction of the Emscher river system, (iii) the emerging energy transition. The ongoing transition of the Ruhr district requires further analysis which the authors plan to undertake. The chapter provides a first attempt at offering an evidence-based narrative of a region as a real-world laboratory for ecological modernization.

Keywords: Regional transition management, North Rhine-Westphalia, sustainability, ecological modernization, Emscher, Innovation City Bottrop, energy transition.

28.1   Introduction

This chapter offers an evidence-based narrative of the process of eco-restructuring of the Ruhr district in Germany between 1970 and 2012 in the form of the cleaning up of a polluted district, the creation of an eco-industry and a shift towards renewable energy in an economy dominated by the coal and steel industry.[4] The authors conceptualize this process of restructuring as a managed sustainability transition because of the use of various programmes and parliamentary acts to achieve an ecological modernization. It was not managed on the basis of a single master plan, but was the result of a number of plans together with sustained efforts by numerous institutional and individual actors.

The chapter documents the eco-restructuring of the Ruhr district with special attention to large projects, for which it offers some insights on the drivers behind the change, the motivating visions and the financing element.

The story of eco-restructuring in the polluted area, dominated by heavy industry, of *North Rhine-Westphalia* (NRW) contributes a case of a purposive transition to the literature of sustainability transitions. The story goes beyond an examination of green innovation in niches, and allows us to draw some conclusions about transition management.

The structure of the chapter is as follows. Section 28.2 states the research questions and explains the

1   Dr. Philipp Schepelmann, Project Coordinator, Wuppertal Institute for Climate, Environment and Energy, Döppersberg 19, 42103 Wuppertal, Germany; *Email*: <philipp.schepelmann@wupperinst.org>.
2   Prof. Dr. René Kemp, United Nations University–UNU MERIT, Keizer Karelplein 19, 6211 TC Maastricht, The Netherlands; *Email*: <rkemp@maastrichtuniversity.nl>.
3   Prof. Dr. Uwe Schneidewind, President of the Wuppertal Institute for Climate, Environment and Energy, Döppersberg 19, 42103 Wuppertal, Germany; Email: *Email*: <uwe.schneidewind@wupperinst.org>.

4   The authors thank the stakeholders for their time and willingness to discuss with us critical events, projects and developments; this helped us to construct a consistent narrative of the Ruhr transition. For the organisation of this process and contributions to improving the text we would like to thank Prof. Dr. Oscar Reutter and Miriam Müller and other colleagues at the Wuppertal Institute. We are also grateful for contributions from Mrs Trenk, Mrs Snowdon and Dr. Beckröge of the Regionalverband Ruhr. Special thanks to Johannes Klement for compiling the factual information and first drafts of a study. Finally, the authors would like to thank PD Dr. Hans Günter Brauch and four anonymous referees for their advice.

Figure 28.1: Germany, the State of North Rhine-Westphalia and the Ruhr Area. **Sources**: Locator map of the Ruhr District, Germany (TUBS Own work. Licensed under Creative Commons Attribution-ShareAlike 3.0 via Wikimedia Commons; at: <http://de. wikipedia.org/wiki/Ruhrgebiet#mediaviewer/File:Locator_map_ RVR_in_Germany.svg>; Ruhr District (Ruhr) with all cities with more than 50,000 inhabitants (Threedots (Daniel Ullrich)—Own work. Licensed Regionalverband under Creative Commons Attribution-ShareAlike 3.0 via Wikimedia Commons.

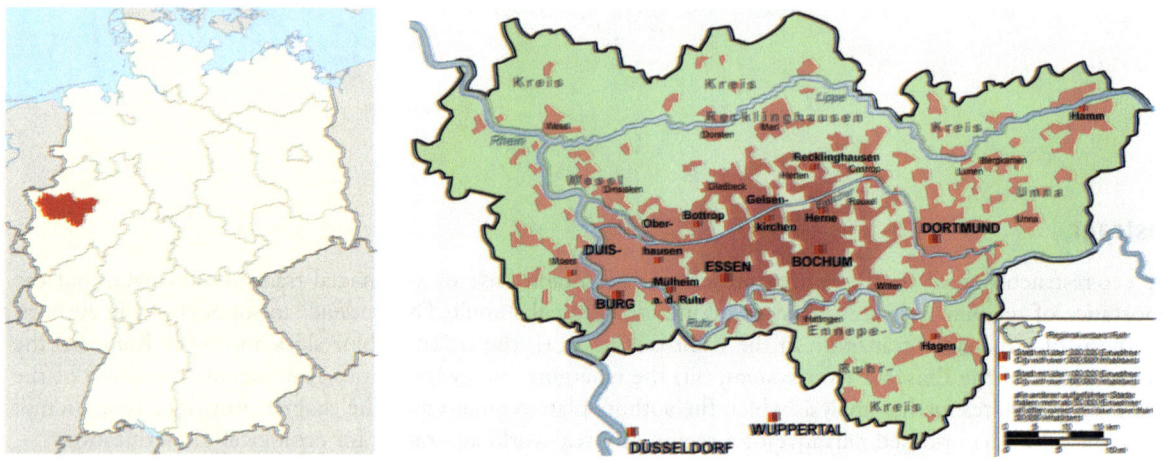

Ruhr Area within the state of   Map of the Ruhr District
North Rhine Westphalia

methodological choices of the research on which the chapter is based. Section 28.3 describes the historical setting of the Ruhr district as the industrial heart of Germany and Europe. Section 28.4 discusses the vision of 'blue sky' above the Ruhr and the increase in government activity for the protection of the environment based on the idea of ecological modernization. Section 28.5 zooms in on an important project forming part of the Ruhr transition to a greener economy: the Emscher Landscape Park initiative (a €4.5 billion 'generation project' for converting the dirtiest part of the Ruhr district into a space for recreation, culture, working and living) together with the low carbon Innovation City Ruhr initiative as two high-profile initiatives based on a partnership between the North Rhine-Westphalian state government, city administrations, business and civil society. Section 6 discusses energy transition as a new political project for the Ruhr district, describing the energy transition (*Energiewende*) towards sustainable energy and the innovation city Bottrop as an important energy transition initiative in the Ruhr district. Section 7 examines the institutional element as well as the interrelationships between the three processes: the introduction of environmental laws giving rise to an eco-industry sector, the clean-up and urban revitalization of the Ruhr district and the emerging energy transition. Section 8 focuses on the local and supra-local element in considering the role of national and international policies

vis-à-vis regional policies. Section 9 draws conclusions about what the case of eco-restructuring in North Rhine-Westphalia can teach us about transition management.

28.2   Research Methodology and Analytical Framework

This chapter offers an evidence-based narrative of the ecological modernization of the Ruhr district. The authors used evidence in grey literature and internet sources as the main sources of factual information. The evidence was compiled, sorted and condensed. This process was supported using secondary sources, experts' views and our own pre-analytical views using a relational, multi-scalar perspective. The major parts of the narrative were validated by central regional stakeholders such as the *Emschergenossenschaft/Lippeverband* (EG/LV), *Regionalverband Ruhr* (RVR), *Wirtschaftsförderung Metropole Ruhr* (wmr), and the *Landesamt für Natur, Umwelt und Verbraucherschutz* (LANUV) NRW, as well as representatives of the cities of Bochum and Essen. The information was presented and discussed in a sequence of conferences and workshops with governmental and non-governmental stakeholders of the Ruhr district. Details of the stakeholder meetings (in terms of place and participants) are described in the annexe.

Figure 28.2: View of the city of Gelsenkirchen, to the right the former coal mine, today centre of the Nordstern Parks— view from a former slag heap in the city of Essen. **Source:** Hans-Jürgen Wiese—Own work. Licensed under Creative Commons Attribution-ShareAlike 3.0 via Wikimedia Commons.

The chapter is based on the proposition that the multilayered process of structural change in the Ruhrgebiet can be understood as a 'managed' process of ecological modernization. It is also based on the proposition that the 'blue sky', 'blue water' and 'green energy' projects had different drivers but also are related in ways whose details this paper seeks to uncover by presenting evidence and its framing in a consistent narrative.

Analytically, the chapter uses the concepts of ecological modernization and institutional entrepreneurs to make sense of what happened. Findings are reused to draw conclusions on transition management as an emerging field of research. To the literature on transition management, the chapter contributes a real case of a 'managed' transition. Special attention is given to the regional element but also to how the transition owed a great deal to developments outside the region, exploited by regional actors. Differently from usual, the chapter is not based on a unified analytical framework applied throughout in the study. The reason is that this chapter gives priority to *describing* the ecological modernization process and to investigating a rather diverse set of themes, namely the politics of transition (Kern/Smith 2008; Meadowcroft 2009; Voß/Smith/Grin 2009; Paredis 2013), the role of institutional entrepreneurs (Maguire/Hardy/Lawrence 2004, Coenen/Benneworth/Truffer 2012), the complex issue of heritage of old industries, and the endogenous nature of green business as additional topics. The chapter helps us to assess the prescriptive elements of transition management as formulated by

Rotmans, Loorbach and Kemp, such as the importance of vision and foresight as strategizing tools, special arenas for discussing and determining transition goals and transition initiatives that act as stepping stones for transitional change, and strategic action linking niche developments with regime and landscape developments (Loorbach 2007; Kemp/Loorbach/Rotmans 2007; Rotmans/Loorbach 2010), and also helps us to describe, for an empirical case, how processes of geographically uneven transition pathways are shaped and mediated by institutional actors and structures at different scales (Coenen/Benneworth/Truffer 2012).

28.3 The Ruhr District and its Socio-Economic Transition

The Ruhr district is Germany's largest urban conglomerate in the federal state of North Rhine-Westphalia, comprising 11 cities and 4 districts (Fig. 28.1 Germany, the state of NRW and the Ruhr area). With a population of 5.2 million (2012), a local GDP of €140 billion (2009) (LIT 2013) and a labour force of 1.7 million (2012) (BA 2012), it is one of Europe's largest industrial clusters.

The Ruhr district is named after the river Ruhr that crosses the south of the region from east to west. In Germany the name itself has become synonymous with the production of coal and steel and later with the crisis of heavy industries and structural change.

Figure 28.3: River Emscher in the Ruhr district near Herne and Herten. **Source**: Arnoldius—Own work. Licensed under Creative Commons Attribution-ShareAlike 2.5 via Wikimedia Commons.

What is now an urban agglomeration was two hundred years ago a rural area of small settlements. The development of the region towards an urbanized industrial region began with German industrialization and the rising demand for energy in the nineteenth century. The huge coal deposits of the region led to the establishment of numerous mines and steelworks along the Ruhr. During the nineteenth century the region became the backbone of Germany's coal and steel industry.

The Ruhr industry not only helped Germany to become one of the most industrialized countries but also was the armaments factory in both world wars. After World War II it was the basis for the rapid German economic recovery (*Wirtschaftswunder*). Its population of about 200,000 in the early nineteenth century rose to four million before World War I and to six million in the 1950s. The massive increase in the labour force promoted a rapid increase in the output of coal and steel. Between the 1920s and the 1940s the output of steel in the Ruhr district exceeded

the total steel production in Great Britain. In the 1930s, every sixth ton of steel consumed worldwide was produced in the Ruhr district (Schlieper/Reinecke/Westholt 1986; Kelly/Matos 2010). Production peaked in the 1950s when more than a million people were employed in the coal and steel industry, roughly seventy per cent of the whole labour force of the Ruhr district.

The economic decline of the Ruhr district began in the 1960s when world market prices for coal and steel fell below the production costs of the Ruhr industry. New mines in the US and eastern Europe supplied the markets with cheap coal mined near the surface, while the average mining depth in the Ruhr in 1960 was 650 metres. Demand also began to shift from coal to oil. The result was large-scale closures of factories and mines. Between 1960 and 1990 more than half a million jobs in the Ruhr were lost in the rapidly declining coal and steel industry. During the 1960s, the massive German economic growth compensated for the job losses. Later, the structural change led to an unemployment rate of fifteen per cent in the region in the 1980s and the massive emigration of its inhabitants (Reicher 2011).

In order to mitigate this development, both the state government and the federal government started to support the mining industry with subsidies. The German coal industry was funded with € 295 billion (at 2008 prices) in total between 1950 and 2008 (Meyer/Eidems 2009). The largest share went to the Ruhr district. Simultaneously, the state started to invest in higher education, establishing five universities and sixteen universities of applied science. As of 2012, there were 223,000 students enrolled in these universities.

The structural change in the Ruhr district's economy coincided with the rise of environmental awareness.

28.4 Blue Sky above the Ruhr

During the federal electoral campaign of 1961, Willy Brandt, the Social Democrat candidate for Chancellor, insisted: "The sky above the Ruhr must turn blue again" (Vierhaus 1994: 86, authors' translation). This idea, which now appears a perfectly reasonable policy goal, was a radical idea at the time, because pollution was viewed as an inescapable element of industrialization. The following historical summary of the rise of environmental awareness in Germany is largely based on Brüggemeier/Scheck/Schepelmann et al. (2012).

When Brandt called for a blue sky above the Ruhr, per year 300,000 tonnes of dust particles were falling from that sky; in some quarters of the Ruhr district it was more than 5 kg per 100 square metres. In 1961 the population of the Ruhr district had to endure 1.5 million tonnes of dust, fly ash and other particles, as well as four million tonnes of sulphur dioxide, in combination with a number of other harmful substances emitted by the booming coal, steel and chemical industry.

Clinical research revealed that children who grew up in the Ruhr district were often smaller, weighed less and suffered more often from rickets than children in other regions in Germany. Their fathers, who often worked in these emitting industries, died more often from cancer. The view that the Ruhr district was a dangerous environment came out of scientific research, and Brandt made reference to it in his speech (Weichelt 1997). Alarming facts also came from the *Siedlungsverband Ruhrkohlebezirk* (SVR), an association of Ruhr municipalities, who said that

the environmental conditions constituted an assault (*Generalangriff*) against the region (Brüggemeier/ Rommelspacher 1992: 63). The traditional symbol of progress was smoking chimneys, but this came more and more into question.

Voices had been raised in concern about the high levels of pollution and the health hazards. Protests began in civil society and shifted to politics, where more and more interest was shown in the issue. In December 1955, the NRW parliament held its first debate about air pollution, and in the following year it was put on the agenda of the German federal parliament, the Bundestag. National media began reporting on air pollution. The Christian Democrat state government of NRW reacted, and in 1960 Franz Meyers, the prime minister of North Rhein-Westphalen, spoke of his concerns at the federal congress of the German Christian Democratic party, but met little response (Weichelt 1993; Hünemörder 2004: 89ff.).

In contrast to Franz Meyers in 1960, the public response to Willy Brandt's speech in 1961 was overwhelming. People in the Ruhr district especially were not willing to tolerate the increasing pollution of their environment much longer and called upon regional and national politicians to take action. In 1962 the NRW parliament enacted the first regional emissions act. In 1964 this was complemented by a directive with thresholds for critical substances (Brüggemeier/ Rommelspacher 1992: 62f.). The breakthrough was complete when Willy Brandt's administration took office as the German federal government in 1969. His administration initiated an immediate action programme for the protection of the environment (Sofortprogramm für den Umweltschutz) and passed a whole series of pollution control measures. The government also created a number of institutions, including an interdepartmental coordination group and a stakeholder forum, as well as the *Federal Agency for Environmental Protection* (UBA).

The 1970s witnessed a significant rise in environmental awareness among the German population. Within a few years it had moved from being a peripheral theme to the centre of political discourse. Hundreds of thousands of Germans organized themselves into environmental citizens' organizations (Wey 1982; Hünemörder 2004). Even the trade unions discovered the environment as an issue on their political agenda (IG Metall 1972).

While pollution control was the predominant mode of environmental policies during the 1970s and 1980s, attention shifted to pollution prevention through the use of cleaner production processes. In 1984, Willy

Brandt supported the position that it was not only important to react to pollution, but that precautionary approaches would be necessary, including technical innovations and production processes. He mentioned that improved energy and resource efficiency and a more careful use of air, water and soil would not only decrease environmental pressure but also save costs. Such a transition would lead to a modernization of society, improve the competitiveness of German industry, and create jobs. All of this came to be known by the name of ecological modernization, a concept first coined by Martin Jänicke and popularized by Joseph Huber (Mol/Jänicke 2009).[5]

During the late 1970s and 1980s the ecological movement in Germany gained more and more momentum. Alarming ecological trends such as the decrease in biodiversity, acidification of soils and waters, and forest die-back (*Waldsterben*), as well as a series of environmental catastrophes (e.g. Seveso, Sandoz, Chernobyl), added to the growing concern of the German population.

One result was that the dispersed anti-nuclear movement joined forces with established nature conservation organizations under the umbrella of new civil society organizations such as Friends of the Earth Germany (*Bund für Umwelt und Naturschutz Deutschland*, BUND). Environmental civil society organizations increased their membership considerably and became large and influential societal forces. Another was that the surge of the Green Party marked a greening of the German political debate, which came to dominate the general political discourse more and more. This increased the pressure on the mainstream political parties as well as federal and state governments, who reacted with an increased number of new programmes and increased legislative activity.

Under the influence of the greening of the (West) German political landscape, the prime minister of North Rhine-Westphalia, Johannes Rau, adopted the increasingly popular idea of ecological modernization. It became the conceptual background against which the Wuppertal Institute for Climate, Environment and Energy and the Energy Agency NRW were

founded in the early 1990s. After losing their absolute majority among NRW voters, the Social Democrat Johannes Rau led the first red-green coalition in 1995. With this political background, ecological modernization became more and more a part of NRWs economic development policies, yet its integration was uneven across the different economic sectors (Schepelmann 2005, 2010).

Ecological modernization as a political programme supported the transformation of "the largest contributors to problems in the Ruhr district into problem solvers" (Kilper 1996: 15, authors' translation). An important facilitating factor was that the energy, chemical, coal and steel industries had the necessary structures and knowledge to develop solutions for protecting the environment. As early as 1992, the NRW eco-industry produced goods and services with a value of about € 10 billion, about one-third of German production in the sector (Nordhause-Janz/Rehfeld 1995: 67).

According to the economic development agency of the Ruhr district (Wirtschaftsförderung Metropole Ruhr, WMR 2012), 'resource efficiency'[6] has a turnover in the Ruhr district of € 63.3 billion. About 96,000 people were employed in about 5,500 companies, representing 6.7 per cent of all jobs in the Ruhr district. Overall, every tenth German job in this leading market is based here.

At a time when production increases were driving pollution to intolerable levels, the Ruhr district became the cradle of modern environmental policy in Germany. Brandt's vision and his promise of "blue skies over the Ruhr" marked the beginning of a change of consciousness and a change of policy in Germany. Environmental end-of-pipe thinking predominated in the Ruhr district for more than thirty years. Strict environmental legislation was passed, and large investments were made in environmental cleanup technologies. Influenced by the awakening of environmental awareness at the international, national and local level, and spurred on by legislation for environmental improvement primarily at the federal level, the NRW state government invested early in the health of its population and their environment. The Ruhr district became a prime mover on the road towards an

---

5   Over time, the meaning of the term 'ecological modernization' became a topic of discussion among social theorists (Spaargaren/Mol 1992; Hajer 1995; Mol/Sonnefeld 2000), but originally "ecological modernization was essentially a political program. It was neither a theory, nor a concept which included the social dimension of this type of modernization" but simply "a practical and normative idea for pressing for far-reaching environmental reform" (Mol/Jänicke 2009).

6   WMR comprises the following economic activities under the lead market 'Resource Efficiency': energy generation and energy-related services, water and waste management, eco-industries, and raw materials extraction and processing, as well as environment-related activities such as engineering, laboratory, monitoring and consultancy services.

environmental goods and services industry.[7] Today, the economic development agency of the Ruhr district perceives eco-industries as highly dynamic and as a pillar of the regional economy (WMR 2012). This perception has been supported by a number of regional studies and assessments (Nordhause-Janz/Rehfeld 1995; Schepelmann 2005; Rehfeld/Schepelmann 2007).

28.5   Reconstruction of the Emscher Region

The ecological modernization process in the Ruhr district included greener production as well as clean-up activities pursued as part of urban revitalization. It goes beyond our ability to give a complete description of this complex multilevel transformational process. Instead of providing an overview of relevant projects, we opt to describe a high-profile project which has already achieved tangible results: the reconstruction of the Emscher region as an exemplary regional development plan and a flagship of the ecological, economic and urban revitalization of the Ruhr district.

In the 1980s the Emscher landscape had been characterized by vacant factories, closed mines and abandoned docks with mining subsidence, large spoil-heaps of mining residue and dams (Figure 28.2: View of the city of Gelsenkirchen). This devastated landscape was deliberately chosen to stimulate and symbolize change by integrating the industrial heritage in the renewal of the Ruhr district. The aim was an ecological, cultural and economic revitalization of the Ruhr district, as well as to increase the attractiveness of its cultural and recreational facilities; this culminated in the *International Building Exhibition*, IBA (1989-1999).

The structural policies symbolized by the IBA Emscher Park focused not only on economic development, but also on environmental protection, nature conservation and increased cooperation in the planning and administration of the region. There was also a desire to preserve the industrial heritage of the district by choosing to convert buildings and to develop industrial factories into tourist attractions.

The conversion of the Ruhr district was to be based on a series of projects that complemented each other. The projects were organized in seven themes (Brüggemeier/Rommelspacher 1992):

- Reconstruction of the landscape: the Emscher Landscape Park
- Ecological improvement of the Emscher river
- Reconstruction of the Rhine-Herne Canal as an experimental space
- Preservation and new use of industrial heritage
- Working in the Park
- New forms of housing and apartments
- New opportunities for social, cultural and sporting activities.

The IBA and the plans that followed it were funded by several public sources and a number of private ones. In 1988 the state of NRW created the Emscher Park Planning Company with an initial funding of € 18 million. In 1993, five years after the establishment of the IBA Park Planning Company, € 1.3 billion was spent on projects, including € 900 million from public sources—the state, the federal government and the European Union (US EPA 2006). As of 2012, private and public funds for the IBA and its successor plans have far exceeded the initial investment.

The basic idea of the Emscher Landscape Park was to convert the abandoned and industrial landscape into a new, future-oriented and clean environment with an excellent quality of life (Figure 28.3: River Emscher in the Ruhr district near Herne and Herten, Figure 28.4: Wastewater treatment plant Deusen at the river Emscher).

The backbone of the Emscher Landscape Park development consists of seven regional green corridors that cut through the Ruhr district from south to north. Originally, these green belts were defined in the 1920s by an association of Ruhr municipalities, the *Siedlungsverband Ruhrkohlenbezirk* (SVR). The green areas were supposed to limit the uncontrolled growth of the increasingly merging cities of the region in order to protect sites for recreation and leisure.

Within the Emscher Landscape Park these green corridors cover an area of 450 km² and have been used to integrate the brownfields of the coal and steel industry into a new system of parks. In 'a landscape of structural change', twenty cities and two counties have initiated 120 projects that contribute to the Emscher Landscape Park. The projects address ten guiding principles:

- Ecology
- Water and rainwater
- Living and working in the Park
- Art in the Park

---

7   It is a story of necessity as the mother of invention, or as the poet Friedrich Hölderlin wrote in his poem "Patmos", "But where danger is, deliverance also grows".

Figure 28.4: Wastewater treatment plant Deusen at the river Emscher. **Source:** Tbachner—Own work. Licensed under Creative Commons Attribution-ShareAlike 3.0 via Wikimedia Commons.

- Park infrastructure
- Leisure and tourism
- Urban development
- Agriculture and forestry
- Industrial heritage
- Landmarks.

The flagship project of the IBA Emscher Park is the revitalization of the Emscher river system, a river which had been canalized and degraded to an open sewer. Owing to ground subsidence, it was not possible to build and use underground sewage systems in the mining area. The settlements and industries of the Ruhr district discharged all wastewater directly into the river Emscher. The Emscher was considered Germany's most polluted river. The abandoning of mining in the Emscher region in the 1980s permitted all wastewater to be discharged underground and the Emscher and its tributaries to be restored. With investments of €4.5 billion over several decades, the Emscher conversion is one of the biggest infrastructure projects in Europe.[8] In the Ruhr district, the Emscher conversion encouraged

several other municipalities to take similar steps in revitalizing waterways in combination with upgrading real estate along the river banks. Historical industrial monuments were given new designations, railway mineral lines were converted into bicycle routes and parks were created.

By giving a new use to abandoned buildings and the creation of parks, the area became a more attractive place to live, work and visit. Techno-cultural highlights such as the Oberhausen Gasometer, the Duisburg-Nord landscape park (Figure 28.5: Heritage of the IBA Emscher Park at night: Furnace No. 5, Figure 28.6: The river Emscher beside the Landscape Park Duisburg), the Duisburg inland port and the UNESCO world heritage site of the Zollverein Coalmine Industrial Complex (Figure 28.7: UNESCO World Heritage Site Zeche Zollverein in Essen) became tourist attractions. The recycling of abandoned land along

---

8   For more information visit <http://www.eglv.de/wasserportal/emscher-umbau.html>.

Figure 28.5: Heritage of the IBA Emscher Park at night: Furnace No. 5 at the Landscape Park Duisburg North. **Source:** Tuxyso—own work. Licensed under Creative Commons Attribution-ShareAlike 2.5 via Wikimedia Commons.

with the provision of reliable and sustainable infrastructure attracted enterprises and created new jobs as well as new perspectives. The ecological revitalization of a river catchment area more and more became the economic, social and cultural transformation of a region. Not only was this the express intention of the makers of the Emscher Park, who had already integrated a number of cultural initiatives into the development of the IBA, it was also accompanied by a number of successful cultural activities such as the Ruhr Triennale.

This three-yearly high-end cultural festival for performing arts started in 2002. The festival events take place in formerly industrial venues transformed by the IBA. In 2010 the Ruhr district managed to obtain the status of European Capital of Culture. The former industrial Beast was more and more turning into a formerly industrial Beauty (Schepelmann 2010). An indicator of this remarkable transformation is the number of overnight stays by foreign and domestic visitors to the Ruhr district: this nearly doubled between 1990 and 2011, rising from 3,588,394 overnight stays in 1990 to 7,026,396 in 2012, an increase of 95 per cent (RVR 2013). The Emscher Landscape Park turned out to be the showcase of a successfully managed transition. Using the Emscher Landscape Park concept, a number of similar plans followed, such as *Konzept Ruhr* (Schwarze-Rodrian/Seltmann 2012) and the *Master Plan Emscher Landschaftspark* 2010. As of December 2012, there were 212 projects that had been realized and 240 more in development (RVR 2013).

28.6   The Emerging Energy Transition

After having discussed the Blue Skies initiative and the Emscher revitalization, we now turn to the third element of eco-restructuring in the Ruhr district: the transformation of the Ruhr district from a centre for fossil industries to a centre for sustainable energy industries. The installed power of renewable energy rose nearly exponentially from 30 MW in 2000 to 588 MW in November 2012 (Amprion 2012). This trend is continuing and is driven by public plans to invest in renewable energy.

The energy transition in the *Ruhrgebiet* (Ruhr district) and NRW is conditioned by the *Energiewende*

(energy transition), a national project of the German federal government with the following elements: to phase out nuclear power by 2022, to achieve a general increase in energy efficiency, and to increase the share of renewables in the energy mix (Bundesregierung 2010).

The phasing out of nuclear power necessitates extra power capacity and creates room for the use of renewables. It means nothing less than an overhaul of the existing system. During the phasing out of nuclear power plants, on the one hand the security of supply in Germany has to be guaranteed, and on the other the network of transmission lines must be adapted to the needs of decentralized sources of renewable energy. As one of the most important energy clusters in Europe, the Ruhr district is of strategic importance, and this makes it eligible for support by national resources in addition to regional policies. It also benefits from international policies, in particular from EU structural funds, which have been utilized to great effect to fund local projects.

Even when the national energy transition necessitates changes in the Ruhrgebiet and offers economic opportunities, *making* a transition to green energy in the Ruhrgebiet is a difficult issue for the regional government because of the coal industry and the resistance of power utilities, who want to make their own investment decisions. Rather than fighting the resistance, the government of NRW orients itself through its innovation and economic development policies to the opportunities that green power affords. In Germany, regions have their own innovation policies, which are oriented towards economic activities in those regions. In NRW, innovation and regional development policies are organized along the value chains of sixteen areas of competence (VDI 2009). Environmental technologies, Energy RTD and logistics are all clusters in which both business actors and knowledge actors from the Ruhr district play an important role. The fact that in the Ruhr district companies traditionally have considerable competencies in energy is beneficial for obtaining innovation support; it is why companies and research institutes have benefitted so much from regional and national innovation policies and from supporting agencies, including regional development agencies as well as thematic agencies such as the Energy Agency NRW.

Research competences in energy are mobilized through the normal science systems and through various innovation policies but especially by *Energiewende in NRW*, an ambitious programme by the NRW state government to shift to low carbon energy and reduce energy use. According to the NRW Ministry for Climate Protection, Environment, Agriculture, Nature Conservation and Consumer Protection, it consists of the following eight measures (MKULNV 2011):

1. The NRW Climate Protection Law: in 2011 the state government established Germany's first Climate Protection Law, which aims at reducing greenhouse gas emissions in NRW by at least 25 per cent by 2020 and by at least 80 per cent by 2050 compared with 1990 levels.

2. The Climate Protection Start Programme: this investment programme will allocate several hundred million euros of loans and grants to support low carbon investments in NRW industry.

3. New Wind Energy Adaption: to stimulate growth, especially in machine and plant engineering, the state government has set a goal of quadrupling the share of wind energy in the electricity supply by 2020.

4. Cogeneration: according to studies commissioned by the NRW state government, 35 per cent of CO_2 emissions and 35 per cent of raw materials can be saved in NRW's power generation through more effective use of co-generation in power plants. The state government has initiated an investment programme, 'Cogeneration', allocating € 250 million over several years.

5. Research: the state encourages research into tapping new sources and locations for renewable energy.

6. Networks and Storage: to strengthen the supply of renewable energy, the state plans to enhance the network of power transition lines by 400 km and to quadruple the capacity of pumped storage power plants.

7. Rhein-Ruhr District Heating: particularly in the densely populated Rhein-Ruhr region, heat sources in existing and planned power stations, as

well as renewable sources of heat and waste heat systems, will be promoted.

8. Resource Efficiency: with the Efficiency and Energy Agencies, the competition 'Ressource.NRW' and an efficiency loan from the public NRW Bank for businesses, the NRW state government intends to promote energy and resource efficiency.

A flagship initiative for the *Energiewende* in NRW is Innovation City Bottrop. In 2010 Bottrop won the Innovation City Ruhr competition for energy efficiency, organized by Initiativkreis Ruhr, a group of private companies. The goal of the Innovation City Ruhr project is to reshape an existing city, with all its industrial facilities, green spaces, and neighbourhoods old and new, along more sustainable lines (Reicher 2011). Sixteen cities participated in the competition which the city of Bottrop won with the blueprint for a low carbon transition process with a substantial element of participation. Interestingly, the fifteen municipalities which took part in the Innovation City competition have now created a network in order to learn from the experiments and solutions developed in Bottrop.

The stated goal of Innovation City Bottrop is to halve CO2 emissions by 2020 in an area of 25 square kilometres and 14,500 buildings (Lechtenbömer/ Fischedick 2012; Reicher 2011). The target area includes several neighbourhoods in the urban south of Bottrop and represents in many ways the rich cultural and social diversity of the Ruhr district. Housing, employment, trade and commerce are intertwined in a small area. By 2020 the city plans an exemplary application of a number of innovations in energy efficiency, decentralized power generation and electric transport.

Unlike other eco-city development schemes such as Masdar City in the United Arab Emirates or Dongtan in China, Bottrop developed its solutions in a bottom-up way, through collaboration between scientific institutions, business, politics and civil society.

The city of Bottrop plans to undertake approximately 100 projects in the following areas:

- Advice and information (e.g. preparation of energy consultancy, information, thermographic imaging of houses)
- City (e.g. greenhouses on roofs, LED street lighting, construction of photovoltaic noise barriers along an autobahn)
- Energy (e.g. a sewage treatment plant as a hybrid power plant, heat recovery from shower water, geothermal and biomass energy from mine water)

- Housing (such as energetic renovation of houses from the 1950s, 60s and 70s, construction of model houses)
- Mobility (charging infrastructure sites, car sharing of electric cars, fuel-cell powered buses)
- Industry and retail work (e.g. implementation of solar energy into production processes, carbon-neutral filling stations and climate-neutral retail).

The measures fall into two categories: measures that municipalities and big organizational actors can take (LED street lightning, hybrid power plants, energy-efficient production processes) and measures which requires different choices by consumers (energy-passive houses, car sharing of electric vehicles). The energy transition thus requires the cooperation of both consumers and business. It also involves important infrastructure issues in the domains of mobility (fast charging points for electric cars) and energy (smart grids, power storage).

28.7 Differences and Interrelationships between the Emscher and Bottrop Showcase Projects

The reconstruction and development of the Emscher river system and Innovation City Bottrop can be interpreted not only as showcase projects but also as real-world experiments ('*realexperiments*'; Groß/Hoffmann-Riem/Krohn 2005) for developing sustainable solutions for local and regional development. They are laboratories for learning about new innovations and forms of cooperation, as well as producing tangible results. For this reason they are of relevance for policy-learning not only for the Ruhr district but also for the state of NRW and for regional policy actors in general. In this section we will compare the different projects.

Both projects relied on positive visions with clearly defined outcomes. IBA Emscher Park promoted the vision of clear water flowing in a natural riverbed for one of the most polluted sewage channels in Europe. Bottrop was based on a vision of halving energy consumption and carbon emissions by increasing the quality of life. Both visions were seen as daring, standing in stark contrast to the negative reality of heavy pollution and unsustainable energy use. Bottrop is a flagship project of energy transition. It should serve as a reference point for what can be achieved through specific innovations, offering lessons about the technology, the usefulness of support action, product imperfections and marketing and distribution (cf. Kemp/Schot/Hoogma 1998).

Both Emscher Park and Innovation City Bottrop are projects of for local ecological transformation. Large areas are being transformed (or are to be transformed) into something greener. They are real-world experiments to investigate and show what is possible. Both projects had specific goals and combined a top-down element of guidance with a bottom-up element of reaching goals, like the eco-cities Graz and Freiburg examined in Rohracher and Späth (2013). In terms of the actors as well as the type of projects there are some important differences. The Emscher project draws on an old structure of cooperation between municipalities, through the *Regionalverband Ruhr* (RVR, Ruhr Regional Association), as well as the Emscher Genossenschaft (Emscher Cooperative). The RVR is responsible for regional planning. For almost a century the RVR and its predecessors (SVR and KVR) have been a permanent platform for cooperation between the different Ruhr communities, while the Emscher Genossenschaft is a cooperative of municipalities for managing water treatment (founded in 1899).

In contrast, Innovation City Bottrop was initiated to large extent by Initiativkreis Ruhr (Ruhr Initiative Group). Initiativkreis Ruhr is an association of sixty-eight companies with a total turnover of € 630 billion and 2.25 million employees worldwide. Established in 1989, it pursues the vision of developing the Ruhr as a strong and attractive location for business. One of the initiatives of Initiativkreis Ruhr is the Zukunft Ruhr 2030 (Future Ruhr 2030) strategy. The strategy is meant to be a regional development plan with a business perspective (Osterhoff 2007).

Even though Initiativkreis Ruhrgebiet represents traditionally important actors in the Ruhr district, the institution as such is a 'new kid on the block'. Ralf Schüle, the deputy director of the Wuppertal Institute's research group for energy, transport and climate policy, observes that the cooperation model of Innovation City Ruhr represents an alternative approach to the traditional formal and informal institutions and networks. By representing primarily corporate decision-makers, it is to some extent a counter-model against established governance structures.

Transition projects must be financed, as no physical area can be converted without spending a substantial sum of money. Where did the finance come from? In the Emscher Park, financing came from various sources: the municipalities, business actors and the state of NRW. Interestingly, the most important

source of funding for the North Rhine-Westphalian structural policies are the EU structural funds. During the period from 2007 to 2013, NRW obtained € 1.3 billion for local development projects in the region. The state government and private donors provided another € 1.2 billion, making a total of € 2.5 billion that will be spent in three priority areas:

- Priority 1: strengthening the entrepreneurial base— € 500 million
- Priority 2: innovation and the knowledge-based economy—€ 1.25 billion
- Priority 3: sustainable urban and regional development—€ 750 million.

The funding activities of the North Rhine-Westphalian energy policies are bundled in the "progress" programme. This offers a wide range of instruments to promote the efficient use of energy and the use of renewable energies in NRW, including solar heating systems, ventilation systems, district heating, combined heat and power generation, biomass facilities, and passive houses. Since 1987, the progress programme and its predecessor the REN programme have funded several thousand projects with a funding volume of more than € 500 million.

For Bottrop we do not have reliable information about money being spent. The overall costs are estimated at € 2.8 billion, of which € 2.3 billion is to come from private partners, € 454 million from the EU and the federal and state governments, and € 42 million from the city of Bottrop itself (Drescher 2012). An important difference between the Emscher and Bottrop cases is the scale of transformation. During the Emscher transformation it was possible to successfully address different levels of infrastructure transformation (Wuppertal Institut 2013):

1. Ecological revitalization
2. Creation of new spaces for broad participation
3. Improvement of the quality of life
4. Increasing the economic value of the location
5. A new image for the location.

The Emscher transformation mainly addressed the transformation of public infrastructure (the Emscher river, parks, monuments, sewage systems etc.). It was generally welcomed by the public, because it mainly affected public infrastructure. In contrast, the low carbon transformation of Bottrop will affect not only public infrastructure, but necessitates the active involvement of private actors and the spending of € 2.5 billion by those actors. Private actors are known for spending money for private matters (fitting in with consumer needs and business profits), not for public matters. Public authorities will spend money for public matters. The dependence on private parties (consumers and companies) puts a constraint on Bottrop. Without private parties willing to spend money on energy conservation projects and low carbon energy systems, Bottrop will not succeed. We should also not expect products requiring lifestyle changes to diffuse quickly.

Energy transition also requires spending from public and private actors. According to the former president of the Wuppertal Institute, Peter Hennicke, support for the *Energiewende* from the traditional energy sector is only marginal. More than forty per cent of the money is from private consumers (civil society) who are investing in renewable energy with the help of government subsidies. Investments by the 'Fat Four' energy utilities (E-ON,[9] RWE, ENBW, Vattenfall) in green power generation amounts to just seven per cent.

In the case of energy, it has proved more difficult to enrol industry. Dieter Rehfeld, head of the research department for Innovation, Space & Culture at the Institute for Work and Technology, observes a general lack of integrated approaches in energy policy, with the explicit management of conflicting objectives in the political discourse.

28.8   Regional and Supraregional Elements

In the transition literature, spatial issues are very much neglected: "Space is only indirectly and implicitly addressed. Both socio-technical regimes and *full term* TIS [technological innovation systems] are implicitly understood as footloose cognitive and institutional structures that influence the activities of different actors largely irrespective of their geographical location" (Truffer/Coenen, 2012, p. 6). In the regional development literature, related variety, differentiated knowledge bases and policy platforms (based on relational and collective types of policy arrangements) are considered key aspects behind the creation of new clusters of innovation (Asheim/Boschma/Cooke

2012). More recently, the concept of institutional entrepreneurship has emerged as an important factor (Coenen/Benneworth/Truffer, 2012). Institutional entrepreneurship refers to 'activities of actors who have an interest in particular institutional arrangements and who leverage resources to create new institutions or to transform existing ones' (Maguire/Hardy/Lawrence 2004, cited in Coenen/Benneworth/Truffer 2012).

The Ruhr transition to a green economy is an interesting case of a managed transition in a geographical area. In this section we will examine more deeply the regional dimension as well as its interaction with national and international factors.

In our discussion of energy policies in NRW, we offered details about the innovation and development policies and specific agencies. The regional policies are important, but according to Johannes Venjacob of the Wuppertal Institute, the *Energiewende* in NRW and Ruhrgebiet is to a large extent determined by national decision-making—the phasing out of nuclear power and the shift to renewables. The national *Energiewende* is a conditioning factor in NRW, which traditionally specialized in energy generation but which is governed by a coalition of Social Democrats and Greens who pursue policy objectives highly compatible with the objectives of the national *Energiewende*.

The NRW state government had played a fruitful role in moderating and financing the activities of the many regional and municipal actors. Nevertheless, support was uneven. This was due to changing priorities, especially under Wolfgang Clement, NRW prime minister from 1998 to 2002, who later publicly turned against the central objectives of the *Energiewende* and left the Social Democratic party in 2008. Ambiguous political leadership and insufficient policy integration among the different NRW ministries got in the way but could not stop the momentum of the transition.

In NRW and particularly in the Ruhr district, regional actors have long collaborated with each other to create green business and to convert an obsolete old industrial landscape into a place for recreation, living and tourism. These changes have been largely led by regional actors who managed to get federal and EU support for their projects.

Institutional entrepreneurship in NRW appears to be a driving factor. The Siedlungsverband Ruhrkohlebezirk and its successor the *Regionalverband Ruhr* (RVR) played an important role in setting the agenda and in implementing and monitoring environmental protection. Other institutional entrepreneurs like IBA

---

9   While this article is going to press, the realities of the energy sector in NRW are in transition. As of December 2014 one of the largest German electricity providers, E-ON, will split into two companies. The larger one will be dedicated to seizing opportunities on the decentralized renewable energy path, while the other will remain on E-ON's traditional fossil and nuclear trajectory.

Emscher Park and the Emscher Genossenschaft (Emscher cooperative) seem to have given the implementation of the transformation a lasting momentum.

In the energy transition, the Ruhr Initiative Group can be viewed as an institutional entrepreneur in drawing attention to the green energy innovation, and in attempting to position the Ruhr district as a competitive centre for industry, commerce, research and services. Among the flagship initiatives of the group are Innovation City Bottrop, EffizienzCluster LogistikRuhr (cluster for regional companies in the logistic sector), the RUHR.2010 presentation of the Ruhr as a European Capital of Culture in 2010, and the Ruhr International School.

Individuals played an important role. Examples of key individuals are SVR's director Sturm Kegel and Willy Brandt, who both managed to put the issue of protecting the human environment on the agenda by highlighting the risk to public health. The seeds of the Ruhr transition were thus sown by a concerned regional decision-maker and a charismatic political leader who was able to translate this concern into a positive, concrete vision of a blue sky over the Ruhr. Another influential individual is the Director of IBA Emscher Park, Karl Ganser. He was not only able to forcefully articulate the vision of a reconstructed Emscher landscape but was also influential in coordinating the financing and implementation projects that provided integrated protection for the cultural and natural heritage of the Ruhr district along the Emscher catchment.

An interesting aspect of the Ruhr transition is that the coal regime plays a certain role as a facilitator. Some low carbon innovations in the region are based on the legacy of the coal mining industry:

- The burning of *coal mine methane* (CMM) from abandoned mines. CMM has a *global warming potential* (GWP) 23 times higher than CO_2. Burning CMM thus helps to reduce greenhouse gas emissions compared to not burning it. The burning is done in co-generation plants. In 1997 the first co-generation plant converted CMM into electricity and heat. In 2012, 148 MW of electricity were produced with CMM co-generation units (Amprion 2012).
- The use of geothermal heat of mine leachate. At many mine sites, leachate has to be pumped out regularly. In Bochum the geothermal heat of mine leachate is used to provide heat for public buildings.
- Hydroelectric storage in abandoned mines. By 2018, Prosper-Haniel in Bottrop, the Ruhr dis-

trict's last working mine, will be decommissioned. Researchers at the Universities of Duisburg-Essen and Bochum have developed plans to use abandoned mines after 2018 as pumped-storage hydroelectric power stations with underground reservoirs. These power plants will serve two purposes. On the one hand, they serve for load balancing the supply of renewable energy. On the other, the geothermal energy of the water, which is heated up to 40 degrees, will be used for district heating.

Clearly, the coal industry is *not* the driving force behind most green energy and energy-saving projects, but in some ways the mining heritage allows for the creation of green energy which is not resisted by the coal and mining actors. Two important drivers behind innovation in green energy are the strong competence in energy research, and *Energiewende* NRW. In total, more than 750 scientists in several research institutes and 250 companies in the Ruhr district work on energy supply, power control and power plant construction.[10] A prominent example of the Ruhr's energy competence is the ef.Ruhr[11] (Energieforschung Ruhr GmbH), which is Germany's largest association of research chairs in the field of energy. More than forty professors from the universities of Bochum, Dortmund and Duisburg-Essen cover the entire spectrum of energy technology. They cooperate closely with global business players of the region such as E.ON, RWE and Evonik. Focal areas of ef.Ruhr are research into reducing CO_2 emissions in power generation, CO_2 transport and storage, in particular *Carbon Capture and Storage* (CCS), and solutions for upgrading coal plants with CO_2 capture. The solutions developed include a mobile plant for CO_2 capture from flue gases, which won the "Hightech.NRW" contest in 2008.

Finally, this section turns to the interlinkages between the green business and nature conservation project, the ecological reconstruction project, and the green energy transition project in the Ruhr district and NRW. Each of the projects started at a different time and had different aims. There is no simple causality from "blue sky" to "blue water" to "green energy". They are all part of the ecological modernization process, often coupled with the belief that dealing with ecological problems creates economic opportunities.

---

10 For more information visit <http://www.ruhrenergy.de>.
11 For more information visit <http://www.efruhr.de>.

An important background factor was that local authorities were responding to problems of pollution, depopulation and the loss of jobs in coal mining and steel. A strong environmental movement and national policies were another. Important policy decisions were taken up by parliament. Flagship projects and investments in education and research also helped to make the transition from dirty industry to a green industry in a greener landscape. The way in which the different transitions are related to or even build on each other is a topic for deeper analysis and will be discussed in further workshops with experts and stakeholders.

28.9   The Lessons for Transition Management

The Ruhr district has been in transition for four decades, ever since the decline of the coal and steel industries was associated with major job losses. The area has been revitalized thanks to investment in the clean-up and projects such as the Emscher Landscape Park. The analysis of the economic development agency for the Ruhr district WMR established that the region obtained economic strengths in resource efficiency and sustainable consumption, as lead markets.[12] In the Ruhr district, the lead market Resource Efficiency constitutes 6.2 per cent of the regional economy, which is considerably above the share for Germany (3.6 per cent) and the state of NRW (3.9 per cent) (see section 4). The strong energy sector is well placed to take advantage of the growth potential for renewable energy and energy efficiency created by the national and regional *Energiewende* goals and programmes. None of the transitions is finished: the share of coal is still higher than the share of renewable and cities are becoming depopulated. The Ruhrgebiet has 600,000 fewer inhabitants than in 1960. The city of Essen now has 140,000 fewer inhabitants (LIT NRW 2013).

It is interesting to investigate the Ruhr transition processes from the point of view of transition man-

agement. Many of the elements of models of transition management as developed by Rotmans, Loorbach and Kemp (Rotmans/Loorback 2010; Loorbach 2007; Kemp/Loorbach/Rotmans 2007; Rotmans/ Loorbach 2010), are being confirmed by the eco-restructuring of the Ruhr district: the importance of an inspiring and concrete vision shared by most actors, platforms for strategic thinking, special innovation projects acting as stepping stones for transition processes, and special transition institutions. Different agendas are being aligned by institutional entrepreneurs: improving air quality for health reasons, environmental amenities, preserving industrial heritage, improving the quality of places, keeping labour in the Ruhrgebiet and stopping depopulation. The Ruhr case also underlines the importance of transition showcases that speak to the imagination of people. It confirms what has been said about real-world experiments: that they help to "build a constituency behind a product—of firms, researchers, public authorities—whose semi-coordinated actions are necessary to bring about a substantial shift in interconnected technologies and practices" (Kemp/Schot/Hoogma 1998: 186). Methodologically, it is difficult to ascertain the precise role of specific projects in the overall transition. Flagships like IBA Emscher or Innovation City Bottrop are necessary but certainly not sufficient. Like Paredis (2013), we found that a lot of contingency is involved in individual cases and that success was not guaranteed.

We consider the proximity of cooperative actors in the state of NRW as a major reason for the success of the eco-transition. While national ministries are known to fight over competing policy choices with little regard for policy coherence, regional actors and city governments are more cooperative and more oriented towards achieving results on the ground. Despite setbacks and conflicts, there has been a long period of cooperation in the Ruhr. Unlike what is predicted by transition theory, the old regime of coal mining and heavy industry played a facilitating role: technical knowledge about pollution and waste helped to establish environmental technology businesses in the NRW area, abandoned quarries were used for storage for renewables, and more than in other states power companies in NRW can take advantage of the regional energy cluster to promote green power. A decisive factor (unique in Europe, but similar to China) is that in Germany, state authorities are responsible for innovation policy, much more than national authorities, leading to and fostering regional specialization.

---

12 A lead market is a regional market that is the first to adopt a global innovation design, thanks to a demand advantage, a price advantage, an export advantage, a transfer advantage and/or a market structure advantage (according to Beise at <http://www.rieb.kobe-u.ac.jp/ academic/ra/dp/English/dp141.pdf>). In NRW and Germany, Resource Efficiency and Sustainable Consumption have obtained the official status of 'lead market'.

Figure 28.8: Main phases and elements of NRW sustainability transition. **Source**: The authors.

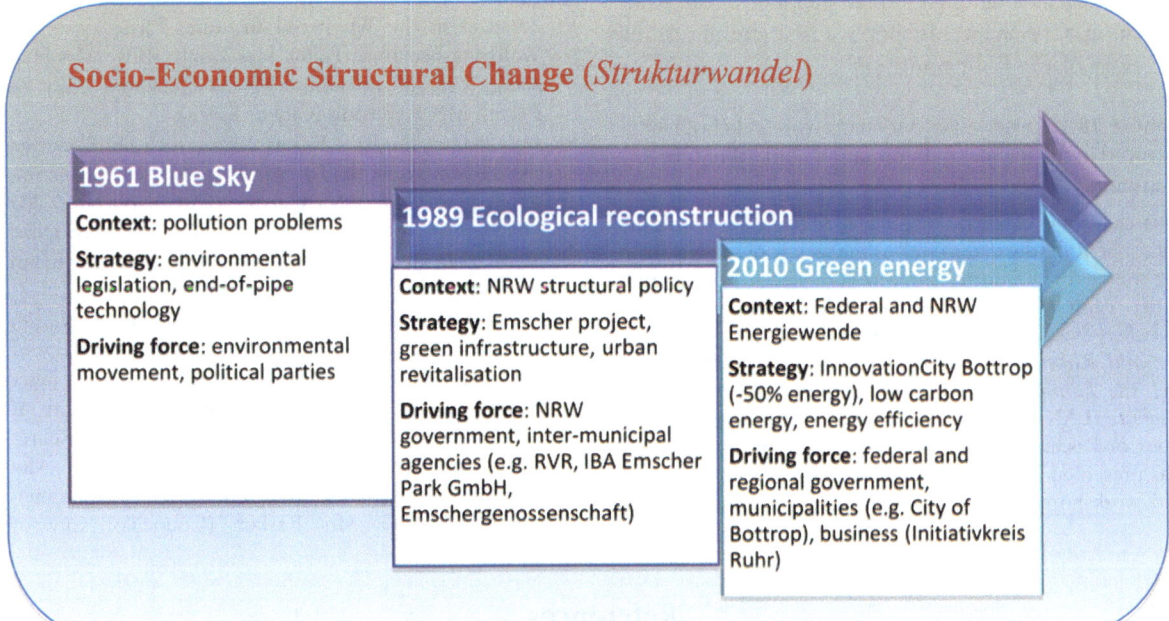

The ecological restructuring of the landscape in the Ruhr district and the energy transition to sustainable energy use project in NRW have top-down as well as a bottom-up elements. The top-down elements consists of a clear sense of direction and making available money for transition projects. Interestingly, the largest funder of the ecological restructuring projects is the European Commission through its structural funds. The Ruhr transition consisted of a three waves of change, which in some ways build on each other (figure 28.8: Main phases and elements of NRW sustainability transition):

1. The greening of dirty industries through pollution control and policies for nature conservation, which helped to establish an eco-industry (1961–1990)
2. The ecological reconstruction, clean-up and urban revitalization of the Ruhr district (1989–2015)
3. The sustainable energy transition (2010 onwards).

Interestingly, the transition frame served as a frame even when the word transition was not used. The overarching transition is the socio-economic *Strukturwandel* (structural change) due to the decline of heavy industries. This coincided with growing environmental awareness by citizens and politicians, which was followed by the implementation of environmental protection. The phase of agenda-setting and the introduction of pollution control and nature conservation was marked in 1961 with Willy Brandt's promise of a blue sky above the Ruhr.

This phase was followed by a (partly overlapping) phase of ecological reconstruction, clean-up and urban revitalization of the Ruhr district, which became manifest by the start of the IBA Emscher Park in 1989.

In 2010 a third transition process began: the transition to sustainable energy. The sustainable energy transition is supported by NRW and national energy transition programmes, as well as regional innovation policies, of which the Innovation City Ruhr contest is an example.

The Ruhr district transformed against the background of wider economic and political developments. Relevant developments for the low carbon energy transition are the prices of fossil fuels, the European Emission Trading system for carbon allowances (of which prices are temporarily low), and German obligations under the UN Framework Convention on Climate Change (UNFCCC).

The eco-restructuring of the Ruhr district presents a remarkable case of a managed transition, bringing out the importance of regional actors and regional factors alongside external stimuli. The Ruhr transition was embedded in a discourse of ecological modernization, which translated into political action and matured into a societal transition. The idea of ecological modernization later became a concept in social theory. The politics of the (sociotechnical) processes and the role of institutions and actors require deeper analysis; it proved difficult to uncover these. Further comparative analysis is needed to determine to which extent there is a unique German element in the Ruhr transition. Our chapter is simply a first attempt to tell the story of the Ruhr transition.

Annexe 28.1: Overview of Meetings with Stakeholders Involved in the Review of Information about the Ruhr Transition

This chapter is based on interaction with major stakeholders of the Ruhr district, as well as the review of various internet sources. The information was compiled in cooperation with central regional stakeholders such as the *Emschergenossenschaft/Lippeverband* (EG/LV), *Regionalverband Ruhr* (RVR), *Wirtschaftsförderung Ruhr* (wmr), and the *Landesamt für Natur, Umwelt und Verbraucherschutz* (LANUV) NRW, as well as representatives of the cities of Bochum, Dortmund and Essen. The information was presented and discussed in a sequence of conferences and workshops with governmental and non-governmental stakeholders of the Ruhr district, involving the following individuals:

- 4.6.2012 at the Wuppertal Institute: Participants: Dr. Wolfgang Beckröge (RVR), Mrs Trenk (RVR), Mrs Herzberg (City of Bochum), Mr Mühlenkamp (City of Essen), Mrs Siepmann (City of Essen).
- 14.6.2012 at the RVR Essen: Participants: Dr. Beckröge (RVR), Mrs Trenk (RVR), Mrs Snowdon (RVR), Mr Fölster (City of Essen), Mr Hartwig (City of Bochum), Mr Herzberg (City of Bochum), Mr Hönig (City of Dortmund), Mr Mühlenkamp (City of Essen), Mrs Siepmann (City of Essen).
- 25.7.2012 at the RVR Essen: Participants: Prof. Dr. Finke, Mr Gendries (RWW), Mrs Heinemann (City of Bottrop), Mr Höing (City of Dortmund), Mr Jäger (wmr), Mrs Mann (RVR), Mr Mühlenkamp (City of Essen), Prof. Dr. Oldengott (EG/LV), Dr. Reuter (Landesarbeitsgemeinschaft Agenda 21 NRW), Mrs Semrau (EG/LV), Mrs Snowdon (RVR), Dr. Sommerhäuser (EG/LV), Mrs Raskob (Councillor City of Essen).

References

Amprion, 2012: "Aktuelle EEG Anlagedaten"; at: <http://www.amprion.net/eeg-anlagenstammdaten-aktuell> (April 2014).

Asheim, Bjorn T.; Boschma, Ron; Cooke, Philip, 2011: "Constructing regional advantage: platform policies based on related variety and differentiated knowledge bases", in: *Regional Studies*, 45: 893-904.

Bundesagentur für Arbeit: "Statistik 2012. Sozialversicherungspflichtig Beschäftigte am Arbeitsort"; at: <http://statistik.arbeitsagentur.de> (April 2014).

Blasé, Dieter, 1997: "Stadtentwicklung im Ruhrgebiet", in: Jan-Pieter Barbian; Leudger Heid (Eds.): *Die Entdeckung des Ruhrgebiets* (Essen: Klartext): 221-245.

BMWI (Bundesministerium für Wirtschaft und Technologie), BMU (Bundesministerium fu r Umwelt, Naturschutz und Reaktorsicherheit), 2010: "Energiekonzept fu r eine umweltschonende, zuverlässige und bezahlbare Energieversorgung; at: <http://www.bmu.de/files/pdfs/allgemein/application/pdf/energiekonzept_bundesregierung.pdf> (April 2014).

Brüggemeier, Franz-Josef; Rommelspacher, Thomas, 1992: *Blauer Himmel über der Ruhr. Geschichte der Umwelt im Ruhrgebiet 1840-1990* (Essen: Klartext).

Brüggemeier, Franz-Josef; Scheck, Hanna; Schepelmann, Philipp; Schneidewind, Uwe, 2012: *Vom 'Blauen Himmel' zur Blue Economy. Fünf Jahrzehnte ökologische Strukturpolitik* (Berlin: Friedrich-Ebert-Stiftung).

Bundesregierung, 2010: *Nationaler Aktionsplan für erneuerbare Energie gemäß der Richtlinie 2009/28/EG zur Förderung der Nutzung von Energie aus erneuerbaren Quellen* (Berlin: Bundesregierung).

Coenen, Lars; Benneworth, Paul; Truffer, Bernhard, 2012: "Toward a spatial perspective on sustainability transitions", in: *Research Policy*, 41: 968-979.

Cooke, Philip, 2010: "Regional innovation systems: development opportunities from the 'green turn'", in: *Technology Analysis and Strategic Management*, 22: 831-844.

Drescher, Burkhard, 2012: "Blauer Himmel. Grüne Stadt. Presentation of the Innovation City Bottrop"; at: <http://www.dnhk.org/fileadmin/ahk_niederlande/Downloads/Veranstaltungen/InnovationCity_Ruhr_Burkhard_Drescher.pdf > (April 2014).

Eikmeier, Bernd; Klatt, Jonas; Sengebusch, Katja; Ludewig, Heidi; Schulz, Wolfgang; Klobasa, Marian; Toro, Felipe; Idrissova, Farikha; Reitze, Felix; Menzler, Gerald, 2011: "Potenzialerhebung von Kraft-Wärme-Kopplung in Nordrhein-Westfalen"; at: <http://www.energieagentur.nrw.de/_database/_data/datainfopool/KWK%20NRW_Entwurf%20Endbericht_Stand%2020040311.pdf> (April 2014).

FES (Friedrich-Ebert-Stiftung), 2010: "Sofortprogramm der Bundesregierung", in: *Archiv der sozialen Demokratie*; at: <http://www.fes.de/archiv/adsd_neu/inhalt/stichwort/sofortprogramm.htm> (April 2014).

Groß, Matthias; Hoffmann-Riem, Holger; Krohn, Wolfgang, 2005: *Realexperimente: ökologische Gestaltungsprozesse in der Wissensgesellschaft* (Bielefeld: Transcript).

Hajer, Maarten Allard, 1995: *The Politics of Environmental Discourse. Ecological Modernization and the Policy Process* (Oxford: Oxford University Press).

Hünemörder, Kai, 2004: *Die Frühgeschichte der globalen Umweltkrise und die Formierung der deutschen Umweltpolitik (1950-1973)* (Stuttgart: Steiner).

IG Metall (Industriegewerkschaft Metall), 1972: *Aufgabe Zukunft—Qualität des Lebens,* Vol. 4 (Frankfurt a. M.: Europäische Verlagsanstalt).

Kelly, Thomas D.; Matos, Grecia R., 2010: "Historical statistics for mineral and material commodities in the United States", in: *U.S. Geological Survey Data Series* 140; at: <http://pubs.usgs.gov/ds/2005/140/> (April 2014).

Kemp, René, 2011: *Ten Themes of Eco-Innovation Policies in Europe. S.A.P.I.E.N.S. (Surveys and Perspectives Integrating Environment & Society)*; at: <http://sapiens.revues.org/1169> (April 2014).

Maguire, Steve; Hardy, Cynthia; Lawrence, Thomas B., 2004: "Institutional entrepreneurship in emerging fields: HIV/AIDS treatment advocacy in Canada", in: *Academy of Management Journal*, 47: 657-679.

Kemp, René; Loorbach, Derk; Rotmans, Jan, 2007: "Transition management as a model for managing processes of co-evolution for sustainable development", in: *The International Journal of Sustainable Development and World Ecology. Special issue on (co)-evolutionary approach to sustainable development*, 14: 78-91.

Kemp, René; Schot, Johan; Hoogma, Remco, 1998: "Regime Shifts to Sustainability through Processes of Niche Formation: The Approach of Strategic Niche Management", in: *Technology Analysis and Strategic Management*, 10,2: 175-195.

Kilper, Heiderose; Lehner, Franz; Rehfeld, Dieter; Schmidt-Bleek, Friedrich, 1996: *Wegweiser in die Zukunft. Perspektiven und Konzepte für den Strukturwandel im Ruhrgebiet* (Essen: Klartext).

Lechtenböhmer, Stefan; Fischedick, Manfred, 2012: "Smart City—Schritte auf dem Weg zu einer CO2-armen Stadt", in: Schneidewind, Uwe; Servatius, Hans-Gerd; Rohlfing, Dirk (Eds.), *Smart Energy: Wandel zu einem nachhaltigen Energiesystem* (Heidelberg: Springer): 395-414.

LIT NRW (Landesbetrieb für Information und Technik Nordrhein-Westfalen), 2013: "Landesdatenbank NRW"; at: <http://www.it.nrw.de/statistik/> (April 2014).

Meyer, Bettina; Schmidt, Sebastian; Eidems, Volker, 2009: *Staatliche Förderungen der Atomenergie im Zeitraum 1950-2008. FÖS-Studie im Auftrag von Greenpeace*; at: <http://www.foes.de/pdf/90903-Subventionen_Atomkraft_Endbericht-3%20li.pdf> (April 2014).

MKULNV (Ministerium für Klimaschutz, Umwelt, Landwirtschaft, Natur- und Verbraucherschutz des Landes Nordrhein-Westfalen), 2011: *Energiewende. Maßnahmen in Nordrhein-Westfalen;* at: <http://www.umwelt.nrw.de/klima/energie/massnahmen/index.php> (April 2014).

Mol, Arthur P. J; Sonnenfeld, David A., 2000: "Ecological Modernization Around the World: An Introduction", in: *Environmental Politics,* 9,1: 3-16.

Mol, Arthur P. J.; Jänicke, Martin, 2009: *The origins and theoretical foundations of ecological modernisation theory* (New York: Routledge).

Nordhause-Janz, Jürgen; Rehfeld, Dieter, 1995: *Umweltschutz 'Made in NRW'. Eine empirische Unter-*suchung der Umweltschutzwirtschaft in Nordrhein-Westfalen (München: Rainer Hampp Verlag).

Osterhoff, Frank, 2007: "Zukunft Ruhr 2030—Regionalstrategien der wirtschaft in einer europäischen Metropolregion", in: *Standort - Zeitschrift für Angewandte Geographie*, 31,4: 179-183.

Paredis, Erik, 2013: *A winding road: transition management, policy change and the search for sustainable development* (PhD Thesis, Ghent University, Faculty of Political and Social Sciences); at: <http://hdl.handle.net/1854/LU-4100031> (April 2014).

Rehfeld, Dieter; Schepelmann, Philipp, 2007: "Wachstumsbranche Umweltwirtschaft", in: Reutter, Oscar (Ed.): *Ressourceneffizienz—der neue Reichtum der Städte. Impulse für eine zukunftsfähige Kommune* (München: oekom): 234-241.

Reicher, Christa; Kunzmann, Klaus R.; Polívka, Jan; Roost, Frank; Utku, Yasmine; Wegener, Michael (Eds.), 2011: *Schichten einer Region—Kartenstücke zur räumlichen Struktur des Ruhrgebietes* (Berlin: Jovis).

Rohracher, Harald; Späth, Philipp, 2014: "The Interplay of Urban Energy Policy and Socio-technical Transitions: the Eco-cities of Graz and Freiburg in Retrospect", in: *Urban Studies*, 51,7: 1415-1431.

Rotmans, Jan; Loorbach, Derk, 2010: "Towards a better understanding of transitions and their governance. A systemic and reflexive approach" in: Grin, John; Rotmans, Jan; Schot, Johann (Eds.): *Transitions to sustainable development: new directions in the study of long term transformative change* (New York: Routledge): 105-222.

RVR (Regionalverband Ruhr), 2013: *Statistik Portal RVR*; at: <http://www.metropoleruhr.de/regionalverband-ruhr/statistik-analysen/statistik-portal.html> (April 2014).

RVR (Regionalverband Ruhr), 2013: *Projektdatenbank—Emscher Landschaftspark*; at: <http://www.metropoleruhr.de/no_cache/regionalverband-ruhr/emscher-landschaftspark/projektdatenbank.html> (April 2014).

Schepelmann, Philipp, 2004: "Querschnittsaufgabe Nachhaltigkeit im Ziel 2-Gebiet", in: Meyer-Stamer, Jörg; Maggi, Claudio; Giese, Michael (Eds.), *Die Strukturkrise der Strukturpolitik. Tendenzen der Mesopolitik in Nordrhein-Westfalen* (Wiesbaden: Springer): 162-171.

Schepelmann, Philipp, 2005: *Die ökologische Wende der EU-Regionalpolitik. Die regionale Resonanz von umweltpolitischen Indikatoren des Lissabon-Prozesses der Europäischen Union* (Hamburg: Verlag Dr. Kovac).

Schepelmann, Philipp, 2010: "From Beast to Beauty? Ecological industry policy in North Rhine-Westphalia", in: *Ekonomiaz*, 75: 104-121.

Schlieper, Andreas; Reinecke, Heike; Westholt, Hans-Joachim, 1986: *150 Jahre Ruhrgebiet: ein Kapitel deutscher Wirtschaftsgeschichte* (Düsseldorf: Schwann).

Schwarze-Rodrian, Michael; Seltmann, Gerhard, 2012: "Konzept Ruhr und Wandel als Chance. Statusbericht 2011/2012"; at: <http://www.konzeptruhr.de/fileadmin/user_upload/metropoleruhr.de/Konzept_Ruhr/Ver-

oeffentlichungen/Konzept_Ruhr_und_Wandel_ als_ Chance_-_Statusbericht_ 2011-2012.pdf > (April 2014).

Smith, Adrian; Stirling, Andy, 2010: "The politics of social-ecological resilience and sustainable socio-technical transitions", in: *Ecology and Society,* 15,1: 11; at: <http://www.ecologyandsociety.org/vol15/iss1/art11/> (April 2014).

Smith, Adrian; Stirling, Andy; Berkhout, Frans, 2005: "The governance of sustainable socio-technical transitions", in: *Research Policy,* 34,10: 1491-1510.

Spaargaren, Gert; Mol, Arthur P., 1992: "Sociology, Environment and Modernity: Ecological Modernisation as a Theory of Social Change", in: *Society and Natural Resources,* 5,4: 323-344.

Truffer, Bernhard; Coenen, Lars, 2012: "Environmental Innovation and Sustainability Transitions in Regional Studies", in: *Regional Studies,* 46,1: 1-21.

US EPA (US Environmental Protection Agency), 2006: "International Brownfields Case Study: Emscher Park, Germany"; at: <http://web.archive.org/web/201010180 54526/ http://www.epa.gov/brownfields/partners/ emscher.html> (April 2014).

VDI (Verein Deutscher Ingenieure), 2009: "Die Landescluster im Überblick", in: *Exzellenz. Das Clustermagazin Nordrhein-Westfalen,* Vol. 1, (Duisburg: WAZ); at: <http://www.exzellenz.nrw.de/exzellenznrw/cluster-magazin> (April 2014).

Vierhaus, Hans-Peter, 1994: *Umweltbewußtsein von oben. Zum Verfassungsgebot demokratischer Willensbildung* (Berlin: Duncker & Humblot).

Voß, Jan-Peter; Smith, Adrian; Grin, John, 2009: "Designing long-term policy: rethinking transition management", in: *Policy Sciences,* 42,4: 275-302.

Weichelt, Rainer, 1993: "Silberstreif am Horizont. Vom langen Weg zum blauen Himmel über der Ruhr. Luftreinhaltepolitik in Nordrhein-Westfalen 1950-1962", in: *Sozialwissenschaftliche Informationen,* 22,3: 169-180.

Weichelt, Rainer,1997: "Der 'verzögerte blaue Himmel' über der Ruhr. Die Entdeckung der Umweltpolitik im Ruhrgebiet aus der Not der Verhältnisse 1949-1975", in: Barbian, Jan-Pieter (Ed.): *Die Entdeckung des Ruhrgebiets. Das Ruhrgebiet in Nordrhein-Westfalen 1946-1996* (Frankfurt am Main: Campus Verlag): 259-284.

Wey, Klaus-Georg, 1982: *Umweltpolitik in Deutschland. Kurze Geschichte des Umweltschutzes in Deutschland seit 1900* (Opladen: Westdeutscher Verl.).

WMR (Wirtschaftsförderung Metropole Ruhr), 2012: "Wirtschaftsförderung Metropole Ruhr. Wirtschaftsbericht Ruhr 2012"; at: <http://business.metropoleruhr.de/fileadmin/user_upload/wmr.de/tmp/Downloads-neu/Wirtschaftsbericht_Ruhr_2012.pdf> (April 2014).

Wuppertal Institut, 2013: *Emscher 3.0. Von Grau zu Blau oder wie der blaue Himmel über der Ruhr in die Emscher fiel* (Bönen: Kettler).

29 Transition towards Sustainable Urbanization in Asia and Africa

Belinda Yuen[1] and Asfaw Kumssa[2]

Abstract

Africa and Asia are two of the world's least urbanized regions. But they are fast urbanizing; eighty per cent of the world's projected urban growth to 2050 is expected to take place in Africa and Asia. Even though cities in Africa and Asia are increasingly recognized as engines of growth, they are sites of extremely high population densities, congestion, informal housing and concomitant expansion of slums and squatter settlements, infrastructure shortages and environmental degradation. As rapid urbanization continues, how Africa and Asia manage the urban transition will not just affect urban economic efficiency but will also define their greenhouse gas footprint. This chapter compares the urbanization of both regions and discusses the challenges and opportunities for transition towards sustainable urbanization.[3]

Keywords: Africa, Asia, urban transition, green growth, sustainable urbanization.

29.1 Introduction

Africa and Asia are among the world's least urbanized regions. But urbanization is occurring at an unprecedented pace; eighty per cent of the world's urban population growth to 2050 is projected to occur in Africa and Asia. The growth of their rural populations is projected to turn negative by 2018 (United Nations 2013a).

The two regions share several common trends in urbanization. The scale and rate of urbanization is unparalleled in terms of urban population size and urban extent. Cities are growing, both large and small. The key factors behind urban growth are several and include internal (rural–urban) and international migration, natural increase and reclassification of urban boundaries (from rural to urban). Even though cities are increasingly recognized as engines of growth, they are sites of extremely high population densities, congestion, informal housing and inadequate basic services such as water supply, sanitation and solid waste management. Many cities are bursting at the seams and do not have the necessary institutions, policies and investments to enable housing and infrastructure service development. Ultimately, human well-being, especially the urban poor's, will suffer.

Despite similarities, there are differences in the two regions' urbanization. The nature of urbanization seems to differ. While Asian urbanization is often associated with economic growth and has led to rising incomes and a lowering of poverty in Asian cities, Africa's urbanization has sometimes been said to occur without growth (Fay/Opal 2000; Yeung 2011). This pattern of 'urbanization without growth' is the result of inappropriate policies that could not cater for properly managed and planned urban development. In developed countries urbanization took root during the industrial revolution, at a time when there was 'redundant' labour in the rural areas and an increase in demand for labour in urban areas. The same cannot be said about the process of urbanization in Africa. This is mainly because African urban economies (weakened by institutional problems such as extreme centralization, misguided policies, rampant corruption and external factors such as unfair global trade practices) have failed to absorb the growing urban populations.

1 Dr. Belinda Yuen, Lee Kuan Yew Centre for Innovative Cities, Singapore University of Technology and Design <belinda_yuen@sutd.edu.sg>.

2 Dr. Asfaw Kumssa, Office of the United Nations Resident Coordinator's Office in Nairobi, Kenya, email: <asfaw.kumssa@undp.org>.

3 The views expressed herein are those of the authors and do not necessarily reflect the views of the United Nations.

© Springer International Publishing Switzerland 2016
H.G. Brauch et al. (eds.), *Handbook on Sustainability Transition and Sustainable Peace*,
Hexagon Series on Human and Environmental Security and Peace 10, DOI 10.1007/978-3-319-43884-9_29

This chapter will compare the two regions' urbanization and discuss the consequences and opportunities for transition towards sustainable urbanization. As rapid urbanization continues, how Africa and Asia manage the urban transition will affect not just urban economic efficiency but also their greenhouse gas footprint.

International evidence points to a strong connection between urban growth and increased energy use for a given unit of economic growth (Kamal-Chaoui/Grazi/Joo et al. 2011). Others have argued that greenhouse gas emissions and energy efficiency are directly and permanently influenced by urban form and density (The World Bank 2010; Suzuki/Cervero/Iuchi 2013). In other words, urban planning, city design and infrastructure investment choices can have a substantial impact on energy demand and emission trends. Recent evidence suggests a clear correlation between investments in energy-efficient infrastructure and economic growth (Ostojic/Bose/Krambeck et al. 2013). Against the upcoming inevitable doubling of urban population in Africa and Asia, it is imperative that cities take full advantage of the potential benefits of sustainable use of urban space. Any failure to transition to sustainable urbanization can but represent a major opportunity lost.

The remainder of this chapter is organized as follows. Section 29.2 offers a brief overview of the urbanization patterns of the two regions under discussion—Africa and Asia, focusing on distinctive characteristics and trends and comparing areas of commonalities and differences. This is followed by a rapid review of their key challenges and the implications for sustainable urbanization (section 29.3). Section 29.4 considers the opportunities for green growth, illustrated by Asian examples, and section 29.5 concludes with lessons and policy implications for sustainable urbanization.

29.2   Urbanization Patterns

Africa and Asia share many similarities. They are the world's two largest regions. Asia, covering 30 per cent of the world's land area (58 countries) and 61 per cent of its population, is the world's largest region (United Nations Economic and Social Commission for Asia and the Pacific 2013). Two countries—China and India—contribute about 60 per cent of Asia's population. Africa is the world's second largest region (55 countries), occupying 20 per cent of the world's land area and about 15 per cent of its population. Africa and Asia's populations are projected to continue to grow. By 2100, Asia is expected to be 2.2 times as populous as Africa (United Nations 2004). Given their vast geographical expanse, Africa and Asia contain diverse cultures, economy, environment and institutions.

In terms of demography, as Zlotnik discusses elsewhere in this volume, Africa is a comparatively 'young' continent with high fertility rates. The average age of its population is 18 years; 40 per cent of Africans are aged 14 years or less and 34 per cent aged between 25 and 59 (World Review n.d.). However, by 2030, it is projected that many African countries will have a growing ageing population. The proportion of Africans who live to age 65 is anticipated to increase from 3.2 per cent in 2010 to 4.5 per cent in 2030. Several middle-income African countries such as Mauritius, Tunisia, Morocco, Algeria, Egypt and South Africa are projected to have 4.5–7.3 per cent of their population aged 65 or older (AfDB 2012a). Understanding this population structure and the changing demographic trends is crucial for urban futures.

Like Africa, Asian population is relatively young (median age in Asia is 30). But population ageing is occurring rapidly (more rapidly than it has in Western countries) and at a much earlier stage of economic development. The proportion of Asian people under age 15 is projected to decline while the number of people aged 65 and over is expected to increase by 314 per cent from 207 million in 2000 to 857 million by 2050 (United Nations 2001). Japan has the most rapidly ageing population in Asia and the world; 36 per cent of its population will be 65 or older by 2050 and median age will rise from 45 in 2010 to 53 in 2050 (United Nations 2013b). India has the youngest population in terms of size (704 million people aged below 30 with a median age of 25.4 in 2011) while Lao People's Democratic Republic is the youngest Asian country in terms of median age (median age of 20).

Africa and Asia are among the world's poorest and least urbanized regions. Asia has the largest number of the world's poor (63 per cent). Estimates indicate that some 750 million people in Asia are living on less than US$1.25 a day, 57 per cent in urban areas (Wan/Sebastian 2011). Many of the urban poor do not have proper housing and access to basic services—one-third have no access to safe water, two-thirds do not have access to adequate sanitation, more than half live in slums. In 2010, Africa was the world's least urbanized region with 40 per cent (414 million) of its population living in urban areas while 42.2 per cent (1.76 billion) of Asia's population lived in urban areas (United

Table 29.1: Urbanization Levels, 1950–2050. **Source:** United Nations (2012).

Major area	Percentage urban					Rate of urbanization (percentage)			
	1950	1970	2011	2030	2050	1950-1970	1970-2011	2011-2030	2030-2050
Africa	14.4	23.5	39.6	47.7	57.7	2.47	1.27	0.98	0.96
Asia	17.5	23.7	45.0	55.5	64.4	1.52	1.57	1.10	0.74
Europe	51.3	62.8	72.9	77.4	82.2	1.02	0.36	0.31	0.30
Latin America & Caribbean	41.4	57.1	79.1	83.4	86.6	1.61	0.80	0.28	0.19
Northern America	63.9	73.8	82.2	85.8	88.6	0.72	0.26	0.22	0.16
Oceania	62.4	71.2	70.7	71.4	73.0	0.66	-0.02	0.05	0.12

Nations 2013a). However, in the coming few decades, it is anticipated that this settlement pattern will change.

29.2.1   Rapid Urbanization Anticipated

Both Africa and Asia are urbanizing rapidly. From 1990 to 2010, Asia's urban population increased from 31.5 per cent to 42.2 per cent (over 754 million people), registering the highest percentage increase among all regions in the world. No other region has experienced such an increase in absolute numbers in such a short time. Every day, an estimated 120,000 people are added to Asian cities (Roberts/Kanaley 2006). Asia's urban population is projected to increase to 63 per cent between 2010 and 2050. Many of the new urban residents are expected to be poor people.

As summarized in table 29.1, Africa and Asia will have the highest urbanization rates in the world in the coming decades. Their urban populations are projected to double between 2000 and 2030. By 2050, about 58 per cent of Africa's population will live in urban areas (United Nations 2012). Africa's urban population will increase from 414 million to over 1.2 billion by 2050 while Asia's urban population will grow from 1.8 billion to 3.3 billion. The largest increases in urban population are expected to take place in China and India in Asia, and in Nigeria in Africa. Between 2014 and 2050, China, India and Nigeria are expected to add 292 million, 404 million and 212 million urban dwellers respectively (United Nations 2014).

The urban expansion has not been uniform but varies substantially across countries and over time. There is a diversity of urbanization patterns. What this means is that there is no single urban transition.

Africa's urban populations are unevenly distributed over the continent's subregions. North Africa is the most urbanized subregion (53.5 per cent), followed by Southern Africa (47.1 per cent) and West Africa (44.1 per cent). East Africa is the least urbanized (24.6 per cent), but the most rapidly urbanizing region of Africa (UN-HABITAT 2008). The process of urbanization in Africa is usually highly influenced by the movement of people displaced by drought, famine, ethnic conflicts, civil strife and war. Acute levels of urban poverty and crime and haphazard urban settlements are common characteristics of urbanization in Africa. The poor live in poor neighbourhoods where they face dire economic and social hardships. Most are uneducated and unskilled workers who have migrated to urban areas, encouraged by the push-and-pull factors associated with rural–urban migration.

Urban centres play a pivotal role in the production of goods and services; they are centres of employment, technology and innovations. Urban areas account for about 55 per cent of Africa's GDP (UN-HABITAT 2008). The common contention is that with urban growth, there is an increase in the use of technology, especially mobile phones, flight connectivity, internet connectivity and a general improvement in information and communication technologies in Africa (Obeng-Odoom 2011). At the same time, others have argued that urbanization in Africa has deviated from an ideal development trajectory, continuing even in periods of negative growth, which the World Bank called 'urbanization without growth' (Fay/Opal 2000; Barrios/Bertinelli/Strobl 2006). Their evidence suggests that 'urbanization without growth' is caused by distorted location policies, misguided policies to stop rural–urban migration, inadequate planning, and corruption, among other factors, all of which have combined to make it difficult for African cities to properly address the ever-increasing population pressure, plan urban infrastructure and development, and grow urban economies to become engines of growth or pivotal centres of industrialization.

Another interesting phenomenon often associated with urbanization in Africa is that a large number of the growing urban population is absorbed by the intermediate and smaller cities of 500,000 inhabitants, and not by larger cities (UN-HABITAT 2008). But smaller cities often have fewer human, financial and technical resources at their disposal to deal with urbanization (United Nations Population Fund 2007).

In Asia, the Pacific subregion, comprising Australia, New Zealand and Pacific Islands, is the most urbanized subregion (70.2 per cent) followed by North and Central Asia (63.9 per cent), East and North-east Asia (50 per cent), South-east Asia (41.8 per cent) and South and South-west Asia (33.3 per cent) (UN-HABITAT 2010a). There are variations within subregions. For instance, Singapore in Southeast Asia is 100 per cent urbanized while Kyrgyzstan, Turkmenistan and Uzbekistan in North and Central Asia have less than 50 per cent of their populations living in cities and towns.

For Asia's two major countries, China's urban population has increased from 27 per cent in 1990 to 36 per cent in 2000 and 51 per cent in 2012 and is projected to reach 70 per cent (1 billion) by 2030. About 80 per cent of this increase can be traced to rural-urban migration and urban area reclassification while the remaining 20 per cent is by natural increase. This urbanization trend is reflected in India though about 60 per cent of urban growth is due to natural increase and less than 20 per cent to rural–urban migration. Even at a relatively low level of 31 per cent urbanization, India's urban population is the second largest in the world. Its urban population is projected to increase from 377 million in 2011 to 600 million (40 per cent) by 2030, an increase of over 200 million in 20 years. On current trends, by 2050, it is projected that India's urban population will constitute nearly 50 per cent of its total population.

The trend points to Asia heading towards a dispersed urbanization pattern with more pronounced expansion in small and medium-sized cities. The majority (60 per cent) of Asia's urban population lives in small (49 per cent) and medium-sized (11 per cent) towns with a population of less than 1 million (UN-HABITAT 2010a). These cities, together with metropolitan cities (with populations of 1-10 million, 29 per cent) and megacities (with over 10 million people, 11 per cent), will drive future urban growth in Asia. Asian cities are increasing in number and size, with many doubling their population every 15-20 years. Estimates

indicate that Bangkok's population has risen from 1 to 8 million within 45 years while other Asian cities increased their population by this amount in half that time—Dhaka 37 years and Seoul 25 years—as compared to 130 years in the case of London (UN-HABITAT 2004).

29.2.2   Growing Cities

The number of large cities, including megacities, with disproportionate urban primacy is growing. Currently, there are two megacities in Africa—Lagos and Cairo, with about 11 million inhabitants each—while Asia has thirteen megacities, half of the world's twenty-five megacities (United Nations 2012). The world's seven largest megacities are in Asia with Tokyo being the largest, followed by Delhi with 23 million inhabitants. The number of Chinese cities with more than 1 million people (million-plus cities) is expected to increase from about 90 to 221, including six megacities, by 2030, while the number of million-plus cities in India is projected to increase to 68 over the same period, of which six will be megacities.

Despite various efforts to restrict the growth of megacities (such as green belt and transmigration schemes), the exurban areas around megacities have continued to grow, engulfing small towns and other settlements on the urban periphery and forming sprawling mega-urban regions. Examples include the Pearl River Delta Region, China, with an estimated 150 million people, and the Jakarta-Bogor-Tangerang-Bekasi Region (Jabodetabek), Indonesia with an estimated 28 million. The planning and governance mechanisms, however, are often fragmented even though there are obvious economic interrelationships among the urban places within the mega-urban region.

That is to say, the peri-urban areas in many large cities are fast expanding, sometimes stretching to 100 kilometres from the urban core. The urban area of Shanghai, China is projected to grow from 410 square kilometres to 1,100 square kilometres (150 per cent) in less than ten years (Martin 2005). Noting the unique feature of Asian urbanization, McGee (2009) has coined the term 'desakota' (village-city) to describe Asian urban agglomerations and indicate their mixed rural-urban characteristics. Asian cities are characterized by mixed land use development and high population densities (e.g. Bangkok, Hong Kong SAR, Seoul, Singapore). Formal and informal activities often take place in the same space while residential developments are located next to commercial activities. Many Asian cities have high population densities, the highest in the world, ranging from 10,000 to over 20,000 people per square kilometre (e.g. Mumbai, Kolkata, Karachi, Seoul). This can be traced to several factors including market forces, planning or other government policies and transportation patterns.

Over the past two decades, many Asian countries have started to recognize cities as engines of economic growth and to reorient national policies to promote urbanization. China has championed an urbanization strategy that develops city clusters across its Central, Western and North-eastern regions into engines of growth, transitioning from a rural, agricultural society to an urban industrial economy and lifting some 660 million people out of poverty. China's Pearl River Delta Region (0.4 per cent of China's land area and 3.5 per cent of its population), for example, contributes 10 per cent of the country's GDP and 30 per cent of its exports. Chinese cities' contribution to GDP is expected to rise from 75 per cent to 95 per cent by 2025 (McKinsey Global Institute 2009).

Similarly, the urban sector in Indian cities contributes about 60 per cent of GDP (Government of India 2011). This contribution is projected to rise to 75 per cent by 2021 and over 70 per cent of new jobs are expected to be created in cities. The suggestion is that cities provide large economies of agglomeration; urbanization and economic growth are often concomitant processes in Asian cities.

29.3   Urban Challenges

While urbanization may represent economic opportunity, it also brings many and varied challenges that threaten the liveability of cities. Much has been written about the challenges and impacts of rapid urbanization (Yeung 2011; UN-HABITAT 2010a; Yuen 2009). The unprecedented scale, size and continued expansion of cities have produced enormous problems, from the decay of inner city areas to urban sprawl, uncontrolled peri-urban area development, ill-regulated land use, urban housing shortages, congestion, and environmental pollution. The demand for urban living, in terms of land, housing, transport, water and energy supplies, social services, etc., has frequently exceeded supply.

In many cities, century-old buildings are being cleared to build modern roads and high-rise buildings (Yuen 2013). Untreated domestic, municipal and industrial waste seriously pollutes rivers and local drinking water supplies and threatens coastal fisheries and livelihoods in several Asian cities (e.g. in China,

Malaysia, Vietnam). Expansion of cities and their built-up areas modifies the local and regional climate mainly by the production of pollution and the urban heat island effect with knock-on effects on agricultural land, biodiversity and ecosystem services (Crutzen 2004; UN-HABITAT 2011; Angel 2012). That is, cities are drivers of climate change.

29.3.1   The Bane of Rapid Urbanization, Africa

As cities grow in size, they become more complex and the ability to manage them becomes a daunting challenge (box 29.1). Rapid urbanization is often followed by urban poverty, with the prospects of a dignified and productive life continuing to elude the poorest among Africans and Asians. As a result, more and more urban poor are pushed into the informal sector and unplanned settlements (Figure 29.1). As many as nine out of ten urban workers in Africa are employed informally, working long hours at very low incomes. Despite African cities generating over 50 per cent of the continent's total GDP, Africa's poor, who earn less than US$2 per day, constitute about 61 per cent of the continent's population (AfDB 2012b).

The other major challenge of urbanization in Africa is related to urban crime and insecurity. The lack of employment opportunities as well as basic social services such as shelter and sanitation services have eroded the dignity of the poor, particularly youth in urban areas. As most of the youth slide into poverty, they join criminal groups and gangs in search of solidarity, identity and empowerment and hence resort to violence and crime to earn a living. In the recent past, inter-state conflict has significantly declined in Africa while urban insecurity, violence and crime have dramatically increased in African cities (UN-HABITAT 2014). Africa has the highest rate of urban crime worldwide (and Asia the lowest).

The higher incidence of Africa's crime has been traced to poverty, inequality, external economic shocks, social exclusion and weak political institutions (Fox/Hoelscher 2012). The prevalence of crime and the escalating urban crime rate can be expected to adversely affect foreign direct investment in urban Africa and reduce employment opportunities for its youth. Such reduction can potentially lead to large-scale social unrest similar to that recently experienced in Northern Africa, dubbed the 'Arab Spring'.

In Africa, inadequate and debilitated infrastructure is adversely affecting the continent's economic growth and reducing private sector productivity (UNECA 2009). Low-level energy access including an

Figure 29.1: A hotel in the slums of Nairobi, Kenya. **Source:** Authors.

Box 29.1: Cairo: Africa's Megacity. **Source:** The author.

With a population of about 11 million (2011), Cairo remains by far Africa's largest city. According to UN-HABITAT (2010b), the population of Cairo will grow by 2 per cent annually between 2010 and 2020. Cairo and Alexandria are the main economic engines of Egypt, with 57 per cent of all manufacturing activities located in the Greater Cairo area. About 67 per cent of the GDP of Egypt comes from the metropolitan area of Cairo. Industries include textiles, cotton and food processing. The service sector and tourism are the main source of revenue and employment for the workforce in Cairo. Like in any other megacity,

rapid urban growth has resulted in serious problems in most aspects of settlements and the life of its population. To address these challenges, the government has embarked on an ambitious plan to steer the urban population to new settlements or areas. In this regard, the policy is to redirect the population pressure from the Nile valley to the adjacent desert plateau. The government has also introduced policies to decentralize population and activities from Cairo to local and regional levels in order to manage its growth and address the challenges of population pressure on this megacity.

---

average rural electrification rate of 10 per cent and a railway network density that ranges from 30 to 50 kilometres per million people pose a major challenge to African development (AfDB 2012b). The proportion of people in Africa still without electricity is higher than in any other continent; the rate of urban electrification is lower than in any other continent. The 2010 data indicates that the urban electrification rate in Africa is 69 per cent compared to 96 per cent

for China and East Asia and 89 per cent for South Asia (International Energy Agency 2010).

Although the challenge of urban transportation in African cities is not new, the situation is getting worse daily (figure 29.2). Rising traffic and congestion levels are having far-reaching negative economic, social and environmental impacts on the cities. The increasing use of personal cars has choked roads, endangering the safety of pedestrians and the health of city residents. Pollution (including noise and automobile

Figure 29.2: Rising traffic congestion in Nairobi city, Kenya. **Source:** Authors.

emissions) has become a serious impediment to urban quality of life. This is made worse by the difficulty in enforcing regulations pertaining to vehicle inspections that address safety and emissions.

Declining urban densities associated with sprawl and combined with weak, fragmented, and underfunded authorities that are unable to maintain or expand existing urban transport services have not only increased travel distances but also pushed up the price of public transport. The poor are often the worst affected, being excluded from work and social services. Facilities for pedestrians are very poor and those for bicycles and other non-motorized transport are non-existent. Improving urban transport in African cities is an urgent priority and will depend on a strategy of coordinated measures to improve infrastructure, traffic management, service quality and network reach.

29.3.2   The Bane of Rapid Urbanization, Asia

Although Asian cities have made some progress in reducing poverty and achieving the Millennium Development Goals, basic urban infrastructure services—water supply, sewerage, solid waste management, storm water drainage, transport—still require major improvement. Over 60 per cent of the world's slum population (505 million people) continues to live in Asia. In many Asian cities, more than 50 per cent of urban residents live, often under hazardous conditions, on land where title is disputed. The living conditions of migrants are poor; they normally live in illegal, overcrowded, unhygienic squatter areas (figure 29.3).

The state of infrastructure in most Asian cities (large and small) is equally beset with problems relating to coverage, quality, operation, persistent underinvestment and poor maintenance (figure 29.4). Even though India's power sector has been expanding in recent years, it still has the problem of providing adequate power to meet the needs of its growing economy (US Energy Information Administration 2013).

Figure 29.3: Squatter settlement in Metro Manila, The Philippines. **Source:** Authors.

Estimates indicate that 25 per cent of India's population lacks basic access to electricity (6 per cent in urban areas and 40 per cent in rural areas), and electricity blackouts are common in electrified areas. In the water sector, 70 per cent of its urban population is covered by individual water connection. Duration of water supply in Indian cities is between 1 and 6 hours. Less than two-thirds of India's urban households are connected to a sewerage system and in some cities only about 20 per cent of sewage generated is treated before disposal. About 13 per cent of India's urban population defecate in the open. These problems are not unique to India. They happen to varying degrees in many other cities in the developing countries of Asia and Africa.

29.3.3   Failure to Act—the Consequences

Failure to meet infrastructure needs could stifle economic growth and, ultimately, human development. Estimates suggest that a rise of 1 per cent of GDP on infrastructure investment could generate an additional 3.4 million direct and indirect jobs in India, while the consequent deterioration in energy, housing, communications, transport and water facilities from the failure to close the infrastructure gap would restrain economic growth by 3 to 4 percentage points of GDP (Tahilyani/Tamhane/Tan 2011; McKinsey Global Institute 2013). Moving in lockstep with urbanization, US$8 trillion is estimated to be required in infrastructure investment (energy, transport, telecommunication, water and sanitation) over the next ten years to redress Asia's historical underinvestment and growing demand.

The projected increase in urban populations to Asian cities calls for unprecedented construction. Estimates suggest that Asia's daily additional 120,000 urban dwellers would require the construction of over 20,000 new houses, 250 km of new roads, and additional infrastructure to provide more than 6 megalitres of potable water (Roberts/Kanaley 2006). On an individual country level, rapidly urbanizing China could require 5 billion square metres of road, 28,000 kilometres of metro rail and 40 billion square metres

Figure 29.4: Lack of proper solid waste disposal creates an environmental problem in Jakarta, Indonesia. **Source:** Authors.

of floor space (20,000–50,000 new buildings of more than 30 floors) over the next twenty years (McKinsey Global Institute 2009). Urban China is projected to require 700–900 gigawatts of new coal-fired power between 2005 and 2025 (50 new power plants a year) to keep pace with urban and manufacturing growth. This will only further increase China's greenhouse gas emissions, strengthening its position as the world's largest emitter of greenhouse gas.

China generates 24 per cent of the world's greenhouse gas emissions (United Nations Environment Programme 2013). It urgently needs to shift towards a model of economic growth focused on low carbon emissions. More than 1 million people living in the coastal areas of China could be displaced and a land area the size of Hong Kong SAR could be submerged by 2050 if the country does not take action to prepare for rising sea levels (ADB 2013). Climate-proofing of China's infrastructure, including roads and drainage, could cost up to US$44 billion per year between 2011 and 2050.

Climate change affects the vulnerability of cities, especially those in low-elevation coastal zones. Africa and Asia (South, South-east and East Asia) are key regional hotspots of coastal vulnerability (IPCC 2014).

Several African cities such as Accra, Durban, Lagos, Luanda and Maputo are located in low-lying coastal areas that are prone to flooding. Climate change is likely to put millions of these people at risk. Changes in the frequency, intensity and duration of climate extremes (droughts, floods, heatwaves, among others) have adversely affected the livelihoods of their urban population, particularly the poor and other vulnerable communities who live in slums and marginalized settlements. Extreme changes in weather patterns have increased incidences of natural disasters and adversely affected all key sectors of the economy including the urban economy, agriculture and forestry, water resources and health.

Yet African cities' ability and readiness to predict and respond to potential complex emergencies, to mitigate natural and other disasters through disaster risk reduction, climate change adaptation planning and ecosystems management are either very weak or altogether lacking. To make matters worse, some African cities are developing new 'ecologically unfriendly' configurations (e.g. using heat-retaining building materials that raise urban temperatures) or 'urban fantasies' of modernist high-rises and highways that are disconnected from the urban core and African land-

---

Box 29.2: Africa's Path to Green Growth. **Source**: The author.

The *African Development Bank* (AfDB 2012b) has argued that green growth will have the desired impact on Africa's two major problems—poverty and climate change—if it is packaged within the following three focal areas:

1. promoting sustainable infrastructure including access to renewable/low carbon energy and energy efficiency; sustainable transport; and sustainable cities;

2. efficient/sustainable management of natural assets including land (agriculture, forests, and other land uses); water (freshwater, marine); and minerals;

3. building resilience of livelihoods including physical/climate; economic; and social.

---

scape, further compromising their resilience to climate change.

Increasingly, international organizations including UN-HABITAT (2009) have advocated proper urban planning to better respond to the growing urban population and urbanization. In the absence of appropriate and effective urban planning and implementation strategies, African and Asian countries will end up with "spiralling poverty, proliferation of slum and squatter settlements, inadequate water and power supply, and degrading environmental conditions" (UN-HABITAT 2009: 26). But with comprehensive urban planning and implementation strategies, urbanization in Africa and in Asia can transition to jump-start sustainable urbanization and promote much-needed political, social and economic development (UN-HABITAT 2010b).

29.4   Opportunities for Green Growth

In part through the effort of international organizations such as the ADB, the OECD, the World Bank and the United Nations, the notion of green growth, first introduced in Asia, has been increasingly offered as a response to cities to forge new ways of acting to reduce greenhouse gas emissions in a climate-changed world. Green growth is defined as a sustainable economic growth "that is either compatible with or driven by reduced emissions, improved efficiencies in the use of natural resources and protection of ecosystems" (AfDB 2012b: 5). As Hallegatte/Heal/Fay et al. (2011: 2) stated, green growth is "about making growth processes resource-efficient, cleaner and more resilient without necessarily slowing them."

Others have argued that haphazard and unsustainable development carries the risk of locking the city into an inferior option of development and may in the longer term result in a loss of competitiveness, retarding economic growth and generating legacy costs for decades (Anas/Timilsina 2009; Corvellec/Campos/Zapata 2013). Even so, escaping lock-ins is possible

and the case for cities to transition towards more sustainable development styles is compelling (Bulkeley/Broto/Hodson et al. 2011; Hallegatte/Heal/Fay et al. 2011; Bridge/Bouzarovski/Bradshaw et al. 2013). Moving towards such a development path would require cities to not only anticipate growth and manage its outcomes more effectively but also to fundamentally change the way they produce and consume energy and resources.

Making the transition is understandably not easy. Many questions remain, especially relating to analysing and understanding the urban complexity and the processes of transition—what are the mechanisms that can be used to effect the transition to green growth? As the effects of climate change intensify and the rising prices of energy are felt, many cities are awakening to the inevitable need to become more proactive in rethinking and repositioning to achieve greater urban sustainability. According to the United Nations Economic and Social Commission for Asia and the Pacific (2012), forty-two million Asians were pushed back into poverty in 2011 due to energy and food price increases.

In the context of Africa, the transition to green growth means "pursuing inclusive economic growth through policies, programmes and projects that invest in sustainable infrastructure, better natural resources, build resilience to natural disasters, and enhance food security" (AfDB 2012b: 5). Pursuing the path of green economy will assist Africa in overcoming the twin challenges of rampant poverty and climate change (box 29.2).

Obviously, not all green growth options are cheap. An economy that is driven by green growth cannot be created overnight. Nonetheless, some African countries have embarked on the path of green growth. Rwanda has committed itself to a long-term green growth strategy. Its *Vision 2020*, adopted in 2002, outlines a long-term national development strategy to achieve economic development and poverty alleviation by modernizing Rwanda from an agrarian economy into a regional service and knowledge-based hub

---

Box 29.3: National Strategy for Green Growth, South Korea. **Source:** The author.

Providing the national agenda and framework for government actions, green growth programmes and laws, the *National Strategy* sets out ten policy directions. These are:

1. effective mitigation of greenhouse gas emissions;
2. reduction of use of fossil fuels and enhancement of energy independence;
3. strengthening the capacity to adapt to climate change;
4. development of green technologies;

5. greening of existing industries and promotion of green industries;
6. advancement of industrial structure;
7. engineering a structural basis for green economy;
8. greening the land and water and building the green transportation infrastructure;
9. bringing the green revolution into daily lives; and
10. becoming a role model for the international community as a green growth leader.

---

with an emphasis on green economy. Similarly, Ethiopia has committed itself to an economy that is based on a green growth strategy to achieve a middle-income country by 2025. In Kenya, where only five per cent of the rural population has access to electricity, the Menengal geothermal energy project is expected to increase energy production by 26 per cent by 2018, thereby providing clean and inexpensive energy for about 500,000 households and 300,000 small businesses (AfDB 2012b).

The unprecedented increase in urban population opens new opportunities to improve urban development. As Roberts and Kanaley (2006) posited, the future prospects of Asian cities will be determined by the management of their sustainability, that is, economic, social and environmental sustainability.

29.4.1 Some Asian Examples

Green growth has been adopted as a regional sustainable development strategy in Asia since 2005 during the Fifth Ministerial Conference on Environment and Development in Asia and the Pacific. South Korea is the first Asian country to introduce a green growth strategy. Championed by its president Lee Myung-bak on the sixtieth anniversary of the founding of South Korea, green growth is put forward as the country's next growth engine with projections to generate additional economic output (206 trillion won) and new jobs (1.8 million) (Kim 2010).

Since August 2008, South Korea has adopted a new national development paradigm based on the concept of low carbon green growth. This is being pursued in a comprehensive way that includes not just strong commitment from its president, as the head of state, to climate change mitigation but also a whole-of-society concerted focus on green innovation for economic growth, green transport, green purchasing and public support for green growth and green living. The purpose is to reduce the country's greenhouse

gas emissions by thirty per cent and to position South Korea among the world's top seven leading countries in green growth by 2020. South Korea offers a case of how strong government leadership and political commitment are brought to bear to act as a catalyst for economic system change and engage businesses and the public to support the transition towards a green economy.

Under the Korean Law for Low Carbon Green Growth, which came into effect in May 2010, the state is obliged to invest at least two per cent of GDP in low carbon production and consumption. In this regard, South Korea has been implementing eco-labelling since 1992. In 2001, its Ministry of Environment implemented the *Environmental Declaration of Products*, a standardized ISO 14025/TR and life cycle assessment-based programme, and in 2005, enacted *The Promotion of Purchase of Environmentally Friendly Products Act*. The green growth effort is being implemented through the National Strategy for Green Growth (2009–2050) and five-year development plans (box 29.3).

In consequence, South Korea has increased its green overseas development assistance. It launched the *Global Green Growth Summit* at the G20 Summit in November 2009 and a *Green Technology Centre* for international green technology cooperation, including cooperating with developing countries in green growth and climate action. Even though still at the implementation stage, the number of venture start-ups in green industries has increased by more than forty per cent since 2009, and the Korean bullet train network is being expanded to connect most major cities nationwide by 2020.

More Korean citizens are becoming socially aware of green living and are participating in green lifestyle choices. The number of households who have joined the carbon point programme increased from 400,000 in 2009 to 2.9 million by July 2011 (Ministry of Environment n.d.). Under the carbon point programme, points, i.e. incentives such as cash, metrocard, and gift coupons, are offered to consumers who save electricity, water and gas in houses and in commercial buildings. The green consumption culture is being strengthened through various efforts to promote green products and green consumption (e.g. green product information system, carbon labelling system, green consumption education). South Korea's government is developing a national carbon cap and trade system, modelled on the European Union Emissions Trading System. Set to begin on 1 January 2015, the proposed South Korean system has been described by Marcacci (2013) as the world's most ambitious cap and trade market with the highest global price on carbon.

Apart from South Korea, a growing number of countries have joined the green growth agenda. In 2010, Cambodia adopted a *National Green Growth Road Map*, while Kazakhstan presented a pioneering *Green Bridge* initiative to link Asia and the Pacific with Europe through green growth during the Sixth Ministerial Conference on Environment and Development in Asia and the Pacific. China, India, Indonesia, Japan, South Korea, Philippines, Singapore, Malaysia and Thailand have started to articulate energy-efficient policies and build energy-efficient and zero-energy buildings as well as to certify green buildings (Wen/ Chiang/ Shapiro et al. 2007). A new paradigm of mainstreaming energy efficiency seems to be emerging.

Energy consumption is a major source of greenhouse gas emissions and the building sector is a major energy consumer in Asia (about 25 per cent). Zero-energy construction has significant implications for the regional as well as the global economy and environment. More than half of the world's new building construction is taking place in China and India. Since 2005, China has introduced a number of measures for green growth including a government decree on mandatory designs to conserve energy used in cooling, heating and lighting in public buildings (United Nations Environment Programme 2013). Both the Eleventh Five-Year Plan (2006-2010) and the Twelfth Five-Year Plan (2011-2015) have emphasized energy and resource efficiency, providing a strong policy framework (energy saving, emission reduction and industrial policies) to support China's transition towards a green economy. China's 2020 emissions reduction target is to reduce its greenhouse gases by 40-50 per cent based on carbon intensity in 2005 and to produce 16 per cent of its primary energy from renewable sources by 2020.

China is actively promoting renewable energy, investing US$67.7 billion in renewable energy in 2012 (United Nations Environment Programme 2013). China is now the world's largest producer of wind power and solar energy equipment, creating millions of green jobs. The renewable (thermal, wind, solar) power generation sector is projected to create 4.4-5.05 million new green jobs between 2005 and 2020 (Wu 2012). Though early days yet, the environment industry represents over 3 per cent of China's GDP. Since 2009, several Chinese cities have started to develop various low carbon action plans (Chinese Society for Urban Studies 2011; Baeumler/Ijjasz-Vasquez/Mehndiratta 2012).

Harbin and Hangzhou, for example, have initiated energy-saving and emission-reduction programmes in communities, mobilizing families, schools, businesses and state organizations to take part in energy conservation and emission reduction. Other cities such as Shanghai and Baoding have joined the *World Wildlife Fund Low Carbon City* initiative, which seeks to find a sustainable development mode for China's urban areas through studying current energy production and utilization patterns and developing new economic approaches for cleaner growth. Other cities, for instance, Wuxi, Shenzhen, Suzhou, Yangzhou, Beijing and Shanghai, are taking action to build low carbon cities. More than eighteen provinces and 150 counties and cities are estimated to have started planning for some form of eco-development. Some of these are being developed in international partnerships, such as the *Sino-Singapore Tianjin Eco-city* (figure 29.5), the *Sino-Swedish Wuxi Low Carbon Eco-city* and the *Sino-Finland Mentougou Eco-Valley* in Beijing.

In 2010, China's National Development and Reform Commission's *National Pilot Programme on Low Carbon Provinces and Cities* was launched to encourage further local low carbon action (People's Daily 2010). Rather than just attempting to reduce carbon emissions from industry, this represents a first comprehensive regional-local government approach to low carbon performance in China, directing five provinces (Guangdong, Liaoning, Hubei, Shaanxi, Yunnan) and eight cities (Tianjin, Chongqing, Shenzhen, Xiamen, Hangzhou, Nanchang, Guiyang, Baoding) to consider the complete built environment as well as the purchasing and consumption habits of those residing, working or visiting these provinces and cities. Together, the five provinces and eight cities contribute 36 per cent of the country's national GDP, 31 per cent of its energy consumption, 27 per cent of population and 27 per cent of energy-related emissions. It

Figure 29.5: Salt pan barren land being transformed into the Sino-Singapore Tianjin eco-city. **Source:** Authors.

represents the largest low carbon city-region development in the world, a critical step in promoting the transformation of local economic development and achieving low carbon performance.

In 2008, India adopted a *National Action Plan on Climate Change* that included a range of measures for renewable energy, energy efficiency, clean technologies, public transport, resource efficiency and tax incentives. It should be noted that low carbon development is not just state action. Non-state actors have an important role. Bangalore, India, for example, has conducted various experiments in smart building technologies, renewable energy and water management. A private company, BCIL, has been developing T-ZED (*Towards Zero-Energy Development*) homes (76 apartments and 15 villas) on a five-acre site since 2004.

The award-winning project has a number of green features including energy efficiency (for example, LED and CFL lighting, renewable energy-based street lighting, solar water heating systems), no import of water (for example, rainwater harvesting, cyclical water use), no export of waste (for example, commu-

nity-led waste management), no industrial floors and no toxic paints (Kikusawa 2010). The company has opted to build only 91 houses instead of the 300–400 apartments allowed by the floor space index, in order to create self-sufficiency in water.

Completed in 2009, the project yields capital savings of about 20,000 tonnes of carbon emissions and operational savings of about 1,500 tonnes of carbon emissions. Its cost and pricing, however, mean that the development is largely accessible only to the upper-middle-class income groups. Many of its residents are IT professionals, doctors, bankers, biologists and the like. In Bangalore, the upper-middle income class constitutes just five per cent of the population. As Kikusawa (2010) argued, the impact of such green building needs to be extended from the current target community of middle-upper income groups to lower-income households.

A number of other Asian cities, including small and medium-sized cities, have started to prepare and implement city development strategies to better strategize urban futures rather than develop without plan-

Figure 29.6: Turning waste to recreation land at Semakau, Singapore. **Source:** Authors.

ning (ADB 2004; Einsiedel 2004). The process would typically identify priority city development issues and formulate medium- and long-term city development strategies through a participatory process. The priority issues include livelihood (e.g. job creation and business development), environmental sustainability, energy efficiency, quality of service delivery, financial resources, governance, spatial form and infrastructure.

Designing eco-efficient cities and infrastructure will require rethinking in urban planning and design. As the United Nations Economic and Social Commission for Asia and the Pacific (2012) acknowledged, this would necessitate a shift towards eco-city development, encompassing eco-efficient transport (change the way people move, from private cars to public transport), green buildings (from energy wasting to energy saving), eco-efficient energy infrastructure (improve efficiency of the energy system and diversify to renewable energy sources), eco-efficient water infrastructure (change the way water resources are managed) and eco-efficient solid waste management (turn waste from a cost into a resource) (figure 29.6).

The 2012 United Nations Conference on Sustainable Development (Rio+20) has recognized and re-emphasized the need for a holistic approach to urban development. Its Resolution notes that, if well planned and managed, cities can promote economically, socially and environmentally sustainable societies. The need to promote an integrated approach to planning, developing and managing sustainable cities and urban settlements as well as sustainable development policies that support inclusive housing and social services and a safe and healthy living environment for all is emphasized. The rationale is simple: cities are the hubs for much of national production and consumption—economic and social progress that generates wealth and opportunities.

29.5   Conclusion

Sustainable urbanization is critical to the future of Africa and Asia. Sustainable urbanization is both more inclusive and resilient as well as more ecologically efficient. The realities of climate change and resource constraints are showing that cities can no longer adopt business-as-usual practices but need to urgently improve their ecological efficiency. That is, African and Asian cities have to act urgently and immediately to close the gap between ecological, social and economic efficiencies. In particular, municipal governments and urban authorities should mainstream policies that are sensitive to energy saving, the creation of fewer sprawls and more climate-resilient infrastructure, especially with respect to drainage, flood control, housing, sanitation, transport and water distribution systems.

While rapid urbanization brings with it many issues, it provides a not-to-be-missed opportunity to pay greater attention to sustainable urbanization. Instead of the conventional 'grow first, clean later' approach to development, cities need to assiduously pursue greener growth and arrive at a green economy where economic prosperity occurs in tandem with ecological sustainability.

Various Asian and African cities have begun this system change towards greater sustainability, from China to Cambodia, Ethiopia, India, Kazakhstan, Kenya, Korea, Malaysia, Rwanda and Singapore. Numerous measures are being used to promote 'green' forms of urban development, including regulation, green building, research (for example, into renewable energy), business and community energy-saving and emission-reduction programmes. Like their counterparts elsewhere, Asian cities are redesigning and experimenting with new 'low carbon' or 'eco' cities. It is important that these be carefully considered and integrated as part of a sustainable development continuum, instead of as isolated phenomena or, worse, as 'fantasy cities' that are disconnected from the everyday needs of the majority of the urban population.

Just as there is no single urban transition, there is no single measure or pathway to sustainable futures. What is at stake is not a simple choice between different approaches to a low carbon future but rather the need to recognize that the transition is a complex and negotiated process and to accept some levels of failure and the opportunity to learn and reinvent our ways towards greater sustainability.

Obviously, Africa and Asia have different political histories and cultural differences in the manner in which both societies are organized and developed. These differences account for some of the major differences in the implementation of urban development policy and the resolution of problems. However, they also have common urban development challenges as discussed above, and this calls for common policy points. At the economic and social levels, to effectively address these challenges, African and Asian cities need to focus on pro-poor urban growth and greener growth strategy that will reduce poverty and, at the same time, create jobs and improve living conditions for the majority of their urban dwellers.

At the physical and spatial levels, African and Asian cities need to invest in infrastructure (including green energy and green transport) that addresses the infrastructure gaps and growth needs. Given the path dependency of infrastructure and spatial locations, there are gains to be made in exploring how such investments could transition out of carbon dependence to focus more on the processes of unlocking new sustainable growth.

Needless to say, there is no 'one size fits all' prescription for implementing sustainable growth. Much depends on the institutional and policy settings and the level of development and resources as well as the specific challenges and opportunities, present and future. The absence of coherent strategies to deal with the urbanization dynamics can impede growth, or worse, exacerbate systemic economic inefficiencies and social and environmental risks.

What is needed is an actionable policy framework that is well designed to mutually reinforce economic growth and the conservation of resources as well as flexible enough to take account of changing urban circumstances and differing stages of development. It will call for policies that promote sustainable infrastructure and efficient sustainable management of natural assets and build the resilience of livelihoods.

It is increasingly apparent that such effort, to succeed, will need to build on a high degree of inter- and intra-collaboration among and between the different urban actors—the layers of government, the private sector, the non-government agencies, and the people—to identify a policy mix that is suitable to local conditions and needs. In many cases, developing appropriate institutional capacity (including vision, the capability to formulate the right agenda, work attitude improvement, coordination of functions, transparency and accountability, eradication of corruption, and improvement in skills and knowledge) will be an essential precondition. Local administrative capacity is generally weak and fragmented in many African and Asian cities.

The challenge is to build on successful experiences already developed in Africa, Asia and elsewhere. Cities would need to identify, learn from and further develop such experiences to catalyse more efficient land, infrastructure, housing and urban development.

Important to these innovation journeys is not just the practical exchange of ideas and experiences between cities but also effective local governance to develop and implement well-targeted policies and investments.

As the United Nations Economic and Social Commission for Asia and the Pacific (2012) observed, the greening of the economy will not happen automatically through the marketplace; governments must lead the systematic transition. This is not to ignore the fact that a low carbon future will not happen without the involvement of the entire society—businesses and communities. Many African and Asian countries are beginning to recognize that policies on green growth need to shift from a purely environmental concern to sustainable development, identifying the synergies with economic, environmental and social sustainability. Promoting greener growth must be an integral part of development processes at both the national and local levels.

References

ADB [Asian Development Bank], 2004: *City Development Strategies to Reduce Poverty* (Manila: Asian Development Bank).

ADB [Asian Development Bank], 2013: *Economics of Climate Change in East Asia* (Manila: Asian Development Bank).

AfDB [African Development Bank], 2012a: *Briefing Notes for AfDB's Long-term Strategy, Briefing Note 4: Africa's Demographic Trends* (Tunisia: African Development Bank).

AfDB [African Development Bank], 2012b: *Facilitating Green Growth in Africa: Perspective from African Development Bank, Discussion Paper* (Tunisia: AfDB).

Anas, Alex; Timilsina, Govinda R., 2009: *Lock-in Effects of Road Expansion on CO2 Emissions: Results from a Core-periphery Model of Beijing, Policy Research Working Paper No. 5017* (Washington, DC: The World Bank).

Angel, Shlomo, 2012: *Planet of Cities* (Cambridge, Massachusetts: Lincoln Institute of Land Policy).

Baeumler, Axel; Ijjasz-Vasquez, Ede; Mehndiratta, Shomik (Eds.), 2012: *Sustainable Low-carbon City Development in China* (Washington, DC: The World Bank).

Barrios, Salvador; Bertinelli, Luisito; Strobl, Eric, 2006: "Climate Change and Rural–Urban Migration: The Case of Sub-Saharan Africa", in: *Journal of Urban Economics*, 60,3: 357-371.

Bridge, Gavin; Bouzarovski, Stefan; Bradshaw, Michael; Eyre, Nick, 2013: "Geographies of Energy Transition: Space, Place and the Low Carbon Economy", in: *Energy Policy*, 53: 331-340.

Bulkeley, Harriet; Broto, Vanesa Castan; Hodson, Mike; Marvin, Simon (Eds.), 2011: *Cities and Low Carbon Transitions* (Abingdon: Routledge).

Chinese Society for Urban Studies (Ed.), 2011: *China Low-carbon Eco-city Development Report 2011* (Beijing: China Building Industry Press).

Corvellec, Herve; Campos, Maria Jose Zapata; Zapata, Patrik, 2013: "Infrastructures, Lock-in and Sustainable Urban Development: The Case of Waste Incineration in the Goteborg Metropolitan Area", in: *Journal of Cleaner Production*, 50,1: 32-39.

Crutzen, Paul, 2004: "New Directions: The Growing Urban Heat and Pollution Island Effect: Impact on Chemistry and Climate", in: *Atmospheric Environment*, 38: 3539-3540.

Einsiedel, Nathaniel von, 2004: "Strengthening Local Government Capacities in Urban Planning and Finance: Lessons from the East Asia City Development Strategies", in: Freire, Mila; Yuen, Belinda (Eds) *Enhancing Urban Management in East Asia* (Aldershot: Ashgate): 159-170.

Fay, Marianne; Opal, Charlotte, 2000: *Urbanisation without Growth: A Not So Uncommon Phenomenon, World Bank, Policy Research Working Paper No. 2412* (Washington, DC: World Bank).

Fox, Sean; Hoelscher, Kristian, 2012: "Political Order, Development and Social Violence", in: *Journal of Peace Research*, 49,3: 431-444.

Government of India, Ministry of Urban Development, 2011: *Strategic Plan of Ministry of Urban Development for 2011-2016* (New Delhi: Ministry of Urban Development).

Hallegatte, Stephane; Heal, Geoffrey; Fay, Marianne; Treguer, David, 2011: *From Growth to Green Growth: A Framework, Policy Research Working Paper 5872* (Washington, DC: The World Bank).

IEA [International Energy Agency], 2010: *World Energy Outlook, 2010* (Paris: OECD); at: <http://www.world energyoutlook.org/media/weo2010.pdf> (2 March 2014).

IPCC [Intergovernmental Panel on Climate Change], 2014: *Fifth Assessment Report*, Working Group II (Geneva: IPCC).

Kamal-Chaoui, Lamia; Grazi, Fabio; Joo, Jongwan; Plouin, Marissa, 2011: "The Implementation of the Korean Green Growth Strategy in Urban Areas", in: *OECD Regional Development Working Papers 2011/02* (Paris: OECD).

Kikusawa, Ikuyo, 2010: "Case Study on Building Zero Energy Development Communities to Mainstream Sustainability ... T-Zed Homes in India", in: *Asia Pacific Forum for Environment and Development 2008 Ryutaro Hashimoto Award Winning Project* (Kanagawa: Asia Pacific Forum for Environment and Development).

Kim, Eugene, 2010: "S Korea sees 'Green Vision' as Growth Engine", in: *Xinhua English News*; at: <http://news.xinhuanet.com/english2010/business/2010-01/29/c_1315 5950.htm> (19 July 2014).

Marcacci, Silvio, 2013: "South Korea may Launch World's Most Ambitious Cap and Trade Market", in: *CleanTechnica*; at: <http://cleantechnica.com/2013/05/18/south-korea-may-launch-worlds-most-ambitious-cap-and-trade-system/> (19 July 2014).

Martin, George, 2005: "The Global Diffusion of Motorised Urban Sprawl: Implications for China and Shanghai", in: Feng, Huan; Yu, Lizhong; Solecki, William (Eds.): *Urban Dimensions of Environmental Change* (Monmouth Junction, NJ: Science Press USA): 122–129.

McGee, Terry, 2009: *The Spatiality of Urbanisation: The Policy Challenges of Mega-urban and Desakota Regions of Southeast Asia, UNU-IAS Working Paper* 161 (Yokohama: UNU-IAS).

McKinsey Global Institute, 2009: *Preparing for China's Urban Billion* (McKinsey & Co.).

McKinsey Global Institute, 2013: *Infrastructure Productivity: How to Save $1 Trillion a Year* (McKinsey & Co.).

Ministry of Environment, South Korea, n.d.: *Practicing Green Lifestyle Korea*; at: <http://eng.me.go.kr/eng/web/index.do?menuId=169> (1 October 2012).

Obeng-Odoom, Franklin, 2011: "Special Issue of African Review of Economics and Finance Editorial: Urbanity, Urbanism, and Urbanization in Africa", in: *African Review of Economics and Finance*, 3,1: 1–7.

Ostojic, Dejan R.; Bose, Ranjan K.; Krambeck, Holly; Lim, Jeanette; Zhang, Yabei, 2013: *Energizing Green Cities in Southeast Asia* (Washington, DC: The World Bank).

People's Daily, 2010: "China Launches Low-carbon Pilot in Select Cities, Provinces", in: *People's Daily*, 19 August 2010; at: <http://english.peopledaily.com.cn/ 90001/90778/90862/7110049.html> (18 July 2014).

Roberts, Brian; Kanaley, Trevor (Eds.), 2006: *Urbanisation and Sustainability in Asia* (Manila: Asian Development Bank).

Suzuki, Hiroaki; Cervero, Robert; Iuchi, Kanako, 2013: *Transforming Cities with Transit: Transit and Land-use Integration for Sustainable Urban Development* (Washington, DC: The World Bank).

Tahilyani, Naveen; Tamhane, Toshan; Tan, Jessica, 2011: *Asia's $1 Trillion Infrastructure Opportunity* (McKinsey & Co.).

The World Bank, 2010: *Winds of Change: East Asia's Sustainable Energy Future* (Washington, DC: The World Bank).

UNECA [United Nations Economic Commission for Africa], 2009: *Workshop on Development Finance Institutions Support to Infrastructure Development: Concept Note* (Addis Ababa: UNECA).

UNESCAP [United Nations Economic and Social Commission for Asia and the Pacific], 2013: *Statistical Yearbook for Asia and the Pacific* 2013 (Bangkok: UNESCAP).

UNESCAP [United Nations Economic and Social Commission for Asia and the Pacific], 2012: *Low Carbon Green Growth Roadmap for Asia and the Pacific* (Bangkok: UNESCAP).

UN-HABITAT, 2004: *The State of the World's Cities 2004/05* (Nairobi: UN-HABITAT).

UN-HABITAT, 2008: *The State of African Cities 2008: A Framework for Addressing Urban Challenges in Africa* (Nairobi: UN-HABITAT).

UN-HABITAT, 2009: *Planning Sustainable Cities: Global Report on Human Settlements* (Nairobi: UN-HABITAT).

UN-HABITAT, 2010a: *The State of Asian Cities 2010/11* (Fukuoka: UN-HABITAT).

UN-HABITAT, 2010b: The State of African Cities 2010: Governance, Inequality and Urban Land Markets (Nairobi: UN-HABITAT).

UN-HABITAT, 2011: *Cities and Climate Change: Global Report on Human Settlements 2011* (London: Earthscan).

UN-HABITAT, 2014: *The Sate of African Cities 2014: Reimagining Sustainable Urban Transitions* (Nairobi: UN-HABITAT).

United Nations Environment Programme, 2013: *China's Green Long March* (New York: United Nations Environment Programme).

United Nations Population Fund (UNFPA), 2007: *State of World Population: Unleashing the Potential of Urban Growth* (New York: UNFPA).

United Nations, 2001: *Living Arrangements of Older Persons, Population Bulletin of the United Nations Special Issue 42/43* (New York: United Nations).

United Nations, 2004: *World Population to 2300* (New York: United Nations Population Division, Department of Economic and Social Affairs).

United Nations, 2012: *World Urbanization Prospects: The 2011 Revision* (New York: United Nations Population Division, Department of Economic and Social Affairs).

United Nations, 2013a: *World Urbanisation Prospects: The 2012 Revision* (New York: United Nations Population Division, Department of Economic and Social Affairs).

United Nations, 2013b: *World Population Ageing 2013* (New York: United Nations Population Division, Department of Economic and Social Affairs).

United Nations, 2014: *World Urbanisation Prospects: Highlights* (New York: United Nations Department of Economic and Social Affairs).

US Energy Information Administration (EIA), 2013: *India* (Washington DC: EIA).

Wan, Guanghua; Sebastian, Iva, 2011: *Poverty in Asia and the Pacific: An Update, ADB Economics Working Paper No. 267* (Manila: Asian Development Bank).

Wen, Hong; Chiang, Madelaine Steller; Shapiro, Ruth A.; Clifford, Mark L., 2007: *Building Energy Efficiency: Why Green Buildings are Key to Asia's Future* (Hong Kong: The Asia Business Council).

World Review (n.d): *Africa's Population Boom*; at: <http://www.worldreview.info/content/africas-population-boom> (25 February 2014).

Wu, Libo, 2012: *Green Jobs in China: Comparative Analysis, Potentials and Prospects* (Washington DC: Friedrich Ebert Stiftung).

Yeung, Yue-mun, 2011: "Rethinking Asian Cities and Urbanization: Four Transformations in Four Decades", in: *Asian Geographer*, 28,1: 65-83.

Yuen, Belinda, 2009: "Regional Study on East Asia, Southeast Asia and the Pacific", Technical Background Paper for *UN-HABITAT Global Report on Human Settlements* 2009 (Nairobi: UN-HABITAT).

Yuen, Belinda, 2013: "Urban Regeneration in Asia: Mega Projects and Heritage Conservation", in: Leary, Michael E.; McCarthy, John (Eds.) *The Routledge Companion to Urban Regeneration* (London: Routledge): 127-137.

30 The Role of University Partnerships in Urban Sustainability Experiments: Evidence from Asia

Gregory Trencher[1] and Xuemei Bai[2]

Abstract

University-driven partnerships and experiments for advancing urban sustainability are flourishing around the world. Responding to drivers such as calls for stakeholder engagement in research, tangible social and economic contributions, and government funding incentives, Asian research universities are also forming cross-sector partnerships and implementing various sociotechnical experiments. In this chapter we examine the role of university partnerships in knowledge co-production and implementation of urban sustainability experiments in industrialized Asian nations. By examining fifteen cases from Singapore, Japan, Hong Kong and Korea, we highlight common attributes (focus areas, actors, motivations and mechanisms) and then investigate the functions, motivations, barriers and significance of roles assumed by differing societal sectors. A detailed case study of an ambitious project from the University of Tokyo then follows to further illustrate these attributes in context.

Key findings are that, overall, university partnerships for urban sustainability in our Asian sample are dominated by technical approaches. Yet the most significant barriers are human aspects such as time restraints, lack of unity, and poor management and leadership, to name several. On key drivers, government funding is playing a major role in enticing partnership formation and influencing particular approaches to urban sustainability. Measures are required to encourage the participation of the social sciences and humanities, and non-technical sustainability experiments. Case study evidence suggests that the ability of partnerships to tackle complex social issues and trigger societal transitions towards sustainability is often constrained by existing research projects and the institutional capacities of universities and their partners.

Keywords: Asia, urban sustainability experiment, transition, transformation, partnership, university, co-design, co-production.

30.1 Background

Sustainability challenges facing cities in the Asian region are significant. They include rising urban populations, associated air and water pollution and other environmental degradation, social inequalities, food and energy security, a growing demand for resources and an expanding carbon footprint (Bai/Imura 2000; Bai/Shi/Liu 2014; Brown 2011). On the other hand, rapid urbanization and economic growth has enhanced the ability to address many of these sustainability problems through technological advancement and improved environmental governance. Concentration of large-scale, dense urban areas and a general cultural enthusiasm for new technologies make industrialized and rapidly industrializing Asia an ideal arena for 'sustainability experiments'.

Experimentation is considered crucial for advancing sustainability (Bai/Roberts/Chen 2010; Konig/Evans 2013). Sustainability experiments are defined as "planned initiatives that embody a highly novel sociotechnical configuration, which typically involve significant learning and are likely to lead to substantial (environmental) sustainability gains" (Berkhout/Verbong/Wieczorek et al. 2010: 262). They are typically

---

1 Ass. Prof. Dr. Gregory Trencher, Visiting Associate Professor, Clark University: Department of International Development, Community, and Environment; email: <gtrencher@clarku.edu>.

2 Prof. Dr. Xuemei Bai, Professor, Australian National University: Fenner School of Environment and Society; email: <xuemei.bai@anu.edu.au>.

characterized by the introduction of a new technology, policy or social arrangement on a small scale; a radical and highly exploratory nature; and involvement of various actors from industry, government, civic society and academia (Bai/Wieczore/Kaneko et al. 2009; Brown/Vergragt/Green et al. 2003). Literature from transitions theory describes the significance of such experiments as, firstly, their ability to function as 'niches' or pockets of protection for emerging forms of social or technical innovation from more established market or societal ways of doing things (i.e. the 'regime') (Geels/Schot 2010), and secondly, their value lies in the potential to trigger wider societal transitions towards sustainability when up-scaled and exported (Bai/Roberts/Chen 2010).

A second body of literature explores the idea of the urban environment functioning as a 'living laboratory', typically focusing on the role of university actors and scientific research (Evans/Karvonen 2011, 2014). However, this topic is not well explored in the sustainability experiments literature.

Cutting-edge experimentation is also prominent in the green campus movement (M'Gonigle/Stark 2009; Konig 2013). Over the past two decades, many universities have taken impressive measures in attempts to become models of urban sustainability for the broader community (ISCN 2014). Tapping into their powerful financial resources and technical expertise, universities are also functioning as test grounds for emerging technologies such as renewable energy production and *electric vehicle* (EV) transport systems, innovative energy, water and waste saving strategies, building design, and campus management. Such on-campus experimentations generate research opportunities and serve as a demonstration site that can entice broader adoption by the society.

A plethora of initiatives emerging around the globe (Trencher/Yarime/Kharazzi 2013; Trencher/Yarime/McCormick et al. 2014; Trencher/Bai/Evans et al. 2014; Zilahy/Huisingh 2009) suggests that university efforts are increasingly extending off-campus into surrounding towns, cities and regions to become key drivers of urban sustainability experiments. Collaboration and partnerships is a defining feature of this trend, which has been recently examined by a number of case studies (Evans/Karvonen 2011, 2014; Liedtke/Welfens/Rohn et al. 2012; Molnar/Ritz/Heller et al. 2011). Several factors are contributing to the rise of university initiatives to engage in research co-production and co-implementation with external stakeholders. There are growing government demands for intensified university-industry relations and outputs

with economic significance such as intellectual property, marketable inventions, spin-off firms and capacity-building for industry and government (Yusuf 2007). The governments of Singapore and Japan, for example, have created generous funding opportunities to encourage cross-sector innovation in areas such as climate change, energy and transport. City governments around the world are also showing increasing willingness to engage with other societal actors in experimental governance and policy initiatives to tackle climate change (Castan Broto/Bulkeley 2013). In addition, there is an increasing awareness within and outside the scientific community that greater innovation can result from collaboration and open networks than in traditional laboratories and closed models of R&D (Shrum/Genuth/Chompalov 2007). Further, there is a growing consensus that the scientific community should better respond to societal needs through the co-design and co-production of knowledge (Lubchenco 1998; Future Earth 2013; Mauser/Klepper/Rice et al. 2013). In the literature, this transition is often described as a shift from 'mode 1' to 'mode 2' type knowledge production, where knowledge is increasingly produced in application and in response to stakeholder needs (Nowotny/Scott/Gibbons et al. 2001).

Of note, literature examining university roles in innovation and experimentation is typically focused on North American or European cases. Very little is known on the characteristics of Asian-based experiments and to what extent cross-sector partnerships are part of such initiatives. This is despite the increasing interests and tendencies of Asian universities to engage with stakeholder networks, building cross-sector R&D platforms, and conduct various forms of technical and social experiments in urban areas.

In this chapter, we explore the role of the university in urban sustainability partnerships and experiments[3] in the context of industrialized (i.e. high-income) Asian nations. With an overall focus on environmental dimensions of urban sustainability, our study is guided by several questions:

1. What are the defining attributes (i.e. focus area, actors, motivations and mechanisms) and com-

---

3   In this study we employ the term 'partnership' to signify the broader actor network serving as an umbrella for a particular implementation project or societal intervention (i.e. the experiment). That is to say, a single partnership may consist of numerous sustainability experiments.

Box 30.1: Method. **Source:** The authors.

The fifteen partnerships and data used in section 30.2 come from a larger (n=70) global survey (Trencher/Bai/Evans et al. 2014) that attempted to identify as many instances as possible of university-driven partnerships for urban sustainability from industrialized (high-income) nations in Europe, Asia and North America. Partnerships were located over the period June 2011 to November 2013 by a combination of i) Internet keyword searches including government funding agency databases, ii) reviews of academic literature and press articles, and iii) referrals from peers and information requests to university actors.

Data analysed in section 30.2 stem from two questionnaires. The first gathered quantitative data describing key partnership characteristics and was administered electronically to one person (in most cases the project leader) from each case. A second questionnaire was then administered to gather both qualitative and quantitative data on key barriers encountered and measures taken to overcome these. This was despatched to several actors in the various societal sectors involved (i.e. government, industry and civil so-

ciety). This led to a total of twenty-six responses from thirteen of the fifteen cases. Both questionnaires asked respondents to assign a numerical score to indicate the relevance of a particular variable (i.e. type of actor, motivation or barrier etc.) to the partnership in question. The first questionnaire on key characteristics offered the following options for each variable: 0 (not at all relevant), 1 (partly relevant), or 2 (extremely relevant). The second survey offered the following scores: 0 (not at all significant), 1 (mildly significant), 2 (very significant), and 3 (extremely significant).

In terms of data calculation, individual scores from each partnership for each variable were tallied and then divided by the total amount of scores possible (i.e. 2 x n for the first questionnaire and 3 x n for the second). This has resulted in a percentage score (observable in all figures in section 30.2) signifying the importance of a particular variable relative to others, from the overall perspective of all partnerships combined.

---

monly encountered barriers in partnership activities and experiments?

2. What are the functions, motivations, barriers and significance of roles assumed by differing societal sectors?

3. What lessons and implications for sustainability transitions research may be drawn from Asian university attempts to co-design and co-produce knowledge and solutions for urban sustainability?

Our study draws some elements (data in section 30.2) from an earlier global study (Trencher/Bai/Evans et al. 2014). Yet this previous work is furthered by focusing specifically on the region of Asia—and in particular the differing roles, motivations and barriers of various societal actors, in addition to the impacts of government policies and funding incentives. Such information is drawn from structured interviews, analysis of survey results, and university and governmental documents, as well as first-hand experience from participating in such partnerships.

The empirical backbone of this study consists of fifteen university partnerships from Singapore, Japan, Hong Kong and Korea (respondents to our survey). All of these nations and regions are characterized on the one hand by a high level of fossil energy imports, per capita *greenhouse gas* (GHG) emissions, dense and mostly urban populations, and limited natural resources such as land, and on the other hand, advanced innovation bases, highly developed university sectors and knowledge-driven economies. Japan is

also facing the dual and interrelated concerns of rapid population ageing and potential socio-economic decline as a result.

Our chapter is structured as follows. Firstly we analyse survey results from the fifteen partnerships to define key trends and barriers in the Asian sample. In the next section we then examine the core roles, motivations and challenges of each societal sector engaging with university attempts to advance urban sustainability. Thirdly, in a detailed case study we examine an attempt by the University of Tokyo to address emerging challenges of climate change and population ageing through research and societal experiments. This project aims to lay a blueprint for a low carbon and elderly-friendly reform of the neighbouring City of Kashiwa. Finally, we reflect on previous sections and conclude with challenges, lessons and implications for sustainability transitions scholarship on Asia.

30.2 Characterizing University Partnerships for Urban Sustainability in Asia

In this section we offer a brief macro-analysis of fifteen university-driven partnerships for advancing urban sustainability in industrialized Asia (listed in table 30.1). Partnerships are scrutinized firstly in terms of urban systems targeted, external and internal actors, and motivational factors and mechanisms, and secondly, of commonly encountered barriers. The

Table 30.1: List of partnerships (n=15). **Source:** The authors.

Name	Lead institution	Target area	Summary	Core disciplines	Imple-mentation	Online resources
CUHK Jockey Club Initiative Gaia	City University of Hong Kong	HONG KONG: Various communities	Initiative consisting of three components: 1) art exhibition showcasing CUHK's research in environment, energy and sustainability; 2) carbon scheme aimed at schools and NGOs to pursue energy efficiency and carbon reductions; and 3) public education and awareness-raising.	• Environment, energy and sustainability • Geography and resource management • Earth system science • Space and earth information science • Architecture	2012–2018 Status: ongoing	Main site: <http://www.gaia.cuhk.edu.hk> Key documents: <http://www.gaia.cuhk.edu.hk/index.php?option=com_content&view=article&id=186&Itemid=85&lang=en>
DHI-NTU Research Centre	Nanyang Technological University	SINGA-PORE: Nationwide	R&D platform to generate new knowledge and strengthen the water and environment industry in Singapore via the development of innovative technologies and training of water and environment professionals.	• Engineering (environmental and civil) • Urban planning	2007–2016 Status: ongoing	Main site: <http://www.dhi-ntu.com.sg> Publications: <http://www.dhi-ntu.com.sg/publications/scientific-publications>
(E2S2) Energy and Environmental Sustainability Solutions for Megacities	Shanghai Jiao Tong University National University of Singapore	CHINA: Shanghai SINGA-PORE: Central Singapore	R&D platform to improve energy recovery from waste and develop system modelling and data management tools to track and mitigate emerging environmental contaminants. Dual test beds set up in several locations across Shanghai and Singapore.	• Engineering (chemical and biomolecular) • Environmental health	2012–2017 Status: ongoing	Main site: <http://www.nus.edu.sg/neri/E2S2.html> Publications: None listed
Hong Kong SME Business Sustainability Index	Hong Kong Polytechnic University	HONG KONG: Nationwide	Platform to promote the understanding and adoption of CSR as a business model to foster sustainability practices of business sector in Hong Kong and encourage reporting of sustainability activities.	• Business management Marketing	2011–n/a Status: ongoing	Main site: None available Publications: None available
Green Society ICT Life Infrastructure	Keio University	JAPAN: Okutama (Tokyo) and Kuribara City (Miyagi-ken)	R&D and testing platform to contribute to the resiliency and sustainability of two semi-rural communities. Involves development of ICT system to boost home energy efficiency and measure climate change impacts on health and agriculture.	• Engineering (information) • Community studies • Social innovation • Geographic information science • Clinical plant science • Agriculture	2010–2015 Status: ongoing	Main site: <http://www.green-lifeinfra.com> (Japanese) Publications: <http://www.green-lifeinfra.com/report/report01.html> (English and Japanese)

Name	Lead institution	Target area	Summary	Core disciplines	Imple-mentation	Online resources
Infrastructure Supporting Project for Wind Power Generation Business in Jeju Region	Jeju National University	KOREA: Jeju Island	R&D effort to drive the development of the wind power industry on Jeju Island, creating jobs, boosting the local economy and building a sustainable energy base.	• Engineering (mechaical, energy)	2004–n/a Status: ongoing	Main site: None available Publications: None available
Low Voltage Direct Current Grid Network	Nanyang Technological University	SINGAPORE: Jalan Bahar (CleanTech One)	Smart grid experiment to use JTC green cluster zone 'CleanTech Park' as a testbed for lighting and smart grids. Direct Current (DC) is used to minimize energy losses from the renewable sources.	• Engineering (electrical)	2010–n/a Status: ongoing	Main site: None available Publications: None available
NUS-JTC Industrial Infrastructure Innovation (NUS-JTC I3) Centre	National University of Singapore	SINGAPORE: Nationwide	R&D and demonstration effort to drive innovation and sustainable development in various areas of industrial zone planning and construction. Focus on solutions to ensure efficient use of space, materials and energy in industrial real estate market.	• Urban planning and design • Real estate • Engineering (mechanical, construction, electrical)	2011–2016 Status: ongoing	Main site: <http://www.sde.nus.edu.sg/nus-jtc> Key documents: <http://www.sde.nus.edu.sg/nus-jtc/news_n_events/PressReleases.htm>
Sustainable Supply Chain Centre Asia Pacific	Singapore National University	1. SINGAPORE 2. Asia-Pacific	Responding to predicted growth of trade and commerce in Asia, collaboration to develop the knowledge and business tools to diffuse green logistics and supply chain innovation.	• Engineering (industrial)	2010–2013 Status: complete	Main site: None available Publications: None available
Sustainable Urban Waste Management for 2020	Nanyang Technological University	SINGAPORE: Western Singapore	R&D and demonstration programme to develop sustainable urban waste management solutions for Singapore based on a decentralized 'waste to resources' concept.	• Waste management • Resource management • Engineering (chemical)	2010–2015 Status: ongoing	Main site: None available Publications: None available
Triple Water Supply (TWS) System	Hong Kong University of Science and Technology	HONG KONG: Tung Chung and Sha Tin	On-going R&D, demonstration and implementation platform to utilize Hong Kong's citywide seawater flushing system to develop energy-efficient and low carbon sewage treatment technologies.	• Engineering (environmental, civil)	2004–n/a Status: ongoing	Main site: None available Publications: None available

Name	Lead institution	Target area	Summary	Core disciplines	Imple-mentation	Online resources
TUM-Create	Technical University of Munich, Nanyang Technological University	SINGA-PORE: Nationwide	Large-scale R&D platform with focus on developing an electric taxi for Singapore, with potential for application in other tropical megacities. Collaboration involves all levels of EV taxi transport: from batteries to car design, extending to citywide infrastructure and traffic control systems.	• Engineering (mechanical, electrical, chemistry, information) • Industrial design	2011–n/a Status: ongoing	Main site: <http://www.tum-create.edu.sg> Publications: <http://www.tum-create.edu.sg/publications2/all>
Underwater Infrastructure and Underwater City of the Future	Nanyang Technological University	SINGA-PORE: Nationwide	R&D and demonstration project to utilize underwater sea space to construct infrastructures such as oil storage facilities or power stations whilst using the topside as reclaimed land.	• Engineering (civil, structural, environmental)	2010–2015 Status: ongoing	Main site: None available Publications: None available
Urban Design Centre Kashiwan-oha	University of Tokyo	JAPAN: Chiba-ken, Kashiwa City	Knowledge exchange, education and research platform addressing issues related to environmental, sociopolitical and urban planning issues in greater Kashiwa City. Brings together academics, citizens, local city authorities and real estate developers.	• Urban planning and design	2006–n/a Status: ongoing	Main site: <http://www.udck.jp/en/> Publications: None available
Urban Reformation Programme for the Realization of a Bright Low Carbon Society	University of Tokyo	JAPAN: Chiba-ken, Kashiwa City	Large-scale applied research initiative to design blueprint for transition to low carbon, elderly-friendly community. Involves extensive demonstrations with technical and social innovation.	• Engineering (mechanical, electrical, information, civil) • Urban planning • Clincal plant science • Agriculture • Sociocultural environmental studies	2010–2015 Status: ongoing	Main site: <http://low-carbon.k.u-tokyo.ac.jp/en> Publications: <http://low-carbon.k.u-tokyo.ac.jp/news_14_05_08.html>

sample consists of eight cases from Singapore, three from Japan, three from Hong Kong and one from Korea. Samples are a subset of a global study (see box 30.1). While this is not sufficient to fully represent university initiatives in industrialized Asia, it can nonetheless highlight some key trends and characteristics. It may also inform attempts from other scholars to more exhaustively survey the region.

30.2.1   Sectors Targeted

Results in figure 30.1 demonstrate that *energy and heating/cooling*, *built environment* and *economy,* *employment and industrial production* are the most commonly targeted urban systems in the fifteen cases. The preoccupation with the built environment and energy and heating/cooling appears to be driven by awareness that these areas represent key opportunities to mitigate climate change through reduced GHG emissions and higher energy efficiency. The overall focus on *economy, employment and industrial production* is indicative of strong collaborations with industry. Overall, a moderately strong resolve to influence government planning and policy can be observed from *governance and planning*. As an overall trend it can be seen that partnerships tend to be

Figure 30.1: Urban systems targeted (n=15 cases). **Source:** The authors.

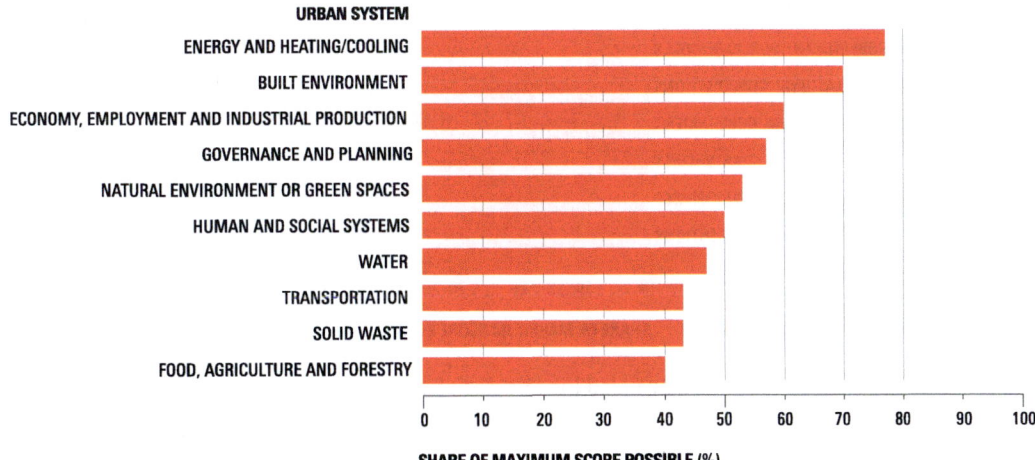

Figure 30.2: Types of internal actors (n=15 cases). **Source:** The authors.

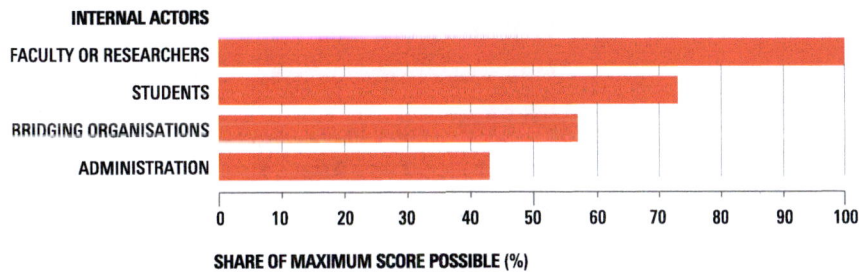

less focused on transforming systems such as *transportation*, *solid waste* and *food, agriculture and forestry*.

Of note, the bulk of partnerships are simultaneously targeting several of the ten urban systems listed in figure 30.1. The choice and quantity of systems targeted appears to be determined by a combination of factors such as the expertise of actors involved, the needs and challenges of the target area, and government funding priorities.

30.2.2   Internal Actors

Figure 30.2 reveals that the main university actors forming, coordinating and implementing urban sustainability partnerships are predominantly *faculty and researchers*. This is followed by postgraduate *students*, usually playing a supporting and secondary role. This demonstrates that sustainability partnerships can generate important educational opportunities. Non-academic actors such as bridging organizations (such as technology transfer offices and community outreach platforms) are also present in several partnerships. This reflects several important supporting roles they

can play in project coordination, establishing external contacts and leveraging university resources. Most partnerships involve more than one of these internal actors.

30.2.3   External Actors

Figure 30.3 below demonstrates that *local or regional government/public services*, *large or multinational corporations* and *state or national government* constitute the most common external partner. The strong links with local or regional government actors may be reflecting the focus on the local or city scale, while the strong overall linkages with corporations appears to reflect the tendency noted above to target the *economy, employment and industrial production* sector. Only about half of the cases involve collaboration between academic institutions (i.e. other universities or research institutes) or the civic sector. This contrasts with findings from North America where civic society actors such as *non-governmental organizations* (NGOs), grassroots organizations and local residents tend to constitute a more frequent external part-

Figure 30.3: Types of external actors (n=15 cases). **Source:** The authors.

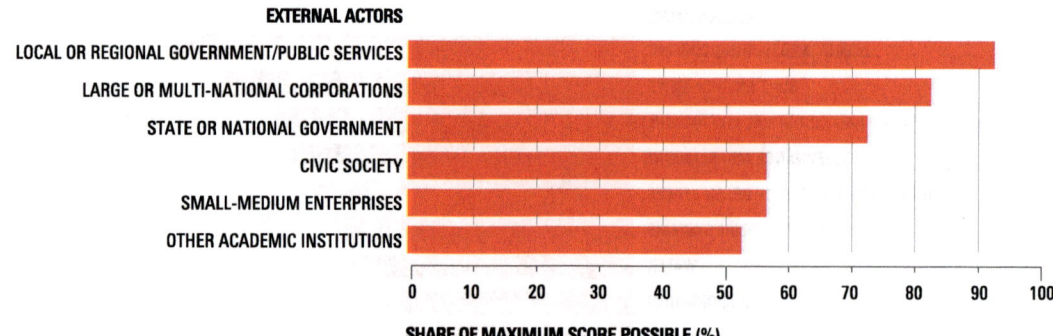

Figure 30.4: Number of external actor types involved (n=15 cases). **Source:** The authors.

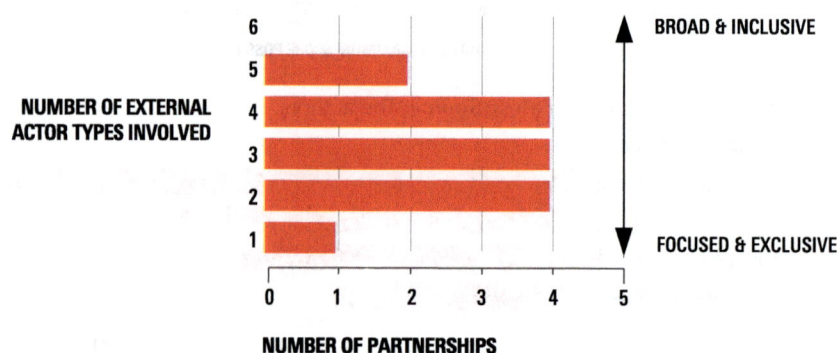

ner for university sustainability initiatives (Trencher/Bai/Evans et al. 2014).

The potential of university partnerships in Asia to integrate multiple societal sectors into the co-design and co-production of urban sustainability is evident in figure 30.4, which shows the number of external actor types from figure 30.3 actively participating in each partnership. It can be seen that the bulk of cases involve two, three, four or even five categories of external actors (specific roles of societal actors are discussed in section 30.3). This structural view suggests that the simplistic analytical lens of 'triple-helix' relations (Etzkowitz/Leydesdorff 2000) between academia, industry and government does not accurately capture the societal inclusiveness of many partnerships.

30.2.4   Motivations and Triggers

Figure 30.5 presents various motivations and triggers behind the university partnership initiatives. Again, for most cases several variables apply simultaneously. Overall, the three most significant motivating factors are *developmental/strategic*, *scientific/scholarly* and *funding*. The developmental/strategic motivation refers to desires to influence development trajectories and transform societal systems in response to external

environmental or socio-economic conditions. With universities described as 'anchor institutions' not having the luxury of being able to relocate (Birch/David/Louis 2013), many university actors are aware that the long-term well-being of their institution is highly dependent on surrounding environmental and socio-economic conditions. The prevalence of the scientific/scholarly motivation shows that virtually all cases were strongly driven by the desire to enhance academic knowledge production by engaging with real-world situations and translating scientific knowledge into tangible and useful outcomes. Lastly, the finding that *funding* also constitutes a crucial motivating factor supports arguments that targeted funding schemes can play a major role in fostering university collaborations for sustainability with external stakeholders (Dedeurwaerdere 2013; Whitmer/Ogden/Lawton et al. 2010).

Interestingly, *entrepreneurial* motivations (i.e. the prospect of generating income for any of the partners involved) least explain the emergence of partnerships for urban sustainability. This is despite earlier noted strong relations with *large or multinational corporations*, and a common focus on *economic, employment and industrial production*. This evidence suggests firstly that the widely documented and

Figure 30.5: Factors motivating partnership formation (n=15 cases). **Source:** The authors.

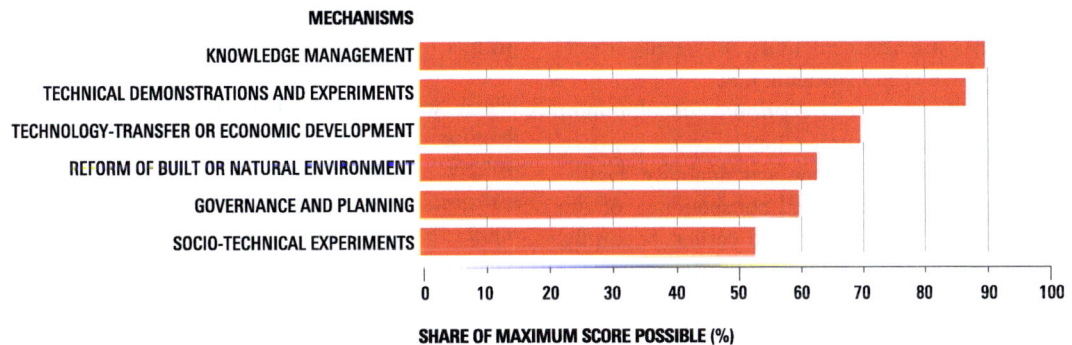

Figure 30.6: Mechanisms (n=15 cases). **Source:** The authors.

promoted entrepreneurial paradigm driving conventional technology transfer practices (Etzkowitz/Webster/Gebhart et al. 2000; Mowery 2007) in other fields such as IT and the life sciences has little relevance for sustainability partnerships in the region. Results from Asia, and other regions around the world (Trencher/Bai/Evans et al. 2014), suggest that other motivational factors (e.g. the desire to influence surrounding development trajectories towards greater sustainability; to enhance research with external knowledge, situations and 'urban laboratories'; and to secure funding as either a means or an end) are pushing university actors to initiate or engage in urban sustainability research and experiments with societal stakeholders.

30.2.5   Mechanisms

Figure 30.6 shows that activities corresponding with *knowledge management* (for example, stakeholder consultations, data production and analysis, publication and communication of results) are the most commonplace mechanism by which university sustainability partnerships seek to co-design and co-produce urban sustainability. This is in agreement with wider trends in Europe and North America (Trencher/Bai/

Evans et al. 2014). Results suggest a prevalence of technical approaches (i.e. *technical demonstrations and experiments*) and relative under exploitation of softer approaches involving social innovation (i.e. *sociotechnical experiments*). This is another defining characteristic that can be retained from our Asian sample.

The overall bias towards technocentric approaches reflects on the one hand a preponderance of R&D-based platforms from faculty and researchers in the fields of engineering and the natural sciences, and on the other government research and funding priorities. In countries such as Singapore, for example, National Science Foundation and government funding programmes have explicitly linked the collaborative pursuit of urban sustainability with areas such as energy- and space-efficient urban development, and disciplines such as the physical sciences and engineering.[4] With technocentric avenues to pursuing urban sustainability shown to be highly limited (Notter/Meyer/Althaus 2013), the integration of social aspects and approaches (i.e. *sociotechnical experiments* and social

4   See URL: <http://www.nrf.gov.sg/research/r-d-ecosystem/research-priorities/physical-sciences-engineering>.

Figure 30.7: Barriers and negative factors (n=13 cases). **Source:** The authors.

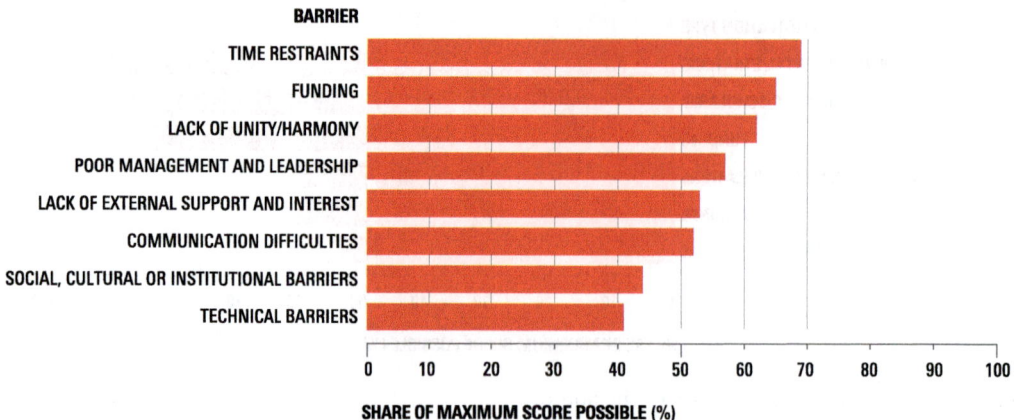

sciences) merits attention in future university efforts in Asia to advance urban sustainability.

Technology transfer and economic development has also emerged as a common means by which partnerships seek to advance urban sustainability. Firstly, this should be seen as a natural consequence of the strong linkages with industry noted earlier. Secondly, this reflects several R&D and demonstration platforms in Singapore fostered through programmes such as the *Campus for Research Excellence and Technological Enterprise* (CREATE) from the National Research Foundation. In such government strategies to enhance national innovation and economic competitiveness, there are explicit expectations for the generation of commercializable intellectual property and economic outcomes such as ventures, employment and technology transfer to industry.[5]

Although not the most significant, mechanisms such as *reform of built or natural environment* and *governance and collaborative planning* are also present. In several cases, interventions on the physical environment (e.g. *Underwater Infrastructure and Underwater City of the Future* and *Low Voltage Direct Current Grid Network*) and activities such as collaborative planning and policy making with government decision-makers, political lobbying and advocacy (e.g. *Urban Design Centre Kashiwanoha* by the University of Tokyo) also represent important avenues by which academic actors seek to translate knowledge into concrete contributions to the advancement of urban sustainability.

30.2.6   Barriers

Figure 30.7 demonstrates that with the exception of *funding*, it is predominantly internal dynamics and project management-related factors such as *time restraints, lack of unity/harmony* and *poor management and leadership* that most commonly hamper partnerships. In connection with the *time restraints* issue, university and societal actors typically engage in cross-sector sustainability initiatives in addition to existing job commitments (Hoover/Harder 2014). The creation of supporting staff and infrastructure such as project offices is vital for addressing these project management-related issues (Hanleybrown/Kania/Kramer et al. 2012). With regard to funding, qualitative responses indicated that the societal impact of university partnerships can be reduced for two reasons. Firstly, funding shortages force actors to divert valuable time from core partnership activities to pursuing funds from other sources, and secondly, decrease the capacity to recruit other necessary partners.

The lack of unity/harmony and poor management and leadership typically impede partnerships when differing values, approaches, goals and expectations can dilute the synergistic effect of cross-sector collaboration. For example, some partnerships cited differing interpretations of timelines and project outcomes between industry and academia as a major challenge, with others suggesting that the cohesiveness and synergy of large-scale sustainability partnerships can deteriorate not only across societal sectors, but also across academic departments and disciplines. With university sustainability partnerships mobilizing numerous partners from differing academic fields and societal sectors, the fostering and maintaining of partnership unity,

---

5   See URL: <http://www.nrf.gov.sg/docs/default-source/new-nrf-factsheets/20131217_create-factsheet-(final).pdf?sfvrsn=2>.

harmony and a common vision and set of expectations is therefore crucial.

Interestingly, despite a preponderance of R&D platforms centred on developing and trialling new and emerging technologies, *technical barriers* have emerged as the least significant, in alignment with other studies examining urban sustainability experiments in Asia (Bai/Roberts/Chen et al. 2010). Evidence from Asia therefore suggests that when collaboratively pursuing urban sustainability, political and human rather than technical barriers are most critical.

30.3 Roles, Motivations and Potential Barriers in Societal Sectors

The following sections summarize potential roles and motivations for each societal sector participating in university partnerships for co-designing and co-producing sustainability knowledge and societal transformations. A summary of this discussion appears in table 30.2.

30.3.1 University

Partnerships examined in this study are mainly from large research-intense universities from various Asian metropolises. The majority originate from engineering and natural sciences, and often from interdisciplinary research institutes with dynamic industry and government linkages, with few including social scientists and other non-engineering disciplines. In all cases, it is university actors who are responsible for the conception, formation and overall direction of each partnership. During formation stages, it is typically senior faculty who apply for research funds, design the overall structure and objectives, and recruit external partners. Coordination and project management roles are typically handled by junior faculty and researchers. Only one project (*CUHK Jockey Club Initiative Gaia*) was formed and coordinated by administration.

Not surprisingly, the chief role for faculty and researchers concerns scientific research. For many partnerships involving R&D in fields such as energy, EVs and *information and communications technology* (ICT) (e.g. *TUM-Create* and *Green Society ICT Life Infrastructure*), such research can be highly fundamental and initially far from the stage of demonstration or application. Nevertheless, such basic and frontier research is vital for innovation. Intense relations with industry and expectations for utility and

concrete outputs to national innovation systems (e.g. technology transfer to industry and venture creation) in several partnerships, especially in Singapore, pose an interesting challenge for the distinct and traditional culture of academic research.

For other partnerships such as *NUS-JTC Industrial Infrastructure Innovation Centre, Sustainable Supply Chain Centre–Asia Pacific* and *Triple Water Supply System*, scientific research can be more mature, applied and driven by ambitions of implementation and diffusion to industry. In such cases, conventional technology transfer channels of patenting, licensing, consulting and personnel transfer to industry (Mowery/Nelson/Sampat et al. 2004) play an important role.

At one point or another, many partnerships serve as testing platforms for unproven technologies or sociotechnical innovation. For example, *Urban Reformation Programme for Realization of Bright Low Carbon Society* involves technical demonstrations of electric mobility solutions for elderly citizens, trials of biofuel production, and attempts to integrate vacant land into citizen led urban agriculture networks. *Low Voltage Direct Current Grid Network* utilizes the technology park (CleanTech One) at Jalan Bahar Singapore as a test bed for emerging low voltage direct current lighting and solar technology. Such demonstrations and pilot projects typically aim to monitor performance and interactions between human and technical systems and assess suitability for, or encourage, wider diffusion. The value of using urban spaces as 'living laboratories' is that such experiments allow a live preview of potential sociotechnical systems for tomorrow and the type of policies and other preconditions required to support, up-scale and export them.

Earlier cited evidence (section 30.2, 'Mechanisms') shows that other common university roles include sharing and communicating research results with societal actors. This can occur through various channels such as publications in academic journals, conference presentations, consulting to industry and government (e.g. *Infrastructure Supporting Project for Wind Power Generation Business in Jeju Region* and *Hong Kong SME Business Sustainability*), and the provision of specialist training to industry and government (e.g. *DHI-NTU Research Centre)* or citizen education and public sector capacity-building efforts (e.g. *CUHK Jockey Club Initiative Gaia*). Evidence from Mowery/Nelson/Sampat et al. (2004) and Philpott/Dooley/O'Reilly et al. (2011) suggests that important transfers of knowledge will also occur through informal meetings and interactions with government and industry partners or stakeholders Other channels for the transferring of research results related to urban sustainability include inputs into government policy and long-term planning. This can also involve the creation and transfer of modelling or visualization tools for aiding government (e.g. *Energy and Environmental Sustainability Solutions for Megacities*) and industry decision-making (e.g. *Sustainable Supply Chain Centre–Asia Pacific*).

30.3.2   Government

Survey results suggest both local or regional government/public services (the most common) and national government participate and play an important role in university partnerships.

Actors from local or regional government are diverse. In Japanese and Korean partnerships, these typically consist of senior members of city planning teams and departments related to transport, industrial development and energy. Short-term roles include the provision of data and urban planning expertise, the provision of land, buildings and test sites for experiments, introduction to other societal stakeholders, public endorsement or promotion of partnership activities, and participation in partnership steering and decision-making.

Participation in cross-sector partnerships for urban sustainability promises much reward for local government. As in Western nations, municipal actors in industrialized Asian countries are faced with various and converging challenges such as pursuing economic development and societal well-being whilst coping with threats such as climate change, ageing and declining populations, energy security, and deterioration of the natural or built environment. With government functions broken up across departments and limitations imposed by existing policies and financial, time and manpower resources, the capacity to single-handedly generate innovative and comprehensive solutions to complex and long-term societal challenges is limited (Hanleybrown/Kania/Kramer et al. 2012; Orr 2013). The opportunity to collaboratively address such problems with universities and other societal experts offers municipalities the precious opportunity to tackle many of these problems simultaneously, and usually with minimal financial contributions. Other ensuing benefits include access to new knowledge and outside best practice, and valuable insights into the type of government policy and planning required to support the collaboratively generated solutions.

Despite potential benefits, several barriers should be expected. These may include tensions between the typically short- to mid-term and implementation-focused priorities of city planners and the long-term and research focus of academics. They may also stem from the radically different cultures of academia and local government where, on one hand, academics are rewarded for risk-taking and innovation, and on the other, government officials are typically encouraged to avoid risks and duplicate established patterns and procedures (Vigoda-Gadot/Shoham/Schwabsky et al. 2005). These differing 'world views' (Trencher 2014: 196) may deter some municipalities from taking a high level of interest in university-driven partnerships and will also influence the ability to assimilate and implement new knowledge and innovation in future policy and planning.

In several cases in Singapore and Hong Kong, participating government actors tend to correspond more with a national level. These include, for instance, officials from national agencies dealing with water supply and drainage, economic and industrial development, and environmental protection. In partnerships such as *Triple Water Supply System* and *DHI-NTU Research Centre,* officials from water-related bodies are assisting with the establishment of pilot plants and testing of emerging technologies (i.e. energy recovery from waste water, energy efficient treatment and recycling, and seawater flushing) in existing government water treatment infrastructures. In partnerships such as *NUS-JTC Industrial Infrastructure Innovation Centre,* actors from national urban planning and engineering divisions play a key role in the design of research agendas, steering and project management. This is in addition to research, marketing and integration of new knowledge into government policy, legislation and construction projects. The value of such contributions is the ability to absorb new knowledge and technologies and contribute to large-scale, rapid societal transformations and 'regime shifts' (Berkhout/Verbong/Wieczorek et al. 2010) through national level policy and legislation.

Motivations for national government actors working in the above contexts appear to be the desire to remain at the forefront of knowledge and innovation, to enhance effectiveness and environmental sustainability of core functions (for example, land planning and water treatment), to accelerate progress towards objectives such as reducing GHG emissions, and to boost national or international competiveness. Barriers to be expected from national level actors concern budgets, political agendas and priorities. Swayed by national political cycles and voters, these largely determine the ability to take up particular technologies and tackle certain societal challenges.

Another important set of government actors operating behind the scenes as 'enablers' in many cross-sector partnerships are national funding agencies. As also described in the triple-helix literature (Etzkowitz/ Leydesdorff 2000), such actors play a crucial role in fostering the formation of cross-sector collaborations by creating institutional conditions and incentives through shaping national research priorities and targeted funding programmes. Singapore is a striking example. One recent measure involved the preparation of a physical infrastructure for university–industry collaboration and sustainability innovation. The National University of Singapore was chosen for construction of the *Campus for Research Excellence and Technological Enterprise* (CREATE)–itself a feat of green building innovation. This *National Research Foundation* (NRF) initiated and financed green technology park is enticing and housing collaborations between prestigious international research universities (e.g. MIT, ETH-Zurich and Munich Technical University), local universities and industry around four themes of human, energy, environmental and urban systems. Fostered in this way are several partnerships such as *TUM-Create* and *Energy and Environmental Sustainability Solutions for Megacities*. Utilitarian motivations behind such generous support are explicit. It is expected that these collaborations generate "positive social and economic outcomes" and lay the knowledge and workforce foundations for a transition to a low carbon, innovation-driven economy in Singapore.[6] Another example is the 'Sustainable Urban Systems' competitive research programme initiated by NRF. This has brought into formation partnerships such as *Sustainable Urban Waste Management for 2020* and *Underwater Infrastructure and Underwater City of the Future*, each allocated SD$ 10 million over five years.

30.3.3   Industry

As noted earlier, large and multinational corporations, rather than *small and medium-sized enterprises* (SMEs) tend to engage with university initiatives for urban sustainability–a trend corresponding with conventional university–industry collaborations in fields

such as life sciences and software engineering. As observed by Laursen and Salter (2004), larger firms with existing capabilities in R&D and open approaches to innovation are most likely to engage in partnerships with universities. Key roles for corporate researchers include participation in research and provision of resources such as company data, expertise and funding. Some cases such as *Sustainable Supply Chain Centre–Asia Pacific* and *Low Voltage Direct Current Grid Network* demonstrate that non-scientific actors from high level corporate management can also become involved, and in the former case, even assume a partnership direction role. As in conventional technology transfer patterns, a chief long-term role for industry in Asian sustainability partnerships is the assimilation of new technologies and knowledge and introduction to the market.

An important motivational factor for firms investing in R&D and commercialization efforts with universities is the prospect of procuring *intellectual property rights* (IPRs) for new inventions and eventually generating income. This entrepreneurial motivation is detectable in several partnerships in Asia, particularly those involving intense R&D collaborations. Another motivation on the industrial side appears to be the prospect of gaining university-won funds from national research bodies. This explains why a plenitude of research funds is so critical in engaging stakeholder support, as reported earlier.

Several factors could potentially impede industry participation in university partnerships for urban sustainability. One would be a limited appreciation of academic topics of enquiry and demands for scientific value and robustness when co-designing projects. Contrary to general belief, industry does not generally place great importance on university research as a source of innovation (Laursen/Salter 2004; Mowery/ Nelson/Sampat et al. 2004). This would particularly be the case for research activities not necessarily connected to near-term income generation. Other potential tensions could arise from differences between academic and industrial decision-making protocols. Industry has a capacity to make rapid and fundamental decisions, with scientific research communities typically driven instead by consensus and slower, deliberative decision-making. Nevertheless, the engagement of large industry is crucial for university attempts to advance urban sustainability through academic research. This is due to expertise in manufacturing, commercializing and marketing, and a unique ability to manifest sustainability impacts on national or inter-

---

6   See URL: <http://www.nrf.gov.sg/docs/default-source/ new-nrf-factsheets/20131217_create-factsheet-(final).pdf?s fvrsn=2>.

national markets though the introduction of new technologies, knowledge or services.

30.3.4   Civic Society

Involvement of civic society in our sample includes NPOs, think tanks, civic associations and individual citizens. Civic society does not feature as prominently in Asian partnerships as in North America (Trencher/Bai/Evans et al. 2014), a core reason being the dominance of R&D platforms. Nonetheless, civil society is playing a key role in several partnerships, both inside and outside Singapore.

As advocated by the Future Earth framework (Mauser/Klepper/Rice et al. 2013), civic society can assume a core role in the co-design of academic research agendas for sustainability by interrogating scientific research questions and activities from the perspective of societal needs and conditions. With less evidence of this occurring in Asia, key roles appear to be more related to implementation and knowledge production activities. Especially for those partnerships targeted at specific locations and sets of community stakeholders, civic society performs functions such as provision of local knowledge (particularly in the three Japanese cases) and co-implementation of various sociotechnical demonstrations and experiments (particularly those requiring human engagement such as *Sustainable Urban Waste Management for 2020, Green Society ICT Life Infrastructure* and *CUHK Jockey Club Initiative Gaia*). The value of citizen participation in scientific demonstrations is that it allows the verification of sociotechnical innovation on living human systems—not only for technical performance, but also for social demand and acceptance. Civic actors can also contribute by mobilizing political and wider community support to either integrate new technologies or social configurations or tackle specific sustainability issues in general. Further, they can also provide academics with valuable connections to other community or political resources and networks.

An important factor motivating civil sector participation in university sustainability partnerships is the prospect of securing capacity-building support for pre-existing commitments and activities. This can entail allocations of university-won funding and enhanced knowledge and learning by engaging with other experts. Other important motivations can include the opportunity to interact with government officials and influence public policy, which might not be possible otherwise.

Barriers that can potentially impede civil society participation include a preponderance of technocentric approaches to urban sustainability, currently dominated by fields of engineering and natural sciences. Without a focus on place-specific human systems and lifestyles, civil society does not form a necessary component for many R&D-based platforms. Other barriers could stem from the undervaluing of time-consuming and uncertain interactions with 'non-experts' from civil society in academic incentive systems (Crow 2011; Yarime/Trencher/Mino et al. 2012). This is particularly relative to exchanges with industry, which can potentially bring economic benefits.

30.4   Case Study: Urban Reformation Programme for the Realization of a Bright Low Carbon Society (University of Tokyo)

To provide a more detailed picture on the particular functions, activities and potential impacts of university-driven sustainability experiments in Asia, this section focuses on a front-runner case from Japan. Data for this case come predominantly from a series of interviews with a total of eight faculty and researchers over the period April 2011 to February 2014, in addition to participation in workshops, steering meetings, conferences and field experiments over the same period .

The *Urban Reformation Programme for the Realization of a Bright Low Carbon Society* (henceforth 'the programme') from the University of Tokyo aims to design the blueprint for a low carbon and elderly-friendly urban transformation of the local City of Kashiwa (30 km from Tokyo) by creating and demonstrating the necessary technologies, and reforming various social systems. The programme is a large-scale sustainability experimentation characterized by a high level of ambition and numerous societal interventions. It brings together various disciplines from the two campuses of Hongo and Kashiwa. The programme has emerged as a dual response to converging challenges of climate change and an ageing society. External partners and stakeholders include local government officials, residents, NPOs, private firms and a corporate think tank. Implementation extends from April 2010 to April 2015, with a total research budget of approximately JP¥ 550 million. This initiative emerged in response to a competitive funding call from the Japan Science and Technology Agency 'Social System Reformation Programme for Adaptation to Climate Change' in 2010.

Table 30.2: Summary of key roles, motivations and potential barriers.

Societal Sector	Common roles	Common motivations	Potential barriers
University	• Conception, design and formation of partnership • Procurement and allocation of funding • Overall direction and project management • Scientific research and implementation of experiments • Technology transfer to industry and government via patenting, licensing, consulting or personnel transfer • Sharing and communication of research results with other actors via publications and conferences • Input to government policy	• Learn from real-world settings, external actors • Increase visibility and impact of research • Influence societal development trajectories • Procure funding (See section 30.2 *Motivations*)	• Academic incentive systems undervaluing time-consuming and uncertain collaborations with societal stakeholders • Poor project management and leadership skills • Funding shortages (See section 30.2 *Barriers*)
Government (National/state)	• Creation of funding and institutional incentives • Assistance with establishment of test beds in government infrastructure • Steering and project management • Co-design of research agendas • Co-production of knowledge • Integration of new knowledge into government policy, legislation and construction projects	• Enhance effectiveness and environmental sustainability of core functions and infrastructure • Boost national or international competiveness through acquisition of new knowledge and innovation • Accelerate progress towards objectives such as reducing GHG emissions etc.	• Budgets, political agendas and priorities (swayed by political cycles and voters)
Government (Local/regional)	• Provision of data, urban planning expertise and resources for experiments such as land, buildings and test sites • Introduction to other societal stakeholders • Public endorsement or promotion of partnership • Partnership steering and decision-making • Integration of results into policy, planning and projects	• Opportunity to tackle multiple societal problems simultaneously with minimal financial contributions (often not possibly with departmentalization) • Access to new knowledge and outside best practices • Insights into type of policies and planning required to support emerging sociotechnical solutions	• Tensions between short to mid-term/implementation-focused priorities of city planners and long-term/research focus of academics • Different cultures in academia and local government (academics rewarded for risk-taking and innovation, government officials often encouraged to avoid risks and duplicate established patterns and procedures)
Industry	• R&D and knowledge co-production • Provision of resources such as data, expertise and funding • Steering and project management • Assimilation of new technologies and knowledge, then introduction to market	• Prospect of acquiring new knowledge via innovation networks • Prospect of procuring intellectual property rights (IPRs) and generating income. • Prospect of gaining university-won funds	• Limited appreciation of academic enquiry and demands for scientific value and robustness in projects • Differences between academic and industrial decision-making protocols.

Table 30.2: Summary of key roles, motivations and potential barriers.

Societal Sector	Common roles	Common motivations	Potential barriers
Civic society	• Co-design of research agendas • Provision of local knowledge • Co-production of knowledge • Co-implementation of demonstrations and experiments • Mobilization of political and wider community support • Provision of connections to other community or political resources and networks	• Prospect of securing capacity-building support (university-won funding and knowledge from other experts) • Opportunity to interact with government officials and influence public policy	• Preponderance of technocentric approaches to urban sustainability (not requiring civil society participation) • Undervaluing of interactions with 'non-experts' from civil society in academic incentive systems (in relation to industry)

Activities are organized into six research groups: *energy* (development and testing of solar-heat-driven heating and cooling systems), *mobility* (development and trials of super-compact electric vehicles and sharing infrastructure), *clinical plant science* (citizen training and capacity-building for diagnosing and alleviating plant diseases), *agriculture and landscape planning* (exploitation of local land and forestry resources for agriculture, greening and biomass production), *urban planning* (creation of mixed-use urban land policies and social networks to prevent urban decay), and lastly, *information systems* (creation and demonstration of supporting ICT infrastructure). As a whole, these groups encompass various disciplines such as engineering, urban design, plant science, environmental design and environmental sciences and involve over forty faculty and researchers. Designed overall to be a catalytic action plan for low carbon urban reform and societal transformation, the bulk of programme activities concern scientific research, both basic and applied. A large proportion of these research projects existed prior to the programme. As such, faculty and researchers engaging in the project are keenly motivated by the desire to advance and increase the visibility of research activities. With procured funding also serving to increase the scale of existing research projects, scientific learning is enhanced by collaborating with other disciplines and external stakeholders, and utilizing the local city as a test bed for emerging technologies and newly conceived 'social experiments' (Yarime/Trencher/Mino et al. 2012). Clearly such motivations correspond with earlier observed factors from section 30.2. A steering committee oversees all activities. Members include the overall project leader, leaders from each research group and external advisors such as senior members of a private sector think tank, the ex-president of the University of Tokyo, the deputy mayor of Kashiwa and senior city planners.

The key challenge addressed by the programme is how to achieve decarbonization in the context of a rapidly aging population. With 23 per cent of the citizens over sixty-five (Statistics Bureau),[7] Japan is a 'hyper-ageing society'—by far exceeding greying trends in any other nation. It is predicted that economic contraction following a shrinking working population could significantly hamper Japan's ability to mitigate and adapt to climate change (University of Tokyo 2014). Although Kashiwa's proportion of elderly sits just below the national average at 19.8 per cent, the working population has already peaked, and resident numbers are forecast to decline as of 2020. With this greying population facing concerns such as access to health care, services and recreation, a patchwork of some 1,000 vacant lots has emerged across the city (University of Tokyo 2014). Any effort to reduce future carbon emissions must therefore contend with this destabilizing decline of urban vitality and socio-environmental conditions.

Regarding external stakeholders, two levels of government are involved: the City of Kashiwa and the surrounding Chiba prefecture. Officials are predominantly senior members from various departments such as city planning, transport, parks and greenery, and agriculture. Immediate roles include official endorsement and decision-making; co-design and implementation of demonstration projects; provision of data, knowledge and test sites; and assistance with legal and political matters. Long-term roles include the integration of research results and the ensuing model for low carbon urban reform into city planning, policy and projects. The civic sector plays a core role. Individual residents and groups provide local

---

7 See URL: <http://www.stat.go.jp/english/data/nenkan/1431-02. htm>.

knowledge and participate in consultations and demonstrations, with a local NPO undertaking commissioned research and assisting in project demonstrations and implementation. Although present, industry plays somewhat a lesser role overall. The think tank Mitsubishi Research Institute provides project management and strategic decision-making support, also mediating stakeholder consultations. Private firms supply technical assistance, data and facilities. Active participation of the civic sector—more so at the level of knowledge production and project implementation—is a defining characteristic of this partnership. It is also a notable deviation from the overall tendency in Asia to collaborate with actors from industry more than civil society (reported earlier in section 30.2).

Table 30.3 offers an overview of the objectives, activities and expected impacts of the six research groups. A defining characteristic of the programme is the prevalence of scientific demonstrations, societal interventions and implementation projects. Being highly experimental in nature, these are often referred to as 'social experiments' (Yarime/Trencher/Mino et al. 2012) and complement traditional scientific activities such as basic research and formal knowledge production. Two factors appear to have contributed to the large quantity of social experiments. Firstly, for several years preceding the programme, the nearby station-front development of Kashiwanoha has sparked a culture of cross-sector innovation between industry, the university and the city and experimental pilot projects such as a rental cycle and EV vehicle network. Secondly, funding requirements from the Japan Science and Technology Agency have explicitly stipulated that qualifying projects involve demonstrations and initiatives to reform urban systems, in addition to basic research and development of new technologies.

Social experiments and implementation projects in the programme are marked by the intimate and ongoing involvement of the civic sector and an array of non-technical approaches and efforts to create new social configurations. For example, trials of *Micro-Electric Vehicles* (MEVs) from the mobility group have heavily involved local residents and NPOs, who are also core actors in field test activities for other groups such as biomass production and forest management experiments of the agriculture and landscape planning group. With activities of the energy and mobility group mostly centred on the creation and trialling of emerging technologies, the social innovation-driven approach of the clinical plant science group is noteworthy. This may be said when recalling from section 30.2 that there appears to be an overall tendency in

Asia to pursue urban sustainability through predominantly technical approaches. Activities in the *clinical plant science* group are focused on capacity-building and the transfer of plant science knowledge to local citizens. The idea is that this knowledge can be used to treat common diseases troubling local farmers and gardeners, thereby reducing wastage and, consequently, GHG emissions (from saved fertilizer and so on). Co-benefits of this approach entail the creation of social insertion opportunities for elderly residents, who through university-led training become certified and functioning 'plant doctors'. A core outcome of this group is the creation of a large-scale and fully functioning 'plant doctor' training and deployment network, which to date has trained approximately 700 local residents (mostly seniors). Experiments are currently under way to establish an on-demand despatch and transport system for trained plant doctors, who would subsequently conduct 'house visits' to farms and gardens in response to Internet reservations. At present, plant science initiatives are heavily planned and directed by university faculty. Yet efforts are being made to make the volunteer plant doctor system an entirely citizen-run and self-sustaining platform, supplemented by revenue from workshops and training courses. Citizen engagement and social innovation-driven approaches are also defining attributes of various initiatives in the urban planning group. At present, one area of activity includes experiments to foster mixed land use and utilize vacant lots for urban agriculture and 'rental gardens'.

From this perspective, outputs from the *Urban Reformation Programme for the Realization of a Bright Low Carbon Society* move far beyond traditional scientific activities such as basic research, publications and conference presentations. Experimental interactions with local government and the civic sector are leading to the co-design and co-production of prototypes of new social systems, which importantly, will continue functioning after the withdrawal of scientific enquiry and completion of the programme. Other outcomes include trials of new technologies and improved scientific learning gained from their insertion into living human systems in the 'urban laboratory'. This is in addition to an enhancement of social capital and provision of valuable policy recommendations for local government concerning, for example, land use and transport. In terms of overall impacts, the various outputs of each group contribute to a wider knowledge base, toolkit and blueprint for urban and social reformation, to counter the dual chal-

Table 30.3: Overview of Urban Reformation Programme for the Realization of a Bright Low Carbon Society. **Source:** The authors.

Focus area	Aim	Core activities	Expected outputs and impacts
Energy	Develop and test solar heat pump driven, super-efficient heating and cooling systems.	• Technological development • Laboratory experiments on CO_2 and energy consumption reduction potential, also using industry on-site data • Field testing	• Prototype creation and testing • Decarbonization and optimization of residential and commercial building heating and cooling systems (potential energy savings of 20 per cent relative to conventional technology)
Mobility	Develop and trial various Micro-Electric Vehicles (MEVs) and sharing network to decarbonize and optimize transport system for elderly citizens.	• Technological development of various models of wireless charging and capacitor-storage MEVs • Field demonstrations, testing with citizens and acceptance studies • Creation of smartphone-enabled sharing network	• Prototype creation and testing of MEVs and supporting infrastructure (charging and sharing network) • Decarbonization and optimization of individual transport system • Social insertion of elderly and activation of lifestyles • Knowledge about public acceptance and demand for MEVs
Clinical plant science	Contribute to decarbonization of agriculture through recruitment and training of elderly plant doctors for diagnosing and alleviating crop diseases	• Creation and implementation of plant science certification and training programme for senior citizens • Deployment of plant doctors throughout gardens and farms in Kashiwa City • Investigation of CO_2 reductions	• Contribution to decarbonization of agriculture by reducing crop loss from disease • Social insertion of elderly and activation of lifestyles • Fully functioning diagnosis and treatment service for plant disease and improved maintenance of parks and gardens • Knowledge base including textbooks, website and plant hospital
Agriculture and landscape planning	Develop citizen-based system for maximizing effective use of urban land and forestry resources for agriculture, greening and biomass production	• Field experiments and investigation of biomass fuel production potential • Study of GHG emissions from rice paddies and agriculture • Design of citizen-run management system for local land resources	• Opportunities for elderly citizens to engage in management of local land resources • Improvement of conservation and resilience of local land and forestry resources against declining population and agricultural activity • Stimulation of urban agriculture and contribution to GHG emissions reduction
Urban planning	Create mixed-use urban land policies and social networks to prevent urban decay and cater to needs of elderly population	• Survey investigation of elderly citizen living conditions, needs and geographical distribution of services • Creation and testing of land rotation system for integrating vacant lots into urban agriculture system	• Policy recommendations and social network prototypes for governance, maximizing effective land use and countering decomposition of city • Quantitative and qualitative understanding on citizen needs, living environment and future changes to urban fabric
Information systems	Study and create various ICT systems for enhancing the integration and decarbonization of various urban elements and functions	• Creation and trials of dynamic sensing and visual feedback technologies for electricity consumption • Creation of storage archive for quantitative and qualitative data from various demonstrations in other groups	• ICT system prototypes for enhancing connectivity and efficiency of various urban elements and functions

lenges of climate change mitigation and an ageing society.

However, an unprecedented and highly experimental nature also means that programme members are confronted by various challenges and uncertainties. Many stem from the large-scale and interdisciplinary structure of the programme. Key issues at stake appear related to communication difficulties. Also a commonly reported barrier for other partnerships (see section 30.2), this is contributing to a loss of unity, harmony and synergy. With the programme uniting over forty faculty and researchers from various disciplines, in addition to countless government and civil society stakeholders, several rifts have occurred between research groups. These appear to be exasperated by differences in activities, approaches and understanding of project goals and expected results. Overall synergy is further hampered by conflicting motivations between groups and individual researchers. Some are seemingly focused on advancing pre-existing research agendas whilst others are more committed to pursuing the goals of the programme. Consequently, at times research groups have functioned as independent research communities, with little cooperation and exchange with others. Although some groups have found areas for collaboration and functioned well together, the overall coordination and integration of the programme has been a key challenge. A recent measure taken to address this is the establishment of an additional project management group and recruitment of a dedicated project coordinator. Other challenges relate to the complexity and locked-in nature of the problems the programme seeks to address, uncertainty regarding results and the lack of an overarching vision for social reform in the community.

Also deserving critical examination is the programme's glowing promise of creating the blueprint for a realizable and exportable model of a 'bright', low carbon and elderly citizen-friendly society in the space of five years. Rapidly approaching the final year of implementation, the mechanisms by which this blueprint will be translated into reality remain unclear, just as the ability to achieve meaningful reduction of GHG emissions with tools such as electric vehicles and improvements in air-conditioning efficiency, plant health and land use. As observed by Bulkeley and Castan Broto (2012: 361), many urban sustainability experiments "promise more on paper" than they actually deliver. The capacity of this programme to achieve its objectives is largely restrained by the scope of existing university research agendas and strengths

(i.e., energy, MEVs, plant science and urban planning). It appears that the relevancy of these fields and research activities to the overarching objective was taken for granted in the design of the programme.

Yet the value of this initiative is nevertheless significant. It lies more so in the new model of knowledge production with stakeholders than in knowledge outcomes *per se*. This case provides an insightful preview into how the modern research university can apply its resources to the co-design and co-production of knowledge, solutions and societal transformations to address complex and place-based societal challenges.

30.5   Concluding Remarks

Several observations can be made about university partnerships attempting to advance urban sustainability through collaborative research and sociotechnical experiments in Asia.

Firstly, the power of government funding programmes to entice the formation of partnerships is evident at various points. Survey results demonstrated that the presence of suitable external funding programmes was a significant motivating factor for many cases. On the other hand, when financial resources were inadequate, funding also constitutes a common barrier. The effectiveness of government funding programmes in spurring cross-sector sustainability innovation was also demonstrated by Singapore and Japan. It was observed that national competitive grant schemes, and the preparation of physical infrastructure in Singapore (CREATE), had successfully brought into fruition several cases. These results suggest that an expansion of such government funding opportunities would have a driving effect on the formation of other partnerships. Such findings give weight to calls from other scholars that government funding programmes can serve as powerful incentives and shapers of university initiatives to initiate or engage in collaborative networks and experiments to advance sustainability (Dedeurwaerdere 2013; Whitmer/Ogden/Lawton et al. 2010).

Secondly, our study has revealed an overall preponderance of technical approaches to urban sustainability in Asia. With numerous R&D platforms and partnerships headed by engineering and physical sciences, it seems that the more human dimensions of urban sustainability and the involvement of the social sciences, the humanities and civil society, which are crucial for the co-design and co-production of knowledge for sustainability transitions, are largely underdeveloped. One influencing factor behind the plenitude of technology-driven platforms in tandem with industry appears to be government funding programmes. In nations such as Singapore and Japan, competitive funding schemes are distinctly privileging fields such as engineering and the physical sciences by explicitly associating urban sustainability with the creation and testing of new technologies. Yet the capacity of predominantly technocentric approaches to reform lifestyles and trigger societal transformations to sustainability has been shown to be limited (Notter/Meyer/Althaus 2013). Measures are therefore required to create more favourable funding opportunities for the social sciences and humanities, and to encourage university actors to engage in social innovation-orientated sustainability experiments in tandem with civil actors.

As argued by Dedeurwaerdere (2013), one pragmatic strategy could involve the 'fine-tuning' of selection criteria and focus areas in existing funding programmes to demand greater collaboration across academic disciplines and engagement with civil actors. Such measures could entice partnerships to move beyond the development and trialling of new technologies to the co-production of new sociotechnical systems (also inclusive of technical elements), social arrangements and services. The case study from the University of Tokyo illustrated to some extent the possibility of such an approach. On the other hand, however, this programme also demonstrated that when large numbers of faculty and researchers from different fields come together, problems of project unity, communication and differences in approach can reduce partnership synergy and alienate certain research groups. Clearly, calls for interdisciplinarity in sustainability partnerships need to take into account these potential tensions and obstacles.

A third point of reflection concerns rising demands from both governments and academic research communities for utility, relevance to stakeholder needs, solutions for societal challenges and economic contributions. How much should be expected from university partnerships for urban sustainability? Can they really prove a 'miracle fix' for converging needs such as achieving innovation-driven growth and responding to complex social and environmental challenges? The case from Japan suggests that progress can be slow, complex and uncertain. Furthermore, the ability of the university to demonstrate relevance or tailor activities to the demands of external stakeholders is also restricted by existing strengths and research projects, and the effective balancing of these with new approaches, activities and stakeholder participation.

These were major factors affecting the ability of the Japanese case (a new programme incorporating many 'old' strengths and research projects) to generate solutions for addressing the designated challenge. Calls for universities to engage in cross-sector sustainability partnerships and experiments should take into account evidence that they can frequently fall short of expectations (Fadeeva 2004; Trencher/Bai/Evans et al. 2014).

In spite of such uncertainty, university partnerships in Asia and the rest of the world are demonstrating a promising capacity to generate important knowledge, technologies, tools and societal transformations for advancing urban sustainability. Also, cases including that from the University of Tokyo suggest that university partnerships are ideal mechanisms for linking broad areas of society and implementing multiple sociotechnical experiments designed to respond to diverse place-specific challenges. That is, they can potentially serve as 'umbrellas' or 'portfolios' for various societal experiments and interventions (i.e. protected spaces or 'niches' of innovation) that can be implemented alongside conventional knowledge production. A wealth of mechanisms highlighted in this study (i.e. knowledge management, governance and planning, technology transfer or economic development, technical demonstrations and experiments, reform of the built and natural environment, and sociotechnical experiments) suggest that university actors in Asia are experimenting with a myriad of ways to co-produce knowledge, co-implement such experiments and magnify societal impacts by facilitating the transfer of lessons elsewhere. Further, there is also evidence to suggest that such activities are increasingly likely to constitute a 'mission' for the university and complement existing, more established functions such as education, basic research and technology transfer or entrepreneurialism (Trencher/Yarime/McCormick et al. 2014).

In closing, in this chapter we have scratched the surface of a potential wealth of university initiatives in Asia to advance the sustainability of particular urban systems or communities, cities and nations. Our contribution to the literature has been our focus on several highly innovative partnerships in Asia that do not typically feature in discussions on sustainability transitions in English-speaking or European settings. Further, our study of fifteen cases (admittedly a modest sample size) has allowed us to extract lessons and common trends. This could not have been possible by a study of one or two cases alone—an approach tending to dominate the existing scholarship. Although our study has shed some light on potential trends of university sustainability partnerships and experiments in industrialized Asia regarding key characteristics, roles, motivations and barriers of each societal sector, further research is required at both a micro- and macro-level. Clearly, a larger survey is needed to better assess the characteristics, roles and impacts of different university partnerships across Asia. As well as encompassing nations such as Korea and Taiwan, it could cover emerging economies such as China and India, which are increasingly significant players in research and innovation in the region. Further understanding is also required at the micro-level. Future studies are required, to examine the potential long-term impacts of different models of partnerships and the processes by which co-designed and co-produced knowledge, technologies, and sociotechnical experiments can be up-scaled to effect wider societal change.

References

Bai, Xuemei; Shi, Peijun; Liu, Yansui, 2014: "Realizing China's urban dream", in: *Nature*, 509 (8 May): 158-160.

Bai, Xuemei; Wieczorek, Anna; Kaneko, Shinji; Lisson, Shaun; Contreras, Antonio. 2009: "Enabling sustainability transitions in Asia: The importance of vertical and horizontal linkages", in: *Technological Forecasting and Social Change*, 76: 255-266.

Bai, Xuemei; Roberts, Brian; Chen, Jing, 2010: "Urban sustainability experiments in Asia: patterns and pathways", in: *Environmental Science and Policy*, 13: 312-325.

Bai, Xuemei; Imura, Hidefumi, 2000: "A comparative study of urban environment in East Asia: Stage model of urban environmental evolution", in: *International Review for Global Environmental Strategies*, 1,1: 135-158.

Birch, Eugenie; Perry, David; Taylor, Henry Louis, 2013: "Universities as anchor institutions", in: *Journal of Higher Education Outreach and Engagement*, 17,3: 7-15.

Berkhourt, Frans; Verbong, Geert; Wieczorek, Anna; Raven, Rob; Lebel, Louis; Bai, Xuemei, 2010: "Sustainability experiments in Asia: innovations shaping alternative development pathways?", in: *Environmental Science and Policy*, 13: 261-271.

Brown, Halina; Vergragt, Philip; Green, Ken; Berchicci, Luca, 2003: "Learning for sustainability transition through bounded socio-technical experiments in personal mobility" in: *Technology Analysis & Strategic Management,* 15,3: 315-291.

Brown, Lester, 2011: *World on Edge: How to Prevent Environmental and Economic Collapse* (New York: WW Norton).

Bulkeley, Harriet; Castan Broto, Vanessa, 2012: "Government by experiment? Global cities and the governing of climate change", in: *Transactions of the Institute of British Geographers*, 38,3: 361-375.

Castan Broto, Vanessa; Bulkeley, Harriet, 2013: "A survey of urban climate change experiments in 100 cities", in: *Global Environmental Change,* 23: 92-102.

Crow, Michel, 2010: "Organizing Teaching and Research to Address the Grand Challenges of Sustainable Development" in: *Bioscience*, 60,7: 488-499.

Dedeurwaerdere, Tom, 2013: "Transdisciplinary Sustainability Science at Higher Education Institutions: Science Policy Tools for Incremental Institutional Change", in: *Sustainability*, 5: 3783-3801.

Etzkowitz, Hendry; Webster, Andrew; Gebhart, Christiane; Cantisano Terra, Branca Regina, 2000: "The future of the university and the university of the future: evolution of ivory tower to entrepreneurial paradigm", in: *Research Policy*, 29: 313-330.

Etzkowitz, Henry; Leydesdorff, Loet, 2000: "The dynamics of innovation: from National Systems and 'Mode 2' to a Triple Helix of university-industry-government relations", in: *Research Policy*, 29: 109-123.

Evans, James; Karvonen, Andrew, 2011: "Living laboratories for sustainability: Exploring the epistemology of urban transition", in: Bulkely, Harriet; Castan Broto, Vanessa; Hudson, Maassen (Eds.): *Cities and Low Carbon Transitions* (Routledge: New York): 126-141.

Evans, James; Karvonen, Andrew, 2014: "Give me a laboratory and I will lower your carbon footprint!—Urban Laboratories and the Governance of Low Carbon Futures", in: *International Journal of Urban and Regional Research*, 38,2: 413-430.

Fadeeva, Zinaida, 2004: "Promise of sustainability collaboration—potential fulfilled?", in: *Journal of Cleaner Production*, 24,13: 165-174.

Future Earth, 2013: "Future Earth Initial Design: Report of the Transition Team" (Paris: International Council for Science).

Geels, Frank; Schot, Johan, 2010: "The dynamics of sociotechnical transitions—a sociotechnical perspective", in: Grin, John; Rotmans, Jan; Schot, Johan (Eds.): *Transitions to Sustainable Development* (New York: Routledge): 9-101.

Hanleybrown, Fay; Kania, John; Kramer, Mark, 2012: "Channelling Change: Making Collective Impact Work", in: *Stanford Social Innovation Review*, 75: 1-8.

Hoover, Elona; Harder, Marie, 2014: "What lies beneath the surface? The hidden complexities of organizational change for sustainability in higher education", in: *Journal of Cleaner Production* (in press); at: <http://dx.doi.org/10.1016/j.jclepro.2014.01.081>.

König, Ariane (Ed.), 2013: *Regenerative sustainable development of universities and cities: the role of living laboratories* (Cheltenham: Edward Elgar).

König, Ariane; Evans, James, 2013: "Experimenting for sustainable development? Living laboratories, social learning, and the role of the university", in: König, A. (Ed.): *Regenerative sustainable development of universities and cities: the role of living laboratories* (Cheltenham: Edward Elgar): 1-27.

International Sustainable Campus Network (ISCN), 2014. "Best Practice in Campus Sustainability"; at: <http://www.international-sustainable-campus-network.org/latest-news/iscn-and-gulf-share-best-practices-in-campus-sustainability-at-the-world-economic-forum.html> (20 October 2014).

Laursen, Keld; Salter, Ammon, 2004: "Searching high and low: what types of firms use universities as a source of innovation?", in: *Research Policy*, 33: 1201-1215.

Liedtke, Christa; Welfens, Maria; Rohn, Holger; Nordmann, Julia, 2012: "LIVING LAB: user-driven innovation for sustainability", in: *International Journal of Sustainability in Higher Education*, 13,2: 106-118.

Lubchenco, Jane, 1988: "Entering the century of the environment: a new social contract for science", in: *Science*, 279,5350: 491-497.

Mauser, Wolfram; Klepper, Gernot; Rice, Martin; Schmalzbauer, Bettina Susanne; Hackmann, Heide; Leemans, Rik; Moore, Howard, 2013: "Transdisciplinary global change research: the co-creation of knowledge for sustainability" in: *Current Opinions in Sustainable Development*, 5,3-5: 420-431.

M'Gonigle, Michael; Starke, Justin, 2006: *Planet U* (Canada: New Society Publishers).

Molnar, Carina; Ritz, Thor; Heller, Benjamin; Solecki, William, 2011: "Using higher education-community partnerships to promote urban sustainability", in: *Environment*, 53,1: 18-28.

Mowery, David; Nelson, Richard; Sampat, Bhaven; Ziedonis, Arvids, 2004: *Ivory tower and industrial innovation: university-industry technology transfer before and after the Bayh-Dole Act in the United States* (Stanford: Stanford University Press).

Notter, Domini; Meyer, Reto; Althaus, Hans-Jo rg, 2013: "The Western Lifestyle and Its Long Way to Sustainability", in: *Environmental Science and Technology*, 47,9: 4014-4021.

Nowotny, Helga; Scott, Peter; Gibbons, Michael T., 2001: *Re-thinking Science: Knowledge Production in an Age of Uncertainty* (Polity: Malden).

Orr, David, 2013: "Governance in the Long Emergency", in: *State of the World* 2013: *Is Sustainability Still Possible?* (Washington DC: Island Press): 279-291.

Philpott, Kevin; Dooley, Lawrence; O'Reilly, Caroline; Lupton, Gary, 2011: "The entrepreneurial university: examining the underlying academic tensions", in: *Technovation*, 31,4: 161-170.

Shrum, Wesley; Genuth, Joel; Chompalov, Ivan, 2007: *Structures of Scientific Collaboration* (MIT Press: Cambridge).

Trencher, Gregory; Masaru, Yarime; Ali Kharazzi: 2013: "Co-creating sustainability: Cross-sector university collaborations for driving sustainable urban transformations", in: *Journal of Cleaner Production*, 50: 40-55.

Trencher, Gregory, Yarime, Masaru; McCormick, Kes; Doll, Christopher; Kraines, Stephen, 2014: "Beyond the third mission: Exploring the emerging university function of co-creation for sustainability", in: *Science and Public Policy*, 41,2: 151-179.

Trencher, Gregory; Bai, Xuemei; Evans, James; Yarime, Masaru; McCormick, Kes, 2014: "University partnerships for co-designing and co-producing urban sustainability", in: *Global Environmental Change,* 28: 153-165.

Trencher, Gregory, 2014: "Co-Creative University Partnerships for Urban Transformations Towards Sustainability: Beyond the Third Mission Through Technology Transfer" (PhD dissertation, University of Tokyo, Graduate Programme in Sustainability Science).

University of Tokyo, 2014: "Urban Reformation Program for Realization of a "Bright" Low-Carbon Society: Progress Report 2013"; at. <http://low-carbon.k.u-tokyo.ac.jp/documents/E%20summargy%20of%20progress%20report.pdf>.

Vigoda-Gadot, Eran; Shoham, Aviv; Schwabsky, Nitza; Ruvio, Ayalla, 2005: "Public Sector Innovation For The Managerial And The Post-Managerial Era: Promises And Realities In A Globalizing Public Administration", *International Journal of Public Sector Management*, 8,1: 57-81.

Whitmer, Ali; Ogden, Laura; Lawton, John; Sturner, Pam; Groffman, Peter; Schneider, Laura; Hart, David; Halpern, Benjamin; Schlesinger, William; Raciti, Steve; Bettez, Neil; Ortega, Sonia; Rustad, Lindsey; Pickett, Steward; Killilea, Mary, 2010: "The engaged university: providing a platform for research that transforms society", in: *Frontiers in Ecology and Environment*, 8: 314-321.

Yarime, Masaru; Trencher, Gregory; Mino, Takashi; Scholz, Roland; Olsson, Lennart; Ness, Barry; Frantzeskaki, Niki; Rotmans, Jan, 2012: "Establishing sustainability science in higher education institutions: towards an integration of academic development, institutionalization, and stakeholder collaborations", in: *Sustainability Science* 7(1), 101-113.

Yusuf, Shahid, 2007: "University-industry links: policy dimensions", in: Yusuf, Shahid; Nabeshima, Kaoru (Eds.): *How universities promote economic growth* (Washington DC: The World Bank): 1-25.

Zilahy, Gyula; Huisingh, Donald, 2009: "The roles of academia in Regional Sustainability Initiatives", in: *Journal of Cleaner Production*, 17: 1057-1066.

Other Literature

Statistics Bureau (of Japan): "Japan Statistical Yearbook"; at: <http://www.stat.go.jp/english/data/nenkan/1431-02.htm> (15 July 2014).

National Research Foundation, "Physical Sciences & Engineering"; at: <http://www.nrf.gov.sg/research/r-d-eco-system/research-priorities/physical-sciences-engineering > (15 June 2014): 1.

National Research Foundation, "Campus for Research Excellence and Technological Enterprise"; at: < http://www.nrf.gov.sg/docs/default-source/new-nrf-fact-sheets/20131217_create-factsheet-(final).pdf?sfvrsn=2> (11 July 2014): 1.

Part VIII Examining Sustainability Transitions in the Water, Food and Health Sectors from Latin American and European Perspectives

31 Future Global Water, Food and Energy Needs

Cecilia Tortajada[1] and Martin Keulertz[2]

Abstract

If present trends continue, it is unlikely that increasingly overused ecosystems subject to deterioration and depletion will be able to meet global water, food and energy needs. Current academic thinking is that scarcity, pollution, mismanagement and misallocation of natural resources will impact every sector on which humankind depends for survival. In a globalized economy with increasingly free movement of commodities and financial and human capital, a poor understanding of the most pressing issues and their interconnectedness and interdependences will cause irreparable damage to the Earth and its billions of people: a clear case of *fait accompli*. This is a challenging context for global development, and to understand and manage the interdependencies between the various sectors and their global impact will require comprehensive planning and policy implementation, institutional resilience, partnerships across economic sectors, and innovation in development in order to sustainably manage resources in the wake of population growth and climate change.

Keywords: Water, food, energy, interconnectedness, globalization.

31.1 Introduction

In an increasingly globalized and interconnected world, societies have become less resilient with respect to water, food and energy resources. Long-term developments such as population growth, urbanization and industrialization in emerging markets, as well as the impending threat of climate change, have increasingly resulted in global impacts on natural resources; consumption of water, energy and food is at risk of becoming unsustainable. At the same time, institutions have not been able to respond to the enormous challenge of implementing policies that address the requirements of the various sectors in an interrelated manner.

Humankind's influence on the global environment and its exploitation of Earth's resources have both escalated to such an extent that Earth is said to have entered a new era, the Anthropocene. This is a human-dominated, geological epoch that has supplemented the Holocene, the warm period of the past ten to twelve millennia (Crutzen 2002). There are clear signs that these changes will affect water resources (Biswas/Tortajada 2009; Rockstrom/ Falkenmark/Allan et al. 2014), food resources (FAO 2011) and energy resources (IEA 2013, 2014) so dramatically that humankind will have to be content with adapting to the new conditions. Consequently, it has become necessary to focus on establishing resilience to an increasing number of stresses, as well as on understanding their impact on the stability of natural, social and economic systems (Rockstrom/Falkenmark/Allan et al. 2014).

In the past, Europe, North America and Oceania were able to meet their water, food and energy security needs in terms of availability and accessibility (affordability and acceptability depend on a number of external factors) through either surplus production or strategic imports. Nonetheless, global economic change is inevitably shifting economic centres from the West to the East. With this economic change have come new demands for food, water, timber, fibre and fuel from populations growing on a scale never before witnessed. For example, a report by McKinsey in 2009 foresaw a forty per cent gap between the demand for

1 Dr. Cecilia Tortajada, Senior Research Fellow, Institute of Water Policy, Lee Kuan Yew School of Public Policy, Singapore; Email: <Cecilia.tortajada@gmail.com>.
2 Dr. Martin Keulertz, Post-Doctoral Research Associate, Department for Agricultural and Biological Engineering, Purdue University, West Lafayette, IN, USA; Email: <mkeulert@purdue.edu>.

© Springer International Publishing Switzerland 2016
H.G. Brauch et al. (eds.), *Handbook on Sustainability Transition and Sustainable Peace*,
Hexagon Series on Human and Environmental Security and Peace 10, DOI 10.1007/978-3-319-43884-9_31

and the supply of water by the year 2030, if practices do not change (Boccaletti 2009). Similarly, the United Nations *Food and Agriculture Organisation* (FAO) concluded that the world will have to grow sixty to seventy per cent more food by 2050 in order to feed nine billion people (FAO 2011). The future of the availability of natural resources to meet the needs of increasingly demanding populations is thus surrounded by complexity and uncertainty, since it depends on global forces that are constantly changing (Hajkowicz/Cook/Littleboy 2012; Ringler/Biswas/Cline 2010).

The consequences of environmental degradation and change are increasingly associated with non-traditional notions of environmental security. While environmental change and degradation, as well as resource scarcity, increase instability and poverty and hold back social development, a discussion of their contribution to insecurity and vulnerability per se is outside the scope of this study. It is important, however, to acknowledge how the intrinsic relationships between societies and their political, human and natural environments are understood (Andrews-Speed 2013; Brauch/Oswald Spring/Grin et al. 2009), as well as the role the environment plays in promoting peace, stability and human security and the potential conflicts and vulnerabilities caused by environmental degradation (Biswas 2011).

In order to understand the complexity surrounding future water, food and energy needs, this chapter will aim to locate the management of these resources in a wider context of development. Most studies either provide a local or regional analysis or emphasize one issue individually, but do not address the three of them in an interrelated manner. Such analyses cannot lead to full comprehension of the crucial role of these resources in twenty-first century development, since they are embedded in complex interdependencies such that attempts to solve problems in one sector may create new problems or exacerbate existing ones.

The rationale for our analysis is that, in the present changing global environment, policy-making and the management, development and governance of natural resources are more and more influenced by numerous social and economic forces and scientific and technological developments; and that these are based on the policies, politics and economic priorities of various actors and sectors who do not take each other into consideration, and who operate concurrently in different geographies. A key concern is that their interconnections, interdependences and feedback loops, as well as their global impacts, are still not fully understood. As a consequence, planning and implementation at national levels are still based on individual sectors without considering their economic, social, physical and political environments, and this frequently makes decisions impracticable.

There is no doubt that things are further complicated by the diversity of views on what should be developed, what sustained, and over what period, particularly when such challenges as the above are involved. Moreover, if current trends in environment and development continue, it is unlikely that the desired state of sustainable development will be achieved (Kates 2001). Although change is inherent in humans, countries have not created systems to cope sustainably with transitions. At a time when our global context is being profoundly transformed, political, economic, social, environmental and technological advances are challenging prevailing assumptions about what was assumed to be a stable state of affairs in the age of the Holocene.

According to the *Millennium Ecosystem Assessment* (MA 2005), there is established but incomplete evidence that changes in ecosystems are increasing the likelihood of non-linear and sometimes irreversible events in the natural environment. As already mentioned, a major aspect of this problem is that the institutional arrangements, policy, legal and regulatory frameworks and management and governance practices necessary to reverse this situation are not currently in hand, and often not even on the drawing boards of decision-makers. This is likely to severely limit the alternative ways in which humankind can provide for global water, food and energy needs, both at present and in the future.

Finally, "needs" is a subjective term. Usually it refers to specific requirements that are essential or very important rather than just desirable. In this chapter, we will discuss both needs and demands, as in many cases this may be a more appropriate phrase given that it refers to "the desire of consumers, clients, employers, etc., for a particular commodity, service or other item".[3] By outlining present and future challenges for the future needs and demands for global water, food and energy, and placing them in a wider framework of development, we aim to broaden the approach from mere resource use to the interconnections and interdependences among a whole series of complex issues, such as population growth, urbaniza-

---

3   See: Oxford Dictionaries, British and World English; at: <http://www.oxforddictionaries.com/> (17 February 2015).

tion, resource scarcity, infrastructure development, agricultural trade, and energy generation, all of which in many cases are driven by geopolitical interests.

31.2 Water, Energy and Food Resource Management

The complexities of water, food and energy sectors need to be analysed in relation to each other as well as within their social, economic, natural, political and cultural environments, and not simply in isolation from each other. Water is a prerequisite resource for global sustainability, with a fundamental role in every other sector, and it has the distinct potential to sustain political, economic, social and environmental systems. Its proper management has an enormous value as it provides an entry point to address broader development objectives that are usually closely linked. On the other hand, its management all over the world does not reflect this value.

Water is also essential for crop production, be this for food, feed, fibre or fuels. Food sustainability, as discussed by Willaarts, Garrido and Llamas (2014), depends on the resilience of related agro-ecosystems, of which water is a fundamental component.

Water, and its proper management, are also indispensable for energy production and power generation. Water is used extensively in energy extraction, refining, processing and transportation; and energy is essential for transporting water over long distances and raising it, treating it and distributing it to end users, as well as for collecting and treating wastewater.

Policies that take into account trade-offs between water, food and energy are not being developed systematically, and much less in a coordinated manner, in spite of the urgent need for them.

As demands for water, food and energy continue to increase, competition for water between the domestic, energy, agriculture, environment and industrial sectors has the potential to have a negative impact on the reliability and security of all of these sectors (Pate/ Hightower/Cameron et al. 2007). If a sustainable future is to be achieved, the management of water resources requires balancing the competing needs for the increasing type and number of uses and users while taking into account the availability of water supplies and storage capacity. Such considerations should include the numerous impacts water policies and regulations have on food and energy supplies and demands and vice versa. This is, however, still far from being implemented anywhere in the developed or developing worlds.

Terms of engagement are also affecting resource utilization policies and practices, contributing to more intense competition and also reshaping how emerging economies and global players are interacting in terms of trade and aid, with commercial and development-oriented policies becoming ever more closely dependent on one another (Lum/Fischer/Gomez-Granger et al. 2009; Tortajada 2014). An example of this is China and India, where competition for natural resources is global in reach, putting pressure on the global environment, and all this within specific political dynamics. The water, energy and food resource needs of both countries are likely to affect the availability of resources for numerous uses and users beyond their borders. In turn, the interlinkages and interdependences of the above sectors are likely to impact on the socio-economic development of different countries (Goldman Sachs 2010).

Within this context, the formulation of policy alternatives able to respond to related changes and challenges and how they could be implemented in a timely, socially and environmentally acceptable and cost-effective manner is becoming a security issue. The unprecedented and accelerating developments at national, regional and global scales and their implications in the individual sectors require in-depth, multidisciplinary and multisectoral policy debates from the strategic viewpoint about how they impact on peace, stability and human security. The resulting analyses will help realistic yet flexible resource-use policy alternatives for the common good to be devised. National, regional and global scenarios need to be formulated, including future-oriented frameworks that are able to respond to upcoming challenges and to further national and international interests.

With regard to water resources, their perceived global scarcity reflects not necessarily their physical availability but rather poor planning, management and governance, and this has resulted in overexploitation and deterioration of both surface and groundwater all over the world, which has had an impact on the food and energy sectors, among others. This sad state of affairs, where the management of water resources has responded to specific agendas rather than contributed to the goal of efficiency, has proved detrimental. In some cases, it has undermined functioning management systems, set back needed agendas for reform, or even become a tool to mask other agendas (see Biswas 2008; Giordano/Shah 2014; Molden/Lautze/Shah et al. 2010).

Global forecasts predict a bleak future where water stress forces river basins and their millions of inhabitants into a transition from a state of water abundance to one of water scarcity. While changes in the urban, energy and environment sectors continue to drive competition between sectors, it is the agricultural sector (already the largest consumer of water globally, with an estimated seventy per cent of total freshwater withdrawals) that will result in further inequities and environmental degradation (Molden/Lautze/Shah et al. 2010). Global water demand (measured as withdrawals) is likely to increase by approximately fifty-five per cent by 2050 because of growing demands from manufacturing (400 per cent), generation of thermal electricity (140 per cent), and domestic use (130 per cent) (UNESCO 2014), not to mention irrigation.

It would be useful to remember at this point that, even though forecasts may give an indication of the global situation, they can be misleading as data can be inaccurate, incomplete or outdated. Previous estimates have been wrong as they have depended on statistics from countries where the systematic collection of water use data has been rare, and where some water uses or needs have been unquantified or unquantifiable. It has been proved that improved management and conservation measures are able to offset population and economic growth (Gleick 2003).

Population growth, rising meat and dairy consumption and expanding biofuel production, among other factors, will continue to put pressure on the agricultural sector. Observations of the yields of major cereal crops have raised the concern that agricultural productivity may not be able to meet demand without major new investments in underperforming regions and without strategies to continue increasing yields in high-performing areas. The focus will be mostly on irrigated agriculture, taking into account the potential impacts of climate change (Scheierling/Treguer/Brooker, et al. 2014). For food and agriculture production and water used for irrigation, the FAO estimates for the 2005/2007–2050 period a growth rate in world consumption of agricultural products of 1.1 per cent per year, and an estimated increase in agricultural production of sixty per cent by 2050. Using optimistic assumptions, water withdrawals for irrigation are projected to increase by six per cent, from 2,761 km3 to 2,926 km3 (Alexandratos/Bruinsma 2012), adding to the already severe competition between uses and users of water.

There is a fragile balance not only between the water, agriculture and energy sectors, but between the users of each of these sectors. With regard to water for human use, in much of the developing world water supply and sanitation facilities are still not adequate or sufficient and untreated solid and liquid wastes are discharged and affect surface and groundwater bodies. In the case of the developed countries, most systems need to be improved as effectively as they were developed in the early twentieth century. According to the European Union, twenty per cent of all surface water in Western Europe is severely polluted; sixty per cent of all cities overexploit their groundwater resources; and fifty per cent of wetlands have "endangered status" because of groundwater overexploitation. Even where wastewater treatment has improved significantly since 1980, the percentage of the population connected to wastewater treatment is still relatively low in Belgium, Ireland, southern Europe and the accession countries. Overall, the wastewater infrastructure needs are so large that wastewater treatment is the largest category of the water industry, with thirty-five per cent of a water market currently worth US$500 billion annually (Stiehler/Vogel 2014).

Depletion of groundwater has lowered water tables in urban and rural areas in developing countries in the semi-arid and arid tropics, as well as in developed countries, causing subsidence in major cities, increasing salinity, and forcing numerous lakes and wetlands to run dry. Groundwater mining for potable use and mostly for irrigation has been documented mostly in arid and semi-arid regions of the world, including China, India, Pakistan, Mexico and Iran (Kumar/Scott/Singh 2013), where very large amounts of energy are required for surface as well as groundwater pumping.

This is the case in India, the largest user of groundwater and the fourth largest energy consumer globally. During the 2003–2004 period alone, around 12.8 million electric pumps with a total of 51.84 gigawatts (GW) of connected load consumed 87.09 billion kilowatt-hours (kWh) of electricity (Kumar/Scott/Singh 2013). Satellite-based estimates of groundwater depletion confirm unsustainable withdrawals in the north-west states of Rajasthan, Punjab and Haryana. Should this situation remain unchanged, some 114 million people in these states may be affected by potable water shortages and a reduction in agricultural outputs, leading, in turn, to extensive socio-economic stress (Rodell/Velicogna/Famiglietti 2010). Figure 31.1 shows groundwater changes during 2002–2008, with losses shown in red and gains in blue, based on GRACE satellite observations. According to the

Figure 31.1: Satellite-based estimates of groundwater depletion in India. **Source**: I. Velicogna, University of California, Irvine; at: <http://www.nasa.gov/topics/earth/features/india_water.html>.

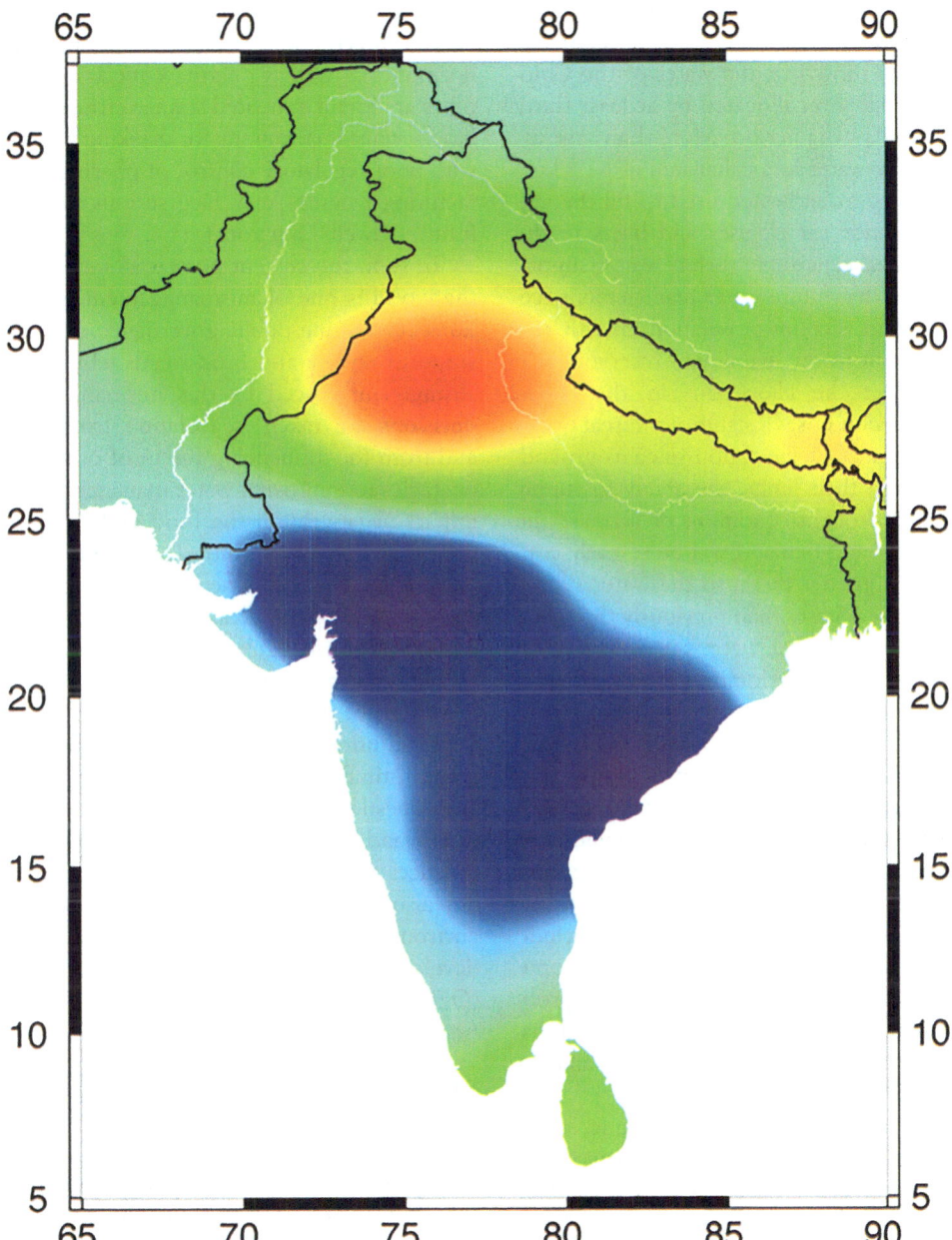

National Aeronautics and Space Administration (NASA),

the estimated rate of depletion of groundwater in northwestern India is 4.0 centimeters of water per year, equivalent to a water table decline of 33 centimeters per year. Increases in groundwater in southern India are due to recent above-average rainfall, whereas rain in northwestern India was close to normal during the study period.[4]

The findings of this study can be contested, however, as they indicate positive groundwater balance in north Gujarat and parts of peninsular India where problems of overexploitation of groundwater and aquifer mining have been studied and documented extensively. These studies are based on monitoring of water level fluctuations pre- and post-monsoon over a

_____

4   See <http://www.nasa.gov/topics/earth/features/india _water.html>.

five-year time period (Central Ground Water Board 2005, 2012), which make them more accurate.

In the case of the United States, studies by NASA and the University of California (Castle/Thomas/Reager et al. 2014) show that the water of the Colorado River Basin was over-allocated by at least thirty per cent in the 2004–2013 period. More than seventy-five per cent of the water loss calculated at 64.8 km3 since late 2004 was groundwater, used to fill the gap between the demands for all uses and users in the basin and the annual renewable surface water supply. The study showed that future water management scenarios that take into account population growth and climate change point to the inability of reservoir storage alone to meet the allocations in the basin. Groundwater depletion poses a significant threat to the long-term development of the region since as groundwater supplies reach their limits, the ability to supply fresh water during drought conditions or increase the gap between supply and demand will be severely constrained, posing "a greater threat to the water supply of the western United States than previously thought." (Castle/Thomas/Reager et al. 2014: 5910).

Ignored for decades, deterioration in water quality has become one of the most pressing problems facing the world, as it further accentuates problems of water scarcity. It has increasingly undermined global economic growth and worsened the physical and environmental health and the quality of life of billions of people because of point and non-point sources of pollution. Concerns about water contaminants that affect human health through drinking water include the presence of arsenic and fluoride that in some places occur naturally in groundwater. The main problem, however, is the billions of litres of untreated discharges that flow into water bodies every year all over the developed and developing world.

One example is China, where the economic, social and environmental costs of water pollution have been estimated at one per cent of *gross domestic product* (GDP) or approximately 150 billion yuan (World Bank 2007). Additionally, subsidence to a depth of several metres has been reported in cities such as Beijing, Tianjin, Shanghai, Taiyuan and Shijiazhuang, and in Shanxi and Hebei provinces. The damage to infrastructure in general caused by subsidence has been estimated to be 1.4 billion yuan. The extent of the potential negative impact of the overexploitation of water and the deterioration in the different sectors in general is so extensive that it threatens to undermine the economic growth of the country as well as the health and quality of life of billions of people.

As analytical procedures have improved, it has been possible to detect an increasing number of so-called emerging contaminants that are the result of human intervention. A range of human and synthetic hormones as well as pharmaceuticals and their metabolites are found in treated sewage effluents and this can have serious implications for drinking water. Countries with poor regulatory control of pharmaceutical manufacturing facilities may face serious problems in the future (Fawell/Ong 2012).

In sum, the current perceived water crisis all over the world is one of mismanagement with strong features concerning public governance. The global landscape of energy and food needs will continue to be strongly influenced by this mismanagement, where decisions are taken in isolation from requirements and from the influencing forces of other sectors. The obstacles to improving water management, which may impact negatively on the food and energy sectors in the future due to their interdependences and interconnectedness, are the same ones that have been identified for several decades: institutional fragmentation, weak legal and regulatory frameworks, poor multilevel governance, limited capacity at local level, weak allocation of roles and responsibilities among sectors and institutions, lack of cooperation between government institutions with common interests, funding gaps and questionable allocation of financial resources. Good and implementable policies and management practices are still needed, not only for the water sector but also for sectors such as energy, agriculture and environment, so that economic, social, and equitable development can be achieved in the long term (OECD 2011).

31.3 Energy Sources in a Changing World

Energy services are provided by converting primary forms of energy (e.g. coal, oil, gas, wood, wind, sunlight, uranium) to intermediate forms (e.g. electricity, liquid fuels) and these to a multitude of services (Wilkinson 2011). As global energy consumption continues to increase, by as much as fifty per cent by 2030, the demand for water supplies and resources will also grow concurrently. This will place the energy sector in greater competition with other water users for already limited resources in many regions of the world. The competition for water resources will impact on future energy development and could have significant impacts on energy reliability and energy security in regions around the globe. The current trends in energy development could significantly increase the water footprint, rather than decrease it (Hightower 2012).

At present, electric power generation is one of the largest water withdrawal and use sectors in countries such as the United States and France, with the energy sector requiring levels beyond forty-five per cent of water withdrawn nationally (NETL 2011). Future energy developments such as biofuels, hydrogen, oil shale development, carbon sequestration and nuclear power could significantly increase water use and consumption and energize economic development. A major concern of gas shale development is the limitation imposed by access to reliable water supplies, as anywhere from eight to twenty-four million litres of water are required for hydraulic fracturing in each well (Mantell 2009). In the United States, the Bureau of Land Management (BLM 2008; Miller 2011) estimates that one to three barrels of water would be needed for each barrel of oil produced by in-situ extraction. Also, the water recovered after fracturing can be extremely high in salinity, presenting concerns not only about water quantity but also about the quality of wastewater (Hightower 2012).

Water availability has also proved to be a limiting resource for thermoelectric generation (NETL 2011; Hightower 2012), and permits to construct thermoelectric stations have been denied because of water-related considerations. Metals and mining, utilities production and supply and manufacturing segments are especially vulnerable to water scarcity as this would mean no electricity and no raw materials. Mining activities have been affected mostly in South America and African countries, where companies have opted to pull out due to high energy costs and competing water demands (Ernest/Young 2014).

Overall, the global landscape of energy consumption indicates nothing but an acceleration in growth. Global energy demand is expected to continue growing by more than one-third over the period to 2035, with China, India and the Middle East accounting for about sixty per cent of the increase and South East Asia emerging as an expanding demand centre. In 2013, energy consumption and production increased for every fuel type except nuclear power, at a rate below the ten-year average of 2.5 per cent, but still resulting in a growth in global CO2 emissions, though this also was below average (BP 2014). Emerging economies dominated global growth with eighty per cent of the global increase in energy consumption.

China once again had the largest growth increment followed by the United States (BP 2014).

Because of economic growth, China and India, together with the ASEAN countries,[5] have transferred the so-called global 'Centre of gravity' from the West to Asia (IEA 2013b: 15). The ten ASEAN member countries, whose combined GDP has increased by around three-quarters since 2000, resulting in increasing energy and water use, have very different patterns of resources use. Their individual policies on energy and water security, affordability and efficiency result from country-specific economic development and endowment with natural resources, as well as from political considerations. In all cases, energy security is driven by increasing reliance on imported energy; reduction of economic costs is linked to rising imports; and sustainability aspects of energy use are driven by concerns about local pollution. The ASEAN Power Grid and Trans-ASEAN Gas Pipeline are examples of collaboration but they are still a long way of implementation due to (for example) a lack of harmonization of technical and regulatory standards, the phasing out of end-user price subsidies, the guarantee of a third-party grid and pipeline access, and the establishment of a regional regulator (IEA 2013b).

The dynamic growth of the Chinese economy and energy demand is reshaping global energy markets with implications of global reach which are not totally understood at present. Together with India, China drives the growing dominance of Asia in global energy demand and trade, and, together, they are building approximately forty per cent of new generation capacity (IEA 2013a). China, the second largest economy in the world, is planning to add 1.2 terawatts (TW) of water-resilient power capacity by 2030, equivalent to 5.9 times India's current installed generation capacity. China is also planning to add 453 GW of coal-fired power capacity by 2020, in many cases in water-stressed areas, further increasing competition for water. By 2030, some eighty-seven per cent of power capacity will still require water for cooling and generation, down from ninety-seven per cent in 2012 (Chan/Robins/Knight 2012). This expansion will place further stress on water resources unless efficiency is greatly improved. Hydropower is also at risk from changing water availability brought about, for

---

5   The International Energy Agency uses the terms *ASEAN* and *South East Asian countries* interchangeably. These include Brunei Darussalam, Cambodia, Indonesia, Lao PDR, Malaysia, Myanmar, Philippines, Singapore, Thailand and Vietnam.

example, by more frequent and more extreme droughts and floods. Energy efficiency is as important as water efficiency because using less energy would reduce demand for power and alleviate some of the water stress. The question remains as to whether there is enough water to fuel China's power expansion.

The rise of unconventional oil and gas and of renewables, both important components of national energy programmes, is transforming our present understanding of the distribution of the world's energy resources. Unconventional oil and gas, or resources with atypical geological locations and that require a specific set or production techniques, include, among others, shale gas. Shale gas has the potential to displace fossil fuels and potentially slow the development of renewable sources. It also has the potential to turn countries that traditionally import natural gas into producers, making them more self-sufficient in domestic supplies. A clear example is the United States, with vast reserves of natural gas commercially available. Shale gas represents at present forty per cent of total natural gas production, a fact reflected in a lower market price for natural gas and that has changed the geopolitics of oil globally. Together with China, the US is expected to join the top tier of producers, allowing both of these major energy consumers to reduce their demand for coal and ultimately for oil (Shell International BV 2013).

Unconventional natural gas and oil resources are not free of criticisms. These relate to the challenges associated with the availability and use of surface water and groundwater for hydraulic fracturing; migration of stray gas and potential effects on overlying aquifers; and contamination of surface and groundwater. Science, technology and policies still have to develop these resources in a manner that is environmentally safe.

There is the question of when or whether the world will achieve an energy system based on one hundred per cent renewable resources. This may be feasible in theory but there are too many societal, political, technological, geographical and market-related concerns that make it improbable and perhaps not even desirable. The first challenge is the geographical location of the renewables resource base, which is frequently a long way from the centres of energy demand, often in different countries or continents. Furthermore, the absence of substantive investment for research and development compared with conventional generation and the fact that such resources still depend on subsidies have significantly reduced their impact (Shell International BV 2013).

Biofuels are not a new concept but the scale of production that is currently proposed is unprecedented. This is expected to more than triple by 2040, representing eight per cent of global demand for road transport fuel, if it is not affected by the cheaper availability of natural gas. Biofuel policies have created and supported a new demand for traditional food crops and this has triggered a cascade of impacts on food, agriculture and energy systems as well as land and water resources, not all of them sustainable. One example is the European Union, whose targets cannot be fully met using only EU domestic biomass. EU biofuel policy has thus triggered the creation of an increasingly globalized market for biofuels and biofuels feedstock, relying on agriculture in developing countries, mostly in Latin America and Asia. At the same time, such production must conform to 'sustainability' criteria (e.g. the Fuel Quality Directive and the *Round Table on Sustainable Biofuels* (RTSB). This is not always the case as it has the potential to influence food prices, further degrade land and water resources and limit access to natural resources for small landholders (HLPE 2013).

Table 31.1 shows the regional production and consumption of total biofuels in 2011. It indicates that the largest consumer and producer of biofuels (including biodiesel and ethanol) is North America (mostly the United States), followed by Central and South America.

Table 31.1: Regional production and consumption of biofuels, 2011. **Source:** Adapted from HLPE (2013). This information is originally from the US Energy Information Administration/ International Energy Statistics, available at: <http://www.eia.gov/>.

Geographical area	Consumption (per cent)	Production (per cent)
North America	52	53
Central and South America	23	28
Europe	19	13
Oceania	6	6
Eurasia[a], Middle East and Africa	0.2	0.2

a) The U.S. Energy Information Administration defines Eurasia as follows, see http://www.eia.gov/tools/glossary/index.cfm?id=E.

A main question is whether biofuels are a possible solution to the challenges posed by demand for fuel.

This is not a simple question and it is unclear what the impacts will be when biofuels become a significant part of agricultural production in any country, as the demand for certain crops will increase whether they are used for food, feed or fuel. This will have consequences for the production of other crops and for the use and distribution of scarce resources all along a diversity of supply chains. Policy and market decisions to stimulate such a large conversion towards biofuels are risky since understanding of the direct and indirect consequences is still uncertain. The interconnectedness of the world markets has shown that decisions taken with regard to agricultural production and use in one area might have enormous impacts on others as they can affect both supply and price (Stattman/Bindraban/Hospes 2008).

Global investment to supply populations with energy is of the order of US$1,600 billion annually. It has more than doubled in real terms since 2000 (IEA 2014). Some additional US$53 billion of investment is required in energy supply and efficiency to move the world on to a 2°C path away from fossil fuels and towards renewables, *carbon, capture and storage* (CCS) and nuclear power. The caveat is that fossil fuels will continue supplying much of the energy used worldwide and will still do so in 2035. This has enormous implications in terms of climate change as electricity demand is expected to increase by 4.2 per cent per year on average until 2035 (Stattman/Bindraban/Hospes 2008).

For energy production, water withdrawals in 2010 were of the order of 580 *billion cubic meters* (bcm), of which 70 bcm were consumed, with the power sector accounting for over ninety per cent of the total amount (IEA 2012). This indicates the strong dependence between water and energy. In the New Policies Scenario of the Organisation of Economic Cooperation and Development (OECD) and of the International Energy Agency (IEA), withdrawals will increase by about twenty per cent between 2010 and 2035 and consumption will rise even more, by eighty-five per cent. These trends will be driven by a shift towards power plants that are more efficient and have more advanced cooling systems where withdrawals will be reduced but consumption per unit of electricity produced will increase, and by expanding production of biofuels (table 31.2).

Given the projected rise in energy demand and the growing burden on natural resources, energy savings can be a very important and cost-effective source of additional energy supply, as well as having a beneficial effect on water resources (Wilkinson 2011). The so-

called *Energy Efficiency* (EE) market is expected to deliver substantial economic gains. For example, investment in EE technology by 2020 could reduce energy consumption in the EU by 250 megatonnes and in the US by 200 megatonnes (Burchardt/Hering/Klose et al. 2014). Challenges that need to be addressed include enhancing national policy coordination and regulatory frameworks, eliminating market distortions, encouraging the financing of energy efficiency projects, improving capacity building and data collection and monitoring, and evaluating the effectiveness of energy efficiency policies (IEA 2013).

The future of energy and thus of water in the world depends on the interplay of economics, social issues, pricing, technology and policies, as well as on political considerations. Demands in Asia, for example, are expected to grow at more than twice the global average, because of strong economic growth and rapid urbanization. Sustained and sustainable growth will require strong policies that are implementable and that encourage the adoption of the best management practices and available technologies to improve energy efficiency and thus improve effective water use (IEA 2013a).

There is a changing energy landscape as the energy future for which we need to plan will happen in an exceptionally different political landscape in the twenty-first century. Trying to think through energy futures while dealing with the pressing problems of energy resources and prices as well as food security and climate change is particularly difficult. To this complexity is added the fact that both developed and developing countries lack energy policy frameworks. Instead, there are energy policy conclusions that are drawn from environmental policies, such as the focus on renewables, and which are not based on environmental considerations but often depend on the requirements of market competition policy (Stattman/Bindraban/Hospes 2008).

Developing an energy system that meets climate targets, sustainable development objectives, and energy, food and water security concerns is a massive challenge. We still have to learn from the individual through to the institutional level that decisions about the ways we produce and use energy cannot be made without regard to wider concerns. Energy system development must be an integral part of sustainable development (Stattman/Bindraban/Hospes 2008).

Table 31.2: Global water use for energy production in the New Policies Scenario by region (bcm). **Source**: OECD/IEA (2012).

	Withdrawal				**Consumption**			
	2010	2020	2035	2010–35 (%)*	2010	2020	2035	2010–35 (%)*
OECD	**307**	**316**	**302**	**-0.1**	**30**	**39**	**46**	**1.7**
Americas	241	253	249	0.1	21	29	38	2.5
United States	206	214	212	0.1	19	26	35	2.5
Europe	61	57	49	-0.9	8	8	6	-1.4
Asia Oceania	5	6	5	0.1	1	2	2	1.9
Non-OECD	**276**	**346**	**388**	**1.4**	**35**	**56**	**76**	**3.1**
E. Europe/Eurasia	95	93	95	0.0	4	5	6	1.5
Asia	157	211	230	1.5	21	34	43	2.9
China	106	134	145	1.3	16	26	30	2.5
India	40	55	58	1.6	4	6	9	3.5
Middle East	3	4	5	1.3	2	3	4	1.5
Africa	5	7	8	1.8	3	4	4	1.6
Latin America	16	31	52	4.9	5	11	19	5.6
World	**583**	**662**	**691**	**0.7**	**66**	**95**	**122**	**2.5**
European Union	66	61	56	-0.7	8	8	6	-1.2

* Compound average annual growth rate.

31.4 The Next Phase of Globalization—The Role of Food

Food security will be one of the most important political topics of the twenty-first century. The FAO has said that food production must increase by seventy per cent in order to feed 9 billion people globally by 2050 (FAO 2011). This increase is needed because of the rising incomes and living standards of the largely urbanized middle classes in Asia. Food security has been a crucial topic throughout the history of humankind. As 'the father of the green revolution', Norman Borlaug, stressed in his Nobel Prize acceptance speech in 1970, "civilization as it is known today could not have evolved, nor can it survive, without an adequate food supply". Yet the development of food security affirms that it encompassed a wider notion of development. While agricultural development was important, food security addressed far wider social, political and economic developments. Since Europe was central to the global economy in the past centuries, the development of food security in Europe will be analysed first.

The pessimist Thomas Malthus predicted as early as 1798 that population growth would outstrip available food supplies, leaving (European) countries with nothing but starvation. Influenced by the dark experiences of the sixteenth and seventeenth centuries when farmers were unable to supply growing urban populations, Malthus foresaw deeper structural problems. However, although Malthus has been repeatedly quoted in history (e.g. by the Club of Rome in the 1970s), his sombre predictions were always nullified by the ability of farmers to produce more for growing populations. This was due to wider economic, political and social developments. European farmers increased their food output by up to 1000 per cent in the past 200 years due to reflexive governance by governments and societies in Europe. Mercantilism and agricultural markets that forced landowners to rethink their strategies replaced the old feudal systems. As a result, newly imported staple food crops such as potatoes from North America were introduced in the seventeenth and eighteenth centuries and allowed farmers to supply growing urban populations with much-needed additional staple foods (Kiple 2000). Another major shift in food security policy occurred in the nineteenth century. At first, agricultural imports from colonies in Asia and Africa permitted the European upper and later middle classes to consume more diversified and affluent diets. The end of colonialism in the twentieth century marked another evolution in food security policy. Agricultural trade increasingly contributed to domestic food security in Europe (Allan 2011). Food security was thus by no means limited to agricultural development but involved wider socio-economic shifts.

However, the most important gains were achieved during the twentieth century. Technological improvements brought about by research and development in Europe and North America increased agricultural outputs. The introduction of fertilizers, improved seeds and irrigation enabled farmers in the rich, Western part of the world to produce more food. In particular, the period after the Second World War saw a period of unprecedented agricultural growth. Governance complemented technological improvements. While America introduced the Farm Bill, and with it agricultural subsidies, during the Great Depression of the 1930s, Europe followed by pooling the agricultural policy of western European states in the 1950s. Europe's cooperation echoed the advent of a liberal functionalist paradigm of food security. This governance shift resulted in an over-supply of crops culminating in the butter mountains and milk lakes of the 1980s and 1990s. North American farmers further expanded their production capacities from the 1940s onwards to supply the world with US staple crops (McMichael 2009).

The successful shift in governance to producing more food also affected wider policy areas. The United States used this outcome of over-supply as part of their bargaining power in foreign policy in the 1970s to contain import-dependent regions such as the Middle East and countries in the former Eastern bloc around the Soviet Union (Woertz 2013). The Carter administration even issued a food export ban on the Soviet Union after the invasion of Afghanistan by the Soviet army in 1980. This had a decisive impact on the US farm sector. American small farms went bankrupt and this led to larger farms operated by fewer farmers (Paarlberg 1980). This laid the foundations for another shift in agricultural governance in the United States. As farm sizes grew, enabling economies of scale, food became increasingly commercialized. As a result of a power shift from small farmers to agribusiness, supermarkets took over increasing parts of the food supply chain. As food supply chains became longer and more sophisticated, supermarket

chains such as WalMart or Carrefour took over from small retailers in providing consumers with fresh food. The enormous outputs of farmers and a strategic subsidization of fodder crops such as wheat and soy in both North America and Latin America allowed the meat and dairy industry to supply more meat and dairy products to consumers (Keulertz 2013). At the same time, eating out in food outlets has increasingly become the norm in most Western countries since the 1980s. In sum, the commercialization of food provided a bullish market for the unprecedented growth of the food sector (McMichael 2009).

The food sector responded to wider shifts in development in the Western world. Farmers increasingly produced cheap food for more diversified economies. Technology played one part, consumer demand and governance of the food system another. Despite stagnating middle-class incomes since the 1970s, economies of scale derived from better technology and improved on-farm and off-farm supply chain management kept food prices low and thus demand growing (Keulertz 2013).

Farmers and agribusiness responded to increasing demand through ever-increasing yields that were quickly absorbed by the market. The size of the global food market was estimated at US$4.21 trillion in 2009 compared to US$3.7 trillion in 20005 (and it is estimated to have grown to US$5 to 7 trillion by 2014). Fresh food accounted for 52.6 per cent of the total value. Packaged foods contributed the rest. The reasons for the rapid expansion of the food market are seen in the economic growth of the first decade of the century alongside population increases that boosted global food consumption. A large part of this growth was driven by increasing prosperity in emerging markets, which have expanded faster than their counterparts in the developed world (Alpen Capital, 2011). However, climate change will severely test the resilience of the food system. As land endowed with water resources is already a very valuable resource, the food system may have to be prepared for systemic shocks that may have a number of effects. Although the role of North America, Australia and Europe remains strong, global economic change in favour of Asia has triggered a new round of food politics specifically related to food trading houses. Traders are of particular importance as they directly liaise with farmers to purchase, store and ship their produce. In 2003, seventy-three per cent of traded agricultural commodities were handled by only four multinational companies: Archers Daniels Midlands, Bunge, Cargill (all from the United States) and Louis Dreyfus (France)

(Keulertz 2013; Murphy 2012). All four companies not only trade the majority of bulk commodities in their home markets but have also acquired market leadership in Latin America and Australia.

31.5   You Can't Build Peace on Empty Stomachs

A pivotal moment for the future reliability of the global food system was the price shock that cast a shadow over the global food system at the end of the first decade of the new millennium. The food price spikes of 2007/08 and 2010/11 reminded decision-makers around the world of the centrality of food in state security. Highly volatile global commodity markets signalled to decision-makers that the food question embedded highly systemic risks for future development (Woertz 2013). Without adequate and reliable access to food, emerging economies could be trapped. As the first general director of FAO, Lord John Boyd Orr, stated, "You can't build peace on empty stomachs" (Borlaug 1970).

It is therefore clear that in a world that will hold 9 billion human beings by 2050, there are significant risks to future food supply and possibly peace *across the globe*. The current and future net exporters are located in water- and land-endowed regions such as Brazil, eastern Europe, Australia and Central Asia. These countries and regions could be pivotal for future peace. The risk areas are those regions affected by high population growth and/or natural resources constraints, such as the Middle East and North Africa, and South and East Asia. The only region that is notably different is sub-Saharan Africa, where food security is uncertain despite the ample availability of land and water resources.

31.6   Countries and Regions with a Positive Outlook

Russia, for example, seeks to utilize its opportunities in this next phase of globalized food. Russia seeks to become the leading wheat producer by 2019 using state capital by fostering the *United Grain Company* (UGC) that invests in farmland along the 'Chernozem belt' (the belt of highly fertile Black Earth soils that stretches from the Ukraine to Siberia). The ambitious goal of this venture is to export 200m tonnes of wheat by 2019. This has enabled Russia to curry favour with import-dependent countries such as Egypt

to which it supplies approximately forty per cent of wheat imports (Woertz/Keulertz forthcoming). Russia also plays a central role in another grain hot spot of the world, the Black Sea region. This region has been a target for investors since the mid-1990s, primarily by Western agribusiness, because it is one of the major agricultural growth zones of the world. With output increases of approximately four per cent per annum, the Black Sea region is expected to export 80m tonnes of grains by 2019 (Marshall 2014). It is expected that the *Middle East and North Africa* (MENA) region will absorb most of these production gains, providing investors with unique market opportunities if both regions achieve political stability.

In Latin America, on the other hand, the agricultural transformation process started during the 1960s and 1970s. Initially largely supported by Western capital (note, for example, that Cargill is the biggest agribusiness company in Argentina because its company policy is to re-invest the profits of its national branches in the market of origin), Latin America is today the biggest soy producer in the world (WWF 2014). Soy is especially important as it is used as animal feed across the world to keep the *proteinization* of diets in place. Europe and Asia are seen as the main destinations of Latin American soy (WWF 2014).

The final continent that has been able to grow and market its food in the global political economy is Australia. China has become one of the key markets for agricultural exports from Australia during the past fifteen years. In addition, Australia enjoys trade opportunities through the *Asia–Pacific Economic Cooperation* (APEC) trade agreement with the United States and the European Union, which has enabled the Australian agricultural sector to utilize opportunities for trade with a region associated with high population and economic growth. As a result, Australia's total agricultural goods and service provisions grew from AUS$3.5 in 1967 to AUS$301.5 billion in 2013 (DAFF 2013). The main commodities that have underpinned growth are wine, fisheries, meat and dairy, staple commodities, fruits and vegetables. At the same time, Australia has become a sought-after target country for agricultural investment from regions faced with water and/or land shortages. The situation of countries that seek farmland in countries like Australia is markedly different. These are seen as the losers in the new round of food globalization.

31.7   Risk regions and countries

The *Middle East and North Africa* (MENA) is a region that since the 1970s has effectively run out of the water needed to achieve self-sufficiency in food (Allan 2003). The tragic political developments in the MENA region in the past decades have also taken their toll with respect to agricultural production. Only 3.9 per cent of MENA arable land is under cultivation, in part due to conflict, in part due to water scarcity (Al Masah Capital 2012). The MENA region imports approximately half of its required food from other world regions (World Bank, 2008). In terms of US dollars, the bill for food imports is projected to reach US\$92.4 billion by 2020 (Al Masah Capital, 2012). Rising food prices and poor public budgets have already contributed to social unrest, as seen during the Arab Spring in 2011. Thus, the MENA region will be hit hardest of all the regions in the world when trying to meet future food security goals.

India and sub-Saharan Africa face similar food security concerns. However, India's cultural preference for 'green proteins' has allowed the country to take a less aggressive stance in the global market. Sub-Saharan Africa has shared a similar fate to the MENA region. High population growth and persistent severe poverty will translate into severe food security concerns in the coming decades. However, in theory, sub-Saharan Africa enjoys readily available water and land resources that could be made more productive to feed the continent (Binswanger 2009).

The other major food importer is China. At the advent of the 'Asian century', it is clear that food controlled by Western agribusiness has the potential to limit Asia's future growth. In an attempt to decouple itself from the Western world, the population giant of the world, China, has traditionally pursued a strategy of food self-sufficiency (Brautigam 2013). However, China's food security strategy has markedly changed during the past ten years. China can no longer feed its population through domestic food production. In particular, industrialization has taken its toll on food production. The Chinese have not only become more affluent, but the natural resources of the country have been decisively directed towards energy. For example, China's focus on coal has led to significant pollution of groundwater resources. It is estimated by the Beijing government that sixty per cent of China's groundwater is polluted. Fifteen per cent of the groundwater cannot be used as potable water (World Bank 2007).

The response by China has been a largely unnoticed paradigm shift. Through the Chinese state food company CofCo, the Republic of China has become increasingly active around the world. For example, China purchased the largest US pork producer, Smithfield, in 2013 (Koons/Venkat 2014). In addition, CofCo has launched a joint venture with agricultural traders in Asia such as the Noble Group, in order to increase its global food footprint (Koons/Venkat 2014). In terms of sales, the Noble Group is already larger than the US trader ADM. Yet Noble still mainly acts as a buyer on the global market. In contrast to Western traders, Noble still needs to establish relationships with farmers across the world. Thus, China's food security will depend on the development of cooperation with farmers around the world.

31.8   The Next Phase of Development in Practice: The Pivotal Role of Africa

Sub-Saharan Africa is the only region in the world where food supply can be significantly increased for both domestic and international consumers. By expanding the area under cultivation, which is currently less than half of its arable land, sub-Saharan Africa could become a crucial region for global peace. For this reason, sub-Saharan Africa is a major region of concern for domestic and international investors as well as for Western and Eastern development agencies seeking to increase food production. However, sub-Saharan Africa's farming sector faces wider development challenges. In particular, social, political, economic and cultural factors make food production in sub-Saharan Africa challenging. With a strong feature of smallholder agriculture dominated largely by female subsistence farmers or pastoralist family farmers, food output has lagged significantly behind other world regions (Keulertz 2013).

In addition, sub-Saharan Africa's infrastructure is in need of multi-billion dollar investments, including both on- and off-farm investments. Potential 'food bowls' such as Nigeria, Ethiopia, the Democratic Republic of Congo, Tanzania, South Sudan and Mali have either experienced long and devastating internal or external wars and conflict or have poor governance systems in place. Nevertheless, sub-Saharan African countries have seen global interest in their farmland resources since the mid-2000s (Deininger/Byerlee/Lindsay et al. 2011; McMichael 2011). Estimates of farmland acquired range from thirty to sixty-two million, but only a tiny proportion of this assumed investment has been made operational. For a while, China was regarded as the major investor in African agriculture. What was by some labelled as 'land grabbing' was deconstructed by analysts with field experience (Brautigam 2013; Woertz 2013). Instead of investing in land and water resources, China has pursued a different, broader development approach.

Although the data is highly opaque, researchers have analysed media reports about China's aid to sub-Saharan African countries. They analysed 1,673 projects that appeared in Chinese, African and international media. In contrast to previous widely-held assumptions, only eighty-three projects were associated with the energy sector, and forty-four were in mining. Instead, the largest Chinese investments were made in the health/government/society sectors and in transport (Strange/Parks/Tierney et al. 2013). This underlines China's wider engagement with Africa, which reflects the fact that Africa is not merely a source of natural resources for China, but also a market for finance, manufacturing, and construction. It is further assumed that thirty per cent of transport projects in Africa are financed by Chinese public and private sector companies (Lancaster 2014). Transportation infrastructure is especially important for the agricultural sector, as it allows farmers to access markets such that their produce remains fresh and edible. Whether or not China will attempt to transform Africa's agricultural sector through direct investment in agriculture remains an open question, but China seems to have learned a lesson from its Western counterparts by understanding how the development of infrastructure is vital to achieving sustainable development on all sides.

A recent initiative by China and its economic allies has signalled a further shift in the global political economy of development. On 15 July 2014, representatives from Brazil, Russia, India, China and South Africa signed an agreement to establish the New Development Bank, seen as a direct countermeasure to the Western-led World Bank. The goal of the new bank is to raise US$100 billion over time to directly fund infrastructure projects in developing countries (Bax 2014). Such a move could tip the global balance of power towards the newly emerging economies in Asia, Africa and Latin America. The lessons of this initiative are directly drawn from the Western model that emerged in the days when the primary sector of the economy was predominantly in the hands of the nobility. It took Europe and America several centuries to produce the immense amounts of food used in foreign political and economic relations. The new Asian century may see a similar model emerging within a

few decades, in which a wider model of development to achieve food security will be pivotal. It could be nothing less than the next phase of globalization, yet this time created by Asian economies instead of by Europe and North America.

31.9 Innovation and Partnerships as the Way Forward

Concerns about the impact that the deterioration of the environment and resource scarcity can have on the livelihoods of billions of people and on all productive sectors have prompted a series of partnerships and innovation initiatives. An increasing number of private sector groups, UN and international organizations, *non-governmental organizations* (NGOs), universities and environmentalists are working towards the common goal of sustainable development and the promotion of peace, stability and human security. Innovative ideas, processes and products focused on conservation and efficiency are being promoted all over the world, reliable data is being collected and cutting-edge technology is being developed, although the necessary policies for their implementation in the longer term are still lagging behind.

Traditional partnerships as well as previously unthinkable ones between the private sector and international organizations and NGOs are being developed by different actors at different scales for the common good. And they are delivering higher outputs than expected. A change of mindset on the part of all actors that encourages them to work together has proved beneficial on the road towards sustainable development (SustainAbility 2014).

A growing number of mostly multinational companies are embracing sustainability policies and practices as integral components of their business models. Water and energy risks are being considered by financiers, investors and companies as a core feature of capital expenditure plans. They are also much more aware of water quotas and pollution targets as their aim is to make operations more efficient, and they are examining the effects of potential water shortages on facilities located in water-scarce provinces.

Companies still aim to achieve the normal goals of continuous profitability and growth but many are also looking to concurrently create social and environmental benefits in the communities within which they operate. Development led by the private sector has been essential in creating numerous communities of interest between producers, factory workers, consumers and stakeholders, with the objective of bringing about sustainable rural development in the developing world. Evidence-based research shows that the private sector can make a major impact in improving the human environment, contributing to social progress through the alleviation of poverty, and increasing the standards of living of millions of people, as well as making a profit (Biswas/Tortajada/Biswas-Tortajada et al. 2014; Biswas-Tortajada/Biswas 2015). As for resource use, many companies are increasingly managing their own water needs and looking for trade-offs along their entire production chain, as well as taking responsibility for the protection and conservation of ecosystems that supply fresh water (see, for example, ILSI 2013).

Decisions taken have the potential to scale across supply chains all over the world multiplying sustainability gains and benefitting large populations. These initiatives are essential for the future water, food and energy needs of the world. Examples include agricultural water management; watershed restoration, protection and payments programmes; eco-compensation models; water funds; green infrastructure investments; alternative management choices for terrestrial, freshwater and marine ecosystems; the Water Risk Atlas (analysis of water risks in terms of quantity, quality, regulatory and reputational risks); development of water markets; water quality trading; and environmental impact bonds (SustainAbility 2014).

In the past decades the water and energy sectors have seen several major technological breakthroughs. Membrane technology has developed to the point where desalination is commercially viable and where wastewater can be treated to convert it to drinking water quality. Anaerobic and biological technologies have greatly improved the sustainability and cost-effectiveness of wastewater treatment. Network monitoring and modelling technologies have reduced water losses in piped water supply systems and allowed the optimization of water resource systems. Drip and micro-irrigation solutions have reduced agricultural water consumption.

The strength of new forms of energy and especially water will be the direct result of strong policies but also of social acceptability. For Singapore, within very strong policy and monitoring frameworks, in addition to communication strategies that have resulted in social acceptability, desalinated water and highly-treated reused wastewater (NEWater) are able to supply up to twenty-five percent and thirty percent of total demand for the city-state respectively for non-direct potable, commercial and industrial uses. There

are plans to augment the capacities of both sources of water to supply up to eighty per cent of Singapore's total water demand by 2060 (Tortajada/Joshi/Biswas 2013). Australia is another country that has planned for the use of desalinated water and recycled wastewater for human consumption, but has not achieved social acceptability for this last one. In the case of desalinated water, one of the largest schemes, the 232,000m3/day Western Corridor Recycled Water Project in Queensland constructed during the Millennium Drought, has now become non operational because it is no longer necessary following the end of the drought. Several cities in the United States as well as Windhoek in Namibia have developed schemes for direct and non-direct potable use of reused water, all with immense energy implications (Lahnsteiner/Lempert 2007; EPA, at: <http://www. twdb.texas.gov/publications/shells/WaterReuse.pdf>). One way or another, cities and countries are looking for additional resources for their continued development.

The rise of unconventional water, oil and gas and of renewables will continue transforming our present understanding of the distribution of the world's energy resources and, most importantly, has the potential to make countries more resilient to the changing environment if policy mixes are developed that take into account the interconnectedness of related sectors. Administrators would do well to consider the water, food and energy sectors among their policy priorities.

31.10   Conclusions

This chapter has framed water, energy and food security in a wider developmental context. It has shown that all three types of security cannot be understood without placing them in a wider conceptual and historical framework of past and future developments. Water, energy and food security will be unquestionably at the heart of the development challenges of the twenty-first century. However, given the interlinkages between the three sectors, only integrated approaches will yield successful management outcomes. Within this challenge of integrated decision-making lies the need for policy-making. All three sectors are decisively characterized by the paramount role of public policy. People in charge of 'statecraft' had for long not worried about natural resources management and in particular water was seen as a 'free' and abundantly available resource. Yet 'business-as-usual' approaches will no longer suffice in the decades ahead. On the contrary, new policy frameworks that promote policy-making outside policy silos will be required to achieve security in all three sectors. The profound environmental and economic changes the world is facing will no longer allow one-dimensional approaches. It will be up to decision-makers to understand that change is inevitable but that it also holds tremendous opportunities. If rightly addressed, those opportunities will yield economic development and prosperity for an unparalleled number of human beings.

The notion of water, energy and food needs, demands and securities within a framework of sustainable development is not new, but it is being reconsidered in the current challenging environment, and invites closer scrutiny. A shared understanding of the challenges is needed as a basis for multi-stakeholder collaboration. This has increasingly been recognized as the most effective way to address global risks and to build resilience against them. However, bringing a very diverse group of stakeholders to the table is not an easy task, as the ill-fated meetings of the Conference of Parties (COP) demonstrate annually. Getting decision-makers to overcome silo thinking is key to achieving new policy outcomes. It is indeed the mega-challenge of our time.

References

Al Masah Capital Limited, 2012: *MENA Food Security: Are We Doing Enough to Feed the Population?* (Dubai: Al Masah Capital Limited).

Alexandratos, Nikos; Bruinsma, Jelle, 2012: *World Agriculture towards 2030/2050. The 2012 Revision.* ESA Working Paper No: 12-03 (Rome: Food and Agriculture Organization of the United Nations).

Allan, John, 2003: *The Middle East Water Question* (London: I. B. Tauris).

Allan, John, 2011: *Virtual Water: Tackling the Threat to our Planet's Most Precious Resource* (London: I. B. Tauris).

Alpen Capital Investment Banking, 2011: *GCC Food Industry* (Dubai: Alpen Capital Investment Banking)

Andrews-Speed, Philip; Bleischwitz, Raimund; Boersma, Tim; Johnson, Corey; Kemp, Geoffrey; Van Deveer, Stacy, 2013: *The Global Resources Nexus. The Struggle for Land, Energy, Food, Water and Minerals* (Washington DC: Transatlantic Academy).

Bax, Tahmeena, 2014: "Poll: Will the BRICS bank shift the balance of power?", in *The Guardian*, 25 July 2014.

Binswanger, H, 2009: *Awakening Africa's Sleeping Giant: Prospects for Commercial Agriculture in the Guinea Savannah Zone and Beyond.* In Directions in Development (Washington, D.C.: World Bank and FAO).

Biswas, Asit, 2008: "Integrated Water Resources Management: Is it Working?", in: *International Journal of Water Resources Development*, 24,1: 5-22.

Biswas, Asit; Tortajada, Cecilia, 2009: "Changing Global Water Management Landscape", in: Biswas, Asit; Tortajada, Cecilia; Izquierdo, Rafael (Eds.): *Water Management in 2020 and Beyond* (Berlin: Springer): 1-34.

Biswas, Asit; Tortajada, Cecilia; Biswas-Tortajada, Andrea; Joshi, Yugal; Gupta, Aishvarya, 2014: *Creating Shared Value. Impacts of Nestle in Moga, India* (Heidelberg: Springer).

Biswas, N. R., 2011: "Is the Environment a Security Threat?", in: *International Affairs Review*, 20,1 (Winter): 1-22.

Biswas-Tortajada, Andrea; Biswas, Asit, 2015: *Sustainability in Coffee Production: Creating Shared Value Chains in Colombia* (London: Routledge).

Boccaletti, G., 2009: *Charting our Water Future.* (London: McKinsey).

Borlaug, N., 1970: *Nobel Lecture: The Green Revolution, Peace and Humanity.* Noble Lecture delivered on 11 December 1970. Noble Prize Committee, Oslo; at: <http://www.nobelprize.org/nobel_prizes/peace/laureates/1970/borlaug-lecture.html> (17 January 2015).

BP (British Petroleum), 2014: *BP Statistical Review of World Energy 2014*; at: <http://www.bp.com/content/dam/bp/pdf/Energy-economics/statistical-review-2014/BP-statistical-review-of-world-energy-2014-full-report.pdf>

Brauch, Hans Günter; Oswald Spring, Úrsula; Grin, John; Mesjasz, Czeslaw; Kameri-Mbote, Patricia; Behera, Navnita Chadha; Chourou, Béchir; Krummenacher, Heinz (Eds.), 2009: *Facing Global Environmental Change: Environmental, Human, Energy, Food, Health and Water Security Concepts* (Berlin–Heidelberg–New York: Springer-Verlag).

Brautigam, Deborah, 2013: "Chinese Engagement in African Agriculture: Fiction and Fact", in: Allan, John; Keulertz, Martin; Sojamo, Suvi; Warner, Jeroen (Eds.): *Handbook of Land and Water Grabs: Foreign Direct Investment and Food and Water Security* (Abingdon: Routledge): 91-103.

Burchardt, Jens; Hering, Gunar; Klose, Frank; Koehn, Jannis; Maciel, Joao; Meures, Nikolai; Rubel, Holger (Eds.), 2014: *The Energy Efficiency Opportunity: Winning Strategies for a High-Growth Market* (Boston: Consulting Group Berlin: BCG).

Castle, Stephanie; Thomas, Brian; Reager, John; Rodell, Matthew; Swenson, Sean; Famiglietti, James, 2014: "Groundwater Depletion during Drought Threatens Future Water Security of the Colorado River Basin", in: *Geophysical Research Letters*, 41,16: 5904-5911.

Central Ground Water Board, 2006: *Ground Water Year Book - India 2005-2006* (Delhi: Government of India)

Central Ground Water Board, 2012: *Ground Water Year Book - India 2011-2012* (Faridabad: Government of India)

Chan, Wai-Shin; Robins, Nick; Knight, Zoe, 2012: *No water, no power. Is There Enough Water to Fuel China's*

Power Expansion?, HSBC Global Research, Climate Change, (September).

Crutzen, Paul J., 2002: "Geology of mankind", in: *Nature*, 415: 23-23.

Deininger, Klaus; Byerlee, Derek; Lindsay, Jonathan; Norton, Andrew; Selod, Harris; Stickler, Mercedes, 2011: *Rising Global Interest in Farmland: Can it Yield Sustainable and Equitable Benefits?* (Washington: World Bank).

DAFF (Department of Agriculture Forestry and Fisheries), 2013: *Australian Food Statistics 2012-13* (Canberra: Government of Australia).

EPA (Environment Protection Agency): *Water for Texas. Water Reuse*; at <http://www.twdb.texas.gov/publications/shells/WaterReuse.pdf>.

Ernest Young, EY's Global Mining & Metals Center; EYGM Limited, 2014: *Business Risks Facing Mining and Metals 2014-2015*; at: <http://www.ey.com/Publication/vwLUAssets/EY-Business-risks-facing-mining-and-metals-2014%E2%80%932015/$File/EY-Business-risks-facing-mining-and-metals-2014%E2%80%932015.pdf>

FAO (Food and Agriculture Organization), 2011: *The State of the World's Land and Water Resources for Food and Agriculture* (Rome: FAO).

Fawell, John; Ong Choon Nam, 2012: "Emerging Contaminants and the Implications for Drinking Water", *in: International Journal of Water Resources Development*, 28,2: 247-263; <DOI: 10.1080/07900627.2012.672394>.

Gleick, Peter, 2003: "Water Use", in: *Annual Review of Environment and Resources*, 28: 275-314.

Giordano, Mark; Shah, Tushaar, 2014: "From IWRM Back to Integrated Water Resources Management", in: *International Journal of Water Resources Development*, 30,3: 364-376; <DOI: 10.1080/07900627.2013.851521>.

Goldman Sachs, 2010: "Is this the 'BRICs decade'?", in: *BRICs Monthly*, Issue No. 10/03 (20 May).

Hajkowicz, Stefan; Cook, Hannah; Littleboy, Anna, 2012: *Our Future World. Global Megatrends that Will Change the Way We Live* (Australia: CSIRO).

Hightower, Mike, 2012: "Water Impacts on Energy Security and Reliability", in: Bigas, Harriet (Ed.): *The Global Water Crisis: Addressing an Urgent Security Issue*. Papers for the InterAction Council, 2011-2012 (Hamilton, Canada: UNU-INWEH): 18-25.

HLPE (High Level Panel of Experts), 2013: *Biofuels and Food Security*. A Report by the High Level Panel of Experts on Food Security and Nutrition of Committee on World Food Security (Rome: Young).

IEA (International Energy Agency), 2012: *World Energy Outlook 2012* (Paris: IEA/OECD).

IEA (International Energy Agency); OECD (Organisation for Economic Co-operation and Development), 2013a: *World Energy Outlook 2013* (Paris: IEA/OECD).

IEA (International Energy Agency); OECD (Organisation for Economic Co-operation and Development), 2013b: *Southeast Asia Energy Outlook* (Paris: IEA/OECD).

IEA (International Energy Agency); OECD (Organisation for Economic Co-operation and Development), 2014: *World Energy Investment Outlook. Special Report* (Paris: IEA/OECD).

ILSI Research Foundation, 2013: *Water Recovery and Reuse: Guideline for Safe Application of Water Conservation Methods in Beverage Production and Food Processing* (Washington: ILSI).

Kates, Robert W., 2001: "Sustainability Transition: Human-Environment Relations", in: Smelser, N.J.; Baltes, Paul, (Eds.): *International Encyclopedia of the Social and Behavioral Sciences* (Oxford: Pergamon): 15325 15329.

Keulertz, Martin, 2013: *Drivers and impacts of farmland investment in Sudan: water and the range of choice in Jordan and Qatar*. In Department of Geography (London: King's College London).

Kiple, Kenneth, 2000: *The Cambridge World History of Food Part II* (Cambridge: Cambridge University Press).

Koons, Cynthia; Venkat, P. R., 2014: "China's Cofco, Hopu to Buy 51% Stake in Noble Agriculture Unit", in: *Wall Street Journal*, 2 April.

Kumar, Dinesh; Scott, Christopher; Singh, P.P., 2013: "Can India Raise Agricultural Productivity while Reducing Groundwater and Energy Use?", in: *International Journal of Water Resources Development*, 29,4: 557-573; <DOI:10.1080/07900627.2012.743957>.

Lahnsteiner, Josef; Lempert, G., 2007: "Water Management in Windhoek/Namibia", in: *Water Science Technology*, 55,1: 441-448.

Lancaster, Kirk, 2014: *China's Investment in Africa* (Chicago: Chicago Council on Foreign Relations).

Lum, Thomas; Fischer, Hanna; Gomez-Granger, Julissa; Leland, Anne, 2009: *China Foreign Aid Activities in Africa, Latin America and Southeast Asia*. CRS Report of Congress (Washington DC: Congressional Research Service).

Mantell, Matthew, 2009: "Deep Shale Natural Gas: Abundant, Affordable, and Surprisingly Water Efficient", Paper presented at the 2009 Ground Water Protection Council Water/Energy Sustainability Symposium, Salt Lake City, 13-26 September. Energy Water Sustainability Symposium.

Marshall, Andrew, 2014: *Black Sea's Rising Grain Tide*. Farmonline; at: <http://www.farmonline.com.au/news/agriculture/cropping/general-news/black-seas-rising-grain-tide/2691044.aspx.>

McMichael, Philip, 2009: "A Food Regime Genealogy", in: *Journal of Peasant Studies*, 36: 139-169.

McMichael, Philip, 2012: "The Land Grab and Corporate Food Regime Restructuring", in: *Journal of Peasant Studies*, 39: 681-701.

McMichael, Philip, 2013: "Land Grabbing as Security Mercantilism in International Relations", in: *Globalizations*, 10, 1: 47-64.

Millennium Ecosystem Assessment, 2005: *Ecosystems and Human Well-Being: Synthesis* (Washington, DC: Island Press).

Miller, Bart, 2011: "Oil Shale and Water", in: Kenney, Douglas; Wilkinson, Robert (Eds.): *The Water-Energy Nexus in the American West* (Cheltenham and Northampton: Edward Elgar): 45-56.

Molden, David; Lautze, Jonathan; Shah, Tushaar; Bin, Dong; Giordano, Mark; Sanford, Luke, 2010: "Governing to Grow Enough Food without Enough Water—Second Best Solutions Show the Way", in: *International Journal of Water Resources Development*, 26,2: 249-263; <DOI: 10.1080/07900621003655643>.

Murphy, S.; Burch, D.; Clapp, J., 2012: *Cereal Secrets: The World's Largest Grain Traders and Global Agriculture*, Oxfam Research Report (Oxford: Oxfam).

NETL (National Energy Technology Laboratory), 2011: *Estimating Freshwater Needs to meet Future Thermoelectric Generation Requirements, 2011 Update* (Albany, OR: US Department of Energy).

OECD (Organisation for Economic Co-operation and Development), 2011: *Water Governance in OECD Countries: A Multi-level Approach* (Paris: OECD).

Paarlberg, Robert, 1980: "Lessons of the Grain Embargo", in: *Foreign Affairs*, 59: 144-162.

Pate, Ron; Hightower, Mike; Cameron, Chris; Einfeld, Wayne, 2007: *Overview of Energy-Water Interdependencies and the Emerging Energy Demands on Water Resources* (Albuquerque: Sandia National Laboratories).

Ringler, Claudia; Biswas, Asit; Cline, Sarah (Eds.), 2010: *Global Change: Impacts on Water and Food Security* (Heidelberg: Springer).

Rockstrom, Johan M.; Falkenmark, Marlin; Allan, Tony; Folke, Carl; Gordon, L.; Jägerskog, Anders; Kummu, M.; Lannerstad, M.; Meybeck, M.; Molden, D.; Postel, S.; Savenije, H.H.G.; Svedin, U.; Turton, Anthony.; Varis, O. 2014: "The unfolding water drama in the Anthropocene: towards a resilience-based perspective on water for global sustainability", in: *Ecohydrology*, 7: 1249-1261; <10.1002/eco.1562>.

Rodell, Matthew; Velicogna, Isabella; Famiglietti, James, 2010: "Satellite-based Estimates of Groundwater Depletion in India", in: *Nature*, 460: 999-1002.

Scheierling, Susanne M.; Treguer, David O.; Brooker, James F.; Decker, Elisabeth, 2014: *How to Assess Agricultural Water Productivity? Looking for Water in the Agricultural Productivity and Efficiency Literature* (Washington, DC: The World Bank).

Shell International BV, 2013: *New Lens Scenarios* (The Hague: Shell).

Stattman, S. L.; Bindraban, P. S.; Hospes, O., 2008: Exploring Biodiesel Production in Brazil, (Wageningen UR: Plant Research International).

Strange, Austin; Parks, Bradley; Tierney, Michael; Fuchs, Andreas; Dreher, Axel; Ramachandran, Vijaya, 2013: *China's Development Finance to Africa: A Media-based Approach to Data Collection* (Geneva: Center for Global Development).

Stiehler, Alexander; Vogel, Sebastian, 2014: *Longer Term Investments*. UBS AG Report (Zürich: UBS).

SustainAbility, 2014: "Evaporating Asset. Water Scarcity and Innovations for the Future"; at: <file:///C:/Users/spphctq/Downloads/evaporating_asset_water_scarcity_and_innovations_sep_2014.pdf>.

Tortajada, Cecilia, 2014: "Water Resources: An Evolving Landscape", in: Currie-Alder, Bruce; Kanbur, Ravi; Malone, David; Medhora, Rohinton (Eds.): *International Development: Ideas, Experience, and Prospects* (Oxford: Oxford University Press): 448-462.

UNESCO (United Nations Educational, Scientific and Cultural Organization), 2014: *Water and Energy*. The United Nations World Water Development Report 2014, Vol. 1 (Paris: UNESCO).

Wilkinson, Robert, 2011: "The Water-Energy Nexus: Methodologies, Challenges and Opportunities", in Kenney, Douglas; Wilkinson, Robert (Eds.): *The Water-Energy Nexus in the American West* (Cheltenham—Northampton: Edward Elgar): 3-17.

Willaarts, Bárbara; Garrido, Alberto; Llamas, Ramón, 2014: *Water for Food Security and Well-Being in Latin America and the Caribbean. Social and Environmental Implications for a Globalized Economy* (Abingdon: Routledge).

Woertz, Eckart, 2013: *Oil for Food. The Global Food Crisis and the Middle East* (Oxford: Oxford University Press).

Woertz, Eckart; Keulertz, Martin: "State Actors in International Agro-Investments: The Role of China, Russia and Gulf", in: *Development Policy*, in press.

World Bank, 2007: *Cost of Pollution in China. Environmental and Social Development Unit, East Asia & Pacific Region* (Washington DC: World Bank).

World Bank, 2008: *Middle East & North Africa: Agriculture and Rural Development* (Washington DC: World Bank).

WWF (World Wildlife Fund), 2014: *Soy* (Geneva: WWF).

32   Sustainability Transition in a Vulnerable River Basin in Mexico

Úrsula Oswald Spring[1]

Abstract

This chapter examines a transition process in a river basin in the central part of Mexico that is highly affected by climate change, social deterioration and the drugs war. The *Yautepec River Basin* is particularly prone to climate impacts because of its abrupt slopes, numerous affluents, and high population density in its floodplain, which is frequently exposed to hurricane impacts. Taking into account laissez-faire policies, illegality, government corruption, and public insecurity in a region with high levels of dual vulnerability (environmental and social), as well as subject to hazard impacts, the chapter reviews the constraints of the process of transition towards sustainability, especially when the goals of the people and of the government are not identical. People organized themselves regionally in order to simultaneously increase public security and freedom from hazard impacts, which both directly affect life and livelihood. Using original survey data, the study examines the risks perceived by the people and the ways in which they reduce these risks of extreme events, as well as how they cooperate with the authorities to improve environmental conditions, to enhance disaster risk reduction activities and to reduce the impact of criminal acts. The transition process requires the integration of numerous social, economic, political, environmental, judicial, cultural and mental factors in order to develop processes and resilience from the bottom up to deal with this complex emergency, as well as to force the different levels of government to enhance all four aspects of the human security of the people: 'freedom from fear', 'freedom from want' 'freedom from hazard impacts', and 'freedom to live in dignity'.

Keywords: Sustainable development, engendered and sustainable peace, Anthropocene, peace concept, patriarchy, civilization, empowerment, equity, values, cultural identity.

32.1   Introduction, Objectives, Research Question, the Scientific Approach and Methods

This chapter examines a transition process in a river basin in the central part of Mexico that is highly affected by climate change, social deterioration and the drugs war.[2] Mexico as a country and the *Yautepec River Basin* (YRB) in particular are particularly prone to climate impacts. They Yautepec River emerges from the glacier of the volcano Popocatepetl at an altitude of 5,450 metres and its basin extends outwards as far as the Federal District of Mexico and the states of Morelos and Mexico (figure 32.1).

The study region depends for its agriculture on a supply of water from the monsoon and from irrigation. The topography consists of high mountains and a wide floodplain, with two dominant ecosystems: pine-oak and low-lying dry deciduous forest, whose trees lose their leaves during the dry season. The climate is influenced by both oceans, and because of the steep mountains the region is extremely vulnerable to flash floods, soil erosion and periodic droughts. People also suffer periodic eruptions of the volcano Popocatepetl and other geophysical hazards such as earthquakes, which are frequent in the region. As well as erosion caused by logging, which is in the hands of organized crime, the mountain part of the basin (Los Altos) is also losing topsoil, and all the affluents are being eroded. Together, these natural and illegal phenomena have caused environmental vulnerability aggravated by poverty, socio-economic stagnation and poor public services (education, health, water, poverty

---

1   Prof. Dr. Ursula Oswald Spring, full time Professor/Researcher at the National University of Mexico (UNAM) in the Regional Multidisciplinary Research Center (CRIM); Email: <uoswald@gmail.com>.

© Springer International Publishing Switzerland 2016
H.G. Brauch et al. (eds.), *Handbook on Sustainability Transition and Sustainable Peace*,
Hexagon Series on Human and Environmental Security and Peace 10, DOI 10.1007/978-3-319-43884-9_32

alleviation, etc.), together with a high degree of government corruption (Oswald Spring 2014a: 151 ff; Bogardi, Oswald Spring, Brauch 2016).

Organized crime has taken further advantage of this situation and involved underemployed and unemployed young people in illegal activities. Drugs are produced and to some extent also consumed in the region, which is also affected by other illegal activities such as drug production, trafficking, kidnapping, extortion and also logging in the national parks. The result is an increasing dual environmental and social vulnerability in a region where law is almost non-existent (98.5 per cent of crimes are not punished, Acosta 2011: 79) because of inadequate investigations and corrupt judges. This has resulted in the loss of the legal right to defence, public insecurity, high homicide rates and a deterioration in environmental and social conditions.[3]

32.1.1   Objective of the Study

The chapter studies possible alternative pathways for a process of transition to sustainability in the face of laissez-faire policies, illegality, government corruption, and public insecurity in a region with a high level of dual vulnerability (environmental and social) and which is subject to extreme events. The social and environmental vulnerability of the people means there is a dual constraint on the process of transition towards sustainability. The goals of the people and of the government in this process are not identical. Government activities are uncoordinated, and this offers opportunities for corruption, where some government depart-

ments protect organized crime. The people want both public security and freedom from hazard impacts, both of which directly affect life and livelihood. Thus a transition process would require major changes to the numerous existing social, economic, political, environmental, cultural and mental constraints imposed by federal government on the people exposed to the problems. Confronted by the lack of action from the top down, and the absence of the rule of law, people are developing processes of resilience from the bottom up to deal with the complex emergency[4] and to force the different levels of government to enhance their human security (Adger, Pulhin, Barnett et al. 2014) as exemplified by 'freedom from fear' and 'freedom from hazard impacts' (Brauch 2005a).

32.1.2   Research Questions

How have the federal, state and municipal governments responded (top-down) to this dual vulnerability in the YRB? How can a bottom-up social organization in cooperation with the three levels of government reduce the hazard impacts of flash floods and droughts, in order to make safe life and livelihood and at the same time find a way to reduce corruption, reduce organized crime and create sustainable jobs? What worries people most? What have been the obstacles and constraints that have prevented and slowed down the implementation process and how could they be overcome? How can an integrated basin management reduce government corruption and organized crime and take advantage of the opportunities offered by the business community to improve the living conditions of the people of the YRB?

32.1.3   Structure of the Chapter

The chapter is divided into seven parts. After this introduction, part 32.1 explains the quantitative and qualitative methodology. Part 32.2 describes the study

---

2   This research project was financed by PAPIIT IN300213 of the National Autonomous University in Mexico (UNAM) for a project on "Integrated water management of a river basin affected by climate change: risks, adaptation and resilience". I am deeply grateful to the Regional Centre of Multidisciplinary Research at the National Autonomous University of Mexico (CRIM-UNAM) together with the sabbatical support of Conacyt and PASPA-DGAPA-UNAM, who supported this research and my sabbatical year, which included a visit to CUSRI at Chulalongkorn University in Bangkok and field research in Thailand. I am deeply indebted to PD Dr. Hans Günter Brauch and the four anonymous reviewers for their careful and detailed comments on the manuscript, and to Michael Headon for his perceptive language editing. I want also to thank MSc. Ariana Estrada Álvarez, Geographer Anahí Bustamante, Geographer Joaquin Cartulina, and Ángel Paredes Rivera for the support in the survey and the design of the maps.

3   Transparencia International (2014); at: <http://www. transparencia.org/>; Human Rights Watch (2014).

---

4   Unicef (1997) defined complex emergency as as "any situation in which the lives and well-being of children are at such risk that extraordinary action, i.e. urgently required action beyond that routinely provided, must be mobilized to ensure their survival, protection and well-being." (See E/ICEF/1997/7 and 15.) IASC (1994) defined it as "a humanitarian crisis in a country, region or society where there is total or considerable breakdown of authority resulting from internal or external conflict and which requires an international response that goes beyond the mandate or capacity of any single agency and/or the ongoing United Nations country program."

area and defines the key scientific concepts of transition goals, dual vulnerability, disaster risk reduction and management, complex emergencies, and resilience. The third part examines the complexity of the river basin and its environmental, social, and political constraints within a contradictory social arena[5] in the region. Part four discusses how social actors cope with these multiple threats, the socio-environmental activities developed by the authorities and the people, the struggle against organized crime, and the resultant agenda for a sustainable transition. Using original survey data, part 32.5 analyses the risks perceived by the people and the way they reduce these risks of extreme events, and how the authorities cooperate with them to improve environmental conditions and to enhance *disaster risk reduction* (DRR) activities. In part 32.6, data on public insecurity in Mexico, the state of Morelos and the river basin are analysed, together with proposals for resilience-building against flood and drought within a context of organized crime and public insecurity. The concluding part (32.7) suggests as a possible outcome an integrated river basin management scheme (IWRM; GWP 2009), together with a reconstruction of the social tissue[6] and the development of policies and practices geared towards a process of sustainable transition.

32.1.4   Scientific Approach and Methods

This chapter is based on original empirical data that were collected and analysed by a multidisciplinary research group at the *National Autonomous University of Mexico* (CRIM/UNAM). This research group compared the activities of different actors; it reviewed the arena of environmental and social threats, assessed the policy agenda in response to dual vulnerability (environmental and social) and the complex emergency in the river basin, and it put together activities carried out by the people to deal with both environmental and social problems.

The scientific background of the research project was multidisciplinary: ecology, anthropology, psychology, geography, demography, and development studies (Adger, Pulhin, Barnett et al. 2014). The research group compiled quantitative and qualitative data using participative observation, semi-structured interviews, focus group discussions, in-depth interviews, and examination of official databases. The survey was conducted in three phases between 2010 and 2013, guided by the research questions.

The full research included three random surveys of 3,955 persons: after the great flood in 2010, later in 2012, and finally in 2013 to assess the impact of the floods in the YRB (table 32.1). In-depth interviews were conducted with local leaders, politicians and key people in the communities and in the state of Morelos. The results from both quantitative and qualitative data together with partial analysis were discussed with the affected communities and authorities. During the fieldwork, participant observation was used, five focus group discussions were conducted, and several participatory rural studies were carried out on water security (Oswald/Brauch 2015), desertification, garbage management, migration, and government support. The research group also explored local social movements and developed comparative regional studies of the processes of adaptation and resilience that women heads of household had developed when, in the absence of their partners, they were confronted with floods, droughts and soil degradation and desertification. During four years of fieldwork, the research

---

5   The term social arena was developed by Pierre Bourdieu (1993) and refers to a field in which agents and their social positions are located. Each particular agent develops activities in the field, which is a result of interactions between existing and created rules of the field, the habitus of the agent and their social, economic, political and cultural capital. In this social arena people manoeuvre to get the necessary resources to achieve their interests. The arena works like a market in which actors compete for resources and benefits. Education and culture together with existing resources, social representations, social classes and economic position play a crucial role in the activities that each agent may develop. Competition and alliances depend on the available types of capital, how they contribute, their history within the arena and their ability to adjust to the rules of this field. Bourdieu (1988) introduced the term habitus, which is to be understood as the subjective system of predispositions and expectations acquired by past experiences which facilitate manoeuvring in the arena and define the activities that strengthen the interests of the actors.

6   Social tissue refers to a set of social relations that increase the cohesion of individuals, consolidate social groups and support social rights. Neo-liberalism has created important contradictions (unemployment among youth, elimination of older workers, increasing poverty, informal economy, etc.), but economic factors do not account entirely for the satisfaction of societies. Improving social relations among individuals may mitigate some of the problems created by dominant and exclusive globalization. Social relations based on confidence and mutual support may improve the ability to live together among social groups with similar interests and create social tissue.

team obtained sensitive information using both quantitative and qualitative methods. After the 2010 floods, corrupt disaster management by the three levels of government obliged the authorities to form the Citizen Council of the Yautepec River Basin, in which several researchers became actively involved. This meant that the research group was not only able to understand adaption and resilience processes but also the sociopolitical limits existing in the whole river basin, as well as the context of organized crime.

Table 32.1: Representative survey in the Yautepec River Basin (YRB). **Source:** Field research (CRIM/UNAM 2010–2013).

Phases of survey	Persons	Families	Percentage of women
First phase	1,440	385	49%
Second phase	2,515	634	51%
Third phase	415	100	57%
Total	4,370	1,119	51%

The communities developed different strategies to deal with the increasing threat of floods and droughts, as well as with deteriorating public security. As a citizen member of the Yautepec River Basin Council, the author had direct access to the representatives of the three levels of government, their reactive and preventive behaviour, and to the data concerning investments from the *National Disaster Fund* (Fonden) as well as government investment in the prevention and mitigation of natural hazards and societal disasters.

32.2 Location and Conceptual Considerations: Dual Vulnerability, Disaster Risk Reduction, Disaster Risk Management, Complex Emergencies and Sustainability Transition

What is the goal of sustainability transition not just in a single community, but also within the whole study region, where institutional, environmental, political and security constraints first of all limit the survival of the people and then limit their attempts to improve their livelihood and to build resilience in order to reduce their dual vulnerability? In this chapter transi-

tion towards the goal of sustainable development is understood as a systemic process with positive and negative feedback, where various actors intervene at different levels. All these actors claim to promote sustainability, but cooperation among the three levels of government and with the people affected is inefficient because of the lack of agreed agendas based on the needs of the people affected and because of government corruption. It is also very hierarchical since most of the investment comes from the federal level in the pursuit of national interests and is not adapted to specific regional conditions. Instead of being used to alleviate the high levels of poverty, the budget is primarily used to maintain an extensive and expensive government bureaucracy in a country where functions are often duplicated at state and local level. The bureaucracy is mainly interested in keeping its jobs, power, and incomes after each election, and so the problems of the people have to wait until the next electoral period. Meanwhile the social, political, and environmental conditions become worse and the interaction between the natural and sociopolitical conditions becomes more and more complex and contradictory. Thus, the transition process is blocked in most sectors and during a disaster its progress is erratic. Public infrastructure is rebuilt with money from the Mexican Disaster Fund (Fonden), and people's livelihoods recover slowly thanks to public and private support and at a rate depending on the severity of the impact. Public security fares worst because of the presence of organized crime and local delinquency. For this reason it is impossible to speak of a change in the landscape (Grin/Rotmans/Schot 2010), but various developments indicate a gradual movement towards change and towards an increase in the social tissue in this contradictory political arena.

The study area of the *Yautepec River Basin* (YRB) is located in central Mexico and represents a bridge between the Atlantic and the Pacific, between the towns of Oaxaca and Toluca, and between Acapulco and Mexico City. This central location has meant an increase in the pressure from organized crime, with different criminal cartels competing with each other following the assassination of the kingpin leader Arturo Beltrán Leyva.

32.2.1 The Study Area

The research area of the YRB is in the centre of Mexico (figure 32.1). It covers the complete areas of fourteen municipalities and parts of five more, giving a total of nineteen municipalities, four in the state of Mexico

Figure 32.1: Location of the study area in the central part of Mexico. **Source:** The author.

(Juchitepec, Atlautla, Ozumba, Tepetixtla, Amecameca), thirteen in Morelos (Totolapan, Tlalnepantla, Tepoztlán, Tlayacapan, Juitepec, Atlatlahucan, Yautepec, Tlatizapan, Tlaquiltenango, Jojutla, Zacatepec, Cuautla, Huitzilac), and a small part of the mountain area of Milpa Alta in Mexico City (figure 32.2). In hydrological terms, the YRB rises from the glacier of the volcano Popocatépetl (5,452 m) and receives hundreds of tributaries and springs from the neovolcanic axis (Chichinautzin 3,476 m; Tepozteco, Ajusco, Sierra de Juchitepec, etc.). Most of the northern part of the basin is located in national parks (Popo-Ixta; Corredor Biológico Chichinautzin, Tepozteco) and the centre and south lies in the Biosphere Sierra de Huautla, El Texcal, Sierra Montenegro, and Las Estacas. Along with dozens of small intermittent tributaries, six sub-basins (Nexpayosutla, Gorge Totolapan, Atongo-Apanquetzalco, Dulce, Tepexi, Apatlaco) converge in the Yautepec floodplain. These rivers recharge the aquifer of Yautepec-Cuautla and supply water to the metropolitan region of Cuautla, the second largest economic region in the state of Morelos (figure 32.3).

32.2.2   Goals of the Transition Process

The governmental goals of the transition process are to get rid of the environmental threats, to overcome the existing poverty, and to reduce the floods in the densely populated floodplain, where the most important agricultural, industrial and service activities are located. During the monsoon the threats to the people are linked to periodic flash floods, and during the

dry season in the agricultural sector they mean lack of water for agricultural and domestic activities.

The people complain about the lack of jobs in the industrial and service sectors, the high cost of living, and especially about threats related to organized crime, where homicides, kidnapping, extortion, prostitution, rape, trafficking of women[7] and drug trafficking are leading to loss of life and livelihoods. This means that the transition process is starting simultaneously on different tracks, since temporary emergencies trigger government actions that limit the implementation of a more stable medium-term strategy. For the people, public insecurity, extreme events, and poverty place limits on a systemic process of development and a transition to sustainability from the bottom up. In this precarious situation, periodic disaster impacts not only constrain a rapid recovery, but also constrain investment for the future and resilience-building. On the other hand, kidnapping and extortion may consume all assets accumulated over a lifetime (savings, homes and means of transport). Frequently kidnapped people are found dead despite having paid the ransom, and so social rage is increasing and people have started to take justice into their own hands.[8] Environmental and social vulnerability are interlinked with public insecurity and threats to life and livelihood. The four pillars of human security (freedom from fear, from want, from hazard impacts and the free-

---

7   The municipality of Yautepec has the highest incidence in Mexico of rape and trafficking of women (Segob, 2014).

Figure 32.2: Location of the study area with municipalities. **Source:** The author.

dom to live in dignity; Brauch 2005a, 2005b) are frequently threatened.

---

8   In Atlautla, basically a peasant community with an incipient urban development, three policemen were lynched, after months of extortion, when they arrested two elderly peasants who were collecting dead wood for cooking. In the mountain area of Guerrero, 132 indigenous communities organized their security through a communitarian police force and built the organization CRAC-PC (Coordinadora Regional de Autoridades Comunitarias-Policías Comunitarias) to protect themselves against organized crime and corrupt police.

32.2.3   Dual Environmental and Social Vulnerability

The dual vulnerability concept (Bohle 2001, 2002) combines environmental and social factors in a scenario of *Global Environmental Change* (GEC) and destructive globalization (Held/McGrew 2007; Calva 2012a; figure 32.4). Environmental vulnerability influences both climate variability and climate change, resulting in floods and drought and the loss of soil fertility by erosion and desertification. Further, the scarcity and pollution of water and the loss of ecosystem services (MA 2005) and biodiversity may trigger disasters, environmentally-induced migration (Oswald Spring 2013; Oswald Spring/Serrano/Flores et al.

Figure 32.3: Complexity of the river basin. **Source:** The author.

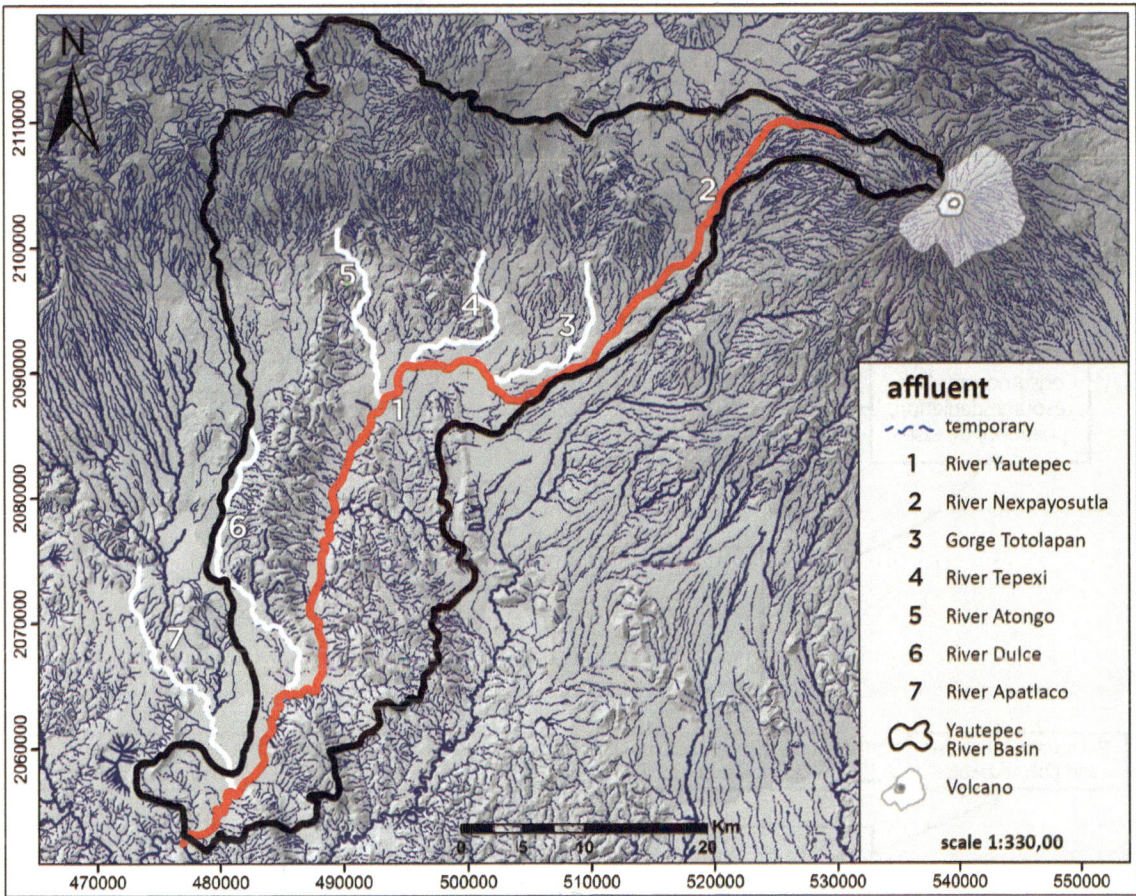

2014), and resource conflicts (Homer-Dixon 1999; Bächler 1998).

Local and regional disasters aggravate the loss of fertile soil, trees and clean water and have increased conflicts over access to natural resources (UNISDR 2013). In the YRB the Spaniards transformed the irrigated floodplain into sugar cane fields, and land grabbing forced peasants to fight, first with Morelos in 1810 against the colonial power, and then in 1910 with the Southern army and Emiliano Zapata against private landlords and sugar mills. During the last two decades, urban landlords have also been converting natural and agricultural land into urban estates in the mountain areas, where the difficult natural conditions caused by the steep-sided volcanoes have caused the fragile soil to further deteriorate and have increased water scarcity during the drought season.

A number of economic crises (1976, 1982, 1988, 1994, 2007–2014; Solis 1973; Sunkel/Paz 1993) and the lack of local jobs and income have confronted an increasing number of families with a survival dilemma

(Brauch 2008; Oswald/Serrano/Flores et al. 2014). In the whole region, the precarious living conditions have forced most families to send one or more members to work in the cities or to migrate to the US. Their remittances have alleviated poverty, and in many cases women who were left behind have become heads of households (Oswald 2013). They have thus become responsible for economic and educational activities and often also for agricultural production, besides taking care of their extended family. This excessive workload has often increased their social vulnerability. In the YRB, twenty-seven per cent of households have a woman as head of household, and these generally experience high levels of poverty because of lack of education and work opportunities.

According the concept of dual vulnerability, environmental vulnerability is the result of physico-chemical changes in the atmosphere produced by intensive use of fossil fuels, which have increased the threat of climate change. Rivers, lakes, seas, groundwater and wetlands become polluted and water from the aqui-

Figure 32.4: Dual vulnerability. **Source:** Adapted by the author from Bohle (2002).

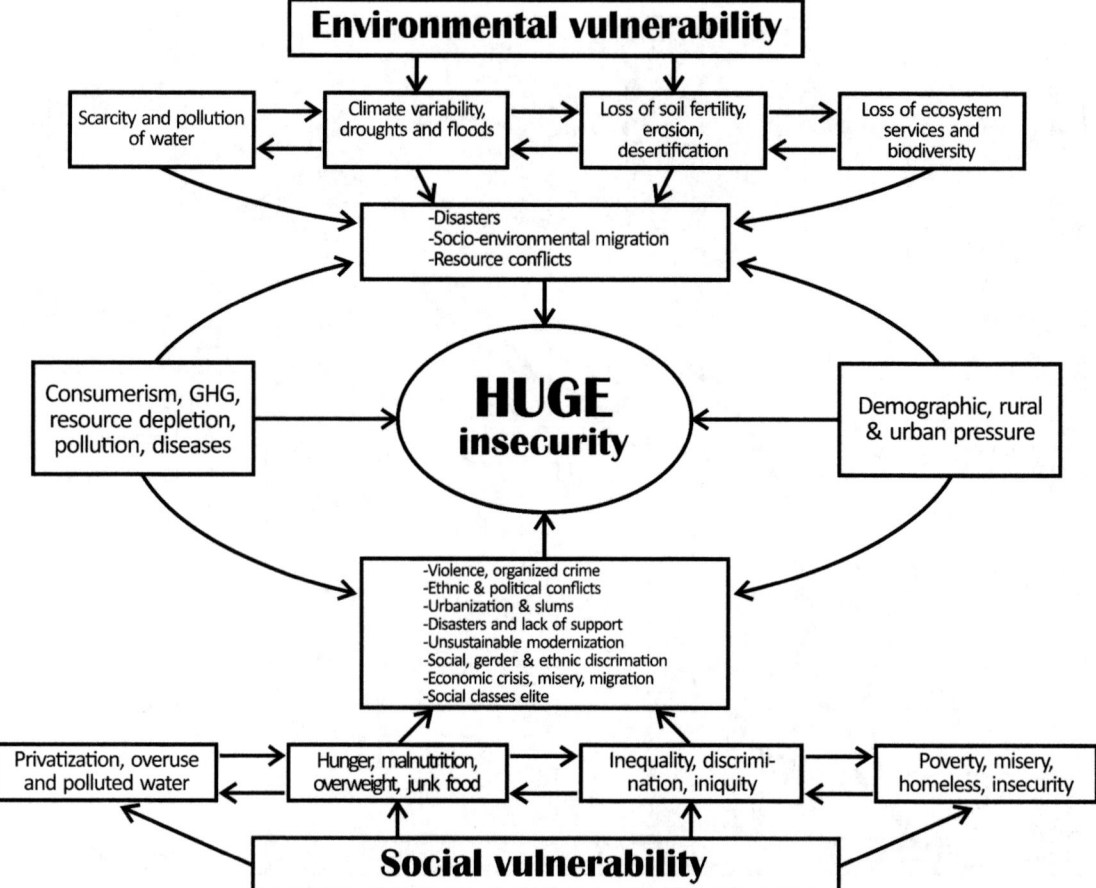

fers and the surface is overexploited. In the lithosphere, shallow soils on steep slopes are easy prey for wind and water erosion and thus lose their potential to sustain the natural vegetation, wildlife and ecosystems, as well as agricultural activities in the highlands, while riverbeds and urban areas have been flooded with mud and sludge. Most of these negative environmental processes affect crop and livestock yields.

These environmental changes have been primarily caused by human intervention into the earth system since the industrial revolution, but especially since the 1950s, through the intensive use of fossil energy, the rapid increase in greenhouse gas emissions into the atmosphere, changes in land use, deforestation, and the pollution and warming of the seas and the diversion of rivers. Taking into account all these anthropogenic interventions into the earth system, Paul Crutzen (2002, 2011) observed a transition in earth history from the Holocene to what he has called the Anthropocene, as a result of anthropogenic *climate change* (CC). According to the Intergovernmental Panel on Climate Change (IPCC 2007, 2012, 2013, 2014), the

direct physical effects of CC include a rise and greater variability in temperature (extreme heat and cold), changes in precipitation, a rise in sea level and an increase in extreme weather events. Higher variability in the monsoon and the interaestival drought in August, together with less rain and more heat, accelerate the process of desertification, resulting in the loss of natural soil fertility and biota. Most ecosystems cannot cope with these fast-changing natural conditions, and once the food chain is affected, wildlife is drastically reduced. The loss of ice in the polar regions and glaciers produces a rise in sea level, which exacerbates coastal erosion and causes seawater intrusion into the coastal aquifers. Warmer seas and sea level rise increase the amount of moisture in the atmosphere and generate more and stronger cyclones and droughts. Both phenomena are becoming more extreme (IPCC 2012, 2014), and in Mexico and Central America extreme weather events have also occurred more frequently (INECC 2012). In the YRB, extreme temperatures, hurricanes, flash floods, drought and hail are the changes related to climate variability, particularly affecting the peasants with rain-fed agriculture, but also traders and people living near the river.

In Mexico and in the YRB, the deterioration and pollution of natural resources has reduced the income and well-being of the people affected. Social vulnerability has increased, due to neo-liberal processes of privatization (Calva 2012a; Stiglitz 2010; Rosas 1996) of water and public services, and the Washington Consensus (Held 2004). The service has rarely improved, but prices have increased. As jobs are rare and often badly paid, people change their diet from nutritious to junk food, often containing sugar, fat, and other carbohydrates (ENSANUT 2013). Health conditions have worsened due to obesity, and Mexico has the second highest incidence of diabetes and the highest incidence of child diabetes mellitus (SS-DGE 2013). Changing environmental conditions are increasing water-borne diseases, especially dengue fever and more recently chikungunya, as well as gastro intestinal illnesses following floods (WHO 2013).

In Latin America, Mexico and Honduras were the sole two countries where poverty was not reduced between 2008 and 2013 (CEPAL 2013) The lack of job opportunities, especially for young people, with eight million young people known as 'ninis', that is, lacking education or jobs, is exploited by organized crime which seeks to involve these unemployed people in criminal activities (Human Rights Watch 2014; Serrano/Oswald Spring/ de la Rúa 2015). In the YRB the situation is worse. Almost thirty per cent of households are headed by women, with high levels of poverty. In 2014, the municipality of Yautepec had the highest rates of rape and women trafficking, the second-highest rate of extortion, and the fourth-highest of kidnapping, and is among the ten municipalities in Mexico with the highest rates of armed robbery (SNSP 2014). These economic, social and security reasons, together with gender inequity, racism and ethnic discrimination, aggravate the difficult socio-economic conditions (Oswald Spring/Moreno/Tena 2014). They have forced people to migrate (Oswald/Serrano/Flores et al. 2014) or to escape poverty through illegal activities (Jaime/Tapia/Goode et al. 2010). Thus social vulnerability is reinforcing environmental destruction, but also social *anomie* (Durkheim 1999), often triggered by disasters.

There is no doubt that environmental and social vulnerability has increased, especially among social groups that depend on natural resources for their survival (in the study area, peasantry with rain-fed land affected by climate change in the mountain regions).

They lack social capital, understood by Bourdieu (1983: 21) as

> the aggregate of the actual or potential resources which are linked to the possession of a durable network of more or less institutionalized relationships of mutual acquaintance and recognition—or in other words, to membership of a group which provides each of its members with the support of collectively-owned capital, a 'credential' which entitles them to credit, in the various senses of the word.

The lack of institutional support and also, because of migration, family support, reduces people's capacity to overcome obstacles (Land/Hannafin 1996), limits the use of other types of capital (Sen 1995), and limits mechanisms for the resolution of conflict (Reychler/Paffenholz 2001; Lederach 1983). Thus, Perona and Rocchi (2008: 1) define social vulnerability as a "status of risk, of difficulty, which disables and overrides, now or in the future, the capacity of affected groups to meet their welfare requirements—subsistence and quality of life—in specific socio-historical and cultural contexts".

Villagrán (2006: 12) analysed dual vulnerability and distinguished between *socio-environmental perspectives* (lack of or inadequate management of the natural environment); *theories of empowerment* (social vulnerability prevents people with little or no education from optimizing existing resources, demanding their rights or improving their living conditions); and *political-economic phenomena* (where inequality, class structure, mechanisms of exploitation, discrimination, and violence rooted in the patriarchal system produce exclusion, exploitation, and conflicts between social groups and the government).

32.2.4 Disaster Risk Reduction and Disaster Risk Management

Disaster risk reduction (DRR) is defined by UNISDR (2009: 10–11) as the "concept and practice of reducing disaster risks through systematic efforts to analyse and manage the causal factors of disasters, including through reduced exposure to hazards, lessened vulnerability of people and property, wise management of land and the environment, and improved preparedness for adverse events". In response to intensive extreme geophysical and climate events, an *international strategy for disaster reduction* (ISDR) has been developed, based on the *Hyogo Framework on Action* (UNISDR 2007), to assist and promote cooperation among governments, civil organizations and private actors in order to better understand extreme events and to stop these natural conditions resulting in disasters through lack of knowledge or the absence of systems for early warning, and through preventive evacuation and mitigation and adaptation to the different natural conditions of the environment. All these processes require knowledge of the risks, systematic monitoring, analysis and forecasting of hazards, and dissemination and mass communication of warnings in order to alert all potentially affected people and to create local capacity to respond effectively to the warnings.

Disaster risk management (DRM) is defined as "the systematic process of using administrative directives, organizations, and operational skills and capacities to implement strategies, policies and improved coping capacities in order to lessen the adverse impacts of hazards and the possibility of disaster" (UNISDR 2009: 10). DRM is oriented towards avoiding, reducing or transferring through insurance the adverse effects of hazards.

The IPCC (2014) has said that besides mitigation, adaptation is crucial in preparing people to proactively adopt measures and actions for the prevention of disasters. Extreme events cannot be avoided, but negative outcomes in terms of loss of life and livelihoods can be reduced through preparedness and early warnings (Villagrán/Pruessner/Breedlove 2013). Often disasters without an early warning can seriously affect people's survival and trigger political responses (Basher 2005). The price hike of basic food in 2007–2008 was instrumental in causing worldwide hunger riots. Physical violence may act as a trigger in existing conflict situations (Schneider 2008), e.g. those existing in the Middle East, or due to the drug war in Mexico (Human Rights Watch 2014). The disappearance of forty-three student teachers in September 2014 led Mexicans to protest in the streets and demand a prompt judicial enquiry by the authorities.

32.2.5 Complex Emergencies

Complex emergencies refer to "a humanitarian crisis in a country, region or society where there is total or considerable breakdown of authority resulting from internal or external conflict and which requires an international response that goes beyond the mandate or capacity of any single agency and/or the on-going United Nations country program" (Oxford Pocket Dictionary 1992).[9] Complex emergencies create massive impacts on human security, increase difficulties for reconstruction and produce extensive violence and loss of life and livelihoods (see earthquake in Haiti; González 2011). As a result, groups of refugees

try to escape the violence, and this means that within the country or in the neighbouring countries large-scale and multifaceted humanitarian assistance is required (Sudan; USAID 2013). The financial, military and political constraints often limit this humanitarian assistance, especially when there are security risks for humanitarian relief workers (e.g. in Syria; IFRC 2013). Emergencies may be created by natural, epidemiological or technological disasters aggravated by conflict, and they disrupt family life and community services and overwhelm the coping capacities of the governments, societies and people affected (Brauch/Oswald Spring/Mesjasz et al. 2015). A complex emergency may initially result from natural or human-made causes, but is outcomes may be magnified by political or military crises.[10]

---

9 The International Federation of Red Cross and Red Crescent Societies (IFRC) [at: <https://www.ifrc.org/en/what-we-do/disaster-management/about-disasters/definition-of-hazard/complex-emergencies/>] states that "some disasters can result from several different hazards or, more often, [from] a complex combination of both natural and man-made causes and different causes of vulnerability. Food insecurity, epidemics, conflicts and displaced populations are examples. A humanitarian crisis in a country, region or society where there is total or considerable breakdown of authority resulting from internal or external conflict and which requires an international response that goes beyond the mandate or capacity of any single agency and/or the on-going UN country program (IASC). Such 'complex emergencies' are typically characterized by:
- extensive violence and loss of life;
- displacements of populations;
- widespread damage to societies and economies;
- the need for large-scale, multi-faceted humanitarian assistance ;
- the hindrance or prevention of humanitarian assistance by political and military constraints;
- significant security risks for humanitarian relief workers in some areas.

The World Health Organization (WHO) defines complex emergencies as "situations of disrupted livelihoods and threats to life produced by warfare, civil disturbance and large-scale movements of people, in which any emergency response has to be conducted in a difficult political and security environment." See: WHO 2002: *Environmental health in emergencies and disasters: a practical guide.* According to WHO: "Complex emergencies combine internal conflict with large-scale displacements of people, mass famine or food shortage, and fragile or failing economic, political, and social institutions. Often, complex emergencies are also exacerbated by natural disasters"; at: <www.who.int/environmental_health_emergencies/complex_emergencies/en>.

32.2.6   Resilience

The concept 'resilience' was initially used in physics and was later adapted by psychologists. The IPCC (2014) uses the definition of the Arctic Council (2013), namely "[t]he capacity of a social-ecological system to cope with a hazardous event or disturbance, responding or reorganizing in ways that maintain its essential function, identity, and structure, while also maintaining the capacity for adaptation, learning, and transformation" (Arctic Council 2013). UNISDR (2009: 24) refers to "[t]he ability of a system, community or society exposed to hazards to resist, absorb, accommodate to and recover from the effects of a hazard in a timely and efficient manner, including through the preservation and restoration of its essential basic structures and functions". In this sense resilience refers to the ability to recover from a shock, whenever the impact of the hazard, and the capacity of the community, internal and external help and the economic situation of the people affected play a crucial role.

32.3   Structural Data on Vulnerable River Basin in Mexico

The state of Morelos has a historically high population density of 364 persons/km² (INEGI 2011), second only to Mexico City and the state of Mexico, the *Metropolitan Valley of Mexico City* (MVMC). This is the result of a favourable climate and favourable environmental conditions that allowed indigenous people to develop irrigated agriculture in the floodplain of the Yautepec as early as three thousand years ago. They produced corn, cotton, chillies, beans, tomatoes and other fruits and vegetables (Maldonado 1984; Mentz 2008), and Cortés introduced sugar cane. This long-standing human impact on the fragile environmental conditions of the mountains, an area of steep slopes, a long dry season, a monsoon from June to October which supplies rain-fed agriculture with water and recharges the aquifers, meltwater from the high volcanoes, and more recently climate variability and climate change, has resulted in a severe deterioration of environmental conditions (Cruz/Delgado/Oswald Spring 2015).

But the floodplain with its deep soils did not experience a loss of soil fertility, and after more than five hundred years of intensive exploitation for sugar cane,

soil fertility is still high, although an increasing area has suffered from land use changes. In 2015, half of the territory of the YRB was used for agriculture, livestock and urban development.

32.3.1   Complexity of River Management Facing Multiple Disasters

Flash floods have increased in the YRB as a result of changes in land use from forest to agriculture and then to urban settlements. Both problems have been further aggravated by logging in the highlands which has had a major impact in the form of the erosion of tributaries and its effect on sedimentation in the valley. Deforestation and change in land use have also reduced the infiltration of rainfall, and with stronger tropical depressions and hurricanes flash floods have increased because of the lack of infiltration of water in the highlands. In 1985, a severe flood caused several deaths and was considered by the authorities and also by older people to be a once-in-a-hundred-years disaster. But in 1998 (a severe Niño year), the YRB and the municipality of Yautepec were flooded once more, and again in 2003. Since 2010, the plain has been flooded every year, and twice in 2010, when at the entrance of the town the river rose twenty-one metres in less than half an hour and flooded the town and the communities downstream (Oswald Spring 2012). During 2013, there was no flood because the river was dredged and some river banks reinforced, despite the double cyclone Ingrid-Manuel, which left 2.5 metres of sediment in the Nexapa derivation dam. But in 2014, due to a lack of preventive behaviour by the three levels of government, five flash floods occurred. All tributaries and the main river together with the four diversion dams were filled with sediment, making them unable to absorb the excessive water from several hurricanes in the Pacific. The YRB faces serious environmental (32.3.1), social (32.3.2) and political (32.3.3) challenges.

32.3.2   Environmental Challenges

In the study region, during the summer the volume of water in the YRB rises due to the melting of the glacier of Popocatepetl and due to tropical depressions and hurricanes. As a consequence, flash floods increasingly threaten life, livelihoods and productive activities, but in addition the natural conditions have worsened, since flash floods tear up trees, rocks and sediment. About 89 per cent of the natural forests has

---

10   Master 2 Action Humanitaire Internationale, accessed 07/12/2014 at: <ahi06.files.wordpress.com/>.

disappeared, and only 11 per cent of the original forest survives (table 32.2).

Table 32.2: Land use changes in the YRB. **Source:** Field research (CRIM/UNAM 2010–2013).

Land use	Area (km²)	Percentage
Agriculture	634.6	50.7
Grassland/livestock	68.3	5.5
Secondary vegetation	372.2	29.7
Forest	112.1	9.0
Dry tropical forest	24.6	2.0
Urban	22.7	1.8
Without vegetation	9.1	0.7
Bodies of water	4.4	0.4
Desert scrub	1.5	0.1

Land use change, deforestation, erosion, extraction of timber and soil, land and water pollution by agrochemicals and human and animal waste have caused the beautiful landscape to deteriorate and have increased environmental vulnerability in the basin (Oswald/Jaramillo 2012). Rain-fed agriculture in the mountains of the north and south of the YRB is highly affected by the variability in rainfall. Agriculture that depends on the monsoon has declined from 10,000 ha in 2009 to less than 9,000 ha in 2012. The pressures caused by flash floods, droughts and more irregular midsummer (interaestival) drought have negatively affected crop yields. Lack of water and irregular rains have also forced subsistence farmers to leave their land, both in the highlands and in the Sierra Madre del Sur. After several years of bad harvests and due to a lack of seeds for the next harvest cycle, whole families have left, in order to escape extreme poverty (Oswald/Serrano/Flores et al. 2014). In the irrigated areas in the central valley, these climatic variations have not affected agriculture, and during the past decade 43,000 ha have been regularly planted each year. Irrigation therefore has become an effective adaptation strategy in a region with high rainfall variability.

What are the concerns of the people who suddenly face changing environmental conditions and how are they responding? According to the survey by Oswald, Serrano, Flores et al., thirteen per cent of the people complained about an excess of water (mostly urban dwellers), while fifteen per cent complained about a shortage of water (basically peasants). The farmers also mentioned longer periods of drought (30 per cent) and irregular rains (42 per cent), in a region where warm temperatures and extreme cold in the

Altos document the impacts of climate change (INECC 2012).

With regard to risk perception, the survey indicated that 62 per cent of the inhabitants feared a lack of water, 26 per cent were afraid of floods and 12 per cent of crop losses. In the mountains, all inhabitants were afraid of a lack of water, while in the river plain over half feared flash floods. In the whole basin 12 per cent mentioned the loss of crops. Thus regional differences and perceptions related to agricultural or service activities influenced their perceptions of risk. When asked about ecological conservation, 45 per cent answered that no natural resources were conserved, 22 per cent mentioned the good quality of air, 20 per cent water and 13 per cent forests and rivers.

In the valley, perceptions differed because the natural ecosystems were different. In the Altos environmental conservation is better, though organized crime is now logging in the protected national parks, where there is limited government control. Thirty-three per cent of the inhabitants also mentioned problems with waste management, 28 per cent the lack of water (Altos), 15 per cent air pollution (around the sugar cane factory in Zacatepec), and 20 per cent other environmental problems, such as the invasion of natural areas, changes in land use, hunting, and clandestine logging; all of them have resulted in a degradation of ecosystem services. This important environmental vulnerability triggers social problems. Lacking clean water and suffering crop failures, families lose their income and food. The absence of income from agriculture increases poverty and social vulnerability.

Environmental and social vulnerability affects the YRB through the loss of biodiversity, which is related to the deforestation of pine-oak forests and especially of the deciduous forest (Maldonado/Ortiz/Dorado 2004; Arias/Dorado/Malddonado et al. 2002; Rzedowski 1973). This environmental deterioration creates additional pressures on social aspects of life, especially with the decline in hunting and gathering of plants, fruits and roots. The loss of corn crops and pasture for livestock in the eroded southern Sierra Madre, together with extreme hydro-meteorological hazards, has forced the people to migrate (Oswald/Serrano/Flores et al. 2014) and in Lorenzo Vázquez, a village in the southern Sierra Madre, two-thirds of the families have at least one emigrant member in the US.

32.3.3   Social Challenges

In terms of social vulnerability, Coneval (2013), the national Mexican institution for poverty evaluation, has stated that nationwide 53.4 million (45.5 per cent) members of the population are living in poverty, and 11.7 million in extreme poverty. ENSANUT (2013) indicates that 53.8 per cent of children and young people between 0 and 17 years do not have enough food. Babies under one year suffered higher levels of deprivation: 27.5 per cent lacked access to health services and 28.2 per cent faced moderate to severe food insecurity. In addition, 54.6 per cent were in poverty, compared with 50.1 per cent of children aged between 12 and 17 (UNICEF 2014). In 2013, of 20.2 million young people, 7.34 million (OECD 2012) were also considered 'ninis' (neither in education nor work). Youth unemployment accounts for 44.7 per cent (ILO 2013). In December 2012, among young people between 20 and 24 years of age, about 1.31 million were without jobs (INEGI-ENOE 2013). The situation among college graduates is particularly difficult: only 70 per cent have found jobs, and of these only four out of ten have a job related to their studies.

The indigenous population in Mexico has the highest poverty rate with 72.3 per cent, compared to the national average (45.5), and rural (61.6) and urban (40.6) populations. They suffer from low income, shortfalls in education, health, social security, housing, food and utilities. Half of the population lacks access to health services, 28 million are hungry, 23.2 million are educationally backward, and 17.1 million lack a home of decent quality. Hence only 23.2 million people (19.8 per cent) suffer no disadvantages (Coneval 2013). In the YRB more than eighty per cent of the rural indigenous population lives in poverty, and in El Pañuelo, a migrant indigenous community from the Mountain of Guerrero, 84 per cent of the women cannot read or write. They are generally 'sold' for marriage at twelve years of age and their life is the most precarious in Mexico, because they live in the poorest municipalities of Mexico.

32.3.4 Political Challenges and a Complex Social Arena

The sugar cane region was historically an active participant in the economic and political struggles of the country. In the state of Morelos, the sugar cane workers were actively involved during the Independence of 1810 and during the revolution of 1910. Social nonconformity is common in the region of the YRB. Environmental vulnerability has also exacerbated local conflicts, especially in the valley, where the control of water and access to irrigated land and wells is crucial

for income and for the export of downstream (Oswald Spring 2012).

Traditional relationships have facilitated corruption and allowed local oligarchies to grab these resources. Nowadays, export agribusiness needs water, while sugar cane still uses water highly inefficiently. Conflicts also exist in the rural villages with limited water services. In several communities empowered women have struggled to control the water service and succeeded in allocating more water to poor people (Oswald 2013). Existing conflicts have increased the uncertainty of access to water and increased the social vulnerability of the disadvantaged, whose lack of income is exacerbated by the country's global socioeconomic situation (Calva 2012b) and the lack of public support.

However, the main political conflict today is related to government agricultural policy, where rural subsidies have been dramatically reduced and basic food prices adjusted to international market prices through NAFTA (the *North American Free Trade Agreement*), and imports of subsidized basic food crops from the US (2013: 16 million tonnes) have increased with no compensatory mechanisms for poor peasants.

Poverty, migration and the absence of conditions for survival have increased among the half of the population with the lowest income, while at the same time the wealth controlled by the top fourteen per cent has increased. The Mexican Study on Household Income and Expenditure (INEGI-ENOE 2008, 2010, 2012) noted important changes in the Gini coefficient in the state of Morelos, which first increased to 0.467 in 2008 (before the effects of the crisis were felt at household level) but then declined to 0.408 in 2010 because of the crisis and because of organized crime, then increased again to 0.425 in 2012. In terms of income, in 2008 the lowest ten per cent of the population earned about 7,950 real pesos (M\$), in 2010 M\$ 5,272 and in 2012 M\$5,562, while the highest strata (ten per cent) spent M\$ 222,047 (2008); M\$ 117, 980 (2010); and M\$ 141,680 (2012) real pesos.[11] The average values of M\$ 53,702 (2008); M\$ 34,426 (2010); and M\$ 38, 666 (2012) therefore do not reflect the living conditions of the people but rather an enormous difference.

In our research area, 44.7 per cent of people live in communities with very high and high levels of mar-

---

11 Real or deflated pesos eliminate the impact of inflation and permit a stable basis of comparison. In 2014 one dollar was equivalent to an average of 12.8 Mexican pesos.

Figure 32.5: Flood over bridge in the centre of Yautepec. **Source:** Photograph taken by Civil Protection, Yautepec, during the 2010 flood.

ginalization, 19.3 per cent with regular levels, and only 13.6 per cent in communites with low standards of poverty. This social inequality has created social resentment, and when the minimum income is not adequate for survival, families explore additional sources of revenue, often in the illegal sector. Therefore, in the YRB the crime rate, especially for kidnapping and extortion, has increased. People no longer trust their neighbours, and they trust the local authorities even less. Crimes are not reported in case local politicians might be involved. Out of every ten kidnappings, a police officer is involved in eight (Segob 2014; INEGI 2014).

32.4   Sustainability Transition from the Top Down and from the Bottom Up

The flood of 2010 in particular affected economic activity in the central market of the municipal capital of Yautepec. River defences, bridges, drinking water pipes, the sewage system and the wall of the market were destroyed by the flood, together with many houses (figure 32.5), and other municipalities were also flooded. Many people lost their income, furniture, domestic animals, water supply and drainage facilities. The authorities managed the emergency situation with a lack of transparency and people lost trust in the authorities. This resulted in bottom-up initiatives to prevent future corruption in the rebuilding process.

32.4.1   Social Actors in the River Basin

After the disaster of 23 August 2010, the victims of the flood in Yautepec city who were protesting against the limited compensation from the official disaster fund (Fonden), peasants organized in the *ejido*[12] of Yaute-

---

12 Ejido is land given collectively and after intensive struggles by agricultural workers after the Mexican Revolution in 1910. Today, ejido land is normally managed individually, wherever in the region the allocation of water and credits "oblige" peasants to produce sugar cane for the sugar mill. With the constitutional reforms of 1992, ejido land is a private good that can be sold or rented, and peasants can form an association with agri-business.

pec, market women, small entrepreneurs, NGOs and academics from different institutions joined together in a first analysis of the damage and the short-term needs in the YRB.

Representatives from the local authorities of the municipalities affected, and later from the state and federal governments, joined to avoid greater unconformity, and based on an informal collective leadership an emergency plan was developed. Specific groups monitored the reconstruction work, budgets were compared and engineers from the informal disaster community organization checked the quality and the prices of the public work. Due to the public unrest the authorities had to accept this citizen control. Public services were re-established very soon and at a lower cost than originally estimated. The key questions raised were how the remaining money could be used in a more sustainable way to avoid new floods, and how this collective work should be organized.

32.4.2 Socio-environmental Activities by Authorities and People

After the informal network of peasant leaders, market representatives, NGOs, municipal authorities, private constructors and academics was created, it held weekly meetings and carried out collective field trips. Conagua, the federal commission in charge of water management, joined the political arena and the system of organization and in November 2010 they organized the first meeting of the Council of the Yautepec River Basin. Federal departments of the environment, water, forestry and agriculture, the ministries of the state of Morelos and the thirteen municipalities of Morelos, together with the water users of the YRB (peasants, enterprises, private users) and academics formed the electoral assembly. The Council elected about fifty-two members, but the large number of people involved meant it was not functional. Therefore an executive group was elected to represent the different interests of stakeholders and to work out a master plan. Conagua and the state water agency (CEA) sent technicians with a appropriate capabilities and an interest in the work. Peasant leaders with local knowledge, engineers and private enterprises were involved, along with academics from the *National Autonomous University of Mexico* (UNAM) and the *Mexican Institute of Water Technology* (IMTA). This constellation offered an effective division of labour and an interdisciplinary working group.

After several months key ideas about sustainable water management were developed, but the group lacked concrete data about the people threatened, their socio-economic conditions, their risk perception, the views of the population of the YRB on waste management, health and education, and detailed knowledge of water management in the river. UNAM supported a multidisciplinary research group that conducted quantitative and qualitative research aiming at short-, middle- and long-term YRB management plans. Demographic, health, socio-economic and educational data were systematized, the investments needed for the restoration of treatment plants, collectors and drinking water facilities were analysed, measures for waste management were proposed, and schools and teachers became involved in the project.

Different sustainable innovation journeys were organized together with universities and technological centres. A photographic exhibition about environmental diversity and the threats to the river basin opened to all citizens. Schoolteachers joined the project and an integrated management was proposed, involving all stakeholders. The narrative explanations with an integrated vision were synthesized in a didactic map (figure 32.6). The short-term activity plan includes: the recovery of the three diversion dams, the reconstruction of drinking and sewage facilities, an association for a collective river management by all ejidos in the basin, the municipal management of solid waste and composting of organic waste, a mathematical model of risk management at municipal level, local reforestation especially in the highlands, and conflict resolution of the Michiate spring. A first proposal for a middle-term management plan was developed collectively, including the restoration of a pre-Hispanic diversion of the river Yautepec, an executive project for the construction of a dam able to retain water for irrigation, municipal urban and territorial planning, and the environmental restoration of the YRB. Some initial proposals for a long-term management plan comprised integrated social, economic, environmental, political and legal actions, together with a socio-environmental museum, that should consolidate the sustainable transition in the river basin (Oswald Spring 2014d, 2015a).

This broad vision clashed with the different interests of governmental officials, mining business and farmers. Conagua favoured the construction of several diversion dams, the municipalities opposed this due to the annual costs of dredging, and the authorities of the *ejidos* wanted to reopen the former diversion to better regulate the amount of water in the

Figure 32.6: Part of the Didactic Map of the Yautepec River Basin. **Source:** The author.

basin, to reduce the risks of flood, and to recover three existing *jagüeyes* (natural water pools) for irrigation in an area with serious problems of desertification. Both proposals were integrated into the didactic map and the map was very widely distributed among different actors, shops, markets and schools.

32.4.3   Agenda for a Sustainability Transition

After the 2012 elections, in December the federal, state and local governments changed and the new officials began to try and understand the tasks facing them. When hurricanes Ingrid (from the Atlantic) and Manuel (from the Pacific) simultaneously merged over Morelos and Guerrero, the YRB was not seriously affected because the sediments from the river and the diversion dams had been dredged, but dams and tributaries were filled with sediment after the extreme event. During the dry season of 2013-2014, the new authorities failed to act on the YRB plan and no meetings of the YRB Council were held.

In 2014, when a new Niño year gradually began to emerge and many floods started to occur, both the authorities and the public were taken by surprise. The change of governments—at the federal level from PAN to PRI and at the state level from PAN to PRD—interrupted the preparation of a broader vision with multiple pathways for short- and middle-term actions, and no budget was available for concrete action during 2013–2014. None of the actions that had previously been agreed between all stakeholders were carried out and from June 2014 five floods demonstrated the lack of prevention measures by federal, state and municipal governments.

The officials reacted to these extreme events too late. The federal government forgot to give permission for the river to be dredged, the state officials did not send the machinery, and the mayor of Yautepec had financial problems in paying for the operators and the gasoline. When the first hurricanes arrived it was too late to dredge the dams and the river.

Subsequent strong flash floods produced several disasters, a consequence of combined factors: the sediments accumulated during the rainy season in 2013 together with waste, rocks and vegetation within the river and its tributaries produced barriers which limited its flow. In the diversion dams, sediments of 2.5 metres in depth had accumulated from previously and the dams could not mitigate the flood; municipal waste was badly managed because the people threw their garbage into the river; and the monsoon began much earlier in summer 2014 with five tropical storms within the first month and twenty-three tropical depressions and hurricanes on the Pacific side during the 2014 monsoon.

All this damage could have been avoided if the actions agreed by all stakeholders had been implemented. These were similar to the first ten steps on the sustainability transition pathway proposed by Dirven, Rotmans and Verkaik (2002). The route towards transition collectively developed before the change of government received organizational and innovative support from different stakeholders. In the YRB, the organization of multi-actor processes had created room for innovation and governance. The processes leading to fewer environmental threats brought about by flash floods and drought, mentioned as transition problems by Dirven, Rotmans and Verkaik (2002), were delineated within a multi-domain, multilevel and multi-actor system (Hillebrand/Heisen/Souwer 2003). Short-, middle- and long-term action plans existed with clearly defined goals. In the transition paths different options (dams vs. diversion of the river) were left open and intermediate goals were adopted related to the regular policy and budget allocations. Instruments and the division of responsibilities were defined, but no intermediate evaluation occurred within a process of learning-by-doing and doing-by-learning. This lack of an intermediate assess-

ment also limited the next transition round and meant restarting the development of new transition arenas, especially after the change of government. Facing these limits and confronted with new disasters, additional inputs into the transition arena were necessary, and the Executive Committee decided to undertake an additional representative survey about the preferred DRR and DRM plans and implementation activities among 415 affected people from 100 families.

32.5 Disaster Risk Reduction and Management from Below

From the analysis of their answers to the questions about DRR and DRM in the YRB, different perceptions of risk emerged. In the highlands (Los Altos) all municipalities mentioned volcanic eruptions as their key threat (between 25.4 to 91.4 per cent), because these communities are permanently exposed to ash and forest fires from the active volcano Popocatepetl, and they have been evacuated several times. Frequent and relatively strong earthquakes (between 6 and 7 on the Richter scale) also explain why six of the fourteen municipalities prioritized this risk, with values between 34 and 51 per cent. Forest fires were mentioned by between 11 and 52 per cent, and landslides were referred to in five municipalities with values between 13 and 18 per cent, distributed across the whole basin. Flash floods were mentioned in four municipalities (Taltizapan, Yautepec, Tlaquiltenango and Zacatepec) which are part of the floodplain, with values between 23 to 37 per cent. Putting together all existing risks in the YRB, only 23.4 per cent of people think that they are not at risk, while the key threats are earthquakes, volcanic eruptions and flash floods (figure 32.7).

Figure 32.7: Multiple risks in the YRB. **Source:** Survey (CRIM/UNAM 2011–2013).

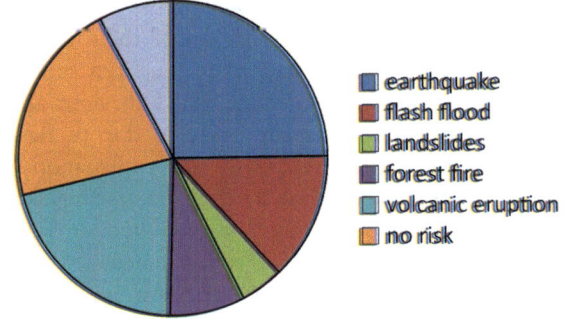

Figure 32.8: Main disasters occurring in the YRB. **Source:** Survey (CRIM/UNAM 2011–2013).

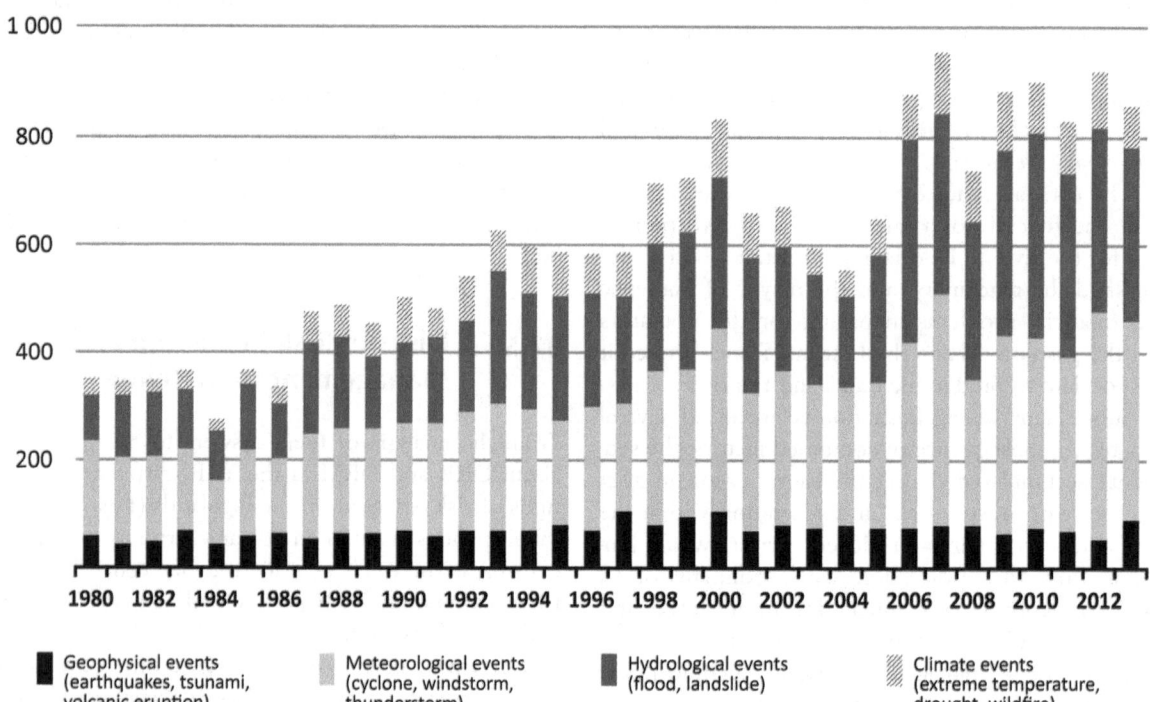

Geophysical events
(earthquakes, tsunami,
volcanic eruption)

Meteorological events
(cyclone, windstorm,
thunderstorm)

Hydrological events
(flood, landslide)

Climate events
(extreme temperature,
drought, wildfire)

32.5.1 Risk Perception by the People

Frequent exposure has created risk perception and consciousness among the people about the threats posed by extreme events. When asked directly which disaster had affected them most, 85.9 per cent mentioned flash floods, followed by drought, earthquake and forest fires (figure 32.8). Houses and furniture were affected according to the answers of 69.3 per cent of the respondents, followed by destruction of infrastructure and services, to which 18.6 per cent of the respondents referred, while 12.1 per cent pointed to the impact on trade and economic activities.

The difference between perception and reality could be related to mass media, films, news, etc., but also to the fact that the people know how to deal with flash floods, after several floods almost every year. During the monsoon season they move their belongings to the upper floor, and when the siren sounds they rush to the nearest refuge. This behaviour also explains why almost nobody has died due to flash floods in the last few years. Another explanation is psychological and linked to the fact that earthquakes and volcanic eruption cannot be prevented. In both cases early warning, building codes, evacuation and training are the ways to reduce negative impacts. In flash floods the lack of dredging, insufficient waste management, deforestation and invasion in the river by blockages are problems for the authorities, so people focus on protecting their own life and furniture. Field research even showed that some disaster victims have learnt how to take advantage of their situation: they put old furniture in the flooded part, cash in the money and buy new furniture which is later sold, and in this way improve their lifestyle. For this reason they do not want to move to a safer place.

Most people relate extreme events to the deterioration of natural resources (with the exception of volcanic eruptions and earthquakes). During the survey, 51 per cent stated that important natural resources were destroyed and 28 per cent claimed that they were somewhat affected, while 17 per cent saw little impact and 4 per cent noted no natural devastation. Figure 32.9 indicates that most of the people interviewed mentioned solid waste, lack of water, and air pollution as the key environmental problems. When asked about other environmental problems, 37.50 per cent mentioned sewage and water pollution.

32.5.2 Reducing the Risks of Extreme Events

When people were asked how to prevent risks increasing, 82 per cent said that human activities were damaging the environment and 80 per cent were con-

Figure 32.9: Environmental destruction in the YRB. **Source:** Survey (CRIM/UNAM 2011–2013).

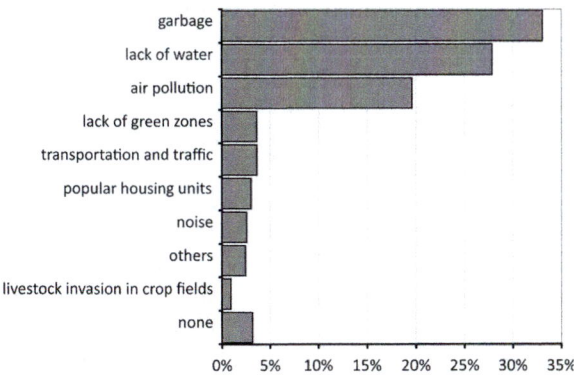

scious that this destruction could be stopped through changes in their lifestyle. About 62 per cent were aware that they had to collaborate with the authorities to recover the affected environment and only 16 per cent believed that it was the exclusive responsibility of the authorities (figure 32.10).

When asked directly what they did after the disaster: 22.2 per cent stated that they went to meetings; 5.4 per cent demanded compensation; 11.9 per cent participated in cleaning up the river and forest; 10.6 per cent reforested, and 49.9 per cent did nothing. This passivity is the result of years of populist policies, where people relied completely on the government. When asked what they had learned from the disaster: 59.9 per cent of those affected mentioned prevention; 29.3 caring for the environment; 10.3 per cent participated in DRM; and only 9.6 per cent did not learn anything. This behaviour indicates that disasters occur at multiple levels and in multi-actor systems, often with contradictory responses.

Figure 32.10: Dealing with increasing risks in the YRB. **Source:** Survey (CRIM/UNAM 2011–2013).

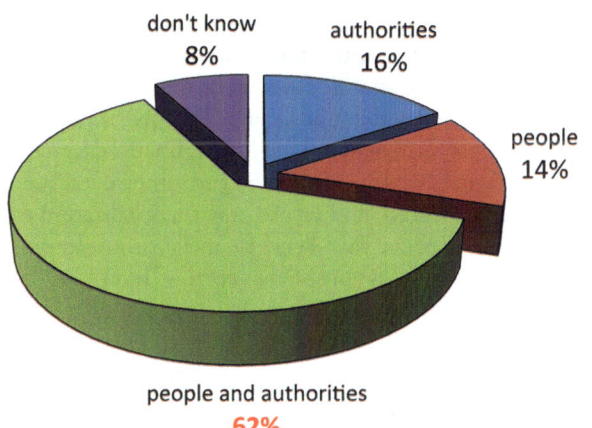

With regard to actions undertaken by all persons interviewed, in first place they mentioned reducing water use and recycling grey water, actions directly related to water scarcity. Then they mentioned recycling organic waste and finally reforesting. This indicates some social learning, bridging individual and collective knowledge, crucial for transition actions capable of dealing with complexity and uncertainty (Wittmayer/Neuteboom 2011), such as the multiple risks existing in the YRB. It has also been called "double loop learning" (Grin/Rotmans/Schot 2010) and demonstrates existing values and conflicting situations, where changes must be carried out through negotiations within the family and community. Grin and Van de Graaf (1996) and Grin and Loeber (2007) insist that surprise, outside views and safe spaces facilitate this second-order learning where the affected are able to go beyond their existing convictions and change their behaviour.

32.5.3 Role of the Authorities in Reducing Disaster Risks

With regard to the role of the authorities in DRR, people noted a lack of preventive actions, a low level of efficiency and a lack of leadership (Villagrán/Pruessner/Breedlove 2013). People claimed that reforestation is a key activity,[13] but the government did their job inadequately. They used seeds from other regions and often from different ecosystems that did not survive in the harsh local environmental conditions, such cold spells and heatwaves. They did not involve the people in the reforestation process and often cows, sheep and goats ate the newly planted trees because the authorities had reforested land belonging to peasants without their permission. The second problem is related to waste management, where municipal authorities do not collect garbage, or only collect it after a long time. So people burn it or throw it into the river. The municipality also destroys the efforts of people who separate their solid waste by mixing it all together, and it does not promote the composting of organic waste, which makes up at least half of all the garbage. A third problem is the lack of communica-

---

13 Additional studies of infiltration of rain and retention of sediments carried out in Los Altos indicated that 72 per cent of water is retained in the upper basin when steep slopes are covered by dense forest, able to produce a healthy soil. This water is slowly infiltrated into the soil and later into the aquifer, thus substantially reducing the threat of flash floods.

tion, training and workshops where the authorities could promote DRR and DRM, improve environmental practices and reduce the risks of extreme events.

Confronted with this lack of government interest, those interviewed proposed campaigns such as the cleaning up and dredging of the YRB and the deviation dams, using the dredge for organic agriculture, fumigating against dengue mosquitos and fruit flies, improving the quality of water, rebuilding drainage and treatment plants, opening up gaps against bush fires, and improving the activities of civil protection. The short period of incumbency of local authorities means that those responsible at the local level can gain no advantage from these proposals, and so they generally ignore them. The people claimed that the authorities were not carrying out the tasks for which they were elected and now they face greater risks from extreme events and feel abandoned. But not all threats in the YRB are related to extreme natural events, and another key concern of those interviewed is public insecurity.

32.6   Complex Emergencies Due to the Drug War and Hydrometeorological Extreme Events

Disasters and organized crime in Mexico and especially in the YRB have created complex emergency scenarios where people with low income and no social security try to create resilience in order to survive under highly precarious conditions.

32.6.1   Public Insecurity in Morelos and the River Basin

With regard to public security, Morelos experiences high crime rates in comparison with the rest of the country and with most parts of the world. The state, one of the smallest in Mexico, has the highest incidence of extortion. In most cases this crime is not reported because political and judicial or municipal authorities are often involved. Morelos has the second highest incidence of kidnapping, which is often not reported because the victim can be killed if the abductors discover the involvement of the police. There are also doubts about which authorities may be involved in this crime. Morelos has also the fourth highest level of homicides (80.01 in 2012; 66.28 in 2013/100,000 inhabitants) and high levels of violent robbery of houses and persons (SNSP 2014).

The violence has increased during the last five years, especially after the drug boss Marcos Arturo Beltrán Leyva was killed in December 2009. Police are frequently involved in crime and the most brutal drug cartel, the Zetas, is composed of highly trained former military and special police forces. The Zetas have widened their illegal activities from drug trafficking to trafficking with humans, arms, women, children, human organs, timber, species threatened with extinction, art and archaeological artefacts, as well as kidnapping and extortion, where migrants to the US are vulnerable victims, including those from Central America. Finally the judicial system is so corrupt that only two per cent of all crimes committed are investigated and only one per cent of the criminals are jailed; often those jailed are innocent victims with no money to pay for a lawyer. This has resulted in a lack of confidence in public authorities.

The precarious socio-economic situation, along with extreme events, public insecurity and the absence of the rule of law, have triggered a situation of *social anomie* (Durkheim 1991, 1999), where each household tries to survive individually. Private security companies seek to supply the public deficiency and in several parts of the YRB the government has lost control of the territory to organized crime. It is the municipality of Yautepec that has the highest figures nationally for rape, directly related to trafficking of girls and prostitution, but the rate of pregnancy among adolescent girls is also very high.

By analysing this public insecurity and linking it to an arena of transition, the first question must be who benefits from these illegal activities. Without doubt most criminals are linked to networks at different levels, but illegal money has to be laundered. According to estimates by the US Department of Homeland Security in 2013, about US$ 400 billion were transferred to the US based on illegal activities in the Americas (INCSR 2014). In Mexico about US$ 30 billion were laundered, permitting a precarious balance between poverty and illegality to be maintained.

Thus a sustainable transition also requires changes at the global level, where Wall Street, financial havens, transnational banks, enterprises and other economic actors are involved, and where only the legalization of drugs could reduce this illegal financial flow and thus the crime rate. As problems at this level can be resolved neither by the Mexican government nor by the authorities of Morelos, the question arises as to what could be done at the local and state level to transform this complex emergency.

At local level, there is a crucial factor related to the theory of social representation and family organization. When we asked all the interviewees about the conditions of their life in the community or colony 66 per cent were highly satisfied or satisfied and 54 per cent were also satisfied with the existing health services. But 35 per cent were highly unsatisfied or unsatisfied with their conditions of life and 46 per cent with the health services. This indicates an important polarization. When the people were interviewed about their emotional links to the colony or village (figure 32.11), almost half of them said that they identified with traditions and nature, while public services and economic facilities were less important. These data indicate a strong identity with traditional values that are deeply anchored in the community and also in urban colonies. Integration into a smaller community is more important than economic facilities or public services. As Moscovici (1984) indicates, the key representation they identify with is significant for its members when they accept norms and values and evaluate their abilities objectivized through cultural processes where identities are permanently being readjusted, and the basic identity of belonging to this social group is anchored (Serrano 2010). A social group, activities and opinions in relation to group expectations, and this permits the group to influence their attitudes and behaviours (Turner 1987: 1). Their identities develop within their primary groups; later they are confronted with different values and they first resist change, but later adapt together with the group to the changing external conditions or reject the new ones. This behaviour is typical for the existing representation of the social group (Duveen 1997, 2001), whether in daily life they adapt collectively or individually to the new necessities.

Figure 32.11: Identity in the YRB (two posible answers). **Source:** Survey (CRIM/UNAM 2011–2013).

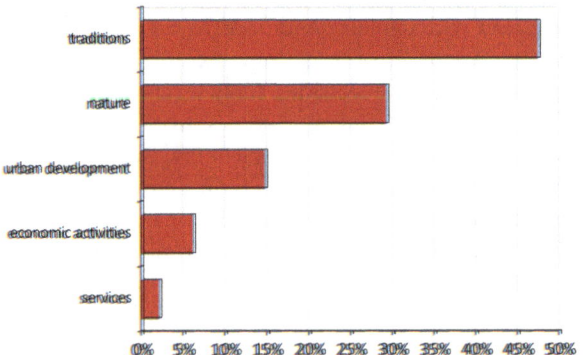

Social representations are crucial to understanding the risks and threats posed by extreme events. But the theory also helps us to understand the links between legality and illegality in the YRB. Three decades of neo-liberal policy with drastic reductions in livelihood, an increase in social inequality, governments becoming more and more corrupt, police involved at local and federal level in illegal activities, and judges letting murders, kidnappers, drug dealers and members of organized crime go free have changed the perception of legality and illegality. The rule of law is permanently broken and the changing reality affects the identity processes. Greed, economic ambitions and social prestige have further increased illegal activities, and the borders between legality and criminality are diluted.

Family units are based on their identity; this identity reinforces the social tissue; organized crime is based on family ties (Segob 2014). The present situation of insecurity creates among members of some groups the necessity of sometimes justifying actions which are considered illegal, such as lynching and producing or selling drugs, and most organized crime is based on extended family structures,[14] called by Garzón (2003) (amoral) *familism*. The term "refers to a model of social organization, based on the prevalence of the family group and its well-being placed against the interests and necessities of each of its members" (Garzón, 2003: 546). To maintain the social tissue within a family, criminal acts are sometimes justified; in other cases part of the family influences another part of the same family without participating in the criminal activities. This complex social process explains why the level of criminality in the region has increased so fast, how existing social tissue has been destroyed and how other social tissues were produced when gangs offered newcomers a social framework, income and social acceptance. Nevertheless, in the longer term these criminal acts create social instability and public insecurity, where the absence of the rule of law leaves people highly exposed. The municipality of Yautepec featured in the international news when in March 2014 forty-four people, including students, were kidnapped and eighteen were killed, even though their families had paid the ransom. Military and state police with a new security scheme called 'mando único' (one sole police controlled by the state authorities), a trained police for all the municipalities

---

14 The cosa nostra in Mexico is basically related to seven families: Arellano Félix, Osiel Cárdenas Guillén, Joaquín Guzmán Loera, Carrillo Fuentes, Armando Valencia, Pedro Díaz Parada and Amezcua Contreras.

Figure 32.12: Income distribution of a peasant household. **Source:** Oswald Spring/Serrano Oswald/Flores Palacio et al. (2014: 343), based on data from SIAP [Statistics from the Ministry of Agriculture].

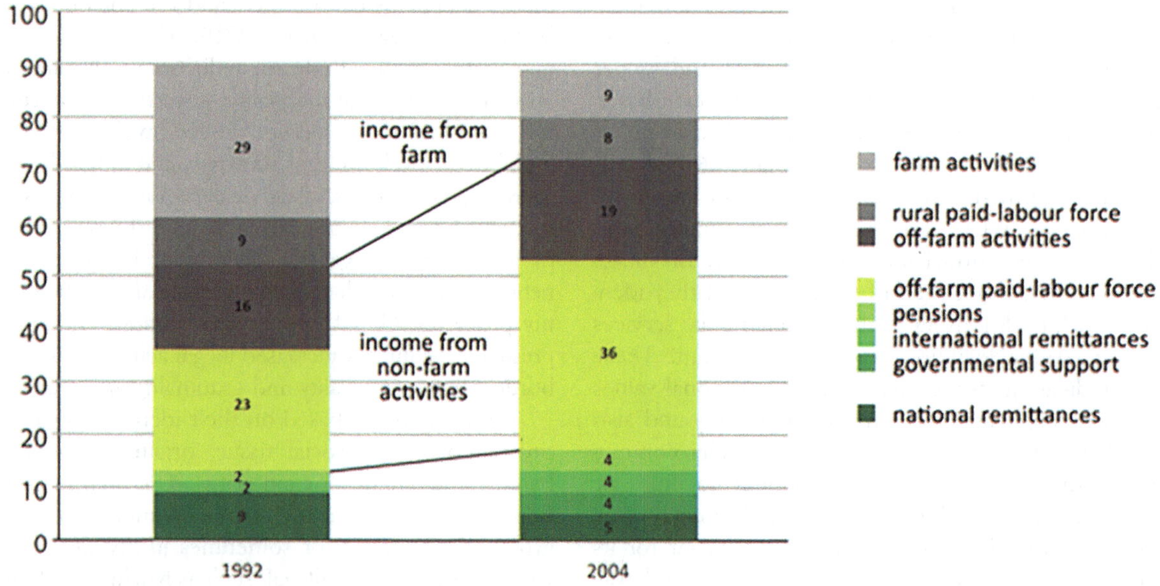

in Morelos, manage to control part of the illegality, but existing family structures and smaller criminal gangs threaten the security of all citizens. The priority given to combatting this high level of crime was also a reason for forgetting preventive behaviour in the YRB, and the start of the Niño year increased the threats of flash floods.

With these multilevel socio-political, security and environmental conditions, how can local people reinforce their resilience and deal with the growing threats of global environmental and climate change (Oswald Spring 2015b) triggered by organized and petty crime? How can especially individual women as heads of household in highly marginal conditions in the YRB increase their resilience?

32.6.2   Resilience Against Floods, Droughts and Crime

All the above three threats were addressed from a bottom-up perspective in cooperation with the three levels of government. People observed that they were unable to deal with the growing threats at family level, which is why they organized themselves in their neighbourhoods. Red buttons in the streets facilitate early alerts against kidnapping, robbery and rape. People generally do not go out after sunset, and festivities take place at home, where those invited normally stay overnight to avoid risks. This type of self-imposed curfew is a rational answer to new threats of crime. As for flash floods the early warning system is working

and people know where their nearest refuge is. As the flash flood comes in very fast, due to the great difference in altitude, the water in the flooded area also disappears very fast, leaving sludge, destruction and diseases behind. Solidarity among the local population helps them to survive the first few days after a flood; later family members and some government support helps them to regain normality. Bigger floods which destroy merchandise and machines sometimes mean it is three years until the small businesspeople regain their former level of income and they are often affected again in the meantime.

With regard to drought the situation is more complicated, because it affects not only the productive cycle but also the livelihood of the people. Drought is especially difficult to manage where it affects rain-fed agriculture in Los Altos, and most families have decided to diversify their economy. At the national level, and with similar conditions to the YRB (figure 32.12), peasant income is complex and only covers ninety per cent of necessities. Only seventeen per cent of this inadequate income comes from agricultural activities (production and agricultural wages), fifty-five per cent comes from off-farm activities, mostly paid labour, nine per cent from national and international remittances, four per cent from government subsidies and another four per cent from pensions. The reduction in income provided by agricultural production from twenty to nine per cent demonstrates the crisis among the peasantry and the absence of a policy of food security of Mexico (Oswald Spring

2009), especially when one takes into account that in 2013 Mexico imported more than sixteen million tonnes of basic grain. Therefore, peasants in regions of rain-fed agriculture have often opted for one of the family members to migrate, and sometimes the whole family has left the region, especially when organized crime wants to make them grow drugs, or the young people are abducted and have to work for the criminal gangs (Oswald Spring/Serrano/Flores et al. 2014).

32.7   Conclusions: Sustainability Transition in a Region Highly Affected by Climate Change and Public Insecurity

The analysis of the complex emergency in the YRB indicates that decentralized governance and citizen participation was possible when the authorities at all three levels were unable to manage the crisis produced by the flash floods. Traditional knowledge held by women and peasants and grass-roots movements against desertification and floods helped the people to deal with these emergencies and to construct step by step a mode of resilience able to deal not only with disasters but also with organized crime and public insecurity; that is, able to deal with complex emergencies.

32.7.1   Integrated River Management

The participative Council of the YRB was able to develop a didactic map (figure 32.6) that let children and adults understand the threats related to global environmental and climate change, and how to manage these threats. The didactic map presented a global understanding of the river basin and its threats from hazards. Later on it was used to illustrate a synthesis of crucial actions where all stakeholders needed to get involved. The most recent survey, in 2013, indicated that people have understood that waste in the river is a threat for everybody not only because of the increase in dengue fever but also because of the waste barriers that increase dangerously when water breaks though, ripping out trees, vegetation, land and sometimes houses. The map also suggests several productive activities, such as composting organic waste and producing safe food for the family in orchards. Conagua proposed a union with the *ejido* that would be able to manage the river water and avoid hydro-conflicts. Today 95 per cent of the surface water and 23 per cent of the groundwater is used in agricultural

activities, in which 16 per cent of the people are employed in generating four per cent of the gross domestic product of the region. This overuse of water is producing conflicts with other productive and service uses, but also removes water needed for ecosystem services.

At the societal level, the collection of seeds from the ecosystem and the reproduction of plants and reforestation is crucial, since 72 per cent of the rain could be retained upstream and infiltrated into the aquifer. This would substantially reduce the number of flash floods, but it will take at least fifteen years before the new trees and the associated recovery of the soil produces this ecosystem service. Additional activities with a more immediate impact must be developed. A waste separation and recycling centre is planned as a combined activity between private enterprise and the municipality, where the regional markets can get involved in separating out the organic waste. Municipal authorities, peasants and NGOs can transform this waste into compost and sell it for organic agriculture. The conflict around the spring of Michiate requires a negotiation process between peasants and local, state and federal authorities. Conagua is responsible for controlling holiday resort companies, who have built infrastructure in the YRB within the protected federal area. It is their responsibility to dismantle these constructions and prevent any new invasion of the YRB.

In the middle term (five to ten years) ecotourism, composting and nurseries using local seeds is society's responsibility. Peasants are committed to transforming 5,000 ha into organic production. Together with the state and federal government, sixty million trees should be planted, especially upstream, and the area should be reforested in order to improve the infiltration of water and reduce flash floods. It is responsibility of the federal and state governments to control the loggers (often linked to organized crime), especially in the national parks in Los Altos. The three levels of government will be involved in the construction of the dam Morelos I which will retain water during the monsoon. The diversion canal, with its archaeological ruins, will restore the traditional river flow and recover four *jagüeyes* and six springs for irrigation in areas of rain-fed agriculture. The building and repair of collectors for sewage water and treatment plants will allow the water quality of the river to be recovered. A regionally established landfill programme and improvement in waste recollection will reduce the amount of garbage in the river. Environmental education in schools, clubs and *ejidos* and the eco-management of the natural flora and fauna of the region will protect endemic plants and animals. The municipalities are committed to thirteen plans for urban and environmental management, centres of waste separation and local recycling, and the rehabilitation of the existing treatment plant. Reforestation activities and ecotourism in the two 'magic cities' (Tepoztlán and Tlayacapan) will help to recover the cultural identity and history of the region, attract tourism, and facilitate the establishment of 2,000 micro-businesses in the areas of handicrafts, souvenirs and food facilities, thus consolidating tourism, income and employment.

In the long term (fifteen to thirty years), education and social activities for youth should reduce adolescent pregnancies, diabetes mellitus, obesity, trafficking of girls, and insecurity. Business communities and society will promote healthy communities and cities, where traditional medicine and indigenous cures can support ecotourism. In economic terms, all these activities will allow small businesses to be consolidated and integrated into chains of support (administration, taxes, legal, labour, market studies, export, etc.) by young professionals who are today unemployed (Cadena 2009). The rational management of the existing natural and social resources in the region can produce 10,000 formal jobs. The restoration of national parks and ecotourism is an additional source of income for poor peasants, together with payment for ecosystem services. An improved socio-economic stability will reduce illegal activities and a better-controlled public administration will offer fewer opportunities for corruption. In environmental terms, seven diversion dams, Morelos 1 and the deviation may protect the region against flash floods and store water for irrigation. In this last phase, about 150 million trees would restore the destroyed ecosystems. Additionally, 20,000 ha of recovered soil thanks to compost, three regional landfills and 5,000 ha of agriculture with efficient irrigation will offer the potential to improve regional food security. Five hundred hectares of fruit trees and 10,000 ha of organic plantations might transform the region into a centre of attraction for tourism, education, archaeology, health and sustainable culture. This model also opens up the potential for international fair-trade markets, where fruit, vegetables and medicinal plants can be exported. At the local level, the recovery of orchards irrigated with recycled grey water and composted waste will improve the availability of healthy food. In the northern area, payments for ecosystem services, the recovery of the national parks and sport and outdoor pursuits in the highlands will offer a variety of ecotourist activities and in the

surrounding regions forestry will offer additional jobs and income.

32.7.2   Reconstruction of the Social Tissue

Ecological education and cleaning up the YRB are two activities where schools and citizens can get involved. The government scholarship-salary for high school students programme opens the door to the active participation of students in communities. Social tissue can only be reconstructed through healthy daily living together where people lose their fears and reclaim public places for social coexistence (Moreno/ Mojica 2013). The reconstruction of social tissue also includes greater transparency in government, the judicial system and the police. The government of Morelos has developed a police force under a single command, and more than 200 corrupt police officers have been dismissed and replaced by professional officers. But it is at the societal level where corruption needs to be stopped, and new roles for the law can be achieved through social pressure.

The consolidation of leadership at the local level is crucial, and teachers, doctors, priests and academics can play a role in reconstructing the social tissue. Unemployment creates not only poverty and a lack of social esteem, but also provides the temptation to become involved in organized crime or to migrate. Migration in adverse condition can be a coping strategy, but may also increase vulnerability at home, especially for women heads of household when the remittances are missing. Integrated labour activities based on concrete action programmes with government support may reduce migration, crises, criminality and conflicts. The concatenation and integration of productive activities at the local level offers jobs for young unemployed high school graduates, reduces the failure of micro-businesses, decreases the cost of professional support in production, marketing and administration, and increases the potential for solid regional development. The consolidation of agricultural, livestock, forestry and off-farm jobs may open up new productive processes, increase financial resources and with local savings can open new development processes. Combined with composting, the recovery of eroded land, integrated water management and a diversity of small local activities, it may create a stable regional economy, able to feed the ple (Oswald Spring 2014c). This may be a way to reconstitute the social tissue and to reduce criminality in the region, especially when unemployed young people have the opportunity to study and find paid jobs.

Figure 32.13: Arenas, agendas, actors and activities of vulnerability and livelihood. **Source:** The author.

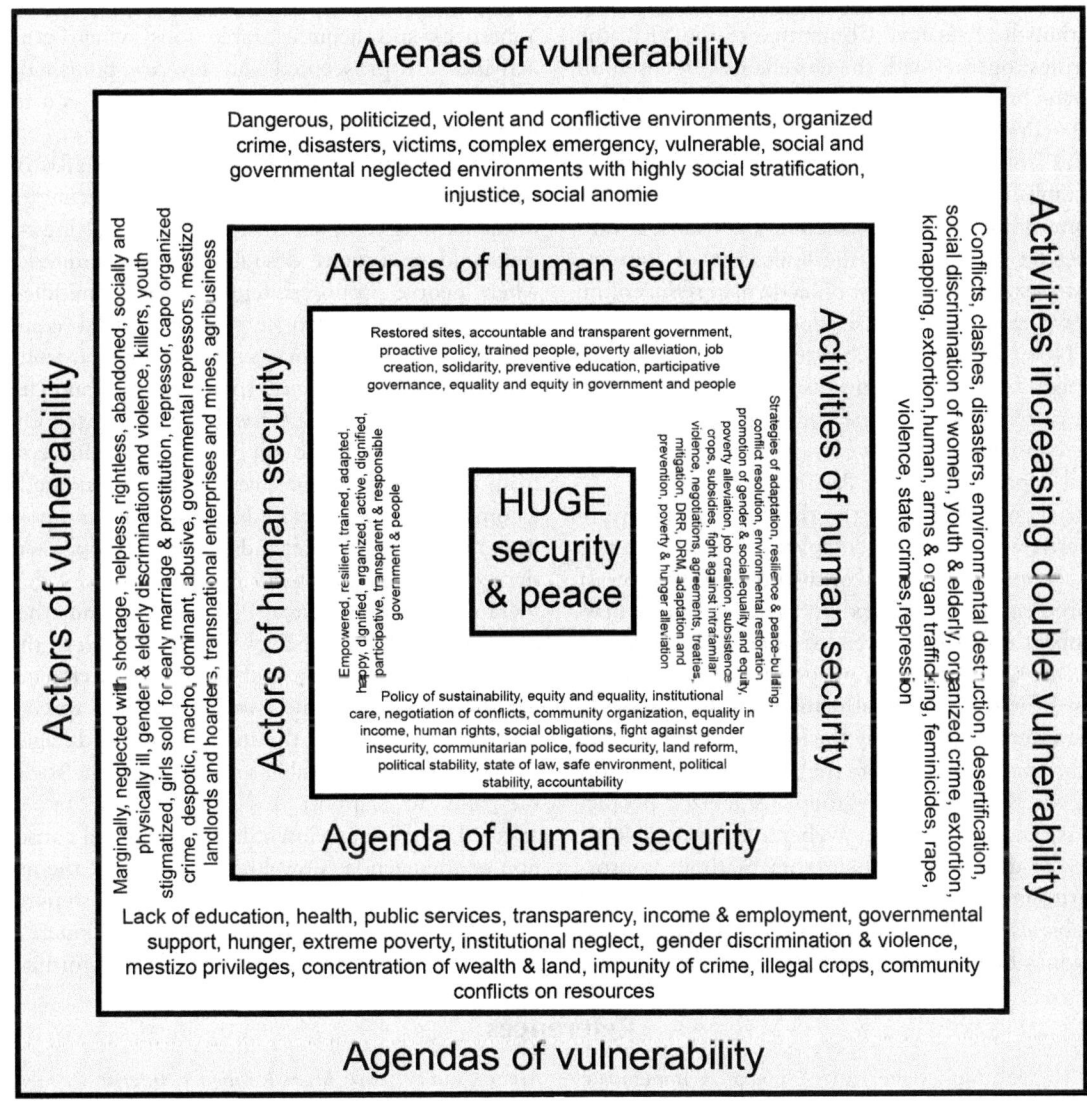

32.7.3 Sustainability Transition: A Long Path Towards a Desirable Future

A sustainable transition explicitly promotes participatory activities, where top-down investment and support from the government and private sectors should consolidate the bottom-up efforts. The development plan for the YRB was jointly developed and is an initial attempt by different stakeholders to understand the present situation of environmental, political and socio-economic deterioration. It also represents the first steps towards designing a framework leading towards a sustainable transition. The integration of a Committee representing different stakeholders and local and regional interests increases the capacity for strategic thinking and innovative experimentation.

Figure 32.13 synthetizes the interaction among arenas, agendas, actors and activities of dual vulnerability and proposes a transition to an agenda of a sustainable socio-economic, political and cultural change towards an integrated HUGE (human, gender and environmental) security and peace (Oswald 2015).

The rapidly deteriorating environmental and sociopolitical conditions have forced people to establish a diagnosis and later a framework for resilience-building to deal with the complex emergency in the YRB. The social mobilization and inconformity has also forced the government to intervene, to collaborate and to recover part of the lost governance. Under the leadership and with the involvement of different stakeholders within the Executive Committee of the YRB, the authorities together with the citizens have focused on increasing freedom from fear and from hazard impacts (Brauch 2005a). It was basically through the activities from the bottom up that a social consensus was established and that initial policies for a sustainability transition were established in a consensual way. The regular meetings and the evaluation of some of the goals established on the didactic map reduced the disaster impacts from flash floods and drought in 2013. Nevertheless, the change of government in December 2012 and the increase in public insecurity in 2014 created new vulnerabilities, triggered by a severe Niño year with twenty-three hurricanes and tropical depressions in the Pacific.

The representatives of the three levels of government are aware that the complex emergency in the YRB can only be managed with the collaboration of organized society. Complex emergencies have a massive impact on various levels of human security, and make the reconstruction or prevention of climate threats difficult. The initial natural and human-made disasters were magnified by the loss of public security and the politicization of the crisis. Cooperation among the three key actor groups (organized people, business, and government; Weber 1987) was able to reduce the disaster impacts in 2013. Without improving environmental and social conditions and creating new jobs, especially among young people, the crime rate cannot be reduced. Only when these young people have the opportunity to study (e.g. at the state university UAEM, which recently opened a branch in Yautepec) and acquire stable jobs, when criminal activities are prosecuted and severely punished and their economic advantages are prevented, is a transition to a safe life and livelihood feasible.

The transition to sustainability in the YRB began with a common problem analysis, made urgent because of the existing complex emergency. By involving crucial stakeholders, a future desirable vision was developed, where people proposed items for short-, middle- and long-term agendas taking into account social (equality), economic (development), gender (equity), environmental (sustainability), political (participation), and cultural (diversity) development (Oswald Spring 2015c). Opportunities came from the appropriate stakeholders and from the three levels of government and merged into a common vision, where collective effort and negotiation allowed a shared agenda to be developed. Activities were planned, but during the change of government several crucial actors disappeared, while the representation of civil society was stable. New threats related to an increase in public insecurity, climate variability, a Niño year and new flash floods forced the authorities to address the integrated agenda again in 2014, when it was possible to reconstruct the social tissue and to support resilience-building by those affected. The lessons learnt include regional consolidation of the agenda (Oswald Spring 2014b), the assigning of a budget for the next five years, the step-by-step evaluation of each activity, and the orientation of activities towards an integrated sustainable transformation.

References

Acosta Urquidi, Marieclaire, 2011: "Superar la Impunidad: hacia una estrategia para asegurar el acceso a la justicia en México, CIDE, Mexico"; at: <http://es.scribd.com/doc/64415702/Superar-la-impunidad-hacia-una-estrategia-para-asegurar-el-acceso-a-la-justicia-en-Mexico-Mariclaire-Acosta-coordinadora>.

Adger, W.N.; Pulhin, J.M.; Barnett, J.; Dabelko, G.D.; Hovelsrud, G.K.; Levy, M.; Oswald Spring Ú.; Vogel, C.H., 2014: "Human security" in: Field, C.B.; Barros, V.R.; Dokken, D.; Mach, K.J.; Mastrandrea, Bilir, M.D.; T.E.; Chatterjee, M.; Ebi, K.L.; Estrada, Y.O.; Genova, R.C.; Girma, B.; Kissel, E.S.; Levy, A.N.; MacCracken, S.; Mastrandrea, P.R.; White, L.L. (Eds.), Climate Change 2014: Impacts, Adaptation, and Vulnerability. Part A: Global and Sectoral Aspects. Contribution of Working Group II to the Fifth Assessment Report of the IntergovernmentalPanel on Climate Change (Cambridge: Cambridge University Press): 755-791.

Arctic Council, 2013: Arctic Resilience, Interim Report 2013 (Stockholm: Stockholm Environment Institute and the Stockholm Resilience Centre).

Arias, Dulce, María; Dorado, Oscar; Maldonado, Belinda, 2002: "Biodiversidad e importancia de la selva baja caducifolia: la reserva de la Biósfera de la Sierra de Huautla", in: Conabio, Biodiversitas, 45: 7-12.

Bächler, Günther, 1998: "Why environmental transformation causes violence", in: Environmental Change and Security Report, No. 4: 24-44.

Basher, Reid, 2005: "Global early warning systems for natural hazards: Systemic and people-centred. Philosophical Transactions of the Royal Society A, vol. 364: 2167-2182.

Bogardi, Janos; Oswald Spring, Úrsula; Brauch, Hans Günter, 2016: "Water Security: Past, Present and Future of a Controversial Concept", in: Pahl-Wostl, Claudia; Gupta Joyeeta; Bhadur, Anik (Eds.), Handbook of Water Security (Swindon: Publisher Edward Elgar), in press.

Bohle, Hans Georg, 2001: "Vulnerability and Criticality: Perspectives from Social Geography", in: *Newsletter of the International Human Dimensions Programme on Global Environmental Change, IHDP Update* 2/2001: 1-7; at: <http://www.ihdp.uni-bonn.de/html/publications/update/update01_02/IHDPUpdate01_02_bohle.html>.

Bohle, Hans Georg, 2002: "Land Degradation and Human Security". In E. Plate (Ed.), *Human Security and Environment. Report on a Workshop held in Preparation for the Creation of a Research and Training Center for the UN* (Bonn).

Bourdieu, Pierre, 1983: "Ökonomisches Kapital, kulturelles Kapital, soziales Kapital". In Kreckel, R. (Ed.): *Soziale Ungleichheiten* (Göttingen: Otto Schwartz & Co.): 183-198.

Bourdieu, Pierre, 1988: La distinción (Madrid: Taurus).

Bourdieu, Pierre: *The Field of Cultural Production* (Cambridge: Polity Press).

Brauch, Hans Günter, 2005a: *Environment and Human Security. Freedom from Hazard Impact*, InterSecTions, 2/2005 (Bonn: UNU-EHS); at: <http://www.ehs.unu.edu/file/get/4031>.

Brauch, Hans Günter, 2005b: *Threats, Challenges, Vulnerabilities and Risks in Environmental Human Security*. Source, 1/2005 (Bonn: UNU-EHS); at: <http://www.ehs.unu.edu/file/get/4040>.

Brauch, Hans Günter, 2008: "From a Security towards a Survival Dilemma", in: Brauch, Hans Günter; Oswald Spring, Úrsula; Mesjasz, Czeslaw; Grin, John; Dunay, Pal; Behera, Navnita Chadha; Chourou, Béchir; Kameri-Mbote, Patricia; Liotta, P. H. (Eds.): *Globalization and Environmental Challenges: Reconceptualizing Security in the 21st Century*. Hexagon Series on Human and Environmental Security and Peace, vol. 3 (Berlin–Heidelberg–New York: Springer-Verlag): 537-552.

Brauch, Hans Günter; Oswald Spring, Úrsula; Mesjasz, Czeslaw; Grin, John; Cheng, Liu (Eds.), 2015: 应对全球环境变化、灾难及安全 —— 威胁、挑战、缺陷和风险 (*Coping with Global Environmental Change, Disasters and Security Threats, Challenges, Vulnerabilities and Risks*) 南京出版社 (Nanjing: Nanjing Publishing Hous).

Cadena Barquín, Félix (Ed.), 2009: *De Foro a Foro. Contribuciones y Perspectivas de la Economía Solidaria en México en el Contexto de Crisis Global* (México, D.F.: Ed. FLASEP, A.C.).

Calva, José Luis, 2012a: *Políticas Macroeconómicas para el Desarrollo Sostenible. Análisis Estratégico para el Desarrollo*, Vol. 4 (Mexico: Consejo Nacional Universitario and Juan Pablos).

Calva, José Luis, 2012b: *Estrategias Económicas Exitosas en Asia y América Latina. Análisis Estratégico para el Desarrollo*, Vol. 2 (Mexico: Consejo Nacional Universitario and Juan Pablos).

Cepal, 2013: *Informe Anual CEPAL* (Santiago de Chile: CEPAL).

Coneval, 2013: *Análisis y medición de la pobreza* (México, D.F.: Coneval).

Coneval, 2014: *National Assessment indicated Social Policy Development* (México, D.F.: Coneval).

Crutzen, Paul J., 2002: "Geology of Mankind", in: *Nature*, 415,3 (January): 23.

Crutzen, Paul J., 2011: "The Anthropocene: Geology by Mankind", in: Brauch, Hans Günter; Oswald Spring, Úrsula; Mesjasz, Czeslaw; Grin, John; Kameri-Mbote, Patricia; Chourou, Béchir; Dunay, Pal; Birkmann, Jörn (Eds.), 2011: *Coping with Global Environmental Change, Disasters and Security–Threats, Challenges, Vulnerabilities and Risks* (Berlin–Heidelberg–New York: Springer-Verlag).

Cruz Núñez, Xóchitl; Delgado Ramos, Gian Carlo; Oswald Spring, Úrsula (Eds.), 2015: *México ante la Urgencia Climática: Ciencia, Política y Sociedad* (México, D.F: CCEIICH, CRIM y PINCC, UNAM).

Dirven, J.; Rotmans, Jan; Verkaik, A.P., 2002: *Society in Transition: an Innovative Viewpoint* (The Hague: Transition Essay).

Durkheim, Émile, 1999: *Le Suicide: étude de sociologie* [Introduced by Serge Paugam] (Paris: Presses Universitaires de France).

Durkheim Émile, 1991: *De la division du travail social* (Paris: PUF).

Duveen, Gerard, 1997: "Psychological Development as a Social Process", in: Smith, Leslie; Dockerell, Julie; Tomlinson, Peter (Eds.): *Piaget, Vygotsky and beyond* (London: Routledge).

Duveen, Gerard, 2001: "Representations, Identities and Resistance", in: Deaux, Kay; Philogén, Gina (Eds.): *Representation of the Social* (Oxford: Blackwell Publishers): 257-271.

Empinotti, Vanessa; De Stefano, Lucia; Oswald, Úrsula; Arrojo, Pedro; Solanes, Miguel; Donoso, Guillermo; Phumpiu Chang, Patricia; Jacobi, Pedro, 2014: "Chapter 14: Civil Society Organizations and their role on water management in Latin America", in: Lucia De Stefano (Ed.), *Water for Food Security and Well-Being in Latin America and the Caribbean. Social and Environmental Implications for a Globalized Economy* (London: Earthscan-Routledge; Fundación Botín): 316-342.

Encuesta Transparencia Internacional, at: <http://www.transparencia.org.es>.

ENSANUT, 2013: *Encuesta Nacional de Salud y Nutrición 2012* (Cuernavaca: SSA-INSP).

Garzón Pérez, Adela, 2003[2]: "Familism", in: Ponetti, J. et al. (eds.), *International Encyclopedia of Marriage and Family*, Mac Millan, Vol 2: 546-549.

González Delgado, Luis David, 2011: *Una visión ampliada de la seguridad humana para el mandato de la MINUSTAH en Haití* (Mexico, D.F.: FCPS-UNAM, MSc Thesis).

Grin, John; Loeber, Anne, 2007: "Theories of learning. Agency, structure and change", in: Fischer, Frank; Miller, Gerard J.; Sidney, Mara S. (Eds.): *Handbook of*

Public Policy Analysis Theory, Politics, and Methods (New York: CRC Press): 201-222.

Grin, John; Van de Graaf, Henk, 1996: "Implementation as communicative action: an interpretive understanding of the interactions between policy makers and target groups", in: *Policy Sciences*, 20,4: 291-319.

Grin, John; Rotmans, Jan; Schot, John, 2010: *Transition to Sustainable Development* (London: Routledge).

GWP [Global Water Partnership], 2009: *A Handbook for Integrated Water Resources Management in Basins* (Elanders: Global Water Partnership [GWP] and the International Network of Basin Organizations [INBO]).

Held, David; McGrew, Anthony (Eds.), 2007: *Globalization Theory. Approaches and controversies* (Cambridge: Polity Press).

Held, David, 2004: *Global Covenant. The Social Democratic Alternative to the Washington Consensus* (Cambridge: Polity Press).

Hernández Oliva, Rocío Citlalli, 2001: *Globalización y privatización: El sector público en México, 1982-1999* (Mexico, DF.: INAP).

Hillebrand, Hans; Hetsen, Hans; Souwer, Maarten, 2003: *System Innovations in Rural Areas* (The Hague: Innovation Network Rural Areas and Agricultural Systems).

Homer-Dixon, Thomas F., 1999: *Environment, Scarcity, and Violence* (Princeton, NJ: Princeton University Press).

Human Rights Watch, 2014: "Informe Mundial: México", at: <http://www.hrw.org/es/world-report/2014>.

IFRC (International Federation of Red Cross and Red Crescent Societies), 2013: "Emergency Appeal n° MDRSYR003 GLIDE n° OT-2011-000025-SYR, 18th November 2013", at: <http://www.ifrc.org/PageFiles/113703/MDRSY003rea3.pdf>.

ILO, 2013: World of Work Report 2013 (Geneva: ILO).

INECC, 2012: *Mexico. Fifth National Communication to the United Nations Framework Convention on Climate Change* (Mexico, D.F.: INECC-SEMARNAT).

INEGI-ENOE, 2005-2013: *Encuesta Nacional de Ocupación y Empleo* (Aguascalientes, INEGI), several years.

INEGI-ENOE, 2008, 2010, 2012: *Encuesta Nacional de Ocupación y Empleo* (Aguascalientes, INEGI).

INEGI, 2011: *Censo Nacional de Población y Vivienda* (Aguascalientes INEGI).

INEGI, 2014: Encuesta Nacional de Seguridad Pública Urbana (ENSU) (Aguascalientes INEGI).

INCSR [United States Department of State, Bureau for International Narcotics and Law Enforcement Affairs], International Narcotics Control, 2014: *Strategy Report, Volume II, Money Laundering and Financial Crimes* (Washington: INCSR, March).

IPCC, 2007: *Climate Change 2007: Climate Change Impacts, Adaptation and Vulnerability* (Cambridge: Cambridge University Press).

IPCC, 2012: *Report on Extreme Events* [SREX] (Cambridge: Cambridge University Press).

IPCC, 2013: *Climate Change 2013. The Physical Science Basis. IPCC Working Group I Contribution to AR5* (Cambridge: Cambridge University Press).

IPCC, 2014: *Climate Change 2014: Impacts, Adaptation, and Vulnerability. IPCC Working Group II Contribution to AR5* (Cambridge: Cambridge University Press).

Jaime, Etna; Tapia, José; Goode, Maralá; García, Gudalupe; Bergman, Marcelo; Cárdenas, Ernesto: Barreda, Luis de la; López-Portillo, Ernesto; Mendoza, Carlos; Sayec, Cecilia; Shirk, David, Zepeda, Guillermo, 2010: SIIS: "Sistema de índices e indicadores de Seguridad Pública"; at: <http://www.mexicoevalua.org/wp-content/uploads/2013/03/SIIS-2010.pdf>.

Land, Susan M.; Hannafin, Michael J., 1996: "A conceptual framework for the development of theories-in-action with open-ended learning environments", in: *Educational Technology Research and Development*, 44,3: 37-53.

Lederach, John-Paul, 1983: *Educar para la Paz* (Barcelona: La Magrana).

MA [Millennium Ecosystem Assessment], 2005: *Ecosystems and Human Wellbeing: Desertification Synthesis* (Washington DC, Island Press).

Maldonado Jiménez, Druzo, 1984: *Cuauhnáhuac y Huaxtepec (Tlahuicas y Xochimilcas en el Morelos Prehispánico)* (Cuernavaca: CRIM-UNAM).

Maldonado, Belinda; Ortiz, Amanda; Dorado, Oscar, 2004; "Preparados galénicos e imágenes de Plantas Medicinales. Una alternativa para promotores de salud en la Reserva de la Biosfera Sierra de Huautla", Cuernavaca: CEAMISH-UAEM.

Mentz, Brigida von, 2008: *Cuauhnáhuac 1450-1675*, (México, D.F.: M.Á. Porrúa).

Moreno, C. and Mojica, F., 2013: "Reconstrucción del tejido social con víctimas de violencia sociopolítica en el Magdalena Media, *Revista de Psicología GEPU* 4(2): 9-29.

Moscovici, Serge, 1984: The Phenomenon of social representations", in: Farr, R.M. and Moscovici, Serge (eds.), Social Representations (Cambridge: Cambridge University Press).

OECD, 2012 : *México. Mejores Políticas para un Desarrollo Incluyente* (Paris: OECD).

Oswald Spring, Úrsula, 1991 : *Estrategias de Supervivencia en la Ciudad de México*, (Cuernavaca, CRIM-UNAM).

Oswald Spring, Úrsula, 2009: "Food as a new human and livelihood security issue", in: Brauch, Hans Günter; Oswald Spring, Úrsula; Grin, John; Mesjasz, Czeslaw; Kameri-Mbote, Patricia; Behera, Navnita Chadha; Chourou, Béchir; Krummenacher, Heinz (Eds.), 2009: *Facing Global Environmental Change: Environmental, Human, Energy, Food, Health and Water Security Concepts*. Hexagon Series on Human and Environmental Security and Peace, vol. 4 (Berlin–Heidelberg–New York: Springer-Verlag): 471-500.

Oswald Spring, Úrsula, 2012: "Cambio ambiental global, desastres y vulnerabilidad social", in Calva, J. L. (Ed.), *Cambio Climático y Políticas de Desarrollo Sustentable* (México: J.P./IIEc-UNAM).

Oswald Spring, Úrsula, 2013: "Dual Vulnerability Among Female Household Heads", in: *Acta Colombiana de Psicología*, 16,2: 19-30.

Oswald Spring, Úrsula, 2014: "Un future sustentable con calidad de vida: ¿una utopia, una realidad o una necesidad en Morelos?", in: Oswald Spring, Úrsula; Serrano Oswald, Serrena Eréndira; Estrada Álvarez, A.; Flores Palacios, Fatima;, Ríos Everardo, M.; Brauch, Hans Günter; Ruíz Pantoja, T.E.; Lemus Ramírez, C.; Estrada Villareal, A.; Cruz, M., 2014: *Vulnerabilidad Social y Género entre Migrantes Ambientales* (Cuernavaca: CRIM, DGAPA-UNAM): 151-168.

Oswald Spring, Úrsula, 2014b: "El Agua como Factor Crítico del Desarrollo Regional", in: Pérez Correa, Fernando (Ed.). *Gestión pública y social del agua en México* (México: UNAM): 78-97.

Oswald Spring, Úrsula, 2014c: "Agua y desarrollo local ante el cambio climático", in: Ocampo Fletes, Ignacio; Villarreal Manzo, Luis (Eds.), *Agua y desarrollo local ante el cambio climático* (Puebla: Colegio de Posgraduados, Campus Puebla): 259-279.

Oswald Spring, Úrsula, 2014d: "Social and Environmental Vulnerability in a River Basin of Mexico", in: Oswald Spring, Úrsula, Brauch, Hans Günte; Tidball, Keith G. (Eds.) *Expanding Peace Ecology: Peace, Security* (Cham: Springer): 85-108.

Oswald Spring, Úrsula, 2015: "Paz positiva, sustentable, culturalmente diversa y engendendrada" (Cuernavaca: CRIM-UNAM and CLAIP-Respuesta para la Paz), in press.

Oswald Spring, Úrsula, 2015a: "México ante el reto de la transición a la sustentabilidad con equidad y desarrollo", in: Cruz Núñez, Xochitl; Delgado Ramos, Gian Carlo y Oswald Spring, Úrsula (Eds.), *México ante la Urgencia Climática: Ciencia, Política y Sociedad* (México, D.F: CCEIICH; CRIM; PINCC-UNAM): 319-338.

Oswald Spring, Úrsula, 2015b: "Managing water resource in Mexico in the context of climate change", in: Shrestha, Sangam; Anal, Anil K.; Salam, P. Abdul; van der Valk, Michael (Eds.), *Managing Water Resources under Climate Uncertainty. Examples from Asia, Europe, Latin America and Australia* (Cham: Springer): 377-404.

Oswald Spring, Úrsula, 2015c: "Paz positiva, sustentable, culturalmente diversa y engendrada", in Serrano Oswald Serena Eréndira; Oswald Spring, Úrsula; de la Rúa, Diana (Eds.). *América Latina en el Camino hacia una Paz Sustentable: Herramientas y Aportes* (Guatemala: FLACSO Guatemala; ARP Asociación respuesta para la paz; CLAIP): 49-68.

Oswald Spring, Úrsula; Brauch, Hans Günter, 2015: "Securitizing Water", in: Anders Jägerskog; Swain, Ashok; Öjendal, Joakim (Eds.), *Water Security- Origin and Foundations*, vol. 1 (London: Sage): 3-35.

Oswald Spring, Úrsula; Moreno, Ana Rosa; Tena, Olivia, 2014. "Cambio Climático, Salud y Género", in: Ímaz Gispert, Mireya, Blazquez, Norma; Velázquez, Margarita et al. (Eds.). *Cambio climático, miradas de género*

(México: UNAM- PUMA; PINCC; CEIICH; CRIM-INAM; PNUD): 85-136.

Oswald Spring, Úrsula; Serrano Oswald, Serrena Eréndira; Estrada Álvarez, A.; Flores Palacios, Fatima; Ríos Everardo, M.; Brauch, Hans Günter; Ruíz Pantoja, T.E.; Lemus Ramírez, C.; Estrada Villareal, A.; Cruz, M., 2014: *Vulnerabilidad Social y Género entre Migrantes Ambientales* (Cuernavaca: CRIM, DGAPA-UNAM).

Oswald Spring, Úrsula; Jaramillo, Fernando, 2012: "Del Holoceno al Antropoceno: evolución del ambiente en Morelos", in: Crespo, Horacio; Morales Moreno, Luis Gerardo (Eds.): *Historiografía de Morelos. Tierra, gente, tiempos del Sur* (Cuernavaca: Congreso del estado de Morelos, UAEM): 325-384.

Oswald Spring, Úrsula; Jaramillo, Fernando, 2012: "Del Holoceno al Antropoceno: evolución del ambiente en Morelos", in: Crespo, Horacio; Morales Moreno, Luis Gerardo (Eds.): *Historiografía de Morelos. Tierra, gente, tiempos del Sur* (Cuernavaca: Congreso del estado de Morelos, UAEM): 325-384.

Oxford Pocket Dictionary, 1992: Dictionary by Oxford Dictionaries (Oxford: Oxford University Press).

Pastrana, A., 2013: "Reconstitución del tejido social"; at: <https://www.dnp.gov.co/Portals/0/archivos/documentos/GCRP/PND/Pastrana2_Compromisos_Fundam.pdf>.

Perona, N.; Rocchi, G.I., 2008: "Vulnerabilidad y exclusión social. Una propuesta metodológica para el estudio de las condiciones de vida de los hogares", in: *Kairos. Revista de temas sociales*, 8 : 1-15.

Reychler, Luc; Pfaffenholz, Tanja (Eds.), 2001: *Peacebuilding. A Field Guide* (London: Routledge).

Rosas, María Cristina, 1996: *México ante los procesos de regionalización económica en el mundo* (México, IIEc-UNAM).

Rzedowski, Jerzy, 1973: "Geographical relations of the flora of Mexican dry regions", in: Graham, Alan (Ed.): *Vegetation and vegetation history of northern Latin America* (Amsterdam, Elsevier): 61-72.

Schneider, Mindi, 2008: "'We are Hungry!': A Summary Report of Food Riots, Government Responses, and State of Democracy in 2008", at: <https://www.academia.edu/238430/_We_are_Hungry_A_Summary_Report_of_Food_Riots_Government_Responses_and_State_of_Democracy_in_2008>

Scott, John, 2010: "Subsidios agrícolas en México. ¿quién gana, y cuánto", in: Fox, Jonathan; Haight, Libby (Eds.): *Subsidios para la desigualdad. Las políticas públicas del maíz en México a partir del libre comercio* (Washington DC: Woodrow Wilson International Center for Scholars).

Segob [Ministry of Interior] 2014: *Incidencia Delictiva del Fuero Común* (November 2014) (Mexico: Segob).

Sen, Amartya, 1995: *Inequality reexamined* (Cambridge, Mass.: Harvard University Press).

Serrano Oswald, Serrena Eréndira, 2010: La Construcción Social y Cutural de la Maternidad en San Martín Tilcajete, Oaxaca, PhD Thesis (Mexico: Institute of Anthropological Research-UNAM).

Serrano Oswald, Serena Eréndira; Oswald Spring, Úrsula; de la Rúa, Diana (Eds.), 2015: *América Latina en el Camino hacia una Paz Sustentable: Herramientas y Aportes* (Guatemala: FLACSO Guatemala; ARP Asociación respuesta para la paz; CLAIP).

SNSP [Sistema Nacional de Seguridad Pública], 2014: "Data bank, May 2014"; at: <http://www.secretariadoejecutivosnsp.gob.mx/work/models/SecretariadoEjecutivo/Resource/131/1/images/CIEISP_mayo14.pdf>.

Solís, Leopoldo, 1973: *La economía mexicana. El Trimestre Económico, Lecturas 4.* México, Ed. FCE.

SS-DGE [Secretaría de Salud, Dirección General de Epidemiología], 2013: *Boletín Epidemiológico Diabetes Mellitus Tipo 2* (México: SS-DGE, Primer Trimestre-2013).

Stiglitz, Joseph, 2010: *Free Fall. America Free Markets, and the Sinking of the World Economy* (New York: W.W. Norton).

Sunkel, Osvaldo; Paz, Pedro, 1993: *El subdesarrollo latinoamericano y la teoría del desarrollo* (México, D.F.: Ed. Siglo XXI).

Transparencia International, 2014: at <http://www.transparencia.org/>.

Turner, John, 1987: *Rediscovering the Social Group: A Self-categorization Theory* (Oxford: Blackwell).

UNICEF, 2014 : *Pobreza y Derechos Sociales de Niños, Niñas y Adolescentes en México, 2010-2012* (Mexico, D.F.: UNICEF).

UNISDR, 2007: *Hyogo Framework for Action 2005-2015: Building the Resilience of Nations and Communities to Disasters. Extract from the final report of the World Conference on Disaster Reduction* (Geneva: UN/UNISDR, A/CONF.206/6).

UNISDR, 2009: 2009 *Terminology on Disaster Risk Reduction* (Geneva: UN/UNISDR).

UNISDR, 2013: "Global Assessment Report on Disaster Risk Reduction (Geneva: UN/UNISDR)"; at: <http://www.preventionweb.net/english/hyogo/gar/2013/en/gar-pdf/GAR2013_EN.pdf>.

USAID, 2013: "South Sudan–Complex Emergency", Fact Sheet #6, Fiscal Year (FY) 2013, 30 September 2013; at: <http://www.usaid.gov/sites/default/files/documents/1866/south_sudan_ce_fs06_09-30-2013.pdf>.

Villagrán de León, Juan Carlos, 2006 : *Vulnerability. A Conceptual and Methodological Review*, UNU-EHS, Source 4, Bonn, UNU-EHS.

Villagrán de León, Juan Carlos; Pruessner, Ines; Breedlove, Harold, 2013: "Alert and Warning Frameworks in the Context of Early Warning System. A Comparative Review", *Intersections* UNU-EHS, No. 12, (Bonn: UNU-EHS, May).

Weber, Max, 1987: *Economía y sociedad* (México: FCE).

WHO, 2013: *Revised Guidelines on Fluid Management of Dengue Fever and Hemorragic Fever 2012*, at: <http://www.ppsstc.com/wp-content/uploads/2013/03/revised-guidelines-fluid-management-oct-20121.pdf>.

Wittmayer, Julia; Neuteboom, J., 2011: *Working Paper. Exploring a transition movement in health care: observations, insights and inspiring examples* (Rotterdam: DRIFT).

33   Sustainability Transition in the Health Sector in Brazil

Monica de Andrade[1]

Abstract

Identifying sustainability transition in the health sector involves analysis of the dynamic and the forces influencing political, sociotechnical and cultural features. The aim of this study is to identify transition in health in a Brazil facing globalization, demographic, migration, environmental and climate change effects and inequities. Brazil has improved its capacity to formulate, implement, monitor and evaluate multi-sectoral and universal public policies in the last twenty years, with the implementation of universal health care and conditional cash transfers. These policies have resulted in the improvement of the health indicators of extreme poverty and hunger, under-five mortality, maternal health, infectious diseases, and primary education coverage. Since the frequency of extreme weather events has been increasing globally and in Brazil, and cardiovascular diseases are a leading cause of death, the consequences for health systems must be considered. Mortality due to traffic accidents has also increased, despite policy efforts and restrictive laws, overwhelming health services and public financial resources.

Keywords: Sustainable transitions, health, environmental health, health policies.

33.1   Introduction

The world population reached 7.2 billion in 2014 and is expected to increase by more than two billion by 2050.[2] Most of the future population growth will occur in the less developed regions. There is considerable diversity in the expected future trajectory of population change across major areas and various countries. More than half of the world's population now lives in urban areas. While the number of large urban agglomerations is increasing, approximately half of all urban dwellers reside in smaller cities and towns. The number of young people has grown rapidly in recent decades and is expected to remain relatively stable over the next thirty-five years. In contrast, both the number and proportion of older persons are expected to continue rising well into the foreseeable future (United Nations 2014).

Good health for all populations has become an accepted international goal and there have been broad gains in life expectancy over the past century. But health inequalities between rich and poor persist, while the prospects for future health depend increasingly on the relative new processes of globalization.

Globalization is causing profound and complex changes in the very nature of society, bringing new opportunities as well as risks. In addition, the effects of globalization are causing a growing concern for human health, and the intergenerational equity implied by sustainable development forces us to think about the right of future generations to a healthy environment and a healthy life (Huynen/Martens/Henk 2005).

Health and development are intimately interconnected. Insufficient development leading to poverty, or inappropriate development resulting in overconsumption, coupled with an expanding world population, can result in severe environmental health problems in both developing and developed nations.

Health security is defined in different ways. In order to distinguish between traditional and socially committed health transition, we use a wider concept of health security,[3] proposed by Leaning (2009). This is a people-centered understanding of health security that integrates underlying aspects such as globaliza-

1   Prof. Dr. Monica de Andrade is Professor of Environment and Health Promotion, Graduate Programme of Health Promotion, University of Franca, Brazil.

2   The author is grateful to Dr Úrsula Oswald Spring and to Dr Hans Günter Brauch for their suggestions and to four anonymous reviewers for their helpful comments on draft versions of this chapter.

tion, demographic effects, the effects of environmental and climate change, the growing disparity between rich and poor nations and between rich and poor people, and migration. Health security is therefore related to the sustainable management of the environment and to the changing conditions posed by climate change, population growth, urbanization and environmental deterioration (Brauch/Oswald Spring/Mesjasz et al. 2008, 2011; Brauch/Oswald Spring/Grin et al. 2009, Oswald Spring 2011).

Sustainable development (SD) as a goal was placed on the policy and scientific agenda by the Brundtland Commission in 1987. This challenge addresses the need for long-term structural changes or transitions in sectors such as energy supply, mobility, agriculture and healthcare. It is increasingly becoming a core focus of many disciplines that link subfields such as urban studies, economics, political studies, ecology and health.

The negotiations at the *World Summit on Sustainable Development* (WSSD 2002) demonstrated a major shift in the perception of sustainable development, away from environmental issues and towards social and economic development. This shift was driven by the needs of the developing countries and strongly influenced by the Millennium Development Goals.[4]

In 2012, the United Nations Conference on Sustainable Development took place in Rio de Janeiro. Its aim was to shape policies to reduce poverty, advance social equity and ensure environmental protection on an ever more crowded planet. One of the main outcomes of the Rio+20 Conference was the agreement by member states to launch a process to develop a set of *Sustainable Development Goals*[5] (SDGs), which will build on the Millennium Development Goals and converge with the post-2015 development agenda. It was decided establish an "inclusive and transparent intergovernmental process open to all stakeholders, with a view to developing global sustainable development goals to be agreed by the General Assembly".

Over the last few years, a large number of scenario-based assessments of global environmental problems and human development have been published. These have used an explorative approach with widely diverging scenarios to assess what might happen in the future. They include the *Intergovernmental Panel on Climate Change* (IPCC) climate assessment, the Global Environment Outlook reports (UNEP 2007, 2002), the *Millennium Ecosystem Assessment* (MA 2005a), the *International Assessment of Agricultural Science and Technology Development* (IAASTD 2009b) and the *World Water Development Reports* (UNESCO 2006, 2009).

In a recent study, a combination of policy options—including the expansion of protected areas into a well-chosen network covering twenty-nine per cent of the world's surface, an increase in agricultural productivity and reduced post-harvest losses, dietary change, improved forest management, and climate mitigation—was put forward, with the aim of achieving a significant restoration of natural areas and reduced biodiversity loss (Ten Brink/van der Esch/Kram et al. 2010).

In order to understand the dramatic changes and transitions in the world's health, it is necessary to access the scenarios using theories that can help to explain the process involved in sustainability transition.

Sustainability transition (ST) is a socio-economic, political and technological process that aims to achieve the goals of sustainable development. Sustainability is an essential part of health during transition and beyond and aligns itself with the sustainability goals agenda.

---

3   Health security is a concept that refers to strategies, policies and measures of sustainable development and preventive behaviour that may contribute to a healthy, participative society and a sufficiency of environmental services.

4   The UN Millennium Declaration (2000) adopted and committed countries to reach eight Millennium Development Goals by 2015. The eight goals included: halving extreme poverty, halting the spread of HIV/AIDS, providing universal primary education, eliminating gender disparity in education, reducing the under-five mortality rate, reducing the maternal mortality rate and achieving universal access to reproductive health, developing a global partnership (to address the needs of the poorest countries, furthering an open non-discriminatory trade system, and dealing with developing country debt); and ensuring environmental sustainability (by integrating sustainable development into national policies and programmes, reducing biodiversity loss, improving access to safe drinking water and sanitation, and improving the lives of slum dwellers) (UN 2010).

5   *Sustainable Development Goals* (SDGs) are under discussion and the proposed SDGs consist of four modules measured through a range of proxy indicators: (i) health (life expectancy; maternal mortality; infant mortality; child mortality); (ii) education (gender ratio in secondary education; years in education); (iii) income (per capita income); (iv) environment (per capita annual carbon emissions).

In order to understand the sustainability transition process, transitions theories[6] have been used as frameworks that allow us to understand the dynamics of ST. The core problem regarding sustainability transitions is how green innovations and sustainable practices (in behaviour and policy) 'struggle' against existing systems or regimes. Incumbent systems in the domains of transport, energy, and agri-food are difficult to dislodge because they are stabilized by various lock-in mechanisms (related to vested interests, low costs, established beliefs, sunk investments, and favourable institutions) that lead to path dependence and entrapment. Green innovations and new practices therefore tend to face an uphill battle, played out in economic, technical, political, scientific, and cultural dimensions (STRN 2010).

Science and technology studies (STS) and innovation studies have developed a 'quasi-evolutionary' approach to studying technological change (van den Bergh/Truffer/Kallis 2011). This innovation process model is characterized as a coupled dynamic of selective pressures and adaptive capacity in the dominant system ('regime'), in which a technology is embedded (Smith/ Stirling/Berkhout 2005).

One approach that has been used to analyse sustainability transition is the *multilevel perspective* (MLP) method, which analyses sociotechnical systems seen as consisting of niches, regimes and landscapes, a nested hierarchy of structuring processes. The multilevel perspective on sociotechnical transitions provides a framework in terms of both organizing analysis and of ordering policy interventions. It identifies three levels within societal systems (such as the health system): *niches*, in which radical innovation emerges; the *regime*, which comprises dominant institutions and technologies; and the *landscape*, which represents macro-level trends and contextual drivers and barriers to change. The main dynamics of change occur within and between the regime and niche levels, which may interact synergistically or antagonistically (Geels/Schot 2007).

Sustainability indicators and the composite index are additional tools that are gaining in importance and increasingly recognized as powerful mechanisms

for policy-making and for public communication to provide information on countries and corporate performance in fields such as environmental, economic, social and technological improvement (Singh/Murty/ Gupta et al. 2012).

The main contribution of this chapter is to provide an interaction between anthropogenic drivers, their impacts on and the policy responses to climate change, their interrelationship with the dominant productive systems of globalization, and their effects on human health. It will attempt to contribute to the understanding the consequences of policies that target the reduction of poverty and hunger, sustainable development, environmental protection, and the reduction of social inequities in Brazil.

The chapter is structured as follows. Section 33.2 addresses the national scale, very important in the delineation of regime contexts of transition studies, as a description of socio-economic, political, and technological scenarios in Brazil. Section 33.3 presents an overview of indicators of improvements in health in the context of sustainability transitions.

33.2 Socio-economic, political, and technological transition scenarios in Brazil

Brazil has the fifth largest population in the world, 197 million people (World Bank 2012), thirty per cent of whom are under the age of 18 (Unicef 2012). Demographic transitions present a major challenge because the population of Brazil increased from 72,775,000 in 1960 to about 202,768,562 in 2014, with 2,938,214 births, corresponding to a crude birth rate of 15.20 per 1,000 inhabitants.

In 2011, Brazil's gross national income was US$11,500 per capita (World Bank 2012). However, it has one of the highest rates of inequality in the world: its 2012 Gini index (51.9) was the sixteenth highest out of 136 countries worldwide (Central Intelligence Agency 2013). Ranking eighty-fifth out of 187 countries on the Human Development Index in 2011, Brazil still faces considerable challenges in meeting the whole population's needs for education, health care, and income.

Brazil is internationally recognized as one of the countries with the widest range of social inequalities. However, in recent years there has been an extraordinary liveliness in social innovation. Declining rates of income inequality have been reported since democratization at the end of the 1980s.

---

6   Transitions theories are theoretical approaches that analyse the development of 'sociotechnical transitions'. 'Sociotechnical' refers to the co-evolution of social and technological relationships, while 'transitions' refer to the dynamics by which fundamental change in these relationships occur (hence the relevance to sustainable consumption).

There have been three main periods of political regimes in Brazil, from military dictatorship (1964-1974) through a transition mix of military dictatorship and political transition called the 'Nova República' (1985-1990) to political (1974-1989) and democratic (1989-2002) consolidation (see table 33.1).

Various social movements emerged during this period, where the union movement in the metropolitan area of São Paulo was of particular importance. In addition, the *Association of Independent Unions* (CUT) was founded (in 1983) as well as the *Labour Party* (PT, in 1980), and these demanded fair wages and working conditions as well as comprehensive social reforms (Lindelow/Araujo 2014).

Urbanization has increased from fifty-four per cent in 1960 to eighty percent in 2010. Improved access to water and sanitation achieved solid economic growth and reduced income inequality.

The Brazilian health situation is going through a process of epidemiological transition in which non-communicable diseases and diseases resulting from external causes (especially violence) are increasing in place of infectious and parasitic diseases.

33.2.1   Social Policy

In Brazil, social policy was seen as a counter-concept to the Bismarckian 'social insurance' model, enabling the creation of a universal welfare regime: education, pensions and healthcare benefits as well as social transfers were universally available to all citizens in urban and rural areas. For the first time, the rural population was included in the welfare system, which has mainly been set up using cash transfers (Leubolt 2014).

The social security system for private workers (general system) is an unfunded defined benefit programme. There is still debate regarding when it began. In 1888 some measures were taken to provide pension benefits for postal workers and employees of the national press. In the following years, retirement benefits were extended to railroad workers, employees of the Ministry of Finance, the Mint and the armed forces. In 1923, the *Lei Eloi Chaves* legislation was approved to regulate social security for both civil servants and private sector workers. This law decentralized the pension system, as each company became responsible for its own employees. The first reform took place in 1933, when the pension funds were structured by professional category. The general pension system was centralized only in 1966, when the House of Representatives approved the Social Security Ordinary Law. The *National Social Security Administration* (INPS) incorporated all the revenues and expenditures from sector-specific programmes as well as its own assets and liabilities. Another major change during this time was that the scheme changed from a capitalization system to PAYGO (Turra/Queiroz 2005).

The last major reform occurred with the 1988 Constitution, which extended mandatory social security coverage to most of the previously excluded groups, including rural workers, without requiring equivalent increases in revenues from contributions.

The main social policy achievements was the establishment of minimum standards of "social security" which can be viewed as an expression of a political equilibrium (or stalemate) between progressive and conservative forces.

The presence of social issues in the 1988 Constitution can be interpreted as a consequence of the 'cultural hegemony' that the left obtained and fortified during the military dictatorship. After that, the group known as *Movimento pela Ética na Política* (Movement for Ethics in Politics) formed an important national social movement against hunger and misery, *Ação da Cidadania contra a Fome, a Miséria e pela Vida* (Citizens' Action against Hunger and Poverty and for Life) in 1993. This further strengthened the national anti-poverty consensus (Leubolt 2014).

The strategies adopted in 2003, such as *conditional cash transfers* (CCTs), can be seen as an instrument of social policy reflecting the widespread belief in Brazil that people are poor through the "fault of an unjust society".

Zero Hunger, Brazil's national strategy for food and nutritional security, consists of more than twenty initiatives on four axes of intervention. Its creation in 2003 was a milestone in the recognition of food and nutritional security as a leading and cross-cutting priority on the political agenda. *Zero Hunger* is a combination of continuity and innovation. It introduced major programmes such as Bolsa Família (Chmielewska/Souza 2011).

Another policy example comes from programmes that link horticultural producers directly to markets. One example is the Family Farming Food Acquisition Programme, part of the Zero Hunger Programme in Brazil. In this programme, the government purchases food produced by family farms. The programme ensures food supplies for poor families, as well as school meals and public hospitals, while creating a market for the small farming sector. There is some evidence that the programme has had a positive nutritional impact in north-eastern Brazil.

Table 33.1: Characterization of sustainability transitions in Brazil: structural changes at local level (landscape level).HD = Hydroelectricity. **Source**: The author.

Structural changes (landscape level)

Timeline	Political Regime	Pop. (million)	Energy	Transport	Social security	Health	Food & Nutrition	Economy	Environment
				Technological innovation (regime)					
1946–1964	Populist Democratic	72.775(1960) 66% rural 44% urban	HD	Fossil fuels	1923 Social security for workers			Agriculture Mineral exportation Industrialization	High deforestation rates
1964–1974	Military dictatorship		HD	Fossil fuels			Subsistence agriculture Cattle ranching	'Economic miracle' Automobile industry	High deforestation rates
1974–1990 "Nova Repúblic a" (1985–1990)	Mix of military dictatorship and political transition		HD +World-renowned technology for exploring deep off shore oil reserves	Fossil fuels +biofuels (ethanol)		Health system exclusively for workers		High inflation (1980) Growth of agribusiness	High deforestation rates
1988 Constitution		80% urban 20% rural			Extended mandatory social security coverage	Universal publicly-funded, rights-based health system (SUS)		Growth of agribusiness	High deforestation rates
1990–2000	1994's Plano Real (Real Plan) Inclusive liberalism	170	HD +World-renowned technology for exploring deep off shore oil reserves				Growth of agribusiness Agriculture Expansion and modernization of traditional cattle ranching		1997 law limiting deforestation to 20% of the private property in the Amazonian region
2001–2010	Developmental welfare		HD Oil + Thermoelectricity + Sugarcane burning	Increased expansion of private transport	Conditional Cash Transfers	Family Health Programme Smiling Brazil Mobile Emergency Services, Popular Pharmacy Programme Health Promotion Policy	(2003) Zero Hunger Programme Bolsa Família (2004) Food and Nutrition Policy (2010)	Automobile industry Growth of agribusiness Mechanization Slow growth of small-scale agriculture	(2006) Decreasing deforestation rates Implementation of Climate Change Policy
2013–2014		202,768,562 86% urban 14% rural	HD Oil + Thermoelectricity + Sugarcane burning Wind power electricity		Unemployment, social exclusion				

33.2.2   The Health System

Brazil was one of the first Latin American countries to establish *universal health care* (UHC) as a fundamental right, based on the principles that health care is a duty of the state and should be free at the point of use. There form in the late 1980s created the *Unified Health System* (*Sistema Único de Saúde*, or SUS) and was based on the principle that health care should be free at the point of use to all Brazilian citizens (Lindelow/Araujo 2014). The goals of universal health coverage are to ensure that all people can access quality health services, to safeguard all people from public health risks, and to protect all people from impoverishment due to illness, whether from out-of-pocket payments for health care or loss of income when a household member falls sick. Brazil's *Unified Health System* (SUS), a universal health care system, was established by the Constitution of 1988, after more than two decades of military dictatorship. It increased access to healthcare for a substantial proportion of the Brazilian population, at a time when the system was becoming increasingly privatized.

The previous system INAMPS (*Instituto Nacional de Assistência Médica da Previdência Social*) had several vertical, independent, and uncoordinated systems, each with its own source of funding, network of facilities, and beneficiary population.

The Brazilian health system is formed by a complex network of public and private institutions devoted to providing, financing and managing health services; surveying, producing, and distributing resources; human resources training; and regulation, legislation and jurisdiction of the system. The *Unified Health System* (SUS) is responsible for exclusive coverage of 78.8 per cent of the Brazilian population, being the main network of public institutions devoted to the provision, financing, and management of health services. The remaining 21.2 per cent of the population, covered by the Supplementary System, also have the right to access the services provided by SUS. SUS is also responsible for the provision of the collective services of health surveillance, disease control and sectorial regulation (WHO 2013a). It has been designed and put in place in an era where neo-liberal reforms elsewhere in the world have driven the marketization of health services, and offers important lessons for future health systems.

The *technological innovation* of the SUS is that institutional mechanisms for popular involvement and accountability are part of the architecture of the governance of the system.

Public involvement has the potential to sustain a compact between state and citizens and to ensure the political momentum required to broaden access to basic health services, while at the same time providing a framework for the emergence of "regulatory partnerships" capable of managing the complex reality of pluralistic provision and multiplying sources of health expertise in a way which ensures that the needs and rights of poor and marginalized citizens are not relegated to the periphery of a segmented health system (Cornwall/Shankland 2008).

The Brazilian health agenda from 2003 to 2008 had as its priorities the Family Health Programme, Smiling Brazil, Mobile Emergency Services, and the Popular Pharmacy Programme (table 33.1). The first is a policy with high institutional density launched by the previous administration, and constitutes an example of path dependence. The other three are innovations in areas where there had been weaknesses in federal government action. The four policy priorities are strategies focused on solving key problems in the Brazilian health system. However, they display important differences in their historical development, political and institutional base, inclusion on the federal agenda, and implications for the principles of the Unified National Health System. Although incremental changes have been introduced, national health policy has been characterized predominantly by continuity (Machado/Baptista/Nogueira 2011).

The main primary health care programme in Brazil, the *Programa Saúde da Família* (Family Health Programme, FHP), is a large-scale national programme, implemented over the past several years. By 2011, it had reached 94 per cent of municipalities, covering 53 per cent of the Brazilian population. FHP aims to broaden access to public health services, especially in deprived areas, by offering free, community-based health care (Programa Saúde da Família 2000).

The National Policy on *Science, Technology and Innovation* (STI) in Health Policy, established in 2004, resulted from a democratic process with inputs from investigators, stakeholders and the public in general. It is guided by six principles: (a) improving the Brazilian population's health conditions in the short, medium and long term; (b) overcoming all forms of inequity and discrimination (regional, social, ethnic, gender-based and others); (c) respecting people's lives and dignity; (d) ensuring the implementation of high ethical standards in health research; (e) respecting methodological and philosophical plurality; and (f) social inclusion, citizen control and respect for the environment and sustainability (Brasil 2008).

The National Agenda of Priorities in Health Research was created at that time to reduce the gap between scientific knowledge and health practice and activities, and aims to contribute to improving Brazilian quality of life. These efforts in guiding health research policy have achieved and legitimated an unprecedented developmental spurt in strategic health research (Pacheco Santos/Moura/BarradasBarata et al. 2011).

33.2.3 Brazilian Environment Policy

Brazil has traditionally been playing an active role in the international climate change arena. It was the first country to sign the United Nations Framework Convention on Climate Change (UNFCCC), in 1992. And during Copenhagen's COP15 run-up, it announced its voluntary pledge of a -36 to -39% reduction in GHG emissions by 2020 compared to a business as usual scenario. In 1997, by signing the Kyoto Protocol, Brazil agreed on limited compulsory greenhouse gas (GHG) reduction emission targets, and in the same year, the country passed law strictly limiting deforestation to twenty per cent of private property in the Amazonian region. However, resistance to the law was very strong until 2005, with the federal government unwilling to enforce the law and strong opposition from most state governments. From 2005 until 2009, the reduction of deforestation in the Amazon was at the core of the federal government's programme.

In 2003, the federal government approved the creation of the Permanent Group for Interministerial Work (*Grupo Permanente de Trabalho Interministerial*, GPTI), which was made up of the heads of thirteen key ministries and led by the Chief of Staff. The group's goal was to propose and coordinate actions aimed at reducing deforestation in the area known as *Amazônia Legal*[7] or Legal Amazon (Casa Civil 2013).

In 2004, the GPTI presented the operational project "Action Plan for the Prevention and Control of Deforestation in the Legal Amazon" (Plano de Ação para a Prevenção e o Controle do Desmatamento na Amazônia Legal, PPCDAm). This action plan is a large set of strategic conservation measures to be implemented and executed as part of a collaborative

effort between federal, state, and municipal governments, alongside specialized organizations and civil society. The launch of the plan in 2004 integrated actions across different government institutions and introduced innovative procedures for monitoring, environmental control, and territorial management (IPEA 2012).

The National Plan on Climate Change (NPCC) was lauched in 2007, which aims to promote actions on both mitigation and adaptation measures to face climate change, involving federal, regional and local governments, and all segments of the national society (Brasil, 2008).

Since 2007, the capacity of the state to control illegal deforestation in large areas has increased so dramatically that a significant part of the remaining deforestation has been reduced to small areas that are more difficult to detect by satellite. It is important to stress that this process was carried out without any negative impact on economic growth (Moutinho 2009).

After 2010, the federal government no longer aims to reduce deforestation, but rather to avoid any new increase. Between 1970 and 2010, approximately eighteen per cent of the Brazilian Amazon was deforested (Baccini/Goetz/Walker al. 2012), with the primary cause being demand for new land for the cultivation of soybeans and expansion of pasture.

With a view to creating a scenario for Brazil, and taking into account MLP as it is applied to particular national systems, to regimes, and to the nationwide creation of conditions for niches, some impacts of policy intervention are presented in table 33.2.

Brazil is a global leader in renewable energy sources, hydropower (designing reservoirs) and ethanol (fuel from sugar cane).The relevant technological innovations are interoperable and interdependent between the agricultural and transport sectors. In 2002, the government announced the *Incentive Programme for Renewable and Alternative Energy* (PROINFA), which supported renewable energy sources, including wind (Dutra/Szklo 2008).

33.3 Impacts of Anthropogenic Drivers on Health Security in Brazil

This section describes the consequences of policies targeting poverty reduction, hunger, sustainable development, environmental protection and the reduction of social inequities in Brazil. By means of universal

---

7   The Legal Amazon of Brazil is defined by law to include the states of Acre, Amapa, Amazonas, Para, Rondonia, Roraima, Mato Grosso, Maranhao, and Tocantins in order to facilitate the economic development planning and integration to other Brazilian regions. Its are represents 61% of Brazilian territory and 13% of population (IBGE 2010).

Table 33.2: Timeline of main policy interventions in Brazil and their impacts on health and the ecosystem, together with their social and economic aspects. **Source:** The author.

Driver	Policy interventions (landscape)	Time	Technological Innovation		Impacts	
			green innovations (niches)	non-environmental	Negative	Positive
Energy supply	Building of hydroelectric power plants	1960 2014	Hydropower		Changes in ecosystem, loss of biodiversity	Increased access to energy
	Incentive Programme for Renewable and Alternative Energy	2002 2014	Solar power			Decreased demand for other energy sources
	Incentive for exploring deep offshore oil reserves	2000 2014		Oil	Increased GHG emissions, environmental pollution	Increased commodities
	Incentive to create thermos-electric power plants	2000 2014		Oil + Thermo-electricity	Increased GHG emissions environmental pollution	
	Building of wind power plant	2010	+Wind power electricity			Decreased demand for energy
	Incentive for bioelectricity	2000	Sugarcane bagasse burning			Decreased demand for other energy sources
Mobility	Incentive for sugarcane mills	1970 2014	Use of ethanol for transport (fuel from sugar cane)		Changes in land use Ecological disequilibria	Decreased GHG emissions
	Public transport	1980 2014		Poor public transport	Increased GHG emissions	Increased job opportunities in industry
	Incentive for acquisition of self-transportation vehicles	2000 2014		Increase expansion of private transport	Increased individual energy expenditure Increased number of deaths from motorcycle accidents (67%)	Increased job opportunities in industry
Industry	Incentives for automobile industry	1970 2014	Cars running on biofuel		Increased GHG emissions	Increased job opportunities in industry
Agriculture	Growth of agribusiness Mechanization	2000 2014		Expansion of culture, habitat fragmentation	Changes in land use Increased unemployment, social exclusion, violence	Improvement of economy
	Expansion of agriculture Modernization of traditional cattle ranching	2000 2014		Intensive use of agro-toxic substances	Deforestation Increased GHG emissions, health issues, use of agro-toxic substances in food production Migration	Food availability
	Food Acquisition Programme (PAA)	2003	Incentive for organic agriculture			Improvement in family agriculture
Healthcare	Unified Health System (SUS)	1988	Popular involvement and accountability		See table 33.3	See table 33.3.3
Poverty Hunger Social inequities	Zero Hunger Strategy	2003	Improvement nutritional status			Reduction in 1. under-5 mortality 2. malnutrition 3. diarrhoea
	Brazil without Extreme Misery	2011				
	Establishment of Bolsa Família	2004	Improvement of income Improvement of education			
Environmental protection	Law limiting deforestation to 20% of private property in the Amazonian region. Policy for solid waste	1997 2006	Decrease in rate of deforestation			Decreased infectious disease transmission related to deforestation Decreased GHG emissions
Sustainable development	National Plan on Climate Change	2008	Stop burning and smoke emissions			Improvement in hospital admissions

Table 33.3: Transition of heath indicators in Brazil. **Source:** The author.

	Indicator	Time	Source
Life expectancy	Life expectancy at birth (years)	67 (1990) 73 (2009) 73.9 (2013)	
Infant mortality	Infant mortality rate (probability of dying by age 1 per 1000 live births)	50 (1990) 31 (2000) 17 (2010)	PNUD 2014
Infant mortality	Under-5 mortality rate	59 (1990) 36 (2000) 19 (2010)	
Adult mortality	Probability of dying between 15 and 60 years old per 1000 population	275 (2000) 219.67 (2011)	World Bank 2014
Extreme poverty eradication	Extreme poverty	15 million (2015)	PNAD, IBGE 2013
External causes	Deaths per 100,000 population	68.7 (2006) 78.4 (2012)	DATASUS 2014
Education	Proportion of people with seven or more years of formal education	19% (1976) 47% (2008)	Paim/Travassos/Almeida et al. 2011
Income inequality	Gini coefficient	0.530 (1960) 0.636 (1989) 0.594 (1999) 0.530 (2012)	IPEA 2012

health coverage, conditional cash transfers and environmental policies, the process of sociotechnical transition has had an impact on quantitative indicators (health indicators), such as life expectancy and mortality among infants and adults (table 33.3).

In Brazil, life expectancy has improved from 67 years in 1990 to 73 years in 2009 and to 73.9 in 2014, according to PNUD or UNDP (2014). The infant mortality rate (the probability of dying by age 1 per 1000 live births) has decreased from 50 in 1990 to 31 in 2000 and 17 in 2010; and the under-5 mortality rate has also decreased from 59 in 1990, to 36 in 2000 and 19 in 2010, according to WHO (2013b). The adult mortality rate (the probability of dying between 15 and 60 years old per 1000 population) has decreased from 275 (2000) to 219.67 (2011) (World Bank 2014).

The leading global risks for mortality in the world are high blood pressure (responsible for 13% of deaths globally), tobacco use (9%), high blood glucose (6%), physical inactivity (6%), and overweight status and obesity (5%). The leading cause of death in Brazil is cardiovascular disease, second is cancer, and external causes are third (Brasil 2014).

The *World Health Organization* (WHO) estimates that the warming and precipitation trends caused by anthropogenic climate change over the past thirtyyears have already claimed over 150,000 lives annually. Many of the prevailing human diseases are

linked to climate fluctuations, from cardiovascular mortality and respiratory illnesses caused by heatwaves, to the altered transmission of infectious diseases and malnutrition from crop failures.

Researchers report (Li/Cheng/Cui et al. 2014, Wang/Lin 2014) that an increase of 1°C in the daily maximum temperature increases mortality and the number of urgent and emergency room visits due to cardiovascular diseases. Considering that extreme weather events have been increasing not only in Brazil but globally, and that cardiovascular disease is the leading cause of death, the consequences for the health systems must be considered.

Another worrying indicator is external causes, the third most common cause of death in Brazil. There was an increase of 14.1 per cent in deaths from external causes, up from 68.7 deaths per 100,000 population in 2006 to 78.4 deaths per 100,000 in 2012.

Among the specific causes of death, homicides increased by 10.5 per cent, from 26.6 deaths per 100,000 population in 2006 to 29.4 deaths per 100,000 in 2012.

Traffic accidents increased by 19 per cent, from 19.9 deaths per 100,000population in 2006 to 23.7 deaths per 100,000 in 2012. The vehicles most frequently involved in traffic accidents were motorcycles, with an increase of 67.8 per cent in the mortality rate (Ortiz 2014).

The implications of these data for understanding the dynamics of sustainability transition are crucial. Policy issues that need to be addressed are the creation of incentives for economic bases such as the automobile industry and sugar cane agribusiness, policies for improved mobility, such as public transport and policies that take into account the forces influencing political, sociotechnical and cultural features.

Deforestation is part of the increasing threat of new illnesses and loss of public health.Measurements carried out in southern Amazonia have demonstrated that exposure to PM2.5 particulates has a positive association with children's respiratory health (Ignotti 2010).

Another study (Smith/Aragão/Sabel et al. 2014) showed a significant increase (from 1.2% to 267%) in hospitalizations forrespiratory diseases in children aged under five in municipalities highly exposed to drought. Aerosols were the primary driver of hospitalizations in drought-affected municipalities during 2005.

The transition to sustainability in Brazil during the last decades has been driven by political regimes and socio-economic and technological processes, and may be summarized as follows. The 2000s were marked by diminishing inequalities. The main impact factors were better employment conditions for the poorer segments of the population and a remarkable expansion in cash transfers. This expansion was accompanied by a slight diminishment in social services and infrastructure until 2005, which corresponded with neo-liberal concepts of social policies. From 2006 onwards, the trend towards 'monetarization' has been reversed, as investments in social services and infrastructure have been raised considerably. The overall strategy has shifted from 'inclusive liberalism' to 'developmental welfare', where the reduction of inequalities through state-induced measures is viewed as an important factor in raising the level of consumption and thereby boosting economic growth (Leubolt 2013).

In a study of the effect of a conditional cash transfer programme on childhood mortality, Rasella, Aquino, Santos et al (2013) show that *Bolsa Familia* had a significant role in reducing under-5 mortality, both overall and from poverty-related causes such as malnutrition and diarrhoea, in Brazilian municipalities between2004 and 2009. The effect was maintained even after adjusting for socio-economic covariates and the FHP. The increase in the duration of *Bolsa Familia* and in the coverage of both the total and target populations strengthened the effect of the programme. The authors concluded that a large-scale programme such as *Bolsa Familia*, combined with an effective primary healthcare system, can strongly reduce childhood mortality, both from poverty-related causes and overall. Mechanisms included the effects on social determinants of health services and the increased use of preventive services for children and pregnant women through programme conditions.

Brazil has not yet embarked on a low carbon green economy paradigm, and lacks serious planning and long-term goals. What is important is the way in which Brazil, particularly in the economic realm, will set the parameters and priorities for alow carbon transition.

The globalization of markets (landscape pressures) is related to international factors of development, and in the case of Brazil it puts pressure on deforestation: Brazil's primary issue being the demand for new land for the cultivation of soybeans and the expansion of pasture.

The growth of agribusiness, together with agricultural expansion and the modernization of traditional cattle ranching, have brought about enormous changes inland use. Land use change and forestry have traditionally been the major drivers of Brazilian greenhouse gas (GHG) emissions.

This has been exerting pressure through demographic trends, with an increase of the population in urban areas from 44 per cent urban to 86 per cent urban in about seven decades.

Agricultural expansion and the modernization and mechanization of large agricultural holdings has resulted in the loss of jobs for rural workers, who then migrate to urban areas.

This migration from rural to urban areas puts pressure on the urban sector and increases the demand for health, education, food and nutrition, social security, energy, and transport, among other things. When there is no planning or policies to meet the demand, there is expansion of human settlement in areas of environmental risk, together with social exclusion and an increase in violence and homicides.

33.4   Conclusions

Transitions are about radical change in systems that are there to meet the needs of societal systems—in this case, healthcare systems.

Brazil has shown how *universal health care* (UHC) can serve as a vital mechanism for improving the health and welfare of its citizens, and can lay the foundations for economic growth and competitiveness grounded on the principles of equity and sustainability. These findings are intended to provide lessons that can be used by countries aspiring to adopt, achieve, and sustain UHC.

The lack of urban planning, the lack of public policies for public transport and to provide the incentive for the acquisition of self-transportation vehicles has increased the number of deaths from traffic accidents by 19.0 per cent. Transport systems in Brazil must be decarbonized, but this depends on increasing energy efficiency and disseminating low carbon energy sources.

The mortality statistics demonstrate how the health infrastructure is under pressure, health services overwhelmed and public financial resources put in danger. The health services need urgently to fit themselves for the future, creating an environment of efficiency, quality and the wise use of resources, and incorporating actions that reduce inequities in the health budgets of municipal, state and federal governments.

The implementation of effective social policies has enabled Brazil to partially reduce absolute poverty and income inequality, thus contributing to decreasing death rates in the population and to the improvement of the health indicators of extreme poverty and hunger, under-5 mortality, maternal health, infectious diseases, and primary education coverage. Reducing income inequality may represent an important step towards improving health and increasing life expectancy.

Despite inter-sectoral collaboration between health, education and social governmental sectors, strong efforts have to be made in the economic area in order to achieve sustainability transition.

References

Assunção, Juliano; Gandour, Clarissa C.; Rocha, Rudi, 2012: "Deforestation Slowdown in the Legal Amazon: Prices or Policies?"; at: <http://climatepolicyinitiative.org/wp-content/uploads/2012/ 03/Deforestation-Prices-or-Policies-Working-Paper.pdf>.

Braubach, Matthias: 2013; "Benefits of environmental inequality assessments for action", in: Journal of the Epidemiology Community Health, 67,8: 661-666; published online first, 1 May 2013; at: <doi:10.1136/jech-2012-201426>.

Baccini, Alessandro; Goetz, Scott J.; Walker, Wayne S.; Laporte, Nadine T.; Sun Mindy; Sulla-Menashe, Damien; Hackler Joseph; Beck, Peter. S. A.; Dubayah Ralph; Friedl Mark A.; Samanta Sudeep; Houghton, Richard A., 2012: "Estimated carbon dioxide emissions from tropical deforestation improved by carbon-density maps", in: Nature Climate Change, 2: 182-185.

Barreto, Mauricio L.; Teixeira, M. Gloria; Bastos, Francisco I.; Ximenes, Ricardo A. A.; Barata, Rita B.; Rodrigues, Laura C., 2011: "Successes and failures in the control of infectious diseases in Brazil: social and environmental context, policies, interventions, and research needs", in: Lancet, 37, 9780:, 1877-1889; at: <doi: 10.1016/S0140-6736(11)60202-X. Epub 2011 May 9>.

Brasil, 2008: National Plan on Climate Change; at: < http://www.mma.gov.br/estruturas/208/ arquivos/national plan_208.pdf>

Brasil, 2008: Ministério da Saúde. Secretaria de Ciência, Tecnologia e Insumos Estratégicos. Departamento de Ciência e Tecnologia. Política Nacional de Ciência, Tecnologia e Inovação em Saúde. 2 ed, 44p, at: < http://livroaberto.ibict.br/handle/1/613>.

Brasil, 2014: Ministério da Saúde. Departamento de Informática do SUS; at: DATASUS <http://www2.datasus.gov.br/datasus/index.php>.

Brauch, Hans Günter; Oswald Spring, Úrsula; Mesjasz, Czeslaw; Grin, John; Dunay, Pal; Behera, Navnita Chadha;

Chourou, Béchir; Kameri-Mbote, Patricia; Liotta, P.H. (Eds.), 2008: Globalization and Environmental Challenges: Reconceptualizing Security in the 21st Century. Hexagon Series on Human and Environmental Security and Peace, vol. 3 (Berlin - Heidelberg - New York: Springer-Verlag).

Brauch, Hans Günter; Oswald Spring, Úrsula; Grin, John; Mesjasz, Czeslaw; Kameri-Mbote, Patricia;Behera, Navnita Chadha; Chourou, Béchir; Krummenacher, Heinz (Eds.) 2009: Facing Global Environmental Change: Environmental, Human, Energy, Food, Healthand Water Security Concepts. Hexagon Series on Human and Environmental Security and Peace, vol. 4 (Berlin - Heidelberg - New York: Springer-Verlag).

Casa Civil (2013). Plano de Ação para a Prevenção e Controle do Desmatamento na Amazônia Legal. Casa Civil da Presidência da República, Brasília at: <http://www. amazonfund.gov.br/FundoAmazonia/export/sites/default/site_pt/Galerias/Arquivos/Publicacoes/PPCDAm _3.pdf>

Chmielewska, Danuta; Souza, Darana, 2011: The food security policy context in Brazil. Country Study (Brasília: International Policy Centre for Inclusive Growth United Nations Development Programme); at: <http://www.ipc-undp.org/pub/IPCCountryStudy22.pdf> (web pdf).

Cornwall, Andrea; Shankland, Alex, 2008: "Engaging citizens: lessons from building Brazil's national health system", in: Social Science and Medicine, 66,10: 2173-2184.

Dutra, Ricardo Marques; Szklo, Alexandre Salem, 2008: "Incentive policies for promoting wind power production in Brazil: Scenarios for the Alternative Energy Sources Incentive Program (PROINFA) under the New Brazilian electric power sector regulation", in: Renewable Energy, 33,1: 65-76.

Galvão, Antonio; Juruá, Mayra; Esteves, L., 2012: "The Amazons and the Use of its Biodiversity". Report for the IDRC funded project Opening up Natural Resource-

Based Industries for Innovation: Exploring New Pathways for Development in Latin America; at: <http://www. ecologyandsociety.org/vol15/iss1/art11/>.

Geels, Frank W; Schot, Johan, 2007: "Typology of sociotechnical transition pathways". Research Policy 36: 399-417.

Goldemberg, Jose, 2007: "Ethanol for a sustainable energy future", in: Science, 315: 808-810.

Huynen, Maud M. T. E.; Martens, Pim; Hilderink, Henk B.M., 2005:"The health impacts of globalisation: a conceptual framework", in: Globalization and Health, 1,14: 1-12.

Ignotti, Eliane, 2010: "Air pollution and hospital admissions for respiratory diseases in the subequatorial Amazon: a time series approach", in: Cad. Saude Publica, 26,4: 747-761.

IBGE, 2010: Censo Populacional, in: <http://www.ibge.gov. br/home/estatistica/populacao/censo2010/default.shtm>

IPAM, 2009: "Evolução na Política para o Controle do Desmatamento na Amazônia Brasileira: O PPCDAm", in: Clima e Floresta, 15.

IPEA (Instituto de Pesquisa Econômica Aplicada), 2012: "A dinâmica recente das transferências públicas de assistência e previdência social", in: Comunicado do Ipea, 138; at: <http://www. ipea.gov.br/portal/images/stories/ PDFs/comunicado/120308_comunicadoipea138.pdf>.

Jerneck, Anne; Olsson, Lennart, 2011:"Breaking out of sustainability impasses: How to apply frame analysis, reframing and transition theory to global health challenges", in: Environmental Innovation and Societal Transitions, 1,2: 255-271.

Leaning, Jennifer, 2009: "Health and Human Security in the 21st Century", in: Brauch, Hans Günter; Oswald Spring, Úrsula; Grin, John; Mesjasz, Czeslaw; Kameri-Mbote, Patricia; Behera, Navnita Chadha; Chourou, Béchir; Krummenacher, Heinz (Eds.): Facing Global Environmental Change: Environmental, Human, Energy, Food, Health and Water Security Concepts. Hexagon Series on Human and Environmental Security and Peace, vol. 4 (Berlin - Heidelberg - New York: Springer-Verlag): 541-552.

Lehtonen, Markku, 2007: Biofuel transitions and global governance: lessons from Brazil. http://www.2007amsterdamconference.org/Downloads/AC2007_Lehtonen. pdf>.

Leubolt, Bernhard, 2014: "Social policies and redistribution in Brazil". Global Labour University working paper; No. 26 (Geneva: International Labour Office, Global Labour University); at: <http://www.global-labour-university.org/ fileadmin/GLU_Working_Papers/GLU_WP_No.26.pdf>.

Li, Yonghong; Cheng, Yibin; Cui, Guoquan; Peng, Chaoqiong; Xu, Yan; Wang, Yulin; Liu, Yingchun; Liu, Jingyi; Li, Chengcheng; Wu, Zhen; Bi, Peng; Jin, Yinlong, 2014: "Association between high temperature and mortality in metropolitan areas of four cities in various climatic zones in China: a time-series study", in: Environmental Health, 7: 13-65.

Lindelow, Magnus; Araujo, Edson C. 2014: Universal Health Coverage for Inclusive and Sustainable Development: Country Summary Report for Brazil (Washington, DC: World Bank); at: <https://openknowledge. worldbank.org/handle/10986/20732>.

Machado, Cristiani. Vieira; Baptista, Tatiana Wargas Faria; Nogueira, Carolina Oliveira, 2011: "Health policies in Brazil in the 2000s: the national priority agenda", in: Cad. Saúde Pública (Rio de Janeiro), 27,3 (March): 521-532.

Markard, Jochen; Raven, Rob; Truffer, Bernhard, 2012: "Sustainability transitions: An emerging field of research and its prospects", in: Research Policy, 41,6: 955-967.

Murray, Joseph; Cerqueira, Daniel Ricardo Castro; Kahn, Tulio, 2013: "Crime and violence in Brazil: Systematic review of time trends, prevalence rates and risk factors", in: Aggressive Violent Behaviour, 18,5 (September): 471-483.

Ortiz, Dennys Samillan, 2014:"Evolution of mortality from external causes in Brazil, 2006-2012" (Master thesis, city: university, Graduate Programme of Health Promotion).

Oswald Spring, Ursula, 2011: "Towards a sustainable health policy in the Anthropocene", in: IHDP Update Issue 1: 19-25.

Pacheco Santos, Leonor Maria; Moura, Erly Catarina; Barradas Barata, Rita de Cássia; Serruya, Suzanne Jacob; Motta, Marcia Luz; Silva Elias, Flávia Tavares; Angulo-Tuesta, Antonia; Paula, Ana Patricia de Paula; de Melo, Gilvania; Guimarães, Reinaldo; Grabois Gadelha, Carlos Augusto, 2011: "Fulfillment of the Brazilian agenda of priorities in health research", Health Research Policy and Systems 2011, 9,35: 1-9 at: <http://www.health-policy-systems.com/content /9/1/35>

Paim, Jairnilson; Travassos, Claudia.; Almeida, Celia.; Bahia, Ligia.; Macinko, James., 2011: "The Brazilian health system: history, advances, and challenges", in: Lancet, 377,9779 (21 May): 1778-1797; at: <doi: 10.1016/S0140-6736(11)60054-8. Epub 2011 May 9>.

Programa Saúde da Família, 2000. Rev. Saúde Pública 34, 3: 316-319; at: <http://www.scielo.br/ scielo.php?script =sci_arttext&pid=S0034-89102000000300018&lng=en> and: <http://dx.doi.org/10.1590/S0034-89102000000300018>.

Rasella, Davide; Aquino, Rosana; Santos, Carlos AT; Paes-Sousa, Rômulo; Lima Barreto, Mauricio, 2013:"Impact of income inequality on life expectancy in a highly unequal developing country: The case of Brazil". The Lancet, 382,9886: 57-64.

Singh, Rajesh Kuma N.; Murty, H. Ramalinga.; Gupta, Suresh Kumar.; Dikshit Anil Kumar, 2012:"An overview of sustainability assessment methodologies", in: Ecological Indicators, 15,1 (April): 281-299.

Smith, Adrian; Stirling, Andy; Berkhout, Frans, 2005: "The governance of sustainable socio-technical transitions", in: Research Policy, 34,10: 1491-1510.

Smith, Lauren T.; Aragão, Luiz E. O. C.; Sabel, Clive E.; Nakaya, Tomoki, 2014:"Drought impacts on children's respiratory health in the Brazilian Amazon", in: *Scientific Reports*, 4,3726: 1-8.

STRN Sustainable Transitions Research Network. "A mission statement and research agenda for the Sustainability Transitions Research Network" 2010, at: <http://www.transitionsnetwork.org/files/STRN_research_agenda_20_August_2010%282%29.pdf>

Ten Brink, Ben.; van der Esch, Stefan; Kram, Tom.; von Oorschot, Mark; Alkemade,Rob; Ahrens, Robert; Bakkenes, Michel; Bakkes, Jan; van den Berg, Maurits, Christensen, Villy; Janse, Jan, Jeuken, Michel, Lucas, Paul, Manders, Ton; van Meijl, Hans; Stehfest, Elke; Tabeau, Andrzej; van Vuuren, Detlef; Wilting, Harry, 2010: *Rethinking Global Biodiversity Strategies: Exploring Structural Changes in Production and Consumption to Reduce Biodiversity Loss* (Bilthoven: Netherlands Environmental Assessment Agency (PBL), 172p.

Turra, Cássio; Queiroz, Bernardo: 2005 *"Before it's too late: demographic transition, labour supply, and social security problems in Brazil"* In: United Nations Expert Group Meeting on Social and Economic Implications of Changing Population Age Structures. United Nations, 104 118. <http://www.un.org/esa/population/meetings/Proceedings_EGM_Mex_2005/turra.pdf>

United Nations General Assembly, 2000:*United Nations Millennium Declaration*, A/RES/55/2 (New York: United Nations).

United Nations, 2010:*The Millennium Development Goals Report* 2010(New York: United Nations).

United Nations, 2014: *The World Population Situation in 2014. A Concise Report* (New York: United Nations, Department of Economic and Social Affairs, Populations Division); at: <http://www.un.org/en/development/desa/population/publications/trends/concise-report2014.shtml>.

van den Bergh, Jeroen C. J. M.; Truffer, Bernhard; Kallis, Giorgos, 2011"Environmental innovation and societal transitions: Introduction and overview", in: *Environmental Innovation and Societal Transitions*, 1,1: 1-23, at: <http://www.sciencedirect. com/science/article/pii/S2210422411000219>

van der Brugge, Rutger; Rotmans, Æ Jan; Loorbach, Derk, 2005: "The transition in Dutch water management", in: *Regional Environmental Change*, 5: 164-176.

Victora, Cesar G.; Barreto, Mauricio L.; do Carmo Leal, Maria; Monteiro, Carlos A., Schmidt, Maria Ines; Paim, Jairnilson; Bastos, Francisco; Almeida, Celia; Bahia, Ligia; Travassos, Claudia; Reichenheim, Michael; Barros, Fernando C., 2011: "Health conditions and health-policy innovations in Brazil: the way forward", in: *Lancet*, 377,9782: 2042-2053 at: <DOI:10.1016/S)140-6736(11)60055-X>.

Wang, Yu-Chun ; Lin, Yu-Kai, 2014: "Association between Temperature and Emergency Room Visits for Cardiorespiratory Diseases: A Case Crossover Study, Metabolic Syndrome-Related Diseases, and Accidents in Metropolitan Taipei" in: *PLOS ONE*, 9,6: 1-9

WHO, 2009: *Health Transition - World Bank Conditional cash transfer. A World Bank Policy Research Report* (Washington, D.C.: World Bank); at: <http://www.who.int/trade/ glossary/story050/en/>.

World Health Organization, 2013a: *Country cooperation strategy at a glance* (Geneva: WHO); at: <http://www.who.int/countryfocus/cooperation_strategy/ccs-brief_bra_en.pdf >

World Health Organization, 2013b: "World Health Statistics", at: <http://www.who.int/gho/ publications/world_health_statistics/EN_WHS2013_Full.pdf>

Part IX   Preparing Sustainability Transitions in the Energy Sector

34 Enabling Environments for Sustainable Energy Transitions: The Diffusion of Technology, Innovation and Investment in Low-Carbon Societies

Jürgen Scheffran[1] and Rebecca Froese[2]

Abstract

A sustainability transition requires innovations, investment and learning to support transformation processes in different fields, including new technologies, products and infrastructures, as well as new social rules, norms and interactions. Greening the economy rests on the rapid and effective dissemination of climate-friendly technologies, and in particular of renewable and efficient energy systems. Substantial financial support and smart governance are required in this process in order to develop the economic, sociopolitical and technological capacities of all countries. Within this progression, the international diffusion of know-how, technologies and investments requires enabling environments to build up local production capacities and demand for low-carbon goods. Further, business, governmental and non-governmental actors rely on social learning to establish cooperation at multiple levels in order to adapt technologies to local contexts within national and global frameworks and to support the transformation towards low-carbon societies. Various mechanisms are analysed and discussed.

Keywords: Diffusion, enabling environments, technology transfer, innovation, knowledge, learning, private investments, low-carbon societies, sustainable energy transition.

34.1 Introduction: The Innovation Challenges for Environment and Development

The world is facing two major challenges which are closely related. On the one hand, atmospheric CO_2 concentration needs to be stabilized at a level that avoids catastrophic climate change. On the other, access to affordable, reliable, and clean energy is required to alleviate poverty and to drive economic development for a growing world population, especially in rural areas of the developing world, where approximately 1.5 billion people are still not connected to the power grid. Within this nexus of environment and development, major innovations in the technical, sociopolitical and financial sectors are required. The main problem is to understand the conditions and linkages of the technical, sociopolitical, economic and financial challenges facing the sustainable energy transition.

34.1.1 The Challenge of Technical Innovation

Replacing fossil-based energy systems with renewable energy sources and the complementary improvement of energy efficiency and other low-carbon energy options can play a key role in simultaneously addressing global warming and access to energy (Santarius/Scheffran/Tricarico 2012).[3] Fast-developing technical innovations and a radical change towards environmentally sound technologies will be essential to solve this

---

1 Prof. Dr. Jürgen Scheffran, Institute of Geography, CliSAP Research Group Climate Change and Security, Center for Earth System Research and Sustainability, University of Hamburg; email: <juergen.scheffran@uni-hamburg.de>. Funded with support by Deutsche Forschungsgemeinschaft (EXC177).

2 Rebecca Froese, MSc Student, University of Hamburg; email: <rebecca.froese@studium.uni-hamburg.de>.

3 This chapter builds on and goes substantially beyond Santarius/Scheffran/Tricarico (2012) and the conference paper Scheffran (2013).

double challenge, especially if global climate disasters are to be prevented. The urgent need for action means that even technology that is not completely 'clean', 'green' or 'climate-friendly' may be acceptable, provided that it is practically available and adequately meets the sustainability criteria.

34.1.2 The Challenges of Sociopolitical Innovation

Innovation is also on the agenda in the social and political realm. The global climate policy debate is no longer about whether to take action but about how, when and where to act, which actions need to be taken and by whom. While the struggle against global climate change ranks high on global political agendas, policymakers are striving to find and agree upon the best policy frameworks. Various conflicting issues need to be mediated, bridged and peacefully integrated in the future: economics and the environment, science and society, international and domestic policies, global and local governance, public and private spheres, governments and civil society, developed and developing countries.[4] Overcoming these differences requires a level of coordination and cooperation across multiple levels and unprecedented in history. The problems can best be solved by collaboration.

34.1.3 The Challenges of Economic and Financial Innovation

Managing the energy revolution requires massive technological and financial resources. Countries worldwide, from China and India to Germany and the United States, have recognized this need and they have already achieved considerable changes through substantial investment in the transformation of their economies' energy base. However, a large gap still remains between current investment levels and the amount of climate investment needed (see table 34.1). Actual climate finance levels in 2011/12 were estimated at €200 billion at global level and €120 billion at EU level. Future investment needs for 2020 are suggested to be €780 billion globally and €280 billion for the EU (EC 2015: 111).

The EU Commission's Impact Assessment of the 2030 Climate and Energy Framework forecasts that €193 billion needs to be invested per year between 2011 and 2030 in order to modernize ageing infra-

Table 34.1: The gap between actual and required climate financing. **Source**: Own representation, data from EC (2015).

	EU	**Globally**
Actual climate finance levels 2011/12	€120 billion	€200 billion
Future investment needed by 2020	€280 billion	€780 billion
Remaining investment gap	€160 billion	€580 billion

structure and to avoid lock-in of inefficient technologies, so that the 2030 climate and energy targets of a forty per cent reduction in emissions can be achieved (EC 2014). A European scenario dealing with the financing of technologies in power production, road transport, and buildings between now and 2020 identified accumulated investment requirements of €350 billion in technology development and €1.65 trillion in technology procurement (Accenture/Barclays 2011).

These figures mean that the accountability of the private sector has to be increased, since public finance is not sufficient to address this global problem. However, public initiatives are essential to leverage private capital and to catalyse it in the most climate-friendly direction (Santarius/Scheffran/Tricarico 2012). Since businesses can be seen as both part of the problem and part of the solution, climate change policies need to encourage business to contribute more constructively to the transformation process. This means that international investment policies are to be incorporated into the climate change framework, with guiding principles for *transnational corporations* (TNCs) and foreign investment. Public and private sectors have to be synergized to galvanize low-carbon investment aimed at climate change mitigation. This process needs time to make an impact and gain a certain visibility. However, an increasing number of TNCs have attempted to integrate sustainable development issues into their strategies, under pressure from consumers and advocacy groups (World Investment Report 2010). At the end of 2014, more than 8,000 corporations had signed up to the United Nations Global Compact, including more than 250 of the world's largest companies (Moon 2014).

Even though some countries are already notable producers and exporters of climate-friendly technologies, a worldwide energy revolution and sustainable economic transformation will only succeed if the economic and technological capacities of all countries are involved. The main task governing the transformation

---

4   In parts of the literature "North" and "South" are used for developed and developing countries respectively.

is to significantly scale up production capacities for clean and energy-efficient technologies all over the world, and in developing countries in particular.

34.2 Context and Framework of Technology Transfer

34.2.1 Shifting Historical Contexts

Technology transfer and related finance mechanisms are key instruments in bridging technology gaps, in particular between developed and developing countries. Technology transfer is widely seen as an important element of development and cooperation, but it can also raise concerns where there is a competitive and conflicting environment that technology may become the object of economic and political power games, because of its role in efficiency increase, path creation and force multiplication. This was a dominant feature of the East-West conflict: *research and development* (R&D) was heavily involved in the arms race, with exploitation of the links between civilian and military science and technology (Altmann/Scheffran 1983; Scheffran 1986). At the same time, comprehensive restraints on technology development were established between the NATO and Warsaw Pact alliances as part of arms control and export control regimes; notable examples were the *Coordinating Committee on Multilateral Export Controls* (COCOM), the Nuclear Suppliers Group, and the *Missile Technology Control Regime* (MTCR). Some of these activities were carried forward into the post-Cold-War era to create instruments of the non-proliferation of weapons of mass destruction in the nuclear, chemical and biological fields of technology, as well as in other hi-tech domains such as missiles, space, and information and communication technologies, with the aim of restricting the transfer of technology to developing countries, notwithstanding the dual-use potential of modern technology (Brauch/Graaf/Grin et al. 1992; Scheffran/Karp 1992; Scheffran/Liebert 1994; Liebert/Rilling/Scheffran 1994).

The focus shifted as part of the push towards conversion of the huge military complexes of the superpowers and their allies, and the growing relevance of addressing environmental problems. Both together led to the idea of transforming the enormous resources in science and technology that existed in the military complex towards environmental purposes. These resources included information and communication technologies, which appeared as particularly promising because

of their dual-use potential and ambivalent role (Scheffran/Vydra 1991, UN 1991; Scheffran 1992a, 1992b). When subsequent attempts to include peace-related issues in the 1992 Earth Summit in Rio de Janeiro failed, the processes and discourses on security/peace and environment/development fell apart, and the opportunity for a peace dividend was missed. This process can be traced through a series of conferences on conversion in Dortmund, Moscow and Hong Kong that led to the foundation of the *Bonn International Center for Conversion* (BICC), and the series of *Conferences of the Parties* (COP) of the *UN Framework Convention on Climate Change* (UNFCCC), whose secretariat is also located in Bonn (for a ten-year review, see Scheffran 2002). While in the first process the dual use of technology became a key dimension of high-tech warfare and its control (most notably in the wars in Iraq and the Balkans, as well as in space and cyber warfare), the second process highlighted technology transfer as an important element in avoiding dangerous climate change and environmental degradation (IPCC 2000). Occasionally the dual role of technology in environment and security was made apparent, e.g. in the use of satellites for Earth observation and reconnaissance (Scheffran 2001).

Following the 1992 Earth Summit and the UNFCCC-COP process, the discourse on technology transfer has been shaped considerably by the merging of calls for sustainable development, for energy transition and for a low-carbon society. R&D is essential for energy supply and distribution as well as for the reduction of greenhouse gas emissions from fossil energy sources that drive climate change, and so science and technology infrastructures are key components of an enabling environment for sustainable energy transitions. In this context, there are three main areas in which the thinking on climate change mitigation has been transformed (see box 34.1 and Lema/Iizuka/Walz 2015, summarizing a special issue of *Innovation and Development*):

- from 'technology' to learning and innovation;
- from international 'transfer' to interactive collaboration;
- from 'diffusion' to systems building.

Cutting across these areas, a comparative assessment of different studies discerns the need for more context-specific and interactive approaches towards the transition to *low-carbon development* (LCD). This includes the idea that the "learning, innovation, and competence-building system lens can help to understand the diverse settings in which 'developing coun-

---

Box 34.1: Aspects of technology transfer and learning in low-carbon development. **Source**: Adapted and condensed from Lema/Iizuka/Walz (2015).

From 'technology' to learning and innovation:
- Going beyond the narrow concept of technology development, distinguish between designs, complete equipment, and installation services (hardware) and skills, knowledge, and expertise for short-term operation and maintenance and long-term change (software).
- Attainment of capabilities (learning) may enable local low-carbon innovation effectively.
- Innovation is a comprehensive and interactive process, which is about breakthrough hi-tech equipment emerging from R&D labs.

From international 'transfer' to interactive collaboration:
- Compared to providing access to hardware technologies, little regard was paid to the facilitation of knowledge exchange and development of local technological capabilities and system building.
- Discussion on trade and intellectual property rights focuses on technology transfer rather than on fitting local circumstances to ensure the absorption, contextualization, and deployment of technology.
- Discussions in UNFCCC took the traditional view, where stimulation of climate-friendly technology transfer was limited to hardware and financing assistance, a strategy widely deemed unsuccessful.

- There is a growing consensus that international action for climate change mitigation and development in the global South must focus on innovation cooperation, i.e. joint action to accelerate the development, adaptation, and deployment of suitable technologies.
- Innovation-cum-cooperation needs to support the deployment of technology, and to identify collaborative patterns for better knowledge creation and diffusion for LCD.

From 'diffusion' to systems building:
- Technology diffusion has been seen as the deployment of low-carbon technologies, e.g. renewable energy technologies or energy efficiency measures.
- Increasing attention is paid to the complexity of the underlying process, which is not simply about rolling out technologies, but also about transforming the relevant sociotechnical setting.
- Effective LCD needs to include organizational and institutional change as well as changes in the realms, instruments, and techniques of policy-making.
- Calling for a systems approach in analysing LCD, we include learning, innovation, and competence-building systems rather than focusing on innovation systems, which is an approach arising from studies in developed countries.

---

tries' are situated and ultimately help to enhance synergies between climate change mitigation and socio-economic development" (Lema/Iizuka/Walz 2015: 183). Finally, the assessment shows that what is required is a more integrated view of trajectories of innovation towards low-carbon societies, including the *water–energy nexus*; it also raises questions about the *political economy* of low-carbon innovation and development; and it identifies new, diverse and changing innovation *pathways* and conditions for low-carbon transformation that are on the horizon in developing countries (Lema/Izuka/Walz 2015). These are "likely to differ markedly among countries because of the diversity in policies, endowments, and technological capabilities" (Lema/Izuka/Walz 2015: 177), and this has implications for "the effectiveness of mitigating climate change and tackling related domestic energy challenges as well as [for] the degree to which low carbon technologies and solutions can become a source of national competitiveness" (Lema/Iizuka/Walz 2015: 177). There is a need for new frameworks to enable LCD to be understood in developing countries.

There is a range of literature on the process of technology transfer that covers the various issues of innovation, development, behavioural change and socio-economic development (see the overviews in

IPCC 2000; Karakosta/Doukas/Psarras 2010). Empirical evidence, theoretical advancements and policy recommendations on low-carbon technology transfer are provided in Ockwell and Mallett (2012), who draw on the literature on adaptive and incremental innovation and sociotechnical transitions to highlight various critical issues: co-evolution of technologies and society; insights from innovation studies and sociotechnical transitions; the context-specific, spatially situated technology needs of different people in different spaces; the way in which innovation capacities are developed via the development of 'innovation capacities' in order to adopt, adapt, develop, deploy and operate technologies effectively within specific contexts (Bell 2009). While there are no overarching theories, in the following sections a conceptual framework of analysis will be considered, applied to the analyses of technology transfer, the role of technology transfer in innovation, and the diffusion of low-carbon energy technologies.

34.2.2   Conceptual Frameworks

34.2.2.1   Adaptive Technology Life Cycle

Technology is made up of processes that create new pathways for action or allow the realization of actions more effectively according to given criteria, such as cost or resource efficiency, the gaining of benefits or the reduction of risk. From a narrow economic viewpoint, technologies are treated as 'objects' or 'hardware' that can be moved around from one location and context to another and sold on markets like goods. In the context of sustainable development, the life cycle approach is more appropriate. In production processes, technology uses input factors (such as material, energy, land, labour or investment) to realize intended output actions. The quality of a technical system is determined by how many input factors are needed to realize a certain action. For instance, a wind turbine converts wind into electricity, and a biorefinery converts biomass into various bioproducts, including bioenergy. Here the quality of a technology is measured by the energy balance and the carbon emissions per energy unit. If investment is the input variable, the quality indicator then is the net financial return throughout the stages of the life cycle (R&D, design, testing, assembling, building, production, maintenances, transportation, utilization, dismantling and waste management). To support the different stages in this life cycle requires an enabling environment and infrastructure that provide the necessary linkages and networks to absorb the input needed to produce the required output, while minimizing side effects and risks. To be effective, the technical system and its natural and social environment must fit to each other, and technology should adapt to local needs and conditions. In particular, an enabling environment and infrastructure are important to facilitate the development of technology in other parts of the world. Without an adequate interface of connections, a technical metabolism and life cycle cannot operate and survive in a new environment. An example is a power plant transferred to a remote area in a developing country without transportation, power lines, water resources, trained employees or the capacity to maintain it under adverse climatic conditions.

34.2.2.2   Enabling Environment and Knowledge

The concept of an 'enabling environment' is not clearly defined and has many different interpretations. According to UNFCCC (2001), an enabling environment "focuses on government actions such as fair trade policies, removal of technical, legal and administrative barriers to technical transfer, sound economic policy, regulatory frameworks and transparency, all of which create an environment conducive to private and public sector technology transfer" (cited in IPCC 2014: 1178). One definition emphasizes government policies for "creating and maintaining an overall macroeconomic environment" (UNCTAD 1998: 24). Another definition interprets an 'enabling environment' as the wider context within which development processes take place, i.e. the role of societal norms, rules, regulations, and systems (Bolger 2000). This environment may either be supportive (enabling) or constraining (IPCC 2014: 1222).

To facilitate technology development, the necessary enabling environment and infrastructure need to be created in the recipient country, following three separate technology flows: a) capital goods and equipment; b) skills and know-how for operating and maintaining equipment, and c) knowledge and expertise for generating and managing technological change in order to adapt technologies to the local context. It is not sufficient to transfer technology without knowledge of how to (re-)produce it and the skills in using it. The knowledge component of technology includes "both codified knowledge (e.g. engineering and manufacturing processes) and tacit knowledge (human-embodied knowledge acquired by doing, e.g. applied engineering and systems integration skills)" (Ockwell/Mallett 2012: 8). This dynamic differentiates between know-how skills (e.g. ability to operate and maintain 'hardware') and know-why skills (ability to understand the principles behind how such 'hardware' works).

For the creation of an enabling environment, particular attention needs to be paid to building the soft elements, such as intellectual property, organizational knowledge and managerial skills, as well as corporate culture, values, norms and standards (Ockwell/Haum/Mallett et al. 2010). Accordingly, Schnepp, von Glinow, Bhambri et al. (1990) define technology transfer as "…a process by which expertise or knowledge related to some aspect of technology is passed from one user to another for the purpose of economic gain". This implies that an infrastructure for establishing these soft skills is required, including education and research facilities for transferring the scientific-technical knowledge about the principles and practical use of technology. When technology is being transferred through an organization such as a manufacturer, there is tacit knowledge associated with the procedures of the organization. For the host developing country, there are knowledge requirements for the

Figure 34.1: The process of technology transfer. **Source**: Adapted from Karakosta/Doukas/Psarras (2010).

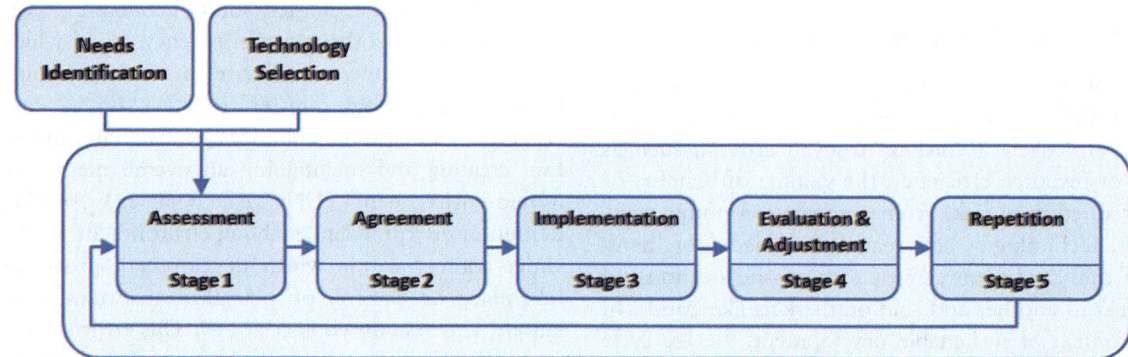

people and institutions who want to interface with the technical systems, including their supply and support chains (Karakosta/Doukas/Psarras 2010).

34.2.2.3   Stages of Technology Transfer

Within the technology transfer process five basic stages can be identified (IPCC 2000; Karakosta/Doukas/Psarras 2010): 1) Assessment; 2) Agreement; 3) Implementation; 4) Evaluation and Adjustment; 5) Repetition and Replication (see figure 34.1). The processes are not necessarily sequential and can occur in parallel, adding to the complexity of technology transfer. The last two stages, for instance, involve the transfer to and adoption in the local energy user market. The Assessment stage is preceded by the identification of needs and the selection of the technology to be assessed. A hierarchical description distinguishes between *vertical transfer* from the laboratory or test bed into commercial use, and *horizontal transfer* from one geographical location to another. The transfer of low-carbon technology between developed and developing countries can also vary depending on industrial sectors, national circumstances and the type of technology. Technology flows can be assisted and supported along various pathways, such as through direct purchases and trade; foreign direct investment; licensing, subcontracting and franchizing. Cooperative activities include joint ventures; cooperative research and co-production (IPCC 2000).

34.2.2.4   Resources and Infrastructures

To be produced, used and maintained, a technical system needs resources and infrastructures for its operation as well as human and social capital that rests on the working capacity and skills of the people who maintain, operate and use it. A viable technology infrastructure requires a "manufacturing capacity, supply chain capacity, end-of-life/waste disposal, institutional

capacity and sustainability of the whole process and the social networks between them" (Karakosta/Doukas/Psarras 2010). Some elements of the infrastructure may exist in a country, others not. Some gaps can be closed domestically by building the capacity needed, other gaps have to be acquired internationally—this may be more feasible for individual components (such as mechanical tools or vehicles) but is more difficult for whole networks and systems (such as power grids or transportation systems). Knowledge is a resource central to successful capacity building in recipient countries, and is potentially transferrable as human beings may relocate in limited numbers. Unlike traditional accumulation theories that see knowledge as a transferable entity following investment flows, assimilation theories of technology transfer consider the impact of the knowledge and the characteristics of suppliers and recipients who assimilate that knowledge (Ockwell/Haum/Mallett et al. 2010). The mutual dependency between consumer behaviour, markets created by new technologies and associated social structures need to be considered in planning for the transfer of the technology and is one of the reasons why technology transfer has to be treated as a complex process in which local organizations and communities take the lead.

34.2.2.5   Technology Development as a Learning Process

Technology development and transfer is a continuous process of learning and decision-making between competing alternative technology pathways which are selected within societal and organizational structures that enable informed choices. To a large extent, the state of the world today is the result of the technological choices of the past. Similarly, the future state of the world will be determined largely by the technologies we choose today (Trindade 1991). The role of learning has been already raised in the definition used by IPCC (2000), where technology transfer involves the process of "learning to understand, utilise and replicate the technology including the ability to decide which technology to transfer and adapt it to local conditions and integrate it with indigenous technologies". According to Archibugi and Michie (1997): "policy to support technology should address the diversity of learning mechanisms and the conditions which enhance the learning capabilities of firms". Schnepp's definition above also emphasizes the economic benefits, which for the transfer of low-carbon technology includes the mitigation of the future costs associated with climate change as well as any financial benefits to the companies involved in the transfer process (Karakosta/Doukas/Psarras 2010). Newer and broader conceptions of technology transfer see it as a process of incremental and cumulative learning by which the results of the initial choice are internalized. Like R&D, it is an essential component of what is described as social learning (Brooks 1995).

34.2.2.6 Technology Transfer in the Context of Climate Change

Preventing dangerous climate change, sustainable development and decarbonization will require radical technological, economic and societal changes in both developed and developing countries, favouring technologies that support adaptation and mitigation strategies. Although most new carbon abatement technologies are being developed in industrialized countries, much of the potential for low-carbon energy production and emission reductions is located in developing countries where fossil fuel consumption is still increasing rapidly. Low-carbon sustainable technologies need to be adopted by both developed and developing countries, and this requires that developing countries avoid being locked into unsustainable practices and technologies and move quickly to environmentally sound and sustainable practices, institutions and technologies. The basic technologies for renewable energy and energy efficiency already exist and promise to address the problem of climate change. The transfer or innovation processes must be rapid enough to prevent dangerous climate change and to make themselves readily available where they are needed most. Economic development is most rapid in developing countries, but it will not be sustainable if these countries follow the historic emission trends of developed countries. Renewable energy investments in the developing world deliver lower carbon abatement costs than in the developed world, while also achieving a broad range of additional social, economic, and environmental benefits. For these reasons, the transition of global energy systems to lower-carbon pathways depends upon "creating an institutional capacity for a national environmental policy" as an important precondition for North–South collaboration (Ipsen/Rösch/Scheffran 2001: 323), and the successful transfer to and absorption of low-carbon technologies within developing country economies (Mallett/Ockwell/Pal et al. 2009). In the absence of strong policy support mechanisms and incentives, as well as for cheap and readily available fossil fuels, public and private funds are unlikely to deliver the necessary technologies at the cost and scale necessary to address climate change. These goals can only be achieved through major changes in the investment framework (WBCSD 2010).

34.3 Implementing the Technology Transfer Framework

34.3.1 Science and Technology Infrastructure

The basis of technology development and transfer depends upon its technology infrastructure, made up of a set of specific, industry-relevant capabilities such as technology centres, research facilities and educational institutions that support the development of technical skills. Establishing a technology infrastructure requires considerable efforts and long lead times; on the other hand it depreciates slowly. Aspects include information relevant for strategic planning and market development and forums for joint industry-government planning and collaboration, as well as the assignment of intellectual property rights.

34.3.1.1 Technology Infrastructure

Technology infrastructure may not be of direct economic value to firms, and this reduces the incentive to build technology infrastructure on their own. To develop a public technology infrastructure, governments can invest public money, stimulate demand and attract technological capabilities that may exist elsewhere but will need to be imported, adapted, and absorbed in the local economy. In addition, it is important to induce institutional change and organizational development in order to adapt the technology infrastructure to various supportive measures. Finally, governments can increase public awareness and spread capability through public networks (Justman/Teubal 1995). National policies and regulations are essential

Box 34.2: Public–Private Partnership for Electricity from Peanut Shells in Senegal. **Source:** Assembled by the authors.

Many villages in remote rural areas of the Senegal lack access to the national power grid. Particularly in the energy development sector, public–private partnerships play an important role in enabling technology transfer and adaptation to the local context. In 2010, Stadtwerke Mainz, a local energy provider for the city of Mainz in Germany, launched a project for rural electrification in collaboration with the German Investment and Development Society (*Deutsche Investitions- und Entwicklungsgesellschaft*, DEG).

The project was implemented in Kalom, a village with 1,200 inhabitants who have no access to the national power grid. Within the framework of the project, a biomass power plant fuelled by peanut shells and millet stalks has been built. Most of the fuel required is purchased from local farmers, providing a secure income for the farmers. The rest is bought from a bigger producer of peanut oil. A foreign private stakeholder has provided the technology for the biomass power plant and supported the project with on-site training for the villagers.

Within the centre of the village, the electricity produced is distributed via underground cables. Households that are not connected to the local grid get electricity access though rechargeable mini solar modules or rental batteries. Some electricity distribution businesses have already been set up in outer parts of the village. Besides the socioeconomic and environmental benefits, these renewable energy sources are steadier and more reliable than the previously used diesel generators or car batteries.

In 2012 the power plant was handed over to a local operating company with the Energy for Africa Foundation and about a dozen villagers as shareholders. The local villagers have been provided with on-site technical and business know-how by the technology provider in order to run and maintain the power plant independently. Since then, seven local employees have been hired by "Kalom Mainz Industries" and run the power plant.

Resources: <http://www.developpp.de/en/content/electrical-power-peanut-shells> and <http://www.energie-fuer-afrika.de/projekte/details/projekt/4-kalom.html>.

for building on the technological capabilities developed by domestic companies and for inducing foreign investment. Domestic progress in low-carbon processes can then be distributed throughout the international network of operations of TNCs, and consequently attract further foreign investment (WIR 2010).

34.3.1.2 Energy Infrastructure

A vital resource within the technical infrastructure is energy which should be integrated into national development strategies. Energy is necessary for preparing food, heating and lighting homes, maintaining schools and hospitals, and driving industry, as well as for connecting people and goods to markets. Public policies and public–private partnerships could focus on improving grid transmission, local electricity generation and energy storage systems, where renewable energy is essential (boxes 34.2 and 34.3). The IEA estimates that about US$5.2 trillion is required in generation investments, and an additional US$6.1 trillion for transmission and distribution networks from now until 2030 (WBCSD 2007).

34.3.1.3 Research and Development Infrastructure

Investing in R&D is an essential precondition for building the technology infrastructure of a country. Historically, governments have played a key role in supporting research and development through national laboratories, universities, and international collabora-

tive ventures. Public funding remains a major source for R&D activities in both industrialized and developing countries, involving either general support for national R&D institutions and laboratories, or direct funding of specific projects according to set government priorities. Effective R&D policies alleviate technical barriers and reduce costs by improving materials, components, system design and tools for installers and users.

34.3.2 Framing Conditions for Innovation

34.3.2.1 Leapfrogging and Diffusion

In addition to the transfer of knowledge and technology, which is in itself not sufficient for effective implementation, a country would be "equipped with a complementary system to enable such acquired knowledge to be diffused and adapted locally. International collaboration should go beyond the technology focus to a capability and system-building focus" (Lema/Iizuka/Walz 2015: 176). If the "political economy is favourable and new actors are allowed to create spaces for innovation", this can be a 'window of opportunity' to leapfrog towards carbon-efficient systems, following multiple and diverse trajectories (Lema/Iizuka/Walz 2015: 185). According to IPCC (2014: p.258), "leapfrogging", or the "skipping of some generations of technology or stages of development, is a useful concept in the climate change mitigation literature for enabling developing countries to avoid the more emissions-intensive stages of development" (Watson/Sauter 2011). The same source notes that "innovation in and diffusion of new technologies also require skills and knowledge from both developers and users, as well as different combinations of enabling policies, institutions, markets, social capital, and financial means depending on the type of technology and the application being considered" (IPCC 2014: 301). Harnessing these kinds of capabilities and processes may require novel mechanisms and institutional forms.

One of the critical phenomena is rapid urbanization in developing countries, associated with radical changes in the provision of large physical infrastructures. At the city scale, a high degree of land use mix can result in significant reductions in vehicle kilometres travelled by increasing the proximity of housing to office developments, business districts, shops, and malls. In service-economy cities with effective air pollution controls, mixed land use can also have a beneficial impact on citizen health and well-being by enabling walking and cycling (IPCC 2014: 955).

34.3.2.2　Societal Structures and Networks in Technology Transfers

Technology development and transfer are embedded in societal structures and at the same time transform them. Technology is interwoven with the social networks of producers and users who depend on the technology. Throughout the process of technology development and transfer, various stakeholders are involved. Their decisions and actions differ depending upon the stages and pathways of technology development, and on the type of technology, especially whether it is bioenergy, wind or solar power. To overcome the various barriers to technology transfer, the interests and influences of stakeholders at each stage need to be considered and incorporated through stakeholder dialogues. In the context of climate change, technology transfer has been defined as a set of processes "covering the flows of know-how, experience and equipment, for mitigating and adapting to climate change amongst different stakeholders such as governments, private-sector entities, financial institutions, non-governmental organizations (NGOs) and research/education institutions" (IPCC 2000). Among the diverse stakeholders in North-South transfers are "project developers, technology owners, technology suppliers, product buyers, recipients, users of the technology, financiers and donors, governments, international organizations, NGOs, and community groups" (IPCC 2000). Not mentioned are institutions for research, trade and education. Technology transfer can take

place between government agencies, within vertically integrated firms, and within partnerships across a network of diverse entities (e.g. information service providers, business consultants, financial firms) (Karakosta/Doukas/Psarras 2010). Networks in technology transfer are critical for market creation and the generation of a social pool of knowledge (Teubal/Yinnon/Zuscovitch 1991). The relevance of the information is different for different stakeholders: governments and end-users need to understand the costs and benefits of a technology; innovators need to understand how to adapt it; firms how to market it; and consumers how to use it.

34.3.2.3　Pathways of Technology Transfer

Different 'pathways' can be distinguished in the actual transfer of technologies, depending on the stakeholder and country context, as well as on technology scale and type (IPCC 2000; Karakosta/Doukas/Psarras 2010). Regarding the type of actors, we differentiate between (1) government-driven pathways of technology transfer to fulfil specific policy objectives, (2) private sector-driven pathways involving commercially oriented entities, and (3) community-driven pathways with a high degree of collective decision-making (see table 34.2). While private-sector driven pathways are still the dominant mode of technology transfer (e.g. *transnational corporations* (TNCs)), multilateral organizations and NGOs are playing an increasing role in promoting technology transfer to support development in a more sustainable and equitable manner (Santarius/Scheffran/Tricarico 2012).

34.3.2.4　Innovative Capacity, Push and Pull Factors

Investments are essential to foster technical innovation and its diffusion. Adequate investment strategies could play a key role in increasing a country's ability to absorb new technologies and improve its innovative capacity. An enabling environment may complement a set of technology-push and demand-pull policies. Technology transfer is part of the innovation chain that ranges from research and development to the commercialization and diffusion of technology. Developing countries may consider three areas of innovative capacity (Tomlinson/Zorlu/Langsley 2008): *disruptive innovation* (radically different from current technologies, e.g. small-scale power generation with thin film solar cells); *orphan areas of research* (neglected by the industrialized world, e.g. small-scale desalination); *adaptive innovation* (e.g. adapting tech-

Table 34.2: Relative importance of particular types of financial flows to technology transfer pathways for different actors. **Source**: Karakosta, Doukas, Psarras (2010).

Technology transfer pathway	Government	Private sector	Community
Cross-border movement of personnel	–	+++	–
Foreign direct investment	+	+++	–
Foreign portfolio equity investment	+	++	+
Government assistance programmes	+++	–	++
Joint ventures	+	+++	–
Licensing	++	+++	–
Loans	++	+++	–
Meetings, workshops, conferences, and other public forums	+	–	+++
NGOs	+	–	+++
Open literature (journals, magazines, books, and articles)	+	+	+++
Trades in goods and services (includes purchases, sales, exports and imports)	+	+++	–

(+) Minor; (++) secondary; (+++) primary component of pathway.

nology to local renewable resources). There is also a need for *incremental innovation*, where user and producer interfaces are important. Innovation incorporates 'push' and 'pull' factors along the innovation chain, with varying levels of public–private finance and policy interventions at different stages (figure 34.2). Push factors drive investors to invest abroad, and include government policies, market conditions, production costs and business conditions, all of which influence investment decisions. Some factors are climate-specific, such as green branding strategies, regulations and pressure from consumers and investors to avoid climate risks or pursue green energy technologies. Pull factors are locational host-country-specific determinants that influence where investors choose to invest, such as policy frameworks, economic conditions, resource endowment and access to skilled labour.

34.3.2.5 Absorption and Capacity Building

Going beyond a narrow focus on technology transfer that predominantly seeks to support access to specific technologies, diffusion of new innovations will require a broad approach for capacity building to enable developing countries to generate their own innovation system. Increases in low-carbon diffusion rates across countries can be achieved at differing development stages through system-wide capacity building to improve internal innovation and absorption. International collaboration will be vital if low-carbon innovations are to be introduced on a commercial scale (Tomlinson/Zorlu/Langsley 2008). Furthermore, international technology dissemination requires the acquisition, mastery, diffusion and indigenization of knowledge, technology and skills in the host country, and these must be not only transferred across borders but also absorbed by local actors (WIR 2010).[5] This approach is vital for the adaptation of technology to the local context and the fostering of the development of endemic technologies.

34.3.2.6 International Collaboration

In responding to climate change and moving towards a low-carbon economy, developing countries face two major challenges: (1) the mobilization of the requisite finance and investment, and (2) the acquisition, generation and dissemination of the relevant technologies and knowledge. In both areas, foreign investment can make a valuable contribution (WIR 2010). While the future international climate change regime—including specific carbon reduction commitments as well as financial and technological support for developing countries—is still to be agreed upon, countries need to

---

5 In this context, 'acquisition' means movement of the technology to local players; and 'mastery' requires that local actors are fully capable of using the knowledge and building on it (i.e. they have the 'absorptive capacity' to do this). 'Indigenization' of technology is a long-term concept, implying that the technology has become part of the national knowledge and innovation system, including diffusion to other enterprises and further research, development and innovation in the host country.

Figure 34.2: Innovation Chain. **Source:** Adapted from Grubb (2004) and Tomlinson, Zorlu, Langsley (2008).

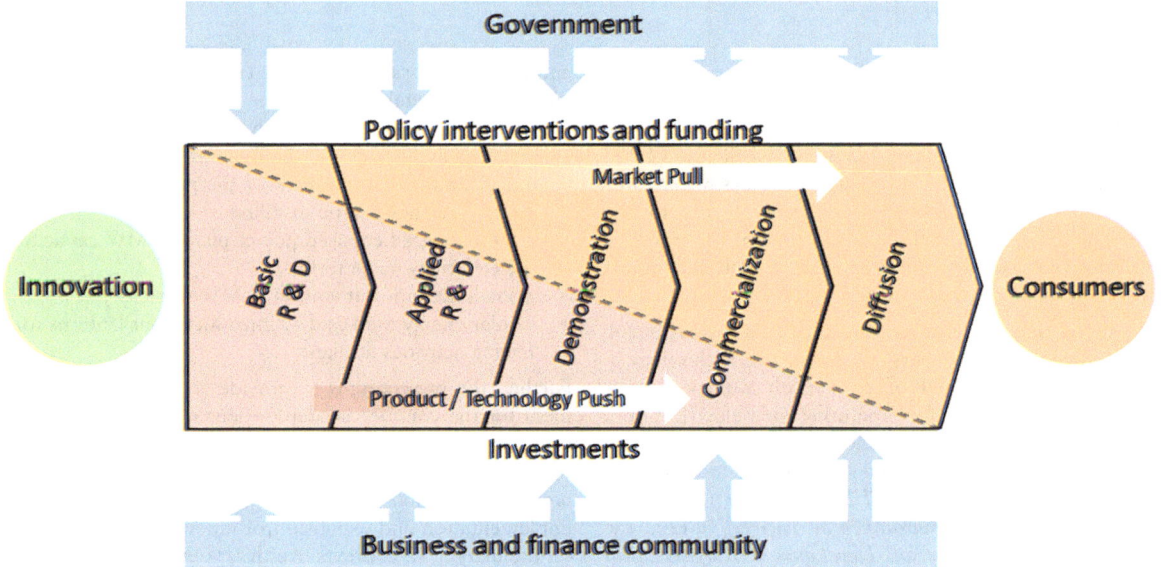

examine how to facilitate low-carbon domestic and foreign investment, including technology transfer from developed to developing countries. Nevertheless, transfers from and within developing countries and especially *triangular cooperation* (TC) should also be considered. The innovation of integrating developed countries into South-South cooperation programmes in order to support these collaborations appears to be a promisingly sustainable approach (UNCTAD 2012). The *United Nations Environment Programme* (UNEP) has described the weaving together of the complementary challenges of investment support in developed countries and investment governance in developing countries as a "Green New Deal" in global climate policy (UNEP 2009). This approach aims at "reviving the global economy and boosting employment, while simultaneously accelerating the fight against climate change, environmental degradation and poverty".[6] Before one can elaborate on and discuss the concept of the "Green New Deal", it is important to analyse the conditions and environments that are needed in developing countries to enable a transfer of low-carbon technologies, and to discuss key areas for future action (see Santarius/Scheffran/Tricarico 2012). In particular, it is recommended that a significant portion of the estimated US$3 trillion that has already been pledged in economic stimulus packages be

invested in five critical areas (Santarius/Scheffran/Tricarico 2012):

1. raising the energy efficiency of old and new buildings;
2. transitioning to renewable energies including wind, solar, geothermal and biomass;
3. increasing reliance on sustainable transport, including hybrid vehicles and high-speed rail and bus rapid transit systems;
4. bolstering the planet's ecological infrastructure, including fresh water, forests, soils and coral reefs;
5. supporting sustainable agriculture, including organic production.

34.3.2.7 Mobilizing Virtuous Cycles through Local Public Policy and Private Investment

While the UN approach suggests that a globally coordinated effort is needed to address the above-mentioned challenges of developing countries, measures at a national or business level could well support technology transfer before global agreements or activities have been achieved. The core idea is to build up public support and unleash private capital for foreign investments while establishing sustainable framework conditions to maximize the mitigation and development potential of these investments on the ground, especially in the energy sector. After establishing such an enabling environment, foreign investments can be a valuable tool to support the transfer of technology as well as related know-how and capacities. Further-

---

6   See: <http://www.unep.org/pdf/A_Global_Green_New_Deal_Policy_Brief.pdf>.

Box 34.3: China–Zambia South–South Cooperation on Renewable Energy Technology Transfer. **Source**: Assembled by the authors.

Worldwide, more than one billion people have no access to modern electricity services, in particular households at the bottom of the income pyramid. Like many other developing countries, Zambia is facing the challenge of the absence of grid coverage in remote rural areas, where only three per cent of the population have access to the national power grid. In these areas, most of the power is provided by diesel generators, resulting in high fuel import costs and inherent environmental impact.

The Government of Zambia has launched a project in collaboration with the Ministry of Science and Technology in China and related UNDP offices in both countries to foster technology transfer for promoting rural electrification. The project is supported and funded by several national and international public and private stakeholders, including the *International Centre on Small Hydro Power* (ICSHP) in China, the *Global Environmental Facility* (GEF), the *United Nations Industrial Development Organization* (UNIDO), UNEP, the *Zambia Electricity Supply Corporation Limited* (ZESCO), the *Development Bank of Zambia* (DBZ) and the *Rural Electrification Authority* (REA). The partnership promotes not only the transfer of renewable energy technology but also its adaptation to the local Zambian context, ensuring that the technology meets the needs of the local population in the optimum way.

The partnership is creating enabling environments for the implementation of isolated mini-grids through the inclusion of national and international know-how. Additionally, the efforts are increasing the share of renewable energy

and consequently supporting the country's socio-economic and environmental objectives in a sustainable way. The provision of clean and sustainably-produced electricity will enhance the livelihood and quality of life of the rural population. Within the framework of the project, the following energy facilities will be installed:

- a biomass gasifier-based power plant (1 MW capacity) for electricity generation;
- a small hydropower station (1 MW capacity);
- a solar energy mini-grid to introduce solar lanterns for fishing activities at night.

Further, the project aims to provide an enabling environment for the commercial deployment of mini-grids based on renewable energy, by contributing to the development of a legal, institutional and policy framework. Such measures facilitate public and private partnership and foster the implementation and expansion of renewable technologies on the market. In contexts like this, South–South cooperation can play a central role in the provision of technology and know-how for the adoption and adaptation of technologies to the local context.

Resources: <http://www.unido.org/where-we-work/africa/selected-projects/zambia-mini-grids.html>; <http://www.unido.org/fileadmin/media/documents/pdf/Energy_Environment/rre_Zambia_factsheet.pdf>; <http://ssc.undp.org/content/dam/ssc/documents/Expo/solutions/2008_to_2012/20132221636470.UNDP-GSSD 2012 -UNIDO-Zambia-v2.pdf>.

---

more, private investments are attracted into such an environment by new opportunities for overcoming limited demand and market size, while the companies' financial risks are reduced by public money, adequate regulation and standards, in order to make foreign investments work for climate protection. To bring such a 'virtuous cycle' into being, complementary steps are required by developed and developing countries, affecting the public and the private sector. A 'virtuous cycle' combines learning in research and development from technology push and market pull, leading to improved products, more market demand and reduced costs, as well as a cycle of policy learning (Jänicke 2012; IPCC 2012; IPCC 2014: WG3).

In this cycle, private investments can be stimulated by five steps: 1) strategic goal-setting and policy alignment; 2) an enabling process and incentives for *low-carbon and climate-resilient* (LCR) investment; 3) financial policies and instruments, 4) harnessing resources and building capacity for an LCR economy, and 5) promoting green business and consumer behaviour (see IPCC 2014: 1041, based on Corfee-Morlot/

Marchal/Kauffmann et al. 2012). Such an approach needs to be broadened to include the demand-side aspects of innovation systems, including their political and institutional contexts, and to ensure that technology development proceeds on a self-determined and needs-based basis (Ockwell/Mallett 2012). While the merits of an integrated approach can best be realized if ambitious mandatory emission reduction targets are set and implemented, this concept is not dependent on global agreement. It is not necessary that national investment support or bilateral investment governance are managed by a global fund. As many decisions as possible can be left to each country's national policies, and these policies will create the framework conditions for technology transfer in a market environment. Global top-down decisions and one-size-fits-all prescriptions can be avoided.

Figure 34.3: Global New Investment in Renewable Power and Fuels, Developed and Developing Countries, 2004–2013. **Source:** Adapted from REN 21 (2014).

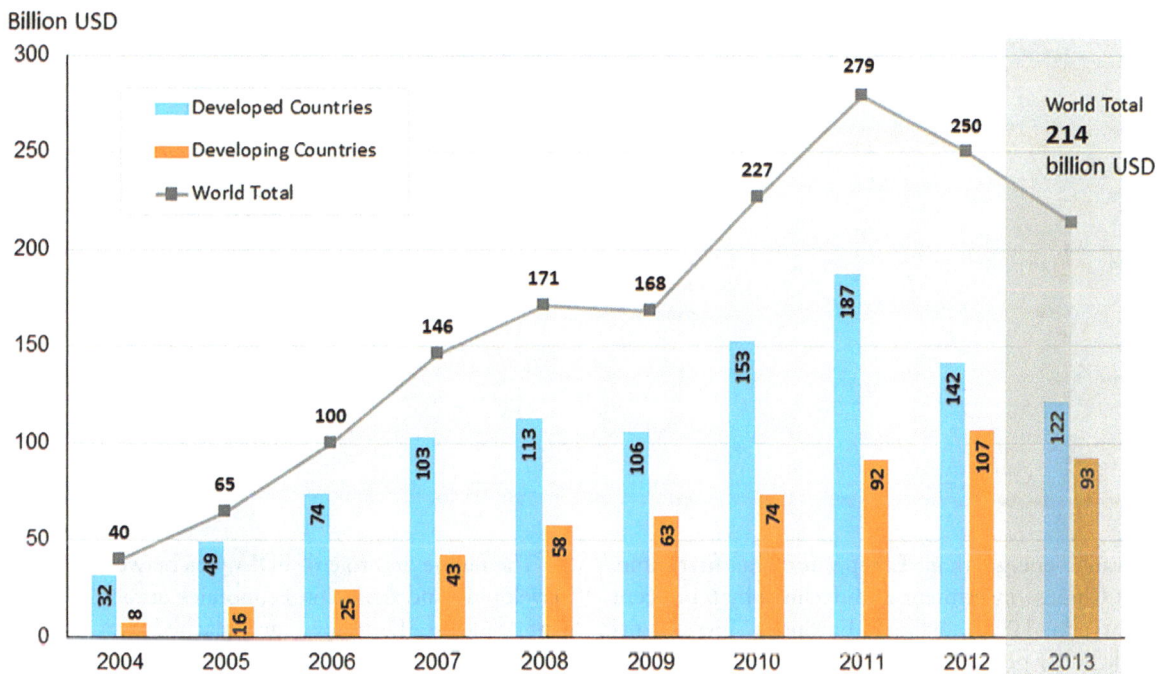

34.4   Global Trends in Sustainable Energy Investment

34.4.1   Trends in Renewable Energies

By the end of 2012, the twenty largest emitting countries were producing seventy per cent of global energy-related CO_2 emissions. This makes them attractive for international private sector investment in low-carbon technologies. Since 2000, significant growth in new investment in sustainable energy has been seen globally, with annual growth rates exceeding fifty per cent. Between 2005 and 2009, the developed countries provided US$2.5 billion of official development assistance to support enabling environments in developing countries (Stadelmann/Michaelowa 2011). With appropriate enabling environments, along with the public sector, the private sector can play an important role in financing mitigation; it represented around 74 per cent and 62 per cent of overall climate finance in 2010 and 2011 on average (IPCC 2014: 1211). In 2011, a record amount of US$279 billion was reached (excluding large hydroelectric projects), followed by a 23 per cent reduction to US$214 billion in 2013, largely in response to a sharp fall in solar system prices and the effect of pol-

icy uncertainty in many countries (Bloomberg 2014). As in previous years, public and private investments in new renewable capacity (including large hydroelectric projects) in 2013 were larger than investment in new fossil fuel capacity.

A decrease of about 14 per cent in 2013 (to US$122 billion) with respect to 2012 was seen not only in developed countries (to US$93 billion) but also in developing economies, interrupting an eight-year trend of increasing investments (figure 34.3). According to the Renewables 2014 Global Status Report, 144 countries had renewable energy targets and 138 countries had renewable energy support policies by early 2014. Although investments weakened in most renewable energy sectors by between 16 and 28 per cent (solar 20 per cent, biofuels 26 per cent, waste-to-energy 28 per cent, small-hydro 16 per cent), investments in the wind sector were quite resilient, falling by only 1 per cent (Bloomberg 2014). However, in 2013 the implementation of onshore wind and solar projects without subsidy support in a growing number of locations around the world was driven by efficiency improvement and cost reductions (Bloomberg 2014).

The most remarkable trend of recent years was the decisive shift towards investments by developing and emerging countries, in particular by China and other Asian countries. In 2013, China invested more in

Figure 34.4: Global New Investment in Renewable Energy by Technology, Developed and Developing Countries, 2013.
Source: Adapted from REN 21 (2014).

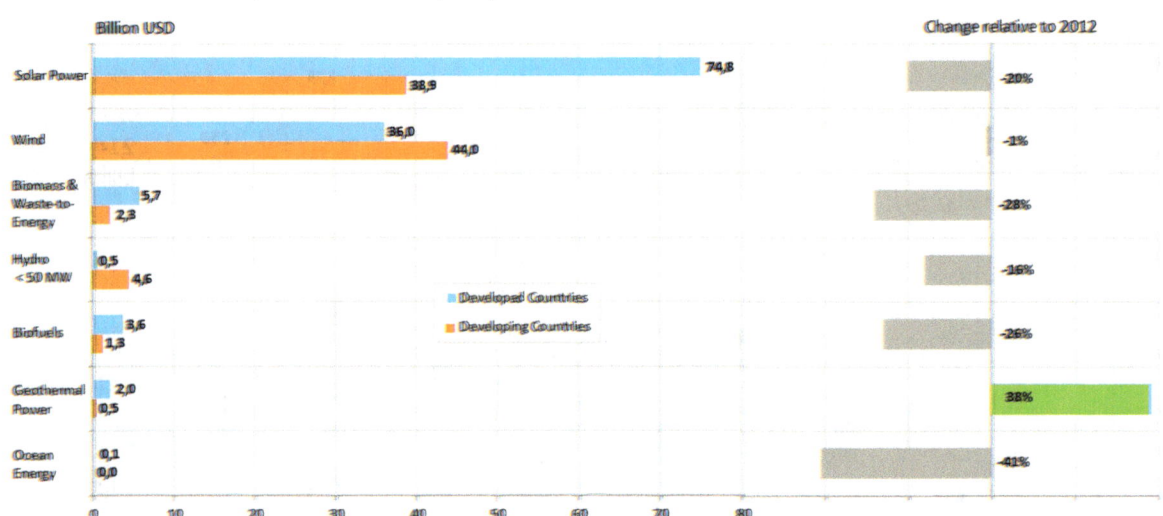

renewable energy than Europe for the first time. While Chinese investment also declined by 6 per cent in 2013 to US$56 billion, European investments shrank by 44 per cent, to US$48 billion. Investments also declined by 10 per cent to US$36 billion in the US, while Indian investments fell by 15 per cent to US$6 billion and in 2013 Brazil reached its lowest investment level since 2005, with a drop of 54 per cent to US$3 billion (Bloomberg 2014). The only regions where investment increased were the Americas (excluding the US and Brazil), with a rise of 26 per cent to US$12 billion in 2013, and Asia and Oceania (excluding India and China), with an increase of 47 per cent to US$43 billion. The largest rise, an 80 per cent increase in renewable energy investments to US$29 billion, was due to the solar boom in Japan after the nuclear catastrophe in Fukushima (Bloomberg 2014).

34.4.2   Foreign Direct Investment

Global *foreign direct investment* (FDI) inflows experienced an 18 per cent drop in 2012; such investment was recovering only slowly due to economic fragility and risks following policy uncertainties and regional instabilities. Inflows increased by only 9 per cent in 2013, to US$1.45 trillion, and are projected to further increase to US$1.6 trillion in 2014, US$1.7 trillion in 2015 and US$1.8 trillion in 2016 (WIR 2014). These projections lie far below the expectations of the *World Investment Report* 2010, where investment was expected to reach between US$1.6 trillion and US$2 trillion in 2012 (WIR 2010).

The nature and role of FDI varies between regions. Developing and transition economies attracted 54 per cent of global FDI inflows, and reached a record level of 39 per cent of global FDI outflows in 2013, due to increasing action by *transnational corporations* (TNCs). Global FDI is mainly shaped by three mega-regional groupings (TPP, TTIP and RCEP),[7] which are currently in negotiation, each accounting for a quarter or more of global FDI flows. The largest regional economic cooperation grouping remains *Asia–Pacific Economic Cooperation* (APEC), with 54 per cent of global inflows (WIR 2014). Africa is seeing the rise of new sources of FDI. Overcoming barriers for attracting FDI remains a key challenge for small, vulnerable and weak economies. *Overseas development assistance* (ODA) can act as a catalyst for boosting the role of FDI in *least developed countries* (LDCs). Focusing on key niche sectors is crucial if *Small Island Developing States* (SIDS) are to succeed in attracting FDI.

Low-carbon FDI reached roughly US$90 billion in 2009 in three key industries: (a) alternative/renewable electricity generation; (b) recycling; and (c) manufacturing of environmental technology products (such as wind turbines, solar panels and biofuels). These industries form the core of new low-carbon business opportunities. The following developments could be seen in these industries (WIR 2010):

---

7   Trans-Pacific Partnership (TPP), Transatlantic Trade and Investment Partnership (TTIP), Regional Comprehensive Economic Partnership (RCEP).

Figure 34.5: FDI inflows, global and by group of economies, 1995–2013 and projections, 2014–2016 in billions of US dollars. **Source**: Adapted from UNCTAD FDI-TNC-GVC Information System, FDI/TNC database; at: <www.unctad.org/fdistatistics> (WIR 2014).

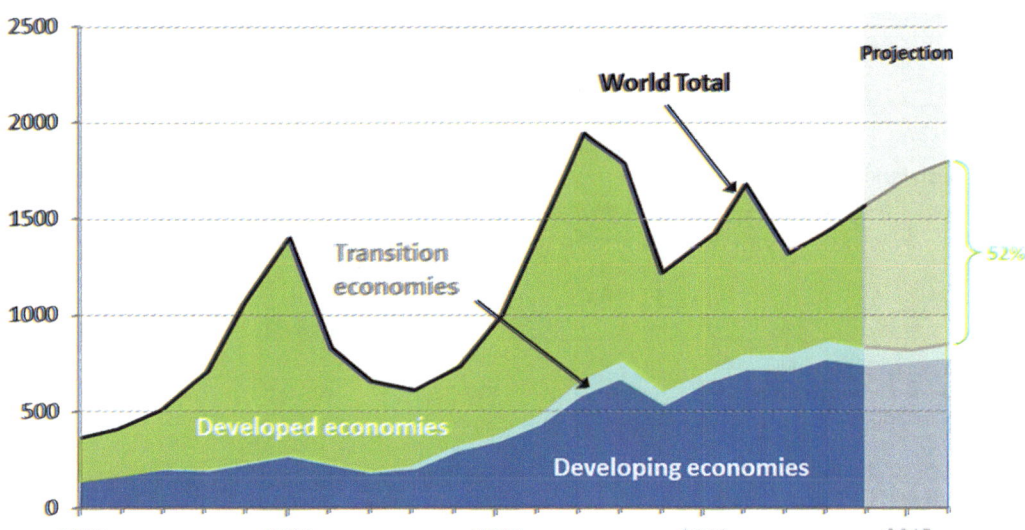

- A rapid increase in low-carbon FDI in recent years, though it declined in 2009 as a result of the financial crisis.
- Around forty per cent of identifiable low-carbon FDI projects by value during the period 2003–2009 were in developing countries.
- Established TNCs are major investors, but new players are emerging, including those from the South. TNCs from other industries are also expanding into the field.
- About ten per cent of identifiable low-carbon FDI projects in 2003–2009 were generated by TNCs from developing and transition economies. The majority of these investments were in other developing countries.

Over time, low-carbon investment is expected to permeate all industries. While foreign investments have recovered and started growing again in recent years, significantly more is required for the level of sustainability transformation envisaged. The global scale of the challenge in reducing *greenhouse gas* (GHG) emissions requires an enormous financial and technological response. Estimates for the period 2015–2030 are that US\$550 billion to US\$850 billion of recurrent additional global investment per year is required to limit GHG emissions to the level needed for a 2°C target to be met, as specified in the Copenhagen Accord (WIR 2010). Current investment in 2014 amounts to US\$170 billion (WIR 2014).

Various studies stress that the financial contribution of the private sector is essential for achieving progress in making economies more climate-friendly, particularly in view of the huge public fiscal deficits worldwide. The average level of the private sector in current mitigation investment is in the area of US\$40 billion in developing countries and US\$90 billion in developed countries (WIR 2014).

TNCs can make significant contributions in the transition towards a low-carbon economy, because they are significant emitters across their vast international operations, but also because they are in a special position to generate and disseminate technology and to finance investments to mitigate GHG emissions (WIR 2010). TNCs are thus not only part of the problem but can also be part of the solution.

34.5 The Context of International Climate Negotiations

34.5.1 UN Conference on Environment and Development (UNCED) 1992

The importance of technology transfer has been realized in the agreements achieved at the 1992 *UN Conference on Environment and Development* (UNCED). In Chapter 34 on the "Transfer of environmentally sound technology, cooperation, and capacity building", Agenda 21 calls for access to scientific and technical information, promotion of technology transfer pro-

Table 34.3: Current investments, investment needs and gaps, and private sector participation in key sustainable development goals (SDG) sectors in developing countries. **Source:** Adapted from WIR 2014; at. <www.unctad.org/fdistatistics>.

Sector	Description	Estimated current investment (latest available year) $ billion A	2015 – 2013 Total investment required Annualized $ billion (constant price) B	2015 – 2013 Investment Gap Annualized $ billion (constant price) C = B - A	Average private sector participation in current investment Developing countries Per cent	Average private sector participation in current investment Developed countries Per cent
Power	Investment in generation, transmission and distribution of electricity	~260	630 - 950	370 – 690	40 - 50	80 – 100
Transport	Investment in roads, airports, ports and rail	~300	350 - 770	50 – 470	30 - 40	60 – 80
Telecommunications	Investment in infrastructure (fixed lines, mobile and internet)	~160	230 – 400	70 – 240	40 – 80	60 – 100
Water and sanitation	Provision of water and sanitation to industry and households	~150	~410	~260	0 – 20	20 – 80
Food security and agriculture	Investment in agriculture, research, rural development, safety nets, etc.	~220	~480	~260	~75	~90
Climate Change mitigation	Investment in relevant infrastructure, renewable energy generation, research and deployment of climate-friendly technologies, etc.	170	550 – 850	380 – 680	~40	~90
Climate Change adaptation	Investment to cope with impact of climate change in agriculture, infrastructure, water management, coastal zones, etc.	~20	80 – 120	60 – 100	0 – 20	0 – 20
Eco-systems/ biodiversity	Investment in conservation and safeguarding ecosystems, marine resource management, sustainable forestry, etc.		70 – 210			
Health	Infrastructural investment, e.g. new hospitals	~70	~210	~140	~20	~40
Education	Infrastructural investment, e.g. new schools	~80	~330	~250	~15	0 - 20

jects, promotion of indigenous and public domain technologies, capacity building, intellectual property rights, and long-term technological partnerships between suppliers and recipients of technology (UN 1993). It points out: "Technology cooperation involves joint efforts by enterprises and governments, both suppliers of technology and its recipients. Therefore, such cooperation entails an interactive process involving government, the private sector, and research and development facilities to ensure the best possible results from transfer of technology" (UN 1993: 305). The document also recommends the utilization of existing technological information, and the promotion and development of research partnerships and assessment networks.

34.5.2 UN Framework Convention on Climate Change (1992)

Technology transfer from the countries and companies that have developed them to other countries in order to reduce GHGs has been a main plank of the UNFCCC. Article 2 of the UNFCCC aims at "preventing dangerous anthropogenic interference with the climate system" by stabilizing greenhouse gas concentrations as an ultimate objective of the Convention. Article 4.1.c requires that the Parties "promote and cooperate in the development, application, diffusion, including transfer, of technologies, practices, and processes that control, reduce, or prevent anthropogenic emissions of greenhouse gases". The promise of access to new technologies was a central incentive for developing nations to support the UNFCCC in 1992 but success has been widely questioned and many developing nations feel frustrated at the lack of technology transfer in practice (Feldmann 1994).

34.5.3 Kyoto Protocol (1997)

The importance of technology transfer was also recognized in the Kyoto Protocol, which in its Article 10c asks all Parties to:

> Cooperate in the promotion of effective modalities for the development, application and diffusion of, and take all possible steps to promote, facilitate and finance, as appropriate, the transfer of, or access to, environmentally sound technologies, know-how, practices and processes pertinent to climate change, in particular to developing countries, including the formulation of policies and programmes for ... the creation of an enabling environment for the private sector.

The Kyoto Protocol has created mechanisms to reduce emissions, including the *Clean Development Mechanism* (CDM), which is focused on low-carbon technology transfer to developing countries. While developing countries are not bound to climate change policies, they are free to choose whether they want to move towards a low-carbon economy, and to what extent. For this purpose, a continuum of options with varying implications, development benefits and costs is available. Although the CDM and its *Certified Emissions Reductions* (CERs) as well as *Joint Implementation* and the *Emissions Reductions Units* (ERUs) were expected to generate foreign investment and technology flows, this expectation has largely not been met. Problems have occurred in the issuing of carbon credits, the determination of emissions baselines, the setting of allowance/credit values or carbon prices, the levying of carbon taxes, and the imposition of rules and procedures for carbon trading.

Because the Protocol's mechanisms were designed for compliance with emission reduction targets at the national level, individual governments were left to decide how best to involve the private sector in the process. Uncertainties about the post-Kyoto framework weakened the private sector's ability and willingness to make decisions in the area of climate change. Strong international and national commitments by governments are nowadays the key conditions to enable and persuade the private sector to take part in the transformation process. In recent years this commitment has been missing and negotiations on these issues have been controversial, with clean technology nearly ignored at the Conference of the Parties (COP) 13 in Bali 2007, because of disagreements between the United States and G77/China.

34.5.4 Green Climate Fund (2009)

New regulations and reformed legislation are needed to enable an environment where climate-friendly innovations can occur and to achieve a breakthrough in this area. This includes innovation in climate finance alongside disagreements on clean technology. One of the critical questions for the nations will be how the necessary financial obligations can be met without taking an excessive amount of money from taxpayers. A possible approach to raising US$100 billion was discussed in the 2010 report by the UN's *Advisory Group on Climate Change Financing* (AGF 2010: 12). It calls this amount "challenging, but feasible", provided that policy would "lay out a clear road-map for making this funding a reality." During the 2009 COP in Copenhagen, Denmark, the *Green Climate Fund* (GCF) was proposed. The aim was to assist developing countries in their efforts to combat climate change through provision of grants and other financing for mitigation and adaptation projects, programmes, policies, and activities. The design of this financial institution, which is linked to the UNFCCC, was agreed on during the 2011 COP in Durban, South Africa, and it came into operation in summer 2014. Contributions from donor countries and other sources, potentially including innovative mechanisms and finance from the private sector, are required to activate the GCF. Over the course of the first year, an immediate activation of US$10 to US$15 billion was requested by the Parties. By December 2014, the GCF had already activated US$10.2 billion and thus fulfilled the call of the Parties (UNFCCC 2015). The GCF contributes to the future targets of financial transfers from the developed to the developing countries; the UNFCCC aims to jointly mobilize US$100 billion per year (UNFCCC 2015). Besides US$100 billion from the public sector, the report suggests that the private sector could mobilize up to US$500 billion a year. With such large amounts of climate finance, an unprecedented scale of investment in energy efficiency, renewable energy and low-carbon technologies can be expected.

34.5.5   Poznan Strategic Programme on Technology Transfer (2010)

The 2010 COP 16 in Cancun approved the *Poznan Strategic Programme on Technology Transfer* (PSP) to support developing countries in moving towards a low-carbon development path through the adoption of environmentally sound technologies. The concept was first presented to COP 14 in Poznan and later developed into a final version. The process of implementing the PSP is structured as four milestones: (1) a *Technology Needs Assessment* (TNA) for the necessary technologies for adaptation to and mitigation of climate change, (2) pilot projects to realize and evaluate the diffusion and transfer of technologies, (3) the dissemination of experiences and (4) long-term implementation (GEF 2012). Here, private investments come into play, especially for reaching the last milestone 'long-term implementation' which seeks to support:

- climate technology centres and climate technology networks;
- pilot priority technology projects to foster innovation and investments;
- public–private partnership for technology transfer;
- technology needs assessments;
- the GEF as a catalyst for technology transfer.

34.5.6   Ad Hoc Working Group on Long-Term Cooperative Action under the Convention (2010)

Building on the PSP, a new chapter was opened at the 2010 COP 16 in Cancun where an agreement adopting a very concrete and practical approach for technology transfer and investment was achieved. The decision made by the "Ad Hoc Working Group on Long-Term Cooperative Action under the Convention" introduced several new elements, including a large section on Finance, Technology and Capacity Building (section 34.4). In this chapter, the working group decided to establish a Technology Executive Committee and a

Climate Technology Centre and Network to guide and facilitate the implementation of a Technology Mechanism. The working group also identified various opportunities for collaboration between governments and private sectors, non-profit organizations and academic and research communities, including the priority areas:[8]

a.) Development and enhancement of endogenous capacities and technologies of developing country Parties, including cooperative research, development and demonstration programmes.

b.) Deployment and diffusion of environmentally sound technologies and know-how in developing country Parties.

c.) Increased public and private investment in technology development, deployment, diffusion and transfer.

d.) Deployment of soft and hard technologies for the implementation of adaptation and mitigation actions.

e.) Improved climate change observation systems and related information management.

f.) Strengthening of national systems of innovation and technology innovation centres.

g.) Development and implementation of national technology plans for mitigation and adaptation.

A conceptual framework was implemented into the more precise and practical Technology Transfer Framework (box 34.4) by the UN Technology Executive Committee, which was initiated in 2010 by the "Ad Hoc Working Group on Long-Term Cooperative Action under the Convention". This Committee has oversight over this framework so that the transfer of environmentally sound technologies and know-how can be increased and improved. The approach describes the relevant conceptual and analytical frameworks comprising the five key themes.

34.5.7   Paris Agreement at COP21 (2015)

In December 2015, the *21st session of the Conference of the Parties in Paris* (COP21) agreed to hold "the increase in the global average temperature to well below 2°C above pre-industrial levels and to pursue efforts to limit the temperature increase to 1.5°C above pre-industrial levels, recognizing that this would significantly weather events and slow onset events, and reduce the risks and impacts of climate change" (Paris

---

8   <https://unfccc.int/files/meetings/cop_16/application/pdf/cop16_lca.pdf>.

Box 34.4: Mechanisms for Technology Transfer. **Source**: Assembled from UNFCCC <http://unfccc.int/ttclear/templates/render_cms_page?TTF_home>.

1. *Technology Needs Assessments:* country-driven activities that identify and determine the mitigation and adaptation technology priorities.
2. *Technology Information:* defines the means to facilitate the flow of information between the different stakeholders to enhance the development and transfer of environmentally sound technologies (including hardware, software and networking).
3. *Enabling Environments:* foster private and public sector technology transfer through focusing on government actions, such as fair trade policies, removal of technical, legal and administrative barriers to technology transfer, sound economic policy or regulatory frameworks and transparency.
4. *Capacity Building:* seeks to build, develop, strengthen, enhance and improve existing scientific and technical skills, capabilities and institutions in developing countries to enable them to assess, adapt, manage and develop environmentally sound technologies.
5. *Mechanisms for Technology Transfer:* facilitates the support of financial, institutional and methodological activities to (a) enhance the coordination of stakeholders in different countries and regions, (b) engage them

in cooperative efforts to accelerate development and diffusion, through technology cooperation and partnerships (public/public, private/public and private/private), (c) to facilitate the development of projects and programmes to support such engagement. The mechanism comprises four sub-themes:
5.1 *Innovative Financing:* aims at improving access to financing for the development and transfer of technology from a wide variety of available public and private sources.
5.2 *International Cooperation:* aims at identifying ways and means to enhance cooperation with relevant conventions and intergovernmental processes involved in the development and transfer of technology.
5.3 *Endogenous Development of Technologies:* aims at identifying actions to promote endogenous development of technology through the provision of financial resources and joint research and development.
5.4 *Collaborative Research and Development:* aims at promoting collaborative research into and development of technologies.

Agreement 2015, Article 2a). The agreement also recognizes "the importance of averting, minimizing and addressing loss and damage associated with the adverse effects of climate change, including extreme the role of sustainable development in reducing the risk of loss and damage" (Article 8). To this end, the Parties address the "urgent need to enhance the provision of finance, technology and capacity-building support by developed country Parties" and formulate the objective of considering "how to enhance linkages and create synergy between, inter alia, mitigation, adaptation, finance, technology transfer and capacity-building, and how to facilitate the implementation of coordination of non-market approaches". Further the Parties recognized the need to incorporate climate-proofing and climate resilience measures into their development assistance and climate finance programmes. To reach these goals they committed themselves to accelerating, encouraging and enabling innovation to reach a long-term global response to climate change, and they called on the established technology framework to

a.) undertake and update *technology needs assessments* (TNA) and technology action plans;
b.) enhance financial and technical support for the implementation of the TNAs;

c.) assess technologies that are ready for transfer; and
d.) enhance enabling environments for and address barriers to the development and transfer of socially and environmentally sound technologies.

These commitments further strengthen the Parties' commitments to sustainable development and create opportunities across institutions and the private sector for participation in the implementation of environmentally sound development.

34.5.8 Other Non-UN Frameworks for Climate Finance and Technology Transfer

Besides the level of the United Nations, various other frameworks have been developed to foster technology cooperation on energy and climate change. For instance, at the Gleneagles summit in July 2005, the G8 highlighted the importance of strengthening technology cooperation in order to develop low-carbon energy options. Many developing countries pressed for a new approach to international cooperation in the area of clean energy technologies. As a follow-up, the UK government and the government of India decided to collaborate on a study to assess the barriers to the transfer of low-carbon energy technology between developed and developing countries (Mallett/Ockwell/Pal et al. 2009). In May 2015 the G7 launched the Hamburg Initiative for Sustainable Energy Security, which supports "the further improvement of the performance and reduction of the cost of technologies such as smart grids, systems optimization, energy storage, electric vehicles, offshore wind energy, and other flexibility options" (G7 2015a: 3). The G7 Declaration at the Summit in Germany on 8 June 2015 took the decision to "foster growth by promoting education and innovation, protecting intellectual property rights, supporting private investment with a business friendly climate especially for small and medium-sized enterprises, ensuring an appropriate level of public investment, promoting quality infrastructure investment to address shortfalls through effective resource mobilization in partnership with the private sector and increasing productivity by further implementing ambitious structural reforms" (G7 2015b: 2). However, the private sector needs a clear, stable and predictable policy framework to reorient its business strategies. Regardless of progress on a global regime, it is left to national governments and private investors to contribute to such a policy framework. It has become clear that creative mechanisms are in demand both at national and international levels to effectively mobilize the private sector's contributions to cross-border capital flows and technology diffusion, especially to developing countries.

34.5.9   Financing Needs for Climate Policies

A challenge in this area is the uncertainty over the financial resources currently available, partly due to the lack of an agreed list of these technologies; there are also no agreed definitions of the costs of technology research, development, deployment, diffusion and transfer. At the time of the 2009 Copenhagen climate summit (COP 15), estimates of the financial resources currently available were classified by UNFCCC (2009) by stage of maturity of the technology they are intended for, whether the resources are from the public or private sector, and whether they are covered by or outside the Convention. The estimates for mitigation technologies, shown in figure 34.5, lie between US$70 billion and US$165 billion per year. As far as adaptation is concerned, R&D is focused on tailoring the technology to the specific site and application. Spending on adaptation projects in developing countries was estimated at about US$1 billion per year, while future spending needs are estimated to be between tens and hundreds of billions of US dollars per year. Financial support for technology

transfer was likely to amount to less than US$2 billion per year (UNFCCC 2009).

Several estimates are available for the additional financing that will be needed for RD&D of mitigation technologies in order to stabilize the levels of GHGs in the atmosphere. The estimates are sensitive to the baseline and to the mitigation scenarios used. As shown in figure 35.6, they indicate that current financing for mitigation technologies needs to increase three- to fourfold. Such increases are consistent with current R&D targets and priorities for developed countries as well as for regions with large R&D budgets. The economic and social benefits of investing in climate change technologies are likely to be greater than the cost of making the investments (UNFCCC 2009). In addition to changing investment in energy-generation sectors, scenarios compatible with the 2°C goal show a rise in annual investments in energy efficiency in the fields of transport, buildings and industry, amounting to US$336 billion (WBGU 2014; IPCC 2014).

34.6   Benefits and Incentives of the Sustainable Energy Transition

To achieve success in the new 'industrial revolution' that has been envisaged and that is required for a climate-smart sustainability transition, as well as for a low-carbon society, it is essential to understand the significance of the task. Energy systems, as key drivers of economic growth and social progress, are the core of the transformation process. They are considered "essential to fuelling industry, powering infrastructure, connecting goods, people and services to markets, and delivering basic services such as heating, lighting and cooking" (WBCSD 2007: 3).

If concerns about climate injustice are to be addressed, countries and communities around the world need to receive their fair share of the fruits and benefits of the technological transformation. Therefore, it is essential to clarify the opportunities, benefits and incentives to encourage support from governments, private sectors and the public in recipient countries. Questions of justice will concern the sharing of benefits from new climate-related investments and financial transfers and the distribution of benefits, costs and risks among the different regions, as well as in the public and private sectors. If these questions are taken seriously, then the groundwork can be laid for a lasting, sustainable and, especially, peaceful energy transition, with concurrent benefits.

Figure 34.6: Estimates of financing for mitigation technologies. **Source**: Adapted from UNFCCC (2009).

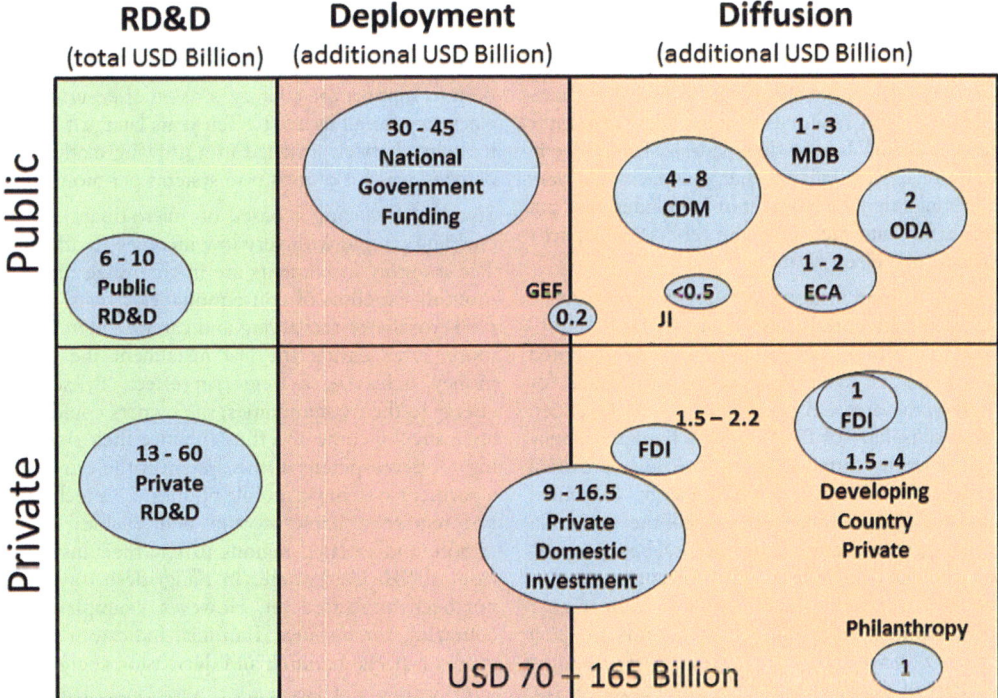

Abbreviations: CDM = clean development mechanism, ECA = export credit agency, FDI = foreign direct investment, GEF = Global Environment Facility, JI = joint implementation, MDB = multilateral development bank, ODA = official development assistance, RD&D = research, development and deployment.

34.6.1 Electrification of Remote Areas in Developing Countries

Investment in energy infrastructures (on- and off-grid) is essential if the billions of people without access to modern energy services are to escape poverty and enter into productive economic activities. One example of successful investment is the combination of small-scale renewable energy systems combined with micro-finance that offers opportunities for less developed countries, as in Bangladesh (see box 34.5). Despite government efforts to promote rural electrification, many remote rural areas in least-developed countries experience shortfalls in energy capacity. Increasing access to electricity by installing off-grid energy generation, such as solar home systems supported by micro-finance institutions, enables low-income communities to enhance their living standards and improve their livelihoods.

34.6.2 Reduction of Future GHG Emissions

With the growing energy demand from developing countries, their share of GHG emissions is also expected to rise, from 39 per cent in the middle of the last decade to 52 per cent by 2030, with China responsible for 29 per cent of the predicted rise (UNDP 2006). India is already the third biggest emitter of CO_2 emissions, yet approximately 25 per cent of its population does not have access to electricity and approximately 68.8 per cent of the population lives on less than US$2 per day (World Bank 2010). Clean energy could possibly help satisfy the growing energy demands of developing countries and at the same time cut their carbon emissions.

34.6.3 Independence from Fluctuating Fossil Fuel Prices

Rising oil prices and dependence on energy imports from the Middle East have increased the demand for alternative energy paths. In response to the OPEC oil embargos and the subsequent price shocks of the 1970s, incentive programmes were implemented to encourage the development of renewable energy. Despite these programmes, and despite increased research efforts, renewable energy production grew only slowly through the 1980s and stagnated during

Box 34.5: Solar Home Systems for Bangladesh. **Source:** Assembled by the authors.

A lack of energy generation, distribution and storage facilities, high upfront costs and insufficient funding sources, coupled with inefficient management, knowledge, manpower, operations and maintenance, has kept economic development potential low in Bangladesh. Especially in remote rural areas, shortfalls in energy capacity are very common, although the Government of Bangladesh has put considerable effort into increasing the generating capacity and promoting rural electrification.

In contrast to government agencies, decentralized private actors and non-profit enterprises can contribute significantly to providing energy infrastructure that are adapted to local needs and that meet the high level of demand for electricity in rural areas. A prominent case is Grameen Shakti (GS), a subsidiary of the Grameen Bank in Bangladesh, which since its foundation in 1996 has provided rural electrification through renewable energy, combined with small-scale loans and micro-credits to the poor. GS is one of the largest and fastest-growing rural-based renewable energy companies in the world, with almost 25,000 solar home systems sold by June 2004 and more than 1 million *solar home system* (SHS) installations in rural areas by end of 2012 with more than 22,250 SHSs installed per month. Part of the success is the combination of the technical system, micro-financing, on-the-spot installation, customer training, and a three-year maintenance warranty. A complete range of services provided, enabling its customers to become independent of central, state-controlled sources of power (Kebir/Philipp/Scheffran et al. 2005).

A complementary approach is the *Infrastructure Development Company Limited* (IDCOL) which was established in 1997 by the Government of Bangladesh and licensed by the Bangladesh Bank as a *non-bank financial institution* (NBFI) in 1998. IDCOL is playing a major role in bridging the financing gap for developing medium to large-scale

infrastructure and renewable energy projects in Bangladesh. In 2003, IDCOL began the installation of SHSs in Bangladesh to support the country's vision of ensuring "Access to electricity for all by 2021". Ten years later, 2.5 million SHSs had already been installed in a growing market where the demand was for over 75,000 systems per month.

Standard financing is based on micro-finance institutions, enabling people with very low incomes to afford an SHS. The monthly instalments are in the range of the average monthly expenses of conventional electricity, as for example kerosene or recharging and carrying batteries. Furthermore, after paying the last instalment the owner saves money, achieving a long-term effect. By selling excess energy to their communities, the owners could further create a small income and thus enhance their standard of living. A development of this system is the establishment of a payment system via mobile phones. This technology does not require a bank account and enables end-users in remote and isolated regions to pay their instalment plan from almost everywhere. In Bangladesh, this system has not been implanted yet. However, examples from other countries, for instance Tanzania, indicate a high success rate in capacity building and decreasing energy poverty.

A German-based company is MicroEnergy International, which develops access to renewable energy technologies and energy efficiency through micro-finance and supporting financial, technical and scientific actors to provide households and micro-entrepreneurs with reliable, affordable and sustainable energy solutions. MicroEnergy International works with micro-finance institutions, energy product and service providers, international development actors and research institutions in order to set up clean energy micro-finance schemes.

Resources: <http://www.gshakti.org>; <http://www.idcol.org/>; and <http://microenergy-project.de>.

the 1990s, but the rate of growth considerably increased in the first decade of this millennium. With increasing oil prices, the competitiveness of renewables will further improve, but it can be slowed down by the current lock-in of high-carbon fossil energy pathways in most sectors. The transportation sector is almost completely dependent on energy from fossil fuels such as gasoline, diesel and kerosene. Developing countries whose oil supply largely relies on imports are particularly dependent, and thus vulnerable.

34.6.4   Reduction of Vulnerability, Long-Term Costs and Risks of Climate Change

Climate change is expected to have many negative impacts in vulnerable developing countries, and these negative impacts could also affect commercial inter-

ests. Reducing carbon emissions contributes to alleviating the long-term economic, social and environmental costs and the risks of climate change, and helps to accomplish the targets for GHG emission reduction as set out in the UNFCCC and the Kyoto Protocol.

Protecting the environment and preserving natural resources can provide additional economic benefits, in particular a reduction in vulnerability to climate change. Although many developing countries are not major GHG emitters, new technology and investment would help them enter low-emission pathways, avoiding the polluting development paths that industrialized countries have pursued in the past. (For a review of pathways of technology cooperation for sustainable energy, see Mallett 2013).

34.6.5   Structural Changes and Employment

The quest for energy security and concerns about global warming is encouraging the search for low-carbon energy alternatives to fossil fuels. Large-scale production of renewables would significantly reduce oil imports and diversify energy sources. Home-grown domestic energy sources offer important development opportunities to structurally weak rural areas and can lead to beneficial structural changes in land use and agricultural practices. Benefits are particularly relevant for the agricultural community, which is expecting fast-expanding future markets for energy and land resources. These will create new sources of income and job opportunities in the agricultural sector, leading to an increase in farmers' income. An example is the production of biofuel resources. According to the UNEP (2009: 20), renewable energy generates more jobs than employment in fossil fuels. Projected investments of US$630 billion by 2030 could translate into at least twenty million additional jobs in the renewable energy sector.

34.6.6   Future Projections

Stabilizing the atmospheric CO_2 concentrations at non-dangerous levels will require a rapid increase in the scale and speed of introduction of low-carbon innovation and technology. To achieve the $2°C$ guard rail with a probability of 67 per cent, by the middle of the twenty-first century no more than a maximum of approximately 750 gigatonnes (Gt; billion tonnes) of CO_2 would have to be released into the atmosphere; for a probability of 75 per cent, the CO_2 budget would have to decline to 600 Gt CO_2 (WBGU 2009: 2). To achieve a 450 ppm (*parts per million*) scenario by 2050, global emissions would need to be reduced by 50 per cent compared with 2005 levels (IEA 2008). The same source also estimates that an additional US$45 trillion in total investment will be required by 2050 for a 450 ppm scenario. With an increase in energy demand in non-OECD countries expected to account for 85 per cent of the increase in world energy demand, a major challenge will be to accelerate the diffusion of low-carbon technologies to developing countries.

34.6.7   Political Incentives Broadening the Economic Basis of Renewables

In both developing and developed countries, there is a significant growth potential for innovative technolo-gies for heating, electricity generation and transportation fuels. Recent years have shown a dramatic shift in policy in many parts of the world; the growing demands and subsidies for renewables have broadened their economic basis (WBCSD 2010). This is expected to lead to a large expansion of sustainable energy and energy efficiency technologies in developing countries over the next decade and beyond. To succeed, renewable energy systems will have to become fully competitive with fossil energy and to avoid some of the current market distortions such as domestic subsidies and import barriers.

34.6.8   Mutual Benefits of Developed and Developing Countries

Although developed and developing countries have different responsibilities and capacities, a smartly designed transformation of the global economy can bring benefits to both. While developed countries are expected to be the main source of the finance needed for the transition, this may become an opportunity rather than a burden for them. Increased foreign investments in environmentally sound technologies will diversify the portfolio of companies, link them to international markets, create new jobs and generate a return on investments. Financial transfers can create social and economic benefits for rural and urban communities in developing countries, including employment opportunities, reduction of poverty, and access to advanced technology. Investments can further strengthen domestic markets and develop future export markets, leading to economic diversification, accelerated commercialization and competitive advantages in developing countries. Successful investment translates into new jobs and infrastructure development and leads to an increase in *gross domestic product* (GDP) and a reduced share of GDP spent on energy imports, either through energy efficiency or through the use of local renewable sources (WBCSD 2010).

34.6.9   Advantages for the Public Sector

The green transformation is significant for the public sector, which in the past has been a main source of political and societal interventions (OECD 2015). Direct governmental expenditure, as well as public choices, regulations and policies, can be drivers of technology development and transfer. Renewable energy systems receive a high level of public support justified by the expected energy, economic and environmental benefits of renewables. Further, a number of co-benefits are included, such as the sectorial effects of rural electrification, energy security through energy diversification and improved efficiency, local environmental benefits, and international funding opportunities.

34.6.10 Advantages for the Private Sector

Major advantages for the private sector to pursue climate change mitigation and the transfer of low-carbon technologies may yield financial benefits for the companies involved in the transfer process. These advantages are likely to improve production processes, including the enhancement of energy efficiency, material efficiency and resource efficiency. Early adopters profit from strengthened productive capacities and enhanced export competitiveness, giving them an edge over competitors. Driving this process would accelerate a developing country's transition and facilitate leapfrogging into a green economy. This is obviously in the best long-term interest of the business sector, assuming that there will be an increasing market and new export opportunities for low-carbon products and services. On the demand side, a growing pool of responsible consumers and the rise of a sustainability-oriented civil society shaping consumer preferences will help to establish such a market (UNCTAD 2010).

34.7 Barriers and Risks to Technology Transfer and Investment

Developing and utilizing the huge potential for low-carbon technology in the developing world means overcoming a number of barriers, obstacles and risks that limit their availability and efficiency. Barriers arise at each stage of the technology transfer and investment process, including infrastructure and ownership, cost and finance, as well as regulation and policy. These factors may vary across developing countries, depending on national context and economic sector, project, pathway and technology, application area, region and geography, as well as the local security situation. Here we can only highlight some of the obstacles and barriers without covering the extensive discussion in the literature (for a list, see box 34.6). Not addressing these issues can undermine the efficiency of technology transfer and investment policies. They can result in social costs, ranging from job losses to the reduced affordability of essential services, and/ or reduced taxes. The consequences are likely to hit

low-developed and other vulnerable countries the hardest. When promoting low-carbon foreign investment, policymakers have to weigh the advantages and disadvantages, both in terms of economic growth on the one hand, and environment, human health and sustainable development on the other.

Box 34.6: Barriers to technology transfer and investment. **Source**: Based on Santarius, Scheffran and Tricarico (2012).

1. Lack of capacity building and domestic infrastructure.
2. Macro-economic conditions and market failure.
3. Lack of access to and sharing of data, information and knowledge.
4. Lock-in due to lack of innovation and diffusion.
5. Failure of R&D spending policy for low-carbon innovation.
6. Tensions and competition undermining the potential for international collaboration.
7. *Intellectual property rights* (IPR).
8. Financial, social and political obstacles for foreign investment.
9. Resource scarcity and environmental impact.

34.7.1 Power Structures and Technology Imperialism

Many of these barriers are subject to vested interests and power games which have not disappeared with the East–West conflict but have re-emerged in the North–South context. The globalized world is driven by an enormous accumulation of capital and concentration of power which shapes technology transfer and financial flows. The large-scale investments necessary for the transitions mentioned above are an expression of the world's power structures and at the same time are transforming them. In the process of technology transfer, the developed countries should act with caution, and take care that they do not establish a 'technology imperialism' by overwhelming their partners and suppressing local companies and economies. The invasion of large companies into local markets is to be avoided. Instead local partnerships, where all stakeholders are at the same level and interdependent, should be established in order to work towards the benefits mentioned above.

34.7.2 Conflict Potential and Ambivalence of Renewable Energy Sources

In a highly competitive environment, the struggle for access to and control of technology can not only strengthen cooperation but also contribute to conflict among the major players, nations as well as corporations. Conflicts may also emerge at a local level, when citizens and non-governmental organizations resist the introduction of mitigation and adaptation measures such as biomass, wind and hydropower (Scheffran/ Cannaday 2013). While renewables are highlighted as being largely carbon neutral, GHG savings can vary significantly for specific production and conversion paths and their fossil fuel inputs. There is an ongoing debate about the carbon balance of biofuels and their impact on land use and food security in developing countries (Scheffran/Summerfield 2009; Hamelinck 2013). If not addressed properly, concerns about the possible adverse implications of renewables may undercut their support. This means that renewable energy production and consumption needs to be established in a sustainable and conflict-sensitive manner in both developed and developing countries.

34.7.3 Policy Frameworks to Address Barriers

To address the barriers and overcome the obstacles it is important to develop a practical framework for policymakers, facilitating technology transfer and foreign investment through public money, regulation and standards, and international cooperation. In each stage of the technological process—R&D, demonstration, deployment, diffusion and commercial maturity—existing barriers need to be overcome in order to develop a commercially mature technology. Adjusting the response to a particular barrier helps to identify the financing vehicles appropriate to it. However, the limiting factors and barriers to technological transformation are not primarily of a technological nature but are rather part of the social, economic, political, and cultural milieux in which technologies are developed, diffused, and used. Market incentives, the structure of regulations, the content and quality of research and education, and social values and preferences all determine technological pathways (Heaton/Repetto/Sobin 1991). To develop effective policy tools, it is important to tailor action to the specific barriers, interests and influences of different stakeholders. Further, requirements such as stable political conditions and an educational system to train and qualify local workers have to be established.

34.8 Policies for Technology Transfer and an Enabling Infrastructure for Innovation

34.8.1 Assessment of Innovation, Investment and Technology Needs

To address the challenges and overcome the barriers described above, it is essential to create the adequate framework conditions with incentives that will cause a large-scale technological shift toward a lower-carbon and more energy efficient economy that can also deliver affordable energy solutions (ICSTD 2012). This shift relies on directing investment flows into the development and deployment of lower-carbon technologies, as well as adapting behaviours and lifestyles to favour these technologies. Deploying low-carbon technologies to developing countries requires innovative mechanisms and a joint effort by the private sectors of developed and developing countries (Byrne/ Zhou/Shen et al. 2007; Watson/Byrne/Ockwell et al. 2014). Previous efforts to support developing countries in creating enabling environments for low-carbon growth have been insufficient, and more efforts need to be made.

34.8.2 National Systems of Innovation

For the transformation of the energy system and a consequent implementation of an enabling environment three steps have to be fulfilled: (1) the true costs of the current emission-intensive energy system have to be determined; (2) targeted support for new technologies has to be provided; and (3) opportunities for refinancing the high investment cost of renewable energy have to be identified (WGBU 2011). However, the transformation process is knowledge-based, involving all social actors and thus attributing a high degree of social responsibility (WGBU 2014). *National Systems of Innovation* (NSI) "integrate the elements of capacity building, access to information and an enabling environment" for investment and technology transfer (IPCC 2000: 6). NSI concepts rest on a complex mixture of institutions, public policies, and business and social relationships, including networks and partnerships with public and private stakeholders in the business, legal, financial and service domains. NSIs can be enhanced by investment and low-carbon technology diffusion through international partnerships.

34.8.3   The Nexus of Public and Private Investments

In this context, the private sector may serve as a major source of innovation, capital and capacity that can potentially make major contributions to a low-carbon global economy. Most low-emission energy technologies will however not be cost-competitive at scale without some combination of investment support mechanisms, technological advances or regulatory regime improvements. A lesson from CDM proposals is that an abundance of potential projects, technologies or investment opportunities will not necessarily mobilize capital flows for implementation and optimization (an analysis of technology transfer in CDM projects can be found in Seres/Haites 2008). Thus, governments can accelerate or guide the process although the private sector could offer innovative market-based solutions without government intervention. However, government involvement can be supportive in some cases, by creating adequate frameworks for investment, including specific regulations tailored to particular technologies and their stage of maturity (for trends in investment and innovation systems in global energy research, development, and demonstration, see Gallagher/Anadon/Kempener et al. 2011).

To facilitate the release of private sector resources, governments need to understand how capital markets work. As a consequence, corporate investment strategies can be incentivized to deliver results consistent with energy and climate policy goals. It should be kept in mind though that quite different mental models and perceptions among policymakers exist, and this could result in inefficient policies, hindering rather than supporting the involvement of the private sector (WBCSD 2007).

34.8.4   Technology Needs Assessments

Many projects in developing countries fail because of inappropriate technology use which is due to lack of capacity building, ill-defined ownership, or a lack of infrastructure. A new technology framework should minimize this risk and ensure efficient diffusion of the appropriate technologies (see section 34.3). Another reason for failing projects is also the lack of reliable statistical data on issues such as energy use, infrastructure, or demand. Solid data will "help policymakers design the most cost-effective policy options and impact assessments, while business can reduce uncertainty and thus risk premiums" (WBCSB 2010).

To overcome information barriers, governments have a key role in creating the necessary information assessment and monitoring capacity. A comprehensive *technology needs assessment* (TNA), by government agencies, think tanks, research institutions, and international and non-governmental organizations, will provide the understanding required for informed decision-making about future technology options and to select the strategies most appropriate to the country, and its capability and technology options. This implies the identification and assessment of specific needs for a technology and how it fits into the domestic environment. To fully undertake TNAs, developing countries might need administrative and technical support by international organizations (UNDP 2010).[9]

Technology roadmaps help to draw future pathways in terms of capabilities, locations and timelines. They prepare the ground for the promotion of technology prototypes, demonstration projects and extension services through linkages between manufacturers, producers and end users. They also facilitate planning for the identification and development of solutions to technical, financial, legal, policy and other barriers. A number of factors might affect host governments' prioritization and targeting of foreign investment to boost prospects for technology dissemination. A government might identify targets by comparing potential growth sectors and assessing the country's natural resources and assets that have been created. For example, Morocco has chosen to enter into renewable power generation and the manufacturing of environmental technologies for a number of reasons (reducing dependence on foreign fossil fuels, supplying and exporting power, encouraging rural electrification), including an assessment of where the technology can best be secured, as well as an analysis of patterns of low-carbon foreign investment in the sector (WIR 2010). Similar approaches are relevant in concentrated solar power projects in North Africa and the Middle East (see box 34.7).

Technology assessment allows synergies to be created from the careful matching, harmonization and utilization of all technological resources and of the relevant actors from private and public sectors. Mechanisms can be put in place to adapt technologies to local needs, or to generate new ones if necessary.

---

9   For TNA guidebooks, see at: <http://www.tech-action. org/Publications/TNA-Guidebooks> and at: <http:// www.unep.org/roap/Activities/ClimateChange/Tech-nologyNeedsAssessmentandTechnologyAction/tabid/ 6847/Default.aspx>.

Attracting low-carbon foreign investment is not only about new and emerging business opportunities, but also about encouraging foreign investments in traditional sectors, with a view to improving their energy, material or resource efficiency.

34.8.5   Building Human, Social and Institutional Capacities

Developing countries not only require finance for energy projects but also capacity building and related funding from public and private sector investors. Capacity building programmes and enabling environments that reduce the risks and restrictions associated with low-carbon technology transfer will increase the flow of technologies close to the commercial margin. Capacity building targets technology acquisition, development of skills, local policies and institutions in order to support the technology transfer process. It focuses on enhancing scientific and technical skills, capabilities, and institutions in developing countries, as a precondition for assessing, adapting, managing, and developing technologies (UNCTAD 1998). But the need for enhanced skills and capabilities can also occur throughout all stages of the technology life cycle. Several studies acknowledge that the needs of capacity building vary greatly from country to country (see Mugabe 1996).

34.8.6   Requirements for Capacity Building

The transfer of green technologies demands a wide range of technical, business and regulatory skills. Human and institutional capacity is needed to assess, select, import, develop and adapt appropriate technologies, including the capacity to optimize and innovate. Not only specific know-how should be transferred, but also related systemic knowledge of the relevant technologies that will enable recipients to add value. Lack of human capital is one of the barriers to the development, acquisition, deployment, and diffusion of technologies required for meeting energy-related CO_2 emissions reduction targets (IRENA 2012). Human capacity is critical in providing a sustainable enabling environment for technology transfer in both the host and recipient countries (IPCC 2014). Previous experiences have demonstrated that the absence of human capacity was a major cause of the many failures of technology transfer. This makes adequate human capacity essential at every stage of the transfer process.

Education and training is important in developing human capital for business, municipalities and governments. The training of employees, engineers as well as operators and users, builds local competency in the use of the technology. Training helps to overcome the lack of knowledge, which is an important barrier to technology diffusion (e.g. for advanced solar technologies). Training should not only focus on the technical aspects but also involve the sales and marketing aspects. Training and support services could be largely public but may involve multinationals and consultancy firms that have extensive training and technical capacity but rarely divert those resources from commercial activities.

Capacity building targets knowledge and gaps in skills, and aims to improve enabling environments by overcoming market, human, and institutional capacity barriers. While frequent and unpredictable policy changes can undermine capability and efficiency, stable, reliable and predictable government policies need well-established legal institutions and the rule of law to reduce uncertainty about the expected return on investment (IPCC 2014).

34.8.7   Outlook for an Improved Integrated Concept of Capacity Building

Future approaches will be more effective if they lay stress on the integration of technology transfer and human resource development, and focus less exclusively on developing technical skills and more on creating improved and accessible competence in associated services, organizational know-how and regulatory management. High-quality training is needed to give the personnel of the receiving firm the skills, knowledge and expertise applicable to particular products and processes. This is an important consideration for developing countries, because the workforce requires continual and cumulative learning. Human capacity building needs to be tackled for large-scale infrastructure projects as well as for small-scale solutions, driven and implemented by local entrepreneurs. For large-scale projects, it is crucial to strengthen the strategic planning capabilities and project management skills of both regulatory authorities and project developers. For small-scale solutions, knowledge of the local market is crucial and needs to be complemented with the strengthening of business skills and the adaptation of technologies to local needs (WBCSD 2007). i

Training skills and learning are an important part of any technology transfer package and can serve as a learning vehicle for the workforce of the recipient

Box 34.7: Clean Power from the Desert—Concentrated Solar Power. **Source**: Assembled by the authors.

The highest solar irradiance can be measured around the equator, roughly between 40°N and 40°S. Therefore, this 'Sun Belt Region' is especially interesting for a new emerging renewable energy technology–*Concentrated Solar Power* (CSP). The idea is simple: incoming sunlight is concentrated through mirrors onto a receiver. The receiver transfers the energy to a heat transfer fluid which can supply the heat when needed—either as end-use heat supply or for electricity generation through steam turbines. The heat can even be stored, to produce energy when sunlight is not available or only to a limited degree, e.g. at nights or on cloudy days. This decoupling can improve the grid integration of CSP-generated electricity and consequently increase economic competitiveness.

The Sun Belt Region bears high potential for the implementation of these technologies, not only, because most of the irradiance can be received here but also because of available space. Many of the world's deserts and less populated areas lie within the subequatorial regions, providing good opportunities to install large-scale CSP. Further, the region's local development can be fostered through the implementation of CSP technology with the jobs that are created for local production of components, operation and maintenance of the power plant.

Due to the low population density and the sparseness of industrial activities, electricity generation from such power plants will exceed the electricity demand in these regions many times over, and this bears high potential for the export of electricity. The idea of large-scale energy export from these regions is also part of the aims of the DESERTEC Foundation. Their project aims to combine and integrate the most efficient and abundant renewable energies in a transnational grid to provide clean energy

from the deserts for ninety per cent of the world's population:
- *Concentrated Solar-Thermal Power* (CSP) in desert regions;
- wind power in coastal areas;
- hydropower in mountain regions with high precipitation;
- photovoltaics in sunny areas;
- biomass and geothermal power where geographic conditions are favourable.

All these technologies already exist and are in use in different parts of the world (IEA 2014). However, a global supergrid as proposed by DESERTEC has so far failed due to the political and socio-economic situation of many of the core countries of the project. CSP plays an integral part in this concept since it has the potential to balance fluctuations emerging from wind or photovoltaic power plants. The power can be transmitted over up to 3,000 kilometres using *High Voltage Direct Current* (HVDC) lines. Since ninety per cent of the world's population lives less than 3,000 km from a desert, DESERTEC aims to expand the idea of clean power from the desert to a global network. To harvest solar power for climate, human and energy security on a larger scale, the future of the concept depends on financial support and favourable political conditions, to be embedded into an integrated framework of environmental, economic and social sustainability (Schinke/Klawitter 2011; Klawitter/Schilling/Scheffran 2010).

Resources: <http://www.irena.org/DocumentDownloads/Publications/IRENA-ETSAP%20Tech%20Brief%20E10%20Concentrating%20Solar%20Power.pdf> and <http://www.desertec.org/concept/>.

firm (Brooks 1995). In those regions of the developing world where existing capabilities are weak in specific technology areas, a basic level of technological capability could be built via the establishment of regional institutes that provide training in technology assessment and management. Experience with implementing the Montreal Protocol provides a useful example for capacity building within enterprises where a multilateral fund, established by the treaty, supports training, research and network building.

34.8.8 Strengthening of Absorptive Capacities

While rural electrification and poverty alleviation is still an issue for emerging economies and small developing countries, a distinct need exists to understand and chart the distribution of capacities for absorbing and working with different low-carbon technologies (Ockwell/Mallett 2012). The absorptive capacities of local firms determine whether they can acquire and

master technologies which depend on technical competencies and commercialization skills. Host developing countries can put in place policies to develop domestic capacities to absorb technology and know-how. To strengthen absorptive capacities, policy seeks to enhance the innovative and competitive performance of firms. Here, innovation and diffusion are two facets of the same process. To reap the benefits of incremental innovations in developing countries, technology transfer should foster technological innovation in recipient firms. This will provide them with the technical capability to generate improved processes and products. In this context, government-driven R&D in green technologies can play an important role, since private investors tend to underinvest in public goods, such as the environment.

The ability to absorb new technology depends on the stages of technology development and transfer, and can be acquired through collaborative initiatives in research, development, demonstration and the deployment of low-carbon technologies. It is important to take into account the potential for regional partnerships to absorb and disseminate information to local business on the best available technologies that suit the local circumstances. To enhance absorptive capacity and build trust between technology developers and users, governments and the private sector could develop local management capabilities and go into partnership with business to find solutions. They could address installation and maintenance as a key element of technology deployment (WBCSD 2010).

34.8.9   Inter-Firm Linkages

Generation of inter-firm linkages through regular local production by foreign operators is also seen as integral to generating knowledge that will result in new technological capacity in recipient countries. Saad and Zawdi (2005) point out how the transfer of plants and equipment to developing countries has often been based on 'turnkey' and 'product-in-hand' contracts that focus on boosting industrial growth rather than fostering innovation. They also highlight the fact that restrictive terms of contracts between TNCs and firms based in developing countries have limited the scope for fostering innovation through 'reverse engineering'.

34.8.10   Creating a Legal and Regulatory Framework

To attract investment in low-carbon infrastructure projects and create new markets for low-carbon products, credible, consistent and non-conflicting regulatory obligations are needed that address the long-term nature and high capital cost of energy infrastructure projects (WBCSD 2007). To operate under the rule of law, business requires stable political and legal systems that will ensure a sound investment environment. Harmonizing regulations and standards are important, especially in sectors that are subject to strong international trade. Open markets, fair trade and rules of competition are essential for enhancing technology diffusion to developing countries.

34.8.11   Institutional Framework Conditions

The quality of a country's enabling environment has a substantial impact on whether private firms invest in new technologies and infrastructures. An effective technology policy will address the framework conditions, including the effectiveness of institutions, regu-

lations and guidelines regarding the private sector, the security of property rights and the credibility of policies, political models, laws, social norms and preferences, individual behaviours, skills, and other characteristics (IPCC 2014: 1178). The enabling environment for low-carbon business activities is "the overall environment including policies, regulations and institutions that drive the business sector to invest in and apply low-carbon technologies and services" (Stadelmann/Michaelowa 2011; Stadelmann 2013). Accordingly, the enabling environment "has three main components: (1) the core business environment, which is relevant for all types of businesses, e.g., tax regime, labour market, and ease of starting and operating a business; (2) the broader investment climate, including education, financial markets, and infrastructure, which is partially low-carbon related, e.g., via climate change education or investments in electricity grids; and (3) targeted policies that encourage the business sector to invest in low-carbon technologies" (IPCC 2014: 1223).

34.8.12   Government Contributions

Governments in developing countries can make significant contributions to build or strengthen regulatory frameworks and institutions to stimulate technology transfer and investments. On the one hand, legal barriers that slow or diminish access to low-carbon technology by local business need to be removed. On the other hand, developing new international standards and regulations is particularly relevant for energy efficiency and could create demand to support the innovation and diffusion of more efficient end-user products. If designed properly, markets will respond to regulatory policies to promote the reduction of GHG emissions.

A number of strategies can create a political and regulatory framework to facilitate the implementation of emission reduction options in the energy sector. These strategies could include but are not limited to:

- Allowing full cost pricing that includes the external costs of fossil energy and recognizes the indirect benefits of renewables such as improvement of the environment, creation of more local jobs, balance of trade, etc.
- Encouraging the implementation and enforcement of energy and environmental standards, including recruitment and training of enforcement personnel together with the necessary tools and administrative support for credible implementation of sanctions. Developing countries can learn from existing environmental standards, e.g. the CAFE

standards for vehicles in the US, the Top Runner Programme of Japan or the EU Eco-Design Directive. *World Trade Organization* (WTO) standards can be influenced to facilitate the trade of products and goods produced from the use of low-carbon technologies.

- Promotion of so-called 'green-labelling' programmes and certification systems for sustainable energy technology to employ trademark or related principles, such as an environmental seal to satisfy and approve certain requirements to products and vendors.
- Establishment of requirements for environmental impact assessment and environmental reporting.

34.8.13   UNFCCC Contributions

Establishing criteria for *measurable, reportable, verifiable* (MRV) action would set out the conditions under which national R&D and development spending would qualify as a contribution to UNFCCC commitments on technology, financing and capacity building support. These conditions would be additional to existing development assistance as well as R&D spending and have a demonstrable link to a developing country's low-carbon development plan. They should meet criteria for enhanced developing country access to new technology and increase the capacity to innovate and adapt.

34.8.14   Protection of Intellectual Property Rights

A major focal point is to regulate the protection of *intellectual property rights* (IPR) in a way that strengthens technology development and transfer in any country (WBCSD 2010). Issues of transferring or purchasing IPR polarize the interests of Parties and prevent progress in global climate negotiations. In some cases companies seem to have strategically withheld or delayed technology from certain markets in order to maximize profits. This contradicts the need to spread technologies into developing countries and is not a sustainable strategy for addressing climate change (Tomlinson/Zorlu/Langley 2008). There is no firm evidence from previous case studies of how IPR affect the diffusion of climate-friendly technologies, and there is no absolute system of IPR protection in any country.

To develop appropriate and effective mechanisms for low-carbon and climate-resilient innovation, the incentives for private innovation and for maximizing

public benefit need to be balanced. To break the deadlock between developed and developing countries over intellectual property, a new agreement for IPR and licensing would provide government-to-government commitments to protect and share low-carbon technologies and encourage joint ventures and public-private partnerships. Access to international R&D funding and credit for national R&D programmes could be made conditional on implementation of agreed protect-and-share principles for IPR. Possible options are improved licensing and parallel markets, as well as 'pay to play' agreements (in exchange for services to engage in certain activities) to meet the climate challenge (Tomlinson/Zorlu/Langley 2008). Support would be made available under a fund to strengthen IPR protection measures in developing countries, consistent with their existing international commitments under the *World Intellectual Property Organization* (WIPO) and World Trade Organization (WTO). Since renewable energy technologies still require some form of government support in many markets, these developments have led to trade-related disputes, via the WTO and domestic trade channels (Lewis 2015). An ambitious example of managing IPR in cross-border clean energy collaboration has been the US–China *Clean Energy Research Center* (CERC) (Lewis 2014a; Lewis 2014b; Watson et al. 2011).

34.8.15   Strengthening Economic Conditions and Incentives

The efficiency of policies for strengthening national systems of innovation depends to a considerable degree on the state of the economy in developing countries. Improving economic conditions is a key factor in creating a long-term viable infrastructure that will absorb low-carbon technology and accelerate development in order to facilitate leapfrogging into a low-carbon future. A diversified economy is also required to handle shocks from climate change and other destabilizing events, and recombine resources in new ways to adjust to a continuous process of change. In the mid- to long term, new fields of economic growth need to be opened, often requiring a different and more skilled workforce; this has implications for education systems and related policies. Already, significant public and private investment in low-carbon innovation in high-income countries is being developed. However, this is often done with a view to creating a national competitive advantage. Competition is a crucial factor in driving innovation but it does not fully capture all

Table 34.4: Summary of actions and incentives for the public and private sectors in the developed and developing countries. **Source:** Adapted from Santarius, Scheffran, Tricarico (2012: 50).

	Public sector	Private sector
Developed country (Investing)	• Reduce the role of current export credit agencies • Build new institutional mechanisms that promote foreign investments to make them work for climate protection and sustainable development • Raise public money to support and leverage climate-friendly foreign investments so as to insure investments against financial and market instabilities, incentivize investments into markets with limited demand, and thus make investments attractive, even where returns on investments are expected to be low	• Support should be offered to companies, e.g. renewable energy companies that aim to invest in countries with low production capacities • The level of support should significantly raise companies' interest in going abroad • Only those investments should be supported that conform to the highest standards at home • Both public and private money raised may be used by countries to fulfil their international climate finance obligations
Developing Country (Recipient)	• Implement strong, stable, transparent, coherent, credible, and ambitious long-term enabling environments • Develop domestic technology roadmaps that identify countries' nationally appropriate mitigation actions • Establish investment policies to effectively govern foreign investment inflows • Design these policies in a way that maximizes the mitigation potentials of foreign investment as well as economic diversification and sustainable development	• Setting the stage for sustained economic growth as 'green growth' and 'energy autonomy' through domestic renewable energies which have more long-term prospects than 'fossil growth' • Provide opportunities for economic diversification by catalysing foreign capital to climate-friendly sectors • Foster ownership and competitive strength of domestic/local companies in the global market by demanding that foreign investors engage in joint ventures and purchase local goods

the global public beneficial aspects of low-carbon technologies (Tomlinson/Zorlu/Langley 2008).

34.9 Conclusions and Outlook

Developing North–South Transitions into Green Economics (Santarius/Scheffran/Tricarico 2012) requires the rapid and effective dissemination of climate-friendly technologies, and in particular of renewable and efficient energy systems. A sustainability transition needs an effective combination of financial mechanisms and smart governance strategies in order to develop the economic and technological capacities of all countries, including products and infrastructures as well as new social rules and norms. Key elements in this process are the international diffusion of know-how, investments and technologies to build up production capacities and the demand for low-carbon goods, and to establish international cooperation between business and government actors to adapt technologies to the local context in a low-carbon society. The different actions and incentives for the public and private sectors in developed and developing countries are summarized in table 34.4.

34.9.1 Innovative Business Concepts Required

New business opportunities will be created for moving to low-carbon economies in developing countries and for establishing new technologies and production modes based on energy efficiency, renewable power generation and low-carbon production pathways. Due to high unit costs, difficult management, lack of markets and infrastructure, new low-carbon products and services can only develop and emerge on a sustainable basis if they are supported by market-creation mechanisms, even if only on a temporary basis (WIR 2010). To overcome initial hurdles, economic incentives are required to compete with existing technologies that are more advanced in their life cycle. As the production costs of new pathways decrease over time and new products become attractive to more people, the need to support emerging markets declines and should eventually be abandoned to avoid market distortions.

34.9.2 Innovative Policy Portfolio Required

To realize the social and environmental benefits of technologies that will not yet diffuse commercially, developing countries can implement policies that combine fiscal and regulatory measures by lowering costs and stimulating demand, thus steering investments into a more desirable direction. Governments can take various measures and tools that will provide incentives for investment in sustainable energy technologies and become a catalyst for establishing new markets. There is a wide range of policies in place that support renewable energy around the world, including mandates and standards, innovation policies, carbon pricing, and others. These include encouraging investment in low-carbon and energy efficient production and transportation systems or accelerated depreciation of existing assets; public procurement of low-carbon products and technologies; energy performance standards or mandatory energy labelling schemes; renewable portfolio standards; blending mandates for alternative fuels; taxes for high-emission technology or tax incentives for green electricity; tendering procedures for green electricity or green certificate systems; and feed-in tariffs to guarantee a preferential price.

34.9.3 Incentive structures and market opportunities

Most of these measures have already been developed and applied in developed countries, in particular in the EU, which has considerable experience with various approaches and mixes (Karakosta/Doukas/Psarras 2010). While developed countries will continue to take the lead in market-creation mechanisms, their experience could be fed into technology support programmes in developing countries, who may wish to adopt market-creation policies in order to build local markets for certain low-carbon products and services. This could support their own export capacity and facilitate the introduction of technologies adapted to their development needs, such as rural electrification using renewable energy sources. Preferences for the various incentive policies vary by country, sector and technology.

While governments have a special role in creating incentive structures, the specific situation of developing countries has to be taken into consideration. Instruments such as 'Feed-in Tariffs' have been established in developed countries where governments have the financial resources to pay for the special treatment required by low-carbon technologies. Most developing countries, however, have limited financial means for setting up market creation programmes to match those of developed economies. This puts them at a disadvantage concerning the attraction of low-carbon foreign investment; it is therefore imperative for more advanced countries to take care not to undermine

efforts being made by poorer countries towards a transition to a low-carbon economy. Furthermore, home countries can help by actively promoting outward low-carbon foreign investment and by avoiding distortions in market mechanisms.

New market opportunities arising from changes in consumer behaviour in the main developed country markets should also be tapped, including markets for organic food, goods produced under responsible practices (fair trade, no child labour, fair treatment of workers), and low-carbon products. Appropriate policies can help protect and promote a host country's economic, social and other interests.

34.9.4 Community Involvement and Micro–Macro Transformations

Due to the failure of top-down and technology-centred approaches, it is now widely recognized that the involvement of community institutions is an essential contribution to environmental projects and is therefore an important factor in successful technology transfer. The involvement of local government agencies, consumer groups, industry associations and NGOs can help to ensure that the technologies being adopted within their particular country/region are consistent with their sustainable development goals. Besides the involvement of such community institutions, lessons from technology-intensive economies teach us that technology increasingly flows through private networks of information and assessment services, management consultants, financial firms, lawyers and accountants as well as through technical specialist groups. Governments can strengthen the growth of such networks for technology transfer (IPCC 2000). This includes a national dialogue of different stakeholders representing various private as well as public interests. Participatory approaches can engage private actors, public agencies, NGOs and grassroots organizations at all levels of environmental policy-making (Brinkerhoff 2004).

The moves towards decentralized and self-organized electricity generation and smart grids may challenge the power structures of large-scale energy producers, e.g. when residents install solar systems in their homes or citizen-driven cooperatives install their own wind power stations. When big utility companies are seeing their value drop substantially, this has been termed the "utility death spiral".[10] Three social chal-

---

10 See at: <http://www.whebgroup.com/the-utility-death-spiral>.

lenges are especially salient as far as the social management of technological systems is concerned: "(1) the size and visibility of transfers and assets created; (2) the predictability of pressure to expand the focus of the policies to broaden the social benefits; and (3) the potential for market incentives and framings of environmental issues to undermine normative motivational systems" (IPCC 2014).

If the micro-macro linkages between mutually impacting action levels are to be bridged, the multi-level interactions across social and technological elements (megatrends, sociotechnical regimes, niche level) need to be involved (Grin/Rotmans/Schot 2010; Geels

2011; Meadowcroft 2011; Foxon 2011; WBGU 2011). Know-how from forerunner actors needs to be tapped and made available at all levels from local to global (Hasselmann/Cremades/Filatova et al 2015). In this context there is a need for understanding policy-business relationships (Scheffran/Leimbach 2006) and organizational learning in technology-society interactions (Engels/Knoll/Huth 2008; Vieira 2015), in order to explain mechanisms such as emissions trading or the substitution of established technologies by new ones, which is at the core of sustainability transition towards the decarbonization of energy systems and society.

References

Accenture/Barclays, 2011: "Carbon Capital–Financing the low-carbon economy"; at: <http://nstore.accenture.com/Geneva/Studie_FINAL.pdf>.

AGF, 2010: Report of the Secretary-General's High-level Advisory Group on Climate Change Financing, 5 November 2010; at: <http://www.un.org/wcm/webdav/site/climatechange/shared/Documents/AGF_reports/AGF_Final_Report.pdf>.

Altmann, Jürgen; Scheffran, Jürgen, 1983: "Ist militärische Überlegenheit erreichbar?–Die neuen Rüstungstechnologien", in: Dürr, Hans-Peter; Harjes, Hans-Peter, Kreck, Matthias; Starlinger, Peter (eds.), Verantwortung für den Frieden- Naturwissenschaftler gegen Atomrüstung (Hamburg: Rowohlt): 138-154.

Archibugi, Daniele; Michie, Jonathan (Eds.), 1997: Technology, Globalisation and Economic Performance (Cambrdge: Cambridge University Press).

Bell, Martin, 2009: "Innovation Capabilities and Directions of Development", STEPS Working Papers. Vol. 33 (Brighton: STEPS Centre).

Bloomberg, 2014: Global Trends in Renewable Energy Investment 2014–Key Findings. Frankfurt School (Nairobi: UNEP Collaborating Centre for Climate & Sustainable Energy Finance. Bloomberg New Energy Finance).

Bolger, Joe, 2000: "Capacity Development: Why, What and How" (Gatineau, Quebec: Canadian International Development Agency); at: <http://portals.wi.wur.nl/files/docs/SPICAD/16.%20Why%20what%20and%20how%20of%20capacity%20development%20-%20CIDA.pdf>.

Brauch, Hans Günter; v.d. Graaf, Henny J.; Grin, John; Smit, Wim (Eds.), 1992: Controlling the Development and Spread of Military Technology (Amsterdam: VU University Press).

Brinkerhoff, Derick W., 2004. "The Enabling Environment for Implementing the Millennium Development Goals: Government Actions to Support NGOs"; Research Triangle Institute, Washington DC; at: <http://www.rti.org/pubs/Brinkerhoff_pub.pdf>.

Brooks, H., 1995: "What We Know and Do Not Know about Technology Transfer: Linking Knowledge to Action", in Marshalling Technology for Development, Proceedings of a symposium organised by the National Research Council and the World Bank, National Academy Press, Washington, DC: 83-96.

Byrne John; Zhou, Aiming; Shen, Bo; Hughes, Kristen, 2007: "Evaluating the potential of small scale renewable energy options to meet rural livelihoods needs: A GIS- and lifecycle cost-based assessment of Western China's options", in: Energy Policy, 35: 4391-4401.

Corfee-Morlot, Jan; Marchal, Virginie; Kauffmann, Céline; Kennedy, Christopher; Stewart, Fiona; Kaminker, Christopher; Ang, Geraldine, 2012: "Towards a Green Investment Policy Framework: The Case of Low-Carbon, Climate-Resilient Infrastructure" (Paris: OECD); at: <http://www.oecd-ilibrary.org/docserver/download/5k8zth7s6s6d.pdf?expires=1456137858&id=id&accname=guest&checksum=A4BE33CD49798D64B5954CD95D5EB2DF>.

EC (European Commission), 2014: "Impact Assessment: A Policy Framework for Climate and Energy in the Period from 2020 up to 2030"; at: <http://eur-lex.europa.eu/legal-content/EN/TXT/PDF/?uri=CELEX:52014DC0015&from=EN>.

EC (European Commission), 2015: "Shifting private finance towards climate-friendly investments" (Brussels: EC, DG Clima, A: CLIMA.A.2/ETU/2013/0035); at: <http://ec.europa.eu/clima/policies/finance/docs/climate-friendly_investments_en.pdf>.

Engels, Anita; Knoll, Lisa; Huth, Martin, 2008: "Preparing for the 'Real' Market: National Patterns of Institutional Learning and Company Behaviour in the European Emissions Trading Scheme (EU ETS)", in: European Environment, 18: 276-297.

Feldman, S., 1994: Market Transformation: Hot Topic or Hot Air? Proceedings of the 1994 ACEEE Summer Study on Energy Efficiency in Buildings (Washington DC: American Council for an Energy-Efficient Economy).

Foxon, Timothy J., 2011: "A coevolutionary framework for analysing a transition to a sustainable low carbon economy", in: *Ecological Economics*, 70: 2258-2267.

G7, 2015a: "G7 Hamburg Initiative for Sustainable Energy Security, G7 Energy Ministerial in Hamburg Communique, Hamburg, May 11/12, 2015"; at: <https://www.bmwi.de/BMWi/Redaktion/PDF/E/energieministertreffen-hambug-kommunique-englische-sprachversion,property=pdf,bereich=bmwi2012,sprache=de,rwb=true.pdf>.

G7, 2015b: "G-7 Leaders' Declaration, Schloss Elmau, Germany, June 8, 2015"; at: <https://www. whitehouse.gov/the-press-office/2015/06/08/g-7-leaders-declaration>.

Gallagher, Kelly Sims; Anadon, Laura Diaz; Kempener, Ruud; Wilson, Charlie, 2011: "Trends in investments in global energy research, development, and demonstration", in: *Wiley Interdisciplinary Reviews: Climate Change*, 2,3: 373-396.

Geels, Frank, 2011: "The multi-level perspective on sustainability transitions: responses to seven criticisms", in: *Journal of Environmental Innovation & Societal Transitions*, 1: 24-40.

GEF, 2012: Implementing the Poznan Strategic and Long-term Program on Technology Transfer. Global Environment Facility; at: <https://www.thegef.org/gef/sites/thegef.org/files/publication/GEF_PoznanTT_lowres%20final.pdf>.

Grin, John; Rotmans, Johan; Schot, Jan, 2010: *Transitions to Sustainable Development. New Directions in the Study of Long Term Transformative Change* (London: Routledge).

Grubb, Michael, 2004: "Technology innovation and climate change policy: An overview of issues and options", in: *Keio economic studies*, 41,2; at <http://seg.fsu.edu/Library/Technology%20Innovation%20and%20Climate%20Policy_%20An%20Overview%20of%20Issues%20and%20Options.pdf>

Hamelinck, Carlo, 2013: "Biofuels and food security—Risks and opportunities" (Utrecht: ECOFYS); at: <http://www.ecofys.com/files/files/ecofys-2013-biofuels-and-food-security.pdf>.

Hasselmann, Klaus; Cremades, Roger; Filatova, Tatiana; Hewitt, Richard; Jaeger, Carlo; Kovalevsky, Dmitry; Voinov, Alexey; Winder, Nick, 2015: "*Free-riders to forerunners. Commentary*", in: *Nature Geoscience*, Advance Online Publication: 1-4.

Heaton, George; Repetto, Robert; Sobin, Rodney, 1991: *Transforming Technology: An Agenda for Environmentally Sustainable Growth in the 21st Century* (Washington DC: World Resources Institute).

ICTSD, 2012: "Ways to promote enabling environment and to address barriers to technology development and transfer", International Centre for Trade and Sustainable Development (ICTSD); at: <http://unfccc.int/ttclear/misc_/StaticFiles/gnwoerk_static/TEM_tec_cfi_ee/7843d4ba5e5e459c99deb4e47b972e83/e5651a5965e0407aae35e7db2e0cae32.pdf>.

IEA, 2008: *Energy Technology Perspectives: Scenarios and Strategies to 2050* (Paris: IEA, OECD).

IEA, 2014: "http://www.iea.org/publications/freepublications/publication/TechnologyRoadmapSolarThermalElectricity_2014edition.pdf>.

IPCC (Intergovernmental Panel on Climate Change), 2000: *Methodological and technological issues in technology transfer* (Cambridge: Cambridge University Press).

IPCC, 2012: *Renewable Energy Sources and Climate Change Mitigation*, Special Report of the Intergovernmental Panel on Climate Change (Cambridge: Cambridge University Press).

IPCC (Intergovernmental Panel on Climate Change), 2014: "Summary for Policymakers", in: IPCC: *Climate Change 2014: Mitigation of Climate Change. Contribution of Working Group III to the Fifth Assessment Report* (Cambridge: Cambridge University Press).

Ipsen, Dirk; Rösch, Roland; Scheffran, Jürgen, 2001: "Cooperation in Global Climate Policy: Potentialities and Limitations", in: *Energy Policy*, 29,4 (January): 315-326.

Jänicke, Martin, 2012: "Dynamic governance of clean-energy markets: how technical innovation could accelerate climate policies", in: *Journal of Cleaner Production*, 22: 50-59.

Justman, Morris; Teubal, Moshe, 1995: "Technological infrastructure policy (TIP): Creating capabilities and building markets", in: *Research Policy*, 24,2: 259-281.

Karakosta, Charikleia; Doukas, Hans; Psarras. John, 2010: "Technology transfer through climate change: Setting a sustainable energy pattern", in: *Renewable and Sustainable Energy Reviews*, 14: 1546-1557.

Kebir, Noara; Philipp, Daniel; Scheffran, Jürgen; Stürmer, Martin, 2005: "Grameen Shakti: An alternative approach to financing energy in rural Bangladesh", in: von Weizsäcker, Ernst Ulrich; Young, Oran R.; Finger, M. (Eds.) *Limits to privatization: how to avoid too much of a good thing*, Report to the Club of Rome (London: Earthscan): page 323.

Klawitter, Jens; Schilling, Janpeter; Scheffran, Jürgen, 2010: "Moving the Desertec Concept Towards Sustainability", DGAP-JREDS Conference "*The Impact of Climate Change on the Middle East*", Amman, Jordan, 25-28 November 2010; at: <http://clisec.zmaw.de/fileadmin/user_upload/fks/publications/conference-papers/Klawitter-et-al-2010_Amman.pdf>.

Lema, Rasmus; Iizuka, Michiko; Walz; Rainer, 2015: "Introduction to low-carbon innovation and development: insights and future challenges for research", in: *Innovation and Development*, 5,2: 173-187.

Lewis, Joanna I., 2014a: "The Rise of Renewable Energy Protectionism: Emerging Trade Conflicts and Implications for Low Carbon Development", in: *Global Environmental Politics*, 14,4: 10-35.

Lewis, Joanna I., 2014b: "Managing intellectual property rights in cross-border clean energy collaboration: The case of the US–China Clean Energy Research Center", in: *Energy Policy*, 69: 546-554.

Lewis, Joanna I., 2015: "A comprehensive look at technology transfer", in: *Climate Policy*, 15,1: 177-179.

Liebert, Wolfgang; Rilling, Rainer; Scheffran, Jürgen (Eds.), 1994: *Die Janusköpfigkeit von Forschung und Technik. Zum Problem der zivil-militärischen Ambivalenz* (Marburg; BdWi-Verlag).

Mallett, Alexandra; Ockwell, David G.; Pal, Prosanto; Kumar, Amit; Abbi, Y.P.; Haum, Rüdiger; MacKerron, Gordon; Watson, Jim; Sethi, Girish, 2009: *UK–India Collaborative Study on the Transfer of Low-Carbon Technology: Phase II Final Report. Report by SPRU and TERI* (New Delhi: Teri).

Mallett, Alexandra, 2013: "Technology cooperation for sustainable energy: a review of pathways", in: *Wiley Interdisciplinary Reviews: Energy and Environment*, 2,2: 234-250.

Meadowcroft, James, 2011: "Engaging with the politics of sustainability transitions", in: *Environmental Innovation and Societal Transitions*, 1: 70-75.

Moon, Jeremy, 2014: *Corporate Social Responsibility–A Very Short Introduction* (Oxford: Oxford University Press).

Mugabe, John, 1996: *Technology and Sustainable Development in Africa–Building policy and institutional capacities for needs assessment. Background paper* (New York, NY: UN, Division for Policy Co-ordination and Sustainable Development).

Ockwell, David G.; Haum, Ruediger; Mallett, Alexandra; Watson, Jim, 2010: "Intellectual property rights and low carbon technology transfer: Conflicting discourses of diffusion and development", in: *Global Environmental Change*, 20,4 (October): 729-738.

Ockwell, David G.; Mallet, Alexandra, 2012: *Low-carbon technology transfer: from rhetoric to reality* (London: Routledge).

OECD, 2015: "Public Interventions and Private Climate Finance Flows: Empirical Evidence from Renewable Energy Financing"; at: <http://www.oecd-ilibrary.org/docserver/download/5js6b1r9lfd4.pdf?expires=1456139240&id=id&accname=guest&checksum=A2543CCEDoECB63F2185300F096B8069>.

REN21, 2014: *Renewables 2014–Global Status Report. Renewable Energy Policy Network for the 21st Century* (Paris: UNEP, REN21).

Saad, Mohammed; Zawdie, Girma, 2005: "From technology transfer to the emergence of a triple helix culture: The experience of Algeria in innovation and technological capability development", in: *Technology Analysis & Strategic Management*, 17: 89-103.

Santarius, Tilman; Scheffran, Jürgen; Tricarico, Antonio, 2012: *North South Transitions to Green Economies–Making Export Support, Technology Transfer, and Foreign Direct Investments Work for Climate Protection* (Berlin: Heinrich Böll Foundation); at: <http://www.santarius.de/wp-content/uploads/2012/08/North-South-Transitions-to-Green-Economies-2012.pdf>.

Scheffran, Jürgen, 1986: "Der Streit um die Hochtechnologieförderung–Kriterien zur Bewertung", in: *Blätter für deutsche und internationale Politik*, 2: 214-228.

Scheffran, Jürgen; Vydra, Jan, 1991: *The Application of Military-Related Resources to Protect the Environment: The Case of Information Technologies*, IANUS Report 3/1991. Report commissioned for the UN Department for Disarmament Affairs.

Scheffran, Jürgen; Karp, Aaron. 1992: "The National Implementation of the Missile Technology Control Regime. The US and German Experiences", in: Brauch, Hans Günter; v. d. Graaf, Henny J.; Grin, John; Smit, Wim (Eds.): *Controlling the Development and Spread of Military Technology* (Amsterdam: VU University Press): 235-255.

Scheffran, Jürgen, 1992a: "Panzer gegen die ökologische Krise?", in: *Spektrum der Wissenschaft,* 10 (October): 128-132.

Scheffran, Jürgen, 1992b: "*Environmental Applications of Military Information and Communication Technologies*", in: Brunn, Anke; Baehr, Lutz; Karpe, Hans-Jürgen (Eds.): *Conversion: Opportunities for Development and Environment* (Berlin–Heidelberg: Springer-Verlag): 122-133.

Scheffran, Jürgen; Liebert, Wolfgang, 1994: "*Ambivalence of Science and Dual-Use of Technology Transfer*", in: Rotblat, Joe (Ed.) *Shaping Our Common Future: Dangers and Opportunities, Proceedings of the 42nd Pugwash Conference on Science and World Affairs, World Scientific* (London: Pugwash): 1117-1134.

Scheffran, Jürgen, 2001: "Peaceful and Sustainable Use of Space–Principles and Criteria", in: Bender, Wolfgang; Hagen, Regina; Kalinowski, Martin; Scheffran, Jürgen (Eds.): *Space Use and Ethics* (Münster: Agenda-Verlag): 49-80.

Scheffran, Jürgen, 2002: "Kein Frieden auf dem Erdgipfel?", in: *Wissenschaft und Frieden*, 20,3: 44-48.

Scheffran, Jürgen; Leimbach, Marian, 2006: "Policy-Business Interaction in Emission Trading between Multiple Regions", in: Antes, Ralf; Hansjürgens, Bernd; Letmathe, Peter (Eds.): *Emissions Trading and Business* (city: Physica-Verlag): 353-367.

Scheffran, Jürgen; Summerfield, G., (Eds.), 2009: "Sustainable Biofuels and Human Security", in: *Swords & Ploughshares*, 27,2: 4-10.

Scheffran, Jürgen, 2013: "The Diffusion of Innovations for the Sustainability Transition", Working Paper for presentation at ISA Annual Convention, San Francisco, CA, 3-6 April 2013; Workshop and Panel on "Sustainability Transition".

Scheffran, Jürgen; Cannaday; Thomas, 2013: "*Resistance against Climate Change Policies: The Conflict Potential of Non-Fossil Energy Paths and Climate Engineering*", in: Maas, Achim; Bodó, Balázs; Comardicea, Irina; Roffey, Roger (Eds.): *Global Environmental Change: New Drivers for Resistance, Crime and Terrorism?* (Baden-Baden: Nomos): 261-292.

Schinke, Boris; Klawitter, Jens, 2011: "Desertec and Human Development at the Local Level in the MENA-Region: A human rights-based and sustainable livelihoods analysis", in: *Germanwatch*, 10/2011; at: <https://germanwatch.org/en/download/6439.pdf>.

Schnepp, O.; von Glinow, M.A.; Bhambri, A., 1990: *United States-China Technology Transfer* (New Jersey: Prentice Hall).

Seres, Stephen; Haites, Erik, 2008: "Analysis of Technology Transfer in CDM Projects", Prepared for the UNFCCC Registration & Issuance Unit CDM/SDM (Toronto, Canada: UNFCCC, December).

Stadelmann, M.; Michaelowa, A., 2011: "How to Enable the Private Sector to Mitigate?", Climate Strategies Working Paper (London, UK: Publisher).

Stadelmann, M., 2013: "The effectiveness of international climate finance in enabling low-carbon development: Comparing public finance and carbon markets", University of Zurich, Zurich; at: < http://opac.nebis.ch/ediss/20131748.pdf>.

Teubal, Martin; Yinnon, Tamar: Zuscovitch, Ehud, 1991: "Networks and Market Creation", in: *Research Policy*, 20: 381-392.

Tomlinson, Shane; Zorlu, Pelin; Langley, Claire, 2008: *Innovation and technology transfer—Framework for a global climate deal*. An E3G report with contributions from Chatham House (London: E3G, November).

Trindade, Sergio C., 1991: "Managing Technological Change in Support of the Climate Change Convention: Framework for Decision-Making." Methodological and Technological Issues, In Technology Transfer: IPCC Special Report on Climate Change. 1.3 1991; at <http://www.grida.no/climate/ipcc/tectran/007.htm>.

UN, 1991: "Potential Uses of Military-Related Resources for Protection of the Environment", Report of the Secretary General, Office for Disarmament Affairs New York; at: <http://www.un.org/disarmament/HomePage/ODA Publications/DisarmamentStudySeries/PDF/SS-25.pdf>.

UN, 1993: *Agenda 21: Program of Action for Sustainable Development* (New York: United Nations).

UNCTAD (United Nations Conference on Trade and Development), 1998: "Report of the Expert Meeting on the Impact of Government Policy and Government/Private Action in Stimulating Inter-Firm Partnerships Regarding Technology, Production and Marketing, United Nations Conference on Trade and Development (Geneva: UNCTAD); at: <http://unctad.org/en/Docs/c3em4d3.en.pdf>.

UNCTAD, 2010: *Trade and Environment Review 2009/10: Promoting Poles of Clean Growth to Foster the Transition to a More Sustainable Energy* (Geneva: UNCTAD).

UNCTAD, 2012: *State of South-South and Triangular Cooperation in the Production, Use and Trade of Sustainable Biofuels* (Geneva: UNCTAD); at: <http://unctad.org/en/PublicationsLibrary/ditcted2011d10_en.pdf>.

UNDP (United Nations Development Programme), 2006: *Beyond scarcity: Power, poverty and the global water crisis. Human Development Report* 2006 (New York: UNDP).

UNDP, 2010: *Handbook for Conducting Technology Needs Assessment for Climate Change* (New York: United Nations Development Programme, November).

UNEP (United Nations Environment Programme), 2009: *Global Green New Deal*, Policy Brief of the UNEP (Geneva: UNEP, March).

UNFCCC, 2001 (United Nations Framework on Climate Change): "Report of the Conference of the Parties on Its Seventh Session", UNFCC, Marrakesh; at: <unfccc.int/resource/docs/cop7/13a01.pdf>.

UNFCCC, 2009: *Advance report on recommendations on future financing options for enhancing the development, deployment, diffusion and transfer of technologies under the Convention*. Note by the Chair of the Expert Group on Technology Transfer, 2F5C CMCa/rSchB/22000099/INF.2 (Bonn: UNFCC).

UNFCCC, 2015: *Report of the Conference of the Parties on its twentieth session, held in Lima from 1 to 14 December 2014* (Lima: UNFCCC).

Vieira do Nascimento, Daniele Maria, 2015: "Interorganisational Situated Learning in Brazil. An Analysis of the Diffusion of the Brazilian Flex-Fuel Vehicle Mitigation Technology". PhD thesis, Universität Hamburg, October 2015.

Watson Jim; Sauter, Raphael, 2011: "Sustainable innovation through leapfrogging: a review of the evidence", in: *International Journal of Technology and Globalisation*, 5, 170-189.

Watson, Jim, Byrne; Rob, Mallett, Alexandra; Stua, Michele: Ockwell, David; Xiliang, Zhang; Da, Zhang; Tianhou, Zhang; Xiaofend, Xunmin, O., 2011: *UK–China collaborative study on low carbon technology transfer* (Brighton: University of Sussex & Tsinghua University).

Watson, Jim; Byrne, Rob; Ockwell, David; Stua, Michele, 2014: "Lessons from China: building technological capabilities for low carbon technology transfer and development", in: *Climatic Change*: 1-13.

WBCSD (World Business Council for Sustainable Development), 2007: Investing in a Low-Carbon Energy Future in the Developing World (Geneva: WBCSD).

WBCSD (World Business Council for Sustainable Development), 2010: *Enabling frameworks for technology diffusion: A business perspective* (Geneva: WBCSD).

WBGU (German Advisory Council on Global Change), 2009: *Solving the climate dilemma: The budget approach*. Special Report (Berlin: WBGU).

WGBU, 2011: *A Social Contract for Sustainability* (Berlin: WBGU).

WGBU, 2014: *Climate Protection as a World Citizen Movement*. Special Report (Berlin: WBGU).

WIR, 2010: *World Investment Report 2010: Investing in a Low-Carbon Economy* (Geneva: UNCTAD).

WIR, 2014: *World Investment Report 2014—Investing in the SDGs: An Action Plan* (Geneva: UNCTAD).

World Bank, 2010: *World View—2010 World Development Indicators* (Washington DC: World Bank).

35 Considering a Structural Adjustment Approach to the Low Carbon Transition

Karlson 'Charlie' Hargroves[1]

Abstract

As the world comes to grips with the need to significantly reduce greenhouse gas emissions, a range of questions are being asked about how to effectively transition economies to low carbon operation over the coming decades. A growing number of pressures are now being felt across a range of sectors to reduce emissions, in particular carbon-related fuel consumption, which is leading to autonomous emissions reduction efforts—typically ad hoc and business-led. However, in order to meet ambitious targets for the reduction of greenhouse gas emissions now set by the world's largest economies a structural adjustment approach may be needed that is effectively underpinned and appropriately expedited at an economy-wide level. This chapter presents an introduction to key lessons from structural adjustment programmes to inform the low carbon transition, and in the absence of conditional lending requirements that have driven structural adjustment programmes the chapter considers how the willingness to adjust structures of the economy to deliver low carbon outcomes can be increased.

Keywords: Structural adjustment, low carbon transition, greenhouse gas emissions, low carbon, carbon structural adjustment.

35.1 Introduction[2]

In the twenty-first century much of the world will experience untold wealth and prosperity that could not be conceived of just a hundred years ago. Much of

---

1 Karlson Hargroves, Senior Research Fellow, Curtin University Sustainability Policy Institute, Curtin University, Perth, Australia; Email: <charlie.hargroves@curtin.edu.au>.

2 I would like to acknowledge and thank my supervisor, mentor, and friend, Professor Peter Newman, both for his extensive contribution to the field of sustainable development internationally, and for his support and patience over the last seven years while I undertook the research to support this chapter. I would also like to acknowledge Dr Cheryl Desha, a long time research collaborator who has brought a great deal of enthusiasm, expertise, and professionalism to the projects that we have worked on together over the last decade that have informed and influenced this chapter. I would also like to thank Professor Frank Geels for his leadership in the field of sustainability transitions and for his mentoring and peer review of my works in this area. I would also like to thank the Sustainability Ninja's for assisting with the publication tasks.

this prosperity has been based on rapid industrialization, mechanization, electrification, and digitization, beginning in the mid-1700s with the Industrial Revolution, as shown in figure 35.1. However as with many of the human civilizations over the last 5,000 years the rapid growth in prosperity has accumulated significant environmental damage that now threatens to result in what esteemed sustainable development expert Lester Brown refers to as *'environmentally induced economic decline'* (Brown 2008).

Governments around the world will face a range of serious global environmental challenges in the twenty-first century, with one of the most pressing due to the fact that the fuel source that has underpinned the industrial revolution and allowed the staggering amounts of development to be undertaken in much of the world, namely fossil fuels, is now widely recognized to have a sinister legacy (Hargroves/Smith 2005; Brown 2008; Stern 2005; OECD 2008). The combustion of fossil fuels such as oil and coal has resulted in the generation of vast quantities of air pollution, a proportion of which that when combined with specific agricultural and industrial emissions are

Figure 35.1: Waves of Innovation. **Source**: Hargroves/Smith (2005).

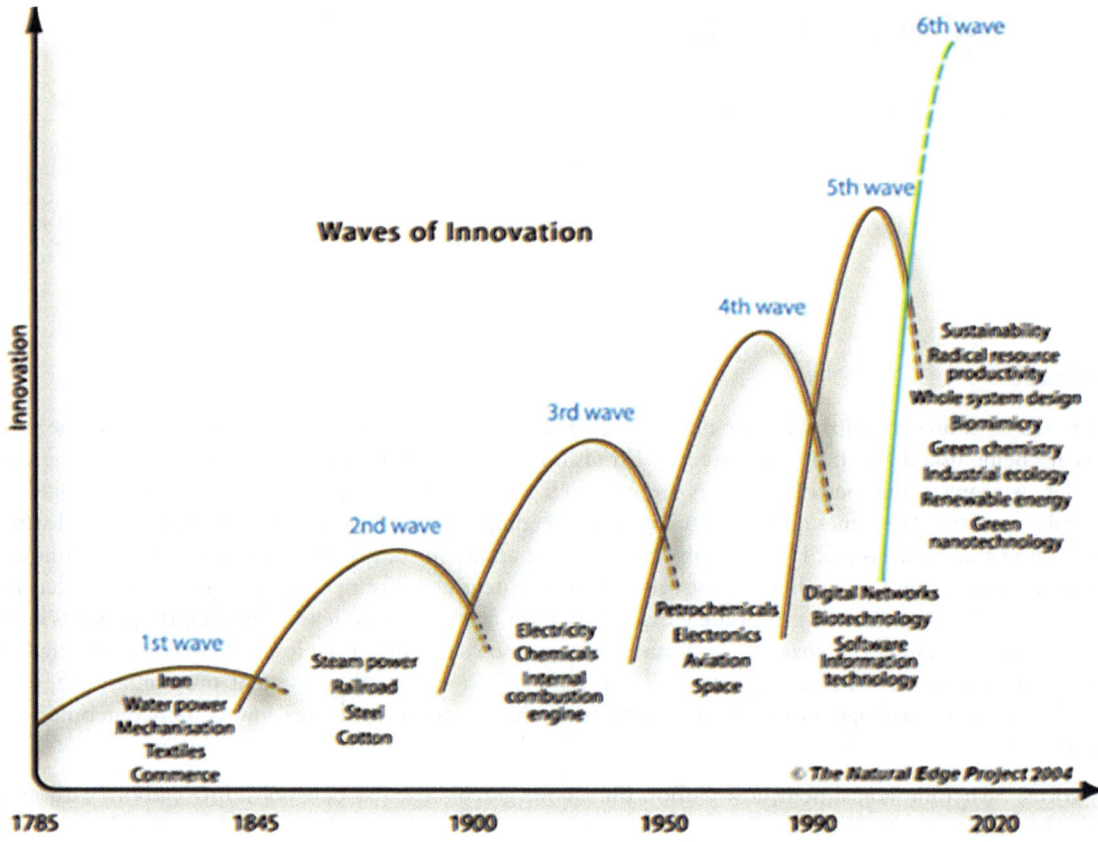

referred to as *'greenhouse gases'* as they increase the atmosphere's capacity to store and release heat. Such emissions are in quantities that now threaten to impact the global heat transfer processes and leading to increases in the average global temperature that can have devastating effects on the planet's biosphere and in turn on our vast developments and settlements. In his 2006 book, *Heat*, George Monbiot laments that "Ours are the most fortunate generations that have ever lived. Ours might also be the most fortunate that ever will" (Monbiot 2006).

Governments around the world are now understanding the threat posed by human-induced climate change and the associated imperative to respond with the world's largest economies setting ambitious greenhouse gas emissions targets. By the end of 2014 the European Union had committed itself to reducing emissions by at least 40 per cent by 2030 (compared to 1990 levels), China committed to 40-45 per cent by 2020 (compared to 2005 levels), India committed to 20-25 per cent by 2020 (compared to 2005 levels), and the United States committed to 26-28 per cent by 2025 (compared to 2005 levels). These ambitious targets will create significant pressure to reduce emis-

sions in the coming decades in a manner that delivers ongoing prosperity, jobs, and profits. The challenge for nations to achieve such targets will be to tackle the temporal issue with the economic impacts of climate change standing to hit hardest in the future (Stern 2008), while political attention and business strategy is firmly focused on performance in the very short-term (Hargroves/Smith 2005). Balancing this mismatch between short- and long-term imperatives will be a crucial aspect of the low carbon transition, one that if not appropriately considered may well lead to significant impacts on economies.

This chapter considers the notion that the scale and pace of economy-wide changes required to significantly reduce greenhouse gas emissions are akin to the results of the application of *structural adjustment programmes* (SAPs) in developing countries in order to improve economic performance as part of conditional lending requirements by financial institutions such as the *International Monetary Fund* (IMF) and the World Bank. According to the IMF, structural adjustment is focused on "changing the way in which an economy is organized in order to raise productive capacity".[3] It may in fact be the case that in the same way that fiscal structural adjustment was implemented largely as a solution to underdevelopment in third world economies in the twentieth century, a new form of structural adjustment, that of *Carbon Structural Adjustment* (CSA), may be needed as a solution to unsustainable development around the world in the twenty-first century. Carbon Structural Adjustment is defined here as "an agenda of swift adjustments to structures across economies to transition to low carbon operation".

A key point of difference however is that unlike SAPs that were enforced as part of the conditions of development loans, Carbon Structural Adjustment programmes will need to be driven by national governments as an economic development strategy. Hence the question considered in this chapter will be, "How can lessons from structural adjustment programmes inform the low carbon transition, and how can a strategic approach be developed to deliver carbon structural adjustment?"

In response to these questions the theoretical approach used in the chapter is based on both 'quantitative' and 'qualitative' approaches; however, the chapter assumes that the most appropriate ontological approach to the questions above is to consider them as largely 'subjective', meaning that it tends to be more qualitative, informed by quantitative aspects including survey and focus group data. Considering the potential for axiological assumptions, although the author has personal values that align to achieving significant reductions in greenhouse gas emissions globally, this bias was separated from the analysis of the literature, surveys and focus groups, and discussions with experts, to allow an 'unbiased and value-free' approach. The chapter takes the methodological approach to the research questions of considering them to be mostly 'deductive' with an 'inductive' component. This means that when considering the various literature and findings the main focus was on a deductive process to consider the 'cause and effect' of various actions and interventions to inform the development of generalizations of the overlying context. Given the growing international interest in achieving a low carbon transition there has been an increasing level of materials released by industry groups, companies, NGOs and other sources that have been used to inform the chapter rather than a purely academic literature approach.

---

3 International Monetary Fund, at: <http://www.imf.org/external/np/exr/glossary/showTerm.asp#136> (13 January 2015).

35.2 Considering the Economics of the Low Carbon Transition

In the 1700s if it were possible to comprehend the future impacts of a fossil-fuel-based system it may have been feasible to take swift action to reduce such pollution with a global population of less than one billion people. However, today in a world with some seven billion people the potential for large-scale change in the basis of the energy system is a seemingly overwhelming task, especially considering that the tools and strategies that have led to the highly successful fossil-fuel-based economy may not be as useful to transition to low carbon operation, and new tools and strategies may be needed. Ironically it may also prove to be the case that without having first harnessed fossil energy to develop industries it would not have been possible to develop the technology required for a transition to large-scale non-fossil-based energy production (such as through solar photovoltaic panels, solar thermal, wind turbines, ocean turbines, fuel cells, geothermal, co-generation equipment, etc.).

Economists have pondered for many years the thought that perhaps if the world stopped focusing on continually increasing economic growth then the unsustainable growth in demands on the environment would slow to a point where it would be sustainable without changing much about the way the economies operate. The challenge is that most of the national economies of the world are built on the prediction that there will be strong economic growth, and further that in its absence the economy will be bailed out by government, as in the 2008 global financial crisis. This assumption underpins many economies, which in reality are an incredibly complex system that for various conditions in various countries leads to a balancing point that allows the economy to run. This fragile balance sets the level of taxation, the price of commodities and products and the spending and saving habits of citizens, and affects levels of investment, insurance, interest rates, public spending, private development, education, research, etc. *The reality is that on any given day the complexity of the average economy is in itself overwhelming, and trying to incorporate future impacts from environmental pressures is an even higher order level of complexity—one that will not be achieved by reducing the level of economic growth alone, or by taking isolated actions in parts of the economy.*

We are now at a point in the development of humankind that offers the most sophisticated and technologically advanced platform upon which to

Figure 35.2: Gross Domestic Product vs. Estimated Environmental Costs (billions) for the United States of America from 1950 to 2004. **Sources:** Smith, Hargroves and Desha (2010), with data reinterpreted by K. Hargroves from Talberth, Cobb and Slattery (2006).

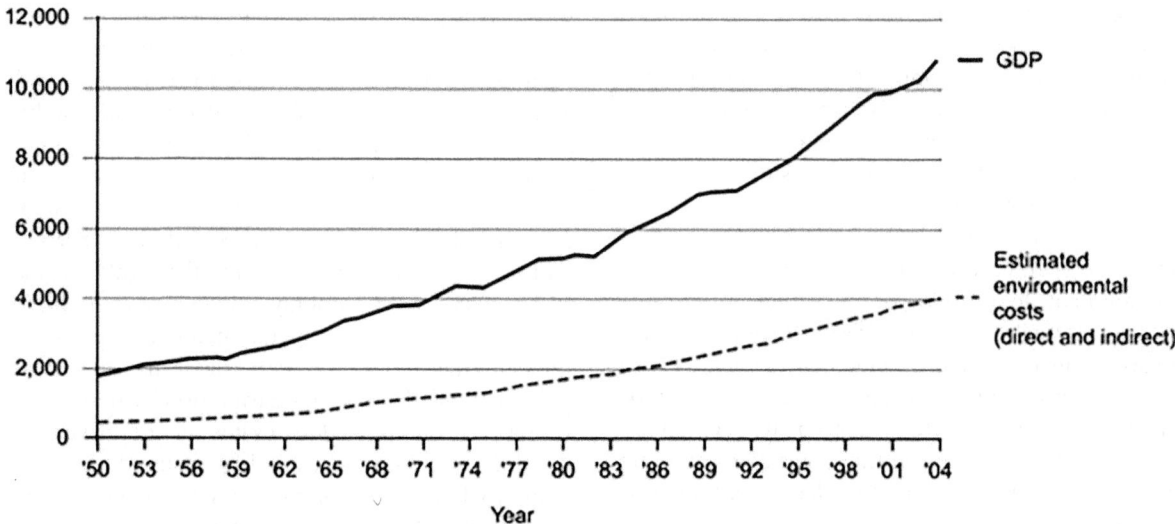

tackle large-scale problems, one that has the potential to underpin a transition in the coming decades to a global society that is able to sustain its preferred living conditions (von Weizsaecker/Hargroves/Smith et al. 2009). However, to achieve such a future this platform needs to be harnessed to address the significant complexities involved in reducing greenhouse gas emissions across economies. Further, the literature shows that it is crucial to understand that even if the impacts of global warming alone may be significantly mitigated or adapted to, when combined with deforestation, the expansion of freshwater intensive modern agriculture and ocean fisheries, and increasing urban waste streams, the overall environmental pressures may still be too much for ecosystems to handle (Brown 2008; OECD 2008).

The longer action is delayed on a meaningful scale the higher the resulting costs from the environmental damage will be to current and future generations, adding to the already high cost of environment-related impacts to the economy. For instance figure 35.2 shows that while US GDP has risen steadily since 1950, the costs related to environmental pressures have also risen, effectively offsetting a growing proportion of GDP rising from 22 per cent in 1950 to 37 per cent in 2004. These costs included household pollution abatement, water pollution, air pollution, loss of wetlands, loss of farmland, loss of primary forests, resource depletion, carbon dioxide emissions damage, and ozone depletion (Talberth/Cobb/Slattery 2006). This simplified example is complicated by a number of factors related to measurement and assumptions

regarding the distribution of costs; however, it provides a clear picture of the trends associated with economic growth and costs related to environmental pressures (Smith/Hargroves/Desha 2010).

Considering the costs from climate change in particular, the *Stern Review* estimated that "the total cost of business-as-usual climate change to equate to an average reduction in global per capita consumption of 5 per cent at a minimum now and forever" (Stern 2006). The *Stern Review* also describes that this cost would increase to as much as twenty per cent per year if the modelling was to take into account the impacts related to human health, various amplifying effects, and the disproportionate burden of climate change on the poor and vulnerable globally. Thus to pay for such costs governments in the future will need to raise taxes, privatize public assets, borrow money, or pass it on to the market. Some argue that future generations will be wealthier and more able to pay, but this assumes that the costs of inaction will not rise, and further that if these costs rise it will be a predictable and manageable increase (Smith/Hargroves/Desha 2010). However, the *Stern Review* is clear that due to the complexity of natural systems and the potential for feedbacks and amplification effects the costs on inaction will continue to rise unless action is taken (Stern 2006).

Those moving early to respond to this challenge are learning that there are in fact a number of ways that reducing pressure on natural systems can underpin continued economic growth—which is possibly the most important finding, or 'convenient truth', of the twenty-first century (von Weizsäcker/A/B et al. 2010). However, the time frame for action is short and the consequences for inaction will be long-lasting (Stern 2006). In order to ensure that the most cost-effective, efficient, and successful transition to low carbon operation is achieved it is crucial that a range of adjustments are made to the various structures of the world's economies—and further that efforts from societies around the world are monitored and quickly learned from to inform further efforts. A number of countries have developed 'National Strategies for Sustainable Development', and although they are largely symbolic this is an encouraging sign of a focus on national-level strategies and frameworks, specifically Cameroon, Canada, Denmark, Finland, France, Germany, Greece, Ireland, Latvia, Mauritius, Mexico, Moldova, Netherlands, Norway, Philippines, South Korea, Sweden, Switzerland, the Czech Republic, the United Kingdom, and the European Union.

Changing an industrial economy to run on non-fossil energy will require significant investments in new energy technology for generation and distribution—such as building vast solar panel arrays, wind farms and ocean turbines, while upgrading electricity distribution systems to be able to manage a network of small-scale energy generation—together with reducing the energy intensity of a range of operations to both reduce overall demand for energy and allow greater levels of economic growth, particularly in developing countries. According to the Pathways to Deep Decarbonization Project, "Directed technological change should not be conceived as picking winners, but as making sure the market has enough winners to pick from to achieve cost-effective low-carbon outcomes" (SDSN/IDDRI 2014).

35.3   Achieving Low Carbon Economic Multipliers

The key to achieving carbon structural adjustment will be the effective implementation of strategies that reduce greenhouse gas emissions in ways that deliver economic multipliers and support economic growth. For instance, investing in improvements to the efficiency and/or productivity of energy, water and other resources can deliver flow on benefits such as reduced input costs, reduced running and maintenance costs, and reduced waste-related costs, and hence can be recovered over a reasonable time frame to then deliver ongoing cost savings that can then be invested to achieve even greater resource productivity improve-

ments. Furthermore, as such investments can lead to reduced levels of consumption of resources, such as water and electricity, this can lead to delays in, or even the avoidance of, costly investments in increasing the capacity of energy and water supply infrastructure, as well as plant and infrastructure in extractive industries (Smith/Hargroves/Desha 2010). Typically such investments are at a local level and can spur jobs growth and economic development, attracting companies and operations keen to be part of such initiatives.

There is strong evidence to show that significant reductions in the energy consumption of our economies can not only be achieved, but can be achieved cost-effectively and in a timely manner. Fortunately unlike some thirty years ago when many of the campaigners for responding to climate change began their important work, such as Amory Lovins (Hawken/Lovins/Lovins 2009), Ernst von Weizsäcker (von Weizsäcker/Lovins/Lovins 1997), Hunter Lovins (Lovins/Cohen 2012), Jim McNeill (World Commission on Environment and Development 1987), Bill McDonough (McDonough/Braungart 2002), Lester Brown (Brown 2008), and others, there are now a vast number of examples of leading practice showing that significant reductions in greenhouse gas emissions can be achieved profitably, such as shown in table 1.1 (von Weizsäcker/Hargroves/Smith et al. 2009). As the *International Energy Agency* (IEA) reported in 2006, "Many processes have very low energy-efficiency and average energy use is much higher than the best available technology would permit" (IEA 2006).

A critical barrier to the achievement of improvements in energy productivity is that the designers, engineers, architects and technicians of today are not versed in a systems-based approach to design (Stasinopoulos/Smith/Hargroves/Desha 2009). The reality that the engineering and design professions now face is that even with significant advances being made by designers across the world, the shift from an incremental approach to a systems approach is in its early stages. This is a problem as our amazing Earth cannot wait for many years of trial and error, research and debate, publications and textbooks, before steps are taken to significantly reduce greenhouse gas emissions. Hence we need to learn from what is being done around the world and rapidly bring this knowledge together. In researching evidence to support the feasibility of carbon structural adjustment it has become clear that in each of the successful examples of such improvements a common set of questions has been considered, namely (von Weizsäcker/Hargroves/Smith et al. 2009):

- Is the current method of delivering the product or service the only way to do so? (Often the first thought when answering this question is 'yes'; however, further investigation often leads to a range of alternatives–from system upgrades, such as energy-efficient motors in an industrial application, to completely new processes, such as shifting to a process to predominantly use scrap metal rather than processing primary resources to make steel.)
- If it is the only way, what are the major areas of energy, water and materials usage, and a) what options are available to reduce the need for such inputs, and b) what alternatives are available to provide these inputs? (The search for such alternative options and inputs can be driven by a requirement to reduce environmental impacts, but also be part of a strategy to improve competitive advantage by reducing input costs, which are inevitably set to increase in the future as availability and impact are factored in.)
- If it is not the only way, what alternatives are there to the system that can be used to profitably deliver the product or service with less resource intensity and environmental pressure? (For instance, geopolymers can be used to create cement with as much as eighty per cent less energy intensity, partly by eliminating the process emissions of greenhouse gases associated with Portland cement production.)

When considering strategies to support such a pragmatic enquiry it is important to understand that once the initial questions as to the best way to meet the design requirement have been answered the conceived system needs to be benchmarked against best practice in order to understand the potential for performance. However, in many cases the new design concept will be part of an emerging wave of innovation and hence there may be little precedent to provide a benchmark. Further, even if there are established examples of the new design, such processes and methodologies are unlikely to have had time to be incorporated into higher education or professional development courses. It is of crucial importance that the feasibility of widespread implementation of low-carbon technologies and processes is established across all sectors of the economy.

It is not possible to consider strategies to reduce greenhouse gas emissions without considering the potential for such achievements to be overwhelmed by growth, referred to as the 'Rebound Effect'. This area of consideration is largely inspired by the 1865 book *The Coal Question* that found that the result of improving the efficiency of coal engines did not lead to an overall reduction in coal consumption, as many assumed it would, but rather a vast increase (Jevons 1865). The book pointed out that those who anticipated that a reduction in the amount of coal needed to produce a ton of iron by over two-thirds in Scotland would lead to a significant reduction in coal demand were mistaken. In reality the lower coal requirement to make iron meant that iron could be made more cheaply, and in a rapidly growing economy that could always use more iron, the overall consumption of coal grew tenfold between the years 1830 and 1863. Applying this logic to modern energy demand reduction programmes and initiatives, many consider these rebound effects to be significant, but they need not be.

Counteracting the rebound effect is based on the observation that, when a more efficient technology, process, or product is introduced into the market, it does not automatically imply that due to the lower consumption of the particular product there will be a reduction in resource consumption overall. This is due to the fact that as the product is more efficient it may be cheaper to purchase and to run, leading to the potential for more people being able to afford to use the product, or to use it more often. In short, as one would expect, a single intervention into a complex system will not necessarily lead to a preferred change in that system. This is why significant economy-wide reductions in greenhouse gas emissions cannot be achieved by a focus on the market-driven uptake of low carbon technologies and process alone, but must be part of an agenda to adjust structures across entire economies.

35.4 Key Aspects and Impacts of Structural Adjustment Programmes

Fundamentally the term 'Structural Adjustment' can be interpreted as the process of making system-wide changes to the very structure of an economy, typically through government policy, to improve its performance. According to the OECD Glossary of Statistical Terms, structural adjustment is a "process of market-oriented economic reform aimed at restoring a sustainable balance of payments, reducing inflation,

Table 35.1: Examples of cost-effective reductions in carbon intensity in various sectors that deliver strong economic multipliers. **Source:** von Weizsäcker, Hargroves, Smith, Desha and Stasinopoulos (2009).

Sector	Best Practice Case Studies
Residential Buildings	Passive house designs have achieved significant reductions in heating requirements in Germany with an eighty per cent improvement over contemporary German standards, and a ninety per cent improvement over the average German building stock (Clinton Climate Initiative 2008). There are now examples of *Passivhaus* design in many OECD countries.
Steel Industry	Leading US steel company, Nucor Steel, is around seventy per cent more energy-efficient than many steel companies around the world (Boyd/Gove 2000), using state-of-the-art electric arc furnace systems, adopting leading practices such as net shape casting, and implementing options such as energy monitoring and systems for energy recovery and distribution between processes (Worrel/Price/Galitsky 2004).
Cement Industry	Ordinary Portland cement manufacture is responsible for between six and eight per cent of global greenhouse emissions and this is rising with demand. The good news is that an Australian company, Zeobond Pty Ltd, based in Melbourne, is now making geo-polymer cement which reduces energy usage and greenhouse gas emissions by over eighty per cent (TNEP 2009). Geo-polymers can be used for most major purposes for which Portland cement is currently used.
Supermarkets	Supermarket chains Tesco (UK) and Whole Foods (USA) are showing that there are numerous ways to significantly reduce electricity usage through, for instance, reducing cooling and heating loads and utilizing more efficient lighting (Faramarzi/Coburn/Sarhadian 2002). They are also experimenting with solar energy and wind micro-turbines (Tesco 2008). Whole Foods Market are set to power an entire store using solar panels and a combined cycle co-generation using fuel cells and heat recovery (Whole Foods Market 2008).
Restaurants	Four profitable restaurants—*Bordeaux Quay*, Bristol UK (The Independent 2006), *Foodorama*, Berlin Germany (Sonnenberg 2009), *The Acorn House*, London UK (Carvalho 2007) and *The Water House* (UK)—demonstrate that restaurants can significantly reduce their energy consumption through building design, energy-efficient lighting and cooking equipment, purchasing their electricity from accredited renewable sources, buying organic fresh local food in season, composting and recycling all waste, and investing in carbon offsets.
Transport— Vehicle Efficiency	Integrating technical advances in light-weighting, hybrid electric engines, batteries, regenerative breaking and aerodynamics is enabling numerous automotive and transport vehicle companies to redesign cars, motorbikes, trucks, trains, ships and aeroplanes to be significantly (fifty to eighty per cent) more fuel-efficient than standard internal combustion vehicles. Plug-in vehicle technologies are opening up the potential for all transportation vehicles to be run on batteries charged by renewable energy (Light Rail Now 2008).
Transport— Efficiency from Modal Shifts. (Passenger)	Shifting transport modes can also lead to significant energy-efficiency gains. One bus with twenty-five passengers can reduce energy and greenhouse gas emissions per capita by over eighty per cent per kilometre compared to twenty-five single-occupant vehicles. Trains are even more efficient. Typically, rail systems in European cities are seven times more energy-efficient than car travel in US cities (Newman/Kenworthy 2007).
Transport Efficiency from Modal Shifts (Freight)	Shifting freight transport from trucks to rail can also lead to large efficiency gains of between seventy-five and eighty-five per cent (Freight on Rai, 2009; Frey/Kuo 2007). Several countries are moving to improve the efficiency of their transport sectors by making large investments in rail freight infrastructure, including improving the modal interfaces. For instance, China has invested US$292 billion to improve and extend its rail network from 78,000 km in 2007, to over 120,000 km by 2020, much of which will be dedicated to freight.

and creating the conditions for sustainable growth in per capita income" (Alexander/Baden 2000). In practice such 'market-oriented economic reform' is typically driven by an intention to improve the economic performance of the economy in a response to a need to reduce debt or position itself to replay development loans. For instance, structural adjustment mechanisms are typically required by financial institutions, such as the IMF or the World Bank, when offering development loans, typically to clear existing debt, or in order to secure lower interest rates on such loans.

In such cases the intention is to enforce structural changes to reduce the risk of the loan by requiring restructuring of economic policy, typically around:

- the level of taxation,
- controls on inflation,
- the provision of subsidies to support selected industries,
- the privatization of government assets and industries,
- levels of government expenditure,
- trade barriers, and
- the de-regulation of markets to encourage greater competition.

The implementation of such adjustments has been heavily criticized for its impact on social and environmental outcomes in the country due to the main focus being on economic performance. Social impacts from this form of restructuring include an increase in the cost of living from a focus on wealth generation, an increase in unemployment from a focus on short-term cost reduction by industry, and a decrease in social welfare expenditure by government (including health care and education). Environmental impacts include increased pressure to rapidly liquidate natural resources for profit (leading to significant environmental degradation) rather than diversification through the development of secondary and tertiary industries. According to Friends of the Earth,

> Exports of natural resources have increased at astonishing rates in many countries under IMF adjustment programs, with no consideration of the environmental sustainability of this approach. Furthermore, the IMF's policies often promote price-sensitive raw resource exports, rather than finished products. Finished products would capture more added value, employ more people in different enterprises, help diversify the economy and disseminate more know-how (Montanye/ Welch 1999).

In order to make certain industries more viable, governments often provide subsidies, particularly to the agricultural sector in response to international competition, and to fossil-fuel-based industries in response to high costs for infrastructure development and maintenance. The intended result is that the product or service is made more affordable to the consumer and economic development is accelerated; however, the unintended result is that the product or service is able to be sold with little consideration of the environmental pressure it may create, and the associated real costs to the economy in the short and long terms. Furthermore, a lower cost not only increases the uptake of the product or service (which is after all the point of a subsidy), it may also lead to its inefficient or wasteful use, further compounding the environmental pressure, and the resulting associated costs.

For example, despite the fact that subsidizing chemical companies to produce fertiliser has led to improved agricultural yields, the low cost has often led to excessive and poorly controlled use, which in turn can lead to contaminated groundwater and waterways. Further examples include the overuse and wastage of water, because of the subsidies for the use of water; wastage of electricity, with appliances, equipment and lights being left on when not needed because of the subsidies for electricity and the absence of a 'carbon price' making it so cheap to households and industry. Further, by making fossil-based energy cheaper, subsidies distort the market and make it harder for renewable and low carbon options to compete. The literature shows that a proportion of the costs that result from the increased environmental pressures from subsidized activities, with such subsidies estimated to be as much as US$650 billion per annum globally, are mostly borne directly by governments—and therefore citizens— through health care costs and public taxes (Roodman 1999). The remaining burden is borne by individuals through health-related impacts, and by the environment itself as the increased levels of toxins and pollution lead to the degradation of ecosystems.

When considering interventions into the market by providing subsidies, governments need to understand the broader implications in the short and long term, and where possible assign a price to these implications. Many experts have campaigned for higher prices for electricity, and/or reductions in government subsidies for the industry, as the increased price would encourage efficiency and reduced consumption, and promote exploration for alternatives to fossil-fuel-based electricity, leading to reduced negative impacts. The challenge here is that any economy exists as part of a complex balance, and increasing the cost of electricity without reducing costs elsewhere may result in an undue burden for citizens—which may then affect voter preferences. As Ernst von Weizsäcker points out in *Factor Five*, "Ultimately, resource consumption should be so expensive that total resource consumption rests in a perfect balance with sustainable supplies of renewable (or recycled) resources, and the resulting ability of the biosphere to assimilate the associated pollution and by-products" (von Weizsäcker/A/B et al. 2010). However, this is not yet the case in our current society, where the major reason environmental degradation occurs is because the current balance reached in many economies around the world makes it cheaper to degrade nature than to appropriately manage its use (Daily/Ellison 2002).

35.5 Harnessing Structural Adjustment Mechanisms for Carbon Management

In order for economies to achieve the 'conditions for sustainable growth in per capita income', advocated by the OECD, the scope of structural adjustment needs to be expanded to include mechanisms that lead to a significant reduction in environmental pressures, and in particular the reduction of greenhouse gas emissions (Smith/Hargroves/Desha 2010). As mentioned above, however, unlike SAPs that are enforced as part of the conditions of large national financial loans, structural adjustments to reduce greenhouse gas emissions will need to be enforced by national governments as part of an agenda of economic development. Such a process would need to be underpinned by effective policies and strategies which address a range of market, institutional, and information failures.

Considering the key mechanisms used in structural adjustment programmes—namely a focus on taxation, inflation, subsidies, government assets, government expenditure, trade, and the regulation of markets—a number of such mechanisms along with a range of others can be used to underpin significant greenhouse gas emissions reductions across economies to achieve *Carbon Structural Adjustment*. As part of a CSA agenda typical structural adjustment mechanisms can be used in a number of ways, including:

- *Taxation*: Pollution taxation is now used by many countries to discourage the generation of greenhouse gas emissions by placing a price on their generation. Revenue from such taxation can be used to provide tax cuts to low carbon-intensive operations and products.
- *Inflation*: Encouraging a transition to low carbon intensive operations can reduce the demand for non-renewable raw materials and avoid future stagflation. For instance a focus on reducing the carbon intensity of the steel industry will see a shift to greater recycled steel affecting the price of iron ore.
- *Subsidies*: Shifting subsidies from fossil-fuel-based energy to renewable and low carbon alternatives will encourage innovation and industry development in low/no carbon alternatives to energy generation.
- *Government Expenditure:* A range of government expenditures can be aligned to support options with lower carbon intensity, such as procurement policies, transport infrastructure spending, and upgrading publically owned built environment infrastructure.
- *Trade*: Encouraging the export of low carbon technologies and expertise will encourage manufacturing and production of products that require less carbon-intensive energy to create and operate in the country to which the trade is destined.
- *Regulation*: Following a focus on pollution regulation many countries are now focused on regulations related to the energy consumption requirements of residential homes (UK Code for Sustainable Homes), banning energy-intensive light bulbs (Cuba and Australia) and requiring levels of renewable energy generation (China's Renewable Energy Laws).

Fundamentally, structural adjustment is about avoiding future negative impacts on the economy by taking steps to achieve system-level change focused on reducing inefficiency and improving performance and stability—while making profits for international money-lenders. Hence, given the imperative to reduce greenhouse gas emissions, it is clear that efforts need to be focused at aligning the structure of economies to significantly reduce carbon intensity over the coming decades, with many options being tested and proven in numerous countries. Such a shift in focus would result in a significant recasting of the various structural adjustment mechanisms, this may include and not be limited to:

- Rather than promoting a reduction in government programmes and spending, *carbon structural adjustment* would include governments investing in research and development of low carbon technologies and processes, along with a strong focus on renewing education and professional development programmes.
- Rather than the deregulation of industry sectors to allow greater wealth generation through reduced environmental charges and requirements, CSA would include regulation and policy designed to drive industry to take innovative approaches to reducing greenhouse gas emissions.
- Rather than focus taxation on employees and businesses, CSA would include a focus on resource taxation to price externalities related to environmental damage that will have future direct impact on economic growth. According to Hawken, Lovins and Lovins. (2009), "Shifting taxes towards

Figure 35.3: Schematic of Key Elements of Carbon Structural Adjustment. **Source:** Hargroves (2014).

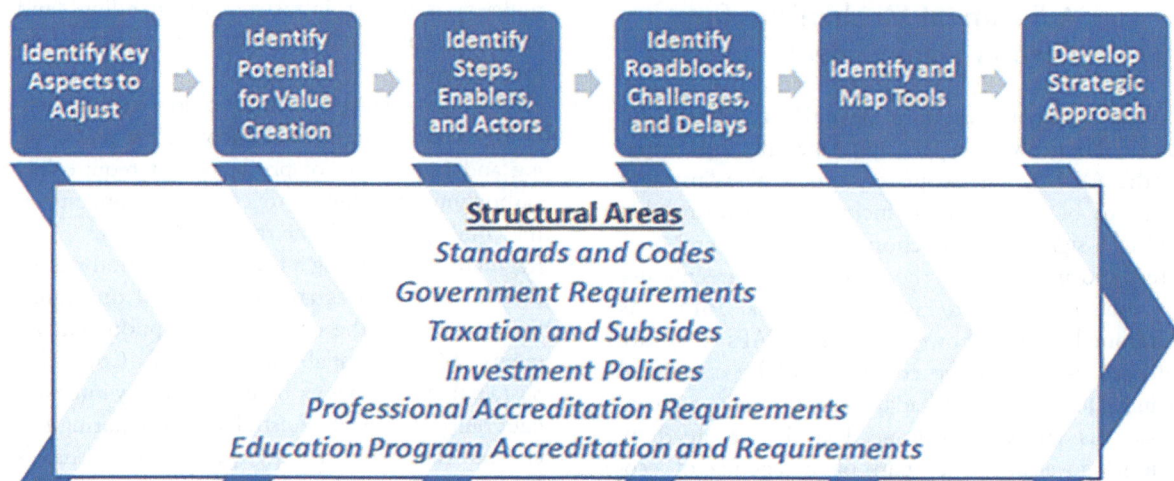

resources creates powerful incentives to use fewer of them".

- Rather than focusing on exporting raw commodities, CSA would focus on creating opportunities to enhance secondary and tertiary industries with such commodities to allow greater value-adding and increased wealth generation through increased exports of low carbon technologies and expertise.

In order to significantly reduce greenhouse gas emissions a focus on reducing the need for energy will need to be complemented by a rapid transition to non-fossil energy. Fortunately our world has an abundance of non-fossil energy due to our sun creating geological energy in the form of underground heat stores, creating kinetic energy in the form of wind that can be harnessed, and providing direct energy through sunlight that we can now transfer into electricity with as much as twenty per cent conversion efficiency.

35.6   A Proposed Carbon Structural Adjustment Road Map[4]

Considering the potential to re-orient structural adjustment programme-type interventions to achieve significant reductions in greenhouse gas emissions as outlined above, the Sustainable Built Environment National Research Centre based at Curtin University in Australia has created a project to propose a carbon structural adjustment road map led by the author and supervised by Professor Peter Newman.[5] The project focuses on six key structural areas that each have a direct impact on the greenhouse gas emissions of

economies. Changes to these structural areas stand to deliver significant reductions in emissions, while building understanding and institution learning outcomes (Hargroves 2014):

1. Standards and codes
2. Government requirements (local, state, federal, and statutory agencies)
3. Taxation and subsides
4. Investment and procurement policies (government and private)
5. Professional accreditation requirements, and
6. Education programme accreditation requirements (higher and vocational).

Building on the growing level of work being done to encourage a transition to sustainable development this project seeks to contribute to nations around the world when responding to inevitable calls in the coming decade(s) for economy-wide structural approaches.

---

4   The following is an edited extract from Hargroves (2014), developed as part of the Sustainable Built Environment National Research Centre (SBEnrc) 'Greening the Built Environment' Programme, led by Professor Peter Newman. The project is supported by the Western Australia Government Department of Treasury, Queensland Government Department of Transport and Main Roads, Main Roads Western Australia, John Holland Group, and the New South Wales Roads and Maritime Services. The research is advised by ClimateWorks Australia and the Clean Energy Finance Commission. The research team is based at the Curtin University Sustainability Policy Institute (CUSP).

5   SBEnrc, at: <http://www.sbenrc.com.au/research-programs/1-24-a-roadmap-for-carbon-structural-adjustment-in-the-built-environment-sector/>.

Figure 35.4: Whole of Society Approach to Sustainable Development. **Source:** Hargroves and Smith (2005).

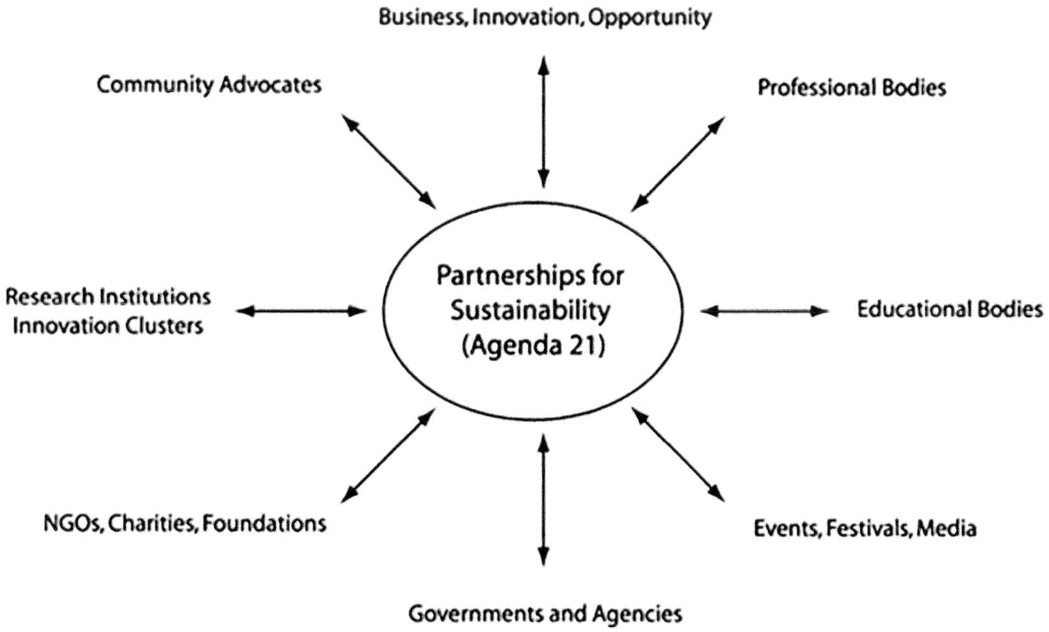

According to the findings of the project, 'When such a call is made it will be important to undertake a strategic process to identify specific areas of the economy to adjust, establish the value in doing so, identify barriers and enablers, select tools and interventions, and develop a strategy specific for each economy and its sectors' (Hargroves 2014), as shown in figure 35.3.

According to the report it is important to understand that each stop will involve a range of actors from across society and that in order to maximize the potential for carbon structural adjustment to be successful a 'whole of society' approach is recommended, as presented in figure 35.4. This is particularly important as unlike SAPs, where there is a strong and enforceable imperative to implement the adjustment measures, a carbon structural adjustment agenda will need to be led by government with organizations, businesses, institutions, and various groups across society actively involved and empowered to contribute to the overall process (Hargroves 2014).

The following provides an overview of each of the steps of the proposed 'Carbon Structural Adjustment Road Map' along with anticipated deliverables (Hargroves 2014).

35.6.1 Identify Key Aspects of Structural Areas to Adjust

The first stop on the road map to achieve carbon structural adjustment is to identify specific aspects of

the structural area being considered, such as 'Standards and Codes' that are deemed to 'need' adjustment in order to ensure greenhouse gas emissions reductions across the sector.

This stop would deliver the following outcomes:

1. The identification of specific aspects of the structural area being considered that contribute to increasing the greenhouse gas emissions of the built environment sector.
2. The provision of a clear and concise summary of the aspects, demonstrating the direct link to greenhouse gas emissions in the sector, and highlighting precedent and evidence for cost-effective emissions reductions.
3. The identification of any current or previous efforts to bring about change in this aspect of the structural area, both those that have gained traction and those that have not. This may include the identification of existing recommendations related to the adjustment of the areas or similar areas internationally.

35.6.2 Identify Potential for Value Creation

Once the specific aspects of the structural areas that require adjustment have been identified, the next stop on the road map involves identifying the potential value that can be created by the adjustment of these aspects. This is important as it identifies potential supporters for the adjustment and demonstrates the value to the economy, sector, and community for taking action to adjust the structural areas to reduce greenhouse gas emissions. Value can be created in a number of direct and indirect areas with the initial focus on both the economic value and the reductions in greenhouse gas emissions. Other areas of value may include job creation, increased trade in services and high value manufactured goods, generation of voter goodwill, along with direct savings to business through reduced energy use.

This stop would deliver the following outcomes:

1. The identification and quantification, where appropriate, of the potential value that would be created through the adjustment of the aspects of the structural area being considered, both direct and indirect.
2. The identification of evidence to support this value creation and precedent of such value being captured. This would include identification of value created for particular parties and stakeholders.

35.6.3   Identify Steps, Enablers, and Actors

Once the specific aspects of the structural areas that require adjustment have been identified, and the value of such adjustments demonstrated, the next stop on the road map involves identifying the main steps required to adjust the aspects. Consideration of the steps includes an investigation into the existence of current enablers to such adjustment that will support the process. Further, as mentioned above it is important to take a whole-of-society approach to the process and at this stage the various actors that should be involved in each step are identified. It may be the case that investigations have been undertaken to identify steps to adjust the selected aspects either in Australia or internationally that can inform the identification of specific steps.

This stop would deliver the following outcomes:

1. The identification of specific steps to be undertaken to achieve the adjustment of the particular aspect of the structural area being considered. (Note: at this stage the steps are identified rather than the actions needed to achieve the steps, which is the focus of stop 5.)
2. The identification of existing enablers to support the adjustment of this aspect of the structural area.
3. The nomination of actors and parties that should be involved in each step, such as business, industry groups, economic think tanks, university research-

ers, government agencies etc. This may involve the consideration of methods to engage and mobilize particular actors that may not be already engaged in the low carbon agenda.

35.6.4   Identify Roadblocks, Challenges, and Delays

Once the steps required to adjust the aspects of the structural area being considered have been identified, the next step in the road map involves the identification of major roadblocks, challenges and areas of potential delay that will be faced should the steps be implemented. Once identified, investigation can be carried out on how to avoid, amend, or remove such barriers.

This stop would deliver the following outcomes:

1. The identification of potential roadblocks, challenges and/or delays to the specific steps to adjust the structural area under consideration, which may include technological, institutional, or market barriers.
2. The identification of possible ways to avoid, amend, or remove the potential roadblocks, challenges or delays, with specific mention of the parties involved.
3. The consideration of the strength of such barriers to carbon structural adjustment and the identification of leverage points to focus tools to reduce such barriers.

35.6.5   Identify and Map Tools

Once the potential for value creation has been established, and the existence of major roadblocks, challenges, and/or delays identified, the next stop in the road map involves identifying and mapping tools that can either manage roadblocks, challenges, and delays, or enhance the value created by the adjustment.

This stop would deliver the following outcomes:

1. The identification of potential tools that may be used to implement the steps. It is anticipated that these tools will be suitable for implementation with involvement across the various actors involved in the process.
2. The mapping of such tools to specific roadblocks, challenges, and/or delays, as well as specific opportunities to enhance value creation.
3. The investigation of perceptions related to the identified tools held by stakeholders in the pro-

Figure 35.5: A method for the prioritization of efforts based on the likely impact on greenhouse gas emissions and the likely willingness to adjust in the area of focus. **Source:** Adapted by K. Hargroves from an adaptation by K. Hargroves and C. Desha of the CBSM Methodology (McKenzie-Mohr/Smith 1999).

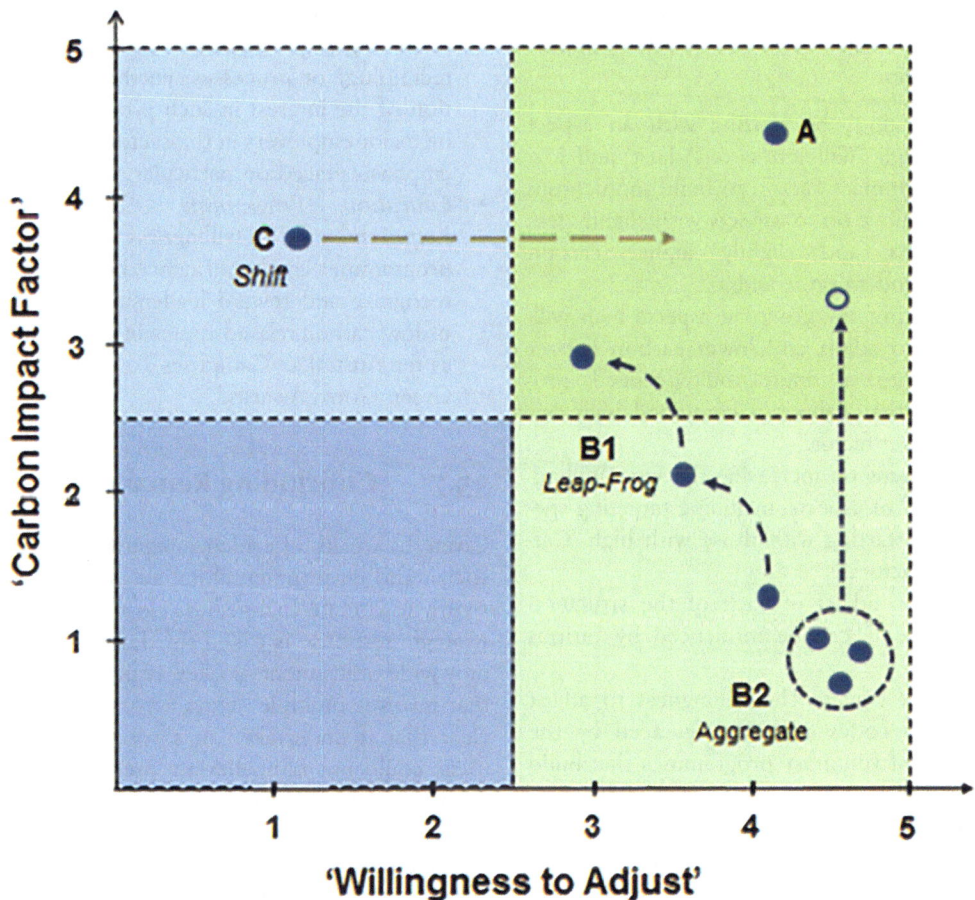

cess as to their suitability and requirements for implementation.

35.6.6 Develop Strategic Approaches

Now that the process of following the road map has created a list of specific aspects of the structural area being considered that are recommended for investigation, an estimate of the potential value that can be created, identification of steps involved and the associated barriers, and a list of possible tools to both reduce the barriers and increase the value creation, the final stop in the road map is to develop strategic approaches. Such approaches will draw on enablers and engage with key actors to create work plans to undertake each step and implement the associated tools. This may include consideration of options for cross-sectorial collaborations. Understanding that it is not feasible for all aspects of each structural area to be adjusted at the start of the carbon structural adjust-

ment process, there needs to be a prioritization process.

In order to identify the priority of aspects to focus on it is important to consider two factors,

a.) the likely 'Carbon Impact Factor', and
b.) the likely *'Willingness to Adjust'*.

The carbon impact factor would take into account the impact on the greenhouse gas emissions in the built environment sector should the aspect identified at stop 1 be adjusted as per the steps at stop 3. The willingness to adjust would take into account the roadblocks, challenges, and delays identified at stop 4, along with a range of other considerations including gauging political will, potential business and industry support, and community views.

Based on figure 35.5, the following three routes can be undertaken:

- *Route A:* A focus on progressing actions with high 'Carbon Impact Factor' and high 'Willingness to Adjust'.
- *Route B:* A focus on progressing parts with high 'Willingness to Adjust' and low 'Carbon Impact Factor' by either:

    B1.) 'Leapfrogging', by starting with an aspect with a high 'Willingness to Adjust' and low 'Carbon Impact Factor' to build momentum to then move on to aspects with slightly less willingness and slightly higher carbon impact, and so on..., and/or

    B2.) 'Aggregating', by grouping aspects high willingness to adjust and lower carbon impact factor into a single multi-pronged programme that delivers a combined high carbon impact factor.

- *Route C:* A focus on increasing the low 'Willingness to Adjust' of aspects, including targeting specific barriers, starting with those with high 'Carbon Impact Factor'.

The willingness to adjust of each of the structural areas along Route C can be influenced by various activities, namely:

- *Standards and Codes:* The willingness to adjust standards and codes can be influenced by the development of voluntary programmes that build industry support and create experience to road-test potential changes. This may be influenced by demonstration projects for new technologies or practices.
- *Government Requirements:* The willingness to adjust government regulations can be influenced by community behaviour change programmes that can build voter support for action towards reducing greenhouse gas emissions.
- *Taxation and Subsides:* The willingness to adjust taxation and subsides can be influenced by providing transparency on the payment of subsides and the impacts on the economy. Furthermore, understanding of the various options for adjusting such structures can provide clarity on the feasibility of using them to encouraging activities that reduce greenhouse gas emissions.
- *Investment and Procurement Policies:* The willingness to adjust investment and procurement policies can be influenced by industry associations through the development of voluntary industry-led sustainability rating tools for infrastructure sustainability that can lead to the inclusion in request for tenders of the nomination of a minimum rating to be achieved by the project.

- *Professional Accreditation Requirements:* The willingness to adjust professional accreditation requirements can be influenced by providing the industry or sector with evidence of the value of ratcheting requirements in line with advances in technology or processes. Furthermore, an indication of the interest in such professional attributes by major employers in the sector can influence the emphasis placed on particular areas.
- *Education Programme Accreditation and Requirements:* The willingness to adjust education programmes can be influenced by programmes to recognize and reward leadership in the coverage of low carbon-related topics in programmes, such as the Australian Campuses Towards Sustainability Green Gown Awards.[6]

35.7   Concluding Remarks

Given the wealth of evidence regarding the economic, social, and environmental impacts of not responding swiftly to climate change, it is clear that a holistic and strategic response is called for that will involve economy-wide adjustments to key structures in a manner that involves multiple actors across society. It is also clear that if undertaken in a robust fashion such a focus could not only alleviate pressures on the environment that stand to impact on economies, but also contribute to delivering sustained economic growth, job creation, and poverty eradication. The costs associated with the environmental pressure created by activities that contribute to GDP are rising and forming a greater proportion of GDP. Further, such costs are set to rise with the impacts of climate change affecting all parts of the world's economies in some way. Hence the twenty-first century is looking to be an era of major transition, one that requires a long-term strategic approach.

Given that the world's economies have experienced many decades of growth based on unbridled fossil fuel consumption, there has been little impetus to act to reduce greenhouse gas. This means that much of the policy and industrial strategy used to achieve development in the past will need to be critically reassessed if it is to continue to deliver progress into the future. Efforts to achieve a low carbon transition will need to be balanced with the level of embedded infrastructure in each economy, as it does not make financial sense to prematurely discard large

---

6   ACTS at: <http://www.acts.asn.au/initiatives/ggaa/>.

amounts of existing fossil-fuel-intensive industries and replace them overnight. Nor does it make sense to do nothing until such infrastructure reaches natural recommissioning or replacement time frames. As has been common in each of the previous waves of innovation, all sectors of the economy will be affected to differing degrees by the imperative to reduce greenhouse gas emissions, some requiring minor changes and others requiring significant overhauls, and even closure.

This presents a significant challenge as if the process takes too long we may fail the ecological imperative, and if it is forced too soon we may fail the economic and social imperatives. The challenge to transition the economy to a low greenhouse gas emissions future will need to be met with sophisticated strategies that build on easily expanded emissions reduction opportunities in the short term to provide a platform for advances in the future. The risk of continued complacency and delayed action is that if greenhouse gas emissions are not significantly reduced it is very likely that the complexity of the impacts on our planet's biosphere may well overwhelm society's ability to find solutions.

There is now a greater level of clarity around actions that can be taken to significantly reduce greenhouse gas emissions across the world's economies. What is missing is a serious agenda to accelerate progress in this area that involves actors from across society dedicated to working in a manner that delivers economic growth and social well-being while achieving preferred greenhouse gas stabilization levels. Such economy-wide changes are akin to the *structural adjustment* of developing countries to improve short-term economic performance. As mentioned in this chapter, it may be the case that in the same way that fiscal structural adjustment was formulated as a solution to underdevelopment in third world economies in the twentieth century, a new form of structural adjustment, such as '*carbon structural adjustment*', may need to be formulated as a solution to unsustain-

able development around the world in the twenty-first century. Hence the scope of traditional structural adjustment needs to be expanded to include mechanisms that lead to a significant reduction in environmental pressures, especially the reduction of greenhouse gas emissions.

In order for such an approach to be successful it will need to draw on lessons learned over the last two decades from efforts to progress the sustainable development and climate change response agendas, and merge this knowledge into an economy-wide structural change agenda. This chapter recommends a focus on six key areas for carbon structural adjustment, namely standards and codes, government requirements, taxation and subsides, investment and procurement policies, professional accreditation requirements, and education programme accreditation and requirements. An important consideration in the undertaking of structural adjustment in these areas is the understanding that they involve a range of actors from across society and that in order to maximize the potential for carbon structural adjustment to be successful a 'whole of society' approach is recommended.

Given that as part of a comprehensive carbon structural adjustment agenda each of the structural areas above would be investigated through the six stops above, and that this would create a comprehensive set of possible actions, this calls for a prioritization process. It is recommended that each possible action generated by the process be considered for its likely 'Carbon Impact Factor', and the likely 'Willingness to Adjust'. Although the Carbon Structural Adjustment Road Map process stands to identify a range of direct interventions for each structural area, the leadership and commitment from government for such an approach has yet to be demonstrated. Hence it is of vital importance that efforts are made to develop complementary activities that increase the willingness to adjust across the economy.

References

Alexander, Patricia; Baden, Sally, 2000: *Glossary on macroeconomics from a gender perspective* (Brighton: University of Sussex, Institute of Development Studies).

Boyd, Brian K.; Gove, Steve, 2000: "Nucor Corporation and the U.S. Steel Industry", in: Hitt, Michael A. Ireland, R. Duane; Hoskisson, Robert E., (Eds.): *Strategic Management: Competitiveness and Globalization*, 4 Edition (Cincinatti: Southwestern Publishing).

Brown, Lester, 2008: *Plan B 3.0: Mobilizing to Save Civilization* (New York: W.W. Norton).

Carvalho, Michael D., 2007: "The Acorn House Bears Fruit", in: *Community Energy* (July 2007).

Clinton Climate Initiative 2008: "C40 Cities: Buildings, Freiburg, Germany" (New York: Clinton Climate Initiative).

Daily, Gretchen C.; Ellison, Katherine, 2002: *The New Economy of Nature: The Quest to Make Conservation Profitable* (Washington DC: Island Press).

Desha, Cheryl J.K.; Hargroves, Karlson J., 2011: "Informing engineering education for sustainable development using a deliberative dynamic model for curriculum renewal", Paper for Research in Engineering Education Symposium 2011, Madrid, Spain.

Desha, Cheryl J.K.; Hargroves, Karlson J., 2012: "Fostering rapid transitions to Education for Sustainable Development through a whole system approach to curriculum and organizational change", Paper for World Symposium on Sustainable Development at Universities, 2012, Hamburg, Germany.

Desha, Cheryl J.K.; Hargroves, Karlson J.; Smith, Michael H.; Stasinopoulos, Peter; Stephens, Renee; Hargroves, Stacey L., 2007: *Energy Transformed: Australian University Survey Summary of Questionnaire Results* (Perth Australia: The Natural Edge Project (TNEP), unpublished).

Faramarzi, Ramin; Coburn, Bruce; Sarhadian, Rafik, 2002: Performance and Energy Impact of Installing Glass Doors on an open Vertical Deli/Dairy Display Case, Associate Member (Atlanta, GA: ASHRAE).

Freight on Rail, 2009: "Useful Facts and Figures", in: Freight on Rail, at: <> (9 April 2009).

Frey, Christopher; Kuo, Po-Yao, 2007: "Assessment of Potential Reduction in Greenhouse Gas (GHG) Emissions in Freight Transportation", Paper for International Emission Inventory Conference, US Environmental Protection Agency, Raleigh, NC, 15-17 May 2007.

Hargroves, Karlson J., 2014: *Roadmap for Carbon Structural Adjustment in the Built Environment Sector–A Report to the Sustainable Built Environment National Research Centre* (Perth: Curtin University, Curtin University Sustainability Policy Institute).

Hargroves, Karlson J.; Desha, Cheryl JK., 2011: *Energy Efficiency Resources for Undergraduate Engineering Education: Energy Efficiency Advisory Group*–Project 2, Report to the Department of Resources, Energy and Tourism, Canberra, August 2011 (unpublished).

Hargroves, Karlson J.; Smith, Michael H. (Eds), 2005: *The Natural Advantage of Nations*, The Natural Edge Project (London: Earthscan).

Hawken, Paul; Lovins, Amory B.; Lovins, L. Hunter, 2009: *Natural Capitalism. Creating the next industrial revolution* (New York: Back Bay Books).

IEA 2006: "Energy Technology Perspectives 2006: Scenarios and strategies to 2050", (Paris: International Energy Agency).

Jevons, William S., 1865: *The Coal Question: an Inquiry Concerning the Progress of the Nation, and the Probable Exhaustion of our Coal-mines* (New York: MacMillan).

Light Rail Now 2008: "CTrain Light Rail growth continues with North East extension", News Release, January 2008 (Austin, Texas: Light Rail Now).

Lovins, Amory B.; Datta, E. Kyle; Bustnes, Odd-Even; Koomey, Johnathan; Glasgow, Nate, 2004: *Winning the Oil Endgame: Innovation for Profits, Jobs, and Security*, Rocky Mountain Institute (London: Earthscan).

Lovins, L. Hunter; Cohen, Boyd, 2012: *Climate Capitalism: Capitalism in the Age of Climate Change* (New York: Hill and Wang).

McDonough, Bill; Braungart, Michael, 2002: *Cradle to Cradle–Remaking The Way We Make Things* (New York: North Point Press).

McKenzie-Mohr, Doug; Smith, William, 1999: *Fostering Sustainable Behavior: An Introduction to Community-Based Social Marketing* (Gabriola Island: New Society Publishers, British Columbia).

Monbiot, Geroge, 2006: *Heat: How to Stop the Planet Burning* (New York: Penguin Books).

Montanye, Dawn; Welch, Carol, 1999: *The IMF: Selling the Environment Short* (Washington, DC: Friends of the Earth).

Newman, Peter; Kenworthy, Jeffery, 2007: "Transportation energy in global cities: sustainability comes in from the cold", in: *Natural Resources Forum*, 25,2: 91-107.

OECD 2008: *OECD Environmental Outlook to 2030* (Paris: OECD).

Roodman, David M., 1999: *The Natural Wealth of Nations: Harnessing the Market and the Environment*, Worldwatch Environment Alert Series (London: Earthscan).

SDSN; IDDRI 2014: "Pathways to Deep Decarbonisation–Interim Report" (New York: Sustainable Development Solutions Network–Paris: Institute for Sustainable Development and International Relations).

Smith, Michael H.; Hargroves, Karlson J.; Desha, Cheryl J.K., 2010: *Cents and Sustainability: Securing Our Common Future through Decoupling Economic Growth from Environmental Pressure* (London: Earthscan).

Sonnenberg, Brittani, 2009: "Berlin's Carbon-Neutral Eatery", *The Associated Press* (21 January 2009).

Stasinopoulos, Peter; Smith, Michael H.; Hargroves, Karlson J.; Desha, Cheryl J.K., 2009: *Whole System Design: An Integrated Approach to Sustainable Engineering* (London: Earthscan,).

Stern, Nicolas, 2006: *The Stern Review: the Economics of Climate Change* (Cambridge: Cambridge University Press).

Talberth, John; Cobb, Clifford; Slattery, Noah, 2006: *The Genuine Progress Indicator 2006: A Tool for Sustainable Development* (Oakland, CA: Redefining Progress); at: <http:// rprogress.org/publications/2007/GPI% 202006.pdf>.

Tesco 2008: *Sustainability Report: More than the weekly shop: Corporate Responsibility Review* (Dundee: Tesco Inc.).

The Independent 2006: "The eco eatery", *The Independent*, 18 May 2006.

TNEP [The Natural Edge Project], 2009: "Factor 5 in Eco-Cements", in: *CSIRO ECOS Magazine*, Issue 149.

Townsville City Council, 2010: "Identification and Assessment of Homeowner Behaviours Related to Reducing Residential Energy Demand: Report to the Townsville

City Solar Community Capacity Building Program" (Townsville: City Council–Brisbane: Griffith University, The Natural Edge Project, Australia).

von Weizsaecker, Ernst; Hargroves, Karlson J.; Smith, Michael H.; Desha, Cheryl J.K.; Stasinopoulos, Peter, 2009: *Factor 5: Transforming the Global Economy through 80% Improvements in Resource Productivity* (London: Earthscan).

von Weizsäcker, Ernst; Lovins, Amory B.; Lovins, L. Hunter, 1997: *Factor Four: Doubling Wealth, Halving Resource Use* (London: Earthscan).

World Commission on Environment and Development, 1987: *Our Common Future* (Oxford: Oxford University Press).

Worrell, Ernst; Price, Lynn; Galitsky, Christina, 2004: "Emerging Energy-Efficient Technologies in Industry: Case Studies of Selected Technologies" (Berkeley, CA: Ernest Orlando Lawrence, Berkeley National Laboratory).

36 Drivers and Barriers to Wind Energy Technology Transitions in India, Brazil and South Africa

Britta Rennkamp[1] and Radhika Perrot[2]

Abstract

This chapter examines the drivers and barriers to innovation in the wind energy sectors in Brazil, India and South Africa. We analyse actors and institutions that have played a role in the development and diffusion of wind energy technologies in these countries from a technological innovation systems (TIS) perspective. We introduce innovation capabilities, the enabling environment and policy-independent strategies as the main drivers for the development of the TIS. This analysis will contribute to improving our understanding of drivers of and barriers to innovation relevant to successful transitions towards cleaner technologies and less carbon-intensive economies. We found that some countries are more successful in developing wind energy technologies because of three factors: stronger technological capability, prime movers, and big market size, and this relates to the enabling environment and policy incentives. We found that in India and Brazil, entrepreneurial activities independent of public policy were critical for the development of the local wind energy industry. In South Africa, policy-independent entrepreneurial activity was scarce and operated in an unfavourable environment which hampered the development of a local wind energy industry.

Keywords: Renewable energy, wind energy, technological innovation systems, innovation, enabling environment, India, South Africa, Brazil.

36.1 Introduction

Why are some countries more successful than others in catching up with creating local renewable energy industries? What drives and hinders innovation in and transitions towards clean technologies? These questions have only partially been answered in the research literature. This chapter examines the drivers and barriers to innovation in and development of local wind energy industries in three developing countries: Brazil, India and South Africa. All three countries are latecomers in developing renewable energy industries, with the intention of catching up in the competition for mature technologies.

The research literature on industrial development provides some explanations for the process of catching up, which is never unique. Success and failure depend on global competition and on the economic, political and cultural circumstances as well as the available skills. The literature often attempts to explain the process of catching up in industrial development as a whole. In this chapter, we will focus on one technology. We propose a framework for the analysis of drivers and barriers in wind energy technology development, based on the concept of *technology innovation systems* (TIS) and the main drivers and barriers identified in the research literature. The technology innovation system approach is a well-established framework, which is useful for explaining technological transitions through seven functions. We apply this framework to new evidence in the wind energy sectors in Brazil, South Africa and India.

Wind energy technologies were first developed in Denmark and Germany in the 1970s. The leading companies[3] in both nations enjoyed the benefits of the first movers in this innovation and continue to lead

1 Dr. Britta Rennkamp, Researcher, Energy Research Centre; Senior Fellow, African Climate and Development Initiative, University of Cape Town (UCT), South Africa; email: <britta.rennkamp@uct.ac.za>.

2 Ms Radhika Perrot, Senior Researcher, Mapungubwe Institute for Strategic Reflection (MISTRA), South Africa; email: <radhikap@mistra.org.za>.

© Springer International Publishing Switzerland 2016
H.G. Brauch et al. (eds.), *Handbook on Sustainability Transition and Sustainable Peace*,
Hexagon Series on Human and Environmental Security and Peace 10, DOI 10.1007/978-3-319-43884-9_36

the global markets. In the early 2000s, the governments of Spain and China helped the industry to catch up through imposing local content requirements, and this supported the creation of now globally operating companies.[4] India and Brazil were the next in line, and attempted to attract global investment to grow local industries. Although these countries had the advantage of a large market, many of the countries' institutional and carbon lock-ins features either prevented or slowed down the diffusion of wind energy technologies. South Africa only started a structured renewable energy programme in 2011, and this did not include the limited manufacturing capability in the country. The programme aimed to attract investment from the struggling industry through local content requirements, which fall short as a stand-alone industrial policy instrument. An analysis of drivers and barriers in all three countries contributes to a better understanding of different approaches to technological catch-up strategies and factors for success and failure.

The main contributions of this chapter are i) the allocation of the technology innovation system (TIS) to new empirical evidence from the wind energy industries in Brazil, India and South Africa and ii) the refining of the technology innovation system framework towards identifying drivers and barriers from the evidence and from additional research literature.

This chapter is structured as follows. In section 36.2 we present the current research literature on drivers and barriers to technological catching up and present the technology innovation system framework based on Bergek, Jacobson and Carlsson (2008) and Perrot (2012). Section 36.3 presents an analysis of technological transition in the wind energy sectors in Brazil, India and South Africa, building on a dataset of 54 interviews, most of which were conducted for a research project on low carbon development in 2012 and 2013.[5] Section 36.4 presents the findings and conclusions.

---

3   Vestas, Enercon, Nordex, Siemens.
4   Gamesa and Acciona in Spain, Goldwind and Sinovel in China.
5   The Brazilian case study was funded by the MAPS programme on *Mitigation Action Plans and Scenarios*. The South African case study was funded by UNITAR. The results of both studies have been previously published in a working paper on local content requirements in Rennkamp and Fortes Westin (2013).

36.2   Towards a Framework of Analysis of Drivers and Barriers to Successful Technological Development

Success factors for the technological and innovative capability of a country depend on its enabling environment or innovation infrastructure, the business environment for innovation in the industrial clusters, and the strength of the linkages between those two (Furman/Porter/Stern 2002), as well as economic policy and taxes, access to resources, foreign technology, and comparative advantages to competitors (Fu/Pietrobelli/Soete 2011; Lundvall/Borrás 2005; Peteraf 1993). The research literature identified these factors quite clearly. It is less clear on successful strategies for technological catching up.

Technological catching up is a difficult and competitive endeavour, which is even more competitive in at-the-frontier technologies (Cimoli/Dosi/Stiglitz 2009). The race for new technologies reflects the inequalities and income gaps between the industrialized and developing worlds (Amsden 2003). A decision-maker needs to be well informed on the prospects of funding a technology before allocating scarce public funds. Choosing strategic sectors at earlier stages will be more successful if decisions are well informed through consultations with local innovators (Lundvall/Borrás 2005). Catching up in a technology race requires strategic policy intervention. These interventions are ideally grounded on a careful evaluation of the factors that drive innovative capability. The history of industrial development shows that most countries have used localization and trade barriers at some point to protect their industries from international competitors (Cimoli/Dosi/Stiglitz 2009). Successful strategies for catching up vary. There is no unique way to pursue it (Fagerberg/Godinho 2005). The timing, the number and quality of global competitors, other countries trying to catch up in the same technology, the interest and demand from local firms and the public, the local skills base and the stage of technology maturity are important to the success and likelihood of catching up in an existing technology. A government needs to assess carefully if it is worth trying if numerous other countries have already tried to attract investment from a global industry.

Studies on innovation have produced the idea that innovation is systemic, and best supported through measures that understand innovation policy as a wider economic policy which focuses on transforming the institutional and organizational settings in which innovation occurs. As opposed to just 'picking winners' and supporting individual firms, innovation policy creates incentives, platforms and network opportunities for businesses to interact with academic research and society to maximize innovative output (Lundvall/Borrás 2005). This understanding goes beyond the neoclassical idea of policy intervention in the case of market failure. The innovation systems approach has been recognized as a useful tool for the analysis of technological transitions. In particular, the technology systems approach has managed to specify concrete functions which help to capture the structural characteristics, performance and dynamics that lead to technological change (Bergek/Jacobsson/Carlsson et al. 2008).

A technology innovation system is a network of agents who interact within a particular institutional infrastructure with the purpose of generating, diffusing, and utilizing technology, according to the definition by Carlsson and Stankiewicz (1991). The unit of analysis is the technology, as opposed to a nation, region or sector. The technology innovation system emphasizes institutions (government, universities, research institutes and firms) that both drive and constrain the performance of a *national innovation system* (NIS) (Winter 1993). The approach has successfully compared the development and diffusion of wind turbine industries in Germany, the Netherlands and Sweden (Bergek/Jacobsson 2003); analysed renewable energy cases in Sweden and discussed the two key processes that formed the TIS—legitimation and the development of positive externalities (Johnson/Jacobson 2001); and analysed the development of biopower in Sweden (Jacobsson 2008). There is a need for further empirical analysis in developing countries that are in the process of catching up with renewable energies.

36.2.1 Functions as Main Components of a Technology System

Bergek, Jacobsson, Carlsson et al. (2008) suggest seven functions in the technology innovation system that influence the development, diffusion and use of new technology or a transition to a low carbon technology:

1. Legitimation refers to creating social acceptance of a technology and compliance with existing institutions. Legitimacy is a crucial consensus of the beliefs and interests of actors in society, political decision-making, research, and private business, which allows for supportive regulation and resource allocation.

2. Entrepreneurial experimentation refers to activities by firms in experimenting, probing and testing technologies under conditions of high uncertainty regarding the outcomes. Experimentation determines the success or failure of a TIS, which otherwise will stagnate and fail to create innovation, business opportunities and wealth.

3. Resource mobilization refers to the allocation of financial, material or human capital.

4. Influence in the direction of search refers to the evolution of the TIS, which suggests a direction which other firms and organizations follow. Sufficient business opportunities or public regulations are necessary to incentivize firms to do so.

5. Knowledge development and diffusion refers to the actions that lead to the exchange of information and learning between actors or networks of actors.

6. Development of positive externalities results from investments or from favourable economic activities beyond the TIS.

7. Market formation refers to activities that lead to the creation of a demand or the provision of protected space for the new technology.

The research literature has acknowledged the seven functions as a useful measure for the development and diffusion of new technologies (e.g. Musiolik/Markard 2011; Suurs/Hekkert/Kieboom et al. 2010). These functions help to identify drivers and barriers to innovation and technological development (Negro/Alkemade/Hekkert 2012). Most papers use the framework to analyse innovation dynamics in industrialized countries and mature markets. The literature on innovation processes in developing countries identifies innovation capabilities and the enabling environment as key determinants for the innovative performance of a developing nation (Musiolik/Markard 2011). We combine the seven functions from the TIS with the concepts of innovation capability, enabling environment and policy-independent firms' strategies. These three concepts determine the success or failure of innovation in developing countries, as we will see in the literature and the analysis in the following sections. We summarize the framework in table 36.1 and group the functions as drivers and barriers in a framework for the analysis of the evolution of the TIS in wind energy in three developing countries.

36.2.2 Innovation Capabilities

Five factors often hinder the diffusion of technologies in developing countries: i) capability failures; ii) short-

Table 36.1: Overview of the framework for analysis of Wind Energy Technology Drivers and Barriers. **Source:** The authors based on Bergek, Jacobsson, Carlsson et al. (2008).

Drivers	TIS Functions	Barriers	Description
Innovation Capability	• Knowledge • Development and Diffusion • Entrepreneurial Activities • Resource mobilization • Influence in the direction of search	• Weak or fragmented innovation capability • Local search processes	Capabilities are weakened or fragmented when there are weak local linkages between institutions and industry. Capabilities drive these four functions of the TIS. Capability depends on pre-existing manufacturing and industrial capacity with skills, design and innovation proximity to the technology and industry under consideration
Enabling Environment	• Market Formation • Influence in the direction of search • Knowledge Development and Diffusion • Resource mobilization • Legitimation	• Unclear, conflicting policies • Poorly articulated demand • Market control by incumbents creating institutional and technological lock-ins • Technology characterized by increasing returns	An enabling environment unlocks institutional lock-ins and drives the five functions of the TIS.
Policy-independent Prime Mover Strategies	• Legitimation • Positive • Externalities • Entrepreneurial Experimentation	• Strategies of firms are stymied due to a non-conducive policy environment • Market control by incumbents creating institutional lock-ins	Policy-independent Prime Mover Strategies are those strategies or initial activities of a firm or firms that prompt entrepreneurial activities and spur the development of an industry. Such strategies induces these three functions of a TIS

falls in the technical skills needed to adopt a new technique; iii) organizational inadequacies which prevent the exploitation of a new technique; iv) deficiencies in business skills; and v) information asymmetry which prevents firms from taking rational decisions (Arnold/Guy 1991). Innovation capability[6] distinguishes between the capability i) to create new configurations of product and process technology, and ii) to absorb and implement changes and improvements to technologies, which are already in use. In the case of low carbon innovation in developing countries, the absorptive capability matters most. Absorbing solar and wind energy technology imports into the national

energy mix, requires a strong base of local innovation capability.

Absorptive innovation capabilities emerge from existing industrial structures and the knowledge base generated in a country. These capabilities are often deeply entrenched in old industry and knowledge structures, behaviours and institutions. For example, finance and banking, and skills and human capital development, all critical for knowledge diffusion, entrepreneurial experimentation and resource mobilization, are centred on old processes and technologies and often in large state owned companies. Market control by large companies and monopoly constellations impede a 'free' choice for customers and bring the search processes of firms to a halt (Johnson/Jacobsson 2001). Power shortages, which cause strong regulatory changes, are often the only way to diversify from the existing energy mix and institutional bases towards innovative energy technologies such as wind.[7]

---

6   Bell and Pavitt (1995) distinguish between production-based and innovation capabilities under technological capability as the capability to carry on producing goods and services with given product technology, and to use and operate given forms of process technology in existing organizational configurations.

36.2.3   Policy-independent Firms Strategies

The literature shows several examples of successful innovation in wind energy technologies through firms that have acted independently of public policy incentives. Early grass-roots innovators of wind turbines and solar collectors in Denmark and Austria evidence entrepreneurial activities that started independently of public policies. Although subsidies for renewable energy from the Danish government created a large home market later on, an important condition for a successful industry was its early environment of learning-by-doing and technological development among the innovators and network of actors within the industry (Brandt/Svendsen 2004).

Such firms and organizations influence the direction of search substantially, because they identify new opportunities. So-called 'infrastructural' organizations have often played an important role in shaping major new technologies and in identifying new opportunities (Bergek/Jacobsson/Carlsson et al. 2008).

In other non-renewable energy cases, firms independently initiated the development of an entire industry (Carlsson 2007):[8] examples are the US-based Texas Instruments' move to Bangalore, India; Olds Motor Works in Detroit, US; and Shockley Semiconductor's move to Silicon Valley from Bell Labs, New Jersey, US. The environment that existed in these cities or regions at that time was not unique, for similar conditions existed elsewhere in the respective countries. Success or failure also depended on the creativity and persistence of the entrepreneur—with a good element of luck (Carlsson 2007).

Public policy can support and sometimes even initiate the development of an industry in its early phase—although spontaneous development and serendipity seem to be more prevalent mechanisms (Carlsson 2007). Carlsson (2007) shows that the frequent role of public policy was to provide support and reinforcement at the late phases of industry formation—once the technology and/or product has been legitimized and positive externalities have been created. The initial activities of one firm or a few firms might prompt entrepreneurial activities among other firms within the same location that share the same advantages and policy environment, which leads us to the next point.

36.2.4   The Enabling Environment

When a technological trajectory is very 'powerful', it will be difficult to switch from one trajectory to an alternative one (Dosi 1982). We find these so-called path dependencies created through specific technological trajectories, especially in energy systems based on fossil fuels, but they also emerge around other technologies, such as hydropower in Brazil. The beneficiaries of the established energy system in developing countries follow their interests in maintaining the status quo and slow the diffusion of renewable energy technologies through their influence on institutions and policies.

Energy based on fossil fuels still forms a techno-institutional complex which is locked-in by mutually reinforcing technological and institutional factors. The techno-institutional complexes evolved historically into 'systems of accumulation', which benefit the incumbent holding most of the capital and for whom it is not in their vested interest to change existing structures. Lagging efforts to transition to renewable energy systems is partly attributed to the path-dependent nature of these countries' institutions, industry, and organizational incumbents. Structural barriers to alternative energy development are the close ties between fossil fuel industries and state-owned or private utilities, which monopolize national energy generation and distribution, as Fine and Rustomjee (1996) explain in the case of South Africa. Energy production and distribution structures are based on fossil fuels that are deeply entrenched and locked-in within fossil-fuels-based technological and industrial paradigms. Moreover, fossil fuels are heavily subsided and this makes them very competitive in comparison to fledgling and unsubsidized renewable energy technologies.

For this reason, a so-called enabling environment matters for renewable energy to take off in energy systems which are locked in to large fossil fuel or hydro powered energy systems. Often energy security and power outages, rather than environmental concerns, are the motivation for support programmes for alternative and renewable energy sources.

The role of public policy in creating an enabling environment has been widely acknowledged in the research literature. The innovation systems literature emphasizes the role of agencies and public support institutions in the interplay with researchers and engineers based in innovating firms and universities (Lun-

---

7   The cases of Brazil and South Africa show that regulatory changes were only applied after severe power shortages in 2001 and 2007 (Rennkamp/Fortes Westin 2013).

8   See Carlsson (2007), which examines the role of individual actors as well as public policy in ten industry clusters around the world.

dvall 1992; Lundvall/Borrás 2005; Nelson 1993). The technology innovation systems (TIS) approach also recognizes the importance of actors, institutions and key policy issues in their influential roles within the seven functions. The role of public policy is especially relevant for the five functions of legitimation, knowledge creation, resource mobilization, market formation, and influencing the direction of search.

The first TIS function, legitimation, refers to the legitimacy of policy processes and determines political outcomes. Legitimacy is a social construct of shared beliefs, logic and interests, which establishes success and failure in policy-making (Wallner 2008). Beliefs in a political order sustain legitimacy and make policy more likely to succeed. Legitimation refers to the process of legitimizing the institutional order (Luckmann/Berger 1991). In unequal developing countries, it is more difficult for policymakers to achieve legitimacy, because beliefs and interests diverge more widely among social groups. These difficulties unfold especially in the field of science, technology and innovation policy, because the interventions are mostly designed on the basis of industrialized countries' experiences and benefit the privileged rather than the poor (Rennkamp, 2011).

This problem translates into the second and third functions of knowledge creation and resource mobilization, because ministries of science and technology battle with small budgets and low R&D expenditure rates of less than one per cent of GDP. In unequal societies where poverty prevails, priorities focus on policy measures with more immediate and visible results that matter to low-income voters. It becomes more difficult to make the case for longer-term investment in industrial development and innovation that provides support to established companies and research institutions where the outcome is uncertain, because other social grants payments, investments in general infrastructure and safety seem more urgent.

The limited resources allocated for innovation often benefit companies in the established technological trajectories rather than benefitting new markets. This problem relates to the fourth function, market creation, which is directly related to the enabling environment. In many developing countries in Asia, Africa and Latin America, governments strictly control their energy markets. Historically, energy and electricity generation counted as a state function with parastatal electricity companies at its core. These governance structures favour the use of the historically predominant large-scale energy technologies, such as coal, nuclear and large hydro. These technologies demand central-

ized energy systems which involve firm governmental regulation. The interplay between actors in the government and the electricity and mining sectors can lead to fixed political economic structures that exclude any alternative trajectory. Once governments decide to promote alternative and renewable energies, there are several choices. Feed-in tariffs fix the price and vary in quantity, whereas quotas and auction systems fix the quantity allocated in the market and allow the price to vary. Both approaches have advantages and disadvantages. International experience has shown that feed-in tariffs can be beneficial in creating a demand that leads to industrial development in first-mover countries such as Germany. Auction and quota systems, in turn, bring market prices down significantly below the levels of the feed-in tariffs. The size of the allocated market then determines whether local content requirements added into the programmes can have positive impacts on industrial development (Lewis/Wiser 2007).

The choice of the incentive system has an immediate impact on the fifth function, the direction of search. The incentive system determines the behaviour of firms and other organizations in developing and diffusing new technologies. The market size and local content levels, for example, shape the competition, visions and expectations in the sector.

36.3 Analysis: Wind Energy Technology Innovation in South Africa, Brazil and India

India, Brazil and South Africa count as emerging markets in the wind energy sector. These markets differ in maturity and size. This comparative analysis of the wind energy sectors reveals similarities and differences in the transitions towards cleaner energy technologies. We apply the technology systems framework introduced in the previous section. We structure the analysis according to the three main factors that influence the ability to catch up: innovation capability, policy-independent firm strategies, and the enabling environment. We find that in all three countries innovation capability was built up over time through a combination of individual firm strategies and government incentives.

36.3.1 Policy-Independent Prime Mover Strategies

In all three countries, we find prime mover companies who have sought to deploy wind energy technologies independently of any public policy incentive. These three strategies differ from each other and have led to very different outcomes.

36.3.1.1 India

Suzlon Energy Limited is an Indian wind power company and the fifth largest wind turbine manufacturer by cumulative installed capacity, with 22,000 megawatts of installations globally. Suzlon was founded by Tulsi Tanti in a move to secure his textile company's energy needs. India's erratic grid power supply and the rising cost of electricity were offsetting any profits the textile company was making. The company was situated in the coastal region of Gujarat and the energy generated using the wind turbines technology made economic sense. It was a few years earlier that the *Danish International Development Agency* (DANIDA), at the request of the *Ministry of New and Renewable Energy* (MNRE), had begun two demonstration projects in the state of Gujarat and was creating interest among the private sector.

Tanti realized that other companies such as his could benefit similarly and so started advising on the use of the technology and introduced a business model that helped its diffusion. Suzlon sought complete ownership of the design and technology of wind turbines while its clients were responsible for twenty-five per cent of the upfront capital investment. Suzlon arranged the remaining seventy-five per cent on loan, which banks were initially very reluctant to fund.[9] In Jolly and Raven (2013), this is an example of "the activities of actors who have interest in particular institutional arrangements and who leverage resources to create new institutions or to transform existing ones" (Maquire/Hardy/Lawrence 2004: 657).

DANIDA was the first foreign development agency to show interest in the potential Indian wind power market back in 1986. DANIDA helped two state agencies[10]–*Tamil Nadu Electricity Board* (TNEB) and the *Gujarat Energy Development Agency* (GEDA)–to

develop demonstration wind farms, and experience was gained in wind farm planning, implementation and management. Three Danish firms worked closely with local partners to develop indigenous technical capacity and skills while Danish contractors manufactured and delivered the turbines and most of the towers (Jolly/Raven 2013).

These demonstration projects in India were very successful, for they legitimized the technology by widening participation and creating interest among the private sector regarding the potential of wind energy. They also influenced the direction of search by establishing technical and economic viability. The programme also instigated new entrepreneurial and industrial activities, including cooperation between Indian and foreign firms (Jolly/Raven 2013). Because of these programmes, Danish-Indian joint ventures between NEPC India Ltd. and Micon and between Vestas and RRB Energy Ltd. were established in 1987.

36.3.1.2 Brazil

As in the cases of Indian and South Africa, there was no direct innovation capability in the Brazilian wind energy sector until the early 1990s. The first wind turbine in Brazil started to rotate its blades in 1992, as test equipment on Fernando de Noronha, a small island 500 km off the shores of Pernambuco in northeastern Brazil. The islands' energy supply depends on shipped petrol. Wind was supposed to diversify this dependence, but visual concerns about polluting the island's landscape made a larger use of wind energy unfeasible.

Two companies entered the wind energy sector in Brazil in the mid-1990s independently of any government programme.

The *original equipment manufacturer* (OEM) Wobben, a subsidiary of the German company Enercon, was the first company to build turbine factories in Brazil. Wobben was founded in 1996. Eighteen years later, the company has three factories in the country, with the main production site in São Paulo and two smaller sites in the north-east.

Enercon's CEO recognized Brazil's market potential for wind energy early and founded the Brazilian subsidiary six years prior to any renewable energy incentives programmes.[11] The rationale for Enercon to invest in local production rather than choosing an import strategy was to build up local capability for the Latin American market and use the advantages of the

---

9 By 2008, about forty to fifty Indian banks were financing wind power projects for Suzlon clients (Nirmalya 2009).

10 The states of Tamil Nadu and Gujarat have long costal shores along the Arabian Sea and the Indian Ocean respectively.

---

11 Interview CEO Wobben Brasil.

prime mover. In the 1990s, investing in Brazil seemed safer than investing in the Asian markets, where cases of industrial spying and reverse engineering were more frequently reported.[12] Registered as a Brazilian company, Wobben can set up its contracts and project finance in local currency and reduce the exchange risk. The company employs exclusively local staff totalling 1,200 employees. German staff is usually only present temporarily for specific training purposes.[13]

The Brazilian wind blade manufacturer Tecsis was founded in 1995 as a spin-off from the aviation industry. The company received its first contract from Wobben/Enercon for blade manufacturing. Wobben was manufacturing blades in-house at the time and wanted to outsource this production line. Tecsis had the necessary know-how from the aviation industry and started to produce 19 m blades. Tecsis managed to establish itself slowly in the market through this opportunity.[14] Today, the company is the leading blade manufacturer in Latin America and continues to supply Wobben for its export market.

Wobben and Tecsis' entrepreneurial experimentation activities contributed significantly to at least three other functions in establishing the wind technology innovation system. Firstly, the investment increased legitimation of wind energy technologies in the countries. The company managed to set up a number of parks before the government's support programme for renewable energies started. This helped to show that wind energy is a functional alternative technology, although the prices at the time were still comparatively high because of the lack of scale. Secondly, resource mobilization into local production and staff helped build innovation capabilities in the firm and in the country. Thirdly, knowledge diffused through subcontracting and the drawing of innovation capabilities from third-tier sectors, such as the aviation sector.

36.3.1.3 South Africa

South Africa is a latecomer to wind energy compared to Brazil and India. Like the other two countries, it took a long time to decide on an incentive programme. In the meantime, some companies developed independent demonstration projects. The Klipheuwel wind farm came on grid in 2003 with 3.2 MW and the Darling wind farm in 2008 with 5.2 MW. The equipment

for these parks was mostly imported, apart from a few towers which were built on site.

One company, IWEC, tried to manufacture wind turbines locally in anticipation of a larger wind energy market from 2006. IWEC entered into a licence agreement for technology and knowledge transfer with the German manufacturer Aerodyn. In 2012, IWEC produced its first full turbines with sixty per cent locally manufactured components. The main challenges for reaching scale in production were restricted finance opportunities for locally manufactured technologies, and restrictive regulation which made it difficult to acquire power purchase agreements. After the main investor pulled out of the business, the market closed up for IWEC and the company went into liquidation in 2013. The *Renewable Energy Independent Power Producer Procurement* (REIPPP) Programme, which opened the market to independent power producers for the first time, initially targeted renewable energy plants that produce more than 5 MW. It also required at least two years of experience and proof of successfully established plants, a condition which was difficult for local newcomers to meet and which advantaged foreign multinational companies.

36.3.1.4 Summary

The Indian joint venture strategy combined a local firm strategy through joint ventures with more mature wind energy companies. This strategy combined existing innovation capability with technology transfer from the more mature companies. The South African case showed a similar case of a South African company, although run by foreigners, partnering with a German company, but only through a licence agreement. The absence of incentives for new independent producers at the time disadvantaged local prime movers in entering the local market. The Brazilian case demonstrated that capability emerged mostly from the strategy of one multinational company which sought to strategically build up domestic capability to position itself for a future market. All three countries demonstrated entrepreneurial experimentation. The success and failure of these efforts depended to some extent on the enabling environment, and this will be explained in more detail in the following section.

36.3.2 Enabling Environment

The regulatory frameworks for renewable energy differ significantly between India on the one hand and Brazil and South Africa on the other. The Indian government engaged in a wider industrial policy reform

---

12 Interview Wobben representative.
13 Interview Enercon representative.
14 Interview CEO Tecsis Brasil, see at: <https://www.youtube.com/watch?v=_1so7ELbJgk>.

whereas the Brazilian and South African administrations only applied local content requirements as their main industrial policy instruments.

36.3.2.1 India

Foreign companies could legally increase their equity shares to above forty per cent after the industrial policy reforms of the early 1990s. Equity rose to forty-nine per cent in most sectors and a hundred per cent in some. This move encouraged private sector participation in many sectors and several Indian companies faced increasing technological and competitive pressures. As a counter-reaction, Indian companies began importing more efficient technologies and increased in-house R&D expenditure (Lall 1987). At the same time, the government of India started to focus its wind energy policy on stronger private sector involvement. This involved extending public finance to private sector wind power projects and providing fiscal and financial incentives to encourage private investment.

The industrial reforms of the 1990s removed many of the restrictions. At the time of the emergence of a wind energy industry in the 1970s, Indian companies lagged behind the technological frontier and innovation capabilities of its international counterparts. Importing older technologies was common practice at the time. Little effort was made in adapting and improving these technologies for domestic use (Kristinsson/Rao 2008). Most Indian firms were observed to make minimal R&D investment in technologies as they already had access to a highly protected and uncontested home market (Lall 1987).

Actual development and growth of the wind energy sector in India began in the 1990s under the Ministry of New and Renewable Energy (MNRE), and by the late 2000s wind power had significantly increased, making India the country with the fifth largest installed wind power capacity in the world. The development and implementation of wind energy technology in India co-evolved with major institutional interventions such as financial support schemes, technology policies, technical standards, grid connection rules, industry organizations and international collaborations (Jolly/Raven 2010). Each state has its own power regulatory, transmission and generation authority, and a range of state-level wind energy programmes and economic incentives associated with the state's renewable energy resource.

Entrepreneurial activities with the Danes further spurred on domestic manufacturing through a series of licensing agreements with other foreign companies such as the German Enercon, Nordex and DeWind.

Many Indian companies diversified to reap the benefits of the economic and technological opportunities which the reforms presented. For example, Shriram had diversified from the financial sector into an energy engineering company by 2001. The company entered into a technological joint venture with an Italian company in 2007 to jointly manufacture and install large-scale wind energy generators. This process of licensing the production of foreign-developed wind turbines resulted in the introduction of somewhat old and small models but it nonetheless allowed for the use of local labour assets and the local production of wind turbines at a lower cost (Neij/Dannemand Anderson 2012).

Suzlon bought two European companies that helped expand its technological capability in gearbox manufacturing. Knowledge development and diffusion processes in India are being enhanced and supported through the development of such international knowledge links and networks. The innovation path in India is characterized by massive support for domestic market formation and an enhanced knowledge development and diffusion process made possible through knowledge spillover from foreign manufacturers (Neij/Dannemand Anderson 2012).

A series of incentives and policies formulated in India influenced the direction of search and market formation. These policy and fiscal incentives successfully assisted the development of the wind energy industry in India (Kristinsson/Rao 2007). Tax exemptions of 100 per cent for some wind turbine components, and an accelerated tax depreciation of 80 per cent, along with feed-in tariff rates for large scale wind technologies, were implemented (GWEC 2011). Since 2011, the Indian government has allowed 100 per cent *foreign direct investment* (FDI) in renewable energy generation projects.

A notable barrier to wind energy in India is the presence of a single utility that monopolizes the entire energy market, including national electricity transmission. This structure hinders the diffusion of renewable energy in many developing countries. The utilities often pursue their interest in fossil fuel technologies, and hinder the development of a successful wind energy industry (Kristinsson/Rao 2008). In India, there is resistance among the state-run utilities[15] to granting 'third party sale'[16] facility to wind power producers (Kristinsson/Rao 2008). Further, there is lack of an appropriate regulatory framework to facilitate

---

15　Each state in India has its own independent power generation and distribution facilities owned and managed by state-run utilities.

the purchase of renewable energy from outside the host states; inadequate grid connectivity; high wheeling costs[17]; and bureaucratic delays in acquiring land and obtaining statutory clearances (GWEC 2011).

36.3.2.2 Brazil

Both in Brazil and South Africa, incentives for renewable energies were direct consequences of shortages in the electric power supply. In Brazil, the shortages resulted from the lack of rain for the hydropower plants, but the blackouts also demonstrated the lack of planning and lack of investment in the electricity supply sector. Hydropower is the main source of electricity production in Brazil.

In 2002, the government announced the *Incentive Programme for Renewable and Alternative Energy* (PROINFA), which supported other renewable energy sources, including wind, in form of a feed-in programme. The programme offered a price of R$ 300 (US$ 128) per MWh of wind energy to power producers under a power purchase agreement for twenty years through the utility Eletrobrás. The aim was to add 3,300 MW generation capacity, of which forty-three per cent was supposed to come from wind. The national development bank (BNDES) offered inexpensive loans for financing up to eighty per cent of the project value. Despite these high-level incentives, the implementation of the programme progressed very slowly.

By 2006, only six of the seventy-five planned parks had been built. The reasons for the delay were the sluggish bureaucracy, which issued environmental permits very slowly, and the local content requirements of sixty per cent. PROINFA intended to attract foreign OEMs to produce in Brazil, but it was difficult for new companies to fulfil these requirements from scratch.[18] Wobben and Tecsis benefitted from the local content requirements and supplied most of the technology to the developers. In 2012, Wobben provided 1 GW of the 2 GW installed wind capacity, most of which still came from contracts under PROINFA. However, the capacity to produce locally was mostly limited to these two companies. The slow implemen-

tation of PROINFA eventually led to a change in the incentive system.

In 2009, the Brazilian government decided to procure additional wind energy capacity through a competitive auction system implemented through the regulator ANEEL, the *Energy Research Enterprise* (EPE), and BNDES. Local content requirements of sixty per cent remained the main industrial policy component, but now these requirements were optional for those companies who wanted to take advantage of the loans of 0.9 per cent from BNDES. Companies with their own financial schemes were free to import, but only a Chinese manufacturer tried this in Aracajú, and failed.

The auction system produced wind energy capacity at a very quick pace. The competitive bidding process brought prices down to a third of the original feed-in tariff under PROINFA. The auction system worked on the opposite principle from the feed-in tariff. It did not set a fixed price but a fixed quantity, which bidders competed for at a flexible price. In 2014, the incentive system produced 3.6 GW installed capacity in 148 wind parks.[19]

The incentive systems both under PROINFA and the auctions increased legitimation for wind technologies and the development of the sector in the country. The auction system attracted twelve international OEMs to do business in the country. After three years, five OEMs managed to set up local manufacturing facilities to fulfil the local content requirements, while others left the market.

Significant resources to support wind energy sector development within the incentive programme came through the loans at 0.9 per cent from BNDES. These resources were compensated to some extent through determining the market size rather than guaranteeing a fixed price, because the costs came down. Market size matters for the successful implementation of local content requirements, because markets that are too small deter investors. The Brazilian government solved this problem by announcing 11 GW of total capacity by 2020 according to the EPE.

36.3.2.3 South Africa

The South African electricity supply system is locked in to coal just as the Brazilian system is locked in to hydropower as the main source. In 2008, the country experienced major blackouts, which raised serious questions about planning and investment in future

---

16 Third-party sale amounts to allowing wind power producers to use the grid infrastructure to sell power to any industrial client at any mutually agreed rate.

17 Wheeling refers to the transfer of electrical power through transmission and distribution lines from one utility's service area to another's.

18 Interview Gamesa representative. Gamesa left the Brazilian market temporarily and only returned once the incentive system changed.

---

19 Brazilian Wind Energy Association Abeeólica; see at: < http://www.portalabeeolica.org.br>.

capacity. The *National Energy Regulator* (NERSA) started a process for a feed-in tariff for renewable energies in 2009. Eventually, the Department of Energy and the National Treasury decided in favour of a competitive bidding programme for a determined market size, rather than paying out feed-in tariffs. The reasoning for this decision was that a feed-in did not comply with the country's procurement rules.

In 2011, the REIPPP Programme started to procure 3.7 GW of renewable energy including wind through three bidding rounds, which concluded in November 2013. In these three rounds, 562 MW, 634 MW and 787 MW respectively were allocated to wind power, which leaves almost 2 GW to be built by 2016. The prices dropped by twenty-two per cent from the first bidding round to the second—from R1140 (US$ 114.27) to R890 (US$ 89) per MWh—and again in the third round to R310 per MWh (US$ 31), which is lower than current new-build coal prices.[20]

As with the Brazilian programme, the REIPPP Programme requires the bidders to fulfil socio-economic development criteria, including local content. The bidding criteria are composed of seventy per cent price determination and thirty per cent local content and social development criteria. The local content requirements are slightly different from the Brazilian ones in that the bidders can provide a range and not a fixed percentage. The targets increased in each bidding round from twenty-five per cent to forty-five per cent and then to sixty-five per cent. Translated into actual technologies, this means that the local content in the first round can be covered through the balance of plant, that is, the infrastructural components. The second round requires locally made towers. In the third round, nacelle boxes and blades would have to be made locally, and this requires a local manufacturer (Rennkamp/Fortes Westin 2013). The local content requirements were met to averages of twenty-one per cent and thirty-nine per cent in the first two rounds. Meeting the local content requirements for the third round will require blade manufacturing. Two local companies and the Danish LM blade manufacturer are currently assessing the possibilities of local manufacturing in South Africa.

Like the Brazilian programme, the South African REIPPP Programme targets knowledge transfer and diffusion through investment from international OEMs in the local market. Both countries have highly

regulated electricity markets in which the government can determine the market space for each technology. The allocated market size for wind, however, is much smaller than in Brazil and does not necessarily translate into profitable investment opportunities.[21] In terms of resource allocation, the South African programme offers just the market opportunity and the power purchase agreement. The programme does not provide for any loans or tax reductions. The developers need to acquire finance from the commercial banks, or the development bank IDC, whose rates are similar to the offers of the commercial bank. The National Treasury claims one per cent of the investment to cover the administration costs. These have now become a surplus and the Treasury is looking into ways of spending these resources. The programme does not rely on support from tax revenue and is currently producing additional revenue. The impact on legitimation of renewable energy was very high, as the market could not take off for a long time without changes in the regulatory framework.

36.3.2.4 Summary

The successful industrial policy reform in India spilled over into the wind energy sector and allowed Indian companies to enter joint ventures and acquire foreign firms, and this increased the R&D activities. State utilities remain obstacles to the market size by objecting to power purchase agreements. In Brazil and South Africa, in turn, industrial policy was rather designed as trade policy, through the narrow local content requirements. In Brazil, these requirements were more fruitfully implemented, because the allocated market size was more significant, and the foreign OEMs could enjoy the incentive of inexpensive loans and existing innovation capability in Wobben and Tecsis. In South Africa, however, local content requirements were not connected to any fiscal incentive or loan scheme. Local firms especially struggled to enter the market because the incentive system narrowed down access and the relatively small market size made the bidding process highly favourable to experienced international OEMs.

36.3.3 Innovation Capabilities

This section analyses the innovation capabilities touched on in the previous sections from a historical perspective. Innovation capabilities were built through

---

20 South African Wind Energy Association; see at: <http://www.greenbusinessguide.co.za/wind-power-cost-drops-even-further-below-coal/>.

21 Interview LM representative.

the sum of independent prime mover strategies and the effects of the enabling environment.

36.3.3.1 India

It was in the early 1900s that Poul Le Cour of Denmark started experimenting with wind turbines and blade designs. At that time, knowledge about aerodynamic flows around blades was underdeveloped and a new aerodynamical theory emerged, largely due to the efforts of three Danish scientists (Kristinsson/Rao 2007). Developing and exchanging the theoretical and applied knowledge of aerodynamic flows became critical to the success of the Danish wind energy industry.

The *Council of Scientific and Industrial Research* (CSIR) established the *National Aerospace Laboratories* (NAL) of India in 1959. Its mandate is to develop civilian aerospace technologies with strong science and design capabilities while its multidisciplinary focus allowed it to initiate one of the first projects on wind energy in India back in the 1960s, when NAL developed the first models of wind turbines for rural applications, used for water pumping and battery charging.

In 1983, India decided to develop a *Light Combat Aircraft* (LCA) for which NAL began to work closely with two other institutions, the *Indian Space Research Organisation* (ISRO)[22] and the *Defence Research and Development Organisation* (DRDO).[23] Collaborations with and research projects with these institutions quickly pushed up NAL's capabilities several notches and by 1990, NAL had become a world-class aerospace laboratory. Its technological capabilities covered advanced computational and experimental fluid dynamics, including the use of smart materials, all capabilities required in the development of wind energy technologies. Around 2006, NAL became involved in the development of a 500 kW low-cost, indigenous, horizontal axis *Wind Energy Generator* (WEG) that claimed to be cheaper and better suited to Indian conditions than the European ones being sold in India at that time. This was a joint venture between the *Structural Engineering Research Centre* (SERC), Sangeet Group of Companies and the *Centre for Wind Energy Technology* (C-WET).

Research and knowledge linkages between the industry and NAL were weak. Public resources for technological development were mobilized in a frag-

mented and silo-like manner. Although innovation capabilities and skills in aero- and fluid dynamics existed in organizations such as NAL, there is no evidence of these capabilities directly spilling over into entrepreneurial activities within the industry.

This fragmentation in resource mobilization and the fracture in knowledge links between the industry and public research organizations came about as a result of the restrictive nature of import-led substitution industrial development policies and the Licence Raj system[24] that was in place for four decades from the 1950s. Though this encouraged industrial and innovation development capabilities in large-scale industries such as petrochemicals, steel and automobile manufacturing, the licence system significantly restricted the wider participation of entrepreneurial firms and activities in high-level and medium-level technology-based areas. It is also for this reason that many of the innovative capabilities built within *public research organizations* (PROs) were unable to translate into applications for the industry. The importance of the criticality of the link between PROs and the industry went under the radar for a long time in India.

One of the essential components of the successful development of the Danish wind energy industry since the 1970s has been the testing and certification[25] of its turbine and blade technologies (Douthwaite 2002). Testing and quality certification programmes and initiatives influence the direction of search by affecting the perceptibility of the requirements of actors and may influence further investments in the technology by improving the reliability of the technology in the market.

From quite early on, there were strong influences on the direction of search in India, and this helped build a wind energy industry through new investments and entrepreneurial activities that improved the perception and reliability of the technology. In 1994, two training institutions were established under a Danish cooperation programme to train regulatory authorities, municipalities and companies about the importance of and need for quality assessment and certification of wind turbines (Kristinsson/Rao 2008). In 1998, the Indian Ministry of New and Renewable Energy

---

22 By then India had launched its first successful space vehicle and its space programme had begun to mature.

23 Soon afterwards, DRDO started its own missile development programme.

---

24 Licences were given to a select few and up to eighty government agencies had to be satisfied before private companies could produce something and, if granted, the government would regulate production.

25 In Denmark, this process was led and institutionalized by Risø Laboratories.

(MNRE)[26] set up the Centre for Wind Energy Technology (C-WET), an R&D institution with similar testing and certification facilities to Risø. Both India and China have developed technology certification programmes to ensure high-quality domestic wind turbines (Kristinsson/Rao 2008; Neij/Dannemand Anderson 2012). Countries such as the US that did not have testing and certification processes for emerging wind turbine models experienced severe market trust problems (Neij/Dannemand Anderson 2012).

In 1995, MNRE (previously MNES) issued guidelines that obliged private wind farm developers to produce *Detailed Project Reports* (DPRs) that contained detailed information on micro-siting,[27] the selection of wind turbine equipment, operation and maintenance data and performance evaluation (Jolly/Raven 2010). Although this was to ensure that incentives provided by the central and state governments were not misused, it contributed to influencing the direction of search and strengthening knowledge links between various actors. State electricity boards, state nodal agencies, manufacturers, developers, and investors became aware of the need for joint and planned development and implementation of wind power projects (Jolly/Raven 2010).

36.3.3.2 Brazil

In Brazil, innovation capabilities in the wind energy sector emerged through a combination of the resource mobilization, entrepreneurial experimentation and knowledge diffusion of Enercon and the spin-off firm Tecsis from the aviation sector. Local capabilities from a related sector matched with the opportunities created through the foreign OEM. Capability at a larger scale only really took off once the auction programme opened the market to the international industry. The content requirements, however, only incentivized the local manufacturing of low to medium technological content, namely the infrastructural components of the balance of plant, the tower, blades, nacelle box and the hub. The gearbox and other micro-electrical components continue to be imported (Rennkamp/Fortes Westin 2013).

Local content requirements were the only instrument to stimulate 'knowledge creation' through foreign direct investment-driven technology transfer in the auction programme. There were no explicit support mechanisms for innovation beyond the technology transfer through the OEM. When 2 GW were installed in 2012, the Wind Energy Association initiated a network to seek public policy support for innovation through a network of business, university researchers and government to fill the gap.[28]

36.3.3.3 South Africa

In the South African situation, the wind energy sector started building innovation capabilities mostly from scratch. The innovation capabilities before the procurement programme were built up tediously through partnerships with international companies. IWEC tried through licensing, the Darling wind farm through imports. Finally, the procurement programme limited market access to only mature OEMs that can present evidence of at least two years of experience, which inadvertently excluded start-up firms. Nationally, innovation capability increased through the REIPPP incentive programme, but it channelled these capabilities through the foreign OEM (Rennkamp/Fortes Westin 2013). As a result, the focus on local content requirements nurtured low technological content and benefited civil construction firms more than wind energy technology companies in the first and second bidding rounds. With growing content, the industry accounts for producing medium technological content including blades in order to fulfil the content requirements.

As in the Brazilian case, the incentive programme influenced the direction of search and entrepreneurial experimentation significantly. The programme attracted significant international investment into the country. So far, there has not been significant resource mobilization through the REIPPP programme to support R&D. Several initiatives fill part of the gap. Research at the Universities of Cape Town and Stellenbosch focuses on smaller scale wind turbines. The Cape Peninsula University of Technology and the Nelson Mandela Metropolitan University host a joint training centre for wind energy engineers with support of the German development cooperation agency. By 2014, the Local content requirements produced two tower factories who manufacture about 140 towers per year. The first manufacturer in Port Elizabeth

---

26 MNRE is a Ministry of the Government of India relating to new and renewable energy. It has its origins in the oil price shock that led to the establishment of a Commission for Additional Sources of Energy in 1981. The Commission was renamed in 2006.

27 Where to place individual wind turbines in a larger wind farm.

---

28 Opening of Brazil Wind Power 2012, Rio de Janeiro.

received most of the contracts from the OEMs for towers in rounds 2 and 3. The second manufacturer in Cape Town opened its gates only towards the end of 2014. Some local companies struggle to meet the current demand on time. The debates on sense and nonsense of local content requirements under the current market size continue between the OEMs and the Department of Trade and Industry. While the DTI aspires to increase the local content requirements to 60–75 per cent and to move to the next step of localizing blade manufacturing, the OEMs worry about the quality, timely delivery and cost of local production.[29]

36.3.3.4 Summary

The three drivers and barriers to successful technological catch-up have had significant impacts on the evolution of wind technology systems in all three countries. Existing innovation capability in India and Brazil facilitated the evolution of the wind sector. Once the incentive systems were put in place, companies could draw on this comparative advantage. In the Indian case, significant innovation capability entered through Suzlon's acquisition of Danish companies. In the Brazilian case, the policy-independent firm strategies of Enercon and later Tecsis built a significant base of innovation capability, which advantaged the companies in complying with local content rules. Their capability also allowed others to be supplied with the necessary local content. The main influence on the direction of search, which sped up the wind energy technology system, was the auction system. The auctions created a significant market size of 11 GW, which attracted the interest and investment of the global wind energy industry. Five international original equipment manufacturers have established themselves in the market over the last five years.

The South African case showed that a lack of an enabling environment can become a barrier to innovative capability built through policy-independent strategies. Once these rules were changed, the technology system could evolve. As in the Brazilian case, the competitive bidding programme attracted the global industry's interest and investment. The market size is not as significant as in Brazil.

All three cases show that innovation capability only builds up over time and requires long-term commitment over time from companies and public support programmes. Capabilities and the knowledge

base lie at the core of the technology system. The Indian case showed that historically-grown capability in the NAL flourished once the enabling environment supported it. In the Brazilian case, the international equipment manufacturer built capability with a long-term vision, which paid off once the market was established. In the South African case, early capabilities and knowledge bases could not unfold in an unfavourable environment. Subsequent initiatives by universities are attempting to establish a knowledge base, and this will build capability in the longer term. Longer-term private sector commitment will be necessary to turn this knowledge into the market, which again will depend on a significant market size. Local content requirements function effectively as a trade barrier, but will have to be complemented through further support in order to establish a knowledge base to support the technology system.

36.4 Conclusion

The three cases above show that different combinations of the seven TIS functions are prevalent in the different countries. Each of these functions can contribute to the success or failure of a wind energy innovation system. In all these countries, knowledge diffusion, entrepreneurial activities, experimentation, and resource mobilization were found to be locked in to each country's existing knowledge, industrial structures and institutions. However, they do in fact induce positive externalities and initiate a process of legitimation within the TIS of each country.

A common feature of these countries is a 'powerful' technological trajectory based on fossil fuels, or large hydropower in the case of Brazil. When trajectories are powerful it is difficult to switch from one trajectory to an alternative one, and elements of path dependencies and institutional lock-ins become commonplace.

We found that Brazil and India have been more successful in catching up so far because of two factors: firstly, in both countries prime moving companies accumulated innovation capabilities independently from public policy incentives. This absorptive capability sped up the development of the wind technology system in both India and Brazil, whereas the South African market needed to build capability almost from scratch.

There is no guarantee that independent firm strategies will succeed or fail. In all three cases, a partnership with a company from a mature wind energy market was necessary to trigger knowledge and technology

---

29 Interviews with OEMs at Windaba 2014; Discussion Panel of OEMs and DTI at Windaba 2014.

Table 36.2: Synthesis of the results. **Source:** The authors.

	Innovation capability	Policy-independent firm strategies	Enabling environment
India	Historical capability in the NAL, built through joint ventures.	Joint ventures and successful technology transfer through firm acquisition.	Tax incentives, renewable energy certificates and feed-in tariffs in various provinces.
Brazil	Significant capability built through multinational and aviation sector.	Technology transfer through multinationals and local spin-offs.	Significant market size nationally allocated in an auction system. High influence on the direction of search, evolution of TIS.
South Africa	Limited capability in local companies and demonstration projects.	Individual efforts which did not succeed because of a lack of an enabling environment.	REIPPP Programme in support of large plants, adjusted later for smaller plants, which targets mature companies and FDI rather than boosting local firms.

transfer. In the Indian case, the Danish cooperation agency supported the joint venture between Danish and Indian firms, which eventually led to the successful transfer of technologies to Indian companies. In Brazil, the well-established OEM operated under a Brazilian name and invested significant resources. The South African case, in turn, was smaller in scale, with fewer resources and support from an overseas company or donor. Another important factor for the failure of IWEC's strategy was the difficult and unsupportive policy environment for smaller-scale, local innovators.

Innovation capabilities emerged through a combination of resource mobilization, entrepreneurial experimentation and knowledge diffusion in Brazil, so the local capabilities from a related sector, the aviation sector, matched the opportunities created through the foreign OEM in the wind energy sector; while in India, although innovation capabilities and skills in aero- and fluid dynamics existed within public research organizations (PROs), there is no evidence of these capabilities spilling over to the industry. So innovation capabilities in India emerged through a combination of entrepreneurial experimentation, influence in the direction of search (testing and certification programmes), and knowledge diffusion (and technology transfer) from foreign joint ventures.

The second critical factor for innovation in the wind energy sector is the enabling environment. The market size and trade barriers influenced the decisions of investors on whether to enter an emerging market or not. In Brazil and South Africa, local content requirements (LCR) are used as an instrument to stimulate knowledge creation and create legitimation of the technology. In India, there are no explicit local content requirements in the wind energy sector, but its restrictions on foreign direct investment, which

require foreign firms to engage in 49:51 joint ventures with local firms, is a similar rule that benefits the local industry. These joint ventures stimulate knowledge creation and entrepreneurial experimentation and influence the direction of search and eventually legitimation. Positive externalities were created as a result because these functions were enabled. In the case of India, an enabling environment was created after the industrial policy reforms of the early 1990s. Foreign companies were allowed to increase their equity shares in the form of joint ventures with local Indian companies. This move encouraged private sector participation in the wind energy sector, thus greatly influencing entrepreneurial activities. A series of incentives and policies were formulated in India and Brazil, which legitimized and influenced the direction of search and formed markets. Market size was found to be critical for the successful implementation of wind energy policies. For example, the Brazilian and South African governments solved the problem of market constraints by announcing wind energy targets. Such a move legitimates the technology in the market place, encouraging further investment. In Brazil, the auctioning process increased legitimation for wind technologies and the development of the sector, and stimulated the expansion of wind energy capacity in the country.

However, in the case of South Africa, the allocated market size for wind is much smaller than in Brazil, and this has translated into fewer profitable investment opportunities and slower and smaller-scale wind technology development.

References

Amsden, Alice, 2003: *The Rise of "The Rest"Challenges to the West from Late-Industrializing Economies* (Oxford: Oxford University Press).

Arnold, Erik; Guy, Ken, 1991: "Diffusion Policies for IT: The Way Forward", Paper prepared for OECD/ICCP Expert Group on the Economic Implications of Information Technologies (Paris: OCED).

Bell, Martin; Pavitt, Keith, 1995: "The Development of Technological Capabilities", in: *Trade, Technology and International Competitiveness*, 22,4831: 69-101.

Bergek, Anna; Jacobsson, Staffan, 2003: "The emergence of a growth industry: a comparative analysis of the German, Dutch and Swedish wind turbine industries", in: *Change, transformation and development*: 197-227.

Bergek, Anna; Jacobsson, Staffan; Carlsson, Bo; Lindmark, Sven; Rickne, Annika, 2008: "Analysing the functional dynamics of technological innovation systems: A scheme of analysis", in: *Research Policy*, 37: 407-429.

Brandt, Urs; Svendsen, Gert, 2004: "Switch Point and First-Mover Advantage: The case of the Wind Energy Industry", Working Paper, 04-02.

Breschi, Stefano; Malerba, Franco, 1997: "Sectoral innovation systems: Technological regimes, Schumpeterian dynamics, and spatial boundaries", in: Edquist, Charles (Ed), *Systems of Innovation: Technologies, Institutions and Organizations* (London: Pinter): 130-156.

Carlsson, Bo; Stankiewicz, Rikard, 1991: "On the nature, function and composition of technological systems", in: *Journal of Evolutionary Economics*, 1,2: 93-118.

Carlsson, Bo, 2007: "The Role of Public Policy in Emerging Clusters", in Braunerhjelm, Pontus and Feldman, Maryann (eds.), *Cluster Genesis*: Technology-Based Industrial Development, Oxford: Oxford University Press

Cimoli, Mario; Dosi, Giovanni; Stiglitz, Joseph, 2009. *Industrial Policy and Development: The Political Economy of Capabilities Accumulation* (Oxford: Oxford University Press).

Dosi, Giovanni, 1982: "Technological paradigms and technological trajectories: A suggested interpretation of the determinants and directions of technical change", in: *Research Policy*, 11,3: 147-162.

Douthwaite, Boru, 2002: "Enabling Innovation: A Practical Guide to Understanding and Fostering", in: *Technological Change* (London: Zed Books).

Fagerberg, Jan; Godinho, Manuel 2005: *Innovation and catching-up, The Oxford Handbook of Innovation* (Oxford: Oxford University Press).

Fine, Ben; Rustomjee, Zavareh, 1996: *The Political Economy of South Africa—From Minerals-Energy complex to Industrialisation* (Johannesburg: Witwatersrand University Press).

Fu, Xiaolan; Pietrobelli, Carlo; Soete, Luc, 2011: "The Role of Foreign Technology and Indigenous Innovation in the Emerging Economies: Technological Change and Catching-up", in: *World Development*, 39: 1204-1212.

Furman, Jeffrey; Porter, Michael; Stern, Scott, 2002: "The determinants of national innovative capacity", in: *Research Policy*, 31, 899-933.

Global Wind Energy Council, 2011, Global Wind Report 2011, GWEC, Brussels.

Jacobsson, Staffan, 2008, "The emergence and troubled growth of a 'biopower' innovation system in Sweden", in: *Energy Policy* 36, 4: 1491-1508.

Johnson, Anna; Jacobsson, Staffan, 2001, "Inducement and blocking mechanisms in the development of a new industry: the case of renewable energy technology in Sweden", in: Technology and the Market, Demand, Users and Innovation, Edward Elgar, 89-111.

Jolly, Suyash; Raven, Rob, 2013, "Collective institutional entrepreneurship and contestations in wind energy in India",Working Paper 13.10, Eindhoven Centre for Innovation Studies, http://cms.tm.tue.nl/Ecis/Files/papers/wp2013/wp1310.pdf

Kristinsson, Kari; Rao, Rekha, 2008, "Interactive Learning or Technology Transfer as a Way to Catch Up? Analysing the Wind Energy Industry in Denmark and India", in: Industry and Innovation 15,3: 297-320.

Lall, Sanjaya, 1987: Learning to Industrialize: The Acquisition of Technological Capability by India (London: Macmillan).

Lewis, Joana; Wiser, Ryan, 2007: "Fostering a renewable energy technology industry: An international comparison of wind industry policy support mechanisms", in: *Energy Policy* 35, 1844-1857.

Luckmann, Thomas; Berger, Peter, 1991: The social construction of reality: A treatise in the sociology of knowledge (London: Penguin UK)

Lundvall, Bengt-Åke, 1992: *National Systems of Innovation Towards a Theory of Innovation and Interactive Learning* (London: Pinter).

Lundvall, Bengt-Åke; Borrás, Susana, 2005: "Science, Technology and Innovation Policy", in: Fagerberg, Jan; Movery, David; Nelson, Richard (Eds.): *The Oxford Handbook of Innovation* (Oxford: Oxford University Press).

Maguire, Steve; Hardy, Cynthia; Lawrence, Thomas, 2004, "Institutional entrepreneurship in emerging fields: HIV/AIDS treatment advocacy in Canada", in: Academy of Management Journal, 47,5: 657-679.

Musiolik, Jörg; Markard, Jochen, 2011: "Creating and shaping innovation systems: Formal networks in the innovation system for stationary fuel cells in Germany", in: *Energy Policy*, 39: 1909-1922.

Negro, Simona; Alkemade, Floortje; Hekkert, Marko, 2012: "Why does renewable energy diffuse so slowly? A review of innovation system problems", in: *Renewable and Sustainable Energy Reviews*, 16: 3836-3846.

Neij, Lena; Dannemand Andersen, Per, 2012: "A Comparative Assessment of Wind Turbine Innovation and Diffusion Policies: Historical Case Studies of Energy Technology Innovation", in: Grubler A. et al. (Ed.): *The Global*

Energy Assessment (Cambridge: Cambridge University Press).

Nelson, Richard, 1993: *National Innovation Systems A comparative Analysis* (New York: Oxford University Press).

Perrot, Radhika, 2012: "The dynamics of renewable energy transition in developing countries: The case of South Africa and India", UNU Working Paper 067–2012 (Maastricht: United Nations University MERIT).

Peteraf, Margaret, 1993: "The cornerstones of competitive advantage: A resource-based view", in: *Strategic Management Journal*, 14: 179-191.

Rennkamp, Britta, 2011: Innovation for all? Legitimizing Science, Technology and Innovation Policy in Unequal Societies (PhD dissertation, Twente University, Department of Science, Technology and Policy Studies (STEPS), Enschede).

Rennkamp, Britta; Fortes Westin, Fernanda, 2013: "Feito no Brasil? Made in South Africa? Boosting technological development through local content policies in the wind energy industry", in: ERC (Ed.): *ERC*, Research Report Series (Cape Town: University of Cape Town, Energy Research Centre).

Sagar, Ambuj, 2009: *Climate Change: Technology Development and Transfer*. Background Paper for Delhi High Level Conference (New York: United Nations, Economic and Social Affairs).

Suurs, Roald; Hekkert, Marko; Kieboom, Sander; Smits, Ruud, 2010: "Understanding the formative stage of technological innovation system development: The case of natural gas as an automotive fuel", in: *Energy Policy*, 38: 419-431.

Wallner, Jennifer, 2008, "Legitimacy and public policy: Seeing beyond effectiveness, efficiency, and performance", in: *Policy Studies Journal* 36,3: 421-443.

37 Sustainability Transitions and the Politics of Electricity Planning in South Africa

Lucy Baker[1]

Abstract

After decades of cheap, abundant coal-fired electricity, from which large international mining and energy conglomerates and wealthy households have benefitted disproportionately, South Africa is experiencing a supply-side crisis. In 2011, the country's first integrated resource plan for electricity (IRP) was promulgated following a prolonged and contested consultation process throughout 2010. This plan anticipates that renewable energy will constitute twenty per cent of installed generation capacity by 2030, which will deliver approximately nine per cent of supply. Coal will retain the greatest share alongside a potential yet currently uncertain nuclear fleet. The objectives of this chapter are twofold: to examine electricity governance in South Africa and the highly politicized policy-making process in relation to IRP in which vested interests have played a major role; and to consider the extent to which the IRP has facilitated a low carbon transition. The chapter finds that despite the creation of a successful renewable energy 'niche', the coal-fired 'regime' is also being reinforced and the electricity mix under analysis is fuelling an unsustainable trajectory of production and consumption. The chapter also considers definitions of sustainability and concepts of a 'just' transition.

Keywords: South Africa, sociotechnical transitions, electricity, integrated resource plan, renewable energy, coal.

37.1 Introduction

South Africa is experiencing a long-term electricity supply-side crisis.[2] One of the cheapest electricity providers in the world until 2010 (Edkins/Marquard/Winkler (2010: 14), following a series of tariff increases, by July 2011 Canada surpassed South Africa as the cheapest provider of electricity.[3] It has had further yearly increases of eight per cent approved between 2013 and 2018.[4] The country's historical dependence on

cheap coal and cheap labour for the production of cheap electricity, from which the mining and minerals beneficiation industry and wealthy households have benefitted disproportionately, is subject to change. The supply-side crisis has been exacerbated by, amongst other factors, rising coal costs, climate change mitigation requirements and a failure to build any new generation capacity since the 1980s. The approval of the *Integrated Resource Plan* for electricity (IRP) in 2011 allows for the construction of 23,500 MW of renewable energy generation capacity by private suppliers by 2030 for a coal-fired electricity sector which until now has been controlled by the monopoly utility Eskom. Electricity is South Africa's largest energy subsector and supplies 29 per cent of its energy demand. About a quarter of the population, 12.3 million people, lack access to electricity (IEA 2011).

---

1 Lucy Baker, School of Global Studies, University of Sussex; email: <L.H.Baker@sussex.ac.uk>.

2 This paper draws from my PhD research, "Power shifts? the political economy of socio-technical transitions in South Africa's electricity sector". Thank you to Peter Newell, Katrina Brown and Heike Schroeder for their supervision. I gratefully acknowledge the support of the UK's *Economic and Social Research Council* (RES-066-27-0005 and ES/J01270X/1). Thank you also to the anonymous reviewers for their constructive and helpful feedback.

3 S. Njobeni: "SA falls further down rankings of cheap electricity producers", in: *Business Day*, 28 June 2012.

---

4 Creamer: "Nersa not expecting a tariff 'reopener' following Eskom bailout declaration", in: *Mining Weekly*, 17 September 2014.

This chapter has two objectives. Firstly, to examine electricity governance in South Africa and the policy-making process in relation to the integrated resource plan (IRP) for electricity as a key regulatory instrument. This objective is guided by the research question: *What is the nature of policy-making in South Africa's electricity sector?* This question is informed by Keeley and Scoones's (2003: 2) explorations of how science and different forms of expertise engage in the policy-making process, the institutional location of the experts in question, "the political context in which decisions are being made", and "the politics of policy-making" (Keeley/Scoones 2003: 17). By demonstrating how any transition to sustainability must involve wider political, economic, social and technological shifts and realignments (Grin/Rotmans/Schot 2010), the chapter seeks to be of value to those trying to understand transitions and/or seeking to influence them in the direction of sustainable development.

In tandem with the Renewable Energy Independent Power Producer Procurement Programme, which has been studied in depth elsewhere (Baker/Newell/Phillips 2014, Eberhard/Kolker/Leigland 2014), the IRP has facilitated significant changes in the country's electricity mix, in particular the rapid introduction of a new renewable energy sector which will constitute over twenty per cent of generation capacity and deliver approximately nine per cent of supply. Exploring the negotiation of the IRP demonstrates how electricity planning, and by implication the governance of sustainability transitions, is an inherently political process and takes place within competing and conflicting economic, political, industrial, environmental and social priorities.

The chapter's second objective is guided by the research question: *To what extent has the Integrated Resource Plan for electricity facilitated a transition to sustainability?* In evaluating the negotiation and implementation of the IRP it can be argued that despite considerable delays in its introduction and criticisms of the negotiation process, on some levels the plan has successfully enabled the emergence of a low carbon niche in South Africa. However, a broader analysis reveals that the gains made by a new renewable energy sector are embedded within a broader and much less sustainable trajectory of demand and supply. Using the multilevel perspective on sociotechnical transitions, I evaluate how low carbon 'niches' are taking place in parallel to the country's continued high carbon 'regime'. This echoes previous findings that low carbon initiatives in South Africa's electricity sector are being integrated into a high carbon industrial

infrastructure, which has limited benefits for the energy-poor and high inequality of access (Baker 2014, Hallowes/Munnik 2007)

In order to obtain rich insights into how national electricity planning is negotiated, this chapter is based on a deep and critical enquiry into the complexity of competing interests and priorities in South Africa. I illustrate the non-linear nature of policy-making following Wildavsky's (1979: 410) argument that policy analysis is concerned with both planning and politics, and that due to the difficulty in defining policy analysis it "should be shown not just defined". The case study of South Africa's IRP shows how the contested and complex nature of policy-making generally constitutes a series or web of decisions that take place over a long period of time and is often carried out by a complex network of actors. It further demonstrates how the policy process is embedded in socio-economic power structures (Dahl 1956, in Grin 2010: 223). Drawing on extensive field research carried out in 2010 and 2013, including approximately sixty qualitative interviews, I evaluate the role of national and international stakeholders and beneficiaries from the public and private spheres including government departments; industrial conglomerates; renewable energy companies; national banks and international investors; bilateral and multi-lateral donors; civil society and grassroots organizations; unions; and academics. Interviews cited in the text have been heavily anonymized in the interests of confidentiality.

In this chapter I firstly consider understandings of 'sustainable' and 'sustainability', particularly in relation to energy and the concept of a 'just transition'. Secondly I put forward the multilevel perspective on sociotechnical transitions as an analytical framework. Thirdly I explore the nature and overarching challenges of electricity governance which I then apply to South Africa by situating the country's electricity sector in a broader political, economic, social and technical context. This sets the context for the case study—that of the negotiation of the country's integrated resource plan for electricity.

37.2 Defining a 'Just' Transition to Sustainability

Motivated by a concern for poverty reduction and social, economic and environmental justice, this chapter should be read in the context of other studies which assert that a fundamental transformation to a low carbon future and industrial structure is needed

(Weischer/Wood/Ballesteros et al. 2011). The starting point is that the world's natural resources are limited (Meadows/Meadows/Randers et al. 1972; Rockström/Steffen/Noone et al. 2009) despite the innovative role that can be played by technological developments. Secondly, global economic dependence on fossil fuels is one of the fundamental obstacles to global environmental change (Adger/Brown/Conway 2010). In this light, the biggest governance challenge is how to manage an equitable low carbon transition that drastically reduces global consumption levels. This challenge includes cross-border dynamics, including the role that national and global institutions have played in driving and perpetuating the high carbon export-based growth of South Africa and other countries. I therefore acknowledge the very complex nature of system innovations and transitions to sustainability (Grin/Rotmans/Schot 2010) and how they may apply to questions of energy.

This study does not assume that all of South Africa's energy needs, however defined, can currently be met with renewable energy and acknowledges the significant technological uncertainties that exist on this point. I also recognize the immense challenges of the entrenched and highly inequitable legacy in the energy sector inherited by the post-apartheid government in 1994 and pay all due respect to the struggle and dedication of the numerous individuals and institutions that have fought to overcome this.

In these times of climate consciousness, 'sustainability' is an oft-used platitude. Despite the origins of the term 'sustainable development' as "development which meets the needs of the present without compromising the ability of future generations to meet their own needs" (WCED 1987), there is no internationally agreed standard or accepted definition of what it should mean. It is a highly contested concept that is a matter of subjective political judgement (Grin 2010: 235). Perhaps for this reason the term has now been co-opted by corporate and financial interests whose core activities may be some distance away from the intentions of its originators (Banerjee 2008). Ambiguities also exist over the term 'low carbon', which is simultaneously applied to various coal technologies, large hydroelectricity and nuclear power.

If the concept of sustainability, in reference to energy at least, is to include a social as well as an environmental dimension it must incorporate the issues of energy poverty, access to energy for the poor, the social consequences of energy exploitation, and the role of labour. This is particularly relevant in light of the failure of grid expansion plans and large infra-

structure projects that meet the needs of economic and industrial centres but often neglect the needs of the rural and urban poor (Prasad 2008). Any sustainable energy transition must also consider ethical debates over how the pressure to decarbonize should be balanced against the need to provide pro-poor energy and/or economic growth on equity and justice principles (Roberts/Parks 2007), possible conflicts that may exist between these two objectives, and potential accusations of being 'anti-development' or 'anti-technology'. A key debate is to what requirements low-income countries should be subjected, in light of their minimal historical carbon emissions and the lack of access of the majority of their populations when compared to developed countries. And what in turn should be the obligations of high-income countries in providing financial and technological assistance: how much, via which channels and to whom (Ockwell/Ely/Mallett et al. 2009: 12)?

With this in mind, the emerging literature on 'just transitions' (Swilling/Annecke 2012; COSATU 2012; Newell/Mulvaney 2013) plays a poignant role in marrying concepts of social justice with the need for a global low carbon shift. In essence, any low carbon energy transition must be based on modes of production and consumption that do not depend on resource depletion, environmental degradation or the exploitation of socio-economic inequalities. Such an argument addresses the tendency of the sociotechnical transitions literature to overemphasize environmental and 'green' concerns in defining sustainability (Geels 2011) with its implicit assumptions that 'radical green niches' will automatically result in concomitant human co-benefits and the realization of social equity. Or in other words, that a 'low carbon' transition will automatically result in a 'just' transition.

We now turn to the literature on sociotechnical transitions and in particular the use of the *multilevel perspective* (MLP) as a framework of analysis with which to explore developments within South Africa's electricity sector.

37.3   Analytical Framework: Sociotechnical Transitions

The MLP analyses sociotechnical systems change from the level of 'landscapes' (macro), 'regimes' (meso), and 'niche-innovations' (micro) (Geels/Schot 2007). It is concerned with the way in which incumbent regimes lose stability and experience low carbon, sociotechnical transitions as a result of coordinated selection pressures from the niche and landscape levels (Byrne/Smith/Watson et al. 2011: 57). As this chapter illustrates, defining the empirical and geographical parameters of the different levels is obviously an incomplete, subjective and temporary exercise. As Smith, Stirling and Berkhout (2005: 1504) explain, "regime membership is neither homogeneous nor clearly bounded" and does not neatly align to the boundaries of the nation state. This lack of conceptual clarity also raises the question as to whether an electricity regime refers only the level of primary fuel such as coal, or encompasses the entire system of generation, transmission and distribution (Berkhout/Smith/Stirling 2004).

With this in mind, I have interpreted the regime concept in this case to refer to South Africa's coal-generated, state-owned, electricity sector under Eskom's monopoly control, supplied by powerful privately-owned coal mining companies and of particular benefit to energy-intensive users. I have interpreted the niche to refer to the renewable energy independent power producers who are supported by national banks and various sources of national and international finance and investment (Baker 2012). The niche and the landscape, so-defined, interact with and are influenced by actors and events at the 'landscape' level, which in this case can include: international trends in renewable energy investment; increasing national and international demands for South Africa's coal resources; and the role of international resource conglomerates. Though not without its own conflicts and tensions, the relative 'stability' of South Africa's coal-fired electricity regime has been characterized by the political, institutional and market dominance of entrenched beneficiaries and stakeholders. This has been supported at the level of national policy and in terms of implicit and explicit subsidies such as cheap coal supply and preferential electricity tariff deals.

In this chapter the MLP framework is employed flexibly rather than rigidly. This is appropriate for a creative interpretation and empirical analysis of the study of complex, interlinked, multidimensional and co-evolving dynamics (Geels 2011: 34; Foxon 2010). I do not test the viability of this model in predicting or determining systems change or measuring the relationship between different variables in a positivist sense, but rather use it as a framework to analyse different stakeholders, institutions and technologies involved in the governance of South Africa's electricity sector. This facilitates "narrative explanations" (Geels/Schot 2007: 414) of the case study in question and allows for an examination of interactions between "technol-

ogy, policy/power/politics, economics/business/markets and culture/discourse/public opinion" (Geels 2011: 25).

Another strand of the sociotechnical transitions literature, Transitions Management, takes a more prescriptive approach and aims to generate low carbon transitions in unsustainable regimes by "steering evolutionary dynamics towards specific visions" using consensus-based decision making (Scrase/Smith 2009: 708). The difference between the MLP and the Transitions Management approach can be related to the difference identified by Ham and Hill (1993: 4) between analysis *of* policy and analysis *for* policy. While the former, which I apply here, is concerned primarily with advanced understanding, the latter refers to applied activity "concerned mainly with contributing to the solution of social problems".

This chapter goes some way towards responding to criticisms of the sociotechnical transitions literature for its inadequate analysis of politics and power relations (Lawhon/Murphy 2012; Smith/Stirling/Berkhout 2005; Meadowcroft 2011). Specifically, I integrate a political economy approach which is crucial for the analysis of relationships between political and economic power, structural change, and the underlying interests of dominant actors and beneficiaries of governance mechanisms (Moe 2007; Büscher 2009; Baker/Newell/Phillips 2014). Such a perspective allows us to examine the interaction between state institutions and private capital in the policy process and the dynamics of global economic expansion. In such a way we avoid reducing a complex debate to a largely technocratic perspective on policy and governance (Torgerson 2003) which is mostly politics-free, ignores the influence of human agency and fails to address "pressing issues of social justice on two fronts: both *within* industrialized nations and *between* highly industrialized and third world nations" (Hajer 1995: 32). As Moe (2009: 1731) explains, technology and economics are insufficient for an analysis of structural change and long-term growth and development. Instead, "*politics*–or the political economy, with its focus on actors and decision-makers, on institutions and regulations, and on past and present interactions–must be included". Following Grin (2010), who places politics and power as an essential aspect of governance in the case of sustainability transitions, decision-making in South Africa's electricity sector cannot be understood without situating it within the complexity in which it is embedded. This complexity refers to a political economy and historical path dependency characterized by the country's minerals-energy complex (Fine/Rustomjee 1996), explored below.

Finally, by applying sociotechnical transitions to South Africa, the chapter adds to limited studies of this literature in relation to a *low- and middle-income countries* (LMICs) context, whose economic growth is based on natural resource extraction and polluting industries within the context of a globalized and financially interdependent world (Lawhon/Murphy 2012: 10).

37.4 The Challenges of Electricity Governance

Electric power systems demanded of their designers, operators, and managers a feel for the purposeful manipulation of things, intellect for the rational analysis of their nature and dynamics, and an ability to deal with the messy economic, political, and social vitality of the production systems that embody the complex objectives of modern men and women (Hughes 1983: 1).

A determining factor in the realization or impediment of sustainable development and climate change mitigation, energy, like water, is a global good but has billions of people without access to it or with access only to poor quality; it is in rapidly growing global demand; it is constrained by natural resources; it varies in regional availability; and it operates in heavily regulated markets (Bazilian/Rogner/Howell et al. 2011: 7897). It is as much a part of a country's development policy (Ljung 2007: 37) as policy on industry, infrastructure development, the environment, foreign relations, trade, climate change and social development. The conflict for policy-making that results from such a complexity of competing and fragmented interests is central to this investigation.

Electricity, defined here as a subset of energy, is a large technological system embedded in broader political, economic and social forces (Hughes 1983; Rip/Kemp 1998). Goldthau and Sovacool (2012: 233) describe electricity as "perhaps one of the most obvious large-scale socio-technical systems involved in converting energy fuels into services", which goes far wider than the arenas of science and engineering. As a technology, electricity is "site-specific" (Hughes 1983: 47), determined by natural and human geography. For all these reasons, electricity is not easily governed or manipulated and is compounded by path-dependency, the role of vested interests and uncertainty in the adoption of new technologies and their integration into the grid.

Electricity as an infrastructural and network technology (Rip/Kemp 1998) and natural monopoly is difficult to separate from the concept of its governance in that it lends itself towards economies of scale. According to Victor and Heller (2007: 1), the treatment of electricity as a commodity like any other, and its separation into generation, transmission, distribution and supply for sale along market principles, is highly flawed. It has proved one of the hardest network industries to reform and it has been difficult to "replace the state with private enterprise because infrastructures usually display strong economies of scale, which arise through network interactions that are prone to natural monopoly".

Yi-chong (2006: 803) classifies electricity as a process rather than a commodity, formed from the conversion of different forms of energy which are then consumed shortly after production. The simultaneous production and consumption process is achieved through "a complex 'coordination' system that integrates a large number of generating facilities dispersed over wide geographic areas to provide a reliable flow of electricity to dispersed demand nodes while adhering to tight physical requirements to maintain network frequency, voltage and stability" (Yi-chong 2006: 804). Smith, Sterling and Berkhout (2005:1493) describe how the electricity-generating regime "is dominated by rules and practices relating to centralized, large-scale (usually thermal) power technology and high voltage alternating current grid infrastructures", which at the national level is "nested within a global energy regime organized primarily around the extraction, trade and combustion of fossil fuels". The national level in turn oversees subordinate regimes, such as "the coal-fired steam turbine, the nuclear fuel cycle... or gas-fired combined cycle turbine systems" (Yi-chong 2006: 1493).

In South Africa there are two different framings of the electrification problem competing for institutional and policy dominance (Bekker/Eberhard/Gaunt et al. 2008). The first one, pioneered by the socially-oriented policy analysts such as the University of Cape Town's Energy and Development Research Centre in the late 1980s and 1990s, was based on an analysis of low-income household energy use which saw electrification as a key intervention in addressing energy poverty. The second framing, inherent in the approach of the national utility Eskom, perceives electrification as a development to be integrated with other service-oriented infrastructure development processes. The first framing aligns with the energy commons approach (cf. Wamukonya 2003: 1284) which perceives energy as a public good rather than as the commodity it is often treated as. Such an approach asserts that concerns of equity and efficiency should be included in energy policy-making and that the natural availability of renewable energy resources "make them potential energy commons". Meanwhile the treatment of energy as a commodity underpins a profit-oriented urban bias in national electrification policy which results in the marginalization of rural communities, overlooks goals of universal access and leans towards technologies based on fossil fuels, in disregard of environmental concerns (Wamukonya 2003: 1284). This brings us back to the concept of 'just transitions' discussed above and debates over how sustainability should be defined.

37.5   South Africa's electricity sector in context

> The energy sector is governed by a monopoly. Huge profits are made by lucky winners. People are trying to put this into the renewable market such as wind. The creation of monopolies is a key theme in energy governance in South Africa. Every player in the market is trying to create their own monopoly. Lack of knowledge and lack of transparency allow this to happen (Energy Specialist, in interview, December 2010).

The structure of South Africa's electricity sector and its coal-dominated sociotechnical 'regime' is characterized by its minerals-energy complex (Fine/Rustomjee 1996). This incorporates national activities organized in and around energy, mining and associated subsectors, and includes the evolving relationship between the state and corporate capital. The minerals-energy complex is central to the country's historical dependence on cheap coal and cheap labour for electricity which in turn serves the interests of export-oriented industry. Central to this is the country's social, political and economic legacy of apartheid, which created an infrastructure to meet the needs of industry and the white minority (Ziramba 2009). According to the IEA, in 2012, 59 per cent of the country's electricity was consumed by industry and 20 per cent by residential consumers.[5]

South Africa has historically been dependent on abundant sources of low-cost coal for ninety per cent of its electricity and this has resulted in an incredibly energy-intensive economy. The country's emissions are higher than the UK's per capita, while its GDP per capita is only a sixth as much. Electricity is South Africa's largest energy subsector, supplying 29 per cent of its energy demand and generating approximately 50 per cent of its carbon emissions compared to 26 per cent globally (CDP 2011). In financial year 2012/2013, the utility Eskom burned 123 million tonnes of coal (Eskom 2013: 85) and emitted 227.9 tonnes of carbon dioxide (Eskom 2013: 34). The country's remaining emissions come from synthetic fuel production, energy-intensive industries (mining, iron and steel, cement), and the transport sector.

South Africa's electricity sector has been characterized by its monopoly parastatal Eskom. The country defied the "standard model" of power sector reform which became universally accepted during 1980s and 1990s (Gratwick/Eberhard 2008). This model, which involves the unbundling of the electricity sector into separate transmission, distribution and generation companies and the introduction of private competition, was unsuccessfully imposed upon other countries in sub-Saharan Africa as part of World Bank-led structural adjustment programmes. Eskom is the sole transmitter of electricity via the country's high-voltage transmission grid, generates 96 per cent of national electricity and is responsible for 60 per cent of distribution, which is consumed by a third of South Africa's customers. Municipal distributors purchase their energy and services from Eskom Distribution and supply about two-thirds of the country's customers, who account for 40 per cent of total sales. This has resulted in a highly fragmented local governance system with poorly performing service delivery departments and the existence of separate local black municipalities (Eberhard 2007: 231).

The historical influence of a small number of large resource-based conglomerates over South Africa's energy policy and economy more generally is a central feature of South Africa's minerals-energy complex. These conglomerates, now internationalized with privileged access to cheap energy, tax breaks and infrastructure, have acute levels of control and concentrated ownership over the country's mining and minerals beneficiation sectors (Fine/Rustomjee 1996) and maintain "enormous collective bargaining power" (Nakhooda 2011: 21). These conglomerates include the country's five largest coal mining companies: BHP Billiton, Glencore-Xstrata, Anglo-American, Sasol and Exxaro. They are in turn members of the country's *Energy Intensive User Group* (EIUG)[6] which has

---

5   See at: <http://www.iea.org/statistics/statisticssearch/report/?year=2012&country=SOUTHAFRIC&product=ElectricityandHeat>.

---

6   See at: <http://www.eiug.org.za/about/membership/>.

thirty-six members and consumes around 44 per cent of the electricity sold in South Africa.

South Africa is one of the most socially unequal countries in the world with 70 per cent of its income going to approximately 20 per cent of its population, 1.6 per cent of its income going to the poorest 20 per cent (The Presidency 2009:23), and an unemployment rate of 31 per cent including 'discouraged work seekers' (Statistics South Africa 2012). Reflecting the country's history of racially determined differentiation (Bekker/Eberhard/Gaunt et al. 2008), one-third of households lack access to electricity. Before the end of apartheid in 1993, only 36 per cent of the population was connected to the electric grid (CURES 2009). While Eskom's unprecedented expansion programme between 1994 and 2000 saw 2.4 million houses connected to the grid and many more connected by local government municipal utilities (McDonald 2009: 15), one-third of the country's population is still without access, particularly in rural areas. Millions of low-income households, who account for no more than five per cent of national electricity consumption, cannot afford to buy electricity even if they are connected to the grid (McDonald 2009: 16), and prioritize paraffin, coal, wood fuel and other sources. Middle- to high-income urban households consume an average of 96,000 kilowatt-hours per annum (McDonald 2009: 15), just below that of the US at 112,000 kWh.

In the course of just two decades South Africa has gone from a period of electricity surplus in the mid-1980s, with some of the lowest electricity prices in the world as a result of the construction of excess generation capacity, to capacity restraints and a severely constrained reserve margin and imminent deficit, resulting in load-shedding in 2008 (Eberhard 2011). The complex reasons for this crisis go beyond the scope of this paper and are the culmination of events over many years (Baker 2012). In policy terms, this crisis can be referred to as a "trigger event" (Keeley/Scoones 2003) in that it has been used simultaneously as a legitimizing discourse for the construction of further coal-fired power stations, the integration of new renewable energy and a potential nuclear fleet. This is now explored in the following section in relation to the Integrated Resource Plan for electricity.

37.6   The Politics of Electricity Planning

South Africa's second *Integrated Resource Plan* (IRP) approved in May 2011 is an electricity master plan cov-

ering total generation requirements from 2010 to 2030. It was initially based on a doubling of national generation capacity from current levels of approximately 41,000 MW to 90,000 MW by 2030. However this project has been under revision since late 2013. The IRP was celebrated in some arenas for diversifying the country's electricity generation mix away from coal by allowing for approximately 23,000 MW of grid-connected renewable energy (of which 9,200 MW wind, 8,400 MW PV [photovoltaics] and 1,200 MW CSP [concentrated solar power]) (DoE 2011a: 14) until 2030, the majority of which will come from independent power producers. However, figure 37.1 reveals that in addition to the construction of new renewable energy, 6,250 MW of new coal will also be introduced. This is in addition to the Medupi and Kusile coal-fired power plants that were approved prior to the IRP and are currently under construction, constituting a combined generating capacity of approximately 9,600 MW. Hence while the proportion of coal in the overall electricity mix decreases from 85 to 46 per cent, in absolute terms it will increase to 41,071 MW. The final plan claimed to be consistent with an emission constraint of 275 million tonnes of carbon dioxide annually after 2024 (DoE 2011a: 6) though the actual emissions associated with it have not been specified.[7] An ambitious nuclear fleet is also planned whose undetermined cost has since been the subject of significant national controversy. A project must be in the IRP in order for the national energy regulator to be able to grant it a licence (Pienaar/Nakhooda 2010), though according to the new generation regulations for electricity the minister holds the right to license generation capacity if s/he deems fit.

This plan has been under revision since late 2013 in keeping with the requirement that it be updated every two years. The revised draft proposes a downward adjustment in the demand forecast which would result in an overall decrease in generation capacity of 6,600 MW. The proposed revisions include a reduced contribution from coal (from 6,250 MW to 2,450 MW) and wind (from 9,200 MW to 4,360 MW), and increased contributions from Solar PV (from 8,400 MW to 9,770 MW) and CSP (from 1,200 to 3,000 MW). There is also a lack of clarity in the draft over the proposed nuclear fleet and a potential increased role for gas, including shale gas. The implications of these revisions, which are some months from being

---

7   H. Winkler, "Will the IRP meet SA's carbon emission target?", in: *Engineering News*, 22 April 2011.

Figure 37.1: Total envisaged capacity 2030 under Policy-adjusted IRP 2010. Source: Author's own using data available in DoE (2011a:17).[a]

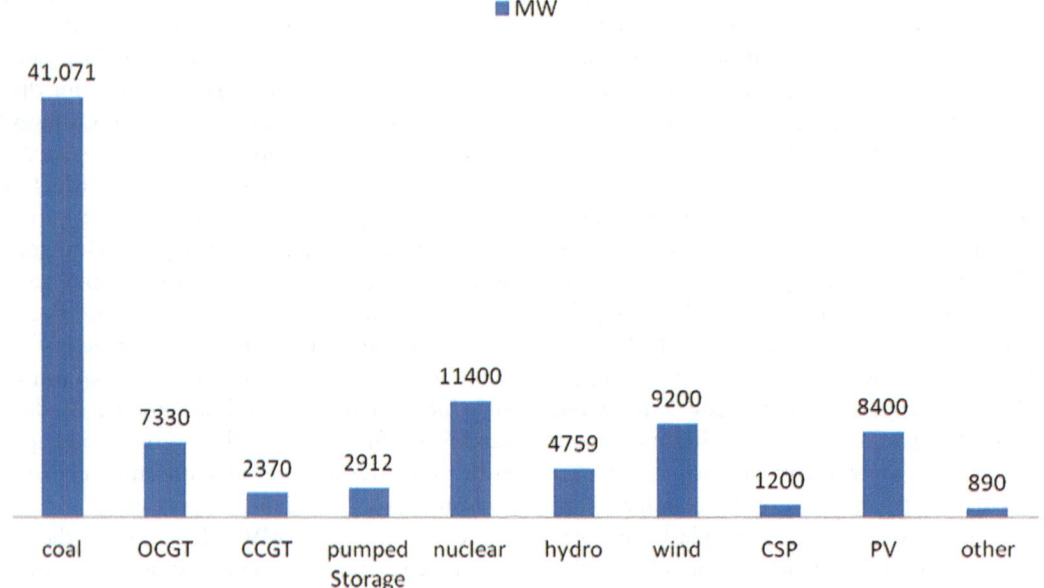

a.   OCGT: Open Cycle Gas Turbine; CCGT: Closed Cycle Gas Turbine; CSP: Concentrated Solar Power; PV: (Solar) Photovoltaics.

finalized, will be uncertain for some time. In brief therefore, despite a decrease in the demand forecast, coal is still set to remain the dominant energy source for the foreseeable future, with minerals extraction and related manufacturing still constituting the bulk of demand.

This section examines the way in which the IRP as a crucial national planning instrument for electricity was negotiated, focusing on participation and transparency; parliamentary oversight; assumptions of demand; the political nature of technical advice; and modelling. In doing so it illustrates how in South Africa the governance of electricity and the nature of sustainability transitions is a highly contested and political process. The IRP claims to "indicate a balance between different government objectives, specifically economic growth, job creation, security of supply and sustainable development" (DoE 2011a: 13). Such a statement encapsulates the nature of conflicting objectives that must be tackled in order for any meaningful transition to take place.

Following various delays, the stakeholder engagement process for IRP took place throughout 2010 and early 2011. The process included representatives of the coal, renewable, and nuclear industries; the country's energy-intensive users; financial stakeholders; and civil society and academics. Government was heavily criticized from all sides for the rushed nature of the plan's

negotiation process; the lack of transparency of critical assumptions; problems with its methodology and input parameters; costs; the overall energy mix; the feasibility of the plan; and its potential impacts on the poor. Despite this, it was considered by some stakeholders as a significant advance on previous electricity planning processes (Hughes 2010, Mainstream Renewable Power 2010b). Nakhooda (2011) argued that while the process was dominated by relatively specialized stakeholders able to engage with the inevitable technical complexity of electricity planning, opening it up to public participation still set an important international precedent.

37.6.1   The Process: Participation and Transparency

Following strong public pressure from civil society, business and the media (Nakhooda 2011: 11), in February 2010 the *Department of Energy* (DoE) announced that a public consultation process would be initiated, with the aim of completing the IRP by June 2010. In the event the final plan, referred to as the 'Policy-Adjusted' IRP (DoE 2011a) was promulgated almost a year later in May 2011. Throughout the consultation civil society groups and academics pushed for an extension to the timeline to ensure sufficient time for public comment (Earthlife Africa 2010). Meanwhile, according to one energy-intensive user, government was keen for the plan to be approved as quickly as possible in the "rush to create more generation capacity" and in the interests of finalizing agreements with international suppliers and contractors for new coal and nuclear build.

Transparency and participation was one of the key concerns cited throughout the process, in particular over a lack of clarity in the IRP's critical assumptions (Hughes 2010). In June, WWF (2010) said that it was "disappointed that the documentation provided to date, as well as the way in which documents have been released, does more to obfuscate the process and issues, than to enable informed engagement in transparent and accountable integrated resource planning". The first 'stakeholder plenary sessions' were held in June 2010 in Pretoria and extended from one to two days at the last minute to accommodate the volume of interest. These heard a total of thirty-two presentations (DoE 2010a) and focused on the input parameters for the IRP modelling (DoE 2011a: 7). According to the DoE (2011b) in its briefing to the parliamentary portfolio committee on energy in January 2011, at the first round of consultations eighty-one submissions were received and "831 specific inputs based on the parameter sheets were captured and analysed", with the major contributors being 67 NGOs and civil society organizations, 63 academics and consultants and 70 organizations.

However, these submissions were not made publicly available and how they contributed to the IRP 2010 was not clarified by the DoE. In October 2010 an NGO report stated that the DoE failed to respond to many of the comments made in relation to its input parameter documents and that there was no recognition as to whether the technical inputs had been included in the modelling or even meaningfully considered (McDaid/Austin/Bragg 2010). On a similar note, in its June 2010 presentation the EIUG (2010) commended the government on its transparency but stated that it should publish a summary of submissions from the first round of consultations and be transparent about the criteria it had used to evaluate scenarios. Sappi, the global pulp and paper giant and EIUG member, also called for greater clarity over the twenty-nine parameters issued and for better structure in the way in which they were presented and the ability of people to comment (Sappi 2010). In the event there was no response to such calls.

On 8 October the DoE put forward a second draft of the plan (DoE 2010b) with a thirty-day comment period to run until 10 November 2010, subsequently

extended to 10 December 2010. Four days of public hearings in Cape Town, Durban and Johannesburg were held at the end of November and in early December 2010.

In the midst of the second comment period, twenty-eight South African faith-based, environmental and social organizations issued a report calling for national electricity planning to be developed within the context of "broader sustainable development goals" and to be connected to "various other processes aimed at addressing these issues arising from our historical legacy" (McDaid/Austin/Bragg 2010: 4). The report, *Power to the People: raising the voice of civil society in electricity planning–IRP 2010 inputs and departmental responses*, outlined key concerns over the nature of the IRP participation process, including its failure to uphold national laws on public participation which are central to the post-apartheid regime's principles of empowerment and democratization, such as the *Protection of Administrative Justice Act* (PAJA) (Nakhooda 2011: 8). The report also pointed to the inability of the general public to be able to participate, firstly given that the main opportunity to do so was via the Internet, which would thereby exclude the majority of South Africans, and secondly in light of the complexity of the process, which would restrict the ability of those with limited education to engage with it. Lastly the report argued that the process had failed to adequately account for rural communities who suffer from poor electricity connections and would be the most affected by rising electricity prices.

Despite the above criticisms from civil society, others welcomed what they saw as the DoE's commitment to participation and transparency. For instance, in November 2010 an energy-intensive user referred to it as "the first consultative electricity plan that South Africa has ever had". In August 2011, Standard Bank's head of energy, utilities & infrastructure, Paul Eardley-Taylor, said that "the IRP is an impressive document, both the speed of the document's promulgation and the transparency of the process".[8] In November 2010 the DoE (2010c) stated "It is clear that the public relished the participation process and its continued use in long-term planning must be ensured".

37.6.2 Parliamentary Oversight

The Parliamentary Portfolio committee on energy has a formal oversight role over the Department of

---

8 L. Pretorius, "Renewable Energy–Moving forward a cause for concern", in: *Financial Mail*, 24 August 2011.

Energy. However the IRP 2010 negotiation process raised doubts over the extent to which the committee complies with this role. For instance, the first parliamentary briefing on IRP 2010 was not held until 3 August 2010 and was criticized for coming too late, after the first hearing and the critical modelling exercises had taken place. Lance Greyling, MP for the Independent Democrats and member of the committee, said "I would go so far as to say that Parliament—and, particularly, the energy committee–has been negligent in its duties at providing proper oversight over the energy sector; and in particular, energy planning" (Greyling 2010). The technical expertise of some of the members of the committee may also be inadequate. For instance at one meeting in June 2010 one member requested clarification over the difference between a wind turbine and a hydroelectric turbine (PC Energy 2010). At the same meeting, a discussion took place about PetroSA's offer to provide committee members with tickets to the World Cup, suggesting that hydrocarbon companies may hold an influence over the committee.

The committee's timing was better in the second round, however, with a second parliamentary hearing on 26 January 2011 (DoE 2011b) which gave a summary of inputs and key debates throughout the consultation process.

37.6.3 Supply and Demand: 'Business as Usual on Steroids'

Despite significant changes that the plan proposes to the generation mix, the IRP was based on assumptions that electricity demand would almost double by 2030, from 260 TWh in 2010 to 454 TWh in 2030. Winkler explained that "the growth in electricity demand outweighs the reduction in the CO2 intensity of electricity" [9] and would result in "GHG emissions from electricity generation increasing from 237 million tons of CO2 in 2010 to 272 million tons in 2030". It will also increase electricity prices "by at least 250 per cent in real terms from their current level by 2020 and by a much higher rate with inflation factored in" (Winkler 2011). This doubling in electricity capacity is based on the moderate energy forecast in the IRP 2010's Revised Balanced Scenario (DoE 2011a: 36) and a projected GDP growth rate of 4.51 per cent (Eskom Systems Operation and Planning 2010: 2). This anticipates a doubling in mining and minerals

beneficiation, something which prompted a representative of the City of Cape Town's climate and energy branch to refer to the plan as "business as usual on steroids".[10]

Energy-intensive users were strong proponents of the doubling of the demand forecast. In June 2010 the EIUG (2010) asserted that it is "essential that we all understand and accept the growth assumptions" of the IRP, which are its "key parameter". This was bolstered by statements in September 2010 by then executive director of Xstrata Alloys, Mike Roussow, who asserted that mining and minerals beneficiation has a major role to play in the economy and nation-building, for which energy is essential.[11] However, despite the EIUG's demands being met in the subsequent demand forecast, their quest to achieve national consensus over such high growth assumptions was challenged in a number of submissions (in DoE 2011a: 10). The Cape Town Chamber of Commerce (2010), for example, stated that the "energy plan may provide for more power than we actually need". Details about how the demand forecast was calculated were not made public, which, Nakhooda (2011: 15) argued, "raises further questions about its reliability, and made it all the more difficult to propose enhancements to the same". Energy analyst (2) stated that the demand forecast was based on a rate of GDP growth which was "politically, not analytically derived".

37.6.4 The Political Nature of Technical Advice

> Essentially then, the department has succeeded in locking all vested-interest groups into the Technical Advisory Panel and even though they claim that it is just about technical expertise, we are not allowed to see the minutes of these meetings; nor are we allowed to see the thinking behind the different energy assumptions that they come up with (Greyling 2010).

The strong influencing role of vested interests in this case reveals how an assumed technical exercise of electricity planning can be inherently political. A technical advisory group set up to provide inputs into the modelling of the plan's input parameters was criticized in the media[12], by civil society (McDaid/Austin/

---

9   H. Winkler, "Will the IRP meet SA's carbon emission target?", in: *Engineering News*, 22 April 2011.

10 L. Donnelly, "Resource plan gets dim reception", in: *Mail and Guardian*, 6 December 2010.

11 T. Creamer, "Mining has major role to play to grow SA's economy–Xstrata", in *Mining Weekly*, 28 September 2010.

12 C. Yelland, "National integrated resource plan, a document to shape our future", in: *Daily Maverick*, 23 April 2010.

Table 37.1: The DoE technical task team. **Source:** McDaid, Austin and Bragg (2010: 5).

NAME	SECTOR	INSTITUTION/ AFFILIATION
Neliswe Magubane	Government	DG Dept of Energy
Ompi Aphane	Government	DoE
Ria Govender	Government	DoE
Thabang Audat	Government	DoE
Kannan Lakmeerharan	State Enterprise	Eskom systems operations and planning
Callie Fabricious	State Enterprise	Eskom planning and market development
Mike Rousouw	Business—coal	Xstrata
Ian Langridge	Business—coal	Anglo American
Brian Day	Business—coal/RE	Exxaro
Piet van Staden	Business—fossil fuels	SASOL
Kevin Morgan	Business—smelters/coal	BHP Billiton
Paul Vermeulen	Local govt-owned company	City Power (Johannesburg)
Doug Kuni	Coal-related business	SA Independent Power Producers Association
Roger Baxter	Mining-related business	Chamber of Mines
Prof. Anton Eberhard	Academic	Graduate School of Business, UCT
Shaun Nel	Business—project manager	Gobodo systems (Eskom is listed [on their website] as one of their clients)

Bragg 2010), and by the renewable energy industry (Mainstream Renewable Power, *in interview*), for consisting largely of representatives from government, Eskom, coal miners and the EIUG. Notably, representatives from the renewable energy industry, civil society and experts from the field of social and environmental impact assessment were not included in the advisory group (see table 37.1). For this reason, concern was expressed that the modelling process on which the committee was advising was likely to reflect the industrial bias of the interests of its members (McDaid/Austin/Bragg 2010: 6). Confirmation of the committee's existence and the names and affiliation of its members was made public in June 2010 in a letter from the Minister of Energy to a request for information from civil society (see Peters 2010).

All committee members signed confidentiality agreements with the DoE and its meetings and minutes were not made publicly available (Peters 2010). Despite this the Minister took "exception to the classification of [its] Technical Task Team as secretive", stating that its members were selected for their technical expertise alone, and that they "are not meant to represent the interests of their constituencies/ employer organizations" (Peters 2010). The company Mainstream Renewable Power (2010a) countered that the "inaccuracies and assumptions in the modelling could be better addressed if the renewable energy

industry were able to participate as part of the technical task team".

The constitution of this team illustrates the privileged access that energy-intensive stakeholders had to decision-makers, and the influential role that regime incumbents continue to play in electricity policy-making in South Africa, despite the incremental steps made by the emerging renewable energy industry. However, a mining industry expert has suggested that, rather than a deliberate conspiracy, the influence of this team was due to the lack of expertise of Eskom's Systems Operator who carried out the IRP on behalf of the DoE: "the chairman of the technical task team of IRP has never built a power station, the Systems Operator has never built a power station either. They are all demand-side specialists and have based their assumptions on EPRI reports. The Systems Operator is a black box. It doesn't distinguish between the needs of different types of plant and it doesn't understand local risks either". This was corroborated by energy analyst (3) who stated, "Eskom writes the IRP, it goes to the DoE who checks it with its little 'in-group', largely coal miners, and then carries out consultation with identified stakeholders. These are people with whom the DoE has a relationship. It is not necessarily a big conspiracy, but there is a serious lack of capacity in the DoE."

37.6.5   Modelling

There were twenty-nine parameters drawn up with inputs from the technical advisory group and made available by the DoE in May 2010. These were divided into matters of: i) demand, such as price elasticity of demand and energy efficiency; ii) supply, such as: reserve margin, discount rate and exchange rate; iii) externalities, such as climate change and carbon tax; and iv) key outputs, such as base scenarios and rate of inflation. As I now discuss, while a number of participants at the hearings felt that publication of the IRP's parameters was a breakthrough in terms of transparency, others found the modelling to be inaccessible and highly specialized and in some cases based on inaccurate assumptions.

For example, the University of Cape Town's Energy Research Centre found that while the report by US-based *Electric Power Research Institute* (EPRI) was a good basis for assessing technology costs, it was difficult to comment at the time of the public hearing in June 2010 in light of its complexity and because the full report had been made available only one day before (Hughes 2010). Hughes (2010) also pointed to major inconsistencies between the previously-released executive summary and the full report. For instance the latter included a tripling of coal costs. Furthermore the modelling process had failed to include technology learning rates, which would make wind competitive with nuclear by 2014. It was also unclear if price elasticity would be incorporated into the modelling. In December 2010, Pienaar and Nakhooda (2010:6) asserted that "the comparative life cycle costs of conventional and renewable energy technologies are arguably the most important input into the IRP 2 modelling process as the emergent scenarios are largely being assessed on the basis of expected costs".

Other examples that contested the accuracy of the model's assumptions include EarthLife Africa (2010), who stated that the modelling had failed to include externalities in the cost of coal, including mono-nitrogen oxides (NOx) and sulphur oxides (SOx). In June 2010 Mainstream Renewable Power (2010a) stated that the key parameters were "informed by myths around RE technologies and their benefits" and that fossil fuel externalities such as price volatility and carbon taxes had not been included. In sum it found that the modelling relies on an "outdated least cost/ levelized cost approach" and would not be able "to accurately model the proper value of a fully diversified supply-side mix".

In an interview, energy analyst (2) described how the IRP's "base plan and selected scenarios" was modelled on an expansion planning software tool named PLEXOS@R (DoE 2010c: 3), which is "very expensive and proprietary" when compared for instance to the IEA's Markal model. S/he added that the Markal model is used extensively internationally, is the official model used in the UK, and is low cost and easy to buy, install and run. Moreover, "Eskom have hardly begun to use PLEXOS@R capabilities" and had not made the model available. In response, WWF-SA commissioned a counter-model called the Sustainable National Accessible Power Planning (SNAPP) tool, developed by the University of Cape Town. Freely available online,[13] this tool allows users to "interrogate proposals for South Africa's new build plan using objective cost analysis, environmental impacts and the reliability of the system" (WWF 2010:22). It allows members of the public to examine the "assumptions and input data and resulting technology choices" of the IRP "without undertaking the complex modelling on which the plan is based". A subsequent report by WWF-SA demonstrated that investing in renewable energy (wind and solar), along with energy efficiency in industry, would provide cheaper electricity for South Africa by 2020 than investing in coal or nuclear power (WWF 2010).

37.7   Conclusion

The case of the IRP's negotiation process illustrates Keeley and Scoones's (2003: 3) claim that policy "is co-constructed across space, through particular networks and connections linking global and local sites", particularly in cases where "national science capacity is weak and under-confident". As we have witnessed, "a lack of interrogation of the nature of expertise in the policy-making process conceals important power dynamics" (Keeley/Scoones 2003: 5). The nature of the plan's negotiation process further illustrates how vested interests are embedded in the technical expertise employed and/or created by government, with the technical task team as the most obvious example. This links to Jasanoff's (2003: 160) claim that "far from being neutral and apolitical, scientific research follows the preferences of those with the power to set research agendas". For this reason, she continues, "expertise has legitimacy only when it is exercised in

---

13 Available at: <http://www.erc.uct.ac.za/Research/Snapp/ snapp.htm>.

ways that make clear its contingent, negotiated character and leave the door open to critical discussion" (Jasanoff 2003: 160).

The IRP process further demonstrates that an assumed technical exercise, such as electricity modelling, can be deeply political, and that technical and political processes are "deeply intertwined" (Keeley/ Scoones 2003: 5). While the public consultation process claimed a relatively high level of participation, the 'technical task team' in which coal miners and energy-intensive users were dominant had the greatest influence over the modelling process, which was itself opaque and subject to contradictory assumptions. This illustrates the privileged access that stakeholders of the country's minerals-energy complex still retain over decision-making processes in electricity policy, as compared to interests in the emerging renewable energy niche, or organizations concerned with social and environmental justice. This relates to Smith, Stirling and Berkhout's (2005: 1505) assertions, who find, in writing about the sociotechnical transition, that "actors more intensively involved than others in system reproduction enjoy quite powerful positions, benefit strongly from the *status quo*, and occupy important gate-keeping positions".

The role of politics in the relationship between energy and industry is central to this case study given that South Africa's energy-intensive users are also the country's major coal miners. They therefore have vested interests in both supply and demand. As Moe (2010: 1733) states, "vested interests tied in with the existing industrial and energy regime is the main factor standing in the way of structural change ... and the state the main actor capable (or incapable) of dealing with these vested interests". However, where the state lacks capacity, vested interests have a wide scope to intervene in policy-making (Hajer 2002). Consequently, powerful actors are likely to undermine the implementation of the 'right' policies (Scrase/Smith 2009:710). As Fine (2009: 38) asserts, "even if the conglomerates know best and have the best capacity, they do not necessarily do best ... those with superior resources may have unacceptable motives and pursue them dysfunctionally for the rest of the population and even for themselves".

Returning to the MLP as a framework for the analysis of sociotechnical change, it can be argued that in order to meet the needs of the country's energy-intensive export-oriented growth trajectory, South Africa's IRP has simultaneously allowed for the entry of a low carbon niche at the same time as supporting the expansion of the incumbent coal-based regime, whilst

facilitating the potential introduction of nuclear energy. Clearly the MLP operates within a much messier, less linear, less predictable and more ambiguous reality than is often implied, and this is often ignored in policy interventions. Leach, Scoones and Stirling (2010) refer to such a reality as "dynamic complexities".

In analytical terms this poses a challenge to the way in which the conceptual and geographical boundaries of a sociotechnical transition within the MLP should be defined, given that what can be considered a sustainability transition at one level may be contributing to a much less sustainable trajectory of supply and demand at another. What is also significant is that this demand has international linkages far beyond the national boundaries of South Africa, which reiterates Smith, Stirling and Berkhout's (2005: 1504) assertion that the regime does not align neatly within the confines of the nation state.

To a certain extent, groups involved in the IRP consultation process succeeded in getting their preferences adopted. Hence, rather than a definitive move away from one technology to another, it leaves the door open for various different technological configurations in which, for time being at least, coal is still set to dominate. At the same time, it can be argued that coal stakeholders have had to compromise a certain level of access, aligning with Meadowcroft's (2011: 72) assertion that political action on sustainability is hampered by uncertainties and that "intervention disrupts established entitlements". As we have witnessed in the case of the IRP, the real politics of sustainability is a "site of conflict" and involves making hard choices and picking winners and losers, particularly among competing technologies and the capital that backs them, as money invested in one particular technological pathway cannot be invested in others (Meadowcroft 2011: 72). For this reason, changes to "the regulatory frameworks within which economic actors conduct their affairs ... are essential to encourage sustainability transitions. And that such changes can only be engineered through political processes, and legitimized and enforced through the institutions of the state. So whatever else they may be, sustainability transitions are inherently political" (Meadowcroft 2011: 71).

Despite plans for significant diversification in the electricity mix, demand is still projected to increase significantly, albeit at a lesser scale than originally planned. As previously discussed, this demand is led by the country's mining and related manufacturing sector which consumes over half the country's electricity supply. Such a trajectory, it can be argued, undermines claims of a transition to sustainability and reveals the fundamental disconnect between the creation of a lower carbon electricity mix and industrial demand. This can be related to Geels, Elzen and Green's (2004: 6) assertion that in order for transitions to take place, they must do so in a co-evolutionary way, involving both supply and demand. On a similar note, Marquard (2010) argues that, "viewing demand as God-given leads to an obsession with low prices resulting in overinvestment, surplus capacity, very low prices, and even faster demand growth, and then incentivization of energy-intensive industries".

Returning now to discussions of a 'just transition' as discussed in 37.2, it is important to note that while the IRP will facilitate a renewable energy niche, its ability to address issues of energy poverty and access to energy for the poor appears to be limited. The plan will contribute to the ongoing increase in electricity prices and fail to reach the thirty per cent of the population who are not connected to the grid. More broadly, it will not tackle the consumption-led model of economic growth or the high levels of national socio-economic inequality in terms of consumption and access to power (Swilling and Annecke 2012). In conclusion, then, any claims of a transition to sustainability and its governance and policy must be evaluated in the most holistic sense, including social, economic, environmental, financial and technological factors.

References

Adger, W. Neill; Brown, Katrina; Conway, Declan, 2010: "Progress in global environmental change", in: *Global Environmental Change*, 20: 547-549.

Baker, Lucy, 2012: "Power shifts? The political economy of socio-technical transitions in South Africa's electricity sector" (PhD thesis, University of East Anglia, School of International Development).

Baker, Lucy, 2014: "Renewable Energy in South Africa's minerals-energy complex: a 'low carbon' transition?", in: *Review of African Political Economy*, 42: 245-261.

Baker, Lucy; Newell, Peter; Phillips, Jon, 2014: "The Political Economy of Energy Transitions: The Case of South Africa", in: *New Political Economy*, published online, January 2014.

Banerjee, Subhabrata Bobby, 2008: "Corporate Social Responsibility: The Good, the Bad and the Ugly", in: *Critical Sociology* 34: 51-79.

Bazilian, Morgan; Rogner, Holger; Howell, Mark; Hermannc, Sebastian; Arentd, Douglas; Gielene, Dolf; Stedutof, Pasquale; Muellerf, Alexander; Komorg, Paul; Tolh, Richard S.J.; Yumkella, Kandeh K., 2011: "Considering the energy, water and food nexus: Towards an integrated modelling approach", in: *Energy Policy*, 39, 7896-7906.

Bekker, Bernard; Eberhard, Anton; Gaunt, Trevor; Marquard, Andrew, 2008: "South Africa's rapid electrification programme: Policy, institutional, planning, financing and technical innovations", in: *Energy Policy*, 36: 3125-3137.

Berkhout, Frans; Smith, Adrian; Stirling, Andrew, 2004: "Socio-technical regimes and transition contexts", in: Elzen, Boelie; Geels, Frank W.; Green, Ken, (Eds): *System Innovation and the Transition to Sustainability. Theory, Evidence and Policy* (Cheltenham: Edward Elgar): 48-75.

Büscher, Bram, 2009: "Connecting political economies of energy in South Africa", in: *Energy Policy* 37,10: 3951-3958.

Byrne, Robert; Smith, Adrian; Watson, Jim; Ockwell, David, 2011: "Energy Pathways in Low Carbon Development: from Technology Transfer to Socio-technical Transformation", STEPS Working Paper 46 (Brighton: STEPS Centre).

Cape Town Chamber of Commerce, 2010: "Comment on IRP 2010, Energy plan may provide for more power than we actually need", 29 November 2010, at: <http://www.doe-irp.co.za/irpCPT/CAPE_CHAMBER_OF_COMMERCE.pdf>.

CDP [Carbon Disclosure Project], 2011: *South Africa JSE 100 Report 2011, Partnering for a low carbon future* (Johannesburg: Carbon Disclosure Project).

COSATU, 2012: *A just transition to a low-carbon and climate resilient economy, COSATU Policy Framework on Climate Change* (Johannesburg: COSATU).

CURES, 2009: *Exploring Energy Poverty in South Africa* (Midrand: CURES).

DoE, 2010a: *Integrated Resource Plan for Electricity, 2010, Revision 2, Report, Draft*, 8 (DoE: October); at: <http://www.energy.gov.za/files/media/pr/INTEGRATED_RESOURCE_PLAN_ELECTRICITY_2010.pdf> (20 October 2014).

DoE, 2010b: *DOE IRP 2010, Power point presentation at the public hearing*, Cape Town (DoE, 29 November); at: <http://www.doe-irp.co.za/irpCPT/DOE_IRP2010_Presnt_Public_Hearings_v2.pdf> (20 October 2014).

DoE, 2010c: Proposed IRP planning process (DoE, 10 May); at: <http://www.doe-irp.co.za/content/Proposed_IRP_Planning_Process.pdf> (20 October 2014).

DoE, 2011a: "Electricity Regulation Act No.4 of 2006, Electricity Regulations on the Integrated Resource Plan 2010-2030 ('Policy-Adjusted IRP')", in: *Government Gazette*, Pretoria, 6 May).

DoE, 2011b: *IRP2: Public hearings and outcomes, Departmental Briefing. Presented to members of the parliamentary portfolio committee on energy* (Parliamentary Monitoring Group, 26 January).

Earthlife Africa, 2010: "IRP2 Presentation, Pretoria", 7 June 2010.

Eberhard, Anton, 2007: "The Political Economy of Power Sector Reform in South Africa", in: Victor, David; Heller, Thomas C. (Eds.): *The Political Economy of Power Sector Reform* (Cambridge: Cambridge University Press): 215-253.

Eberhard, Anton, 2011: "The future of South African coal: market, investment, and policy challenges", Paper for the Programme on Energy and Sustainable Development, Stanford University, Working paper #100 (Stanford: Stanford University, January).

Eberhard, Anton; Kolker, Joel; Leigland, James, 2014: "South Africa's Renewable Energy IPP Procurement Program: Success Factors and Lessons" (Washington, DC: World Bank, Public-Private Infrastructure Advisory Facility (PPIAF), May).

Edkins, Max; Marquard, Andrew; Winkler, Harald, 2010: *Assessing the effectiveness of national solar and wind energy policies in South Africa* (Cape Town: University of Cape Town, Energy Research Centre, June).

EIUG (Energy Intensive User's Group), 2010: "Comments on the IRP2010 Input Planning Parameters", Pretoria, 8 June 2010.

Eskom, 2010: *On the path to recovery, Integrated Report 2010* (Johannesburg: Eskom).

Eskom, 2013: *Integrated Report for the year ended 31 March 2013* (Johannesburg: Eskom).

Eskom Systems Operation and Planning, 2010: *IRP 2010 Energy Forecast Revision 2 Report* (Johannesburg: Eskom, July).

Fine, Ben; Rustomjee, Zavareh, 1996: *The Political Economy of South Africa: From Minerals-Energy-Complex to Industrialisation* (London: C. Hurst).

Fine, Ben, 2009: "Engaging the MEC: Or a few of my views on a few things", in: *Transformation:* 71: 26-49.

Foxon, Timothy, 2010: "A Co-evolutionary framework for analysing a transition to a sustainable low carbon economy", Centre for Climate Change Economics and Policy, Working Paper No. 31 (Leeds–London: Centre for Climate Change Economics and Policy).

Geels, Frank W.; Elzen, Boelie; Green, Ken, 2004: "General introduction: system innovation and transitions to sustainability", in: Elzen, G; Geels, Frank W; Green, K. (Eds.): *System innovation and the transition to sustainability: theory evidence and policy* (Cheltenham: Edward Elgar).

Geels, Frank W.; Schot, Johan, 2007: "Typology of socio-technical transition pathways", in: *Research Policy*, 36,3: 399-417.

Geels, Frank W., 2011: "The multi-level perspective on sustainability transitions: Responses to seven criticisms", in: *Environmental Innovation and Societal Transitions*, 1,1: 24-40.

Goldthau, Andreas; Sovacool, Benjamin K., 2012: "The uniqueness of the energy security, justice, and governance problem", in: *Energy Policy*, 41: 232-240.

Gratwick, Katherine Nawal; Eberhard, Anton, 2008: "Demise of the standard model for power sector reform and the emergence of hybrid power markets", in: *Energy Policy*, 36,10: 3948-3960.

Greyling, Lance, 2010: "The role of Parliament in energy oversight, Lance Greyling, MP, Independent Democrats", in: Baker, Lucy (Ed.), 2011: *Sustainable energy solutions for South Africa: Ensuring public participation and improved accountability in policy processes.* Conference report of the South African Civil Society Energy Caucus in Cape Town on 14 and 15 September 2010" (Pretoria: Institute for Security Studies).

Grin, John; Rotmans, Jan; Schot, Johan, 2010: "Introduction from persistent problems to system innovation and transitions", in: Grin, John; Rotmans, Jan; Schot, Johan: *Transitions to Sustainable Development, New Directions in the Study of Long Term Transformative Change* (London: Routledge): 1-11.

Grin, John, 2010: "Understanding transitions from a governance perspective", in: Grin, John; Rotmans, Jan; Schot, Johan: *Transitions to Sustainable Development, New Directions in the Study of Long Term Transformative Change* (London: Routledge): 223-319.

Hajer, Maarten, 2002: "Discourse analysis and the study of policy making", in: *European political science,* 2,1; 61-65.

Hallowes, David; Munnik, Victor, 2007: "Peak Poison: The Elite Energy Crisis and Environmental Justice" (Pietermaritzburg: groundwork).

Ham, Christopher; Hill, Michael, 1993: *The Policy Process in the Modern Capitalist State* (London: Harvester Wheatsheaf).

Hughes, Thomas Parke, 1983: *Networks of Power: Electrification in Western Society 1880-1930* (Baltimore, MD: Johns Hopkins University Press).

Hughes, Alison, 2010: "IRP 2010 Assumptions" (Cape Town: University of Cape Town, Energy Research Centre, Energy modelling and analysis group, June).

IEA (International Energy Association), 2011: *World Energy Outlook* (Paris: ORCD/IEA); at: <http://www.worldenergyoutlook.org/resources/energydevelopment/accesstoelectricity/> (20 June 2014).

Jasanoff, Sheila, 2003: "Accountability: (No?) Accounting for Expertise", *Science and Public Policy*, 30, 152-162.

Keeley, James; Scoones, Ian, 2003: "Knowledge, Power and Politics: Environmental Policy Processes in Africa", in: Keeley, James; Scoones, Ian (Eds.): *Understanding Environmental Policy Processes: Cases from Africa* (London: Earthscan).

Lawhon, Mary; Murphy, James T., 2012: "Socio-technical regimes and sustainability transitions: Insights from political ecology", in: *Progress in Human Geography*, 36,3: 354-378.

Leach, Melissa; Scoones, Ian; Stirling, Andrew, 2010: *Dynamic Sustainabilities: Technology, Environment, Social Justice* (London: Earthscan).

Ljung, Per, 2007: *Energy sector reform: strategies for growth, equity and sustainability*. (Stockholm, Sida).

McDaid, Liziwe; Austin, Brenda; Bragg, Christy (Eds.), 2010: "Power to the people: raising the voice of civil society in electricity planning–Integrated Resources Plan 2010 inputs and departmental responses" (Cape Town: WWF, South African Faith Communities Environment Institute, Institute for Security Studies, October).

Mainstream Renewable Power, 2010a: "IRP 2 Preliminary Comments, 1st Stakeholder Workshop" by Davin Chown, Director, Pretoria, 7 June 2010 (Mainstream Renewable Power); at: <http://www.doe-irp.co.za/hearing1/MAINSTREAM_REN_POWER.pdf> (20 October 2014).

Mainstream Renewable Power, 2010b: "IRP 2010 comments and inputs, presented to Department of Energy IRP2010 Hearings", Johannesburg, 2 December 2010 (Mainstream Renewable Power). at: <http://www.doe-irp.co.za/irp-JHB/MAINSTREAM_RENEWABLE_POWER.pdf> (20 October 2014).

McDonald, David (Ed.), 2009: *Electric Capitalism: Recolonising Africa on the Power Grid* (Cape Town: HSRC Press).

Meadowcroft, James, 2011: "Engaging with the politics of sustainability transitions", in: *Environmental Innovation and Societal Transitions*, 1: 70-75.

Meadows, Donella H.; Meadows, Dennis; Randers, Jørgen; Behrens III, William W., 1972: *The Limits to Growth: A Report for the Club of Rome's Project on the Predicament of Mankind* (New York: Universe).

Moe, Espen, 2007: *Governance, Growth and Global Leadership* (Aldershot: Ashgate).

Moe, Espen, 2009: "Mancur Olson and structural economic change", in: *Review of International Political Economy*, 16,2: 202-230.

Moe, Espen, 2010: "Energy, industry and politics: Energy, vested interests, and long-term economic growth and development", in: *Energy*, 35: 1730-1740.

Nakhooda, Smita, 2011: "Empowering a Sustainability Transition? Electricity Planning in a Carbon Constrained South Africa" (MSc thesis, London School of Economics and Political Science).

Newell, Peter; Mulvaney, Dustin, 2013: "The political economy of the 'just transition'", in: *The Geographical Journal*, 179,2: 132-140.

Ockwell, David; Ely, Adrian; Mallett, Alexandra; Johnson, Oliver, 2009: "Low Carbon Development: the Role of Local Innovative Capabilities", STEPS Working Paper 31 (Brighton: STEPS Centre & Sussex Energy Group, SPRU, University of Sussex).

Pienaar, Gary; Nakhooda, Smita, 2010: "The great policy disconnect", Heinrich Boell Stiftung Southern Africa, Cape Town.

PC Energy (Parliamentary Portfolio Committee on Energy), 2010: *Meeting of the Energy Portfolio Committee: Committee Programme & Outstanding Reports*, Parliamentary Monitoring Group, 1 June 2010.

Peters, Dipuo, 2010: "Letter from the Minister of Energy to Ms. Samantha Bailey of 350.org South Africa, regarding the Integrated Resource Plan for Electricity", 21 June 2010.

Prasad, Gisela, 2008: "Energy sector reform, energy transitions and the poor in Africa", in: *Energy Policy*, 36,8: 2806-2811.

Rip, Arie; Kemp, René, 1998: "Technological Change", in: Rayner, S.; Malone, E.; Columbus, L., in: *Human Choice and Climate Change volume 2: Resources and Technology* (city: OH, Battelle Press).

Roberts, J. Timmons; Parks, Bradley C, 2007: *A Climate of Injustice* (London: The MIT Press).

Rockström, Johan; Steffen, Will; Noone, Kevin; Persson, Åsa; Chapin, F. Stuart; Lambin, Eric F.; Lenton, Timothy M.; Scheffer, Marten; Folke, Carl; Schellnhuber, Hans Joachim; Nykvist, Björn; de Wit, Cynthia A.; Hughes, Terry; van der Leeuw, Sander; Rodhe, Henning; Sörlin, Sverker; Snyder, Peter K.; Costanza, Robert; Svedin, Uno; Falkenmark, Malin; Karlberg, Louise; Corell, Robert W.; Fabry, Victoria J.; Hansen, James; Walker, Brian; Liverman, Diana; Richardson, Katherine; Crutzen, Paul; Foley, Jonathan A., 2009b: "A safe operating space for humanity", in: *Nature*, 461,24 (September): 472-475.

Sappi, 2010: Sappi Presentation at the IRP 2010 Stakeholder function, 7 June 2010.

Scrase, Ivan; Smith, Adrian, 2009: "The (non-)politics of managing low carbon socio-technical transitions", in: *Environmental Politics*, 18: 707-726.

Smith, Adrian; Stirling, Andrew; Berkhout, Frans, 2005: "The Governance of Sustainable Socio-Technical Transitions" in: *Research Policy*, 34: 1491-1510.

Swilling, Mark; Annecke, Eve, 2012: *Just Transitions: Explorations of Sustainability in an Unfair World* (South Africa: UCT Press).

Torgerson, Douglas, 2003: "Democracy through policy discourse", in: Hajer, Maarten A.; Wagenaar, Hendrik (Ed.): *Deliberative Policy Analysis: Understanding Governance in the Network Society* (Cambridge: Cambridge University Press).

Victor, David; Heller, Thomas C. (Ed.), 2007: *The Political Economy of Power Sector Reform*, (New York–Cambridge: Cambridge University Press).

Wamukonya, Njeri, 2003: "Power sector reform in developing countries: mismatched agendas", in: *Energy Policy*, 31,2: 1273-1289.

WCED (World Commission on Environment and Development), 1987: *Our Common Future* (Oxford University Press).

Weischer, Lutz; Wood, Davida; Ballesteros, Athena; Fu-Bertaux, Xing, 2011: *Grounding green power, bottom-up perspectives on smart renewable energy policy in developing countries*. Climate and Energy Paper series (Washington DC: German Marshall Fund of the United States–World Resources Institute–Heinrich Böll Stiftung).

Wildavsky, Aaron, 1979: *Speaking truth to power: the art and craft of policy analysis*, (Canada: Little Brown and Company).

WWF South Africa, 2010: *50% renewable energy for a just transition to sustainable electricity supply through low carbon re-industrialisation*. Living Planet Unit Climate Change Programme, IRP2 input, 7 June 2010.

Yi-chong, Xu, 2006: "The myth of the single solution: electricity reforms and the World Bank", in: *Energy* 31,6-7: 802-814.

Ziramba, Emmanuel, 2009: "Disaggregate energy consumption and industrial production in South Africa", in: *Energy Policy*, 37,6: 2214-2220.

Other Literature

Statistics South Africa, 2012: *Quarterly Labour Force Survey, Quarter 4, 2012*, 7 February 2012; at: <http://www.statssa.gov.za/publications/P0211/P02114thQuarter2011.pdf> (20 March 2012).

The Presidency, Republic of South Africa, 2009: *Development Indicators 2009*; at: <http:// www.thepresidency.gov.za/learning/me/indicators/2009/indicators.pdf> (22 August 2011).

38   Low Carbon Green Economy: Brazilian Policies and Politics of Energy, 2003–2014

Eduardo Viola[1] and Larissa Basso[2]

Abstract

Achieving sustainability is a complex issue because of the lack of a precise definition of the term. Very recently, science has acknowledged that the human species has become the main driver of transformations in bio-geo-physical systems, and has identified boundaries to be respected if a systemic planetary disruption is to be avoided. A low carbon green economy—a development model in which environmental concerns are at the same level as economic and social concerns—is a good paradigm for guiding policy-making that aims at respecting these boundaries; and it includes the objective of the decarbonization of energy systems in order to mitigate climate change, the most studied and best understood of the boundaries. Having these parameters in mind, this chapter shows how Brazilian policies and politics of energy evolved in the period between 2003 and 2014. It identifies an upward and later downward trend in the production of energy from low carbon sources, and a stable trend in energy efficiency. It concludes that, despite the higher proportion of low carbon energy sources in the Brazilian energy matrix compared with other countries, Brazil has not yet embraced the low carbon green economy or the decarbonization paradigms: in spite of significant forces promoting both these, important political and cultural features are preventing them from being adopted.

Keywords: Energy governance, low carbon green economy, decarbonization, policy and politics, Brazil.

38.1   Introduction

The transition to sustainability has been on the international agenda since the 1970s and it still generates more heat than light. Complexity and the lack of a precise definition of sustainable development has led to the use of new terms for guiding domestic and international policy-making, including 'low carbon development'—a pattern that sets global climate stability as a condition for national and global economic growth—

and 'green economy'—an economy that results in improved human well-being and social equity, while significantly reducing environmental risks and ecological scarcities. However, science has proved that the world has entered a new geological epoch, the Anthropocene, in which human beings are the key drivers of Earth's system transformations; therefore, a low carbon green economy—one that respects the nine planetary boundaries that need to be monitored to prevent a systemic disruption of Earth's systems (see chapter 3 above by Dalby)—is a better guideline for policies that face up to current sustainability challenges.

Applying the 'low carbon green economy' paradigm means changing the way in which policy-making is thought about and enforced. Environmental concerns must be at the same level as economic and social ones. As far as energy is concerned, for instance, reliable energy sources and access to energy for all are not enough: energy production and consumption need to decrease their impact on the environment; in order to respect climate stability, the most studied and best

---

1   Prof. Dr. Eduardo Viola is full professor at the Institute of International Relations and coordinator of the Climate Change and International Relations Research Programme, University of Brasilia; and senior researcher at the Brazilian Council for Scientific and Technological Development (CNPq). Email: <eduviola@gmail.com>.

2   Larissa Basso is a PhD candidate at the Institute of International Relations and member the Climate Change and International Relations Research Programme, University of Brasília; and researcher of Brazilian Federal Agency for Support and Evaluation of Graduate Education (CAPES). E-mail: <larissabasso@gmail.com>

© Springer International Publishing Switzerland 2016   811
H.G. Brauch et al. (eds.), *Handbook on Sustainability Transition and Sustainable Peace*,
Hexagon Series on Human and Environmental Security and Peace 10, DOI 10.1007/978-3-319-43884-9_38

understood of the planetary boundaries, energy must be decarbonized. Investing in energy efficiency and employing low carbon sources–especially renewable sources–to produce energy is a must.

Brazil has a relatively low carbon energy matrix, compared to other countries: 46 per cent of its total primary energy production and 42.4 per cent of its total primary energy supply[3] come from low carbon sources (EPE 2013: 21–22). In fact, *land use, land use change and forestry* (LULUCF) have traditionally been the major drivers of Brazilian greenhouse gas (GHG) emissions. Deforestation has been successfully tackled in recent decades: by about 2008, LULUCF was no longer the main source of Brazil's emissions. The year marked a new upward trend in emissions induced by the effects of enhanced economic development:[4] emissions from energy, agribusiness, industry, transportation and waste treatment had increased (Brasil 2010; Brasil 2013; Lapola/Martinelli/Peres et al. 2014).

When it comes to the challenge of embracing the 'low carbon green economy' paradigm and the central role of energy in achieving an embedded challenge, that of decarbonization, given that energy accounts for more than 60 per cent of global GHG emissions[5] it is very relevant to understand how energy policy and politics evolve. There are hardly any academic papers about energy in the transition to a low carbon green economy, and only a few analyse the evolution of low carbon energy sources and energy efficiency in Brazil. This chapter aims to address this topic. Further reflections on how energy policy-making should be designed to respect the other eight planetary boundaries are needed.

The chapter focuses on how Brazilian energy policy and politics evolved between 2003 and 2014. The topic will be explored in sections. First, a reflection on the definition of low carbon green economy, and how decarbonization is embedded in it. Next, an analysis of the Brazilian energy matrix transformations and energy policy and politics between 2003 and 2014. This analysis will show (i) two trends in the production of energy by low carbon sources: first an upward trend, where these sources in the energy

matrix increased, and then a downward trend, due to the discovery of deep offshore oil reserves and the use of oil prices as a heterodox policy tool to artificially maintain greater economic growth rates; and (ii) a stable trend in energy efficiency: no consistent development of policy and politics during the decade. Finally, some traits of Brazilian culture that hinder the implementation of a low carbon green economy in general, and of decarbonization in particular, will be described, focusing on 'crony capitalism' and short-term thinking. The conclusion argues that Brazil has not yet embarked on a low carbon green economy paradigm, and serious planning and long-term goals, including tackling deep-rooted cultural traits, are needed.

38.2   Low Carbon Green Economy

Sustainability has been on the international agenda since the *United Nations Conference on the Human Environment* (1972), when it was stated that development should consider its impact on the environment. In 1987, the *Brundtland Report* was released; it defined sustainable development as "development that meets the needs of the present without compromising the ability of future generations to meet their own needs" (WCED 1987). This concept was adopted by the *United Nations Conference on Environment and Development* (UNCED 1992), and has become the leading guideline for international and national policies directed towards sustainability. However, the term is very loose. It does not translate the impact of human development patterns in the environment, nor the action necessary to mitigate them. For this reason, other definitions have been developed, among which are 'low carbon development' and 'green economy'.

Low carbon development aims at a consistent reduction of GHG emissions per GDP unit (beyond the 'normal' reduction derived from business-as-usual economic development). By reducing carbon emissions, economic growth and climate stability become structurally compatible (UNDESA 2012: 51). The concept was developed within the scope of the *United Nations Framework Convention on Climate Change* (UNFCCC) and gives, in fact, a clear direction to policy-making.

Still, it is not enough, because the world has entered a new geological epoch. Humanity has developed during the Holocene, the era that started approximately 11,000 years ago, when the last glaciation ended and the climate became warmer and more stable. During this period, human beings changed

---

3   Excluding and including energy imports, respectively; 2012 data.

4   While LULUCF accounted for approximately 57 per cent of total emissions in 2005, its share was approximately 23 per cent in 2010.

5   Data are available at <http://www.grida.no/graphicslib/detail/world-greenhouse-gas-emissions-by-sector_6658>, (15 Dec 2013).

from a few groups of hunter-gatherers to a complex population of seven billion people, most of whom require an unprecedented amount of Earth's resources to live. The expansion of human population—both in numbers and in per capita exploitation of the Earth's resources (Crutzen/Stoermer 2000)—was such that it became the driver of the transition to a new geological epoch, the Anthropocene. This is unprecedented: from the evidence currently available, all previous changes in Earth systems were caused either by Earth's natural dynamics or the dynamics of the cosmos; instead, in the Anthropocene, human beings are the main drivers of transformations in Earth systems (Crutzen/Stoermer 2000; Viola/Franchini/Ribeiro 2013; see chapter 3 by Dalby in this volume).

The concept of a 'green economy' was developed in order to deal with this new reality. There is no official definition of a green economy, but the one most frequently used refers to an "economy that results in improved human well being and social equity, while significantly reducing environmental risks and ecological scarcities" (UNEP 2011). The concept was discussed during the *United Nations Conference on Sustainable Development* (UNCSD 2012), but there was no agreement to make it more prescriptive—the final report of Rio+20 states that a 'green economy' is "one of the important tools available for achieving sustainable development and that it could provide options for policy-making but should not be a rigid set of rules" (UNCSD 2012). Even if it represents an evolution from the loose concept of sustainable development, this more recent definition of a green economy[6] remains almost as elusive as the former definition, unable to guide effective policy-making.

A more precise concept should acknowledge the nine biophysical planetary boundaries science has identified whose transgression would lead to a planetary disruption of Earth systems, a disruption which would threaten life on Earth, especially human life: (i) climate change, (ii) biodiversity loss, (iii) the nitrogen cycle and the phosphorus cycles, (iv) stratospheric ozone depletion, (v) ocean acidification, (vi) global freshwater use, (vii) change in land use, (viii) aerosol pollution, and (ix) chemical contamination (Rockström/Steffen/Noone et al. 2009). Three of the boundaries have already been already trespassed on. First, the nitrogen cycle, due to the massive use of

nitrogenized fertilizers since the 1950s. Second, biodiversity loss, especially since the 1970s, when an expanded human population started impacting natural ecosystems and other species to the point of exhausting them. Third, climate stability, due to the accumulation of GHG in the atmosphere—currently, 400 part per million (ppm),[7] when the safe boundary is 350 ppm. Trespassing on the boundaries indicates that change is needed, and urgently: development must take into account its effects on Earth's equilibrium and on human survival in the long term; yet policy-making is stuck in an outdated framework, incapable of establishing a new development pattern.

Take, for instance, climate change. Of the nine planetary boundaries, climate change is the most studied and best understood by science; yet tackling it is still a great challenge, because of the discrepancy between the nature of the issue and the characteristics of current political models. Climate change has been defined as a truly wicked or diabolical problem because its complexity defies standard problem-solving (Prins/Galiana/Green et al. 2010; Steffen 2011). First, it is intrinsically global: it is led by changes in the concentration of GHG in the atmosphere, a global common good. However, there are no institutions for direct global policy-making: policy-making is still mostly carried out by nation states that only occasionally cede small shares of their sovereignty to international institutions. Second, climate change is not linear (see chapter 11 by Schellnhuber et al. in this volume): vulnerabilities to and responsibilities for the problem are skewed—the poorest in every society usually contribute the least to the problem, but suffer its effects the most. Third, climate change operates on a timescale beyond human daily experience, so the appeal of passing the burden to future generations is always present. Climate is a key civilizational driver, together with the expansion and deepening of globalization and the diffusion of democracy (Viola/Franchini/Ribeiro 2013). Yet policy-making has only marginally faced up to the challenge of its stabilization, especially in the international arena.

After signing the UNFCCC, in 1992, countries have tried to build a multilateral framework to tackle climate change. In 1997, by signing the Kyoto Protocol, countries agreed on limited compulsory GHG reduction emission targets (seven per cent reduction

---

6 The European Union proposed a more scientifically based concept of green economy, which was vetoed by the BRICS countries, the United States and most developing countries.

7 The concentration of GHG in the atmosphere was 401.85 ppm on 12 June 2014, according to the Carbon Dioxide Information Analysis Center. Data are available at <http://cdiac.ornl.gov>, accessed 19 June 2014.

in twenty years) for developed countries (Annex I countries) and on mechanisms for their implementation; emerging economies rejected any commitment to reduce the curve of emissions growth (moving out from business-as-usual scenarios). For this reason, and fearing what it considered unfair competition in global markets, the United States withdrew from the protocol in 2001 (Viola 2002). This was the first setback for the protocol: in 1997, mandatory commitments covered countries responsible for 65 per cent of global GHG emissions, but the amount was reduced to around 45 per cent when the United States left. The second setback occurred in 2012, when the period in which the protocol was to be in force expired. At that moment, Japan, Canada and Russia decided not to sign the extension of the protocol,[8] so countries currently obliged to reduce GHG emissions—the European Union, Switzerland, Norway and New Zealand—account for only 13 per cent of the global amount.[9]

Discussions to establish a new legally binding agreement have advanced by the presentation of *Intended National Determined Contributions* (INDCs) by the members of UNFCCC prior to the Paris Conference, in December 2015. There has been some advance toward an agreement, even if it will be short of what science identifies as needed to guarantee climate stability. Only European Union's pledge to reduce emissions is in tandem with what science defines as needed to avoid serious climate change. The United States, Canada and Japan pledges have 2005 as baseline, and none means substantial reduction of GHG emissions when translated into 1990 baseline. China committed to peak its emissions by 2030, but Chinese emissions are expected to reach 35% to 40% of total global emissions by then. India did not commit to peak emissions. Mexico promises reductions regarding a business as usual scenario. Russian commitments are poor considering that the baseline is 1990 and after that Russian emissions were reduced in more than 50% due to economic collapse. Brazil pledged to cut 37% of its GHG emissions until 2025, compared to 2005 levels, and 43% until 2030, an important advance compared to previous pledges.

Yet, commitments for specific sectors are either not ambitious (energy, easily met before 2025, raising concerns over the increase of fossil fuels in the energy matrix), or too broad (agriculture, transport and industry), or shameful (LULUCF, as Brazil pledges to zero illegal deforestation only by 2030).

It is clear that current policy-making patterns are not going to achieve the results that are needed to mitigate climate change. This is so because mitigation requires the implementation of measures that interfere with core issues of the contemporary lifestyle, such as energy sources and use, institutions, governance, economic organization and values (Steffen 2011: 21; Jamieson 2011: 38-42). It has been said that it "will only be successfully achieved as a benefit contingent upon other goals which are politically attractive and relentlessly pragmatic" (Prins/Galiana/Green et al. 2010). However, a clearer paradigm guiding national and international policy-making may be helpful.

The concept of low carbon green economy aims at doing this.[10] It has three main assumptions: (i) the nine planetary boundaries identified by science should be respected to prevent a systemic disruption of Earth systems—starting with climate change, since science has already produced clear indicators on how to achieve climate stability; (ii) economic growth should be guided by equity; (iii) population dynamics should be considered a central issue (Viola/Franchini 2012b). If this concept is applied, development will involve a balance of economic, social and environmental aspects (Viola/Franchini 2012b). Environmental concerns become embedded: they should be incorporated in any policy-making as one of its pillars.

Under the low carbon green economy paradigm, energy policies must take into account the environmental impacts of producing and using energy. It is not enough to guarantee energy security and energy access; the processes by which energy is produced and distributed are relevant. For example, if climate change is to be mitigated, stabilizing the concentration of GHG in the atmosphere is necessary. This can

---

8 In 2012, countries agreed on an extension of the Kyoto Protocol from 2013 to 2019. In 2015, it is expected that they will have agreed on the terms for a legally binding agreement to replace the Kyoto Protocol from 2020 onwards.

9 2013 figure, data from the US Environmental Protection Agency, available at <http://www.epa.gov/climatechange/ghgemissions/global.html>, accessed 9 July 2014.

10 The paradigm was inspired by the European Commission communication "Rio+20: towards the green economy and better governance", released ahead of the UNCSD 2012 Conference, and the UNDP Report "Green Economy in action: articles and excerpts that illustrate green economy and sustainable development efforts"; but by acknowledging the need to respect the nine planetary boundaries before a systemic planetary disruption, as well as equity in economic growth and population dynamics, it goes further as a concrete policy guideline.

be done either by reducing emissions or capturing and storing carbon (CCS) (UNFCCC 1992); but since CCS is still a limited option (IEA 2013a), it is imperative to reduce emissions. Energy supply and use accounts for more than sixty per cent of the world's GHG emissions, and it is likely that this share will increase even more once emerging economies achieve higher levels of development. Energy systems must be decarbonized, and decarbonization depends to a large degree on increasing energy efficiency and disseminating low carbon energy sources (Viola/Franchini/Ribeiro 2013), especially among the biggest energy producers and consumers. Brazil is one of these.

38.3   The Brazilian Energy Matrix and its Transformations (2003–2014)

Brazil is a big country, in all aspects: it has a territory of 8,514,215 km2, the fifth largest in the world; as one of the emerging countries, Brazil is boosting its international power assets: it is among the ten biggest economies—economic growth averaged 3.6 per cent a year between 2003 and 2012, compared with 2.8 per cent during the previous decade,[11] and it has considerably reduced inequality-from a Gini score of 57.78 (2003) to 54.70 (2009).[12] It has limited military power, despite its increased role in peacekeeping and human rights missions, but has important soft power assets (Nye 2011): it is a pacific, multi-ethnic, multicultural and multi-religious country, with magnificent nature and extraordinary tourism potential. Between 2003 and 2014, Brazil's enhanced international role could be seen in several different arenas, including energy governance in the transition to a low carbon green economy: it is among the eight largest GHG emitters; with 190,732,694 inhabitants (IBGE 2010), it is the eleventh largest energy producer and seventh largest energy consumer;[13] and it already occupies the fifth place among producers of low carbon technology[14] (WWF/Roland Berger 2012; The Pew Charitable Trusts 2012).

Brazil is generally self-sufficient in energy: it imports natural gas from Bolivia, and oil and coal from several countries, but in amounts that are small compared with its domestic production. The same is true of hydroelectricity, where a small amount is imported from Paraguay. It has developed world-renowned technology for exploring deep offshore oil reserves, and it is a world leader in hydropower technology (especially in designing reservoirs) and in producing electricity and fuel from sugar cane.

With its large territory located in a warm climate, its relief system of plains and plateaux, and its extensive drainage basins, Brazil has enormous potential for developing renewable energy. In fact, it is safe to say that, given the currently available technology, Brazil has one of the world's biggest potentials for renewable energy resources. The reality, however, does not match the potential. From 2003 to 2012, both Brazilian energy production and domestic supply increased by approximately 40 per cent; however, during the same period, energy production and energy supply from non-renewable sources increased by 42.5 per cent and 45 per cent respectively, and from renewable sources by only 37 per cent and 36 per cent respectively. The data show that non-renewable energy sources answered for a greater share of energy production and supply in Brazil during the decade from 2003 to 2012.

When it comes to energy efficiency, the picture is even gloomier. Energy efficiency is crucial for a low carbon green economy. In fact, due to (i) the time span required for research and development into low carbon energy sources, and (ii) the intrinsic need to reduce energy consumption so as to assimilate a low carbon way of living, and given that saving energy reduces costs and enables resources to be used for something else, investing in energy efficiency is the most rational choice for achieving low carbon development. Yet in Brazil, energy efficiency plays second fiddle to the expansion of energy supply: from 1984 to 2004,[15] the amount of energy effectively employed in final use (in relation to total energy input) increased from 46.9 per cent to merely 57.8 per cent (EPE 2013, 193); performances in many sectors were even poorer, for example, from 31.4 per cent to 37.5 per cent in transport (EPE 2013, 193) ; compared with the US,

---

11 On GDP growth, data from the World Bank (1993-2002, 2003-2012) are available at: <http://data.worldbank.org/indicator/NY.GDP.MKTP.KD.ZG?page=4> (25 January 2014).

12 World Bank, available at: <http://iresearch.worldbank.org/PovcalNet/index.htm?2> (28 July 2013).

13 World Bank data (2010) are available at: <http://data.worldbank.org/indicator> (28 July 2013).

14 The definition covers manufacturing inputs such as silicon and specialized machinery, intermediate products such as solar cells, and end products such as wind turbines, heat pumps and biofuels. Brazil is a leader in technology for the exploration of offshore deep oil, reservoir-building and obtaining ethanol from biomass (sugar cane).

15 Latest available data.

Table 38.1: Domestic energy production and domestic energy supply, by source 2003–2012 (103 tep (toe)). **Source:** Author's elaboration based on: *Balanço Energético Nacional–séries completas 1970-2012*; at: <https://ben.epe.gov.br/BENSeriesCompletas.aspx> (12 January 2014).

Energy source	Energy production 2003	Energy production 2012	Variation (%)	Energy supply 2003	Energy supply 2012	Variation (%)
Oil	77,225	107,017	38.5%	80,688	111,193	37.8%
Natural gas	15,681	25,574	63%	15,512	32,598	110%
Coal + Peat	1,823	2,517	38%	12,848	15,287	19%
Uranium	2,745	3,881	41.3%	3,621	4,286	18%
Total non-renewable	**97,474**	**138,989**	**42.5%**	**112,669**	**163,364**	**45%**
Hydro	26,283	35,719	36%	29,477	39,181	33%
Wood	25,965	25,735	-1%	25,973	25,735	-1%
Sugar cane	28,357	45,132	60%	27,093	43,572	60.8%
Other renewable (*)	5,663	11,723	107%	5,663	11,754	107.5%
Total renewable	**86,268**	**118,309**	**37%**	**88,206**	**120,242**	**36%**
Total	**183,742**	**257,298**	**40%**	**200,875**	**283,606**	**41%**

(*) Includes wind and solar.

Brazilian truck freight consumes 85 per cent more fuel (MME 2011).

Why has energy production from non-renewable sources surpassed renewable sources between 2003 and 2014, in a context of increasing national and international debates about the importance of a low carbon green economy and the need to commit to it? And why is energy efficiency such a marginal topic in Brazil? In order to answer these questions, it is necessary to analyse how energy policy and politics evolved in Brazil over the decade.

38.4   Brazilian Energy Policy and Politics (2003–2014)

Energy policy and politics of the last decade are part of the big picture of Brazilian recent economic and political developments and their consequences for domestic politics and international affairs. From 2003 to 2010, Brazil's economy grew faster than the average world economy, due to three factors: (i) a big change in the relative prices of industrial products and primary commodities, favouring the latter–and Brazil is a big exporter of primary commodities–; (ii) the effects of the pro-market macroeconomic reforms carried out during the second half of the 1990s; and (iii)

high rates of *foreign direct investment* (FDI). The redistribution of income was initiated during the Cardoso administration (1995-2002) and deepened in Lula da Silva's administration (2003-2010). This virtuous cycle of economic growth ended around 2010. The consequences of the lack of pension, labour, fiscal, tax and political system reforms–reforms that should have been undertaken in order to finish the process initiated in the 1990s–began to be felt during the Rousseff administration (2011-present): economic growth declined to less than 2 per cent a year; inflation was always significantly above the target defined by the Central Bank (target 4.5 per cent, real inflation around 6 per cent); FDI, compared with the period to 2010, stagnated–although it remains among the highest in the world–as a result of uncertainty generated by federal government interventionism into the economy. Tackling inequality became more difficult with lower growth rates and higher inflation.

Nevertheless, the effects of a stronger economy were felt politically. Very powerful and competitive media and a more attentive society were more perceptive of corruption–corruption was actually lower during this decade compared with previous ones, but there was much more information about it. Brazilian courts demonstrated a level of independence rare in non-OECD countries: core members of the first Lula administration (except the former president himself) were convicted of corruption offences in 2012. In June 2013, massive demonstrations against corruption and the state of public transport, education and health care surprised political actors and analysts.

Brazil has not yet incorporated the low carbon green economy paradigm; not even a clear decarbonization outcome was envisaged. During the last decade, two trends in low carbon energy sources can be observed: from 2003 to 2007, an upward trend in which progressive forces—committed to low carbon development—were stronger, and the proportion of renewable sources in the energy matrix increased; from 2007 to 2014, a downward trend in which regressive forces—which favour fossil energy sources— were empowered due to the discovery of deep offshore oil reserves and the use of oil prices as a heterodox policy tool to artificially maintain greater economic growth rates and fight inflation. As for energy efficiency, however, the trend was stable: there was no consistent development of policy and politics during the decade.

38.4.1 Energy Sources

38.4.1.1 Fossil Fuels: Oil, Gas and Coal

Oil remains the main source of fuel employed in transportation in Brazil. The Brazilian transport system is not adequate for the country's characteristics and needs: a large country with mostly plain relief, important river basins and an extensive coastline could make a better use of those assets in transportation. Instead, its transport system is highly dependent on roads—terribly built and poorly maintained—, and its cargo fleet employs diesel as fuel.

Brazilian diesel had traditionally been of low quality because of the high level of sulphur in its mix, but recent regulation has attempted to improve it (though very slowly).[16] Despite some delays and implementa-

tion problems,[17] the new regulation contributes to the provision of cleaner air. It has minimal impact on reducing carbon emissions, though: other measures— such as increasing the use of biodiesel or other biofuels in transportation and adopting railways and coastal and inland navigation systems—would bring greater benefits in this area.

The competition between oil/derivatives and renewable fuel prices is not a fair one. Until 2006, the domestic prices of oil/derivatives followed international prices; after that, however, due to the discovery of the deep offshore reserves—and the illusion that Brazil would rapidly become a big producer and exporter of oil—the federal government was misled into using domestic prices as a heterodox economic tool. In 2007, the Brazilian government subsidized oil prices to maintain economic growth rates at a high level, changing the relative prices of gasoline and ethanol and undermining the competiveness of the latter. After the economic crisis that followed the collapse of Lehman Brothers, some contracyclical measures were adopted, including tax exemptions for the automobile industry. The exemptions led to an increase in car ownership and a dramatic increase in fuel demand, while ethanol prices were still not competitive. These heterodox policies enhanced short-term economic growth but increased long-term macroeconomic imbalance and penalized both Petrobras[18]—who faced several important losses—and the ethanol production chain.

General expectations of economic gains from the deep offshore reserves were high between 2008 and 2010. Action by environmentalists was of a very low profile: there was no significant environmental movement to defend exploring low carbon energy sources rather than the new oil reserves. Nevertheless, it was a victory without a battle: while it was expected that exploitation would start soon after the discovery, disputes between state governments over the distribution of revenues from the reserves postponed the auctions (of the reserves) for a couple of years; in this period, economic production of the technology to exploit shale gas was achieved in the US—supposedly the big-

---

16 There are two types of diesel: metropolitan diesel (used in larger cities with poor air quality) and interior diesel (used in smaller cities, roads and other parts of the country). From 2001 onwards, sulphur concentration in the mix began to be limited: 2000 mg/kg for metropolitan diesel and 3500 mg/kg for interior diesel (2001); 500 mg/kg for metropolitan and 1800 mg/kg for interior diesel (2005-2006); 50 mg/kg for metropolitan and 500 mg/kg for interior diesel (2009, coming into force in 2013); finally, 10 mg/kg for metropolitan (2013). See National Oil Agency (in Portuguese, ANP) Resolutions nr. 310/2001, 12/2005, 15/2006, 31/2009, 42/2009, 65/2011 and 50/2013.

17 E.g. the 500 mg/kg sulphur concentration goal in metropolitan diesel, determined in 2006, was not achieved until 2012, due to technical problems at refineries and in the automotive industry; the switch to S500, determined in 2009, was delayed in some cities by up to fifteen months.

18 Petrobras is the most famous Brazilian energy corporation. It is worldly known for its vanguard technology to explore oil in offshore platforms.

gest market for the Brazilian deep offshore oil–and Mexico passed a bill that would end PEMEX monopoly over oil, facts that appealed to American companies such as Chevron and Exxon Mobile; Petrobras, whose participation in the auction was obligatory due to the requirement to include national companies in the exploitation, had lost the financial capacity to act due to the continued losses it faced through subsidizing oil derivatives. In 2013, when only one consortium presented itself at the auction for the exploitation of the Libra field, enthusiasm for deep offshore oil reached its lowest point. Subsidies for gasoline peaked in 2014–in fact, estimates indicate that gasoline would cost on average thirty per cent more without the subsidies.

The use of natural gas in Brazil more than doubled during the decade. This is a progressive trend, especially when natural gas is employed in place of gasoline or diesel (in part of the metropolitan transport system and taxis) and coal (in industries, especially in Sao Paulo state after the building of the gas pipe from Bolivia). Nevertheless, the share of gas in the Brazilian energy matrix is still very small; it could be bigger if there were incentives to employ natural gas in the thermal power plants that currently rely on oil or coal. Nowadays, some old practices, such as gas flaring on offshore oil platforms, are beginning to be condemned by (a minority in) Brazilian society; it is expected that this trend will increase and a more environmentally-friendly use of this gas will become mandatory in Brazil. The rational use of the flared gas is also an argument against the exploitation of shale gas: Brazil has important reserves, but it lacks the technology, infrastructure (pipes and transport) and policies (incentives for its production and the creation of a market) that are vital for its exploitation.

Coal has never been a relevant energy source in Brazil. The quality of Brazilian coal is low; it is employed mostly in producer regions, such as the south of the country. Brazilian society is against the use of coal, and groups in favour of it have very limited appeal. Recently, the steel industry, the largest consumer of coal in Brazil, has started to employ vegetable charcoal from planted forests to replace the charcoal from native forests that was used previously,[19] an important advance towards sustainability and a low carbon green economy. The greatest backwards step involving coal is its inclusion among the accepted

sources for electricity generation in Brazil: in August 2013, the federal government authorized coal's participation in future auctions for electricity, along with hydropower, natural gas and biomass.

38.4.1.2 Hydropower

Hydropower is the main source of Brazilian electricity. In 2012, 70.1 per cent of Brazilian electricity production and 76.9 per cent of the Brazilian electricity supply[20] came from hydro (EPE 2013: 16). From 1920 to the 1970s, relying on hydropower for electricity generation was a 'natural' process in a country with little coal and oil; after 1973, the process gained strategic importance due to concerns about energy security in the context of world oil crises. Brazil has one of the world's greatest hydropower potentials, but most of its southern river basins are already largely exploited; 70 per cent of the remaining potential is located in Amazonian basins (Eletrobras 2012),[21] a region with complex ecosystems and where it is of great importance to maintain the local and regional climate and biodiversity.

Most of Brazilian hydroelectricity comes from hydropower plants that have an installed capacity of at least 30MW and a reservoir larger than 3 km2, called by the Brazilian government and in this article *Usinas Hidrelétricas* (UHEs). Most UHEs were built between 1968 and 1984, when lower environmental standards were in force and financial resources, both national and foreign, were available. Since 1985, it has become much more difficult to license a new UHE; many of those projected had their construction postponed or disrupted due to further environmental demands and the fiscal crisis of the Brazilian state. It was mainly after the 2001 crisis in the electricity supply[22] that the federal government resumed efforts to build new hydropower plants, including several in the Amazon region–not without controversy (table 38.2). Two of the biggest sources of controversy are run-of-the-river technology and the impact of UHEs on the indigenous populations of the Amazon forest.

---

19 The Climate Change law prescribes the substitution of mineral coal with charcoal from planted forests, but this prescription has not been implemented yet.

20 Excluding and including electricity imports, respectively.
21 According to official data, the rivers of the Amazon basin have 34,000 MW of hydropower potential not yet explored.
22 In 2001, the Brazilian electricity supply was disrupted after an abnormal dry summer reduced water levels at key dams; the government rationed electricity, and fossil-fuel thermal power plants were put into operation as a backup system.

Table 38.2: UHEs in the Amazon region (Amazon basin and Tocantins-Araguaia basin). **Source**: Author's elaboration, based on data from Agência Nacional de Energia Elétrica—Banco de Informações sobre Geração; at: <http://www.aneel.gov.br/aplicacoes/capacidadebrasil/GeracaoTipoFase.asp?tipo=1&fase=3>, and Programa de Aceleração do Crescimento—PAC; at: <http://www.pac.gov.br/infraestrutura-energetica/geracao-de-energia-eletrica> (26 July 2014). PCHs and UHEs under study or not found on any of the government websites above are excluded from the table.

UHE name	River	Generating (MW)	Capacity (MW)	Status
Balbina	Uatumã	249.75	250.00	In operation since 1989
Belo Monte	Xingu	-	11,233.00	Under construction
Bem Querer	Branco	-	708.40	Under construction
Cachoeira dos Patos	Tapajós	-	528.00	Preparing to construct
Cana Brava	Tocantins	450.00	450.00	In operation since 2002
Castanheira	Arinos	-	192.00	Under construction
Chacorão	Tapajós	-	3,336.00	Preparing to construct
Colíder	Teles Pires	-	300.00	Under construction
Dardanelos	Aripuanã	261.00	261.00	In operation since 2011
Estreito	Tocantins	1,087.00	1,087.00	In operation since 2011
Ferreira Gomes	Araguari	-	252.00	Under construction
Jamanxim	Tapajós	-	881.00	Preparing to construct
Jatobá	Tapajós	-	2,338.00	Preparing to construct
Jirau	Madeira	750.00	3,750.00	In operation since 2013
Luís Eduardo Magalhães	Tocantins	902.50	902.50	In operation since 2009
Paredão A	Mujacaí	-	199.30	Preparing to construct
Peixe Angical	Tocantins	498.75	498.75	In operation since 2006
Prainha	Aripuanã	-	796.40	Under construction
Rondon II	Comemoração	73.50	73.50	In operation since 2011
Salto Augusto de Baixo	Juruena	-	1,461.00	Preparing to construct
Santo Antonio	Madeira	1,927.03	3,150.40	In operation since 2012
São Luiz do Tapajós	Tapajós	-	6,133.00	Preparing to construct
São Manoel	Teles Pires	-	746.00	Preparing to construct
São Salvador	Tocantins	243.20	243.20	In operation since 2010
São Simão do Alto	Juruena	-	3,509.00	Preparing to construct
Serra da Mesa	Tocantins	1,275.00	1,275.00	In operation since 1998
Serra Quebrada	Tocantins	-	1,328.00	Preparing to construct
Sinop	Teles Pires	-	461.00	Preparing to construct
Sumaúma	Aripuanã	-	458.20	Preparing to construct
Tabajara	Machado	-	350.00	Under construction
Teles Pires	Teles Pires	-	1.820,00	Under construction
Tucumã	Juruena	-	510.00	Preparing to construct
Tucuruí I and II	Tocantins	8,535.00	8,535.00	In operation since 1984

Run-of-the-river technology was developed to accommodate demands from both environmentalists and defenders of development at all costs. This technology allows for smaller reservoirs to be created, so, in theory, it diminishes the impact of hydropower on the environment—especially on biodiversity. Nevertheless, the results obtained from the Amazon UHEs are, to date, ambiguous.[23] For hydropower to be efficient, the amount of water that goes into the turbines must remain constant over time. River flows change over the course of the year, so either dams or backup systems must be built to maintain electricity production at a stable rate. Amazonian rivers present great hydrological variation over the seasons; by applying run-of-the-river technology to the basin, the choice of backup system becomes ever more relevant—and, sadly, in recent years, this role has been played by fossil-fuel thermal power plants. Therefore, when the aggregate impacts—impacts from the plant plus impacts from the backup system—are taken into account, it is not a straightforward conclusion that a run-of-the-river UHE in the Amazon region would cause less environmental impact than a UHE with a large reservoir built in the same or another region of the country.

The impact of UHEs on indigenous populations is a great source of conflict in Brazil. Take, for instance, the UHE Belo Monte. It is a controversial project. First attempts at implementation—including the building of a large dam—date from the second half of the 1980s, but these failed in the face of strong opposition from both environmentalists and indigenous people, as well as the fiscal collapse of the Brazilian state. In the second half of the 2000s, the project was redesigned to apply run-of-the-river technology. In 2010, the construction contract was signed, and in 2011 the environmental licence was issued. The construction is advancing; it has, however, faced several legal battles, including disputes between the Brazilian federal state and the Inter-American Court of Human Rights.[24] Public opinion is mostly against the project, because of both the environmental impact and the impact on

indigenous populations. UHE projects in the Tapajos River face the same controversy.

Adding to the controversy, the federal government has changed some environmental laws to speed up the construction of new UHEs in the Amazon region. In 2011, the boundaries of several national parks and national forests were modified, and areas judged to be flooded were excluded from the obligation to be preserved.[25] The government is also postponing the demarcation of indigenous lands—a potential retaliation to demands from indigenous populations to be consulted about the projects, since they will be built on land that the indigenous populations have historically occupied. On several occasions in 2013, the indigenous population of the Tapajos River area demanded prompt and diligent action regarding this matter from government officials.

Plans to construct hydropower plants in Peru in order to export electricity to Brazil were also developed. In 2010, Brazil and Peru signed an Energy Agreement[26] specifying the construction of six hydropower plants in the Peruvian Amazon: the plants would be built with Brazilian money (both private and public); the electricity produced would be directed to the Peruvian market, but the surplus would be exported to Brazil; and, after thirty years, the ownership of the plants would be Peruvian. This measure is also highly controversial: part of the Peruvian population, in the manner of a traditional South American ideological dispute, argue that the country has

23 Of many works on the environmental impacts of dams, three sets of international guidelines are especially recommended: World Commission on Dams: *Dams and Development: A New Framework for Decision-Making* (2000); at: <http://www.internationalrivers.org/files/attached-files/world_commission_on_dams_final_report.df>; International Hydropower Association: *IHA Sustainability Guidelines* (2004): at: <http://www.ydropower.org/downloads/IHA%20Sustainability%20Guidelines Feb04.pdf> (7 November 2013); IEA Hydropower Agreement, *Annex III—Hydropower and the environment: present context and guidelines for future action*; at: <http://www.ieahydro.org/reports/HyA3S5V2. pdf>, revised in 2010: *Update of Recommendations for Hydro-power and the Environment*; at: <http://www.ieahydro.org/uploads/files/finalannexxii_task2_briefingdocument_ oct2010.pdf> (7 November 2013).

24 In 2011, the Inter-American Court of Human Rights, following a claim presented by NGOs that the UHE Belo Monte would create social and environmental impacts not covered by the environmental licensing in process, asked Brazil to suspend the licensing. The request came after the Brazilian government had answered several demands from the Court on the same issue, and was not welcomed. The relationships between the country and the Court reached its lowest ebb; Brazil has occasionally supported the pledge by Venezuela, Ecuador and Bolivia to diminish the Court's powers concerning interventions on matters of human rights.

25 Law 12.678/2012, available at <http://www.planalto.gov.br/ccivil_03/_Ato2011-2014/2012/Lei/L12678.htm> (13 January 2014).

26 Agreement between the government of the Federal Republic of Brazil and the government of the Republic of Peru for the supply of electricity to Peru and exports of surplus to Brazil, available in Portuguese at <http://daimre.serpro.gov.br/atos-internacionais/bilaterais/2010/acordo-entre-o-governo-da-republica-federativa-do-brasil-e-o-governo-da-republica-do-peru-para-fornecimentos-de-energia-eletrica-ao-peru-e-exportacao-de-excedentes-ao-brasil> (12 January 2014).

become the subject of Brazilian imperialism, since there is an imbalance between the large social and environmental impacts of the hydropower plants on local communities and the small benefits for the Peruvian population (whose energy demand is low and could be met without the new projects). Peruvian indigenous populations are challenging the Agreement by arguing that they should have been consulted before the government signed it, since their interests are at stake. The treaty is not yet in force and its ratification was delayed, although some of the projects are at an advanced stage of planning.

An alternative is building hydropower plants with an installed capacity between 1 and 30 MW and reservoirs smaller than 3 km², referred to as *Pequenas Centrais Hidrelétricas* (PCHs).[27] PCHs are an important alternative source of hydroelectricity: they can be located near consumer regions, reducing transmission costs; their smaller dams also reduce the environmental impact; their employment on a large scale increases the redundancy of electric systems, reducing the risks of blackouts during peak demand hours. They were included among the renewable energy options in the *Alternative Energy Sources Incentive Programme* (in Portuguese, PROINFA).[28] Between 2004 and 2011, PCHs sold around 1,800 MW in electricity auctions (Nogueira/Costa 2012); nevertheless, they produce only 3.63 per cent[29] of Brazilian electricity. Unlike other renewable sources, PCHs are not acknowledged by public opinion; the political strength of groups related to UHEs prevents the development of policies that would benefit this energy source.

Another alternative would be to diversify Brazil's renewable electricity sources, by making serious investments in wind and solar power and using hydropower as a reserve source. The complementarity between wind and hydroelectric power, for example, is scientifically proven (Amarante/Schultz/Bittencourt et al.

2001; Jaramillo/Borja/Huacuz 2004; Dutra/Szklo 2008; Ricosti/Sauer 2013). Brazil has great wind and solar power potential, and their intermittency can be compensated for by hydropower, saving the environmental and economic costs of accumulators. The world already has cases of successful joint exploration of the sources.[30] Joint exploitation in Brazil is not wishful thinking: the country already has the *National Interconnected System* (in Portuguese, SIN), an extensive electricity grid that connects the whole country;[31] however, massive investment would be needed in order to make the grid a smart one.

38.4.1.3 Ethanol

Brazil is the leading producer of ethanol from sugar cane. The high price of oil during the crises of the 1970s and reasons of energy security played important roles in encouraging its use as fuel—back then, reasons of sustainability were absent. With Brazilian government incentives and subsidies, large sugar cane plantations were established, mainly at São Paulo and in the north-eastern states of Pernambuco and Alagoas. In the 1990s, due to a severe crisis in supply, ethanol's role as a biofuel faded; it came back into the spotlight in 2003, after the development of flexible fuel technology.

Brazilian ethanol does not compete with food production. Brazil has more than 200 million hectares of land occupied by cattle grazing, almost half of which is degraded land no longer in use or used but with very low productivity; sugar cane plantations occupy around six million hectares. It is also important to note that Brazilian sugar cane production is currently concentrated in São Paulo state and the surrounding areas, and there is no prospect of expanding it into the Amazon region (Goldemberg 2008), so ethanol cannot be blamed for Amazonian deforestation, though it could be for the conversion of the Cerrado savannah.[32]

Ethanol has faced unfair competition with gasoline since 2007, when the federal government began

---

27 ANEEL Resolution 304/98.
28 The policy was enacted by the federal government in 2002 and was regulated in 2004. It specifies that ten per cent of Brazil's energy supply in the 2020s should to be produced by alternative renewable sources, and lists small hydropower plants, wind and biomass among them. After 2004, these sources took part in the federal government's auctions for electricity, sometimes selling important amounts, but still facing difficulties in competing with traditional sources due to the minimum cost criteria employed in the auctions.
29 Data from <http://www.aneel.gov.br/aplicacoes/capacidadebrasil/capacidadebrasil.cfm>, accessed 12 February 2014.

---

30 For example, the partnership between Denmark, a big producer of wind energy, and Norway, a big producer of hydroelectric power.
31 There are some isolated systems in the north of the country.
32 Until 2007, the expansion of the ethanol industry was an indirect driver of deforestation in the Amazon, because some areas in which there was substitution of cattle raising and soybeans were pushed in the Amazon by ethanol. Since 2008, it has not happened any more.

to subsidize oil prices, reducing the competitiveness of ethanol. Its competitiveness is also affected by the transportation system: since Brazil has not developed a piped distribution system and other transport options are underdeveloped, ethanol travels around Brazil on trucks fuelled with diesel. It is a true paradox: oil prices affect ethanol prices even when there is no direct manipulation by the federal government, just by adding the freight costs to the price of the biofuel.

From 2005 to 2007, ethanol played an important role in Brazilian diplomacy. Brazil supported the creation of a global market for ethanol, and tried to make it an international commodity. It has exported technology and established partnerships in order to develop ethanol markets in various countries such as Colombia, El Salvador, Honduras, Argentina and some African countries; multinational companies (Shell, Petrobras and Chevron) bought assets from ethanol companies during this period. This situation suited the Brazilian national interest but not the positions of China, India and Indonesia, who were Brazil's allies in climate negotiations, and it did not last. By the time the deep offshore oil was discovered, ethanol vanished from Brazilian official speeches, as did investment in ethanol companies.

It is uncertain how ethanol will develop over the next few years. Brazilian public opinion favours the biofuel, but few groups are mobilized or understand the big picture. A clear example is the positive connotation the federal government subsidies for gasoline had for the average Brazilian. If there is no revolutionary change in the next few years (creating strategic reserves to avoid price volatility and maintain reliability of supply; investing in the production of pure ethanol vehicles), it is likely that ethanol will remain a fuel of secondary importance in Brazil. Other important obstacles to greater ethanol production are the influence of oil and gas groups in the government and the misconception that investing in ethanol implies a further commoditization of the Brazilian economy and another step towards the Dutch disease.[33] In the international market, competition with ethanol from corn produced by the US, and a possible ethanol from cellulose produced in the US and Europe, undermines the possibility of the export of significant quantities of Brazilian ethanol.

38.4.1.4 Biodiesel

Biodiesel production is still in its infancy in Brazil. Created in 2005, the National Biodiesel Programme aims at increasing the uptake of biodiesel nationally. Following its creation, several regulatory developments[34] established a minimum percentage of biodiesel in the diesel mix; there is no pure biodiesel in Brazil.[35] Biodiesel currently being produced is mixed with the diesel used by the freight sector: type B diesel has eight to ten per cent biodiesel in its mix.[36]

Soybeans are the main source of Brazilian biodiesel; despite the proven possibilities of making biodiesel out of other vegetable species,[37] these crops are mostly produced on small family farms, making it difficult for the crops to be commoditized and sold on a large scale. Just like ethanol, biodiesel production cannot be blamed for deforestation: the share of soybeans employed in biodiesel production is almost insignificant when compared with the share exported for human and animal consumption. The situation could change if the federal government's predicted incentives and subsidies for biodiesel production should materialize in the next few years.

In order to become a competitive biofuel in Brazil, biodiesel needs to go through a technological and economic transformation revolution, and this is unlikely to happen in the near future.[38]

---

33 Dutch disease refers to the relative deindustrialization of the Brazilian economy due to expanded agriculture production and exports and energy commodities (deep offshore oil reserves).

34 Law 11.097/2005 and ANP Resolution 15/2006 and 14/2012.

35 The current restrictions that prevent engines from being fuelled by biodiesel could be overcome with technology innovation, and an official programme in support of it would be highly beneficial (Dias, 2007).

36 Besides the different sulphur concentrations, Brazilian diesel is classified according to the absence (type A) or presence (type B) of biodiesel in the mix. Therefore, following ANP Resolution 50/2013, there are four types of diesel in Brazil: S10A, S10B, S50A and S50B.

37 Studies were mandated by ANP Resolution 14/2012 and were carried out.

38 Soybean producers are currently more interested in exporting their crops for human and animal food production; other crops must be produced intensively in order to support biofuel production, and no relevant movements towards this can be seen in Brazil; in addition, there are significant vested interests in exploiting deep offshore oil reserves instead of investing in biofuel production.

38.4.1.5   Nuclear

Despite being non-renewable,[39] nuclear power is a low carbon energy source. Nuclear power was developed in Brazil by the military regime (1964–1985). It was seen as both an alternative source of energy and a strategic technology to be dominated by Brazilian scientists. Brazil has two nuclear power plants: Angra 1, built between 1972 and 1984, in operation since January 1985, and Angra 2, whose construction started in 1976, was paralyzed between 1983 and 1995 due to lack of financial resources, and was resumed in 1996; the power plant began operation in 2001. A third nuclear power plant, Angra 3, had its foundation laid at the beginning of the 1980s, but construction never got off the ground due to a lack of finance. Building began in earnest in 2010, and it is predicted that the plant will become active in 2018. Plans for developing four new nuclear power plants in Brazil were postponed after the Fukushima accident.

Nuclear power is an important low carbon energy source for a great part of the world, and could be most successfully applied as a transitional source; since it is highly effective in generating electricity, it can replace fossil fuels until technology for renewable sources is developed (Carvalho 2012). Nevertheless, it is important to bear in mind that nuclear energy involves risks that are more obvious than those of other energy sources, since the consequences of nuclear accidents spread out in space and time; so its externalities and requirements for safe operation have to be fully understood before it is employed on a large scale.

Brazilian public opinion does not favour nuclear energy. Interestingly, it is one of the energy sources with the strongest opposition in the country—even most of the academic community is against it. The lobby in favour of nuclear power (nuclear engineers, the military, some diplomats) is losing ground, especially since nuclear engineers, its fiercest defenders and the most politically engaged group, are starting to retire. Nuclear power will never be an important source of energy in Brazil, although it will remain among the sources linked with technological know-how (Brazil is one of the few countries in the world able to enrich uranium) and with the fact that the country has huge uranium reserves. And maybe this is for the best, since Brazil has other options for low carbon energy sources and nuclear power involves important considerations

about dealing with nuclear waste and managing evacuation exercises.

38.4.1.6   Wind and Solar Power

Wind and solar power, together with small hydropower plants, are classified as alternative renewable energy sources in Brazil. Brazil has enormous potential for wind and solar power, and a very small amount is currently employed in energy generation. Besides low carbon reasons, wind and solar energy development would (i) enable the decentralization of energy production in Brazil, once places with high potential for both are located in areas at the borders of the grid (wind) or with a precarious connection to the grid (solar); (ii) result in technological gains, since Brazil still imports most of the components for both types of energy system (though it exports wind blades) and could develop the national industry; and (iii) increase employment opportunities in both sectors. Wind power, in particular, has even more in its favour: the Brazilian wind regime is complementary to important Brazilian river basin regimes (highs/lows), and the connection of both energy systems to the same grid would guarantee renewable electricity throughout the year.

Wind power has expanded greatly in the last decade, due to PROINFA. In the first auction for electricity from alternative sources (2004), wind sold 1,422.9 MW; between 2005 and 2008, however, there were eight auctions in which alternative renewable energy sources competed with other sources, such as the UHEs of Jirau, Santo Antonio and Belo Monte and several fossil-fuel thermal power plants—which, in the end, sold 73 per cent of the total auctioned (Nogueira/Costa 2012). At the beginning of 2007, the federal government established specific auctions for each alternative source, and there were impressive results for wind power in 2009. In 2010 and 2011, wind sold 2,047.8 MW and 1,928.8 MW respectively (Nogueira/Costa 2012).

Despite the great expansion in Brazilian wind power, beyond earliest expectations, there is neither an official policy to establish long-term incentives, quotas and minimum prices for alternative energy sources, nor is there a legal obligation to auction alternative energy production periodically. The fact that there are no regional auctions for electricity (all auctions are national, and controlled by the federal government) is an important obstacle to the best use of different regional renewable electricity potential. Wind power is gaining momentum in Brazil, but a large part of the Brazilian electricity bureaucracy does

---

39   Nuclear energy production is limited by the availability of uranium ore, or similar elements.

not seriously consider it an alternative to other energy developments.

Solar power faces even more challenges. It was not included in PROINFA, because the federal government argued that the costs of production were too high. In fact, lack of competitiveness when compared to other energy sources is still the argument of most of those who dismiss solar energy. Besides, top technical decision-makers in the Brazilian energy matrix make some valid arguments against the development of solar power in the present decade: they consider that technological innovation in solar power is advancing very fast and a choice of technology at this point could be a trap. On top of that, the taxation system does not favour solar energy. Since 2013, solar energy has been included as a possible energy source in the electricity auctions, but it has still not sold anything.

38.4.1.7 Biomass for Electricity Generation

In addition to producing fuel, biomass is a source of Brazilian electricity. Production started in ethanol plants, where sugar cane pomace is used for electricity generation—the co-generation process. It has helped ethanol plants to become self-sufficient in electricity: co-generation is employed at all stages of the ethanol production chain, especially in the state of São Paulo, and has increased the efficiency of the sector in energy conversion, carbon intensiveness and economic productivity. Besides this, PROINFA authorized biomass to take part in the electricity auctions, and the results were significant: from 2004 to 2011, it sold around 2,500 MW (Nogueira/Costa 2012).

Currently, electricity from biomass comes mostly from sugar cane. Several other raw materials are being tested, as well as waste—a useful destination for one of the biggest outputs of the modern lifestyle. Initial projects using waste received certified carbon credits through the *Clean Development Mechanism* (CDM) of the Kyoto Protocol, but its use on a large scale requires further research. Public opinion favours this energy source, but it is unlikely that it will be extensively developed in the next few years.

38.4.2 Energy Efficiency, the Ignored Issue

The topic of energy efficiency has been on the agenda since the 1980s. Brazil has developed a regulatory framework and mandatory programmes of energy efficiency: the *National Electrical Energy Conservation Programme* (in Portuguese, PROCEL) aims at reducing the consumption of electricity; the *National*

Programme of Rational Use of Oil and Natural Gas By-products (in Portuguese, CONPET) targets the use of oil and derivatives; the *Brazilian Labelling Programme* (in Portuguese, PBE) classifies domestic appliances, devices and light utility vehicles according to their energy use. Law no. 9991/2000 requires electricity companies to invest a specified minimum amount in energy efficiency research and development. However, public debate about energy efficiency and public response to the initiatives were minimal.

The picture changed during the 2001 electricity supply crisis. Electricity was rationed, and surprisingly, Brazilian society was capable of a rapid and efficient response to electricity scarcity. Following this, the National Policy for Conservation and Rational Use of Energy (Law no. 10295/2001) was enacted, and an inter-ministerial committee was set up to manage and enhance initiatives related to energy efficiency; labelling became much more stringent; industries were obliged to increase energy efficiency so as not to face economic losses. After the crisis, though, energy waste resumed its high levels.

Electricity transmission also performs very poorly. Brazilian electricity is mostly transmitted through the SIN. An integrated transmission system has supply security advantages, as long as the transmission lines are quality-built and well maintained; unfortunately, the criterion in Brazil is short-term cost, not reliability and efficiency: lines are of inferior quality and undergo little maintenance—even cheap electronic leak detectors are not yet in place in the country. In addition, the expansion of the electricity supply and grid improvements are not coupled together: for instance, UHE Belo Monte's connection to the SIN will be concluded two years after it starts producing electricity; several wind farms are disconnected as well. The lack of a smart grid prevents consumers from becoming energy 'prosumers' by injecting electricity produced by solar panels and wind farms into the grid, and prevents the intelligent use of renewable energy potential utilizing the characteristics of the Brazilian regions, as well as hindering the joint exploitation of complementary renewable energy sources (solar and wind with hydropower).

Energy inefficiency is also a concern in the automotive industry. Since the 1950s, this has been one of the main sectors of the Brazilian economy; the federal government has always encouraged car production, for domestic consumption or to be exported to markets that are undemanding in terms of carbon emissions. The automotive industry accepts only vague energy efficiency labelling: Brazilian branches of European and US companies lobby against strict energy efficiency labelling and sell in Brazil outdated and inefficient models that are no longer commercially viable in their home countries.[40] Most consumers do not understand the concept of energy efficiency, and still buy cars according to short-term costs only. Even flex-fuel technology was not developed though purely environmental concerns, but as a means of employing a boosted ethanol production.

The *Brazilian Association for Energy Efficiency* (in Portuguese, ABEE) and other NGOs are in favour of energy efficiency and a smart grid in Brazil, but they are not strong enough to change the picture. Given that energy access is no longer a major issue in the country—electricity is available to 99 per cent of the Brazilian population; LPG for cooking, to 94 per cent (IEA 2013b)—and the recent supply crises,[41] quality in production and transmission of energy should be a priority. However, government propaganda is always based on quantity only; recent policies—reducing prices for electricity, based on a distorted calculation that benefited inefficient companies who do not invest in productivity and efficiency were clearly energy populist measures targeting the 2014 elections. If Brazil is to become an energy efficient country, quality, defined in economic, social and environmental terms, needs to become the criterion for energy production and consumption. This is a major transformation, and will not be achieved through increasing economic interventionism and the reduction of macroeconomic predictability.

38.5   Crony Capitalism and Short-term Thinking: a Major Stumbling Block to a Low Carbon Green Economy

Every society has deeply-rooted cultural traits that shape social relations at every level. Some of the traits are positive, others negative. Understanding policy and politics without keeping in mind these traits and their

impact on social relations is counterproductive and will lead to superficial analysis (Hochstetler/Viola 2012).

When countries face the challenge of becoming low carbon green economies, it is necessary to check which cultural traits might be behind the difficulties of putting environmental concerns on the same level as economic and social concerns. In Brazil, this is a double challenge, since the struggle between economic and social concerns is still far from being resolved. Big inequalities in income translated into a high Gini index, a low level of education, inequality of opportunities and a bureaucratic culture are big stumbling blocks to setting up a low carbon green economy, but two cultural features play a pivotal role: crony capitalism and short-term thinking.

Crony capitalism has existed in Brazil from the time it was a Portuguese colony: personal relationships call the tune of almost every venture that takes place in the country. Following the democratization of the 1980s and the macroeconomic reforms of the 1990s, crony capitalism had been on a downward trend; unfortunately, the trend was reversed in the late 2000s. Unlike mostly social democratic Europe, Brazil is a plutocratic democracy, even more so than the US: big money and in particular vested interests have a major influence on government decision-making and the political system. When hardship hits, instead of investing in productivity and efficiency, big companies look for more concessions from the federal government to solve their problems; in exchange, they fund election campaigns. In absolute terms Brazilian electoral campaigns are the second most expensive in the world (after the US); by per capita comparison, they are the most expensive. Around seventy per cent of the spending in electoral campaigns comes from companies and very rich individuals—with big providers of government services at the top—; thirty per cent comes from large trade unions and public funding. There is almost no tradition of small individual contributions.

Crony capitalism is spread all over the Brazilian economy, but its effects are stronger in sectors whose activities require large initial capital and substantial investment. This is the case with energy production and distribution. Long-established energy groups are powerful in Brazil, and they exert great pressure on federal government decisions concerning both energy sources and energy efficiency. The developers of UHEs, for instance, have special access to decision-making and public financial resources. This has an impact on electricity policies: because they defend the expansion of investment in UHEs over investment in PCHs, wind and solar (insisting on Brazil's UHE

---

40 Japanese and Korean companies are exceptions to this rule.

41 In February 2014, due to an unusually hot summer and the increased use of cooling devices, there were several cuts in the electricity supply due to the incapacity of the system to answer the demand; given that the summer was also unusually dry and Brazil relies on summer rains to refill the reservoirs of UHEs, either electricity will be rationed or fossil-fuel thermal power plants will be further employed over the winter.

potential and technological know-how and dismissing the social and environmental impacts of UHEs—Basso/Viola, forthcoming), diversification of the renewable matrix, investment in a smart grid and the decentralization of electricity production are endlessly postponed. It is true that Brazil will not be able to maintain its low carbon electricity profile for the next ten years without UHEs, and it should not abandon them; still, current UHE projects have problems and should be examined.

The oil sector is also very powerful. The federal government is the major shareholder in Petrobras, so company interests and state interests become entangled; the use of subsidized oil as a heterodox economic measure to increase consumption and maintain an artificial economic growth, imposing sequenced financial losses on the company, is an example of this. Vested interests in oil block bigger investment in natural gas and renewable fuels.

Crony capitalism in the energy sector is best understood through examining the general business culture of the country. In Brazil, in general (there are exceptions, of course, but they are still a minority, especially among the big companies), the business sector lacks an entrepreneurial culture: companies are mostly focused on rent-seeking. The manoeuvres of business, supported by governmental measures,[42] distort the market and create uncertainty for foreign investors. This is not inevitable, and the picture is different in the energy sectors of other countries: in Scandinavia and Germany, markets work well and the state interferes to accelerate investment in renewable energy and energy efficiency; in the US, markets work quite well, and the state interferes, in a limited way and directly, to fix market failures; in Brazil, however, not only are energy markets distorted, but the state interferes to solve the short-term problems of specific companies or to enhance its electoral prospects, creating a self-perpetuating vicious circle that deepens the problem and takes Brazil further away from low carbon green development.

Short-term thinking does not bring any relief. Brazil has more organized forces that oppose traditional energy sources (fossil fuels, UHEs) than it has forces that are in favour of new renewable sources or energy efficiency. This is due to the reactive nature of civil society movements: civil society reacts to government

action concerning a specific venture or energy source, for instance, the movement of people affected by dams associated with UHEs or the movement in favour of indigenous peoples' rights over forests that would be affected by UHEs, and these lobbies later attract NGOs and academia and form opposition to the venture or energy source. But there are no consistent proactive movements directed towards promoting an energy source or encouraging strategic energy assessment in Brazil. In order to change this state of affairs, think tanks are needed and research must be conducted to understand the big picture concerning energy production and consumption. The large number of published papers on strategic environmental and energy assessment in China and the negligible number in Brazil is evidence of the different situation in the two countries.

Furthermore, energy efficiency is hardly part of the Brazilian mindset. Brazilian culture favours short-term benefits over long-term gains, and most consumer decisions only consider the immediate costs; concerns about maintenance, the cost of electricity or fuel consumption, waste generated, damage to public health or harm caused to the environment are not taken into account by the majority of the population, rich or poor. In fact, once they are given the opportunity,[43] the actions of the poorer population mirror the irresponsible consumerism of the richer population. This is not to say that the poor should not be given the opportunity to consume or that socio-economic policies are wrong: they have been very effective and have helped reduce the abnormal inequality of the country. But it is important to note that the environment, renewable energy, energy efficiency or any other improvement that would mean that the links between economic, social and environmental issues are understood are not present in social claims—an example is provided by the demands made during the demonstrations of June 2013. The absence of long-term thinking will only be lessened through education, through transparent rules that tackle privileges and dissolve crony capitalism, and through accountability, creating a sense of community and belonging that is still lacking in Brazil.

And even if crony capitalism and short-term thinking have started to be tackled by some groups in Brazilian society, this approach will never spread and become the national rationale if the Brazilian political system is not reformed. In Brazil, politics is not the

---

42 E.g. the large subsidies received from BNDES by Eike Batista for risky projects that lacked consistency without guarantees being requested from him; the Bank incurred great losses that are unlikely to be recovered.

43 Due to the socio-economic policies of the last decade, for example.

game of achieving national consensus about what is needed in order to further the development of its population; it is the game in which parties fight each other to have their particular interests met, whatever the consequences for the country. Patronage practices are the main strategy for gaining votes; states fight each other for resources and investments in a predatory fashion; the federal government has a large, compartmentalized and, in general, inefficient bureaucracy, privileged with job stability and the right to receive a full-salary pension, unlike private workers. The system must be reformed so that fragmentation is decreased and meritocracy and universal policies are promoted.

38.6   Conclusions and Perspectives

Brazil has so far not adopted the low carbon green economy paradigm. This type of transformation cannot be achieved without planning, without long-, medium- and short-term goals, and without consistent implementation policies. Even the attempt at the goal of decarbonization, a smaller share of the paradigm, lacks consistency in Brazil.

It is true that low carbon energy sources form a higher proportion of the Brazilian energy matrix than the world average. Nevertheless, the share of non-renewable energy sources increased during the last decade, due to both heterodox economic measures that distorted the relative prices of oil/derivatives, and the lack of long-term policies to promote renewables. Two trends can be seen in the energy policy and politics of the last decade for low carbon energy sources. From 2003 to 2007, there was an upward trend in which progressive forces, committed to low carbon development, were stronger, and the participation of renewable sources in the energy matrix increased. From 2007 to 2014, there was a downward trend in which regressive forces, which favour non-renewables, especially fossil energy sources, were empowered, due to the discovery of deep offshore oil reserves and the use of oil prices as a heterodox policy tool to artificially maintain greater economic growth rates. As for energy efficiency, however, practices are underdeveloped and the trend during the last decade was stable: there was no consistent development of energy efficiency policy and politics.

There is no clear direction towards decarbonization in Brazil. A clear sign would be sent if the federal government opted for: (i) an intelligent use of the Brazilian potential for renewable energy, by offering incentives for exploiting solar and wind energy and making the SIN a smart grid–actions that would allow consumers to become 'prosumers' and fewer hydropower plants to be built, reducing their impact on the environment and the local population–; (ii) serious policies leading towards systemic energy efficiency in the Brazilian economy, not just by labelling domestic appliances but also by promoting better efficiency targets for vehicles, by supporting green building and by investing heavily in collective transport; (iii) a change in the criteria for electricity auctions, from the lowest price possible to the greenest and least carbon-intensive energy; (iv) the removal of all subsidies for oil and derivatives, and the enforcement of only limited exploitation of the deep offshore oil reserves, prioritizing that exploitation where it can be proved that natural gas is also present; (v) the use of natural gas, and not oil or coal, in thermal power plants, and the phasing out of oil and coal over the next decades; (vi) changing the incentives for the use of fuel by taxing fuel by its carbon emission and other impacts of the cycle of production and use on the planetary boundaries; (vii) supporting public transport and limiting the promotion of the automobile industry; and (viii) intermodal transportation, through a combination of railroads and river navigation.

It is difficult to predict how energy policy and politics will develop from 2014 onwards. It is clear that without undertaking consistent tax, pension, labour, fiscal, political system and energy policy reforms, Brazilian reality will not match up to its low carbon potential. National and foreign economic agents are almost as concerned about Brazil's economic prospects as they were before the 2002 elections, when the expected victory of a party that supported increased economic interventionism by the state shook economic expectations.

The public demonstrations of June 2013 did not have enough impetus to being about significant change in the general election of 2014. Marina Silva, the candidate of the Socialist Party/Sustainability Network alliance, whose manifesto laid emphasis on the transition to a low carbon economy, gained 22 per cent of the vote in the first round. This was a small increase on 2010, when she received 19 per cent of the vote, but still not enough for her to make the second round. In the second round, Ms Silva supported the Social Democratic Party's candidate Aécio Neves, on the condition that he laid more stress on sustainability; her support was contested by a significant part of Ms Silva's Sustainability Network, however—they decided to cast a "white" vote (an unmarked ballot paper; voting is compulsory in Brazil). President Dilma Rousseff was re-elected by the lowest margin in Brazilian history in the second round (51.6 per cent of the valid votes, with abstentions, white and null votes totalling more than 30 per cent, meaning that more than 60 per cent of the electorate did not support her re-election). In an extremely fragmented multiparty system (twenty-eight parties are represented in the new Congress), the parliamentary election was overwhelmingly won by traditional parties. In this beginning of second mandate of President Rousseff, the status quo of an underdeveloped low carbon economy is being maintained. The federal government is trying to implement partial economic reforms but the disagreements between the Presidency and the Parliament have slowed the process. The emergence of serious corruption scandals, linking high officers of state, politicians and large Brazilian corporations to the contracts and operations of Petrobras and other major companies of the energy area has put the country in the middle of the major crisis it has faced after the redemocratization. In this scenario, in which the government is focused in surviving, low carbon energy situation is becoming even worse.

Above all, it is vital to acknowledge that Brazil has not embraced decarbonization, let alone become a low carbon green economy, and that deep-rooted features of Brazilian culture, especially crony capitalism and short-term thinking, are great barriers to the implementation of these goals. Crony capitalism has been on an upward trend during the last few years and this has undermined investment in energy efficiency and new renewable energy sources. Federal government manoeuvres to reduce electricity prices at the expense of quality and reliability led to more inefficiency in the electricity system and to crises of supply in February 2014, and this led to the operation of more fossil-fuel thermal power plants in order to prevent the situation recurring during the winter of 2014. The lack of incentives for producing energy from renewable sources, such as higher prices in electricity auctions, undermines the possibility of their being employed on a large scale. In a country in which energy access is almost universal, energy security depends less on the expansion of energy production at distorted prices and more on cost-effective long-term quality and reliability of the supply and the transmission systems.

As well as tackling crony capitalism, minimizing short-term thinking should be taken very seriously. Brazil lacks strategic planning at all levels, from government policies to household budgets; measures take into account specific situations, usually urgent ones, and often create contradictory situations when seen in succession. Recent socio-economic policies that have been successful in reducing economic inequality spread irresponsible consumerism—encouraged by the government—throughout social groups, revealing that the pattern lies much deeper in Brazilian culture. There is no easy or quick fix: only with quality education, transparency, accountability and rules that tackle privileges in Brazilian society will there develop the sense of community and belonging—a sense that lies at the basis of a low carbon green economy—that it still lacks.

References

Amarante, Odilon A. C. do; Schultz, Dario J.; Bittencourt, Rogério M.; Rocha, Nelson A., 2001: "Wind/Hydro Complementary Seasonal Regimes in Brazil", in: *DEWI* (German Wind Energy Institute) *Magazin*, no. 19: at: <http://www.dewi.de/dewi_res/fileadmin/pdf/publications/Magazin_19/13.pdf> (12 July 2014).

Basso, Larissa; Viola, Eduardo (forthcoming): *Sustentabilidade da política brasileira para hidreletricidade: estudo da análise socioambiental do PDE 2021* (Paper presented at REGSA International Conference—Promoting Renewable Electricity Generation in South America, at UNISUL, Florianópolis, Brazil from 6 May 2014 to 8 May 2014).

Brasil, 2010: *Segunda Comunicação Nacional do Brasil à Convenção Quadro das Nações Unidas*; at: <http://www.mct.gov.br/upd_blob/0213/213909.pdf> (12 January 2014).

Brasil, 2013: *Estimativas anuais de emissões de gases de efeito estufa no Brasil*; at: <http://www.mct.gov.br/upd_blob/0226/226591.pdf> (12 January 2014).

Carvalho, Joaquim Francisco de, 2012: "O espaço da energia nuclear no Brasil", in: *Estudos Avançados*, 26,74: 193-307.

Crutzen, Paul J.; Stoermer, E. F., 2000: "The 'Anthropocene'", in: *Global Change Newsletter*, 41: 17-18.

Dias, Guilherme Leite da Silva, 2007: "Biodiesel: a new challenge", in: *Estudos Avançados*, 21,59: 179-183.

Dutra, Ricardo; Szklo, Alexandre, 2008: "Assessing long-term incentive programs for implementing wind power in Brazil using GIS rule-based methods", in: *Renewable Energy*, 33,12: 2507-2515.

Eletrobras, 2012: "Potencial hidrelétrico brasileiro por bacia—Dezembro 2012"; at: <http://www.eletrobras.com/elb/data/Pages/LUMIS21D128D3PTBRIE.htm> (17 January 2014).

EPE [Empresa De Pesquisa Energetica], 2013: *Balanço Energético Nacional–BEN* 2013; at: <https://ben. epe. gov.br/downloads/Relatorio_Final_BEN_2013.pdf> (17 December 2013).

Goldemberg, Jose, 2008: "The Brazilian biofuels industry", in: *Biotechnology for Biofuels*, 1,6.

Hochstetler, Kathryn; Viola, Eduardo, 2012: "Brazil and the Politics of Climate Change: Beyond the Global Commons", in: *Environmental Politics*, 21,05: 753-771.

IBGE [Instituto Brasileiro De Geografia E Estatistica], 2010): *Censo 2010,* at: <http://censo2010.ibge.gov.br/resultados>.

IEA, 2013a: *Technology roadmap carbon capture and storage*; at: <http://www.iea.org/publications/freepublications/publication/TechnologyRoadmapCarbonCaptureandStorage.pdf> (15 November 2013).

IEA, 2013b: *World Energy Outlook* 2013, energy access database; at: <http://www.worldenergyoutlook.org/resources/energydevelopment/energyaccessdatabase/> (25 February 2014).

Jamieson, Dale, 2011: "The nature of the problem", in: Dryzek, John S.; Norgaard, Richard B.; Schlosberg, David (Eds.): *The Oxford Handbook of Climate Change and Society* (Oxford: Oxford University Press): 38-54.

Jaramillo, Oscar Alfredo; Borja, Miguel Angel; Huacuz, Jorge M., 2004: "Using hydropower to complement wind energy: a hybrid system to provide firm power", in: *Renewable Energy*, 29,11: 1887-1909.

Lapola, David M.; Martinelli, Luiz A.; Peres, Carlos A.; Ometto, Jean P. H. B.; Ferreira, Manuel E. Nobre, Carlos A.; Aguiar, Ana Paula; Bustamante, Mercedes M. C.; Cardoso, Manoel F.; Costa, Marcos H.; Joly, Carlos A.; Leite, Christiane C.; Moutinho, Paulo; Sampaio, Gilvan; Strassburg, Bernardo B. N.; Vieira, Ima C. G., 2014: "Pervasive transition of the Brazilian land-use system", in: *Nature Climate Change*, 4 (January): 27-35.

MME [Ministry of Mines and Energy], 2011: *Plano Nacional de Eficiência Energética*; at: <http://www.orcamentofederal.gov.br/projeto-esplanada-sustentavel/pasta-para-arquivar-dados-do-pes/Plano_Nacional_de_Eficiencia_Energetica.pdf> (12 February 2014).

Nogueira, Luiz A. Horta; Costa, Jonas Carvalheira, 2012: *Opções tecnológicas em energia: uma visão brasileira,* FBDS [Fundação Brasileira para o Desenvolvimento Sustentável]; <http://fbds.org.br/fbds/IMG/pdf/doc-531.pdf>, accessed 17 Dec 2013

Nye, Joseph, 2011: *The future of power* (New York: Public Affairs).

Prins, Gwyn; Galiana, Isabel; Green, Christopher; Grundmann, Reiner; Korhola, Atte; Laird, Frank; Nordhaus, Ted; Pielke Jnr Roger; Rayner, Steve; Sarewitz, Daniel; Shellenberger, Michael; Stehr, Nico; Tezuko, Hiroyuki, 2010: *The Hartwell Paper: a new direction for climate policy after the crash of* 2009 (London: London School of Economics and Political Science–Oxford: University of Oxford); at: <http://eprints.lse.ac.uk/27939/1/HartwellPaper_English_version. pdf> (2 August 2013).

Ricosti, Juliana F. Chader; Sauer, Ildo, 2013: "An assessment of wind power prospects in the Brazilian hydrothermal system", in: *Renewable and Sustainable Energy Reviews,* 19: 742-753.

Rockström, Johan; Steffen, Will; Noone, Kevin; Persson, Åsa; Chapin, F. Stuart; Lambin, Eric F.; Lenton, Timothy M.; Scheffer, Marten; Folke, Carl; Schellnhuber, Hans Joachim; Nykvist, Björn; de Wit, Cynthia A.; Hughes, Terry; van der Leeuw, Sander; Rodhe, Henning; Sörlin, Sverker; Snyder, Peter K.; Costanza, Robert; Svedin, Uno; Falkenmark, Malin; Karlberg, Louise; Corell, Robert W.; Fabry, Victoria J.; Hansen, James; Walker, Brian; Liverman, Diana; Richardson, Katherine; Crutzen, Paul; Foley, Jonathan A., 2009: "A safe operating space for humanity", in: *Nature*, 461,24 (September): 472-475; 2009; at: <http://www.nature.com/nature/journal/v461/n7263/full/461472a.html> (1 July 2013).

Steffen, Will, 2011: "A truly complex and diabolical policy problem", in: Dryzek, John S.; Norgaard, Richard B.; Schlosberg, David (Eds.); *The Oxford Handbook of Climate Change and Society* (Oxford: Oxford University Press): 21-37.

The Pew Charitable Trusts, 2012: "Who is winning the energy race"; at: <http://www.pewenvironment.org/newsroom/reports/whos-winning-the-clean-energy-race-2012-edition-85899468949> (12 July 2013).

UNDESA, 2012: *A guidebook to the green economy*; at: <http://www.uncsd2012.org/content/documents/528Green%20Economy%20Guidebook_100912_FINAL.pdf> (28 June 2013).

UNEP, 2011: "Towards a Green Economy: Pathways to Sustainable Development and Poverty Eradication"; at: <http://www.unep.org/publications/contents/pub_details_search.asp?ID=4188> (1 July 2013).

UNCSD [United Nations Conference on Sustainable Development], 2012: *The future we want*; United Nations General Assembly Resolution A/RES/66/288; at: <http://www.un.org/en/ga/66/resolutions.shtml> (1 July 2013).

UNFCCC [United Nations Framework Convention On Climate Change], 1992; at: <http://unfccc.int/files/essential_background/background_publications_htmlpdf/application/pdf/conveng.pdf> (1 July 2013).

Viola, Eduardo, 2002: "O Regime Internacional da Mudança Climática e o Brasil", in: *Revista Brasileira de Ciências Sociais*, 17,50: 25-46.

Viola, Eduardo; Franchini, Matías, 2012a: "Os limiares planetários, a Rio+20 e o papel do Brasil", in: *Cadernos EBAPE*, 10,3: 470-491.

Viola, Eduardo; Franchini, Matías, 2012b: "O Brasil na transic a o para a economia verde de baixo carbono", in: *O Brasil e a agenda da sustentabilidade: desafios e oportunidades para o Estado, o setor privado e a sociedade civil* (Rio de Janeiro: CEBRI [Centro Brasileiro de Relações Internacionais]): 31-53.

Viola, Eduardo; Franchini, Matías; Ribeiro, Thais Lemos, 2012: "Climate Governance in an international system under conservative hegemony: the role of major powers", in: *Revista Brasileira de Política Internacional*, 55 (special edition): 9-29.

Viola, Eduardo; Franchini, Matías; Ribeiro, Thais Lemos, 2013: *Sistema internacional de hegemonia conservadora – governança global e democracia na era da crise climática* (São Paulo: Annablume).

WCED [World Commission On Environment And Development], 1987: *Our Common Future* (Brundtland Report); annex of United Nations General Secretary document A/42/427; at: <http://www.un-documents.net/wced-ocf.htm> (1 July 2013).

WWF; Roland Berger; *Clean Economy, Living Planet: The Race to the Top of Global Clean Energy Technology Manufacturing* 2012; at: <http://www.rolandberger.com/media/publications/2012-06-06-rbsc-pub-Clean_Economy_Living_Planet.html> (10 July 2013).

39 Sustainable Electricity Transition in Thailand and the Role of Civil Society

Carl Middleton[1,2]

Abstract

The chapter explains the creation and resistance to change of Thailand's centralized and fossil-fuel intensive electricity regime through a Sustainability Transition and Multilevel Perspective lens, with an emphasis on the sector's political economy. The incumbent electricity industry has evolved from a state-owned monopoly to a partially-privatized industry structure dominated by the state utility and several large independent power producers. The analysis demonstrates how important global landscape shifts articulate with the sector's domestic political economy, including a shifting global development paradigm from developmentalist state to liberal market principles, as well as the impact of waves of global economic crisis. The chapter highlights the role played by civil society coalitions in unsettling the incumbent electricity regime since the late 1970s, despite significant power asymmetries, through opposing problematic projects, advocating for progressive policy, and proposing alternative plans, values and visions for Thailand's electricity sector. Important but small steps towards sustainability transition are identified, including greater energy conservation and distributed renewable energy generation, the creation of an independent regulator, and a small increase in public participation and accountability in the power planning process. The chapter argues that civil society has been—and will continue to be—important in shaping the incumbent electricity regime and often acts as a catalyst for transition towards sustainability.

Keywords: Thailand, electricity, civil society, sustainability transition, multilevel perspective.

39.1 Introduction

Since the 1960s and more rapidly since the 1980s, Thailand's economy, society and environment have witnessed a major transformation. Throughout this period of high economic growth, as the country rapidly industrialized and urbanized, a centralized electricity system was established led by the state's utility, the *Electricity Generating Authority of Thailand* (EGAT). At first a monopoly, several waves of privatization since the 1980s have created a partially-privatized electricity regime that remains dominated by fossil fuels and centralized generation owned by EGAT and a small number of *independent power producers* (IPPs).

Whilst largely meeting the rapid growth in demand for electricity, the environmental and social costs have been high. Since the 1970s, an increasingly established civil society has emerged within Thailand's often precarious democracy (Phongpaichit/ Baker 2002), which has included project-affected communities, their wider social movements, and various *non-governmental organizations* (NGOs) (Foran 2006). Nowadays, Thailand's electricity policy is an active arena of policy deliberation, including on the impacts of domestic projects and power-import projects from neighbouring countries, as well as on the transparency and accountability of the electricity plan-

1 Carl Middleton is a lecturer on the *MA in International Development Studies* (MAIDS) programme and Deputy Director for International Research in the *Center for Social Development Studies* (CSDS) in the Faculty of Political Science, Chulalongkorn University, Bangkok, Thailand; email: <Carl.Chulalongkorn@gmail.com>.

2 The author is indebted to PD Dr. Hans Günter Brauch for his patience and encouragement in the production of the chapter. In formulating this chapter, the author valued time spent with Dr. Brauch and Professor Dr. Úrsula Oswald Spring during their visiting professorship at Chulalongkorn University, hosted by Professor Surichai Wun'gaeo. The author also gratefully appreciates the suggestions of four anonymous peer reviewers.

H.G. Brauch et al. (eds.), *Handbook on Sustainability Transition and Sustainable Peace*,
Hexagon Series on Human and Environmental Security and Peace 10, DOI 10.1007/978-3-319-43884-9_39

ning process as a whole. There are wide power asymmetries between actors in decision-making, however, and this remains largely one-sided and favours the established electricity industry. Despite these power asymmetries, there have been several important reforms, including some towards energy conservation, and the introduction of small-power producer and very-small-power producer legislation that has allowed a growing proportion of renewable energy generation within Thailand's electricity system.

Thailand's electricity system faces a number of contemporary challenges. Whilst on paper there may be a commitment to low-carbon economic growth (EPPO 2012), in practice higher on the policy agenda is ensuring energy security for Thailand's sustained economic growth. Meanwhile, policies on electricity production are fragmented and often contradictory in relation to other policy issues such as water resource management and food production (Middleton/Dore 2015). Finally, Thailand's electricity generation is also heavily dependent on natural gas, yet domestic reserves are rapidly depleting and growing dependence on imports (together with hydroelectricity) has raised concerns about national energy security.

Drawing upon the concepts of *sustainability transition* (ST) and the *multilevel perspective* (MLP) and adopting a political economy approach (actors, interests, power), the purpose of this chapter is to analyse the emergence and embedding of Thailand's incumbent electricity regime and the role played by civil society within it (Verbong/Loorback 2012). Synthesizing diverse literature on Thailand's power sector through an ST and MLP lens, the chapter addresses a general gap in ST literature on *newly industrialized countries* (NICs) in South East Asia, and on strategies taken by actors in incumbent electricity regimes to resist change, together with the role played by civil society in catalysing alternative pathways. The next section outlines the conceptual framework of ST and the MLP as relevant to electricity system transitions in NICs. Section 39.3 maps out the emergence and embedding of Thailand's electricity regime, differentiated into two phases: the construction of a centralized national power system (1950–1980); and the partial privatization of this system (1980–present). The section shows how globalization landscape pressures reformulated the regime over the two periods, and how the partially-privatized regime has emerged as a stable industry formation. Section 39.4 demonstrates how civil society has challenged and shaped the incumbent regime, including through transforming the landscape in which the regime exists, unsettling

the regime directly through resistance and cooperation, imagining a new regime, and catalysing new innovative niches. Section 39.5 draws conclusions and suggests strategic directions for transitions towards sustainability in Thailand's electricity sector.

39.2 Sustainable Electricity Transition and the Role of Civil Society

39.2.1 Transitions Studies and the Multilevel Perspective

Sustainability Transition (ST) explores how sociotechnical systems change over time. It is a mid-range theory that sees technology and innovation as emerging from, embedded within, and shaping society. The scope of ST aims to incorporate "multi-level dynamics, multi-actor networks, radical innovation and uncertainty, and the impossibility of full control" (Verborg/Loorback 2012: 7). ST adopts a multi-disciplinary perspective, drawing on evolutionary economics, science and technology studies, structuration theory and neo-institutional theory (Geels 2011).

The *Multilevel Perspective* (MLP) conceptualizes processes of sociotechnological change as occurring at and between three levels: niche, regime and landscape (Geels 2002). 'Regime' refers to the existing dominant and dynamically stable sociotechnical system with its associated scientific knowledge, policy, industrial networks, markets and user practices, technology, infrastructure, and cultural meaning (Geels 2002: 1263). Markard/Raven/Truffer (2012: 956) note that a regime:

> consists of (networks of) actors (individuals, firms, and other organizations, collective actors) and institutions (societal and technical norms, regulations, standards of good practice), as well as material artefacts and knowledge.

Geels (2011) warns against reifying the regime, emphasizing that it is the actors within the regime who hold agency, intentionality and strategy.

'Niches', meanwhile, are spaces where radical sociotechnical innovation can occur and that are protected from regime selection criteria, for example by subsidized research and pilot projects, or by culturally-held values. Finally, 'landscape' refers to the exogenous environment in which regimes evolve and niches also exist; it includes broad geographical factors, and political, economic and societal trends. In response to recent critiques of MLP's conceptualization of scale, which implies that micro-meso-macro levels are nested and hierarchical (e.g. Shove/Walker 2010), Geels (2011: 37) has proposed that levels may refer to "different degrees of stability, which are not necessarily hierarchical".

The transition from one sociotechnical regime to another is a set of non-linear process entailing a radical change in configuration of technologies, infrastructures, institutions, governance and actors. Existing sociotechnical regimes may become unsettled when: broad landscape-level changes create pressure (macro-economic environment, society values, and so on); there are growing processes towards and within the regime itself that destabilizes it; and niche-level experiments offer alternatives (Geels 2002; Grin/Rotmans/Schot 2010) and gain momentum, for example through learning processes, price/performance improvements, and support from powerful groups (Geels 2014: 23). Other have noted that over long periods, incremental changes in regime can amount to significant transformation; As Smith/Voß/Grin (2010: 440) put it: "In a Kuhnian vein, regimes tend to produce 'normal' innovation patterns, whilst 'revolutionary' change originates in 'niches'" (see also Pavitt, 1984).

The incumbent regime may be resistant to significant rapid change for a range of economic, institutional, cultural, and technical reasons, including the resistance of existing actors that benefit from it. The regime can be analysed in terms of its lock-in, path dependence, and inertia; Geels (2014: 35) emphasizes, however, that an incumbent regime is not monolithic, and its stability is not guaranteed. Smith/Voß/Grin (2010:433) observe that "radical niche practice can be co-opted by a regime without unduly unsettling and transforming it" (Smith/Voß/Grin 2010: 443). Recent literature has considered in more detail how existing regimes exhibit resistance to change, and thus how sustainability transitions are inherently a political process (Smith/Stirling/Berkhout 2005; Meadowcroft 2011). One approach has been to adopt a political economy perspective, giving explicit consideration to various forms of power, agency and resistance (Geels, 2014; see also Schreuer/Rohracher/Späth 2010). This perspective rebalances previous literature that placed too greater emphasis on green niche innovation leading to ST, without also considering how the incumbent regime might also be resistant to being destabilized; as Geels (2014: 25) puts it, "we should better understand the 'destruction' part of Schumpeter's 'creative destruction' concept".

Studies of ST, by explaining how sociotechnical transitions occur, ask how they can be steered (or 'managed' or 'governed') towards sustainability, and ideally how transition can be accelerated given the urgency of the sustainability challenge faced (Smith/Voß/Grin 2010). This approach is normative in that it recognizes that social and environmental values, alongside economic efficiency, are—and must be—accounted for in policy and practice. Thus the focus is not only upon the technical regime and its associated innovation, but also its relationship with broader social arrangements (Smith/Stirling/Berkhout 2005).

39.2.2 Sustainability Transition in South East Asia

ST as a concept has been developed and applied predominantly in *Organisation for Economic Co-operation and Development* (OECD) country contexts, with a particular focus on Europe (Markard/Raven/Truffer 2012). The concerns of geography matter to ST, however, ranging from the role of place and associated spatial relationships, to the transnational flows of capital, actors and ideas, to the availability of domestic resources and the implications of sovereignty (Smith/Voß/Grin 2010). Despite this, a literature on ST in developing Asia is only now beginning to emerge (Berkhout/Angel/Wieczorek 2009; Smits 2012; Sawdon 2014).

For Asia's NICs, the spatial dynamics and flows of knowledge, technology and investment as well as export-market driven growth have held significant implications for economy, society and politics (Nevins/Peluso 2008). To understand the sociotechnical regimes of NICs in Asia, therefore, it is important to account for the international context (Berkhout/Angel/Wieczorek 2009: 225). At the same time, global diffusion of technology, knowledge and policies are not transplanted off-the-shelf, but articulate with local context (Dobbins/Simmons/Garrett 2007). Regarding sustainability transformations in Asia's NICs, Angel/Rock (2009) propose that whilst initial development was at the sacrifice of environment quality, there has been a second wave of environmental governance reform building on pragmatic policy innovation, institutional strengths of the development state, and the changing norms of globalization, although impacts on sustainability have actualized unevenly.

39.2.3 Electricity System Transition

Geert/Loorbach (2012: 5) emphasize that "current energy systems are deeply entrenched in our economy, consumption patterns, regulations and infrastructure". In most cases around the world, electricity provision was initially provided by the private sector, but during the end of the nineteenth century and first half of the twentieth century its provision was taken over by municipal and later national state electricity suppliers, at which point electricity increasingly became considered as a public service.[3] The liberal wave of privatization policies in the 1980s that began in Northern countries and spread to developing countries restructured the electricity industry and reintroduced the private sector, but often resulted in only partial privatizations (Victor/Heller 2007).

Electricity services, like other network industry services such as water supply and telecommunications, have been of great political interest to governments and politicians, for reasons ranging from the importance of ensuring the provision of these services to the public (i.e. a form of social contract), to the importance of electricity security to economic growth, and the high capital investment costs and associated possibility of job creation, political patronage and corruption (Victor/Heller 2007: 1). In the context of Asian NICs, construction and maintenance of electricity systems have been promoted by the government as a priority for national economic development. Yet, this 'national' economic development has disproportionately benefited politically-connected industrial elites and their networks (Hildyard/Lohmann/Sexton 2012). Actors in the incumbent electricity regime, who are bound together by mutual dependencies, act to maintain regime stability and resist change. They include established business and state enterprises, supportive state agencies, policymakers, unions and politicians. Aside from considerations of politics and vested interests, there are a number of attributes of the electricity sector that lock in the incumbent regime, including high capital costs, political visibility, network monopoly effects, technological stasis, and daunting regulatory tasks (Victor/Heller 2007: 2).

Finally, in ST studies there has recently been an effort to unpack in greater detail how technical systems have implications for the practices of everyday life and its reproduction (Shove/Walker 2010). In the context of electricity systems, this means consideration not only of sociotechnical practices of supply, but also those of demand. Shove/Walker (2014: 41) observe "Whereas theories of practice highlight basic questions about what energy is for, these issues are routinely and perhaps necessarily obscured by those who see energy as an abstract resource that structures

or that is structured by a range of interlocking social systems". Thus, everyday practices of electricity use are important considerations to understand how regimes and niches become embedded (or entrenched), and how (or whether) energy transitions occur (Smith/Voß/Grin 2010: 443).

39.2.4   Role of Civil Society in Electricity Transitions

Amongst state agencies and enterprises, firms and civil society groups, networks exist that promote different visions for electricity systems. These partnerships—of differing degree of affiliation and asymmetrical power relations—may either seek to maintain and stabilize the incumbent regime, or promote alternative pathways. For example, Hess (2014) explores the formation, strategies and power asymmetries of broad coalitions of actors and how they shape the US energy sector, namely: the incumbent regime seeking to maintain its position; (often) grassroots coalitions, including social movements, seeking transition to a more sustainable regime; a 'countervailing industry mobilization' that support grassroots coalitions with financial and political resources around shared interests; and the multifaceted role of the state (Hess 2014).

Smith (2012) proposes that the MLP perspective helps social movement theory broaden its scope of analysis from political systems to sociotechnical regimes (table 39.1). Smith (2012: 180) asks "What visions do civil associations hold, and what roles do they play in transitions processes?" Smith (2012: 190) stresses, however, that civil society—whether unsettling regimes, nurturing niches, or shaping societal values—are "not amendable to herding and corralling" and thus it is unrealistic to anticipate diverse civil society activity as constituting a "coordinated transition management". Instead, he emphasizes that it is the diversity of civil society that is one of its strengths, including ensuring that electricity system transitions do not lose sight of principles of justice.

39.2.5   Conceptual Contribution

As a heuristic framework (Geels 2011: 34), ST and MLP can offer new conceptual insights into the emergence of Thailand's electricity production-consumption system, and opportunities for a transition to sustainability. A significant number of studies have analysed electricity systems from an ST and MLP perspective but mainly in an OECD-country context (Markard/Raven/Truffer 2012). Few have examined

---

3   The author appreciates the insight of an anonymous reviewer in raising this point.

Table 39.1: Civil society activity in relation to sustainable electricity transitions. **Source:** The author, adapted from Smith (2012: 189).

MLP domain	Types of civil society activity
Sociotechnical landscape	Awareness raising; social pressure
Unsettling existing sociotechnical regime	Consumer boycotts; protest and lobbying; standards and counter-expertise
Aspired-to sociotechnical regime	Community aspirations; plural visions in civil society
Technological niches	Grassroots innovation; green consumption; citizen science

electricity systems in Asian NICs, an area to which this chapter contributes (Geert/Loorbach 2012). Responding to knowledge gaps identified by Geels (2011; 2014) and Smith/Voß/Grin (2010: 445), the chapter also highlights the active resistance of the incumbent regime to change, and the role and strategies of civil society groups in destabilizing it, whilst at the same time promoting alternative pathways (Smith 2012). Finally, in exploring the interaction between the incumbent electricity regime and its actors—and those who would like to change it—, the chapter adopts a political economy approach and explicitly considers the actors involved within decision-taking processes and the relative distribution of political power, so contributing towards a nascent political economy turn in MLP literature (Geels 2014).

39.3   Emergence and Embedding of Thailand's Incumbent Electricity Regime

Since the 1960s, Thailand has been transformed from a predominantly agrarian to an industrializing, urbanizing society, where until the 1980s the country was a weak developmentalist state which subsequently increasingly liberalized. Thailand is presently mainland South East Asia's largest economy, and has been classified as an upper-middle-income country by the World Bank since 2011. Rapid economic growth came in the 1980s as the country shifted towards an export-orientated economy, infused with *foreign direct investment* (FDI) from Japan. The country's economic growth has been dependent upon—and as a political priority has shaped—the construction of a national electricity system fuelled predominantly by large-scale fossil-fuelled technologies, and to a lesser extent by large hydropower dams. Reflecting Thailand's development pathway, however, the country has experienced escalating challenges, including high levels of economic, social and political inequality, an oligarchic political structure, and widespread environmental degradation (UNEP/TEI 2007; UNDP 2010).

Thailand has regularly faced political crisis, and this has once again escalated since 2005 leading to the most recent military coup d'état in May 2014.

This section maps out the emergence and progressive embeddedness of Thailand's electricity regime. Two broad phases are identified: the construction of a centralized national power system (1950-1980; 39.3.1); and the partial privatization of this system (1980-present; 39.3.2).[4] Several researchers have described the evolution of Thailand's electricity sector in detail, including Wattana/Sharma/Vaiyavuth (2008), Greacen/Greacen (2004), Foran (2006; 2013) and Sulistiyanto/Xun (2004). This section furthers these analyses by examining the electricity system's emergence through the lens of ST and MLP with a focus on landscape and regime scales and guided by the electricity regime typology of Smith (2012:193-195) (table 39.2). Particular attention is paid to the politics of resistance of the incumbent regime during the second phase. The role played by civil society groups is also introduced, and then expanded upon in section 39.4.

39.3.1   Establishment of a Development State-Led Sociotechnical Regime (1950–1980)

In 1932, Thailand transformed from an absolute to a constitutional monarchy, and this was followed by almost five decades of predominantly military and elite bureaucratic rule with only brief periods of representative democracy (Phonpaichit/Baker 2012). During this period, the aristocratic, senior bureaucratic and merchant oligarchic elites from the time of the absolute monarchy yielded to accommodate rising military and new business elites. The foundation of Thailand's national electricity system was laid in the

---

4   Wattana/Sharma/Vaiyavuth (2008) provide a useful detailed chronology of Thailand's electricity sector's evolution as: early days (1884-1949); industry establishment (1950-79); foundation for privatization (1980-89); first steps in electricity reform (1990-97); and proposal for market-oriented reform (1998-2006).

Table 39.2: Evolution of Thailand's electricity regime. **Source:** Table structure adapted from Smith (2012: 193–195).

Sociotechnical dimension[a]	Establishment of a development state-led sociotechnical regime (1950–1980)	Partial privatization of sociotechnical regime (1980–present)
'Guiding principles'	Electricity is a public good produced by the state utility for socio-economic development; electrification linked to nation-building, modernity, and bringing the benefits of 'development'	Electricity produced according to regulated market principles by the state utility and private enterprises; electrification and energy security required for rapid export-orientated economic growth
'Favoured technologies'	Large-scale technologies (lignite coal-fired plants; hydropower dams); centralized production with transmission through a national grid	Increased dependency on technologies fuelled by imported natural gas (combined-cycle gas turbines); still general commitment to large-scale projects and centralized system; some niche spaces emerge for energy conservation and renewable energy generation
'Industrial structure'	Vertically integrated. Dominated by *state-owned enterprises* (SOEs).	Remains vertically integrated. Increased private sector role in generation at large, small and very small scales; independent regulator since 2008; increasing electricity imports from neighbouring countries
'User relations and markets'	SOEs as monopoly generator and distributor of electricity, guided by a weak developmentalist state. Electricity consumers as passive recipients. Cost-plus arrangement in terms of EGAT investment, yet electricity prices low.	Electricity consumers generally remain passive, but increasingly aware of issues around energy including the *Power Development Plan* (PDP) and renewable energy. Cost-plus arrangements for EGAT remain in place, but limited competition introduced with *independent power producers* (IPPs).
'Policy and regulations'	Utilities essentially self-regulating. Policies on electricity and energy fragmented. Little concern for broader social, environmental and governance issues.	A period of policy flux from state monopoly to *enhanced single-buyer* (ESB) market regulations. Creation of an energy regulatory commission in 2008. Economic efficiency, social, environmental and governance issues rising up the policy agenda.
'Knowledge'	Expert engineering knowledge dominates policy development, planning, and operation	Expert engineering knowledge continues to dominate, but challenged by counter-knowledge related to social and environmental impacts of projects, and governance principles for electricity planning.
'Culture'	Cheap and reliable electricity perceived as important to economic development and Thailand's modernization. The electricity system also as a source of patronage and corruption.	Thailand's modernization and urbanization underpinned by cheap and reliable electricity. Ownership of electricity production increasingly contested, including between centralized versus decentralized models and role of state versus private sector.

a) Sociotechnical dimensions are flagged in the analysis by being place in 'single quotation marks'.

1950s and 1960s, shaped by this landscape (Greacen 2004: 129).

As the cold war escalated, so Thailand's relationship with Western aid agencies grew closer, in particular with the World Bank and the United States. In 1961, during a period of martial law, the government announced its first five-year National Economic Development Plan with advice from the World Bank aimed at transforming Thailand from an agricultural to an industrial country under a market economy; at this time the dominant development paradigm informed by the success of the Marshall Plan in Europe and Rostow's notion of economic take-off was for large-scale state-led development. Thailand pursued a weak development state model where the role of the government was to invest public expenditure in basic infrastructure to support the private sector—including in large-scale electricity generation and rural electrification (Wattana/Sharma/Vaiyavuth 2008: 42). Some, however, have suggested that Thailand was closer to an authoritarian state rather than a developmental one (Siriprachai 2012).

The establishment of a national electricity infrastructure entailed significant 'industrial restructuring'. With World Bank guidance, as concessions granted to over 200 small cooperative-, municipal- and privately-owned concessions ended in the 1950s, the government did not renew them. Instead, the Government brought the concessions under their control. With borrowing from private-sector lenders and concessionary lending and technical advice from USAID and the World Bank, Thailand built a series of large dams and lignite-fired power plants, together with a national high-voltage transmission network (Foran 2006:13-15). These 'favoured technologies' (lignite-fired power stations, hydropower) reflected the domestic availability of resources, before natural gas that could be used as a fuel was discovered in the early 1980s. In 1968, EGAT, a *state-owned enterprise* (SOE), was established to be responsible for electricity generation and transmission,[5] alongside the earlier-created *Provincial Electricity Authority* (PEA) (1960) and the *Metropolitan Electricity Authority* (MEA) (1958) which are SOEs responsible for rural and urban distribution respectively. Overall, this was a move to consolidate, vertically integrate, and standardize the 'industry's structure', and also to bring the industry under the control of the state. These early infrastructure investments and institutional arrangements established the centralized system and cost-plus[6] organizational model that shapes the electricity system regime to this day.

Whilst in the early 1960s less than two per cent of Thailand's population had access to electricity, this had increased to over eighty per cent by 2000 (Grea-

cen 2004: 127, 144). As a 'guiding principle', not only was electricity central to the country's industrialization and economic growth strategy, but it was also viewed as a public good. Universal and affordable electricity became a promise of modernity by the state to the people, which, given the massive public investment and for some the social and environmental cost, also served as a justification for expectations of self-sacrifice (Williams/Dubash 2004). Electricity generation and provision delivered by the state was integral to the process of nation-building and was a symbol of development delivered. In terms of 'knowledge' (and its associated power relations), electricity provision was largely viewed as an administrative and technical exercise to be undertaken by the state, reflecting the prevailing global development paradigm of electricity system design, construction and operation.

During this period, EGAT in particular, but also PEA and MEA, became very powerful politically. These SOEs were 'effectively self-regulating' towards attaining the ends of meeting electricity demand for economic growth (Greacen/Greacen 2004: 519; Foran 2006: 17). Little concern was paid to "competition and profitability, environmental and social constraints, and governance issues such as transparency, accountability and public participation" (Williams/Dubash 2004: 413). There was a general atmosphere of 'grow now and clean up later', with implications for electricity production (Angel/Rock 2009: 232). More broadly, there was accelerating consumption of resources and production of pollution and waste. Environmental concerns—often raised by civil society groups and affected communities—emerged gradually in the late 1970s and grew in the late 1980s. Significant environmental legislation was first created in Thailand in 1975 and reformed in the early 1990s (Harashima 2000), although it was often weakly enforced. Contested projects have included the Mae Moh lignite-fired power station in Lampang province, northern Thailand (Boonlong 2011), and various dams such as the Sirindhorn dam in Ubon Ratchatani province, north-east Thailand (Blake 2013; Missingham 2003).

During this period of rapid growth, EGAT was located within the prime minister's office.[7] With privileged access in the government structure, EGAT emerged as the country's largest SOE with access to large amounts of development funds (Smith 2003). Thus, EGAT was also "regarded by politicians as a vehi-

---

5   Greacen (2004: 130) notes that the earliest large generation projects in the 1950s and 1960s were initially "set up as 'independent' state-owned enterprises at the behest of the World Bank in order to ensure that World Bank loans would not be deposited directly into the Thai national Treasury".

6   A cost-plus regulatory model allows for a fixed rate of return (i.e. profit) based on expenditure, and globally has been a common model for regulated monopoly utilities. It provides a strong incentive for the rapid expansion of infrastructure, but also carries the risk of over-investment given that all costs—including excessive ones—are ultimately passed on to the consumer (Greacen/Footner 2006: 23).

7   The PEA and MEA, meanwhile, were established under the Ministry of the Interior.

cle for patronage in appointments and awarding contracts" (Smith 2003: 281). EGAT's long-established well-connected political influence shaped its ability to muster resistance against later attempts at privatization. Meanwhile, at least until the 1970s, during this period of fledgling democracy in Thailand, military rather than elected government was more common, and civil society groups were little active on the electricity sector.

39.3.2 Partial Privatization of Electricity Regime (1980–Present)

From the mid-1970s, following protests by students, workers and farmers, there were lengthening periods of elected government–still punctuated by military coups–as democratic institutions became increasingly embedded.[8] Business owners became elected as *members of parliament* (MPs), displacing bureaucrats and generals from Parliament; in the process politics–via oligarchic networking- became more directly linked to the interests of private business and power relations between these elite groups partly shifted towards the latter (Phongpaichit/Benyaapikul 2013: 35).

In the late 1980s and early 1990s, Thailand's economy boomed, growing at eight to nine per cent a year as a result of massive FDI inflows, in particular from Japan, and growth in export of manufactured goods. At the same time, the government increasingly liberalized Thailand's economy. The Asian financial crisis severely affected Thailand in 1997, and recovery only began in 2000 although at a lower rate of growth than before (Phongpaichit/Baker 2008). The global financial crisis (2008-9), a major flood in 2011, and political unrest in 2010 and since 2013 have all affected growth since, and hence also the electricity sector coupled to it. Some economists consider Thailand caught in a 'middle-income trap', where it faces a significant challenge to transform from an industrializing economy based on cheap labour, exploitation of natural resources, and imported technology via FDI and transnational corporations, to an economy that can compete with advanced economies through producing value via knowledge creation and innovation

(Phongpaichit/Benyaapikul 2013); indicatively, in contrast to Korea and Japan, as well as China and India, very few globally competitive multinational companies have emerged from Thailand. The pathway set by Thailand's economic model of development has also been related to the country's contemporary environmental, social and political challenges.

Meanwhile, in 1997, following growing protests by social movements and now with an established and vocal NGO sector, Thailand passed a new 'people's' constitution. These social movements, Phonpaichit/Baker (2012: 85) write, "focused on specific issues, particularly growing competition over resources of land and water, declining agricultural prices, corruption and over-centralization". At the same time, the Asian financial crisis challenged the old oligarchic elites, and opened the door to Thaksin Shinawatra's *Thai Rak Thai* (TRT) party and a new era of populist politics in Thailand (Phonpaichit/Baker 2012; Walker 2012). Ultimately, this led from 2005 onwards to an intense 'politics of colour' and entrenched political deadlock, and ultimately to two military coups in September 2006 and May 2014.

Thailand's incumbent electricity regime has been fundamentally shaped by this economic, social and political 'landscape' and reconfigured power relations. With rapid economic growth, electricity consumption rose, and this was further bolstered by 'policy principles' for a commitment to nationwide electrification, and a pricing system that made electricity attractive over other fuels to the manufacturing and services sectors (Wattana/Sharma/Vaiyavuth 2008: 44). Thailand's electricity system peak capacity grew elevenfold from 2,838 *megawatts* (MW) in 1982 to 32,600 MW in 2012 (Greacen/Greacen 2004; EPPO 2013). Yet, as democratic space expanded, controversy also erupted with increasing regularity over the environmental and social costs of individual projects and the procedures for electricity 'policy-making and planning.'

The 'landscape' for the incumbent regime's partial privatization was laid in the 1980s with a series of pro-business governments. Thailand had experienced a public sector debt crisis from 1978 to 1981 during the second global oil price shock,[9] compounded by inflationary pressures coupled with stagnant economic growth (i.e. stagflation) in OECD countries affecting

---

8 Thailand has witnessed since 1932 twelve successful military coups and nine attempted ones, and even during periods of democracy the military continues to play a pivotal, if often covert, role in Thai politics: Nicholas Farrelly, "Counting Thailand's coups", at: <http://asia-pacific.anu.edu.au/newmandala/2011/03/08/counting-thailands-coups/> (31 May 2015).

---

9 The second oil price shock tripled Thailand's fuel import bill, constituting thirty per cent of all imports by 1982, even as natural gas was at that time coming online (Foran 2006, citing Phongpaichit/Baker 1995; Greacen/Greacen 2004).

export demand. Thailand's electricity sector particularly struggled, as its rapid expansion in the late 1960s and 1970s had been debt-financed; furthermore, an emphasis on supply-side-led expansion (rather than energy conservation) also created perceptions of capital shortage (Foran 2006: 41). The government entered into a structural adjustment programme from 1981 to 1985 with the World Bank and *International Monetary Fund* (IMF), who had realigned themselves towards a neo-liberal development paradigm; loan conditions of the World Bank and the IMF included commitments to privatize SOEs, and for the electricity sector to raise prices (ultimately by two and a half times as much). Yet EGAT and its union resisted and ultimately defeated their privatization (Greacen/Greacen 2004).

By the late 1980s, Thailand's economy had recovered, and the country began to experience power shortages. EGAT, however, was still burdened by high levels of foreign-sourced debt, servicing of which constituted over half of its budget (Greacen/Greacen 2004). In 1992, the pro-market National Energy Policy Council chaired by the prime minister, and its secretariat the *National Energy Policy Office* (NEPO), tried to bring together a fragmented series of energy 'policies' within one entity (Wattana/Sharma/Vaiyavuth 2008). It also sought to address power shortages through 'restructuring the industry' by launching an *Independent Power Producers* (IPPs) programme in 1994. There were subsequent plans for a competitive power pool electricity market for generation and privatization of retail distribution, which remained unfulfilled. As EGAT also strengthened its technical capacity ('knowledge') for planning, it introduced least-cost planning practices. Thailand's *National Economic and Social Development Board* (NESDB), meanwhile, increased its 'regulatory' scrutiny of EGAT's *Power Development Plans* (PDPs) that it prepared (Foran 2006: 19). Power shortages also sparked EGAT's first concern for energy conservation, catalysed by the NGO *International Institute for Energy Conservation* (IIEC), signifying a shift in planning 'culture' (see Foran 2006; 39.4.2).

NEPO permitted EGAT to sign several *Power Purchase Agreements* (PPAs) with IPPs on a 'take-or-pay' basis, thus allowing entry of large private-sector actors into Thailand's electricity generation. EGAT established a subsidiary, the *Electricity Generating Company* (EGCO), to operate two of its most profitable plants as IPPs whilst maintaining a forty-five per cent share in the company, and began negotiating contracts with other IPPs. These 'policies' were reinforced by pressure from the World Bank and IMF, alongside the apparent availability of international private capital seeking profitable returns in developing countries. Greacen (2007) notes that the IPP programme, whilst contributing to meeting Thailand's electricity demand, has presented a number of risks to IPPs, including delays due to electricity demand gluts, community opposition, and political and policy/regulatory uncertainty, as well as risks to electricity consumers in Thailand who have had to pay for unused generation capacity (39.3.3).

At the same time, the *Small Power Producers* (SPP) programme, launched in 1992, allowed an additional role for the private sector, either selling to EGAT or directly to nearby industry. The SPP programme purchases electricity from either *combined heat and power* (CHP) or renewable private sector generation projects of up to 90 MW. Thailand was the first country in Asia to adopt such a programme,[10] which was modelled on the US *Public Utilities Policies Act* (PURPA) 'regulations' (Greacen 2007). As of 2009, the capacity of SPPs was 1,962 MW, representing 6.7 per cent of Thailand's total capacity (EGAT 2010).

As EGAT resisted NEPO's attempt to unbundle generation, transmission and distribution, in 1997 Thailand was hit by the Asian financial crisis (Smith 2003). The crash of the economy reduced electricity demand, whilst the collapse of the Thai baht left EGAT struggling with its foreign-denominated debt. The 'take-or-pay' IPP contracts signed by EGAT were also left in a precarious position, and the government subsequently renegotiated these (Greacen 2007). An economic adjustment package offered by the IMF required further 'policy change' towards privatization; EGAT was required to sell assets, including a sixty-five per cent share in its just-built, profitable Ratchaburi gas-fired plant (Smith 2003).

Following the Asian financial crisis, Thaksin Shinawatra's *Thai Rak Thai* (TRT) government which came to power in 2001 redefined the direction of Thailand's electricity 'industry restructuring' (as well as many other functionings of the government). Having established a new Ministry of Energy in 2002 and redesignated NEPO as the *Energy Policy and Planning Office* (EPPO) with significantly reduced powers, TRT

---

10 World Bank "Retoolkit Case Study: Small Power Producers in Thailand", at: <http://siteresources.worldbank.org/EXTRENENERGYTK/Resources/5138246-1238175210723/Thailand0Small0Power0Producer0Program0.pdf> (31 May 2015).

shelved plans to create a competitive power-pool electricity market and replaced it with an *enhanced single-buyer* (ESB) model, in which EGAT owned approximately fifty per cent of the generation capacity and a hundred per cent of the transmission. Wattana/Sharma/Vaiyavuth (2008: 47) observe that the ESB model was quite similar to the previous single-buyer model, thus maintaining many traits of the existing electricity regime. TRT's approach de-emphasized market competition within the electricity sector, and promoted EGAT as a 'national champion.' A greater priority for TRT was to corporatize EGAT as a public company on the *Stock Exchange of Thailand* (SET), reflecting TRT's economic policy of expanding the SET, building Thailand's domestic market, and projecting influence into neighbouring countries. The SET listing, however, was blocked in Thailand's Supreme Administrative court in 2006 by a coalition formed of EGAT's labour union and civil society consumer groups concerned about the absence of an independent energy regulator. Subsequent to the court case, the Energy Industry Act, B.E. 2550 (2007) was passed (under a military-appointed government), which established the *Energy Regulatory Commission* (ERC) reshaping 'user relations' (Wattana/Sharma/Vaiyavuth 2008; Wisuttisak 2012).

As a result of these waves of partial-privatization, EGAT's share in generation dropped from eighty-nine per cent in the late 1980s to forty-six per cent in 2012 (EPPO 2013). Meanwhile, 'favoured technology' had shifted towards combined-cycle gas turbines fuelled by natural gas, which had become available domestically and via import from Myanmar. The claimed aims of the IPP programme were to reduce EGAT's investment burden and the cost of power generation. Wattana/Sharma/Vaiyavuth (2008: 45) conclude, however, that although IPP bidding was competitive, it appeared that cartels were formed to push up bidding prices, and thus any reduced costs in production were not passed on to consumers but rather remained with the companies and their shareholders. Conflicts of interest were also noted; Wattana/Sharma/Vaiyavuth (2008: 50) write "Some of the business-orientated politicians with dual roles—as citizens' representatives and as executive directors of companies—played a part in promoting the privatization of the industry".

The corollary of the partial privatization of Thailand's electricity sector is the growing role of the private sector in decision-making on power projects; this includes not only IPPs but also construction companies and commercial banks, among others (Middleton/Matthews/Mirumachi 2015). Many of Thailand's

IPPs and SPPs, as well as the major commercial banks that fund them, are listed on the Stock Exchange of Thailand (SET). For these actors, return on investment and minimization (or redistribution) of investment risks are key criteria by which power projects are financed and thus built. For example, the controversial 1,285 MW, US$3.5 billion Xayaburi Dam on the Mekong River's main stream in northern Laos (see also 39.4.2.1), now under construction, is owned by a predominantly Thai consortium with financing from Thai commercial banks. The project signed a PPA with EGAT to export ninety-five per cent of its power to Thailand. When the Laos government announced its final approval of the project in November 2012, the share price of the lead developer, the Thai construction company Ch. Karnchang, hit a twenty-one-month high, unsurprisingly revealing the stock market incentives for listed IPPs.[11] Merme/Ahlers/Gupta (2014) warn that such strict market logic may undermine a power project's long-term commitment to environmental sustainability and local livelihoods, or displace these costs from the private developer to the state, as is highly likely to be the case with the Xayaburi Dam (Matthews 2012; Middleton/Matthews/Mirumachi 2015). Meanwhile, Phongpaichit/Benyaapikul (2013) flag up that EGAT's role as both state utility and major investor in private affiliates, such as Ratchaburi and EGCO (which also holds a 12.5 per cent share in the Xayaburi Dam), creates a potential conflict of interest as the organization becomes a blurred semi-public, semi-private entity.

In the 1990s, some initial domestic concern (reflecting also a global concern) for renewable energy, energy efficiency, and demand-side management moved up the 'policy' agenda within the incumbent electricity regime (see 39.3.3 and 39.4). This was catalysed by both collaborative and adversarial relationships with NGOs (see 39.4.2). An increasingly vocal civil society also sought to challenge individual projects, including fossil-fuel-fired projects and hydropower dams that still remained as 'favoured technologies', as well as to influence EGAT's planning process. In total, this reflects the emergence of a "small sustainability agenda" in pursuit of cost effectiveness and environmental consideration, although Foran (2006: 4) argues that "Thailand's electricity planning, and its overall industry structure, impede sustainable energy

---

11 "Ch Karnchang hits 21-mth high on Xayaburi dam nod"; in: *Reuters,* 6 November 2014, at: <http://www.reuters.com/article/markets-thailand-stocksnews-idUSL3E8M61E520121106> (31 May 2015).

futures". Foran (2006: 4) further observes that the privatization debate in the 1990s and 2000s insufficiently addressed who plans and strategizes the electricity system, and that:

> Of all the agencies involved, EGAT plays a major role in shaping the details of what appears in the PDP, particularly plant size, fuel source, and location. These conditions prevailed in the 1980s, at the time Pak Mun was identified as a potential addition to the Thai power system. They prevail today. The continuity surrounding the PDP process is remarkable considering the dynamism surrounding EGAT (Foran 2006:5).

Whilst preparation of the PDP has been a closed-door process, since 2007 a slight (but imperfect) increase in transparency and public participation has emerged, with public hearings held during the 2010 PDP preparation process (Foran 2013) and more limited ones for the 2015 PDP. Thailand's PDP process has been criticized by civil society groups and critical academics for emphasizing supply-side options, in particular large-scale centralized technologies, downplaying the potential for renewable small-scale technologies and distributed systems, and not integrating demand-side planning, including energy efficiency and demand-side management (Greacen/Greacen 2012). Thus, even as countervailing 'knowledge' of civil society advocacy coalitions increasingly challenged mainstream discourses, EGAT's technical 'knowledge' continued to predominate, and constructive discussion about any actual technical and economic constraints and how they might be overcome has been limited.

It is of note that the circumstances, processes, industry structure, and outcomes of partial electricity sector reform in Thailand parallel the experience of other emerging economies that have incumbent electricity regimes. These include: the significant role of debt and financial crisis as a context for (partial) reform; the creation of an (enhanced) single-buyer model rather than full privatization of distribution; the prioritization of a limited number of (politically connected) IPPs to meet immediate generation demand that seems politically and technically easier, and the use of long-term PPAs rather than a competitive generation market, guaranteeing revenues; the relatively weak role of independent regulation; and, overall, the continued strong presence of the state (Victor/Heller 2007).

Thailand has also experienced the emergence of what Victor/Heller (2007: xvii) term 'dual firms' that are owned jointly between SOEs and private-sector actors. In other emerging economies, including China, India and South Africa, Victor/Heller (2007:

xvii) observe that dual firms "thrive in the murky middle ground between the old state-dominated system and a fully open and competitive private marketplace," and this captures the character of EGCO and Ratchaburi. They are well connected politically, yet relatively efficiently managed, and act to protect the partially privatized regime and their privileged status within it (Victor/Heller 2007: 289-290). The limited number of actors within the regime facilitates policy negotiation internal to the regime, which outsiders— including critical NGOs and civil society groups find difficult to influence, reflecting overall power relations between these actors.

39.3.3 Thailand's Current Electricity System and Challenges

In terms of the current structure of the industry, of the total installed capacity of 32,600 MW in 2012, EGAT generates 46 per cent; IPPs generate 39 per cent; SPPs generate 8 per cent; and 7 per cent is imported from Lao PDR and exchanged with Malaysia (EPPO 2013: 89). The maximum peak demand in 2012 was 26,121 MW. The figures for fuel type are: natural gas fuels 67 per cent; coal/lignite fuels 20 per cent; domestic hydropower fuels 5 per cent; oil fuels 1 per cent; and electricity import/exchange (principally hydropower) fuels 7 per cent. By sector, the largest consumers of electricity were industry (45 per cent), followed by residential (23 per cent) and commercial (17 per cent) (EPPO 2013: 93).

Thailand faces a particular challenge in terms of long-term fuel supply. At present, seventy per cent of Thailand's electricity generation is fuelled by natural gas, of which (as of 2006) twenty-seven per cent was imported from Myanmar (Nakawiro/Bhattacharyya/Limmeechokchai 2008). As Thailand's domestic gas supplies could potentially be exhausted by 2025 (Sutabutr 2010, cited in Meerow/Baud 2012: 21), and opposition to further coal-fired power stations and hydropower remains staunch, as does proposals for nuclear power, EGAT faces a difficult dilemma in continuing a business-as-usual model without encountering civil society and community opposition (Nakawiro/Bhattacharyya 2010). Recognizing these challenges, and based on Thailand's 2010 Power Development Plan, Kamsamrong/Sorapipatana (2014) argue that a renewable energy scenario for Thailand is possible, which would strengthen energy security through dependence on domestic fuel sources, whilst also reducing CO_2 emission intensity (see also Greacen/

Greacen 2012). However, the cost per unit of electricity in the 2010 PDP would increase by fourteen per cent.

On the other hand, at present Thailand's electricity system has an excess of generation capacity; in 2015, the reserve margin[12] was twenty-five per cent, and with several new projects contracted to come online in the next couple of years this could increase to as much as thirty-five per cent over the next decade.[13] EPPO and EGAT have claimed that an unexpectedly weak economy has led to unanticipated low growth in electricity demand, whilst civil society groups, frustrated that these costs are ultimately passed on to consumers, claim that the creation of overcapacity is a systemic flaw in the incentives and oversight of the current cost-plus electricity regulations (39.3.1) and power planning model (39.4.2.1).

Climate change (as a 'landscape' factor) has also risen up the 'policy' agenda, creating pressure for change in the electricity system. For civil society groups, this has led to a push for energy conservation and renewable energy (see 39.1.4). EGAT, meanwhile, whilst incorporating these concerns into its so-called 'Green PDP' (EGAT 2010), has also responded by promoting 'clean coal technology', more large-scale hydropower dams, and a nuclear power station.

Another challenge is created by the institutional disjunctures, not unique to Thailand, between electricity planning and those agencies related to water management and food production (Middleton/Dore 2015). Since 2008, there has been growing discussion about the water–energy–food nexus globally and in South East Asia as a policy and research agenda, although to date this is yet to be extensively translated into national policy and practice (Middleton/Allouche/Gyawali et al. 2015).

At the time of writing, Thailand's most recent Power Development Plan is the PDP 2010 revision 3, approved in December 2011 (EPPO 2012).[14] It anticipates a total system capacity growth of 52,256 MW by 2030, almost double that of 2010, although this growth is contested by civil society groups (Greacen/Greacen 2012). Guided by an Alternative Energy Development Plan (2012– 2021), the PDP proposes that "total capacity of renewable energy will be

around 20,546.3 MW (or 29 percent of total generating capacity in the power system)", although this includes a significant proportion of large hydropower plants whose 'renewable' credentials (in the sense of broad-based sustainability and social justice) are contested. This strategy reflects a concern for energy security, in particular fuel diversification, and a reduction in dependence on natural gas. Meanwhile, the Twenty-Year Energy Efficiency Development Plan 2011–2030 proposes a twenty-five per cent reduction in energy intensity within twenty years.

39.4 Role of Civil Society in Sustainable Electricity Transition in Thailand

This section discusses how civil society has acted to transform the 'landscape' in which the incumbent electricity regime exists (39.4.1), to unsettle the existing 'regime' (39.4.2) and to imagine a new one (39.4.3), and how new 'niches' have been created (39.4.4). The analysis is structured according to Smith's (2012: 189) typology of civil society activity in relation to sustainable energy transitions (see table 39.1).

39.4.1 Landscape Transformation

Thailand's 1997 People's Constitution—since replaced, following a military coup, by a new Constitution in 2007[15]—was a fundamental shift in landscape (39.3.2) that reconfigured Thailand's political system, including recognizing many human rights, and creating a Constitutional Court, an Administrative Court, a National Counter Corruption Commission, and a National Human Rights Commission. Articles[16]

---

12 'Reserve margin' reflects the percentage of excess capacity relative to maximum annual peak demand in the system.

13 "Officials to tackle surplus in future electricity supply"; in: Bangkok Post, 29 April 2015, at: <http://www.bangkokpost.com/business/news/544859/officials-to-tackle-surplus-in-future-electricity-supply> (31 May 2015).

14 On 14 May 2015, Thailand's National Energy Policy Committee approved the PDP 2015. Full details were not available at the time of writing, but the plan anticipates a growth in generation capacity from 37,612 MW in 2015 to 70,410 MW by 2036 ("Public hearing held on the Energy Ministry's PDP"; in: Thai PBS, 28 April 2015, at: <http://englishnews.thaipbs.or.th/public-hearing-held-on-the-energy-ministrys-pdp> (31 May 2015)); and a reduction in the proportion of natural gas as a fuel from sixty-seven per cent to forty per cent, to be replaced by coal, hydropower and renewable sources ("National Energy Policy Committee approves Thailand's power development plan (PDP 2015)", in: media release by the Royal Thai Government, 14 May 2015, at: <http://www.thaigov.go.th/index.php?option=com_k2 &view=item& id=91997:91997&Itemid=398&lang=en> (31 May 2015)).

allowed for the right to access information and to hold public hearings (Articles 56 and 57), and the right to public participation (Article 67); the latter is significant given that Thailand's *Enhancement and Conservation of the National Environmental Quality Act, NEQA* 1992, that predates the 1997 and 2007 constitutions and provides the law on *Environmental Impact Assessment* (EIA), does not detail requirements for public participation.

These mechanisms to counterbalance the power of the state have been framed by often contentious debate about democracy and development in Thailand, incorporating public interest issues such as economic, social and political (in)equality, and environmental protection versus development. These 'landscape' shifts have challenged (and been shaped by) the incumbent electricity regime, including in terms of transparency, participation and accountability, and they also provide some opportunity for emergent niches (Smith/Voß/Grin 2010: 441). For example, article 67 states that people have the right to ask for and participate in a *Health Impact Assessment* (HIA), which was translated into rules and procedures following a constitutional court ruling—brought by civil society—in December 2009; power development plans and large power projects are subject to a HIA (National Health Commission Office Thailand 2010: 12). In another example, as mentioned above (39.3.2), opponents of EGAT's listing on SET won their case in Thailand's Supreme Administrative Court in 2006. These entitlements under the law are legitimized and reinforced as they are exercised with regard to cases in the electricity sector.

Some civil society groups have also conducted awareness-raising campaigns about the impact of consumerist lifestyles in Thailand, perhaps the most high-profile of which have related to climate change, for example the activities of the Thai Climate Justice Network.[17] Whilst Thailand has made important steps in reducing material poverty and increasing *gross domestic product* (GDP), its ecological footprint is rising;[18]

Thailand's carbon intensity—CO_2 emissions per unit of GDP—are significantly higher than neighbouring Cambodia and Laos, as well as greater than some other large industrial countries in Asia such as Japan.[19] According to the International Energy Association, for the period 2011 to 2035, carbon intensity will decrease by 1.4 per cent per year on average, whilst per-capita emissions will rise from 36 per cent to 90 per cent of the OECD average over the same period (IEA 2013).

39.4.2   Unsettling the Incumbent Electricity Regime

Since the 1990s, a diverse range of civil society groups has sought to unsettle Thailand's existing electricity regime at scales ranging from individual projects to the PDP itself (Foran 2013). Not all civil society, however, is seeking to unsettle the incumbent regime. EGAT's union has maintained a close alignment with EGAT, staunchly resisting efforts to privatize EGAT in order to protect members' jobs and the benefits that EGAT provides for its employees. On the other hand, when interests have been aligned, EGAT's union has also partnered with consumer groups and NGOs, most notably when the Thai Rak Thai party sought to corporatization EGAT in 2006 (see 39.3.2); whilst EGAT's union opposed the corporatization in general and sought to maintain a minimum generation capacity of at least fifty per cent of the total system for EGAT, consumer groups and NGOs disagreed with the corporatization in the absence of an independent regulator.

39.4.2.1   Challenging the Incumbent Regime Through Resistance

Since the 1990s, community protests—supported by a range of NGOs—emerged around numerous projects proposed by EGAT and IPPs, including the Pak Mun and Nam Choan hydropower dams (Foran/Manorom 2009; Hirsch 1998) and coal-fired power plants in Prachub Khiri Khan province and the Mae Moh project in Lampang province (Hildyard/Lohmann/Sexton 2012;

---

15   The new constitution was approved in October 2006. At the time of writing (May 2015), the 2006 constitution had been repealed following another military coup in May 2014. In July 2014 an interim constitution was enacted, with a new constitution under preparation. Meanwhile, martial law was mostly lifted in April 2015, although many restrictions on political freedoms remain in place.

16   Carried over from the 1997 into the 2006 constitution.

17   See Thai Climate Justice website, at: <http://www.thai-climatejustice.org/> [in Thai] (28 May 2015).

18   See at: <http://www.gms-eoc.org/gms-statistics/overview/ecological-footprint> (28 May 2015).

19   According to World Bank Indicators (see at: <http://data.worldbank.org/indicator/EN.ATM.CO2E.PP.GD> (28 May 2015)), CO_2 emissions (kg per PPP $ of GDP) for 2010-2014 are: 0.4 in Thailand; 0.1 in Cambodia and Laos; 0.2 in Indonesia and the Philippines; 0.4 in Vietnam, Malaysia and South Korea; and 0.3 in Japan.

Greenpeace 2005). The case of Mae Moh, for example, has been the subject of a decade-long court case for remedy and restitution.[20] Whilst it has been claimed that EGAT was a relatively early adopter of EIA, partly due to the strong influence of the World Bank (Shepard/Ortolano 1997), numerous other researchers have highlighted the shortcomings of such assessment processes so far in Thailand (ADB 2010; Boonlong/Farbotko/Parfondry et al. 2011; Friend/Pradubsuk/Badenoch/ et al. 2011).

In the case of the proposed private coal-fired power plants in Prachub Khiri Khan in the 1990s, for example, EGAT signed a PPA before holding hearings or conducting an EIA. International NGOs supported local opposition "by pointing to the conflict of interest between the government's investment in private projects, and its regulation of those projects' rates of return" (Ryder 1997, cited in Foran 2006). Meanwhile, others pointed out that the power was surplus to requirements due to the impact of the Asian financial crisis. As Thailand has shifted to power-import projects from neighbouring countries, transnational campaigns have also emerged, for example around the Nam Theun 2 and Theun Hinboun Dams in Laos, and the Yandana gas pipeline and proposed dams on the Salween River in Myanmar (Middleton 2012). These have challenged EGAT as the electricity buyer, as well as the host project government, project developers and financiers.

A recent example of a transnational campaign is the Xayaburi Dam (see 39.3.2). Project proponents argue that the Xayaburi Dam would contribute towards Thailand's energy security and generate cheap electricity, and that the FDI and project revenues would bring development to Laos. Those opposing the project, including various civil society groups, emphasize that the project will displace 2,100 people and that more than 200,000 people located near the dam would experience a negative impact on their livelihoods, both within Laos and in neighbouring countries. Civil society strategies to challenge the project have included: challenging the intergovernmental process hosted by the Mekong River Commission; direct protests including peace walks along the Mekong River, in front of the lead company's headquarters, and at various government agencies; consumer boycotts against Thai banks; public petitions; and media coverage (International Rivers 2014; Matthews 2012). A range of counter-expertise has been invoked, including legal interpretation,[21] scientific assessment of project documents (Hirsch/Hogan/Lanza et al. 2011), and alternative ways of valuing the river, ranging from ecological economic assessments (Costanza/Kubiszewski/Paquet et al. 2011) to local knowledge (Herbertson 2012). The project is currently a case before the Supreme Administrative Court in Thailand brought by Thai riparian communities against the Thai government agencies involved, including the Ministry of Energy, benefiting from the 1997 'landscape' changes—and seeking to expand them, given that the project is located in neighbouring Laos.[22]

From the late 1990s, Thai NGO civil society groups have also sought to challenge the power planning process itself, seeking to make it more transparent and accountable (Greacen/Palettu 2007; Foran 2013). Electricity planning in Thailand has been enshrouded in a discourse of expert knowledge—and the power relations that that entails—which in turn has supported the incumbent electricity regime (Foran 2006). Since the 1980s, EGAT's power planning has become more technically sophisticated, including least-cost planning, load forecasting, and determining reserve capacity (Foran 2006). Yet counter-expertise has challenged assumptions within EGAT's PDP (see 39.4.3). For example, a 2006 study by Greenpeace showed that twelve of the last thirteen power-demand estimates by EGAT had been overestimates, resulting in over-investment in generation capacity (Greacen/Footner 2006). A combination of factors is likely to have contributed to this track record, including: a tendency to over forecast Thailand's GDP growth rate, which is a fundamental assumption in the long-term PDP, and hence demand growth; risk perception of planners towards demand-side management and energy efficiency measures in the face of the need to maintain a stable electricity supply; the moral hazard of cost-plus regulations;[23] and the existing interests and politics of the incumbent electricity regime.

Civil society has sought also to open up the electricity planning process to more participation, transpar-

20 "Victory for Mae Moh victims", in: Bangkok Post, 25 February 2015, at: <http://www.bangkokpost.com/news/general/483656/victory-for-mae-moh-victims> (31 May 2015).

21 Perkins Coei, "Letter to International Rivers and Environmental Defenders Law Center Re: PNPCA Process for Xayaburi Dam", dated 5 July 2011, at: <http://www.internationalrivers.org/files/attached-files/xayaburipnpcaprocess.pdf > (31 May 2015).

22 "Thai Court Takes Villagers' Case against Power Firm, Laos Dam", in: Reuters, 24 June 2014, at: <http://uk.reuters.com/article/2014/06/24/thailand-laos-lawsuit-dam-idUKL4N0P51PN20140624> (31 May 2015).

ency and accountability, citing the broader 'landscape changes' and associated entitlements. For example, under a wider World Resources Institute initiative, a consortium of Thai organizations researched a benchmarking report that assessed electricity governance performance with regard to policy process, regulatory process, and environmental and social considerations, and placed Thailand in a comparative perspective against nine other emerging economy countries (Sukkumnoed/Greacen/Limstit et al. 2006). The report and policy work surrounding it both contributed to blocking EGAT's corporatization in 2006, and helped shape the subsequent law that established the Energy Regulatory Commission (WRI/EGI/Prayas Energy Group n.d.).

39.4.2.2 Challenging the Incumbent Regime Through Cooperation

Some civil society groups have sought to cooperate with EGAT to reform its practices. For example, the rise of Thailand's energy conservation agenda is linked to the catalytic work of the *International Institute for Energy Conservation* (IIEC), a technically-orientated NGO with its headquarters in the US. In the early 1990s, IIEC proposed a US$179 million energy conservation programme to EGAT, PEA and MEA, that was taken up and implemented. IIEC formulated its proposal on the basis that investment in *demand-side management* (DSM) and *energy efficiency* (EE) was the least-cost option for Thailand. In tracing the origin of the programme and its impact, Foran (2006: 28) observes that there was a "good rapport" between EGAT and IIEC.

Yet energy conservation also presented a number of challenges to EGAT, such that its full potential has not been exploited (Foran 2006: 27–32; Foran/du Pont/Parinya/ et al. 2010). IIEC proposed a new approach to PDP preparation based on the principles of *Integrated Resources Planning* (IRP), namely least-cost planning that optimizes both the supply side and the demand side. EGAT's senior management and planners, however, perceived interventions to change user behaviour as risky with regard to ensuring electricity system reliability in contrast to building new

supply, and thus continued to pursue a business-as-usual approach that privileged the expansion of capacity. Such nationally-scaled framing of electricity systems—including of aggregated demand—renders invisible local-scaled practices surrounding electricity consumption, including its micro-politics (Smits 2012), and how the preferences of users are shaped through broader consumer society preferences and practices (Shove/Walker 2014) (39.2.3).

39.4.3 Imagined Alternative Sociotechnical Regime

Civil society has also unsettled the incumbent electricity regime through imagining and practising alternatives. At the community scale, for example, groups resisting EGAT's plans for two large-scale coal-fired power plants in Prachub Khiri Khan province sought to reframe the concept of development. The power stations were part of the Southern Seaboard development plan that proposes industrial steel production in the region. They envision

> a just provincial level programme for defence of local subsistence and prosperity through rice, coconut and pineapple cultivation, local marketing, small fisheries, tourism, and wind and other non-fossil energy sources (Sureerat 2010, cited in Lohmann/Hildyard 2013:14).

They reimagine energy production as locally produced and at an appropriate scale for their area, rather than meeting 'national' energy demand; this reflects an often-found 'politics of scale' tension between local- and national-level priorities and the discourses (and power inequalities) that frame them (Sneddon 2003). In another example, the *Appropriate Technology Association* at its energy ashram in Nakorn Ratchasima Province researches, promotes and puts into practice appropriate technologies related to rural development.[24] It was created by an early-retired engineering professor from Chulalongkorn University in Bangkok, and those visiting the ashram can learn about a range of methods by which to produce energy sustainably.

Other civil society groups have worked to create different PDPs from EGAT's plans. In 2004, in the context of the opposition to the corporatization of EGAT, civil society groups under the auspices of the National Economic and Social Advisory Council developed an "Alternative PDP" that was presented to the Senate Committee on Public Participation (Permpongsachar-

---

23 It has been pointed out by numerous critics that EGAT's cost-plus investment arrangement is a disincentive to investment in electricity-saving measures, as profit cannot be made by reducing electricity sales; in contrast, it incentivizes over-investment, as well as creating unnecessary social and environmental costs (Foran 2006: 31; Greacen/Footner 2007).

24 Appropriate Technology Association website; at: <http://www.ata.or.th/th/index.php> (28 May 2015).

oen 2004). Since then, two more complete studies have been prepared, in 2006 (Greacen/Footner 2006) and in 2012 (Greacen/Greacen 2012). These studies highlight the limitations of existing electricity planning and its governance, and propose plans whereby Thailand could meet its energy needs through energy conservation, renewable energy, and refurbishing existing power stations. They also propose improved practices for planning, namely IRP, and additional decision-making criteria that internalize environmental and societal costs. Whilst EGAT has not officially responded to these plans, as credible counter-expertise they have empowered community groups to challenge the business-as-usual approach.

39.4.4 Electricity Niche Development

Civil society groups also play a role in nurturing niches (Seyfang/Smith 2007; Smith 2012). As an example in Thailand, this section focuses on the emergence of the *Very Small Power Producer* (VSPP) programme and the role of civil society, approved by the Thai Cabinet in 2002 and expanded in 2006. The internationally-applauded programme has allowed for the distributed production of electricity from renewable energy systems, and has been argued to increase the resilience of Thailand's electricity system (Meerow/Baud 2012).

The origin of the VSPP programme can be traced back to a community micro-hydropower project in Mae Kampong village, Mae On district, Chiang Mai province, northern Thailand (Smits 2012; Greacen 2004), and the activities of a small number of researchers, civil society groups and progressive government staff in EPPO and EGAT (Brouwer 2012). Briefly, three micro-hydro schemes of twenty to forty kilowatts have been built in Mae Kampong village in 1983, 1988 and 1994. The projects were built as a partnership between the community and the government's *Department of Alternative Energy Development and Efficiency* (DEDE), resulting in a strong sense of community ownership (Smits 2012). Importantly, even when PEA's grid electricity reached the village in 2000, the community preferred to maintain the micro-hydro system as a cooperative alongside it, partly because it enhanced their image as an 'environmentally friendly forest community' and thus supported an ecotourism project also implemented by the villagers (Smits 2012). The micro-hydro projects, in other words, are linked to the village's culturally-valued ecological identity (cf. Smith 2010).

In 2001, a PhD researcher working with the community, Christopher Greacen, began to consider how it might be technically possible to connect one of the micro-hydro projects to PEA's grid. At the same time, Cheunchom Sangarasri Greacen, a staff member at EPPO, explored how a regulatory framework might enable this arrangement.[25] According to Brouwer (2012: 4), over 2001:

> Working closely with the utility companies and regulators, and operating without public pressure, lobbying, op-eds or media coverage, Chris and Chom saw Cabinet approval for the VSPP rules in less than a year. Their strategy was to work collaboratively with the authorities (government regulators), and to establish allies in the sector with the biggest influence on the process (utilities).

Thus, catalysed by this groundbreaking work emerging from Mae Kampong, in 2007 one of the village's micro-hydro projects was synchronized to sell to PEA's grid, benefiting from the VSPP programme it helped create and earning the community approximately US$1,000 per month.[26]

Thailand's VSPP regulations are drawn from net-metering rules that were already in operation in the US at the time of drafting, and are also based on the existing Small Power Producers regulations passed in Thailand in 1992 (Greacen 2007). The first-phase VSPP regulations, approved in 2002, allowed projects of up to 1 MW to connect to the grid, thus assuaging the concerns of EGAT's engineers regarding distributed generation sources creating grid instability. The 2006 VSPP modifications included an adder feed-in tariff (i.e. a subsidy) and allowed for projects of up to 10 MW, including co-generation units (Greacen 2007).

The VSPP projects have challenged Thailand's business-as-usual electricity regime, in that most contributing renewable energy systems are owned by small- and medium-scale businesses[27] rather than EGAT or large IPP companies, thus diversifying the ownership of gen-

---

25 Christopher and Cheunchom Greacen went on to create the NGO energy think tank Palang Thai in 2002 (Foran 2006), which continues to promote the VSPP programme.

26 Unfortunately, after six months operation, sales to PEA were halted due to a disagreement over the sharing of revenues between the village and the sub-district authority (Smits 2012).

27 Since 2014, community groups—in hybrid governance arrangements with the state and private sector—have sought to benefit from the VSPP programme, albeit on a small scale at present (On 2015).

eration capacity. On the other hand, at the time of the VSPP programme's development, EGAT was facing pressure to privatize, and thus EGAT's permitting a greater (but still comparatively small) role for the wider private sector could also be viewed as a minor compromise that actually bolstered the incumbent electricity regime. As Brouwer (2012) writes, interviewing Christopher Greacen:

> the VSPP regulations were politically very 'nicely aligned, because they were consistent with private sector participation in the power sector in the sense that they would enable more customer-owned generators. The utilities could give back a little bit by making a concession on this while they were fighting against [utility privatization]'.

Despite the apparent success of the VSPP programme and Thailand's Ten-Year Alternative Energy Development Plan (39.3.3), challenges remain for the widespread scaling-up of renewable energy. Tongsopit/Greacen (2013: 440) point out that one "major impediment" is that Thailand has six separate long-term national energy plans which are neither integrated nor coordinated. Meanwhile, whilst the adder feed-in tariff programme as first implemented between 2007 and 2010 was, according to Tongsopit/Greacen (2013: 441), "systematic and transparent", since then loopholes have been exploited whereby private companies have bid for solar VSPP PPAs with the intent to resell the PPA rather than develop the capacity themselves. The government's response has been to create a new Management Committee and evolve rules since 2010 which have been criticized as lacking transparency and also as slowing down investment in renewable energy (see also Meerow/Baud 2013). Other controversies have also emerged, for example the local pollution impacts of some poorly-managed biomass projects under the VSPP programme (Yoo 2013).

39.5   Conclusion: Towards Decentralizing and Democratizing Electricity

This chapter has applied a Sustainability Transition and Multilevel Perspective lens to explain the creation and resistance to change of Thailand's incumbent electricity regime, which was first a state monopoly and since the 1980s has evolved into a partially-privatized structure. This chapter has conceptualized the incumbent regime not as monolithic but as a coalition of actors (pursuing their interests) including the state utility, its labour unions, and various 'dual-firm' IPPs,

thus adopting an explicitly political economy approach. The chapter has contributed to sparse literature on sustainability transition in Asia's NICs, and in particular on how Thailand's political economy has shaped the electricity regime's establishment and embedding, and its resistance to change that combines economic, institutional, cultural and political factors, with technical lock-ins and historical pathway dependencies.

Changes in the structure of the incumbent electricity regime have been shaped by shifts in 'landscape'. These shifts include those at the global scale (changing development paradigm from developmentalist state to neo-liberalism, global economic crisis and Thailand's exposure to it as an export-orientated economy, and climate change) and at the national scale (increasingly embedded democratic institutions, and a growing role for civil society). Regarding the latter, these political spaces have been widened as civil society groups have sought to claim, utilize, and thus legitimize them—and have been closed as elites have sought to reduce the role of progressive civil society groups. It may be noted that not only has reform of the electricity sector regime been shaped by these landscape level shifts, but the electricity sector has also constituted an important theme within broader political movements seeking landscape shifts in the first place. Incremental reforms, some of which have originated in niches, have been accommodated by the incumbent regime, including the growing contribution of distributed renewable energy generation and energy conservation, the creation of an independent regulator, and a small increase in civil society participation and accountability of the power planning process.

The chapter has emphasized in particular the role of civil society in shaping the incumbent electricity regime. Civil society groups have adopted a range of strategies to unsettle the regime, including opposing problematic projects, advocating for progressive policy, and proposing alternative plans, values and visions for Thailand's electricity sector. In the past two decades there has been a significant growth in contestation and deliberation about how electricity should be produced, how decisions should be taken around it, and how implications for ecological sustainability and justice should be internalized. Whilst significant power asymmetries exist, there is now somewhat more debate between government agencies with civil society. They have drawn upon independent state agencies, such as the justice system and the Thai National Human Rights Commission, to counterbalance the power of the state (Middleton 2012), and their existence reflects 'landscape' level shifts. Civil society is evidently a source of reflexivity in Thailand's electricity sector (Grin/Rotmans/Schot 2010, cited by Smith 2012).

The ST approach and MLP has helped explain why the structure of the incumbent electricity regime has remained relatively intransient, despite the argued-for benefits of electricity sector reform (Greacen/ Greacen 2012), and from the perspective of ongoing broader environmental, social and political challenges in Thailand (Phongpaichit/Benyaapikul 2013). Transformation towards sustainability in Thailand, therefore, is a long-term challenge. Drawing insight from sustainability transition management literature (Verbong/Loorbach 2012), some strategies and policies to steer Thailand's incumbent electricity regime in the direction of sustainability include:

- The importance of building long-term and broad coalitions formed of civil society, government reformers, academics, and media (amongst others) that can counterbalance the political influence of the incumbent electricity regime, and that can support the emergence of green niches.
- Expand and maintain political space (or 'transition arenas') and support more deliberative processes that increase the responsiveness of the incumbent

regime, and in particular government agencies (EGAT; Ministry of Energy), to public interest concerns. Adopting IRP practices can integrate supply-side and demand-side multi-criteria least-cost planning. Governance and planning tools such as strategic environmental assessment and (transboundary) environmental impact assessment can also be employed, through which deliberation can occur (Middleton/Dore 2015).

- Build upon landscape pressures to destabilize unsustainable aspects of the incumbent electricity regime. Landscape pressures range from international expectations for sustainability transition towards renewable energy and energy conservation in the context of climate change to utilizing (and thus reinforcing) relatively recent independent state agencies such as the Administrative Courts, the Thai National Human Rights Commission, and the Energy Regulatory Commission.
- Acknowledge and support the legitimate role of civil society in its diverse forms, ranging from local community groups to progressive NGOs, as regime watchdogs and advocates, and as niche innovators.

References

ADB (Asian Development Bank), 2010: *Analysis of EIA/ EMP in the Greater Mekong Subregion (GMS) Countries and Identification of Gaps, Needs, and Areas for Capacity Development* (Manila: Asian Development Bank).

Angel, David; Rock, Michael T., 2009: "Environmental Rationalities and the Development State in East Asia: Prospects for a Sustainability Transition", in: *Technological Forecasting and Social Change*, 76,2: 229-40.

Berkhout, Frans; Angel, David; Wieczorek, Anna J., 2009: "Asian Development Pathways and Sustainable Socio-Technical Regimes", in: *Technological Forecasting and Social Change*, 76,2: 218-28.

Blake, David J. H., 2013: "Thai Dam-Affected Villagers Demand Fair Compensation", in: *World Rivers Review*, 28: 5.

Boonlong, Raine, 2011: "Energy Issues in Thailand", in: Boonlong, Raine; Farbotko, Carol; Parfondry, Claire; Graham, Colum; Macer, Darryl (Eds.), 2011: *Representation and Decision-Making in Environment Planning with Emphasis on Energy Technologies Ethics and Climate Change in Asia and the Pacific* (Bangkok: UNESCAP).

Boonlong, Raine; Farbotko, Carol; Parfondry, Claire; Graham, Colum; Macer, Darryl (Eds.), 2011: *Representation and Decision-Making in Environment Planning with*

Emphasis on Energy Technologies Ethics and Climate Change in Asia and the Pacific (Bangkok: UNESCAP).

Brouwer, Ben, 2012: "Institutional Aikido: Bringing Power to the People of Thailand" (San Francisco: The Democracy Center).

Costanza, Robert; Kubiszewski, Ida; Paquet, Peter; King, Jeffrey; Halimi, Shpresa; Sanguanngoi, Hansa; Nguyen, Luong Bach; Frankel, Richard; Ganaseni, Jiragorn; Intralawan, Apisom; Morell, David, 2011: *Planning Approaches for Water Resources Development in the Lower Mekong Basin* (Portland and Chiang Rai: Portland State University and Mae Fah Luang University).

Dobbin, Frank; Simmons, Beth; Garrett, Geoffrey, 2007: "The Global Diffusion of Public Policies: Social Construction, Coercion, Competition, or Learning?", in: *Annual Review of Sociology*, 33,1: 449-72.

Doner, Richard F., 2009: *The Politics of Uneven Development: Thailand's Economic Growth in Comparative Perspective* (Cambridge: Cambridge University Press).

EGAT (Electricity Generating Authority of Thailand), 2010: *Thailand's Power Development Plan (2010-2030)* (Bangkok: EGAT).

EPPO (Energy Policy and Planning Office, Ministry of Energy), 2013: *Energy Statistic of Thailand 2013* (Bangkok: EPPO).

EPPO (Energy Policy and Planning Office, Ministry of Energy), 2012: *Summary of the Thailand Power Development Plan 2012-2030 (PDP2010: Revision 3)* (Bangkok: EPPO).

Foran, Tira, 2006: *Thailand's Politics of Power System Planning and Reform* (Chiang Mai: Mekong Program on Water Environment and Resilience (M-POWER)).

Foran, Tira, 2013: "Action Research to Improve Electricity Planning: Experience from Thailand", in: Daniel, Rajesh; Lebel, Louis; Manorom, Kanokwan (Eds.), 2013: *Governing the Mekong: Engaging in the Politics of Knowledge* (Selangor: SIRDC Press): 49-69.

Foran, Tira; du Pont, Peter T.; Parinya, Panom; Phumaraphand, Napaporn, 2010: "Securing Energy Efficiency as a High Priority: Scenarios for Common Appliance Electricity Consumption in Thailand", in: *Energy Efficiency* 3: 347-64.

Foran, Tira; Manorom, Kanokwan, 2009: "Pak Mun Dam: Perpetually Contested?", in Molle, François; Foran, Tira; Käkönen, Mira (Eds.), *Contested Waterscapes in the Mekong Region: Hydropower, Livelihoods and Governance* (London, Sterling, VA: Earthscan): 55-80.

Friend, Richard; Pradubsuk, Suphasuk; Badenoch, Nathan; Limpiyawon, Patcharapol (Eds.), 2011: *Environmental Governance in Asia: Independent Assessments of National Implementation of Rio Declaration's Principle 10* (Bangkok: Thailand Environment Institute (TEI)).

Geels, Frank W., 2014: "Regime Resistance against Low-Carbon Transitions: Introducing Politics and Power into the Multi-Level Perspective", in: *Theory, Culture & Society*, 31,5: 21-40.

Geels, Frank W., 2011: "The Multi-Level Perspective on Sustainability Transitions: Responses to Seven Criticisms", in: *Environmental Innovation and Societal Transitions*, 1,1: 24-40.

Geels, Frank W., 2002: "Technological Transitions as Evolutionary Reconfiguration Processes: A Multi-Level Perspective and a Case-Study", in: *Research Policy*, 31,8-9: 1257-74.

Greacen, Christopher E., 2004: "The Marginalization of 'Small Is Beautiful': Micro-Hydroelectricity, Common Property and the Politics of Rural Electricity Provision in Thailand" (PhD dissertation, University of California, Energy and Resources Group).

Greacen, Chris, 2007: "An Emerging Light: Thailand Gives the Go-Ahead to Distributed Energy", in: *Cogeneration & On-Site Power Production Magazine* March/April: 65-73.

Greacen, Chris; Footner, James, 2006: *Decentralizing Thai Power: Towards a Sustainable Energy System* (Bangkok: Greenpeace Southeast Asia).

Greacen, Chris; Palettu, Apsara, 2007: "Electricity Sector Planning and Hydropower", in: Lebel, Louis; Dore, John; Daniel, Rajesh; Koma, Yang Saing (Eds.), 2007: *Democratizing Water Governance in the Mekong Region* (Chiang Mai: Mekong Press): 93-126.

Greacen, Chuenchom Sangarasri; Greacen, Chris, 2012: *Proposed Power Development Plan (PDP) 2012 and a Framework for Improving Accountability and Performance of Power Sector Planning* (Bangkok: Palang Thai).

Greacen, Chuenchom Sangarasri; Greacen, Chris, 2004: "Thailand's Electricity Reforms: Privatization of Benefits and Socialization of Costs and Risks", in: *Pacific Affairs*, 77,3: 717-541.

Greenpeace Southeast Asia, 2005: *All Emission, No Solution: Energy Hypocrisy and the Asian Development Bank in Southeast Asia* (Bangkok: Greenpeace Southeast Asia)

Grin, John; Rotmans, Jan; Schot, Johan (Eds.), 2010: *Transitions to Sustainable Development: New Directions in the Study of Long Term Transformative Change* (New York and London: Routledge).

Harashima, Yohei, 2000: "Environmental Governance in Selected Asian Developing Countries", in: *International Review for Environmental Strategies*, 1,1: 193-207.

Hausman, William J.; Hertner, Peter; Wilkins, Mira, 2008: *Global Electrification: Multinational Enterprise and International Finance in the History of Light and Power 1878-2007* (Cambridge, UK: Cambridge University Press).

Herbertson, Kirk, 2012: "Citizen Science Supports a Healthy Mekong", in: *World Rivers Review*, 28, 2: 1,7.

Hess, David J., 2014: "Sustainability Transitions: A Political Coalition Perspective", in: *Research Policy*, 43, 2: 278-83.

Hildyard, Nicholas; Lohmann, Larry; Sexton, Sarah, 2012: *Energy Security: For Whom? For What?* (Sturminster Newton: The Corner House).

Hirsch, Philip; Hogan, Zeb S.; Lanza, Guy R.; Blake, David J. H., 2011: *Technical Review of the Xayaburi Environmental Impact Assessment* (Berkeley: International Rivers).

Hirsch, Philip, 1998: "Dams, Resources and the Politics of Environment in Mainland Southeast Asia", in: Hirsch, Philip; Warren, Carol (Eds.), 1998: *The Politics of Environment in Southeast Asia: Resources and Resistance* (London: Routledge): 55-70.

IEA (International Energy Association), 2013: *Southeast Asia Energy Outlook* (Paris, IEA).

Jomo, Kwame Sundaram, 2001: "Introduction: Growth and Structural Change in the Second Tier Southeast Asian NICS", in: Jomo, Kwame Sundaram (Ed.), 2001: *Southeast Asia's Industrialization: Industrial Policy, Capabilities and Sustainability* (New York: Palgrave Macmillan): 1-29.

Kamsamrong, Jirapa; Sorapipatana, Chumnong, 2014: "An Assessment of Energy Security in Thailand's Power Generation", in: *Sustainable Energy Technologies and Assessments*, 7: 45-54.

Lohmann, Larry; Hildyard, Nicholas, 2013: *Energy Alternatives: Surveying the Territory* (Sturminster Newton: The Corner House).

Markard, Jochen; Raven, Rob; Truffer, Bernhard, 2012: "Sustainability Transitions: An Emerging Field of Research and Its Prospects", in: *Research Policy*, 41,6: 955-67.

Matthews, Nathanial, 2012: "Water Grabbing in the Mekong Basin–an Analysis of the Winners and Losers of Thailand's Hydropower Development in Lao PDR", in: *Water Alternatives*, 5,2: 392-411.

Meadowcroft, James, 2011: "Engaging with the Politics of Sustainability Transitions", in: *Environmental Innovation and Societal Transitions*, 1: 70-75.

Meerow, Sara A.; Baud, Isa, 2012: "Generating Resilience: Exploring the Contribution of the Small Power Producer and Very Small Power Producer Programs to the Resilience of Thailand's Power Sector", in: *International Journal of Urban Sustainable Development*, 4,1: 20-38.

Merme, Vincent; Ahlers, Rhodante; Gupta, Joyeeta, 2014: "Private equity, public affair: Hydropower financing in the Mekong Basin", in: *Global Environmental Change*, 24: 20-29.

Middleton, Carl; Matthews, Nathanial; Mirumachi, Naho, 2015: "Whose Risky Business?: Public-Private Partnerships (PPP), Build-Operate-Transfer (BOT) and Large Hydropower Dams in the Mekong Region", in: Matthews, Nathanial; Geheb, Kim (Eds.), 2015 *Hydropower Development in the Mekong Region: Political, Socio-economic and Environmental Perspectives* (London: Earthscan): 127-152.

Middleton, Carl; Dore, John, 2015: "Transboundary Water and Electricity Governance in Mainland Southeast Asia: Linkages, Disjunctures and Implications", in: *International Journal of Water Governance* 3,1: 93-120.

Middleton, Carl; Garcia, Jelson; Foran, Tira, 2009: "Old and New Hydropower Players in the Mekong Region: Agendas and Strategies", in: Molle, François; Foran, Tira; Käkönen, Mira (Eds.), 2009 *Contested Waterscapes in the Mekong Region: Hydropower, Livelihoods and Governance* (London, Sterling, VA: Earthscan): 23-54.

Middleton, Carl; Allouche, Jeremy; Gyawali, Dipak; Allen, Sarah, 2015: "The Rise and Implications of the Water-Energy-Food Nexus in Southeast Asia through an Environmental Justice Lens", in: *Water Alternatives*, 8,1: 627-54.

Middleton, Carl, 2012: "Transborder Environmental Justice in Regional Energy Trade in Mainland South-East Asia", in: *Austrian Journal of Southeast Asian Studies*, 5,2: 292-315.

Missingham, Bruce, 2003: *The Assembly of the Poor: From Local Struggle to National Social Movement* (Chiang Mai: Silkworm Books).

Nakawiro, Thanawat; Bhattacharyya, Subhes C., 2010: "Security of Supply Concerns and Environmental Impacts of Electricity Capacity Expansion in Thailand", in: *International Energy Journal*, 11,4: 181-92.

Nakawiro, Thanawat; Bhattacharyya, Subhes C; Limmeechokchai, Bundit, 2008: "Electricity Capacity Expansion

in Thailand: An Analysis of Gas Dependence and Fuel Import Reliance", in: *Energy*, 33, 5: 712-23.

Nevins, Joseph; Peluso, Nancy Lee, 2008: *Taking Southeast Asia to Market* (Ithaca and London: Cornell University Press).

On, Thita, 2015: "Hybrid Environmental Governance in Very Small Power Producers (VSPPs): Comparative Case Studies in Three Communities Micro-Hydropower Projects, Thailand" (MA dissertation, Chulalongkorn University, MA in International Development Studies Programme).

Pavitt, Keith, 1984: "Sectoral patterns of technical change: Towards a taxonomy and a theory", in: *Research Policy*, 13: 343-73.

Phongpaichit, Pasuk; Baker, Chris, 2012: "Populist Challenge to the Establishment: Thaksin Shinawatra and the Transformation of Thai Politics", in: Robison, Richard (Ed.), 2012: *Routledge Handbook of Southeast Asian Politics* (London and New York: Routledge): 83-96.

Phongpaichit, Pasuk; Baker, Chris (Eds.), 2008: *Thai Capital after the 1997 Crisis* (Chiang Mai: Silkworm Books).

Phongpaichit, Pasuk; Baker, Chris, 2002: *Thailand Economy and Politics (2nd edition)* (Oxford and New York: Oxford University Press).

Phongpaichit, Pasuk; Benyaapikul, Pornthep, 2013: *Political Economy Dimension of a Middle Income Trap: Challenges and Opportunities for Policy Reform: Thailand* (Bangkok: Faculty of Economics, Chulalongkorn University).

Sawdon, John, 2014: "The Political Economy of Environmental Technological Change with a Case Study of the Power Sector in Vietnam" (PhD Dissertation, University of East Anglia, Norwich, UK, School of International Development).

Schreuer, Anna; Rohracher, Harald; Späth, Philipp, 2010: "Transforming the Energy System: The Role of Institutions, Interests and Ideas", in: *Technology Analysis & Strategic Management*, 22,6: 649-52.

Seyfang, Gill; Smith, Adrian, 2007: "Grassroots Innovations for Sustainable Development: Towards a New Research and Policy Agenda", in: *Environmental Politics*, 16,4: 584-603.

Shepherd, Anne; Ortolano, Leonard, 1997: "Organizational Change and Environmental Impact Assessment at the Electricity Generating Authority of Thailand: 1972-1988", in: *Environmental Impact Assessment Review*, 17,5: 329-56.

Shove, Elizabeth; Walker, Gordon, 2010: "Governing Transitions in the Sustainability of Everyday Life", in: *Research Policy*, 39,4: 471-76.

Shove, Elizabeth; Walker, Gordon, 2014: "What Is Energy For? Social Practice and Energy Demand", in: *Theory, Culture & Society*, 31,5: 41-58.

Siriprachai, Somboon, 2012: *Industrialization with a Weak State: Thailand's Development in Historical Perspective* (Singapore and Kyoto: NUS Press in association with Kyoto University Press).

Smith, Adrian; Stirling, Andy; Berkhout, Frans, 2005: "The Governance of Sustainable Sociotechnical Transitions", in: *Research Policy*, 34,10: 1491-1510.

Smith, Adrian, 2012: "Civil Society in Sustainable Energy Transitions", in: Verbong, Geert; Loorback, Derk (Eds.), 2012: *Governing the Energy Transition* (London and New York: Routledge): 180-202.

Smith, Adrian; Voß, Jan-Peter; Grin, John, 2010: "Innovation Studies and Sustainability Transitions: The Allure of the Multi-Level Perspective and Its Challenges", in: *Research Policy*, 39,4: 435-48.

Smith, Thomas. B., 2003: "Privatising Electric Power in Malaysia and Thailand: Politics and Infrastructure Development Policy", in: *Public Administration and Development*, 23,3: 273-83.

Smits, Mattijs, 2012: "The Benefits and Complexities of Distributed Generation: Two Energy Trajectories in Laos and Thailand", in: *Forum for Development Studies*, 39,2: 185-208.

Sneddon, Chris, 2003: "Reconfiguring Scale and Power: The Khong-Chi-Mun Project in Northeast Thailand", in: *Environment and Planning A*, 35,12: 2229-50.

Sukkumnoed, Decharut; Greacen, Chuenchom Sangarasri; Limstit, Paisan; Bureekul, Thawilwadee; Thongplon, Sairung; Nuntavorakarn, Suphakij, 2006: *Governing the Power Sector: An Assessment of Electricity Governance in Thailand* (Washington DC and Pune: World Resources Institute, National Institute of Public Finance and Policy, and Prayas Energy Group).

Sulistiyanto, Priyambudi; Xun, Wu, 2004: "The Political Economy of Power Sector Restructuring in Southeast Asia", Paper for the Conference on Regulation, Deregulation and Re-regulation in Globalizing Asia, National University of Singapore, March 22-24.

UNDP (United Nations Development Program), 2009: *Human Security Today and Tomorrow: Thailand's Human Development Report* 2009 (Bangkok: United Nations Development Programme).

UNEP (United Nations Environment Programme) and TEI (Thailand Environment Institute), 2007: *Greater Mekong Environmental Outlook* (Bangkok: United Nations Environment Program and Thailand Environment Institute (TEI)).

Verbong, Geert; Loorback, Derk, 2012: "Introduction", in: Verbong, Geert; Loorback, Derk, 2012: *Governing the Energy Transition* (London and New York: Routledge): 1-23.

Victor, David G; Heller, Thomas C. (Eds.), 2007: *The Political Economy of Power Sector Reform: The Experiences of Five Major Developing Countries* (Cambridge, New York, Melbourne, Madrid, Cape Town, Singapore, São Paulo: Cambridge University Press).

Walker, Andrew, 2012: *Thailand's Political Peasants: Power in the Modern Rural Economy*. (Madison: University of Wisconsin Press).

Wattana, Supannika; Sharma, Deepak; Vaiyavuth, Ronnakorn, 2008: "Electricity Industry Reforms in Thailand: A Historical Review", in: *GMSARN International Journal* 2,2: 41-52.

Williams, James H; Dubash, Navroz K., 2004: "Asian Electricity Reform in Historical Perspective", in: *Pacific Affairs*, 77,3: 411-36.

Wisuttisak, Pornchai, 2012: "Regulation and Competition Issues in Thai Electricity Sector", in: *Energy Policy*, 44: 185-98.

WRI (World Resources Institute), EGI (Electricty Governance Initiative), and Prayas Energy Group (n.d.): *Shining a Light on Electricity Governance* (Washington DC: World Resources Institute (WRI)).

Yoo, Yeji, 2013: "Renewable Energy Development and Environmental Justice in Thailand: Case Studies of Biomass Energy Projects in Roi-Et and Suphanburi Provinces" (MA dissertation, Chulalongkorn University, MA in International Development Studies Programme).

Part X   **Relying on Transnational, International, Regional and National Governance for Strategies and Policies Towards Sustainability Transition**

Chapter 40   **Governance of Sustainable Development in Knowledge Democracies— Its Consequences for Science**
Roeland Jaap in 't Veld

Chapter 41   **Discourse and Practice of Transitions in International Policy-making on Resource Efficiency in the EU**
Sander Happaerts

40 Governance of Sustainable Development in Knowledge Democracies—Its Consequences for Science

Roeland Jaap in 't Veld

Abstract

Governance is the way in which a society organizes decision-making. Advanced societies turn into knowledge democracies where the relationships between politics, media and science intensify and change continuously. The concept of knowledge democracy embraces participatory democracy besides representative democracy, the rise of social media besides corporate media, the emergence of transdisciplinary trajectories besides classical disciplinary science with complex interactions that also change. The quest for sustainable development takes place within knowledge democracies, where sustainability covers economic, social and ecological issues, whose governance is complex. The future direction and content of sustainable development depend on uncertain future determinants such as technological innovation and the evolution of social values. The precautionary principle as a powerful moral imperative has its limitations. The multidimensional nature of sustainability requires integration and a recognition of a multilevel, multiscale, multidisciplinary character. Development refers to change, transitions and transformations. Governance of sustainable development has to cope with complex dynamics. Governance that furthers transitions focuses on the interaction between representative and participatory democracy, and the optimalization of the contribution of science. This chapter addresses the specific consequences for science, in particular the organization of science, relating both to disciplinary research and transdisciplinarity.

Keywords: Sustainable development, knowledge democracy, science, transdisciplinarity.

40.1 Introduction

Sustainable development can be described as a normative notion presenting the preferred evolution of society. The realization of sustainable development demands change in many domains. These changes can be described as transformations or transitions. Governance presents in a normative manner how a society organizes its decision-making. Governance of sustainable development has specific characteristics dependent on its complexity. This chapter concentrates on the necessary changes in the organization of science as an essential component of the governance of sustainable development.

Representative parliamentary democracy is an essential part of governance. Representation gradually became the predominant mechanism by which the population at large, through elections, provided a body with a general authorization to take decisions in all public domains for a certain period of time. Representative parliamentary democracy became the icon of advanced nation states.

The recent decline of representative parliamentary democracy has been noted by many authors (Castells 1996, 2009; Dahrendorf 2002). Media politics destroys the original meaning of representation. The evolution of political parties to marketeers in the political realm destroys their capacity for designing consistent broad political strategies. Like willow trees they move with the winds of the voters' supposed preferences. Volatility therefore has increased. *This is dramatic, because important challenges like sustainable development demand consistent policies over a longer time period.*

The debate on the future of democracy has not yet led to major innovations in advanced societies in Europe. Established political actors try to tackle populism with trusted resources: a combination of anti-populist rhetoric and adoption of the populist agenda. Some media have tried to become "more

© Springer International Publishing Switzerland 2016
H.G. Brauch et al. (eds.), *Handbook on Sustainability Transition and Sustainable Peace*,
Hexagon Series on Human and Environmental Security and Peace 10, DOI 10.1007/978-3-319-43884-9_40

populist than populists themselves", almost always at the expense of analytical depth. The development in different parts of the world points to various directions: in parts of South America city government is characterized by remarkable citizens' participation, while in Asia the rule of law is introduced without classical democracy in influential nations. Meanwhile, the Internet and the evolution of social media provide for a drastic change in the rules of the game. A better-educated public has wider access to information, and selects it increasingly itself, instead of relying on media filters by classical media. Moreover, citizens themselves have become prosumers (simultaneously producing and consuming): they utilize social media as consumers but also produce YouTube pictures at home.

At the national level, representative democracy must find new ways of interacting with various forms of bottom-up participatory democracy. The relationships between corporate, top-down media and politics have considerably changed due to the rise of social media because politicians may utilize social media to directly communicate with voters, reducing their dependence on the top-down media. The corporate media are no longer the sole necessary intermediaries between politicians and voters.

But people get also tired of social media, because the latter produce information of different quality, increasing the costs of finding trustworthy information and producing confusion and ambiguity. In most social media no editor exists. There is no selection of information on the basis of quality to enable consumers to minimize their costs of searching.

The crucial combination of a network society and media politics (Castells 1996, 2009) provides new problems and tensions. The political agenda is increasingly filled with so-called wicked problems, characterized by the absence of consensus both on the relevant values and on the necessary knowledge and information. Uncertainty and complexity prevail. Wicked problems cannot be 'solved' in a classical way but it is possible to live with them (Hajer 2003, 2005; Hoppe 2008, 2010; Meuleman 2013). *Sustainable development is characterized by the presence of major wicked problems.*

In general contemporary societies experience an increasing intensity and speed of reflexive mechanisms (In 't Veld, 2009, 2010, 2012). In a more or less lenient political environment reflexive mechanisms cause overwhelming volatility of bodies of knowledge about social systems. As all available knowledge is utilized to facilitate reflexive processes, their result might establish new relationships undermining existing knowledge. Social reality has thus become unpredicta-

ble in principle. Both knowledge and information are increasingly volatile and are surrounded by uncertainty.

In their introductory chapter to *Reflexive Governance to Sustainable Development*, Voß and Kemp (2006) deal with reflexivity and distinguish between first- and second-order reflexivity. First-order reflexivity "refers to how modernity deals with its own implications and side effects, the mechanism by which modern societies grow in cycles of producing problems and solutions to these problems that produce new problems. The reality of modern society is thus a result of self-confrontation" (Voß/Kemp 2006: 6). Second-order reflexivity concerns "the cognitive reconstruction of this cycle". It "entails the application of modern rational analysis not only to the self-induced problems but also to its own working, conditions and effects".

The relationships between science and politics demand new designs in an environment of media politics, wicked problems and reflexivity. Jasanoff's (2003, 2004) classical theory on boundary to cope with existing gaps between science and politics is now widely accepted among experts. The underlying insight is that scientific knowledge by its very structure never directly relates to action, because it is fragmented, partial, conditional and immunized. This observation applies to both mono- and multidisciplinary knowledge. Translation activities are always necessary to utilize scientific knowledge for policy purposes.

The literature on transdisciplinary research is dominated by process-directed normative studies. The key concept of transdisciplinarity should be defined as the trajectory in a multi-actor environment from a political agenda and existing expertise to a robust, plausible perspective for action. Thus, in the terminology of Voß and Kemp we mainly deal with second-order reflexivity.

We try to explore the relationships between disciplinary research and transdisciplinarity further, and to come up with several recommendations for the future organization of science.

40.2   Sustainable Development

The concept of sustainability deals with three key dimensions of human societies, the economic, social and ecological dimensions, as the three P's: people, planet, profit. This implies that multiple changes are relevant regarding sustainable development, not only ecological dynamics.

Statistics indicate that on average human beings live longer and in better health than ever before, but the pursuit of happiness relates to more than statistics. Our values concerning distributive justice urge us to pay attention to differences. Many normative environmental perspectives are formulated in terms of threats that require immediate action. While increasing wealth appears to reduce the willingness to accept risks of wealthy people (Beck 2003, 2009), these threats are shaped as extreme risks.

Humankind is now able to cause irreversible change that partially diminishes the options of future generations. The normative insight derived from this principle is formulated as the precautionary principle, which leads to the norm that we should abstain from action that reduces the choice among valuable future options.

The reconciliatory character of sustainable development raises specific questions for judgment about changes that lead to improvement in two dimensions but to a deterioration in the third. Until now we have lacked a satisfactory interdimensional measuring rod in order to evaluate this type of changes. This deficiency is serious because as a consequence we are unable to provide convincing criteria to judge policy options in a comparative manner. In real-life situations Pareto-optimal solutions are exceptions. The remembrance of the slow death of welfare economics as an academic subdiscipline should warn us: the inability of welfare economics to formulate convincing policy recommendations in cases of non-Pareto-optimal situations appeared to be the main cause of its death (In 't Veld 1975).

Many different dialogues on sustainable development occur simultaneously. It functions as a unifying concept because its vagueness breeds a consensus that might be later used. It is an asset if it triggers action. On the other hand, if sustainable development is everything, maybe it is nothing. Although—or maybe because—the concept is so vague, it has overwhelming appeal on political agendas and in programmes and dialogues.

Sustainable development is a container notion. The use of the singular form fits holistic viewpoints. The supporters of these viewpoints speak about *the* climate, *the* earth, *the* emissions, *the* planetary boundaries (Meuleman 2013; In 't Veld 2011), which are all at stake, and disasters threaten. Such constructs enable us subsequently to deal with a global challenge that should be addressed in a well-coordinated manner. Thus, the normative construction of the problematique leads to a specific line of argumentation on governance. Once again it appears that framing is normative. The supporters of this view may be found in international organizations that make continuous efforts to produce consensus on internationally binding agreements to prevent disasters. Basic metaphors like the exhaustion of the earth and planetary boundaries are then very useful.

However, people do not experience *the* climate but a climate in the neighbourhood. They pursue a good life according to their own values and in many cases try to find a satisfactory relationship with the surrounding nature. Their visible world is not abstract or systemic but specific and concrete. Likewise, until a few years ago, climatologists distinguished many different climates.

Entrepreneurs make attempts to design and apply more sustainable technologies. They act in a specific environment too, not in an abstract universe. Not just perceptions are context- bound but also acceptable ways of dealing with problematic issues.

Thus, major discrepancies may exist here between the systemic world and daily life. Those who recognize the twofold framing of sustainable development will prefer multiple politics and policies on different scales, with both top-down and bottom-up influence.

The Western world has developed environmental policies during the last half century. In the international realm younger nation states, often former colonies, more recently also become aware of the disagreeable side effects of economic growth. They want to counterbalance these effects in their own manner. However, in the diplomatic arena they are continuously confronted with urgent calls to participate in bargaining processes on treaties with the former colonial powers. These partners now urge dramatic reductions of emissions and the like. Quotas for a certain future year are symbols of urgency. The young nation that is coping with the need for reduction of backwardness in technologies and is just starting to think about clean technologies will not feel inspired by the short-term limits set by others. It will experience those as unnatural.

Cultural diversity should be recognized both as a component of sustainability and as a complicating factor that prohibits progress in reaching consensus on collective action. A society needs a certain cohesion that is produced as a moral order, based on consensus about some fundamental values and norms. Culture within a society also implies the sharing of some common substantial and relational values. A society consists of configurations that possess a specific culture but—as observed earlier—this leads to outside walls (In 't Veld, 2009; Teisman 2009). Thus, tensions arise. In particular the tensions between emerging identities on one side, accompanied necessarily by outer walls, and the need for cohesion and collective action on the other will never disappear. Thus, shaping governance is walking a high wire.

We should argue that both biodiversity and cultural diversity are components of sustainability. We may mourn the loss of a language somewhere on this globe and the loss of a species. But our general attitude towards cultural diversity in daily practice is far more critical than towards biodiversity. We do not believe that each culture is intrinsically good. On the contrary, some cultures are horrifying to many. As sustainability also implies the economic and social dimension, we realise that "diversity always is a bedfellow of inequality" (Van Londen in De Ruijter 2011: 14). Inequality might be a threat to sustainable development. So our attitude towards cultural diversity is ambiguous. In other terms, we do not embrace the precautionary principle for culture. *The full recognition of cultural diversity as a major component of sustainability delivers a strong argument against uniform global governance of transitions and transformations.*

According to the concept of second modernity (Beck 1997, 2006), it is probable that from the tense relations between emerging opposites variety increases. Striving at sustainable development urges us to take these tensions fully into account when dealing with governance.

Because sustainable development is a long-range trajectory, with considerable uncertainty and lack of forecasting options, the notion of resilience is crucial: like Noah we can act sensibly without any certainty about future events by answering the question of how to avoid a disaster, *in casu* by building an Ark. Nowadays, for instance, it is uncertain which theory about climate change is the right one, but once the theory that allies climate change to carbon emissions is there, the justification of measures to reduce emissions can be based on the resilience norm: in order to avoid disasters we have to take into account the feasible theoretical viewpoints irrespective of our beliefs.

We should realise in accordance with the view of Grunwald (2004), Grin (2006) and others that the plurality of notions of sustainable development and their normative origins and connotations lead to the necessity to consider the recommendable knowledge-producing and policy-making processes as reflexive. In Grunwald's terminology: "The normative character of the imperative of sustainability, its inseparable connection with deep-rooted societal structures and val-

ues, the long-term nature of many relevant developments, as well as the often necessary inclusion of societal groups and actors, result in specific demands on scientific problem-solving contributions. Research for sustainable development is a particularly marked type of post-normal science (Funtowitz/Ravetz 1993: 151)". Therefore we argue that dealing with reflexivity and transdisciplinarity are necessary once we strive for sustainable development.

40.3   Knowledge Democracy

Representative democracy has been a successful governance concept for societies (In 't Veld, 2010). During the last two centuries, several forms of representative constitutional democracy developed at the national level and democracy is recommended by Western and most Southern political leaders as the preferred system of rule.

In the twentieth and twenty-first centuries parliamentary democracy, politics and media have become more interdependent; policies are increasingly funded in science, but at the same time science relies on public resources, which is why the linkages between politics and science have intensified. There have been large cognitive and emotional investments in current democratic institutions. As a consequence the stability of these institutions is embraced. Resistance to change has been considerable. But exogenous and endogenous developments threaten the continued success of representative parliamentary democracy.

The recent decline in acceptance, legitimacy and effectiveness of representative parliamentary democracy has been noticed by many authors (In 't Veld 2009, 2010). Three intertwining simultaneous developments have taken place on the macro-, meso- and micro-level of societies, with important effects. On the micro-level of the individual citizen, the classical assumption of a consistent individual position, based on an ideologically-based consistent value pattern, has disappeared. Separate values prevail but the glue of a focal ideological principle often no longer exists. Fragmentation of values has resulted in individualization and uniqueness but also in the impossibility of being represented for all different purposes by a single actor such as a member of parliament. None of the values cherished by an individual may be unique, but the combination probably is. The preference of individuals to be partially represented by a *non-governmental organization* (NGO) per value-domain is therefore no mistake, but a logical evolution. On the meso-level the development of political parties to mar-keteers who try to optimize their future votes in the political realm destroys their capacity to design broad and consistent political strategies. And on the macro-level media politics dominate. As a consequence the epicentre of politics has shifted from parliament to the media. Media can handle personalities better than pro-grammes.

Personalities instead of programmes become the major focus for selection and therefore the voters choose personalities. In the attempt to maximize their votes, political parties are keen to use the media, as it is only possible to actually 'sell' personalities through mass media. This of course significantly increases the structural dependence of politicians on the mass media. The classical function of democracy of protect-ing the people against tyranny and random or arbitrary action by rulers is endangered by stressing personalities over programmes. But media politics destroy the origi-nal meaning of representation.

We now envisage a world where representative democracy is enhanced with participatory democracy, in which social media are added to classical corporate top-down media, and in which disciplinary science is increasingly accompanied by transdisciplinary trajec-tories. The evolutionary patterns at each corner of the triangle are not without tensions: the institutions towards the centre of the triangle feel threatened by the younger ones at the edges. Each corner in the tri-angle is prone to profound change, as indicated in the second-order relationships:

- the bottom-up media not only supplement the classical media, but compete with them;
- participatory democracy is complementary to rep-resentative democracy but is also seen as a threat to it;
- transdisciplinary research designs are not only a bridge between classical science and the real world but also produce deviant knowledge and insights, which may challenge the disciplinary viewpoints.

The Internet, better education and other societal changes have made knowledge accessible to many more people than in the past. This leads to an abun-dance of knowledge and information that needs to be interpreted. It also leads to different types of knowl-edge: not only scientific knowledge appears to be rel-evant, but also citizens' knowledge. This is a huge challenge for policymakers, for scientists and for the media. Politics is not just about how knowledge can be selected for political decisions, but also about how

democratic decision-making processes should change in order to incorporate the different types of knowl-edge adequately.

Moreover citizens themselves have become pro-sumers: any citizen may produce a YouTube picture that is world-famous within two days.

The classical media suffer from the new ones: not only in a commercial sense, but also because of the influence of the new media. We call the new media the bottom-up media in order to distinguish them from the classical top-down media. This distinction does not imply that the top is more powerful than the bottom. An increasing series of empirical counter-proofs is available. Many new media do not have an editor: nobody accepts the obligation to select the rubbish from the trustworthy materials. The develop-ments within and with the media are confusing. Our capacity to observe and to interpret adequately seems to be deficient. Information and knowledge of very different origins are available within a second but it is hard to judge quality. As usual in second modernity the top-down media do not disappear altogether but develop innovative strategies, accepting Internet options and modes of cooperation with social media. The social media are in the process of discovering their own deficiencies, and in some cases organize a revival of editorial functions. The 'wicked' character of many problems on the political agenda sheds a fas-cinating light on the complexities caused by the inter-action of top-down and bottom-up media.

Both contribute to the agenda-setting of politics. The top-down media operate in structural interde-pendency with politics. The expression 'media poli-tics' refers to this interdependency. The bottom-up media are to a considerable degree independent from both the top-down media and politics. Participation in decision preparation and -making may be invited by public authorities, but uninvited participation also occurs, in particular with support from bottom-up media. It is too early to draw consolidated conclu-sions on this development: it is fluid, fast and reflex-ive, and also unpredictable.

The current global economic situation raises new, very challenging questions that are mainly about the institutional frameworks of today's societies. It is therefore time for a transition to a new concept that concentrates on institutional and functional innova-tion. As the industrial economy has been combined with mass democracy through universal suffrage and later by the rise of mass media, one might suggest that the logical successor is a new type of governance, to be called 'knowledge democracy'.

Which challenges and threats are we facing? How will parliamentary and new direct forms of democracy mix, and which roles will knowledge play in the transition towards a knowledge democracy? The crucial combination of a network society and media politics provides new problems and tensions. Today policymaking in many instances is evidence- or knowledge-based, providing both legitimacy and effectiveness, according to its supporters. Effectiveness is assured as the knowledge concerns true statements about the relationships between political interventions and their societal effects. The argument runs thus: according to them, legitimacy is enhanced when the policies are based on 'objective' truth. It is easy to undermine this belief.

Scientific research is a specific form of research, aimed at the creation or accumulation of scientific knowledge. Classical scientific research is performed within disciplines, specialized branches of science with specific theories and methodologies. A discipline studies an aspect of reality, not reality itself. A disciplinary methodology consists of several approved conditions under which truth claims are accepted.

This mono-disciplinary knowledge is methodologically formalized in a particular way: for example it is subject to peer review. It is often put into a rule-based form, such as: 'A implies B' in a particular set of circumstances, whenever these circumstances occur. Such an assertion is known as a hypothesis. 'The more a child participates in sports, the less likely the child is to turn to drugs', is a statement which could originate from empirical research and which probably applies to white families in European cities since 1990, but not to rural areas in Colombia. And why should this statement be true in the future? Scientific knowledge is therefore by definition both fragmented and conditional. Its scientific value depends on the correct application of the agreed methodology. Scientific knowledge requires validity, and methodology protects against criticism. This is called 'normal research'.

It is difficult to integrate different areas of scientific knowledge because scientific knowledge is by its very nature fragmented. And its conditional character means that in order to apply the knowledge in real-world situations, it is necessary to verify whether the relevant characteristics of the situation in which the original study was performed have been complied with. Regarding the future, this question can never be definitively answered. This means that every application of social scientific knowledge for the purpose of policy implies an element of risk.

If a policymaker—in the course of preparing policy proposals—wishes to apply an assertion which is based on a rule, such as 'for every X, under condition Y: A implies B', she/he first has to verify:

- 'Is the X that I am talking about the same X as in the assumption?'
- 'Are the conditions which I am faced with the same as the Y in the assumption?'
- 'Is there really an A in my situation?'
- 'Will the implication still apply at the time when the policy is implemented?'

In particular the last question is a nasty one because the consciousness of reflexivity urges us to wonder whether the drug dealers might have reflected upon the research results too, and might have ensured for themselves a position on the boards of the sports clubs.

This implies that applying scientific knowledge in policy does not always and should not follow the accepted route of meeting the methodological requirements which applied when the knowledge in question was developed. The application of scientific knowledge in a political and governmental context is an exercise in uncertainty, partly based on suppositions, and it also requires competences other than scientific ones, such as social intelligence and well-developed social intuition. It appears necessary to link scientific knowledge to other types of insights without detracting from its relevance and usefulness. Combining knowledge from different scientific disciplines and mixing it with other insights is an opportunity to try to maintain the relevance and usefulness of such knowledge in the relevant application. Multi-, inter- and transdisciplinary developments in research are in full swing. Anyone who realizes this cannot fail to be impressed by the speculative nature of many elements of the methods used. The precision of a great deal of scientific knowledge very soon gets lost in these methods. Robust concepts are often unrefined. As Silvio Funtowicz has repeatedly explained, this image of evidence-based policies based on 'sound' knowledge is not adequate according to the advanced science model (Funtowicz/Ravetz 1991, 1992, 1993). Let us now state that knowledge about social systems is by definition volatile as a consequence of the reflexivity we discuss below. The predominant position of 'wicked problems' on political agendas as indicated earlier is the main reason that linear problem-solving strategies cannot be used. Wicked problems cannot be solved, but they can be managed. In many cases interactive processes are part of effective management.

Today's societies are characterized by the increasing intensity and speed of reflexive mechanisms. We define reflexive mechanisms as events and arrangements that bring about a redefinition of the action perspectives, the focal strategies of the groups and people involved, as a consequence of mindful or thoughtful considerations concerning the frames, identities, and the underlying structures of themselves as well as other relevant stakeholders. Defined in this manner, reflexivity has to do with a particular kind of learning potential. Reflexive systems have the ability to reorient themselves and adapt accordingly based on available self-knowledge.

Reflexive mechanisms in a more or less lenient political environment cause the overwhelming volatility of bodies of knowledge related to social systems. As all available knowledge is utilized to facilitate reflexive processes, the result of such processes might establish new relationships that undermine the existing knowledge. Social reality has then become unpredictable in principle. The efficacy of reflexive mechanisms is furthered by institutional arrangements that enable individual liberty and tolerance.

It is necessary to develop this notion of reflexive learning further because it is of utmost importance for the design of an advanced way of thinking on policy-making: we should realize that a social theory of any kind may never be used to create policy measures without an additional research effort on the specific issue. Such an effort should include the question of whether it is probable or plausible that the theory is already undermined by reflexive reactions in or around the target group of the measure. This latter effort will never deliver results with an absolute truth claim. So uncertainty is overwhelmingly present there too. The policy dialogue will then be characterized by different layers of uncertainty, and so by a discussion about the impact of the different layers of uncertainty too.

Evidence-based policy-making as a normative concept probably bears some relevance when it concerns the application of a physical, chemical or biological scientific theory. But it becomes a hazardous pretention if the decision support comes from a theory in the social sciences for the reasons just explained. Thus, the fashionable approach towards evidence-based policies in social domains should be moderated in a more modest and thoughtful framework.

Knowledge democracy could become an emerging concept with political, ideological and persuasive meaning. The concept of knowledge democracy is meant to enable a new focus on the relationships between knowledge production and dissemination,

the functioning of the media and our democratic institutions.

The media are far from neutral or passive. The illusion that they are a neutral mirror of reality belongs to a forgotten past. We have already shed light on the relationships between politics and media. Media create realities; they also produce knowledge, and moreover report on citizens' knowledge. They are the reporters on scientific findings but also competitors of scientists. The same goes for the relationships between media and citizens.

Moreover, citizens utilize social media independently from authorities either in order to mobilize support for ideas, or to attack existing policy theories. Science is involved in fierce competition, in continuous marketing efforts in order to gain support for viewpoints based upon research and aiming at the acquisition of public resources for further research.

Advocacy coalitions between the proponents of a certain policy theory, the scientific representatives of related scientific theoretical viewpoints, and sympathetic NGOs and citizens' initiatives are born, live and disassemble later on.

40.4   Transdisciplinarity

Much valuable scientific work has been performed on the relationships between science and politics, in order to answer the last question at least partially. Jasanoff (1990, 2003, 2004, 2005) and others have argued that it would be wise to design an independent boundary function in order to foster the quality of the translation (Jasanoff/Martello 2004). The classical theory of boundary work in order to master the existing gaps between science and politics is nowadays widely accepted among experts. The underlying insight is that scientific knowledge by its very structure never directly relates to action, because it is fragmented, partial, conditional and immunized. This observation is valid for both mono- and multidisciplinary knowledge. So translation activities are always necessary in order to utilize scientific knowledge for policy purposes. Pohl, Scholz, Nowotny, Regeer, Bunders and many others have explored this vast domain and developed the concept of transdisciplinarity in a number of variations. The literature on transdisciplinary research is dominated by process-directed normative studies (see Bunders in In 't Veld 2010).

Many authors suggest that transdisciplinary research is just a specific category of scientific research, characterized by the acceptance of some normative

Figure 40.1: The emergence of the knowledge democracy concept. **Source:** The author.

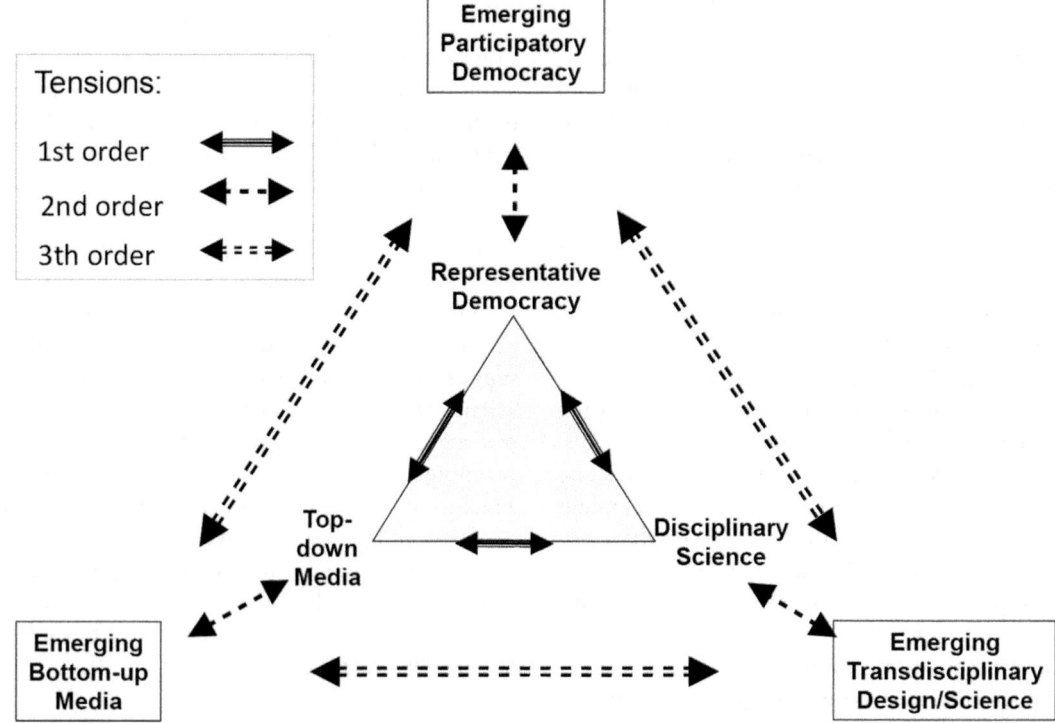

base for scientific reasoning. Here another viewpoint is defended: it appears clear that the core concept of transdisciplinarity is to be defined as the trajectory in a multi-actor environment, a trajectory that in due course leads from two sources: a political agenda and existing scientific expertise, to an acceptable robust, plausible perspective for action. This trajectory bears the character of a communicative and argumentative process (In 't Veld 2013). Its character is design. Funtowicz's later models contain both solutions and caveats on this thorny road (Funtowicz/Ravetz 1991, 1992, 1993).

Figure 40.1 illustrates the twofold relationships between the corners of the triangle. The original, inner institutional framework was fit for the application of the fruits of disciplinary science, in order to solve rather simple policy problems within the framework of representative democracy. Society was ordered clearly in terms of ideological patterns, and classical top-down media fulfilled their roles. The first-order relationships show this picture. The second-order relationships describe the evolution of each corner. As a consequence of that evolution we are confronted with tensions, threats and opportunities around the outer corners of the triangle that are indicated in the third-order relationships.

As we may observe, the outer points of the extended triangle also strengthen and stimulate each other. Transdisciplinarity nears participatory democracy, and social media play crucial roles in large-scale communication processes. So the tensions relate mainly to the inside–outside relations in the triangle while the stimuli relate to the outer point of the corners. Hardly any empirical research is available yet.

Figure 40.2 shows some of the relations between each inner and each outer corner. This type of relation also has far-reaching consequences for the governance of sustainable development in knowledge democracies. These fourth-order relations might prove to be very diversified: for instance, bottom-up media might be utilized by representative democracy but also cause conflicts as shown in the case study on vaccination. Citizens initiatives might internalize the fruits of disciplinary science, but this might also cause application problems. Top-down media might organize transdisciplinary trajectories, but they could prove to be boomerangs for those media themselves. In any society, a wide diversity of actors possesses relevant knowledge concerning important societal problems. In a knowledge democracy both dominant and non-dominant actors could and maybe should have equal access and ability to put this knowledge forward in the process of solving societal problems. We have

Figure 40.2: Old and new forms coexist and influence each other. **Source:** The author.

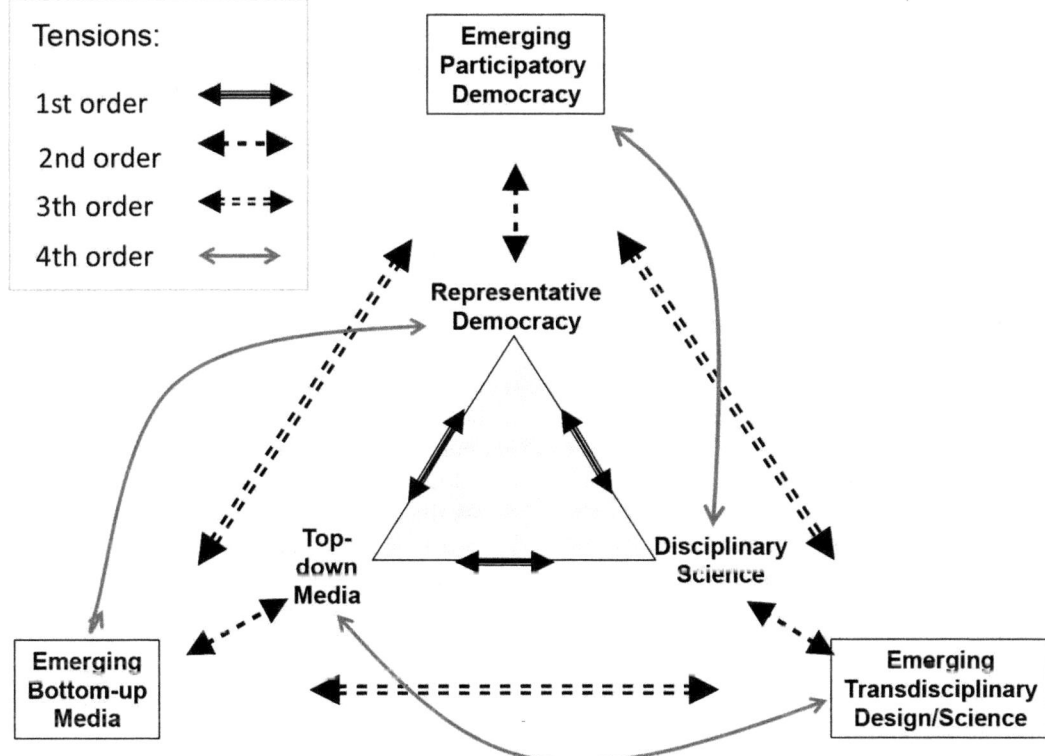

already explained why disciplinary knowledge on its own is not fit to solve broader societal problems.

During the past centuries the tendency towards specialization has dominated in science, destroying the practical meaning of the *homo universale*, and leading to more and more disciplines and subdisciplines. Sometimes innovation was brought about by new combinations of those, called multidisciplinary or interdisciplinary cooperation or even mergers.

According to the earlier terminology, transdisciplinary research developed during the 1980s and early 1990s. Multidisciplinary and interdisciplinary research then can be placed on a continuum between monodisciplinary research and transdisciplinary research. Thompson Klein, Grossenbacher-Mansuy, Häberli et al. (2001: 7) at the start of this century defined transdisciplinarity as:

> A new form of learning and problem-solving involving co-operation between different parts of society and science in order to meet complex challenges of society. Transdisciplinary research starts from tangible, real-world problems. Solutions are devised in collaboration with multiple stakeholders.

So she already states that cooperation and mutual learning are key notions in transdisciplinary trajectories.

We observe, following Bunders and Regeer (2010), that in the scholarly literature the core of transdisciplinary research is most often presented as a shared set of principles. Principles differ from theories, methods, tools and conditions because they refer to the attitudes of the researcher-participant; the researcher is said to perform genuine transdisciplinary research as long as he or she acknowledges and acts in accordance with the intention of these principles. These principles relate to process demands such as joint problem definition, orientation towards robust action perspectives, and so on. As such, a set of principles describes the intentions that guide the researcher in choices he or she has to make for the design of the project or programme, that is, the choice of methods, tools and the sequence of these. In other words, 'the approach' is the manner in which the issue at stake is addressed. This is in line with the widespread convention of labelling specific realizations of transdisciplinary research as 'approaches'.

If one concentrates on the essentials of transdisciplinarity as communication and argumentation, the demands for specific attitudes and even principles concerning the other participants besides researchers are as crucial. The policymakers will tend to accept those scientific viewpoints that are closely related to the predominant policy theory, if present. They should, however, develop a certain willingness to open up to other scientific insights because the aim of the exercise could be to end up with resilient proposals, having answered the question of how to avoid disasters. This demands a sophisticated degree of reflexivity on their part.

Once all participants are touched by the need for mutual adapting, learning and the common goal of a resilient design, the transdisciplinary process could really be successful. Considering the existing literature, one might observe that these conditions are seldom fulfilled.

40.5   Agenda for the Future Organization of Science

Transitions and transformations are major changes that may contribute to sustainable development. There is an impressive number of studies available, also represented in this volume, on the character of these phenomena, under different names such as transformation theory, transition theory, etc. Above we argued that transdisciplinarity is a promising approach to tackling the wicked problems that emerge as we aim for sustainable development. Our analysis of knowledge democracies leads to specific questions about the organization of science in society in order to enable the coexistence of disciplinary research and transdisciplinarity.

In general, assuring the independence of scientific disciplinary research becomes more complicated as the interdependencies between politics, media and science become so intense and numerous. Reflexive mechanisms also considerably complicate the reactions of the public at large to scientific findings: the *Intergovernmental Panel on Climate Change* (IPCC) consensus on the causes of climate change irritated a large proportion of political actors and media so intensely that the support for sustainable policies declined. Each actor is involved in the interdependencies and turmoil of knowledge democracies: there is no outsider position available!

So consciousness of incompatible relations should increase. Scientific activities are performed by numerous organizations, by government and enterprises, by NGOs and by individual researchers, and most of all by specialized organizations devoted to research such as universities. One might say that society at large becomes a research institute.

Governments stimulate and finance research in order to further national competitive positions and innovation, and to contribute to an attractive cultural climate. The belief is widespread that expenditure on research and development causes economic progress without causing damage to other dimensions of societal evolution. Inside government the allocation of scarce resources to specific fields and objectives often leads to conflicts between the innovators on the one hand and the gatekeepers of academic excellence on the other. The latter defend the ideal of meaningful accumulation of knowledge while the former accentuate the need for applicability of scientific insights. Systematically of course both are complementary in nature, but the practice of budgeting in most cases is not adapted to that nature. These conflicts are reproduced nowadays within universities.

As the problematique of sustainable development bears a long-term, multilevel, complex character, the consistency and stability of science policies are of the utmost importance, but so also is free access to scientific sources for local and regional public authorities and citizens' initiatives. This is an essential characteristic of the governance of sustainable development.

In Europe, the scientific infrastructure is primarily connected to national decision-makers. Innovation, however, mainly originates in local and regional communities, so free access is necessary in order to strengthen the nature of participatory democracy.

Growing tensions surround the well-established prototype of the Humboldtian university with its accent on funding disciplinary research—at a distance from society—which is executed in absolute autonomy and 'disinterest'. This university will not show much sympathy for involvement in transdisciplinary trajectories. Academic excellence is here defined as excellence according to academic peers. The idea that there is more 'between heaven and earth' than disciplinary science alone, even in the realm of knowledge production, may be hard to digest for Humboldtian scientists.

The tradition of the American university—our reference concerns among others the University of California, Harvard and MIT—demonstrates a quite different picture: here the assignment of the university as an institution that considers itself obliged to render services to society—besides performing basic research—is rooted in centuries of institutional value formation.

European nation states have developed advisory bodies and semi-autonomous agencies in order to perform boundary work between the worlds of science and politics. If knowledge democracy is accompanied by an increasing degree of populism, political hostility to these intermediaries grows fast, as can be seen now in Denmark, the Netherlands and France.

The wicked character of the sustainability problem necessitates transdisciplinarity, but political conflicts are often fought out here by playing two-level games. As the political controversy may lead to a dead end, an attractive strategy might be to accuse the knowledge source of the opponent of lack of trustworthiness. As the scientific experts are not present in the political arena, this type of indirect accusation will not stimulate the willingness of scientists to participate in transdisiplinary trajectories.

We foresee that in advanced societies the quantitative dimensions of research activities will still increase considerably, because a large proportion of the population will be capable of creating new knowledge, and productive activities will become still more knowledge-intense. As a consequence, transdisciplinarity

and participatory democracy will become more closely interwoven.

The competences of scientists who are successful in transdisciplinarity are concentrated around empathy, and more on specific insights into the specific rationality of politicians.

Adequate governance arrangements in order to ensure acceptable equilibria between different modes of knowledge production demand redesign of the external and internal organization of scientific institutions.

More fundamentally, one could come up with an organizational logic that stimulates peaceful coexistence between the different research orientations in universities and the like that have just been mentioned. The organization of science is a focal issue in the pursuit of sustainable development.

References

Bunders, Joske; Regeer, Barbara, 2009: *Knowledge co-creation: Interaction between science and society* (The Hague: RMNO).

Castells, Manuel, 1996: *The Rise of the Network Society*, vol. 1: *The Information Age: Economy, Society and Culture* (Cambridge, Mass.–Oxford: Blackwell).

Castells, Manuel, 2009: *Communication Power* (Oxford: Oxford University Press).

De Bruijn Hans, 2006: "One fight, one team: the 9/11 commission report on intelligence, fragmentation and information", in: *Public Administration*, 84,2: 267-287.

De Bruijn, Hans, 22007: *Managing Performance in the Public Sector* (London–New York–Melbourne: Routledge).

De Bruijn, Hans; Ten Heuvelhof, Ernst, 1999: "Scientific expertise in complex decision-making processes", in: *Science and Public Policy*, 26,3: 179-184.

De Zeeuw, Aart; In 't Veld, Roel; Van Soest, Daan; Meuleman, Louis; Hoogewoning, Paul, 2008: *Social Cost-Benefit Analyses for Environmental Policy-making* (The Hague: RMNO).

Funtowicz, Silvio; Ravetz, Jerome, 1991: "A new scientific methodology for global environmental issues", in: Constanza, R. (Ed.): *Ecological economics: the science and management of sustainability* (New York: Columbia University Press): 137-152.

Funtowicz, Silvio; Ravetz, Jerome, 1992: "Three types of risk assessment and the emergence of post-normal science", in: Krimsky, Sheldon; Golding, David (Eds): *Social theories of risk* (Westport: Praeger): 211-232.

Funtowicz, Silvio; Ravetz, Jerome, 1993: "Science for the post-normal age", in: *Futures*, 25: 739-755.

Gee, David, 2008: "Costs of Inaction (or Delayed Action) to Reduce Exposures to Hazardous Agents: Some Lessons

from History". Paper to the SCBA Conference in The Hague, 17 January 2008.

Gee, David, 2009: "Evaluating and Communicating Scientific Evidence on Environment and on Health". Presentation to EEAC, 12 June 2009, EEA, Copenhagen.

Gibbons, Marc; Limoges, Camille; Nowotny, Helga.; Schwartzman, Simon, Scott, Peter; Trow, Martin, 1994: *The New Production of Knowledge: The Dynamics of Science and Research in Contemporary Societies* (London–Thousand Oaks–New Delhi: Sage).

Grin, John; Rotmans, Jan; Schot, Johan, 2010: *Transitions to Sustainable Development* (New York: Routledge).

Hajer, Maarten, 2003: "Policy without Polity: Policy Analysis and the Institutional Void", in: *Policy Sciences*, 36,2: 175-195.

Hajer, Maarten, 2005: "Setting the Stage, a Dramaturgy of Policy Deliberation", in: *Administration and Society*, 36: 624-647.

Hajer, Maarten; Wagenaar, Henk, 2003: *Deliberative policy analysis: understanding governance in the network society* (Cambridge: Cambridge University Press).

Hoogeveen, Hans; Verkooijen, Patrick, 2010: *Transforming Sustainable Development Diplomacy* (Wageningen).

Hoppe, Rob, 2008: "Scientific advice and public policy: expert advisers' and policy-makers' discourses on boundary work", in: *Poièsis and Praxis: International Journal of Technology Assessment and Ethics of Science*, 6,3-4: 235-263.

Hoppe, Rob, 2010: *The Governance of Problems. Puzzling, Powering, and Participation* (Bristol: Policy Press).

Hoppe, Rob (in press): "Lost in translation? A boundary work perspective on making climate change governable", in: Driessen, Peter; Leroy, Pieter; Van Viersen, Wim (Eds.): *From Climate Change to Social Change:*

Perspectives on Science-Policy Interactions (London: Earthscan).

In 't Veld, Roeland, 2009: "Towards Knowledge Democracy. Consequences for science, politics and the media". Paper for the international conference Towards Knowledge Democracy, 25-27 August, Leiden.

In 't Veld, Roeland (Ed.), 2000, 2009: *Willingly and knowingly. The roles of knowledge about nature and the environment in policy processes* (The Hague: RMNO).

In 't Veld, Roeland (Ed.), 2001, 2008: *The rehabilitation of Cassandra. A methodological discourse on future research for environmental and spatial policy* (The Hague: WRR/RMNO/NRLO); English version available at: <www.rmno.nl>.

In 't Veld, Roeland; Verhey, Tanja, 2000, 2009: "Willingly and Knowingly: about the relationship between values, knowledge production and use of knowledge in environmental policy", in: In 't Veld, R.J. (Ed.): *Willingly and Knowingly: the roles of knowledge about nature and environment in policy processes* (The Hague: RMNO): 105-145.

In 't Veld, Roeland; Maassen van den Brink, Henriette; Morin, Pierre; Van Rij, Victor; Van der Veen, Hans; et al. (Eds.), 2007: *Horizon Scan Report 2007. Towards a Future Oriented Policy and Knowledge Agenda* (The Hague: COS).

In 't Veld, Roeland, 2010: *Kennisdemocratie* (The Hague, SDU).

In 't Veld, Roeland (Ed.), 2010a: *Knowledge Democracy* (City: Springer Verlag).

In 't Veld, Roeland; de Bruijn, Hans; ten Heuvelhof, Ernst, 2010: *Process management* (Heidelberg: Springer Verlag).

Jasanoff, Sheila, 1990: *The fifth branch: advisers as policy makers* (Cambridge. Mass.: Harvard University Press).

Jasanoff, Sheila, 2003: Technologies of humility: Citizen participation in governing science. Minerva, 41, 223-244.

Jasanoff, Sheila (Ed.), 2004: *States of Knowledge: The Co-Production of Science and Social Order* (London–New York: Routledge).

Jasanoff, Sheila, 2005: *Designs on Nature. Science and Democracy in in Europe and the United States* (Princeton: Princeton University Press).

Jasanoff, Sheila; Martello, Marybeth (Eds.), 2004: *Earthly Politics. Local and Global in Environmental Governance* (Cambridge, Mass.–London: MIT Press).

Kickert, Walter; Koppenjan, Joop; Klijn, Eric-Hans, 1997: *Managing complex networks: strategies for the public sector* (London: Sage).

Lindblom, Charles; Cohen, David, 1979: *Usable knowledge: social science and social problem solving* (New Haven: Yale University Press).

Meuleman, Louis, 2008: *Public management and the metagovernance of hierarchies, networks and markets* (Heidelberg: Springer).

Meuleman, Louis, 2010a: "The Cultural Dimension of Metagovernance: Why Governance Doctrines May Fail", in: *Public Organisation Review*; online, 12 August 2009; <DOI 10.1007/s11115-009-0088-5>.

Meuleman, Louis, 2010b: Climate governance conference, 2010, Yale University, New Haven CT, USA.

Meuleman, Louis, 2011: "Metagoverning governance styles: Increasing the metagovernors' toolbox", in: RMNO cahiers 2009.

Meuleman, Louis; In 't Veld, Roel J., 2009: *Sustainable development and the Governance of Long-term Decisions* (The Hague: RMNO/EEAC).

Meuleman, Louis (Ed.), 2013: *Transgovernance* (Heidelberg: Springer Verlag).

Nowotny, Helga; Scott, Peter; Gibbons, Marc, 2002: *Rethinking science: knowledge and the public in an age of uncertainty* (Cambridge: Polity Press).

Petschow, Ulrich; Rosenau, J.; von Weizsäcker, Ernst Ulrich, 2005: *Governance and Sustainability* (Sheffield: Greenleaf Publishing).

Pohl, Christian; Hirsch Hagedorn, Georg, 2008: "Methodological challenges of transdisciplinary research", in: *Natures Sciences Sociétés*, 16,1: 111-121.

Pollitt, Chris; Bouckaert, Geert, 2000: *Public Management Reform. A comparative analysis* (Oxford: Oxford University Press).

Regeer, Barbara; Mager, Sara; Van Oorsouw, Yvonne, 2011: *Licence to Grow* (Amsterdam, VU University Press).

Scholz, Roland W., 2010: *Environmental literacy in science and society: From knowledge to decision* (Cambridge: Cambridge University Press).

Scholz, Roland W.; Stauffacher, Michael, 2007: "Managing transition in clusters: area development negotiations as a tool for sustaining traditional industries in a Swiss pre-alpine region", in: *Environment and Planning A*, 39: 2518-2539.

Selin, Henrik; Najam, Adil (Ed.), 2011: *Beyond Rio + 20: Governance for a Green Economy* (Boston, Mass.: Boston University).

Schwarz, Michiel; Elffers, Joop, 2011: *Sustainism is the New Modernism* (New York: DAP).

Thompson Klein, Julie; W.; Grossenbacher-Mansuy, Walter; Häberli, Rudolf; Bill, Alain; Scholz, Rolnd W.; Welti, Myrtha (Eds.), 2001: *Transdisciplinarity: Joint problem solving among science, technology, and society* (Basel: Birkhäuser).

Teisman, Geert; Van Buuren, Arwin; Gerrits, Lasse, 2009: *Managing complex governance systems* (New York: Routledge).

Voß, Jan-Peter; Bauknecht, Dierk; Kemp, Rene, 2006: *Reflexive Governance for Sustainable Development* (Cheltenham: Edward Elgar).

WBGU, 2011: *World in Transition – A Social Contract for Sustainability* (Berlin: German Advisory Council on Global Change, July 2011).

Weick, Karl; Sutcliffe, Kathleen, 2001: *Managing the Unexpected: Assuring High Performance in an Age of Complexity* (San Francisco: Jossey-Bass).

Wynne, Brian, 1991: "Knowledges in context: science", in: *Technology, and Human Values*, 16: 111-121.

Wynne, Brian, 1996: "May the Sheep Safely Graze? A Reflexive View of the Expert-Lay Knowledge Divide", in: Lash, Scott; Szerszynski, Läronislaw; Wynne, B. (Eds.): *Risk, Environment and Modernity, Towards a New Ecology* (London–Thousand Oaks–New Delhi: Sage).

Wynne, Brian, 2006: "Public Engagement as a Means of Restoring Public Trust in Science—Hitting the Notes, but Missing the Music?", in: *Community Genetics*, 9: 211-220.

Wynne, Brian, 2007: "Risky delusions: Misunderstanding science and misperforming publics in the GE crops issue", in: Taylor, Shelly; Barrett, Katherine (Eds.): *Genetic engineering: Decision making under uncertainty* (Vancouver: University of British Columbia Press).

41   Discourse and Practice of Transitions in International Policy-making on Resource Efficiency in the EU

Sander Happaerts[1]

Abstract

For a number of years now, 'transitions' has been the new buzzword in international policy-making. The emergence of an international transitions discourse is linked to debates on the green economy and a low carbon society, and is manifested within the UN system as well as in EU discussions. Does the apparent political appeal of a concept that introduces fundamental changes mean more than the superficial use of the word 'transition'? And how far is it also translated into policy? This chapter analyses international policy initiatives from the perspective of sustainability transitions, with the aim of moving towards a better understanding of the potential role of international policy-making in current and future transitions. A case study of the European Commission's policy initiatives on resource efficiency demonstrates a remarkably strong awareness of transitions thinking, consistently integrated into the discourse with the intention of creating a sense of urgency and convincing other actors of the need for fundamental change. Furthermore, transitions thinking has served as the inspiration for a number of principles, goals and instruments, such as the European Resource Efficiency Platform.

Keywords: EU, green economy, international organizations, public policy, resource efficiency, sustainability transitions.

41.1   Introduction

For some years now, 'transitions' has been the new buzzword in international policy and politics. The concept refers to a strand of thinking that advocates fundamental changes in the main societal systems (food, mobility, housing, etc.) in order to overcome persistent problems such as climate change or resource scarcity. It is explicitly linked to the overall goal of sustainable development, and can be considered as a new way of thinking in sustainability science (Dedeurwaerdere 2013). Transitions thinking assumes

that persistent problems, rooted as they are in our prevailing modes of production and consumption, can only be overcome when deep innovations are realized in societal systems ('system innovations'), by means of profound changes to the dominant structures, practices, technologies, policies, lifestyles, etc. (Kemp/Rotmans 2005).

Recently, various gradations of a 'transitions discourse' have been permeating international policy documents (Audet 2014; Happaerts 2014). Such a discourse, that borrows frames and assumptions from the academic transitions literature to promote certain transformative 'pathways' out of contemporary crises, is for instance observed in policy initiatives by international agencies in the framework of the 'green economy' (e.g. OECD 2010; UNCSD 2012; UNEP 2011; 2014). Transitions have also become an essential part of the language of recent EU initiatives in the field of climate policy and related issues. In these EU initiatives, the transitions discourse is strongly linked to the systemic character and the long-term time horizon of the societal transformations needed to overcome per-

---

1   Sander Happaerts (sander.happaerts@kuleuven.be) was until 2014 a research manager in the field of environmental policy and sustainable development at HIVA (Research Institute for Work and Society), KU Leuven. This research was funded by the Flemish government in the framework of the Policy Research Centre on Sustainable Materials Management (SuMMa). The goal of the project was to map and analyse the global and European context of the transition towards sustainable materials management, in order to support the Flemish government's ambitions in that field.

sistent problems (e.g. Council of the EU 2009; European Commission 2011b; EEA 2014).

The emergence of such an international discourse inevitably means that certain actors will appropriate it in order to influence policies and practices. Consequently, Audet (2014: 47) indicates that scholars should focus on how the appropriation of transition in international discourses might influence future transitions. In a similar vein, this chapter starts from the observation that a transitions discourse is emerging at the level of international agencies. The chapter aims at, first, gaining a better understanding of what such a discourse means in specific cases and, second, assessing whether it also influences the policy-making practices of those agencies. This could contribute to an evaluation of the potential role of international policy-making in current and future transitions and, ultimately, to better substantiating the multi-scalarity of transitions (see Hansen/Coenen 2013).

This chapter moves towards better knowledge of the role of international policy-making by conducting in-depth, exploratory research in one case study, proceeding in two steps. First, the discourse of the case is analysed by looking into how sustainability transitions are framed. Policy framing refers to the process of interpreting a concept and of giving meaning to a problem. It involves the use of available knowledge and information in order to select, name, emphasize or organize certain aspects of a policy problem (Daviter 2007; Schön/Rein 1994). In a policy context, framing takes the form of 'problem-setting stories', which are told to persuade others in policy debates, and to shape the preferred policy solutions (Hajer/Versteeg 2005; Mazey/Richardson 1997; Peters/Hoornbeek 2005). The importance of policy framing cannot be overestimated (Hajer 2000: 287). The particular way in which a policy problem is framed has important consequences for the solutions that can be chosen for it and can already imply who will be responsible for that solution, thus limiting the policy choices that actors can make (Hajer/Versteeg 2005: 178; Peters/Hoornbeek 2005: 82-85; Rochefort/Cobb 1994: 4). As a second step, the chapter looks at whether and how the transitions discourse influences policy-making, by studying the precise policy content. In policy analysis, policy content is commonly broken down into different components (Heichel/Pape/Sommerer 2005; Howlett 2009). This chapter focuses on three such components: the stated principles that guide policy-making, the policy goals, and the instruments chosen to attain those goals. The case study is guided by two research questions: how are transitions framed in the

discourse of international policy initiatives, and to what extent does transitions discourse influence policy-making at the level of principles, goals and instruments?

The next section offers theoretical and conceptual background. In order to explore what sort of influence transitions can be expected to have on policy-making, the most significant policy implications of transitions thinking are elaborated. Before turning to the case study, a subsequent section is dedicated to the link between sustainability transitions and the green economy, as it has been observed that the emergence of a transitions discourse in international policy-making is closely related to that issue. Afterwards, the in-depth analysis of one case study is presented, namely, the EU's recent policy initiatives on resource efficiency. Findings and a future research agenda are laid out in a final section.

41.2 Sustainability Transitions and International Policy-making

41.2.1 The Concept of Sustainability Transitions

In the theoretical literature on sustainability transitions (building upon earlier traditions such as innovation and technology studies and complex systems theory), transitions are presented as a new solution to the problems of sustainable development. Those problems are said to be 'persistent', mainly because they are deeply embedded in societal structures, and for this reason are difficult to manage or steer (Loorbach 2007).[2] Examples are climate change, the scarcity of certain resources, and the housing and food needs of a growing world population. Because of their specific characteristics, persistent problems will not be solved by regular or short-term policy-making, incremental institutionalism or market mechanisms (Geels 2010; Meadowcroft 2011: 71). As persistent problems are rooted in existing structures, they require a fundamental change in the dominant institutions, policies and lifestyles that shape societal systems. Such fundamental changes are called *transitions*—or *sustainability transitions* when their fundamental goal of achieving

---

2 The characterization of persistent problems bears strong similarities to the 'wicked problems' defined in the policy literature (Rittel/Webber 1973), but it adds the specific perspective of sustainable development and the different layers of complexity that distinguish that policy issue from others (cf Lafferty 2004: 12).

sustainable development is underlined. The assumption that such fundamental changes are needed distinguishes transitions thinking from other paradigms such as ecological modernization theory, which hold to a more incremental type of change based upon technological innovation and increased efficiency within existing institutions.

Most transition theorists use the *multilevel perspective* (MLP) as a common tool to analyse changes in societal systems. Their research has demonstrated that transitions are the result of the interaction (or 'co-evolution') of phenomena taking place at three levels: the *landscape* (the exogenous societal and material environment that encompasses social, political and economic values and trends, shocks, etc.), the *regime* (the dominant setting of the societal system, containing elements of technology, industry, markets, consumer preferences, science, culture and policy) and the level of *niches* (where new rules, technologies, behaviour and so on are developed that deviate from the regime) (Geels/Schot 2010; Geels 2011). Developments at the landscape level can put pressure on the regime and trigger innovations in niches which ultimately cause the regime to adjust. But many other patterns of change (or inertia) are envisaged (Geels/Schot 2007).

The trajectory of transitions can be idealized as an S-curve with four consecutive phases: *predevelopment* (in which only minor changes occur or are prepared), *take-off* (where the change process gains momentum), *acceleration* (in which the change becomes structural and visible) and *stabilization* (where a new dynamic equilibrium of the societal system is reached) (Rotmans/Loorbach 2010: 126). But not all initiated transitions end in a new dynamic equilibrium; other outcomes are possible too. Authors describe the risks of a *lock-in* (when attention is paid only to making existing structures more efficient, rather than changing them), a *backlash* (when the exclusive focus on one part of the solution causes unexpected side-effects) or a *system breakdown* (when no fundamental changes are initiated and the persistent problems eventually cause the sociotechnical system to collapse) (van der Brugge/Rotmans 2007: 254–56).

41.2.2   Policy Implications of Sustainability Transitions

Transitions have been observed to occur spontaneously. The endeavour to deliberately influence or steer them, usually in a sustainable direction, is called *transition governance* (Paredis 2013). It involves a large-scale societal effort, in which the role of governmental policies is limited but essential. Little has been written about the consequences of sustainability transitions for policies (Paredis 2013).[3] This is surprising, both because sustainability transitions are unlikely to proceed in a desirable direction if regular policies are not at least altered, and because the implementation of a transitions approach will have a great impact on policies. From a study of the transitions literature, at least five major policy implications of sustainability transitions can be deduced.

First, a transitions perspective implies that the policy focus moves away from traditional policy sectors towards the core societal systems that shape contemporary production and consumption patterns. In the literature these are called 'sociotechnical systems'. They are organized around essential societal needs (such as feeding and sheltering people, moving around goods and persons, caring for people, etc.) and consist of three main components. The *structure* is the material infrastructure, the technologies, the different kinds of institutions and the associated economic reality of a sociotechnical system. A system's *culture* contains the principal images, values and paradigms. And the *practices* denote the 'normal' behaviour and routines within a system (Rotmans/Loorbach 2010). To achieve a transition, changes need to happen in all these components. This requires specific consultation and coordination instruments but also demands a systemic approach that goes far beyond the principle of horizontal policy integration advocated by the sustainable development paradigm (Happaerts/Bruyninckx 2014).

Sustainability transitions ultimately aim at fundamental changes within the core sociotechnical systems, which is a significant policy challenge. Not only do changes need to happen at the combined level of structure, culture and practices, they also need to be deep and fundamental (e.g. Loorbach/Lijnis Huffenreuter 2013). This stands in contrast with regular policy (Paredis 2013), which is usually of an incremental nature: policies are usually aimed at making existing systems more efficient or at improving specific aspects of their functioning. However, with system innovation being the purpose of sustainability transitions, an incremental improvement of an incumbent

---

3   Looking at specific approaches to transition governance, the picture is more nuanced. For example, see Loorbach (2007) on the policy implications of Transition Management, and Schot and Geels (2008) for the policy implications of Strategic Niche Management.

system is unlikely to lead to a fundamentally new system. Rather, policies will need to destabilize the existing system and stimulate a different constellation, in order to eventually "replac[e] established ways of doing things, as well as their structural embedment" (Voß/Smith/Grin 2009: 278). But it is not because incrementalism is the wrong approach for transitions that radical revolutions are the only alternative. Of essential importance is the difference between *incremental* change and *gradual* change. A transition can materialize in a stepwise approach, as long as each step, no matter how small, is aimed at destabilizing the incumbent regime and constructing a new one (or, in less radical language, fundamentally changing that regime). The reason is that transition policies must avoid a lock-in. Regular, incremental policies are based on the pre-established structure, culture and practices of the regime—including infrastructure, financial mechanisms, and networks of people and organizations—and continuously reinforce them, making innovation ever more difficult and thus 'trapping' society in suboptimal outcomes (Meadowcroft 2009: 329). According to Geels (2013), policies and institutional changes in support of transitions are especially required in the take-off phase, when fundamental changes have been prepared in niches and need to be supported and scaled up.

Third, one of the essential features of transitions thinking is its long-term perspective. Again, this clashes with regular policies, which have a short-term orientation (Loorbach 2007). Similarly, in policy studies only scarce attention has been given to long-term policy design (Voß/Smith/Grin 2009). Rather, experts in policy studies have described why long-term policies are rare. Because of electoral cycles and uncertainty, among other things, it is assumed that long-term policies are difficult to implement in democratic societies (EEA 2011; Hendriks 2009: 334; Lempert/Popper/Min et al. 2009; Sprinz 2012). The popularity of long-term planning has, however, been increasing again (Voß/Smith/Grin 2009). Especially within the field of environmental policy and policy analysis, reflections can be found on how to tackle long-term policy challenges such as climate change (Sprinz 2009). Knowledge of different policy designs and of facilitating and impeding factors is growing (EEA 2011; Lempert/Scheffran/Sprinz 2009; Underdal 2010). Some years ago the European Commission issued a number of 'road maps' in climate policy and related areas, that define 2050 as their long-term time horizon (see box 41.1). These initiatives are punctuated with a transitions discourse (see section 40.1). It

is thus argued that the renewed popularity of long-term policy planning is associated with the salience of a transitions discourse. Similarly, Voß, Smith and Grin (2009: 294) claim that the transitions framework offers opportunities for long-term policy design, and Meadowcroft (2009: 325) mentions the adoption of a long-term perspective as one of the most promising features of transitions thinking. As change processes leading from one system state to an entirely new system state, transitions typically take a generation or longer (Loorbach/Frantzeskaki/Thissen 2011: 76). If one wants to manipulate the direction or the sustainable character of such transitions, a vision is needed of what the intended system state looks like. Moreover, transition paths to achieve a systemic change should be evaluated against that long-term vision. If not, they are likely to lead to a lock-in of the system. The design of a long-term policy should thus structure the current, short-term and mid-term activities of political actors (Voß/Smith/Grin 2009: 291).

A fourth policy implication is the 'reflexive' character of long-term transition policies, as opposed to a rational and unidirectional conceptualization of planning (Voß/Smith/Grin 2009). Reflexive governance processes accept the uncertainty of policy choices and their effects. A reflexive perspective furthermore sees a constant interaction between policies and broader societal changes (Voß/Smith/Grin 2009). The different policy focus that emerges as a consequence of this reflexivity also assumes a more reflexive view on policy monitoring. Transition policies therefore accept experimenting and learning as principal drivers of change, as systematic learning activities should continuously lead to policy adjustments (Bussels/Happaerts/Bruyninckx 2013).

Making policy processes more reflexive is not necessarily the exclusive task of governments. Other actors, such as research institutes or civil society organizations should also contribute to knowledge-building and reflexivity (Meadowcroft 2009: 336). This points to a more general policy implication of transitions, the involvement of a broad group of stakeholders. Not only do policies account for merely a fraction of transition governance (see above), but the paradigm also demands that such governmental policies are developed in close cooperation with different types of non-governmental stakeholders, in the participatory tradition of sustainable development (Meadowcroft 2009: 325; van der Brugge/Rotmans 2007: 261). This involvement is not limited to the implementation phase but should be incorporated in policy formulation and during the formation of long-term

---

Box 41.1: Transitions in the European Commission's Roadmaps. **Source**: The author.

In 2011, the European Commission published a number of 'road maps' in the environmental domain. They defined 2050 as a long-term time horizon, something new in EU policy-making. The overview of the road maps given below demonstrates how this policy innovation followed on from the EU's climate agenda.

- In order to reach the global goal of limiting global warming to 2° Celsius, the European Council decided in October 2009 that the EU should aim for an 80 to 95 per cent reduction of greenhouse gases by 2050 (compared to 1990 levels). This decision supplemented the previously agreed EU climate and energy objectives for 2020 with a long-term dimension. The Commission's newly created Directorate-General (DG) Climate Action then elaborated the long-term goal by issuing the *Roadmap for Moving to a Competitive Low Carbon Economy in* 2050. It proposed a way of sharing efforts to reduce emissions among different sectors and policy domains between 2011 and 2050 (European Commission 2011b).

- Other DGs followed with similar road maps for the same time horizon. DG Environment prepared the *Roadmap to a Resource Efficient Europe*, which is the main case study of this chapter (European Commission 2011a).

- DG Energy drafted the *Energy Roadmap* 2050, in which different scenarios for Europe's energy use are

elaborated based upon different assumption of decarbonization, and thus only partly based on DG Climate Action's road map (European Commission 2011d).

These road maps are not legally binding decisions or legislative proposals. They appear under the form of Communications, non-binding documents adopted by the European Commission without a Treaty basis. They are intended to communicate the Commission's position on a particular issue to the other institutions, the public and other stakeholders, and as such are not different from documents issued by other international agencies. Communications also promote the policy options that the Commission considers desirable to deal with that issue (Krämer 2012). With these documents, the Commission actually anticipates legislative proposals that it might formulate in following years. Communications are thus meant to develop societal and political support for its intentions with regard to future policies (Delreux/Happaerts 2016).

Although not always bearing the name of 'road map', other Commission Communications adopted in the same period have the same characteristics. For example, the *White Paper on Transport* also looks to 2050 and refers to long-term climate targets (European Commission 2011e). Likewise, the EU's *Seventh Environment Action Programme* frames the EU's environmental challenges in a 2050 vision (European Commission 2012c).

---

visions that will guide transition policies. The difficulty of this final implication for policy-making is illustrated by the critiques of the democratic deficit of specific transition processes (Hendriks 2009).

41.2.3   Transitions at an International Scale

It has been observed by previous scholars that a significant part of the transitions literature has a 'bottom-up bias': a built-in tendency to focus on bottom-up dynamics, on local experiments, or on processes taking place within economic or technological niches (Jørgensen 2012: 999; Lauridsen/Jørgensen 2010: 487). This bias is currently being challenged by a growing field of studies on the geography of transitions looking more closely at the spatial context of transitions and at the interactions between multiple scales (Hansen/Coenen 2013; Raven/Schot/Berkhout 2012), but it is still in need of remedy. One of the remaining issues is that few analyses of transition governance pay substantive attention to international politics and policy (exceptions include Lauridsen/Jørgensen 2010; Nilsson 2012; van der Brugge/Rotmans 2007). Most case studies use a national or subnational level of observation (Raven/Schot/Berkhout 2012). If they do include

international policy-making in their analyses, it is often underexposed or too easily categorized as an exogenous (stimulating or inhibiting) factor or as a 'kitchen sink' variable of change (Happaerts/Bruyninckx 2012).

Transitions are nevertheless essentially international phenomena. Indeed, the persistent problems of sustainable development have a transboundary character and, as a consequence, the systemic changes that are needed to overcome them are international challenges, manifested at a regional or even a global scale. While niches can be localized (Hansen/Coenen 2013), the sociotechnical systems and the patterns of production and consumption that shape them are increasingly globalized. Transition studies therefore need to take account of the globalized character of economic, technological and regulatory structures (Meadowcroft 2005: 490–91). At the same time, the literature's lack of attention to international policy initiatives stands in contrast with the burgeoning transitions discourse that permeates those initiatives (see section 41.1), making them additionally interesting to investigate from a transitions perspective.

The transitions literature is thus in need of, on the one hand, a better conceptualization of the international scale and, on the other, more attention to international policy initiatives (both by adopting an international level of observation and by involving such initiatives more systematically in case studies in general). This chapter hopes to contribute to that second challenge by conducting an exploratory case study of one particular steering initiative. The intention of the case study is to increase the scholarly community's understanding of the potential role of international policy-making in transitions, which, ultimately, should also bring us closer to a better conceptualization of the international scale in transition studies.

41.3   Sustainability Transitions and the Green Economy

Many of the instances where a transitions discourse can be seen emerging are international initiatives in the context of the green economy (Audet 2014). This is a concept that became popular in the sustainable development debate of the end of the 2000s (Happaerts/Bruyninckx 2014), and especially after the eruption of the global financial and economic crisis in 2008. In an attempt to redefine the crisis as an opportunity for sustainable development, UNEP's Green Economy Report proposed investing heavily in the greening process of a number of key sectors, in order to create jobs, eradicate poverty and support the transition to a low carbon, resource-efficient and socially inclusive economy (UNEP 2011). Other actors launched similar proposals, such as the OECD's Green Growth strategy (OECD 2011). A number of studies have analysed the content of different green economy proposals (Althaus 2013; Bina 2013; Ferguson 2014; Jänicke 2012). When compared to sustainable development—the concept that dominated environmental and development debates in previous decades—the green economy at first glance seems to offer some opportunities for a bigger impact on policy. It has a certain appeal for governments, because it justifies the priority they intuitively accord to economic development. Optimistically, it could thus be more easily integrated into governments' main economic policies and lead to concrete improvements (Ferguson 2014). But, on the other hand, a number of the important policy principles of sustainable development appear to have been overlooked by applications of the green economy concept, such as North-South equity and intergenerational solidarity (Althaus 2013; Onestini 2012). Another risk is that a purely instrumental view of the environment is adopted, as a 'capital

stock' which offers resources and ecosystem services, and which absorbs waste (Happaerts/Bruyninckx 2014).

Although the choice of the green economy as one of the main themes of the Rio+20 conference (the UN Conference on Sustainable Development held in 2012 in Rio de Janeiro, Brazil) appeared to denote a consensus among states to put the economy at the centre of attention at a time of global economic crisis, some transition scholars invoke that same crisis as a favourable occasion for sustainability transitions. Loorbach and Lijnis Huffenreuter (2013) claim that the landscape shocks brought about by that crisis inevitably led to structural systemic changes. Those changes, of course, do not necessarily head in a sustainable direction. That is why the green economy should be seized as an opportunity to propel sustainability transitions (Loorbach/Lijnis Huffenreuter 2013). Geels (2013) agrees, although he deduces from public discourse and investments that the window of opportunity for sustainability transitions is by now much smaller than at the beginning of the crisis, so before the green economy had gained its most popularity.

If the crisis is not seized as an opportunity for sustainability transitions, it could be because behind the green economy, despite its transitions discourse, are ideas closer to ecological modernization theory than to transitions thinking. Indeed, the emphasis on technological innovation as a driver of change and on the decoupling of economic growth from environmental impact (UNEP 2014), to name just two examples, corresponds with the key assumptions of ecological modernization (cf Bina 2013). One of the core differences between the two theoretical paradigms is the depth of change envisaged. The path to a green economy promoted by international organizations such as UNEP and OECD is situated somewhere between radical change and business as usual (Bina 2013: 1033-34), and embraces the idea of transitions while steering clear of its fundamental ambitions. Bina (2013) demonstrates "an uneasy balance" between wanting to change and effectively reconfirming the incumbent sociotechnical regimes, as green economy policies target incremental, technology-driven changes towards a more efficient and cleaner economy. Moreover, the dominant economic paradigm, with its focus on sustained growth, is not questioned, while it is acknowledged that the market mechanisms, so prominent in that paradigm, need fixing (Bina 2013: 1034-37). Based upon these horizontal findings, the transitions discourse of the green economy at best suggests tentative support for transitions in international policy-making.

But more detailed analysis, for which the case study presented in this chapter is a first step, is needed in this field.

41.4   The Case of Resource Efficiency in the EU

The case study that was selected for this study consists of a recent steering initiative at the level of the EU in the area of resource efficiency. Although the EU is not a global organization, non-legislative initiatives by the European Commission can be taken as a case of policy-making initiatives by international agencies that potentially play a role in stimulating transitions and influencing other levels of governance and other scales of societal action. In other words, although the EU is hardly comparable to other international organizations and obviously has a direct influence on national policy-making in various ways, it also produces a body of soft law, general guidance and other types of steering initiatives (e.g. Commission Communications, see box 41.1) which have a potential impact equal to that of the OECD, the IEA, UN bodies or other international agencies. The initiative that serves as our case study, the *Roadmap to a Resource Efficient Europe*, falls in that category, and touches upon a topic that is extremely relevant for sustainability transitions.

The international transitions discourse is manifested in a number of thematic areas, including natural resource use (e.g. UNEP 2014). Resources are central to the major global production and consumption patterns, which are at the core of many persistent problems of sustainability. As the extraction, production and consumption of materials are necessary for the provision of buildings, transport, energy and food, materials have a vital link to the main sociotechnical systems. At the global level, initiatives for a more sustainable use of natural resources have tried to address the multiple environmental impacts that materials generate throughout their life cycle. UNEP, the OECD and G8 have taken the issue to heart. A horizontal analysis indicates that those international initiatives were boosted by two recent events (Happaerts 2014: 32–33). First, sharp rises in commodity prices in the second half of the 2000s, a phenomenon closely linked to the outbreak of the economic crisis, pitchforked the issue of raw materials and their perceived scarcity to the highest ranks of geopolitics. It fundamentally heightened the level of governmental intervention concerning raw materials, and signified a

break with an era in which the efficient or more sustainable use of natural resources was purely an environmental affair. Second, the economic advantages of sustainable resource use are prominently promoted in green economy strategies (see above), as it offers both ecological gains and economic savings in win-win scenarios for various sectors

In this context, the concept of 'resource efficiency' was endorsed by several actors in 2010 and gave a new impetus to the debate (Happaerts 2014). This concept attempts to link the economic and geopolitical concerns about commodity prices and resource scarcity with an environmentally motivated concern for the responsible use of natural resources. At the same time, it presents itself as an integrating concept: a more efficient use of resources allows connections with a broad spectrum of environmental problems (such as climate change, biodiversity and waste) and provides an effective solution to several forms of environmental impact (such as air, soil and water pollution).

Also in the EU, resource efficiency became a central policy concept. While the EU has been taking action on natural resources since the beginning of the 2000s, mainly from an environmental policy point of view (with waste and product policies as the predecessors of a genuine resource policy), the economic crisis brought the EU to focus on the economic governance of raw materials. Subsequently, the defining moment for resource efficiency was the formulation of seven 'flagship initiatives' within Europe 2020 (European Commission 2011f), the EU's main socio-economic strategy and the successor of the Lisbon Strategy. Resource efficiency is now the number one priority of the EU's environmental policy (Potočnik 2010, 2012). Moreover, with its inclusion in Europe 2020, it achieved a political position in the EU unprecedented by any other initiative proposed by DG Environment (see box 41.2). The Commission first issued a general Communication about the flagship initiative (European Commission 2011f), and then established an interdepartmental task force led by DG Environment to draft the *Roadmap to a Resource Efficient Europe* (European Commission 2011a).

The *Roadmap*, its accompanying documents, and the EU discourse on the *Roadmap* form the case study of this chapter. The analysis is based on a thorough document analysis, on nine interviews with European and national policy officials and on non-participant observation at two informal workshops of EU member states on resource efficiency (April 2012, Brussels and November 2012, Berlin).

Box 41.2: Resource efficiency and Europe **2020**: the back story. **Source**: The author.

The development of Europe 2020 coincided with the start of the second Barroso Commission and with a significant institutional shift in EU environmental policy-making. Since 2010, climate change has come under the newly created DG Climate Action within the European Commission (Schoenefeld 2014). Previously, climate change was arguably the chief policy-making issue of DG Environment, for instance because the EU took a leadership role in global climate change discussions until 2009 (Bäckstrand/Elgström 2013). As a consequence, when Janez Potočnik took office as the new European Commissioner for the Environment in 2010, he had to look for a new 'grand story' if he wanted to play a prominent role within the Commission,

after his 'loss' of climate change. It is suggested that Potočnik connected several policy streams to create a window of opportunity for 'resource efficiency' to become a major policy priority of the European Commission (Happaerts 2014). Those streams were the existing waste and product policies of the EU, in which natural resources had become increasingly prominent since the early 2000s, and the focus on raw materials within the industrial policy. Potočnik was explicitly looking for an issue that had the chance to weigh on the EU's economic agenda in times of crisis. Linked to the perception of scarcity of raw materials and to the problem of European competitiveness, resource efficiency was such an issue.

41.4.1   Policy Framing

Following the first research question, this chapter assesses how and under which conditions the concept 'transitions' is used in the discourse of the *Roadmap*, and the *Roadmap*'s policy framing is studied. The three main European institutions frequently use the word 'transition' in the context of the *Roadmap*. The concept either stands alone or appears in combination with 'resource efficiency' (Council of the EU 2011; European Commission 2010, 2011f, 2011a; European Parliament 2012), 'green economy' (Council of the EU 2011: 4; European Commission 2011a: 3, 8, 22), or 'low carbon economy' (European Commission 2011c: 11, 22; 2011a: 8, 14). The Commission defines a transition as "a fundamental transformation within a generation—in energy, industry, agriculture, fisheries and transport systems, and in producer and consumer behaviour" (European Commission 2011a: 2), and the terms 'transition' and 'transformation' are used interchangeably in the policy discourse (European Commission 2011c: 24; 2011a: 4). The *Roadmap* systematically stresses its long-term perspective and that it will take a generation to achieve the transition (European Commission 2011a: 2).

Sociotechnical systems and system innovation are also referred to in the EU's resource efficiency discourse. The Commission calls for "a significant transition in energy, industrial, agricultural and transport systems" (European Commission 2011f: 3). The Council states that the shift towards resource efficiency "will require, in addition to technological innovation, innovation at the level of our socio-economic system, i.e. new governance models, new business and education models, new consumption patterns, and lifestyles geared towards the sustainable management of resources" (Council of the EU 2011: 5). Awareness of

the other possible outcomes of a transition trajectory is equally manifested in the discourse. The Commission adopts the explicit wording of "lock-ins" of socio-technical systems when it refers to market distortions, to Europe's technological and knowledge systems, to prevailing institutions or to dominant production and consumption patterns (European Commission 2011c: 20-23; 2011g: 14-16, 18, 28, 34, 54, 56; 2011a: 4, 10). In a single instance, the Commission warns against the risk of "systemic collapse" when current resource use persists (European Commission 2011c: 6).

Another manifestation of the transition discourse is the explicit reference to transition governance. The *Roadmap* expresses the ambition to "[p]repar[e] th[e] transformation in a timely, predictable and controlled manner" (European Commission 2011a: 2) and mentions the use of 'transition paths' (European Commission 2010). Referring to the historical development of transition governance in the Netherlands (see above), the Commission advances the example of the "Dutch government['s] … 'transition management' as national policy" (European Commission 2011c: 28).[4]

Although the Commission defines a transition as a fundamental transformation, and although interviewees stress the Commission's awareness that incremental changes are not enough for a transition, it is unclear from the discourse how fundamental the proposed

----

4   The concept of sustainability transitions and the ensuing 'transition management' approach were developed in the Netherlands in the context of the Fourth National Environmental Policy Plan (Ministerie van VROM 2001). As a policy concept, it was afterwards diffused to other jurisdictions, with the Flanders region in Belgium as the first testing ground for transition management (Paredis 2008).

changes are considered to be, as Commission documents have a tendency to be cautious and avoid radical wording. This is illustrated by the references to the concept of decoupling, an important principle of the *Roadmap*. While the goal of decoupling economic growth from resource use is frequently mentioned (European Commission 2011f: 5; 2011a: 2; European Parliament 2012: §7), it is unclear whether the aim is to achieve absolute decoupling (i.e. economic growth continues while resource use decreases) or relative decoupling (i.e. resource use continues to increase, but less intensively than economic growth). The Commission only mentions absolute decoupling in a staff working document that contains background analyses and trends of resource use (see also European Commission 2010, 2011g: 96), but interviews demonstrate that a lack of consensus within the Commission prevents the explicit promotion of absolute decoupling of resource use. Although it is never formally confirmed, the Commission's discourse on 'efficiency' and 'doing more with less' goes much more in the direction of relative decoupling. The Council of Environment Ministers is in that regard more ambitious than the Commission, when it calls for "absolute decoupling of growth from resource use and negative environmental impacts" (Council of the EU 2011: 6; see also Council of the EU 2012: 4, 6).

In the policy framing of resource efficiency, the EU's resource problem is very clearly an economic one. The *Roadmap* stresses different forms of scarcity that follow a rising demand for a number of resources at a global scale and a volatility of resource prices (European Commission 2011c: 3-4; 2011f: 2; 2011a: 2-3). Besides global economic trends, the *Roadmap* deals with market distortions or systemic construction errors within Europe. These are situated in three domains. First, the EU addresses the problem of externalities (such as pollution or waste) that are not reflected in the price of resources. As a consequence, signals of scarcity are not perceived, and unsustainable exploitation and inefficient consumption are promoted (European Commission 2011c: 20; 2011a: 2, 4-5, 9-11, 23). Second, the *Roadmap* deprecates the exclusive orientation of financial markets towards short-term interests, whereas the long-term gains of resource efficiency are not sufficiently rewarded (European Commission 2011c: 19; 2011a: 20). In a working document of the European Parliament, it is argued that this "vested interest focussed on short-term economic gains needs to be defied" (European Parliament 2011). Third, a related distortion is the unfamiliarity of investors with the returns of resource

efficiency, which functions as an obstacle for sustainable investments (European Commission 2011a: 9, 12, 20). Besides an economic problem, the transition towards resource efficiency is also presented as a policy problem, and explained as the result of policy failures and inconsistencies, e.g. inefficient subsidies (European Commission 2011c: 20-22; 2011a: 9-11; Potočnik 2011). Finally, although the *Roadmap* is an initiative of DG Environment, environmental problems take a back seat in the framing of economic policy. Only limited attention is given to the pressure of the EU's extensive resource use on ecosystems and to its contribution to climate change (European Commission 2011c: 4-6; 2011a: 2). The same goes for social and demographic changes (European Commission 2011c: 2). Moreover, when problems related to environment, energy or health are invoked, they are framed as economic problems and expressed in, for instance, the loss of working days (e.g. European Commission 2011a: 14).

The solution to the resource problem is also framed as an economic story. The EU sees a significant role for itself in what it calls "transforming the economy" (European Commission 2011a: 23). The *Roadmap* proposes a number of adjustments that can be made in the system's policy framework (of which the EU controls a significant part) in order to eliminate the distortions and construction errors of the market framework. The solutions for a new policy framework mostly revolve around the principle of 'getting the prices right'. Policies must create economic opportunities, offer the right incentives for investments, reward innovation and resource efficiency, and ensure security of supply. The way to do this is on the one hand by implementing policies that promote product redesign and increased reuse, recycling and substitution of materials, and on the other by improving coherence among existing policies which "shape our economy and our lifestyles" (European Commission 2011a: 2). An improved policy framework, in short, offers long-term certainty to businesses and investors about the future policy direction of Europe, which is aimed at resource efficiency. The Commission systematically emphasizes the economic advantages of resource efficiency, such as less dependence on imports and increased competitiveness (European Commission 2011a: 4, 6, 8, 10-11, 19-20), more growth, profit and jobs (European Commission 2011a: 2, 4-5, 8, 10, 20), opportunities for businesses to capitalize on the commercialization of their innovations and to use waste as a resource, and savings for consumers (European Commission 2011a:

2, 4-6). With all these economic benefits, the *Roadmap to a Resource Efficient Europe* is presented as a guidebook to a way out of the economic crisis.

To conclude, the appearance of a discourse on transitions in the EU's resource efficiency policy is more than the sporadic mention of the word 'transition', but a systematic use of several concepts central to transitions thinking. Interviews confirm that some of those who wielded the pen in the writing of the *Roadmap* (including Dutch and Flemish EU officials, see footnote 4) are very aware of the framework of sustainability transitions, and deliberately positioned the EU's efforts for resource efficiency within that framework.[5] They are conscious of the importance of such a discourse and indicate that they see it as an important strategy for creating awareness and pushing for a sense of urgency and the idea that fundamental changes are necessary. Furthermore, the *Roadmap*'s framing of economic policy suggests a firm belief in the market's ability to steer itself, provided that policies offer the right incentives. The EU's confidence in the functioning of improved market mechanisms should come as no surprise and can partly be explained by the fact that the EU's weight in and influence on national policies is strongest with regard to the internal market. In several domains, the European Commission tends to invoke internal market tools as the legal basis of policy proposals (e.g. Pollack/Shaffer 2005: 331). As early as the recession of the late 1970s, the EU's environmental policies were framed as economic measures (Delreux/Happaerts 2016, Lenschow 2005: 307, 12).

41.4.2   Principles

Following the second research question, the case study now assesses to what extent transitions thinking permeates the content of the EU's resource efficiency policy, measured at the level of principles, goals and instruments. An analysis of the *Roadmap* shows that a number of principles that are rooted in transitions thinking are strongly embedded in the policy initiative. Those include a fundamental observation about the unsustainability of our current system, a systemic approach, a long-term approach and the new paradigm of a circular economy. However, a number of

---

5   The interviewees also indicate that those same officials intended to integrate the discourse on transitions in the 2005 Thematic Strategy on the Sustainable Use of Natural Resources (European Commission 2005), but that it was impossible at that time.

other principles that give shape to the *Roadmap*—decoupling, getting the prices right, monitoring and participation—appear focused on the optimization of the sociotechnical system instead of on a fundamental renewal of it.

As a first fundamental principle, the *Roadmap* departs from the observation that current patterns of resource use are unsustainable (European Commission 2011a: 2), thus implying that business as usual is not an option. The *Roadmap* then advances a systemic approach to suggest solutions to alter the EU's resource use. It does so by consistently coupling resource efficiency to other major domains and European policies (e.g. energy, agriculture, industry, transport). Even though concrete measures in those domains are not always elaborated, the *Roadmap* proposes to take land use impacts into account in all policies, to move from energy-efficient policies (e.g. for buildings) to resource-efficient policies, and to integrate resource efficiency into the European Semester, among other examples. Besides its systemic approach, the long-term approach embedded in the *Roadmap* also fits well with a transitions perspective. The strategy is built on the idea that current practices must be aligned with the long-term gains of sustainable resource use, even though that means short-term investments and costs incurred by or clashes with normal market practices (European Commission 2011a: 6, 20). The most visible expression is the formulation of the vision for 2050. A final principle that highlights a systemic transition is the idea of a circular economy. Although not a new concept, this has recently become a popular buzzword in the governance of resources (e.g. Ellen MacArthur Foundation 2012) and is increasingly used by the EU to frame the transition towards sustainable materials management (European Commission 2012b, 2014, 2015). The principle is used in the *Roadmap* to advance certain ideas and approaches that suggest a move towards a new materials system based on closed material loops, e.g. turning waste into a resource or adopting life-cycle approaches.

The *Roadmap* rests on four other principles that suggest a change in the EU's resource use, but which appear to avoid rather than support a fundamental renewal of the system. The first one is decoupling. As stated above, the principle is interpreted as relative decoupling, meaning that the main focus is on efficiency and that decreasing resource use is a secondary issue. The second principle is 'getting the prices right', the dominant image of the economic policy framing of the *Roadmap* (see above). It is linked to the idea that consumers will make more resource-efficient choices if they are properly informed and if taxes and product prices reflect the real environmental costs and the true value of natural capital. The two principles of decoupling and correct prices are oriented towards optimizing the current system of resource use, which according to the EU is characterized by market failures, rather than towards a real innovation of that system. Another principle that is strongly present in the *Roadmap* is monitoring. While transitions thinking promotes a reflexive form of monitoring, the *Roadmap* focuses narrowly on the development of indicators to measure the EU's progress towards resource productivity. A similar assessment is made with regard to participation, a final principle of the *Roadmap*. While the EU strongly invests in the involvement of all relevant stakeholders (member states, business, civil society, etc.), this participation is only invoked for the development of the resource efficiency indicators, a very narrow aspect of the governance of the initiative.

In conclusion, a number of principles suggest that the EU makes an attempt at intervening in the culture of the sociotechnical system, i.e. the prevailing images, values and paradigms of resource use in the EU. A point of departure assuming that the current system is unsustainable, the emphasis on a long-term, systemic approach and the new paradigm of a circular economy are clearly intended to achieve a necessary "shift in mindset" (European Commission 2011a: 20). However, the elaboration of some other principles, such as decoupling and the internalization of external costs, shows that the *Roadmap* tends to aim at an optimization of the current system, i.e. making current resource use much more efficient, rather than striving for a fundamental renewal of that system.

41.4.3   Goals

As a Commission Communication, the *Roadmap to a Resource Efficient Europe* sets out policy goals that are aimed at shaping EU policies in the years to come (see box 41.1). In the *Roadmap*, goals are formulated at three levels: long-term (2050), medium-term (2020) and short-term. The long-term vision contains the EU's long-term goals for resource efficiency in a condensed form (European Commission 2011a: 3). It is relatively vague and certainly not ambitious enough for some, but its main merit is that it exists, and thus integrates a long-term perspective into the policy process. The *Roadmap* is the only one of Europe 2020's flagships that looks beyond 2020, but it joins a recent

trend of other Commission road maps that depart from a long-term vision (see box 41.1).

The vision is translated into eighteen 'milestones' or mid-term goals to be reached by 2020. The milestones relate to the 'key issues' and governance aspects identified by the *Roadmap*.[6] Some correspond to 'hard goals' (e.g. environmentally harmful subsidies are phased out by 2020) or figured targets (e.g. water abstraction remains below twenty per cent of available resources), while others remain very vague (e.g. transport will use less and cleaner energy). Interviewees reveal that most milestones were formulated according to different resources rather than sectors to avoid creating the perception that DG Environment would encroach upon other domains (e.g. agriculture, fisheries, transport, construction, etc.). Nevertheless, the Commission found that the milestones were the object of the bulk of the criticism that the *Roadmap* received from member states and other actors, as they were too vague for some or too ambitious for others.

Short-term actions are formulated for each milestone. Those actions, around a hundred in total, are proposals for what either the Commission or the member states (or a combination of the two) will or should do between 2012 and 2020. About half of the actions involve the implementation, evaluation or (partial) improvement of existing policies, while only a fifth deal with new policy measures (the rest consisting of vague goals related to research or the promotion of certain issues). The Commission defends the inclusion of so many actions related to existing policies, arguing that many subsectors already have targets, and that the actions formulated by the *Roadmap* can help those targets to be reached by exploiting synergies between subsystems (European Commission 2011c: 24). Interviews show that the formulation of the actions was even harder than the definition of the milestones, which is why they are the most similar to incremental policy-making.

---

6 These are: 'sustainable consumption and production' (two separate milestones), 'turning waste into a resource', 'supporting research and innovation', 'environmentally harmful subsidies and getting the prices right' (containing one milestone on subsidies and one on taxation), 'ecosystem services', 'biodiversity', 'water', 'air', 'land and soils', 'marine resources', 'addressing food', 'improving buildings', 'ensuring efficient mobility', 'new pathways to action on resource efficiency', 'supporting resource efficiency internationally' and 'improving the delivery of benefits from EU environmental measures'.

The analysis of the policy goals reveals that, following the strong discourse on transitions in the *Roadmap*, the long-term and systemic approach is translated into a long-term vision for a resource-efficient Europe and a number of strategic goals covering a wide array of fields important for resource use. However, the more concrete and immediate the policy goals become, the more they deviate from fundamental transformations and suggest an approach based on incrementalism and improved business as usual. The *Roadmap* thus demonstrates a firm sense of urgency but little willingness to make fundamental choices right away.

41.4.4   Instruments

As a broad strategy proposed by the Commission, the *Roadmap* does not create radically new instruments. It does, however, lay out a spectrum of governance strategies that should be used to achieve a more resource-efficient Europe. Those strategies include addressing markets and prices, promoting new business models and production patterns, stimulating innovation, integrating life-cycle thinking, adequately informing consumers, boosting knowledge and information-sharing, and developing indicators.

Three types of instruments are particularly promoted. First, the Commission attributes a significant role to the instruments pertaining to the internal market and the EU's new competences with regard to 'economic governance'. Exactly those instruments give the EU the "sufficient scale of influence" needed for the transition (European Commission 2010). For instance, the Commission wants to integrate resource efficiency objectives into the European Semester (European Commission 2011a: 11, 19, 21). Its preference for using market-based instruments is explained by the weight of the EU's internal market tools (see above).

A second type is information instruments. This refers to the work on indicators, but also to the proposed measures with regard to product footprinting and other informational tools (e.g. European Commission 2011a: 7). The preference for information instruments demonstrates that the Commission realizes that it is the best placed institution for the monitoring of the transition towards resource efficiency in the EU, and that it can partially influence the transition by developing indicators. The focus on indicators can also be explained by the common criticism of the Lisbon Strategy, Europe 2020's predecessor, concerning the lack of concrete targets and indicators (Natali 2010; Steurer/Berger 2011). This time, the Commis-

sion wants to actively monitor the progress of Europe 2020. Furthermore, some authors have stated previously that working on indicators is popular because it is less threatening than actually intervening for change (Kemp/Parto/Gibson 2005: 21).

While the formulation of the *Roadmap* rested on unprecedented intra-Commission consultation, a third type of instrument is used to encourage the cooperation of specific stakeholders during the implementation process. Several of the short-term actions call for the establishment of platforms and partnerships for specific topics. The most prominent proposal was the launch of an *EU Resource Efficiency Transition Platform* (European Commission 2011a: 21). This is an explicit reference to the 'transition arenas' that the Dutch government used for its transition policy (European Commission 2011c: 28). The platform was formally established in June 2012, under the name *European Resource Efficiency Platform* (EREP).[7] It was a high-level consultation group, composed of Commissioners, MEPs, environment ministers, international organizations and representatives of different stakeholder groups, that presented recommendations, after a series of meetings, on the implementation of the *Roadmap* and on further action. These recommendations focused, for instance, on setting targets for resource productivity, promoting non-financial company reporting, phasing out environmentally harmful subsidies, and developing instruments for SMEs (EREP 2013, 2014).

The *Roadmap* thus contains at least one instrument that is explicitly derived from transitions thinking. But in general, the *Roadmap* mostly relies on market-based instruments and on information instruments. Rather than developing new instruments, it refers to a significant degree to already existing policies and measures, thus supporting an incremental approach.

41.5   Conclusions

The strong transitions discourse in the EU's resource efficiency policy, a case of a broader trend of emerging transitions thinking in international policy initiatives for the green economy, suggests the political appeal of a concept that demands fundamental

---

7   According to interviews, the choice was eventually made to give the platform a shorter name for communication purposes, and because the concept of 'transition arenas' is not well known in government circles apart from in the Netherlands and Belgium.

changes in our way of life. Indeed, the case study points out that the adoption of a transitions discourse was a deliberate choice by policy officials, who saw in transitions thinking the appropriate framework to strengthen the awareness that deeper societal changes are needed to achieve a shift towards a more sustainable use of natural resources. In the broader context of international green economy policies, it is hypothesized that the attraction of transitions thinking lies in its strong wording of deep change, in its comprehensive systems approach and in its focus on a number of principles and assumptions that differentiate regular policy-making from a more radical approach. However, the analysis of the case's policy framing indicates that the choice for an economic framing limits the scope of policy choices, focusing almost exclusively on improved market dynamics. The case also confirms the tendency to adopt a very instrumental view of the environment, as was observed with regard to green economy policies.

The analysis furthermore points out that the EU's resource efficiency policy experiences a real impact from transitions thinking, demonstrating that the adoption of the transitions discourse is more than a change of terminology. A number of policy components informed by this discourse are in line with transitions thinking. The long-term horizon is the most significant manifestation of such influence, and could be seen as a very meaningful policy innovation in the EU. However, while these policy components are symptoms of a significant change in the mindset of EU policy-making circles, the analysis confirms that the changes envisaged fall short of a true transition. Confirming previous observations concerning international green economy policies, the *Roadmap* seems situated between radical change and business as usual.

In conclusion, the underlying theoretical assumption about the nature of change in the policy initiative analysed is not fully in line with transitions thinking. The adoption of a long-term, systemic perspective and the awareness of the need for fundamental changes clash with the organizations and routines of EU policy-making, which are not at all oriented towards profound changes, whether radical or gradual. Rather, the EU's resource efficiency policy is built predominantly on incremental policy-making, guided by the conviction—which is at odds with transitions thinking—that an improvement in market-oriented policies will be enough to realize a fundamental societal change. The findings from this case study are in agreement with Meadowcroft (2009: 336), who states that transition governance in practice logically looks more like regular policy-making than like theoretical transition governance. The real innovation is that these international policy initiatives have embraced the concept of transitions in the first place. From the perspective of the transitions framework, that could mean two things. A first possibility is that the transition towards sustainable materials management in Europe in still at the predevelopment phase, where the adoption of a transitions discourse is applied to facilitate upcoming changes, but where policies and institutional modifications in support of fundamental changes have not yet been put in place. A second possibility is that, despite the adoption of a number of the key principles of sustainability transitions, the policies conducted under that denominator are actually leading to a lock-in of the current system.

Future research should focus on at least three questions in order to deepen the understanding of the potential role of international policy-making in sustainability transitions. First, scholars should assess to what extent transitions thinking and related principles are permeating EU policies, including in a less visible way and independently of a transitions discourse. The example of regional policy, where in the context of Europe 2020 regions are encouraged to aim for transformative innovation rather than reactive or incremental innovation (European Commission 2012a), suggests that other EU policy processes are also moving in the direction of fundamental and systemic changes. Second, a detailed analysis of additional cases, exceeding the scope of the EU, is needed to corroborate the findings of this case study and to integrate them into a more universal framework. Third, from the perspective of multilevel governance, research needs to delve into the question of how international policy initiatives oriented towards transitions can be linked to bottom-up processes of innovation. Some of those processes are propelled by ambitious national and subnational governments, and such front-runners are likely to become a model for diffusion when their ideas are picked up at the international level (cf Jänicke 2013).

References

Althaus, Lisa-Marie, 2013: *Green Transformation towards Sustainable Development? A Comparative Analysis of the Green Transformation Concepts by UNEP, OECD, and WBGU through the Lens of Sustainable Development* (Duisburg/Bochum: UAMR Graduate Centre for Development Studies).

Audet, René, 2014: "The double hermeneutic of sustainability transitions", in: *Environmental Innovation and Societal Transitions*, 11: 46-49.

Bäckstrand, Karin; Elgström, Ole, 2013: "The EU's role in climate change negotiations: from leader to 'leadiator'", in: *Journal of European Public Policy*; at: <DOI:10.1080/13501763.2013.781781>.

Bina, Olivia, 2013: "The green economy and sustainable development: an uneasy balance?", in: *Environment and Planning C: Government and Policy*, 31: 1023-1047.

Bussels, Matthias; Happaerts, Sander; Bruyninckx, Hans, 2013: *Evaluating and monitoring transition initiatives. Lessons from a field scan* (Leuven: Steunpunt TRADO).

Council of the EU, 2009: *An integrated approach to a competitive and sustainable industrial policy in the European Union* (Brussels: Competitiveness Council).

Council of the EU, 2011: *Roadmap to a resource-efficient Europe–Draft Council conclusions* (Brussels: Council of the European Union).

Council of the EU, 2012: *Conclusions on setting the framework for a Seventh EU Environment Action Programme* (Luxembourg: Council of the European Union).

Daviter, Falk, 2007: "Policy Framing in the European Union", in: *Journal of European Public Policy*, 14,4: 654-666.

Dedeurwaerdere, Tom, 2013: *Sustainability Science for Strong Sustainability* (Louvain-la-Neuve: Université catholique de Louvain).

Delreux, Tom; Happaerts, Sander, 2016: *Environmental Policy and Politics in the European Union* (London: Palgrave).

EEA [European Environment Agency], 2011: *BLOSSOM–Bridging long-term scenario and strategy analysis: organisation and methods. A cross-country analysis* (Copenhagen: EEA).

EEA [European Environment Agency], 2014: *Multiannual Work Programme 2014-2018. Expanding the knowledge base for policy implementation and long-term transitions* (Copenhagen: EEA).

Ellen MacArthur Foundation, 2013: "Circular Economy"; at: <http://www.ellenmacarthurfoundation.org/circular-economy> (19 March 2014).

European Commission, 2005: *Thematic strategy on the sustainable use of natural resources* (Brussels: Commission of the European Communities).

European Commission, 2010: *Roadmap to a Resource Efficient Europe (Impact Assessment roadmap)* (Brussels: European Commission).

European Commission, 2011a: *Roadmap to a Resource Efficient Europe* (Brussels: European Commission).

European Commission, 2011b: *A Roadmap for moving to a competitive low carbon economy in 2050* (Brussels: European Commission).

European Commission, 2011c: *Analysis associated with the Roadmap to a Resource Efficient Europe. Part I* (Brussels: European Commission).

European Commission, 2011d: *Energy Roadmap 2050* (Brussels: European Commission).

European Commission, 2011e: *White Paper. Roadmap to a Single European Transport Area–Towards a competitive and resource efficient transport system* (Brussels: European Commission).

European Commission, 2011f: *A resource-efficient Europe–Flagship initiative under the Europe 2020 Strategy* (Brussels: European Commission).

European Commission, 2011g: *Analysis associated with the Roadmap to a Resource Efficient Europe. Part II* (Brussels: European Commission).

European Commission, 2012a: *Connecting Smart and Sustainable Growth through Smart Specialisation* (Brussels: European Commission, Directorate-General for Regional and Urban Policy).

European Commission, 2012b: *Manifesto for a resource-efficient Europe* (Brussels: European Resource Efficiency Platform).

European Commission, 2012c: *Living well, within the limits of our planet. Proposal for a general Union Environment Action Programme* (Brussels: European Commission, DG Environment).

European Commission, 2014: *Towards a circular economy: A zero waste programme for Europe* (Brussels: European Commission).

European Commission, 2015: *Closing the loop - An EU action plan for the circular economy* (Brussels: European Commission).

European Parliament, 2011: *Working Document on the Roadmap for a Resource Efficient Europe* (Brussels: European Parliament, Committee on the Environment, Public Health and Food Safety).

European Parliament, 2012: *Motion for a European Parliament Resolution on a resource-efficient Europe* (Brussels: European Parliament).

European Resource Efficiency Platform (EREP), 2013: *Action for a resource efficient Europe* (Brussels: European Resource Efficiency Platform).

European Resource Efficiency Platform, 2014: *Manifesto & Policy Recommendations* (Brussels: European Commission, DG Environment).

Ferguson, Peter, 2014: "The green economy agenda: business as usual or transformational discourse?", in: *Environmental Politics*; at: <DOI: 10.1080/09644016.2014.919748>.

Geels, Frank W., 2010: "Ontologies, socio-technical transitions (to sustainability), and the multi-level perspective", in: *Research Policy*, 39: 495-510.

Geels, Frank W., 2011: "The multi-level perspective on sustainability transitions: Responses to seven criticisms", in: *Environmental Innovation and Societal Transitions,* 1: 24-40.

Geels, Frank W., 2013: "The impact of the financial-economic crisis on sustainability transitions: Financial investment, governance and public discourse", in: *Environmental Innovation and Societal Transitions,* 6: 67-95.

Geels, Frank W.; Schot, Johan, 2007: "Typology of socio-technical transition pathways", in: *Research Policy,* 36: 399-417.

Geels, Frank W.; Schot, Johan, 2010: "The Dynamics of Transitions. A Socio-Technical Perspective", in: Grin, John; Rotmans, Jan; Schot, Johan (Eds.): *Transitions to Sustainable Development. New Directions in the Study of Long Term Transformative Change* (New York: Routledge).

Hajer, Maarten, 2000: *The Politics of Environmental Discourse. Ecological Modernization and the Policy Process* (Oxford: Clarendon Press).

Hajer, Maarten; Versteeg, Wytske, 2005: "A Decade of Discourse Analysis of Environmental Politics: Achievements, Challenges, Perspectives", in: *Journal of Environmental Policy & Planning,* 7,3: 175-84.

Hansen, Teis; Coenen, Lars, 2013: *The Geography of Sustainability Transitions: A Literature Review* (Lund University: Centre for Innovation, Research and Competence in the Learning Economy (CIRCLE)).

Happaerts, Sander, 2014: *International Discourses and Practices of Sustainable Materials Management* (Leuven: Policy Research Centre on Sustainable Materials Management).

Happaerts, Sander; Bruyninckx, Hans (2012), 'Upscaling transition governance. An exploratory analysis of the Roadmap to a Resource Efficient Europe', *the International Conference on Sustainability Transitions* (Copenhagen).

Happaerts, Sander; Bruyninckx, Hans, 2014: "Sustainable development: the institutionalization of a contested policy concept", in: Betsill, Michele M.; Hochstetler, Kathryn; Stevis, Dimitris (Eds.): *Advances in international environmental politics* (Basingstoke: Palgrave): 300-327.

Heichel, Stephan; Pape, Jessica; Sommerer, Thomas, 2005: "Is there convergence in convergence research? An overview of empirical studies on policy convergence", in: *Journal of European Public Policy,* 12,5: 817-840.

Hendriks, Carolyn M., 2009: "Policy design without democracy? Making democratic sense of transition management", in: *Policy Sciences,* 42,4: 341-368.

Howlett, Michael, 2009: "Governance modes, policy regimes and operational plans: A multi-level nested model of policy instrument choice and policy design", in: *Policy Sciences,* 42: 73-89.

Jänicke, Martin, 2012: ""Green growth": From a growing eco-industry to economic sustainability", in: *Energy Policy,* 48: 13-21.

Jänicke, Martin, 2013: "Accelerators of Global Energy Transformation: Horizontal and Vertical Reinforcement in Multi-Level Climate Governance", *The Thematic Workshop of the Multi-Level Climate Governance Research Network* (Kolkata).

Jørgensen, Ulrik, 2012: "Mapping and navigating transitions—The multi-level perspective compared with arenas of development", in: *Research Policy,* 41: 996-1010.

Kemp, René; Rotmans, Jan, 2005: "The Management of the Co-evolution of Technical, Environmental and Social Systems", in: Weber, M.; Hemmelskamp, J. (Eds.): *Towards Environmental Innovation Systems* (Heidelberg: Springer).

Kemp, René; Parto, Saeed; Gibson, Robert B., 2005: "Governance for sustainable development: moving from theory to practice", in: *International Journal of Sustainable Development,* 8,1-2: 12-30.

Krämer, Ludwig, 2012: *EU Environmental Law* (London: Sweet & Maxwell).

Lafferty, William M., 2004: "Introduction: form and function in governance for sustainable development", in: Lafferty, William M. (Ed.): *Governance for Sustainable Development. The Challenge of Adapting Form to Function* (Cheltenham & Northampton: Edward Elgar), 1-31.

Lauridsen, Erik Hagelskjær; Jørgensen, Ulrik, 2010: "Sustainable transition of electronic products through waste policy", in: *Research Policy,* 39: 486-494.

Lempert, Robert J.; Scheffran, Jürgen; Sprinz, Detlef F., 2009: "Methods for Long-Term Environmental Policy Challenges", in: *Global Environmental Politics,* 9,3: 106-133.

Lempert, Robert J.; Popper, Steven W.; Min, Endy Y.; Dewar, James A., 2009: *Shaping Tomorrow Today. Near-Term Steps Towards Long-Term Goals* (Santa Monica, Arlington & Pittsburgh: RAND).

Lenschow, Andrea, 2005: "Environmental Policy. Contending Dynamics of Policy Change", in: Wallace, Helen; Wallace, William; Pollack, Mark A. (Eds.): *Policy-Making in the European Union* (New York: Oxford University Press), 305-327.

Loorbach, Derk, 2007: *Transition Management. New mode of governance for sustainable development* (Utrecht: International Books).

Loorbach, Derk; Lijnis Huffenreuter, Rueben., 2013: "Exploring the economic crisis from a transition management perspective", in: *Environmental Innovation and Societal Transitions,* 6: 35-46.

Loorbach, Derk; Frantzeskaki, Niki; Thissen, Wil, 2011: "A Transition Research Perspective on Governance for Sustainability", in: Jaeger, Carlo C.; Tàbara, J. David; Jaeger, Julia (Eds.): *European Research on Sustainable Development* (Berlin: Springer), 73-89.

Mazey, Sonia; Richardson, Jeremy, 1997: "Policy Framing: Interest Groups and the lead up to 1996 Inter-Governmental Conference", in: *West European Politics,* 20,3: 111-133.

Meadowcroft, James, 2005: "Environmental Political Economy, Technological Transitions and the State", in: *New Political Economy,* 10,4: 479-498.

Meadowcroft, James, 2009: "What about the politics? Sustainable development, transition management, and long term energy transitions", in: *Policy Sciences,* 42,4: 323-340.

Meadowcroft, James, 2011: "Engaging with the *politics* of sustainability transitions", in: *Environmental Innovation and Societal Transitions,* 1: 70-75.

Ministerie van VROM [Volkshuisvesting, Ruimtelijke Ordening en Milieubeheer], 2001: *Een wereld en een wil. Werken aan duurzaamheid. Nationaal Milieubeleidsplan 4* (Den Haag: Ministerie van VROM).

Natali, David, 2010: "The Lisbon Strategy, Europe 2020 and the Crisis in Between", in: Marlier, Eric; Natali, David (Eds.): *Europe 2020: Towards a More Social EU?* (Brussels: P.I.E. Peter Lang).

Nilsson, Måns, 2012: "Energy Governance in the European Union. Enabling Conditions for a Low Carbon Transition", in: Verbong, Geert; Loorbach, Derk (Eds.): *Governing the Energy Transition. Reality, Illusion or Necessity?* (New York–London: Routledge): 296-316.

OECD, 2010: *Transition to a Low-carbon Economy. Public goals and corporate practices* (Paris: OECD).

OECD, 2011: *Towards Green Growth* (Paris: OECD).

Onestini, Maria, 2012: "Latin America and the Winding Road to Rio+20: From Sustainable Development to Green Economy Discourse", in: *The Journal of Environment & Development,* 21,1: 32-35.

Paredis, Erik, 2008: *Transition management in Flanders. Policy context, first results and surfacing tensions* (Leuven: Steunpunt Duurzame Ontwikkeling).

Paredis, Erik, 2013: *A winding road. Transition management, policy change and the search for sustainable development* (PhD thesis, Gent: Universiteit Gent).

Peters, B. Guy; Hoornbeek, John A., 2005: "The Problem of Policy Problems", in: Eliadis, Pearl; Hill, Margaret M.; Howlett, Michael (Eds.): *Designing Government. From Instruments to Governance* (Montreal & Kingston: McGill-Queen's University Press): 77-105.

Pollack, Mark A.; Shaffer, Gregory C., 2005: "Biotechnology Policy. Between National Fears and Global Disciplines", in: Wallace, Helen; Wallace, William; Pollack, Mark A. (Eds.): *Policy-Making in the European Union* (New York: Oxford University Press). 329-351.

Potočnik, Janez, 2010: "Opening remarks of Janez Potočnik, Commissioner-Designate for Environment" (European Parliament Hearing, 13 January 2010).

Potočnik, Janez, 2011: "Without public action, some resources will never have a price nor a market" (Speech made at the Conference of Confederation of British Industry and the Green Alliance, London, 12 December 2011).

Potočnik, Janez, 2012: "The role of resource efficiency for Europe's future" (Speech made at the European Resources Forum at Berlin, 13 November 2012).

Raven, Rob; Schot, Johan; Berkhout, Frans, 2012: "Space and scale in socio-technical transitions", in: *Environmental Innovation and Societal Transitions,* 4: 63-78.

Rittel, H. W. J.; Webber, M. M., 1973: "Dilemmas in a General Theory of Planning", in: *Policy Sciences,* 4,2: 155-69.

Rochefort, David A.; Cobb, Roger W., 1994: "Problem Definition: An Emerging Perspective", in: Rochefort, David A.; Cobb, Roger W. (Eds.): *The Politics of Problem Definition. Shaping the Policy Agenda* (Lawrence: University Press of Kansas), 1-31.

Rotmans, Jan; Loorbach, Derk, 2010: "Towards a Better Understanding of Transitions and Their Governance. A Systemic and Reflexive Approach", in: Grin, John; Rotmans, Jan; Schot, Johan (Eds.): *Transitions to Sustainable Development. New Directions in the Study of Long Term Transformative Change* (New York: Routledge).

Schoenefeld, Jonas, 2014: "The Politics of the Rise of DG Climate Action"; at: <http://environmentaleurope.ideasoneurope.eu/2014/02/24/the-politics-of-the-rise-of-dg-climate-action/> (24 July 2014).

Schön, Donald A.; Rein, Martin, 1994: *Frame Reflection. Toward the Resolution of Intractable Policy Controversies* (New York: Basic Books).

Schot, Johan; Geels, Frank W., 2008: "Strategic niche management and sustainable innovation journeys: theory, findings, research agenda, and policy", in: *Technology Analysis & Strategic Management,* 20,5: 537-554.

Sprinz, Detlef F., 2009: "Long-Term Environmental Policy: Definition, Knowledge, Future Research", in: *Global Environmental Politics,* 9,3: 1-8.

Sprinz, Detlef F., 2012: "Long-Term Environmental Policy: Challenges for Research", in: *The Journal of Environment Development,* 21,1: 67-70.

Steurer, Reinhard; Berger, Gerald, 2011: "The EU's double-track pursuit of sustainable development in the 2000s: how Lisbon and sustainable development strategies ran past each other", in: *International Journal of Sustainable Development & World Ecology,* 18,2: 99-108.

Underdal, Arild, 2010: "Complexity and challenges of long-term environmental governance", in: *Global Environmental Change,* 20: 386-393.

UNCSD [United Nations Conference on Sustainable Development], 2012: *The Future We Want* (Rio de Janeiro: United Nations).

UNEP [United Nations Environment Programme], 2011: *Towards a Green Economy: Pathways to Sustainable Development and Poverty Eradication* (UNEP).

UNEP [United Nations Environment Programme], 2014: *Decoupling 2: technologies, opportunities and policy options.*

van der Brugge, Rutger; Rotmans, Jan, 2007: "Towards transition management of European water resources", in: *Water Resource Management,* 21: 249-267.

Voß, Jan-Peter; Smith, Adrian; Grin, John, 2009: "Designing long-term policy: rethinking transition management", in: *Policy Sciences,* 42,4: 275-302.

Part XI   Conclusions and Mapping
Future Research Needs

**Chapter 42   Sustainability Transition with Sustainable
Peace: Key Messages and Scientific
Outlook**
*Úrsula Oswald Spring, Hans Günter Brauch,
and Jürgen Scheffran*

42 Sustainability Transition with Sustainable Peace: Key Messages and Scientific Outlook

Úrsula Oswald Spring[1], Hans Günter Brauch[2], and Jürgen Scheffran[3]

Abstract

This chapter presents the key messages of this *Handbook on Sustainability Transition and Sustainable Peace found in the previous texts by the sixty authors, arranged into ten parts. They focus on I)* moving towards sustainability transition; II) aiming for sustainable peace; III) meeting the challenges of the twenty-first century: demographic imbalances, temperature rise and the climate–conflict nexus; IV) initiating research on global environmental change, the limits to growth, and the decoupling of growth and resource needs; V) developing theoretical approaches to sustainability and transitions; VI) analysing national debates about sustainability in North America; VII) preparing transitions towards a sustainable economy and society, production and consumption and urbanization; VIII) examining sustainability transitions in the water, food and health sectors from Latin American and European perspectives; IX) preparing sustainability transitions in the energy sector; and X) relying on international, regional and national governance for strategies and policies leading towards sustainability transition.

This chapter proposes moving from disciplinary perspectives towards a transdisciplinary and anticipatory transformative approach. It points to research deficits and maps future research needs on 'sustainability transition', on 'sustainable peace', and on the linkages between both discourses, so that we can move from knowledge to action, and towards governance strategies, policies and measures aiming at *Sustainability Transition with Sustainable Peace*. Four examples are used to briefly illustrate this transformative scientific approach towards proactive policies. The first examines the sustainable energy transition achievable by moving from fossil fuels to enhancing energy efficiency and to renewables; this would grant access to energy for up to twelve billion people by 2100, while GHG emissions would be reduced. The second proposes a shift from resource- and carbon-intensive agriculture and a high degree of waste in the food sector to *climate-smart agriculture* with less waste. The third and fourth examples address proposed changes to different lifestyles in industrialized countries, and a shift in values as suggested, for example, by the Kingdom of Bhutan (Gross Happiness Index) and by indigenous people in Bolivia (Pachamama) and Chiapas. These alternatives may not be globally acceptable but they indicate that new viable pathways are needed that will lead towards a sustainable and peaceful world, and enable us to move beyond a continuation of the unsustainable Western way of life based on abundance and waste in consumption and production.

Keywords: Anthropocene, business-as-usual, climate-smart agriculture, complexity, decarbonization, global environmental change, interdisciplinary, lifestyles, multidisciplinary, sustainable development, sustainable peace, sustainability transition, transdisciplinary, transformative.

42.1 Introduction

This concluding chapter summarizes the key messages of the sixty authors in the ten parts (42.2). It goes on to stress the need to move from a disciplinary perspective towards a transdisciplinary and anticipatory and transformative approach (42.3), by pointing to research deficits and mapping future research needs on 'sustainability transition', on 'sustainable peace', and on the linkages between both discourses (42.4). In this way it will be possible to move from knowledge to action, and towards strategies, policies and measures aiming at *Sustainability Transition with Sustainable Peace* (42.5).

---

1 Prof. Dr. Úrsula Oswald Spring, full-time Professor/Researcher at the *National Autonomous University of Mexico* (UNAM) in the *Regional Multidisciplinary Research Center* (CRIM); email: <uoswald@gmail.com>.

© Springer International Publishing Switzerland 2016
H.G. Brauch et al. (eds.), *Handbook on Sustainability Transition and Sustainable Peace*,
Hexagon Series on Human and Environmental Security and Peace 10, DOI 10.1007/978-3-319-43884-9_42

42.2   Key Messages of the Book in the Ten Parts

The *Handbook on Sustainability Transition and Sustainable Peace* examines in ten parts: moving towards sustainability transition (42.2.1); aiming for sustainable peace (42.2.2); meeting the challenges of the twenty-first century: demographic imbalances, temperature rise and the climate–conflict nexus (42.2.3); initiating research on global environmental change, the limits to growth, and the decoupling of growth and resource needs (42.2.4); developing theoretical approaches to sustainability and transitions (42.2.5); analysing national debates about sustainability in North America (42.2.6); preparing transitions towards a sustainable economy and society, production and consumption and urbanization (42.2.7); examining sustainability transitions in the water, food and health sectors from Latin American and European perspectives (42.2.8); preparing sustainability transitions in the energy sector (42.2.9); and relying on transnational, international, regional and national governance for strategies and policies leading towards sustainability transition (42.2.10).

42.2.1   Moving towards Sustainability Transition

In the introductory chapter, *Hans Günter Brauch* (Germany) and *Úrsula Oswald Spring* (Mexico) addressed the global context, the scientific concepts, the evolution of a new research paradigm, and the dimensions of the debate on sustainability transition and their linkages with sustainable peace. The Handbook links research on sustainable development, human security and sustainable peace with 'sustainability transition', focusing on the long-term transformative change from a carbon-intensive development path by addressing the causes of global environmental and climate change. The texts foster research on longer-term proactive strategies and policies in order to realize 'sustainable development' with 'sustainable peace', resulting from a transition of the systems of production, consumption, and governance. The Handbook addresses the key questions of whether business-as-usual policies and the growing number of climate-induced natural hazards pose threats to international peace and security; whether anticipative learning about long-term transformative changes might contribute to sustainable development and prevent new security dangers; and what lessons may be drawn from the violent consequences of the industrial revolution for a long-term transformative change towards sustainable development. After a brief sketch of opposite visions, the purpose and objectives of the Handbook were highlighted and a survey reviewed the challenges posed by global environmental change driven by population growth: climate change, biodiversity loss, soil erosion and desertification, water scarcity and stress, food scarcity and hunger, and gender. The survey integrated the results of global research programmes, of their linkages and their assessment by the IPCC, and on the nexus debates between water, food and energy security. Three key concepts—sustainable development, sustainability transition and sustainable peace—were introduced, the evolution of different approaches to sustainability transition were reviewed, the debates on green growth and decarbonization were noted, and six dimensions of the research on 'sustainability transition' were presented.

Simon Dalby (Canada), in chapter 2, "Contextual Changes in Earth History: From the Holocene to the Anthropocene: Implications for Sustainable Development and for Strategies of Sustainable Transition", recontextualized peace and sustainability for the Anthropocene. Taking an earth systems science perspective, he examined ecological phase shifts, boundaries, thresholds and tipping points, and multiple system stressors in the context of planetary boundaries. He placed the debate on 'sustainable transitions' in a rapidly changing future caused by global environmental change and reviewed the implications for contextualizing sustainable peace beyond national security by taking into account geopolitical contexts from the perspective of political geoecology and of the creation of the Anthropocene. Dalby's key conclusions were: 1. Production is a key consideration in understanding and constructing our future and integral to a

---

2   PD Dr. Hans Günter Brauch is chairman of *Peace Research and European Security Studies* (AFES-PRESS), an international scientific NGO (Germany) and editor of five English language book series published by Springer: the Hexagon series <http://www.afes-press-books.de/html/hexagon.htm>, of Springer Briefs series on ESDP <http://www.afes-press-books.de/html/SpringerBriefs_ESDP.htm>, and PSP <http://www.afes-press-books.de/html/SpringerBriefs_PSP.htm>; APESS <http://www.afes-press-books.de/html/APESS.htm> and PAHSEP <http://www.afes-press-books.de/html/PAHSEP.htm>, email: <brauch@afes-press.de>.

3   Prof. Dr. Jürgen Scheffran, Institute of Geography, CliSAP Research Group Climate Change and Security, Center for Earth System Research and Sustainability, University of Hamburg; email: <juergen.scheffran@uni-hamburg.de>.

sustainable transition that will be shaped by decisions of community executives, where global cooperation is a key component. 2. Territorial strategies to 'green' some societies by outsourcing the production of energy- and pollution-intensive industries to others do not solve problems. 3. Authority over ecological processes necessitates deeper action than state territorial strategies of rule. 4. This new geography of connection in the globalized economy, and the possibilities of commodity chain governance, suggest that traditional assumptions of state-based national security are insufficient frameworks for governance. 5. This becomes more urgent if the earth system discussion is engaged, precisely because assumptions of a stable ecological context for humanity can no longer be taken for granted. 6. The future struggle for peace will be carried out in rapidly changing ecological and social conditions. Any transition from consumption-based extractive modes of economy to a sustainable one will take place during dramatic ecological change. 7. Strategies linking development and sustainability must consider ecological phase shifts if earth system boundaries are further transgressed. 8. Planning is more difficult if humanity is making the future of the planetary system as well as its own economic and social future. 9. This requires a transition to new ways of thinking about economics and politics if peaceful human societies are to be created as the next phase of the Anthropocene.

Carolyn Stephenson (US) examined, in chapter 3, "Paradigm and Praxis Shifts: Transitions to Sustainable Environmental and Sustainable Peace Praxis", transitions in paradigms and praxis, a sustainable environment in environmental studies, sustainable peace and peace studies, and the beginning of convergence between sustainable environment and sustainable peace in the 1970s, with a special focus on Kenneth Boulding (1966, 1970, 1978, 1983). She focused on conceptualizing peace in the context of UNESCO's 'culture of peace' and the rethinking of peace by the Nobel Peace Prize committee, on concepts of security, especially of human security, and on the UN debate that led to the Sustainable Development Goals. Stephenson's key concluding messages were: 1. Sustainability transitions may offer important contributions to peace and environmental studies in systems thinking. 2. While the environment has changed over millennia, transitions to environmental sustainability must acknowledge the importance of the environment for humankind. 3. Human beings have played a detrimental role in the health of the environment, and in environmental protection. 4. A key maxim is the

precautionary principle, where threats of serious damage and lack of full scientific certainty should be no reason for postponing measures to prevent environmental degradation. This principle could be important for peace and human rights. 5. A transition to an expanded concept of our human community would improve our work on both peace and the environment.

In chapter 4 on "Transition Studies: Basic Ideas and Analytical Approaches" John Grin (The Netherlands) offered a concise introduction of different approaches to (i) understanding and (ii) shaping transition dynamics: 1) A *socio-technical approach*, with the multi-level perspective as its main concept, and strategic niche management as its governance concept; 2) A *complex (adaptive) system based approach*, using the concept of transition patterns, with transition management as its governance concept; 3) A *governance approach*, which has elaborated the politics of transition dynamics, with reflexive design as the core of its approach to shaping transition and he reviewed the literature based on the social theory of practice. Grin concluded that the *sociotechnical approach* has yielded transition pathways that enable discussion of possible patterns of interaction between these levels. The *complex systems approach* offers more analytical insight into the underlying mechanisms, summarized in various types of transition patterns that underlie the sociotechnical pathways. The *governance theory based* approach has added an understanding of how the 'politics' of transitions are part of transition dynamics—thus yielding new insight into both transition dynamics and its politics. He argued that *social practice theory* has contributed insight into the agency in the embedded practices which shape transition dynamics, including the under-emphasized agency of users and consumers. Further development of these approaches and their relations may be both used in and informed by emerging research on new objects and units of analysis, which has expanded from societal domains, such as energy, agrifood, mobility, water management and health care, to cities and regions. Recently transition research has become aware of the role of place and space in all the approaches, which shapes regime-niche dynamics for all units of analysis; the governance and social practices approaches have highlighted the importance of proper analysis of the interaction between local, embedded, practices and globalization; and place and space are obviously particularly important when studying regions or cities. A final trend concerns the increasing attention paid to non-developed countries; this may help develop the theoretical approaches, as novel contexts mean a more varied empirical basis: comparison of European with non-European cases may shed new light on the former as they may reveal how far regime dynamics is intertwined with the development of the capitalist welfare state, or how (reflexive) modernization takes different forms in different places.

In chapter 5, "Transformative Science for Sustainability Transitions", *Uwe Schneidewind*, *Mandy Singer-Brodowski*, and *Karoline Augenstein* (Germany) reviewed the need for and definition of 'transformative science', the methodological challenges of transformative research given the status quo of transdisciplinary science. They suggested moving from transdisciplinary to transformative research, and discussed the institutional challenges of a transformative science that could achieve institutional self-transformation and a 'new governance of science' by shifting from science policy to governance of science if civil society were given a larger role. Their main messages were: 1. As humanity is facing massive challenges, the key role of science in sustainability transitions has hardly been recognized. 2. 'Transformative science' has catalysed necessary processes through suitable forms of knowledge production. Transformative science is based on debates about transdisciplinary/transformative research and places emphasis on the aspirations of scientists to intervene in complex systems and to carry out research in real-world laboratories. It focuses on the problem dimensions of sustainability science and aims for an institutional change as the framework condition for sustainability science. Transformative science focuses on the whole science system, which faces massive transformations. 3. Building on the 'governance of science' approach, they argued for non-hierarchical forms of organization in science, where external actors play an important role. 4. In the context of sustainability transitions, science system transformations require reflection on the institutional conditions for a broadening and a quality enhancement of sustainability sciences as a whole. 5. They sketched out such change processes for the German science system and showed how processes of reform have prepared the ground. Although this case cannot be generalized, structural similarities may exist in other cases. 6. This was illustrated for global initiatives, such as the Future Earth programme and the global change research agenda of the International Social Science Council.

42.2.2 Aiming for Sustainable Peace

The four chapters in this second part examined sustainable peace from the perspective of peace psychology (chapters 6 and 7) and with a focus on patriarchy (chapter 8) and on the linkages between peace and nature, especially in the context of the emerging concepts of a political geoecology and peace psychology (chapter 9).

Morton Deutsch and *Peter T. Coleman* (US), in chapter 6, encouraged other psychologists to sensitize citizens, scientists and public officials to the psychological components of peace. They proposed thirteen elements in a sustainable peace process: 1. *Effective cooperation* on pressing international issues and at the interpersonal level on religious concerns, sexual relations, political views, lifestyles, etc.; 2. *Constructive Conflict Resolution* at the interpersonal, intergroup and international level; 3. *Social Justice* that distinguishes between injustice and oppression in order to enhance constructive conflict resolution; 4. *Power and Equality* in order to raise awareness of hidden discriminative relationships and to overcome confrontation by negotiation, in order to promote a successful peace process; 5. *Human Needs and Emotions*, in order to fulfil basic human needs and overcome apathy, fear, depression, humiliation, rage, and anger; 6. *The Psychodynamics of Peace*, to be achieved by emphasizing the interdependence between internal and external conflicts; 7. *Creative Problem-Solving* to bring in imagination, freedom and abilities to create new possibilities and capacities for problem-solving; 8. *Complex Thinking*, achieved by overcoming dichotomies between different regional experiences and integrating opposed alternatives to give unexpected results; 9. *Persuasion and Dialogue* as mutual processes where the parties listen to one another with mutual respect and equality; 10. *Reconciliation* of conflicting parties by encouraging them to cooperate with the support of third parties; 11. *Education*, by transmitting and discussing the psychological components of peace through cooperative learning, training in conflict resolution, constructive use of controversy, creation of dispute resolution centres and the development of awareness of human rights and social justice; 12. *Norms for Policy*, as being crucial for sustainable peace, where psychological principles and activism help change norms and policies towards processes of a sustainable peace; and 13. *The Practice of Sustainable Peace*, fostered by cognitive, affective, behavioural, structural, institutional, spiritual, and cultural elements. The authors concluded that "decision-making within the individual as within the nation can entail a struggle among different interests and values for control over action. Internal structure and internal process, while less observable in individuals than in groups, are characteristic of all social units".

In chapter 7, *Peter T. Coleman* (US) analysed the essence of sustainable peace, highlighting the basic commonalities and synthesizing its main components by using a model of dynamic systems theory and minimalism. He outlined an agenda for future study and education that moves beyond destructive conflict, violence or war towards sustainable peace solutions. Instead of a simple, linear model of cause-and-effect he proposed a more complex, holistic model of sustainable peace that would integrate ecological, biological, psychological, social, economic, and other structural forces, promote transdisciplinary collaboration and foster sustainable peace through multiple perspectives and approaches. These processes go beyond short-term outcomes (peace treaties, agreements, etc.) and establish positive conditions for communities. Through communication and local, regional, and global networks, partnerships between both science and policy/practice should emerge to support these processes. His proposed initiatives included developing basic theory and research into sustainable peace with its fundamental conditions and processes. Applied frameworks should include empirically-tested theoretical models to foster peace and to enable the teaching of graduate, multidisciplinary, theory-and-practice courses on sustainable peace, using transdisciplinary approaches. He suggested a data-based index for annual reporting on state and regional levels of sustainable peace. The business and academic communities could develop comprehensive methodologies beyond early warning and violence prevention, and measure processes of sustainable peace. Coleman proposed an annual theory–practice–policy forum, where policymakers, peace practitioners and scholars would meet and cooperate on translating research into practice. This would require a common language with an integrative platform for transdisciplinary communication and coordination that would contribute to a new paradigm across scientific disciplines.

Úrsula Oswald Spring addressed a sustainable-engendered peace as a challenge in the Anthropocene. She examined different peace concepts, from the Neolithic and agricultural revolutions to the twenty-first century, such as negative, positive, structural, sustainable, and cultural peace, noting that none included long-term violence, authoritarianism, exclusion, discrimination or exploitation, all of which are deeply embedded in the patriarchal system. In agreement with Betty Reardon she argued that a sustainable peace requires overcoming patriarchy, militarism and injustice through: 1. the adoption of a holistic, gender-equal and feminist perspective; 2. a fundamental change in the occidental world view to include feminist values at all levels of society, in the public domain and in government; 3. a shift from national security towards human and gender security; and 4. an increasing global self-awareness of the people, especially the most vulnerable. She argued that this sustainable-engendered peace includes 'power-from-within', leading to a preference for change and to a commitment to action. Such a peace starts from a positive understanding of power and encourages new forms of action to overcome all types of oppression. This concept encourages a process of empowerment in order to overcome systemic violence, economic oppression, socio-economic marginalization, lack of autonomy in decision-making, cultural imperialism and environmental destruction. This peace is based on diversity in a world with numerous ethnic, religious, cultural, and social groups. The integration of traditional and modern knowledge may offer new approaches for dealing with conflicts and new ways of promoting equality and solidarity. From a radical feminist perspective, she criticized the current world view arising from greed, exploitation, violence, and oppression. Her peace paradigm is based on a different world view that is inclusive, just, participative, and equal, and focuses on a holistic, cosmopolitan, ethically oriented approach with the clean management of natural resources, ecosystem services, and safe food. This transformative approach may overcome the patriarchal system through education, peace-building, mediation, and negotiation aimed at an ethnically-oriented, gender-sensitive, religiously diverse and just society.

In chapter 9, "Sustainable Peace in the Anthropocene: Towards Political Geoecology and Peace Ecology", *Hans Günter Brauch* (Germany) argued that policymakers favouring a transformation towards a low-carbon society will face opposition. A sustainability transition in the energy and transport sectors would reduce energy costs and dependence on fossil energy imports from unstable regions. At least two peaceful outcomes of such a transition process may be assumed: the probability of energy resource conflicts would be reduced; and climate wars, dangerous climate change and tipping points would become unlikely. Preventing these conflicts would become a diplomatic strategy towards sustainable peace. Brauch's messages were: 1. Humankind has interfered in nature and the earth system, causing it to move towards the Anthropocene. 2. This transition requires a scientific revolution for sustainability. 3. The social sciences must reconceptualize research programmes on peace, security, development and environment. 4. Irrespective of multilateral global climate diplomacy, strategies for a 'sustainability transition' must have an economic rationale. 5. Such strategies may contribute to 'sustainable development' with 'human security' and 'sustainable peace'. 6. Two research programmes, 'political geoecology' and 'peace ecology,' may contribute to a new scientific paradigm. 7. 'Political geoecology' adds a 'political' dimension to physical geography. 8. 'Peace ecology' links environmental and peace studies. 9. 'Sustainable peace in the Anthropocene' requires a rethinking of peace, focusing on 'peace with nature'. 10. The 'sustainability revolution' requires a *transformation* of global cultural, environmental, economic and political relations, replacing the dominant strategy of business-as-usual in a Hobbesian world that may lead to a major crisis for humankind.

42.2.3   Meeting the Challenges of the Twenty-First Century: Demographic Imbalances, Temperature Rise and the Climate–Conflict Nexus

The four chapters of the third part (chapters 10–13) assessed the key challenges and obstacles for sustainability transitions and possible ways of addressing them. These include demographic and population imbalances (chapter 10), the consequences of global warming in a 4°C world (chapter 11), the climate–conflict nexus (chapter 12), and the transformation from complex climate risk cascades to viability and sustainable peace in the Anthropocene (chapter 13).

In chapter 10, "Population imbalances: Will they continue over the rest of the century?", *Hania Zlotnik* (Mexico/US) analysed variations among countries in current population trends, which are expected to continue in the medium term. This will maintain today's population imbalances and produce a high potential for marked population growth during this century, especially in some of the world's poorest countries. According to the most recent estimates of fertility levels, 62 countries (19 per cent of the world population in 2014) had high fertility; another 64 countries (35 per cent of the world population) had intermediate fertility, and 75 countries (46 per cent of the world population) had low fertility. Zlotnik's key messages were: 1. Current population imbalances result from different trends in the reduction of fertility. Clear dif-

ferences exist between the countries of high fertility, intermediate fertility and low fertility in terms of contraceptive use, which is lowest in the high-fertility countries and highest in the low-fertility countries. 2. Given the differences in levels and trends of fertility that already exist between the countries, their influence cannot be erased in the medium term. 3. If the high-fertility countries could achieve the reductions of fertility projected by the low variant while all other countries followed the medium variant, the world might witness the beginning of population stabilization by the end of this century. 4. If the world population is to remain below ten billion by the end of this century, there is an urgent need to accelerate the reduction of fertility in the countries that still have high fertility levels, and that countries with intermediate levels of fertility continue to reduce them. 5. The use of modern contraception is the major immediate means of reducing fertility, especially in the high-fertility countries, where the unmet need for contraception remains high, especially among low-income groups, but also in the intermediate-fertility countries. 6. Increasing use of modern contraception seems feasible but requires sustained support for family planning, especially considering that the number of women of reproductive age in intermediate-fertility countries will increase over the next two decades. 7. In both the high-fertility and the intermediate-fertility countries, the unmet need for contraception is projected to decline between 2010 and 2030, while in the intermediate-fertility countries, the decline in unmet need is slower, but for the high-fertility countries it is faster than in the past. 8. Groups of countries will still leave a substantial proportion of women of reproductive age with an unmet need for contraception in 2030, especially in the high-fertility countries where, as fertility declines and increasing numbers of women wish to reduce their fertility, the demand for contraception will likely grow faster than the capacity to meet it.

In chapter 11, *Hans Joachim Schellnhuber, Olivia Maria Serdeczny, Sophie Adams, Claudia Köhler, Ilona Magdalena Otto* and *Carl-Friedrich Schleussner* (Adams Australia; all others Germany) focused on "The Challenge of a 4°C World by 2100". Recognizing the limitations of our understanding, they addressed processes that are scarcely or not at all reflected in current assessments of global warming, including critical thresholds and tipping points in the earth system, suggesting that the risk of crossing such thresholds might be much greater than previously thought. The overall effect of climate impacts has been determined in large part by the patterns of capital and commodity flows and the degree of complexity in social networks. Another decisive factor driving the nature and scale of impacts and impact cascades and human vulnerability is future socio-economic development. The authors' main messages were: 1. Climate change acts as a multiplier on other environmental and social stressors, further increasing the vulnerability of the poor and those already disadvantaged. In a 4°C world the negative impacts of increasing temperatures and changing rainfall patterns on human sectors interact in non-linear and complex ways and have the potential for cascading impacts that affect the stability and well-being of societies. 2. Different degrees of vulnerability need to be taken into account across and within nation states when the risk of complex and abrupt impacts combines with the direct effects of unprecedented heat extremes and asymmetric sea-level rise. 3. Cumulative effects can be mediated and shaped by societal developments, including population and economic growth, which in turn affect and are affected by changes in climate. Climate change will likely widen the divide between rich and poor, or between more and less resilient populations. 4. Impacts projected for ecosystems, agriculture, and water supply in the twenty-first century could lead to large-scale displacement of populations, with manifold consequences for human security, health and economic and trade systems. 5. Given the intricate complexity of the earth system, impacts may also occur in a non-linear manner, significantly exacerbating the challenge for precautionary risk reduction and other adaptation measures. 6. Adaptation policy-making and planning needs to find ways of minimizing the risks associated with the impacts of climate change, both expected and unexpected. 7. The safest course for humanity is to avoid the unexpected rather than to prepare for it, and mitigating climate change remains the safest bet for the lives and well-being of present and future generations.

In chapter 12, *Tobias Ide, P. Michael Link, Jürgen Scheffran* and *Janpeter Schilling* (Germany) investigated "The climate–conflict nexus: Pathways, regional links and case studies". They reviewed the most recent quantitative and qualitative literature, assessed empirical, theoretical and conceptual issues, and analysed case studies. Their key conclusions were: 1. Research on the climate-vulnerability-conflict nexus has produced few consensual findings and remains disputed. 2. Depending on the spatial and temporal scales of data, large-N studies differ regarding the impacts of increasing temperatures, precipitation variability and extremes, lower freshwater availability, soil degradation, and natural disasters on the onset of violent conflict. 3. Qualitative studies also provide no clear picture: climate change in combination with other political, social, and economic trends seems to influence the intensity of pastoralist violence in northwestern Kenya, unlike the Nile Basin where coordinated efforts by the major riparian states have been able to avoid increasing conflict. 4. In Israel and Palestine, confrontational discourses and asymmetric power structures are more important for the course of the conflict than availability of water, which is predicted to decline due to climate change. 5. Possible reasons for the lack of scientific consensus may include difficulties in capturing multiple complex links in theoretical and empirical research on the dependent and independent variables. 6. Shortcomings also exist regarding methodology and the availability and quality of data sets, e.g. the onset of civil war at the country level with an annual temporal resolution is more likely driven by other factors than the persistence of communal violence at administrative unit levels with a monthly temporal resolution. 7. As climate change progresses, the relevance of conflict over adaptation, mitigation and geoengineering increases, while violent conflict in (peri-)urban areas may become more likely. 8. The potential impacts of climate change on societal stability could erode institutions, and this might trigger a downward spiral of violent conflict and environmental degradation, while violent conflict affects the environment negatively. 9. The debate on climate change and (violent) conflict is relevant for the debate on sustainability transitions and sustainable peace: to realize a sustainable world characterized by low CO_2 emissions and small ecological footprints, stable, reliable and accountable institutions are needed. 10. Research is needed on the climate-cooperation nexus, e.g. how climate change can facilitate cooperation and reconciliation between rival groups who adapt to common threats such as storms, floods or droughts.

In chapter 13, "From a Climate of Complexity to Sustainable Peace: Viability Transformations in the Anthropocene", *Jürgen Scheffran* (Germany) investigated major challenges in international relations caused by an increasingly complex and interconnected world. Occurring across different scales, this demands the stabilization of human-environment interactions and the integration of complexity science with global governance. Scheffran discussed cases of complex crises where climate change can be considered as a 'risk multiplier' that disturbs the balance between natural and social systems and amplifies the consequences through complex impact chains. He used an integrative framework of human–environment interaction to analyse destabilizing developments, tipping elements and cascading risks, as well as concepts of resilience, viability and sustainable peace. His conclusions were: 1. Climate change can act as a 'risk multiplier' where the consequences are amplified through complex impact chains that affect the functioning of critical infrastructures and supply networks; intensify the nexus of water, energy and food; lead to production losses, price increases and financial crises in other regions through global markets; undermine human security, social living conditions and political stability; and trigger or aggravate migration movements and conflict situations. 2. Affected systems need to develop strategies to tackle the problems, including climate mitigation and adaptation; building resilient societal structures; cultivation of diversity and flexibility; social networks that facilitate the exchange of knowledge, income and other resources; new capabilities for disaster management, crisis prevention and conflict resolution; arms control, non-proliferation and disarmament; regional security concepts, crisis prevention, conflict resolution and confidence-building; and innovative institutional frameworks and legal mechanisms. 3. To avoid a cycle of violence and support, a transition towards a cycle of cooperation with new concepts of anticipative and adaptive governance is required, based on the concepts of resilience, viability and sustainable peace. To influence critical decision points and adjust actions towards human security, social resilience and societal stability, 'win-win' solutions can help in resource sharing and risk management; new social and political structures and institutions; and rules and regulations. The viability concept can serve as a guiding framework for transition and transformation processes towards a viable world.

42.2.4 Research on Global Environmental Change, The Limits to Growth, and Decoupling and Resource Needs

In the fourth part, Lourdes Arizpe, Martin F. Price and Sir Robert Worcester, together with Eckart Ehlers, reviewed the emergence of the human dimension of global environmental change, Marit Sjovaag analysed four reports by the authors of *The Limits of Growth* (1972-2012), and Mark Swilling assessed the work of the *United Nations Environment Programme's* (UNEP) *International Resource Panel*.

In chapter 14 on "The first decade of initiatives for research on the Human Dimensions of Global (Environmental) Change (1986-1995)", *Lourdes Arizpe* (Mexico), *Martin F. Price* (UK), and *Sir Robert Worcester* (UK) analysed the institutional framework, the emerging social science issues on global change, and the initial steps towards a research programme on the human dimension of global (environmental) change between 1990 and 1995. Their main messages were: 1. Between 1986 and 1995 scientists from the natural and social sciences, politicians and members of the public recognized the rapid changes in the biophysical and human systems of our planet, with climate change and development as key drivers. 2. From the scientific perspective a critical need for new ways of thinking, data, and approaches to research emerged, requiring collaboration across disciplines with a new conceptualization of research problems, agreements on measurement, and time to learn to work together. 3. New research strategies developed on the *human dimension of global environmental change* (HDGEC), within the ISSC's (International Social Science Council) research framework and in its subsequent implementation. 3. This was one of many international initiatives concerning HDGEC, while much research was carried out by individuals and groups within and across disciplines at the local, subnational and national levels. 4. In the next two decades, substantial advances occurred as a result of these initiatives.

Eckart Ehlers (Germany), in chapter 15, examined the "Evolution of the International Human Dimensions of Global Environmental Change Programme (1996-2014)" by moving from HDP to IHDP (1995-1996) and offering an overview of the activities of the IHDP (1996-2014) in its search for stability and sustainability. He assessed IHDP's new mission and organizational restructuring and its transformation into a scientific programme. In an outlook, he examined the GEC research within the new 'Future Earth' initiative. Ehlers' key messages were: 1. IHDP fulfilled its mission and paved the way for a new understanding of the interactions between nature and societies. 2. The results from IPCC and the postulate of the Anthropocene are milestones of a new interpretation of the interactions between society and nature. 3. IHDP participated in developing these new paradigms. 4. Humans have changed the Earth in fundamental ways, and as a result stronger commitments are needed from the social science community to cope with these threats to Planet Earth. 5. IHDP and its projects have raised awareness and expanded the consciousness of the causes and consequences of climate and environmental change through supporting a new 'sustainability science' and highlighting its responsibility for society and its well-being. 6. While for millennia, humanity faced the threat of nature, now nature is threatened by humankind. 7. A Nobel Prize Laureate symposium on global sustainability called for new ways of thinking and a new alliance between nature and society, and for new forms of social-ecological systems of research into global change. 8. The symposium proposed a 'new social contract for global sustainability', with a shift of perception from people and nature to interdependent social-ecological systems. 9. Humankind needs a 'social-ecological' innovation and technologies that work more directly for social justice, the alleviation of poverty, environmental sustainability, and democracy. 10. IHDP has contributed to meeting these challenges.

In chapter 16, "From The Limits to Growth to 2052", *Marit Sjovaag* (Norway) reviewed the debate on *The Limits to Growth* and its updates from 1972 to 2012. She assessed *The Limits to Growth—A System Dynamic Analysis of Planet Earth* (1972) and analysed its key concepts and methodology, its scientific approach and messages, and the immediate reactions to its publication. She examined the update *Beyond the Limits* (1993) and the 30-Year Update (2002), including the reactions to these reports, as well as the fourth report, *2052—A Global Forecast for the Next Forty Years* (2012); she also reviewed further resources. Sjovaag's key messages were: 1. Major industrial interests saw their wealth and fortunes threatened. Economists and politicians found their world view questioned. The population saw economic growth as the safest option for meeting their needs. 2. In 1972, the report's messages were radical and its plea for a sustainable development path threatened the established economic, political and industrial elites. 3. The report was not the only environmental message that caused hostility. 4. The updates were presented in different lights, emphasizing 'hope' and 'love' (1992), and 'doom and gloom' (2004). The authors chose different strategies to engage the world in a debate about where we are heading. 5. Three decades later, Graham Turner showed that the model offered a good description of reality. 6. The global UN-based governance system has been unable to produce a globally binding GHG emission reduction treaty in the forty years since 1972. 7. The four books concentrated on the 'physical' variables of human development and its global effect, while discussions on security and social development were absent. The reports assumed that resource shortages might lead to social unrest, but did not carry out a closer causal

analysis. 8. The '2052' book included superficial discussions on quality of life. There was no discussion of how the political process can solve *social* problems caused by climate change. 9. *The Limits to Growth* pointed to two stories about how global governance failed to take action to prevent the challenges highlighted in the 1970s, and it incited debate, raised ethical questions and spurred on a search for deeper knowledge of our world.

In chapter 17, "Preparing for Global Transition: Implications of the Work of the International Resource Panel" (IRP), *Mark Swilling* (South Africa) contextualized the activities of the IRP, gave an overview of its work, reviewed global resource perspectives, addressed the nexus themes by discussing city-level decoupling, GHG mitigation technologies, food systems and trade, examined specific resource challenges posed by metals, land-use and soils, and water, and discussed the implications of the IRP's work for global transition thinking. Swilling's key messages were: 1. In this assessment long-wave theories of transition were used for understanding its relevance for anticipatory science. 2. The IRP's work suggests that it is unlikely that the industrial epoch can continue relying on the continuous increase in the consumption of natural resources. 3. Elements in its reports could be woven into a more robust and systematic conception of transition: the types of decoupling in the *Decoupling 2 Report*, the dietary and land-use changes in the *Land and Soils Report*, the key role of cities in the *City-Decoupling Report*, and the unintended consequences of a transition to clean energy, to give a few examples. 4. Three broad conclusions flow from this analysis: a) The IRP's reports support the notion that future well-being must be decoupled from rising rates of resource use. This requires a fundamental restructuring of prevailing modes of production and consumption. b) The IRP's work reveals the futility of naïve assumptions on what is attainable in a sustainable world of over nine billion people, mostly living in cities. Anticipatory science must show that this cannot continue. A world of over nine billion people without poverty may well need what was extracted in 2000. c) The IRP's work on resource limits and potentials must be integrated into a wider holistic theory of economic development.

42.2.5 Developing Theoretical Approaches to Sustainability and Transitions

In the fifth part, in chapter 18 Czeslaw Mesjasz addressed linkages between sustainability and complexity, deriving lessons from modern systems think-ing. In chapter 19, Bonno Pel, Flor Avelino and S. S. Jhagroe developed critical approaches to transitions theory, while in chapter 20, Mike Hodson, Simon Marvin and Philipp Späth investigated inter-scalar dynamics in the differentiated geographies of low-carbon transitions, with examples from the UK.

In chapter 18, *Czeslaw Mesjasz* (Poland) addressed "Sustainability and Complexity: A Few Lessons from Modern Systems Thinking". He identified and assessed the applications of concepts derived from complexity studies in the discourse on transition to sustainability and sustainable development, which are associated with the growing complexity of ecological and socio-political systems, and of their interactions. This includes a clarification of the terminology of complex-ity and related narratives to provide new insights into causal links as well as prediction, the normative approach, and influence upon societal phenomena. The author came to several conclusions and recom-mendations concerning sustainability, sustainable development and transition management: 1. Complex-ity cannot be defined in a precise way, and thus its meaning emerges in intersubjective exchange. Applica-tions in sustainability and sustainable development may vary, beginning with loosely defined metaphors and analogies, and ending with various attempts at operationalization. 2. In many cases, the terms associ-ated with complexity theory are misused in discus-sions about sustainability, where their vague meanings are applied as a kind of decoration, adding a 'scien-tific value' to the discourse. 3. Well-thought-out use of those concepts would contribute to a better under-standing of the fast-changing modern world, and should lead to the development of more effective pol-icy measures. 4. Sustainability transition and transition management are based upon ideas drawn from com-plexity theory and complexity science which relate to phenomena, processes, names and mathematical models for studying the complexity of natural and social phenomena with a set of simple descriptions, explanations and predictions, including ideas such as adaptation, the edge of chaos, learning, and self-organization. 5. It is a challenge to identify and stimu-late the process of self-organization among the reflex-ive and self-reflexive members of an *autopoietic* social system which is described metaphorically as being at the edge of chaos. 6. A methodological toolbox should be open, but at the same time, any ideas and methods, before inclusion, should be tested for their conceptual and cognitive validity as well as for their applicability. The development of sets of indicators

for sustainable development is the first step in this direction

In chapter 19, *Bonno Pel* (Belgium), *Flor Avelino* (The Netherlands), and *S. S. Jhagroe* (The Nether-lands) developed "Critical Approaches to Transitions Theory". The authors showed how transitions theory has started to gain currency among both policymakers and researchers, and a process of institutionalization and mainstreaming is increasingly being established. At the same time, transitions theory remains contro-versial and questions have been raised about the theo-rization of agency and transformation dynamics, par-ticularly about the normative assumptions underlying intervention strategies. Key issues discussed were: 1. Critical approaches to transitions theory are guided by a constructive attitude that reflects distinctly 'criti-cal' elements and avoids polemical juxtapositions between critical and uncritical transitions approaches. 2. Key questions are addressed through a historical comparison with the critical-theoretical project initi-ated by Marx, Horkheimer, Adorno and others to diagnose the social problems of its time and to artic-ulate corresponding remedial strategies. 3. Critical transitions theory runs into various internal contradic-tions, provides useful insights, and is well equipped to deal with critical-theoretical paradoxes, but also dis-plays tendencies towards relapsing into pitfalls. 4. The relapse into objectivism and ivory tower critiques, into fortification and scientization, and the negligence of the normative assumptions underlying transitions diagnoses and remedies, seems to be the flip side of the aspiration towards transitions science, where tran-sitions theory is somewhat uprooted from its origins in the humanities. 5. There have been considerable developments towards a more emphatically critical transitions theory, with many recent advances that take direct issue with the paradoxes discussed. 6. The existence of repertoires such as transition manage-ment, strategic niche management and the technolog-ical innovation systems framework shows how transi-tions theory is strongly dedicated to the development of remedies. 7. The critical-theoretical legacy reminds us that transitions theory too is shaped by the prevail-ing social conditions of its time.

In chapter 20, *Mike Hodson* (UK), *Simon Marvin* (UK) and *Philipp Späth* (Germany) investigated "Sub-national, Inter-Scalar Dynamics: The Differentiated Geographies of Governing Low-Carbon Transitions–With Examples from the UK". Taking as a starting point the *Multilevel Perspective* (MLP), and distin-guishing niche, regime and landscape levels, the authors improved the analytical understanding of low-carbon transitions at and in between multiple geographical scales, particularly 'below' a national level. They set out a range of spatial levels relevant to transition activity, and explicitly identified different, yet coexisting and interacting, scales of low-carbon transition activity in the UK. Their key results were: 1. The MLP approach offers tools for thinking through institutional and technological conditions and rules through which regimes reproduce or change over time. 2. Activities that contribute to a transition (electricity systems, heat and buildings, mobility systems, food systems, etc.) have effects of different spatial reach, and are connected with a wide range of socially constructed spaces. 3. From a geographical perspective, both spaces (e.g. cities, regions, neighbourhoods) and scales (the city scale, the regional scale) are continually being made, negotiated and remade; transition activities are defined on national, urban, hyperlocal and even smaller scales. 4. Coalitions are tightly linked to the space of the national state, promoting a view of the primacy of markets in low-carbon transitions and the need for effective political coordination of institutions to shape low-carbon transitions. 5. Their chapter illustrated the ways in which political-economic institutional configurations have shaped a particular low-carbon response at UK national level, mediated landscape pressures and set conditions for regime reproduction/reconfiguration and niche activities. 6. The literature on sociotechnical niches shows how, why and the extent to which hyperlocal spaces are 'protected' and the limits of geographical protection where national and local priorities may differ. 7. Scalar transition activity works on one dimension to promote top-down low-carbon activity or alternatively bottom-up activity.

42.2.6 Analysing National Debates about Sustainability in North America

In the North American geopolitical context Twig Johnson, Simon Dalby and Juan Antonio Le Clercq analysed for the US, Canada and Mexico the constraints for sustainability transition.

In chapter 21, "Policy, Politics and the Impact of Transition Studies", *Twig Johnson* (US) addressed US interests in linking research and political action as a key challenge for transition studies. Following the publication of *Our Common Journey: A Transition toward Sustainability* (NRC 1999), a contested presidential election and the terrorist attacks of 9/11 totally changed the political arena for a sustainability transition. Several universities and research centres

working on sustainability studies shifted to interdisciplinary research and teaching, while others established new courses, launched a transdisciplinary approach, and founded new academic journals. The award of the Nobel Prize for Chemistry for work on the depletion of the ozone layer (1995) and of the Nobel Peace Prize to the IPCC and Al Gore (2007), as well as the increasing number of disasters from extreme weather events, put pressure on the US government, which failed to react to this but rather securitized terrorism. The high level of support for environmental issues by the Clinton administration was replaced during the Bush administration with hostility and the suppression of research results. During the Obama administration, multinational insurance companies maintained the pressure because of the increasing costs of disaster payments, but the Republican-controlled Congress together with many lobbyists continued to block a radical decarbonization. The problem-solving policies oriented towards sustainability have shifted to the states, municipalities, the people affected, scientists, engineers and educational institutions, as well as to *non-governmental organizations* (NGOs) and some corporations, who pointed out the potential for job creation and innovation in a decarbonized and sustainable economy, where water, soil, biodiversity, health, food and population growth were all addressed.

In chapter 22, "Geopolitics, Ecology and Stephen Harper's Reinvention of Canada", *Simon Dalby* (Canada) examined the consequences of Harper's policies. Harper promoted a geopolitical foreign and domestic policy approach based on energy exports, with the aim of making Canada an energy superpower. This neo-liberal approach prioritized the market, since its key objectives were wealth and job creation. His conservative government withdrew from the Kyoto Protocol, dropped some regulations concerning environmental standards, and pushed for a carbon-intensive economy. Dalby found that six lessons could be learnt: 1) Activists for a sustainable world in Canada and elsewhere generally ignore politics. 2) They primarily focus on narrow technical environmental matters and assume that legislation can protect nature in the long term. 3) They forget that "power has also slipped away from states as they become enmeshed in complicated international trade agreements where foreign companies have the rights to sue governments trying to initiate social and environmental regulations deemed a challenge to corporate profits". 4) Dalby argued that governance driven by the market is often only a chimera resulting in business-as-usual, rising inequality and social unrest. Profit-driven policies cause long-term market failures and threaten the future of Earth. The biosphere and human well-being should not be exclusively managed by market forces. 5) In a globalized world economy, environmental and climate governance of countries, provinces, cities, social organizations, corporations and international institutions, which are often based on competing values, require cooperation and negotiation of common interests towards sustainability. 6) Dalby questioned the ontological premise of competition as the sole way of reaching sustainable futures, and stressed inter- and intra-generational equity, cooperation and care about nature and humans as a different geopolitical postulate for promoting sustainability globally.

In chapter 23, "Regime Change, Transition to Sustainability and Climate Change Law in México", *Juan Antonio Le Clercq* (Mexico) argued that "the environmental governance perspective points to a relationship between the quality of institutional design, regime interplay and effectiveness in dealing with global challenges". He asked why parliament had opted for this model of regulation for a governance regime, and whether this law would have an impact on Mexico's environmental policy and lead to a sustainable transition. The law required the government to establish an inter-ministerial commission whose objectives should reflect the goals of the National Development Plan (2007–2012) and be linked to a sustainable human development approach for mitigating GHG. The law required an *Inter-Ministerial Commission on Climate Change* (ICCC) to develop binding strategic guidelines including a diagnosis of environmental challenges; to formulate a hundred objectives and reportable goals on adaptation and mitigation; and to promote a voluntary register of emissions from the private sector. The decision in favour of a climate regime permitted the creation of a national climate policy with only minimal modifications of the existing environmental law. Climate policy instruments alone did not grant an effective mitigation and adaptation process, and after the government changed, climate policy was downgraded. The government of Peña Nieto adopted a constitutional change that allowed the privatization of hydrocarbons, and during the first three years of his government GHG emissions increased by twenty-eight per cent. In December 2015 (during COP 21 in Paris) the Law of Energy Transition was approved by Congress, and this facilitated investment in renewable energy and moves towards a more sustainable energy future in Mexico. This adoption of an energy regime not only weakened the Law on Cli-

mate Change but slowed down Mexico's transition toward a sustainable future.

42.2.7 Preparing Transitions towards a Sustainable Economy and Society, Production and Consumption and Urbanization

Part seven deals with the economic and societal dimensions of preparing sustainability transitions. Its seven chapters cover the discourse-institutional perspective (chapter 24); new business models relating to transactions and interactions (chapter 25); and sustainable consumption (chapter 26). Case studies focus on sustainable consumption and production in China (chapter 27); the eco-restructuring of the Ruhr as an example of a managed transition (chapter 28); the transition towards sustainable urbanization in Asia and Africa (chapter 29); and university partnerships in urban sustainability experiments, with evidence from Asia (chapter 30).

Audley Genus (UK) provided, in chapter 24, a review of "Sustainability Transitions: A Discourse-Institutional Perspective", based on diverse contributions concerning the role of language and institutional arrangements. Addressing the complex web of activities and actors in the transition process, the author presented a framework for a cross-disciplinary conversation on sustainability transitions, neo-institutional theory, and critical discourse analysis. Potential insights were considered from a discourse-institutional perspective in an attempt to improve understanding of the problems involved, the issues at stake, and the implications for policy and practice. The author's key conclusions were: 1. It is essential to embed or reproduce pro-environmental or unsustainable values and actions; the role of language in substantiating the status quo and in 'carrying' institutional change; and the role of firms, citizens/consumers and academic researchers as agents of institutional change or inertia, or as catalytic intermediaries. 2. Some contributors have developed combined discourse-institutional frameworks, highlighting how the discourse of 'ecology' has 'disciplined' the debate about nature in the Netherlands, leading to new and 'enduring policy practices'. 3. Several contributions highlighted different types of factors and actors in (un)sustainable patterns of production and consumption and the (in)effective governance of environmental sustainability-related science and technology; others highlighted the potential or marginalization of certain actors, such as NGOs and citizens, in grass-roots social movements. 4. Conceptual issues apply to the definition of the types of institutions that may be significant to transition governance and innovation as well as to a more explicit analysis of the carriers, mechanisms and processes of institutionalization, non-institutionalization or de-institutionalization of arrangements, norms and cultures. 5. Earlier contributions have lacked a comprehensive treatment of specific *texts*, practices and their social conditions which together (re)produce the institutional reality of governance in a sustainability transition.

In chapter 25, *Jan Jonker* (The Netherlands/France) and *Linda O'Riardon* (Ireland/Germany) studied "New Business Models: Examining the Role of Principles Relating to Transactions and Interactions". Recognizing the signals that current economic ideas no longer function, they found that new ways of organizing are emerging with relevance for sustainability. The results of exploratory research, based on a series of interviews, demonstrate the relevance of *new business models* (NBMs) to create 'multiple value(s)' regarding the task of organization and the organizing entities. The empirical evidence is combined with theoretical underpinnings to explore the nature, features and functioning of NBMs. The authors' main results may be summarized as: 1. The world lives in a time of great transition, which is rapidly changing the conventional practice of merely economic business models. 2. Early NBMs can be generally categorized into different streams based on the practice of sharing, trading, and creating. 3. At the interface of connecting, community-building and cooperation, renewal, innovation, and thus NBMs come into existence. 4. The ability to connect increasing social and economic values, such as nature, care, attention, and money. 5. These connections create new consortia and constituent configurations of parties. 6. While money no longer predominates as the only means of trade, economic traffic is based on 'exchanging and satisfying' needs; having 'access to' the means of production becomes more important than actually owning them. 7. It is only possible to organize such an economy cooperatively based on long-term commitment where securing trust in relations via collaboration is a necessary condition. 8. Almost all of the models under investigation included activities to create an experience or a community or both that has to do with combining a product or products with a service or services.

Chapter 26, "Sustainable Consumption", by *Sylvia Lorek* (Germany) drew key arguments from the literature on sustainability research. As an early concept, 'environmental space' embedded sustainable con-

sumption within natural and social boundaries, and considered floor and ceiling as well as the space itself, its size and how to share it. She described various possible paths of transition to environmental space starting from overconsumption and under-consumption. Providing a more detailed analysis of the concepts of both 'green growth' and 'de-growth', the chapter distinguished between strong and weak sustainable consumption, and outlined some enabling mechanisms for sustainable consumption. The author's key arguments were: 1. Human consumption is embedded in the boundaries of natural systems but shaped by economic, societal and cultural structures. 2. The concept of environmental space helps us to visualize how the consumption level of countries, regions or individuals can be unsustainable if it exceeds upper and lower limits. 2. The limits often have not been clearly formulated, nor are there concrete suggestions about how to reach the safe operating space and what the best pathways are to achieving sustainable consumption; this is debated. 3. Green growth pathways rely on the transformative capacity of the economic system and respect consumer sovereignty. 4. (Global) consumption patterns that recognize these conditions do not necessarily remain within environmental limits and so characterize a weak sustainable consumption path. 5. De-growth paths demand shrinking economic activities, investment in research, and policies that lead to a change of systemic structures, 6. A development favouring sustainable consumption and 'sustainable shrinking' would require enabling mechanisms, framing conditions for the economy, strong sustainable consumption governance, and convincing research to show that it does not lead to adverse consequences, such as the collapse of economic or social security systems, and that it is completely different from austerity or recession situations.

In chapter 27, *Hongmin Chen* (China) explored "Sustainable Consumption and Production in China" as a primary way of addressing the constraints on economic development caused by a large population, severe resource scarcity, environmental pollution and ecological degradation. China's consumption has grown rapidly, in particular among the middle class and because of the shift to a consumption-driven economy. This increases the need for *sustainable consumption and production* (SCP) as a national strategy for economic and social development that can decouple economic development from environmental degradation. The key messages were: 1. SCP is still facing crucial challenges as extensive economic growth, uneven development, imbalanced expenditure between consumption and investment, and rising inequality in income, accompanied by overconsumption, are challenging the coordination of development and protection and is thus aggravating problems of resources and environment. 2. Weak capacities for scientific and technological innovation and underdevelopment of non-governmental organizations in China leave the government as the main player in SCP, although long-term capacity for the improvement of SCP lies mainly in markets, independent technical innovation in enterprises, and extensive social participation by civil society. 3. SC has not received as much attention as SP from the government and the public because of the relatively low proportion of consumption in the GDP expenditure structure. 4. China's further reforms are expected to lift institutional barriers, and to include the optimization of fiscal and tax systems, adjustment of the performance evaluation system, transition of government functions, expansion of market spaces and social forces, and the construction of the rule of law. 5. With great efforts, China has made achievements in SP in some regions, especially improvements to production efficiency where the pace of improvements appears inadequate to offset the rapid increases in consumption. 6. SCP has not been clearly established as a development strategy in China, where it should be conducted more systematically and efficiently and better integrated in order to reduce duplication of effort and minimize conflicts. 7. Energy-saving and environmental protection policies and practices include the adjustment of the industrial structure, promotion of a circular economy and clean production, the Energy Conservation and Emission Reduction Project, green procurement, and progress in waste management.

In chapter 28, *Philipp Schepelmann* (Germany), *René Kemp* (The Netherlands) and *Uwe Schneidewind* (Germany) presented "The Eco-restructuring of the Ruhr Area as an Example of a Managed Transition", demonstrating the importance of regional actors and factors, alongside external stimuli. Since the decline of the coal and steel industries was associated with major job losses, the Ruhr district has been in transition for four decades. The chapter provides a narrative of a real-world laboratory for ecological modernization which became revitalized by investments and projects. The chapter offered relevant lessons for transition management: 1. The managed transition consisted of three phases: a) After Willy Brandt's promise of blue skies, pollution control and policies for nature conservation helped to establish an eco-industry (1961–1990). b) Starting with the IBA

Emscher Park, came the ecological reconstruction, clean-up and urban revitalization of the Ruhr district (1989-2015). c) The sustainable energy transition initiated regional innovation policies, supported by state and national energy programmes (2010 onwards). 2. The transition frame is the yet unfinished socio-economic structural change away from coal mining and heavy industries, together with growing environmental awareness and protection, building on existing regional energy clusters and infrastructures. 3. The region achieved economic strengths in the leading markets of resource efficiency and sustainable consumption, above the share for Germany and the state of North Rhine-Westphalia, taking advantage of the German *Energiewende*. 4. The eco-restructuring confirms many elements of transition management: a vision shared by most actors, platforms for strategic thinking, special innovation projects acting as stepping stones for transition processes, and special transition institutions. 5. Institutional entrepreneurs follow different agendas: air quality for health reasons, environmental amenities, preserving industrial heritage, quality of places, labour in the Ruhr area and halting depopulation. 6. The importance of transition showcases and real-world experiments to build a constituency behind a product (firms, researchers, public authorities) was noted. 7. It is difficult to determine the role of flagship projects in the transition when success depends on factors like the proximity of cooperative actors in the state. 8. Regional actors and city governments are more cooperative and result-oriented than national ministries. 9. Ecological restructuring and energy transition have top-down and bottom-up elements, where state authorities support innovation policy for regional specialization, while national authorities and the European Commission are funders of ecological restructuring projects.

Belinda Yuen (Singapore, China) and *Asfaw Kumssa* (Kenya, Sweden) analysed, in chapter 29, "Transition towards Sustainable Urbanization in Asia and Africa", two of the world's least urbanized but fastest urbanizing regions, where cities are engines of growth, with high population densities, congestion, expansion of slums, infrastructure shortages and environmental degradation. As rapid urbanization continues, how Africa and Asia manage the urban transition affects urban economic efficiency and greenhouse gas footprints. This chapter compared the urbanization patterns and transitions of both regions and discussed challenges and opportunities for sustainable urbanization. Major lessons highlighted were: 1. To abandon business-as-usual practices that impede economic growth and efficiencies, lead to social and environmental risks, and ignore the realities of climate change and resource scarcity, African and Asian cities have to act towards sustainable urbanization in a way that is more inclusive, resilient and ecologically efficient. 2. Various Asian and African cities have begun sustainable urbanization with greener growth and economic prosperity, including regulation; green building, research and technology; and business and community energy-saving and emission-reduction. 3. Cities need to invest in infrastructure and to focus on pro-poor and greener growth strategies to reduce poverty, create jobs and improve living conditions. 4. Municipal governments and urban authorities should mainstream policies leading towards energy saving, fewer sprawls and climate-resilient infrastructure, that address drainage, flood control, housing, sanitation, transport and water distribution systems. 5. Rather than focusing on single pathways to sustainable futures, the transition is a complex and negotiated process allowing for some levels of failure and opportunities to learn. 6. An institutional and policy framework will mutually reinforce sustainable infrastructure and the management of natural assets, economic growth, the conservation of resources and the resilience of livelihoods in changing urban circumstances and at different stages of development. 7. Such effort needs to build collaboration among different urban actors (government, private sector, NGOs, people) to identify policies suitable to local conditions and needs.

In chapter 30, *Gregory Trencher* (USA, Canada) and *Xuemei Bai* (Australia, China) studied "The Role of University Partnerships in Urban Sustainability Experiments: Evidence from Asia". Starting from the observation that Asian research universities are forming cross-sector partnerships and implementing sociotechnical experiments for urban sustainability, the authors examined fifteen cases from Singapore, Japan, Hong Kong and Korea for knowledge co-production and urban sustainability experiments. They found common attributes (focus areas, actors, motivations and mechanisms) and investigated functions, motivations, barriers and significance of roles by differing societal sectors. Their key findings were: 1. University partnerships addressing urban sustainability are dominated by technical approaches, with numerous R&D platforms and competitive funding schemes privileging the engineering and physical sciences, although the capacity of technocentric approaches to reform lifestyles and trigger societal transformations to sustainability is limited. 2. Since human aspects are

significant barriers in sustainability experiments (for example, time restraints, lack of unity, poor management and leadership), and since the human and societal dimensions are crucial for the co-design and co-production of knowledge, more funding opportunities for the social sciences and humanities are needed to engage university actors and civil actors in social innovation-orientated sustainability experiments. 3. The ability of partnerships to tackle complex social issues and trigger societal transitions is often constrained by existing research projects and institutional capacities, and calls for interdisciplinarity in sustainability partnerships to address potential tensions and obstacles. 4. University partnerships demonstrate a promising capacity to generate knowledge, technologies, tools and societal transformations for advancing urban sustainability, and potentially serve as 'umbrellas' or 'portfolios' for societal experiments and interventions (i.e. protected spaces or 'niches' of innovation). 5. There is a wealth of mechanisms (e.g. knowledge management, governance and planning, technology transfer or economic development, technical demonstrations and experiments, reform of the built and natural environment, sociotechnical experiments) to co-produce knowledge, co-implement experiments and magnify societal impacts by the transfer of lessons.

42.2.8 Examining Sustainability Transitions in the Water, Food and Health Sectors from Latin American and European Perspectives

In the twentieth century, international politics focused on hydrocarbons. During the summit in Rio de Janeiro in 1992 an understanding emerged of the complex linkages among energy, soil, water, biodiversity and ecosystem services and population growth and land use changes due to rural, urban and productive processes.

Cecilia Tortajada (Mexico/Singapore) and *Martin Keulertz* (Germany/US) argued in chapter 31, "Future global water, food and energy needs", that "scarcity, pollution, mismanagement and misallocation of natural resources will impact every sector on which humankind depends for survival. In a globalized economy with increasingly free movement of commodities and financial and human capital, a poor understanding of the most pressing issues and their interconnectedness and interdependences will cause irreparable damage to the Earth and its billions of people: a clear case of *fait accompli*." The Millennium

Ecosystem Assessment (MA 2005) claimed that the changes in ecosystem services could produce non-linear, chaotic, unpredictable and often irreversible tipping points in the environment. Water is crucial for life and most productive processes, especially for agriculture and food production, which still consume seventy per cent of global water. The nexus between water, food and energy links the productive, service and domestic sectors, promotes or inhibits socio-economic development, and creates well-being and health globally. As there can be no peaceful development with empty stomachs and chronic undernourishment, both authors argued that water, energy and food security is "at the heart of the development challenges of the twenty-first century". In the past these resources were globally abundant and their use was considered free. At present both drastic environmental deterioration and economic globalization are threatening not only these three resources and their security, but could also directly affect the survival of humankind through polluted water and food scarcity. Population growth, increased demands for hygiene and new productive processes require more water, food and energy and thus put pressure on ecosystem services. Scarce and polluted resources also require different forms of international and local collaboration that involve all stakeholders. The authors concluded that "getting decision-makers to overcome silo thinking is key to achieving new policy outcomes. It is indeed the mega-challenge of our time."

In chapter 32, "Sustainability Transition in a Vulnerable River Basin in Mexico", *Úrsula Oswald Spring* (Mexico) analysed climate impacts in the context of increased organized crime in Central Mexico. Abrupt slopes, numerous affluents, high population density in the floodplain, laissez-faire politics, illegality, governmental corruption, and public insecurity have produced high levels of dual environmental and social vulnerability. Using survey data, she examined the risks perceived by the people and how they reduced their socio-economic and security risks. She explored cooperation between people and different levels of government in extreme events, and asked how the management of disaster risk reduction could be enhanced, the social tissue be reinforced, and the impact of criminal acts be avoided. For dealing with these complex emergencies, she suggested that a sustainable transition process requires the linking of socio-economic, political, environmental, judicial, cultural and psychological factors to develop adaptation and resilience from the bottom up. She proposed a new governance model, where the political arena

needs to be changed at the local level from vulnerability to human and gender security. Human security for the people requires freedom from fear, want and hazard impacts, as well as freedom to live in dignity. Empowerment, equality, and equity are key mechanisms for establishing this new arena. She proposed that all stakeholders should develop a common agenda, with new activities and agents to change this complex emergency into sustainable livelihood and well-being for the most vulnerable people, including women and girls. Oswald Spring concluded that the transition to sustainability in the Yautepec river basin began with a common problem analysis involving crucial stakeholders and a desirable future vision including short-, middle- and long-term agendas for social (equality), economic (development), and gender (equity) issues.

In chapter 33, *Monica de Adrade* (Brazil) examined "Sustainability Transition in the Health Sector in Brazil". Brazil faces demographic, environmental and climate problems aggravated by migration and globalization. To address these challenges, over the past two decades the government of Brazil has developed multisectorial and universal public policies, whose objectives are universal health care, zero hunger, family health, and conditional cash transfer for extremely poor people. This has improved the indicators of extreme poverty, hunger, under-five mortality, maternal health, infectious and vector diseases, and primary education. During the last two decades Brazil has changed: the traditional epidemiological pattern of a poor country has given way to the modern diseases of industrialized countries, and cardiovascular diseases, cancer and traffic accidents now dominate. She linked environment policies to health issues, such as biofuel, hydro-energy, solar power plants, and the reduction of deforestation in the Amazon Basin, with a reduction in forest burnings and smoke emissions. These policies are combined with better urban planning and the promotion of public transport, and this has improved air quality. She showed that the health infrastructure and services are insufficient because of inadequate financial support. Efficiency, quality and the optimal use of investments may consolidate the achievements in the health sector and promote the transition process in Brazil. This requires a drastic reduction of income inequality and intersectorial collaboration between health, education and social development. This would optimize existing resources and support a sustainability transition through new investment in the Brazilian health and social sector.

42.2.9 Preparing Sustainability Transitions in the Energy Sector

Chapters 34 to 39 in part nine discuss the issues of sustainable energy transformation (Jürgen Scheffran and Rebecca Froese), low-carbon transition (Karlson Hargroves), wind energy technology transitions in India, Brazil and South Africa (Britta Rennkamp and Radhika Perrot), sustainability transitions and the politics of electricity planning in South Africa (Lucy Baker), Brazilian policies and the politics of energy and a low-carbon green economy (Eduardo Viola and Larissa Basso), and sustainable electricity transition in Thailand and the role of civil society (Carl Middleton).

In chapter 34, *Jürgen Scheffran* and *Rebecca Froese* (Germany) focused on "Enabling Environments for a Sustainable Energy Transformation: The Diffusion of Know-How, Technologies and Innovations in Low-Carbon Societies". The chapter started from the assumption that a sustainability transition addressing global warming and access to energy requires the diffusion of innovations, investment and learning to support transformation processes, including new technologies, products and infrastructures, as well as new social rules, norms and interactions. Their major conclusions included: 1. Developing North–South transitions into green economies requires an enabling environment for the innovation and dissemination of climate-friendly energy systems, including financial support, smart governance, and new social rules and norms in order to develop the economic, sociopolitical and technological capacities of all countries. 2. Key elements are the international diffusion of know-how, investments and technologies to establish market-creation mechanisms, including production capacities, the demand for low-carbon goods and international cooperation between business and government to adapt technologies to local contexts in a low-carbon society. 3. New market opportunities arising from changes in consumer behaviour could be tapped, including markets for organic food, goods produced under responsible practices (fair trade, no child labour, fair treatment of workers), and low-carbon products. 4. Governments can take up various policies to encourage investment in low-carbon and energy efficient production and transportation systems, including public procurement of low-carbon products and technologies; energy performance standards or mandatory energy labelling schemes; renewable portfolio standards; blending mandates for alternative fuels; taxes for high-emission technology or tax incentives for green electricity; tendering procedures for green electricity or green certificate systems; and feed-in tariffs. 5. Technology increasingly flows through private networks of information and assessment services, management consultants, financial firms, lawyers and accountants as well as through technical specialist groups. 6. Due to the failure of top-down and technology-centred approaches, participatory approaches involving community institutions, local public agencies, consumer groups, industry associations and NGOs are essential for successful technology transfer. 7. Multi-level interactions, social learning and institutions can help adapt technologies to local contexts.

In chapter 35, "Considering a Structural Adjustment Approach to the Low-Carbon Transition", *Karlson Hargroves* (Australia) argued "that the scale and pace of economy-wide changes required to significantly reducing greenhouse gas emissions are akin to … the application of *structural adjustment programmes* (SAPs) in developing countries". His key question was: "How can lessons from structural adjustment programmes inform the low-carbon transition, and how can a strategic approach be developed to deliver carbon structural adjustment?" After discussing the low-carbon transition (35.2) and efforts towards creating low-carbon economic multipliers (35.3), Hargroves examined the impacts of structural adjustment programmes (35.4), efforts towards harnessing structural adjustment mechanisms for carbon management (35.5), and a proposed carbon structural adjustment road map (35.6). His key messages were that "economy-wide adjustments to key structures [involving] multiple actors across society" are needed that could deliver "sustained economic growth, job creation, and poverty eradication". He proposed a reassessment "of the policy and industrial strategy used to achieve development", arguing that "efforts to achieve a low-carbon transition will need to be balanced with the level of embedded infrastructure in each economy", as a transition of "the economy to a low greenhouse gas emissions future [requires] sophisticated strategies". He proposed that a '*carbon structural adjustment*' may be needed "as a solution to unsustainable development", and that we should draw "on lessons learned … to progress the sustainable development and climate change response agendas, and merge this knowledge into an economy-wide structural change agenda". He recommended "six key areas for carbon structural adjustment, namely standards and codes, government requirements, taxation and subsides, investment and procurement policies, professional accreditation requirements, and education programme accreditation and requirements". He suggested that for each action we should consider "its likely 'Carbon Impact Factor', and the likely 'Willingness to Adjust'".

In chapter 36, "Drivers and Barriers to Wind Energy Technology Transitions in India, Brazil and South Africa", *Britta Rennkamp* (Germany/South Africa) and *Radhika Perrot* (South Africa) provided a framework for the analysis of drivers and barriers to successful technological development (36.2), and applied it to wind energy technology innovation in three countries (36.3). They concluded that different combinations of the seven TIS (*technological innovation systems approach*) functions predominated in these countries, where each function might be crucial for the success or failure of a wind energy innovation system. Knowledge diffusion, entrepreneurial activities, experimentation, and resource mobilization were locked in to each country's existing knowledge, industrial structures and institutions. This induced positive externalities and initiated a process of legitimation within each country. They had a 'powerful' technological trajectory in common, based on fossil fuels, or large-scale hydropower in the case of Brazil, and this impeded switching from one trajectory to an alternative. The authors concluded that Brazil and India had succeeded in catching up because major companies had accumulated innovation capabilities independently of public policy incentives, thus speeding up the development of wind technology systems, while South Africa had had to start almost from scratch. They described a partnership with a company in a mature wind energy market that could trigger the transfer of knowledge and technology. While in Brazil a combination of resource mobilization, entrepreneurial experimentation and knowledge diffusion had led to innovation capabilities, in India there was a combination of entrepreneurial experimentation, influence in the direction of search, and knowledge diffusion from foreign joint ventures. Another crucial innovative factor was the enabling environment, for example, market size and trade barriers, where joint ventures stimulated knowledge creation and entrepreneurial experimentation and influenced the direction of search and eventually legitimation.

In chapter 37, "Sustainability Transitions and the Politics of Electricity Planning in South Africa", *Lucy Baker* (UK) defined a 'just' transition to sustainability (37.2), offered an analytical framework for sociotechnical transitions (37.3), examined the challenges of electricity governance (37.4), reviewed South Africa's electricity sector in context (37.5), and discussed the politics of electricity planning (37.6). She concluded that South Africa's minerals–energy complex still retains a key role in decision-making processes in electricity policy, and that the role of politics in the relationship between energy and industry is central because energy-intensive users are also the country's major coal mining enterprises and have vested interests in both supply and demand. She further argued that in order to meet the needs of the country's energy-intensive export-oriented growth trajectory, South Africa's IRP (*Integrated Resource Plan* for electricity) has simultaneously permitted a low-carbon niche, supported the expansion of the incumbent coal-based regime, and facilitated the potential introduction of nuclear energy. Despite plans for significant diversification in the electricity mix, Baker noted that demand is still projected to increase significantly, although much more slowly than projected; this reduces the possibility of a transition to sustainability due to a fundamental disconnect between the creation of a lower-carbon electricity mix and industrial demand. She further claimed that while the IRP will facilitate a renewable energy niche, its ability to address issues of energy poverty and access to energy for the poor is limited. She stated that this plan will not deal with the consumption-led model of economic growth or the high levels of national socio-economic inequality in consumption and access to power, and she concluded that any claims of a transition to sustainability and its governance and policy must be evaluated from a holistic perspective, including social, economic, environmental, financial and technological factors.

In chapter 38, "Low-Carbon Green Economy: Brazilian Policies and the Politics of Energy, 2003-2014", *Eduardo Viola* and *Larissa Basso* (Brazil) reviewed the debate on a low-carbon green economy (38.2), discussed the Brazilian energy matrix and its transformations (38.3), examined Brazilian energy policy and politics (38.4), and criticized crony capitalism and short-term thinking as major stumbling blocks to a low-carbon green economy (38.5). They concluded that Brazil had not so far adopted the low-carbon green economy model, and that the share of non-renewable energy sources had increased during the last decade, because of a lack of long-term policies towards renewables. Two opposite trends dominated Brazil's energy policy: while from 2003 to 2007 the share of renewables increased, from 2007 to 2014 it declined due to new deep offshore oil reserves and the use of oil prices as a policy tool to maintain higher rates of economic growth. Energy efficiency remained underdeveloped because a lack of policy, with no clear direction towards decarbonization because crony capitalism opposed investment in energy efficiency and renewable energy sources. In Brazil, energy security depends on cost-effective long-term quality and reliability of the supply and the transmission systems. The authors claimed that Brazil lacks strategic planning from government policies down to household budgets and that measures often simply respond to urgent problems and create contradictory situations. The authors concluded that only with quality education, transparency, accountability and rules that challenge privileges in Brazilian society can a sense of community and belonging, still absent in Brazilian society, develop.

In chapter 39, "Sustainable Electricity Transition in Thailand and the Role of Civil Society", *Carl Middleton* (UK/Thailand) examined the role of civil society in Thailand's sustainable electricity transition (39.2), analysed the emergence and embedding of Thailand's incumbent electricity regime (39.3) and the role of civil society in sustainable electricity transition (39.4), and discussed efforts towards decentralizing and democratizing electricity (39.5). Taking a multilevel perspective, Middleton explained the emergence and opposition to change of Thailand's electricity regime, which since the 1980s has been transformed from a state monopoly into a partially-privatized structure. Taking a political economy approach, he analysed Thailand's energy regime as a coalition of actors, including the state utility, its labour unions, and various independent power producers. Structural changes had been caused by shifts in 'landscape' at the global and national scales. While civil society groups supported the changes, elites had tried to limit their role. Incremental reforms were accommodated by the regime, including the growing contribution of distributed renewable energy generation and energy conservation, the creation of an independent regulator, and a small increase in civil society participation and accountability of the power planning process. Middleton argued that civil society groups had challenged the regime, opposed problematic projects, advocated a progressive policy, and proposed alternative plans, values and visions for Thailand's electricity sector. In his view, "the ST approach and MLP has helped explain why the structure of the incumbent electricity regime has remained relatively intransient, despite the argued-for benefits of electricity sector reform, and from the perspective of ongoing broader environmental, social and political challenges in Thailand". Reviewing the sustainability transition management literature, he pointed to strategies and policies that could transform Thailand's incumbent electricity regime in the direction of sustainability.

42.2.10 Relying on Transnational, International, Regional and National Governance for Strategies and Policies Leading Towards Sustainability Transition

In chapter 40, "Governance of Sustainable Development in Knowledge Democracies–Its Consequences for Science", *Roeland J. in 't Veld* (The Netherlands) examined sustainable development, knowledge democracy, transdisciplinarity, and an agenda for future science organization. He argued that transdisciplinarity contributes to tackling the 'wicked' problems of sustainable development, and that knowledge democracies address the organization of science in society. Independent disciplinary scientific research becomes more complicated given the growing interdependencies between politics, media and science. Reflexive mechanisms complicate the reactions of the public to scientific findings. Scientific activities are performed by many organizations, governments and enterprises, by NGOs and individual researchers, and by specialized research organizations, such as universities. Governments stimulate and finance research to further national competitive positions and innovation, and to contribute to an attractive cultural climate. As sustainable development is long-term, multilevel and complex, consistent and stable science policies have free access to scientific sources for local and regional public authorities and citizens' initiatives. Innovation originates in local and regional communities, so free access is necessary for participatory democracy. The Humboldtian university is not interested in transdisciplinary trajectories, while the American university has rendered services to society based on centuries of institutional value formation. European nation states developed advisory bodies and semi-autonomous agencies to perform boundary work between science and politics. If a knowledge democracy is accompanied by populist tendencies, political hostility to these intermediaries grows fast. Sustainability problems necessitate transdisciplinarity but political conflicts are fought by playing two-level games. As the political controversy may lead to a dead end, an attractive strategy is to accuse the knowledge source of the opponent of a lack of trustworthiness. This indirect accusation will not stimulate the willingness of scientists to participate in transdisciplinary trajectories. In advanced societies the quantitative dimensions of research activities will increase, as large proportions of the population create new knowledge and productive activities become more knowledge-intense. Transdisciplinarity and participatory democracy are more closely interwoven. Scientists who are successful in transdisciplinarity have insights into the rationality of politicians. Adequate governance arrangements for acceptable equilibria between different modes of knowledge production make necessary a redesign of the external and internal organization of scientific institutions. Peaceful coexistence between different research orientations in universities, and the organization of science, are focal issues for sustainable development.

Finally, in chapter 41, "Discourse and Practice of Transitions in International Policy-Making on Resource Efficiency in the EU", *Sander Happaerts* (Belgium) examined sustainability transitions and international policy-making, focusing on policy implications, the green economy, and resource efficiency in the EU. Happaerts' key messages were: The transitions discourse in the EU's resource efficiency policy demands changes in our way of life. With this discourse policy officials wanted to strengthen awareness that deeper societal changes are needed to achieve a more sustainable use of natural resources. In the context of international green economy policies, transitions thinking puts the emphasis on deep change and a comprehensive systems approach, and its principles and assumptions differ from a radical approach. Its economic framing limits the policy choices and focus on improved market dynamics. This discourse on the EU's resource efficiency policy is more than a change of terminology, and many policy components show that transitions thinking as symptoms of a change in the mindset of EU policy-making circles remains below the level of a true transition. The adoption of a long-term, systemic perspective and the awareness of the need for fundamental changes clash with the organization and routines of EU policy-making, which resist radical change. EU policy is built predominantly on incremental policy-making. Transition governance looks more like regular policy-making than theoretical transition governance. In the transition towards sustainable materials management in Europe, policies and institutional modifications for fundamental changes are not yet in place. Future research should address three questions about international policy-making in sustainability transitions. Scholars should assess whether transitions thinking is permeating EU policies. The example of regional policy suggests that other EU policy processes are also moving towards fundamental and systemic changes. Additional cases beyond the EU are needed to integrate the author's

case study into a universal framework. From the perspective of multilevel governance, research needs to address how such international policy can be linked to bottom-up processes of innovation. Some are propelled by ambitious national and subnational governments, and they are likely to become a model for diffusion when their ideas are picked up at the international level.

42.3   From Disciplinary, Multi-, Inter- and Transdisciplinary Approaches towards Transformative and Anticipatory Science

The linkages between sustainability transition and sustainable peace require bridge-building between different scientific disciplines in the natural and social sciences and different research programmes of political science: on the one hand between environmental and development studies, especially those that focus on sustainable development, and on the other between peace and security studies. This requires a fundamental shift from narrow disciplinary and programme-specific approaches to *multi- and interdisciplinary perspectives* as well as *transdisciplinary and transformative research designs and policy proposals*.

Each discipline has its specific epistemology, its premises and its methods of generating new knowledge. As the problems and issues that need to be examined scientifically become more complex, multidisciplinarity offers a first step in analysing complex problems from different disciplinary perspectives. These multidisciplinary studies rely on the methodologies of their respective disciplines.

The Swiss scholar Jean Piaget[4] who worked simultaneously in different disciplines, e.g. in developmental psychology, cognitive theory and genetic epistemology, pioneered a new transdisciplinary scientific approach, which resulted in educational reforms. Piaget promoted communication among different disciplines, and in the 1960s he proposed using the term 'interdisciplinary'. He applied interdisciplinarity to pedagogic units or modules in order to integrate knowledge from different disciplines. This interdisciplinary approach was taken up by new approaches and fields, such as bioengineering and brain sciences.

Brauch, Dalby and Oswald Spring (2011) suggested the creation of a 'political geo-ecology' for the Anthropocene by linking the 'geoecology' approach of physical geography (Huggett 1995) with the 'political' dimension of the human security approach within political science and peace and security studies. The goal was to sensitize social scientists to take the results of the natural and earth sciences on global environmental change into account and to ask natural scientists to bring the political and security dimensions into their research paradigms and teaching on *Earth Systems Science* (ESS) and *Earth Systems Analysis* (ESA). Within the social and political sciences, Oswald Spring, Brauch and Tidball (2014) proposed a 'peace ecology' approach that would encourage bridge-building research and discourse between two parallel research programmes, each of which often ignored the results of the other (see chapter 3 by Stephenson).

Given the complexity of the Anthropocene, of global environmental change and of resource scarcity, several research centres and think tanks proposed transdisciplinarity as a new scientific approach to overcome the narrow disciplinary boundaries of specialized subfields and epistemic schools of knowledge creation. Darbellay (2015: 165) has argued that

> the increasing interest in ITD [inter- and transdisciplinarity] reflects a growing awareness of the multidimensional complexity of research contexts and objects and, hence also, of societal issues that require the greater synergization of institutionalized disciplinary skills. Today, the implementation of this type of approach responds to a need within the community of researchers, who must provide answers on a daily basis to theoretical and practical questions that are highly complex and not reducible to a single disciplinary point of view.

For Hirsch Hadorn, Hoffmann-Riem, Biber-Klemm et al. (2008) and Jaeger and Scheringer (1998), transdisciplinarity refers to "the cause of the present problems and their future development (*system knowledge*)"; to

the "values and norms ... [to] be used to form goals of the problem-solving process (*target knowledge*)"; and to "how a problematic situation can be transformed and improved (*transformation knowledge*)". They argue that "transdisciplinarity requires adequate [ways of] addressing ... the complexity of problems and the diversity of perceptions of them, that abstract and case-specific knowledge are linked, and that practices promote the common good" (Hirsch Hadorn/Hoffmann-Riem/Biber-Klemm et al. 2008; Pohl/Hirsch Hadorn 2007; Jaeger/Scheringer 1998; Mittelstrass 2003; Brand/Schaller/Völker 2004).

While "*multidisciplinarity* draws on knowledge from different disciplines but stays within their boundaries" (Choi/Pak, 2006), a definition of transdisciplinary and interdisciplinary research[5] states:

> *Transdisciplinary Research* is defined as research efforts conducted by investigators from different disciplines working jointly to create new conceptual, theoretical, methodological, and translational innovations that integrate and move beyond discipline-specific approaches to address a common problem. *Interdisciplinary Research* is any study or group of studies undertaken by scholars from two or more distinct scientific disciplines. The research is based upon a conceptual model that links or integrates theoretical frameworks from those disciplines, uses study design and methodology that is not limited to any one field, and requires the use of perspectives and skills of the involved disciplines throughout multiple phases of the research process.

In short, *transdisciplinarity* refers to a research strategy that establishes a common research objective that crosses disciplinary boundaries. The goal is to create a holistic approach by addressing complex problems that require close cooperation between several disciplines, such as brain or cancer research or issues of global environmental change, where medical, behavioural, environmental, economic and political sciences work together. Funtowicz and Ravetz (1993) argued that "transdisciplinarity can help determine the most relevant problems and research questions involved" (Funtowicz/Ravetz 1993, 2003).

Holistic system analysis (Güvenen 2015) also contributed to *transdisciplinary* research, which includes all possible aspects and focuses on the interaction among different elements. *Transdisciplinarity* takes a structural approach (Nicolescu no/year) and distinguishes between different levels of analysis. The surrounding conditions facilitate dynamic adjustment of undesirable disturbers. Of particular interest is a sys-

---

4　Jean Piaget and Barbara Inhelder (1969) argued that meta-thinking and behaviour are not simply acquired from the outer world but emerge from within, in relation to the outside. His constructivism is embedded in social, economic, environmental, political, gender and cultural structures. According to Piaget's epistemology the development of a child's knowledge came about through understanding of the chronological sequences and the structural interrelations among each of these processes, their advances and sometimes also their regressions. He used empirical data to show how a simple structure becomes more complex, then global, and how this generalizable structure may permit a sustainable transition process. From this epistemology comes the idea of complex structures, processes, and social interactions, where a sustainable-engendered peace may help overcome greed, social differentiation, discrimination, and violence.

5　"Definitions"; at: <http://www.hsph.harvard.edu/trec/about-us/definitions/> (4 February 2016).

temic dissipative and self-regulating approach, based originally on Ilya Prigogine and the thermodynamic understanding of processes (Prigogine/Stengers 1977: 184) and Haken's (1983) *Synergetics*.

These 'dissipative and open systems' operate outside and mostly far from thermodynamic equilibrium in a setting where energy and matter are exchanged. Prigogine characterized the 'dissipative structures' with the spontaneous appearance of anisotropy[6] or symmetry-embracing processes, where complex and often chaotic structures interact and create unpredictable new system formations. These dissipative systems are part of a permanent dynamic process which creates a new equilibrium among the existing structures and substructures, but also among the flows at different levels. The outcomes are permanently changing processes and new structures, which are far from equilibrium but able to maintain some dynamic functionality within the global system.

Niklas Luhmann (1991) applied dynamic system analysis to sociology and used the term 'autopoiesis', which originally described and explained the nature of a living system (Maturana 1996, e.g. a biological cell consisting of various biochemical components such as nucleic acids and proteins which are organized and bound in a cell nucleus). Luhmann's term 'autopoiesis' refers to the complexity of dynamic systems which interact with the complexity of the environment. Luhmann insisted on the radical nature of the concept and assessed five key characteristics: autonomy, emergency, operative closure, self-structuration and autopoietic reproduction. These elements are essential for the analysis of new risks and uncertainties caused by changes in the environment and social behaviour in the Anthropocene.

In chapter 5, Uwe Schneidewind, Mandy Singer-Brodowski, and Karoline Augenstein proposed moving from a 'transdisciplinary' approach to a 'transformative science', while in chapter 17 Swilling suggested an 'anticipatory science'. The concept of

'transformative research' or 'science' has been used since the 2000s for a new approach that cuts across the dominant scientific paradigms. The US National Science Board (2007) adopted the following working definition of 'transformative research': "[it] involves ideas, discoveries, or tools that radically change our understanding of an important existing scientific or engineering concept or educational practice or leads to the creation of a new paradigm or field of science, engineering, or education. Such research challenges current understanding or provides pathways to new frontiers".

Building on this approach, in *World in Transition–A Social Contract for Sustainability*, the WBGU (2011: 21-23, 321-356) referred to "four transformative pillars of the knowledge society": 'transformation research' and 'transformation education', as well as 'transformative research' and 'transformative education'. It proposed (2011: 21) that 'transformation research' should "specifically addresses the future challenge of transformation realisation" by exploring "transitory processes in order to come to conclusions on the factors and causal relations of transformation processes" and should "draw conclusions for the transformation to sustainability based on an understanding of the decisive dynamics of such processes, their conditions and interdependencies. ... Transformative research supports transformation processes with specific innovations in the relevant sectors and it should encompass, for example, "new business models such as the shared use of resource-intensive infrastructures, and research for technological innovations like efficiency technologies" by aiming at a "wider transformative impact". Uwe Schneidewind and Mandy Singer-Brodowski (2013) and Maja Göpel (2017) have developed this transformative approach further for climate policy and for research on sustainability transition.

The International Social Science Council (ISSC 2012: 21-22) in its report on the *Transformative Cornerstones of Social Science Research for Global Change* identified six cornerstones: 1) historical and contextual complexities; 2) consequences; 3) conditions and visions for change; 4) interpretation and subjective sense-making; 5) responsibilities; and 6) governance and decision-making. The report concluded that

> the transformative cornerstones framework speaks to the full spectrum of social science disciplines, interests and approaches—theoretical and empirical, basic and applied, quantitative and qualitative. By not fashioning a global change research agenda around a substantive

---

6   According to *Encyclopedia Britannica* at: <http://global. britannica.com/science/anisotropy>, '*anisotropy*' refers in physics to "the quality of exhibiting properties with different values when measured along axes in different directions. Anisotropy is most easily observed in single crystals of solid elements or compounds, in which atoms, ions, or molecules are arranged in regular lattices. In contrast, the random distribution of particles in liquids, and especially in gases, causes them rarely, if ever, to be anisotropic. A familiar example of anisotropy is the difference in the speed of light along different axes of crystals of the mineral calcite."

focus on concrete topics—water, food, energy, migration, development, and the like—the cornerstones are not only inclusive of many social science voices but, perhaps most importantly, show that climate change and broader processes of global environmental change are organic to the social sciences, integral to social science preoccupations, domains par excellence of social science disciplines. … The transformative cornerstones of social science function not only as a framework for understanding what the social sciences can and must contribute to global change research. They function as a charter for the social sciences, a common understanding of what it is that the social sciences can and must do to take the lead in developing a new integrated, transformative science of global change.

The seventh conference of the *Sustainability Transitions Research Network* (STRN) in September 2016 will take as its theme "Exploring Transition Research as Transformative Science". Contributions have been invited that "[focus] on the conceptual and methodological challenges of transformative science, i.e. research that is actively involved in societal transformation processes". The speakers are invited "to reflect on the challenges and lessons learned in concrete research projects, on theoretical contributions advancing our understanding of transitions, and on the role of science and scientists involved in transitions to sustainability".[7]

Various initiatives by the US National Science Board (2007), the ISCC (2012), and the STRN (2016) have called for a new scientific paradigm in research into both global environmental change and sustainability transitions. The policy dimension should be included in the research design, by moving from knowledge creation to action, to policy initiatives, development and implementation.

These efforts are still highly dependent on the top-down efforts of governments and multinational enterprises. In the South, meanwhile, many excluded stakeholders have for decades put into practice a transformative education, e.g. by implementing the pedagogy of liberation inspired by Paulo Freire (1970, 1975). These excluded social groups promote transformative processes from their daily situation of marginalization, violence and exclusion, and promote sustainable livelihoods not for elites, but for wider social groups (Benavenides 2015; Silvia/Tavares 2015).

42.4 Need for Research: Deficits and Mapping

There have been few systematic intellectual exchanges between the early US pioneers on sustainability transitions such as the Tellus Institute in Boston, the suggestions by Speth (1992) and the Report of the NRC (1999), and the *Dutch Knowledge Network on Systems Innovations and Sustainability Transition* (2005–2010) that has since influenced the sociotechnological debate in the social sciences in Europe. Both approaches largely excluded the demand side, the values, preferences and behaviour of human beings as customers and citizens.

We need to integrate the contributions of the different disciplines on the spatial, scientific, economic, societal and cultural dimensions of sustainability transition into a wider and more holistic approach. The report on a *Social Contract for Sustainability* (WBGU 2011) took the earth systems and the global economic and social megatrends (chapter 1) into account as well as changing values (chapter 2), stages of the great transformation (chapter 3), and its technical and economic feasibility (chapter 4), as well as transformation governance (chapter 5) and the agents of transformation (chapter 6), including 'transformative science' and education (chapter 7). This approach was taken up by UNDP and was discussed by scholars from developing countries in June 2012 in Rio de Janeiro. So far, most approaches have been top-down and not bottom-up and have often involved only policymakers, opinion leaders in society, scientists and a few businesspeople. However, there have also been many grassroots movements calling for slow growth or 'degrowth' strategies (<http://www.degrowth.org/>), primarily for the industrialized countries.

In both the top-down and bottom-up approaches and initiatives, there has been no relationship between the call for sustainability transition and the normative goal of a sustainable peace and processes of conflict resolution and negotiation. The root causes of exploitation, discrimination, violence, neo-colonialism, and authoritarian leadership have been ignored. Greed has prevailed and since the 1970s has resulted in a rising level of social inequality and the accumulation of wealth by the top one per cent of the population, the super-rich. So far, only a few modest efforts by governments and social movements in Africa, Asia, Latin America, Europe and Australia have begun to link the transition process with sustainable peace initiatives.

---

7 See at: <http://www.ist2016.uni-wuppertal.de/fileadmin/ist2016/Call_for_Papers_IST_2016.pdf> (6 February 2016).

Research on 'sustainable peace' exists in the context of peace psychology (see chapters 6 and 7 by Coleman and Deutsch), and of conflict prevention and conflict and post-conflict peace-building, but it scarcely exists in the context of environmental problems, nor is there a focus on 'peace with nature'. An attempt to widen this narrow approach was made by Oswald Spring in chapter 8 by bringing in the debates on the root causes of violence with gender and patriarchy, and in chapter 9 by Brauch, who discussed the concept in the context of GEC in the Anthropocene. Since the 1990s the ecumenical movement has been calling for justice, peace and the preservation of creation, and more recently Pope Francis issued *Laudato Si* (2015), an encyclical on the environment. This is the first papal statement on the issue directed not just to the Catholic church but to "all people of good will".

So far hardly any research results exist on linking the process of 'sustainability transitions' with the normative goal of a 'sustainable peace'. The emerging debate on climate change and security (Gleditsch 2012; Scheffran/Brzoska/Brauch et al. 2012) in the Anthropocene (Dalby 2007, 2009, 2013, 2014, 2015, 2016; Brauch 2009) and on the nexus between water, food, energy and biodiversity security (Oswald Spring 2016) has crossed the boundaries between researchers working on GEC and those who work on security, both from a narrow military perspective and on human security from a liberal security and peace research perspective.

Only a few scholars have addressed the linkages between environmental and peace studies (Kenneth Boulding in the 1960s and 1970s; Conca 2002; Amster 2014; Oswald Spring/Brauch/Tidball 2014 and the other authors of the IPRA's *Ecology and Peace Commission* (EPC)). However, so far there is no systematic conceptual, theoretical, empirical and scenario research that addresses the possible consequences and impacts of strategies, policies and measures aiming at 'sustainability transition' on international relations, on international and national security or on 'sustainable peace'.

Robert W. Kates and Thomas M. Parris (2003: 8062) discussed "Long-term Trends and a Sustainability Transition" and found "10 classes of trends, which makes them complex, contradictory, and often poorly understood". They argued that "each class includes trends that make a sustainability transition more feasible as well as trends that make it more difficult. Taken in their entirety, they serve as a checklist for the consideration of global trends that impact place-based sustainability studies."

Using the NRC (1999) report, Kates and Parris (2003: 8062) added "material related to peace, security, and globalization" and described "10 major classes and 26 long-term global and regional trends that make a sustainability transition more feasible as well as more difficult". They indicated a "paradox of sustainable development [that] lies in the need to meet human needs through growing affluence and economy and the threat to the planet that such affluence poses through rapid and growing consumption". In their view, sustainability transition "reflects a process of globalization or interconnectedness, emerging forms of governance, and changing institutions and values" (Kates/Parris 2003: 8063). They argued that "favourable shifts in investment, income, and job opportunities in some parts of the interconnected world are accompanied by the loss of jobs elsewhere, unpredictable withdrawals of capital, and a deepening divide in needed skills and innovation, all adding new sources of instability for the world."

The post-development agenda called the *Sustainable Development Goals* (SDG) promotes, in its goal 16, 'Peace and Justice': peaceful and inclusive societies with access to justice for all and the building of institutions that are accountable, inclusive, effective and equal for all citizens, regardless of gender, race, caste, social class or age. Pisano, Lepuschitz and Berger (2014: 23) noted that the UN Secretary-General's High-Level Panel of Eminent Persons on the Post-2015 Development Agenda proposed

> that two out of 'five transformative shifts' for the Post-2015 agenda should consider to (a) *build peace and effective, open and accountable institutions for all* (with responsive and legitimate institutions that encourage the rule of law, property rights, freedom of speech and the media, open political choice, access to justice, and accountable government and public institutions), and (b) *forge a new global partnership* that involves governments but also other stakeholders, including people living in poverty, civil society and indigenous and local communities, multilateral institutions, local and national government, the business community, and academia.

Without taking up this suggestion, they reviewed the "sustainability transition initiatives undertaken in six EU member states (Austria, Belgium, Finland, France, Germany, and [The] Netherlands)" but without discussing the prospective impacts of these sustainability transitions initiatives on European and international peace and security. Nevertheless, all the attempts mentioned basically addressed processes of decarbonization but did not go further to processes of dematerialization, the recovery of ecosystem services, or resource recovery. The peace dividend was overlooked.

42.5   From Knowledge to Action: Sustainability Transition with Sustainable Peace

There is little or no scientific knowledge about the possible consequences and impacts of global and national strategies, policies and measures dealing with national and international peace and security or achieving the goal of sustainable peace with increased global equity and social justice. Policy-relevant considerations on the linkages between "Sustainability Transition with Sustainable Peace" are even scarcer.

How can we move from knowledge to action to achieve 'sustainability transition' with 'sustainable peace' in the Anthropocene? 'Anticipatory' and 'transformative' research and science mean integrating a 'proactive policy perspective' into research design. There must be a shift away from the dominant policy perspective of business-as-usual. This leads to ignoring the challenges and postponing action (or to non-action) and calls for 'reactive policies' of adaptation by technical and military means. In this way the aim is to contain the causes instead of addressing them in a sustainable scientific way and developing a process of 'anticipatory' sustainability transition.

However, this alternative paradigm and vision has almost totally ignored the dimension of international peace and security. A proactive policy perspective requires a systematic analysis of the constraints and opposed political and societal actors at national and local levels. This should include, for example, businesses and workers involved in the fossil fuel industries, especially in coal, natural gas and oil and in the fossil fuel and nuclear energy sectors.

Apart from global climate change, losses in biodiversity and increased levels of soil degradation and desertification and degradation, scarcity and stress of water have received less scientific attention; similarly, there is less public awareness of these issues and they are accorded a much lower political priority. For all these sectors both 'anticipative' scientific research and 'proactive' policy activities are needed.

The systemic model shown in figure 42.1 presents an analysis of natural and anthropogenic impacts on the earth system. There are several doubts about the impacts of chemical pollution and so new research is needed in order to obtain safe scientific data. The greatest impact on the earth system, following data provided by Rockström, Steffen, Noone et al. (2009), is loss of biodiversity, and this affects flora and fauna, ecosystem services, soil composition and food production, but also the mitigation for climate change.

The nitrogen flow is another threat. Together with other national and international institutions the FAO (2013) has promoted a soft approach to agriculture. They encourage the composting of organic waste and nitrogen fixation of microorganisms from the air to the soil (*Azospirillum, Rhizobium*), as well as a association of crops and mixed farming.

The melting of glaciers is a further danger. Schellnhuber and Martin (2014) argued that the ice shields of the Antarctic and of Greenland, as well as other glaciers, could trigger dangerous tipping points, while in the Indian subcontinent, in parts of China and in the Andes progressive melting of glaciers (Painter 2007) has occurred. This may create threats from glacier break-ups, as well as leading to a lack of drinking water for people who depend exclusively on meltwater.

As discussed in the introductory chapter, climate change (IPCC 2013, 2014a, 2014b), soil erosion, loss of natural fertility and desertification, and loss of phosphorous may seriously impact on food production (FAO 2016). This poses a threat to an ever-increasing population with changing dietary habits and calling for more meat and dairy products.

Finally, lack of water and increasing pollution has become another serious threat. Some conflicts are related to drought, lack of pasture for pastoralists, and lack of water facilities (e.g. in the Sahel countries, as also in Kenya, Sudan, etc.). During the last century, the world's population increased threefold and water use sixfold, partly because of changing habits of hygiene, but also because of massive water pollution from mining and oil extraction. In dryland areas, electricity production has caused serious threats to survival, further exacerbated by climate change and extreme events. The anthropogenic impacts of global environmental change and extreme events, together with business-as-usual behaviour with respect to natural resources, have put pressure on the natural cycles of nitrogen, phosphorous, sulphur, and dioxide.

Resource management has become more complex because of GEC. It now includes alternative approaches to managing water and energy, the restoration of ecosystems, and the recovery of depleted soils (figure 42.2). If 'transformative' science and progressively 'dematerialized' technologies can be implemented, effective strategies, policies and measures of mitigation and adaptation may emerge, and sufficient financial support, effective governmental and societal institutions, and active participation by citizens and stakeholders may come about. Social and environmental vulnerability could be reduced, and the resil-

Figure 42.1: System approach to natural and anthropogenic factors of global environmental change. **Source:** Oswald Spring (2016), based on data from Rockström, Steffen, Noone et al. (2009).

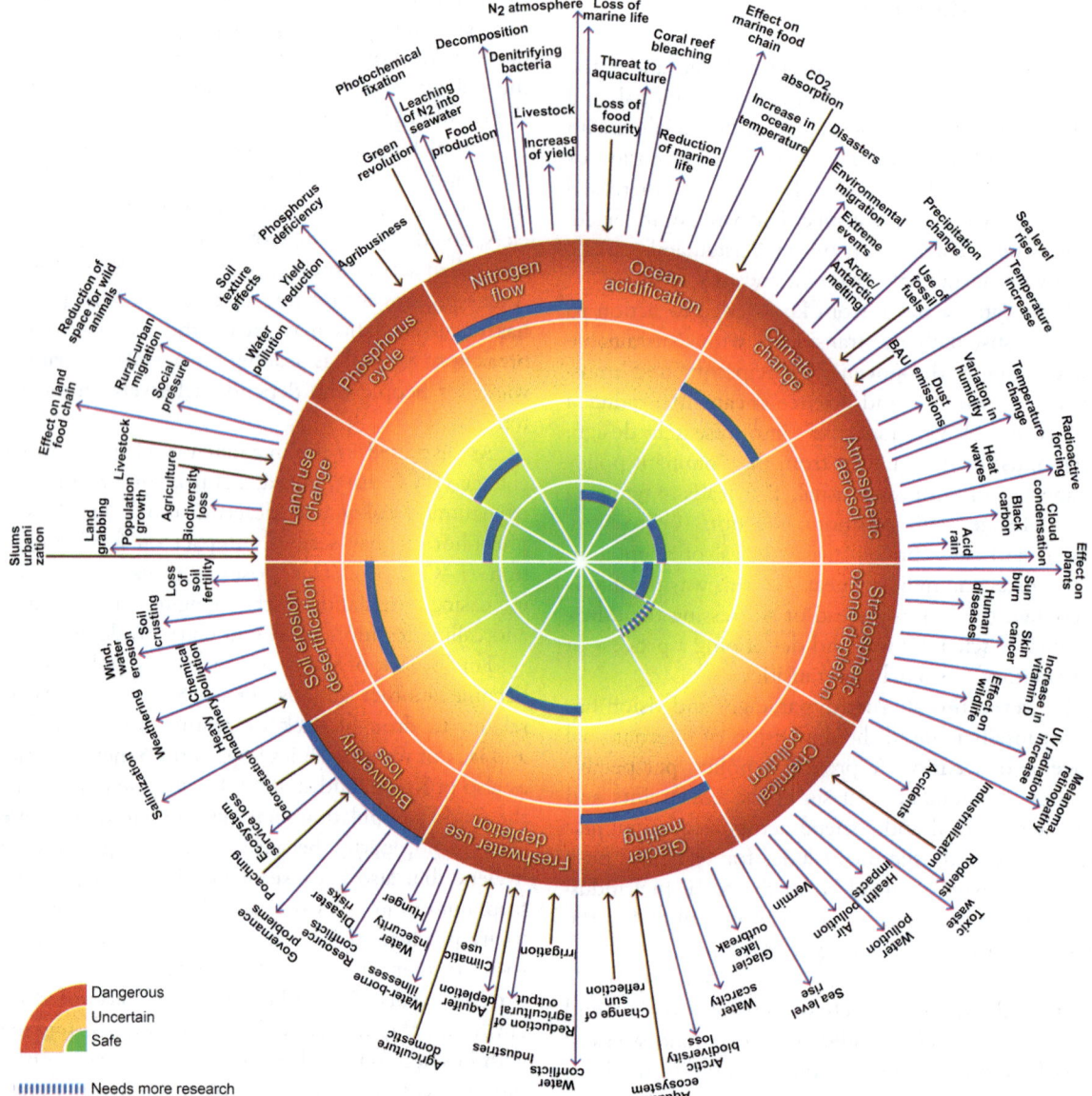

ience and the well-being of all social groups could be strengthened. While GEC and CC have already become 'threat multipliers', extreme weather events and disasters could destroy the development path of many poor countries (IPCC 2012), creating hunger, causing death, and impelling affected people to migrate (Oswald Spring 2012; Oswald Spring/Serrano Oswald/Estrada Álvarez et al. 2014).

Nevertheless, with a solid development path which is human-oriented and takes the most vulnerable into account, new threats could be better managed through disaster risk reduction ((McBean/

Ajibade 2009), and water, food, energy, health and sanitary security. It will be possible to restore and maintain the social tissue, create employment, consolidate transparent institutions, and maintain or recover essential ecosystem services.

Finally, there is the question of how 'sustainability transition' with 'sustainable peace' may be achieved in the Anthropocene by moving from knowledge to action. The above analysis has not specifically addressed the international, security and peace aspects. It was argued in the introduction that a process of sustainability transition may make two conflict

Figure 42.2: Transition to sustainability. **Source:** Oswald Spring (2016).

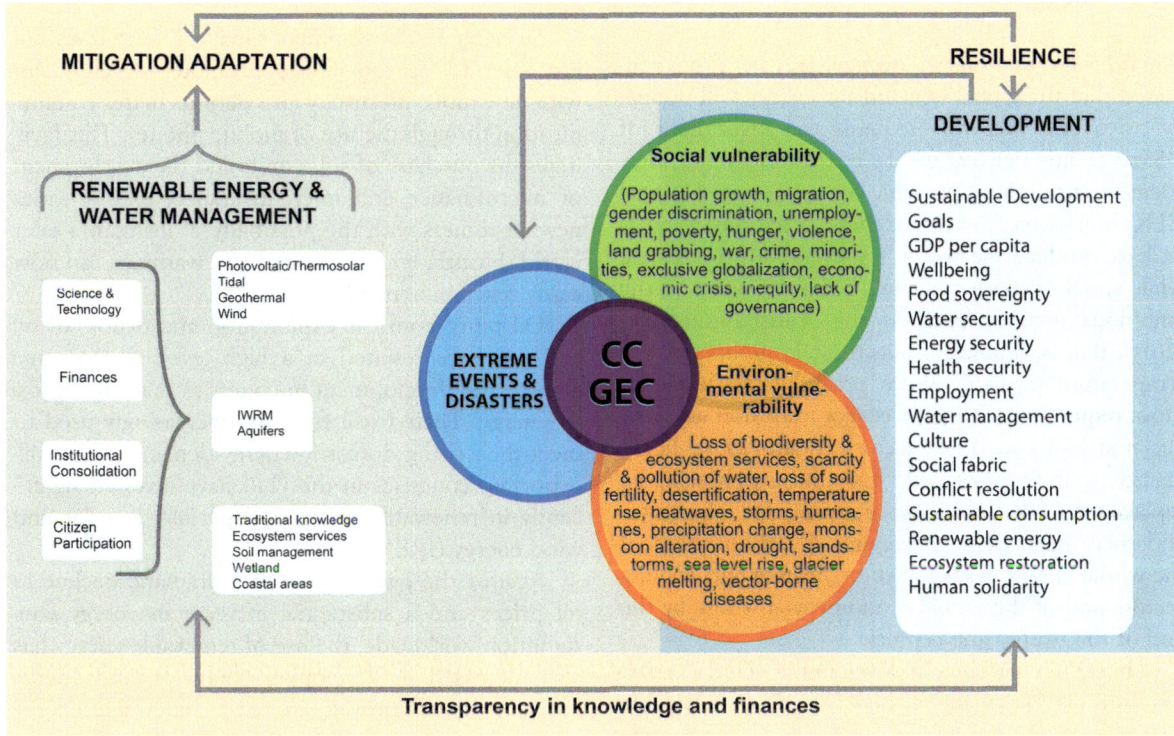

constellations less likely: resource conflicts and climate-induced violent confrontations.

Peace is not a stable process and there are different approaches to a sustainable-engendered peace. There cannot be worldwide peaceful behaviour if water, food, energy and well-being are not available to the 9.5 billion of people forecast for 2050 and the twelve billion in 2100. However, the SDGs adopted in 2015 as a global pathway to achieving these goals will remain a paper tiger unless governments can persuade their elites to implement them as targets at national and local levels. This requires effective national policies to reduce poverty and social inequality. Mahatma Gandhi said "the world has enough for everyone's need, but not enough for everyone's greed".

A major change is needed in the prevailing mindset, from favouring a business-as-usual approach towards supporting more sustainable behaviour and culture. It requires, first of all, peaceful negotiations at international, national and local levels on the use and efficient management of scarce resources to give reduced environmental footprint. This may result during this century in a decarbonized, dematerialized sustainable world with constructive human relations and care and solidarity for the most vulnerable. However, this means that scientists need to integrate 'transformative' perspectives in their research designs that will

contribute to 'proactive' knowledge, and policymakers need to have the courage to move from mere policy declarations to proactive policies that will produce the desired results.

This transformative scientific approach towards proactive policies cannot be systematically developed here. Rather, just four examples will be offered for a sustainable energy transition: firstly, moving from fossil fuels to enhancing energy efficiency and to renewables so there can be access to energy for up to twelve billion people by 2100 while GHG emissions can be reduced (42.5.1); secondly, shifting from resource- and carbon-intensive agriculture and a high degree of waste in the food sector to climate-smart agriculture with a lower level of waste (42.5.2). The third and fourth examples address proposed changes to totally different lifestyles in industrialized countries (42.5.3), and a shift in values as suggested, for example, by the kingdom of Bhutan (Gross Happiness Index) and by indigenous people in Bolivia (Pachamama) and Chiapas (42.5.4). These alternatives will not be globally acceptable but they indicate that a continuation of the Western way of life based on abundance and waste in consumption is not sustainable for a global population reaching twelve billion people in the year 2100.

42.5.1   From Fossil Fuels to Renewable Energy Sources

The *International Energy Agency* (IEA 2014) has estimated that the global demand for energy will double by 2050 and that this requires the investment of US$2.5 trillion per year for the next twenty years. Electricity is currently responsible for a quarter of global GHG emissions. To achieve these development goals and to produce electricity for everybody on Earth, while simultaneously reducing GHG emissions to the ambitious goal adopted in the 'Paris Agreement' (2015), that is, limiting the increase in global average temperature to 1.5°C above pre-industrial levels by 2100, requires a sustainable energy transition and similar transitions in other sectors (figure 42.3). Fossil energy sources (coal, oil, gas) must gradually be replaced by renewables (solar, wind, etc.) by 2050, and there must be a progressive decarbonization of the whole energy, transportation, heating and cooling sectors and of the global economy as a whole by the end of the twenty-first century.

The UN Environment Programme (UNEP 2015) has forecast the many co-benefits of a sustainable energy transition towards renewable and clean energy; impacts on ecosystems will be reduced, especially eutrophication and acidification. Human health will also benefit from the limiting of particles, toxicity, and GHG emissions. Furthermore, the use of many resources such as concrete, metals, energy intensity, water use, and land use may be reduced and the resources reused; all this will contribute to a process of dematerialization.

The year 2014 was the first year in four decades that saw a 'decoupling' of global economic growth from an increase in carbon dioxide emissions (UNEP 2011, 2014; see also chapter 17 by Swilling). This encouraging sign may be a turning point in countering the worst outcomes of GEC and CC. The rapid increase in renewable energy sources in developing countries (led by China, India et al.) may offer the 1.3 billion people without access to electricity the possibility of obtaining energy in a decentralized way that is cheaper than through the traditional centralized electricity grid. There are still ninety million African children who have no access to electricity at school and who cannot do their homework after sunset. The economic, social and personal benefits of electrification based on clean energy may overcome regional inequalities and promote health and education across gender and ethnic barriers, as well as increasing income and well-being and thus contributing to social equity.

Sustainable Development Goal (SDG) 7 does not just refer to energy; it also promotes development with new tools. Electricity also permits better communication through the use of mobile phones. This facilitates the purchase of solar units and the development of microfinance and micro-insurance, and provides new customers with the possibility of living in a safer world. Health, emergency and risk warnings can now easily reach billions of people.

Oil rents from the exploitation and export of oil and gas have resulted in a high level of economic growth and development in countries that export fossil energy. Their fossil fuels are increasingly used to meet their rising domestic energy demand. Many oil-exporting countries in the Gulf have invested significantly in renewable energies, especially in solar and wind energy (IRENA 2016).[8]

Against the background of a dramatic decline in oil prices and a substantial increase in energy consumption worldwide, the use of renewable energy has also increased in developing countries, and energy efficiency has improved. In 2013, "renewable energy provided an estimated 19.1% of global final energy consumption. … Traditional biomass, used primarily for cooking and heating in remote and rural areas of developing countries, accounted for about 9%, and modern renewables increased their share slightly over 2012 to approximately 10.1%". From the 10.1 per cent of modern renewables: 4.1 per cent comes from biomass, geothermal and solar heat; 3.9 per cent from hydroelectricity; 1.3 per cent from biomass, solar, geothermal and wind power; and 0.8 per cent from biofuel. Most of the global final energy consumption (78.3 per cent) came from fossil energy, with 2.6 per cent from nuclear (REN21 2015: 27).[9]

About 58.5 per cent of net additions to global power capacity came from wind, solar *photovoltaic* (PV) and hydropower (REN21 2015: 30). In late 2014, "renewables comprised an estimated 27.7 per cent of the world's power generating capacity, enough to supply an estimated 22.8 per cent of global electricity, with hydropower providing about 16.6 per cent" (REN21 2015: 30). Energy support policies and subsi-

---

8   IRENA, 2016: *Renewable Energy Market Analysis: The GCC Region* (Abu Dhabi: IRENA); at: <http://www.irena.org/DocumentDownloads/Publications/IRENA_Market_GCC_2016.pdf>.

9   REN 21, 2015: *Renewables 2015–Global Status Report* (Paris: UNEP).

Figure 42.3: From diagnosis to remedy: Sustainability Transition and Sustainable Peace: **Source:** Hans Günter Brauch (2016), with many suggestions from Úrsula Oswald Spring.

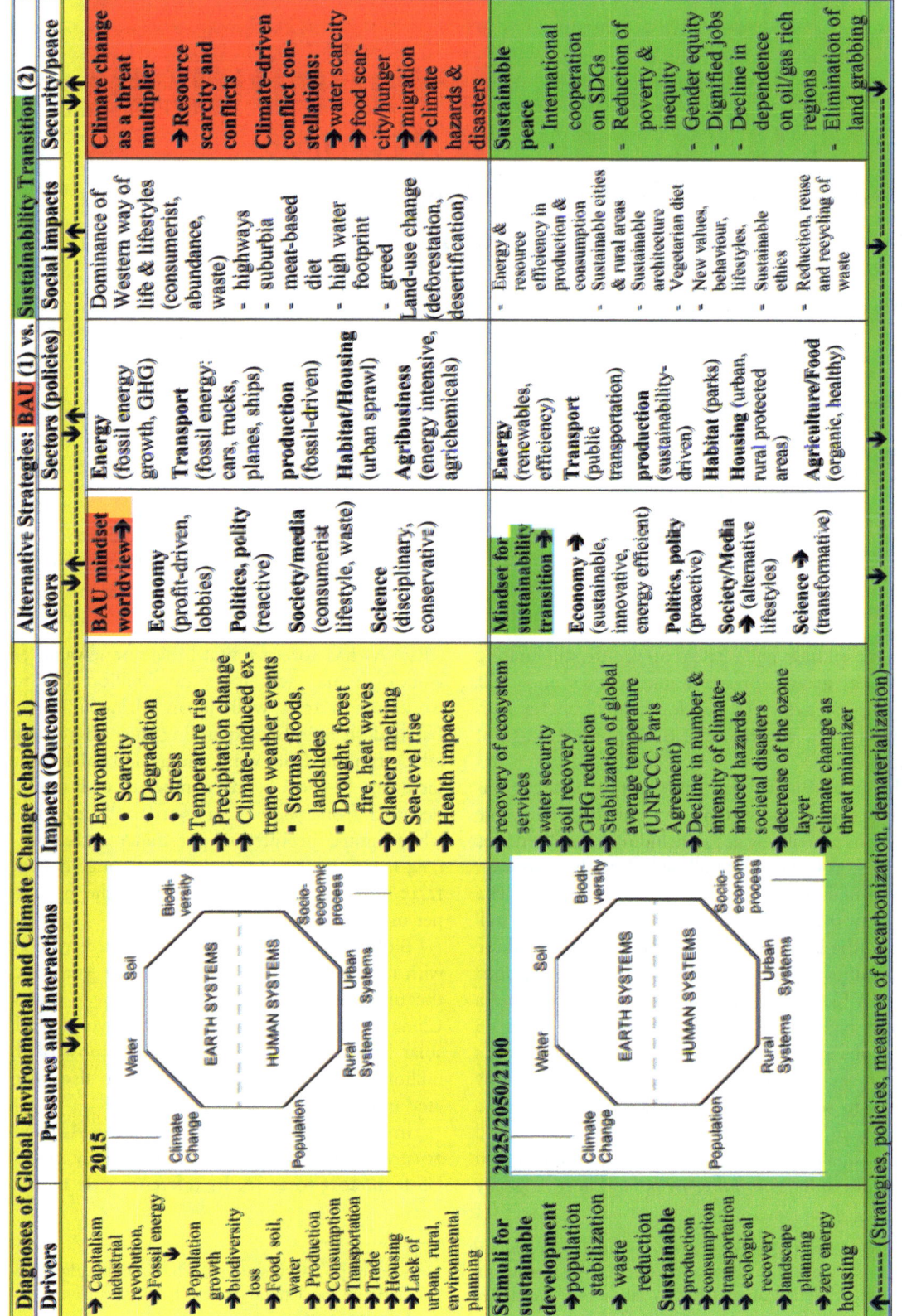

Figure 42.4: :Renewable Energy Employment by Technology. **Source**: IRENA: *Renewable Energy and Jobs-Annual Review 2016*; at: <http://www.irena.org/DocumentDownloads/Publications/IRENA_RE_Jobs_Annual_review_2016.pdf>.

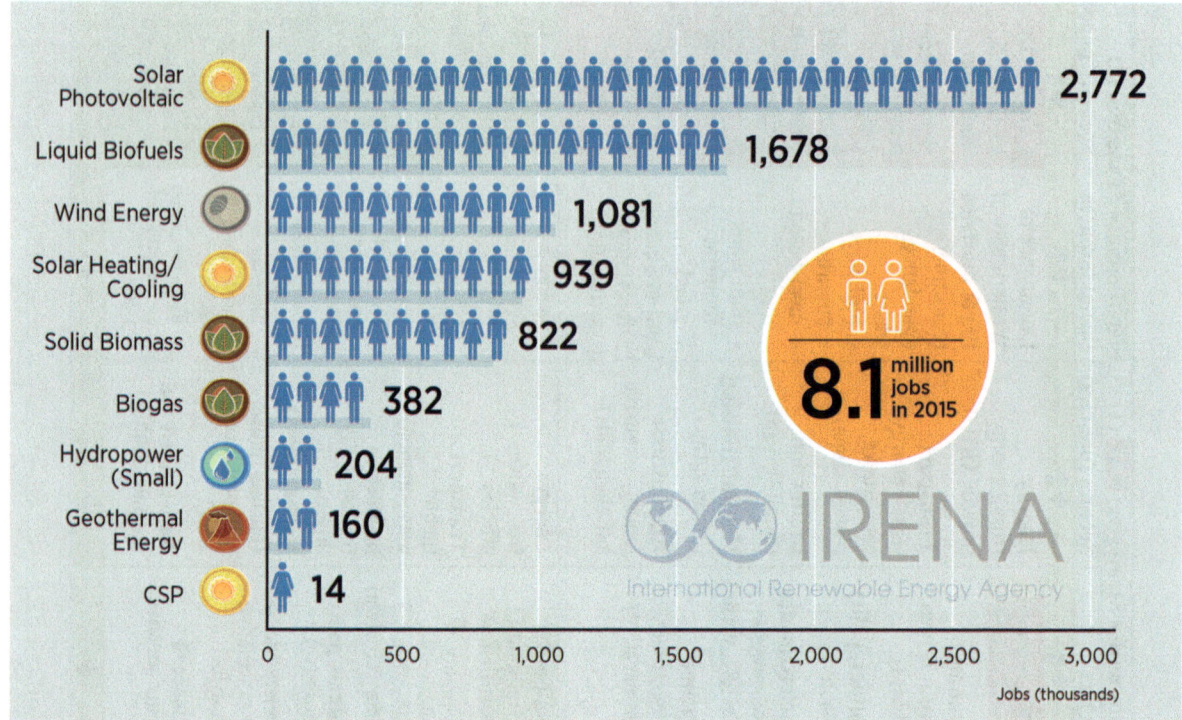

dies for fossil fuels and nuclear power are still limiting the present growth of renewable sources and their cost-competiveness. The development of energy storage systems for all sectors also supports the use of renewables .

The transition to renewable energy sources has already created more than 7.7 million renewable energy jobs globally by 2014, including 2.5 million jobs in solar photovoltaics, 1.5 million in large hydropower, and one million in wind energy (figure 42.4).[10] The majority of the jobs have moved from Europe (1.2 million) and the US (724.000) to Asia, especially to three BRIC countries: China (3.390.000), Brazil (934.000) and India (427.000), two of which China and India have very high GHG and CO₂ growth rates. With 1.6 million jobs in the solar photovoltaic industry, China's share of the world's solar manufacturing capacity amounts to seventy per cent, and it has become the world leader for jobs in wind power, solar heating and cooling, small and large hydropower, biomass and biogas. The *International Renewable Energy Agency*

(IRENA) has forecast that India's renewable energy workforce may add about one million jobs to the workforce by 2022, while Japan, Malaysia, Korea, and Bangladesh are also expanding their share of green jobs, primarily from wind energy and solar PV manufacturing. In the US, the number of coal miners has declined from 784,621 in 1920 to 80,209 in 2013, while annual production per miner increased from 0.84 (1920) to 15.01 (2000) and has since declined to 12.25 (2013).[11] From 1920 to 2000, the productivity per miner rose eighteen fold.

Biofuel jobs are concentrated in feedstock supply, with 1.8 million (Brazil, US, South East Asia), and half the one million jobs in wind have been created in China. China has also become the world leader in solar heating and small hydropower; most of the 1.5 million jobs in large hydropower have also been created in China.

In the US, the *Energy Information Agency* (EIA 2015: 40) has projected the future energy mix of the US from 2013 to 2040. Its reference case model pro-

---

10 IRENA, 2016a: *Renewable Energy and Jobs-Annual Review* 2016; at: <http://www.irena.org/Document-Downloads/Publications/IRENA_RE_Jobs_Annual_Review_2016.pdf.

11 See: <http://www.sourcewatch.org/index.php/Coal_and_jobs_in_the_United_States#Total_coal-related_jobs> (6 February 2016).

Figure 42.5: Renewable electricity generation in the US by fuel type in the reference case, 2000–2040 (billion kilowatt hours). **Source:** EIA (2015: 41).

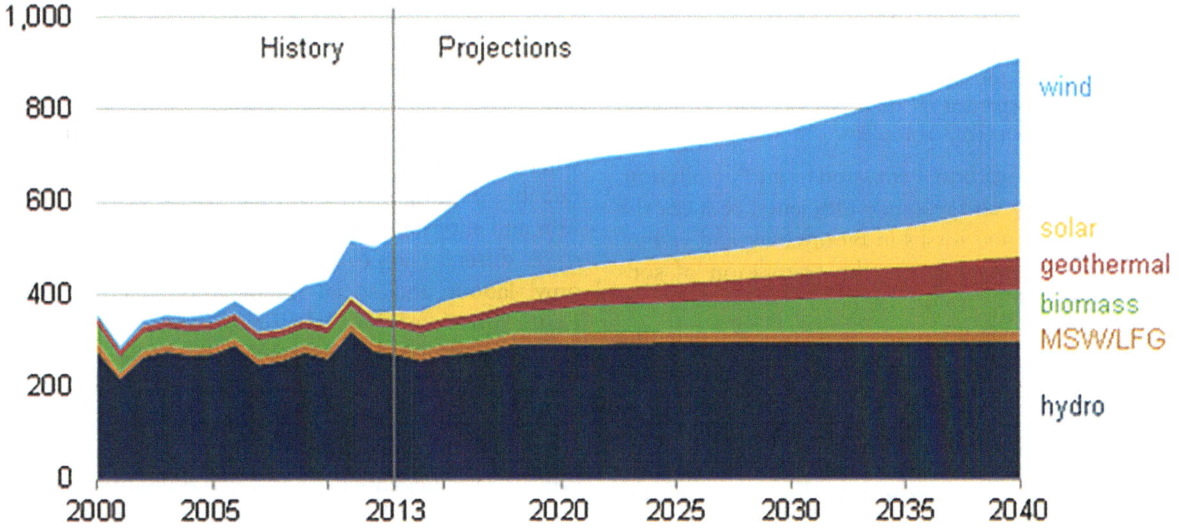

MSW/LFG stands for municipal solid waste and landfill gas.

jects that the share of natural gas will increase from 27 per cent to 31 per cent and of renewables from 13 to 18 per cent, while the share of nuclear energy is projected to decline from 19 to 16 per cent, and of coal from 39 to 31 per cent; other liquid fuels may account for one per cent (figure 42.5). According to this official US assessment, renewables would account only for a modest share of the energy mix. The EIA (2015) has also analysed the evolution of renewables in the generation of electricity. Figure 42.5 projects an important increase in wind and solar energy by 2040, less for geothermal and biomass, and a steady figure in the case of hydropower.

Renewables also have an important impact on human and environmental health. During 2015 improvements were achieved in the storage and sequestration of GHG emissions: chemical absorption of CO_2 from flue gas (post-combustion), physical adsorption from synthesis gas (pre-combustion), and the combustion of fossil fuels with pure oxygen producing CO_2 and water are mechanisms for capturing and storing carbon dioxide. Life-cycle GHG assessments have shown that PV and wind energy have comparatively the lowest impacts of GHG emissions, and these emissions will continue to decline between now and 2050. The impacts of ionizing radiation, photochemical oxidant formation, particulate matter formation, human toxicity, and ozone depletion are also the lowest in renewable energies. This offers co-benefits for human health and health care (REN 21 2015:24), as well as for the environment and for ecosystem ser-

vices. Renewables, a carbon-free and dematerialized economy, waste reduction and recycling help realize a different way of life based on a different world view, one that calls for deep changes in culture and its relationship with the environment and with ecosystem services. A transformation is also needed in the relationships between human beings; greed, exploitation, violence and discrimination need to be replaced by solidarity, confidence, care, commitment and mutual responsibility for others, especially for the vulnerable, but also for the environment.

42.5.2 From Agribusiness and Hunger to a Sustainable and Equitable Livelihood

Despite the MDGs of 2000, by 2015 the level of global hunger had remained quite high (GHI 2105). The FAO (2013), IPCC (2014a) and other international and national institutions have called for an agriculture with drastic reductions in the carbon, water, soil, biological and ecosystem service footprints (Torquebiau 2016). Confronted with the complexity of climate change, FAO (2011a) developed the concept of *'climate-smart agriculture'* (CSA). This refers to "agriculture that sustainably increases productivity, resilience (adaptation), reduces/removes greenhouse gases (mitigation), and enhances the achievement of national food security and development goals" (FAO, 2016a: 548). CSA is not a new agricultural system or a different set of practices.

CSA shares with sustainable development and green economy objectives and guiding principles. It aims also for food security and contributes to preserve natural resources. As such, it has close links with the concept of sustainable intensification, which has been fully developed by FAO for crop production (FAO 2011b) and is now being extended to other sectors and to a food chain approach (FAO 2013a: 27).

CSA reduces carbon emissions and pollution, enhances energy and resource efficiency, reduces the water footprint and losses in biodiversity and ecosystem services, and promotes the restoration of soils and forests. FAO (2016b) widened the concept and integrated the whole food chain into it, including waste management and losses in field, harvest, staple and market.

This new approach to a safe food system also needs to be regionally consolidated to diminish the carbon footprint for transport. It should use organic production processes to restore soils that have suffered a drastic decline through the use of agrochemicals. It should promote local market structures and food transformation. The IPCC (2014a) also stated that half of the world's food is produced by women or small-scale farmers, usually in orchards, while highly technological agriculture uses massive amounts of fossil energy, depletes soil and water, and produces basically grains for biofuel and livestock. A climate-smart agriculture can not only reduce up to thirty-two per cent of GHG emissions related to modern agriculture and change in land use, but it may also improve the local food system, restore soils, compost organic waste, reduce transport, limit the carbon footprint nationally and locally, recover ecosystems and improve the livelihood of poor people.

Pretty (2006) suggested integrating biological and ecological processes, with the goal of improving the nutrient cycle. Organic agriculture can increase crop yields by up to seventy-nine per cent. When combined with agroforestry, aquaculture, mixed farming, rainwater harvesting, water retention during the rainy season and soil conservation, these practices can restore natural environments, recover human health, restore ecosystem services and improve food quality for consumers, at affordable prices. Biofertilizers naturally fix nitrogen from the air to the soil and regenerate impoverished soils, while biopesticides efficiently combat pests. Nitrogen fixation from the air to the soil (*Azospirillum, Rhizobium*), organic fertilization, mixed agriculture, rotation of crops and soils, and biopesticides not only reduce production costs, but also open new markets for organic food products and help the restoration of soils and the recovery of eco-

system services. These interlinked processes minimize non-renewable inputs that harm the environment, conserve and save water, and reduce production costs. Fewer agrochemicals also mean that natural and human health is protected. Links between science and modern technology, together with traditional knowledge and practices, facilitate adaptation everywhere, improve environmental security, promote sustainable development, and limit the need for migration and survival strategies.

A different approach to the environment, economy, labour and social interactions challenges the present processes of globalization and the powerful role of multinational enterprises (Scholte 2005). It aims at human security instead of greed and financial accumulation. Hunger, undernourishment and obesity create a vicious circle along with ignorance, poverty, disease, dependency and cultural discrimination.

Locally-based food production, also called 'food sovereignty', reduces the dependence of national governments on the world economy and on short-term investments, and inhibits speculation on food staples. Food sovereignty increases national savings, creates local jobs and develops creativity, science and technology; in addition, it has positive effects on health and well-being. A skilled labour force finds local jobs, and this stimulates the internal market and reduces foreign investment and restrictions on social development imposed by the IMF, the World Bank and the *World Trade Organization* (WTO). Productive systems that are stable at the local and regional level reduce dependence on international trade and prevent the cultural homogenization of consumerism. They reinforce local productive niches and increase regional food security and food sovereignty (Oswald Spring 2009). A healthy economy produces savings and reduces international debt and high-interest payments. This may also allow the use of public funds for the development of infrastructure, for education, and also for adaptation and resilience-building measures.

Positive linkages between a safe environment, restored ecosystem services, a healthy population and cultural diversity also allows the creation of new political arenas, in which a participative society can contribute to an agenda of sustainable well-being. Corrupt public actors can be viewed critically, and corruption itself countered more effectively, as public activities are reoriented to the well-being of the people, including the mitigation of the impacts of GEC. Such postmodern governance (see chapter 41 by in t' Veld.) may reorient the relationships between science, the mass media, and the government.

No longer will a small global oligarchy of the super-rich and powerful national economic and political elites benefit most. This has been widely documented as a reason for the rapid increase in social inequality since the 1970s, as wealth is concentrated in fewer and fewer hands. People and opinion leaders need to develop a totally different political and economic mindset and a new societal and scientific world view (Oswald Spring/Brauch 2011). Society needs to develop, adopt and implement new goals inspired by the SDGs and should aim at a new praxis based on an understanding of the underlying processes driving the present processes of GEC and CC.

42.5.3 Alternative Lifestyle Activities and Projects

In the industrialized countries a number of authors, social movements and NGOs have put the case for alternative values and promoted a more sustainable lifestyle. In the context of its *Education for Sustainable Development* (ESD), UNESCO focused on sustainable lifestyles alongside ten other themes.[12] In the framework of its programme on sustainable consumption, UNEP has also addressed the theme of sustainable lifestyles, which it defines as

> rethinking our ways of living, how we buy, what we consume and how we organize our daily lives. It is about transforming our societies and living in balance with our natural environment. All our choices and actions ... —on energy use, transport, food, waste and communication— contribute to sustainable lifestyles. For sustainable lifestyles to be part of our cultures and societies and become part of our everyday lives, they must be enabled and developed at all levels, through the social and technical systems and institutions that surround us. These include efficient infrastructures, services and products, such as efficient public transport systems, and by individual choices and actions to help minimize the use of natural resources, emissions, wastes and pollution while supporting equitable socio-economic development and conserving the Earth's life support systems within the planet's ecological carrying capacity.[13]

With the support of Sweden, UNEP has launched several projects to examine and promote sustainable lifestyles in the context of the Marrakech *Task Force on Sustainable Lifestyles*,[14] including the *Global Survey on Sustainable Lifestyles*, among other key projects.

The European 'social platform' on *Sustainable Lifestyles* 2050 (SPREAD), a project funded by the EU (2011-2012), addressed four key questions: "What is a sustainable lifestyle? What will a sustainable future mean for the way we live, move, and consume? How do we know if our lifestyles are sustainable or not? How can our aspirations for continuous life improvements be enabled sustainably (within one planet)?"[15] The SPREAD project gave rise to four working groups who considered: a) sustainable moving, b) sustainable living, c) sustainable consuming and d) sustainable society. It developed a "roadmap of action strategies" for "different societal actors, including 2012-2050 pathways to enable sustainable living across Europe by 2050". SPREAD distinguished four visions: i) economies for dense communities; ii) the convenience of trust; iii) entrepreneurial and self-aware society, connected wealth; and iv) happy sharing communities.[16] The project developed four scenarios for sustainable lifestyles (2012-2050): 1) Singular Super Champions; 2) Governing the Commons; 3) Local Loops; and 4) Empathetic Communities.[17]

These four alternative scenarios for a gradual transition of lifestyles in Europe between 2012 and 2050 (figure 42.6) were stimulated by a conceptual

---

12 See UNESCO, "Sustainable Lifestyles"; at: <http://www.unesco.org/new/en/education/themes/leading-the-international-agenda/education-for-sustainable-development/sustainable-lifestyles> (5 March 2016): "Together with UNEP, UNESCO has been active in raising awareness and providing educational resources in relation to sustainable consumption via its YouthXChange Project. ... YXC involves a participatory process based on interaction and cooperation between teachers and youth, discussion and learning from experiences."

13 See UNEP, "Sustainable Lifestyles"; at: <http://www.unep.org/resourceefficiency/Consumption/EducationLifestylesandYouth/SustainableLifestyles/tabid/101304/Default.aspx> (6 March 2016).

14 Through its *Marrakech Task Force on Sustainable Lifestyles*, UNEP supported nine key projects, including 1. *Communicating Sustainability*; 2. *Creative Communities*; 3. *Global Survey on Sustainable Lifestyles*; 4. *Intercultural Sister Classrooms*; 5. *Literature Review on Sustainable Lifestyles*; 6. *Making the business case for Sustainable Lifestyles*; 6. *Sustainable Entrepreneurship in African Universities*; at: <http://www.unep.org/resourceefficiency/Consumption/EducationLifestylesandYouth/SustainableLifestyles/KeyProjectsTaskForceonSustainableLifestyles/tabid/101314/Default.aspx> (6 March 2016).

15 See "The SPREAD Sustainable Lifestyles 2050 Project"; at: <http://www.sustainable-lifestyles.eu/> (6 March 2016).

16 See SPREAD, "European Policy Brief: Emerging Visions for Future Sustainable Lifestyles"; at: <http://www.sustainable-lifestyles.eu/fileadmin/images/content/D5.2_PolicyBrief_Roadmap_01.pdf>.

Figure 42.6: Matrix of lifestyle differences in the four scenarios: **Source**: SPREAD [Leppänen, Juha; Neuvonen, Aleksi; Ritola, Maria; Ahola, Inka; Hirvonen, Sini; Hyötyläinen, Mika; Kaskinen; Tuuli; Kauppinen, Tommi; Kuittinen, Outi; Kärki, Kaisa; Lettenmeier, Michael; Mokka, Roope, all from Demos Helsinki] (2013): 56–57.

	Singular Super Champions	Governing the Commons	Local Loops	Empathetic Communities
Education	Embedded into everyday life and practice, lifelong instead of short cycles in the beginning of life. Individualized and commoditied. Basis of welfare provision.	The main focus of education is on peer-to-peer self-sharing. Learning by doing is emphasized in iterative ways. Pandemic technology enables instant feedback loops, which accelerate sharing of knowledge. Focus on informal education where classrooms do not exist.	Education is about transmitting and sharing skills. Craftsmanship and specialisation are promoted through mutual teaching and problem-based learning.	Education and learning are problem-based and collaboration-driven.
Work	Human resources highlight work. Talent is concentrated in global organisations. Entrepreneur vs. super talented multinational class	People's sources of income have fragmented. Comprehensive use of skills are valued. Everyone has something to offer to society. Key words describing work include micro-tasks, crowdsourcing and being useful to one's peers.	Work is characterised by engagement with issues and collaboration within and among guilds. Needs met by applying design thinking and formulating local solutions.	Work happens collaboratively in hubs and people learn through asking for input from colleagues. Hands-on work is highly valued. Work is neighbourhood-based and aims at contributing to the community.
City	10–15 highly urbanised metropolises in Europe. Extremely dense. Lots of new infrastructure. New specialised areas of excellence.	Cities are based on already existing infrastructure. Office and school buildings have been converted into flats and public spaces. Urban experiences are enriched by augmented reality. People find personalised solutions to fulfil their needs and aspirations on both the physical and virtual layers.	Cities are multicentred and formed into their own loops. Guilds working and living in the loops lay their own strong characteristics on their loops.	Village infill from sprawl to farm village. Parking lots are turned into places of food production. The public space gains great significance. Village within cities are key elements of the urban fabric.
Health	Preventative public healthcare. Rational diets. Self-diagnosis.	Peer-to-peer network support is characteristic of both preventative and reactive health care. Public funding is provided for health-care cooperatives. A wide variety of healthy lifestyles have become routinised. Digital feedback tools are used by everyone.	Work places provide health-care and skilled doctors. There are basic rights that all regions agree to prioritise with regard to health-care.	Paradigm of quality over quantity characterises the health-care system. This means that the meaningfulness of a person's lifespan is seen as more important than the amount of years lived. Local administration prioritises health-care and healthy living. Every municipality has a hospital. Healthy-living-circles share preventative knowledge locally.
Living	Location compensates size of the flat. New materials and design. Price drives clarity.	People live in small flats and work in the office lofts. The digital layer is key to enable people with quality in their lives. Smart homes, austere furniture and digital services characterise domesticity.	Living in the loops is characterised by shared spaces, existing infrastructure and co-working spaces. Guild members often live in the same neighbourhood.	Farming opportunities raise property values. People live in shared apartments and make use of shared spaces.
Food	Price and health efficient diet. Large scale organic production.	Food production and distribution are managed by global food systems and smart food storage mechanisms are in place. A multitude of diets are offered and energy intake is reduced. Vegetable choices and synthetic meals form differentiations in diets.	Food production and distribution are marked by locality, minimised transportation and neighbourhood canteens. Energy used for food production is optimised.	Growing food in urban farming circles meets local food demand. Food transportation needs are very low. In addition to production, high importance is also placed on food quality, and distribution.
Mobility	New rail systems within and between metropoles. Personalized rapid transport systems. Smart mobility solutions. High prices.	Mobility is greatly reduced by the use of digital tools. Commuting is minimised and the construction of new traffic infrastructure has become unnecessary. Smart public transit and car and ride sharing are the main forms of transport	Transportation is about walkability and cycling. Existing infrastructure is optimised. Intercity mobility is needed less and services are home-delivered. Local tourism and long vacations are favoured by people.	Local mobility is emphasised and less road space is devoted for private vehicles. Old and new infrastructure is adapted to cycling.
Consuming	Meanings and symbols get consumed more than products. Education and self-projected me. Price mechanism.	3D-printing personalises consumption. Material consumption is reduced by using modular appliances, which enable do-it-yourself (DIY) repair and upgrade of products. High degrees of appliance personalisation, virtual consumption and recyclable generic materials form new design and producer cultures, helping to reduce the overall number of appliances.	Consumption drivers include a mass quest to reduce the overall volume of appliances needed through sharing schemes and replacement services. Products are made with high-quality local materials and design. Availability of foreign goods is limited. All products are repairable.	Consumption is geared towards meeting people's basic needs. Sharing, swapping and renting succeed private ownership.
Economy	Large multinational firms. Efficiency. Competition. Eco-industrial revolution. Thought leaders.	Micro-tasks characterise economic organisation. New businesses are created in and by data-rich environments. Open source, open data and free distribution of information drive new innovation. Personal optimisation, DIY, peer services and manufacturing are drivers of the new economy.	The economy is based around local user-centric adaptations and efficient local clustering.	The economy is organized around the self-sufficiency of small units. Food production is prioritised. Experimentation happens on the local level and high value is given to community activities.
Sense of security	From technological progress. Transparency. Surveillance. Individual choices. Thought leaders.	Sense of security is heightened by membership in peer-to-peer communities. Democratised data empowers people. Easy access to services, products and global knowledge-bases promotes equality. Personalised appliances and direct participation increase a sense of belonging and security.	Sense of security is generated through guilds and understanding of how the system, i.e. the closed cycles, work. People identify themselves primarily as part of their work communities.	Sense of security is generated by communities, closed circles, PPP-systems guarantee participation and sense of ownership in public and social affairs.
Leisure Time	Investing in own education and training.	Leisure time is formed around a multitude of digital interactions. Home consumption, high quality household capabilities and digital crowd experiences are the main ways of spending free time.	People have outsourced their housework in order to be able to maximise their inputs in work communities. Leisure time is mostly spent with guild members.	Leisure time is mostly public and used for social activities, e.g. in gardening circles. Vacations provide time for self-reflection.

Figure 42.7: How to spread sustainable lifestyles. **Source:** SPREAD: *Scenarios for Sustainable Lifestyles–From Global Champions to Local Loops* (Wuppertal: 2013), based on Geels (2002).

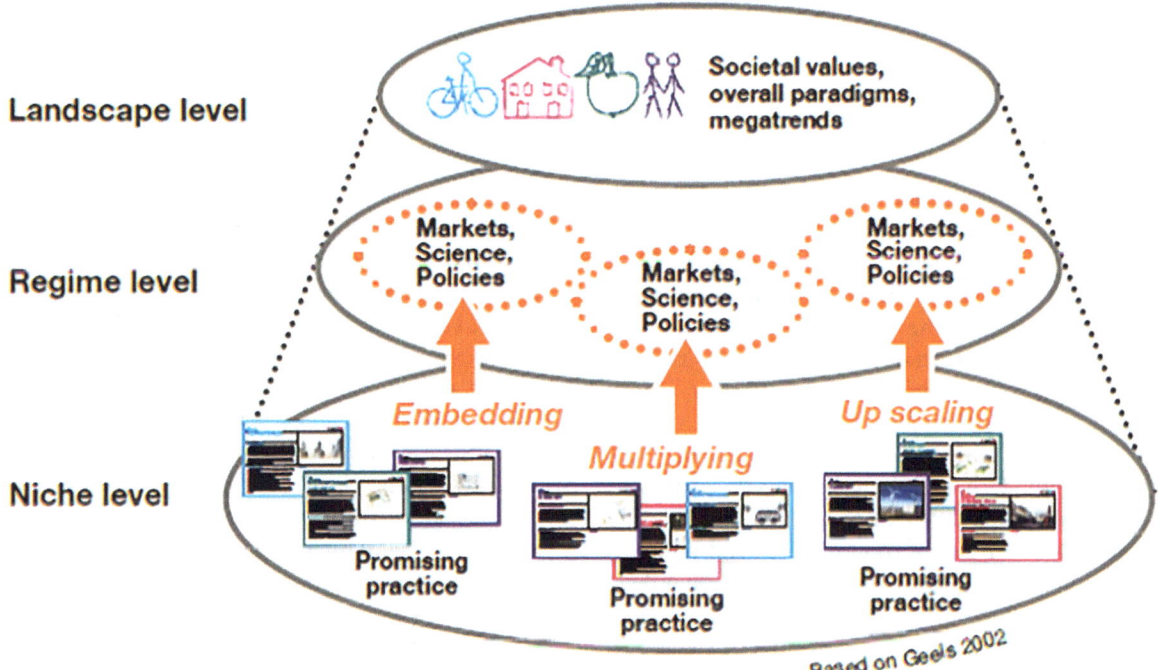

model based on Geels (2002; see figure 42.7) that distinguished the three levels of niche, regime and landscape change (see figure 1.11 in chapter 1 by Brauch/Oswald Spring); it was deeply influenced by the Dutch approach to sustainability transition (Grin/Rotmans/Schot 2010).

Many grass-roots initiatives and research groups in the *Organisation for Economic Co-operation and Development* (OECD) countries have promoted alternative sustainable lifestyles since the end of the Cold War, for example, in the US (*The Global Stewards*[18]), the UK (Sustainable Lifestyle Research Group[19]), France (Voluntary Simplicity Movement[20]), Germany

(De-Growth Movement[21]), Spain (Research & Degrowth[22]). These have been supplemented by economic movements (e.g. in Mexico, on the economics of solidarity (Cadena 2009) and the Peasant University of the South (UNICAM, on subsistence agriculture[23]) and in developing countries (Bangladesh: Yunus Centre;[24] India: Ashok Khosla;[25] Brazil: MST,[26]

---

17 SPREAD: *Scenarios for Sustainable Lifestyles2050: From Global Champions to Local Loops* (Wuppertal: UNEP/Wuppertal Institute Collaborating Centre on Sustainable Consumption and Production (CSCP)– Brussels: EU Commission, n.d.); at: <http://www.sustainable-lifestyles.eu/fileadmin/images/content/D4.1_FourFutureScenarios.pdf> (6 March 2016).

18 See at: <http://www.globalstewards.org/index.htm> (6 March 2016) for many eco tips; for specific proposals on "How to Live a Sustainable Lifestyle: Three Essential Steps for Creating a Sustainable Life"; see at: <http://www.globalstewards.org/sustainable-lifestyle.htm> (6 March 2016).

19 See at: <http://www.sustainablelifestyles.ac.uk/> (6 March 2016).

20 See the MA thesis by Perrine Vandecastèle (France): *How sustainable lifestyles spread social change: An analysis of the Voluntary Simplicity Movement* (Berlin: Otto-Suhr Institute, Free University of Berlin, Spring semester 2012).

21 See: "Degrowth–eine kapitalismuskritische Bewegung?!"; at: <http://www.degrowth.de/de/2015/01/degrowth-eine-kapitalismuskritische-bewegung/> (6 March 2016).

22 See: "Research & Degrowth: Research and actions to consume less and share more"; for its publications, see at: <http://www.degrowth.org/publications> (6 March 2016).

23 Universidad Campesina del Sur (UNICAM del Sur), at: <http://www.unicamsur.org.mx/index.php/component/content/article/17-multimedia/24-tv-unicamsur>; for its publications see at: <http://www. unicamsur.org.mx/index.php/publicaciones> (6 March 2016).

24 For the Yunus Centre, which was founded by Nobel Peace Laureate (2006) Muhammad Yunus, see at: <http://muhammadyunus.org/> (6 March 2016).

as well as intellectual leaders (Samir Amin (2014, 2014a)[27]). Completely different ways of life have been developed in the kingdom of Bhutan and proposed by indigenous people in Bolivia and in Chiapas (Mexico).

42.5.4 Alternative Indigenous World Views and Mindsets

Both biocentric world views and the dominant socio-political and economic mindset that promotes business-as-usual development paths and strategies reflect a 'cornucopian' (Gleditsch 2003) understanding of the economy, society, the environment and culture. In the Anthropocene, human beings, especially the top one per cent of the moneyed elite, exploit all the existing natural, human, political and cultural resources, often understood as natural, human, political and cultural capital (Bourdieu 1983). Earth and society may face challenges to their survival during this century as a result of this dominant and homogenized mindset, a mindset that is reinforced on a daily basis by a mass media that promotes consumerism and ignores the potentially catastrophic societal outcomes of this cornucopian world view. But not everybody has adopted the business-as-usual ideology.

For the people of Bhutan a 'gross national happiness index'[28] expresses a different approach and a guiding principle for a good life, where nature and human beings live together in harmony and peace, and where diverse traditions and rules open the way to cultural diversity, the defence of Mother Earth and the protection of the 'Pacha Mama'.[29]

Alternative lifestyles have at their centre the goal of a sustainable peace within a sustainability transition process that aims for a decent and good life and not just a better consumerist way of life. These lifestyles

combine ideas of complementarity and try to counter the destruction of nature and humankind. They accept unity in diversity, prioritize life within community, respect the natural cycle of nature and promote the recovery of the wisdom of the elderly. This postmodern understanding recovers the deep indigenous knowledge of 'cosmovisions'[30] and their traditions, where the Western paradigm of development is decolonized and the diversity of pluricultural integration of society permits the understanding of a biocentric world view, where harmony with nature and its ecosystem services respect simultaneously the intrinsic rights of nature and social rights.

In Mexico, the indigenous Zapatista movements proposed a shell or 'caracol'[31] model of sustainability, where they organized their way of life based on values, norms, institutions and productive processes which include the development of modern science and technology. Through formal and informal processes they transmit their knowledge and wisdom and integrate new scientific achievements. They reinforce their own structures of power, where society controls the mechanisms of power, and where they rotate authority and obey the community ('mandando obediciendo': directing obeying) for their necessities. Their period of authority is limited to one year and they are not paid for their social service, but the community provides the daily necessities for them and their families.

In analytical terms this cultural product of an indigenous society, which was exposed during hundreds of years of colonialization to exploitation, discrimination and violence, has developed new structures of power sharing since 1994, which are deeply internalized and perceived by the people as natural. Their cognition process legitimizes thoughtful structures of beliefs and behaviours, where nature and

25 Ashok Khosla is the founding director of "Development Alternatives: eco-solutions for people and the planet" in New Delhi, India, see at: <http://www.devalt.org/GoverningPersonDetail.aspx?Gid=1> and co-chair of UNEP's resource panel; see at: <http://www.unep.org/resourcepanel/PanelMembers/Ashok/tabid/794042/Default.aspx> (6 March 2016).

26 See in English on the Friends of MST, at: <http://www.mstbrazil.org/>.

27 Samir Amin has been the director of the Third World Forum in Dakar since 1981; see on his publications at: <http://www.afes-press-books.de/html/SpringerBriefs_PSP_Amin.htm> (6 March 2016).

28 The Kingdom of Bhutan has developed a *Gross National Happiness Index* (GNHI) as an alternative to GNP or GSP; see at: <http://www.grossnationalhappiness.com/> (23 February 2016).

29 TranSMART, Penny, 1992: "Pachamama: The Inka Earth Mother of the Long Sweeping Garment", in: Barnes, Ruth; Eicher, Joanne B. (Eds.): *Dress and Gender: Making and Meaning* (New York/Oxford: Berg): 145-163; Molinie, Antoinette, 2004: "The Resurrection of the Inca: The Role of Indian Representations in the Invention of the Peruvian Nation", in: *History and Anthropology*, 15,3: 233-250.

30 See: Adamson, Joni: "Indigenous Literature", in: Garrard, Greg (Ed.), 2014: *The Oxford Handbook of Ecocriticism* (Oxford: Oxford University Press).

31 The Zapatista movement proposed the shell model on 15 February 2010; see <http://enlacezapatista.ezln.org.mx/2010/02/15/gobierno-autonomo-zapatista-caracteristicas-antisistema-politico-mexicano/> or its general website < http://enlacezapatista.ezln.org.mx/>.

community interact in a complex interdependence and where individuals and social actors, institutions and regimes have transformed a business-as-usual approach or a cornucopian world view to a complex sustainable interaction between the biological and human systems.

This is just an example of a totally different and slow way of moving towards a cultural transformation within small groups of indigenous people. These changes resulted not simply from social nonconformity, but also from a longer learning process, where these social groups analysed all the different types of development processes proposed by governments, enterprises and other social groups—indigenous people, mestizos, *non-governmental organizations* (NGOs), and *civil society organizations* (CSOs). In this process identity patterns and social representations were recreated, and the deep sustainable biological-human system was anchored into their culture, later objectivized and then assimilated and transferred into local or regional policy.

A key process was the sharing of cultural and material products, which consolidated the social cohesion and at the same time limited internal social stratification. Besides this, extreme hydro-meteorological events have reinforced the process of solidarity among the affected and their capacity for sharing scarce resources among the most vulnerable communities. Their new world view is based on a way of life which is mostly carbon-free and dematerialized, where waste is recycled and consumption oriented towards renewable production and consumption.

This fundamentally different indigenous world view links the transition to sustainability with the recovery of crucial ecosystem services and the environment. The changes in mindset from a patriarchal business-as-usual approach is embedded in a daily transformation of the understanding and exercise of non-authoritarian authority, economy, society and environment. A critical approach to incipient social stratification processes and mechanisms has the function of limiting it. Interest-driven political and economic interests are controlled by participative governance and collective decision-making to protect the most vulnerable and the environment. It includes a sustainable peace-building process that is focused on a sustainable culture in their radically alternative world view, mindset and system of governance.

These different approaches from Bhutan, Bolivia and the Zapatista movement in Chiapas in Mexico are deeply culturally embedded in their respective cosmovisions and religious beliefs. However, they do not offer a solution for highly industrialized modern societies in the OECD world and for the elites and upper and middle classes in many economic threshold countries and also in most parts of the developing world.

References

Adamson, Joni, 2014: "Indigenous Literature", in: Garrard, Greg (Ed.): *The Oxford Handbook of Ecocriticism* (Oxford: Oxford University Press).

Amin, Samir, 2014: *Pioneer on the Rise of the South*—Presented by Dieter Senghaas. SpringerBriefs on Pioneers in Science and Practice No. 16 (Cham—Heidelberg—New York—Dordrecht—London: Springer-Verlag).

Amin, Samir, 2014a: *Theory is History*. SpringerBriefs on Pioneers in Science and Practice No. 17. Texts and Protocols No. 9 (Cham—Heidelberg—New York—Dordrecht—London: Springer-Verlag).

Amster, Randall, 2014, 2015: *Peace Ecology* (Boulder, CO: Paradigm).

Benavenides de Pérez, Raquel Amada, 2015: "Estrategia para el desarrollo integral adolescente y juvenil", in: Serrano, S. Eréndira, Oswald Spring, Úrsula, de la Rúa, Diana (Eds.), *América Latina en el camino hacia una paz sustentable: herramientas y aportes* (Guatemala: ARP, FLACSO, CLAIP): 142-162.

Boulding, Kenneth E., 1966: "The Economics of the Coming Spaceship Earth", in: Jarrett, Henry (Ed.): *Environmental Quality in a Growing Economy, Essays from the Sixth RFF Forum on Environmental Quality held in Washington, March 8 and 9, 1966* (Baltimore, Johns Hopkins Press): 3-14.

Boulding, Kenneth E., 1970: "The Economics of the Coming Spaceship Earth", in: Boulding, Kenneth E., (Ed.): *Beyond Economics: Essays on Society, Religion, and Ethics* (Ann Arbor: University of Michigan Press): 275-287.

Boulding, Kenneth E., 1978: *Ecodynamics: A new theory of societal evolution* (Beverly Hills: Sage).

Boulding, Kenneth E., 1983: "Ecodynamics", in: *Interdisciplinary Science Reviews*, 8,2 (June): 108-113.

Bourdieu, Pierre, 1983: *The Forms of Capital*, originally published as: "Ökonomisches Kapital, kulturelles Kapital, soziales Kapital", in: Kreckel, Reinhard (Ed.): *Soziale Ungleichheiten* (Göttingen: Otto Schwartz & Co.): 183-198.

Brand, Frank; Schaller, Franz; Völker, Harald (Eds.), 2004: *Transdisziplinarität. Bestandsaufnahme und Perspektiven. Beiträge zur THESIS-Arbeitstagung im Oktober 2003 in Göttingen* (Göttingen: Universitätsverlag).

Brauch, Hans Günter, 2009: "Securitzing Global Environmental Change", in: Brauch, Hans Günter; Oswald Spring, Úrsula; Grin, John; Mesjasz, Czeslaw; Kameri-Mbote, Patricia; Behera, Navnita Chadha; Chourou, Béchir; Krummenacher, Heinz (Eds.), 2009: *Facing Global Environmental Change: Environmental, Human, Energy, Food, Health and Water Security Concepts* (Berlin–Heidelberg–New York: Springer-Verlag): 65-102.

Brauch, Hans Günter; Dalby, Simon; Oswald Spring, Úrsula, 2011: "Political Geoecology for the Anthropocene", in: Brauch, Hans Günter; Oswald Spring, Úrsula; Mesjasz, Czeslaw; Grin, John; Kameri-Mbote, Patricia; Chourou, Béchir; Dunay, Pal; Birkmann, Jörn (Eds.), 2011: *Coping with Global Environmental Change, Disasters and Security–Threats, Challenges, Vulnerabilities and Risks* (Berlin–Heidelberg–New York: Springer-Verlag): 1453-1486.

Cadena, Félix (Ed.), 2009: *De Foro a Foro. Contribuciones y perspectivas de la economía solidaria en México, en contexto de crisis global* (Mexico: FLASEP).

Choi, B. C.; Pak, A. W., 2006: "Multidisciplinarity, interdisciplinarity and transdisciplinarity in health research, services, education and policy: 1. Definitions, objectives, and evidence of effectiveness", in: *Clinical and Investigative Medicine*; 29,6 (December): 351-364; at: <http://www.ncbi.nlm.nih.gov/pubmed/17330451>.

Conca, Ken, 2002: "The Case for Environmental Peacemaking", in: Conca, Ken; Dabelko, Geoffrey (Eds.): *Environmental Peacemaking* (Baltimore: Johns Hopkins University Press): 1-22.

Dalby, Simon, 2007: "Ecology, Security, and Change in the Anthropocene", in: *Brown Journal of World Affairs*, 13,2: 155-164.

Dalby, Simon, 2009: *Security and Environmental Change* (Cambridge: Polity).

Dalby, Simon, 2009a: "Peacebuilding and Environmental Security in the Anthropocene", in: Péclard, Didier (Ed.): *Environmental Peacebuilding: Managing Natural Resource Conflicts in a Changing World*. Conference Paper 2009-1 (Bern: Swisspeace): 8-21.

Dalby, Simon, 2013: "Human Security in the Anthropocene: The Implications of Earth Systems Analysis", in: O'Brien, Karen; Wolf, Johanna; Sygna, Linda (Eds.): *The Changing Environment for Human Security: Transformative Approaches to Research, Policy, and Action* (London: Earthscan): 27-33.

Dalby, Simon, 2014: "Rethinking Geopolitics: Climate Security in the Anthropocene", in: *Global Policy*, 5,1 (February): 1-9.

Dalby, Simon, 2015: "Anthropocene Formations: Environmental Security, Geopolitics and Disaster", in: *Theory, Culture and Society*. Special issue on *Geosocial Formations and the Anthropocene*: in: *OnlineFirst* (August); at: <http://tcs.sagepub.com/content/early/recent>.

Dalby, Simon, 2016: "Climate Security in the Anthropocene: 'Scaling up' the Human Niche", in: Wepner, Paul; Lever,

Hill (Eds.): *Reimagining Climate Change* (New York: Routledge): 29-48.

Darbellay, Féderic, 2015: "Rethinking inter- and transdisciplinarity: Undisciplined knowledge and the emergence of a new thought style", in: *Futures*, 65: 163-174.

Dransart, Penny, 1992: "Pachamama: The Inka Earth Mother of the Long Sweeping Garment", in: Barnes, Ruth; Eicher, Joanne B. (Eds.): *Dress and Gender: Making and Meaning* (New York–Oxford: Berg): 145-163.

EIA, 2015: *Annual Energy Outlook 2015 with Projections to 2040–April 2015* (Washington DC: US EIA).

FAO, 2011a: *Climate-Smart Agriculture: Smallholder Adoption and Implications for Climate Change Adaptation and Mitigation* (Rome: FAO).

FAO, 2011b: *Save and grow. A policymaker's guide to the sustainable intensification of smallholder crop production* (Rome: FAO).

FAO, 2013a: *Climate-smart agriculture sourcebook* (Rome: FAO).

FAO, 2013b: *Food wasting footprint. Impacts on natural resources* (Rome: FAO).

FAO, 2015: *Global guidelines for the restoration of degraded forests and landscapes in drylands: building resilience and benefiting livelihoods,* Forestry Paper No. 175 (Rome: FAO).

FAO, 2016a: *Climate change and food security: risks and responses* (Rome: FAO).

FAO, 2016b: *Status of the World's Soil Resources* (Rome: FAO).

Freire, Paulo, 1975: *Conscientization* (Geneva, World Council of Churches).

Funtowicz, S.; Ravetz, J. R., 1993: "Science for the Post-Normal Age", in: *Future*, 25:735-755.

Funtowicz, S.; Ravetz, J. R., 2003: "Post-Normal Science", in: International Society for Ecological Economics (Ed.): *Internet Encyclopaedia of Ecological Economics*; at: <http://isecoeco.org/pdf/pstnormsc.pdf> (6 February 2016).

Geels, Frank W., 2002: "Technological Transitions as evolutionary reconfiguration processes: A multi-level perspective and a case study", in: *Research Policy*, 31,8-9: 1257-1274.

GHI [Global Hunger Index], 2015: 2015 *Global Hunger Index: Armed Conflict and the Challenge of Hunger* (Bonn: Welthungerhilfe; Washington DC: International Food Policy Research Institute).

Gleditsch, Nils Petter, 2003: "Environmental Conflict: Neomalthusians vs. Cornucopians", in: Brauch, Hans Günter; Liotta, P. H; Marquina, Antonio; Rogers, Paul; Selim, Mohammed El-Sayed (Eds.): *Security and Environment in the Mediterranean. Conceptualising Security and Environmental Conflicts* (Berlin–Heidelberg: Springer): 477-486.

Gleditsch, Nils-Petter, 2012: "Whither the weather? Climate change and conflict", in: *Journal of Peace Research, special issue: Climate change and conflict,* 49,1 (January-February): 9-18.

Göpel, Maja, 2017: *The Great Mindshift—Why We Need a New Economic Paradigm for Sustainability Transformations* (Cham–New York–Heidelberg–Dordrecht–London: Springer).

Grin, John; Rotmanns, Jan; Schot, Johan, 2010: *Transitions to Sustainable Development. New Directions in the Study of Long Term Transformative Change* (New York, NY–London: Routledge).

Haken, Hermann, [3]1983: *Synergetics: An Introduction. Nonequilibrium Phase Transition and Self-Organization in Physics, Chemistry, and Biology* (Berlin–Heidelberg: Springer-Verlag).

Hirsch Hadorn, Gertrude; Hoffmann-Riem, Holger; Biber-Klemm, Susette; Grossenbacher-Mansuy, Walter; Joye, Dominique; Pohl, Christian; Wiesmann, Urs; Zemp, Elisabeth (Eds.), 2008: *Handbook of Transdisciplinary Research* (Berlin–Heidelberg: Springer);

Huggett, Richard John, 1995: *Geoecology. An Evolutionary Approach* (London–New York: Routledge).

IEA (International Energy Agency), 2014: *World Energy Outlook 2014* (Paris: OECD/IEA).

IPCC, 2012: *Special Report on Managing the Risks of Extreme Events and Disasters to Advance Climate Change Adaptation* (SREX) (Cambridge–New York: Cambridge University Press); at: <http://ipcc-wg2.gov/SREX/report/> (7 March 2016).

IPCC, 2013: *Climate Change 2014—The Physical Science Basis. Working Group I Contribution to the Fifth Assessment Report of the Intergovernmental Panel on Climate Change* (Cambridge–New York: Cambridge University Press).

IPCC, 2014a: "Human Security", in: *Climate Change 2014—Impacts, Adaptation, and Vulnerability—Part A: Global and Sectoral Aspects. Working Group II Contribution to the Fifth Assessment Report of the Intergovernmental Panel on Climate Change* (Cambridge–New York: Cambridge University Press). 755-701, at: <http://www.ipcc.ch/pdf/assessment-report/ar5/wg2/WGIIAR5-Chap12_FINAL.pdf>.

IPCC, 2014b: *Climate Change 2014—Impacts, Adaptation, and Vulnerability—Part A: Global and Sectoral Aspects. Working Group II Contribution to the Fifth Assessment Report of the Intergovernmental Panel on Climate Change* (Cambridge–New York: Cambridge University Press).

IPCC, 2014c: *Climate Change 2014—Impacts, Adaptation, and Vulnerability—Part B .... Working Group II Contribution to the Fifth Assessment Report of the Intergovernmental Panel on Climate Change* (Cambridge–New York: Cambridge University Press).

IRENA, 2016: *Renewable Energy Market Analysis: The GCC Region* (Abu Dhabi: IRENA); at: <http://www.irena.org/DocumentDownloads/Publications/IRENA_Market_GCC_2016.pdf.

IRENA, 2016a: *Renewable Energy and Jobs-Annual Review 2016*; at: <http://www.irena.org/DocumentDownloads/Publications/IRENA_RE_Jobs_Annual_Review_2016.pdf.>.

ISSC, 2012: *Transformative Cornerstones of Social Science Research for Global Change* (Paris: ISSC).

Jaeger, Jochen; Scheringer, Martin, 1998: "Transdisziplinarität: Problemorientierung ohne Methodenzwang", in: *GAIA*, 7,1: 10-25.

Kates, Robert W.; Parris, Thomas M., 2003: "Long-term trends and a sustainability transition", in: *PNAS*, 100,14 (8 July): 8062-8067.

Luhmann, Niklas, 1991: *Soziale Systeme* (Frankfurt a. M: Suhrkamp).

Maturana, Umberto II, 1996: "*Fundamentos biológicos del conocimiento*", in: *Objectiva o construida?* (Mexico, D.F.: Anthropos/Universidad Iberoamericana/Iteso).

McBean, Gordon; Ajibade, Idowu, 2009: "Climate change, related hazards and human settlements", in: *Current Opinion in Environmental Sustainability*, 1: 179-186.

Mittelstrass, Jürgen, 2003: *Transdisziplinarität—wissenschaftliche Zukunft und institutionelle Wirklichkeit* (Konstanz: Universitätsverlag Konstanz).

Molinie, Antoinette, 2004: "The Resurrection of the Inca: The Role of Indian Representations in the Invention of the Peruvian Nation", in: *History and Anthropology*, 15,3: 233-250.

National Science Board, 2007: *Enhancing Support of Transformative Research at the National Science Foundation* (Washington DC: National Science Foundation), at: <https://www.nsf.gov/about/transformative_research/definition.jsp> (6 February 2016).

NRC (National Research Council), 1999: *Global Environmental Change: Research Pathways for the Next Decade* (Washington DC: National Academy Press).

Oswald Spring, Úrsula, 2009: "Sustainable Development", in: De Rivera, Joe (Ed.): *Handbook on Building Cultures of Peace* (New York: Springer): 211-227.

Oswald Spring, Úrsula, 2009a: "Food as a new human and livelihood security issue", in: Brauch, Hans Günter; Oswald Spring, Úrsula; Grin, John; Mesjasz, Czeslaw; Kameri-Mbote, Patricia; Behera, Navnita Chadha; Chourou, Béchir; Krummenacher, Heinz (Eds.), 2009: *Facing Global Environmental Change: Environmental, Human, Energy, Food, Health and Water Security Concepts* (Berlin–Heidelberg–New York: Springer-Verlag): 471-500.

Oswald Spring, Úrsula, 2012: "Climate-Induced Migration as a Security Risk and an Additional Threat for Conflict in Mexico", in: Jürgen Scheffran, Michael Brzoska, Hans Günter Brauch, Peter Michael Link, Janpeter Schilling (Eds.): *Climate Change, Human Security and Violent Conflict: Challenges for Societal Stability* (Berlin: Springer-Verlag): 315-350.

Oswald Spring, Úrsula, 2016: "The Water, Energy, Food and Biodiversity Nexus: New Security Issues in the Case of Mexico", in: Brauch, Hans Günter; Oswald Spring, Úrsula; Bennett, Juliet; Serrano Oswald, Serena Eréndira (Eds.): *Addressing Global Environmental Challenges from a Peace Ecology Perspective* (Cham–Heidelberg–New York–Dordrecht–London: Springer International Publishing, forthcoming).

Oswald Spring, Úrsula; Brauch, Hans Günter, 2011: "Coping with Global Environmental Change–Sustainability Revolution and Sustainable Peace", in: Brauch, Hans Günter; Oswald Spring, Úrsula; Mesjasz, Czeslaw; Grin, John; Kameri-Mbote, Patricia; Chourou, Béchir; Dunay, Pal; Birkmann, Jörn, 2010: *Coping with Global Environmental Change, Disasters and Security–Threats, Challenges, Vulnerabilities and Risks* (Berlin–Heidelberg–New York: Springer-Verlag): 1487-1504.

Oswald Spring, Úrsula; Brauch, Hans Günter; Tidball, Keith G. (Eds.), 2014: *Expanding Peace Ecology: Security, Sustainability, Equity and Peace: Perspectives of IPRA's Ecology and Peace Commission* 1 (Cham–Heidelberg–New York: Springer).

Oswald Spring, Úrsula; Serrano Oswald, Serena Eréndira; Estrada Álvarez, Adriana; Flores Palacios, Fátima; Ríos Everardo, Maribel; Brauch, Hans Günter; Ruíz Pantoja, Teresita E.; Lemus Ramírez, Carlos; Estrada Villareal, Ariana; Cruz, Mónica, 2013: *Vulnerabilidad Social y Género entre Migrantes Ambientales* (Cuernavaca: CRIM, DGAPA-PAPIIT-UNAM, RETAC-Conacyt).

Painter, James, 2007: *UN Human Development Report 2007/2008. Deglaciation in the Andean Region* (New York: UNDP).

Pisano, Umberto; Lepuschitz, Katrin; Gerald Berger, 2014: *Sustainability transitions at the international, European and national level: Approaches, objectives and tools for sustainable development governance.* Sustainability transitions ESDN Quarterly Report 33, July 2015 (Vienna: Vienna University of Economics and Business).

Pohl, Christian; Hirsch Hadorn, Gertrude, 2007: *Principles for Designing Transdisciplinary Research–proposed by the Swiss Academies of Arts and Sciences* (München: oekom Verlag).

Pope Francis, 2015: *Laudato Si*; at: < https://laudatosi.com/watch> (7 March 2016).

Pretty, Jules, 2006: "Agroecological Approaches to Agricultural Development"; at: <https://openknowledge.worldbank.org/bitstream/handle/10986/9044/WDR2008_0031.pdf?sequence=1>.

Prigogine, Ilya; Stengers, Isabelle, 1984: *Order out of Chaos: Man's new dialogue with nature* (London: Flamingo).

Prigogine, Ilya; Stengers, Isabelle, 1997: *The End of Certainty* (Glencoe: The Free Press).

REN 21, 2015: *Renewables 2015–Global Status Report* (Paris: UNEP).

Rockström, Johan; Steffen, Will; Noone, Kevin; Persson, Åsa; Chapin, III, F. Stuart; Lambin, Eric; Lenton, Timothy M.; Scheffer, Marten; Folke, Carl; Schellnhuber, Hans Joachim; Nykvist, Björn; De Wit, Cynthia A.; Hughes, Terry; Leeuw, Sander van der; Rodhe, Henning; Sörlin, Sverker; Snyder, Peter K.; Costanza, Robert; Svedin, Uno; Falkenmark, Malin; Karlberg, L.; Corell, Robert W.; Fabry, Victoria J.; Hansen, James;

Walker, Brian; Liverman, Diana M.; Richardson, Katherine; Crutzen, Paul; Foley, Jonathan A., 2009a: "Planetary boundaries: exploring the safe operating space for humanity", in: *Ecology and Society*, 14,2: 32; at: <http://www.ecologyandsociety.org/vol14/iss2/art32/> (13 March 2106).

Scheffran, Jürgen; Brzoska, Michael; Brauch, Hans Günter: Link, Michael; Schilling, Janpeter (Eds.), 2012: *Climate Change, Human Security and Violent Conflict: Challenges for Societal Stability* (Berlin–Heidelberg: Springer-Verlag).

Schellnhuber, Hans Joachim; Martin, Maria A., 2014: "Sustainable Humanity, Sustainable Nature: Our Responsibility", in: Pontifical Academy of Sciences, Extra Series 41, Vatican City 2014, Pontifical Academy of Social Sciences, *Acta* 19; at: <http://www.pas.va/content/dam/accademia/pdf/es41/es41-schellnhuber.pdf>.

Schneidewind, Uwe; Singer-Brodowski, Mandy, 2013: *Transformative Wissenschaft – Klimawandel im deutschen Wissenschafts- und Hochschulsystem* (Marburg: Metropolis).

Scholte, Jan Aart, ²2005: *Globalization: A Critical Introduction* (London: Macmillan Palgrave).

Silva Diniz, Barbara, Tavares Beleza, Flávia, 2015: "La mediación social en la escuela: espacio de construcción de paz", in: Serrano, S. Eréndira, Oswald Spring, Úrsula, de la Rúa, Diana (Eds.): *América Latina en el camino hacia una paz sustentable: herramientas y aportes* (Guatemala: ARP, FLACSO, CLAIP): 181-200.

Speth, James Gustave, 1992: "The transition to a sustainable society", in: *Proceedings of the National Academy of Sciences of the United States of America*, 89: 870-872.

SPREAD [Leppänen, Juha; Neuvonen, Aleksi; Ritola, Maria; Ahola, Inka; Hirvonen, Sini; Hyötyläinen, Mika; Kaskinen, Tuuli; Kauppinen, Tommi; Kuittinen, Outi; Kärki, Kaisa; Lettenmeier, Michael; Mokka, Roope, all from Demos Helsinki], 2013: *Scenarios for Sustainable Lifestyles 2050: From Global Champions to Local Loops* (Wuppertal: UNEP/Wuppertal Institute Collaborating Centre on Sustainable Consumption and Production (CSCP)–Brussels: EU Commission); at: <http://www.sustainable-lifestyles.eu/fileadmin/images/content/D4.1_FourFutureScenarios.pdf> (6 March 2016).

Torquebiau, Emmanuel (Ed.), 2016: *Climate Change and Agriculture Worldwide* (Dordrecht–Heidelberg–New York–London: Springer).

UNEP, 2015: *Our Planet: Global Climate Action* (Nairobi: UNEP).

UNEP, International Resource Panel, 2011: *Decoupling Natural Resource Use and Environmental Impacts from Economic Growth* (Paris: UNEP).

UNEP, International Resource Panel, 2014: *Decoupling 2: Technologies, Opportunities and Policy Options* (Paris: UNEP).

US National Science Board, 2007: *Enhancing Support for Transformative Research at the National Science Foun-*

dation (Washington DC: NSF): at: <https://www.nsf.gov/nsb/documents/2007/tr_report.pdf>.

Vergara, Walter; Deeb, Alejandro M.; Valencia, Adriana M.; Bradley, Raymond S.; Francou, Bernard; Zarzar, Alonso; Grünwaldt, Alfred; Haeussling, Seraphine M., 2007:

"Economic Impacts of Rapid Glacier Retreat in the Andes", in: *Eos*, 88,25 (19 June): 261-268.

WBGU, 2011: *World in Transition—A Social Contract for Sustainability* (Berlin: German Advisory Council on Global Change, July 2011).

Abbreviations

10YFP	10-Year Framework of Programmes
AA	Auswärtiges Amt (Federal Ministry for Foreign Affairs, Germany)
ABEE	Associação Brasileira para Eficiência Energética (Brazilian Association for Energy Efficiency)
AC^4	Advanced Consortium on Cooperation, Conflict and Complexity
AD	Anno Domini
ADB	Asian Development Bank
AfDB	African Development Bank
AFES-PRESS	Peace Research and European Security Studies
AFOLU	Agriculture, Forestry and Other Land Use
AGF	Advisory Group on Climate Change Financing
AHDR	American Human Development Report
ANEEL	Agência Nacional de Energia Elétrica (Brazilian Electricity Regulatory Agency)
ANT	actor network theory
APEC	Asia-Pacific Economic Cooperation
APESS	The Anthropocene: Politik–Economics–Society–Practice (book series)
APPRA	Asia-Pacific Peace Research Association
APREA	African Peace Research and Education Association
AR5	Fifth Assessment Report of the IPPC
ASEAN	Association of Southeast Asian Nations
BAU	business-as-usual
BC	Before Christ
BCE	Before Common Era
Bha	billion hectares
BICC	Bonn International Center for Conversion
BTL	Beyond the Limits
BNDES	Brazilian Development Bank
BP	Before Present
BP	A supermajor oil and gas companies with headquarters in London
BRIC	Brazil, Russia, India, China
BRICS	Brazil, Russia, India, China, South Africa
BRISE	the 14 largest 'emerging' economies
BS	biodiversity security
BSD	Board on Sustainable Development
Bt	billion tonnes
BTL	Beyond the Limits
BUND	Bund für Umwelt und Naturschutz Deutschland (Friends of the Earth Germany)
C3	Consejo Consultivo de Cambio Climático (Advisory Council on Climate Change)
CAFE	Corporate Average Fuel Economy

CAPES	Coordenação de Aperfeiçoamento de Pessoal de Nível Superior (Brazilian Federal Agency for Support and Evaluation of Graduate Education)
CAS	complex adaptive systems
CBD	UN Convention on Biodiversity
CC	climate change
CCL	Ley General de Cambio Climático (Climate Change Law), Mexico
CCS	carbon capture and storage
CCSP	Climate Change Science Programme
CCT	conditional cash transfer
CDA	critical discourse analysis
CDM	clean development mechanism
CE	circular economy
CE	Common Era
CEA	Comisión de Agua (State Water Agency, Morelos, Mexico)
CEN	Center for Earth System Research and Sustainability
CEO	Chief Executive Officer
CER	Certified Emissions Reductions
CERC	US-China Clean Energy Research Center
cf	confer (compare)
CFCs	chlorofluorocarbons
CFL	compact fluorescent lamp
CFSP	European Foreign and Security Policy
CH$_4$	methane
CHP	combined heat and power
CHS	Commission on Human Security
CICC	Comisión Intersecretarial de Cambio Climático (Interministerial Commission on Climate Change), Mexico
CIDA	Canadian International Development Authority
CIESIN	Centre for International Earth Science Information Network
CIGI	Centre for International Governance Innovation (Waterloo, Ontario, Canada)
CIM	climate-induced migration
CISEN	Centro de Investigación y Seguridad Nacional (Centre for Investigation and National Security), Mexico
CLAIP	Latin American Council on Peace Research
CliSAP	Cluster of Excellence, Hamburg University, Germany
CLISEC	Research Group Climate and Security
CMM	coal mine methane
CMS	Centre for Mountain Studies, Perth College, Scotland
CNA	Center for Naval Analyses

CNPq	Brazilian Council for Scientific and Technological Development
CO_2	carbon dioxide
CO_2e	carbon dioxide equivalent
COCOM	Coordinating Committee on Multilateral Export Controls
CONAGUA	Comisión Nacional de Agua (National Water Commission, Mexico)
CONPET	Programa Nacional da Racionalização do Uso dos Derivados do Petróleo e do Gás Natural (National Programme of Rational Use of Oil and Natural Gas By-products), Mexico
COP	Conference of the Parties
COP 15	Conference of the Parties of the UNFCCC, Copenhagen (2009)
COP 19	Conference of the Parties of the UNFCCC, Warsaw (2013)
COP 21	Conference of the Parties of the UNFCCC, Paris (2015)
COPRED	Consortium on Peace Research Education and Development
COPRI	Copenhagen Peace Research Institute
CP	Cleaner Production
CPA	Cleaner Production Audit
CREATE	Campus for Research Excellence and Technological Enterprise
CRED	Centre for Research on the Epidemiology of Disasters (Université Catholique de Louvain, Belgium)
CRIM	Center for Regional Multidisciplinary Studies (of UNAM, in Mexico)
CSA	climate-smart agriculture
CSA	Carbon Structural Adjustment
CSD	United Nations Commission on Sustainable Development
CSDS	Center for Social Development Studies, Faculty of Political Science, Chulalongkorn University, Bangkok, Thailand.
CSIR	Council for Scientific and Industrial Research)
CSO	civil society organization
CSP	concentrated solar power
CSR	corporate social responsibility
CT	critical theory
CUHK	City University of Hong Kong
CUT	Central Única dos Trabalhadores (Association of Independent Trade Unions), Brazil
CvJO	Centre for Young Entrepreneurship
C-WET	Centre for Wind Energy Technology
DAFF	Department of Agriculture Forestry and Fisheries
DANIDA	Danish International Development Agency
DBZ	Development Bank of Zambia
DC	direct current
DDT	dichloro-diphenyl-trichloroethane
DECC	Department of Energy & Climate Change (UK)
DEDE	Department of Alternative Energy Development and Efficiency, Thailand
DEFRA	Department for Environment, Food and Rural Affairs (UK)
DEG	Deutsche Investitions- und Entwicklungsgesellschaft (German Investment and Development Society)
DG	Directorate-General
DHI	DHI Water & Environment (Private) Limited
Diversitas	(international research programme on biodiversity science)
DLDD	desertification, land degradation and drought
DNA	deoxyribonucleic acid
DoE	Department of Energy (South Africa)
DPKO	Department for Peacekeeping Operations of the United Nations
DPR	Detailed Project Report
DRDO	Defence Research and Development Organization
DRIFT	Dutch Research Institute for Transitions
DRM	disaster risk management
DRR	disaster risk reduction
DSM	demand-side management
DTU	Danmarks Tekniske Universitet (The Technical University of Denmark)
EC	European Commission
ECA	Export Credit Agency
ECER	energy conservation and emissions reduction
ECO	Energy Company Obligation
EE	energy efficiency
EEA	European Environment Agency
EF	ecological footprint
EFM	environmentally-forced migration
EG/LV	Emschergenossenschaft/Lippeverband (planning association of the region of Emscher and Lippe in North Rhine-Westphalia, Germany)
EGAT	Electricity Generating Authority of Thailand
EGCO	Electricity Generating Company
EHS	environment and human security
EIA	environmental impact assessment
EIA	US Energy Information Agency
EIB	European Investment Bank
EIST	Environmental Innovation and Sustainability Transitions
EIUG	Energy Intensive Users Group
EMC	environmentally-weighted material consumption
Em-DAT	Emergency Disasters Database (of the Centre for Research on the Epidemiology of Disasters at the Catholic University of Louvain, Brussels)

ENACC	Estrategia Nacional de Cambio Climático (National Strategy on Climate Change), Mexico
ENIGH	Encuesta Nacional de Ingresos y Gastos de los Hogares (National Survey of Household Income and Expenditure), Mexico
ENS	energy security
ENSO	El Niño Southern Oscillation
EoLRR	end-of-life recycling rate
EOM	Equal Opportunity Model
EOP	end-of-pipe
EPA	Environmental Protection Agency
EPC	Ecology and Peace Commission of IPRA
EPC	Energy Performance Certificate
EPE	Empresa de Pesquisa Energética (Brazilian Federal Energy Planning Company)
EPPO	Energy Policy and Planning Office
ERC	Energy Regulatory Commission
EREP	European Resource Efficiency Platform
ERU	emission reduction unit
ES	environmental security
ES	environmental space
ESA	Earth Systems Analysis
ESB	enhanced single-buyer
ESD	Education for Sustainable Development
ESDP	European Security and Defence Policy
ESDP	European Spatial Development Perspective
ESG	Earth System Governance
ESGP	Earth System Governance Project
ESS	Earth System Science
ESSC	Earth System Science Committee, NASA
ESSP	Earth System Science Partnership
ETS	emissions trading system
EU	European Union
EU-28	European Union (with its 28 member states)
EUPRA	European Peace Research Association
EV	electric vehicle
EW-MFA	economy-wide material flow analysis
FAG	Forest Advisory Group
FAO	Food and Agriculture Organization of the United Nations
FAOSTAT	Statistics Division of the Food and Agriculture Organization
FAST	Frühanalyse von Spannungen und Tatsachenermittlung (Early Analysis of Tensions and Fact Finding)
FDI	foreign direct investment
FE	Future Earth
FHP	Family Health Programme (Programa Saúde da Família), Brazil
FiT	Feed-in Tariff
FRONTEX	European Agency for the Management of Operational Cooperation at the External Borders
FS	food security
FYP	Five-Year Plan

G20	Group of twenty (most important industrial and threshold states)
G7	Group of seven (major industrialized countries: Canada, France, Germany, Italy, Japan, UK, US)
G8	Group of eight (major industrialized countries: Canada, France, Germany, Italy, Japan, Russia, UK, US)
GACGC	German Advisory Committee on Global Change
GBO	global biological outlook
GCC	global climate change
GCCP	natural gas combined cycle power
GCF	Green Climate Fund
GCM	General Circulation Model
gCO$_2$e/kWh	grams of carbon dioxide equivalent per kilowatt-hour
GCP	Global Carbon Project
GCRP	Global Change Research Program
GCTE	Global Change and Terrestrial Ecosystems
GDI	Gender-Related Development Index
GDP	gross domestic product
GEC	global environmental change
GECAFS	Global Environmental Change and Food Systems
GECHH	Global Environmental Change and Human Health
GECHS	Global Environmental Change and Human Security
GEDA	Gujarat Energy Development Agency
GEF	Global Environmental Facility
GEO	Global Environmental Outlook
GEO 4	Global Environment Outlook 4
Gg	gigagrams
Gg/y	gigagrams per year
Gha	global hectares
GHG	greenhouse gases
GHI	Global Hunger Index
GIM	Global Impact Model
GLADA	Global Assessment of Land Degradation and Improvement
GLASOD	Global Assessment of Human-Induced Soil Degradation
GLP	Global Land Project
GMP	genetically modified organism
GNH	Gross National Happiness
GNI	gross national income
GNP	gross national product
GONGO	government-organized non-governmental organization
GOP	Grand Old Party (Republican Party of the USA)
GPI	genuine progress indicator
GPP	green public rocurement
GPTI	Grupo Permanente de Trabalho Interministerial (Permanent Group for Interministerial Work), Brazil

GRF–SPaC	Global Research Forum on Sustainable Production and Consumption		IDC	International Development Cooperation
GRIT	graduated reduction in tension		IDCOL	Infrastructure Development Company Limited
GTC	global technology company		IDDRI	Institute for Sustainable Development and International Relations
GTI	Great Technological Institute			
GTI	Great Transition Initiative		IDGEC	Institutional Dimensions of Global Environmental Change
GTS	geological time scale			
GTZ	Agency of German Technical Cooperation		IDMC	Internal Displacement Monitoring Centre
GVC	global value chain		IDRC	International Development Research Centre
GW	gigawatt			
GWEC	Global Wind Energy Council		IEA	International Energy Agency
GWP	global warming potential		IFIAS	International Federation of Institutes for Advanced Study
GWSA	Global Warming Solutions Act			
GWSP	Global Water Systems Project		IGBP	International Geosphere-Biosphere Programme
ha	hectares			
HD	hydroelectricity		IGO	international governmental organization
HDGCP	Human Dimensions of Global Change Programme		IHDP	International Human Dimensions Programme on Global Environmental Change
HDGEC	Human Dimensions of Global Environmental Change			
HDI	Human Development Index		IHOPE	Integrated History of People on Earth
HDP	Human Dimensions of Global Environmental Change Programme		IIASA	International Institute for Applied Systems Analysis
HDR	Human Development Report		IIEC	International Institute for Energy Conservation
HFCs	hydrofluorocarbons			
HIA	Health Impact Assessment		IISD	International Institute for Sustainable Development
HIV/AIDS	Human immunodeficiency virus/Acquired immune deficiency syndrome			
			ILO	International Labour Organization
HIVA	Onderzoeksinstituut voor Arbeid en Samenleving (Research Institute for Work and Society)		IMF	International Monetary Fund
			IMTA	Instituto Mexicano de Tecnología del Agua (Mexican Institute of Water Technology)
HLPE	High Level Panel of Experts		INAMPS	Instituto Nacional de Assistência Médica da Previdência Social (National Administration for Medical Assistance and Social Welfare), Brazil
HNR	Human-Nature Relationships			
HPI	Happy Planet Index			
HS	human security			
HSN	Human Security Network			
HUGE	human, gender and environmental (security)		INDC	Intended Nationally Determined Contributions
			INE	Instituto Nacional de Ecología (National Institute of Ecology, Mexico)
HVDC	high voltage direct current			
IAD	Institutional Analysis and Development (framework)		INECC	Instituto Nacional de Ecología y Cambio Climático (National Institute of Ecology and Climate Change), Mexico
IAEA	International Atomic Energy Agency		INEGI	Instituto Nacional de Estadísticas y Geografía e Informática (National Institute for Statistics and Geography), Mexico
IBA	Internationale Bauausstellung (International Building Exhibition)			
			INPS	Instituto Nacional de Previdência Social (National Social Security Administration), Brazil
ICCC	Interministerial Commission on Climate Change			
ICIS	International Centre for Integrated Assessment and Sustainable Development, Maastricht University, The Netherlands		IOM	International Organization for Migration
			IPCC	Intergovernmental Panel on Climate Change
ICNC	International Center on Nonviolent Conflict		IPP	Independent Power Producer
			IPR	intellectual property rights
ICSHP	International Centre on Small Hydro Power		IPRA	International Peace Research Association
ICSU	International Council of Scientific Unions (now International Council for Science)		IRENA	International Renewable Energy Agency
			IRG	integrated risk governance
			IRP	integrated resources planning
ICT	information and communication technology		IRP	International Resource Panel
			ISA	International Sociological Association
ICT	integrated circuit technology		ISA	International Studies Association

ISDR	international strategy for disaster reduction	LUCC	Land Use and Land Cover Change
ISEW	Index of Sustainable Economic Welfare	LULUCF	Land Use, Land Use Change and Forestry
ISFH	Institut für Friedensforschung und Sicherheitspolitik an der Universität Hamburg (Institute for Peace Research and Security Studies at the University of Hamburg)	MA	Millennium Ecosystem Assessment
		MAB	Man and the Biosphere Programme, UNESCO
		MAD	mutually assured destruction
ISI-MIP	Intersectoral Impact Model Intercomparison	MAIRS	Monsoon Asia Integrated Regional Study
		MAS	Mutually Assured Survival
ISRO	Indian Space Research Organisation	MDB	Multilateral Development Bank
ISS	Institute for Security Studies in Pretoria (South Africa)	MDGs	Millennium Development Goals
		MD-ICCCR	Morton Deutsch International Center for Cooperation and Conflict Resolution
ISSC	International Social Science Council	MEA	Metropolitan Electricity Authority
IT	industrial transformation	MENA	Middle East and North Africa
IT	information technology	MEP	Member of the European Parliament
ITD	inter- and transdisciplinarity	MFPRC	Ministry of Finance of the People's Republic of China
IUAES	International Union of Anthropological Sciences.		
		MEVs	micro-electric vehicles
IUCN	International Union for the Conservation of Nature	Mha	million hectares
		MIIT	Ministry of Industry and Information Technology
I-WEC	Isivunguvungu Wind Energy Converter		
		MIoIR	Manchester Institute of Innovation Research
JI	Joint Implementation		
JPOI	Johannesburg Plan of Implementation	MISTRA	Mapungubwe Institute for Strategic Reflection
JTC	JTC Corporation		
		MIT	Massachusetts Institute of Technology
kg	kilogram	MJ	megajoules
kg/cap	kilogram per capita	MLG	multilevel governance
KLSC	Knowledge, Learning and Societal Change	MLP	multilevel perspective (on sociotechnical transitions)
km/l	kilometres per litre		
km²	square kilometre	MME	Ministério de Minas e Energia (Ministry of Mines and Energy), Brazil
KP	Kyoto Protocol		
KSI	Dutch Knowledge Network on Systems Innovations and Transitions	MNES	Ministry of Non-Conventional Energy Sources
kt	kilotonne	MNRE	Ministry of New and Renewable Energy
KU Leuven	Katholieke Universiteit Leuven (Catholic University of Leuven)	MNRES	Ministry of New and Non-Conventional Energy Sources
LANUV	Landesamt für Natur, Umwelt und Verbraucherschutz NRW (environment agency of North Rhine-Westphalia, Germany)	MP	Member of Parliament
		MRI	Mountain Research Initiative
		MRV	monitoring, reporting, verification
		MST	Movimento dos Trabalhadores Rurais Sem Terra (Landless Workers' Movement), Brazil
LCA	light combat aircraft		
LCD	low-carbon development	MSW	municipal solid waste
LCEA	Low-Carbon Economic Area	Mt	megatonne
LCHRS	Low-Carbon Housing Retrofit Strategy	MtCO₂	megatonnes of carbon dioxide
LCR	Local Content Requirement	MTCR	Nuclear Suppliers Group or the Missile Technology Control Regime
LCR	low-carbon and climate-resilient		
LDC	least developed country	MVMC	Metropolitan Valley of Mexico City
LED	light-emitting diode	MW	megawatt
LEGEEPA	Ley General del Equilibrio Ecológico y La Protección al Ambiente (General Law on Ecological Equilibrium and Protection of the Environment), México	MWh	megawatt hour
		N₂O	Nitrous Oxide
		NABU	Naturschutzbund Deutschland (Nature and Biodiversity Conservation Union)
LETS	Local Exchange Trading System		
LM	Lunderskov Moebelfabrik	NAFTA	North American Free Trade Agreement
LOICZ	Land-Ocean Interactions in the Coastal Zone	NAL	National Aerospace Laboratories
		NAMA	Nationally Appropriate Mitigation Actions (in the context of UNFCCC)
LTG	Limits to Growth		

NAS	National Academy of Sciences
NASA	US National Aeronautics and Space Administration
NATO	North Atlantic Treaty Organization
NBFI	non-bank financial institution
NBI	Nile Basin Initiative
NBM	new business model
NBS	National Bureau of Statistics
NBSAP	National Biodiversity Strategies and Action Plans
NDRC	National Development and Reform Commission
NEPAD	New Partnership for Africa's Development
NEPO	National Energy Policy Office
NERSA	National Energy Regulator South Africa
NESDB	National Economic and Social Development Board
NETL	National Energy Technology Laboratory
NGO	non-governmental organization
NHDR	National Human Development Report
NIC	National Intelligence Council
NIC	newly industrialized country
NIMBY	not in my backyard
NOAA	US National Oceanic and Atmospheric Administration
NPC	The National People's Congress
NPO	not-for-profit organization
NRC	US National Research Council
NRDC	Natural Resources Defense Council
NRF	National Research Foundation
NRR	net reproduction rate
NRW	North Rhine–Westphalia
NSI	national system of innovation
NSM	Nijmegen School of Management
NTU	Nanyang Technological University
NUS	National University Singapore
OAS	Organization of American States
ODI	overseas development assistance
OECD	Organisation for Economic Co-operation and Development
OEM	original equipment manufacturer
OIC	observed irreducible complexity
OPEC	Organization of the Petroleum Exporting Countries
OSCE	Organization for Security and Co-operation in Europe
PAHSEP	Pioneers in Arts, Humanities, Science, Engineering, Practice (book series)
PAN	Partido Acción Nacional (National Action Party) country?
PAWSS	Peace and World Security Studies
PAYGO	pay-as-you-go
PBE	Programa Brasileiro de Etiquetagem (Brazilian Labelling Programme)
PCH	Pequena Central Hidrelétrica (Small Hydroelectric Plant)
PDP	Power Development Plan

PDSI	Palmer Drought Severity Index
PEA	Provincial Electricity Authority
PEC	Peace Education Commission of IPRA
PECC	Programa Especial de Cambio Climático (Climate Change Special Programme), Mexico
PEISOR	pressure, effect, impact, societal outcome and response model
PERN	Population–Environment Research Network
PIK	Potsdam Institute for Climate Impact Research and Climate Analytics
PJSA	Peace and Justice Studies Association
PLO	Palestinian Liberation Organization
PND	Programa Nacional de Desarrollo (National Development Programme), Mexico
PPA	Power Purchase Agreement
PPCDAm	Plano de Ação para a Prevenção e o Controle do Desmatamento na Amazônia Legal (Action Plan for the Prevention and Control of Deforestation in the Legal Amazon), Brazil
ppm	parts per million
PPP	purchasing power parity
PRD	Partido de la Revolución Democrática (Revolutionary Democratic Party), Mexico
PRI	Partido Revolucionario Institucional (Institutional Revolutionary Party), Mexico
PRIF	Peace Research Institute Frankfurt
PRIO	International Peace Research Institute in Oslo (Norway)
PRO	public research organization
PROCEL	Programa Nacional de Conservação de Energia Elétrica (National Electrical Energy Conservation Programme), Brazil
PROINFA	Programa de Incentivo às Fontes Alternativas de Energia Elétrica (Alternative Energy Sources Incentive Programme), Brazil
PSP	Poznan Strategic Program on Technology Transfer
PSP	Springer Briefs in Science and Practices (book series)
PSR	pressure-state-response
PT	Partido dos Trabalhadores (Labour Party, Brazil)
PT	Partido del Trabajo (Labour Party, Mexico)
PURPA	Public Utilities Regulatory Policies Act
PV	photovoltaics
PVEM	Partido Verde Ecologista de México (Green Ecological Party of Mexico)
R&D	research and development
RAST	resource abundance and scarcity threshold
RCEP	Regional Comprehensive Economic Partnership
RCP	representative concentration pathways
RD&D	research, development and deployment
RE	renewable energy
REA	Rural Electrification Authority

REDD+	Reducing Emissions from Deforestation and Forest Degradation in Developing Countries and the Conservation, Sustainable Management of Forests and Enhancement of Forest Carbon Stocks policy		SNM	strategic niche management
			SOE	state-owned enterprise
			SOM	climate model simulation
			SOM	soil organic matter
			SP	sustainable production
REIPPPP	Renewable Energy Independent Power Producer Procurement Programme		SPP	small power producer
			SPREAD	European 'social platform' on Sustainable Lifestyles 2050
REN 21	Renewable Energy Network for the 21st Century (c/o UNESCO)		SPREAD	Scenarios for Sustainable Lifestyles–From Global Champions to Local Loops
RHI	Renewable Heat Incentive		SREX	Special Report on Managing the Risks of Extreme Events and Disasters to Advance Climate Change Adaptation (IPCC)
ROW	the rest of the world			
RTD	Research and Technological Development			
RVR	Regionalverband Ruhr (regional association of municipalities of the Ruhr district, Germany)		SRREN	Special Report on Renewable Energy Sources and Climate Change Mitigation (IPCC)
			ST	sustainability transition
SAP	Structural Adjustment Programme		START	System for Analysis, Research and Training
SAR	Special Administrative Region		STI	science, technology and innovation
SC	Scientific Committee		STRN	Sustainability Transition Research Network
SC	State Council		STS	science and technology studies
SC	Sustainable Consumption		SuMMa	Policy Research Centre on Sustainable Materials Management
SCI	KompetenzCentrum for Corporate Social Responsibility, University of Applied Sciences in Essen, Germany		SUS	Sistema Único de Saúde (Unified Health System), Brazil
SCI	Sustainable Consumption Institute, University of Manchester		SYR	Synthesis Report
SCOPE	Scientific Committee on Problems of the Environment, ICSU		t/cap	tonnes per annum per capita
			TAMA	there are many alternatives
SCORAI	Sustainable Consumption Research and Action Initiative		TAPRI	Tampere Peace Research Institute
			TC	triangular cooperation
SCP	sustainable consumption and production		tce	tonnes of coal equivalent
SD	sustainable development		tep	tonelada equivalente de petróleo (tons of oil equivalent, in Portuguese)
SDG	Sustainable Development Goal			
SDSN	Sustainable Development Solutions Network		TINA	there is no alternative
			TIS	technological innovation systems approach
SEGOB	Secretaría de Gobernación (Ministry of the Interior), Mexico		TM	transition management
			TMR	total material requirements
SEMARNAT	Secretaría de Medio Ambiente y Recursos Naturales (Mexican Ministry of the Environment and Natural Resources)		TNA	technology needs assessment
			TNC	transnational corporation
			TNEB	Tamil Nadu Electricity Board (India)
			TNEP	The Natural Edge Project
SERC	Structural Engineering Research Centre		TNO	Nederlandse Organisatie voor Toegepast Natuurwetenschappelijk Onderzoek (Netherlands Organisation for Applied Scientific Research)
SERI	Sustainable Europe Research Institute			
SES	social-ecological systems theory			
SET	Stock Exchange of Thailand			
SHS	solar home systems			
SIAP	Sistema de Información Agropecuario (System of Information for Agriculture and Livestock), Mexico		toe	tonnes of oil equivalent
			TPP	Trans-Pacific Partnership
			TRIP	trade-related aspects of intellectual property rights
SIDS	small island developing states			
SIL	Sistema de Información Legislativa (Legislative Information System) Mexico		TRT	Thai Rak Thai
			TSB	Technology Strategy Board
SIN	Sistema Interligado Nacional (National Interconnected System), Brazil		TT	transitions theory
			TTIP	Transatlantic Trade and Investment Partnership
SIPRI	Stockholm International Peace Research Institute			
			TUM	Technical University of Munich
SME	small and medium-sized enterprise		TUPADO	Turkana Pastoralist Organization
SNCC	Sistema Nacional de Cambio Climático (National Climate Change System) Mexico		TWA	tolerable windows approach

TWAS	Third World Academy of Science	UNWR	United Nations Water Report
UAEM	Universidad Autónoma del Estado de Morelos (Autonomous University of the State of Morelos, Mexico)	UPEACE	University for Peace
		US	United States of America
		USA	United States of America
UCDP	Uppsala Conflict Data Programme	USAID	United States Agency for International Development
UDLAP	Universidad de las Américas, Puebla		
UGEC	Urbanization and Global Environmental Change	USD	United States dollar
		USIP	United States Institute of Peace
UHC	universal health care	USSR	Union of Soviet Socialist Republics
UHE	Usina Hidrelétrica (Hydroelectric Plant)	UV	ultraviolet
UHI	University of the Highlands and Islands, Scotland, UK	VIABLE	Values and Investments in Agent-Based Interaction and Learning for Environmental Systems
UK	United Kingdom		
UN	United Nations	VKT	vehicle kilometres travelled
UNAM	Universidad Nacional Autónoma de México (National Autonomous University of Mexico)	VSPP	very small power producer
		WAPOR	World Association for Public Opinion Research
UNCCD	United Nations Convention to Combat Desertification	WASH	water access, sanitation and hygiene
UNCED	United Nations Conference on Environment and Development	WB	World Bank
		WBCSD	World Business Council for Sustainable Development
UNCSD	United Nations Commission on Sustainable Development	WBGU	Wissenschaftlicher Beirat Globale Umweltveränderungen (German Advisory Council on Global Change)
UNCSD	United Nations Conference on Sustainable Development (or Rio+20) in June 2012		
UNCTAD	United Nations Conference on Trade and Development	WCED	World Commission on Environment and Development
UNDESA	United Nations Department of Economic and Social Affairs	WCR	World Climate Research Programme
		WEF	World Economic Forum
UNDP	United Nations Development Programme	WEG	wind energy generator
UNECA	United Nations Economic Commission for Africa	WFE	water, food, and energy security nexus
		WFE	World Federation of Engineering Organizations
UNEP	United Nations Environment Programme	WG	Working Group (of the IPCC)
UNESCAP	United Nations Economic and Social Commission for Asia and the Pacific	WHO	World Health Organization
		WIPO	World Intellectual Property Organization
UNESCO	United Nations Educational, Scientific and Cultural Organization	WIR	World Investment Report
		WMO	World Meteorological Organization
UNFCCC	United Nations Framework Convention on Climate Change	WMR	Wirtschaftsförderung Ruhr (economic development agency of the Ruhr district, Germany)
UNGA	United Nations General Assembly		
UNHCR	United Nations High Commissioner for Refugees	WS	water security
		WSSD	World Summit on Sustainable Development, Johannesburg (2002)
UNICAM	Peasant University of the South		
UNICEF	United Nations Children's Fund	WTO	World Trade Organization
UNIDO	United Nations Industrial Development Organization	WVS	World Value Survey
		WWF	Worldwide Fund for Nature (also known as World Wildlife Fund)
UNODC	United Nations Office on Drugs and Crime		
UNPD	United Nations Population Division		
UNSC	United Nations Security Council	YHDR	Young Human Dimensions Researchers
UNSG	United Nations Secretary-General	YRB	Yautepec River Basin (Mexico)
UNU	United Nations University		
UNU-EHS	United Nations University–Environment and Human Security Institute	ZESCO	Zambia Electricity Supply Corporation Limited
UNU-MERIT	United Nations University–Maastricht Economic and Social Research Institute on Innovation and Technology		

Biographies of Editors

Hans Günter Brauch (Germany), Dr. phil. habil., Adj. Prof. (Privatdozent) at the Faculty of Political and Social Sciences, Free University of Berlin (1990-2012); since 1987 chairman of *Peace Research and European Security Studies* (AFES-PRESS). He is editor of the *Hexagon Book Series on Human and Environmental Security and Peace* (HESP), of the *Springer Briefs in Environment, Security, Development and Peace* (ESDP), of the *SpringerBriefs on Pioneers in Science and Practice* (PSP); of *Pioneers in Arts, Humanities, Science, Engineering, Practice (PAHSEH)* and of *The Anthropocene: Politik–Economics–Society–Science (APESS)* with Springer International Publishing. He has been visiting professor of international relations at the universities of Frankfurt am Main, Leipzig, Greifswald, and Erfurt; research associate at Heidelberg and Stuttgart universities, and a research fellow at Harvard and Stanford Universities. He has taught at the Free University of Berlin (1990-2012), at Sciences Po (Paris, 2010-2012), at the *European Peace University* (EPU, Schlaining, Austria), at the *National Autonomous University of Mexico* (UNAM), at the *National University of Malaysia* (UKM), Malaysia (2010, 2012), and in autumn and winter 2013/2014 he was a visiting professor at Chulalongkorn University in Bangkok, Thailand. He has published on security, armament, climate, energy, and migration, on Mediterranean issues and on sustainability transition in English and German, and his work has been translated into Chinese, Spanish, Greek, French, Danish, Finnish, Russian, Japanese, Portuguese, Serbo-Croat, and Turkish. *Publications*: His books include: (co-author with H. v. d. Graaf, J. Grin, W. Smit): *Militärtechnikfolgenabschätzung und präventive Rüstungskontrolle*, 1997; *Klimapolitik der Schwellenstaaten: Südkorea, Mexiko und Brasilien; Osterweiterung der Europäischen Union. Umwelt- und Energiepolitik der Tschechischen Republik*, 2000. Books in English, Spanish, Greek, Turkish and Chinese: (co-ed. with D. L. Clark): *Decision-making for Arms Limitation–Assessments and Prospects*, 1983; (ed.): *Star Wars and European Defence–Implications for Europe: Perceptions and Assessments*, 1987; (co-author with R. Bulkeley): *The Anti-Ballistic Missile Treaty and World Security*, 1988; (ed.): *Military Technology, Armaments Dynamics and Disarmament*, 1989; (co-ed. with R. Kennedy): *Alternative Conventional Defense Postures in the European Theater*, vol. 1: *The Military Balance and Domestic Constraints*, 1990; Vol. 2: *Political Change in Europe: Military Strategy and Technology*, 1992; Vol. 3: *Military Alternatives for Europe after the Cold War*, 1993; (co-ed. with H. J. v. d. Graaf, J. Grin; W. Smit): *Controlling the Development and Spread of Military Technology*, 1992; (co-ed. with A. Marquina): *Confidence Building and Partnership in the Western Mediterranean. Tasks for Preventive Diplomacy and Conflict Avoidance*, 1994; *Energy Policy in North Africa (1950-2050). From Hydrocarbon to Renewables*, 1997; (co-ed. with A. Marquina, A. Biad): *Euro-Mediterranean Partnership for the 21st Century*, 2000; (co-ed. with A. Marquina): *Political Stability and Energy Cooperation in the Mediterranean* (2000); *Liberalisation of the Energy Market for Electricity and Gas in the European Union: a Survey and implications for the Czech Republic*, 2002; (co-ed. with P. Liotta, A, Marquina, P. Rogers, M. Selim): *Security and Environment in the Mediterranean. Conceptualising Security and Environmental Conflicts*, 2003; *Environmental Dimension of Human Security: Freedom from Hazard Impact*, 2005; *Threats, Challenges, Vulnerabilities and Risks in Environmental and Human Security*, 2005; (co-ed. with Ú. Oswald Spring, C. Mesjasz, J. Grin, P. Dunay, N. Chadha Behera, B. Chourou, P. Kameri-Mbote, P. H. Liotta): *Globalization and Environmental Challenges: Reconceptualizing Security in the 21st Century*, 2008; (co-ed. with Ú. Oswald Spring, J. Grin, C. Mesjasz, P. Kameri-Mbote, N. Chadha Behera, B. Chourou, H. Krummenacher): *Facing Global Environmental Change: Environmental, Human, Energy, Food, Health and Water Security Concepts* (2009); (co-ed. with Ú. Oswald Spring): *Reconceptualizar la Seguridad en el Siglo XXI* (2009); (co-author with Ú. Oswald Spring: *Securitizing the Ground–Grounding Security* and: *Seguritizar la Tierra–Aterrizar la Seguridad* (Bonn: UNCCD, 2009); (guest co-ed. with Ú. Oswald Spring and M. Aydin of a special issue of: *Uluslararasi Iliskiler / International Relations*, 5,18 (Summer) Special Issue on "Security" (2009): (co-ed. with Ú.Oswald Spring, C. Tsardanidis and Y. Kinnas: Greek translation of 7 chapters, vol. 3: *Globalization and Environmental Challenges*, in: *Agora*, Spring 2010; *Climate Change and Mediterranean Security: International, National, Environmental and Human Security Impacts for the Euro-Mediterranean Region during the 21st Century–Proposals and Perspectives*. Papers IEMed No. 9 (Barcelona: IE, 2010); (co-ed. with Oswald Spring, Mesjasz, Grin, Kameri-Mbote, Chourou, Dunay, Birkmann), *Coping with Global Environmental Change, Disasters and Security–Threats,*

© Springer International Publishing Switzerland 2016   937
H.G. Brauch et al. (eds.), *Handbook on Sustainability Transition and Sustainable Peace,*
Hexagon Series on Human and Environmental Security and Peace 10, DOI 10.1007/978-3-319-43884-9

Challenges, Vulnerabilities and Risks (2011); (co-ed with Scheffran, Brzoska, Link, Schilling), *Climate Change, Human Security and Violent Conflict: Challenges for Societal Stability* (2012); with Aydin, Celikpata, Oswald Spring, Polat (Eds.): *Uluslararasi Iliskilerde Catismadam Güvenige* (2012); co-ed. with Oswald Spring, Tidball (Eds.): *Expanding Peace Ecology: Security, Sustainability, Equity and Peace: Perspectives of IPRA's Ecology and Peace Commission* (2014); co-ed. with Oswald Spring, Serrano Oswald, Estrada Álvarez, Flores Palacios, Ríos Everardo, Ruíz Pantoja, Lemus Ramírez, Estrada Villareal, Cruz: *Vulnerabilidad Social y Género entre Migrantes Ambientales* (Cuernavaca: CRIM, DGAPA-PAPIIT-UNAM, RETAC-Conacyt, 2014); co-ed. with Grimwood (Ed.): *Jonathan Dean: Pioneer in Détente in Europe, Global Cooperative Security, Arms Control and Disarmament* (2015); co-ed. with Oswald Spring, Mesjasz, Grin, Cheng (Eds.): [in Chinese] *Globalization and Environmental Challenges: Reconceptualizing Security in the 21st Century* (2015); co-ed. with Oswald Spring, Mesjasz, Grin, Cheng (Eds.): [in Chinese]: *Facing Global Environmental Change: Environmental, Human, Energy, Food, Health and Water Security Concepts* (2015); co-ed. with Oswald Spring, Mesjasz, Grin, Cheng (Eds.): [in Chinese] *Coping with Global Environmental Change, Disasters and Security—Threats, Challenges, Vulnerabilities and Risks* (2015); co-ed. with Oswald Spring, Bennett, Serrano: *Addressing Global Environmental Challenges from a Peace Ecology Perspective* (2016); co-ed. with Oswald Spring, Serrano, Bennett: *Regional Ecological Challenges for Peace in Africa, the Middle East, Latin America and Asia Pacific* (2016).

Address: PD Dr. Hans Günter Brauch, Alte Bergsteige 47, 74821 Mosbach, Germany.
Email: <brauch@afes-press.de>.
Website: <http://www.afes-press.de> and <http://www.afes-press-books.de/>.

Úrsula Oswald Spring (Mexico), full time Professor/Researcher at the *National University of Mexico* (UNAM) in the *Regional Multidisciplinary Research Center* (CRIM), she was national coordinator of water research for the *National Council of Science and Technology* (RETAC-CONACYT), first Chair on Social Vulnerability at the *United National University Institute for Environment and Human Security* (UNU-EHS); founding Secretary-General of El Colegio de Tlaxcala; General Attorney of Ecology in the State of Morelos (1992-1994), National Delegate of the Federal General Attorney of Environment (1994-1995); Minister of Ecological Development in the State of Morelos (1994-1998). She was President of the *International Peace Research Association* (IPRA, 1998-2000), and General Secretary of the Latin-American Council for

Peace Research (2002-2006). She studied medicine, clinical psychology, philosophy, anthropology, ecology, and classical and modern languages. She obtained her PhD from the University of Zürich (1978). For her scientific work she received the Premio Sor Juana Inés de la Cruz (2005), the Environmental Merit in Tlaxcala, Mexico (2005, 2006), and the UN Development Prize. She was recognized as Women Academic of UNAM (1990 and 2000); and Women of the Year (2000). She works on nonviolence and sustainable agriculture with groups of peasants and women and is President of the Advisory Council of the Peasant University. She has written fifty-two books and more than 379 scientific articles and book chapters on sustainability, water, gender, development, poverty, drug consumption, brain damage due to undernourishment, peasantry, social vulnerability, genetically modified organisms, bioethics, and human, gender, and environmental security, adaptation, resilience, climate-induced migration, peace and conflict resolution, democracy and processes of negotiation. Among her major publications are: (co-author with Rudolf Strahm): *Por esto somos tan pobre* (translated into 17 languages, 1.5 million copies); *Unterentwicklung als Folge von Abhängigkeit* (Berne: Lang, 1978); *Mercado y Dependencia* (México, D.F.: Ed. Nueva Imagen, 1979); *Piedras en el Surco* (México, D.F.: UAM-X, 1983); *Campesinos Protagonistas de su Historia: la Coalición de los Ejidos Colectivos de los Valles del Yaqui y Mayo, una Salida a la Cultura de la Pobreza* (México, D.F.: UAM-X, 1986); *Estrategias de Supervivencia en la Ciudad de México* (Cuernavaca: CRIM/UNAM, 1991); *Fuenteovejuna o Caos Ecológico* (Cuernavaca: CRIM/UNAM, 1999); *Peace Studies from a Global Perspective: Human Needs in a Cooperative World* (ed.) (New Delhi: Mbooks, 2000); (co-author with Mario Salinas): *Gestión de Paz, Democracia y Seguridad en América Latina* (México, D.F.: UNAM-CRIM/Coltlax, Böll, 2002); *El recurso agua en el Alto Balsas* (ed.) (México: IGF, CRIM/UNAM, 2003); *Soberanía y Desarrollo Regional. El México que queremos* (ed.) (México, D.F.: UNAM, 2003); *Resolución noviolenta de conflictos en sociedades indígenas y minorías* (ed.) (México, D.F.: CLAIP, IPRA & Böll Fundation, COLTLAX, 2004); *El valor del agua: una visión socioeconómica de un conflicto ambiental* (COLTLAX, CONACYT, 2005); *International Security, Peace, Development, and Environment*, Book 39: *Encyclopaedia on Life Support Systems* (ed.) (Paris: UNESCO–EOLSS, UK, online); *Gender and Disasters* (Bonn: UNU-EHS, 2008); (co-ed. with H. G. Brauch, C. Mesjasz, J. Grin, P. Dunay, N. Chadha Behera, B. Chourou, P. Kameri-Mbote, P. H. Liotta), 2008: *Globalization and Environmental Challenges: Reconceptualizing Security in the 21st Century* (translated into Chinese); (co-ed. with H. G. Brauch, J. Grin, C. Mesjasz, P. Kameri-Mbote, N. Chadha Behera, B. Chourou, H. Krummenacher (translated into Chinese), 2009: *Facing Global Environmental Change: Environmental, Human, Energy, Food, Health and Water Security Concepts*; (co-ed. with H.G. Brauch): *Reconceptualizar la Seguridad en el Siglo XXI*, 2009; (co-author with H. G. Brauch: *Securitizing the Ground—Grounding Security* and *Seguritizar la Tierra—Aterrizar la Seguridad* (Bonn: UNCCD), 2009; (guest co-ed. with H. G. Brauch and M. Aydin of a special issue of *Uluslararasi Iliskiler / International Relations*, 5,18 (Summer) Special Issue on "Security", 2009: (co-ed. with H. G. Brauch, C. Tsardanidis and Y. Kinnas: Greek translations of 7 chapters, vol. 3: *Globalization and Environmental Challenges*, in: *Agora*, Spring 2010; (co-ed. with H. G. Brauch, C. Tsardanidis and Y. Kinnas): Greek translations of chapters, vol. 4: Facing Global Environmental Change, in: *Agora*, Summer, 2010; *Retos de la investigación del agua en México* (ed.), 2011, CRIM-UNAM, CONACYT, México; *Water Research in Mexico,* Springer Verlag, Berlin; *Coping with Global Environmental Change: Disaster and Security* (co-ed. with H. G. Brauch, C. Mesjasz, J. Grin, P. Dunay, N. Chadha Behera, B. Chourou, P. Kameri-Mbote, P. Dunay, J. Birkman) 2011 (translated into Chinese: 应对全球环境变化、灾难及安全 —— 威胁、挑战、缺陷和风险); "Can health be Securitized?", *Human evolution* 2012, 27 (1-3), 2015, pp. 21-29; "Vulnerabilidad social en eventos hidrometeorológios extremos: una comparación entre los huracanes Stan y Wilma", *SocioTam*, 2012, 22 (2) (July-Dec), pp. 125-145.; "Forced migration, climate change, mitigation and adaptive policies in Mexico. Some functional relationships", *International Migration*, 2013 (co-author with I. Sánchez, G. Díaz et al.); "Water security and national water law in Mexico", *Earth Perspectives*, 2013: vol.1, núm. 7: 1-15. "Dual Vulnerability among Female Household Heads", *Acta Colombiana de Psicología,* 2013: 16(2), pp. 19-30; *Vulnerabilidad Social y Género entre Migrantes Ambientales,* (co-author with S. E. Serrano, Adriana. Estrada, F. Flores, M. Ríos, H. G. Brauch, T. E. Ruíz, C. Lemus, Ariana Estrada, M. Cruz), 2014, CRIM, DGAPA-UNAM, Cuernavaca; *Expanding Peace Ecology: Security, Sustainability, Equity and Peace*, 2015, (co-ed. with H. G. Brauch, K. G. Tidball) Berlin, Springer Verlag; "Human Security", 2014, (co-author with N. Adger et al.), *Fifth Assessment Report, IPCC, WG 2*, Cambridge University Press, Cambridge, pp. 755-791; "Cambio Climático, Salud y Género", 2014, (co-author with A. R. Moreno, O. Tena, in: M. Ímaz et al. (eds.). *Cambio climático, miradas de género*, México, UNAM, pp. 85-136; *América Latina en el Camino hacia una Paz Sustentable: Herramientas y Aportes*, 2015: (co-ed. with S. E. Serrano, D. de la Rúa) ARP, FLACSO, CLAIP, Guatemala; *México ante la Urgencia Climática: Ciencia, Política y Sociedad* (co-ed. with X. Cruz, G. C. Delgado), 2015, CCEIICH, CRIM, PINCC, UNAM; "Water Security: Past, Present and Future of a Controversial Concept", 2015, (co-author with J. Bogardi, H. G. Brauch), in: *Handbook on Water Security*, C. Pahl-Wostl, A. Bhaduri, J. Gupta (eds.), Edward Elgar, Cheltenham, pp. 38-58; "Seguridad humana", in C. Gay, C. Ruiz (eds.), 2015: *Reporte Mexicano de Cambio Climático GRUPO II Impactos, vulnerabilidad y adaptación*, PINCC-UNAM, pp. 183-210.

Address: Prof. Dr. Úrsula Oswald Spring, Priv. Río Bravo Núm.1, Col. Vistahermosa, Cuernavaca, Morelos, 62290 México.

Email: <uoswald@gmail.com>.

Website: <http://www.afes-press.de/html/download_oswald. html>.

John Grin (The Netherlands) is a Full Professor in Policy Science, in particular of System Innovation, Dept. of Political Science, University of Amsterdam, and Co-Director (with Marieke de Goede) of the Programme Group Transnational Configurations, Conflicts and Governance, Amsterdam Institute for Social Science Research (AISSR). A physicist by training (MSc 1986, VU University Amsterdam) he gained a PhD (1990) on defence technology assessment. He is driven by a profound, life-long interest in the relationships between knowledge, technology, society and policy. He is particularly interested in understanding the 'politics' of knowledge and technology and ways to analyse them; and in the intertwining between the two. He co-established and was co-director of the transdisciplinary *Dutch Knowledge Network for System Innovations and Transitions* (KSI; <www.ksinetwork.nl>), in which some hundred researchers of ten different universities cooperate on major changes towards a sustainable society. From 2006 to 2009 he was scientific director of the Amsterdam School for Social Science Research (ASSR; at: <www.assr.nl>). He co-edited special issues of the journals *Policy Sciences* (vol. 43[2009], no. 4) and *Research Policy* (vol. 39 [2010], no. 4) on transitions, and he published (co-author with Jan Rotmans, Johan Schot): *Transitions to Sustainable Development. New Directions in the Study of Long Term Structural Change (*New York: Routledge, 2010). Earlier (co-)authored books include (with Hans Günter Brauch, Henk van de Graaf, Wim Smit): *Militärtechnikfolgenabschätzung und Präventive Rüstungskontrolle. Institutionen, Verfahren und Instrumente* (Münster: LIT, 1997); and *Military-technological choices and political implications. Command and control in established NATO posture and a non-provocative defence* (Amsterdam: VU University Press—New York: St. Martin's Press, 1990); (co-ed. with Wim A. Smit, Lev Voronkov): *Military-technological innovation and stability in a changing world. Politically assessing and influencing weapon innovation and military research and development* (Amsterdam: VU University Press, 1992); (with Henk van de Graaf, Rob Hoppe): *Technology assessment through interaction: A guide* (Den Haag: SDU, 1997). His edited volumes include (with Armin Grunwald): *Vision Assessment: Shaping Technology in 21st century society. Towards a repertoire for Technology Assessment* (Heidelberg: Springer, 2000); (co-ed. with Brauch, Oswald Spring, Mesjasz, Dunay, Chadha Behera, Chourou, Kameri-Mbote, Liotta, 2008): *Globalization and Environmental Challenges: Reconceptualizing Security in the 21st Century* (Springer); (co-ed. with Brauch, Oswald Spring, Mesjasz, P. Kameri-Mbote, Chadha Behera, Chourou, Krummenacher, 2009): *Facing Global Environmental Change: Environmental, Human, Energy, Food, Health and Water Security Concepts* (Springer), (co-ed. with Brauch, Oswald Spring, Mesjasz, Kameri-Mbote, Chourou, Dunay, Birkmann: *Coping*

with Global Environmental Change, Disasters and Security—Threats, Challenges, Vulnerabilities and Risks, 2011); (co-ed. with Brauch, Oswald Spring, Scheffran (Eds.): *Handbook on Sustainability Transition and Sustainable Peace*, 2014). Some key journal articles and book chapters on transitions are Niki Frantzeskaki, Wil Thissen, John Grin: "Drifting between transitions. Lessons from the environmental transition around the river Acheloos Diversion project in Greece", in: *Technological Forecasting and Social Change,* 102 (January 2016): 275-286; Andrew Switzer, Luca Bertolini, John Grin, 2015: "Understanding transitions in the regional transport and land-use system: Munich 1945-2013", in: *Town Planning Review*, 86,6: 699-723; Vanja Karadzic, Paula Antunes, John Grin, 2014: "Adapting to environmental and market change: Insights from Fish Producer Organizations in Portugal", in: *Ocean & Coastal Management,* 102, Part A: 364-374 ; Jan Hassink, Wim Hulsink, John Grin, 2014: "Farming with care: the evolution of care farming in the Netherlands", in: *NJAS–Wageningen Journal of Life Sciences*, 68,7: 1-11; Vanja Karadzic, John Grin, Paula Antunes, Marija Banovic, 2014: "Social learning in fish producers' organizations: how fishers perceive their membership experience and what they learn from it", in: *Marine Policy*, 44: 427-437; Andrew Switzer, Luca Bertolini, John Grin, 2013: "Transitions of Mobility Systems in Urban Regions: A Heuristic Framework", in: *J. Environmental Policy and Planning*, 15,2: 141-160 (JIF 1,421; SCI 1,316); Jan Hassink, John Grin, Wim Hulsink, 2013: "Multifunctional Agriculture Meets Health Care: Applying the Multi-Level Transition Sciences Perspective to Care Farming in the Netherlands", in: *Sociologia Ruralis*, 53,2: 223-245; John Grin, 2012: "The politics of transition governance in Dutch agriculture. Conceptual understanding and implications for transition management", in: *Int. J. Sustainable Development*, 15, 2:72-89 (SCI 0,297); Vanja Karadzic, Paula Antunes, John Grin, 2012: "'How to learn to be adaptive?' An analytical framework for organizational adaptivity and its application to a fish producers organization in Portugal", in: *Journal of Cleaner Production*, 45: 29-37; Boelie Elzen, Marc Barbier, Marianne Cerf, John Grin, 2012: "Stimulating transitions towards sustainable farming systems", in: Ika Darnhofer, David Gibbon, Benoit Dedieu (Eds.) (2012). *Farming Systems Research into the 21st century: The new dynamic* (Berlin etc.: Springer): 431-455; John Grin, 2012: "Between governments, kitchens, firms and farms: the governance of transitions between societal practices and supply systems", in: Spaargaren, Gert, Anne Loeber, and Peter Oosterveer (Eds.), 2012: *Food Practices in Transition. Changing Food Consumption, Retail and Production in the Age of Reflexive Modernity.* (London: Routledge): 35-56; John Grin, Johan Schot, Jan Rotmans, 2011: "On patterns and agency in transition dynamics: Some key insights from the KSI programme", in: *Environmental Innovation and Societal Transitions*, 1,1: 76-81; John Grin, Esther Marijnen, 2010: "Global Threats, Global Changes and Connected Communities in the Agrofood system", in: (co-ed. with Brauch, Oswald Spring, Mesjasz, Grin, Kameri-Mbote, Chourou, Dunay, Birkmann, 2011): *Coping with Global Environmental Change, Disasters and Security—Threats, Challenges, Vulnerabilities and Risks*): 1005-1018.

Address: Prof. Dr. John Grin, Dept. of Political Science, University of Amsterdam P.O. Box 15578, 1001 NB Amstedam, Netherlands.
E-mail: <j.grin@uva.nl>.
Website: www.uva.nl/profiel/j.grin and at: <http://www.afes-press.de/html/ grin_en.html>.

Jürgen Scheffran (Germany) is Professor at the Institute of Geography and the *Center for Earth System Research and Sustainability* (CEN) of the University of Hamburg in Germany and head of the *Research Group Climate Change and Security* (CLISEC) in the Cluster of Excellence *Integrated Climate System Analysis and Prediction (CliSAP)*. Until summer 2009 he held faculty positions at the Departments of Political Science and Atmospheric Sciences at the *University of Illinois at Urbana-Champaign* (UIUC) where he also was a researcher in the *Program in Arms Control, Disarmament and International Security* (ACDIS), the *Center for Advanced BioEnergy Research* (CABER) and the *Energy Biosciences Institute* (EBI). After his PhD in physics at Marburg University he held various research positions in the *Interdisciplinary Research Group Science, Technology and Security* (IANUS) at Technical University Darmstadt, the *Potsdam Institute for Climate Impact Research* (PIK) and as a Visiting Professor at the University of Paris (Pantheon-Sorbonne). At all these locations he taught more than fifty different courses on a wide range of topics, and attracted project funding from a number of funding organizations, e.g. *Deutsche Forschungs-Gemeinschaft* (DFG), *German Academic Exchange Service* (DAAD), *Federal Ministry for Education and Research* (BMBF), European Commission, COST Action, VW Foundation, Berghof Foundation, Heinrich Böll Foundation, Federal Environmental Agency, US Department of Energy, UIUC Environmental Council, Critical Initiatives for Research and Scholarship, MacArthur Foundation, Ploughshares Fund, Vattenfall Europe, BP. Besides CliSAP and CLISEC, participation in recent projects includes the ConflictSpace project, the Renewable Energy Initiative and related projects on bioenergy at UIUC, the EuTRACE assessment of climate engineering, and the COST Action on climate change and migration. Research and teaching interests include: climate change and energy security; natural resources, land use, environmental conflict and human migration; rural-urban interaction and the water–energy–food nexus; complex systems analysis, agent-based modelling and human-environment interactions; transformation and sustainability science; technology assessment, and arms control and international security. He has served as a consultant to the United Nations, the Technology Assessment Bureau of the German Parliament, and the Federal Environmental Agency, and he was part of the German delegation to the climate negotiations in New Delhi in 2002

(COP-8). He has published more than twenty books and more than 300 articles in journals and books, is co-editor of *Wissenschaft und Frieden (Science and Peace)*, and has co-edited special issues of the journals *Complexity, Foreign Policy Analysis, Earth System Dynamics, Migration and Development*. Recent books include: (co-ed. with Brzoska, Brauch, Link, Schilling): *Climate Change, Human Security and Violent Conflict: Challenges for Societal Stability* (2012); (co-ed. with Khanna, M., Zilberman, D.): *Handbook of Bioenergy Economics and Policy* (Heidelberg etc.: Springer, 2010); (co-ed. with Blaschek, H., Ezeji, T.): *Biofuels from Agricultural Wastes and Byproducts* (Ames, Iowa: Wiley/Blackwell, 2010); (co-ed. with Mascia, P. N., Widholm, J.): *Plant Biotechnology for Sustainable Production of Energy and Co-Products* (Heidelberg etc.: Springer, 2010); (co-ed. with Kropp, J.): *Advanced Methods for Decision Making and Risk Management in Sustainability Science* (New York: Nova Science, 2007); (co-ed. with Billari, F., Fent, T., Prskawetz, A.): *Agent-Based Computational Modelling in Demography, Economic and Environmental Sciences* (Heidelberg etc.: Springer/Physica, 2006); (co-ed. with Johnke, B., Soyez, K.): *Abfall, Energie und Klima. Wege und Konzepte für eine integrierte Ressourcennutzung* (Erich Schmidt Verlag, Berlin: 2004); (co-author with Ott, K; Klepper, G. M., Lingner, S., Schäfer, A., Sprinz, D.): *Reasoning Goals of Climate Protection. Specification of Article 2 UNFCCC* (Berlin: Umweltbundesamt, 2004); (co-ed. with Hummel, M., Simon, H.-R.): *Konfliktfeld Biodiversität* (Münster: agenda, 2002); (co-ed. with Bender, W., Hagen, R., Kalinowski, M.): *Space Use And Ethics* (Münster: agenda, 2001); (co-ed. with Zoll, R., Ackermann, H., Melsheimer, O.): *Energiekonflikte: Problemübersicht und empirische Analysen zur Akzeptanz von Windkraftanlagen* (Münster: LIT Verlag); (co-author with Abid, M., Schilling, J., Zulfiqar, F.): "Climate change vulnerability, adaptation and risk perceptions at farm level in Punjab, Pakistan", in: *Science of The Total Environment*, 547 (2/2016): 447-460; (co-author with Shu, K., Schneider, U.): "Bioenergy and Food Supply: A Spatial-Agent Dynamic Model of Agricultural Land Use for Jiangsu Province in China", in: *Energies*, 8,11 (11/2015): 13284-13307; (co-author with Gioli, G., Hu-

go, G., Máñez Costa, M.): "Human mobility, climate adaptation, and development. Introduction to special issue", in: *Migration and Development* (2015); (co-author with Schilling, J., Locham, R, Weinzierl, T., Vivekanada, J.): "The nexus of oil, conflict, and climate change vulnerability of pastoral communities in northwest Kenya", in: *Earth System Dynamics*, 6,2 (11/2015): 703-717; (co-author with Schäfer, M. S.; Penniket, L.): "Securitization of media reporting on climate change?", in: *Security Dialogue*, 10/2015; (co-author with Rodriguez Lopez, J. M., Rosso, P., Delgado, D.): "Remote Sensing of Sustainable Rural-Urban Land Use in Mexico City", in: *Journal Interdisciplina*, 3,7 (2015): 1-20; (co-author with Link, P. M., Brücher, T., Claussen, M., Link, J. S. A.): "The Nexus of Climate Change, Land Use, and Conflict: Complex Human-Environment Interactions in Northern Africa", in: *Bulletin of the American Meteorological Society* (September 2015): 1561-1564); (co-author with Yang, L., Qin, H., You, Q.): "Climate-related Flood Risks and Urban Responses in the Pearl River Delta, China", in: *Regional Environmental Change*, 15,2 (2015): 379-391; (co-author with Ide, T., Schilling, J., Link, J. S. A., Ngaruiya, G., Weinzierl, T.: "On exposure, vulnerability and violence: Spatial distribution of risk factors for climate change and violent conflict across Kenya and Uganda", in: *Political Geography*, 43,1 (2014): 68-81); (co-author with Scheffran, J., Marmer, E., Sow, P.): "Migration as a contribution to resilience and innovation in climate adaptation: Social networks and co-development in Northwest Africa", in: *Applied Geography*, 33: 119-127); (co-author by Scheffran, J., Brzoska, M., Kominek, J., Link, P. M., Schilling, J.): "Climate change and violent conflict", in: *Science*, 336: 869-871.

Address: Prof. Dr. Jürgen Scheffran, CLiSAP/CEN, Grindelberg 7, 20144 Hamburg, Germany.
Email: <juergen.scheffran@uni-hamburg.de>.
Website: <http://clisec.zmaw.de/Prof-Dr-Juergen-Scheffran.867.0.html>; <https://www.geo.uni-hamburg.de/geographie/mitarbeiterverzeichnis/scheffran.html>.

Biographies of the Authors of Forewords

Heide Hackmann (a dual citizen of South Africa and Germany) has been Executive Director of the *International Council for Science* (ICSU) since March 2015, following eight years as Executive Director of the *International Social Science Council* (ISSC) from 2007 to 2015. She was born in South Africa, read for an MPhil in contemporary social theory at the University of Cambridge, UK, and completed her PhD in Science and Technology Studies at the University of Twente, Netherlands in 2003. She has worked as a science policymaker, researcher and consultant in the Netherlands, Germany, the United Kingdom and South Africa. Her career in science policy dates back to the early 1990s, when she worked at the Human Sciences Research Council in South Africa. From 2004 until 2007 she was Head of the Department of International Relations and National Quality Assurance and Director of CO-REACH (an EU-funded multilateral initiative for the Coordination of Research between Europe and China) at the *Royal Netherlands Academy of Arts and Sciences* (KNAW). She is a member of several international advisory committees, including the Scientific Advisory Board of the Potsdam Institute for Climate Impact Research in Germany, and co-chairs the UN's ten-member group supporting the *Technology Facilitation Mechanism* (TFM) on the Sustainable Development Goals.

ICSU is a non-governmental organization with a global membership of national scientific bodies (121 members, representing 141 countries) and International Scientific Unions (31 members). It mobilizes the knowledge and resources of the international scientific community to strengthen international science for the benefit of society. The ISSC was established by the United Nations Educational, Scientific and Cultural Organization (UNESCO) in 1952. The ISSC represents the social, behavioural, and economic sciences at a global level, and aims to bring the best social science to bear on the biggest social challenges of our times. As Executive Director of the International Social Science Council she strengthened ISSC's activity profile, membership base and financial position, and forged strong links with the International Council for Science through key partnerships. These include the Integrated Research on Disaster Risk programme and the Science and Technology Alliance for Global Sustainability, the consortium of international organizations that founded Future Earth, the new global research initiative on global sustainability that also coordinates inputs from the international scientific community on key policy processes at the United Nations. Hackmann also led the launch of the World Social Science Forums and spearheaded the development of a new series of World Social Science Reports. She initiated a new global social science research funding and coordination pro-

gramme on Transformations to Sustainability, which was launched in March 2014 as a major contribution to Future Earth. Her major co-authored publications include: Heide Hackmann, Susanne C. Moser, Asuncion Lera St. Clair: "The social heart of global environmental change", in: Nature Climate Change, 4 (2014): 653-655; Peter P. J. Driessen, Jelle Behagel, Dries Hegger, Heleen Mees, Lisa Almesjö, Steinar Andresen, Fabio Eboli, Sebastian Helgenberger, Kirsten Hollaender, Linn Jacobsen, Heide Hackmann, Jörg Knieling, Corrine Larrue, Björn Ola Linnér, Orla Martin, Karen O'Brien, Saffron O'Neill, Marleen van Rijswick, Bernd Siebenhuener, Catrien Termeer, Aviel Verbruggen: "Societal transformations in the face of climate change", Working Paper, December 2013; Wolfram Mauser, Gernot Klepper, Martin Rice, Bettina Susanne Schmalzbauer, Heide Hackmann, Rik Leemans, Howard Moore. "Transdisciplinary Global Change Research: the co-creation of knowledge for sustainability", in: Current Opinion in Environmental Sustainability (July 2013); Gisli Palsson, Bronislaw Szerszynski, Sverker Sörlin, John Marks, Bernard Avril, Carole Crumley, Heide Hackmann, Poul Holm, John Ingram, Alan Kirman, Mercedes Pardo Buendía, Rifka Weehuizen: "Reconceptualizing the 'Anthropos' in the Anthropocene: Integrating the Social Sciences and Humanities in Global Environmental Change Research", in: Environmental Science & Policy (April 2013); Karen O'Brien, Jonathan Reams, Anne Caspari, Andrew Dugmore, Maryam Faghihimani, Ioan Fazey, Heide Hackmann, David Manuel-Navarrete, John Marks, Riel Miller, Kari Raivio, Patricia Romero Lankao, Hassan Virji, Coleen Vogel, Verena Winiwarter: "You say you want a revolution? Transforming education and capacity building in response to global change", in: Environmental Science & Policy (April 2013); Gisli Palsson, Bronislaw Szerszynski, Sverker Sorlin, John Marks, Bernard Avril, Carole Crumley, Heide Hackmann, Poul Holm, John Ingram, Alan Kirman, Mercedes Pardo Buendia, Rifka Weehuizen: "Challenges of the Anthropocene: Contributions from Social Sciences and Humanities for the Changing Human Condition", in: Environmental Science & Policy (January 2013); W. V. Reid, D. Chen, L. Goldfarb, H. Hackmann, Y. T. Lee, K. Mokhele, E. Ostrom, K. Raivio, J. Rockström, H. J. Schellnhuber, A. Whyte: "Earth System Science for Global Sustainability: Grand Challenges", in: Science (November 2010); Heide Hackmann, Arie Rip: "Priorities and quality incentives for university research: A brief international survey: Countries studied: Australia, Belgium, Germany, Switzerland, UK" (January 2000); H. Hackmann, P. J. D. Drenth, J. J. F. Schroots: "Evaluating for Science: Processes & Protocols" (Amsterdam: ALLEA, 2004).

H.G. Brauch et al. (eds.), *Handbook on Sustainability Transition and Sustainable Peace*,
Hexagon Series on Human and Environmental Security and Peace 10, DOI 10.1007/978-3-319-43884-9

Address: Dr. Heide Hackmann, Executive Director, International Council for Science (ICSU), 5 rue Auguste Vacquerie, 75116 Paris, France.
Email: <heide.hackmann@icsu.org>.
Website: <http://www.icsu.org/about-icsu/about-us/our-staff> and <http://www.icsu.org/news-centre/news/international-council-for-science-announces-dr.-heide-hackmann-to-be-executive-director-and-dr.-lucilla-spini-head-of-science-programmes>.

Hoesung Lee (Republic of Korea) is the Chair of the Intergovernmental Panel on Climate Change (IPCC). He is a professor of economics of climate change, energy and sustainable development at Korea University's Graduate School of Energy and Environment in the Republic of Korea. He serves on various boards including as executive member of the Korean Academy of Environmental Sciences; a member of the Asia Development Bank President's advisory board, a council member of the Global Green Growth Institute and an editorial board member of UK-based journal *Climate Policy*. Hoesung Lee has published extensively in the field of energy and climate change. He was the founding president of the Korea Energy Economic Institute and the former president of the International Association for Energy Economics. Professor Lee has been involved in the IPCC in various capacities including as Vice-Chair and Working Group III Co-Chair since its Second Assessment Report of 1992. He obtained a bachelor's degree in economics at Seoul National University in 1969 and received his PhD in economics at Rutgers University (USA) in 1975. Among his major publications are: (co-authored with Jin-Gyu Oh): "Mitigation initiatives: Korea's experiences", in: *Climate Policy*, 2,2-3 (2002); (with Dennis Eklof): "Refining in the Far East: Its Potential and Constraints", in: *The Energy Journal*, 15 (1994); (Ed.): *The Role of Technology Transfer in Coping with Global Warming: The Case of Korea* (Seoul: Korea Energy Economics Institute, 1993); "Energy outlook and environmental implications for Korea", in: *Energy*, 16,11-12 (1991); "Korea's energy outlook", in: *Energy*, 6,8 (1981); (with Paul Davidson, Laurence H. Falk): "Oil: Its Time Allocation and Project Independence", in: *Brookings Papers on Economic Activity*, 2 (1974): 411-448; (with E. Haites): "Review of IPCC Working Group III Findings on the Impacts of Energy Policy Options on GHG Emissions", in: *Wiley Interdisciplinary Review: Energy and Environment*, 1,1 (2012); (with J. Oh): "Integrating Climate Change Policy with a Green Growth Strategy: the Case of Korea", in Bryce Wakefield (Ed.): *Green Tigers: The Politics and Policy of Climate Change in Northeast Asian Democracies*, Asia Program Special Report (Washington DC: Woodrow Wilson International Center for Scholars, 2010); (with J. Byrne et al.): "Electricity Reform at a Crossroads: Problems in South Korea's Power Liberalization Strategy", in: *Pacific Affairs*, 77,3 (Fall 2004); (with F. Toth, M. Mwandosya, C. Cararro et al.): "Decision-making Frameworks", in: IPCC [B. Metz and O. Davidson] (Eds.): *Climate Change 2001: Mitigation, IPCC Third Assessment* Report, Working Group III (Cambridge: Cambridge University Press, 2001): 601-688; (with Robert T. Watson, John A. Dixon, Steven P. Hamburg, Anthony C. Janetos, Richard H.

Moss et al.): *Protecting Our Planet—Securing Our Future* (United Nations Environment Programme, US National Aeronautics and Space Administration, World Bank, 1998); "Approaches to differentiation for advanced developing countries", in: M. Peterson, M. Grubb (Eds.): *Sharing the Effort: Options for Differentiating Commitments on Climate Change* (London: Royal Institute of International Affairs, 1996): 77-82; IPCC [P. Bruce, E. Haites] (Ed.): *Economics and Social Dimensions of Climate Change: IPCC Second Assessment Report* 1995 (Cambridge: Cambridge University Press, 1996); (with Z. Dadi, Y. Jung, J. Wisniewski, J. Sathaye): "Greenhouse Gas Emissions Inventory and Mitigation for Asia and Pacific Countries", in: *Ambio*, 25,4 (June 1996): 220-228.
Address: Prof. Dr. Hoesung Lee, Chairman, IPCC, c/o World Meteorological Organization, 7bis Avenue de la Paix, C. P. 2300, CH-1211 Geneva 2, Switzerland.
Email: <hoesung@korea.ac.kr>.
Website: <https://www.ipcc.ch/nominations/cv/cv_hoesung_lee.pdf>.

Paul Raskin (United States of America) is the founding President of the Tellus Institute, Boston, Massachusetts, and Director of the Great Transition Initiative. Born in Chicago in 1942, he was raised in California, receiving a BA in physics and philosophy in 1964 from the University of California, Berkeley, where his senior thesis was supervised by philosopher of science Paul Feyerabend. He earned a PhD in theoretical physics from Columbia University in 1970, and taught at university level, becoming Chair of an interdisciplinary department at the State University of New York at Albany in 1974. In 1976, he co-founded the Tellus Institute, a non-profit research and policy organization, where he has directed an interdisciplinary team working on environmental, resource, and development issues throughout the world. He also founded the US Centre of the Stockholm Environment Institute in 1989, the Global Scenario Group in 1995, and the Great Transition Initiative (GTI) in 2003. The overarching theme of Dr. Raskin's research has been the development of scientifically-based visions and strategies for a transformation to resilient and equitable forms of social development. Toward this larger aim, his research has spanned issues (energy, water, climate change, ecosystems, and sustainable development) and spatial scales (local, national, and global). He has conceived and built widely-used models for integrated scenario planning for energy (LEAP), fresh water (WEAP), and sustainability (PoleStar). His recent research and writing has centred on formulating and analysing long-range, integrated global and regional scenarios, and, in particular, the requirements for a 'Great Transition' to a sustainable, just, and liveable future. Dr. Raskin has published widely, and served as a lead author for the US National Academy of Science's Board on Sustainability, the Intergovernmental Panel on Climate Change, the Millennium Ecosystem Assessment, the Earth Charter, and UNEP's Global Environment Outlook. In 1995, he convened the Global Scenario Group (GSG) to explore alternative futures and their implications for contemporary values, choices, and actions. The GSG's 2002 valedictory essay—*Great Transition: The Promise and Lure of* *the Times Ahead*—became the point of departure for the Great Transition Initiative that Dr. Raskin launched in 2003 and continues to direct, which engages an eminent network of hundreds of international scholars and activists in research, publications, and public outreach. His publications include: *Great Transition: The Promise and Lure of the Times Ahead* (Boston: Stockholm Environment Institute, 2002); "Contours of a Resilient Global Future", in: *Sustainability*, 6 (2014): 123-135; "The Century Ahead: Searching for Sustainability", in: *Sustainability*, 2,8 (2010): 2626-2651; "World Lines: A Framework for Exploring Global Pathways", in: *Ecological Economics*, 65,3 (2008): 461-470; "Scenes from the Great Transition", in: *Solutions*, 3,4 (2012): 11-17; "Global Scenarios: Background Review for the Millennium Ecosystem Assessment", in: *Ecosystems*, 8 (2005): 133-142; "Imagine All the People: Advancing a Global Citizens Movement", in: *Development*, 54 (2011): 287-290; "Future", in: Willis Jenkins (Ed.): *Encyclopedia of Sustainability: The Spirit of Sustainability* (Great Barrington, MA: Berkshire, 2010); "Planetary Praxis: Rhyming Hope and History", in: Stephen R. Kellert; James Gustave Speth (Eds.): *The Coming Transformation: Values to Sustain Human and Natural Communities* (New Haven: Yale University, 2009); "The Problem of the Future: Sustainability Science and Scenario Analysis", in: *Global Environmental*

Change, 14 (2004): 137-146; *Global Environmental Outlook Scenario Framework* (Nairobi: United Nations Environment Programme, 2002); *Global Sustainability: Bending the Curve* (London: Routledge, 2002); *Halfway to the Future: A Reflection on the Global Condition* (Boston: Tellus Institute, 2002); *Bending the Curve: Toward Global Sustainability*. Second Report of the Global Scenario Group (Stockholm: Stockholm Environment Institute, 1998); "Windows on the Future: Global Scenarios and Sustainability", in: *Environment Magazine*, 40,3 (April 1998): 6-11; "Global Energy, Sustainability and the Conventional Development Paradigm", in: *Energy Sources*, 20 (1998): 363-383; *Water Futures: Assessment of Long-Range Patterns and Problems Perspectives* (Stockholm: Stockholm Environment Institute—United Nations, 1997); *Branch Points: Global Scenarios and Human Choice* (Stockholm: Stockholm Environment Institute, 1997). See his recent Tellus publications at: <http://www.tellus.org/tellus/author/paul-raskin>.

Address: Dr. Paul Raskin, President, Tellus Institute, 11 Arlington Street, Boston, MA 02116, USA.
Email: <praskin@tellus.org>.
Website: <http://www.tellus.org/tellus/author/paul-raskin#sthash.xof4OyCV.dpuf> and <https://en.wikipedia.org/wiki/Paul_Raskin>.

Biographies of Contributors

Sophie Adams (Australia) has a background in Development Studies and works with the German organization *Climate Analytics*. She is currently based in Sydney, Australia, and is completing doctoral research on community participation in climate change policy-making.

Address: Ms Sophie Adams, School of Social Sciences, Morven Brown Building, The University of New South Wales, Sydney, Australia.
Email: <s.m.adams@unsw.edu.au>.

Monica de Andrade (Brazil) holds a PhD and a Master's in Ecology and Natural Resources and diplomas in Biological Sciences. She is based at the Health Promotion Graduate Programme, University of Franca, Brazil, responsible for research projects about the role of environmental hazards on human health, including correlation between deforestation and emergent diseases, extreme weather and water-related diseases and cardiovascular disease. Since 2006, she has been advisor on more than twenty Master's theses. She has also worked as a consultant for national organizations and is a Visiting Professor at the University of Costa Rica. Recent publications include: "The role of land use and land cover changes in epidemiology of Hantaviruses in Minas Gerais, Brazil"; "The role of land use and land cover changes in epidemiology of Brazilian Spotted Fever in São Paulo, Brazil"; "The impacts of natural disasters on Brazilian public health. Dry eye as an environmental disease".

Address: Dr. Monica de Andrade, Unifran, Av Dr Armando Salles Oliveira, 201, Franca-SP, BRASIL CEP 14.404-600.
Email: <mmonicandrade@gmail.com> and >monica.andrade@unifran.edu.br>.
Website: <http://www.unifran.edu.br/pos-graduacao-pesquisa-extensao/mestrado-e-doutorado/saude/corpo-docente//>.

Lourdes Arizpe (Mexico) received an MA from the National School of History and Anthropology in Mexico in 1970; a PhD in Anthropology from the London School of Economics and Political Science, UK, in 1975; and an Honorary Doctorate from the University of Florida at Gainesville in 2010. She is a professor at the Regional Center for Multidisciplinary Research of the National University of Mexico. Her early anthropological studies were focused on migration, gender, rural development, and the human dimensions of environmental change. Professor Arizpe received a Fulbright grant in 1979 and a John F. Guggenheim grant in 1981. Elected President of the International Union of Anthropological and Ethnological Sciences in 1988, she became a member of the United Nations Commission on Culture and Development in 1990 and soon afterwards was designated Assistant Director General for Culture at UNES-

CO 1994-98. As President of the International Social Science Council 2004-2008, and the Academic Faculty of the Global Economic Forum at Davos, Switzerland 2000-2004, she participated in projects to further policy-oriented studies on culture and sustainability. At the United Nations Institute for Research on Social Development, she was Chair of the Board 2005-2011 and a member of the United Nations Committee for Development Policy of the Economic and Social Council (ECOSOC). Her major studies were published in three volumes by Springer-Verlag in 2014-15 and may be consulted at <http://www.afes-pressbooks.de/html/SpringerBriefs_ ESDP06.htm> and <http://www.afes-press-books.de/html/SpringerBriefs_PSP_Arizpe. htm>.

Address: Prof. Dr. Lourdes Arizpe, Centro Regional de Investigaciones Multidisciplinarias—UNAM, Av Universidad s/n Circuito II, campus UAEM, Col. Chamilpa, CP 62210 Cuernavaca, Morelos, México.
Email: <la2012@correo.crim.unam.mx>.
Website: <www.crim.unam.mx>.

Karoline Augenstein is a Research Fellow at the Wuppertal Institute for Climate, Environment and Energy. Her background is in political science and economics. She completed her BA in European Studies at the University of Maastricht and her MA in Sustainability Economics and Management at the University of Oldenburg. Since 2010, she has worked as scientific assistant to the management board at the Wuppertal Institute and in 2014 completed her PhD at the University of Wuppertal. Her main research interest is in sustainable system innovations in various fields, narrative approaches in sustainability transitions research, and concepts of transdisciplinary research for sustainability. Among her publications on this theme are: Schneidewind, Uwe; Singer-Brodowski, Mandy; Augenstein, Karoline, 2016: "Sustainability and Science Policy", in: Heinrichs, Harald et al. (Eds.): *Sustainability Science*, Springer Science, Dordrecht, 149-160; Schneidewind, Uwe; Augenstein, Karoline, 2012: "Analyzing a transition to a sustainability-oriented science system in Germany", in: *Environmental Innovation and Societal Transitions*, 3: 16-28.

Address: Dr. Karoline Augenstein, Wuppertal Institute for Climate, Environment and Energy, Döppersberg 19, 42103 Wuppertal, Germany.
Email: <karoline.augenstein@wupperinst.org>.
Website: <http://wupperinst.org/home/>.

Flor Avelino (The Netherlands) works at DRIFT, Erasmus University Rotterdam, as a researcher and lecturer. Her research focus is on the role of power and empowerment in sustainability transitions. As the academic director of the

© Springer International Publishing Switzerland 2016

H.G. Brauch et al. (eds.), *Handbook on Sustainability Transition and Sustainable Peace*,
Hexagon Series on Human and Environmental Security and Peace 10, DOI 10.1007/978-3-319-43884-9

Transition Academy, she strives to co-create new learning environments to challenge people to think and act for radical change. As scientific coordinator of the TRANSIT (*Transformative Social Innovation Theory*) -project, she is currently involved in empirically and theoretically investigating social innovation and transformation, including case-studies on social entrepreneurship and the ecovillage movement. Among her major publications are: with Wittmayer, J. (2015): "Shifting Power Relations in Sustainability Transitions: A Multi-actor Perspective", in: *Journal of Environmental Policy & Planning*, at: <http://www.tandfonline.com/doi/abs/ 0.1080/1523908X.2015.1112259>; with Rotmans, J. (2011): "A dynamic conceptualization of power for sustainability research", in: *Journal of Cleaner Production*, 19,8: 796-804; (2009): "Empowerment and the challenge of applying transition management to ongoing projects", in: *Policy Sciences*, 42,4: 369-390.

Address: Dr. Flor Avelino, DRIFT/ Erasmus University of Rotterdam, P.O. 1738, 3000 DR Rotterdam, The Netherlands
Email: <avelino@drift.eur.nl>.
Website: <htp://transitionacademy.nl/people/dr-flor-avelino/>.

Xuemei Bai (Australia/China) is a Professor of Urban Environment and Human Ecology at Fenner School of Environment and Society, Australian National University. She obtained her bachelor's degree in science from Peking University, and her Master's in engineering and PhD from the University of Tokyo. Prior to joining ANU, she worked at ecology and environmental research institutes in Japan and at CSIRO in Australia, and was a Visiting Professor at Yale University for two years. Her current research interests include: drivers and impacts of urbanization, cities as complex social ecological systems, evolution of urban environment and urban systems, urban metabolism, cities and climate change, and system innovation and sustainability transition/transformation, as well as broader Anthropocene futures and the sustainability agenda. She is a member of the Science Committee of Future Earth, and served as a vice chair of the International Human Dimensional Programme of Global Change (IHDP). She serves on the editorial board of several international journals including *Current Opinion of Environmental Sustainability*; *Sustainability Science*; *Environment, Computer, Environment and Urban Systems*; etc. Selected academic journal publications include: "Realizing China's Urban Dream" (2014), in: *Nature*, 509: 158-160; "Landscape Urbanization and Economic Growth" (2012), in: *Environmental Science & Technology*, 46,1: 132-139; "Plausible and Desirable Futures in the Anthropocene" (2015), in: *Global Environmental Change*, <doi:1 0.1016/j.gloenvcha.2015.09.017>; "Urban Sustainability Experiments: Patterns and Pathways" (2012), in: *Environmental Science & Policy*, 13: 312-325; "Global Change and the Ecology of Cities" (2008), in: *Science*, 319: 756-760.

Address: Prof. Dr. Xuemei Bai, Fenner School of Environment and Society, College of Medicine, Biology and Environment, Frank Fenner Building, The Australian National University, Canberra 2601, Australia.

Email: <xuemei.bai@anu.edu.au>.
Website: <https://researchers.anu.edu.au/researchers/bai-x>.

Lucy Baker (United Kingdom) is a Research Fellow in the Science Policy Research Unit at the University of Sussex and a Visiting Fellow at the Energy Research Centre, University of Cape Town. Between 2013 and 2015 Lucy was a Research Fellow in the School of Global Studies, University of Sussex on an ESRC-funded project, *Rising powers and the low carbon transition in Southern Africa*. Lucy's areas of research include: the political economy of energy; electricity policy in South Africa; sociotechnical transitions; and low-carbon development in low- and middle-income countries. Lucy holds a PhD in *The political economy of sociotechnical transitions in South Africa's electricity sector* from the University of East Anglia, UK, an MSc in Development Studies from the School of Oriental and African Studies, and an MA in Hispanic Studies and French from the University of Edinburgh. Prior to joining academia she worked for ten years in the fields of environment, development and human rights as a policy officer and campaigner with non-governmental organizations Amnesty International, Oxfam and the Bretton Woods Project. A native speaker of English, she is fluent in Spanish, Portuguese and French.

Address: Dr. Lucy Baker, Science Policy Research Unit, University of Sussex, Sussex House, Falmer, Brighton, BN1 9RH.
Contact details: <L.H.Baker@sussex.ac.uk>.
Website: <http://www.sussex.ac.uk/profiles/326431>.

Larissa Basso (Brazil) is currently a PhD candidate in International Relations at the University of Brasília and a researcher at the *Brazilian Federal Agency for Support and Evaluation of Graduate Education* (CAPES). She is also a member of the *Brazilian Research Network on International Relations and Climate Change* (Redeclim). She holds Bachelor of Laws—LLB (2003) and Master's in International Law—LLM (2008) degrees from the University of Sao Paulo, and an MPhil in Environmental Policy (2010) from the University of Cambridge. She is also a Graduate Teaching Assistant in the Institute of International Relations, University of Brasília, and an Attorney-at-Law since 2004 (legal consulting and litigation). She has researched extensively on international trade and development; since 2009, her main research interests are sustainable development, environmental governance, and climate change and global energy governance in the transition to low-carbon development.

Address: Ms Larissa Basso, Institute of International Relations, University of Brasília (Campus Darcy Ribeiro, Caixa Postal: 04306, CEP: 70904-970, Brasília - DF, Brazil).
Email: <larissabasso@gmail.com>.
Website: <http://lattes.cnpq.br/6094310548211358>.

Hans Günter Brauch (Germany), Privatdozent (Adj. Prof.) at the Faculty of Political Science and Social Sciences, Free University of Berlin (ret.), since 1987 chairman of *Peace Research and European Security Studies* (AFES-PRESS). See *Biographies of editors*.

Hongmin Chen (China) holds a PhD in population, resources and environmental economics. She is an associate professor in the Department of Environmental Science and Engineering at Fudan University and a researcher in the Fudan Urban Environmental Management Research Center and Fudan Tyndall Center. Her research interests focus on the transformation towards a low-carbon society. Her current research covers a range of topics including trade-embodied CO_2 emissions and high-emission groups and their role in accelerating the transition towards a low-carbon society. Recent publications include: *Energy and Environmental Impacts of China's International Trade: A Research on Embodied Flows* (in Chinese, Shanghai: Fudan University Press, 2011); (co-authored with Xingyi Dai et al.): 2050 *Shanghai Low Carbon Development Roadmap Report* (in Chinese, Beijing: Science Press, 2011); "Development and Prospect of Personal Carbon Trading Research", in: *China Population, Resources and Environment*, 24, 9 (2014): 30-36; (with Yanni Zhu and Miaojing Lu): "Identification and Analysis of High-Emission Commuter Groups in Shanghai" (in Chinese), in: *Resources Science*, 36,7 (2014): 1469-1477; (with Miaojing Lu): "Review of the Inequality of Carbon Emissions" (in Chinese), in: *Resources Science*, 35,8 (2013): 1617-1624; (with Jing Chen): "Energy Consumption Differences in Different Income Groups: The Case Study of Shanghai" (in Chinese), in: *Population and Development*, 18,4 (2012): 30-38.

Address: Dr. Hongmin Chen, 220 Handan Road, Department of Environmental Science and Engineering, Fudan University, 200433, Shanghai, China.
Email: <swingboat77@gmail.com>.

Peter T. Coleman (USA) holds a PhD in Social/Organizational Psychology from Columbia University and is a Professor of Psychology and Education at Columbia University with a joint appointment at Teachers College and The Earth Institute and teaches courses in Conflict Resolution, Social Psychology, and Social Science Research; Director of the *Morton Deutsch International Center for Cooperation and Conflict Resolution* (MD-ICCCR) at Teachers College, Columbia University; Chair of Columbia University's *Advanced Consortium on Cooperation, Conflict, and Complexity* (AC4); research affiliate of the *International Center for Complexity and Conflict*, The Warsaw School for Social Psychology in Warsaw, Poland. He currently conducts research on optimality of motivational dynamics in conflict, power asymmetries and conflict, intractable conflict, multicultural conflict, justice and conflict, environmental conflict, mediation dynamics, and sustainable peace. In 2003, he became the first recipient of the Early Career Award from the American Psychological Association, Division 48: Society for the Study of Peace, Conflict, and Violence. He has edited: *Handbook of Conflict Resolution: Theory and Practice* (2000; 2006; 2014); *The Five Percent: Finding Solutions to Seemingly Impossible Conflicts* (2011); *Conflict, Justice, and Interdependence: The Legacy of Morton Deutsch* (2011); *Psychological Components of Sustainable Peace* (2012); *Attracted to Conflict: Dynamic Foundations of Destructive Social Relations* (2013); and *Making Conflict Work: Harnessing the Power of Disagreement* (2014). He has authored over seventy journal articles and chapters, and is a member of the United Nation Mediation Support Unit's Academic Advisory Council, a founding board member of the Leymah Gbowee Peace Foundation USA, and a New York State certified mediator and experienced consultant.

Address: Prof. Peter T. Coleman, Director, *Morton Deutsch International Center for Cooperation and Conflict Resolution*; Professor of Psychology and Education. Program in Social-Organizational Psychology, Department of Organization and Leadership, Teachers College, Columbia University, Box 53, 525 West 120th Street, New York, NY 10027, USA; Co-Director, Advanced Consortium for Cooperation, Conflict, and Complexity (The Earth Institute at Columbia University), New York, NY, USA.
Email: <coleman@exchange.tc.columbia.edu>.
Website: <http://icccr.tc.columbia.edu/> and <http://www.makingconflictwork.com/>.

Simon Dalby (Canada/United Kingdom/Ireland), Professor, formerly at Carleton University, is now CIGI Chair in the Political Economy of Climate Change at the Balsillie School of International Affairs and Professor of Geography and Environmental Studies at Wilfrid Laurier University, Waterloo, Ontario. He is author of *Creating the Second Cold War* (Pinter 1990), *Environmental Security* (University of Minnesota Press, 2002) and *Security and Environmental Change* (Polity, 2009). His articles have appeared in various scholarly journals including *Alternatives, Antipode, Australian Journal of International Affairs, Contemporary Security Policy, Geoforum, Geopolitics, Global Environmental Politics, Global Policy, Intelligence and National Security, International Politics, Political Geography, Society and Space* and *Studies in Political Economy*.

Address: Prof. Simon Dalby, Balsillie School of International Affairs, 67 Erb Street West, Waterloo, ON N2L 6C2, Canada.
Email: <sdalby@gmail.com>.
Website: <http://www.balsillieschool.ca/people/simon-dalby>.

Morton Deutsch (USA) is E. L. Thorndike Professor and Director Emeritus of the *Morton Deutsch International Center for Cooperation and Conflict Resolution* (MD-ICCCR) at Teachers College, Columbia University. He studied with Kurt Lewin at MIT's Research Center for Group Dynamics, where he obtained his PhD in 1948. He is well-known for his pioneering studies in intergroup relations, cooperation-competition, conflict resolution, social conformity, and the social psychology of justice. His books include *Interracial Housing, Research Methods in Social Relations, Preventing World War III: Some Proposals, Theories in Social Psychology, The Resolution of Conflict, Applying Social Psychology*, and *Distributive Justice*. His work has been widely honoured by the Kurt Lewin Memorial Award, the G. W. Allport Prize, the Carl Hovland Memorial Award, the AAAS Socio-Psychological Prize, APA's Distinguished Scientific Contribution Award, SESP's Distinguished Research Scientist Award, and the Nevitt Sanford Award. He is a William James Fellow of APS. He has also received lifetime achievement awards for his work on conflict management, cooperative learning, peace psychology, and applications of psychology to social issues. In addition, he has received the Teachers College Medal for his contributions to education, the Helsinki University medal for his contributions to psychology, and the doctorate of humane letters from the City University of New York. He has been president of the Society for the Psychological Study of Social Issues, the International Society of Political Psychology, the Eastern Psychological Association, the New York State Psychological Association, and several divisions of the American Psychological Association. It is not widely known, but after postdoctoral training, Deutsch received a certificate in psychoanalysis in 1958 and conducted a limited practice of psychoanalytic psychotherapy for more than twenty-five years.

Address: Prof. Morton Deutsch, *Morton Deutsch International Center for Cooperation and Conflict Resolution.* Professor of Psychology and Education. Program in Social-Organizational Psychology, Department of Organization and Leadership, Teachers College, Columbia University, Box 53, 525 West 120th Street, New York, NY 10027, USA.
Email: <deutsch@tc.columbia.edu >.
Website: <http://icccr.tc.columbia.edu/>.

Eckart Ehlers (Germany) received his PhD (1965) and his Habilitation (1970) in geography from the University of Tübingen. After appointments as Professor of Geography in Gießen (1970-1972) and Chair of the Department of Geography in Marburg (1972-1986), he accepted an offer from the University in Bonn in 1986, where he retired as Professor Emeritus in 2004. In 1996 the *Deutsche Forschungsgemeinschaft* (DFG) appointed him as chair of the newly-created *Nationales Kommittee für Global Change Forschung* (NKGCF), a position he held until 2002. Internationally, he served as Secretary General and Treasurer of the *International Geographical Union* IGU (1992-2000), as Chair of the *International Human Dimensions of Global Environmental Change Programme* IHDP (1996-1999), and as a member of the Scientific Steering Committee of START between 2007 and 2013. Since 2007 he has been on the Board of Trustees of the *International Foundation for Science* (IFS) in Stockholm. He is a member of *Academia Europaea* and Senior Fellow at the *Center for Development Research/Zentrum für Entwicklungsforschung* (ZEF) at the University of Bonn. In 2008, he published: *Das Anthropozän. Die Erde im Zeitalter des Menschen*, summarizing his plea for interdisciplinarity and integrative research with a specific focus on the interrelationships between nature and society.

Address: Prof. Dr. Eckart Ehlers (Em.), ZEF, Walter-Flex-Str. 3, 53113 Bonn, Germany.
Email: <eckart.ehlers@t-online.de>.
Website: <https://www.geographie.uni-bonn.de/das-institut/personal/emeritierte-pensionierte-professoren/ehlers_e> and <http://www.zef.de/index.php?id=2232&tx_zefportal_staff[ref]=2252&tx_zefportal_staff[uid]=658&no_cache=1>.

Rebecca Froese (Germany) is a Research Assistant in the *Research Group on Climate Change and Security* (CLISEC) at Hamburg University and a Master's student in the programme "Integrated Climate System Sciences" at the Cluster of Excellence "Integrated Climate System Analysis and Prediction" (CliSAP) at Hamburg University. She com-

pleted her bachelor's degree in Geoscience at the University of Bremen and at the University of Victoria, Canada, focusing on geochemistry, oceanography and climatology. After graduating in 2013 she spent six months in Brazil, working as an intern for the GIZ (German Society for International Development). Currently she is writing her Master's thesis on the topic of climate-proof development cooperation.

Address: Ms Rebecca Froese, Grindelberg 7, 20144 Hamburg, Germany.
Email: <rebecca.froese@studium.uni-hamburg.de>.
Website: <http://clisec.zmaw.de/Rebecca-Froese.3000.0.html>.

Audley Genus (United Kingdom), Professor at the Small Business Research Centre, Kingston University, Kingston Hill, Kingston upon Thames, has research interests in the areas of innovation and entrepreneurship, technology policy, and new approaches for stimulating 'green' innovation in firms and communities. Current projects include an investigation into the nature of institutional innovation and entrepreneurship in community energy initiatives and the emergence of permaculture-inspired small firms. Audley publishes regularly in high-impact journals such as *Research Policy* and *Technological Forecasting and Social Change* and has published three single-authored books. He is currently working on a book entitled *Sustainable Consumption: Design, Innovation and Practice*, to be published by Springer in 2016. Audley is a member of the editorial board of *Technology Analysis and Strategic Management* and a member of the *European Association for the Study of Science and Technology* (EASST), the *British Academy of Management*, and the *Sustainable Consumption Action Research Initiative*.

Address: Prof. Audley Genus, YTL Professor of Innovation and Technology Management, Small Business Research Centre, KHBS410 Kingston Business School, Kingston University, Kingston Hill, Kingston upon Thames, Surrey, KT2 7LB, UK.
Email: <a.genus@kingston.ac.uk>.
Website: <http://business.kingston.ac.uk/staff/professor-audley-genus>.

John Grin (The Netherlands) is Professor at the Department of Political Science of the University of Amsterdam and was scientific director to the *Amsterdam School for Social Science Research* (ASSR). See *Biographies of editors*.

Sander Happaerts (Belgium) was until September 2014 research manager for environmental policy and sustainable development at HIVA (*Research Institute for Work and Society*), University of Leuven, Belgium. He is also a part-time Assistant Professor at Leuven International and European Studies, University of Leuven, Belgium. His expertise is situated in the fields of sustainable development, transitions and sustainable materials management. Between 2007 and 2011, Sander conducted PhD research into governance for sustainable development. His PhD was awarded in the Thesis Competition of the EU Committee of the Regions. He has published extensively on the sustainable development policies of subnational governments, on their embedding in national and international governance, and on symbolic politics. In 2010, he held a Visiting Fellowship at the *Université du Québec à Montréal* (UQÀM, Canada), where he became the Social Responsibility and Sustainable Development Research Chair. After completing his chapter, he took up a position at the European Commission. The information and views set out in this chapter are those of the author and do not reflect the official opinion of the European Commission.

Address: KU Leuven, Leuven International and European Studies, Parkstraat 45 box 3600, 3000 Leuven, Belgium
Email: <sander.happaerts@kuleuven.be>.
Website: <http://soc.kuleuven.be/lines>.

Karlson 'Charlie' Hargroves (Australia) holds a Bachelor of Engineering degree and has a PhD in low-carbon transitions and carbon structural adjustment. He is an internationally renowned sustainability transitions researcher, strategist and author, who has co-authored five books, numerous chapters and papers, and delivered over fifty keynote presentations and guest lectures around the world. His first book, 'The Natural Advantage of Nations', won the Australian Banksia Award for Environmental Leadership, Education and Training in 2005, and the two published in 2010 were ranked among the 'Top Forty Sustainability Books' in the world that year by the Cambridge Programme for Sustainability Leadership (with 'Cents and Sustainability' ranked fifth and 'Factor 5' ranked twelfth). He is a Senior Research Fellow at the *Curtin University Sustainability Policy* (CUSP) Institute, working with Professors Peter Newman and Dora Marinova, and is the Sustainable Development Fellow at the University of Adelaide. He has a focus on low-carbon transitions and works on a series of national and international projects and publications. He is a full member of the Club of Rome. He lives in Adelaide and he and his wife, Stacey, have two children, Grace and Tyson. Recent related publications include: (with Desha): *Higher Education & Sustainable Development: A Model for Curriculum Renewal* (London: Earthscan/Routledge Publishing 2014); (with Smith and Desha): *Cents and Sustainability: Securing Our Common Future through Decoupling Economic Growth from Environmental Pressure* (London: Earthscan/Routledge Publishing 2010); (with von Weizsäcker, Smith, Desha and Stasinopolous): *Factor 5: Transforming the Global Economy through 80% Increase in Resource Productivity* (London: Earthscan/Routledge Publishing 2010); (with von Weizsäcker, de Larderel, Hudson, Smith, and Rodrigues): *Decoupling 2: Technologies, opportunities and policy options* (Nairobi/Paris: UNEP 2014).

Address: Karlson 'Charlie' Hargroves, Curtin University, Perth, Australia.
Website: <http://www.curtin.edu.au/research/cusp/>.
Email: <charlie.hargroves@curtin.edu.au>.

Mike Hodson (United Kingdom) is Research Fellow at the University of Manchester, where he is based jointly in the *Sustainable Consumption Institute* (SCI) and the *Manchester Institute of Innovation Research* (MIoIR). Mike was previously Senior Research Fellow at Salford University, where he spent a decade in the *Centre for Sustainable Urban and Regional Futures* (SURF) working in the area of urban and regional governance and transitions. He has published and presented widely on this research agenda. His developing research interests are at the interface of systemic transitions and territorial transitions and the ways in which relationships between the two are, are not and can be organized.

Address: Dr. Mike Hodson, Sustainable Consumption Institute (SCI) & Manchester Institute of Innovation Research (MIoIR), Harold Hankins Building, Room 8.15, University of Manchester, Manchester M13 9PL, United Kingdom.
Email: <michael.hodson@mbs.ac.uk>
Website: <http://www.sci.manchester.ac.uk/people/dr-mike-hodson>.

Tobias Ide (Germany) is Head of the research field Peace and Conflict at the *Georg Eckert Institute* in Braunschweig and associated researcher in the *Research Group Climate Change and Security*, University of Hamburg. His research on the links between environmental change, resource scarcity, conflict and cooperation has been published in *Global Environmental Change* and *Political Geography*, among other places. He has received various grants from the German Research Foundation (DFG) and the German Environmental Foundation (DBU) for his research.

Address: Dr. Tobias Ide, Celler Straße 3, 38114 Braunschweig, Germany.
Email: <ide@gei.de>.
Website: <http://www.gei.de/en/staff/dr-tobias-ide.html>.

Roeland J. in't Veld (The Netherlands) is UNESCO Professor of Governance and Sustainability at Tilburg University and teaches in postgraduate education. Previously, he was professor at eight European universities (1977-2010). He chaired the National Council on Spatial, Nature and Environmental research in the Netherlands (1999-2009), as well as a project at the *Institute for Advanced Sustainability Studies* (IASS) in Potsdam (2010-2011). He has served as a World Bank and OECD consultant, and as a Director-General for Higher Education. He was Chief Strategist in the Airport Region Haarlemmermeer (2012-2013), and carried out scenario studies on the future sustainable development of the region. He is dedicated to efforts to further social innovation in the Tilburg Region, in particular in the fields of participatory democracy and private-public partnerships. Some of his past positions include supervision both in the private sector (IBM) and the public sector (National Railways Infraprovider). His research focuses on democracy theory, the interface of science and politics, and foresight studies. He is involved in transdisciplinary research on sustainable development and argues that fundamental change towards sustainability demands improvements in the functioning of democratic institutions. He is author, co-author and editor of about thirty books and 110 papers and articles on public choice theory, educational economics, cost-benefit analysis, boundary work and political theory. Recently he has published *Kennisdemocratie* (Den Haag: SDU, 2010), *Knowledge Democracy* (Heidelberg: Springer,

2011) and two chapters in *Transgovernance* (Heidelberg: Springer, 2013).

Address: Prof. Dr. R. J. in't Veld, Waterbieskreek 40, 2353 jh, The Netherlands.
Email: <roelintveld@hotmail.com>.
Website: <www.roelintveld.nl>.

Shivant Jhagroe (The Netherlands) is a researcher at the Eindhoven University of Technology. Shivant graduated in Public Administration and History, and completed his doctoral thesis at the Erasmus University Rotterdam (DRIFT). His (forthcoming) dissertation examines the politics of urban sustainability transitions. His current project studies the transformation of domestic energy practices in the context of smart homes and energy technologies. Shivant's research interests include sustainability transitions, environmental politics, critical urban studies, political philosophy and STS.

Address: Shivant Jhagroe, MSc, MA, Faculty of Industrial Engineering and Innovation Science, Eindhoven University of Technology (TIS). De Rondom 70, 5612 AP Eindhoven, The Netherlands. Room IPO 2.07.
Email: <s.s.jhagroe@tue.nl>.

Twig Johnson (USA) received his MA in 1971, his MPhil in 1974 and his PhD in 1977 in Anthropology from Columbia University, taught in the Anthropology Departments of Queens College, City University of New York and the University of Maine, and spent two years as a Visiting Scientist at *Massachusetts Institute of Technology* (MIT). He has served as a Peace Corps Brazil Volunteer (1964-66) and as its Country Director (1977-78). US Governmental service included Chief of the Studies Division in the US Agency for International Development's Policy and Program Coordination Bureau and Director of Environment and Natural Resources in its Global Bureau. At the UN, he was the Chief of Evaluation for UNICEF. With the *Worldwide Fund for Nature* (WWF) he served as Regional Director for Latin America and as Vice President of WWF/US for Latin America. He also served on the Brazil/World Bank/G8 International Advisory Group for the Amazon. At Earthwatch he was Director of the Center for Field Research. He served on US delegations to UNCED and WCSD as well as taking part in numerous preparatory meetings. In the Policy and Global Affairs Division of the *National Research Council* (NRC), he was responsible for developing a programme on Science, Technology and Sustainability.

Address: Twig Johnson, Writer/Consultant, 29 Wapping Road, Kingston, Massachusetts, 02364 USA.
Email: <twigjohnson@gmail.com>.

J. (Jan) Jonker (The Netherlands/France) is a professor of Sustainable Entrepreneurship at the Nijmegen School of Management at Radboud University Nijmegen. Since 2014 he has also held the Chaire d'Excellence Pierre de Fermat at the Toulouse Business School in Toulouse (France). His research focuses on the interface of sustainability, (new) business models, and transitions. He is the author of the 'green' bestseller *Duurzaam Denken Doen* [*Thinking and Acting in a Sustainable Way*] (Deventer: Kluwer, 2011) and

Werken aan de WEconomy [*Working on the WEconomy*] (Deventer: Kluwer, 2013), among other books, and has recently published *Nieuwe Business Modellen* [*New Business Models*] (The Hague: Academic Service, 2014). As a follow-up to these previous projects, he is now working on a system of Hybrid Banking (see the Dutch website <www.nieuwebusinessmodellen.info>). Dr. Jonker ranked five years in a row (between 2010 and 2014) in the Dutch daily paper *Trouw*'s Top 100 as one of the most influential 'green' people in the Netherlands. In order to keep in touch with everyday practice, he also works as a management consultant, his motto being 'practise what you preach'. According to Dr. Jonker, over the next few years sustainable and responsible entrepreneurship and identifying how to realize fundamental change by organizing in different ways will continue to increase in significance.

Address: Prof. Dr. J. Jonker, Nijmegen School of Management, Radboud University, PO Box 9108, 6500 HK Nijmegen, The Netherlands.
Email: <j.jonker@fm.ru.nl>.
Website: <www.ru.nl/fm>.

René Kemp (The Netherlands) is Professor of Innovation and Sustainable Development at the *International Centre for Integrated Assessment and Sustainable Development* (ICIS) of Maastricht University and a Professorial Fellow of UNU-MERIT. He is a multidisciplinary researcher interested in real-life problems. He has published in innovation journals, ecological economics journals, policy journals, transport and energy journals, and sustainable development journals. He is advisory editor of *Research Policy*, editor of *Sustainability Science* and editor of *Environmental Innovation and Societal Transitions*. His work on technological regime shifts, strategic niche management and sustainability transitions is foundational to the field of sustainability transition studies. He currently works on resource efficiency, transformative social innovation and urban labs.

Address: Prof. Dr. René Kemp, ICIS, Maastricht University, and UNU-MERIT, Boschstraat 24, 6211 AX Maastricht, The Netherlands.
Email: <rkemp@maastrichtuniversity.nl> and <Kemp@merit.unu.edu>.
Website: <http://unu.edu/experts/rene-kemp.html#profile>.

Martin Keulertz (Germany/USA) is a postdoctoral research associate in the Department for Agricultural and Biological Engineering at Purdue University in Indiana (United States). He is part of the core water-energy-food nexus team around Prof. Rabi Mohtar (Texas A&M). Prior to joining Purdue, he received his PhD from King's College, London; in it he analysed the role of water resources in foreign direct investment in African agriculture. During his PhD research, he carried out fieldwork in fifteen MENA and SSA countries to empirically investigate the risks and opportunities of FDI in agriculture. Martin has published several peer-reviewed articles in journals, books and conference proceedings. In 2012, Water International selected his co-authored article on the role of corporates in global 'virtual water' trade as paper of the year. He further led the editing process of the *Handbook of Land and Water Grabs in Af-*

rica: foreign direct investment and food and water security published by Routledge in 2013 and the special issue on The water-food-energy-climate nexus in global drylands in the *International Journal of Water Resources Development* in 2015. His research interests centre around the water-energy-food nexus with reference to the West Asia/North Africa region, 'virtual water', the impact of global economic change on global agricultural trade and food systems, and sub-Saharan African agriculture and water accounting methods.

Address: Dr. Martin Keulertz, Department for Biological and Agricultural Engineering, 225 South University Street, West Lafayette, IN 47907, USA.
Email: <mkeulert@purdue.edu>.
Website: <http://wefnexus.tamu.edu/511-2/>.

Claudia Köhler (Germany) studied sociology and philosophy and earned her MA at Potsdam University in 2004. Since 2008 she has been employed at the *Potsdam Institute for Climate Impact Research* (PIK). Her previous positions were at the Max Planck Institute and the German Institute for International and Security Affairs.

Address: Ms Claudia Köhler MA, Potsdam Institute for Climate Impact Research, Telegraphenberg A 31, P.O. Box 601203, D-14412 Potsdam, Germany.
Email: <koehler@pik-potsdam.de>.
Website: <https://www.pik-potsdam.de/members/ckoehler>.

Dr. Asfaw Kumssa (Sweden) is a Chief Technical Advisor at the Office of the United Nations Resident Coordinator's Office in Nairobi, Kenya. Dr. Kumssa has over twenty years of experience in the field of development studies. Prior to his appointment with the UN Resident Coordinator's Office, Dr. Kumssa, served for sixteen years as the Coordinator of the United Nations Centre for Regional Development Africa Office (UNDESA's Project Office) in Nairobi, Kenya. While in UNCRD, he undertook training-cum-research activities, advisory services, and information exchange related to local and regional development, and assisted African countries in designing and implementing effective and innovative local and regional development policies to address their socio-economic needs and problems. Prior to that, Dr. Kumssa worked as a Regional Economic Development Planner in Nagoya, Japan for four years. From 1992 to 1994, Dr. Kumssa served as an Adjunct Professor of Global Political Economy and Development Economics at the University of Denver, Colorado, USA. Dr. Kumssa holds a PhD and an MA in International Studies, specializing in International Economics (University of Denver, CO, USA); and an MSc in National Economic Planning (Odessa Institute of National Economy, Ukraine). From 1984-1986 he also studied economics at Stockholm University, Sweden. He has published a number of books and articles in the field of urban and regional planning; climate change; globalization; capacity-building and distance training; human security and conflict; refugee issues; police corruption; social development; transitional economies; and other development related subjects.

Address: Dr. Asfaw Kumssa, Office of the United Nations Resident Coordinator's Office in Nairobi, Kenya.

Email: <asfaw.kumssa@undp.org>.
Website: For his CV please search at: <www.academia.edu/attachments/.../download>.

Juan Antonio Le Clercq (Mexico) has a doctorate in Political and Social Sciences from the Facultad de Ciencias Políticas y Sociales, UNAM. He is a full-time professor at the International and Political Sciences Department at the *Universidad de las Américas Puebla* (UDLAP), and his recent publications include "Cambio Climático: políticas nacionales y bases institucionales", in: *Diálogo Político* 3/2011, Konrad Adenauer Stiftung, Argentina (2011): 97–115, and "Las consecuencias del cambio climático, la responsabilidad del daño y la protección de los derechos humanos, una relación problemática", in: José Pablo Abreu and Juan Antonio Le Clercq: *La Reforma Humanista. Derechos Humanos y cambio Constitucional en México* (Mexico, D.F.: Miguel Angel Porrúa, 2011): 381–407; he is also co-author of *The Global Impunity Index* (UDLAP, 2015).

Address: Dr. Juan Antonio Le Clercq, Professor of International Relations and Political Science, Universidad de las Américas Puebla (UDLAP); Sta. Catarina Mártir. Cholula, Puebla. C.P. 72810. México.
Email: <juan.leclercq@udlap.mx>.o
Website: <http://www.udlap.mx/ofertaacademica/profesores.aspx?cveCarrera=LPI&profesor=0019873&extracto=8>.

Peter Michael Link (Germany) is a postdoctoral scientist at the *Research Group Climate Change and Security* (CLISEC) at the Institute of Geography of the University of Hamburg, Germany. His main research interest is the numeric modelling of societal and economic impacts of climate change in different regions of the world. He has published numerous journal articles and several edited volumes on the climate change and conflict nexus and on coastal and marine geography. Michael Link is also Speaker of the Working Group on Coastal and Marine Geography of the German Geographical Society and currently a board member of the Coastal and Marine Specialty Group of the American Association of Geography.

Address: Dr. Peter Michael Link, Grindelberg 7, 20144 Hamburg, Germany.
Email: <michael.link@uni-hamburg.de>.
Website: <http://clisec.zmaw.de/Dr-Peter-Michael-Link.868.0.html>

Sylvia Lorek (Germany) holds a PhD in consumer economics diploma in household economics and nutrition (Oecotrophologie) and in economics. The combination of these different disciplines provides her with the tools - the individual micro-economic and the societal macroeconomic perspective - for a well-founded analysis of the contexts in which the scientific and societal discourses about sustainable consumption take place. She is based at the *Sustainable Europe Research Institute* (SERI) and is head of SERI Germany e.V. Here she has worked on studies and as consultant for national and international organizations und institutes. She is also engaged in civil society organization (CSO) activities dealing with sustainable consumption at national, European and global levels. Among other roles, she is a leading functionary at the *Sustainable Consumption Research and Action Initiative* (SCORAI) in Europe and the Global Research Forum on Sustainable Production and Consumption. Recent publications include: *Towards Strong Sustainable Consumption Governance* (LAP Publishing, Saarbrücken; 2010); *Sustainable Production and Consumption Systems* (co-edited with Lebel and Daniel; Springer, Dordrecht 2010); *Strong Sustainable Consumption Governance–Precondition For A Degrowth Path?* (with Fuchs, 2013); and *Sustainable consumption within a sustainable economy–beyond green growth and green economies* (with Spangenberg, 2014), both published in the *Journal of Cleaner Production*.

Address: Dr. Sylvia Lorek, SERI, Schwimmbadstr. 2e, 51491 Overath, Germany.
Email: <sylvia.lorek@seri.de>.
Website: <http://www.seri.de>.

Simon Marvin (United Kingdom) is Director of the Urban Institute and a Professor in the Department of Geography, Sheffield University. He is interested in developing both conceptual and empirical understanding of the changing relations between sociotechnical networks and urban contexts. To date, he has played major roles within urban research towards addressing important questions surrounding telecommunications, infrastructure and mobility, sustainability and infrastructure, smart, interdisciplinary urban research, and, most recently, cities, systemic transitions, energy and low carbon, climate change and ecological security.

Address: Professor Simon Marvin, Urban Institute, The University of Sheffield, 219 Portobello, S1 4DP, United Kingdom.
Email: <s.marvin@sheffield.ac.uk>.
Website: <http://www.sheffield.ac.uk/urbaninstitute/our-people>.

Czeslaw Mesjasz (Poland), Dr habil., Associate Professor, Faculty of Management, Cracow University of Economics, Cracow, Poland. His research interests include applications of systems approach in management and in international relations, conflict resolution and negotiation, corporate governance, international management, and the links between economics and security. From 1992 to 1996 he was the convener of the Defence and Disarmament Commission of IPRA (*International Peace Research Association*). In 1991-1992 he received a NATO Democratic Institutions Fellowship. In 1992-1993 he was a Visiting Research Fellow at the *Centre for Peace and Conflict Research* in Copenhagen (later COPRI). From 2010 to 2014 he was Secretary of the Research Committee Sociocybernetics (RC51) of the International Sociological Association. He has published some 250 items–two books in Polish, and papers and book chapters in Polish, English, Chinese and Turkish on management, security and related areas. Among his major publications are: "Applications of Systems Modelling in Peace Research", in: *Journal of Peace Research*, 25,3, 1988. He is a co-editor and co-author of the Hexagon Series. His most recent publications are: (co-ed. with Brauch, Oswald Spring, Grin, Kameri-Mbote, Chourou, Dunay, Birkmann): *Coping with Global Environmental Change, Disasters and*

Security–Threats, Challenges, Vulnerabilities and Risks, (Berlin–Heidelberg–New York: Springer, 2011); and "Complex Systems Studies and Terrorism", in: Fellman, Bar-Yam, Minai (Eds.): *Conflict and Complexity. Countering Terrorism, Insurgency, Ethnic and Regional Violence* (New York: Springer, 2015).

Address: Assoc. Prof. dr habil. Czeslaw Mesjasz, Cracow University of Economics, Pl-31-510 Kraków, ul Rakowicka 27, Poland.
Email: <mesjaszc@uek.krakow.pl>
Website: <http://kpz.uek.krakow.pl/portal/pl/user/43> and <http://www.afes-press.de/html/mesjasz_en.html>.

Carl Middleton (United Kingdom/Thailand) is a lecturer on the *MA in International Development Studies* (MAIDS) programme and Deputy Director for International Research in the *Center for Social Development Studies* (CSDS) in the Faculty of Political Science, Chulalongkorn University, Bangkok, Thailand. Before joining the MAIDS programme in 2009, he spent eight years working with international and local civil society organizations throughout the Mekong Region. His research interests are the politics and policy of the environment in South East Asia, with a particular focus on environmental justice and the political ecology of water and energy. He graduated from the University of Manchester, UK, with a bachelor's degree in Civil Engineering, and a Doctorate in Environmental Chemistry. Among his major publications are: Middleton, Carl; Elmhirst, Rebecca; Chantavanich, Supang (Eds.), 2016: *Living With Floods in a Mobile Southeast Asia: A Political Ecology of Vulnerability, Migration and Environmental Change* (London: Earthscan); Middleton, Carl; Dore, John, 2015: "Transboundary Water and Electricity Governance in Mainland Southeast Asia: Linkages, Disjunctures and Implications", in: *International Journal of Water Governance* 3,1: 93-120; Middleton, Carl; Allouche, Jeremy; Gyawali, Dipak; Allen, Sarah, 2015: "The Rise and Implications of the Water-Energy-Food Nexus in Southeast Asia through an Environmental Justice Lens", in: *Water Alternatives* 8,1: 627-54; Middleton, Carl; Matthews, Nathanial; Mirumachi, Naho, 2015: "Whose Risky Business?: Public-Private Partnerships (PPP), Build-Operate-Transfer (BOT) and Large Hydropower Dams in the Mekong Region", in: Matthews, Nathanial; Geheb, Kim (Eds.), 2015: *Hydropower Development in the Mekong Region: Political, Socio-economic and Environmental Perspectives* (London: Earthscan): 127-152; Smits, Mattijs; Middleton, Carl, 2014: "New arenas of engagement at the water governance–climate finance nexus? An analysis of the boom and bust of hydropower CDM projects in Vietnam", *Water Alternatives* 7,3: 561-583.

Address: Carl Middleton, Deputy Director for International Research, *Center for Social Development Studies* (CSDS), Faculty of Political Science, Chulalongkorn University, Henri Dunant Road, Bangkok, 10330, Thailand.
Email: <carl.Chulalongkorn@gmail.com>.
Website: <www.csds-chula.org; www.maids-chula.org>.

L. (Linda) O'Riordan (Germany/Ireland) is a Professor of Business Studies and International Management and Director of the *KompetenzCentrum for Corporate Social Re-*

sponsibility (KCC) at the FOM University of Applied Sciences in Germany. Dr. O'Riordan researches and writes with a focus on responsible management and sustainable entrepreneurship via an inclusive approach to organizing business in society. She has led an edited book on CSR, *"New Perspectives on Corporate Social Responsibility: Locating the Missing Link"*, and her work has been published in peer-reviewed international journals such as the *Journal of Business Ethics*. Dr. O'Riordan lectures mainly at MBA level but also to undergraduate students in subjects such as Business Strategy, Corporate Governance and CSR, International Management, Management Consulting and Problem-Solving. She is a general manager by background, with a PhD from the University of Bradford, UK. Previous to her academic role, she gained business and consultancy experience from working in industry. Some of her past employers include Accenture, UCB–Schwarz Pharma, and the Government of Ireland (Bord Bia), as well as apprenticeship positions during her studies in Ireland at the Irish Telecom Company, Dublin (now Eircom), and the Bank of Ireland, Cork. Her qualifications include a bachelor's degree in Business Studies from the University of Limerick, Ireland, as well as an MBA and a Master's in Research from the University of Bradford in the UK.

Address: Professor Dr. Linda O'Riordan, FOM University of Applied Sciences, Leimkugelstraße 6, 45141 Essen, Germany.
Email: <linda.oriordan@fom.de>.
Website: <http://www.fom.de/forschung/kompetenzcentren/kcc. html>.

Úrsula Oswald Spring (Mexico), Research Professor at the National University of Mexico (UNAM), in the Regional Multidisciplinary Research Center (CRIM) in Cuernavaca. See *Biographies of editors*.

Ilona Magdalena Otto (Germany) is a researcher at the *Potsdam Institute for Climate Impact Research* (PIK). She obtained a PhD in resource economics at Humboldt University, Berlin, where she teaches climate management and advanced research methods. In her research she investigates coordination mechanisms for the provision of public and common resources such as biodiversity and water. She is interested in combining different research methods and approaches to assess the socio-economic impacts of climate change.

Address: Dr. Ilona Magdalena Otto, Potsdam Institute for Climate Impact Research, Telegraphenberg A 31, P.O. Box 601203, D-14412 Potsdam, Germany.
Email: <ilona.otto@pik-potsdam.de>.
Website: <https://www.pik-potsdam.de/members/banaszak/>.

Bonno Pel (The Netherlands) is a postdoctoral researcher at the *Centre d'Etudes du Développement Durable* of the *Université Libre de Bruxelles* (Belgium). After graduating in environmental planning and sociopolitical philosophy at the University of Amsterdam, he completed a dissertation on system innovation in-the-making at the Erasmus University of Rotterdam in 2012. Bonno specializes in the governance and politics of system innovations and transitions, with a specific interest in mobility issues. Currently he is researching social innovation processes in the context of societal transformations.

Address: Dr. Bonno Pel, Institut de Gestion de l'Environnement et d'Aménagement du Territoire (IGEAT), Room DB.6.250, Université Libre de Bruxelles, 50 avenue FD Roosevelt, 1050 Bruxelles, Belgique.
Email: <Bonno.Pel@ulb.ac.be>.
Website: <http://igeat.ulb.ac.be/fr/equipe/details/person/bonno-pel/>.

Radhika Perrot (South Africa) is senior researcher at the *Mapungubwe Institute for Strategic Reflection* (MISTRA) in South Africa. She is currently based in Johannesburg, and is completing doctoral research on firm innovation in alternative energy systems, from UNU-MERIT, the Netherlands. She has researched and worked on various issues and topics around renewable energy, including feed-in policies and sociotechnological factors behind solar PV, wind and hydrogen-fuel cells innovation, and understanding market competition and firm strategies in industry and global markets. Radhika has authored and co-authored a number of articles, book chapters, and media articles on innovation, green economy, sustainable energy and technologies, and climate change.

Address: Ms Radhika Perrot, MISTRA, PostNet Suite 586, Private Bag X29, Gallo Manor, Johannesburg 2052, South Africa.
Email: <radhika.perrot@gmail.com>.
Website: <www.mistra.org.za>.

Martin F. Price (United Kingdom) has been Director of the *Centre for Mountain Studies* (CMS) at Perth College, University of the Highlands and Islands (UHI), UK since 2000; since 2009, Chairholder of the UNESCO Chair in Sustainable Mountain Development; and since 2013, Adjunct Professor at the University of Bergen, Norway. He has a PhD in Geography from the University of Colorado at Boulder and was appointed Professor of Mountain Studies by UHI in 2005. He worked previously at the University of Oxford; the University of Bern, Switzerland; and the National Center for Atmospheric Research, USA. The primary emphasis of his research has been on mountain people and environments; he has acted as a consultant on mountain issues to many international organizations. In 2012, the King Albert I Memorial Foundation awarded him the King Albert Mountain Award: the citation states that "Martin Price, with his exceptional knowledge and his editorial competence, has played a vital role for the mountains of the world". Further emphases of his work have been on interdisciplinary research and the human dimensions of global (environmental) change, beginning with his experience as the Secretary of the International Social Science Council's Standing Committee on the Human Dimensions on Global Change in 1989–90, which produced the first global framework for research on this topic. See his publications at: <http://www.perth.uhi.ac.uk/subject-areas/centre-for-mountain-studies/staff/professor-martin-price>.

Address: Prof. Martin F. Price, Director, Centre for Mountain Studies, Perth College, University of the Highlands and

Islands (UHI), Crieff Road, Perth, PH1 2NX, UK.
Email: < Martin.Price.perth@uhi.ac.uk >.
Website: <http://www.perth.uhi.ac.uk/subject-areas/centre-for-mountain-studies/staff/pro fessor-martin-price>.

Britta Rennkamp (Germany/South Africa) is a researcher at the Energy Research Centre and Fellow at the African Climate and Development Initiative, University of Cape Town. She holds a PhD in Science and Technology Policy from the University of Twente, the Netherlands, and a Master's Degree in Latin American Studies from University of Cologne, Germany. Prior to joining the University of Cape Town, Britta worked at the interface of technological innovation and social development in international cooperation for *German International Cooperation* (GIZ) in Brazil. After her return to Germany, she worked as a researcher at one of Europe's leading think tanks on international development, the *German Development Institute* (DIE) in Bonn, for four years. Her research focuses on policy analysis of climate technology, climate change mitigation, and poverty eradication in developing countries. Previous work has analysed renewable energy and nuclear programmes, carbon taxation, and green industrial and innovation policies. She has published various papers and book chapters on the overall question of integrating policies on emissions reductions, energy supply and development in Africa and Latin America.

Address: Britta Rennkamp, PhD, Energy Research Centre, University of Cape Town, Private Bag X3, 7701 Cape Town.
Email: <britta.rennkamp@uct.ac.za>.
Website: <http://www.erc.uct.ac.za>.

Jürgen Scheffran (Germany) is Professor at the Institute of Geography and the *Center for Earth System Research and Sustainability* (CEN), Germany, and Head of the *Research Group Climate Change and Security* (CLISEC) in the CliSAP Cluster of Excellence, Hamburg University. See *Biographies of Editors*.

Hans Joachim Schellnhuber (Germany) has been Director of the *Potsdam Institute for Climate Impact Research* (PIK) since he founded the institute in 1992. He is Professor of Theoretical Physics at the University of Potsdam and External Professor at the Santa Fe Institute, USA, and is Co-Chair of the *German Advisory Council on Global Change* (WBGU). Schellnhuber has authored, co-authored or edited around 300 articles and over fifty books in the fields of condensed matter physics, complex systems dynamics, climate change research, Earth System analysis, and sustainability science. Among his major publications are: (with Coumou, D.; Petoukhov, V.; Rahmstorf, S.; Petri, S., 2014): "Quasi-resonant circulation regimes and hemispheric synchronization of extreme weather in boreal summer", in: *Proceedings of the National Academy of Sciences of the United States of America (PNAS)*, 111, 12331; (with Menck, P.; Heitzig, J.; Kurths, J., 2014): "How Dead Ends Undermine Power Grid Stability", in: *Nature Communications*, 5; (with Rahmstorf, S., 2006): *Der Klimawandel. Diagnose, Prognose, Therapie* (München: C.H. Beck); (with Lenton, T.M., 2007): "Tipping the scales. Nature Reports", in: *Cli-*

mate Change, 1, 97; (with Lenton, T.; Held, H.; Kriegler, E.; Hall, J.; Lucht, W.; Rahmstorf S., 2008): "Tipping elements in the Earth's climate system", in: *PNAS*, 105: 1786-1793; (2008): "Global warming: Stop worrying, start panicking?", in: *PNAS*, 105: 14239-14240; (with Hall, J.; Held, H.; Dawson, R.; Kriegler, E., 2009): "Imprecise probability assessment of tipping points in the climate system", in: *PNAS*, 106: 5041-5046; (with Hofmann, M., 2009): "Ocean acidification affects marine carbon pump and triggers extended marine oxygen holes", in: *PNAS*, 106: 3017-3022; (Ed.): "*Tipping Elements in Earth Systems Special Feature*", in: *PNAS*, 106, 20561-20621.

Address: Prof. Dr. Dr. h.c. Hans Joachim Schellnhuber, Potsdam Institute for Climate Impact Research, Telegraphenberg A 31, P.O. Box 601203, D-14412 Potsdam, Germany.
Email: <director@pik-potsdam.de>.
Website: <http://www.pik-potsdam.de/institute/director>.

Philipp Schepelmann (Germany) is Acting Director of the Research Group for Material Flows and Resource Management at the Wuppertal Institute for Climate, Environment and Energy. His academic work focuses on societal and political conditions for successful ecological transition in multilevel governance systems. In particular he explores the use of scientific evidence for steering social and technological innovations in the right direction such as narratives, heuristics, indicators, and appraisal procedures.

Address: Dr. Philipp Schepelmann, Wuppertal Institute for Climate, Environment and Energy, Döppersberg 19, 42103 Wuppertal, Germany.
Email: <philipp.schepelmann@wupperinst.org>.
Website: <http://wupperinst.org/kontakt/details/wi/c/s/cd/89/>.

Janpeter Schilling (Germany) is a Postdoctoral Fellow in the research group Climate Change and Security (CLISEC) at Hamburg University. He is also an Associate at International Alert in London. His previous roles include: Programme Officer in Environment, Climate Change and Security at International Alert, Postdoctoral Research Fellow in Environmental Security at Colgate University in New York, and Research Associate at Hamburg University, where he received his PhD in Geography. Janpeter Schilling has published several journal articles and book chapters on climate change, vulnerability, adaptation, resilience and conflict.

Address: Dr. Janpeter Schilling, Grindelberg 7, 20144 Hamburg, Germany.
Email: <janpeter.schilling@uni-hamburg.de>.
Website: http://clisec.zmaw.de/index.php?id=869

Carl-Friedrich Schleussner (Germany) is a climate scientist currently with *Climate Analytics*, Berlin and the *Potsdam Institute for Climate Impact Research*. He earned a PhD in climate physics, under the supervision of Anders Levermann at the Potsdam Institute for Climate Impact Research, on the stability of the North Atlantic Ocean current system, and he is working on a variety of climate-related topics including stability analysis of tipping elements, cou-

Biographies of Contributors

pled socio-ecosphere systems and the Amazonian rainforest.

Address: Dr. Carl-Friedrich Schleussner, Climate Analytics, Friedrichstr 231 - Haus B, 10969 Berlin, Germany.
Email: <carl.schleussner@climateanalytics.org>.
Website: <http://www.climateanalytics.org/the-team/dr-carl-friedrich-schleussner>.

Prof. Dr. Uwe Schneidewind (Germany) is President of the Wuppertal Institute for Climate, Environment, Energy, and Professor of Sustainable Transition Management at the University of Wuppertal. He is a member of the Club of Rome and of the Steering Group of the Sustainability Transitions Research Network. He has been a member of the German Advisory Council on Global Change (WBGU) since 2013. His latest book, "Transformative Wissenschaft" (Transformative Science), analyses the status of sustainability science in the German science system. Among his major recent publications on this theme are: Schneidewind, Uwe; Singer-Brodowski, Mandy; Augenstein, Karoline, 2016: "Sustainability and Science Policy", in: Heinrichs, Harald et al. (Eds.): *Sustainability Science*, Springer Science, Dordrecht, 149-160; Schneidewind, Uwe; Singer-Brodowski, Mandy, 2014: *Transformative Wissenschaft. Klimawandel im deutschen Wissenschafts- und Hochschulsystem* (Marburg: Metropolis); Schneidewind, U. (2013); "Transformative Literacy. Rahmen für den wissensbasierten Umgang mit der 'Großen Transformation'"; in: *GAIA* 22:2, 82-86; Schneidewind, Uwe; Augenstein, Karoline, 2012: "Analyzing a transition to a sustainability-oriented science system in Germany", in: *Environmental Innovation and Societal Transitions,* 3: 16-28.

Address: Prof. Dr. Uwe Schneidewind, President of the Wuppertal Institute for Climate, Environment and Energy, Döppersberg 19, 42103 Wuppertal, Germany.
Email: <uwe.schneidewind@wupperinst.org>.
Website: <http://wupperinst.org/home/>.

Olivia Maria Serdeczny (Germany) earned her MA in Philosophy at the Freie Universität Berlin and currently works as a research analyst at *Climate Analytics*. From 2010 to 2013 she worked as a research analyst for Professor Schellnhuber at the German Advisory Council on Global Change to the Federal Government and at the Potsdam Institute for Climate Impact Research. In 2012 and 2013 she coordinated and co-authored two reports commissioned by the World Bank on climate impacts in a 4 °C world.

Address: Olivia Serdeczny, Climate Analytics, Friedrichstr 231 - Haus B, 10969 Berlin, Germany (affiliated to Potsdam Institute for Climate Impact Research, Telegraphenberg A 31, P.O. Box 601203, D-14412 Potsdam, Germany).
Email: <olivia.serdeczny@climateanalytics.org>.
Website: <http://www.climateanalytics.org/the-team/olivia-serdeczny>.

Mandy Singer-Brodowski (Germany) studied educational science at the University of Erfurt and was a founding member of the nationwide network for student initiatives in sustainability in Germany. She was also a member of the German national committee for the implementation of the UN "Decade of Education for Sustainable Development".

She has worked at the Wuppertal Institute for Climate, Environment and Energy from 2012 till 2015. In 2015 she completed her PhD at the Leuphana University of Lüneburg, where she is also a lecturer. At the moment she is scientific coordinator of the Center for Transformation Research and Sustainability at the University of Wuppertal. Her main interests in research are the development of student competency in sustainability projects and transformative learning. Among her major recent publications on this theme are: Schneidewind, Uwe; Singer-Brodowski, Mandy; Augenstein, Karoline, 2016: "Sustainability and Science Policy", in: Heinrichs, Harald et al. (Eds.): *Sustainability Science*, Springer Science, Dordrecht, 149-160; Schneidewind, Uwe; Singer-Brodowski, Mandy, 2014: *Transformative Wissenschaft. Klimawandel im deutschen Wissenschafts- und Hochschulsystem* (Marburg: Metropolis).

Address: Dr. Mandy Singer-Brodowski, Center for Transformation Research and Sustainability (TransZent), University of Wuppertal, Bendahler Str. 31, 42285 Wuppertal, Germany.
Email: <singer@uni-wuppertal.de>.
Website: <http://www.transzent.uni-wuppertal.de/home.html>.

Marit Sjøvaag (Norway) is a political scientist whose main research interest is climate policy both in Norway and internationally. She holds a PhD from the London School of Economics and Political Science. She has also studied mathematics and natural sciences in Oslo, Tromsø and Caen, France, and management and social sciences at the Norwegian Business School and at Sciences-Po, Paris. In the period 2008-2011 she was responsible for the annual Status Report on the implementation of climate policies in Norway at the Norwegian Business School's Center for Climate Strategy. She has written and taught on climate policy, regulatory regimes and democratic processes, and aims to contribute to the transition to a low-carbon society. She currently works with energy and climate issues both in Norway and in an international context, for the environmental organization Bellona. Among her major publications are: (with Nils Erik Bjørgo, Torgeir Ericson, Per Arild Garnåsjordet, Håkon Karlsen, Jørgen Randers, Daniel Rees): *People's opinion of climate policy—Popular support for climate policy alternatives in Norway*, CICERO working paper 2012: 03; *The Study of the European Union from Outside: European Integration Studies in Norway and Iceland 1990-2010*; (with Kjell A. Eliassen, Eirikur Bergmann): in: Eliassen, K.A.; Bindi, F. (Eds.): *Analyzing European Union Politics* (Bologna: Il Mulino).

Address: Dr. Marit Sjøvaag Marino, Bellona, Maridalsveien 17B, 0175 Oslo; and: Boks 2141 Grünerløkka, 0505 Oslo, Norway.
Email: <marit@bellona.no> and <sjovaag@aol.com>.
Website: <http://bellona.org/employee/marit-sjovaag-marino>.

Philipp Späth (Germany) was Assistant Professor at the chair group of environmental governance in the Faculty of the Environment and Natural Resources at Albert-Ludwigs University, Freiburg. He now heads a research project on Smart EcoCities funded by the German Research Foundation (DFG) in the same institute. Trained as a geographer and political scientist, he obtained a PhD in Science and Technology Studies in 2009. His research is on urban environmental governance, with a particular focus on the governance of sociotechnical change in multilevel governance systems, with the German Energiewende being a primary example. From 2003 to 2009 he was senior researcher at the Inter-University Research Centre for Technology, Work & Culture (IFZ) in Graz, Austria. From 1997 to 2003 he worked as a practitioner in renewable energy projects in Freiburg, Germany. He has published in *Urban Studies, Research Policy, Energy Policy, Transportation Reviews* etc.

Address: Dr. Philipp Späth, Institute of Environmental Social Sciences and Geography, Albert-Ludwigs University Freiburg, Tennenbacherstr. 4, D-79106 Freiburg, Germany.
Email: <spaeth@envgov.uni-freiburg.de>.
Website: <https://www.envgov.uni-freiburg.de/mitarbeiterinnen/philippspaeth-en/>.

Carolyn M. Stephenson (USA) is Associate Professor of Political Science at the University of Hawaii at Manoa, and affiliate faculty member in its Matsunaga Institute for Peace. With a BA from Mt. Holyoke and an MA and PhD from Ohio State, she taught previously at Colgate, where she was Director of Peace Studies. Her current research is on alternative security systems, the role of NGOs at the UN on environment and disarmament, and the UN role in the development of women's rights. Among her major publications are: "Alternative International Security Systems: An Introduction", and "Alternative Methods for International Security: A Review of the Literature"; in: Stephenson, Carolyn (Ed.), 1982: *Alternative Methods for International Security* (Washington DC: University Press of America); "NGOs and the Principal Organs of the United Nations"; in: Taylor, Paul; Groom, A. J. R. (Eds.), 2000: *The United Nations at the Millennium: The Principal Organs* (London: Continuum): 271–294; "Gender Equality and a Culture of Peace"; in: de Rivera, Joseph (Ed.), 2008: *Handbook for Building Cultures of Peace* (New York: Springer): 123–137; "Peace Research/Peace Studies: a Twentieth Century Intellectual History"; in: Denemark, Robert E. (Ed.), 2010: *The International Studies Encyclopedia* (Oxford, UK: Blackwell): 5579–5603; "Women Leaders in the Peace/Anti-War Movements"; in: O'Connor, Karen (Ed.), 2010: *Gender and Women's Leadership*, Volume I (Thousand Oaks, CA: Sage): 279–289.

Address: Prof. Carolyn M. Stephenson, Political Science Department, Saunders 640, 2424 Maile Way, University of Hawaii at Manoa, Honolulu, HI 96825, USA.
Email: <cstephen@hawaii.edu>.

Mark Swilling (South Africa) is Professor of Sustainable Development in the School of Public Leadership, University of Stellenbosch <www.sopmp.sun.ac.za> and Academic Director of the Sustainability Institute <www.sustainabilityinstitute.net>. He is responsible for a Master's and doctoral programme in Sustainable Development delivered at the Sustainability Institute in Stellenbosch, South Africa. He also heads up the TSAMA Hub, a new Centre for the transdisciplinary study of sustainability and complexity at Stellenbosch University <www.tsama.org.za>. The TSAMA Hub hosts a transdisciplinary doctoral programme that involves collaboration between seven of Stellenbosch University's Faculties. Professor Swilling obtained his PhD from the University of Warwick in 1994. He co-authored with Eve Annecke: *Just Transitions: Explorations of Sustainability in an Unfair World* (Tokyo: United Nations University Press, 2012), runner-up in 2013 for the Harold and Margaret Sprout Award for best environmental governance book. He is also a member of UNEP's International Resource Panel, which addresses the challenge of global transition to more sustainable production and consumption, and he has specific responsibility for coordinating the Cities Working Group.

Address: Professor Mark Swilling, School of Public Leadership, Stellenbosch University, Private Bag XI, Matieland, 7602, South Africa.
Email: <Mark.Swilling@spl.sun.ac.za>.
Website: <www.sopmp.sun.ac.za>.

Cecilia Tortajada (Mexico/Singapore) is a Senior Research Fellow, Institute of Water Policy, Lee Kuan Yew School of Public Policy, National University of Singapore, Singapore. The main focus of her work at present is on the future of the world's water, especially in terms of water, food, energy and environmental security through coordinated policies. She has been an advisor to major international institutions such as FAO, UNDP, JICA, ADB, OECD and IDRC, and has worked in countries in Africa, Asia, North and South America and Europe on water and environment-related policies. She is a member of the OECD Initiative in Water Governance and is engaged in independent studies on the Corporate Social Responsibility of major multinational corporations. She is a past President of the International Water Resources Association (2007–2009) and Editor-in-Chief of the *International Journal of Water Resources Development*. Recent publications include: (co-edited with Biswas, 2015): "Water Security, Climate Change and Sustainable Development" (Heidelberg: Springer); (co-authored with Hering, J. G., D. L. Sedlak, A. K. Biswas, C. Niwagaba, and T. Breu, 2015): "The need for local perspectives", in: *Science*; (co-authored with Muller, M., A. K. Biswas, and R. Martin-Hurtado, 2015): "Building Water Security: The Role of Infrastructure in the Anthropocene", in *Science*; (co-edited with Wegerich and Warner, 2014): "'Dark' Side of Water Governance", in: *International Journal of Water Governance*; (co-authored with Joshi and Biswas, 2013): *The Singapore Water Story, Sustainable Development in an Urban City State* (London: Routledge).

Address: Dr. Cecilia Tortajada, IWP, LKYSPP, 469C Bukit Timah Road, Oei Tiong Ham Building, Singapore 259772.
Email: <cecilia.tortajada@gmail.com>.
Website: <http://lkyspp.nus.edu.sg/iwp/>.

Gregory Trencher (USA) holds a PhD in Sustainability Science and is currently an Assistant Professor in Environmental Science and Policy at Clark University. His main field of research examines innovative cases of collaboration between universities, government, industry and civil society to drive societal transformations towards sustainability. In parallel, his research also analyses approaches, impacts and barriers of innovative policies implemented in various C40 cities to advance the energy efficiency of existing commercial buildings. This is conducted with the Tokyo Metropolitan Government Bureau of Environment and C40 Cities Climate Leadership Group London. He recently held a Visiting Assistant Professor appointment at the University of Tokyo Graduate School of Frontier Sciences. Recent publications include: "Policy innovation for building energy efficiency and retrofitting: Approaches, impacts and challenges in ten C40 cities" (submitted to *Local Environment*); "Implementing sustainability co-creation between universities and society: Potential models, challenges and strategies" (submitted to *Sustainability Science*); "The role of students in the co-creation of knowledge and societal transformations towards sustainability: Experiences from Lund, Tokyo and Ohio", in: Leal Filho, W.; Brandli, L. (Eds.): *Engaging Stakeholders in Education for Sustainable Development at the University Level* (Berlin: Springer, 2016); "*University partnerships for co-designing and co-producing urban sustainability*" in: *Global Environmental Change* (2014), 28: 153-165; "Beyond the third mission: Exploring the emerging university function of co-creation for sustainability", in: *Science and Public Policy* (2014), 41,2: 151-179.

Address: Ass. Prof. Dr. Gregory Trencher, Clark University, Department of International Development and Environment. 950 Main Street, Worcester, 01610 Massachusetts, USA.
Email: <gtrencher@clarku.edu>.
Website: <https://www.clarku.edu/faculty/facultybio.cfm?id=992>.

Eduardo Viola (Brazil) has been a Full Professor of International Relations at the University of Brasília since 1993 and senior researcher of the Brazilian Council for Scientific Research (CNPQ) since 1986. He holds a PhD in Political Science from the University of Sao Paulo (1982) and has postdoctoral training in international political economy from the University of Colorado at Boulder. He has also been the coordinator of the *Brazilian Research Network on International Relations and Climate Change* (Redeclim) since 2009. Dr. Viola has been Visiting Professor in several first-rank international universities, among them Stanford, Colorado at Boulder, Notre Dame, Texas at Austin, Amsterdam, Campinas, San Martin and Buenos Aires. He has also taught for some years at the Rio Branco Institute, the Brazilian school for diplomatic training. Prof. Viola has published five books, more than seventy articles in journals and more than forty book chapters on issues of globalization and governance, democracy and democratization in South America, Brazilian foreign policy, international environmental policy and politics and international political economy of climate change. He has published in Portuguese, English, Spanish, French and Italian in thirteen countries. He has more than 2,700 citations in Google Scholar.

Address: Prof. Dr. Eduardo Viola, Institute of International Relations, University of Brasília (Campus Darcy Ribeiro, Caixa Postal: 04306, CEP: 70904-970, Brasília – DF, Brazil).
Email: eduviola@gmail.com.
Website: <http://irel.unb.br/professores/>; CV: <http://lattes.cnpq.br/2685286492991791>.

Sir Robert Worcester (United Kingdom) is Emeritus Chancellor of the University of Kent, and is Visiting Professor and Honorary Fellow at the London School of Economics, Visiting Professor and Fellow of King's College London, Honorary Professor at Warwick University and Adjunct Professor at the University of Kansas. He was Senior Vice President of the International Social Science Council, UNESCO and is Past President of the World Association of Public Opinion Research. Sir Robert was the co-editor and co-founder of the *International Journal of Public Opinion Research* with Professors Elisabeth Noelle-Neumann and Seymour Martin Lipset. In 2015 he was Chairman of the 800th Anniversary Commemoration Committee of the Magna Carta Trust.

Address: Sir Robert Worcester, Allington Castle, Maidstone, Kent, ME16 0NB, UK.
Email: <rmworcester@yahoo.com>.
Website: <http://www.kcl.ac.uk/sspp/departments/icbh/people/worcester/index.aspx>; <http://magnacarta800th.com/magna-carta-today/membership-of-the-magna-carta-800th-committee/>; <http://www.kent.ac.uk/politics/about-us/staff/members/worcester.html>.

Belinda Yuen (Singapore) is with the Lee Kuan Yew Centre for Innovative Cities, Singapore University of Technology and Design. She is a qualified urban planner. Her academic qualifications include: BA (Hons), MA (Town & Regional Planning) and PhD (Environmental Planning). Aside from master planning practice, Belinda has extensive research and publication in spatial planning and urban policy analysis, most recently on planning liveable, sustainable cities. She has been elected President, Singapore Institute of Planners (2005-08) and Vice-President, Commonwealth Association of Planners (South East Asia) (2006-08; 2010-12). Belinda has served on various Singapore urban planning committees, including as Planning Appeals Inspector, and on international advisory committees including the International Advisory Board of the United Nations Global Research Network on Human Settlements (2008-11); United Nations Commission on Legal Empowerment of the Poor Working Group; and on the editorial boards of international journals including *Asia Pacific Planning Review*; *Regional Development Studies*; *Cities*; *Journal of Planning History*; and *Journal of Comparative Asian Development*. She has been a Juror, Shaikh Khalifa Bin Salman Al Khalifa Habitat Award, 2008-2009, and science reviewer of various international multidisciplinary research programmes.

Address: Singapore University of Technology and Design, Singapore.
Email: <Belinda_yuen@sutd.edu.sg>.

Hania Zlotnik (Mexico/US) is currently an independent consultant. She retired in early 2012 from the post of Director of the Population Division of the United Nations. She joined the Division in 1982 and worked previously as Chief of the Mortality and Migration Section and Chief of the Population Estimates and Projections Section. Before joining the United Nations, she served as research associate for the Committee on Population and Demography of the US National Research Council. She holds a PhD in Statistics and Demography from Princeton University and is a graduate of the *Universidad Nacional Autónoma de México*

(UNAM). Her work has spanned the field of demography, covering the analysis of fertility, mortality and migration with especial emphasis on their quantitative aspects. She has prepared manuals on demographic estimation techniques and on the collection of international migration statistics. She has edited or written reports published by the United Nations on a variety of topics, including international migration and development, female migration, levels and trends in urbanization, population distribution and migration, population ageing, health and mortality, and population estimates and projections. Recent publications include: "Population ageing in the context of globalization", in: Diego Sánchez González and Vicente Rodríguez Rodríguez (Eds.): *Environmental Gerontology in Europe and Latin America* (New York: Springer, 2016): 45-72; (with David Bloom; Emmanuel Jimenez): *Seven Billion and Growing: A Twenty-First Century Perspective on Population* (Geneva: World Economic Forum, 2012); "International migration and population ageing", in: *Global Population Ageing: Peril and Promise* (Geneva: World Economic Forum, 2012): 97-102; "Does population matter for climate change?", in: José Miguel Guzmán et al (Eds.): *Population Dynamics and Climate Change* (New York: UNFPA–London: IIED, 2009): 31-44. She has contributed texts to volumes 1 and 5 of the Hexagon Book Series.

Address: 78-06 Kew Forest Lane, Forest Hills, NY 11375, USA.
Email: <hania.zlotnik@hotmail.com>.

Index

4°C world 51, 267, 269, 270, 279, 892, 893

9/11 terrorist attacks (World Trade Center and Pentagon) 54, 92–94, 204, 307, 481, 485, 486, 897

A

Abbott, Anthony John (Tony) 494
ability of future generations 31, 68, 165, 202, 324, 425, 572, 795, 812
abolition of the military and of warfare 177
absence of structural violence 93, 163, 195
absence of violence 324
absence of war 93, 163, 324
absorption 385, 584, 724, 727, 730, 917
absorptive capacities 748
abundance 21, 72, 75, 287, 319, 660, 746, 766, 859, 913
academics 157, 370, 385, 493, 501, 636, 642, 644, 645, 689, 794, 800, 801, 841, 848
access to irrigated land and wells 687
access to material and immaterial resources 163
accountability 110, 627, 710, 712, 722, 826, 828, 831, 837, 843, 845, 847, 905
acculturation 141
achievement 17, 19, 20, 112, 255, 323, 372, 386, 530, 571, 573, 575, 586, 708, 761, 762, 900, 902, 917, 922
acid rain 482
activities and actors in the transition process 898
activities of vulnerability 699
ad hominem politics 498
Adams, Sophie 48, 51
adapt technologies 721, 725, 746, 751, 903
adaptation 13, 16, 17, 22, 24, 25, 27, 39, 42, 47, 67, 69, 71, 81, 82, 84, 109, 111, 112, 156, 162, 167, 200, 212, 272, 273, 276, 279, 285, 295, 297, 305, 307, 310, 312–315, 329, 340, 352, 365, 367, 368, 370, 431, 433, 443, 446, 501, 508–511, 513, 514, 520, 527, 584, 677, 684, 685, 724, 728, 730, 732,

738, 739, 740, 747, 894, 896, 898, 911, 917, 918
adaptation and mitigation strategies 4, 209, 727
adaptation and resilience-building measures 918
adaptation mandates 519
adaptation measures 711, 745, 893
adaptation of patriarchy 166, 170
adaptation planning 16, 17, 621
adaptation policies 16, 506, 515
adaptation policy-making and planning 279
adaptation pressure 270, 271
adaptation programmes 516
adaptation projects 737
adaptation response 17
adaptation strategies 317, 505, 686
adaptation to climate change 209, 317, 318, 644
adaptation to risks 170
adaptation, mitigation and sustainable development 15
adaptive capacity 17, 36, 287, 295, 308, 311, 315, 707
adaptive control 326–328, 334
adaptive governance 327, 329, 340, 894
adaptive innovation 729
adaptive management 35, 327, 328
adaptive networks 328, 329
adaptive technology life cycle 725
additional sources of revenue 688
adequate governance 865, 906
adjustment 55, 177, 294, 501, 571, 726, 761, 764, 767, 768, 799, 839, 872, 877, 900, 907
adjustment mechanisms 765, 903
adjustment of industry structure 572, 586
adjustment rules 339
administrative approaches 575
Adorno, Theodor 53, 451–453, 455, 460, 896
adult mortality 713
adultery 168
Advanced Consortium on Cooperation, Conflict and Complexity 211
advanced technology 574
aerosol pollution 813

aerosols 77, 714
aerospace technologies 786
affluents 56, 675, 902
Afghanistan 10, 204, 205, 306, 307, 318, 486, 495, 496, 498, 667
Africa 10, 14, 20, 25, 48, 55, 75, 141, 143, 167, 169, 171, 174, 198, 199, 208, 243, 244, 271, 272, 307, 311, 318, 354, 393, 401, 406, 409, 484, 555, 613, 614–618, 621, 627, 628, 663, 664, 666–670, 734, 780, 822, 898, 900, 901, 909, 914
African Development Bank (AfDB) 622
African Peace Research and Education Association 93
African Union 208
ageing 31, 239, 241, 261, 262, 614, 633, 642, 644, 646, 649, 722
Agenda 21 31, 111, 176, 202, 203, 383, 559, 572, 577, 735
agenda for peace 93, 213
agent-based model 271, 305, 328, 329
agribusiness 22, 667–669, 687, 709, 712, 714, 812, 917
agricultural activities 20, 77, 167, 399, 682, 696, 697
agricultural and industrial revolutions 192
agricultural change 13
agricultural exports 668
agricultural management 45
agricultural policy 112, 113, 667, 687
agricultural pollution 15
agricultural practices 743
agricultural revolution 5, 167, 170, 175, 891
agricultural sector 45, 218, 269, 272, 273, 276, 552, 660, 668, 670, 679, 743, 764
agricultural subsidies 667
agricultural technologies 170
agricultural trade 659, 667
agriculture 10, 12, 18, 23, 24, 27, 31, 40, 45, 47, 69, 78, 109, 113, 163, 170, 180, 200
Agriculture, Forestry and Other Land Use (AFOLU) 47
agrifood 115, 889
agrochemicals 15, 21, 24, 27, 686, 918

Hexagon Series on Human and Environmental Security and Peace (HESP)

This book series includes monographs and edited volumes that cross scientific disciplines and develop common ground among scientists from the natural and social sciences, as well as from North and South, who are addressing common challenges and risks for humankind in the 21st century.

The 'hexagon' represents six key factors contributing to global environmental change—three *nature-induced* or supply factors: *soil*, *water* and *air* (atmosphere and climate), and three *human-induced* or demand factors: *population* (growth), *urban systems* (habitat, pollution) and *rural systems* (agriculture, food). Throughout the history of the earth and of *Homo sapiens* these six factors have interacted. The supply factors have created the preconditions for life, while human behaviour and economic consumption patterns have contributed to its challenges (increase in extreme weather events) and led to fatal outcomes for human beings and society. The series covers the complex interactions among these six factors and their often extreme and sometimes fatal outcomes (hazards/disasters, internal displacement and migrations, crises and conflicts), as well as the crucial social science concepts relevant for their analysis.

Further issues relate to three basic areas of research: the approaches and schools of environment, security, and peace, especially in the environmental security realm and from a human security perspective, will be addressed. The aim of this book series is to contribute to a fourth phase of research on environmental security from the perspective of normative peace research and human security. In this series, the editor welcomes books by natural and social scientists, as well as by multidisciplinary teams of authors. The material should address issues of global change (including climate change, desertification, deforestation), and its impacts on humankind (natural hazards and disasters), on environmentally-induced migration, on crises and conflicts, together with cooperative strategies to cope with these challenges either locally or within the framework of international organizations and regimes.

From a human-centred perspective, this book series offers a platform for scientific communities dealing with global environmental and climate change, disaster reduction, human, environmental and gender security, and peace and conflict research, as well as for the humanitarian aid community and the policy community in national governments and international organizations.

The series editor welcomes brief concept outlines and original manuscripts as proposals. If they are considered of relevance, these proposals will be peer-reviewed by specialists in the field from the natural and the social sciences. Inclusion in this series will also require a positive decision by the publisher's international editorial conference. Prior to publication, the manuscripts will be assessed by the series editor and external peer reviewers.

Mosbach, Germany, January 2016

Hans Günter Brauch,
AFES-PRESS, Chairman

Hexagon Series on
Human and Environmental Security and Peace (HESP)

Edited by Hans Günter Brauch, AFES-PRESS, Chairman, Mosbach, Germany

Vol. 1: Hans Günter Brauch, P. H. Liotta, Antonio Marquina, Paul Rogers, Mohammad El-Sayed Selim (Eds.): *Security and Environment in the Mediterranean - Conceptualising Security and Environmental Conflicts. With Forewords by the Hon. Lord Robertson, Secretary General of NATO, and the Hon. Amre Moussa, Secretary General of the League of Arab States* (Berlin – Heidelberg – New York: Springer, 2003).
ISBN: 978-3-540-40107-0 (Print) / ISBN: 978-3-642-55854–2 (Online)
DOI 10.1007/978-3-642-55854-2

Vol. 2: Hillel Shuval, Hassan Dweik (Eds.): *Water Resources in the Middle East: Israel-Palestinian Water Issues - from Conflict to Cooperation* (Berlin – Heidelberg – New York: Springer-Verlag, 2007).
ISBN: 978-3-540-69508-0 (Print) / ISBN: 978-3-540-69509-7 (Online)
DOI 10.1007/978-3-540-69509-7

Vol. 3: Hans Günter Brauch, Úrsula Oswald Spring, Czeslaw Mesjasz, John Grin, Pál Dunay, Navnita Chadha Behera, Béchir Chourou, Patricia Kameri-Mbote, P.H. Liotta (Eds.): *Globalization and Environmental Challenges: Reconceptualizing Security in the 21st Century* (Berlin – Heidelberg – New York: Springer-Verlag, 2008).
ISBN : 978-3-540-75976-8 (Print) / ISBN: 978-3-540-75977-5 (Online)
DOI 10.1007/978-3-540-75977-5

Vol. 4: Hans Günter Brauch, Úrsula Oswald Spring, John Grin, Czeslaw Mesjasz, Patricia Kameri-Mbote, Navnita Chadha Behera, Béchir Chourou, Heinz Krummenacher (Eds.): *Facing Global Environmental Change: Environmental, Human, Energy, Food, Health and Water Security Concepts* (Berlin – Heidelberg – New York: Springer-Verlag, 2009).
ISBN: 978-3-540-68487-9 (Print) / ISBN: 978-3-540-68488-6 (Online)
DOI 10.1007/978-3-540-68488-6

Vol. 5: Hans Günter Brauch, Úrsula Oswald Spring, Czeslaw Mesjasz, John Grin, Patricia Kameri-Mbote, Béchir Chourou, Pal Dunay, Jörn Birkmann (Eds.): *Coping with Global Environmental Change, Disasters and Security - Threats, Challenges, Vulnerabilities and Risks* (Berlin - Heidelberg - New York: Springer-Verlag, 2011).
ISBN: 978-3-642-17775-0 (Print) / ISBN: 978-3-642-17776-7 (Online)
DOI 10.1007/978-3-642-17776-7

Vol. 6: Thanh-Dam Truong, Des Gasper (Eds.): *Transnational Migration and Human Security: The Migration - Development - Security Nexus.* Hexagon Series on Human and Environmental Security and Peace, vol. 6 (Heidelberg – Dordrecht – London – New York: Springer, 2011).
ISBN 978-3-642-12756-4 (Print) / ISBN 978-3-642-12757-1 (Online)
DOI 10.1007/978-3-642-12757-1

Vol. 7: Úrsula Oswald Spring (Ed.): *Water Resources in Mexico. Scarcity, Degradation, Stress, Conflicts, Management, and Policy.* Hexagon Series on Human and Environmental Security and Peace, vol. 7 (Heidelberg – Dordrecht – London – New York: Springer, 2011).
ISBN: 978-3-642-05431-0 (Print) / ISBN 978-3-642-05432-7 (Online)
DOI 10.1007/978-3-642-05432-7

In Preparation

Authors or editors who would like to have their publication project considered for inclusion in this series should contact both the series editor:

PD Dr. phil. habil. Hans Günter Brauch, Alte Bergsteige 47, 74821 Mosbach, Germany
Phone: 49-6261-12912 FAX: 49-6261-15695
Email afes@afes-press.de
http://www.afes-press.de and http://www.afes-press-books.de/html/hexagon.htm

and the publisher:
Dr. Johanna Schwarz, Senior Publishing Editor, Springer Nature, Tiergartenstrase 17, 69121 Heidelberg, Germany,
Phone: +49 6221 487 8614,
Email: <johanna.schwarz@springer.com> and <http://www.springer.com>

Springer: <http://www.springer.com/series/8090>
AFES-PRESS: <http://www.afes-press-books.de/html/hexagon.htm>

SpringerBriefs in Environment, Security, Development and Peace

Editor:
Hans Günter Brauch, *AFES-PRESS, Mosbach, Germany*

SpringerBriefs in Environment, Security, Development and Peace (ESDP) present concise summaries of cutting-edge research as well as innovative policy perspectives. The series focuses on the interconnection of new and nontraditional global environmental and development challenges facing humankind that may pose dangers for peace and security in the Anthropocene era of earth history. SpringerBriefs in ESDP publish monographs as well as edited volumes of topical workshops that are peer-reviewed by scholars from many disciplines and from all parts of the world. SpringerBriefs in ESDP will give more "voice" and "visibility" to scientists and innovative political thinkers in developing countries.

The books will be published 8 to 12 weeks after acceptance. Featuring compact volumes of 50 to 125 pages (approx. 20,000–55,000 words), the series covers a wide scope of policy-relevant issues from policy-oriented, to professional and academic perspectives.

SpringerBriefs in ESDP address the "conceptual quartet" among the four research research programmes in the social sciences focusing on *Environment, Security, Development and Peace* as well as the "consilience" between the natural and the social sciences that try to initiate debates and to provide multi-, inter- and transdisciplinary knowledge relevant for coping with the multiple projected impacts of global environmental change.

- Timely reports of state-of-the art analyses (e.g. from earth systems science or analysis, geoecology) and policy assessments dealing with the global challenges facing humankind in the 21st century
- SpringerBriefs in ESDP bridges between new research results offering snapshots of hot and/or emerging topics, literature reviews and in-depth case studies

Briefs will be published as part of Springer's eBook collection, with millions of users worldwide. In addition, Briefs will be available for individual print and electronic purchase. Briefs are characterized by fast, global electronic dissemination, standard publishing contracts, easy-to-use manuscript preparation and formatting guidelines, and expedited production schedules.

Interested? Please send your proposal to:

PD Dr. habil. Hans Günter Brauch
Alte Bergsteige 47, 74821 Mosbach, Germany
hg.brauch@onlinehome.de

Your proposal should consist of the following items: Title, Author(s) information, Abstract, 5 Keywords (would be used when searching your topic on Google or Amazon), Manuscript delivery date, estimated number of pages, concise outline (ca. 200 words), Table of contents. Please allow 1–2 months for the peer review process.

SpringerBriefs in Environment, Security, Development and Peace

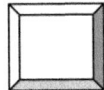

Edited by
Hans Günter Brauch, AFES-PRESS Chairman, Mosbach, Germany

Vol. 1: Mely Caballero-Anthony, Youngho Chang and Nur Azha Putra (Eds.): *Energy and Non-Traditional Security (NTS) in Asia.* SpringerBriefs in Environment, Security, Development and Peace, vol. 1 (Heidelberg - Dordrecht - London - New York: Springer-Verlag, 2012).

Vol. 2: Mely Caballero-Anthony, Youngho Chang and Nur Azha Putra (Eds.): *Rethinking Energy Security in Asia: A Non-Traditional View of Human Security.* SpringerBriefs in Environment, Security, Development and Peace, vol. 2 (Heidelberg – Dordrecht - London - New York: Springer-Verlag, 2012).

Vol. 3: Philip Jan Schäfer: *Human and Water Security in Israel and Jordan.* SpringerBriefs in Environment, Security, Development and Peace, vol. 3 (Heidelberg - Dordrecht - London - New York: Springer-Verlag, 2013).

Vol. 4: Gamal M. Selim: *Euro-American Approaches to Arms Control and Confidence-Building Measures in the Middle East: A Critical Assessment from the South.* SpringerBriefs in Environment, Security, Development and Peace, vol. 4 – Mediterranean Studies Subseries No. 1 (Heidelberg - Dordrecht - London - New York: Springer-Verlag, 2013).

Vol. 5: Mansoureh Ebrahimi: *The British Role in Iranian Domestic Politics (1951-1953* (Cham- New York – Heidelberg – Dordrecht – London: Springer International Publishing, 2016).

Vol. 6: Lourdes Arizpe, Cristina Amescua: *Anthropological Perspectives on Intangible Cultural Heritage.* (Heidelberg – Dordrecht – London – New York: Springer-Verlag, 2013).

Vol. 7: Ebru Gencer: *Natural Disasters and Risk Management in Urban Areas: A Case Study of the Istanbul Metropolitan Area* (Heidelberg – Dordrecht – London – New York: Springer-Verlag, 2013).

Vol. 8: Selim Kapur, Sabit Er ahin (Eds.): *Soil Security for Eco-system Management. Mediterranean Soil Ecosystems* 1 (Heidelberg – Dordrecht – London – New York: Springer-Verlag, 2013).

Vol. 9: Ayesah Uy Abubakar: *Peacebuilding and Sustainable Human Development: A Case Study from the Bangsamoro, Philippines* (Cham- New York - Heidelberg - Dordrecht - London: Springer International Publishing, 2017).

Vol. 10: Nur Azha Putra, Aulalia Han (Eds.): *Governments Responses to Climate Change: Selected Examples from Asia-Pacific* (Heidelberg – Dordrecht – London – New York: Springer-Verlag, 2014).

Vol. 11: Sara Hellmüller, Martina Santschi (Eds.): *Is Local Beautiful? Peacebuilding between International Interventions and Locally Led Initiatives* (Heidelberg – Dordrecht – London – New York: Springer-Verlag, 2014).

Vol. 12: Úrsula Oswald Spring, Hans Günter Brauch, Keith G. Tidball (Eds.): *Expanding Peace Ecology: Security, Sustainability, Equity and Peace: Perspectives of IPRA's Ecology and Peace Commission* (Heidelberg – Dordrecht – London – New York: Springer-Verlag, 2014).

Vol. 13: Mansoureh Ebrahimi (Eds.): *The Dynamics of Iranian Borders: Issues of Contention* (Cham- New York - Heidelberg - Dordrecht - London: Springer International Publishing, 2017).

Vol. 14: Liliana Rivera-Sánchez, Fernando Lozano-Ascencio (Eds.): *The Practice of Research on Migration and Mobilities.* SpringerBriefs in Environment, Security, Development and Peace, vol. 14 – Migration Studies Subseries No. 1 (Heidelberg – Dordrecht – London – New York: Springer-Verlag, 2014).

Vol. 15: Yongyuth Chalamwong - Naruemon Thabchumpon (Eds.): *Livelihood Opportunities, Labour Market, Social Welfare and Social Security in Temporary Sheltered and Surrounding Communities* (Heidelberg – Dordrecht – London – New York: Springer-Verlag, 2014).

Vol. 16: Suwattana Thadaniti (Ed.): *The Impact of Displaced People's Temporary Shelters on their Surrounding Environment .* SpringerBriefs in Environment, Security, Development and Peace, vol. 16 – Migration Studies Subseries No. 3 (Heidelberg – Dordrecht – London – New York: Springer-Verlag, 2014).

Vol. 17: Premjai Vungsiriphisal and Dares Chusri (Eds.): *Royal Thai Government Policy and Donor, INGO/NGO and UN Agency Delivery of Humanitarian Assistance for Displaced Persons from Myanmar* (Heidelberg – Dordrecht – London – New York: Springer-Verlag, 2014).

Springer: <http://www.springer.com/series/10357>
AFES-PRESS: <http://www.afes-press-books.de/html/SpringerBriefs_ESDP.htm>

Authors or editors who would like to have their publication project considered for inclusion in this series should contact both the series editor:

PD Dr. phil. habil. Hans Günter Brauch, Alte Bergsteige 47, 74821 Mosbach, Germany
Phone: 49-6261-12912 – FAX: 49-6261-15695 – Email: afes@afes-press.de
http://www.afes-press.de and http://www.afes-press-books.de/html/hexagon.htm
and the publisher:

Dr. Johanna Schwarz, Senior Publishing Editor, Springer, Tiergartenstraße 17
69121 Heidelberg, Germany, Phone: +49 6221 487 8614
Email: <johanna.schwarz@springer.com> and <http://www.springer.com>

Springer Briefs on Pioneers in Science and Practice (PSP)

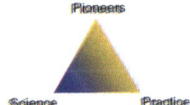

Edited by
Hans Günter Brauch, AFES-PRESS Chairman, Mosbach, Germany

Vol. 1: Arthur H. Westing: Arthur H. Westing: *Pioneer on the Environmental Impact of War* (Heidelberg – New York – Dordrecht – London: Springer-Verlag, 2013).

Vol. 2: Rodolfo Stavenhagen: *Pioneer on Indigenous Rights* (Heidelberg – Dordrecht – London – New York: Springer-Verlag, 2013).

Vol. 3: Rodolfo Stavenhagen: *The Emergence of Indigenous Peoples* (Heidelberg – New York – Dordrecht – London: Springer-Verlag, 2013).

Vol. 4: Rodolfo Stavenhagen: *Peasants, Culture and Indigenous Peoples: Critical Issues* (Heidelberg – New York – Dordrecht – London: Springer-Verlag, 2013).

Vol. 5: Johan Galtung Dietrich Fischer: *Pioneer of Peace Research* (Heidelberg – New York – Dordrecht – London: Springer-Verlag, 2013).

Vol. 6: Dieter Senghaas: *Pioneer of Peace and Development Research* (Heidelberg – New York – Dordrecht – London: Springer-Verlag, 2013).

Vol. 7: Chadwick Alger: *Pioneer in the Study of the Political Process and on NGO Participation in the United Nations* (Heidelberg – New York – Dordrecht – London: Springer-Verlag, 2014).

Vol. 8: Chadwick F. Alger: *The UN System and Cities in Global Governance* (Heidelberg – New York – Dordrecht – London: Springer-Verlag, 2014).

Vol. 9: Chadwick F. Alger: *Peace Research and Peacebuilding* (Heidelberg – New York – Dordrecht – London: Springer-Verlag, 2014).

Vol. 10: Lourdes Arizpe: *Lourdes Arizpe Schlosser: A Mexican Pioneer in Anthropology* (Heidelberg – New York – Dordrecht – London: Springer-Verlag, 2014).

Vol. 11: Lourdes Arizpe: *Migration, Women and Social Development: Key Issues* (Heidelberg – New York – Dordrecht – London: Springer-Verlag, 2014).

Vol. 12: Lourdes Arizpe: *Culture, Diversity and Heritage: Major Studies* (Heidelberg – New York – Dordrecht – London: Springer-Verlag, 2014).

Vol. 13: Arthur H. Westing: *Texts on Environmental and Comprehensive Security* (Heidelberg – New York – Dordrecht – London: Springer-Verlag, 2014).

Vol. 14: Klaus von Beyme: *Pioneer in the Study of Political Theory and Comparative Politics* (Heidelberg – New York – Dordrecht – London: Springer-Verlag, 2013).

Vol. 15: Klaus von Beyme: *On Political Culture, Culture, Art and Politics* (Heidelberg – New York – Dordrecht – London: Springer-Verlag, 2014).

Vol. 16: Samir Amin: *Pioneer on the Rise of the South* – Presented by Dieter Senghaas (Heidelberg – New York – Dordrecht – London: Springer-Verlag, 2014).

Vol. 17: Samir Amin: *Theory is History* (Cham – Heidelberg – New York – Dordrecht – London: Springer, 2014).

Vol. 18: Ulrich Beck (Ed.): *Ulrich Beck: A Pioneer in Cosmopolitan Sociology and Risk Society* – Presented by John Urry (Cham – Heidelberg – New York – Dordrecht – London: Springer, 2014).

Vol. 19: Hans Günter Brauch – Teri Grimwood (Eds.): *Jonathan Dean: Pioneer in Détente in Europe, Global Cooperative Security Arms Control and Disarmament* (Cham – Heidelberg – New York – Dordrecht – London: Springer, 2014).

Vol. 20: Hartmut Soell (Ed.): *Helmut Schmidt: Pioneer of International Economic and Financial Cooperation* (Cham – Heidelberg – New York – Dordrecht – London: Springer, 2014).

Vol. 21: Ulrich Bartosch (Ed.): *Carl Friedrich von Weizsäcker: Pioneer of Physics, Philosophy, Religion, Politics and Peace Research* (Cham – Heidelberg – New York – Dordrecht – London: Springer, 2015).

Vol. 22: Michael Drieschner (Ed.): *Carl Friedrich von Weizsäcker: Major Texts in Physics* (Cham – Heidelberg – New York – Dordrecht – London: Springer-Verlag, 2014).

Vol. 23: Michael Drieschner (Ed.): *Carl Friedrich von Weizsäcker: Major Texts in Philosophy* (Cham – Heidelberg – New York – Dordrecht – London: Springer, 2014).

Vol. 24: Konrad Raiser (Ed.): *Carl Friedrich von Weizsäcker: Major Texts on Religion* (Cham – Heidelberg – New York – Dordrecht – London: Springer, 2014).

Vol. 25: Ulrich Bartosch (Ed.): *Carl Friedrich von Weizsäcker: Major Texts on Politics and Peace Research* (Cham – Heidelberg – New York – Dordrecht – London: Springer, 2014).

Vol. 26: Betty A. Reardon; Dale Snauwaert (Eds.): *Betty A. Reardon: A Pioneer in Education for Peace and Human Rights* (Cham – Heidelberg – New York – Dordrecht – London: Springer-Verlag, 2015).

Vol. 27: Betty A. Reardon; Dale Snauwaert: *Key Texts in Gender and Peace* (Cham – Heidelberg – New York – Dordrecht – London: Springer, 2015).

Vol. 28: Ernst Ulrich von Weizsäcker (Ed.): *Ernst Ulrich von Weizsäcker: A Pioneer on Environmental, Climate and Energy Policy* (Cham – Heidelberg – New York – Dordrecht – London: Springer, 2014).

Vol. 29: Nils Petter Gleditsch: *Nils Petter Gleditsch: Pioneer in the Analysis of War and Peace* (Cham – Heidelberg – New York – Dordrecht – London: Springer, 2015).

Vol. 30: Morton Deutsch and Peter T. Coleman: *Morton Deutsch: A Pioneer in Developing Peace Psychology* (Cham – Heidelberg –New York – Dordrecht – London: Springer, 2015).

Vol. 31: Morton Deutsch and Peter T. Coleman: *Morton Deutsch: Major Texts on Peace Psychology* (Cham – Heidelberg –New York – Dordrecht – London: Springer, 2015).

Vol. 32: Herbert C. Kelman and Ronald J. Fisher: *Herbert C. Kelman: Pioneer in the Social Psychology of Conflict Analysis and Resolution* (Cham – Heidelberg – New York – Dordrecht – London: Springer International Publishing, 2016).

Vol. 33: Ronald J. Fisher: *Ronald J. Fisher: A North American Pioneer in Interactive Conflict Resolution* (Cham – Heidelberg – New York – Dordrecht – London: Springer International Publishing, 2016).

Vol. 34: Harvey Starr (Ed.): *Bruce M. Russett: Pioneer in the Scientific and Normative Study of War, Peace, and Policy* – Presented by Harvey Starr (Cham – Heidelberg – New York – Dordrecht – London: Springer, 2015).

Vol. 35: Bruce M. Russett (Ed.): *Karl W. Deutsch: Pioneer in the Theory of International Relations* – Presented by Bruce M. Russett (Cham – Heidelberg – New York – Dordrecht – London: Springer International Publishing, 2016).

Vol. 36: Naresh Dadhich (Ed.): *Mahatma Gandhi; Pioneer of Nonviolent Action –Presented by Naresh Dadhich* (Cham – Heidelberg – New York – Dordrecht – London: Springer International Publishing, 2016).

Vol. 37 Nils-Petter Gleditsch (Ed.): *R. J. Rummel - An Assessment of His Many Contributions* (Cham- New York - Heidelberg - Dordrecht - London: Springer International Publishing, 2017).

Vol. 38: Lucas De Melo Melgaço (Ed.): *Milton Santos: A Pioneer in Geography from Latin America - On World Time and Space, Globalization, Citizenship, Reason, and Knowledge* (Cham- New York - Heidelberg - Dordrecht - London: Springer International Publishing, 2017).

Vol. 39: Lucas De Melo Melgaço (Ed.): *Milton Santos: Towards Another Globalization – From a Unique Idea to a Universal Conscience* (Cham- New York - Heidelberg - Dordrecht - London: Springer International Publishing, 2017).

Vol. 41: Kalevi Holsti: *Kalevi Holsti: A Pioneer in International Relations Theory, Foreign Policy Analysis, History of International Order, and Security Studies* - To be presented by Barry Buzan (Cham – Heidelberg –New York – Dordrecht – London: Springer International Publishing, 2016).

Vol. 42: Kalevi Holsti: *Kalevi Holsti: Major Texts on War, Peace, and International Order* (Cham – Heidelberg –New York – Dordrecht – London: Springer International Publishing, 2015).

Vol. 50: Paul J. Crutzen, Hans Günter Brauch (Eds.): *Paul J. Crutzen: A Pioneer on Atmospheric Chemistry and Climate Change in the Anthropocene* (Cham – Heidelberg – New York – Dordrecht – London: Springer International Publishing, 2016).

Vol. 52: Richard D. Knowles, Céline Rozenblat (Eds.): *Sir Peter Hall: Pioneer in Regional Planning, Transport and Urban Geography* (Cham – Heidelberg – New York – Dordrecht – London: Springer International Publishing, 2016).

Springer: <http://www.springer.com/series/10970>
AFES-PRESS: <http://www.afes-press-books.de/html/SpringerBriefs_PSP.htm>.

Two New Book Series Starting in 2016

Pioneers in Arts, Humanities, Science, Engineering, Practice (PAHSEP)

Vol. 1: Louis Kriesberg: *Louis Kriesberg: Pioneer in Peace and Constructive Conflict Resolution Studies* (Cham – Heidelberg – New York – Dordrecht – London: Springer International Publishing, 2016).

Vol. 2: Richard Ned Lebow (Ed.): *Richard Ned Lebow: A Pioneer in International Relations Theory, History, Political Philosophy and Psychology* – Presented by Simon Reich (Cham – Heidelberg –New York – Dordrecht – London: Springer International Publishing, 2016).

Vol. 3: Richard Ned Lebow: *Richard Ned Lebow: Major Texts on Methods and Philosophy of Science* (Cham – Heidelberg –New York – Dordrecht – London: Springer International Publishing, 2016).

Vol. 4: Richard Ned Lebow: *Richard Ned Lebow: Key Texts in Political Psychology and International Relations Theory* (Cham – Heidelberg –New York – Dordrecht – London: Springer International Publishing, 2016).

Vol. 5: Richard Ned Lebow: *Richard Ned Lebow: Essential Texts on Classics and History and Ethics and International Relations* (Cham – Heidelberg –New York – Dordrecht – London: Springer International Publishing, 2016).

Vol. 6: Russell Boulding (Ed.): *Elise Boulding: A Pioneer in Peace Research, Peacemaking, Feminism and the Family: From a Quaker Perspective* (Cham – Heidelberg –New York – Dordrecht – London: Springer International Publishing, 2016).

Vol. 7: Russell Boulding (Ed.): *Elise Boulding: Major Texts in Peace Research and Peacemaking* (Cham – Heidelberg –New York – Dordrecht – London: Springer International Publishing, 2016).

Vol. 8: Russell Boulding (Ed.): *Elise Boulding: Key Texts on Feminism and the Family* (Cham – Heidelberg –New York – Dordrecht – London: Springer International Publishing, 2016).

Vol. 9: Russell Boulding (Ed.): *Elise Boulding: Autobiographical Writings and Selections from Unpublished Journals and Letters* (Cham – Heidelberg –New York – Dordrecht – London: Springer International Publishing, 2016).

Vol. 10: Úrsula Oswald Spring (Ed.): *Úrsula Oswald Spring: A Pioneer on Gender, Ecology, Peace, Security, and Development* (Cham – Heidelberg –New York – Dordrecht – London: Springer International Publishing, 2017).

Vol. 11: Úrsula Oswald Spring: *Úrsula Oswald Spring: Text on Peasantry, Food Systems and Development and Development with a Focus on Mexico* (Cham – Heidelberg –New York – Dordrecht – London: Springer International Publishing, 2018).

Vol. 12: Úrsula Oswald Spring: *Úrsula Oswald Spring: Texts on Ecology, Biodiversity, Water, Energy, on Climate Change, Environmental Migration, and as Minister of Ecological Development in Morelos* (Cham – Heidelberg –New York – Dordrecht – London: Springer International Publishing, 2018).

Vol. 13: Úrsula Oswald Spring: *Úrsula Oswald Spring: Texts on Peace, Peace Research, on Human, Gender and Environmental Security, on Sustainable-Engendered Peace, and Health and Disaster from a Gender Perspective* (Cham – Heidelberg –New York – Dordrecht – London: Springer International Publishing, 2019).

Springer: <http://www.springer.com/series/15230>
AFES-PRESS: <http://www.afes-press-books.de/html/PAHSEP.htm>.

The Anthropocene:
Politik – Economics – Society – Science
(APESS)

Springer: <http://www.springer.com/series/15232>
AFES-PRESS: <http://www.afes-press-books.de/html/APESS.htm>.

Authors or editors who would like to have their publication project considered for inclusion in this Springer book series should contact both the series editor:
PD Dr. Hans Günter Brauch, Alte Bergsteige 47, 74821 Mosbach, Germany
Email: <afes@afes-press.de>
and the publisher:
Dr. Johanna Schwarz, Editor, Springer, 69121 Heidelberg, Germany
Email: <johanna.schwarz@springer.com>.